D1083688

THE INFRARED HANDBOOK
revised edition

Environmental Research Institute of Michigan. Infrared Information and Analysis (IRIA) Center.

<u>Editors</u>

William L. Wolfe
Professor of Optical Sciences, University of Arizona.
Consultant, Infrared Information Analysis (IRIA) Center
Environmental Research Institute of Michigan.

George J. Zissis
Director Emeritus, Infrared Information Analysis (IRIA) Center
Environmental Research Institute of Michigan.

Prepared by

Σ The Infrared Information Analysis (IRIA) Center,
Environmental Research Institute of Michigan.

for the

Office of Naval Research, Department of the Navy,
Washington, DC

First Edition 1978
Revised Edition 1985
3rd Printing 1989

NOTICES AND DISCLAIMERS

Prepared by the Infrared Information and Analysis (IRIA) Center — A Defense Logistics Agency administered Department of Defense Information Analysis Center — of the Environmental Research Institute of Michigan (ERIM) under Contract Numbers N00014-73-A-0321-0002, N00014-74-C-285, N00014-76-C-0607, N00014-77-C-0125, and N00014-84-C-0493 with the Office of Naval Research. Scientific Officer for the contract is the Director, Physics Program of the Office of Naval Research, with funding and administrative management provided by the Defense Logistics Agency of the Department of Defense (DLA-SCT). Pre-printing preparation for the first edition was by the Naval Research Laboratory (NRL) with printing by the United States Government Printing Office. For the second edition, preparation was again by NRL with printing by ERIM.

Library of Congress Catalog Card No: 77-90786

ISBN: 0-9603590-1-X

Preface to the Third Printing

This third printing of 2500 copies of *The Infrared Handbook* was undertaken by the Infrared Information Analysis Center, the Environmental Research Institute of Michigan to bridge the gap between this final printing and the introduction of a completely revised sourcebook series scheduled for release in mid 1991. The support for this printing was secured through the Defense Logistics Agency, Defense Electronics Supply Center, under contract DLA900-88-D-0392 under the technical direction of Dr. John M. MacCallum, Office of the Deputy Undersecretary of Defense, Research and Advanced Technology. The 20,000 copies of *The Infrared Handbook* printed since 1979 stand as testimony to its general success and as a lasting tribute to editors George J. Zissis and William L. Wolfe.

Joseph S. Accetta
Director, Infrared Information
  Analysis Center
Ann Arbor, MI
April 1989

PREFACE

Origins

The Infrared Handbook has been prepared by the Infrared Information and Analysis (IRIA) Center as a replacement for the Handbook of Military Infrared Technology* which was published in 1965. The old handbook has served the infrared community as a ready reference for data, techniques and equations. It is a venerable publication, but has become obsolete in the decade since its publication. New detectors have been invented, new materials discovered, and many instrumentation techniques undergone vast improvement since 1965. There are other deficiencies as well. For example, relatively little attention was paid to infrared systems, and too much emphasis was placed on control theory (which is well documented elsewhere).

Discussions in 1975 between one of the two editors of The Infrared Handbook, George Zissis, the Director of IRIA, and William Condell, ONR Physics, Washington, DC, crystallized the recognition of the obsolescence of the first Handbook and set into motion the chain of events that lead to the existence of this publication. Subsequently, William Wolfe, Professor of Optical Sciences, University of Arizona, and editor of the previous Handbook, agreed to become a consultant to IRIA and join George Zissis as coeditor of The Infrared Handbook.

Our first step was to make a survey of a sample population of the infrared community. We sought to determine the parts of the 1965 Handbook which were most useful and the parts which were least used together with any criticisms of commission and omission. The results of the survey in summary were: the chapters on radiation, detectors, optical design, optical materials, and backgrounds were used a great deal; the chapters on control systems, thermal control, and reticles were little used; the treatment of infrared systems was inadequate. The results agreed well with our own estimates of the strong and weak points of the 1965 Handbook. We, therefore, contacted the previous authors of chapters on radiation theory and sources, detectors, optical materials, optics, and backgrounds to obtain updated and expanded replacement chapters. We deleted the less useful chapters and converted the single, unsatisfactory chapter on system design into four chapters on selected, important aspects of infrared system design. We also included a chapter on displays because they have become recognized as a critical element in the overall design and performance of a system. The advent of the pyroelectric sensing element has renewed the hope that one can design and construct an infrared vidicon completely analogous to the standard televison tube. Accordingly, we added a chapter on infrared tubes, with vidicons and pyroelectric materials emphasized. A major technological advance in semiconductor electronics is the charge-coupled device (CCD), which promises to make the

---

*The Handbook of Military Infrared Technology was prepared by the IRIA Center in the University of Michigan's Willow Run Laboratories, which became a separate non-profit institute, the Environmental Research Institute of Michigan, in 1973.

v

two-dimensional array of detectors a practical reality. Although infrared CCDs are still in the developmental stage and, as such, not really ready for handbook treatment, we have chosen to include the theoretical discussion and data on a few of the experimental models which are now in existence. The reader should expect rapid changes in realizable CCD structures. Active systems, too, have started to take their place among the armamentarium of infrared technology, and this volume includes some discussion and a few data on direct and coherent detection of laser signals as used in communication, direction-finding and ranging systems.

Preparation

Part of the preparation of a well organized handbook is the establishment of uniform nomenclature, a consistent set of symbols, and identical units. The last of these turned out to be easiest, but not easy. We simply adopted the Système International (SI) set of units (basically metric) as *the* system for the Handbook. This seemed natural for all but the chapters on aerodynamic influences and on coolers. Here the work has been done almost entirely by engineers who work in the English system, and thus the references and the data are in that system. Nomenclature uniformity was more difficult to obtain. Our first rule, of course, was to define the terms as they are used. The most troublesome technical word was "intensity." Most astronomers use "intensity" or "specific intensity" as a term referring to the distribution of flux (or radiant power) with respect to area and solid angle. We use "radiance" for this. Workers in the fields of electromagnetic theory often use "intensity" when they refer to the distribution of flux with respect to area alone. We use "irradiance" or "exitance" for this. We use "intensity" only for referring to the distribution of flux with respect to solid angle. Signal-to-noise ratio had an ambiguity that we attempted to resolve by definition. First, unless otherwise specified the rms signal-to-rms noise is meant. When the ratio of peak-to-peak signal-to-rms noise is meant, that is specified. When it makes no difference, nothing is said.

Next we considered the potential confusion inherent in the many frequencies associated with infrared technology. In this Handbook we use ν for the very large optical-radiation frequencies (e.g., $\nu \sim 10^{12}$ to 10^{14} Hz). Modulating frequencies (as may be provided by a chopper) are associated with a modulation or temporal frequency symbolized by f. Finally, radiation patterns often have spatial distributions with which can be associated "spatial frequencies." These we symbolize by f_x or f_y or f_x, as may be most appropriate.

In the text we indicate vectors by bold-face type; however, in figures, it was necessary to use normal symbols with an arrow above them, e.g., $\vec{x}$. In addition, symbols are italicized in the text, but could not be so displayed in the figures.

There are not enough English and Greek symbols to establish a one-to-one correspondence between symbols and physical quantities. We started symbol designation by using the SI set of symbols for atomic constants and radiometric terms. The Table on the inside front cover represents the symbols that are used universally but not uniquely throughout the book. For example, the speed of light in a vacuum is always c, but c sometimes represents some other constant. Each chapter has its own set of symbols, nomenclature and units. These are given in the first table of each chapter. We believe that uniformity was impossible, but hope that ambiguity has been eliminated.

Organization

We have attempted to arrange the chapters in a logical sequence to assist the reader in finding a particular subject. The sequence is traditional: sources, atmosphere, optics, detectors, electronics, displays and systems. Chapter 1 emphasizes the general calculational procedures with blackbody radiators. We have chosen to give several example pocket calculator programs for the distributions and integrals so that the reader can quickly obtain a value based on the most recent values of the atomic constants. Chapters 2 and 3 provide information on most laboratory and field sources. A catalog of all the natural sources is not feasible, so models are presented so that natural scenes can be constructed from relatively few data. Models for scattering, absorption, and turbulence effects by the atmosphere are given in Chapters 4, 5, and 6, together with references for the reader who wishes to be more accurate or thorough. Chapters 7, 8, 9 and 10 should allow the reader to pick a material and make a "first cut" at an optical design. More careful design should be done with one of the several ray-trace and optical design programs available and referenced. The final choice of optical materials and design should be made in conjunction with a vendor and specific measurements on the real thing. Chapters 11, 12, 13 and 14 should allow the reader to pick a generic detector: element, array, CCD, tube or photographic film. Again the specific choice should be made after more detailed analyses. Chapters 15 and 16 give information on the cooling of and signal extraction from detectors. Chapters 17 and 18 describe methods for the processing or display of the information. Chapter 17 gives a detailed treatment of two-dimensional Fourier analysis and of techniques which have not been widely utilized in the design or analysis of infrared devices. Chapters 19 through 24 deal with systems; imaging, radiometric, tracking, warning, communication, ranging and simulation. The equations and data are about as specific to the design of a system as we could make them without using illustrative examples or very special applications. Finally, many of the constants presented in Chapter 25 are different (and improved) from those contained in the 1965 Handbook and other publications. They are the recently recognized values of the International Physical Constants of 1973 produced by a Task Group of the Committee on Data for Science and Technology (CODATA) under the International Council of Scientific Unions (ICSU). Significant differences will be found between the radiometric values calculated here, using the given values of the first and second radiation constants, and those with the older, less exact values.

Acknowledgments

Many have contributed to this undertaking in diverse ways. We thank them all. Clearly the most significant contribution was by Gary Gatien. Throughout the entire project, he was our colleague, chief copy editor and organizer. He read manuscripts, logged them, clarified text and coordinated the corrections of the two coeditors. Corliss Hugg assisted in much of this and performed yeoman efforts in figure sizing and clarification. Marie Nichols and Rose Coleman of the IRIA Staff provided the bulk of the necessary peerless typing, and the triumvirate of William Condell, John Ivory (ONR, Chicago), and Joseph

L. Blue (Defense Logistics Agency) kept the project moving at a timely and steady pace toward the date of publication.

Vitally important contributions came from the ERIM Publications Department from James Cooper, Robert DiGiovanni, Marcella Dunton, Alice Otto, Judith Steeh, and Walter Ellis and the staff of the ERIM Art Department. The project benefited by assistance from Mildred Denecke, Manager of IRIA, the index preparation efforts of Ruth Chatton, and copy-editing contributions by Janice McKimmy. Three members of the ERIM scientific staff, i.e., Anthony J. LaRocca, the late Leo Larsen, and Joseph Mudar, acted as technical editors for several chapters. Others offered useful reviews of several specific sections within chapters.

Of special note is the most cooperative attitude and critical contributions of John Berezansky, Jr. and Kenneth T. Lassiter, Eastman Kodak Company, and of Michael N. Ernstoff and Gerald K. Slocum, Hughes Aircraft Company, who would not rest until we had the data correct on their publications and their companies' products in Chapters 7, 14 and 18. We are grateful to Robert Ansell, Bulova Watch Company, for similar assistance for Chapter 10, and to the Society of Information Display (SID) and its publications Chairman, Thomas V. Curran, for granting special blanket permission to reproduce in Chapter 18 information contained in SID publications. Chapter 22 profited considerably from the efforts of Lawrence C. Caplan, John M. Fitts, Jack M. Sacks, Reo Yoshitani and Albert Zalon, all of Hughes Aircraft Company, while Chapter 23 received similar contributions from A. R. Kraemer, P. J. Titterton, and S. C. Morford, all of GTE Sylvania. Acknowledgments for the use of specific, copyrighted material are listed on the following pages.

Finally, the authors should receive applause for the work they did, the corrections they made, and the patience they had with the editors. The honoraria they received for their efforts may have paid partially for the pen and ink they used.

Editors:
George J. Zissis
William L. Wolfe
Ann Arbor, MI

Special Figure, Table and
Text Acknowledgments

Chapter 1

Fig. 1-16: MIT Press.

Chapter 2

Figs. 2-24, 2-26, 2-27, 2-70, 2-71, 2-72, 2-73 and Tables 2-30, 2-31 and 2-32: John Wiley and Sons, Inc.

Figs. 2-54 and 2-55: General Electric.

Figs. 2-56 through 2-60 and Table 2-27 and Text p. 2-69: GTE Sylvania, Inc.

Figs. 2-74, 2-75, and 2-76: Headquarters, U.S. Army Materiel Development and Readiness Command.

Table 2-5: Electro-Optical Industries.

Text pp. 2-2 and 2-35: McGraw-Hill Book Company.

Chapter 3

Figs. 3-8, 3-9 and Table 3-2: American Association for the Advancement of Science.

Fig. 3-95: National Research Council of Canada.

Table 3-22: Prentice-Hall.

Tables 3-28, 3-29, and 3-30: USDA-ARS.

Chapter 4

Fig. 4-29 and Table 4-1: American Geophysical Union.

Table 4-12: Academic Press.

Chapter 5

Fig. 5-50 and Table 5-22: American Geophysical Union.

Table 5-19: Pergamon Press.

Tables 5-21 and 5-26: Ohio State University.

Table 5-26: American Chemical Society.

Chapter 7

Fig. 7-8: Purdue Research Foundation.

Fig. 7-15 and Tables 7-30 and 7-42: The Institute of Physics.

Fig. 7-21: The Royal Society of London.

Fig. 7-28: A. Hadni, E. Decamp, and C. Janot.

Fig. 7-29: Pergamon Press.

Fig. 7-53: Army Night Vision Laboratory.

Figs. 7-64 and 7-65: W.H. Freeman and Company.

Chapter 8

Figs. 8-10, 8-11, 8-16, 8-17, 8-18, 8-22, and 8-24 through 8-28: McGraw-Hill Book Company.

Chapter 9

Figs. 9-4, 9-6, 9-7, 9-8, 9-11, 9-12, 9-13, 9-14, 9-29, 9-30, 9-32 through 9-36, 9-38, and 9-39: McGraw-Hill Book Company.

Figs. 9-39 and 9-40: W.H. Freeman and Company.

Chapter 11

Text pp. 11-36 and 11-37: Butterworth & Co. (Publishers) Ltd.

Chapter 12

Figs. 12-21, 12-22 and Text pp. 12-11, 12-16, 12-24, 12-30, and 12-37: Institute of Electrical and Electronics Engineers.

Chapter 13

Table 13-7: The Electrochemical Society, Inc.

Chapter 14

Figs. 14-1 through 14-31 inclusive: Eastman Kodak Company.

Figs. 14-32 and 14-33: American Institute of Physics.

Chapter 15

Figs. 15-32 and 15-33: American Institute of Aeronautics and Astronautics.

Figs. 15-43 and 15-44: United States Energy Research and Development Administration.

Fig. 15-45 and Table 15-17: McGraw-Hill Book Company.

Chapter 18

Figs. 18-17 through 18-23 and Tables 18-4, 18-9, 18-11, 18-12, 18-13, 18-14 and Text pp. 18-25, 18-26, and 18-29 through 18-44: Society for Information Display.

Table 18-7: The Electrochemical Society, Inc.

Chapter 20

Figs. 20-29, 20-37, and 20-39: McGraw-Hill Book Company.

Fig. 20-43: W.H. Freeman and Company.

Figs. 20-45 and 20-46: Academic Press.

Fig. 20-50: American Institute of Physics.

Chapter 21

Figs. 21-1, 21-2, and 21-3: McGraw-Hill Book Company.

Chapter 22

Fig. 22-49: American Institute of Aeronautics and Astronautics.
Text pp. 22-28, 22-29, and 22-42: McGraw-Hill Book Company.

Chapter 23

Figs. 23-33 and 23-34: Society of Photo-Optical Instrumentation Engineers (SPIE).

Figs. 23-36, 23-37, 23-40 and Tables 23-13, 23-14, 23-15 and Text pp. 23-60, 23-61, 23-62, 23-64, 23-66, and 23-67: U.S. Army Missile Research and Development Command.

Table 23-10: Chemical Rubber Publishing Company.

Text pp. 23-10, 23-13, 23-14, 23-15, 23-55, 23-56, 23-57 and Table 23-3: John Wiley & Sons, Inc.

Text pp. 23-16, 23-17, 23-34, 23-36, and 23-37: McGraw-Hill Book Company.

Chapter 25

Table 25-2: American Institute of Physics.

Tables 25-6, 25-7, 25-8, 25-10 through 25-14, 25-17, 25-18, and 25-19: CRC Press, Inc.

Chapter 1

RADIATION THEORY

William L. Wolfe
University of Arizona

CONTENTS

1. Radiation Theory

1.1. Introduction

1.1.1. Symbols, Nomenclature and Units.
The symbols and quantities used here are based largely on the SI system. Table 1-1 lists the symbols, nomenclature and units for each of the quantities used in this chapter (exclusive of those for radiometric quantities). Table 1-2 gives symbols, units and formulas for radiometric quantities.

The photometric quantities are based on visual response. Thus,

$$\int K(\lambda)S(\lambda)d\lambda = K_m S \int V(\lambda)s(\lambda)d\lambda \tag{1-1}$$

where K_m = maximum, luminous efficacy, 683 lm W^{-1}
S = maximum source output, W
$V(\lambda)$ = relative, spectral, luminous efficiency for the CIE-standard photometric observer
$s(\lambda)$ = relative spectral output of source

The basic unit is the candela or international standard candle. It is defined as the luminous intensity of $1/60$ cm^2 of a blackbody at the temperature of freezing platinum. (See Chapter 2.) Table 1-3 gives the units, dimensions and definitions of most photometric quantities. Table 1-4 gives conversions among photometric units.

1.1.2. Fluometry and Other Proposed Systems.
Jones [1-1] pointed out that most concepts in radiometry have to do only with the geometry of a flux, e.g., flux geometry or "fluometry". (Jones proposed the prefix "phlu-".) In this section, Q is the quantity whose flux is of interest. The names Jones suggested and some alternatives suggested by Nicodemus are given in Table 1-5.

1.1.3. Chinese Restaurant Nomenclature [1-2].
With this system, one picks the quantity and the modifiers. It's like choosing the different entrees in a Chinese restaurant. The system is specific and descriptive, but wordy. For example, one would choose (initially) one word each from columns A, B, C and D of Table 1-6 and then, perhaps, use a shorter word in the sequel.

1.2. Blackbody (Planck) Functions

A blackbody can be defined as a perfect radiator, i.e., one that radiates the maximum number of photons in a unit time from a unit area in a specified spectral interval into a hemisphere that any body in thermodynamic equilibrium at the same temperature can radiate. The number of photons, per unit volume of such a blackbody with energy between $\hbar ck$ and $\hbar c(k + dk)$ is

$$n_k dk = \pi^{-2} k^2 (e^x - 1)^{-1} dk \tag{1-2}$$

where n_k = number of photons per unit k-interval per unit volume
$k = 2\pi/\lambda$
λ = wavelength
$x = c_2/\lambda T = h\nu/k_B T = \hbar ck/k_B T$

Table 1-1. Symbols, Nomenclature and Units

Symbols	Nomenclature	Units
A	Area	m^2
c	Speed of light	$m\ sec^{-1}$
c_1	First radiation constant	$W\ m^2$
c_2	Second radiation constant	$m\ K$
D	Density of states function	Varies
E_q	Photon irradiance	$sec^{-1}\ m^{-2}$
h	Planck's constant	$J\ sec$
$\hbar$	$h/2\pi$	$J\ sec$
K_m	Maximum luminous efficacy	$683\ lm\ W^{-1}$
k	Angular or radian wavenumber, $2\pi/\lambda$	$rad\ m^{-1}$
k_B	Boltzmann constant	$J\ K^{-1}$
M_q	Photon exitance (flux density)	$sec^{-1}\ m^{-2}$
N	Number	—
$\dot{N}$	Number rate	sec^{-1}
n	Number density; also, an integer	m^{-3}
R	General radiometric quantity	Varies
RD	Relative difference	—
RE	Relative error	—
T	Temperature	K
$V(\lambda)$	Relative, spectral, luminous efficiency for the CIE-standard photometric observer	—
V	Volume	
x	Normalized radiation variable, $x = h\nu/k_B T$	
α, β	Angles	rad
β	$1/k_B T$	J^{-1}
γ	Coherence factor	—
$\zeta(\ )$	Zeta function	—
θ	Angle between line of sight and normal	rad
λ	Wavelength	$m, \mu m, nm$
ν	Frequency	Hz
$\tilde{\nu}$	Wavenumber, $1/\lambda$	m^{-1}, cm^{-1}
σ	Stefan-Boltzmann constant	$W\ m^{-2}\ K^{-4}$
ω	Frequency, angular, $\omega = 2\pi\nu$	$rad\ sec^{-1}$
Ω	Solid angle	sr
Ω'	Projected solid angle, $\Omega' = \Omega \cos\theta$	sr
Subscripts		
e	Radiometric quantities	—
q	Photon number	—
v	Visual quantities, photometric	—
λ, ν, k	Distribution with respect to spectral variable indicated	—

Table 1-2. Radiometric Quantities, Symbols and Units [1-3]

Quantity	Symbol	Defining Equation	Units
Energy	Q_e	–	J
Energy (volume) density	w_e	$\dfrac{\partial Q_e}{\partial V}$	J m^{-3}
Flux (power)	Φ_e	$\dfrac{\partial Q_e}{\partial t}$	W
Flux (area) density	–	$\dfrac{\partial \Phi_e}{\partial A}$	W m^{-2}
Radiant exitance	M_e	$\dfrac{\partial \Phi_e}{\partial A}$	W m^{-2}
Irradiance	E_e	$\dfrac{\partial \Phi_e}{\partial A}$	W m^{-2}
Radiance	L_e	$\dfrac{\partial^2 \Phi_e}{\partial A \cos\theta \, \partial\Omega}$	W m^{-2}sr^{-1}

Table 1-3. Photometric Quantities, Symbols and Units [1-4]

Quantity	Symbol	Formula or Value	Unit Names	Unit Symbols
Luminous energy (quantity of light)	Q_v	$\displaystyle\int_{380}^{760} k(\lambda)Q_{e\lambda}d\lambda$	Lumen hour Lumen second (Talbot)	lm h lm sec
Luminous (energy) density	w_v	$\partial Q_v/\partial V$	Lumen second per cubic meter (Talbot per cubic meter)	lm sec m^{-3}
Luminous flux (light watt)	Φ_v	$\partial Q_v/\partial t$	Lumen (Talbot/second)	lm
Luminous flux density	Φ_v/A	$\partial \Phi_v/\partial A$	Lumen per square meter (lux)	lm m^{-2} (lx)
Luminous exitance	M_v	$\partial \Phi_v/\partial A$	Lumen per square centimeter (phot)	lm cm^{-2} (ph)

Table 1-3. Photometric Quantities, Symbols and Units [1-4] *(Continued)*

Quantity	Symbol	Formula or Value	Unit Names	Unit Symbols
Illuminance	E_v	$\partial\Phi_v/\partial A$	Lumen per square foot (foot candle)	lm ft^{-2} (ft c)
Luminous intensity	I_v	$\partial\Phi_v/\partial\Omega$	Lumen per steradian (candela)	lm sr^{-1} (cd)
Luminance (photometric brightness)	L_v	$\dfrac{\partial^2\Phi_v}{\partial A\,\partial\Omega\cos\theta}$	Lumen per steradian and square meter (nit)	lm sr^{-1} m^{-2} (nt)
			Candela per square meter	cd m^{-2}
			Candela per square centimeter (stilb)	cd cm^{-2} (sb)
			*foot lambert	ft L (π^{-1} cd ft^{-2})
			*lambert	L (π^{-2} cd cm^{-2})
			*apostilb	asb (π^{-1} cd m^{-2})

*The units with a factor of π^{-1} are generally applicable only to diffuse materials.

This is the product of D, the density of states ($\pi^{-2}k^2$ in this case), and the average occupancy of a given mode, $(e^x - 1)^{-1}$. The density of states can be written in different ways depending on whether it is expressed in terms of k, ω, ν, λ, $\tilde{\nu}$ or some other radiation variable; the expression is also dependent on whether the states are expressed in terms of intervals of dk, $d\omega$, $d\nu$, $d\lambda$, $d\tilde{\nu}$, etc. (See Table 1-7.)

1.2.1 Radiometric Quantities and Conversions. The Planck function for n_k, the number of photons per unit k-interval per unit volume, is (Equation (1-2) above)

$$n_k = k^2\pi^{-2}(e^x - 1)^{-1} \tag{1-3}$$

The energy per unit volume in the same spectral interval is

$$w_k = h\nu n_k = \hbar c k n_k \tag{1-4}$$

The energy is always the number of photons times the energy per photon, monochromatically.

The number of photons per unit area per unit time radiated into a hemisphere by a blackbody is given by

$$M_q = \frac{\partial^2 N}{\partial t\,\partial A} = \frac{c}{4}\,n \tag{1-5}$$

Table 1-4. Photometric Conversions [1-5]

	Nit (Candela/m^2)	Stilb	Boucie Hectométre Carré	Apostilb	Milli-apostilb
1 Nit (nt) =	1	10^{-4}	10^4	3.14	3.14×10^3
1 Stilb (sb) =	10^4	1	10^8	3.14×10^4	3.14×10^7
1 Bougie Hectomètre Carré	10^{-4}	10^{-8}	1	3.14×10^{-4}	3.14×10^{-1}
1 Apostilb (asb) =	3.183×10^{-1}	3.183×10^{-5}	3.183×10^3	1	10^3
1 Milli-apostilb = (m asb)	3.183×10^{-4}	3.183×10^{-8}	3.183	10^{-3}	1
1 Micro-apostilb = (μ asb)	3.183×10^{-7}	3.183×10^{-11}	3.183×10^{-3}	10^{-6}	10^{-3}
1 Lambert (L) =	3.183×10^3	3.183×10	3.183×10	10^4	10^7
1 Milli-Lambert = (mL)	3.183	3.183×10^{-4}	3.183×10^4	10	10^4
1 Micro-lambert = (μmL)	3.183×10^{-3}	3.183×10^{-7}	3.183×10	10^{-2}	10
1 Foot-lambert = (ft L)	3.426	3.426×10^{-4}	3.426×10^4	10,764	1.0764×10^4
1 Candle per Sq ft =	1.0764×10	1.0764×10^{-3}	1.0764×10^5	3.382×10	3.382×10^4
1 Candle per Sq in. =	1.55×10^3	1.55×10^{-1}	1.55×10^{-5}	4.869×10^3	4.869×10^6

In terms of energy, the relationship is

$$M_k = \frac{\partial^3 Q_e}{\partial k \, \partial t \, \partial A} = \frac{c}{4} w_k \tag{1-6}$$

The number of photons per unit area per unit time per unit solid angle radiated from a blackbody is

$$L_q = \frac{\partial^3 N}{\partial t \, \partial A \, \cos\theta \, \partial \Omega} = \frac{c}{4\pi} n \tag{1-7}$$

Table 1-4. Photometric Conversions [1-5] (*Continued*)

Micro-apostilb	Lambert	Milli-lambert	Micro-lambert	Foot-lambert	Candle per Sq. ft	Candle per Sq. in.
3.14×10^6	3.14×10^{-4}	3.14×10^{-1}	3.14×10^2	2.919×10^{-1}	9.29×10^{-2}	6.452×10
3.14×10^{10}	3.14	3.14×10^3	3.14×10^6	2.919×10^3	9.29×10^2	6.452
3.14×10^{-2}	3.14×10^{-8}	3.14×10^{-5}	3.14×10^{-2}	2.919×10^{-5}	9.29×10^{-6}	6.452×10^8
10^6	10^{-4}	10^{-1}	10^2	9.29×10^{-2}	2.957×10^{-2}	2.054×10^4
10^3	10^{-7}	10^{-4}	10^{-1}	9.29×10^{-5}	2.957×10^{-5}	2.054×10
1	10^{-10}	10^{-7}	10^{-4}	9.29×10^{-8}	2.957×10^{-8}	2.054×10^{10}
10^{10}	1	10^3	10^6	9.29×10^2	2.957×10^2	2.054
10^7	10^{-3}	1	10^3	9.29×10^{-1}	2.957×10^{-1}	2.054×10^{-3}
10^4	10^{-6}	10^{-3}	1	9.29×10^{-4}	2.957×10^4	2.054×10^{-6}
1.0764×10^7	1.0764×10^{-3}	1.0764	1.0764×10^3	1	0.3183	2.14×10^{-3}
3.382×10^7	3.382×10^{-3}	3.382	3.382×10^3	3.14	1	6.944×10^{-3}
4.869×10^9	4.869×10^{-1}	4.869×10^2	4.869×10^5	4.524×10^2	1.44×10^2	1

Again, for energy the relationship is

$$L_k = \frac{\partial^4 Q_e}{\partial k\, \partial t\, \partial A\, \cos\theta\, \partial\Omega} = \frac{c}{4\pi}\, w_k \qquad (1\text{-}8)$$

These different forms are summarized in Table 1-7.

1.2.2. Radiometric Quantities (Spectral Scale Conversions). Table 1-8 gives the expressions for the different radiometric functions in different spectral scales. The independent variables are k, ν, $\tilde{\nu}$, x, λ, and ω, all indicated in general by the variable y.

Table 1-5. Fluometry Quantities [1-6]

Name	Symbol	Definition	Equation	Alternates
Quantity	Q	—	—	—
Flux	Φ	Time rate	$\partial Q/\partial t$	—
Flux density	—	Time rate per unit area normal to flow	$\partial Q/\partial t\,\partial A\,\cos\theta$	Areance
Exitance	M	Flux density emitted	—	Areance
Incidance	E	Flux density received	—	Areance
Intensity	I	Flux per unit solid angle from a small source	$\partial Q/\partial t\,\partial\Omega$	Pointance
Sterance	L	Flux density per unit solid angle per unit area	—	Sterance
Sterisent	—	Change in radiance per unit pathlength	—	Sterisent
Fluence	—	—	—	Fluence
Exposure	—	Flux density resulting from integration of a projected solid angle over a hemisphere	—	Exposure

Table 1-6. Chinese Restaurant Nomenclature

A	B	C	D
Incident	Total	Energy	Through a surface
Scattered	Spectral	Entropy	Radiance
Reflected	Weighted	Power	Irradiance
Absorbed	Photopic	Photon	Intensity
Transmitted	Scotopic	Photon Rate	Radiance per unit length
Emitted	Erythemal	Momentum	—

1.2.3. Generalized Planck Functions. One useful form for radiant exitance is

$$M_x = 2\pi\,c^2 h\left(\frac{k_B T}{ch}\right)^4 x^3 (e^x - 1)^{-1} \tag{1-9}$$

This is of the form

$$R(x, T) = CT^l x^m (e^x - 1)^{-1} \tag{1-10}$$

The general radiometric function, R, is a function of temperature, T, and a spectral

Table 1-7. Density of States Function, D, in the Planck Function. The independent variable is given as y. The functions are number density n_y, energy density w_y, photon exitance M_{qy}, radiant exitance M_y, photon radiance, radiance. Each entry is D the multiplier of $(e^x - 1)^{-1}$ in the Planck expression. (See Section 1.2.)

$y =$ Function	k	$\tilde{\nu}$	ν	$x = h\nu/kT$	λ	ω
$n_y = N_y/V$	k^2/π^2	$8\pi\tilde{\nu}^2$	$8\pi\nu^2/c^3$	$8\pi\left(\dfrac{kT}{ch}\right)^3 x^2$	$8\pi\lambda^{-4}$	$\omega^2/(c^3\pi^2)$
$\omega_y = h\nu n_y$	$chk^3/2\pi^3$	$8\pi ch\nu^3$	$8\pi h\nu^3/c^3$	$8\pi ch\left(\dfrac{kT}{ch}\right)^4 x^3$	$8\pi ch\lambda^{-5}$	$h\omega^3/(2\pi^3 c^3)$
$M_{qy} = nc/4$	$ck^2/4\pi^2$	$2\pi c\tilde{\nu}^2$	$2\pi\nu^2/c^2$	$2\pi c\left(\dfrac{kT}{ch}\right)^3 x^2$	$2\pi c\lambda^{-4}$	$\omega^2/(4\pi^2 c^2)$
$M_y = \omega_y c/4$	$c^2 hk^3/8\pi^3$	$2\pi c^2 h\nu^3$	$2\pi h\nu^3/c^2$	$2\pi c^2 h\left(\dfrac{kT}{ch}\right)^4 x^3$	$2\pi c^2 h\lambda^{-5} = c_1\lambda^{-5}$	$h\omega^3/(8\pi^3 c^2)$
$L_{qy} = M_{qy}/\pi$	$ck^2/4\pi^3$	$2c\tilde{\nu}^2$	$2\nu^2/c^2$	$2c\left(\dfrac{kT}{ch}\right)^3 x^2$	$2c\lambda^{-4}$	$\omega^2/(4\pi^3 c^2)$
$L_y = M_y/\pi$	$c^2 hk^3/8\pi^4$	$2c^2 h\tilde{\nu}^3$	$2h\nu^3/c^2$	$2c^2 h\left(\dfrac{kT}{ch}\right)^4 x^3$	$2c^2 h\lambda^{-5}$	$h\omega^3/(8\pi^4 c^2)$

Table 1-8. Relationships Among Spectral Variables

$$\nu = c\tilde{\nu} = (2\pi)^{-1}\omega = (c/2\pi)k = (kT/h)x = c\lambda^{-1}$$

$$\lambda = c\nu^{-1} = \tilde{\nu}^{-1} = (2\pi c)\omega^{-1} = 2\pi k^{-1} = (hc/kT)x$$

$$d\nu = cd\tilde{\nu} = (2\pi)^{-1}d\omega = (c/2\pi)dk = (kT/h)dx = -c\lambda^{-2}d\lambda$$

$$d\lambda = -\frac{d\tilde{\nu}}{\tilde{\nu}^2} = -2\pi c\frac{d\omega}{\omega^2} = -2\pi\frac{dk}{k^2} = -c\frac{d\nu}{\nu^2}$$

$$\frac{d\lambda}{\lambda} = -\frac{d\tilde{\nu}}{\tilde{\nu}} = -\frac{d\omega}{\omega} = -\frac{dk}{k} = -\frac{dx}{x} = -\frac{d\nu}{\nu}$$

variable that is proportional to either frequency or wavelength:

$$R = R(p\nu, T)$$
$$R = R(l\lambda, T) \tag{1-11}$$

where p and l are constants of proportionality. A change in R can result from a change in the spectral variable or in the temperature, or both.

$$dR = \frac{\partial R}{\partial(p\nu)}d(p\nu) + \frac{\partial R}{\partial T}dT \tag{1-12}$$

or

$$dR = \frac{\partial R}{\partial (l\lambda)} d(l\lambda) + \frac{\partial R}{\partial T} dT \tag{1-13}$$

The second derivatives are

$$\frac{\partial^2 R}{\partial T^2}, \frac{\partial^2 R}{\partial (l\lambda)^2}, \frac{\partial^2 R}{\partial (pv)^2}, \frac{\partial^2 R}{\partial T \partial (l\lambda)}, \frac{\partial^2 R}{\partial T \partial (pv)} \tag{1-14}$$

The signs of any of the first three determine the nature of the inflection points whereas setting either of the third or fourth equal to zero will determine the spectral variable at which the change with temperature is an extremum.

The term $R(l\lambda, T)$ can be written in general (for photon flux density, energy flux density, energy or photon volume density, etc.) as

$$R = \text{constant } y^{\pm m} (e^x - 1)^{-1} \tag{1-15}$$

Then

$$\frac{dR}{R} = \left[\pm m \frac{dy}{y} + \frac{xe^x}{(e^x - 1)} \left(\frac{dT}{T} \mp \frac{dy}{y} \right) \right] \tag{1-16}$$

where $y = l\lambda$ corresponds to the upper signs (i.e., plus and minus)
$\quad\quad y = pv$ corresponds to the lower signs (i.e., minus and plus)

(See References [1-7, 1-8].) This gives the generalized expression for the Wien distribution law. Table 1-9 gives the values of the maxima for the different distributions.

The total integrals of the functions can also be considered in a general way

$$I_m = \int_0^\infty CT^l x^m (e^x - 1)^{-1} dx \tag{1-17}$$

and, if T is constant,

$$I_m = CT^l \int_0^\infty x^m (e^x - 1)^{-1} dx \tag{1-18}$$

These definite integrals are identifiable in terms of zeta functions:

$$\int_0^\infty x^m (e^x - 1)^{-1} dx = m! \zeta(m + 1) \tag{1-19}$$

Table 1-10 gives the values of $\zeta(m + 1)$ and $m! \zeta (m + 1)$ for m from 1 to 5. These values, multiplied by the constants for the appropriate distributions, are given in Table 1-11. The total integral is independent of the spectral scale. Thus, the total photon volume density is independent of scale. The energy density is independent of scale but different from the photon density. The photon exitance is $c/4$ times the density; the photon radiance is π^{-1} times the exitance. The results are summarized in Table 1-11.

Table 1-9. Maxima of the Dependent and Independent Variables for
Different Isothermal Planck Spectral Distributions

| Function | | m | x_{max} | R_{max} |
Dependent Variable	Independent Variable			
Photons	$\tilde{\nu}$	2	1.593624260	0.6476
Power	ν	3	2.821439372	1.4214
Photons	λ	4	3.920690395	4.7796
Power	λ	5	4.96511423	21.2036
Power contrast	λ	6	5.96940917	115.9359

Table 1-10. Values of $\zeta(m + 1)$ and $m!\zeta(m + 1)$

m	1	2	3	4	5
$\zeta(m + 1)$	$\pi^2/6$	1.2021	$\pi^4/90$	1.0369	$\pi^4/945$
$m!\zeta(m + 1)$	$\pi^2/6$	2.4041	$\pi^4/15$	24.9863	$8\pi^6/63$

Indefinite integrals are more complicated though just as important. For constant temperature, the function can be written as

$$R^m = Cx^m(e^x - 1)^{-1} = Cx^m \sum_{n=1}^{\infty} e^{-nx} \qquad (1\text{-}20)$$

Integration is done by parts:

$$I_m = \sum_{n=1}^{\infty} e^{-nx} x^m \sum_{l=1}^{n} \left[(mx)^l(n-1)!\right]^{-1} \qquad (1\text{-}21)$$

Table 1-12 gives the appropriate information for the different functions. The functions can be developed from an iterative formula

$$I_l = -\frac{x^l e^{-mx}}{m} + \frac{l}{m} I_{l-1} \qquad (1\text{-}22)$$

The maximum for any isothermal spectral scale is given by

$$\frac{xe^x}{e^x - 1} = m \qquad (1\text{-}23)$$

Table 1-11. Values of the Total Integrals of some Planck Functions

Distribution	Variable	Equation	Value
Photon density	n	$\int_0^\infty n_y dy$	$\dfrac{4}{c} M_q$
Energy density	w	$\int_0^\infty w_y dy$	$\dfrac{4}{c} M$
Photon exitance	M_q	$\int_0^\infty M_{qy} dy$	$\dfrac{2\pi k_B^3 T^3}{c^2 h^3}(2.4041)$
			$= 1.5202 \times 10^{11} T^3 \simeq \dfrac{\sigma T^3}{2.75 kT}$
Radiant exitance	M	$\int_0^\infty M_y dy$	$\dfrac{2\pi^5 k_B^4}{15 c^2 h^3} T^4 = \sigma T^4$
Photon sterance	L_q	$\int_0^\infty L_{qy} dy$	$\pi^{-1} M_q$
Radiance	L	$\int_0^\infty L_y dy$	$\pi^{-1} M$
Contrast	$\dfrac{1}{R}\dfrac{\partial R}{\partial T} dT$	$\dfrac{dT}{T}\int_0^\infty \dfrac{x e^x}{e^x - 1} dx = 1!\zeta(2)\dfrac{dT}{T} = \dfrac{\pi^2}{6}\dfrac{dT}{T}$	

Table 1-12. Solutions of the Integral of $x^l e^{-mx}$

l	I_l	Formula
0	I_0	$-\dfrac{e^{-mx}}{m}$
1	I_1	$-\dfrac{e^{-mx}}{m^2} = -\dfrac{x e^{-mx}}{m} + \dfrac{1}{m} I_0$
2	I_2	$-\dfrac{x e^{-mx}}{m} - \dfrac{2}{m}\dfrac{x e^{-mx}}{m} + \dfrac{2}{m}\dfrac{e^{-mx}}{m} = -\dfrac{x e^{-mx}}{m} + \dfrac{2}{m} I_0$
3	I_3	$-\dfrac{x e^{-mx}}{m} + \dfrac{3}{m} I_2$
4	I_4	$-\dfrac{x e^{-mx}}{m} + \dfrac{4}{m} I_3$

This is the generalized expression for the Wien law. The isospectral curves have a slope given by $(R/T) x e^x (e^x -1)^{-1}$.

In a similar way, it can be shown that all the mixed derivatives have a maximum determined by

$$\frac{x e^x}{e^x - 1} = \frac{x + m + 1}{2} \tag{1-24}$$

A good approximation is

$$\frac{x e^x}{e^x - 1} = m + 1 \tag{1-25}$$

The spectral maximum of the temperature change is related to the spectral maximum by

$$\frac{x + m + 1}{2m} \approx 1 + m^{-1} \tag{1-26}$$

For example, the radiant exitance expression has $m = 5$. Therefore $1 + m^{-1}$ is 1.2 and the maximum of the contrast is at a wavelength of approximately $(1.2)^{-1} \doteq 0.83$ of the wavelength of the maximum. A general treatment of this information is given (in part) in References [1-7, 1-8]. Table 1-13 gives values for the contrast maxima.

Table 1-13. Contrast Maxima Functions

Function		$m + 1$	x_{md}	x_{md}/x_{max}
Dependent variable	Independent variable			
$\dfrac{\partial M_{qv}}{\partial T}$	v	3	2.575678910	1.50
$\dfrac{\partial M_v}{\partial T}$	v	4	3.830016096	1.33
$\dfrac{\partial M_{q\lambda}}{\partial T}$	λ	5	4.928119359	1.25
$\dfrac{\partial M_\lambda}{\partial T}$	λ	6	5.969409172	1.20

1.2.4. Approximations of the Planck Function. Table 1-14 gives two approximations often used for the Planck law and its variation.

Table 1-14. Approximations of the Planck Function

	Large x (i.e., $x \gg 1$) $c_1 \lambda^{-5} e^x$ (Wien Law)	Small x (i.e., $x \ll 1$) $2\pi c k_B T \lambda^{-4}$ (Rayleigh-Jeans Law)
M_λ		
RE (relative error)	$\left[1 - \left(\dfrac{e^x - 1}{e^x} \right) \right]$	$\left[1 - \left(\dfrac{e^x - 1}{x} \right) \right]$

Other techniques can be used as approximating functions. The binomial or power series expansion which is used for obtaining the integrals in the pocket calculator portion of this chapter (Section 1.2.7) is accurate to about 0.1% for ten terms of the series even for values of $1/\lambda T$ greater than 10^8. The calculators are quite fast under these conditions. Other, more complicated procedures may be used with larger computers. Reference [1-9] is a good place to begin to investigate these techniques.

1.2.5. Fluctuations in the Planck Function [1-10]. It can be shown that for coherent radiation the fluctuations from the average in a mean square sense are

$$\overline{(R - \bar{R})^2} = D \frac{e^x}{(e^x - 1)^2} = D \sum m e^{-mx} \tag{1-27}$$

For noncoherent radiation, the mean square deviation is just equal to the mean

$$\overline{(R - \bar{R})^2} = D(e^x - 1)^{-1} = D \sum e^{-mx} \tag{1-28}$$

The maxima of these functions can be found in the usual way for the coherent case. The difference relative to the coherent case, RD, is

$$RD = 1 - \frac{(e^x - 1)^{-1}}{e^x(e^x - 1)^{-2}} = 1 - \frac{e^x - 1}{e^x} = e^{-x} \tag{1-29}$$

A general formulation can be written in terms of the degree of coherency, γ.

$$\overline{(R - \bar{R})^2} = D \frac{e^x - 1 + \gamma}{(e^x - 1)^2} \tag{1-30}$$

For coherent radiation $\gamma = 1$ and for incoherent radiation $\gamma = 0$.

1.2.6. Planck Curves. Visual impressions are often very useful. This section includes curves of various Planck and related functions. Figures 1-1, 1-2 and 1-3 provide information about the spectral distribution of blackbody radiation. The first of these is a linear curve with wavelength as the abscissa and M_λ as the ordinate. The second is semilogarithmic but for a different range of temperatures. The third is a curve plotted on a logarithmic wavelength scale and linear exitance scale in terms of the variable λT. Figure 1-4 is a log-log plot in which all Planck functions are the same shape and the Wien distribution law is a straight line. Thus a "do-it-yourself" slide rule can be constructed by putting an overlay on this figure, tracing the curve and the line and placing an index marker at 6000 K, the temperature of the top curve. Then, by keeping the lines overlapped and setting the index marker at the desired temperature, one finds that the template becomes the blackbody curve for that temperature. Figure 1-5 is the same sort of curve but for a different temperature region. The next group of figures includes the following: the integral of radiant exitance divided by total radiant exitance (Figure 1-6); the same function for photon exitance (Figure 1-7); the integral of the change of radiant exitance with temperature

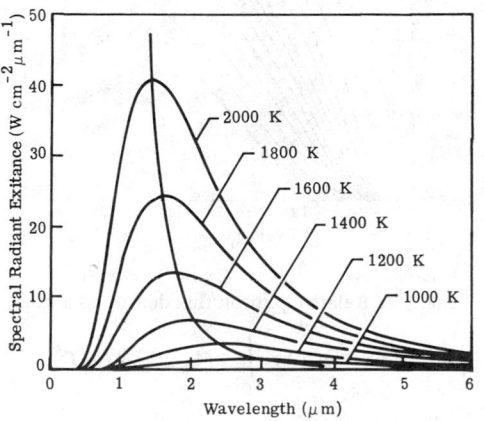

Fig. 1-1. Blackbody curves, 1000 to 2000 K.

Fig. 1-3. $M_\lambda/M_{\lambda_{max}}$ versus T for a blackbody.

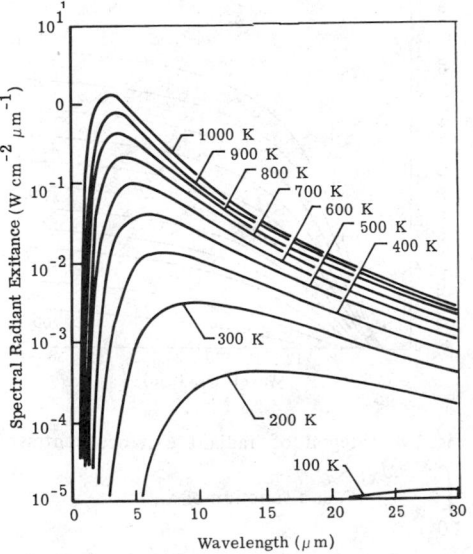

Fig. 1-2. Blackbody curves, 100 to 1000 K.

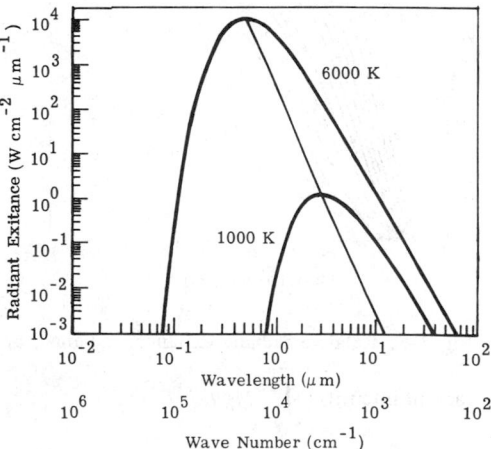

Fig. 1-4. M_λ versus λ and ν.

(Figure 1-8); and the integral of the change in photon exitance with temperature (Figure 1-9).

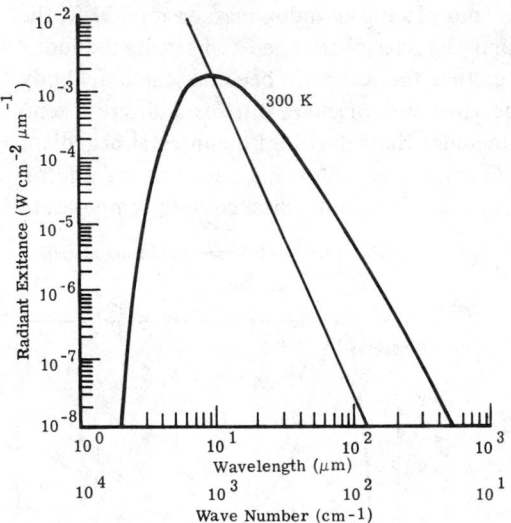

Fig. 1-5. M_λ versus λ and ν.

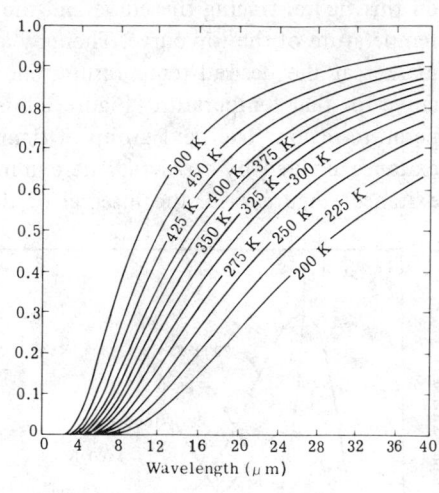

Fig. 1-7. Relative photon flux density as a function of wavelength: $\int_0^\lambda M_{q,\lambda}(\lambda)d\lambda/\sigma_q T^3$.

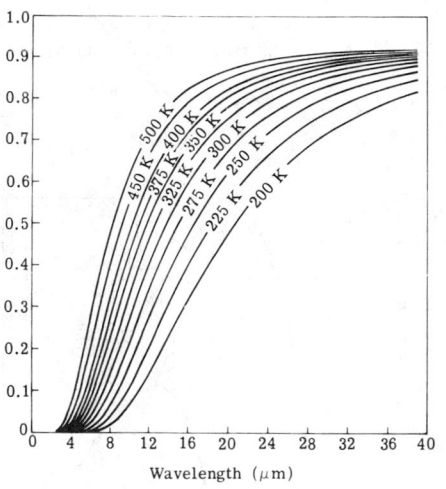

Fig. 1-6. Relative radiant exitance, radiance, or radiant intensity: $\int_0^\lambda M_\lambda d\lambda/\sigma T^4$.

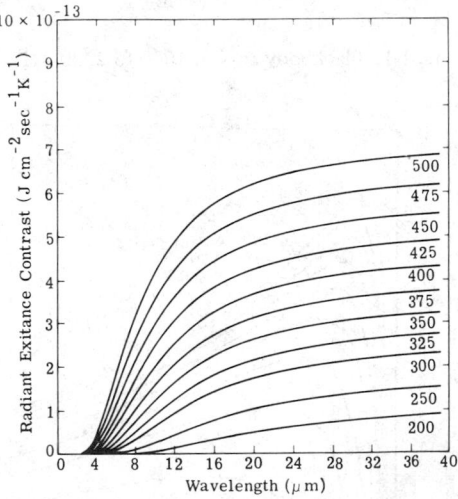

Fig. 1-8. Integral of radiant exitance contrast $\int_0^\lambda \frac{\partial M_\lambda}{\partial T} d\lambda$ as a function of λ.

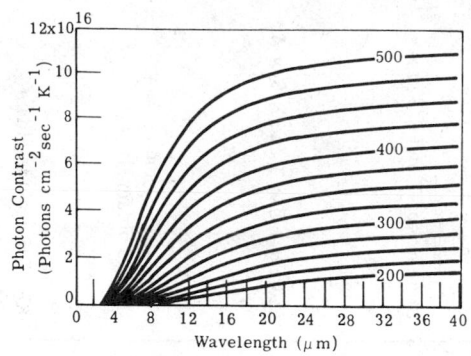

Fig. 1-9. The integral of photon contrast

$$\int_0^\lambda \frac{\partial M_{q,\lambda}}{\partial T} \, d\lambda \text{ as a function of } \lambda.$$

1.2.7. Pocket Calculator Programs. Several pocket calculator programs are presented here. They were written for the Hewlett-Packard HP-25 and the Texas Instruments SR-56 and SR-52 calculators. The programs reflect differences in programming between Reverse Polish Notation (RPN) and Algebraic Operation (AO) systems, as well as differences resulting from calculators which can and cannot store programs on cards. Table 1-15 gives the equations, units, a catalog of the programs, and initialization and starting procedure. In most cases some of the storage registers need to be filled at the start with terms indicated by quantities *not* in parentheses. Those registers which show a parenthesized expression are used in the calculation, but do not need to be initialized. In the programs for integrals, a value of ϵ, the allowed fractional value of the next term of the series, needs to be entered. For most problems and their required accuracies, 10^{-9} is an appropriate value for this.

For additional information on programming pocket calculators, see References [1-11, 1-12].

Table 1-15
Pocket Calculator Programs

Equation	Units	For Program for:			Instructions
		HP-25 TABLE	SR-56† TABLE	SR-52*† TABLE	
$M = \sigma T^4$ $= 5.67032 \times 10^{-12}\, T^4$	[W cm^{-2}]	1-16	1-17	—	Store σ. Key T and R/S or A
$M_q = \sigma_q T^3$ $= 1.5204 \times 10^{11}\, T^3$	[sec^{-1} cm^{-2}]	1-18	1-19	—	Store σ_q. Key T and R/S or A
$M_\lambda = c_1 \lambda^{-5} (e^x - 1)^{-1}$ $= 3.741832 \times 10^{-12}\, \lambda^{-5} (e^x - 1)^{-1}$	[W cm^{-3}]	1-20	1-21	—	Store λ, T, c_2 and c_1. Key R/S or A to obtain M_λ.
$\dfrac{\partial M_\lambda}{\partial T} = x e^x (e^x - 1)^{-1} T^{-1} M_\lambda$	[W cm^{-3} K^{-1}]	1-20	—	—	Key R/S again to obtain $\left(\partial M_\lambda / \partial T\right)$
$M_{q\lambda} = c_1 (ch)^{-1} \lambda^{-4} (e^x - 1)^{-1}$ $= 1.883652 \times 10^{11}\, \lambda^{-4} (e^x - 1)^{-1}$	[sec^{-1} cm^{-3}]	1-22	—	—	Store λ, T, c_2 and c. Key R/S to obtain $M_{q\lambda}$.
$\dfrac{\partial M_{q\lambda}}{\partial T} = x e^x (e^x - 1)^{-1} T^{-1} M_{q\lambda}$	[sec^{-1} cm^{-3} K^{-1}]	1-22	—	—	Key R/S again to obtain $\left(\partial M_{q\lambda} / \partial T\right)$

Equation	Units			Instructions
$M_\nu = c_1 c^{-4} \nu^3 (e^x - 1)^{-1}$ $= 4.632351 \times 10^{-54} \nu^3 (e^x - 1)^{-1}$	[W cm^{-2} Hz^{-1}]	1-23	—	Store ν, T, c_2, c, and c_1. Key R/S to obtain M_ν.
$\dfrac{\partial M_\nu}{\partial T} = x e^x (e^x - 1)^{-1} M_\nu$	[W cm^{-2} Hz^{-1} K^{-1}]	1-23	—	Key R/S again to obtain $(\partial M_\nu / \partial T)$
$M_{q\nu} = c_1 c^{-4} h^{-1} \nu^2 (e^x - 1)^{-1}$ $= 6.990987 \times 10^{-21} \nu^2 (e^x - 1)^{-1}$	[sec^{-1} cm^{-2} Hz^{-1}]	1-24	—	Store ν, T, c_2 and c. Key R/S to obtain $M_{q\nu}$
$\dfrac{\partial M_{q\nu}}{\partial T} = x e^x (e^x - 1)^{-1} T^{-1} M_{q\nu}$	[sec^{-1} cm^{-2} Hz^{-1} K^{-1}]	1-24	—	Key R/S again to obtain $(\partial M_{q\nu} / \partial T)$
$M_{\Delta y} = \displaystyle\int_{y_1}^{y_2} M_y \, dy = c_1 (T/c_2)^4 \sum_1$ $= 8.7317134 \times 10^{-13} T^4 \sum_1$	[W cm^{-2}]	1-25	—	Calculate and store: x_1, x_2, $(c_1 c_2^{-4})$, and ϵ. Key T and R/S to obtain $M_{\Delta y}$
$\dfrac{\partial M_{\Delta y}}{\partial T} = \displaystyle\int_{y_1}^{y_2} \dfrac{\partial M_y}{\partial T} \, dy = c_1 (T/c_2)^4 T^{-1} \sum_3$ $= 8.7317134 \times 10^{-13} T^3 \sum_3$	[W cm^{-2} K^{-1}]	1-26	—	Calculate and Store: x_1, x_2, $(c_1 T^3 c_2^{-4})$, and ϵ. Key R/S and, after program-run stops, Key [−], [RCL 1] and [×] to obtain $(\partial M_{\Delta y}/\partial T)$
$L_{\Delta \lambda} = \displaystyle\int_{\lambda_1}^{\lambda_2} L_\lambda \, d\lambda = \dfrac{c_1 T^4}{\pi \; c_2^4} \sum_1$ $= 2.7794 \times 10^{-13} T^4 \sum_1$	[W cm^{-2} sr^{-1}]	—	1-27	Key T, A, λ_1, B, λ_2, C and D to obtain $L_{\Delta\lambda}$. Card-stored constants are limited by the selected value, $c_2 = 14388$. λ is in μm.

*Programs stored on magnetic cards.

†Programs for the TI SR calculators were compiled by Anthony J. LaRocca and George H. Lindquist, Environmental Research Institute of Michigan.

Table 1-15 (Continued)
Pocket Calculator Programs

Equation	Units	For Program for:			Instructions
		HP-25	SR-56	SR-52*	
		TABLE	TABLE	TABLE	
$\dfrac{\partial L_{\Delta\lambda}}{\partial T} = \displaystyle\int_{\lambda_1}^{\lambda_2} (\partial L_\lambda/\partial T)d\lambda = \dfrac{c_1}{\pi}\left(\dfrac{T}{c_2}\right)^4 T^{-1} \sum_3$ $= 2.7794 \times 10^{-13} T^3 \sum_3$	[W cm^{-1} sr^{-1} K^{-1}]	—	—	1-27	Key T, A, λ, B, λ$_2$, C and *D' to obtain $\partial L_{\Delta\lambda}/\partial T$.
$M_{q\Delta y} = \displaystyle\int_{y_1}^{y_2} M_{qy}\, dy = 2\pi c(T/c_2)^3 \sum_2$ $= 6.324287 \times 10^{10} T^3 \sum_2$	[sec^{-1} cm^{-2}]	1-28	—	—	Calculate and Store: $x_1, x_2, (2\pi c\, c_2^{-3})$, and ϵ. Key T and R/S to obtain $M_{q\Delta y}$.
$M_{q\Delta\lambda} = \displaystyle\int_{\lambda_1}^{\lambda_2} M_{q\lambda}\, d\lambda = 2\pi c(T/c_2)^3 \sum_2$ $= 6.324287 \times 10^{10} T^3 \sum_2$	[sec^{-1} cm^{-2}]	—	—	1-29	Key T, A, λ$_1$, B, λ$_2$, C, ϵ and D to obtain $M_{q\Delta\lambda}$. Card stored constants are c = 2.99793 × 10^{10} and c$_2$ = 1.4388. λ is in μm.
$\dfrac{\partial M_{q\Delta y}}{\partial T} = \displaystyle\int_{y_1}^{y_2} \dfrac{\partial M_{qy}}{\partial T} dy = 2\pi c(T/c_2)^3 T^{-1} \sum_4$ $= 6.324287 \times 10^{10} T^2 \sum_4$	[sec^{-1} cm^{-2} K^{-1}]	Modified 1-25	—	—	Modify Table 1-25 as follows:

Line	Key Entry	Registers
02	2	R$_1$: $2\pi c\, c_2^{-3}$
27	3	

Store $x_1, x_2, (2\pi c\, c_2^{-3})$ and ϵ. Key T and R/S to obtain $M_{q\Delta y}$

$$\Sigma_1 = \sum_{m=1}^{8} e^{-u_2} m^{-4}(u_2^3 + 3u_2^2 + 6u_2 + 6) - \sum_{m=1}^{8} e^{-u_1} m^{-4}(u_1^3 + 3u_1^2 + 6u_1 + 6)$$

$$\Sigma_2 = \sum_{m=1}^{8} e^{-u_2} m^{-3}(u_2^2 + 2u_2 + 2) - \sum_{m=1}^{8} e^{-u_1} m^{-3}(u_1^2 + 2u_1 + 2)$$

$$\Sigma_3 = \sum_{m=1}^{8} e^{-u_2} m^{-4}(u_2^4 + 4u_2^3 + 12u_2^2 + 24u_2 + 24) - \sum_{m=1}^{8} e^{-u_1} m^{-4}(u_1^4 + 4u_1^3 + 12u_1^2 + 24u_1 + 24)$$

$$\Sigma_4 = \sum_{m=1}^{8} e^{-u_2} m^{-3}(u_2^3 + 3u_2^2 + 6u_2 + 6) - \sum_{m=1}^{8} e^{-u_1} m^{-3}(u_1^3 + 3u_1^2 + 6u_1 + 6)$$

$u_i = mx_i; i = 1, 2$

$x = c_2/(\lambda T) = c_2 \nu/(cT)$**

λ = wavelength in cm (except for SR-52 programs)

ν = frequency in Hz = c/λ

T = Temperature in K

$c_1 = 3.741832 \times 10^{-12} = 2\pi c^2 h$ [W cm²]

$c_2 = 1.438786 = hc/k_B$ [cm K]

$k_B = 1.380662 \times 10^{-23}$ [J K^{-1}]

$h = 6.626176 \times 10^{-34}$ [J sec]

$c = 2.997924580 \times 10^{10}$ [cm sec^{-1}]

$$\sigma = 5.67033 \times 10^{-12} = \frac{2\pi^5 k_B^4}{15 h^3 c^2} \quad [\text{W cm}^{-2}\,\text{K}^{-4}]$$

$$\sigma_q = \frac{2\pi k_B^3}{h^3 c^2}(2.4041) = 1.5204 \times 10^{11} \quad [\text{sec}^{-1}\,\text{cm}^{-2}\,\text{K}^{-3}]$$

*Programs stored on magnetic cards.

**For values of x > 100, the series converges slowly and is subject to rounding errors.

Table 1-16
Stefan-Boltzmann Law
HP-25, RPN

Line	Key Entry	Registers
01	↑	
02	4	$R_0 : \sigma$
03	y^x	
04	RCL 0	
05	×	

Table 1-17
Stefan-Boltzmann Law
SR-56, AO

LOC	KEY	Registers
000	*LBL	
001	A	$R_{00} :$
002	y^x	
003	4	$R_{01} : \sigma$
004	×	
005	RCL	
006	0	
007	1	
008	=	
009	HLT	

*Denotes second function key.

Table 1-18
Stefan-Boltzmann Law –
Photons:
HP-25, RPN

Line	Key Entry	Registers
01	↑	
02	3	$R_0 : \sigma_q$
03	y^x	
04	RCL 0	
05	×	

Table 1-19
Stefan-Boltzmann Law –
Photons
SR-56, AO

LOC	KEY	Registers
000	*LBL	
001	A	$R_{00} :$
002	y^x	
003	3	$R_{01} : \sigma_q$
004	×	
005	RCL	
006	0	
007	1	
008	=	
009	HLT	

*Denotes second function key.

Table 1-20
M_λ and $\partial M_\lambda / \partial T$
HP-25, RPN

Line	Key Entry	Registers
01	RCL 3	$R_0 : \lambda$
02	RCL 0	
03	5	$R_1 : T$
04	y^x	
05	÷	$R_2 : c_2$
06	RCL 2	
07	RCL 0	$R_3 : c_1$
08	RCL 1	
09	×	$R_4 : (x)$
10	÷	
11	STO 4	
12	e^x	
13	1	
14	–	
15	÷	
16	R/S	
17	RCL 4	
18	RCL 1	
19	÷	
20	RCL 4	
21	CHS	
22	e^x	
23	CHS	
24	1	
25	+	
26	÷	
27	×	

Table 1-21
M_λ
SR-56, AO

LOC	KEY	Registers
000	*LBL	
001	A	$R_{00} :$
002	RCL	
003	0	$R_{01} : T$
004	3	
005	×	$R_{02} : \lambda$
006	RCL	
007	0	$R_{03} : c_1$
008	2	
009	y^x	$R_{04} : c_2$
010	5	
011	+/–	$R_{05} :$
012	=	
013	÷	
014	(	
015	(	
016	RCL	
017	0	

Table 1-21 (Continued)

018	4
019	÷
020	RCL
021	0
022	2
023	÷
024	RCL
025	0
026	1
027	)
028	INV
029	ln x
030	–
031	1
032	=
033	HLT

*Denotes second function key.

Table 1-22
$M_{q\lambda}$ and $\partial M_{q\lambda} / \partial T$
HP-25, RPN

Line	Key Entry	Registers
01	RCL 3	$R_0 : \lambda$
02	RCL 0	
03	4	$R_1 : T$
04	y^x	
05	÷	$R_2 : c_2$
06	2	
07	×	$R_3 : c$
08	π	
09	×	$R_4 : (x)$
10	RCL 2	
11	RCL 0	
12	÷	
13	RCL 1	
14	÷	
15	STO 4	
16	e^x	
17	1	
18	–	
19	÷	
20	R/S	
21	RCL 4	
22	RCL 1	
23	÷	
24	RCL 4	
25	CHS	
26	e^x	
27	CHS	
28	1	
29	+	
30	÷	
31	×	

Table 1-23
M_ν and $\partial M_\nu/\partial T$
HP-25, RPN

Line	Key Entry	Registers
01	RCL 4	$R_0{:}\nu$
02	RCL 3	
03	4	$R_1{:}T$
04	y^x	
05	÷	$R_2{:}c_2$
06	RCL 0	
07	3	$R_3{:}c$
08	y^x	
09	×	$R_4{:}c_1$
10	RCL 2	
11	RCL 3	$R_5{:}(x)$
12	÷	
13	RCL 0	
14	×	
15	RCL 1	
16	÷	
17	STO 5	
18	e^x	
19	1	
20	−	
21	÷	
22	R/S	
23	RCL 5	
24	RCL 1	
25	÷	
26	RCL 5	
27	CHS	
28	e^x	
29	CHS	
30	1	
31	+	
32	÷	
33	×	

Table 1-24
$M_{q\nu}$ and $\partial M_{q\nu}/\partial T$
HP-25, RPN

Line	Key Entry	Registers
01	2	$R_0{:}\nu$
02	π	
03	×	$R_1{:}T$
04	RCL 3	
05	x^2	$R_2{:}c_2$
06	÷	
07	RCL 0	$R_3{:}c$
08	x^2	

Table 1-24 (Continued)

Line	Key Entry	Registers
09	×	
10	RCL 0	$R_4{:}(x)$
11	RCL 2	
12	×	
13	RCL 3	
14	÷	
15	RCL 1	
16	÷	
17	STO 4	
18	e^x	
19	1	
20	−	
21	÷	
22	R/S	
23	RCL 4	
24	RCL 1	
25	÷	
26	RCL 5	
27	CHS	
28	e^x	
29	CHS	
30	1	
31	+	
32	÷	
33	×	

Table 1-25
$$\int_{y_1}^{y_2} M_y \, dy$$
HP-25, RPN

Line	Key Entry	Registers
01	↑	$R_0{:}x_1$
02	4	
03	y^x	$R_1{:}c_1 c_2^{-4}$
04	STO × 1	
05	RCL 0	$R_2{:}\epsilon$
06	1	
07	RCL 7	$R_3{:}x_2$
08	+	
09	STO 7	$R_4{:}(\Sigma)$
10	×	
11	STO 6	$R_5{:}(\Sigma)$
12	x^2	
13	RCL 6	$R_6{:}(u)$
14	3	
15	+	$R_7{:}(m)$
16	×	
17	RCL 6	

Table 1-25 (Continued)

Line	Key Entry
18	1
19	+
20	6
21	×
22	+
23	RCL 6
24	e^x
25	÷
26	RCL 7
27	4
28	y^x
29	÷
30	STO + 5
31	RCL 5
32	÷
33	RCL 2
34	x < y
35	GTO 05
36	RCL 4
37	x ≠ 0
38	GTO 46
39	RCL 5
40	STO 4
41	0
42	STO 5
43	STO 7
44	RCL 3
45	GTO 06
46	RCL 5
47	−
48	RCL 1
49	×

Table 1-26
$$\frac{\partial M_{\Delta y}}{\partial T} = \int_{y_1}^{y_2} (\partial M_y/\partial T) \, dy$$
HP-25, RPN

Line	Key Entry	Registers
01	RCL 0	$R_0{:}x_1$
02	1	$R_1{:}c_1 T^3 c_2^{-4}$
03	RCL 7	
04	+	$R_2{:}\epsilon$
05	STO 7	
06	×	$R_3{:}x_2$
07	STO 6	
08	4	$R_4{:}(\Sigma)$
09	+	

Table 1-26. (Continued)

Table 1-26 (Continued)

LOC	KEY	Registers
10	RCL 6	$R_5:(\Sigma)$
11	3	
12	y^x	$R_6:(u)$
13	×	
14	RCL 6	$R_7:(m)$
15	2	
16	+	
17	RCL 6	
18	×	
19	2	
20	×	
21	4	
22	+	
23	6	
24	×	
25	+	
26	RCL 6	
27	e^x	
28	÷	
29	RCL 7	
30	4	
31	.y^x	
32	÷	
33	STO + 5	
34	RCL 5	
35	÷	
36	RCL 2	
37	x < y	
38	GTO 01	
39	RCL 4	
40	x ≠ 0	
41	GTO 49	
42	RCL 5	
43	STO 4	
44	0	
45	STO 5	
46	STO 7	
47	RCL 3	
48	GTO 02	
49	RCL 5	
After Run Stops:		
(49+1)	-	
(49+2)	RCL 1	
(49+3)	×	

Table 1-27†

$$\int_{\lambda_1}^{\lambda_2} L_\lambda \, d\lambda \quad \text{and} \quad \int_{\lambda_1}^{\lambda_2} (\partial L_\lambda/\partial T) \, d\lambda$$

SR-52, AO (card)

LOC	KEY	Registers
000	*LBL	R_{00}(# of terms)
001	A	
002	STO	$R_{01}:T$
003	0	
004	1	$R_{02}:\lambda_1$
005	HLT	
006	*LBL	$R_{03}:(c_2/\lambda_1 T)$
007	B	
008	STO	$R_{04}:\lambda_2$
009	0	
010	2	$R_{05}:$
011	HLT	
012	*LBL	R_{06}
013	C	
014	STO	R_{07}
015	0	
016	4	R_{08}
017	HLT	
018	*LBL	R_{09}
019	D	
020	*set flg	R_{10}
021	0	
022	*LBL	R_{11}
023	*8'	
024	RCL	
025	0	R_{12}
026	2	
027	SBR	R_{13}
028	*4'	
029	STO	R_{14} (mx)
030	0	
031	6	R_{15}
032	RCL	
033	0	$R_{16}:(\Sigma)$
034	4	
035	SBR	
036	*4'	
037	-	
038	RCL	
039	0	
040	6	
041	=	
042	×	
043	RCL	
044	0	
045	1	
046	y^x	
047	4	
048	×	
049	2	

LOC	KEY
050	7
051	7
052	9
053	4
054	EE
055	1
056	7
057	+/-
058	=
059	HLT
060	*LBL
061	*D'
062	INV
063	*set flg
064	0
065	GTO
066	*8'
067	*LBL
068	*4'
069	*if zro
070	*3'
071	*1/x
072	×
073	1
074	4
075	3
076	8
077	8
078	÷
079	RCL
080	0
081	1
082	=
083	*LBL
084	E
085	STO
086	0
087	3
088	-
089	1
090	0
091	0
092	=
093	*if pos
094	*3'
095	0
096	STO
097	1
098	6
099	STO
100	1

LOC	KEY
101	4
102	3
103	5
104	STO
105	0
106	0
107	*LBL
108	*1'
109	RCL
110	0
111	3
112	SUM
113	1
114	4
115	INV
116	*dsz
117	*2'
118	RCL
119	1
120	4
121	*1/x
122	+
123	1
124	=
125	×
126	2
127	÷
128	RCL
129	1
130	4
131	+
132	1
133	=
134	×
135	3
136	÷
137	RCL
138	1
139	4
140	+
141	1
142	=
143	÷
144	RCL
145	1
146	4
147	×
148	RCL
149	1
150	4
151	+/-
152	INV

Table 1-27 (Continued)

FOOTNOTES TO TABLE 1-27:

*Denotes second function key.

†The program in Table 1-27 includes a fixed value of $\epsilon = 10^{-6}$ and limits the series (Σ) to 35 terms. Thus, calculation is accurate to at least four significant numbers for x < 0.1 and five for 0.1 < x < 100. Convergence of (Σ) can be checked by use of a sub-routine namely: Key[*st flg], 0, x_1, or x_2, and E. The number of terms to obtain the displayed value is given by 35 minus the accumulated contents of R_{00}.

Table 1-27. (Continued)

153	ln x
154	=
155	SUM
156	1
157	6
158	÷
159	RCL
160	1
161	6
162	-
163	1
164	EE
165	6
166	+/-
167	=
168	'if pos'
169	*1'
170	*LBL
171	*2'
172	RCL
173	1
174	6
175	*if flg
176	0
177	*6'
178	x
179	4
180	+
181	(
182	RCL
183	0
184	3
185	INV
186	ln x
187	-
188	1
189	)
190	*1/x
191	=
192	÷
193	RCL
194	0
195	1
196	=
197	*LBL
198	*6'
199	x
200	RCL
201	0
202	3
203	y^x
204	4
205	=
206	*RTN
207	*LBL
208	*3'
209	0
210	*RTN

Table 1-28

$$M_{q\Delta y} = \int_{y_1}^{y_2} M_{qy}\, dy$$

HP-25, RPN

Line	Key Entry	Registers
01	↑	$R_0{:}x_1$
02	3	
03	y^x	$R_1 \cdot \dfrac{2\pi c}{c_2^3}$
04	STO × 1	
05	RCL 0	$R_2{:}\epsilon$
06	1	
07	RCL 7	$R_3{:}x_2$
08	+	
09	STO 7	$R_4{:}(\Sigma)$
10	×	
11	STO 6	$R_5{:}(\Sigma)$
12	x^2	
13	RCL 6	$R_6{:}(u)$
14	↑	
15	+	$R_7{:}(m)$
16	+	
17	2	
18	+	
19	RCL 6	
20	e^x	
21	÷	
22	RCL 7	
23	3	
24	y^x	
25	÷	
26	STO + 5	
27	RCL 5	
28	÷	
29	RCL 2	
30	x < y	
31	GTO 05	
32	RCL 4	
33	x ≠ 0	
34	GTO 42	
35	RCL 5	
36	STO 4	
37	0	
38	STO 5	
39	STO 7	
40	RCL 3	
41	STO 0	
42	GTO 06	
43	RCL 5	
44	-	
45	RCL 1	
46	×	

Table 1-29

$$M_{q\Delta\lambda} = \int_{\lambda_1}^{\lambda_2} M_{q\lambda}\, d\lambda$$

SR-52, AO (Card)

LOC	KEY	Registers
000	*LBL	R_{00}
001	A	
002	STO	R_{01}
003	0	
004	3	$R_{02}{:}(c_2)$
005	HLT	
006	*LBL	$R_{03}{:}T$
007	B	
008	×	$R_{04}{:}\lambda_1$
009	1	
010	EE	$R_{05}{:}\lambda_2$
011	+/-	
012	4	$R_{06}{:}\epsilon$
013	=	
014	STO	R_{07}
015	0	
016	4	R_{08}
017	HLT	
018	*LBL	$R_{09}{:}(c_2/T)$
019	C	
020	×	$R_{10}{:}(x)$
021	1	
022	EE	$R_{11}{:}(2\pi c(T/c_2)^3)$
023	+/-	
024	4	$R_{12}{:}(m)$
025	=	
026	STO	$R_{13}{:}(u)$
027	0	
028	5	$R_{14}{:}(\Sigma)$
029	HLT	
030	*LBL	$R_{15}{:}\left(\int_0^\lambda M_{q\lambda}\, d\lambda\right)$
031	D	
032	STO	
033	0	
034	6	
035	1	
036	.	
037	4	
038	3	
039	8	
040	8	

*Denotes second function key.
Table 1-29. (Continued)

Table 1-29. (Continued)

041	STO	085	5	129	2	173	1
042	0	086	RCL	130	STO	174	2
043	2	087	0	131	1	175	y^x
044	÷	088	4	132	4	176	3
045	RCL	089	*if zro	133	SBR	177	+/-
046	0	090	1	134	*5'	178	×
047	3	091	1	135	-	179	(
048	=	092	1	136	RCL	180	RCL
049	STO	093	*1/x	137	1	181	1
050	0	094	×	138	5	182	3
051	9	095	RCL	139	=	183	y^x
052	2	096	0	140	HLT	184	2
053	.	097	9	141	0	185	+
054	9	098	=	142	0	186	2
055	9	099	STO	143	0	187	×
956	7	100	1	144	0	188	RCL
057	9	101	0	145	0	189	1
058	3	102	0	146	0	190	3
059	EE	103	STO	147	0	191	+
060	1	104	1	148	0	192	2
061	0	105	2	149	*LBL	193	)
062	×	106	STO	150	*5'	194	=
063	*π	107	1	151	*LBL	195	SUM
064	×	108	4	152	E	196	1
065	2	109	SBR	153	1	197	4
066	×	110	*5'	154	SUM	198	÷
067	(	111	STO	155	1	199	RCL
068	RCL	112	1	156	2	200	1
069	0	113	5	157	RCL	201	4
070	3	114	RCL	158	1	202	-
071	÷	115	0	159	2	203	RCL
072	RCL	116	5	160	×	204	0
073	0	117	*1/x	161	RCL	205	6
074	2	118	×	162	1	206	=
075	)	119	RCL	163	0	207	*if pos
076	y^x	120	0	164	=	208	E
077	3	121	9	165	STO	209	RCL
078	=	122	=	166	1	210	1
079	STO	123	STO	167	3	211	4
080	1	124	1	168	+/-	212	×
081	1	125	0	169	INV	213	RCL
082	0	126	0	170	ln x	214	1
083	STO	127	STO	171	×	215	1
084	1	128	1	172	RCL	216	=
						217	*rtn

1.2.8. Radiation Slide Rules. Rules have been devised for rapid, fairly accurate calculations of radiometric quantities.

The General Electric Rule. This rule is designated GEN-15C*. (See Figure 1-10.) Calculations which can be made on the rule are as follows:

(1) Conversions of temperatures among Celsius, Kelvin, Fahrenheit, and Rankine by setting the temperature on one scale and reading it on another. Use scales ABKL.

(2) Multiplication by the use of standard C and D log scales.

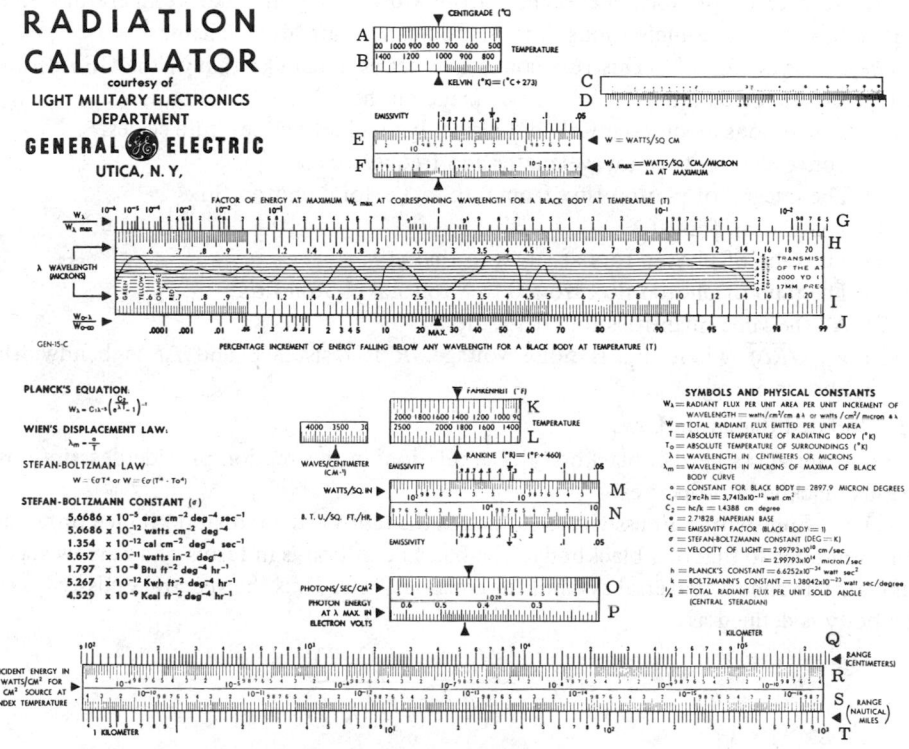

Fig. 1-10. The GE slide rule.

(3) Total blackbody radiant exitance by setting the temperature of the blackbody source on a temperature scale and reading on the E scale (W cm^{-2}). An emissivity scale associated with the E scale permits direct calculation for graybodies; the value on the E scale is read under the appropriate emissivity.

(4) Incremental blackbody radiant exitance $W_{\Delta\lambda}$ at maximum. The power density for 1 μm bandpass can be read directly from the $W_{\lambda max}$ or F scale.

(5) The ratio of W_λ at any wavelength, λ, to that at λ_{max}, $W_\lambda/W_{\lambda max}$. The temperature scale; then $W_\lambda/W_{\lambda max}$ is read from the G scale opposite the desired λ on the H scale. Thus one can find $W_{\lambda max}$ for a given λ on the $W_{\lambda max}$ scale and then calculate the value of W_λ at any wavelength on the $W_\lambda/W_{\lambda max}$ scale.

---

*Available from the General Electric Company, Schenectady, NY, for about four dollars.

(6) The blackbody radiation in any spectral interval. The temperature scale is set at the appropriate temperature. Then on the $W_{0-\lambda}/W_{0-\infty}$, or J, scale, one reads the percentage radiation that lies below a particular wavelength, λ_1, on the I scale. The same is done for λ_2, with subtraction.

(7) Conversion of range in nautical miles to range in centimeters with the aid of a straight edge, and vacuum calculation of irradiance. These can be made with the QRST scales.

(8) Conversion from W in.$^{-2}$ to Btu ft^{-2} h^{-1}.

(9) Number of photons sec^{-1} cm^{-2} from a blackbody at index temperature. Useful constants and other combinations of these calculations are also available.

The "Cussen" Rule*. This rule consists of scales much like that of the GE rule but is more accurate. Calculations which can be made on the rule are as follows:

(1) Conversions among the Kelvin, Celsius, Fahrenheit temperature scales.
(2) Conversions between wavelengths and frequencies .
(3) The integral of photon flux from 0 to ∞, i.e. total photon flux.
(4) The photon flux at λ_{max}.
(5) The ratio of photon flux at any λ to that at λ_{max}.
(6) The ratio of photon flux from 0 to λ to total photon flux.
(7) All the same quantities for energy flux.
(8) $V_n/\sqrt{R\Delta f}$ where V_n is noise voltage, R is resistance and Δf is bandwidth.
(9) Photon energy.

1.3. Related Radiation Laws

No material is a perfect blackbody. The relations in this section provide descriptions of various blackbody properties.

1.3.1. Emissivity.
Emissivity is defined as the ratio of the radiant exitance or radiance of a given body to that of a blackbody. The basic definition is in terms of a narrow spectral interval. The almost universal symbol for emissivity is ϵ. Thus the spectral emissivity, $\epsilon(\lambda)$, of a body is defined as

$$\epsilon(\lambda) \equiv \frac{M_\lambda}{M_\lambda^{BB}} = \frac{M_\lambda(\lambda, T)}{M_\lambda^{BB}(\lambda, T)} \tag{1-31}$$

where λ and T must be alike for both the numerator and denominator. The directional spectral emissivity is the ratio of radiances.

$$\epsilon(\lambda, \theta, \phi) = \frac{L_\lambda(\theta, \phi)}{L_\lambda^{BB}(\theta, \phi)} \tag{1-32}$$

Over a spectral interval the emissivity is

$$\bar{\epsilon} \equiv \frac{\int_{\Delta\lambda} L_\lambda d_\lambda}{\int_{\Delta\lambda} L_\lambda^{BB} d_\lambda} = \frac{\int_{\Delta\lambda} \epsilon(\lambda) L_\lambda^{BB} d\lambda}{\int_{\Delta\lambda} L_\lambda^{BB} d\lambda} \tag{1-33}$$

---

*Available from Electro-Optical Industries, Inc., Santa Barbara, CA, for about fifty dollars.

Over the entire spectral band

$$\epsilon \equiv \frac{\int_0^\infty \epsilon(\lambda) L_\lambda^{BB}\, d\lambda}{\pi^{-1}\sigma T^4} \qquad (1\text{-}34)$$

The same definitions apply to photon distributions

$$\epsilon_q \equiv \frac{\int \epsilon(\lambda) L_{q\lambda}^{BB}\, d\lambda}{\sigma T^4 (2.75 k_B T)^{-1}} = \frac{\int \epsilon(\lambda) L_{q\lambda}^{BB}\, d\lambda}{1.5202 \times 10^{11}\, T^3} \qquad (1\text{-}35)$$

$$\overline{\epsilon}_q(\lambda) = \frac{\int \epsilon(\lambda) \lambda^{-4} (e^x - 1)^{-1}\, d\lambda}{\int \lambda^{-4}(e^x - 1)^{-1}\, d\lambda} \qquad (1\text{-}36)$$

For any material, these two different emissivities are not the same.

A material which has an emissivity that is independent of wavelength is often called a *gray body*. Bodies which have an emissivity which varies with wavelength are often called *spectral* or *colored bodies*.

1.3.2. Total Power Law. When radiation is incident upon a body, some of it is transmitted, some absorbed, and some is reflected. Thus, the ratios of each of these to the incident power must add up to unity:

$$\alpha + \rho + \tau = 1 \qquad (1\text{-}37)$$

or

$$\Phi_{\text{absorbed}} + \Phi_{\text{reflected}} + \Phi_{\text{transmitted}} = \Phi_{\text{incident}} \qquad (1\text{-}38)$$

where $\alpha = \Phi_{\text{absorbed}}/\Phi_{\text{incident}}$ = absorptivity
$\rho = \Phi_{\text{reflected}}/\Phi_{\text{incident}}$ = reflectivity
$\tau = \Phi_{\text{transmitted}}/\Phi_{\text{incident}}$ = transmissivity

It is also true that the spectral quantities add up to one

$$\alpha(\lambda) + \rho(\lambda) + \tau(\lambda) = 1 \qquad (1\text{-}39)$$

The interpretation is simple for a relatively transparent, smooth, plane parallel plate in which the absorbed flux is small enough that the plate does not increase significantly in temperature. For bodies of irregular shape, the reflected radiation must be considered as all the radiation that is scattered or reflected in the entire sphere. Under any circumstances, the absorbed radiation can increase the temperature of the body thereby increasing the radiation in all spectral bands. For total radiation, these components will act in such a way that the total power remains constant.

The conditions do not apply generally for two different components of polarization.

1.3.3. Kirchhoff's Law [1-13]. By considering two bodies in thermal equilibrium, one a blackbody and one arbitrary, one can show that

$$\alpha = \int_0^\infty \alpha(\lambda)\, d\lambda = \int_0^\infty \epsilon(\lambda)\, d\lambda = \epsilon \qquad (1\text{-}40)$$

It can also be shown that

$$\alpha(\lambda) = \epsilon(\lambda) \qquad (1\text{-}41)$$

$$\int_{\Delta\lambda} \alpha(\lambda)\, d\lambda = \int_{\Delta\lambda} \epsilon(\lambda)\, d\lambda \qquad (1\text{-}42)$$

where the temperature of both bodies is the same and the spectral region of consideration is the same for both α and ϵ. Since these depend upon the total power law, they also depend upon thermal equilibrium.

1.3.4. Reflectivity and the Bidirectional Reflectance Distribution Function (BRDF) [1-14]. The differential element of flux reflected from a surface (Figure 1-11) can be written as

$$d\Phi_r \equiv \lim \Delta\Phi_r = L_r \lim \{\Delta A_s \cos\theta_r\, \Delta\Omega_r\} \qquad (1\text{-}43)$$

where L_r = reflected radiance
  ΔA_s = a differential area of sample surface
  θ_r = the angle of reflection
  $\Delta\Omega_r$ = a differential solid angle in reflection space

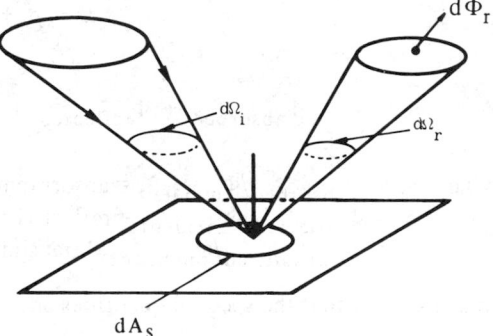

and where the limit is taken as ΔA_s and $\Delta\Omega_r$ approach 0. In the limit, this can be written as

$$d\Phi_r = L_r\, dA_s \cos\theta_r\, d\Omega_r \qquad (1\text{-}44)$$

The incident flux can be written

$$d\Phi_i = L_i\, dA_s \cos\theta_i\, d\Omega_i \qquad (1\text{-}45)$$

where the subscript i represents incident quantities. Then, in general, any element of area and solid angle has a reflectivity factor such that

$$d\Phi_r = \rho(\theta_i, \phi_i; \theta_r, \phi_r; A_s)\, d\Phi_i \qquad (1\text{-}46)$$

Fig.1-11. Differential element of flux reflected from a surface.

and

$$\Phi_r = \rho_{\text{total}} \Phi_i \qquad (1\text{-}47)$$

The first of these is a definition; the second is a normalization. In general,

$$d\Phi = L\, dA \cos\theta\, d\Omega \qquad (1\text{-}48)$$

Therefore, reflectivity can be defined as the ratio of reflected flux to incident flux

$$\rho = \frac{L_r \, dA_s \cos\theta \, d\Omega_r}{L_i \, dA_s \cos\theta \, d\Omega_i} = \frac{L_r \, d\Omega_r}{L_i \, d\Omega_i} \tag{1-49}$$

There is ambiguity unless the solid angles of incidence and reflectance are specified. The BRDF is defined so that

$$d\Phi_r = f\Phi_i \, d\Omega_r \tag{1-50}$$

$$\frac{d\Phi_r}{\Phi_i} = f \, d\Omega_r \tag{1-51}$$

$$f \equiv \frac{L_r \, dA_s}{L_i \, d\Omega_i \, dA_s} = \frac{L_r}{L_i \, d\Omega_i} = \frac{L_i \, dA_s}{d\Phi_i} \tag{1-52}$$

The use of ρ requires that the solid angles be specified; BRDF is independent of that specification. Like most differentially defined quantities, BRDF accuracy depends upon the approximation to the derivative. This is most noticeable in dealing with the beam profile. BRDF will, in general, be a function of all four angles (of incidence and reflection) and the area of the sample that is chosen.

The two extreme cases are Lambertian and specular surfaces. By definition, the former surface has no dependence on angle, and a flat surface therefore has a BRDF of $\rho\pi^{-1}$. It is the integral of the projected solid angle over a hemisphere. A purely specular surface has a nonzero value only for $\theta_i = \theta_r$ and $\phi_i = \phi_r$ so that the BRDF $= \rho\delta(\theta_i - \theta_r)\delta(\phi_i - \phi_r)$. The hemispherical reflectance in each case is ρ.

An alternate definition for reflection is

$$\rho = \frac{d\Phi_r}{d\Phi_i} = \frac{L_r \, dA_s' \, d\Omega_r}{L_i \, dA_s' \, d\Omega_i} = \frac{E \, dA_s}{d\Phi_i} \tag{1-53}$$

The return flux is given by

$$d\Phi_r = \rho \, d\Phi_i = \rho \, dA_s' \, E\frac{dA_s'}{dA_s'} = (\rho \, dA_s') E \tag{1-54}$$

where $\rho \, dA_s'$ is called the cross-section.

1.4. Radiation Geometry

The geometric transfer of radiation is independent of spectral characteristics and based on the equation of transfer in a nonabsorbing, nonscattering medium.

1.4.1. Transfer Equation. For a surface, A_1, with radiance L_1 and a body, A_2, with radiance L_2 joined by a general ray of length r_{12} (as shown in Figure 1-12) the net radiative change between them is given by

$$\Delta\Phi = \Phi_{2-1} - \Phi_{1-2} =$$

$$\int (L_{12} - L_{21}) \, dA_1 \cos\theta_2 \, dA_2 \cos\theta_2 r_{12}^{-2}$$

$$= \int (\Delta L) \, r_{12}^{-2} \cos\theta_1 \cos\theta_2 \, dA_1 \, dA_2$$

$$\tag{1-55}$$

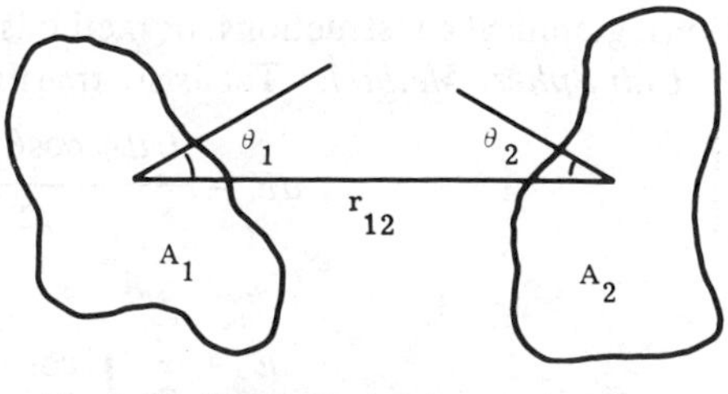

Fig. 1-12. The radiation interaction factor.

Often, the radiance of each surface is independent of angle so that

$$\Delta\Phi = \Delta L \int r_{12}^{-2} \cos\theta_1 \cos\theta_2 \, dA_1 dA_2 \qquad (1\text{-}56)$$

The integral in this expression is useful for many radiation interaction problems and has been given a number of different names: projected solid angle, radiation interaction factor, and configuration factor. Some authors substitute $M\pi^{-1}$ for L and consider π part of the definition of the configuration factor.

1.4.2. Lambertian Surfaces. A Lambertian surface can be defined as one whose radiance is independent of angle. For the differential area of a Lambertian surface, the flux density (flux per unit area) radiated into a hemisphere is given by

$$M = \pi L \qquad (1\text{-}57)$$

The flux density radiated into a solid angle defined by a cone of half angle ϕ is given by

$$M(\theta) = \pi L \sin^2\theta = \frac{\pi}{2}(1 - \cos 2\theta) L \qquad (1\text{-}58)$$

This can be interpreted as the interaction factor or projected solid angle between a differential element of area and a segment of a sphere.

1.4.3. Surfaces with Cosine Radiance Distributions. Surfaces with nonconstant angular distributions of emission or reflection can often be characterized as a power of $\cos\theta$: $L = L_0 \cos^n\theta$. Then

$$M_\theta = \frac{2\pi L_0}{n+2}\cos^{n+2}\theta = \frac{2\pi}{n+2} L(\theta) \cos^2\theta \qquad (1\text{-}59)$$

1.4.4. Lambertian Disc. It can be shown that a Lambertian disc of radius R a distance r_{12} away from an on-axis normal differential element of surface has irradiance given by

$$E = 2\pi\left(\frac{R}{r_{12}}\right)^2 \left[1 - \left(\frac{R}{r_{12}}\right)^2 + \left(\frac{R}{r_{12}}\right)^4 - \ldots\right] \qquad (1\text{-}60)$$

This shows how the inverse square law is an approximation that only applies when $r_{12} \gg R$, i.e., the distance away from the source is much larger than the dimension of the source. (This is only shown here for a disc but applies more generally for two-dimensional objects that have an aspect ratio of about one).

1.4.5. Short-Cuts in Calculations [1-15]. These short-cuts are either approximations, graphical constructions, or methods that apply in special situations.

Unit Sphere Method. The basic transfer equation can be written

$$dE = \frac{L \, dA \cos\theta_1 \cos\theta_2}{r^2} = \frac{M}{\pi} d\Omega \cos\theta_1 \qquad (1\text{-}61)$$

$$E = \frac{M}{\pi} \int \cos\theta_1 \, d\Omega = \frac{M}{\pi} \Omega' \qquad (1\text{-}62)$$

This is sometimes called the unit sphere method because it can be interpreted as the area on a unit sphere intercepted by the contour of the true area, as shown in Figure 1-13. The disc problem can be solved quickly this way. The area shown (in Figure 1-14) shaded on the plane P is $\pi R^2 (R^2 + r^2_{12})^{-2}$. Therefore

$$E = M R^2 (R^2 + r^2_{12})^{-2} = M \sin^2\theta = \frac{M}{2} (1 - \cos 2\theta) \qquad (1\text{-}63)$$

This is the same result that a hemispherical (or any other shape) radiator with the same boundary would give. A disc with radiant exitance proportional to its radius yields an irradiance given by

$$M \{r_{12} \tan^{-1} (R/r_{12}) - R [1 + (R/r_{12})^2]^{-1}\} \qquad (1\text{-}64)$$

Sumpner Method. On a sphere of radius r, a cap, ANB, can be constructed as shown in Figure 1-15. The irradiance at any point on the sphere from this cap will be the same as that from a disc with the same boundary (by virtue of the unit sphere method, above). The angle γ subtended by AB is independent of the position P. The angle 2θ is equal to γ and the irradiance is everywhere the same on the inside of the sphere wall. Therefore the irradiance at P on the wall is

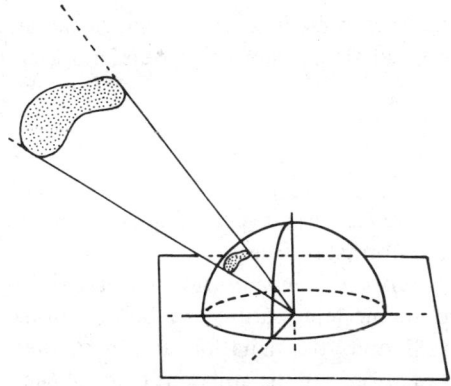

Fig. 1-13. The unit sphere method.

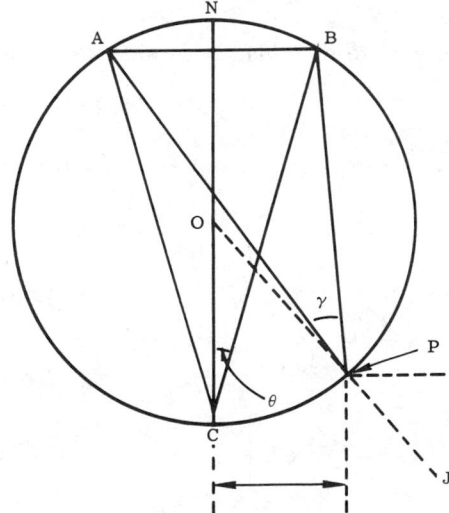

Fig. 1-15. The Sumpner method.

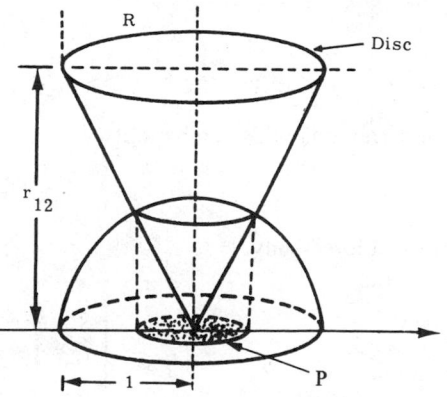

Fig. 1-14. Disc problem solved by the unit sphere method.

$$E_w = \frac{M}{2}(1 - \cos 2\theta) = M\left[1 + \left(\frac{r_{12}}{r}\right)^2\right]^{-1} \tag{1-65}$$

Contour Integration. The irradiance can be calculated for a diffuse surface of constant radiance as follows:

$$E = \frac{M}{\pi} \oint d\hat{\alpha} \tag{1-66}$$

where $d\hat{\alpha}$ is the differential angle that is intercepted by a differential element of the boundary. For surfaces with polygonal boundaries

$$E = \frac{M}{2\pi}\sum_i \hat{\alpha}_i \gamma_i \tag{1-67}$$

where $\hat{\alpha}_i$ = the unit vector of the outward drawn normal
$\quad\gamma_i$ = the angle subtended by the ith side

Vector Method. One can calculate the irradiance for inclined surfaces on a plane normal to the line of centers and then develop the irradiance on the inclined surfaces:

$$E = \frac{1}{\pi} \int \mathbf{r}\frac{M\cos\theta}{r_{12}} dA \tag{1-68}$$

where $\mathbf{r}$ is the unit vector from the element of area in question to the point where r_{12} meets the normal surface.

Optical Temperatures [1-7, 1-16]. Many methods exist for determining the thermodynamic temperature of a body radiometrically. These include measurements of the total radiation, the radiation at a particular wavelength, the wavelength distribution of the radiation, and the apparent color of the radiation.

Radiation temperature is defined as the temperature of a blackbody that gives the same total radiance or radiant exitance:

$$T_R = \left(\frac{M}{\sigma}\right)^{1/4} \tag{1-69}$$

For a gray body this can be written

$$T_R = \epsilon^{1/4} T \tag{1-70}$$

For a colored body

$$T_R = \left[\sigma^{-1} \int_0^\infty \epsilon c_1 \lambda^{-5} (e^x - 1)^{-1} d\lambda\right]^{1/4} \tag{1-71}$$

The relative error for the gray body can be written simply as

$$RE = 1 - \left(\frac{T_R}{T}\right) = 1 - \epsilon^{1/4} \tag{1-72}$$

For a colored body it is

$$RE = 1 - T\left[\sigma^{-1} \int_0^\infty \epsilon(\lambda)\, c_1\, \lambda^{-5}\, (e^x - 1)^{-1}\, d\lambda\right]^{-1/4} \tag{1-73}$$

The analysis requires specific values for $\epsilon(\lambda)$.

Brightness temperature is defined as the temperature of a blackbody that gives the same radiance in a narrow spectral band as the body in question. Then

$$e^{c_2/\lambda T_B} - 1 = \epsilon^{-1}\, (e^{c_2/\lambda T} - 1) \tag{1-74}$$

In the region for which $e^{c_2/\lambda T}$ is much greater than 1, this can be written

$$T_B = c_2\left(\frac{c_2}{T} - \lambda \ln \epsilon\right)^{-1} \tag{1-75}$$

Otherwise

$$T_B = c_2\, \{\lambda \ln\, [\epsilon^{-1}\, (e^{c_2/\lambda T} - 1) + 1]\}^{-1} \tag{1-76}$$

The relative errors in these two cases are

$$RE = 1 - \left(1 - \frac{\lambda T}{c_2}\, \ln \epsilon\right)^{-1} \tag{1-77}$$

$$RE = 1 - c_2\, \{\lambda T \ln\, [\epsilon^{-1}\, (e^{c_2/\lambda T} - 1) + 1]\}^{-1} \tag{1-78}$$

Distribution or Ratio Temperature. The distribution temperature is the temperature of the blackbody that best matches the spectral distribution of the body in question. Often two or three narrow bands are used for this determination. If two wavelengths are used, then the ratio is

$$\frac{M_{\lambda_1}}{M_{\lambda_2}} = \frac{\epsilon_1\, \lambda_1^{-5}(e^{c_2/\lambda_1 T} - 1)^{-1}}{\epsilon_2\, \lambda_2^{-5}(e^{c_2/\lambda_2 T} - 1)^{-1}} \tag{1-79}$$

For the Wien approximation, the temperature is given by

$$T_D = \frac{c_2(\lambda_1 - \lambda_2)}{\lambda_1 \lambda_2 \ln\left[\dfrac{M_{\lambda_1}}{M_{\lambda_2}} \dfrac{\epsilon_2}{\epsilon_1}\left(\dfrac{\lambda_1}{\lambda_2}\right)^5\right]} \tag{1-80}$$

If the Wien expression is not applicable, then

$$\frac{e^{c_2/\lambda_2 T_D} - 1}{e^{c_2/\lambda_1 T_D} - 1} = \left[\frac{M_{\lambda_1}}{M_{\lambda_2}} \frac{\epsilon_2}{\epsilon_1}\left(\frac{\lambda_1}{\lambda_2}\right)^5\right] \tag{1-81}$$

$$(e^{c_2/\lambda_2 T_D} - 1)(e^{-c_2/\lambda_1 T_D} + e^{-2c_2/\lambda_1 T_D} + \ldots) = \frac{M_{\lambda_1}}{M_{\lambda_2}} \frac{\epsilon_2}{\epsilon_1}\left(\frac{\lambda_1}{\lambda_2}\right)^5 \tag{1-82}$$

Color Temperature. The color temperature is the temperature of a blackbody that has the same chromaticity coordinates as the body in question. (See Figure 1-16.)

If the body has a given set of normalized chromaticity coordinates x, y, then it has a set of relative radiances such that

$$\frac{M_{\lambda_1}}{M_{\lambda_2}} = \frac{\lambda_1^{-5}(e^{c_2/\lambda_1 T_c} - 1)^{-1}}{\lambda_2^{-5}(e^{c_2/\lambda_2 T_c} - 1)^{-1}} = \frac{\lambda_2^5(e^{c_2/\lambda_2 T_c} - 1)}{\lambda_1^5(e^{c_2/\lambda_1 T_c} - 1)} \tag{1-83}$$

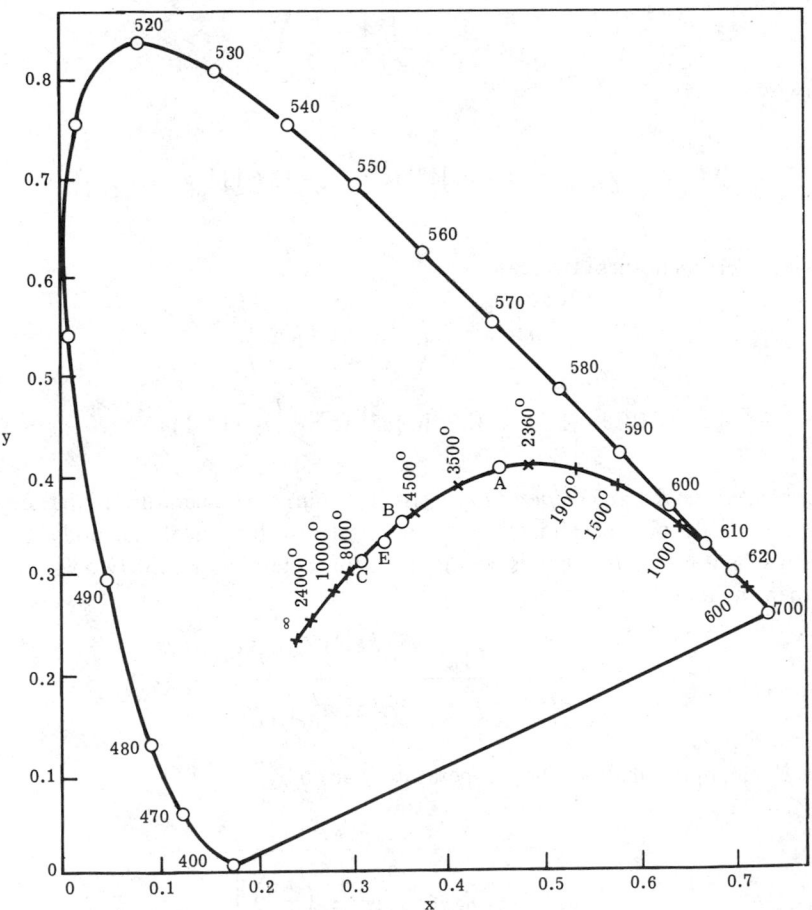

Fig. 1-16. Planckian locus [1-16].

If ϵ_1 and ϵ_2 are the emissivities at these wavelengths, then the color temperature is found by solving for T_c

$$\frac{e^{c_2/\lambda_2 T_c} - 1}{e^{c_2/\lambda_1 T_c} - 1} = \frac{\epsilon_1 \left(e^{c_2/\lambda_2 T} - 1\right)}{\epsilon_2 \left(e^{c_2/\lambda_1 T} - 1\right)} \qquad (1\text{-}84)$$

Table 1-30 gives the chromaticity coordinates of blackbodies with temperatures from 1000 K to infinity. Figure 1-16 shows this locus from zero Kelvin to infinity.

Table 1-30. Colorimetry Chromaticity Coordinates of Blackbodies
[1-15] (c_2 = 1.438 cm K)

T(K)	x	y	T(K)	x	y
1000	0.652	0.344	3400	0.410	0.393
1500	0.585	0.393	3500	0.405	0.390
1600	0.573	0.399	3600	0.399	0.387
1700	0.560	0.404	3700	0.394	0.384
1800	0.549	0.408	3800	0.389	0.382
1900	0.537	0.411	3900	0.384	0.379
2000	0.526	0.413	4000	0.380	0.376
2100	0.515	0.414	4500	0.360	0.363
2200	0.505	0.415	5000	0.345	0.351
2300	0.495	0.415	5500	0.332	0.341
2400	0.486	0.414	6000	0.322	0.331
2500	0.476	0.413	6500	0.313	0.323
2600	0.468	0.412	7000	0.306	0.316
2700	0.459	0.410	7500	0.300	0.310
2800	0.451	0.408	8000	0.295	0.304
2900	0.444	0.406	8500	0.290	0.299
3000	0.436	0.404	9000	0.286	0.295
3100	0.429	0.401	9500	0.283	0.291
3200	0.423	0.398	10000	0.280	0.288
3300	0.417	0.396	20000	0.256	0.257
			∞	0.239	0.234

1.5. References and Bibliography

1.5.1. References

[1-1]   R. C. Jones, "Terminology in Photometry and Radiometry," *Journal of the Optical Society of America*, Optical Society of America, Washington, DC, Vol. 53, 1963, p. 1314.

[1-2]   J. Geist and F. Zalewski, "Chinese Restaurant Nomenclature for Radiometry, *Applied Optics*, Optical Society of America, Washington, DC, Vol. 12, 1973, p. 435.

[1-3]   G. A. Melchty, "The International System of Units, Physical Constants and Conversion Factors," National Aeronautics and Space Administration Special Publication 7012, United States Government Printing Office, Washington, DC, 1973, Second Revision, pp. iii, 2-6, 11-20.

[1-4]   J. J. Muray, F. E. Nicodemus, and I. Wunderman, "Proposed Supplement to the Systeme International d'Unites Nomenclature for Radiometry and Photometry," *Applied Optics*, Optical Society of America, Washington, DC, Vol. 10, 1971, pp. 1465-1468; see also: National Bureau of Standards, *The International System of Units (SI)*, U.S. Department of Commerce, Washington, DC, NBS Special Publication 330, 1972.

[1-5]   J. Mayer-Arendt, Foothills College, Los Angeles, CA, Private Communication.

[1-6]   J. J. Spiro, "Radiometry and Photometry," *Optical Engineering*, Society of Photo-Optical Instrumentation Engineers, Palos Verdes Estates, CA, Vol. 13, 1974, pp. G 183-187, Vol. 15, 1976, pp. SR-7.

[1-7]   M. A. Bramson, *Infrared: A Handbook for Applications*, Plenum Press, New York, NY, 1966, Chapter 1.

[1-8]   M. W. Makowski, *Review of Scientific Instruments*, American Institute of Physics, New York, NY, Vol. 20, 1945, p. 876.

[1-9]   R. B. Johnson and E. E. Branstetter, "Integration of Planck's Equation by the Laguerre-Gauss Quadrature Method," *Journal of the Optical Society of America*, Optical Society of America, Washington, DC, Vol. 64, 1974, p. 1445.

[1-10]   M. Ross, *Laser Receivers*, John Wiley and Sons, New York, 1966, p. 18.

[1-11]   Robert M. Eisberg, *Applied Mathematical Physics with Programmable Pocket Calculators*, McGraw-Hill, New York, NY, 1976, p. 176.

[1-12]   Jon M. Smith, *Scientific Analysis on the Pocket Calculator*, John Wiley and Sons, New York, NY, 1975.

[1-13]   F. E. Nicodemus, "Directional Reflectance and Emissivity of an Opaque Surface," *Applied Optics*, Optical Society of America, Washington, DC, Vol. 4, 1965, p. 767.

[1-14]   G. Kirchhoff, *Philosophical Magazine and Journal of Science,* Fourth Series, Vol. 20, No. 130, July 1860 (translated from *Poggendorf Annalen*, Vol. CXX, p. 275).

[1-15]   Parry Moon, *The Scientific Basis of Illumination Engineering*, Dover, New York, NY, 1936.

[1-16]   A. C. Hardy, *Handbook of Colorimetry*, Technology Press, Massachusetts Institute of Technology, Cambridge, MA, 1936.

1.5.2. Bibliography

Bennett, H. E., and J. O. Porteus, "Relation Between Surface Roughness and Specular Reflectance at Normal Incidence," *Journal of the Optical Society of America*, Optical Society of America, Washington, DC, Vol. 51, February 1961, pp. 123-129.

Blau, Henry H., and Heinz Fischer (eds.), *Radiative Transfer from Solid Materials*, Macmillan, New York, 1962.

Blau, Henry H., Jr., John L. Miles, and Leland E. Ashman, *Thermal Radiation Characteristics of Solid Materials—a Review*, Arthur D. Little, Inc., Contract No. AF19(604)-2639, Scientific Report No. 1, AFCRC-TN-58-132, AD 146 883, 1958.

Born, Max, and Emil Wolf, *Principles of Optics*, Pergamon Press, New York, NY, 1959.

Bramson, M. A., *Infrared: A Handbook for Applications*, Plenum Press, New York, NY, 1966.

Canada, A. H., *General Electric Review*, Vol. 51, 1948, pp. 50-54.

Chandos, Ray, "A Program for Blackbody Radiation Calculations," Electro-Optical Industries, Santa Barbara, CA, No. L2227, pp. 10-17.

Czerny, M., and A. Walther, *Tables of the Fractional Function for the Planck Radiation Law*, Springer-Verlag, Berlin, 1961.

DeBell, A. G., "Rocketdyne Research Report 59-32," Rocketdyne Division, North American Aviation, Canoga Park, CA, 1959.

DeHoop, A. T., "A Reciprocity Theorem for the Electromagnetic Field Scattered by an Obstacle," *Applied Scientific Research*, Nijhowf, The Hague, Netherlands, Sec. B, Vol. 8, 1960, p. 135.

DeVos, J. C., "Evaluation of the Quality of a Blackbody," *Physica*, North Holland, Amsterdam, Netherlands, Vol. 20, 1954, p. 669.

Edwards, D. K., J. T. Gier, K. E. Nelson, and R. D. Roddick, "Integrating Sphere for Imperfectly Diffuse Samples," *Journal of the Optical Society of America*, Optical Society of America, Washington, DC, Vol. 51, 1961, p. 1279.

Eisberg, Robert M., *Applied Mathematical Physics with Programmable Pocket Calculators*, McGraw-Hill, New York, NY, 1976, p. 176.

Geist, J., and E. Zalewski, "Chinese Restaurant Nomenclature for Radiometry," *Applied Optics*, Optical Society of America, Washington, DC, Vol. 12, 1973, p. 435.

Golden, S. A., "Spectral and Integrated Blackbody Radiation Functions," Rocketdyne Division, North American Aviation, Canoga Park, CA, Report 60-23, 1960.

Hapke, Bruce W., "A Theoretical Photometric Function for the Lunar Surface," *Journal of Geophysical Research*, American Geophysical Union, Washington, DC, Vol. 68, No. 15, 1 August 1963, p. 4571.

Hapke, Bruce W., and Hugh Van Horn, "Photometric Studies of Complex Surfaces with Applications to the Moon," *Journal of Geophysical Research*, American Geophysical Union, Washington, DC, Vol. 68, No. 15, 1 August 1963, p. 4545.

Hardy, A. C., *Handbook of Colorimetry*, Technology Press, Massachusetts Institute of Technology, Cambridge, MA, 1936.

Helmholtz, H. von, "Treatise on Physiological Optics," James P. C. Southhall (ed.), *Journal of the Optical Society of America*, Optical Society of America, Washington, DC, Vol. 1, 1924, p. 231.

Holter, M. et al., *Fundamentals of Infrared Technology*, MacMillan, New York, NY, 1963.

Jahnke, E., and F. Emde, *Tables of Functions*, Dover, New York, NY, 1945.

Johnson, R. B., and E. E. Branstetter, "Integration of Planck's Equation by the Laguerre-Gauss Quadrature Method," *Journal of the Optical Society of America*, Optical Society of America, Washington, DC, Vol. 64, 1974, p. 1445.

Jones, R. Clark, "Immersed Radiation Detectors," *Applied Optics*, Optical Society of America, Washington, DC, Vol. 1, 1962, p. 607.

Jones, R. Clark, "Terminology in Photometry and Radiometry," *Journal of the Optical Society of America*, Optical Society of America, Washington, DC, Vol. 53, 1963, p. 1314.

Kerr, D. D., "Application of the Lorentz Reciprocity Theorem to Scattering," *Propagation of Short Radio Waves*, Massachusetts Institute of Technology Radiation Laboratory Series, McGraw-Hill, New York, Vol. 13, Appendix A, 1951, First Edition.

Kirchhoff, G., *Philosophical Magazine and Journal of Science*, Fourth Series, Vol. 20, No. 130, July 1860 (translated from *Poggendorf Annalen*, Vol. CXX, p. 275).

Lowan, A. N., and G. Blanch, *Journal of the Optical Society of America*, Optical Society of America, Washington, DC. Vol. 30, 1940, p. 70.

MacAdam, D. L., "OSA Nomenclature," *Journal of the Optical Society of America*, Optical Society of America, Washington, DC, Vol. 57, 1967, p. 854.

Makowski, M. W., *Review of Scientific Instruments*, American Institute of Physics, New York, NY, Vol. 20, 1945, p. 876.

McNicholas, H. J., "Absolute Methods in Reflectometry," *U.S. National Bureau of Standards Journal of Research*, Vol. 1, p. 29, (RP-3); PhD dissertation, The Johns Hopkins University, 1928.

Melchtly, G. A., *The International System of Units, Physical Constants and Conversions*, National Aeronautics and Space Administration Special Publication 7012, United States Government Printing Office, Washington, DC, 1973, Second Revision, pp. iii, 2-6, 11-20.

Merritt, T. P., and F. F. Hall, Jr., "Blackbody Radiation," *Proceedings of the IRE*, Institute of Radio Engineers, New York, NY, Vol. 47, No. 9, September 1959, pp. 1435-1441.

Moon, Parry, "A Table of Planckian Information," *Journal of the Optical Society of America*, Optical Society of America, Washington, DC, Vol. 38, 1948, p. 291.

Moon, Parry, *The Scientific Basis of Illumination Engineering*, Dover, New York, NY, 1936.

Muray, J. J., F. E. Nicodemus, and I. Wunderman, "Proposed Supplement to the Systeme International d'Unites Nomenclature for Radiometry and Photometry," *Applied Optics*, Optical Society of America, Washington, DC, Vol. 10, 1971, pp. 1465-1468.

National Bureau of Standards, *The International System of Units (SI)*, U.S. Department of Commerce, Washington, DC, NBS Special Publication 330, 1972.

Nicodemus, F. E., "Directional Reflectance and Emissivity of an Opaque Surface," *Applied Optics*, Optical Society of America, Washington, DC, Vol. 4, 1965, p. 767.

Nicodemus, F. E., "Normalization in Radiometry," *Applied Optics*, Optical Society of America, Washington, DC, Vol. 12, 1973, p. 2960.

Nicodemus, F. E., "Radiance," *American Journal of Physics*, American Institute of Physics, New York, NY, Vol. 31, No. 5, 1963.

Nicodemus, F. E., *Radiometry*, American Association of Physics Teachers, State University of New York, Stoney Brook, NY.

Nicodemus, F. E., "Radiometry," *Applied Optics and Optical Engineering*, R. Kingslake (ed.), Academic Press, New York, NY, Vol. 4, Chapter 8, 1967.

Optical Society of America Committee on Colorimetry, *The Science of Color*, Crowell, New York, NY, 1954.

Penner, S. S., *Quantitative Molecular Spectroscopy and Gas Emissivities*, Addison Wesley, Reading, MA, 1959,

Pivovonsky, M., and M. Nagel, *Tables of Blackbody Radiation Functions*, Macmillan, New York, 1961.

Planck, M., *Theory of Heat*, H. L. Brose (trans.), Macmillan, New York, NY, 1957.

Ross, M., *Laser Receivers*, John Wiley and Sons, New York, 1966, p. 18.

Smith, Jon M., *Scientific Analysis on the Pocket Calculator*, John Wiley and Sons, New York, NY, 1975.

Sparrow, E. M., and R. D. Cess, *Radiation Heat Transfer*, Brooks/Cole, Belmont, CA, 1970.

Spiro, J. J., "Radiometry and Photometry," *Optical Engineering*, Society of Photo-Optical Instrumentation Engineers, Palos Verdes Estates, CA, Vol. 13, 1974, pp. G 183-187, Vol. 15, 1976, pp. SR-7.

Svet, D. Y., *Thermal Radiation*, Consultants Bureau, New York, NY, 1965, (now Plenum Publishing Company).

Twersky, V., "On Scattering and Reflection of Electromagnetic Waves by Rough Surfaces," *IRE Transactions—Antennas and Propagation*, Institute of Radio Engineers, New York, NY, AP-5, 1957, p. 81.

Twersky, V., "On Multiple Scattering of Waves," *National Bureau of Standards Journal of Research,* Vol. 64D, 1960, pp. 715-730.

Twersky, V., "Multiple Scattering of Waves and Optical Phenomena," *Journal of the Optical Society of America*, Optical Society of America, Washington, DC, Vol. 52, 1962, p. 145.

University of Michigan, *Report of the Working Group on Infrared Backgrounds,* "Part II: Concepts and Units for the Presentation of Infrared Background Information," The University of Michigan Institute of Science and Technology, Ann Arbor, MI, Report No. 2389-3-S, AD 123 097, 1956.

Worthing, G., and D. Halliday, *Heat*, John Wiley and Sons, New York, NY, 1948.

Chapter 2

ARTIFICIAL SOURCES

Anthony J. LaRocca
Environmental Research Institute of Michigan

CONTENTS

2. Artificial Sources

2.1. Introduction

This chapter deals with artificial sources of radiation as subdivided into two classes: laboratory and field sources. Normally, laboratory sources are used in some standard capacity and field sources are used as targets. Both varieties appear to be limitless.

The sources in this chapter were chosen arbitrarily, often depending upon manufacturer response to requests for information. The purpose of this chapter is to consolidate much of this information to assist the optical-systems designer in making reasonable choices.

Regarding the selection of a source, Worthing [2-1] suggests that one ask the following questions:

(1) Does it supply energy at such a rate or in such an amount as to make measurements possible?

(2) Does it yield an irradiation that is generally constant or that may be varied with time as desired?

(3) Is it reproducible?

(4) Does it yield irradiations of the desired magnitudes over areas of the desired extent?

(5) Has it the desired spectral distribution?

(6) Has it the necessary operating life?

(7) Has it sufficient ruggedness for the proposed problem?

(8) Is it sufficiently easy to obtain and replace, or is its purchase price or its construction cost reasonable?

2.1.1. Symbols, Nomenclature and Units.
Table 2-1 lists the symbols, nomenclature and units used in this chapter.

2.2. Laboratory Sources

2.2.1. Standard Sources

Blackbody Cavity Theory. Radiation levels can be standarized by the use of a source that will emit a quantity of radiation that is both reproducible and predictable. Cavity configurations can be produced to yield radiation theoretically sufficiently close to Planckian (Chapter 1) that it is necessary only to determine what the imprecision is. Several theories have been expounded over the years to calculate the quality of a blackbody simulator.* Two of the older, most straightforward, widespread, and demonstrable theories are those of Gouffé and DeVos.

The Method of Gouffé [2-2]. For the total emissivity of the cavity forming a blackbody (disregarding temperature variation) Gouffé gives

$$\epsilon_0 = \epsilon'_0 (1 + k) \tag{2-1}$$

---

*Generically used to describe those sources designed to produce radiation that is nearly Planckian.

Table 2-1. Symbols, Nomenclature and Units*

Symbols	Nomenclature	Units
a	Ratio: l/r	—
B	Subscript used for blackbody radiation	—
c_2	Blackbody radiation constant	cm K
I	Radiant intensity	W sr^{-1}
k	Emissivity corrective factor	—
L	Radiance	W cm^{-2} sr^{-1}
l	Length (or depth) of cavity	cm
M	Radiant exitance	W cm^{-2}
m	Mach number	—
R	Cavity radius	cm
r	Aperture radius	cm
r_a^{bc}	Partial reflectivity of DeVos	sr^{-1}
S	Interior surface area	cm^2
s	Aperture area	cm^2
T	Temperature	K, °C
x	Variable length	cm
y	Ratio: x/r	—
ϵ	Emissivity	—
θ	Angle of ray from surface normal	rad
λ	Wavelength	μm, Å
σ	Boltzmann constant	W cm^{-2} K^{-4}
Φ	Power, flux	W

*Most of the units, symbols and nomenclature in this chapter are shown in the tables and graphs corresponding to particular sources.

where

$$\epsilon_0' = \frac{\epsilon}{\epsilon \left[1 - \dfrac{s}{S}\right] + \left(\dfrac{s}{S}\right)} \qquad (2\text{-}2)$$

and $k = (1 - \epsilon) [(s/S) - (s/S_0)]$, and is always nearly zero—it can be either positive or negative

ϵ = emissivity of materials forming the blackbody surface

s = area of aperture

S = area of interior surface

S_0 = surface of a sphere of the same depth as the cavity in the direction normal to the aperture

Figure 2-1 is a graph for determining the emissivities of cavities with simple geometric shapes. In the lower section, the value of the ratio s/S as a function of the ratio l/r is read. The value of ϵ_0' is found by reading up from this value to the value of the intrinsic emissivity of the cavity material. The emissivity of the cavity is found by multiplying ϵ_0' by the factor $(1 + k)$.

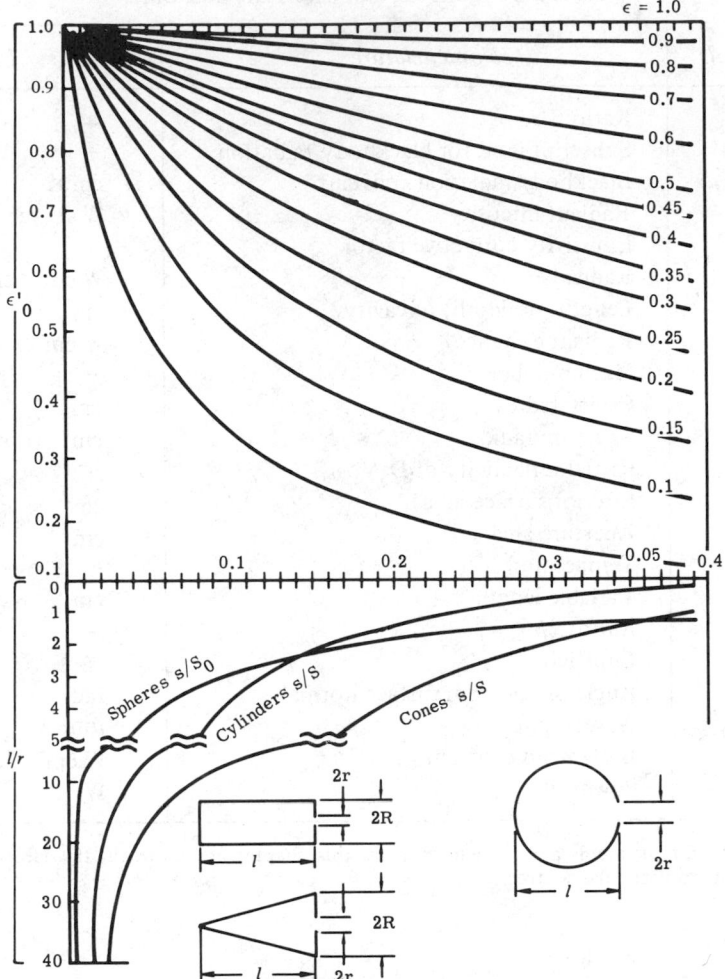

Fig. 2-1. Emissivities of conical, spherical, and cylindrical cavities.

When the aperture diameter is smaller than the interior diameter of the cylindrical cavity, or the base diameter of a conical cavity, it is necessary to multiply the values of s/S determined from the graph by $(r/R)^2$, which is the ratio of the squares of the aperture and cavity radii (Figure 2-1).

The Method of DeVos [2-3]. DeVos considers a cavity of arbitrary shape, with opaque walls, in a nonattenuating medium, initially at a uniform steady temperature, with one small opening. He adds additional openings and temperature variations along the cavity walls and indicates several practical approximations necessary for the calculation of numerical values. (See Figure 2-2.)

The power emitted from the opening $d0$ is

$$em\Phi_w^0 = \epsilon_w^0(\lambda, T)L_{\lambda, B}(\lambda, T)\, dw \cos\theta_w^0\, d\Omega_w^0 \qquad (2\text{-}3)$$

where $\epsilon_w^0(\lambda, T)$ = the spectral emissivity of dw in the direction of $d0$ (indicated by subscripts and superscripts throughout) for temperature T at wavelength λ

$L_{\lambda,B}(\lambda, T)$ = the spectral radiance of a blackbody for temperature T and wavelength λ given by either modification of Equation (1-9) or, approximately, by the Wien law

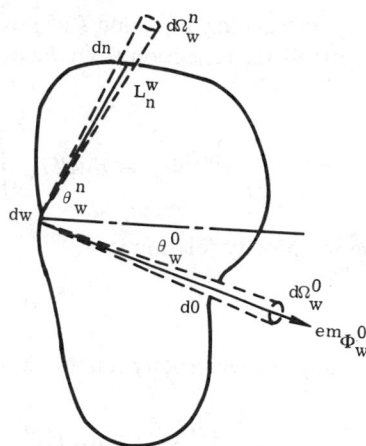

$$L_{\lambda,B} = (\text{constant})e^{-c_2/\lambda T} \qquad (2\text{-}4)$$

dw = area of the emitting infinitesimal element

Fig. 2-2. Definition of terms for the Devos method.

θ_w^0 = the angle of the direction from dw to $d0$ with respect to the normal to dw

$d\Omega_w^0$ = the solid angle subtended by $d0$, the hole, as seen from dw

The power from dw through $d0$ which is due to the reflection of the power received at dw from some arbitrary elemental wall area, dn, is

$$^{refl}\Phi_w^{n0} = L_{\lambda,n}^w(\lambda, T)d\Omega_w^n dw \cos \theta_w^n r_w^{n0}(\lambda, T)d\Omega_w^0 \qquad (2\text{-}5)$$

where Φ_w^{n0} = the power from dn to $d0$ via dw

$L_{\lambda,n}^w(\lambda, T)$ = the spectral radiance from dn to dw for a temperature, T, and at a wavelength, λ

$d\Omega_w^n$ = the solid angle subtended by dn as seen from dw

θ_w^n = the angle to the normal to dw made by the direction from dw to dn

$r_w^{n0}(\lambda, T)$ = the partial reflectivity of dw for radiation from dn at θ_w^n reflected from dw toward $d0$ at θ_w^0, at a wavelength, λ, for a temperature, T

Partial reflectivity* can be defined as follows:

$$r_w^{n0} = \frac{L_w^{n0} \cos \theta_w^0}{L_n^w \cos \theta_w^n d\Omega_w^n} \qquad (2\text{-}6)$$

or

$$r_w^{n0} = \frac{L_w^{n0}}{E_w^n} \cos \theta_w^0 \qquad (2\text{-}7)$$

---

*This terminology is used slightly differently in other literature. The method of DeVos is retained here.

By integrating Equation (2-5) over the walls excluding $d0$ one obtains the power from dw due to the reflection from dw of the radiation from all parts of the cavity walls except $d0$:

$$^{\text{refl}}\Phi_w^0 = dw d\Omega_w^0 \int_{\text{all } dn} L_{\lambda,n}^w(\lambda, T) \cos \theta_w^n r_w^{n0}(\lambda, T) d\Omega_w^n \qquad (2\text{-}8)$$

The reciprocity relation is

$$r^{ab} \cos \theta^a = r^{ba} \cos \theta^b \qquad (2\text{-}9)$$

By using the reciprocity relation in Equation (2-9)

$$^{\text{refl}}\Phi_w^0 = dw d\Omega_w^0 \cos \theta_w^0 \int_{\text{all } dn} L_{\lambda,n}^w(\lambda, T) r_w^{0n}(\lambda, T) d\Omega_w^n \qquad (2\text{-}10)$$

To a first-order approximation the following relationship is true:

$$L_{\lambda,n}^w(\lambda, T) = L_{\lambda,B}(\lambda, T)$$

The hole can thus be considered to have an emissivity given by

$$\epsilon_0 = 1 - r_w^{00} d\Omega_w^{h0} \qquad (2\text{-}11)$$

to a first approximation neglecting temperature variations. If additional holes exist, then the reflected contributions of these elements must also be excluded. This leads to

$$\epsilon_0 = 1 - \sum_h r_w^{0h} d\Omega_w^h \qquad (2\text{-}12)$$

for the emissivity of $d0$ in the direction from dw, when there exist several holes numbered from $h = 0$ to some finite integer.

For the second-order approximation, neglecting temperature variations, DeVos considers the use of a value for $L_{\lambda,n}^w$ which is not $L_{\lambda,B}$, but which is calculated by considering the effects of the holes on this spectral radiance (from each element dn). The second-order approximation for the quantity $^{\text{total}}\Phi_w^0$ is

$$^{\text{total}}\Phi_w^0 = L_{\lambda,B}(\lambda, T) \, dw \cos \theta_w^0 d\Omega_w^0 \left(1 - \sum_h r_w^{0h} d\Omega_w^h - \sum_h \int r_n^{wh} d\Omega_n^h r_w^{0n} d\Omega_w^n \right)$$

$$(2\text{-}13)$$

where the integration is over the entire surface excluding the holes.

DeVos applied this theory to the V-wedge, cylindrical (closed at one end), spherical, and tubular shapes. He did not treat the cone, a combination of cones, or a cone-cylinder.

For a cylinder of a radius, r, and a length, l (Figure 2-1), the value of ϵ_0 to a first-order approximation is

$$\epsilon_0 \simeq \left(1 - r_w^{00}\frac{\pi r^2}{l^2}\right) \simeq \left(1 - r_w^{00}\frac{\pi}{a^2}\right) \tag{2-14}$$

where $a = l/r$.

For the second-order approximation (neglecting temperature gradients), one needs $d\Omega_w^n$ and $d\Omega_w^0$. If dn is an annulus of the cylinder with a length dx, then

$$d\Omega_w^n = d\Omega_w^x = \frac{2\pi r^2 dx}{[(l-x)^2 + r^2]^{3/2}} \tag{2-15}$$

and

$$d\Omega_x^0 = \frac{\pi r^2}{x^2 + r^2} \tag{2-16}$$

If $x/r = y$, then

$$\epsilon_0 = 1 - r_w^{00}\frac{\pi}{a^2} - 2\pi^2 \int_0^a \frac{r_w^{0y}r_y^{w0}}{(y^2 + 1)[(y-a)^2 + 1]^{3/2}}dy \tag{2-17}$$

DeVos evaluated this expression by numerical integration. His values, corrected by Edwards [2-4] for a small numerical error, are given in Table 2-2. For a similar calculation for a sphere, DeVos obtained the (corrected) results in Table 2-3.

Table 2-2. DeVos' Emissivities, Cylindrical Blackbody[a] [2-4]

a	(d)	(s_1)	(s_2)	(s_3)
6	0.970	0.954	0.865	0.668
10	0.990	0.985	0.953	0.864
15	0.995	0.994	0.980	0.947
20	0.997	0.997	0.989	0.972
30	0.999	0.999	0.996	0.988

[a]Emissivity values for a cylindrical blackbody with second-order corrections for various values of a (= l/r = depth of cylinder/radius of hole) and surfaces of different smoothness. These are DeVos' values corrected for a numerical error.

Table 2-3. DeVos' Emissivities, Spherical Blackbody[a] [2-4]

a	(d)	(s_1)	(s_2)	(s_3)
10	0.992	0.989	0.963	0.894
20	0.998	0.993	0.991	0.976

[a]Emissivity values for a spherical blackbody with second-order corrections for various values of a (= l/r = diameter of sphere/radius of hole) and surfaces of different smoothness. These are DeVos' values corrected for a numerical error.

There is an attempt by DeVos to examine the effect of temperature gradients in a cavity. This factor is the most important in determining the quality of a blackbody since it is not very difficult to achieve emissivities as near to unity as desired. Manufacturers of blackbody simulators strive to achieve uniform heating of the cavity because it is only under this condition that the radiation is Planckian. The ultimate determination of a radiator that is to be used as the standard is the quality of the radiation that it emits.

Another review of methods for calculating blackbody effectiveness (including the so-called Sparrow method) is presented by Bedford [2-47].

There has been a division historically between the standards of photometry and those used to establish thermal radiation and the thermodynamic temperature scale. Thus, in photometry the standard has changed from the use of candles, the Carcel lamp, the Harcourt pentane lamp, and the Hefner lamp to the present primary standard of light adopted in January 1918 [2-5].

Primary Standard of Light. The primary standard now established is the candela, cd, corresponding to 1.0000 lm sr^{-1}, 0.10^4 Carcel unit (approximately), 1 pentane candle (approximately), 1 English sperm candle (approximately), and 1.11 Hefner unit (approximately). The construction of the primary standard of light is shown in Figure 2-3. The radiator itself consists of a small cylinder of pure fused thoria, about 45 mm long, with an internal diameter of about 2.5 mm and a wall thickness of 0.2 to 0.3 mm. This cylinder, the bottom of which is packed with powdered fused thoria to a depth of 10 to 15 mm, is supported vertically in a fused-thoria crucible of about 20 mm internal diameter nearly filled with pure platinum, as shown in Figure 2-3. The crucible has a lid with a small hole in the center, about 1.5 mm in diameter; this hole, which is the source of light, is surrounded above by a funnel-shaped sheath forming part of the crucible lid. The crucible is embedded in powdered fused thoria in a larger refractory container and is heated by enclosing it in a high-frequency induction furnace. The power required to melt the platinum is about 7 kW at a frequency of the order of 1 MHz. With this arrangement, it is possible to regulate the temperature so closely that the period required for the solidification of the platinum may exceed 20 minutes. The purity of the platinum is controlled by taking samples before and after use and determining the ratio of the electric resistance at 100 and 0°C. A minimum value of 1.390 for this ratio is required to ensure that the temperature of solidification is sensibly the same as that of pure platinum. This corresponds to an impurity of less than 3 parts in 100,000. The luminance (radiance) of the hole is given the value of 60 cd cm^{-2}.

To use the standard, one must first raise the temperature of the crucible and its container well above the melting point of platinum. Both are then allowed to cool slowly until the period of solidification of platinum is reached. (This period is used in preference

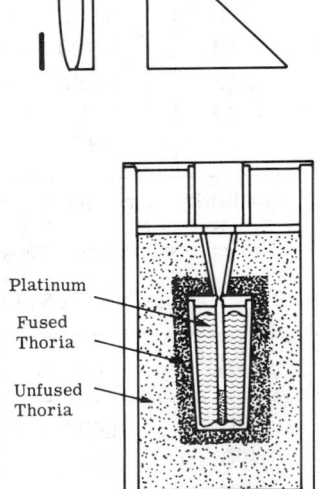

Fig. 2-3. The primary standard of light [2-5].

to the period of fusion because it gives more consistent results.) At this temperature, a change of 10 K would vary the luminous intensity by about 5 to 6%. The freezing point of platinum was recently measured as 1769.5 ± 0.6°C. An excellent treatment of the history of photometric standards especially with regard to standard lamps, is found in Jones and Preston [2-48].

Thermal Radiation and Radiometry. Two approaches have been taken to realize standards for thermal radiation and radiometry. One has been to achieve a "black" detector with extremely stable response characteristics and to use it in an arrangement by which electrical heating is introduced to balance the heat gained by incoming radiation. The other has been to produce a "black" source or blackbody radiator.

The attainment of a black detector or source involves

$$M = \epsilon \sigma T^4 \ \mathrm{W cm^{-2}} \qquad (2\text{-}18)$$

where ϵ = emissivity
   T = absolute temperature
   σ = Stefan-Boltzmann constant
   M = exitance

The National Bureau of Standards' standard of radiant energy or total thermal radiation was first established by Coblentz in 1914 with a ceramic wire-wound furnace used between about 1000 and 1150°C. This source was used to calibrate the carbon-filament lamp which was then issued by the Bureau. The NBS instructions on use of this now-obsolete standard carefully define the required geometry of the setup. The instructions, as revised in 1960, included statements to the effect that the values of radiant flux from this standard were based upon direct measurements and upon comparison to a blackbody using a value of 5.7 pW cm^{-2} K^{-4} for the Stefan-Boltzmann constant [2-6, 2-7, 2-8]. Results were believed to be accurate to about 1% for the absolute values of the primary standard and within about 1/2% for a compared secondary standard.

The source used by Coblentz was dismantled. Subsequently, another design was assembled with an oxidized nickel-chromium alloy cavity (80% Ni, 20% Cr) and used up to 1400 K. The cavity had an aperture of 0.95 cm diameter and a depth of about 12.5 or 14 mm. A calculation by the DeVos method gives a value of $\epsilon = 0.999$ for the cavity. At the operating temperature of 1300 K, the uncertainty is ± 0.2% or ± 1/2 K.

Using this blackbody for the 0.7 to 2.6 μm region with a higher-temperature cylindrical graphite source (estimated emissivity by the Gouffé method of 0.996) for the 0.25 to 0.75 μm region), Stair et al. calibrated tungsten strip lamps (GE 30A/T24/17) to be used as working standards. The uncertainties ranged from 8% at the short wavelengths to 3% at the long wavelengths [2-9].

For spectral irradiance from 0.25 to 2.6 μm, Stair et al. prepared (as working standards) tungsten-quartz-iodine 200 W lamps (GE 6.6A/T4Q 1 CL) and similar 1000 W lamps. Again the uncertainties were estimated to range from 8 to 3% [2-10].

Issued originally in 1973, the DXW spectral irradiance standard lamp has been replaced by the type FEL* lamp, the latter being easier to align and free from absorption bands and emission lines. The following material is quoted from the NBS description of the type FEL lamp.

---

*Sylvania-type FEL or GE equivalent.

II. Lamp Description

Type FEL lamps are 1,000 watt, clear bulb, quartz-halogen, tungsten coiled-coil filament (CC-8) lamps. They have a rated life of 500 hours at 120 volts. The lamps are munufactured with a two-pin base. Before calibration, the lamp base is converted to a medium bipost base and the base structure encapsulated in an epoxy compound [Figure 2-4]. The posts that form the medium bipost base are 1/4 inch diameter cylindrical stainless steel rods that extend 13/16 inch from the bottom of the epoxy block. The posts are spaced 7/8 inch between centers. A metal plate bearing the lamp identification number and indicating the electrical polarity is attached to the rear (side away from the radiometer) surface of the epoxy block.

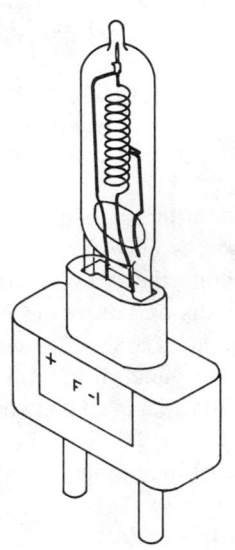

Fig. 2-4. FEL lamp.

III. Preparation and Screening

Before calibration all type FEL lamps are seasoned on direct current for 40 hours at 120 volts. Lamp output is then monitored at 654.6 mm for a 24-hour period to determine its drift rate. Only lamps exhibiting a drift of less than 0.5% for this period are selected for calibration (about 40% of the lamps fail this test).

All type FEL lamps are spectrally scanned from 250 to 800 nm to check for emission lines and absorption bands (0.02 - 0.03 nm bandpass). None of the approximately 80 type FEL lamps checked to date has exhibited either emission lines or absorption bands.

At the working distance of 50 cm, the irradiance field from type FEL lamps has been observed to be non-uniform to a small extent. The amount of non-uniformity varies from lamp to lamp. . .

IV. Orientation

During calibration, type FEL lamps are oriented as follows. The lamp is positioned base down with its identification plate facing away from the measuring instrument and with the base posts vertical. These posts are made perpendicular to and equidistant from the optical axis of the measuring instrument. The lower end of the base's positive post is 3.75 inches (approximately 9.53 cm) below the horizontal plane containing the optical axis. The plane tangent to the side of both posts nearest the measuring instrument is set 50 cm from the limiting aperture of the measuring instrument. This alignment fixes the lamp base posts (*not* the filament) relative to the optical axis of the measuring instrument. It is possible for the center of the filament to be serval millimeters off the optical axis with the lamp correctly aligned.

V. Operation

Type FEL lamps are calibrated while operating on direct current with the designated contact at positive potential. The exact operating current (set to the nearest 0.1 amp) is determined for each test lamp by matching its radiant output to the working standard lamps at 654.6 mm. Typically currents in the range 7.6 - 8.0 amps are used. Electrical measurements of the lamp operating current are made potentiometrically to an accuracy of 0.02%. After positioning and alignment, the lamps are slowly (30 seconds) brought up to the designated electrical operating point and allowed to stabilize for at least 10 minutes before irradiance measurements are made [2-11].

The NBS routinely supplies, with the exception generally of blackbodies, radiometric and photometric standards.*

... The Optical Radiation Section provides radiometric and photometric calibrations of three classes: Basic, Gage, and Special. Basic calibrations are those considered fundamental to all work in radiometry and photometry, and for which documentation of uncertainties relative to the International System of Measurements exists. Gage calibrations are those routinely available calibrations for which uncertainties have only been documented relative to NBS standards. Basic and gage calibrations are offered as fixed fee items, ... and require only a purchase order identifying the item number. Details of the calibration procedure and a summary of the documentation of the uncertainties are provided in each calibration report. The listed Basic and Gage calibrations are performed under rigid, predetermined conditions and are, therefore, restricted as to lamp type, measuring geometry, wavelength points, etc.; and requests for departure from these conditions will be considered as special calibrations.

... Special calibrations are those having unique requirements not satisfied by the listed Basic and Gage items, and are considered as small research efforts. They are accepted on a limited basis, in order to avoid serious disruption of the long term standards research effort. Fees are charged on an actual cost basis, with an estimate of cost, delivery time, and uncertainty being provided after receipt of description of desired test, and before actual work commences. The request for a special calibration should include the following information:

(1) Detailed description of desired calibration

(2) Uncertainty required (SI units, NBS standards)

(3) Manner in which calibration will be used

(4) The consequences of this calibration not being provided by NBS

221.123 Basic Radiometric Calibrations

(a) Spectral radiance standard, ribbon filament lamp 30A/T24/13, calibrated at 33 wavelengths from 225 to 2400 nm, target [sic] area 0.6 mm wide by .8 mm high, at a radiance temperature of about 2675 K at 225 nm, 2495 K at 650 nm, 2415 K at 800 nm and 1620 K at 2400 nm, with approximate uncertainties relative to SI units of spectral radiance of 4 1/2% at 225 nm, 1% at 650 and 800 nm, and 1 1/2% at 2400 nm. Lamp requires about 40 amperes DC at 12 volts. Interpolation formula allows calculation at all wavelengths except in regions of absorption bands. Lamp normally provided by NBS.

(b) Spectral radiance standard calibrated as ... (a) above for 20 wavelengths from 225 to 800 nm.

(c) Spectral radiance standard calibrated as ... (a) above for 17 wavelengths from 650 to 2400 nm.

(d) Spectral radiance standard, fixed temperature blackbody (500 - 700 K) sub-

---

*This quotation is edited and abridged. The values quoted are valid up to 1975. Beginning in June, 1975, the modified FEL lamp was used for irradiance calibrations. The values in square brackets represent post-1975 ones.

mitted for calibration at eight wavelengths from 1.5 to 14 μm, for a target size no larger than 2 mm wide by 4 mm high, one aperture between f/8 and f/23, physical dimensions no larger than 10 inches wide by 20 inches long by 6 inches high (to center of cavity aperture), with approximate uncertainties relative to SI units of spectral radiance of 2% at 1.5 μm to 1% at 14 μm.

(e) Spectral radiance standard, blackbody calibrated as . . . (d) above for four wavelengths in either 1.5 to 4 μm region or in 4 to 14 μm region.

(f) Spectral irradiance standard, quartz-halogen 1000 watt DXW lamp, calibrated at 24 wavelengths from 250 to 1600 nm, at a distance of 50 cm, at a spectral irradiance of about 0.2 watts/cm^3 at 250 nm, 220 [120] watts/cm^3 at 1600 nm, with approximate uncertainties relative to SI units of 5% [2.6%] at 250 nm, 1 1/2% [1.2%] at 650 nm, and 2% [1.2%] at 1600 nm. Lamp normally supplied by NBS, requires about 8 amperes DC at 110 [120] volts. Interpolation formula allows calculation of value at any wavelength except in regions of absorption bands.

(g) Spectral irradiance standard, lamp calibrated as . . . (f) above, from 250 to 750 [or 800] nm.

(h) Spectral irradiance standard, lamp calibrated as . . . (f) above, from 600 [or 650] to 1600 nm.

(i) Irradiance standard, quartz-halogen, 1000 watt DXW [modified type FEL] lamp in reflector submitted for calibration at distance of 40 cm, irradiance level about 130 mw/cm^2, approximate uncertainty 1/2% relative to SI units. Lamp requires about 8 amperes DC at 110 volts.

(j) Irradiance standard, quartz-halogen 1000 watt DXW [modified type FEL] lamp calibrated at distance of 50 cm, irradiance level about 30 mw/cm^2, approximate uncertainty 1% relative to SI units. Lamp normally supplied by NBS, requires about 8 amperes DC at 110 volts.

(k) Irradiance standard, airway beacon lamp 500 T 20/13 calibrated at distance of 100 cm, irradiance level about 3 mw/cm^2, approximate uncertainty 2% relative to SI units. Lamp normally supplied by NBS, requires 110 volts DC.

(1) Irradiance standard, microscope illuminator lamp 100 T 8 1/2 9 submitted for calibration at distance of 100 cm, irradiance level of 0.6 mw/cm^2, approximate uncertainties 2% relative to SI units. Lamp requires 110 volts DC.

221.124 Basic Photometric Calibrations

(a) Luminous intensity standard, 100 watt tungsten lamp c 13 B filament, inside frosted bulb, medium bipost base calibrated for luminous intensity at about 90 candelas, approximate color temperature 2700 K, approximate uncertainty 4% of luminous intensity relative to SI units, 1.5% relative to NBS standards. Lamp normally provided by NBS, requires 110 volts.

(b) Luminous intensity standard calibrated as . . . (a) above, with calibration performed at lamp current required to produce color temperature of 2700 K, color temperature uncertainty approximately 9 K relative to NBS standards.

(c) Luminous intensity standard calibrated as . . . (b) above, with calibration performed at lamp current required to produce color temperature of illuminant A (2856 K), approximate luminous intensity 140 candelas, lamp requires about 127 volts. Color temperature uncertainty approximately 9 K relative to NBS standards.

(d) Luminous intensity standard, 500 watt tungsten lamp c 13 B filament, inside frosted bulb, medium bipost base, calibrated for luminous intensity at about 700 candelas, approximate color temperature 2856 K, approximate uncertainty 4% of luminous intensity relative to SI units, 1.5% relative to NBS standards. Lamp normally provided by NBS, requires 110 volts.

(e) Luminous intensity standard calibrated as . . . (d) above, with calibration performed at current required to produce color temperature of illuminant *A* (2856 K), color temperature uncertainty approximately 9 K relative to NBS standards.

(f) Luminous intensity standard, 1000 watt tungsten lamp *C* 13 *B* filament, inside frosted bulb, medium bipost base, calibrated for luminous intensity at about 1400 candelas, approximate color temperature 2856 K, approximate uncertainty 4% of luminous intensity relative to SI units, 1.5% relative to NBS standards. Lamp normally provide by NBS, requires 110 volts.

(g) Luminous intensity standard calibrated as . . . (f) above, with calibration performed at current required to produce color temperature of illuminant *A* (2856 K), color temperature uncertainty approximately 9 K relative to NBS standards.

(h) Luminous flux standard (geometrically total), 25 watt vacuum lamp, base-up burning submitted for calibration for luminous flux at about 270 lumens; approximate color temperature 2500 K, approximate uncertainty 4.5% of luminous flux relative to SI units, 2% relative to NBS standards. Lamp requires 120 volts.

(i) Luminous flux standard (geometrically total), 60 watt gas-filled lamp, base-up burning, submitted for calibration for luminous flux of about 870 lumens, approximate color temperature 2800 K, approximate uncertainty 4.5% of luminous flux relative to SI units, 2% relative to NBS standards. Lamp requires 120 volts.

(j) Luminous flux standard (geometrically total), 100 watt gas filled lamp, base-up burning, submitted for calibration for luminous flux of about 1600 lumens, approximate color temperature 2900 K, approximate uncertainty 4.5% of luminous flux relative to SI units, 1.5% relative to NBS standards. Lamp requires 120 volts.

(k) Luminous flux standard (geometrically total) 200 watt gas-filled lamp, base-up burning submitted for calibration for luminous flux of about 3300 lumens. approximate color temperature 3000 K, approximate uncertainty 4.5% of luminous flux relative to SI units, 1.5% relative to NBS standards. Lamp requires 120 volts.

(l) Luminous flux standard (geometrically total), 500 watt gas filled lamp, base-up burning, submitted for calibration for luminous flux of about 10,000 lumens, approximate color temperature 3000 K, approximate uncertainty 4.5% of luminous flux relative to SI units, 1.5% relative to NBS standards. Lamp requires 120 volts . . .

232.071 Near and Vacuum Ultraviolet

(z) Near and vacuum ultraviolet spectral radiance standard, deuterium arc lamp, calibrated in wavelength range from 165 nm to 300 nm, with approximate uncertainties of 15% at 165 nm and 10% above 200 nm. Target region approximately 0.3 nm diameter; lamps are commercially available. Fee is charged dependent on details of the test [2-12].

Working standard samples of various NBS-type standard lamps can be obtained from companies such as The Eppley Laboratory, Inc., Newport, Rhode Island, and Optronic Laboratories, Inc., Silver Spring, Maryland.

Deuterium Lamp Standards of Spectral Irradiance

...Modified deuterium lamps manufactured by Cathodeon (model C70. 3V.H) have been selected for issuance. Care has been [is] taken to insuere [sic] maximum interchangeability with the type FEL incandescent lamp standards of spectral irradiance presently available from NBS.

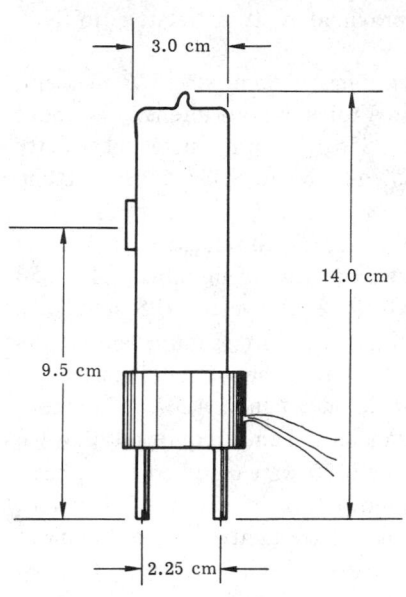

Fig. 2-5. Lamp modified for calibration.

II. Lamp Description

The lamp as modified for calibration is shown in Figure [2-5]. Mechanically, but *not* electrically, the lamps have been mounted in a black anodized medium bipost base. The exact orientation of the lamp envelope with respect to the pins of the medium bipost base is determined as follows. The lamp is set loosely in the base and the spectral irradiance produced at 250 nm along the optical axis, described below, is examined. The lamp is rotated (pitch, roll, and yaw) until a peak output is observed. The lamp is then cemented into the bipost base at this orientation. Thus, lamp envelopes may have a skew orientation with respect to the base pins.

III. Lamp Operation

A. Electrical

The lamps are calibrated while operating at a dc current of about 300 ma (0.1% regulation; operating voltage approximately 100 volts). A current limiting device, such as a constant current dc power supply or external resistor, must be used. In order to start the lamps, the filament is first heated (about 4 volts, 5 amps) for about 5 second. A starting voltage of 400 volts is then applied to the lamp. After the lamp has started, the heater circuit is turned off and the lamp allowed to stabilize at 300 ma for 20 minutes before spectral irradiance measurements are made.

B. Optical

Spectral irradiance measurements are made at a position which is specified with respect ot the posts of the medium bipost base. As shown in Figure [2-5], the optical axis of the measuring instrument intersects the lamp at a point 9.5 cm above the bottom of the posts. The posts are made perpendicular to and equidistant from the optical axis of the measuring instrument. The plane tangent to the side of both posts nearest the measuring instruments is set 50 cm from the limitng aperture of the measuring instrument.

Values of spectral irradiance 50 cm from the tangent plane of the base posts described above, will be reported at 10 nm intervals for the spectral range 200 to 350 nm. The spectral irradiance produced under these conditions is a monotonically increasing function starting at 0.07 watts/cm^3 at 350 nm and increasing to 0,55 watts/cm^3 at 200 nm. At 260 nm the spectral irradiance produced by the deuterium

Table 2-4. Miniature Lamp Standards of Luminous Flux* [2-45]

Lamp Type	Current (set) (A)	Voltage (V)	Color Temperature (K)	Luminous Flux (lm)	(MS Cp)**
1183	5.80	5.0	2950	400.0	32.0
1183	5.55	4.7	2856	300.0	24.0
1133	3.50	5.3	2856	220.0	18.0
87	1.80	6.1	2856	130.0	10.0
81	1.00	6.1	2856	66.0	5.2
63	0.63	7.4	2750	44.0	3.5
51	0.197	6.3	2400	6.2	0.49

*In response to many requests from industry, the Optical Radiation Section has established a calibration service for miniature lamp standards of luminous flux (6 to 400 lumens). Beginning December 1, 1973, six types of calibrated lamps were available on a routine basis. To promote operating stability, all of the issued lamps . . . have white electrical leads soldered to their bases and all calibrations are performed with the lamps operating at a set current.
**MS Cp, Mean Spherical Candle power.

lamp (0.24 watts/cm^3) is approximately equal to that produced by the incandescent type FEL lamps. . . [2-11].

Miniature lamp standards are listed in Table 2-4.

A series of blackbodies using the melting and freezing points of several metals has been made by the NBS for use in a continuing research program by Kostkowski, Richmond, and their coworkers to increase the precision and accuracy of radiation standards [2-49, 2-50]. In particular, the goldpoint blackbody (Figure 2-6) represents a close approach to a blackbody.

Work by Kostkowski, Erminy, and Hattenberg at NBS required a blackbody usable to 3000 K, stable to within 0.1% in temperature for 30 to 45 minutes, and with a uniformity of 5°C or better. A graphite cylinder used for this source is believed to have achieved these values with an ϵ = 0.999. The temperature measurement required direct photoelectric pyrometry. The pyrometer was calibrated with respect to the International Practical Temperature Scale, IPTS, using the gold-point blackbody. In the vicinity of 6545 Å, the radiation is believed to be within 0.01% of that from a blackbody at the temperature of freezing gold. The resulting total uncertainty of the spectral-radiance determination was believed to be 0.35% relative to IPTS at 6500 Å and 1.5% at 2100 Å. These uncertainties are changed to 1 and 3.4% respectively, for the thermodynamic temperature scale, TTS, due to present uncertainties of the temperature for freezing gold on the TTS. The gold-point source is limited by the temperature value 1064.43°C assigned to the melting point of gold.

Wavelength Standard. One of the most important standard sources became important in 1960, when the international standard of length became the wavelength in vacuum of a spectral line, the orange line of ^{86}Kr. Thus, the meter was defined as 1,650,763.73

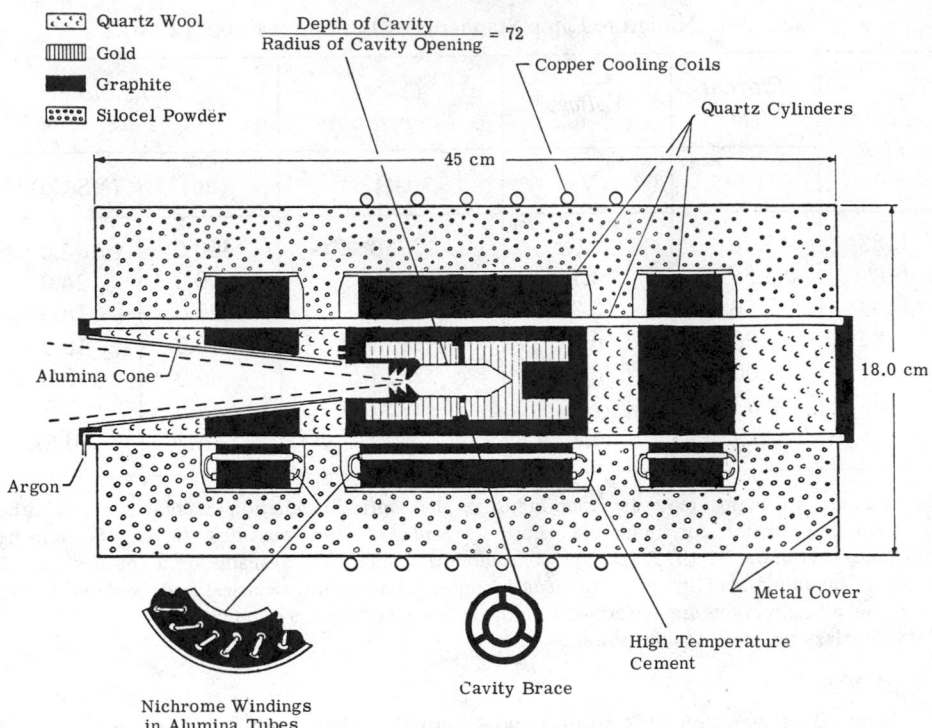

Fig. 2-6. Cross section of the National Bureau of Standards' horizontal gold point blackbody and furnace [2-13].

wavelengths of the radiation from the transitions between the $2p_{10}$ and $5d_5$ levels in that isotope of krypton. The vacuum wavelength was stated as

$$\lambda_{vac} = 6057.80210_5 \text{ Å} \tag{2-19}$$

2.3. Commercial Sources

2.3.1. Blackbodies. Virtually any cavity can be used to produce radiation of high quality, but practicality limits the shapes to a few. The most popular shapes are cones and cylinders, the former being more popular. Spheres, combinations of shapes, and even flat-plate radiators are used occasionally. Blackbodies can be bought rather inexpensively, but there is a fairly direct correlation between cost and quality (i.e., the higher the cost the better the quality).

Few manufacturers specialize in blackbody construction. Some, whose products are specifically described here, have been specializing in blackbody construction for many years. Any other companies of this description which exist did not appear as such in the 1976 "Optical Industry and Systems Directory."

The largest known selection of standard (or blackbody) radiators is offered by Electro-Optical Industries, Inc. (EOI), Santa Barbara, California. For example, Table 2-5 shows a summary of the various, more-or-less conventional types offered by EOI, along with some not-so-conventional ones. As depicted in Figure 2-7 most blackbodies can be characterized

Table 2-5. Summary of Families of Available Blackbodies

Series	Features	Temperature Range	No. of Models
Primary standard series	Metal freezing points	−78 − 1772°C	8
Secondary standard	Very high quality reverse cone and spherical cavities	50 − 1000°C	6
Working standard	High quality simple cavities	10 − 3300 K	126
MS series	Lower quality competitive series	50 − 1000°C	8
Aerospace series	Minimum size controllers small blackbodies	200 − 1200 K	24
LT series	Cryogenic vacuum operation	10 − 400 K	14
V series	Vacuum operation	100 − 1200 K	24
Extended area	Flat plates, V grooves, and Intersecting cones	100 − 1073 K	45 24 24
Thermoelectric cooled sources	Heat and cool electrically up to 12 inches square or larger	0 − 60°C −30 − 80°C	36
Refrigerated source	Using conventional freon compressor	−50 − 600°C	8
Differential targets	High resolution and uniformity	−30 − 100°C	18
Gradient target	Adjustable gradient	Gradient 0 − 50°C range ambient to 200°C	2
Rate-table blackbodies	Special purpose remote control	50 − 1200°C	12
High speed blackbodies	Fast warm up and cool down	50 − 800°C	6
Gas fired	Flat plates and cavities	1000 − 2000 K	4

as one of the following: primary, secondary, and working standard. The output of the primary must, of course, be checked with those standards retained at NBS.

Table 2-6 demonstrates the various working standards obtainable from EOI blackbodies. Figure 2-8 pictures blackbody no. 153 and its controller. Table 2-7 is a similar listing of the various working standards offered by Barnes Engineering. Figure 2-9 pictures blackbody no. 11-200T and its controller. Finally, Table 2-8 shows the different working

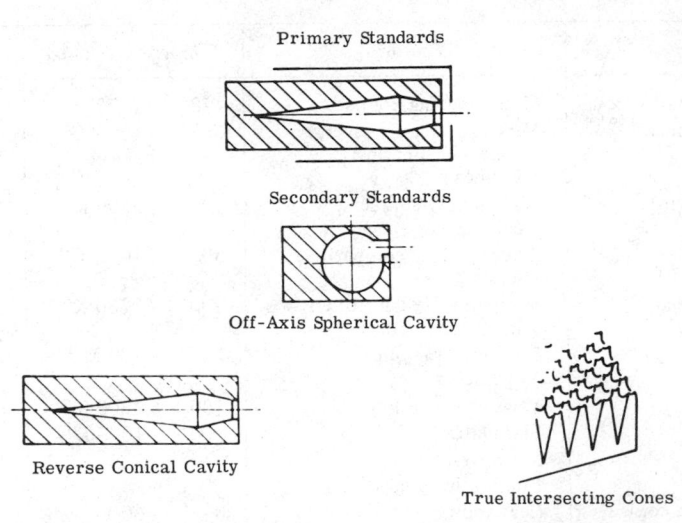

(*a*) Blackbody sources. Primary standard: freezing point sources using lead, tin, zinc, aluminum, silver, gold, copper, or platinum. Uniformity of temperature $\sim 0.01°$C, $\epsilon = 0.999$. Secondary standards: uniformity of temperature $\sim 0.05°$C, $\epsilon = 0.999$.

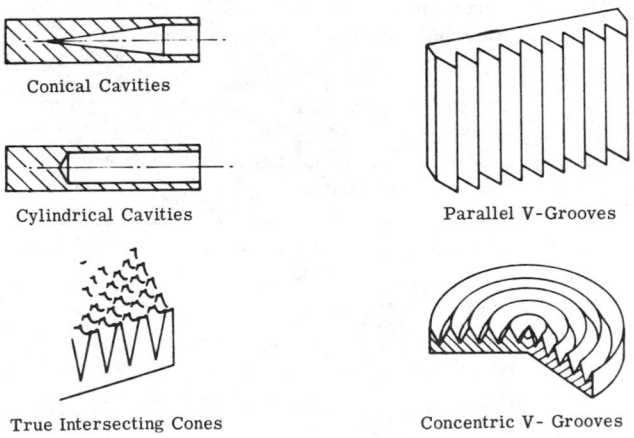

(*b*) Blackbody working standards. Uniformity of temperature $\sim 1°$C. $\epsilon = 0.99$ to 0.999 depending on temperature range and configuration

Fig. 2-7. Blackbody standards and sources.

Table 2-6. Specifications of Working Standards Series from EOI [2-14]

Specification Categories	For Models up to 600°C	For Models above 600°C	For 1900 K Model
Control accuracy	±1°C	±1°C	±1°C
Stability (Long term)	0.1°C	0.5°C	0.05%
Stability (Short term)	0.02°C	0.25°C	0.25°C
Sensing element	Platinum resistance thermometer	Platinum resistance thermometer	Silicon detector
Type of control	Linear proportional, true null	Linear proportional, true null	Linear proportional, true null
Cavity	Recessed cone 15°	Recessed cone 15°	Recessed cone 15°
Cavity emissivity	0.99 ± .01	0.99 ± .01	0.99 ± .01
Source housing temperature	Less than 10°C above ambient	Less than 10°C above ambient	Less than 10°C above ambient
Ambient temperature range	-40 to 60°C	-40 to 60°C	-40 to 60°C

Model	Temperature Range	Cavity (in.)	Aperture Set Included	Field of View[a]	Thermo-couple	Temperature Controller Model	Control Power (W)	Maximum Input (W)	Warm-up Time (min)	Dimensions H × W × D (in.)	Weight (lb)
111	50-600°C	0.040	none	10°	no	210	3	10	1.5	3/8 dia. × 1/2 long	0.5 oz
113	50-1000°C	0.040	none	10°	no	210	6	25	3.0	2.50 × 2 × 3	1.5
114	500-1500 K	0.040	none	10°	no	210	8	40	8	4.25 × 4 × 6.50	5
115	1000-1900 K	0.040	none	5°	no	210B	10	45	10	4.25 × 4 × 6.50	5
121	50-600°C	0.080	.015" dia.	10°	no	210	4	10	1.5	1.25 × 0.75 × 1.75	4 oz
123	50-1000°C	0.080	.015" dia.	10°	no	210	7	25	3	3 × 3 × 4.75	3
125	1000-1900 K	0.080	.015" dia.	5°	no	210B	12	45	10	4.25 × 4 × 6.50	5
131	50-600°C	0.25	AS-1	15°	no	210B	18	40	20	4.25 × 4 × 6.50	6
133	50-1000°C	0.25	AS-1	15°	yes[b]	210B	50	100	15	7.25 × 6.50 × 11.50	17
135	1000-1900 K	0.25	AS-1	10°	no	210C	200	400	60	8.25 × 7.25 × 11.50	25
141	50-600°C	0.50	AS-2 or 5	15°	yes[c]	210B	60	120	20	7.25 × 6.50 × 11.50	15
142	325-1000 K	0.50	AS-2 or 5	15°	yes[c]	210C	90	200	45	7.25 × 6.50 × 11.50	17
143	50-1000°C	0.50	AS-2 or 5	15°	yes[b]	215B	120	300	75	8.25 × 7.25 × 11.50	25

[a] Nominal Value, Cosine Distribution.
[b] Platinum vs Platinum-10% Rhodium Thermocouple.
[c] Chromel/Alumel Thermocouple.

Table 2-6. Specifications of Working Standards Series from EOI [2-14] (Continued)

Model	Temperature Range	Cavity (in.)	Aperture Set Included	Field of View[a]	Thermo-couple	Temperature Controller Model	Control Power (W)	Maximum Input[a] (W)	Warm-up Time (min)	Dimensions H × W × D (in.)	Weight (lb)
144	500–1500 K	0.50	AS-2 or 5	15°	no	215B	180	350	60	8.25 × 7.25 × 11.50	25
145	1000–1900 K	0.50	AS-2 or 5	10°	no	216	400	1000	90	10.50 × 9.50 × 14	30
146	1000–3000°C	0.50	AS-2	5°	no	204	—	5000	30–90	50 × 32 × 59	900
147	400–1400°C	0.50	AS-2 or 5	10°	no	205	350	1000	60	11.13 × 10.25 × 14.25	30
152	325–1000 K	1.0	AS-3	30°	yes[c]	215B	200	400	60	8.25 × 7.50 × 14.25	25
153	50–1000°C	1.0	AS-3	30°	yes[b]	205	300	1000	90	10.50 × 9.50 × 14	30
154	500–1500 K	1.0	AS-3	25°	no	205	300	1000	90	11.13 × 10.25 × 14.25	30
155	1000–1900 K	1.0	AS-3	15°	no	216	500	1000	120	13 × 12 × 20	35
157	400–1400°C	1.0	AS-3	20°	no	205	300	1000	120	11.13 × 10.25 × 14.25	30
161	50–600°C	2.0	AS-9	45°	yes[c]	215C	300	600	60	14 × 13 × 18	60
163	50–1000°C	2.0	AS-9	40°	yes[b]	205	500	1000	120	14 × 13 × 18	60
165	1000–1900 K	2.0	AS-9	20°	no	216B	1000	2000	180	14 × 13 × 21.75	70
167	400–1400°C	2.0	AS-9	25°	no	205B	800	1500	150	14 × 13 × 21.75	70
173	50–1000°C	3.0	AS-4	15°	yes[b]	205B	900	2000	180	21 × 19 × 30	350
181	50–600°C	4.0 (sq.)	none	90°	no	205		1000	75	12 × 13 × 8	35

Refrigerated Blackbody Sources

Model	Temperature Range	Cavity (in.)	Aperture Set Included	Field of View[a]	Thermo-couple	Temperature Controller Model	Warm-up Time (min)	Cool-down Time (min)	Dimensions H × W × D (in.)	Weight (lb)
1951	–20 to 100°C	1.0	AS-3	30°	yes[c]	215B	10	10	13 × 12 × 20	50
1952	–50 to 100°C	1.0	AS-3	30°	yes[c]	215B	10	10	13 × 12 × 20	50
1953	–20 to 600°C	1.0	AS-3	30°	yes[c]	215B	10	20	13 × 12 × 20	50
1954	–50 to 600°C	1.0	AS-3	30°	yes[c]	215B	10	20	13 × 12 × 20	50

[a] Nominal Value, Cosine Distribution.
[b] Platinum vs Platinum-10% Rhodium Thermocouple.
[c] Chromel/Alumel Thermocouple.

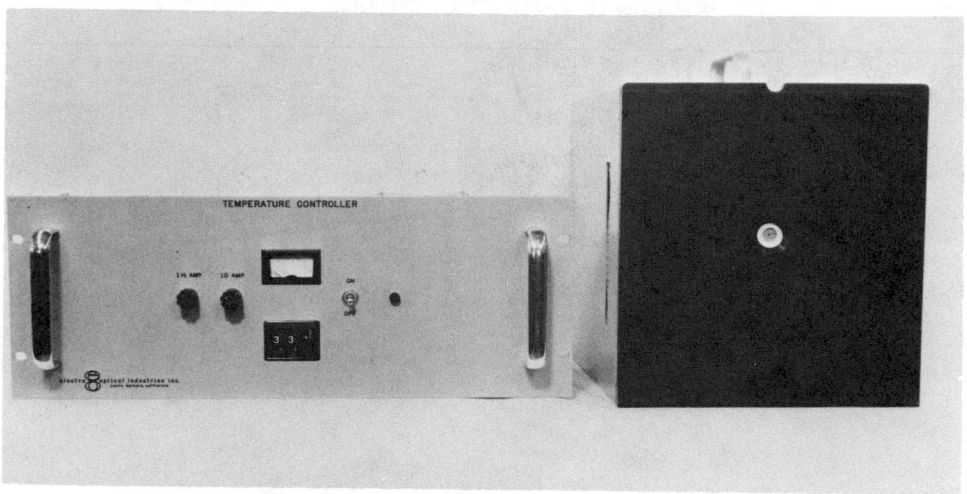

Fig. 2-8. EOI WS #153 and its controller with a one-inch cavity and an operating range of 50 to 1000°C.

Table 2-7. Barnes Engineering Radiation Reference Sources [2-15]

Model No.	Temperature Range (°C)	Maximum Aperture Diameter (in.)	Maximum Power Required (at 115V, 50-60 cps) (W)	Blackbody Simulator Case Style	Controller Case Style
11-100T	60-230°	0.625	25	C	B
11-101T	0-230°	0.625	30	D	A
11-120a	200-600°	0.008-0.140	55	F	A
FCS-1b	100-700°F	3.0	325	E	N/A
11-200T	50-1000°	0.50	275	G	A
11-201T	50-1000°	1.00	500	G	A
RM-121	60-230°	0.020	20	H	B
11-140T	0-230°	12 × 12	1250	J	A

aNow designed for use with interchangeable apertures. Typical standard apertures are .0086, 0.015, and 0.040 in.
bIntended primarily for use with the infrared microscope.

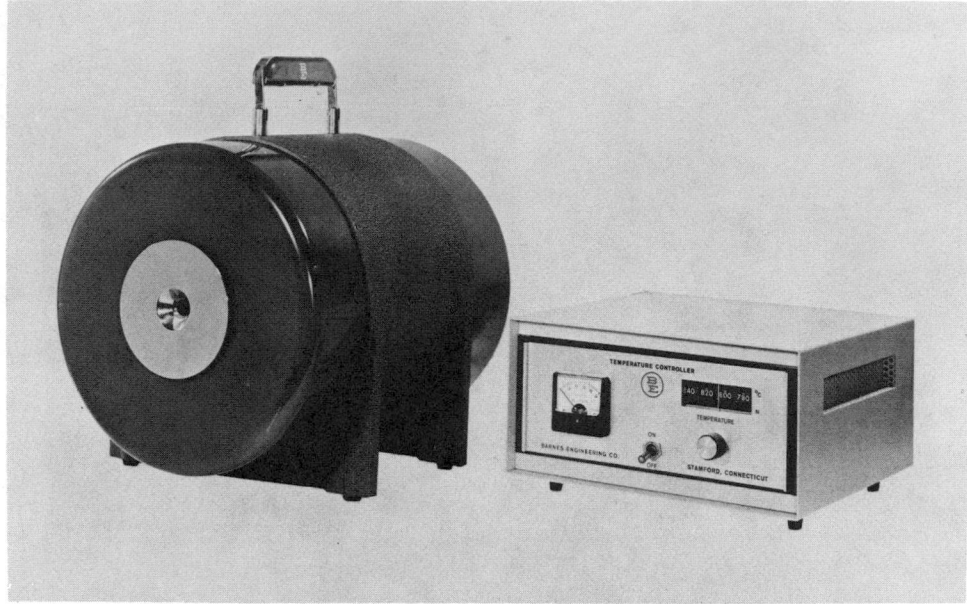

Fig. 2-9. Barnes Engineering #11-200T and its controller.

standards obtainable from Infrared Industries. Figure 2-10 shows an example of their Series 400. All of the companies sell separate apertures (some of which are water cooled) for controlling the radiation output of the radiators. Another piece of auxiliary equipment which can be purchased is a multispeed chopper.

Figure 2-11 demonstrates two of the less-conventional working standards manufactured by EOI. The radiator on the left consists of parallel V-grooves, the one on the right of an array of intersecting conical cavities. Figure 2-12 is EOI's Model 19165 radiator. It is 12 in. in diameter and 9 in. deep. The base is an array of intersecting conical cavities whose walls form a hexagonal honeycomb. The operating temperature range is 175 to 340 K.

Information on blackbody simulators was also obtained from the following companies:

(1) DBA Systems, Inc., Melbourne, Florida
(2) Advanced Kinetics, Inc., Costa Mesa, California
(3) Ircon, Inc., Skokie, Illinois
(4) Optronic Laboratories, Inc., Silver Spring, Maryland
(5) The Eppley Laboratory, Inc., Newport, Rhode Island

DBA Systems issues both Series 202 and 210 Ambient Range Sources, which have multiple elements that are investment castings of four overlapping 15-degree conical cavities to ensure lambertian emission ± 10 degrees from 0 to 99.9°C. The model 202 series incorporates sources with from 2 to 9 cavity units, depending on the temperature range desired. The Model 210 Ambient Range Source uses a single aperture cavity element similar to the one described above, operating at controlled temperatures over the range 0 to 99.9°C.

Table 2-8. Infrared Industries Blackbody Radiation Sources Series 400 [2-16]

Model	Temperature Range	Cavity (in.)	Apertures (dia.; in.)	Emissivity	Field of View[a]	Warm-Up Time (Min)	Controller[b]	Maximum Input (W)	Thermo-Couple	Dimensions H × W × D (in.)	Shipping Weight (lb)
407A	50-600°C	0.08	0.015[c]	0.99 ± 0.01	20°	15	101B	35	Available	3-1/2 × 1-3/4 × 3-1/4	13 oz
408	50-600°C	0.25	0.1[c]	0.99 ± 0.01	50°	10	101B	60	Yes[d]	5-1/2 × 3 × 4-3/8	2
413	500 K	0.5	0.2[c]	0.99 ± 0.01	12°	20	Built in	100	No	7-1/2 × 6 × 3-1/2	5
414	50-250°C	0.5	0.2[c]	0.99 ± 0.01	12°	20	Built in	100	Available	8 × 6-3/8 × 4	5
427	50-600°C	0.08	–	0.99 ± 0.01	–	15	101B	15	Available	5/8 dia. × 3-1/2 lg.	10 oz
440	50-600°C	0.08	–	0.99 ± 0.01	–	5	None	35	No	2-1/4 dia. × 1-1/4 lg.	8 oz
436	400-1500°C	0.5	0.2-0.0125	0.99 ± 0.01	12°	3(h)	Manual or 101B	300	Yes[d]	10 × 8 × 11-3/8	15
461	325-1000 K	1.	0.6-0.0125	0.99 ± 0.01	20°	60	101B	460	Yes[d]	10 × 8 × 11-3/8	17
463	50-1000°C	1.	0.6-0.0125	0.99 ± 0.01	20°	60	101B	460	Yes[d]	10 × 8 × 11-3/8	17
464	50-1200°C	0.5	0.2-0.0125	0.99 ± 0.01	12°	60	101B	400	Yes[d]	10 × 8 × 11-3/8	15

[a] With small aperture.
[b] Temperature Controller Specifications.
[c] Other sizes available.
[d] Platinum versus platinum–10% rhodium thermocouple.

Fig. 2-10. Infrared Industries Series 400.

(a) Parallel V-Grooves   (b) An array of intersecting conical cavities.

Fig. 2-11. Two emitting large-area surfaces by EOI.

Fig. 2-12. EOI model 19165. This model is 12 in. in diameter and 9 in. deep. The base is an array of intersecting conical cavities. The walls are hex-honeycomb and the temperature range is 175 to 340 K.

Advanced Kinetics manufactures a Model BB Radiation Source with a 10 mm standard cavity aperture, operating in a range of 300 to 500 K (Model BB-500) and 300 to 1000 K (Model BB-1000). The Ircon Source, Model BC-15, operates between ambient and 815°C. The aperture diameter 0.50 in. Optronic Laboratories issues a copper-point source used primarily for calibrating and monitoring the long-term stability of automatic and visual optical pyrometers at 1083.3 ± 0.1°C.

The Eppley Laboratory issues a low-temperature source for use generally in precision radiometry and in the calibration of satellite spectrometers. An array of blackened copper cavities of hexagonal cross-section (honeycomb) constitutes the radiator with cell wall thickness of 0.1 mm. Operating ranges of –100 to + 70°C have been achieved.

2.3.2. Incandescent Nongaseous Sources (Exclusive of High-Temperature Blackbodies).

Nernst Glower. The Nernst glower is usually constructed in the form of a cylindrical rod or tube from refractory materials (usually zirconia, yttria, beria, and thoria) in various sizes. Platinum leads at the ends of the tube conduct power to the glower from the source. Since the resistivity of the material at room temperature is quite high, the working voltage is insufficient to get the glower started. Once started, its negative temperature coefficient-of-resistance tends to increase current, which would cause its destruction, so that a ballast is required in the circuit. Starting is effected by applying external heat, either with a flame or an adjacent electrically-heated wire, until the glower begins to radiate.

Data from Infrared Industries for specific-sized glower and operating conditions are as follows:

(1) Power requirements: 117 V, 50 to 60 A, 200 W
(2) Color temperature range: 1500 to 1950 K
(3) Dimension: 0.05 in. diameter by 0.3 in.

The spectral characteristics of a Nernst glower in terms of the ratio of its output to that of a 900°C blackbody are shown in Figure 2-13.

The life of the Nernst glower diminishes as the operating temperature is increased. Beyond a certain point, depending on the particular glower, no great advantage is gained by increasing the current through the element. The glower is fragile, with low tensile strength, but can be maintained intact with rigid support. The life of the glower depends on the operating temperature, care in handling, and the like. Lifetimes of 200 to 1000 hours are claimed by various manufacturers.

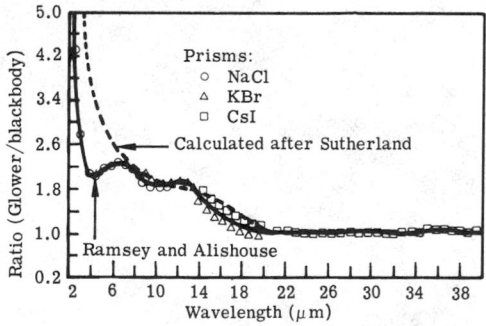

Fig. 2-13. The ratio of a Nernst glower to a 900°C blackbody versus wavelength [2-17].

Since the Nernst glower is made in the form of a long thin cylinder, it is particularly useful for illuminating spectrometer slits. Its useful spectral range is from the visible region to almost 30 μm, although its usefulness compared with other sources diminishes beyond about 15 μm. As a rough estimate, the radiance of a glower is nearly that of a graybody at the operating temperature with an emissivity in excess of 75%, especially below about 15 μm.

The relatively low cost of the glower makes it a desirable source of moderate radiant power for optical uses in the laboratory.

The makers of spectroscopic equipment constitute the usual source of supply of glowers (or of information about suppliers).

Globar. The globar is a rod of bonded silicon carbide usually capped with metallic caps which serve as electrodes for the conduction of current through the globar from the power source. The passage of current causes the globar to heat, yielding radiation at a temperature above 1000°C. A flow of water through the housing that contains the rod is needed to cool the electrodes (usually silver). This complexity makes the globar less convenient to use than the Nernst glower and necessarily more expensive. This source can be obtained already mounted from a number of manufacturers of spectroscopic equipment. Feedback in the controlled power source makes it possible to obtain high radiation output.

Ramsey and Alishouse [2-17] provide information on a particular sample globar as follows:

(1) Power consumption: 200 W, 6 A
(2) Color temperature: 1470 K

They also provide the spectral characteristics of the globar in terms of the ratio of its output to that of a 900°C blackbody. This ratio is plotted as a function of wavelength in Figure 2-14.

Figure 2-15 is a representation of the spectral emissivity of a globar as a function of wavelength. The emissivity values are only representative and can be expected to change considerably with use.

Gas Mantle. The Welsbach mantle is typified by the kind found in high-intensity gasoline lamps used where electricity is not available. The mantle is composed of thorium oxide with

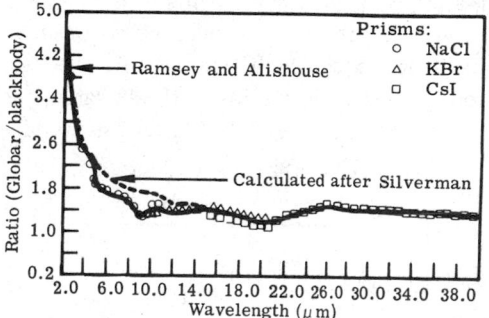

Fig. 2-14. The ratio of a Globar to a 900°C black-body versus wavelength [2-17].

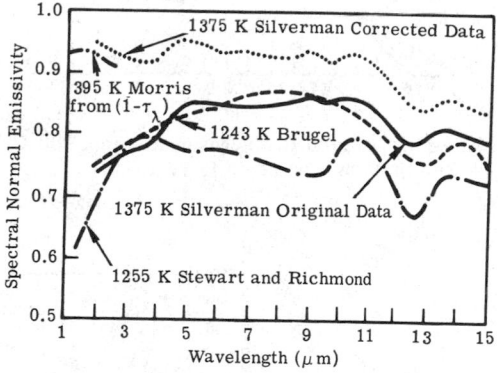

Fig. 2-15. The spectral emissivity of a Globar [2-18].

some additive to increase its efficiency in the visible region. Its near-infrared emissivity is quite small, except for regions exemplified by gaseous emission, but increases considerably beyond 10 μm.

Ramsey and Alishouse [2-17] provide information on a propane-heated sample from an experiment in which a comparison of several sources is made:

(1) Color temperature: 1670 K

(2) Dimensions: 25.4 by 38.1 mm

The spectral characteristics of the mantle in terms of the ratio of its output to that of a 900°C blackbody are shown in Figure 2-16.

Pfund modified the gas mantle so that it became more a laboratory experimental source than an ordinary radiator. By playing a gas flame on an electrically heated mantle, he was able to increase its radiation over that from the gas mantle itself [2-19]. Figure 2-17 shows a comparison of the gas mantles and the gas, electrically heated mantle with a Nernst glower. Strong [2-20] points out that playing a flame against the mantle at an angle produces an elongated area of intense radiation useful for illuminating the slits of a spectrometer.

Comparison of Nernst Glower, Globar, and Gas Mantle. Figure 2-18 compares these three types of sources, omitting a consideration of differences in the instrumentation used in making measurements of the radiation from the sources.

Availability, convenience, and cost usually influence a choice of sources. At the very long wavelength regions in the infrared, the gas mantle and the globar have a slight edge

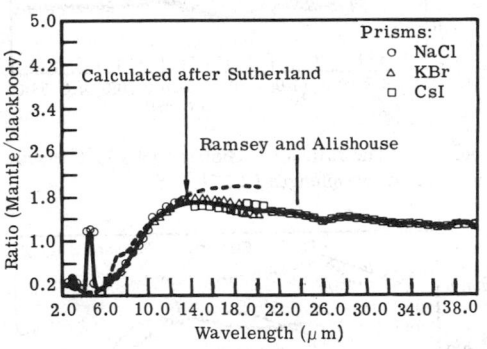

Fig. 2-16. The ratio of the gas mantle to a 900°C blackbody versus wavelength [2-17].

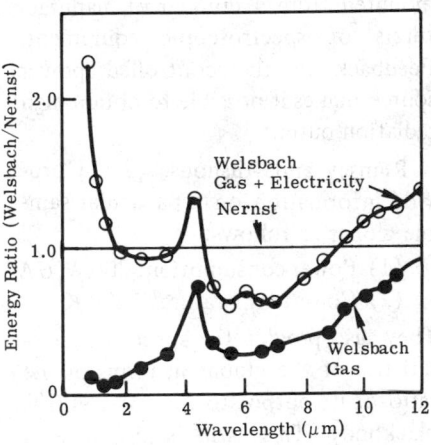

Fig. 2-17. Emission relative to that of a Nernst glower (2240 K) of the gas-heated mantle (lower curve) and that of the mantle heated by gas plus electricity (upper curve) [2-19].

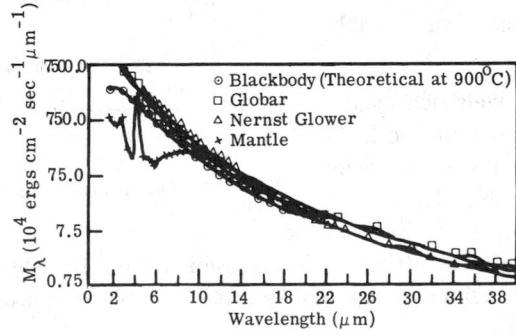

Fig. 2-18. The spectral radiant emittances of a Globar, Nernst glower, 900°C blackbody, and gas mantle versus wavelength [2-17].

over the Nernst glower because the Nernst glower—a convenient, small, and inexpensive source—does not have the power of the gas mantle and globar.

Tungsten-Filament Lamps*. A comprehensive discussion of tungsten-filament lamps is given by Carlson and Clark [2-21]. Figures 2-19 to 2-21 show the configurations of lamp housings and filaments. Tables 2-9 to 2-11 give pertinent characteristics of tungsten and photoflood lamps. In addition to those shown in the tables, a variety of miniature and subminiature lamps can be obtained from various manufacturers.

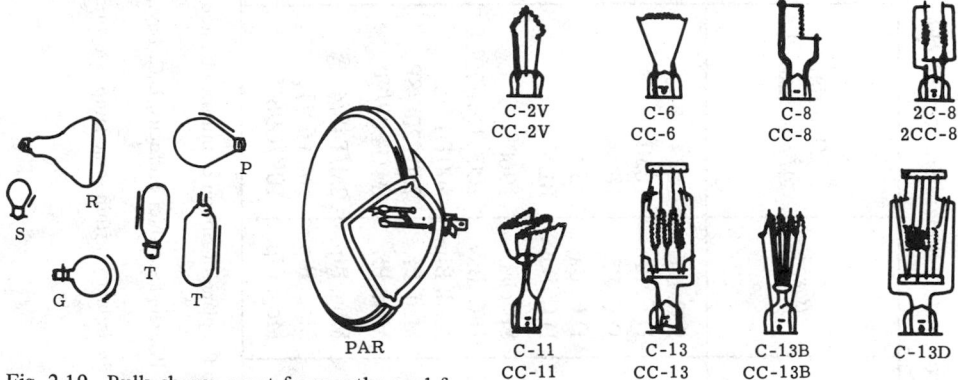

Fig. 2-19. Bulk shapes most frequently used for lamps in optical devices. Letter designations are for particular shapes [2-21].

Fig. 2-20. Most commonly used filament forms. Letters designate the type of filament [2-21].

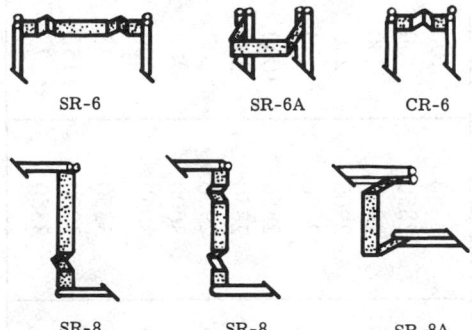

Fig. 2-21. Ribbon-type tungsten filaments. Type designations are by number [2-21].

Tungsten lamps have been designed for a variety of applications; few lamps are directed toward scientific research, but some bear directly or indirectly on scientific pursuits insofar as they can provide steady sources of numerous types of radiation. One set of sources cited here, particularly for what the manufacturer calls their scientific usefulness, is described in Reference [2-22]. Their filament structures are similar to those already described, but their designs reduce extraneous radiation and ensure the quality and stability of the desired radiation. The lamps can be obtained with a certification of their calibration values.

---

*See also Section 2.2.1

Table 2-9. Typical Tungsten Filament Lamps with Approximately Square Sources [2-21]

Watts Nominal	Approximate Source Size[a,b] (mm)		Filament Form	Approximate Average Color Temperature Through Life (K)	Average Life at Design Voltage(h)	Bulb	Base[c]	Ordering Abbreviation[d]	Approximate Initial Lumens at Design Voltage
	Width	Height							
10	1.0	1.0	C-6	2850	100	T-5	S.C. Bay.	1874	140
17.9	2.0	1.1	C-6	2850	100	S-8	D.C. Pf.	1630	290
20	2.0	1.7	CC-6	2900	200	S-8	D.C. Pf.	1634	300
30	2.5	1.6	C-6	2900	250	S-8	D.C. Bay.	1594	450
50	7.0	7.0	CC-2V	2850	50	S-11	D.C. Bay.	BLX	780
60	3.8	2.2	CC-8	2850	1000	T-4	Special	1960[e]	880
100	5.8	5.8	CC-13	2850	200	G-16½	D.C. Bay.	100G16½/29DC	1660
500	9.8	10.2	C-13D[f]	2950	200	T-14	Med. Bip.	500T 14/8	11000
750	14.2	13.7	C-13	3000	200	T-20	Med. Pf.	750T20P/SP	17000
1000	10.9	13.4	C-13D	3050	200	T-20	Med. Pf.	IM/T20P/SP	23500
2000	24	20	C-13	3350	25	T-48	Mog. Bip.	2M/T48/4	65000
5000	36	27	C-13	3200	150	T-64	Mog. Bip.	5M/T64/1	161500
10000	52	34	C-13	3350	75	G-96	Mog. Bip.	10K/G96	335000

[a]Viewed in the direction perpendicular to the long axis of straight or coiled filaments and to the plane of ribbon or multisegmented coiled filaments.

[b]Source-aspect ratios (width:height) between 2:1 and 1:2.

[c]Abbreviations: S.C. Bay., single-contact bayonet; D.C. Pf., double-contact prefocus; S.C. Pf., single-contact prefocus; D.C. Bay., double-contact bayonet; Med. Pf., medium prefocus; Med. Bip., medium bipost; Mog. Bip., Mogul bipost.

[d]All-letter and all-numeral groups are ANSI standard identification codes used throughout the industry within some lamp categories; some manufactures use different identifications for similar lamps.

[e]This lamp employs the iodine regenerative cycle.

[f]Designed for base-up operation.

Table 2-10. Photoflood Lamps[a] [2-21]

Lamp Identification	Approximate (W)	Average Life (h)	Illumination (lm) or (cd)[b]	Approximate Beam Spread[c] (deg)	Bulb Shape and size
			Nonreflectorized Bulbs		
BBA	250	3	8500 lm	–	A-21
EBV	500	6	17000 lm	–	PS-25
BWY/FAB[d]	650	25	20000 lm	–	T-4
DXR	1000	10	31000 lm	–	PS-35
			Reflectorized Bulbs		
BEP	300	4	11000 cd	20-30	R-30
EBR	375	4	14000 cd	30-40	R-30
BFA	375	4	12000 cd	30-40	R-40
DXC	500	6	6000 cd	90	R-40
DXB	500	6	38000 cd	20	R-40
DWE[d]	650	100	24000 cd	Medium	PAR-36
DXK[d]	650	16	30000 cd	30 X 40	PAR-36
FBJ[d]	650	16	75000 cd	15 X 22	PAR-36
FBE/FGK[d,e]	650	20	35000 cd	22 X 30	PAR-36

[a]3400 K at nominal 120 V.
[b]Approximate mean candlepower in 10° axial cone.
[c]To one-half maximum candlepower.
[d]Iodine regenerative getter.
[e]5000 K.

The physical descriptions of some of these sources are given in Figure 2-22. Applications (according to the manufacturer) are photometry, pyrometry, optical radiometry, sensitometry, spectroscopy, spectrometry, polarimetry, saccharimetry, spectrophotometry, colorimetry, microscopy, microphotography, microprojection, and stroboscopy.

Quartz envelope lamps are particularly useful as standards because they are longer-lasting (due to action of iodine in the quartz-iodine series), can be heated to higher temperatures, are sturdier, and can transmit radiation to longer wavelengths in the infrared than glass-envelope lamps. Studer and Van Beers [2-23] have shown the spectral deviation to be expected of lamps containing no iodine. The deviation, when known, is readily acceptable in lieu of the degradation in the lamp caused by the absence of iodine.

The particular tungsten-quartz-iodine lamps used in accordance with the NBS are described above. Others can be obtained in a variety of sizes and wattages.

Carbon Arc. The carbon arc has been passed down from early lighting applications in three forms: low-intensity arc; flame; and high-intensity arc. The low- and high-intensity arcs are usually operated on direct current; the flame type adapts to either

Table 2-11. Typical Tungsten Filament Lamps With Elongated
Sources [2-21]

Watts Nominal	Approximate Source Size[a] (mm)		Filament Form	Approximate Average Color Temperature Through Life (K)	Design Values (V or A)
	Width	Height			
Source-Aspect Ratios (Width:Height) Greater than 10:1					
11	8.2	0.25	C-6	2900	11V
22	16	0.25	C-6	2950	22V
500	60	1.4	C-8[d]	3000	120V
1000	65	2	C-8[d]	3200	120V
10500	530	2.5	C-8[d]	3200	120V
Source-Aspect Ratios (Width:Height) Between 10:1 and 5:1					
6.4	2	0.4	C-6	2900	5.8V
13.1	4.5	0.5	C-6	2900	6.7V
25	4.5	0.8	C-6	3000	6 V
100	10	1.5	C-8	2900	6.6A
2000			C-13		
Source-Aspect Ratios (Width:Height) Between 5:1 and 2:1					
7.2	1.8	0.6	C-6	2850	8 V
18	3.4	0.7	C-6	2850	6.5V
25	2.7	1.2	C-6	2950	6.1V
45	4	1.5	C-8	2800	6.6A
125	15.3	6.2	C-13B	2800	120 V
150	11.2	2.2	CC-8	3050	28 V
150	8	2	CC-6[e]	3150	120 V
150	5.3	2.2	CC-6[f]	3300	21 V
200	10	3	CC-8	3050	6.6A
1000	22	6	CC-8	3200	120 V
Source-Aspect Ratios (Width:Height) Between 1:2 and 1:5					
25	1.2	2.8	C-8	2950	6.1V
75	1.8	5.2	C-8	3100	7.5A
108	1.9	5.5	C-8	3075	18 A
Source-Aspect Ratios (Width:Height) Between 1:5 and 1:10					
25	0.8	4.5	C-8	3000	6V
108	2	14	SR-8A	3000	18A
Source-Aspect Ratios (Width-Height) Less than 1:10					
6.5	0.08	15	S-8	2850	3.6V
180	3	33	SR-8	2300	30 A

[a]Viewed in the direction perpendicular to the long axis of straight or coiled filaments and to the plane of ribbon or multisegmented coiled filaments.

[b]Abbreviations: D.C. Pf., double-contact prefocus; S.C. Pf., single-contact prefocus; S.C. Bay., single-contact bayonet; Med. Pf., medium prefocus; L.S. Med., long-screw medium; Mog. Bip., mogul bipost; R.S.C., recessed single contacts on double ended lamp.

Table 2-11. Typical Tungsten Filament Lamps with Elongated Sources
[2-21] (Continued)

Average Life at Design Volts or Amperes (h)	Bulb	Base[b]	Ordering Abbreviation[c,d]	Approximate initial lumens at design Volts or Amperes
\multicolumn{5}{c}{Source-Aspect Ratios (Width:Height) Greater than 10:1}				
200	T-8	D.C. Pf.	1926	165
200	T-8	D.C. Pf.	1936	350
2000	T-3	R.S.C.	Q500T3/CL	10950
500	T-3	R.S.C.	Q1000T3/4CL	28000
400	T-4	R.S.C.	Q1500T4/4CL	41200
\multicolumn{5}{c}{Source-Aspect Ratios (Width:Height) Between 10:1 and 5:1}				
50	G-6	Min. Sc.	153	100
500	S-8	S.C. Bay.	1619	190
50	S-11	L.S. Med.	25S11/7	460
1000	T-3	R.S.C.	Q6.6A/T3/CL	1900
750	T-30	Mog. Bip.	Q2000T30/4CL	58500
\multicolumn{5}{c}{Source-Aspect Ratios (Width:Height) Between 5:1 and 2:1}				
50	T-4½	Min. Bay.	872	115
125	T-5	S.C. Bay.	1462	315
125	S-11	S.C. Pf.	1763	400
1000	T-2½	R.S.C.	Q6.6A/T2½/CL	630
500	T-10	Med. Pf.	125T10P	1820
300	T-4	Special	1958	3300
15	T-12	4-Pin	DFA	–
15	T-12	4-Pin	DLS	–
500	T-4	R.S.C.	Q6.6A/T4/CL	4400
150	T-5	R.S.C.	DXW	28000
\multicolumn{5}{c}{Source-Aspect Ratios (Width:Height) Between 1:2 and 1:5}				
125	S-11	S.C. Pf.	1759	400
100	T-8	S.C. Pf.	7.5A/T8/94(10V)	1575
50	T-10	Med. Pf.	CPR	2250
\multicolumn{5}{c}{Source-Aspect Ratios (Width:Height) Between 1:5 and 1:10}				
50	T-8	S.C. Pf.	25T8/SCP	460
370	T-10	Med. Pf.	18A/T10/1P(6V)	–
\multicolumn{5}{c}{Source-Aspect Ratios (Width:Height) Less than 1:10}				
100	T-5	S.C. Bay.	1874	140
–	T-24	Mog. Bip.	30A/T24/6(6V)	–

[c]All-letter and all numeral groups are ANSI standard identification codes used throughout the industry within some lamp categories. (Occasionally, a suffix letter is added to the ANSI numerical codes.) Letter-numeral combinations are the ordering abbreviations in common use for other categories; some manufacturers use different identifications for similar lamps.

[d]Lamps 1958 and DXW and all lamps that include Q in the ordering abbreviation employ the iodine regenerative cycle.

[e]Internal reflector, designed for f/1.5 lens.

[f]Internal dichroic reflector, designed for f/1.2 lens.

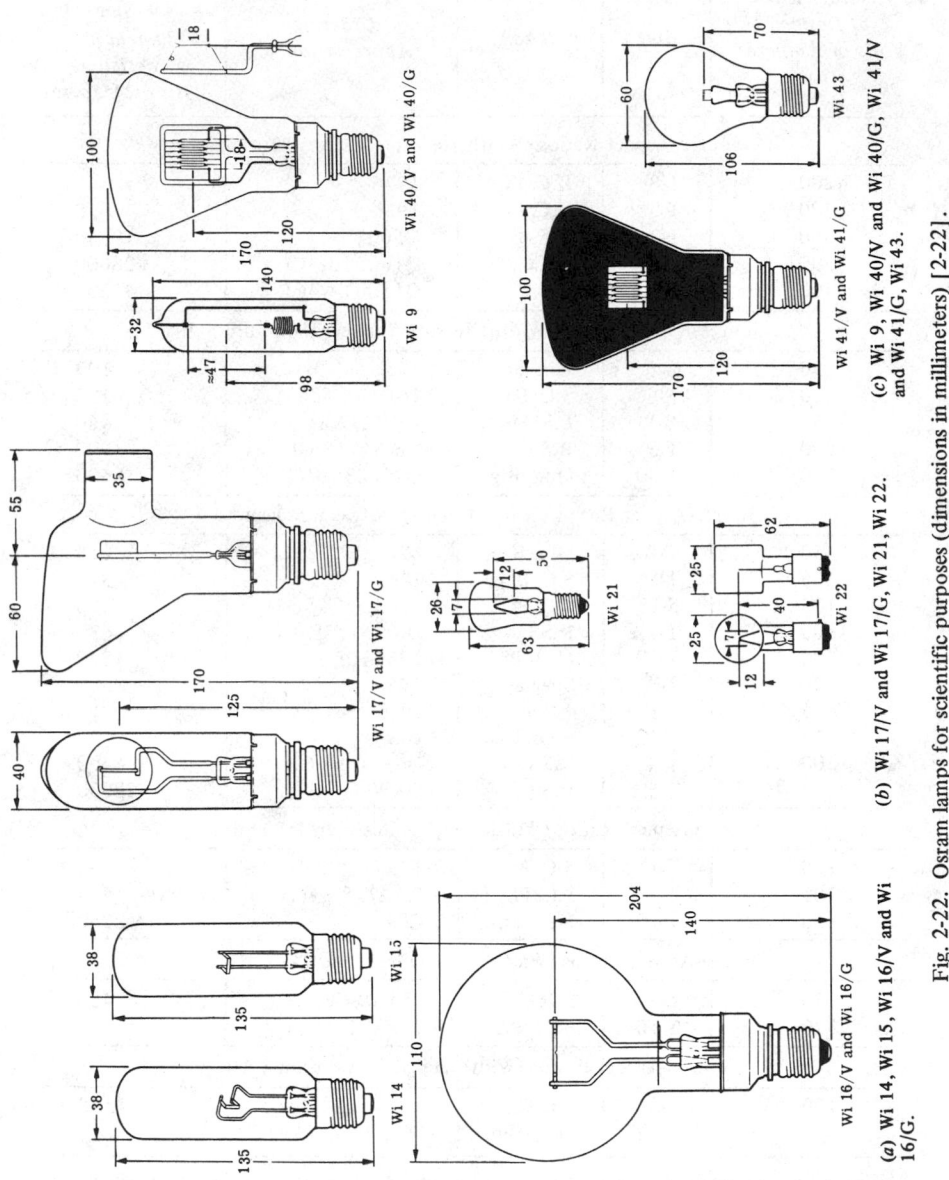

(a) **Wi 14, Wi 15, Wi 16/V and Wi 16/G.**

(b) **Wi 17/V and Wi 17/G, Wi 21, Wi 22.**

(c) **Wi 9, Wi 40/V and Wi 40/G, Wi 41/V and Wi 41/G, Wi 43.**

Fig. 2-22. Osram lamps for scientific purposes (dimensions in millimeters) [2-22].

direct or alternating current. In all cases, a ballast must be used. "In the alternating-current arc, the combined radiation from the two terminals is less than that from the positive crater of the direct-current arc of the same wattage" [2-1].

Spatial variation in the amount of light energy across the crater of dc arcs for different currents is shown in Figure 2-23.

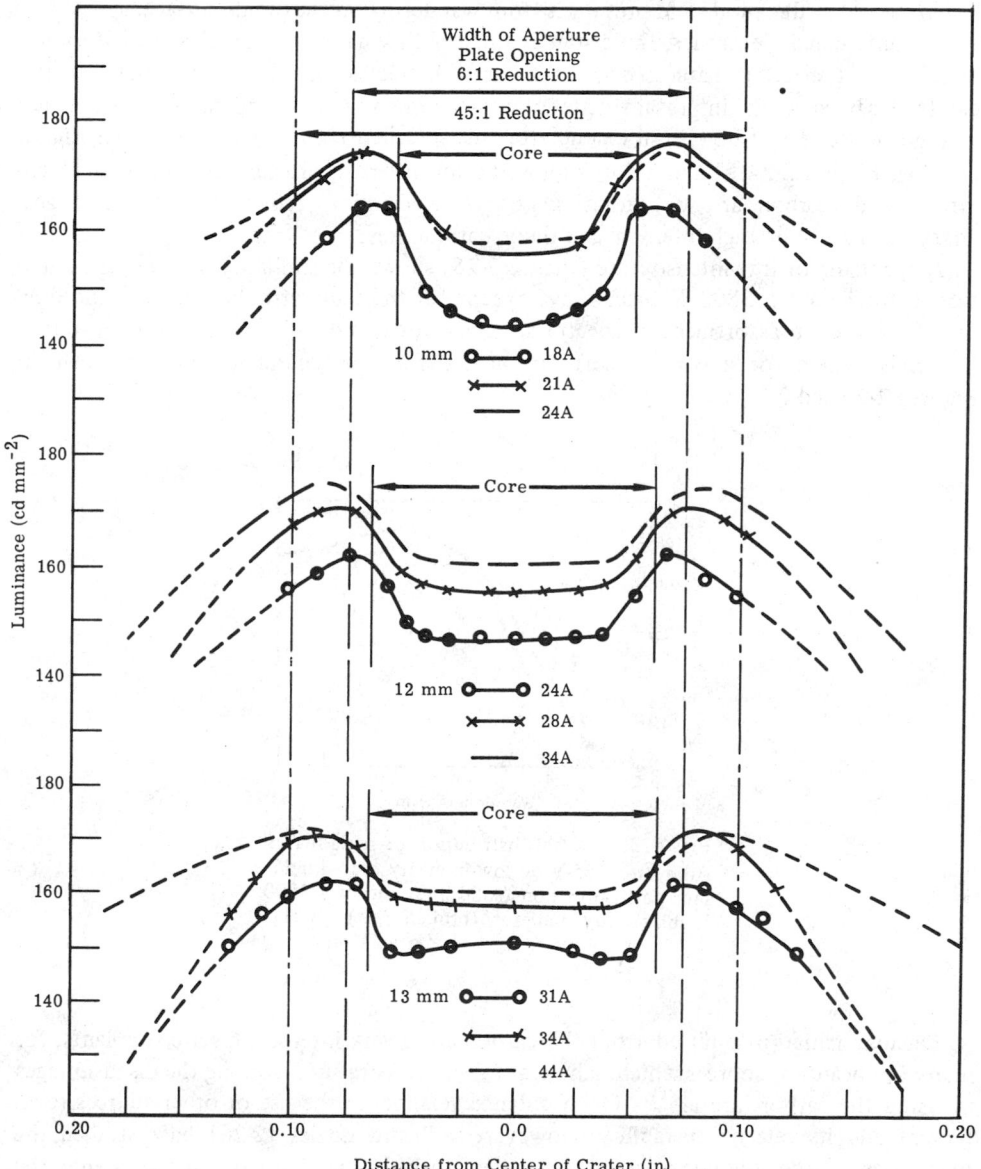

Fig. 2-23. Variations in brightness across the craters of 10-, 12-, and 13-mm positive carbons of dc plain arcs operated at different currents in the regions of recommended operation [2-1].

The carbon arc is a good example of an open arc, widely used because of its very high radiation and color temperatures (from approximately 3800 to 6500 K, or higher). The rate at which the material is consumed and expended during burning (5 to 30 cm h^{-1}) depends on the intensity of the arc. The arc is discharged between two electrodes that are moved to compensate for the rate of consumption of the material. The anode forms a crater of decomposing material which provides a center of very high luminosity. Some electrodes are hollowed out and filled with a softer carbon material which helps keep the arc fixed in the anode and prevent it from wandering on the anode surface.

In some cored electrodes, the center is filled with whatever material is needed to produce desired spectral characteristics in the arc. In such devices, the flame between the electrodes becomes the important center of luminosity and color temperatures reach values as high as 8000 K [2-24]. An example of this so-called *flaming arc* is shown in Figure 2-24(a). Figures 2-24(b) and 2-24(c) show the low-intensity dc carbon arc and the high-intensity dc carbon arc with rotating positive electrodes. Tables 2-12 and 2-13 give characteristics of dc high-intensity and flame carbon arcs.

A spectrum of low-intensity arc (Figure 2-25) shows the similarity between the radiation from it and a 3800 K blackbody, except for the band structure at 0.25 and 0.39 μm. In Koller an assortment of spectra are given for cored carbons containing different materials. Those for a core of soft carbon and for a polymetallic core are shown in Figures 2-26 and 2-27.

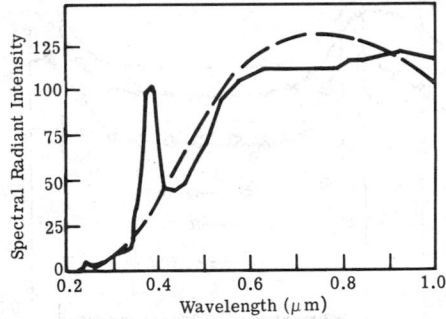

Fig. 2-25. Spectral distribution of radiant flux from 30-A, 55-V dc low-intensity arc with 12-mm positive carbon (solid line) and of a 3800 K blackbody radiator (broken line) [2-21].

Because radiation emitted from the carbon arc is very intense, this arc supplants, for many applications, sources which radiate at lower temperatures. Among the disadvantages in using the carbon arc are its inconvenience relative to the use of other sources (e.g., lamps) and its relative instability. However, Null and Lozier [2-26] have studied the properties of the low-intensity carbon arc extensively and have found that under the proper operating conditions the carbon arc can be made quite stable; in fact, they recommend its use as a standard of radiation at high temperatures.

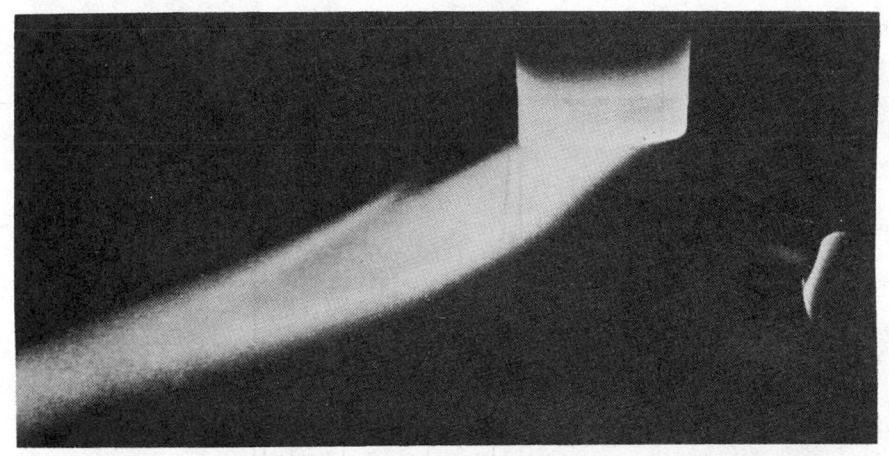

(c) High-intensity dc arc with rotating positive carbon.

(b) Low-Intensity dc arc.

Fig. 2-24. Various types of carbon arc [2-24].

(a) Flame type.

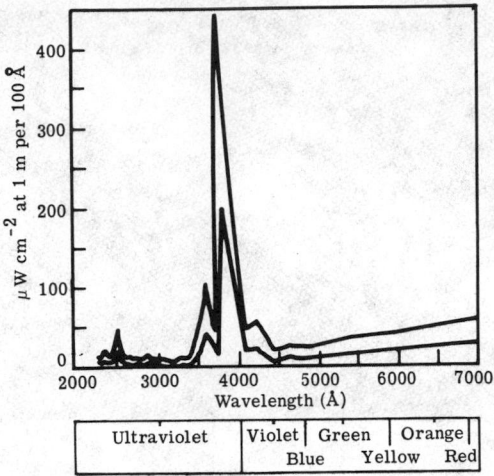

Fig. 2-26. Spectral energy distribution of carbon arc with core of soft carbon: upper curve: 60-A ac 50-V across the arc; lower curve: 30-A ac 50-V across the arc [2-24].

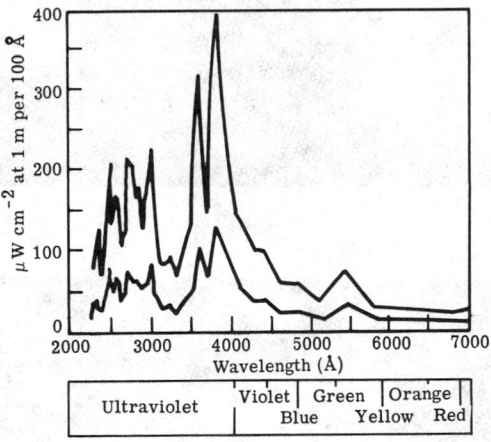

Fig. 2-27. Spectral energy distribution of carbon arc with polymetallic-cored carbons: upper curve: 60-A ac, 50-V across arc; lower curve: 30-A ac, 50-V across arc [2-24].

Table 2-12. dc Carbon Arcs [2-25]

	Low Intensity	Nonrotating High Intensity		Rotating High Intensity						
Application Number[a]	1	2	3	4	5	6	7	8	9	10
Type of carbon	Microscope	Projector	Projector	Projector	Projector	Projector	Projector	Searchlight	Studio	
Positive carbon:										
Diameter (mm)	5	7	8	10	11	13.6	13.6	16	16	16
Length (in.)	8	12-14	12-14	20	20	22	22	22	22	22-30
Negative carbon:										
Diameter	6 mm	6 mm	7 mm	11/32 in.	3/8 in.	0.5 in.	0.5 in.	11 mm	17/32 in.	7/16 in.
Length (in.)	4.5	9	9	9	9	9	9	12	9	12-48
Arc current (A)	5	50	70	105	120	160	180	150	225	400
Arc volts (dc)	59	40	42	59	57	66	74	78	70	80
Arc power (W)	295	2000	2940	6200	6840	10600	13300	11700	15800	32000
Burning rate (in. h^{-1})										
Positive carbon	4.5	11.6	13.6	21.5	16.5	17	21.5	8.9	20.2	55
Negative carbon	2.1	4.3	4.3	2.9	2.4	2.2	2.5	3.9	2.2	3.5
Approximate crater diameter (in.)	0.12	0.23	0.28	0.36	0.39	0.5	0.5	0.55	0.59	0.59
Maximum luminance of crater (cd cm^{-2})	15000	55000	83000	90000	85000	96000	95000	65000	68000	45000
Forward crater										
candlepower	975	10500	22000	36000	44000	63000	78000	68000	99000	185000
Crater lumens[b]	3100	36800	77000	126000	154000	221000	273000	250000	347000	660000
Total lumens[c]	3100	55000	115000	189000	231000	368000	410000	374000	521000	999000
Total lumens per arc watt	10.4	29.7	39.1	30.5	33.8	34.7	30.8	32	33	30.9
Color temperature (K)[d]	3600	5950	5500-6500	5500-6500	5500-6500	5500-6500	5500-6500	5400	4100	5800-6100

[a]Typical applications: 1, microscope illumination and projection; 2 to 7, motion-picture projection; 8, searchlight projection; 9, motion-picture-set lighting and motion-picture and television background projection.
[b]Includes light radiated in forward hemisphere.
[c]Includes light from crater and arc flame in forward hemisphere.
[d]Crater radiation only.

Table 2-13. Flame-Type Carbon Arcs [2-25]

	Application Number[a]			
	1	3	3	4
Type of carbon	C	E	Sunshine	Sunshine
Flame materials	Polymetallic	Strontium	Rare earth	Rare earth
Burning position[g]	Vertical	Vertical	Vertical	Vertical
Upper carbon[d]				
Diameter	22 mm	22 mm	22 mm	22 mm
Length (in.)	12	12	12	12
Lower carbon[d]				
Diameter	13 mm	13 mm	13 mm	13 mm
Length (in.)	12	12	12	12
Arc current (A)	60	60	60	80
Arc voltage (ac)[h]	50	50	50	50
Arc power (kW)	3	3	3	4
Candlepower[i]	2100	6300	9100	10000
Lumens	23000	69000	100000	110000
Lumens per arc watt	7.6	23	33.3	27.5
Color temperature (K)			12800[j]	24000[j]
Spectral intensity (μW cm^{-2})				
1 m from arc axis:				
Below 270 nm	540.0	180.0	102	140
270-320 nm	540.0	150.0	186	244
320-400 nm	1800.0	1200.0	2046	2816
400-450 nm	300.0	1100.0	1704	2306
450-700 nm	600.0	4050.0	3210	3520
700-1125 nm	1580.0	2480.0	3032	3500
Above 1125 nm	9480.0	10290.0	9820	11420
Total	14930	19460	20100	24000
Spectral radiation (percent of input power):				
Below 270 nm	1.8	0.6	0.34	0.35
270-320 nm	1.8	0.5	0.62	0.61
320-400 nm	6.0	4.0	6.82	7.04
400-450 nm	1.3	3.7	5.68	5.90
450-700 nm	2.0	13.5	10.7	8.80
700-1125 nm	5.27	8.27	10.1	8.75
Above 1125 nm	31.6	34.3	32.7	28.55
Total	49.77	64.87	67.00	60.00

[a]Typical applications: 1 to 5 and 8, photochemical, therapeutic, accelerated exposure testing, or accelerated plant growth; 6, 7, and 9 blueprinting diazo printing, photo copying, and graphic arts; 10, motion-picture and television studio lighting.
[b]Photographic white-flame carbons.
[c]High intensity copper-coated sunshine carbons.
[d]Both carbons are same in horizontal, coaxial ac arcs.

[e]High-intensity photo carbons.
[f]Motion-picture-studio carbons.
[g]All combinations shown are operated coaxially.
[h]All operated on alternating current except item 10.
[i]Horizontal candlepower, transverse to arc axis.
[j]Deviates enough from blackbody colors to make color temperature of doubtful meaning.

Table 2-13. Flame-Type Carbon Arcs [2-25] (*Continued*)

5	6	7[b]	8[c,d]	9[d,e]	10[f]
W	Enclosed arc	Photo	Sunshine	Photo	Studio
Polymetallic	None	Rare earth	Rare earth	Rare earth	Rare earth
Vertical	Vertical	Vertical	Horizontal	Horizontal	Vertical
22 mm	1/2 in.	1/2 in.	6 mm	9 mm	8 mm
12	3-16	12	6.5	8	12
13 mm	1/2 in.	1/2 in.	6 mm	9 mm	7 mm
12	3-16	12	6.5	8	9
80	16	38	40	95	40
50	138	50	24	30	37 dc
4	2.2	1.9	1	2.85	1.5
8400	1170	6700	4830	14200	11000
92000	13000	74000	53000	156000	110000
23	5.9	39.8	53	54.8	73.5
		7420[j]	6590	8150	4700
1020		95	11		12
1860		76	49	100	48
3120	1700	684	415	1590	464
1480	177	722	405	844	726
2600	442	2223	1602	3671	3965
3220	1681	1264	1368	5632	2123
14500	6600	5189	3290	8763	4593
27800	10600	10253	7140	20600	11930
2.55		0.5	0.11		0.08
4.65		0.4	0.49	0.35	0.32
7.80	7.7	3.6	4.15	5.59	3.09
3.70	0.8	3.8	4.05	2.96	4.84
6.50	2.0	11.7	16.02	12.86	26.43
8.05	7.6	6.7	13.68	10.75	14.15
36.25	29.9	27.3	32.90	30.60	30.62
69.50	48.0	54.0	71.40	72.20	79.53

2.3.3. Enclosed Arc and Discharge Sources (High Pressure). Koller [2-24] states that the carbon arc is generally desired if a high intensity is required from a single unit but that it is less efficient than the mercury arc. Other disadvantages are the short life of the carbon with respect to mercury, and combustion products which may be undesirable.

Worthing [2-1] describes a number of the older, enclosed, metallic arc sources, many of which can be built in the laboratory for laboratory use. Today, however, it is rarely necessary to build one's own source unless it is highly specialized.

Uviarc®*. This lamp is an efficient radiator of ultraviolet radiation. The energy distribution of one type is given in Figure 2-28. Since the pressure of this mercury-vapor lamp is intermediate between the usual high- and the low-pressure lamps, little background (or continuum) radiation is present.

In the truly high-pressure lamp, considerable continuum radiation results from greater molecular interaction. Figure 2-29 shows the dependence on pressure of the amount of continuum in mercury lamps of differing pressure. Bulb shapes and sizes are shown in Figure 2-30.

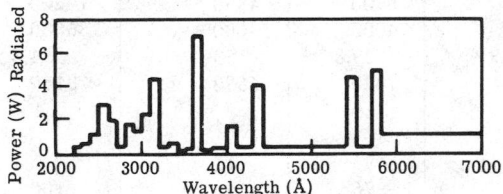

Fig. 2-28. Intensity distribution of UA-2 interme-diate-pressure lamp [2-24].

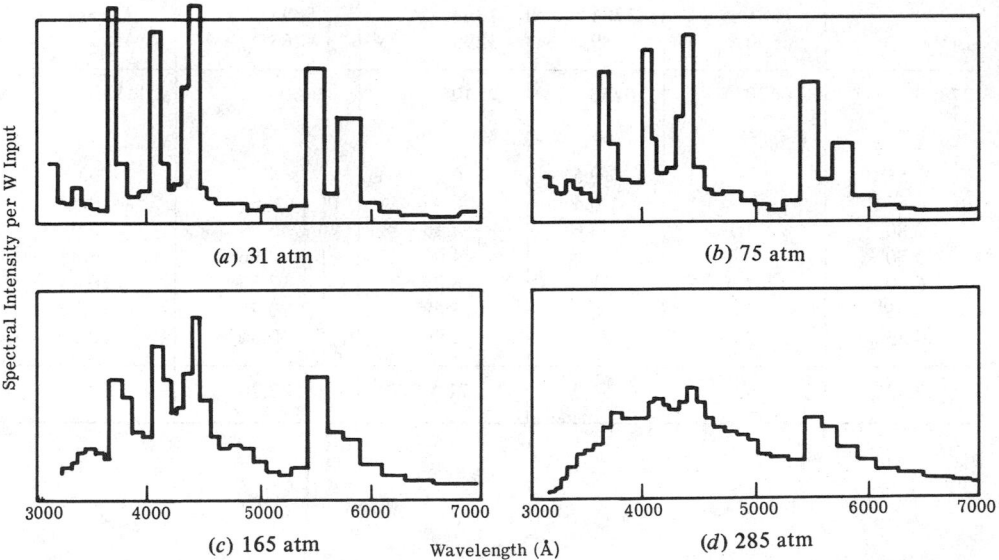

Fig. 2-29. Emission spectrum of high-pressure mercury-arc lamps showing continuum background [2-27].

---

*Registered trademark of General Electric.

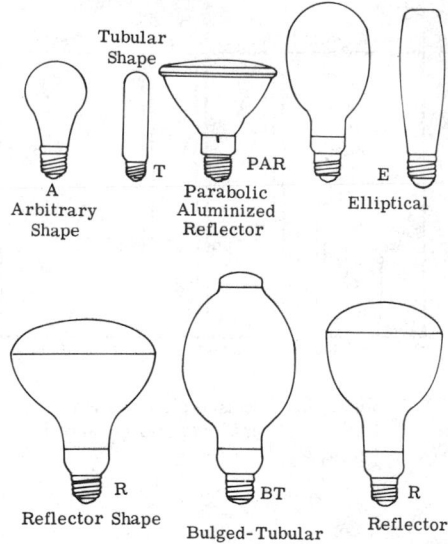

Fig. 2-30. Bulb shapes and sizes (not to scale) [2-28].

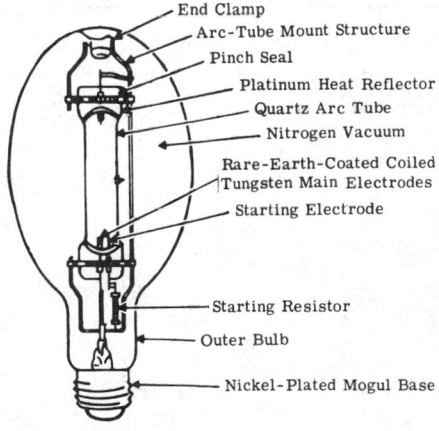

Fig. 2-31. High-pressure mercury lamp showing various components [2-28].

Mercury Arcs. A widely used type of high-pressure, mercury-arc lamp and the components necessary for its successful operation are shown in Figure 2-31. The coiled tungsten cathode is coated with a rare-earth material (e.g., thorium). The auxiliary electrode is used to help in starting. A high resistance limits the starting current. Once the arc is started, the operating current is limited by ballast supplied by the high reactance of the power transformer. Spectral data for clear, 400 W mercury lamps of this type are given in Figure 2-32.

Multivapor Arcs. In these lamps, argon and mercury provide the starting action. Then sodium iodide, thallium iodide, and indium iodide vaporize and dissociate to yield the bulk of the lamp radiation. The physical appearance is like that of mercury lamps of the same general nature. Ballasts are similar to their counterparts for the mercury lamp. Sample lamp performance data are given in Table 2-14. Spectral characteristics of these lamps are given in Figure 2-33.

Lucalox®* Lamps. The chief characteristics of this lamp are high-pressure sodium discharge and a high-temperature-withstanding ceramic, Lucalox (translucent aluminum oxide), to yield performance typified in Table 2-15. The spectral output of the 400 W Lucalox lamp is shown in Figure 2-34. Ballasts for this lamp are described in Reference [2-28].

Capillary Mercury-Arc Lamps [2-24]. As the pressure of the arc increases, cooling is required to avoid catastrophic effects on the tube. The AH6 tube (Figure 2-35) is constructed with a quartz bulb wall and a quartz outer jacket, to allow 2800 Å radiation to pass, or a Pyrex®** outer jacket to eliminate ultraviolet. Pure water is forced through at a rapid rate, while the tube is maintained at a potential of 840 V.

Table 2-16 shows the characteristics of tubes manufactured by Illumination Industries, Inc., Sunnyvale, California. Figure 2-36 shows spectral characteristics.

---

*Registered trademark of General Electric.
**Registered trademark of Corning Glass Works.

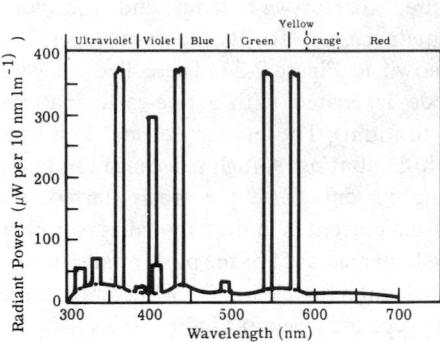

Fig. 2-32. Spectral energy distribution for clear mercury-arc lamp [2-28].

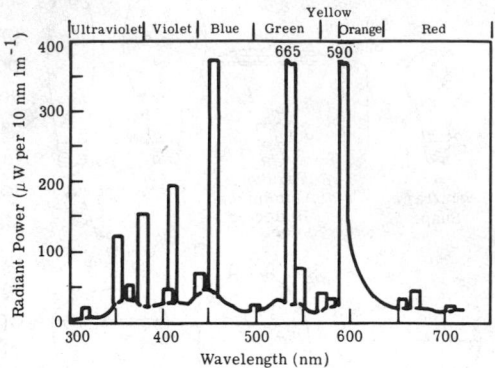

Fig. 2-33. Spectral energy distribution of multivapor-arc lamp [2-28].

Table 2-14. Lamp Performance Data [2-28]

Specification	GE *ordering code*[a]	
	MV–400/BU/I *MV–400/BD/I*	*MV–1000/BU/I*
Minimum open-circuit rms starting voltage with ballast that meets waveform requirements:		
To –20°F	280	400
To +50°F	225	–
Operating voltages	135	265
Finish	Clear	Clear
Bulb	E-37	BT-56
Base	Mogul	Mogul
Rated life (h)	15,000	10,000
Approximate initial lumen		
Vertical	34,000	88,000
Average mean lumen		
Vertical	26,500	70,400
Light center length (in.)	7	9 3/8
Maximum overall length (in.)	11 5/16	15 1/16

[a]Burning position of BU lamps is base up to within ± 15° from vertical of BD lamps it is base down to 15° above horizontal.

Table 2-15. Lamp Performance Data[a] [2-28]

GE Ordering Code	Watts	Lamp Rated Voltage (V)	Average Rated Life (h)[b]	Initial (lm)[c]	Mean (lm)	Maximum Overall Length (in.)	Light Center Length (in.)
LU–400/BD or BU	400	100	20,000	50,000	45,000	9 3/4	5 3/4
LU–1000/BD or BU	1000	250	15,000	140,000	127,400	5 1/16	8 3/4

[a]All lamps have clear finish and mogul base.
[b]At 10 or more burning hours per start.
[c]Any burning position.

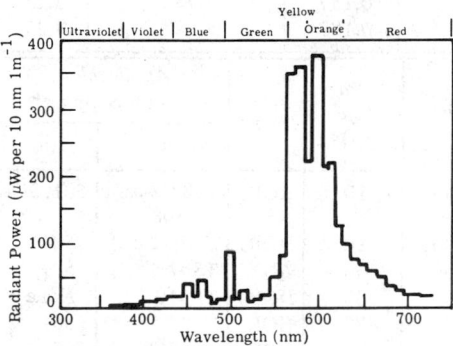

Fig. 2-34. Spectral output of 400-W <u>Lucalox</u> lamp [2-28].

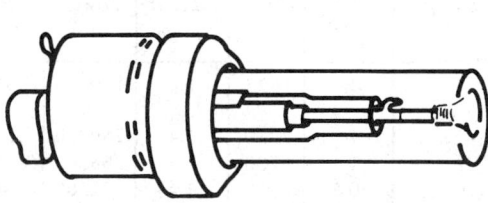

Fig. 2-35. Water-cooled high-pressure (110 atm) mercury arc lamp showing lamp in water jacket [2-24].

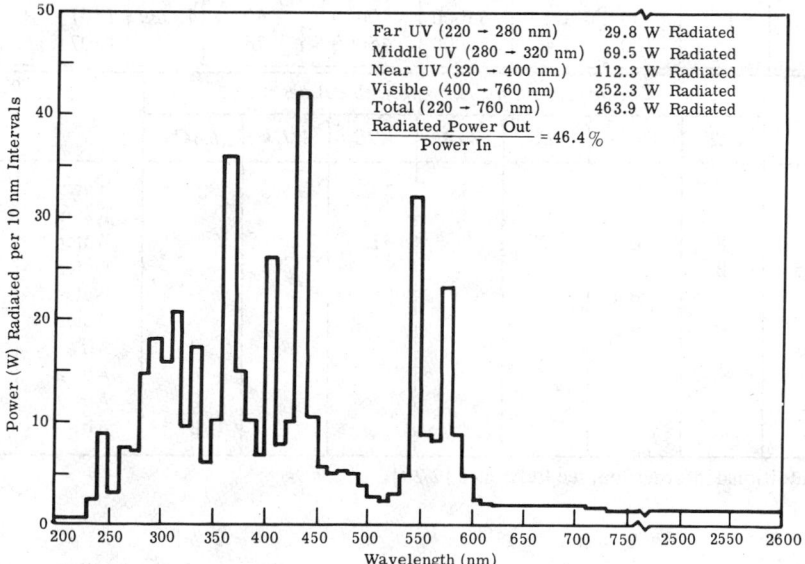

Far UV (220 → 280 nm)   29.8 W Radiated
Middle UV (280 → 320 nm)   69.5 W Radiated
Near UV (320 → 400 nm)   112.3 W Radiated
Visible (400 → 760 nm)   252.3 W Radiated
Total (220 → 760 nm)   463.9 W Radiated

$$\frac{\text{Radiated Power Out}}{\text{Power In}} = 46.4\%$$

Fig. 2-36. Spectral energy distribution of type BH6-1 mercury capillary lamp [2-29].

Table 2-16. Radiation from Water- and Air-Cooled Capillary Lamps [2-29]

Lamp Type Number	Base Code (see below)	Actual Luminous Length (in.)	Bore Dimension (mm)	Dimensions (in.)		
				(A) Base Diameter	(B) Exposed Envelope Length	(C) Overall Length
A-1	B, L	1.125	1	0.125	2.0	3.25
A-2	B, L	2.0	1	0.125	2.875	4.125
C-1	B, L	1.125	1.5	0.187	1.75	3.25
AH6-.5	B	0.5	2	0.187	1.5	1.75
AH6-1	B	1.0	2	0.187	1.75	3.25
AH6-2	B	2.0	2	0.187	2.75	4.25
BH6-.5	B	0.5	2	0.187	1.5	2.75
BH6-.8	B	0.8	2	0.187	1.5	3.25
BH6-1	B	1.0	2	0.187	1.75	3.25
PA-1	B	0.68	2	0.187	1.62	3.25

Lamp	Power (kVA)	Volts	Current	Output Power (lm)	Average Bright (cd cm^{-2})	Average Life (h)	Mtg. Pos.	Applicable Power Supplies*	
								ac	dc
A-1	2	1425	1.4	130000	90000	10	HOR.	T507(2 ea.), T508	501, 504
A-2	4	2850	1.4	260000	90000	10	HOR.	T508(2 ea.)	504
C-1	1	700	1.4	65000	40000	60	ANY	T507	501
AH6-.5	0.5	350	1.4	32500	40000	80	HOR.	TL510	None
AH6-1	1	700	1.4	65000	40000	80	HOR.	T507	501
AH6-2	2	1425	1.4	130000	40000	80	HOR.	T507(2 ea.), T508	501
BH6-.5	0.5	350	1.4	32500	40000	60	HOR.	TL510	None
BH6-.8	0.8	550	1.4	52000	40000	60	HOR.	None	None
BH6-1	1	700	1.4	65000	40000	60	HOR.	T507	501
PA-1	1	700	1.4	75000	45000	75	HOR.	T507	501

Lamp Type	Cooling Assembly and Applicable Bases*							Coolant*	Volume*
	SEB	DEL	DES	WJA	WJB	WJD	BAC		
A-1			B	L				Water	1.50 GPM
A-2			B	L				Water	1.50 GPM
C-1	B	B				L		Water	1.20 GPM
AH6-.5	B	B						Water	1.20 GPM
AH6-1	B	B						Water	1.20 GPM
AH6-2	B	B						Water	1.20 GPM
BH6-.5							B	Air	5.50 CFM
BH6-.8							B	Air	5.50 CFM
BH6-1							B	Air	5.50 CFM
PA-1							B	Air	5.50 CFM

*For additional information, see Reference [2-29].

Compact-Source Arcs [2-21, 2-30]. Some common characteristics of currently available compact-source arc lamps are as follows:

(1) A clear quartz bulb of roughly spherical shape with extensions at opposite ends constituting the electrode terminals. In some cases, the quartz bulb is then sealed within a larger glass bulb, which is filled with an inert gas.

(2) A pair of electrodes with relatively close spacing (from less than 1 mm to about 1 cm)—hence the sometimes-used term *short-arc lamps*.

(3) A filling of gas or vapor through which the arc discharge takes place.

(4) Extreme electrical loading of the arc gap, which results in very high luminance, internal pressures of many atmospheres, and bulb temperatures as high as 900°C. Precautions are necessary to protect people and equipment in case the lamps should fail violently.

(5) The need for a momentary high-voltage ignition pulse, and a ballast or other auxiliary equipment to limit current during operation.

(6) Clean, attention-free operation for long periods of time.

Table 2-17 gives characteristics of commonly available compact-source arc lamps. These lamps are designated by the chief radiating gases enclosed as mercury, mercury xenon, and xenon lamps.

Figure 2-37 shows the compact-source arc lamp in a housing purchasable from one supplier. The housing shown is for lamps up to 1000 W. Typical lamp constructions are shown in Figure 2-38. Since starting may be a problem, some lamps (Figure 2-39) are constructed with a third (i.e., a starting) electrode, to which a momentary high voltage is applied for starting (and especially restarting) while hot. The usual ballast is required for compact-source arcs. For stability, these arcs, particularly mercury and mercury-xenon, should be operated near rated power on a well-regulated power supply [2-30].

The spatial distribution of luminance from these lamps is reported in the literature already cited, and typical contours are shown in Figure 2-40. Polar distributions are similar to those shown in Figure 2-41.

Spectral distributions are given in Figures 2-42 through 2-44 for a 1000 W ac mercury lamp, a 5 kW dc xenon lamp, and 1000 W dc mercury-xenon lamp. Lamps are available at considerably less wattage.

Cann [2-30] reports on some interesting special lamps tested by Jet Propulsion Laboratories for the purpose of obtaining a good spectral match to the solar distribution. The types of lamps tested were: Xe; Xe-Zn; Xe-Cd; Hg-Xe-Zn; Hg-Xe-Cd; Kr; Kr-Zn; Kr-Cd; Hg-Kr-Zn; Hg-Kr-Cd; Ar; Ne; and Hg-Xe with variable mercury-vapor pressure. For details, see Reference [2-21].

A special design of a short-arc lamp manufactured by Varian [2-32] is shown in Figure 2-45. Aside from its compactness and parabolic selector, it has a sapphire window which allows a greater amount of IR energy to be emitted. It is operated either dc or pulsed, but the user should obtain complete specifications, because the reflector can become contaminated, with a resultant decrease in output. A similar type of lamp with a sapphire window is manufactured by Optitron, Inc., Torrance, California.

2.3.4. Enclosed Arc and Discharge Sources (Low Pressure) [2-24]. With pressure reduction in a tube filled with mercury vapor, the 2537 Å line becomes predominant so that low-pressure mercury tubes are usually selected for their ability to emit ultraviolet radiation.

Table 2-17. Compact-Source Arc-Lamp Data

Mfgr. No. Lamp Type[a]	Rated Power (W)	dc (V)	Amperes	Luminous Flux (lm)	Luminance (cd mm^{-2})	Gap Length (mm)	Arc Life (h)	Cooling
Typical Mercury Lamps								
PEK H–50[b]	50	39	1.30–1.6	2000	300	0.4	100	Convection
PEK 110[b]	100	20	4.20–6.2	2200	1400	0.4	100	Convection
PEK 200–2[b]	200	57	3.10–4	10000	330	2.0	400	Convection
PEK 350[b]	350	60–75	7–5	17500	300	3.0	400	
Typical Mercury-Xenon Lamps								
H 900B0011[c]	200	20–25	8.0–9.5	4500	222	1.9	1000	
H 941B0010[c]	600	20–25	26	23000	700	3.3	1000	
H 528B0010[c]	1000	65	15.5	50000	250	6.1	1000	
H 929B0010[c]	2500	50	50	122000	760	5.0	1000	
H 932B0010[c]	5000	60	83	265000	869	7.2	1000	
H L5181–000[c]	7000	65–70	100	400000	260	12.0	500	
Typical Xenon Lamps								
PEK X–35[b]	35	12	2.50–3.5	315	400	0.3	200	Convection
PEK X–75[b]	75	14	4.60–6.2	1125	800	0.5	300	Convection
H L–5122–000[c]	150	20	7.5	2200	470	1.9	1000	Forced air
PEK X–500[b]	500	20	14–30	15000	350	3.0	1200	
H L–5179–000[c]	1000	23	43.5	30000	400	4.0	1000	
H 966C–0010[c]	5000	32–36	147	210000	600	10.0	1000	
H L–5341–000[c]	7000	46	150	300000	700	14.0	500	

[a]Lamps are operated vertically.
[b]PEK = PEK Brand of Illumination Industries, Inc.
[c]H = Conrad-Hanovia, Inc.

Fig. 2-37. Lamp housing for lamps up to 1000 W showing lamp, rear reflector, beam part, and reflective thermal shielding [2-31].

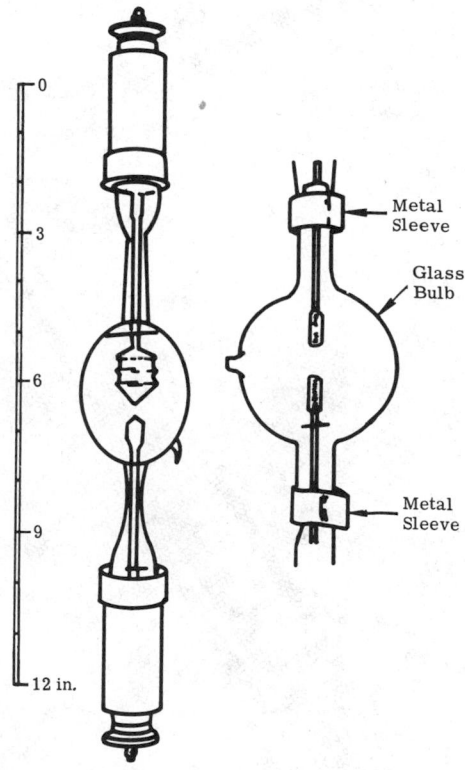

Fig. 2-38. Construction of different lamps showing differences in relative sizes of electrodes for dc (left) and ac (right) operation [2-21].

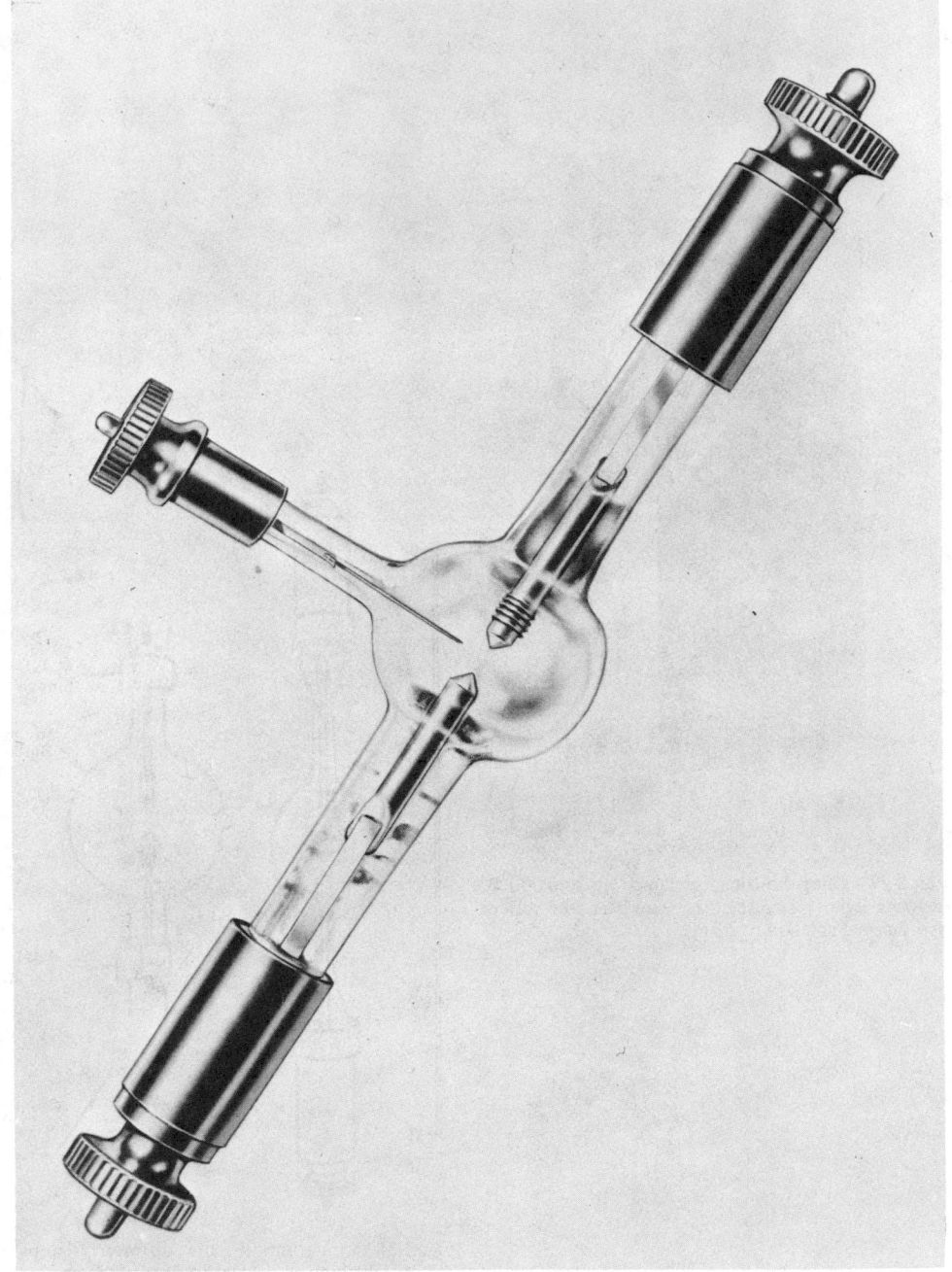

Fig. 2-39. Construction of a lamp with a third, starting electrode [2-22].

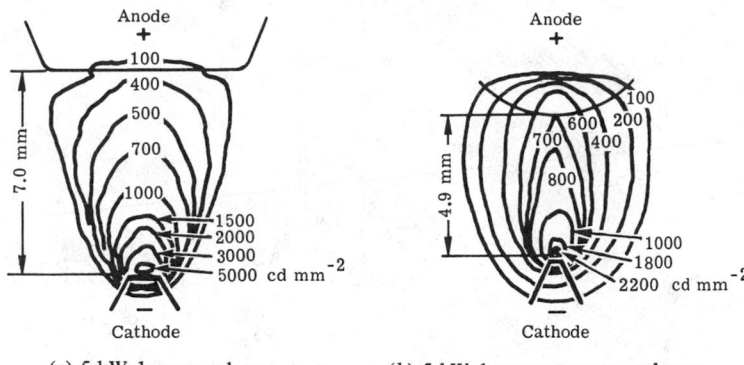

(a) 5-kW dc xenon lamp   (b) 5-kW dc mercury-xenon lamp

Fig. 2-40. Spatial luminance distribution, of compact-arc lamps [2-21].

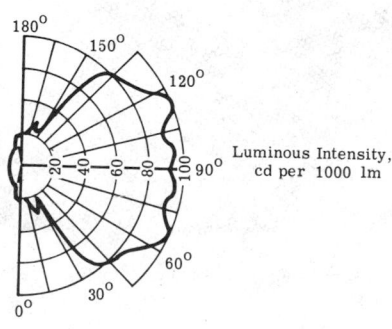

(a) 7.5-kW ac mercury lamp

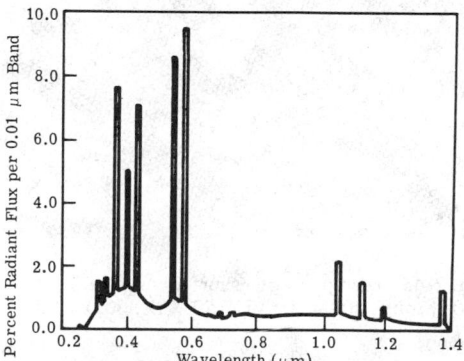

Fig. 2-42. Spectral distribution of radiant intensity from a 1000-W ac mercury lamp perpendicular to the lamp axis [2-21].

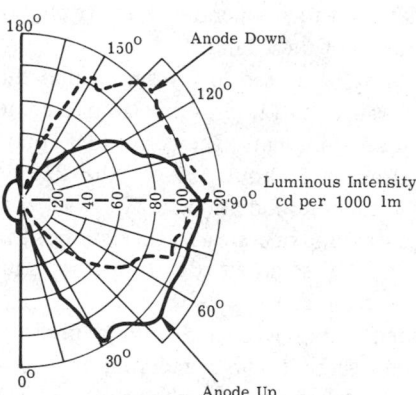

(b) 2.5-kW dc mercury-xenon lamp

Fig. 2-41. Polar distribution of radiation in planes that include arc axis. Asymmetry in (b) is due to unequal size of electrodes [2-21].

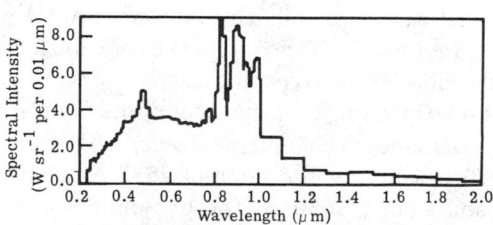

Fig. 2-43. Spectral distribution of radiant intensity from a 5-kW dc xenon lamp perpendicular to the lamp axis with electrode and bulb radiation excluded [2-21].

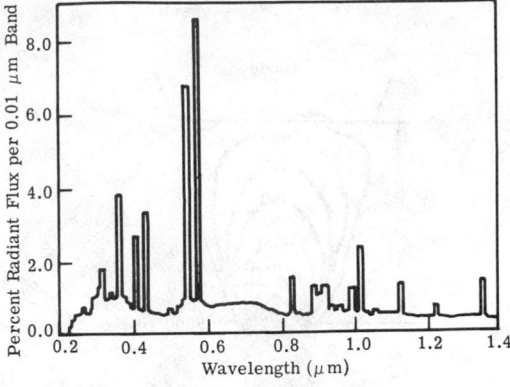

Fig. 2-44. Spectral distribution of radiant flux from a 1000-W dc mercury-xenon lamp [2-21].

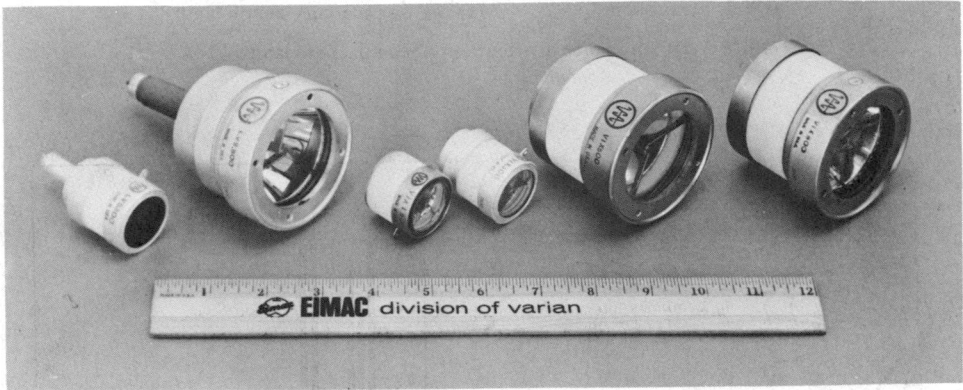

Fig. 2-45. High-pressure, short-arc xenon illuminators with sapphire windows. Low starting voltage, 150 through 800 watt; VIX150, VIX300, VIX500, VIX 800 [2-32].

Germicidal Lamps. These are hot-cathode lamps which operate at relatively low voltages. Tables 2-18 and 2-19 give some characteristics of these lamps.

These differ from ordinary fluorescent lamps which are used in lighting in that they are designed to transmit ultraviolet, whereas the wall of the fluorescent lamp is coated with a material that absorbs ultraviolet and reemits visible light. The germicidal lamp is constructed of glass of 1-mm thickness which transmits about 65% of the 2537 Å radiation and virtually cuts off shorter-wavelength ultraviolet radiation.

*Sterilamp® * Types.* These cold cathode lamps start and operate at higher voltages than the hot-cathode type and can be obtained in relatively small sizes as shown in Figure 2-46. Operating characteristics of the Sterilamps are given in Table 2-20.

Black-light Fluorescent Lamps. This fluorescent lamp is coated with a phosphor efficient in the absorption of 2537 Å radiation, emitting ultraviolet radiation in a broad band around 3650 Å. The phosphor is a cerium-activated calcium phosphate, and the glass bulb is impervious to shorter-wavelength ultraviolet radiation. The characteristics of

---

*Registered trademark of Westinghouse Electric.

Table 2-18. Radiation From Low-Pressure Mercury Lamps [2-24]

Normal (W)	Length (in.)	Diameter (in.)	Ultraviolet Output at 2537 Å (W)	Ultraviolet Irradiance (μW cm^{-2} at 1 m)
30	36	1	8.3	85
15	18	1	3.6	38
8	12	5/8	1.3	14
4	5 3/4	1/2	0.5	7

Table 2-19. T-6 Slimline-Type Germicidal Lamp [2-24]

Lamp (W)	Lamp Operating (V)	Lamp Operating Current	Ultraviolet (W)	Ultraviolet Irradiance (μW cm^{-2} at 1 m)
16	150	120	6	65
23	135	200	8	90
30	115	300	10	110
36	105	420	11	120

Table 2-20. Sterilamp Operating Characteristics[a] [2-24]

Lamp Designation	782L-30	782-20	782H-10	793
Rated Watts	17	14	12	3.5
Overall-Length (in.)	34-3/4	24-3/4	14-3/4	8-1/8
Starting ac voltage	750	575	400	200
Operating ac voltage	410	325	240	90
Operating current (A)	0.05	0.005	0.06	0.04
Ultraviolet output (2537 Å) (W)	5.2	2	2	0.13
Ultraviolet (μW cm^{-2} at 1 m (2537 Å))	46	20	20	1.3

[a]These lamps are made in a number of glasses having the same transmission for 2537 Å but having a transmission for 1849 Å of 10, 1.5, and 0.1 percent, respectively. Thus the quantity of ozone produced varies from a fairly large to a negligible amount.

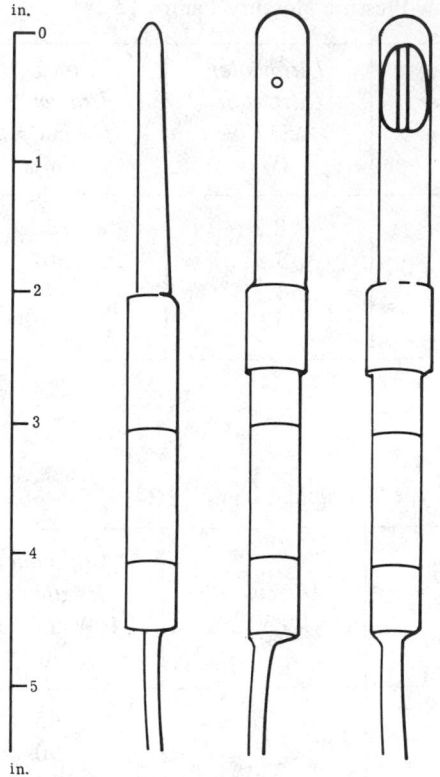

Fig. 2-46. Pen-Ray low-pressure lamp [2-24].
Pen-Ray® is a registered trademark of Ultraviolet Products, Inc.

one type are given in Table 2-21.

Hollow Cathode Lamps. A device described early in this century and used for many years by spectroscopists is the hollow-cathode tube. The one used by Paschen [2-1] consisted of a hollow metal cylinder and contained a small quantity of inert gas, yielding an intense cathode-glow characteristic of the cathode constituents. Materials that vaporize easily can be incorporated into the tube so that their spectral characteristics predominate [2-1].

A number of companies sell hollow-cathode lamps which do not differ significantly from those constructed in early laboratories. The external appearance of these modern tubes shows the marks of mass production and emphasis on convenience. They come with a large number of vaporizable elements, singly or in multiples, and with Pyrex® or quartz windows. A partial list of the characteristics of the lamps available from two manufacturers is given in Table 2-22. Their physical appearance is shown in Figure 2-47. A schematic of the different elements obtainable in various lamps is shown in Figure 2-48. Some sample spectra from those of another manufacturer are shown in Figure 2-49 and Table 2-23.

Table 2-21. Spectral Energy Distribution for Black-Light (360 BL) Lamps [2-24]

(W)	Length (in.)	3200-3800 Å		Total Ultra-violet below 3800 Å		Total Visible (W)	3800-7600 Å %[a]	Erythemal Flux
		(W)	%[a]	(W)	%[a]			
6	9	0.55	9.1	0.56	9.4	0.1	1.7	250
15	18	2.10	14.0	2.20	14.6	0.4	2.7	950
30	36	4.60	15.3	4.70	15.8	0.9	3.0	2100
40	48	6.70	16.8	6.90	17.3	1.5	3.8	3000

[a]Percentage of input power.

Table 2-22. Single-Element and Multiple-Element Hollow-Cathode Lamps (Tubes listed in this table are issued by Fisher Scientific and produced by Westinghouse Electric.) [2-46]

Single-Element 22000 Series

Element	Window[a]	Gas Fill[b]	Size[c]	Analytical Line (Å)	Catalog Number[d]
Aluminum	Q	A	B	3092	WL22804
	P	N	A	3092	JA-45-452
	Q	A	A	3092	WL22870
	Q	N	B	3092	WL22929
	Q	A	A	3092	WL22954
Antimony	Q	A	B	2311	WL22840
	Q	A	A	2311	WL22872
	Q	N	A	2311	JA-45-461
	Q	N	A	2311	WL22956
Arsenic	Q	N	B	1937	WL22873
	Q	N	A	1937	JA-45-315
	Q	N	A	1937	JA-45-315
Barium	P	N	A	5536	JA-45-480
Beryllium	Q	N	B	2349	WL23407
Bismuth	Q	A	B	3068	WL22841
	Q	A	A	3068	WL22874
	Q	N	B	3068	JA-45-469
	Q	N	A	3068	WL22957
Boron	Q	A	B	2497	JA-45-568
	Q	A	A	2497	WL22917
Cadmium	Q	A	B	3261	WL22816
	Q	A	A	3261	WL22875
	Q	A	A	3261	JA-45-462
	Q	N	A	3261	WL22958
Calcium	P	N	A	4227	JA-45-440
Cerium	Q	N	B	—	JA-45-569
Cesium	Q	N	A	—	WL22978
	Q	A	A	4556	WL22817
	P	A	A	4556	JA-45-441
Chromium	P	A	A	3579	WL22812
	P	A	B	3579	WL22821
	Q	A	A	3579	WL22877
	Q	A	B	3579	JA-45-454
	Q	N	A	3579	WL22959
Cobalt	P	A	A	3454	WL22813
	Q	A	A	3454	WL22814
	Q	A	B	3454	WL22878
	Q	N	A	3454	JA-45-456
	Q	N	A	3454	WL22953

Element	Window[a]	Gas Fill[b]	Size[c]	Analytical Line (Å)	Catalog Number[d]
Copper	P	N	A	3247	JA-45-458
	Q	A	B	3247	WL22606
	Q	A	A	3247	WL22879
	Q	A	B	3247	JA-45-490
	Q	N	A	3247	WL23042
Dysprosium	Q	N	B	4212	JA-45-595
	Q	N	A	4212	WL22880
Erbium	Q	A	B	4008	JA-45-571
	Q	A	A	4008	WL22881
Europium	Q	N	B	4594	JA-45-572
	Q	N	A	4594	WL22882
Gadolinium	P	A	A	4079	WL22975
	Q	N	A	4079	JA-45-573
	Q	N	B	4079	WL22986
Gallium	Q	A	A	4172	JA-45-470
	Q	A	B	4172	WL22884
Germanium	Q	N	A	2651	JA-45-575
	Q	N	B	2651	JA-45-313
Gold	Q	A	A	2676	WL22839
	Q	N	B	2676	WL22883
	Q	A	A	2676	JA-45-467
	Q	N	B	2676	WL22960
Hafnium	Q	N	A	3072	JA-45-303
Holmium	Q	N	B	4104	WL22885
	Q	A	A	4104	JA-45-576
Indium	Q	A	B	3040	WL22867
	Q	A	A	3040	WL22915
Iridium	Q	N	A	3040	JA-45-471
	Q	N	B	2850	JA-45-577
Iron	P	A	A	3270	WL22602
	Q	A	B	3270	WL22611
	Q	A	B	3720	JA-45-455
Iron, high-purity	P	A	A	3720	WL22820
	Q	N	A	3720	WL22886
	Q	A	A	3720	WL22887
	Q	A	B	3720	WL22837
	Q	A	A	3720	WL22888
Lanthanum	P	N	B	5501	WL22846
	Q	A	A	5501	WL22889
	Q	A	B	5501	JA-45-495

[a]P = Pyrex, Q = quartz; [b]N = neon, A = argon; [c]A = 1½ in. diameter, B = 1 in. diameter, C = 2 in. diameter; [d]WL = Westinghouse, JA = Jarrell-Ash.

Table 2-22.　Single-Element and Multiple-Element Hollow-Cathode Lamps (Tubes listed in this table are issued by Fisher Scientific and produced by Westinghouse Electric.) [2-46] (Continued)

Element	Window[a]	Gas Fill[b]	Size[c]	Analytical Line (Å)	Catalog Number[d]
Lead	Q	A	B	2833	WL22838
	Q	A	A	2833	WL22890
	Q	N	B	2833	JA-45-468
	Q	A	A	2833	WL22952
Lithium 6	P	N	A	6708	JA-45-579
	P	A	A	6708	WL22925
Lithium 7	P	N	A	6708	JA-45-580
	P	A	A	6708	WL22926
Lithium, natural	P	A	A	6708	WL22825
	P	A	A	6708	JA-45-444
	P	N	B	6708	WL23115
Lutetium	Q	A	B	3282	JA-45-581
	Q	A	A	3282	WL23010
Magnesium	Q	A	B	2852	WL22609
	Q	A	A	2852	WL22891
	Q	N	A	2852	WL22951
	Q	A	B	2852	JA-45-451
Manganese	Q	A	B	2795	WL22608
	P	A	A	2795	WL22815
	Q	N	B	2795	JA-45-472
	Q	A	A	2795	WL22961
	Q	A	A	2795	WL22876
Mercury	Q	A	B	2537	JA-45-493
	Q	A	A	2537	WL22892
Molybdenum	Q	A	B	3133	WL22805
	Q	N	B	3133	WL22893
	Q	A	B	3133	JA-45-460
	Q	A	A	3133	WL22962
Neodymium	Q	A	B	4925	JA-45-582
	P	N	A	4925	WL22980
Nickel	Q	A	A	3415	WL22605
	Q	A	B	3415	WL22663
	Q	A	B	3415	JA-45-457
	Q	A	A	3415	WL22894
Niobium	Q	N	A	4059	WL22895
	Q	N	B	4059	JA-45-486
	Q	A	A	4059	WL22912
Osmium	Q	A	B	2909	JA-45-584

Element	Window[a]	Gas Fill[b]	Size[c]	Analytical Line (Å)	Catalog Number[d]
Palladium	Q	A	B	3404	WL22857
	Q	A	A	3404	WL22911
	Q	A	B	3404	JA-45-475
	Q	A	A	3404	WL22970
Phosphorus	Q	N	B	2136	JA-45-449
	Q	N	A	2136	WL22990
Platinum	Q	A	B	2659	WL22851
	Q	A	A	2659	WL22896
	Q	A	B	2659	JA-45-466
Potassium	P	N	A	4044	JA-45-484
Praseodymium	Q	N	B	4951	JA-45-585
	Q	N	A	4951	WL22982
Rhenium	Q	N	B	3460	JA-45-489
	Q	A	A	3460	WL22967
Rhodium	Q	A	B	3435	WL22850
	Q	A	A	3435	WL22897
	Q	A	B	3435	JA-45-476
Rubidium	Q	A	A	7800	WL23046
	P	A	B	7800	JA-45-586
Ruthenium	Q	A	B	3499	JA-45-587
Samarium	Q	N	A	4760	WL22899
	Q	A	B	4760	JA-45-309
Scandium	Q	A	B	3912	WL22843
Selenium	Q	A	B	1960	WL22898
	Q	A	A	1960	JA-45-477
	Q	N	B	1960	WL22963
Silicon	Q	A	A	2516	WL22832
	Q	N	B	2516	WL22900
	Q	N	A	2516	JA-45-479
	Q	A	B	2516	WL22964
Silver	Q	A	A	3281	WL22901
	P	A	A	3281	WL22864
Sodium	P	N	A	5890	JA-45-485
Strontium	P	N	A	5890	JA-45-481
	Q	N	B	4607	JA-45-588
Sulphur	Q	N	A	—	

Table 2-22. Single-Element and Multiple-Element Hollow-Cathode Lamps (Tubes listed in this table are issued by Fisher Scientific and produced by Westinghouse Electric.) [2-46] (Continued)

Element	Window[a]	Gas Fill[b]	Size[c]	Analytical Line (Å)	Catalog Number[d]
Tantalum	Q	A	B	2714	JA-45-488
	Q	A	A	2714	WL22913
	Q	N	B	2714	WL22971
	Q	N	A	2714	WL22972
Tellurium	Q	A	B	2143	WL22842
	Q	A	A	2143	WL22902
	Q	N	A	2143	JA-45-473
Terbium	Q	N	B	4326	WL22965
	Q	N	A	4326	JA-45-589
	Q	N	B	4326	WL22903
Thallium	Q	N	A	3776	WL23408
Thorium	Q	N	B	3245	WL23028
	Q	N	A	3245	JA-45-590
Thulium	Q	N	B	4105	JA-45-591
	Q	N	A	4105	WL23008
Tin	Q	A	B	2863	WL22822
	Q	A	A	2863	WL22904
	Q	N	B	2863	JA-45-463
Titanium	Q	A	A	2863	WL22966
	Q	N	B	3643	JA-45-592
	Q	N	A	3643	WL22992
Tungsten	Q	N	B	4009	JA-45-465
	Q	N	A	4009	WL22849
	Q	N	B	4009	WL22905
Uranium	Q	A	A	4009	WL22906
	Q	N	B	5027	JA-45-447
	Q	N	A	5027	WL22907
Vanadium	Q	A	B	3184	WL22856
	Q	N	A	3184	WL22910
	Q	N	B	3184	JA-45-453
Ytterbium	Q	N	A	3184	WL22974
	Q	A	B	3988	JA-45-593
	Q	A	A	3988	WL22984
Yttrium	P	A	A	4102	WL22976
	Q	N	B	4102	JA-45-594
	Q	N	A	4102	WL22988
Zinc	Q	A	B	2139	WL22607
	Q	A	A	2139	JA-45-459
	Q	N	B	2139	WL22908
	Q	N	A	2139	WL22909

Element	Window[a]	Gas Fill[b]	Size[c]	Analytical Line (Å)	Catalog Number[d]
Zirconium	Q	A	B	3601	JA-45-482
	Q	A	A	3601	WL22914
	Q	N	B	3601	WL22998
Single-Element 36000 Series					
Aluminum	P	N	C	3092	JA-45-36009
Antimony	Q	N	C	2311	JA-45-36010
Arsenic	P	A	C	1937	JA-45-36011
Barium	P	N	C	5536	JA-45-36012
Beryllium	Q	N	C	2349	JA-45-36013
Bismuth	Q	N	C	3068	JA-45-36014
Boron	Q	A	C	2497	JA-45-36015
Cadmium	Q	N	C	3261	JA-45-36016
Calcium	P	N	C	4227	JA-45-36017
Cerium	Q	N	C	—	JA-45-36019
Cesium	P	N	C	4556	JA-45-36020
Chromium	P	N	C	3579	JA-45-36021
Cobalt	Q	N	C	3454	JA-45-36022
Copper	P	N	C	3247	JA-45-36024
Dysprosium	P	N	C	4212	JA-45-36025
Erbium	P	N	C	4008	JA-45-36026
Europium	P	N	C	4594	JA-45-36027
Gadolinium	P	N	C	4079	JA-45-36028
Gallium	Q	N	C	4172	JA-45-36029
Germanium	Q	N	C	2651	JA-45-36030
Gold	Q	N	C	2676	JA-45-36031
Hafnium	P	N	C	3072	JA-45-36032
Holmium	P	N	C	4104	JA-45-36033
Indium	P	N	C	3040	JA-45-36034
Iridium	Q	N	C	2850	JA-45-36036
Iron	P	N	C	3720	JA-45-36037
Lanthanum	Q	N	C	5501	JA-45-36038
Lead	P	N	C	2833	JA-45-36039
Lithium 6	P	N	C	6708	JA-45-36090
Lithium 7	P	N	C	6708	JA-45-36091
Lithium, natural	P	N	C	6708	JA-45-36040
Lutetium	P	N	C	3282	JA-45-36041
Magnesium	Q	N	C	2852	JA-45-36042
Manganese	Q	N	C	2795	JA-45-36043

[a] P = Pyrex, Q = quartz; [b] N = neon, A = argon; [c] A = 1½ in. diameter, B = 1 in. diameter, C = 2 in. diameter; [d] WL = Westinghouse, JA = Jarrell-Ash.

Table 2-22. Single-Element and Multiple-Element Hollow-Cathode Lamps (Tubes listed in this table are issued by Fisher Scientific and produced by Westinghouse Electric.) [2-46] (Continued)

Element	Window[a]	Gas Fill[b]	Size[c]	Analytical Line (Å)	Catalog Number[d]
Mercury	Q	A	C	2537	JA-45-36044
Molybdenum	Q	N	C	3133	JA-45-36045
Neodymium	P	N	C	4925	JA-45-36046
Nickel	Q	N	C	3415	JA-45-36047
Niobium	P	A	C	4059	JA-45-36023
Osmium	Q	N	C	2909	JA-45-36048
Palladium	Q	N	C	3404	JA-45-36049
Phosphorus	Q	N	C	2136	JA-45-36050
Platinum	Q	N	C	2659	JA-45-36051
Potassium	P	N	C	4044	JA-45-36052
Praseodymium	P	N	C	4951	JA-45-36053
Rhenium	P	N	C	3460	JA-45-36056
Rhodium	P	N	C	3435	JA-45-36057
Rubidium	P	N	C	7800	JA-45-36058
Ruthenium	P	A	C	3499	JA-45-36059
Samarium	P	N	C	4760	JA-45-36060
Scandium	P	N	C	3912	JA-45-36061
Selenium	Q	N	C	1960	JA-45-36062
Silicon	P	N	C	2516	JA-45-36063
Silver	Q	A	C	3281	JA-45-36064
Sodium	P	N	C	5890	JA-45-36065
Strontium	P	N	C	4607	JA-45-36066
Sulphur	Q	N	C		JA-45-36067
Tantalum	Q	A	C	2714	JA-45-36068
Tellurium	O	N	C	2143	JA-45-36069
Terbium	P	N	C	4326	JA-45-36070
Thallium	Q	N	C	3776	JA-45-36071
Thorium	O	N	C	3245	JA-45-36072
Thulium	P	N	C	4105	JA-45-36073
Tin	P	N	C	2863	JA-45-36074
Titanium	P	N	C	3643	JA-45-36075
Tungsten	O	N	C	4009	JA-45-36076
Uranium	P	N	C	5027	JA-45-36077
Vanadium	P	N	C	3184	JA-45-36078
Ytterbium	P	A	C	3988	JA-45-36079
Yttrium	P	N	C	4102	JA-45-36080
Zinc	O	N	C	2139	JA-45-36081
Zirconium	P	A	C	3601	JA-45-36082

Multiple-Element 22000 Series

Element	Window[a]	Gas Fill	Size[c]	Analytical Line (Å)	Catalog Number[d]
Aluminum, calcium	Q	N	B	-	WL23246
Aluminum, calcium, magnesium	Q	A	B	-	WL22604
Aluminum, calcium, magnesium	Q	A	A	-	WL22871
Aluminum, calcium magnesium	Q	N	B	-	JA-45-450
Aluminum, calcium, magnesium	Q	N	A	-	WL22955
Aluminum, calcium, magnesium, iron	Q	N	B	-	JA-45-310
Aluminum, calcium, magnesium, lithium	Q	N	B	-	JA-45-436
Aluminum, calcium, magnesium, lithium	Q	A	A	-	WL23036
Aluminum, calcium, strontium	P	N	A	-	WL23403
Antimony, arsenic, bismuth	Q	N	B	-	WL23147
Arsenic, nickel	Q	N	B	-	JA-45-434
Arsenic, selenium, tellurium	Q	N	B	-	JA-45-598
Barium, calcium, strontium	P	N	A	-	JA-45-437
Barium, calcium, silicon, magnesium	O	N	B	-	JA-45-478
Cadmium, copper, zinc, lead	O	N	B	-	JA-45-597
Cadmium, silver, zinc, lead	O	N	B	-	JA-45-308
Calcium, magnesium, strontium	O	N	B	-	WL23605
Calcium, magnesium, zinc	O	N	B	-	JA-45-311
Calcium, magnesium, aluminum, lithium	O	A	B	-	WL23158
Calcium, zinc	O	N	B	-	JA-45-304
Chromium, iron, manganese, nickel	O	N	B	-	JA-45-442
Chromium, cobalt, nickel	O	N	B	-	WL23174
Chromium, copper	O	N	B	-	JA-45-306

Table 2-22. Single-Element and Multiple-Element Hollow-Cathode Lamps (Tubes listed in this table are issued by Fisher Scientific and produced by Westinghouse Electric.) [2-46] (Continued)

Element	Window[a]	Gas Fill[b]	Size[c]	Analytical Line (Å)	Catalog Number[d]
Chromium, manganese	Q	N	B	—	WL23499
Chromium, cobalt, copper, manganese, nickel	Q	N	B	.	WL23601
Chromium, cobalt, copper, iron, manganese, nickel	Q	N	B	—	JA-45-599
Cobalt, copper	Q	N	B	—	JA-45-305
Cobalt, copper, gold, nickel	Q	N	B	—	WL23295
Cobalt, copper, zinc, molybdenum	Q	N	B	—	JA-45-596
Cobalt, iron	Q	N	B	—	WL23291
Cobalt, nickel	Q	N	B	—	JA-45-426
Copper, gallium	Q	N	B	—	JA-45-431
Copper, iron	Q	N	B	—	JA-45-312
Copper, iron, manganese	Q	N	B	—	JA-45-435
Copper, iron, molybdenum	Q	N	B	—	JA-45-301
Copper, iron, gold, nickel	Q	N	B	—	JA-45-307
Copper, iron, manganese, zinc	Q	N	B	—	JA-45-492
Copper, manganese	Q	N	B	—	JA-45-491
Copper, nickel	Q	N	B	—	WL23441A
Copper, nickel, zinc	Q	N	B	—	JA-45-405
Copper, zinc, molybdenum	Q	N	B	—	JA-45-496
Copper, zinc, lead, silver	Q	N	B	—	JA-45-448
Copper, zinc, lead, tin	Q	N	B	—	JA-45-438
Gold, nickel	Q	N	B	—	JA-45-433
Gold, silver	Q	N	B	—	WL23269
Indium, silver	Q	N	B	—	WL23294
Lead, silver, zinc	Q	N	B	—	WL23171
Mangesium, zinc	Q	N	B	—	WL23455
Sodium, potassium	P	A	A	—	JA-45-439
Sodium, potassium	P	A	A	—	WL23230
Zinc, lead, tin	Q	N	B	—	WL23404

Multiple-Element 36000 Series

Element	Window[a]	Gas Fill[b]	Size[c]	Analytical Line (Å)	Catalog Number[d]
Aluminum, calcium, magnesium	Q	N	C	—	JA-45-36099
Aluminum, calcium, magnesium lithium	Q	N	C	—	JA-45-36250
Antimony, arsenic, bismuth	Q	N	C	—	JA-45-36203
Barium, calcium, strontium, magnesium	O	N	C	—	JA-45-36228
Cadmium, silver, zinc, lead	O	N	C	—	JA-45-36205
Cadmium, copper, zinc, lead	O	N	C	—	JA-45-36227
Calcium, magnesium	O	N	C	—	JA-45-36092
Calcium, magnesium, zinc	O	N	C	—	JA-45-36097
Calcium, zinc	O	N	C	—	JA-45-36093
Chromium, iron, manganese, nickel	O	N	C	—	JA-45-36201
Chromium, cobalt, copper, manganese, nickel	O	N	C	—	JA-45-36094
Chromium, cobalt, copper, manganese, nickel	Q	N	C	—	JA-45-36103
Chromium, copper, nickel, silver	O	N	C	—	JA-45-36096
Chromium, copper, iron, nickel, silver	O	N	C	—	JA-45-36108
Cobalt, copper, iron, manganese, molybdenum	O	N	C	—	JA-45-36102
Copper, zinc, lead, tin	O	N	C	—	JA-45-36202
Copper, iron	O	N	C	—	JA-45-36200
Copper, iron, nickel	O	N	C	—	JA-45-36101
Copper, iron, lead, nickel, zinc	O	N	C	—	JA-45-36204
Copper, iron, manganese, zinc	O	N	C	—	JA-45-36105
Sodium, potassium	P	A	C	—	JA-45-36095

[a]P = Pyrex, Q = quartz; [b]N = neon, A = argon; [c]A = 1½ in. diameter, B = 1 in. diameter, C = 2 in. diameter; [d]WL = Westinghouse, JA = Jarrell-Ash.

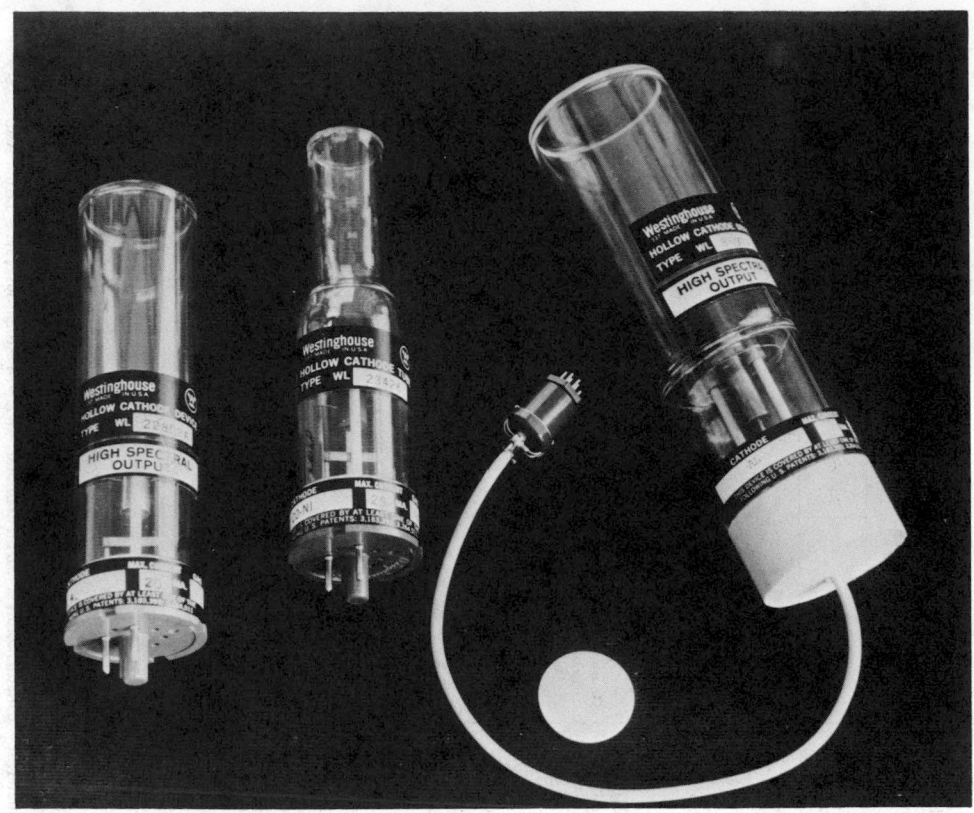

Fig. 2-47. Hollow-cathode spectral tubes described in Table 2-22.

		Transition Elements															
Li	Be											B					
Na	Mg				Group 8							Al	Si	P	S		
K	Ca	Sc	Ti	V	Cr	Mn	Fe	Co	Ni	Cu	Zn	Ga	Ge	As	Se		
Rb	Sr	Y	Zr	Nb	Mo		Ru	Rh	Pd	Ag	Cd	In	Sn	Sb	Te		
Cs	Ba	La	Hf	Ta	W	Re	Os	Ir	Pt	Au	Hg	Tl	Pb	Bi			

Lantha-nides	Ce	Pr	Nd		Sm	Eu	Gd	Tb	Dy	Ho	Er	Tm	Yb	Lu
Actinides	Th		U											

Fig. 2-48. Periodic table showing the prevalence of elements obtainable in hollow-cathode tubes.

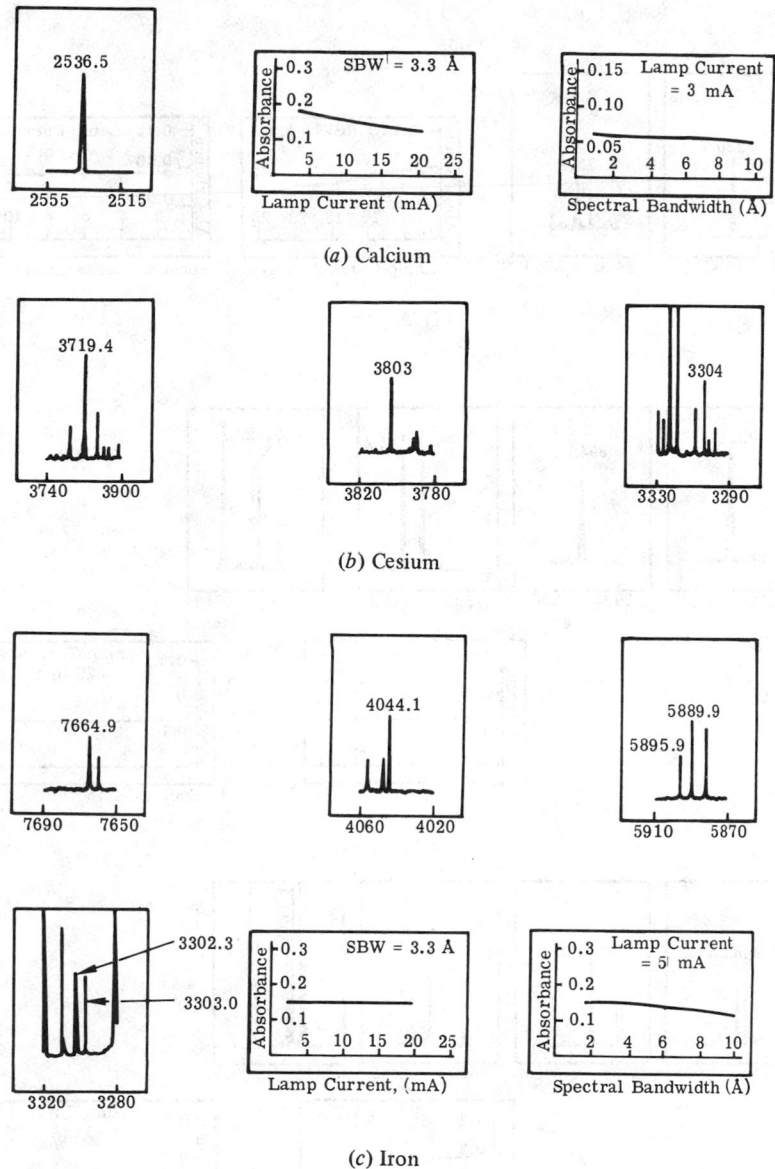

(a) Calcium

(b) Cesium

(c) Iron

Fig. 2-49. Spectral characteristics of some hollow-cathode lamps. The data in these curves are identical with data in the 1973 *Varian Techtron Catalog*. [2-32].

Electrodeless Discharge Lamps [2-33, 2-34, 2-35]. The electrodeless lamp gained popularity when Meggers used it in his attempt to produce a highly precise standard of radiation. Simplicity of design makes laboratory construction of this type of lamp easy. Some of the simplest lamps consist of a tube, containing the radiation-producing element, and a microwave generator for producing the electric field (within the tube) which in turn excites the elemental spectra.

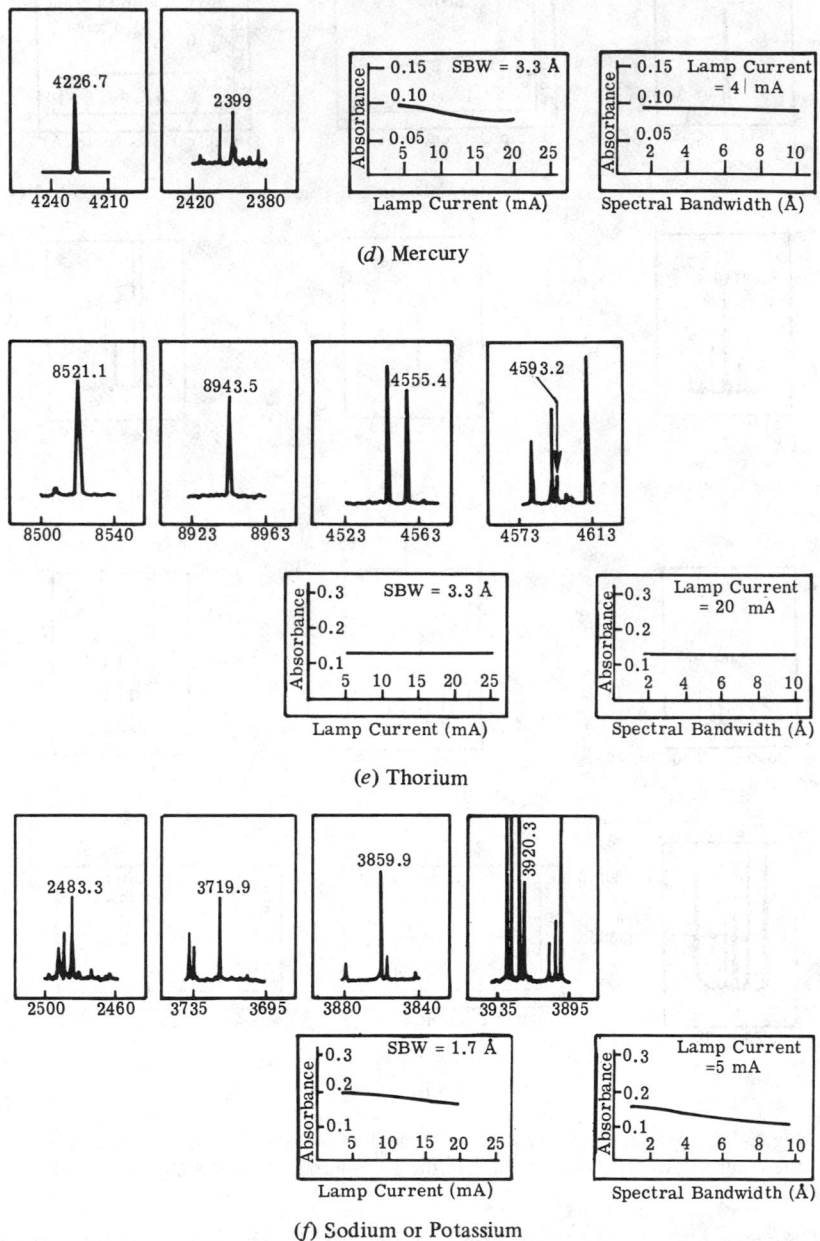

(d) Mercury

(e) Thorium

(f) Sodium or Potassium

Fig. 2-49 (Continued). Spectral characteristics of some hollow-cathode lamps. The data in these curves are identical with data in the 1973 *Varian Techtron Catalog.* [2-32].

Table 2-23. Spectral Characteristics of Varian Hollow-Cathode Tubes

Lamp Specifications	Calcium	Cesium	Iron	Mercury	Thorium	Sodium or Potassium
Fill Gas	Neon	Argon	Neon	Neon	Neon	Neon
Window	Quartz	Pyrex	Quartz	Quartz	Quartz	Pyrex
Operating current (mA)	4	20	5	3	10	5
Strike voltage (V)	260	–	260	310	270	280
Operating voltage (V)	150	–	160	230	140	210
Most sensitive line (Å)	4226.7	8521.1	2483.3	2536.5	a	5889.9
Spectral bandwidth (Å)	3.3	9.9	1.7	3.3	0.8	3.3
Sensitivity (ppm)	0.03[b]	0.16[c,d]	0.08[e]	2[e]	a	0.01[c]

Alternative Absorption Lines[f]

Element	Wavelength (Å)	Sensitivity (ppm)	Relative Sensitivity	Relative Signal Intensity	Recommended Spectral Bandwidth (Å)
Calcium	4226.7	0.03	1[g]	100[h]	3.3
	2398.6	6	200.0[g]	0.25[h]	0.8
Cesium	8521.1	0.16	1[i]	100[j,d]	9.9
	8943.5	0.22	1.2[i]	47[i,d]	9.9
	4555.4	10	5.3[i]	100[k]	3.3
	4593.2	40	21[i]	26[k,l]	0.8
Iron	2483.3	0.08	1[m]	1[n]	1.7
	3719.9	0.76	9[m]	100[n]	0.8
	3859.9	1.2	15.0[m]	53[n]	1.7
	3920.3	22	270[m]	4[n]	0.8
Thorium	3719.4	–	–	100[o]	0.8
	3803.1	–	–	57[o]	0.8
	3304.2	–	–	32[o]	0.8
Sodium or potassium	5889.9	0.1	1.0[p]	100[q]	3.3
	5895.9	0.02	2.1[p]	56[q]	3.3
	3302.3 3302.0	2.9	360[p]	3.7[q]	3.3

[a]No absorption has been detected for this element with the Techtron burner system.
[b]With $N_2O-C_2H_2$ flame.
[c]With air-propane flame.
[d]With R196 photomultiplier tube.
[e]With air-C_2H_2 flame.
[f]No alternative lines for mercury have been examined.
[g]Measured relative to the 4226.7 Å line.
[h]Measured at 4 mA lamp current with spectral bandwidth of 3.3 Å relative to the 4226.7 Å line.
[i]Measured relative to the 8521.1 Å line.
[j]Measured at 20 mA lamp current with spectral bandwidth of 3.3 Å.
[k]With R213 photomultiplier tube.
[l]Measured relative to 4555.4 Å.
[m]Measured relative to 2483.3 Å line.
[n]Measured at 5 mA lamp current with spectral bandwidth of 0.8 Å relative to 3719.9 Å line.
[o]Measured at 20 mA lamp current with spectral bandwidth of 0.8 Å relative to 3719.9 Å line.
[p]Measured relative to 5889.9 Å line.
[q]Measured at 5 mA lamp current with spectral bandwidth of 0.8 Å relative to 5889.9 Å line.

Lamps of this type can be purchased with specially designed microwave cavities for greater efficiency in coupling. Those made of fused quartz can transmit from ultraviolet to near infrared. The electrodeless lamp is better able than the arc lamp to produce stable radiation of sharp spectral lines; this makes it useful in spectroscopy and interferometry. The Hg 198 lamp makes a suitable secondary standard of radiation, particularly at the following selected (vacuum) wavelengths (in Å):

2537.2687	3022.3796	4047.7145
2652.8324	3126.5761	4359.5622
2753.5968	3342.4422	5462.2704
2894.4464	3651.1966	5771.1982
2968.1498		

Electrodeless UV lamps with variable output power and a choice of bulbs with different UV spectra are manufactured by Fusion Systems Corporation, Rockville, Maryland. Quoted technical specifications for a 10-inch lamp are as follows:

UV output power: ~ 1000 W
Lamp input power: 3000 W
Bulb lifetime: up to 3000 h
Cooling: water or forced air
IR power: 1000 W maximum

Spectral Lamps. Some manufacturers produce groups of arc sources, which are similar in construction and filled with different elements and rare gases, and which yield discontinuous or monochromatic radiation throughout most of the ultraviolet and visible spectrum. They are called spectral lamps. The envelopes of these lamps are constructed of glass or quartz, depending on the part of the spectrum desired. Thus, discrete radiation can be obtained from around 2300 Å into the near infrared.

Figure 2-50 represents the various atomic lines observable from Osram spectral lamps. Figure 2-51 gives a physical description of various spectral lamps obtainable from Philips. Tables 2-24 and 2-25 list the characteristics of the various types of lamps obtainable from Osram and Philips, respectively.

Pluecker Spectrum Tubes. These are inexpensive tubes made of glass (Figure 2-52) with an overall length of 25 cm and capillary portion 8.5 to 10 cm long. They operate from an ordinary supply with a special transformer which supports the tubes in a vertical position and maintains the voltage and current values adequate to operate the discharge and regulate the spectral intensity. Table 2-26 lists the various gases in available tubes.

2.3.5. Concentrated Arc Lamps

Zirconium Arc. The cathodes of these lamps are made of a hollow refractory metal containing zirconium oxide. The anode, a disk of metal with an aperture, resides directly above the cathode with the normal to the aperture coincident with the longitudinal axis of the cathode. Argon gas fills the tube. The arc discharge causes the zirconium to heat (to about 3000 K) and produce an intense, very small source of light.

These lamps can be obtained from the Cenco Company in a number of wattages (from 2 to 300). The end of the bulb through which the radiation passes comes with ordinary curvature or (for a slight increase in price) flat. Examples are shown in Figure 2-53.

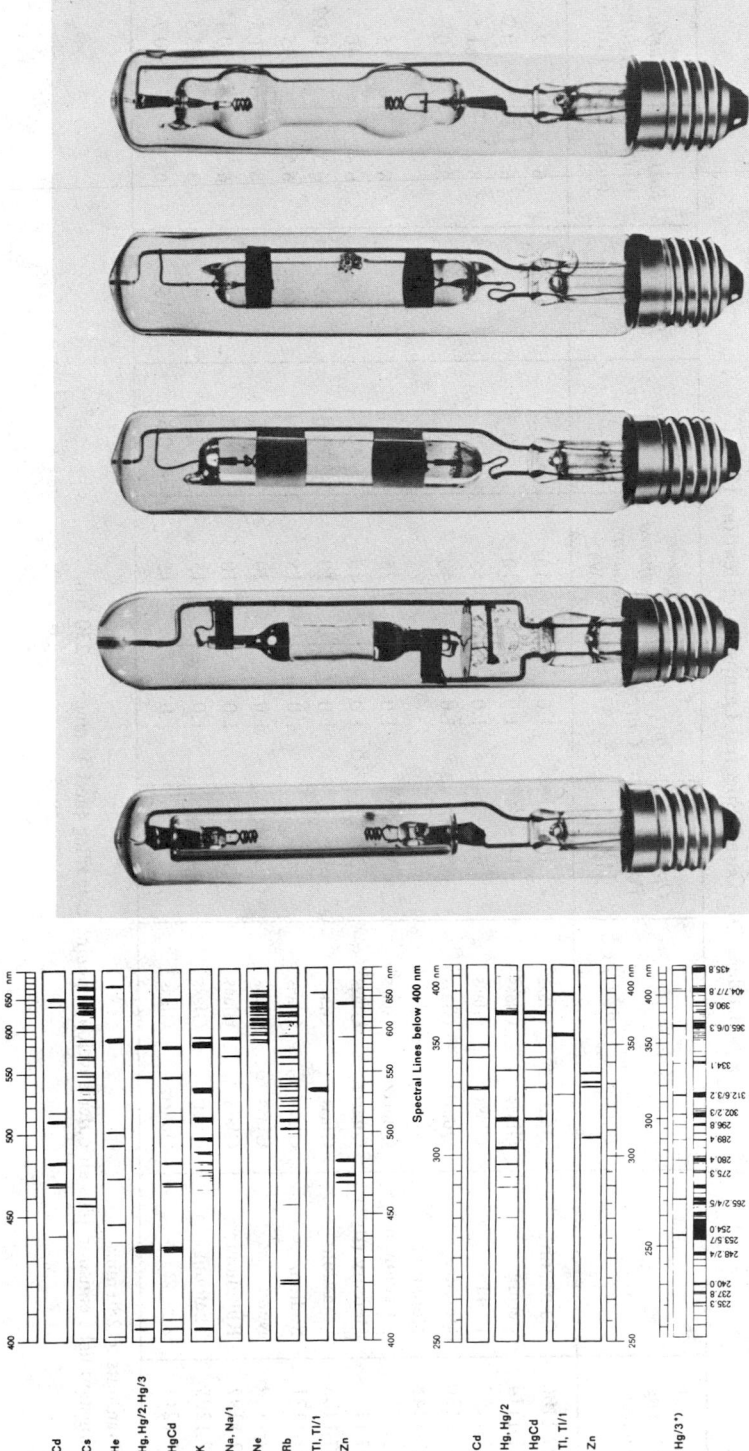

Fig. 2-51. Examples of Philips spectral lamps [2-36].

Fig. 2-50. Spectograms of Ostram spectral lamps [2-22].

Table 2-24. Osram Spectral Lamps Specification [2-36]

Catalog Number	Type	Material		Operating Potential (V)	Operating Current (A)	Power Dissipated in Lamp (W)	Light Intensity (cd)	Emitting Height (mm)	Surface Width (mm)	Surface Brightness (Stilb)
		Burner	Envelope							
27-1106	Cadmium	Quartz	Glass[a]	15	1.0	15	1.2	15	6	2
27-1114	Cesium	Glass	Glass	10	1.0	10	0.3	15	6	0.2
27-1122	Helium	Glass	Glass	60	1.0	55	2	15	8	1.5
27-1130	Mercury	Quartz	Glass[a]	50	1.0	40	90	20	8	110
27-1148	Mercury & Cadmium	Quartz	Glass[a]	30	1.0	25	10	20	8	15
27-1155	Mercury (Hg3)	Quartz	Glass[b]	50	1.0	40	40	25	6	40
27-1163	Potassium	Glass	Glass	10	1.0	10	0.04	15	6.5	0.02
27-1171	Sodium	Glass	Glass	15	1.0	15	40	15	6.5	15
27-1189	Neon	Glass	Glass	30	1.0	30	3.5	15	8	1.5
27-1197	Rubidium	Glass	Glass	10	1.0	10	0.2	15	6	0.1
27-1205	Thallium	Quartz	Glass[a]	15	1.0	15	1.5	8	3	3.5
27-1213	Zinc	Quartz	Glass[a]	15	1.0	15	0.5	15	6	0.7

[a]Transmits to 280 nm.
[b]Constant light emission plus an outer envelope opening extending the UV range to 230 nm.

Table 2-25. Specifications of Philips Spectral Lamps [2-36]

Catalog Number	Symbols	Type	Material		Operating Current (A)	Wattage	Arc Length (mm)
			Burner	Envelope			
26–2709	Hg	Mercury (low-pressure)	Quartz	Glass	0.9	15	40
26–2717	Hg	Mercury (high-pressure)	Quartz	Glass	0.9	90	30
26–2725	Cd	Cadmium	Quartz	Glass	0.9	25	30
26–2733	Zn	Zinc	Quartz	Glass	0.9	25	30
26–2741	Hg, Cd, Zn	Mercury, cadmium, and zinc	Quartz	Glass	0.9	90	30
26–2758	He	Helium	Glass	Glass	0.9	45	32
26–2766	Ne	Neon	Glass	Glass	0.9	25	40
26–2774	A	Argon	Glass	Glass	0.9	15	40
26–2782	Kr	Krypton	Glass	Glass	0.9	15	40
26–2790	Xe	Xenon	Glass	Glass	0.9	10	40
26–2808	Na	Sodium	Glass	Glass	0.9	15	40
26–2816	Rb	Rubidium	Glass	Glass	0.9	15	40
26–2824	Cs	Caesium	Glass	Glass	0.9	10	40
26–2832	K	Potassium	Glass	Glass	0.9	10	40
26–2857	Hg	Mercury (low-pressure)	Quartz	Quartz	0.9	15	40
26–2865	Hg	Mercury (high-pressure)	Quartz	Quartz	0.9	90	30
26–2873	Cd	Cadmium	Quartz	Quartz	0.9	25	30
26–2881	Zn	Zinc	Quartz	Quartz	0.9	25	30
26–2899	Hg, Cd, Zn	Mercury, cadmium, and zinc	Quartz	Quartz	0.9	90	30
26–2907	In	Indium[a]	Quartz	Quartz	0.9	25	25
26–2915	Tl	Thallium	Quartz	Quartz	0.9	20	30
26–2923	Ga	Gallium	Quartz	Quartz	0.9	20	30

[a]Requires a Tesla coil to cause it to strike initially

Table 2-26. Gas Fills in Pluecker Tubes[a] [2-37]

Cenco Number	Type
87210	Argon Gas
87215	Helium Gas
87220	Neon Gas
87225	Carbonic Acid Gas
87230	Chlorine Gas
87235	Hydrogen Gas
87240	Nitrogen Gas
87242	Air
87245	Oxygen Gas
87255	Iodine Vapor
87256	Krypton Gas
87258	Xenon
87260	Mercury Vapor
87265	Water Vapor

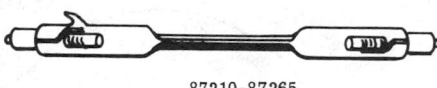

87210-87265

Fig. 2-52. Physical construction of Pluecker spectrum tubes [2-37].

[a]Consists of glass tube with overall length of 25 cm with capillary portion about 8.5 to 10 cm long. Glass-to-metal seal wires are welded in metal caps with loops for wire connection are firmly sealed to the ends. Power supply no. 87208 is recommended as a source of excitation.

Fig. 2-53. Physical construction of some zirconium arc lamps [2-37]. Two 2 W lamps are available, but not shown here [2-28].

Tungsten-Arc (Photomicrographic) Lamp. The essential elements of this discharge-type lamp (Figure 2-54) are a ring electrode and a pellet electrode, both made of tungsten. The arc forms between these electrodes, causing the pellet to heat incandescently. The ring also incandesces but to a lesser extent. Thus, the hot pellet (approximately 3100 K) provides an intense source of small-area radiation. A plot of the spectral variation of this radiation is given in Figure 2-55.

As with all tungsten sources, evaporation causes a steady erosion of the pellet surface with the introduction of gradients, which is not serious if the pellet is used as a point source.

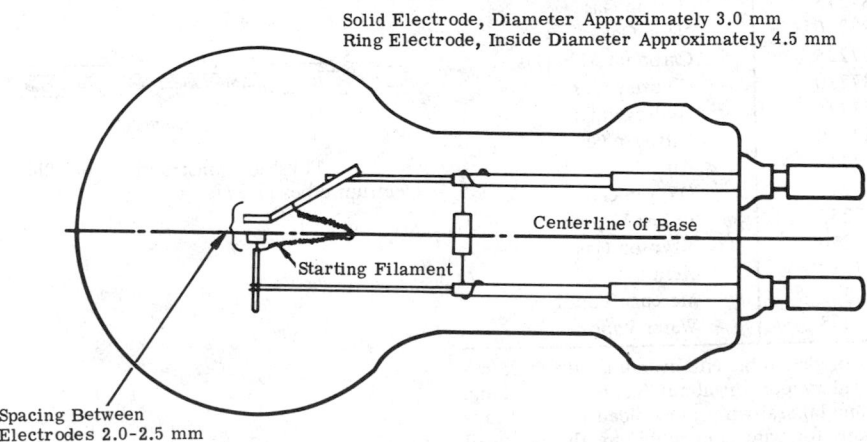

Fig. 2-54. Construction of tungsten arc lamp. The lamp must be operated base-up in a well-ventilated housing and using a special high-current socket which does not distort the position of the posts [2-28].

General Electric, manufacturer of the 30A/PS22 photomicrographic lamp—which uses a 30 Å operating current, states that this lamp requires a special heavy-duty socket obtainable through certain manufacturers suggested in their brochure.

2.3.6. Glow Modulator Tubes. According to technical data supplied by Sylvania, these are cold-cathode light sources uniquely adaptable to high-frequency modulation. Pictures of two types are shown in Figure 2-56. The cathode is a small hollow cylinder and the

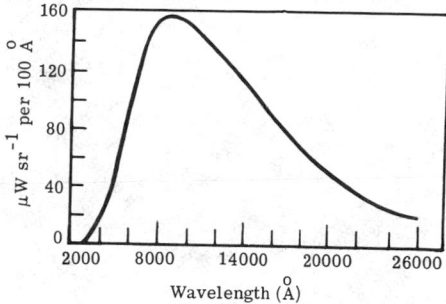

Fig. 2-55. Spectral distribution of a 30A PS 22 photomicrographic lamp [2-28].

high ionization density in the region of the cathode provides an intense source of radiation.

Figure 2-57 is a graph of the light output as a function of tube current. Figure 2-58 is a graph depicting the response of the tube to a modulating input. The spectral outputs of a variety of tubes are shown in Figure 2-59. Table 2-27 gives some of the glow-modulator specifications.

2.3.7. Hydrogen and Deuterium Arcs. For applications requiring a strong continuum in the ultraviolet region, the hydrogen arc at a few millimeters pressure provides a useful source. It can be operated with a cold or hot cathode. One hot-cathode type is shown in Figure 2-60. Koller [2-24] plots a distribution for this lamp down to about 200 Å.

Deuterium lamps (Figure 2-61) provide a continuum in the ultraviolet with increased intensity over the hydrogen arc. Both lamps have quartz envelopes. The one on the left is designed for operation down to 2000 Å: the one on the right is provided with a Suprasil® * window to increase the ultraviolet range down to 1650 Å.

NBS is offering a deuterium lamp standard of spectral irradiance between 200 and 350 nm. The lamp output at 50 cm from its medium bipost base is about 0.7 W cm^{-3} at 200 nm and drops off smoothly to 0.3 W cm^{-3} at 250 nm and 0.07 W cm^{-3} at 350 nm. A working standard of the deuterium lamp can be obtained also, for example, from Optronic Laboratories, Incorporated, Silver Springs, Maryland.

2.3.8. Other Commercial Sources

Activated-Phosphor Sources. Of particular importance and convenience in the use of photometers are sources composed of a phosphor activated by radioactive substances. Readily available, and not subject to Atomic Energy Commission licensing with small quantities of radioactive material, are the ^{14}C-activated phosphor light sources. These are relatively stable sources of low intensity, losing about 0.02% per year due to the half-life of ^{14}C and the destruction of phosphor centers.

Light Emitting Diodes (LEDs). These sources fit most readily into the category of display devices. However, for some special purposes they are often useful since they are small, simple to integrate into a system, and can be made highly directional. They emit very narrow-band radiation, usually in the visible, but some are designed to operate in the infrared. The performance characteristics of different types of LEDs are shown in Table 2-28. Comments about LEDs are given in Table 2-29.

---

*Registered trademark of Heraeus-Amersil.

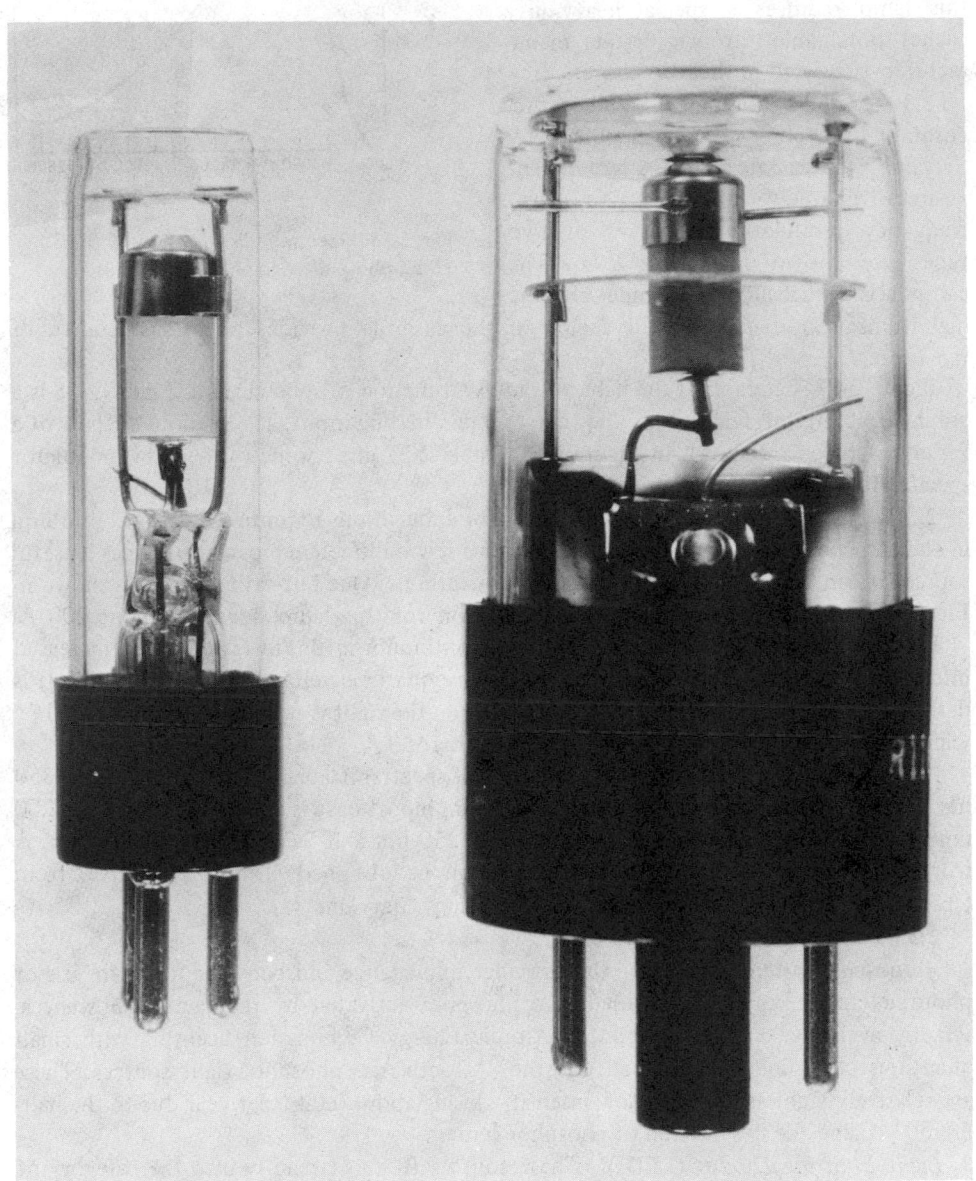

Fig. 2-56. Construction of two glow modulator tubes [2-38].

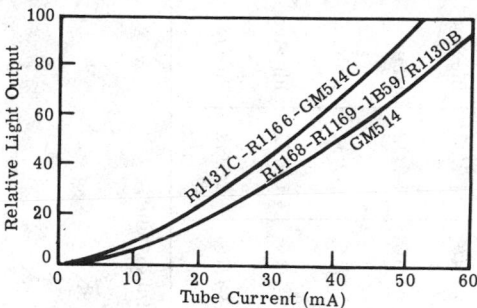

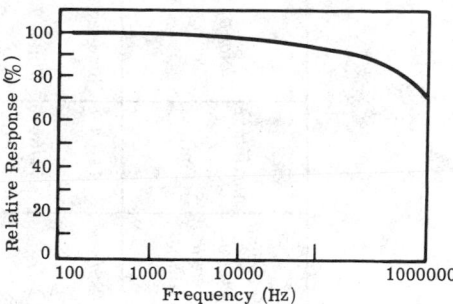

Fig. 2-57. Variations of the light output from a glow modulator tube as a function of tube current [2-38].

Fig. 2-58. Response of the glow modulator tube to a modulating input [2-38].

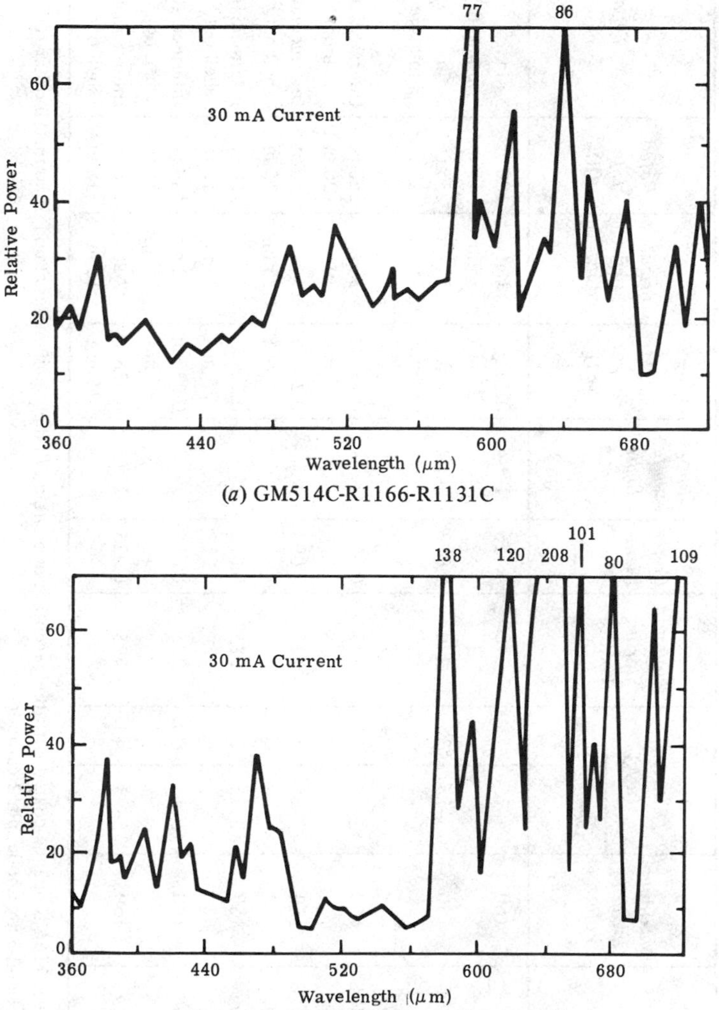

(a) GM514C-R1166-R1131C

(b) R1168-R1169-1B59/R1130D-GM514

Fig. 2-59. Spectral variation of the output of glow modulator tubes [2-38].

Table 2-27. Glow-Modulator Specifications [2-38]

No.[a]	Maximum Operating Voltage	Current (mA) Average	Current (mA) Peak	Minimum Starting Voltage (V)	Crater Diameter (in.)	Approximate Light Center Length (in.)	Light Output (cd)	Brightness (cd in.$^{-2}$)	Rated Life (h)	Base Type	Bulb Type	Maximum Overall Length (in.)	Maximum Diameter (in.)	Color of Discharge
GM-514	160	5–25	55	240	0.056	1-3/4	0.1 at 25 mA	41 at 25 mA	100 at 15 mA	3-pin miniature[b]	T-4½	2-5/8	41/64	Blue-red
GM-514C	160	5–15	35	240	0.093	1-3/4	0.1 at 15 mA	15 at 15 mA	25 at 10 mA	3-pin miniature[b]	T-4½	2-5/8	41/64	White
1B59/ R-1130B	150	5–35	75	225	0.056	2	0.13 at 30 mA	43 at 30 mA	250 at 20 mA	Intermediate shell oct.[c]	T-9	3-1/16	1-9/32	Blue-red
R-1131C	150	3–25	55	225	0.093	2	0.2 at 25 mA	29 at 25 mA	150 at 15 mA	Intermediate shell oct.[c]	T-9	3-1/16	1-9/32	White
R-1166	150	3–25	55	225	0.093	2	0.2 at 25 mA	29 at 25 mA	150 at 15 mA	Intermediate shell oct.[c]	T-9	3-1/16	1-9/32	White
R-1168	150	5–15	30	225	0.015	2	0.023 at 15 mA	132 at 15 mA	150 at 15 mA	Intermediate shell oct.[c]	T-9	3-1/16	1-9/32	Blue-red
R-1169	150	5–25	45	225	0.025	2	0.036 at 15 mA	72 at 15 mA	250 at 15 mA	Intermediate shell oct.[c]	T-9	3-1/16	1-9/32	Blue-red

[a] Type R-1166 is opaque-coated with the exception of a circle 3/8 inch in diameter at end of lamp. All other types have clear-finish bulb.
[b] Pins 1 and 3 are anode; pin 2 cathode.
[c] Pin 7 anode; pin 3 cathode.

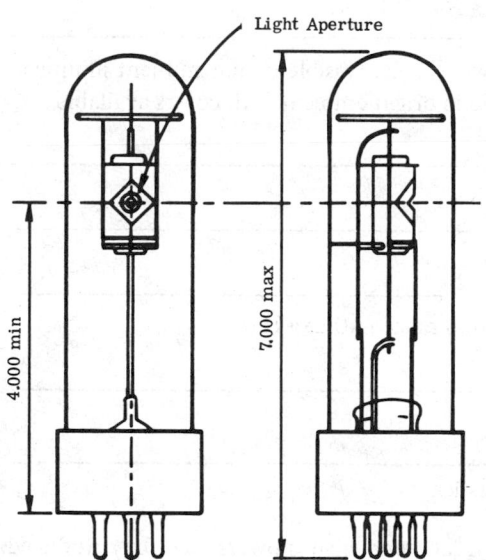

Fig. 2-60. Hydrogen arc lamp [2-38].

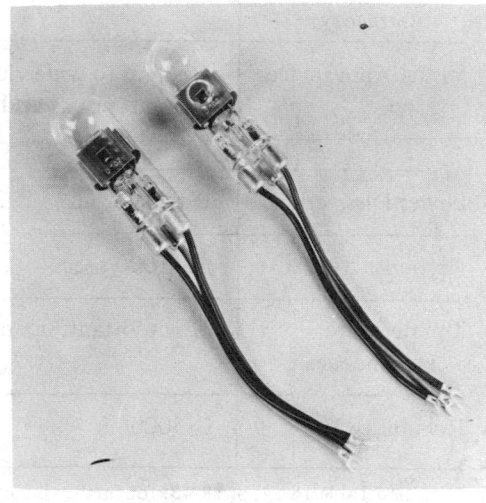

Fig. 2-61. Two types of deuterium arc lamps [2-39].

Table 2-28. Performance Characteristic of Different Types of LEDs[a] [2-43]

Type	Color	Peak Emission Wavelength (Å)	Quantum Efficiency		Luminous Power Efficiency (lm W^{-1})	
			Research	Commercial Performance	Research Result	Commercial Performance
GaAs$_{0.6}$P$_{0.4}$	Red	6490	5×10^{-3}	2×10^{-3}	0.33	0.15
GaAs$_{0.35}$P$_{0.65}$:N	Red-Orange	6320	4×10^{-3}		0.76	–
LPE: GaP:Zn, O	Red	6925	15×10^{-2}	$0.5-2 \times 10^{-2}$	3.0	0.1 –0.4
VPE GaP:N (Zn diffusion)	Green	5700	1×10^{-3}	$0.3-0.8 \times 10^{-3}$	0.6	0.18-0.48
PE GaP:N (grown junction)	Green	5700	3×10^{-3}	$0.5-2.0 \times 10^{-4}$	1.8	0.03-0.12
GaAs$_{0.15}$P$_{0.85}$:N	Yellow	5890	0.8×10^{-3}	$0.5-0.8 \times 10^{-3}$	0.36	0.23-0.36
VPE GaP:N (N > 10^{20} cm^{-2})	Yellow	5900	1×10^{-3}	–	0.45	–

[a]The value given in this table are for dc current densities less than 20 A cm^{-2}.

Table 2-29. Summary of LED Characteristics*

Category	LEDs
Visual Appearance	Medium to wide viewing angle. Visible in dim ambient illumination but not as visible in bright ambient. All colors available except blue.
Power Dissipation	0.1–10 W cm^{-2} at 2 V
Response Times	10–1000 nsec
Temperature Dependence	Unimportant; operation range –40 to +100°C
Reliability	> 50000 h

*See also Table 18-14, p. 18-43, for further information.

Other (High-Energy) Sources. Radiation at very high powers can be produced. Sources are synchrotrons, plasmatrons, arcs, sparks, exploding wires, shock tubes, and atomic and molecular beams, to name but a few. Among these, one can purchase in convenient, usuable form, precisely controlled spark-sources for yielding many joules of energy in a time interval of the order of microseconds. Also, consumable plasma tubes can be purchased as radiation sources emitting effectively at many thousands of degrees, yielding radiation rich in ultraviolet energy.

Other Special Sources. An enormous number of special-purpose sources are obtainable from manufacturers and scientific instrument suppliers. One source that remains to be mentioned is the so-called miniature, sub- and microminiature lamps. These are small, even tiny, incandescent bulbs of glass or quartz, containing tungsten filaments. They serve excellently in certain applications where small, intense radiators of visible and near-infrared radiation are needed.

2.4. Other Artificial Sources (Man-Made Targets.) (See also Chapter 3.)

2.4.1. **Surface and Subsurface Vehicles.** In this group are included: surface land vehicles, (i.e., trucks, tanks, personnel carriers, automobiles, and motorcycles, trains, hovercraft, and others) and ships (i.e., aircraft carriers, missile ships, cargo vessels, hydrofoil craft, small boats (wooden and metal), and submarines). Specific radiation characteristics of these vehicles are seldom, if ever, reported in the open literature. For the sake of visualization, however, a thermal image of an Army Shop Van is given in Figure 2-62. Whereas specific radiative patterns may be of use in certain instances, many of the target data are calculated by the use of predictive models. Figures 2-63 and 2-64 show faceted models (i.e., they are made of many facets placed so as to approximate the actual shape of whatever target is to be calculated). The faces or facets, are composed of radiating material with appropriate emissivities, emitting at temperatures which have been derived from the conditions under which the vehicle has been operating. Because these conditions vary considerably, it is often not feasible to report specific radiative properties except in special cases suited to special purposes.

Fig. 2-62. Thermal image of an Army 6 × 6 shop van [2-42].

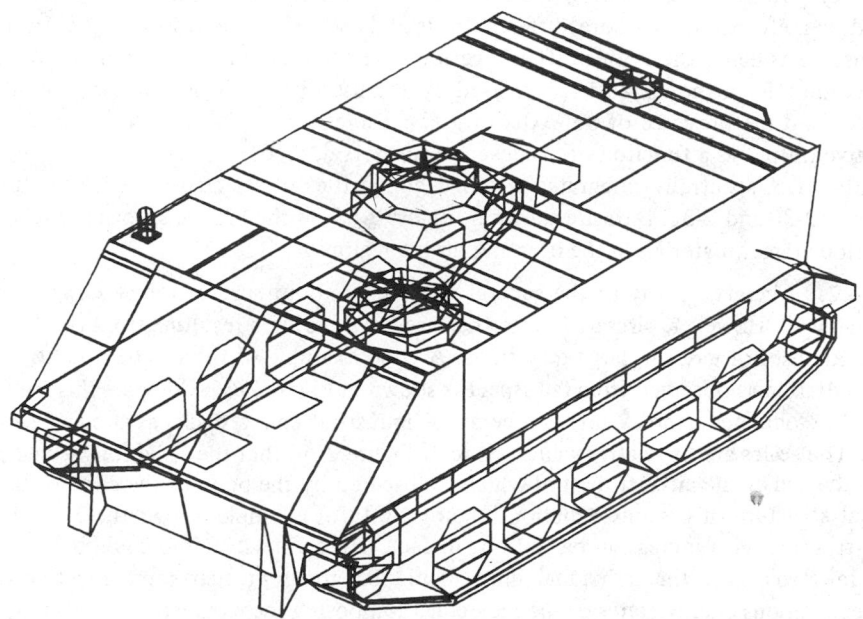

Fig. 2-63. Faceted model of an APC.

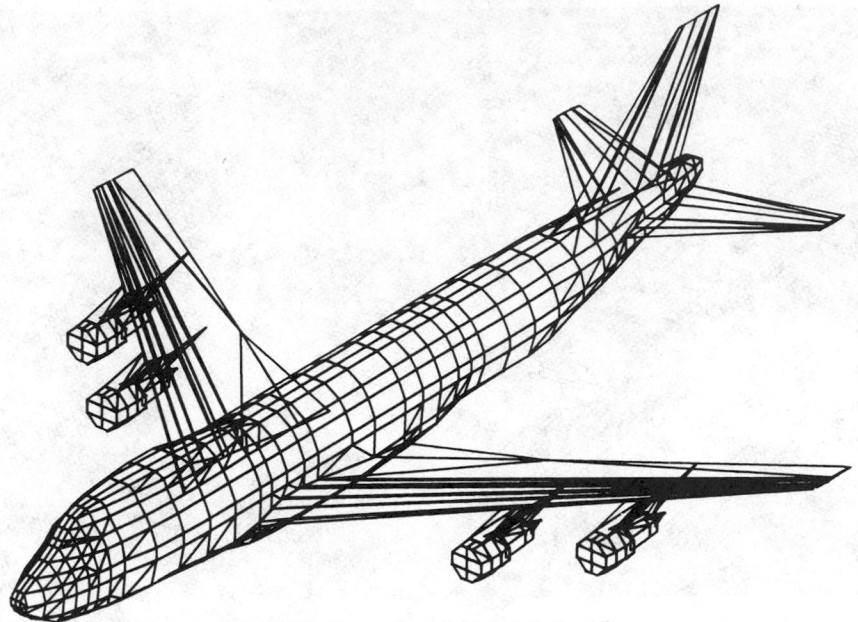

Fig. 2-64. Faceted model of a jet aircraft.

Painted surfaces usually have emissivities in the infrared of about 0.9. This can change in reality because of weathering and general deterioration. Dust and dirt can accumulate to add their effect. Furthermore, different parts of the vehicle are at different temperatures, exhaust parts being the hottest. The uncertainty, however, of the exact amount of radiation coming from the exhaust is increased by the use of suppression devices on military vehicles and the presence of hot exhaust gases. Therefore, models are used to predict the radiative output as a function of whatever parameters desired.

Total (i.e., spectrally integrated) normal emissivities of selected materials are shown in Tables 2-30 and 2-31. Data are shown in Figure 2-65 on the diffuse spectral reflectances of various target materials in the near and visible regions.

2.4.2. Aircraft. Many of the general statements about surface vehicles apply to aircraft as well. However, aircraft have rather special signatures resulting from the fact that they exhaust hot gases in large quantities. A good example of the spectral quality of aircraft radiation as observed from tail aspect is shown in Figure 2-66. Compare this spectrum with that obtained from a high-temperature industrial smokestack, as shown in Figure 2-67. The scales are of relative values. Note in Figure 2-68 that the radiation viewed away from the tail of the aircraft, part of which is obscured by the body of the aircraft, has the typical structure of gaseous radiation. Figure 2-69, for example, shows the evolution of fine structure with increasing resolution. In fact, if the fine structure could be seen with infinite resolution, the individual lines would be apparent that represent transitions between various energy states of the molecules composing the exhaust.

For hydrocarbon fuels, the exhaust consists mainly of H_2O vapor and CO_2. Because of these, one finds a lot of radiation in certain spectral regions, especially around 2.7 and 4.3 μm, at which the emissivities of these gases are quite high. However, the atmosphere, composed also of H_2O vapor and CO_2, tends to absorb quite readily at these wavelengths.

Table 2-30. Emissivity (Total Normal) of Various Common Materials [2-40]

Material	Temperature (°C)	Emissivity
METALS AND THEIR OXIDES		
Aluminum: polished sheet	100	0.05
sheet as received	100	0.09
anodized sheet, chromic acid process	100	0.55
vacuum deposited	20	0.04
Brass: highly polished	100	0.03
rubbed with 80-grit emery	20	0.20
oxidized	100	0.61
Copper: polished	100	0.05
heavily oxidized	20	0.78
Gold: highly polished	100	0.02
Iron: cast, polished	40	0.21
cast, oxidized	100	0.64
sheet, heavily rusted	20	0.69
Magnesium: polished	20	0.07
Nickel: electroplated, polished	20	0.05
electroplated, no polish	20	0.11
oxidized	200	0.37
Silver: polished	100	0.03
Stainless steel: type 18–8, buffed	20	0.16
type 18–8, oxidized at 800°C	60	0.85
Steel: polished	100	0.07
oxidized	200	0.79
Tin: commercial tin-plated sheet iron	100	0.07
OTHER MATERIALS		
Brick: red common	20	0.93
Carbon: candle soot	20	0.95
graphite, filed surface	20	0.98
Concrete[a]	20	0.92
Glass: polished plate	20	0.94
Lacquer: white	100	0.92
matte black	100	0.97
Oil, lubricating (thin film on nickel base):		
nickel base alone	20	0.05
film thickness of 0.001, 0.002, 0.005 in.	20	0.27, 0.46, 0.72
thick coating	20	0.82
Paint, oil: average of 16 colors	100	0.94
Paper: white bond	20	0.93
Plaster: rough coat	20	0.91
Sand	20	0.90
Skin, human	32	0.98
Soil: dry	20	0.92
saturated with water	20	0.95
Water: distilled	20	0.96
ice, smooth	−10	0.96
frost crystals	−10	0.98
snow	−10	0.85
Wood: planed oak	20	0.90

[a]See also Chapter 3.

Table 2-31. Solar Absorptance, α, and Low-Temperature (300 K) Emissivity, ϵ, for Spacecraft Materials [2-40]

Material	α	ϵ	α/ϵ
Aluminum:			
polished and degreased	0.387	0.027	14.35
foil, dull side, crinkled and smoothed	0.223	0.030	7.43
foil, shiny side	0.192	0.036	5.33
sandblasted	0.420	0.210	2.00
oxide, flame sprayed, 0.001 inch thick	0.422	0.765	0.55
anodized	0.150	0.770	0.19
Fiberglass:	0.850	0.750	1.13
Gold: plated on stainless steel and polished	0.301	0.028	10.77
Magnesium: polished	0.300	0.070	4.30
Paints:[a]			
Aquadag, 4 coats on copper aluminum	0.782	0.490	1.60
aluminum	0.540	0.450	1.20
Microbond, 4 coats on magnesium	0.936	0.844	1.11
TiO_2, gray	0.870	0.870	1.00
TiO_2, white	0.190	0.940	0.20
Rokide A	0.150	0.770	0.20
Stainless steel: type 18-8, sandblasted	0.780	0.440	1.77

[a]See also Chapter 3.

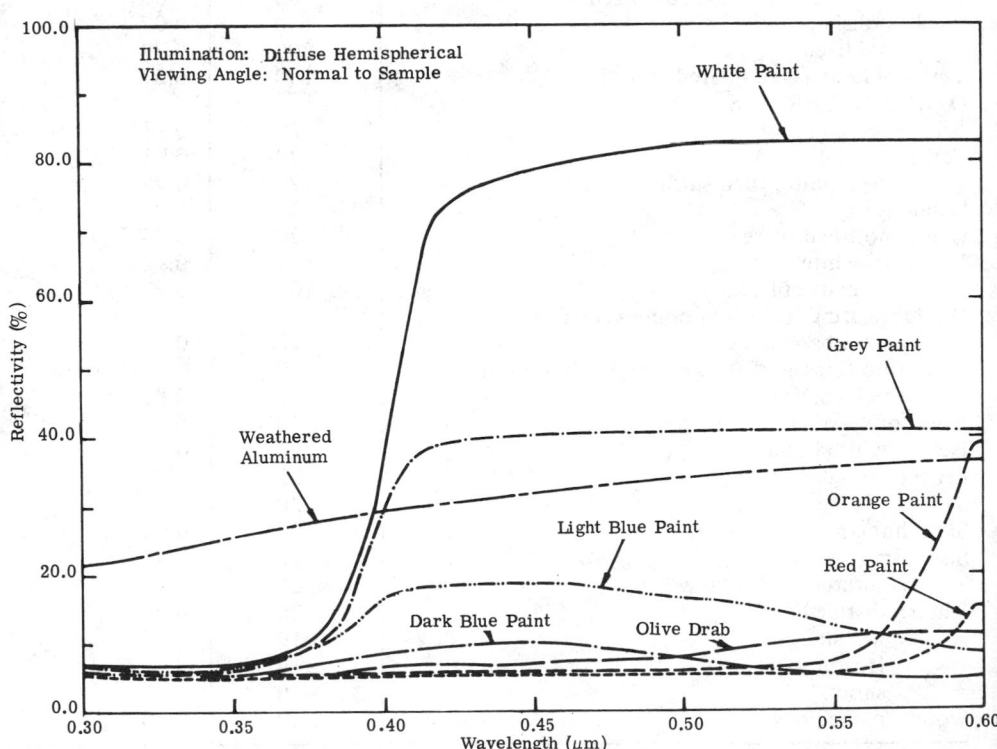

Fig. 2-65. Diffuse spectral reflectance of various materials in the near UV/VIS [2-51].

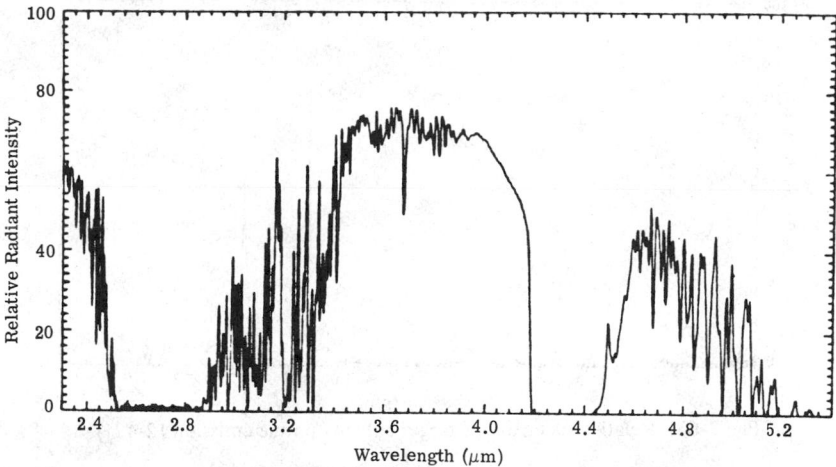

Fig. 2-66. Relative spectrum of jet aircraft as seen from tail aspect at a range of 1.5 mi [2-42].

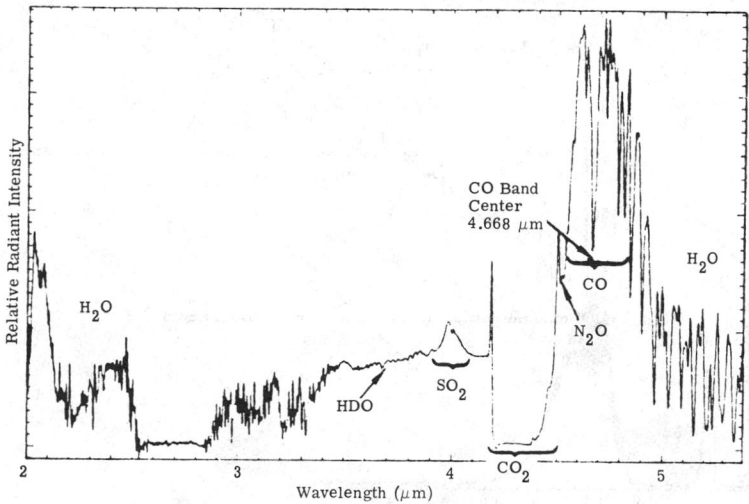

Fig. 2-67. Emission from a high temperature industrial smokestack [2-42].

Since the gases are hotter than the atmosphere, some molecular lines radiate outside of the region of strong atmospheric absorption and propagate with much less attenuation than in the atmospheric bands. This effect is pronounced in the 4.3 μm region of CO_2 absorption.

As the observation angle changes to the tail, the continuum radiation emitted from the tail pipe is clearly seen. The radiation from the hot gases is, of course, still present.

Below are calculations of the radiation emitted from a Boeing 707 jet transport [2-40]. The features of a turbojet engine are shown in Figure 2-70 with temperatures and pressures given in the lower portion of the figure. The higher temperature in afterburning gives rise to more radiation than obtained otherwise. But the calculation would be

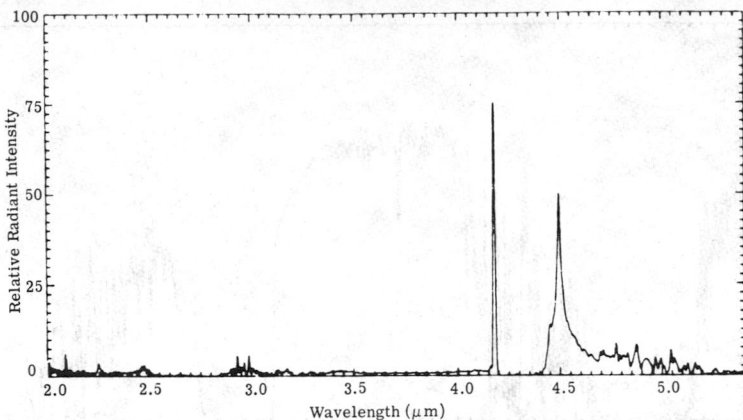

Fig. 2-68. Relative spectrum of target exhaust plume emission [2-42].

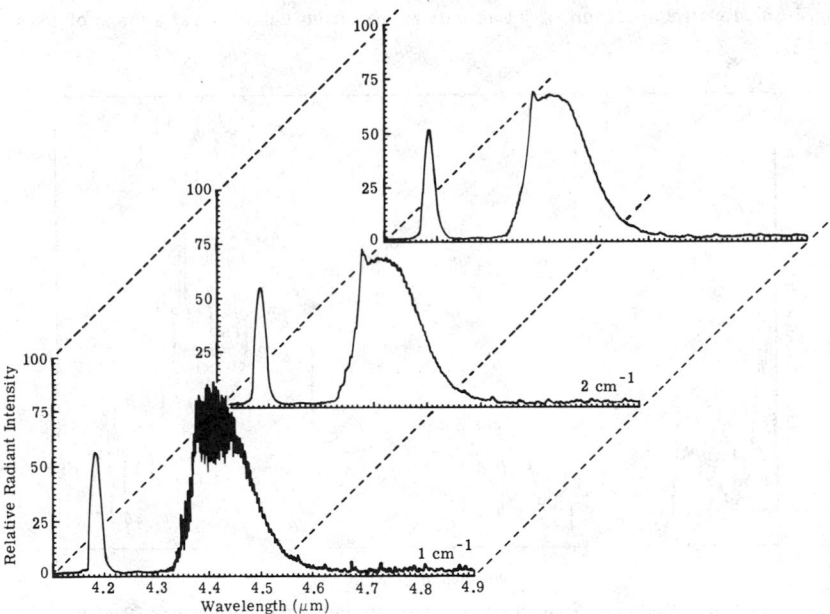

Fig. 2-69. Aircraft exhaust plume as seen through 200 ft of atmosphere and measured with three different spectral resolutions [2-42].

complicated, furthermore, by the fact that large amounts of hot gases would be emitted with an accompanying increase in molecular band radiation. In other words, the exhaust plume radiation dominates that from the hot tailpipe.

   The turbofan exhaust of a 707 jet transport is cooler than that of the turbojet. The characteristics of each are given in Table 2-32. The total radiance from the tailpipe of the aircraft can be calculated from a knowledge of the emissivities and the temperatures of the emitting surfaces. The Gouffé results earlier in this chapter make it reasonable to assume a cavity emissivity better than 0.9. Using this value and a uniform temperature equal to the exhaust gas temperature for maximum cruise (from Table 2-32), the radiant

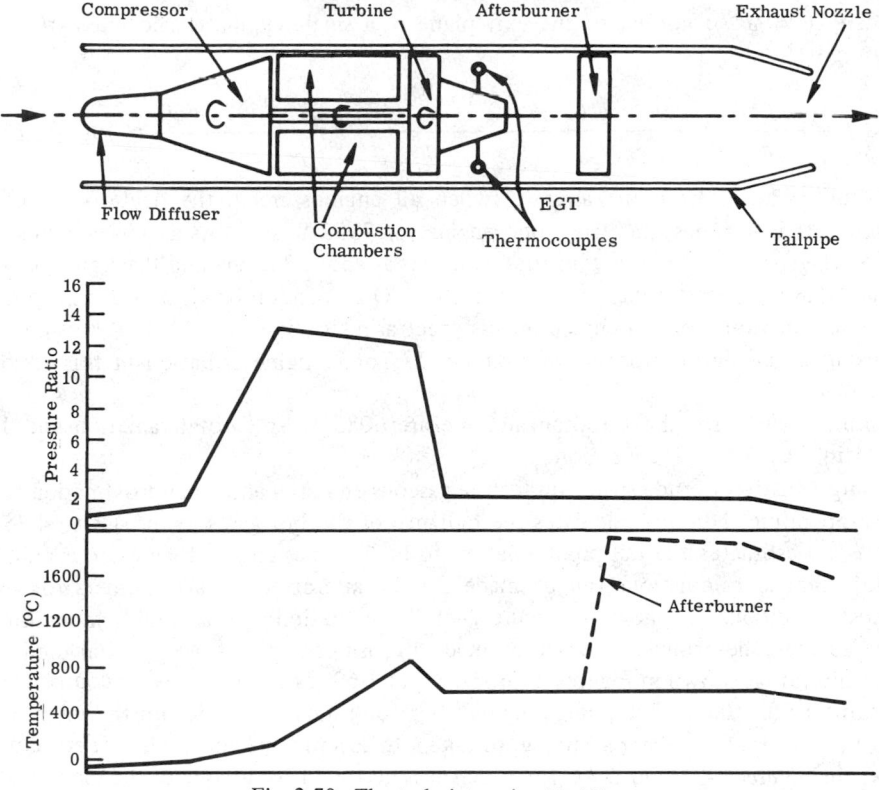

Fig. 2-70. The turbojet engine [2-40].

Table 2-32. Characteristics of the Boeing 707 Intercontinental Jet Transports [2-40]

Model	707-320	707-320B
Engines	4-P&WA JT4A-9	4-P&WA JT3D-3
Type of engine	Turbojet	Turbofan
Maximum rated thrust (lb per engine)	16800	18000
Area of engine exhaust nozzle (cm²)	3660	3502
Exhaust gas temperatures:		
Maximum allowable takeoff (°C)	635	555
Maximum continuous thrust (°C)	515	490
Maximum cruise thrust (°C)	485	445
Fuel flow (per engine):		
Sea level, Mach 0.4 (lb h⁻¹)	10814	9068
35 000 ft. Mach 0.8 (lb h⁻¹)	4819	3962
Spacing between engines		
Inner engines (ft)	66	66
Outer engines (ft)	104	104
Maximum speed (mph)	585	592

intensity (W sr^{-1}) normal to the exit plane of a single engine is calculated from (See Chapter 1.):

$$I = \frac{\epsilon\sigma T^4 A}{\pi} = \frac{0.9 \times 5.67 \times 10^{-12} \times (485 + 273)^4 \times 3660}{\pi} = 1963 \quad (2\text{-}20)$$

Total radiation from the aircraft (when all engines are in the field-of-view of the instrument) is 4 times that from one engine, or 7852 W sr^{-1}. As a radiation slide-rule shows, the spectrum for this temperature peaks at about 3.8 μm, and the region between 3.2 and 4.8 μm passes radiation between the H_2O absorption bands between 2.7 and 6.3 μm. The radiation from an engine in this spectral region is thus 580 W sr^{-1}, as obtained by using a calculator program of Chapter 1. Total engine radiation in this region is 2320 W sr^{-1}.

Similar values for the turbofan airplane are 6085 W sr^{-1} total radiation and 1500 W sr^{-1} in the 3.2 to 4.8 μm region.

Using a relatively crude spectrum for the gaseous emission and an approximation to the gas temperature, Hudson calculates the radiance of the hot gases in the 4.3 to 4.55 μm region and estimates the tail pipe radiation to be 25 times greater than gaseous radiation [2-40]. Similar calculations can be made for the aircraft using afterburners for which exhaust conditions are shown in Figure 2-71. The radiation is considerably more intense. More accurate determinations can be made by using more precise spectral data from actual aircraft as shown in Figures 2-66, 2-68 and 2-69. The conditions, of course, are not the same as for the 707 and adjustments, therefore, must be made. Furthermore, it becomes necessary, for greater accuracy, to take into account the actual aircraft structure as shown in Figures 2-71 and 2-72, for which it is pertinent to use one of the mathematical models cited earlier.

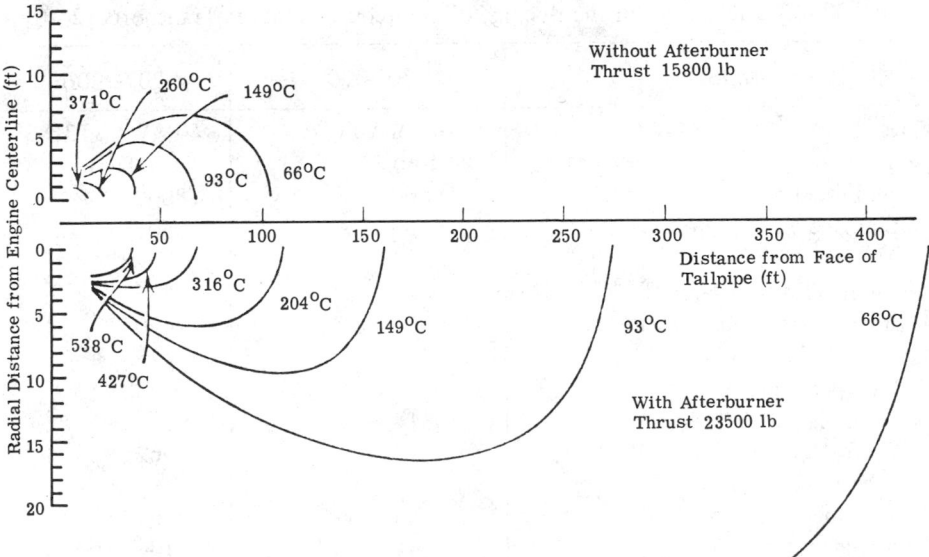

Fig. 2-71. Exhaust temperature contours for the Pratt and Whitney JT4a turbojet engines at maximum sea level thrust with and without afterburner [2-40].

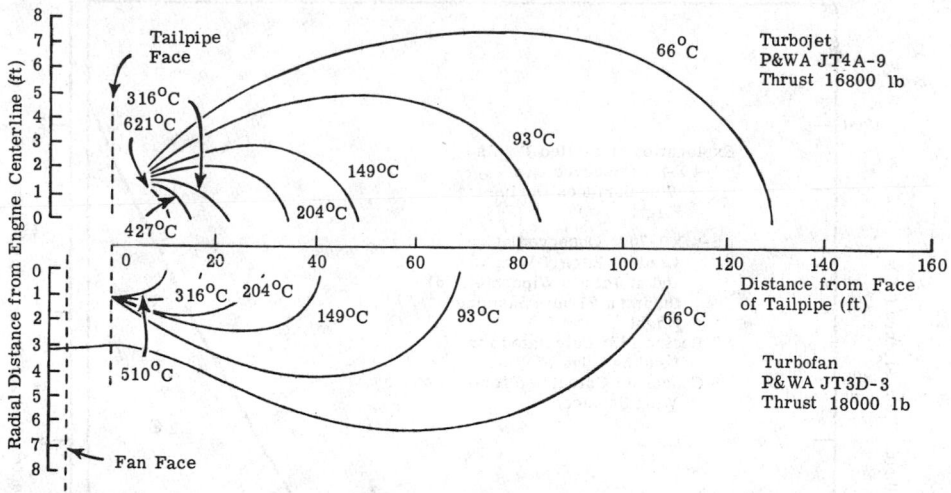

Fig. 2-72. Exhaust temperature contours for the turbojet and turbofan engines used on the Boeing 707 [2-40].

As the speed of an aircraft increases the temperature of the skin reaches high enough temperatures to produce radiation which becomes readily observed at large distances, often competitive with other sources from the aircraft. Figure 2-73 represents centigrade temperatures as a function of Mach number, M, for laminar flow over a surface above 37,000 feet altitude. The equation of this curve is:

$$T(K) = 216.7(1 + 0.164 M^2) \tag{2-21}$$

which is compared to some data obtained as shown in the legend to the figure. Hudson explains that the low values for the X-15 can be explained by the short flight time and by the high emissivity of the blackened surface, reducing the likelihood of equilibrium.

2.4.3. **Muzzle Flash.** The hot gases ejected from the muzzle of a gun contain, among other things, a large amount of combustible constituents, such as CO, and H_2, along with CO_2, N_2, and H_2O at a sufficiently high temperature to cause the accompanying particulate matter to radiate visible radiation. In addition, considerable IR radiation is emitted from these same hot gases which accumulate near the muzzle exit. The radiation from this volume is called primary flash. A gaseous flow pattern, shown schematically in Figure 2-74, is set up in front of the gun. As the gases flow through the normal shock they are heated and radiate in what is called the intermediate flash. On mixing with the atmosphere the hot gases ignite and burn with a large luminous flame, giving rise to what is known as secondary flash.

To suppress the secondary flash, which emits large quantities of radiation, the gun is often fitted with a device attached to the muzzle which thwarts the formation of a shock wave. In other cases, suppressants are added chemically to the propellant to prevent ignition of the muzzle combustible gases.

Reference [2-41] describes muzzle flash in all its aspects for a range of weapon types and sizes. A well-controlled experiment on a special .50 caliber test gun (with 20 mm chamber) was made by the Franklin Institute to obtain spectra in the UV, visible, and IR

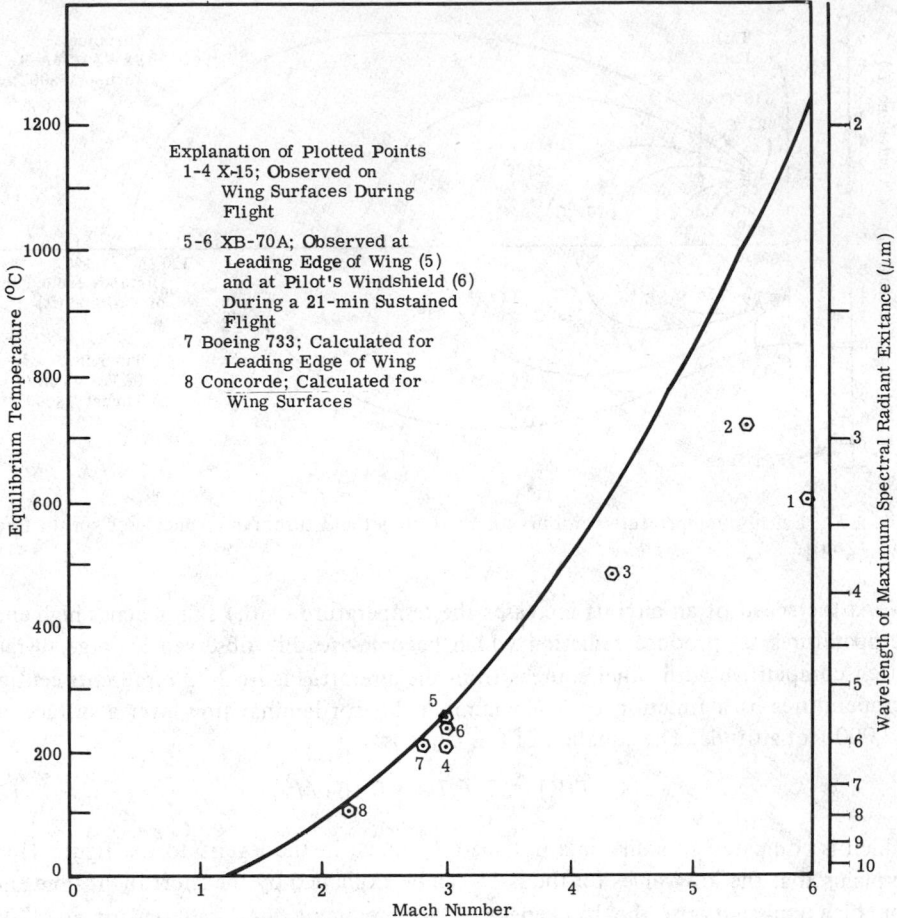

Fig. 2-73. Equilibrium surface temperature caused by aerodynamic heating (for altitudes above 37000 ft and laminar flow) [2-40].

regions. Results showed that a rather weak UV spectrum of secondary radiation developed yielding a continuum on which was superimposed strong lines of atomic potassium (4044 to 4047 Å), medium-strength lines of atomic copper (3248 to 3274 Å), strong bands of CuCl (4300 to 4900 Å), relatively strong bands of CuH (4005 and 4280 Å), weak OH bands (2811 and 3064 Å), and very weak NH bands (3360 and 3370 Å).

In the visible region of the spectrum, the intermediate (suppressed) flash consists of a continuum with a superposition of radiation of Na (5890 Å) and CuOH (6230 Å). Incandescent solids contribute to the continuum. Copper compounds are also present, although when the copper jacketed projectiles are replaced by steel projectiles with steel rotating bands, the radiation from the copper compounds disappears.

Infrared radiation is copious in muzzle flash and IR radiation from large-caliber weapons has been measured. The relative spectral intensity of the secondary flash from a 155 mm artillery weapon at different observation distances is reported in Reference [2-41]. The barrel of the gun was 40 calibers long and the projectile was a dummy HE (M101) weighing 95 lbs. The muzzle velocity was 2100 ft sec^{-1}. A relative spectrum for a 200 foot observation distance is shown in Figure 2-75.

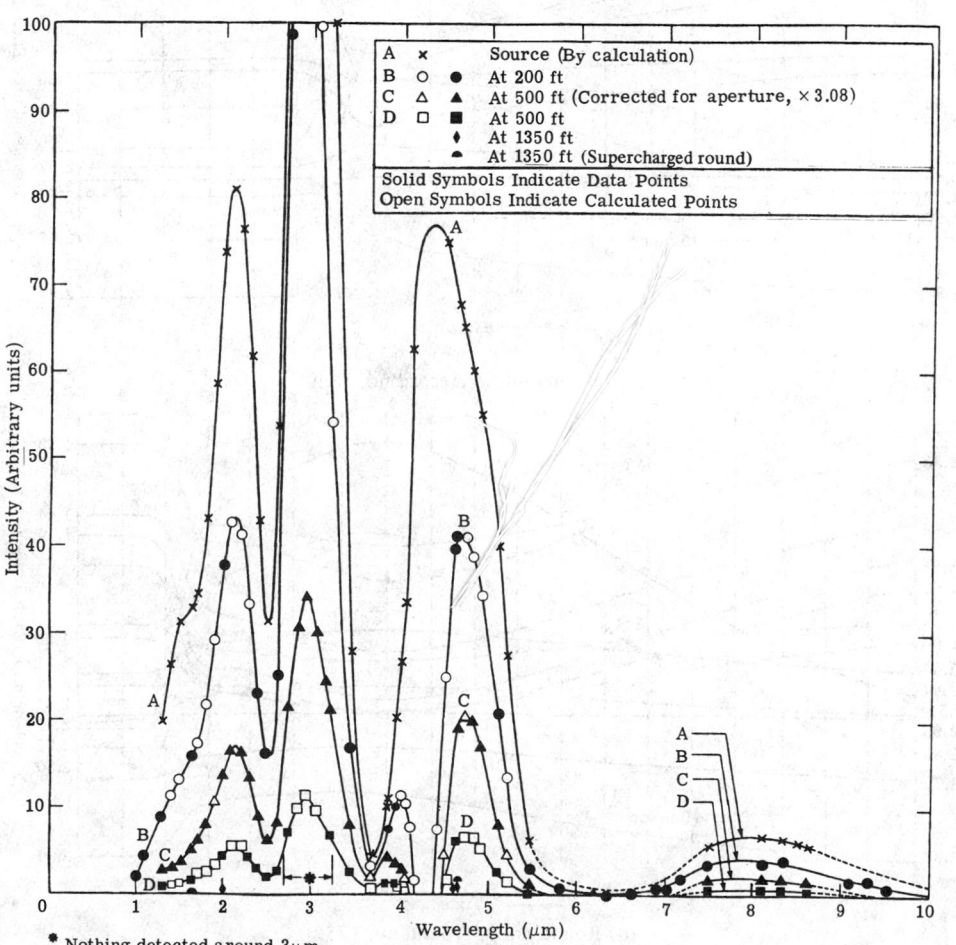

Fig. 2-74. Flow pattern at a gun muzzle (dashed lines indicate composition boundaries) [2-41]. By permission of Headquarters, U.S. Army Materiel Development and Readiness Command.

Fig. 2-75. Variation of spectral distribution of radiation from secondary flash with distance from the gun. Weapon: 155-mm, M2 artillery gun [2-41]. By permission of Headquarters, U.S. Army Materiel Development and Readiness Command.

Time records of the relative radiant intensity from a 280 mm gun are given in Figure 2-76. These are given, however, only as specially chosen, representative examples. In observing other records given, for example, in Reference [2-41] huge variations in the time histories are evident, depending on a large number of factors including gun size, propellant type and charge, firing history of the gun, etc. These factors often determine whether or not secondary flash will occur, causing multiple peaks in the time record.

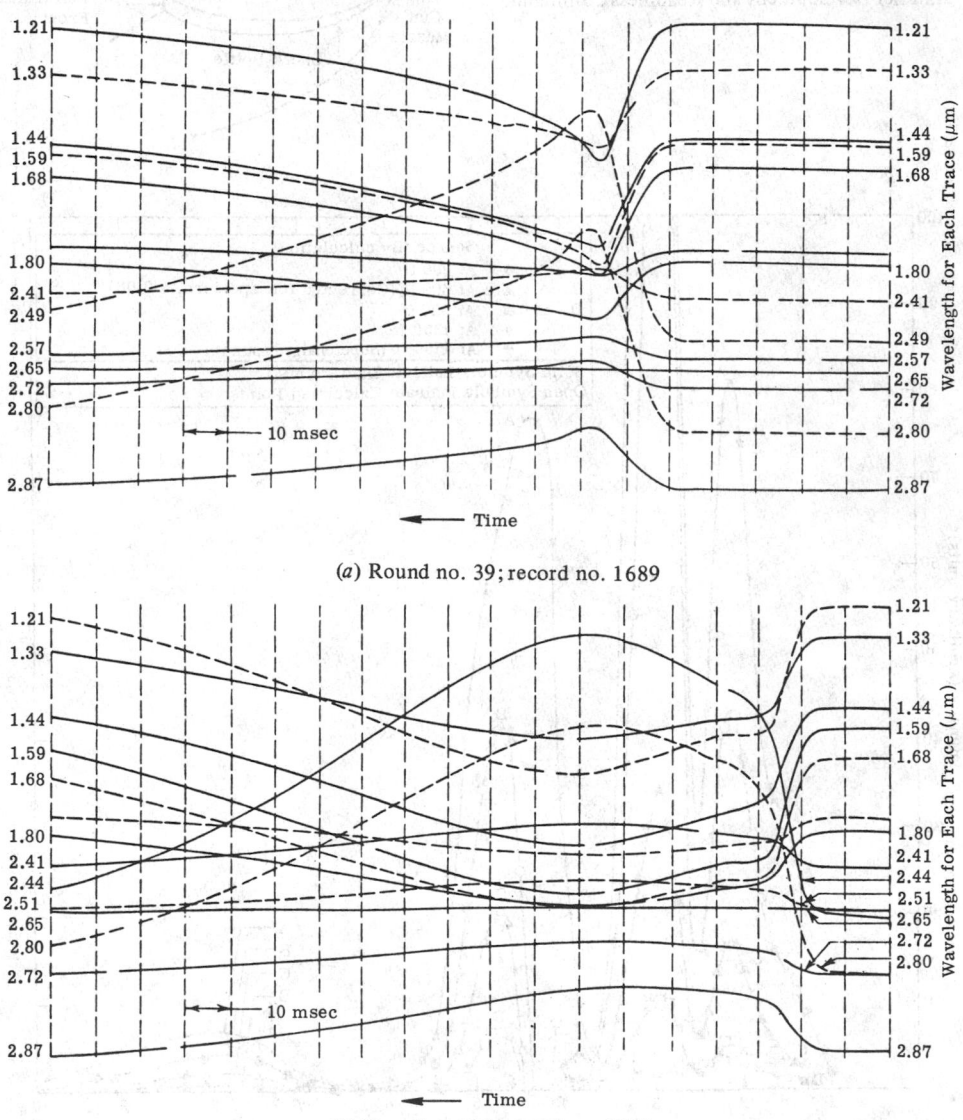

(a) Round no. 39; record no. 1689

(b) Round no. 95; round no. 1759

Fig. 2-76. Spectrograms of flash from 280-mm gun. Both solid and dotted curves are used for clarity. Positive deflection is downward for traces having wavelengths between 1.21 and 1.80 μm [2-41]. By permission of Headquarters, U.S. Army Materiel Development and Readiness Command.

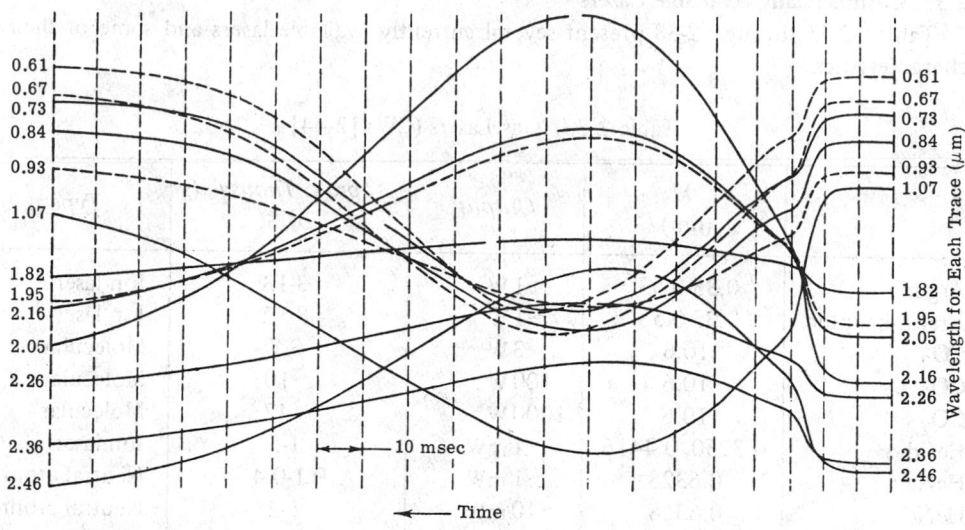

(c) Round no. 40; record no. 1690

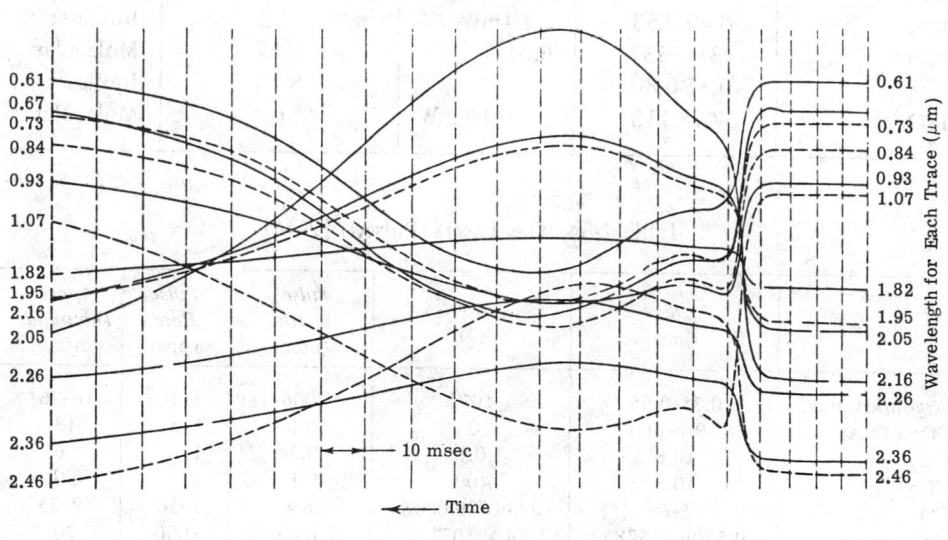

(d) Round no. 92; round no. 1756

Fig. 2-76. (Continued). Spectrograms of flash from 280-mm gun. Both solid and dotted curves are used for clarity. Positive deflection is downward for traces having wavelengths between 0.61 and 1.07 μm [2-41]. By permission of Headquarters, U.S. Army Materiel Development and Readiness Command.

2.5. Commercially Available Lasers

Tables 2-33 through 2-38 present several currently available lasers and some of their characteristics.

Table 2-33. Gas Lasers (CW) [2-44]

	λ (μm)	Output	Approx. Thousands of $	Type
Argon	0.35-0.53	~1 W	3-13	Ion laser
Argon-krypton	0.35-0.53	~1 W	8-12	Ion laser
CO_2	10.6	3 W	3.7	Molecular
CO_2	10.6	100 W	~10	Molecular
CO_2	10.6	1000 W	47	Molecular
HeCd	0.3250, 0.4416	~1 mW	1-5	Ion laser
HeNe	0.6328	~1 mW	0.1-0.4	Neutral atom
HeNe	0.6328	~10 mW	1-2	Neutral atom
HeNe	0.6328	~80 mW	5	Neutral atom
HeNe	1.15	~1 mW	0.6	Neutral atom
HeNe	3.39	~1 mW	0.6	Neutral atom
HeSe	0.49-0.53	10 mW	3	Ion laser
HCN	311, 337	6.01-1	15	Molecular
Kr	0.35-0.80	-1	8-13	Ion laser
H_2O	27,78,118	1-10 mW	16	Molecular

Table 2-34. Gas Lasers (Pulsed) [2-44]

	λ (μm)	Output (J)	Pulse Width (μsec)	Pulse Rate (pps)	Approx. Thousands of $
Argon/krypton	0.35–0.65	10^{-6}	0.0006–1	$1-10^8$	16–26
CO_2 (TEA)	9.1–11	3	0.1	1–5	13
CO_2	10.6	0.1	0.2	1	6
CO_2	10.6	400	0.1	1	90
CO	5–6	0.0007–0.0008	1–4	1–50	9–25
Cu	0.5106, 0.5782	4×10^{-4}	0.02	1000	20
DF	3.5–4	0.2	0.5	1	15
HCN	311, 337	$10^{-3}-10^{-4}$	30	1–30	15
HF	2.8–5	0.3	1	1–3	15
Ne	0.5401	3×10^{-5}	3	2–100	14
H_2	0.3371	$10^{-2}-10^{-3}$	0.01	0–100	3–24
N_2O	10.5–11	0.3	0.1	1–5	15
H_2O	28, 78, 118	10^{-5}	30	1–10	15

Table 2-35. Optically Pumped Solid State Lasers (CW) [2-44]

	λ (μm)	Output (W)	Approx. Thousands of $
Nd: YAG	1.06	1	~15
Nd: YAG	1.06	1000	~100
Nd: YAG	0.53	2	20 doubled
Nd: YAG	0.53	0.25	6 doubled
Nd: YAG	0.67	0.05	6 doubled 1.34 μm
Nd: YAG	1.06	0.1-10	2-20

Table 2-36. Optically Pumped Solid State Lasers (Pulsed) [2-44]

	λ (μm)	Output (J)	Pulse Width (μsec)	Pulse Rate (pps)	Approx. Thousands of $
Nd: glass	1.060	1-45.00	~500.000	0.07-30	1-15
Nd: glass	0.265	~0.05	~0.010	4-60	20-40 doubled twice
Nd: glass	0.530	2.00	0.030	1-30	40-60 doubled
Nd: glass	1.060	100.00	0.005	1-240	80
Nd: YAG	0.265	0.02	0.015	20	40-50 doubled twice
Nd: YAG	0.530	0.10	0.020	30	40-60 doubled
Nd: YAG	1.060	0.1-20.00	0.01-100.00	2-50	5-55
Nd: YAG	Variable	2.50×10^{-4}	0.1-0.4	30-3000	35 Parametric oscillator
Ruby	0.6943	0.02-100.00	$10^{-2}-10^3$	1-60	7-50

Table 2-37. Semiconductor or Diode Lasers (Pulsed) [2-44]

	λ (μm)	Peak Power (W)	Rep Rate (Hz)	Pulse Width (μsec)	Approx. Thousands of $
GaAs (diode)	0.85	3-1800	10^3-10^4	10^{-3}-10^{-4}	0.02-0.05
GaAs (diode)	0.9	-1-10	10^3-10^6	10^{-5}-1	0.09-1.00
PbSnTe	4.3-30 (not covered by a single diode)	0.01	2×10^7	10^{-8}	3.50

Table 2-38. Dye Lasers (Tunable) [2-44]

λ (μm)	Output Power (W)	Rep Rate (pps)	Pulse Width (nsec)	Approx. Thousands of $
0.2300-1.06	10^2-10^7	1-100	1-1000	2.5-20
0.440-0.560	0.1	cw	Ar + Dye	27
0.570-0.640	1-5	cw	Ar + Dye	13-24
0.640-0.760	0.1	cw	Ar + Dye	13

2.6. References and Bibliography

2.6.1. References

[2-1] A. G. Worthing, "Sources of Radiant Energy," *Measurement of Radiant Energy*, W. E. Forsythe (ed.), McGraw-Hill, New York, NY, 1937, Chapter 2.

[2-2] Andre Gouffé, "Corrections d'Ouverture des Corps-noir Artificiels Compte Tenu des Diffusions Multiples Internes," *Revue d'Optique*, Masson, Paris, France, Vol. 24, 1945, p. 1.

[2-3] J. C. DeVos, "Evaluation of the Quality of a Blackbody," *Physica*, North Holland Publishing, Amsterdam, Netherlands, Vol. 20, 1954, p. 669.

[2-4] David F. Edwards, "The Emissivity of a Conical Blackbody," The University of Michigan Engineering Research Institute, Ann Arbor, MI, 2144-105-T, 1956.

[2-5] J. W. T. Walsh, *Photometry*, Dover, New York, NY, 1965, Third Edition.

[2-6] J. A. Van Allen, "Dynamics Composition of Geometrically-Trapped Corpuscular Radiation," *Transactions of the International Astronomical Union*, Academic Press, New York, NY, Vol. 11 A, 1962, pp. 99-136.

[2-7] National Bureau of Standards, Washington, DC, *NBS Bulletin*, Vol. 11, No. 227, 1914, p. 87.

[2-8] National Bureau of Standards, Washington, DC, *NBS Journal of Research*, Vol. 11, No. 578, 1933, p. 79.

[2-9] Ralph Stair, Russel G. Johnston, and E. W. Halbach, "Standard of Spectral Radiance for the Region of 0.25 to 2.6 Microns," *NBS Journal of Research*, National Bureau of Standards, Washington, DC, Vol. 64A, No. 4, July-August 1960.

[2-10] Ralph Stair, William E. Schneider, and John K. Jackson, "A New Standard of Spectral Irradiance," *Applied Optics,* Optical Society of America, Washington, DC, Vol. 2, No. 11, November 1963, p. 1151.

[2-11] D. A. Sparron, National Bureau of Standards, Washington, DC, Private Communication, 1976.

[2-12] National Bureau of Standards, Washington, DC, *NBS Measurement Users Bulletin,* No. 4, February 1973.

[2-13] H. J. Kostkowski and R. D. Lee, "Theory and Methods of Optical Pyrometry," *NBS Monograph,* National Bureau of Standards, Washington, DC, Vol. 41, March 1962.

[2-14] Electro-Optical Industries, Santa Barbara, CA, "Working Standard Series Blackbody Sources," 1976, p. 4.

[2-15] Barnes Engineering, Stamford, CT, "Radiation Reference Sources," Bulletin 11-035, 1976, p. 2.

[2-16] Infrared Industries, Santa Barbara, CA, "Blackbody Radiation Sources, Series 400," Data Sheet, 1976, p. 2.

[2-17] W. Y. Ramsey and J. C. Alishouse, "A Comparison of Infrared Sources," *Infrared Physics,* Pergamon Publishing, Elmsford, NY, Vol. 8, 1968, p. 143.

[2-18] J. C. Morris, "Comments on the Measurement of Emittance of the Globar Radiation Source," *Journal of the Optical Society of America,* Optical Society of America, Washington, DC, Vol. 51, July 1961, p. 798.

[2-19] A. H. Pfund, "The Electric Welsbach Lamp," *Journal of the Optical Society of America,* Optical Society of America, Washington, DC, Vol. 26, December 1936, p. 439.

[2-20] J. Strong, *Procedures in Experimental Physics,* Prentice-Hall, New York, NY, 1938, p. 346.

[2-21] F. E. Carlson and C. N. Clark, "Light Sources for Optical Devices," R. Kingslake (ed.), *Applied Optics and Optical Engineering,* Academic Press, New York, NY, Vol. 1, 1965, p. 80.

[2-22] Osram Gmbh, "Lamps for Scientific Purposes," Munchen, West Germany, 1966.

[2-23] F. J. Studer and R. F. Van Beers, "Modification of Spectrum of Tungsten Filament Quartz-Iodine Lamps Due to Iodine Vapor," *Journal of the Optical Society of America,* Optical Society of America, Washington, DC, Vol. 54, No. 7, July 1964, p. 945.

[2-24] L. R. Koller, *Ultraviolet Radiation,* John Wiley and Sons, New York, NY, 1965, Second Edition.

[2-25] J. E. Kaufman (ed.), *IES Lighting Handbook,* Illuminating Engineering Society, New York, NY, 1972, Fifth Edition.

[2-26] M. R. Null and W. W. Lozier, "Carbon Arc as a Radiation Standard," *Journal of the Optical Society of America,* Optical Society of America, Washington, DC, Vol. 52, No. 10, October 1962, pp. 1156-1162.

[2-27] E. B. Noel, "Radiation from High Pressure Mercury Arcs," *Illuminating Engineering,* Vol. 36, 1941, p. 243.

[2-28] General Electric, Cleveland, OH, Bulletin TP-109R, 1975.

[2-29] Illumination Industries, Sunnyvale, CA, Catalog No. 108-6-72-3M, 1972.

[2-30] M. W. P. Cann, "Light Sources for Remote Sensing Systems," NASA-CR-854, August 1967.

[2-31] Oriel Corporation, Stamford, CT, "Oriel Corporation of America Catalog," 1976.

[2-32]   Varian Associates, Palo Alto, CA, "Hollow Cathode Lamp Technical Data,"
          1969.

[2-33]   E. F. Worden, R. G. Gutmacher, and J. F. Conway, "Use of Electrodeless
          Discharge Lamps in the Analysis of Atomic Spectra," *Applied Optics,* Optical
          Society of America, Washington, DC, Vol. 2, No. 7, July 1963, pp. 707-713.

[2-34]   W. F. Meggers and F. O. Westfall, "Lamps and Wavelengths of Mercury 198,"
          NBS Journal of Research, National Bureau of Standards, Washington, DC, Vol.
          44, 1950, pp. 447-455.

[2-35]   W. F. Meggers, "Present Experimental Status of Rare Earth Spectra," *Journal of
          the Optical Society of America,* Optical Society of America, Washington, DC,
          Vol. 50, 1960, p. 405.

[2-36]   Ealing Corporation, South Natick, MA, "Ealing Catalog," Optical Components
          Section, 1976-1977.

[2-37]   Central Scientific, Chicago, IL, "Cenco Scientific Educational Catalog," Physics-
          Light Section, 1975, p. 602.

[2-38]   GTE Sylvania, Lighting Products Group, Danvers, MA, "Special Purpose
          Lamps," TR-29R, 1 May 1966.

[2-39]   Ealing Corporation, South Natick, MA, "Optical Services Supplement," No. 1,
          1969-1970, p. 26; See also, Oriel Corporation, Stamford, CT, "Oriel Deuterium
          Catalog, 1975, p. 62, Lamp no. 6316, and GTE Sylvania, Salem, MA, "Sylvania
          Deuterium Lamps," L-524.

[2-40]   R. D. Hudson, *Infrared System Engineering,* John Wiley and Sons, New York,
          NY, 1969, Chapter 2.

[2-41]   S. P. Carfagno, "Spectral Characteristics of Muzzle Flash," *Engineering Design
          Handbook,* Headquarters, U. S. Army Materiel Command, Alexandria, VA,
          AMCP-706-255, June 1967. Available through NTIS, AD-818 532/PDM.

[2-42]   General Dynamics, Applied Research Laboratories, Electro Dynamics Division,
          Pomona, CA, "Optical Radiation: Measurement and Analysis Capability," 1975.

[2-43]   L. A. Goodman, "The Relative Merits of LEDs and LCDs," *Proceedings of the
          Society of Information Display,* Society for Information Display, Montvale, NJ,
          Vol. 16, No. 1, First Quarter, 1975.

[2-44]   S. Jacobs, University of Arizona, Tucson, AZ, Private Communication, 1976.

[2-45]   National Bureau of Standards, Washington, DC, "Optical Radiation News,"
          January 1974.

[2-46]   Fisher Scientific, Pittsburgh, PA, "Special Catalog to Spectrophotometer Users,"
          1972.

[2-47]   Bedford, R. E., "A Low Temperature Standard for Total Radiation," *Canadian
          Journal of Physics,* Vol. 38, 1960.

[2-48]   O. C. Jones and J. S. Preston, "Photometric Standards and the Unit of Light,"
          Notes on Applied Science, National Physics Laboratory of London, London,
          England, Vol. 24, 1969, Second Edition.

[2-49]   L. J. Nivert, M. Mulbrandon, and J. C. Richmond, "Primary Reference Standards
          of Spectral Radiance in the Infrared," *Journal of the Optical Society of
          America,* Optical Society of America, Washington, DC, Vol. 57, 1967, p. 1417.

[2-50]   R. D. Lee, "Construction and Operation of a Simple High Precision Copper
          Point Blackbody and Furnace," *NBS Technical Note,* National Bureau of
          Standards, Washington, DC, Note 483, May 1969.

[2-51]   Dennis Blay, General Dynamics, Pamona, CA, Private Communication, 1977.

2.6.2. Bibliography

Allen, C. W., *Astrophysical Quantities,* Athlone Press, London, 1963, Second Edition.

American Institute of Physics, New York, NY, *Review of Scientific Instruments,* Vol. 11, 1940, p. 429.

Barnes Engineering Company, Stamford, CT, "Radiation Reference Sources," Bulletin 11-035.

Bedford, R. E., "A Low Temperature Standard of Total Radiation," *Canadian Journal of Physics,* Ottawa, Ontario, Canada, Vol. 38, 1960, p. 1256.

Bedford, R. E., and C. K. Ma, "Emissivities of Diffuse Cavities, Isothermal and Nonisothermal Cones and Cylinders," *Journal of The Optical Society of America,* Optical Society of America, Washington, DC, Vol. 64, No. 3, March 1974, p. 339.

Bedford, R. E., and C. K. Ma, "International Comparison of Measurements of Irradiance," *Metrologia,* Springer-Verlag, New York, NY, Vol. 4, No. 3, 1968.

Bell, E. E., et al., "Spectral Radiance of Sky and Terrain at Wavelengths Between 1 and 20 microns, II," *Journal of the Optical Society of America,* Optical Society of America, Washington, DC, Vol. 50, No. 12, December 1960, p. 1313.

Benedict, W. S., et al., "The Water Vapor Laser," *IEEE Journal of Quantum Electronics,* Institute of Electrical and Electronics Engineers, New York, NY, Vol. QE-1, 1965, p. 66.

Bennett, W. R., Jr., "Chemical Lasers," Supplement 2, "Inversion Mechanisms in Gas Lasers," *Applied Optics,* Optical Society of America, Washington, DC, 1965, p. 3.

Bevolo, A. J., and W. A. Barker, "Laser Emission Lines and Materials," *Applied Optics,* Optical Society of America, Washington, DC, Vol. 4, 1965, p. 531.

Bloom, A. L., *Gas Lasers,* John Wiley and Sons, New York, NY, 1968.

Bloom, A. L. and J. P. Goldsborough, "New CW Laser Transitions in Cadmium and Zinc Ion," *IEEE Journal of Quantum Electronics,* Institute of Electrical and Electronics Engineers, New York, NY, Vol. QE-6, 1970, p. 164.

Bradley, D. J., et al., "Characteristics of Organic Dye Lasers as Tunable Frequency Sources for Nanosecond Absorption Spectroscopy," *IEEE Journal of Quantum Electronics,* Institute of Electrical and Electronics Engineers, New York, NY, Vol. QE-4, 1969, p. 707.

Bridges, W. B., and A. H. Chester, "Spectroscopy of Ion Lasers," *IEEE Journal of Quantum Electronics,* Institute of Electrical and Electronics Engineers, New York, NY, Vol. QE-1, 1965, p. 66.

Campen, C. F., Jr., et al., (eds.), *Handbook of Geophysics,* Macmillan, New York, NY, 1961.

Cann, M. W. P., "Light Sources for Remote Sensing Systems," NASA CR-854, August 1967.

Carfagno, S. P., "Spectral Characteristics of Muzzle Flash," *Engineering Design Handbook,* Headquarters, U. S. Army Materiel Command, Alexandria, VA, AMCP-706-255, June 1967. Available through NTIS, AD-818 532/PDM.

Carlson, F. E., and C. N. Clark, "Light Sources for Optical Devices," *Applied Optics and Optical Engineering,* R. Kingslake (ed.), Academic Press, New York, NY, Vol. 1, 1965, p. 80.

DeVos, J. C., "Evaluation of the Quality of a Blackbody," *Physica,* North Holland Publishing, Amsterdam, Netherlands, Vol. 20, 1954, p. 669.

Edwards, David F., "The Emissivity of a Conical Blackbody," The University of Michigan Engineering Research Institute, Ann Arbor, MI, 2144-105-T, 1956.

Electro Optical Industries, Inc., Santa Barbara, CA, "Working Standard Series Blackbody Sources," 1976.

Finkelnberg, W., "The High Current Carbon Arc and Its Mechanism," *Journal of Applied Physics*, American Institute of Physics, New York, NY, Vol. 20, 1949, p. 468.

Frederickson, W. R., et al., "Infrared Spectral Emissivity of Terrain," Syracuse University Research Institute, Syracuse, NY, Final Report AD155552, 1958.

Frederickson, W. R., et al., "Infrared Spectral Emissivity of Terrain," Syracuse University Research Institute, Syracuse, NY, Int. Dev. Report 2, AF 33 (616)-5034, 1 August 1957.

Furumoto, H. W., and H. L. Ceccon, "Ultraviolet Dye Lasers," *IEEE Journal of Quantum Electronics*, Institute of Electrical and Electronics Engineers, New York, NY, Vol. QE-6, 1970.

General Dynamics, Applied Research Laboratories, Electro Dynamics Division, Pomona, CA, "Optical Radiation: Measurement and Analysis Capability," 1975.

General Electric, Cleveland, OH, Bulletin TP-109R, 1975.

Gillham, E. J., "The Measurement of Optical Radiation," *Research Applied in Industry*, Vol. 12, 1959, p. 404.

Gillham, E. J., "Radiometric Standards and Measurements," *Notes on Applied Science*, National Physics Laboratory of London, London, England, Vol. 23, 1961.

Gillham, E. J., "Recent Investigation in Absolute Radiometry," *Proceedings of the Royal Society*, London, England, Series A, Vol. 269, 1962, p. 240.

Goodman, L. A., "The Realtive Merits of LEDs and LCDs," *Proceedings of the Society for Information Display*, Society for Information Display, Montvale, NJ, Vol. 16, No. 1, First Quarter, 1975.

Gouffé, Andre, "Corrections d'Ouverture des Corps-noir Artificiels Compte Tenu des Diffusions Multiples Internes," *Revue d'Optique*, Masson, Paris, France, Vol. 24, No. 1, 1945.

GTE Sylvania, Salem, MA, "Sylvania Deuterium Lamps," L-524.

Harrison, W. N., Joseph C. Richmond, Earle K. Plyler, Ralph Stair, and Harold K. Skramstad, "Standardization of Thermal Emittance Measurements," National Bureau of Standards, Washington, DC, WADC TR 59-510, AD 238 918, 1960.

Hayashi, I., M. B. Panish, and P. W. Foy, "A Low-Threshold Room-Temperature Injection Laser," *IEEE Journal of Quantum Electronics*, Institute of Electrical and Electronics Engineers, New York, NY, Vol. QE-5, 1969, p. 211.

Heavens, O. S., *Optical Masers*, John Wiley and Sons, New York, NY, 1964.

Hornton, J. R., et al., "Properties of Neodymium Laser Materials," *Applied Optics*, Optical Society of America, Washington, DC, Vol. 8, 1969.

Hudson, R. D., *Infrared System Engineering*, John Wiley and Sons, New York, NY, 1969, Chapter 2.

Humphreys, C. J., and E. Paul, Jr., "Interferometric Observations in the Spectra of 86 Kr," *Journal of the Optical Society of America*, Optical Society of America, Washington, DC, Vol. 60, 1970.

Illumination Industries, Inc., Sunnyvale, CA, Catalog No. 108-6-72-3M.

Infrared Industries, Santa Barbara, CA, "Blackbody Radiation Sources, Series 400," Data Sheet, 1976, p. 2.

Johnson, F. S., *Journal of Meteorology*, Vol. 11, 1954.

Jones, O. C., and J. S. Preston, "Photometric Standards and the Unit of Light," *Notes on Applied Science*, National Physics Laboratory of London, London, England, Vol. 24, 1969, Second Edition.

Jones, T. P., and J. Tapping, "The Freezing Point of Platinum," *Metrologia*, Springer-Verlag, New York, NY, Vol. 12, No. 1, 1976, p. 19.

Kagan, M. R., G. L. Farmer and B. Huth, "Organic Dye Lasers," *Laser Focus*, Advanced Technology Publications, Newtonville, MA, Vol. 4, September 1968, p. 26.

Karoli, A. R., J. R. Hickey, and R. E. Nelson, "An Absolute Calibration Source for Laboratory and Satellite Infrared Spectrometers," *Applied Optics*, Optical Society of America, Washington, DC, Vol. 6, 1967, p. 1183.

Kaufman, J. E. (ed.), *IES Lighting Handbook*, Illuminating Engineering Society, New York, NY, 1972, Fifth Edition.

Kiss, Z. J., and R. J. Pressley, "Crystalline Solid Lasers," *Proceedings of the IEEE*, Institute of Electrical and Electronics Engineers, New York, NY, Vol. 51, 1966, p. 1236.

Koller, L. R., *Ultraviolet Radiation*, John Wiley and Sons, New York, NY, 1965, Second Edition.

Kostowski, H. J., and R. D. Lee, "Theory and Methods of Optical Pyrometry," National Bureau of Standards Monograph, National Bureau of Standards, Washington, DC, Vol. 41, March 1962.

Kruse, P. W., L. D. McGlauchlin, and R. B. McQuistan, *Elements of Infrared Technology-Generation, Transmission, and Detection*, John Wiley and Sons, New York, NY, 1962.

LaRocca, A. J., and G. J. Zissis, "Field Sources of Blackbody Radiation," *Review of Scientific Instruments*, American Institute of Physics, New York, NY, Vol. 30, 1959, p. 200.

Lee, R. D., "Construction and Operation of a Simple High Precision Copper Point Blackbody and Furnace," *NBS Technical Note*, National Bureau of Standards, Washington, DC, Note 483, May 1969.

Mathias, L. E. S., A. Crocker, and M. S. Wills, "Spectroscopic Measurements on the Laser Emission from Discharges in Compounds of Hydrogen, Carbon, and Nitrogen," *Journal of Quantum Electronics*, Institute of Electrical and Electronics Engineers, New York, NY, Vol. QE-6, 1968, p. 205.

Meggers, W. F., "Present Experimental Status of Rare Earth Spectra," *Journal of the Optical Society of America*, Optical Society of America, Washington, DC, Vol. 50, 1960, p. 405.

Meggers, W. F., and F. O. Westfall, "Lamps and Wave-lengths of Mercury," 198, *NBS Journal of Research*, National Bureau of Standards, Washington, DC, Vol. 44, 1950, p. 447.

Morris, J. C., "Comments on the Measurement of Emittance of the Globar Radiation Source," *Journal of the Optical Society of America*, Optical Society of America, Washington DC, Vol. 51, July 1961, p. 798.

Murcray, D. G., A. Goldman, A. Csoeke-Poeckh, et al., "Nitric Acid Distribution in the Stratosphere," *Journal of Geophysical Research*, American Geophysical Union, Washington, DC, Vol. 78, No. 30, 1973, p. 7033.

Murcray, D. G., F. H. Murcray, W. T. Williams, T. G. Kyle, and A. Goldman, "Variation of the Infrared Solar Spectrum Between 700 cm^{-1} and 2240 cm^{-1} with Altitude," *Applied Optics*, Optical Society of America, Washington, DC, Vol. 8, No. 12, December 1969, p. 2519.

Nathan, M. I., "Semiconductor Lasers," *Proceedings of the IEEE*, Institute of Electrical and Electronics Engineers, New York, NY, Vol. 54, 1966, p. 1276.

National Bureau of Standards, Washington, DC, *NBS Bulletin*, Vol. 11, No. 227, 1914, p. 87.

National Bureau of Standards, Washington, DC, *NBS Journal of Research*, Vol. 11, No. 578, 1933, p. 79.

National Bureau of Standards, Washington, DC, *NBS Journal of Research*, Vol. 53, 1954.

National Bureau of Standards, Washington, DC, *NBS Measurement Users Bulletin*, No. 4, February 1973.

National Bureau of Standards, Washington, DC, *NBS Technical News Bulletin*, Vol. 54, No. 3, March 1970.

Nester, J. F., "Dynamic Optical Properties of CW Nd:YA1G Lasers," *IEEE Journal of Quantum Electronics*, Institute of Electrical and Electronics Engineers, New York, NY, Vol. QE-6, 1970, p. 97.

Nivert, L. J., M. Mulbrandon, and J. C. Richmond, "Primary Reference Standards of Spectral Radiance in the Infrared," *Journal of the Optical Society of America*, Optical Society of America, Washington, DC, Vol. 57, 1967, p. 1417.

Noel, E. B., "Radiation from High Pressure Mercury Arcs," *Illuminating Engineering*, Vol. 36, 1941, p. 243.

Null, M. R., and W. W. Lozier, "Carbon Arc as a Radiation Standard," *Journal of the Optical Society of America*, Optical Society of America, Washington, DC, Vol. 52, No. 10, October 1962, p. 1156.

Oriel Corporation, Stamford, CT, "Oriel Deuterium Source for the Ultraviolet, 180 to 400 nm," *Oriel Corporation of America Catalog*, 1975, p. 62, Lamp No. 6316.

Osgood, R. M., Jr., et al., "An Investigation of the High Power CO Laser," *Journal of Quantum Electronics*, Institute of Electrical and Electronics Engineers, New York, NY, Vol. QE-6, 1970, p. 145.

Osram Gmbh, "Lamps For Scientific Purposes," Munchen, West Germany, 1966.

Paschen, F., "Bohrs Heliumlinien," *Annalen der Physik*, Johann Ambrosius Barth, Leipsig, East Germany, Vol. 50, 1916, p. 901.

Pfund, A. H., "The Electric Welsbach Lamp," *Journal of the Optical Society of America*, Optical Society of America, Washington, DC, Vol. 26, December 1936, p. 439.

Plyler, E. K., D. J. C. Yates, and H. A. Gebbie, "Radiant Energy from Sources in the Far Infrared," *Journal of the Optical Society of America*, Optical Society of America, Washington, DC, Vol. 52, 1962, p. 859.

Ramsay, R. C., "Spectral Irradiance from Stars and Planets, Above the Atmosphere, from 0.1 to 100.0 Microns," *Applied Optics*, Optical Society of America, Washington, DC, Vol. 1, No. 4, July 1962, p. 465.

Ramsay, W. Y. and J. C. Alishouse, "A Comparison of Infrared Sources," *Infrared Physics*, Pergamon Publishing, Elmsford, NY, Vol. 8, 1968, p. 143.

Ripper, J. E. (ed.), "Special Issues on Semi-Conductor Lasers," *IEEE Journal of Quantum Physics*, Institute of Electrical and Electronics Engineers, New York, NY, Vol. QE-6, No. 6, June 1970.

Silverman, S., "The Emissivity of Globar," *Journal of the Optical Society of America*, Optical Society of America, Washington, DC, Vol. 38, No. 11, November 1948, p. 989.

Simmons, F., A. G. DeBell and Q. S. Anderson, "A 2000°C Slit Aperture Blackbody Source," *Review of Scientific Instruments*, American Institute of Physics, New York, NY, Vol. 32, 1961, p. 1265.

Smith, R. A., F. F. Jones, and R. P. Chasmar, *The Detection and Measurement of Infra-Red Radiation*, Oxford University Press, Oxford, 1957; See also, S. Sliverman, *Journal of the Optical Society of America*, Optical Society of America, Washington, DC, Vol. 38, 1948.

Snitzer, E., "Glass Lasers," *Proceedings of the IEEE* Institute of Electrical and Electronics Engineers, New York, NY, Vol. 54, 1966, p. 1249.

Sparrow, E. M., L. V. Albers, "The Apparent Emissivity and Heat Transfer in a Long Cylindrical Hole," *Journal of Heat Transfer Transactions*, American Society of Mechanical Engineers, New York, NY, Series C, Vol. 82, 1960, pp. 253-255.

Stair, R., R. G. Johnston, and E. W. Halbach, "Standard of Spectral Radiance for the

Region of 0.25 to 2.6 Microns," *NBS Journal of Research*, National Bureau of Standards, Washington, DC, Vol. 64A, No. 4, July-August, 1960.

Stair, R., W. E. Schneider, and J. K. Jackson, "A New Standard of Spectral Irradiance," *Applied Optics*, Optical Society of America, Washington, DC, Vol. 2, No. 11, November 1963, p. 1151.

Stimson, H. F., *NBS Journal of Research*, National Bureau of Standards, Washington, DC, Vol. 42, 1949, p. 209.

Stimson, H. F., "International Practical Temperature Scale of 1948, Text Revision of 1960," *NBS Journal of Research*, National Bureau of Standards, Washington, DC, Vol. 65A, 1961, p. 139.

Strong, J., *Procedures in Experimental Physics*, Prentice-Hall, New York, NY, 1938, 1939.

Studer, F. J., and R. F. Van Beers, "Modification of Spectrum of Tungsten Filament Quartz-Iodine Lamps Due to Iodine Vapor," *Journal of the Optical Society of America*, Optical Society of America, Washington, DC, Vol. 54, No. 7, July 1964, p. 945.

Taylor, H. H., C. S. Ruppert, and J. Strong, "An Incandescent Tungsten Source for Infrared Spectroscopy," *Journal of the Optical Society of America*, Optical Society of America, Washington, DC, Vol. 41, 1951, p. 626.

Thekaekara, M. P., R. Kruger, and C. H. Duncan, "Solar Irradiance Measurements from a Research Aircraft," *Applied Optics*, Optical Society of America, Washington, DC, Vol. 8, No. 8, August 1969, p. 1713.

Thornton, J. R., et al., "Properties of Neodymium Laser Materials," *Applied Optics*, Optical Society of America, Washington, DC, Vol. 8, 1969, p. 1087.

Tomiyasu, K., "Bibliography of Laser Devices," *IEEE Journal of Quantum Electronics*, Institute of Electrical and Electronics Engineers, New York, NY, Vol. QE-4, 1968, p. 274.

Tomiyasu, K., "Bibliography of Laser Devices," *IEEE Journal of Quantum Electronics*, Institute of Electrical and Electronics Engineers, New York, NY, Vol. QE-4, 1968, p. 674.

Truenfels, E. W., "Emissivity of Isothermal Cavities," *Journal of the Optical Society of America*, Optical Society of America, Washington, DC, Vol. 53, 1963, p. 1162.

Valley, S. L. (ed.), *Handbook of Geophysics and Space Environments*, Air Force Cambridge Research Laboratories, Bedford, MA, 1965.

Van Allen, J. A., "Dynamics Composition of Geometrically-Trapped Corpuscular Radiation," *Transactions of the International Astronomical Union*, Academic Press, New York, NY, Vol. 11A, 1962, pp. 99-136.

Varian Techtron, Palo Alto, CA, "Hollow Cathode Lamp Technical Data," 1969.

Walsh, J. W. T., *Photometry*, Dover, New York, NY, 1965, Third Edition.

Williams, C. S., "Discussion of the Theories of Cavity-Type Sources of Radiant Energy," *Journal of the Optical Society of America*, Optical Society of America, Washington, DC, Vol. 51, 1961, p. 564.

Worden, E. F., R. G. Gutmacher, and J. F. Conway, "Use of Electrodeless Discharge Lamps in the Analysis of Atomic Spectra," *Applied Optics*, Optical Society of America, Washington, DC, Vol. 2, No. 7, July 1963, p. 707.

Worthing, A. G., "Sources of Radiant Energy," *Measurement of Radiant Energy*, W. E. Forsythe (ed.), McGraw-Hill, New York, NY, 1937, Chapter 2.

Yariv, A., and J. P. Gordon, "The Laser," *Proceedings of the IEEE*, Institute of Electrical and Electronics Engineers, New York, NY, Vol. 51, 1963.

Zissis, G. J., "Precision Radiometry-Theory," *Special Topics in Infrared Technology*, Engineering Summer Conferences, The University of Michigan, Ann Arbor, MI, 1963.

Chapter 3

NATURAL SOURCES

Gwynn H. Suits

Environmental Research Institute of Michigan

CONTENTS

3. Natural Sources

3.1. Introduction

Natural sources constitute a large and diverse class of radiators including terrestrial, atmospheric, and astronomical materials. Included in terrestrial materials are some construction materials which commonly appear in terrestrial landscapes.

3.1.1. Symbols, Nomenclature and Units. Table 3-1 lists the symbols, nomenclature and units used in this chapter.

3.1.2. Definitions of Terms. The common radiometric terms used in this section are defined in Chapter 1. Several other important and widely used specialized terms are defined below.

The *apparent temperature* of an extended natural source is the temperature of that blackbody placed at the aperture of the receiver which will produce the same output signal as that produced by the extended natural source. Apparent temperature is a substitute measure of radiance. Consequently, if the spectral variation of radiance of a natural source differs from the spectral radiance of a blackbody, the apparent temperature obtained by narrow-band radiometry will be wavelength-dependent. The common term used in radio-astronomy for apparent temperature is *brightness temperature.*

Distribution or color temperature is the blackbody temperature for which the spectral exitance of the source, $M_\lambda(\lambda, T)$, is spectrally best fit to the spectral irradiance at the receiver, $E_\lambda(\lambda)$, in the spectral range of interest where $M_\lambda(\lambda, T)$ is given by Planck's radiation law.

Spectral apparent temperature is the same as apparent temperature where narrow radiometer bandwidth is assumed. It is the temperature of a blackbody at the radiometric aperture which would have the same spectral radiance at one wavelength as was observed from the target.

Bulk reflectance is the reflectance due to the return of internal scattered radiation within a body.

Surface reflectance is the reflectance due to the change of optical properties at the boundary between two media.

Bidirectional reflectance, ρ', is defined (See also Chapter 1.) as

$$\rho' = \frac{L(\theta_r, \psi_r)}{E(\theta_i, \psi_i)} \qquad (3\text{-}1)$$

where θ_r, ψ_r = reflected polar and azimuthal angles of view, respectively

θ_i, ψ_i = polar and azimuthal angles of collimated incident radiation, respectively

Thermal flux density is the flow rate of heat energy through a surface area.

The *heat transfer coefficient* is the ratio of thermal flux density through a plane section of material to the temperature difference causing the flow across the material. Heat transfer by thermal conductance is h_k; heat transfer by convection is h_c.

Table 3-1. Symbols, Nomenclature and Units

Symbols	Nomenclature	Units
A	Area	m^2 or cm^2
a	Detector area	m^2 or cm^2
A_B	Altitude of a background point	m
A_0	Area of entrance pupil	m^2 or cm^2
A-0 to A-9	Stellar spectral classifications for 11,000 K stars	—
a_i	Constant for i^{th} class of source	—
$a(\lambda)$	Spectral absorption coefficient	m^{-1}
b	Galactic latitude	degree
B-0 to B-9	Stellar spectral classifications for 20,000 K stars	—
C	Relative concentration	—
c	Velocity of light in vacuum	$m\ sec^{-1}$
C_h	Specific heat	$J\ kg^{-1}\ K^{-1}$
C_0	Altitude of an observer	m
C_i	Constant for i^{th} class of source	—
D	Density	$kg\ m^{-3}$
d	Thickness or depth	m
$D(0-\lambda)$	Percent of total irradiance in a spectral band from 0 to λ μm	—
E	Irradiance	$W\ m^{-2}$
E (sky)	Total irradiance from the sky	$W\ m^{-2}$
E (sun)	Total solar irradiance	$W\ m^{-2}$
E_0	Initial irradiance at top	$W\ m^{-2}$
E_v	Illuminance	$lm\ m^{-2}$
$E_v(m)$	Exoatmospheric illuminance for a magnitude, m, star	$lm\ m^{-2}$
$E_v(1)$	Exoatmospheric illuminance from stellar object number 1	$lm\ m^{-2}$
$E_v(2)$	Exoatmospheric illuminance from stellar object number 2	$lm\ m^{-2}$
E_λ	Spectral irradiance	$W\ m^{-2}\ \mu m$
$E_\lambda(lim)$	Limiting spectral irradiance at the earth	$W\ m^{-2}\ \mu m^{-1}$
E_λ (sky)	Spectral irradiance from the sky	$W\ m^{-2}\ \mu m$
$E_\lambda(sun)$	Spectral irradiance from the sun	$W\ m^{-2}\ \mu m^{-1}$
$E_\lambda(x)$	Spectral irradiance on horizontal plane at x	$W\ m^{-2}\ \mu m^{-1}$
$E(0-\lambda)$	Irradiance in spectral band	$W\ m^2$
F	Focal length	m
f	Fraction in a mixture	—
F-0 to F-9	Stellar spectral classifications for 7,500 K stars	—
G-0 to G-9	Stellar spectral classifications for 6,000 K stars	—
h	Planck's constant	$J\ sec$

Table 3-1. Symbols Nomenclature and Units *(Continued)*

Symbols	Nomenclature	Units
H_0	The Hubble constant, 2.4×10^{-18} (1970);	sec^{-1}
	also, satellite orbital altitude	nmi or km
h_c	Convective heat transfer coefficient	$W\ m^{-2}\ K^{-1}$
$\bar{h}_c$	Time average of h_c	$W\ m^{-2}\ K^{-1}$
h_k	Conductive heat transfer coefficient	$W\ m^{-2}\ K$
i	Running index; also $\sqrt{-1}$	—
IBC	International brightness coefficient	—
$I_{\lambda,i}$	Mean spectral radiant intensity of ith class of	$W\ sr^{-1}\ \mu m^{-1}$
	celestial object	
j_c	Thermal flux density lost by convection	$W\ m^{-2}$
j_k	Thermal flux density due to conduction	$W\ m^{-2}$
j_s	Thermal flux density income from sun	$W\ m^{-2}$
K	Attenuation coefficient	$W\ m\ K^{-1}$
k	Thermal conductivity of a material	—
K-0 to K-9	Stellar spectral classification for 5,000 K stars	—
k_a	Absorption coefficient	cm^{-1}
$k(\lambda)$	Spectral attenuation (extinction) coefficient	—
L	Radiance	$W\ m^{-2}\ sr^{-1}$
l	Galactic longitude	degree
$L_{BB,\lambda}(225)$	Blackbody spectral radiance for $T = 225$ K	$W\ m^{-2}\ sr^{-1}$
$L_{BB,11}(225)$	Blackbody spectral radiance for $T = 225$ K at 11 μm	$W\ m^{-2}\ sr^{-1}$
L_q	Photon radiance	photon sec^{-1} $cm^{-2}\ sr^{-1}$
$L_s(o)$	Sky radiance at the zenith	$W\ m^{-2}\ sr^{-1}$
$L_v(\theta)$	Luminance at angle θ	$lm\ m^{-2}\ sr^{-1}\ \mu m^{-1}$
$L_{\Delta\lambda_1}$	Narrow band radiance, bandwidth $\Delta\lambda_1$ centered	$W\ m^{-2}\ sr^{-1}$
	at λ_1	
$L_{\Delta\lambda_2}$	Narrow band radiance, bandwidth $\Delta\lambda_2$ centered	$W\ m^{-2}\ sr^{-1}$
	at λ_2	
L_λ	Spectral radiance	$W\ m^{-2}\ sr^{-1}\ \mu m^{-1}$
$L_\lambda(A), (B), (C)$	Spectral radiance of materials A, B or C	$W\ m^{-2}\ sr^{-1}\ \mu m^{-1}$
$L_{\lambda,G}$	Diffuse-background, spectral radiance due to	$W\ m^{-2}\ sr^{-1}\ \mu m^{-1}$
	extragalactic sources	
l_c	Characteristic length	m
m	Visual stellar magnitude	—
$M(A), (B), (C)$	Exitance of materials A, B or C	$W\ m^{-2}$
$M_{BB,\lambda}(\lambda, T)$	Spectral exitance of blackbody surface at	$W\ m^{-2}\ \mu m$
	temperature, T	
msl	Mean sea level	m

Table 3-1. Symbols Nomenclature and Units *(Continued)*

Symbols	Nomenclature	Units
$M(T)$	Total radiant exitance	W m^{-2}
M_V	Absolute visual magnitude	—
$M_v(T)$	Luminous exitance for temperature, T	lm m^{-2}
M_λ	Spectral exitance	W m^{-2} μm^{-1}
M_λ (rock, T)	Spectral exitance from a rock sample with surface contact temperature, T	W m^{-2} μm^{-1}
$M_\lambda(\lambda, T)$	Radiant spectral exitance at temperature, T	W m^{-2} μm^{-1}
M-0 to M-9	Stellar spectral classifications for 3,500 K stars (also indicated as the letter subscripted by the number)	—
m_a	Number of atmospheric air masses	—
m_V	Apparent visual magnitude	—
m_1	Visual stellar magnitude of object number 1	—
m_2	Visual stellar magnitude atmospheric air masses of object number 2	—
n	Index of refraction	—
$N_{l,b}$	Number of infrared sources per square degree	degree^{-2}
$N_{\lambda, G}$	Number of extragalactic sources observed per square degree with limiting spectral irradiance, E_λ	degree^{-2}
$N_{\lambda,a}$	Number of asteroids per square degree observed to a limiting spectral irradiance, E_λ	degree^{-2}
n_g	Number density of galaxies	m^{-3}
n_i	Mean space density of ith class at galactic center	cm^{-3}
p	Object distance	m
R_E	Radius of earth, or reference earth sphere	nmi, or km
R_H	Range from exoatmospheric point and earth's horizon	nmi, or km
$\mathcal{R}_p$	Responsivity for photon flux	signal (sec^{-1})$^{-1}$
Ry	Symbol used for rayleigh unit	—
R_1, R_2	Thermal resistances	W m^{-2} K^{-1}
S	Signal in arbitrary signal units	—
S_{ph}	Scattering coefficient for phytoplankton	m^{-1}
T	Temperature	K
$T(A), (B)$	Temperature of material A or B	K
T_a	Local air temperature	°C
T_0	Average diurnal temperature	°C
T(steady state)	Steady state temperature	°C
V_a	Air velocity	m sec^{-1}

Table 3-1. Symbols Nomenclature and Units *(Continued)*

Symbols	Nomenclature	Units
$V(\lambda)$	Spectral luminous efficiency	—
x	Distance	m
α	Attenuation coefficient; also, angle from nadir at a satellite as vertex to background point	cm^{-1}; rad
$\alpha(L)$	Absorptance for long wavelength thermal radiation	—
α_{max}	Angle between nadir and earth horizon for exoatmospheric point	degree, rad
$\alpha(s)$	Absorptance for solar radiation	—
$\alpha(\lambda)$	Spectral absorptance	—
β	Scattering angle between solar flux direction and direction from background point to satellite with background point as vertex	—
β_e	Ecliptic latitude	degree
γ	Angle from nadir from an exoatmospheric point to a point on the earth's surface; also satellite elevation angle from point on earth's surface	degree
γ_e	Elevation angle, from local horizontal at background point as vertex, to satellite	—
γ'	The polar angle, from a point on the earth's surface, to an exoatmospheric point	—
Δ	Latitude of the sun's prime ray	—
δ	Zenith angle of satellite, from a point on the earth's surface	degree
ϵ	Emissivity	—
$\epsilon(L)$	Long wavelength thermal emissivity	—
$\epsilon(\lambda)$	Spectral emissivity	—
θ	Phase angle	degree
θ_i	Polar angle of incidence	degree
θ_r	Polar angle of reflectance	degree
θ_s	Polar angle of sun	degree
λ	Wavelength	μm
ξ	Latitude of a background point	degree
$\rho(A), (B), (C)$	Reflectance of material, A, B or C	—
$\rho(dry)$	Reflectance of dry surface	—
$\rho(wet)$	Reflectance of wet surface	—
$\rho(\lambda)$	Spectral hemispherical reflectance	—
$\bar{\rho}$	Hemispherical reflectance for Lambertian source	—
ρ'	Bidirectional reflectance	sr^{-1}

Table 3-1. Symbols Nomenclature and Units *(Continued)*

Symbols	Nomenclature	Units
$\rho'(A), (B), (C)$	Bidirectional reflectance of material, A, B or C	sr^{-1}
σ	Stefan Boltzmann constant	$W\,m^{-2}\,K^{-4}$
σ_i	The dispersion of ith class normal to galactic plane	cm
σ_s	Measure of sea state	—
τ	Transmittance	—
τ_c	Time constant	sec
$\tau(A), (B), (C)$	Transmittance of material A, B or C	—
$\tau(\lambda)$	Spectral transmittance	—
ϕ	Scanner azimuth angle from the direction of the sun	rad
Φ_λ	Spectral flux	$W\,\mu m^{-1}$
ψ	Solar zenith angle	degree
ψ_i	Azimuthal angle of incidence	degree
ψ_r	Azimuthal angle of reflectance	degree rad
ω	Angular frequency	$rad\,sec^{-1}$
Ω_i	Fraction of galaxies with i^{th} radiant flux $\Phi_{i,\lambda}$	—

Thermal conductivity is the ratio of the thermal flux density through a surface within a material to the temperature gradient across the surface.

The *absorptance for solar radiation*, $\alpha(s)$, is

$$\alpha(s) = \frac{\int_0^{3.5\,\mu m} \alpha(\lambda) E_\lambda(\text{sun}) d\lambda}{\int_0^{3.5\,\mu m} E_\lambda(\text{sun}) d\lambda} \tag{3-2}$$

The *absorptance for thermal sky radiation*, $\alpha(L)$, is

$$\alpha(L) = \frac{\int_{3.5\,\mu m}^{\infty} \alpha(\lambda) E_\lambda(\text{sky}) d\lambda}{\int_{3.5\,\mu m}^{\infty} E_\lambda(\text{sky}) d\lambda} \tag{3-3}$$

Aurorae and airglow brightness are usually measured in rayleighs. The rayleigh measure is derived from the observed photon radiance but is not equivalent to the radiance concept. Each unit volume of an optically thin cloud of radiating gas is assumed to radiate isotropically without foreground obscuration by the rest of the cloud.

$$R\,\text{rayleigh} = L_q \times 4\pi \times 10^{-6} \tag{3-4}$$

where L_q is expressed in units of photon $sec^{-1}\,cm^{-2}\,sr^{-1}$.

Some confusion may arise when brightness (which implies the concept of radiance) is specified in rayleighs. Usually the author has made an irradiance measurement in a narrow wavelength band and converted it numerically to rayleighs.

3.1.3 Radiative Transfer Modeling. Natural sources of radiation are almost always composites of a number of constituents of more common materials at different temperatures in a variety of spatial arrangements. Consequently, no handbook is large enough to contain even a significant portion of the radiative properties of natural sources of interest. However, there is a very useful alternative: to make use of the known laws of nature to calculate the radiative properties of a composite from the properties and geometric arrangement of its more common constituent materials. This procedure requires that the composite be conceived as an ensemble or model of a known but idealized arrangement of constituents, the spectral properties of which are known.

The following simple models are commonly used to extend the limited amount of data available to predict the spectral properties of composites using the spectral properties of common constituents.

The Plane Mixtures Model. An example is a composite of two types of plane sources, type A and type B, covering a section of a plane surface without overlap (Figure 3-1). The spectral radiance of the composite section, $L_\lambda(C)$, is the area-weighted average of the spectral radiance of A, $L_\lambda(A)$, and B, $L_\lambda(B)$. Thus

$$L_\lambda(C) = fL_\lambda(A) + (1-f)L_\lambda(B) \tag{3-5}$$

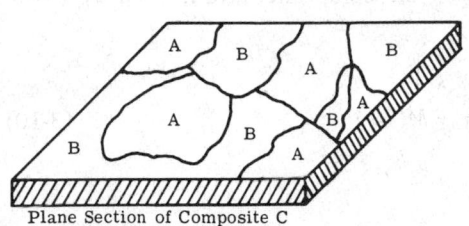

Plane Section of Composite C

Fig. 3-1. Plane Mixture of A and B materials comprising a composite plane section.

where f is the fraction of the section covered by type A sources. The extension to any number of types of plane sources is straightforward. If the spectral radiance of the sources is due to reflectance under uniform spectral irradiance, E_λ, then the composite, spectral bidirectional reflectance, $\rho'(C)$, can be determined if the spectral bidirectional reflectances, $\rho'(A)$, of the constituents are known. That is

$$\frac{L_\lambda(C)}{E_\lambda} = \frac{fL_\lambda(A)}{E_\lambda} + \frac{(1-f)L_\lambda(B)}{E_\lambda} \tag{3-6}$$

so that

$$\rho'(C) = f\rho'(A) + (1-f)\rho'(B) \tag{3-7}$$

where $\rho'(C) = L_\lambda(C)/E_\lambda$
$\rho'(A) = L_\lambda(A)/E_\lambda$
$\rho'(B) = L_\lambda(B)/E_\lambda$

The same mixtures model applies in the case of exitance and hemispherical reflectance of the composite and constituents.

The Plane Stacking Reflectance and Transmittance Model. An example is a composite of two perfectly diffuse plane materials, A and B where material A is stacked (not in optical contact with B) as a layer over material B (e.g., a leaf resting upon a concrete road). Analysis will show that the hemispherical reflectance, $\rho(C)$, and transmittance, $\tau(C)$, of the stack is derivable from the hemispherical reflectances*, $\rho_t(A)$, $\rho_b(A)$ and $\rho(B)$, and the hemispherical transmittances, $\tau(A)$ and $\tau(B)$, of the constituent materials by

$$\rho(C) = \rho_t(A) + \frac{\tau^2(A)\rho(B)}{1 - \rho_b(A)\rho(B)} \tag{3-8}$$

and

$$\tau(C) = \frac{\tau(A)\tau(B)}{1 - \rho_b(A)\rho(B)} \tag{3-9}$$

The factor, $(1 - \rho(A)\rho(B))$, accounts for the multiple reflection between layers. The transmittance of the composite is independent of the order of stacking, but the reflectance does change with stacking order. These results can be compared to the transmittance and reflectance of a plane-parallel plate in Chapter 7 (that is not diffuse).

The Plane Stacking Reflectance-Thermal Emission Model. A composite of two diffuse plane materials, A and B, as in the case of the plane stacking reflectance and transmittance model, with material A at temperature $T(A)$ and opaque material B at temperature $T(B)$, has an exitance from the top of the composite, $M(C)$, given by (with the top and bottom reflectances of A the same)

$$M(C) = \rho(A)E_0 + M(A) + \tau(A)\{\rho(B)[Q] + M(B)\} \tag{3-10}$$

where E_0 = the irradiance on A

$\quad\quad M(A)$ = the thermal exitance of A

$\quad\quad\quad\quad$ = $(1 - \tau(A) - \rho(A))M_{BB}(T(A))$

$\quad\quad M(B)$ = the thermal exitance of B

$\quad\quad\quad\quad$ = $(1 - \rho(B))M_{BB}(T(B))$

$\rho(A), (B), (C)$ = the reflectance of material, A, B or C

$\tau(A), (B), (C)$ = the transmittance of material, A, B or C

$$Q = \frac{\rho(A)M(B) + M(A) + E_0\tau(A)}{1 - \rho(A)\rho(B)}$$

---

*The subscripts t and b on $\rho(A)$ refer respectively to the top and bottom reflectance of material A.

This relation can not be extended by iteration. Where $M(A)$, $M(B) = 0$, the relation reduces to the previous case by letting $\rho(C) = M(C)/E_0$. The analytical method of deriving the above relation can be solved for more than two layers, but the resulting relation becomes progressively more complex as additional layers are added.

The Wet-Dry Plane Stacking Reflectance Model. In the above plane stacking models, the layers were not in perfect optical contact. A common condition occurring in nature is a thin plane of liquid (usually water) or a wax-like material covering and wetting an opaque, diffuse, reflecting surface. The wet or waxed material almost always reflects less than the dry material. There are two principal reasons for this reduction of reflectance upon wetting with a liquid. One is that the thin liquid film tends to form an impedance-matching layer between air and the opaque solid, so that greater penetration of radiation is permitted by reducing the impedance mismatch. Fresnel's equations for ordinary dielectrics specify the reflection resulting from the impedance mismatch. The second cause, which appears to be dominant in natural circumstances, is the reduction in reflectance resulting from multiple internal reflections within the liquid layer between the diffuse surface and the specular liquid surface. Radiation entering a medium with a high index of refraction, such as water, from one with a low index of refraction, such as air, is refracted towards the normal to the surface, according to Snell's law. Thus radiation reaching the solid diffusing surface cannot arrive at all angles. After reflection from the diffusing surface, the radiation may leave at all angles, so that some radiation arrives back at the liquid surface at angles greater than the critical angle for internal reflection. Consequently, that part of the radiation is reflected back to the solid surface to be diffusely reflected again. With each reflection, some radiation is lost by absorption.

When the multiple internal reflection effect is dominant, an approximate relation between wet and dry reflectance is

$$\rho(\text{wet}) \cong \frac{(1 - \bar{\rho})\tau^2 \rho(\text{dry}) \left[1 - \left(\frac{n-1}{n+1} \right)^2 \right]}{n^2 - \rho(\text{dry})\tau^2 (n^2 - 1 + \bar{\rho})} \tag{3-11}$$

where $\bar{\rho}$ = the hemispherical reflectance of the liquid for a Lambertian source

τ = the transmittance of the liquid layer

n = the index of refraction of the liquid relative to air

The dry material is assumed to be Lambertian and opaque so that $\rho(\text{dry})$ is the hemispherical reflectance of the dry material. The reflectance of the wet material is for near normal illumination and $\rho(\text{wet})$ is the relative directional reflectance $\pi L/E$ which does not include the specular reflection of the illuminating source from the liquid surface.

For a thin layer of clear water in the visible spectral range, $n \cong 1.33$, $\bar{\rho} \approx 0.08$, and $\tau \approx 1.0$, so that

$$\rho(\text{wet}) \cong \frac{0.90\rho(\text{dry})}{1.77 - 0.85\rho(\text{dry})} \tag{3-12}$$

In the infrared, the refractive index, n, and average hemispherical reflectance, $\bar{\rho}$, are different. No data could be found on $\bar{\rho}$ but one may calculate values for $\rho(\text{wet})$ using

$\bar{\rho} = 0.08$ and proper values of n for the wavelength of interest. The above relation for water is shown in Figure 3-2 for the visible range.

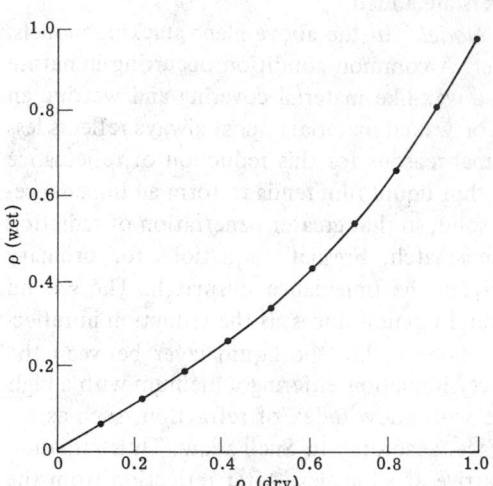

Fig. 3-2. The illustration of the relation for water: $\rho(\text{wet}) \cong 0.90\rho(\text{dry})/[1.77-0.85\rho(\text{dry})]$.

Some surfaces may already be covered by thin dielectric films in the form of oils, waxes, or glossy paint finishes so that no further reduction in reflectance resulting from wetting with water can be expected.

Lambertian Approximation for Rough Surfaces. Many natural surfaces reflect and emit radiation more or less diffusely. For purposes of estimating the expected radiance from such surfaces, such surfaces are usually considered Lambertian reflectors and emitters to a first approximation, so that

$$L_\lambda \cong \frac{M_\lambda}{\pi}, \text{W m}^{-2} \text{ sr}^{-1} \quad (3\text{-}13)$$

Specular Approximation for Smooth Surfaces. A ripple-free, liquid surface forms a partially reflecting plane mirror. The radiance contribution of the surface of such a liquid depends upon the radiance of the objects which are seen by reflection in the surface. Thus

$$L_\lambda(\text{image in surface}) = \rho(\lambda)L_\lambda(\text{object}) \quad (3\text{-}14)$$

where $\rho(\lambda)$ is given by Fresnel's equations. Natural water bodies are rarely ripple free. However, the above relation is a good first-order estimate for low-sea states. The surface reflectance contributes additively to the bulk and bottom reflectance of a water body.

Bulk and Surface Reflectance. The reflectance of radiation from solids and liquids consists of two components. One component, surface reflectance, is produced at the interface of the material with air. The change of the impedance of the space in which the waves propagate causes this reflectance, as specified by Fresnel's equations. The reflected wave penetrates the surface to a nominal depth of one-quarter wavelength and returns. The reflected wave is little attenuated by the absorption process in the body of the material. The second component, bulk reflectance, is produced by the internal back-scatter of radiation that was transmitted into the material. This component is largely altered by the spectral absorption properties of the body of the material. Consequently, the bulk spectral reflectance contains features similar to the spectral transmittance for thin sections of the material.

Surface reflectance tends to be single surface reflectance, so that the degree of polarization of incident radiation is largely preserved. Surface reflectance also tends to produce most of the non-Lambertian character of the bidirectional reflectance near the specular angle. Bulk reflectance tends to be randomly polarized and more nearly Lambertian in goniometric distribution.

When the incident radiation frequencies approach the resonant absorption frequencies of a uniform material, the electromagnetic impedance of the material rapidly changes and increases the impedance mismatch between air and the material. Consequently, the surface reflectance increases sharply near the reststrahlen frequency. The remaining penetrating radiation is absorbed just below the surface of the material because the absorption coefficient also increases near resonance. Consequently, bulk reflectance decreases sharply. The penetrating radiation is converted to thermal energy before backscattering can be effective. Thus, surface reflectance is the dominant component near resonance. Even though the absorption coefficient of the material is very large for penetrating radiation, the amount of penetrating radiation is reduced by the increase in surface reflectance. The absorptance of the material for incident radiation from the air medium is reduced due to the increased reflectance, so that the spectral emissivity must also be less at resonance in accordance with Kirchhoff's law.

Most natural materials are mixtures, aggregates, and solutions. Minerals are generally solid mixtures with traces of impurities. Rocks are aggregates of these minerals. Natural water bodies are generally solutions of water with salts as solutes and with particulate suspensions. The surface reflectance of these natural materials is largely determined by the character of the solvent. The bulk reflectances, on the other hand, are largely determined by the solutes, the absorption coefficient of the solvent, and particulate suspensions which can absorb and scatter penetrating radiation.

The state of division of an aggregate must affect its spectral reflectance. If there is no division at all and the material is optically homogeneous, there can be no bulk reflectance other than Rayleigh scattering by molecules. If the material is powdered into extremely fine particles, the multiple scattering caused by the many surfaces can return radiation to the surface before large distances have been traversed through the material. Bulk reflectance will tend spectrally to resemble surface reflectance. However, if the aggregate particle dimensions are about equal to $1/a(\lambda)$, then bulk reflectance should exhibit the spectral absorption coefficient spectral properties of the material.

Spectral properties of various samples of natural materials are presented in this chapter. The reader should be aware of the inherent variations of spectral properties which must occur within any natural material category, caused by the state of division and surface preparation.

A Vegetative Canopy Reflectance Model. The calculation of the bidirectional reflectance of a uniform vegetative canopy is somewhat more complex than the simple mixtures and stacking models. Not only are the spectral reflectance and transmittance properties of the components (such as leaves and stalks) involved, but also the statistical geometrical arrangement of these components. The latter is a significant influence on the determination of the final reflectance of the canopy.

Most vegetative canopies tend to exhibit a layered structure with particular botanical features occupying the top, middle and lower layers. Soil forms the lower boundary. The materials at the top tend to provide the dominant influence on bidirectional reflectance because the line of sight from sun to top layer and top layer to observer is more certain. Vertically standing opaque parts such as stems cast shadows on other components and the soil. Such shadows are not necessarily resolved at large distances. Thus, the reflectance of a canopy as viewed remotely will be a summation of directly illuminated components, shadowed components, and components indirectly illuminated by multiple scattering within the canopy.

Thermal Modeling. The thermal emission of a natural source depends upon the temperature of the material of the source and the effectiveness of that material in converting thermal energy into radiant energy as compared to the ideal blackbody emitter. The conversion effectiveness of the ideal blackbody emitter is the maximum permitted by thermodynamic laws. Generally, a natural material *in situ* may not have one temperature, but rather may sustain a temperature gradient from the interior of the material to its surface caused by the time rate of change in the thermal exchange between the material and its environment.

The effective temperature for emission of a semi-infinite layer can be approximated by the internal temperature at a distance $1/a(\lambda)$ below the surface, where $a(\lambda)$ is the spectral absorption coefficient (in m^{-1}) of the material for diffuse flux. The spectral exitance from the surface, M_λ, for a body with a uniform temperature gradient below the surface, is approximated by

$$M_\lambda \cong (1 - \rho(\lambda))M_{BB,\lambda}(\lambda, T) \tag{3-15}$$

where $\rho(\lambda)$ = the spectral hemispherical reflectance
$\quad T$ = the temperature at $1/a(\lambda)$ below the surface

The effective body temperature may be wavelength-dependent. The emission spectrum will exhibit bulk absorption properties as well as surface reflectance properties if a large internal thermal gradient exists in the material. When no temperature gradient exists, the relation reduces to the familiar relation

$$M_\lambda = \epsilon(\lambda)M_{BB,\lambda}(\lambda, T) \tag{3-16}$$

where $\epsilon(\lambda) = 1 - \rho(\lambda)$
$\quad T$ = uniform body temperature

The thermal exchange relations in some natural circumstances can be greatly complicated by convection both at the surface, by wind currents, and by matter transport within the material. The flow of soil moisture through plants results in evaporation of moisture from the leaves and a cooling of the plant leaves. The thermal gradients within moist soils produce a vapor transport of heat from warm to cool soil layers. Deep, geothermal heat is most frequently made manifest on the surface by the seepage of hot gases and liquids through fissures leading to the surface rather than by thermal conduction through rock layers, which is a comparatively slow process.

Some natural materials such as dry dust, sand, and wood are capable of sustaining very large temperature gradients under solar irradiation and that such materials may also have a moderately low, diffuse, absorption coefficient for radiation in the thermal range. The altered thermal emission caused by thermal gradients may be significant.

Heat Transfer Relations. It is often necessary in infrared system studies to estimate probable signal levels from terrestrial materials. Elementary heat transfer relations may be used to provide estimates of contact temperature under various environmental conditions.

The following relations provide estimates of the thermal exchange between plane surfaces with low heat capacity per unit area where the heat transfer may be considered to be in steady-state.

The surface-to-air thermal flux density, j_c, is given by Newton's law of cooling for convection

$$j_c = h_c(T_a - T) \tag{3-17}$$

where h_c = the convective heat transfer coefficient, $\text{W m}^{-2}\,^\circ\text{C}^{-1}$
T_a = the local air temperature
T = surface temperature

The value of h_c depends upon geometry, the temperature difference between surface and air, and the air velocity. For plane surfaces, h_c is approximately

$$h_c \cong (1.7|T - T_a|^{1/3} + 6 V_a^{0.8} l_c^{-0.2}) \tag{3-18}$$

where V_a = air velocity, m sec^{-1}
l_c = a characteristic length of the air path across the plane surface with h_c in W m^{-2} $^\circ\text{C}^{-1}$ and T and T_a in $^\circ\text{C}$.

The first term accounts for free convection and the second for forced convection. Turbulent flow is assumed in both cases.

Terrestrial conditions allow for further approximations, since the forced convection term will usually be dominant and $|T - T_a|$ is frequently of the order of 10°C. In addition, common regions of interest are of the order of 10 m in dimension so that

$$h_c \approx (4 + 3.8 V_a^{0.8}),\ \text{W m}^{-2}\,^\circ\text{C}^{-1} \tag{3-19}$$

The total solar irradiance, $E(\text{sun})$, normal to the sun at sea level on a clear day may be approximated from Table 3-2, as interpolated from Gates. The solar, thermal flux density, j_s, incoming to the surface is

$$j_s = \alpha(s)\ E(\text{sun}) \cos \theta_i \tag{3-20}$$

where $\alpha(s)$ = the absorptance of the surface for solar radiation
θ_i = the angle of incidence of the sun relative to surface normal

The long wavelength thermal irradiance, $E(\text{sky})$, incident upon a plane horizontal surface at sea level from a clear sky may be estimated using the Idso-Jackson formula [3-1]

$$E(\text{sky}) = \sigma T_a^4 \{1 - 0.261 \exp[-7.77 \times 10^{-4}(273 - T_a)^2]\} \tag{3-21}$$

where T_a = the meteorological air temperature, K
$\sigma = 5.6686 \times 10^{-8},\ \text{W m}^{-2}\,\text{K}^{-4}$

Table 3-2. Calculated Clear Day Solar Irradiance at Sea
Level [3-2]

Normal to rays for particulate concentration of 200 cm^{-3}
and 10 mm *precipitable water.*

Zenith Angle (deg)	E(sun) (W m^{-2})
0	865
10	862
20	852
30	832
40	796
50	735
60	640
70	488
80	260
85	100

The thermal flux density entering a isothermal supporting material is

$$j_k = h_k(T - T_0) \tag{3-22}$$

where $h_k = k/d$ = the conductive heat transfer coefficient of the surface material
k = the thermal conductivity of the surface material
d = the thickness of the surface material
T_0 = the isothermal support temperature assuming infinite heat capacity of the support

Steady-state conditions require that thermal input be equal to thermal output.

$$\text{Input} = \alpha(s)E(\text{sun}) \cos \theta_i + \alpha(L)E(\text{sky}) \tag{3-23}$$

$$\text{Output} = M(\text{surface}) + h_c(T - T_a) + h_k(T - T_0) \tag{3-24}$$

where $M(\text{surface}) = \epsilon(L)\sigma T^4$
$\epsilon(L)$ = the long wavelength thermal emissivity of the surface material

Again, considering common terrestrial conditions, a further approximation helps to simplify the solution for surface contact temperature in °C.

$$M(\text{surface}) \approx \epsilon(L)(316 + 4.62T), \text{ W m}^{-2} \tag{3-25}$$

The steady-state surface contact temperature is then

$$T\text{(steady-state)} \cong \frac{\alpha(s)E\text{(sun)} \cos \theta_i + \epsilon(L)[E\text{(sky)} - 316] + h_c T_a + h_k T_0}{4.62\epsilon(L) + h_c + h_k} \quad (3\text{-}26)$$

where all quantities are expressed in MKS and temperature is in $^\circ$C. If a surface is insulated on the underside, then $h_k = 0$.

The surface temperature variation of thick, massive substances such as concrete and compacted soils is the result of a dynamic thermal exchange process and is not steady-state. Thermal flux moves into the interior in slowly moving, damped waves. The solution of the dynamic flow involves Fourier series descriptions of the input and output. The diurnal surface temperature variation may be estimated using steady-state results if one assumes that the maximum depth of material involved in the diurnal heat transfer is that depth were the diurnal thermal wave falls to $1/e$ of its surface value and that the thermal path is one-half of this depth. The effective depth of penetration of a sinusoidal thermal wave is

$$d = \sqrt{\frac{2k}{\omega C_h D}} \quad (3\text{-}27)$$

where d = the depth of attenuation to $1/e$ of the surface value
k = the thermal conductivity of the material
ω = 2π times the frequency of the thermal wave
C_h = the specific heat of the material
D = the density of the material

The surface dynamic temperature is then related to the steady-state temperature for infinite, internal heat capacity at temperature T_0 through the lumped constant electric circuit analogy shown in Figure 3-3. The temperature, T_0, is taken as the average diurnal air temperature. Thus, the dynamic surface temperature is the real part of the following:

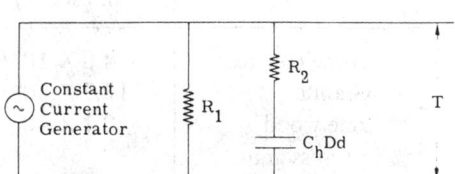

Fig. 3-3. Electrical analog of surfaces of massive materials.

$$T \cong T\text{(steady state)} \left[\frac{\omega^2 \tau^2 + \dfrac{(R_1 + R_2)}{R_2} - \dfrac{iR_1\omega\tau}{R_2}}{1 + \omega^2 \tau^2} \right] \quad (3\text{-}28)$$

where R_1 = $[4.62\epsilon(L) + \bar{h}_c]^{-1}$
$\bar{h}_c$ = average value of h_c
R_2 = $d/2k = h_k^{-1}$
τ = $(R_1 + R_2)C_h D d$
ω = $2\pi f$

The quantities τ and R_2 are frequency dependent through d for thick materials. The phase angle of the dynamic surface temperature relative to the steady-state temperature is

$$\theta = \arctan \frac{Y}{X} \tag{3-29}$$

where $Y = -\omega\tau R_1/R_2$
$X = [\omega^2\tau^2 + (R_1 + R_2)/R_2]$

For materials which are much thinner than the depth of penetration of the diurnal cycle, d is set equal to the actual thickness of the material and R_2 is given by

$$R_2 = \frac{d}{2k} + \left(\frac{d}{2k} + \frac{1}{h'}\right) \tag{3-30}$$

where h' is the heat transfer coefficient of the junction between the thin material and its isothermal support at temperature, T_0. Table 3-3 lists the approximate values k, $k/(C_hD)$, and d for a number of common materials where d is computed for the frequency of the diurnal cycle.

Table 3-3. Thermal Properties of Common Materials [3-3, 3-4]

Material	k/c_hD $(m^2 \ sec^{-1})$	k $(W \ m^{-1} \ °C)$	Diurnal Depth, d (m)
Stone concrete	4.8×10^{-7}	0.92	0.115
Granite	12.7×10^{-7}	1.9	0.187
Pine wood (cross grain)	0.7×10^{-7}	0.1	0.043
Lime stone	8.1×10^{-7}	0.7	0.149
Ice	11.2×10^{-7}	2.2	0.176
Damp soil	5.0×10^{-7}	2.6	0.116
Dry soil	3.1×10^{-7}	0.35	0.093
Building brick	4.4×10^{-7}	0.63	0.11
Cast iron	121.0×10^{-7}	57.0	1.73
Aluminum	860.0×10^{-7}	203.0	4.48

Shallow water bodies undergo a small diurnal temperature fluctuation. The electrical analogy is the same as shown in Figure 3-3, but with $R_2 \rightarrow 0$. One assumes thermal mixing in the photic zone. The depth, d, is taken as the depth of the euphotic zone or the actual depth, whichever is the smaller. The bottom is assumed to form an insulating barrier, i.e. the bottom material has a much poorer thermal heat transfer coefficient compared to that afforded by thermal convection within the water body. The water is isothermal and thermally separated from the bottom. The steady-state temperature is calculated with $h_k = 0$. The dynamic temperature is then the real part of

$$T \cong \frac{T(\text{steady state})(1 - i\omega\tau)}{(1 + \omega^2\tau^2)} \tag{3-31}$$

The phase angle is

$$\theta = \text{arc tan}(-\omega\tau) \tag{3-32}$$

The thermal exchange caused by evaporation is not taken into account. If the loss rate caused by evaporation is relatively constant, the effect will be to alter the average temperature of the water body rather than alter the magnitude of the diurnal variation.

3.2. The Celestial Background in the Visible Spectral Range

3.2.1. Stellar Magnitudes [3-5]. The brightness of celestial bodies is usually measured in magnitudes. The scale of magnitudes is adjusted so that a star of magnitude $+ 1.00$ (the first magnitude) gives an illuminance of 0.832×10^{-10} lm cm^{-2} at a point outside the atmosphere of the earth.

The relation between the visible light received from two stars and their magnitudes is expressed by the formula

$$\log \frac{E_v(1)}{E_v(2)} = 0.4(m_2 - m_1) \tag{3-33}$$

where E_v = illuminance
  m = magnitude

3.2.2. Stellar Spectral Classes. Under the Harvard system of classification, the principal types of spectra are designated by the letters B, A, F, G, K, and M. Stars intermediate to these designations are designated by suffixed numbers from 0 to 9.

The apparent temperatures corresponding to the various spectral classes are not always the same, but vary according to the methods used to measure or calculate the temperature. The following list should be considered only an approximation for main sequence stars:

Spectral Classification	Surface Temperature of Star (K)
B-0	20,000
A-0	11,000
F-0	7,500
G-0	6,000
K-0	5,000
M-0	3,500

3.2.3. Numbers of Stars. Table 3-4 shows the estimated number of stars brighter than a given magnitude for both photographic and visual magnitudes. From magnitude 0 to 18.5, the figures are based on direct observation; the values from magnitude 18.5 to 21 are extrapolated.

Table 3-4. Estimated Total Number of Stars Brighter than
a Given Magnitude [3-6]

Magnitude	Number of Stars	
	Photographic	Visual
−1	–	1
0	–	2
1	–	12
2	–	40
3	–	140
4	360	530
5	1,030	1,620
6	2,940	4,850
7	8,200	14,300
8	22,800	41,000
9	62,000	117,000
10	166,000	324,000
11	431,000	870,000
12	1,100,000	2,270,000
13	2,720,000	5,700,000
14	6,500,000	13,800,000
15	15,000,000	32,000,000
16	33,000,000	71,000,000
17	70,000,000	150,000,000
18	143,000,000	296,000,000
19	275,000,000	560,000,000
20	505,000,000	1,000,000,000
21	890,000,000	–

The photographic results are based on all available material such as photographs, star charts, and catalogs. The data for visual magnitudes are derived from the photographic results by allowing for the color of the stars. Very few stars are bluer than class A-0, for which the visual and photographic magnitudes are equal; but many stars are redder and have color indices of +1 magnitude or more.* A list of stars brighter visually than the tenth magnitude, for example, will contain many red stars which are photographically of the eleventh magnitude or fainter, and a great many which are photographically fainter than the tenth magnitude. On the other hand, a list of stars to the tenth photographic magnitude will contain a few blue stars which are visually below the tenth magnitude, but

---

*A color index of +1 means that the photographic magnitude can be obtained by adding +1 to the visual magnitude.

not many. The difference in the numbers in the two columns is thus explained. As seen by the table, this effect increases for the fainter stars, which are generally redder than the brighter ones. Table 3-5 shows the percentage of stars in the six principal spectral classes for various ranges of magnitudes.

Table 3-5. Percentage of Stars of Various Spectral Classes [3-6]

Visual Magnitude	B-0 to B-5	B-8 to A-3	A-5 to F-2	F-5 to G-0	G-5 to K-2	K-5 to M-8
<2.24	28	28	7	10	15	12
2.25 to 3.24	25	19	10	12	22	12
3.25 to 4.24	16	22	7	12	35	8
4.25 to 5.24	9	27	12	12	30	10
5.25 to 6.24	5	38	13	10	28	6
6.26 to 7.25	5	30	11	14	32	7
7.26 to 8.25	2	26	11	16	37	7
8.5 to 9.4	2	18	13	20	36	12
9.5 to 10.4	1	16	12	24	38	9
For all Magnitudes	2	29	9	21	33	6

Photographic Magnitude	B-0 to B-5	B-6 to A-4	A-5 to F-4	F-5 to G-4	G-5 to K-4	K-5 to M-8
8.5 to 9.5	2	31	16	24	24	3
9.5 to 10.5	1	24	16	31	26	3
10.5 to 11.5	1	17	13	40	27	3
11.5 to 12.5	0	10	13	47	26	3
12.5 to 13.5	0	3	10	58	26	2

3.2.4. Galactic Concentration of Stars.

The Number of Stars (Galactic Concentration) in Different Parts of the Sky.   Table 3-6 shows the number of stars per square degree brighter than a given photographic magnitude for different galactic latitudes.

Galactic Concentration of Stars of Various Spectral Classes.   An approximation of the number of stars of a certain spectral class and magnitude range can be obtained by applying the data of Table 3-5 to Table 3-4, since Table 3-4 gives the estimated number of stars brighter than a given magnitude for each magnitude. For example, by interpolation of Table 3-4, one can estimate the number of stars brighter than magnitudes 7.25 and 8.25. The number of stars in the magnitude range 7.25 to 8.25 can be determined by subtraction. The percentage of stars of the six principal spectral classes for this range of magnitudes, as shown in Table 3-5, can be used to obtain the approximate number of stars in these spectral classes for this range of magnitudes.

Table 3-6. Number of Stars Per Square Degree Brighter than Photographic Magnitude as a Function of Galactic Latitudes [3-6]

Photographic Magnitude	+90°	+40°	+20°	+10°	0°	-10°	-20°	-40°	-90°
5.0	0.014	0.0175	0.023	0.031	0.059	0.045	0.032	0.0178	0.012
6.0	0.039	0.053	0.071	0.089	0.166	0.126	0.087	0.051	0.042
7.0	0.015	0.151	0.20	0.257	0.436	0.323	0.224	0.144	0.123
8.0	0.275	0.42	0.59	0.741	1.230	0.851	0.617	0.398	0.316
9.0	0.724	1.12	1.62	2.14	3.55	2.34	1.69	1.10	0.832
10.0	1.78	2.95	4.50	5.89	10.5	6.61	4.68	2.95	2.09
11.0	4.3	7.4	12.0	16.2	30.9	18.2	12.8	7.76	5.25
12.0	10.2	18.2	32.0	43.6	89.1	50.1	34.7	19.50	13.2
13.0	24.0	43.0	79.0	112.0	245.0	138.0	89.1	47.8	30.2
14.0	50.0	93.0	190.0	282.0	661.0	371.0	218.0	107.0	60.3
15.0	95.0	200.0	457.0	708.0	1660.0	977.0	525.0	218.0	104.0
16.0	182.0	407.0	1047.0	1778.0	3981.0	2455.0	1175.0	436.0	182.0
17.0	338.0	794.0	2291.0	4365.0	9120.0	5754.0	2512.0	832.0	302.0
18.0	616.0	1413.0	4677.0	9330.0	20890.0	2590.0	4786.0	1514.0	501.0
19.0	770.0	2180.0	6860.0	—	—	—	—	—	—
20.0	—	—	—	—	—	—	—	—	—
21.0	1670.0	5000.0	21200.0	—	—	—	—	—	—

Table 3-7 shows the average number of stars per 100 square degrees, near the galactic equator and in regions remote from it, for the six principal spectral classes.

Table 3-7. Galactic Concentration of Stars of the Principal Spectral Classes in 100 Square Degrees near Galactic Equator [3-6]

Stellar Magnitudes	Galactic Latitudes	B	A	F	G	K	M	Total
Above 7^m.0								
	40° - 90°	0.2	6.6	3.0	3.4	10.2	1.5	24.9
	0°	10.8	21.1	5.1	5.1	15.1	3.9	61.1
7^m.0 to 8^m.25								
	40° - 90°	0.1	6.6	9.5	16.4	32.8	6.1	71.5
	0°	18.9	75.8	13.6	20.9	53.9	13.6	196.7

Table 3-8 gives more detailed information of the distribution of stars by spectral class and magnitude. There are differences in the data of Tables 3-7 and 3-8 because somewhat different areas of the sky were considered in preparing the tables. For example, Table 3-7 considers the latitude from 40 to 90°, whereas Table 3-8 considers the latitude from 60 to 90°, in arriving at an average galactic distribution. The most important difference is that Table 3-8 has been prepared by selecting narrower ranges of stellar magnitude.

Galactic distribution is not presented for stars of magnitudes less than 5 because the total number of these stars is not large enough to make the concept of the number of stars per square degree meaningful.

Table 3-9, an index of apparent galactic concentration, has been prepared from Table 3-8 by taking the ratios of numbers of stars in low latitudes to numbers of stars in high latitudes. For a given spectral class, more stars are concentrated in the lower galactic latitudes as the index number becomes higher.

3.2.5. Spectral Distribution of Stellar Radiation. Figure 3-4 shows the relative spectral distribution of stellar radiation as a function of star classes and surface temperature.

The terrestrial exoatmospheric spectral irradiance, $E_\lambda(\lambda)$ of a celestial body may be calculated if its visual magnitude, m, and its distribution temperature (sometimes called color temperature), T, are known. (See Chapter 1.) In this case, distribution temperature is the blackbody temperature for which $M_\lambda(\lambda, T)$ is spectrally best fit to $E_\lambda(\lambda)$ in the spectral range of interest where $M_\lambda(\lambda, T)$ is given by Planck's radiation law. The defining relation between exoatmospheric illuminance and visual magnitude is

$$m = -2.5 \log_{10}\left(\frac{E_v(m)}{E_v(m = 0)}\right) \tag{3-34}$$

where $E_v(m = 0)$ for zero magnitude is 2.089×10^{-6} lm m^{-2} .

Table 3-8. Galactic Concentration of Stars of Various Spectral Classes [3-6]

Spectrum Visual Magnitude	Galactic Latitude 0° to 5°					
	B	A	F	G	K	M
<6.0	4.5	6.0	1.7	2.1	3.5	1.3
6.0 to 7.0	6.3	15.0	3.4	3.0	12.0	2.6
7.0 to 8.25	19.0	76.0	14.0	21.0	54.0	14.0
8.5 to 9.4	46.0	190.0	85.0	96.0	200.0	57.0
9.5 to 10.4	82.0	610.0	240.0	310.0	490.0	150.0
Photographic Magnitude						
9.5 to 10.5	38.0	510.0	150.0	220.0	180.0	19.0
10.5 to 11.5	87.0	970.0	430.0	720.0	460.0	42.0
11.5 to 12.5	100.0	1390.0	1200.0	1960.0	940.0	140.0
Visual Magnitude	Galactic Latitude 60° to 90°					
<6.0	0.2	2.6	0.8	1.0	2.9	0.7
6.0 to 7.0	0.0	3.8	1.8	2.4	7.5	0.7
7.0 to 8.25	0.0	7.4	9.2	16.0	32.0	6.3
8.5 to 9.4	0.0	8.0	20.0	83.0	75.0	0.0
9.5 to 10.4	0.0	8.0	20.0	170.0	210.0	16.0
Photographic Magnitude						
9.5 to 10.5	0.0	9.0	32.0	120.0	75.0	9.0
10.5 to 11.5	0.0	10.0	27.0	290.0	160.0	12.0
11.5 to 12.5	0.9	14.0	34.0	680.0	270.0	26.0

Table 3-9. Index of Apparent Galactic Concentration* [3-6]

Visual Magnitude	B	A	F	G	K	M
<6.0	22.0	2.8	2.3	2.1	1.2	1.9
6.0 to 7.0	—	4.0	1.9	1.2	1.5	3.7
7.0 to 8.25	—	10.0	1.5	1.3	1.7	2.2
8.5 to 9.4	—	24.0	4.2	1.2	2.7	—
9.5 to 10.4	—	76.0	12.0	1.8	2.3	0.9
Photographic Magnitude						
9.5 to 10.5	—	56.0	4.8	1.8	2.4	2.1
10.5 to 11.5	—	97.0	16.0	2.5	2.9	3.5
11.5 to 12.5	—	99.0	35.0	2.9	3.5	5.5

*The irregularities here are attributable in part to inadequate sampling.

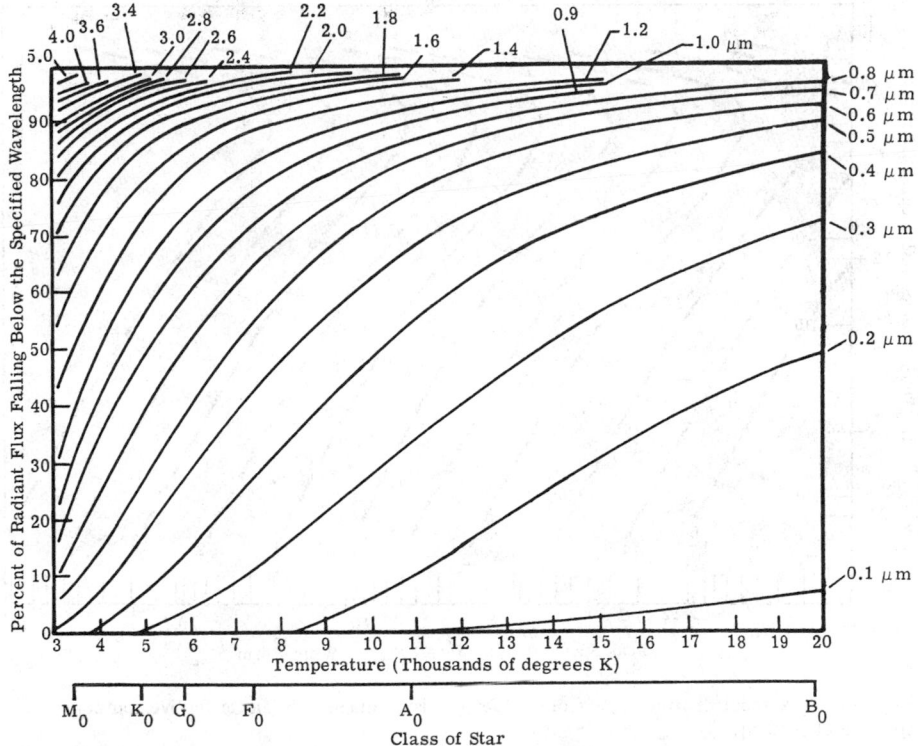

Fig. 3-4. Relative spectral distribution of stellar radiation as a function of star classes [3-6].

The luminous exitance, $M_v(T)$, and total radiant exitance, $M(T)$, are found by using Planck's radiation law and the color temperature.

$$M_v(T), = 680 \int_0^\infty M_\lambda(\lambda, T) V(\lambda) d\lambda \qquad (3\text{-}35)$$

and

$$M(T) = \int_0^\infty M_\lambda(\lambda, T) d\lambda \qquad (3\text{-}36)$$

Assuming no spectrally selective absorption in the path between celestial body and earth, one may relate luminous and radiant quantities by

$$\frac{E_v}{E} = \frac{M_v(T)}{M(T)} \qquad (3\text{-}37)$$

and also relate spectral quantities to total by

$$\frac{E_\lambda(\lambda)}{E} = \frac{M_\lambda(\lambda, T)}{M(T)} \qquad (3\text{-}38)$$

because only the geometric factors of size and distance are involved.

Therefore,

$$E_\lambda(\lambda) = M_\lambda(\lambda, T) \frac{E_v(m)}{M_v(T)} \qquad (3\text{-}39)$$

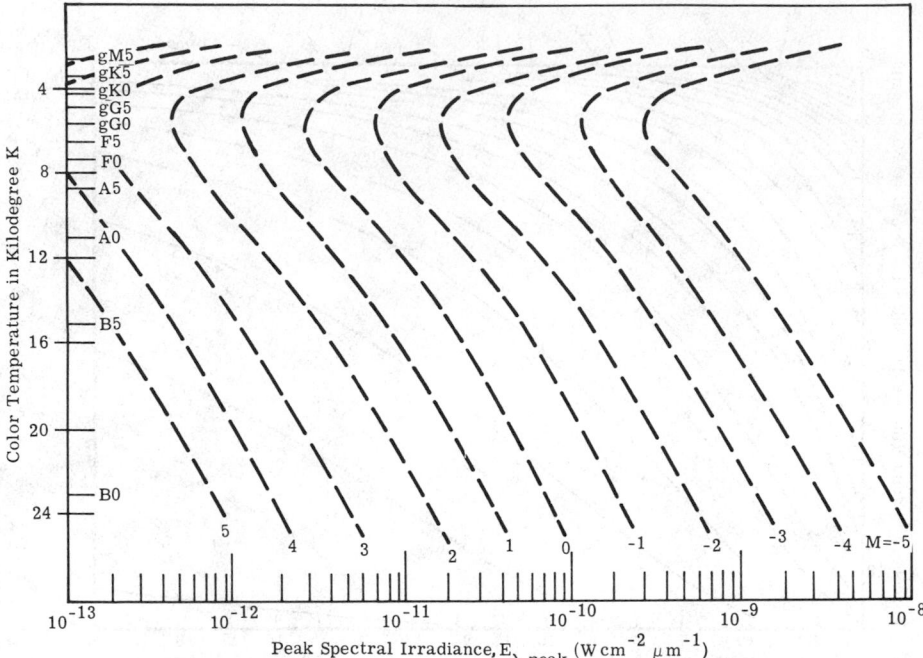

Fig. 3-5. Peak spectral irradiance from values of visual magnitude and effective temperature or spectral class [3-5].

where $M_\lambda(\lambda, T)$ is the spectral exitance given by Planck's radiation law. $E_v(m)$ is derived from Equation (3-34), and $M_v(T)$ is derived from Equation (3-35).

The relationship between peak spectral irradiance, color temperature and visual magnitude is shown in Figure 3-5.

The relations used pertain to irradiance received above the atmosphere. Values for absorption by the atmosphere in the various spectral regions can be readily applied to the chart values.

Table 3-10 shows the visual magnitude and color temperature for the brightest celestial bodies and also for the important red stars. The list contains all the stars in Schlesinger's *Catalogue of Bright Stars* which give an irradiance of at least 10^{-12} W cm^{-2} in either the PbS region, 1 to 3 μm, or the bolometer region, 0.3 to 13.5 μm. (See also the revised edition of *Catalogue of Bright Stars* by Hoffleit.) Using Figure 3-5 and Planck's law, one may obtain the spectral irradiance curves for any star or planet. In determining the spectral irradiance of the moon and planets, one may assume an effective temperature of 5900 K for reflected sunlight. Thermal emission must be added.

One may have to use the visual magnitudes of bright stars to compute their infrared irradiance simply because the enormous job of cataloging the infrared emission from all the stars has just begun. Computations using visual magnitudes will not adequately predict the irradiance from massive cool stars whose infrared magnitude might far exceed their visual magnitude [3-7]. For additional information, see References [3-8] through [3-15].

Table 3-10. Visual Magnitudes and Color Temperature of Planets
and the Brightest Visual and Red Stars [3-5]

Name	Visual Magnitude, m_V	Color Temperature, T (K)
1. Moon (full)	−12.2	5,900
Planets		
2. Venus (at brightest)	− 4.28	5,900
3. Mars (at brightest)	− 2.25	5,900
4. Jupiter (at brightest)	− 2.25	5,900
5. Mercury (at brightest)	− 1.8	5,900
6. Saturn (at brightest)	− 0.93	5,900
Stars		
1. Sirius	− 1.60	11,200
2. Canopus	− 0.82	6,200
3. Rigel Kent (double)	0.01	4,700
4. Vega	0.14	11,200
5. Capella	0.21	4,700
6. Arcturus	0.24	3,750
7. Rigel	0.34	13,000
8. Procyon	0.48	5,450
9. Achernar	0.60	15,000
10. β Centauri	0.86	23,000
11. Altair	0.89	7,500
12. Betelguex (variable)	0.92	2,810
13. Aldebaran	1.06	3,130
14. Pollux	1.21	3,750
15. Antares	1.22	2,900
16. α Crucis	1.61	2,810
17. Mira (variable)	1.70	2,390
18. β Gruis	2.24	2,810
19. R. Hydrae (variable)	3.60	2,250

3.3. The Celestial Background in the Spectral Range from 4.0 to 30.0 μm*

Infrared emission from solid material (dust, ice, metal condensates, etc,) distributed throughout our galaxy becomes increasingly important at wavelengths longer than about 4 μm. While it is true that the infrared flux from most main sequence stars (A through M dwarfs) is fairly well represented by the spectrum of a blackbody at the star's distribution temperature, many giant and supergiant stars show large amounts of infrared flux in

*Section 3.3 was contributed by Russell G. Walker, NASA, Ames Research Center, Moffett Field, Calif. Note that the data in this section represent the best data available as of the Spring of 1976; errors can be great as new, more accurate measurements are made.

excess of that predicted on the basis of the star's distribution temperature. Furthermore, the spectrum is nonthermal in shape, showing emission peaks characteristic of particulate materials (such as carbon and silicates) in a cloud or shell surrounding the star. Stars of early spectral type showing emission lines in the visible and ultraviolet, (e.g. Be stars) are often found to have strong infrared excesses due to free-free transitions of electrons and molecules in their extended atmospheres.

At wavelengths longer than 10 μm emission nebulae, such as regions of ionized hydrogen gas, compact nebulae, and planetary nebulae, become strong sources of infrared radiation. The spectral envelopes of these sources are characteristic of the emission from dust mixed with a highly excited gas and often show emission lines such as that of Ne^+ at 12.8 μm.

In the following discussion, the wavelength λ is restricted to the range $4 \leqslant \lambda \leqslant 30$ μm.

3.3.1. The Density of Galactic Infrared Sources.

Infrared surveys of infrared celestial sources have been completed to a level of irradiance of about 10^{-16} W cm^{-2} μm^{-1} in the wavelength region from 3 to 26 μm [3-16]. These data have been used to develop a model for estimating the density of infrared sources within our galaxy which would be observed at faint irradiance levels [3-17].

Let $N = N_{l,b}(E_\lambda(\text{lim}))$ be the number of infrared sources per square degree observed at galactic longitude l and galactic latitude b to a limiting spectral irradiance E_λ, in W cm^{-2} μm^{-1}. If $I_{\lambda,i}$ is the mean spectral radiant intensity of the ith class of celestial objects, n_i is the mean space density of these sources at the galactic center, and σ_i is the dispersion of the ith class normal to the galactic plane, then

$$N = k \sum_{i=1}^{m} n_i I_{\lambda,i}^{3/2} \int_{E_\lambda(\text{lim})}^{\infty} \left(\frac{R_c^2}{R_c^2 + R^2}\right) E_\lambda^{-5/2} \exp\left[-\frac{I_{\lambda,i}}{E_\lambda}\left(\frac{\sin b}{\sigma_i}\right)^2\right] dE_\lambda \quad (3\text{-}40)$$

where $R^2 = I_{\lambda,i} E_\lambda^{-1} \cos^2 b + R_0^2 - 2R_0 I_{\lambda,i}^{1/2} E_\lambda^{-1/2} \cos b \cos l$
$R_0 = 3.09 \times 10^{22}$ cm
$R_c = 9.27 \times 10^{21}$ cm
$I_{\lambda,i} = a_i \lambda^{c_i}$

Constants for the above model, derived from sky survey and other astronomical data, are given in Table 3-11. The source counts caused by emission nebulae have been lumped with the supergiants because they share similar space distributions. Source densities estimated by means of Equation (3-40) are given in Tables 3-12, 3-13, and 3-14 for wavelengths of 4.2, 11.0, and 19.8 μm.

Table 3-11. Constants of the Galactic Source Model* [3-17, 3-18, 3-19, 3-20]

Class	σ_i (cm)	n_i (Stars cm^{-3})	a_i (W sr^{-1} μm^{-1})	c_i
Dwarfs	4.94E+20	2.63E-56	2.77E+24	-3.693
Giants	1.39E+21	9.46E-60	1.07E+28	-1.915
Supergiants	3.09E+20	7.90E-62	1.09E+29	-1.108

*Exponential notation is used in computer format; e.g., $500 \equiv 5 \times 10^2 \equiv 5.00E + 02$.

Table 3-12. Number of Sources Per Square Degree at 4.2 μm*

Spectral Irradiance	Galactic Latitude									Galactic Longitude
	0.0	2.0	5.0	10.0	20.0	30.0	45.0	60.0	90.0	
1.0E-14	4.9E-03	4.8E-03	4.5E-03	3.8E-03	2.4E-03	1.8E-03	1.6E-03	1.5E-03	1.5E-03	0.0
	4.8E-03	4.7E-03	4.5E-03	3.7E-03	2.3E-03	1.8E-03	1.6E-03	1.6E-03	1.5E-03	45.0
	4.6E-03	4.6E-03	4.3E-03	3.6E-03	2.3E-03	1.8E-03	1.6E-03	1.6E-03	1.5E-03	90.0
	4.4E-03	4.4E-03	4.1E-03	3.5E-03	2.3E-03	1.8E-03	1.6E-03	1.5E-03	1.5E-03	180.0
1.0E-15	1.7E-01	1.5E-01	9.4E-02	5.6E-02	4.8E-02	4.6E-02	4.3E-02	4.1E-02	3.9E-02	0.0
	1.6E-01	1.4E-01	9.1E-02	5.6E-02	4.8E-02	4.6E-02	4.3E-02	4.1E-02	3.9E-02	45.0
	1.5E-01	1.3E-01	8.4E-02	5.4E-02	4.7E-02	4.5E-02	4.2E-02	4.0E-02	3.9E-02	90.0
	1.3E-01	1.1E-01	7.7E-02	5.1E-02	4.5E-02	4.3E-02	4.1E-02	3.9E-02	3.9E-02	180.0
1.0E-16	8.8E+00	3.1E+00	1.7E+00	1.6E+00	1.3E+00	9.8E-01	6.0E-01	3.9E-01	2.7E-01	0.0
	6.4E+00	2.7E+00	1.7E+00	1.5E+00	1.2E+00	9.5E-01	5.9E-01	3.9E-01	2.7E-01	45.0
	4.3E+00	2.2E+00	1.5E+00	1.4E+00	1.2E+00	8.8E-01	5.6E-01	3.7E-01	2.7E-01	90.0
	3.1E+00	1.8E+00	1.3E+00	1.2E+00	1.0E+00	8.1E-01	5.2E-01	3.6E-01	2.7E-01	180.0
1.0E-17	7.6E+02	7.1E+01	6.0E+01	3.7E+01	8.9E+00	2.7E+00	8.9E-01	4.7E-01	2.9E-01	0.0
	1.9E+02	6.0E+01	5.1E+01	3.2E+01	8.1E+00	2.5E+00	8.6E-01	4.6E-01	2.9E-01	45.0
	9.3E+01	4.5E+01	3.8E+01	2.4E+01	6.7E+00	2.5E+00	8.0E-01	4.4E-01	2.9E-01	90.0
	5.9E+01	3.3E+01	2.9E+01	1.9E+01	5.4E+00	1.9E+00	7.4E-01	4.2E-01	2.9E-01	180.0
1.0E-18	1.0E+04	6.3E+03	1.5E+03	1.1E+02	9.7E+00	3.0E+00	1.2E+00	7.9E-01	6.1E-01	0.0
	3.3E+03	2.0E+03	6.2E+02	7.7E+01	8.8E+00	2.9E+00	1.2E+00	7.8E-01	6.1E-01	45.0
	1.4E+03	8.4E+02	3.0E+02	5.0E+01	7.2E+00	2.6E+00	1.1E+00	7.6E-01	6.1E-01	90.0
	8.0E+02	4.7E+02	1.8E+02	3.5E+01	5.9E+00	2.3E+00	1.1E+00	7.4E-01	6.1E-01	180.0
1.0E-19	9.4E+04	3.1E+04	2.1E+03	1.2E+02	2.0E+01	1.2E+01	9.9E+00	8.8E+00	8.1E+00	0.0
	2.9E+04	6.9E+03	7.2E+02	3.7E+01	1.9E+01	1.2E+01	9.8E+00	8.7E+00	8.1E+00	45.0
	1.3E+04	2.6E+03	3.3E+02	8.0E+01	1.7E+01	1.2E+01	9.7E+00	8.7E+00	8.1E+00	90.0
	7.3E+03	1.4E+03	2.0E+02	4.4E+01	1.5E+01	1.1E+01	9.5E+00	8.6E+00	8.1E+00	180.0
1.0E-20	1.7E+05	3.2E+04	2.4E+03	4.2E+02	2.5E+02	1.7E+02	8.8E+01	5.3E+01	3.5E+01	0.0
	9.1E+04	7.3E+03	1.0E+03	3.9E+02	2.4E+02	1.6E+02	8.8E+01	5.3E+01	3.5E+01	45.0
	5.9E+04	3.0E+03	6.4E+02	3.5E+02	2.4E+02	1.6E+02	8.6E+01	5.2E+01	3.5E+01	90.0
	4.2E+04	1.7E+03	5.0E+02	3.2E+02	2.2E+02	1.5E+02	8.3E+01	5.1E+01	3.5E+01	180.0

*Exponential notation is used in computer format; e.g., $500 = 5 \times 10^2 \equiv 5.00E + 02$.

Table 3-13. Number of Sources Per Square Degree at 11.0 μm

Spectral Irradiance	Galactic Latitude									Galactic Longitude
	0.0	2.0	5.0	10.0	20.0	30.0	45.0	60.0	90.0	
1.0E-14	4.4E-04	4.4E-04	4.3E-04	3.9E-04	2.8E-04	1.9E-04	1.2E-04	9.3E-05	8.1E-05	0.0
	4.4E-04	4.3E-04	4.2E-04	3.9E-04	2.8E-04	1.9E-04	1.2E-04	9.3E-05	8.1E-05	45.0
	4.3E-04	4.2E-04	4.1E-04	3.8E-04	2.8E-04	1.9E-04	1.2E-04	9.2E-05	8.1E-05	90.0
	4.1E-04	4.1E-04	4.0E-04	3.7E-04	2.7E-04	1.8E-04	1.2E-04	9.2E-05	8.1E-05	180.0
1.0E-15	1.5E-02	1.4E-02	1.1E-02	5.6E-03	2.5E-03	2.0E-03	1.9E-03	1.9E-03	1.8E-03	0.0
	1.5E-02	1.4E-02	1.1E-02	5.5E-03	2.4E-03	2.0E-03	1.9E-03	1.9E-03	1.8E-03	45.0
	1.3E-02	1.3E-02	9.9E-03	5.2E-03	2.4E-03	2.0E-03	1.9E-03	1.9E-03	1.8E-03	90.0
	1.2E-02	1.1E-02	9.0E-03	4.9E-03	2.3E-03	2.0E-03	1.9E-03	1.8E-03	1.8E-03	180.0
1.0E-16	6.3E-01	3.7E-01	1.0E-01	6.4E-02	5.8E-02	5.6E-02	5.2E-02	4.9E-02	4.6E-02	0.0
	5.4E-01	3.3E-01	9.8E-02	6.3E-02	5.8E-02	5.5E-02	5.2E-02	4.8E-02	4.6E-02	45.0
	4.1E-01	2.6E-01	9.0E-02	6.1E-02	5.5E-02	5.4E-02	5.1E-02	4.8E-02	4.6E-02	90.0
	3.2E-01	2.1E-01	8.1E-02	5.9E-02	5.4E-02	5.2E-02	4.9E-02	4.7E-02	4.6E-02	180.0
1.0E-17	7.4E+01	3.2E+00	2.1E+00	1.9E+00	1.5E+00	1.1E+00	6.5E-01	4.1E-01	2.7E-01	0.0
	2.3E+01	2.7E+00	2.0E+00	1.8E+00	1.5E+00	1.1E+00	6.4E-01	4.1E-01	2.7E-01	45.0
	1.0E+01	2.3E+00	1.8E+00	1.7E+00	1.4E+00	1.0E+00	6.0E-01	3.9E-01	2.7E-01	90.0
	6.2E+00	1.9E+00	1.6E+00	1.5E+00	1.2E+00	9.1E-01	5.6E-01	3.7E-01	2.7E-01	180.0
1.0E-18	7.5E+02	8.8E+01	7.4E+01	4.3E+01	9.1E+00	2.7E+00	8.8E-01	4.6E-01	2.9E-01	0.0
	2.8E+02	7.4E+01	6.2E+01	3.7E+01	8.2E+00	2.5E+00	8.5E-01	4.5E-01	2.9E-01	45.0
	1.4E+02	5.4E+01	4.6E+01	2.7E+01	6.8E+00	2.2E+00	7.9E-01	4.4E-01	2.9E-01	90.0
	8.7E+01	3.9E+01	3.3E+01	2.1E+01	5.5E+00	1.9E+00	7.8E-01	4.1E-01	2.9E-01	180.0
1.0E-19	1.4E+04	8.4E+03	1.7E+03	1.0E+02	9.4E+00	2.7E+00	9.4E-01	5.2E-01	3.5E-01	0.0
	3.9E+03	2.4E+03	6.5E+02	7.6E+01	8.5E+00	2.6E+00	9.1E-01	5.1E-01	3.5E-01	45.0
	1.7E+03	9.6E+02	3.0E+02	5.0E+01	7.0E+00	2.3E+00	8.6E-01	5.0E-01	3.5E-01	90.0
	9.9E+02	5.3E+02	1.8E+02	3.4E+01	5.6E+00	2.0E+00	7.9E-01	4.7E-01	3.5E-01	180.0
1.0E-20	9.7E+04	3.1E+04	2.1E+03	1.1E+02	1.1E+01	4.7E+00	2.8E+00	2.3E+00	2.1E+00	0.0
	3.1E+04	7.0E+03	7.1E+02	7.8E+01	1.0E+01	4.5E+00	2.8E+00	2.3E+00	2.1E+00	45.0
	1.4E+04	2.6E+03	3.3E+02	5.2E+01	8.9E+00	4.2E+00	2.7E+00	2.3E+00	2.1E+00	90.0
	8.3E+03	1.4E+03	1.9E+02	3.6E+01	7.6E+00	3.9E+00	2.6E+00	2.3E+00	2.1E+00	180.0

Table 3-14. Number of Sources Per Square Degree at 19.8 μm

Spectral Irradiance	Galactic Latitude									Galactic Longitude
	0.0	2.0	5.0	10.0	20.0	30.0	45.0	60.0	90.0	
1.0E-14	1.4E-04	1.4E-04	1.4E-04	1.3E-04	1.1E-04	8.6E-05	5.9E-05	4.4E-05	3.6E-05	0.0
	1.4E-04	1.4E-04	1.4E-04	1.3E-04	1.1E-04	8.6E-05	5.8E-05	4.4E-05	3.6E-05	45.0
	1.4E-04	1.4E-04	1.4E-04	1.3E-04	1.1E-04	8.5E-05	5.8E-05	4.4E-05	3.6E-05	90.0
	1.4E-04	1.4E-04	1.3E-04	1.3E-04	1.1E-04	8.3E-05	5.7E-05	4.3E-05	3.6E-05	180.0
1.0E-15	4.8E-03	4.5E-03	4.0E-03	2.6E-03	1.0E-03	7.1E-04	6.1E-04	5.8E-04	5.7E-04	0.0
	4.7E-03	4.4E-03	3.9E-03	2.5E-03	1.0E-03	7.1E-04	6.1E-04	5.8E-04	5.7E-04	45.0
	4.4E-03	4.2E-03	3.7E-03	2.4E-03	1.0E-03	7.0E-04	6.0E-04	5.8E-04	5.7E-04	90.0
	4.1E-03	3.9E-03	3.4E-03	2.3E-03	9.8E-04	6.9E-04	6.0E-04	5.8E-04	5.7E-04	180.0
1.0E-16	1.8E-01	1.4E-01	5.0E-02	2.2E-02	1.8E-02	1.7E-02	1.7E-02	1.6E-02	1.6E-02	0.0
	1.7E-01	1.2E-01	4.7E-02	2.1E-02	1.8E-02	1.7E-02	1.6E-02	1.6E-02	1.6E-02	45.0
	1.4E-01	1.0E-01	4.2E-02	2.1E-02	1.7E-02	1.7E-02	1.6E-02	1.6E-02	1.6E-02	90.0
	1.1E-01	8.5E-02	3.8E-02	2.0E-02	1.7E-02	1.6E-02	1.6E-02	1.6E-02	1.6E-02	180.0
1.0E-17	1.3E+01	1.5E+00	6.2E-01	5.7E-01	5.2E-01	4.4E-01	3.4E-01	2.6E-01	2.0E-01	0.0
	7.1E+00	1.2E+00	6.0E-01	5.6E-01	5.0E-01	4.3E-01	3.3E-01	2.5E-01	2.0E-01	45.0
	3.7E+00	9.3E-01	5.6E-01	5.3E-01	4.8E-01	4.1E-01	3.2E-01	2.5E-01	2.0E-01	90.0
	2.4E+00	7.6E-01	5.2E-01	4.9E-01	4.4E-01	3.8E-01	3.0E-01	2.4E-01	2.0E-01	180.0
1.0E-18	4.8E+02	2.4E+01	2.1E+01	1.6E+01	6.9E+00	2.6E+00	8.8E-01	4.6E-01	2.8E-01	0.0
	1.3E+02	2.1E+01	1.9E+01	1.5E+01	6.4E+00	2.5E+00	8.5E-01	4.5E-01	2.8E-01	45.0
	5.9E+01	1.7E+01	1.5E+01	1.2E+01	5.4E+00	2.2E+00	7.9E-01	4.3E-01	2.8E-01	90.0
	3.6E+01	1.3E+01	1.2E+01	9.8E+00	4.5E+00	1.9E-01	7.3E-01	4.1E-01	2.8E-01	180.0
1.0E-19	2.6E+03	1.4E+03	6.7E+02	1.0E+02	9.3E+00	2.7E+00	8.8E-01	4.6E-01	2.9E-01	0.0
	1.3E+03	7.8E+02	4.0E+02	7.5E+01	8.4E+00	2.5E+00	8.5E-01	4.5E-01	2.9E-01	45.0
	6.9E+02	4.0E+02	2.1E+02	4.9E+01	6.9E+00	2.2E+00	8.0E-01	4.3E-01	2.9E-01	90.0
	4.3E+02	2.4E+02	1.3E+02	3.4E+01	5.5E+00	1.9E+00	7.3E-01	4.1E-01	2.9E-01	180.0
1.0E-20	7.5E+04	3.1E+03	2.1E+03	1.1E+02	9.4E+00	2.7E+00	9.4E-01	5.2E-01	3.4E-01	0.0
	1.8E+04	6.5E+03	7.0E+02	7.6E+01	8.5E+00	2.6E+00	9.1E-01	5.1E-01	3.4E-01	45.0
	7.2E+03	2.4E+03	3.2E+02	5.0E+01	6.9E+00	2.3E+00	8.5E-01	4.9E-01	3.4E-01	90.0
	4.1E+03	1.3E+03	1.9E+02	3.4E+01	5.6E+00	2.0E+00	7.9E-01	4.7E-01	3.4E-01	180.0

3.3.2. The Infrared Radiance of the Galaxy. The radiance L_λ of the galaxy resulting from the integrated effect of all the sources considered in 3.2.1 can be estimated from Equation (3-40) by noting that $2kdL_\lambda = E_\lambda dN$, where $k = 1.52 \times 10^{-4}$ deg^2 sr^{-1}, which leads to

$$L_\lambda = \frac{1}{2} \sum_{i=1}^{m} n_i I_{\lambda,i}^{3/2} \int_0^\infty \left(\frac{R_c^2}{R_c^2 + R^2} \right) E_\lambda^{-3/2} \exp\left[-\frac{I_{\lambda,i}}{E_\lambda} \left(\frac{\sin b}{\sigma_i} \right)^2 \right] dE_\lambda, \quad (3-41)$$

W cm^{-2} μm^{-1} sr^{-1}

Radiance values estimated by Equation (3-41) are given in Table 3-15 for a line of sight directed toward the galactic pole ($b = 90°$) and along the galactic plane ($b = 0°$). Equation (3-41) may underestimate the radiance in the galactic plane at long wavelengths since it fails to account for the general dust distribution. At higher galactic latitudes, more relevant values of L_λ may be obtained by replacing the upper limit of the integral (Equation 3-4) with that value of $E_\lambda(\text{lim})$ at which the infrared system fails to resolve individual sources.

Table 3-15. Infrared Radiance of the Sky Due to Galactic Point Sources

		L_λ, (W cm^{-2}sr$^{-1}\mu$m^{-1})		
	l	$4\ \mu$m	$11\ \mu$m	$20\ \mu$m
Case I, b=0°	0°	2.6×10^{-10}	3.5×10^{-11}	1.6×10^{-11}
	10°	2.2	3.0	1.3
	20°	1.6	2.2	9.7×10^{-12}
	30°	1.2	1.6	7.2
	45°	8.3×10^{-11}	1.1	5.0
	60°	6.3	8.4×10^{-12}	3.7
	90°	4.2	5.6	2.5
	135°	3.0	4.0	1.8
	150°	2.8	3.8	1.7
	180°	2.7×10^{-11}	3.6×10^{-12}	1.6×10^{-12}
Case II, b=90°		7.8×10^{-13}	9.5×10^{-14}	4.2×10^{-14}

3.3.3. Density of Extragalactic Infrared Sources. All-sky infrared surveys have not yet been performed at irradiance levels faint enough to detect a significant number of extragalactic objects. Infrared radiometry of a representative sample of galaxies has been used to estimate the mean radiant flux of those galaxies which contribute to the infrared background [3-21]. These data, when coupled with other basic astronomical data [3-18], lead to the following expression approximating the number of extragalactic sources observed per square degree $N_{\lambda,G}$ at a limiting irradiance $E_\lambda(\text{lim})$ [3-17]:

$$N_{\lambda,G} = 6 \times 10^{-28} \lambda^{-1.27} E_\lambda^{-3/2}(\text{lim}) \quad (3-42)$$

which assumes that 3% of the total extragalactic population are significant infrared emitters. The effects resulting from red shift less than a factor of 2 for $E_\lambda(\text{lim}) > 10^{-20}$ W cm^{-2} μm^{-1}.

3.3.4. The Diffuse Background Radiance Caused by Extragalactic Sources. The background radiance, $L_{\lambda, G}$, caused by the integrated light from all extragalactic sources is isotropic. Taking into account the uncertainties of the basic quantities involved [3-22], background radiance is given by

$$L_{\lambda, G} = \frac{n_g c}{4 \pi H_0} \sum_{i=1}^{m} \Omega_i \Phi_{i, \lambda} \qquad (3\text{-}43)$$

where n_g = the number density of galaxies
H_0 = the Hubble constant*
c = the velocity of light
Ω_i = the fraction of galaxies with radiant flux $\Phi_{i, \lambda}$

Results of Equation (3-43) are given in Table 3-16.

Table 3-16. Extragalactic Background Radiance [3-17, 3-18, 3-21]

λ (μm)	3	5	10	15	20	25	30	
$L_{\lambda, G}$ (W cm^{-2}sr$^{-1}\mu$m^{-1})	11.0	7.2	3.8	2.6	2.0	1.7	1.4	$\times 10^{-14}$

3.3.5. Density of Asteroids. Asteroids are solid bodies orbiting the sun between the orbits of Mars and Jupiter. Their surface temperatures are normally in the range 180 to 280 K, with a weighted mean spectral temperature of about 255 K. A number of asteroids have been observed in the infrared [3-23]. Statistical treatment of these and other data available on about 1800 known asteroids [3-24, 3-17] leads to the relation

$$N_{\lambda, a} = 4.6 \times 10^{-19} \frac{L_{BB, \lambda}(225)}{L_{BB, 11}(225)} E_\lambda^{-1}(\text{lim}) \exp(-0.14 \beta_e) \qquad (3\text{-}44)$$

for the number of asteroids per square degree observed to a limiting irradiance $E_\lambda(\text{lim})$, in W cm^{-2} μm^{-1}, at ecliptic latitude β_e degrees. $N_{\lambda, a}$ is independent of ecliptic longitude. $L_{BB, \lambda}(225)$ is the Planck function evaluated at $T = 225$ K.

3.3.6. Zodiacal Radiance. Thermal emission from dust particles distributed throughout the solar system produces a significant diffuse background radiance in the 5 to 30 μm wavelength region. A number of models have been proposed to estimate the zodiacal radiance [3-25, 3-26]; none, however, consider details of the spectral properties of individual grains. The single measurement available [3-27] places upper limits to the radiance in the ecliptic plane ($\beta_e = 0°$) at 160 degrees from the sun of: 3×10^{-11}

---

*The Hubble Constant is 2.4×10^{-18} sec^{-1} (as of 1970) and has continually changed with new measurements.

W cm^{-2} sr^{-1} μm^{-1} in the region 5 to 6 μm; 6 × 10^{-11} in the region from 12 to 14 μm; and 2.5 × 10^{-11} in the region 16 to 23 μm. The radiance may be expected to decrease toward the ecliptic pole ($\beta_e = 90°$) and increase toward the sun.

3.4. The Sun

The sun is a G class star with a mean radius of about 695,000 km. The surface temperature is approximately 5900 K by best-fit blackbody curve, or about 5770 K for the temperature of a blackbody source which is the size and distance of the sun and which would produce an exoatmospheric total irradiance of 1353 W m^{-2} at the earth.

The surface is largely a heated plasma which should radiate very nearly as a blackbody. However, the cooler atomic gases of the solar atmosphere and the large temperature gradient below the surface, coupled with the nonisothermal character of transient surface features, lead to deviations from blackbody radiation.

Table 3-17 lists a proposed, standard solar spectral irradiance ($m = 0$) at the mean distance of the earth from the sun, where the mean distance is 149.68 × 10^6 km or 149.68 Gm. Figure 3-6 shows the spectral irradiance of exoatmospheric solar radiation, the comparable blackbody spectrum for a temperature of 5900 K, and the approximate solar spectral irradiance at sea level through one atmospheric air mass ($m = 1$).

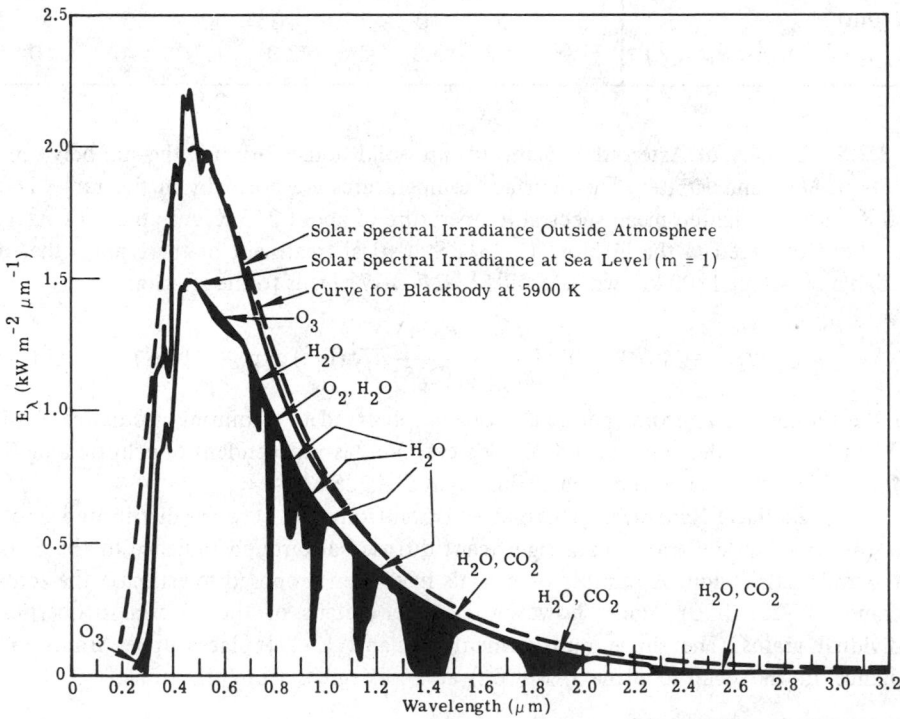

Fig. 3-6. Spectral distribution curves related to the sun. The shaded areas indicate absorption at sea level due to the atmospheric constituents shown.

Table 3-17. Solar Spectral Irradiance $(m = 0)$ — Proposed Standard Curve [3-28]

λ - Wavelength in micrometers
$E_\lambda (\lambda)$ - Solar spectral irradiance averaged over small bandwidth centered at λ, in W m^{-2} μm^{-1}
$D(0-\lambda)$ - Percentage of the solar constant associated with wavelengths shorter than λ
Solar Constant - 1353 W m^{-2}.

λ	$E_\lambda (\lambda)$	$D(0-\lambda)$	λ	$E_\lambda (\lambda)$	$D(0-\lambda)$	λ	$E_\lambda (\lambda)$	$D(0-\lambda)$
0.120	0.100	0.00044	0.375	1157.0	6.582	0.575	1719.0	32.541
0.140	0.030	0.00053	0.380	1120.0	7.003	0.580	1715.0	33.176
0.150	0.07	0.00057	0.385	1098.0	7.413	0.585	1712.0	33.809
0.160	0.23	0.00068	0.390	1098.0	7.819	0.590	1700.0	34.439
0.170	0.63	0.00100	0.395	1189.0	8.241	0.595	1682.0	35.064
0.180	1.25	0.00169	0.400	1429.0	8.725	0.600	1666.0	35.683
0.190	2.71	0.00316	0.405	1644.0	9.293	0.605	1647.0	36.295
0.200	10.7	0.00811	0.410	1751.0	9.920	0.610	1635.0	36.902
0.210	22.9	0.02053	0.415	1774.0	10.571	0.620	1602.0	38.098
0.220	57.5	0.05024	0.420	1747.0	11.222	0.630	1570.0	39.270
0.225	64.9	0.0728	0.425	1693.0	11.858	0.640	1544.0	40.421
0.230	66.7	0.0971	0.430	1639.0	12.473	0.650	1511.0	41.550
0.235	59.3	0.1204	0.435	1663.0	13.083	0.660	1486.0	42.657
0.240	63.0	0.1430	0.440	1810.0	13.725	0.670	1456.0	43.744
0.245	72.3	0.1680	0.445	1922.0	14.415	0.680	1427.0	44.810
0.250	70.4	0.1944	0.450	2006.0	15.140	0.690	1402.0	45.855
0.255	104.0	0.2266	0.455	2057.0	15.891	0.700	1369.0	46.879
0.260	130.0	0.2698	0.460	2066.0	16.653	0.710	1344.0	47.882
0.265	185.0	0.3280	0.465	2048.0	17.413	0.720	1314.0	48.864
0.270	232.0	0.4051	0.470	2033.0	18.167	0.730	1290.0	49.826
0.275	204.0	0.4857	0.475	2044.0	18.921	0.740	1260.0	50.769
0.280	222.0	0.5644	0.480	2074.0	19.681	0.750	1235.0	51.691
0.285	315.0	0.6636	0.485	1976.0	20.430	0.760	1211.0	52.595
0.290	482.0	0.8109	0.490	1950.0	31.155	0.770	1185.0	53.480
0.295	584.0	1.0078	0.495	1960.0	21.878	0.780	1159.0	54.346
0.300	514.0	1.2107	0.500	1942.0	22.599	0.790	1134.0	55.194
0.305	603.0	1.4171	0.505	1920.0	23.312	0.800	1109.0	56.023
0.310	689.0	1.6558	0.510	1882.0	24.015	0.810	1085.0	56.834
0.315	764.0	1.9243	0.515	1833.0	24.701	0.820	1060.0	57.627
0.320	830.0	2.2188	0.520	1833.0	25.379	0.830	1036.0	58.401
0.325	975.0	2.552	0.525	1852.0	26.059	0.840	1013.0	59.158
0.330	1059.0	2.928	0.530	1842.0	26.742	0.850	990.0	59.899
0.335	1081.0	3.323	0.535	1818.0	27.418	0.860	968.0	60.622
0.340	1074.0	3.721	0.540	1783.0	28.084	0.870	947.0	61.330
0.345	1069.0	4.117	0.545	1754.0	28.737	0.880	926.0	62.022
0.350	1093.0	4.517	0.550	1725.0	29.380	0.890	908.0	62.700
0.355	1083.0	4.919	0.555	1720.0	30.017	0.900	891.0	63.365
0.360	1068.0	5.316	0.560	1695.0	30.648	0.910	880.0	64.019
0.365	1132.0	5.723	0.565	1705.0	31.276	0.920	869.0	64.665
0.370	1181.0	6.150	0.570	1712.0	31.907	0.930	858.0	65.304

Table 3-17. Solar Spectral Irradiance—Proposed Standard Curve [3-28] *(Continued)*

λ   - Wavelength in micrometers
$E_\lambda(\lambda)$ - Solar spectral irradiance averaged over small bandwidth centered at λ, W m^{-1} μm^{-1}
$D(0-\lambda)$ - Percentage of the solar constant associated with wavelengths shorter than λ
Solar Constant - 1353 W m^{-2}

λ	$E_\lambda(\lambda)$	$D(0-\lambda)$	λ	$E_\lambda(\lambda)$	$D(0-\lambda)$	λ	$E_\lambda(\lambda)$	$D(0-\lambda)$
0.940	847.0	65.934	2.40	62.0	95.8580	9.0	0.380	99.913939
0.950	837.0	66.556	2.50	55.0	96.2903	10.0	0.250	99.937221
0.960	820.0	67.168	2.60	48.0	96.6710	11.0	0.170	99.952742
0.970	803.0	67.768	2.70	43.0	97.0073	12.0	0.120	99.963459
0.980	785.0	68.355	2.80	39.0	97.3103	13.0	0.087	99.971108
0.990	767.0	68.928	2.90	35.0	97.5838	14.0	0.055	99.976356
1.000	748.0	69.488	3.00	31.0	97.8277	15.0	0.049	99.980199
1.050	668.0	72.105	3.10	26.0	98.0383	16.0	0.038	99.983414
1.100	593.0	74.435	3.20	22.6	98.2179	17.0	0.031	99.985964
1.150	535.0	76.519	3.30	19.2	98.3724	18.0	0.024	99.987997
1.200	485.0	78.404	3.40	16.6	98.5047	19.0	0.02000	99.989623
1.250	438.0	80.109	3.50	14.6	98.6200	20.0	0.01600	99.990953
1.300	397.0	81.652	3.60	13.5	98.7238	25.0	0.00610	99.995037
1.350	358.0	83.047	3.70	12.3	98.8192	30.0	0.00300	99.996718
1.400	337.0	84.331	3.80	11.1	98.9056	35.0	0.00160	99.997568
1.450	312.0	85.530	3.90	10.3	98.9847	40.0	0.00094	99.998037
1.500	288.0	86.639	4.00	9.5	99.0579	50.0	0.00038	99.998525
1.550	267.0	87.665	4.10	8.7	99.1252	60.0	0.00019	99.998736
1.600	245.0	88.611	4.20	7.8	99.1861	80.0	0.00007	99.998928
1.650	223.0	89.475	4.30	7.1	99.2412	100.0	0.00003	99.999002
1.70	202.0	90.261	4.40	6.5	99.291507	1000.0	0.00000	100.000000
1.75	180.0	90.967	4.50	5.9	99.337331	–	–	–
1.80	159.0	91.593	4.60	5.3	99.378721	–	–	–
1.85	142.0	92.149	4.70	4.8	99.416045	–	–	–
1.90	126.0	92.644	4.80	4.5	99.450413	–	–	–
1.95	114.0	93.088	4.90	4.1	99.482195	–	–	–
2.00	103.0	93.489	5.00	3.83	99.511500	–	–	–
2.10	90.0	94.202	6.00	1.75	99.717708	–	–	–
2.20	79.0	94.826	7.00	0.99	99.818965	–	–	–
2.30	69.0	95.373	8.00	0.60	99.877723	–	–	–

Table 3-18 shows the spectral irradiance at sea level for solar radiation after passing through two atmospheric air masses (m_a = 2) or a zenith sun angle of 60°. For these data, the total exoatmospheric solar irradiance, E_0, is taken to be 1322 W m^{-2}.

Figures 3-7 and 3-8 show the spectral variation of spectral irradiance, caused by the penetration of solar radiation through various air masses. Figure 3-9 shows the global and direct solar spectral irradiance on a horizontal surface for various slant paths.

Table 3-18. Solar Irradiance at Sea Level on an Area Normal to the Sun for m = 2,
E_0 = 1322 W m^{-2} [3-29]

λ (μm)	$E_\lambda(\lambda)$ (W m^{-2}μm^{-1})	λ (μm)	$E_\lambda(\lambda)$ (W m^{-2} μm^{-1})	λ (μm)	$E_\lambda(\lambda)$ (W m^{-2} μm^{-1})	λ (μm)	$E_\lambda(\lambda)$ (W m^{-2} μm^{-1})
0.301	0.177	0.510	1206.0	0.91	375.0	1.31	203.0
0.302	0.342	0.520	1199.0	0.92	258.0	1.32	168.0
0.303	0.647	0.530	1188.0	0.93	169.0	1.33	115.0
0.304	1.16	0.540	1198.0	0.94	278.0	1.34	58.1
0.305	1.91	0.550	1190.0	0.95	487.0	1.35	18.1
0.306	2.89	0.560	1182.0	0.96	584.0	1.36	0.660
0.307	4.15	0.570	1178.0	0.97	633.0	1.37	–
0.308	6.11	0.580	1168.0	0.98	645.0	1.38	–
0.309	8.38	0.590	1161.0	0.99	643.0	1.39	–
0.310	11.0	0.600	1167.0	1.00	630.0	1.40	–
0.311	13.9	0.61	1168.0	1.01	620.0	1.41	1.91
0.312	17.2	0.62	1165.0	1.02	610.0	1.42	3.72
0.313	21.0	0.63	1176.0	1.03	601.0	1.43	7.53
0.314	25.4	0.64	1175.0	1.04	592.0	1.44	13.7
0.315	30.0	0.65	1173.0	1.05	551.0	1.45	23.8
0.316	34.8	0.66	1166.0	1.06	526.0	1.46	30.5
0.317	39.8	0.67	1160.0	1.07	519.0	1.47	45.1
0.318	44.9	0.68	1149.0	1.08	512.0	1.48	83.7
0.319	49.5	0.69	978.0	1.09	514.0	1.49	128.0
0.320	54.0	0.70	1108.0	1.10	252.0	1.50	157.0
–	–	0.71	1070.0	1.11	126.0	1.51	187.0
–	–	0.72	832.0	1.12	69.9	1.52	209.0
0.330	101.0	0.73	965.0	1.13	98.3	1.53	217.0
0.340	151.0	0.74	1041.0	1.14	164.0	1.54	226.0
0.350	188.0	0.75	867.0	1.15	216.0	1.55	221.0
0.360	238.0	0.76	566.0	1.16	271.0	1.56	217.0
0.370	279.0	0.77	968.0	1.17	328.0	1.57	213.0
0.380	336.0	0.78	907.0	1.18	346.0	1.58	209.0
0.390	397.0	0.79	923.0	1.19	344.0	1.59	205.0
0.400	470.0	0.80	857.0	1.20	373.0	1.60	202.0
0.410	672.0	0.81	698.0	1.21	402.0	1.61	198.0
0.420	733.0	0.82	801.0	1.22	431.0	1.62	194.0
0.430	787.0	0.83	863.0	1.23	420.0	1.63	189.0
0.440	911.0	0.84	858.0	1.24	387.0	1.64	184.0
0.450	1006.0	0.85	839.0	1.25	328.0	1.65	173.0
0.460	1080.0	0.86	813.0	1.26	311.0	1.66	163.0
0.470	1138.0	0.87	798.0	1.27	331.0	1.67	159.0
0.480	1183.0	0.88	614.0	1.28	382.0	1.68	145.0
0.490	1210.0	0.89	517.0	1.29	346.0	1.69	139.0
0.500	1215.0	0.90	480.0	1.30	264.0	1.70	132.0

Table 3-18. Solar Irradiance at Sea Level on an Area Normal to the Sun for m = 2,
$E_0 = 1322$ W m^{-2} [3-29] *(Continued)*

λ (μm)	$E_\lambda(\lambda)$ (W m$^{-2}\mu$m^{-1})	λ (μm)	$E_\lambda(\lambda)$ (W m^{-2} μm^{-1})	λ (μm)	$E_\lambda(\lambda)$ (W m^{-2} μm^{-1})	λ (μm)	$E_\lambda(\lambda)$ (W m^{-2} μm^{-1})
1.71	124.0	1.82	–	1.93	3.68	2.04	54.70
1.72	115.0	1.83	–	1.94	5.30	2.05	38.3
1.73	105.0	1.84	–	1.95	17.7	2.06	56.2
1.74	97.1	1.85	–	1.96	31.7	2.07	77.0
1.75	80.2	1.86	–	1.97	37.7	2.08	88.0
1.76	58.9	1.87	–	1.98	22.6	2.09	86.8
1.77	38.8	1.88	–	1.99	1.58	2.10	85.6
1.78	18.4	1.89	–	2.00	2.66	2.11	84.4
1.79	5.70	1.90	–	2.01	19.50	2.12	83.2
1.80	0.920	1.91	0.705	2.02	47.60	2.13	20.7
1.81	–	1.92	2.34	2.03	55.40	2.14	–

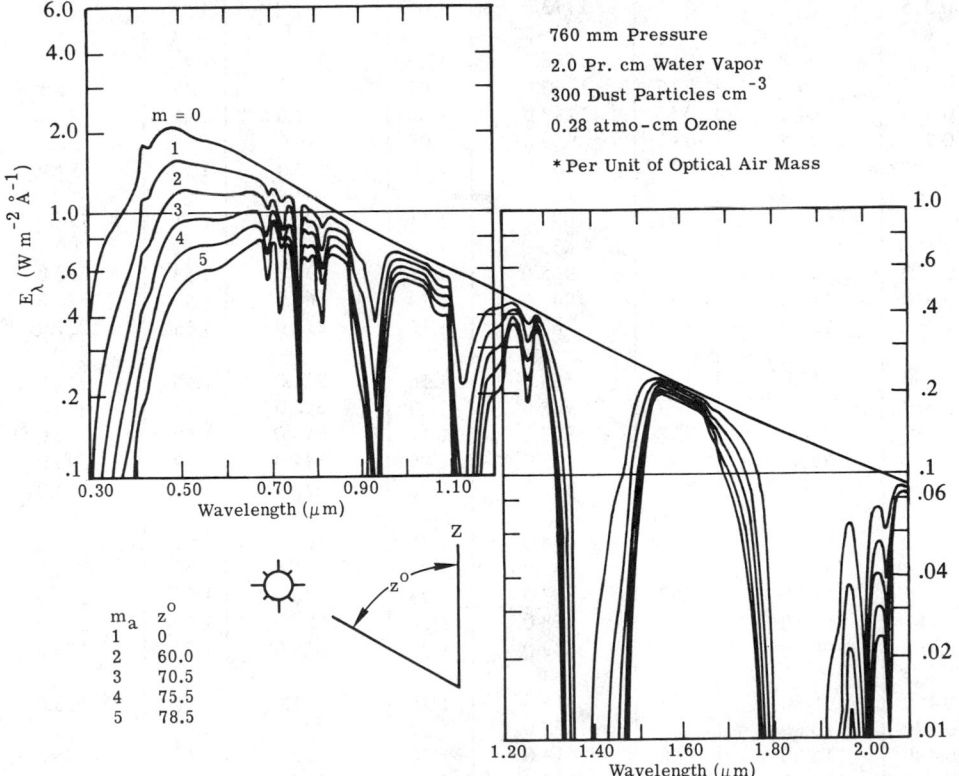

760 mm Pressure
2.0 Pr. cm Water Vapor
300 Dust Particles cm^{-3}
0.28 atmo-cm Ozone

*Per Unit of Optical Air Mass

m_a	z^0
1	0
2	60.0
3	70.5
4	75.5
5	78.5

Fig. 3-7. Solar spectral irradiance curves at sea level for various optical air masses. The value of the solar constant used in this calculation was 1322 W m^{-2} [3-29].

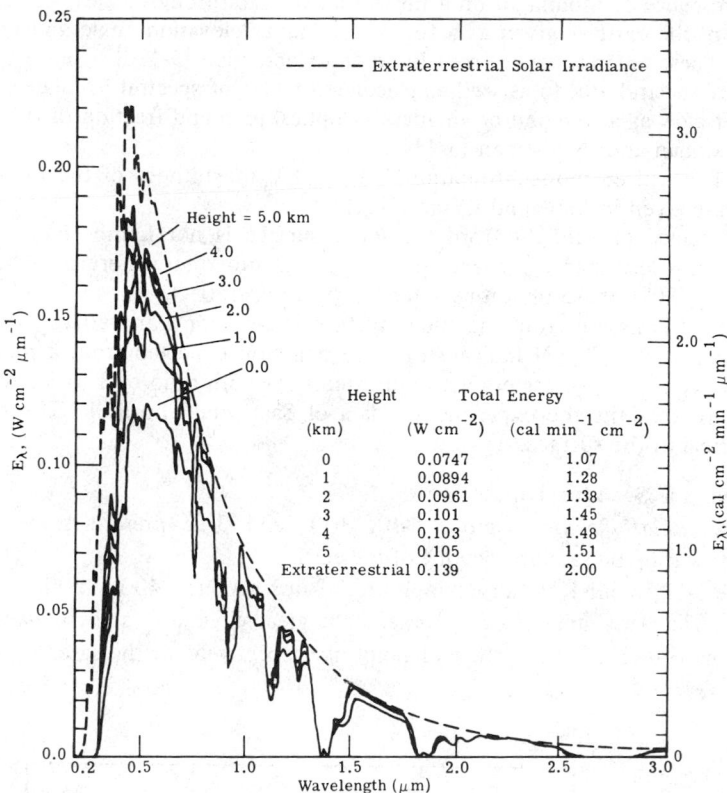

Fig. 3-8. Variation with altitude of spectral irradiance versus wavelength for direct perpendicular to the sun's rays for an air mass of 1.5 [3-2].

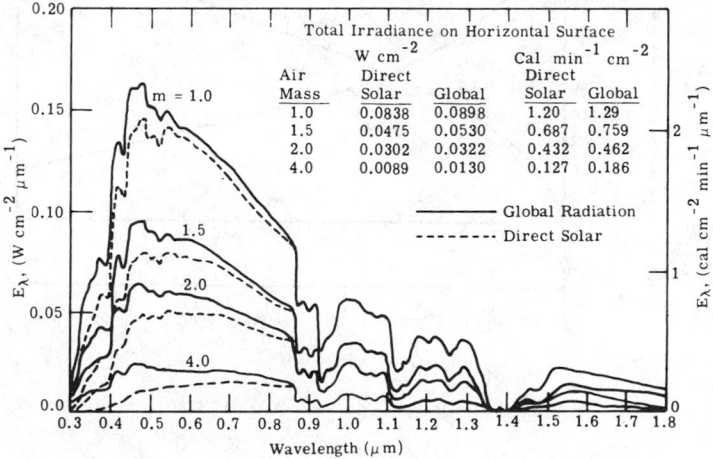

Fig. 3-9. Spectral distribution as a function of wavelength of the global and direct solar radiation incident at sea level on a horizontal surface for various slant paths corresponding to air mass 1.0, 1.5, 2.0, and 4.0 [3-2].

The illuminance of moonlight on a horizontal surface through a clear atmosphere on the surface of the earth is given as a function of lunar elevation angle and phase angle in [3-109]. These data are taken from Bond and Henderson [3-110].

The lunar spectral albedo as well as calculated tables of spectral irradiance of moonlight plus air glow as a function of air mass of optical path and fraction of full moon are given in Biberman and Nudleman [3-111].

Tables of measured monochromatic, U, B, and V magnitudes of the moon in various phases are given in Lane and Irvine [3-112].

The reflectance of sunlight from the lunar surface is not Lambertian. The lunar phase function is less than the phase function of a Lambertian sphere by about a factor of 2 at quarter (90°) phase indicating a tendency to retroreflect.

The thermal emission from the lunar surface is also non-Lambertian. Calculations given in Sexl, et al. [3-113] indicate that the emission expected from a model of the lunar surface consisting of craters with a diameter-to-depth ratio of 3 to 1 covering 50% of the surface area fits the experimental data of Saari and Shorthill. The radiance is greatest normal to the surface.

3.5. The Earth as Seen from Space

Geometric Relationships. Figures 3-10, 3-11, and 3-12 present some important relationships bearing on satellite viewing of the earth.

In Figure 3-12, point P (as an example) represents a vehicle 400 nmi high at an elevation of 35°. The slant range is 655 nmi, and the great circle arc (angle between vectors located at the center of the earth and pointing respectively to the satellite and to the ground point viewed) is 7°.

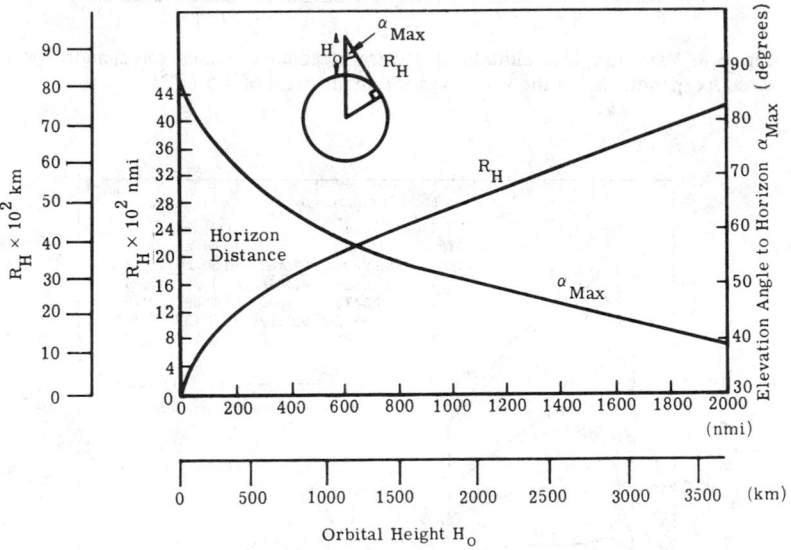

Fig. 3-10. Ranges and View Angles [3-30].

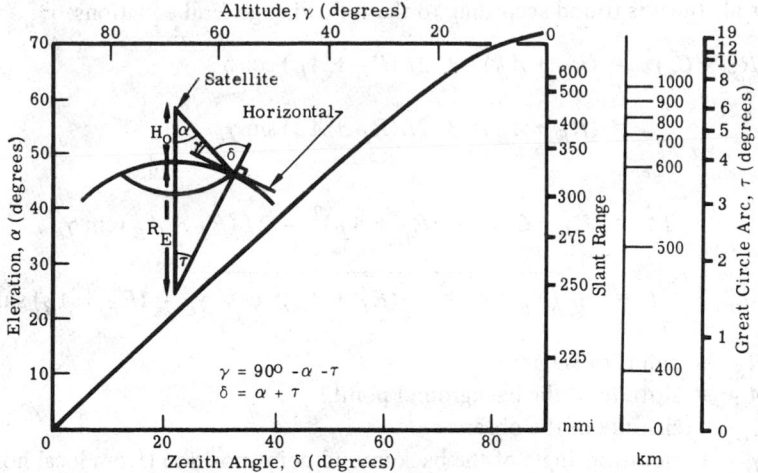

Fig. 3-11. View angles for 200 nmi orbit [3-30].

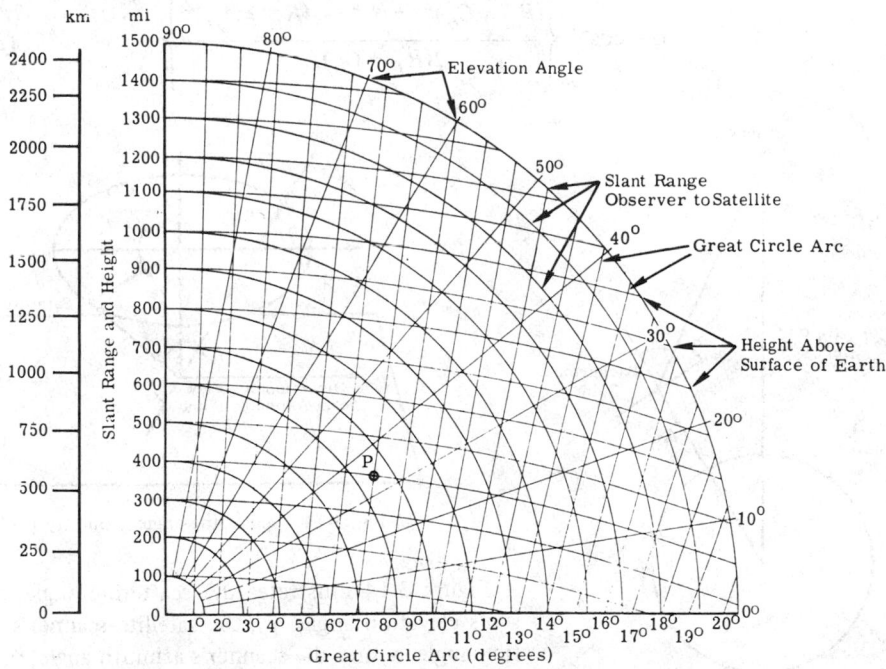

Fig. 3-12. Satellite coordinate conversion [3-30].

Path Lengths (See Figure 3-13.). The length, L, of a line between any two points at different altitudes is found according to the following general equations:

$$(R_E + C_o)^2 = (R_E + A_B)^2 + 2L(R_E + A_B) \cos \delta$$

$$= (R_E + A_B)^2 + 2L(R_E + A_B) \sin \gamma_e$$

(3-45)

$$L^2 = (R_E + C_o)^2 - (R_E + A_B)^2 - 2L(R_E + A_B) \sin \gamma_e$$

$$L = \sqrt{(R_E + C_o)^2 - (R_E + A_B)^2 \cos^2 \gamma_e} - (R_E + A_B) \sin \gamma_e$$

where R_E = radius of sphere
     A_B = altitude of the background point
     C_o = altitude of the observer
     γ_e = elevation angle of the background point position (from local horizontal)
     α = elevation angle from nadir at observer's point

The angle α is computed by the following relationship:

$$\alpha = \cos^{-1} \left\{ \frac{(R_E + C_o)^2 + L^2 - (R_E + A_B)^2}{2(R_E + C_o)L} \right\}$$

(3-46)

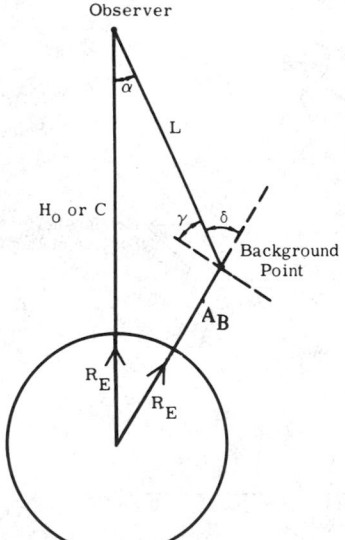

Fig. 3-13. Path length diagram.

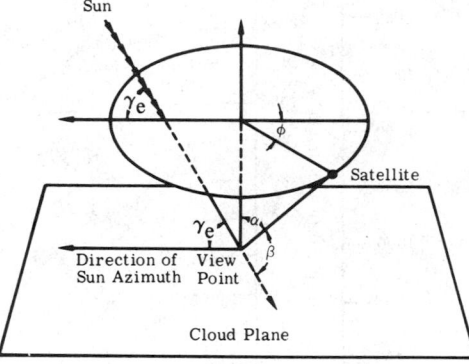

Fig. 3-14. Scattering-angle geometry [3-30].

Figure 3-14 illustrates the scattering angle, β, the sun's elevation angle, γ_e, the satellite scanner's elevation angle, α, and the scanner's azimuth angle, ϕ, from the direction of the sun. The scattering angle is:

$$\beta = \cos^{-1} (\cos\gamma \sin\alpha \cos\phi - \sin\gamma \cos\alpha) \qquad (3\text{-}47)$$

Figure 3-15 illustrates these angles for a spherical earth. Equation (3-47) becomes

$$\beta = \cos^{-1} [\cos (90° - (\xi - \Delta)) \sin\alpha \cos\phi - \sin [90° - (\xi - \Delta)]] \qquad (3\text{-}48)$$

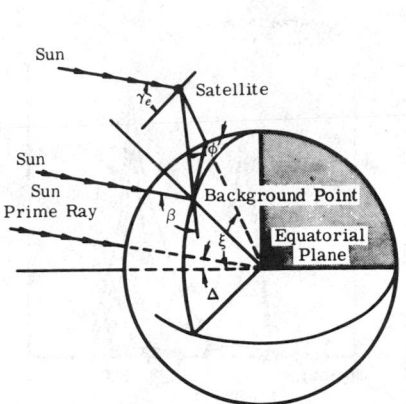

Sun's Position	Sun's Altitude at Noon
Equinox	$(90^\circ$ - Lat.$)$
Summer Solstice	$(90^\circ$ - Lat.$) + 23\ 1/2^\circ$
Winter Solstice	$(90^\circ$ - Lat.$) - 23\ 1/2^\circ$

Fig. 3-15. Solar scattering angle [3-30].   Fig. 3-16. Solar declination to the equator [3-30].

where ξ = the elevation angle (latitude) of the observed point from the earth's equatorial plane

Δ = the latitude of the sun's prime ray

Figure 3-16 shows the range of the sun's declination angle. (Both ξ and Δ are taken positive for a northern latitude.)

3.5.1. Albedo

Spectra in the Scattering Region. The daytime spectra of objects at ambient temperatures show minima around 3 to 4 μm. In the more transparent regions of the spectrum, between 3 and 5 μm in the daytime, the sky radiates less a few degrees above the horizon than the ground a few degrees below the horizon. In the scattering region of the spectrum, the sky and the ground often show radiances of comparable magnitude. Usually the ground radiance is somewhat higher than that of the sky and it is frequently at a minimum near the horizon.

Figures 3-17 and 3-18 are elevation scans of the spectral radiance near the horizon at different wavelengths.

The scan covers alternating patches of shaded and sunlit ground, trees, mountains (the northern slope of Pike's Peak in Colorado), as well as the clear sky up to 15 to 20° near the horizon.

Average spectral reflectances of common terrains are shown in Figure 3-19 [3-32]. Figures 3-20, 3-21, and 3-22 provide information on spectral reflectance, luminance, and variations in reflectance respectively. Figures 3-23 through 3-26 show the upwelling, reflected, spectral radiance for an exoatmospheric observer as calculated with Turner's atmospheric scattering model [3-33]. (See Chapter 4.)

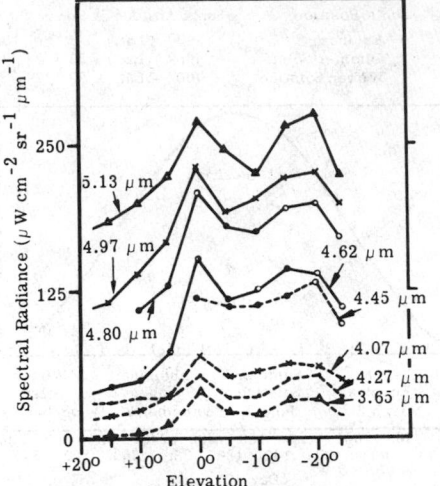

Fig. 3-17. Elevation scans across mountainous terrain at fixed wavelengths in the interval 1.8 to 3.2 μm [3-31].

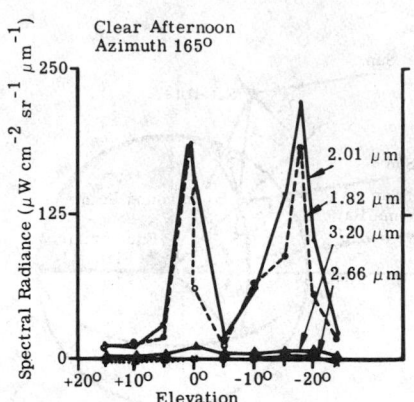

Fig. 3-18. Elevation scans across terrain at the same time as in Figure 3-17 at fixed wavelengths in the intervals 3.5 to 5.2 μm [3-31].

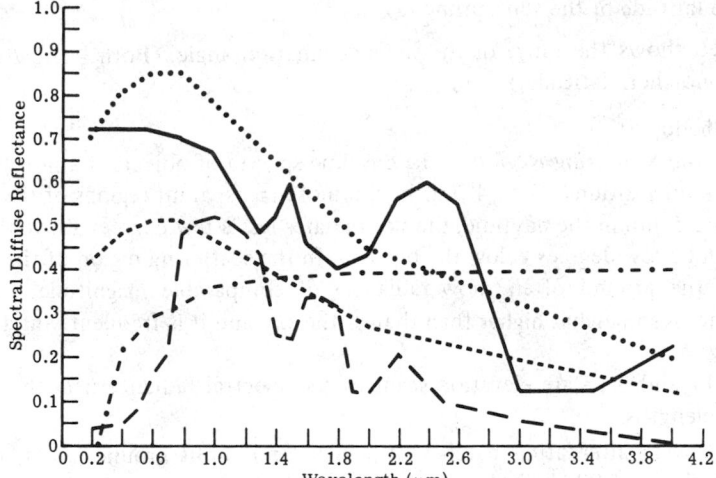

———————Clouds. Data are directional reflectance of a middle layer cloud.

• • • • • • • • Winter Snow and Ice. Data are directional reflectance of dry snow.

• • • • • • • • • • • • Summer Ice. Data are directional reflectance of summer Arctic ice.

• — • — • — Soil and Rocks. Data represent the average value of the bidirectional reflectance, ρ_λ (45°, 0, 0, 0), of gravel, wet clay, dry clay, tuff bedrock, and sandy loam.

— — — Vegetation. Data represent the average value of directional reflectance of many types of vegetation (from the ERIM Data File).

Fig. 3-19. Spectral diffuse reflectance of earth-atmosphere constituents [3-34, 3-35, 3-36, 3-37, 3-32].

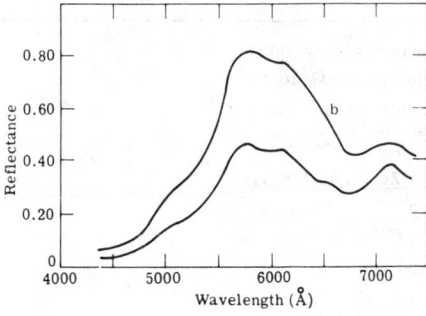

Fig. 3-20. Spectral reflectance, measured at 90,000 ft. (27.5 km) altitude above summertime cultivated farmland at 50° solar zenith angle. Curve b shows the effect of the intrusion of 0.6 cumular cloud structure [3-32].

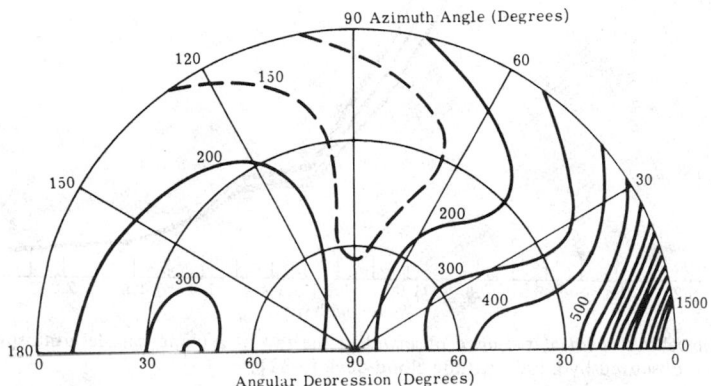

Fig. 3-21. Luminance (cd ft^{-2}) over the downward hemisphere. It is measured with balloon-borne equipment over mountains area in early autumn at an altitude of 98,000 ft (29.8 km) in a relatively turbid atmosphere. (For results in cd m^{-2}, multiply cd ft^{-2} by 10.764.) [3-38].

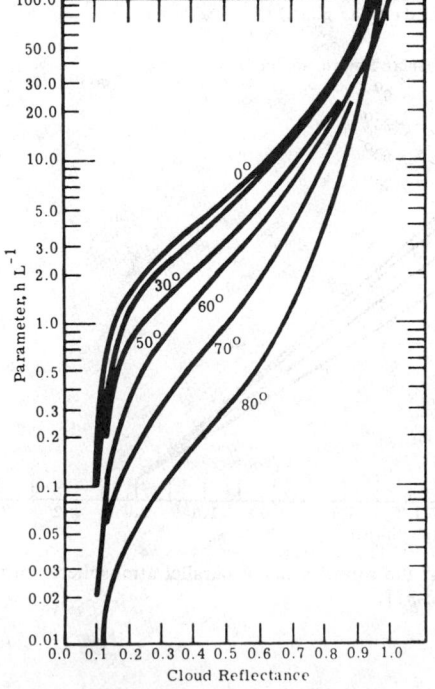

Fig. 3-22. The variation in reflectance as a function of cloud thickness parameter h L^{-1} for different solar zenith angles. The cloud thickness is h and the mean free path of a light ray is L [3-38].

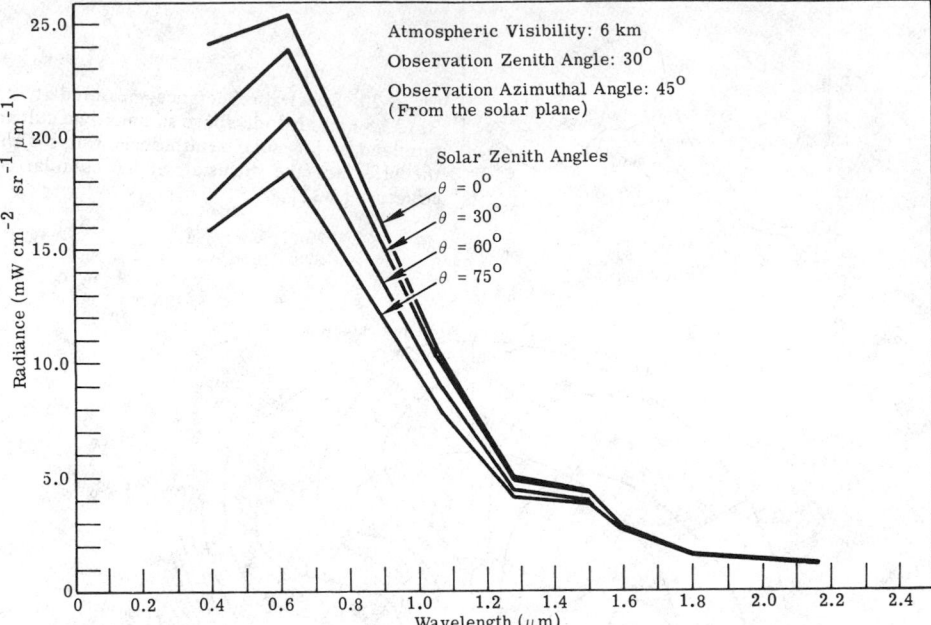

Fig. 3-23. Upwelling spectral radiance observed at the top of a plane parallel atmosphere with the earth's surface obscured by a low-altitude cloud deck [3-32].

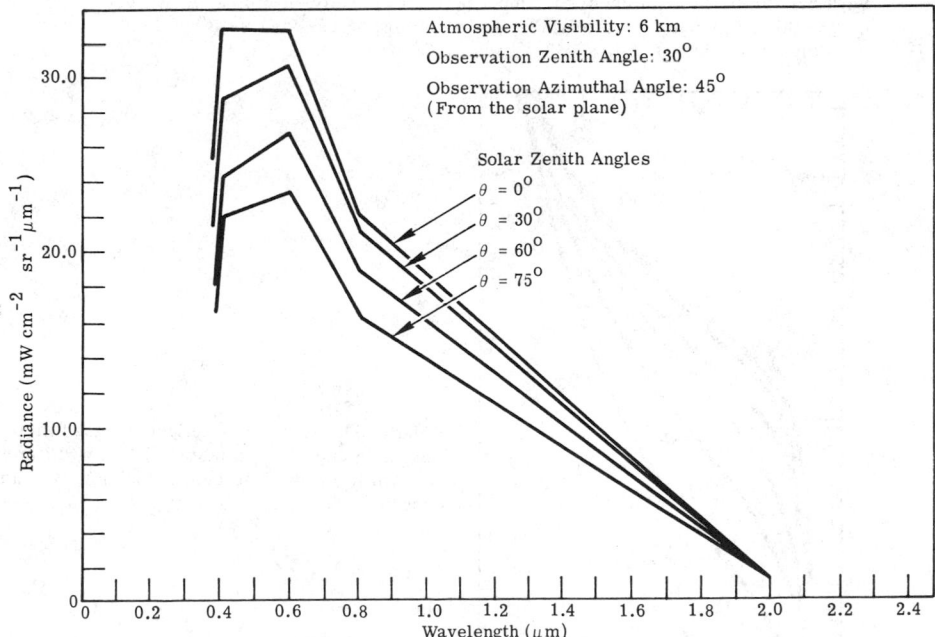

Fig. 3-24. Upwelling spectral radiance observed at the top of a plane parallel atmosphere with the earth's surface composed of winter snow and ice [3-32].

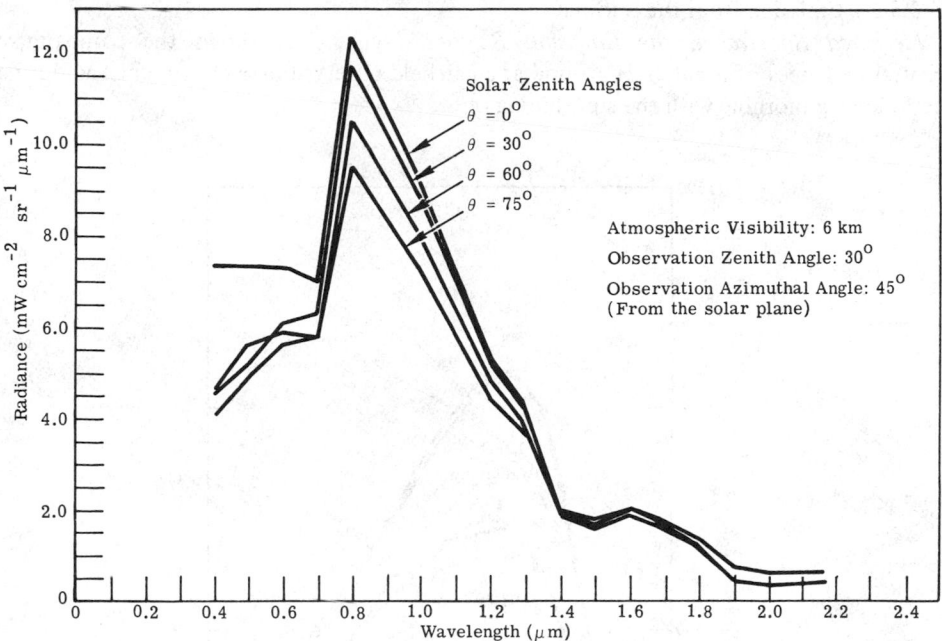

Fig. 3-25. Upwelling spectral radiance observed at the top of a plane parallel atmosphere with the earth's surface composed of vegetation [3-32].

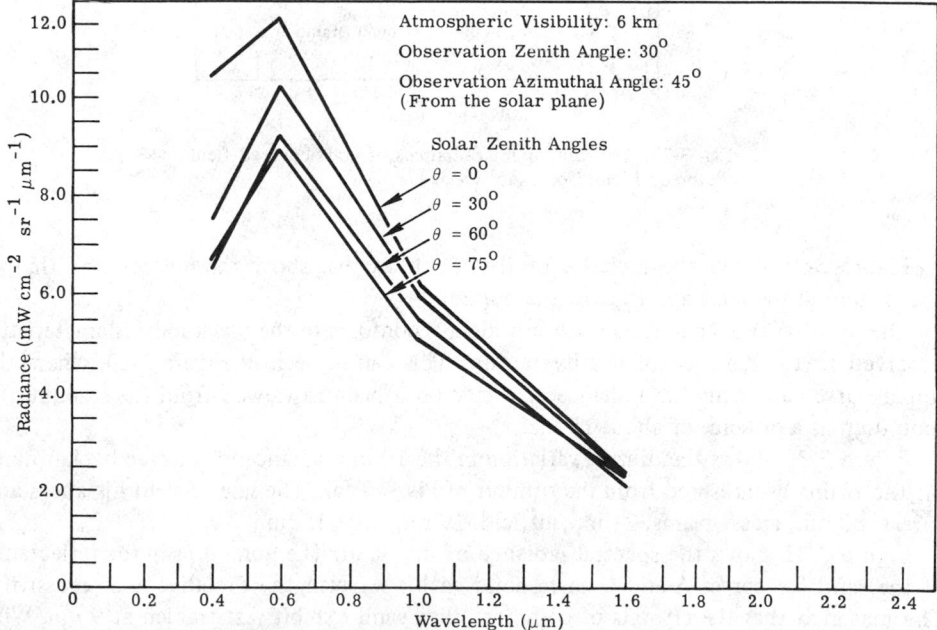

Fig. 3-26. Upwelling spectral radiance observed at the top of a plane parallel atmosphere with the earth's surface composed of soil and rocks [3-32].

3.5.2. Emission from the Surface

Measured Spectra in the Emission Region. Figure 3-27 shows the comparative spectral radiances of a patch of ground at an airfield observed on a clear night and during the following morning with the sun shining on it.

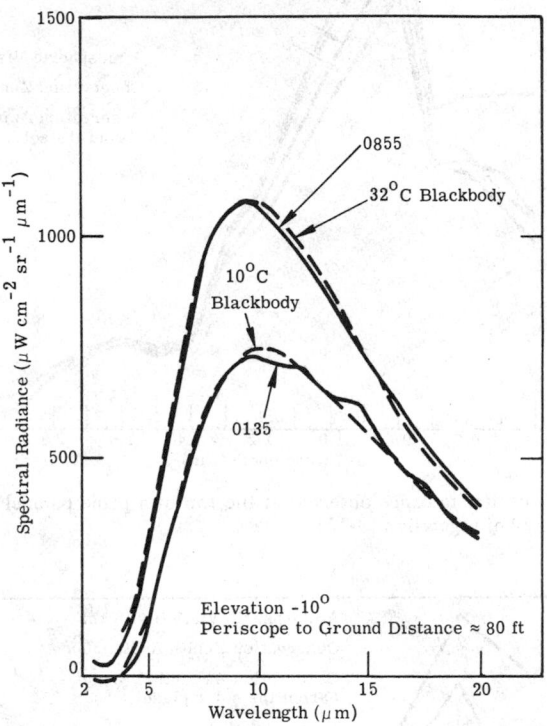

Fig. 3-27. Day and night radiances of grass-covered field (Peterson Field, Colorado) [3-31].

Figure 3-28 shows the radiance of the night sky just above the horizon and that of the ground at the same angle below the horizon.

The spectra of distant terrain do not always conform to the blackbody characteristics observed in the radiance of nearby terrain. This can be seen in Figure 3-29, where the upper curve represents the radiance of a city on a plain as viewed from the summit of a mountain at a distance of about 15 mi.

Figure 3-30 shows the diurnal variation in the 10 μm radiance of selected backgrounds on the plains as measured from the summit of Pike's Peak. The line-of-sight distances are: forest, 30 mi.; grassy plains, 21 mi.; airfield, 19 mi.; city, 15 mi.

Figure 3-31 shows the spectral radiance of dry sand. The 9 μm dips in the reflectance of the sand for curves A and C correspond with a wavelength of relative poor emissivity. The reason is that the crystals of common silica sand exhibit reststrahlen at 9 μm. With overcast sky (curve B), the added sky radiance reflected at this wavelength just compensates for the loss of emissivity.

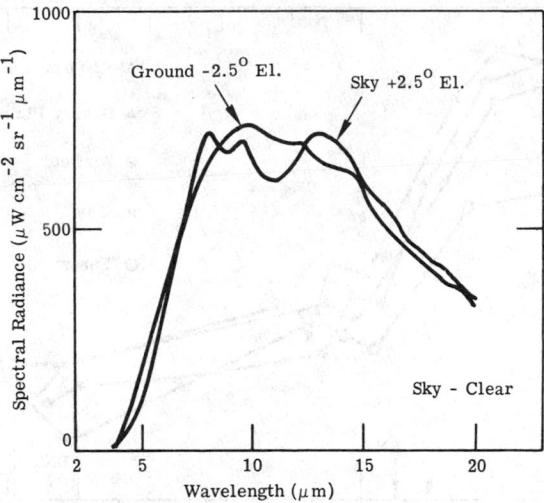

Fig. 3-28. Comparative spectra of the ground and sky near the horizon (Peterson Field, Colorado) [3-31].

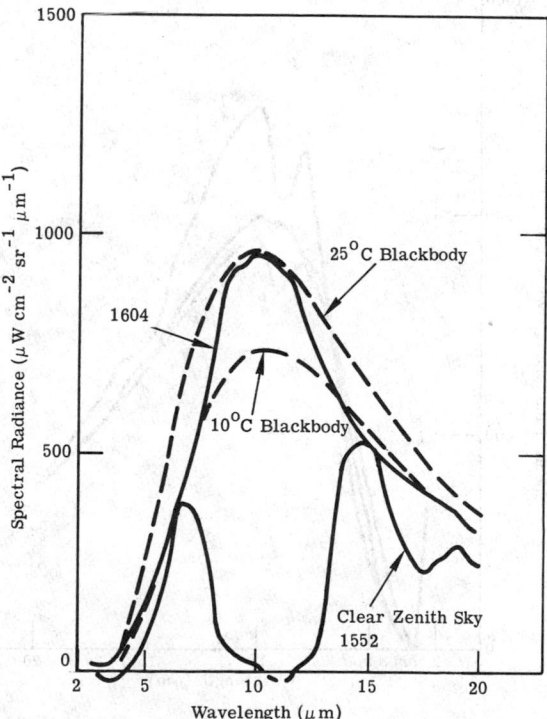

Fig. 3-29. Radiance of an urban area and of clear zenith sky (Colorado Springs from Pikes Peak) [3-31].

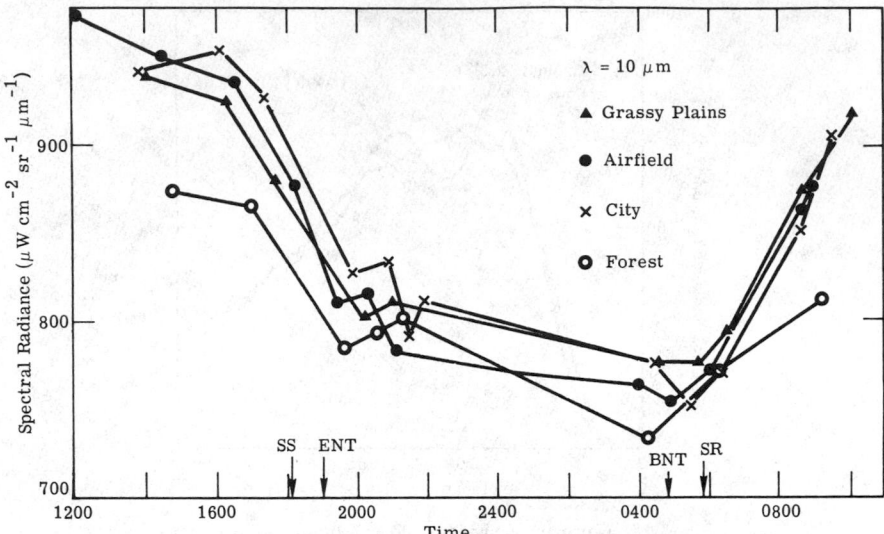

Fig. 3-30. Diurnal variation in the 10 μm Radiance of selected backgrounds. SS = sunset; SR = sunrise; ENT = end of nautical twilight; BNT = beginning of nautical twilight [3-31].

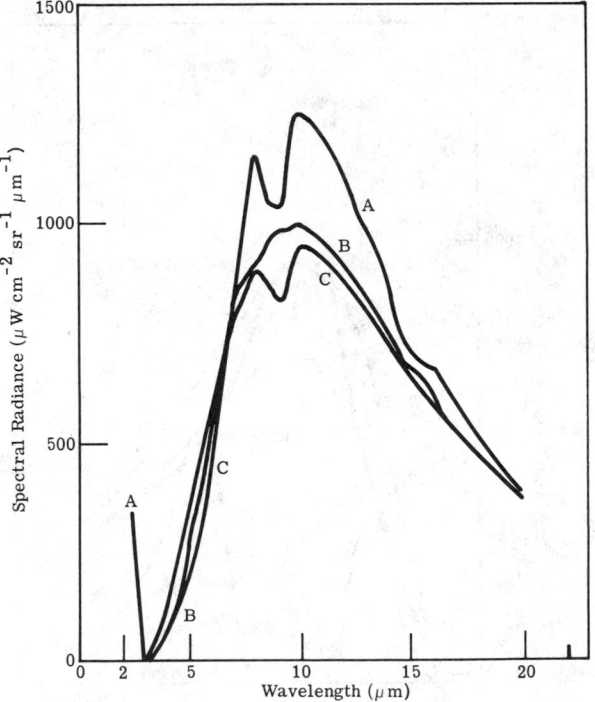

Fig. 3-31. Spectral radiance of dry sand (Cocoa Beach, Florida). A = sunlit sand; B = sand under a cloudy night sky; C = sand on a clear night [3-31].

The effect of moisture on the radiance of sand is shown in Figure 3-32.

Data on the infrared spectral radiance of the night horizon and sky are reported in Reference [3-39]. The data, accumulated by sounding rockets, show that the 4.3 μm night sky is measurably bright at altitudes up to 50 km. Definite horizon limb brightening, at this wavelength of strong carbon dioxide emission–absorption, was observed. A peak limb spectral radiance of 3.2×10^{-5} W cm^{-2} sr^{-1} μm^{-1} at 4.3 μm was measured.

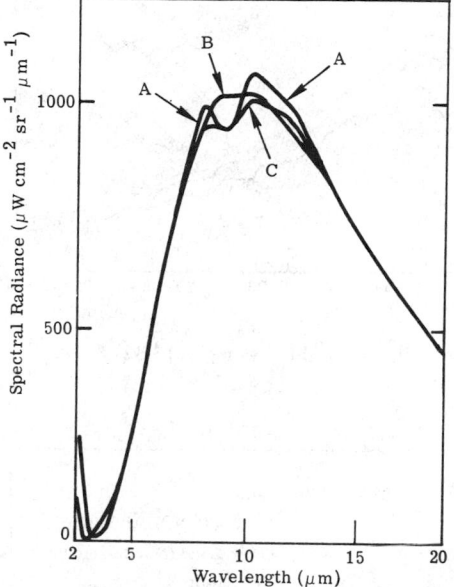

Fig. 3-32. Spectral radiance of moist sand (Cocoa Beach, Florida). A = dry sand; B = extremely wet sand; C = moist sand [3-31].

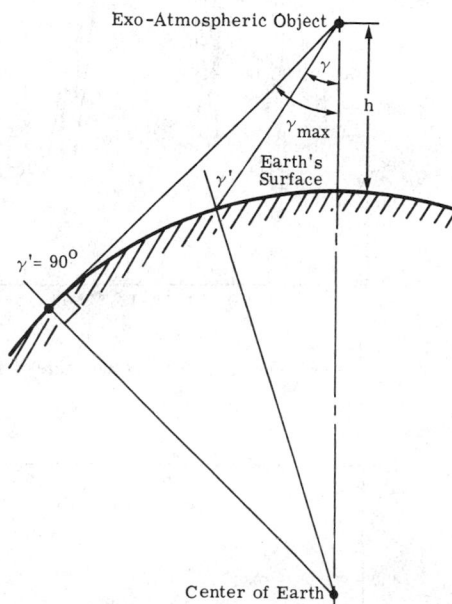

Fig. 3-33. Cross-section view of the geometry pertaining to the calculation of the thermal radiance field at an exoatmospheric object.

Calculated Spectra in the Emission Region. Figure 3-33 shows the cross-section view of the geometry used in calculating thermal spectra from the surface by modeling. Figure 3-34 illustrates the meaning of the terms limb darkening and limb brightening. Figures 3-35 to 3-42 show calculated spectral radiances to be expected by an exoatmospheric observer over various portions of North America for different seasons of the year. The prominent absorption band between 9 and 10 μm is due to ozone.

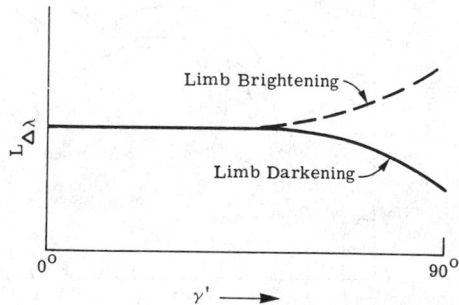

Fig. 3-34. Typical plots of upwelling radiance versus nadir angle.

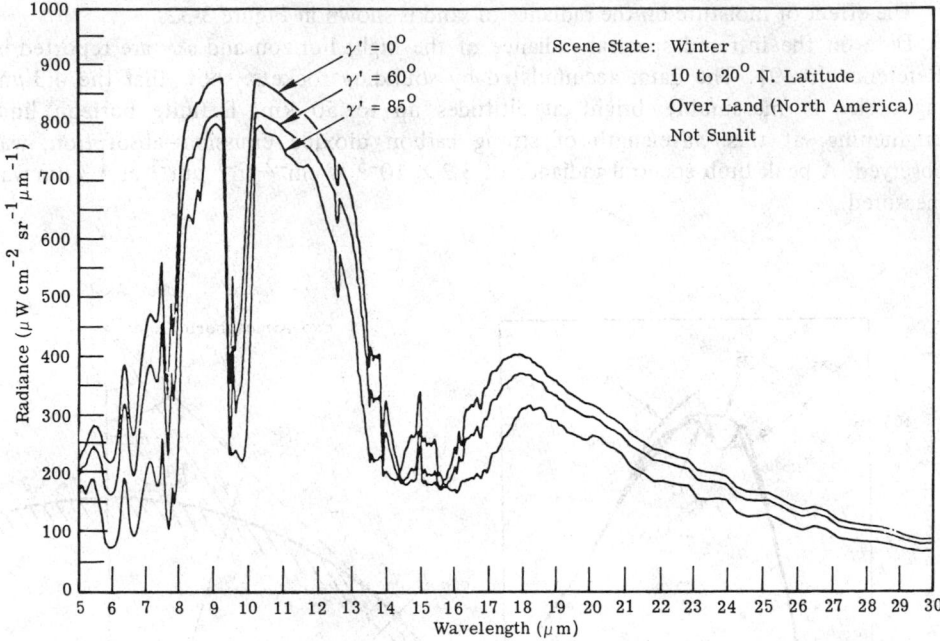

Fig. 3-35. The expected value of spectral radiance for three γ′ angles [3-32].

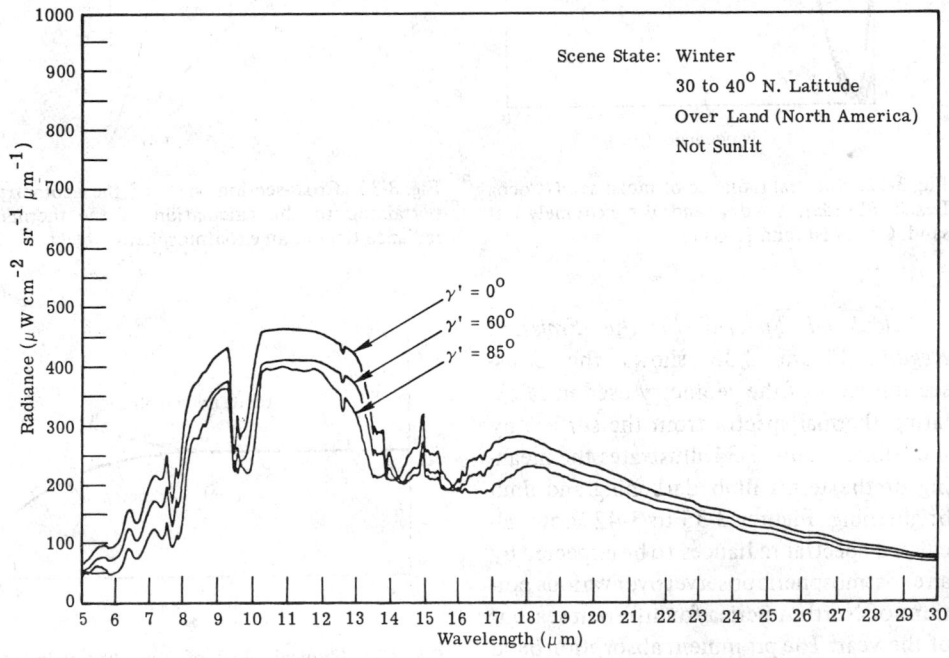

Fig. 3-36. The expected value of spectral radiance for three γ′ angles [3-32].

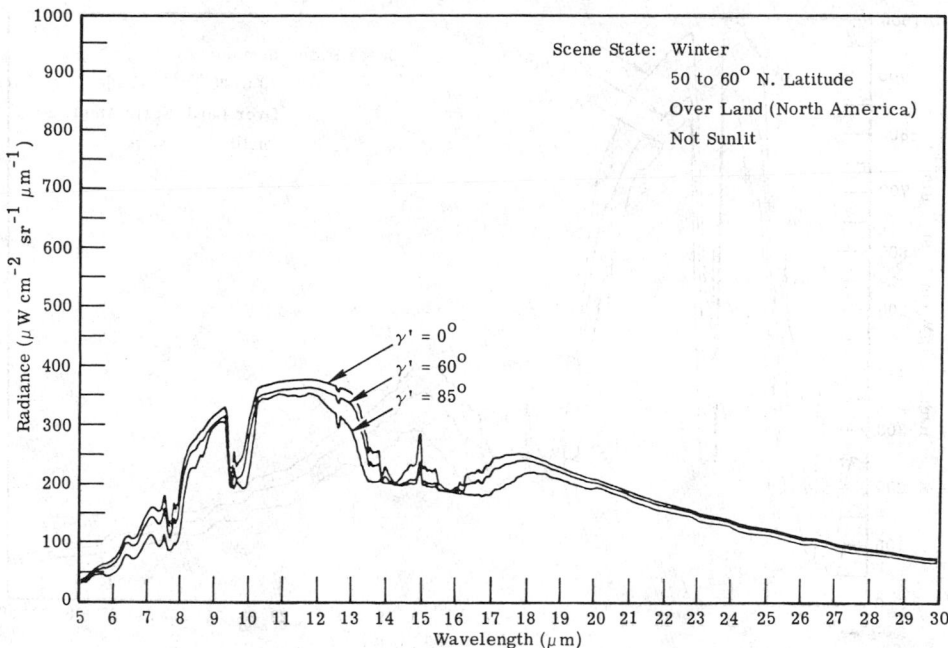

Fig. 3-37. The expected value of spectral radiance for three γ' angles [3-32].

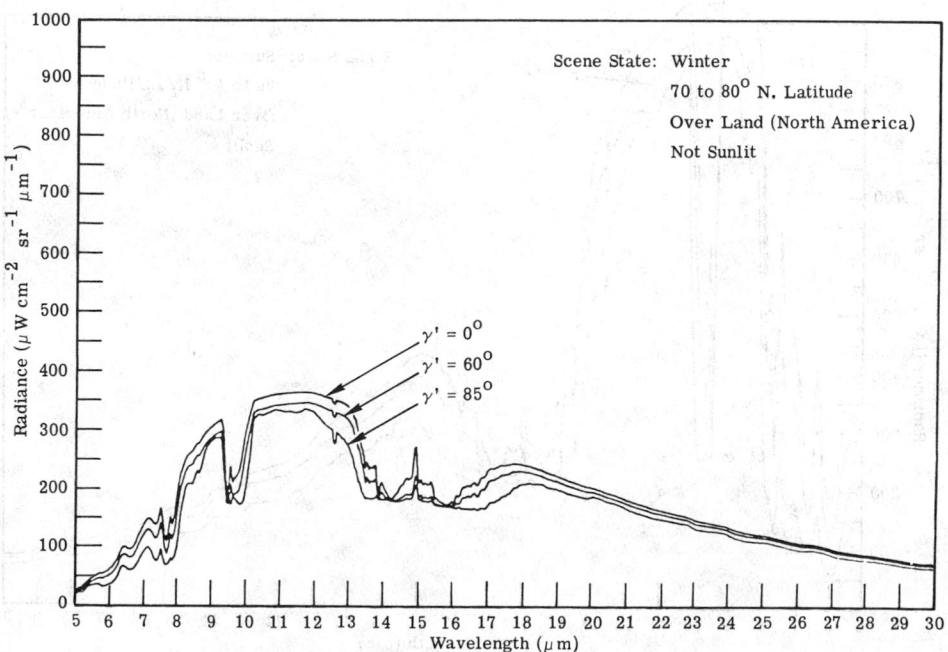

Fig. 3-38. The expected value of spectral radiance for three γ' angles [3-32].

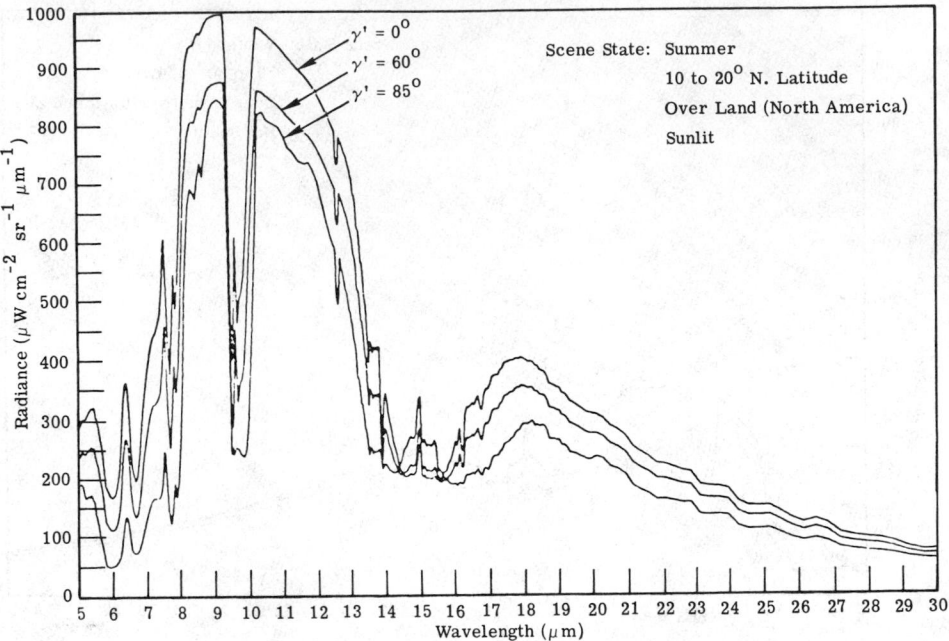

Fig. 3-39. The expected value of spectral radiance for three γ' angles [3-32].

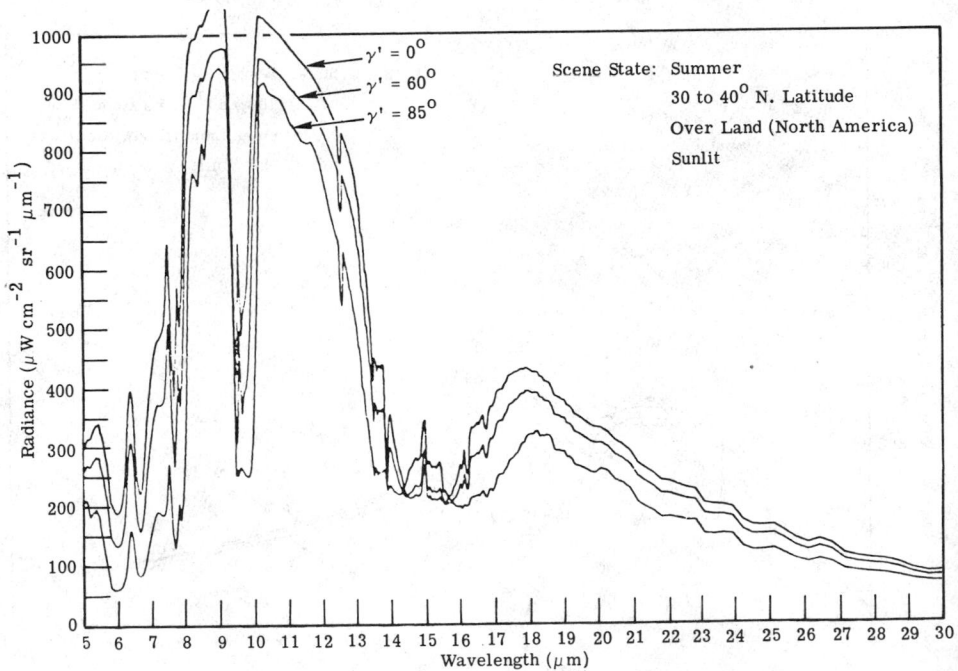

Fig. 3-40. The expected value of spectral radiance for three γ' angles [3-32].

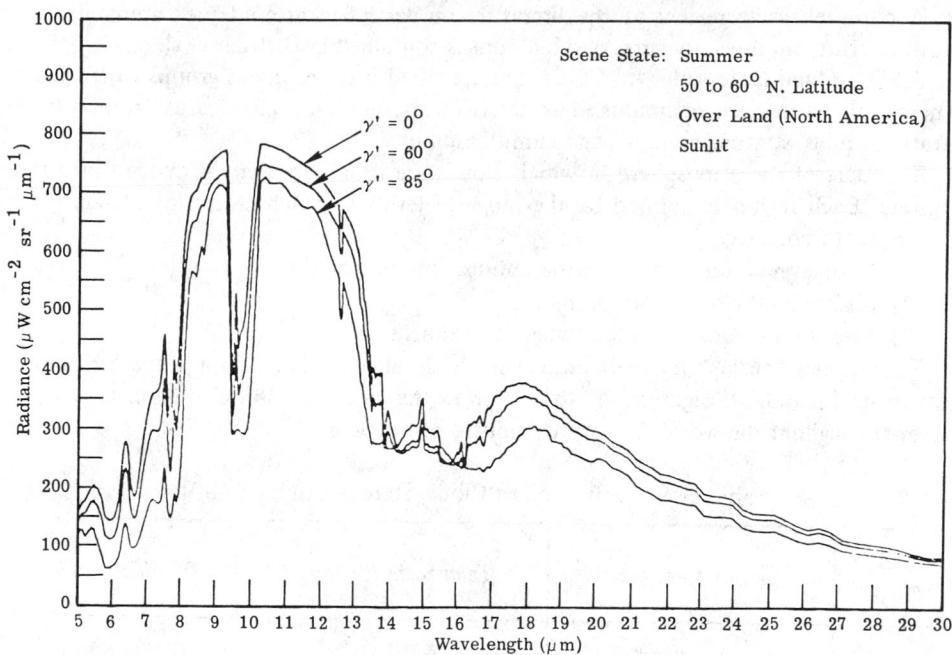

Fig. 3-41. The expected value of spectral radiance for three γ' angles [3-32].

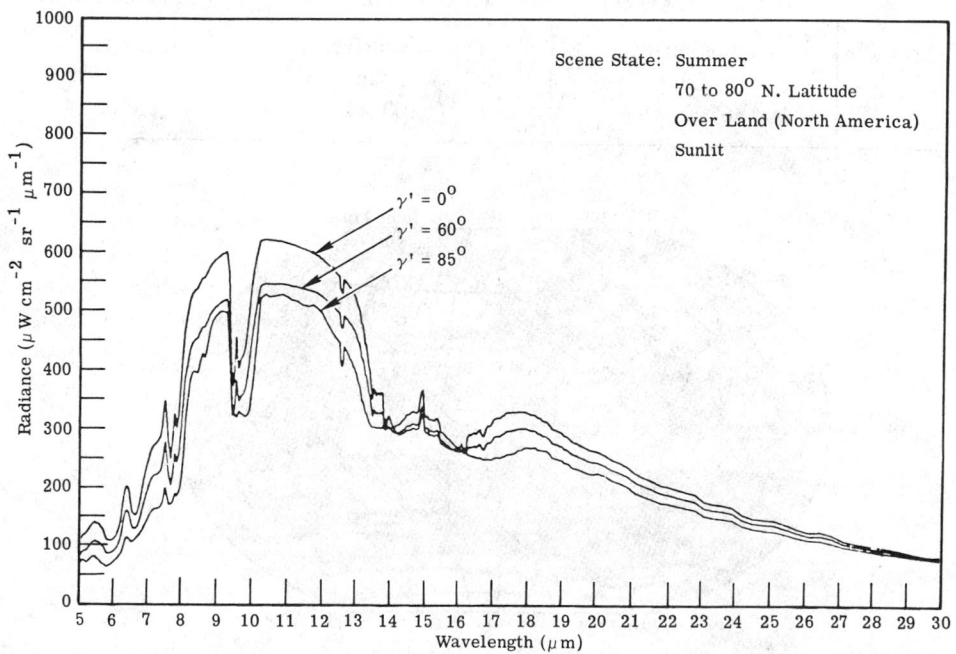

Fig. 3-42. The expected value of spectral radiance for three γ' angles [3-32].

A comprehensive review of the literature on earth-background measurements taken from aircraft, satellites, rockets, and balloons is contained in Dittmar et al.

3.5.3. Cloud Meteorology. Clouds are classified into ten main groups called genera. These are cirrus, cirrocumulus, cirrostratus, altocumulus, altostratus, nimbostratus, stratocumulus, stratus, cumulus, and cumulonimbus.

The part of the atmosphere in which clouds are usually present is divided into three regions. Each region is defined by the range of levels at which clouds of certain genera occur most frequently.

(1) High-level clouds - cirrus, cirrocumulus, and cirrostratus.

(2) Middle-level clouds - altocumulus.

(3) Low-level clouds - stratocumulus and stratus.

The regions overlap and their limits vary with latitude. Their approximate ranges are shown in Table 3-19. Figures 3-43 to 3-48 show the mean cloudiness in percentage of sky cover throughout the world for various months of the year.

Table 3-19. Definition of Cloud State Altitudes [3-30]

Cloud Level	Polar Regions	Temperate Regions	Tropical Regions
High	3–8 km (10,000–25,000 ft)	5–13 km (16,500–45,000 ft)	6–18 km (20,000–60,000 ft)
Middle	2–4 km (6500–13,000 ft)	2–7 km (6500–23,000 ft)	2–8 km (6500–25,000 ft)
Low	Earth's surface to 2 km (6500 ft)	Earth's surface to 2 km (6500 ft)	Earth's surface to 2 km (6500 ft)

Fig. 3-43. Mean cloudiness in percentage of sky cover for the month of January [3-40].

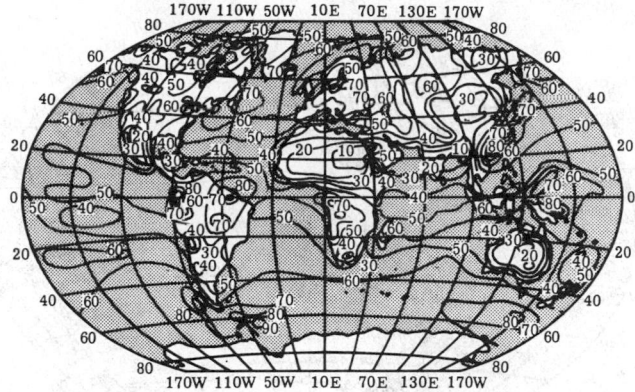

Fig. 3-44. Mean cloudiness in percentage of sky cover for the month of March [3-40].

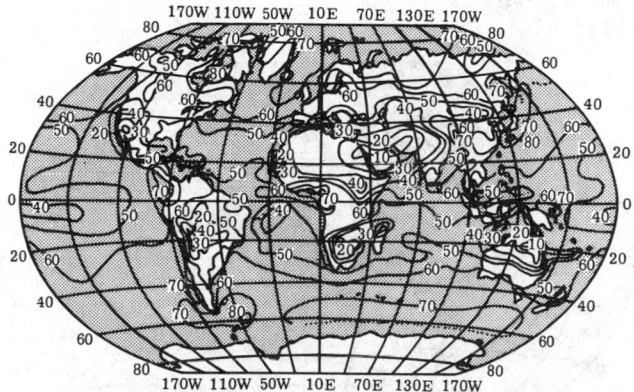

Fig. 3-45. Mean cloudiness in percentage of sky cover for the month of May [3-40].

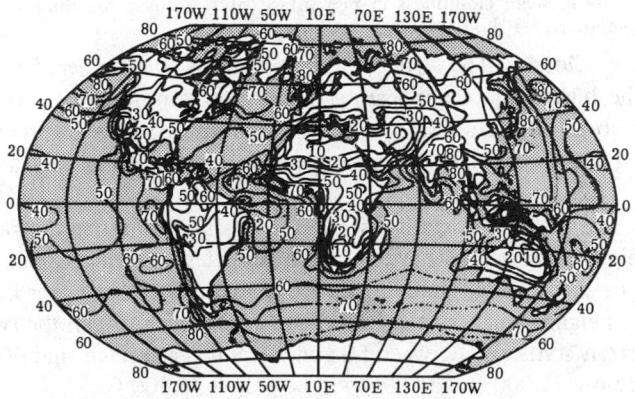

Fig. 3-46. Mean cloudiness in percentage of sky cover for the month of July [3-40].

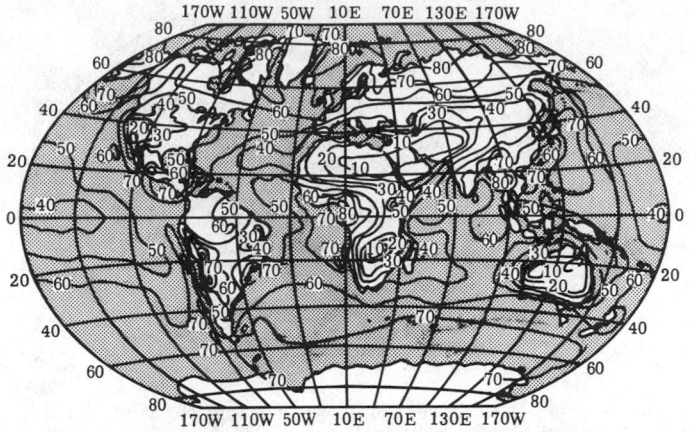

Fig. 3-47. Mean cloudiness in percentage of sky cover for the month of September [3-40].

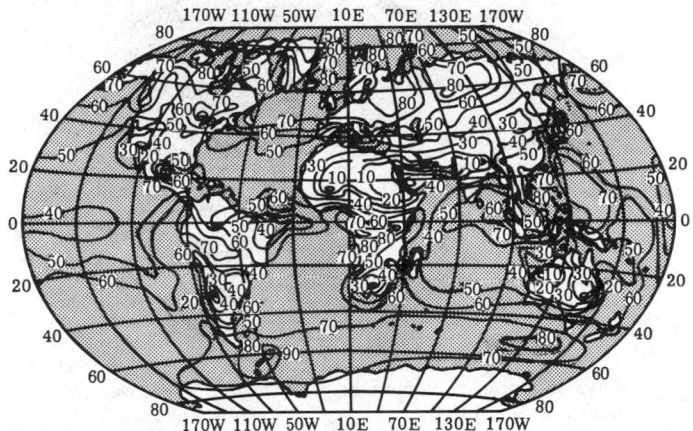

Fig. 3-48. Mean cloudiness in percentage of sky cover for the month of November [3-40].

3.5.4. Cirrus Clouds. The tropopause represents the upper limit of the cloud atmosphere. The highest clouds appearing within the troposphere are composed of large ice crystals of about 100 μm. Frequently these particles become oriented in the same direction, giving rise to unusual visible (and possibly infrared) effects such as haloes and arcs.

Tropopause and cloud top statistics are not available for the central Eurasian land mass. Cirrus height observations have not been reported anywhere north of 55° latitude. Inferences can be made about the annual tropopause distribution over Eurasia, and from this a cirrus top height model constructed. The correlation between the two parameters is based on American statistics. Between 50 and 70°N it is expected that 90% of the annual clouds will be below 32,000 ft, and 99% will be below 36,000 ft.

Figure 3-49 shows cloud top and tropopause heights based on a collation of cirrus and tropopause data averaged on a yearly basis for the entire United States. Figure 3-50 and

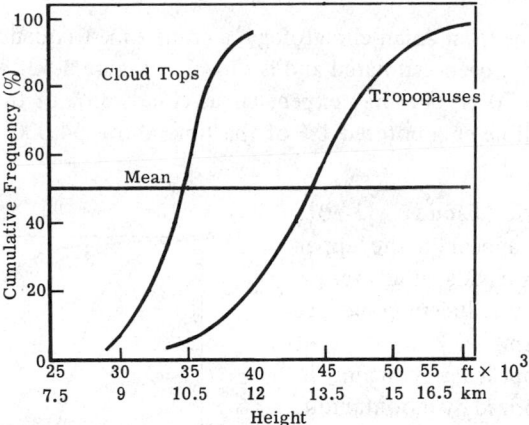

Fig. 3-49. Distribution of cloud top and tropopause heights (United States average) [3-30].

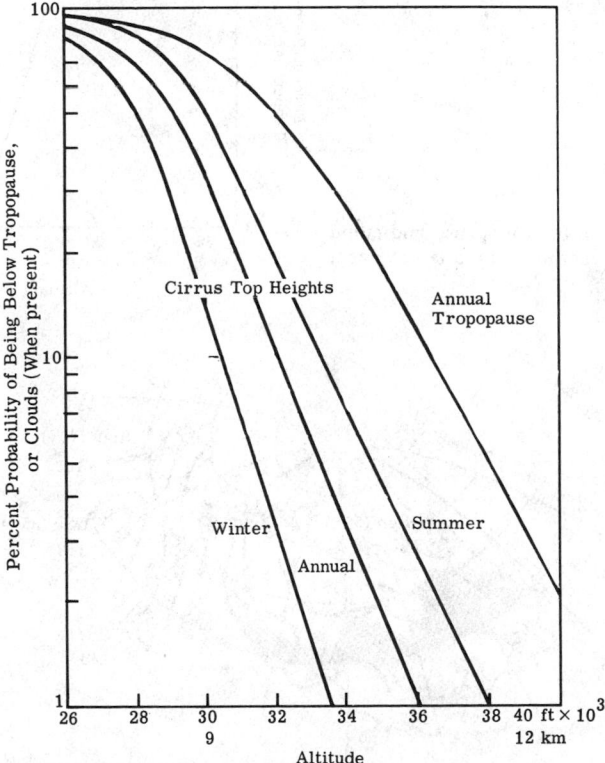

Fig. 3-50. Distribution of tropopause and cloud tops (50 to 70° N. latitude, Eurasian average) [3-30].

Figure 3-51 represent the distribution of tropopause and cloud heights between 50 and 90° latitude.

Based on deductions from Asian climatology, a crude time-frequency occurrence chart for cirriform clouds has been estimated and is shown in Figure 3-52. Averaging the entire year between 50 and 70°N, one may expect cirrus clouds 35% of the time. This means that cirrus clouds will be encountered 1% of the time above 34,000 ft, and 10% of the time above 30,000 ft.

3.5.5. Stratospheric Clouds [3-30].

Two types of clouds appear in the upper stratosphere: nacreous clouds, at an average height of 24 km, and noctilucent clouds, at a height of about 82 km.

Nacreous clouds appear rarely, mainly in high latitudes characterized by mountainous terrain. They are generally observed in the direction of the sun during sunset or sunrise

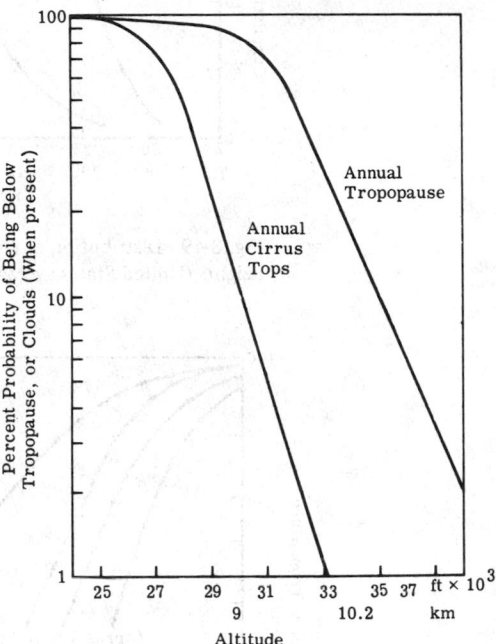

Fig. 3-51. Distribution of tropopause and cloud tops (65 to 90° N. Latitude, Eurasian average) [3-30].

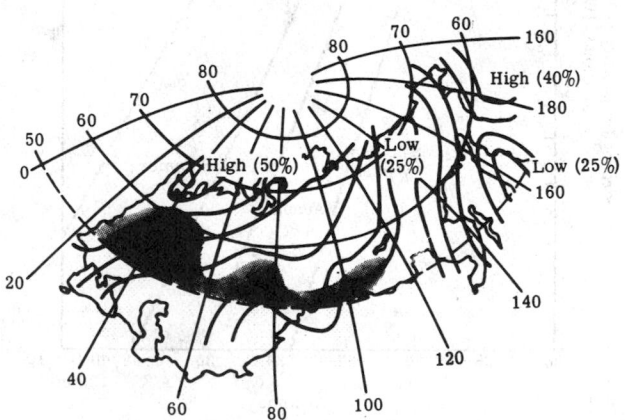

Fig. 3-52. The estimated annual temporal frequency of cirriform clouds. The dotted portion delineates area where twenty or more thunderstorms per year are reported (overall average = 35%) [3-30].

and are irridescent. Characteristic synoptic conditions that exist with these clouds are strong and consistent northwest winds extending to great heights with below average stratospheric temperatures. Theoretical considerations of water-droplet and ice-crystal growth in nacreous clouds suggest that the radii are less than 1.2 μm, with a very narrow size spectrum of about 0.1 μm. The particle concentration should be essentially that of the available condensation nuclei, about cm^{-3}. The liquid water content would be therefore between 10^{-12} to 10^{-11} g cm^{-3}. Such liquid water content is lower by about a factor of 10^4 than those observed in the tropospheric clouds.

Noctilucent clouds are visible against the nightime sky when the upper levels of the atmosphere are still illuminated by sunlight. These clouds have generally been reported only in the northern hemisphere during summer (August through October) within a restricted zone of latitudes extending from about 45 to 63°N.

Sunlight scattered from noctilucent clouds exhibits a spectrum and a degree of polarization which can be attributed to the scattering of sunlight by dielectric particles with predominant radii of around 0.1 μm and not greater than 0.2 to 0.4 μm. The observed brightness of the clouds suggests that the corresponding concentrations and matter content should be between 1 to 10^{-2} particles cm^{-3} and between 10^{-17} to 10^{-16} g cm^{-3}, respectively. Such cloud-particle concentrations are about five orders of magnitude less than those given for nacreous clouds.

3.5.6. Probability of Coverage at Various Altitudes. Figures 3-53 through 3-68 are charts showing, for the northern hemisphere, altitudes above which the probabilities of less than 0.1 sky coverage are 95, 90, 80, and 60%. Charts are presented for the midseason months January, April, July, and October. The criterion of less than 0.1 sky cover (actually less than 0.05 sky cover) can be taken as essentially no interference by clouds for air-to-air operation.

3.5.7. Particulate Statistics

Stratospheric Aerosols [3-30, 3-42 *through* 3-47]. A uniform distribution of the stratospheric aerosol content tends to decrease infrared gradients. Apparently the stratosphere will contribute to background noise because of the tendency of particles to form clouds which become arranged in periodic structures. Intensities will be high at small scattering angles, and atmospheric attenuation is negligible at these high altitudes. Table 3-20 presents a summary of information on the particle content of the stratosphere.

Stratospheric particulate matter may be divided into two classes: dust particles and condensed water.

Catastrophic volcanic eruptions and forest fires have deposited vast quantities of dust in the upper atmosphere. These can indirectly increase the upward intensity of reflected sunlight by acting as nucleating agents for ice. Unusual concentrations of clouds might result.

The earth is surrounded by belts of dust, smoke, and ice particles. Encounters with the dust by the earth's gravitational field cause an accretion of 10 to 50 lb of matter per square mile per year, based on an estimate of 24,000,000 visible meteors per day [3-40]. Some of this dust is concentrated into two extreme outer shells: the lighter smoke between altitudes of 2000 to 4000 mi., and dust from 600 to 1000 mi.

In general, the total particle concentrations just above the tropopause are between 10 and 100 cm^{-3}, but decrease to 1 cm^{-3} or less above about 20 km.

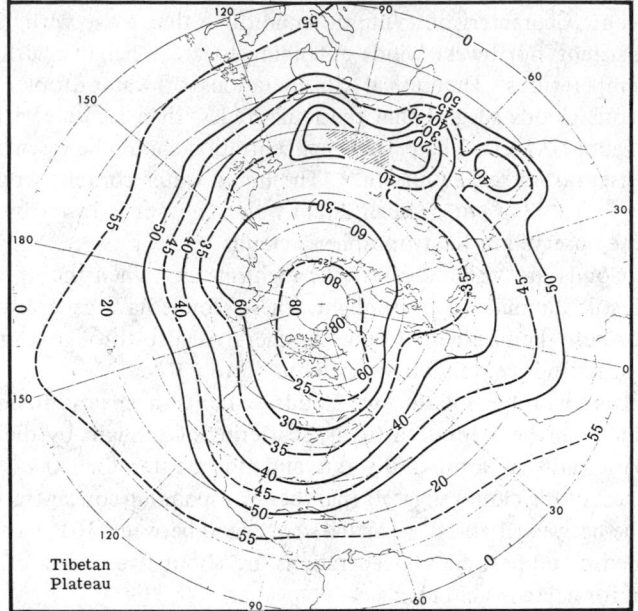

Fig. 3-53. Altitudes (thousands of feet msl) above which there is 95% probability of having less than 0.1 sky cover for the month of January [3-41].

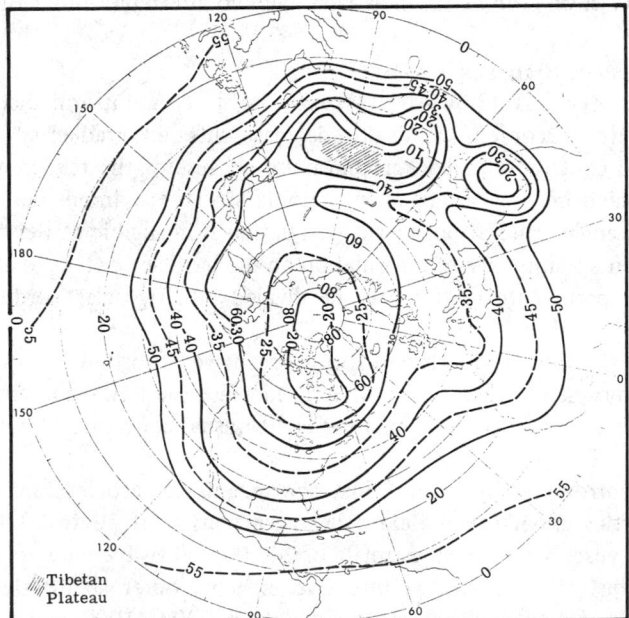

Fig. 3-54. Altitudes (thousands of feet msl) above which there is 90% probability of having less than 0.1 sky cover for the month of January [3-41].

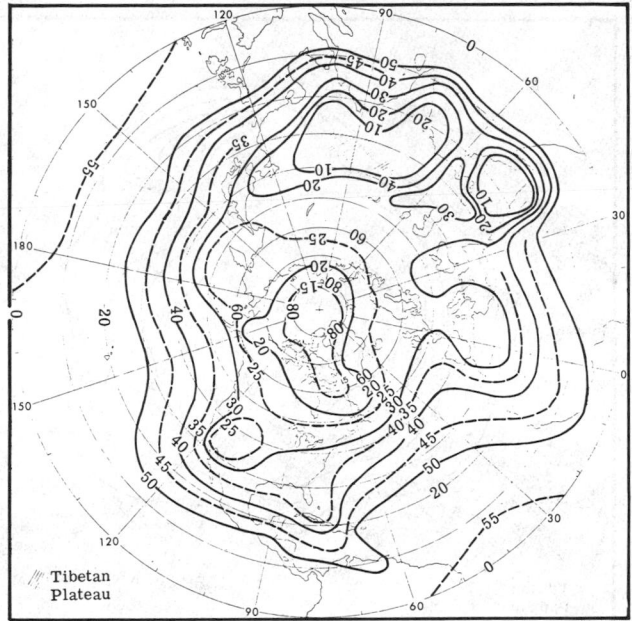

Fig. 3-55. Altitudes (thousands of feet ms1) above which there is 80% probability of having less than 0.1 sky cover for the month of January [3-41].

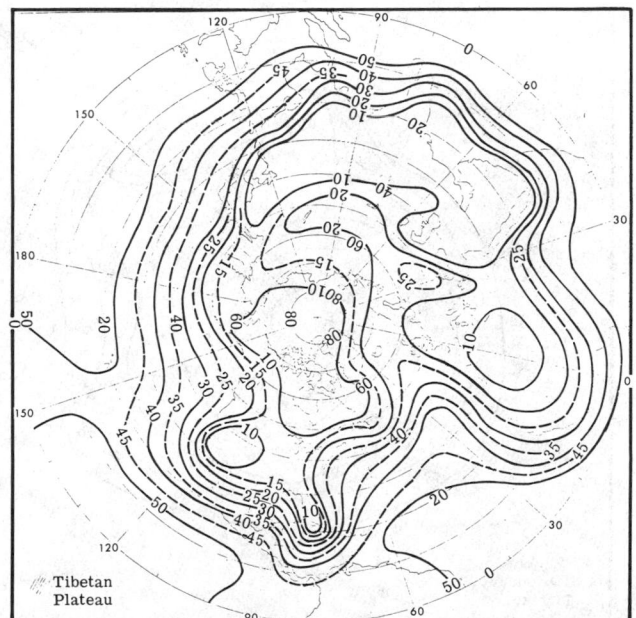

Fig. 3-56. Altitudes (thousands of feet ms1) above which there is 60% probability of having less than 0.1 sky cover for the month of January [3-41].

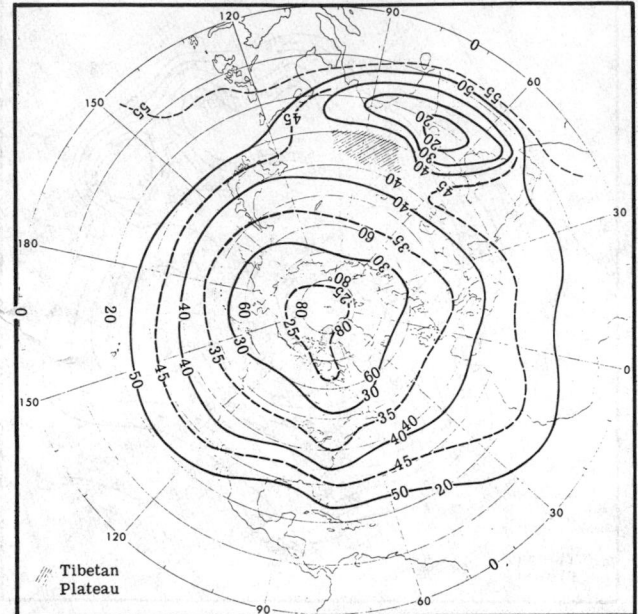

Fig. 3-57. Altitudes (thousands of feet msl) above which there is 95% probability of having less than 0.1 sky cover for the month of April [3-41].

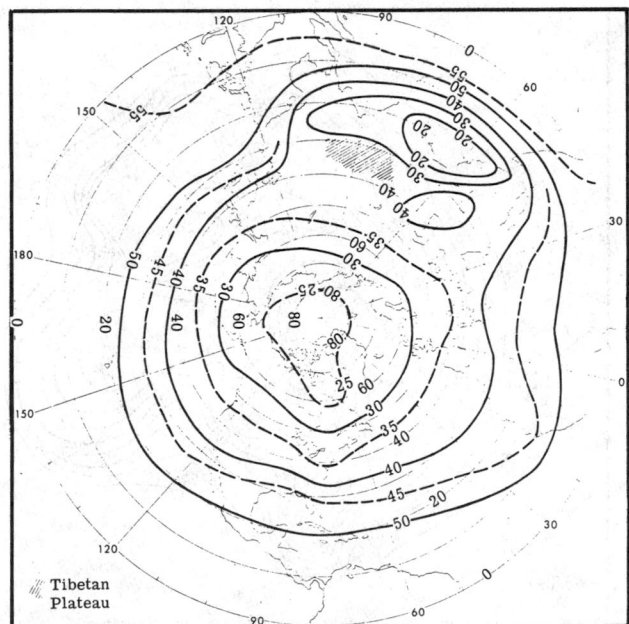

Fig. 3-58. Altitudes (thousands of feet msl) above which there is 90% probability of having less than 0.1 sky cover for the month of April [3-41].

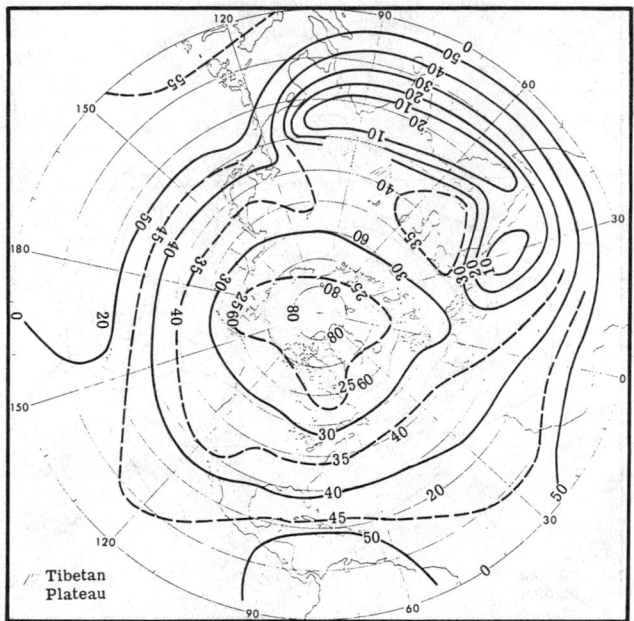

Fig. 3-59. Altitudes (thousands of feet msl) above which there is 80% probability of having less than 0.1 sky cover for the month of April [3-41].

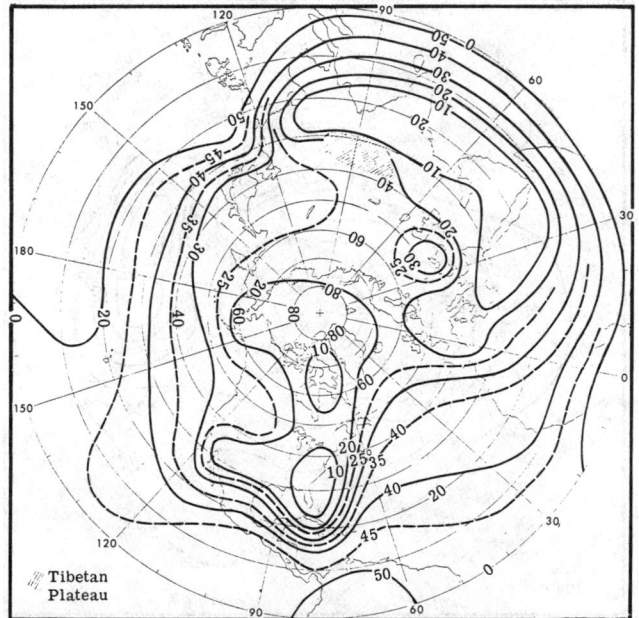

Fig. 3-60. Altitudes (thousands of feet msl) above which there is 60% probability of having less than 0.1 sky cover for the month of April [3-41].

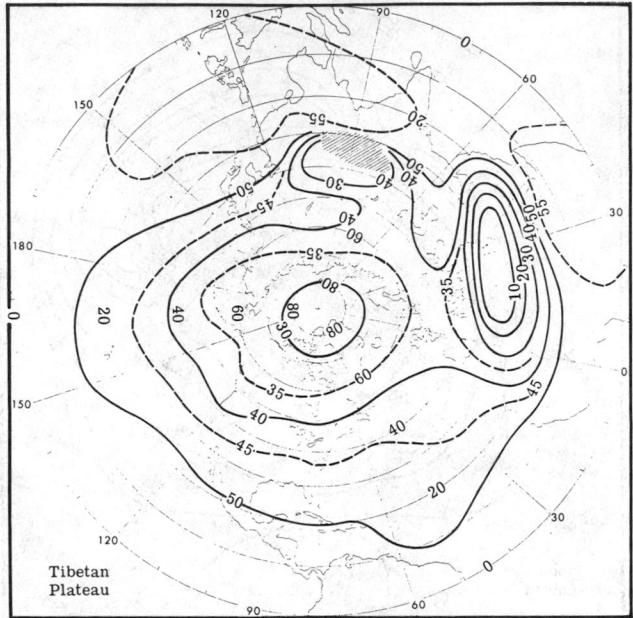

Fig. 3-61. Altitudes (thousands of feet msl) above which there is 95% probability of having less than 0.1 sky cover for the month of July [3-41].

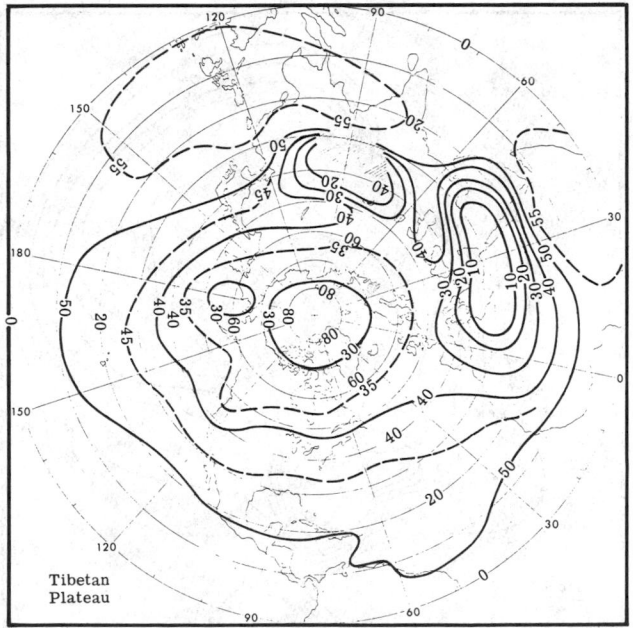

Fig. 3-62. Altitudes (thousands of feet msl) above which there is 90% probability of having less than 0.1 sky cover for the month of July [3-41].

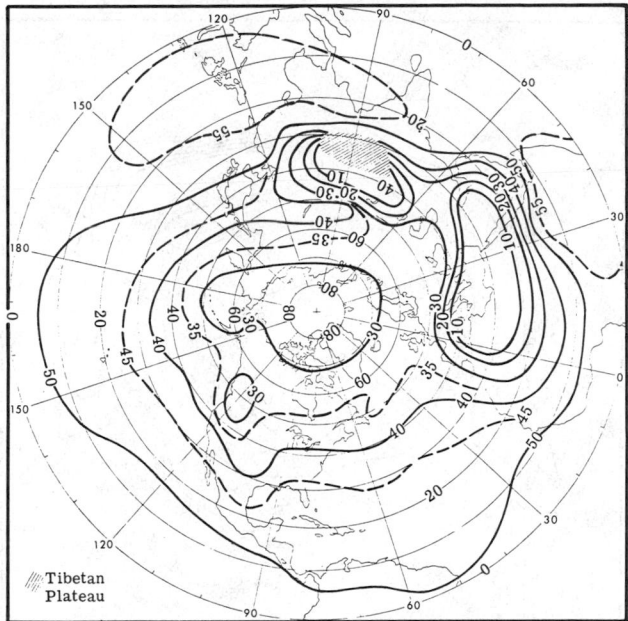

Fig. 3-63. Altitudes (thousands of feet ms1) above which there is 80% probability of having less than 0.1 sky cover for the month of July [3-41].

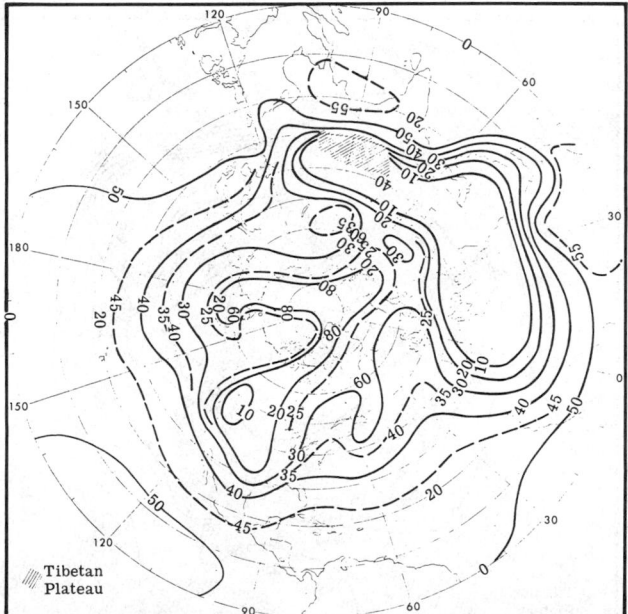

Fig. 3-64. Altitudes (thousands of feet ms1) above which there is 60% probability of having less than 0.1 sky cover for the month of July [3-41].

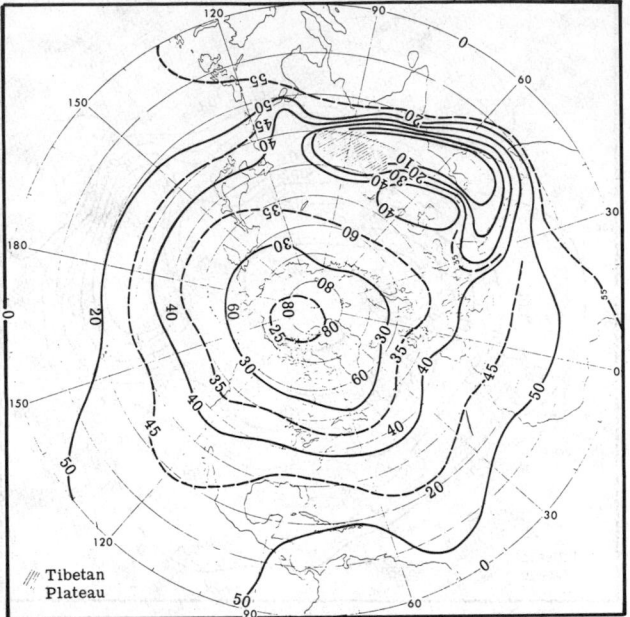

Fig. 3-65. Altitudes (thousands of feet msl) above which there is 95% probability of having less than 0.1 sky cover for the month of October [3-41].

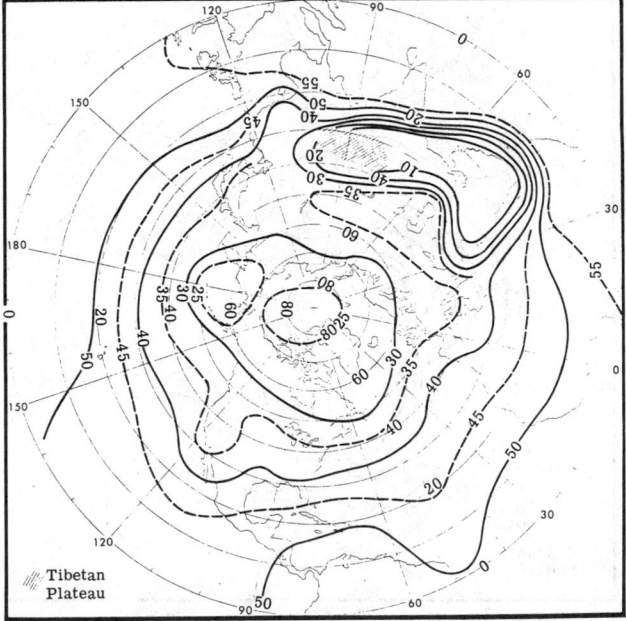

Fig. 3-66. Altitudes (thousands of feet msl) above which there is 90% probability of having less than 0.1 sky cover for the month of October [3-41].

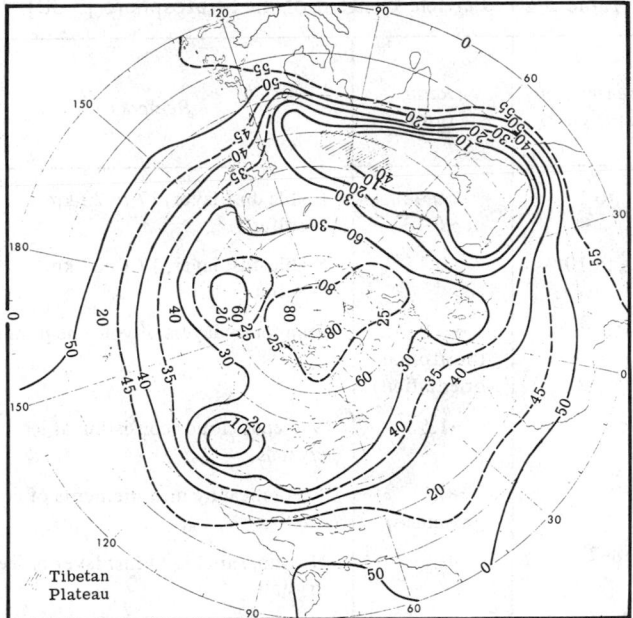

Fig. 3-67. Altitudes (thousands of feet msl) above which there is 80% probability of having less than 0.1 sky cover for the month of October [3-41].

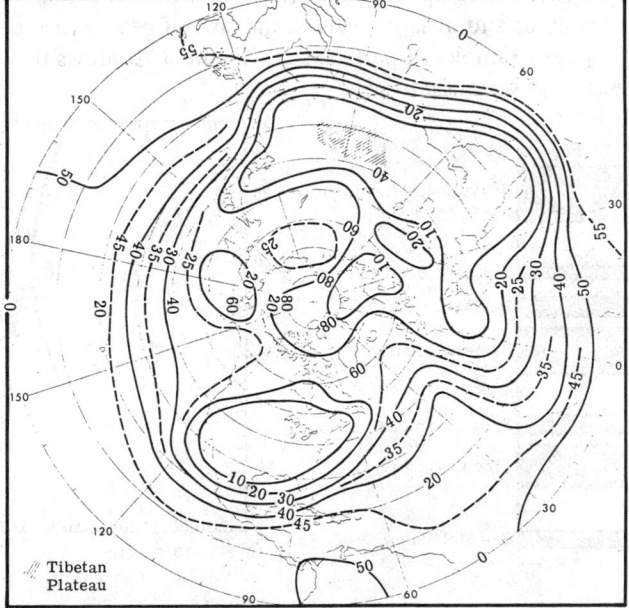

Fig. 3-68. Altitudes (thousands of feet msl) above which there is 60% probability of having less than 0.1 sky cover for the month of October [3-41].

Table 3-20. Particle Content of the Stratosphere [3-30]

Altitude (km)	Concentrations (no. per cm³)	Radii (µm)	Remarks	Typical Band Spacings (km)
10-30	10^{-1} to 1	>0.08 or ~0.10	Stable dust layer, 17 to 22 km – sulfur	–
10-30	10^{-2} to 10^{-1}	~0.15	Stable dust layer, 17 to 22 km – sulfur	–
10-30	~<10^3	~0.8 (horizontal orientation)	Temporary layers of volcanic pumice	1
17-31	~<1	~1.5	Nacreous clouds consisting of ice crystals	40
30-80	1	~0.1 (assumed)	Theoretical by measurements of conductivity	–
74-92	10^{-2}	~0.1	Noctilucent cloud (dust layer or ice crystal)	10 and 60
80	10^{-4} to 10^{-1}	0.1 (assumed)	Theoretical interplanetary dust sources, 10^{-23} to 10^{-20} g cm⁻³	–

The manner in which a particle scatters light depends on the ratio of its radius to the wavelength of light. For ratios up to about 0.08, Rayleigh's laws hold; between 0.08 and 3 the Mie theory is used; and at larger values, the laws of geometrical optics are satisfactory. Figure 3-69 gives examples of particle sizes. Figure 3-70 shows the concentration of different particle sizes at various altitudes.

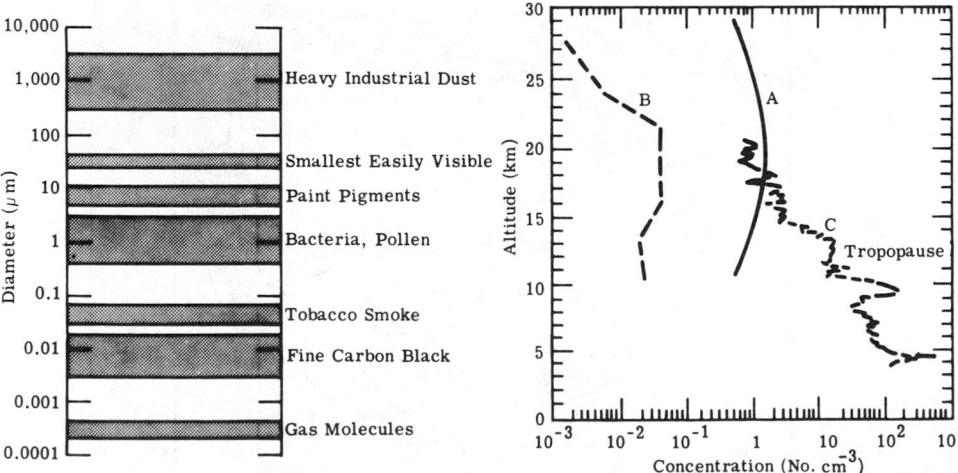

Fig. 3-69. Examples of particle sizes [3-30].

Fig. 3-70. Vertical profiles of particle concentrations. Curve A = radii greater than 0.08 µm; curve B = radii greater than 0.1 to 0.3 µm; curve C = Aitken nuclei, 0.01 to 0.1 µm radii [3-30].

3.6. Radiation of the Sky As Seen from the Earth

3.6.1. Sky Spectral Radiance.

There is a moderate amount of literature on the spatial and temporal fluctuations of the sky background, [3-48 to 3-54].

Sky-background radiation in the infrared is caused by scattering of the sun's radiation and by emission from atmospheric constituents. Figure 3-71 illustrates the separation of the spectrum into two regions: the solar scattering region short of 3 μm, and the thermal emission region beyond 4 μm. Solar scattering is represented by reflection from a bright sunlit cloud and, alternatively, by a curve for clear-air scattering. The thermal region is represented by a 300 K blackbody. Figure 3-72 shows blackbody curves for temperatures ranging from 0 to 40°C. This simple model is modified by a number of factors: in the solar region there are absorption bands of water vapor at 0.94, 1.1, 1.4, 1.9, and 2.7 μm, and of carbon dioxide at 2.7 μm. The effect of these bands is shown in Figure 3-73.

In the thermal region, curves for those bands which have strong absorption (and thus strong emission) will approach the blackbody curve appropriate to the temperature of the atmosphere. Less strongly emitting regions may contribute only a small fraction of the radiation of a blackbody at the temperature of the atmosphere. The bottom curve in Figure 3-74 is a good example. This zenith measurement, taken from a high, dry location, shows low emission except in the strong band of CO_2 at 15 μm and of H_2O at 6.3 μm. There is also a weak emission peak, due to ozone, at 9.6 μm. The low-level continuum is due to the wings of the strong bands of H_2O and CO_2. The effect of increased humidity and air mass can be seen by comparing the bottom curves of Figures 3-74 and 3-75. Figure 3-75 shows measurements taken at a humid, sea-level location.

The effect of increasing air mass alone can be seen in both Figures 3-74 and 3-75 by comparing curves taken from the same altitude at various elevation angles. The emission shows a systematic decrease with increasing elevation angle. The direction of look also has an effect in the solar scattering region, as seen in Figure 3-73, where, for a clear sky, the sun's position is fixed and the spectral radiance is plotted for several observer angles.

The position of the sun has a strong effect on the scattered radiation in the solar region, as shown in Figure 3-76. Here, the observer looks at the zenith. The elevation angle of the sun is varied but has little effect on the radiation in the thermal region. The temperature of the atmosphere, on the other hand, has a strong effect on the radiation

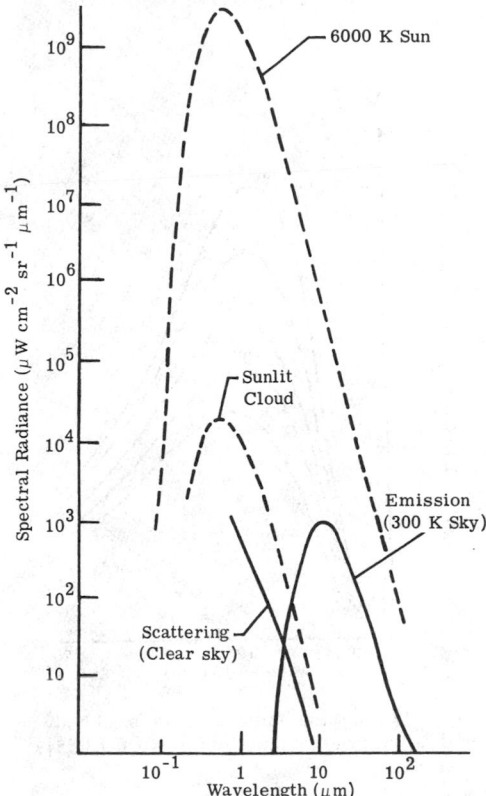

Fig. 3-71. Contributions from scattering and atmospheric emission to background radiation [3-55].

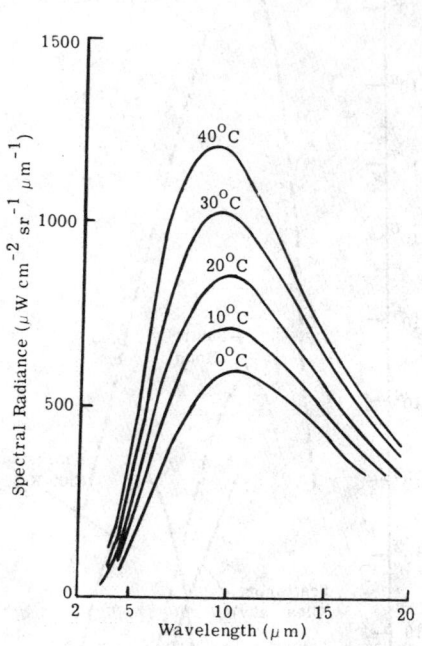

Fig. 3-72. Spectral radiance of a blackbody with a temperature range of 0 to 40°C [3-55].

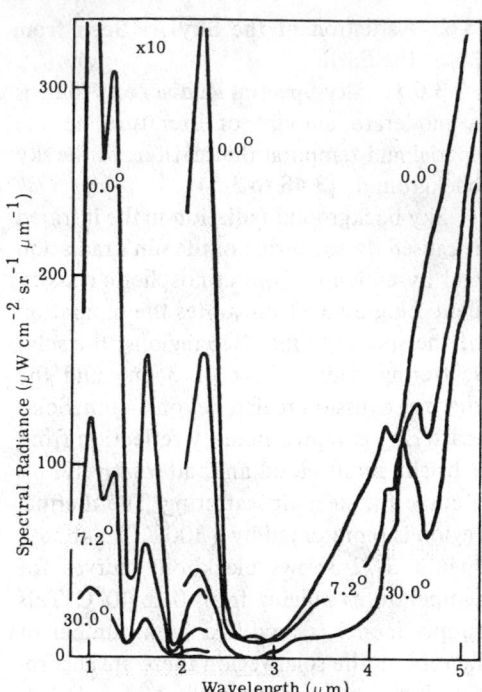

Fig. 3-73. The spectral radiance of a clear daytime sky [3-55].

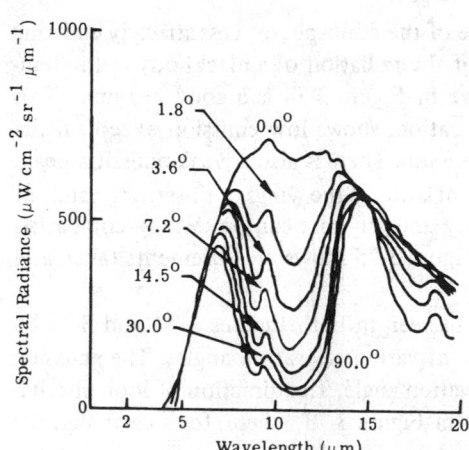

Fig. 3-74. The spectral radiance of a clear nighttime sky. It is for several angles of elevation above the horizon (Elk Park Station, Colorado) [3-55].

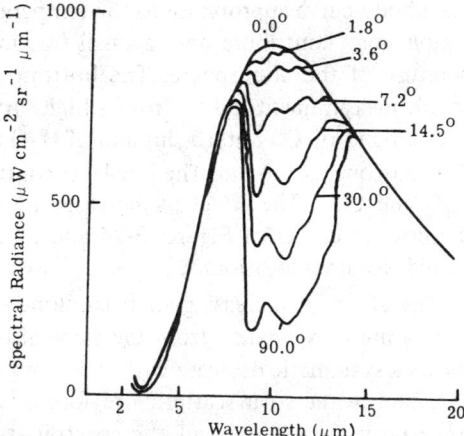

Fig. 3-75. The spectral radiance of a clear nighttime sky. It is for several angles of elevation above the horizon (Cocoa Beach, Florida) [3-55].

in the thermal region but little effect in the solar region. The presence of clouds will affect both near-infrared solar scattering and thermal-region emission.

Near-infrared radiation exhibits strong forward scattering in clouds. Thus the relative positions of sun, observer, and cloud cover become especially important. For a heavy, overcast sky, multiple scattering reduces the strong forward scattering effect.

Thick clouds are good blackbodies. Emission from clouds is in the 8 to 13 μm region and is, of course, dependent on the cloud temperature. Because of the emission and absorption bands of the atmosphere at 6.3 and 15.0 μm, a cloud may not be visible in these regions and the radiation here is determined by the temperature of the atmosphere. A striking example is given in Figure 3-77. Here, the atmospheric temperature is +10°C and the radiation in the emission bands at 6.3 and 15.0 μm approaches a value appropriate to that temperature. The underside of the cloud has a temperature of –10°C, and the radiation in the 8 to 13 μm window approaches that of a blackbody at –10°C.

Figure 3-78 shows the variation of sky radiance as a function of elevation angle. Figure 3-79 shows the variation with respect to variations of ambient air temperature. Figure 3-80 shows seasonal variations. Figures 3-81 and 3-82 show recorder traces of sky spectra under several different conditions. The curves may be calibrated by the dash-dot curves for the blackbody reference.

3.6.2. Overcast Sky Luminance. An approximate relation for the overcast sky luminance as viewed from ground level is reported as

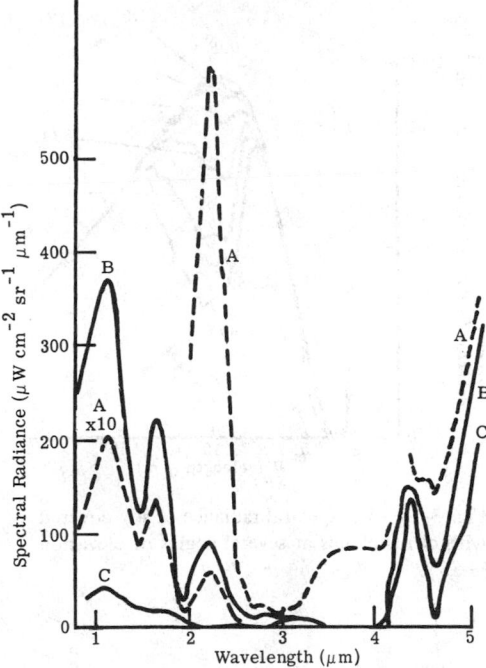

Fig. 3-76. The spectral radiance of a clear zenith sky as a function of the sun position. Curve A = sun elevation 77°, temperature 30°C; curve B = sun elevation 41°, temperature 25.5°C; curve C = sun elevation 15°, temperature 26.5 °C [3-55].

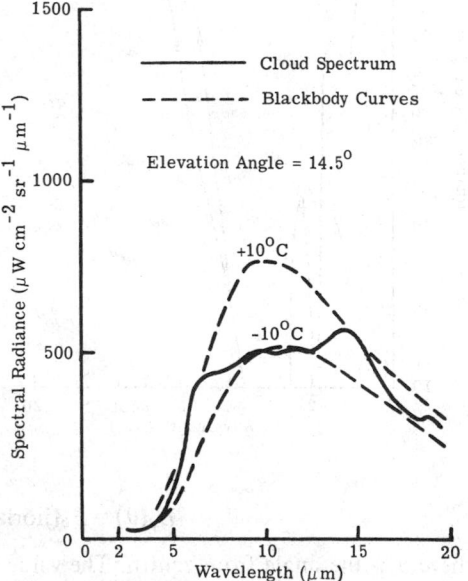

Fig. 3-77. The spectral radiance of the underside of a dark cumulus cloud [3-55].

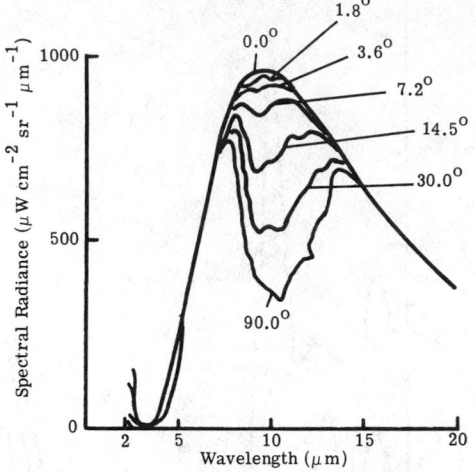

Fig. 3-78. The spectral radiance of sky covered with cirrus clouds at several angles of elevation [3-55].

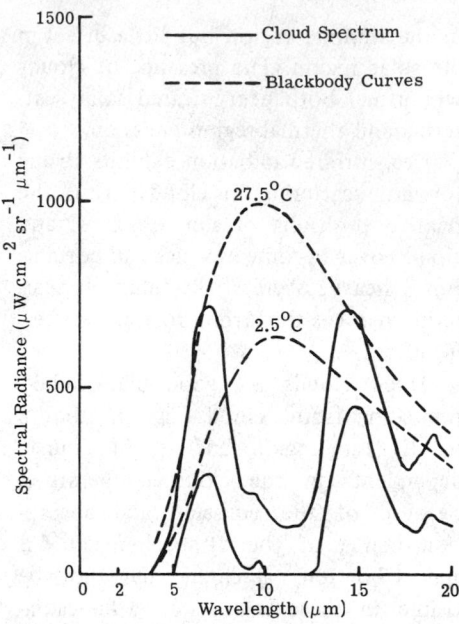

Fig. 3-79. Zenith sky spectral radiance showing the large variation with ambient air temperature [3-55].

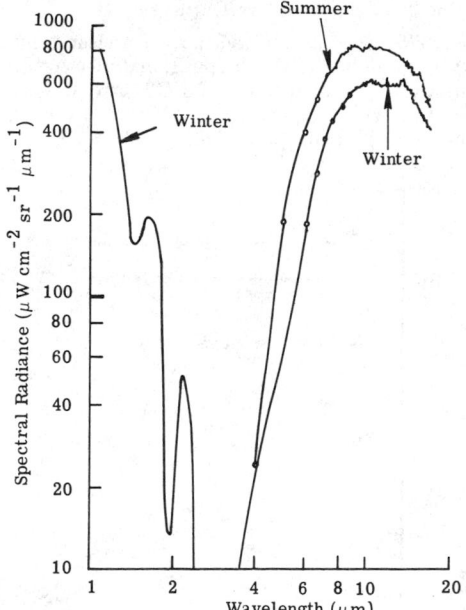

Fig. 3-80. Spectral radiance of overcast skies in winter and summer [3-56].

$$L_V(\theta) = L_V(\text{horizon})(1 + A \cos \theta) \qquad (3\text{-}49)$$

where θ is the angle from zenith. The value of A is 2.0 according to Moon and Spencer. However, Hood and also Kasten and Moller have found A to be about 1.0 for arctic skies and skies over snow [3-58].

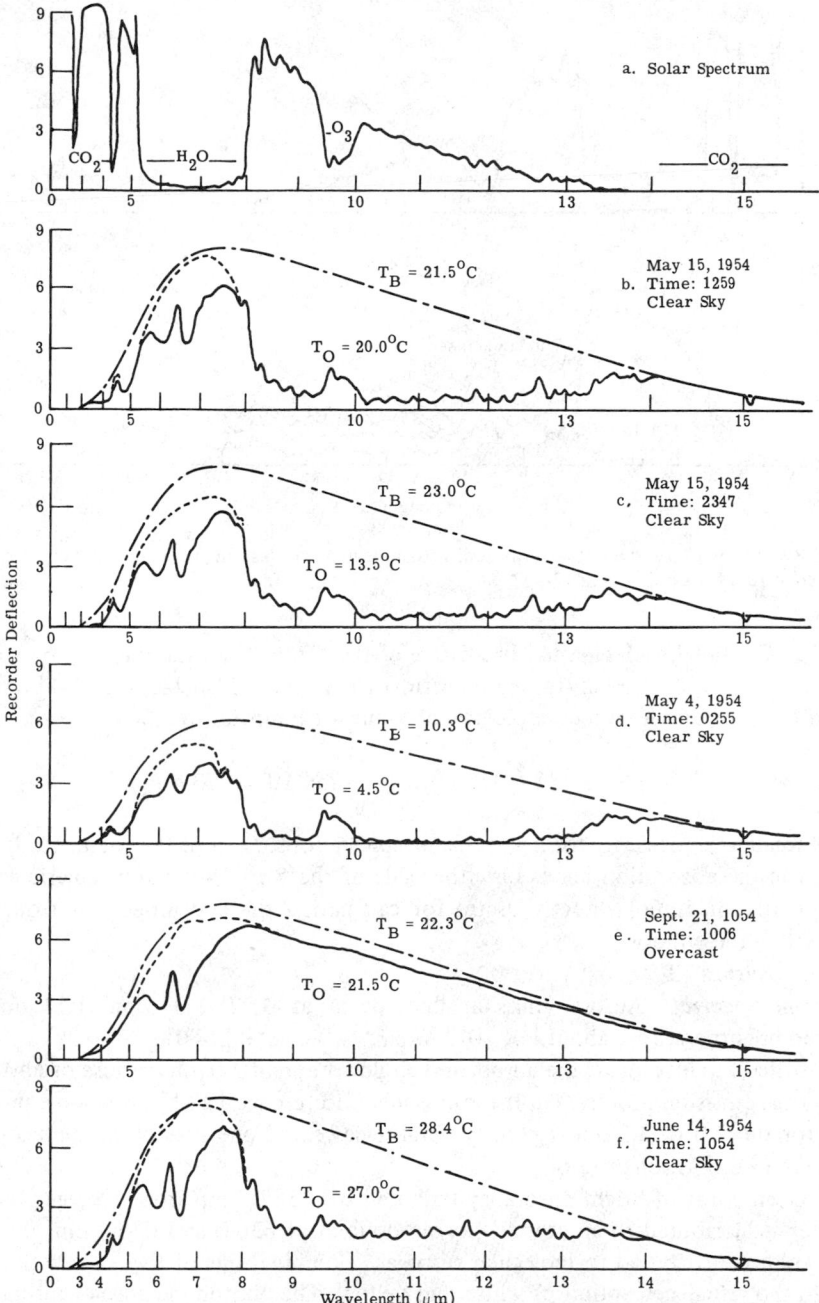

Fig. 3-81. Typical zenith sky spectra with a solar spectrum for comparison. The dashed-dot curve above each tracing was obtained with a laboratory blackbody of temperature T_B. The ground temperature, T_0, was measured by automatic recording in a standard screen. Water vapor concentrations (g m^{-3}) six feet above ground level have the following values for various parts of the figure: (b) 9.3, (c) 8.0, (d) 5.9, (e) 15.2, and (f) 20.3 [3-57].

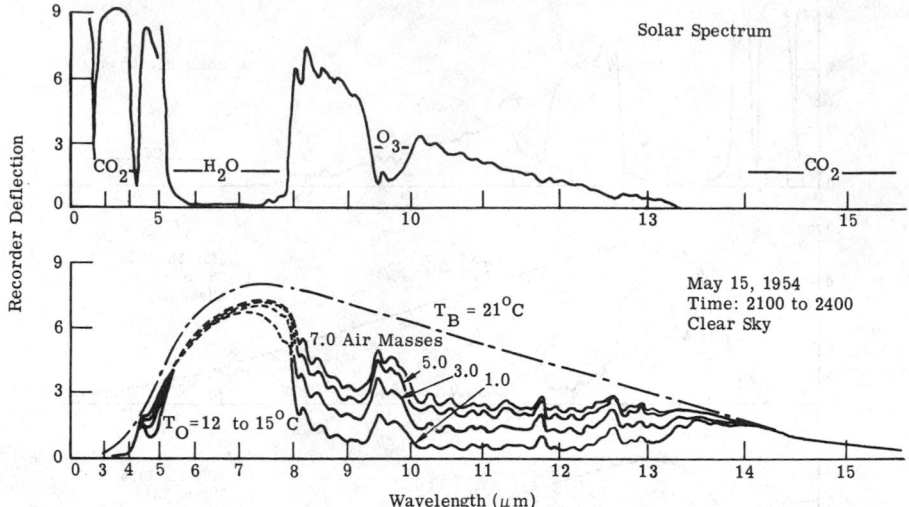

Fig. 3-82. Spectra observed at various zenith distances on a clear night. The upper curve gives a solar spectrum for comparison [3-57].

3.6.3. Ground Level Thermal Irradiance of the Sky. The total thermal irradiance of a clear sky at sea level, $E(\text{sky})$, may be estimated from the Idso-Jackson [3-1] empirical relation using ground-level meteorological absolute air temperature, T_a,

$$E(\text{sky}) = \sigma T_a^4 \{1 - 0.261 \exp[-7.77 \times 10^{-4}(273 - T_a)^2]\} \qquad (3-50)$$

This relation is primarily for heat transfer applications. Most of the irradiance is due to emission in the absorption bands on either side of the 8 to 14 μm atmospheric window. Consequently, it is not directly useful for calculating the sky reflections from metal plates within the window.

3.6.4. Aurora [3-30, 3-59]

Auroral Spectral. Aurora emission lines occur at 0.92, 1.04, and 1.11 μm. The measured brightnesses are about 6×10^{-8} W cm^{-2}sr^{-1}line^{-1} [3-60].

It is difficult to investigate the aurora and airglow beyond 2.0 μm because of absorption and thermal emission processes in the atmosphere. Reference [3-61] gives some predicted values for the 2.0 to 3.5 μm region. General reviews and one case of an application are covered in [3-62] through [3-65].

The green color of bright aurora is attributed to a 557.7 nm atomic oxygen line. The red color is attributed to an atomic oxygen doublet at 630.0 and 636.4 nm; the bluish purple color is attributed to molecular nitrogen. Combinations of these excitations may result in the visual perception of white and yellow. The photon radiance of auroral range from 10^3 to 10^6 Ry. (Ry is the unit symbol used here for a rayleigh where 1 Ry corresponds to a photon radiance of $1/4\pi \times 10^6$ photons sec^{-1} cm^{-2} sr^{-1}.)

Figure 3-83 shows the auroral spectrum between 0.9 and 1.2 μm. This reproduction was obtained by averaging a number of individual spectra. Figure 3-84 shows auroral spectra between 1.4 and 1.65 μm. The dotted curve is the airglow spectrum fitted to the

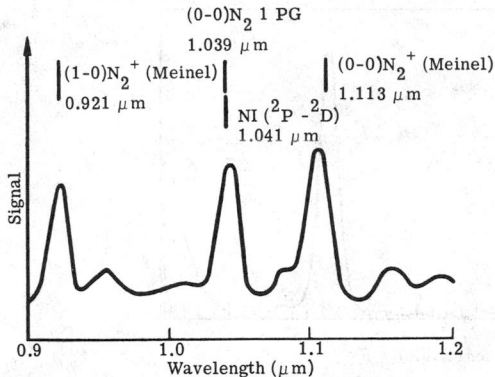

Fig. 3-83. Auroral spectrum, 0.9 to 1.2 μm. It was obtained with a lead sulfide spectrometer; projected slit width 100 Å [3-60].

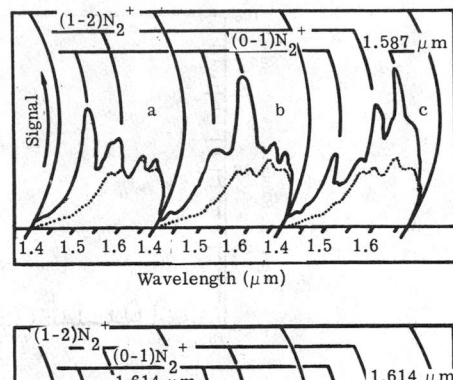

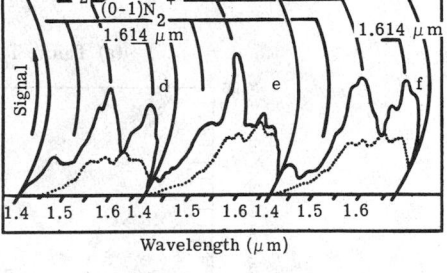

Fig. 3-84. Auroral spectra, 1.4 to 1.65 μm. It was obtained with a lead sulfide spectrometer; projected slit width 200 Å [3-66].

auroral spectrum in a region where the auroral emission appears feeble. Spectra a, b, and c were made in consecutive scans, with a total time of 3 min. The relative intensities of features on a single scan are not significant since the aurora fluctuates in brightness during the scanning period.

High-altitude, rocket-borne spectrometers were used to obtain aurora spectra in 1973 [3-67]. Figures 3-85 and 3-86 show the spectral radiance in the short-wavelength, infrared range, SWIR, from 1.6 to 5.6 μm and in the long wavelength infrared, LWIR, from 7.0 to 23.0 μm. Figure 3-87 shows a profile of spectral radiance at 4.3 μm as a function of altitude. The unit MRy is a megarayleigh.

Auroral Zones [3-59]. Auroral zones are divided into 3 areas: the north and south auroral regions extending from geomagnetic latitudes 60° to the poles, the subauroral belts between 45 and 60°, and the minauroral belt between 45°N and 45°S. The auroral regions include the auroral zones (the regions of maximum occurrence) and the auroral caps (the polar regions within the auroral zones).

Although aurora occur primarily in the auroral regions, large displays may occur in quite low latitudes. However, in tropical and even low temperature latitudes they are extremely rare.

The frequency of auroral occurrences is maximum 20 to 25° from the geomagnetic poles. Figure 3-88 shows the geographic distribution of the frequency of aurora in the northern hemisphere. The isochasms refer to the number of nights during the year in which an aurora might be seen at some time during the night, in any part of the sky, if clouds and other factors affecting visual detection of aurora do not interfere. Figure 3-89 shows the zone of maximum auroral frequency in the southern hemisphere.

Periodic Variations. The number of aurora observed from a particular point over the course of a year may vary widely and is strongly correlated with solar activity. Minimum auroral activity corresponds with minimum solar activity. Maximum auroral activity usually occurs about two years after sunspot maximum.

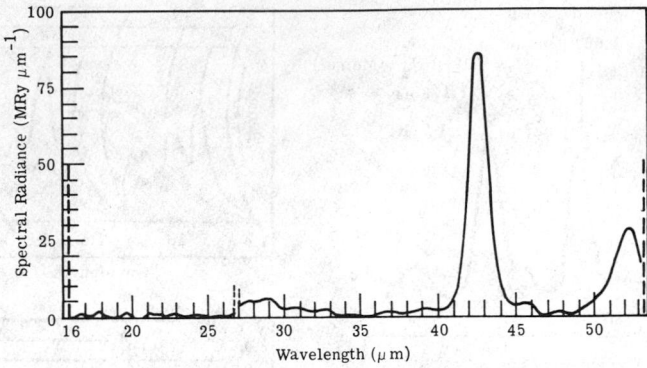

(a) Paiute Tomahawk 10.205-2 97 km.

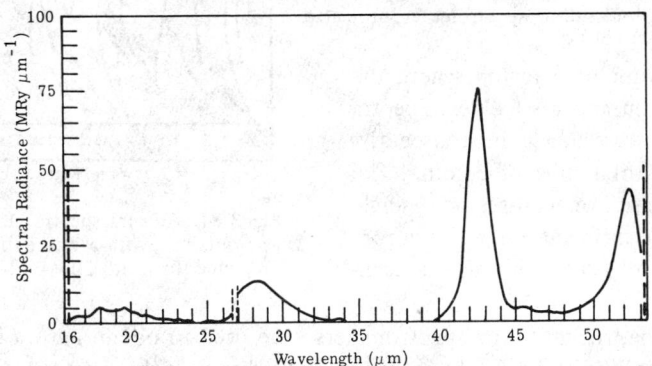

(b) Paiute Tomahawk 10.205-2 86 km.

Fig. 3-85. Sample spectra scans from a SWIR spectrometer aboard a Paiute-Tomahawk rocket launched from Poker Flat, Alaska, 24 March 73. Although uncorrected for actual rocket aspect, the data approximate the zenith radiance.

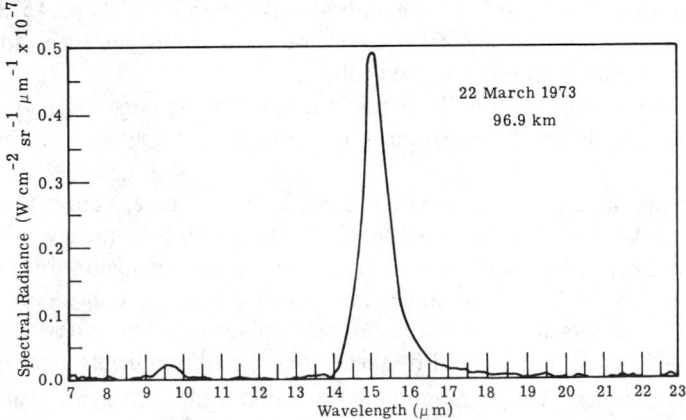

Fig. 3-86. Sample spectrum scan (vertical) from a LWIR spectrometer aboard a Black Brant rocket flown from Poker Flat, Alaska, 22 March 73. The rocket altitude is 96.9 km on rocket descent.

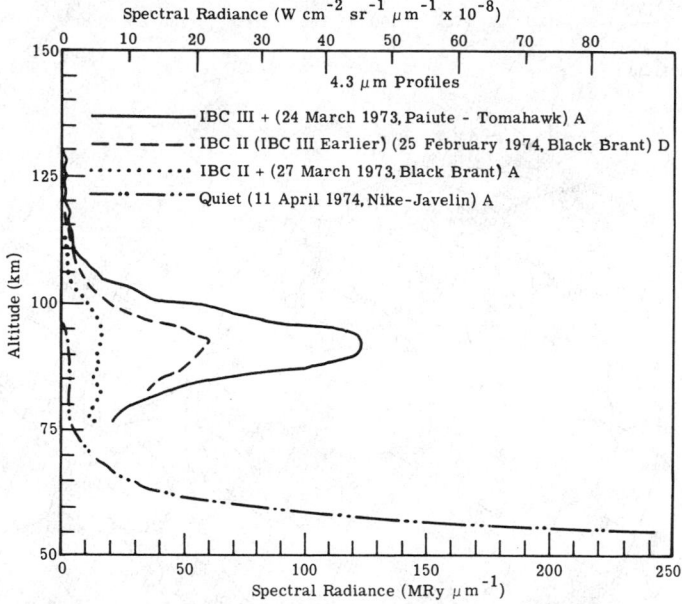

Fig. 3-87. Zenith peak spectral radiance at 4.3 μm measured with SWIR spectrometers flown on four different rockets under various auroral conditions.

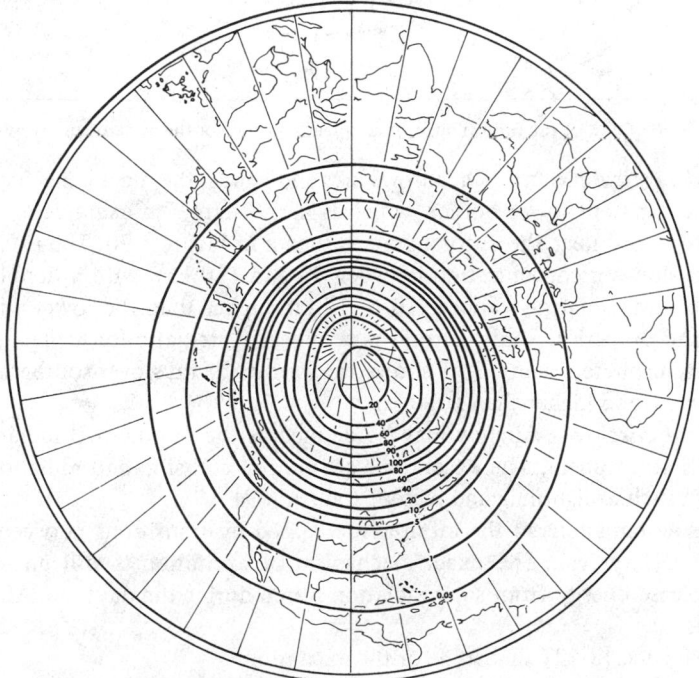

Fig. 3-88. Geographic distribution of the frequency of aurora in the northern hemisphere [3-68].

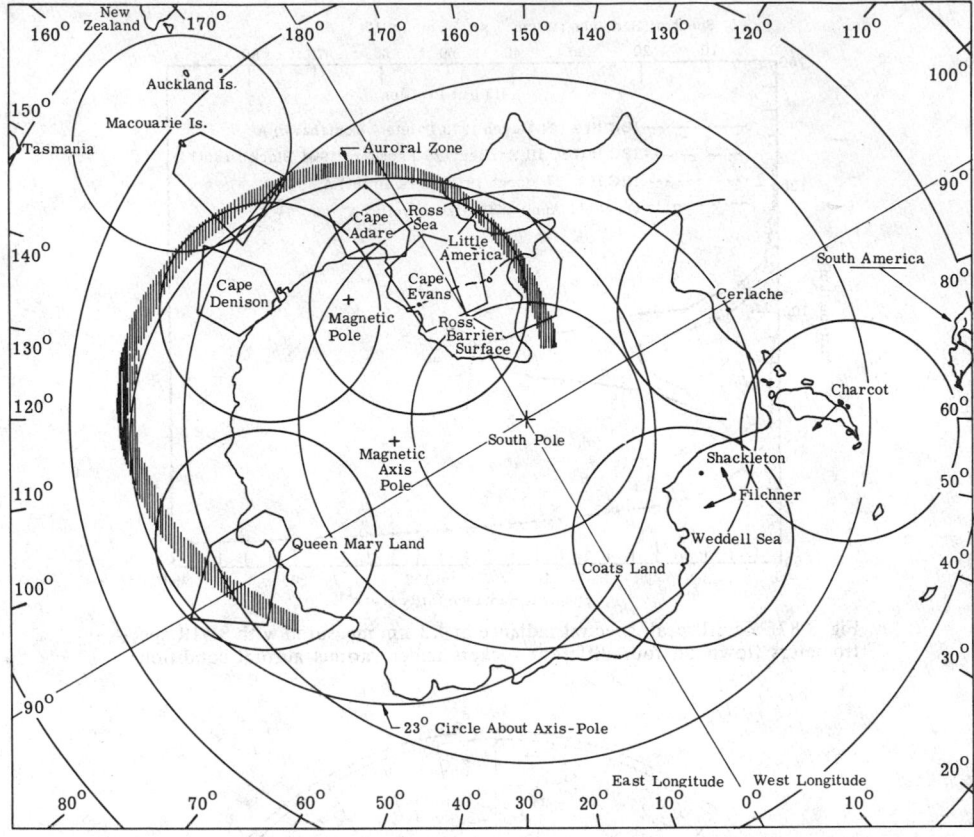

Fig. 3-89. Zone of maximum auroral frequency in the southern hemisphere [3-69].

Height and Vertical Extent. On auroral arcs and bands, the most convenient height to measure is the apparent lower border, which is fairly sharp. An example of a set of such measurements in and near the auroral zone is shown in Figure 3-90. The total number of measurements shows a concentration between 95 and 110 km, with a double peak. The lower limits of individual rays appear 10 or 15 km higher than the lower edges of most arcs, bands, and draperies. Sunlit auroral rays appear systematically higher than displays in the dark atmosphere. Figure 3-91 shows the heights of rays over southern Norway. A few sunlit rays extend higher than 1000 km.

3.6.5. Night Airglow. Airglow may be defined as the nonthermal radiation emitted by the earth's atmosphere. The exceptions are auroral emission and radiation of a cataclysmic origin, such as lightning and meteor trails [3-59].

Night airglow emissions in the infrared are caused by transitions between vibrational states of the OH^- radical. The exact mechanism of excitation is still unclear, but the effect is to release energy from solar radiation stored during the daytime. Airglow occurs at all latitudes.

There is evidence [3-72] that some of the excitation is

$$H + O_3 \rightarrow OH + O_2$$
$$OH + O \rightarrow O_2 + H$$

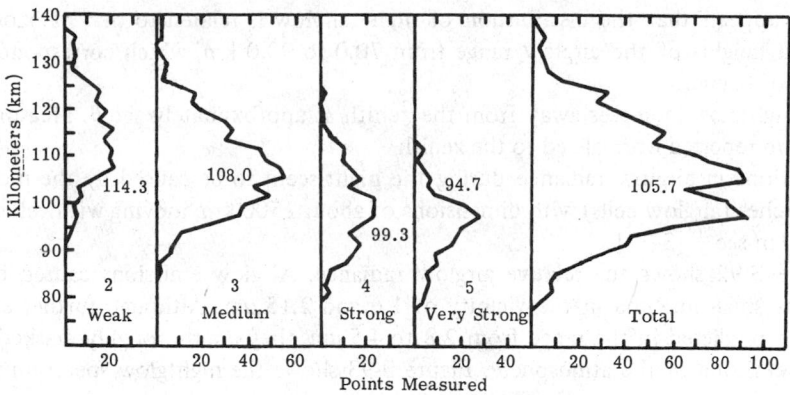

Fig. 3-90. Frequency distribution of heights of lower borders of auroral arcs [3-70].

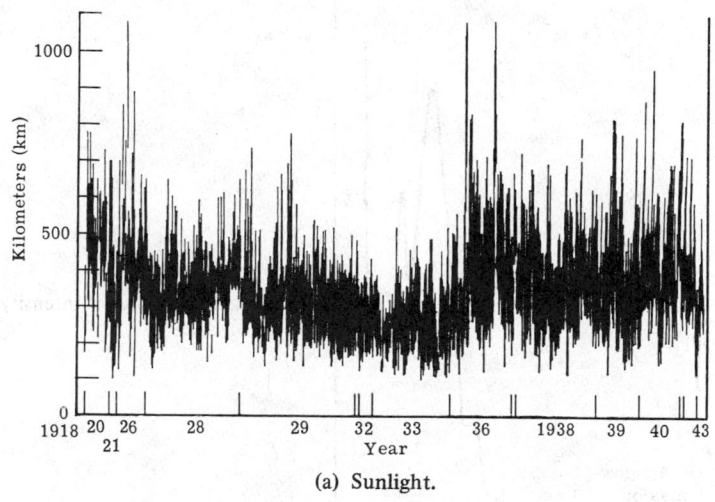

(a) Sunlight.

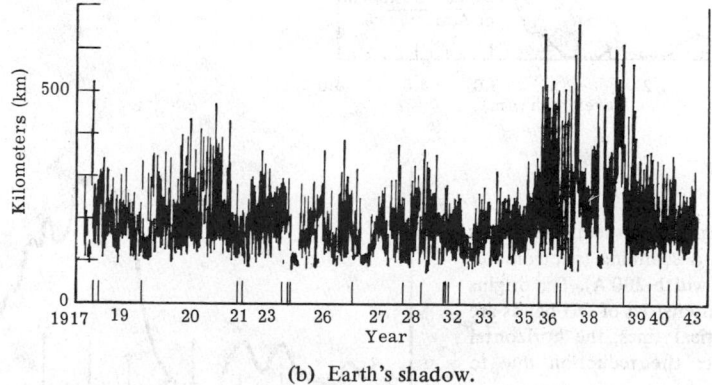

(b) Earth's shadow.

Fig. 3-91. Length and position in the atmosphere of the vertical projections of auroral rays (1917 to 1943) [3-71].

Thus, it appears that the distribution of night airglow is related to that of ozone. The measured heights of the airglow range from 70.0 to 90.0 km, which correspond to the location of ozone.

The nightglow increases away from the zenith at approximately sec θ. Measurements usually are reported normalized to the zenith.

Variations in airglow radiance during the night seem to be caused by the motion of large patches (airglow cells) with dimensions of about 2500 km, moving with velocities of about 70 m sec^{-1} [3-73].

Figure 3-92 shows the relative airglow radiance. Airglow emissions caused by OH⁻ appear as small maxima in the vicinity of 1.6 and 2.15 μm. Although further emission bands are predicted in the range from 2.8 to 4.5 μm, they are thoroughly masked by the thermal emission of the atmosphere. Figure 3-93 shows the nightglow spectrum in the 1 to 2 μm region.

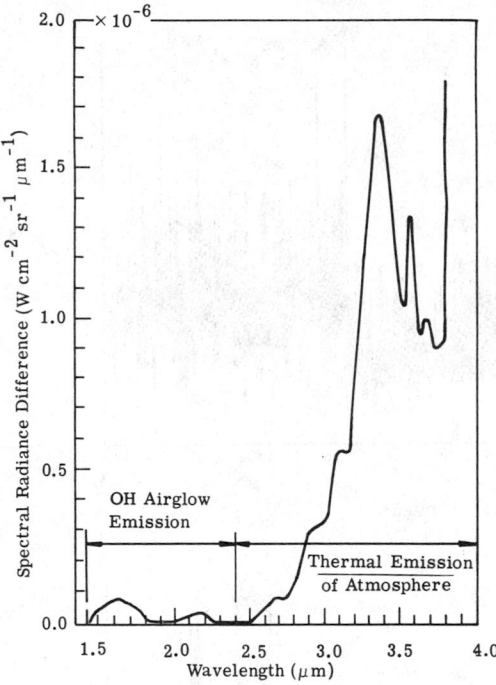

Fig. 3-92. Airglow intensity [3-74].

Fig. 3-93. Nightglow spectrum. It is obtained with a scanning spectrometer (projecting slit width 200 A). The origins and expected intensities of OH bands are shown by vertical lines; the horizontal strokes indicate the reduction due to water vapor [3-75].

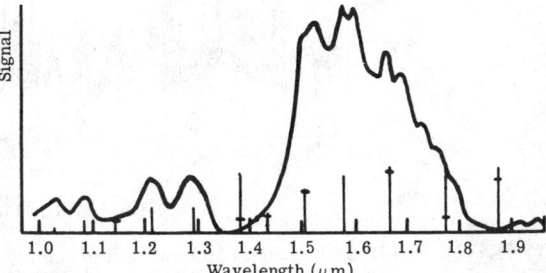

If one looks straight down from a satellite, the atmospheric spectrum should be very similar to that shown in Figures 3-92 and 3-93. Table 3-21 compares the approximate rates of emission for various airglow and auroral lines. The references in the footnotes should be consulted for further details. Note that for the airglow all results are given for the zenith itself rather than for the angles at which observations are usually made.

Table 3-21. Comparison of Aurora and Airglow Photon Emission Rates [3-59]

Source	Emission	Airglow Brightness
Aurora, IBC* I	$[OI]_{32}$ 5577 A	1 kRy
II	—	10 kRy
III	—	100 kRy
IV	—	1000 kRy
Night airglow**	$[OI]_{32}$ 5577 A	250 Ry
(in the zenith)	$[OI]_{21}$ 6300 A	50–100 Ry
	Na 5893 A	—
	summer	<30 Ry
	winter	200 Ry
	H α 6563 A	5–20 Ry
	Ly α 1215 A	2.5 kRy
	O_2 Atmospheric (0–1) 8645 A	1.5 kRy
	O_2 Herzberg (observable range)	430 Ry
	OH (4–2) 1.58 μm	175 kRy
	OH (estimated total)	4500 kRy
Twilight airglow†	N_2^+ 3914 A	1 kRy
(referred to the	(quiet magnetic conditions)	—
zenith)	Na I 5893 A	—
	summer	1 kRy
	winter	5 kRy
	$[OI]_{21}$ 6300 A	1 kRy
	Ca II 3933 A	150 Ry
	Li I 6708 A	200 Ry
	$[NI]_{21}$ 5199 A	10 Ry
	O_2 IR Atmospheric (0–1) 1.58 μm	20 kRy
Day airglow††	Na 5893 A	—
(referred to the	summer	2 kRy
zenith)	winter	15 kRy
	$[OI]_{21}$ 6300 A	50 kRy
	OI 8446 A	0.5 kRy
	OI 11,290 A	0.5 kRy
	N_2^+ 3914 A	<70 kRy
		> 1 kRy

*Recommended as definitions of the International Brightness Coefficients (IBC) [3-76, 3-77].
**Average values.
†Approximate values of the maximum emission rates that are observed during twilight. These values are often governed by the time after sunset when observation first become possible.
††Values predicted from theory [3-78] to [3-81].

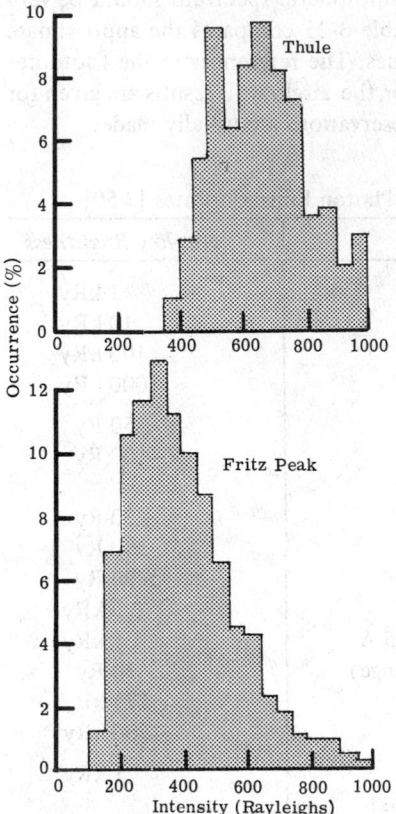

Fig. 3-94. Frequency distribution of airglow [3-70].

Figure 3-94 shows the frequency distribution of airglow and weak auroral radiance near the geomagnetic pole (Thule, Greenland) and at a subauroral station (Fritz Peak, Colorado).

There is some evidence that suggests a general increasing radiance of airglow emissions towards higher latitudes and a bright belt at middle latitudes.

Figure 3-95 shows spectrum of night airglow observed from high altitude.

Meteoroid Radiation [3-38] Elements contributing to meteor spectra are

Al, Ca, Cr, Fe, H. Mg, Mn, Na, Ni, Si, Sr, O, N.

The absolute visual magnitude, M_V, of a meteor for an observer on the ground is the magnitude it would have at 100 km through 1 air mass. The symbol M_V, in this section should not be confused with luminous exitance. Apparent visual magnitude, m_V, requires a correction for τ/R^2

$$M_V = 6.8 - 2.5 \log_{10} \left[\frac{(\int V(\lambda)\Phi_\lambda d\lambda)}{1 \text{ W}} \right] \quad (3-51)$$

The maximum luminous intensity is reached in the range, 78 to 95 km.

3.7. Properties of Terrestrial Materials

3.7.1. Soils.
Soil reflectance measurements of U.S. soils were made by Condit [3-83] in the 0.32 to 1.0 μm range. He observed that there were three main types of soil

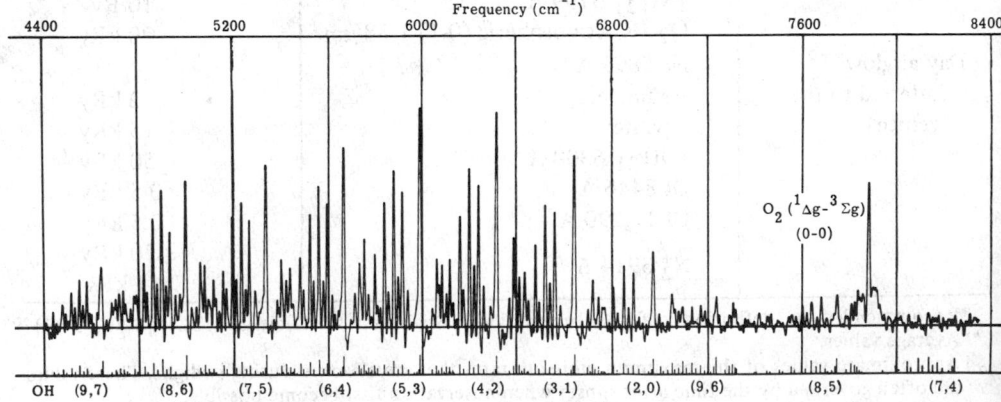

Fig. 3-95. Spectrum of night airglow observed from an altitude of about 80,000 feet. The total recording time of the twelve interferograms from which this spectrum is calculated equals 141 minutes [3-82].

spectra as exemplified in Figure 3-96 (a), (b), (c). The application of characteristic vector analysis to these spectra showed that, along with the average spectral reflectances of all soil samples, the spectral reflectance values at only five different wavelengths (0.44, 0.54, 0.64, 0.74 and 0.86 μm) were needed to replicate closely the entire spectral reflectance curve from 0.32 to 1.0 μm. His spectral reflectance curves are shown in Figure 3-96 where the circles are predicted values and the solid lines are experimental values.

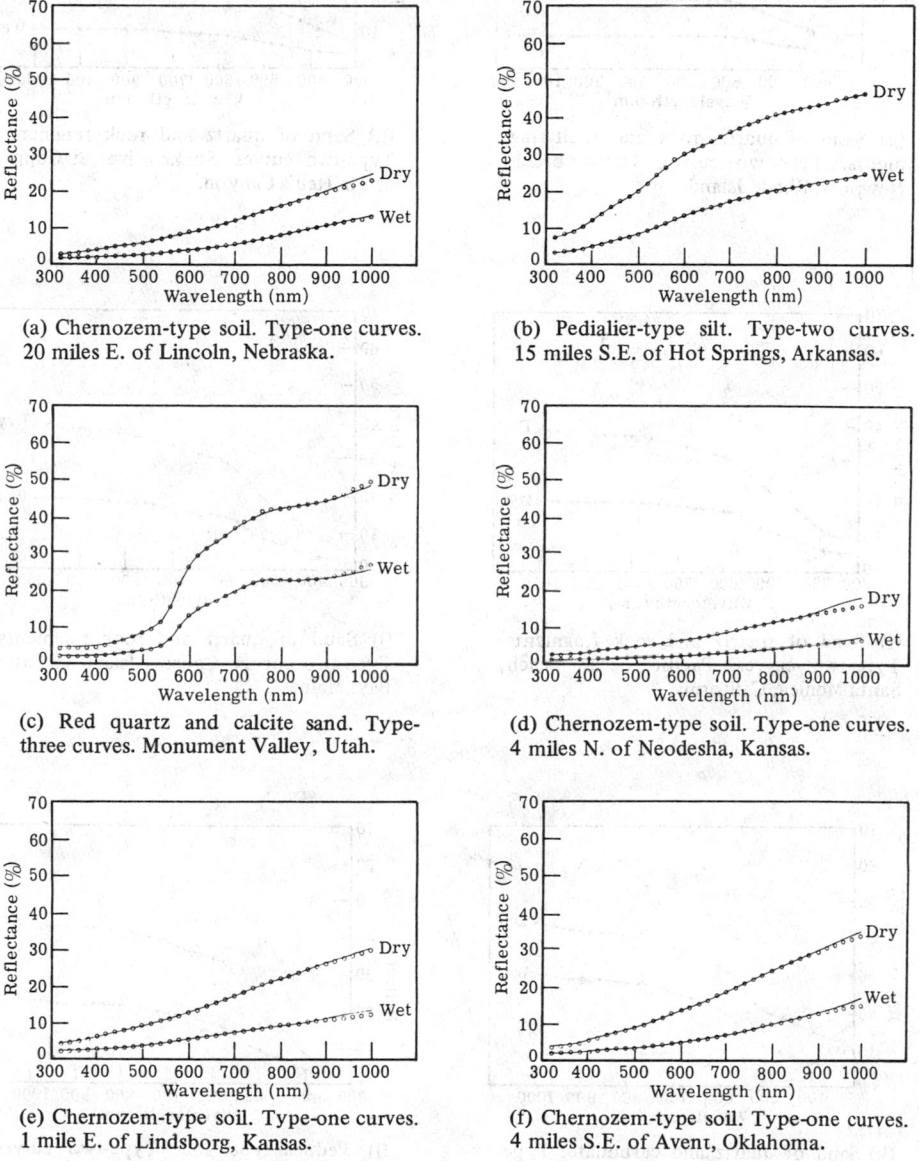

(a) Chernozem-type soil. Type-one curves. 20 miles E. of Lincoln, Nebraska.

(b) Pedialier-type silt. Type-two curves. 15 miles S.E. of Hot Springs, Arkansas.

(c) Red quartz and calcite sand. Type-three curves. Monument Valley, Utah.

(d) Chernozem-type soil. Type-one curves. 4 miles N. of Neodesha, Kansas.

(e) Chernozem-type soil. Type-one curves. 1 mile E. of Lindsborg, Kansas.

(f) Chernozem-type soil. Type-one curves. 4 miles S.E. of Avent, Oklahoma.

Fig. 3-96. Spectral reflectance curves. Measured values are shown by lines and predicted values by a series of circles.

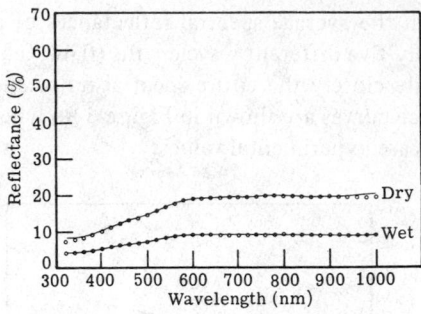

(g) Sand of quartz, rock and shell fragments. Type-two curves. Public beach, Newport, Rhode Island.

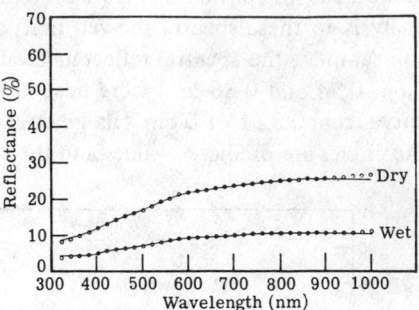

(h) Sand of quartz and rock fragments. Type-two curves. Snake River at Copper Mt. in Hell's Canyon.

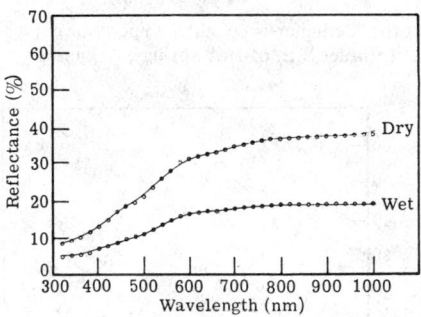

(i) Sand of quartz and rock fragments. Type-two curves. Pacific ocean beach, Santa Monica, California.

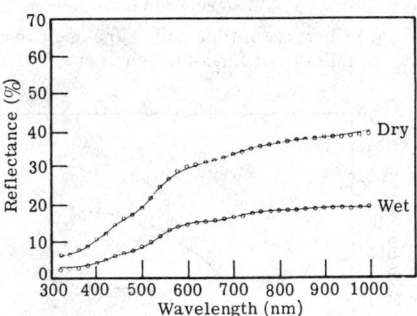

(j) Sand of quartz and rock fragments. Type-two curves. Cousins Island in Casco Bay, Maine.

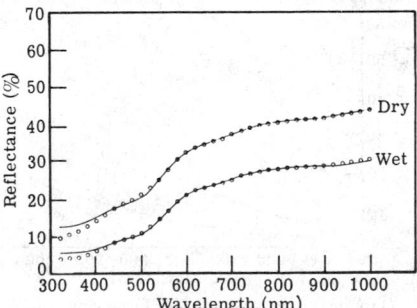

(k) Sand of quartz and carbonate. Type-two curves. Miami Beach, Florida.

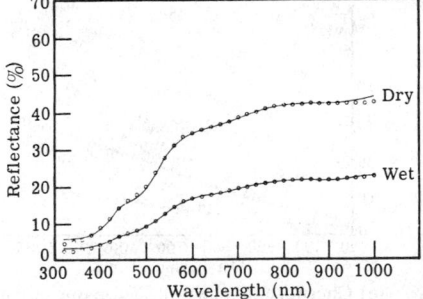

(l) Pedocal-type soil. Type-two curves. 1 mile S. of Sagamore Hills, Ohio.

Fig. 3-96. (Continued) Spectral reflectance curves. Measured values are shown by lines and predicted values by a series of circles.

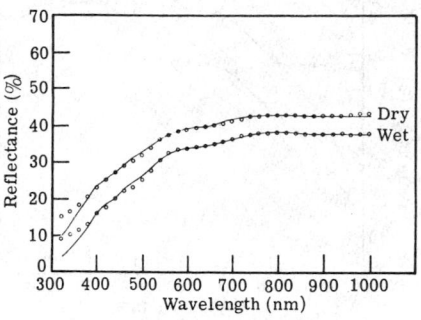

(m) Pedocal-type soil. Type-two curves. 10 miles S. of Lyman, Nebraska.

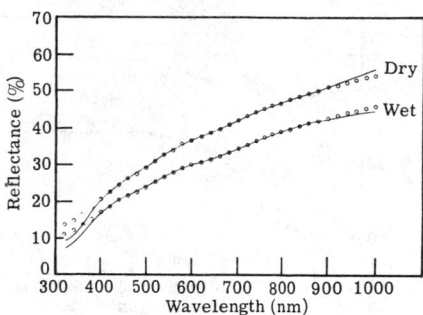

(n) Clay. Type-two curves. 3 miles E. of Paris, Missouri.

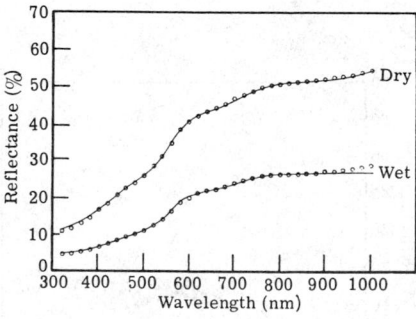

(o) Quartz sand. Type-two curves. 20 miles N. of Coos Bay, Oregon.

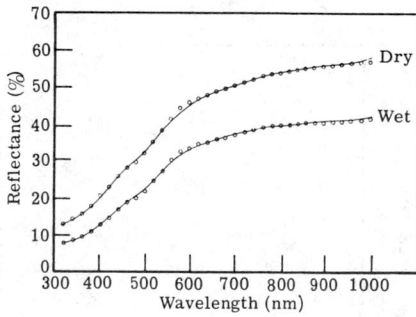

(p) Carbonate sand. Type-two curves. Waikiki Beach, Honolulu, Hawaii.

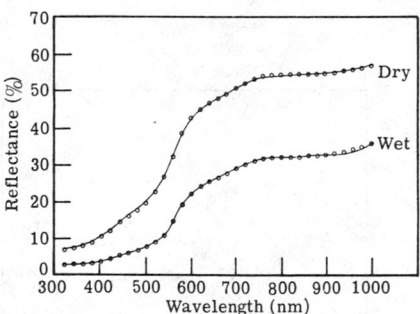

(q) Pedalfer-type Soil. Type-two curves. 3 miles E. of Mountain View, Missouri.

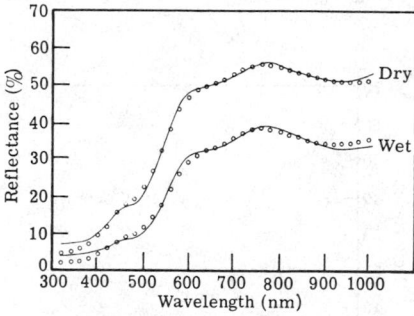

(r) Pedalfer-type soil. Type-two curves. 12 miles N. of Dalton, Georgia.

Fig. 3-96. (Continued) Spectral reflectance curves. Measured values are shown by lines and predicted values by a series of circles.

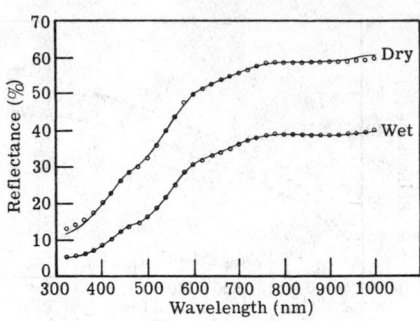

(s) Pedocal-type soil. Type-two curves. 3 miles W. of Phillipsburg, Missouri.

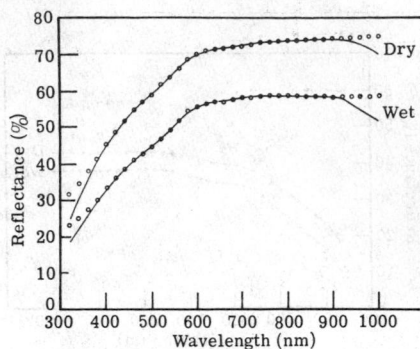

(t) Gypsum sand. Type-two curves. White sands National Monument, New Mexico.

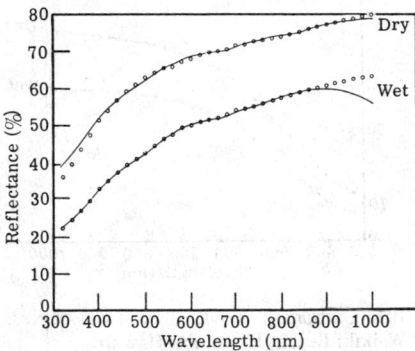

(u) Quartz sand. Type-two curves. Ft. Walton Beach, Florida.

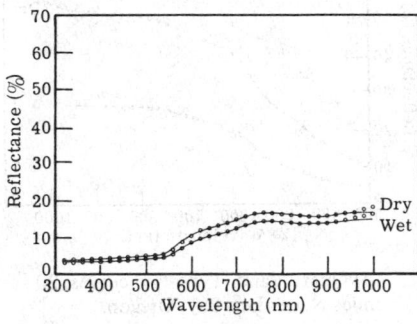

(v) Pedalfer-type soil. Type-three curves. 5 miles S.E. of Rowe, New Mexico.

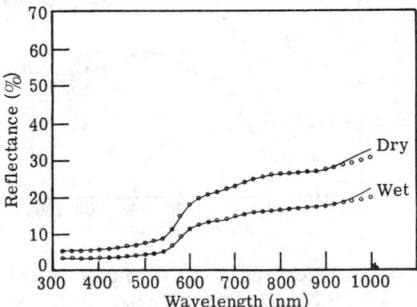

(w) Pedalfer-type soil. Type-three curves. Garden of the Gods, Colorado.

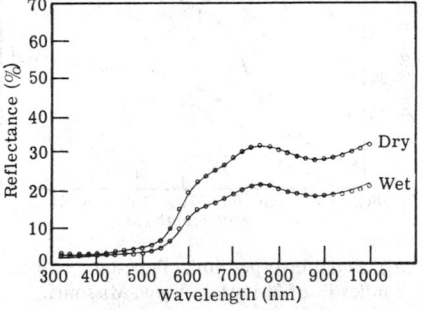

(x) Laterite-type soil. Type-three curves. 2 miles N.E. of Lexington, North Carolina.

Fig. 3-96. (Continued) Spectral reflectance curves. Measured values are shown by lines and predicted values by a series of circles.

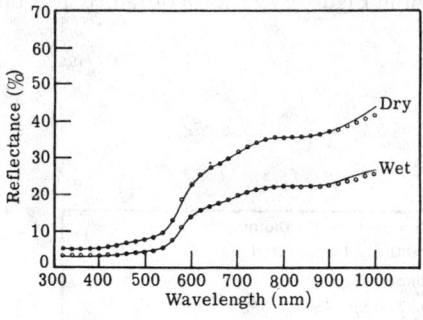

(y) Pedalfer soil. Type-three curves. Woodlawn, Colorado.

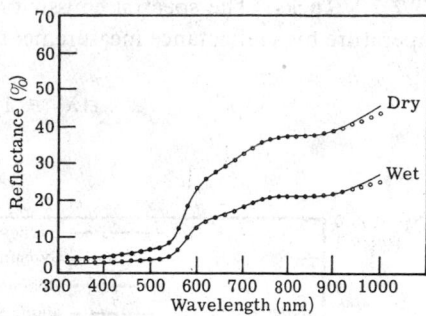

(z) Pedocal-type soil. Type-three curves. 12 miles W. of Elk City, Oklahoma.

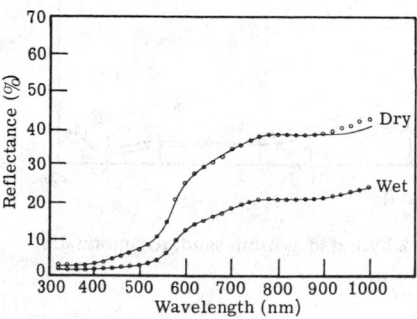

(aa) Pedalfer-type soil. Type-three curves. 6 miles N. of Del Rio, Texas.

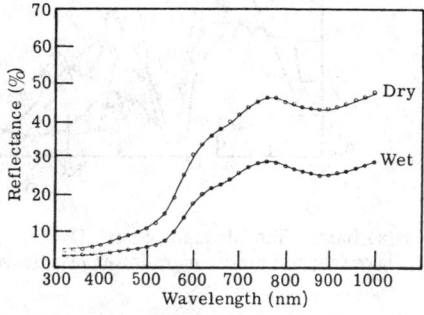

(bb) Pedalfer-type soil. Type-three curves. 4 miles N. of Griffin, Georgia.

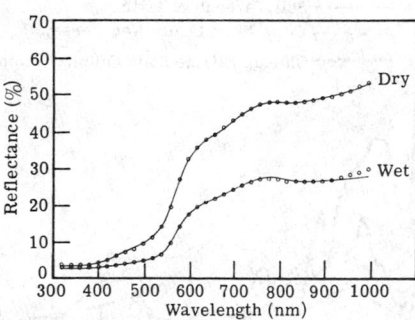

(cc) Quartz sand with hematite stain. Type-three curves. Bok Tower, Florida.

Fig. 3-96. (Continued) Spectral reflectance curves. Measured values are shown by lines and predicted values by a series of circles.

Spectral reflectances extending into the thermal infrared range are shown in Figure 3-97.

3.7.2. Rocks. The spectral emissivities shown in Figure 3-98 were obtained at room temperature by a reflectance measurement where

$$\epsilon(\lambda) = 1 - \rho(\lambda) \tag{3-52}$$

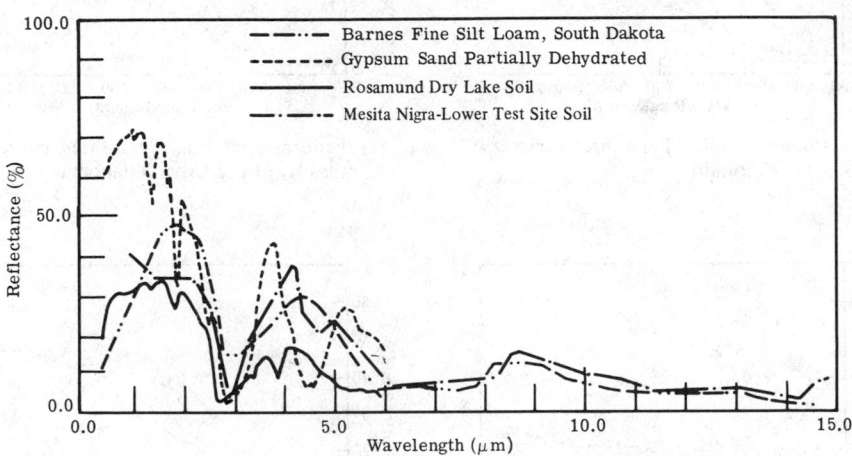

(a) Barnes fine silt loam, South Dakota; partially dehydrated gypsum sand; rosamond dry lake soil; and mesita nigra-lower test site soil.

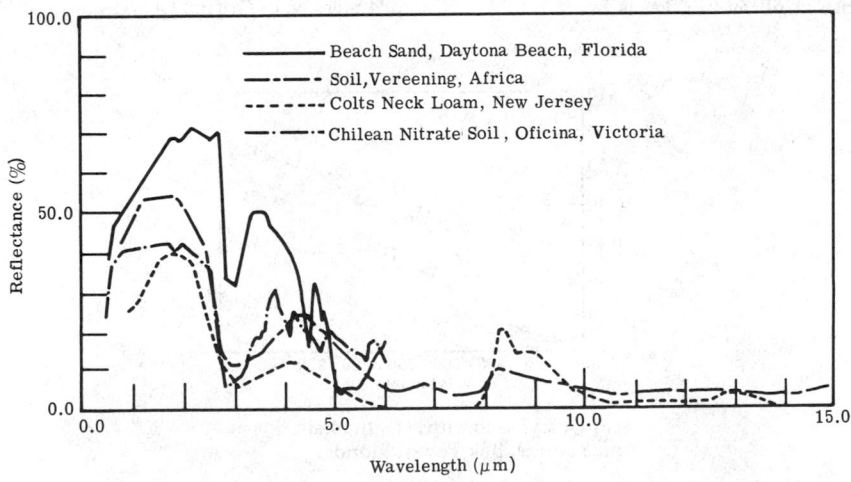

(b) Beach sand, Daytona Beach, Florida, soil, Vereening, Africa; colts neck loam, New Jersey; and Chilean nitrate soil, Oficiana, Victoria.

Fig. 3-97. Spectral reflectances extending into the thermal infrared range [3-84].

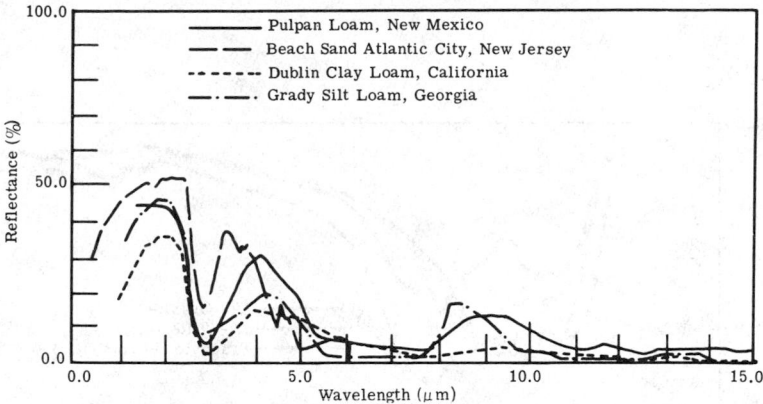

(c) Pulpan loam, New Mexico; beach sand, Atlantic City, New Jersey; Dublin clay loam, California; and Grady silt loam, Georgia.

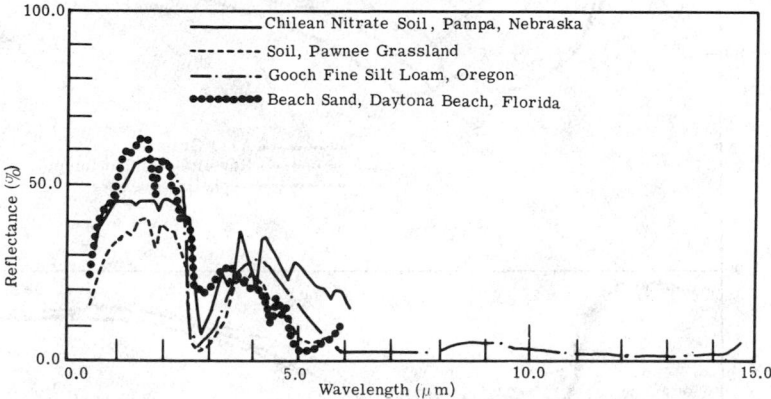

(d) Chilean nitrate soil, Pampa, Nebraska; soil Pawnee grassland; Gooch fine silt loam, Oregon; and beach sand, Daytona Beach, Florida.

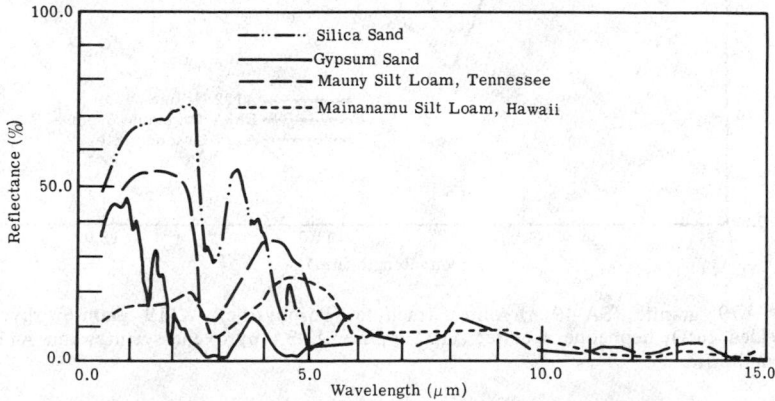

(e) Silica sand; natural gypsum sand; Mouny silt loam, Tennessee; and Mainanamu silt loam, Hawaii.

Fig. 3-97. (Continued) Spectral reflectances extending into the thermal infrared range [3-84].

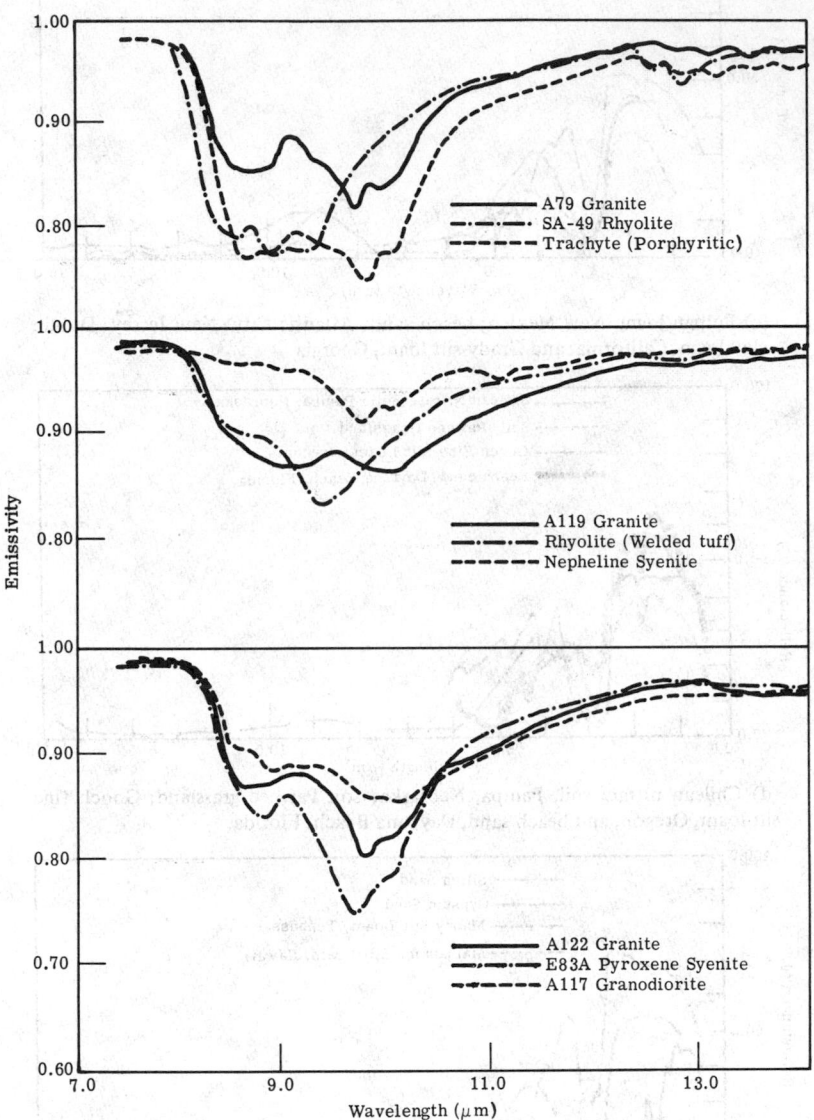

(a) A79 granite; SA-49 rhyolite; trachyte (Porphyritic); A119 granite; rhyolite (welded tuff); nepheline syenite; A122 granite; E83A pyroxene syenite; and A117 granodiorite.

Fig. 3-98. Infrared spectra of igneous silicate rocks [3-85].

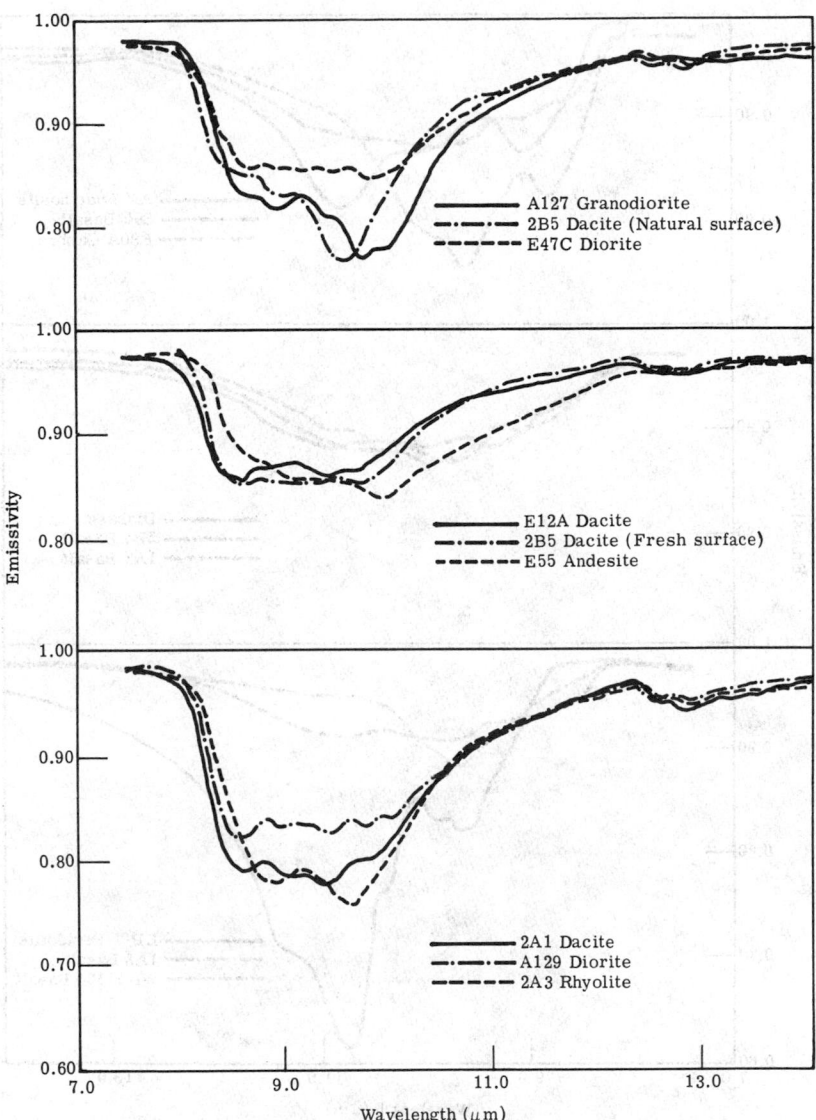

(b) A127 granodiorite; 2B5 dacite (natural surface); E47C diorite; E12A dacite; 2B5 dacite (fresh surface); E55 andesite; 2A1 dacite; A129 diorite; and 2A3 rhyolite.

Fig. 3-98. (Continued) Infrared spectra of igneous silicate rocks [3-85].

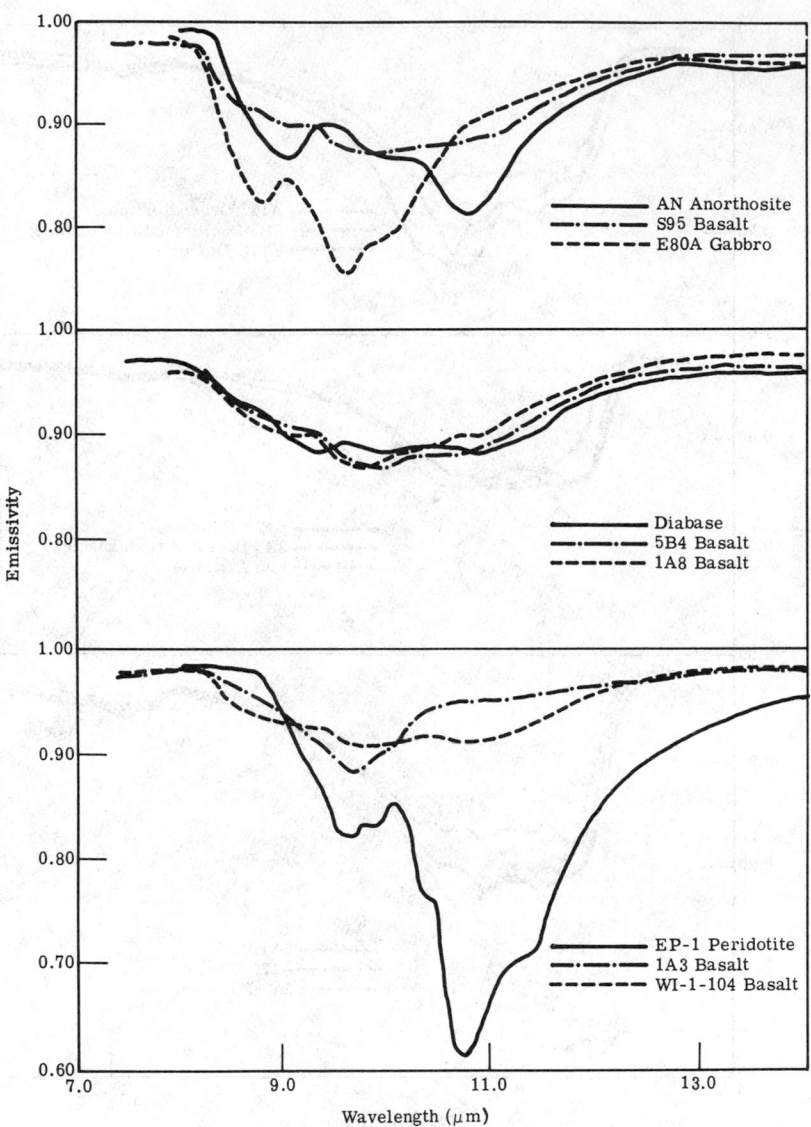

(c) An anorthosite; S95 basalt; E 80A gabbro; diabase; 5B4 basalt; 1A8 basalt; EP-1 peridotite; 1A3 basalt; and WI-1-104 basalt.

Fig. 3-98. (Continued) Infrared spectra of igneous silicate rocks [3-85].

The spectrometer field of view on the samples was about 0.5 × 12.0 mm. Coarse grained rock, such as granite, yielded variable results depending upon sample position. Figure 3-99 illustrates the degree of variation with sample position for three granite samples.

Under daylight conditions where convective heat transfer to the atmosphere is very low, some rock surfaces may sustain a large temperature gradient. This condition will alter the apparent emissivity because the effective-radiation temperature is the temperature at a depth of $1/a(\lambda)$ below the surface where $a(\lambda)$ is the spectral absorption coefficient of the rock. Figure 3-100 shows a comparison between spectral emissivity derived from reflectance measurements and a direct spectral emissivity measurement,

$$\epsilon(\lambda) = \frac{M_\lambda(\text{rock}, \lambda, T)}{M_{BB,\lambda}(\lambda, T)}$$

(3-53)

with the rock sample holder held at room temperature and the rock surface exposed to about 100 K radiation from a liquid nitrogen container. Rock surface contact temperature was measured and was used as the temperature of the reference blackbody surface.

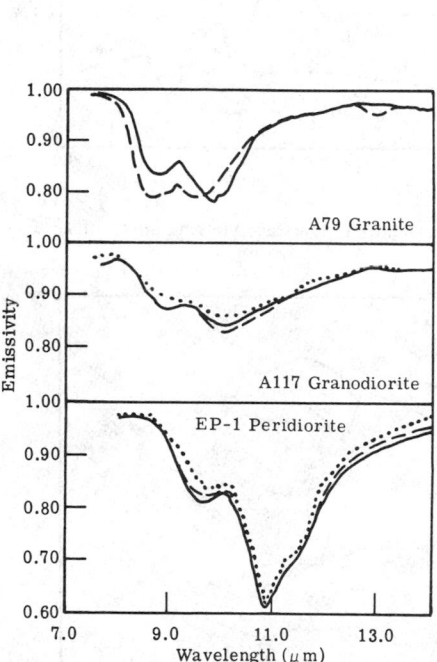

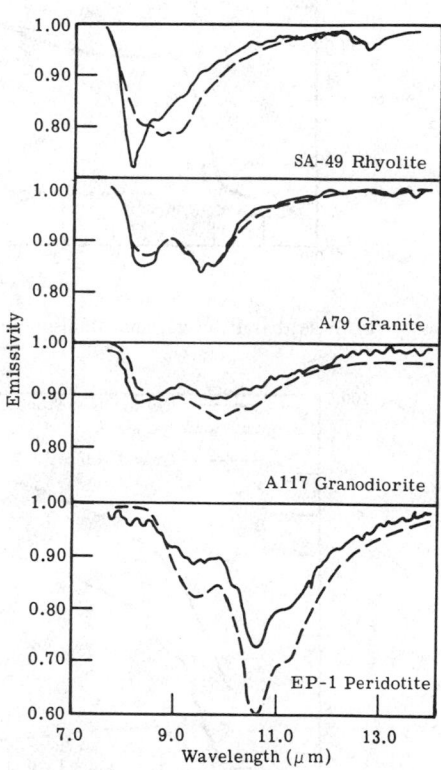

Fig. 3-99. Spectral variations resulting from measurements taken at different locations on the samples.

Fig. 3-100. Comparison of direct spectral emissivity measurements (solid line) with spectral emissivity curves derived from Kirchhoff's law and reflectance measurements (dashed line).

Figure 3-101 shows the spectral emissivity of iron rust. Iron rust stain, hematite, occurs on rocks frequently.

Spectral reflectances and emissivities of other common rock types are shown in Figure 3-102.

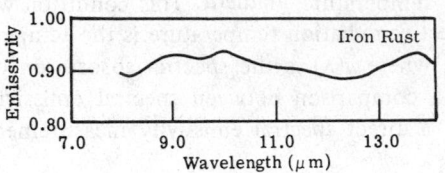

Fig. 3-101. Spectral emissivity of iron rust.

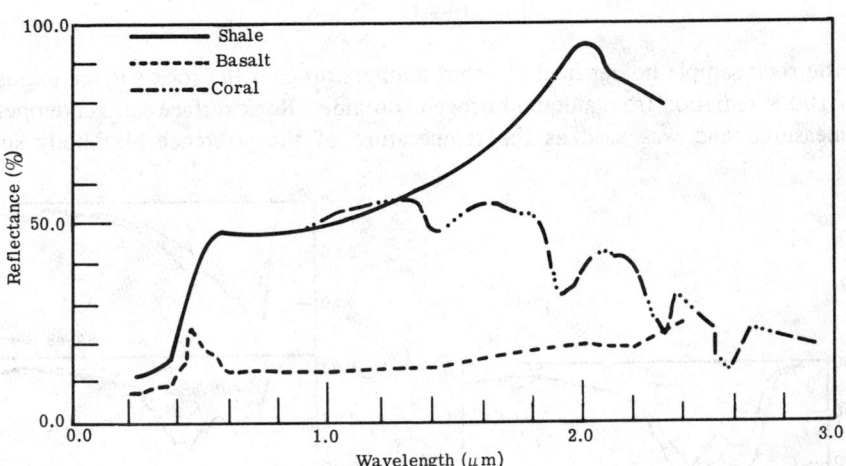

(a) Shale; basalt; coral; and Manitou limestone, weathered surface viewing angle of 15°.

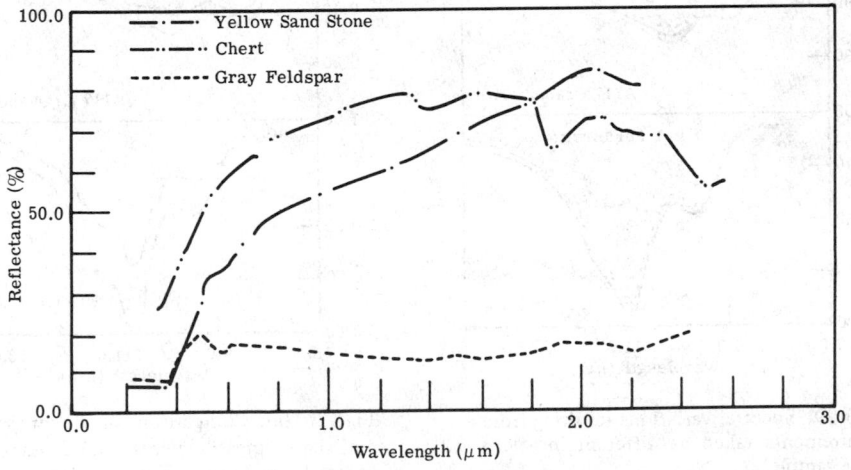

(b) Yellow sand stone; chert; and gray feldspar.

Fig. 3-102. Spectral reflectances and emissivities of common rock types [3-84].

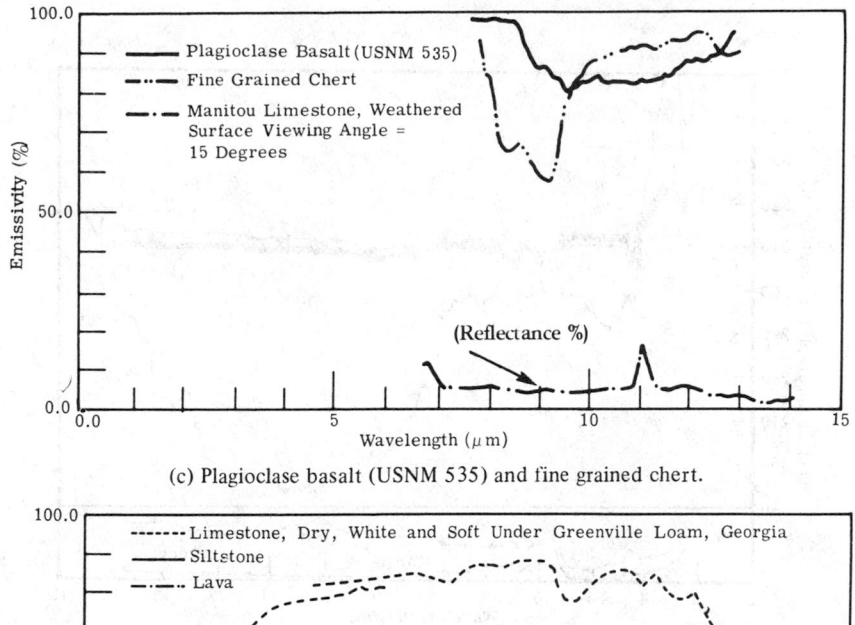

(c) Plagioclase basalt (USNM 535) and fine grained chert.

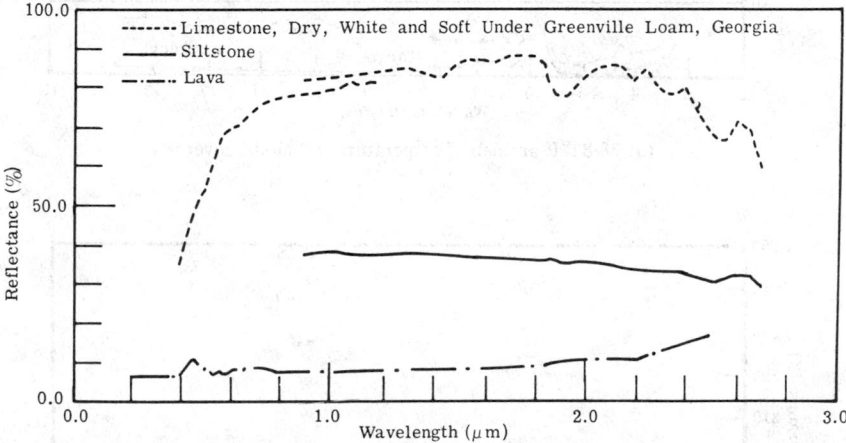

(d) Limestone, dry white and soft under Greenville loam, Georgia; siltstone; and lava.

Fig. 3-102. (Continued) Spectral reflectances and emissivities of common rock types [3-84].

3.7.3. Road Construction Materials (See also Chapter 2.). Figures 3-103, 3-104, and 3-105 show the spectral apparent temperature of some common natural and construction materials *in situ*. Spectral apparent temperature (brightness temperature) is the temperature of a blackbody surface at the radiometer aperture which will have the same spectral radiance as was observed originally. The contact temperature of each subject was not taken, so a spectral emissivity can not be inferred from these data. One may observe temperature variations of several degrees Celsius on stone concrete only a few millimeters apart when the concrete is under direct solar illumination due to differential heating of the components. The same should be true of a gravel road.

A grey body would plot as a horizontal line, except in the 3 to 4 μm atmospheric window where scattered solar radiation may increase the surface radiance by reflection. Under overcast skys, the reflection of cloud radiation tends to remove spectral details produced by emission.

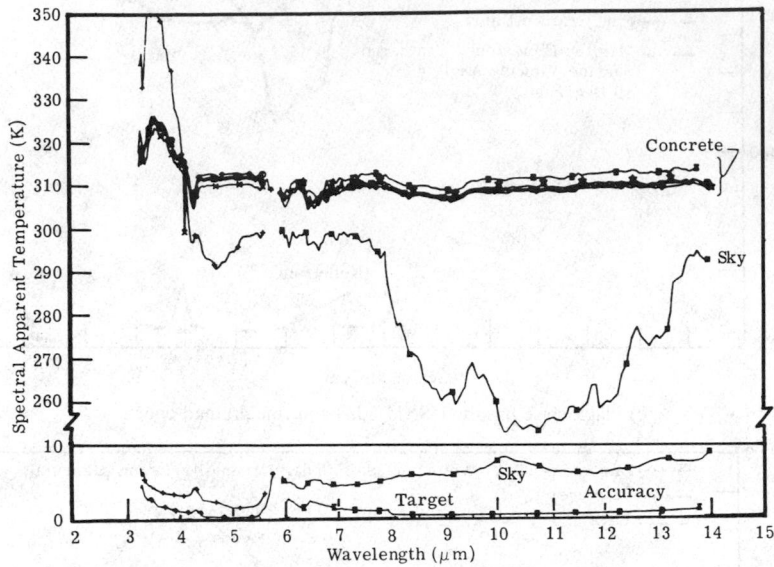

(a) 78-81°F ambient Temperature 0.0 cloud cover.

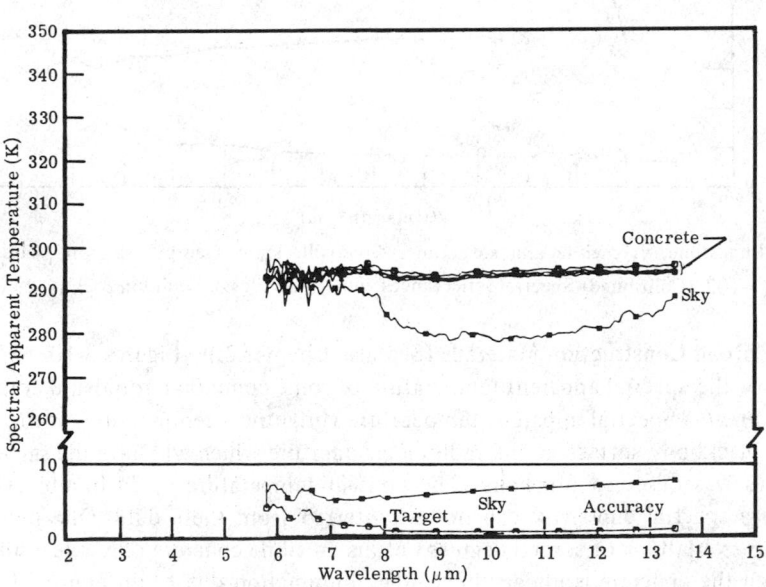

(b) 72-73°F ambient temperature 1.0 cloud cover.

Fig. 3-103. Concrete runway and sky spectral apparent temperature [3-86].

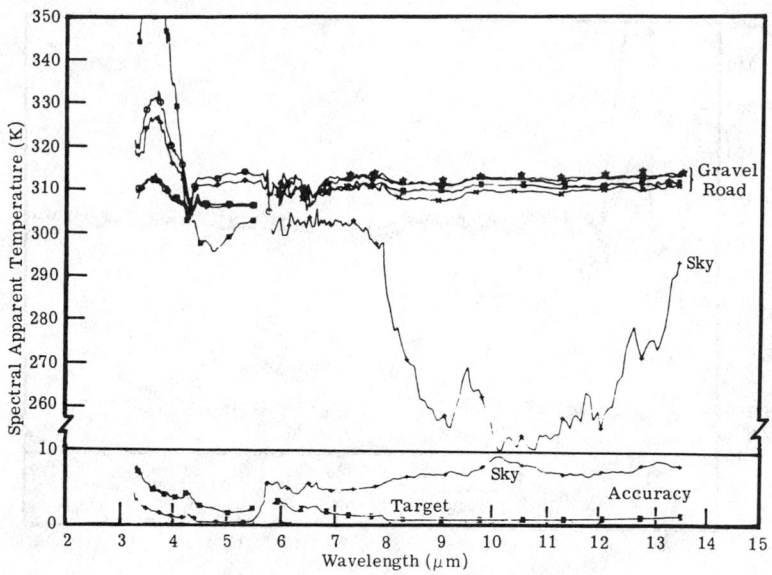

(a) 83-87°F ambient temperature 0.0 cloud cover.

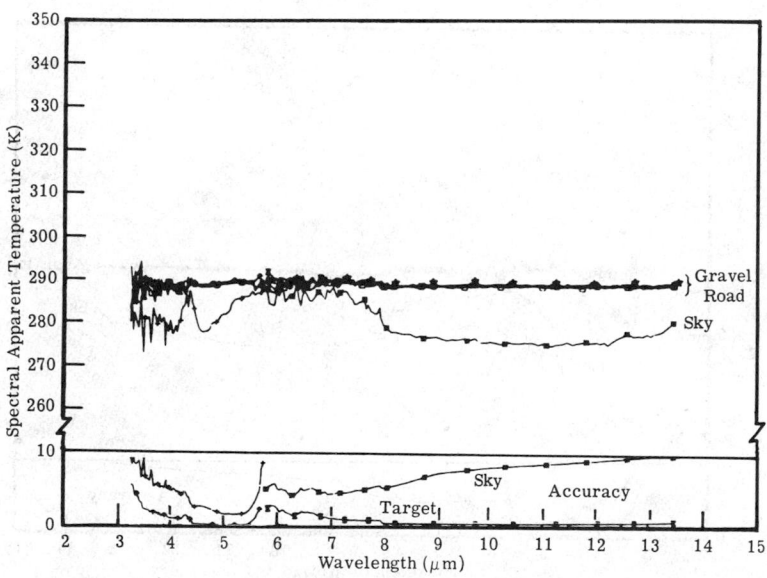

(b) 57-58°F ambient temperature 1.0 cloud cover.

Fig. 3-104. Heavily used gravel road and sky spectral apparent temperature [3-86].

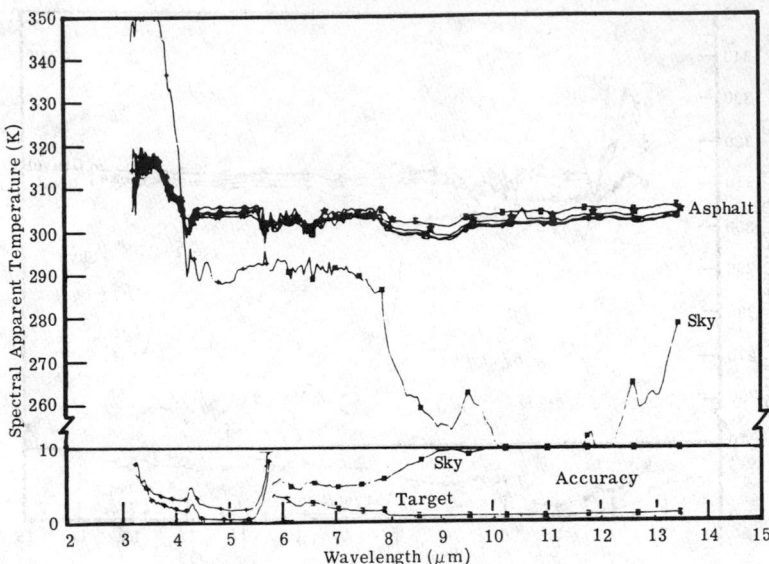

(a) 48-51°F ambient temperature 0.3 cloud cover.

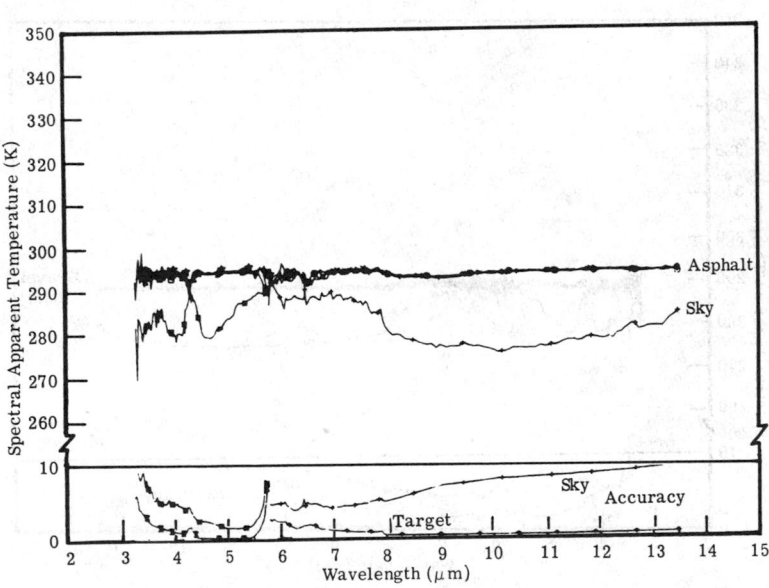

(b) 54-56°F ambient temperature 1.0 cloud cover.

Fig. 3-105. Asphalt and sky spectral apparent temperature [3-86].

The curves at the bottom of each graph indicate the expected accuracy of the measurements. The precision is indicated by the replication of data in each figure.

Spectral reflectance curves of common construction materials are shown in Figure 3-106. Observe that a large wavelength scale change occurs at 1 μm in some of these figures.

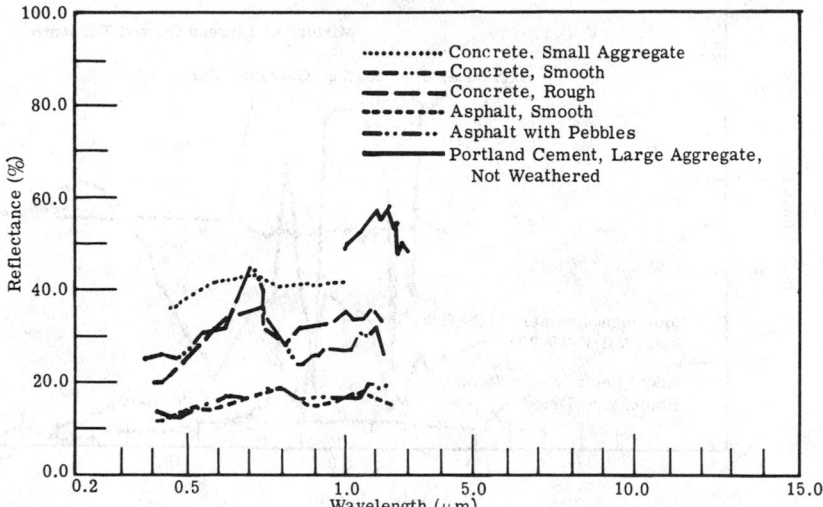

(a) Concrete, small aggregate; concrete, smooth; concrete, rough; asphalt, smooth; asphalt with pebbles; and Portland cement, large aggregate, not weathered.

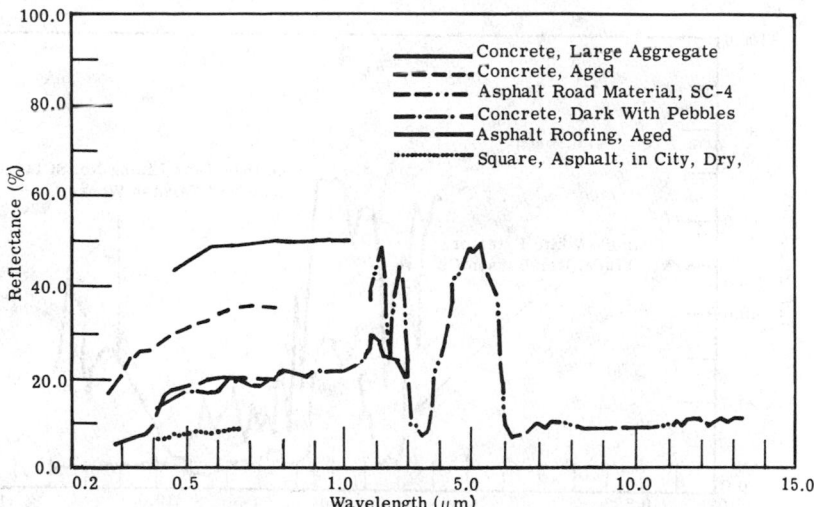

(b) Concrete large aggregate; concrete aged; asphalt road material, SC-4; concrete, dark with pebbles; asphalt roofing, aged; square, asphalt in city, dry.

Fig. 3-106. Spectral reflectance curves of common construction materials [3-87, 3-88].

3.7.4. Painted Surfaces. Many exposed surfaces of buildings are covered with a paint or stain for purposes of weather protection. A popular binder for the pigments in paints is linseed oil. The reflectance spectrum of linseed oil is shown in Figure 3-107. There are absorption bands near 3.5 and 5.9 μm and a broad band near 9 μm. Paints using

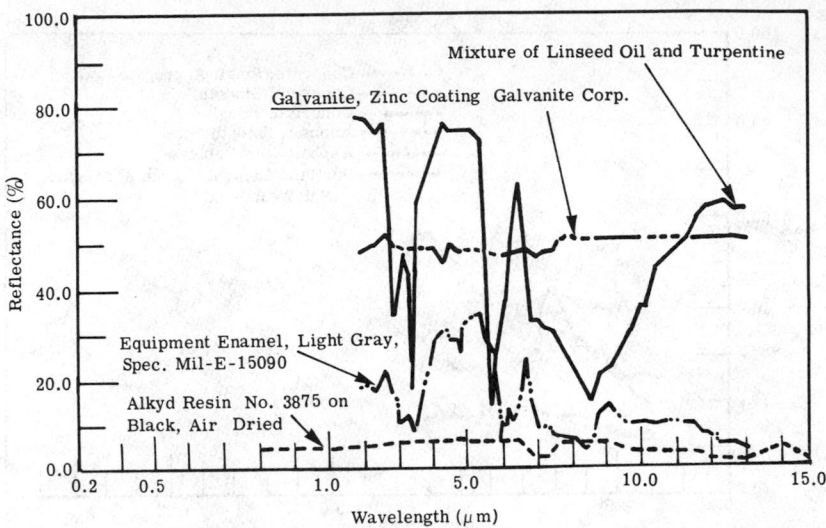

(a) Mixture of linseed oil and turpentine; Galvanite® (Registered trademark of Otley Paint Mfg.), zinc coating, Galvanite Corp.; equipment enamel, light gray, spec.mil-E-15090; and alkyd resin no. 3875 on black, air dried.

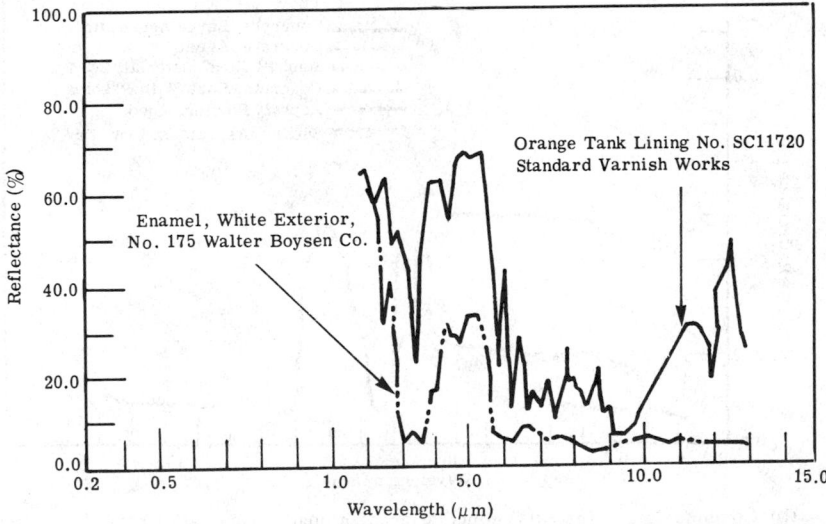

(b) Orange tank lining no. SC11720, standard varnish works; and enamel, white exterior, no. 175, Walter Boysen Co.

Fig. 3-107. Spectral reflectance curves of common paints and stains [3-87, 3-88].

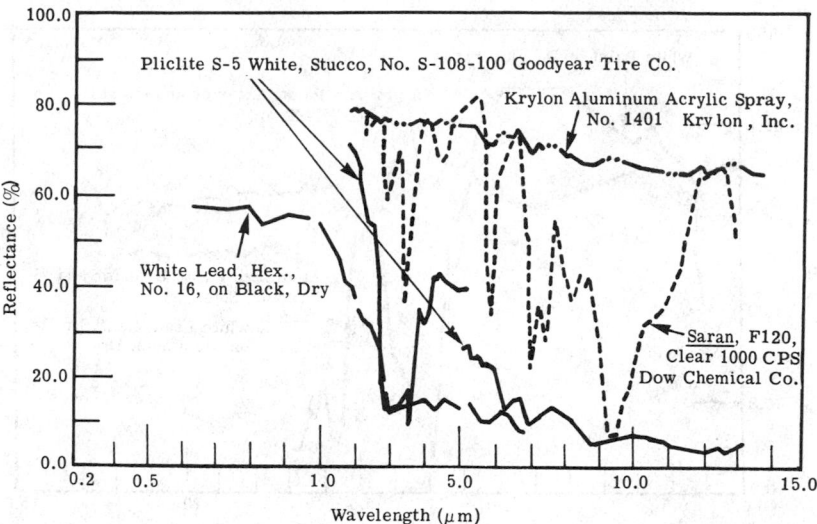

(c) Pliclite S-5 white, Stucco no. S-108-100, Goodyear Tire Co.; Krylon aluminum acrylic spray, no. 1401 Krylon; white lead, hex., no. 16 on black, dry; and Saran®, F120, clear 1000 CPS, Dow Chemical Co. (Saran® is a registered trademark of Dow Chemical Co.)

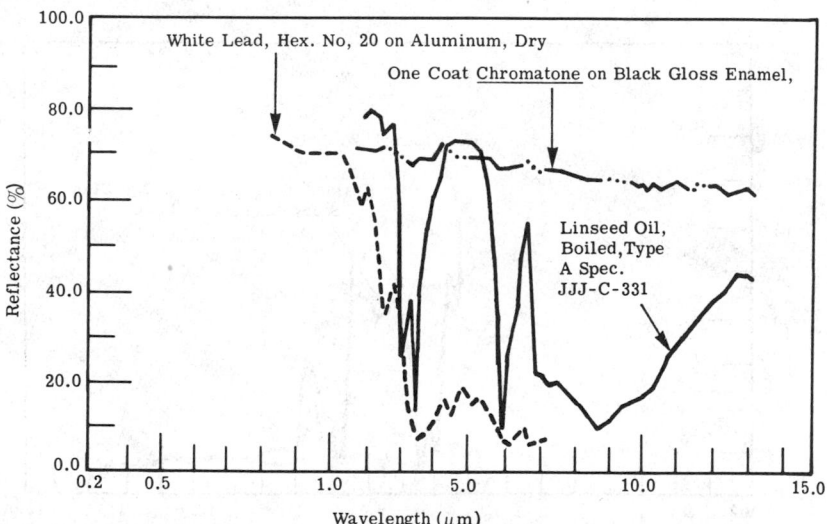

(d) White lead, hex., no. 20 on aluminum; dry; one coat Chromatone® on black gloss enamel (Chromatone® is a registered trademark of Alumatone Corp.); and linseed oil, boiled, type A spec. JJJ-C-33.

Fig. 3-107. (Continued) Spectral reflectance curves of common paints and stains [3-87, 3-88].

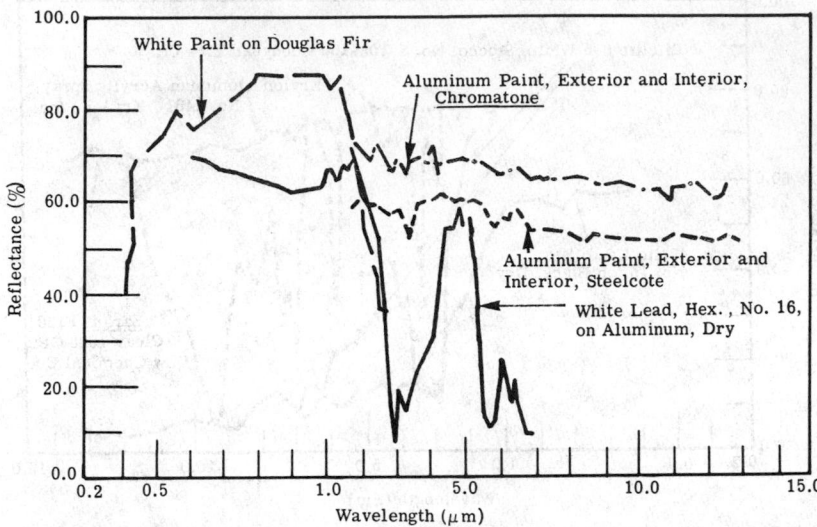

(e) White paint on douglas fir; aluminum paint, exterior and interior, Chromatone®, aluminum paint, exterior and interior steelcote; and white lead, hex., no. 16, on aluminum, dry.

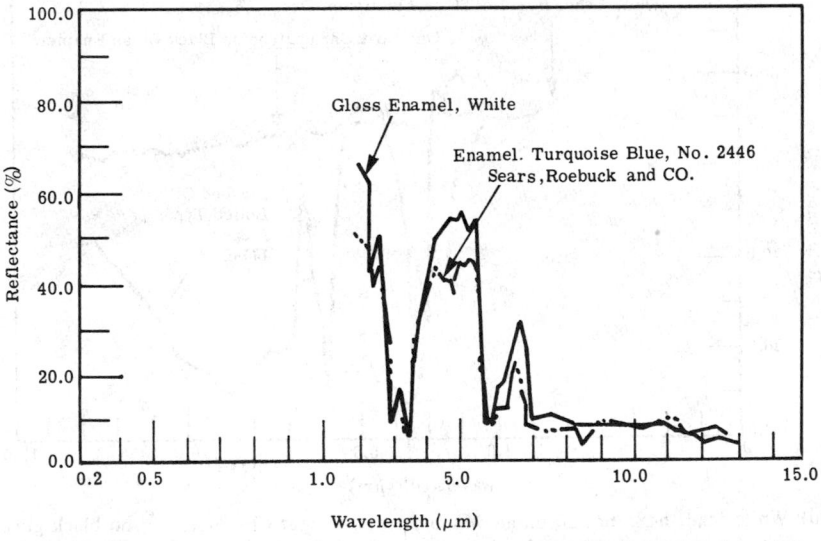

(f) Gloss enamel, white; and enamel, turquoise blue, no. 2446, Sears, Roebuck and Co.

Fig. 3-107. (Continued) Spectral reflectance curves of common paints and stains [3-87, 3-88].

linseed oil generally exhibit these same bands if the substrate is reflective and the pigments are not opaque at these wavelengths. These absorption bands are not apparent in aluminum paints, as shown in Figure 3-107. The paint coatings exhibit considerable spectral detail in the thermal infrared range so that grey body approximations are likely to lead to significant error. (Note the large wavelength scale change at 1 μm.)

3.7.5. Marine Backgrounds. The radiance of the sea surface is the sum of its thermal emission and reflected incident radiation. Factors that determine the character of the marine background are the following:

(1) the optical properties of water.

(2) surface geometry and wave-slope distribution.

(3) surface temperature distribution.

(4) bottom material properties.

Atmospheric scattering, transmission and emission in the optical path from scene to observing instrument contribute significantly to observed radiances.

Optical Properties of Water in the Thermal Range. Water is essentially opaque to infrared radiation longer than 3 μm. Few liquids have absorption coefficients of the same order of magnitude. Consequently, the water surface, which is 0.01 cm thick, determines the thermal radiance of the water. Subsurface scattering of sky radiation is absent in the thermal range. The optical influence in the thermal range of thin layers of surface contamination is negligible except for the suppression of capillary waves by surface tension changes—causing slicks and the alteration of heat exchange by evaporation. There is no significant difference in the transmittance of sea and distilled water for these thin layers in the 2 to 15 μm region.

The infrared transmittance, reflectance, emissivity and indices of refraction for water are shown in Figures 3-108 through 3-111. The spectral absorption coefficient for sea water is shown in Figures 3-112 and 3-113.

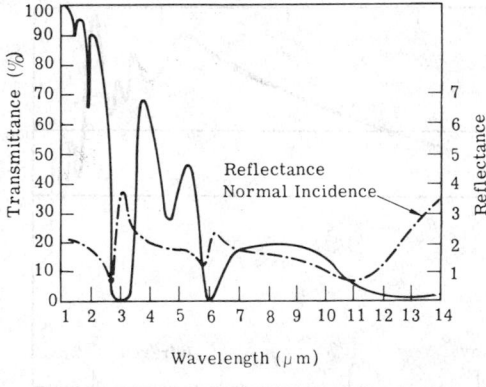

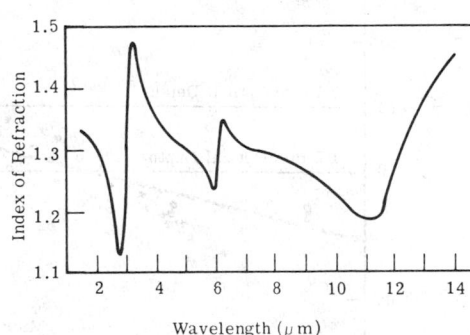

Fig. 3-108. Transmittance of 0.002 cm of sea water and reflectance of a free sea-water surface [3-89].

Fig. 3-109. Indices of refraction of water calculated from reflectivity data in Figure 3-108.

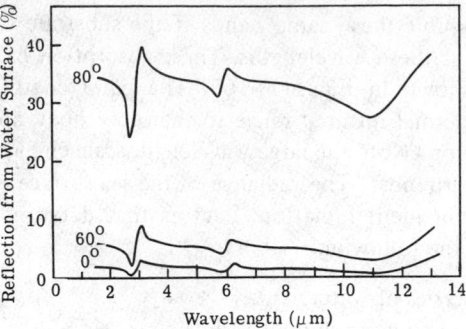

Fig. 3-110. Reflection from a water surface at 0°, 60°, and 80° angle of incidence calculated from data in Figure 3-109.

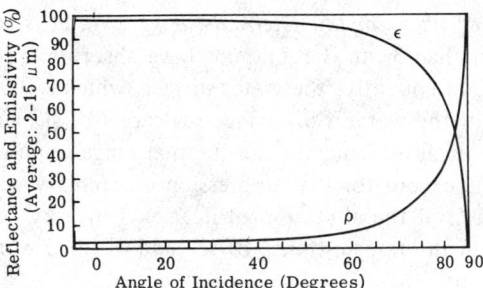

Fig. 3-111. Reflectance and emissivity of water (2 to 15 μm average) versus angle of incidence, calculated from averaged data of Figure 3-109. (Note the scale change.) [3-90].

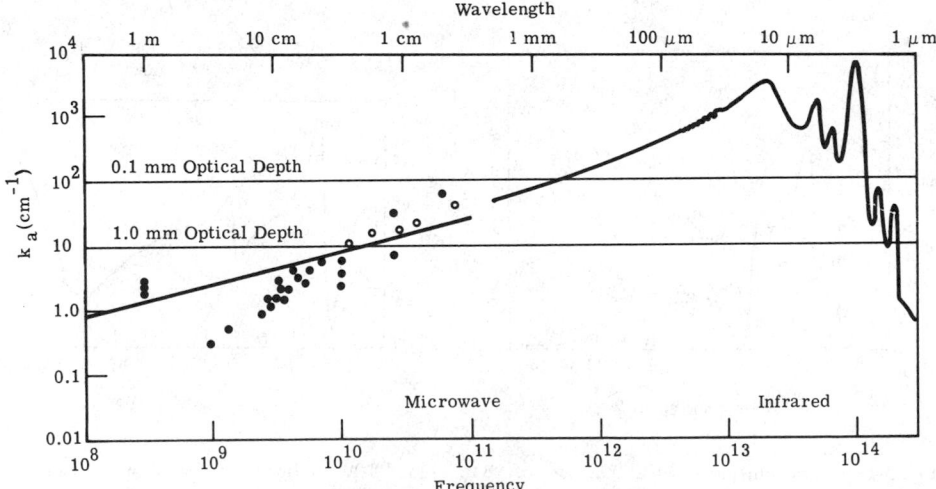

Fig. 3-112. Absorption coefficient, k_a, of sea water versus wavelength [3-91].

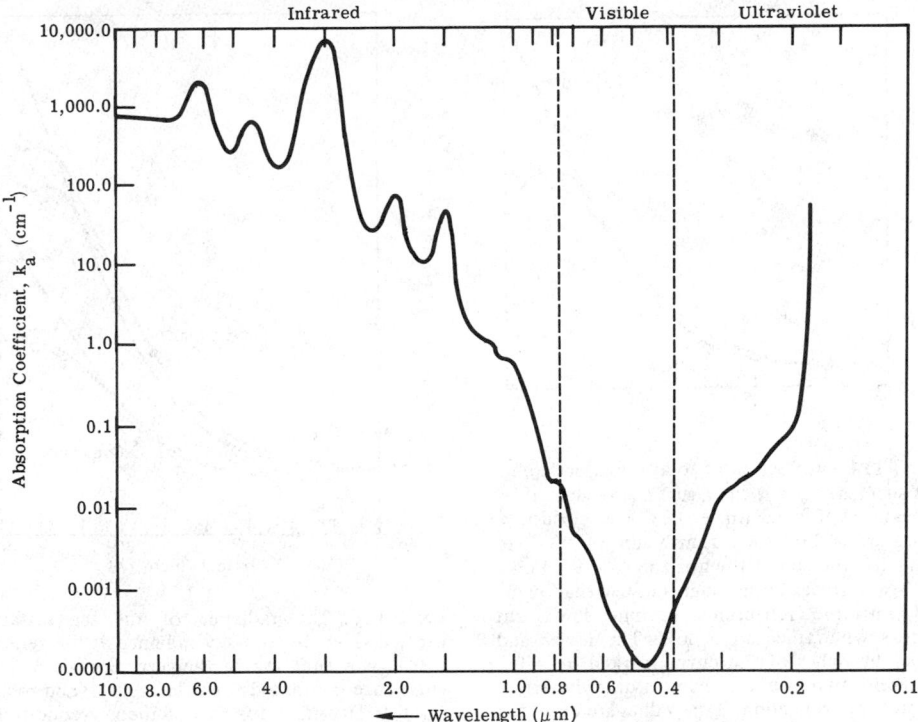

Fig. 3-113. Absorption of radiation by sea water [3-91].

Sea-Surface Geometry. The effect of wave slope on the reflectance of a sea surface roughened by a Beaufort 4 wind (11 to 16 knots, white caps) is seen in Figure 3-114. Here, for an average rough sea, the reflectance approaches 20% near the horizon. Consequently, the emissivity remains at 80% or higher.

The radiance of the sea surface along an azimuth $90°$ from that of the sun (in daylight, for clear and for overcast conditions) is shown in Figure 3-115.

Information is lacking on similar observations for the radiance of the sea surface at night. However, the variation of sky radiance with zenith angle is similar day and night and the photographic reflectance is about equal to the average for the infrared from 2 to 15 μm (Figure 3-111). Consequently, the curves in Figure 3-115 are instructive because they show the general shape of that part of the radiance of the sea surface at night caused by the reflection of sky radiation. To these curves must be added the infrared radiance of the sea surface because of its thermal emission. Examples of the spectral radiance of the sea for day and after sundown are shown in Figures 3-116, 3-117 and 3-118 [3-55].

For further data on sea-surface geometry see Reference [3-93].

Sea-Surface Temperature Distribution. The temperature of the sea surface determines the contribution of emission to its total radiance. In arctic regions, this temperature is near $0°$ C; near the equator it rises to $29°$ C. Currents, such as the warm water of the Gulf Stream, produce anomalies of several degrees Celsius as it flows into colder areas. However, in most infrared scenes of marine interest, it is the radiance variation from point-to-point that determines the background against which a target is seen. Recent improvements

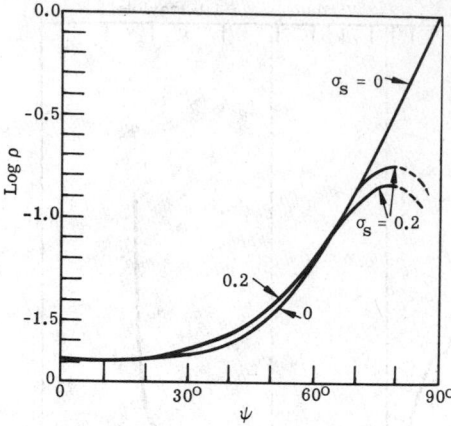

Fig. 3-114. Reflection of solar radiation from a flat surface ($\sigma_S = 0$) and a surface roughened by a beaufort 4 wind ($\sigma_S = 0.2$). The albedo, ρ, varies from 0.02 for a zenith sun, ($\psi = 0°$) to unity for the sun at the horizon ($\psi = 90°$) on a flat sea surface. For a rough surface, shadowing and multiple reflections become important factors when the sun is low. The lower and upper branches of the curve marked $\sigma_S = 0.2$ represent two assumptions regarding the effect of multiple reflection. True values are expected to lie between the indicated limits [3-92].

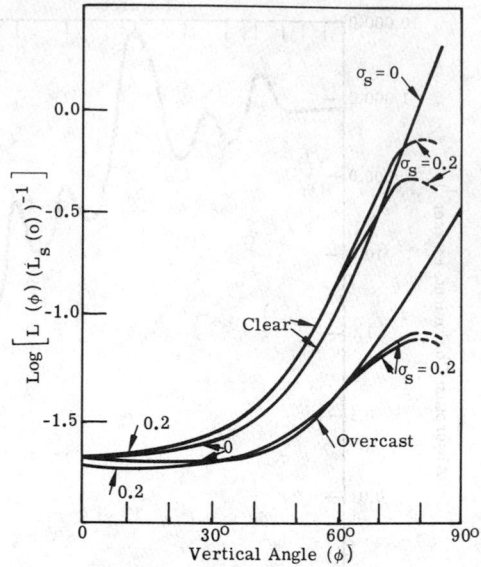

Fig. 3-115. The radiance of the sea surface, $L(\phi)$, divided by the sky radiance at the zenith, $L_S(o)$, as a function of the vertical angle ϕ. The curves are computed for a flat ($\sigma_S = 0$) and rough ($\sigma_S = 0.2$) surface for two of the sky conditions illustrated in Figure 3-14 [3-92].

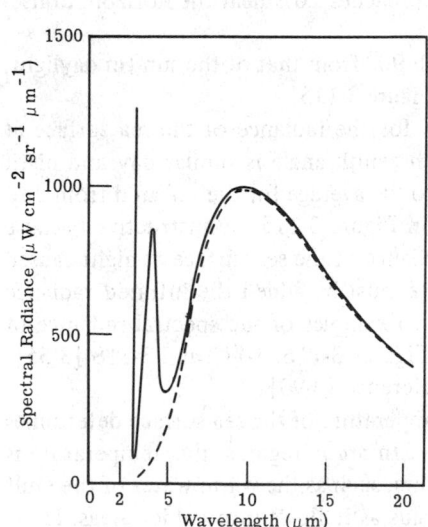

Fig. 3-116. Spectral radiance of the Banana River at Cocoa Beach, Florida [3-31].

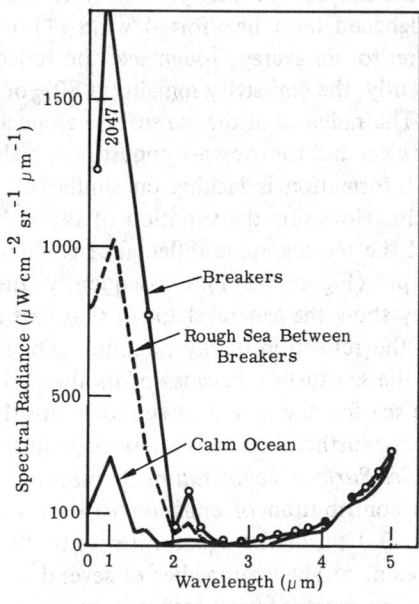

Fig. 3-117. Spectral radiance of the Ocean [3-31].

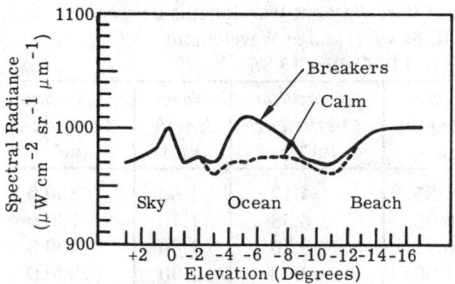

Fig. 3-118. Spectral radiance of the ocean versus the elevation angle [3-31].

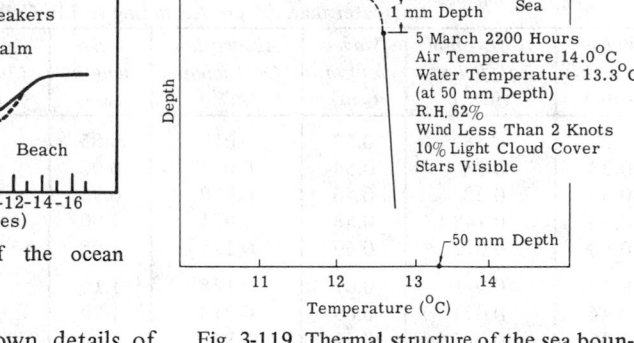

Fig. 3-119. Thermal structure of the sea boundary layer. Previous conditions: 12 h cool (12 to 15°C), no rain. The data were taken during the passage of a warm front.

in thermal mappers have shown details of this variation which is usually caused by temperature differences over the sea surface, but under some conditions reflected sky radiance predominates.

The temperature of the upper 0.1 mm of the sea surface under evaporative conditions has been measured as 0.6°C colder than water a few centimeters below [3-94]. The sharpest gradient is in the upper 1.0 mm [3-93]. Measurements typical for the conditions noted are shown in Figure 3-119.

The temperature of this layer with low heat capacity is determined by the rate of evaporation, radiation exchange, and the flow of heat from the air and from below. It has been found experimentally that the presence of surface contaminations reduces (slightly) the flow of heat from below so that a slick (a region in the sea with enough surface contamination to alter surface tension) appears colder than adjacent areas outside the slick. Finally, the flow of heat from below is also influenced by the convective activity of the water layer above the thermocline.

Optical Properties of Water in the Solar Range–0.35 to 3.0 μm. The spectral absorption coefficient for pure water falls to very low values in the 0.4 to 0.7 μm range, so radiation can penetrate deeply into the water body and scatter from suspended particulates and from the bottom. The spectral absorption coefficient, $a(\lambda)$, defined by the relation

$$\tau(\lambda) = e^{-a(\lambda)x} \tag{3-54}$$

for pure water of depth x is given in Table 3-22. Figure 3-112 shows the spectral absorption coefficient of sea water over the infrared and microwave spectral regions.

Natural water bodies contain both solutes and particulates in suspension. The spectral extinction or attenuation coefficient, $k(\lambda)$, is a measure of the combined effects of the absorption of the solution and particulates (as well as the backscattering due to the particulates) in reducing the downwelling radiation in natural water so that

$$E_\lambda(x) = E_\lambda(0)e^{-k(\lambda)x} \tag{3-55}$$

where $E_\lambda(x)$ = the spectral irradiance on the upper surface of a horizontal plane
$\quad\quad x$ = depth of plane

Table 3-22. Absorption Coefficients Per Meter of Pure Water at Wavelengths Between
0.32 μm and 0.65 μm According to W.R. Sawyer; and at Wavelengths
Greater than .65 μm According to J.R. Collins [3-95]

Wave-length (μm)	Absorption Coefficient (m^{-1})	Wave-length (μm)	Absorption Coefficient (m^{-1})	Wave-length (μm)	Absorption Coefficient (m^{-1})	Wave-length (μm)	Absorption Coefficient (m^{-1})
0.32	0.58	0.52	0.019	0.85	4.12	1.60	800.0
0.34	0.38	0.54	0.024	0.90	6.55	1.70	730.0
0.36	0.28	0.56	0.030	0.95	28.80	1.80	1700.0
0.38	0.148	0.58	0.055	1.00	39.70	1.90	7300.0
0.40	0.072	0.60	0.125	1.05	17.70	2.00	8500.0
0.42	0.041	0.62	0.178	1.10	20.30	2.10	3900.0
0.44	0.023	0.65	0.210	1.20	123.20	2.20	2100.0
0.46	0.015	0.70	0.84	1.30	150.00	2.30	2400.0
0.48	0.015	0.75	2.72	1.40	1600.00	2.40	4200.0
0.50	0.016	0.80	2.40	1.50	1940.00	2.50	8500.0

Because of multiple scattering and particulate absorption, $k(\lambda)$ is not necessarily the simple sum of the absorption coefficients of the solution and backscattering coefficient of the particulates. Figure 3-120 shows the spectral attenuation coefficient for natural water bodies. The attenuation coefficient for fresh water falls below that for distilled water. This difference may be an artifact of the experimental methods which were used. The downwelling spectral irradiance varies greatly with water composition. Figures 3-121 and 3-122 present actual measurements.

The reflectance of water bodies has a surface reflectance component given by Fresnel's equations, and a bulk reflectance component which depends upon specific water composition. The calculated reflectance of a deep ocean, including surface reflection and the effects of a turbid atmosphere, is shown in Figure 3-123 with mg m^{-3} of chlorophyll as a parameter. The calculated bulk spectral reflectance of a 64-m-deep sand bottom ocean with 16-m-deep photic zone is shown in Figure 3-124, with the phytoplankton scattering coefficient, $S = \sigma n$, as a parameter where σ is the average plankton cross-section and n is the number per unit volume. Note that in both Figures 3-123 and 3-124 the curves tend to cross near 0.5 μm. This crossover point is sometimes called the hinge point. The location of the hinge point changes somewhat with the introduction of other materials in the water. Airborne spectral reflectance data for sea surfaces are shown in Figure 3-125. No correction was made for path radiance of the atmosphere or surface reflectance of the sky. Curves B, C, and E appear to exhibit the hinge point.

The spectral attenuation coefficients of prepared algae suspensions are presented in Table 3-23. The algae concentrations were measured as the weight per liter of filtered dry solids dried at 103°C for 90 minutes. Chlorella and Selenastrum are green algae. Anabaena and Microcystis are blue-green algae. The spectral reflectances of prepared columns of algae suspensions are shown in Figures 3-126 and 3-127*.

---

*These spectra were made using a specialized optical arrangement that included reflected radiation from the container walls so that these spectra will not correspond quantitatively to spectra made by other arrangements.

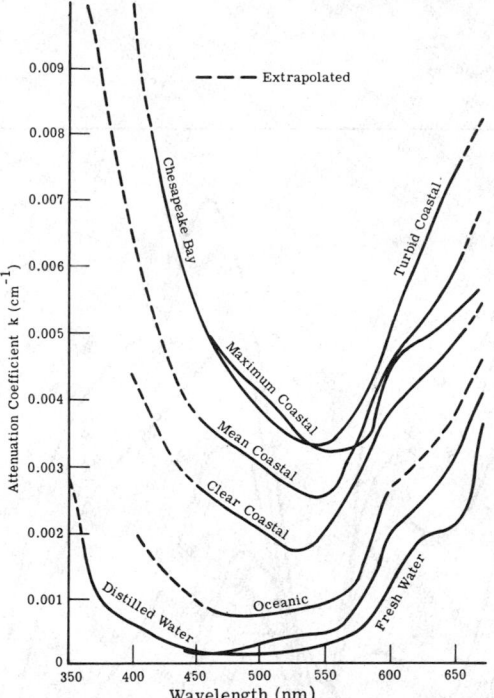

Fig. 3-120. Attenuation coefficient, k, versus wavelength for distilled, fresh, and sea water [3-96].

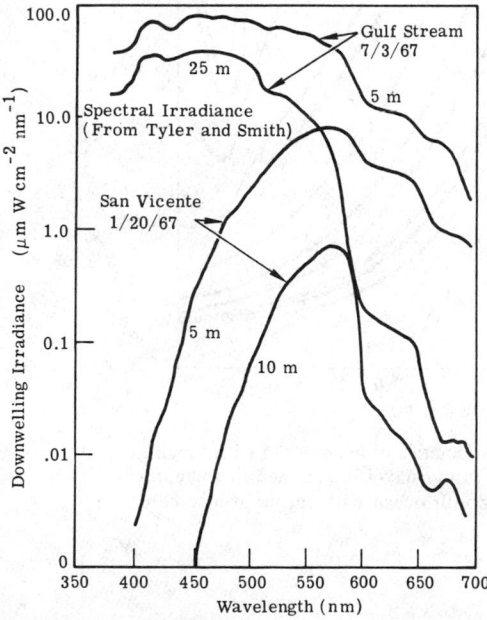

Fig. 3-121. Downwelling spectral irradiance measured by Tyler and Smith [3-97].

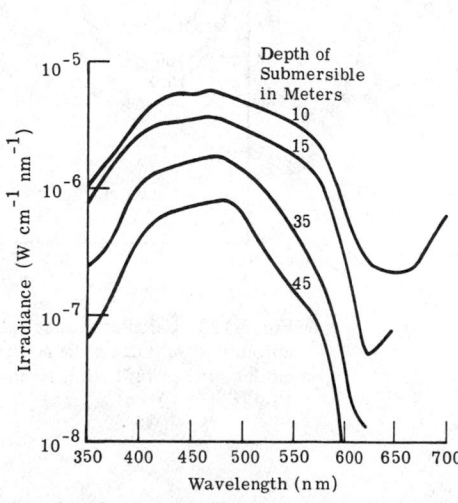

Fig. 3-122. Downwelling spectral irradiance measured in the Ben Franklin experiment [3-97].

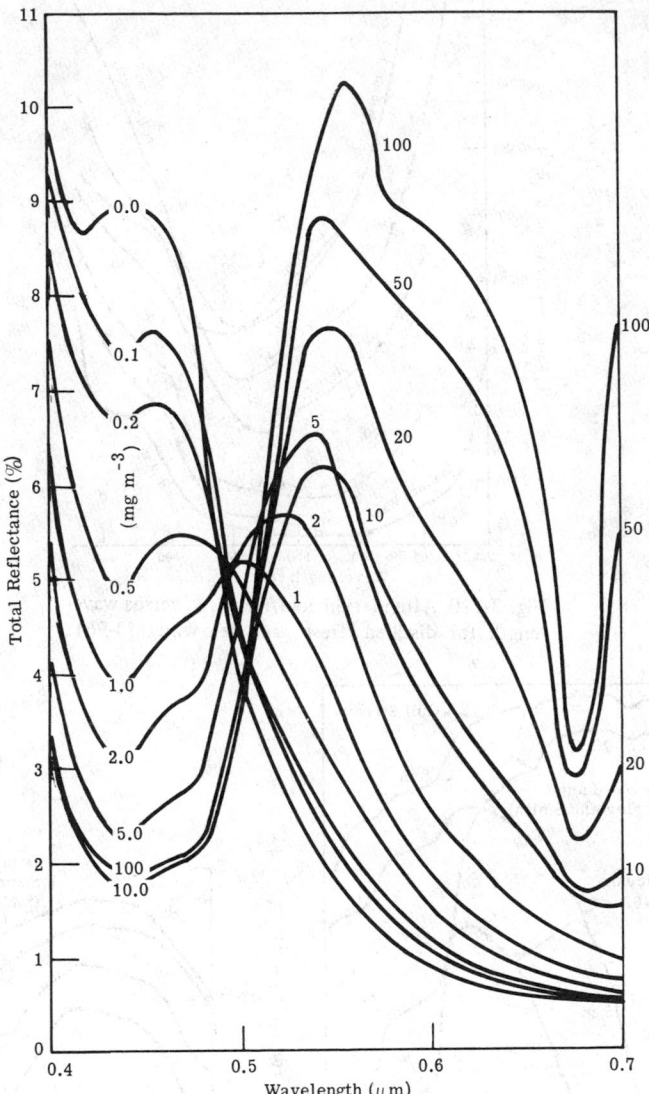

Fig. 3-123. Calculated spectral reflectance of deep ocean with varying amounts of chlorophyll. A clear sunny day-45° solar zenith angle, includes surface reflection from smooth ocean with turbid atmosphere [3-98].

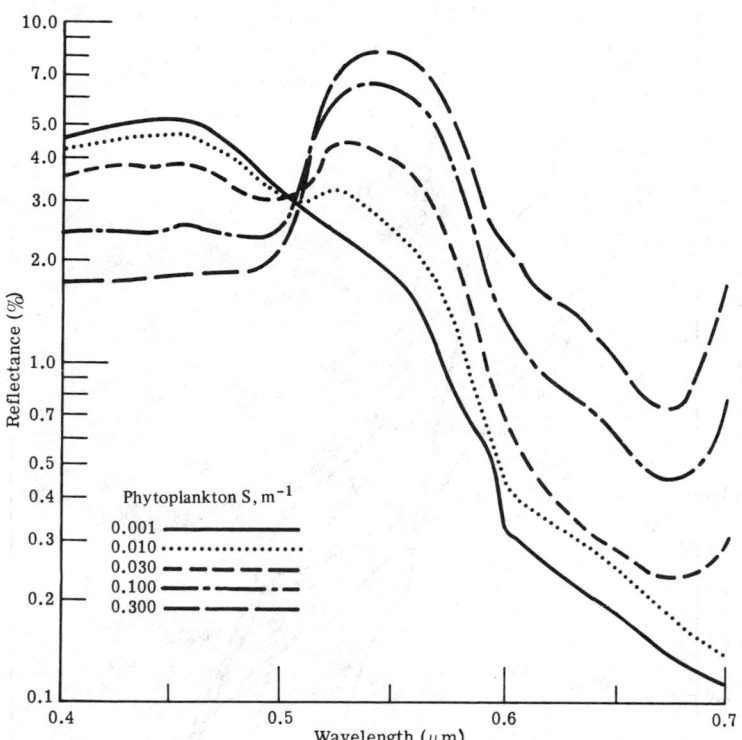

Fig. 3-124. Calculated change in bulk reflectance of ocean water with increasing concentration of phytoplankton [3-99].

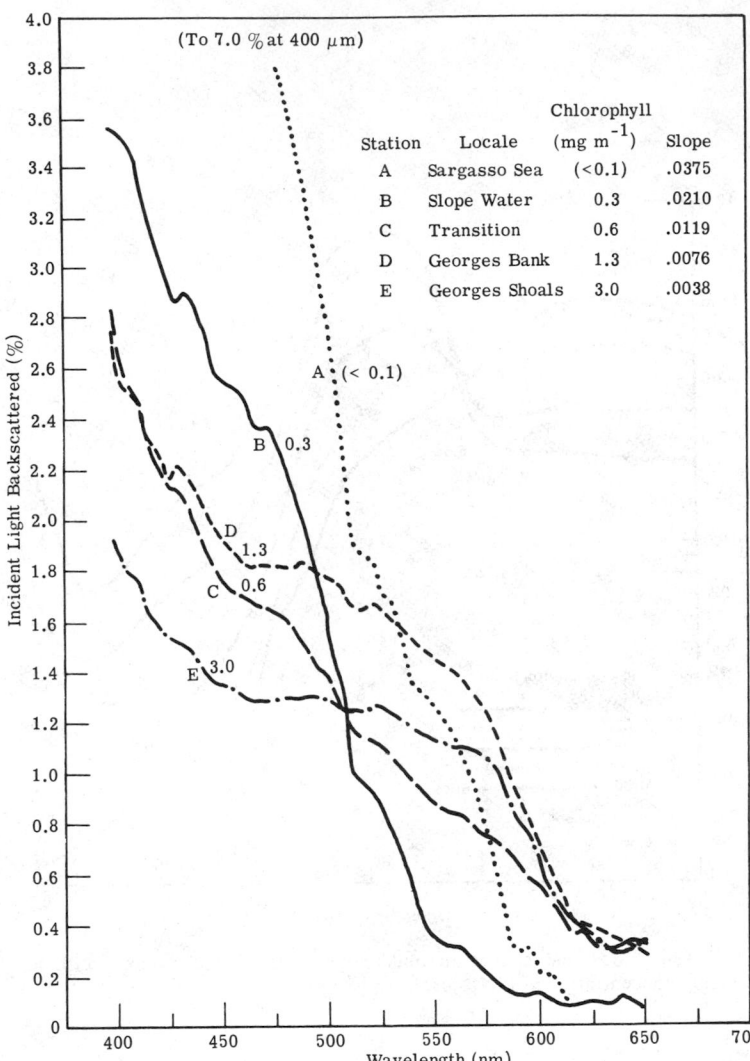

Fig. 3-125. Spectra of backscattered light, measured from the aircraft at 305 m on 27 August 1968 at the following stations and times (all E.D.T.): Station A, 1238 hours; Station B, 1421 hours; Station C, 1428.5 hours; Station D, 1445 hours; Station E, 1315 hours. The spectrometer with polarizing filter was mounted at a 53° tilt and directed away from the sun. Concentrations of chlorophyll a were measured from shipboard: Station A, 1238 hours (27 August); Station B, 0600 hours (28 August); Station C, 0730 hours (28 August); Station D, 1230 hours (28 August) [3-98].

Table 3-23. Attenuation Coefficients* at Selected Wavelengths
for the Four Experimental Algae [3-100]

| λ (μm) | Anabaena | | Microcystis | | Chlorella | | Selenastrum | | Dist. |
	4 mg l^{-1}	8 mg l^{-1}	4 mg l^{-1}	8 mg l^{-1}	4 mg l^{-1}	8 mg l^{-1}	4 mg l^{-1}	8 mg l^{-1}	H$_2$O**
375	2.09	4.24	1.66	3.24	2.48	4.80	2.24	4.39	0.017
400	2.30	4.36	1.83	3.31	2.73	5.13	2.52	4.81	0.046
425	2.45	4.80	1.97	3.84	3.04	5.86	2.80	5.46	0.050
450	2.26	4.31	1.97	3.58	2.99	5.72	2.72	5.32	0.045
475	1.76	3.17	1.67	2.96	2.79	5.32	2.53	4.73	0.051
500	1.65	2.95	1.57	2.74	2.50	4.47	2.17	3.83	0.059
525	1.43	2.56	1.19	2.20	1.61	2.59	1.46	2.39	0.075
550	1.28	2.42	1.07	1.95	1.48	2.29	1.33	2.07	0.098
575	1.59	2.85	1.17	2.11	1.56	2.49	1.40	2.24	0.117
600	2.04	3.52	1.56	2.56	1.85	2.85	1.63	2.54	0.262
625	2.40	4.12	1.77	2.78	2.09	3.22	1.92	2.96	0.316
650	2.23	3.64	1.76	2.61	2.44	4.08	2.08	3.38	0.341
675	2.55	4.08	2.16	3.26	3.34	5.71	3.06	5.27	0.408
700	2.24	3.11	2.08	2.76	2.60	3.61	2.37	3.32	0.612
725	2.57	3.28	2.73	3.32	3.07	3.60	2.86	3.46	1.183
750	4.07	4.85	4.21	4.91	4.55	5.22	4.40	5.05	2.576
775	4.08	4.79	4.12	4.77	4.28	5.01	4.15	4.85	2.195
800	3.62	4.54	3.59	4.35	3.83	4.73	3.65	4.51	1.723

*Attenuation coefficients expressed per meter, calculated from 30 inch column data.
**Distilled water plus 75 mg l^{-1} of MgSO$_4$ and 20 mg l^{-1} of Na$_2$CO$_3$.

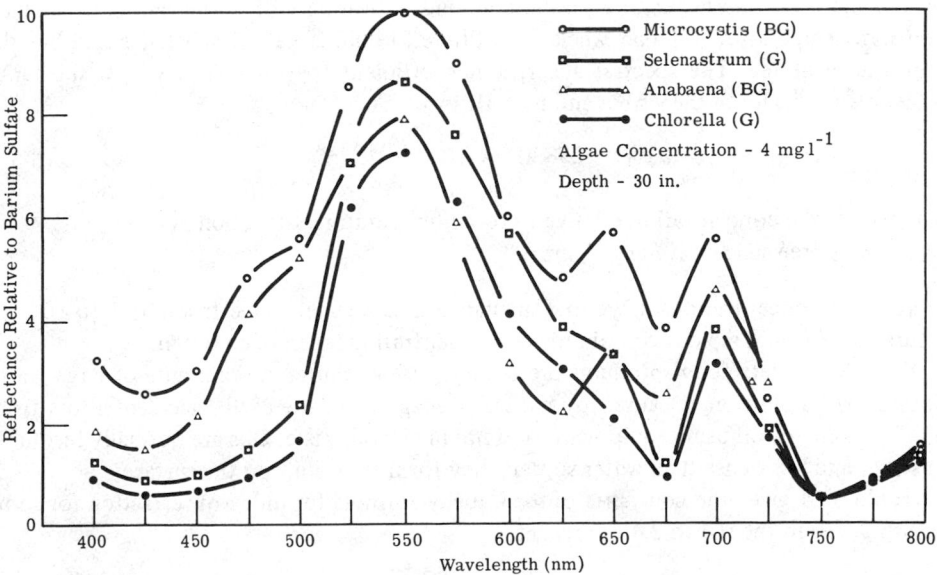

Fig. 3-126. Relative spectral reflectance curves for two green and two blue-green algae at 4 mg l^{-1} concentration [3-100].

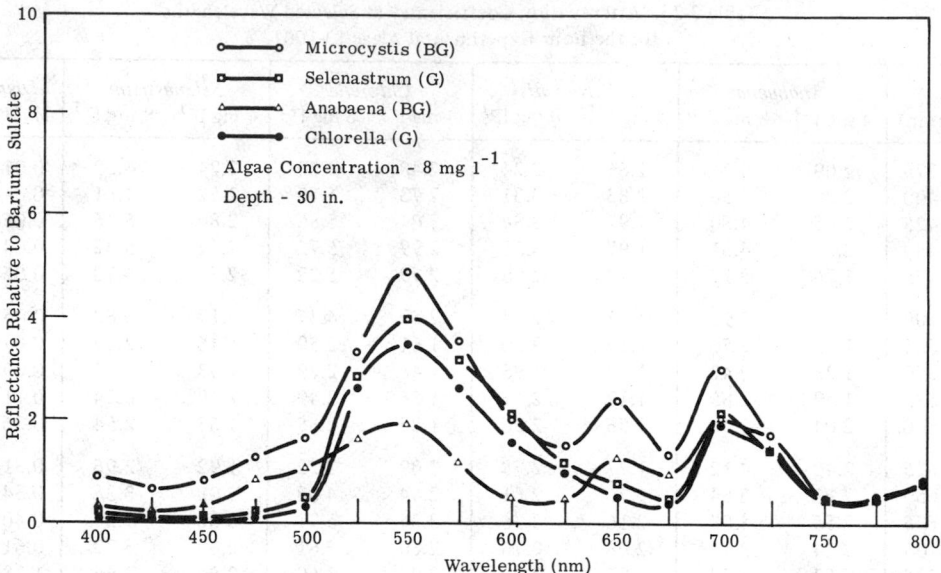

Fig. 3-127. Relative spectral reflectance curves for two green and two blue-green algae at 8 mg 1^{-1} concentration [3-100].

Yellow Substance. According to Kalle [3-101], sea water may contain a yellow soluble material which he calls yellow substance. This material is presumed to be related to humic acids from decaying vegetation. Water run-off from forested areas may be a source of yellow substance in lake and river water. Yellow substance is easily prepared by placing dead tree leaves in a plastic container, covering with water, and letting stand at room temperature for one week. The filtered liquid is yellow substance and has the appearance of tea. The spectral absorption coefficient (per meter) of yellow substance is given empirically in the visible and near-IR by

$$a(\lambda) = 25330 C \times 10^{-0.004935\lambda} \qquad (3\text{-}56)$$

where C = the concentration relative to room temperature saturation
λ = free space wavelength, nm

Relative concentrations of yellow substance in sea water range from 10^{-5} to 10^{-3} of saturation. Coastal regions contain higher concentrations than open ocean.

Oil Films. Petroleum products are commonly encountered pollutants of large water bodies. The pollution is caused by natural seepage from the earth, accidental loss from oil-processing or oil-using equipment, or dumping of oil waste. Oils are normally insoluble in water and less dense than water so that they form thin films on the surface.

Table 3-24 gives the constants of the Cauchy formula for index of refraction for some common oils in the 0.4 to 2.4 μm range.

$$n = A + \frac{B}{\lambda^2} + \frac{C}{\lambda^4} \qquad (3\text{-}57)$$

Table 3-24. Constants for Cauchy Relation for Index of Refraction of Oils
0.4 to 2.4 μm [3-103]

$$(n = A + B/\lambda^2 + C/\lambda^4)$$

Oil	Sample No.	A	B (μm^2)	C (μm^4)
Bradford crude oil	A1654	1.417	-0.00161	0.00359
Bradford residual fuel oil	A1655	1.494	0.01394	-0.00073
Bradford medium lube	A1656	1.432	0.00284	0.00146
Bradford diesel fuel	A1657	1.410	0.00558	-0.00030
Hastings crude oil	A1658	1.433	-0.00964	0.00843
Wilmington crude oil	A1659	1.456	-0.00653	0.00725
SAE 30 lube oil	A1660	1.436	-0.00748	0.00779
Used SAE 30 lube oil	A1661	1.438	0.00425	0.00149
Menhaden fish oil	A1662	1.428	0.00205	0.00211

The constants were derived from experimental data utilizing direct and indirect index of refraction measurements [3-102]. Only two to three decimal-place accuracy can be expected from these data.

Figures 3-128 to 3-130 show the index of refraction of oils in the 1.5 to 10.0 μm range.

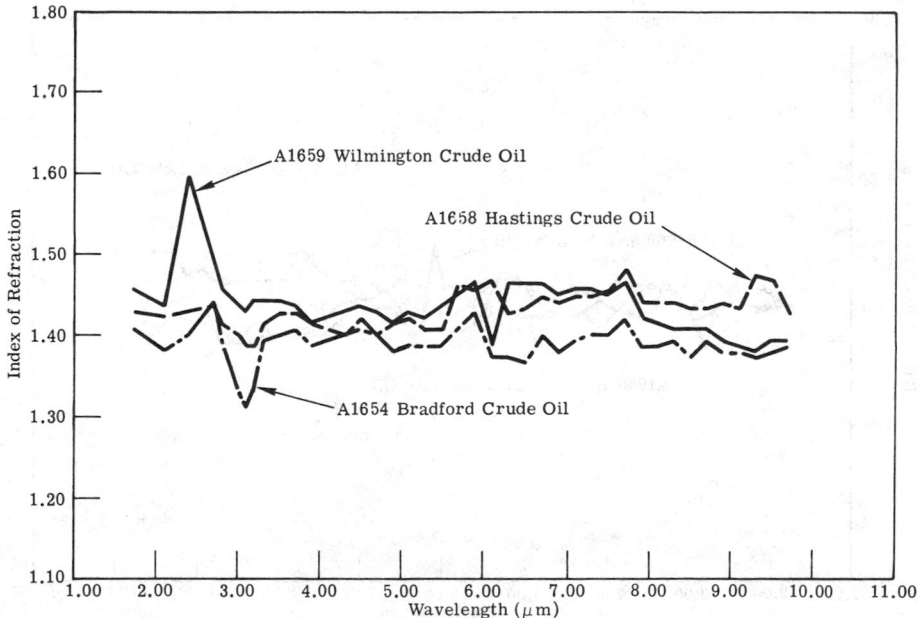

Fig. 3-128. The index of refraction A1659 Wilmington crude oil, A1658 Hastings crude oil, and A1654 Bradford crude oil [3-102].

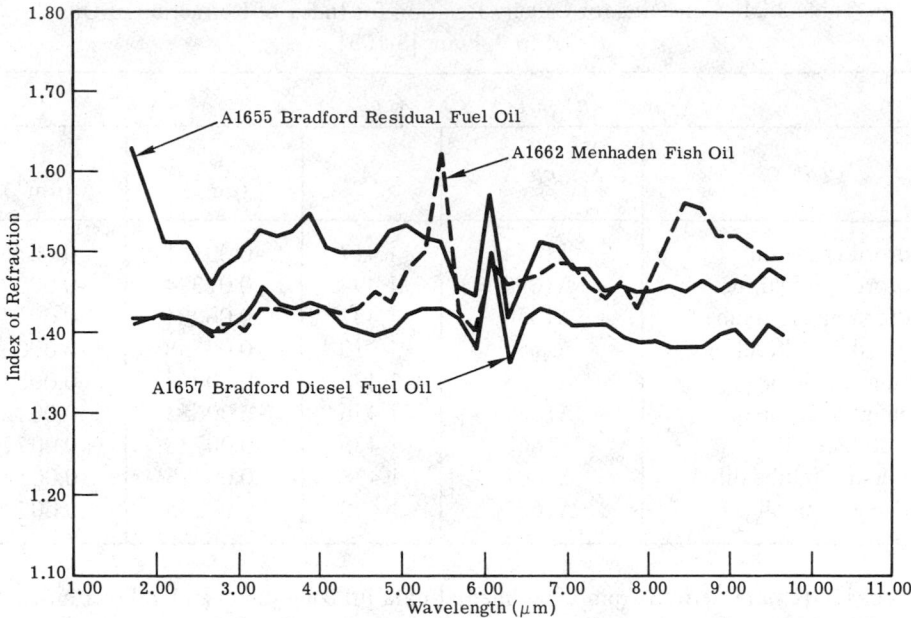

Fig. 3-129. The index of refraction of A1655 Bradford residual fuel oil, A1662 Menhaden fish oil, and A1657 Bradford diesel fuel oil [3-102].

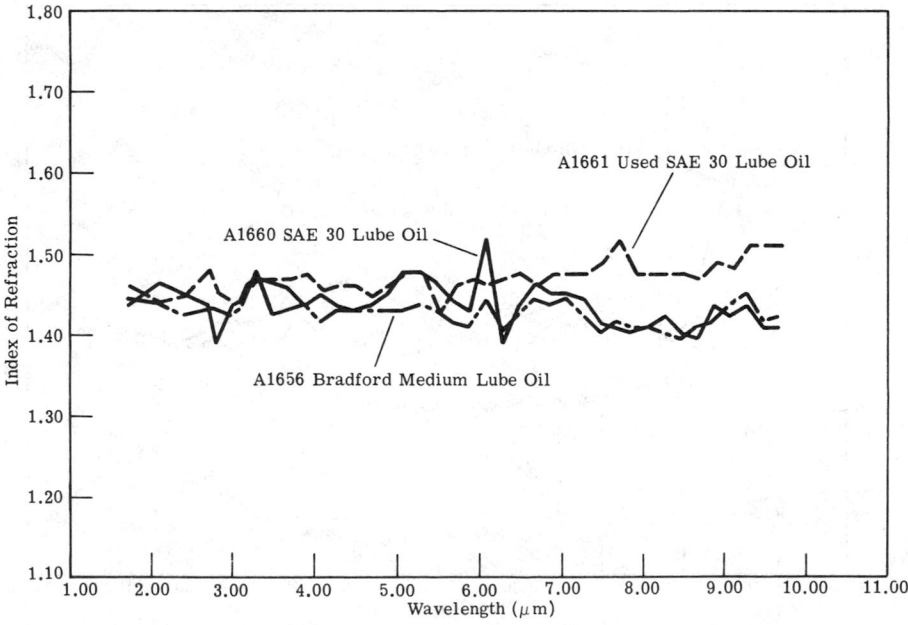

Fig. 3-130. The index of refraction of A1661 used SAE 30 lube oil, A1660 SAE 30 lube oil, and A1665 Bradford medium lube oil [3-102].

Figures 3-131 and 3-132 show the spectral extinction coefficients of oils. The spectral details of the spectral extinction coefficient generally correspond to similar spectral variations in index of refraction.

Figures 3-133 to 3-135 present the calculated spectral radiance of oil slicks on sea water under solar illumination viewed at the nonspecular angle.

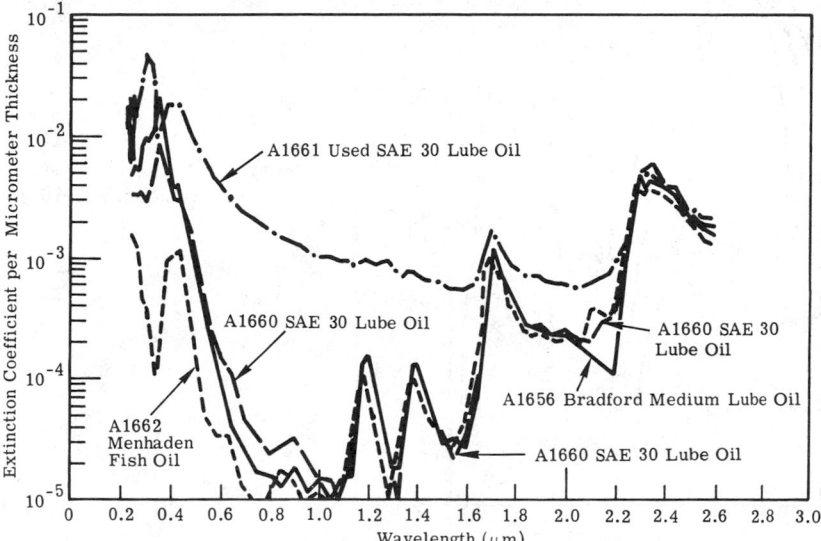

Fig. 3-131. The spectral extinction coefficients of A1661 used SAE 30 lube oil, A1660 SAE 30 lube oil, A1656 Bradford medium lube oil, and A1662 Menhaden fish oil [3-102].

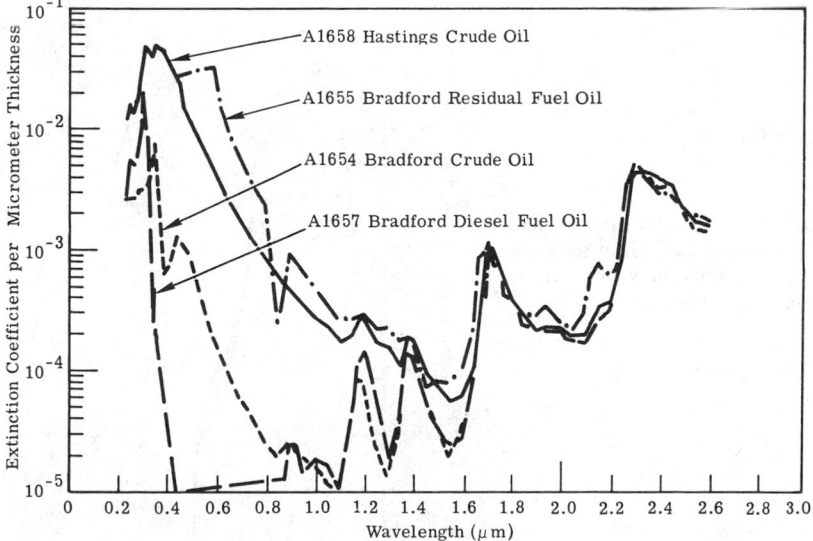

Fig. 3-132. The spectral extinction coefficients of A1658 Hastings crude oil, A1655 Bradford residual fuel oil, A1654 Bradford crude oil, A1657 Bradford diesel fuel oil [3-102].

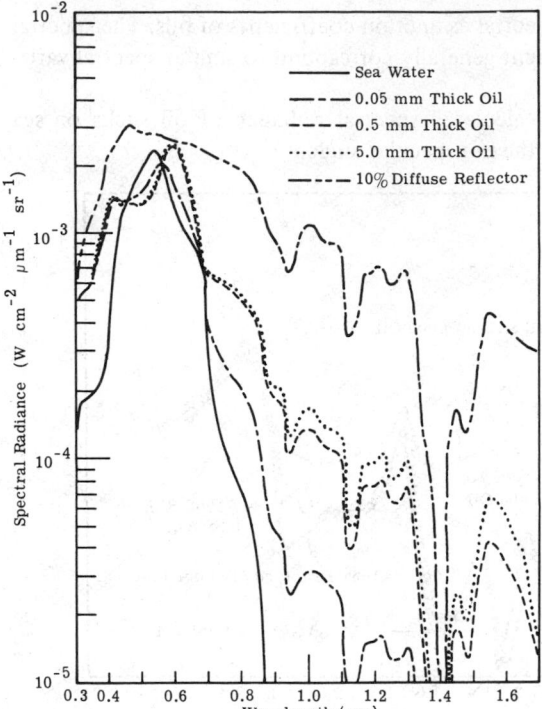

Fig. 3-133. The calculated spectral radiance of used SAE 30 lube oil on sea water under solar illumination viewed at the nonspecular angle [3-102].

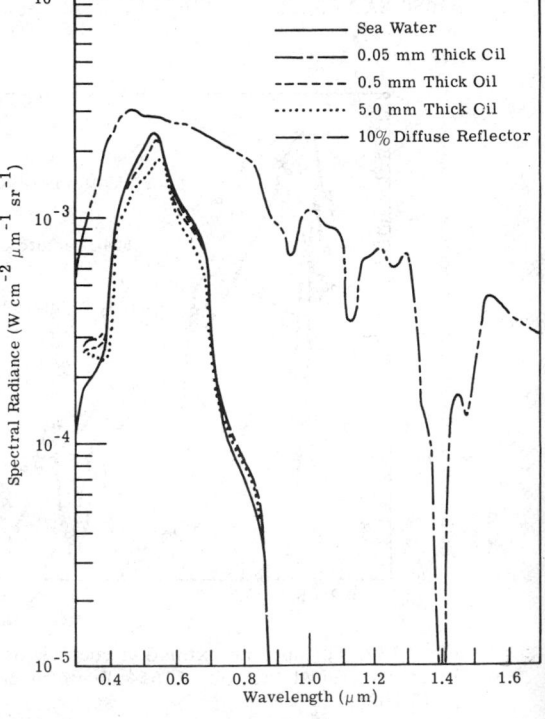

Fig. 3-134. The calculated spectral radiance of Bradford diesel oil on sea water under solar illumination viewed at the nonspecular angle [3-102].

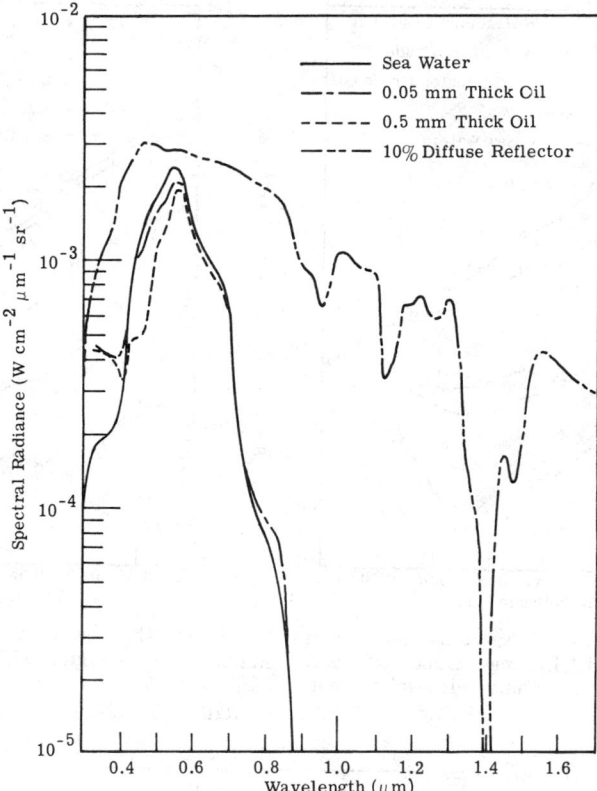

Fig. 3-135. The calculated spectral radiance of Menhaden oil on sea water under solar illumination viewed at the nonspecular angle [3-102].

Figures 3-136 to 3-138 show the measured spectral radiance of oil samples under ultraviolet lamp excitation. The magnesium oxide is not fluorescent.

3.7.6. Snow. Snow is a common surface cover of terrain in the high latitudes. High altitude clouds are frequently composed of small ice crystals, as is snow, so that the spectral distinction between terrestrial snow cover and high altitude clouds is not reliable. The clearest distinguishing attribute of cloud cover is the motion of the cloud pattern. Snow patterns on terrain do not exhibit lateral motion in time. Snow differs from place to place and from time to time by differences in the crystal sizes, the state of division of flakes, the compactness or density of snow cover, and the amount of free water within the snow cover.

New-fallen snow is close to Lambertian in character. However, old snow may develop a compacted crust which exhibits marked specular properties.

The spectral reflectance of snow under various conditions is shown in Figures 3-139 to 3-150.

3.7.7. Botanical Materials. The major spectrally significant constituents of terrestrial botanical materials are water, cellulose, and chlorophyll. For this reason, botanical materials appear very similar both by reflected solar radiation and by thermal emission.

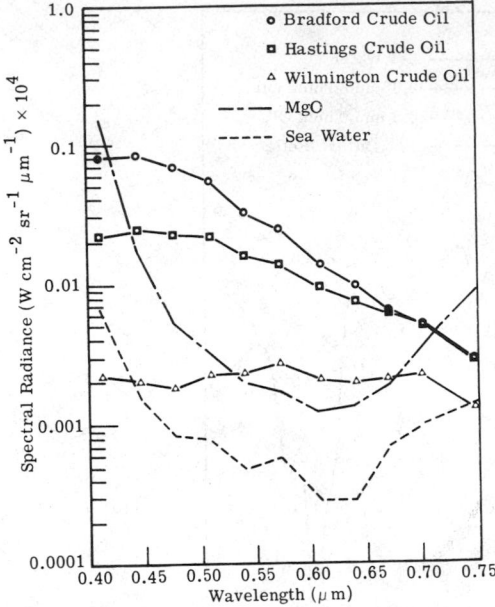

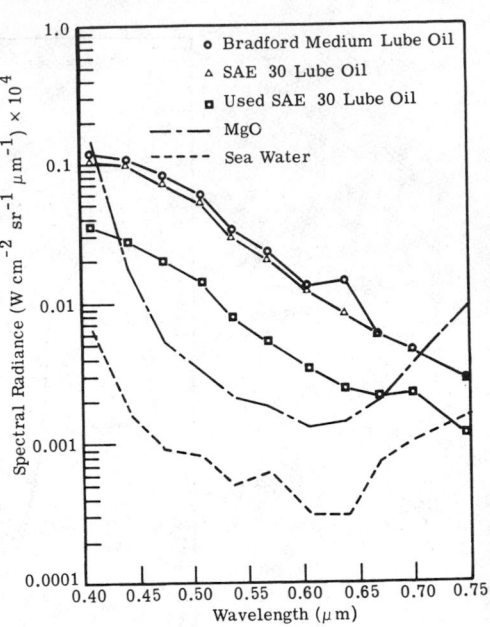

Fig. 3-136. The measured spectral radiance of Bradford crude oil, Hastings crude oil, and Wilmington crude oil under ultraviolet lamp excitation [3-102].

Fig. 3-137. The measured spectral radiance of Bradford medium lube oil, SAE 30 lube oil and used SAE 30 lube oil under ultraviolet lamp excitation [3-102].

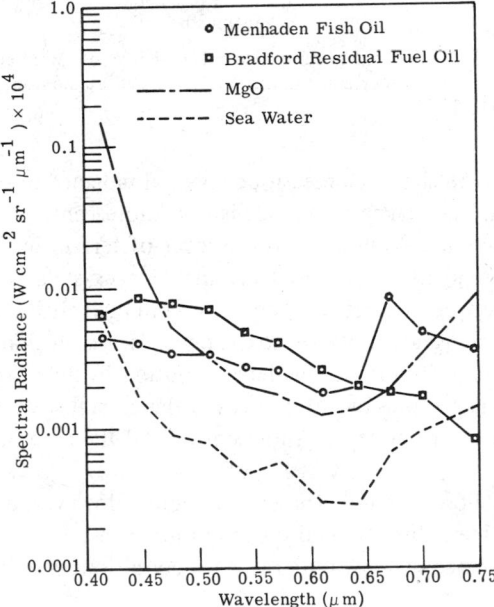

Fig. 3-138. This measured spectral radiance of Menhaden fish oil and Bradford residual fuel oil under ultraviolet lamp excitation [3-102].

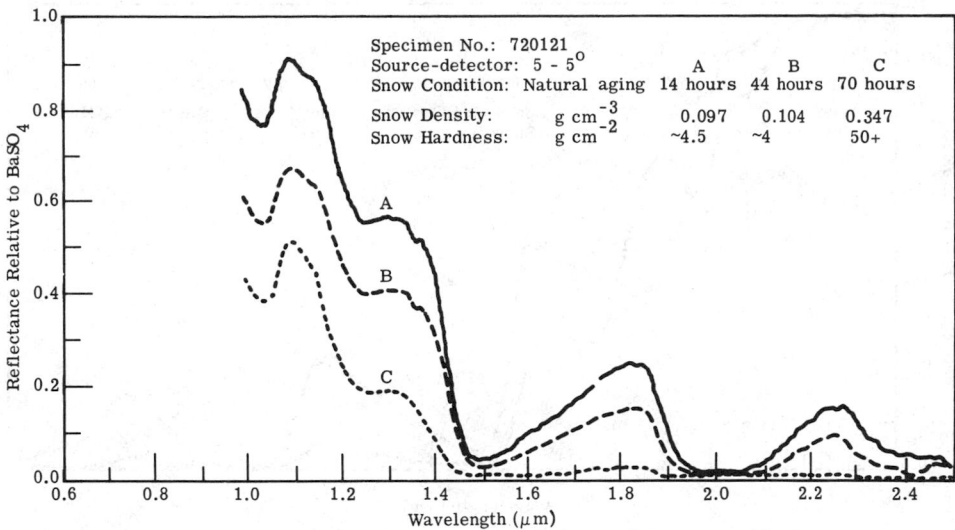

Fig. 3-139. Changes in snow reflectance with natural aging (5-5° source-detector) [3-103].

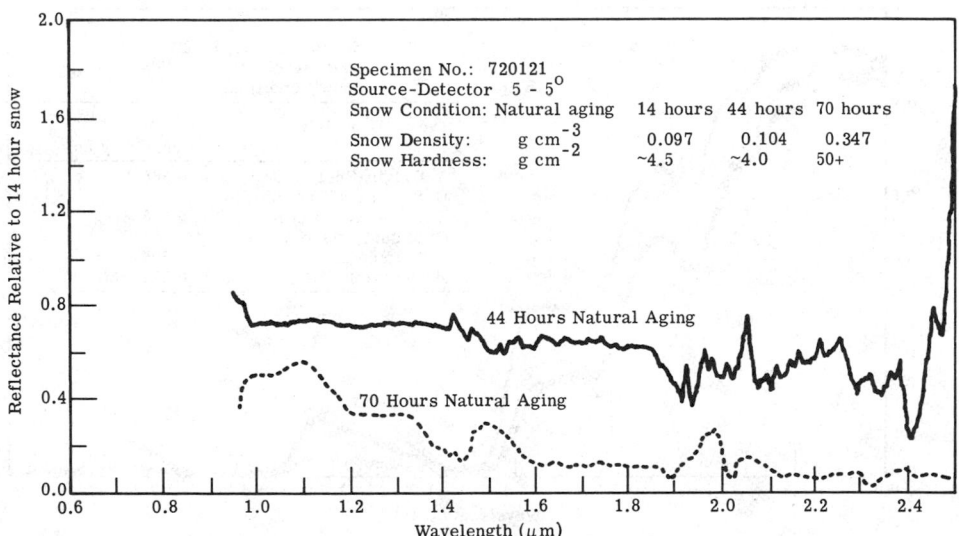

Fig. 3-140. Reflectance of aging snow relative to fresher (14 hour) snow (5-5° source-detector) [3-103].

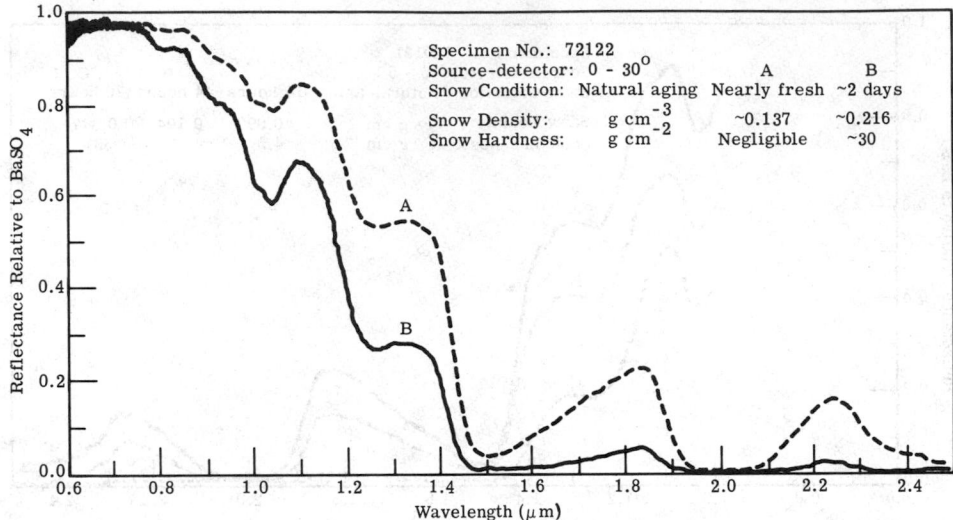

Fig. 3-141. Changes in snow reflectance with natural aging (0-30° source-detector) [3-103].

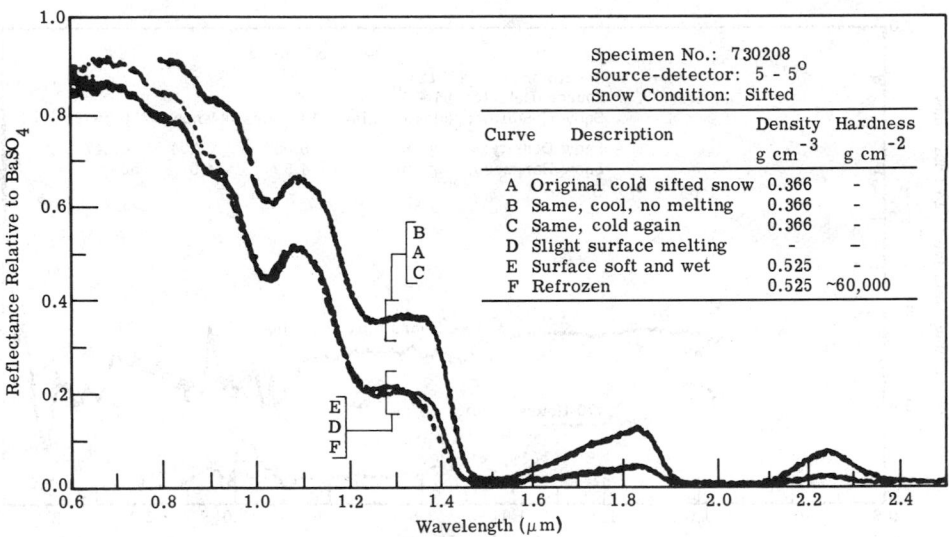

Fig. 3-142. Effects of temperature conditions on sifted snow reflectance (5-5° source-detector) [3-103].

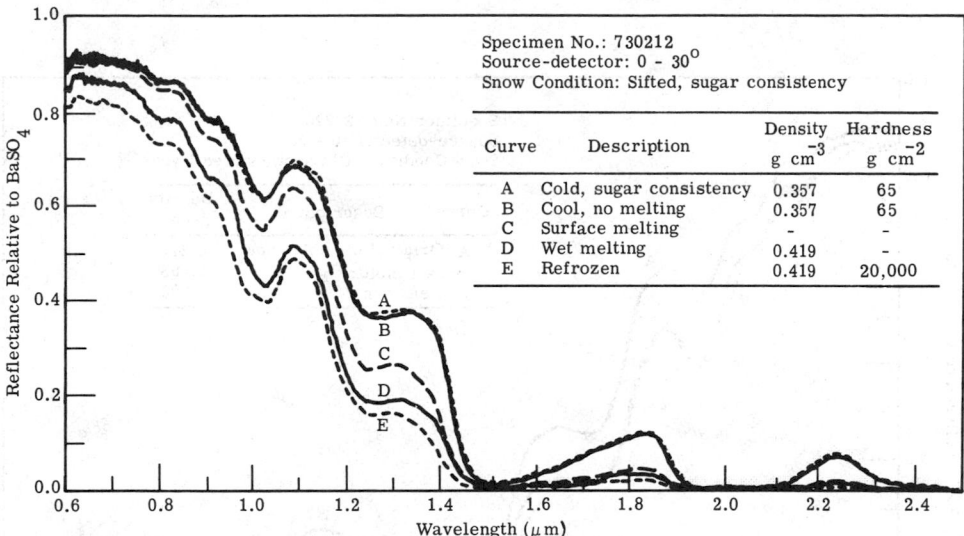

Fig. 3-143. Effects of temperature conditions on sifted snow reflectance (0-30° source-detector) [3-103].

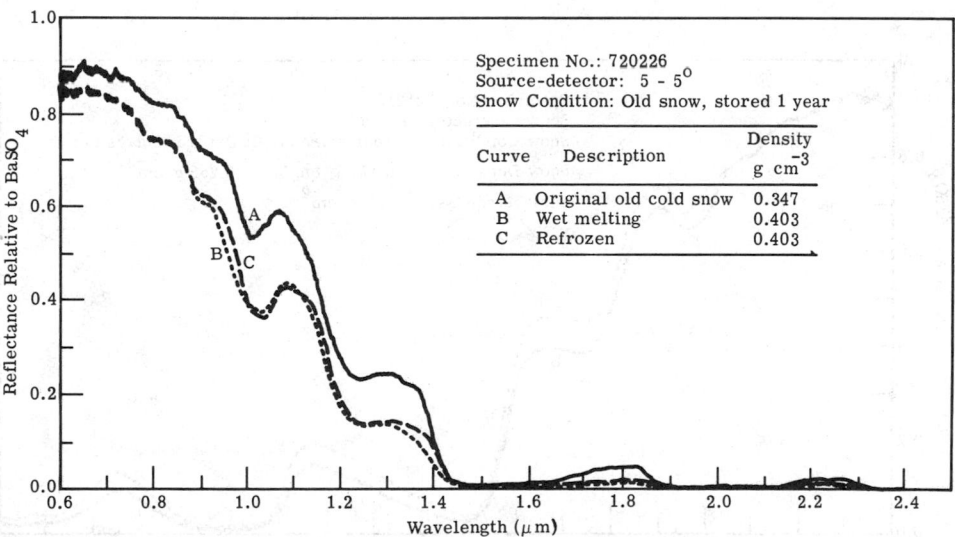

Fig. 3-144. Effects of temperature conditions on old snow reflectance (5-5° source-detector) [3-103].

NATURAL SOURCES

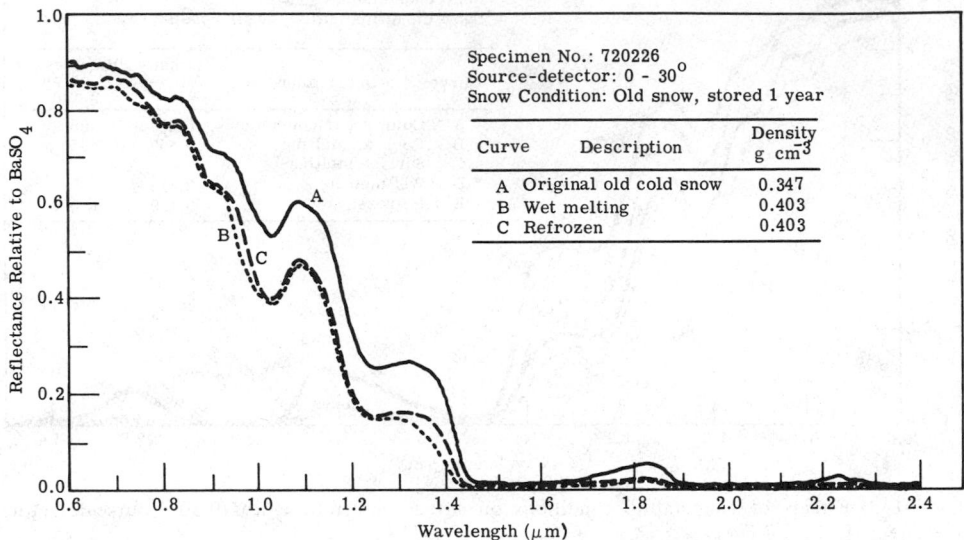

Fig. 3-145. Effects of temperature conditions on old snow reflectance (0-30° source-detector) [3-103].

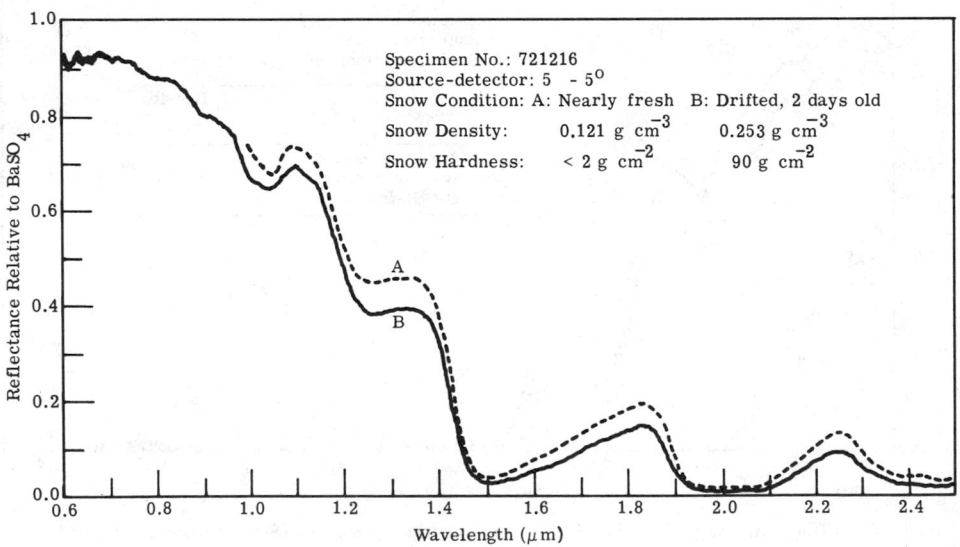

Fig. 3-146. Effect of snow drifting on snow reflectance [3-103].

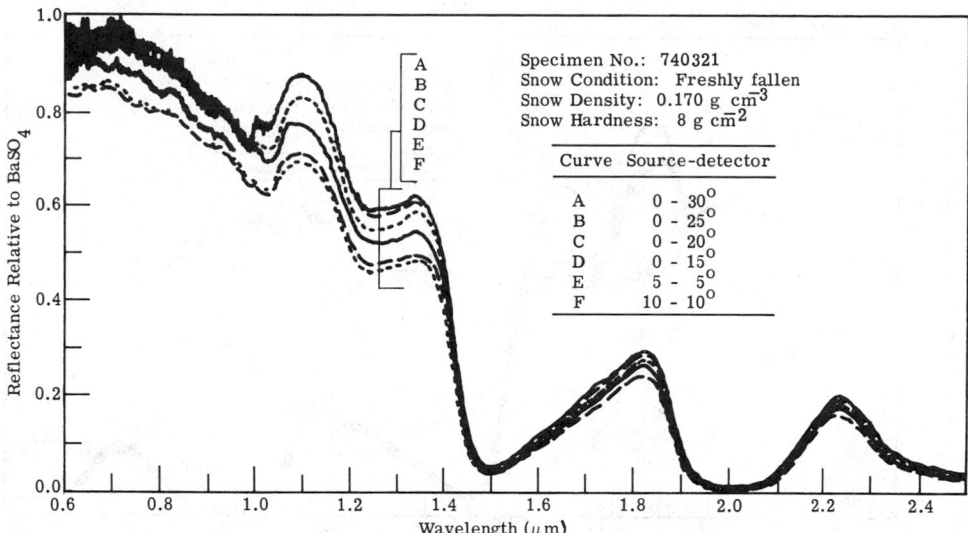

Fig. 3-147. Comparison of relative reflectance of freshly fallen snow at various source-detector angles [3-103].

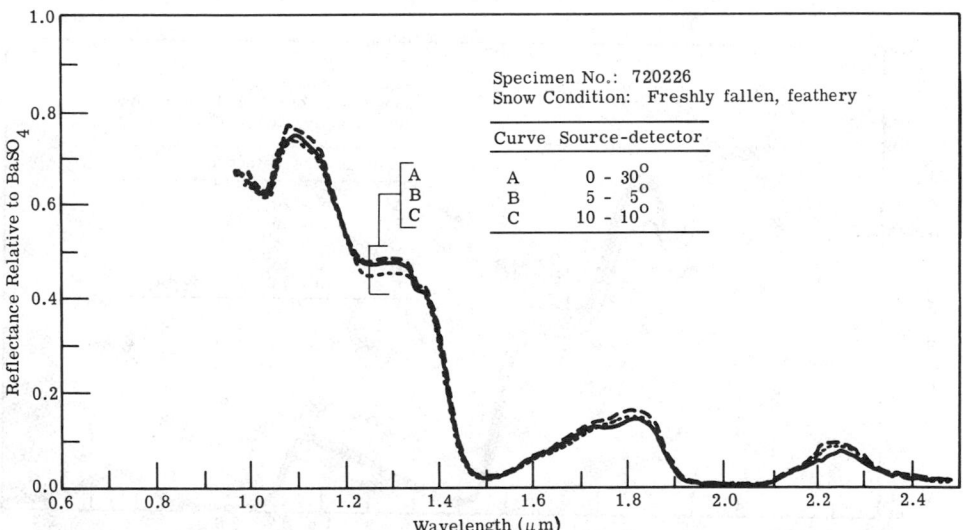

Fig. 3-148. Comparison of relative reflectance of freshly fallen, feathery snow at various source-detector angles [3-103].

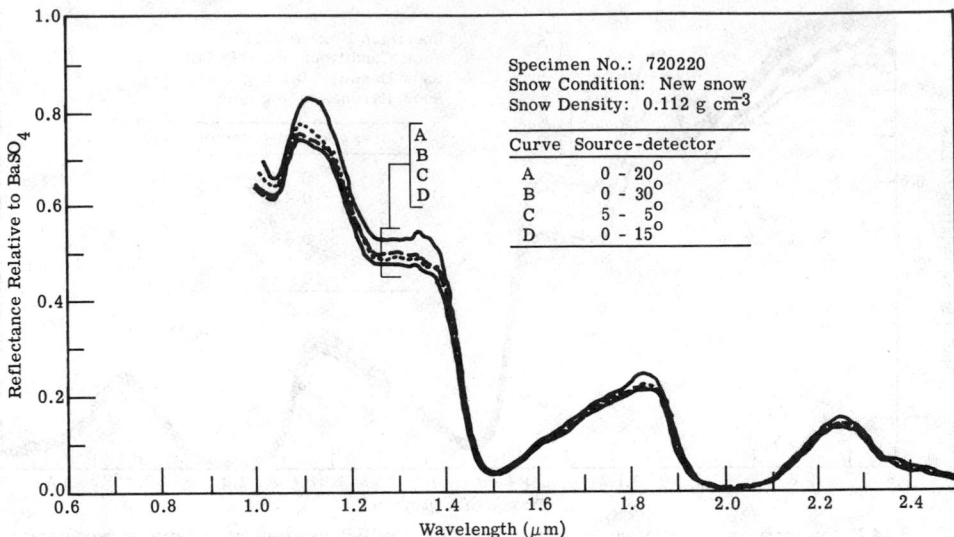

Fig. 3-149. Comparison of relative reflectance of new snow at various source-detector angles [3-103].

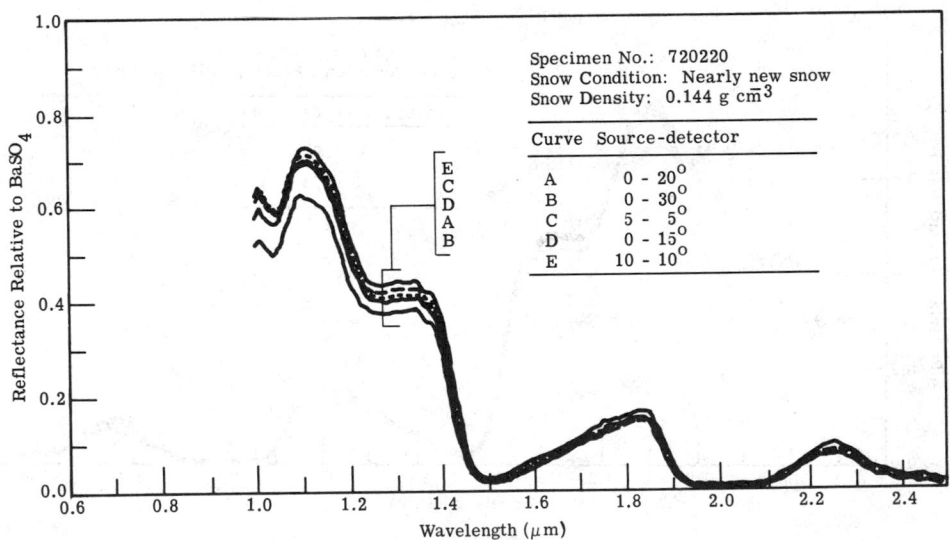

Fig. 3-150. Comparison of relative reflectance of nearly new snow at various source-detector angles [3-103].

According to Nobel [3-104], carotenoids and phycobilins also occur commonly in botanical materials as accessory pigments. Carotenoids are found in essentially all green plants but are largely masked by the predominant spectral absorptance properties of chlorophyll. Carotenoids absorb strongly in the 0.425 to 0.490 μm spectral band and, to a somewhat lesser extent, in the 0.49 to 0.56 μm band; they are primarily responsible for the yellow and orange fall colors of trees. Phycobilins generally have their major absorption bands in the 0.5 to 0.65 μm band with a minor band in the ultraviolet. One of the two main phycobilins, phycoerythrin, has an absorption band between 0.53 and 0.57 μm, so that the visual appearance of phycoerythrin alone would be magenta. The other phycobilin, phycocyanin, has an absorption band maximum between 0.61 and 0.66 μm and should visually appear cyan.

There are several kinds of chlorophyll, all of which absorb in the UV, but more strongly in the 0.4 to 0.45 μm band and in the 0.64 to 0.69 μm band. One type of chlorophyll has an absorption maximum near 0.67 μm while another has a maximum near 0.68 μm. The absorption of both decreases sharply at 0.70 μm producing the characteristic sharp rise to high values of spectral reflectance of green botanical materials in the 0.72 to 1.3 μm range. The spectral absorptance falls to low values of 5 to 15% in the near infrared range for many types of green leaves. Multiple scattering becomes a major contributor to canopy reflectance. Thick leaves tend to exhibit high reflectance and low transmittance while thin leaves exhibit low reflectance and high transmittance.

The high spectral absorption coefficient of the water within healthy tissue produces deep reflectance and transmittance minima near 1.4 and 2.0 μm. Since healthy tissue containing active chlorophyll must also contain some water to permit photosynthesis, the concurrent appearance of a chlorophyll absorption band near 0.68 μm and the water absorption bands near 1.4 and 2.0 μm is generally expected. Frequently the change of leaf spectra resulting from plant stress is first made manifest by disruption of the photosynthetic process. The disruption is caused by the destruction of chlorophyll before water has been completely lost by the leaf. Consequently, the water absorption bands may still be present in leaf spectra after the leaf is dead.

Figure 3-151 shows the spectral reflectances of particular samples of various vegetative canopies. The reflectances include the spectral properties of the understory or soil, the

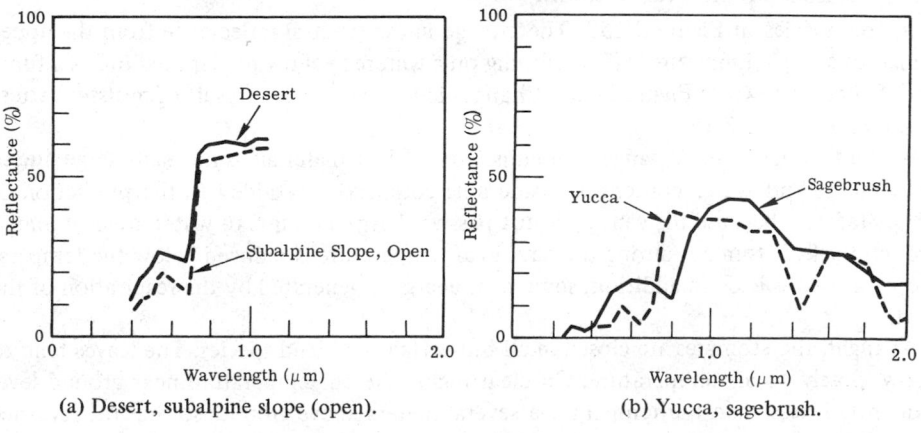

(a) Desert, subalpine slope (open).　　　(b) Yucca, sagebrush.

Fig. 3-151. Spectral reflectances of various vegetative canopies [3-84].

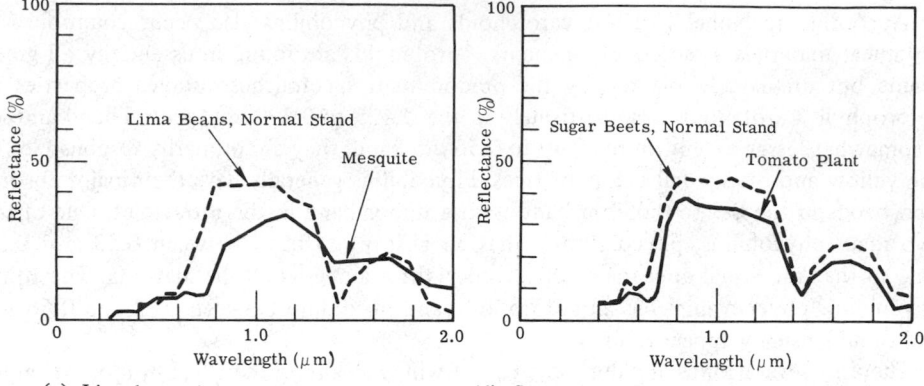

(c) Lima beans (normal stand), mesquite. (d) Sugar beets (normal stand), tomato plant.

percentage of cover, the effects of shadow cast by opaque stems and woody parts, the effects of leaf orientation, and sun and view angles. Changes in any of these will produce a change in canopy bidirectional reflectance so that one can not expect these sample spectral reflectances to be entirely characteristic of the canopy type.

Table 3-25 lists the description of plant leaf materials which produce the reflectance and transmittance spectra listed in Tables 3-26 and 3-27. Table 3-28 lists the descriptions of plant leaf materials which produce the spectral reflectances listed in Table 3-29 and the transmittances in Table 3-30.

The dependence of leaf spectral reflectance on seasons of the year is shown for

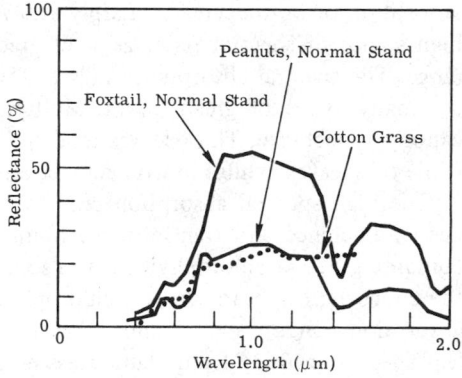

(e) Foxtail (normal stand), peanuts (normal stand), and cotton grass.

Fig. 3-151. (Continued) Spectral reflectances of various vegetative canopies [3-84].

seven tree species in Figure 3-152. The change in the spectral reflectance from the upper surface of a single immature leaf developing on a watered yellow poplar seedling as a function of time is shown in Figure 3-153. Changes in leaf spectra as a result of moisture stress are shown in Figure 3-154.

In the thermal infrared range, the emissivity of leaf materials is normally large due to the cellulose and water content. Considerable complexity is added to the prediction of canopy radiance because the canopy is not passive. Large amounts of water are evaporated through the leaf stomates during the day. Leaf temperature is reduced below the temperature of a passive leaf. In addition, some heat energy is generated by the respiration of the plant.

At night, the stomates are closed in all but certain arid land species. The leaves tend to follow closely to air temperature. On clear nights the air temperature near ground level frequently is less than the temperature several meters above ground, so that the radiance of low lying grass canopies may be less than the radiance of tree tops.

Table 3-25. Leaf Mesophyll Arrangement and other Structural Characteristics of Plant Leaves Used in this Study. (Common names are used in the text.) [3-105]

Common Name	Latin Name*	Mesophyll Arrangement	Additional Structural Characteristics
Corn	*Zea mays* L.	Compact	Bulliform cells on adaxial surface.
Banana	*Musa acuminata* Colla (*M. cavendishii* Lamb.)	Dorsiventral	Adaxial and abaxial hypodermal layers, palisade layer.
Begonia	*Begonia cucullata* Willd. (*B. sempertlorens* Link & Otto)	Succulent, central chlorenchyma	Malacophyllous-type xerophytic leaf; large thin-walled storage cells on each side of central chlorenchyma
Eucalyptus	*Eucalyptus camaldulensis* Dehnh. (*E. restrata* Schlecht.)	Isolateral	Thick adaxial cuticle; no spongy parenchyma cells.
Rose	*Rosa* var. unknown	Dorsiventral	Multiple palisade layers.
Hyacinth	*Eichhornia crassipes* (Mart.) Solms	Dorsiventral	Multiple palisade layers; large air chambers characteristic of hydrophytes.
Sedum	*Sedum spectabile* Boreau	Succulent	Well-differentiated cellular structure.
Ficus	*Ficus elastica* Roxb. ex Hornem.	Dorsiventral	Thick adaxial cuticle; multiseriate adaxial epidermis; and multiple palisade layer.
Oleander	*Nerium oleander* L.	Dorsiventral	Thick adaxial cuticle; multiseriate epidermis; multiple palisade layer abaxial grooves.
Ligustrum	*Ligustrum lucidum* Ait.	Dorsiventral	Thick adaxial cuticle; multiple palisade layer.
Crinum	*Crinum fimbriatulum* Baker	Dorsiventral	Poorly differentiated and lobed palisade parenchyma cells; large central air spaces.

*Names are those used by New Crops Research Branch (Dr. Edward E. Terrell), ARS, USDA, Beltsville, Maryland 20705.

Table 3-26. Percent Diffuse Reflectance of Top (T) and Bottom (B) Leaf Surfaces of 11 Plant
Genera at Seven Wavelengths [3-105]

Plant Genera	550 nm		800 nm		1000 nm		1450 nm		1650 nm		1950 nm		2200 nm	
	T	B	T	B	T	B	T	B	T	B	T	B	T	B
Ficus	8.1	19.2	54.2	54.1	52.4	53.1	7.8	17.2	27.0	34.4	3.7	7.2	10.2	20.6
Ligustrum	10.2	26.9	55.5	53.8	54.1	53.4	9.0	20.8	27.7	37.7	4.1	8.7	11.1	24.2
Rose	10.4	20.5	48.8	43.2	48.4	42.6	23.2	24.6	38.0	34.6	9.4	13.8	25.1	25.8
Banana	10.5	16.9	44.0	41.3	43.3	40.7	12.0	15.9	29.0	29.4	5.1	8.2	16.0	19.0
Oleander	10.7	18.1	54.3	53.0	54.0	52.9	13.0	20.1	32.8	37.1	5.5	9.1	16.1	23.4
Hyacinth	12.1	19.0	49.8	49.4	49.6	49.1	11.7	18.4	31.0	34.8	4.8	8.2	15.3	21.9
Eucalyptus	12.6	15.8	45.9	46.0	45.4	45.6	16.0	18.2	29.7	31.0	7.0	8.3	15.6	17.6
Begonia	12.9	20.1	45.3	39.9	43.4	38.2	6.2	9.5	21.6	21.6	3.8	4.4	8.4	11.1
Corn	15.4	15.6	42.6	43.9	42.4	43.2	17.2	19.1	31.6	32.8	7.2	8.5	19.6	21.1
Crinum	15.6	23.1	55.8	54.6	54.1	53.2	10.0	14.8	29.3	33.4	5.2	6.8	13.6	19.4
Sedum	20.1	27.1	54.2	52.2	50.9	50.0	5.2	10.1	18.4	26.1	3.2	4.2	6.2	12.8

Table 3-27. Percent Transmittance of Top (T) and Bottom (B) Leaf Surface of 11 Plant
Genera at Seven Wavelengths [3-105]

Plant Genera	550 nm		800 nm		1000 nm		1450 nm		1650 nm		1950 nm		2200 nm	
	T	B	T	B	T	B	T	B	T	B	T	B	T	B
Ficus	1.2	1.5	40.1	41.5	39.8	41.2	2.8	2.8	20.6	21.2	0.5	0.5	6.6	6.9
Ligustrum	3.9	3.9	38.1	40.1	38.4	40.0	4.2	4.4	21.4	22.0	0.5	0.5	8.2	8.4
Rose	9.3	11.2	48.6	54.1	49.5	55.2	27.1	31.1	44.1	49.3	10.9	13.0	33.8	38.1
Banana	12.2	13.2	52.1	54.8	52.5	55.4	17.1	18.5	41.4	43.8	2.6	2.9	26.9	28.6
Oleander	1.6	1.8	34.1	35.8	35.0	36.8	4.4	4.7	20.9	21.8	0.5	0.5	8.4	8.8
Hyacinth	8.5	8.8	45.2	47.0	45.9	47.6	9.1	10.2	31.6	33.6	0.7	0.9	16.6	18.4
Eucalyptus	7.5	7.3	48.3	48.3	49.8	49.8	17.9	18.2	36.2	34.2	3.8	3.8	22.4	22.3
Begonia	15.5	21.0	51.7	56.6	51.4	56.3	6.4	7.6	31.6	36.9	0.5	0.5	13.9	16.4
Corn	12.9	12.6	53.1	52.6	54.3	54.0	26.2	25.8	48.1	47.7	8.3	8.2	36.5	35.9
Crinum	6.1	6.6	38.1	39.4	37.7	38.8	1.9	2.1	18.5	19.2	0.5	0.5	6.2	6.4
Sedum	10.2	10.9	42.6	44.2	40.3	42.1	0.5	0.6	14.7	15.6	0.5	0.5	3.0	3.3

Table 3-28. Common, Scientific, and Family Names; Leaf Mesophyll Arrangements; and Structural Characteristics of Plant Leaves used in this Study. (Common names are used in the text.) [3-106]

Common Name	Scientific Name*	Family Name	Mesophyll Arrangement**	Additional Structural Characteristics
Avocado	*Persea americana* Mill.	*Lauraceae*	Dorsiventral	Thick cuticle, multiple palisade layers, long and narrow palisade cells.
Bean	*Phaseolus* vulgaris L.	*Leguminosae*	Dorsiventral	Very porous mesophyll
Cantaloupe	*Cucumis melo* L. var. *cantalupensis* Naud.	*Cucurbitaceae*	Dorsiventral	Multiple palisade layers, hairs lower epidermis.
Corn	*Zea mays* L.	*Gramineae*	Compact	Bulliform cells, hairs upper epidermis
Cotton	*Gossypium hirsutum* L.	*Malvaceae*	Dorsiventral	Glandular hairs, nectaries, lysigenous glands.
Lettuce	*Lactuca sativa* L.	*Compositae*	Compact	Large cells, porous mesophyll.
Okra	*Hibiscus esculentus* L.	*Malvaceae*	Dorsiventral	Well differentiated, porous mesophyll.
Onion	*Allium cepa* L.	*Amaryllidaceae*	Dorsiventral	Tubular leaves.
Orange	*Citrus sinensis* (L.) Osbeck	*Rutaceae*	Dorsiventral	Thick cuticle with wax layers, multiple palisade layers, lysigenous cavities.
Peach	*Prunus persica* (L.) Batsch	*Rosaceae*	Dorsiventral	Multiple palisade layers, porus mesophyll
Pepper	*Capsicum annuum* L. and other spp.	*Solanaceae*	Dorsiventral	Druse crystals
Pigweed	*Amaranthus cetroflexus* L.	*Amaranthaceae*	Compact	Druse crystals, veins surrounded by large, cubical, parenchymatous cells
Pumpkin	*Cucurbita pepo* L.	*Cucurbitaceae*	Dorsiventral	Multiple palisade layers, hairs upper, and lower epidermis
Sorghum	*Sorghum bicolor* L. Moench	*Gramineae*	Compact	Bulliform cells
Soybean	*Glycine max* (L.) Merr.	*Leguminosae*	Dorsiventral	Porous mesophyll
Sugarcane	*Saccharum officinarum* L.	*Gramineae*	Compact	Bulliform cells
Sunflower	Helianthus annuus L.	Compositae	Isolateral	Hairs upper and lower epidermis
Tomato	Lycopersicon esculentum Mill.	Solanaceae	Dorsiventral	Hairs upper and lower epidermis, glandular hairs lower surface
Watermelon	Citrullus lanatus (Thumb.) Mansf.	Cucurbitaceae	Dorsiventral	Multiple palisade layers, glandular hairs lower surface
Wheat	Triticum aestivum L.	Gramineae	Compact	Bulliform cells

*Names are those used by New Crops Research Branch (Dr. Edward E. Terrell), ARS, USDA, Beltsville, Maryland.

**Arbitrary definitions of mesophyll arrangements used herein are: dorsiventral, a usually porous (many intercellular air spaces) mesophyll with palisade parenchyma cells in its upper and spongy parenchyma cells in its lower part; compact mesophyll with intercellular air space and no differentiation into palisade and spongy parenchyma cells; isolateral, tending to have long narrow cells through a porous mesophyll.

Table 3-29. Average Percent Reflectances of Top Leaf Surfaces of 10 Leaves for Each of 20 Crops for 41 WL* (nm) Over the 500- to 2500-nm WLI** [3-106]

Crop	500	550	600	650	700	750	800	850	900	950	1000	1050	1100	1150
Avocado	8.2	8.9	6.8	7.2	26.6	47.9	50.4	50.3	50.1	49.4	49.7	49.7	49.3	47.1
Bean	15.2	18.5	12.0	10.7	37.3	55.7	56.9	56.9	56.5	55.8	56.2	56.6	56.0	53.6
Cantaloupe	11.6	12.7	10.0	9.9	28.6	46.1	47.7	47.7	47.5	46.8	47.3	47.6	47.0	44.6
Corn	12.7	16.2	12.0	9.3	24.8	45.4	46.3	46.4	46.2	45.5	45.7	46.0	45.5	43.3
Cotton	9.8	11.8	8.0	7.7	28.6	45.8	47.2	47.2	46.9	46.2	46.6	47.0	46.4	44.2
Lettuce	27.6	30.3	26.8	23.6	33.7	37.6	37.6	37.5	36.7	34.6	35.3	36.3	35.0	30.3
Okra	10.8	12.9	9.5	9.2	29.0	47.2	49.0	49.2	49.0	48.4	48.7	49.0	48.5	46.6
Onion	10.1	11.6	8.5	8.1	25.0	39.4	40.5	40.4	39.6	37.7	38.5	39.4	38.2	33.3
Orange	8.9	10.2	7.2	7.1	28.9	53.2	55.8	55.9	55.7	55.2	55.6	55.7	55.4	53.1
Peach	9.6	10.9	8.3	8.6	29.1	47.7	49.5	49.5	49.3	49.0	49.3	49.4	49.1	47.7
Pepper	12.8	16.8	11.0	9.3	32.8	50.5	51.6	51.6	51.4	50.7	51.0	51.4	40.8	48.5
Pigweed	10.9	12.4	9.3	9.0	26.6	43.9	45.7	45.5	45.4	44.8	45.1	45.1	44.6	42.8
Pumpkin	10.2	11.8	8.9	10.6	29.1	44.9	46.4	46.3	46.2	45.8	46.7	46.2	45.7	44.2
Sorghum	15.0	17.2	13.3	11.3	28.2	45.8	47.3	47.4	47.3	46.9	47.0	47.0	46.8	45.5
Soybean	10.9	13.1	8.7	7.9	28.8	45.6	46.6	46.5	46.3	45.9	46.0	46.2	45.8	44.5
Sugarcane	15.9	18.6	13.4	11.4	29.9	45.8	46.9	46.8	46.4	45.6	45.7	46.0	45.4	42.9
Sunflower	9.6	11.0	8.4	8.5	27.5	45.4	47.3	47.3	47.1	46.5	46.9	47.2	46.6	44.1
Tomato	10.0	11.1	8.6	8.6	25.9	46.6	48.4	48.6	48.5	47.8	48.3	48.6	48.0	45.4
Watermelon	11.9	14.4	10.7	9.9	30.4	45.6	46.8	47.0	47.0	46.3	46.8	47.2	46.6	44.5
Wheat	10.3	13.4	9.6	7.7	27.3	50.2	51.5	51.7	51.4	51.0	51.2	51.5	51.0	48.9

Crop	1200	1250	1300	1350	1400	1450	1500	1550	1600	1650	1700	1750	1800	1850
Avocado	46.8	47.1	45.2	41.0	26.3	19.2	23.1	29.0	32.5	34.1	33.2	31.2	30.3	23.1
Bean	53.5	53.6	50.8	44.9	25.6	18.5	24.6	33.1	38.4	40.9	40.6	37.5	35.2	24.2
Cantaloupe	44.3	44.5	41.9	36.7	20.6	14.8	19.1	25.5	29.9	32.0	31.5	28.9	27.4	19.4
Corn	43.2	43.5	41.8	38.3	23.4	16.8	21.0	27.1	31.0	32.9	32.6	30.1	28.8	23.1
Cotton	44.0	44.2	42.0	37.5	21.7	15.2	19.6	26.2	30.4	32.3	31.9	29.4	27.9	19.9
Lettuce	29.6	29.8	26.4	21.4	11.8	9.1	10.4	13.0	15.4	16.8	16.8	15.0	13.8	10.6
Okra	46.2	46.4	44.5	40.4	25.6	18.1	22.3	28.8	33.0	35.0	34.5	32.3	30.8	23.0
Onion	32.5	32.9	29.0	23.0	10.3	6.8	8.4	12.0	15.1	17.2	17.0	14.6	13.1	9.4
Orange	52.8	53.0	51.2	47.1	31.2	22.3	26.6	33.3	37.6	39.8	39.0	36.6	35.4	27.8
Peach	47.7	47.8	46.5	43.0	30.3	24.3	28.8	34.3	37.5	38.9	38.0	38.4	35.6	27.4
Pepper	48.4	48.6	46.4	41.7	25.0	17.6	22.6	30.0	34.7	36.9	36.6	33.9	32.2	23.4

*WL = wavelengths　　　**WLI = wavelength interval

Table 3-29. Average Percent Reflectances of Top Leaf Surfaces of 10 Leaves for Each of 20 Crops for 41 WL* (nm) Over the 500- to 2500-nm WLI** [3-106] (Continued)

Crop	1200	1250	1300	1350	1400	1450	1500	1550	1600	1650	1700	1750	1800	1850
Pigweed	42.5	42.6	40.6	36.2	21.5	15.6	19.9	26.1	30.0	31.8	31.3	29.1	27.6	19.5
Pumpkin	44.0	44.0	42.1	37.4	24.6	19.0	23.6	29.2	32.6	34.6	33.1	31.3	29.5	21.6
Sorghum	45.3	45.4	44.3	41.7	30.9	24.7	28.2	33.2	36.1	37.4	36.9	35.3	34.2	28.2
Soybean	44.5	44.4	43.1	40.1	27.7	21.8	26.1	31.9	35.2	36.6	36.3	34.5	33.3	25.5
Sugarcane	42.6	42.7	40.5	35.9	20.7	14.4	18.3	24.2	28.0	30.4	30.0	27.5	25.9	18.8
Sunflower	44.0	44.2	41.7	36.4	20.4	14.3	18.4	24.9	29.3	31.3	30.5	28.1	26.6	18.9
Tomato	45.2	45.4	42.7	37.3	20.5	14.4	18.9	25.6	30.0	32.1	31.7	28.9	27.3	19.1
Watermelon	44.4	44.5	42.2	37.5	22.0	16.6	21.2	27.4	31.2	33.0	32.4	29.9	28.7	20.5
Wheat	48.8	49.2	47.2	43.5	27.7	21.7	26.5	32.7	36.4	38.2	37.4	35.2	34.3	27.3

Crop	1900	1950	2000	2050	2100	2150	2200	2250	2300	2350	2400	2450	2500
Avocado	9.7	7.5	10.2	13.2	15.7	18.1	19.5	17.4	14.2	11.6	9.5	7.8	7.0
Bean	8.0	6.0	9.4	14.1	18.9	22.6	24.0	21.5	17.2	12.8	9.5	7.2	5.9
Cantaloupe	8.1	6.9	8.6	11.1	14.2	16.5	17.5	15.7	12.6	9.9	8.0	6.6	6.0
Corn	7.9	7.2	9.7	12.6	15.8	18.3	19.8	17.6	14.4	11.6	9.3	7.5	6.7
Cotton	7.6	6.0	7.9	10.8	14.1	16.7	16.8	15.8	12.5	9.8	7.5	6.0	5.3
Lettuce	6.2	5.6	6.4	7.4	8.4	9.2	9.4	8.8	7.7	6.6	5.8	5.2	4.9
Okra	9.4	7.0	9.4	12.8	16.3	19.0	20.2	18.3	14.9	11.8	9.3	7.3	6.5
Onion	4.9	4.4	4.9	5.6	6.6	7.6	8.0	7.4	6.3	5.4	4.8	4.6	4.5
Orange	11.4	8.6	12.0	15.8	19.2	22.1	23.6	21.2	17.4	14.1	11.1	9.0	7.8
Peach	12.5	10.5	14.4	18.3	21.6	24.3	25.7	23.1	19.3	16.0	13.2	10.7	9.5
Pepper	8.5	6.6	9.4	13.2	17.1	20.2	21.5	19.3	15.4	11.7	8.9	6.8	5.7
Pigweed	7.7	5.8	8.0	11.0	14.3	16.8	17.8	15.9	12.9	9.9	7.6	5.9	5.1
Pumpkin	9.0	7.1	10.6	14.0	17.2	19.5	20.9	18.2	14.9	12.1	9.6	7.6	7.0
Sorghum	14.1	12.0	15.6	19.1	22.1	24.5	25.8	23.7	20.4	17.4	14.7	12.4	11.3
Soybean	10.2	8.1	12.1	16.6	20.6	23.5	24.8	22.7	19.1	15.4	12.1	9.5	8.2
Sugarcane	7.6	6.2	8.2	10.5	13.1	15.5	16.4	14.5	11.8	9.5	7.8	6.5	6.0
Sunflower	8.0	6.5	8.1	10.4	13.2	15.4	16.2	14.4	11.6	9.3	7.6	6.5	6.0
Tomato	7.3	6.0	7.9	10.7	13.7	16.3	17.3	15.3	12.2	9.5	7.4	6.0	5.4
Watermelon	8.0	6.9	9.1	12.1	15.3	17.7	18.8	16.8	13.5	10.8	8.5	6.9	6.2
Wheat	9.7	9.0	12.8	16.6	20.2	22.6	24.4	21.7	18.2	15.0	12.2	9.7	8.5

*WL = wavelengths   *WLI = wavelength interval

Table 3-30. Average Percent Transmittances of Top Leaf Surfaces of 10 Leaves for Each of 20 Crops for 41 WL* (nm) Over the 500- to 2500-nm WLI** [3-106]

Crop	500	550	600	650	700	750	800	850	900	950	1000	1050	1100	1150
Avocado	2.3	4.1	1.4	3.1	24.9	42.4	44.8	45.4	45.5	45.5	46.1	46.6	46.3	45.0
Bean	6.9	10.9	5.5	3.6	26.6	40.9	42.0	42.2	42.0	41.5	42.2	42.4	41.9	39.9
Cantaloupe	4.9	8.7	3.9	2.4	27.5	46.3	48.1	48.6	48.6	48.0	48.8	49.5	49.0	46.5
Corn	8.1	9.8	3.7	0.7	22.6	48.9	50.5	50.9	51.1	50.7	51.2	51.7	51.6	49.7
Cotton	8.1	13.1	7.0	4.2	30.6	47.8	49.1	49.4	49.3	39.0	49.4	49.9	49.6	47.8
Lettuce	38.4	44.3	39.5	34.0	49.5	55.3	55.6	55.5	54.8	52.6	53.7	54.9	53.7	48.2
Okra	5.9	14.8	5.8	4.1	27.1	44.6	46.4	46.7	46.9	46.7	47.3	47.8	47.6	46.0
Onion	11.7	18.8	10.8	6.6	35.8	54.3	55.7	55.7	55.0	52.9	54.0	55.4	54.1	48.2
Orange	0.7	1.9	0.5	0.5	17.6	36.0	38.2	38.6	38.6	38.4	38.9	39.5	39.3	37.7
Peach	3.5	6.2	2.6	2.8	27.1	45.5	47.3	47.6	47.7	47.6	47.9	48.3	48.1	47.1
Pepper	6.9	12.6	6.4	3.1	28.4	44.8	46.2	46.5	46.4	46.0	46.5	47.0	46.7	44.9
Pigweed	5.4	9.5	3.7	2.7	28.6	49.2	51.6	52.0	52.0	51.9	52.4	52.9	52.6	51.0
Pumpkin	5.6	8.8	4.3	5.6	30.0	47.1	48.9	49.4	49.6	49.5	50.1	50.6	50.4	49.1
Sorghum	5.0	9.0	4.2	2.1	24.4	46.7	49.1	49.6	49.8	49.9	50.3	50.8	50.7	49.8
Soybean	10.0	15.6	8.7	5.4	32.5	50.0	51.4	51.8	51.9	51.8	52.2	52.6	52.4	51.4
Sugarcane	7.5	12.2	6.9	4.1	26.7	45.0	46.9	47.2	47.3	46.9	47.6	48.1	47.9	46.0
Sunflower	6.3	9.1	5.7	5.1	27.8	46.4	48.4	48.8	48.8	48.4	49.1	49.7	49.2	46.8
Tomato	2.6	5.5	1.5	0.9	23.6	41.9	43.8	44.3	44.4	44.0	44.7	45.3	44.9	42.6
Watermelon	5.2	9.6	4.3	2.0	28.7	45.2	46.6	47.1	47.4	47.2	47.9	48.5	48.2	46.3
Wheat	1.9	5.8	2.1	0.7	20.3	41.8	43.4	43.9	44.1	43.9	44.6	45.2	45.1	43.4

Crop	1200	1250	1300	1350	1400	1450	1500	1550	1600	1650	1700	1750	1800	1850
Avocado	45.1	45.6	44.0	39.4	26.1	20.5	25.6	32.0	35.8	37.6	37.0	35.4	34.1	25.1
Bean	40.0	40.2	38.1	33.5	17.3	11.8	17.3	24.9	29.6	32.2	32.2	29.5	27.9	18.5
Cantaloupe	46.6	47.0	44.5	39.2	20.6	14.6	19.7	28.2	33.7	36.6	36.5	33.6	32.0	21.8
Corn	49.8	50.5	49.0	45.9	28.8	20.5	26.8	35.1	40.2	43.0	43.1	40.6	39.6	32.0
Cotton	47.9	48.3	46.6	42.7	26.7	19.6	25.4	33.2	38.0	40.4	40.3	38.1	37.0	27.1
Lettuce	47.4	48.0	43.7	35.9	14.6	6.2	11.1	19.9	26.6	30.5	31.0	27.0	24.4	15.2
Okra	46.1	46.5	45.0	41.5	26.8	19.3	24.5	32.0	36.6	39.2	39.1	37.0	35.9	27.3
Onion	47.4	48.1	43.4	35.1	12.5	4.1	8.7	17.5	24.3	28.4	28.8	24.7	22.0	13.1
Orange	37.6	38.2	36.9	33.7	20.1	13.0	17.2	23.5	27.6	30.0	29.6	27.6	26.9	20.2
Peach	47.3	47.7	46.7	43.9	31.5	26.2	31.3	37.4	40.9	42.8	42.3	40.9	40.6	31.7
Pepper	45.0	45.4	43.7	39.8	23.9	16.9	22.7	30.4	35.3	37.8	37.8	35.4	34.0	25.0

*WL = wavelengths   *WLI = wavelength interval

Table 3-30. Average Percent Transmittances of Top Leaf Surfaces of 10 Leaves for Each of 20 Crops for 41 WL* (nm) Over the 500- to 2500-nm WLI** [3-106] (Continued)

Crop	1200	1250	1300	1350	1400	1450	1500	1550	1600	1650	1700	1750	1800	1850
Pigweed	51.2	51.6	49.9	45.8	29.9	23.1	29.1	37.1	41.9	44.5	44.4	42.2	41.0	30.3
Pumpkin	49.3	49.7	48.2	43.7	29.5	23.8	29.7	36.8	41.2	43.5	43.1	41.1	39.5	29.2
Sorghum	50.0	50.4	49.6	47.3	35.1	28.2	33.2	39.9	44.0	46.2	46.3	44.8	44.1	36.6
Soybean	51.6	51.9	50.8	48.0	34.9	28.7	34.3	41.3	45.3	47.4	47.5	45.8	44.8	35.5
Sugarcane	46.0	46.5	44.9	40.8	24.6	17.3	23.0	30.7	35.7	38.5	38.4	36.0	34.8	25.1
Sunflower	46.8	47.3	45.1	40.0	22.2	15.0	21.0	29.1	34.4	37.0	36.6	34.0	32.7	22.5
Tomato	42.6	43.0	40.7	35.9	18.6	12.3	17.9	25.7	30.8	33.4	33.3	30.5	29.1	19.0
Watermelon	46.5	47.0	45.0	40.7	24.1	18.3	24.3	31.8	36.5	38.8	38.6	36.2	35.3	25.3
Wheat	43.6	44.2	42.8	39.7	24.3	18.5	23.9	30.7	34.7	36.8	36.3	34.3	33.7	26.7

Crop	1900	1950	2000	2050	2100	2150	2200	2250	2300	2350	2400	2450	2500
Avocado	8.8	6.7	12.3	17.3	21.2	24.0	25.2	23.3	19.8	16.2	12.1	9.8	6.9
Bean	3.7	1.9	5.4	10.3	15.3	18.6	19.7	18.4	15.2	11.3	7.8	4.9	3.5
Cantaloupe	4.2	2.1	6.0	11.5	16.9	20.5	21.8	20.1	16.4	12.1	8.1	5.0	3.4
Corn	6.5	5.0	11.8	18.6	24.6	28.5	30.3	28.3	24.4	19.9	14.8	9.7	7.0
Cotton	7.4	4.5	10.2	16.8	22.6	26.2	27.7	26.1	22.5	17.9	12.9	8.8	6.6
Lettuce	2.1	0.5	1.7	4.5	8.8	12.2	13.5	12.2	9.0	5.6	2.9	1.4	0.8
Okra	8.6	5.2	10.7	16.8	22.2	25.7	27.1	25.7	22.2	18.1	13.6	9.7	7.5
Onion	1.2	0.5	0.6	2.5	6.0	9.0	10.2	8.8	6.0	3.1	1.2	0.5	0.5
Orange	5.3	2.6	6.2	10.3	14.1	16.8	18.1	16.5	13.6	10.7	7.8	5.1	3.8
Peach	12.6	10.4	17.3	23.4	28.1	31.2	32.5	30.6	27.2	23.6	19.1	14.7	12.2
Pepper	6.3	3.8	8.9	15.0	20.5	24.2	25.6	24.2	20.8	16.4	11.9	8.1	6.0
Pigweed	9.9	6.9	13.5	20.7	26.8	30.7	32.2	30.6	26.9	22.2	17.0	12.4	9.6
Pumpkin	10.2	8.3	14.9	21.4	26.8	30.1	31.3	29.4	25.8	21.5	16.7	11.9	10.2
Sorghum	15.4	12.2	19.9	26.7	31.9	35.3	36.9	35.4	32.1	28.2	23.6	18.4	15.6
Soybean	14.6	11.7	19.3	26.7	32.7	36.3	37.7	36.3	33.0	28.6	23.5	18.5	15.8
Sugarcane	6.7	4.0	9.3	15.1	20.0	23.7	25.0	23.0	19.3	15.1	10.8	7.0	4.9
Sunflower	6.0	2.3	6.5	11.9	17.1	20.5	21.6	19.7	16.1	12.1	8.2	5.0	3.3
Tomato	3.7	1.8	5.4	10.2	15.2	18.6	19.8	18.2	14.8	10.8	7.2	4.3	3.0
Watermelon	6.1	4.6	10.1	16.1	21.3	24.7	26.1	24.4	20.7	16.6	12.2	8.2	6.3
Wheat	6.0	5.2	10.7	15.9	20.4	23.3	24.7	22.8	19.6	16.2	12.3	8.6	6.5

*WLI = wavelength interval   *WL = wavelengths

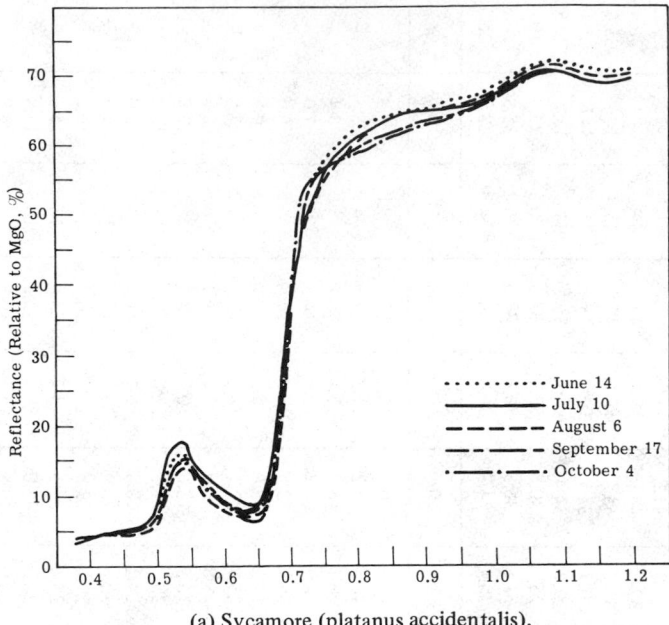

(a) Sycamore (platanus accidentalis).

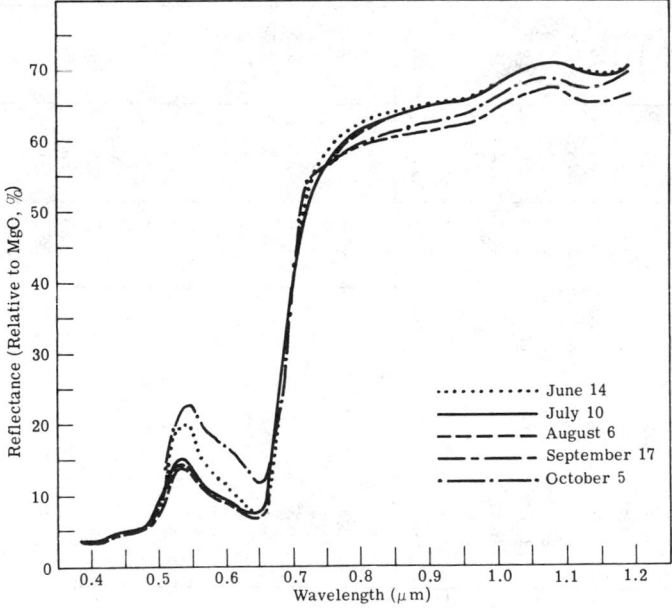

(b) Silver maple (acer saccharinum).

Fig. 3-152. Selected average reflectance curves for the upper surface of leaves from various trees [3-107].

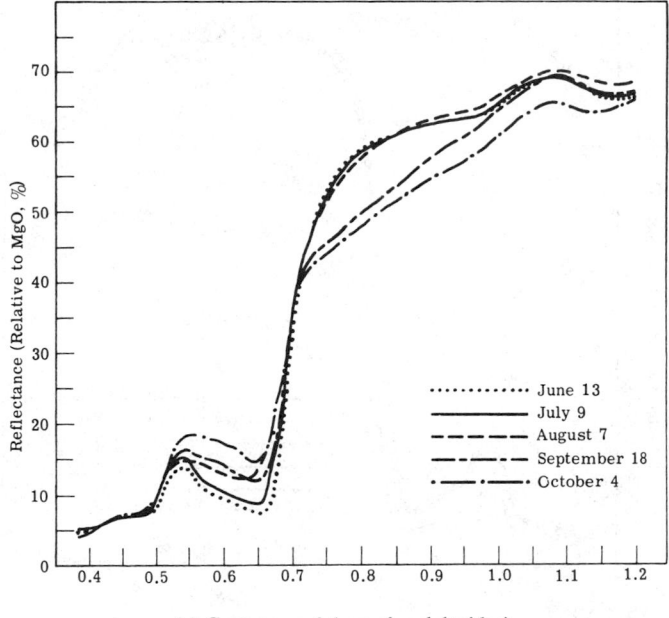

(c) Cottonwood (populus deltoides).

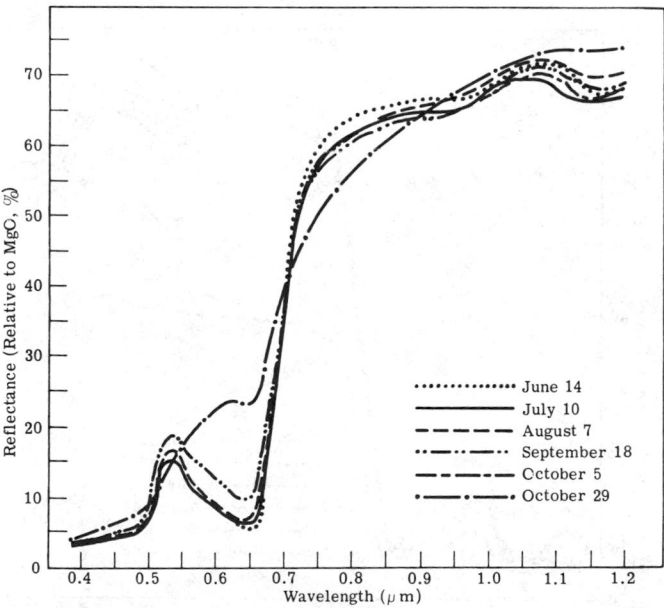

(d) Yellow poplar (liriodendron tulipifera).

Fig. 3-152. (Continued) Selected average reflectance curves for the upper surface of leaves from various trees [3-107].

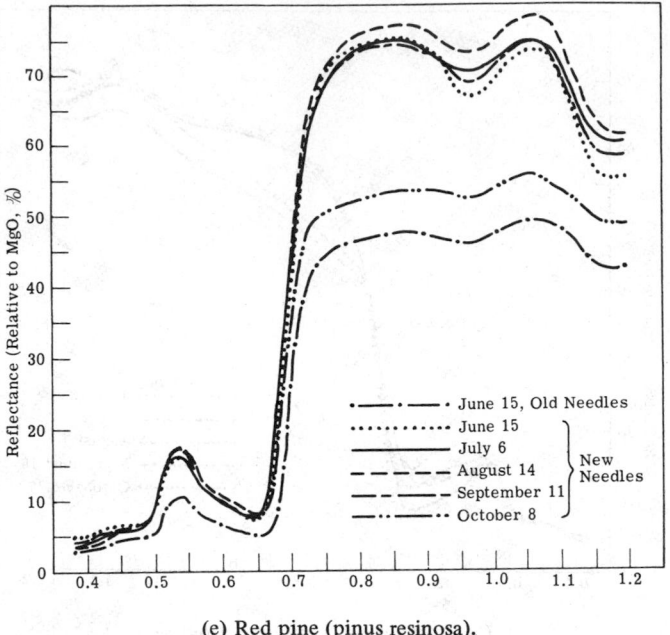

(e) Red pine (pinus resinosa).

(f) White pine (pinus strobus).

Fig. 3-152. (Continued) Selected average reflectance curves for the upper surface of leaves from various trees [3-107].

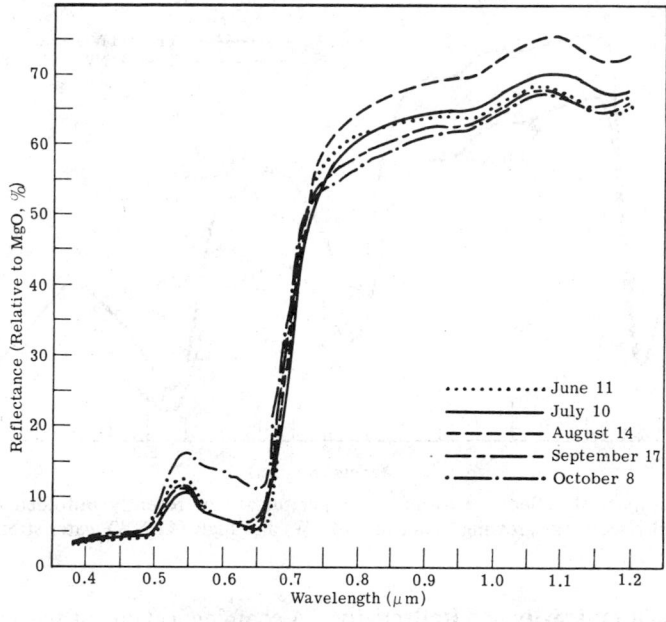

(g) Bur oak (quercus macrocarpa).

Fig. 3-152. (Continued) Selected average reflectance curves for the upper surface of leaves from various trees [3-107].

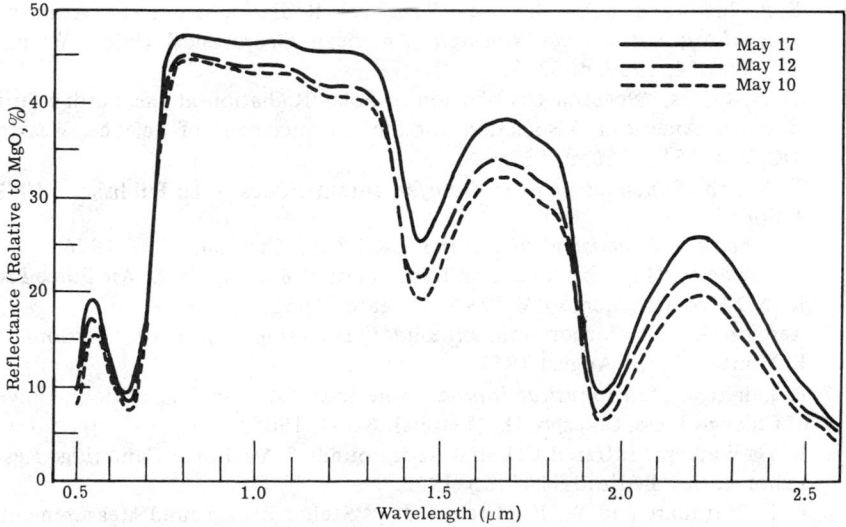

Fig. 3-153. Change in spectral reflectance from the upper surface of a single immature leaf developing on a watered yellow poplar seedling as a function of time [3-107].

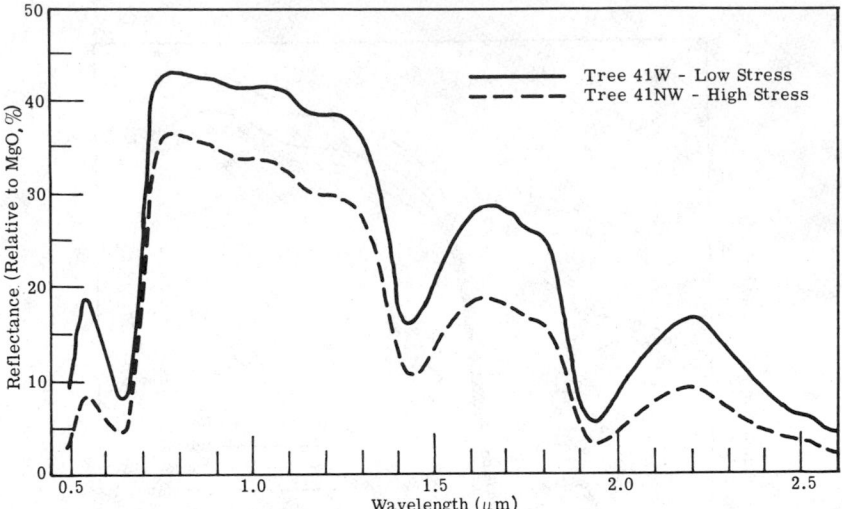

Fig. 3-154. Spectral reflectance from the upper surface of recently unfolded leaves on yellow poplar seedlings growing under low (41 W) and high (41 NW) water stress regimes [3-107].

3.7.8. Terrain Emissivity and Reflectivity. A complete catalog of spectral reflectance data of terrain from all available sources, reduced to a standard format of presentation, is now available [3-108]. The report itself is very bulky and has had limited circulation. However the data are on file at the NASA Johnson Space Center, Houston, TX.

3.8. References and Bibliography

3.8.1. References

[3-1]   S. B. Idso and R. D. Jackson, "Thermal Radiation from the Atmosphere," *Journal of Geophysical Research,* American Geophysical Union, Washington, DC, Vol. 74, 1969, p. 5397.

[3-2]   D. M. Gates, "Spectral Distribution of Solar Radiation at the Earth's Surface," *Science,* American Association for the Advancement of Science, Washington, DC, Vol. 151, 1966, p. 523.

[3-3]   F. Kreith, *Principles of Heat Transfer,* Intext Educational Publishers, 1973, 3rd Edition.

[3-4]   *Handbook of Chemistry and Physics,* CRC Press, Cleveland, OH, 1976.

[3-5]   L. Larmore, "Infrared Radiation from Celestial Bodies," U. S. Air Force Project, RAND Research Memo RM-793-1, 17 March 1952.

[3-6]   Aerojet-General Corporation, *Infrared Engineering Handbook,* Astrionics Div., El Monte, CA, 25 August 1961.

[3-7]   R. Leighton, *Astrophysical Journal,* American Astronomical Society, University of Chicago Press, Chicago, IL, (Letters), No. 4, 1965.

[3-8]   R. G. Walker, "Infrared Celestial Backgrounds," Air Force Cambridge Research Laboratories, Bedford, MA, July 1962.

[3-9]   P. E. Barnhart and W. E. Mitchell, Jr., "Stellar Background Measurement Program," The Ohio State University, Columbus Research Foundation, Columbus, OH, July 1964.

[3-10] Eastman Kodak Company, Rochester, NY, "Space Background Study for Project DEFENDER," AD 403 780, April 1963.

[3-11] ACF Industries Inc., *Space Handbook No. 1*, ACF Electronics Division, Riverdale, MD, November 1962.

[3-12] Air Force Cambridge Research Labs., Bedford, MA, "Celestial Background Radiation," AD 602 616, March 1964.

[3-13] L. L. Collins and R. B. Freund, "Celestial Background Simulation Techniques," Northrop Space Laboratories, Hawthrone, CA, AD 282 788, 1961.

[3-14] F. J. Low and H. L. Johnson, "Stellar Photometry at 10 μ," *Journal of Applied Physics*, Vol. 139, 1964, p. 1130.

[3-15] H. L. Johnson, R. I. Mitchell, B. Iriarte, and W. K. Wisniewski, "Magnitudes and Colors of 1300 Bright Stars," *Sky and Telescope*, Sky Publishing Corporation, Cambridge, MA, Vol. 30, No. 1, 24 July 1965.

[3-16] R. G. Walker and S. D. Price, "AFCRL Infrared Sky Survey," AFCRL-TR-75-0373, Vol. 1, July 1975.

[3-17] R. G. Walker and S. D. Price, "A Model for the Infrared Celestial Background," manuscript in preparation, June, 1977.

[3-18] C. W. Allen, "Astrophysical Quantities," University of London, Athlone Press, 1974, 3rd Edition.

[3-19] B. W. Bopp and R. D. Gehrz, and J. A. Hackwell, PASP, Vol. 86, No. 514, December, 1974.

[3-20] L. N. Mavridis, *Structure and Evolution of the Galaxy*, Springer-Verlag, New York, 1971.

[3-21] G. H. Rieke, and F. J. Low, *Astrophysical Journal*, American Astronomical Society, University of Chicago Press, Chicago, IL, Vol. 176, L95, 1972.

[3-22] D. W. Sciama, *Modern Cosmology*, Cambridge University Press, New York, NY, 1971.

[3-23] D. Morrison, *Astrophysical Journal*, American Astronomical Society, University of Chicago Press, Chicago, IL, Vol. 194, 1974, p. 203.

[3-24] T. L. Murdock, "Contribution of Asteroids to the Infrared Astronomical Sky Survey," AFCRL-TR-73-0154, 1973.

[3-25] R. B. Partridge and P. J. E. Peebles *Astrophysical Journal*, American Astromonical Society, University of Chicago Press, Chicago, IL, Vol. 148, 1967, p. 377.

[3-26] A. W. Peterson, *Astrophysical Journal*, American Astronomical Society, University of Chicago Press, Chicago, IL, Vol. 138, 1963, p. 1218.

[3-27] B. T. Soifer, J. R. Houck, and M. Harwit, *Astrophysical Journal*, American Astronomical Society, University of Chicago Press, Chicago, IL, Vol. 168, L73, 1971.

[3-28] M. P. Thekakara, "Evaluating the Light from the Sun," *Optical Spectra*, Optical Publishing, Pittsfield, MA, Vol. 6, No. 3, March 1972, pp. 32-35.

[3-29] After P. Moon, *Journal of the Franklin Institute*, Vol. 230, No. 5, 1940.

[3-30] The Boeing Company, "Infrared Satellite Backgrounds, Part I: Atmospheric Radiative Processes," Aero-Space Division, Seattle, WA, AFCRL 1069 (I), 30 September 1961.

[3-31] E. E. Bell, L. Eisner, J. Young, and R. A. Oetjen, *Journal of the Optical Society of America*, Vol. 52, February 1962, pp. 201-209.

[3-32] H. Rose, et al., *The Handbook of Albedo and Thermal Earthshine*, Environmental Research Institute of Michigan, Ann Arbor, MI, Report No. 190201-1-T, 1973.

[3-33] D. Anding, R. Kauth, and R. Turner, "Atmospheric Effects on Multispectral Sensing of Sea Surface Temperature from Space," Willow Run Laboratories of the Institute of Science and Technology, The University of Michigan, Ann Arbor, MI, Report No. 26760-6-T, December 1970.

[3-34] Y. P. Novoseltsev, "Spectral Reflectivity of Clouds," Translation of Spektrolnaya Otrazhatelnaya Oblakov, Trudy Glavnoy, Geofizicheskoy Observatoiri Uneni A. I. Voyykova, NASA TT F-328, No. 152, 1964, pp. 186-191.

[3-35] K. Y. Kondratyev, Z. F. Mironova, and A. N. Otto, "Spectral Albedo of Natural Surfaces," *Pure and Applied Geophysics*, Birkhauser Verlag, Switzerland, Vol. 59, No. 3, 1964, pp. 207-216.

[3-36] K. Y. Kondratyev, "Actinometry," NASA TT F-9712, Washington, DC, November 1965 (translation of "Aktinometriya," Gidrometerologicheskoye Izdatel'stvo, Leningrad, 1965).

[3-37] D. G. Orr, S. E. Dwamik, and L. M. Young, "Reflectance Curves of Soils, Rocks, Vegetation and Pavement," U.S. Army Engineer Research and Development Laboratories, Fort Belvoir, Report No. 7746-RR, 22 April 1963.

[3-38] S. L. Valley (ed.), *Handbook of Geophysics and Space Environments,* Air Force Cambridge Research Laboratories, Hanscom Air Force Base, Bedford, MA, 1965.

[3-39] G. A. Wilkins and J. A. Hoyem, "The Terrestrial Night Horizon and Sky, 4.3 Micron Radiance Data Generated by the HITAB-TRIS Experiments," U.S. Naval Ordnance Test Station, China Lake, CA, NAVWEPS Report 8552, June 1964.

[3-40] F. A. Berry, et al., *Handbook of Meteorology,* McGraw-Hill Book Company, New York, NY, 1945.

[3-41] I. Soloman, "Estimates of Altitudes with Specified Probabilities of Being Above All Clouds," Air Weather Service (MATS), U.S. Air Force, Technical Report 159, October, 1961.

[3-42] A. Adel, "Observations of Atmospheric Scattering Near the Sun's Limb," Arizona State College, Flagstaff, Ariz., AD 273 599, January 1961.

[3-43] R. C. Jones and A. M. Nagvi, "Satellite Navigation of Terrestrial Occulations of Stars; III: Interference due to Brightness of the Earth's Atmosphere," AD 287 869, Geophysics Corporation of America, Boston, MA, October 1962.

[3-44] A. M. Nagvi, "Satellite Navigation by Terrestrial Occulations of Stars— Considerations Relating to Refraction and Extinction," Geophysics Corporation of America, Bedford, MA, AD 287 868, October 1962.

[3-45] G. Newkirk and J. Eddy, "Light Scattering by Particles in the Upper Atmosphere," University of Colorado, Boulder, CO, May 1963.

[3-46] R. K. McDonald, Boeing Co., Seattle, WA, Private Communication.

[3-47] N. P. Laverty and W. M. Clark, Boeing Co., Seattle, WA, Private Communication.

[3-48] "Measurement of Infrared Radiation Gradients in the Sky," Midwest Research Institute, Kansas City, MO, AD 206 453, 1953.

[3-49] R. C. Jones, "Sky Noise—Analysis of Circular Scanning," Polaroid Corporation, Cambridge, MA, November 1953.

[3-50] R. C. Jones, "Sky Noise—Its Nature and Analysis," Polaroid Corporation, Cambridge, MA, September 1953.

[3-51] R. E. Eisele, "Infrared Background Investigation," Ramo-Wooldridge Division of Thompson-Ramo Wooldridge, Inc., Los Angeles, CA, Report AFCRC-TN59-843, June 1959.

[3-52] R. E. Eisele, "Infrared Background Investigation," Thompson-Ramo Wooldridge, Inc., Canoga Park, CA, AD 236913, March 1960.

[3-53] H. E. Bennett, J. M. Bennett, and M. R. Nagel, "The Spatial Distribution of Infrared Radiation from the Clear Sky Including Sequences of Sky Maps at Various Elevations," U.S. Naval Ordnance Test Station, China Lake, CA, NAVORD Report 6577, September 1959.

[3-54] H. E. Bennett, J. M. Bennett, and M. R. Nagel, "Measurements of Infrared and Total Radiance of the Clear Winter Sky of Wright-Patterson Air Force Base, Ohio," Wright Air Development Center, USAF Air Research and Development Command, Wright-Patterson AFB, OH, AD 118 127, March, 1957.

[3-55] E. E. Bell, L. Eisner, J. Young, and R. A. Oetjen, "Spectral Radiance of Sky and Terrain at Wavelengths between 1 and 20 Microns. II. Sky Measurements," *Journal of the Optical Society of America,* Vol. 50, December 1960, pp. 1313-1320.

[3-56] E. E. Bell, I. L. Eisner, and R. A. Oetjen, "The Spectral Distribution of the Infrared Radiation from the Sky," *Proc. of the Symposium on Infrared Backgrounds,* Engineering Research Institute, The University of Michigan, Nonr-1224 (12), AD 121010, March 1956.

[3-57] R. Sloan, J. Shaw, and D. Williams, "Infrared Emission Spectrum of the Atmosphere," *Journal of the Optical Society of America,* Vol. 45, No. 6, June 1955, pp. 455, 459.

[3-58] J. Gordon and P. Church, "Overcast Sky Luminances and Directional Luminous Reflectances of Objects and Backgrounds under Overcast Skies," *Applied Optics,* Vol. 5, 919, June 1966.

[3-59] J. W. Chamberlain, *Physics of the Aurora and Airglow,* Academic Press, 1961.

[3-60] A. W. Harrison and A. V. Jones, *Journal of Atmospheric and Terrestrial Physics,* Pergamon Press, New York, NY, Vol. 11, 1957, pp. 192-199.

[3-61] A. V. Jones, "Possible Methods for Studying the Auroral Spectrum in the 2.0 to 3.5 Micron Region," Saskatchewan University, Saskatoon, Canada, November, 1959.

[3-62] D. M. Hunten, "Optics of the Upper Atmosphere," *Applied Optics,*Vol. 3, No. 2, February 1964.

[3-63] National Aeronautics and Space Adminiatration, *Aurorae and Airglow,* Washington, D.C., April 1964.

[3-64] I. Sellin, "Auroral Radiations in the Infrared," Laboratories for Applied Sciences, University of Chicago, October, 1961.

[3-65] R. Chapman, R. Jones, A. Dalgarno, and D. Beining, "Investigation of Auroral, Airglow and Night Emissions as Related to Space-Based Defense Systems," Geophysics Corporation of America, Bedford, MA, June 1962.

[3-66] A. W. Harrison and A. V. Jones, *Journal of Atmospheric and Terrestrial Physics,* Pergamon Press, New York, NY, Vol. 13, 1957, pp. 291-294.

[3-67] A. T. Stair, Jr., J. C. Ulwick, K. D. Baker, and D. J. Baker, "Rocketborne Observations of Atmospheric Infrared Emissions in the Auroral Region," *Atmospheres of Earth and Planets,* Reidel Publishing Company, Boston, MA, 1975, pp. 335-346.

[3-68] E. H. Vestine, "Terrestrial Magnetism," *Journal of Geophysical Research,* Vol. 49, June 1944, pp. 77-102.

[3-69] F. W. G. White and M. Geddes, "Terrestrial Magnetism," *Journal of Geophysical Research,* Vol. 44, December 1939, pp. 367-377.

[3-70] L. Harang, *The Aurorae,* John Wiley & Sons, Inc., New York, NY, 1951.

[3-71] C. Störmer, *The Polar Aurora,* Clarendon Press, Oxford, 1955.

[3-72] E. L. Krinov, "Spectra Reflectance Properties of Natural Formations,"

(Originally published in Russian) 1947. Translation: National Research Council of Canada. Tech. Trans. TT439, Ottawa, Canada, 1953.

[3-73] F. F. Roach, Proceedings of the Institute of Radio Engineers, Vol. 47, 1959, p. 267.

[3-74] J. F. Noxon, A. Harrison, and A. V. Jones, "The Infrared Spectrum of the Night Airglow 1-4 μ to 4.0 μ," *Journal of Atmospheric and Terrestrial Physics*, Pergamon Press, New York, NY, Vol. 16, 1959, pp. 246-251.

[3-75] A. Vallance Jones and H. Gush, *Nature*, Vol. 172, 12 September 1953, p. 496.

[3-76] D. M. Hunten, *Journal of Atmospheric and Terrestrial Physics*, Pergamon Press, New York, NY, Vol. 7, 1955, pp. 141-151.

[3-77] M. J. Seaton, *Journal of Atmospheric and Terrestrial Physics*, Pergamon Press, New York, NY, Vol. 4, 1954, pp. 285-294.

[3-78] J. C. Brandt and J. W. Chamberlain, "Resonance Scattering by Atmospheric Sodium-V, Theory of the Day Airglow," *Journal of Atmospheric and Terrestrial Physics*, Pergamon Press, New York, NY, Vol. 13, December 1958, pp. 90-98.

[3-79] J. C. Brandt, *Astrophysical Journal*, American Astronomical Society, University of Chicago Press, Chicago, IL, Vol. 128, 1958, pp. 118-123.

[3-80] J. C. Brandt, *Astrophysical Journal*, American Astronomical Society, University of Chicago Press, Chicago, IL, Vol. 130, July 1959, pp. 228-240.

[3-81] J. W. Chamberlain and C. Sagan, *Planetary and Space Science*, Vol. 2, 1960, pp. 157-164.

[3-82] H. P. Gush and H. L. Buijs, "The Near Infrared Spectrum of Night Airglow Observed from High Altitude," *Canadian Journal of Physics*, National Research Council of Canada, Ottawa, Quebec, Canada, Vol. 42, 1964, p. 1037.

[3-83] H. R. Condit, "The Spectral Reflectance of American Soils," *Photogrammetric Engineering*, American Society of Photogrammetry, Falls Church, VA, Vol. 36, No. 9, September 1970.

[3-84] V. Leeman, D. Earing, R. Vincent, S. Ladd, "The NASA Earth Resources Spectral Information System: A Data Compilation," Institute of Science and Technology, The University of Michigan, Ann Arbor, MI, Report No. NASA CR-31650-24-T, May 1971.

[3-85] R. K. Vincent, "A Thermal Infrared Ratio Imaging Method for Mapping Compositional Variations Among Silicate Rock Types," PhD Thesis, The University of Michigan, 1973.

[3-86] D. Faulkner, R. Horvath, J. Ulrich and E. Work, "Spectral and Polarization Characteristics of Selected Targets and Backgrounds: Instrumentation and Measured Results (3.3-14.0 μm)," Willow Run Laboratories, Institute of Science and Technology, The University of Michigan, Ann Arbor, MI, Report No. 3692-23-T, AD 886 916 L, 1971.

[3-87] D. Earing, "Target Signature Analysis Center: Data Compilation, Second Supplement," Willow Run Laboratories, Institute of Science and Technology, The University of Michigan, Ann Arbor, MI, Report No. 8492-5-B, AD 819 712, July 1967.

[3-88] D. Carmer, "Target Signature Analysis Center: Data Compilation Seventh Supplement," Willow Run Laboratories, Institute of Science and Technology, The University of Michigan, Ann Arbor, MI, Report No. 8492-35-B, AD 856 343, January 1969.

[3-89] E. D. McAlister, University of California, Private Communication, 1951-1952.

[3-90] H. O. McMahon, "Thermal Radiation from Partially Transparent Reflecting Bodies," *Journal of the Optical Society of America*, Vol. 40, June 1950, pp. 376-380.

[3-91] G. C. Ewing (ed.), *Oceanography from Space*, Woods Hole Oceanographic Institution, Woods Hole, MA, WHOI Ref. No. 65-10, April 1965.

[3-92] C. Cox and W. Munk, *Bulletin of the Scripps Institution of Oceanography*, University of California, Vol. 6, 1956, pp. 401-488.

[3-93] E. D. McAlister, "Application of Infrared-Optical Techniques to Oceanography," *Journal of the Optical Society of America*, Vol. 52, May 1962, p. 607.

[3-94] G. C. Ewing and E. D. McAlister, *Science*, Vol. 131, May 6, 1960, pp. 1374-1376.

[3-95] H. U. Sverdrup, M. Johnson, and R. Fleming, *The Oceans*, Prentice Hall, 1972.

[3-96] R. D. Watson, et al., "Prediction of the Fraunhofer Line Detectivity of Luminescent Materials," *Proceedings of the Ninth Symposium on Remote Sensing of Environment*, Environmental Research Institute of Michigan, Ann Arbor, MI, April 1974, p. 1969.

[3-97] M. R. Specht, D. Needler, and N. L. Fritz, "New Color Film for Water-Photography Penetration," *Photogrammetric Engineering*, American Society of Photogrammetry, Falls Church, VA, Vol. 39, April 1973, p. 359.

[3-98] V. E. Derr, *Remote Sensing of the Troposphere*, U.S. Dept. of Commerce, NOAA, Catalog No. C55.602 T75, Stock No. C323-0011, August 1972.

[3-99] C. T. Wezernak, "The Use of Remote Sensing in Limnological Studies," *Proceedings of the Ninth Symposium on Remote Sensing of Environment*, Environmental Research Institute of Michigan, Ann Arbor, MI, April 1974, p. 973.

[3-100] L. C. Gramms and W. C. Boyle, "Reflectance and Transmittance Characteristics of Several Selected Green and Blue-Green Unialgae," *Proceedings of the Seventh Symposium on Remote Sensing of Environment*, Environmental Research Institute of Michigan, Ann Arbor, MI, May 1971, pp. 1637, 1642, 1643.

[3-101] K. Kalle, "Zurn Problem der Meereswasserfarbe," *Ann. d. Hydrogr. und Mar. Meteor. Bd.*, Vol. 66, S1-13, 1938.

[3-102] R. Horvath, W. L. Morgan, and R. Spellicy, "Measurements Program for Oil Slick Characteristics," Institute of Science and Technology, The University of Michigan, Report No. 2766-7-F, 1970. p. 17, 19, 21, 23, 24, 28, 30, 32, 83-91, 144-146, 154-156.

[3-103] H. W. O'Brien, et al., "Red and Near Infrared Spectral Reflectance of Snow," U.S. Army Cold Region Research and Engineering Laboratory, Hanover, NH, CREEL (AD-A007732), 1975.

[3-104] P. S. Nobel, *Introduction to Biophysical Plant Physiology*, W. H. Freeman and Company, 1974.

[3-105] H. W. Gausman, et al., "Leaf Light Reflectance, Transmittance, Absorptance, and Optical and Geometrical Parameters for Eleven Plant Genera with Different Leaf Mesophyll Arrangements," *Proceedings of the Seventh International Symposium on Remote Sensing of Environment*, Environmental Research Institute of Michigan, Ann Arbor, MI, May 1971, pp. 1606-1608.

[3-106] H. W. Gausman, et al., "The Leaf Mesophylls of Twenty Crops, Their Light Spectra, and Optical and Geometrical Parameters," U.S. Department of Agriculture, Washington, DC, U.S.D.A. Technical Bulletin No. 1465, 1973.

[3-107] C. E. Olson, Jr., "Seasonal Change in Foliar Reflectance of Five Broadleaved Forest Tree Species," PhD Thesis, The University of Michigan, 1969, p. 76.

[3-108] "Target Signatures Study Interim Report, Volume V: Catalog of Spectral Reflectance Data," The University of Michigan, Ann Arbor, MI, Report No. 5698-22-T(V), October 1964.

[3-109] *Electro-Optics Handbook*, RCA, Commercial Engineering, Harrison, NJ, Technical Series EOH-11, 1974.

[3-110] D. S. Bond and F. P. Henderson, "The Conquest of Darkness," AD 346 297, 1963.

[3-111] L. M. Biberman and S. Nudleman (eds.), *Photoelectronic Imaging Devices*, Plenum Press, NY, 1971.

[3-112] A. P. Lane and W. M. Irvine, "Monochromatic Phase Curves and Albedos for the Lunar Disk," *The Astronomical Journal*, Vol. 78, No. 3, April 1973.

[3-113] R. U. Sexl, et al., "The Directional Characteristics of Lunar Infrared Radiation," *The Moon*, Vol. 3, No. 2, August 1971.

3.8.2. Bibliography

Adel, A., "Observations of Atmospheric Scattering Near the Sun's Limb," Arizona State College, Flagstaff, AZ, AD 273 599, January 1961.

Aerojet-General Corporation, *Infrared Engineering Handbook*, Astrionics Division, 25 August 1961.

ACF Industries, Inc., *Space Handbook No. 1*, ACF Electronics Division, Riverdale, MD, November 1962.

Air Force Cambridge Research Laboratories, "Celestial Background Radiation," AFCRL, Hanscom Air Force Base, Bedford, MA, AD 602 616, March 1964.

Allen, C. W., *Astrophysical Quantities*, University of London, Athlone Press, 1974, 2nd Edition.

Allen, David A., *Infrared–The New Astronomy*, John Wiley and Sons, 1975.

Anding, D., "Band Model Methods for Computing Atmospheric Slant Path Molecular Absorption," Willow Run Laboratories of the Institute of Science and Technology, The University of Michigan, Ann Arbor, MI, Report No. 7142-21-T, February 1967.

Anding, D., R. Kauth, and R. Turner, "Atmospheric Effects of Multispectral Sensing of Sea Surface Temperature from Space," Willow Run Laboratories of the Institute of Science and Technology, The University of Michigan, Ann Arbor, MI, Report No. 26760-6-T, December 1970.

Barnhart, P. E., and W. E. Mitchell, Jr., "Stellar Background Measurement Program," Ohio State University, Columbus Research Foundation, Columbus, OH, July 1964.

Bell, E. E., I. L. Eisner, and R. A. Oetjen, "The Spectral Distribution of the Infrared Radiation from the Sky," *Proceedings of the Symposium on Infrared Backgrounds*, Engineering Research Institute, The University of Michigan, Ann Arbor, MI, Nonr-1224(12), AD 121 010, March 1956.

Bell, E. E., L. Eisner, J. Young, and R. A. Oetjen, "Spectral Radiance of Sky and Terrain at Wavelengths between 1 and 20 Microns. II. Sky Measurements," *Journal of the Optical Society of America*, Vol. 50, No. 12, December 1960, pp. 1313-1320.

Bell, E. E., L. Eisner, J. Young, and R. A. Oetjen, "Spectral Radiance of Sky and Terrain of Wavelengths between 1 and 20 μ. III. Terrain Measurements," *Journal of the Optical Society of America*, Vol. 52, No. 2, February 1962, pp. 201-209.

Benford, Frank, *Journal of the Illuminating Engineering Society*, Vol. 42, May 1947.

Bennett, H. E., J. M. Bennett, and M. R. Nagel, "Measurements of Infrared and Total Radiance of the Clear Winter Sky of Wright-Patterson Air Force Base, Ohio," Wright Air Development Center, USAF, Air Research and Development Command, Wright-Patterson AFB, OH, AD 118 127, March 1957.

Bennett, H. E., J. M. Bennett, and M. R. Nagel, "The Spatial Distribution of Infrared Radiation from the Clear Sky Including Sequences of Sky Maps at Various Elevations," U.S. Naval Ordnance Test Station, China Lake, CA, NAVORD, Report No. 6577. September 1959.

Berry, F. A., et al., *Handbook of Meteorology*, McGraw-Hill Book Company, New York, NY, 1945.

Biberman, L. L., and S. Nudleman (eds.), *Photoelectronic Imaging Devices*, Plenum Press, NY, 1971.

Boeing Corporation, "Infrared Satellite Backgrounds, Part I: Atmospheric Radiative Processes," Aero-Space Division, Seattle, WA, AFCRL 1069(I), 30 September 1961.

Bond, D. S., and F. P. Henderson, "The Conquest of Darkness," AD 346 297, 1963.

Bopp, B. W., R. D. Gehrz, and J. A. Hackwell, PASP, Vol. 86, No. 514, December 1974.

Brandt, J. C., *Astrophysical Journal*, American Astronomical Society, University of Chicago Press, Chicago, IL, Vol. 128, 1958, pp. 118-123.

Brandt, J. C., *Astrophysical Journal*, American Astronomical Society, University of Chicago Press, Chicago, IL, Vol. 130, July 1959, pp. 228-240.

Brandt, J. C., and J. W. Chamberlain, "Resonance Scattering by Atmospheric Sodium-V, Theory of the Day Airglow," *Journal of Atmospheric and Terrestrial Physics*, Pergamon Press, New York, NY, Vol. 13, December 1958, pp. 90-98.

Brooks, F. A., *Journal of Meteorology*, Vol. 9, 1952.

Carmer, D., "Target Signature Analysis Center: Data Compilation, Seventh Supplement," Willow Run Laboratories, Institute of Science and Technology, The University of Michigan, Ann Arbor, MI, Report No. 8492-35-B, AD 856 343, January 1969.

Centerno, M., *Journal of the Optical Society of America*, Vol. 31, 1941.

Chamberlain, J. W., *Physics of the Aurora and Airglow*, Academic Press, New York, NY, 1961.

Chamberlain, J. W., and C. Sagan, *Planetary and Space Science*, Pergamon Press, New York, NY, Vol. 2, 1960, pp. 157-164.

Chang, Jen-Hu, *Ground Temperature*, Blue Hill Meteorological Observatory, Harvard University Press, Cambridge, MA, 1958.

Chapman, R., R. Jones, A. Dalgarno, and D. Beining, "Investigation of Auroral, Airglow and Night Emissions as Related to Space-Based Defense Systems," Geophysics Corporation of America, Bedford, MA, June 1962.

Collins, L. L., and R. B. Freund, "Celestial Background Simulation Techniques," Northrop Space Laboratories, Hawthrone, CA, AD 282 788, 1961.

Condit, H. R., "The Spectral Reflectance of American Soils," *Photogrammetric Engineering*, American Society of Photogrammetry, Falls Church, VA, Vol. 36, No. 9, September 1970.

Cox, C. and W. Munk, *Bulletin of Scripps Institition of Oceanography*, University of California, Vol. 6, 1956.

Derr, V. E., "Remote Sensing of the Troposphere," U.S. Department of Commerce, NOAA, Catalog No. C55, 602:T75, Stock No. C323-0011, August 1972.

Diermendjian, P., *Electromagnetic Scattering on Spherical Polydispersions*, American Elsevier Publishing Company, New York, NY, 1969.

Dittmar, N., F. Farley, and J. Boyse, "Earth Background Measurements: A Survey of the Unclassified Literature," Willow Run Laboratories, Institute of Science and Technology, The University of Michigan, Ann Arbor, MI, Report No. 6054-16-X, June 1969.

Earing, D., "Target Signature Analysis Center: Data Compilation, Second Supplement," Willow Run Laboratories, Institute of Science and Technology, The University of Michigan, Ann Arbor, MI, Report No. 8492-5-B, AD 819 712, July 1967.

Eastman Kodak Company, "Space Background Study for Project DEFENDER," Rochester, NY, AD 403 780, April 1963.

Eisele, R. E., "Infrared Background Investigation," Ramo-Wooldridge Division of Thompson-Ramo Wooldridge, Inc., Los Angeles, CA, Report AFCRC-TN59-843, June 1959.

Eisele, R. E., "Infrared Background Investigation," Ramo-Wooldridge Division of Thompson-Ramo Wooldridge, Inc., Canoga Park, CA, AD 236 913, March 1960.

Elterman, L., U. V., "Visible and IR Attenuation for Altitudes to 50 km," Air Force Cambridge Research Laboratories, Bedford, MA, Report No. AFCRL-68-0153, 1968.

Elterman, L., "Vertical-Attenuation Model with Eight Surface Meteorological Ranges 2 to 13 Kilometers," Air Force Cambridge Research Laboratories, Bedford, MA, Report No. 70-0200, March 1970.

Ewing, G. C., (ed.), "Oceanography from Space," Woods Hole Oceanographic Institution, Woods Hole, MA, WHOI Ref. No. 65-10, April 1965.

Ewing, G. C., and E. D. McAlister, Science, American Association for the Advancement of Science, Washington, DC, Vol. 131, 6 May 1960, pp. 1374-1376.

Faulkner, O., R. Horvath, J. Ulrich, and E. Work, "Spectral and Polarization Characteristics of Selected Targets and Backgrounds: Instrumentation and Measured Results (3.3-14.0 μm)," Willow Run Laboratories, Institute of Science and Technology, The University of Michigan, Ann Arbor, MI, Report No. 3692-23-T, AD 886 916 L, 1971.

Fredrickson, W. R., N. Ginsburg, and R. Paulson, "Infrared Spectral Emissivity to Terrain," Final Report, Syracuse University Research Institute, Syracuse, NY, AD 155 552.

Fredrickson, W. R., H. Ginsburg, R. Paulson, and D. L. Stierwalt, "Infrared Spectral Emissivity of Terrain," Syracuse University Research Institute, Syracuse, NY, Int. Dev. Report No. 2, AF33(616)-5034, August 1957.

Gates, D. M., "Spectral Distribution of Solar Radiation at the Earth's Surface," Science, Vol. 151, 1966.

Gates, D. M., and W. Tantraporn, Science, American Association for the Advancement of Science, Washington, DC, Vol. 115, 1952.

Gausman, H. W., et al., "Leaf Light Reflectance, Transmittance, Absorptance, and Optical and Geometrical Parameters for Eleven Plant Genera with Different Leaf Mesophyll Arrangements," Proceedings of the Seventh International Symposium on Remote Sensing of Environment, Environmental Research Institute of Michigan, Ann Arbor, MI, May 1971.

Gausman, H. W., et al., "The Leaf Mesophylls of Twenty Crops, Their Light Spectra, and Optical and Geometrical Parameters," U.S. Department of Agriculture, Washington, DC, USDA Technical Bulletin No. 1465, 1973.

Gordon, J., and P. Church, "Overcast Sky Luminances and Directional Luminous Reflectances of Objects and Backgrounds Under Overcast Skies," Applied Optics, Vol. 5, No. 6, June 1966, p. 919.

Gramms, L. C., and W. C. Boyle, "Reflectance and Transmittance Characteristics of Several Selected Green and Blue-Green Unialgae, Proceedings of the Seventh Symposium on Remote Sensing of Environment, Environmental Research Institue of Michigan, Ann Arbor, MI, May 1971, pp. 1637, 1642, 1643.

Gush, H. P., and H. L. Buys, "The Near Infrared Spectrum of Night Airglow Observed from High Altitude," Canadian Journal of Physics, National Research Council of Canada, Ottawa, Quebec, Canada, Vol. 42, 1964.

Handbook of Chemistry and Physics, CRC Press, Cleveland, OH, 1976.

Harang, L., The Aurorae, John Wiley and Sons, Inc., New York, NY, 1951.

Harrison, A. W., and A. V. Jones, Journal of Atmospheric and Terrestrial Physics, Pergamon Press, New York, NY, Vol. 11, 1957, pp. 192-199.

Harrison, A. W., and A. V. Jones, Journal of Atmospheric and Terrestrial Physics, Pergamon Press, New York, NY, Vol. 13, 1957, pp. 291-294.

Herin, W., and T. Borden, Jr., "Ozonesonde Observations Over North America," Air Force Cambridge Research Laboratories, Bedford, MA, Environmental Research Paper No. 38, Report No. AFAL-64-30, July 1964.

Hoffleit, D., *Catalogue of Bright Stars*, Tuttle, Morehouse, and Taylor, New Haven, CT, 1964, third revised edition.

Horvath, R., W. L. Morgan, and R. Spellicy, "Measurements Program for Oil Slick Characteristics," Institute of Science and Technology, The University of Michigan, Ann Arbor, MI, Report No. 2766-7-F, 1970, pp. 17, 19, 21, 23, 24, 28, 30, 32, 83-91, 144-146, 154-156.

Hunten, D. M., *Journal of Atmospheric and Terrestrial Physics*, Pergamon Press, New York, NY, Vol. 7, 1955, pp. 141-151.

Hunten, D. M., "Optics of the Upper Atmosphere," *Applied Optics*, Vol. 3, No. 2, February 1964.

Idso, S. B., and R. D. Jackson, "Thermal Radiation from the Atmosphere," *Journal of Geophysical Research*, American Geophysical Union, Washington, DC, Vol. 74, 1969.

Johnson, H. L., R. I. Mitchell, B. Iriarte, and W. K. Wisniewski, "Magnitudes and Colors of 1300 Bright Stars," *Sky and Telescope*, Sky Publishing Corporation, Cambridge, MA, Vol. 30, No. 1, July 1965, p. 24.

Jones, A. V., "Possible Methods for Studying the Auroral Spectrum in the 2.0 to 3.5 Micron Region," Saskatchewan University, Saskatoon, Canada, November 1959.

Jones, A. V., and H. Gush, *Nature*, Macmillan and Co., London, England, Vol. 172, 12 September 1953, p. 496.

Jones, R. C., "Sky Noise—Analysis of Circular Scanning," Polaroid Corporation, Cambridge, MA, November 1953.

Jones, R. C., "Sky Noise—Its Nature and Analysis," Polaroid Corporation, Cambridge, MA, September 1953.

Jones, R. C., and A. M. Nagvi, "Satellite Navigation of Terrestrial Occulations of Stars, III: Interference due to brightness of the Earth's Atmosphere," Geophysics Corporation of America, Boston, MA, AD 287 869, October 1962.

Kalle, K., "Zurn Problem der Meereswasserfarbe," Ann. d. Hydrogr. und mar. Meteor. BD, Vol. 66, S1-13, 1938.

Kaschmieder, H., *Beitr. Phys. freien Afm*, Vol. 12, 1924.

Kauth, R., and J. Penquite, "The Probability of Clear Lines of Sight through a Cloudy Atmosphere," *Journal of Applied Meteorology*, American Meteorological Society, Boston, MA, Vol. 6, No. 6, December 1967.

Kondratyev, K. Y., "Actinometry," NASA TTF-9712, Washington, DC, November 1965 (Translation of "Aktinometriya," Gidrometerologicheskoye Izadtel' stvo, Leningrad, 1965).

Kondratyev, K. Y., Z. F. Miconova, and A. N. Otto, "Spectral Albedo of Natural Surfaces," *Pure and Applied Geophysics*, Birkhauser Verlag, Switzerland, Vol. 59, No. 3, 1964, pp. 207-216.

Kreith, F., *Principles of Heat Transfer*, Intext Educational Publishers, 1973, 3rd Edition.

Krinov, E. L., "Spectra Reflectance Properties of Natural Formations," (originally Published in Russian) 1947, Tech. Trans. TT439, National Research Council of Canada, Ottawa, Canada, 1953.

Lane, A. P., and W. M. Irvine, "Monochromatic Phase Curves and Albedos for the Lunar Disk," *The Astronomical Journal*, Vol. 78, No. 3, April 1973.

Larmore, L., "Infrared Radiation from Celestial Bodies," U.S. Air Force Project, RAND Research Memo, Rand Corp., Santa Monica, CA, RM-793-1, 17 March 1952.

Leeman, V., D. Earing, R. Vincent and S. Ladd, "The NASA Earth Resources Spectral Information System: A Data Compilation," Willow Run Laboratories, Institute of Science and Technology, The University of Michigan, Ann Arbor, MI, Report No. NASA CR-31650-24-T, May 1971.

Leighton, R., *Astrophysical Journal*, American Astronomical Society, University of Chicago Press, Chicago, IL, (Letters), No. 4, 1965.

Low, F. J., and H. L. Johnson, "Stellar Photometry at 10 μ, " *Journal of Applied Physics*, Vol. 139, 1964, p. 1130.

Mavridis, L. N., *Structure and Evolution of the Galaxy*, Springer-Verlag New York, NY, 1971.

McAlister, E. D., "Application of Infrared-Optical Techniques to Oceanography," *Journal of the Optical Society of America*, Vol. 52, May 1962, p. 607.

McMahon, H. O., "Thermal Radiation from Partially Transparent Reflecting Bodies," *Journal of the Optical Society of America*, Vol. 40, No. 6, June 1950, pp. 376-380.

Midwest Research Institute, "Measurement of Infrared Radiation Gradients in the Sky," Kansas City, MO, AD 206 453, 1953.

Moon, P., *Journal of the Franklin Institute*, Vol. 230, No. 5, 1940.

Morrison, D., *Astrophysical Journal*, American Astronomical Society, University of Chicago Press, Chicago, IL, Vol. 194, 1974, p. 203.

Murdock, T. L., "Contribution of Asteroids to the Infrared Astronomical Sky Survey," AFCRL-TR-73-0154, 1973.

Nagvi, A. M., "Satellite Navigation by Terrestrial Occulations of Stars—Considerations Relating to Refraction and Extinction," Geophysics Corporation of America, Bedford, MA, AD 287 868, October 1962.

National Aeronautics Space Administration, "Aurorae and Airglow," NASA, Washington, DC, April 1964.

Nelgner, H. D., and J. R. Thompson, "Airborne Spectral Radiance Measurements of Terrain and Clouds," Emerson Electric Manufacturing Company, St. Louis, MO, Report 1323, April 1962.

Newkirk, G., and J. Eddy, "Light Scattering by Particles in the Upper Atmosphere," University of Colorado, Boulder, CO, May 1963.

Nobel, P. S., *Introduction of Biophysical Plant Physiology*, W. H. Freeman and Company, San Francisco, CA, 1974.

Novoseltsev, Y. P., "Spectral Reflectivity of Clouds," Translation of "Specktrolnaya Otrazhatelnaya Oblakov," Trudy Glavnoy, Geofizicheskoy Observatoiri Uneni, A. I. Voyykova, NASA TT F-328, No. 152, 1964, pp. 186-191.

Noxon, J. F., A. Harrison, and A. V. Jones, "The Infrared Spectrum of the Night Airglow 1-4 μ to 4.0 μ," *Journal of Atmospheric and Terrestrial Physics*, Pergamon Press, New York, NY, Vol. 16, 1959, pp. 246-251.

O'Brien, H. W., et al., "Red and Near Infrared Spectral Reflectance of Snow," U.S. Army Cold Region Research and Engineering Laboratory, Hanover, NH, CRREL (AD-A007732), 1975.

Olson, Jr., C. E., "Seasonal Change in Foliar Reflectance of Five Broadleaved Forest Tree Species," PhD Thesis, University of Michigan, Ann Arbor, MI, 1969.

Orr, D. G., S. E. Dwamik, and L. M. Young, "Reflectance Curves of Soils, Rocks, Vegetation and Pavement," U.S. Army Engineer Research and Development Laboratories, Fort Belvoir, VA, Report No. 7746-RR, April 1963, p. 22.

Partridge, R. B., and P. J. E. Peebles, *Astrophysical Journal*, American Astronomical Society, University of Chicago Press, Chicago, IL, Vol. 148, 1967, p. 377.

Peterson, A. W., *Astrophysical Journal*, American Astronomical Society, University of Chicago Press, Chicago, IL, Vol. 138, 1963, p. 1218.

RCA, *Electro-Optics Handbook*, Commerical Engineering, Harrison, NJ, Technical Series EOH-11.

Ramsey, R. C., "Spectral Irradiance from Stars and Planets above the Atmosphere from 0.1 to 100.0 microns," *Applied Optics, VI*, Vol. 4, July 1962.

Rasool, S. I., Latitudinal Distribution of Cloud Cover from TIROS III Photographs, *Science*, February 1964.

Rieke, G. H., and F. J. Low, *Astrophysical Journal*, American Astronomical Society, University of Chicago Press, Chicago, IL, Vol. 176, L95, 1972.

Roach, F. F., *Proceedings of the IRE*, J. D. Ryder (ed.), The Institute of Radio Engineers, Inc., New York, NY, Vol. 47, 1959, p. 267.

Rose, H., et al., "The Handbook of Albedo and Thermal Earthshine," Environmental Research Institute of Michigan, Ann Arbor, MI, Report No. 190201-1-T, 1973.

Safir, G., G. Suits, and M. Wiese, "Application of a Directional Reflectance Model to Wheat Canopies Under Stress," *Proceedings of the International Conference on Remote Sensing in Arid Lands*, Tucson, AZ, November 1972.

Schimpf, R., and C. Aschenbrenner, *Z Phot Wiss Tech*, Vol. 2, 1940.

Schlesinger, F., *Catalogue of Bright Stars*, Tuttle, Morehouse, and Taylor, New Haven, CT, 1930.

Sciama, D. W., *Modern Cosmology*, Cambridge University Press, New York, NY, 1971.

Seaton, M. J., *Journal of Atmospheric and Terrestrial Physics*, Pergamon Press, New York, NY, Vol. 4, 1954, p. 285-294.

Sellin, I., "Auroral Radiations in the Infrared," Laboratories for Applied Sciences, University of Chicago, Chicago, Ill. October 1961.

Sexl, R. U., et al., "The Directional Characteristics of Lunar Infrared Radiation," *The Moon*, Vol. 3, No. 2, August 1971.

Sloan, R., J. Shaw, and D. Williams, "Infrared Emission Spectrum of the Atmosphere," *Journal of the Optical Society of America*, Vol. 45, No. 6, June 1955, pp. 455-459.

Soifer, B. T., J. R. Houck, and M. Harwit, *Astrophysical Journal*, American Astronomical Society, University of Chicago Press, Chicago, IL, Vol. 168, L73, 1971.

Solomon, I., "Estimates of Altitudes with Specified Probabilities of Being Above all Clouds," Air Weather Service (MATS), U.S. Air Force, Tech. Report 159, October 1961.

Space Handbook No. 1, ACF Industries, Inc., ACF Electronics Division, Riverdale, MD. November 1962.

Specht, M. R., D. Needler, and N. L. Fritz, "New Color Film for Water-Photography Penetration," *Photogrammetric Engineering*, American Society of Photogrammetry, Falls Church, VA, Vol. 39, April 1973, p. 359.

Spiegler, D., and J. Greaves, "Development of Four-Dimensional Atmospheric Models (Worldwide)," Allied Research Associates, Inc., Concord, MA, NASA CR-61362, August 1971.

Stair, A. T., Jr., J. C. Ulwick, K. D. Baker, and D. J. Baker, "Rocketborne Observations of Atmospheric Infrared Emissions in the Auroral Region," *Atmospheres of Earth and Planets*, B. M. McCormac (ed.), Reidel Publishing Company, Boston, MA, 1975, pp. 335-346.

Störmer, C., *The Polar Aurora*, Clarendon Press, Oxford, 1955.

Suits, G., "The Calculation of the Directional Reflectance of a Vegetative Canopy," *Remote Sensing of Environment*, Vol. 2, 1972.

Suits, G., "The Cause of Azimuthal Variations in Directional Reflectance of Vegetative Canopies," *Remote Sensing of Environment*, Vol. 2, 1972.

Suits, G., and G. Safir, "Verification of a Reflectance Model for Mature Corn with Applications to Corn Blight Detection," Vol. 2, 1972.

Sverdrup, H. U., M. Johnson, and R. Fleming, *The Oceans*, Prentice Hall, Inc., Englewood, NJ, 1972.

Thekaekara, M. P., "Evaluating the Light from the Sun," *Optical Spectra*, Vol. 6, No. 3, March 1972.

University of Michigan, "Target Signature Study Interim Report, Volume V: Catalog of Spectral Reflectance Data," Ann Arbor MI, Report No. 5698-22-T(V), October 1964.

Valley, S. L., (ed.), *Handbook of Geophysics and Space Environments,* Air Force Cambridge Research Laboratories, Bedford, MA, 1965.

Vestine, E. H., "Terrestrial Magnetism," *Journal of Geophysical Research*, Vol. 49, June 1944, p. 77-102.

Vincent, R. K., "A Thermal Infrared Ratio Imaging Method for Mapping Compositional Variations Among Silicate Book Types," PhD Thesis, The University of Michigan, Ann Arbor, MI, 1973.

Walker, R. G., "Infrared Celestial Backgrounds," Air Force Cambridge Research Laboratories, Bedford, MA, July 1962.

Walker, R. G., and S. D. Price, "AFCRL Infrared Sky Survey," AFCRL-TR-75-0373, Vol. 1, July 1975.

Walker, R. G. and S. D. Price, "A Model for the Infrared Celestial Background," manuscript in preparation, June, 1977.

Watson, R. D., et al., "Prediction of the Fraunhofer Line Detectivity of Luminescent Materials," *Proceedings of the Ninth Symposium on Remote Sensing of Environment,* Environmental Research Institute of Michigan, Ann Arbor, MI, April 1974.

Wezernak, C. T., "The Use of Remote Sensing in Limnological Studies," *Proceedings of the Ninth Symposium on Remote Sensing Environment,* Environmental Research Institute of Michigan, Ann Arbor, MI, April 1974.

White, F. W. G., and M. Geddes, "Terrestrial Magnetism," *Journal of Geophysical Research,* Vol. 44, December 1939.

Wilkins, G. A., and J. A. Hoyem, "The Terrestrial Night Horizon and Sky, 4.3 Micron Radiance Data Generated by the HITABTRIS Experiments," U.S. Naval Ordnance Test Station, China Lake, CA, NAVWEPS Report 8552, June 1964.

Willow Run Laboratories, "Optical Sensor Technology for Threat Detection and Identification, (Final Report)," The Institute of Science and Technology, The University of Michigan, Ann Arbor, MI, Report No. 2647-8-F, April 1970.

Chapter 4

ATMOSPHERIC SCATTERING

Benjamin Herman*
University of Arizona

Anthony J. LaRocca**
Robert E. Turner***
Environmental Research Institute of Michigan

CONTENTS

---

*Section 4.7.
**Sections 4.1, 4.2, and 4.3, based upon material provided by Benjamin Herman.
***Sections 4.4, 4.5, and 4.6.

4. Atmospheric Scattering

4.1. Introduction

This and the following chapter treat the processes by which the Earth's atmosphere produces and transfers radiation. Here, the radiative transfer equation is described, and its solution is discussed for scattering atmospheres. Absorption processes are discussed in Chapter 5.

In general, the solution to the radiative transfer equation for scattering atmospheres involves very complex mathematical techniques which, except for very special and often less important cases, will not yield closed-form solutions. Compromises in the establishment of boundary conditions or approximations are made to effect a mathematical formulation amenable to reasonably handy solutions. Principal among these is the assumption of single scattering. (See Section 4.2.) This often leads to a valid, relatively simple, viable solution to the transfer equation, with accuracy commensurate with the use of much of the currently available, atmospheric input-data, itself of uncertain accuracy. Many problems, however, are characterized in such a way that multiple scattering is important in the transfer of radiation.

Methods of solving the radiation transfer equation are described in Section 4.4 starting with the so-called exact solutions, i.e., those with no approximations in the basic mathematical formulations of the transfer equation for ideal atmospheres. Because the exact methods are laborious and the number of physical problems which can be solved is limited, approximations to the exact solution have been developed. These are described in Section 4.4.3. The limitation to accuracy in these solutions is the number of terms in a series or the amount of computer time allowed. When high accuracy is not necessary, especially when the pertinent input parameters are not well enough known, speed and efficiency can be preserved by resorting to direct approximate methods, described, with examples, in Section 4.4.4.

Throughout this chapter, monochromatic radiation is assumed. Since the scattering coefficient is wavelength-dependent, it is necessary to calculate this effect on scattering in any cases in which wide-spectral-band results are required. Scattering by molecules is evident mainly in the visible region. The wavelength dependence is strong. On the other hand, scattering by aerosols, depending on their sizes, is normally less strongly dependent on wavelength, and can be effective in both the visible and infrared regions. This chapter treats scattering chiefly from 0.3 μm to approximately 2 μm for an atmosphere with a particulate density consistent with so-called clear-sky or near-clear-sky conditions.

Sections 4.2 and 4.4 are devoted to various applications of scattering formulae. The albedo problem is covered in Section 4.5 and illustrative results of scattering calculations are given in Section 4.6. Section 4.7 gives a description of the optical properties of the atmosphere pertinent to scattering phenomena.

4.1.1. Symbols, Nomenclature and Units. Table 4-1 lists symbols, nomenclature and units used in this chapter.

4.1.2. Radiative Transfer in the Atmosphere. The extinction of radiation which traverses a medium (Figure 4-1) is proportional to the initial radiance, to the density of

Table 4-1. Symbols, Nomenclature and Units

Symbols	Nomenclature	Units
A	Area	cm^2
a	Number density; also, a subscript used to designate absorptance	cm^{-3}
b	Shaping parameter	—
c	Speed of light	$cm\ sec^{-1}$
E_s	Solar irradiance at a point with optical depth, q	$W\ cm^{-2}$
E_λ	Spectral irradiance	$W\ cm^{-2}\ \mu m^{-1}$
$e_\lambda\ (\lambda,\ s)$	Spectral radiant intensity per unit mass emitted within the medium at s	$W\ g^{-1}\ \mu m^{-1}\ sr^{-1}$
h	Planck's constant	$W\ sec^2$
I_λ	Spectral radiant intensity	$W\ sr^{-1}\ \mu m^{-1}$
J_λ	Source function	$W\ cm^{-2}\ sr^{-1}\ \mu m^{-1}$
K_S	Volume scattering coefficient	$cm^{-1}\ (or\ km^{-1})$
$k(\lambda)$	Spectral mass extinction coefficient	$g^{-1}\ cm^2$
$k_a(\lambda)$	Spectral mass absorption coefficient	$g^{-1}\ cm^2$
$k_S(\lambda)$	Spectral mass scattering coefficient	$g^{-1}\ cm^2$
L_G	Surface (or ground) radiance	$W\ cm^{-2}\ sr^{-1}$
L_λ	Spectral radiance	$W\ cm^{-2}\ sr^{-1}\ \mu m^{-1}$
L_P	Path radiance	$W\ cm^{-2}\ sr^{-1}$
L_T	Total radiance for downward looking observer	$W\ cm^{-2}\ sr^{-1}$
m_m	Mass of an air molecule	g
$N(r)$	Particle count	—
N_{STP}	Molecular number density at STP	cm^{-3}
$N(z)$	Molecular number density at z	cm^{-3}
$n(\lambda)$	Real index of refraction at λ	—
$n_i(\lambda)$	Imaginary portion of index of refraction at λ	—
$P(\Theta)$	Scattering cross-section per unit volume per steradian at angle Θ	$cm^{-1}\ sr^{-1}$
p	Pressure	Torr, mm Hg
$p(\Theta)$	Scattering phase function	—
q	Optical depth $(= q(\lambda,\ z) = \int_z^\infty k(\lambda,\ z')\ \rho\,dz')$	—
q_a	Analogous to q, for $k_a(\lambda)$	—
q_o	Optical depth of total atmosphere	—
q_S	Analogous to q, for $k_S(\lambda)$	—
R	Distance	cm or m
r	Particle radius	μm or cm
S	Subscript used to designate scattering	—
s	Coordinates $(x,\ y,\ z)$ of a point, or distance	m
T	Absolute temperature	K
V	Scattering volume	cm^3
$(x,\ y,\ z)$	Cartesian coordinates (z directed upward)	m

Table 4-1. Symbols, Nomenclature and Units (*Continued*)

Symbols	Nomenclature	Units
α, β	Shaping constants	—
γ	Scattering parameter	—
δ	Depolarization factor	—
$\delta(\mu - \mu')$	Dirac delta function	—
η	Fraction of radiation scattered into a forward hemisphere	—
Θ	Scattering angle; i.e., angle between incident and scattered beam	rad or degree
(θ, ϕ)	Angular coordinates in a cylindrical coordinate system; θ is the positive angle with the zenith; ϕ is the azimuth angle in the x-y plane related to any convenient direction; usually the plane containing the sun and the center of the system.	rad or degree
k	Boltzmann's constant	W sec K^{-1}
λ	Wavelength	μm, Å
μ	Positive value of the cos θ	—
μ_o	cos θ_o corresponding to sun angle	—
ξ	General angle coordinate	rad or degree
ρ	Density; also, reflectance	g cm^{-3}
τ	Transmittance	—
ϕ_o	Sun's azimuth angle	rad or degree
Ω	Solid angle in direction (θ, ϕ)	sr
ω_o	Single scattering albedo $(= \omega_o(\lambda) = k_S(\lambda)/k(\lambda))$	—

the attenuating medium, and to the distance traversed, ds, so that

$$dL_\lambda(\lambda,s) = -k(\lambda,s)L_\lambda(\lambda,s)\rho\, ds \quad (4\text{-}1)$$

where $L_\lambda(\lambda,s)$ = spectral radiance at a point s with coordinates (x,y,z).

ρ = density of the medium, g cm^{-3}

$k(\lambda,s)$ = spectral mass extinction coefficient

$\qquad = k_a(\lambda,s) + k_S(\lambda,s)$

$k_a(\lambda,s)$ = spectral mass absorption coefficient

$k_S(\lambda,s)$ = spectral mass scattering coefficient

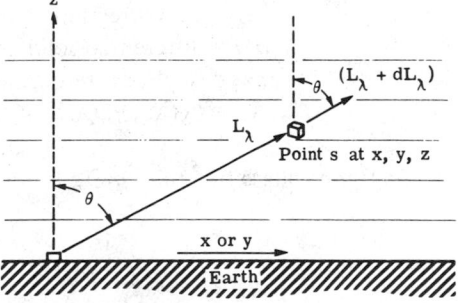

Fig. 4-1. Radiance extinction.

Dimensionally, the mass extinction coefficient is in (Length)^2(Mass)$^{-1}$, and is thus an extinction cross-section per unit mass. The usual assumption is made that the atmosphere is structured in plane-parallel, homogeneous layers, so that the only spatial variations in k and ρ are vertical, in the z-direction. Thus the radiative transfer equation following from Equation (4-1) is

$$-\mu \frac{dL_\lambda}{dq(\lambda,z)} = J_\lambda(\lambda,z) - L_\lambda(\lambda,z) \tag{4-2}$$

where
$$\mu = \cos\theta$$
$$\theta = \text{zenith angle}$$
$$q(\lambda,z) = \text{optical depth}$$

$$= \int_z^\infty k(\lambda,z')\rho\,dz'$$

$$= q_a + q_S = \int_z^\infty [k_a(\lambda,z') + k_S(\lambda,z')]\rho\,dz' \tag{4-3}$$

(Here primes symbolize integration variables.)
q_a = optical depth resulting only from absorption
q_S = optical depth resulting only from scattering
$J_\lambda(\lambda,z)$ = source function

$$= \frac{\omega_0}{4\pi} \int_0^{4\pi} p(\Theta)L_\lambda(\lambda,z,\theta,\phi)d\Omega + \frac{e_\lambda(\lambda,z)}{k(\lambda,z)} \tag{4-4}$$

$e_\lambda(\lambda,z)/k(\lambda,z)$ = source term due to internal emission.*
$p(\Theta)$ = scattering phase function
Θ = scattering angle between the direction of the scattered beam and each direction of the incident beams
$d\Omega$ = differential solid angle in the direction of each incident beam
ω_0 = single scattering albedo
$$= k_S(\lambda,z)/k(\lambda,z) \leqslant 1$$

The scattering phase function, $p(\Theta)$, is normalized so that

$$\int_0^{4\pi} p(\Theta)d\Omega = 4\pi \text{ sr} \tag{4-5}$$

The simplest example of a phase function is that for isotropic scattering for which

$$p(\Theta) = 1 \tag{4-6}$$

---

*Under conditions of thermal equilibrium, $e_\lambda = k_a(\lambda, z)L_\lambda{}^*$, or $k_a(\lambda, z)L_{\lambda,BB}$, where $L_\lambda{}^*$ or $L_{\lambda,BB}$ is the Planck function. (See Chapter 1.)

The next most common phase function is for Rayleigh scattering in which

$$p_{Ray}(\Theta) = \frac{3}{4}(1 + \cos^2\Theta) \tag{4-7}$$

In general, it is assumed that the phase function may be expanded in a series of Legendre polynomials of the form

$$p(\Theta) = \sum_{l=0}^{\infty} A_l P_l(\cos\Theta) \tag{4-8}$$

Integration of Equation (4-2) yields the formal solution of the radiative transfer equation. The angles (θ, ϕ) are general zenith and azimuth angles, with $\theta = 0$ at the zenith. Often ϕ is selected so that the value of zero corresponds to the plane containing the sun and the local vertical. Since negative values of $\mu = \cos\theta$ would result for $\theta > 90°$, a sign convention is adopted in this chapter in which μ is replaced by $-\mu$ throughout all equations for downward propagating radiation. Thus upward propagating radiation is designated by $L_\lambda(\lambda,z,\mu,\phi)$ and downward propagating radiation by $L_\lambda(\lambda,z,-\mu,\phi)$. With these conventions, the formal solution to Equation (4-2), with $0 < \mu \leqslant 1$, is as follows:

(1) For downward propagating radiation, i.e., observer at z, looking upward,

$$L_\lambda(\lambda,q,-\mu,\phi) = L_\lambda(\lambda,q_1,-\mu,\phi)e^{-(q-q_1)/\mu} + \int_{q_1}^{q} e^{-(q-q')/\mu} J_\lambda(\lambda,q',-\mu,\phi)\frac{dq'}{\mu} \tag{4-9}$$

where $\qquad q_1$ = optical depth from any point z_1 to the top of the atmosphere, $z = \infty$
$\qquad q$ = optical depth from an observer at $z < z_1$, so that $q > q_1$

$$J_\lambda(\lambda,q',-\mu,\phi) = \frac{\omega_0}{4\pi}\int_0^{4\pi} p(\Theta)L_\lambda(\lambda,q',-\mu,\phi)d\Omega + (1 - \omega_0(\lambda))L_\lambda^*(\lambda, T(z'))$$

$$+ \frac{\omega_0}{4\pi}p(\Theta)E_\lambda(\lambda,0,-\mu_0,\phi_0)e^{-q'/\mu_0} \tag{4-10}$$

with $\quad L_\lambda^*(\lambda, T(z'))$ = Planck function at z' corresponding to q'
$\qquad T(z')$ = temperature at z'
$E_\lambda(\lambda,0,-\mu_0,\phi_0)$ = spectral irradiance as a result of the sun at the top of the atmosphere, $(q = 0)$, from the direction (θ_0,ϕ_0)

A special case of Equation (4-9) is for an observation through the whole atmosphere, in which case $q_1 = 0$ (i.e., $z_1 = \infty$). For this case, the first term on the right side of Equation (4-9) is zero.

(2) For upward propagating radiation, i.e., observer at z looking downward,

$$L_\lambda(\lambda,q,\mu,\phi) = L_\lambda(\lambda,q_2,\mu,\phi)e^{-(q_2-q)/\mu} + \int_{q}^{q_2} e^{-(q'-q)/\mu} J_\lambda(\lambda,q',\mu,\phi)\frac{dq'}{\mu} \tag{4-11}$$

where q_2 = optical depth from any point z_2 to the top of the atmosphere, $z = \infty$
$\quad q$ = optical depth from an observer at $z > z_2$, so that $q_2 > q$

and $J_\lambda(\lambda, q', \mu, \phi)$ is as given in Equation (4-10) after replacement of $-\mu$ by μ. The first term on the right side of Equation (4-11) represents the radiation from an opaque object at $z = z_2$. When no object intervenes, the observation penetrates to the earth's surface at $z = 0$;

$$q(0) = \int_0^\infty k(\lambda)\rho(z)dz \tag{4-12}$$

This represents the total optical depth of the atmosphere.

4.2. Single Scattering (See also Chapter 5.)

Radiative transfer through optically thin, scattering media (i.e., $q \leqslant 0.05$) may always be approximated reasonably accurately with a single scattering transfer equation. Optically denser media, in which there is, in addition to scattering, absorption, may also be sometimes approximated with single scattering only. In this latter set of problems, the validity of the single scattering approximation is not always obvious, and individual problems must be examined separately. In Chapter 5, under the Aggregate and LOW-TRAN methods, single scattering is calculated for an aerosol model corresponding to 5 and 23 km visibility.

4.2.1. Single Scattering Transfer Equation for Incident Sunlight.
In many problems of practical interest, the incident radiation may be considered to be infinite, plane-parallel wave-fronts (e.g., sunlight illuminating a planetary atmosphere). The single-scattering equations for plane-parallel radiation incident on a horizontally homogeneous atmosphere, are, from Equation (4-9) before,

$$L_\lambda(\lambda, q, -\mu, \phi) = \frac{\omega_0}{4\pi} \int_0^q \exp\left[\frac{-(q-q')}{\mu}\right] p(\Theta) E_\lambda(\lambda, 0, -\mu_0, \phi_0) \exp\left[\frac{-q'}{\mu_0}\right] \frac{dq'}{\mu} \tag{4-13}$$

and

$$L_\lambda(\lambda, q, \mu, \phi) = \frac{\omega_0}{4\pi} \int_q^{q(0)} \exp\left[\frac{-(q'-q)}{\mu}\right] p(\Theta) E_\lambda(\lambda, 0, -\mu_0, \phi_0) \exp\left[\frac{-q'}{\mu_0}\right] \frac{dq'}{\mu} \tag{4-14}$$

Often, $p(\Theta)$ is constant with q. An example is a purely molecular atmosphere, where $p(\Theta)$ becomes the Rayleigh phase function given by

$$p(\Theta) \equiv P_{Ray}(\Theta) = \frac{3}{4}(1 + \cos^2\Theta) \tag{4-15}$$

For a turbid atmosphere, $p(\Theta)$ may be considered as constant for the aerosols if the size distribution of the particulates is constant with q, i.e., with height. If, additionally, the relative amounts of scattering by both aerosols and the molecules remain constant, then in Equations (4-13) and (4-14),

$$p(\Theta) = \frac{q(0)_{aer}}{q(0)} p(\Theta)_{aer} + \frac{q(0)_{Ray}}{q(0)} p(\Theta)_{Ray} \qquad (4\text{-}16)$$

where $q(0)_{aer}$ = total optical depth due to aerosols
$\quad\quad\ q(0)_{Ray}$ = total optical depth due to molecules

Equations (4-13) and (4-14) may be integrated, yielding

$$L_\lambda(\lambda,q,-\mu,\phi) = \frac{\omega_0}{4\pi} p(\Theta)E_\lambda(\lambda,0,-\mu_0,\phi_0)\left(\frac{\mu_0}{\mu-\mu_0}\right)\left\{1 - \exp\left[\frac{-q}{\mu}\left(\frac{\mu-\mu_0}{\mu_0}\right)\right]\right\}\exp\left[\frac{-q}{\mu}\right]$$
$$(4\text{-}17)$$

and

$$L_\lambda(\lambda,q,\mu,\phi) = \frac{\omega_0}{4\pi} p(\Theta)E_\lambda(\lambda,0,-\mu_0,\phi_0)\left(\frac{\mu_0}{\mu+\mu_0}\right)\left\{\exp\left[\frac{-q}{\mu}\left(\frac{\mu+\mu_0}{\mu_0}\right)\right]\right.$$
$$\left.- \exp\left[\frac{-q(0)}{\mu}\left(\frac{\mu+\mu_0}{\mu_0}\right)\right]\right\}\exp\left[\frac{q}{\mu}\right] \qquad (4\text{-}18)$$

4.2.2. Finite Wave-Front Sources.
Another slightly different class of problems in which single scattering approximations are generally employed involves the illumination of the atmosphere by radiation of a finite cross-section. Laser, lidar, or search-light probes fall into this category. Because of the nonuniform illumination of the atmosphere along a horizontal plane, this class of problems differs somewhat from that described above.

The spectral irradiance at the scattering volume, V, for such a source with spectral radiant intensity, $I_\lambda(\lambda, 0)$, and receiver throughput, $A_r\Omega$ (Figure 4-2), is

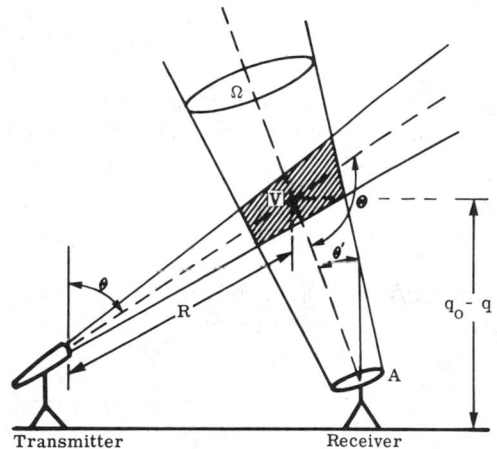

Fig. 4-2. Transmitter-receiver geometry.

$$E_\lambda = \frac{I_\lambda(\lambda,0)}{R^2} e^{-(q(0)-q(z))/\mu} \qquad (4\text{-}19)$$

where $I_\lambda(\lambda,0)$ = spectral radiant intensity of the transmitter at the Earth's surface
$\quad\quad R$ = distance between transmitter and scattering volume
$\quad\quad \mu = \cos\theta$

If one assumes only single scattering within the scattering volume, the spectral radiant intensity of the beam scattered toward the receiver from V is

$$I_\lambda(V) = \frac{I_\lambda(\lambda,0)}{R^2} e^{-(q(0)-q(z))/\mu} V P(\Theta) \qquad (4\text{-}20)$$

where V = scattering volume
        $P(\Theta)$ = scattering cross-section per unit volume per unit solid angle
        Θ = scattering angle

The spectral power at the receiver is

$$\Phi_R = I_\lambda(V)e^{-(q(0)-q(z))/\mu'}A_r R'^{-2} \tag{4-21}$$

or

$$\Phi_R = \frac{I_\lambda(\lambda,0)}{R^2}[e^{-(q(0)-q(z))/\mu}]\,V P(\Theta)[e^{-(q(0)-q(z))/\mu'}]A_r[R']^{-2} \tag{4-22}$$

where A_r = receiver aperture area
     R' = distance from V to the receiver
     μ' = cos θ' (See Figure 4-2).

If the transmitter and receiver are located side by side, then $R = R'$, $\Theta = 180°$; $\mu = \mu'$, and

$$\Phi_R = \frac{I_\lambda(\lambda,0)}{R^4}[e^{-2(q(0)-q(z))/\mu}]\,V P(180°)A_r \tag{4-23}$$

or

$$\Phi_R = L_\lambda(\lambda,0)[e^{-2(q(0)-q(z))/\mu}]VP(180°)A_t A_r R^{-4} \tag{4-24}$$

where A_t is the transmitter aperture area.

Figure 4-3 shows a typical plot of a computed $p(\Theta)$ in relative units for an atmospheric aerosol distribution, given by

$$\frac{dN(r)}{d(\ln r)} = cr^{-\beta} \tag{4-25}$$

where $dN(r)$ = number of particles with radius between r and $r + dr$
     r = particle radius
     β = shaping constant

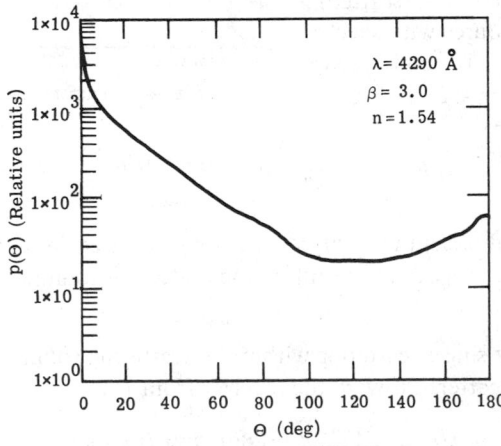

Fig. 4-3. Typical plot of a computed $p(\Theta)$ in relative units for an atmospheric aerosol distribution.

The wavelength, λ, assumed for these calculations was 4290 Å, and the real portion of the index of refraction was assumed to be 1.54. The particles were assumed to be present in the range of sizes between $r = 0.04$ and 10.0 μm to be distributed according to Equation (4-25) in this range, and to be absent outside this range. This shape is, in general, quite typical of scattering by particles outside the Rayleigh range (i.e., for particles whose parameter $2\pi r/\lambda > 1$) as a result of the presence of diffracted energy scattered around the edges of particles. At a fixed wavelength, the diffracted energy concentrates into a smaller and smaller solid angle about the 0° scattering angle as the size of the particles gets larger. This feature may be more readily seen from Figures 4-4, 4-5, and 4-6. In these figures $p(\Theta)$ is plotted as a function of Θ for discrete-sized particles.

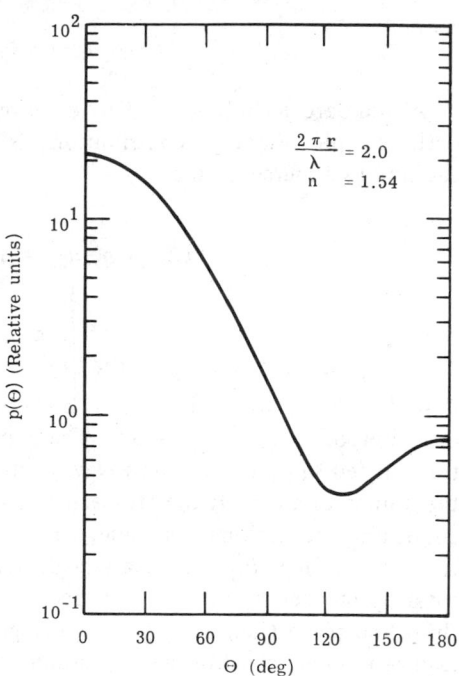

Fig. 4-4. Plot of $p(\Theta)$ as a function of Θ for discrete-sized particles.

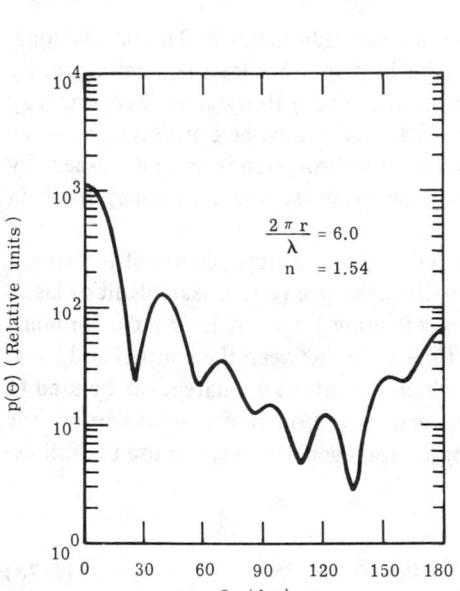

Fig. 4-5. Plot of $p(\Theta)$ as a function of Θ for discrete-sized particles.

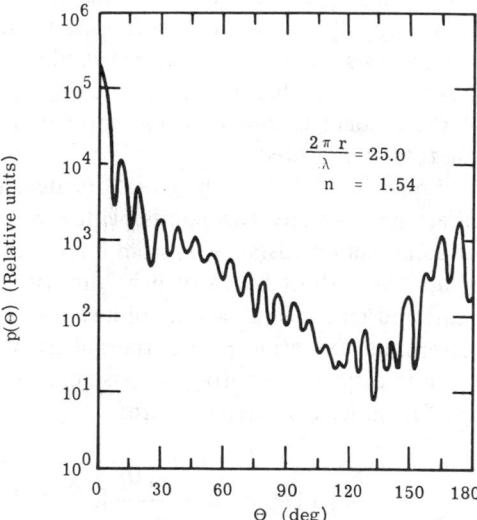

Fig. 4-6. Plot of $p(\Theta)$ as a function of Θ for discrete-sized particles.

4.3. Atmospheric Transmission Through a Turbid Atmosphere

The equation of transfer appropriate for atmospheric transmission is one with no sources, i.e., with J_λ set equal to zero. For a homogeneous atmosphere this becomes

$$E_\lambda(\lambda,q,\mu,\phi) = E_\lambda(\lambda,0,\mu,\phi)\,e^{-q/\mu} \tag{4-26}$$

A standard technique used to determine the total optical depth of the atmosphere with Mie plus Rayleigh contributions, is to write Equation (4-26) logarithmically, with the sum as a source, so that

$$\ln E_\lambda(\lambda,q,-\mu_0) = \ln E_\lambda(\lambda,0,-\mu_0) - q/\mu_0 \tag{4-27}$$

where $\mu_0 = \cos\theta_0$

θ_0 = the zenith angle of the sun

A plot of $\ln E_\lambda$ versus $1/\mu_0$ from Equation (4-27) is a straight line of slope $(-q)$ and y-intercept of $\ln E_\lambda$ $(\lambda, 0, -\mu_0)$. The experimental procedure is to measure the directly transmitted sunlight at a given wavelength over a wide range of solar zenith angles during the course of a day. If the atmosphere is sufficiently homogeneous in the horizontal and conditions are stationary in time, a plot of the logarithm of the transmitted solar radiation versus $1/\cos\theta_0$ will be a straight line. The slope of this line is the negative of the total optical depth, and the y-intercept is a measure of the incident sunlight at the top of the atmosphere (i.e., at $q = 0$) at the given wavelength. With an absolute instrumental calibration, one can thus also determine the monochromatic solar constant. (See Chapter 3.)

Figure 4-7 gives examples of data taken with a solar radiometer at Tucson, Arizona. The straight lines through the data points are the best fits in a least squares sense. By subtracting the known Rayleigh optical depth from the total q thus determined, one may readily obtain the optical depth caused by particulates. One must be cautious in choosing wavelengths for this technique. If molecular or gaseous absorption is present —especially if the amount is unknown and correction cannot be made—serious errors may result in the total q measured.

Equation (4-26) can be used for calculating the horizontal transmission of the atmosphere between any two points, with a known artificial source (e.g., a searchlight or laser) at some known distance, s, from a receiver or a reflecting target. A laser radar (or lidar) which emits short bursts of light and uses the time delay between the emitted and back-scattered return as a measure of distance from which the return orignated, can be used to determine the atmospheric transmissivity between two points or, equivalently, the volume extinction coefficient. Assuming horizontal homogeneity, one can use the following for the back-scattered return:

$$\Phi_R(180°) = \frac{I_\lambda(\lambda, 0)}{R^4}\,[e^{-2k(\lambda,s)\rho R}]\,VP(180°)A_r \tag{4-28}$$

where $2k(\lambda,s)\rho R = q/\mu$

R = distance between the Lidar and the scattering volume

$= (t/2)c$

t = time delay between emitted and received pulses

c = speed of light

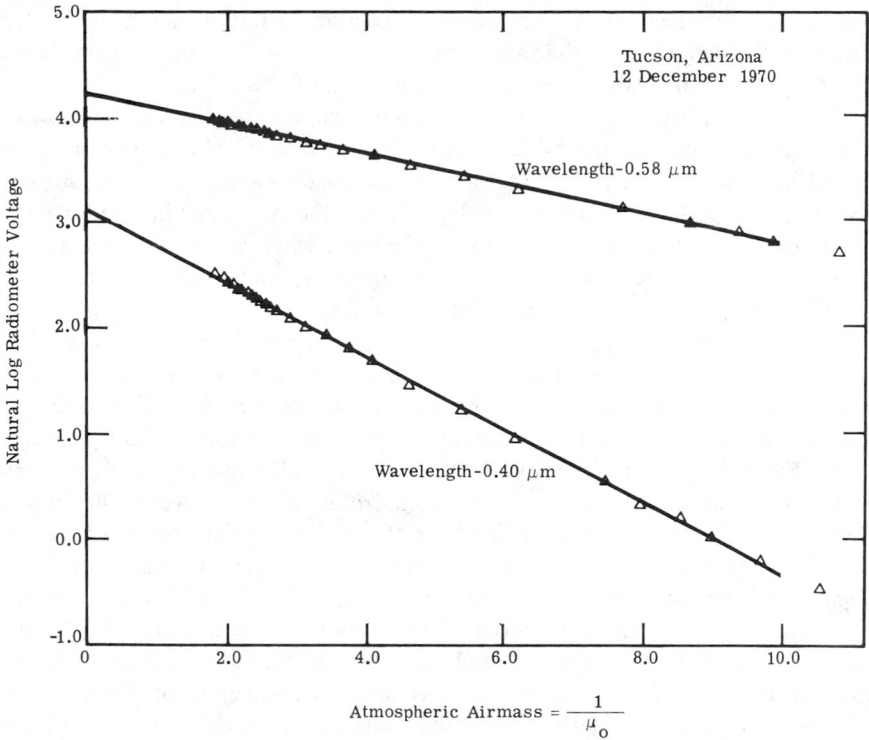

Fig. 4-7. Examples of data taken with a solar radiometer.

Equation (4-28) contains two unknowns, $k(\lambda,s)$ and $P(180°)$. Under the assumed condition of horizontal homogeneity, measurements of $P_R(180°)$ for two values of R suffice to solve for the unknowns. Similar experiments may be performed with pulsed lasers firing in the vertical to obtain the vertical distribution of the volume extinction coefficient. In this case, however, the assumption of homogeneity along the pathlength is no longer valid. As a result, either certain assumptions must be made relating the values of $P(180°)$ to $k(\lambda,s)$ or additional measurements must be taken.

Another, less direct technique for determining the atmospheric volume extinction coefficient is to determine the size distribution function, $f(r)$, of the particulates. This may be done either by direct measurement or by remote sensing utilizing inversion techniques. Once the size distribution is inferred, with a known or assumed index of refraction, and once the further assumption is made that the particulates may be treated as spherical particles, the volume extinction coefficient may be computed from Mie theory.

4.4. Methods of Calculating Radiative-Transfer for Multiple Scattering

4.4.1. Introduction. Scattering occurs from aerosols as well as from molecules in the atmosphere. Compared to aerosol scattering, molecular scattering is negligible outside of the visible part of the spectrum. Significantly, the sun's emission spectrum peaks in the center of the visible region. Thus, scattering by molecules can be considered confined mainly to the visible region, the sun being the chief emitter of radiation scattered from molecules, since the peak of the atmospheric radiation, emitted mainly by molecules, occurs beyond 10 μm. On the other hand, scattering by aerosols, depending on their sizes, is less strongly dependent on wavelength and, in cases of heavy haze, can be effective in both the visible and infrared regions. This section is concerned primarily with an atmosphere with a particulate density which is consistent with so-called clear-sky or near-clear-sky conditions, so that aerosol scattering in the long wavelength infrared region can usually be neglected. In the region of overlap between the visible and infrared parts of the spectrum, or in the long wavelength regions for cases when aerosol scattering is significant, the mathematical representations of scattering and molecular absorption are considered to be completely separable.

This section considers only the effects of scattering by gases, and scattering and absorption by aerosols, in the spectral region $0.3 \gtrsim \lambda \gtrsim 2\ \mu$m. All methods described are not given equal space, especially if they have adequate exposure in the literature. Discussion concentrates on those methods which form the basis for practical applications.

4.4.2. Exact Solutions. The origins of modern radiative-transfer theory can be traced to the classic works of Chandrasekhar and Ambartsumian in which they developed the fundamental mathematics for the analysis of radiation in plane-parallel atmospheres. The mathematical complexities of radiative-transfer theory present major difficulties to investigators who want to model the natural or artificial radiation field in a scattering medium. The problem in the determination of the radiation field is basically due to both the uncertainty in the knowledge of the physical state of the medium (as a result of one's inability to measure enough state parameters), and the complexity of the mathematical analysis (and the resultant length of computer time needed to obtain significant results).

This section briefly discusses exact solutions to radiative-transfer problems, i.e., those for which there are no approximations in the basic mathematical formulations of the radiative-transfer equation for ideal atmospheres. Chandrasekhar was able to derive a set of nonlinear equations which could be solved to determine the radiation field for a homogeneous plane-parallel atmosphere which is illuminated by solar radiation. He also considered polarization. Results based on his analysis for the case of a pure Rayleigh atmosphere are given by Coulson, et al. [4-2] for the polarized radiation field. The computations are laborious and are limited to small optical thicknesses and the radiant energy emerging from the top and bottom of the atmosphere. A later, extended mathematical study by Busbridge [4-3, 4-4], Mullikin [4-5, 4-6, 4-7, 4-8], and Sekera [4-9] showed that one can also use the nonlinear equations to determine the radiation field within the atmosphere and for very large as well as small optical thicknesses. The solutions are exact for *any* optical thickness. For the case of inhomogeneous atmospheres, the analysis is less well developed but in recent years investigators have made progress on an exact solution for inhomogeneous atmospheres [4-10, 4-11, 4-12]. It should be pointed out, however, that this is a special form of inhomogeneity for which exact analytic solutions are possible; it does not necessarily correspond to realistic atmospheric inhomogeneities.

A powerful mathematical method which has been used in recent years to find a rigorous solution to the radiative-transfer equation is the normal-mode expansion technique developed by Case [4-13, 4-14]. It is basically an attempt to formulate a solution to the linear transport equation by using singular eigenfunctions, whose unknown, expansion coefficients are determined by constraining the solution to fit the boundary conditions. In this way, this technique is similar to the classical methods of solving partial differential equations in mathematical physics. Two advantages of this method are that it allows one to understand the nature of the solutions, and that the method can easily be adapted to approximation procedures. Details of the application areas of this approach can be found in References [4-15, 4-16, 4-17, 4-18, 4-19], and [4-20].

The normal-mode expansion technique has been applied to a number of problems involving heat transfer, scattering in atmospheres, and neutron transport theory. It is difficult, however, to carry out the numerical procedures needed to obtain the final solution. Therefore, this technique is not necessarily competitive with the simpler but less elegant techniques dealt with in the next sections.

Another method, which was devised by Wiener and Hopf [4-21], is to make use of the Fourier transform analysis of the integral form of the radiative transfer equation. The Wiener-Hopf technique is applicable to the Milne problem of astrophysics, i.e., that of finding the radiation field for a semi-infinite atmosphere such as that approximated by a star. For finite atmospheres, Chandrasekhar derived an integral relation for the radiation field emerging from the top and bottom of a uniform homogeneous plane-parallel atmosphere. Based upon the general analysis, Coulson et al. [4-2] computed the radiation field for seven atmospheric optical thicknesses, three surface reflectances, and several sun angles and view angles for a Rayleigh scattering law. The analysis has also been extended to consider cases of large optical thicknesses and inhomogeneous atmospheres. A more complete survey is provided by Sobolev.

4.4.3. Adaptations to Exact Solutions.

Iterative Method. The equation for total radiance in a homogeneous atmosphere in short-wavelength spectral regions for which thermal emission is considered negligible (or separable), is written in general form

$$L(q) = L(q_0)\tau + \omega_0 HL(q) \tag{4-29}$$

where $\tau = e^{-(q_0-q)/\mu}$

H = operator

The operator is defined by

$$HL(q) = \frac{1}{4\pi\mu} \int_0^{2\pi} \int_{-1}^{1} \int_q^{q_0} p(q',\mu,\phi,\mu',\phi') e^{-(q'-q)/\mu} L(q',\mu',\phi') dq' d\mu' d\phi' \tag{4-30}$$

where $p = p(\Theta) = p(q',\mu,\phi,\mu',\phi')$.

Equation (4-29) can formally be written as

$$(I - \omega_0 H)L(q) = L(q_0)\tau \tag{4-31}$$

and the solution is

$$L(q) = (I - \omega_0 H)^{-1} L(q_0)\tau \tag{4-32}$$

where $(I - \omega_0 H)^{-1}$ denotes the inverse operator and I is a unit operator. Thus,

$$L(q) = (I + \omega_0 H + \omega_0^2 H^2 + \ldots + \omega_0^n H^n) L(q_0)\tau \tag{4-33}$$

$$= L(q_0)\tau + \omega_0 HL(q_0)\tau + \omega_0^2 H^2 L(q_0)\tau + \ldots + \omega_0^n H^n L(q_0)\tau \tag{4-34}$$

which will converge to the exact solution as $n \to \infty$, provided certain conditions hold true for ω_0 and the operator, H. For this Neumann series to converge,

$$|\omega_0| < \frac{1}{M(q_o - q)} \tag{4-35}$$

where M is the maximum value of the kernel in the radiative transfer integral equation. Physically, the first term on the right-hand side of Equation (4-34) is the directly attenuated radiance, the second term is the singly scattered radiance, and so on up to the n-th term, which represents the scattering of a photon n times in the atmosphere. It can be seen from Equation (4-34) that if the single-scattering albedo, ω_0, is small (i.e., if there is little scattering or much absorption), the series can converge rapidly and only a few terms will provide a reasonable solution. Irvine (1968, 1968) has applied the Neumann series method to the solution of radiative transfer problems, and Herman and Browning [4-22] have used the Gauss-Seidel method of iteration. Some results of the Herman and Browning method are illustrated in Figure 4-8, with polarization included. Here, they calculated the radiance emerging from the top and bottom of a homogeneous, plane-parallel atmosphere with Rayleigh scattering and compared the results with those of Coulson, et al. [4-2]. Herman et al. [4-23] extended this analysis to include aerosol scattering in more realistic atmospheres. Figure 4-9 illustrates typical results of their calculations for a fit to experimental data on the optical thickness of the atmosphere in the Tucson, Arizona area. The normalized radiance is shown in the solar plane for a solar zenith angle of ~22.5°. The quantity q_{aer} is the aerosol optical thickness, and q_o is the total (aerosol plus Rayleigh). The iteration technique can be quite time consuming on a computer, especially if large optical thicknesses and vertical inhomogeneity is considered. But the computer time can be drastically cut if polarization is neglected.

Spherical Harmonics Method. The scattering phase function, p, is represented by an expansion in Legendre polynomials, P_l, i.e.,

$$p(\xi) = \sum_{l=0}^{N} A_l P_l(\xi) \tag{4-36}$$

where the orthogonality property of the polynomials is used to obtain the expansion

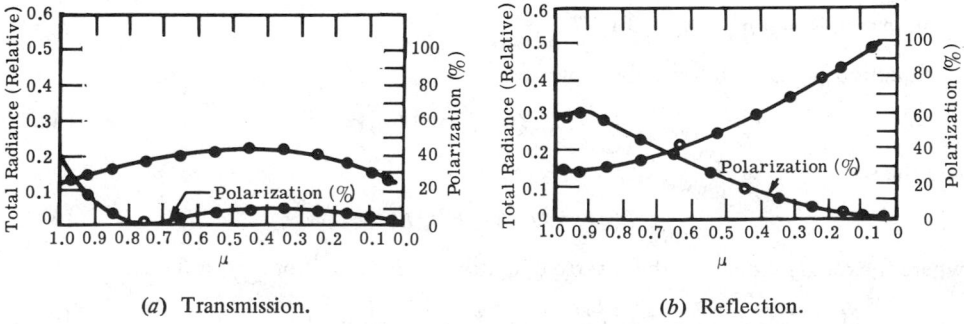

(a) Transmission.   (b) Reflection.

Fig. 4-8. Total radiance (relative), percent polarization ($q = 1.0$; $\lambda = 0.4$; and $\phi = 0$) [4-2, 4-22].

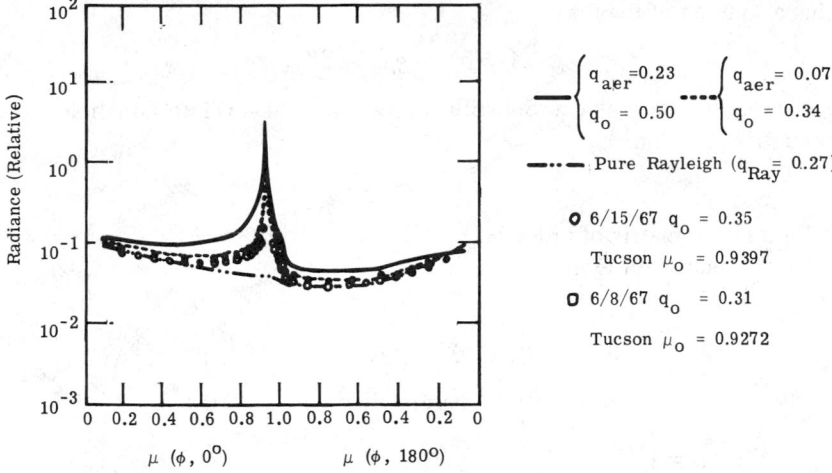

Fig. 4-9. Measured and theoretical transmitted radiances [4-23].

coefficients, A_l, i.e.,

$$A_l = \frac{2l+1}{2} \int_{-1}^{1} p(\xi)P_l(\xi)d\xi \qquad (4\text{-}37)$$

In practice, the representation of a typical, atmospheric, haze-type, polydisperse phase function will require up to 200 terms. Using the additive properties of the spherical harmonics, one gets

$$p(\mu,\phi,\mu',\phi') = \sum_{l=0}^{N} A_l \left[P_l(\mu)P_l(\mu') + 2\sum_{m=1}^{l} \frac{(l-m)!}{(l+m)!} P_l^m(\mu)P_l^m(\mu') \cos m(\phi'-\phi) \right] \qquad (4\text{-}38)$$

The radiance is also expanded in a set of spherical harmonics as

$$L(q,\mu,\phi) = \sum_{m=0}^{\infty} \sum_{l=m}^{N} A_{lm}(q)P_l^m(\mu) \cos m(\phi_0 - \phi) \qquad (4\text{-}39)$$

and inserted into the transfer equation to get a system of differential equations:

$$\frac{(l-m)}{(2l-1)} \frac{dA_{l-1,m}}{dq} + \frac{(l+m+1)}{(2l+3)} \frac{dA_{l+1,m}}{dq}$$

$$= \left(1 - \frac{\omega_0 A_l}{2l+1} \right) A_{lm} - \frac{\omega_0 E_s(q)(2-\delta_{0m})b_{lm} (-1)^{m+l}P_l^m(\mu_0)}{4\pi} \qquad (4\text{-}40)$$

where

$$b_{lm} = A_l \frac{(l-m)!}{(l+m)!} \qquad (4\text{-}41)$$

with the conditions $\qquad l = m, \ldots, N; \ 0 \leqslant m \leqslant N \qquad (4\text{-}42)$

Assuming a solution of the form

$$A_{lm}(q) = g_{lm}(\xi)e^{\xi q} \tag{4-43}$$

one gets an eigenvalue problem. Substituting Equation (4-43) into Equation (4-40), one gets the matrix equation

$$\tilde{\tilde{A}}\,\tilde{V}_m = \lambda \tilde{V}_m \tag{4-44}$$

where $\tilde{\tilde{A}}$ = a square matrix of order $2r$
$\quad\quad r$ = the number of terms chosen
$\quad\quad \lambda = 1/\xi$

$$\tilde{\tilde{A}} = \begin{pmatrix} 0 & \beta_m & 0 & 0 & \cdots & 0 \\ \alpha_{m+1} & 0 & \beta_{m+1} & 0 & \cdots & 0 \\ \cdot & \cdot & \cdot & \cdot & \cdots & \cdot \\ \cdot & \cdot & \cdot & \cdot & \cdots & \cdot \\ \cdot & \cdot & \cdot & \cdot & \cdots & \beta_{2r-2+m} \\ 0 & 0 & \cdot & \cdot & \alpha_{2r-1+m} & 0 \end{pmatrix} \tag{4-45}$$

and the matrix elements are given by

$$\alpha_{m+1} = \frac{(l+m)(2l+1)}{(2l-1)(2l+1-\omega_0 A_l)} \tag{4-46}$$

$$\beta_{m+1} = \frac{(l+m+1)(2l+1)}{(2l+3)(2l+1-\omega_0 A_l)} \tag{4-47}$$

The eigenvalues are found by taking the determinant of the matrix equation:

$$det\,(\tilde{\tilde{A}} - \lambda \tilde{\tilde{I}}) = 0 \tag{4-48}$$

Knowing the eigenvalues $\lambda(=1/\xi)$ allows one to use Equations (4-43) and (4-39), along with the appropriate boundary conditions, to determine the coefficients $A_{lm}(q)$. One of the problems encountered in using this method is the propagation of roundoff errors. The matrices are ill-conditioned in some instances and the system is numerically unstable. Canosa and Penafiel, however, were able to produce well-conditioned matrices and elimi-nate most of the numerical instability by using various transformations on matrices [4-24]. Under certain conditions, for example a Rayleigh-type atmosphere, the spherical harmonics method is much faster and more accurate than the iterative technique; but for realistic atmospheres with strongly anisotropic aerosol scattering, the advantage over the iterative technique is not that great. This is because many coefficients (A_l) are required in Equation (4-36) to represent the scattering phase function, which increases the size of the matrix (Equation (4-45)), whereas the iterative method is independent of the degree of anisotropy. Because of recent advancements in matrix computer analysis, the spherical harmonics method can be made more efficient than the iterative technique for thick atmospheres. Typical computer times are given in Table 4-2 for the calculation of fluxes in homogeneous atmospheres (i.e., integrals over the angles θ and ϕ). Table 4-3 gives the storage requirements using double precision arithmetic.

Table 4-2. Net Fluxes for a Mie Atmosphere of One Optical Thickness.
All cases solved using 64 layers. [4-24]

Cosine of the Sun's Zenith Angle (μ_O)	Haze*	Wavelength (μm)	Net Flux (Relative)					No. of Conditioning Points	Computation Time (sec) IBM 360/195#	
			Integral Equation Iteration Method		Spherical Harmonics (P_l) Method**				t_i	t_{sh}
			Top	Bottom	Top	Bottom	l			
1.0	L	2.45	2.7198	2.7193	2.7193	2.7193	9	0	18.41	—
0.5	L	2.45	1.0612	1.0605	1.0599	1.0600	9	0	20.84	0.17†
1.0	L	0.595	2.8130	2.8176	2.8130	2.8130	21	2	21.17	—
0.5	L	0.595	1.1402	1.1387	1.1394	1.1395	21	2	21.17	0.86†
1.0	L	0.3025	—	—	2.8078	2.8078	41	4	—	—
0.5	L	0.3025	1.1511	1.1501	1.1506	1.1507	41	4	24.15	4.19‡
0.5	M	0.3025	1.1825	1.1818	1.1821	1.1822	67	8	24.77	16.78†
1.0	M	0.3025	—	—	2.8466	2.8466	67	8	—	—
0.8	M	0.3025	—	—	2.1747	2.1748	67	8	—	—
0.6	M	0.3025	—	—	1.5076	1.5077	67	8	—	—
0.4	M	0.3025	—	—	0.8693	0.8694	67	8	—	—
0.2	M	0.3025	—	—	0.3275	0.3278	67	8	—	19.28***

*L = terrestrial haze; M = marine haze

**l + 1 = No. of terms in expansion

†This timing is only for the value μ_O = 0.5.

†This timing is for the two values of μ_O shown for the given wavelength, which are solved in one pass.

‡This time is for the five values of μ_O shown for Haze M, λ = 0.3025 μm which are solved in one pass.

#Using FORTRAN IV with an H compiler.

Table 4-3. Storage Requirements for Double Precision [4-24]

Approximation, l	No. of Layers	No. of Condition Points	Storage (10^3)
7	64	3	115.0
7	128	3	120.0
15	128	17	130.0
21	64	3	250.0
41	64	5	306.3
67	64	9	585.0
7	512	33	156.0

Discrete Ordinates Methods. One can make discrete the angular variables, the space variables, or both to arrive at a set of difference equations [4-25]. This method is, therefore, more easily adapted to computer methods [4-26].

Thus, for azimuthal symmetry, the angular integral is replaced by a summation

$$\int_{-1}^{1} p(\mu,\mu')L(q,\mu')d\mu' \approx \sum_{i=1}^{N} p(\mu,\mu_i)L(q,\mu_i)w_i \qquad (4\text{-}49)$$

where μ_i = zeroes of the Legendre polynomials $P_l(\mu)$
w_i = Christoffel numbers or relative weights

Another variation is the S_N method, in which the angular interval $-1 \leqslant \mu \leqslant 1$ is divided into N subintervals $[\mu_{i-1},\mu_i]$, $i = 1,2,\ldots,N$, and the radiance is assumed to vary linearly with μ:

$$L(q,\mu) = \frac{\mu - \mu_{i-1}}{\mu_i - \mu_{i-1}} L(q,\mu_i) + \frac{\mu_i - \mu}{\mu_i - \mu_{i-1}} L(q,\mu_{i-1}) \qquad (4\text{-}50)$$

When Equation (4-50) is substituted into the radiative-transfer equation, a set of N equations results, known as the S_N set. This procedure has the advantage that it can be applied to a large number of geometric configurations of both boundaries and sources. Some disadvantages are that negative fluxes may result and a very fine mesh is needed for high accuracy. This procedure is useful in solving nuclear reactor problems, so many programs and options have been developed over the years. A summary of some of these programs is given by Lathrop and by Carlson and Lathrop, and the basic ideas of the S_N method are described by Lee. For investigations in radiative-transfer, these programs are somewhat cumbersome to use. Nevertheless, the Oak Ridge National Laboratory can provide an almost complete compilation.

The discrete-ordinates approach to solving the radiative-transfer equation is regarded by some as a more fundamental method of solving the equation rather than as an approximation to an exact formulation. Thus, there has emerged a rigorous basis for the so-called discrete space theory by Preisendorfer. These ideas have culminated in the practical application of discrete space theory to the solution of the radiative-transfer equation by Grant and Hunt (1969a, 1969b). This method has been applied to a number of practical problems, but limitations do exist. So far, only a Lambertian surface has been considered and multidimensional problems have not been investigated in detail. Also, the computer time can be significant, especially for a large number of parameter values.

Invariant Imbedding. The invariant imbedding method, which uses the accumulation of thin layers of scattering media, has been successfully used in the analysis of transport problems. Many of the ideas of the invariant imbedding approach were formulated by Ambartzumian and Chandrasekhar based upon fundamental principles of invariance. However, the first application of invariant imbedding to the practical solution of radiative-transfer problems was by Bellman and Kalaba (1956) with numerical results by both Bellman, Kalaba, and Prestrud (1963), and Bellman, Kagiwada, Kalaba, and Prestrud (1964).

Application of the method results in a nonlinear integro-differential equation subject to an initial value condition instead of the usual linear transport equation with a two-point boundary condition. The advantage of using this method is that an initial value problem can be solved by means of a simple iteration procedure on a computer, whereas the classical approach (using the two-point boundary condition) usually involves the solving of a large system of linear equations. In recent years, the invariant imbedding approach has been applied to a great number of problems, including anisotropic scattering (Kagiwada and Kalaba), spherical shell atmospheres (Ueno, Kagiwada and Kalaba), and time dependence (Bellman, Kagiwada and Kalaba (1966)). Comparisons have been made between Chandrasekhar's results and those based on invariant imbedding and the agreement is good. Some disadvantages of this technique are the long computer time, especially for optically thick media, and that each problem requires a whole new formulation of equations.

Doubling Technique. Van de Hulst and Grossman [4-27] considered the simplification of the invariant imbedding method by adding layers, each of which was twice the optical thickness of the preceding one. Thus, even though one starts the computational process with a very thin layer with a total optical thickness q_o, after 10 cycles (for example), a thickness of $2^{10}q_o$ is reached, resulting in a rapid computational procedure. Some of the disadvantages are that only homogeneous atmospheres can be dealt with efficiently and that overall accuracy is difficult to estimate. Nevertheless, Hansen (1969(a), 1969(b)) has found some interesting results by using this method in the analysis of clouds.

Moment Methods. Another mathematical method to solve radiative-transfer problems is the moments method, originally devised by Krook, and later developed by Sherman and Liner. The method is simple, and essentially equivalent to the discrete ordinates method.

Monte Carlo. The statistical procedure called Monte Carlo consists basically of a particle counting method with probabilities associated with the physical processes involved. Much of the early work dealt with neutron transport in complicated geometrical systems. In recent years the method has been applied to a variety of problems, including photon transport in the atmosphere.

The basic description of a Monte Carlo process is as follows. A computer program is devised in which the physics of the problem is described in a probabilistic manner. When one has a well-defined coordinate system and boundary conditions, the particles are released from a source and followed as they undergo scattering and absorption. The processing continues until reasonable statistical estimates have been obtained. In this general way, Collins and Wells, Collins, and Wells and Marshall have calculated the visible radiation field within the atmosphere; Plass and Kattawar (1968), and Kattawar and Plass (1969) have done similar calculations for clouds.

The main advantage of the Monte Carlo procedure is that very complicated geometries can be considered. The main disadvantage is the excessive amount of computer time expended to obtain reasonably accurate values. The accuracy is proportional to the square root of the number of particles counted.

4.4.4. Approximate Methods

Schuster-Schwarzschild Method. Early in the twentieth century, Schuster [4-28] and Schwarzschild [4-29] introduced a simple method of directional averaging. The radiation field is assumed to be nearly isotropic, and averages of radiance upward and downward are taken over the upward and downward hemispheres:

$$L_+(q) \equiv \int_0^{2\pi} \int_0^1 L(q, \mu, \phi) d\mu d\phi \qquad (4\text{-}51)$$

$$L_-(q) \equiv \int_0^{2\pi} \int_{-1}^0 L(q, \mu, \phi) d\mu d\phi \qquad (4\text{-}52)$$

Also a weighted average (i.e., the flux density upward or downward) is taken:

$$E_+(q) \equiv \int_0^{2\pi} \int_{-1}^0 \mu L(q, \mu, \phi) d\mu d\phi \qquad (4\text{-}53)$$

$$E_-(q) \equiv \int_0^{2\pi} \int_{-1}^0 \mu L(q, \mu, \phi) d\mu d\phi \qquad (4\text{-}54)$$

Then, with the assumption of a nearly isotropic field,

$$E_+(q) \cong \frac{1}{2} L_+(q)$$

$$E_-(q) \cong \frac{1}{2} L_-(q) \qquad (4\text{-}55)$$

If one uses these formulae in the radiative-transfer equation with an isotropic scattering law

$$\mu \frac{dL}{dq} = L(q, \mu, \phi) - \frac{\omega_o}{4\pi} \int_0^{2\pi} \int_{-1}^1 L(q, \mu', \phi') d\mu' d\phi' \qquad (4\text{-}56)$$

one gets two differential equations instead of the integro-differential equation:

$$\frac{dE_+(q)}{dq} = (2 - \omega_o) E_+(q) - \omega_o E_-(q) \qquad (4\text{-}57)$$

$$\frac{dE_-(q)}{dq} = \omega_o E_+(q) - (2 - \omega_o)E_-(q) \tag{4-58}$$

When these coupled equations are solved using the appropriate boundary conditions, one obtains the irradiance (upward and downward) at any point in the medium. A model based upon the assumptions made about the directionality of the field (Equation (4-55)) is taken to be valid deep within an atmosphere rather than near the boundaries. Note, however, that radiance is not calculable.

Eddington Method. In 1926, Eddington [4-30] introduced another averaging method which used the average over all space rather than over hemispheres, i.e.,

$$\bar{L} = \frac{1}{4\pi} \int_0^{2\pi} \int_{-1}^{1} L(q,\mu,\phi)d\mu d\phi \tag{4-59}$$

The result (using Eddington's assumption that the field is almost isotropic) is

$$\frac{d^2 L}{dq^2} = 3(1 - \omega_o)L(q) \tag{4-60}$$

an approximation expected to hold for very thick atmospheres.

Irvine [4-1] has made comparisons between the Schuster-Schwarzschild (two-stream) model and the Eddington model. The conclusion is that the two-stream model is better when low-order scattering dominates, but in general that the Eddington approximation and the two-stream approximation are quite similar in terms of their regions of validity. The Eddington approximation has been applied to inhomogeneous atmospheres by Shettle and Weinman.

Romanova's Method. One of the major difficulties involved in the solution of the radiative-transfer equation is how to deal with the high degree of anisotropy of the scattering phase function. Romanova [4-31, 4-32, 4-33] has developed a method based upon a small-angle approximation similar to that used by Wang and Guth [4-34] in nuclear physics. If one were to expand the phase function in a series of Legendre polynomials, many terms would be required. This leads to the rather difficult eigenvalue problem in the spherical harmonics method. Romanova's method consists of expressing the radiance as the sum of two terms, one, L_{sa}, for small angles (i.e., the highly anisotropic part of the phase function) and another, $\tilde{L}$, for the remaining part of the phase function:

$$L(q,\mu,\phi) = L_{sa}(q,\mu,\phi) + \tilde{L}(q,\mu,\phi) \tag{4-61}$$

The $L_{sa}(q,\mu,\phi)$ solution is found by replacing μ by μ_o in the radiative-transfer equation and setting new boundary conditions, i.e.,

$$\begin{aligned} \tilde{L}(q,\mu,\phi) &= 0 & \mu &> 0 \\ \tilde{L}(q,\mu,\phi) &= -L_{sa}(q,\mu,\phi) & \mu &< 0 \end{aligned} \tag{4-62}$$

$\tilde{L}(q,\mu,\phi)$ can then be expanded in a series of Legendre polynomials which presumably will require fewer terms since most of the anisotropy is accounted for by the $L_{sa}(q,\mu,\phi)$

term. Irvine (1968) has made comparisons between Romanova's method and the doubling method and found excellent agreement over a wide range of angles and optical thicknesses. No estimate is given for the amount of computer time, but it is considerably reduced over the other, more conventional techniques.

The Turner Method. Turner [4-35, 4-36, 4-37, 4-38] has developed a modified two-stream iterative model especially designed for hazy atmospheres.

The one-dimensional, radiative-transfer equation for a wavelength, λ, is

$$\mu \frac{dL}{dq} = L(q,\mu,\phi) - \frac{\omega_o}{4\pi} \int_0^{2\pi} \int_{-1}^{1} p(\mu,\phi,\mu',\phi')L(q,\mu',\phi',)d\mu'd\phi'$$
$$- \frac{\omega_o E_s(q)}{4\pi} p(\mu,\phi,-\mu_o,\phi_o) \tag{4-63}$$

where $E_s(q)$ is the solar irradiance at a point with optical depth q, and the scattering phase function is approximated as

$$p(\mu,\phi,\mu',\phi') = 4\pi\eta\delta(\mu'-\mu)\delta(\phi'-\phi) + 4\pi(1-\eta)\delta(\mu'+\mu)\delta(\phi'-\pi-\phi) \tag{4-64}$$

Here $\delta(\mu'-\mu)$ and similar quantities are the Dirac delta function and η represents the fraction of the radiation scattered into a forward hemisphere, i.e., under conditions of azimuthal symmetry,

$$\eta = \frac{1}{2} \int_0^1 p(\xi)d\xi \tag{4-65}$$

For Rayleigh scattering, $\eta = 0.5$; whereas for aerosol scattering in the visible region, $\eta \sim 0.95$. Using Equation (4-64) in Equation (4-63) and making a similar assumption regarding the radiation field, i.e.,

$$L(q,\mu,\phi) = \frac{1}{\mu_o} [E'_+(q)\delta(\mu-\mu_o)\delta(\phi-\pi-\phi_o) + E'_-(q)\delta(\mu+\mu_o)\delta(\phi-\phi_o)] \tag{4-66}$$

(where $E'_+(q)$ and $E'_-(q)$ are irradiances in the upward and downward directions, respectively), one gets two, linear, differential equations for $E'_+(q)$ and $E'_-(q)$:

$$\frac{dE'_+(q)}{dq} = \frac{1-\omega_o\eta}{\mu_o} E'_+(q) - \frac{\omega_o(1-\eta)}{\mu_o} E'_-(q) - \omega_o(1-\eta)E_\Delta(q) \tag{4-67}$$

$$-\frac{dE'_-(q)}{dq} = \frac{1-\omega_o\eta}{\mu_o} E'_-(q) - \frac{\omega_o(1-\eta)}{\mu_o} E'_+(q) - \omega_o\eta E_\Delta(q) \tag{4-68}$$

The $1/\mu_o$ dependence is included to normalize the diffuse irradiance field. The primed quantities indicate that the field is calculated for a surface albedo of zero. The solution to these equations and similar ones for an isotropic radiation reflected from a Lambertian surface is given in functional form by

$$E_+(q) = \mu_o E_o f_+(\eta,\gamma,\omega_o,\mu_o,\rho,q,q_o,\lambda) \tag{4-69}$$

$$E_-(q) = \mu_o E_o f_-(\eta, \gamma, \omega_o, \mu_o, \rho, q, q_o, \lambda) \tag{4-70}$$

$$\tilde{E}_-(q) = \mu_o E_o [\eta, \gamma, \omega_o, \mu_o, \rho, q, q_o, \lambda) + e^{-q/\mu_o}] \tag{4-71}$$

where $E_+(q)$, $E_-(q)$ and $\tilde{E}_-(q)$ are the upward diffuse, downward diffuse, and total downward irradiances, respectively, and

$$\gamma = \sqrt{(1 - \omega_o)(1 + \omega_o - 2\eta\omega_o)}/\mu_o \tag{4-72}$$

One may calculate these irradiance components as a function of wavelength, solar zenith angle, single-scattering albedo, surface albedo, altitude, visual range, and optical depth. Figure 4-10 illustrates the diffuse downward component for $\eta = 0.5$ and $\theta_o = 0°$, thus simulating a Rayleigh atmosphere. A comparison is made in Figure 4-10 with calculations using the exact method. It should be noted that a homogeneous atmosphere is assumed and polarization is not included in the approximate method. A similar comparison, carried out for the diffuse radiation emitted by the atmosphere, is shown in Figure 4-11.

Using a relation similar to Equation (4-66) and the irradiances as determined above, one can then find the spectral radiances. These are given formally as

$$L_{Sky} = L_{Sky}(\eta, \gamma, \omega_o, \mu_o, \mu, \phi, \bar{\rho}, q, q_o, \lambda) \tag{4-73}$$

$$L_P = L_P(\eta, \gamma, \omega_o, \mu_o, \mu, \phi, \bar{\rho}, q, q_o, \lambda) \tag{4-74}$$

$$L_T = L_T(\eta, \gamma, \omega_o, \mu_o, \mu, \phi, \rho, \bar{\rho}, q, q_o, \lambda) \tag{4-75}$$

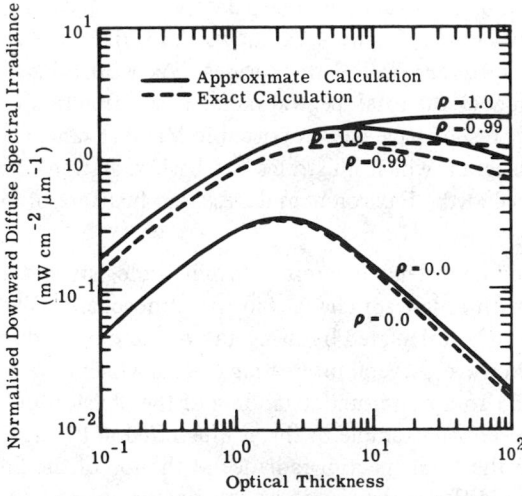

Fig. 4-10. Normalized downward diffuse spectral irradiance versus optical thickness for various surface reflectances. ρ is surface reflectance; λ is 0.55 μm; $0°$ is the solar zenith angle; no absorption; and Rayleigh atmosphere [4-39].

where L_{Sky} = spectral sky radiance at any point in the medium (i.e., the radiance for an observer looking into the upper hemisphere)

L_P = path radiance (i.e., the radiance for an observer looking into the lower hemisphere at a ground target with zero reflectance)

L_T = the total spectral radiance for a downward looking observer

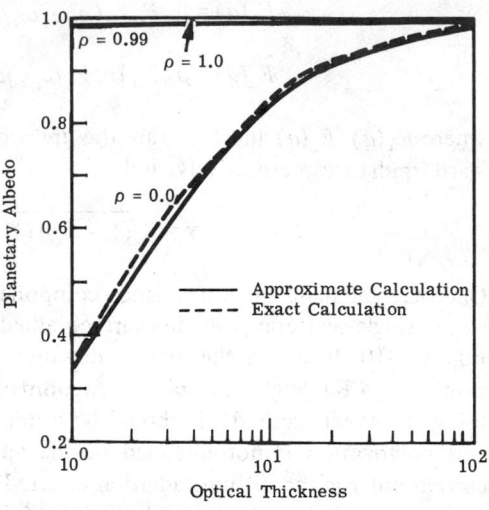

Fig. 4-11. Planetary albedo versus optical thickness for various surface reflectances. $0°$ is the solar zenith angle; no absorption; θ equals $180°$; and Rayleigh atmosphere [4-39].

L_T is not only a function of the background albedo, $\bar{\rho}$, but also of the target reflectance, ρ. In general, then,

$$L_T = L_G \tau + L_P \qquad (4\text{-}76)$$

where L_G = the radiance from the ground surface
τ = the transmittance from the surface to the observer

The computational accuracy of this model was tested by comparing it with exact results. Figure 4-12 illustrates the comparison between the Turner method calculations and those in Coulson, et al. [4-2]. Figure 4-13 depicts the path radiance as a function of q_o for the sun angles shown. Figure 4-14 shows the variation of sky radiance with solar zenith angle for two different optical thicknesses. As a final illustration of the accuracy of the approximate model, the total spectral radiance as a function of nadir angle is shown in Figure 4-15. Other results have been reported in Malila, Crane, and Turner.

Figure 4-16 is a diagram which illustrates the basic radiation-transfer multiple scattering model as developed at the Environmental Research Institute of Michigan.

4.5. The Albedo Problem

This problem is of considerable interest to climatologists and atmospheric scientists who are concerned with long-term changes in the atmosphere. The total albedo (atmosphere plus surface) can be calculated by using any of the models discussed in the previous sections. There are, however, several interesting effects which have not been made evident before and which lead to a better understanding of the physical processes which occur in a scattering atmosphere. An example of this is illustrated in Figure 4-17, which shows the relationship between the total spectral radiance at the top of the Earth's atmosphere and the total (direct plus diffuse) irradiance at the bottom of the atmosphere for various transmittances and surface albedos. For low albedos, the atmospheric transmittance increases and the radiance decreases. On the other hand, radiance decreases for extremely high surface albedos and then increases. Many other radiometric relationships have also been investigated for satellite studies for realistic atmospheres in Turner (1976).

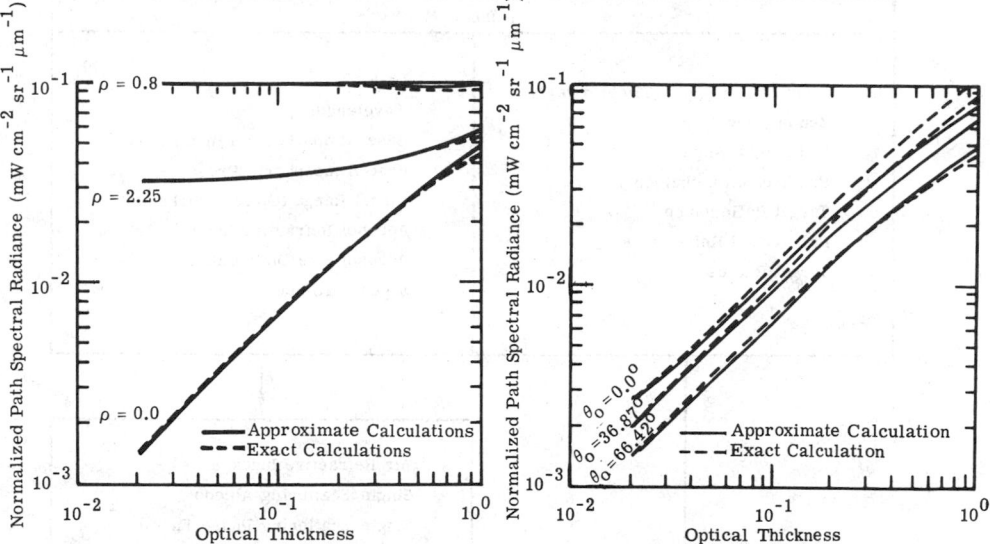

Fig. 4-12. Normalized path radiance at the top of the atmosphere versus optical thickness for several surface reflectances. ρ is surface reflectance; λ is 0.55 μm; $0°$ is the solar zenith angle; no absorption; and Rayleigh atmosphere [4-39].

Fig. 4-13. Normalized path radiance at the top of the atmosphere versus optical thickness for various solar zenith angles. 0.0 is surface reflectance; λ is 0.55 μm; θ equals $180°$; no absorption; and Rayleigh atmosphere [4-39].

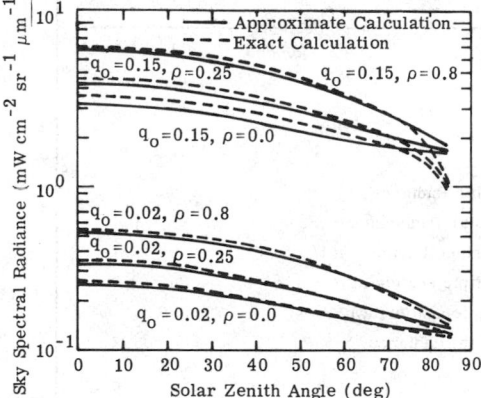

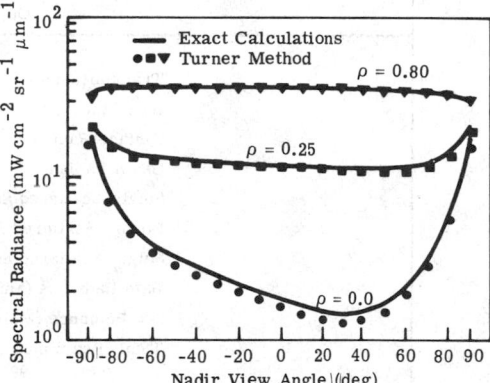

Fig. 4-14. Sky radiance versus solar zenith angle. q is optical thickness; ρ is surface reflectance; λ is 0.55 μm; θ equals $0°$; and Rayleigh atmosphere [4-39].

Fig. 4-15. Total spectral radiance in the solar plane at the top of a Rayleigh atmosphere bounded by a surface with Lambertian reflectance ρ. λ is 0.546 μm; the solar zenith angle is $36.87°$; and optical thickness is 0.1 [4-39].

4.6. Some Illustrative Sample Results

Below are illustrations of the results of calculations based upon radiative-transfer models. The diagrams are for the wavelength of 0.55 μm, but an extrapolation to other wavelengths is possible by using the standard extraterrestrial solar spectrum.

Figure 4-18 illustrates the total radiance, i.e., the radiance as detected by an extra-terrestrial sensor at the top of the atmosphere looking toward the Earth's surface. Also depicted is the sky radiance, i.e., that radiance detected by a sensor looking upward. The

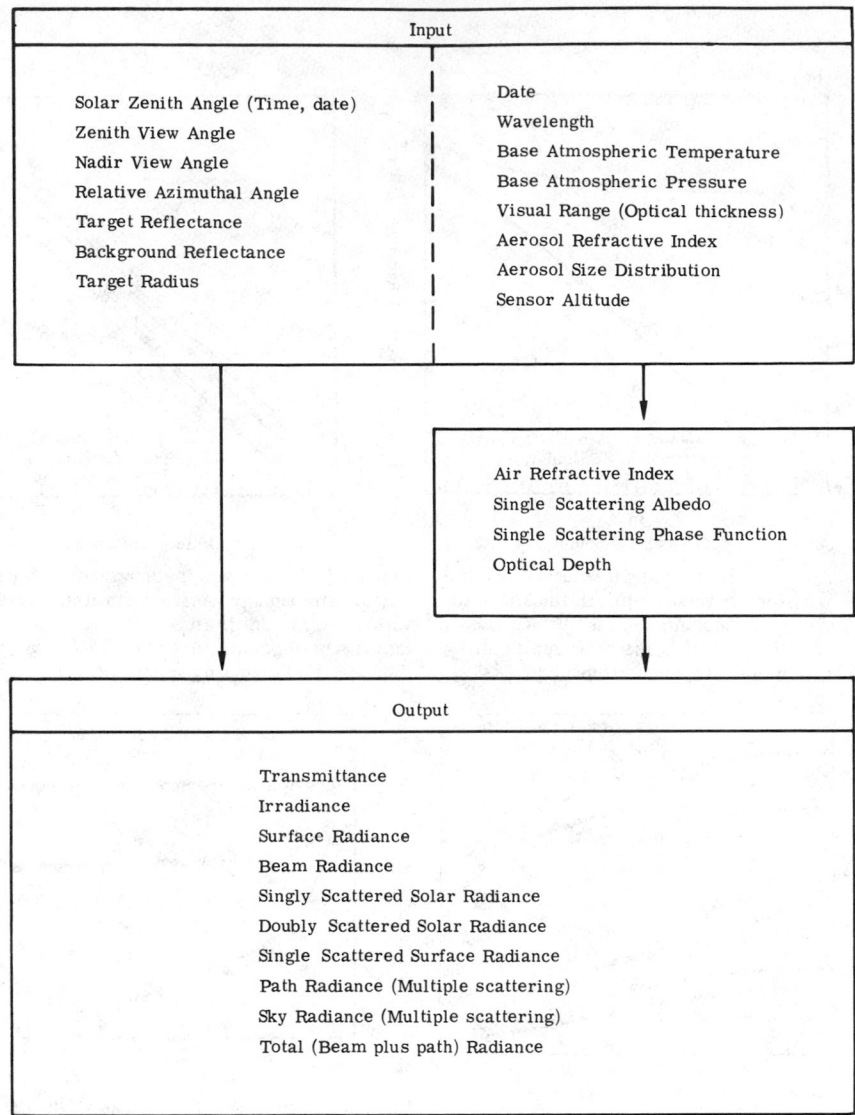

Fig. 4-16. Turner radiative transfer model [4-39].

atmosphere is devoid of particulate matter and hence is referred to as a Rayleigh atmosphere. The radiation field is illustrated on the solar plane, i.e., as a function of zenith and/or nadir view angle in the plane of the sun.

Figure 4-19 illustrates the same radiation field except that the optical thickness is ten times greater than before. In this case, the dipole scattered radiation becomes obvious for a Rayleigh atmosphere.

Figure 4-20 depicts the radiation field for a Rayleigh plus a small aerosol component atmosphere. This pattern is almost identical to that in Figure 4-18 because the Rayleigh optical thickness of Earth's atmosphere is actually 0.098 at a wavelength of 0.55 μm. In addition, all the illustrations beyond Figure 4-19 are based upon the Turner-modified two-stream model. Figures 4-18 and 4-19 were based upon Chandrasekhar's exact calculations for a Rayleigh atmosphere.

A more realistic situation is illustrated in Figure 4-21. Here the optical thickness is 0.2, which corresponds to an atmosphere with a visual range of 250 km at sea level. Even this small aerosol component changes the scattering phase function drastically and the bright region around the sun becomes obvious in the sky radiance. The total radiance, on the other hand, is nearly isotropic because the multiple scattering tends to eliminate the anisotropy characteristic of the phase function. For the case of a black surface, however, the total radiance is not affected by the surface and the distribution of radiance is anisotropic.

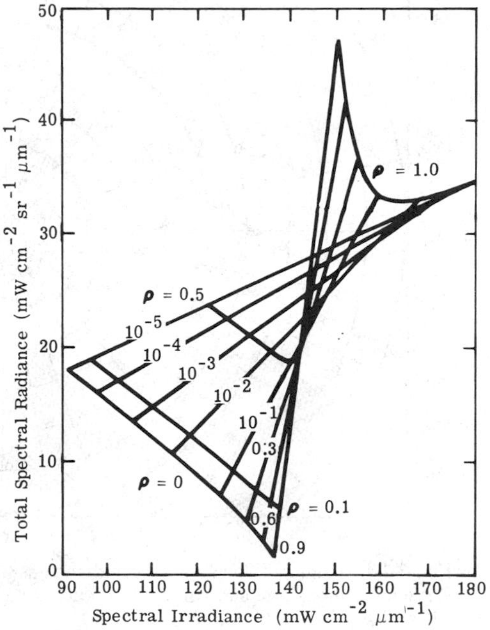

In Figure 4-22, the same situation is depicted as in Figure 4-22 except that the optical thickness is 1.0, which corresponds to a visual range of $\sim$ 4.5 km (a very hazy day). A strong, well-defined back-scattering peak develops opposite the sun in the total radiance and the sky radiance has a strong aureole or bright region near the sun.

Fig. 4-17. Dependence of total radiance on total irradiance for various transmittances and surface albedos. The solar zenith angle is 30°; the nadir view angle is 0°; and λ is 0.55 μm [4-39].

The total and path radiances for all optical thicknesses and all reflecting surfaces can now be illustrated simultaneously. Figure 4-23 depicts the relationship connecting the two radiances for a sun angle of 45° and for an observer looking straight down at the surface from space. The optical thickness of 0.1 corresponds to a visual range of 336 km, 0.2 is 250 km, 0.5 is 14 km, 1.0 is 4.5 km, and 10 corresponds to a dense fog. Intermediate values can easily be scaled from the graph.

A relationship between the total spectral radiance and path spectral radiance can also be established for any point within the medium. In Figure 4-24, this relationship is illustrated for an atmosphere with an optical thickness of 0.2. The actual altitude which corresponds to a given optical depth depends upon the density profile within the atmosphere. Figure 4-25 depicts the same relationship for an atmosphere with an optical thickness of 1.0.

It is interesting to look at the angular distribution of the multiply-scattered sky radiation as a function of optical depth. This is illustrated in Figure 4-26 for an atmosphere with an optical thickness of 1.0 and a surface reflectance of 0.2. The solar peak is obvious at an angle of 45°. The minimum occurs at an angle approximately 90° away

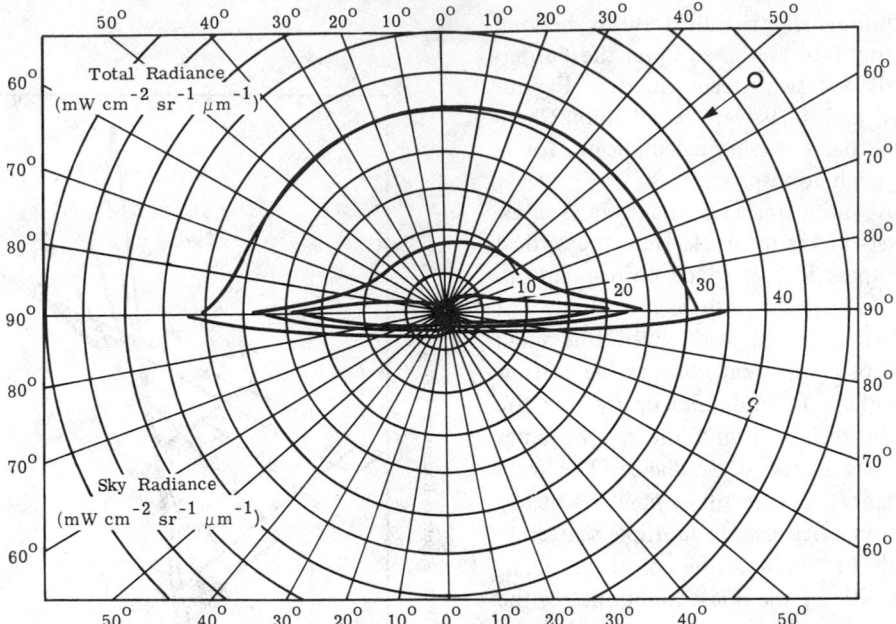

Fig. 4-18. Total radiance and sky radiance as a function of view angle in the solar plane for a Rayleigh atmosphere and three reflectances. θ_O is $53°8'$; q_O is 0.1; λ is 0.55 μm; the day is June 21; and ρ is 0.0, 0.25, and 0.80.

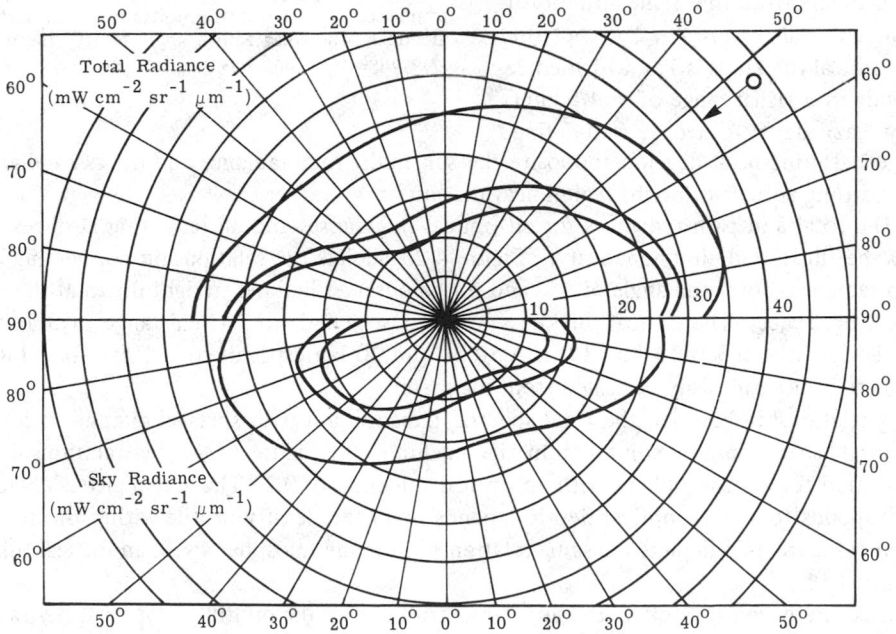

Fig. 4-19. Total radiance and sky radiance as a function of view angle in the solar plane for a Rayleigh atmosphere and three reflectances. θ_O is $53°8'$; q_O is 1.0; λ is 0.55 μm; the day is June 21; and ρ is 0.0, 0.25, and 0.80.

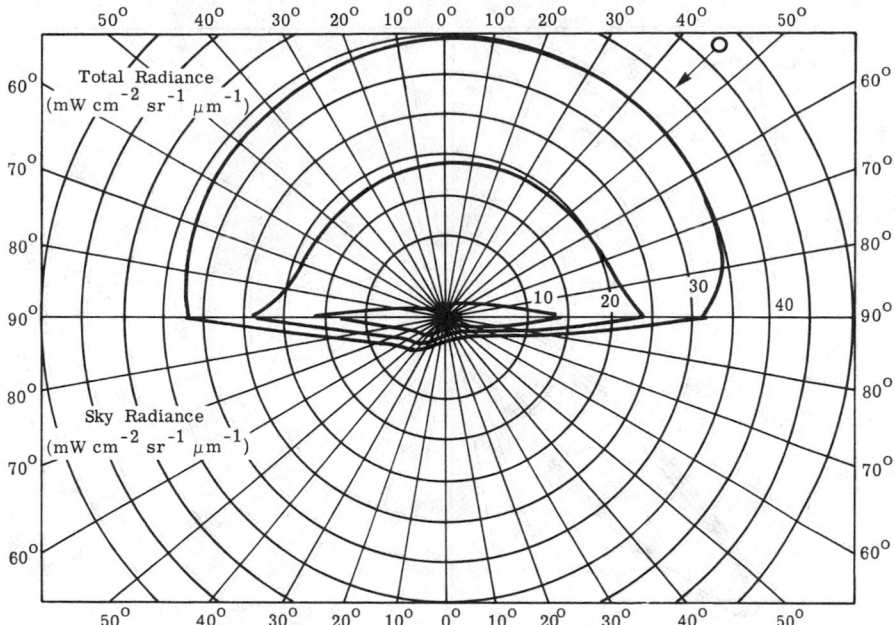

Fig. 4-20. Total radiance and sky radiance as a function of view angle in the solar plane for three reflectances. θ_O is 45°; q_O is 0.10; λ is 0.55 μm; the day is June 21; and ρ is 0.0, 0.5, and 0.9.

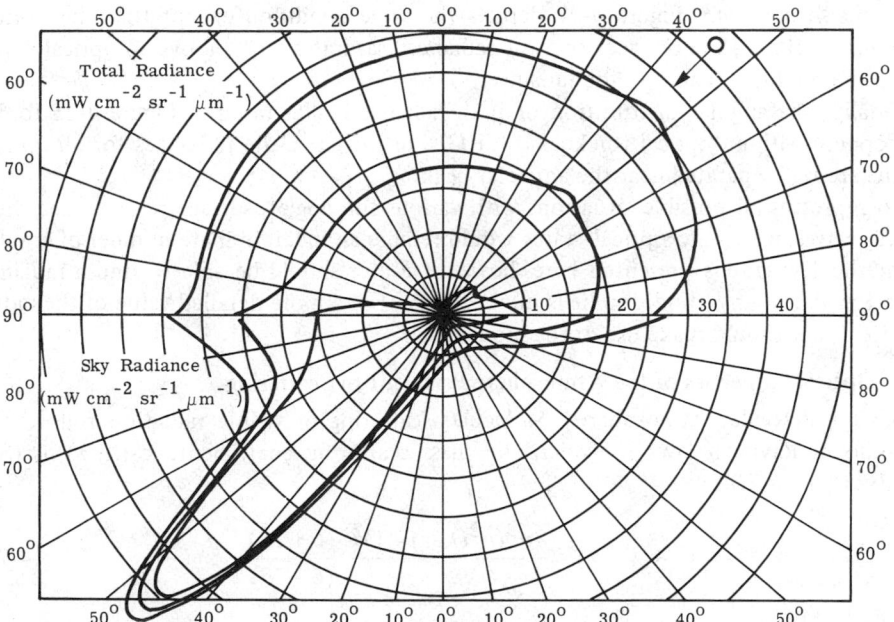

Fig. 4-21. Total radiance and sky radiance as a function of view angle in the solar plane for three reflectances. θ_O is 45°; q_O is 0.2; λ is 0.55 μm; the day is June 21; and ρ is 0.0, 0.5, and 0.9.

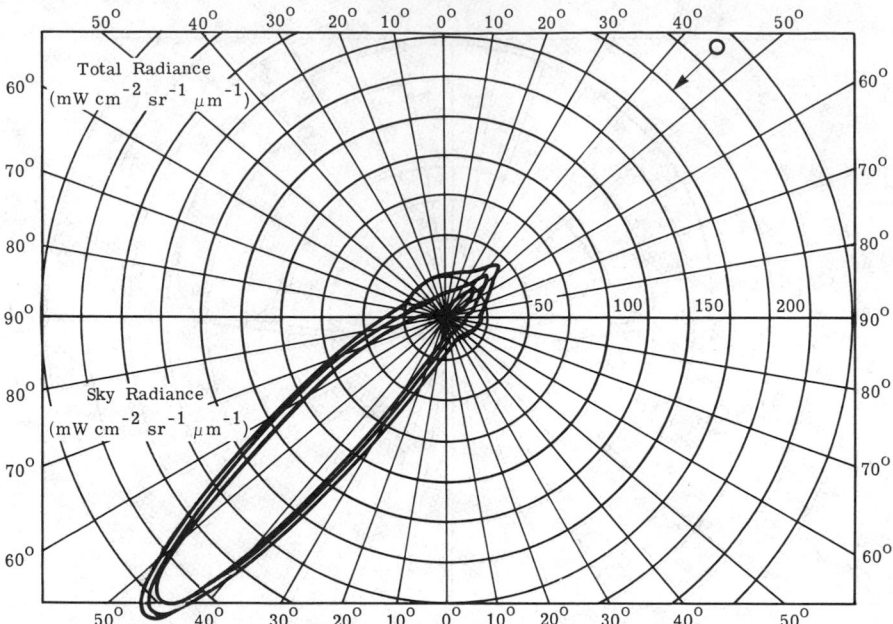

Fig. 4-22. Total radiance and sky radiance as a function of view angle in the solar plane for three reflectances. θ_O is 45°; q_O is 1.0; λ is 0.55 μm; the day is June 21 and ρ is 0.0, 0.5, and 0.9.

from the maximum. Figure 4-27 depicts the same condition except that the optical thickness is 10.0. Here one sees that the radiance "saturates," i.e., above an optical depth of 1.0 there is little change in the radiance.

Finally, the angular distribution of path radiance is illustrated in Figure 4-28 for an atmosphere with an optical thickness of 10.0. In this case, the radiances for all optical depths are well separated near the antisolar peak.

To represent all possible situations with various sun angles, surface reflectances, view angles, wavelengths, and optical states would require an indeterminate number of graphs. From the few graphs presented here, however, a user should be able to find a radiance value for almost any condition he is interested in. For a more detailed value of the radiation field, one should make use of the model itself.

4.7. Optical Properties of the Atmosphere Pertinent to Scattering

4.7.1. Molecular Atmosphere. Molecular scattering of visible radiation follows the well-known Rayleigh laws in which the mass scattering coefficient, $k_S(\lambda, z)$, is (See Goody, Chapter 7.):

$$k_S(\lambda, z) = \frac{8\pi^3 (n^2(\lambda, z) - 1)^2}{3\lambda^4 \, N_{STP}^2 m_m} \frac{(6 + 3\delta)}{(6 - 7\delta)} \tag{4-77}$$

where δ = the depolarization factor due to molecular anisotropy = 0.035

$n(\lambda, z)$ = real index of refraction for air at wavelength, λ, for standard temperature and pressure

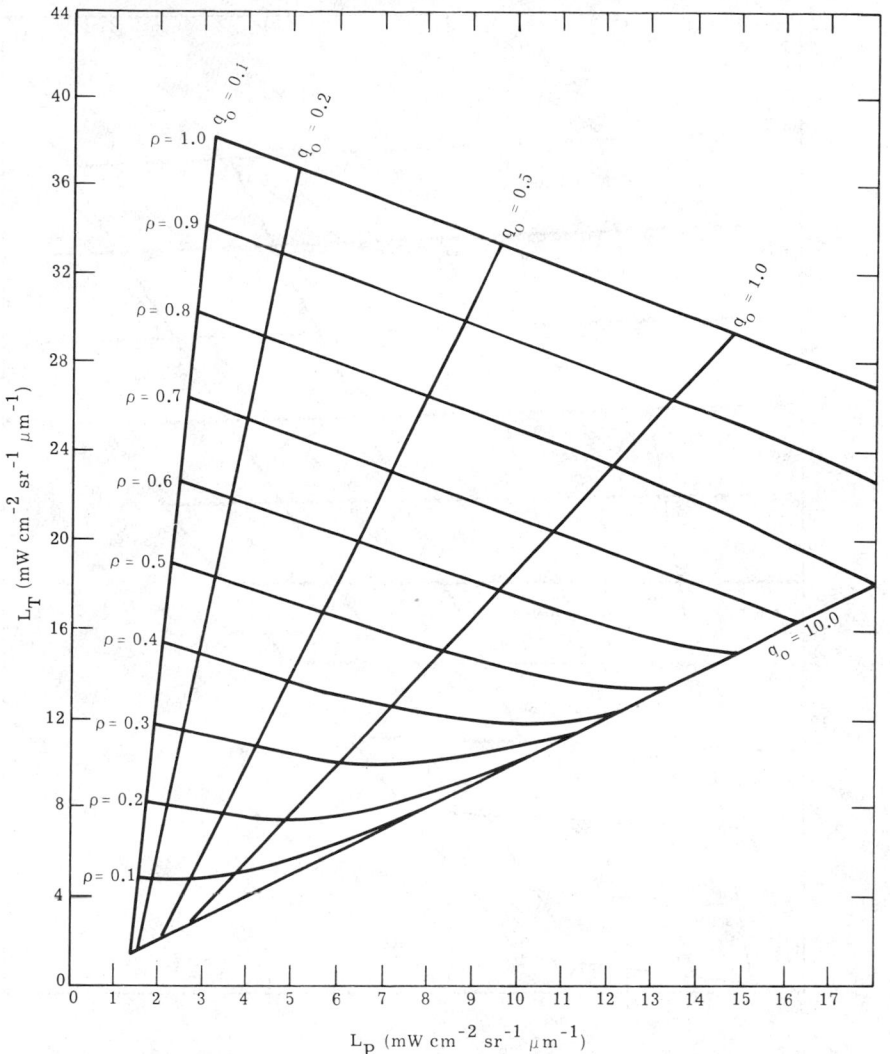

Fig. 4-23. Total radiance versus path radiance in the solar plane for various reflectances and optical thicknesses. θ_o is 45°; θ is 0°; λ is 0.55 μm; and the day is June 21.

N_{STP} = molecular number density at standard temperature and pressure
m_m = mass of an air molecule

The volume scattering coefficient, $K_S(\lambda, z)$, is

$$K_S(\lambda, z) = \frac{8\pi^3(n^2(\lambda, z) - 1)^2 N(z)}{3\lambda^4 N_{STP}^2} \frac{(6 + 3\delta)}{(6 - 7\delta)} \qquad (4\text{-}78)$$

where $N(z)$ is the molecular number density at height z.

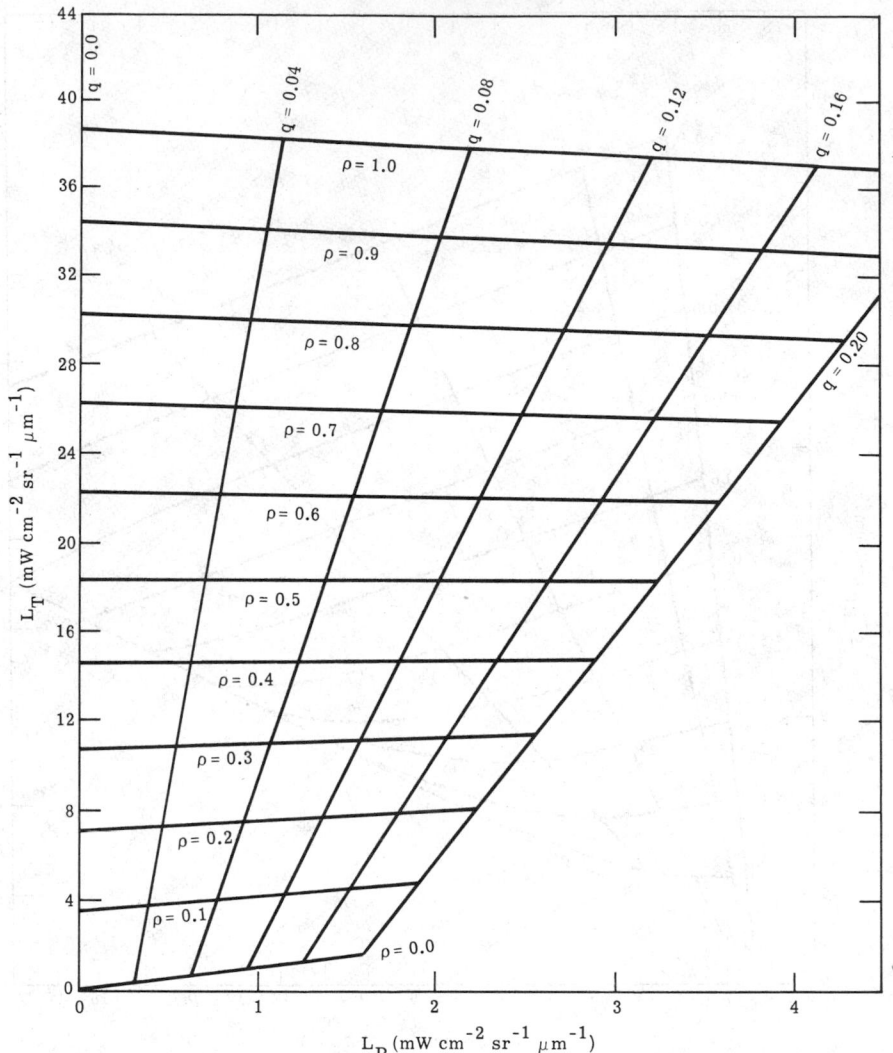

Fig. 4-24. Total radiance versus path radiance in the solar plane for various reflectances and optical depths. θ_o is 45°; θ is 0°; q_o is 0.2; λ is 0.55 μm; and the day is June 21.

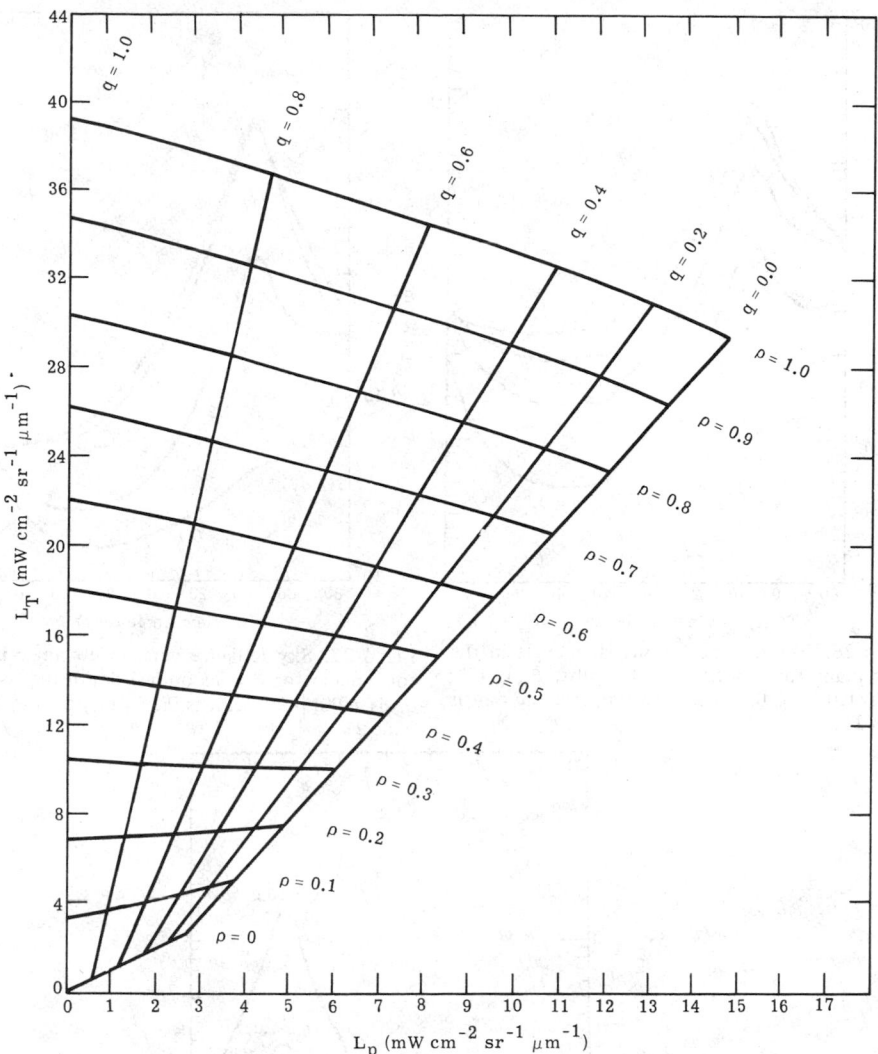

Fig. 4-25. Total radiance versus path radiance in the solar plane for various reflectances and optical depths. θ_O is 45°; θ is 0°; q_O is 1.0; λ is 0.55 μm; and the day is June 21.

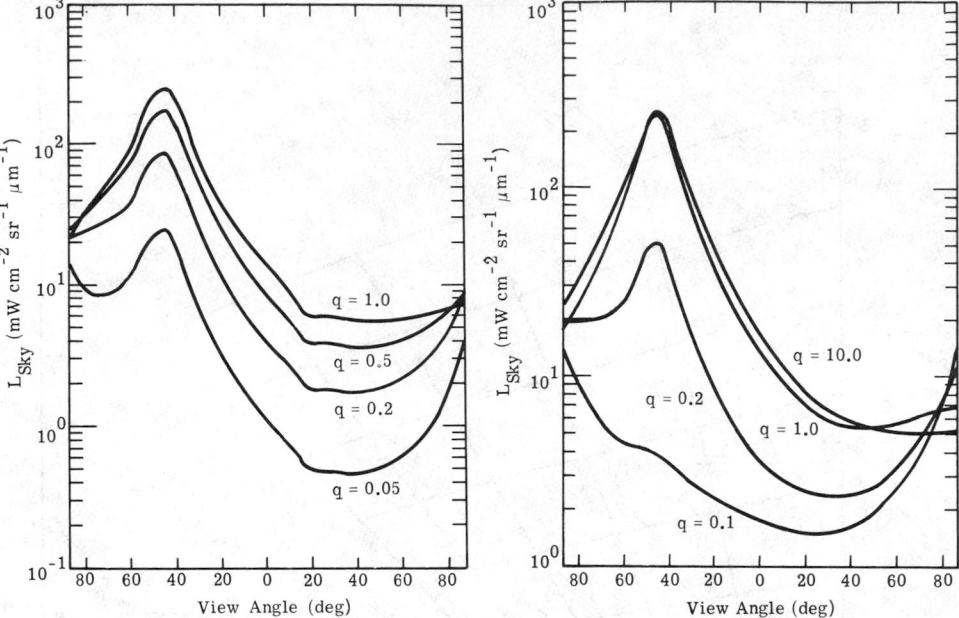

Fig. 4-26. Sky radiance versus view angle in the solar plane for various optical depths. θ_o is 45°; q_o is 1.0; ρ is 0.2; λ is 0.55 μm, and the day is June 21.

Fig. 4-27. Sky radiance versus view angle in the solar plane for various optical depths. θ_o is 45°, q_o is 10.0; ρ is 0.2; λ is 0.55 μm; and the day is June 21.

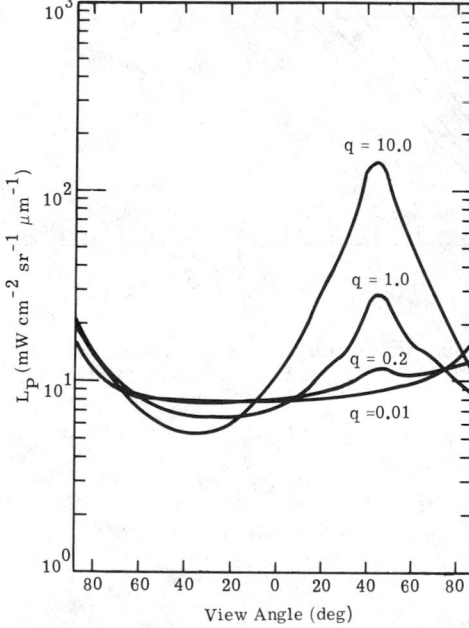

Fig. 4-28. Path radiance versus view angle in the solar plane for various optical depths. θ_o is 45°; q_o is 10.0; ρ is 0.2; λ is 0.55 μm; and the day is June 21.

The wavelength relation of $n(\lambda, z)$ is given by Edlen [4-40] as

$$(n - 1) \times 10^8 = 6432.8 + \frac{2949810}{146 - \left(\frac{1}{\lambda^2}\right)} + \frac{25540}{41 - \left(\frac{1}{\lambda^2}\right)} \tag{4-79}$$

Elterman [4-41] published a set of tables listing Rayleigh volume extinction coefficients and optical depths at selected wavelengths from the near UV to the near IR for altitudes from sea level to 50 km. Table 4-4 is a partial listing of Elterman's results.

The molecular optical depth, q_S, caused by scattering varies directly with the pressure. (See Equation (4-3).) The volume extinction coefficient resulting from molecular constituents varies with N, the molecular number density, and thus directly with air density, ρ. The values listed in Table 4-4 are based upon a standard atmospheric distribution of temperature and pressure with height.

Since the index of refraction variation with wavelength is quite small, and since $K_S(\lambda)$, $k_S(\lambda)$, and q_S all may vary with λ^{-4}, values of these parameters at wavelengths other than those listed in Table 4-4 may be easily computed from the expression

$$\frac{Q_o}{Q_1} = \frac{\lambda_1^4}{\lambda_o^4} \tag{4-80}$$

where Q_o is the known value of $K_s(\lambda)$, $k_s(\lambda)$, or q_s at λ_o, and Q_1 is the desired value at λ_1.

4.7.2. Aerosol Properties. Aerosols are a normal constituent of the atmosphere, with mass-mixing ratios near the surface varying from a few $\mu g\ m^{-3}$ in very clear air to over 100 $\mu g\ m^{-3}$ in polluted air. Aerosols may be introduced into the atmosphere as wind-raised dust and sea salt, products of combustion such as soot, ash, condensed organics, and products formed by chemical reactions within the atmosphere involving gaseous materials such as sulphates, nitrates, H_2S, NH_3, terpenes, etc. Particulates are removed from the atmosphere principally by gravitational fallout, condensation on the particle followed by subsequent rain-out, and capture by falling precipitation. Because of the nature of these removal processes, residence times may vary widely, from minutes to weeks in the troposphere, up to years in the stratosphere, where the removal processes resulting from condensation and precipitation are normally absent.

In the stratospheric regions of the atmosphere, an important component of aerosols is sulfuric acid; at higher levels, above about 30 km, meteoric and cometary dust may be a major constituent. The contribution of these aerosols to the total columnar loading is, however, very small.

To compute the radiative transfer properties or the optical properties of the atmosphere, one must know the properties of the aerosols as well as the molecular constituents. Such properties as particulate size and height distribution, chemical composition, complex index of refraction given by the real (n) and imaginary (n_i) parts in the expression $n - in_i$, etc. are thus of considerable importance. Because of the variety of source and sink mechanisms and the variability of their effectiveness on both time and space, it follows that there will be wide spatial and temporal variations in these properties. Furthermore, only in recent years has enough data on aerosol properties been obtained to begin to gain some insight into their average or typical properties.

Table 4-4. Values of $K_S(\lambda)$ in Units of km^{-1}, and $q_S(\lambda)$ for a Purely Molecular Atmosphere at Selected Wavelengths and Heights* [4-41]

λ (μm)	0.3	0.34	0.40	0.45	0.50	0.55	0.60	0.70	0.80	0.90	1.06
$K_S(\lambda)$ (km^{-1}); z = 0 km	0.1446	0.0849	0.0430	0.0264	0.0172	0.0116	0.0082	0.0044	0.0025	0.0016	0.0008
2 km	0.1188	0.0698	0.0354	0.0217	0.0141	0.0096	0.0067	0.0036	0.0021	0.0013	0.0007
5 km	0.0870	0.0511	0.0259	0.0159	0.0103	0.0070	0.0049	0.0026	0.0015	0.0010	0.0005
10 km	0.0488	0.0287	0.0145	0.0089	0.0058	0.0034	0.0027	0.0015	0.0009	0.0005	0.0003
20 km	0.0105	0.0002	0.0031	0.0019	0.0012	0.0008	0.0006	0.0003	0.0002	0.0001	0.0001
$q_S(\lambda)$; z = 0 km	1.222	0.717	0.354	0.223	0.145	0.098	0.069	0.037	0.021	0.013	0.007
2 km	0.959	0.563	0.285	0.175	0.114	0.077	0.054	0.029	0.017	0.010	0.005
5 km	0.652	0.383	0.194	0.119	0.077	0.052	0.037	0.020	0.011	0.005	0.004
10 km	0.320	0.188	0.095	0.059	0.038	0.026	0.018	0.010	0.006	0.004	0.002
20 km	0.067	0.039	0.020	0.012	0.008	0.005	0.004	0.002	0.001	0.001	0.000

*The scattering coefficient used in this table is $K_S(\lambda)$, related to $k_S(\lambda)$ by: $K_S(\lambda) = k_S(\lambda) \cdot \rho$.

4.7.3. Atmospheric Aerosol Models.

Of the parameters required to describe the interaction of atmospheric aerosols with electromagnetic radiation, particle size distribution has probably received the most attention. The literature on the subject is quite lengthy, providing useful input to certain theoretical models that have been developed to fit the bulk of the observations. At the end of this chapter, references to some of the experimental work are given.

Junge's model [4-42] is

$$\frac{dN(r)}{dr} = c\, r^{-(\beta+1)} \qquad (4\text{-}81)$$

where $dN(r)$ = number of particles per unit volume of radius between r and $r + dr$

c = normalizing constant to adjust the total number of particles per unit volume

β = shaping constant

Most measured size distributions can best be fit by values of β in the range $2 \leqslant \beta \leqslant 4$, for aerosols whose radii lie between 0.1 and 10 μm. Experimentally, it is known that aerosol numbers drop off rapidly below some radii less than 0.1 μm. (See, for example, Figure 4-29.) A plot of $dN(r)/d(\ln r)$ versus r on a log scale from Equation (4-81) yields a straight line of slope $-\beta$.

To allow for the dropoff of particle number density at small radii, and also to allow for more versatility, Deirmendjian [4-43] introduced the so-called modified gamma function of the form:

$$\frac{dN(r)}{dr} = ar^{\alpha}\,\exp(-br^{\beta}) \qquad (4\text{-}82)$$

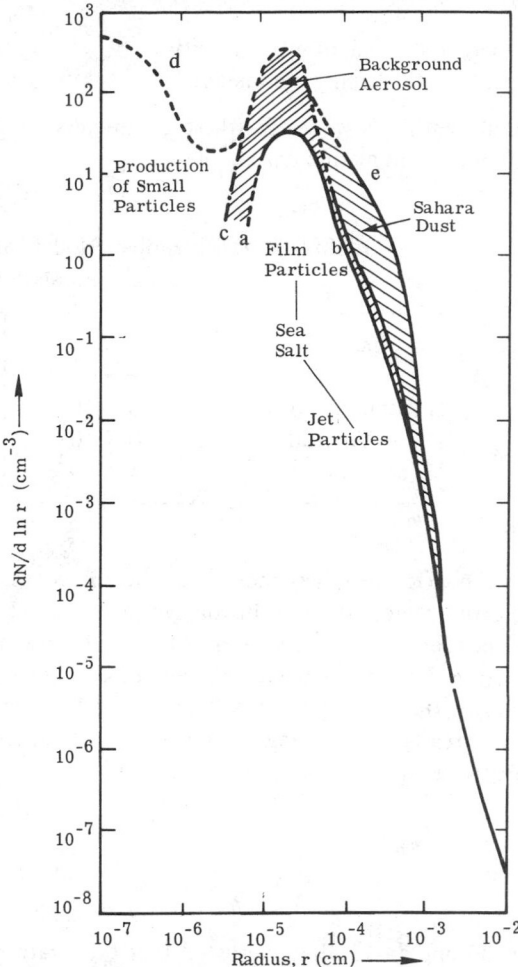

Fig. 4-29. Idealized diagram of the various components of the undisturbed marine aerosol. The diagram combines the information from several sources and tries to represent the present status of our knowledge. Below line (a) is the sea spray component separated by line (b) into film and jet particles. Line (b) is very uncertain and tries to demonstrate the fact that there is considerable overlapping of the two production mechanisms. Between line (c) and line (a) is the background aerosol, some part of which at least is of continental origin. Above about 10^{-4} cm the aerosol consists of about 15% of insoluble particles (at least over the Atlantic). Below 10^{-4} cm it consists most likely of sulfate, perhaps $(NH_4)_2SO_4$. Between lines (c) and (d) there is some indication of a component, which if confirmed may represent a steady-state distribution due to continuous production of very small particles. In air masses originating in Sahara dust storms the additional dust component is represented by line (e). All curves below about 10^{-4} cm are broken because they are rather uncertain [4-42].

where a = total number density
α, b, β = shaping parameters

Deirmendjian's values for these parameters are given in Table 4-5 for stratospheric, continental, and marine hazes.

Table 4-5. Deirmendjian Model Parameters for Stratospheric,
Continental, and Marine Hazes [4-43]

Hazes	a	α	b	β
H (Stratosphere)	4.0×10^5	2	20.0	1
L (Continental)	5.0×10^6	2	15.1	1/2
M (Marine)	5.3×10^4	1	8.9	1/2

The Deirmendjian model has the advantage of proper behavior as r approaches zero. Furthermore, the continental haze model (Table 4-5) takes on an approximate r^{-4} dependence for radii greater than about 0.6 μm, in agreement with the Junge power law. Figure 4-30 shows curves of these three aerosol models. Kuriyan and Sekera [4-44] have shown that, by letting α = 2.0 and β = 1.0, the continental and marine distributions are not greatly changed regarding their effects on atmospheric scattering. Therefore, all three distributions may be approximated:

$$\frac{dN(r)}{dr} = ar^2 \exp(-br) \tag{4-83}$$

Friend [4-45] has suggested that the stratospheric aerosol may best be described by a so-called log-normal, or Zold, distribution function given by

$$\frac{dN(r)}{dr} = \frac{c}{\sigma r \sqrt{2\pi}} e^{-(\ln r - \ln \bar{r})^2/2\sigma^2} \tag{4-84}$$

where $\bar{r}$ = mean radius
σ = natural logarithm of the standard deviation

Friend determined values of $\bar{r}$ = 0.3 μm and σ = 1.3 to give a best fit to his measurements. More recently, Toon and Pollack [4-46] (based upon the experimental work of Ferry and Lem, Kondratiev, et al., Zuyev, et al., and Bigg, et al.) determined that a best fit to the data is σ = 2.0 and $\bar{r}$ = 0.035 μm. Shettle and Fenn [4-47], on the other hand, have chosen a modified gamma-function to represent the stratospheric aerosol, although with different values of the parameters than those used by Deirmendjian. The work of Shettle and Fenn is described in more detail below. Figure 4-31 shows the modified gamma-function, Haze H model of Deirmendjian [4-43] and the log normal distribution of Toon and Pollack [4-46]. The two disagree considerably below 0.3 μm, a point discussed in the above referenced work of Toon and Pollack.

To utilize the extensive measurements which are now available, Shettle and Fenn [4-47] have recently constructed a series of aerosol models for different environmental conditions and seasons. They have divided their aerosol models into four altitude regimes:

(1) The boundary layer below 2 km in which 10 models are described for several surface visibilities and for rural, urban, and maritime environments.

(2) The troposphere, above, 2 km, in which spring-summer and fall-winter models are described.

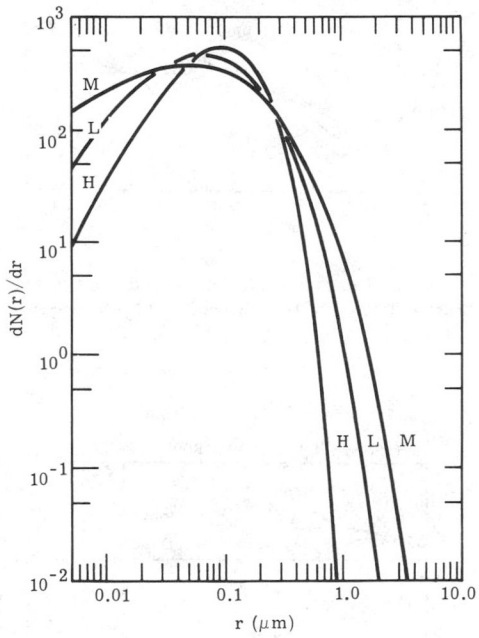

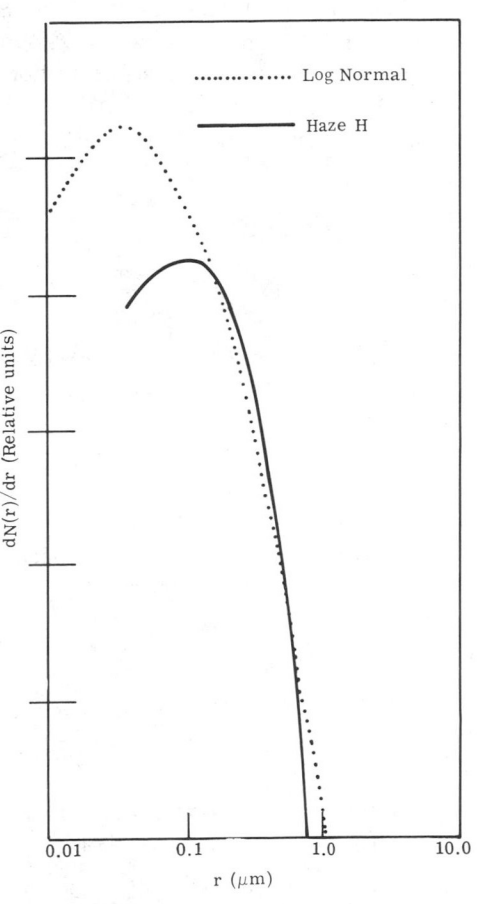

Fig. 4-30. Curves of $dN(r)/dr$ versus r for the modified gamma function for marine continental, and stratosphere aerosol models [4-43].

Fig. 4-31. Plots of $dN(r)/dr$ versus r for Haze H stratospheric aerosol model, and the log normal, or zold distribution for the stratosphere [4-43, 4-46].

(3) The stratosphere, up to 30 km, in which background conditions, as well as moderate, high, and extreme volcanic conditions, for the two seasons as above are described.

(4) Altitudes above 30 km, in which two models are described, one for background conditions and one for high aerosol concentrations.

In the boundary layer, the "rural model" is intended to represent conditions in clean, continental areas. Based on measurements by Volz [4-48], Shettle and Fenn [4-47] chose an example in this layer composed of 70% water-soluble substances (i.e., ammonium and calcium sulfate and organic compounds) and 30% dustlike aerosols. Figures 4-32, 4-33, and 4-34 show the real and imaginary parts of the index of refraction for these substances. The different curves for similar substances in these figures represent measurements taken from different samples. Shettle and Fenn are not clear as to how a composite index was determined for a given substance, but presumably some sort of average was used when more than one set of data was available. Since the data presented by Volz (Figures 4-32 through 4-34) extend only down to 2.5 μm, it is also not clear what values of the indices were used in the near IR and visible portions of the spectrum.

For their size distribution, Shettle and Fenn chose a function which is the sum of two log-normal distributions given by

$$\frac{dN(r)}{dr} =$$

$$\sum_{i=1}^{2} \left\{ \frac{N_i}{\sigma_i r \sqrt{2\pi}} \exp\left[-\frac{(\ln r - \ln \bar{r}_c)^2}{2\sigma_i^2} \right] \right\}$$

$$(4\text{-}85)$$

where the terms are as earlier defined for the log-normal distribution. This model was used to reproduce experimental evi-

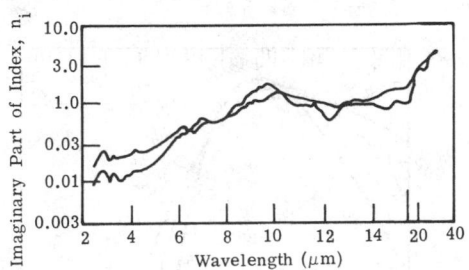

Fig. 4-32. Imaginary part of the index of refraction versus λ for two dust-like aerosols [4-48].

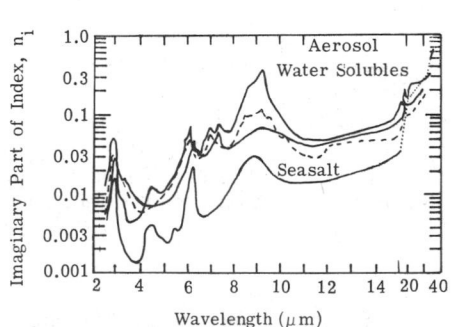

Fig. 4-33. Imaginary part of the index of refraction versus λ for water soluble substances and sea salts [4-48].

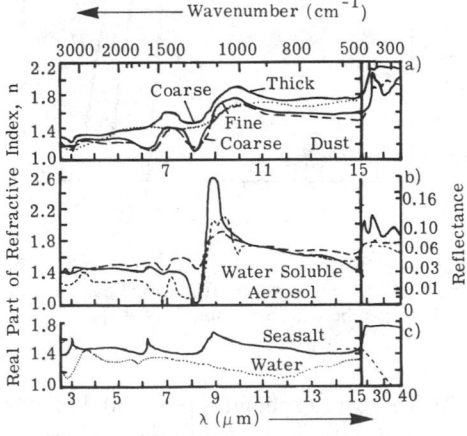

Fig. 4-34. Real part of the index of refraction versus λ of dust-like aerosols, water soluble substances, and sea salt [4-48].

dence (Whitby, et al.) of a bimodel structure in $dV/(d \ln r)$ versus $\log r$, where V is the cumulative particle volume density, with one broad peak occurring for particle diameters between 0.1 and 1.0 μm and another between 5 and 100 μm.

For their urban model, Shettle and Fenn added combustion and industrial aerosols to the rural model in the lowest 2 km. The added aerosols were assumed to have the same size distribution as the rural model and are mixed in the proportion of 35% urban to 65% rural aerosol. The index of refraction of the added urban aerosols was based on the data of Twitty and Weinman [4-49] who give an index (1.8 – 0.5i) for the spectral range 0.15 to 15.0 μm.

Finally, for their maritime model, Shettle and Fenn assumed a mixture of evaporation products of sea spray and continental-produced aerosols. The sea-spray component was assumed to exist at humidities of 80%, with appropriately weighed indices of refraction for water and sea salt (Figure 4-33 and Table 4-6). Table 4-7 gives the parameters used for the size distributions, given by Equation (4-85) for the three boundary-layer-aerosol models.

In the troposphere, above the boundary layer, the rural aerosol model is used, but without the large particle distribution (i.e., $N_2 = 0$) which is assumed to have settled out. Therefore, the tropospheric parameters are $N_1 = 1.0$, $r_1 = 0.005$ μm, $\sigma_1 = 0.457$, and $N_2 = 0$.

Stratospheric aerosol measurements indicate that there is a more-or-less steady background of aerosols, occasionally perturbed by factors of 100 or more by volcanic eruptions. The background aerosol was taken to be a 75% solution of sulfuric acid (See also

Table 4-6. Optical Constants of Water in the 0.4 to 14.0 μm Range [4-50]

$\lambda(\mu m)$	n	n_i	$\lambda(\mu m)$	n	n_i
0.400	1.339	1.86×10^{-9}	0.775	1.330	1.48×10^{-7}
0.425	1.338	1.30×10^{-9}	0.800	1.329	1.25×10^{-7}
0.450	1.337	1.02×10^{-9}	0.825	1.329	1.82×10^{-7}
0.475	1.336	9.35×10^{-10}	0.850	1.329	2.93×10^{-7}
0.500	1.335	1.00×10^{-9}	0.875	1.328	3.91×10^{-7}
0.525	1.334	1.32×10^{-9}	0.900	1.328	4.86×10^{-7}
0.550	1.330	1.96×10^{-9}	0.925	1.328	1.06×10^{-6}
0.575	1.333	3.60×10^{-9}	0.950	1.327	2.93×10^{-6}
0.600	1.332	1.09×10^{-8}	0.975	1.327	3.48×10^{-6}
0.625	1.332	1.39×10^{-8}	1.000	1.327	2.89×10^{-6}
0.650	1.331	1.64×10^{-8}	1.20	1.324	9.89×10^{-6}
0.675	1.331	2.23×10^{-8}	1.40	1.321	1.38×10^{-4}
0.700	1.331	3.35×10^{-8}	1.60	1.317	8.55×10^{-5}
0.725	1.330	9.15×10^{-8}	1.80	1.312	1.15×10^{-4}
0.750	1.330	1.56×10^{-7}	2.00	1.306	1.10×10^{-3}

Table 4-6. Optical Constants of Water in the 0.4 to 14.0 μm Range [4-50] (*Continued*)

$\lambda(\mu m)$	n	n_i	$\lambda(\mu m)$	n	n_i
2.20	1.296	2.89×10^{-4}	5.00	1.325	0.0124
2.40	1.279	9.56×10^{-4}	5.10	1.322	0.0111
2.60	1.242	3.17×10^{-3}	5.20	1.317	0.0101
2.65	1.219	6.70×10^{-3}	5.30	1.312	0.0098
2.70	1.188	0.019	5.40	1.305	0.0103
2.75	1.157	0.059	5.50	1.298	0.0116
2.80	1.142	0.115	5.60	1.289	0.0142
2.85	1.149	0.185	5.70	1.277	0.0203
2.90	1.201	0.268	5.80	1.262	0.0330
2.95	1.292	0.298	5.90	1.248	0.0622
3.00	1.371	0.272	6.00	1.265	0.107
3.05	1.426	0.240	6.10	1.319	0.131
3.10	1.467	0.192	6.20	1.363	0.0880
3.15	1.483	0.135	6.30	1.357	0.0570
3.20	1.478	0.0924	6.40	1.347	0.0449
3.25	1.467	0.0610	6.50	1.339	0.0392
3.30	1.450	0.0368	6.60	1.334	0.0356
3.35	1.432	0.0261	6.70	1.329	0.0337
3.40	1.420	0.0195	6.80	1.324	0.0327
3.45	1.410	0.0132	6.90	1.321	0.0322
3.50	1.400	0.0094	7.00	1.317	0.0320
3.60	1.385	0.00515	7.10	1.314	0.0320
3.70	1.374	0.00360	7.20	1.312	0.0321
3.80	1.364	0.00340	7.30	1.309	0.0322
3.90	1.357	0.00380	7.40	1.307	0.0324
4.00	1.351	0.00460	7.50	1.304	0.0326
4.10	1.346	0.00562	7.60	1.302	0.0328
4.20	1.342	0.00688	7.70	1.299	0.0331
4.30	1.338	0.00845	7.80	1.297	0.0335
4.40	1.334	0.0103	7.90	1.294	0.0339
4.50	1.332	0.0134	8.00	1.291	0.0343
4.60	1.330	0.0147	8.20	1.286	0.0351
4.70	1.330	0.0157	8.40	1.281	0.0361
4.80	1.330	0.0150	8.60	1.275	0.0372
4.90	1.328	0.0137	8.80	1.269	0.0385

Table 4-6. Optical Constants of Water in the 0.4 to 14.0 μm Range [4-50] (*Continued*)

$\lambda(\mu m)$	n	n_i	$\lambda(\mu m)$	n	n_i
9.00	1.262	0.0399	12.5	1.123	0.259
9.20	1.255	0.0415	13.0	1.146	0.305
9.40	1.247	0.0433	13.5	1.177	0.343
9.60	1.239	0.0454	14.0	1.210	0.370
9.80	1.229	0.0479	–	–	–
10.0	1.218	0.0508	–	–	–
10.5	1.185	0.0662	–	–	–
11.0	1.153	0.0968	–	–	–
11.5	1.126	0.142	–	–	–
12.0	1.111	0.199	–	–	–

Table 4-7. Size Distribution Parameters (Normalized to 1 Particle/cm^3) [4-47]

Type of Aerosol	N_1^*	r_1 (μm)	σ_1	N_2^*	r_2 (μm)	σ_2
Rural	0.9999975	0.005	0.475	2.5×10^{-6}	0.5	0.475
Urban	0.9999975	0.005	0.475	2.5×10^{-6}	0.5	0.475
Maritime (continental origin)	1.0	0.005	0.475	0	–	–
Maritime (sea spray origin)	1.0	0.3	0.4	0	–	–

*$N_1 + N_2 = 1$

Rosen (1971) and Toon and Pollack (1973.), whose complex index of refraction was based upon the work of Remsberg (1971), (1973) and Palmer and Williams [4-51]). The latter data are shown in Table 4-10. The refractive index of the volcanic models, based upon the work of Volz [4-52], is shown in Figure 4-35. The size distribution chosen for the stratospheric aerosol is a modified gamma distribution, based upon the work of Mossop, given by Equation (4-82).

Table 4-8 gives the parameters used for background, fresh volcanic and aged, volcanic, stratospheric, aerosol distributions.

The final classification of Shettle and Fenn is for the upper atmosphere, above 30 km. These aerosols, which represent only a very small fraction of the total, are considered to be primarily meteoric dust (Newkirk and Eddy (1964) Rosen (1969)). The refractive index for these particles was based on the work of Volz and Shettle, and the size distribution was taken as log normal with parameters $N_1 = 1.0$, $r_1 = 0.03$ μm, $\sigma_1 = 0.5$, and $N_2 = 1$. The upper atmosphere region is represented with a normal aerosol distribution with height, and an extreme model. The extreme model is used to represent those occasion's when this region is invaded with micrometeoric dust or noctilucent clouds.

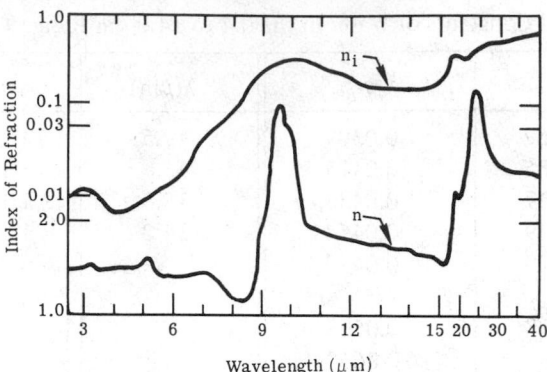

Fig. 4-35. Real, n, and imaginary, n_i, parts of the index of refraction of volcanic dust versus λ [4-52].

Table 4-8. Size Distributions for Stratospheric Aerosols [4-47]

	a	α	β	b
Background stratospheric	324.0	1	1	18
Fresh volcanic	341.33	1	1/2	8
Aged volcanic	5461.33	1	1/2	16

Figures 4-36 and 4-37 show the extinction coefficient at $\lambda = 0.55$ μm as a function of height for the various models for the spring-summer and fall-winter seasons, respectively. Also shown in these figures is the Elterman [4-41] model for comparison. This model, as readily can be seen, is more typical of moderate volcanic conditions in the 10 to 30 km region, which indeed existed during the period when most of the measurements which went into Elterman's model were made. In the lowest levels, five distributions are given, representing visibilities of 50, 23, 10, 5, and 2 km at the surface.

Figures 4-38 through 4-41 show aerosol extinction, absorption, and scattering coefficients as functions of wavelength for the urban, maritime, rural, and tropospheric aerosol models. These cross-sections were computed from Mie theory, using the size distributions and indices of refraction as previously described for each model. In each case, the total number of particles per unit volume was normalized to yield an extinction coefficient of 0.155 km^{-1} at $\lambda = 0.55$ μm, which corresponds to a 23 km visibility. Figure 4-42 gives the total extinction cross-sections per unit volume (normalized as before as a function of wavelength) for background stratosphere aerosols, fresh volcanic aerosols, and aged volcanic aerosols, after some settling out has occurred. There are significant differences in the wavelength dependencies of the different models. These differences are caused by the differing chemical compositions of the aerosols (thus giving different index or refraction variations with wavelength), as well as by the different size distributions used for the different models. It thus becomes a matter of some concern to choose a representative model for calculating atmospheric optical properties. To further

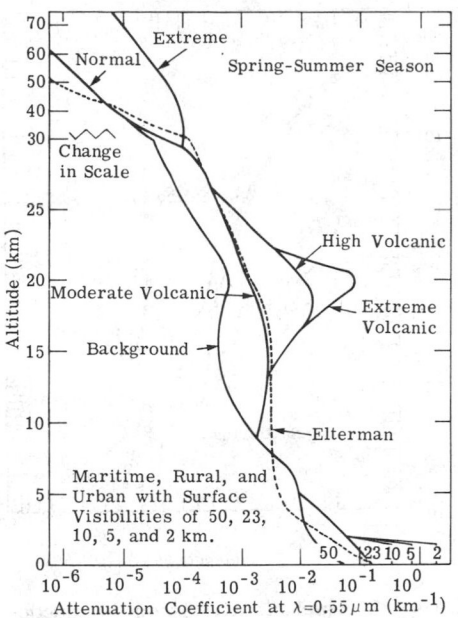

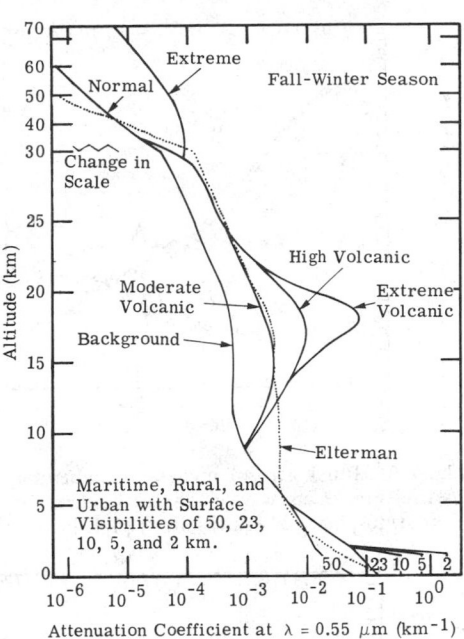

Fig. 4-36. Extinction coefficient as a function of height for the various models for the spring-summer season [4-47]

Fig. 4-37. Extinction coefficient as a function of height for the various models for the fall-winter season [4-47].

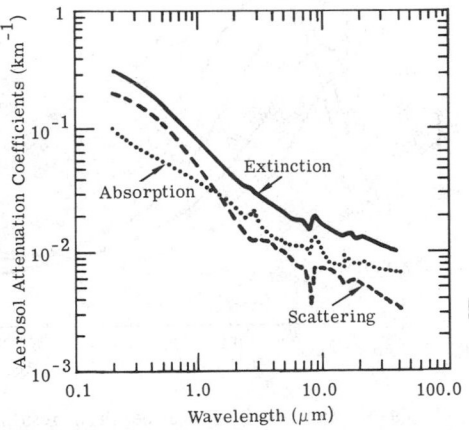

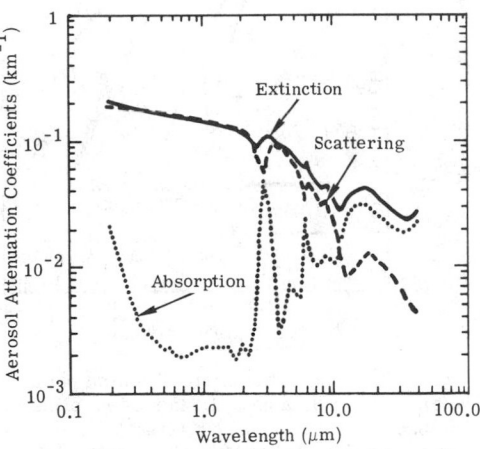

Fig. 4-38. Urban aerosol model. Attenuation coefficients versus wavelength for moderately clear atmosphere (23 km visibility) [4-47].

Fig. 4-39. Maritime aerosol model. Attenuation coefficients versus wavelength for moderately clear atmosphere (23 km visibility) [4-47].

demonstrate this point, Figure 4-43 shows the increase in optical depth resulting from an increase of 1 μg m^{-3} of particulates through a 10 km thick stratospheric layer. The calculations were made assuming the particulates have a log-normal size distribution given by Equation (4-84). Total extinction cross-sections per unit volume were computed at

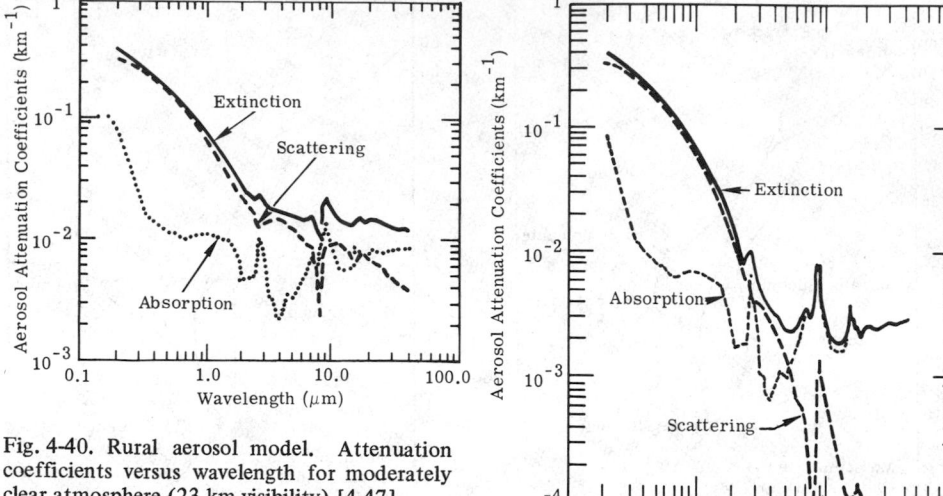

Fig. 4-40. Rural aerosol model. Attenuation coefficients versus wavelength for moderately clear atmosphere (23 km visibility) [4-47].

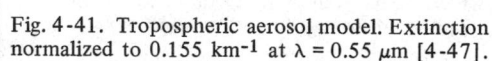

Fig. 4-41. Tropospheric aerosol model. Extinction normalized to 0.155 km^{-1} at $\lambda = 0.55$ μm [4-47].

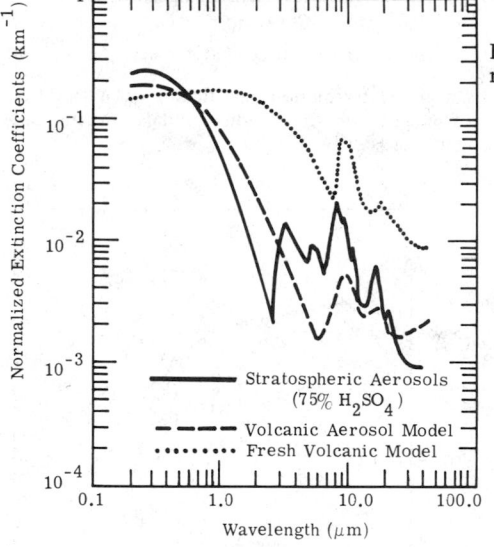

Fig. 4-42. Total extinction cross-sections per unit volume (normalized as a function of wavelength) for background stratosphere aerosols, fresh volcanic aerosols and aged volcanic aerosols [4-53].

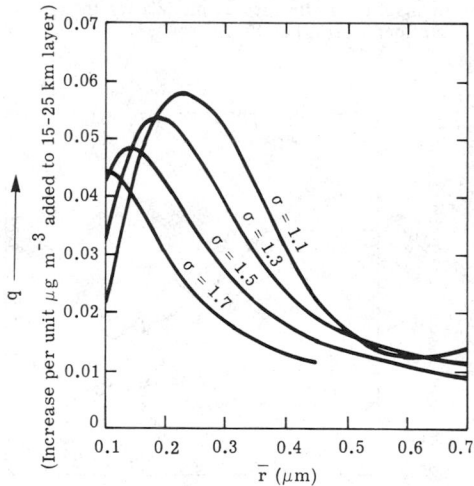

Fig. 4-43. Increase in optical depth, q, resulting from an increase of 1 μg m^{-3} of particles through a 10 km thick stratospheric layer for different values of σ, the natural logarithm of the standard deviation [4-53].

small intervals of radius and summed over the size distribution function from $r = 0.01$ to 10.0 μm, and for $\lambda = 5,000$ Å. The abscissa is the value of $\bar{r}$, while the different curves are for the value of σ used in the distribution function. This figure demonstrates the wide range of resulting q, all for the same mass of aerosols, which can result from different size distribution functions.

The recent work of Toon and Pollack [4-46] presents a global average model for atmospheric aerosols to be used for calculating radiative transfer. Their model is intended to be a composite average based on available data. Table 4-9 presents the pertinent features of their model. The second column gives the optical depth down to the corresponding height in column 1, based upon their model, at a wavelength of $\lambda = 0.55\ \mu m$. For their height profile they have adopted a slightly modified form given by Elterman, et al. The indices of refraction were based upon the compositions listed in column 3 and are presented in the form of single scattering albedos. Tables 4-6 and 4-10 through 4-14 list experimental values for some common atmospheric aerosol substances. Table 4-11 lists experimental values for dust as well as determinations of optical constants of quartz. In this latter set of data, values for both ordinary and extraordinary rays are listed with the subscripts o and e, respectively. Table 4-12 lists data on the variation with relative humidity of the optical constants of both urban and maritime aerosols.

Table 4-9. A Global Average Model of Atmospheric Aerosols [4-46]

h (km)	q ($\lambda = 0.55\ \mu m$)	Composition	Size Distribution
0 1 2 3	0.125 0.083 0.058 0.042	15% sea salt (NaCl) 35% soil (basalt) 50% sulfate $[(NH_4)_2SO_4]$	$dN(r)/dr \propto r^{-(\beta+1)}$ $r < 0.1\ \mu m,\ \beta = -1;\ 0.1 < r < 5\ \mu m,\ \beta = 2$ $5\ \mu m < r < 50\ \mu m,\ \beta = 4;\ 50\ \mu m < r,\ dN/dr = 0$
4 5 6 7 8 9 10 11 12	0.034 0.028 0.024 0.021 0.017 0.014 0.011 0.008 5.0×10^{-3}	40% soil (basalt) 60% sulfate $[(NH_4)_2SO_4]$	$dN(r)/dr \propto r^{-(\beta+1)}$ $r < 0.045\ \mu m,\ \beta = -1;$ $0.045 < r < 5\ \mu m,\ \beta = 2.6;$ $5 < r < 30\ \mu m,\ \beta = 4.6;$ $30\ \mu m < r,\ dN(r)/dr = 0.$
15 18 21 24 27 30 33 36 39 42 45	3.3×10^{-3} 1.8×10^{-3} 8.3×10^{-4} 4.2×10^{-4} 2.0×10^{-4} 8.7×10^{-5} 3.8×10^{-5} 1.6×10^{-5} 6.4×10^{-6} 2.0×10^{-6} 0.0	100% sulfate (75% H_2SO_4, 25% H_2O liquid solution)	$dN(r)/dr \propto \exp\{[-\ln^2(r/r_m)]/(2\ln^2\sigma)\}$ $\sigma = 2, r_m = 0.035\ \mu m$

Table 4-10. Optical Constants of Sulfuric Acid From Near Infrared to Ultraviolet [4-51]

ν (cm⁻¹)	25% n	25% n_i	38% n	38% n_i	50% n	50% n_i	75% n	75% n_i	84.5% n	84.5% n_i	95.6% n	95.6% n_i	λ (µm)
4000	1.286	—	1.300	—	1.311	—	1.344	3.76×10^{-3}	1.358	3.38×10^{-3}	1.368	2.11×10^{-3}	2.500
4100	1.296	—	1.309	—	1.321	—	1.352	2.97	1.363	2.79	1.372	1.86	2.439
4200	1.304	—	1.316	—	1.328	—	1.358	2.41	1.368	2.43	1.376	1.71	2.381
4300	1.310	—	1.321	—	1.333	—	1.362	2.09	1.374	2.13	1.379	1.62	2.326
4400	1.315	—	1.326	—	1.338	—	1.367	1.86	1.378	1.94	1.383	1.41	2.273
4500	1.319	—	1.330	—	1.342	—	1.370	1.67	1.382	1.69	1.385	1.29	2.222
4600	1.323	—	1.333	—	1.346	—	1.374	1.54	1.384	1.56	1.388	1.15	2.174
4700	1.326	—	1.336	—	1.348	—	1.377	1.43	1.388	1.44	1.391	1.02	2.128
4800	1.328	—	1.339	—	1.351	—	1.380	1.35	1.390	1.34	1.393	9.47×10^{-4}	2.083
4900	1.329	—	1.341	—	1.353	—	1.382	1.30	1.392	1.28	1.394	8.93	2.041
5000	1.331	—	1.343	—	1.355	—	1.384	1.26	1.392	1.19	1.396	8.37	2.000
5100	1.332	—	1.345	—	1.357	—	1.386	1.24	1.394	1.10	1.398	7.41	1.961
5200	1.334	—	1.346	—	1.358	—	1.388	1.11	1.396	9.61×10^{-4}	1.399	6.58	1.923
5300	1.335	—	1.347	—	1.359	—	1.389	7.96×10^{-4}	1.397	7.58	1.400	5.86	1.887
5400	1.336	—	1.348	—	1.361	—	1.391	5.95	1.398	6.26	1.401	5.33	1.852
5500	1.338	2.50×10^{-4}	1.350	—	1.362	—	1.392	5.37	1.399	5.66	1.403	4.93	1.818
5600	1.339	2.37	1.351	2.87×10^{-4}	1.364	—	1.393	4.86	1.400	4.99	1.404	4.52	1.786
5700	1.340	2.09	1.352	2.68	1.365	3.00×10^{-4}	1.394	4.24	1.401	4.44	1.404	4.02	1.754
5800	1.341	1.87	1.353	2.37	1.366	2.65	1.396	3.61	1.402	3.76	1.405	3.48	1.724
5900	1.342	1.71	1.355	2.16	1.367	2.37	1.397	3.14	1.403	3.21	1.406	2.99	1.695
6000	1.342	1.58	1.356	1.98	1.368	2.18	1.398	2.72	1.404	2.76	1.407	2.55	1.667
6100	1.343	1.50	1.357	1.85	1.369	2.01	1.398	2.34	1.404	2.32	1.408	2.14	1.639
6200	1.344	1.48	1.358	1.73	1.370	1.87	1.399	2.02	1.405	1.95	1.409	1.81	1.613
6300	1.345	1.49	1.358	1.69	1.371	1.76	1.400	1.76	1.406	1.67	1.410	1.52	1.587
6400	1.346	1.55	1.359	1.72	1.372	1.68	1.402	1.55	1.407	1.41	1.410	1.24	1.563
6500	1.346	1.69	1.360	1.80	1.373	1.74	1.403	1.38	1.408	1.22	1.411	9.88×10^{-5}	1.538
6600	1.346	1.87	1.360	1.94	1.373	1.82	1.403	1.25	1.409	1.09	1.412	7.85	1.515
6700	1.347	2.17	1.361	2.14	1.374	1.91	1.404	1.17	1.410	9.41×10^{-5}	1.413	6.21	1.493
6800	1.348	2.43	1.362	2.28	1.375	1.99	1.405	1.10	1.410	8.13	1.414	4.53	1.471
6900	1.348	2.54	1.363	2.19	1.376	1.83	1.406	1.02	1.411	6.95	1.415	3.40	1.449

Table 4-10. Optical Constants of Sulfuric Acid From Near Infrared to Ultraviolet [4-51] (Continued)

ν (cm^{-1})	25% n	25% n_i	38% n	38% n_i	50% n	50% n_i	75% n	75% n_i	84.5% n	84.5% n_i	95.6% n	95.6% n_i	λ (μm)
7000	1.349	2.36	1.363	1.72	1.377	1.47	1.406	8.78×10^{-5}	1.411	5.62	1.416	2.54	1.429
7100	1.349	1.60	1.364	1.04	1.377	8.80×10^{-5}	1.407	6.16	1.412	3.80	1.416	1.98	1.408
7200	1.350	7.52×10^{-5}	1.364	6.61×10^{-5}	1.377	5.35	1.408	3.89	1.413	2.61	1.416	1.65	1.389
7300	1.351	4.88	1.365	4.41	1.378	3.80	1.409	2.59	1.413	2.01	1.417	1.38	1.370
7400	1.351	3.37	1.366	3.06	1.378	2.69	1.410	1.98	1.414	1.53	1.418	1.15	1.351
7500	1.351	2.24	1.366	2.10	1.379	1.97	1.410	1.59	1.415	1.36	1.419	9.44×10^{-6}	1.333
7600	1.352	1.66	1.367	1.60	1.380	1.54	1.411	1.27	1.416	1.12	1.420	7.85	1.316
7700	1.352	1.33	1.367	1.29	1.380	1.26	1.411	1.05	1.416	9.50×10^{-6}	1.420	6.67	1.299
7800	1.353	1.13	1.368	1.10	1.381	1.07	1.412	8.93×10^{-6}	1.416	8.06	1.421	5.61	1.282
7900	1.353	1.03	1.368	9.74×10^{-6}	1.382	9.48×10^{-6}	1.413	7.75	1.417	6.88	1.422	4.68	1.266
8000	1.354	1.00	1.368	9.23	1.382	8.85	1.413	6.94	1.418	6.00	1.422	3.95	1.250
8200	1.355	1.02	1.369	8.98	1.383	8.22	1.415	5.63	1.419	4.72	1.422	2.91	1.220
8400	1.355	9.95×10^{-6}	1.369	8.83	1.384	7.90	1.416	4.95	1.420	3.85	1.423	2.29	1.190
8600	1.356	8.41	1.371	8.02	1.384	7.25	1.416	4.05	1.422	3.01	1.423	1.94	1.163
8800	1.357	4.46	1.371	4.04	1.385	3.73	1.417	2.46	1.422	2.17	1.424	1.69	1.136
9000	1.357	1.92	1.372	1.85	1.386	1.87	1.418	1.84	1.423	1.77	1.425	1.51	1.111
9200	1.358	1.44	1.373	1.43	1.387	1.45	1.419	1.60	1.424	1.60	1.425	1.44	1.087
9400	1.358	1.30	1.373	1.29	1.387	1.30	1.420	1.50	1.425	1.52	1.426	1.40	1.064
9600	1.358	1.49	1.374	1.39	1.388	1.37	1.421	1.48	1.425	1.46	1.426	1.32	1.042
9800	1.358	2.01	1.375	1.75	1.389	1.62	1.421	1.52	1.426	1.37	1.427	1.11	1.020
10000	1.359	2.75	1.375	2.36	1.389	2.09	1.422	1.53	1.427	1.19	1.427	8.67×10^{-7}	1.000
10200	1.359	3.23	1.376	2.86	1.390	2.40	1.422	1.41	1.427	9.67×10^{-7}	1.427	6.20	0.980
10400	1.360	2.91	1.377	2.51	1.390	1.97	1.423	1.03	1.428	7.51	1.427	4.53	0.962
10600	1.360	1.39	1.377	1.24	1.391	9.38×10^{-7}	1.423	6.05×10^{-7}	1.428	4.97	1.427	3.27	0.943
10800	1.360	7.88×10^{-7}	1.377	6.90×10^{-7}	1.391	5.25	1.424	3.62	1.429	3.49	1.428	2.39	0.926
11000	1.361	5.20	1.377	4.98	1.391	3.53	1.424	2.84	1.429	2.73	1.428	1.97	0.909
11200	1.361	4.01	1.378	3.79	1.392	2.65	1.425	2.33	1.430	2.15	1.428	1.71	0.893
11400	1.361	3.21	1.378	2.92	1.392	2.24	1.425	2.02	1.430	1.70	1.429	1.52	0.877
11600	1.361	2.72	1.379	2.50	1.392	2.07	1.425	1.83	1.431	1.48	1.429	1.34	0.862
11800	1.361	2.44	1.379	2.27	1.392	1.87	1.426	1.58	1.432	1.29	1.430	1.17	0.847

Table 4-10. Optical Constants of Sulfuric Acid From Near Infrared to Ultraviolet [4-51] (Continued)

ν (cm⁻¹)	λ (μm)	25% n	25% n_i	38% n	38% n_i	50% n	50% n_i	75% n	75% n_i	84.5% n	84.5% n_i	95.6% n	95.6% n_i
12000	0.833	1.362	1.92	1.380	1.73	1.392	1.42	1.426	1.24	1.432	1.07	1.430	9.95×10^{-8}
12200	0.820	1.362	1.25	1.380	1.14	1.392	1.04	1.427	9.98×10^{-8}	1.432	8.94×10^{-8}	1.430	8.35
12400	0.806	1.362	1.17	1.380	1.01	1.392	9.18×10^{-8}	1.427	8.79	1.432	7.51	1.430	7.06
12600	0.794	1.362	1.24	1.380	1.07	1.393	9.28	1.427	8.46	1.433	6.57	1.430	5.87
12800	0.781	1.362	1.32	1.380	1.14	1.393	9.51	1.427	8.39	1.433	6.09	1.431	4.85
13000	0.769	1.362	1.43	1.381	1.21	1.393	9.92	1.427	8.20	1.434	4.90	1.431	3.92
13200	0.758	1.362	1.55	1.381	1.25	1.393	1.04×10^{-7}	1.427	7.84	1.434	3.92	1.431	3.13
13400	0.746	1.362	1.45	1.381	1.18	1.393	9.86×10^{-8}	1.427	6.83	1.434	3.09	1.431	2.26
13600	0.735	1.362	1.09	1.381	8.89×10^{-8}	1.393	7.02	1.427	4.80	1.435	2.34	1.431	—
13800	0.725	1.363	6.86×10^{-8}	1.381	5.94	1.394	4.61	1.427	3.58	1.435	1.73	1.432	—
14000	0.714	1.363	4.72	1.381	3.87	1.394	3.13	1.427	2.79	1.435	1.14	1.432	—
14250	0.702	1.363	3.02	1.382	2.46	1.394	2.07	1.428	2.07	1.436	—	1.432	—
18000	0.556	1.366	—	1.384	—	1.397	—	1.431	—	1.438	—	1.434	—
22250	0.449	1.369	—	1.387	—	1.402	—	1.432	—	1.442	—	1.438	—
24500	0.408	1.373	—	1.392	—	1.408	—	1.438	—	1.448	—	1.443	—
27800	0.360	1.383	—	1.407	—	1.421	—	1.452	—	1.463	—	1.459	—
1390	7.194	1.304	0.102	1.297	0.135	1.285	0.161	1.222	0.173	1.197	0.161	1.057	0.334
1410	7.092	1.309	0.097	1.305	0.127	1.296	0.150	1.249	0.158	1.219	0.138	1.024	0.226
1430	6.993	1.313	0.092	1.313	0.121	1.307	0.143	1.272	0.143	1.246	0.116	1.092	0.100
1450	6.897	1.319	0.089	1.322	0.117	1.315	0.136	1.297	0.143	1.273	0.102	1.159	0.067
1470	6.803	1.325	0.086	1.327	0.112	1.325	0.130	1.308	0.137	1.299	0.094	1.209	0.058
1490	6.711	1.332	0.084	1.335	0.108	1.335	0.125	1.323	0.130	1.323	0.088	1.237	0.052
1500	6.667	1.335	0.083	1.339	0.107	1.340	0.122	1.331	0.126	1.335	0.087	1.252	0.047
1510	6.623	1.339	0.083	1.344	0.106	1.346	0.121	1.340	0.122	1.347	0.087	1.270	0.045
1520	6.579	1.342	0.083	1.349	0.105	1.352	0.120	1.351	0.121	1.360	0.088	1.285	0.048
1530	6.536	1.346	0.083	1.354	0.104	1.357	0.120	1.361	0.122	1.371	0.093	1.297	0.052
1540	6.494	1.351	0.085	1.360	0.105	1.363	0.120	1.368	0.123	1.378	0.098	1.305	0.056
1560	6.410	1.359	0.088	1.371	0.111	1.375	0.122	1.384	0.125	1.389	0.105	1.318	0.065
1580	6.329	1.368	0.096	1.380	0.120	1.386	0.128	1.399	0.130	1.400	0.111	1.322	0.070
1600	6.250	1.378	0.112	1.385	0.135	1.398	0.138	1.413	0.138	1.410	0.116	1.326	0.069

Table 4.10. Optical Constants of Sulfuric Acid From Near Infrared to Ultraviolet [4-51] (Continued)

v (cm^{-1})	25% n	25% n_i	38% n	38% n_i	50% n	50% n_i	75% n	75% n_i	84.5% n	84.5% n_i	95.6% n	95.6% n_i	λ (μm)
1610	1.379	0.125	1.387	0.143	1.403	0.148	1.422	0.144	1.416	0.120	1.329	0.069	6.211
1620	1.374	0.139	1.386	0.154	1.406	0.159	1.428	0.152	1.422	0.126	1.331	0.068	6.173
1630	1.365	0.156	1.381	0.170	1.402	0.172	1.434	0.164	1.427	0.133	1.333	0.067	6.135
1640	1.344	0.165	1.361	0.174	1.395	0.185	1.433	0.175	1.430	0.143	1.336	0.065	6.098
1650	1.321	0.159	1.349	0.170	1.379	0.187	1.430	0.184	1.429	0.153	1.339	0.064	6.061
1660	1.310	0.148	1.343	0.165	1.371	0.185	1.427	0.191	1.425	0.161	1.341	0.062	6.024
1680	1.299	0.130	1.335	0.157	1.361	0.181	1.420	0.203	1.413	0.169	1.347	0.057	5.952
1700	1.296	0.116	1.330	0.151	1.356	0.177	1.410	0.215	1.404	0.173	1.357	0.054	5.882
1720	1.298	0.105	1.327	0.147	1.352	0.176	1.392	0.225	1.394	0.173	1.365	0.057	5.814
1740	1.300	0.099	1.323	0.143	1.347	0.177	1.371	0.220	1.385	0.169	1.369	0.058	5.747
1760	1.300	0.094	1.317	0.139	1.340	0.181	1.361	0.212	1.377	0.162	1.371	0.058	5.682
1770	1.299	0.091	1.315	0.136	1.330	0.182	1.356	0.209	1.375	0.158	1.373	0.057	5.650
1780	1.299	0.088	1.314	0.133	1.319	0.173	1.350	0.206	1.374	0.154	1.375	0.056	5.618
1800	1.299	0.082	1.311	0.126	1.316	0.160	1.341	0.194	1.373	0.146	1.379	0.055	5.556
1820	1.301	0.076	1.310	0.119	1.316	0.151	1.337	0.182	1.373	0.140	1.384	0.053	5.495
1840	1.303	0.071	1.309	0.114	1.314	0.143	1.336	0.171	1.374	0.134	1.387	0.052	5.435
1860	1.306	0.068	1.310	0.107	1.315	0.136	1.336	0.160	1.375	0.127	1.393	0.051	5.376
1880	1.309	0.064	1.310	0.103	1.315	0.130	1.339	0.151	1.378	0.119	1.397	0.051	5.319
1900	1.311	0.061	1.312	0.097	1.316	0.123	1.342	0.144	1.385	0.113	1.403	0.050	5.263
1930	1.315	0.058	1.314	0.090	1.318	0.116	1.347	0.135	1.393	0.110	1.409	0.052	5.181
1960	1.317	0.055	1.318	0.085	1.321	0.108	1.353	0.128	1.401	0.108	1.416	0.053	5.102
2020	1.324	0.049	1.324	0.076	1.329	0.096	1.366	0.118	1.414	0.110	1.426	0.058	4.950
2080	1.330	0.046	1.330	0.071	1.338	0.089	1.379	0.116	1.423	0.116	1.434	0.064	4.808
2120	1.333	0.044	1.335	0.068	1.343	0.087	1.384	0.117	1.426	0.122	1.438	0.069	4.717
2180	1.337	0.043	1.340	0.065	1.348	0.086	1.386	0.121	1.422	0.131	1.441	0.074	4.587
2240	1.340	0.042	1.343	0.063	1.350	0.085	1.384	0.119	1.416	0.127	1.446	0.075	4.464
2290	1.341	0.040	1.345	0.060	1.352	0.081	1.386	0.113	1.419	0.122	1.457	0.077	4.367
2330	1.343	0.036	1.349	0.057	1.357	0.079	1.395	0.109	1.425	0.122	1.470	0.092	4.292
2340	1.344	0.036	1.350	0.057	1.357	0.079	1.397	0.110	1.428	0.122	1.472	0.099	4.274
2370	1.347	0.035	1.354	0.058	1.359	0.080	1.405	0.117	1.436	0.131	1.461	0.116	4.219

Table 4-10. Optical Constants of Sulfuric Acid From Near Infrared to Ultraviolet [4-51] (Continued)

ν (cm^{-1})	2.5% n	2.5% n_i	38% n	38% n_i	50% n	50% n_i	75% n	75% n_i	84.5% n	84.5% n_i	95.6% n	95.6% n_i	λ (μm)
2410	1.350	0.034	1.356	0.059	1.361	0.080	1.399	0.121	1.430	0.142	1.446	0.120	4.149
2450	1.353	0.033	1.356	0.058	1.362	0.080	1.400	0.124	1.425	0.146	1.438	0.118	4.082
2500	1.355	0.032	1.358	0.056	1.363	0.080	1.398	0.126	1.418	0.149	1.435	0.114	4.000
2530	1.357	0.030	1.359	0.056	1.363	0.080	1.396	0.127	1.416	0.150	1.436	0.113	3.953
2560	1.361	0.030	1.361	0.054	1.364	0.078	1.395	0.127	1.414	0.152	1.438	0.115	3.906
2590	1.364	0.029	1.363	0.053	1.365	0.078	1.395	0.127	1.412	0.156	1.437	0.118	3.861
2600	1.364	0.028	1.364	0.053	1.365	0.077	1.395	0.127	1.410	0.157	1.437	0.117	3.846
2620	1.367	0.029	1.366	0.052	1.366	0.077	1.396	0.128	1.406	0.157	1.438	0.116	3.817
2660	1.372	0.029	1.370	0.052	1.369	0.076	1.396	0.130	1.403	0.156	1.445	0.122	3.759
2710	1.377	0.030	1.375	0.054	1.374	0.077	1.397	0.136	1.402	0.159	1.448	0.137	3.690
2760	1.384	0.031	1.379	0.056	1.378	0.081	1.394	0.143	1.399	0.168	1.444	0.155	3.623
2810	1.391	0.034	1.383	0.059	1.379	0.086	1.388	0.153	1.388	0.178	1.431	0.173	3.559
2880	1.400	0.040	1.389	0.064	1.377	0.092	1.370	0.161	1.361	0.189	1.403	0.192	3.472
2930	1.408	0.047	1.393	0.070	1.375	0.095	1.357	0.159	1.341	0.181	1.377	0.197	3.413
2990	1.418	0.058	1.397	0.078	1.373	0.099	1.341	0.159	1.321	0.171	1.347	0.195	3.344
3050	1.428	0.075	1.401	0.090	1.371	0.102	1.325	0.150	1.306	0.159	1.315	0.182	3.279
3150	1.431	0.118	1.400	0.117	1.369	0.113	1.306	0.131	1.283	0.131	1.274	0.143	3.175
3250	1.408	0.166	1.380	0.149	1.357	0.130	1.296	0.109	1.273	0.098	1.260	0.092	3.077
3310	1.380	0.193	1.359	0.165	1.343	0.141	1.294	0.099	1.272	0.079	1.263	0.064	3.021
3350	1.354	0.207	1.342	0.175	1.330	0.148	1.292	0.093	1.273	0.067	1.267	0.048	2.985
3400	1.313	0.218	1.312	0.184	1.308	0.154	1.288	0.086	1.276	0.053	1.276	0.030	2.941
3430	1.285	0.219	1.290	0.185	1.291	0.153	1.284	0.082	1.277	0.045	1.283	0.022	2.915
3470	1.243	0.206	1.256	0.176	1.267	0.147	1.277	0.073	1.282	0.034	1.291	0.012	2.882
3520	1.204	0.174	1.222	0.150	1.238	0.125	1.273	0.056	1.289	0.022	1.303	—	2.841
3530	1.198	0.166	1.216	0.144	1.233	0.119	1.272	0.052	1.290	0.019	1.305	—	2.833
3610	1.170	0.089	1.191	0.077	1.215	0.062	1.277	0.023	1.304	—	1.327	—	2.770
3620	1.171	0.079	1.190	0.068	1.216	0.054	1.279	0.019	1.307	—	1.329	—	2.762
3670	1.183	0.031	1.207	0.025	1.231	0.022	1.293	0.006	1.320	—	1.339	—	2.725
3720	1.216	—	1.234	—	1.252	—	1.308	—	1.330	—	1.345	—	2.688
3800	1.250	—	1.266	—	1.282	—	1.320	—	1.341	—	1.353	—	2.632

Table 4-10. Optical Constants of Sulfuric Acid From Near Infrared to Ultraviolet [4-51] (Continued)

ν	25%		38%		50%		75%		84.5%		95.6%		λ
(cm^{-1})	n	n_i	n	n_i	n	n_i	n	n_i	n	n_i	n	n_i	(μm)
3900	1.271	—	1.284	—	1.299	—	1.332	—	1.348	—	1.360	—	2.564
4000	1.286	—	1.300	—	1.311	—	1.344	—	1.358	—	1.368	—	2.500
400	1.700	0.303	1.749	0.327	1.806	0.319	1.930	0.200	1.938	0.099	1.896	0.212	25.000
410	1.696	0.305	1.744	0.328	1.808	0.328	1.939	0.226	1.954	0.144	1.880	0.245	24.390
430	1.692	0.329	1.736	0.356	1.783	0.371	1.918	0.300	1.905	0.219	1.822	0.274	23.256
440	1.676	0.343	1.719	0.372	1.758	0.380	1.881	0.320	1.874	0.231	1.781	0.248	22.727
450	1.657	0.351	1.696	0.378	1.734	0.384	1.848	0.329	1.846	0.229	1.785	0.218	22.222
470	1.627	0.354	1.658	0.374	1.689	0.361	1.781	0.290	1.807	0.199	1.826	0.193	21.277
480	1.615	0.353	1.645	0.368	1.690	0.350	1.782	0.257	1.804	0.169	1.848	0.194	20.833
490	1.605	0.351	1.635	0.363	1.690	0.345	1.804	0.240	1.833	0.145	1.874	0.197	20.408
500	1.596	0.349	1.627	0.357	1.690	0.344	1.823	0.235	1.873	0.146	1.913	0.209	20.000
510	1.590	0.349	1.622	0.349	1.690	0.344	1.842	0.238	1.903	0.161	1.961	0.241	19.608
530	1.580	0.352	1.627	0.348	1.692	0.353	1.892	0.261	1.981	0.216	2.057	0.405	18.868
540	1.576	0.355	1.628	0.354	1.694	0.360	1.926	0.299	2.011	0.291	2.045	0.569	18.519
550	1.572	0.360	1.629	0.364	1.700	0.375	1.946	0.362	2.011	0.391	1.912	0.740	18.182
560	1.569	0.368	1.629	0.379	1.703	0.402	1.939	0.457	1.955	0.520	1.680	0.802	17.857
570	1.564	0.383	1.625	0.404	1.693	0.443	1.869	0.554	1.800	0.575	1.477	0.699	17.544
580	1.550	0.412	1.603	0.443	1.653	0.496	1.741	0.594	1.671	0.549	1.389	0.540	17.241
590	1.509	0.422	1.551	0.460	1.576	0.509	1.621	0.564	1.584	0.468	1.382	0.415	16.949
600	1.473	0.415	1.498	0.448	1.506	0.479	1.542	0.479	1.552	0.400	1.410	0.340	16.667
620	1.433	0.374	1.438	0.381	1.441	0.379	1.512	0.352	1.530	0.290	1.466	0.253	16.129
630	1.427	0.358	1.437	0.350	1.450	0.338	1.512	0.299	1.538	0.244	1.488	0.232	15.873
650	1.420	0.340	1.445	0.322	1.472	0.299	1.551	0.221	1.578	0.175	1.520	0.203	15.385
670	1.407	0.329	1.444	0.304	1.483	0.277	1.596	0.191	1.617	0.151	1.543	0.183	14.925
680	1.400	0.322	1.443	0.295	1.488	0.268	1.613	0.183	1.632	0.144	1.552	0.176	14.706
700	1.388	0.309	1.441	0.280	1.496	0.254	1.643	0.173	1.657	0.134	1.567	0.160	14.286
720	1.377	0.294	1.440	0.266	1.503	0.242	1.663	0.171	1.677	0.125	1.584	0.143	13.889
740	1.367	0.279	1.440	0.253	1.511	0.230	1.681	0.165	1.689	0.116	1.604	0.126	13.514
760	1.358	0.262	1.440	0.240	1.520	0.221	1.701	0.160	1.722	0.108	1.628	0.110	13.158
780	1.353	0.242	1.443	0.226	1.532	0.213	1.726	0.157	1.751	0.100	1.663	0.090	12.821

Table 4-10. Optical Constants of Sulfuric Acid From Near Infrared to Ultraviolet [4-51] (Continued)

ν (cm^{-1})	25% n	25% n_i	38% n	38% n_i	50% n	50% n_i	75% n	75% n_i	84.5% n	84.5% n_i	95.6% n	95.6% n_i	λ (μm)
790	1.353	0.233	1.447	0.218	1.541	0.210	1.741	0.157	1.771	0.099	1.693	0.090	12.658
800	1.354	0.224	1.455	0.211	1.549	0.210	1.757	0.158	1.793	0.100	1.710	0.094	12.500
820	1.357	0.206	1.471	0.209	1.568	0.215	1.796	0.168	1.839	0.112	1.751	0.096	12.195
840	1.373	0.191	1.483	0.212	1.588	0.227	1.844	0.194	1.896	0.130	1.812	0.107	11.905
850	1.382	0.190	1.491	0.216	1.599	0.238	1.869	0.216	1.934	0.158	1.848	0.121	11.765
870	1.392	0.205	1.512	0.240	1.617	0.303	1.916	0.313	1.993	0.247	1.940	0.181	11.494
874	1.385	0.208	1.515	0.259	1.607	0.338	1.911	0.341	1.998	0.277	1.956	0.206	11.442
880	1.381	0.209	1.491	0.275	1.579	0.345	1.904	0.386	2.005	0.321	1.978	0.243	11.364
890	1.372	0.207	1.461	0.277	1.519	0.338	1.842	0.464	1.984	0.416	2.007	0.356	11.236
900	1.365	0.200	1.436	0.269	1.484	0.311	1.739	0.463	1.884	0.483	1.937	0.458	11.111
910	1.360	0.188	1.416	0.247	1.471	0.279	1.676	0.410	1.789	0.466	1.841	0.460	10.989
920	1.360	0.173	1.407	0.207	1.463	0.228	1.663	0.351	1.735	0.402	1.808	0.407	10.870
930	1.368	0.160	1.431	0.184	1.502	0.200	1.678	0.301	1.742	0.348	1.870	0.391	10.753
940	1.382	0.152	1.454	0.173	1.535	1.717	1.717	0.275	1.775	0.323	1.953	0.468	10.638
950	1.396	0.151	1.475	0.172	1.563	0.199	1.756	0.271	1.807	0.328	1.967	0.627	10.526
960	1.407	0.154	1.495	0.176	1.582	0.210	1.788	0.277	1.820	0.346	1.856	0.795	10.417
963	1.412	0.157	1.501	0.181	1.587	0.215	1.807	0.282	1.820	0.349	1.789	0.833	10.384
970	1.417	0.158	1.510	0.186	1.597	0.222	1.822	0.292	1.818	0.347	1.634	0.869	10.309
980	1.425	0.165	1.523	0.197	1.613	0.235	1.849	0.311	1.839	0.341	1.415	0.809	10.204
990	1.432	0.173	1.535	0.212	1.631	0.255	1.882	0.338	1.877	0.353	1.301	0.624	10.101
1010	1.441	0.195	1.549	0.247	1.650	0.314	1.947	0.453	1.954	0.443	1.372	0.350	9.901
1020	1.442	0.209	1.556	0.272	1.649	0.349	1.944	0.538	1.968	0.528	1.457	0.311	9.804
1030	1.438	0.226	1.557	0.314	1.644	0.402	1.907	0.637	1.937	0.630	1.527	0.326	9.709
1040	1.431	0.246	1.525	0.364	1.594	0.463	1.807	0.708	1.848	0.707	1.548	0.352	9.615
1050	1.398	0.273	1.460	0.376	1.509	0.462	1.702	0.711	1.749	0.724	1.548	0.373	9.524
1060	1.350	0.245	1.407	0.339	1.456	0.428	1.624	0.668	1.666	0.693	1.537	0.366	9.434
1070	1.346	0.213	1.396	0.297	1.433	0.370	1.589	0.612	1.625	0.639	1.548	0.347	9.346
1080	1.362	0.194	1.419	0.266	1.450	0.337	1.590	0.560	1.623	0.594	1.578	0.342	9.259
1090	1.377	0.193	1.434	0.265	1.471	0.314	1.626	0.540	1.647	0.574	1.617	0.360	9.174
1100	1.383	0.197	1.442	0.267	1.502	0.318	1.655	0.556	1.682	0.603	1.639	0.391	9.091

Table 4-10. Optical Constants of Sulfuric Acid From Near Infrared to Ultraviolet [4-51] (Continued)

ν (cm⁻¹)	25% n	25% n_i	38% n	38% n_i	50% n	50% n_i	75% n	75% n_i	84.5% n	84.5% n_i	95.6% n	95.6% n_i	λ (μm)
1110	1.387	0.201	1.452	0.269	1.523	0.335	1.669	0.590	1.682	0.634	1.654	0.439	9.009
1120	1.389	0.206	1.464	0.280	1.529	0.360	1.666	0.634	1.670	0.686	1.645	0.481	8.929
1130	1.391	0.211	1.473	0.302	1.529	0.384	1.643	0.681	1.633	0.731	1.632	0.524	8.850
1150	1.395	0.229	1.451	0.344	1.507	0.436	1.545	0.755	1.515	0.777	1.572	0.615	8.696
1160	1.393	0.246	1.435	0.360	1.483	0.465	1.479	0.761	1.456	0.773	1.516	0.651	8.621
1170	1.382	0.268	1.417	0.376	1.448	0.485	1.421	0.758	1.403	0.764	1.447	0.669	8.547
1190	1.329	0.285	1.353	0.400	1.369	0.498	1.320	0.719	1.308	0.726	1.300	0.643	8.403
1210	1.277	0.270	1.283	0.381	1.280	0.473	1.241	0.663	1.230	0.666	1.230	0.532	8.264
1230	1.238	0.228	1.222	0.322	1.218	0.409	1.179	0.593	1.179	0.590	1.217	0.453	8.130
1240	1.230	0.199	1.212	0.285	1.203	0.370	1.161	0.547	1.166	0.557	1.219	0.427	8.065
1250	1.233	0.173	1.214	0.252	1.197	0.329	1.151	0.513	1.153	0.527	1.218	0.400	8.000
1270	1.254	0.143	1.236	0.206	1.212	0.268	1.145	0.445	1.132	0.471	1.225	0.358	7.874
1290	1.271	0.127	1.258	0.184	1.239	0.239	1.144	0.397	1.116	0.412	1.244	0.321	7.752
1310	1.284	0.121	1.269	0.172	1.251	0.222	1.136	0.351	1.102	0.351	1.276	0.308	7.634
1320	1.288	0.119	1.275	0.166	1.255	0.214	1.133	0.323	1.100	0.318	1.292	0.313	7.576
1340	1.293	0.115	1.284	0.157	1.263	0.197	1.142	0.262	1.116	0.242	1.314	0.362	7.463
1360	1.297	0.109	1.290	0.152	1.272	0.182	1.173	0.211	1.156	0.201	1.245	0.431	7.353
1370	1.300	0.107	1.390	0.146	1.275	0.175	1.192	0.195	1.171	0.187	1.180	0.437	7.299

Table 4-11. Experimental Values for Dust/Optical Constants of Quartz

$\lambda(\mu m)$	n	n_i	n	n_i	(1969) Quartz			
					n_o^*	n_{io}^*	n_e^*	n_{ie}^*
0.40	1.65	0.005	—	—	—	—	—	—
0.45	1.65	0.005	—	—	—	—	—	—
0.50	1.65	0.005	—	—	—	—	—	—
0.55	1.65	0.005	—	—	—	—	—	—
0.60	1.65	0.005	—	—	—	—	—	—
0.65	1.65	0.005	—	—	—	—	—	—
0.70	1.65	0.005	—	—	—	—	—	—
0.75	1.65	0.005	—	—	—	—	—	—
0.768	—	—	—	—	1.53903	—	1.54794	—
0.80	1.65	0.005	—	—	—	—	—	—
0.85	1.649	0.005	—	—	—	—	—	—
0.90	1.648	0.005	—	—	—	—	—	—
0.95	1.648	0.005	—	—	—	—	—	—
1.00	1.647	0.0051	—	—	—	—	—	—
1.10	1.647	0.0051	—	—	—	—	—	—
1.159	—	—	—	—	1.53283	—	1.54152	—
1.20	1.646	0.0052	—	—	—	—	—	—
1.30	1.646	0.0058	—	—	—	—	—	—
1.40	1.645	0.0068	—	—	—	—	—	—
1.50	1.645	0.0070	—	—	—	—	—	—
1.60	1.644	0.0076	—	—	—	—	—	—
1.681	—	—	—	—	1.52583	(10^{-7})	1.53422	—
1.70	1.644	0.0082	—	—	—	—	—	—
1.80	1.643	0.0088	—	—	—	—	—	—
1.90	1.643	0.0094	—	—	—	—	—	—
2.00	1.642	0.010	—	—	—	—	—	—
2.10	1.642	0.0110	—	—	—	—	—	—
2.20	1.641	0.0120	—	—	—	—	—	—
2.30	1.641	0.013	—	—	—	—	—	—
2.40	1.640	0.0140	—	—	—	—	—	—
Ref.	[4-54]		[4-48, 4-52]		[4-55]			

*The subscripts o and e refer to the ordinary and extraordinary rays in the quartz crystal.

Table 4-11. Experimental Values for Dust/Optical Constants of Quartz (*Continued*)

$\lambda(\mu m)$	n	n_i	n	n_i	(1969) Quartz n_o^*	n_{io}^*	n_e^*	n_{ie}^*
2.50	1.64	0.15	1.45	0.01	–	–	–	–
2.60	1.630	0.018	–	–	–	–	–	–
2.70	1.618	0.021	–	–	–	–	–	–
2.80	1.608	0.0440	–	–	–	–	–	–
2.90	1.622	0.73	1.47	0.04	–	–	–	–
3.00	1.646	0.076	–	–	–	–	–	–
3.10	1.673	0.068	1.48	0.025	–	–	–	–
3.20	1.665	0.051	–	–	–	–	–	–
3.30	1.689	0.040	–	–	–	–	–	–
3.40	1.678	0.021	–	–	–	–	–	–
3.50	1.655	0.020	–	–	1.48451	0.00001	–	–
3.60	1.651	0.019	–	–	–	–	–	–
3.70	1.649	0.018	–	–	–	–	–	–
3.80	1.645	0.018	–	–	–	–	–	–
3.90	1.640	0.018	–	–	–	–	–	–
4.00	1.637	0.018	1.47	0.0045	–	–	–	–
4.10	1.633	0.018	–	–	–	–	–	–
4.20	1.630	0.018	–	–	–	–	–	–
4.30	1.626	0.018	–	–	–	–	–	–
4.40	1.620	0.018	–	–	–	–	–	–
4.50	1.620	0.018	–	–	–	–	–	–
4.60	1.612	0.018	–	–	–	–	–	–
4.70	1.605	0.018	–	–	–	–	–	–
4.80	1.600	0.018	–	–	–	–	–	–
4.90	1.598	0.018	–	–	–	–	–	–
5.00	1.592	0.018	1.50	0.013	1.41249	0.00079	1.41887	0.00091
5.10	1.588	0.018	–	–	–	–	–	–
5.20	1.580	0.018	–	–	–	–	–	–
5.30	1.572	0.018	–	–	–	–	–	–
5.40	1.567	0.018	–	–	–	–	–	–
Ref.	[4-54]		[4-48, 4-52]		[4-55]			

Table 4-11. Experimental Values for Dust/Optical Constants of Quartz (*Continued*)

λ(μm)	n	n_i	n	n_i	(1969) Quartz n_o^*	n_{io}^*	n_e^*	n_{ie}^*
5.50	1.558	0.018	–	–	–	–	–	–
5.60	1.551	0.020	–	–	–	–	–	–
5.70	1.551	0.024	–	–	–	–	–	–
5.80	1.546	0.029	–	–	–	–	–	–
5.90	1.535	0.036	–	–	–	–	–	–
6.00	1.515	0.045	1.43	0.045	–	–	–	–
6.10	1.494	0.054	–	–	–	–	–	–
6.20	1.482	0.052	–	–	–	–	–	–
6.30	1.471	0.041	1.42	0.05	1.27875	0.00323	1.28246	0.00380
6.40	1.443	0.038	–	–	–	–	–	–
6.50	1.402	0.042	–	–	–	–	–	–
6.60	1.350	0.071	–	–	–	–	–	–
6.70	1.270	0.135	–	–	–	–	–	–
6.75	1.212	0.237	–	–	–	–	–	–
6.80	1.280	0.363	–	–	–	–	–	–
6.90	1.420	0.445	–	–	–	–	–	–
7.00	1.528	0.434	1.45	0.11	–	–	–	–
7.10	1.611	0.415	–	–	–	–	–	–
7.20	1.723	0.360	–	–	–	–	–	–
7.25	1.780	0.240	–	–	–	–·	–	–
7.30	1.740	0.150	–	–	–	–	–	–
7.40	1.635	0.102	–	–	–	–	–	–
7.50	1.560	0.071	1.37	0.071	–	–	–	–
7.60	1.475	0.061	–	–	0.86663	0.03104	0.86023	0.03412
7.70	1.415	0.075	–	–	–	–	–	–
7.80	1.366	0.103	–	–	0.70229	0.05774	0.68949	0.05973
7.90	1.318	0.143	–	–	–	–	–	–
8.00	1.269	0.178	1.18	0.085	0.42984	0.13825	0.39076	0.14379
8.10	1.220	0.263	–	–	–	–	–	–
8.20	1.200	0.330	–	–	0.14154	0.53495	0.12328	0.58352
Ref.	[4-54]		[4-48, 4-52]		[4-55]			

*The subscripts o and e refers to the ordinary and extraordinary rays in the quartz crystal.

Table 4-11. Experimental Values for Dust/Optical Constants of Quartz (*Continued*)

					(1969)			
					Quartz			
$\lambda(\mu m)$	n	n_i	n	n_i	n_o^*	n_{io}^*	n_e^*	n_{ie}^*
8.30	1.210	0.432	–	–	–	–	–	–
8.40	1.200	0.515	–	–	0.09642	0.98098	0.08965	1.00302
8.50	1.186	0.600	1.07	0.19	–	–	–	–
8.60	1.221	0.650	1.07	0.23	0.65842	1.44755	0.08715	1.44831
8.70	1.218	0.745	–	–	–	–	–	–
8.80	1.249	0.945	1.65	0.33	0.11009	1.75810	0.11421	2.03496
8.90	1.390	1.130	–	–	–	–	–	–
9.00	1.650	1.240	1.87	0.44	0.17463	2.59701	0.22995	3.04158
9.10	1.940	1.248	–	–	–	–	–	–
9.20	2.170	1.090	2.17	0.52	0.62206	4.51698	1.66428	6.37770
9.30	2.268	0.89	–	–	3.43313	7.51737	7.48857	2.54419
9.40	2.327	0.76	2.80	0.61	6.38644	1.37131	4.87845	0.50951
9.50	2.342	0.600	–	–	4.51517	0.39770	3.90448	0.23041
9.60	2.335	0.460	3.07	0.70	–	–	–	–
9.70	2.303	0.362	–	–	–	–	–	–
9.80	2.262	0.255	–	–	–	–	–	–
9.90	2.205	0.185	2.75	0.80	–	–	–	–
10.0	2.140	0.126	2.59	0.92	2.66527	0.05190	2.57228	0.04402
10.10	2.078	0.101	–	–	–	–	–	–
10.20	2.031	0.091	2.77	0.88	–	–	–	–
10.30	1.990	0.082	–	–	–	–	–	–
10.40	1.948	0.080	1.82	0.70	–	–	–	–
10.50	1.904	0.078	–	–	–	–	–	–
10.60	1.869	0.078	1.70	0.60	–	–	–	–
10.70	1.841	0.079	–	–	–	–	–	–
10.80	1.808	0.088	1.79	0.37	–	–	–	–
11.0	1.750	0.118	1.84	0.31	–	–	–	–
11.1	1.721	0.143	–	–	–	–	–	–
11.2	1.690	0.169	–	–	–	–	–	–
11.3	1.680	0.235	–	–	1.90613	0.01706	1.90783	0.01490
Ref.	[4-54]		[4-48, 4-52]		[4-55]			

Table 4-11. Experimental Values for Dust/Optical Constants of Quartz (*Continued*)

λ(μm)	n	n_i	n	n_i	(1969) Quartz			
					n_o^*	n_{io}^*	n_e^*	n_{ie}^*
11.4	1.726	0.320	—	—	—	—	—	—
11.5	1.784	0.331	—	—	—	—	—	—
11.6	1.819	0.256	—	—	—	—	—	—
11.7	1.798	0.222	—	—	—	—	—	—
11.8	1.776	0.218	—	—	—	—	—	—
11.9	1.764	0.222	—	—	—	—	—	—
12.0	1.756	0.230	1.80	0.18	—	—	—	—
12.1	1.750	0.238	—	—	1.45911	0.07039	1.63688	0.02861
12.2	1.740	0.246	—	—	—	—	—	—
12.3	1.734	0.254	—	—	—	—	—	—
12.4	1.725	0.263	—	—	0.72462	1.04211	1.41016	0.07535
12.5	1.713	0.270	—	—	1.52652	2.23992	1.25602	0.13312
12.6	1.701	0.280	—	—	3.24575	1.06551	0.98837	0.31437
12.7	1.711	0.285	—	—	2.72881	0.28899	0.77628	0.99290
12.8	1.722	0.291	—	—	2.43777	0.13172	1.54157	2.01706
12.9	1.752	0.295	—	—	2.27734	0.07753	3.00871	1.04450
13.0	1.790	0.296	1.74	0.18	—	—	—	—
13.1	1.828	0.291	—	—	—	—	—	—
13.3	1.887	0.270	—	—	—	—	—	—
13.4	1.889	0.256	—	—	—	—	—	—
13.5	1.880	0.243	—	—	—	—	—	—
13.6	1.850	0.233	—	—	0.89958	0.02029	1.93921	0.02676
13.7	1.794	0.229	—	—	—	—	—	—
13.8	1.745	0.232	—	—	—	—	—	—
13.9	1.710	0.263	—	—	—	—	—	—
14.0	1.754	0.435	1.64	0.20	—	—	—	—
Ref.	[4-54]		[4-48, 4-52]		[4-55]			

*The subscripts o and e refers to the ordinary and extraordinary rays in the quartz crystal.

Table 4-12. Variation of the Optical Constants of Aerosols with Fractional
Relative Humidity, F, at a Fixed Wavelength of 0.55 μm [4-56]

	Maritime (over Atlantic 13-16 April, 1969)			Urban (at Mainz in January, 1970)		
	F	n	n_i	F	n	n_i
increasing	0.2	1.546	0.054	0.2	1.545	0.047
F	0.4	1.541	0.053	0.4	1.539	0.046
	0.6	1.522	0.048	0.6	1.503	0.038
	0.7	1.457	0.031	0.7	1.461	0.028
	0.8	1.383	0.012	0.8	1.404	0.016
	0.9	1.364	0.008	0.9	1.380	0.010
	0.95	1.352	0.004	0.95	1.355	0.005
	0.975	1.344	0.002	0.975	1.344	0.002
	0.99	1.338	0.001	0.99	1.338	0.001
decreasing	0.8	–	–	0.8	1.395	0.014
F	0.7	1.397	0.016	0.7	1.435	0.022
	0.6	1.496	0.041	0.6	1.478	0.032
	0.4	1.531	0.050	0.4	1.522	0.042
	0.2	1.543	0.053	0.2	1.541	0.046

Table 4-13. Optical Constants of Ammonium Sulfate
at Various Humidities [4-57]

Frequency	n_i	n	Frequency	n_i	n
		21% $(NH_4)_2SO_4$			
1296.6	0.03505	1.31906	1150.3	0.12816	1.15447
1279.4	0.03580	1.30978	1139.5	0.18741	1.14945
1261.8	0.03596	1.29978	1127.8	0.27266	1.16388
1245.3	0.03702	1.29010	1117.6	0.33626	1.21397
1225.6	0.03770	1.27499	1104.5	0.37796	1.31509
1211.0	0.03977	1.26153	1100.2	0.37863	1.34957
1191.5	0.04599	1.23843	1093.9	0.36788	1.39932
1178.2	0.05753	1.21545	1088.7	0.35537	1.43614
1170.4	0.06798	1.19791	1081.6	0.31812	1.47667
1160.2	0.08960	1.17425	1071.6	0.24586	1.51320

Table 4-13. Optical Constants of Ammonium Sulfate
at Various Humidities [4-57] (*Continued*)

Frequency	n_i	n	Frequency	n_i	n
21% $(NH_4)_2SO_4$ *(Continued)*			30% $(NH_4)_2SO_4$ *(Continued)*		
1063.8	0.19470	1.51542	1078.4	0.45448	1.60129
1056.3	0.15107	1.49959	1071.5	0.37688	1.63820
1047.9	0.11783	1.47064	1064.7	0.30372	1.64214
1037.1	0.09073	1.44069	1057.0	0.23328	1.61406
1026.6	0.07727	1.41307	1049.6	0.16430	1.60800
1019.8	0.07351	1.39897	1041.3	0.13061	1.55314
1009.1	0.06685	1.37888	1032.5	0.10375	1.52153
996.3	0.06577	1.36017	1024.6	0.08900	1.49806
984.8	0.06625	1.34631	1015.3	0.07771	1.47766
976.6	0.06921	1.34019	1006.3	0.07094	1.45998
973.6	0.06799	1.33724	996.7	0.06723	1.44458
			984.4	0.06781	1.42097
30% $(NH_4)_2SO_4$			977.6	0.07252	1.40960
			973.2	0.06899	1.41057
1295.8	0.04162	1.31950			
1271.8	0.04062	1.30374	39% $(NH_4)_2SO_4$		
1250.7	0.04103	1.28872			
1232.1	0.04103	1.27568	2253.9	0.01885	1.38826
1214.4	0.04408	1.25629	2208.9	0.02141	1.38617
1197.4	0.05444	1.22382	2142.3	0.02565	1.38324
1185.1	0.06334	1.20256	2088.7	0.02743	1.38264
1170.6	0.09178	1.15504	2045.9	0.02565	1.38256
1160.4	0.13131	1.32206	1980.0	0.02128	1.37501
1150.5	0.18754	1.09417	1897.8	0.02335	1.36142
1139.6	0.27516	1.09384	1825.0	0.02855	1.35077
1130.2	0.36280	1.12661	1755.8	0.03325	1.33047
1121.1	0.44642	1.18501	1697.8	0.07250	1.30412
1114.3	0.49015	1.24303	1648.9	0.12911	1.33852
1106.7	0.52554	1.32780	1624.3	0.11221	1.37688
1103.4	0.54080	1.35902	1607.4	0.08776	1.37884
1100.2	0.55154	1.39353	1571.9	0.05738	1.35568
1095.9	0.53946	1.44399	1521.6	0.05631	1.32394
1088.6	0.52753	1.52109	1496.8	0.09054	1.29171
1084.5	0.48938	1.56896	1475.8	0.17194	1.28432

Table 4-13. Optical Constants of Ammonium Sulfate
at Various Humidities [4-57] (*Continued*)

Frequency	n_i	n	Frequency	n_i	n
\multicolumn{6}{c}{39% $(NH_4)_2SO_4$ (*Continued*)}					
1455.7	0.25253	1.35778	778.1	0.22604	1.23878
1443.5	0.25317	1.42103	761.8	0.23866	1.25573
1431.7	0.21654	1.45926	753.0	0.25094	1.25414
1415.6	0.13395	1.47682	747.5	0.24003	1.27832
1391.3	0.06913	1.43265	\multicolumn{3}{c}{41% $(NH_4)_2SO_4$}		
1368.3	0.04676	1.41051	1298.1	0.04067	1.32910
1316.4	0.04139	1.35720	1280.7	0.04065	1.31316
1246.3	0.04011	1.28296	1264.0	0.04134	1.29678
1197.8	0.05381	1.20999	1248.0	0.04214	1.28022
1176.3	0.08802	1.13824	1232.5	0.04083	1.26821
1159.6	0.14841	1.07411	1217.7	0.04306	1.25099
1144.9	0.28067	1.03229	1203.3	0.05138	1.21830
1137.7	0.36978	1.03490	1188.1	0.06265	1.18267
1125.0	0.54788	1.08337	1174.8	0.09393	1.12871
1118.2	0.61667	1.15026	1162.0	0.14906	1.07339
1103.9	0.60829	1.34741	1148.3	0.27589	1.02268
1095.4	0.69451	1.47968	1136.3	0.41161	1.03468
1089.1	0.67187	1.56863	1124.7	0.54821	1.10846
1080.0	0.58075	1.69107	1113.4	0.65259	1.23246
1066.2	0.40110	1.78119	1102.5	0.70549	1.39092
1058.5	0.30972	1.76905	1091.9	0.70839	1.55477
1041.9	0.16497	1.67226	1081.6	0.63823	1.69239
1021.0	0.10784	1.57122	1071.6	0.50452	1.78736
1006.9	0.09311	1.51689	1061.9	0.37023	1.79980
983.5	0.09003	1.45607	1051.5	0.23499	1.75627
973.8	0.09297	1.44425	1041.5	0.17134	1.67597
935.0	0.08700	1.39000	1032.6	0.13493	1.62778
896.8	0.08236	1.33918	1023.9	0.10846	1.59427
875.8	0.09914	1.30979	1015.4	0.09392	1.56869
856.3	0.11679	1.29634	1007.2	0.08523	1.54286
838.9	0.13824	1.27604	999.2	0.07984	1.52124
822.7	0.15602	1.27268	991.3	0.07894	1.49956
802.0	0.19400	1.24312	983.7	0.08239	1.48057
790.6	0.19878	1.25768	976.2	0.08417	1.47753

Table 4-14. The Optical Constants
of NaCl [4-46]

ν (cm^{-1})	λ (μm)	n	n_i	ν (cm^{-1})	λ (μm)	n	n_i
50000	0.20	1.79	1.0-7*	370	27.0	1.21	7.0-3
45455	0.22	1.72	–	357	28.0	1.17	9.2-3
41667	0.24	1.67	–	345	29.0	1.13	1.2-2
35714	0.28	1.62	–	333	30.0	1.09	1.5-2
33333	0.30	1.61	–	313	32.0	0.97	4.0-2
28571	0.35	1.58	–	294	34.0	0.82	8.0-2
25000	0.40	1.57	–	278	36.0	0.61	1.8-1
20000	0.50	1.55	–	263	38.0	0.45	4.8-1
10000	1.00	1.53	–	250	40.0	0.52	7.0-1
3333	3.00	1.52	–	238	42.0	0.46	7.3-1
1429	7.00	1.51	–	227	44.0	0.27	9.4-1
1111	9.00	1.50	–	217	46.0	0.18	1.3-0
1000	10.0	1.49	1.0-7	208	48.0	0.17	1.6-0
909	11.0	1.49	1.9-7	200	50.0	0.18	2.0-0
833	12.0	1.48	8.0-7	192	52.0	0.22	2.5-0
769	13.0	1.47	2.7-6	185	54.0	0.31	3.8-0
714	14.0	1.46	7.8-6	179	56.0	0.49	4.3-0
667	15.0	1.45	2.0-5	172	58.0	1.02	5.1-0
625	11.0	1.44	4.4-5	170	59.0	1.73	6.0-0
588	17.0	1.43	9.0-5	167	60.0	3.49	7.2-0
556	18.0	1.41	1.7-4	164	61.0	6.81	6.3-0
526	19.0	1.40	3.0-4	161	62.0	7.46	3.2-0
500	20.0	1.38	5.1-4	159	63.0	6.54	1.6-0
476	21.0	1.37	8.2-4	156	64.0	5.77	9.9-1
454	22.0	1.34	1.3-3	154	65.0	5.22	6.8-1
435	23.0	1.31	1.9-3	143	70.0	3.93	2-1
417	24.0	1.28	2.7-3	125	80.0	3.16	1-1
400	25.0	1.25	3.8-3	111	90.0	2.88	7-2
385	26.0	1.23	5.2-3	100	100.0	2.74	6-2

*The second number in the n_i column is the power of ten following the value of n_i. For example, $1 - 7 = 1 \times 10^{-7}$.

4.8 References and Bibliography

4.8.1. References

[4-1]   W. M. Irvine, "Multiple Scattering by Large Particles, II: Optically Thick Layers," *Astrophysical Journal,* American Astronomical Society, University of Chicago Press, Chicago, IL, Vol. 152, June 1968.

[4-2]   K. L. Coulson, J. V. Dave, and Z. Sekera, "Tables Related to Radiation Emerging from a Planetary Atmosphere with Rayleigh Scattering," University of California Press, Berkeley, CA, 1960.

[4-3]   I. W. Busbridge, "On the X- and Y- Functions of S. Chandrasekhar," *Astrophysical Journal,* Vol. 122, 1955, p. 327.

[4-4]   I. W. Busbridge, *The Mathematics of Radiative Transfer,* Cambridge University Press, New York, NY, 1960.

[4-5]   T. W. Mullikin, "Radiative Transfer in Finite Homogeneous Atmospheres with Anisotropic Scattering, II: The Uniqueness Problem for Chandrasekhar's ψ_l and θ_l Equations," *Astrophysical Journal,* Vol. 139, No. 4, 1964 (a), pp. 1267-1289.

[4-6]   T. W. Mullikin, "Radiative Transfer in Finite Homogeneous Atmospheres with Anisotropic Scattering, I: Linear Singular Equations," *Astrophysical Journal,* Vol. 139, No. 1, 1964(b), pp. 379-396.

[4-7]   T. W. Mullikin, "Chandrasekhar's X and Y Equations," *Transactions of the American Mathematical Society,* American Mathematical Society, Providence, RI, Vol. 113, No. 2, 1964(c).

[4-8]   T. W. Mullikin, "The Complete Rayleigh-Scattered Field Within a Homogeneous Plane Parallel Atmosphere," *Astrophysical Journal,* Vol. 145, No. 3, 1966, pp. 886-931.

[4-9]   Z. Sekera, "Reduction of Equations of Radiative Transfer in a Planetary Plane-Parallel Atmosphere," RM-4951-PR and RM-5056-PR, The RAND Corporation, Santa Monica, CA, 1966.

[4-10] Z. Sekera, "Radiative Transfer in a Planetary Atmosphere with Imperfect Scattering," The RAND Corporation, Santa Monica, CA, RAND Publication R-413-PR, 1963.

[4-11] J. W. Chamberlain and M. B. McElroy, "Diffused Reflection by an Inhomogeneous Planetary Atmosphere," *Astrophysical Journal,* Vol. 144, 1966, p. 1148.

[4-12] A. L. Fymat and K. D. Abhyankar, "Theory of Radiative Transfer in Inhomogeneous Atmospheres, I. Perturbation Method," *Astrophysical Journal,* No. 158, 1969, pp. 325-335.

[4-13] K. M. Case, "Elementary Solutions of the Transport Equation and Their Applications," *Annalen der Physik,* Vol. 9, 1960, pp. 1-23.

[4-14] K. M. Case, "Recent Developments in Neutron Transport Theory," *Lectures Presented at the Neutron Physics Conference,* Michigan Memorial Phoenix Project, University of Michigan, Ann Arbor, 1961.

[4-15] K. M. Case and P. F. Zweifel, *Linear Transport Theory,* Addison-Wesley Publishing Company, Reading, MA, 1967.

[4-16] E. Inonu and P. F. Zweifel (eds.), *Developments in Transport Theory,* Academic Press, New York, NY, 1967.

[4-17] N. J. McCormick and I. Kuscer, "Half-Space Neutron Transport with Linearly Anisotropic Scattering," *Journal of Mathematical Physics,* Vol. 6, 1965, pp. 1939-1945.

[4-18] K. M. Case, "On the Boundary Value Problems on Linear Transport Theory," *Transport Theory, Proceedings of a Symposium in Applied Mathematics of the American Mathematical Society and the Society for Industrial and Applied*

Mathematics, Providence, RI, R. Bellman, G. Birkhoff and I. Abu-Shumays, (eds.) Vol. I, 1969.

[4-19] M. N. Ozisik, *Radiative Transfer and Interactions with Conduction and Convection*, John Wiley & Sons, NY, 1973.

[4-20] B. W. Roos, *Analytic Functions and Distributions in Physics and Engineering*, John Wiley & Sons, NY, 1969.

[4-21] E. Hopf, *Mathematical Problems of Radiative Equilibrium*, Cambridge University Press, 1934.

[4 22] B. M. Herman and S. R. Browning, "A Numerical Solution to the Equation of Radiative Transfer," *Journal of Atmospheric Sciences*, Vol. 22, No. 5, 1965, pp. 559–566.

[4-23] B. M. Herman, S. R. Browning, and R. J. Curran, "The Effect of Atmospheric Aerosols on Scattered Sunlight," *Journal of Atmospheric Sciences*, Vol. 28, No. 3, 1971, pp. 419–428.

[4-24] J. Canosa and H. R. Penafiel, "A Direct Solution of the Radiation Transfer Equation: Application to Rayleigh and Mie Atmospheres," *Journal of Quantitative Spectroscopy and Radiative Transfer*, Vol. 13, 1973, pp. 21-39.

[4-25] G. C. Wick, "Über ebene Diffusions Probleme," *Zeitschrift Fur Physik*, Vol. 121, 1943, p. 702.

[4-26] C. Whitney, "Implications of a Quadratic Stream Definition in Radiative Transfer Theory, " *Journal of Atmospheric Sciences*, Vol. 29, No. 8, November 1972, pp. 1520-1530.

[4-27] H. C. van de Hulst and K. Grossman, "Multiple Light Scattering in Planetary Atmospheres," *The Atmospheres of Venus and Mars*, J. C. Brandt and M. B. McElroy (eds.), Gordon and Breach, NY, 1968.

[4-28] A. Schuster, "Radiation Through a Foggy Atmosphere," *AstroPhysical Journal*, Vol. 21, No. 1, 1905.

[4-29] K. Schwarzschild, "Uber das Gleichgewicht der Sonnenatmosphere," *Gottinger Nachrichten*, Vol. 41, 1906.

[4-30] A. S. Eddington, *The Interval Constitution of the Stars*, Cambridge University Press, 1926.

[4 31] L. M. Romanova, "The Solution of the Radiative-Transfer Equation for the Case When the Indicatrix of Scattering Greatly Differs from the Spherical One, I," *Optika I Spektroskopiia*, Vol. 13, 1962 (a), p. 429.

[4-32] L. M. Romanova, "Solution of the Radiative-Transfer Equation for the Case of a Highly Nonspherical Scattering Index, II," *Optika I Spektroskopiia*, Vol. 13, 1962(b), p. 819.

[4-33] L. M. Romanova, "Radiation Field in Plane Layers of a Turbid Medium with Highly Anisotropic Scattering," *Optika I Spektroskopiia*, Vol. 14, 1963, p. 262.

[4-34] M. C. Wang and E. Guth, "On the Theory of Multiple Scattering, Particularly of Charged Particles," *Physical Review*, Vol. 84, 1951, p. 1092.

[4-35] R. E. Turner, "Remote Sensing in Hazy Atmospheres," *Proceedings of American Congress on Surveying and Mapping*, American Society of Photogrammetry, Meeting in Washington, DC, March 1972, American Society of Photogrammetry, Falls Church, VA.

[4-36] R. E. Turner, "Atmospheric Effects in Remote Sensing," *Remote Sensing of Earth Resources*, F. Shahrokhi (ed.), University of Tennessee, Vol. II, 1973.

[4 37] R. E. Turner, "Contaminated Atmospheres and Remote Sensing," *Remote Sensing of Earth Resources*, F. Shahrokhi (ed.), University of Tennessee, Vol. III, 1974.

[4-38] R. E. Turner, "Radiative Transfer on Real Atmospheres," Environmental Research Institute of Michigan, Ann Arbor, MI, Report No. 190100-24-T, July 1974.

[4-39] Anthony J. LaRocca and Robert E. Turner, "Atmospheric Transmittance and Radiance: Methods of Calculation," *Infrared Information and Analysis Center State-of-the-Art Report*, Infrared and Optics Division, Environmental Research Institute of Michigan, Ann Arbor, MI, Contract Nos. N00014-73-A-0321-0002 and N00014-74-C-0285, June 1975, pp. 57, 76, 78, 79, 81, 82, 83.

[4-40] B. Edlen, "The Dispersion of Standard Air," *Journal of Optical Society of America*, Vol. 43, 1953, p. 339.

[4-41] L. Elterman, "UV, Visible, and IR Attenuation for Altitudes to 50 km," Air Force Cambridge Research Laboratories, Bedford, MA, Report No. AFCRL-68-0153, 1968.

[4-42] C. E. Junge, "Our Knowledge of the Physics-Chemistry of Aerosols in the Undisturbed Marine Environment," *Journal of Geophysical Research*, Vol. 77, No. 27, 1972, pp. 5183-5200.

[4-43] D. Deirmendjian, *Electromagnetic Scattering On Spherical Polydispersions*, American Elsevier Publishing Company, New York, NY, 1969.

[4-44] J. Kuriyan and Z. Sekera, "Scattering in Liquid Haze—Analytic Approximations, *Quarterly Journal of the Royal Meteorological Society*, Vol. 100, 1974, pp. 67–75.

[4-45] J. P. Friend, "Properties of the Stratosphere Aerosol," *Tellus*, Vol. 18, 1966, pp. 465–473.

[4-46] O. B. Toon and J. P. Pollack, "A Global Average Model of Atmospheric Aerosols for Radiative Transfer Calculations," *Journal of Applied Meteorology*, Vol. 15, No. 3, 1976, pp. 225-246.

[4-47] E. P. Shettle and R. W. Fenn, "Models of Atmospheric Aerosols and Their Optical Properties," Paper presented at Electromagnetic Wave Propagation Panel of the Advisory Group for Aerospace Research and Development, 22nd Technical Meeting on *Optical Propagation in the Atmosphere*, The Technical University of Denmark, Lyngby, Denmark, 27-31 October 1975.

[4-48] F. E. Volz, "Infrared Refractive Index of Atmospheric Aerosol Substances," *Applied Optics*, Vol. 11, 1972, pp. 755-759.

[4-49] J. T. Twitty and J. A. Weinman, "Radiative Properties of Carbonaceous Aerosols," *Journal of Applied Meteorology*, Vol. 10, 1971, pp. 725-731.

[4-50] G. M. Hale and M. R. Querry, "Optical Constants of Water in the 200 nm to 200 μm Wavelength Region," *Applied Optics*, Vol. 12, 1973, pp. 555–563.

[4-51] K. F. Palmer and D. Williams, "Optical Constants of Sulphuric Acid: Applications to the Clouds of Venus," *Applied Optics*, Vol. 14, 1975, pp. 208-219.

[4-52] F. E. Volz, "Infrared Optical Constants of Ammonium Sulphate, Sahara Dust, Volcanic Pumice and Flyash," *Applied Optics*, Vol. 12, 1973, pp. 564-568.

[4-53] B. M. Herman, S. R. Browning, and R. Rabenoff, *The Change in Earth-Atmosphere Albedo Due to Stratospheric Pollution*, manuscript in preparation, 1976.

[4-54] L. S. Ivlev and S. I. Popova, "The Complex Index of Refraction of the Matter of the Disperse Phase of the Atmospheric Aerosol," *Atmospheric Oceanic Physics*, Vol. 9, No. 10, 1973, pp. 587-591.

[4-55] J. T. Peterson and J. A. Weinman, "Optical Properties of Quartz Dust Particles at IR Wavelengths," *Journal of Geophysical Research*, Vol. 74, 1969, pp. 6947-6952.

[4-56]　G. Hänel, "The Properties of Atomspheric Aerosol Particles as Functions of the Relative Humidity at Thermodynamic Equilibrium with the Surrounding Moist Air," *Advances in Geophysics,* Academic Press, New York, NY, Vol. 19, 1976, pp. 73-188.

[4-57]　E. E. Remsberg, *Radiative Properties of Several Probable Constituents of Atmospheric Aerosols,* Ph.D. Thesis, Department of Meteorology, University of Wisconsin, Madison, WI, 1971.

4.8.2 Bibliography

Ambartsumian, V. A., "Diffuse Reflection of Light by a Foggy Medium," *Comptes Rendus,* Doklady (English Translation), Plenum Publishing Company, New York, NY, Vol. 38, 1943.

Bellman, R. E., R. E. Kalaba, and M. Prestrud, *Invariant Imbedding and Radiative Transfer in Slabs of Finite Thickness,* American Elsevier Publishing, New York, NY, 1963.

Bellman, R. E., H. H. Kagiwada, R. E. Kalaba, and M. Prestrud, *Invariant Imbedding and Time Dependent Processes,* American Elsevier Publishing, New York, NY, 1964.

Bellman, R. E., H. H. Kagiwada, and R. E. Kalaba, "Time-Dependent Diffuse Reflection From Slabs with Multiple Scattering," The RAND Corporation, Santa Monica, CA, Memorandum RM-5070-PR, July 1966.

Bellman, R. E. and R. E. Kalaba, "On the Principle of Invariant Imbedding and Propagation Through Inhomogeneous Media," *Proceedings of the National Academy of Sciences,* National Research Council, Washington, DC, Vol. 42, 1956.

Bigg, E. K., and A. Ono, "Size Distribution and Nature of the Stratospheric Aerosols," *Proceedings, International Conference on Structure, Composition and General Circulation of the Upper and Lower Atmospheres and Possible Anthropogenic Perturbations,* International Association of Meteorology and Atmospheric Physics, Atmospheric Environment Service, Downsview, Ontario, Canada, 1974.

Busbridge, I. W., "On the X- and Y- Functions of S. Chandrasekhar," *Astrophysical Journal,* American Astronomical Society, University of Chicago Press, Chicago, IL, Vol. 122, 1955, p. 327.

Busbridge, I. W., *The Mathematics of Radiative Transfer,* Cambridge University Press, New York, NY, 1960.

Canosa, J., and H. R. Penafiel, "A Direct Solution of the Radiation Transfer Equation: Application to Rayleigh and Mie Atmospheres," *Journal of Quantitative Spectroscopy and Radiative Transfer,* Pergamon Press, Elmsford, NY, Vol. 13, 1973, pp. 21-39.

Carlson, B. G., and K. D. Lathrop, "Transport Theory, The Method of Discrete Ordinates," *Computing Methods in Reactor Physics,* H. Greenspan, C. N. Kelber and D. Okrent (eds.), Gordon and Breach, New York, NY, 1968.

Case, K. M., "Elementary Solutions of the Transport Equation and Their Applications," *Annalen der Physik,* Springer Verlag, New York, NY, Vol. 9, 1960, pp. 1-23.

Case, K. M., "Recent Developments in Neutron Transport Theory," *Lectures Presented at the Neutron Physics Conference,* Michigan Memorial Phoenix Project, University of Michigan, Ann Arbor, MI, 1961.

Case, K. M. and P. F. Zweifel, *Linear Transport Theory,* Addison-Wesley Publishing Company, Reading, MA, 1967.

Case, K. M., "On the Boundary Value Problems of Linear Transport Theory," Transport Theory, *Proceedings of a Symposium in Applied Mathematics of the American Mathematical Society and The Society for Industrial and Applied Mathematics,* R. Bellman, G. Birkhoff and I. Abu-Shumays (eds.), Providence, RI, Vol. 1, 1969.

Chamberlain, J. W. and M. B. McElroy, "Diffused Reflection by an Inhomogeneous Planetary Atmosphere," *Astrophysical Journal,* American Astronomical Society, University of Chicago Press, Chicago, IL, Vol. 144, 1966, p. 1148.

Chandrasekhar, S., *Radiative Transfer,* Oxford University Press, New York, NY, 1950.

Collins, D. G., "Atmospheric Path Radiance Calculations for a Model Atmosphere," Air Force Cambridge Research Laboratories, Bedford, MA, Report No. AFCRL-68-0124, 1968.

Collins, D. G., and M. B. Wells, *Monte Carlo Codes for Study of Light Transport in the Atmosphere,* Vol. 1, *Description of Codes,* Vol. II, *Utilization,* U. S. Army ECOM, Ft. Monmouth, NJ, ECOM-00240-FI, ECOM-00240-FII, 1965.

Coulson, K. L., J. V. Dave, and Z. Sekera, "Tables Related to Radiation Emerging from a Planetary Atmosphere with Rayleigh Scattering," University of California Press, Berkeley, CA, 1960.

Davison, B., *Neutron Transport Theory,* Oxford University Press, New York, NY, 1957.

Deirmendjian, D., *Electromagnetic Scattering on Spherical Polydispersions,* American Elsevier Publishing, New York, NY, 1969.

Eddington, A. S., *The Internal Constitution of the Stars,* Cambridge University Press, New York, NY, 1926.

Edlen, B., "The Dispersion of Standard Air," *Journal of the Optical Society of America,* Optical Society of America, Washington, DC, Vol. 43, 1953, p. 339.

Elterman, L., "UV, Visible, and IR Attenuation for Altitudes up to 59 km," Air Force Cambridge Research Laboratories, Bedford, MA, Report No. AFCRL-68-0153, 1968.

Elterman, L., R. Wexler, and D. T. Chang, "Features of Tropospheric and Stratospheric Dust," *Applied Optics,* Optical Society of America, Washington, DC, Vol. 8, 1969, pp. 893-902.

Feigelson, E. M., M. S. Malkevich, S. Ya. Kogan, T. D. Koronatova, K. S. Glazova, and M. A. Kuznetsova, *Calculation of the Brightness of Light in the Case of Anisotropic Scattering,* Consultants Bureau, New York, NY, 1960.

Ferry, G. V., and H. Y. Lem, "Aerosols in the Stratosphere," *Proceedings, Third Conference of Climatic Impact Assessment Program,* A. Broderick and T. Hard (eds.), National Technical Information Service, Springfield, VA, DOT-TSC-OST-74-15, 1974.

Friend, J. P., "Properties of the Stratosphere Aerosol," *Tellus, A Quarterly Journal of Geophysics,* Swedish Geophysical Society, Stockholm, Sweden, Vol. 18, 1966, pp. 465-473.

Fymat, A. L., and K. D. Abhyankar, "Theory of Radiative Transfer in Inhomogeneous Atmospheres, I. Perturbation Method," *Astrophysical Journal,* American Astronomical Society, University of Chicago Press, Chicago, IL, Vol. 158, 1969, pp. 325-335.

Gelbard, E. M., "Spherical Harmonics Methods: P_l and Double P_l Approximations," *Computing Methods in Reactor Physics,* H. Greenspan, C. N. Kelber and D. Okrent (eds.), Gordon and Breach, New York, NY, 1968.

Goody, R. M., *Atmospheric Radiation, I,* Oxford University Press, New York, NY, 1964.

Grant, I. P., and G. E. Hunt, "Discrete Space Theory of Radiative Transfer, I: Fundamentals," *Proceedings, Royal Society of London,* London, England, A., Vol. 313, 1969 (a).

Grant, I. P., and G. E. Hunt, "Discrete Space Theory of Radiative Transfer, II: Stability and Non-Negativity," *Proceedings, Royal Society of London,* London, England, A., Vol. 313, 1969 (b).

Gucker, F. T., and S. Basu, "Right-Angle Molecular Light Scattering from Gases," University of Indiana, Bloomington, IN, Scientific Report No. 1, Contract AF 10(122)-400, 1953.

Guillemot, "Contribution a L'Etude du Transfert de Rayonnement dans les Nuages par la Methods des Harmonique Spheriques," *Revue d'Optique*, Paris, France, Vol. 46, No. 6, 1967.

Hale, G. M., and M. R. Querry, "Optical Constants of Water in the 200 nm to 200 μm Wavelength Region," *Applied Optics*, Optical Society of America, Washington, DC, Vol. 12, 1973, pp. 555-563.

Hänel, G. "The Properties of Atmospheric Aerosol Particles as Functions of the Relative Humidity at Thermodynamic Equilibrium with the Surrounding Moist Air," *Advances in Geophysics*, Academic Press, New York, NY, Vol. 19, 1976, pp. 73-188.

Hänel, G., "Ratio of the Extinction Coefficient to the Mass of Atmospheric Aerosol Particles as a Function of the Relative Humidity," *Journal of Aerosol Science*, Pergamon Press, Oxford, England, Vol. 3, 1972.

Hansen, J. E., "Exact and Approximate Solutions for Multiple Scattering by Cloudy and Hazy Planetary Atmospheres," *Journal of Atmospheric Sciences*, American Meteorological Society, Boston, MA, Vol. 26, No. 3, May 1969 (a), pp. 478-487.

Hansen, J. E., "Radiative Transfer by Doubling Very Thin Layers," *Astrophysical Journal*, American Astronomical Society, University of Chicago Press, Chicago, IL, Vol. 155, February 1969(b).

Herman, B. M., and S. R. Browning, "A Numerical Solution to the Equation of Radiative Transfer," *Journal of Atmospheric Sciences*, American Meteorological Society, Boston, MA, Vol. 22, No. 5, 1965, pp. 559-566.

Herman, B. M., S. R. Browning, and R. J. Curran, "The Effect of Atmospheric Aerosols on Scattered Sunlight," *Journal of Atmospheric Sciences*, American Meteorological Society, Boston, MA, Vol. 28, No. 3, 1971, pp. 419-428.

Herman, B. M., S. R. Browning, and R. Rabenoff, *The Change in Earth-Atmosphere Albedo Due to Stratospheric Pollution*, manuscript in preparation, 1976.

Hildebrand, F. B., *Introduction to Numerical Analysis*, McGraw-Hill, New York, NY, 1956.

Hildebrand, F. B., *Methods of Applied Mathematics*, Prentice-Hall, New York, NY, 1952.

Hopf, E., *Mathematical Problems of Radiative Equilibrium*, Cambridge University Press, New York, NY, 1934.

Hunt, G. E., and I. P. Grant, "Discrete Space Theory of Radiative Transfer and Its Application to Problems in Planetary Atmospheres," *Journal of Atmospheric Sciences*, American Meteorological Society, Boston, Ma, Vol. 26, September 1969, pp. 963-972.

Inonu, E., and P. F. Zweifel (eds.), *Developments in Transport Theory*, Academic Press, New York, NY, 1967.

Irvine, W. M., "Multiple Scattering by Large Particles, II: Optically Thick Layers," *Astrophysical Journal*, American Astronomical Society, University of Chicago Press, Chicago, IL. Vol. 152, June 1968.

Irvine, W. M., "An Evaluation of Romanova's Method in the Theory of Radiative Transfer," *The Atmospheres of Venus and Mars*, J. C. Brandt and M. B. McElroy (eds.), Gordon and Breach, New York, NY, 1968.

Ivlev, L. S., and S. I. Popova, "The Complex Index of Refraction of the Matter of the Disperse Phase of the Atmospheric Aerosol," *Atmospheric Oceanic Physics*, American Geophysical Union, Washington, DC, Vol. 9, No. 10, 1973, pp. 587-591.

Junge, C. E., "Atmospheric Chemistry," *Advances in Geophysics*, H. E. Landsberg (ed.), Academic Press, New York, NY, Vol. 4, 1958.

Junge, C. E., "Our Knowledge of the Physics-Chemistry of Aerosols in the Undisturbed Marine Environment," *Journal of Geophysical Research*, American Geophysical

Union, Washington, DC, Vol. 77, No. 27, 1972, pp. 5183-5200.

Kagiwada, H. H., and R. E. Kalaba, "Estimation of Local Anisotropic Scattering Properties Using Measurements of Multiply Scattered Radiation," *Journal of Quantitative Spectroscopy and Radiative Transfer*, Pergamon Press, Elmsford, NY, Vol. 7, 1967.

Kattawar, G. W., and G. N. Plass, "Infrared Cloud Radiance," *Applied Optics*, Optical Society of America, Washington, DC, Vol. 8, No. 6, June 1969, pp. 1169-1178.

Kofink, W., "Recent Developments in the Spherical Harmonics Method and New Integral Solution of the Boltzmann Equation in Spherical Geometry," *Developments in Transport Theory*, E. Inonu and P. F. Zweifel (eds.), Academic Press, New York, NY, 1967.

Kondratiev, K. Ya., L. S. Ivlev and G. W. Nikolsky, "Complex Investigations of the Stratospheric Aerosols," *Proceedings, Third Conference of Climatic Impact Assessment Program*, A. Broderick and T. Hard (eds.), National Technical Information Service, Springfield, VA, DOT-TSC-OST-74-15, 1974.

Kourganoff, V., *Basic Methods in Transfer Problems*, Dover Publications, New York, NY, 1963.

Krook, M., "On the Solution of Equation of Transfer," *Astrophysical Journal*, American Astronomical Society, University of Chicago Press, Chicago, IL, Vol. 122, 1955, p.488.

Kuriyan, J., and Z. Sekera, "Scattering in Liquid Haze – Analytic Approximations," *Quarterly Journal of the Royal Meteorological Society*, Royal Meteorological Society, Berkshire, England, Vol. 100, 1974, pp. 67-75.

LaRocca, Anthony J. and Robert E. Turner, "Atmospheric Transmittance and Radiance: Methods of Calculation," *Infrared Information and Analysis Center State-of-the-Art Report*, Infrared and Optics Division, Environmental Research Institute of Michigan, Ann Arbor, MI, Contract Nos. N00014-73-A-0321-0002 and N00014-74-C-0285, June 1975, pp. 57, 76, 78, 79, 81, 82, 83.

Lathrop, K. D., "Discrete Ordinates Methods for the Numerical Solution of the Transport Equation," *Reactor Technology*, U. S. Government Printing Office, Washington, DC, Vol. 15, No. 2, 1972.

Lee, C. E., "The Discrete S_N Approximation for Transport Theory," Los Alamos Scientific Laboratory for the University of California, Los Alamos, NM, Report No. LA-2595, 1962.

Malila, W. A., R. B. Crane, and R. E. Turner, "Information Extraction Techniques for Multispectral Scanner Data," Willow Run Laboratories of the Institute of Science and Technology, University of Michigan, Ann Arbor, MI, Report No. 31650-74-T, 1972.

Marengo, J., "Application Numerique de la Methode des Harmoniques Spheriques," *Nouvelle Revue d'Optique Appliquie*, Saint-Germain, Paris, France, Vol. 1, No. 3, 1970.

McCormick, N. J. and I. Kuscer, "Half-Space Neutron Transport with Linearly Anisotropic Scattering," *Journal of Mathematical Physics*, American Institute of Physics, New York, NY, Vol. 6, 1965, pp. 1939-1945.

Mossop, I. C., "Volcanic Dust Collected at an Altitude of 20 km," *Nature*, Macmillan Journals, Ltd., Washington, DC, Vol. 203, 1964.

Mullikin, T. W., "Radiative Transfer in Finite Homogeneous Atmospheres with Anisotropic Scattering, II: The Uniqueness Problem for Chandrasekhar's ψ_l and ϕ_l Equations," *Astrophysical Journal*, American Astronomical Society, University of Chicago Press, Chicago, IL, Vol. 139, No. 4, 1964 (a), pp. 1267-1289.

Mullikin, T. W., "Radiative Transfer in Finite Homogeneous Atmospheres with Anisotropic Scattering, I: Linear Singular Equations," *Astrophysical Journal*, American

Astronomical Society, University of Chicago Press, Chicago, IL, Vol. 139, No. 1, 1964(b), pp. 379-396.

Mullikin, T. W., "Chandrasekhar's X and Y Equations," *Transactions of the American Mathematical Society*, American Mathematical Society, Providence, RI, Vol. 113, No. 2, 1964(c).

Mullikin, T. W., "The Complete Rayleigh-Scattered Field Within a Homogeneous Plane Parallel Atmosphere," *Astrophysical Journal*, American Astronomical Society, University of Chicago Press, Chicago, IL, Vol. 145, No. 3, 1966, pp. 886-931.

Newkirk, G., and J. A. Eddy, "Light Scattering by Particles in the Upper Atmosphere," *Journal of Atmospheric Sciences*, American Meteorological Society, Boston, MA, Vol. 21, 1964.

Ozisik, M. N., *Radiative Transfer and Interactions with Conduction and Convection*, John Wiley & Sons, New York, NY, 1973.

Palmer, K. F., and D. Williams, "Optical Constants of Sulphuric Acid: Applications to the Clouds of Venus," *Applied Optics*, Optical Society of America, Washington, DC, Vol. 14, 1975, pp. 208-219.

Peterson, J. T., and J. A. Weinman, "Optical Properties of Quartz Dust Particles at IR Wavelengths," *Journal of Geophysical Research*, American Geophysical Union, Washington, DC, Vol. 74, 1969, pp. 6947-6952.

Plass, G. N., and G. W. Kattawar, "Influence of Single Scattering Albedo on Reflected and Transmitted Light from Clouds," *Applied Optics*, Optical Society of America, Washington, DC, Vol. 7, No. 2, February 1968, pp. 361-367.

Preisendorfer, R., *Radiative Transfer on Discrete Spaces*, Pergamon Press, Oxford, England, 1965.

Remsberg, E. E., *Radiative Properties of Several Probable Constituents of Atmospheric Aerosols*, Ph.D. Thesis, Department of Meteorology, University of Wisconsin, Madison, WI, 1971.

Remsberg, E. E., "Stratospheric Aerosol Properties and Their Effects on Infrared Radiation," *Journal of Geophysical Research*, American Geophysical Union, Washington, DC, Vol. 78, 1973.

Romanova, L. M., "The Solution of the Radiative-Transfer Equation for the Case When the Indicatrix of Scattering Greatly Differs from the Spherical One, I," *Optika I Spektroskopiia*, Johnson Reprint Corporation, New York, NY, Vol. 13, 1962(a), p. 429.

Romanova, L. M., "Solution of the Radiative-Transfer Equation for the Case of a Highly Nonspherical Scattering Index, II," *Optika I Spektroskopiia*, Johnson Reprint Corporation, New York, NY, Vol. 13, 1962(b), p. 819.

Romanova, L. M., "Radiation Field in Plane Layers of a Turbid Medium with Highly Anisotropic Scattering," *Optika I Spektroskopiia*, Johnson Reprint Corporation, New York, NY, Vol. 14, 1963, p. 262; Also, English Translation: *Optics and Spectroscopy*, Optical Society of America, Washington, DC, Vol. 14, pp. 135-138.

Roos, B. W., *Analytic Functions and Distributions in Physics and Engineering*, John Wiley & Sons, New York, NY, 1969.

Rosen, J. M., "The Boiling Point of Stratospheric Aerosols," *Journal of Applied Meteorology*, American Meteorological Society, Boston, MA, Vol. 10, 1971.

Rosen, J. M., "Stratospheric Dust and Its Relationship to the Meteoric Influx," *Space Science Reviews*, D. Reidel Publishing, Boston, MA, Vol. 9, 1969.

Schuster, A., "Radiation Through a Foggy Atmosphere," *AstroPhysical Journal*, American Astronomical Society, University of Chicago Press, Chicago, IL, Vol. 21, No. 1, 1905.

Schwarzschild, "Uber das Gleichgewicht der Sonnenatmosphere," *Gottinger Nachrichten*, Vol. 41, 1906.

Sekera, Z., "Radiative Transfer in a Planetary Atmosphere with Imperfect Scattering," The RAND Corporation, Santa Monica, CA, RAND Publication R-413-PR, 1963.

Sekera, Z., "Reduction of Equations of Radiative Transfer in a Planetary Plane-Parallel Atmosphere, The RAND Corporation, Santa Monica, CA, RM-4951-PR and RM-5056-PR, 1966.

Sherman, M. P., "Moment Methods in Radiative Transfer Problems," *Journal of Quantitative Spectroscopy and Radiative Transfer,* Pergamon Press, Elmsford, NY, Vol. 7, 1967.

Shettle, E. P., and R. W. Fenn, "Models of Atmospheric Aerosols and Their Optical Properties, " Paper presented at Electromagnetic Wave Propagation Panel of the Advisory Group for Aerospace Research and Development, 22nd Technical Meeting on *Optical Propagation in the Atmosphere,* The Technical University of Denmark, Lyngby, Denmark, 27-31 October 1975.

Shettle, E. P., and J. A. Weinman, "The Transfer of Solar Irradiance through Inhomogeneous Turbid Atmosphere Evaluated in Eddington's Approximation," *Journal of Atmospheric Sciences,* American Meteorological Society, Boston, MA, Vol. 22, No. 1048, 1971.

Sobolev, V. V., *Light Scattering in Planetary Atmospheres,* Pergamon Press, New York, NY, 1975.

Toon, O. B., and J. B. Pollack, "A Global Average Model of Atmospheric Aerosols for Radiative Transfer Calculations," *Journal of Applied Meteorology,* American Meteorological Society, Boston, MA, Vol. 15, No. 3, 1976, pp. 225-246.

Toon, O. B., and J. B. Pollack, "Physical Properties of the Stratospheric Aerosol," *Journal of Geophysical Research,* American Geophysical Union, Washington, DC, Vol. 78, 1973.

Turner, R. E., "Atmospheric Effects in Multispectral Remote Sensor Data," Environmental Research Institute of Michigan, Ann Arbor, MI, Final Report No. 109600-15-F, 1975.

Turner, R. E., "Atmospheric Effects in Remote Sensing," *Remote Sensing of Earth Resources,* F. Shahrokhi (ed.), University of Tennessee, Vol. II, 1973.

Turner, R. E. "Contaminated Atmospheres and Remote Sensing," *Remote Sensing of Earth Resources,* F. Shahrokhi (ed.), University of Tennessee, Vol. III, 1974.

Turner, R. E., "Investigation of Earth's Aerosol Albedo Using Skylab Data," Environmental Research Institute of Michigan, Ann Arbor, MI, Final Report, ERIM 102200-20-F, 1976.

Turner, R. E., "Radiative Transfer on Real Atmospheres," Environmental Research Institute of Michigan, Ann Arbor, MI, Report No. 190100-24-T, July 1974.

Turner, R. E., "Remote Sensing in Hazy Atmospheres," *Proceedings of American Congress on Surveying and Mapping, American Society of Photogrammetry,* American Society of Photogrammetry meeting in Washington, DC, March 1972, American Society of Photogrammetry, Falls Church, VA.

Twitty, J. T., and J. A. Weinman, "Radiative Properties of Carbonaceous Aerosols," *Journal of Applied Meteorology,* American Meteorological Society, Boston, MA, Vol. 10, 1971, pp. 725-731.

Ueno, S., H. Kagiwada, and R. Kalaba, "Radiative Transfer in Spherical Shell Atmospheres with Radial Symmetry," *Journal of Mathematical Physics,* American Institute of Physics, New York, NY, Vol. 12, No. 6, 1971.

van de Hulst, H. C., and K. Grossman, "Multiple Light Scattering in Planetary Atmospheres," *The Atmospheres of Venus and Mars,* J. C. Brandt and M. B. McElroy (eds.), Gordon and Breach, New York, NY, 1968.

Volz, F. E., "Infrared Refractive Index of Atmospheric Aerosol Substances," *Applied Optics,* Optical Society of America, Washington, DC, Vol. 11, 1972, pp. 755-759.

Volz, F. E., "Infrared Optical Constants of Ammonium Sulphate, Sahara Dust, Volcanic Pumice and Flyash," *Applied Optics,* Optical Society of America, Washington, DC, Vol. 12, 1973, pp. 564-568.

Wang, M. C., and E. Guth, "On the Theory of Multiple Scattering, Particularly of Charged Particles," *Physical Review,* American Institute of Physics, New York, NY, Vo. 84, 1951, p. 1092.

Wells, M. B., and J. D. Marshall, "Monochromatic Light Intensities Above the Atmosphere Resulting from Atmospheric Scattering and Terrestrial Reflection," Radiative Research Association, Ft. Worth, TX, 1968.

Wick, G. C., "Über ebene Diffusions Probleme," *Zeitschrift Für Physik,* New York, NY, Vol. 121, 1943, p. 702.

Williams, M. M. R., *Mathematical Methods in Particle Transport Theory,* Wiley-Interscience, New York, NY, 1971.

Whitby, K. T., R. B. Husan, and B. Y. H. Liu, "The Aerosol Size Distribution of Los Angeles Smog," *Journal of Colloid and Interface Science,* Academic Press, New York, NY, Vol. 39, 1972.

Whitney, C., "Implications of a Quadratic Stream Definition in Radiative Transfer Theory," *Journal of Atmospheric Sciences,* American Meteorological Society, Boston, MA, Vol. 29, No. 8, November 1972, pp. 1520-1530.

Yvon, J., "La Diffusion Macroscopique des Neutrons: Une Methode d' Approximation," *Journal of Nuclear Energy,* Pergamon Press Elmsford, New York, NY, Vol. 4, 1957.

Zuyev, V. Ye., L. S. Ivlev, and K. Ya. Kondratiev, "Recent Results from Studies of Atmospheric Aerosols," *Atmospheric Oceanic Physics,* American Geophysical Union, Washington, DC, Vol. 9, 1973.

Chapter 5

ATMOSPHERIC ABSORPTION

Anthony J. LaRocca
Environmental Research Institute of Michigan

CONTENTS

5. Atmospheric Absorption

5.1. Introduction

This chapter presents simple, practical means for calculating atmospheric spectral transmittance and radiance to an accuracy of 5 to 10%. The theory of molecular band absorption is presented, briefly, to provide insight into the use of the mathematical models developed later for the performance of transmittance calculations in narrow regions of the spectrum.

A short description of molecular line structure is presented to show the mechanism for making precise calculations; however, emphasis is on methods using models and empirical techniques. Only the Aggregate and LOWTRAN methods are considered generally useful for the wide variety of problems needing knowledge of spectral transmittance from approximately 1 to 30 μm [5-1].

The *Aggregate method* is a compilation of various models. Each model is applied to the spectral regions for which it will give the highest accuracy. The specific models used for given spectral intervals, together with their formulae, are listed in the text. To calculate spectral transmittance, one introduces the appropriate atmospheric parameters into the formulae; once the spectral transmittance is known, the spectral radiance can be calculated.

All calculations assume a flat earth and standardized, horizontally-uniform, atmospheric conditions. Adaptation to slant-path conditions is made through the calculation of a so-called *equivalent absorber path*, using the same formulae as for uniform conditions. Most calculations are considered for short paths which make angles with the zenith or nadir of less than about 80°. Horizontal atmospheric paths (90°) will often be sufficiently short that the path can be assumed to be uniform. Within the acceptable 80° limit, the slant path is obtained by using the calculation for vertical paths and dividing by $\mu = \cos\theta$, where θ is the zenith angle. For cases in which accuracies between 5 and 10% are not essential it is feasible to carry out the calculation using the $\cos\theta$ approximation to as far as 87°. However, for angles between 80° and 90° the user is advised to consult Reference [5-2], in which angles in this range are considered, or use the flat-earth techniques in this chapter with some risk.

The *LOWTRAN method* (generated by AFGL*) is more empirically derived and simpler to use than the Aggregate method, although its output is slightly less accurate. Determination of transmittance using LOWTRAN is done in this chapter by use of AFGL graphs [5-2]. One graph is used to determine the equivalent horizontal or slant-path absorber amount and four others to calculate spectral transmittance due to absorption by atmospheric gases. Scattering is calculated by the use of yet another graph. As explained above, slant paths are obtained using the $\cos\theta$. Reference [5-2] should be consulted for cases not covered using the approximation.

---

*U.S. Air Force Geophysics Laboratory, formerly the U.S. Air Force Cambridge Research Laboratories.

So far (as of November, 1976), LOWTRAN has evolved through four major stages. LOWTRAN 3B is available as a printout [5-3]; however, graphs from LOWTRAN 3B calculations have not yet appeared. Thus, LOWTRAN 2 graphs are used here in the designated calculations. The compromise is not serious, particularly for absorption, since the authors claim that changes are slight. For scattering, the changes are more significant, but even then the compromise is acceptable because atmospheric models for scattering represent less predictable conditions than do absorption models.

Much of the rest of this chapter is designed to provide reference material for quick answers to a host of problems involving atmospheric phenomena, particularly absorption. A final section is devoted to a quantitative description of atmospheric constituents. The values used in the calculations in the earlier sections of this chapter are derived from the material of this last section.

5.1.1. Symbols, Nomenclature and Units. The symbols, nomenclature and units used in this chapter are summarized in Table 5-1.

Table 5-1. Symbols, Nomenclature and Units

Symbol	Nomenclature	Units
A	Absorptance (absorption)	—
A_1	Absorptance for a single line	—
a	Number density ratio	—
B	Self-broadening coefficient	—
b	Line shape factor	cm
C	Band model exponent for O_3; also H_2O continuum parameter	—
c	Velocity of light	cm sec^{-1}
D	Wavenumber interval for the quasi-random model; also particle density	—; cm^{-3}
d	Mean spacing between spectral lines	cm^{-1}
e, f	H_2O continuum constants	Molecule^{-1} cm^2 atm^{-1}
G	LOWTRAN spectral parameter	—
I_1, I_0	Modified Bessel functions	—
i	Integer or running index	—
J_ν	Source function	W cm^{-2} sr^{-1} cm
j	Integer or running index	—
K_1	Band model coefficient for H_2O	—
K_2	Band model coefficient for CO_2	—
K_3	Band model coefficient for O_3	—
K_4	Band model coefficient for O_3	—
K_5	Band model coefficient for N_2O	—
k	Mass extinction coefficient	cm^2 g^{-1}
k_a	Mass absorption coefficient	cm^2 g^{-1}
k_s	Mass scattering coefficient	cm^2 g^{-1}

Table 5-1. Symbols, Nomenclature and Units (*Continued*)

Symbol	Nomenclature	Units
L^*	Radiance of a Planck emitter	W cm^{-2} sr^{-1}
L_λ	Spectral radiance (in terms of wavelength)	W cm^{-2} sr^{-1} μm^{-1}
L_ν	Spectral radiance (in terms of wavenumber)	W cm^{-2} sr^{-1} cm
M, M'	Mixing ratio	g kg^{-1}, g g^{-1}
m	Molecular mass	g molecule^{-1}
N	Number of randomly superposed Elsasser bands	—
N_r	Particle size distribution	cm^{-3} μm^{-1}
n	Integer or running index	—
P	Total pressure	atm, mb, etc.
$P(S)$	Line strength distribution	—
p	Partial pressure	atm, mb
q	Optical depth	—
r	Particle radius	μm
S	Line strength	cm g^{-1}
s	Path length	cm, km
T	Temperature	K
t	Variable of integration	—
w	Absorber amount	g cm^{-2}
w^*	Equivalent absorber amount	g cm^{-2}
Y	Band model parameter	—
z	Vertical distance (height)	cm, km
α	Lorentz-line half-width	cm^{-1}
α_D	Doppler-line half width	cm^{-1}
α_o'	Line half-width per unit pressure, at standard conditions	cm^{-1} mb^{-1}
γ	H$_2$O continuum constant	cm
ϵ	Emissivity	—
κ	Boltzmann constant	erg K^{-1}
ζ	Spectral slit function (normalized)	—
λ	Wavelength	μm or cm
ν	Wavenumber	cm^{-1}
ρ	Density	kg m^{-3}, g cm^{-3}
Ψ	Band model parameter	—
τ	Transmittance	—
ξ	Slit function	—

5.2. The Formal Solution for Absorption Only

The formal solution for radiative transfer in the atmosphere is given in Chapter 4. This chapter concentrates on the effect of atmospheric absorption, with implicit emphasis on the spectral region beyond the visible.

5.2.1. Exact Solution. For simplicity, the radiative transfer equation (Equations (4-9) and (4-11)) is considered for vertically propagated radiation, in which case the value of μ becomes unity. The solution of the differential equation, for a flat earth is*

$$L_\nu(\nu,q_o) = L_\nu(\nu,q)\, e^{-(q_o-q)} + \int_q^{q_o} e^{-(q_o-q')} J_\nu(\nu,q')\, dq' \qquad (5\text{-}1)$$

where ν = wavenumber = $1/\lambda$, cm^{-1}
λ = wavelength, cm
$L_\nu(\nu,q_o)$ = spectral radiance, W cm^{-2} sr^{-1} cm, at observer station ($z=0$)
$L_\nu(\nu,q)$ = spectral radiance, W cm^{-2} sr^{-1} cm, at target station ($z=z$)
(q_o-q) = optical depth,† defined by

$$\int_0^z k(\nu,z')\rho\, dz' \quad \text{(See Chapter 4.)‡}$$

$k(\nu,z')$ = spectral mass extinction coefficient
ρ = density of absorber, g cm^{-3}
$J_\nu(\nu,q)$ = $L_\nu^*(\nu,q) = L_\nu^*(\nu,T(z))$ LTE
$J_\nu(\nu,q)$ = source function
$L_\nu^*(\nu,T(z))$ = Planck function (See Chapter 1.)
$T(z)$ = temperature, K, at altitude z

The source function differs from that in Chapter 4, Equation (4-10), in that the sun is neglected as a source. Scattering is considered later in this chapter as a secondary cause of extinction. Scattering into the path from sources external to the path is excluded.

By definition, the transmittance in a path between an initial point, z_1, and a final point, z_2, above z_1, for which the optical depths are q_1 and q_2, is given by

---

*Primes are used to designate variables of integration and the explicit spectral dependence is shown in ν rather than λ.

†The quantity optical depth is often defined without explanation, sometimes leading to confusion. By definition, the optical depth, $q(z)$, corresponding to some altitude, z, is:

$$q(z) = \int_z^\infty k(z')\rho(z')\, dz'$$

In Equation (5-1) the path is from the observer at altitude $z = 0$, to the target at $z = z$, that is (q_o-q), which is

$$(q_o-q) = \int_0^z k(z')\rho(z')\, dz'$$

There need be no confusion regarding the sign of (q_o-q), that is, whether it should be (q_o-q) or $(q-q_o)$. The sign of the integral must be positive, and the limits of integration, $0 \rightarrow z$ or $z \rightarrow 0$, must be fixed to make it so.

‡The observer, or detector, and the target can be at any desired location, z.

$$\frac{L_\nu(\nu,q_2)}{L_\nu(\nu,q_1)} = \tau(\nu,q_1,q_2) = \exp\{-(q_1-q_2)\} \tag{5-2}$$

where $\tau(\nu,q_1,q_2)$ is the spectral transmittance of the medium between q_1 and q_2. It follows that the expression within the integral in Equation (5-1) is equivalent to $L_\nu^*(\nu,q')\,d\tau(\nu,q')$ yielding

$$L_\nu(\nu,q_o) = L_\nu(\nu,q)\exp\{-(q_o-q)\} + \int_{\tau(q_o)}^{\tau(q)} L_\nu^*(\nu,q')\,d\tau(\nu,q') \tag{5-3}$$

5.2.2. Numerical Integration. Because the integral form of Equation (5-3) is almost never amenable to a closed-form solution, a numerical integration is performed. In the following equation, L_ν(Obs) is the radiance at the observer station at altitude z_o. The earth is considered to be flat and homogeneous in layers as shown in Figure 5-1. The case shown in the figure is for $z_o = 0$, although this need not be. The layers are separated by shells $z_o, z_1, \ldots z_n$. The observer is considered to be looking vertically upward at a target at shell z_n. Slant paths are considered later in the chapter by multiplying the optical path between observer and target by $\sec\theta$, where θ is the elevation angle between the vertical and the line-of-sight. Considerations for earth curvature are discussed later.

The numerical form of Equation (5-3) (referring to Figure 5-1) is written [5-1]:

$$L_\nu(\text{Obs}) = \epsilon_{\text{Targ}}(\nu)\,L_\nu^*(\text{Targ})\,\tau_n(\nu) + \sum_{i=1}^{n} L_\nu^*(\nu,T_i)\left[\tau_{i-1}(\nu) - \frac{\tau_i(\nu)}{\tau_{S_i}'(\nu)}\right] \tag{5-4}$$

where $\epsilon_{\text{Targ}}(\nu)$ = spectral emissivity of the target

$L_\nu^*(\text{Targ})$ = spectral blackbody radiance corresponding to the target temperature

$\tau_n(\nu)$ = spectral transmittance of the atmosphere between target and observer

$L_\nu^*(\nu,T_i)$ = spectral blackbody radiance corresponding to the average temperature, T_i, of the ith layer of the atmosphere

$\tau_{i-1}(\nu)$ = spectral transmittance of the atmosphere from the bottom of the ith layer to the observer

$\tau_i(\nu)$ = spectral transmittance of the atmosphere from the top of the ith layer to the observer

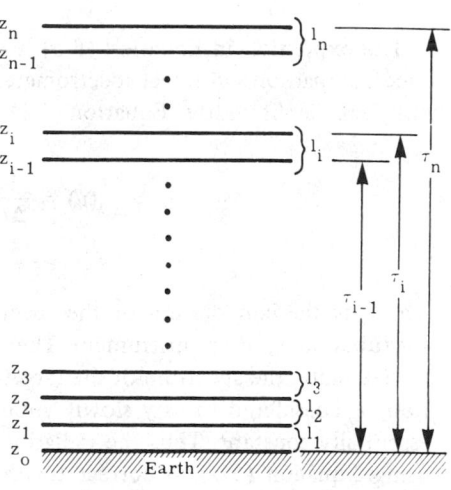

Fig. 5-1. Schematic and layered atmosphere for transmission and radiance calculations. Note: $\tau_o = 1$. z_i is the altitude of the ith layer. Observer is in the oth shell. Target is at the top of the nth shell.

$\tau'_{S_i}(\nu)$ = spectral transmittance of the ith layer accounting only for scattering (See below.)

The second term on the right side of Equation (5-4) represents the atmospheric radiation at the detector as emitted from every layer, modified by the transmittance of the atmosphere between the detector and the lower bound of the layer from which it originated. The term $\tau_i(\nu)$ is the transmittance through both absorbing and scattering media so that $\tau_i(\nu) = \tau_{a_i} \tau_{S_i}$, where τ_{a_i} is the transmittance accounting for absorption and τ_{S_i} is the transmittance by virtue of scattering. As used in this chapter, absorption refers to molecular absorption. The value τ'_{S_i} is the transmittance of just the ith layer as affected only by scattering. It appears in Equation (5-4) to account for the fact that the source function in the ith layer is a result of only emitted radiation, neglecting any scattering into the layer from outside the layer. Note, however, that the transmittance through the atmosphere below the emitting layer is affected by both absorption and scattering.

The solution of Equation (5-4) for a given value of ν is relatively simple once the values of $\tau_i(\nu)$ are known. The values of τ_i are obtained by methods described later in Section 5.5. To ascertain what might be obtained in measurements of target radiance with a spectrometer with a given spectral slit width (i.e., a spectral region of effective width, $\Delta\nu$), one must calculate $\bar{\tau}_{\Delta\nu}$. The value of $\Delta\nu$ is often of the order of 10 cm^{-1} or larger. In this chapter, $\bar{\tau}_{\Delta\nu}$ is represented by

$$\bar{\tau}_{\Delta\nu} = \frac{1}{\Delta\nu} \int_{\Delta\nu} \tau(\nu)d\nu \tag{5-5}$$

The expression in Equation (5-5) is an approximation from two points of view. First, since comparisons with real spectrometers require that spectral transmittance be modified with realistic slit widths, Equation (5-5) should actually be written as

$$\bar{\tau}_{\Delta\nu}(\nu) = \frac{1}{\Delta\nu} \int_{\Delta\nu} \tau(\nu')\xi(\nu'-\nu)d\nu' \tag{5-6}$$

which is the convolution of the spectral transmittance with the (normalized) spectral slit function, ξ, of the instrument. The center (or centroid) of the interval is ν.

Secondly, the spectrum of the target-source, whose radiation is affected by the absorption, is considered to vary slowly enough over the interval $\Delta\nu$ that it can be considered essentially constant. Thus the radiation from greybodies ($\epsilon(\nu)$ = constant) can be treated using Equation (5-5) to correct for the absorption in the atmosphere. Use of Equation (5-5) in conjunction with the radiation from hot gases (whose spectral structure varies rapidly and generally correlates very closely with that of the atmosphere) could lead to serious errors. The source generating the radiation is usually assumed to be constant in ν over the interval $\Delta\nu$, unless simplifying assumptions about the source spectra can be made which allow the use of Equation (5-5) as an approximation.

Thus all expressions which evolve are eventually averaged over $\Delta\nu$ and all results are dependent on the calculation (in absorption) of

$$\bar{\tau}_{\Delta\nu}\Delta\nu = \int_{\Delta\nu} e^{-q(\nu)}\,d\nu \tag{5-7}$$

where

$$q(\nu) = \int_{path} k_a(\nu,z)\rho\,dz \tag{5-8}$$

An example of the possible complexity of the determination of $\bar{\tau}$ in Equation (5-7) is shown in Figure 5-2, wherein the implied structure of the absorption coefficient is given for a certain spectral region. The total extent of the curve in Figure 5-2 represents a region of only about 0.025 μm. Thus, $\bar{\tau}$ is found for single lines and for various assumed groupings of lines (i.e., bands).

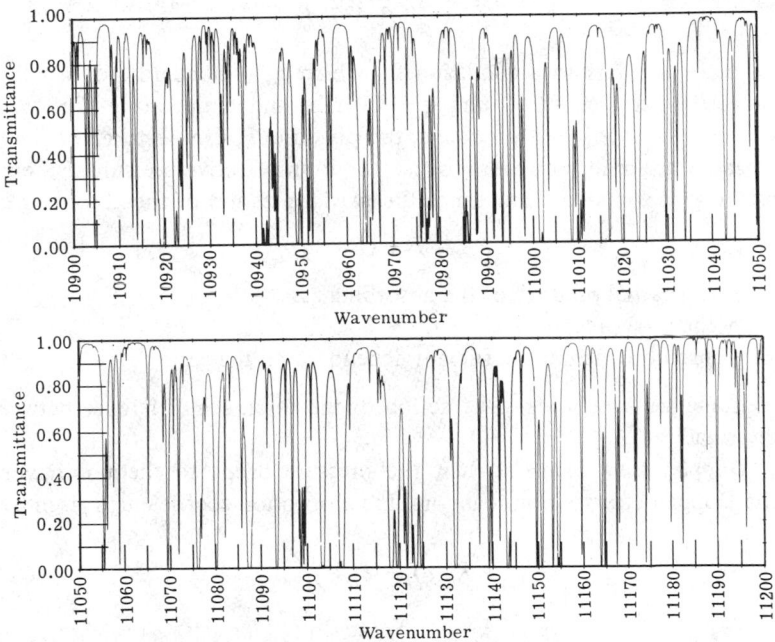

Fig. 5-2. Atmospheric transmittance due to molecular absorption through a 10 km, horizontal path at sea level [5-35].

5.3. Absorption by a Single Line

A single-pressure-broadened, molecular absorption line is shown in Figure 5-3, where α is called the half-width and S is the line strength. Actually, such lines are broadened by a combination of the pressure (Lorentz broadening) of the surrounding medium, and the motion (Doppler broadening) of the molecule. (See Goody, 1964(a), and Penner, 1959.) Above a pressure of 10 mb, Lorentz broadening dominates, while Doppler broadening dominates below 10 mb.

5.3.1. Lorentz Line Shape.

The absorption coefficient for a Lorentz line shape is

$$k_a(\nu) = \frac{S}{\pi} \frac{\alpha}{(\nu-\nu_o)^2 + \alpha^2} \qquad (5\text{-}9)$$

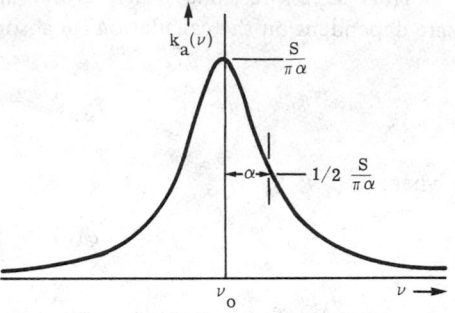

Fig. 5-3. Features of a Lorentz line.

where S = line strength, cm g^{-1}
ν_o = frequency at the center of the line, cm^{-1}
α = Lorentz half-width as defined by the structure of Figure 5-3, cm^{-1}

According to kinetic theory, the half-width depends on the pressure and absolute temperature as follows:

$$\alpha = \alpha_o \frac{P}{P_o} \left(\frac{T_o}{T}\right)^{1/2} \qquad (5\text{-}10)$$

where the index o refers to a condition for which α_o is standardized, usually at standard pressure (P_o = 1 atm = 760 mm = 1000 mb) and temperature (T_o = 273.16 K). For most atmospheric infrared problems, the pressure, P, can be taken as the total pressure. It is more accurate, however, to use for P an effective pressure, P_e, equal to the total pressure plus some constant times the partial pressure of the absorbing gas [5-4]:

$$P_e = P + (B - 1)p_a \qquad (5\text{-}11)$$

where p_a = the partial pressure of the absorbing gas
P = the total pressure
B = a self-broadening coefficient dependent on the gas

If the absorbing gas is only a small fraction of the total, the difference between P_e and P is usually small.

5.3.2. Doppler Line Shape.

When the pressure drops to the point where $\alpha \ll (\nu-\nu_o)$, the Doppler effect is dominant and the absorption coefficient is given by

$$k_a(\nu) = k_o e^{-y^2} \qquad (5\text{-}12)$$

where

$$k_o = \frac{S}{\alpha_D} \left(\frac{\ln 2}{\pi}\right)^{1/2} \qquad (5\text{-}13)$$

$$y = \frac{(\nu-\nu_o)}{\alpha_D} (\ln 2)^{1/2} \qquad (5\text{-}14)$$

and α_D, the Doppler half-width, in cm^{-1}, is given by

$$\alpha_D = \frac{\nu_o}{c} \left[2 \kappa \left(\ln 2\right) \frac{T}{m}\right]^{1/2} \qquad (5\text{-}15)$$

where c = velocity of light, cm sec^{-1}
      κ = Boltzmann constant $(1.380662 \times 10^{-23}\ J\ K^{-1})$
      m = molecular mass, gm molecule^{-1}

5.3.3. Mixed Line Shape. At intermediate altitudes, the Doppler effect convolves with the pressure broadened line to yield the mixed Doppler-Lorentz line shape [5-17]. The absorption coefficient for the mixed line shape is given by

$$k_a(\nu) = \frac{k_o u}{\pi} \int_{-\infty}^{\infty} \frac{e^{-t^2}}{u^2 + (y-t)^2}\ dt \tag{5-16}$$

where t is the variable of integration (arbitrary). The symbols k_o and y are defined as in Equations (5-13) and (5-14), respectively, and

$$u = \frac{\alpha}{\alpha_D} (\ln 2)^{1/2} \tag{5-17}$$

The line shape that is most simply applied to the band-model method (described subsequently) is the Lorentz shape. This shape is recommended for the first-order calculations evolving from the discussion in this chapter. In making calculations using the so-called *line-by-line technique* (Section 5.6.1), one can use any line shape that is appropriate. However, the Lorentz line shape is the only one which does not involve extensive numerical manipulation. A method described by Gille and Ellingson [5-5] for obtaining the Lorentz shape equivalent to the mixed profile gives a more reasonable approximation than the straight Lorentz line shape.

5.4. Band-Model Methods of Atmospheric Transmittance Calculations

5.4.1. The Band-Model Concept. The most practical approach to computing atmospheric absorption is to use an approximate, mathematically workable model of the band structure. This assumes that line positions and strengths are distributed in a way that can be represented by a simple mathematical model. The most commonly used band models are listed below.

(1) The *Elsasser*, or *regular model* [5-6] assumes spectral lines of equal strength, equal spacing, and identical half-widths. The transmission function is averaged over an interval equal to the spacing between the line centers.

(2) The *statistical*, or *random model*, originally developed for water vapor, assumes that the positions and strengths of the lines are given by a probability function. The statistical model was worked out by Mayer [5-7] and (independently) Goody [5-8].

(3) The *random-Elsasser model* [5-9] is a generalization of the Elsasser and the statistical models. It assumes a random superposition of any number of Elsasser bands of different strengths, spacings, and half-widths.

(4) The most accurate, presently-available model is the *quasi-random model* [5-10], provided the averaging interval can be made sufficiently small. It requires the greatest amount of computation of all the models.

5.4.2. A Single Lorentz Line. Assuming that the shape of a single spectral line for a homogeneous path in a single absorbing gas is represented by the Lorentz formula, the absorption is given by

$$A \Delta \nu = \int_{\Delta \nu} \left\{ 1 - \exp\left[-\frac{1}{\pi} \int_{path} \frac{S \alpha \rho}{(\nu - \nu_o)^2 + \alpha^2} dz \right] \right\} d\nu \qquad (5\text{-}18)$$

For a homogeneous path, S, α and ρ are constant; Equation (5-18) then further reduces to

$$A \Delta \nu = \int_{\Delta \nu} \left\{ 1 - \exp -\left[\frac{S}{\pi} \frac{\alpha w}{(\nu - \nu_o)^2 + \alpha^2} \right] \right\} d\nu \qquad (5\text{-}19)$$

where $w = \int_{path} \rho \, dz = \rho z$, and is defined as the absorber amount in g cm^{-2}.*

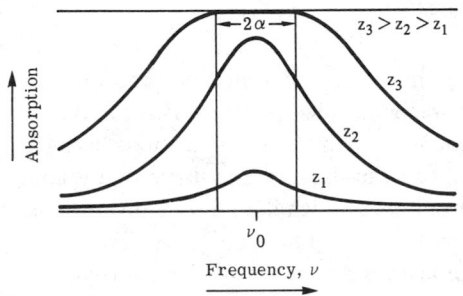

Fig. 5-4. Absorption versus frequency for a single line and for different absorber amounts.

A plot of absorption versus frequency is shown in Figure 5-4 for different path lengths, or for different values of w. For an optical path of length z_1, the absorption is small even at the line center. For a path of length z_3, the center of the line is completely absorbed and any further increase in path length would only change the absorption in the wings of the line. Absorption by paths of length equal to or greater than z_3 are considered *strong-line absorption*.

If, in Equation (5-19), one assumes that the interval $\Delta \nu$ is such that substantially the entire line is included, then the limits of integration can be taken from $-\infty$ to ∞ without introducing significant error. When these limits are used, Equation (5-19) can be solved exactly for the total absorption. Ladenburg and Reiche [5-11] have solved the integral to obtain

$$A \Delta \nu = 2\pi \alpha \psi e^{-\psi} \left[I_0(\psi) + I_1(\psi) \right] \qquad (5\text{-}20)$$

where $\psi = Sw/2\pi\alpha$ and I_0 and I_1 are modified Bessel functions. For *weak-line absorption* ($\psi \ll 1$) Equation (5-20) reduces to

$$A \Delta \nu = 2\pi \alpha \psi = Sw \qquad (5\text{-}21)$$

and absorption is linear with the amount of absorber w. Under conditions of *strong-line absorption*, ψ is large and Equation (5-20) reduces to

---

*The basic unit for the absorber amount is g cm^{-2}, or alternatively, molecules cm^{-2}. These units refer to the amount of absorber in a transmission path confined within an imaginary cylindrical volume, one square centimeter in cross-section. Other units are used in the case of water vapor where one considers the resulting depth of water if the vapor in the cylinder were all condensed. The units are "precipitable centimeters," pr cm. In the case of the atmospheric compressible gases, the amount of absorber is treated as if it were compressed within the cylinder until sea-level atmospheric pressure is reached. The length of the resulting gas volume would then be designated "atmospheric centimeters," atm cm.

$$A \Delta v = \sqrt{S \alpha w} \qquad (5\text{-}22)$$

known as the square-root approximation. The above formulations for a single spectral line are also valid for absorption when many spectral lines are present but do not overlap.

5.4.3. Elsasser Model. The Elsasser model of an absorption band is formed by repeating a single Lorentz line periodically throughout the interval Δv. This gives rise to a series of lines that are equally spaced and that have a constant strength and half-width throughout the interval. This arrangement of spectral lines was first proposed by Elsasser in 1938 [5-6].

The general expression for absorption by an Elsasser band is

$$A = (\sinh \beta) \int_0^Y I_0(Y) \exp(-Y \cosh \beta) \, dY \qquad (5\text{-}23)$$

where $\beta = 2\pi\alpha/d$
  $Y = \beta\psi/\sinh \beta$
  $\psi = Sw/2\pi\alpha$
  d = mean spacing between spectral lines
  I_0 = Bessel function of imaginary argument

A plot of this function for various values of β is given in Figure 5-5. Because the function in its present form is difficult to evaluate, considerable effort has been expended comparing approximate formulas and evaluating the integral. Kaplan [5-12] has expanded the integral into a series which is convergent only for values of β less than 1.76. An algorithm for the Elsasser integral which is convergent for all values of β and ψ is written in a computer program used in the calculation of spectral transmittance with

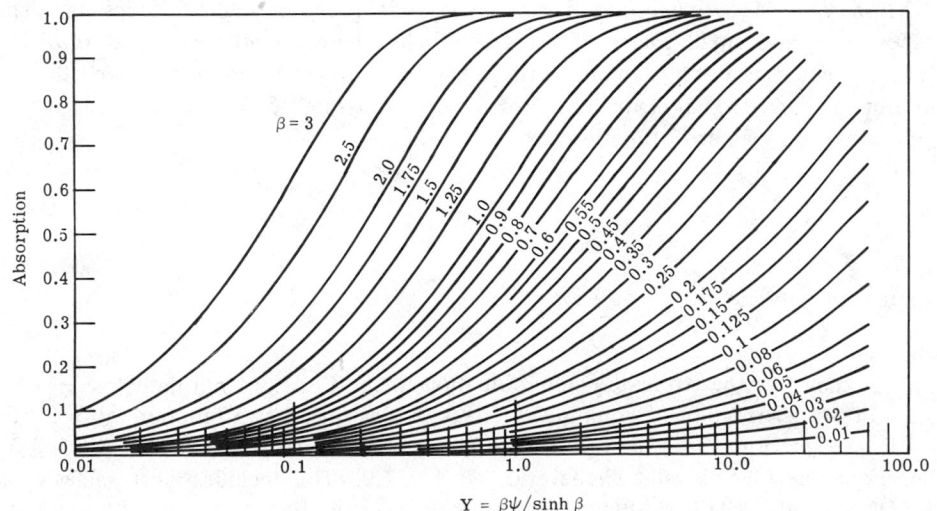

Fig. 5-5. Absorption by a single Elsasser band.

the Aggregate method (Section 5.5). However, it is frequently desirable to work with approximations to the function which are valid for certain conditions.

Weak-Line Approximation. In Figure 5-6, the absorption given by Equation (5-23) is plotted as a function of the product $\beta\psi = Sw/d$ for four values of β. For $\beta \gtrsim 1$, the absorption curves become nearly superimposed for all values of ψ. Since the parameter β measures the ratio of line width to the distance between neighboring lines, $\beta \gtrsim 1$ implies that the spectral lines

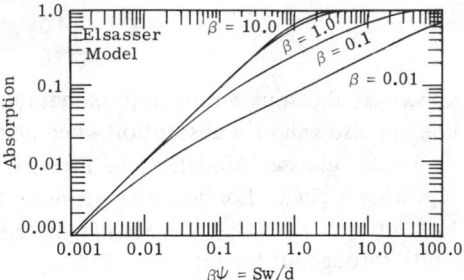

Fig. 5-6. Absorption as a function of $\beta\psi = Sw/d$ for the Elsasser model. The weak-line approximation is the uppermost curve [5-112].

are strongly overlapping and spectral line structure is not observable. This condition corresponds to large pressures which would be realistic for atmospheric paths at low altitudes. For $\beta \gtrsim 1$, Equation (5-23) can be approximated by

$$A = 1 - e^{-\beta\psi} \tag{5-24}$$

Equation (5-24) is also a good approximation to Equation (5-23) whenever the absorption is small at the line centers (small ψ) regardless of the value of β. This approximation is referred to as the weak-line approximation, and is independent of the position of the spectral lines within the band. Table 5-2 summarizes the regions of β and ψ for which the weak-line approximation is valid with an error of less than 10%. This approximation is particularly useful in extrapolating the absorption to small values of ψ and to large values of pressure. The weak-line approximation reduces to the linear approximation when the absorption is small even if the lines overlap; i.e., when $\beta\psi$ is small, all terms in the expansion of the exponential of Equation (5-24) are neglected except the linear one.

Strong-Line Approximation. Long atmospheric paths at high altitudes have large values of w and small values of pressure. Under these conditions, the absorption at the line centers is usually complete (large ψ), the half-widths are narrow, and the lines do not overlap strongly (small β). For large ψ and small β, Equation (5-23) may be approximated by an error function:

$$A = \mathrm{erf}\left(\frac{1}{2}\beta^2\psi\right)^{1/2} \tag{5-25}$$

where $\mathrm{erf}(a) = \dfrac{2}{\sqrt{\pi}}\displaystyle\int_0^a e^{-t^2}\,dt$.

This is known as the strong-line approximation to the Elsasser band model.

Figure 5-7 is a plot of Equation (5-23) with absorption as a function of $\beta^2\psi$. For $\beta \lesssim 0.01$, Equations (5-23) and (5-25) are nearly equal when $\beta^2\psi > 0.003$. If $\beta \lesssim 0.1$, then Equation (5-25) is valid whenever $0.1 \lesssim A \lesssim 1.0$. This includes most values of absorption that are usually of interest. This case differs from the square-root approximation in that it does not require nonoverlapping. For overlapping spectral lines (larger β), the values of ψ for which the approximation is valid are simply restricted to large values of ψ.

Table 5-2. Regions of Validity of Various Approximations
for Band Absorption* [5-49]

Approximation	$\beta = 2\pi\alpha/d$	Elsasser Model	Statistical Model; All Lines Equally Intense	Statistical Model; Exponential Line Intensity Distribution
Strong-line approximation:	0.001	$\psi > 1.63$	$\psi > 1.63$	$\psi_0 > 2.40$
	0.01	$\psi > 1.63$	$\psi > 1.63$	$\psi_0 > 2.40$
	0.1	$\psi > 1.63$	$\psi > 1.63$	$\psi_0 > 2.30$
	1.0	$\psi > 1.35$	$\psi > 1.10$	$\psi_0 > 1.40$
	10.0	$\psi > 0.24$	$\psi > 0.24$	$\psi_0 > 0.27$
	100.0	$\psi > 0.024$	$\psi > 0.024$	$\psi_0 > 0.24$
Weak-line approximation:	0.001	$\psi < 0.20$	$\psi < 0.20$	$\psi_0 < 0.10$
	0.01	$\psi < 0.20$	$\psi < 0.20$	$\psi_0 < 0.10$
	0.1	$\psi < 0.20$	$\psi < 0.20$	$\psi_0 < 0.10$
	1.0	$\psi < \infty$	$\psi < 0.23$	$\psi_0 < 0.11$
	10.0	$\psi < \infty$	$\psi < \infty$	$\psi_0 < \infty$
	100.0	$\psi < \infty$	$\psi < \infty$	$\psi_0 < \infty$
Nonoverlapping-line approximation:	0.001	$\psi < 600000$	$\psi < 63000$	$\psi_0 < 80000$
	0.01	$\psi < 6000$	$\psi < 630$	$\psi_0 < 800$
	0.1	$\psi < 60$	$\psi < 6.30$	$\psi_0 < 8.00$
	1.0	$\psi < 0.7$	$\psi < 0.22$	$\psi_0 < 0.23$
	10.0	$\psi < 0.02$	$\psi < 0.020$	$\psi_0 < 0.020$
	100.0	$\psi < 0.002$	$\psi < 0.0020$	$\psi_0 < 0.0020$

*When $\psi = Sw/2\pi\alpha$ satisfies the given inequalities, the indicated approximation for the absorption is valid with an error of less than 10%. For the exponential line intensity distribution, $\psi_0 = S_0 w/2\pi\alpha$, where S_0 is defined by $P(S) = S_0^{-1} \exp(-S/S_0)$. (See Section 5.4.4.)

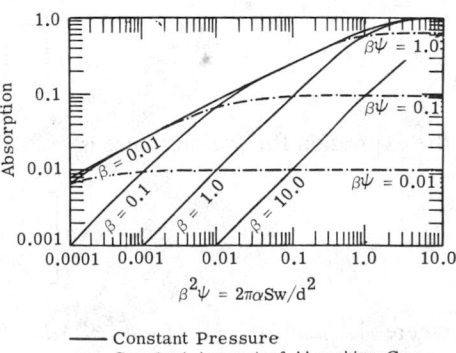

Fig. 5-7. Absorption of a function of $\beta^2\psi = 2\pi\alpha\, Sw/d^2$ for the Elsasser model. The curves are shown for constant pressure (β = constant) and for the constant amount of the absorbing gas ($\beta\psi$ = constant). The strong-line approximation is the uppermost curve [5-112].

Constant Pressure
Constant Amount of Absorbing Gas

Specific regions of validity are given in Table 5-2.

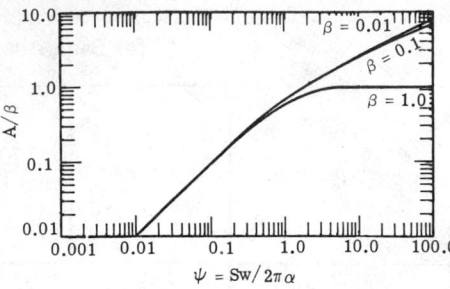

Nonoverlapping Approximation. The third approximation to the Elsasser band model is known as the nonoverlapping approximation. It is particularly useful for extrapolating the absorption to small values of w and small values of pressure which correspond to short paths at high altitudes. Under these conditions, Equation (5-23) reduces to

Fig. 5-8. Absorption divided by β as a function of $\psi = Sw/2\pi\alpha$ for the Elsasser model. The nonoverlapping-line approximation is the uppermost curve [5-112].

$$A = \beta\psi e^{-\psi} \left[I_0(\psi) + I_1(\psi)\right] \qquad (5\text{-}26)$$

This is exactly the same expression as that obtained for the absorption by a single spectral line.

In Figure 5-8, A/β is given as a function of ψ. The uppermost curve is the nonoverlapping approximation. For $\psi \ll 1$, the curve has a slope of 1 (i.e., a region where the weak-line approximation is valid) and for $\psi \gg 1$ the curve has a slope of one-half (i.e., a region where the strong-line approximation is valid). The regions of validity for various values of β and ψ are given in Table 5-2.

The general expression for absorption by an Elsasser band given by Equation (5-23) and the strong-line approximation given by Equation (5-25) are useful for determining absorption by CO_2, since the bands consist of fairly regularly spaced lines. However, the bands of H_2O and O_3 have a highly irregular, fine-structure and cannot be well described by Equation (5-23). Statistical methods must be used to develop an analytical expression for the transmissivity function for H_2O and O_3.

Golden [5-13] has described an Elsasser-like model using the Doppler shape instead of a Lorentz line-shape.

5.4.4. Statistical Band Model.

With a spectral interval

$$\Delta\nu = nd \qquad (5\text{-}27)$$

where there are n lines of mean wavenumber-spacing d, and with $P(S)dS$ the normalized probability that a line will have a strength between S and $S + dS$, so that

$$\int_0^\infty P(S)\,dS = 1 \qquad (5\text{-}28)$$

the expression for transmittance is

$$\tau = \exp\left[-\frac{1}{d}\int_0^\infty [A_{sl}\Delta\nu]\,P(S)\,dS\right] \qquad (5\text{-}29)$$

where $[A_{sl}\Delta\nu] = \int_{\Delta\nu} (1 - e^{-k_a w})\,d\nu$

A_{sl} = the absorptivity of a single line integrated over the interval $\Delta\nu$

Equal-Intensity Lines. Equation (5-29) can be evaluated for two special cases. When all the lines have equal intensities, $P(S) = \delta(S-S_o)$, in which $\delta(x)$ is the dirac delta function, and Equation (5-29) reduces to

$$\tau = e^{-A_{sl}\Delta\nu/d} = \exp\{-\beta\psi\, e^{-\psi}\, [I_0(\psi)+I_1(\psi)]\,\} \qquad (5\text{-}30)$$

In terms of absorption,

$$A = 1 - \exp\{-\beta\psi e^{-\psi}\, [I_0(\psi)+I_1(\psi)]\} \qquad (5\text{-}31)$$

If each of the lines absorbs *weakly*, so that ψ is small, then Equation (5-24) reduces to

$$A = 1 - \exp(-\beta\psi) \qquad (5\text{-}32)$$

If the lines absorb *strongly*, then Equation (5-31) reduces to

$$A = 1 - \exp\left(-2\,\sqrt{\frac{S\alpha w}{d}}\,\right) \qquad (5\text{-}33)$$

In terms of β and ψ,

$$A = 1 - \exp\left[-\left(\frac{2}{\pi}\beta^2\psi\right)^{1/2}\right] \qquad (5\text{-}34)$$

The nonoverlapping approximation to Equation (5-31) is obtained from the first term in the expansion of the exponential, so that

$$A = \beta\psi e^{-\psi}\, [I_0(\psi)+I_1(\psi)] \qquad (5\text{-}35)$$

This is the same expression as that obtained for the nonoverlapping approximation to the Elsasser model, Equation (5-26).

Line Strength with an Exponential Probability Distribution. When lines are of different strength and the distribution for the probability of their strength is an exponential distribution, namely,

$$P(S) = \frac{1}{S_o}\, e^{-S/S_o} \qquad (5\text{-}36)$$

where S_o is a mean line strength, then

$$\tau = \exp\left[-\,\frac{wS_o\alpha}{d\,\sqrt{\alpha^2 + \dfrac{(wS_o\alpha)}{\pi}}}\right] \qquad (5\text{-}37)$$

In terms of absorption, and β and ψ,

$$A = 1 - \exp\left[\frac{-\beta\psi_o}{(1 + 2\psi_o)^{1/2}}\right] \qquad (5\text{-}38)$$

where $\psi_o = S_o w/2\pi\alpha$. This is the formula developed by Goody and is therefore referred to as the Goody band model.

The weak-line approximation to Equation (5-38) is obtained when $\psi_o \ll 1$. Under these conditions,

$$A = 1 - \exp(-\beta\psi_o) \qquad (5\text{-}39)$$

The strong-line approximation to the statistical model with an exponential distribution of line strengths is obtained when $\psi_o \gg 1$. Under these conditions, Equation (5-38) becomes

$$A = 1 - \exp\left[-\left(\frac{1}{2}\beta^2\psi_o\right)^{1/2}\right] \qquad (5\text{-}40)$$

The nonoverlapping approximation is obtained from Equation (5-38) when the exponent is small and is therefore given by the first two terms of the expansion, or

$$A = \frac{\beta\psi_o}{(1 + 2\psi_o)^{1/2}} \qquad (5\text{-}41)$$

Strong-Line, Weak-Line, and Nonoverlapping Approximations. These three approximations to the statistical models are discussed concurrently because they are so closely related. For the weak-line approximation, absorption is given by $A = 1 - \exp(-\beta\psi)$, similar to that for the Elsasser model. When an exponential distribution of line strengths is assumed, ψ is replaced by ψ_o. This confirms that weak-line absorption is independent of the arrangement of the spectral lines within the band. Absorption versus $\beta\psi$ is plotted for Equation (5-31) in Figure 5-9. The solid curves give the absorption for the statistical model for the case in which all lines are equally strong. The dashed curves give the absorption for the statistical model with an exponential distribution of line strengths. The uppermost solid and dashed curves represent the weak-line approximations for those strength distributions. The regions of validity are given in Table 5-2.

When all lines are equally strong, the weak-line approximation is always valid within 10% when $\psi < 0.2$. It is valid for the exponential strength distribution when $\psi < 0.1$.

The strong-line approximations to the statistical model for all lines of equal strength and for an exponential distribution of line strengths are given respectively by Equation (5-34) and (5-40). The absorption for these models as a function of $\beta^2\psi$ is shown in Figure 5-10. The strong-line approximation is the uppermost curve in the figure. The distribution of the line strengths in a band only slightly influences the shape of the absorption curve. As with the strong-line approximation to the Elsasser model, the strong-line approximation to the statistical

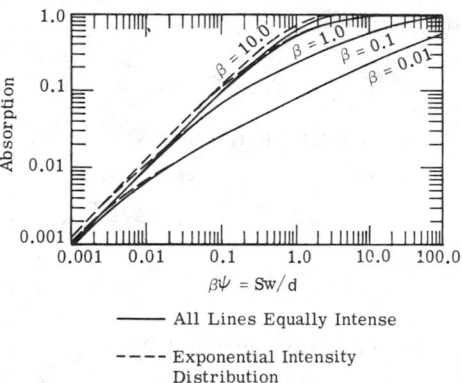

Fig. 5-9. Absorption as a function of $\beta\psi = Sw/d$ for the statistical model. The absorption for a model in which the spectral lines are all of equal intensity is compared with that for a model in which the spectral lines have an exponential intensity-distribution function. The weak-line approximation is the uppermost curve [5-112].

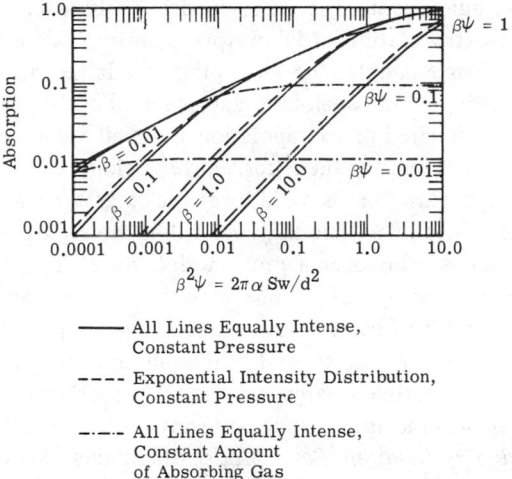

Absorption (vertical axis)

$\beta^2 \psi = 2\pi\alpha\, Sw/d^2$

—— All Lines Equally Intense,
Constant Pressure

- - - - Exponential Intensity Distribution,
Constant Pressure

—·—·— All Lines Equally Intense,
Constant Amount
of Absorbing Gas

Fig. 5-10. Absorption as a function of $\beta^2\psi = 2\pi\alpha Sw/d^2$ for the statistical model. Curves are shown for constant pressure (β = constant) and for the constant amount of the absorbing gas ($\beta\psi$ = constant). The absorption is shown when all the spectral lines have equal intensity and when there is an exponential intensity-distribution function. The strong-line approximation is the uppermost curve [5-112].

model for either distribution of the line strengths is always valid where $\beta \leqq 0.1$ and $0.1 \leqq A \leqq 1$. The complete regions of validity are given in Table 5-2.

For all lines of equal strength, the absorption is given by the expression used for the nonoverlapping approximation to the Elsasser model, namely

$$A = \beta\psi e^{-\psi}\left[I_0(\psi) + I_1(\psi)\right] \qquad (5\text{-}42)$$

For an exponential distribution of line strengths, the absorption is given by Equation (5-41). Therefore, the strength distribution, but not the spacing of the spectral lines, influences the absorption curve in this approximation. In Figure 5-11, A/β is plotted as a function of ψ for the statistical model. The nonoverlapping approximation to the Elsasser model has a considerably larger region

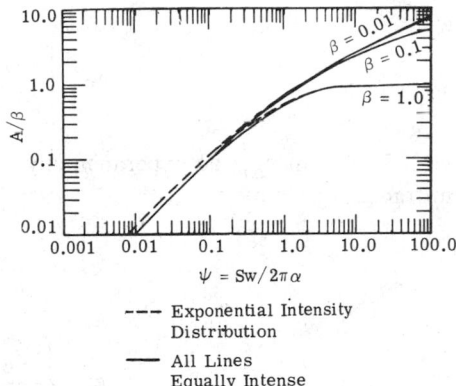

$\psi = Sw/2\pi\alpha$

- - - Exponential Intensity
Distribution

—— All Lines
Equally Intense

Fig. 5-11. Absorption divided by β as a function of $\psi = Sw/2\pi\alpha$ for the statistical model. The absorption for a model in which the spectral lines are all of equal intensity is compared with that for a model in which the spectral lines have an exponential intensity-distribution function. The nonoverlapping-line approximation is the uppermost curve [5-112].

of validity because the spectral lines begin to overlap at considerably larger path lengths for the Elsasser model than for the statistical model (Table 5-2).

These three approximations to the band models of Elsasser and Goody provide a reliable means for the extrapolation of laboratory absorption data to values of the pressure and path length. For example, the absorption for large values of pressure can be obtained from the strong- and weak-line approximations, depending upon whether w is relatively large or small. For extrapolation to small values of pressure, one may use all three approximations in their respective regions of validity; however, the nonoverlapping-line approximation is valid over the largest range of values of w. For extrapolation to large values of w, one may use either the strong- or weak-line approximation, but the former is valid over a much wider range of pressure than the latter. For extrapolation to small values of w, one may use all three approximations in their respective regions of validity; however, the nonoverlapping-line approximation is valid over a wider range of pressure. In general, atmospheric slant paths of interest to the systems engineer contain relatively large amounts of absorber, and the range of pressures is such that the strong-line approximation to any of the models is applicable.

Exponential-Tailed S^{-1} Random Band Model. Malkmus [5-14] described a model comprising a line strength distribution which is defined by

$$P(S) \propto S^{-1} \exp\left\{-\frac{1}{\pi}\frac{S}{S_E}\right\} \qquad (5\text{-}43)$$

where S_E is fixed in the expressions

$$\lim_{w\to 0}\left[\frac{\overline{W}/d}{w}\right] = \frac{S_E}{d_E} \qquad (5\text{-}44)$$

and

$$\lim_{w\to\infty}\left[\frac{\overline{W}/d}{w^{1/2}}\right] = \frac{2(S_E\alpha)^{1/2}}{d_E} \qquad (5\text{-}45)$$

with $\overline{W}/d = -\ln\tau_{\Delta\nu}$, an experimentally determined value. The resultant calculated transmittance is given by

$$\tau_{\Delta\nu} = \exp\left\{-\frac{2\alpha}{d_E}\left[\left(1+\frac{S_E w}{\alpha}\right)^{1/2}-1\right]\right\} \qquad (5\text{-}46)$$

$$= \exp\left\{-\frac{\beta_E}{\pi}\left[\left(1+2\pi\psi_E\right)^{1/2}-1\right]\right\} \qquad (5\text{-}47)$$

5.4.5. Random Elsasser Band Model. For some gases and spectral regions, absorption can be represented more accurately by the random Elsasser band model than by either the statistical or Elsasser model alone. This is a natural generalization of the original models which assumes that the absorption can be represented by the random superposition of Elsasser bands. The individual bands may have different line spacings, half-widths, and strengths. As the number of superposed Elsasser bands becomes large, the predicted absorption approaches that of the usual statistical model.

The absorption for N, randomly-superposed, Elsasser bands with arbitrary strengths, half-widths, and line spacings is given by

$$A = 1 - \prod_{i=1}^{N} [1 - A_{E,i}(\psi_i, \beta_i)] \tag{5-48}$$

where
$$\psi_i = S_i w / 2\pi\alpha_i$$
$$\beta_i = 2\pi\alpha_i / d_i$$
S_i and α_i = the line strength and half-width of the ith Elssaer band
d_i = the line spacing of the ith band

The derivation of this general equation for different strengths, half-widths, and the line spacings is discussed in detail by Plass [5-9].

5.4.6. Quasi-Random Model. The quasi-random model is the most accurate and, necessarily, the most complicated of the band models. It is especially useful when the absorptance is required over a wide range of path lengths and pressures.

The spectral lines in an actual band are arranged neither as regularly as required by the Elsasser band nor in as random a fashion as the statistical model; but there is some order in their arrangement. In the quasi-random model, the absorptance is calculated first for a frequency interval that is much smaller than the interval size of interest. This localizes the stronger lines to a narrow interval around their actual positions and prevents the introduction of spurious overlapping effects. The absorptance of this narrow interval is calculated from the equation for the single-line absorptance over a finite interval. The absorptance for each of the N spectral lines in the interval is calculated separately and the results combined by assuming a random placing of the spectral lines within the small interval. The absorption from the wings of lines in neighboring intervals is included in the calculation. The results are averaged for at least two different arrangements of the mesh that divides the spectrum into frequency intervals. Finally, the absorptance values for all of the small intervals that fill the larger interval of interest are averaged to obtain the final value for the absorptance. A computer is commonly used to calculate results for this model when many spectral lines are involved. The many, weak, spectral lines and their relative spacing are accurately taken into account by this model.

The absorptance for the quasi-random model is given by

$$A = \frac{1}{n} \sum_{j=1}^{n} A_j \tag{5-49}$$

where A_j is the absorptance of each of the n, smaller, wave-number intervals of width D, into which the original interval $\Delta\nu$ is subdivided. The absorptance is calculated from

$$A_j = 1 - \prod_{i=1}^{M} [1 - A_{sl}(\psi_{i,j}; \beta_{i,j})] \tag{5-50}$$

where A_{sl} = single line absorptance of the ith line over the jth finite interval, D
M = the number of lines in the jth frequency interval, D [5-10]

References [5-15 and 5-16] give absorptance tables for H_2O and CO_2 based on the quasi-random model calculations. In these calculations, the lines in the small frequency interval were grouped by line-strength decades. The average strength and number of lines in each of these decades were calculated and used in Equation (5-50). All lines (from all isotopic species) having strengths greater than 10^{-8} of the strongest line in an absorption region were included in the calculation.

5.4.7. Temperature and Frequency Dependence of Band-Model Parameters. Table 5-3 is a summary of the band models that yield closed-form expressions for spectral-band absorption. All of these expressions are functions of two parameters, β and ψ, which in turn are functions of temperature, pressure, absorber concentration, and wave-number or wavelength (or frequency).

Since the variation of line strength with temperature is different for each spectral line, it would be impossible to include the effect of temperature on line strength and still retain the band-model expressions in closed form. For this reason, the band-model expressions in their present form can be used only to predict absorption for homogeneous paths at standard temperature. Since Drayson [5-17] has shown that the effect of temperature on line half-width has a secondary effect on absorption, this dependence is neglected also. Therefore, if $T = T_o$, the expressions for β and ψ become

$$\beta = \left(\frac{2\pi\alpha_o}{d}\frac{P}{P_o}\right) = \frac{2\pi\alpha_o'}{d}P \tag{5-51}$$

$$\psi = \left(\frac{S\,w}{2\pi\alpha_o}\frac{P_o}{P}\right) = \frac{S}{2\pi\alpha_o'}\frac{w}{P} \tag{5-52}$$

where α_o' is the half-width at standard conditions, per unit pressure.

The expressions listed in Table 5-3 are of the following general form:

(1) $A = A(\beta,\psi)$ when no approximation of the model is assumed.
(2) $A = A(\beta^2\,\psi)$ for the strong-line approximation.
(3) $A = A(\beta\psi)$ for the weak-line approximation.

The first expression is a function of two frequency-dependent parameters, $2\pi\alpha_o'/d$ and $S/2\pi\alpha_o'$, and two path-parameters, w and P. The second expression gives absorption as a function of one frequency-dependent parameter, $2\pi\alpha_o'S/d^2$, and two path-parameters, w and P. The last is a function of S/d and w, being independent of pressure.

The next problem in completely specifying the absorption expressions is that of evaluating the frequency-dependent parameters by empirically fitting the respective equations to laboratory, homogeneous-absorption spectra. The empirical procedure is most involved when no approximations to the model are assumed, since in this case two parameters must be evaluated. After the frequency-dependent parameters have been specified and the value of w and P have been determined for a given slant path, it becomes a simple matter to generate absorption spectra.

5.5. Practical Methods for Calculating Spectral Transmittance

There are two practical methods for calculating spectral transmittance: The Aggregate and LOWTRAN methods [5-1]. The Aggregate method, so-called because it uses a

Table 5-3. Summary of Closed-Form Expressions for Spectral Absorption

Band Model	Approximation	Equation*
Single Lorentz line	None	$A = \dfrac{1}{\Delta\nu} 2\pi\alpha \, \psi \, e^{-\psi} \, [I_0(\psi) + I_1(\psi)]$
Single Lorentz line	Linear	$A = Sw/\Delta\nu$
Single Lorentz line	Square Root	$A = 2\sqrt{S\alpha \, w/\Delta\nu}$
Elsasser band	None	$A = \sinh\beta \displaystyle\int_0^Y I_0(Y) \exp(-Y\cosh\beta) \, dY$
Elsasser band	Weak	$A = 1 - e^{-\beta\psi}$
Elsasser band	Strong	$A = \mathrm{erf}\left(\dfrac{1}{2}\beta^2\psi\right)^{1/2}$
Statistical band (Exponential)	None	$A = 1 - \exp\left[-\beta\psi_o/(1 + 2\psi_o)^{1/2}\right]$
Statistical band	None	$A = 1 - \exp - \left\{\beta\psi e^{-\psi}\,[I_0(\psi) + I_1(\psi)]\right\}$
Statistical band (Exponential and equal)	Weak	$A = 1 - e^{-\beta\psi}$
Statistical band (Exponential)	Strong	$A = 1 - \exp\left[-\left(\dfrac{1}{2}\beta^2\,\psi_o\right)^{1/2}\right]$
Statistical band (equal)	Strong	$A = 1 - \exp\left[-\left(\dfrac{2}{\pi}\beta^2\,\psi\right)^{1/2}\right]$
Exponential tailed S^{-1} Statistical band	Strong	$A = 1 - \exp\left\{-\dfrac{\beta_E}{\pi}\left[(1 + 2\pi\psi_E)^{1/2} - 1\right]\right\}$
Random Elsasser band (equal line strengths, half-widths, and spacings)	Strong	$A = 1 - \left[1 - \mathrm{erf}\left(\dfrac{1}{2}\beta^2\psi/N^2\right)^{1/2}\right]^N$

*$\beta = 2\pi\alpha/d$; $\psi = Sw/2\pi\alpha$; where S is the line strength (S_o is mean strength), α is the Lorentz-line half-width, d is the (mean) line spacing, and w is the absorber amount. (For ψ_E, β_E, see Section 5.4.4.)

selected collection of the conventional two-parameter models described earlier in this chapter, is perhaps more adaptable to a wide variety of different atmospheric conditions than LOWTRAN, which is strictly empirical and depends essentially on a single adjustable parameter. However, for transmittances between roughly 10 and 90%, the values calculated using either method are not greatly different, and the simplicity of the use of LOWTRAN makes it much more attractive for routine computations. Neither method perfectly reproduces the results of field measurements. Direct line-by-line calculations are required to achieve accuracies consistent with that of the basic data.

5.5.1. Aggregate Method [5-1, 5-20]. Table 5-4 and Figure 5-12* show which models are used for different spectral regions for the various atmospheric molecules. (The models and data for their use are given below.)

For H_2O:

(1) Strong-Line Goody Model (1.0 to 2.0 μm and 4.3 to 15.0 μm). The transmittance is calculated from

$$\tau(\lambda) = \exp\{-[w^*K_1(\lambda)]^{1/2}\} \tag{5-53}$$

where w^* = equivalent absorber amount, given by

$$w^* = \rho_o \int_{path} M(z)\left(\frac{P(z)}{P_o}\right)^2\left(\frac{T_o}{T(z)}\right)^{1.5} dz, \text{ pr cm} \tag{5-54}$$

$K_1(\lambda)$ = spectral coefficient
ρ_o = air density, kg cm^{-3}, at standard temperature and pressure
$M(z)$ = mixing ratio at altitude, z, g H_2O kg^{-1} air
$P(z)$ = pressure at z, mm Hg
P_o = standard pressure ($= 760.0$ mm Hg)
$T(z)$ = temperature at z, K
T_o = standard temperature ($= 273.16$ K)

If measured data are not available, M, P, and T can be determined from Table 5-27 for model atmospheres. In this case, in Equations (5-54) and (5-56) below, the product

$$\rho_o M(z)\frac{P(z)}{P_o}\frac{T_o}{T(z)} = \rho_{H_2O}$$

is the H_2O vapor density found in column 5 of Table 5-27, converted to units of g cm^{-3}. Values for $K_1(\lambda)$ are given in Table 5-5.

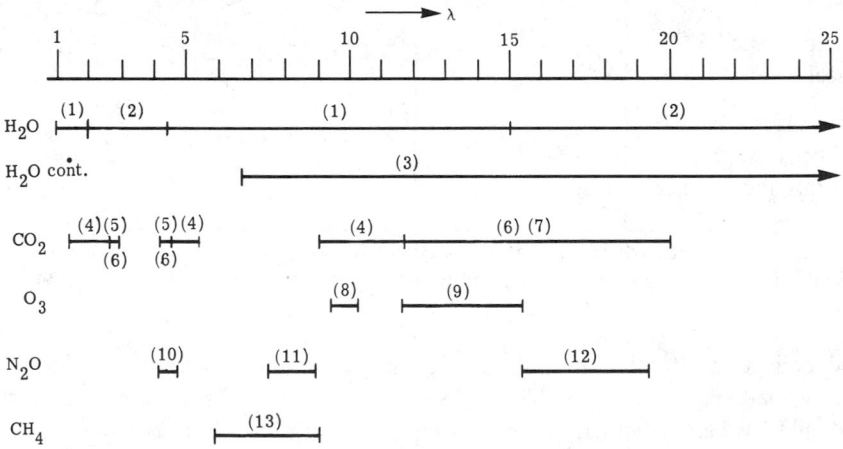

Fig. 5-12. Models used for different spectral regions for the various atmospheric molecules.

---

*The numbers in parentheses in Figure 5-12 correspond to the following numbered models, and the molecule to which they apply.

Table 5.4. Summary of Band Models for the Aggregate Method [5-20]

Gas	Spectral Region (μm)	Approximate Resolution (μm)	Model	Coefficient Acquisition Procedure	Source of Data	Reference
H_2O	1 to 2	0.10	Strong-line Goody*	Empirical fit to lab data	Howard, et al., (1955)	[5-21]
	2 to 4.3	0.05	Goody*	Empirical fit to lab data	Burch, et al., (1962)	[5-22]
	4.3 to 15	0.50	Strong-line Goody*	Empirical fit to lab data	Howard, et al., (1955)	[5-21]
	7.0 to 30	–	H_2O continuum	Empirical fit to lab data	Burch (1970) & Bignell (1970)	[5-23, 5-24]
	15 to 30	1.00	Goody*	Direct from line parameters	McClatchey, et al., (1973)	[5-24]
CO_2	1.37 to 2.64	0.20	Empirical	Empirical fit to lab data	Howard, et al., (1955)	[5-21]
	2.64 to 2.88	0.01	Classical Elsasser	Empirical fit to lab data	Burch, et al., (1962)	[5-22]
	4.184 to 4.454	0.02	Classical Elsasser	Empirical fit to lab data	Bradford, et al., (1963)	[5-26]
	4.465 to 5.355	0.50	Empirical	Empirical fit to lab data	Howard, et al., (1955)	[5-21]
	9.13 to 11.67	0.10	Temperature dependent classical Elsasser	Empirical fit to line by line spectra	Drayson and Young (1967)	[5-27]
	11.67 to 19.92					
O_3	9.398 to 10.19	0.10	Modified classical Elsasser	Empirical fit to lab data	Walshaw (1957)	[5-29]
	11.7 to 15.4	0.50	Goody*	Empirical fit to lab data	McCaa and Shaw (1968)	[5-30]
N_2O	4.228 to 4.73	0.50	Strong-line Elsasser	Empirical fit to lab data	Plyler and Barker (1931)	[5-28]
	7.53 to 8.91	0.50	Classical Elsasser	Empirical fit to lab data	Burch, et al., (1962)	[5-22]
	15.4 to 19.3	0.50	Goody*	Empirical fit to lab data	Burch, et al., (1962)	[5-22]
CH_4	5.91 to 9.1	0.10	Classical Elsasser	Empirical fit to lab data	Burch, et al., (1962)	[5-22]

*The Goody model is the statistical model with an exponential line strength distribution.

Table 5-5.* H_2O Vapor Parameters Corresponding to the Strong-Line Goody Model

λ	$K_1(\lambda)$	λ	$K_1(\lambda)$	λ	$K_1(\lambda)$
1.050	1.01E-04	1.470	8.39E-01	1.920	9.50E 00
1.070	2.19E-04	1.480	5.60E-01	1.930	7.50E 00
1.090	3.64E-03	1.485	3.56E-01	1.940	4.49E 00
1.100	1.23E-02	1.490	2.61E-01	1.950	2.59E 00
1.110	3.46E-02	1.497	1.85E-01	1.955	1.93E 00
1.120	8.24E-02	1.505	1.27E-01	1.963	1.44E 00
1.125	8.88E-02	1.510	8.12E-02	1.974	1.10E 00
1.130	1.27E-01	1.522	4.20E-02	1.977	8.39E-01
1.134	1.85E-01	1.530	4.04E-02	1.980	5.60E-01
1.140	2.61E-01	1.550	9.50E-03	1.985	4.80E-01
1.150	2.97E-01	1.570	6.50E-03	1.990	3.56E-01
1.165	1.72E-01	1.580	2.50E-03	1.995	2.61E-01
1.170	1.27E-01	1.590	1.50E-03	2.000	1.27E-01
1.180	9.36E-02	1.610	8.40E-04	2.010	4.97E-02
1.190	3.46E-02	1.620	7.40E-04	2.030	2.50E-02
1.200	1.00E-02	1.630	7.02E-04	2.040	1.60E-02
1.220	1.88E-03	1.650	8.76E-04	2.045	1.18E-02
1.240	1.12E-03	1.700	4.61E-03	2.052	6.40E-03
1.280	6.99E-03	1.750	1.74E-02	2.057	4.20E-03
1.300	1.93E-02	1.760	3.03E-02	2.062	2.50E-03
1.310	3.96E-02	1.770	5.57E-02	2.065	1.85E-03
1.320	8.12E-02	1.780	1.08E-01	2.069	1.30E-03
1.330	1.85E-01	1.790	2.13E-01	2.072	1.00E-03
1.335	2.75E-01	1.800	4.54E-01	2.075	7.60E-04
1.340	3.79E-01	1.810	9.92E-01	2.078	5.70E-04
1.345	7.12E-01	1.820	2.04E 00	2.081	4.20E-04
1.350	1.04E 00	1.830	3.59E 00	2.084	3.30E-04
1.355	1.44E 00	1.840	5.29E 00	2.089	1.90E-04
1.360	2.16E 00	1.850	8.00E 00	2.100	7.40E-05
1.370	3.59E 00	1.860	9.70E 00	2.120	1.00E-05
1.375	3.50E 00	1.880	1.30E 01	4.274	1.70E-03
1.380	7.05E 00	1.890	1.36E 01	4.283	1.48E-03
1.400	1.12E 01	1.900	1.29E 01	4.292	1.30E-03
1.410	9.09E 00	1.910	1.09E 01	4.301	1.12E-03
1.430	5.29E 00	1.915	1.01E 01	4.310	1.00E-03
1.460	1.10E 00	1.918	9.80E 00	4.319	9.00E-04

*In this and similar Tables, exponential notation is used in computer format; e.g., $0.05 \equiv 5 \times 10^{-2} \equiv 5.00E-02$. λ is in μm.

Table 5-5. H_2O Vapor Parameters Corresponding to the
Strong-Line Goody Model (*Continued*)

λ	$K_1(\lambda)$	λ	$K_1(\lambda)$	λ	$K_1(\lambda)$
4.329	7.90E-04	4.925	2.40E-01	6.088	1.49E 02
4.338	7.20E-04	4.940	2.66E-01	6.126	1.37E 02
4.348	6.30E-04	4.950	2.85E-01	6.163	1.22E 02
4.357	5.90E-04	4.965	3.12E-01	6.202	1.02E 02
4.367	5.30E-04	4.970	3.22E-01	6.240	7.47E 01
4.376	4.90E-04	4.975	3.35E-01	6.279	4.99E 01
4.386	4.60E-04	4.990	3.65E-01	6.319	3.60E 01
4.396	4.35E-04	5.000	3.89E-01	6.359	2.96E 01
4.405	4.25E-04	5.025	4.70E-01	6.400	3.02E 01
4.415	4.40E-04	5.075	6.51E-01	6.441	3.64E 01
4.425	4.80E-04	5.105	7.53E-01	6.483	5.54E 01
4.435	5.30E-04	5.130	8.82E-01	6.525	1.11E 02
4.444	5.90E-04	5.145	9.56E-01	6.568	1.58E 02
4.454	6.40E-04	5.160	1.06E 00	6.612	1.94E 02
4.465	7.50E-04	5.180	1.17E 00	6.656	1.28E 02
4.470	7.80E-04	5.220	1.49E 00	6.700	2.12E 02
4.475	8.20E-04	5.260	1.86E 00	6.745	1.77E 02
4.490	1.00E-03	5.290	2.25E 00	6.791	1.40E 02
4.495	1.05E-03	5.320	2.61E 00	6.838	1.04E 02
4.505	1.22E-03	5.355	3.36E 00	6.885	8.00E 01
4.520	1.50E-03	5.380	3.56E 00	6.932	6.57E 01
4.526	1.62E-03	5.430	4.80E 00	6.981	5.60E 01
4.540	2.00E-03	5.470	5.60E 00	7.030	4.90E 01
4.548	2.30E-03	5.500	8.39E 00	7.080	4.32E 01
4.575	3.40E-03	5.540	1.10E 01	7.130	3.81E 01
4.600	5.00E-03	5.580	1.44E 01	7.181	3.32E 01
4.660	1.36E-02	5.610	1.93E 01	7.233	2.79E 01
4.700	2.45E-02	5.650	2.59E 01	7.286	2.03E 01
4.730	3.70E-02	5.700	3.59E 01	7.339	1.31E 01
4.750	4.70E-02	5.800	9.00E 01	7.394	9.10E 00
4.770	5.90E-02	5.900	1.22E 02	7.449	6.59E 00
4.800	7.20E-02	5.908	1.26E 02	7.505	4.63E 00
4.830	1.10E-01	5.944	1.38E 02	7.533	3.91E 00
4.860	1.45E-01	5.979	1.47E 02	7.561	3.39E 00
4.890	1.83E-01	6.015	1.52E 02	7.619	2.68E 00
4.910	2.15E-01	6.051	1.53E 02	7.678	2.31E 00

Table 5-5.* H_2O Vapor Parameters Corresponding to the
Strong-Line Goody Model (*Continued*)

λ	$K_1(\lambda)$	λ	$K_1(\lambda)$	λ	$K_1(\lambda)$
7.737	2.00E 00	10.290	0.00109	13.230	0.053
7.797	1.74E 00	10.440	0.00129	13.260	0.055
7.859	1.56E 00	10.550	0.00145	13.300	0.058
7.921	1.77E 00	10.900	0.0022	13.330	0.060
7.984	1.54E 00	11.050	0.0026	13.370	0.063
8.048	8.30E-01	11.173	0.0032	13.400	0.066
8.114	0.0990	11.268	0.0034	13.440	0.069
8.180	0.0270	11.364	0.0038	13.480	0.073
8.247	0.0080	11.500	0.0046	13.510	0.076
8.316	0.0038	11.670	0.0056	13.550	0.080
8.386	0.0023	11.790	0.0066	13.590	0.084
8.457	0.0017	11.900	0.0076	13.620	0.087
8.529	0.00147	12.050	0.0095	13.660	0.092
8.602	0.00128	12.200	0.012	13.700	0.096
8.677	0.0011	12.350	0.0145	13.740	0.103
8.753	0.00099	12.500	0.0184	13.770	0.107
8.830	0.0009	12.530	0.019	13.810	0.112
8.909	0.00083	12.560	0.020	13.850	0.117
8.989	0.00076	12.590	0.021	13.890	0.124
9.070	0.0007	12.630	0.022	13.950	0.133
9.130	0.00068	12.660	0.023	13.990	0.143
9.220	0.00064	12.690	0.0246	14.030	0.145
9.310	0.00062	12.720	0.0253	14.060	0.148
9.335	0.00061	12.760	0.027	14.100	0.150
9.398	0.0006	12.790	0.028	14.140	0.152
9.463	0.00059	12.820	0.029	14.180	0.155
9.494	0.00058	12.850	0.031	14.220	0.157
9.526	0.00058	12.890	0.033	14.270	0.160
9.590	0.00059	12.920	0.035	14.310	0.162
9.652	0.00061	12.950	0.036	14.350	0.165
9.713	0.00063	12.990	0.038	14.390	0.168
9.773	0.00066	13.020	0.040	14.430	0.170
9.834	0.00069	13.050	0.042	14.470	0.173
9.893	0.00072	13.090	0.044	14.510	0.176
9.953	0.00075	13.120	0.045	14.560	0.180
10.070	0.00086	13.160	0.048	14.600	0.183
10.190	0.00098	13.190	0.050	14.640	0.186

*In this and similar Tables, exponential notation is used in computer format; e.g., $0.05 \equiv 5 \times 10^{-2} \equiv 5.00E-02$. λ is in μm.

(2) Goody Model (2.0 to 4.3 μm and 15.0 to 30.0 μm). The transmittance is calculated from the following equations:

$$\tau(\lambda) = \exp\left\{-\frac{S}{d} \frac{w}{\left[1 + \frac{2}{P}\frac{Sw}{2\pi\alpha_o'}\right]^{1/2}}\right\}$$ (5-55)

$$w = \rho_o \int_{\text{path}} M(z) \frac{P(z)T_o}{P_o T(z)}\, dz, \text{ pr cm}$$ (5-56)

$$\bar{P} = \frac{P_o}{w} w^*, \ (P_o = 760.0 \text{ mm Hg})$$ (5-57)

where α_o is the Lorentz half-width at P_o. Values of S/d and $S/[2\pi\alpha_o']$ are given in Table 5-6, and

$$\alpha_o' = \frac{\alpha_o}{P_o}$$ (5-58)

(3) H_2O Continuum (7 to 30 μm). The H_2O continuum is a region influenced by H_2O and H_2O dimers and by the wings of many lines whose centers are well outside of the continuum region. This is in addition to the effect of local lines whose centers are in the region (as given by one of the other models). Since the continuum region cannot be modeled accurately, the computation is derived from a set of empirical data [5-24 and 5-32].

Roberts et al. [5-18] have reported work with results which supersede those used in the original Aggregate method. Roberts' results are used here.

The transmittance is calculated from the following equation:

$$\tau(\lambda) = \exp\left(-k'_{H_2O \text{ cont.}}\, s\right)$$ (5-59)

where s = horizontal* path length, km

$$k'_{H_2O \text{ cont}} = C(\lambda, T)\, w'_{H_2O} P_{H_2O}$$ (5-60)

with

$$w'_{H_2O} = 2.69 \times 10^{24}\, p_{H_2O}, \text{ molecules cm}^{-2} \text{ km}^{-1}$$ (5-61)

$$p_{H_2O} = \text{partial pressure of } H_2O \text{ vapor, atm}$$

$$C(\lambda, T) = C_o(\lambda) \exp\left[1800\left(\frac{1}{T} - \frac{1}{296}\right)\right]$$ (5-62)

where

$$C_o(\lambda) = e + f \exp\left(-g\frac{10^4}{\lambda}\right)$$ (5-63)

---

*For vertical paths an integration must be performed using z-dependent parameters. Slant paths are achieved by dividing the result from vertical integration by $\cos\theta$. See Section 5.1.

Table 5-6. H_2O Vapor Parameters Corresponding to the Goody Model

λ	S/d	$S/2\pi\alpha'_O$	λ	S/d	$S/2\pi\alpha'_O$	λ	S/d	$S/2\pi\alpha'_O$
2.160	7.03E-04	8.67E-01	2.577	2.32E 02	1.15E 05	2.855	8.03E 01	6.88E 04
2.180	9.34E-04	1.16E 00	2.584	8.13E 02	1.30E 06	2.862	3.87E 01	3.49E 04
2.200	1.68E-03	2.08E 00	2.589	5.98E 02	7.88E 05	2.869	2.73E 01	5.20E 04
2.240	7.59E-04	9.36E-01	2.592	1.02E 03	1.31E 06	2.875	1.50E 01	1.80E 04
2.280	3.07E-02	3.79E 01	2.597	2.88E 02	1.31E 05	2.878	1.49E 01	1.60E 04
2.331	5.65E-02	6.79E 01	2.602	4.95E 02	2.02E 05	2.881	1.50E 01	1.78E 04
2.336	5.66E-02	6.80E 01	2.609	2.94E 02	1.27E 05	2.886	1.54E 01	7.93E 03
2.342	7.93E-02	8.21E 01	2.614	4.07E 02	1.36E 05	2.894	5.91E 00	8.93E 03
2.347	7.98E-02	7.88E 01	2.622	2.51E 02	1.74E 05	2.903	3.89E 01	5.70E 04
2.353	8.39E-02	8.24E 01	2.629	2.38E 02	6.22E 04	2.910	8.03E 00	1.13E 04
2.358	8.82E-02	8.64E 01	2.634	9.65E 01	4.90E 04	2.915	8.73E 00	1.07E 04
2.364	9.31E-02	8.67E 01	2.640	1.26E 02	5.67E 04	2.916	8.16E 00	1.55E 04
2.370	9.89E-02	9.27E 01	2.643	1.13E 02	7.03E 04	2.924	2.60E 01	2.48E 04
2.375	1.04E-01	9.91E 01	2.650	1.93E 02	7.88E 04	2.931	4.14E 00	4.92E 03
2.381	1.09E-01	9.92E 01	2.667	6.18E 02	1.53E 05	2.933	3.81E 00	9.71E 03
2.387	1.15E-01	1.03E 02	2.672	4.89E 02	1.19E 05	2.941	7.17E 00	6.61E 03
2.392	1.21E-01	1.05E 02	2.674	6.09E 02	1.85E 05	2.948	1.56E 01	2.02E 04
2.398	1.25E-01	1.05E 02	2.680	3.05E 02	8.04E 04	2.951	1.21E 01	5.43E 04
2.404	9.58E-02	5.26E 01	2.683	3.09E 02	8.64E 04	2.955	1.15E 01	1.72E 04
2.410	1.37E-01	9.79E 01	2.688	2.15E 02	7.07E 04	2.961	2.32E 00	2.47E 03
2.415	1.50E-01	1.03E 02	2.691	2.59E 02	7.98E 04	2.972	1.57E 01	1.24E 04
2.421	1.62E-01	9.81E 01	2.694	3.07E 02	8.88E 04	2.976	1.34E 01	2.13E 04
2.427	1.86E-01	1.06E 02	2.697	1.47E 02	4.34E 04	2.989	2.05E 00	6.27E 03
2.433	2.18E-01	1.32E 02	2.703	1.31E 02	4.83E 04	2.994	8.18E 00	1.63E 04
2.439	2.49E-01	1.69E 02	2.709	2.49E 02	3.05E 04	3.003	6.85E 00	3.00E 04
2.444	2.91E-01	1.76E 02	2.714	2.13E 02	1.20E 05	3.007	8.01E 00	8.37E 04
2.451	3.97E-01	2.49E 02	2.721	5.79E 02	3.09E 05	3.014	4.74E 00	1.50E 04
2.457	6.46E-01	4.39E 02	2.732	1.24E 02	7.18E 04	3.021	1.12E 01	9.92E 03
2.463	7.49E-01	4.85E 02	2.740	4.95E 02	2.02E 05	3.025	7.72E 00	5.22E 04
2.470	4.38E 00	5.11E 04	2.747	1.39E 02	6.15E 04	3.030	8.14E 00	1.49E 04
2.474	2.53E 00	2.32E 04	2.755	2.78E 02	6.89E 04	3.036	9.81E 00	6.65E 03
2.480	6.65E 00	2.11E 04	2.758	2.72E 02	1.10E 05	3.039	8.43E 00	2.45E 04
2.483	5.28E 00	3.72E 04	2.760	2.66E 02	1.52E 05	3.044	1.01E 01	2.39E 04
2.488	8.63E 00	3.14E 04	2.766	2.94E 02	6.69E 04	3.049	1.67E 01	1.77E 04
2.494	1.00E 01	2.15E 04	2.768	2.50E 02	6.64E 04	3.053	7.60E 00	3.21E 04
2.501	9.86E 00	1.67E 04	2.774	1.71E 02	6.46E 04	3.060	8.28E 00	1.94E 04
2.505	1.03E 01	5.09E 04	2.778	1.90E 02	7.20E 04	3.067	9.68E 00	1.19E 04
2.509	1.59E 01	3.63E 04	2.780	1.80E 02	7.63E 04	3.072	7.19E 00	4.48E 04
2.511	1.39E 01	3.07E 04	2.785	2.30E 02	8.46E 04	3.078	1.28E 01	2.19E 04
2.519	2.60E 01	2.48E 04	2.792	1.13E 02	6.27E 04	3.084	7.03E 00	1.05E 04
2.524	3.45E 01	1.73E 05	2.794	8.74E 01	4.97E 04	3.088	8.86E 00	7.01E 03
2.528	5.16E 01	9.29E 04	2.803	2.55E 02	1.32E 05	3.094	8.63E 00	3.13E 04
2.536	9.20E 01	1.15E 05	2.813	1.00E 02	1.15E 05	3.099	6.25E 00	4.90E 03
2.539	7.22E 01	1.03E 05	2.815	1.05E 02	9.17E 04	3.107	9.86E 00	1.66E 04
2.544	1.44E 02	1.52E 05	2.818	1.36E 02	9.23E 04	3.116	6.43E 00	1.15E 05
2.550	9.06E 01	5.65E 05	2.824	8.17E 01	1.30E 05	3.121	9.78E 00	1.44E 04
2.556	1.12E 02	2.70E 05	2.828	6.12E 01	9.38E 04	3.125	8.12E 00	9.55E 03
2.563	1.41E 03	6.69E 06	2.834	1.07E 02	6.00E 04	3.130	2.20E 00	2.48E 03
2.566	4.57E 02	8.50E 05	2.843	3.74E 01	6.24E 04	3.135	1.42E 00	1.47E 03
2.571	2.59E 02	1.31E 05	2.851	6.00E 01	5.74E 04	3.140	7.67E 00	9.38E 03

Table 5-6. H_2O Vapor Parameters Corresponding to the Goody Model (*Continued*)

λ	S/d	$S/2\pi\alpha_o'$	λ	S/d	$S/2\pi\alpha_o'$	λ	S/d	$S/2\pi\alpha_o'$
3.145	7.10E 00	7.25E 03	3.419	7.17E-01	8.84E 01	3.738	5.10E-02	4.26E 01
3.150	4.28E 00	3.54E 03	3.425	3.32E-01	2.80E 02	3.745	5.89E-02	1.23E 02
3.155	9.24E-01	8.14E 02	3.430	2.03E-01	1.33E 02	3.752	6.22E-02	5.76E 01
3.160	4.57E-01	2.30E 02	3.436	3.06E-01	1.98E 02	3.759	5.88E-02	4.92E 01
3.165	2.70E-01	1.26E 02	3.442	5.29E-01	5.51E 02	3.766	2.32E-02	3.89E 01
3.170	2.29E-01	1.02E 02	3.448	2.09E-01	1.59E 02	3.774	4.42E-02	4.39E 01
3.175	4.26E-01	3.30E 02	3.454	1.72E-01	9.16E 01	3.781	7.12E-02	5.25E 01
3.180	1.30E 00	1.31E 03	3.460	1.13E-01	5.26E 01	3.788	2.71E-02	5.42E 01
3.185	1.56E 00	1.28E 03	3.466	1.11E-01	8.36E 01	3.795	2.29E-02	2.65E 01
3.190	7.34E 00	8.01E 03	3.472	2.28E-01	1.60E 02	3.802	5.94E-02	4.74E 01
3.195	1.13E 01	1.38E 04	3.478	1.19E-01	5.21E 01	3.809	4.12E-02	4.13E 01
3.200	1.10E 01	7.69E 03	3.484	5.31E-02	3.61E 01	3.817	2.19E-01	1.76E 01
3.205	1.00E 01	8.07E 03	3.490	8.09E-02	4.46E 01	3.824	3.48E-02	3.41E 01
3.210	1.29E 01	1.10E 04	3.496	1.13E-01	7.35E 01	3.831	4.78E-01	3.55E 01
3.215	1.11E 01	1.00E 04	3.503	9.14E-02	7.05E 01	3.839	2.03E-02	2.97E 01
3.221	1.29E 01	1.07E 04	3.509	8.84E-02	3.83E 01	3.846	2.07E-02	2.98E 01
3.226	1.45E 01	7.96E 03	3.515	8.62E-02	3.52E 01	3.854	3.79E-02	3.00E 01
3.231	1.11E 01	6.42E 03	3.521	1.29E-01	4.77E 01	3.861	2.90E-02	2.92E 01
3.236	8.78E-01	5.26E 02	3.527	8.30E-02	5.23E 01	3.868	1.94E-02	3.00E 01
3.241	4.87E 00	8.41E 03	3.534	8.39E-02	5.40E 01	3.876	2.38E-02	3.28E 01
3.247	1.02E 01	1.40E 04	3.540	8.38E-02	5.27E 01	3.883	3.18E-02	2.62E 01
3.252	6.47E-01	5.07E 02	3.546	5.25E-02	3.78E 01	3.891	2.40E-02	3.35E 01
3.257	4.35E 00	9.14E 03	3.552	1.49E-01	6.22E 01	3.899	1.94E-02	4.07E 01
3.263	1.50E 01	4.61E 04	3.559	7.63E-02	4.83E 01	3.906	2.22E-02	2.83E 01
3.268	3.33E 00	4.67E 03	3.565	4.39E-02	2.31E 01	3.914	2.27E-02	2.45E 01
3.273	6.40E-01	4.15E 02	3.571	1.55E-01	5.31E 01	3.922	2.70E-02	3.51E 01
3.279	1.21E 00	1.07E 03	3.578	1.30E-01	7.10E 01	3.929	1.99E-02	5.50E 01
3.284	6.47E-01	5.82E 02	3.584	1.14E-01	7.83E 01	3.937	2.12E-02	2.77E 01
3.289	1.94E 00	3.61E 03	3.591	1.90E-01	1.27E 02	3.945	2.01E-02	3.18E 01
3.295	1.48E 01	1.59E 04	3.597	3.28E-02	3.70E 01	3.953	2.23E-02	5.29E 01
3.300	8.47E 00	8.88E 03	3.604	4.26E-02	4.60E 01	3.960	1.96E-02	3.40E 01
3.306	5.16E 00	4.64E 03	3.610	7.68E-02	4.67E 01	3.968	2.12E-02	3.42E 01
3.311	3.83E 00	4.09E 03	3.617	3.83E-02	2.62E 01	3.976	1.94E-02	3.87E 01
3.317	2.01E 00	3.86E 03	3.623	3.47E-02	4.36E 01	3.984	2.34E-02	5.75E 01
3.322	1.37E 00	1.56E 03	3.630	5.92E-02	5.41E 01	3.992	2.18E-02	3.87E 01
3.328	1.71E-01	9.10E 01	3.636	4.99E-02	4.06E 01	4.000	2.44E-02	3.93E 01
3.333	1.09E-01	6.10E 01	3.643	3.82E-02	2.83E 01	4.008	2.11E-02	4.20E 01
3.339	4.56E-01	5.05E 02	3.650	5.97E-02	3.14E 01	4.016	2.00E-02	4.12E 01
3.344	1.87E 00	2.78E 03	3.656	4.70E-02	3.65E 01	4.024	2.09E-02	4.16E 01
3.350	2.88E-01	1.67E 02	3.663	5.57E-02	6.29E 01	4.032	3.40E-02	6.32E 01
3.356	5.69E-01	5.05E 02	3.670	3.78E-01	2.02E 02	4.040	2.57E-02	4.68E 01
3.361	5.30E 00	9.43E 03	3.676	3.92E-01	1.07E 02	4.049	2.15E-02	4.19E 01
3.367	1.12E 00	1.15E 03	3.683	1.42E-01	4.18E 01	4.057	2.26E-02	4.09E 01
3.373	1.13E 00	1.09E 03	3.690	8.36E-02	2-24E 01	4.065	2.20E-02	4.00E 01
3.378	2.02E-01	1.31E 02	3.697	5.33E-02	1.73E 01	4.073	4.97E-02	7.81E 01
3.384	8.97E-01	1.22E 03	3.704	3.54E-02	2.59E 01	4.082	3.28E-02	5.28E 01
3.390	1.79E 00	2.22E 03	3.711	5.58E-02	4.03E 01	4.089	2.20E-02	3.91E 01
3.396	1.73E 00	2.15E 03	3.717	3.53E-02	3.11E 01	4.098	2.61E-02	3.88E 01
3.401	2.48E-01	1.64E 02	3.724	2.32E-02	3.74E 01	4.107	4.97E-02	7.32E 01
3.407	1.68E-01	1.03E 02	3.731	6.80E-02	5.99E 01	4.115	9.70E-02	1.66E 02
3.413	8.31E-01	1.05E 03	–	–	–	4.124	2.49E-02	3.87E 01

Table 5-6. H_2O Vapor Parameters Corresponding to the Goody Model (*Continued*)

λ	S/d	$S/2\pi\alpha'_O$	λ	S/d	$S/2\pi\alpha'_O$
4.132	2.34E-02	4.16E 01	16.340	0.1103E 02	0.7795E 05
4.141	3.19E-02	4.77E 01	16.393	0.1103E 02	0.7702E 05
4.149	2.74E-02	3.87E 01	16.447	0.1818E 02	0.1101E 06
4.158	2.38E-02	3.77E 01	16.502	0.1556E 02	0.1034E 06
4.167	2.56E-02	4.81E 01	16.556	0.1557E 02	0.1017E 06
4.175	3.55E-02	5.82E 01	16.611	0.1712E 02	0.1145E 06
4.184	4.99E-02	6.11E 01	16.667	0.1710E 02	0.1166E 06
4.193	2.75E-02	3.91E 01	16.722	0.1733E 02	0.1056E 06
4.202	2.80E-02	3.57E 01	16.779	0.1979E 02	0.1127E 06
4.211	2.53E-02	4.05E 01	16.835	0.1954E 02	0.1194E 06
4.219	5.36E-02	6.97E 01	16.892	0.2690E 02	0.1609E 06
4.228	1.17E-01	2.07E 02	16.949	0.2690E 02	0.1614E 06
4.237	3.48E-02	3.70E 01	17.007	0.2784E 02	0.1531E 06
4.246	3.36E-02	3.36E 01	17.065	0.3056E 02	0.1600E 06
4.255	2.98E-02	3.36E 01	17.123	0.3325E 02	0.1583E 06
4.264	3.19E-02	3.49E 01	17.182	0.2995E 02	0.1514E 06
14.684	0.5182E 01	0.2918E 05	17.241	0.3056E 02	0.1425E 06
14.728	0.4486E 01	0.2593E 05	17.301	0.3056E 02	0.1425E 06
14.771	0.4999E 01	0.2875E 05	17.361	0.2954E 02	0.1454E 06
14.815	0.4956E 01	0.3093E 05	17.422	0.2238E 02	0.1193E 06
14.859	0.4956E 01	0.3093E 05	17.483	0.2332E 02	0.1109E 06
14.903	0.3568E 01	0.2397E 05	17.544	0.2331E 02	0.1120E 06
14.948	0.3206E 01	0.2184E 05	17.606	0.2353E 02	0.1081E 06
14.993	0.3206E 01	0.2184E 05	17.668	0.2411E 02	0.1096E 06
15.038	0.3206E 01	0.2184E 05	17.730	0.2615E 02	0.1113E 06
15.083	0.3247E 01	0.2077E 05	17.794	0.2596E 02	0.1155E 06
15.129	0.3744E 01	0.2440E 05	17.857	0.2601E 02	0.1127E 06
15.175	0.3896E 01	0.2326E 05	17.921	0.1914E 02	0.7320E 05
15.221	0.4992E 01	0.2876E 05	17.986	0.1931E 02	0.7372E 05
15.267	0.5339E 01	0.2984E 05	18.051	0.1950E 02	0.6942E 05
15.314	0.5223E 01	0.2966E 05	18.116	0.1991E 02	0.6875E 05
15.361	0.7378E 01	0.5491E 05	18.182	0.1701E 02	0.5829E 05
15.408	0.8425E 01	0.5594E 05	18.248	0.1432E 02	0.5118E 05
15.456	0.8436E 01	0.5487E 05	18.315	0.1427E 02	0.5039E 05
15.504	0.8197E 01	0.5875E 05	18.382	0.1375E 02	0.5092E 05
15.522	0.8475E 01	0.5639E 05	18.450	0.3140E 02	0.1632E 06
15.601	0.9847E 01	0.7567E 05	18.519	0.3136E 02	0.1646E 06
15.649	0.9847E 01	0.7567E 05	18.587	0.3119E 02	0.1738E 06
15.699	0.1078E 02	0.7245E 05	18.657	0.3957E 02	0.2190E 06
15.748	0.1081E 02	0.7031E 05	18.727	0.5714E 02	0.2732E 06
15.798	0.1238E 02	0.7396E 05	18.797	0.5475E 02	0.2633E 06
15.848	0.1238E 02	0.7305E 05	18.868	0.5341E 02	0.2677E 06
15.898	0.1263E 02	0.6868E 05	18.939	0.5198E 02	0.2512E 06
15.949	0.1260E 02	0.6996E 05	19.011	0.5143E 02	0.2649E 06
16.000	0.1188E 02	0.7116E 05	19.084	0.6114E 02	0.2821E 06
16.051	0.1173E 02	0.7455E 05	19.157	0.6046E 02	0.3001E 06
16.129	0.1079E 02	0.7307E 05	19.231	0.6153E 02	0.2875E 06
16.181	0.1066E 02	0.7604E 05	19.305	0.8327E 02	0.4012E 06
16.234	0.1130E 02	0.7937E 05	19.380	0.8295E 02	0.4075E 06
16.287	0.9990E 01	0.7982E 05	19.455	0.8293E 02	0.4109E 06

Table 5-6. H_2O Vapor Parameters Corresponding to the Goody Model (*Continued*)

λ	S/d	$S/2\pi\alpha_o'$	λ	S/d	$S/2\pi\alpha_o'$
19.531	0.8293E 02	0.4129E 06	23.348	0.3501E 03	0.1571E 07
19.608	0.8379E 02	0.4095E 06	23.614	0.3147E 03	0.1710E 07
19.685	0.7445E 02	0.3193E 06	23.906	0.2867E 03	0.1610E 07
19.763	0.7472E 02	0.3129E 06	24.195	0.5240E 03	0.2060E 07
19.841	0.7472E 02	0.3147E 06	24.492	0.6525E 03	0.2683E 07
19.920	0.6945E 02	0.3005E 06	24.795	0.5966E 03	0.2971E 07
20.068	0.6306E 02	0.2773E 06	25.107	0.4370E 03	0.2349E 07
20.272	0.6651E 02	0.3185E 06	25.426	0.5587E 03	0.2692E 07
20.479	0.8812E 02	0.4054E 06	25.753	0.8262E 03	0.3746E 07
20.691	0.7649E 02	0.3138E 06	26.089	0.8704E 03	0.3380E 07
20.907	0.8768E 02	0.3567E 06	26.434	0.6253E 03	0.3806E 07
21.128	0.1441E 03	0.5227E 06	26.788	0.6156E 03	0.3016E 07
21.354	0.1533E 03	0.5154E 06	27.152	0.1221E 04	0.4662E 07
21.584	0.1567E 03	0.5035E 06	27.525	0.1624E 04	0.7594E 07
21.820	0.1871E 03	0.5503E 06	27.910	0.1548E 04	0.7538E 07
22.060	0.1821E 03	0.6477E 06	28.305	0.1333E 04	0.6586E 07
22.306	0.1785E 03	0.6553E 06	28.711	0.1438E 04	0.6578E 07
22.558	0.1687E 03	0.6117E 06	29.129	0.2238E 04	0.9170E 07
22.815	0.1593E 03	0.6694E 06	29.560	0.2593E 04	0.9800E 07
23.079	0.3643E 03	0.1470E 07	30.003	0.1990E 04	0.1012E 08

with $e = 1.25 \times 10^{-22}$, molecules^{-1} cm^2 atm^{-1}

$f = 2.34 \times 10^{-19}$, molecules^{-1} cm^2 atm^{-1}

$g = 8.30 \times 10^{-3}$, cm

For CO_2:

(4) Empirical Model (1.37 to 2.64 μm, 4.65 to 5.355 μm, and 9.13 to 11.67 μm)
[5-33]. The transmittance is calculated from the following equation

$$\tau(\lambda) = \tau[w_p^* K_2(\lambda)] \qquad (5\text{-}64)$$

$$w_p^* = \int_{path} M'(z)\left(\frac{P(z)}{P_o}\right)^2\left(\frac{T_o}{T(z)}\right)^{1.5} dz, \text{ atm cm} \qquad (5\text{-}65)$$

where $K_2(\lambda)$ = spectral coefficient given in Table 5-7
$M'(z)$ = ratio of partial pressure of absorber to total pressure (See Table 5-25.)

Table 5-8 lists a set of values of $w_p^* K_2(\lambda)$. For each one of these values, a value of τ is given; for values of $w_p^* K_2(\lambda)$ between those shown in the table, τ must be interpolated linearly.

(5) Elsasser Model (2.64 to 2.88 μm and 4.184 to 4.454 μm). The transmittance is

$$\tau(\lambda) = 1 - \sinh\beta \int_0^Y I_0(y) \exp(-y \cosh\beta)dy \qquad (5\text{-}66)$$

Table 5-7. CO_2 Parameters for the Empirical Model

λ	$K_2(\lambda)$	λ	$K_2(\lambda)$	λ	$K_2(\lambda)$	λ	$K_2(\lambda)$
1.375	1.24E-07	1.977	9.40E-05	2.634	1.50E-06	5.145	1.50E-05
1.380	2.50E-07	1.980	9.94E-05	2.640	2.00E-05	5.160	2.34E-05
1.400	2.21E-06	1.985	1.15E-04	4.465	9.94E-03	5.180	3.94E-05
1.410	4.50E-06	1.990	1.33E-04	4.470	5.69E-03	5.220	5.66E-05
1.430	9.28E-06	1.995	1.60E-04	4.475	3.94E-03	5.260	3.94E-05
1.460	4.50E-06	2.000	1.64E-04	4.490	2.34E-03	5.290	2.34E-05
1.470	3.70E-06	2.010	1.75E-04	4.495	1.50E-03	5.320	1.50E-06
1.480	3.15E-06	2.030	2.48E-04	4.505	5.59E-04	5.355	5.59E-06
1.485	2.90E-06	2.040	1.95E-04	4.520	1.64E-04	9.130	1.24E-07
1.490	2.70E-06	2.045	1.64E-04	4.526	9.94E-05	9.220	2.05E-06
1.497	2.50E-06	2.052	1.33E-04	4.540	6.90E-05	9.310	6.25E-06
1.505	2.25E-06	2.057	9.98E-05	4.548	5.69E-05	9.335	6.25E-06
1.510	2.15E-06	2.062	7.84E-05	4.575	3.94E-05	9.398	4.00E-06
1.522	1.90E-06	2.065	5.69E-05	4.600	2.34E-05	9.463	2.16E-06
1.530	1.75E-06	2.069	4.87E-05	4.660	1.64E-05	9.494	4.00E-06
1.550	1.64E-06	2.072	3.94E-05	4.700	2.34E-05	9.526	5.00E-06
1.570	3.50E-06	2.075	3.13E-05	4.730	3.94E-05	9.590	5.00E-06
1.580	9.22E-06	2.078	2.34E-05	4.750	5.69E-05	9.652	2.18E-06
1.590	8.00E-06	2.081	1.77E-05	4.770	9.94E-05	9.713	6.22E-07
1.610	5.00E-06	2.084	1.50E-05	4.800	1.94E-04	9.773	2.67E-07
1.620	3.50E-06	2.089	5.59E-06	4.830	2.48E-04	9.834	1.00E-08
1.630	2.20E-06	2.100	2.21E-06	4.860	4.06E-04	9.893	1.00E-08
1.650	6.25E-07	2.120	1.24E-07	4.890	2.48E-04	9.953	1.00E-08
1.700	1.40E-07	2.563	1.30E-07	4.910	1.64E-04	10.070	3.40E-08
1.890	1.24E-07	2.566	1.80E-07	4.925	9.94E-05	10.190	2.05E-06
1.900	1.24E-06	2.571	2.00E-07	4.940	5.69E-05	10.290	6.06E-06
1.910	9.88E-06	2.577	3.50E-07	4.950	3.94E-05	10.440	1.94E-06
1.915	1.50E-05	2.584	5.00E-07	4.965	2.34E-05	10.550	6.03E-06
1.918	1.77E-05	2.589	7.00E-07	4.970	1.50E-05	10.900	1.24E-07
1.920	2.34E-05	2.592	7.20E-07	4.975	9.88E-06	11.050	2.80E-07
1.930	4.86E-05	2.597	1.10E-06	4.990	5.59E-06	11.173	5.40E-07
1.940	5.50E-05	2.602	1.60E-06	5.000	5.59E-06	11.268	8.80E-07
1.950	5.69E-05	2.609	2.50E-06	5.025	2.05E-06	11.364	1.50E-07
1.955	6.00E-05	2.614	4.00E-06	5.075	8.00E-07	11.500	3.13E-06
1.963	7.00E-05	2.622	6.40E-06	5.105	2.05E-06	11.670	4.36E-06
1.974	8.40E-05	2.629	1.00E-05	5.130	5.59E-06	—	

Table 5-8. Transmittance, τ, as a Function of $w_p^* K_2(\lambda)$

$w_p^* K_2(\lambda)$	$\tau(w_p^* K_2(\lambda))$	$w_p^* K_2(\lambda)$	$\tau(w_p^* K_2(\lambda))$
0.398E −03	0.100E +01	0.141E +00	0.575E +00
0.178E −02	0.980E +00	0.150E +00	0.560E +00
0.199E −02	0.970E +00	0.162E +00	0.540E +00
0.630E −02	0.955E +00	0.178E +00	0.525E +00
0.707E −02	0.940E +00	0.182E +00	0.500E +00
0.100E −01	0.930E +00	0.200E +00	0.480E +00
0.126E −01	0.920E +00	0.251E +00	0.450E +00
0.159E −01	0.910E +00	0.282E +00	0.425E +00
0.200E −01	0.890E +00	0.318E +00	0.400E +00
0.252E −01	0.875E +00	0.355E +00	0.375E +00
0.282E −01	0.867E +00	0.426E +00	0.350E +00
0.316E −01	0.850E +00	0.446E +00	0.330E +00
0.355E −01	0.840E +00	0.526E +00	0.300E +00
0.390E −01	0.830E +00	0.562E +00	0.275E +00
0.398E −01	0.820E +00	0.603E +00	0.270E +00
0.447E −01	0.810E +00	0.620E +00	0.260E +00
0.479E −01	0.800E +00	0.630E +00	0.240E +00
0.526E −01	0.780E +00	0.708E +00	0.225E +00
0.565E −01	0.750E +00	0.795E +00	0.200E +00
0.631E −01	0.740E +00	0.100E +01	0.175E +00
0.640E −01	0.725E +00	0.111E +01	0.150E +00
0.750E −01	0.700E +00	0.159E +01	0.125E +00
0.795E −01	0.675E +00	0.178E +01	0.100E +00
0.100E +00	0.650E +00	0.251E +01	0.750E −01
0.112E +00	0.625E +00	0.426E +01	0.500E −01
0.126E +00	0.600E +00	0.741E +01	0.250E −01
—	—	0.224E +02	0.0
—	—	0.300E +02	0.0

where

$$\beta = \frac{2\pi\alpha_o'}{d}\,\bar{P} \tag{5-67}$$

$$y = \frac{\dfrac{s}{d}}{\sinh\beta}\,w_p \tag{5-68}$$

$$\bar{P} = \frac{P_o}{w_p}\,w_p^*, \quad P_o = 760 \text{ mm Hg} \tag{5-69}$$

$$w_p = \int_{\text{path}} M'(z)\left(\frac{P(z)}{P_o}\right)\frac{T_o}{T(z)}\,dz, \text{ atm cm} \tag{5-70}$$

Y = value of y at end of path

Table 5-9. CO_2 Parameters for the Elsasser Model

λ	S/d	2πα'_o/d	λ	S/d	2πα'_o/d
2.643	3.30E-04	4.55E-06	2.851	3.30E-03	5.00E-04
2.650	1.60E-03	3.75E-05	2.855	2.40E-03	5.42E-04
2.667	3.70E-01	1.35E-04	2.862	1.00E-03	4.50E-04
2.672	5.00E-01	5.00E-04	2.869	3.00E-05	3.23E-03
2.674	5.00E-01	6.30E-04	4.158	2.60E-04	1.00E-06
2.680	5.50E-01	1.00E-03	4.167	5.90E-03	5.00E-06
2.683	8.80E-01	6.14E-04	4.175	2.40E-01	8.00E-05
2.688	8.60E-01	3.26E-04	4.184	9.94E-02	1.81E-03
2.691	5.40E-01	3.61E-04	4.193	9.80E-01	7.37E-04
2.694	8.80E-01	2.95E-04	4.202	5.05E 00	7.63E-04
2.697	5.20E-01	6.25E-04	4.211	1.46E 01	8.03E-04
2.703	7.00E-01	4.93E-04	4.219	2.73E 01	1.16E-03
2.709	7.40E-01	3.78E-04	4.228	3.36E 01	1.50E-03
2.714	6.20E-01	2.18E-04	4.237	3.36E 01	1.55E-03
2.721	4.40E-01	8.86E-05	4.246	2.56E 01	1.60E-03
2.732	1.50E-01	6.67E-05	4.255	1.34E 01	1.68E-03
2.740	2.50E-02	6.00E-04	4.264	2.32E 01	1.58E-03
2.747	7.60E-02	9.87E-04	4.274	2.59E 01	1.42E-03
2.755	3.60E-01	7.22E-04	4.283	2.47E 01	1.35E-03
2.758	5.50E-01	5.45E-04	4.292	2.07E 01	1.21E-03
2.760	5.80E-01	4.40E-04	4.301	1.70E 01	1.43E-03
2.766	5.30E-01	2.45E-04	4.310	1.22E 01	1.26E-03
2.768	3.30E-01	2.45E-04	4.319	7.30E 00	1.32E-03
2.774	2.60E-01	9.23E-04	4.329	5.00E 00	1.32E-03
2.778	4.20E-01	6.67E-04	4.338	2.91E 00	1.45E-03
2.780	4.70E-01	5.74E-04	4.348	2.02E 00	1.34E-03
2.785	4.40E-01	3.75E-04	4.357	1.30E 00	1.45E-03
2.792	2.90E-01	2.41E-04	4.367	6.86E-01	1.29E-03
2.794	1.60E-01	3.50E-04	4.376	3.76E-01	1.30E-03
2.803	1.30E-01	2.69E-04	4.386	4.60E-01	8.29E-04
2.813	7.00E-02	1.20E-04	4.396	6.09E-01	5.53E-04
2.815	3.70E-02	1.57E-04	4.405	5.89E-01	5.00E-04
2.818	2.70E-02	1.44E-04	4.415	5.61E-01	4.34E-04
2.824	1.70E-02	1.18E-04	4.425	3.83E-01	4.80E-04
2.828	5.00E-03	3.30E-04	4.435	2.50E-01	6.85E-04
2.834	3.00E-03	5.17E-04	4.444	1.85E-01	9.22E-04
2.843	3.90E-03	5.05E-04	4.454	8.62E-02	1.01E-03

The quantity w_p^* is defined in Equation (5-65) and $2\pi\alpha'_o/d$ and S/d are given in Table 5-9. Equation (5-66) is solved graphically with the use of the family of curves in Figure 5-5.

(6) Strong-Line Elsasser Model (2.64 to 2.88 μm and 4.184 to 4.454 μm). Since the exact solution converges slowly for large values of $\psi = Sw_p/(2\pi\alpha)$, the strong-line approximation is used for $\psi > 20$, so that

$$\tau(\lambda) = 1 - \text{erf} \, [0.5 \, \beta^2 \psi)^{1/2}]$$ (5-71)

where

$$\text{erf}(x) = \frac{2}{\sqrt{\pi}} \int_0^x \exp(-t^2) \, dt$$ (5-72)

The quantities $\bar{P}$ and w_p are calculated as above.

A tabulation of error functions, $\text{erf}(\zeta)$, is given in Table 5-10.

(7) *Temperature-Dependent Elasser Model (11.67 to 19.92 μm)*. The transmittance is given by Equation (5-66) with the following band-model parameters:

$$\frac{S}{d} = \frac{A_1 \exp \left(-\dfrac{A_2}{T_h} \right)}{T_h^2}$$ (5-73)

$$\frac{2\pi\alpha_o' S}{d^2} = \frac{A_3 \exp \left(-\dfrac{A_4}{T_h} \right)}{T_h^2}$$ (5-74)

A_1, A_2, A_3, and A_4 are tabulated in Table 5-11. The value of S/d is used from Equation (5-73). The parameter $2\pi\alpha_o'/d$ is calculated from the quotient of Equations (5-73) and (5-74). The quantity T_h is the homogeneous temperature which has an effect on the transmittance equivalent to that from an atmosphere in which the temperature changes in the path. If the temperature is constant, then T_h is used exactly. Otherwise it must be obtained by iteration.

For O_3:

(8) *Modified Elasser (9.398 to 10.19 μm)* [5-114]. For this spectral region, the transmittance is given by Equation (5-66). The β-parameters are given by

$$\beta' = \left(\frac{2\pi\alpha_o}{d} \right) \left(\frac{\bar{P}}{P_o} \right)^{C(\lambda)} = \beta_o \left(\frac{\bar{P}}{P_o} \right)^{C(\lambda)}$$ (5-75)

The family of curves in Figure 5-5 is used as for the Elsasser model, except that β' is substituted for β. Table 5-12 gives a tabulation of β_o, $C(\lambda)$ and S/d (from which Y in Equation 5-66) is calculated. The quantity $\bar{P}$ is calculated as in Equation (5-69) and w_p is calculated as in the Elsasser model for CO_2. In this case, $M'(s) = 0.6 \, \rho_{\text{ozone}}/\rho_{\text{air}}$ and the densities are obtained from Table 5-27. If a vertical path is used, the integration is performed numerically. The value 0.6 is the ratio of the molecular weights of air and ozone.

(9) *Goody Model (11.7 to 15.4 μm)*. A special form of the Goody model is given by

$$\tau(\lambda) = \exp \left\{ \frac{-w \, K_3(\lambda)}{\left[1 + \dfrac{K_3(\lambda) \, w}{4\bar{P} \, K_4(\lambda)} \right]^{1/2}} \right\}$$ (5-76)

Table 5-10. Error Function

x	erf x	x	erf x	x	erf x	x	erf x
0.00	0.00000	0.50	0.52049	1.00	0.84270	1.50	0.96610
0.01	0.01128	0.51	0.52924	1.01	0.84681	1.51	0.96727
0.02	0.02256	0.52	0.53789	1.02	0.85083	1.52	0.96841
0.03	0.03384	0.53	0.54646	1.03	0.85478	1.53	0.96951
0.04	0.04511	0.54	0.55493	1.04	0.85864	1.54	0.97058
0.05	0.05637	0.55	0.56332	1.05	0.86243	1.55	0.97162
0.06	0.06762	0.56	0.57161	1.06	0.86614	1.56	0.97262
0.07	0.07885	0.57	0.57981	1.07	0.86977	1.57	0.97360
0.08	0.09007	0.58	0.58792	1.08	0.87332	1.58	0.97454
0.09	0.10128	0.59	0.59593	1.09	0.87680	1.59	0.97546
0.10	0.11246	0.60	0.60385	1.10	0.88020	1.60	0.97634
0.11	0.12362	0.61	0.61168	1.11	0.88353	1.61	0.97720
0.12	0.13475	0.62	0.61941	1.12	0.88678	1.62	0.97803
0.13	0.14586	0.63	0.62704	1.13	0.88997	1.63	0.97884
0.14	0.15694	0.64	0.63458	1.14	0.89308	1.64	0.97962
0.15	0.16799	0.65	0.64202	1.15	0.89612	1.65	0.98037
0.16	0.17901	0.66	0.64937	1.16	0.89909	1.66	0.98110
0.17	0.18999	0.67	0.65662	1.17	0.90200	1.67	0.98181
0.18	0.20093	0.68	0.66378	1.18	0.90483	1.68	0.98249
0.19	0.21183	0.69	0.67084	1.19	0.90760	1.69	0.98315
0.20	0.22270	0.70	0.67780	1.20	0.91031	1.70	0.98379
0.21	0.23352	0.71	0.68466	1.21	0.91295	1.71	0.98440
0.22	0.24429	0.72	0.69143	1.22	0.91553	1.72	0.98500
0.23	0.25502	0.73	0.69810	1.23	0.91805	1.73	0.98557
0.24	0.26570	0.74	0.70467	1.24	0.92050	1.74	0.98613
0.25	0.27632	0.75	0.71115	1.25	0.92290	1.75	0.98667
0.26	0.28689	0.76	0.71753	1.26	0.92523	1.76	0.98719
0.27	0.29741	0.77	0.72382	1.27	0.92751	1.77	0.98769
0.28	0.30788	0.78	0.73001	1.28	0.92973	1.78	0.98817
0.29	0.31828	0.79	0.73610	1.29	0.93189	1.79	0.98864
0.30	0.32862	0.80	0.74210	1.30	0.93400	1.80	0.98909
0.31	0.33890	0.81	0.74800	1.31	0.93606	1.81	0.98952
0.32	0.34912	0.82	0.75381	1.32	0.93806	1.82	0.98994
0.33	0.35927	0.83	0.75952	1.33	0.94001	1.83	0.99034
0.34	0.36936	0.84	0.76514	1.34	0.94191	1.84	0.99073
0.35	0.37938	0.85	0.77066	1.35	0.94376	1.85	0.99111
0.36	0.38932	0.86	0.77610	1.36	0.94556	1.86	0.99147
0.37	0.39920	0.87	0.78143	1.37	0.94731	1.87	0.99182
0.38	0.40900	0.88	0.78668	1.38	0.94901	1.88	0.99215
0.39	0.41873	0.89	0.79184	1.39	0.95067	1.89	0.99247
0.40	0.42839	0.90	0.79690	1.40	0.95228	1.90	0.99279
0.41	0.43796	0.91	0.80188	1.41	0.95385	1.91	0.99308
0.42	0.44746	0.92	0.80676	1.42	0.95537	1.92	0.99337
0.43	0.45688	0.93	0.81156	1.43	0.95685	1.93	0.99365
0.44	0.46622	0.94	0.81627	1.44	0.95829	1.94	0.99392
0.45	0.47548	0.95	0.82089	1.45	0.95969	1.95	0.99417
0.46	0.48465	0.96	0.82542	1.46	0.96105	1.96	0.99442
0.47	0.49374	0.97	0.82987	1.47	0.96237	1.97	0.99466
0.48	0.50274	0.98	0.83423	1.48	0.96365	1.98	0.99489
0.49	0.51166	0.99	0.83850	1.49	0.96489	1.99	0.99511

Table 5-11. CO_2 Parameters Corresponding to the Temperature-Dependent
Elsasser Model

λ	A_1	A_2	A_3	A_4
11.790	0.0496E 02	5.1198E 02	0.0521E 00	0.1371E 04
11.900	0.0868E 02	5.1198E 02	0.3049E 00	0.1418E 04
12.050	0.2281E 02	5.1198E 02	0.6201E 00	0.1631E 04
12.200	1.0398E 02	0.7055E 03	0.3694E 01	0.1646E 04
12.350	2.9232E 02	8.3624E 02	0.4218E 01	0.1496E 04
12.500	8.3536E 02	9.9206E 02	0.3968E 01	0.1418E 04
12.530	0.1134E 04	0.1063E 04	0.3932E 01	0.1406E 04
12.560	0.1581E 04	0.1131E 04	0.1134E 02	0.1629E 04
12.590	0.2657E 06	0.2131E 04	0.5443E 02	0.1753E 04
12.630	0.2886E 06	0.2127E 04	0.6953E 02	0.1769E 04
12.660	0.5547E 10	0.4303E 04	0.1056E 02	0.1372E 04
12.690	0.4121E 03	0.8242E 03	0.4582E 01	0.1422E 04
12.720	0.9680E 03	0.9952E 03	0.8368E 01	0.1520E 04
12.760	0.1208E 04	0.1034E 04	0.1004E 02	0.1543E 04
12.790	0.1576E 04	0.1083E 04	0.1342E 02	0.1590E 04
12.820	0.2138E 04	0.1140E 04	0.2241E 02	0.1686E 04
12.850	0.3828E 04	0.1258E 04	0.4513E 02	0.1819E 04
12.890	0.4000E 04	0.1262E 04	0.4791E 02	0.1826E 04
12.920	0.6420E 04	0.1356E 04	0.9884E 02	0.1966E 04
12.950	0.1156E 05	0.1471E 04	0.1776E 03	0.2066E 04
12.990	0.1779E 05	0.1554E 04	0.1533E 03	0.2013E 04
13.020	0.2833E 05	0.1639E 04	0.1526E 03	0.1981E 04
13.050	0.3799E 05	0.1685E 04	0.1220E 03	0.1902E 04
13.090	0.1563E 06	0.1949E 04	0.1826E 03	0.1912E 04
13.120	0.1121E 06	0.1852E 04	0.1225E 03	0.1801E 04
13.160	0.1529E 06	0.1880E 04	0.1956E 03	0.1849E 04
13.190	0.2323E 06	0.1887E 04	0.5823E 03	0.1983E 04
13.230	0.1396E 06	0.1850E 04	0.4183E 03	0.1894E 04
13.260	0.2016E 06	0.1777E 04	0.2805E 03	0.1742E 04
13.300	0.2388E 06	0.1708E 04	0.2968E 03	0.1652E 04
13.330	0.1046E 06	0.1504E 04	0.2820E 03	0.1647E 04
13.370	0.5863E 05	0.1316E 04	0.4374E 03	0.1678E 04
13.400	0.2818E 05	0.1023E 04	0.3298E 03	0.1522E 04
13.440	0.7802E 05	0.1246E 04	0.2798E 03	0.1442E 04
13.480	0.1341E 06	0.1360E 04	0.8617E 03	0.1571E 04
13.510	0.3475E 06	0.1528E 04	0.1488E 04	0.1588E 04
13.550	0.3454E 05	0.9799E 03	0.8273E 03	0.1568E 04
13.590	0.1602E 05	0.7234E 03	0.4963E 03	0.1433E 04
13.620	0.2796E 05	0.8623E 03	0.2765E 03	0.1241E 04

Table 5-11. CO_2 Parameters Corresponding to the Temperature-Dependent
Elsasser Model (*Continued*)

λ	A_1	A_2	A_3	A_4
13.660	0.1876E 05	0.7702E 03	0.2419E 03	0.1247E 04
13.700	0.1719E 05	0.7776E 03	0.1590E 03	0.1262E 04
13.740	0.3462E 04	0.3097E 03	0.3093E 03	0.1304E 04
13.770	0.1281E 04	0.4217E 02	0.3339E 03	0.1350E 04
13.810	0.1229E 04	-.3377E 00	0.4200E 03	0.1412E 04
13.850	0.3928E 03	-.3496E 03	0.7854E 11	0.5328E 04
13.890	0.1847E 06	0.2947E 03	0.3052E 05	0.1857E 04
13.950	0.2866E 04	0.1781E 03	0.4001E 04	0.1699E 04
13.990	0.9503E 04	0.5409E 03	0.2722E 04	0.1626E 04
14.030	0.9309E 04	0.4656E 03	0.3134E 04	0.1616E 04
14.060	0.2199E 05	0.6026E 03	0.4106E 04	0.1624E 04
14.100	0.2451E 05	0.5670E 03	0.3670E 04	0.1551E 04
14.140	0.7930E 04	0.2646E 03	0.7788E 04	0.1686E 04
14.180	0.4181E 04	0.3927E 02	0.1300E 05	0.1744E 04
14.220	0.3320E 04	-.1113E 03	0.1322E 05	0.1680E 04
14.270	0.7648E 04	0.7435E 02	0.1013E 05	0.1569E 04
14.310	0.2042E 04	-.2913E 03	0.1056E 05	0.1526E 04
14.350	0.5341E 03	-.6627E 03	0.1562E 05	0.1535E 04
14.390	0.4605E 03	-.7497E 03	0.7459E 04	0.1299E 04
14.430	0.5586E 03	-.7104E 03	0.8157E 04	0.1275E 04
14.470	0.6195E 03	-.7126E 03	0.9928E 04	0.1246E 04
14.510	0.6513E 03	-.7033E 03	0.8860E 04	0.1139E 04
14.560	0.5009E 03	-.8024E 03	0.1799E 04	0.7323E 03
14.600	0.2645E 03	-.9942E 03	0.2513E 04	0.7932E 03
14.640	0.3811E 03	-.8926E 03	0.2483E 04	0.7391E 03
14.684	0.2964E 03	-.9596E 03	0.3039E 04	0.7337E 03
14.728	0.3596E 03	-.9178E 03	0.1042E 04	0.4670E 03
14.771	0.2856E 03	-.9768E 03	0.1364E 04	0.5543E 03
14.815	0.7554E 03	-.7286E 03	0.1794E 04	0.6056E 03
14.859	0.7240E 03	-.7466E 03	0.1069E 05	0.9695E 03
14.903	0.2785E 04	-.4781E 03	0.5989E 06	0.1726E 04
14.948	0.2061E 04	-.6501E 03	0.1234E 07	0.1716E 04
14.993	0.4158E 04	-.5511E 03	0.1326E 08	0.2253E 04
15.038	0.2406E 04	-.3470E 03	0.2558E 04	0.6745E 03
15.083	0.7086E 03	-.6943E 03	0.2048E 04	0.6550E 03
15.129	0.5428E 03	-.7764E 03	0.5274E 03	0.3265E 03
15.175	0.8198E 03	-.7070E 03	0.1450E 04	0.6131E 03
15.221	0.9123E 03	-.6771E 03	0.9524E 03	0.4844E 03
15.267	0.5597E 03	-.7968E 03	0.7001E 03	0.4015E 03

Table 5-11. CO_2 Parameters Corresponding to the Temperature-Dependent
Elsasser Model (*Continued*)

λ	A_1	A_2	A_3	A_4
15.314	0.7001E 03	-.7238E 03	0.1610E 04	0.6188E 03
15.361	0.1007E 04	-.6039E 03	0.2147E 05	0.1184E 04
15.408	0.1023E 04	-.5910E 03	0.1086E 05	0.9644E 03
15.456	0.7357E 03	-.6335E 03	0.1292E 05	0.1094E 04
15.504	0.6620E 03	-.6161E 03	0.2224E 05	0.1507E 04
15.552	0.8717E 03	-.5512E 03	0.1442E 04	0.8306E 03
15.601	0.8254E 03	-.5590E 03	0.2645E 04	0.1012E 04
15.649	0.2784E 04	-.1832E 03	0.1331E 05	0.1465E 04
15.699	0.1947E 04	-.2642E 03	0.6661E 04	0.1332E 04
15.748	0.2217E 04	-.1996E 03	0.4236E 04	0.1248E 04
15.798	0.4612E 04	0.4628E 02	0.1233E 05	0.1582E 04
15.848	0.4473E 04	0.6327E 02	0.3717E 04	0.1341E 04
15.898	0.6447E 04	0.2145E 03	0.3651E 04	0.1385E 04
15.949	0.2368E 04	0.1793E 02	0.2638E 04	0.1384E 04
16.000	0.1196E 04	-.1171E 03	0.1823E 04	0.1372E 04
16.051	0.4167E 03	-.3686E 03	0.2667E 03	0.8802E 03
16.129	0.1495E 07	0.1765E 04	0.4081E 03	0.1119E 04
16.181	0.1440E 12	0.4279E 04	0.1041E 04	0.1073E 04
16.234	0.1730E 12	0.4334E 04	0.1366E 04	0.1115E 04
16.287	0.1373E 06	0.1271E 04	0.9795E 03	0.1450E 04
16.340	0.3874E 05	0.9882E 03	0.4113E 03	0.1310E 04
16.393	0.3590E 05	0.9714E 03	0.3965E 03	0.1305E 04
16.447	0.2904E 05	0.9440E 03	0.4175E 03	0.1351E 04
16.502	0.2409E 05	0.8957E 03	0.3030E 03	0.1309E 04
16.556	0.2806E 05	0.9308E 03	0.2406E 03	0.1248E 04
16.611	0.2389E 05	0.9042E 03	0.2028E 03	0.1273E 04
16.667	0.5346E 05	0.1099E 04	0.1641E 03	0.1271E 04
16.722	0.1513E 06	0.1394E 04	0.6753E 03	0.1502E 04
16.779	0.8236E 05	0.1280E 04	0.1396E 04	0.1659E 04
16.835	0.6057E 05	0.1282E 04	0.6766E 03	0.1658E 04
16.892	0.9037E 05	0.1408E 04	0.2069E 03	0.1491E 04
16.949	0.1730E 06	0.1592E 04	0.1997E 03	0.1520E 04
17.007	0.1675E 06	0.1626E 04	0.2119E 03	0.1556E 04
17.065	0.1305E 06	0.1650E 04	0.1728E 03	0.1584E 04
17.123	0.1385E 06	0.1705E 04	0.1876E 03	0.1650E 04
17.182	0.1376E 06	0.1742E 04	0.2395E 03	0.1740E 04
17.241	0.1669E 06	0.1836E 04	0.2560E 03	0.1807E 04
17.301	0.9789E 05	0.1776E 04	0.1094E 03	0.1770E 04
17.361	0.8987E 05	0.1790E 04	0.1124E 03	0.1749E 04
17.422	0.7901E 05	0.1794E 04	0.9238E 02	0.1753E 04

Table 5-11. CO_2 Parameters Corresponding to the Temperature-Dependent
Elsasser Model (*Continued*)

λ	A_1	A_2	A_3	A_4
17.483	0.3691E 05	0.1662E 04	0.1141E 03	0.1856E 04
17.544	0.1833E 05	0.1535E 04	0.1071E 03	0.1889E 04
17.606	0.1127E 05	0.1449E 04	0.9019E 02	0.1890E 04
17.668	0.7836E 04	0.1385E 04	0.9961E 02	0.1941E 04
17.730	0.3778E 04	0.1243E 04	0.5891E 02	0.1861E 04
17.794	0.1855E 04	0.1101E 04	0.3837E 02	0.1795E 04
17.857	0.1400E 04	0.1049E 04	0.2995E 02	0.1759E 04
17.921	0.8305E 03	0.9456E 03	0.1453E 02	0.1620E 04
17.986	0.6014E 03	0.8841E 03	0.9487E 01	0.1543E 04
18.051	0.4370E 03	0.8227E 03	0.6361E 01	0.1471E 04
18.116	0.3425E 03	0.7780E 03	0.5063E 01	0.1436E 04
18.182	0.2392E 03	0.7083E 03	0.3146E 01	0.1348E 04
18.248	0.5335E 03	0.8883E 03	0.1070E 02	0.1622E 04
18.315	0.4398E 04	0.1332E 04	0.8373E 02	0.2020E 04
18.382	0.3669E 04	0.1300E 04	0.4536E 02	0.1901E 04
18.450	0.3762E 03	0.8294E 03	0.1862E 01	0.1267E 04
18.519	0.1092E 03	0.5745E 03	0.1133E 01	0.1194E 04
18.587	0.1169E 03	0.5958E 03	0.1239E 01	0.1224E 04
18.657	0.1289E 03	0.6231E 03	0.1426E 01	0.1264E 04
18.727	0.1454E 03	0.6558E 03	0.1694E 01	0.1312E 04
18.797	0.1387E 03	0.6527E 03	0.1599E 01	0.1312E 04
18.868	0.1377E 03	0.6578E 03	0.1660E 01	0.1332E 04
18.939	0.1592E 03	0.6960E 03	0.2170E 01	0.1401E 04
19.011	0.1179E 03	0.6380E 03	0.1445E 01	0.1328E 04
19.084	0.1087E 03	0.6269E 03	0.1354E 01	0.1326E 04
19.157	0.1122E 03	0.6406E 03	0.1539E 01	0.1365E 04
19.231	0.8112E 02	0.5768E 03	0.1005E 01	0.1286E 04
19.305	0.7337E 02	0.5611E 03	0.9072E 00	0.1275E 04
19.380	0.7022E 02	0.5577E 03	0.8894E 00	0.1283E 04
19.455	0.5596E 02	0.5148E 03	0.6276E 00	0.1217E 04
19.531	0.5220E 02	0.5054E 03	0.6226E 00	0.1227E 04
19.608	0.4822E 02	0.4938E 03	0.5122E 00	0.1195E 04
19.685	0.4317E 02	0.4753E 03	0.4272E 00	0.1166E 04
19.763	0.4096E 02	0.4693E 03	0.4289E 00	0.1177E 04
19.841	0.3796E 02	0.4589E 03	0.3730E 00	0.1155E 04
19.920	0.3632E 02	0.4546E 03	0.3659E 00	0.1161E 04

Table 5-12. O_3 Parameters Corresponding to the Modified Elsasser Model

λ	S/d	β_o	$C(\lambda)$	λ	S/d	β_o	$C(\lambda)$
9.398	4.76E-01	7.12E-01	4.70E-01	9.773	8.00E 00	1.67E 00	6.90E-01
9.463	1.00E 01	1.70E 00	6.50E-01	9.834	7.14E 00	1.86E 00	7.50E-01
9.494	1.43E 01	1.48E 00	6.40E-01	9.893	5.20E 00	1.74E 00	7.20E-01
9.526	9.43E 00	2.07E 00	6.90E-01	9.953	3.22E 00	1.82E 00	7.40E-01
9.590	4.58E 00	1.36E 00	6.00E-01	10.070	9.44E-01	2.27E 00	7.40E-01
9.652	7.15E 00	2.03E 00	6.00E-01	10.190	3.18E-01	5.54E-01	3.00E-01
9.713	8.34E 00	1.49E 00	6.40E-01	–	–	–	–

The quantities $K_3(\lambda)$ and $K_4(\lambda)$ are given in Table 5-13, and $\bar{P}$ (in atm) is calculated as in Equation (5-69). Finally, w (in atm cm) is calculated from

$$w = \int_{\text{path}} M'(z) \frac{P(z)}{P_o} \frac{T_o}{T(z)} \, dz \qquad (5\text{-}77)$$

For N_2O:

(10) Strong-Line Elsasser Model (4.228 to 4.73 μm).

$$\tau(\lambda) = 1 - \text{erf} \, [w_N^* \, K_5(\lambda)]^{1/2} \qquad (5\text{-}78)$$

where

$$w_N^* = \int_{\text{path}} M'(z) \left(\frac{P(z)}{P_o}\right)^2 \left(\frac{T_o}{T(z)}\right)^{1.5} \, dz \qquad (5\text{-}79)$$

$M'(z)$ = ratio of partial pressure of N_2O to total. (See Table 5-25.)

$K_5(\lambda)$ is obtained from Table 5-14.

(11) Elsasser (7.53 to 8.91 μm). See Equations (5-66) and (5.71) for CO_2. The values for $2\pi\alpha_o'/d$ and S/d are given in Table 5-15.

(12) Goody Model (15.4 to 19.3 μm). The special form of the Goody model given for ozone is given in Equation (5-76). The values for $K_3(\lambda)$ and $K_4(\lambda)$ are in Table 5-16.

For CH_4:

(13) Elsasser Model (5.91 to 9.1 μm). The Elsasser model is used for methane so that Equations (5-66) and (5-71) for CO_2 are valid. The values for $2\pi\alpha_o'/d$ and S/d are given in Table 5-17. $\bar{P}$ and w_p are calculated as for the CO_2 molecule using Elsasser. In this case, $M'(z)$ is 1.1×10^{-6}.

Absorption due to gaseous HNO_3, nitric acid, is included in some more complete calculations but is of minor importance for the first-order calculation described in this chapter. Calculations for the N_2 absorption continuum and for extinction by scattering are described in Section 5.5.2 on the LOWTRAN method.

Table 5-13. O_3 Parameters Corresponding to the Goody Model

λ	$K_3(\lambda)$	$K_4(\lambda)$	λ	$K_3(\lambda)$	$K_4(\lambda)$	λ	$K_3(\lambda)$	$K_4(\lambda)$
11.790	.0010	1.14	13.230	.059	1.14	14.270	0.83	1.14
11.900	.0015	1.14	13.260	.064	1.14	14.310	.094	1.14
12.050	.0025	1.14	13.300	.068	1.14	14.350	.104	1.14
12.200	.0039	1.14	13.330	.073	1.14	14.390	.114	1.14
12.350	.0063	1.14	13.370	.078	1.14	14.430	.128	1.14
12.500	.0100	1.14	13.400	.083	1.14	14.470	.142	1.14
12.530	.0109	1.14	13.440	.089	1.14	14.510	.157	1.14
12.560	.0120	1.14	13.480	.096	1.14	14.560	.166	1.14
12.590	.0133	1.14	13.510	.102	1.14	14.600	.171	1.14
12.630	.0144	1.14	13.550	.109	1.14	14.640	.169	1.14
12.660	.0161	1.14	13.590	.117	1.14	14.684	.163	1.14
12.690	.0176	1.14	13.620	.125	1.14	14.728	.148	1.14
12.720	.0192	1.14	13.660	.133	1.14	14.771	.148	1.14
12.760	.0210	1.14	13.700	.142	1.14	14.815	.141	1.14
12.790	.0230	1.14	13.740	.152	1.14	14.859	.133	1.14
12.820	.0251	1.14	13.770	.159	1.14	14.903	.125	1.14
12.850	.0271	1.14	13.810	.167	1.14	14.948	.118	1.14
12.890	.0292	1.14	13.850	.174	1.14	14.993	.110	1.14
12.920	.0315	1.14	13.890	.181	1.14	15.038	.103	1.14
12.950	.0338	1.14	13.950	.190	1.14	15.083	.096	1.14
12.990	.0365	1.14	13.990	.185	1.14	15.129	.089	1.14
13.020	.0390	1.14	14.030	.176	1.14	15.175	.082	1.14
13.050	.0418	1.14	14.060	.162	1.14	15.221	.077	1.14
13.090	.0447	1.14	14.100	.150	1.14	15.267	.072	1.14
13.120	.0479	1.14	14.140	.133	1.14	15.314	.067	1.14
13.160	.051	1.14	14.180	.114	1.14	15.361	.062	1.14
13.190	.055	1.14	14.220	.095	1.14	—	—	—

Table 5-14. N_2O Parameters Corresponding to the Strong-Line Elsasser Model

λ	$K_5(\lambda)$	λ	$K_5(\lambda)$	λ	$K_5(\lambda)$	λ	$K_5(\lambda)$
4.228	1.00E–04	4.319	2.20E–03	4.415	2.10E–02	4.505	4.90E 00
4.237	5.00E–04	4.329	2.65E–03	4.425	5.40E–02	4.520	4.60E 00
4.246	9.40E–04	4.338	2.20E–03	4.435	1.50E–01	4.526	4.80E 00
4.255	1.40E–03	4.348	1.35E–03	4.444	3.50E–01	4.540	5.70E 00
4.264	1.80E–03	4.357	9.20E–04	4.454	1.00E 00	4.548	5.00E 00
4.274	2.40E–03	4.367	7.20E–04	4.465	2.70E 00	4.575	1.60E 00
4.283	2.60E–03	4.376	9.00E–04	4.470	4.20E 00	4.600	4.15E–01
4.292	1.90E–03	4.386	1.50E–03	4.475	7.30E 00	4.660	1.05E–02
4.301	1.30E–03	4.396	3.00E–03	4.490	6.20E 00	4.700	9.29E–04
4.310	1.55E–03	4.405	7.20E–03	4.495	5.60E 00	4.730	1.50E–04

Table 5-15. N_2O Parameters Corresponding to the Elsasser Model

λ	S/d	$2\pi\alpha_o'/d$	λ	S/d	$2\pi\alpha_o'/d$
7.533	.190	1.32E-03	8.180	.013	1.32E-03
7.561	.600	1.32E-03	8.247	.009	1.32E-03
7.619	2.370	1.32E-03	8.316	.033	1.32E-03
7.678	5.400	1.32E-03	8.386	.096	1.32E-03
7.737	5.120	1.32E-03	8.457	.173	1.32E-03
7.797	1.900	1.32E-03	8.529	.128	1.32E-03
7.859	5.020	1.32E-03	8.602	.106	1.32E-03
7.921	3.080	1.32E-03	8.677	.177	1.32E-03
7.984	1.140	1.32E-03	8.753	.101	1.32E-03
8.048	.270	1.32E-03	8.830	.034	1.32E-03
8.114	.055	1.32E-03	8.909	.007	1.32E-03

Table 5-16. N_2O Parameters Corresponding to the Goody Model

λ	$K_3(\lambda)$	$K_4(\lambda)$	λ	$K_3(\lambda)$	$K_4(\lambda)$
15.408	.00030	.3	17.182	.869	.3
15.456	.00039	.3	17.241	1.030	.3
15.504	.00051	.3	17.301	.975	.3
15.552	.00075	.3	17.361	.925	.3
15.601	.00111	.3	17.422	.712	.3
15.649	.00168	.3	17.483	.660	.3
15.699	.00265	.3	17.544	.661	.3
15.748	.00418	.3	17.606	.520	.3
15.798	.00632	.3	17.668	.442	.3
15.848	.00955	.3	17.730	.237	.3
15.898	.0143	.3	17.794	.205	.3
15.949	.0214	.3	17.857	.177	.3
16.000	.0321	.3	17.921	.122	.3
16.501	.0521	.3	17.986	.0845	.3
16.129	.108	.3	18.051	.0628	.3
16.181	.141	.3	18.116	.0502	.3
16.234	.185	.3	18.182	.0401	.3
16.287	.240	.3	18.248	.0278	.3
16.340	.309	.3	18.315	.0193	.3
16.393	.398	.3	18.382	.0120	.3
16.447	.492	.3	18.450	.00675	.3
16.502	.608	.3	18.519	.00378	.3
16.556	.711	.3	18.587	.00228	.3
16.611	.787	.3	18.657	.00138	.3
16.667	.872	.3	18.727	.00090	.3
16.722	.885	.3	18.797	.00063	.3
16.779	.898	.3	18.868	.00044	.3
16.835	.992	.4	18.939	.00030	.3
16.892	1.192	.8	19.011	.00021	.3
16.949	1.434	1.5	19.084	.00015	.3
17.007	1.060	.8	19.157	.00012	.3
17.065	.785	.4	19.231	.00010	.3
17.123	.733	.3			

Table 5-17. CH_4 Parameters Corresponding to the Elsasser Model

λ	S/d	$2\pi\alpha_o'/d$	λ	S/d	$2\pi\alpha_o'/d$
5.908	1.270E 00	8.493E-04	7.233	5.506E 01	1.058E-01
5.944	1.579E 00	1.287E-03	7.286	6.718E 01	1.927E-01
5.979	2.030E 00	1.621E-03	7.339	7.149E 01	2.803E-01
6.015	2.333E 00	1.965E-03	7.394	7.904E 01	3.328E-01
6.051	2.279E 00	2.411E-03	7.449	8.371E 01	2.933E-01
6.088	1.125E 00	2.820E-03	7.505	6.814E 01	1.815E-01
6.126	2.269E 00	3.001E-03	7.533	5.503E 01	1.472E-01
6.163	2.706E 00	3.031E-03	7.561	1.646E 02	3.013E-01
6.202	3.336E 00	2.758E-03	7.619	9.383E 01	4.427E-01
6.240	3.631E 00	2.246E-03	7.678	7.703E 01	5.420E-01
6.279	3.994E 00	1.740E-03	7.737	6.335E 01	3.009E-01
6.319	2.223E 00	1.461E-03	7.797	6.889E 01	2.554E-01
6.359	1.396E 00	2.222E-03	7.859	8.934E 01	2.679E-01
6.400	1.623E 00	5.437E-03	7.921	1.050E 02	2.582E-01
6.441	1.805E 00	9.401E-03	7.984	8.854E 01	1.913E-01
6.483	1.810E 00	8.250E-03	8.048	8.237E 01	1.462E-01
6.525	1.445E 00	3.352E-03	8.114	1.009E 02	1.139E-01
6.568	1.460E 00	1.219E-03	8.180	8.215E 01	7.593E-02
6.612	1.525E 00	1.551E-03	8.247	3.499E 01	3.492E-02
6.656	1.878E 00	2.477E-03	8.316	1.171E 01	1.717E-02
6.700	2.437E 00	3.548E-03	8.386	1.071E 01	1.299E-02
6.745	2.822E 00	4.497E-03	8.457	1.045E 01	8.246E-03
6.791	2.924E 00	5.358E-03	8.529	4.038E 00	4.308E-03
6.838	2.863E 00	6.086E-03	8.602	5.726E-01	2.267E-03
6.885	2.798E 00	6.675E-03	8.677	1.856E-01	1.478E-03
6.932	2.841E 00	7.130E-03	8.753	1.633E-01	9.756E-04
6.981	2.935E 00	7.449E-03	8.830	4.226E-01	7.445E-04
7.030	3.212E 00	7.771E-03	8.909	7.121E-01	5.811E-04
7.080	3.728E 00	8.835E-03	8.989	7.207E-01	4.008E-04
7.130	1.089E 01	1.759E-02	9.070	6.525E-01	1.852E-04
7.181	3.026E 01	4.604E-02	—	—	—

5.5.2. LOWTRAN Method (See also Section 5.1.). This method is empirical, yielding a form for the transmittance given by

$$\tau(\lambda) = f[G(\lambda), w, P^n] = f[G(\lambda), w^*] \tag{5-80}$$

where $G(\lambda)$ = the parameters derived for each wavelength
  w = the quantity of absorber
  P = the quantity of pressure
  w^* = the equivalent absorber amount

The functional form of τ has been determined by a fit of a specialized curve to a plot of actual data. The value of n is determined for each absorber.

This method of atmospheric transmittance calculation was formulated at the Air Force Cambridge Research Laboratories* [5-34] and computerized with a card deck which is obtainable from them. For quick calculations, LOWTRAN computation is graphically presented and fully described by McClatchey et al. [5-2].

The LOWTRAN method might be considered slightly less accurate than the Aggregate method, especially in the extremes, but its ease and versatility in use easily compensate for this possible shortcoming, especially for calculations made on a first-order basis.

5.5.3. **Determination of Attenuator Amount for LOWTRAN.** First one must calculate the equivalent amount of absorber to apply to the transmittance curves. For model atmospheres tabulated for the so-called standard seasonal conditions, these values are obtained directly from the curves of Figures 5-13(a) through (j), which are in turn obtained from the values of model atmosphere parameters tabulated in Table 5-27. For any other atmospheric conditions, chiefly for the variable gases H_2O and O_3, the values are calculable from the following formulae.

For Water Vapor (H_2O):

If the path is horizontal, then, at the altitude, z, for which pressure is P,

$$w_h^* = w_h(z)\left(\frac{P}{P_o}\right)^{0.9} \tag{5-81}$$

or, using values similar to those in Table 5-27,

$$w_h^* = \rho(z)\left(\frac{P}{P_o}\right)^{0.9} \times 10^{-1} \tag{5-82}$$

Equation (5-82) is used to calculate the equivalent amount of absorber.[†] The H_2O vapor concentration, normalized to 1 km, is given by $w_h(z)$, in g cm^{-2} km^{-1}. The range, R, is in km and is multiplied by the appropriate value plotted in Figure 5-13(a). The equivalent absorber amount in a slant path is determined by using the curve in Figure 5-13(b), obtained from

$$w_v^* = \int_z^\infty w(z')\left(\frac{P}{P_o}\right)^{0.9} dz' \tag{5-83}$$

or, using values similar to those in Table 5-27,

$$w_v^* = 10^{-1} \int_z^\infty \rho(z')\left(\frac{P}{P_o}\right)^{0.9} dz' \tag{5-84}$$

---

*Now the Air Force Geophysics Laboratory.

†For horizontal paths, in LOWTRAN, the absorber amount is stated as a gradient, g cm^{-2} km^{-1}. This is done for convenient use of the graphs in Figure 5-13. The symbol w_h^* is consistently used for this quantity with these units as given in Reference [5-2].

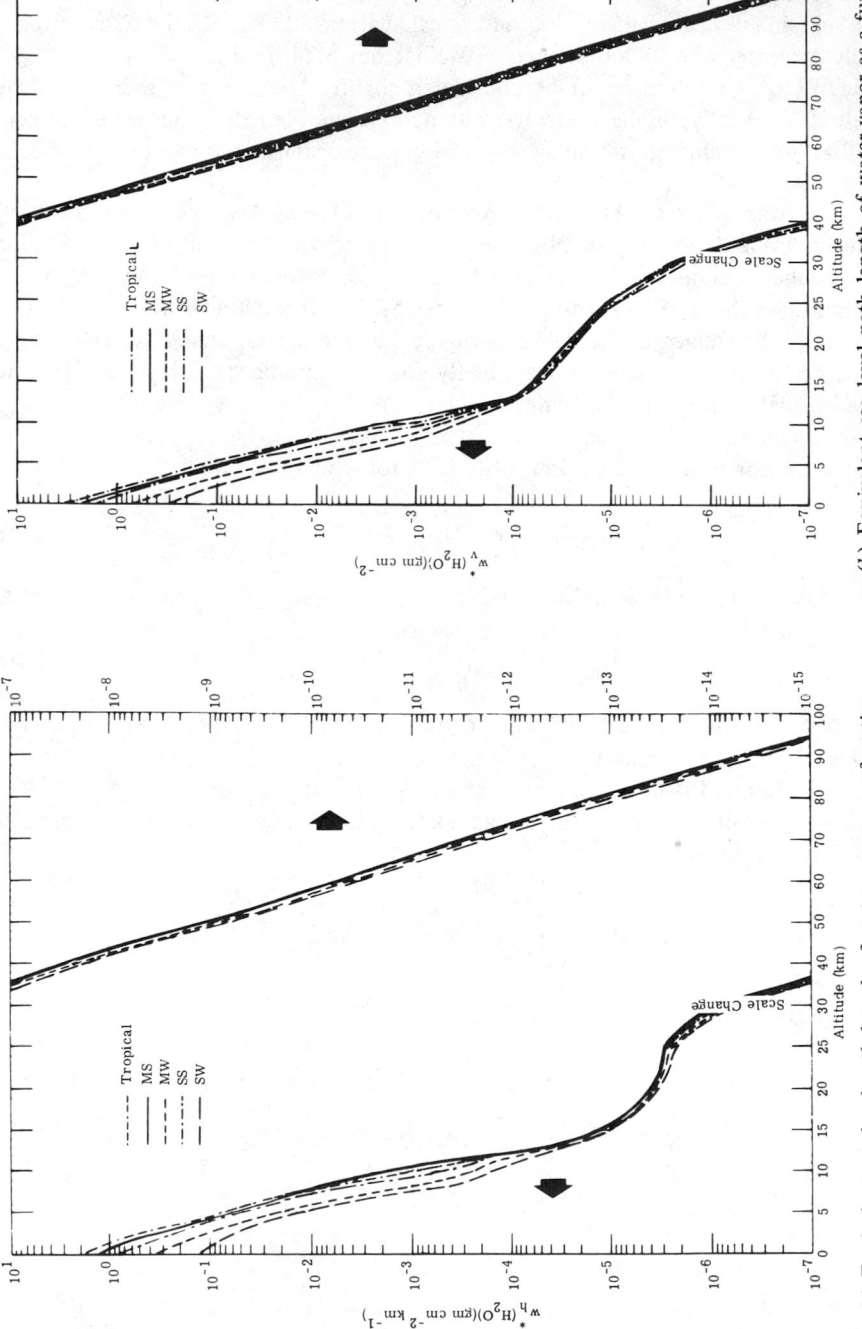

(a) Equivalent sea level path length of water vapor as a function of altitude for horizontal atmospheric paths.

(b) Equivalent sea level path length of water vapor as a function of altitude for vertical atmospheric paths. For slant paths multiply by secant θ, for $\theta < 80°$.

Fig. 5-13. Values obtained for model atmospheres tabulated for the so-called standard seasonal conditions.

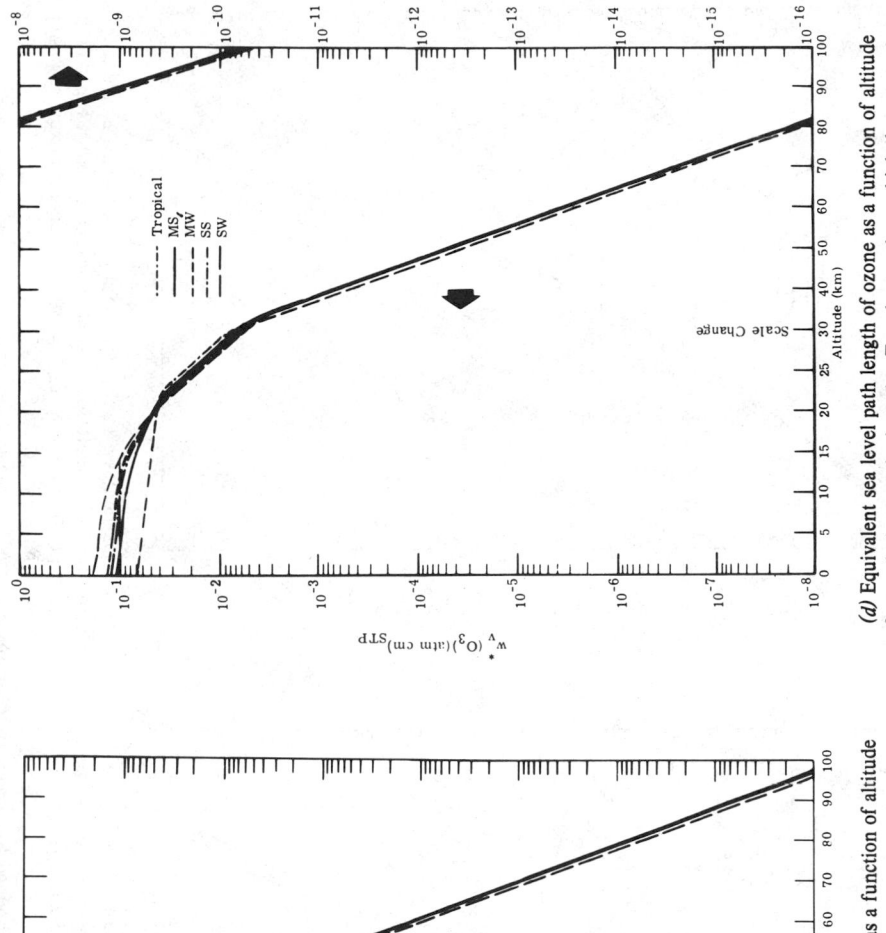

(c) Equivalent sea level path length of ozone as a function of altitude for horizontal atmospheric paths.

(d) Equivalent sea level path length of ozone as a function of altitude for vertical atmospheric paths. For slant paths multiply by secant θ, for $\theta < 80°$.

Fig. 5-13 (Continued). Values obtained for model atmospheres tabulated for the so-called standard seasonal conditions.

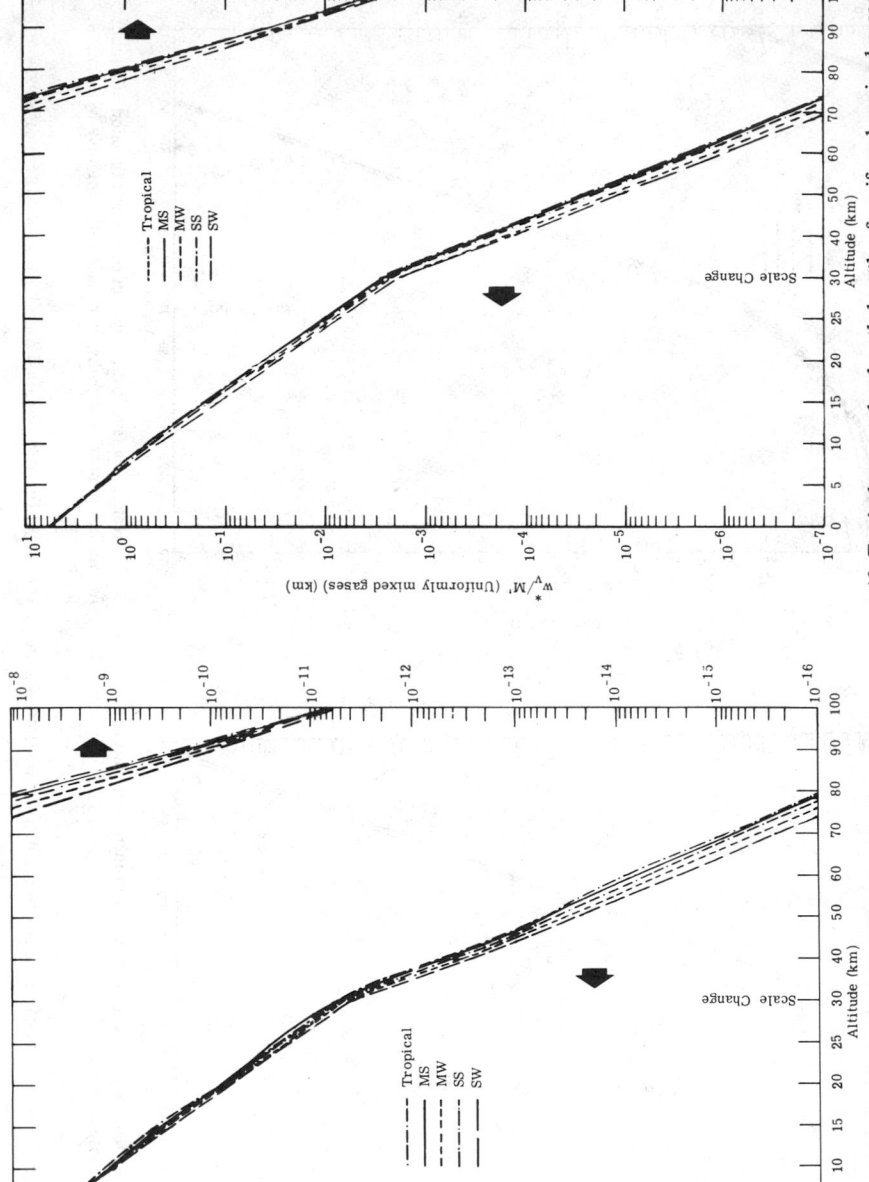

(e) Equivalent sea level path length of uniformly mixed gases as a function of altitude for horizontal atmospheric paths. (M' equals mixing ratio in parts by volume: See Table 5-25.)

(f) Equivalent sea level path length of uniformly mixed gases as a function of altitude for vertical atmospheric paths. For slant paths multiply by secant θ, for $\theta < 80°$. (M' equals mixing ratio in parts by volume: See Table 5-25.)

Fig. 5-13 (Continued). Values obtained for model atmospheres tabulated for the so-called standard seasonal conditions.

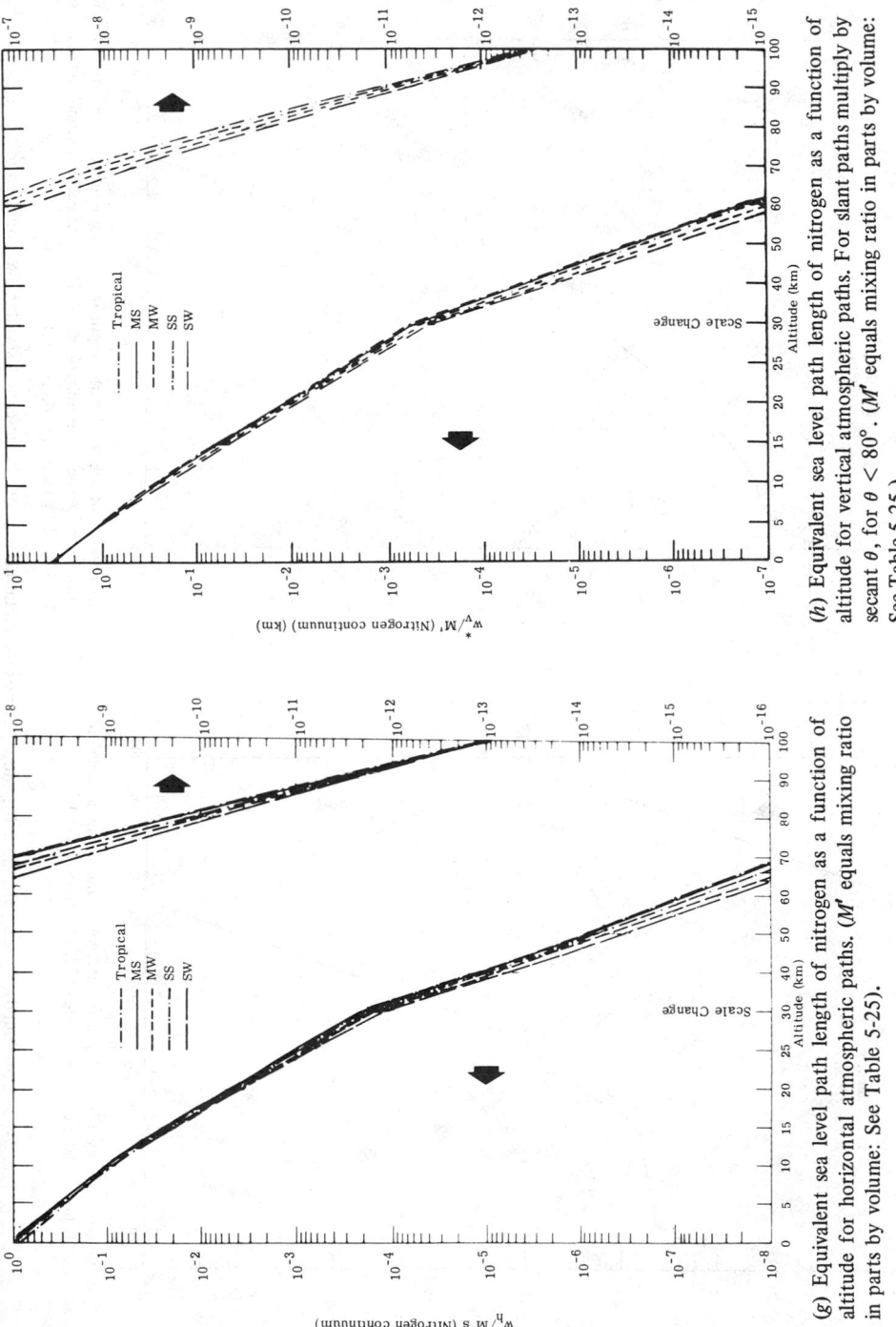

(h) Equivalent sea level path length of nitrogen as a function of altitude for vertical atmospheric paths. For slant paths multiply by secant θ, for $\theta < 80°$. (M' equals mixing ratio in parts by volume: See Table 5-25.)

(g) Equivalent sea level path length of nitrogen as a function of altitude for horizontal atmospheric paths. (M' equals mixing ratio in parts by volume: See Table 5-25).

Fig. 5-13 (Continued). Values obtained for model atmospheres tabulated for the so-called standard seasonal conditions.

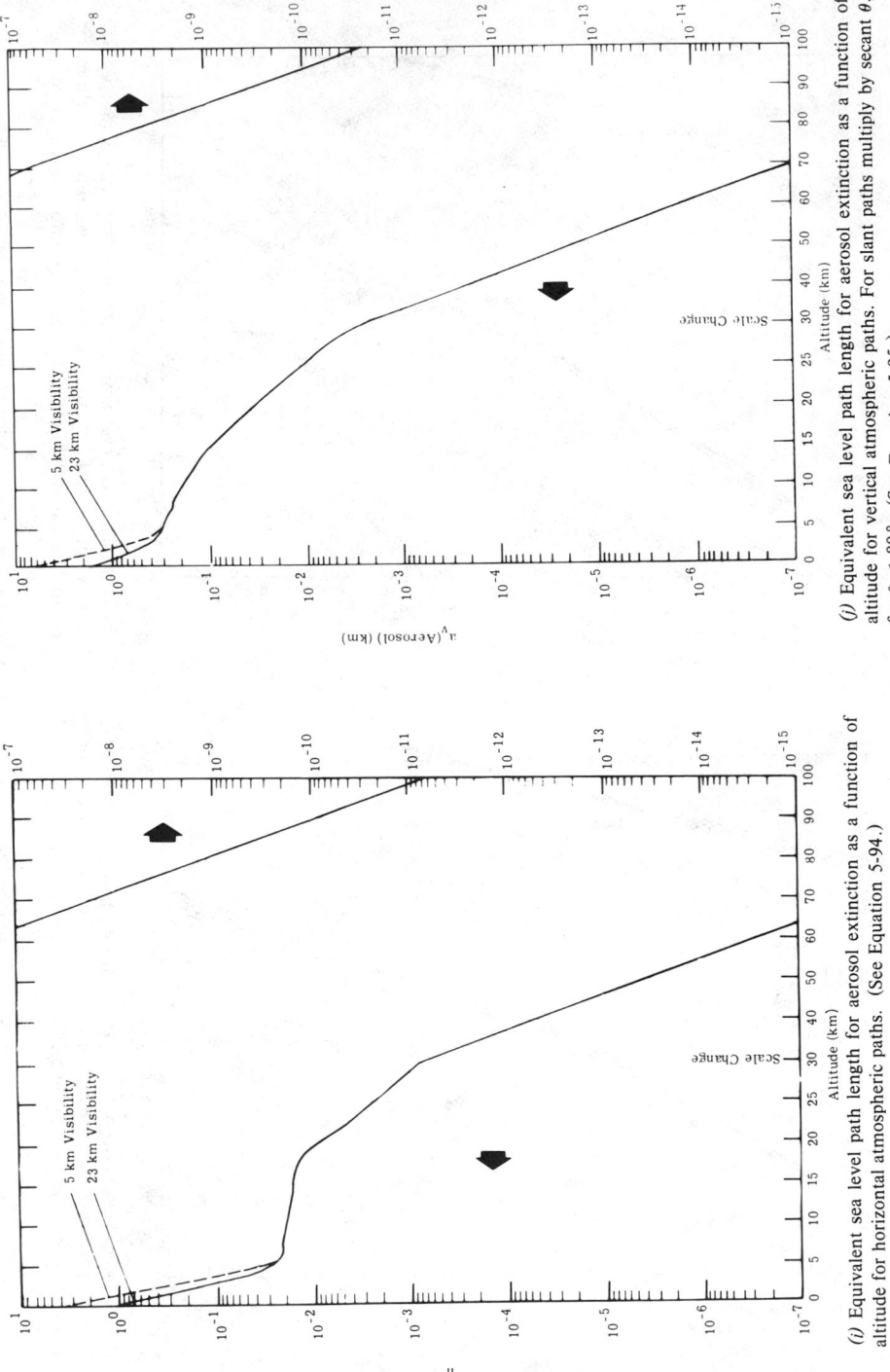

(j) Equivalent sea level path length for aerosol extinction as a function of altitude for vertical atmospheric paths. For slant paths multiply by secant θ, for $\theta < 80°$. (See Equation 5-95.)

(i) Equivalent sea level path length for aerosol extinction as a function of altitude for horizontal atmospheric paths. (See Equation 5-94.)

Fig. 5-13 (Continued). Values obtained for model atmospheres tabulated for the so-called standard seasonal conditions.

From this curve one obtains the value in a vertical path between the altitude shown on the abscissa and infinity. The value in a slant path between two altitudes is determined by the difference in the values obtained at the two altitudes multiplied by the secant of the angle between the upward vertical and the slant path. Corrections for refraction and earth curvature are given in Reference [5-2] for angles greater than $80°$. Each curve in the family of curves in this figure represents a model atmosphere.

For Ozone (O_3):

$$w_h^* = w_h(z)\left(\frac{P}{P_o}\right)^{0.4} \tag{5-85}$$

or, using values similar to those in Table 5-27, one has

$$w_h^* = 47\, \rho(z)\left(\frac{P}{P_o}\right)^{0.4} \tag{5-86}$$

for the horizontal path, with $w_h(z)$ (normalized) in atm cm km^{-1}. For the slant path, the vertical component is calculated from

$$w_v^* = \int_z^\infty w(z')\left(\frac{P}{P_o}\right)^{0.4} dz' \tag{5-87}$$

or, using values similar to those in Table 5-27, one has

$$w_v^* = 47 \int_z^\infty \rho(z')\left(\frac{P}{P_o}\right)^{0.4} dz' \tag{5-88}$$

The curves representing the quantities

$$w_h(z)\left(\frac{P}{P_o}\right)^{0.4}$$

and

$$\int_z^\infty w(z')\left(\frac{P}{P_o}\right)^{0.4} dz'$$

respectively are found in Figures 5-13(c) and 5-13(d).

For the spectral region between 0.25 to 0.75 μm, the absorption coefficient is independent of pressure and the absorber curves take on a slightly different character. See Reference [5-2].

For Uniformly Mixed Gases (CO_2, N_2O, CO, CH_4, O_2):

For the uniformly mixed gases the amount of absorber reduced to standard conditions is proportional to

$$M's\left(\frac{P}{P_o}\right)\left(\frac{T_o}{T}\right)$$

where M' = fractional concentration of gas by volume
       s = transmission path length

In Reference [5-2], the quantity above is designated as the absorber concentration reduced to *STP* conditions and is given the symbol:

$$\Delta L_o = M's \left(\frac{P}{P_o}\right)\left(\frac{T_o}{T}\right) \tag{5-89}$$

The units are atm cm. Thus, the justification for designating $\Delta L_o \equiv w$. The value of n used for determining the normalized equivalent amount of absorber is 0.75 for the uniformly mixed gases. For the horizontal path (at the altitude for which the calculation is made for the equivalent sea level length per km) the curve is plotted in Figure 5-13(e) as a function of altitude for

$$\frac{w_h^*}{M's} = \left(\frac{P}{P_o}\right)^{1.75}\left(\frac{T_o}{T}\right) \tag{5-90}$$

Figure 5-13(e) is normalized by the value of M' to keep it applicable to any of the uniformly mixed gases regardless of their concentrations. For vertical paths the formula is given by

$$\frac{w_v^*}{M'} = \int_z^\infty \left(\frac{P}{P_o}\right)^{1.75}\left(\frac{T_o}{T}\right) dz' \tag{5-91}$$

which is plotted as a function of altitude in Figure 5-13(f). In the horizontal-path case, the equivalent path-length is obtained by multiplying the value from the appropriate curve (for a chosen point of the abscissa) by the actual number of kilometers in the path. For the vertical case, the number of equivalent kilometers upward from the chosen altitude is obtained from the ordinate corresponding to a prescribed value on the abscissa. The slant path is obtained as before.

For N_2 Continuum:

Absorption by the nitrogen collision-induced band is taken proportional to P^2. Thus, for a concentration M', analogous to the case of the uniformly mixed gases

$$\frac{w_h^*}{M's} = \left(\frac{P}{P_o}\right)^2\left(\frac{T_o}{T}\right) \tag{5-92}$$

for the horizontal case; and

$$\frac{w_v^*}{M'} = \int_z^\infty \left(\frac{P}{P_o}\right)^2\left(\frac{T_o}{T}\right) dz' \tag{5-93}$$

for the vertical case. The figures representing these cases are shown in Figures 5-13(g) and 5-13(h).

For H_2O Continuum:

See Section 5.5.1, H_2O continuum.

For Aerosol Extinction:

Determinations of aerosol extinction are made with the use of Figures 5-13(i) and 5-13(j). The first of these is a plot, as a function of altitude, of the ratio

$$a_h = \frac{D(z)}{D_o} \qquad (5\text{-}94)$$

where $D(z)$ = number density of particles (given in Section 5.10.1) at altitude, z

D_o = number density at sea level for a visibility of 23 km

The value used in determining the horizontal path extinction in Section 5.5.4 is obtained by multiplying the value obtained from Figure 5-13(i) at altitude, z, by the actual transmission path.

In Figure 5-13(j) is plotted the effective vertical path length designated by

$$a_v = \int_z^\infty \frac{D(z')}{D_o} \, dz' \qquad (5\text{-}95)$$

In this case, as in the cases for the atmospheric gases discussed above, the vertical extinction path length obtained on the ordinate of Figure 5-13(j) is the effective height above the corresponding value on the abscissa.

The effective vertical distance between two altitudes is obtained by subtracting the ordinates corresponding to the two altitudes. For a slant path, the vertical distance, as explained previously, is multiplied by sec θ.

5.5.4. **Use of the LOWTRAN Method.** The values of equivalent absorber amount are computed in the LOWTRAN program. They can also be used with the scaling factors provided by referring to Figures 5-14(a) through (h). These figures are convenient tools for determining rapidly and accurately the value of transmittance at any frequency (wavelength) and for any path conditions, provided one is willing to accept one of the standard model atmospheres as representative of the conditions pertinent to the prescribed situation. The figures are simple to use in conjunction with the graphs of Figure 5-13 from which the equivalent absorber amounts are determined. The scale need only be moved, as demonstrated below, to the properly determined absorber amount in the appropriate dimensions. The transmittance is read on the accompanying transmittance scale from the curve.

The accuracy of the LOWTRAN technique decreases as the transmittance approaches unity. When conditions are more closely consistent with the weak- and strong-line representations of the model, then the LOWTRAN models will tend, respectively, to underestimate and overestimate the actual value of transmittance. On the other hand, one should refrain from using the model for the calculation of path radiance under conditions of high transmittance, since the uncertainty in atmospheric emissivity increases greatly. Errors also occur from a neglect of the temperature dependence of line parameters.

An example of the use of the graphs in the LOWTRAN method is presented here for a vertical transmission path from sea level to an altitude of 25 km, under midaltitude-summer conditions. For a slant path between the same two altitudes, the absorber amount is obtained as for the vertical path and multiplied by sec θ, if $\theta < 80°$. Here, θ is the zenith angle for the line of sight between source and receiver. For nearly horizontal paths, i.e., $\theta > 80°$, correction must be made for the sphericity of the earth. For more accurate calculations under all circumstances, the index of refraction of the atmosphere must be considered. (See Reference [5-2].) The latter correction, however, is usually less than the error inherent in band model calculations, and is ignored.

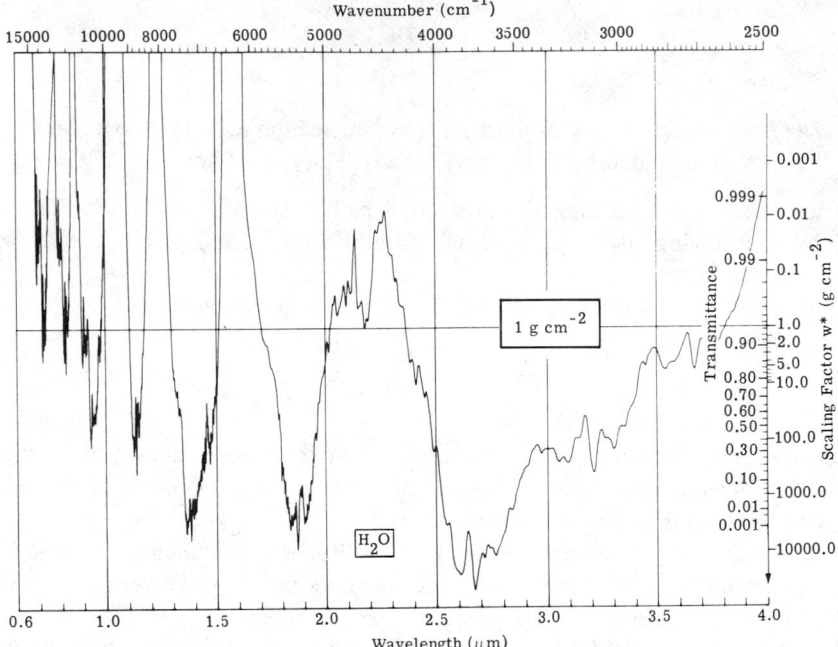

(a) Transmittance of water vapor (0.6 to 4.0 μm).

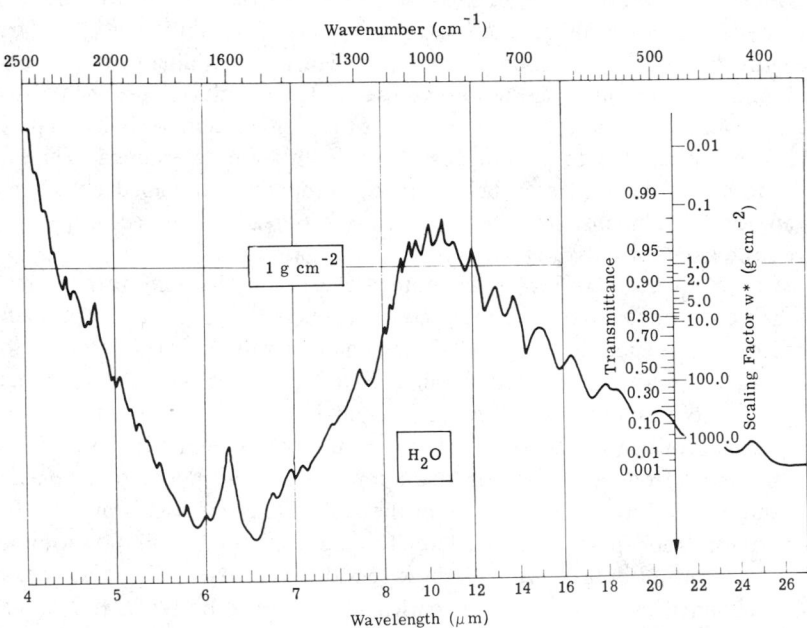

(b) Transmittance of water vapor (4.0 to 26.0 μm).

Fig. 5-14. Scaling factors for determining the value of transmittance at any wavelength and path condition using standard model atmospheres as representative of conditions pertinent to the prescribed situation.

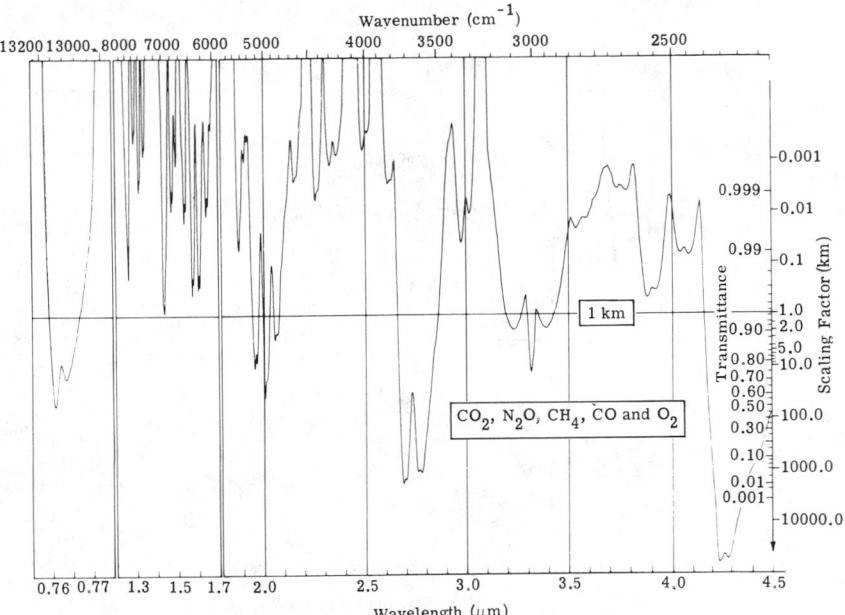

(c) Transmittance of uniformly mixed gases (CO_2, N_2O, CO, CH_4, O_2) (0.76 to 4.5 μm).

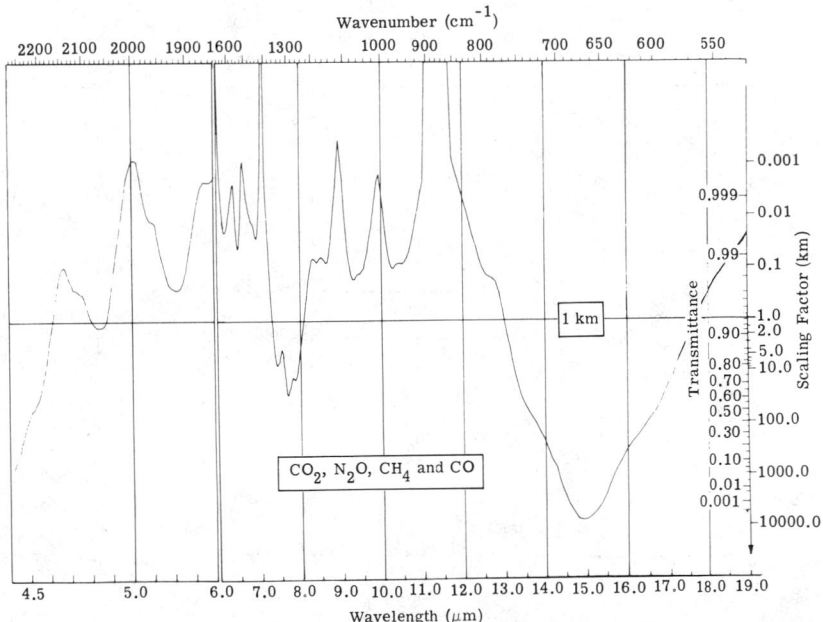

(d) Transmittance of uniformly mixed gases (CO_2, N_2O, CO, CH_4, O_2) (4.5 to 19.0 μm).

Fig. 5-14 (Continued). Scaling factors for determining the value of transmittance at any wavelength and path condition using standard model atmospheres as representative of conditions pertinent to the prescribed situation.

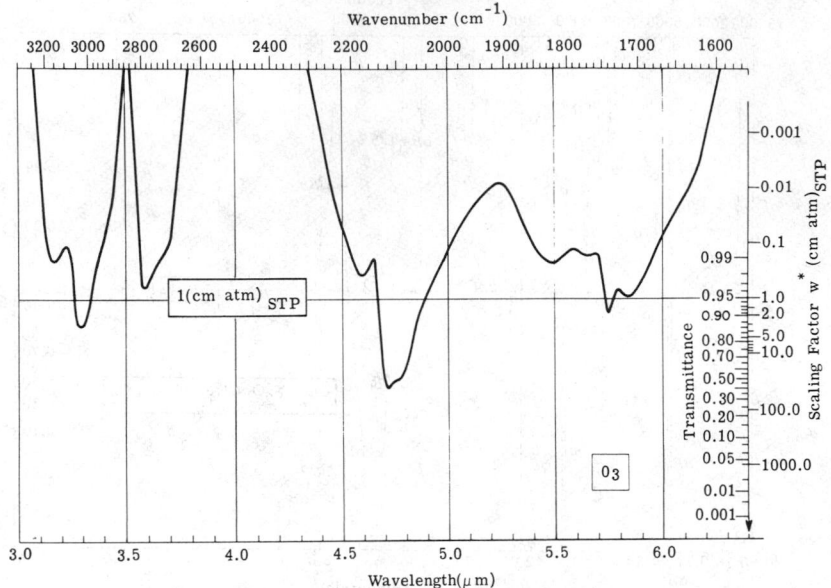

(e) Transmittance of ozone (3.0 to 6.4 μm).

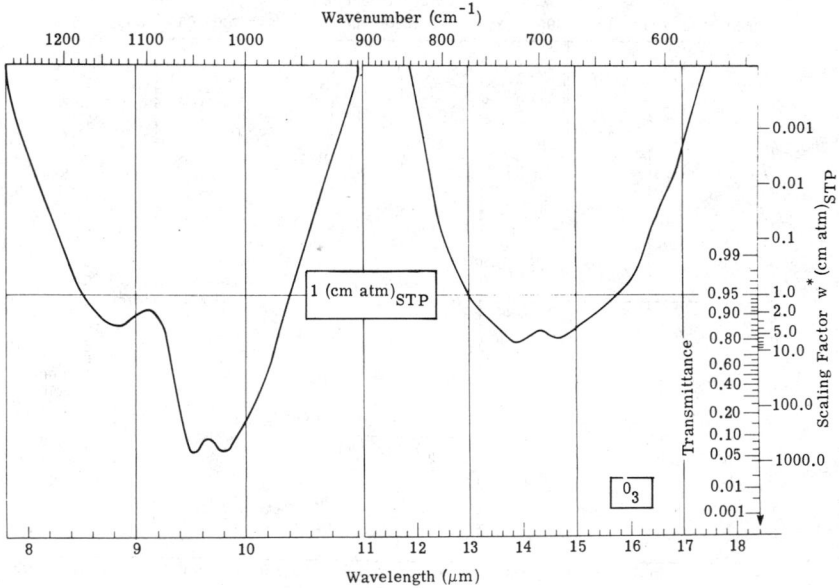

(f) Transmittance of ozone (8.0 to 18.0 μm).

Fig. 5-14 (Continued). Scaling factors for determining the value of transmittance at any wavelength and path condition using standard model atmospheres as representative of conditions pertinent to the prescribed situation.

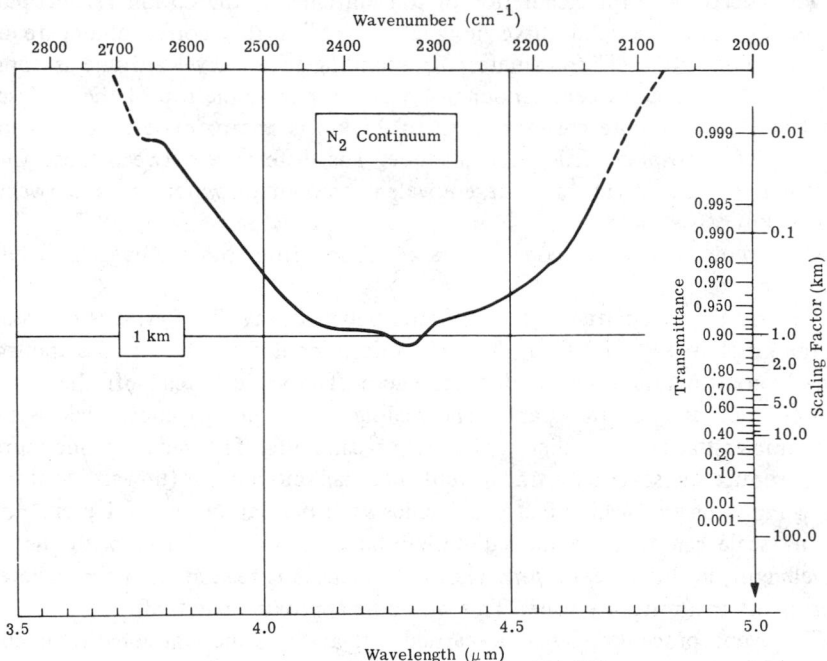

(g) Transmittance of nitrogen.

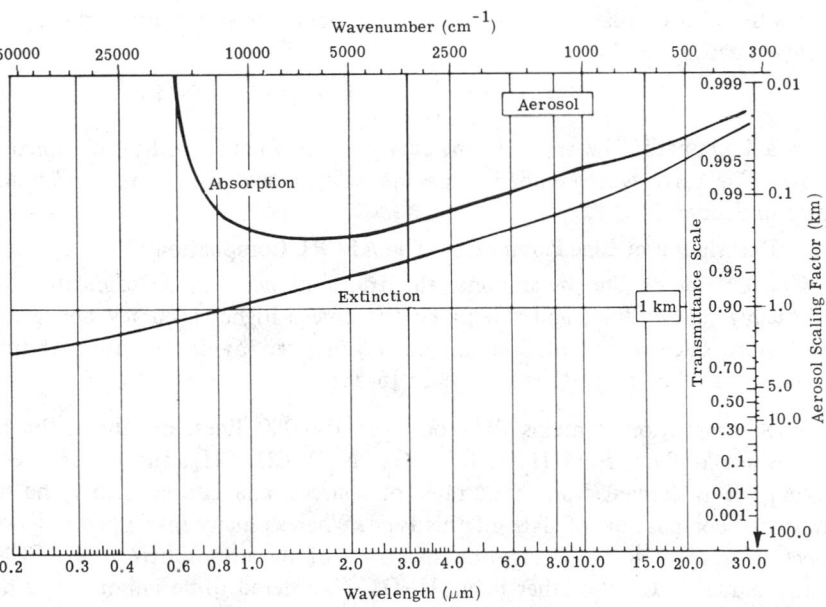

(h) Transmittance of aerosol (scattering and absorption).

Fig. 5-14 (Continued). Scaling factors for determining the value of transmittance at any wavelength and path condition using standard model atmospheres as representative of conditions pertinent to the prescribed situation.

To proceed with the calculation of transmittance in the 25 km vertical path, refer to Figure 5-13(b). The solid curve designated "MS" is that corresponding to midlatitude summer conditions. The ordinate corresponding to sea level altitude is approximately 2.2 g cm^{-2}, the equivalent amount of water vapor to the top of the atmosphere from sea level. The ordinate corresponding to 25 km is approximately 10^{-5} g cm^{-2} to the top of the atmosphere from that altitude. The difference between these values, which is effectively 2.2 g cm^{-2}, is the equivalent amount of water vapor between sea level and 25 km.

The amounts of other constituents are found from other curves in Figure 5-13 by the same procedure.

To find the transmittance due to water vapor at, say, 3.5 μm, it is necessary to refer to the graph of Figure 5-14(a). This curve is computed to indicate the transmittance of a path containing 1 g cm^{-2} of water vapor. The value is read off the left-most scale on the right side of the graph. The scaling factor on right-most scale is used to obtain transmittance for other water vapor amounts. The recommended procedure is to reproduce the scale on a transparent material, and move it (upward, in this case) until 2.2 g cm^{-2} on the scaling factor coincides with the datum line at 1 g cm^{-2}. The transparent scale can then be moved horizontally until it coincides with the appropriate wavelength, in this case 3.5 μm. The transmittance is read at the point where the curve crosses the transmittance scale. This would be approximately 0.80.

The same procedure must be carried out at the same wavelength for all other extinction components. The product of all of these is then the resultant transmittance at the desired wavelength.

In the region where water vapor continuum is effective, the procedure described in Section 5.5.1 must be used. Aerosol extinction is found using the curve of Figure 5-14(h), which is a plot of

$$\tau(\nu) = \exp\left[-3.745 \times 10^{-5}\, \nu^{0.8543}\right] \qquad (5\text{-}96)$$

from a 1 km path. The value on the curve for other path lengths is obtained as described above. The curve was obtained using the particle size distribution and number density given in Section 5.10.1.

5.6. Description of Line Parameters—The AFCRL Compilation*

Given a set of line parameters, the transmittance can be calculated directly (and tediously), line-by-line, and integrated to achieve higher accuracy and greater spectral resolution. Such a compilation has been generated by investigators at the Air Force Geophysics Laboratory. (See Reference [5-25].)

The compilation contains data on over 100,000 lines, mainly in the infrared and microwave regions, from H_2O, CO_2, O_3, N_2O, CO, CH_4, and O_2 in the atmosphere [5-25]. It is derived from a number of sources, and is presumably the most nearly-complete compilation of data of this type. Whereas many investigators ignore the presence of oxygen in the absorption spectrum of the atmosphere, the AFCRL compilation includes it. On the other hand, HNO_3, considered to be important at high altitudes,

---

*According to L. S. Rothman ("Atmospheric Optics," OSA Technical Group Meeting, Tucson, AZ, 19 October 1976, *Applied Optics*, Optical Society of America, Washington, DC, Vol. 16, No. 2, February 1977, p. 277) the compilation is available from NOAA National Climatic Center, Federal Building, Asheville, NC 28801 for $60.00.

is not included. The compilation also contains information on the line strengths and half-widths corresponding to a temperature of 296 K and a pressure of 760 mm Hg (i.e., one standard atmosphere). For ease of use in generalized atmospheric calculations, the compilation also includes the energy of the ground state for each line, as well as several sets of identification numbers. The strengths are derived on the basis of the number of molecules of the isotopes in their normal abundance. The isotopic abundances used are shown in Table 5-18.

Determinations of the values compiled in Reference [5-25] are an intricate combination of theoretical and experimental results, each being used as a check on the other. Included in the report is a set of nomenclature which forms the basis for the calculation of the more than 100,000 lines.

Table 5-18. Isotopic Abundances

Molecule	Isotope	Abundance	Molecule	Isotope	Abundance
H_2O	161	0.99729	CH_4	211	0.98815
	162	0.000300		311	0.01110
	181	0.00204	CH_3D	212	0.00060
	171	0.000370		–	–
CO_2	626	0.98414	O_2	66	0.99519
	636	0.01105	–	68	0.00407
	628	0.00402	–	67	0.00074
	627	0.000730	–	–	–
	638	0.0000452	–	–	–
	637	0.00000820	–	–	–
	828	0.00000412	–	–	–
O_3	666	0.99279	–	–	–
	668	0.00406	–	–	–
	686	0.00203	–	–	–
N_2O	446	0.99022	–	–	–
	456	0.00368	–	–	–
	546	0.00368	–	–	–
	448	0.00202	–	–	–
	447	0.00037	–	–	–
CO	26	0.98652	–	–	–
	36	0.01107	–	–	–
	28	0.00202	–	–	–
	27	0.000369	–	–	–

5.6.1. Calculations Using Line-by-Line Techniques. The tape containing the line parameters is available from its originators, along with the specific instructions that have been compiled for its use. The compilation is thoroughly described in Reference

[5-25] and a computer program listing is given for reading the tape. The tape is designed for use on a CDC 6600 computer, as is the program given in the appendix of Reference [5-25], for performing a line-by-line calculation.

Other methods, more efficient than the one prescribed in Reference [5-25], exist for performing a line-by-line calculation. The costs are still high, however, and the programs are more complicated to use. A description of these methods can be found in Reference [5-1].

5.6.2. Laser Propagation. Detailed spectra of the transmittance have been published using the line parameters discussed above [5-35]. Thus, it is possible to determine the effect of the atmosphere on different laser wavelengths and, in fact, to choose lasers with wavelengths which lie in a propitious spectral region with respect to atmospheric absorption.

Reference [5-35] contains a set of tables of absorption coefficients corresponding to certain emission lines for CO, HF, DF and CO_2 lasers. These entries are those for which the absorption coefficients (per km) are the lowest. Absorption coefficients have been tabulated only for tropical, midlatitude-winter and subarctic-winter atmospheric models at sea level, and for a midlatitude-winter model at 12 km altitude. These absorption coefficients are given in Table 5-19. The effect of scattering can be determined by methods presented in Section 5.5.4.

5.7. Total Absorption in Selected Spectral Bands* [5-21] and [5-108 through 5-110]

Total absorption data provide a means to predict absorption for known paths through known absorbing gases and to test the validity of theories describing absorption phenomena. The total absorption of an absorption band is the area under the curve obtained when the fractional absorption at a given frequency is plotted against frequency. Usually, the integral

$$\int_{\nu_1}^{\nu_2} A(\nu)\, d\nu$$

is called the total absorption of an absorption band, defined by the limits ν_1 and ν_2, and expressed in units of cm^{-1}. Sometimes, however, the integral is referred to as the equivalent bandwidth of the absorption because the same integral can be considered as applying to an equivalent band having complete absorption over a frequency interval

$$\Delta\nu = \int_{\nu_1}^{\nu_2} A(\nu)\, d\nu \tag{5-97}$$

The former definition is used in this section.

Data presented in this section show the functional relationship between total absorption, $\int A(\nu)\, d\nu$, absorber concentration, w, partial pressure of the absorbing gas, p, total pressure, P, (which includes the partial pressure of absorbing and nonabsorbing gases), and the absolute temperature of the gas, T. This relationship is expressed as

$$\int_{\nu_1}^{\nu_2} A(\nu)\, d\nu = f(w, p, P, T) \tag{5-98}$$

and is described for various wavelength regions of high characteristic absorption.

*This section essentially reproduces material prepared by Gilbert N. Plass, Texas A and M University, College Station, TX, and Harold Yates, National Environmental Satellite Center, NOAA, Washington, DC, and originally published in the 1965 *Handbook of Military Infrared Technology*.

Table 5-19. Attenuation Coefficients for Laser Frequencies [5-35]

CO Laser Parameters			Atmospheric Absorption Coefficients (km^{-1})			
			z = 0 km, Sea Level			z = 12 km
Band	Rot. ID	$v(cm^{-1})$	k_{trop}	k_{mw}	k_{sw}	k_{mw}
(a) 1-0	P2	2135.549	0.661	0.249	0.224	0.266
	P14	2086.325	0.409	0.202	0.176	0.141
	P17	2073.267	0.608	0.159	0.104	0.0511
	P18	2068.849	0.268	0.101	0.0792	0.0352
	P21	2055.402	0.141	0.0750	0.0654	0.0112
	P22	2050.856	0.152	0.0522	0.0392	0.00630
	P25	2037.027	0.411	0.0765	0.0369	0.00574
	P26	2032.354	0.178	0.0292	0.0124	0.000813
	P27	2027.651	0.757	0.137	0.0477	0.000650
	P30	2013.353	0.548	0.0784	0.0230	0.000077
(a) 2-1	P1	2112.977	0.0935	0.0144	0.00665	0.00035
	P2	2109.132	0.0525	0.0168	0.0126	0.00902
	P3	2105.256	0.120	0.0264	0.0125	0.0038
	P4	2101.342	0.122	0.0246	0.0127	0.0055
	P7	2089.393	1.52	0.191	0.0527	0.00671
	P8	2085.343	0.186	0.0346	0.0218	0.00196
	P9	2081.258	0.151	0.0276	0.0140	0.00109
	P11	2072.987	0.366	0.0733	0.0332	0.00268
	P12	2068.802	0.240	0.0761	0.0563	0.00427
	P15	2056.046	0.144	0.0218	0.0118	0.000605
	P16	2051.729	1.09	0.0846	0.0283	0.000769
	P17	2047.379	0.350	0.0718	0.0413	0.00118
	P19	2038.582	0.365	0.0542	0.0190	0.000178
	P21	2029.656	0.213	0.0314	0.00956	0.000032
	P22	2025.145	0.537	0.0746	0.0221	0.000079
	P25	2011.423	0.407	0.0577	0.0167	0.000014
	P26	2006.786	0.801	0.108	0.0300	0.000020
	P27	2002.118	0.320	0.0504	0.0156	0.000016
	P28	1997.419	0.938	0.157	0.0501	0.000045
3-2	P1	2086.594	0.479	0.0565	0.0263	0.00305
	P2	2082.784	0.114	0.0181	0.00920	0.00084
	P3	2078.940	0.630	0.171	0.125	0.045
	P4	2075.061	0.333	0.0558	0.0216	0.0064

(a) Laser frequencies calculated using molecular constants of Young [5-60].

Table 5-19. Attenuation Coefficients for Laser Frequencies [5-35] (*Continued*)

CO Laser Parameters			Atmospheric Absorption Coefficients (km^{-1})			
			z = 0 km, *Sea Level*			z = 12 km
Band	Rot. ID	$\nu(cm^{-1})$	k_{trop}	k_{mw}	k_{sw}	k_{mw}
	P5	2071.148	0.123	0.0235	0.0125	0.000861
	P6	2067.200	0.679	0.122	0.0508	0.00181
	P7	2063.218	0.801	0.130	0.0561	0.00152
	P8	2059.203	0.571	0.0937	0.0365	0.000655
	P10	2051.071	0.414	0.0581	0.0236	0.000598
	P11	2046.954	0.851	0.104	0.0292	0.000119
	P12	2042.804	1.49	0.225	0.0735	0.000429
	P13	2038.621	0.367	0.0525	0.0174	0.000122
	P14	2034.405	0.882	0.0896	0.0217	0.000239
	P15	2030.157	0.317	0.0406	0.0116	0.000073
	P16	2025.875	1.13	0.166	0.0513	0.000365
	P17	2021.561	0.734	0.098	0.0277	0.000066
	P19	2012.835	0.739	0.102	0.0290	0.000077
	P20	2008.424	1.68	0.231	0.0654	0.000044
	P21	2003.981	0.299	0.0416	0.0127	0.000117
	P25	1985.891	1.06	0.155	0.0455	0.000030
	P26	1981.290	0.843	0.0773	0.0188	0.000011
	P27	1976.658	1.15	0.214	0.0735	0.000094
	P28	1971.995	0.607	0.0944	0.0290	0.000040
	P30	1962.577	1.37	0.216	0.0660	0.000058
(a) 4-3	P2	2056.506	0.127	0.0568	0.0497	0.00233
	P3	2052.697	0.0955	0.0198	0.0114	0.000392
	P4	2048.853	0.283	0.0616	0.0406	0.00151
	P5	2044.975	0.779	0.125	0.0407	0.000133
	P7	2037.116	0.568	0.0802	0.0305	0.00110
	P8	2033.135	0.172	0.0215	0.00596	0.000012
	P9	2029.121	0.180	0.0284	0.00939	0.000049
	P10	2025.074	0.503	0.0708	0.0214	0.000069
	P11	2020.993	0.859	0.119	0.0338	0.000050
	P13	2012.731	0.581	0.0816	0.0234	0.000022
	P14	2008.550	1.43	0.203	0.0590	0.000053
	P15	2004.337	0.302	0.0406	0.0117	0.000001
	P17	1995.812	1.12	0.170	0.0513	0.000039
	P20	1982.783	0.507	0.0753	0.0225	0.000017

(a) Laser frequencies calculated using molecular constants of Young [5-60].

Table 5-19. Attenuation Coefficients for Laser Frequencies [5-35] (*Continued*)

CO Laser Parameters			Atmospheric Absorption Coefficients (km^{-1})			
			z = 0 km, Sea Level			z = 12 km
Band	Rot. ID	$\nu(cm^{-1})$	k_{trop}	k_{mw}	k_{sw}	k_{mw}
	P21	1978.375	0.281	0.0446	0.0141	0.000048
	P22	1973.936	0.386	0.0607	0.0187	0.000016
(a) 5-4	P2	2030.297	0.186	0.0236	0.00682	0.000011
	P6	2014.993	1.62	0.229	0.0666	0.000117
	P7	2011.082	1.02	0.138	0.0392	0.000120
	P8	2007.137	1.70	0.225	0.0623	0 000219
	P9	2003.158	0.373	0.0502	0.0144	0.000018
	P11	1995.100	1.61	0.243	0.0731	0.000075
	P14	1982.764	0.496	0.0730	0.0217	0.000017
	P15	1978.586	0.266	0.0416	0.0129	0.000016
	P16	1974.376	0.412	0.0631	0.0194	0.000016
	P21	1952.838	0.900	0.145	0.0453	0.000046
	P25	1935.035	1.29	0.205	0.0681	0.001500
	P26	1930.506	1.13	0.180	0.0563	0.000071
(a) 6-5	P2	2004.155	0.588	0.0587	0.0151	0.000026
	P3	2000.415	0.783	0.134	0.0434	0.000040
	P4	1996.641	1.089	0.155	0.0464	0.00039
	P7	1985.115	0.738	0.108	0.0319	0.000024
	P8	1981.205	1.55	0.119	0.0257	0.000013
	P9	1977.261	0.437	0.0737	0.0238	0.000023
	P10	1973.284	0.432	0.0669	0.0205	0.000022
	P15	1952.901	0.917	0.147	0.0459	0.000044
(b) 7-6	P3	1974.409	0.424	0.0641	0.0196	0.000016
	P4	1970.670	1.16	0.176	0.0529	0.000042
	P6	1963.089	1.26	0.195	0.0594	0.000052
	P7	1959.247	0.969	0.152	0.0469	0.000048
	P14	1931.380	1.36	0.212	0.0653	0.000106

(a) Laser frequencies calculated using molecular constants of Young [5-60].
(b) Laser frequencies calculated using molecular constants of Mantz [5-61].

Table 5-19. Attenuation Coefficients for Laser Frequencies [5-35] (*Continued*)

HF Laser Parameters				Atmospheric Absorption Coefficients (km^{-1})			
				z = 0 km, Sea Level			z = 12 km
Band	Rot. ID	$\nu(cm^{-1})$		k_{trop}	k_{mw}	k_{sw}	k_{mw}
(c) 1-0	P11	3436.12		2.21	0.221	0.0542	0.0000287
(c)	P12	3381.50		0.496	0.0751	0.0231	0.000022
(c) 2-1	P8	3435.17		2.01	0.209	0.0512	0.0000267
(c) 3-2	P6	3373.46		0.364	0.0537	0.0168	0.000029
(d) 4-3	P8	3130.09		0.801	0.148	0.0554	0.000295
	P9	3083.83		1.12	0.211	0.0808	0.000806
(d) 5-4	P4	3150.67		0.498	0.126	0.0736	0.00229
(d) 6-5	P6	2921.74		0.586	0.0453	0.0103	0.000077
	P7	2880.70		0.0430	0.00424	0.00121	0.000006
	P8	2838.59		0.369	0.0654	0.0218	0.000044

DF Laser Parameters				Atmospheric Absorption Coefficients (km^{-1})			
				z = 0 km, Sea Level			z = 12 km
Band	Rot. ID	$\nu(cm^{-1})$		k_{trop}	k_{mw}	k_{sw}	k_{mw}
(e) 1-0	P1	2884.934		0.414	0.123	0.0772	0.00316
(e)	P2	2862.652		0.0540	0.0115	0.00485	0.00316
(e)	P3	2839.779		0.0386	0.00725	0.00266	0.000038
(e)	P4	2816.362		0.0837	0.0190	0.0104	0.00108
(e)	P5	2792.437		0.0471	0.0106	0.00496	0.000157
(e)	P6	2767.914		0.0719	0.0184	0.00952	0.000672
(e)	P7	2743.028		0.0352	0.00801	0.00352	0.000043
(e)	P8	2717.536		0.114	0.0204	0.00718	0.000034

(c) Measured, Deutsch [5-62].
(d) Calculated, Basov, et al. [5-63].
(e) Measured, Spanbauer [5-64].

Table 5-19. Attenuation Coefficients for Laser Frequencies [5-35] (*Continued*)

DF Laser Parameters			Atmospheric Absorption Coefficients (km^{-1})			
			z = 0 km, *Sea Level*			z = 12 km
Band	Rot. ID	$\nu(cm^{-1})$	k_{trop}	k_{mw}	k_{sw}	k_{mw}
(e)		P9	2691.409	0.0248	0.00485	0.00252	0.000053
(e)		P10	2665.20	0.0237	0.00752	0.00489	0.000307
(e)		P11	2638.396	0.337	0.0664	0.0247	0.000187
(e)		P12	2611.125	0.0133	0.00394	0.00302	0.000090
(f)		P13	2584.91	0.0145	0.0102	0.00981	0.00390
(f)		P14	2557.09	0.0176	0.0180	0.0185	0.00335
(c)		P15	2527.06	0.0145	0.0155	0.0161	0.000565
(c)		P16	2498.02	0.0261	0.0282	0.0295	0.00103
(c)	2-1	P3	2750.05	0.0401	0.00898	0.00403	0.000074
(c)		P4	2727.38	0.0378	0.00653	0.00272	0.000033
(c)		P5	2703.98	0.00528	0.00171	0.00118	0.0000307
(c)		P6	2680.28	0.0600	0.0139	0.00611	0.000069
(c)		P7	2655.97	0.0535	0.0134	0.00667	0.000733
(c)		P8	2631.09	0.00950	0.00348	0.00293	0.000761
(c)		P9	2605.87	0.0311	0.00776	0.00455	0.000110
(c)		P10	2580.16	0.282	0.0295	0.0311	0.00180
(c)		P11	2553.97	0.0144	0.0163	0.0177	0.000883
(c)		P12	2527.47	0.0140	0.0152	0.0158	0.000554
(c)		P13	2500.32	0.0240	0.0265	0.0278	0.000072
(c)		P16	2417.27	0.0811	0.0901	0.0943	0.00330
(c)	3-2	P3	2662.17	0.354	0.00790	0.00361	0.000047
(c)		P4	2640.04	0.0437	0.00914	0.00424	0.000075
(c)		P5	2617.41	0.00490	0.00276	0.00253	0.000090
(c)		P6	2594.23	0.0118	0.00557	0.00480	0.000152
(c)		P7	2570.51	0.0507	0.0560	0.0613	0.00557
(c)		P8	2546.37	0.0322	0.0356	0.0379	0.00228
(c)		P9	2521.81	0.0150	0.0164	0.0171	0.00599
(c)		P10	2496.61	0.0319	0.0298	0.0307	0.00107
(c)		P11	2471.34	0.0509	0.0491	0.0508	0.00184
(c)		P12	2445.29	0.0659	0.0728	0.0756	0.00266

(c) Measured, Deutsch [5-62].
(e) Measured, Spanbauer [5-64].
(f) Calculated, Spanbauer [5-64].

Table 5-19. Attenuation Coefficients for Laser Frequencies [5-35] (*Continued*)

DF Laser Parameters			Atmospheric Absorption Coefficients (km^{-1})			
			$z = 0$ km, *Sea Level*			$z = 12$ km
Band	Rot. ID	$\nu(cm^{-1})$	k_{trop}	k_{mw}	k_{sw}	k_{mw}
(c)		P13	2419.02	0.0797	0.0885	0.0927	0.00325
(c)		P14	2392.46	0.141	0.199	0.115	0.00369
(c)	4-3	P5	2532.50	0.0134	0.0143	0.0148	0.000528
(c)		P6	2500.86	0.0199	0.0218	0.0228	0.000795
(c)		P7	2486.83	0.0318	0.0349	0.0356	0.00129
(d)		P8	2463.25	0.0681	0.0563	0.0571	0.00198
(d)		P9	2439.29	0.0686	0.0758	0.0794	0.00279
(d)		P10	2414.89	0.0829	0.0921	0.0964	0.00338
(d)	5-4	P7	2404.63	0.0878	0.0965	0.101	0.00354
	7-6	P8	2222.68	0.251	0.233	0.226	0.0102
		P10	2177.99	0.123	0.0979	0.0867	0.00297
		P11	2155.03	0.186	0.0344	0.0225	0.000846
		P12	2131.68	0.272	0.187	0.195	0.0311
(d)	8-7	P7	2165.93	0.0698	0.0459	0.0466	0.00258
(d)		P8	2144.80	1.34	0.129	0.0349	0.000357
(d)		P9	2123.24	0.187	0.0296	0.0169	0.00410
(d)		P10	2101.27	0.144	0.0322	0.0180	0.00599
(d)		P12	2056.14	0.114	0.0222	0.0131	0.000494
(d)		P13	2033.01	0.153	0.0198	0.00580	0.000100
(d)	9-8	P6	2108.48	0.0603	0.0172	0.0119	0.00969
(d)		P7	2088.34	0.444	0.0567	0.0188	0.00663
(d)		P8	2067.76	0.791	0.112	0.0554	0.00259
(d)		P10	2025.36	0.646	0.0864	0.0253	0.000025
(d)		P11	2003.56	0.367	0.0480	0.0138	0.000085
(d)		P12	1981.38	0.476	0.0557	0.0152	0.000010

(c) Measured, Deutsch [5-62].
(d) Calculated, Basov, et al. [5-63].

Table 5-19. Attenuation Coefficients for Laser Frequencies [5-35] (*Continued*)

CO_2 Laser Parameters			Atmospheric Absorption Coefficients (km^{-1})			
			$z = 0$ km, *Sea Level*			$z = 12$ km
Band	Rot. ID	$\nu(cm^{-1})$	k_{trop}	k_{mw}	k_{sw}	k_{mw}
	P40	924.970	0.514	0.0359	0.0112	0.000812
	P38	927.004	0.521	0.0423	0.0154	0.00164
	P36	929.013	0.744	0.0581	0.0190	0.00211
	P34	930.997	0.538	0.0536	0.0227	0.00311
	P32	932.956	0.557	0.0650	0.0302	0.00520
	P30	934.890	0.572	0.0737	0.0360	0.00677
	P28	936.800	0.588	0.0852	0.0440	0.00887
	P26	938.684	0.583	0.0853	0.0447	0.00955
	P24	940.544	0.603	0.0955	0.0517	0.0118
	P22	942.380	0.606	0.1021	0.0569	0.0136
	P20	944.190	0.609	0.0958	0.0521	0.0125
	P18	945.976	0.635	0.1223	0.0717	0.0186
	P16	947.738	0.572	0.0747	0.0378	0.00897
	P14	949.476	0.607	0.1101	0.0642	0.0173
	P12	951.189	0.591	0.1058	0.0619	0.0171
	P10	952.877	0.596	0.1008	0.0580	0.0161
	P8	954.541	0.553	0.0817	0.0452	0.0123
	P6	956.181	0.513	0.0615	0.0314	0.00810
	P4	957.797	0.484	0.0498	0.0236	0.00573
	P2	959.388	0.978	0.0753	0.0282	0.00609
	R0	961.729	0.456	0.0347	0.0130	0.00234
	R2	963.260	0.461	0.0401	0.0170	0.00367
	R4	964.765	0.478	0.0502	0.0241	0.00590
	R6	966.247	0.519	0.0614	0.0308	0.00783
	R8	967.704	0.505	0.0663	0.0352	0.00931
	R10	969.136	0.510	0.0714	0.0389	0.0104
	R12	970.544	0.578	0.0788	0.0418	0.0109
	R14	971.927	0.556	0.0796	0.0427	0.0110
	R16	973.285	0.554	0.0799	0.0425	0.0106
	R18	974.618	0.522	0.0755	0.0405	0.0101
	R20	975.927	0.194	0.2140	0.0740	0.0109
	R22	977.210	0.674	0.0871	0.0398	0.00803
	R24	978.468	0.503	0.0641	0.0318	0.00699
	R26	979.701	0.484	0.0579	0.0280	0.00585
	R28	980.909	0.474	0.0529	0.0245	0.00471
	R30	982.091	0.552	0.0587	0.0240	0.00378
	R32	983.248	0.454	0.0436	0.0183	0.00324
	R34	984.379	0.455	0.0439	0.0158	0.00229
	R36	985.484	0.436	0.0357	0.0133	0.00176
	R38	986.563	0.428	0.0328	0.0114	0.00138
	R40	987.616	0.423	0.0306	0.0102	0.00121

The data were measured using multiple-traversal cells containing absorbing gas, the partial pressure of which could be varied. High-altitude conditions were simulated at appropriate pressures by adding broadening gases such as nitrogen and oxygen to the cell. Path lengths from 1.5 to 4.8 m were achieved by successive reflections of radiation back and forth through the cell.

5.7.1. CO_2 Absorption. Strong absorption by CO_2 exists in the 2.7 μm (3660 cm^{-1}) region, the 4.3 μm (2350 cm^{-1}) region and the region between 11.4 μm (875 cm^{-1}) and 20 μm (495 cm^{-1}). Weaker absorption bands are present at 1.4 μm (6975 cm^{-1}), 1.6 μm (6230 cm^{-1}), 2.0 μm (4983 cm^{-1}), 4.8 μm (2075 cm^{-1}), 5.2 μm (1930 cm^{-1}), 9.4 μm (1064 cm^{-1}), and 10.4 μm (961 cm^{-1}). In Figures 5-15 through 5-30, P_e is equivalent pressure and P is total pressure.

The 2.7 μm (3660 cm^{-1}) Region. The CO_2 absorption in the 2.7 μm region is caused primarily by two strong absoption bands, the $2\nu_2 + \nu_3$ band centered at 2.77 μm (3609 cm^{-1}) and the $\nu_1 + \nu_3$ band centered at 2.69 μm (3716 cm^{-1}). Total absorption for the 2.77 μm band is shown in Figure 5-15, and total absorption for the 2.69 μm band in Figure 5-16. Total absorption for the entire 2.7 μm (3660 cm^{-1}) region, i.e., the 2.77 μm band plus the 2.69 μm band, is shown in Figure 5-17. For a total absorption of more than 10 cm^{-1}, the curves in all of the illustrations are estimated to be accurate within ± 5%; between 3 and 10 cm^{-1}, the estimated accuracy is ± 10%. For total absorption values less than 3 cm^{-1}, the estimated accuracy is ± 20%.

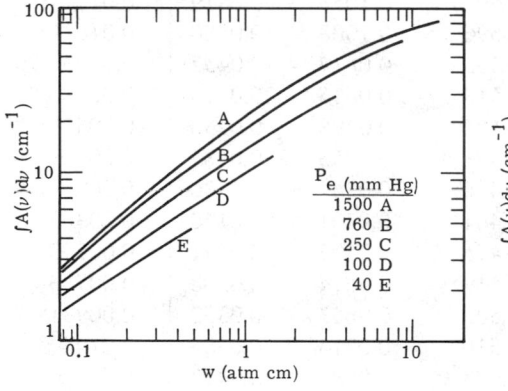

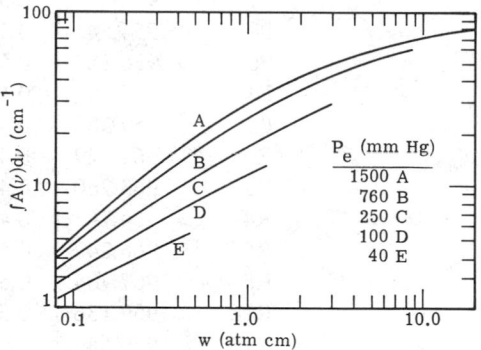

Fig. 5-15. Total absorption versus absorber concentration of the 2.77 μm (3609 cm^{-1}) CO_2 band.

Fig. 5-16. Total absorption versus absorber concentration for the 2.69 μm (3716 cm^{-1}) CO_2 band.

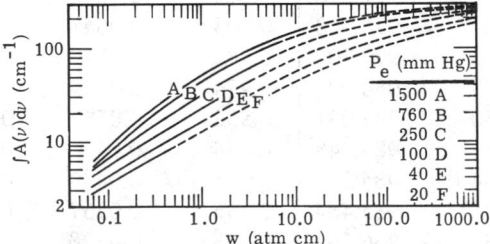

Fig. 5-17. Total absorption versus absorber concentration for the 2.7 μm (3660 cm^{-1}) CO_2 region (2.77 μm band plus 2.69 μm band).

The 4.3 μm (2350 cm⁻¹), 4.8 μm (2075 cm⁻¹) and 5.2 μm (1930 cm⁻¹) Bands. The 4.3 μm (2350 cm⁻¹) CO_2 band causes almost complete absorption between about 4.19 μm (2386 cm⁻¹) and 4.45 μm (2250 cm⁻¹). Total absorption curves, which are estimated to be accurate within ± 5% above 10 cm⁻¹ total absorption, ± 10% between 3 and 10 cm⁻¹ total absorption, and 20% below 3 cm⁻¹ total absorption, are shown in Figure 5-18.

Total absorption curves for the 4.8 μm (2075 cm⁻¹) and the 5.2 μm (1930 cm⁻¹) bands are shown in Figure 5-19. These bands are very weak and are significant only for high values of absorber concentration.

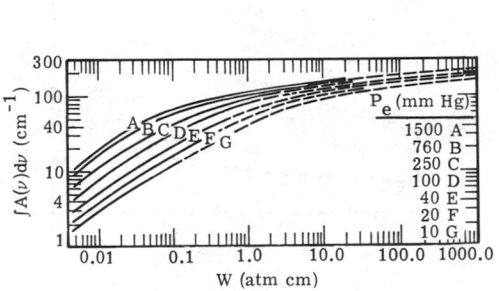

Fig. 5-18. Total absorption versus absorber concentration for the 4.3 μm (2350 cm⁻¹) CO_2 band.

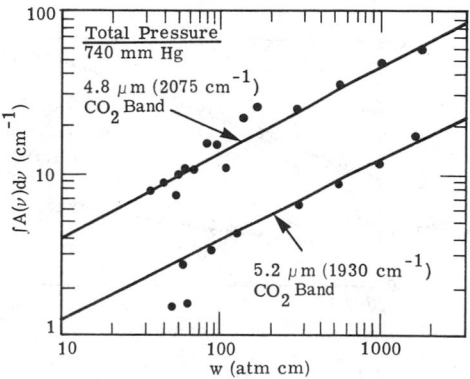

Fig. 5-19. Total absorption versus absorber concentration for the 4.8 μm (2075 cm⁻¹) CO_2 band and the 5.2 μm (1930 cm⁻¹) CO_2 band.

The 9.4 μm (1064 cm⁻¹) and 10.4 μm (961 cm⁻¹) Bands. Total absorption of the weak CO_2 bands at 9.4 μm (1064 cm⁻¹) and 10.4 μm (961 cm⁻¹) are shown in Figures 5-20 and 5-21 respectively. Total absorption of these bands is strongly dependent upon temperature; thus, values of total absorption are temperature-corrected. The curves shown in Figures 5-20 and 5-21 are for a temperature of 26°C, and are estimated to be accurate within ± 5% for more than 10 cm⁻¹ total absorption, ± 10% between 3 and 10 cm⁻¹ total absorption, and ± 20% below 3 cm⁻¹ total absorption.

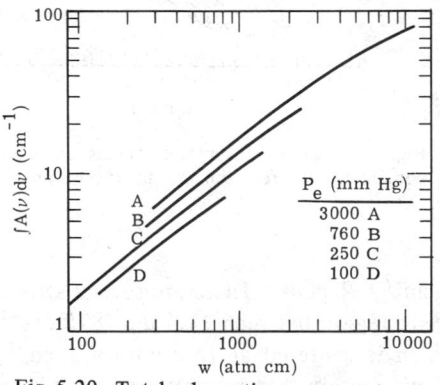

Fig. 5-20. Total absorption versus absorber concentration for the 9.4 μm (1064 cm⁻¹) CO_2 band.

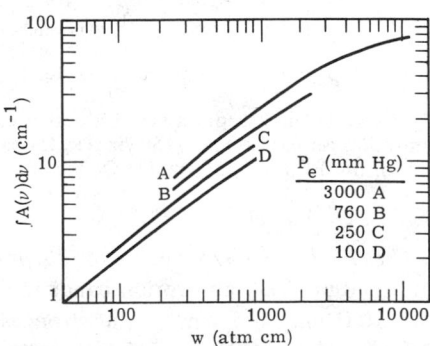

Fig. 5-21. Total absorption versus absorber concentration for the 10.4 μm (961 cm⁻¹) CO_2 band.

The temperature dependence of both bands is illustrated in Figure 5-22. From the curves, it can be seen that total absorption increases with temperature.

The 1.4 μm (7150 cm^{-1}), 1.6 μm (6250 cm^{-1}), and 2.0 μm (5000 cm^{-1}) Bands. Total absorption of the weak CO_2 bands at 1.4 μm (7150 cm^{-1}), 1.6 μm (6250 cm^{-1}) and 2.0 μm (5000 cm^{-1}) are shown in Figures 5-23 and 5-24. The band at 1.6 μm is actually a group of very weak bands centered at 1.645 μm (6077 cm^{-1}), 1.604 μm (6231 cm^{-1}), 1.574 μm (6351 cm^{-1}), and 1.536 μm (6510 cm^{-1}). The CO_2 band near 2 μm (5000 cm^{-1}) consists primarily of three weak absorption bands centered at 2.057 μm (4861 cm^{-1}), 2.006 μm (4983.5 cm^{-1}), and 1.957 μm (5109 cm^{-1}).

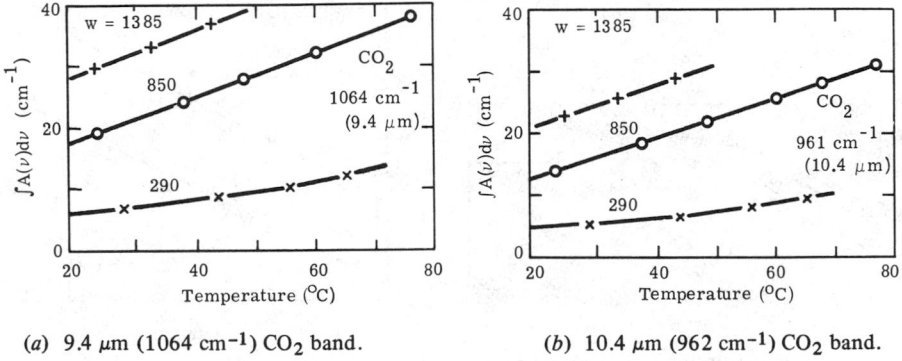

(a) 9.4 μm (1064 cm^{-1}) CO_2 band.   (b) 10.4 μm (962 cm^{-1}) CO_2 band.

Fig. 5-22. Effects of temperature on total absorption.

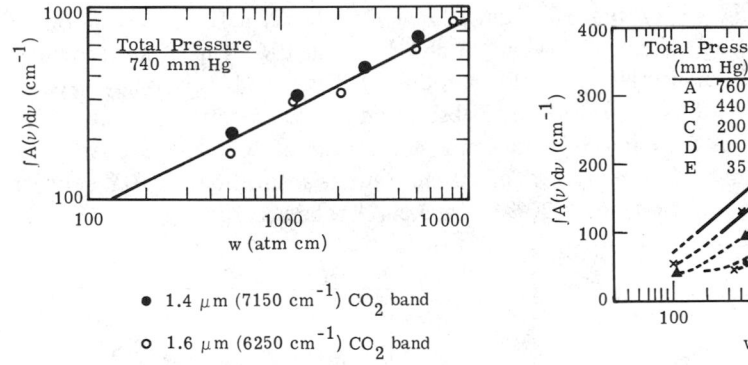

● 1.4 μm (7150 cm^{-1}) CO_2 band

○ 1.6 μm (6250 cm^{-1}) CO_2 band

Fig. 5-23. Total absorption versus absorber concentration for the 1.4 μm (7150 cm^{-1}) CO_2 band and for the 1.6 μm (6250 cm^{-1}) CO_2 band.

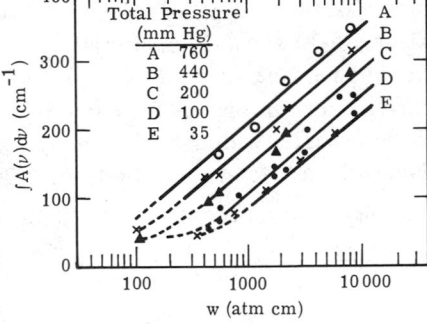

Fig. 5-24. Total absorption versus absorber concentration for the 2 μm (5000 cm^{-1}) CO_2 band.

The 11.4 μm (875 cm^{-1}) to 20 μm (495 cm^{-1}) Region. There are several strong and medium CO_2 absorption bands in the spectral region between 11.4 μm (875 cm^{-1}) and 20.0 μm (495 cm^{-1}), the strongest of which is centered at 15.0 μm (667 cm^{-1}). Because of the wide spectral region covered in this section, the spectral region is divided into smaller subregions, and absorber concentration is plotted against mean fractional absorption rather than against total absorption (which usually refers to an entire band).

Total absorption, $\int A(\nu)\,d\nu$, of an entire band is independent of spectral slit-width, provided that there is no absorption beyond the limits of integration. It follows that the total absorption of a subregion will be independent of spectral slit-width if the subregions are divided at frequencies where the absorption is zero. Practically, however, no frequencies exist where the absorption is zero, and optimum frequencies, i.e., frequencies at which absorption is very slight, are chosen to divide the subregions.

Mean fractional absorption is related to total absorption by the following equation:

$$\bar{A}\,(\nu_2 - \nu_1) = \frac{1}{(\nu_2 - \nu_1)} \int_{\nu_1}^{\nu_2} A(\nu)\,d\nu \tag{5-99}$$

which represents the mean fractional absorption in the spectral region $\nu_1 \nu_2$.

Figure 5-25 shows mean fractional absorption for each of the five subregions of the 11.4 μm (875 cm^{-1}) to 20.2 μm (495 cm^{-1}) CO_2 region. The curves are based upon a temperature of 26°C and are estimated to be accurate within ± 5% for a mean fractional absorption greater than 0.10 cm^{-1} and increase to approximately ± 20% for smaller values of total absorption.

The effect of temperature on mean fractional absorption is shown in Figure 5-26 for four of the five subregions. No data are given for the fifth subregion, which extends from 18.3 μm (545 cm^{-1}) to 20 μm (495 cm^{-1}). From the curves in Figure 5-26, it can be seen that mean fractional absorption increases with temperature.

5.7.2. H_2O Absorption. Total absorption of the 6.27 μm (1595 cm^{-1}), 2.70 μm (3700 cm^{-1}), and 1.87 μm (5332 cm^{-1}) H_2O bands are shown in Figures 5-27(a) through (c). The absorption in the 2.70 μm region is caused primarily by two absorption bands with centers at 2.73 μm (3657 cm^{-1}) and 2.66 μm (3756 cm^{-1}). The curves in Figure 5-27 are estimated to be accurate within ± 6%.

Minor H_2O absorption bands at 3.2 μm (3150 cm^{-1}), 1.38 μm (7250 cm^{-1}), 1.1 μm (8807 cm^{-1}), and 0.94 μm (10,613 cm^{-1}) are shown in Figures 5-27 (d) through (h). These curves are estimated to be accurate within about ± 10%.

5.7.3. N_2O Absorption. Figure 5-28(a) shows total absorption as a function of absorber concentration and equivalent pressure for the ν_3 fundamental band of N_2O centered at 4.5 μm (2224 cm^{-1}). This band occurs in the atmospheric window between the strong 4.3 μm (2350 cm^{-1}) CO_2 band and the 6.3 μm (1595 cm^{-1}) H_2O band. Thus, it gives rise to the major portion of the atmospheric absorption in the 4.3 to 6.2 μm window. It is estimated that the accuracy of the total absorption values given in Figure 5-28(a) is within ± 5% above 10 cm^{-1} total absorption.

Figures 5-28(b) through (f) show total absorption of N_2O in the 3.9 μm (2563 cm^{-1}), 4.06 μm (2461 cm^{-1}), 7.8 μm (1285 cm^{-1}), 8.6 μm (1167 cm^{-1}), and 17.0 μm (589 cm^{-1}) bands. These curves show total absorption within an estimated accuracy of ± 5% above a total absorption of 30 cm^{-1}. Below 30 cm^{-1}, the accuracy is somewhat less.

5.7.4. CO Absorption. Total absorption of the 4.7 μm (2143 cm^{-1}) CO band is shown in Figure 5-29(a). The data are estimated to be accurate within ± 5% above 10 cm^{-1} total absorption and within ± 10% below 10 cm^{-1} total absorption.

Figure 5-29(b) shows total absorption for the 2.3 μm (4260 cm^{-1}) CO band. The curves in Figure 5-29(b) do not represent nearly as wide a range of absorber concentration and equivalent pressure as those of Figure 5-29(a) because the largest values of

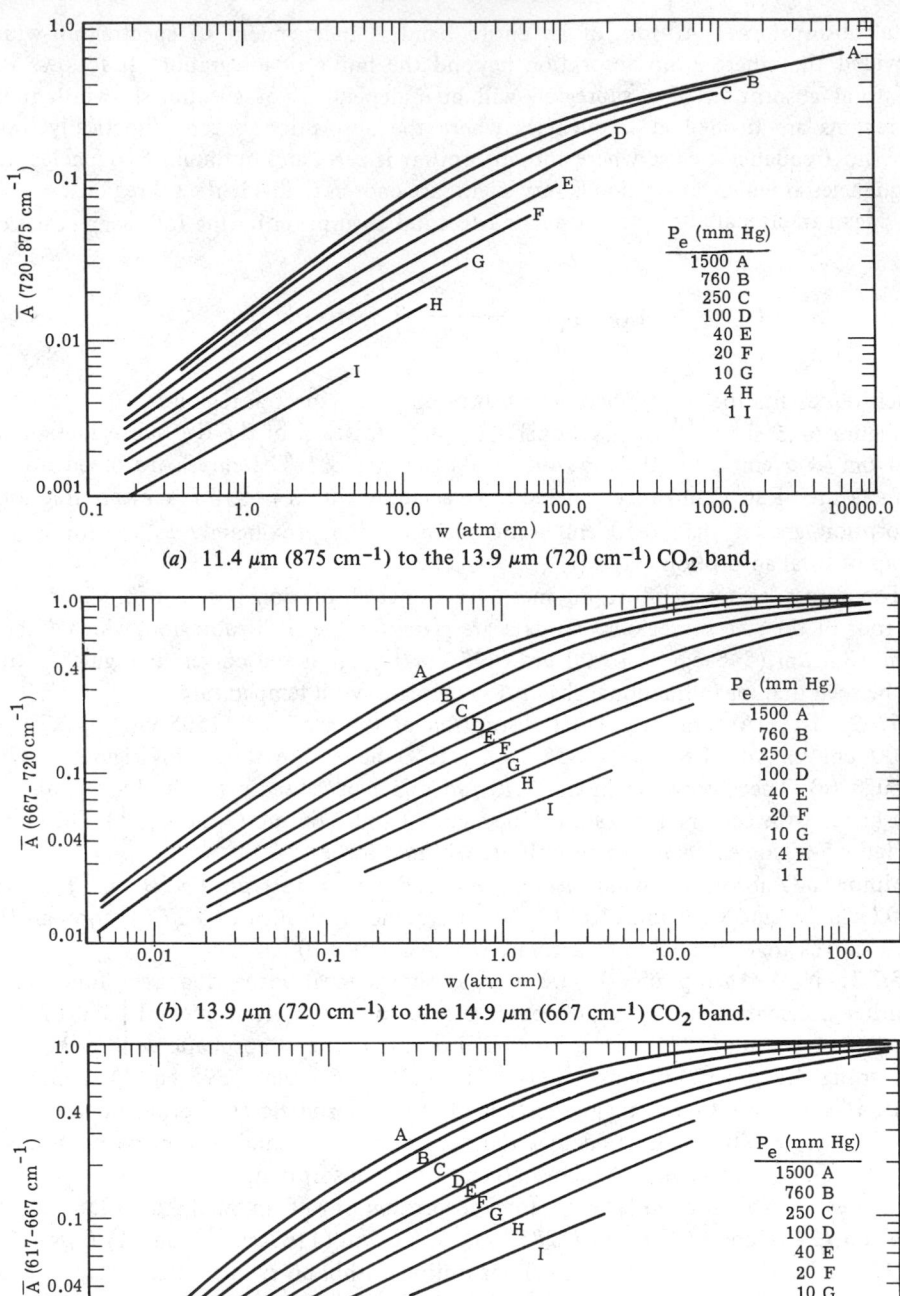

(a) 11.4 μm (875 cm^{-1}) to the 13.9 μm (720 cm^{-1}) CO_2 band.

(b) 13.9 μm (720 cm^{-1}) to the 14.9 μm (667 cm^{-1}) CO_2 band.

(c) 14.9 μm (667 cm^{-1}) to the 16.2 μm (617 cm^{-1}) CO_2 band.

Fig. 5-25. Mean fractional absorption versus absorber concentration.

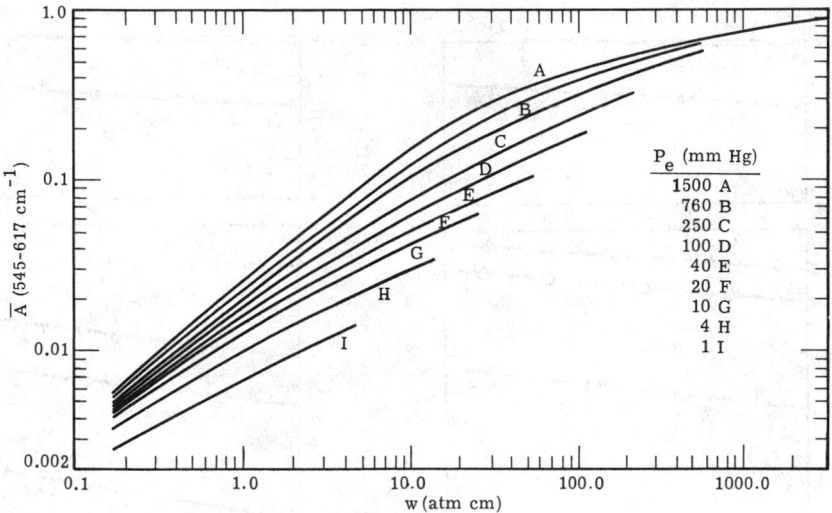

(d) 16.2 μm (617 cm^{-1}) to the 18.3 μm (545 cm^{-1}) CO_2 band.

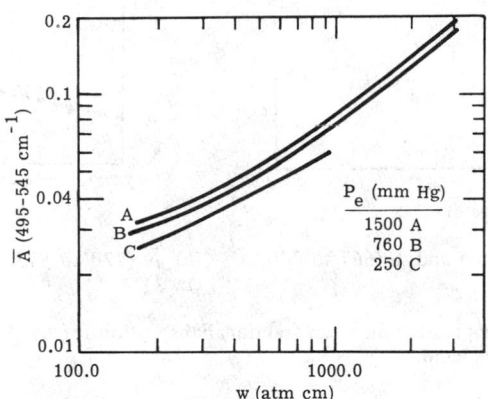

(e) 18.3 μm (545 cm^{-1}) to the 20.0 μm (495 cm^{-1}) CO_2 band.

Fig. 5-25 (Continued). Mean fractional absorption versus absorber concentration.

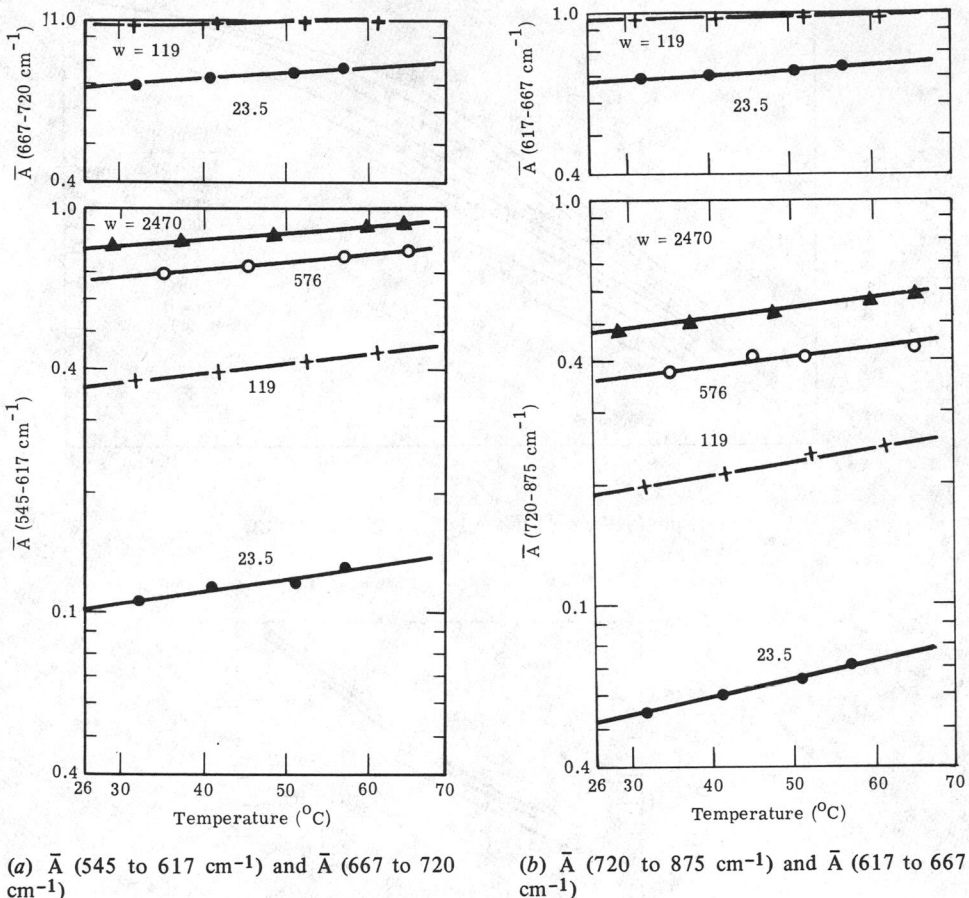

(a) $\bar{A}$ (545 to 617 cm^{-1}) and $\bar{A}$ (667 to 720 cm^{-1})

(b) $\bar{A}$ (720 to 875 cm^{-1}) and $\bar{A}$ (617 to 667 cm^{-1})

Fig. 5-26. Effects of temperature on mean fractional absorption for the 11.4 μm (875 cm^{-1}) to the 18.3 μm (545 cm^{-1}) CO_2 band.

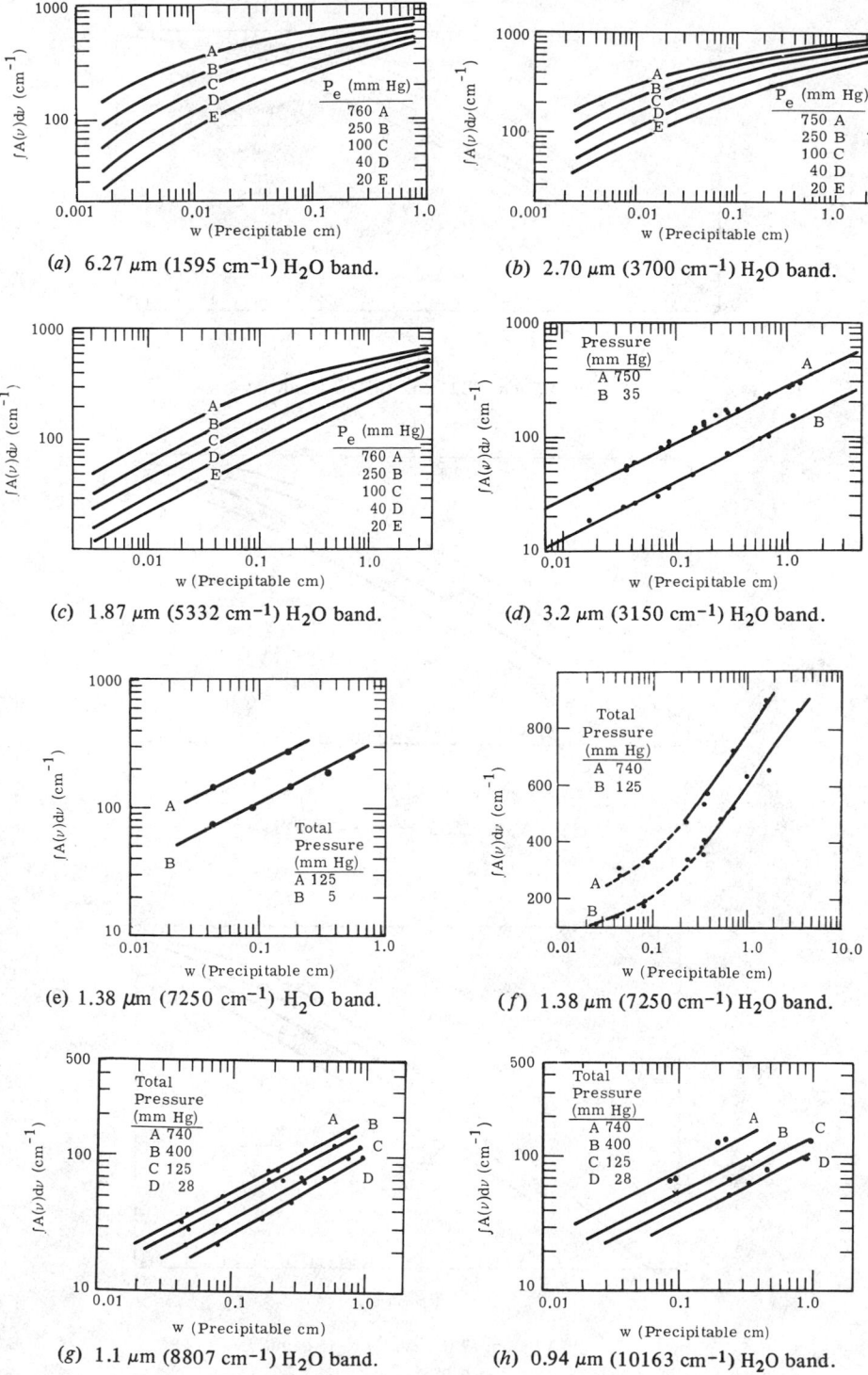

(a) 6.27 μm (1595 cm⁻¹) H₂O band.

(b) 2.70 μm (3700 cm⁻¹) H₂O band.

(c) 1.87 μm (5332 cm⁻¹) H₂O band.

(d) 3.2 μm (3150 cm⁻¹) H₂O band.

(e) 1.38 μm (7250 cm⁻¹) H₂O band.

(f) 1.38 μm (7250 cm⁻¹) H₂O band.

(g) 1.1 μm (8807 cm⁻¹) H₂O band.

(h) 0.94 μm (10163 cm⁻¹) H₂O band.

Fig. 5-27. Total absorption versus absorber concentration for H₂O bands.

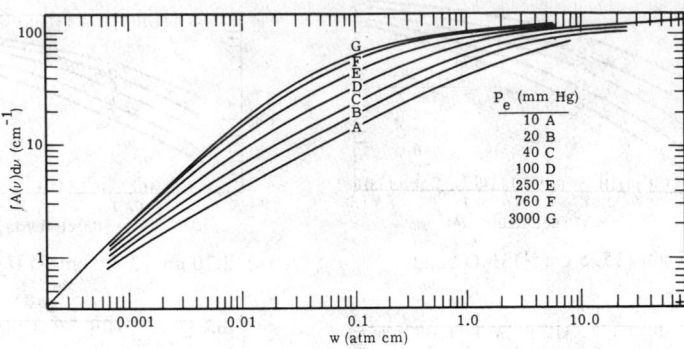

(a) 4.5 μm (2224 cm⁻¹) N₂O band.

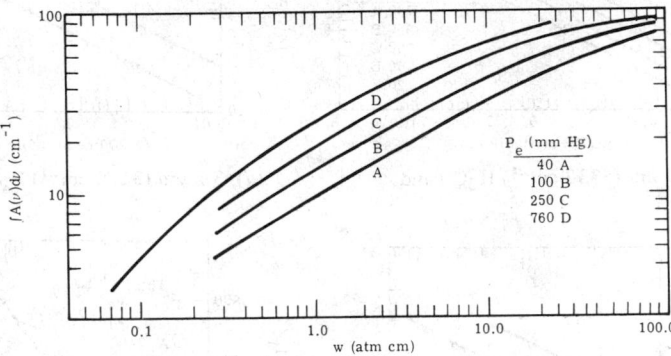

(b) 3.9 μm (2563 cm⁻¹) N₂O band.

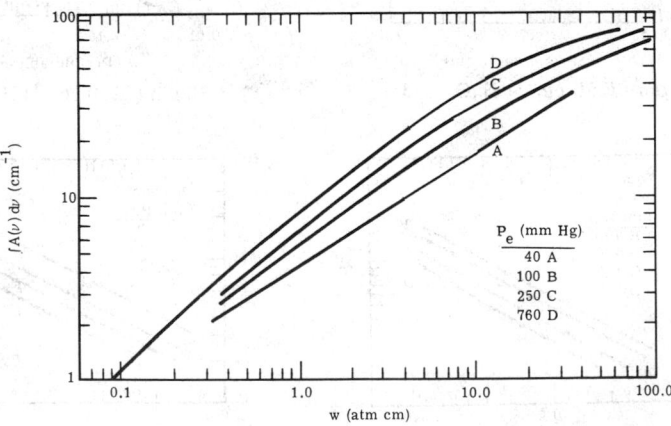

(c) 4.05 μm (2461 cm⁻¹) N₂O band.

Fig. 5-28. Total absorption versus absorber concentration for N₂O bands.

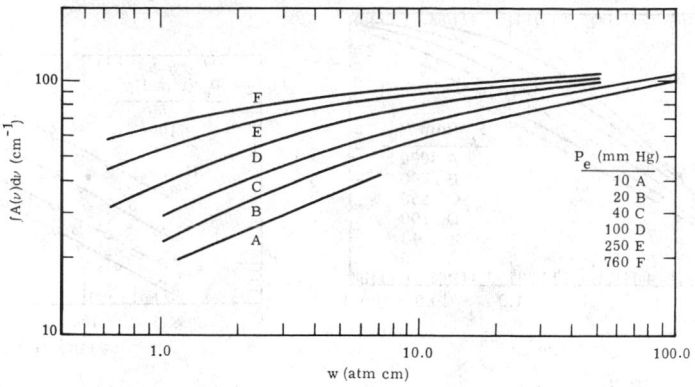

(d) 7.7 μm (1285 cm⁻¹) N₂O band.

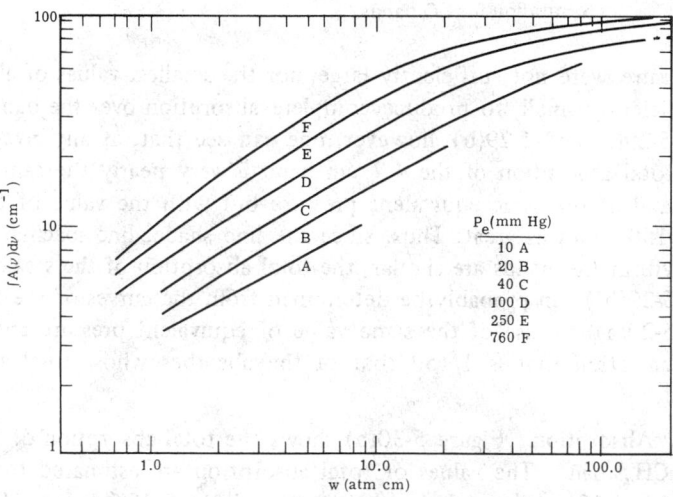

(e) 8.6 μm (1167 cm⁻¹) N₂O band.

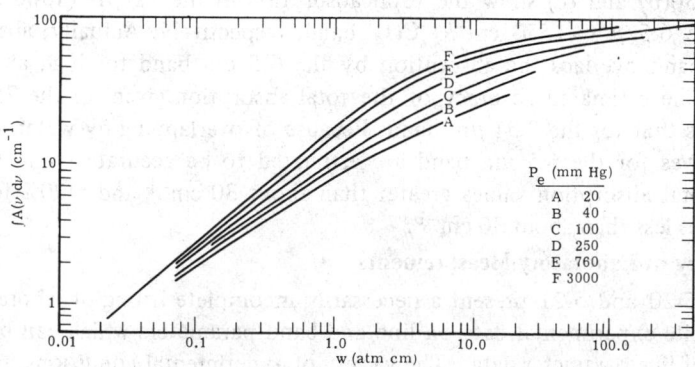

(f) 17.1 μm (589 cm⁻¹) N₂O band.

Fig. 5-28 (Continued). Total absorption versus absorber concentration for N₂O bands.

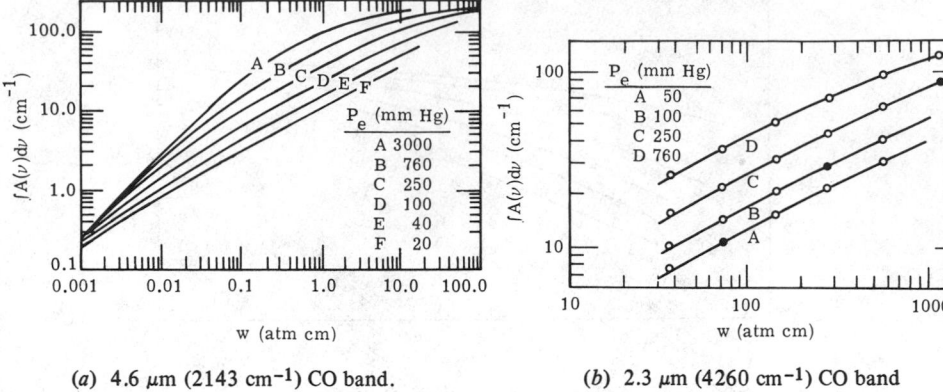

(a) 4.6 μm (2143 cm⁻¹) CO band.   (b) 2.3 μm (4260 cm⁻¹) CO band

Fig. 5-29. Total absorption versus absorber concentration for CO bands.

equivalent pressure were not sufficiently large, nor the smallest values of absorber concentration sufficiently small, to produce complete absorption over the bands. By comparing Figure 5-29(a) and 5-29(b), however, one can see that, at any given equivalent pressure, the total absorption of the 4.7 μm band is very nearly the same as that of the 2.3 μm band at the same equivalent pressure but with the value of the absorber concentration 150 times as great. Thus, since the line shape, line spacing, and relative line strength within the bands are similar, the total absorption of the weaker over-tone band (Figure 5-29(b)) can probably be determined from the curves of the fundamental band (Figure 5-29(a)) by use of the same value of equivalent pressure and a value of absorber concentration that is 1/150 that of the absorber whose total absorption is being measured.

5.7.5. CH₄ Absorption. Figure 5-30(a) shows the total absorption of the 3.31 μm (3020 cm⁻¹) CH₄ band. The values of total absorption are estimated to be accurate within ± 5% above 10 cm⁻¹ total absorption and within ± 10% below 10 cm⁻¹ total absorption.

Figures 5-30(b) and (c) show the total absorption of the 7.6 μm (1306 cm⁻¹) CH₄ band and the 6.5 μm (1550 cm⁻¹) CH₄ band, respectively. Actually, absorption by the 7.6 μm band overlaps the absorption by the 6.5 μm band for high absorber concentrations. The estimated accuracy of the total absorption given for the 7.6 μm band is the same as that for the 3.31 μm band. Because of overlapping by water vapor, however, the curves for the 6.5 μm band are estimated to be accurate to no better than ± 10% for total absorption values greater than about 30 cm⁻¹ and ± 20% for total absorption values less than about 30 cm⁻¹.

5.8. Summary of Laboratory Measurements

Tables 5-20 and 5-21 present a necessarily incomplete listing of laboratory measurements to cite experimental data on line and band parameters which can be used in a compilation of line parameter data. The variety of experimental conditions under which measurements were made are listed to provide a reliable data base for empirically fitting data to band-model functions and ultimately for determining the reliability of band-model parameters.

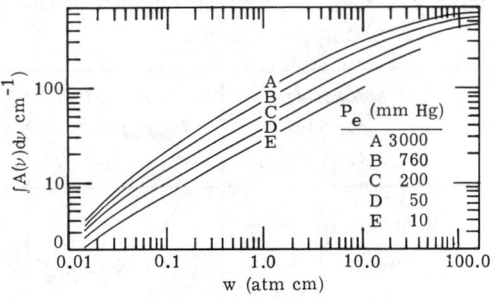

(a) 3.3 μm (3020 cm⁻¹) CH₄ band.

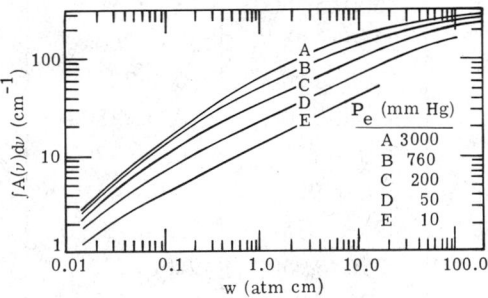

(b) 7.6 μm (1306 cm⁻¹) CH₄ band.

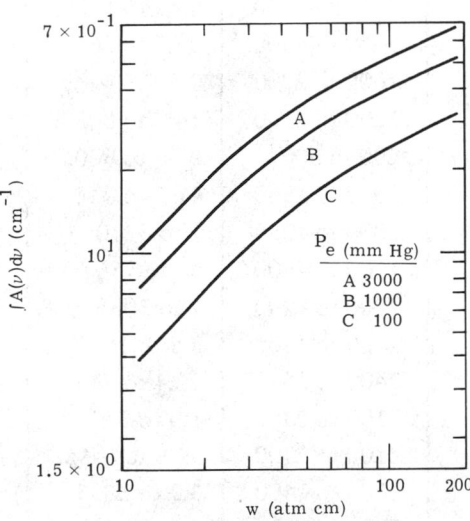

(c) 6.5 μm (1550 cm⁻¹) CH₄ band.

Fig. 5-30. Total absorption versus absorber concentration for CH₄ bands.

Table 5-20. Summary of Laboratory Measurements of Homogeneous-Path
Absorption Spectra [5-1]

Gas	Band (μm)	Observed Interval (cm^{-1})	Range of Pressures* (mm Hg)	Range of Absorber Amount, w[†]
CO_2	1.6 & 1.4	6000 to 7200	75 to 760	540 to 8100
	2.0	4600 to 5400	10 to 760	108 to 8630
	2.7	3300 to 4100	1 to 755	11 to 1619
	2.7	3450 to 3850	14.2 to 1065	0.164 to 24.4
	4.3	2200 to 2450	3.8 to 2115	0.0108 to 22.8
	4.3 & 4.8	2000 to 2500	1 to 735	9 to 1570
	4.3	2250 to 2450	38 to 760	1.0 to 300
	5.2	1800 to 2000	10 to 735	104 to 1570
	9.398 & 10.41	900 to 1100	103 to 3800	48 to 11.200
	–	720 to 875	103 to 3800	305 to 11.200
	15	500 to 900	20 to 745	1 to 863
	–	495 to 875	0.26 to 3190	0.0118 to 2470
H_2O	1.1	8250 to 9500	9.8 to 740	0.03 to 1.93
	1.38	6500 to 8000	3 to 740	0.026 to 3.85
	1.875	4000 to 6000	3 to 740	0.026 to 3.85
	1.875	4950 to 5800	27.5 to 862	0.0033 to 0.101
	2.7 & 3.2	2800 to 4400	2 to 750	0.017 to 2.1
	2.7	3000 to 4300	27.5 to 862	0.0033 to 0.101
	6.3	1000 to 2200	2.5 to 742	0.021 to 1.49
	6.3	1200 to 2200	14.0 to 805	0.0041
H_2O	–	200 to 500	0.76 to 600	0.0041 to 0.143
O_3	9.6	1000 to 2560	11.2 to 744	0.00278 to 1.968
CO	4.666	2000 to 2250	1 to 3210	0.00096 to 22.2
	2.347	4100 to 4400	54 to 756	36.6 to 1140
CH_4	3.311	2700 to 3200	2 to 3085	0.015 to 188
	6.452 & 7.657	1100 to 1800	3.8 to 3050	0.026 to 188
N_2O	4.0	2400 to 2650	22.4 to 746	0.38 to 18.6
	4.5	2100 to 2300	1.0 to 3120	0.00016 to 76.4
	4.5	2100 to 2300	99.8 to 849	0.01 to 2.3
	7.78 & 8.57	1100 to 1400	3.2 to 3035	1.37 to 46.7
	14.45 & 16.98	500 to 800	4.5 to 851	1.89 to 359

*Pressure designated is generally the equivalent pressure.

[†] w is expressed in precipitable centimeters for H_2O and in atmospheric centimeters for the other gases.

Table 5-20. Summary of Laboratory Measurements of Homogeneous-Path
Absorption Spectra [5-1] (*Continued*)

No. of Curves	Resolution‡ (approx.)	Researchers	Reference No.
13	0.12 μm	Howard et al. (1955)	[5-21]
32	0.9 μm	Howard et al. (1955)	[5-21]
67	0.07 μm	Howard et al. (1955)	[5-21]
32	10 to 15 cm^{-1}	Burch et al. (1962)	[5-4]
65	5 to 10 cm^{-1}	Burch et al. (1962)	[5-4]
43	0.1 μm	Howard et al. (1955)	[5-21]
10	10 cm^{-1}	Bradford et al. (1963)	[5-26]
9	0.1 μm	Howard et al. (1955)	[5-21]
25	5 to 10 cm^{-1}	Burch et al. (1962)	[5-4]
14	5 to 10 cm^{-1}	Burch et al. (1962)	[5-4]
37	0.5 μm	Howard et al. (1955)	[5-21]
30	5 to 10 cm^{-1}	Burch et al. (1962)	[5-4]
41	0.13 μm	Howard et al. (1955)	[5-21]
62	0.12 μm	Howard et al. (1955)	[5-21]
62	0.1 μm	Howard et al. (1955)	[5-21]
14	20 cm^{-1}	Burch et al. (1962)	[5-4]
114	0.07 μm	Howard et al. (1955)	[5-21]
5	20 cm^{-1}	Burch et al. (1962)	[5-4]
69	0.4 μm	Howard et al. (1955)	[5-21]
15	6 cm^{-1}	Burch et al. (1962)	[5-4]
	5 to 10 cm^{-1}	Palmer (1957)	[5-65]
	7 cm^{-1}	Walshaw (1957)	[5-29]
147	25 cm^{-1}	Burch et al. (1962)	[5-4]
26	15 to 20 cm^{-1}	Burch et al. (1962)	[5-4]
88	25 cm^{-1}	Burch et al. (1962)	[5-4]
86	10 cm^{-1}	Burch et al. (1962)	[5-4]
8	20 cm^{-1}	Burch et al. (1962)	[5-4]
17	25 cm^{-1}	Burch et al. (1962)	[5-4]
65	—	Abels (1962)	[5-66]
11	10 cm^{-1}	Burch et al. (1962)	[5-4]
7	6 cm^{-1}	Burch et al. (1962)	[5-4]

‡Resolution specified is ordinarily center-band resolution and is used only to give an approximate designation.

Table 5-21. Summary of Laboratory Transmittance
Measurements [5-106]

Absorber	Wavelength (μm)	Resolution (cm^{-1})	Absorber Concentration	Path Length (m)
CO_2	1.4, 1.6, 2.0, 2.7, 4.3, 5.2, 15	50 to 200	1 to 1000 atm cm	88 to 1400
CO_2	9.4, 10.4, 12.64	2 to 5	3 to 320 atm cm	3.28
CO_2	1.4, 1.6, 2.7, 4.3, 4.8, 5.2, 9.4, 10.4, 15	20 to 250	0.5 to 1225 atm cm	0.388, 1.29
CO_2	2.7, 4.3, 10, 15	~5	0.01 to 11,200 atm cm	0.0155 to 32
CO_2	1 to 1.4, 1.6, 2, 2.7, 4.3	0.3 to 2.5	0.08 to 8.4 $\times$ 10^4 atm cm	4 to 933
H_2O	0.94, 1.1, 1.38, 1.87	~100	0.001 to 3.8 pr cm	88 to 1936
H_2O	4.2 to 23	–	–	–
H_2O	22 to 200	2.5	4 $\times$ 10^{-5} to 8 $\times$ 10^{-3}	0.125 to 9.06
H_2O	20 to 2500	19 to 3	0.007 to 0.0079 pr cm	7.5
H_2O	14.5 to 21.05	–	–	–
H_2O	20 to 40	3 to 8	0.0041 to 0.143 pr cm	196
H_2O	1.87, 2.7, 6.3	6 to 20	0.0017 to 0.109 pr cm	0.0155 to 48.75
H_2O	0.69 to 1.98, 2.7, 4.4 to 6.3	0.4 to 1.75	0.0017 to 3.1 pr cm	4 to 933
N_2O	3.9, 4.06, 4.5, 7.8, 8.6, 7.8, 14.5, 17	6 to 20	1.6 $\times$ 10^{-4} to 359 atm cm	0.0155 to 16
N_2O	1.5 to 2.5, 4.2 to 13.2	0.2 to 2.2	0.04 to 605 atm cm	0.001 to 900
CO	4.7, 2.35	15 to 20	9.6 $\times$ 10^{-4} to 1140 atm cm	0.0155 to 48.75
CH_4	3.3, 6.45, 7.7	10 to 25	0.015 to 188 atm cm	0.0635 to 16
O_3	9.6	~25	0.02 to 0.5 atm cm	0.90
O_3	4.5 and 9.6	~10	2.7 atm cm	0 to 1, 0.4
O_3	9.6	~10	0.27 atm cm	0.04
O_3	9.6	6.5	0.03 to 1.5 atm cm	0.065, 0.35
O_3	3 to 15	5 to 40	0.009 to 36.7 atm cm	0.4, 32
H_2O cont. (T=296K)	8.5 to 12	–	5.48 $\times$ 10^{22} mol cm^{-2}	1185
H_2O, CO_2	2.5 to 40	~1	–	92

Table 5-21. Summary of Laboratory Transmittance
Measurements [5-106] (*Continued*)

Pressure		Broadening Gas	Originator	Reference
Partial (mm Hg)	*Total* (mm Hg)			
1 to 50	1 to 740	N_2	Howard, et al., (1955)	[5-21]
7 to 740	370 to 740	N_2	Kostkowski (1955)	[5-67]
38 to 760	380 to 7600	N_2	Edwards (1960)	[5-68]
0.26 to 2920	0.26 to 3800	N_2	Burch, et al., (1962)	[5-4]
0.005 to 1920	0.005 to 11,000	N_2	Burch, Gryvnak, and Patty (1964 to 1968)	[5-69]
2 to 28	2 to 740	N_2	Howard, et al., (1955)	[5-21]
–	–	–	Daw (1956)	[5-70]
3.8 to 10	760	Air	Bell (1956)	[5-71]
10	760	Air	Yaroslavskii (1958)	[5-72]
–	–	–	Izatt (1960)	[5-73]
0.015 to 0.7	0.015 to 61	N_2	Palmer (1957)	[5-65]
2.8 to 22.5	2.8 to 775	N_2	Burch, et al., (1962)	[5-4]
4.16 to 1467	134 to 7350	N_2	Burch, Gryvnak, and Patty (1965 to 1973)	[5-74]
0.09 to 760	1 to 3035	N_2	Burch, et al., (1962)	[5-4]
0.07 to 10,000	47 to 10,000	N_2	Burch, et al., (1971)	[5-22]
3 to 3000	3 to 3000	N_2	Burch, et al., (1962)	[5-4]
1.52 to 100	3.to 3000	N_2	Burch, et al., (1962)	[5-4]
0.15 to 4.2	6 to 760	O_2	Summerfield (1941)	[5-77]
100 to 200	100 to 760	O_2	Gutowsky and Peterson (1950)	[5-79]
50 to 760	50 to 760	O_2	Wilson and Ogg (1950)	[5-78]
0.5 to 33	10 to 760	Dry Air	Walshaw (1957)	[5-29]
0.002 to 800	15 to 1600	O_2	McCaa and Shaw (1968)	[5-30]
14.2	14.2	–	Burch (1970)	[5-75]
–	–	Air	Lovell (1969)	[5-76]

5.9 Field Measurements of Atmospheric Transmittance

5.9.1. Horizontal Paths. Several sets of measurements have been made of atmospheric transmittance, mainly at sea-level. The majority of those published since 1950 are delineated in Table 5-22. Although approximately 20 years old and subject to some question in certain spectral regions, the Taylor and Yates curves are still widely used to roughly check the effect of the atmosphere by experiment.* Figure 5-31 shows the effect of moderately small path, specifying the spectral resolution of the instrument for the various regions and the absorbing constituents. Figure 5-32 shows nearly the same spectral range of data, and demonstrates the effect of increases in path length with commensurate increases in absorber amounts.

5.9.2 Solar Spectrum Measurements (See also Chapter 3). [5-37 through 5-50].[†] A solar spectrum is a transmission spectrum of the earth's atmosphere, with the sun used as the source of radiation. Many Fraunhofer lines are observed in the visible and ultraviolet portions of the solar spectrum. In the infrared, however, very few Fraunhofer lines are present and the sun emits approximately as a uniform 6000 K blackbody. Spectral details observed in the infrared solar spectrum are almost entirely due to the absorption of solar radiation by the molecules present in the earth's atmosphere.

The length of an absorption path through the atmosphere is dependent upon the elevation of the sun in the sky. Thus, when the sun is at the zenith, the solar radiation traverses one air mass of atmosphere. At any angle from the zenith called the solar altitude or zenith distance, the radiation traverses longer paths through the atmosphere.

Table 5-23 gives the equivalent air mass for a sea-level observer as the solar altitude varies from 0° to 90° [5-36]. To obtain the same type of table for an observer at a different reference altitude, one must multiply the values in Table 5-23 by the ratio of the pressure at the new reference level to that at sea level. This approximation holds well up to 85° angles and 100 km altitude.

Figure 5-33 shows a low-resolution solar spectrum for the region from 1 to approximately 15 μm. The other curves show the position and approximate relative strengths of the infrared absorption bands for various molecules in the atmosphere.

A number of high-resolution measurements of the solar spectrum have been made [5-37 to 5-50]. Figures 5-34 through 5-48 show the spectrum from approximately 2.80 to 14.2 μm [5-37 to 5-40]. The measurements were made from the Jungfraujoch, Switzerland, at an altitude of approximately 12,000 ft. Other figures in the references contain more detail.

---

*An extensive program for the field measurement of atmospheric transmittance is currently under way at the U. S. Naval Research Laboratory in Washington, D. C. Results are just being obtained at the writing of this chapter. They are expected to extend the output of the earlier Taylor and Yates measurements with laser extinction data and high resolution, long-path atmospheric transmission spectra. For a descriptive paper, see Reference [5-111].

†This section essentially reproduces material prepared by Gilbert N. Plass, Texas A&M University, College Station, TX, and Harold Yates, National Environmental Satellite Center, NOAA, Washington, DC, and originally published in the 1965 *Handbook of Military Infrared Technology*.

Table 5-22. Summary of Atmospheric Transmission Field Measurements

Wavelength Range (μm)	Resolution (cm⁻¹)	Path Length (km)	Water Vapor Content (pr cm)	Altitude (km)	Notes	Originator	Reference
1 to 14	5	2.26, 4.4	1.7	0.03	Visual Range 14.5 km	Gebbie, et al. (1951)	[5-80]
0.35 to 10	30	0.05 to 1.2	—	0	Fog, haze transmittance	Arnulf, et al. (1957)	[5-81]
0.5 to 15	3 to 60	0.3, 5.5, 16.25	11 to 38	0	Over water	Taylor and Yates (1957)	[5-82]
0.5 to 15	3 to 60	27.7	8 to 20	0.3	Horizontal paths	Yates (1957)	[5-83]
0.5 to 26	1 to 50	0.3	0.22 to 0.57	0 and 3	Horizontal paths	Yates and Taylor (1960)	[5-84]
3.7 to 5.5	5	0.77, 1.9, 8.4	0.24 to 1.9	5.2	Horizontal paths	Farmer, et al. (1963)	[5-85]
0.54 to 0.85	0.4	16.25	22	0	Relative Transmittance	Curcio, et al. (1964)	[5-86]
0.56 to 10.7	50	25	21.5 to 43.3	0	—	Streete (1968)	[5-87]
0.59 to 12	2.6 to 6	1.3, 2.6	0.047 to 3.5	0	Visual range 1.5 to 20 km	Filippov, et al. (1969)	[5-88]
1.9 to 5.4, 4 to 14	2.5 to 6	0.057 to 11	0.01 to 10	0 and 0.1	No visual range data	Ashley, et al. (1971 to 1972)	[5-89]

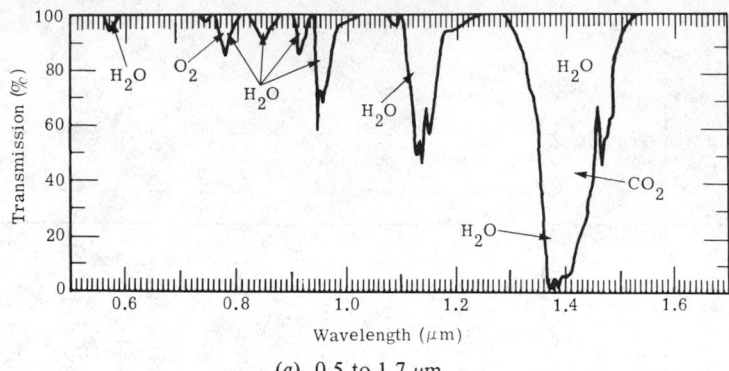

(*a*)　0.5 to 1.7 μm.

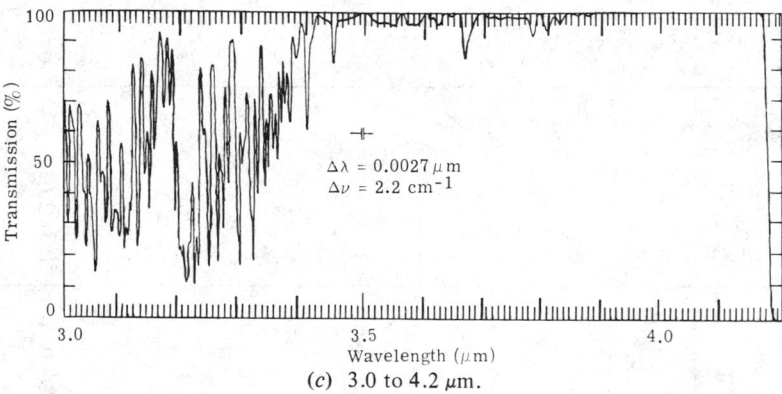

(*b*)　1.7 to 3.0 μm.

$\Delta\lambda = 0.0027\,\mu m$
$\Delta\nu = 2.2\ cm^{-1}$

(*c*)　3.0 to 4.2 μm.

Fig. 5-31.　Atmospheric transmission at sea level over a 0.3 km path.

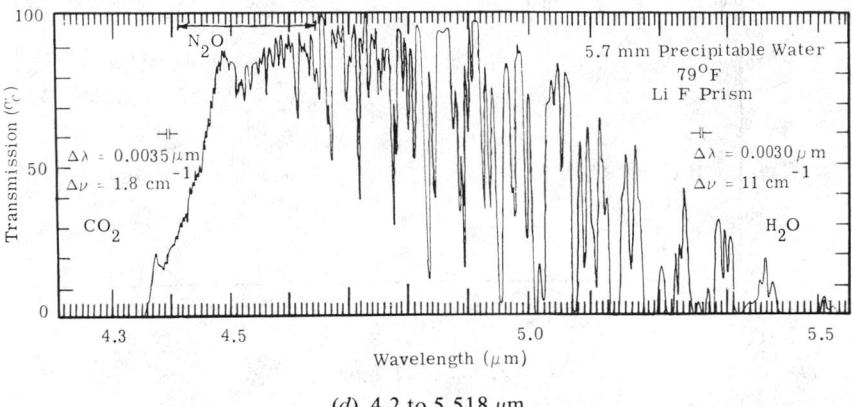

(*d*) 4.2 to 5.518 μm.

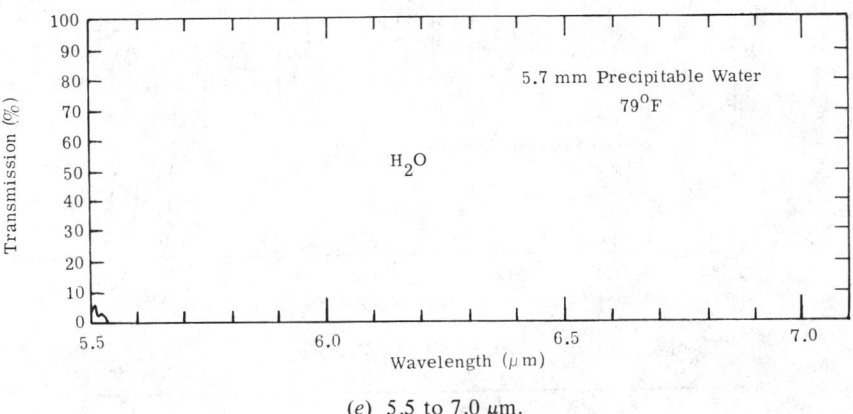

(*e*) 5.5 to 7.0 μm.

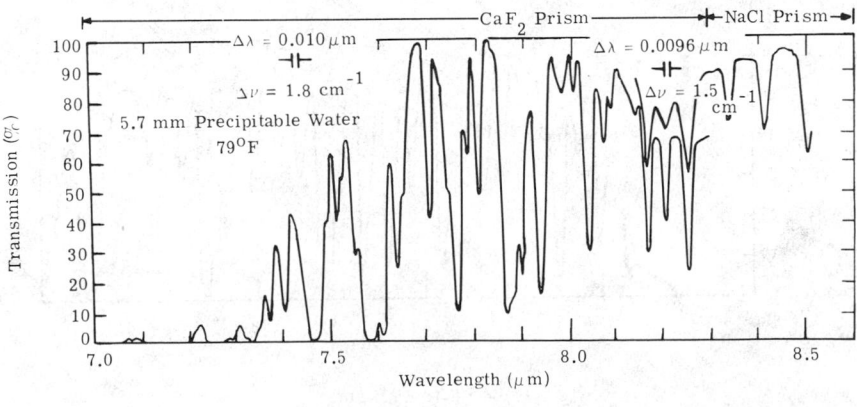

(*f*) 7.0 to 8.5 μm.

Fig. 5-31 (Continued). Atmospheric transmission at sea level over a 0.3 km path.

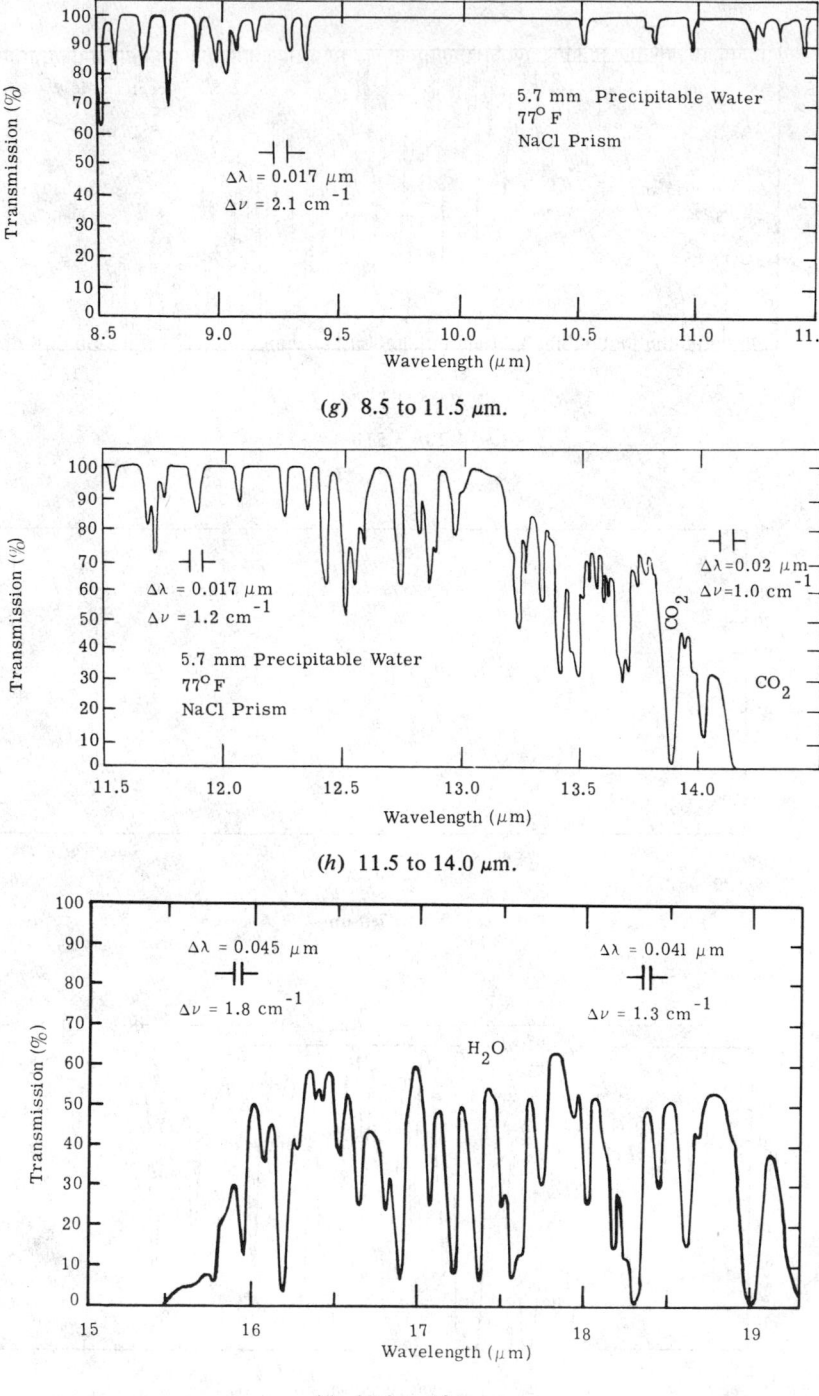

(g) 8.5 to 11.5 μm.

(h) 11.5 to 14.0 μm.

(i) 15.0 to 19.0 μm.

Fig. 5-31 (Continued). Atmospheric transmission at sea level over a 0.3 km path.

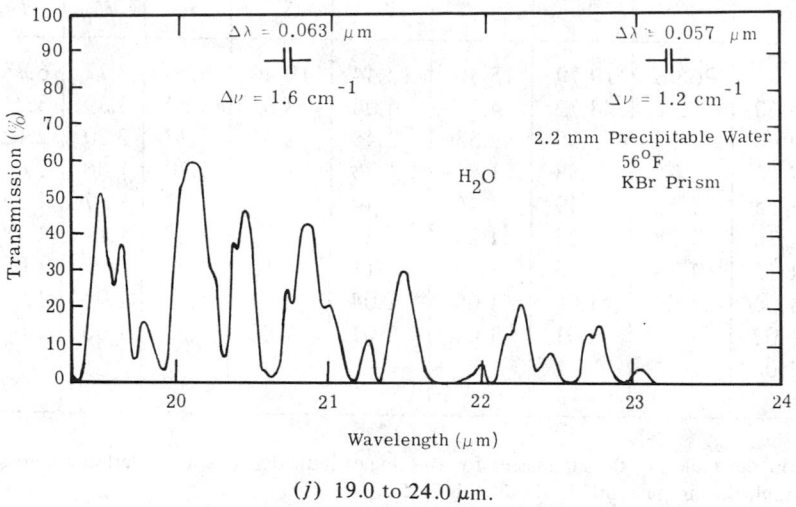

(*j*) 19.0 to 24.0 μm.

Fig. 5-31 (Continued). Atmospheric transmission at sea level over a 0.3 km path.

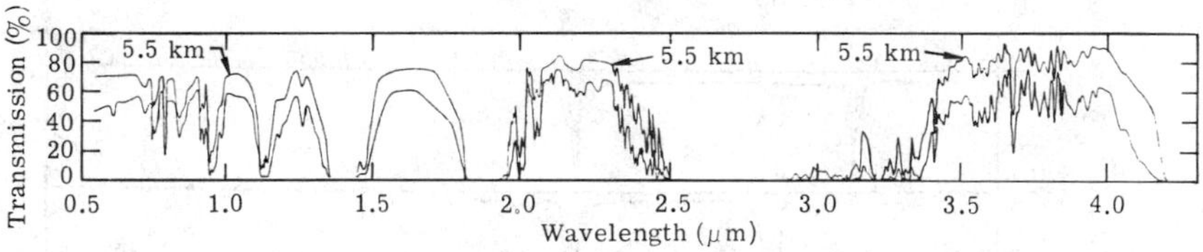

(*a*) 0.5 to 4.0 μm.

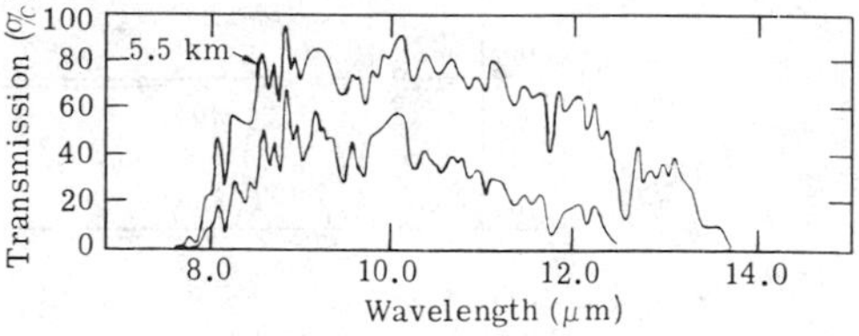

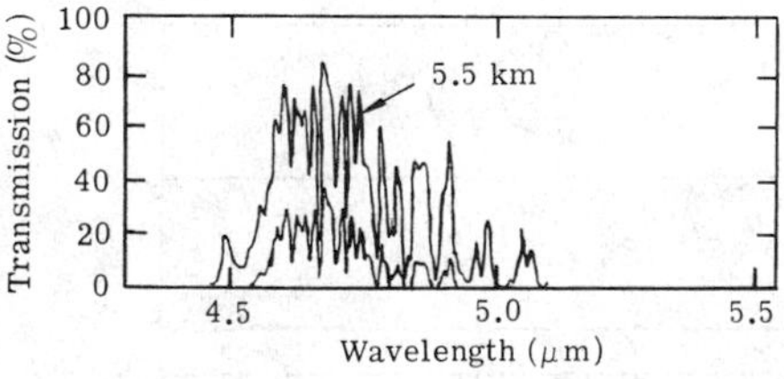

	5.5 km	16.25 km
R. H. (%)	51	53
Temp. (°F)	64	68
H_2O in Path (cm)	4.18	15.1
Transmission at 0.55 μm (%)	70	43

(*b*) 4.5 to 5.5 μm.

(*c*) 8.0 to 14.0 μm.

Fig. 5-32. Atmospheric transmission at sea level over 5.5 and 16.25 km paths.

Table 5-23. Equivalent Air Masses for Solar Altitudes 0-90°* [5-36]

	0°	1°	2°	3°	4°	5°	6°	7°	8°	9°
0°	–	26.96	19.79	15.36	12.44	10.40	8.90	7.77	6.88	6.18
10°	5.60	5.12	4.72	4.37	4.08	3.82	3.59	3.39	3.21	3.05
20°	2.90	2.77	2.65	2.55	2.45	2.36	2.27	2.20	2.12	2.06
30°	2.00	1.94	1.88	1.83	1.78	1.74	1.70	1.66	1.62	1.59
40°	1.55	1.52	1.49	1.46	1.44	1.41	1.39	1.37	1.34	1.32
50°	1.30	1.28	1.27	1.25	1.24	1.22	1.20	1.19	1.18	1.17
60°	1.15	1.14	1.13	1.12	1.11	1.10	1.09	1.09	1.08	1.07
70°	1.06	1.06	1.05	1.05	1.04	1.04	1.03	1.03	1.02	1.02
80°	1.02	1.01	1.01	1.01	1.01	1.00	1.00	1.00	1.00	1.00
90°	1.00	–	–	–	–	–	–	–	–	–

*Entries in the table are the air masses for the angles indicated down the left and across the top. For example the air mass for 22° elevation is 2.65.

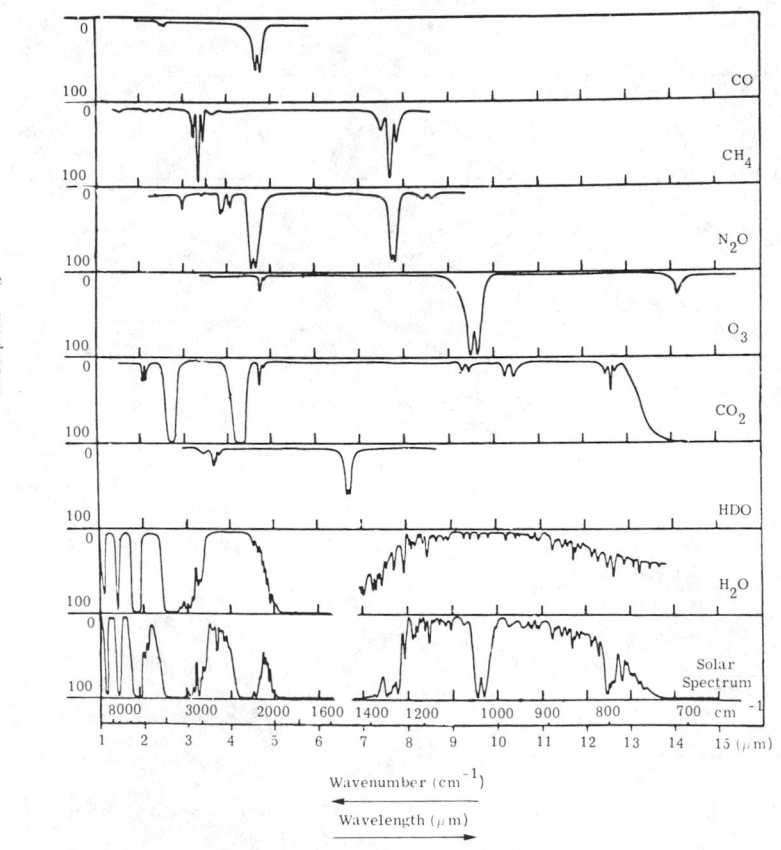

Fig. 5-33. Low resolution solar spectrum from 1.0 to 15.0 μm.

In the region from 2.80 to 3.15 μm (Figure 5-34) the absorption is due mainly to H_2O, although lines of a 13 CO_2 band, centered at 2.834 μm, of a $\nu_1 + \nu_3$ band of N_2O centered at 2.87 μm, and of a $2\nu_2 + \nu_3$ band of N_2O centered at 2.97 μm, are present. In the region of 3.15 to 3.50 μm, Figure 5-35, the strong fundamental band ν_3 of CH_4, centered at 3.31 μm, is present. The remaining absorption is due to the weaker $2\nu_2$ band of H_2O.

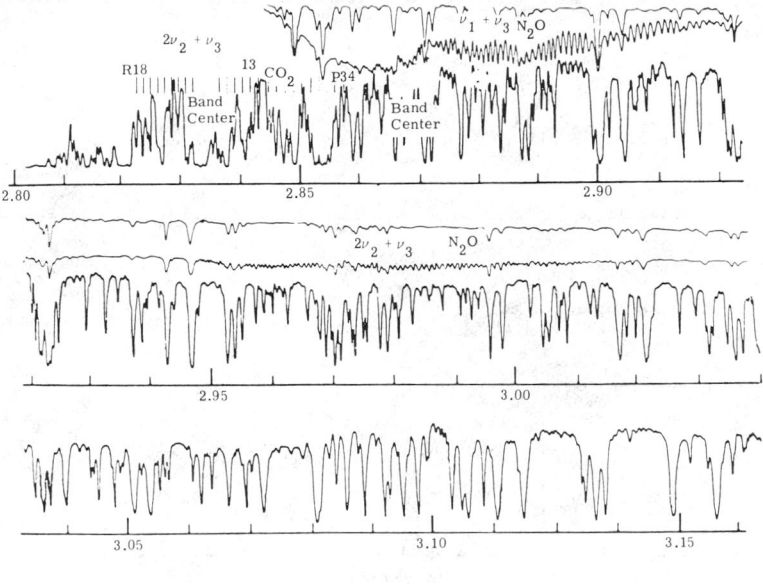

Fig. 5-34. Solar spectrum from 2.8 to 3.15 μm (lowest curve); laboratory spectrum of H_2O (top curve); and laboratory spectrum of N_2O (middle curve).

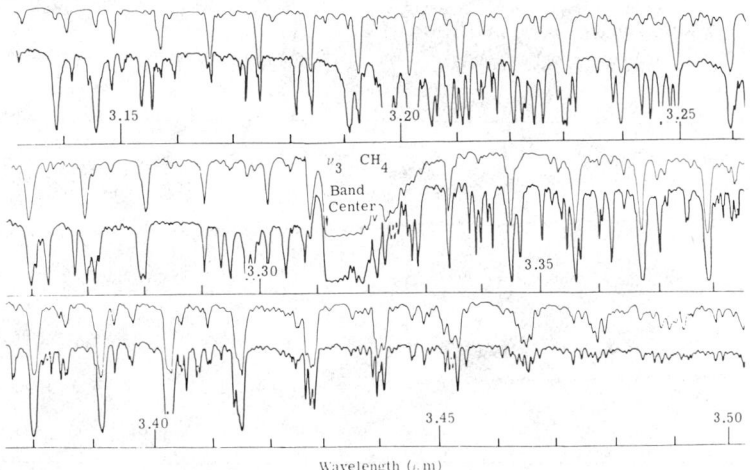

Fig. 5-35. Solar spectrum from 3.15 to 3.5 μm (lower curve) and laboratory spectrum of CH_4 (upper curve).

From 3.50 to 3.85 μm (Figure 5-36), many strong lines of the ν_1 fundamental band of HDO are present, in addition to some CH_4 absorption ($\nu_2 + \nu_4$ band at 3.55 μm) and a weak N_2O combination band ($\nu_2 + \nu_3$ at 3.57 μm). The Q-branch of the HDO band appears as a weak cluster of lines near 3.67 μm. All lines marked X in Figure 5-36 are due to HDO.

The region from 3.85 to 4.20 μm (Figure 5-37) contains $2\nu_1$ bands of N_2O at 3.91 μm and $\nu_1 + \nu_2$ bands of N_2 at 4.06 μm. It also contains the ν_3 fundamental of CO_2 beginning at 4.18 μm.

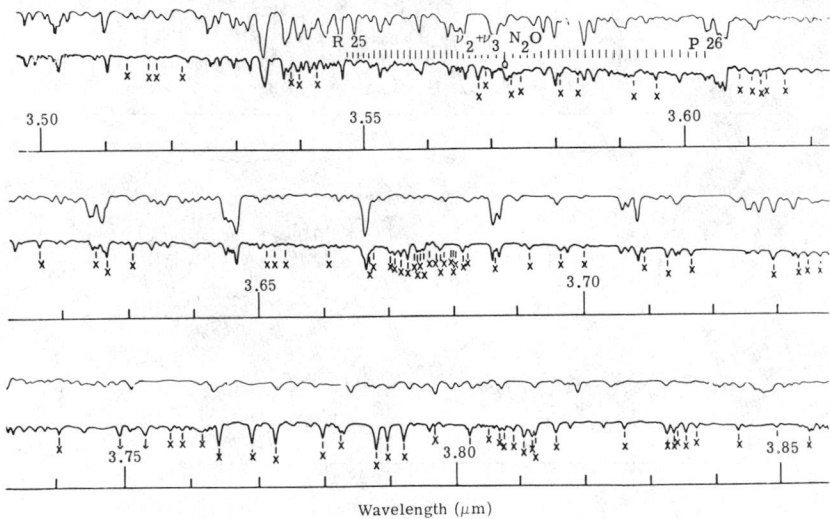

Fig. 5-36. Solar spectrum from 3.5 to 3.85 μm (lower curve) and laboratory spectrum of CH_4 (upper curve).

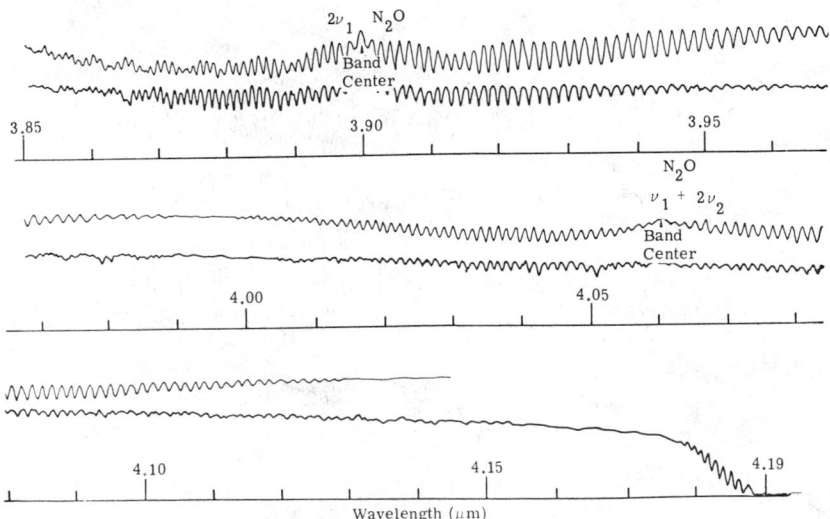

Fig. 5-37. Solar spectrum from 3.85 to 4.19 μm (lower curve) and laboratory spectrum of N_2O (upper curve).

Virtually complete absorption by CO_2 occurs in the region from 4.19 μm (Figure 5-37) to 4.45 μm (Figure 5-38). From 4.43 to 4.48 μm are some high-J lines in the P-branch of ν_3 band of $^{13}CO_2$, centered at 4.38 μm. The intense ν_3 fundamental band of N_2O is at 4.49 μm, with a weaker band $(\nu_3 + \nu_2) - \nu_2$ of N_2O centered near 4.52 μm.

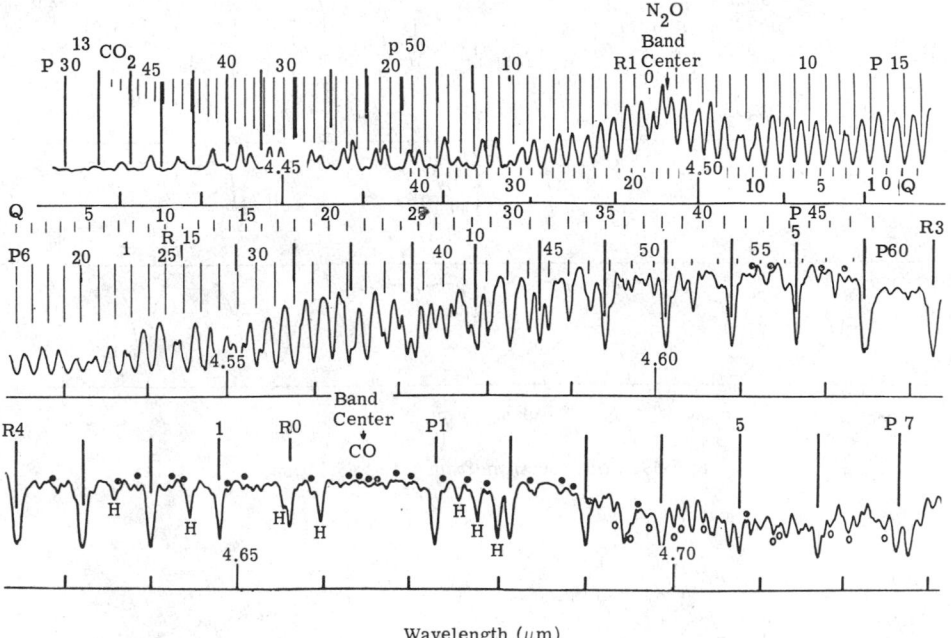

Wavelength (μm)

Fig. 5-38. Solar spectrum from 4.43 to 4.73 μm. The absorption structures of CO_2, N_2O and CO are shown schematically.

Near 4.66 μm (Figure 5-38) is the fundamental band of CO. Circles above the lines in Figure 5-38 indicate CO transitions of solar origin. The absorption at approximately 4.7 μm is due to ozone. The ozone absorption lines are indicated by circles below the spectrum. H_2O absorption lines between about 4.64 and 4.68 μm are indicated by H.

A weak CO_2 band is present near 4.8 μm (Figure 5-39) and a very weak CO_2 band near 5.2 μm (Figure 5-40). Beyond the 5.2 μm there is strong absorption from the edge of the ν_2 fundamental band of H_2O which is centered at 6.2 μm. Absorption by water vapor is complete between 5.5 and 6.9 μm. From 6.9 to 7.65 μm (Figure 5-41), the primary absorption is the edge of the 6.2 μm H_2O band, although absorption due to the ν_4 vibration-rotation band of CH_4, centered at 7.65 μm, is present. The region between 7.65 μm (Figure 5-42) and 9.0 μm (Figure 5-43) possesses relatively high transmission, except for the overtone band $2\nu_2$ of N_2O centered at 8.56 μm.

Figure 5-44 shows very intense absorption due to the ν_3 band of O_3 centered at 9.60 μm. The very faint structure observed between 8.90 μm (Figure 5-43) and 9.15 μm (Figure 5-44) is due to the very weak ν_1 band of O_3.

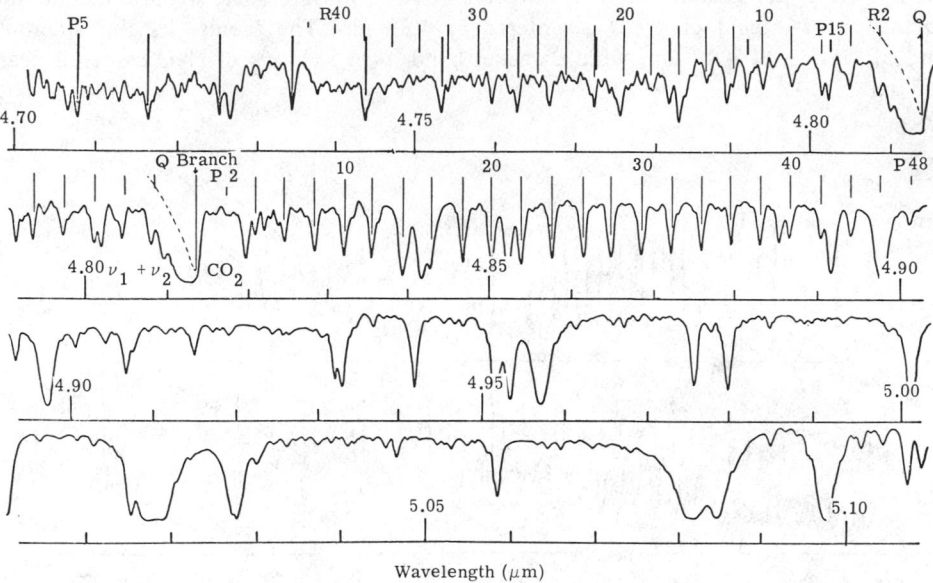

Fig. 5-39. Solar spectrum from 4.70 to 5.11 μm.

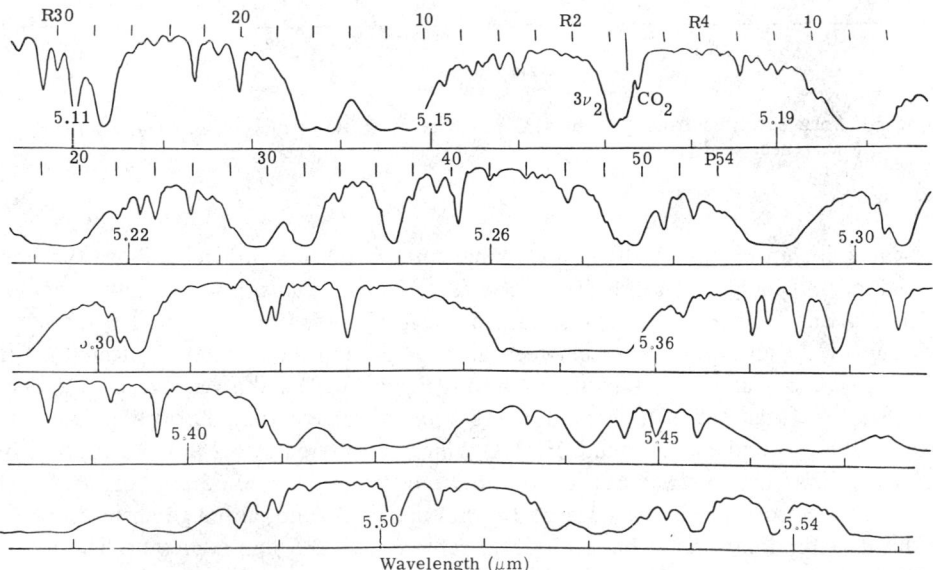

Fig. 5-40. Solar spectrum from 5.11 to 5.55 μm.

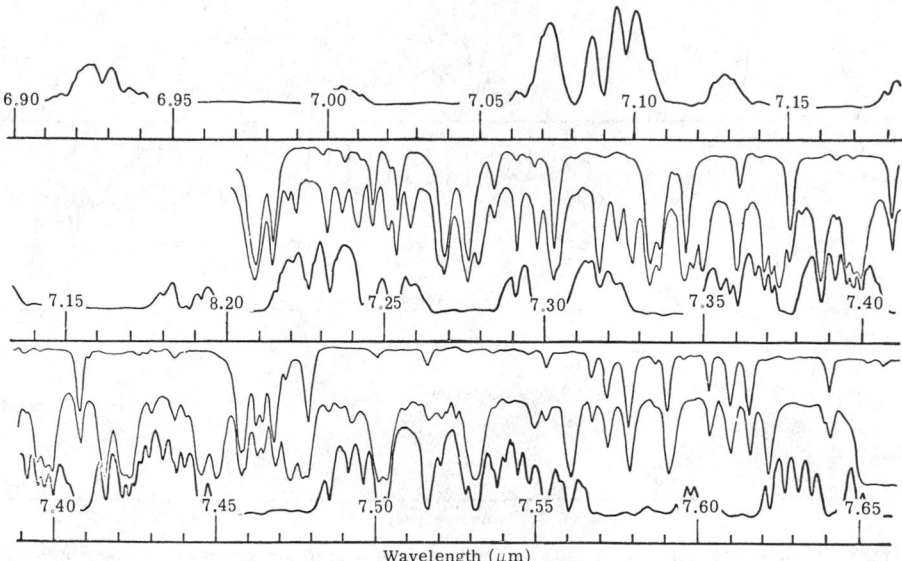

Fig. 5-41. Solar spectrum from 6.90 to 7.65 μm (lowest curve); laboratory spectrum of H_2O (top curve); and laboratory spectrum of CH_4 (middle curve).

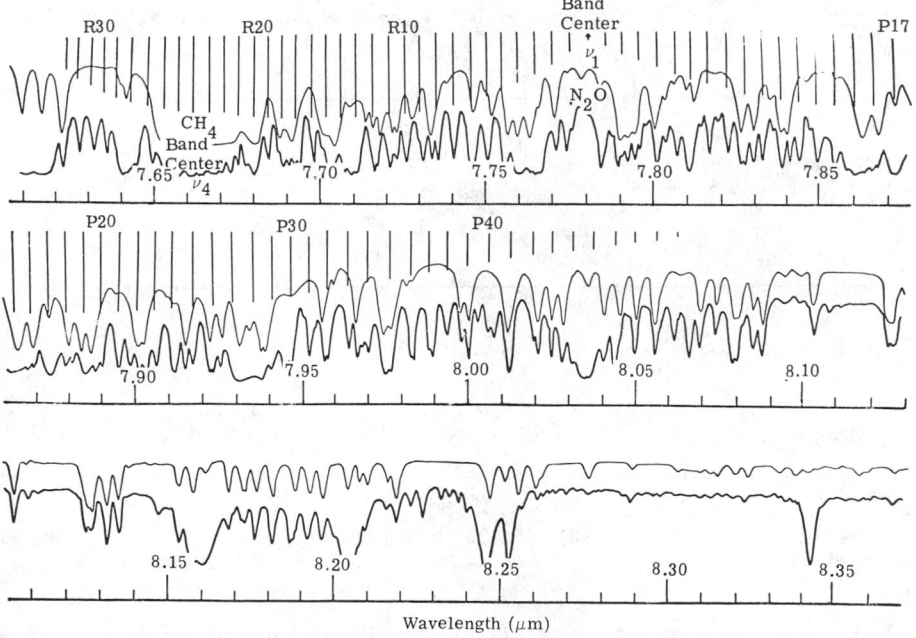

Fig. 5-42. Solar spectrum from 7.65 to 8.35 μm (lower curve); laboratory spectrum of CH_4 (upper curve); and absorption structure of N_2O indicated schematically.

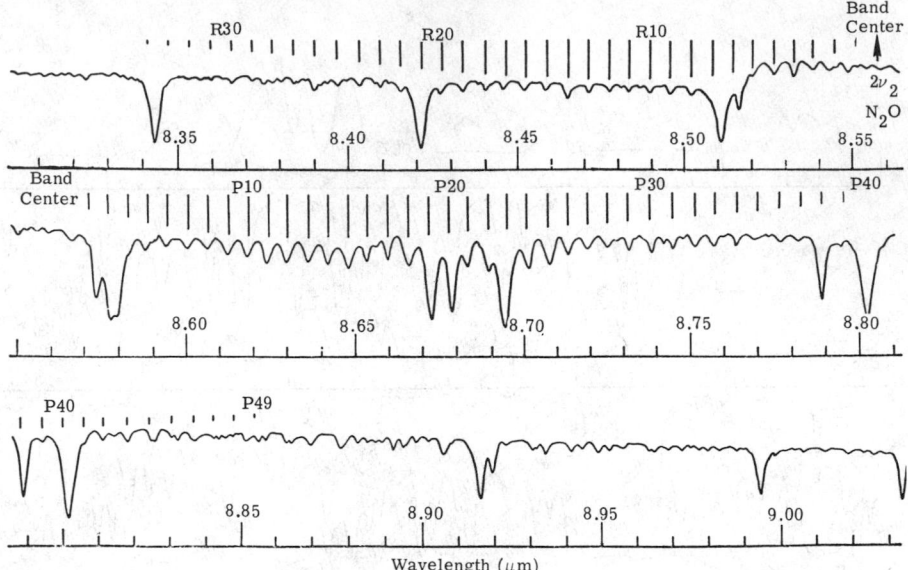

Fig. 5-43. Solar spectrum from 8.35 to 9.03 μm and absorption structure of N_2O indicated schematically.

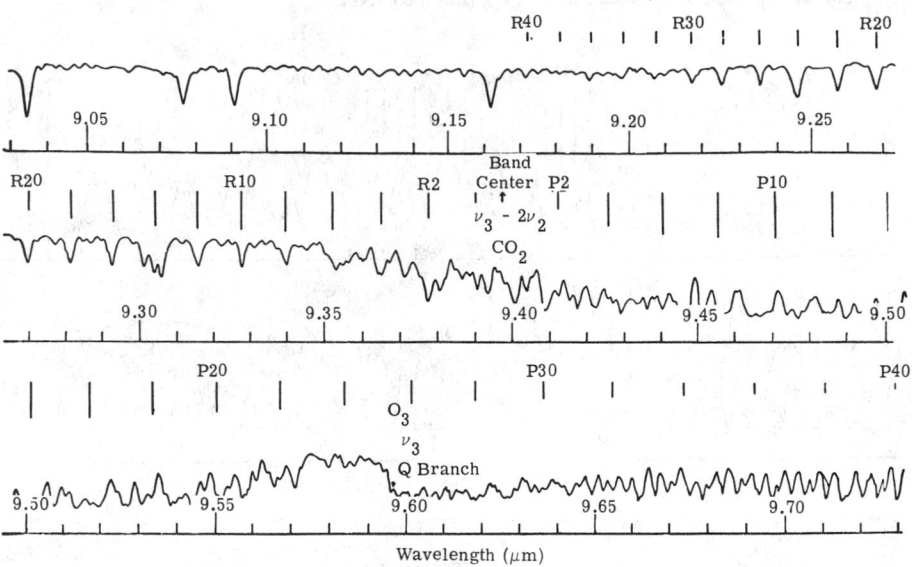

Fig. 5-44. Solar spectrum from 9.03 to 9.73 μm and absorption structure of CO_2 indicated schematically.

The $\nu_3 - 2\nu_2$ difference band of CO_2, centered at 9.4 μm, is also shown in Figure 5-44. In a difference band, the individual absorption line is not caused by a quantum transition from the ground state of the molecule to an excited level but rather to a transition from an excited level to a higher level. Because the strength of an absorption line depends very strongly on the population of the energy level from which the transition originates, and since this population decreases as temperature decreases, difference bands fade out very rapidly as temperature decreases. Thus, a band such as the 9.40 μm CO_2 band, although it may cause significant absorption near ground level where the ambient temperature may be 300 K, may have very little strength near 100,000 ft where the temperature is about 200 K.

The region from 9.75 to 10.6 μm (Figure 5-45) shows more of the structure of the 9.6 μm ozone band. It also shows another difference band of CO_2, $\nu_3 - \nu_1$, centered at 10.4 μm.

Relatively high transmission is present in the region from 12.2 to 13.25 μm (Figures 5-46 and 5-47). Figure 5-47 shows a CO_2 difference band, $\nu_1 - \nu_2$, centered near 13.2 μm and a much weaker CO_2 difference band, $(\nu_1 + \nu_2) - 2\nu_2$, centered near 12.6 μm.

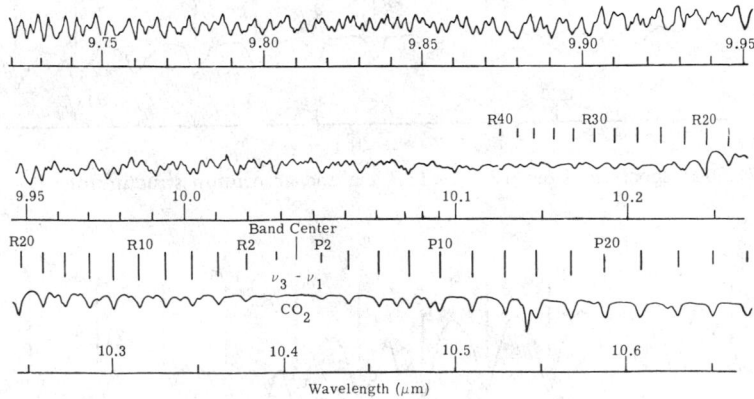

Fig. 5-45. Solar spectrum from 9.72 to 10.7 μm and absorption structure of CO_2 indicated schematically.

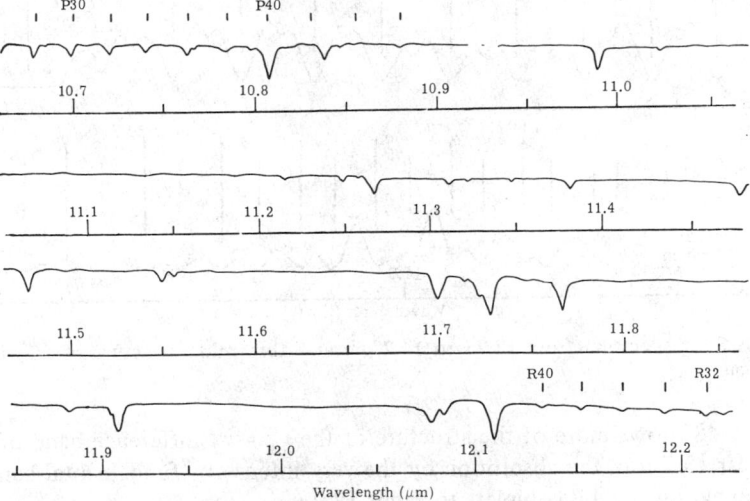

Fig. 5-46. Solar spectrum from 10.7 to 12.2 μm and absorption structure of CO_2 indicated schematically.

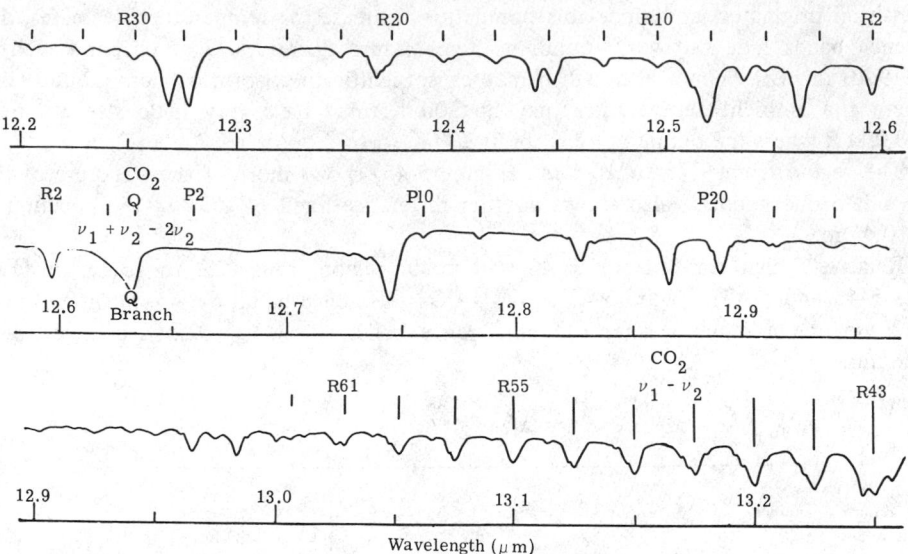

Fig. 5-47. Solar spectrum from 12.2 to 13.3 μm and absorption structure of CO_2 indicated schematically.

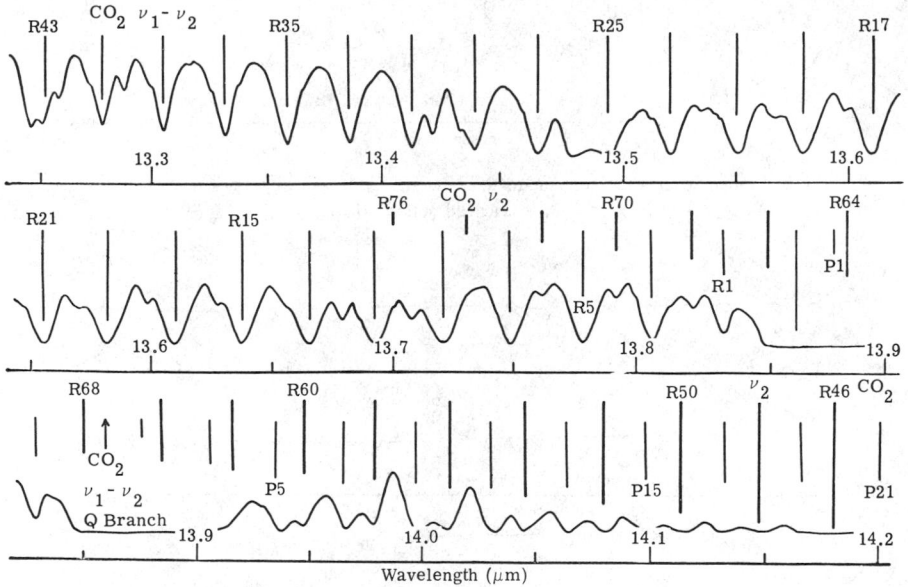

Fig. 5-48. Solar spectrum from 13.25 to 14.2 μm and absorption structure of CO_2 indicated schematically.

Figure 5-48 shows more of the structure of the $\nu_1 - \nu_2$ difference band of CO_2 and, starting near 13.7 μm, the absorption by the very intense ν_2 fundamental band of CO_2. Absorption by this band is complete to about 17.0 μm.

5.9.3. Satellite Data. Field data of terrestrial-atmospheric radiation are available from the results of the measurements obtained with the interferometer-spectrometer of Nimbus satellites. Data tapes are available from the NASA Goddard Space Flight Center.* With weather data obtained from the U.S. Air Force Air Weather Service, it was possible for Selby [5-106] to make calculations which compare with Nimbus data for the same or similar conditions. One such comparison with a LOWTRAN calculation is shown in Figure 5-49. The variability of the spectra obtained with Nimbus is sufficiently large, depending on variability in weather, that it is essential to know the exact weather conditions as input to a model calculation.

An example of the comparison between Nimbus and the calculation with the line-by-line method is shown in Figure 5-50.

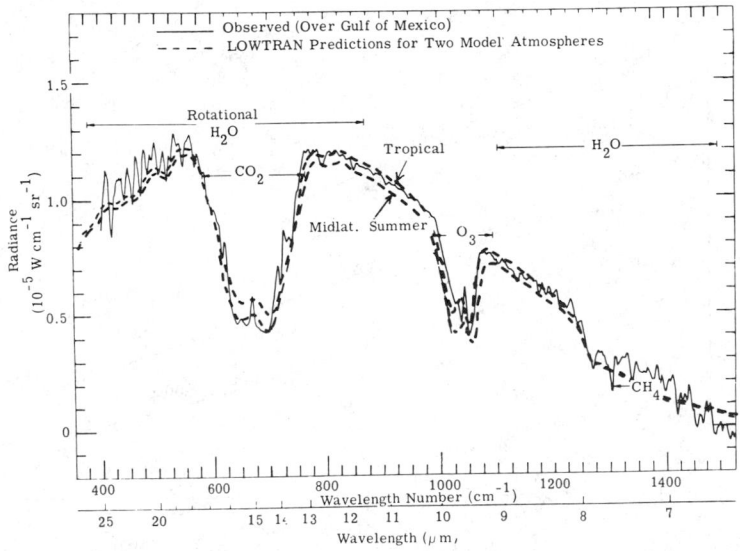

Fig. 5-49. Comparison of LOWTRAN 2 calculation with Nimbus data [5-106].

5.10. Physical Properties and Constituents of the Atmosphere

The two important factors to consider when making atmospheric transmittance calculations are what parameters affect the computation and how they vary in a given slant path. The solution to Equation (5-7) demands that the concentration of the absorber as a function of the path through which it acts and the path dependence of the absorption coefficient be known. The latter factor imposes the supplemental requirement of a knowledge of atmospheric temperature and pressure. For horizontal path calculations, all these quantities are usually taken to be constant with the values measured (or otherwise/ascertained) at the presumed location of the detector. For slant-path calculations, the quantities are usually assumed to follow the profile of some model atmosphere previously generated and published for general use.

5.10.1. Principal Constituents for Absorption and Scattering

Carbon Dioxide (CO_2). Carbon dioxide is chief among the fixed gases in absorption and emission of atmospheric radiation. Table 5-24 depicts the values of the concentrations

---

*Contact the Project Manager, Landsat/Nimbus Project, Goddard Space Flight Center, Greenbelt, MD.

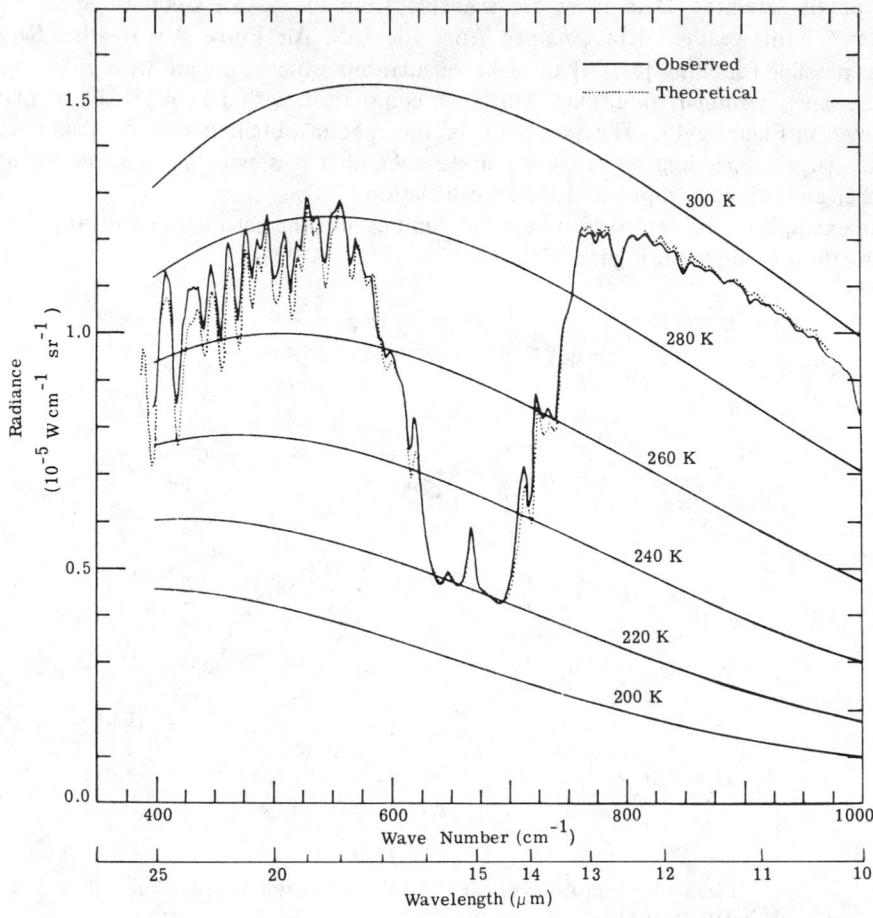

Fig. 5-50. 400 to 1000 cm^{-1} portion of the Nimbus spectrum. The dashed curve is calculated with line-by-line method [5-107].

Table 5-24. Concentration of Carbon Dioxide

Source	CO_2 Content (ppm)	Reference
Valley	314	[5-19]
Keeling	314	[5-90]
Callendar	320	[5-91]
Glueckauff	330	[5-92]
Fonselius, et al.	321 to 329	[5-93]
Bray	320	[5-94]
Average	321 ± 5	

of CO_2 obtained by various investigators. The concentration chosen for the Aggregate method is 320 ppm. The quoted value is about 325 ppm, uniformly mixed up to the stratopause where it is presumably destroyed by photochemical decomposition. The value used in the LOWTRAN model is shown in Table 5-25, along with the values for other fixed gases used in that model.

Table 5-25. Concentrations of Uniformly Mixed Gases [5-2, 5-19, 5-95 through 5-98]

Constituent	Molecular wt.	ppm by vol.	Vertical Path from Sea Level (cm atm)STP	Horizontal Path at Sea Level (cm atm)STP/km	$w^\dagger$ (g cm^{-2} mb^{-1})
Air	28.97	10^6	8×10^5	10^5	1.02
CO_2	44	330	264	33	5.11×10^{-4}
N_2O	44	0.28	0.22	0.028	4.34×10^{-7}
CO	28	0.075	0.06	0.0075	7.39×10^{-8}
CH_4	16	1.6	1.28	0.16	9.01×10^{-7}
O_2	32	2.095×10^5	1.68×10^5	2.095×10^4	0.236

$\dagger$ Absorber amount in a vertical increment of one millibar.

Nitrous Oxide (N_2O). N_2O is considered to have a fractional volume abundance of 2.5×10^{-7}. Table 5-26 shows the values obtained by different investigators. The value chosen for both the Aggregate method and the LOWTRAN method is 0.28 ppm (Table 5-25). The measurements giving these results were made at ground level; the mean value for a range of altitudes from 5.0 to 13.0 km is 0.14 ± 0.04 ppm [5-51].

Table 5-26. Concentration of Nitrous Oxide [5-1]

Date	Location	Content (ppm)	Author	Reference
1941	Arizona	0.38	Adel	[5-99]
1948	England	1.25	Shaw, Sutherland, and Wormell	[5-100]
1948	Michigan	0.5	McMath and Goldberg	[5-101]
1950	Texas	0.5	Slobod and Krogh	[5-102]
1957	Chesapeake Bay	0.43	Birkeland, Burch, and Shaw	[5-103]
1957	Ohio	0.28	Birkeland	[5-104]
1959	Ohio	0.28	Bowman	[5-105]
1976	–	0.27	U.S. Standard Atmosphere	[5-113]

Carbon Monoxide (CO). As with the other fixed gases, CO is considered to be uniformaly mixed vertically up to tropopause, above which it is oxidized by CO_2. The concentration then increases higher in the stratosphere and in the mesosphere. There is no strong indication that there are significant long-term increases in CO, partially because short-term temporal and spatial variations tend to conceal whatever long-term increases might occur. Laulainen [5-52] suggests that natural production and destruction of CO are sufficient to override manproduced causes. The variability from different regions causes a range of concentrations from 0.01 to 0.20 ppm in remote locations and from 0.4 to 2.2 ppm in urban areas. The fixed value in the Aggregate method is 0.12 ppm, whereas, as shown in Table 5-25, LOWTRAN uses 0.075 ppm.

Methane (CH_4). The volume mixing ratio of methane is almost 1.6 ppm [5-52] which is approximately constant with altitude up to the tropopause. It then decreases rapidly in the lower stratosphere. The value used in the Aggregate method is 1.1 ppm and in the LOWTRAN method, Table 5-25, is 1.6 ppm.

Nitric Acid (HNO_3). The values used in the Aggregate method for nitric acid, discovered recently in the stratosphere [5-53], are shown as a curve of mixing ratio versus altitude in Figure 5-51. More recent balloon-flight measurements show distributions with altitude produced as in Figure 5-52 [5-54] where the absorber amounts corresponding to the measured data are shown with the curves.

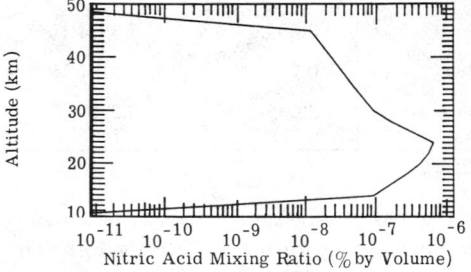

Fig. 5-51. Model atmospheres—nitric acid mixing ratio [5-20].

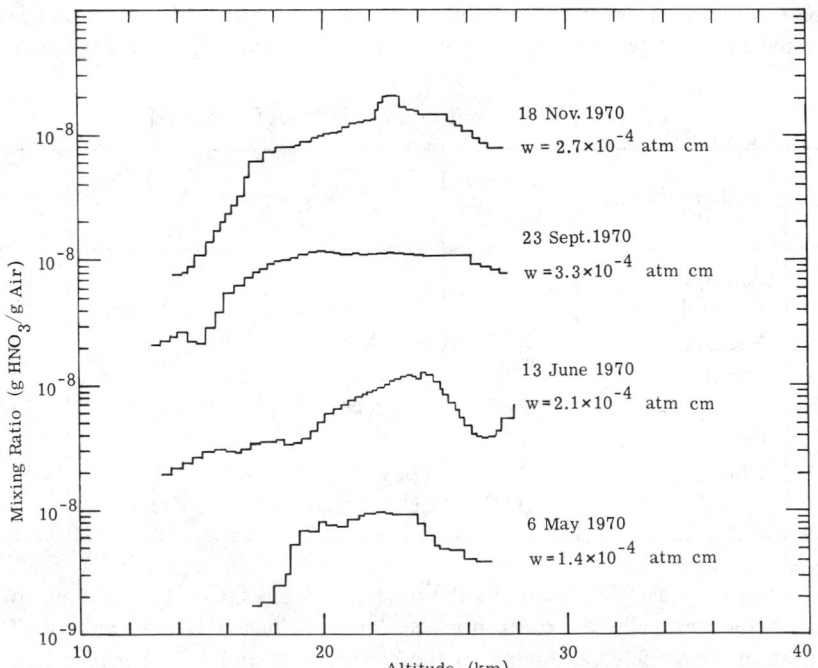

Fig. 5-52. Distribution of HNO_3 with altitude as determined from atmospheric emission data [5-55].

Water Vapor (H$_2$O). Water vapor is the atmosphere's most variable gaseous constituent from the standpoint of atmospheric transmittance. Laulainen [5-52] quotes the variability as 10^{-2} to 10^{-5} in the mixing ratio, with a diminution as a result of condensation and precipitation of about 1/3 for every 2.0 km, reaching the value of roughly 3×10^{-6} in the stratosphere. Table 5-27 gives the profiles of water vapor in conjunction with the regional-temporal profiles of pressure, temperature, and air density, corresponding to the following climates: tropical, midlatitude summer, midlatitude winter, subarctic summer, and subarctic winter, as functions of altitude. Values for the water vapor profile are shown in the fifth column. Results of some measurements with balloon-borne instrumentation are shown in Figure 5-53. A summary article comparing the values of various measurements has concluded that the 30 mbar H$_2$O-vapor mixing-ratio is approximately 6.0 to 8.0 ppm in the high latitudes of the northern hemisphere over populated areas [5-56].

Ozone (O$_3$). Ozone is another highly variable atmospheric constituent, with respect to both the time of year and geographical location. The concentration peaks at about 30 km, where the mixing ratio is better than 10.0 ppm [5-52].

The Aggregate and LOWTRAN methods use the tabulated values of the separate regional climates from Reference [5-19]. These are shown as the sixth column in Table 5-27. The results of rocket soundings made from latitudes 58°S to 64°N are reproduced in Figure 5-54, showing variabilities in the measurements [5-57].

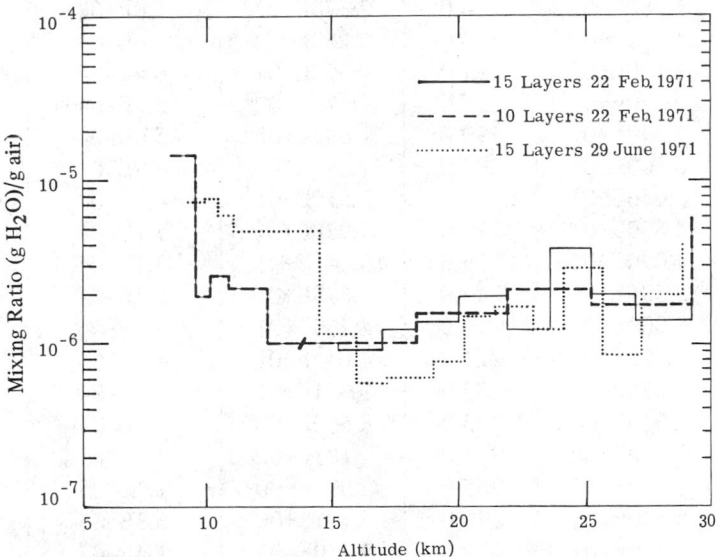

Fig. 5-53. Mixing ratio of water vapor as derived from a 15-layer and a 10-layer calculation using the 25 μm line group for the 22 February 1971 balloon flight. The slash near 14 km indicates the boundary between two layers with similar mixing ratio. The mixing ratio derived from a 15 layer calculation using the 25 μm line group for the 29 June 1971 balloon flight is shown by the dotted line [5-55].

Table 5-27. Model Atmospheres Used as a Basis of the
Computation of Atmospheric Optical Properties [5-2]

Ht (km)	Pressure (mb)	Temp (K)	Density (g m^{-3})	Water Vapor (g m^{-3})	Ozone (g m^{-3})
(a) Tropical					
0	1.013E+03	300.0	1.167E+03	1.9E+01	5.6E−05
1	9.040E+02	294.0	1.064E+03	1.3E+01	5.6E−05
2	8.050E+02	288.0	9.689E+02	9.3E+00	5.4E−05
3	7.150E+02	284.0	8.756E+02	4.7E+00	5.1E−05
4	6.330E+02	277.0	7.951E+02	2.2E+00	4.7E−05
5	5.590E+02	270.0	7.199E+02	1.5E+00	4.5E−05
6	4.920E+02	264.0	6.501E+02	8.5E−01	4.3E−05
7	4.320E+02	257.0	5.855E+02	4.7E−01	4.1E−05
8	3.780E+02	250.0	5.258E+02	2.5E−01	3.9E−05
9	3.290E+02	244.0	4.708E+02	1.2E−01	3.9E−05
10	2.860E+02	237.0	4.202E+02	5.0E−02	3.9E−05
11	2.470E+02	230.0	3.740E+02	1.7E−02	4.1E−05
12	2.130E+02	224.0	3.316E+02	6.0E−03	4.3E−05
13	1.820E+02	217.0	2.929E+02	1.8E−03	4.5E−05
14	1.560E+02	210.0	2.578E+02	1.0E−03	4.5E−05
15	1.320E+02	204.0	2.260E+02	7.6E−04	4.7E−05
16	1.110E+02	197.0	1.972E+02	6.4E−04	4.7E−05
17	9.370E+01	195.0	1.676E+02	5.6E−04	6.9E−05
18	7.890E+01	199.0	1.382E+02	5.0E−04	9.0E−05
19	6.660E+01	203.0	1.145E+02	4.9E−04	1.4E−04
20	5.650E+01	207.0	9.515E+01	4.5E−04	1.9E−04
21	4.800E+01	211.0	7.938E+01	5.1E−04	2.4E−04
22	4.090E+01	215.0	6.645E+01	5.1E−04	2.8E−04
23	3.500E+01	217.0	5.618E+01	5.4E−04	3.2E−04
24	3.000E+01	219.0	4.763E+01	6.0E−04	3.4E−04
25	2.570E+01	221.0	4.045E+01	6.7E−04	3.4E−04
30	1.220E+01	232.0	1.831E+01	3.6E−04	2.4E−04
35	6.000E+00	243.0	8.600E+00	1.1E−04	9.2E−05
40	3.050E+00	254.0	4.181E+00	4.3E−05	4.1E−05
45	1.590E+00	265.0	2.097E+00	1.9E−05	1.3E−05
50	8.540E−01	270.0	1.101E+00	6.3E−06	4.3E−06
70	5.790E−01	219.0	9.210E−02	1.4E−07	8.6E−08
100	3.000E−04	210.0	5.000E−04	1.0E−09	4.3E−11

Table 5-27. Model Atmospheres Used as a Basis of the
Computation of Atmospheric Optical Properties [5-2]
(*Continued*)

Ht (km)	Pressure (mb)	Temp (K)	Density (g m^{-3})	Water Vapor (g m^{-3})	Ozone (g m^{-3})
		(b) Midlatitude Summer			
0	1.013E+03	294.0	1.191E+03	1.4E+01	6.0E-05
1	9.020E+02	290.0	1.080E+03	9.3E+00	6.0E-05
2	8.020E+02	285.0	9.757E+02	5.9E+00	6.0E-05
3	7.100E+02	279.0	8.846E+02	3.3E+00	6.2E-05
4	6.280E+02	273.0	7.998E+02	1.9E+00	6.4E-05
5	5.540E+02	267.0	7.211E+02	1.0E+00	6.6E-05
6	4.870E+02	261.0	6.487E+02	6.1E-01	6.9E-05
7	4.260E+02	255.0	5.830E+02	3.7E-01	7.5E-05
8	3.720E+02	248.0	5.225E+02	2.1E-01	7.9E-05
9	3.240E+02	242.0	5.669E+02	1.2E-01	8.6E-05
10	2.810E+02	235.0	4.159E+02	6.4E-02	9.0E-05
11	2.430E+02	229.0	3.693E+02	2.2E-02	1.1E-04
12	2.090E+02	222.0	3.269E+02	6.0E-03	1.2E-04
13	1.790E+02	216.0	2.882E+02	1.8E-03	1.5E-04
14	1.530E+02	216.0	2.464E+02	1.0E-03	1.8E-04
15	1.300E+02	216.0	2.104E+02	7.6E-04	1.9E-04
16	1.110E+02	216.0	1.797E+02	6.4E-04	2.1E-04
17	9.500E+01	216.0	1.535E+02	5.6E-04	2.4E-04
18	8.120E+01	216.0	1.305E+02	5.0E-04	2.8E-04
19	6.950E+01	217.0	1.110E+02	4.9E-04	3.2E-04
20	5.950E+01	218.0	9.453E+01	4.5E-04	3.4E-04
21	5.100E+01	219.0	8.056E+01	5.1E-04	3.6E-04
22	4.370E+01	220.0	6.872E+01	5.1E-04	3.6E-04
23	3.760E+01	222.0	5.867E+01	5.4E-04	3.4E-04
24	3.220E+01	223.0	5.014E+01	6.0E-04	3.2E-04
25	2.770E+01	224.0	4.288E+01	6.7E-04	3.0E-04
30	1.320E+01	234.0	1.971E+01	3.6E-04	2.0E-04
35	6.520E+00	245.0	9.264E+00	1.1E-04	9.2E-05
40	3.330E+00	258.0	4.505E+00 *	4.3E-05	4.1E-05
45	1.760E+00	270.0	2.268E+00	1.9E-05	1.3E-05
50	9.510E-01	276.0	1.202E+00	6.3E-06	4.3E-06
70	6.710E-02	218.0	1.071E-01	1.4E-07	8.6E-08
100	3.000E-04	210.0	5.000E-04	1.0E-09	4.3E-11

*Original numbers have been corrected.

Table 5-27. Model Atmospheres Used as a Basis of the
Computation of Atmospheric Optical Properties [5-2]
(*Continued*)

Ht (km)	Pressure (mb)	Temp (K)	Density (g m^{-3})	Water Vapor (g m^{-3})	Ozone (g m^{-3})
(c) Subarctic Winter					
0	1.013E+03	257.1	1.372E+03	1.2E+00	4.1E-05
1	8.878E+02	259.1	1.193E+03	1.2E+00	4.1E-05
2	7.775E+02	255.9	1.058E+03	9.4E-01	4.1E-05
3	6.798E+02	252.7	9.366E+02	6.8E-01	4.3E-05
4	5.932E+02	247.7	8.339E+02	4.1E-01	4.5E-05
5	5.158E+02	240.9	7.457E+02	2.0E-01	4.7E-05
6	4.467E+02	234.1	6.646E+02	9.8E-02	4.9E-05
7	3.853E+02	227.3	5.904E+02	5.4E-02	7.1E-05
8	3.308E+02	220.6	5.226E+02	1.1E-02	9.0E-05
9	2.829E+02	217.2	4.538E+02	8.4E-03	1.6E-04
10	2.418E+02	217.2	3.879E+02	5.5E-03	2.4E-04
11	2.067E+02	217.2	3.315E+02	3.8E-03	3.2E-04
12	1.766E+02	217.2	2.834E+02	2.6E-03	4.3E-04
13	1.510E+02	217.2	2.422E+02	1.8E-03	4.8E-04
14	1.291E+02	217.2	2.071E+02	1.0E-03	4.9E-04
15	1.103E+02	217.2	1.770E+02	7.6E-04	5.6E-04
16	9.431E+01	216.6	1.517E+02	6.4E-04	6.2E-04
17	8.058E+01	216.0	1.300E+02	5.6E-04	6.2E-04
18	6.882E+01	215.4	1.113E+02	5.0E-04	6.2E-04
19	5.875E+01	214.8	9.529E+01	4.9E-04	6.0E-04
20	5.014E+01	214.1	8.155E+01	4.5E-04	5.6E-04
21	4.277E+01	213.6	6.976E+01	5.1E-04	5.1E-04
22	3.647E+01	213.0	5.966E+01	5.1E-04	4.7E-04
23	3.109E+01	212.4	5.100E+01	5.4E-04	4.3E-04
24	2.649E+01	211.8	4.358E+01	6.0E-04	3.6E-04
25	2.256E+01	211.2	3.722E+01	6.7E-04	3.2E-04
30	1.020E+01	216.0	1.645E+01	3.6E-04	1.5E-04
35	4.701E+00	222.2	7.368E+00	1.1E-04	9.2E-05
40	2.243E+00	234.7	3.330E+00	4.3E-05	4.1E-05
45	1.113E+00	247.0	1.569E+00	1.9E-05	1.3E-05
50	5.719E-01	259.3	7.682E-01	6.3E-06	4.3E-06
70	4.016E-02	245.7	5.695E-02	1.4E-07	8.6E-08
100	3.000E-04	210.0	5.000E-04	1.0E-09	4.3E-11

Table 5-27. Model Atmospheres Used as a Basis of the
Computation of Atmospheric Optical Properties [5-2]
(*Continued*)

Ht (km)	Pressure (mb)	Temp (K)	Density (g m^{-3})	Water Vapor (g m^{-3})	Ozone (g m^{-3})
(d) Subarctic Summer					
0	1.010E+03	287.0	1.220E+03	9.1E+00	4.9E−05
1	8.960E+02	282.0	1.110E+03	6.0E+00	5.4E−05
2	7.929E+02	276.0	9.971E+02	4.2E+00	5.6E−05
3	7.000E+02	271.0	8.985E+02	2.7E+00	5.8E−05
4	6.160E+02	266.0	8.077E+02	1.7E+00	6.0E−05
5	5.410E+02	260.0	7.224E+02	1.0E+00	6.4E−05
6	4.730E+02	253.0	6.519E+02	5.4E−01	7.1E−05
7	4.130E+02	246.0	5.849E+02	2.9E−01	7.5E−05
8	3.590E+02	239.0	5.231E+02	1.3E−02	7.9E−05
9	3.107E+02	232.0	4.663E+02	4.2E−02	1.1E−04
10	2.677E+02	225.0	4.142E+02	1.5E−02	1.3E−04
11	2.300E+02	225.0	3.559E+02	9.4E−03	1.8E−04
12	1.977E+02	225.0	3.059E+02	6.0E−03	2.1E−04
13	1.700E+02	225.0	2.630E+02	1.8E−02	2.6E−04
14	1.460E+02	225.0	2.260E+02	1.0E−03	2.8E−04
15	1.250E+02	225.0	1.943E+02	7.6E−04	3.2E−04
16	1.080E+02	225.0	1.671E+02	6.4E−04	3.4E−04
17	9.280E+01	225.0	1.436E+02	5.6E−04	3.9E−04
18	7.980E+01	225.0	1.235E+02	5.0E−04	4.1E−04
19	6.860E+01	225.0	1.062E+02	4.9E−04	4.1E−04
20	5.890E+01	225.0	9.128E+01	4.5E−04	3.9E−04
21	5.070E+01	225.0	7.849E+01	5.1E−04	3.6E−04
22	4.360E+01	225.0	6.750E+01	5.1E−04	3.2E−04
23	3.750E+01	225.0	5.805E+01	5.4E−04	3.0E−04
24	3.227E+01	226.0	4.963E+01	6.0E−04	2.8E−04
25	2.780E+01	228.0	4.247E+01	6.7E−04	2.6E−04
30	1.340E+01	235.0	1.979E+01	3.6E−04	1.4E−04
35	6.610E+00	247.0	9.320E+00	1.1E−04	9.2E−05
40	3.400E+00	262.0	4.526E+00	4.3E−05	4.1E−05
45	1.810E+00	274.0	2.314E+00 *	1.9E−05	1.3E−05
50	9.870E−01	277.0	1.240E+00	6.3E−06	4.3E−06
70	7.070E−02	216.0	1.137E−01	1.4E−07	8.6E−08
100	3.000E−04	210.0	5.000E−04	1.0E−09	4.3E−11

*Original numbers have been corrected.

Table 5-27. Model Atmospheres Used as a Basis of the
Computation of Atmospheric Optical Properties [5-2]
(*Continued*)

Ht (km)	Pressure (mb)	Temp (K)	Density (g m^{-3})	Water Vapor (g m^{-3})	Ozone (g m^{-3})
(e) Midlatitude Winter					
0	1.018E+03	272.2	1.301E+03	3.5E+00	6.0E-05
1	8.973E+02	268.7	1.162E+03	2.5E+00	5.4E-05
2	7.897E+02	265.2	1.037E+03	1.8E+00	4.9E-05
3	6.938E+02	261.7	9.230E+02	1.2E+00	4.9E-05
4	6.081E+02	255.7	8.282E+02	6.6E-01	4.9E-05
5	5.313E+02	249.7	7.411E+02	3.8E-01	5.8E-05
6	4.627E+02	243.7	6.614E+02	2.1E-01	6.4E-05
7	4.016E+02	237.7	5.886E+02	8.5E-02	7.7E-05
8	3.473E+02	231.7	5.222E+02	3.5E-02	9.0E-05
9	2.992E+02	225.7	4.619E+02	1.6E-02	1.2E-04
10	2.568E+02	219.7	4.072E+02	7.5E-03	1.6E-04
11	2.199E+02	219.2	3.496E+02	6.9E-03	2.1E-04
12	1.882E+02	218.7	2.999E+02	6.0E-03	2.6E-04
13	1.610E+02	218.2	2.572E+02	1.8E-03	3.0E-04
14	1.378E+02	217.7	2.206E+02	1.0E-03	3.2E-04
15	1.178E+02	217.2	1.890E+02	7.6E-04	3.4E-04
16	1.007E+02	216.7	1.620E+02	6.4E-04	3.6E-04
17	8.610E+01	216.2	1.388E+02	5.6E-04	3.9E-04
18	7.350E+01	215.7	1.188E+02	5.0E-04	4.1E-04
19	6.280E+01	215.2	1.017E+02	4.9E-04	4.3E-04
20	5.370E+01	215.2	8.690E+01	4.5E-04	4.5E-04
21	4.580E+01	215.2	7.421E+01	5.1E-04	4.3E-04
22	3.910E+01	215.2	6.338E+01	5.1E-04	4.3E-04
23	3.340E+01	215.2	5.415E+01	5.4E-04	3.9E-04
24	2.860E+01	215.2	4.624E+01	6.0E-04	3.6E-04
25	2.430E+01	215.2	3.950E+01	6.7E-04	3.4E-04
30	1.110E+01	217.4	1.783E+01	3.6E-04	1.9E-04
35	5.180E+00	227.8	7.924E+00	1.1E-04	9.2E-05
40	2.530E+00	243.2	3.625E+00	4.3E-05	4.1E-05
45	1.290E+00	258.5	1.741E+00	1.9E-05	1.3E-05
50	6.820E-01	265.7	8.954E-01	6.3E-06	4.3E-06
70	4.670E-02	230.7	7.051E-02	1.4E-07	8.6E-08
100	3.000E-04	210.2	5.000E-04	1.0E-09	4.3E-11

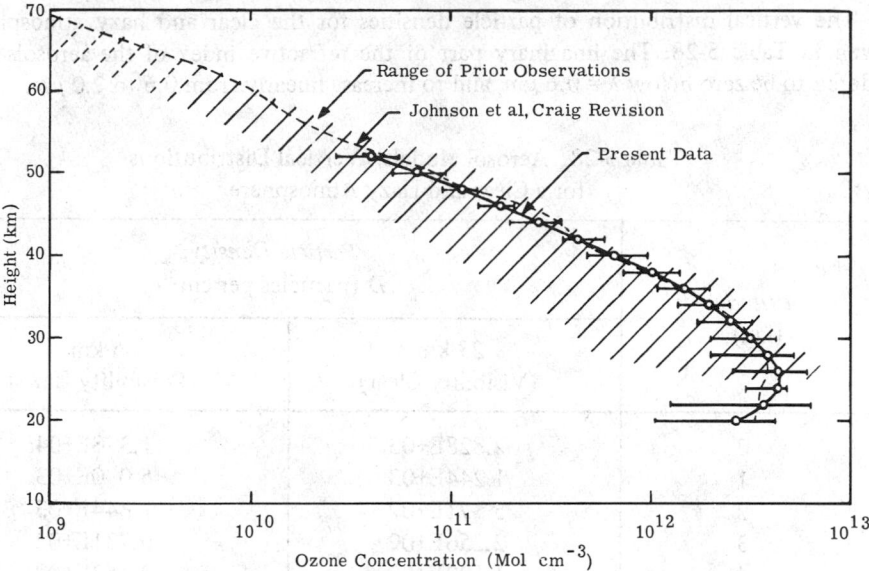

Fig. 5-54. Average and extreme range of vertical ozone distribution data [5-57].

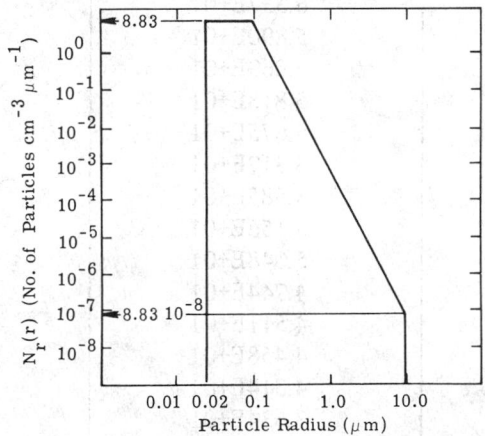

Fig. 5-55. Normalized particle size distribution for aerosol models [5-2].

Scattering Constituents. The LOWTRAN model considers two aerosol models describing a clear and hazy atmosphere corresponding to visibilities of 23.0 and 5.0 km respectively at ground level. The size distribution, considered to be the same for all altitudes, is that suggested by Diermendjian [5-31] for continental haze. It is shown in Figure 5-55 with the regions given by

$$N_r(r) = 8.33 \times 10^{-4}\, r^{-4} \text{ for } 0.1\ \mu m \leqslant r < 10\ \mu m$$

$$N_r(r) = 8.83 \text{ for } 0.02\ \mu m < r < 0.1\ \mu m$$

$$N_r(r) = 0 \text{ for } r < 0.02\ \mu m \text{ and } r > 10.0\ \mu m \qquad (5\text{-}100)$$

where r is the particle radius.

The vertical distribution of particle densities for the clear and hazy atmospheres is given in Table 5-28. The imaginary part of the refractive index of the aerosols is considered to be zero below $\lambda = 0.6$ μm, and to increase linearly from 0.6 to 2.0 μm.

Table 5-28. Aerosol Models: Vertical Distributions
for a Clear and Hazy Atmosphere

Altitude (km)	Particle Density, D (particles per cm^3)	
	23 km (Visibility Clear)	5 km (Visibility Hazy)
0	2.828E+03	1.378E+04
1	1.244E+03	5.030E+03
2	5.371E+02	1.844E+03
3	2.256E+02	6.731E+02
4	1.192E+02	2.453E+02
5	8.987E+01	8.987E+01
6	6.337E+01	6.337E+01
7	5.890E+01	5.890E+01
8	6.069E+01	6.069E+01
9	5.818E+01	5.818E+01
10	5.675E+01	5.675E+01
11	5.317E+01	5.317E+01
12	5.585E+01	5.585E+01
13	5.156E+01	5.156E+01
14	5.048E+01	5.048E+01
15	4.744E+01	4.744E+01
16	4.511E+01	4.511E+01
17	4.458E+01	4.458E+01
18	4.314E+01	4.314E+01
19	3.634E+01	3.634E+01
20	2.667E+01	2.667E+01
21	1.933E+01	1.933E+01
22	1.455E+01	1.455E+01
23	1.113E+01	1.113E+01
24	8.826E+00	8.826E+00
25	7.429E+00	7.429E+00
30	2.238E+00	2.238E+00
35	5.890E-01	5.890E-01
40	1.550E-01	1.550E-01
45	4.083E-02	4.082E-02
50	1.078E-02	1.078E-02
70	5.550E-05	5.550E-05
100	1.969E-08	1.969E-08

The values in Table 5-28 are adjusted to match the Elterman coefficients at 0.55 μm wavelength [5-58, 5-59]. The effective ground-level visible range is 23.0 km. The hazy model is the same as the clear model above 5.0 km, but it is increased exponentially below 5.0 km to yield as effective visible range at ground level to 5.0 km.

There is always an aerosol component in the atmosphere, even on very clear days, but the Rayleigh limit, with $k'_s = 1.162 \times 10^{-2}$ km^{-1} (where $k'_s = \rho k_s$), corresponds to a horizontal visual range of 336.0 km. On the other hand, for large extinction coefficients such as those for a fog, the visual range is only about 1.2 km.

5.11. References and Bibliography

5.11.1. References

[5-1]   A. J. LaRocca and R. E. Turner, "Atmospheric Transmittance and Radiance: Methods of Calculation," Environmental Research Institute of Michigan, Ann Arbor, MI, Report No. 107600-10-T.

[5-2]   R. A. McClatchey, et al., "Optical Properties of the Atmosphere," AFCRL-72-0497, Air Force Cambridge Research Laboratories, Bedford, MA, August 1972, Third Edition.

[5-3]   J. E. A. Selby, E. P. Shettle, and R. A. McClatchey, "Atmospheric Transmittance from 0.25 to 0.28 μm: Supplement LOWTRAN 3B (1976)," Air Force Geophysics Laboratory, Bedford, MA, AFGL-TR-76-0258, 1 November 1976.

[5-4]   D. E. Burch, D. A. Gryvnak, and E. B. Singleton et al., "Absorption by Carbon Dioxide, Water Vapor, and Minor Atmospheric Constituents," Ohio State University, OH, Report No. AFCRL 62-298, 1962.

[5-5]   J. C. Gille and R. G. Ellingson, "Correction of Random Exponential Band Transmissions for Doppler Effects," *Applied Optics*, Optical Society of America, Washington, DC, Vol. 7, No. 3, 1968, pp. 471-474.

[5-6]   W. M. Elsasser, "Mean Absorption and Equivalent Absorption Coefficient of a Band Spectrum," *Physical Review*, American Institute of Physics, New York, NY, Vol. 54, 1938, p. 126.

[5-7]   H. Mayer, Methods of Opacity Calculations, V. Effect of Lines on Opacity, Methods for Treating Line Contributions, Los Alamos Scientific Laboratory, Los Alamos, CA, Report No. AECD-1870, 1947.

[5-8]   R. M. Goody, "A Statistical Model for Water-Vapour Absorption," *Quarterly Journal of Royal Meteorological Society*, Royal Meteorological Society, Berkshire, England, Vol. 58, 1952, pp. 165-169.

[5-9]   G. N. Plass, "Models for Spectral Band Absorption," *Journal of Optical Society of America*, Optical Society of America, Washington, DC, Vol. 48, 1958, pp. 690-703.

[5-10]   P. J. Wyatt, V. R. Stull and G. N. Plass, "Quasi-Random Model of Band Absorption," *Journal of Optical Society of America*, Optical Society of America, Washington, DC, Vol. 52, No. 11, 1962, p. 1209.

[5-11]   R. Ladenburg and F. Reiche, "Über Selektive Absorption," *Annalen der Physik*, Johann Ambrosius Barth Verlag, Leipzig, E. Germany, Vol. 42, 1913, p. 181.

[5-12]   L. D. Kaplan, "Regions of Validity of Various Absorption-Coefficient Approximations," *Journal of Meteorology*, American Meteorological Society, Boston, MA, Vol. 10, 1953, pp. 100-104.

[5-13]   S. A. Golden, "The Doppler Analog of the Elsasser Band Model," *Journal of Quantitative Spectroscopy and Radiative Transfer*, Pergamon Press, Oxford, England, Vol. 7, 1967, pp. 483-494.

[5-14]　W. Malkmus, "Random Lorentz Band Model with Exponential-Tailed S^{-1} Line Intensity Distribution Function," *Journal of Optical Society of America*, Optical Society of America, Washington, DC, Vol. 57, No. 3, 1967, pp. 323-329.

[5-15]　P. J. Wyatt, V. R. Stull and G. N. Plass, "The Infrared Transmittance of Water Vapor," *Applied Optics,* Optical Society of America, Washington, DC, Vol. 3, No. 2, 1964; Aeronutronic Report U-1717, Aeronutronic Systems, Newport Beach, CA, 1962.

[5-16]　V. R. Stull, P. J. Wyatt, and G. N. Plass, "The Infrared Transmittance of Water Vapor," *Applied Optics*, Optical Society of America, Washington, DC, Vol. 3, No. 2, February 1964; Aeronutronic Report U-1718, Aeronutronic Systems, Newport Beach, CA, 1962.

[5-17]　S. R. Drayson, "Atmospheric Slant-Path Transmission in the 15μ CO_2 Band," The Institute of Science and Technology, The University of Michigan, Ann Arbor, MI, Report No. 05863-6-T, 1964.

[5-18]　R. E. Roberts, L. M. Biberman and J. E. A. Selby, "Infrared Continuum Absorption of Atmospheric Water Vapor in the 8-12 μm Window," Paper P-1184, Institute of Defense Analyses, Arlington, VA, April 1976.

[5-19]　S. L. Valley, *Handbook of Geophysics and Space Environments,* Air Force Cambridge Research Laboratories, L. G. Hanscom Field, Bedford, MA, 1965.

[5-20]　J. N. Hamilton, J. A. Rowe and D. Anding, "Atmospheric Transmission and Emission Program," Aerospace Corporation, El Segundo, CA, Report No. TOR-0073 (3050-02)-3, 1973.

[5-21]　J. N. Howard, D. Burch and D. Williams, "Near-Infrared Transmission Through Synthetic Atmospheres," Ohio State University Research Foundation, Columbus, OH, Geophysics Research Paper No. 40, Report No. AFCRL-TR-55-213, 1955.

[5-22]　D. E. Burch, D. A. Gryvnak, and J. D. Pembrook, "Investigation of Infrared Absorption by Nitrous Oxide from 4000 to 6000 cm^{-1} (2.5 to 1.5 μm)," Aeronutronics Report U-4393, AFCRL-71-0536, June 1971; "760 to 2380 cm^{-1} (13.2 to 4.2 μm)," Aeronutronics Report U-4995, AFCRL-71-0620, December 1971; Philco-Ford Corporation, Newport Beach, CA.

[5-23]　D. E. Burch, "Investigation of the Absorption of Infrared Radiation by Atmospheric Gases," Philco-Ford Corporation, Newport Beach, CA, Philco Report No. U-4784, 31 January 1970.

[5-24]　K. Bignell, The Water Vapor Infrared Continuum," *Quarterly Journal of the Royal Meteorological Society*, Royal Meteorological Society, Berkshire, England, Vol. 96, 1970, pp. 390-403.

[5-25]　R. A. McClatchey, et al., "AFCRL Atmospheric Absorption Line Parameters Compilation," Air Force Research Laboratories, L. G. Hanscom Field, Bedford, MA, Report No. AFCRL-TR-73-0096, 1973.

[5-26]　W. R. Bradford, T. M. McCormick and J. A. Selby, "Laboratory Representation of Atmospheric Paths for Infrared Absorption," EMI Electronics, Hayes, Middlesex, England, Report No. DMP 1431, 1963.

[5-27]　S. R. Drayson and C. Young, "The Frequencies and Intensities of Carbon Dioxide Absorption Lines Between 12 to 18 Microns," The Institute of Science and Technology, The University of Michigan, Ann Arbor, MI, Report No. 08183-1-T, 1967.

[5-28]　E. Plyler and E. Barker, "Infrared Spectrum and Molecular Configuration on N_2O," *Physical Review,* American Institute of Physics, New York, NY, Vol. 38, 1931, p. 1827.

[5-29] C. Walshaw, "Integrated Absorption by the 9.6 μm Band of Ozone," *Quarterly Journal of Royal Meteorological Society*, Royal Meteorological Society, Berkshire, England, Vol. 83, 1957, pp. 315-321.

[5-30] D. J. McCaa and J. H. Shaw, "The Infrared Spectrum of Ozone," *Journal of Molecular Spectroscopy*, Academic Press, New York, NY, Vol. 25, No. 3, March 1968, pp. 374-397.

[5-31] D. Deirmendjian, "Scattering and Polarization Properties of Polydispersed Suspensions with Partial Absorption," *Proceedings of the Interdisciplinary Conference on Electromagnetic Scattering*, Potsdam, Pergamon Press, New York, NY, 1963.

[5-32] D. E. Burch, D. A. Gryvnak, and D. D. Pembrook, "Investigation of the Absorption of Infrared Radiation by Atmospheric Gases," Philco Ford Corporation, Newport Beach, CA, Philco Report No. U-4829, 1970.

[5-33] T. L. Altshuler, "Infrared Transmission and Background Radiation by Clear Atmospheres," General Electric Company, Philadelphia, PA, No. 61SD199, 1961, p. 140.

[5-34] J. E. A. Selby and R. M. McClatchey, "Atmospheric Transmittance from 0.25 to 2.85 μm: Computer Code LOWTRAN 2," Air Force Research Laboratories, L. G. Hanscom Field, Bedford, MA, Report No. AFCRL-72-0745, 1972.

[5-35] R. A. McClatchey and J. E. A. Selby, "Atmospheric Attenuation of Laser Radiation from 0.76 to 31.25 μm," Air Force Research Laboratories, L. G. Hanscom Field, Bedford, MA, Report No. TR-74-0003, January 1974.

[5-36] F. Benford, "Duration of Intensity of Sunshine—Part I—General Considerations and Corrections," *Illuminating Engineering*, General Electric Company, Philadelphia, PA, Vol. 42, 1947, p. 527.

[5-37] M. Migeotte, L. Neven and J. Swensson, *The Solar Spectrum from 2.8 to 23.7 Microns, Part I, Photometric Atlas*, University of Liege, Belgium, Contract AF 61 (514)-432, Phase A, Part I, Geophysics Research Directorate, AFCRC, Cambridge, MA, ASTIA AD 210043.

[5-38] M. Migeotte, L. Neven, and J. Swensson, *The Solar Spectrum from 2.8 to 23.7 Microns, Part II, Measures and Identification*, University of Liege, Belgium, Contract AF 61 (514)-432, Phase A, Part II, Geophysics Research Directorate, AFCRC, Cambridge MA, ASTIA, AD 210044.

[5-39] M. Migeotte, L. Neven and J. Swensson, *An Atlas of Nitrous Oxide, Methane and Ozone Infrared Absorption Bands, Part I, The Photometric Records*, University of Liege, Contract AF 61 (614)-432, Phase B, Part I, Geophysics Research Directorate, AFCRC, Cambridge, MA, ASTIA AD 210045.

[5-40] M. Migeotte, L. Neven and J. Swensson, *An Atlas of Nitrous Oxide, Methane and Ozone Infrared Absorption Bands, Part II, Measures and Identifications*, University of Liege, Contract AF 61 (514)-432, Phase B, Part II, Geophysics Research Directorate, AFCRC, Cambridge, MA, ASTIA AD 210046.

[5-41] J. N. Howard and J. S. Garing, *Infrared Atmospheric Transmission: Some Source Papers on the Solar Spectrum from 3 to 15 Microns*, Air Force Surveys in Geophysics, No. 142, Geophysics Directorate, Air Force Research Laboratories, L. G. Hanscom Field, Bedford, MA, AFCRL Report No. 1098, December 1961.

[5-42] J. N. Howard, Air Force Geophysics Laboratory, L. G. Hanscom Field, Bedford, MA, Private Communication.

[5-43]   J. N. Howard, "Atmospheric Transmission in the 8 to 13 Micron Region," *Proceedings of the Symposium on Optical Radiation from Military Airborne Targets*, Final Report No. AFCRL-TR-58-146, AFCRL, Cambridge, MA, Contract No. AF19(604)-2451, Haller, Raymond and Brown, State College, PA, ASTIA AD 152411; *see also* J. Yarnell and R. M. Goody, "Infrared Solar Spectroscopy in a High-altitude Aircraft," *Journal of Scientific Instruments* (Journal of Physics E: Scientific Instruments), Institute of Physics, Bristol, England, Vol. 29, 1952, p. 352.

[5-44]   O. C. Mohler, A. K. Pierce, P. R. McMath, and L. Goldberg, *Atlas of the Solar Spectrum from 0.84 to 2.52 Microns*, University of Michigan Press, Ann Arbor, MI, 1950.

[5-45]   O. C. Mohler, *Table of Solar Spectrum Wavelengths from 1.20 to 2.55 Microns*, Univeristy of Michigan Press, Ann Arbor, MI, 1955.

[5-46]   J. H. Shaw, R. M. Chapman, J. N. Howard, and M. L. Oxholm, "A Grating Map of the Solar Spectrum from 3.0 to 5.0 Microns," *Astrophysical Journal*, University of Chicago Press, Chicago, IL, Vol. 113, No. 2, 1951; *see also* J. Yarnell and R. M. Goody, "Infrared Solar Spectroscopy in a High-altitude Aircraft," *Journal of Scientific Instruments* (Journal of Physics E: Scientific Instruments), Institute of Physics, Bristol, England, Vol. 29, 1952, p. 352.

[5-47]   J. H. Shaw, M. L. Oxholm and H. H. Classen, "The Solar Spectrum from 7 to 13 Microns," *Astrophysical Journal*, University of Chicago Press, Chicago, IL, Vol. 116, No. 3, 1952; *see also* J. Yarnell and R. M. Goody, "Infrared Solar Spectroscopy in High-Altitude Aircraft," *Journal of Scientific Instruments*, (Journal of Physics E: Scientific Instruments), Institute of Physics, Bristol, England, Vol. 29, 1952, p. 352.

[5-48]   W. W. Talbert, H. A. Templin, and R. E. Morrison, *Quantitative Solar Spectral Measurements at Mt. Chacaltaya (17,100 ft)*, U.S. Naval Ordnance Laboratory, White Oak, MD: *see also Journal of the Optical Society of America*, Optical Society of America, Washington, DC, Vol. 47, 1957, p. 1056.

[5-49]   J. E. Seeley, J. T. Houghton, T. S. Moss, and N. D. Hughes, "Solar Spectrum from 1 to 6.5 Microns at Altitudes up to 15 KM," *Philosophical Transactions of the Royal Society*, Royal Society of London, London, England.

[5-50]   C. B. Farmer and S. J. Todd, "Reduced Solar Spectra 3.5 to 5.5 Microns", EMI Electronics, Hayes, Middlesex, England, Report DP 927, 1961.

[5-51]   A. Goldman, D. G. Murcray and F. H. Murcray, et al., "Abundance of N_2O in the Atmosphere between 4.5 and 13.5 km," *Journal of the Optical Society of America*, Optical Society of America, Washington, DC, Vol. 60, No. 11, 1970, pp. 1466-1468.

[5-52]   N. Laulainen, *Minor Gases in the Earth's Atmosphere: A Review and Bibliography of Their Spectra*, University of Washington, Seattle, WA, ASTRA Publication No. 18, 1972.

[5-53]   D. G. Murcray, T. G. Kyle, F. H. Murcray, and W. J. Williams, "Nitric Acid and Nitric Oxide in the Lower Stratosphere," *Nature*, Macmillan Journals, Ltd., London, England, Vol. 218, 1968, p. 78.

[5-54]   D. G. Murcray, A. Goldman, and A. Csoeke-Poeckh, et al., "Nitric Acid Distribution in the Stratosphere," *Journal of Geophysical Research*, American Geophysical Union, Washington, DC, Vol. 78, No. 30, 1973, pp. 7033-7038.

[5-55]   J. N. Brooks, A. Goldman, J. J. Kosters, D. G. Murcray, F. H. Murcray, and W. J. Williams, "Balloon-Borne Infrared Measurements," *Physics and Chemistry*

of Upper Atmospheres, B. N. McCormack (ed)., Proceedings of a Symposium organized by the Summer Advanced Study Institute held at the University of Orleans, France, July 31–August 11, 1972, D. Reidel Publishing Company, Dordrecht, Holland, 1973, pp. 278-285.

[5-56] J. L. Stanford, "Stratospheric Water-Vapor Upper Limits Inferred from Upper-Air Observations, Part I: Northern Hemisphere," *Bulletin of the American Meteorological Society,* American Meteorological Society, Boston, MA, Vol. 55, No. 3, 1974, p. 194.

[5-57] A. J. Krueger, "The Mean Ozone Distribution from Several Series of Rocket Soundings to 52 km Latitudes from 54°S to 64°N," Goddard Space Flight Center, Greenbelt, MD, NASA Report No. X-651-73-67-1973.

[5-58] L. Elterman, "UV, Visible, and IR Attenuation for Altitudes to 50 km," Air Force Research Laboratories, L. G. Hanscom Field, Bedford, MA, Report No. AFCRL-68-0153, 1970.

[5-59] L. Elterman, "Vertical Attenuation Model with Eight Meteorological Ranges 2 to 13 Kilometers," Air Force Research Laboratories, L. G. Hanscom Field, Bedford, MA, Report AFCRL-70-0200-1970.

[5-60] L. A. Young, "Infrared Spectra" *Journal of Quantitative Spectroscopy and Radiative Transfer,* Pergamon Press, Oxford, England, Vol. 8, No. 2, February 1968, pp. 693-716.

[5-61] A. W. Mantz, E. R. Nichols, D. B. Alpert, and K. N. Rao, "CO Laser Spectra Studied with a 10-Meter Vacuum Infrared Grating Spectrogram," *Journal of Molecular Spectroscopy,* Academic Press, New York, NY, Vol. 35, 1970, p. 325.

[5-62] T. F. Deutsch, "Molecular Laser Action in Hydrogen and Deuterium Halides," *Applied Physics Letters,* American Institute of Physics, New York, NY, Vol. 10, 1971, p. 234.

[5-63] N. G. Basov, V. T. Galochkin, V. I. Igoshin, L. V. Kulakov, E. P. Martin, A. J. Nitikin, and A. N. Oraevsky, "Spectra of Stimulated Emission in the Hydrogen-Fluorine Reaction Process and Energy Transfer from DF to CO_2," *Applied Optics,* Optical Society of America, Washington, DC, Vol. 10, 1971, p. 1814.

[5-64] R. N. Spanbauer, K. N. Rao, and L. H. Jones, *Journal Molecular Spectroscopy,* Academic Press, New York, NY, Vol. 16, No. 1, May 1965, p. 100.

[5-65] C. H. Palmer, "Long Path Water Vapor Spectra with Pressure Broadening, I: 20–21.7," *Journal of the Optical Society of America,* Optical Society of America, Washington, DC, Vol. 47, 1957, p. 1024.

[5-66] L. A. Abels, "A Study of Total Absorption near 4.5 by Two Samples of N_2O As Their Total Pressure and N_2O Concentrations were Independently Varied," Ohio State University Research Foundation, Columbus, OH, Scientific Report No. 3, AFCRL-62-236, January 1962.

[5-67] H. J. Kostkowski, "Half Widths and Intensities from the Infrared Transmission of Thermally Excited CO_2," Progress Report, Office of Naval Research Contract 248(01), Johns Hopkins University, Baltimore, MD, October 1955.

[5-68] D. K. Edwards, et al., "Absorption by Infrared Bands of Carbon Dioxide Gas at Elevated Pressures and Temperatures," *Journal of the Optical Society of America,* Optical Society of America, Washington, DC, Vol. 50, 1960, pp. 130, 617.

[5-69] D. E. Burch, D. A. Gryvnak, and R. R. Patty, "Absorption of CO_2 Between 4500 and 54000 cm^{-1} (2 μm Region," Aeronutronic Report U-2955 (1964); "6600 and 7125 cm^{-1} (1.4 μm Region)," Aeronutronic Report U-3127 (1965); "8000 and 10,000 cm (1 to 1.25 μm Region)," Aeronutronic Report U-3200 (1965); "5400 and 6600 cm (1.6 μm Region)," Aeronutronic Report U-3201 (1965); "1800 and 2850 cm (3.5 - 5.6 μm Region)," Aeronutronic Report U-3857 (1966); "7125 to 8000 cm (1.25 to 1.4 μm Region)," Aeronutronic Report U-3930 (1967); "3100 and 4100 cm^{-1} (2.44 to 3.22 μm Region)," Aeronutronic Report U-4132 (1968); Philco-Ford Corporation, Newport Beach, CA.

[5-70] H. A. Daw, "Transmission of Radiation through Water Vapor Subject to Pressure Broadening in the Region 4.2 Microns to 23 Microns," University of Utah, Salt Lake City, UT, Technical Report No. 10, 1956.

[5-71] E. E. Bell, "Infrared Techniques and Measurements," Ohio State University Research Foundation, Columbus, OH (Interim Engineering Report for Period July-September 1956 on Contract AF 33 (616)-3312), 1956.

[5-72] N. G. Yaroslavskii, and A. E. Stanevich, "The Long Wavelength Infrared Spectrum of H_2O Vapor and the Absorption Spectrum of Atmospheric Air in the Region 20-2500 μ (500-4 cm^{-1})," *Optics and Spectroscopy*, Optical Society of America, Washington, DC, Vol. 7, November 1959, p. 380; also Optika Spektroskopiia, Vol. 5, 1958, p. 382.

[5-73] J. R. Izatt, "Office of Naval Research Progress Report," Johns Hopkins University, Baltimore, MD. Contract No. 248(01), 1960.

[5-74] D. E. Burch, D. A. Gryvnak and R. R. Patty, "Absorption of H_2O Between 2800 and 4500 cm^{-1} (2.7 μm Region)," Aeronutronic Report U-3203 (1965); "5045 and 14,485 cm (0.69 to 1.98 μm Region)," Aeronutronic Report U-3704 (1966); "1630 and 2245 cm^{-1} (6.13 to 4.44 μm Region)," Aeronutronic Report U-5090 (1973); Philco-Ford Corporation, Newport Beach, CA.

[5-75] D. E. Burch, "Investigation of the Absorption of Infrared Radiation by Atmospheric Gases," Air Force Cambridge Research Laboratories, Philco Report No. U-4784, Philco-Ford Corporation, Newport Beach, CA, 1970.

[5-76] D. J. Lovell, *An Atlas of Air Absorption in the Infrared*, Interim Report, University of Massachusetts, Amherst, MA, June, 1969.

[5-77] M. Summerfield, "Pressure Dependence of the Absorption in the 9.6 Micron Band of Ozone," thesis, California Institute of Technology, Pasadena, CA, 1941.

[5-78] M. K. Wilson and R. A. Ogg, "The Infrared Spectrum and Structure of Ozone," *Journal of Chemical Physics*, American Institute of Physics, New York, NY, Vol. 18, 1950, p. 766.

[5-79] H. S. Gutowsky and E. M. Petersen, "The Infrared Spectrum and Structure of Ozone," *Journal of Chemical Physics*, American Institute of Physics, New York, NY, Vol. 18, 1950, p. 564.

[5-80] H. A. Gebbie, W. Harding, C. Hilsum, A. Pryce, and V. Roberts, "Atmospheric Transmission in the 1-14 μm Region," *Proceedings of the Royal Society*, Royal Society of London, London, England, Vol. 206A, 1951, p. 87.

[5-81] A. Arnulf, J. Bricard, E. Cure, and C. Veret, "Transmissions by Haze and Fog in the Spectral Region 0.35 to 10 Microns," *Journal of the Optical Society of America*, Optical Society of America, Washington, DC, Vol. 47, 1957, p. 491.

[5-82] J. H. Taylor and H. W. Yates, "Atmospheric Transmission in Infrared," *Journal of the Optical Society of America*, Optical Society of America, Washington, DC, Vol. 47, No. 3, 1957, pp. 223-226.

[5-83] H. W. Yates, "The Absorption Spectrum from 0.5 to 25 microns of a 1000 ft. Atmospheric Path at Sea Level," Naval Research Laboratory, Washington, DC, NRL Report 5033, September 1957.

[5-84] H. W. Yates and J. H. Taylor, "Infrared Transmission of the Atmosphere," U.S. Naval Research Laboratory, Washington, DC, NRL Report 5453, 1960.

[5-85] C. B. Farmer, P. J. Berry and D. B. Lloyd, "Atmospheric Transmission Measurements in the 3.5 – 5.5 micron band at 5.200 m. Altitude," EMI Electronics, Hayes, Middlesex, England, EMI Electronics Report DMP 1578, 1963.

[5-86] J. Curcio, L. Drummeter, and G. Knestrick, "An Atlas of the Absorption Spectrum of the Lower Atmosphere from 5400 Å to 8520 Å," *Applied Optics*, Optical Society of America, Washington, DC, Vol. 3, No. 12, December 1964, pp. 1401-1410.

[5-87] J. L. Streete, "Infrared Measurements of Atmospheric Transmission at Sea Level," *Applied Optics*, Optical Society of America, Washington, DC, Vol. 7, No. 8, 1968, pp. 1545-1549.

[5-88] V. L. Filippov, L. M. Artem'yeva, S. O. Mirumyants, *Bull. Izv. Acad. Sci. U.S.S.R. Atmos. and Oceanic Physics*, Academy of Sciences of the U.S.S.R. Bulletin (Izvestiya), Atmospheric and Oceanic Physics Services (English Edition), American Geophysical Union, Washington, DC, Vol. 5, 1969.

[5-89] G. Ashley, L. Gastineau and D. Blay, General Dynamics, Pomona, CA, Private Communication, 1971-1972.

[5-90] C. D. Keeling, "The Concentration and Isotopic Abundance of Carbon Dioxide in the Atmosphere," *Tellus*, Swedish Geophysical Society, Stockholm, Sweden, Vol. 12, 1960, pp. 200-203.

[5-91] G. S. Callendar, "On the Amount of Carbon Dioxide in the Atmosphere," *Tellus*, Swedish Geophysical Society, Stockholm, Sweden, Vol. 10, 1958, pp. 243-248.

[5-92] E. G. Glueckauff, "CO_2 Content of the Atmosphere, *Nature,* Macmillan Journals, Ltd., London, England, Vol. 153, 1944, pp. 620-621.

[5-93] S. Fonselius, F. Koroleff and K. Burch, "Microdetermination of CO_2 in the Air with Current Data for Scandinavia," *Tellus*, Swedish Geophysical Society, Stockholm, Sweden, Vol. 7, 1955, pp. 258-265.

[5-94] J. R. Bray, "An Analysis of the Possible Recent Change in Atmospheric Carbon Dioxide," *Tellus*, Swedish Geophysical Society, Stockholm, Sweden, Vol. 11, 1959, pp. 220-230.

[5-95] U. Fink, D. H. Rank, and T. A. Wiggins, "Abundance of Methane in the Earth's Atmosphere," *Journal of the Optical Society of America*, Optical Society of America, Washington, DC, Vol. 54, 1964, p. 472.

[5-96] J. W. Birkeland and J. H. Shaw, "Abundance of Nitrous Oxide in Ground-Level Air," *Journal of the Optical Society of America*, Optical Society of America, Washington, DC, Vol. 49, 1959, p. 637.

[5-97] J. H. Shaw, "Monthly Report on Infrared Temperature Sounding," Ohio State University Research Foundation, Columbus, OH, RF Project 2469, Report No. 16, October 1968.

[5-98] J. H. Shaw, "A Determination of the Abundance of N_2O, CO, and CH_4 in Ground Level Air at Several Locations Near Columbus, OH," Air Force Research Laboratories, L. G. Hanscom Field, Bedford, MA, Scientific Report No. 1, Contract AF19(604)-2259.

[5-99] A. Adel, "Equivalent Thickness of the Atmospheric Nitrous Oxide Layer,"

[5-100] J. H. Shaw, G. B. B. M. Sutherland and T. W. Wormell, "Nitrous Oxide in the Earth's Atmosphere," *Physical Review*, American Institute of Physics, New York, NY, Vol. 74, 1958, p. 978.

[5-101] R. McMath and L. Goldberg, "The Abundance and Temperature of Methane in the Earth's Atmosphere," *Proceedings of the American Physical Society*, American Physical Society, American Institute of Physics, New York, NY, Vol. 74, 1958, p. 623.

[5-102] R. L. Slobod and M. E. Krogh, "Nitrous Oxide as a Constituent of the Atmosphere," *Journal of the American Chemical Society*, American Chemical Society, Washington, DC, Vol. 72, 1950, pp. 1175-1177.

[5-103] J. W. Birkeland, D. E. Burch and J. H. Shaw, "Some Comments on Two Articles by Taylor and Yates," *Journal of the Optical Society of America*, Optical Society of America, Washington, DC, Vol. 47, January 1957, p. 441.

[5-104] J. W. Birkeland, "Determination of Ground Level N_2O," M.S. thesis, Ohio State University, Columbus, OH, 1957.

[5-105] A. L. Bowman, "A Determination of the Abundance of Nitrous Oxide, Carbon Monoxide and Methane in Ground Level Air at Several Locations near Columbus, Ohio," Ohio State University, Columbus, OH, Scientific Report No. 1, Contract AF19(604)-2259, 1959.

[5-106] J. Selby, "Atmospheric Transmittance, I.," Notes for Advanced IR Technology, University of Michigan Engineering Summer Conference, Ann Arbor, MI, 1974.

[5-107] B. J. Conrath, R. Hanel, V. Kunde and C. Prabhahara, "The Infrared Interferometer Experiment on Nimbus 3," *Journal of Geophysical Research*, American Geophysical Union, Washington, DC, Vol. 75, No. 30, 1970, pp. 5831-5857.

[5-108] D. E. Burch and D. Williams, "Infrared Absorption by Minor Atmospheric Constituents," The Ohio State University Research Foundation, Columbus, OH, Scientific Report No. 1, Contract AF 19(604)-2633, Geophysics Research Directorate, AFCRL Report No. TN-60-674, AFCRL, Cambridge, MA (1960), AD 246921.

[5-109] D. E. Burch, D. Gryvnak, and D. Williams, "Infrared Absorption by Carbon Dioxide," The Ohio State University Research Foundation, Columbus, OH, Scientific Report No. 11, Contract No. AF 19(604)-2632, Geophysics Research Directorate, AFCRL Report No. 255, AFCRL, Cambridge, MA (1960), AD 253435.

[5-110] D. E. Burch, E. B. Singleton, W. L. France, and D. Williams, "Infrared Absorption by Minor Atmospheric Constituents," The Ohio State University Research Foundation, Columbus, OH, Final Report, Contract No. AF 19(604)-2633, Geophysics Research Directorate, AFCRL Report No. 412, AFCRL, Cambridge, MA (1960), AD 256952.

[5-111] J. A. Dowling, et al., "*Atmospheric Transmission Field Experiments Using IR Lasers, Fourier Transform Spectroscopy and Gas Filter Correlation Techinques,*" Office of the Director of Defense Research and Engineering, Workshop Paper Delivered December 1976, Private Communication.

[5-112] G. N. Plass, "Useful Representations for Measurements of Spectral Band Absorption," *Journal of the Optical Society of America*, Optical Society of America, Washington, DC, Vol. 50, No. 9, September 1960, pp. 868-875.

[5-113] National Oceanic and Atmospheric Administration, Washington, DC, "U. S. Standard Atmosphere 1976," NOAA-S/T 76-1562, November 1976.

[5-114] D. Anding, "Band Model Methods for Computing Atmospheric Slant-Path Molecular Absorption," Willow Run Laboratories, University of Michigan, Ann Arbor, MI, IRIA State-of-the-Art Report No. 7142-21-T, 1967.

5.11.2. Bibliography

Abels, L. A., "A Study of Total Absorption near 4.5 by Two Samples of N_2O, as their Total Pressure and N_2O Concentrations were Independently Varied," Ohio State University Research Foundation, Columbus, OH, Scientific Report No. 3, AFCRL-52-236, January 1962.

Abramowitz, M., and I. A. Stegun, *Handbook of Mathematical Functions*, National Bureau of Standards, Washington, DC, Applied Math Series No. 55, 1964.

Adel, A., "Equivalent Thickness of the Atmospheric Nitrous Oxide Layer, *Physical Review*, American Institute of Physics, New York, NY, Vol. 59, 1941, p. 944.

Adel, A., "Identification of Carbon Monoxide in Atmosphere above Flagstaff, AZ," *Astrophysical Journal*, University of Chicago, IL, Vol. 116, September 1952, pp. 442-443.

Adel, A., "Note on the Atmospheric Oxides of Nitrogen," *Astrophysical Journal*, University of Chicago, IL, Vol. 90, November 1939, p. 627.

Adel, A., and C. O. Lampland, "A New Band of the Absorption Spectrum of the Earth's Atmosphere," *Astrophysical Journal*, University of Chicago, IL, Vol. 87, March 1938, pp. 198-203.

Altshuler, T. L., "Infrared Transmission and Background Radiation by Clear Atmosphere," General Electric Company, Philadelphia, PA, No. 61SD199, 1961, p. 140.

Anding, D., "Band Model Methods for Computing Atmospheric Slant-Path Molecular Absorption," Willow Run Laboratories, University of Michigan, Ann Arbor, MI, IRIA State-of-the-Art Report, No. 7142-21-T, 1967.

Arking, A., and K. Grossman, "The Influence of Line Shape and Band Structure on Temperatures in Planetary Atmosphere," *Journal of Atmospheric Sciences*, American Meteorological Society, Boston, MA, Vol. 29, July 1972, p. 937-949.

Armstrong, B. H., "Analysis of the Curtis-Godson Approximation and Radiation Transmission Through Inhomogeneous Atmospheres," *Journal of Atmospheric Sciences*, American Meteorological Society, Boston, MA, Vol. 26, 1969, pp. 312-322.

Armstrong, B. H., "Exponential Integral Approximations," *Journal of Quantitative Spectroscopy and Radiative Transfer*, Pergamon Press, Oxford, England, Vol. 9, 1969, pp. 1390-1040.

Armstrong, B. H., "Spectrum Line Profiles: The Voight Function," *Journal of Quantitative Spectroscopy and Radiative Transfer*, Pergamon Press, Oxford, England, Vol. 7, 1967, pp. 483-494.

Armstrong, B. H., and J. Dave, "Gaussian Quadratures of the Exponential Integral," IBM Palo Alto Science Center, Palo Alto, CA, No. 320-3250, 1969.

Arnulf, A. J., J. Bricard, E. Cure, and C. Veret, "Transmissions by Haze and Fog in the Spectral Region 0.35 to 10 Microns," *Journal of the Optical Society of America*, Optical Society of America, Washington, DC, Vol. 47, June 1957, p. 491.

Basov, N. G., V. T. Galochkin, V. I. Igoshin, L. V. Kulakov, E. P. Martin, A. I. Nitikin, and A. N. Oraevsky, "Spectra of Stimulated Emission in the Hydrogen-Fluorine Reaction Process and Energy Transfer from DF to CO_2," *Applied Optics*, Optical Society of America, Washington, DC, Vol. 10, 1976, p. 1814.

Baumeister, and Marquardt, "Least Squares Estimation of Non-Linear Parameters," IBM SHARE Program No. 1427 FORTRAN Program.

Bell, E. E., "Infrared Techniques and Measurements," Ohio State University Research Foundation, Columbus, OH (Interim Engineering Report for July-September 1956 on Contract AF 33(616)-3312), 1956.

Benesch, W., M. V. Migeotte, and L. Neven, "Investigations of Atmospheric CO at the

Jungfraujoch," *Journal of the Optical Society of America*, Optical Society of America, Washington, DC, Vol. 43, November 1953, pp. 1119-1123.

Benford, F., "Duration of Intensity of Sunshine—Part I-Geneeral Considerations and Corrections," *Illuminating Engineering*, General Electric Company, Philadelphia, PA, Vol. 42, 1947, p. 527.

Bignell, K., "The Water Vapor Infrared Continuum," *Quarterly Journal of the Royal Meteorological Society*, Royal Meteorological Society, Berkshire, England, Vol. 96, 1970, pp. 390-403.

Birkeland, J. W., "Determination of Ground Level N_2O," M.S. Thesis, Ohio State University, Columbus, OH, 1957.

Birkeland, J. W., D. E. Burch, and J. H. Shaw, "Some Comments on Two Articles by Taylor and Yates," *Journal of the Optical Society of America*, Optical Society of America, Washington, DC, Vol. 47, 25 January 1957, p. 441.

Birkeland, J. W., and J. H. Shaw, "Abundance of Nitrous Oxide in Ground-Level Air," *Journal of the Optical Society of America*, Optical Society of America, Washington, DC, Vol. 49, 10 October 1959, p. 637.

Bowman, A. L., "A Determination of the Abundance of Nitrous Oxide, Carbon Monoxide and Methane in Ground Level Air at Several Locations near Columbus, OH," Ohio State University, Columbus, OH, Scientific Report No. 1, Contract AF 19(604)-2259, 1959.

Bradford, W. R., T. M. McCormick, and J. A. Selby, "Laboratory Representation of Atmospheric Paths for Infrared Absorption," EMI Electronics, Hayes, Middlesex, England, Report No. DMP 1431, 1963.

Bray, J. R., "An Analysis of the Possible Recent Changes in Atmospheric Carbon Dioxide," *Tellus, A Quarterly Journal of Geophysics*, Swedish Geophysical Society, Stockholm, Sweden, Vol. 11, 1959, pp. 220-230.

Brichard, J., "Etude de la Constitution des Nuages au sommet du Puyde-Dame," *La Meteorologie*, Societe Meteorologique de France, 73-77 rue de Sevres 92100 Boulogie-sur-Seine, France, Vol. 15.

Brooks, J. N., A. Goldman, J. J. Kosters, D. G. Murcray, F. H. Murcray, and W. J. Williams, "Balloon-Borne Infrared Measurements," B. N. McCormack (ed.), *Physics and Chemistry of Upper Atmospheres, Proceedings of a Symposium Organized by the Summer Advanced Study Institute*, University of Orleans, France, D. Reidel Publishing Co., Dordrecht, Holland, 1973, pp. 278-285.

Burch, D. E., "Investigation of the Absorption of Infrared Radiation by Atmospheric Gases," Philco-Ford Corporation, Newport Beach, CA, Philco Report No. U-4784, 31 January 1971.

Burch, D. E., and D. A. Gryvnak, "Infrared Radiation Emitted by Hot Gases and Its Transmission Through Synthetic Atmospheres," Philco-Ford Corporation, Newport Beach, CA, Philco Report No. U-1929, 1962.

Burch, D. E., and D. A. Gryvnak, "Strengths, Widths, and Shapes of the Lines of the $3\nu CO$ Band," Philco-Ford Corporation, Newport Beach, CA, Philco Report No. U-3972, 1972.

Burch, D. E., D. A. Gryvnak, and R. R. Patty, "Absorption of CO_2 Between 4500 and 5400 cm^{-1} (2 μm Region)," Aeronutronic Report No. U-2955 (1964); "6600 and 7125 cm^{-1} (1.4 μm Region)," Aeronutronic Report No. U-3127 (1965); "8000 and 10,000 cm^{-1} (1 to 1.25 μm Region)," Aeronutronic Report No. U-3200 (1965); "5400 and 6600 cm^{-1} (1.6 μm Region)," Aeronutronic Report No. U-3201 (1965); "1800 and 2850 cm^{-1} (3.5 - 5.6 μm Region)," Aeronutronic Report No. U-3857 (1966); "7125 and 8000 cm^{-1} (1.25 and 1.4 μm Region)," Aeronutronic Report

No. U-3930 (1967); "3100 and 4100 cm^{-1} (2.44 to 3.22 μm Region)," Aeronutronic Report No. U-4132 (1968); Philco-Ford Corporation, Newport Beach, CA.

Burch, D. E., D. A. Gryvnak, and R. R. Patty, "Absorption by H_2O Between 2800 and 4500 cm^{-1} (2.7 μm Region)," Aeronutronic Report No. U-3202 (1965); "5045 and 14,485 cm^{-1} (0.69 to 1.98 μm Region)," Aeronutronic Report No. U-3704 (1966); "1630 and 2245 cm^{-1} (6.13 to 4.44 μm Region)," Aeronutronic Report No. U-5090 (1973); Philco-Ford Corporation, Newport Beach, CA.

Burch, D. E., D. A. Gryvnak, and J. D. Pembrook, "Investigation of the Absorption of Infrared Radiation by Atmospheric Gases," Philco-Ford Corporation, Newport Beach, CA, Philco Report No. U-4829, 1970.

Burch, D. E., D. A. Gryvnak, and J. D. Pembrook, "Investigation of Infrared Absorption by Nitrous Oxide from 4000 to 6700 cm^{-1} (2.5 to 1.5 μm)," Aeronutronic Report No. U-4943, AFCRL-71-0536, 1971; "760 to 2380 cm^{-1} (13.2 to 4.2 μm)," Aeronutronic Report No. U-4995, AFCRL-71-0620, Philco-Ford Corporation, Newport Beach, CA, 1971.

Burch, D. E., D. A. Gryvnak, and E. B. Singleton, et al., "Absorption by Carbon Monoxide, Water Vapor, and Minor Atmospheric Constituents," Ohio State University, Columbus, OH, No. AFCRL 62-298, 1962, pp. 359-363.

Burch, D. E., D. A. Gryvnak, and D. Williams, Ohio State University, Columbus, OH, No. 2, AF 19(604)-2633, January 1961.

Burch, D. E. E. B. Singleton, and D. Williams, "Absorption Line Broadening in the Infrared," Applied Optics, Optical Society of America, Washington, DC, Vol. 1, No. 3, May 1962, pp. 359-363.

Calfee, R. F., and W. S. Benedict, "Carbon Dioxide Spectral Line Positions and Intensities Calculated for the 2.05 and 2.7 Micron Regions," National Bureau of Standards, Washington, DC, Technical Note 332, 1966.

Callendar, G. S., "On the Amount of Carbon Dioxide in the Atmosphere," Tellus, A Quarterly Journal of Geophysics, Swedish Geophysical Society, Stockholm, Sweden, Vol. 10, 1958, pp. 243-248.

Chaney, L. W., "High Resolution Spectroscopic Measurements of Carbon Dioxide and Carbon Monoxide," The Institute of Science and Technology, The University of Michigan, Ann Arbor, MI, No. 036350-3T, 1972.

Clough, S. A., and F. X. Kneizys, "Ozone Absorption in the 9.0 Micron Region," Air Force Cambridge Research Laboratories, Bedford, MA, No. AFCRL-65-862, 1965.

Cody, W., and H. Thacker, Jr., "Rational Chebyshev Approximations for the Exponential Integral E (χ), Mathematics of Computation, Vol. 22, July 1968, p. 641.

Connes, J., P. Connes, and J. P. Maillard, "Near Infrared Spectra of Venus, Mars, Jupiter, and Saturn," and "Atlas of Near Infrared Spectra of Venus, Mars, Jupiter, and Saturn," Centre National de la Recherche Scientifique, Paris, France, 1969.

Conrath, B. J., R. Hanel, V. Kunde, and C. Prabhahara, "The Infrared Interferometer Experiment on Nimbus 3," Journal of Geophysical Research, American Geophysical Union, Washington, DC, Vol. 75, No. 30, 1970, pp. 5831-5857.

Curcio, J., L. Drummeter, and G. Knestrick, "An Atlas of the Absorption Spectrum of the Lower Atmosphere from 5400 Å to 8520 Å," Applied Optics, Optical Society of America, Washington, DC, Vol. 3, No. 12, December 1964, pp. 1401-1410.

Curcio, J., R. Eckhardt, C. Acton, and T. H. Cosden, "An Atlas of the Absorption of the Atmosphere from 8512 to 11,600 Å," U.S. Naval Research Laboratory, Washington, DC, No. 6352, 1965.

Curtis, A. R., "Discussion of Goody's 'A Statistical Model for Water-Vapour Absorption,'" Quarterly Journal of the Royal Meteorological Society, Royal Meteorological Society, Berkshire, England, Vol. 58, 1952, pp. 165-169.

Dang-Nhu, M., Thesis, Universite de Paris, Paris, France, 1968.

Daniels, G., "A Computer Program for Atmospheric Infrared Transmission and Background Calculations," *Optical Engineering*, The Society of Photo-Optical Instrumentation Engineers, Palos Verdes Estates, CA, Vol. 13, No. 2, 1974, pp. 92-97.

Dave, J., "A Subroutine for Evaluation of the Exponential Integral with Fifteen Significant Figure Accuracy," IBM Palo Alto Science Center, Palo Alto, CA, No. 320-3251, 1968.

Daw, H. A., "Transmission of Radiation through Water Vapor Subject to Pressure Broadening in the Region 4.2 Microns to 23 Microns," University of Utah, Salt Lake City, UT, Technical Report No. 10, 1956.

Deirmendjian, D., "Scattering and Polarization Properties of Polydispersed Suspensions with Partial Absorption," *Proceedings of the Interdisciplinary Conference on Electromagnetic Scattering*, Potsdam, Pergamon Press, New York, NY, 1963.

De Lucia, F. C., P. Helminger, R. L. Cook, and W. Gordy, "Millimeter and Submillimeter Wave Rotational Spectrum and Centrifugal Distortion Effects of HDO," *Journal of Chemical Physics*, American Institute of Physics, New York, NY, Vol. 55, No. 11, December 1971, pp. 5334-5339.

DeLucia, F. C., P. Helminger, R. L. Cook and W. Gordy, *Physical Review*, American Institute of Physics, New York, NY, A 3, Vol. 5, 1972, p. 487.

Deutsch, T. F., "Molecular Laser Action in Hydrogen and Deuterium Halides," *Applied Physics Letters*, American Institute of Physics, New York, NY, Vol. 10, 1971, p. 234.

Dionne, J., "Atmospheric Spectra from 9.1 to 11.6 μ," Thesis, Universite de Paris, 1972.

Drayson, S. R., "Atmospheric Slant-Path Transmission in the 15μ CO_2 Band," Report No. 05863-6-T, University of Michigan, Ann Arbor, MI, 1964.

Drayson, S. R., "The Calculation of Long-Wave Radiative Transfer in Planetary Atmospheres," Report No. 07584-1-T, University of Michigan, Ann Arbor, MI, 1967.

Drayson, S. R., "A Listing of Wave Numbers and Intensities of Carbon Dioxide Absorption Lines between 12 and 20 μm," Report No. 036350-4-T, University of Michigan, Ann Arbor, MI, 1973.

Drayson, S. R., "Polynomial Approximations of Exponential Integrals," *Journal of Quantitative Spectroscopy and Radiative Transfer*," Pergamon Press, Oxford, England, Vol. 8, 1968.

Drayson, S. R., "Transmittance for Use in Remote Soundings of the Atmosphere," Space Research XI, Akademie-Verlag, Berlin, 1971.

Drayson, S. R., and C. Young, "The Frequencies and Intensities of Carbon Dioxide Absorption Lines Between 12 to 18 Microns," University of Michigan, Ann Arbor, MI, Report No. 08183-1-T, 1967.

Drayson, S. R., and C. Young, "Theoretical Investigations of Carbon Dioxide Radiative Transfer." University of Michigan, Ann Arbor, MI, Report No. 07349-1-F, 1966.

Drummod, A. J., *Advances in Geophysics*, H. E. Landsberg and J. Van Miegham (eds.), Academic Press, New York, NY, Vol. 14, 1970, p. 25.

Dutsch, H. U., "Current Problems of the Photochemical Theory of Atmospheric Ozone," paper given at Inter Symposium on Chem. Reac. in the Lower and Upper Atmosphere, Stanford Research Institute, Stanford, CA, April 1961

Edlen, B., "The Refractive Index of Air," *Metrologia*, Springer-Verlag, New York, NY, Vol. 2, No. 2, 1966, p. 71.

Edwards, D. K., et al., "Absorption by Infrared Bands of Carbon Dioxide Gas at Elevated Pressures and Temperatures," *Journal of the Optical Society of America*, Optical Society of America, Washington, DC, Vol 50, 1960, pp. 130, 617.

Edwards, D. K., et al., "Apparatus for the Determination of the Band Absorption of

Gases at Elevated Pressures and Temperatures," *Journal of the Optical Society of America*, Optical Society of America, Washington, DC, Vol. 50, February 1960, pp. 130-136.

Ellingson, R. G., "A New Long-Wave Radiative Transfer Model: Calibration and Application to the Tropical Atmosphere," Florida State University, Tallahassee, FL, Report No. 72-4, 1972.

Elsasser, W. M., "Heat Transfer by Infrared Radiation in the Atmosphere," Harvard University Press, Harvard University, Cambridge, MA, Report No. 6, 1942.

Elsasser, W. M., "Mean Absorption and Equivalent Absorption Coefficient of a Band Spectrum," *Physical Review*, American Institute of Physics, New York, NY, Vol. 54, 1938, p. 126.

Elterman, L., "Parameters for Attenuation in the Atmosphere Windows for Fifteen Wavelengths," *Applied Optics*, Optical Society of America, Washington, DC, Vol. 3, No. 6, June 1964, pp. 745-750.

Elterman, L., "UV, Visible, and IR Attenuation for Altitudes to 50 km," Air Force Cambridge Research Laboratories, L. G. Hanscom Field, Bedford, MA, Report No. AFCRL-68-0153, 1968.

Elterman, L., "Vertical Attenuation Model with Eight Surface Meteorological Ranges 2 to 13 kilometers," Air Force Cambridge Research Laboratories, L. G. Hanscom Field, Bedford, MA, Report No. AFCRL-70-0200, 1970.

Ely, R., and T. McCubbin, "The Temperature Dependence of the Self-Broadened Half Width of the P-20 Line in the 001-100 Band of CO_2," *Applied Optics*, Optical Society of America, Washington, DC, Vol. 9, No. 5, May 1970, pp. 1230-1232.

Erdelyi, A., et al., *Higher Transcendental Functions*, McGraw-Hill, New York, NY, Vol. 2, 1953.

Farmer, C. B., P. J. Berry, and D. B. Lloyd, "Atmospheric Transmission Measurements in the 3.5 – 5.5 Micron Band at 5.200 m Altitude," EMI Electronics, Hayes, Middlesex, England, Report DMP 1578, 1963.

Farmer, C. B., and S. J. Todd, "Reduced Solar Spectra 3.5 to 5.5 Microns," EMI Electronics, Hayes, Middlesex, England, Report DP 927, 1961.

Feddeyeva, V., and N. Tarentev, *Tables of the Probability Integral for Complex Argument*, Pergamon Press, New York, NY, 1961.

Filippov, V. L., L. M. Artem'yeva, S. O. Mirumyants, *Bull. Iz. Acad. Sci. U.S.S.R. Atmos. and Oceanic Physics*, Academy of Sciences of the U.S.S.R. Bulletin (Izvestiya), Atmospheric and Oceanic Physics Series (Eighth Edition), American Geophysical Union, Washington, DC, Vol. 5, 1969.

Fink, U., D. H. Rank, and T. A. Wiggins, "Abundance of Methane in the Earth's Atmosphere," *Journal of the Optical Society of America*, Optical Society of America, Washington, DC, Vol. 54, 1964, p. 472.

Fonselius, S., F. Koroleff and K. Burch, "Microdetermination of CO_2 in the Air with Current Data for Scandinavia," *Tellus, A Quarterly Journal of Geophysics*, Swedish Geophysical Society, Stockholm, Sweden, Vol. 7, 1955, pp. 258-265.

Gebbie, H. A., W. Harding, C. Hilsum, A. Pryce and V. Roberts, "Atmospheric Transmission in the 1-14 μm Region," *Proceedings of the Royal Society*, Royal Society of London, London, England, Vol. 206A, 1951, p. 87.

Gille, J. C., and R. G. Ellingson, "Correction of Random Exponential Band Transmissions for Doppler Effects," *Applied Optics*, Optical Society of America, Washington, DC, Vol. 7, No. 3, March 1968, pp. 471-474.

Glueckauff, E. G., "CO_2 Content of the Atmosphere," *Nature*, Macmillan Journals, Ltd., London, England, Vol. 153, 1944, pp. 620-621.

Golden, S. A., "The Doppler Analog of the Elsasser Band Model," *Journal of Quantitative Spectroscopy and Radiative Transfer*, Pergamon Press, Oxford, England, Vol. 7, 1967, pp. 483-494.

Goldman, A., "Distribution of Water Vapor in the Stratosphere as Determined from Balloon Measurements of Atmospheric Emission Spectra in the 24 to 29 m Region," University of Denver, Denver, CO, Report No. AFCRL-72-0077, 1972.

Goldman, A., T. G. Kyle, and F. S. Bonomo, "Statistical Band Model Parameters and Integrated Intensities for the 5.9 μ, 7.5 μ, and 11.3 μ Bands of HNO_3 Vapor," *Applied Optics*, Optical Society of America, Washington, DC, Vol. 10, No. 1, January 1971, pp. 65-73.

Goldman, A., D. G. Murcray, and F. H. Murcray, et al., "Abundance of N_2O in the Atmosphere Between 4.5 and 13.5 km," *Journal of the Optical Society of America*, Optical Society of America, Washington, DC, Vol. 60, No. 11, November 1970, pp. 1466-1468.

Goody, R. M., *Atmospheric Radiation*, Oxford University Press, New York, NY, 1964(a).

Goody, R. M., "A Statistical Model for Water-Vapour Absorption," *Quarterly Journal of the Royal Meteorological Society*, American Meteorological Society, Boston, MA, Vol. 78, 1952.

Goody, R. M., "Transmission of Radiation Through an Inhomogeneous Atmosphere," *Journal of Atmospheric Sciences*, Boston, MA, Vol. 21, No. 6, 1964(b).

Goody, R. M., and W. T. Roach, "Determinations of the Vertical Distribution of Ozone From Emission Spectra," *Quarterly Journal of the Royal Meteorological Society*, Royal Meteorological Society, Berkshire, England, Vol. 82, No. 352, 217, 1956.

Green, A. E. S., C. Lindenmeyer, and M. Griggs, "Molecular Absorption in Planetary Atmosphere," *Journal of Geophysical Research*, American Geophysical Union, Washington, DC, Vol. 69, No. 3, 1964.

Gutnick, M., "How Dry is the Sky?," *Journal of Geophysical Research*, American Geophysical Union, Washington, DC, Vol. 66, 1961.

Gutnick, M., "Mean Atmospheric Moisture Profiles to 31 km for Middle Latitudes," *Applied Optics*, Optical Society of America, Washington, DC, Vol. 1, September 1962, pp. 670-672.

Gutowsky, H. S., and E. M. Petersen, "The Infrared Spectrum and Structure of Ozone," *Journal of Chemical Physics*, American Institute of Physics, New York, NY, Vol. 18, 1950, p. 564.

Hall, D. N. B., "Observations of the Infrared Sunspot Spectrum Between 11340A and 24778A, Thesis, Harvard University, Cambridge, MA, 1970.

Hall, R. T., and J. M. Dowling, "Pure Rotational Spectrum of Water Vapor," *Journal of Chemical Physics*, American Institute of Physics, New York, NY, Vol. 47, No. 7, October 1967, pp. 2454-2461.

Hamilton, J. N., J. A. Rowe, and D. Anding, "Atmospheric Transmission and Emission Program," Aerospace Corporation, El Segundo, CA, Report No. TOR-0073 (3050-02)-3, 1973.

Haurwitz, F. D., "The Distribution of Tropospheric Infrared Radiative Fluxes and Associated Radiative Fluxes and Associated Heating and Cooling Rates in the Southern Hemisphere," University of Michigan, Ann Arbor, MI, 1972.

Howard, J. N., "Atmospheric Transmission in the 8 to 13 Micron Region," *Proceedings of the Symposium on Optical Radiation from Military Airborne Targets*, Final Report No. AFCRL-TR-58-146, AFCRL, Cambridge, MA, Contract No. AF 19(604)-2451, Haller, Raymond and Brown, Inc., State College, PA, ASTIA AD 152411; *see also* J. Yarnell and R. M. Goody, "Infrared Solar Spectroscopy in a High-altitude Aircraft," *Journal of Scientific Instruments*, Institute of Physics, Bristol, England, Vol. 29, 1952.

Howard, J. N., and J. S. Garing, *Infrared Atmospheric Transmission: Some Source Papers on the Solar Spectrum from 3 to 15 Microns*, Air Force Surveys in Geophysics No. 142, Geophysics Research Directorate, AFCRL, Cambridge, MA, AFCRL, Report No. 1098, December 1961.

Howard, J. N., D. Burch, and D. Williams, "Near Infrared Transmission Through Synthetic Atmospheres," Ohio State University Research Foundation, Columbus, OH, Geophysics Research Paper No. 40, Report No. AFCRL-TR-55-213, 1955.

Howard, J. N., and J. S. Garing, "The Transmission of the Atmosphere in the Infrared," GRD, Air Force Research Laboratories, L. G. Hanscom Field, Bedford, MA, 1962.

Hurlburt, E. O., "Physics of the Upper Atmosphere," *Meteorological Research Rev.*, Vol. 3, No. 17, 1957.

Izatt, J. R., "Office of the Naval Research Progress Report," Johns Hopkins University, Baltimore, MD, Contract No. 248(01).

Johansen, H., *On the Relation Between Meteorological Conditions and Total Amount of Ozone Over Tromso*, Polar Atmosphere Symposium, Part I, Pergamon Press, London, 1958.

Junge, C. E., "Atmospheric Composition," *Handbook of Geophysics for Air Force Designers*, AFCRC, Cambridge, MA, 1957.

Junge, C. E., "Our Knowledge of the Physics-Chemistry of Aerosols in the Undisturbed Marine Environment," *Journal of Geophysical Research*, American Geophysical Union, Washington, DC, Vol. 77, No. 27, 1972, pp. 5183-5200.

Kaplan, L. D., "A Quasi-Statistical Approach to the Calculation of Atmospheric Transmission," *Proceedings of Toronto Meteorological Conference* 1953 (b), pp. 43-48.

Kaplan, L. D., "Regions of Validity of Various Absorption-Coefficient Approximations," *Journal of Meteorology*, American Meteorological Society, Boston, MA, Vol. 10, 1953, pp. 100-104.

Keeling, C. D., "The Concentration and Isotopic Abundance of Carbon Dioxide in the Atmosphere," *Tellus, A Quarterly Journal of Geophysics*, Swedish Geophysical Society, Stockholm, Sweden, Vol. 12, 1960.

Kostkowski, H. J., "Half Widths and Intensities from the Infrared Transmission of Thermally Excited CO_2," Office of Naval Research Contract 248(01), Johns Hopkins University, Baltimore, MD, Progress Report, October 1955.

Kruger, A. J., "The Mean Ozone Distribution from Several Series of Rocket Soundings to 52 km Latitudes from 54°S to 64°N," Goddard Space Flight Center, Greenbelt, MD, NASA Report No. X-651-73-67, 1973.

Kruse, P., L. McGlaughlin, and R. McQuistan, *Elements of Infrared Technology*, John Wiley and Sons, New York, NY, 1962.

Ladenberg, R., and F. Reiche, "Uber Selektive Absorption," *Annalen der Physik*, Leipsig, E. German, Vol. 42, 1913, p. 181.

LaRocca, A. J., and R. E. Turner, "Atmospheric Transmittance and Radiance: Methods of Calculation," Environmental Research Institute of Michigan, Ann Arbor, MI, Report No. 107600-10-T.

Laulainen, N., "Project ASTRA, Astronomical and Space Techniques for Research on the Atmosphere," University of Washington, Seattle, WA, ASTRA Publication No. 18, 1972.

Laulainen, N., *Minor Gases in the Earth's Atmosphere: A Review and Bibliography of Their Spectra*, University of Washington, Seattle, WA, ASTRA Publication No. 18, 1972.

Locke, J. L., and L. Herberg, "The Absorption due to Carbon Monoxide in the Infrared Solar Spectrum," *Canadian Journal of Physics*, National Research Council of Canada, Ottawa, Canada, Vol. 31, 1953, p. 504.

Lovell, D. J., *An Atlas of Air Absorptions in the Infrared*, University of Massachusetts, Amherst, MA, Interim Report, June 1969.

Malila, W. A., R. B. Crane, and R. E. Turner, "Information Extraction Techniques for Multispectral Scanner Data," Willow Run Laboratories for the Institute of Science and Technology, University of Michigan, Ann Arbor, MI, Report No. 31650-74-T, 1972.

Malkmus, W., "Random Lorentz Band Model with Exponential-Tailed S^{-1} Line-Intensity Distribution Function," *Journal of the Optical Society of America*, Optical Society of America, Washington, DC, Vol. 57, No. 3, 1967, pp. 323-329.

Mantz, A. W., E. R. Nichols, D. B. Alpert, and K. N. Rao, "CO Laser Spectra Studied with a 10-Meter Vacuum Infrared Grating Spectrogram," *Journal of Molecular Spectroscopy*, Academic Press, New York, NY, Vol. 35, 1970, p. 325.

Mateer, C. L., and W. L. Godson, "The Vertical Distribution of Atmospheric Ozone Over Canadian Stations from Umkehr Observations," *Quarterly Journal of the Royal Meteorological Society*, Berkshire, England, Vol. 18, No. 3, 1960.

Mayer, H., "Methods of Opacity Calculations," Los Alamos, CA, LA-647, 1947.

Mayer, H., "Methods of Opacity Calculations, V, Effect of Lines on Opacity, Methods for Treating Line Contribution," Los Alamos Scientific Laboratory, Los Alamos, CA, Report No. ACED-1870, 1947.

McCaa, D. J., and J. H. Shaw, "The Infrared Spectrum of Ozone," *Journal of Molecular Spectroscopy*, Academic Press, New York, NY, Vol. 25, No. 3, March 1968, pp. 374-397.

McClatchey, R. A., and J. E. A. Selby, "Atmospheric Attenuation of Laser Radiation from 0.76 to 31.25 μm," Air Force Cambridge Research Laboratories, Bedford, MA, Report No. TR-74-0003, January, 1974.

McClatchey, R. A., et al., "Atmospheric Absorption Line Parameters Compilation," Air Force Systems Command, Andrews Air Force Base, Washington, DC, Report No. AFCRL-TR-73-0096, 1973.

McClatchey, R. A., et al., "Optical Properties of the Atmosphere," Air Force Cambridge Research Laboratories, Bedford, MA, Report No. AFCRL-72-0497, Third Edition, 1972.

McDowell, R. S., "The ν_3 Infrared Bands of $C^{12} H_4$ and $C^{13} H_4$," *Journal of Molecular Spectroscopy*, Academic Press, New York, NY, Vol. 21, No. 3, November 1966, p. 280.

McMath, R., and L. Goldberg, "The Abundance and Temperature of Methane in the Earth's Atmosphere," *Proceedings of the American Physical Society*, American Physical Society, New York, NY, Vol. 74, 1948, p. 623.

Migeotte, M. V., "The Fundamental Band of Carbon Monoxide at 4.7 μ in the Solar Spectrum," *Physical Review*, American Physical Society, New York, NY, Vol. 75, 1949, p. 1108.

Migeotte, M., L. Neven and J. Swensson, *An Atlas of Nitrous Oxide, Methane and Ozone Infrared Absorption Bands, Part I, The Photometric Records,* University of Liege, Contract AF 61(614)-432, Phase B, Part I, Geophysics Research Directorate, AFCRC, Cambridge, MA, AD 210045.

Migeotte, M., L. Neven and J. Swensson, *An Atlas of Nitrous Oxide, Methane and Ozone Infrared Absorption Bands, Part I, The Photometric Records,* University of Liege, Contract AF 61(514)-432, Phase B, Part II, Geophysics Research Directorate, AFCRC, Cambridge, MA, AD 210046.

Migeotte, M., L. Neven and J. Swensson, "The Solar Spectrum from 2.8 to 23.7 Microns, Measures and Identifications," *Societe Royal Des Sciences De Liege, Memoires*, Special Vol. 2, 1957.

Migeotte, M., L. Neven, and J. Swensson, *The Solar Spectrum from 2.8 to 23.7 Microns, Part I, Photometric Atlas*, University of Liege, Contract AF 61(514)-432, Phase A, Part I, Geophysics Research Directorate, AFCRC, Cambridge, MA, AD 210043.

Migeotte, M., L. Neven and J. Swensson, *The Solar Spectrum from 2.8 to 23.7 Microns, Part II, Measures and Identifications*, University of Liege, Contract AF 61(514)-432, Phase A, Part II, Geophysics Research Directorate, AFCRC, Cambridge, MA, AD 210044.

Mohler, O. C., *Table of Solar Spectrum Wavelengths from 1.20 to 2.55 Microns*, University of Michigan Press, Ann Arbor, MI, 1955.

Mohler, O. C., A. K. Pierce, P. R. McMath, and L. Goldberg, *Atlas of the Solar Spectrum from 0.84 to 2.52 Microns*, University of Michigan Press, Ann Arbor, MI, 1950.

Moyers, J. L., and R. A. Duce, "Gaseous and Particulate Iodine in the Marine Atmosphere," *Journal of Geophysical Research*, American Geophysical Union, Washington, DC, Vol. 77, No. 27, 1972, pp. 5529-5238.

Murcray, D. G., A. Goldman, and A. Csoeke-Poeckh, et al., "Nitric Acid Distribution in the Stratosphere," *Journal of Geophysical Research*, American Geophysical Union, Washington, DC, Vol. 78, No. 30, 1973, pp. 7033-7038.

Murcray, D. G., T. G. Kyle, F. H. Murcray, and W. J. Williams, "Nitric Acid and Nitric Oxide in the Lower Stratosphere," *Nature*, Macmillan Journals, Ltd., London, England, Vol. 218, 1968, p. 78.

National Oceanic and Atmospheric Administration, Washington, DC, "U. S. Standard Atmosphere 1976," NOAA-S/T 76-1562, November 1976.

Palmer, C. H., "Long Path Water Vapor Spectra with Pressure Broadening, 1: 20–21.7," *Journal of the Optical Society of America*, Optical Society of America, Washington, DC, Vol. 47, 1957, p. 1024.

Penner, S. S., *Quantitative Molecular Spectroscopy and Gas Emissivities*, Addison-Wesley Publishing Co., Inc., Reading, MA, 1959.

Plass, G. N., "Models for Spectral Band Absorption," *Journal of the Optical Society of America*, Optical Society of America, Washington, DC, Vol. 48, October 1958, pp. 690-703.

Plass, G. N., "Useful Representations for Measurements of Spectral Band Absorption," *Journal of the Optical Society of America*, Optical Society of America, Washington, DC, Vol. 50, No. 9, September 1960, pp. 868-875.

Plass, G. N., and D. I. Fivel, *Astrophysical Journal*, American Astronomical Society, University of Chicago, IL, Vol. 117, 1953.

Plyler, E., and E. Barker, "Infrared Spectrum and Molecular Configuration of N_2O," *Physical Review*, American Institute of Physics, New York, NY, Vol. 38, 1931, p. 1827.

Ramanthan, K. R., and R. N. Kulkarni, "Mean Meridional Distribution of Ozone in Different Seasons," Calculated from Umkehr Observations and Probable Vertical Transport Mechanisms, *Quarterly Journal of the Royal Meteorological Society*, Berkshire, England, Vol. 86, No. 368, 1960, p. 144.

Roberts, R. E., L. M. Biberman and J. E. A. Selby, "Infrared Continuum Absorption by Atmospheric Water Vapor in the 8-12 μm Window," Paper P-1184, Institute for Defense Analyses, Arlington, VA, April 1976.

Seeley, J. E., J. T. Houghton, T. S. Moss, and N. D. Hughes, "Solar Spectrum from 1 to 6.5 Microns at Altitudes up to 15 KM," *Philosophical Transactions of the Royal Society*, Royal Philosophical Society of London, London, England.

Selby, J., "Atmospheric Transmittance, I.," Notes for Advanced IR Technology, University of Michigan Engineering Summer Conference, Ann Arbor, MI, 1974.

Selby, J. E. A., and R. M. McClatchey, "Atmospheric Transmittance from 0.25 to 2.85 μm: Computer Code LOWTRAN 2," Air Force Research Laboratories, Bedford, MA, Report No. AFCRL-72-0745, 1972.

Selby, J. E. A., E. P. Shettle, and R. A. McClatchey, "Atmospheric Transmittance from 0.25 to 0.28 μm: Supplement LOWTRAN 3B (1976)," Air Force Geophysics Laboratory, Bedford, MA, AFGL-TR-76-0258, 1 November 1976.

Shaw, J. H., "A Determination of the Abundance of N_2O, CO, and CH_4 in Ground Level Air at Several Locations Near Columbus, Ohio," Air Force Research Laboratories, Bedford, MA, Scientific Report No. 1 Contract AF 19(604)-2259.

Shaw, J. H., "Monthly Report on Infrared Temperature Sounding," Ohio State University Research Center, Columbus, OH, RF Project 2469, Report No. 16, October 1968.

Shaw, J. H., "The Abundance of Atmospheric CO above Columbus, Ohio," Ohio State University Research Foundation, Columbus, OH (Contract AF 19(604)-1003), Report No. AFCRC TN 57-212, 1957.

Shaw, J. H., R. M. Chapman, J. N. Howard, and M. L. Oxholm, "A Grating Map of the Solar Spectrum from 3.0 to 5.0 Microns," *Astrophysical Journal*, University of Chicago Press, Chicago, IL, Vol. 113, No. 2, 1951; *see also* J. Yarnell and R. M. Goody, "Infrared Solar Spectroscopy in a High-altitude Aircraft," *Journal of Scientific Instruments*, Institute of Physics, Bristol, England, Vol. 29, 1952, p. 352.

Shaw, J. H., and J. N. Howard, "A Quantitative Determination of the Abundance of Telluric CO above Columbus, Ohio," *Physical Review*, American Physical Society, Washington, DC, Vol. 87, April 1957, p. 380.

Shaw, J. H., and J. N. Howard, "Absorption of Telluric CO in the 23 μ Region," *Physical Review*, American Physical Society, Washington, DC, Vol. 87, June 1952, p. 679.

Shaw, J. H., and H. H. Nielson, "Infrared Studies of the Atmosphere," Ohio State University Research Foundation, Columbus, OH, Final Report on Contract AF 19(122)-65, 1954.

Shaw, J. H., M. L. Oxholm and H. H. Classen, "The Solar Spectrum from 7 to 13 Microns," *Astrophysical Journal*, University of Chicago Press, Chicago, IL, Vol. 116, No. 3, 1952; *see also* J. Yarnell and R. M. Goody, "Infrared Solar Spectroscopy in a High-altitude Aircraft," *Journal of Scientific Instruments*, Institute of Physics, Bristol, England, Vol. 29, 1952.

Shaw, J. H., G. B. B. M. Sutherland and T. W. Wormell, "Nitrous Oxide in the Earth's Atmosphere," *Physical Review*, American Physical Society, Washington, DC, Vol. 74, June 1948, p. 978.

Sissenwine, N., D. Grantham, and N. S. Salmela, "Humidity Up to the Mesopause," Air Force Research Laboratories, Bedford, MA, Report No. AFCRL-68-0550, 1968.

Slobod, R. L., and M. E. Krogh, "Nitrous Oxide as a Constituent of the Atmosphere," *Journal of the American Chemical Society*, American Chemical Society, Washington, DC, Vol. 72, May 1950, pp. 1175-1177.

Spanbauer, R. N., K. N. Rao, and L. H. Jones, *Journal Molecular Spectroscopy*, Academic Press, New York, NY, Vol. 16, No. 1, May 1965, p. 100.

Stanford, J. L., "Stratospheric Water-Vapor Upper Limits Inferred from Upper-Air Observations, Part I: Northern Hemisphere," *Bulletin of American Meteorological Society*, American Meteorological Society, Boston, MA, Vol. 55, No. 3, 1974, p. 194.

Stratton, J. A., and H. G. Houghton, "A Theoretical Investigation of the Transmission of Light Through Fog," *Physical Review*, American Physical Society, Washington, DC, Vol. 38, No. 159, 1931.

Stratton, J. S., *Electromagnetic Theory*, McGraw-Hill, New York, NY, 1941.

Stewart, H. S., and J. A. Curcio, "The Influence of Field-of-View on Measurements of Atmospheric Transmission," *Journal of the Optical Society of America*, Optical Society of America, Washington, DC, Vol. 42, No. 801, 1952.

Streete, J. L., "Infrared Measurements of Atmospheric Transmission at Sea Level," *Applied Optics*, Optical Society of America, Washington, DC, Vol. 7, No. 8, 1968, pp. 1545-1549.

Stull, V. R., P. J. Wyatt, and G. W. Plass, "The Infrared Transmittance of Water Vapor," *Applied Optics*, Optical Society of America, Washington, DC, Vol. 3, No. 2, February 1964, pp. 229-241. Aeronutronic Report U-1718, Aeronutronic Systems, Inc., Newport Beach, CA, 1962.

Summerfield, M., "Pressure Dependence of the Absorption in the 9.6 Micron Band of Ozone," Thesis, California Institute of Technology, Pasadena, CA, 1941.

Talbert, W. W., H. A. Templin, and R. E. Morrison, *Quantitative Solar Spectral Measurements at Mt. Chacaltaya (17,000 ft.)*, U. S. Naval Ordnance Laboratory, White Oak, MD; *see also Journal of the Optical Society of America*, Optical Society of America, Washington, DC, Vol. 47, 1957, p. 1056.

Taylor, J. H., and H. W. Yates, "Atmospheric Transmission in Infrared," *Journal of the Optical Society of America*, Optical Society of America, Washington, DC, Vol. 47, No. 3, March 1957, pp. 223-226.

Valley, S. L., *Handbook of Geophysics and Space Environments*, Air Force Research Laboratories, Bedford, MA, 1965.

Van de Hulst, H. C., "Theory of Absorption Lines in the Atmosphere of the Earth," *Annales d'Astrophysique*, Paris, Vol. 8, 1945, pp. 21-34.

Verkateswaren, S., J. G. More, and A. J. Krueger, "Determination of the Vertical Distribution of Ozone by Satellite Photometry," *Journal of Geophysical Research*, American Geophysical Union, Washington, DC, Vol. 66, 1961.

Walshaw, C., "Integrated Absorption by the 9.6 μm Band of Ozone," *Quarterly Journal of the Royal Meteorological Society*, Berkshire, England, Vol. 83, 1957, pp. 315-321.

Wark, D. Q., and M. Wolk, *Monthly Weather Review*, Government Printing Office, Washington, DC, Vol. 88, 1960.

Weickman, H. J., and J. J. Afum Kampe, "Physical Properties of Cumulus Clouds," *Journal of Meteorology*, American Meteorological Society, Boston, MA, Vol. 10, No. 204, 1953.

Weinreb, M. P., and A. C. Neuendorffer, "Method to Apply Homogeneous-Path Transmittance Models to Inhomogeneous Atmospheres," *Journal of Atmospheric Sciences*, American Meteorological Society, Boston, MA, Vol. 39, 1973, p. 662.

Williamson, J. G., K. N. Rao, and L. H. Jones, "High Resolution Infrared Spectra of Water Vapor ν_2 Band of $H_2{}^{18}O$," *Journal of Molecular Spectroscopy*, Academic Press, New York, NY, Vol. 40, No. 2, November 1971.

Wilson, M, K., and R. A. Ogg, "The Infrared Spectrum and Structure of Ozone," *Journal of Chemical Physics*, American Institute of Physics, New York, NY, Vol. 18, 1950, p. 766.

Winters, B. H., et al., "Line Shape in the Wing Beyond the Band Head of the 4.3 μ Band of CO_2," *Journal of Quantitative Spectroscopy and Radiative Transfer*, Pergamon Press, New York, NY, Vol. 4, 1964, p. 527-537.

Wyatt, P. J., V. R. Stull, and G. N. Plass, "The Infrared Transmittance of Water Vapor," *Applied Optics*, Optical Society of America, Washington, DC, Vol. 3, No. 2, February 1964, pp. 229-241. Aeronutronic Report U-1717, Aeronutronic Systems, Inc., Newport Beach, CA, 1962.

Wyatt, P. J., V. R. Stull, and G. N. Plass, "Infrared Transmission Studies, Vol. 2: The Infrared Absorption of Water Vapor," Ford Motor Company, Report No. SSD-TDR-62-127, Vol. 2, 1962(a).

Wyatt, P. J., V. R. Stull, and G. N. Plass, "Quasi-Random Model of Band Absorption," *Journal of the Optical Society of America*, Optical Society of America, Washington, DC, Vol. 52, No. 11, November 1962 (b), p. 1209.

Yaroslavskii, N. G., and A. E. Stanevich, "The Long Wavelength Infrared Spectrum of H_2O Vapor and the Absorption Spectrum of Atmospheric Air in The Region 20-2500 μ (500-4 cm^{-1})," *Optics and Spectroscopy*, Optical Society of America, Washington, DC, Vol. 7, November 1959, p. 380, also *Optika Spektroskopiia*, Leningrad, U.S.S.R., Vol. 5, 1958, p. 382.

Yates, H. W., "The Absorption Spectrum from 0.5 to 26 microns of a 1000 ft. Atmospheric Path at Sea Level," Naval Research Laboratory, Washington, DC, NRL Report 5033, September 1957.

Yates, H. W., and J. H. Taylor, "Infrared Transmission of the Atmosphere," U. S. Naval Research Laboratory, Washington, DC, NRL Report 5453, 1960.

Young, L. A., "Infrared Spectra," *Journal of Quantitative Spectroscopy and Radiative Transfer*, Pergamon Press, New York, NY, Vol. 8, No. 2, February 1968, pp. 693-716.

Chapter 6

PROPAGATION THROUGH
ATMOSPHERIC TURBULENCE

Robert E. Hufnagel

The Perkin-Elmer Corporation

CONTENTS

6. Propagation Through Atmospheric Turbulence

6.1 Introduction

6.1.1. Symbols, Nomenclature and Units.
Table 6-1 lists the symbols, nomenclature and units used in this chapter.

Table 6-1. Symbols, Nomenclature and Units

Symbols	Nomenclature	Units
A	System aperture area	m^2
a	Entrance pupil radius	m
$b(z)$	Beam radius parameter at distance z	m
$b_0(z)$	Vacuum propagation beam radius parameter	m
C^2	Structure function for subscripted quantity	(Subscript units)^2 $m^{-2/3}$
c	Speed of light in vacuum	m sec^{-1}
D	Structure function for subscripted quantity	(Subscript units)^2
E	Irradiance	W m^{-2}
$\mathcal{E}$	Wave field strength	W$^{1/2}$ m^{-1}
e	2.718 ...	–
F	Focal length or distance to focus	m
f	Temporal frequency $0 < f < \infty$	Hz
$\mathbf{f}$	Spatial frequency $-\infty < \mathbf{f} < \infty$	Cycles m^{-1}
G	Weighted atmospheric integral	(m)$^{q+1/3}$
g	Acceleration of gravity	m sec^{-2}
H	Aperture filter function	–
$\mathcal{H}$	Beam profile filter function	–
h	Altitude above local terrain	m
$\tilde{h}$	Altitude above mean sea level	m
h_0	Altitude of lower end of path	m
$\mathcal{G}$	Normalized point spread function	m^{-2}
i	$\sqrt{-1}$	–
J_1	First order Bessel function	–
K	Spatial frequency	rad m^{-1}
k	Wave vector $k \equiv 2\pi/\lambda$	m^{-1}
L	Path length	m
L_0	Outer scale of turbulence	m

Table 6-1. Symbols, Nomenclature and Units *(Continued)*

Symbols	Nomenclature	Units
ℓ_0	Inner scale of turbulence	m
$M(\mathbf{f})$	Image modulation transfer function	–
$N.$	Number of degrees of freedom	Integer
n	Index of refraction	–
$\mathcal{P}$	Local atmospheric pressure	Millibars
$P(R)$	Probability density function for variable R	$(R\text{ units})^{-1}$
p, p	Independent two-dimensional position vectors	m
Q, Q_0	Radiation flux densities	W m^{-2}
q	Altitude weighing parameter	–
R	Generalized random variable	Arbitrary
Ri	Richardson's number	–
$\mathbf{r}$	Independent position or distance vector	m
r_0	Coherence diameter	m
S	Normalized collected power	–
s	Subscript for short exposure	–
T	Optical transfer function (OTF)	–
$\mathcal{T}$	Local absolute temperature	K
T_0	OTF of system with no turbulence	–
t	Time	sec
U	Wind parameter	m sec^{-1}
$\overline{\mathbf{U}}$	Horizontal component of local average wind	m sec^{-1}
$\mathbf{v}$	Wind or pseudo wind velocity	m sec^{-1}
$\mathbf{v}_\perp$	Projection of $\mathbf{v}$ on a plane normal to the line of sight	m sec^{-1}
$W(f)$	One-sided temporal Wiener spectral density function for subscripted variable	(Subscript units)^2 Hz^{-1}
x, y, z	Cartesian coordinate or coordinate subscripts	m
α	Image centroid shift	m
β	Beam centroid shift	m
ζ	Propagation zenith angle	degrees
$\bar{\zeta}$	Sighting zenith angle	degrees
η	Intensity fluctuation parameter	–
Θ	Complex part of generalized Gaussian random variable	–
θ	Potential temperature	K
λ	Wavelength of light	m
μ	Source angular radius	rad
ν	Number of dimensions in vector space	Integer
ξ	Solar elevation angle	degrees
π	3.14159 ...	–

Table 6-1. Symbols, Nomenclature and Units (*Continued*)

Symbols	Nomenclature	Units
ρ	Normalized autocorrelation function of subscripted variable	—
σ	Standard deviation of subscripted variable	(Subscripted variable units)
σ^2	Variance of subscripted variable	(σ units)^2
Υ	Generalized log-normal random variable	—
Φ	Spatial Wiener spectrum	(Subscripted variable units)^2 $(\text{rad m}^{-1})^{-\nu}$
ϕ	Relative wavefront phase	rad
χ	Log_e amplitude	Nepers
Ψ	Generalized Gaussian random variable	—
Ω	Fresnel parameter	—
ω	Electric field oscillation frequency	sec^{-1}
∇	$(\partial/\partial x,\ \partial/\partial y,\ \partial/\partial z)$	m^{-1}
∇^2	$\partial^2/\partial x^2 + \partial^2/\partial y^2 + \partial^2/\partial z^2$	m^{-2}

6.1.2. Notation. Vector quantities are indicated by bold-face characters, and their corresponding scalar magnitudes by the corresponding light-face characters. Vector components are indicated by x, y, and z subscripts. Thus, in three-dimensional space

$$\mathbf{K} \equiv (K_x, K_y, K_z) \text{ and } K \equiv (K_x{}^2 + K_y{}^2 + K_z{}^2)^{1/2} \equiv |\mathbf{K}| \qquad (6\text{-}1)$$

Area and volume integrals are written in vector notation. In three dimensions, for example,

$$\iiint \Phi(\mathbf{K})\, d\mathbf{K} \equiv \iiint \Phi(K_x,\ K_y,\ K_z)\, dK_x dK_y dK_z \qquad (6\text{-}2)$$

The single set of limits – ∞ to + ∞ in a ν-dimensional integral indicate an integration over the entire ν-dimensional space.

Vector products are always dot products. Thus, in three dimensions,

$$\mathbf{K} \cdot \mathbf{r} \equiv xK_x + yK_y + zK_z \qquad (6\text{-}3)$$

When Cartesian coordinates are used, the positive z axis is aligned along the average direction of light propagation. The source of light lies in the plane $z = 0$, and the target or receiver is located in the plane $z = L$.

A superscript asterisk indicates the complex conjugate of a complex quantity. Vertical bars indicate the magnitude of the enclosed quantity. Angle brackets indicate an ensemble average of the enclosed quantity. In most cases, a time average may be substituted without changing the results.

Greek and Roman letter subscripts other than o, s, x, y, and z indicate that the symbol describes the subscripted quantity. Thus, σ_E is the standard deviation of the irradiance E.

6.1.3. Qualitative Description. Turbulence induces random irregularities in the atmosphere's index of refraction. In passing through these irregularities, optical wavefronts become distorted. As radiation with a distorted wavefront continues to propagate, its local irradiance also must vary under the focusing and spreading effects of that wavefront. These effects are random and complicated, but by dealing with statistically averaged quantities one can quantitatively describe the effects of turbulence quite well.

Distortions of wavefronts are the primary cause of image motion, image distortion, and added blur in optical imaging systems. Wavefront distortions also are the primary cause of beam wander and added spreading in the transmission of narrow beams of radiation.

The term *scintillation* is used to describe the turbulence-induced, random variations of local irradiance. This includes the twinkling of starlight, fading in optical communications channels, and some speckle effects in radiation distributions at targets.

6.1.4. Definition of Geometries. Four basic geometries can be used to describe most practical situations:

(1) *Plane wave propagation* is for radiation from stars and other exo-atmospheric sources. The radiation must be essentially collimated before interacting with the turbulent medium.

(2) *Spherical wave propagation* is the propagation of light from sources which are themselves in or near the turbulence. This geometry can describe the imaging of objects in the atmosphere, and also the propagation of light radiating from point sources or from strongly divergent optics. In this chapter, the sources of spherical waves is always at the location $z = 0$.

(3) *Beam wave propagation* is the propagation of focused or collimated beams of light in which the lateral extent of the beam is not large compared to the turbulence-induced lateral distortions of the radiation. This situation is often encountered in the propagation of laser beams.

(4) *Folded paths* occur when the radiation is reflected off a mirror or a retro-reflector and passes through the same turbulent region two or more times. In this case, the random disturbances to the inward and outward beams become correlated, and formulae derived for cases 1, 2, or 3 above must be modified. Equations for folded paths are not given in this chapter. Both Smith, and Smith and Pries, give examples of folded path analysis techniques.

6.1.5. Methodology for Solving Problems. The analysis of problems concerning propagation through turbulent media can be conveniently segmented into three parts.

First, statistics describing the random irregularities of the atmospheric index of refraction along the propagation path must be quantitatively determined. This is discussed in Section 6.2 of this chapter.

Second, one must determine the statistics for the propagated radiation. This involves choosing the appropriate solution to the wave equation, and substituting into it the statistics for the atmospheric path found in Section 6.2. This step is discussed in Section 6.3.

Third, the statistics of the radiation must be translated into appropriate statistics which describe overall system performance. This is discussed in Sections 6.4 through 6.8.

Section 6.9 contains a brief mathematical appendix for general use.

6.2. Atmospheric Turbulence

6.2.1. Index of Refraction. The index of refraction for the Earth's atmosphere (below 40 km) depends on temperature, pressure, humidity, and optical wavelength [6-1].

Except for supersonic air flow, the effects of turbulence-induced pressure variations are negligible compared to those of temperature or humidity fluctuations. The effect of humidity fluctuations upon the index of refraction becomes stronger as the optical wavelength increases beyond each major water vapor absorption band. There is very little literature on this subject. (See Friehe, et al.) However, for most optical engineering applications the effect is small enough to be ignored, except perhaps in very humid atmospheres or in very critical applications.

The dependence of the index upon wavelength can usually be ignored when considering fluctuations of the index. However, the wavelength dependence of the average (non-fluctuating) part of the index is often important. Section 6.4.4 illustrates this latter point.

With these simplifications, index of refraction fluctuations depend only on fluctuations in temperature. The variation of index of refraction, n, with temperature is

$$\frac{dn}{d\mathcal{T}} = \frac{78\mathcal{P}}{\mathcal{T}^2} \times 10^{-6} \tag{6-4}$$

where $\mathcal{P}$ = local air pressure, mb

$\mathcal{T}$ = absolute temperature, K

Table 6-2 gives values of $|dn/d\mathcal{T}|$ versus altitude for the 1962 U.S. standard model atmosphere. Below 10 km, $|dn/d\mathcal{T}|$ decays exponentially with altitude with a 12.6 km (e^{-1}) scale height; above 10 km the exponential scale height is 6.6 km.

A convenient rule of thumb for use near sea level is that $|dn/d\mathcal{T}| \cong 10^{-6}$.

6.2.2. Atmospheric Spatial Statistics.

Because most optical phenomena of interest depend on differential rather than absolute optical path lengths, the spatial statistics of random index variations and of random wavefronts are given in terms of *structure functions* rather than the more conventional autocovariance or autocorrelation functions. The structure function, $D_R(\mathbf{r})$, for an arbitrary spatially distributed random variable, $R(\mathbf{r})$, is defined as

$$D_R(\mathbf{r}) = \langle [R(\mathbf{r}_1) - R(\mathbf{r}_2)]^2 \rangle \tag{6-5}$$

where $\mathbf{r} = \mathbf{r}_1 - \mathbf{r}_2$, and the angle brackets indicate an ensemble average of the enclosed quantity. If the random process generating R is isotropic, then $D_R(\mathbf{r})$ is a function only of $r = |\mathbf{r}|$. The subscript under the D is a reminder that the random variable is R. Thus the structure function for temperature fluctuations is $D_\mathcal{T}(\mathbf{r})$, etc.

The structure function is related to the more familiar statistics, $\sigma_R{}^2$, the variance of R, and the autocorrelation function, $\rho_R(r)$, by the relationship

$$D_R(\mathbf{r}) = 2\sigma_R{}^2 [1 - \rho_R(\mathbf{r})] \tag{6-6}$$

Table 6-2. Nominal* Values of the Ratio of Index of Refraction** Fluctuations to Temperature Fluctuations for Dry Air Versus Altitude

Altitude Above Sea Level, $\tilde{h}$ (km)	$-\partial n/\partial \mathcal{T}$ (K^{-1})
0	0.95×10^{-6}
1	0.88×10^{-6}
2	0.82×10^{-6}
3	0.76×10^{-6}
4	0.70×10^{-6}
5	0.65×10^{-6}
6	0.59×10^{-6}
8	0.50×10^{-6}
10	0.42×10^{-6}
12	0.32×10^{-6}
14	0.24×10^{-6}
16	0.17×10^{-6}
18	0.13×10^{-6}
20	0.09×10^{-6}
22	0.07×10^{-6}
24	0.05×10^{-6}

*Based on 1962 U.S. Standard Atmosphere
**For infrared wavelengths and dry air only. See Section 6.2.1.

The ν-dimensional Fourier transform of $\sigma_R{}^2\rho_R(\mathbf{r})$ is the Wiener Spectrum, $\Phi_R(\mathbf{K})$, of the random variable, $R(\mathbf{r})$, where ν is the dimensionality of the space occupied by the vector $\mathbf{r}$. In this chapter spatial frequency, $\mathbf{K}$ (also a ν-dimensional vector), will be expressed in rad m^{-1}. The Fourier transformation used is*

$$\Phi_R(\mathbf{K}) = \frac{\sigma_R{}^2}{(2\pi)^\nu} \int\!\!\!\int\!\!\!\int_{-\infty}^{+\infty} \rho_R(\mathbf{r})\, e^{-i\mathbf{K}\cdot\mathbf{r}}\, d\mathbf{r} \tag{6-7}$$

If ρ_R is a function of $r = |\mathbf{r}|$ only, Φ_R will be a function of $K = |\mathbf{K}|$ only. The corresponding inverse transform is

$$\sigma_R{}^2\rho_R(\mathbf{r}) = \int\!\!\!\int\!\!\!\int_{-\infty}^{+\infty} \Phi_R(\mathbf{K})\, e^{i\mathbf{K}\cdot\mathbf{r}}\, d\mathbf{K} \tag{6-8}$$

Since by definition $\rho_R(o) \equiv 1$, Equations (6-6) and (6-8) can be combined to yield the relationship

$$D_R(\mathbf{r}) = 2 \int\!\!\!\int\!\!\!\int_{-\infty}^{+\infty} \Phi_R(\mathbf{K})[1 - e^{i\mathbf{K}\cdot\mathbf{r}}]\, d\mathbf{K} \tag{6-9}$$

which has the useful property that D_R will often be finite even when the integral of Φ_R is infinite or undefined. This is another principal reason for using structure functions rather than correlation functions. When the integral of Φ_R is finite, $\sigma_R{}^2$ can be computed by evaluating Equation (6-8) with $\mathbf{r} = 0$.

6.2.3. **Atmospheric Temporal Statistics.** For many problems, it is necessary to know about temporal changes in the radiation disturbed by turbulence. For this, Taylor's hypothesis usually works quite well.

In this hypothesis, one may assume that the shape of the turbulent field of index fluctuations is "frozen" and that the whole field moves in its frozen form with the local mean wind. In reality, of course, the fluctuations change as they move along, just as clouds change their shape as they move. The point of the hypothesis is that these shape changes are very slow compared to the gross movement past a stationary observer, and that for many statistical purposes they can be ignored.

Only that component of the local wind velocity which is perpendicular to the line of sight causes temporal changes in the received light field. The symbol $v_\perp$ is often used to denote this component.

---

*There is an alternative definition for Fourier transformations in which spatial frequencies have the units of cycles/unit length rather than radians/unit length. When discussing the atmosphere, the transform definitions given above will be used and spatial frequencies will always be described by the symbol $\mathbf{K}$. When describing images, we use the alternative definitions, given by Equation (6-56); in this case spatial frequencies, in units of cycles/meter, are always described by the symbol $\mathbf{f}$. For electronic engineers, a third alternative (described in the next section) is used for temporal signals.

When an optical system is tracking a moving object, or when the system itself is moving, one should use the relative velocity between each portion of the line of sight and the local mean wind. This relative velocity is called the *pseudowind.*

With Taylor's hypothesis, the relationship between the one-sided temporal Wiener spectrum, $W_R(f)$, and the corresponding two-dimensional spatial spectrum with $\Phi_R(\mathbf{K}) \equiv \Phi_R(K_x, K_y)$ becomes

$$W_R(f) = \frac{4\pi}{v_\perp} \int\limits_{-\infty}^{+\infty} \Phi_R\left(\frac{2\pi f}{v_\perp}, K_y\right) dK_y \tag{6-10}$$

where, for simplicity, it has been assumed that $v_\perp$ lies in the direction of the x axis. As defined here, the temporal frequency, f, in Hertz, assumes only positive values. That is, W_R obeys the relationship

$$\sigma_R{}^2 = \int\limits_0^\infty W_R(f)\,df \tag{6-11}$$

If Φ_R is formed by a summation (or integration) over several regions of turbulence, each with a different value of $v_\perp$, then Equation (6-10) must be changed in a corresponding way to a summation (or integration) of integrals where each term summed is the contribution to Φ_R corresponding to each value of $v_\perp$.

Taylor's hypothesis fails when $v_\perp$ is less than the turbulent fluctuations in wind velocity. This will happen, for example, when the average wind direction is parallel to the line of sight.

6.2.4. Kolmogorov Theory of Turbulence. In 1941 Kolmogorov [6-2] proposed a mathematical model for fluid velocity turbulence. Key to this model is the hypothesis that the kinetic energy associated with the larger eddies is redistributed without loss to successively smaller and smaller eddies until finally dissipated by viscosity.

In the open atmosphere, solar heating generates atmospheric kinetic energy over scale sizes that extend from a few meters to global scales*.

Kinetic energy leaves the air by frictional generation of heat over scales that, near the earth's surface, are smaller than one centimeter. The intervening range (typically from a few millimeters to a few meters) is called the *inertial sub range.*

Over this sub range the structure function for temperature fluctuations obeys an isotropic two-thirds power law:

$$D_{\mathcal{T}}(\mathbf{r}) \simeq C_{\mathcal{T}}{}^2\, r^{2/3} \tag{6-12}$$

where $C_{\mathcal{T}}{}^2$ is a parameter of proportionality**.

This equation is valid only over the inertial sub range, $\ell_o \ll r \ll L_o$, where ℓ_o and L_o are called, respectively, the inner and outer scales of turbulence. For $r \ll \ell_o$, $D_{\mathcal{T}}(\mathbf{r})$ varies as r^2.

Within a few meters of the ground, $L_o \simeq h/2$, where h is the distance from the ground. Estimates of ℓ_o and L_o and their effects on optical propagation are given in Tatarskii (1961), Reinhardt and Collins, and Hufnagel (July 1966).

---

*Infrared radiation exchange processes, gravity wave effects (Section 6.2.6), and wind interactions with the ground also cause transfers of kinetic energy. These are also typically at large scales.
**Because of an effect known as *intermittency,* the exponent of r in Equation (6-12) departs slightly from 2/3. For most engineering purposes, this effect may be ignored. (See Van Atta.)

The parameter of proportionality in Equation (6-12) is called the structure parameter, subscripted by the fluctuating quantity. Thus $C_{\mathcal{T}}{}^2$ is the *temperature structure parameter,* and $C_n{}^2$, defined in a similar manner, is the *index (of refraction) structure parameter*.*

In optical calculations, the scales of engineering interest will usually fall within the inertial subrange. Because of this, $C_n{}^2$ *is the single most important parameter for an optical engineer in describing the turbulent atmosphere.*

An alternative, equally useful, and completely equivalent method of describing the statistics of a turbulent quantity is by its three-dimensional Wiener spectrum. The spectrum corresponding to the structure function in Equation (6-12) is

$$\Phi_n(\mathbf{K}) = 0.033 C_n{}^2 K^{-11/3} \qquad (6-13)$$

where $\mathbf{K}$ is spatial frequency measured in rad m^{-1}. For an isotropic medium, $\mathbf{K}$ becomes a scalar.

In MKS units, $C_{\mathcal{T}}{}^2$ has the units $\mathrm{K}^2\mathrm{m}^{-2/3}$, and $C_n{}^2$ has the units $\mathrm{m}^{-2/3}$. $C_n{}^2$ is related to $C_{\mathcal{T}}{}^2$ by the formula

$$C_n{}^2 = |\partial n/\partial \mathcal{T}|^2 C_{\mathcal{T}}{}^2 \qquad (6-14)$$

where, for dry air and optical wavelengths, $|\partial n/\partial \mathcal{T}|$ is given by Equation (6-4) or Table 6-2. In a similar manner

$$\Phi_n(\mathbf{K}) = |\partial n/\partial \mathcal{T}|^2 \Phi_{\mathcal{T}}(\mathbf{K}) \qquad (6-15)$$

6.2.5. Turbulence Near the Ground. The ground affects the air in two ways. First, it shears the wind through a frictional interaction with the free stream air flow. Second, it serves as a source (or sink) of thermal energy which can be imparted to the air.

In typical, sunny, daytime conditions the ground becomes hotter than the air and, therefore, becomes a source of heat. When this happens, thermal convection takes place and the conditions are said to be unstable or *active.* At night (especially with dry, cloudless skies) the ground cools by radiation, and becomes colder than the air above it. It thus becomes a heat sink with respect to the air, but convection is inhibited since the cold air is below the less dense warm air. When this happens, conditions are said to be stable or *passive.* When the ground and the air are the same temperature, conditions are said to be *neutral.*

To quantify these conditions, one introduces a dimensionless stability parameter. Called the Richardson number, *Ri*, it is defined by

$$Ri = \frac{g}{\overline{\mathcal{T}}} \frac{\partial \overline{\theta}/\partial h}{|\partial \overline{\mathbf{U}}/\partial h|^2} \qquad (6-16)$$

where g = acceleration of gravity
$\overline{\mathcal{T}}$ = mean absolute air temperature
h = altitude
$\partial \overline{\theta}/\partial h$ = $(\partial \overline{\mathcal{T}}/\partial h) + (9.8 \times 10^{-3})$
$\partial \overline{\mathbf{U}}/\partial h$ = vertical gradient of mean horizontal wind

---

*Some authors call $C_{\mathcal{T}}$ rather than its square, $C_{\mathcal{T}}{}^2$ the temperature structure parameter.

The factor 9.8×10^{-3} (in K m^{-1}) used above is the adiabatic lapse rate of a parcel of dry air [6-3]. Negative values of Ri correspond to active conditions, and positive to passive ones.

The structure parameter for temperature fluctuations near the ground is given by

$$C_{\mathcal{T}}^2(h) = h^{4/3}(\partial\bar{\theta}/\partial h)^2 f_3(Ri) \quad (6\text{-}17)$$

where $f_3(Ri)$ is plotted as in Figure 6-1. In using this formula, one must be

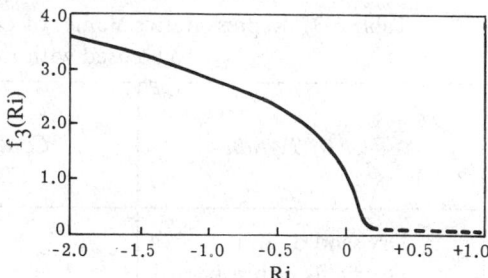

Fig. 6-1. Dimensionless turbulence strength parameter versus the Richardson number [6-4].

very careful to measure both $d\bar{\mathcal{T}}/dh$ and Ri at the same height, h, since they vary rapidly with h. In addition, the thermometers used to measure $\bar{\mathcal{T}}$ must be shielded from solar radiation.

When $Ri \gtrsim 0.2$, the air is so stably stratified that the development of turbulence is inhibited, but not completely. There is no satisfactory theory for turbulence under these conditions, and $f_3(Ri)$ is shown as a dotted line at this point in the figure, indicating that significant departures from these values can occur.

An alternative formulation for $C_{\mathcal{T}}(h)$ valid for active conditions only [6-4] is

$$C_{\mathcal{T}}^2(h) = \frac{2 \times 10^{-3} Q^{4/3} h^{-4/3}}{\left[1 + \dfrac{14000 u_*^3}{hQ}\right]^{2/3}} \quad (6\text{-}18)$$

where Q = the upward convective heat flux, W m^{-2}

h = height, m

u_* = a characteristic friction velocity, m sec^{-1}; u_* may be approximated by the relationship $u_* \cong 0.35\, h|\partial U/\partial h|$, valid for heights where $h(\partial\bar{u}/\partial h)$ is independent of h.

Typically, u_* is one order of magnitude smaller than local wind speed. The numerical coefficients in Equation (6-18) will differ from the values given here if the ground is not within 2 km of sea level.

In Equation (6-18), the height dependence of $C_{\mathcal{T}}^2(h)$ can be seen more clearly. For low winds and/or large values of h, $C_{\mathcal{T}}^2 \sim h^{-4/3}$; but for low values of h or high winds, $C^2 \sim h^{-2/3}$. Note that increasing winds tend to decrease $C_{\mathcal{T}}^2$.

The value of Q depends on the amount of sunlight falling on the ground, as well as on the wind and the properties of the ground itself.

As a rough guide, one may take

$$Q = (Q_0 \sin \xi) - 50, \quad \text{W m}^{-2} \quad (6\text{-}19)$$

where ξ is the solar elevation angle and Q_0 is as given in Table 6-3.

A typical value for C_n^2 at a point located one meter above dry, strongly sunlit, grassy, level ground is 10^{-12} m$^{-2/3}$. When sun illumination is strong, the transition between the $h^{-2/3}$ and the $h^{-4/3}$ dependence of C_n^2 will typically occur at an altitude of a few meters.

Figure 6-2 shows a plot of C_n^2, derived from measured values of $C_{\mathcal{T}}^2$ versus time of

Table 6-3. Representative Values of Q_0 Versus Terrain and Cloud Cover,
to be used with Equation (6-19)

Type of Terrain	Q_0 Continuously Clear Sky	Q_0 Overcast Sky
Dry sand or lava	500	200
Dry fields or brush	400	150
Wet fields	200	70

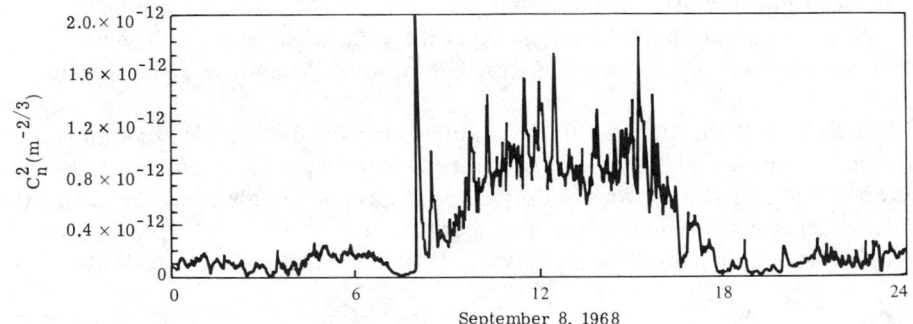

September 8, 1968

Fig. 6-2. Refractive index structure parameter versus time of day. The measurements were derived from temperature-structure-function measurements made with vertical spacings of 0.01 m at an elevation of 2 m [6-6].

day, two meters above a grassy plain near Boulder, Colorado. The sky was clear until 1630, followed by a partial covering by cumulus clouds. The increase in turbulence associated with midday solar heating is clearly evident. The minima near dawn and dusk occurred when the atmosphere passed through neutral conditions.

Associated with strong active convection is a phenomenon known as pluming, in which the heated air rises in the form of giant bubbles or plumes. Cool air descends between these plumes to replace the hot air. The turbulent temperature fluctuations in the hot plumes are greater than those in the descending cool air. Thus, there are spatial and temporal variations in C_T^2 even at the same altitude [6-5].

Changes in solar heating caused by clouds or variations in the properties of the underlying terrain also cause variations in C_T^2.

Equations (6-17) and (6-18) are valid from the ground up to near the first temperature inversion, which in a hot summer afternoon may be over 1000 m high. In the early morning, however, the first inversion may occur at a few tens of meters.

At the first inversion there is often a local peak in the strength of $C_T^2(h)$, which is typically an order of magnitude larger than $C_T^2(h)$ in the region before the inversion.

Figure 6-3 shows a typical profile which illustrates this phenomenon. On the left, C_n^2 is plotted on a logarithmic scale, and the local average temperature is plotted on a linear scale on the right. These measurements were made with an airplane flying over open water. The first inversion began at 250 m. The atmosphere below this region was active.

At night, under conditions of ground radiative cooling, a strong inversion starts at the ground and extends upward for several tens of meters. Equation (6-17) can be used to

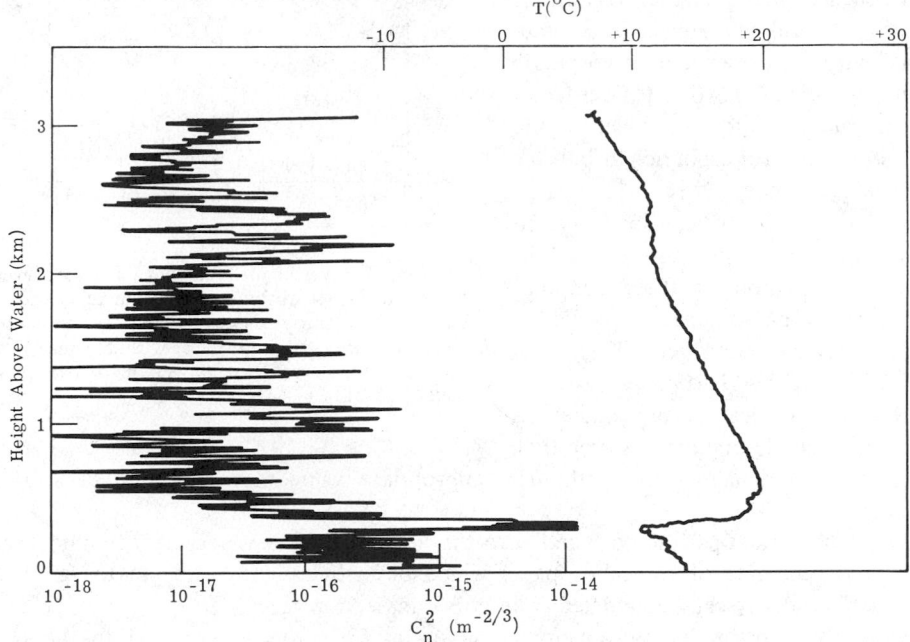

Fig. 6-3. C_n^2 versus altitude for the first 3000 m over the ocean. The local air temperature is shown on the right, with the scale on the upper right. (Note the increase in C_n^2 at the base of the strong temperature inversion.) [6-7].

estimate $C_{\mathcal{T}}^2$ in this region, but the average temperature gradient, $\partial \mathcal{T}/\partial h$, varies rapidly in space and time and cannot be easily estimated. A typical value of C_n^2 in this region is 2×10^{-14} m$^{-2/3}$ [6-7]. The dependence of $C_{\mathcal{T}}^2$ on altitude is irregular, but for most analytical purposes one can assume that above two meters, C_n^2 is constant throughout the strong ground inversion region, with a rapid drop above this region to the free atmospheric value (described in the next section).

6.2.6. Turbulence in the Free Atmosphere. In and above the first inversion (where $Ri > 0.2$) the dynamics of the turbulence become considerably more complex. There are no simple formulas to predict $C_{\mathcal{T}}^2$ or C_n^2. The available quantitative data are empirical, with the available theory providing some qualitative insight into the processes observed.

Direct *in situ* measurements of $C_{\mathcal{T}}^2$ have been made with aircraft [6-7, 6-8, 6-9], and balloon-borne, fast-responding temperature sensors [6-10, 6-11, 6-12]. Remote sensing of C_n^2 has been achieved with radars [6-13, 6-14] and acoustic sounders [6-13, 6-15]. Weighted path integrals of $C_n^2(h)$ have also been deduced by analyzing the effects of turbulence on starlight [6-16, 6-17, 6-18, 6-10].

In the stratosphere, a good correlation has been observed between wind (velocity) turbulence and the strength of temperature turbulence [6-19], thus allowing some use of an expanded base of observations.

Figure 6-4 shows a typical measured profile of $C_n^2(\tilde{h})$ versus $\tilde{h}$. Here $\tilde{h}$ is altitude above sea level (as compared to h, which is measured from the local terrain). As can be seen, there is a considerable fine structure in the profile. The turbulence distribution appears to be layered with narrow zones of greatly increased C_n^2. These zones are often associated with the base of temperature inversions, especially when accompanied by high wind shears [6-20]. Other regions of increased turbulence appear to be associated with the presence of atmospheric gravity waves* [6-22, 6-23]. Over larger regions there is good correlation between upper tropospheric turbulence and average wind speed [6-24].

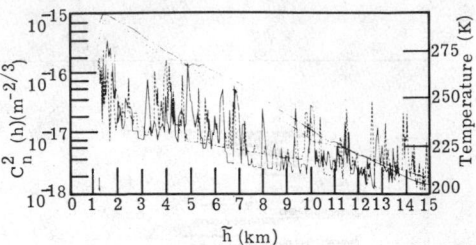

Fig. 6-4. C_n^2 versus altitude above mean sea level, measured on two different dates from an ascending balloon. The smooth lines are the corresponding local air temperature profiles, with the scales on the right. The measurements were made over New Mexico [6-21].

In addition to having spatial variations, C_n^2 also varies considerably in time. A discussion of these temporal variations and a mathematical model for them are given in Hufnagel (1974).

6.2.7. Model Turbulent Atmospheres. For optical paths parallel to the local ground, one may assume that C_n^2 is a constant along the path. The appropriate value may be found in Section 6.2.5.

When the optical path is not parallel to the local ground, it is necessary to model the altitude dependence of C_n^2. For paths which nevertheless stay close to the ground, altitude dependence can be modeled as described in Section 6.2.5.

An extremely simple atmosphere model useful for paths extending into the free atmosphere or beyond is given by the following formula (in MKS units):

$$C_n^2(h) = \begin{cases} \dfrac{1.5 \times 10^{-13}}{h} & \text{below 20,000 m above sea level} \\ 0 & \text{above 20,000 m above sea level} \end{cases} \qquad (6\text{-}20)$$

Because of its simplicity, this model is very handy in evaluating some of the more complex path integrals which occur in propagation calculations. In this formula, h (in meters) is altitude above local ground (assumed to be less than 2500 m above sea level), but the cutoff relates to elevation above sea level.

The simple model given above obviously does not match the complex details of real atmospheres as exemplified by Figure 6-4. It is also time invariant, in contradiction to reality; but it may be considered to represent a median of the real world.

A slightly more complex model, valid only above the first strong inversion layer, is given by the formula

$$C_n^2(\tilde{h}) = 8.2 \times 10^{-56}\, U^2\, \tilde{h}^{10} e^{-(\tilde{h}/1000)} + 2.7 \times 10^{-16} e^{-(\tilde{h}/1500)} \qquad (6\text{-}21)$$

where $\tilde{h}$ is measured in meters above sea level. Here U is a root-mean-squared wind speed, averaged over the 5 to 20 km altitude interval [6-17], and may be set equal to 27 m sec^{-1} for a fixed model. To obtain a random model atmosphere, one may allow U to be a Gaussian random variable which varies from day to day about a 27 m sec^{-1} mean value with a standard deviation of 9 m sec^{-1}. A value of U computed from radiosonde measurements made on a particular day can be used to estimate the turbulence conditions on that

---

*A gravitationally stable interface between two fluids can support a wave-like propagation of disturbances. A common example of this is the generation of ripples caused by a stone thrown upon the air/water interface of a pond. These disturbances are called *gravity waves*, and are present with various wavelengths and amplitudes throughout most of the earth's stably stratified atmosphere.

day. Equation (6-21) does not match the observed fine structure of Figure 6-4, but it is more accurate than Equation (6-20). It also manages to account for much of the total day-to-day variation in upper air turbulence—if U is allowed to vary as specified.

A still more complex model, complete with fine-structured layering, is given in Reference [6-17].

In some models, all the upper altitude turbulence is lumped into a single layer at the tropopause. Although it is not very realistic, this kind of model is simple to use and can give acceptable accuracy. The strength of this single layer can be made variable with specified statistics to match reality [6-26].

In later sections of this chapter, C_n^2 will enter into the final results in the form of weighted integrals along the path. These integrals can usually be expressed in the form

$$G(q, h_0) = \int_{h_0}^{\infty} C_n^2(h)(h - h_0)^q \, dh \qquad (6-22)$$

where q = parameter determined by the problem of interest

  h_0 = height above ground of the lower end of the light path

Values of q of particular interest to turbulence are 0, 5/6, 5/3, and 2. Integrals of this form have been calculated from *in situ* measured profiles of C_n^2 and are reported in Reference [6-10].

Table 6-4 gives the median value of $G(q, h_0)$ for various values of q and h_0, where it is assumed that the ground level is close to sea level. Since C_n^2 varies randomly from day to day, G varies accordingly. This variation is of particular interest to those designers interested in worst- or best-case situations. For example, the reliability of an optical communications channel would depend on the statistics of G. Experimentally, it has been found that nighttime values of $G(0, h_0)$ and $G(2, h_0)$ are both log-normally distributed* [6-27, 6-26, 6-17]. The standard deviation of $\log_e G(0, 10)$ is 0.5, and the standard deviation of $\log_e G(2, 0)$ is 0.7. Presumably, G also has a log-normal probability distribution for intermediate values of q with correspondingly intermediate standard deviations. Estimates of these intermediate values are given in Table 6-4.

The following problem illustrates the method of estimating path statistics: what is the 95th percentile value of $G(2, 0)$ (i.e., the value below which $G(2, 0)$ will fall 95% of the time)? From standard Gaussian probability tables, one finds that the 95th percentile point is 1.645 standard deviation above the mean. The standard deviation of $\log_e G(2, 0)$ is 0.7, which means that the 95th percentile level is a factor exp [(1.645) (0.7)] = 3.2 greater than the median value given in Table 6-3. Thus, the 95th percentile level of $G(2, 0)$ is 8×10^{-5} m$^{7/3}$.

On the other hand, the 5th percentile point of a cumulative Gaussian distribution falls 1.645 standard deviations below the mean, and the 5th percentile level is a factor of 3.2 smaller than the median, that is at 8×10^{-6} m$^{7/3}$.

Other properties of log-normal probability distributions are given in Section 6.9. It is

---

*The standard deviation of $\log_e G(0, 10)$ was measured from two good mountain-top observatory sites only, and may not be true for all sites. The standard deviation of $\log_e G(2, 0)$ was measured from several sites near sea level and appears to be the same for many (low altitude) sites throughout the world. These are for nighttime measurements only. For combined day/night periods the corresponding standard deviations may be different than those given here.

Table 6-4. Statistics of Atmospheric Turbulence Integrals*

Weighting Parameter, q (Dimensionless)	Height of Lower End of Path, h_0 (m)	Median Value of $G(q, h_0)$	Units of G	Standard Deviation of $\log_e G$ (Dimensionless)
0	1	2.0×10^{-12}	$m^{1/3}$	0.5
0	10	1.6×10^{-12}	$m^{1/3}$	0.5
0	100	8.0×10^{-13}	$m^{1/3}$	0.5
5/6	0	6.0×10^{-10}	$m^{7/6}$	0.6
5/3	0	1.1×10^{-6}	m^2	0.7
2	0	2.5×10^{-5}	$m^{7/3}$	0.7

*Equation (6-19) defines G. These values are valid for non-overcast nights only, although the values for $q \geqslant 5/6$ are probably valid for days too. The median values are typical values for ordinary open sites located at an elevation not more than 1 km from sea level. For $q \geqslant 5/6$, $G(q, h_0) \cong G(q, 0)$ when $0 \leqslant h_0 < 100$ m.

especially important to note that the average value of a log-normal random variable is greater than its median.

Beginning with Section 6.3, the calculation of optical statistics will involve weighted integrals along the line of sight. For light paths directed upward at an angle, ζ, with respect to zenith, the distance along the path is related to the altitude, h, by the formula (based on a flat earth approximation)

$$z = (h - h_0) \text{ secant } \zeta \qquad (6-23)$$

where h_0 is the altitude of the light source at the lower end of the path.

For downward-directed light beams, where the propagation direction is at an angle, ζ, with respect to zenith ($\zeta > 90°$), the corresponding relationship is

$$L - z = (h - h_0) \text{ secant } (180° - \zeta) \qquad (6-24)$$

Here the receiver or target is at the lower end of the optical path at $z = L$, which is at an altitude h_0. It will often be more convenient to express slant angles in terms of the line of sight rather than the direction of propagation, where these two directions are exactly 180° apart. For this purpose, one may define the sighting zenith angle, $\bar{\zeta}$, which is given by

$$\bar{\zeta} \equiv 180° - \zeta \qquad (6-25)$$

Substitution of the appropriate transformation into the path integral will yield the predicted zenith angle dependence for the optical quantity of interest. Examples are given in Section 6.4.4.

6.3. Propagation Through Turbulence

6.3.1. Wave Propagation. Several excellent works deal with the propagation of electromagnetic radiation through random media. For further information, see Tatarskii (1961 and 1971), Lawrence and Strohbehn, Fante (1975), Strohbehn (1971), and Shapiro (to be published). Accordingly, no attempt is made in this chapter to present the mathematical details which lead to the final results presented later in this chapter.

The starting point for mathematical analyses is the wave equation. A key assumption, supported by empirical evidence, is that the index of refraction, $n(x, y, z, t)$, does not vary rapidly in either space or time. As a result, turbulent fluctuations in the index of refraction cause only small angle refraction and diffraction of the radiation and do not introduce any depolarization in the radiation [6-28]. The vector wave equation, therefore, may be replaced by its scalar equivalent.

Unfortunately, even with these simplifications, the wave equation cannot be solved exactly in closed form. Further simplifying assumptions must be made, which then yield only approximate solutions to the wave equation.

One of the most widely used of these approximations is called the *Rytov approximation*. Two others, called the *near field approximation* and the *Born approximation*, are special cases of the Rytov approximation. Additional methods of solving the wave equation, which overcome some of the limitations of the Rytov solution, are described in Section 6.3.4.

Results obtained with the Rytov approximation are presented in Section 6.3.3, but first it is necessary to describe the notation used.

6.3.2. Wave Descriptors. Electromagnetic radiation can be described in terms of a wave field designated by the symbol $\&(\mathbf{r}, t)$. In the absence of turbulence, a propagating monochromatic* field can be represented in the form

$$\& = [E(\mathbf{r})]^{1/2} \exp[i\mathbf{k}\cdot\mathbf{r} - i\omega t] \tag{6-26}$$

where ω = temporal frequency of electromagnetic field

$\mathbf{k}$ = vector of magnitude $2\pi/\lambda$ pointing in the direction of wave propagation

λ = is the wavelength of the light

The irradiance, $E(\mathbf{r})$, measured in units of W m^{-2}, is uniform for plane wave propagation, but would obey the familiar inverse square law for radiation from a point source.

The effect of the turbulence is to change both the phase and intensity of the radiation. For reasons of mathematical elegance and simplicity it is desirable to describe the intensity changes by changes in the natural logarithm of $E^{1/2}$. Thus, in the presence of turbulence, $\&$ is represented by the form

$$\& = <E(\mathbf{r})>^{1/2} \exp[i\mathbf{k}\cdot\mathbf{r} - i\omega t + \chi(\mathbf{r}, t) + i\phi(\mathbf{r}, t)] \tag{6-27}$$

Here $<E(\mathbf{r})>$ is the average (nonfluctuating) part of the irradiance while $\chi(\mathbf{r}, t)$ describes the randomly fluctuating part. The pure imaginary term $i\phi(\mathbf{r}, t)$ describes the turbulence-induced, phase angle variation in $\&$. Alternatively, ϕ may be considered to be the wave-front retardation, measured in units of $\lambda/2\pi$.

6.3.3. Rytov Approximations. The solution of the wave equation under the Rytov approximation presents χ and ϕ as weighted linear sums of contributions from all the random index fluctuations in the semi-infinite volume between source and receiver. By

---

*The equations presented in this chapter will work well enough for most engineering applications with polychromatic radiation, provided the effective spectral range spans less than an octave. A spectrally averaged wavelength should be used. The reader is cautioned, however, on atmospheric dispersion effects over long slant paths (as discussed in Section 6.4.4).

the Law of Large Numbers [6-29], both X and ϕ will become random variables with Gaussian probability distributions. Their descriptions must be in statistical terms (mean values, variances, and correlation functions).

The Rytov approximation predicts that the mean values $<X>$ and $<\phi>$ are both zero. In fact, neither is exactly true, as discussed in Sections 6.4.1 and 6.9.

As shown in Section 6.2.2, the variances and correlation functions of random variables can be related to their spatial Wiener spectra. The spectra also are useful in calculating system outputs, including signal-fading statistics and angle-of-arrival fluctuations.

The following four equations present the Wiener spectra for X and ϕ as derived by the Rytov approximation under the additional simplifying assumptions that (1) the turbulence obeys Kolmogorov's law, (2) the inner scale of turbulence, ℓ_0, is negligibly small, and (3) the outer scale of turbulence, L_0, is sufficiently large. Equation (6-13) was used for $\Phi_n(\mathbf{K})$. Lawrence and Strohbehn (1970) and Reinhardt and Collins (1972) contain corresponding formulae under less restrictive conditions.

For plane wave propagation the equations are*

$$\Phi_\phi(\mathbf{K}) = 2\pi(0.033)k^2 K^{-11/3} \int_0^L C_n^2(z) \cos^2\left[\frac{K^2(L-z)}{2k}\right] dz \qquad (6\text{-}28)$$

$$\Phi_X(\mathbf{K}) = 2\pi(0.033)k^2 K^{-11/3} \int_0^L C_n^2(z) \sin^2\left[\frac{K^2(L-z)}{2k}\right] dz \qquad (6\text{-}29)$$

where the spatial frequency $\mathbf{K}$ (in rad m^{-1}) is a two-dimensional vector located in the plane $z = L$. The corresponding equations for spherical wave propagation are

$$\Phi_\phi(\mathbf{K}) = 2\pi(0.033)k^2 K^{-11/3} \int_0^L C_n^2(z)(z/L)^{5/3} \cos^2\left[\frac{K^2 L(L-z)}{2zk}\right] dz \qquad (6\text{-}30)$$

$$\Phi_X(\mathbf{K}) = 2\pi(0.033)k^2 K^{-11/3} \int_0^L C_n^2(z)(z/L)^{5/3} \sin^2\left[\frac{K^2 L(L-z)}{2zk}\right] dz \qquad (6\text{-}31)$$

where $\mathbf{K}$ lies on the spherical surface passing through $z = L$ with its center of curvature at the radiation point source located in the plane $z = 0$. The variance of X, which is σ_X^2, may by computed from Equation (6-8) with $r = 0$. With isotropic turbulence, the azimuthal part of the two-dimensional spatial frequency integration is easily performed, leaving the result

$$\sigma_X^2 = \int_0^\infty 2\pi K \, \Phi_X(\mathbf{K}) \, dK \qquad (6\text{-}32)$$

Integration yields the following useful results. For plane wave propagation, the Rytov

---

*Not presented here is the cross-spectrum related to the cross-correlation between X and ϕ. The cross-correlation is usually not needed in system analyses because it cancels out after symmetry considerations. However, it is wise to remember that ϕ and X are correlated [6-30].

theory predicts

$$\sigma_\chi^2 = 0.56\, k^{7/6} \int_0^L C_n^2(z)\,(L-z)^{5/6}\, dz \tag{6-33}$$

For spherical wave propagation, the Rytov theory predicts

$$\sigma_\chi^2 = 0.56\, k^{7/6} \int_0^L C_n^2(z)\,(z/L)^{5/6}(L-z)^{5/6}\, dz \tag{6-34}$$

The integration is along the light path from $z = 0$ to $z = L$. The integrals appearing here can be evaluated with the aid of Table 6-4. These equations assume that $\ell_0^2 \ll \lambda L \ll L_0^2$.

If the optical path is parallel to the local ground, then $C_n^2(z)$ is usually independent of z, and Equations (6-33) and (6-34) may be easily integrated to yield

$$\sigma_\chi^2 = 0.31\, C_n^2\, k^{7/6}\, L^{11/6} \tag{6-35}$$

for plane wave sources, and

$$\sigma_\chi^2 = 0.124\, C_n^2\, k^{7/6}\, L^{11/6} \tag{6-36}$$

for spherical wave sources. These equations for Φ_χ and σ_χ^2 are valid only if they predict that $\sigma_\chi^2 \lesssim 0.3$. This restriction is discussed further in Section 6.3.4. The relationship between intensity fluctuations, signal fading, and fluctuations of χ are presented in Section 6.4.

The autocorrelation functions for χ evaluated through Equation (6-8) for horizontal paths are shown in Figures 6-5 and 6-6. The agreement with experiment is good when $\sigma_\chi^2 < 0.3$.

The variance of ϕ and its autocorrelation function do not mathematically exist under the assumption of an infinitely large outer scale of turbulence. However the structure function, $D_\phi(r)$, does exist and has been measured. (See Breckenridge.) For most engineering problems, the phase structure function, D_ϕ, can be replaced, with little loss of

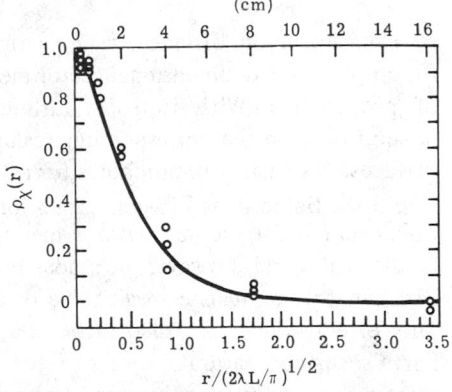

Fig. 6-6. Log-amplitude autocorrelation function versus normalized spatial separation, for spherical wave propagation geometry [6-45]. Other assumptions are the same as for Figure 6-5. The circles are experimental data points for a high horizontal path, 5.5 km long, using a 0.63 μm wavelength laser. An unnormalized correlation distance scale for the experiment is shown on the top [6-31].

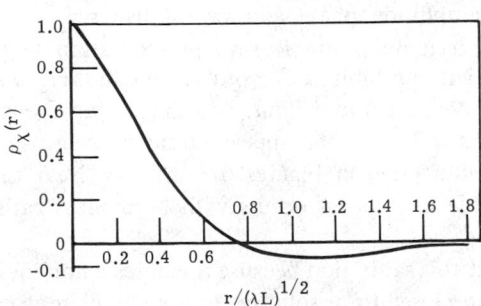

Fig. 6-5. Log-amplitude autocorrelation function versus normalized spatial separation, for plane wave propagation geometry. This curve is computed by the Rytov approximation with the added assumptions that $\ell_0 \ll (\lambda L)^{1/2} \ll L_0$ and that the turbulence is uniform along the path [6-25].

accuracy, by a quantity called the *wave structure function*. In addition, the wave structure function, $D_w(\mathbf{r})$, which is defined as

$$D_w(\mathbf{r}) \equiv D_\phi(\mathbf{r}) + D_\chi(\mathbf{r}) \tag{6-37}$$

has a direct physical significance in image-blurring and beam-spreading phenomena. The wave structure function can be evaluated by substituting $\Phi_\phi(\mathbf{K}) + \Phi_\chi(\mathbf{K})$ into Equation (6-9).

Because $\cos^2$ [arg] $+ \sin^2$ [arg] $\equiv 1$, the expression for $\Phi_\phi(\mathbf{K}) + \Phi_\chi(\mathbf{K})$ is simpler than either component separately. For plane wave propagation, the wave structure function, $D_w(\mathbf{r})$, is

$$D_w(\mathbf{r}) = 2.914 \, k^2 \, r^{5/3} \int_0^L C_n{}^2(z) \, dz \tag{6-38}$$

while for spherical wave propagation, it is

$$D_w(\mathbf{r}) = 2.914 \, k^2 \, r^{5/3} \int_0^L C_n{}^2(z) \, (z/L)^{5/3} \, dz \tag{6-39}$$

The wave structure function is identical to the phase structure function calculated with the near field approximation. The wave structure function will be used in Sections 6.5 and 6.6. In the special case when $C_n{}^2(z)$ has a uniform value along the entire optical path $o < z < L$, Equation (6-38) simplifies to

$$D_w(\mathbf{r}) = 2.914 \, k^2 \, r^{5/3} \, C_n{}^2 \, L \tag{6-40}$$

and Equation (6-39) for spherical waves simplifies to

$$D_w(\mathbf{r}) = 1.093 \, k^2 \, r^{5/3} \, C_n{}^2 \, L \tag{6-41}$$

The separation distance, $\mathbf{r}$, appearing in the phase, amplitude, and wave structure functions is a two-dimensional vector measured in a surface normal to the average direction of propagation. With isotropic turbulence, as assumed here, the structure functions depend only on the corresponding scalar separation distances measured in along the same surfaces. A similar situation holds for other amplitude, phase, and wave statistics.

6.3.4. Saturation Effects. Equations based upon the Rytov approximation (e.g., Equation (6-35)), predict that $\sigma_\chi{}^2$ increases without limit as $C_n{}^2$ or path length increases. In the real world, however, $\sigma_\chi{}^2$ does not increase without limit. With increasing $C_n{}^2$ or path length, $\sigma_\chi{}^2$ reaches peak value of about 0.3, and then appears to slowly decrease*. This effect, which is called saturation, is illustrated in Figures 6-7 and 6-8. Near the Earth's surface, saturation of $\sigma_\chi{}^2$ for visible light often occurs with horizontal paths greater than about a few hundred meters long.

The Rytov approximation fails to predict this saturation because it ignores a nonlinear term in the wave equation for χ and ϕ. No closed-form solution to the complete wave equation has been found, but several attempts at alternate approximate solutions have been made, (See Tatarskii (1971), de Wolf, and Strohbehn (1971).) The success of these alternative methods has been limited, but helpful in understanding the problem.

---

*Slightly higher peak values of $\sigma_\chi{}^2$ are reached with beam-wave propagation.

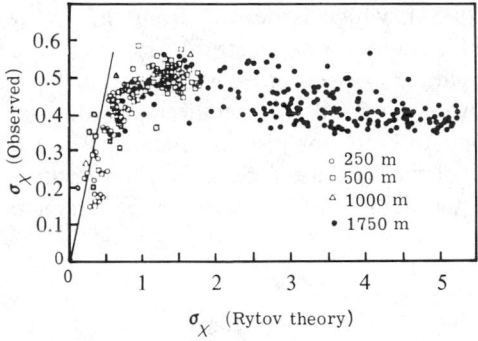

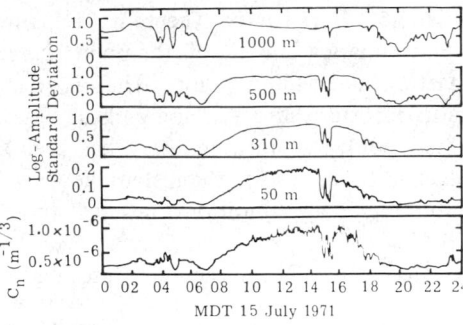

Fig. 6-7. Observed strength of scintillation versus scintillation predicted by Rytov approximation, for various horizontal path lengths, with evident saturation [6-32]. The original data were taken by Soviet workers [6-90].

Fig. 6-8. The variation of the square root of the logamplitude variance at each of four path lengths compared to the refractive index structure parameter, C_n, for a 24 h period. All paths greater than 50 m show evidence of saturation. (Note the effect of clouds in the afternoon at 15 h.) [6-33].

In qualitative terms, saturation arises because the wavefront distortions cause an initially single source of light to appear as extended multiple sources of light, each of which scintillates somewhat independently of the others. Thus an averaging effect takes place, such that the net overall intensity scintillation becomes limited [6-34, 6-35].

The effect of saturation upon the statistics of the wavefront, ϕ, is relatively minor. For most engineering purposes, the Rytov approximation—and often the near field approximation—may be used to compute statistics of ϕ, even in the presence of saturation.

The Rytov solution for the wave structure function given by Equations (6-38) and (6-39) may be considered valid for all engineering purposes, provided that Equation (6-12) is valid.

Although it has not been possible to obtain exact solutions for X and ϕ separately, in many problems the quantities of interest involve special combinations of X and ϕ such that more exact solutions are possible. Computations of average image blurring and beam spreading involve just such combinations.

More generally, many quantities of physical interest can be expressed in terms of second- and higher-order coherence functions. (See Born and Wolf.) Very good solutions have been obtained for coherence functions valid over a larger domain than for the Rytov approximation [6-36, 6-37, 6-38].

The key to the solution of the coherence equations has been the assumption that local electromagnetic-field fluctuations are independent of local index of refraction fluctuations. Mathematically, this is known as the assumption of a Markov process [6-29]. In fact, the local field is both correlated and dependent on local index fluctuations. In coherence equations, however, these correlations cancel because of symmetry considerations, and the Markov assumption works.

This assumption, coupled with a generalization of the classical Huygens-Fresnel principle, is now the primary method of analyzing beam-wave propagation problems. Further details are given in References [6-39, 6-40] and [6-41]. Sections 6.4, 6.5, and 6.6 make use of these results.

6.3.5. Reciprocity Theorem. Equation (6-34), which is derived from the Rytov approximation and useful for point sources of radiation, is symmetrical with respect to starting and ending points. That is, the amplitude statistic, σ_χ^2, is unchanged if the radiation starts at $z = L$ and ends at $z = 0$. This result has been generalized in two ways. First, point-to-point symmetry has been shown to exist not just for a statistic of the electric field, but for the instantaneous values of the field itself. Secondly, this result is valid under all conditions, not just those in the domain of the Rytov approximation [6-42].

More generally the theorem states that

> for two optical systems, each consisting of a coherent transmitter and receiver sharing a common antenna aperture through a beamsplitter, the effect of atmospheric turbulence on the signal received by the first unit from the second will be identical instant-by-instant to the effect of atmospheric turbulence on the signal received by the second unit from the first. [6-43]

This theorem has successfully undergone experimental testing [6-43]. It is useful in optical communications, as well as in other fields where it is desirable to maintain a concentrated beam of radiation on a single target through the use of fast adaptive optics in laser transmitters. Specifically, it means that radiation from a target may be used to sense instantaneous path distortions; this information, in turn, may be used to introduce a negative compensating distortion in the outgoing beam, as described in Section 6.7.

6.4. Scintillation

6.4.1. Irradiance Fluctuations. Instantaneous irradiance, $E(r, t)$, is proportional to the square of the absolute value of the field amplitude, $\mathcal{E}$, as defined by Equation (6-27). Thus,

$$E(r, t) = <E(r)> \exp\left[2\chi(r, t)\right] \qquad (6-42)$$

where the angle brackets indicate an average of the enclosed quantity.

The Rytov approximation predicts that χ has a Gaussian (normal) probability density distribution. From Equation (6-42) it follows, by definition, that E has a log-normal probability distribution.

Experimentally, it is observed that E has approximately a log-normal distribution for essentially all conditions of engineering interest, even beyond the valid range of the Rytov approximation. Section 6.9.1 describes useful properties of log-normal random variables.

In particular, it is shown there that the average value of χ cannot be zero. Of more interest is the relationship between the variance of the irradiance and the variance of χ.

This relationship is

$$\frac{\sigma_E^2}{<E>^2} \equiv \frac{<(E - <E>)^2>}{<E>^2} = (\exp 4\sigma_\chi^2) - 1 \qquad (6-43)$$

The normalized standard deviation, $<E>^{-1}\,\sigma_E$, is just the square root of the normalized variance, and it is a measure of irradiance modulation.

Although Equation (6-43) is widely used, its accuracy is significantly reduced when the probability distribution of X departs even slightly from an exact Gaussian distribution. Experimental evidence for this effect is given in Reference [6-44].

6.4.2. Receiver Aperture Averaging.

In many applications, radiation is collected by an optical system with an entrance aperture of finite diameter. Turbulence-induced fluctuations in incident radiation will result in corresponding fluctuations in total collected signal strength. If the aperture is large compared to the lateral dimensions of the irradiance fluctuations, then an averaging effect will take place.

To quantify this effect, let S represent the collected signal power, normalized so that S has an average value of unity. Thus, S is the normalized integral of E where the integration is taken over the system entrance pupil. For an unobscured pupil of radius a, one has

$$S(t) = (\pi a^2)^{-1} \int \exp\left[2X(x, y, t)\right] \, dx \, dy \qquad (6\text{-}44)$$

where the integration is taken over the region $x^2 + y^2 < a^2$.

When a is so small that no aperture averaging takes place, the variance of S becomes $\sigma_E^2/<E>^2$.

References [6-45] and [6-46] give expressions for σ_S^2 versus a and σ_x^2 for a limited number of path geometries. Figure 6-9 shows a plot of this relationship in normalized form for plane wave propagation along a path of constant turbulence strength. The calculations required to obtain these results are somewhat involved.

If $\sigma_x^2 \ll 1$, the exponent in Equation (6-44) can be expanded in a power series and the following simplified result obtained:

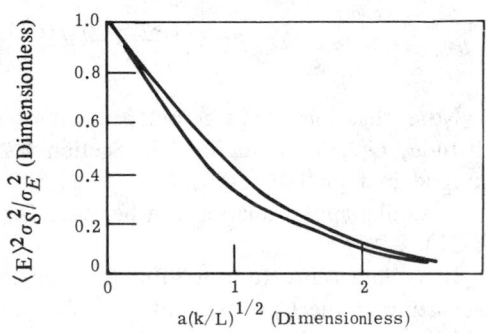

Fig. 6-9. Normalized, relative, collected power fluctuation versus normalized collector pupil radius. The top curve is for σ_x^2 near zero and the bottom curve is for $\sigma_x^2 = 0.25$. (Other assumptions are the same as for Figure 6-5.) [6-45].

$$\sigma_S^2 = 4 \int\limits_{-\infty}^{+\infty} |H(\mathbf{K})|^2 \, \Phi_x(\mathbf{K}) \, d\mathbf{K} \qquad (6\text{-}45)$$

Here $H(\mathbf{K})$ is the Fourier transform of the pupil irradiance transmission function, normalized to a value of unity at $\mathbf{K} = 0$.

For a system with a clear aperture of radius a,

$$H(\mathbf{K}) = 2J_1(aK)/aK \qquad (6\text{-}46)$$

where J_1 is a first-order Bessel function of the first kind. Azimuthal spatial frequency integration can be carried out to yield

$$\sigma_S^2(a) = 32\pi a^{-2} \int\limits_0^\infty J_1^2(aK) \, \Phi_x(\mathbf{K}) \, K^{-1} \, dK \qquad (6\text{-}47)$$

For large values of a, $J_1{}^2(aK)$ is small enough at large spatial frequencies that it is possible to greatly simplify the evaluation of this integral. If, for example, Equation (6-29) is used for $\Phi_\chi(\mathbf{K})$, the approximation

$$\sin^2\left[\frac{K^2(L-z)}{2k}\right] \cong \frac{K^4(L-z)^2}{4k^2} \tag{6-48}$$

may be made when K is sufficiently small. With this approximation, Equations (6-47) and (6-29) become

$$\sigma_S{}^2(a) \cong 16\pi^2(0.033)\,a^{-2} \int_0^\infty J_1{}^2(aK)\,K^{-2/3}\,dK \int_0^L (L-z)^2 C_n{}^2(z)\,dz \tag{6-49}$$

The K integration can be performed first, yielding both for plane wave propagation and for large values of a,

$$\sigma_S{}^2(a) \cong 3.44\,a^{-7/3} \int_0^L (L-z)^2 C_n{}^2(z)\,dz \tag{6-50}$$

Note that this result is independent of wavelength. When $L - z$ can be related to altitude, $G(2, 0)$, as defined in Section 6.2.7, can be used to estimate the remaining integral in Equation (6-50).

A similar approximation can be made for spherical wave propagation using Equation (6-31).

It is important to note that the receiver aperture averaging changes the relative weighting of the $C_n{}^2$ path integral. The weighting factor in Equation (6-33) for $\sigma_\chi{}^2$, which corresponds to very small receiver, varies as $(L - z)^{5/6}$; but for large receivers (and weak scintillation) the corresponding weighting factor will vary approximately as $(L - z)^2$. For zenith visible starlight, the transition between these two domains occurs for aperture diameters between 2 and 20 cm. For other wavelengths the transition aperture diameter scales as $\lambda^{1/2}$.

6.4.3. Scintillation from Extended Sources. The scintillation of the total radiation from multiple independent sources will be less than the scintillation from a single source because of averaging effects similar to those described in the previous section.

For a disk of angular radius μ (as viewed from the location $z = L$), Equations (6-29) and (6-31) for $\Phi_\chi(\mathbf{K})$ are modified to become

$$\Phi_\chi(\mathbf{K}) = 8\pi(0.033)\,k^2 K^{-11/3} \int_0^L C_n{}^2(z)\sin^2\left[\frac{K^2(L-z)}{2k}\right]\frac{J_1{}^2[\mu K(L-z)]}{[\mu K(L-z)]^2}\,dz \tag{6-51}$$

for extended plane wave sources (e.g., planets or the sun), and

$$\Phi_\chi(\mathbf{K}) = 8\pi(0.033)\,k^2 K^{-11/3} \int_0^L C_n{}^2(z)\left(\frac{z}{L}\right)^{5/3}\sin^2\left[\frac{K^2 L(L-z)}{2kz}\right]\frac{J_1{}^2[\mu K(L-z)]}{[\mu K(L-z)]^2}\,dz \tag{6-52}$$

for incoherent extended sources at a finite distance.

As before, σ_χ^2 may be found by substituting these equations into Equation (6-32). As was shown in Section 6.4.2, the evaluation of the resulting double integral may be considerably eased when μ is large by making the approximation

$$\sin^2\left[\frac{K^2(L-z)}{2k}\right] \cong \left[\frac{K^2(L-z)}{2k}\right]^2 \tag{6-53}$$

A similar approximation may be applied for Equation (6-52).

With this approximation in Equation (6-51), Equation (6-32) yields a simple result for plane wave propagation:

$$\sigma_\chi^2 = 0.86\,\mu^{-7/3} \int_0^L (L-z)^{-1/3} C_n^2(z)\, dz \tag{6-54}$$

This approximation method works quite well in explaining both the variance and the correlation function of turbulence-induced, solar-flux scintillations, as reported in Reference [6-47].

As a rough rule, for $\lambda < 10\ \mu m$, μ must be larger than about 2×10^{-4} rad for this approximation to be valid.

Because of the heavy weighting of the path integral near the observer in Equations (6-51) and (6-52), the lateral correlation distance of intensity fluctuations will be quite small—often only a few millimeters.

Therefore, receiver aperture averaging effects, as described in Section 6.5.2, become important even for small aperture sizes.

Receiver aperture averaging effects can be included by multiplying Equation (6-51) or (6-52) by $|H(K)|^2$ before substitution into Equation (6-32). ($H(K)$ was defined in Section 6.4.2.)

Reference [6-48] describes the effects of source averaging when the Rytov approximation is not valid

6.4.4. Scintillation of Starlight. For stars not too far from the zenith, the simple Rytov approximation works rather well. Equation (6-33) for σ_χ^2 can be evaluated directly, using the value of $G(5/6, 0)$ given in Table 6-3. For stars at a sighting zenith angle, $\bar{\zeta}$, the substitution $h = (L - z) \cos \bar{\zeta}$ from Equation (6-24) is used. For $\lambda = 0.5 \times 10^{-6}$ m and $\bar{\zeta} = 0$, Table 6-3 predicts, for example, that the median value of σ_χ^2 is 0.064.

According to Equations (6-24) and (6-33), the zenith angle dependence of σ_χ^2 should be of the form $\sigma_\chi^2 \sim (\text{secant } \bar{\zeta})^{11/6}$. In practice, this dependence is observed for visible white starlight only for $\bar{\zeta} < 40°$. Beyond $40°$, σ_χ^2 grows less rapidly than predicted, then saturates, and for $\bar{\zeta} > 70°$, decreases with $\bar{\zeta}$.

Two effects cause this saturation. One of these is the breakdown of the Rytov approximation, as described in Section 6.3.4. The other effect is color (wavelength) averaging caused by atmospheric dispersion. Light of different wavelengths traverse different paths in the atmosphere and, therefore, scintillate somewhat uncorrelated from each other. This effect is akin to the source averaging effects described in Section 6.4.3, where here the source width is proportional to the spectral width of the measured radiation and the $\bar{\zeta}$-dependent atmospheric dispersion [6-49].

References [6-50] and [6-34] give quantitative analyses of these two sources of scintillation reduction. The color dispersion effect is not present, of course, for monochromatic radiation but the Rytov approximation failure remains.

The results given above are observed only for small aperture light collectors. With very large apertures, when Equation (6-50) is valid, a (secant $\bar{\zeta}$)^3 dependence on σ_S^2 is predicted and is observed for zenith angles out to about 65°.

6.4.5. Temporal Variations in Collected Power. As shown in Section 6.2.3, an application of Taylor's hypothesis allows the computation of a temporal Wiener spectrum from the spatial Wiener spectrum.

The temporal spectrum of intensity fluctuations has been calculated in the literature for a variety of cases with the dual assumption of a uniform cross wind and a uniform value of C_n^2 along the entire optical path.

Figure 6-10 shows the spectrum in normalized form for plane wave propagation, uniform cross wind, uniform C_n^2, under the assumption that $\sigma_\chi^2 \ll 1$. The effect of receiver aperture averaging is also shown.

Figure 6-11, shown for comparison, gives measured data for the scintillation of starlight. Even though neither C_n^2 nor the winds were constant along the path, there is a good match between the two figures.

For spherical wave propagation, the total temporal spectrum of intensity fluctuations is smaller in strength but extends outward toward higher frequencies when compared to the corresponding plane wave spectrum. Examples of spherical wave scintillation spectra are given in Reference [6-51].

Measurements of the temporal spectrum have been made under conditions of σ_χ^2 saturation; they are presented in dimensionless form in Reference [6-36]. Yura (March 1974) gives an analysis of intensity temporal spectra for beam waves in the saturation regime.

6.5. Image Degradation

6.5.1. Image Properties. Turbulence in an optical path distorts images in two ways: each portion of the image moves in relationship to other portions of the image; and the image itself becomes blurred. At times the image may even appear to be broken up.

In analyzing an imaging system, one must first decide whether the image motion will contribute to image degradation. For photographic systems with exposure times longer

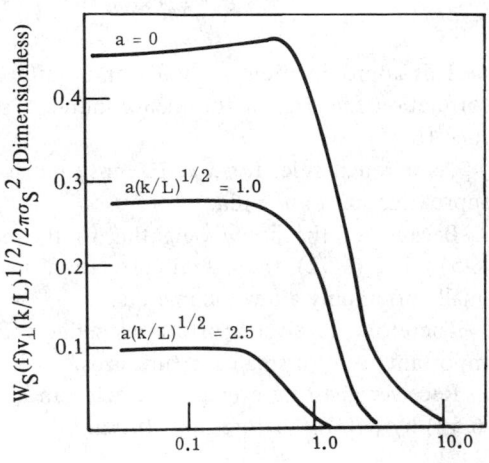

Fig. 6-10. Theoretical shape of temporal Wiener spectrum for fluctuations of collected radiation versus normalized radius of collector aperture. Plane wave geometry, uniform turbulence strength, constant cross winds along the path, and weak scintillation are all assumed [6-25].

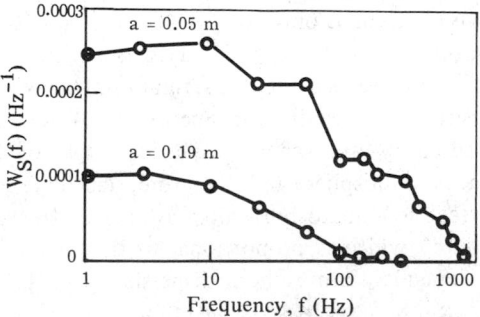

Fig. 6-11. Experimental measurements of temporal Wiener spectrum of fluctuations in collected starlight for two telescope radii. Observations were made within 18° of the zenith in the visible portion of the spectrum [6-24].

than several milliseconds, there is usually sufficient motion during the exposure that it contributes significantly to the overall blur. On the other hand, with very short exposures, the motion will yield local geometric distortions but will not contribute to image blur. The two limiting cases are described as *long exposures* and *short exposures*. The analysis of intermediate cases are not given here, but Roddier and Roddier describe the type of analyses required.

In the normal situation, where the object is considered to be an incoherent source of radiation, the image consists of a linear superposition of images from each point in the object. In this case, image quality can be described in terms of the image from a single point object. This image of a point, which is called the *point spread function,* is usually normalized to a total integrated flux of unity. (See Chapter 8.)

Let **p** be a position vector in the system image plane with the origin of the coordinate system arranged so that, in the absence of turbulence, the point image is centered at **p** = 0. Let $\mathscr{J}$ (**p**) be the normalized point spread function.

The instantaneous centroid of $\mathscr{J}$(**p**) is defined as

$$\alpha \equiv (\alpha_x, \alpha_y) = \int\int_{-\infty}^{+\infty} \mathscr{J}(\mathbf{p})\, \mathbf{p}\, d\mathbf{p} \tag{6-55}$$

and may be taken as a measure of turbulence-induced, image-position shift. In general, $\mathscr{J}$ (**p**), and therefore α, will be time-varying random functions.

In the presence of turbulence, the point spread function may assume quite complex shapes. Indeed, for large aperture systems it will contain a speckle-like pattern rather similar to the appearance of a laser-illuminated surface. It is convenient and useful to compute average values for this random function.

The ensemble average $<\mathscr{J}$ (**p**) $>$ may be taken as a measure of long-exposure system blur. In a corresponding way, $<\mathscr{J}$ (**p** - α) $>$ may be taken as a measure of the short-exposure blur. In this latter case, the image blurring is continuously measured about the instantaneous (random) image centroid.

Because of diffraction from the system pupil and the possible presence of fixed aberrations, $<\mathscr{J}$ (**p**) $>$ will have some spread even in the absence of turbulence. To isolate the effect of turbulence, one will find it convenient to deal not with point spread functions but with their two-dimensional Fourier transforms, which are the optical transfer functions.

When $\mathscr{J}$ (**p**) is normalized to a total integrated flux of unity, the corresponding optical transfer function is defined as

$$T(\mathbf{f}) = \int\int_{-\infty}^{+\infty} \mathscr{J}\ (\mathbf{p}) \exp(-2\pi i \mathbf{f} \cdot \mathbf{p})\, d\mathbf{p} \tag{6-56}$$

which differs from Equation (6-7) by a normalizing factor and in the definition of spatial frequency. Here $\mathbf{f} = (f_x, f_y)$ is spatial frequency in the image plane expressed in units of cycles per unit length.

The long-exposure, average optical transfer function then becomes

$$<T(\mathbf{f})> = \int\int_{-\infty}^{+\infty} <\mathscr{J}\ (\mathbf{p}) > \exp(-2\pi\, \mathbf{f} \cdot \mathbf{p})\, d\mathbf{p} \tag{6-57}$$

and the ensemble-averaged, short-exposure optical transfer function becomes

$$<T_s(\mathbf{f})> = \int\int\limits_{-\infty}^{+\infty} <\mathfrak{s}(\mathbf{p} - \boldsymbol{\alpha})> \exp(-2\pi i \mathbf{f} \cdot \mathbf{p}) \, d\mathbf{p} \qquad (6\text{-}58)$$

where T_s is the short-exposure, instantaneous optical transfer function.*

Having now expressed the system blur in terms of optical transfer functions, one may divide out the optical transfer function, $T_0(\mathbf{f})$, corresponding to pupil diffraction and all other nonturbulence-related contributions to blur [6-52].

The resulting optical transfer functions given by the expressions**

$$<M(\mathbf{f})> = <T(\mathbf{f})>/T_0(\mathbf{f}) \qquad (6\text{-}59)$$

$$<M_s(\mathbf{f})> = <T_s(\mathbf{f})>/T_0(\mathbf{f}) \qquad (6\text{-}60)$$

describe the blurring effects of only atmospheric turbulence. In the absence of path turbulence, $<M> = 1$ and $<M_s> = 1$ for all $\mathbf{f}$.

Formulas for evaluating $<M>$ and $<M_s>$ are given in Sections 6.5.3 and 6.5.6, respectively.

6.5.2. Isoplanatism. An isoplanatic region is that area over which a system point spread function remains essentially unchanged. The extent of the area depends upon the exact definition of what is meant by "essentially unchanged." One useful definition appropriate for estimating maximum allowable exposure times and for studying various image compensation schemes involves a cross correlation between the system's optical transfer functions measured at two positions in the field of view [6-54, 6-55]. If one is looking upward through the entire Earth's atmosphere, the typical diameter of the isoplanatic region corresponds to several arc seconds. Reference [6-56] shows that the isoplanatic region diameter will be smaller than the long-exposure, point spread function diameter whenever the corresponding value of $\sigma_\chi{}^2$ is saturated.

6.5.3. Long Exposure Blur. Reference [6-52] shows that $<M(\mathbf{f})>$ is identical to a spatial coherence function for the radiation from an initially spatial coherent source. The specific relationship is

$$<M(\mathbf{f})> = <E>^{-1} <\mathcal{E}(\mathbf{r} + F\lambda\mathbf{f})\mathcal{E}^*(\mathbf{r})> \qquad (6\text{-}61)$$

where $\mathcal{E}$ is given by Equation (6-27). This expression may be evaluated with the aid of Equations (6-5), (6-37), (6-110), and (6-115) to yield

$$<M(\mathbf{f})> = \exp\left[-\left(\frac{1}{2}\right)D_w(F\lambda\mathbf{f})\right] \qquad (6\text{-}62)$$

---

*The limited size of the isoplanatic patch (Section 6.5.2) may prevent $T(\mathbf{f})$ and $T_s(\mathbf{f})$ from being rigorously useful as multiplicative operators, but this restriction will usually not apply to their ensemble averages.

**Since $T_0(\mathbf{f})$ is zero for $\mathbf{f}$ beyond the diffraction limit, $<M(\mathbf{f})>$ is not defined beyond this limit. On the other hand, since $<M(\mathbf{f})>$ is a property of the atmosphere independent of $T_0(\mathbf{f})$, one can pass to the limit of an infinitely large aperture system without a limiting value of $\mathbf{f}$.

Alternatively, Equation (6-62) can be derived from the coherence propagation equations. The wave structure function for both plane and spherical wave propagation is evaluated in Equations (6-38) and (6-39). When these substitutions are made, the final results are

$$<M(f)> = \exp\left[- 2\pi^2(2.914)\lambda^{-1/3}(Ff)^{5/3} \int_0^L C_n^2(z)\,dz\right] \qquad (6-63)$$

for plane-wave geometry and

$$<M(f)> = \exp\left[-2\pi^2(2.914)\lambda^{-1/3}(Ff)^{5/3} \int_0^L C_n^2(z)(z/L)^{5/3}\,dz\right] \qquad (6-64)$$

for spherical-wave geometry. These two equations are valid for all problems of engineering interest provided that the scales of interest lie well within the inertial subrange. Reference [6-57] gives an example of experimental agreement with these results.

If the combination $\lambda^{-1/3}f^{5/3}$ is rewritten as $(\lambda^{-1/5}f)^{5/3}$, it becomes clear that resolution varies as $\lambda^{-1/5}$ (i.e., "seeing" is slightly better at longer wavelengths).

One measure of image quality is *Strehl definition*, which is the relative intensity of the point spread function measured at the centroid location of the undisturbed point spread function. In the limit of a large telescope aperture, the Strehl definition for $< \mathcal{I}(p) >$ is the same as that for a diffraction-limited telescope with aperture diameter = r_0, where

$$r_0 = \left[\frac{D_w(r)}{6.884\,r^{5/3}}\right]^{-3/5} \qquad (6-65)$$

The quantity r_0, thus defined, is called the *coherence diameter* [6-58], and is a convenient measure of the lateral distance over which the radiation remains coherent. Equations (6-38) through (6-41), as appropriate, should be substituted for $D_w(r)$ in Equation (6-65). The factor $r^{5/3}$ automatically drops out of the final expression. For example, for horizontal plane wave propagation, substitution of Equation (6-40) into (6-65) would yield

$$r_0 = [0.4233\,C_n^2 k^2 L]^{-3/5} \qquad (6-66)$$

Equations (6-63) and (6-64) may be expressed in terms of the coherence diameter. The result is

$$<M(f)> = \exp[-3.44\,(F\lambda f/r_0)^{5/3}] \qquad (6-67)$$

for either spherical wave or plane wave propagation. For $\lambda = 0.5 \times 10^{-6}$ m, an r_0 value of 0.1 m corresponds approximately to a 1 arc second "resolution limit." For the same geometry and turbulence conditions, but for $\lambda = 10^{-5}$ m, r_0 would be equal to 3.6 m, corresponding to a 1/2 arc second resolution limit.

6.5.4. Optical Superheterodyne Reception. The quantities $< \mathcal{I}(p) >$, $<M(f)>$, and r_0 are also relevant to the analysis of the effects of turbulence upon optical superheterodyne reception. Attaining a maximized heterodyne signal requires that the collected radiation be coherent (i.e., in phase) over the entire system pupil. This will not be possible for pupil diameters greater than r_0.

In the absence of turbulence, the output signal-to-noise ratio of an optical superheterodyne receiver could increase without limit as receiver aperture area is increased. With turbulence, there is a limiting maximum signal-to-noise ratio which can be achieved. This limit is reached when the diameter of the system's entrance pupil exceeds r_0.

References [6-59, 6-60, and 6-61] give theoretical and experimental details. Some signal-to-noise improvement can be gained with rapid tracking of the signal angle of arrival. (See Section 6.7.)

6.5.5. Image Motion. In terms of the radiation at the system entrance pupil, the image centroid position, α, is given by the equation

$$\alpha = \frac{F}{k} \frac{\iint E(\mathbf{r})\nabla\phi(\mathbf{r})d\mathbf{r}}{\iint E(\mathbf{r})d\mathbf{r}} \tag{6-68}$$

where the integration is taken over the entrance pupil area and where F is the system's effective focal length. Here $\nabla\phi$ is the lateral gradient of the wavefront phase, and E is the irradiance—both evaluated just prior to entering the imaging system.

The physical interpretation of Equation (6-68) is that α is proportional to the wavefront slope averaged over the system pupil [6-62, 6-16]. The focal length, F, converts wavefront slope into position at the focal plane.

The easiest way to solve for the variance of α is to use a Fourier transform. The spatial Wiener spectrum for $(\partial/\partial x)\,\phi(r)$ is $K_x^2\,\Phi_\phi(K)$, whereas for $(\partial/\partial y)\,\phi(r)$ it is $K_y^2\,\Phi_\phi(K)$.

If the relative random irradiance fluctuations are small, then

$$<\alpha_x^2> \cong (F/k)^2 \int\!\!\int_{-\infty}^{+\infty} K_x^2\,\Phi_\phi(\mathbf{K}) <|H(\mathbf{K})|^2>\, d\mathbf{K} \tag{6-69}$$

and similarly for $<\alpha_y^2>$. Here $H(\mathbf{K})$ is the two-dimensional Fourier transform of the product of the irradiance at the pupil times the pupil transmission function, all normalized to a value of unity at $\mathbf{K} = 0$. If scintillation is ignored, the averaging brackets in Equation (6-69) can be omitted, and $H(\mathbf{K})$ becomes the normalized Fourier transform of the pupil irradiance transmission function. If the pupil is a clear circular aperture of radius a, then Equation (6-46) may be used for H.

For small irradiance fluctuations,

$$\Phi_\phi \cong \Phi_\phi + \Phi_\chi = 2\pi k^2(0.033)K^{-11/3} \int_0^L C_n^{\,2}(z)\,dz \tag{6-70}$$

for plane wave geometry. Equations (6-69), (6-46), and (6-70) combine to yield an expression for $<\alpha\cdot\alpha>=<\alpha_x^2 + \alpha_y^2>$:

$$<\alpha \cdot \alpha> = F^2 a^{-2} 16\pi^2(0.033) \int_0^L C_n^{\,2}(z)\,dz \int_0^\infty J_1^{\,2}(aK)\,K^{-8/3}\,dK \tag{6-71}$$

where the azimuthal K integration has already been carried out. The second integral above is equal to $0.8644\, a^{5/3}$, so that

$$<\alpha \cdot \alpha> \cong 4.50\, F^2 a^{-1/3} \int_0^L C_n^{\,2}(z)\, dz \qquad (6\text{-}72)$$

This result is valid for plane wave propagation in the near-field limit. This result may also be expressed in the form

$$\frac{1}{2} <\alpha \cdot \alpha> = <\alpha_x^{\,2}> = <\alpha_y^{\,2}> \cong 0.975\, (F/k)^2\, [D_\phi(2a)]\, /(2a)^2 \quad (6\text{-}73)$$

where the pupil diameter is 2a. $D_\phi\,(r)$ is the phase structure function for plane-wave propagation given approximately by

$$D_\phi(\mathbf{r}) = 2.91 k^2 r^{5/3} \int_0^L C_n^{\,2}(z)\, dz \qquad (6\text{-}74)$$

For spherical waves

$$D_\phi(\mathbf{r}) = 2.91 k^2 r^{5/3} \int_0^L C_n^{\,2}(z)(z/L)^{5/3}\, dz \qquad (6\text{-}75)$$

As a common approximation, the coefficient 0.975 in Equation (6-73) is replaced by unity [6-30].

Diffraction decreases Φ_ϕ, but this effect upon $<\alpha \cdot \alpha>$ is opposed by the effect of irradiance fluctuations. The net balance depends on the ratio r_0/a, but for most purposes it is sufficient to replace D_ϕ in Equation (6-73) by D_w, the wave structure function defined by Equation (6-37), as was done here.

6.5.6. Short Exposure Blur. The mathematically rigorous evaluation of $<T_s(\mathbf{f})>$ is rather involved, but fortunately an approximation can be made which yields nearly correct results with little effort. By the shifting theorem for Fourier transforms [6-63], $T_s(\mathbf{f})$, which is the Fourier transform of $\mathcal{g}\,(\mathbf{p} - \alpha)$, is related to $T(\mathbf{f})$, the transform of $\mathcal{g}\,(\mathbf{p})$, by the equation

$$T(\mathbf{f}) = T_s(\mathbf{f})\, \exp(-\,2\pi i \alpha \cdot \mathbf{f}) \qquad (6\text{-}76)$$

The approximation involves the assumption that the random variations in $T_s(\mathbf{f})$ are uncorrelated with the random variations in $\exp(-2\pi i \alpha \cdot \mathbf{f})$. With this approximation, an ensemble average of the right-hand side of Equation (6-76) can be written as the product of two averages. That is,

$$<T(\mathbf{f})> \cong <T_s(\mathbf{f})> \quad <\exp(-\,2\pi i \alpha \cdot \mathbf{f})> \qquad (6\text{-}77)$$

Since α_x and α_y are random variables with Gaussian distributions, the results discussed in Section 6.9 can be used to obtain the following formula:

$$<\exp(-\,2\pi i \alpha \cdot \mathbf{f})> = \exp\left[-\,2\pi^2 <(\alpha \cdot \mathbf{f})^2>\right]$$

$$= \exp\left[-\,\pi^2 f^2 <\alpha \cdot \alpha>\right] \qquad (6\text{-}78)$$

It follows that

$$<T_s(f)> \cong <T(f)> \exp(\pi^2 f^2 <\alpha \cdot \alpha>) \qquad (6\text{-}79)$$

For a system with a clear circular aperture or radius a, Equations (6-59), (6-65), and (6-67) for $<T(f)>$ and Equation (6-73) for $<\alpha \cdot \alpha>$ can be substituted into the equations above to yield the following result:

$$<T_s(f)> \cong T_0(f) \exp\left\{-\left(\frac{1}{2}\right)D_w(F\lambda f)\left[1 - (f/f_0)^{1/3}\right]\right\} \qquad (6\text{-}80)$$

where $f_0 = 2a/\lambda F$ is the spatial frequency at the diffraction limit. With the appropriate substitution for D_w, Equation (6-80) is valid for both plane wave and spherical wave propagation. Although derived under the near-field approximation, Equation (6-80) is reasonably accurate in the presence of modest scintillation. It may be useful over an even larger domain.*

Reference [6-64] gives Wiener spectra for the temporal variations in optical wavefront distortions. (See also Noll, and Greenwood and Fried.)

6.5.7. Speckle Imaging. The averaging operators applied to $\mathcal{I}(p)$ and $\mathcal{I}(p - \alpha)$ in Equations (6-57) and (6-58) smooth out any fine, speckle-like pattern which may have been present in the instantaneous point spread function. However, the statistics of this speckle pattern are vital to a quantitative understanding of the various forms of speckle imaging techniques, as described in References [6-65] and [6-66].

Since the cross correlation function $< \mathcal{I}(p) \mathcal{I}(p + p') >$ does contain the appropriate information on the speckle statistics it is of interest. In addition, this function is a useful measure of isoplanatism. (See Section 6.5.2.)

From Equation (6-56) one may derive the relationship

$$<T^*(f)T(f)> = \int\!\!\int_{-\infty}^{+\infty} <\mathcal{I}(p)\,\mathcal{I}(p + p')> \exp(-2\pi i f \cdot p')dp' \qquad (6\text{-}81)$$

where T^* is the complex conjugate of T, and $T^*(f)\,T(f) \equiv |T(f)|^2$.

Reference [6-67] gives a detailed description of the methods of evaluating this quantity.

For medium-to-high spatial frequencies, use of the Rytov approximation yields

$$<|T(f)|^2> \cong 0.15\,(r_0/a)^2 T_0(f) \qquad (6\text{-}82)$$

where $T_0(f)$ is the transfer function for the system optics (assumed here to be diffraction limited) in the absence of turbulence. In this section, r_0 is the coherence diameter defined in Section 6.5.3, and a is the radius of the system entrance pupil.

Conventional speckle imaging can yield information about the magnitude, but not the phase angles, of the various object spatial spectral components. Various schemes have been proposed to retrieve phase angle information as well. Among these is the Knox-Thompson algorithm [6-68], which makes use of the fact that the phase angle of $T(f_1)$ is approximately equal to the phase angle of $T(f_2)$ when $|f_2 - f_1| < r_0/F\lambda$.

---

*Equation (6-60) can be used to derive $<M_s>$.

6.6. Beam Wave Propagation

6.6.1. Beam Properties. Equations regarding beam wave propagation differ from those involving plane wave or spherical wave propagation in that the lateral beam dimension plays an important role in the determination of the radiation field at the receiver or target. There is a close, qualitative similarity between beam wave propagation and image degradation as described in Section 6.5 where the size of the system pupil also plays an important role. For example, beam wander corresponds to image motion, and time-averaged, beam intensity profiles correspond to long-exposure, image point spread functions.

Beam paths of interest are often long enough to cause scintillation saturation. Except where explicitly noted, the equations presented in Section 6.6 are valid in the saturation domain, beyond the limitations of the Rytov approximation.

Historically, for reasons of mathematical convenience, most papers concerning beam waves have assumed that the beam intensity is launched with a Gaussian-shaped intensity profile, and that the radiation is perfectly coherent on a spherical surface centered about a point of focus.* These assumptions will be used here also, except where explicitly noted otherwise.

Beam irradiance, $E(x, y, z)$, and its statistics are the primary quantities of interest. As elsewhere in this chapter, x and y will be coordinates perpendicular to the average propagation path. The beam is assumed to start at the plane $z = 0$, with an irradiance profile given by

$$E(x, y, 0) = E_0(0) \exp[-(x^2 + y^2)/b_0^2(0)] \qquad (6\text{-}83)$$

where $E_0(0)$ = peak irradiance at center of starting beam

$b_0(0)$ = effective radius of starting beam

The area integral of the irradiance profile is the total beam power. For the profile given in Equation (6-83), this becomes

$$\int\!\!\!\int_{-\infty}^{+\infty} E(x, y, 0)dx\, dy = \pi E_0(0) b_0^2(0) \qquad (6\text{-}84)$$

In the absence of turbulence, the profile of the beam will maintain its initial Gaussian shape as it propagates along the z axis, but with a modified beam-radius parameter. Thus, in the absence of turbulence

$$E(x, y, z) = E_0(z) \exp[-(x^2 + y^2)/b_0^2(z)] \qquad (6\text{-}85)$$

where $b_0(z)$ is a measure of the beam radius in the absence of turbulence.

If the beam is focused at a point $z = F$ on the z axis,** then in the absence of turbulence the beam radius, $b_0(z)$, would obey the following relationship [6-69]:

$$b_0^2(z) = b_0^2(0)[1 - (z/F)]^2 + z^2/k^2 b_0^2(0) \qquad (6\text{-}86)$$

The first part of the right side of Equation (6-86) expresses the geometric contraction

---

*For collimated beams, the corresponding assumption is that the radiation is perfectly coherent over a planar surface.

**A negative value of F indicates a beam diverging from a virtual focal point located behind the transmitter aperture.

of the beam-width near focus. The second part gives the added contribution of diffraction to the total beam spread.

Since the total power in the beam is unchanged* along the propagation path, $E_0(z)$ and $b_0{}^2 (z)$ are related by the condition

$$\pi E_0(z) b_0{}^2(z) = \pi E_0(0) b_0{}^2(0) \qquad (6\text{-}87)$$

which is a generalization of Equation (6-84). Alternatively, one can write

$$b_0{}^2(z) = b_0{}^2(0) E_0(0)/E_0(z) \qquad (6\text{-}88)$$

a result which will be of later use.

6.6.2. Time-Averaged Beam Profile. Reference [6-69] summarizes the methodology for computing the time-averaged beam profile. A second-order coherence function is evaluated with the aid of the modified Huygens-Fresnel principle [6-39]. In this section we give a simplified and approximate version of these results, accurate to better than 20%, provided that the beam diameter falls well within the inertial subrange of the Kolmogorov turbulence spectrum.

The time average of $E(x, y, L)$ is designated by $< E(x, y, L) >$, where, as usual, the angle brackets indicate the averaging operation. In the presence of turbulence, the average intensity of the center of the beam is given approximately by

$$<E(0, 0, L)> \cong E_0(0)[b_0{}^2 (0)/b_0{}^2 (L)] \{1 + 0.46\, D_w\, [2L/kb_0(L)]\}^{-6/5} \qquad (6\text{-}89)$$

where $D_w [2L/kb_0(L)]$ is the wave structure function computed as though there were a point source of radiation located in the plane $z = L$, and evaluated for a lateral separation distance equal to $2L/kb_0(L)$ at the $z = 0$ end of the path. Analogous to Equation (6-39), this wave structure function is given by the expression

$$D_w [2L/kb_0(L)] = 2.91\, k^2 [2L/kb_0(L)]^{5/3} \int_0^L C_n{}^2(z)[1 - z/L]^{5/3} dz$$

$$= 9.25\, k^{1/3} [b_0(L)]^{-5/3} \int_0^L C_n{}^2(Z)(L - z)^{5/3}\, dz \qquad (6\text{-}90)$$

For a horizontal path with a uniform value of $C_n{}^2$, Equation (6-90) reduces to

$$D_w [2L/kb_0(L)] = 3.47\, C_n{}^2 k^{1/3} L^{8/3} [b_0(L)]^{-5/3} \qquad (6\text{-}91)$$

A special case of interest occurs when the beam is focussed on the plane $z = L$, that is when $L = F$ and $F > 0$. Equation (6-86) then gives $b_0(L) = L/kb_0(0)$, which may be substituted into Equation (6-89). If, in addition, $C_n{}^2$ is constant along the path so that Equation (6-91) applies, then Equation (6-89) becomes

$$<E(0, 0, F)> \cong E_0(0)k^2 F^{-2} [b_0(0)]^4 \{1 + 1.60\, C_n{}^2 k^2 F[b_0(0)]^{5/3}\}^{-6/5} \qquad (6\text{-}92)$$

---

*As discussed in Chapters 4 and 5, various atmospheric constituents will, in fact, cause absorption. These equations assume that corrections for this effect will be performed after the calculation of turbulence effects.

Even in the presence of turbulence, the transverse profile of $<E(x, y, L)>$ maintains an approximately Gaussian shape.* That is,

$$<E(x, y, L)> \simeq <E(0, 0, L)> \exp[-(x^2 + y^2)/b^2(L)] \qquad (6\text{-}93)$$

Here, $b(L)$ may be considered to be a measure of the radius of the local averaged beam, as contrasted with $b_0(L)$, which was the corresponding radius measure in the absence of turbulence.

Because the turbulence does not change the total power carried by the beam, Equation (6-88) may be generalized to yield

$$b(L) = b_0(0) \left[\frac{E_0(0)}{<E(0, 0, L)>} \right]^{1/2} \qquad (6\text{-}94)$$

When Equation (6-89) is substituted into Equation (6-94), the result is

$$b(L) \simeq b_0(L)\{1 + 0.46\, D_w\, [2L/kb_0(L)]\}^{3/5} \qquad (6\text{-}95)$$

When C_n^2 is uniform along the path and when the beam is focussed on the plane of interest, then Equation (6-92) may be substituted into Equation (6-94) to yield

$$b(F) \simeq [F/kb_0(0)]\{1 + 1.60\, C_n^2 k^2 F[b_0(0)]^{5/3}\}^{3/5} \qquad (6\text{-}96)$$

For long paths with strong uniform turbulence, this expression becomes asymptotic to

$$b(F) \approx 1.26\, C_n^{6/5} k^{1/5} F^{8/5} \qquad (6\text{-}97)$$

An experimental check on these predictions is given in Reference [6-70].

CAUTION must be used in applying the equation just presented since there are two additional effects which can invalidate the strict applicability of these results. The first of these is *thermal blooming,* which occurs when the radiation is strong enough to significantly heat the propagation medium. A discussion of this effect is beyond the scope of this chapter, but it can be noted that in pure air thermal blooming usually occurs only with high-power laser beams.

The second effect, which may at times be quite significant, is *undulant beam wander.* *Undulance* is the word used to describe slow, quasi-periodic, temporal and spatial variations in the atmosphere often associated with atmospheric gravity waves. (See Section 6.2.6.) Unfortunately a quantitative predictive theory for undulance does not yet exist, but its effect will usually be most noticeable under nighttime, stable atmospheric conditions.

A more detailed description of undulant beam wander is given in Section 6.6.3, but here it is important to note that this is a source of time-averaged radiation spreading in addition to diffraction and turbulence effects.

Even when the beam intensity profile at the transmitter does not have a Gaussian shape, the equations given in this section will yield reasonably correct predictions if the appropriate value of $b_0(0)$ is used. Equation (6-84) should be used to determine $b_0(0)$. For example, if the initial profile is uniform over a circular pupil of radius a, then that radius should be substituted for $b_0(0)$.

---

*The actual profile will have slightly stronger tails than an exact Gaussian profile. Details are given in References [6-69] and [6-36].

When the radiation is not spatially coherent as it leaves the transmitting aperture, the generalized time-averaged beam profile shape may be estimated as a spatial convolution of a vacuum propagation beam profile and a Gaussian-like profile with a radius parameter as given by Equation (6-95).

6.6.3. Beam Motion. The instantaneous centroid of the beam in the plane $z = L$ is defined* as

$$\beta = (\beta_x, \beta_y) = \frac{\iint \mathbf{r} E(\mathbf{r}, L) d\mathbf{r}}{\iint E(\mathbf{r}, L) d\mathbf{r}} \tag{6-98}$$

where $E(\mathbf{r}, L) = E(x, y, L)$ is the instantaneous, beam intensity profile in the plane $z = L$. Here $\mathbf{r} = (x, y)$, and the area integration is over the infinite plane $z = L$. Equation (6-98) is analogous to Equation (6-55). It will be assumed that the (x, y) coordinate system axes are centered on $<\beta>$ so that $<\beta> \equiv 0$.

The exact evaluation of $<\beta \cdot \beta> \equiv <\beta_x{}^2> + <\beta_y{}^2>$ requires $<E(\mathbf{r}, L) E(\mathbf{r}', L)>$ be known. It is related to the fourth-order coherence function. Reference [6-71] solves for $<\beta \cdot \beta>$ using an approximate form of the required coherence function. Reference [6-72] gives a method of solution which is valid when turbulence is not too strong. Both references include the effect of scale lengths outside the inertial subrange.

Described below is a simple derivation which yields results equivalent to those in the two references just described. The starting point is Equation (6-69), which describes image motion. In that equation it is assumed that all optical wavefront distortions take place before radiation is focussed by the optical system. The focal length, F, which appears outside the integral in Equation (6-69) is a "lever arm" which converts angular deviations into position deviations.

The beam-wander equation is quite analogous, except that the lever-arm term is weighted proportionally to the strength of the turbulence along the path. The basic expression is

$$<\beta \cdot \beta> = 2\pi(0.033) \int_0^L C_n{}^2(z) (L - z)^2 dz \int_0^\infty K^{-5/3} <|\mathcal{H}(\mathbf{K}, z)|^2> d\mathbf{K} \tag{6-99}$$

where $L - z$ is the lever arm, and Equation (6-70) was used for $\Phi_\phi(\mathbf{K})$. Here $\mathcal{H}(\mathbf{K}, z)$, which is analogous to $H(\mathbf{K})$ in Section 6.5.5, is the normalized, two-dimensional Fourier transform of the instantaneous, beam-irradiance profile across a plane at position z. If one ignores scintillation and assumes that the beam has a fixed, Gaussian-shaped profile with a radius parameter $b(z)$, then

$$\mathcal{H}(\mathbf{K}, z) = \exp[-b^2(z) K^2/4] \tag{6-100}$$

and the averaging brackets in Equation (6-99) can be dropped.

When Equation (6-100) is substituted into Equation (6-99), the result becomes

$$<\beta \cdot \beta> = 4.070 \int_0^L C_n{}^2(z) (L - z)^2 [b(z)]^{-1/3} dz \tag{6-101}$$

---

*Note that β is a linear measure of beam position. Some authors report beam wander in angular units, equivalent to β/L.

Numerical integration is usually needed to solve this integral, but it is simple enough to be done on a modest, programmable pocket calculator.

Of special interest is the case where the image motion is measured in the focal plane of the beam and where C_n^2 is uniform along the path. The required integration has been carried out, and the numerical results may be approximated (10% accuracy) by the following analytic expression:

$$<\beta \cdot \beta> \cong \frac{1.53\, C_n^2 F^3 [b_0(0)]^{-1/3}}{\{1 + 0.2\, [F/kb_0^2(0)] + 0.1\, [C_n^{6/5} k^{1/5} F^{8/5} b_0^{-1}(0)]\,\}^{1/3}} \qquad (6\text{-}102)$$

To obtain the numerical results, one substitutes Equation (6-91) with $L = z$ into Equation (6-95) to yield $b(z)$, where Equation (6-86) is used for $b_0(z)$.

The corresponding results for an initially collimated beam are very similar and will be identical to those for a focused beam when $L/kb_0^2(0) \gg 1$.

The effect of scintillation, which was ignored to get this result, is to increase $<\beta \cdot \beta>$ slightly from the results given here [6-71].

The effects of the inner and outer scales of turbulence are to decrease $<\beta \cdot \beta>$ from the results given here. This decrease will be negligible if $b(z)$ is well within the inertial subrange for $z < L$ [6-72].

The dimensionless quantity $kb_0^2(0)/L$, often designated by the symbol Ω, is called the *Fresnel number;* it is a measure of whether a point in the plane $z = L$ is in the near or far field of the effective transmitter aperture. The state $\Omega \gg 1$ designates a near-field condition, whereas $\Omega \ll 1$ designates a far-field condition.

Both the vertical and horizontal components, (β_y, β_x), of the turbulence-induced beam motion are random variables with a Gaussian-probability distribution. The standard deviation of both β_x and β_y is $[(1/2) <\beta \cdot \beta>]^{1/2}$.

If, instead of a Gaussian-shaped intensity profile at the transmitter, the profile were uniform in a pupil of radius a, then Equation (6-101) should have the radius a substituted for $b_0(0)$, and a coefficient of 4.50 replacing the coefficient 4.07.

Figure 6-12, shows experimentally determined, beam-wander data for both vertical and horizontal motions. The rapid fluctuations are caused by turbulence. The added slower fluctuation (most predominant for vertical motion) is caused by large-scale, atmospheric refractivity changes. These changes, which may be called undulance, are probably related to atmospheric gravity waves. The equations for beam wander presented in this section do not include these slow, undulant beam wanderings.

In the absence of undulance, the temporal spectrum of the turbulence-induced, beam-position deviations obeys an approximate $f^{-2/3}$ power law in the $0.1 < f < 10$ Hz frequency range. Undulance will cause a departure from this power law relationship at low frequencies.

The average vertical gradient of the atmosphere's index of refraction will, of course, cause a corresponding, average, vertical, bending of the beam. An additional source of beam wander is equipment motion or vibration at the transmitter site.

6.6.4. Beam Irradiance Fluctuations. To understand beam-wave irradiance fluctuations, one must recognize their two relatively distinct causes. The first is the ordinary generation of a speckle pattern superimposed on the beam irradiance profile. This process is very similar to scintillation for unbounded plane waves or spherical wave propagation, as discussed in Sections 6.3.3.

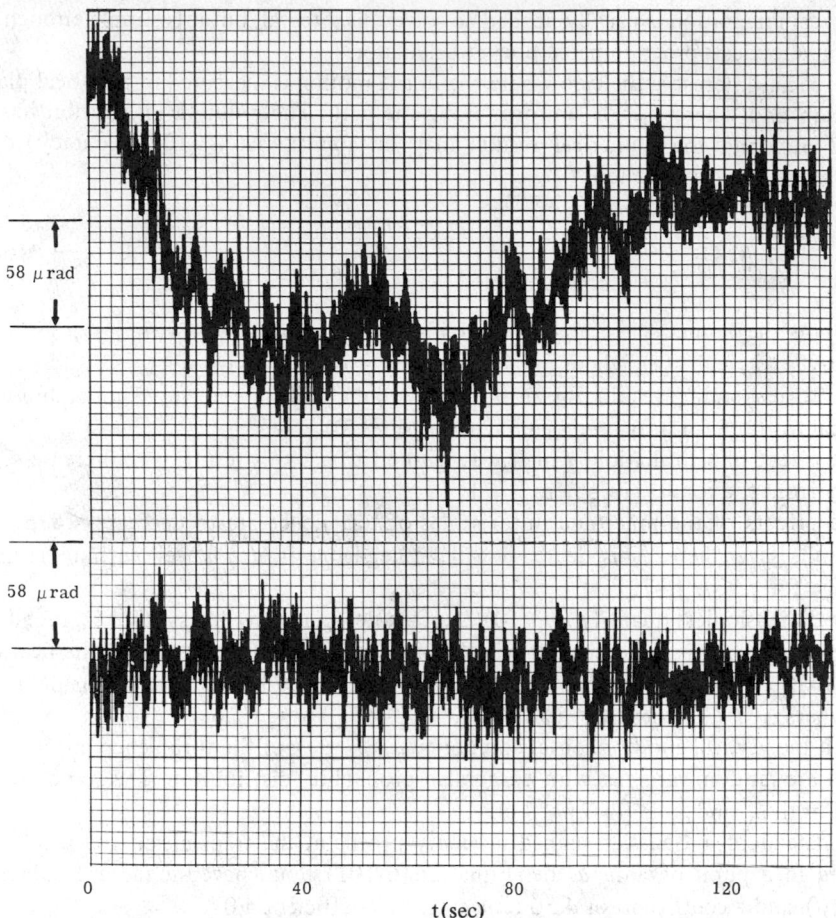

Fig. 6-12. Laser beam wander versus time for horizontal path propagation at night under low-wind conditions. The vertical wander signal is shown at the top and the horizontal wander signal on the bottom [6-73].

The second process, unique to beam waves, is the effect of beam wander and related, gross, beam-shape changes upon the received irradiance. If, as an extreme example, a beam with a finite width were to be moved on and off the receiving location, total modulation of the received irradiance would result. In the absence of saturation, this effect will be strongest near the edges of the beam where, because the average, lateral-irradiance gradients are strongest, even small beam motions cause significant irradiance changes.

Not surprisingly, dual parameters are needed for an analytical description of the beam-wave, irradiance-fluctuation process [6-36]. One parameter, partially describing the speckle pattern, will be of the form $C_n{}^2 \, k^{7/6} \, L^{11/6}$, following Equations (6-35) and (6-36).

There are several possible parameters for describing beam-motion effects. Which is best is the subject of current research. Among the candidates are $C_n{}^2 \, k^2 \, L \, [b_0(0)]^{5/3}$, which is a wave structure function for a separation distance $b_0(0)$, and the Fresnel parameter $\Omega = b_0{}^2(0) \, k/L$. However, one should note that

$$C_n^2 k^2 L [b_0(0)]^{5/3} = [C_n^2 k^{7/6} L^{11/6}] \Omega^{5/6} \qquad (6\text{-}103)$$

so that either of the latter two parameters, when combined with the speckle parameter, is usable, provided C_n^2 is uniform along the path.

Under the Rytov approximation, the normalized irradiance variance in the center of a collimated beam will be approximately

$$<E>^{-2} \sigma_E^2 \simeq 1.24 \, C_n^2 \, k^{7/6} L^{11/6} \left[\frac{\Omega}{2.5 + \Omega} + \frac{0.4}{1 + 8\Omega} \right] \qquad (6\text{-}104)$$

This equation is an analytic approximation (10% accuracy) of the prediction presented in tabular form in Reference [6-74]. In the limit $\Omega \to \infty$, the result is equivalent to that for plane wave propagation. In the limit $\Omega \to 0$, the result is equivalent to that for spherical wave propagation. For intermediate values of Ω, the quantity is square brackets has a minimum value when $\Omega \approx 0.3$. This prediction agrees with observation when $<E>^{-2} \sigma_E^2 < 0.8$ [6-91]. The irradiance-fluctuation variance also has been predicted for positions off the center of the beam [6-76].

Equation (6-104) assumes that C_n^2 is uniform along the path. References [6-75] and [6-77] give more general formulas and experimental data.

The spatial autocorrelation function for irradiance fluctuations varies with the average position with respect to the beam center. Examples of calculated autocorrelation functions are given in Reference [6-37]. Reference [6-36] shows that the irradiance autocovariance function, when averaged over the entire, beam cross-section, is identical to the corresponding autocovariance function for laterally unbounded beams.

Figure 6-13 shows typical, measured, irradiance-autocorrelation functions under conditions of saturated scintillation. This figure should be compared to Figure 6-6, which is valid only for conditions of weak scintillation. Two changes are evident. With saturation the initial decay of correlation from the origin occurs at distances small compared to $(\lambda L)^{1/2}$.

Various theories have been proposed to explain these two, new, characteristic correlation distances. According to some [6-37], the characteristic distance for the initial decay is r_0, the coherence diameter defined by Equation (6-65). These same theories also predict that the tail of the autocorrelation function extends to a distance proportional to $C_n^{6/5} k^{1/5} L^{8/5}$, which may be interpreted through Equation (6-97) as an effective beam-spreading parameter. Although agreeing reasonably with experiment, these and other theories have not yet been tested critically.

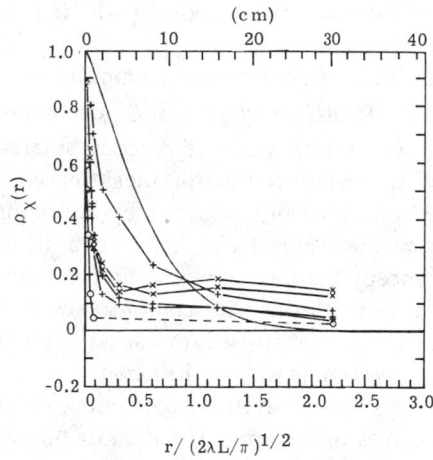

Fig. 6-13. Experimentally observed log-amplitude autocorrelation functions under conditions of irradiance fluctuation saturation. The measurements were made over a 45 km long path with 0.63 μm wavelength laser radiation. An unnormalized correlation distance scale is shown at the top with a normalized scale at the bottom. The smooth solid line is the autocorrelation function predicted by the Rytov approximation, and is shown for comparison. (Compare with Figure 6-6.) [6-31].

The long tail on the autocorrelation function limits the averaging effects of moderately large, receiver apertures. Figure 6-9, derived for weak scintillation, is not valid under conditions of saturated scintillation. However, calculations for aperture averaging would be valid if the correct irradiance-autocorrelation function were substituted.

The temporal spectrum of irradiance fluctuations under conditions of saturation contains two characteristic frequencies corresponding (through the wind velocity) to the two characteristic scales in the irradiance-autocorrelation functions. Typical spectra are given in Gurvich and Tatarskii.

The Rytov approximation has been used to predict log-amplitude fluctuation in the focus of a beam [6-74]. Agreement with observation is poor, however, presumably because beam-wander effects are not accounted for in that approximation [6-78].

Reference [6-70] claims that the single quantity $C_n^2 k^2 [b_0(0)]^{5/3}$ is a sufficient measure to determine the irradiance-fluctuation in the center of a focused beam. Both the experimental data and the analytical predictions in Reference [6-70] can be approximated quite well by the simple formula

$$<E>^{-1} \sigma_E = \frac{(\eta + \eta^2)}{(1 + \eta^2)} \tag{6-105}$$

where

$$\eta = 0.35 \, C_n k L^{1/2} [b_0(0)]^{5/6} \tag{6-106}$$

The quantity $<E>^{-1} \sigma_E$ is, of course, the standard deviation of the irradiance E, normalized by the average irradiance. This equation is valid only when C_n^2 is reasonably uniform along the optical path. It is valid, however, both below and in the saturation regime.

Figure 6-14 shows an example of an instantaneous beam profile.

6.7. Real-Time Wavefront Compensation

For several cases of practical interest, it is possible to construct a feedback control system which can perform real-time correction of turbulence-induced wavefronts. Correction is generally achieved by reflecting the optical radiation either off a single, fast-response, deformable mirror element, or off an array of smaller movable elements. This concept has been explored both for correction of blur in an image-forming system, and for beam-flux-density concentration in a laser-transmitter system. With a transmitter, the beam is predistorted to cancel approximately the atmospheric perturbation the beam receives on its way to the target.

For good wavefront correction, the deformable mirror must be controlled with many degrees of freedom. The number of degrees of freedom, N, is related to the area of the system aperture, A, and the desired image Strehl definition*, $\mathcal{S}$, by the approximate relationship.

$$\mathcal{S} \leqslant \exp[-0.3 \, A^{5/6} \, N^{-5/6} \, r_0^{-5/3}] \tag{6-107}$$

where r_0 is the coherence diameter given by Equation (6-65) or (6-66) [6-80]. For visual astronomy, $r_0 \sim 0.1$ m; thus, at least 100 degrees of freedom per square meter of telescope aperture is required for a 70% Strehl definition. Since r_0 is proportional to $\lambda^{6/5}$,

*Defined in Section 6.5.3.

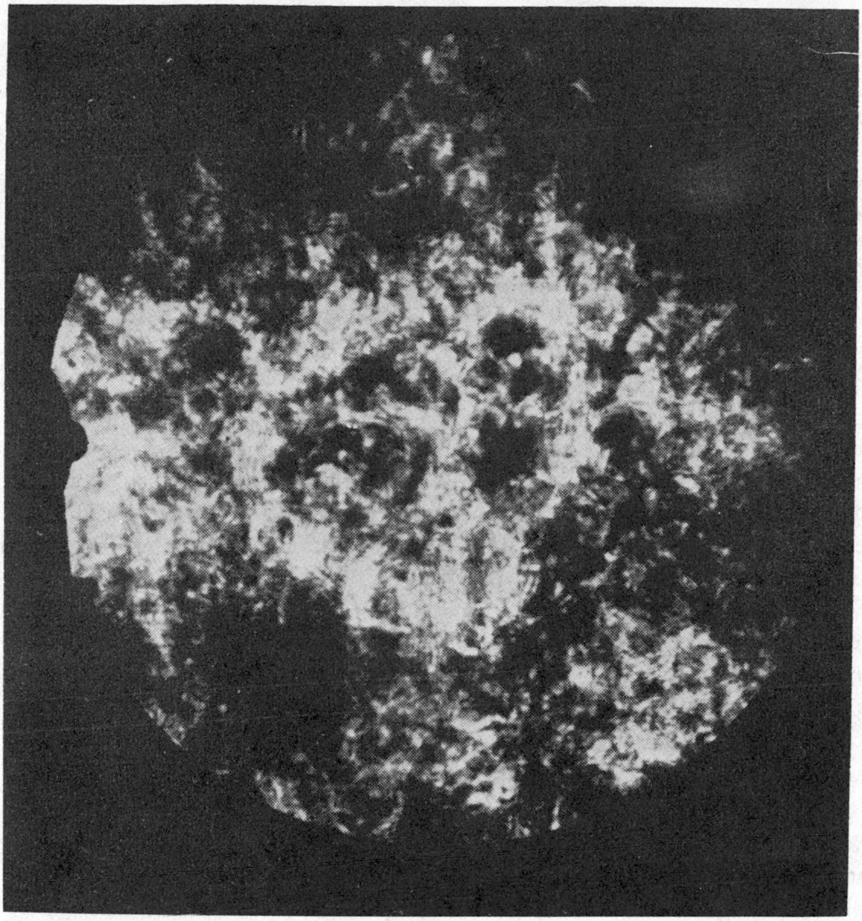

(a) Photograph of a typical laser beam profile after propagation through a long turbulent path.

(b) A single scan of irradiance versus lateral distance for the profile shown in part (a), showing positive spikes characteristic of a log-normal irradiance distribution. The measurements were made with a pulsed ruby laser operating at a 0.69 μm wavelength. The path was 1000 m long and 2 m above ground with strong turbulence; measurement field is 610 mm in diameter.

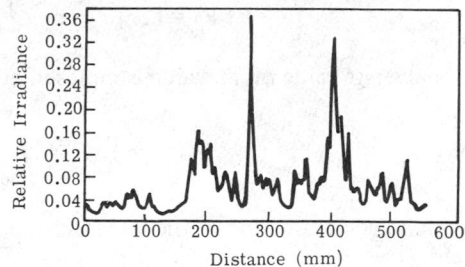

Fig. 6-14. Instantaneous beam profile [6-79].

considerably fewer degrees of freedom per unit of aperture area are required at infrared wavelengths than for visual-wavelength systems.

A key problem for the designers of real-time, compensation systems is the method of obtaining error signals which must be used to drive the wavefront-correcting element. For astronomical applications, it is often possible to analyze the wavefront error in real time with a shearing interferometer or an electronic Hartmann sensor. Nevertheless, some measurement noise will always be present, which will degrade the Strehl resolution from the equality limit in Equation (6-105). Great care in signal processing must be taken to achieve adequate system results.

The finite size of the isoplanatic region (Section 6.5.2) implies that wavefront compensation will normally be achieved over a restricted field of view. Multiple wavefront-correction planes are needed if system operation over a wide field is required.

The simplest form of real-time, wavefront compensation is image-motion or beam-motion compensation. Such a system has only two degrees of freedom (corresponding to x-motion and y-motion), but with system pupil diameters of proper size significant performance gains can be achieved.

Details are given in References [6-81, 6-82, 6-83, 6-84] and [6-85].

6.8. Remote Sensing

In previous sections, the statistics of output signals from optical systems are given in terms of the spatial distribution of the turbulence and the cross-track components of the wind velocities. For systems with single, simple, clear circular apertures, it is difficult with imperfect data to work the statistics backward (i.e., to determine the turbulence and/or wind profiles from the output signal statistics).

The system apertures and geometrics can be modified, however, to allow reliable remote probing of the atmosphere. Typically, systems with multiple apertures or with grating-like apertures are used. References [6-16, 6-18, 6-86] and [6-87] give examples of such remote-sensing systems.

6.9. Log-Normal Random Variables

Several quantities of physical interest have log-normal statistics. In this section, some properties of log-normal random variables are described.

A random variable, Υ, is said to have a log-normal distribution if it can be represented in the form $\Upsilon \equiv \exp \Psi$, where Ψ is a random variable with a normal (Gaussian)-, probability density function. Let $\sigma_\Psi^2 \equiv \langle(\Psi - \langle\Psi\rangle)^2\rangle$ be the variance of Ψ. The probability density function for Ψ then becomes

$$P(\Psi) = (2\pi)^{-1/2} \sigma_\Psi^{-1} \exp[-(\Psi - \langle\Psi\rangle)^2/2\sigma_\Psi^2] \qquad (6\text{-}108)$$

The average value of Υ, which is represented by $\langle\Upsilon\rangle$, is

$$\langle\Upsilon\rangle = \int_{-\infty}^{+\infty} d\Psi\, P(\Psi) \exp \Psi \qquad (6\text{-}109)$$

which upon evaluation of the integral becomes

$$\langle\Upsilon\rangle = \exp\left[\langle\Psi\rangle + \left(\frac{1}{2}\right)\sigma_\Psi^2\right] \qquad (6\text{-}110)$$

As an illustration in the use of this result, consider the irradiance E for plane wave propagation, where $E = \exp 2X$. Here X, which is defined by Equation (6-27), has a normal probability density distribution with a mean value of $\langle X\rangle$ and a variance σ_X^2.

By identifying Υ with E, and Ψ with $2X$, one finds from Equation (6-110) that

$$\langle E\rangle = \exp[2\langle X\rangle + 2\sigma_X^2] \qquad (6\text{-}111)$$

Since turbulence redistributes energy, but does not absorb it, it follows that $<E>$ should be a constant independent of the strength of the turbulence. Thus, if $E = 1$ in the absence of turbulence, it follows that

$$<\chi> = -\sigma_\chi^2 \qquad (6\text{-}112)$$

In a similar way, one can find that

$$<E^2> = <\exp 4\chi> = \exp[4<\chi> + 8\sigma_\chi^2] \qquad (6\text{-}113)$$

It follows that if $<E> = 1$, then

$$\sigma_E^2 \equiv <E^2> - <E>^2 = (\exp 4\sigma_\chi^2) - 1 \qquad (6\text{-}114)$$

This result is helpful in relating experimentally observed, irradiance-fluctuation statistics σ_χ, given in Equation (6-43). A variant of this result is used to evaluate aperture averaging effects.

Equation (6-110) is also correct if Ψ is an imaginary quantity. Let $\Psi \equiv i\Theta$, then

$$<\Upsilon> = \exp\left[i<\Theta> - \frac{1}{2}\sigma_\Theta^2\right] \qquad (6\text{-}115)$$

This result is useful where Θ becomes identified with image motion or the difference in optical phases, as in Section 6.5.

A very interesting property of log-normal distributions is that the probability distribution of the sum of a number of log-normal random variables is itself very close to a log-normal distribution [6-88]. Even after aperture averaging, therefore, the total captured-signal flux also has a nearly log-normal, probability distribution.

Some other properties of log-normal variables are covered in Section 6.2.7 and in Reference [6-89].

6.10. References and Bibliography

6.10.1. References

[6-1]   *Handbook of Geophysics*, United States Air Force, The Macmillan Company, New York, NY, 1960, Revised Edition, pp. 13-1 and 13-2.

[6-2]   A. N. Kolmogorov, "The Local Structure of Turbulence in Incompressible Viscous Fluid for Very Large Reynolds Numbers," *Doklady Akad. Nauk. USSR.*, Academy of Sciences of USSR, New York, NY, Vol. 30, 1941, p. 301.

[6-3]   John L. Lumley and Hans A. Panofsky, *The Structure of Atmospheric Turbulence*, John Wiley and Sons, New York, NY, 1964, pp. 62, 72.

[6-4]   John C. Wyngaard, Y. Izumi, and Stuart A. Collins, Jr., "Behavior of the Refractive-Index-Structure Parameter Near the Ground," *Journal of the Optical Society of America*, Optical Society of America, Washington, DC, Vol. 61, No. 12, December 1971, pp. 1646-1650.

[6-5]   E. K. Webb, "Daytime Thermal Fluctuations in the Lower Atmosphere," *Applied Optics*, Optical Society of America, Washington, DC, Vol. 3, No. 12, December 1964, pp. 1329-1336.

[6-6]   Robert S. Lawrence, Gerald R. Ochs, and Steven F. Clifford, "Measurements
          of Atmospheric Turbulence Relevant to Optical Propagation," *Journal of the
          Optical Society of America*, Optical Society of America, Washington, DC,
          Vol. 60, No. 6, June 1970, pp. 826-830.

[6-7]   Gerald R. Ochs and Robert S. Lawrence, "Temperature and C_n^2 Profiles Meas-
          ured Over Land and Surface," United States Government Printing Office,
          Washington, DC, National Oceanic and Atmospheric Agency Technical Report
          ERL 251-WPL22, October 1972.

[6-8]   F. Ya. Voyt, Ye. Ye. Korniyenko, V. P. Kukharets, S. B. Khusid, and L. R.
          Tsvang, "Structural Characteristics of the Temperature Field in the Surface
          Layer of the Atmosphere," *Izvestia*, Academy of Sciences, USSR, *Atmospheric
          and Oceanic Physics*, (English Edition), American Geophysical Union, Washing-
          ton, DC, Vol. 9, No. 5, May 1973, pp. 251-255.

[6-9]   V. M. Koprov and L. R. Tsvang, "Characteristics of Very Small-Scale Turbulence
          in a Stratified Boundary Layer," *Izv. Atmospheric and Oceanic Physics*, English
          Edition, American Geophysical Union, Washington, DC, Vol. 2, No. 11, pp.
          705-709.

[6-10]   Jack L. Bufton, "Comparison of Vertical Profile Turbulence Structure with
          Stellar Observations," *Applied Optics*, Optical Society of America, Washington,
          DC, Vol. 12, No. 8, August 1973, pp. 1785-1793.

[6-11]   C. E. Coulman, "Vertical Profiles of Small-Scale Temperature Structure in the
          Atmosphere," *Boundary-Layer Meteorology*, Dordrecht, Netherlands, Vol. 4,
          April 1973, pp. 169-177.

[6-12]   R. Barletti, G. Ceppatelli, E. Moroder, L. Paterno, and A. Righini, "A Vertical
          Profile of Turbulence in the Atlantic Air Mass Measured by Balloon-Borne
          Radiosondes," *Journal of Geophysical Research*, American Geophysical Union,
          Washington, DC, Vol. 79, No. 30, October 1974, pp. 4545-4549.

[6-13]   Hans Ottersten, Kenneth R. Hardy, and C. Gordon Little, "Radar and Sodar
          Probing of Waves and Turbulence in Statically Stable Clear-Air Layers,"
          Boundary-Layer Meteorology, Dordrecht, Netherlands, Vol. 4, April 1973,
          pp. 47-89.

[6-14]   Kenneth R. Hardy and Isadore Katz, "Probing the Clear Atmosphere with High
          Power, High Resolution Radars," *Proceedings of the IEEE*, Institute of Elec-
          trical and Electronics Engineers, New York, NY, Vol. 57, No. 4, April 1969,
          pp. 468-480.

[6-15]   C. Gordon Little, "Acoustic Methods for the Remote Probing of the Lower
          Atmosphere," *Proceedings of the IEEE*, Institute of Electrical and Electronics
          Engineers, New York, NY, Vol. 57, No. 4, April 1969, pp. 571-578.

[6-16]   Jack L. Bufton, "An Investigation of Atmospheric Turbulence by Stellar Obser-
          vations," National Aeronautics and Space Administration Technical Report,
          Washington, DC, NASA TR R-369, August 1971.

[6-17]   Robert E. Hufnagel, "Variations of Atmospheric Turbulence," *Digest of Tech-
          nical Papers*, Topical Meeting on Optical Propagation Through Turbulence,
          Optical Society of America, Washington, DC, July 9-11, 1974.

[6-18]   J. Vernin and F. Roddier, "Experimental Determination of Two-Dimensional
          Spatiotemporal Power Spectra of Stellar Light Scintillation, Evidence for a
          Multilayer Strucutre of the Air Turbulence in the Upper Troposphere," *Journal
          of the Optical Society of America*, Optical Society of America, Washington, DC,
          Vol. 63, No. 3, March 1973, pp. 270-273.

[6-19] Nikolay Vinnichenko and John A. Dutton, "Empirical Studies of Atmosphere Structure and Spectra in the Free Atmosphere," *Radio Science*, American Geophysical Union, Washington, DC, Vol. 4, No. 12, December 1969, pp. 1115-1126.

[6-20] Jack L. Bufton, "Correlation of Microthermal Turbulence Data with Meteorological Soundings in the Troposphere," *Journal of the Atmospheric Sciences*, American Meteorological Society, Boston, MA, Vol. 30, No. 1, January 1973, pp. 83-87.

[6-21] Jack L. Bufton, Peter Minott, and Michael W. Fitzmaurice, "Measurements of Turbulence Profiles in the Troposphere," *Journal of the Optical Society of America*, Optical Society of America, Washington, DC, Vol. 62, No. 9, September 1972, pp. 1068-1070.

[6-22] Linday G. McAllister, John R. Pollard, Allan R. Mahoney, and Peter J. R. Shaw, "Acoustic Sounding—A New Approach to the Study of Atmospheric Structure," *Proceedings of the IEEE*, Institute of Electrical and Electronics Engineers, New York, NY, Vol. 57, No. 4, April 1969, pp. 579-587.

[6-23] C. O. Hines, "Gravity Waves in the Atmosphere," *Nature*, Macmillan Journals, Washington, DC, Vol. 239, September 8, 1972, pp. 73-78.

[6-24] A. H. Mikesell, "The Scintillation of Starlight," U.S. Naval Observatory, Washington, DC, Vol. 17, Second Series, Part IV, 1955.

[6-25] V. I. Tatarskii, *Wave Propagation in a Turbulent Medium*, R. A. Silverman (trans.), McGraw-Hill Book Company, New York, NY, 1961.

[6-26] R. E. Hufnagel, "An Improved Model Turbulent Atmosphere," *Restoration of Atmospherically Degraded Images*, Woods Hole Summer Study, National Academy of Sciences, Washington, DC, Vol. 2, Appendix 3, July 1966.

[6-27] David L. Fried and Gus E. Mevers, "Evaluation of r_o for Propagation Down Through the Atmosphere," *Applied Optics*, Optical Society of America, Washington, DC, Vol. 13, No. 11, November 1974, pp. 2620-2622; Errata in Vol. 14, No. 11, November 1975, p. 2567; Second Errata in Vol. 16, No. 3, 1977.

[6-28] A. A. M. Saleh, "An Investigation of Laser Wave Depolarization Due to Atmospheric Transmission," *IEEE Journal of Quantum Electronics*, Institute of Electrical and Electronics Engineers, New York, NY, Vol. QE-3, November 1967, pp. 540-543.

[6-29] William Feller, *Probability Theory and Its Applications*, Vol. 1, John Wiley and Sons, New York, NY, 1950.

[6-30] V. I. Tatarskii, *The Effects of the Turbulent Atmosphere on Wave Propagation*, (Translated from Russian), Israel Program for Scientific Translations, Jerusalem, 1971.

[6-31] G. R. Ochs, R. R. Bergman, and J. R. Snyder, "Laser-Beam Scintillation Over Horizontal Paths from 5.5 to 145 Kilometers," *Journal of the Optical Society of America*, Optical Society of America, Washington, DC, Vol. 59, No. 2, February 1969, pp. 231-234.

[6-32] John W. Strohbehn, "Line-of-Sight Wave Propagation Through the Turbulent Atmosphere," *Proceedings of the IEEE*, Institute of Electrical and Electronics Engineers, New York, NY, Vol. 56, No. 8, August 1968, pp. 1301-1318.

[6-33] S. F. Clifford, G. R. Ochs, and R. S. Lawrence, "Saturation of Optical Scintillation by Strong Turbulence," *Journal of the Optical Society of America*, Optical Society of America, Washington, DC, Vol. 64, No. 2, February 1974, pp. 148-154.

[6-34]   Andrew T. Young, "Aperture Filtering and Saturation of Scintillation," *Journal of the Optical Society of America*, Optical Society of America, Washington, DC, Vol. 60, No. 2, February 1970, pp. 248-250.

[6-35]   H. T. Yura, "Physical Model for Strong Optical-Amplitude Fluctuations in a Turbulent Medium," *Journal of the Optical Society of America*, Optical Society of America, Washington, DC, Vol. 64, No. 1, January 1974, p. 59.

[6-36]   A. S. Gurvich and V. I. Tatarskii, "Coherence and Intensity Fluctuations of Light in the Turbulent Atmosphere," *Radio Science*, American Geographical Union, Washington, DC, Vol. 10, No. 1, January 1975, pp. 3-14.

[6-37]   Ronald L. Fante, "Electromagnetic Beam Propagation in Turbulent Media," *Proceedings of the IEEE*, Institute of Electrical and Electronics Engineers, New York, NY. Vol. 63, No. 12, December 1975, pp. 1660-1688.

[6-38]   S. F. Clifford and H. T. Yura, "Equivalence of Two Theories of Strong Optical Scintillation," *Journal of the Optical Society of America*, Optical Society of America, Washington, DC, Vol. 64, No. 12, December 1974, pp. 1641-1644.

[6-39]   R. F. Lutomirski and H. T. Yura, "Propagation of a Finite Optical Beam in an Inhomogeneous Medium," *Applied Optics*, Optical Society of America, Washington, DC, Vol. 10, No. 7, July 1971, pp. 1652-1658.

[6-40]   R. L. Fante, "Mutual Coherence Function and Frequency Spectrum of a Laser Beam Propagating Through Atmospheric Turbulence," *Journal of the Optical Society of America*, Optical Society of America, Washington, DC, Vol. 64, No. 5, May 1974, pp. 592-598.

[6-41]   R. L. Fante and J. Leon Poirier, "Mutual Coherence Function of a Finite Optical Beam in a Turbulent Medium," *Applied Optics*, Optical Society of America, Washington, DC, Vol. 12, No. 10, October 1973, p. 2247.

[6-42]   D. L. Fried and H. T. Yura, "Telescope-Performance Reciprocity for Propagation in a Turbulent Medium," *Journal of the Optical Society of America*, Optical Society of America, Washington, DC, Vol. 62, No. 4, April 1972, pp. 600-602.

[6-43]   H. V. Hance and D. L. Fried, "Experimental Test of Optical Antenna-Gain Reciprocity," *Journal of the Optical Society of America*, Optical Society of America, Washington, DC, Vol. 63, No. 8, August 1973, pp. 1015-1016.

[6-44]   Michael W. Fitzmaurice and Jack L. Bufton, "Measurement of Log-Amplitude Variance," *Journal of the Optical Society of America*, Optical Society of America, Washington, DC, Vol. 59, No. 4, April 1969, pp. 462-463.

[6-45]   D. L. Fried, "Aperture Averaging of Scintillation," *Journal of the Optical Society of America*, Optical Society of America, Washington, DC, Vol. 57, No. 2, February 1967, pp. 169-175.

[6-46]   R. F. Lutomirski and H. T. Yura, "Aperture-Averaging Factor of a Light Signal," *Journal of the Optical Society of America*, Optical Society of America, Washington, DC, Vol. 59, No. 9, September 1969, pp. 1247-1249.

[6-47]   Harry W. Wessely and McLaren P. Mitchell, "Solar-Scintillation Measurements," *Journal of the Optical Society of America*, Optical Society of America, Washington, DC, Vol. 61, No. 2, February 1971, pp. 242-247.

[6-48]   V. H. Rumsey, "Intensity Fluctuations Due to an Incoherent Extended Source Seen Through Extended Strong Turbulence," *Radio Science*, American Geophysical Union, Washington, DC, Vol. 11, No. 6, June 1976, pp. 545-549.

[6-49]   Andrew T. Young, "Saturation of Scintillation," *Journal of the Optical Society of America*, Optical Society of America, Washington, DC, Vol. 60, No. 11, November 1970, pp. 1495-1500.

[6-50]   Andrew T. Young, "Photometric Error Analysis VIII. The Temporal Power Spectrum of Scintillation," *Applied Optics*, Optical Society of America, Washington, DC, Vol. 8, No. 5, May 1969, pp. 869-885.

[6-51]   S. F. Clifford, "Temporal-Frequency Spectra for a Spherical Wave Propagating Through Atmospheric Turbulence," *Journal of the Optical Society of America*, Optical Society of America, Washington, DC, Vol. 61, No. 10, October 1971, pp. 1285-1292.

[6-52]   R. E. Hufnagel and N. R. Stanley, "Modulation Transfer Function Associated with Image Transmission Through Turbulent Media," *Journal of the Optical Society of America*, Optical Society of America, Washington, DC, Vol. 54, No. 1, January 1964, pp. 52-61.

[6-53]   David M. Chase, "Coherence Function for Waves in Random Media," *Journal of the Optical Society of America*, Optical Society of America, Washington, DC, Vol. 55, No. 11, November 1965, pp. 1559-1560.

[6-54]   R. E. Hufnagel, "Correlation Functions for Optical Transfer Functions," *Restoration of Atmospherically Degraded Images*, Woods Hole Summer Study, National Academy of Sciences, Washington, DC, Vol. 2, July 1966.

[6-55]   D. Korff, G. Dryden, and R. P. Leavitt, "Isoplanicity: The Translation Invariance of the Atmospheric Green's Function," *Journal of the Optical Society of America*, Optical Society of America, Washington, DC, Vol. 65, No. 11, November 1975, pp. 1321-1330.

[6-56]   Jeffrey H. Shapiro, "Propagation-Medium Limitations on Phase-Compensated Atmospheric Imaging," *Journal of the Optical Society of America*, Optical Society of America, Washington, DC, Vol. 66, No. 5, May 1976, pp. 460-469.

[6-57]   F. C. Dainty and R. F. Scaddan, "A Coherence Interferometer for Direct Measurement of the Atmospheric Transfer Function," *Monthly Notices*, Royal Astronomical Society, London, England, Vol. 167, 1974, pp. 69-73.

[6-58]   D. L. Fried, "Statistics of a Geometric Representation of Wavefront Distortion," *Journal of the Optical Society of America*, Optical Society of America, Washington, DC, Vol. 55, No. 11, November 1965, pp. 1427-1435.

[6-59]   R. D. Rosner, "Performances of an Optical Heterodyne Receiver for Various Receiving Apertures," *IEEE Transactions, Antennas and Propagation*, Institute of Electrical and Electronics Engineers, New York, NY, Vol. AP-17, No. 3, May 1969, pp. 324-331.

[6-60]   I. Goldstein, P. A. Miles, and A. Chabot, "Heterodyne Measurements of Light Propagation Through Atmospheric Turbulence," *Proceedings of the IEEE*, Institute of Electrical and Electronics Engineers, New York, NY, Vol. 53, No. 9, September 1965, pp. 1172-1180.

[6-61]   D. L. Fried, "Optical Heterodyne Detection of an Atmospherically Distorted Signal Wave Front," *Proceedings of the IEEE*, Institute of Electrical and Electronics Engineers, New York, NY, Vol. 55, No. 1, January 1967, pp. 57-67.

[6-62]   R. E. Hufnagel, "Optical Propagation Study," Technical Report RADC-TR-65-511, ASTIA No. AD 476244, Perkin-Elmer Corporation, Norwalk, CT, January 1966.

[6-63]   Athanasios Papoulis, *The Fourier Integral and Its Applications*, McGraw-Hill, New York, NY, 1962, pp. 15-16.

[6-64]   G. Barry Hogge and R. Russell Butts, "Frequency Spectra for the Geometric Representation of Wavefront Distortions Due to Atmospheric Turbulence," *IEEE Transactions, Antennas and Propagation*, Institute of Electrical and Electronics Engineers, New York, NY, Vol. AP-24, No. 2, March 1976, pp. 144-154.

[6-65]   D. Y. Gezari, A. Labeyrie, and R. V. Stachnik, "Speckle Interferometry: Diffraction-Limited Measurements of Nine Stars with the 200-Inch Telescope," *Astrophysical Journal*, American Astronomical Society, University of Chicago Press, Chicago, IL, Vol. 173, April 1, 1972, L1-L5.

[6-66]   A. Labeyrie, "Speckle Interferometry Observations at Mount Palomar," *Nouvelle Revue D'Optique*, Saint Germain, Paris, France, Vol. 5, No. 3, 1974, pp. 141-151.

[6-67]   D. Korff, "Analysis of a Method for Obtaining Near-Diffraction-Limited Information in the Presence of Atmospheric Turbulence," *Journal of the Optical Society of America*, Optical Society of America, Washington, DC, Vol. 63, No. 8, August 1973, pp. 971-980.

[6-68]   Keith T. Knox and Brian J. Thompson, "Recovery of Images from Atmospherically Degraded Short-Exposure Photographs," *Astrophysical Journal*, American Astronomical Society, University of Chicago Press, Chicago, IL, Vol. 193, October 1973, L45-L48.

[6-69]   A. M. Prokhorov, F. V. Bunkin, K. S. Gochelashvily, and V. I. Shishov, "Laser Irradiance Propagation in Turbulent Media," *Proceedings of the IEEE*, Institute of Electrical and Electronics Engineers, New York, NY, Vol. 63, No. 5, May 1975, pp. 790-811.

[6-70]   V. A. Banakh, G. M. Krekov, V. L. Mironov, S. S. Khmelevtsov, and R. Sh. Tsvik, "Focused-Laser-Beam Scintillations in the Turbulent Atmosphere," *Journal of the Optical Society of America*, Optical Society of America, Washington, DC, Vol. 64, No. 4, April 1974, pp. 516-518.

[6-71]   V. L. Mironov and V. V. Nosov, "On the Theory of Spatially Limited Light Beam Displacements in a Randomly Inhomogeneous Medium," *Journal of the Optical Society of America*, Optical Society of America, Washington, DC, Vol. 67, No. 8, August 1977, pp. 1073-1080.

[6-72]   Richard J. Cook, "Beam Wander in a Turbulent Medium: An Application of Ehrenfest's Theorem," *Journal of the Optical Society of America*, Optical Society of America, Washington, DC, Vol. 65, No. 8, August 1975, pp. 942-948.

[6-73]   J. Richard Kerr and James R. Dunphy, "Propagation of Multi Wavelength Laser Radiation Through Atmospheric Turbulence," Oregon Graduate Center, Beaverton, Oregon, Technical Report RADC-TR-74-183, May 31, 1974.

[6-74]   D. L. Fried and J. B. Seidman, "Laser-Beam Scintillation in the Atmosphere," *Journal of the Optical Society of America*, Optical Society of America, Washington, DC, Vol. 57, No. 2, February 1967, pp. 181-185.

[6-75]   R. S. Lawrence and J. W. Strohbehn, "A Survey of Clean-Air Propagation Effects Relevant to Optical Communications," *Proceedings of the IEEE*, Institute of Electrical and Electronics Engineers, New York, NY, Vol. 58, No. 10, October 1970, pp. 1523-1546.

[6-76]   T. L. Ho, "Log-Amplitude Fluctuations of Laser Beam in a Turbulent Atmosphere," *Journal of the Optical Society of America*, Optical Society of America, Washington, DC, Vol. 59, No. 4, April 1969, pp. 385-390.

[6-77]   J. Richard Kerr and James R. Dunphy, "Experimental Effects of Finite Transmitter Apertures on Scintillations," *Journal of the Optical Society of America*, Optical Society of America, Washington, DC, Vol. 63, No. 1, January 1973, pp. 1-8.

[6-78]   Paul J. Titterton, "Power Reduction and Fluctuations Caused by Narrow Laser Beam Motion in the Far Field," *Applied Optics*, Optical Society of America, Washington, DC, February 1973, pp. 423-425.

[6-79]   Paul H. Deitz and Neal J. Wright, "Saturation of Scintillation Magnitude in Near-Earth Optical Propagation," *Journal of the Optical Society of America*, Optical Society of America, Washington, DC, Vol. 59, No. 5, May 1969, pp. 527-535.

[6-80] Robert J. Noll, "Zernike Polynomials and Atmospheric Turbulence," *Journal of the Optical Society of America*, Optical Society of America, Washington, DC, Vol. 66, No. 3, March 1976, pp. 207-211.

[6-81] Darryl P. Greenwood and David L. Fried, "Power Spectra Requirements for Wave-Front-Compensation Systems," *Journal of the Optical Society of America*, Optical Society of America, Washington, DC, Vol. 66, No. 3, March 1976, pp. 193-207.

[6-82] James E. Pearson, "Atmospheric Turbulence Compensation Using Coherent Optical Adaptive Techniques," *Applied Optics*, Optical Society of America, Washington, DC, Vol. 15, No. 3, March 1976, pp. 573-826.

[6-83] J. C. Wyant, "Use of an AC Heterodyne Lateral Shear Interferometer with Real-time Wavefront Correction Systems," *Applied Optics*, Optical Society of America, Washington, DC, Vol. 14, No. 11, November 1975, pp. 2622-2626.

[6-84] Jeffrey H. Shapiro, "Imaging and Optical Communication through Atmospheric Turbulence," to be published in *Laser Beam Propagation Through the Atmosphere*, J. W. Strohbehn (ed.), Springer Verlag, New York, NY.

[6-85] Optical Society of America, Washington, DC, *Journal of the Optical Society of America*, Vol. 67, No. 3, 1977. This is a special issue featuring adaptive optics.

[6-86] G. R. Ochs, S. F. Clifford, and Ting-i Wang, "Laser Wind Sensing: The Effects of Saturation of Scintillation," *Applied Optics*, Optical Society of America, Washington, DC, Vol. 15, No. 2, February 1976, pp. 403-408.

[6-87] A. Rocca, F. Roddier, and J. Vernin, "Detection of Atmospheric Turbulent Layers by Spatiotemporal and Spatioangular Correlation Measurements of Stellar-Light Scintillation," *Journal of the Optical Society of America*, Optical Society of America, Washington, DC, Vol. 64, No. 7, July 1974, pp. 1000-1004.

[6-88] R. L. Mitchell, "Permanence of the Log-Normal Distribution," *Journal of the Optical Society of America*, Optical Society of America, Washington, DC, Vol. 58, No. 9, September 1968, pp. 1267-1272.

[6-89] Richard Barakat, "Sums of Independent Lognormally Distributed Random Variables," *Journal of the Optical Society of America*, Optical Society of America, Washington, DC, Vol. 66, No. 3, March 1976, pp. 211-220.

[6-90] M. E. Gracheva, "Research into the Statistical Properties of the Strong Fluctuation of Light when Propagated in the Lower Layer of the Atmosphere," *Izv. Vuz. Radio fiz.*, Moscow, USSR, Vol. 10, 1967, pp. 775-787.

[6-91] S. S. Khmelevtsov, "Propagation of Laser Radiation in a Turbulent Atmosphere," *Applied Optics*, Optical Society of America, Washington, DC, Vol. 12, No. 10, October 1973, pp. 2421-2433.

6.10.2. Bibliography

Banakh, V. A., G. M. Krekov, V. L. Mironov, S. S. Khmelevtsov, and R. Sh. Tsvik, "Focused-Laser-Beam Scintillations in the Turbulent Atmosphere," *Journal of the Optical Society of America*, Optical Society of America, Washington, DC, Vol. 64, No. 4, April 1974, pp. 516-518.

Barakat, Richard, "Sums of Independent Lognormally Distributed Random Variables," *Journal of the Optical Society of America*, Optical Society of America, Washington, DC, Vol. 66, No. 3, March 1976, pp. 211-220.

Barletti, R., G. Ceppatelli, E. Moroder, L. Paterno, and A. Righini, "A Vertical Profile of Turbulence in the Atlantic Air Mass Measured by Balloon-Borne Radiosondes," *Journal of Geophysical Research*, American Geophysical Union, Washington, DC, Vol. 79, No. 30, October 1974, pp. 4545-4549.

Born, Max, and Emil Wolf, *Principles of Optics*, Pergamon Press, Oxford, England, 1975, Fifth Edition.

Breckenridge, J. B., "Measurement of the Amplitude of Phase Excursions in the Earth's Atmosphere," *Journal of the Optical Society of America*, Optical Society of America, Washington, DC, Vol. 66, No. 2, February 1976, pp. 143-144.

Bufton, Jack L., "An Investigation of Atmospheric Turbulence by Stellar Observations," National Aeronautics and Space Administration Technical Report, NASA TR R-369, Washington, DC, August 1971.

Bufton, Jack L., "Comparison of Vertical Profile Turbulence Structure with Stellar Observations," *Applied Optics*, Optical Society of America, Washington, DC, Vol. 12, No. 8, August 1973, pp. 1785-1793.

Bufton, Jack L., "Correlation of Microthermal Turbulence Data with Meteorological Soundings in the Troposphere," *Journal of Atmospheric Sciences*, American Meteorological Society, Washington, DC, Vol. 30, No. 1, January 1973, pp. 83-87.

Bufton, Jack L., Peter Minott, and Michael W. Fitzmaurice, "Measurements of Turbulence Profiles in the Troposphere," *Journal of the Optical Society of America*, Optical Society of America, Washington, DC, Vol. 62, No. 9, September 1972, pp. 1068-1070.

Chase, David M., "Coherence Function for Waves in Random Media," *Journal of the Optical Society of America*, Optical Society of America, Washington, DC, Vol. 55, No. 11, November 1965, pp. 1559-1560.

Clifford, S. F., "Temporal-Frequency Spectra for a Spherical Wave Propagating Through Atmospheric Turbulence," *Journal of the Optical Society of America*, Optical Society of America, Washington, DC, Vol. 61, No. 10, October 1971, pp. 1285-1292.

Clifford, S. F., and H. T. Yura, "Equivalence of Two Theories of Strong Optical Scintillation," *Journal of the Optical Society of America*, Optical Society of America, Washington, DC, Vol. 64, No. 12, December 1974, pp. 1641-1644.

Clifford, S. F., G. R. Ochs, and R. S. Lawrence, "Saturation of Optical Scintillation by Strong Turbulence," *Journal of the Optical Society of America*, Optical Society of America, Washington, DC, Vol. 64, No. 2, February 1974, pp. 148-154.

Cook, Richard J., "Beam Wander in a Turbulent Medium: An Application of Ehrenfest's Theorem," *Journal of the Optical Society of America*, Optical Society of America, Washington, DC, Vol. 65, No. 8, August 1975, pp. 942-948.

Coulman, C. E., "Vertical Profiles of Small-Scale Temperature Structure in the Atmosphere," *Boundary-Layer Meteorology*, Dordrecht, Netherlands, Vol. 4, April 1973, pp. 169-177.

Dainty, F. C., and R. F. Scaddan, "A Coherence Interferometer for Direct Measurement of the Atmospheric Transfer Function," *Monthly Notices*, Royal Astronomical Society, London, England, Vol. 167, 1974, pp. 69-73.

Deitz, Paul H., and Neal J. Wright, "Saturation of Scintillation Magnitude in Near-Earth Optical Propagation," *Journal of the Optical Society of America*, Optical Society of America, Washington, DC, Vol. 59, No. 5, May 1969, pp. 527-535.

Fante, Ronald L., "Electromagnetic Beam Propagation in Turbulent Media," *Proceedings of the IEEE*, Institute of Electrical and Electronics Engineers, New York, NY, Vol. 63, No. 12, December 1975, pp. 1660-1688.

Fante, R. L., "Mutual Coherence Function and Frequency Spectrum of a Laser Beam Propagating Through Atmospheric Turbulence," *Journal of the Optical Society of America*, Optical Society of America, Washington, DC, Vol. 64, No. 5, May 1974, pp. 592-598.

Fante, R. L., and J. Leon Poirier, "Mutual Coherence Function of a Finite Optical Beam in a Turbulent Medium," *Applied Optics*, Optical Society of America, Washington, DC, Vol. 12, No. 10, October 1973, p. 2247.

Feller, William, *Probability Theory and Its Applications*, John Wiley and Sons, New York, NY, Vol. 1, 1950.

Fitzmaurice, Michael W., and Jack L. Bufton, "Measurement of Log-Amplitude Variance," *Journal of the Optical Society of America*, Optical Society of America, Washington, DC, Vol. 59, No. 4, April 1969, pp. 462-463.

Fried, D. L., "Aperture Averaging of Scintillation," *Journal of the Optical Society of America*, Optical Society of America, Washington, DC, Vol. 57, No. 2, February 1967, pp. 169-175.

Fried, D. L., "Optical Heterodyne Detection of an Atmospherically Distorted Signal Wave Front," *Proceedings of the IEEE*, Institute of Electrical and Electronics Engineers, New York, NY, Vol. 55, No. 1, January 1967, pp. 57-67.

Fried, D. L., "Optical Resolution Through a Randomly Inhomogeneous Medium For Very Long and Very Short Exposures," *Journal of the Optical Society of America*, Optical Society of America, Washington, DC, Vol. 56, No. 10, October 1966, pp. 1372-1379.

Fried, D. L., "Propagation of a Spherical Wave in a Turbulent Medium," *Journal of the Optical Society of America*, Optical Society of America, Washington, DC, Vol. 57, No. 2, February 1967, pp. 175-180.

Fried, D. L., "Statistics of a Geometric Representation of Wavefront Distortion," *Journal of the Optical Society of America*, Optical Society of America, Washington, DC, Vol. 55, No. 11, November 1965, pp. 1427-1435.

Fried, David L., and Gus E. Mevers, "Evaluation of r_o for Propagation Down Through the Atmosphere," *Applied Optics*, Optical Society of America, Washington, DC, Vol. 13, No. 11, November 1974, pp. 2620-2622; Errata in Vol. 14, No. 11, November 1975, p. 2567; Second Errata in Vol. 16, No. 3, 1977.

Fried, D. L., and J. B. Seidman, "Laser-Beam Scintillation in the Atmosphere," *Journal of the Optical Society of America*, Optical Society of America, Washington, DC, Vol. 57, No. 2, February 1967, pp. 181-185.

Fried, D. L., and H. T. Yura, "Telescope-Performance Reciprocity for Propagation in a Turbulent Medium," *Journal of the Optical Society of America*, Optical Society of America, Washington, DC, Vol. 62, No. 4, April 1972, pp. 600-602.

Friehe, Carl A., John C. LaRue, F. H. Champagne, C. H. Gibson, and G. F. Dreyer, "Effects of Temperature and Humidity Fluctuations on the Optical Refractive Index in the Marine Boundary Layer," *Journal of the Optical Society of America*, Optical Society of America, Washington, DC, Vol. 65, No. 12, December 1975, pp. 1502-1511.

Gezari, D. Y., A. Labeyrie, and R. V. Stachnik, "Speckle Interferometry: Diffraction-Limited Measurements of Nine Stars with the 200-Inch Telescope," *Astrophysical Journal*, American Astronomical Society, University of Chicago Press, Chicago, IL, Vol. 173, April 1, 1972, L1-L5.

Goldstein, I., P. A. Miles, and A. Chabot, "Heterodyne Measurements of Light Propagation Through Atmospheric Turbulence," *Proceedings of the IEEE*, Institute of Electrical and Electronics Engineers, New York, NY, Vol. 53, No. 9, September 1965, pp. 1172-1180.

Goodman, Joseph W., *Introduction to Fourier Optics*, McGraw-Hill, New York, NY, 1968.

Gracheva, M. E., "Research into the Statistical Properties of the Strong Fluctuation of Light when Propagated in the Lower Layer of the Atmosphere, *Izv. Vuz. Radio fiz.*, Moscow, USSR, Vol. 10, 1967, pp. 775-787.

Greenwood, Darryl P., and David L. Fried, "Power Spectra Requirements for Wave-Front-Compensative Systems," *Journal of the Optical Society of America*, Optical Society of America, Washington, DC, Vol. 66, No. 3, March 1976, pp. 103-207.

Gurvich, A. S., and V. I. Tatarskii, "Coherence and Intensity Fluctuations of Light in the Turbulent Atmosphere," *Radio Science*, American Geophysical Union, Washington, DC, Vol. 10, No. 1, January 1975, pp. 3-14.

Hance, H. V., and D. L. Fried, "Experimental Test of Optical Antenna-Gain Reciprocity," *Journal of the Optical Society of America*, Optical Society of America, Washington, DC, Vol. 63, No. 8, August 1973, pp. 1015-1016.

Handbook of Geophysics, Revised Edition, United States Air Force, The Macmillan Company, New York, NY, 1960, pp. 13-1 and 13-2.

Hardy, Kenneth R., and Isadore Katz, "Probing the Clear Atmsophere with High Power, High Resolution Radars," *Proceedings of the IEEE*, Institute of Electrical and Electronics Engineers, New York, NY, Vol. 57, No. 4, April 1969, pp. 468-480.

Hines, C. O., "Gravity Waves in the Atmosphere," *Nature*, Macmillan Journals, Washington, DC, Vol. 239, September 8, 1972, pp. 73-78.

Ho, T. L., "Log-Amplitude Fluctuations of Laser Beam in a Turbulent Atmosphere," *Journal of the Optical Society of America*, Optical Society of America, Washington, DC, Vol. 59, No. 4, April 1969, pp. 385-390.

Hogge, G. Barry, and R. Russell Butts, "Frequency Spectra for the Geometric Representation of Wavefront Distortions Due to Atmospheric Turbulence," *IEEE Transactions, Antennas and Propagation*, Institute of Electrical and Electronics Engineers, New York, NY, Vol. AP-24, No. 2, March 1976, pp. 144-154.

Hufnagel, R. E., "An Improved Model Turbulent Atmosphere," *Restoration of Atmospherically Degraded Images*, Woods Hole Summer Study, National Academy of Sciences, Washington, DC, Vol. 2, July 1966, Appendix 3.

Hufnagel, R. E., "Correlation Functions for Optical Transfer Functions," *Restoration of Atmospherically Degraded Images*, Woods Hole Summer Study, National Academy of Sciences, Washington, DC, Vol. 2, July 1966.

Hufnagel, R. E., "Optical Propagation Study," Technical Report (RADC-TR-65-511, ASTIA No. AD 476244), Perkin-Elmer Corporation, Norwalk, CT, January 1966.

Hufnagel, Robert E., "Variations of Atmospheric Turbulence," *Digest of Technical Papers*, Topical Meeting on Optical Propagation Through Turbulence, Optical Society of America, Washington, DC, July 9-11, 1974.

Hufnagel, R. E., and N. R. Stanley, "Modulation Transfer Function Associated with Image Transmission Through Turbulent Media," *Journal of the Optical Society of America*, Optical Society of America, Washington, DC, Vol. 54, No. 1, January 1964, pp. 52-61.

Kerr, J. Richard, and James R. Dunphy, "Experimental Effects of Finite Transmitter Apertures on Scintillations," *Journal of the Optical Society of America*, Optical Society of America, Washington, DC, Vol. 63, No. 1, January 1973, pp. 1-8.

Kerr, J. Richard, and James R. Dunphy, "Propagation of Multi Wavelength Laser Radiation Through Atmospheric Turbulence," Oregon Graduate Center, Beaverton, Oregon, Technical Report RADC-TR-74-183, May 31, 1974.

Khmelevtsov, S. S., "Propagation of Laser Radiation in a Turbulent Atmosphere," *Applied Optics*, Optical Society of America, Washington, DC, Vol. 12, No. 10, October 1973, pp. 2421-2433.

Knox, Keith T., and Brian J. Thompson, "Recovery of Images from Atmospherically Degraded Short-Exposure Photographs," *Astrophysical Journal*, American Astronomical Society, University of Chicago Press, Chicago, IL, Vol. 193, October 1973, L45-L48.

Kolmogorov, A. N., "The Local Structure of Turbulence in Incompressible Viscous Fluid for Very Large Reynolds Numbers," *Doklady Akad. Nauk.*, USSR, Academy of Sciences of USSR, New York, NY, Vol. 30, 1941, p. 301.

Koprov, V. M., and L. R. Tsvang, "Characteristics of Very Small-Scale Turbulence in a Stratified Boundary Layer," *Izv. Atmospheric and Oceanic Physics*, English Edition, American Geophysical Union, Washington, DC, Vol. 2, No. 11, pp. 705-709.

Korff, D., "Analysis of a Method for Obtaining Near-Diffraction-Limited Information in the Presence of Atmospheric Turbulence," *Journal of the Optical Society of America*, Optical Society of America, Washington, DC, Vol. 63, No. 8, August 1973, pp. 971-980.

Korff, D., G. Dryden, and R. P. Leavitt, "Isoplanicity: The Translation Invariance of the Atmospheric Green's Function," *Journal of the Optical Society of America*, Optical Society of America, Washington, DC, Vol. 65, No. 11, November 1975, pp. 1321-1330.

Labeyrie, A., "Speckle Interferometry Observations at Mount Palomar," *Nouvelle Revue D'Optique*, Saint Germain, Paris, France, Vol. 5, No. 3, 1974, pp. 141-151.

Lawrence, R. S., and J. W. Strohbehn, "A Survey of Clear-Air Propagation Effects Relevant to Optical Communications," *Proceedings of the IEEE*, Institute of Electrical and Electronics Engineers, New York, NY, Vol. 58, No. 10, October 1970, pp. 1523-1546.

Lawrence, Robert S., Gerald R. Ochs, and Steven F. Clifford, "Measurements of Atmospheric Turbulence Relevant to Optical Propagation," *Journal of the Optical Society of America*, Optical Society of America, Washington, DC, Vol. 60, No. 6, June 1970, pp. 826-830.

Little, C. Gordon, "Acoustic Methods for the Remote Probing of the Lower Atmosphere," *Proceedings of the IEEE*, Institute of Electrical and Electronics Engineers, New York, NY, Vol. 57, No. 4, April 1969, pp. 571-578.

Lumley, John L., and Hans A. Panofsky, *The Structure of Atmospheric Turbulence*, John Wiley and Sons, New York, NY, 1964, pp. 62, 72.

Lutomirski, R. F., and H. T. Yura, "Aperture-Averaging Factor of a Light Signal," *Journal of the Optical Society of America*, Optical Society of America, Washington, DC, Vol. 59, No. 9, September 1969, pp. 1247-1249.

Lutomirski, R. F., and H. T. Yura, "Propagation of a Finite Optical Beam in an Inhomogeneous Medium," *Applied Optics*, Optical Society of America, Washington, DC, Vol. 10, No. 7, July 1971, pp. 1652-1658.

McAllister, Linday G., John R. Pollard, Allan R. Mahoney, and Peter J. R. Shaw, "Acoustic Sounding—A New Approach to the Study of Atmospheric Structure," *Proceedings of the IEEE*, Institute of Electrical and Electronics Engineers, New York, NY, Vol. 57, No. 4, April 1969, pp. 579-587.

Mikesell, A. H., "The Scintillation of Starlight," Publications of U.S. Naval Observatory, Washington, DC, Vol. 17, Second Series, Part IV, 1955.

Mironov, V. L., and V. V. Nosov, "On the Theory of Spatially Limited Light Beam Displacements in a Randomly Inhomogeneous Medium," *Journal of the Optical Society of America*, Optical Society of America, Washington, DC, Vol. 67, No. 8, August 1977, pp. 1073-1080.

Mitchell, R. L., "Permanence of the Log-Normal Distribution," *Journal of the Optical Society of America*, Optical Society of America, Washington, DC, Vol. 58, No. 9, September 1968, pp. 1267-1272.

Noll, Robert J., "Zernike Polynomials and Atmospheric Turbulence," *Journal of the Optical Society of America*, Optical Society of America, Washington, DC, Vol. 66, No. 3, March 1976, pp. 207-211.

Ochs, Gerald R., and Robert S. Lawrence, "Temperature and C_n^2 Profiles Measured Over Land and Surface," United States Government Printing Office, Washington, DC, National Oceanic and Atmospheric Agency Technical Report ERL 251-WPL22, October 1972.

Ochs, G. R., R. R. Bergman, and J. R. Snyder, "Laser-Beam Scintillation Over Horizontal Paths from 5.5 to 145 Kilometers," *Journal of the Optical Society of America*, Optical Society of America, Washington, DC, Vol. 59, No. 2, February 1969, pp. 231-234.

Ochs, G. R., S. F. Clifford, and Ting-i Wang, "Laser Wind Sensing: The Effects of Saturation of Scintillation," *Applied Optics*, Optical Society of America, Washington, DC, Vol. 15, No. 2, February 1976, pp. 403-408.

Optical Society of America, Washington, DC, *Journal of the Optical Society of America*, Vol. 67, No. 3, 1977. This is a special issue featuring adaptive optics.

Ottersten, Hans, Kenneth R. Hardy, and C. Gordon Little, "Radar and Sodar Probing of Waves and Turbulence in Statically Stable Clear-Air Layers," *Boundary-Layer Meteorology*, Dordrecht, Netherlands, Vol. 4, April 1973, pp. 47-89.

Papoulis, Athanasios, *The Fourier Integral and Its Applications*, McGraw-Hill, New York, NY, 1962.

Pearson, James E., "Atmospheric Turbulence Compensation Using Coherent Optical Adaptive Techniques," *Applied Optics*, Optical Society of America, Washington, DC, Vol. 15, No. 3, March 1976, pp. 573-826.

Prokhorov, A. M., F. V. Bunkin, K. S. Gochelashvily, and V. I. Shishov, "Laser Irradiance Propagation in Turbulent Media," *Proceedings of the IEEE*, Institute of Electrical and Electronics Engineers, New York, NY, Vol. 63, No. 5, May 1975, pp. 790-811.

Reinhardt, G. W., and S. A. Collins, Jr., "Outer-Scale Effects in Turbulence-Degraded Light-Beam Spectra," *Journal of the Optical Society of America*, Optical Society of America, Washington, DC, Vol. 62, No. 12, December 1972, pp. 1526-1530.

Rocca, A., F. Roddier, and J. Vernin, "Detection of Atmospheric Turbulent Layers by Spatiotemporal and Spatioangular Correlation Measurements of Stellar-Light Scintillation," *Journal of the Optical Society of America*, Optical Society of America, Washington, DC, Vol. 64, No. 7, July 1974, pp. 1000-1004.

Roddier, C., and F. Roddier, "Influence of Exposure Time on Spectral Properties of Turbulence-Degraded Astronomical Images," *Journal of the Optical Society of America*, Optical Society of America, Washington, DC, Vol. 65, No. 6, June 1975.

Rosner, R. D., "Performance of an Optical Heterodyne Receiver for Various Receiving Apertures," *IEEE Transactions, Antennas and Propagation*, Institute of Electrical and Electronics Engineers, New York, NY, Vol. AP-17, No. 3, May 1969, pp. 324-331.

Rumsey, V. H., "Intensity Fluctuations Due to an Incoherent Extended Source Seen Through Extended Strong Turbulence," *Radio Science*, American Geophysical Union, Washington, DC, Vol. 11, No. 6, June 1976, pp. 545-549.

Saleh, A. A. M., "An Investigation of Laser Wave Depolarization Due to Atmospheric Transmission," *IEEE Journal of Quantum Electronics*, Institute of Electrical and Electronics Engineers, New York, NY, Vol. QE-3, November, 1967, pp. 540-543.

Shapiro, Jeffrey H., "Imaging and Optical Communication through Atmospheric Turbulence," to be published in *Laser Beam Propagation Through the Atmosphere*, J. W. Strohbehn (ed.), Springer Verlag, New York, NY.

Shapiro, Jeffrey H., "Propagation-Medium Limitations on Phase-Compensated Atmospheric Imaging," *Journal of the Optical Society of America*, Optical Society of America, Washington, DC, Vol. 66, No. 5, May 1976, pp. 460-469.

Smith, Jack, "Folded-Path Weighting Function for a High-Frequency Spherical Wave," *Journal of the Optical Society of America*, Optical Society of America, Washington, DC, Vol. 63, No. 9, September 1973, pp. 1095-1097.

Smith, Jack, and Thomas H. Pries, "Temporal-Frequency Spectra For Waves Propagating over Straight and Folded Paths: A Comparison," *Applied Optics*, Optical Society of America, Washington, DC, Vol. 14, No. 5, May 1975, pp. 1161-1168.

Strohbehn, John W., "Line-of-Sight Wave Propagation Through the Turbulent Atmosphere," *Proceedings of the IEEE*, Institute of Electrical and Electronics Engineers, New York, NY, Vol. 56, No. 8, August 1968, pp. 1301-1318.

Strohbehn, John W., "Optical Propagation Through the Turbulent Atmosphere," *Progress in Optics*, North Holland Publishing Company, Amsterdam, Netherlands, Vol. 9, 1971.

Tatarskii, V. I., *The Effects of the Turbulent Atmosphere on Wave Propagation*, (Translated from Russian), Israel Program for Scientific Translations, Jerusalem, 1971.

Tatarskii, V. I., *Wave Propagation in a Turbulent Medium*, R. A. Silverman (trans.), McGraw-Hill Book Company, New York, NY, 1961.

Titterton, Paul J., "Power Reduction and Fluctuations Caused by Narrow Laser Beam Motion in the Far Field," *Applied Optics*, Optical Society of America, Washington, DC, February 1973, pp. 423-425.

Van Atta, C. W., "Influence of Fluctuations in Local Dissipation Rates on Turbulent Scalar Characteristics in the Inertial Subrange," *The Physics of Fluids*, Vol. 14, No. 8, August 1971, Erratum is in Vol. 16, No. 4, April 1973.

Vernin, J., and F. Roddier, "Experimental Determination of Two-Dimensional Spatiotemporal Power Spectra of Stellar Light Scintillation, Evidence for a Multilayer Structure of the Air Turbulence in the Upper Troposphere," *Journal of the Optical Society of America*, Optical Society of America, Washington, DC, Vol. 63, No. 3, March 1973, pp. 270-273.

Vinnichenko, Nikolay, and John A. Dutton, "Empirical Studies of Atmospheric Structure and Spectra in the Free Atmosphere," *Radio Science*, American Geophysical Union, Washington, DC, Vol. 4, No. 12, December 1969, pp. 1115-1126.

Voyt, F. Ya., Ye. Ye. Korniyenko, V. P. Kukharets, S. B. Khusid, and L. R. Tsvang, "Structural Characteristics of the Temperature Field in the Surface Layer of the Atmosphere," *Izvestia*, Academy of Sciences, USSR, *Atmospheric and Oceanic Physics*, English Edition, American Geophysical Union, Washington, DC, Vol. 9, No. 5, May 1973, pp. 251-255.

Webb, E. K., "Daytime Thermal Fluctuations in the Lower Atmosphere," *Applied Optics*, Optical Society of America, Washington, DC, Vol. 3, No. 12, December 1964, pp. 1329-1336.

Wessely, Harry W., and McLaren P. Mitchell, "Solar-Scintillation Measurements," *Journal of the Optical Society of America*, Washington, DC, Vol. 61, No. 2, February 1971, pp. 242-247.

de Wolf, David A., "Are Strong Irradiance Fluctuations Log Normal or Rayleigh Distributed?" *Journal of the Optical Society of America*, Optical Society of America, Washington, DC, Vol. 59, No. 11, November 1969, pp. 1455-1460.

Wyant, J. C., "Use of an AC Heterodyne Lateral Shear Interferometer with Realtime Wavefront Correction Systems," *Applied Optics*, Optical Society of America, Washington, DC, Vol. 14, No. 11, November 1975, pp. 2622-2626.

Wyngaard, John C., Y. Izumi, and Stuart A. Collins, Jr., "Behavior of the Refractive-Index-Structure Parameter Near the Ground," *Journal of the Optical Society of America*, Optical Society of America, Washington, DC, Vol. 61, No. 12, December 1971, pp. 1646-1650.

Young, Andrew T., "Aperture Filtering and Saturation of Scintillation," *Journal of the Optical Society of America*, Optical Society of America, Washington, DC, Vol. 60, No. 2, February 1970, pp. 248-250.

Young, Andrew T., "Photometric Error Analysis VIII. The Temporal Power Spectrum of Scintillation," *Applied Optics*, Optical Society of America, Washington, DC, Vol. 8, No. 5, May 1969, pp. 869-885.

Young, Andrew T., "Saturation of Scintillation," *Journal of the Optical Society of America*, Optical Society of America, Washington, DC, Vol. 60, No. 11, November 1970, pp. 1495-1500.

Yura, H. T., "Physical Model for Strong Optical-Amplitude Fluctuations in a Turbulent Medium," *Journal of the Optical Society of America*, Optical Society of America, Washington, DC, Vol. 64, No. 1, January 1974, p. 59.

Yura, H. T., "Temporal-Frequency Spectrum of an Optical Wave Propagating Under Saturation Conditions," *Journal of the Optical Society of America*, Optical Society of America, Washington, DC, Vol. 64, March 1974, pp. 357-359.

Chapter 7

OPTICAL MATERIALS

William L. Wolfe
University of Arizona

CONTENTS

7. Optical Materials

7.1. Introduction

7.1.1. Symbols, Nomenclature, and Units.
Table 7-1 gives the symbols, descriptions and identifying equations, where appropriate, for the terms that are used in this chapter.

Table 7-1. Symbols, Nomenclature, Units

Symbols	Nomenclature	Units
A	Area; also, anisotropy factor	m^2; —
BRDF	Bidirectional reflectance distribution function	sr^{-1}
c	Speed of light	$m\ sec^{-1}$
c_L	Longitudinal velocity of acoustic waves	$m\ sec^{-1}$
c_T	Transverse velocity of acoustic waves	$m\ sec^{-1}$
c_p	Specific heat at constant pressure	$J\ g^{-1}\ K^{-1}$
c_v	Specific heat at constant volume	$J\ g^{-1}\ K^{-1}$
d	Plate thickness	m
dW	An increment of work	$N\ m$
E_p	Amplitude of the incident electric field in the direction parallel to the plane of incidence	$V\ m^{-1}$
E_s	Amplitude of the incident electric field in the direction perpendicular to the plane of incidence	$V\ m^{-1}$
e	Electronic charge	C
g	Damping constant	sec^{-1}
h	Planck's constant	$J\ sec$
j	$\sqrt{-1}$	—
k_a	Linear absorption coefficient	m^{-1}
k_{at}	Atomic absorption coefficient	—
k_B	Boltzmann constant	W
k_c	Concentration absorption coefficient	$(atm\ cm)^{-1}$
k_m	Mass absorption coefficient	m^{-3}
k_{mol}	Molecular absorption coefficient	—
k_T	Isothermal compression	$m\ K^{-1}$
l	Length	m
m	Reduced mass	g
N	Number of oscillators per unit cell	—
n	Refractive index	—

Table 7-1. Symbols, Nomenclature, Units (*Continued*)

Symbols	Nomenclature	Units
N_A	Avogadro's number	mol^{-1}
N_m	Number of atoms	—
n_e, n_o	Refractive indices for extraordinary and ordinary rays	—
n_i	Refractive index of the material i	—
$n(v)$	Number of modes per unit volume	m^{-3}
P_i	Power at position i	W
Q	Heat	J
R	Universal gas constant; also, electrical resistance	J K^{-1} mol^{-1}; ohms
r	Number of atoms per mole	mol^{-1}
R_p	Amplitude of the reflected electric field in the direction parallel to the plane of incidence	V m^{-1}
R_s	Amplitude of the reflected electric field in the direction perpendicular to the plane of incidence	V m^{-1}
r_p	$R_p/E_p \equiv$ Amplitude reflectivity for radiation polarized in the parallel direction	—
r_s	$R_s/E_s \equiv$ Amplitude reflectivity for radiation polarized in the perpendicular direction	—
S	Entropy	W
T	Temperature	K, °C
T_p	Amplitude of the transmitted electric field in the direction parallel to the plane of incidence	V m^{-1}
T_s	Amplitude of the transmitted electric field in the direction perpendicular to the plane of incidence	V m^{-1}
t_g	Thickness of gap or layer	m
t_p	T_p/E_p	—
t_s	T_s/E_s	—
u	Enthalpy or internal energy	J
V	Volume	m^3
V_c	Primitive (unit) cell volume	m^3
X	Distance along a beam	m
x	$h\nu/k_B T$	—
x_{ij}	Applied stress	N m^{-2}
z	Atomic number	—
α	Thermal expansion, also, polarizability	m K^{-1}
β	Volume expansion	m^3 K^{-1}
γ	Poisson's ratio	—

Table 7-1. Symbols, Nomenclature, Units (*Continued*)

Symbols	Nomenclature	Units
δ	Phase relationship or angle; also, deformation	—; m
ϵ_0	Dielectric constant of vacuum	F
ϵ_0	Low frequency dielectric constant	F
ϵ_∞	High frequency dielectric constant	F
θ_B	Polarizing or Brewster angle	rad
θ	Angle of incidence	rad
θ_D	Debye temperature	K
θ_c	Critical angle of total reflection	rad
θ_r	Angle of reflection	rad
θ''	Angle of refraction	rad
κ	Extinction coefficient	—
λ	Wavelength of the radiation	m
μ	Dipole moment	C m
ν	Frequency	$\sec^{-1}$, Hz
$\tilde{\nu}$	Wavenumber, $1/\lambda$	—
ν_D	Highest allowed acoustic frequency	$\sec^{-1}$, Hz
ρ_p	$\lvert\tilde{r}_p\rvert^2 = r_p r_p^* =$ power or intensity of reflectivity for radiation polarized in the parallel direction	—
ρ_s	$\lvert\tilde{r}_s\rvert^2 = r_s r_s^* =$ power or intensity of reflectivity for radiation polarized in the perpendicular direction	—
ρ_o	Resistivity	ohm cm
σ	Engineering stress; also, thermal conductivity	$N\,m^{-2}$; $W\,m^{-1}\,{}^\circ C^{-1}$
σ_o	Low frequency electrical conductivity	$mho\,m^{-1}$
σ_∞	High frequency electrical conductivity	$mho\,m^{-1}$
τ_p	$\lvert\tilde{t}_p\rvert^2 = t_p t_p^*$	—
τ_s	$\lvert\tilde{t}_p\rvert^2 = t_s t_s^*$; also, substrate transmission	—
Φ	Flux, radiant power	W
ψ	Optical phase	rad
ω	Radian frequency	$\sec^{-1}$
ω_e	Radian frequency of longitudinal lattice mode	$\sec^{-1}$
ω_n	Natural lattice frequency	$\sec^{-1}$
ω_0	Radian frequency of transverse lattice mode	$\sec^{-1}$

7.2. Description of Properties

The properties especially pertinent to optical materials and their use are given in this section. Tables of data are given in Section 7.3.

7.2.1. Reflection and Transmission of Nonabsorbing Materials.

The reflection of radiation in general involves both polarization and the phase of the field components. This section gives expressions for the transmitted and reflected components for polarized and unpolarized light in terms of amplitude and flux quantities.

Figure 7-1 illustrates reflection and transmission at a single surface. The relationships among the quantities are given in Tables 7-2, 7-3, 7-4, and Figure 7-2. The reflective, absorptive and transmissive properties of a plane parallel plate in a single medium (usually air) are included. See also Figure 7-3. A polarized beam is assumed to be incident on the plate. The reflectivities and transmissivities are given in Table 7-5.

Table 7-2. Relations for Single-Surface Reflection for Dielectrics

Type	Formula	Remarks
Angular relations	$\theta_r = \theta$	Snell's Law
	$n \sin \theta = n_t \sin \theta''$	Snell's Law
Reflected wave amplitude ratios	$\tilde{r}_s = |r_s| e^{j\delta s} = \dfrac{\tilde{R}_s}{\tilde{E}_s}$	Definition
	$\tilde{r}_p = |r_p| e^{j\delta p} = \dfrac{\tilde{R}_p}{\tilde{E}_p}$	Definition
Reflected wave amplitude ratios	$\dfrac{\tilde{R}_s}{\tilde{E}_s} = \dfrac{\sin(\theta - \theta'')}{\sin(\theta + \theta'')}$	Fresnel's Law for dielectrics
	$\dfrac{R_p}{E_p} = \dfrac{\tan(\theta - \theta'')}{\tan(\theta + \theta'')}$	Fresnel's Law for dielectrics
Transmitted wave amplitude ratios	$\tilde{t}_s = |t_s| e^{j\delta s} = \dfrac{\tilde{T}_s}{\tilde{E}_s}$	Definition
	$\tilde{t}_p = |t_p| e^{j\delta p} = \dfrac{\tilde{T}_p}{\tilde{E}_p}$	Definition
Transmitted wave amplitude ratios	$\dfrac{\tilde{T}_s}{\tilde{E}_s} = \dfrac{2 \sin \theta'' \cos \theta}{\sin(\theta + \theta'')}$	Fresnel's Law for dielectrics
	$= \dfrac{2 \sin \theta'' \cos \theta}{\sin(\theta + \theta'') \cos(\theta - \theta'')}$	Fresnel's Law for dielectrics
Reflected wave amplitude ratio for perpendicular polarization	$\tilde{r}_s = \dfrac{(n_t^2 - n^2 \sin^2 \theta)^{1/2} - n \cos \theta}{(n_t^2 - n^2 \sin^2 \theta)^{1/2} - n \cos \theta}$	Fresnel's Law in terms of incidence angle only
Transmitted wave amplitude ratio for perpendicular polarization	$\tilde{t}_s = \dfrac{2n \sin \theta}{n \sin \theta + (n_t^2 - n^2 \sin^2 \theta)^{1/2}}$	Fresnel's Law in terms of incidence angle only
Reflected wave amplitude ratio for parallel polarization	$\tilde{r}_p = \dfrac{(n_t^2 - n^2 \sin \theta)^{1/2} \sin \theta - n \sin \theta \cos \theta}{(n_t^2 - n^2 \sin \theta)^{1/2} \sin \theta - n \sin \theta \cos \theta}$	Fresnel's Law in terms of incidence angle only
	$= \dfrac{(n_t^2 - n^2 \sin \theta)^{1/2} \cos \theta - n \sin \theta}{(n_t^2 - n^2 \sin \theta)^{1/2} \cos \theta - n \sin \theta}$	Fresnel's Law in terms of incidence angle only
Transmitted wave amplitude ratio for parallel polarization	$\tilde{t}_p = 2nn_t \sin \theta \times$ $[n \sin \theta + (n_t^2 - n^2 \sin \theta)^{1/2}]^{-1} \times$ $[n \sin^2 \theta + (n_t^2 - n^2 \sin^2 \theta)^{1/2}]^{-1}$	Fresnel's Law in terms of incidence angle only

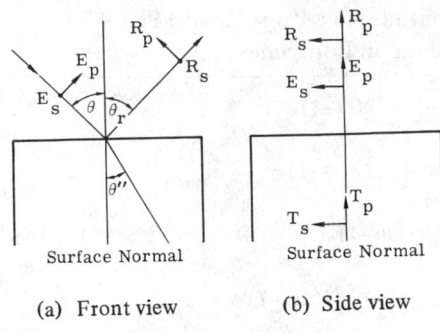

(a) Front view (b) Side view

Fig. 7-1. Geometry of reflection and refraction of light. Front and side views are given to show the plane of incidence (coincident with the paper in the front view) and the various orientations of the electric vector before and after transmission and reflection.

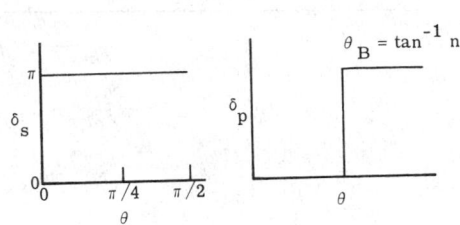

Fig. 7-2. Phase relation between the reflected and incident electromagnetic fields. The Brewster angle is designated as θ_B.

Table 7-3. Intensity Relations for Reflection and Transmission at a Single Surface

Type	Formula	Remarks
Reflection ratio for power	$\rho_s = |\tilde{r}_s|^2 = \tilde{r}_s\tilde{r}_s^*$	Definition
Reflection ratio for power	$\rho_p = |\tilde{r}_p|^2 = \tilde{r}_p\tilde{r}_p^*$	Definition
Transmission ratio for power	$\tau_s = |\tilde{t}_s|^2 = \tilde{t}_s\tilde{t}_s^* = -\dfrac{(1-\rho_p)\cos\theta}{n\cos\theta_t}$	Definition
Transmission ratio for power	$\tau_p = |\tilde{t}_p|^2 = \tilde{t}_p\tilde{t}_p^* = \dfrac{(1-\rho_s)\cos\theta}{n\cos\theta_t}$	Definition

Table 7-4. Phase Relations for Dielectric Reflections and Transmission at a Single Surface

$\delta_s = 0°, \quad \theta < \theta_C$

$\delta_s = \dfrac{2\tan^{-1}(n^2\sin^2\theta - 1)^{1/2}}{n\cos\theta}, \theta > \theta_C$

$\delta_p = 180°, \quad \theta < \theta_B$

$\delta_p = 0°, \qquad \theta_B < \theta < \theta_C$

$\delta_p = \dfrac{2\tan^{-1} n(n^2\sin^2\theta - 1)^{1/2}}{\cos\theta}$

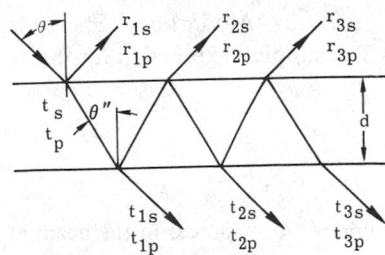

Fig. 7-3. Multiple reflections in a plane parallel plate. Radiation is incident at an angle θ and refracts at θ_t. The first reflection is the same as for a single surface and does not bear an index number. No attempt has been made to show the orientation of the electric vectors.

Table 7-5. Expressions for Amplitudes in a Plane Parallel Plate
With the Same Medium on Both Sides

$$\tilde{r}_{ms} = -t_s t_{sr} r_s^{2(m-1)} e^{j2(m-1)\phi}$$

$$\tilde{r}_{mp} = -t_p t_{pr} r_p^{2(m-1)} e^{j2(m-1)\phi}$$

$$\tilde{t}_{ms} = t_s t_{sr} r^{2m-1} e^{j(2m-1)\phi}$$

$$\tilde{t}_{mp} = t_{ms} [\cos(\theta - \theta'')]^{-2} e^{j(2m-1)\phi}$$

where t_s and t_p = the single surface transmission
t_{sr} and t_{pr} = the single surface transmissions t_s and t_p, but with θ and θ'' interchanged
m = the order of reflection or transmission starting with 2
ϕ = optical phase = $(2\pi n d \sec \theta'')\lambda^{-1}$

The total reflections or transmissions can be found by summing. Both r_s and r_p have the same form. The equations are given in Table 7-6.

Table 7-6. Amplitude Reflections and Transmissions for
the Sum of Multiple Reflections in a Plane Parallel Plate
Immersed in a Single Medium

$$\sum \tilde{r}_s = \tilde{r}_s - \frac{t_s t_{sr} r_s}{1 - r_s^2 e^{2j\phi}} \qquad \sum \tilde{t}_s = \frac{t_s t_{sr} e^{j\phi}}{1 - r_s^2 e^{2j\phi}}$$

$$\sum \tilde{r}_p = \tilde{r}_p - \frac{t_p t_{pr} r_p}{1 - r_p^2 e^{2j\phi}} \qquad \sum \tilde{t}_p = \frac{\tilde{t}_s}{\cos^2(\theta - \theta'')}$$

7.2.2. Absorption. Energy is lost from a beam in part by absorption and reflection. The simplest expression for this is the exponential law

$$\frac{d\Phi}{\Phi} = -k_a dx \qquad\qquad (7\text{-}1)$$

$$\Phi = \Phi_0 \exp(-k_a x) \qquad\qquad (7\text{-}2)$$

where Φ = power in the beam at x
x = distance from the beginning of the beam under consideration
Φ_0 = power at the beginning of the beam
k_a = absorption coefficient

Absorption is a function of wavelength. For most purposes, k_a is found experimentally as a function of wavelength from a sample of the material. The units cited are cm^{-1}, $(atm\ cm)^{-1}$, $g\ l^{-1}$, $dB\ km^{-1}$, and $ppm\ cm^{-1}$. The loss is also expressed in terms of the extinction coefficient, κ, as defined in Equation (7-18).

(1) The mass absorption coefficient, k_m, is equal to the linear (ordinary) absorption coefficient divided by the mass of material. It refers to energy absorbed from a unit cross-section of beam for 1 g of material.

(2) The atomic absorption coefficient, k_{at}, is equal to α_m times the mass of an atom of the absorber (=at. wt. divided by Avogadro's number, 6.02×10^{23}).

(3) The molecular absorption coefficient, k_{mol}, is equal to α_m times the mass of a molecule of absorber (=molecular weight divided by Avogadro's number).

(4) The concentration absorption coefficient, k_c, is generally given in $(atm\ cm)^{-1}$ to indicate the absorption per unit path length and partial pressure.

7.2.3. Scattering.
Mechanisms of scattering are discussed in Chapter 4. For optical materials, scattering can usually be calculated by an exponential expression

$$\Phi = \Phi_0 \exp(-k_s x) \tag{7-3}$$

All terms are identical to Equation (7-2) except k_s, which is the scattering coefficient in cm^{-1}

7.2.4. Transmission with Reflecting and Absorbing Materials.
Figure 7-4 is a general representation of collimated unpolarized light of power Φ_0, incident on a plane parallel plate. The beam is shown incident, but only a ray representation is given for all subsequent paths of the beam. In general, all of the incident power is reflected, absorbed, or transmitted. No account is taken of any polarizing properties. The Φ_i values indicate the power in the beam in the places shown. Then, for a single pass through the surface or for a single reflection, the reflection coefficient for power is given by

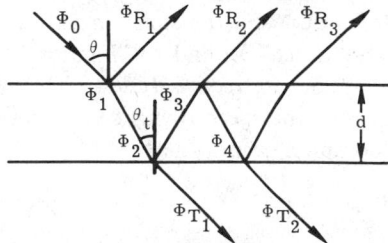

$$\rho_1 = \frac{\Phi_{R_1}}{\Phi_0} = \text{single surface reflectance} \tag{7-4}$$

Fig. 7-4. Geometrical and power relations for a beam incident at an angle on a plane parallel plate.

$$\tau_1 = \frac{\Phi_{T_1}}{\Phi_0} = \text{single surface (power) transmittance} \tag{7-5}$$

$$\tau_i = \frac{\Phi_2}{\Phi_1} = e^{-k_a} = \text{internal transmittance (single pass)} \tag{7-6}$$

$$\tau_{ext} = \frac{\Phi_{T_1}}{\Phi_0} = (1 - \rho_1)^2 \tau_1 = \text{external transmittance} \tag{7-7}$$

For an infinite number of passes, the expressions are

$$\tau_\infty = \frac{(1 - \rho_1)^2 \tau_i}{(1 - \rho_1^2)\tau_i^2} \tag{7-8}$$

$$\rho_\infty = \rho_1 + \frac{\rho_1 \tau_1 \tau_i (1 - \rho_1)^2}{1 - \rho_1^2 \tau_i^2} \tag{7-9}$$

$$\alpha_\infty = \rho_1 + \frac{\tau_1 \tau_i}{1 - \rho_1^2 \tau_i^2} \tag{7-10}$$

If there is no absorption, transmission becomes

$$\tau_\infty = \frac{(1 - \rho_1)^2 \tau_i}{1 - \rho_2 \tau_i^2} \tag{7-11}$$

For normal incidence and no absorption, the appropriate quantities are

$$\rho_1 = \left(\frac{n - 1}{n + 1}\right)^2 \tag{7-12}$$

$$\tau_1 = \frac{4n}{(n + 1)^2} \tag{7-13}$$

$$\tau_\infty = \frac{2n}{n^2 + 1} \tag{7-14}$$

Reflection and transmission data for many common infrared optical materials are given in Sections 7.3.1 and 7.4, Figures 7-5 and 7-8 through 7-29 respectively.

7.2.5. Refractive Index. The refractive index of a material can be defined as the ratio of the speed of light in a vacuum to that in the material. It can also be defined as a complex quantity.

$$\hat{n} \equiv \frac{c}{v} \quad \tilde{n} \equiv n - j\kappa \tag{7-15}$$

Refractive index values change with wavelength and temperature. Both are sensitive changes. The refractive index also changes with pressure, but this is a very mild function for every solid material. The basic equation for a change in the value of the refractive index is derived from the differential equation for the damped, harmonic motion of a dipole or similar charge distribution [7-1]. The result is

$$n^2 - \kappa^2 = 1 + \frac{Ne^2}{m\epsilon_0}\left[\frac{\omega_n^2 - \omega^2}{(\omega_n^2 - \omega^2)^2 + \omega^2 g^2}\right] \tag{7-16}$$

$$2n\kappa = \frac{Ne^2}{m\epsilon_0} \frac{\omega g}{(\omega_n^2 - \omega^2)^2 + \omega^2 g^2} \tag{7-17}$$

where N = number of oscillators per unit cell
n = real refractive index
κ = extinction coefficient (due to absorption)
ω_n = natural lattice vibration (angular) frequency
ω = optical (radian) frequency = $2\pi\nu$ rad sec^{-1}

ϵ_0 = dielectric constant of vacuum

g = damping constant

m = reduced mass of the dipoles

This simplified treatment shows that the absorption by free carriers is proportional to the square of the wavelength. Bound charges exhibit resonances when the optical frequency, ω, approaches the natural frequency, ω_n. At that point, the extinction coefficient for bound carriers is given by

$$2n\kappa = \frac{Ne^2}{m\epsilon_0} \frac{1}{\omega g} \kappa^2 = -\frac{1}{2} \pm \frac{1}{2}\left[1 - \left(\frac{Ne^2}{m\epsilon_0} \frac{1}{\omega g}\right)^2\right]^{1/2} \qquad (7\text{-}18)$$

Authors have used a variety of equations to represent the variation of n with λ. These are given with the material descriptions where available.

No similar attempt has been made to obtain regression-approximation equations for index versus temperature.

The complex index can be used in most of the equations for reflectance and transmittance. For instance, the reflectance at normal incidence is given by

$$\rho = \left|\frac{\tilde{n} - 1}{\tilde{n} + 1}\right|^2 = \frac{(n-1)^2 + \kappa^2}{(n+1)^2 + \kappa^2} \qquad (7\text{-}19)$$

The restrahlen frequencies, which are about equal to the natural resonances of the crystals, are close to the resonance frequencies described by the dispersion equation. If ω_0 and ω_l are the frequencies of the transverse and longitudinal optical mode, then

$$\omega_0{}^2 = \omega_l{}^2\left(\frac{\epsilon_\infty}{\epsilon_0}\right)\left(\frac{\epsilon_0 + 2}{\epsilon_\infty + 2}\right) - \frac{4\pi(Ze)^2}{3mV_c}\left(\frac{\epsilon_\infty + 2}{3}\right) \qquad (7\text{-}20)$$

where Ze = effective charge

m = reduced mass

$\epsilon_\infty, \epsilon_0$ = high and low frequency dielectric constant respectively

V_c = primitive (unit) cell volume

Refractive index data and dispersion equations are given in Figures 7-6 and 7-7 in Sections 7.3.2 and 7.3.3, respectively, and Figures 7-30 through 7-37 and Tables 7-15 through 7-52 in Section 7.4.2.

7.2.6. Thermal Properties. Although almost every physical property of a material is a function of temperature, specific heat, thermal conductivity, and thermal expansion bear more directly on the reaction of a material to changes in temperature or heat. Other properties logically included are melting or softening temperature, the temperature at which some other change of state takes place, and the heats associated with these changes.

Specific heat is closely related to heat capacity. The latter is the amount of heat required to raise a unit mass of material one unit of temperature. It is usually in cal g^{-1} K^{-1}, or Btu lb^{-1} $°F^{-1}$. The specific heat of a material is the ratio of its heat capacity to that of water and is therefore dimensionless.

The coefficient of linear thermal expansion is the increase in length per unit temperature change and per unit length of the bar. The dimensions are reciprocal degrees. Because the expansion is usually a function of temperature, it can be given as polynomial

$$\alpha = A + BT + CT^2 \qquad (7\text{-}21)$$

Three terms are usually a sufficient approximation. Area and volume expansion can be shown to be just two and three times this value.

Thermal conductivity is a measure of the ease with which heat is conducted along a body. The heat flow is proportional to the area, the temperature difference, and the time, and is inversely proportional to the length. The thermal conductivity, σ, is defined by the heat rate equation

$$\dot{q} = \frac{\sigma A}{l} \Delta T \qquad (7\text{-}22)$$

where $\dot{q} = \partial Q/\partial t$ = heat rate or power
$\qquad A$ = area
$\qquad l$ = length
$\qquad \sigma$ = thermal conductivity
$\qquad \Delta T$ = temperature difference

Note that $\sigma A l^{-1}$ is called the thermal conductance. The dimensions of σ are cal cm^{-2} cm $°C^{-1}$ sec^{-1}, or cal cm^{-1} $°C^{-1}$ sec^{-1}, or Btu ft^{-1} $°F^{-1}$ sec^{-1}, or W m^{-1} $°C^{-1}$.

Specific heat, thermal conductivity and thermal expansion data are given in Sections 7.3.5 and 7.3.6, Tables 7-8 and 7-9, respectively.

7.2.7. Debye Temperature. The Debye characteristic temperature, θ_D, can be defined such that

$$\theta_D = \frac{h\nu_D}{k_B} \qquad (7\text{-}23)$$

Debye temperatures for many common infrared optical materials are given in Section 7.3.7, Table 7-10.

7.2.8. Elastic Moduli. Among the important measures of the mechanical properties are the fundamental ones of the elastic constants c_{ij} and s_{ij} and the practical ones that include Young's modulus E, the compressibility, the rupture modulus, and various expressions for hardness like the Moh scale, Knoop, Rockwell and Vickers hardness values.

The values of Young's modulus quoted in the data enumeration sections were determined in one of several ways. The preferred value is that obtained from measurements made in either tension of compression, where the experimental procedure conforms most closely to the elongation system indicated by the definition of Young's modulus. However, for most optical materials, such data are rare and difficult to obtain because these materials cleave when subjected to either tension or compression. Young's modulus may also be determined from measurements made with the sample bent in flexure. The values obtained for optical materials are usually from flexure and, in general, are lower than those obtained in either tension or compression. The data can also be inferred from sound-velocity measurements, but none obtained that way are reported here.

The bulk modulus is the reciprocal of the compressibility and is generally determined from compressibility measurements. In the absence of experimental values, the bulk modulus was computed from the elastic coefficients.

In general, crystals are anisotropic with respect to their elastic properties. That is, the values of these moduli differ with direction in the crystal. A measure of the anisotropy of a cubic crystal is given by the anisotropy factor, A. It is defined as $2c_{44}(c_{11} - c_{12})^{-1}$. This factor equals unity for elastically isotropic materials and, for elastically anisotropic crystals, has values either greater than or less than unity. For those crystals with $A > 1$, such as germanium and silicon, Young's modulus has its maximum value along $<100>$ directions and a minimum value along $<111>$ directions. For crystals with $A < 1$, such as sodium chloride, Young's modulus has its maximum value along $<111>$ directions and its minimum value along $<100>$ directions. The variation in elastic properties with direction may be as great as 30% and should be considered in design problems.

The discussion of elastic properties given above is applicable only when the deformation of the crystal is small, that is, less than 1% in research studies and less than 2% for practical applications. When a crystal is deformed to such an extent that Hooke's law is no longer obeyed, the crystal is said to be plastically deformed. The elastic limit is used to indicate the stress above which plastic deformation occurs and below which Hooke's law is obeyed. For metals, the stress-strain curve shows an abrupt change. For most optical materials the stress-strain curve gradually changes from a straight line to a curve with decreasing slope. In the absence of a unique departure from Hooke's law, an apparent elastic limit must be defined. For some crystals that were tested in flexure, an apparent elastic limit is quoted. This limit is taken as that stress on the stress-strain curve where the slope is half the slope at the origin of the stress-strain curve.

The modulus of rupture value is strongly dependent upon the history of the particular sample and should be used only for approximate calculations. The presence of small cracks or barely macroscopic cleavages will change the apparent rupture strength by at least an order of magnitude. There are several materials that show a markedly different performance at stresses where rupture might be anticipated. The stress-strain curves of thallium bromide-iodide, thallium bromide-chloride, and silver chloride show rapid increases in slope at large deformations and appreciable cold flow, that is, a marked increase in deformation with time.

Elastic coefficients and moduli data are given in Sections 7.3.8 and 7.3.9, Tables 7-11 and 7-12 respectively.

7.2.9. Engineering Mechanical Properties. The engineering stress is the ratio of the instantaneous load, P, on the specimen to its original cross sectional, A_0. The central portion of the specimen between two reference marks is the gauge length, l_0. The engineering strain is the ratio of the change in length of this section, $\Delta l = l - l_0$, to the original length. The relation of stress to strain is linear up to the elastic limit, and the constant of proportionality is Young's modulus E, that is

$$E = \frac{P/A_0}{\Delta l/l_0} \tag{7-24}$$

Sometimes the transition from elastic to plastic deformation is sharp and one can define the yield stress. Otherwise a microyield stress can be defined as the stress required for a yield of 10^{-5}. An arbitrary, practical measure is the proof stress or offset-yield stress, defined as the stress required to produce a certain degree of plastic strain, e.g., 0.1%, 0.2%, or 0.5%. Then one specifies, for instance, the 0.2% proof stress.

Since most metals can work-harden and the stress needed to produce further plastic deformation increases as the strain increases, other measures are useful. The strain-hardening rate, or the rate at which a material hardens with increasing strain, decreases with increasing strain until it is zero at the ultimate tensile strength, UTS, or tensile strength, TS, for short. Fracture occurs at the fracture stress, which is lower than the UTS. The reduction in area is the ratio of the specimen area after it has necked (at or near the UTS) to the original area. The fracture strain, ductility, or fracture elongation all describe the amount produced before fracture (and this is almost equal to the total strain). The true stress is the ratio of the load to the instantaneous area. This describes the stress in the necked region of the specimen and is only of academic interest. However, it leads to the definition of true fracture stress which is the true stress at fracture. True strain or elongation is the definite integral of the ratio of incremental charge in length to the instantaneous length

$$\int_{l_0}^{l_i} \frac{dl}{l} = \ln \frac{l_i}{l_0} \qquad (7\text{-}25)$$

This definition is valid until necking occurs, at which point a second definition is needed. The true strain is then $\ln A_i/A_0$ where A_i and A_0 are the instantaneous and initial cross-sectional areas respectively. The true stress-strain curve is often called a flow curve. Then one has the quantities of flow stress, flow strain, and initial-flow stress which is synonymous with yield stress. The toughness of a material is a measure of one energy it absorbs prior to fracture. It is the area under the flow curve.

Engineering mechanical properties of Young's moduli, rupture moduli, and hardness are given in Sections 7.3.9 and 7.3.10, Tables 7-12 and 7-13 respectively.

7.2.10. Thermodynamic Relationships

First Law $\qquad dQ = dU + dW$ $\qquad\qquad\qquad\qquad\qquad\qquad (7\text{-}26)$

Entropy $\qquad dQ = TdS$ $\qquad\qquad\qquad\qquad\qquad\qquad\qquad (7\text{-}27)$

Specific Heats $\qquad mc_p = \left(\frac{\partial Q}{\partial T}\right)_p = T\left(\frac{\partial S}{\partial T}\right)_p$ $\qquad\qquad\qquad (7\text{-}28)$

$$mc_v = \left(\frac{\partial Q}{\partial T}\right)_V = T\left(\frac{\partial S}{\partial T}\right)_V \qquad (7\text{-}29)$$

Internal Energy $\qquad dU = TdS - PdV$ $\qquad\qquad\qquad\qquad\qquad (7\text{-}30)$

where Q = heat
$\qquad\quad W$ = work
$\qquad\quad U$ = internal energy

Maxwell's Relations

$$\left(\frac{\partial T}{\partial V}\right)_S = \left(\frac{\partial P}{\partial S}\right)_V \qquad \left(\frac{\partial T}{\partial P}\right)_S = \left(\frac{\partial V}{\partial S}\right)_P$$

$$\left(\frac{\partial S}{\partial V}\right)_T = \left(\frac{\partial P}{\partial T}\right)_V \qquad \left(\frac{\partial S}{\partial P}\right)_T = \left(\frac{\partial V}{\partial T}\right)_P \tag{7-31}$$

Difference in Specific Heats

$$c_p - c_v = \frac{T}{m}\left(\frac{\partial S}{\partial T}\right)_p - \frac{T}{m}\left(\frac{\partial S}{\partial T}\right)_V = \frac{T}{m}\left(\frac{\partial P}{\partial T}\right)_V \left(\frac{\partial V}{\partial T}\right)_p \tag{7-32}$$

If $PV = RT$, then $c_p - c_v = R/m$; for molar volume, V, volume expansion, β, and isothermal compressibility, k_T, $c_p - c_v = TV\beta^2 \, k_T^{-1}m^{-1}$. The Nernst-Lindeman relation is

$$c_p - c_v = A \, c_p^{\,2} \frac{T}{m} \tag{7-33}$$

where A is a constant.

7.2.11. Dielectric Constant and Loss Tangent. Theoretical relationships exist which predict the dielectric constant for materials in different states.

(1) Dilute Gases (Claussius Mosotti)

$$\frac{\epsilon_0 - 1}{\epsilon_0 + 2} = \frac{4}{3} \, \pi N\left(\alpha + \frac{\mu^2}{3kT}\right) \tag{7-34}$$

where N = number of molecules per unit volume
α = polarizability
μ = dipole moment

(2) Condensed Medium (Onsager)

$$\frac{9(\epsilon_0 - \epsilon_\infty)(2\epsilon_0 + \epsilon_\infty)}{(\epsilon_\infty + 2)^2(3\epsilon_0 + 2)} = \frac{4\pi N\mu^2}{3k_B T} \tag{7-35}$$

The dielectric constant can be written as follows

$$\tilde{\epsilon} = \epsilon_\infty + \frac{\epsilon_0 - \epsilon_\infty}{1 - j} = \epsilon_r + j\epsilon_i \tag{7-36}$$

where $\epsilon_r = \epsilon_\infty + \dfrac{\epsilon_0 - \epsilon_\infty}{1 + \omega^2\gamma^2}$ \hfill (7-37)

$$\epsilon_i = \frac{(\epsilon_0 - \epsilon_\infty)\omega}{1 + \omega^2\gamma^2} \tag{7-38}$$

Dielectric-constant data are given in Table 7-7.

7.2.12. Hardness and Microhardness. To a large extent, the measures of hardness are empirical and comparative. The simplest, least-quantitative measurement scale is the Moh scale—a simple listing of ten materials in order of increasing hardness. Comparisons are made on the basis of which materials scratch each other. The steps in the scale are not equal. Below are the ten materials and their given values.

10 diamond, C	5 apatite $Ca_5(PO_4)_3(F,Cl,OH)$
9 corundum, Al_2O_3	4 fluorite, CaF_2
8 topaz, $Al_2SiO_4(OH,F)_2$	3 calcite, $CaCO_3$
7 quartz, SiO_2	2 gypsum, $CaSO_4 \cdot 2H_2O$
6 orthoclase, $KAlSi_3O_8$	1 talc, $Mg_3Si_4O_{10}(OH)_2$

Most other methods of measuring hardness are based on pressing a material with a specified force with an indenter of a prescribed shape. The load is usually specified (as a mass in grams). The shape determines whether the method is called Brinell, Vickers, or Knoop. Hardness measurements can vary with applied load, duration of load and speed of application and release. The variability depends on material properties of creep, cold flow, and stress relief by minute cracks. Hardness data are given in Section 7.3.10, Table 7-13.

7.2.13. Chemical and Related Properties. These properties include gram molecular weight, specific gravity, and solubility. The first of these is the molecular weight of the substance expressed in grams if this is applicable. For example, the gram molecular weight of table salt, sodium chloride, is 28 g. Specific gravity is a measure of density, but rather than being in specific units like g cm^{-3}, it is the density of a substance related to the density of water at 4°C. Thus, since water at these temperatures has a density of 1 g cm^{-3} or 62.5 lb ft^{-3}, the density of any substance is its specific gravity multiplied by either of these dimensional, numeric quantities.

Solubility data are usually given in terms of the number of grams of a substance that under equilibrium conditions are dissolved in 100 ml of water at the temperature specified (or normal room temperature if not specified). Thus, it is roughly a percentage solubility and somewhat indicative of the weathering properties of each material. These properties are all given in Section 7.3.11, Table 7-14.

7.3. Summary Data

7.3.1. Transmission Region. Figure 7-5 shows transmission regions of most infrared optical materials. The white bars represent the wavelength region in which a particular material transmits appreciably.

The limiting wavelengths chosen are those at which a sample of 2 mm thickness has 10% external transmittance. For some cases this criterion is insufficient, as in the consideration of semiconducting materials where purity and temperature must also be specified.

Several different endings are used for the bars of the chart in Figure 7-5. A bar with a straight vertical ending indicates that the cutoff exists at the wavelength represented by the end of the bar exactly as defined above; a bar with an S-shape ending represents a material which cuts off at approximately that wavelength; a bar ending in an angle

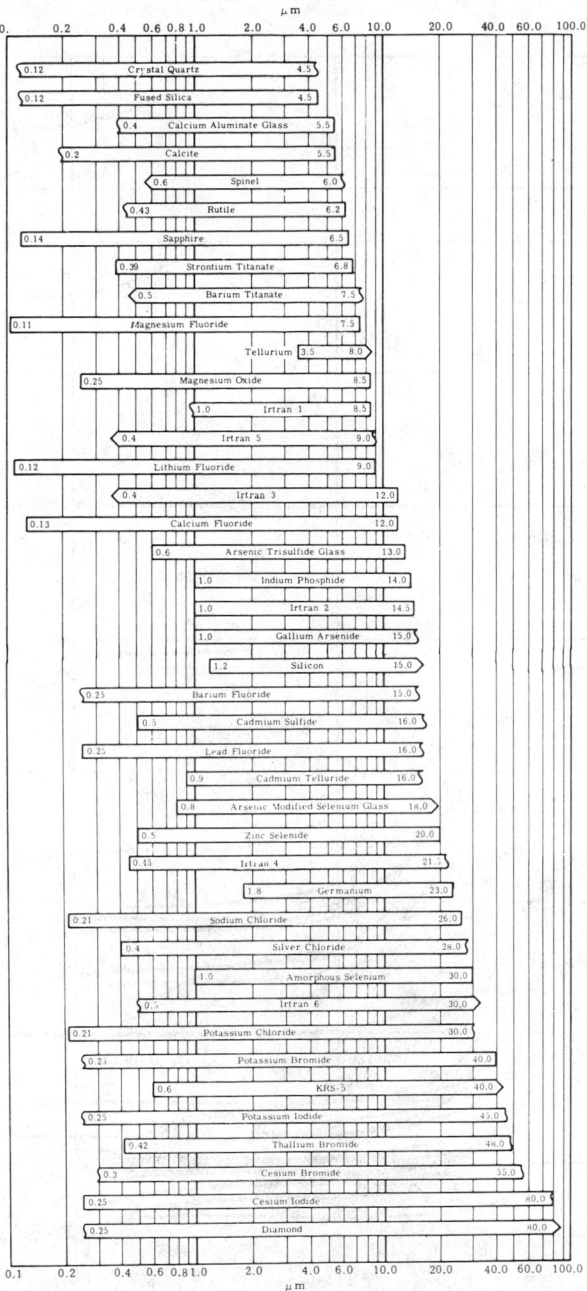

Fig. 7-5. Transmission regions of optical materials (2 mm thickness). The cutoff is defined as 10 percent external transmittance. Irtran 6 is no longer manufactured by the Eastman Kodak Co. Irtran® is a registered trademark of the Eastman Kodak Co.

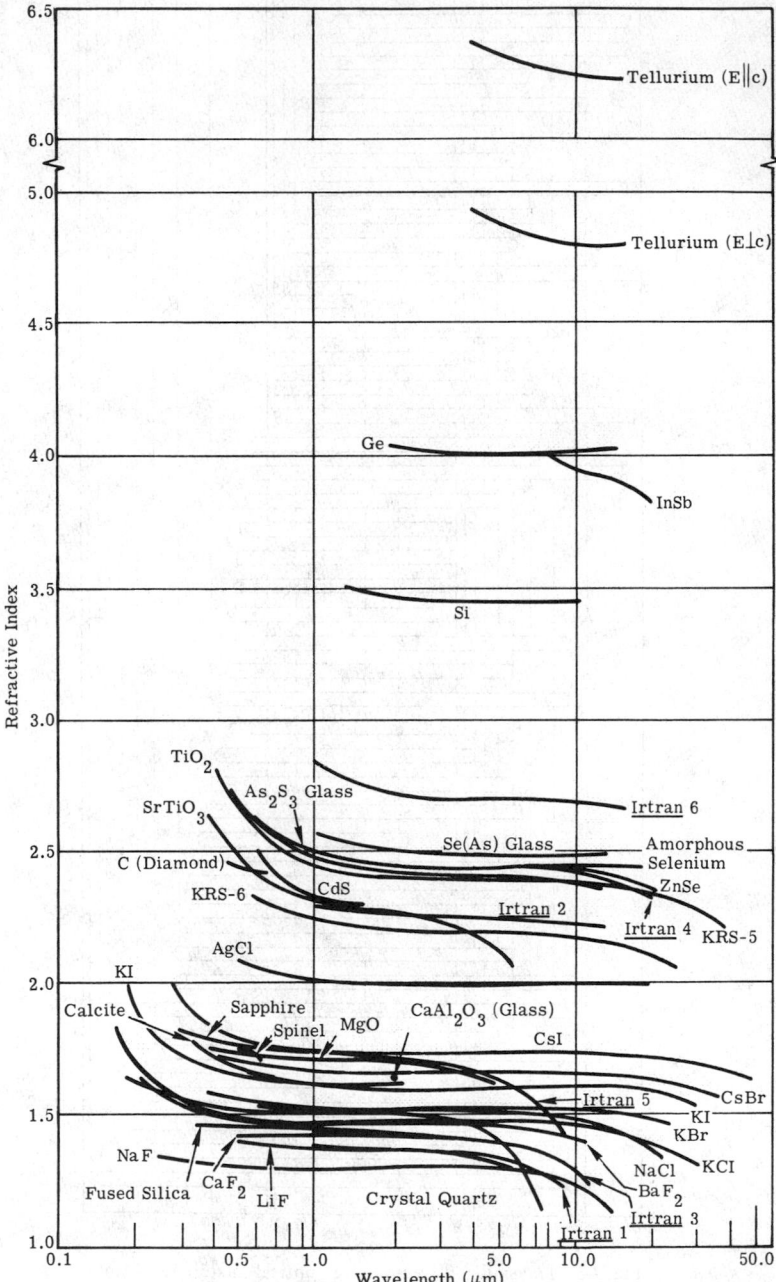

Fig. 7-6. Refractive index values. <u>Irtran</u> 6 is no longer manufactured by the Eastman Kodak Co. Irtran® is a registered trademark of the Eastman Kodak Co.

indicates that the material transmits at least to that wavelength, and probably further. Measurements made on materials in this last group have not been made to sufficiently long or sufficiently short wavelengths to determine the cutoff.

7.3.2. Refractive Index. Figure 7-6 presents curves of refractive index versus wavelength and illustrates the range of the refractive indices of many of the materials. For a crystal whose refractive index varies with direction, only the refractive index corresponding to the ordinary ray has been plotted.

7.3.3. Dispersion. The data of Figure 7-6 are plotted in Figure 7-7 to show the rate of change of the refractive index versus wavelength.

7.3.4. Dielectric Constant. Values for the dielectric constant as a function of frequency and temperature are given in Table 7-7. These values are the relative dielectric constants of materials, that is, the ratios of the dielectric constants of the material to that of a vacuum. They include measurements taken at microwave frequencies and

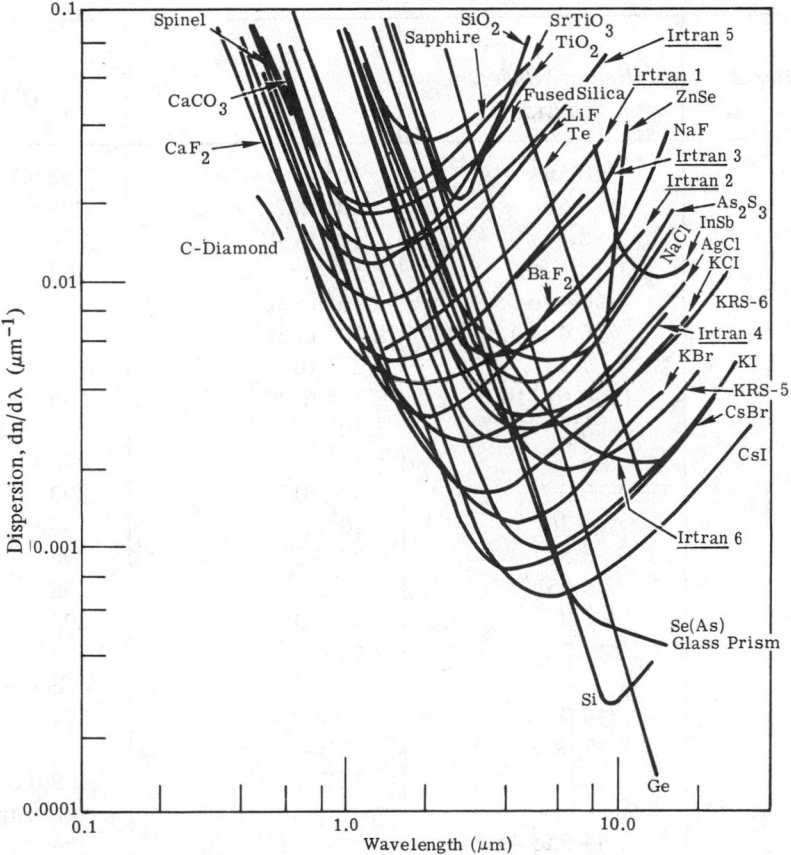

Fig. 7-7. $dn/d\lambda$ versus λ for selected materials. <u>Irtran</u> 6 is no longer manufactured by the Eastman Kodak Co. Irtran® is a registered trademark of the Eastman Kodak Co.

indicate such peculiarities as variation with orientation. Since dielectric properties depend upon purity, particularly in semiconductors, the purity of the sample measured is given, where available.

When the dielectric constant is measured with the electric field parallel to the c axis (the optic axis), the measurement is identified with a superscript p; when the electric field is perpendicular to the optic axis, the measurement is identified with a superscript s.

7.3.5. Melting Temperature, Specific Heat, and Thermal Conductivity. Table 7-8 lists the following: the melting temperature of crystals and the softening temperature

Table 7-7. Permittivity [7-2]
(Dielectric constant)

Material	ϵ_r (Relative dielectric constant)	f (Hz)	T (K)
Al_2O_3	10.55^p	10^2 to 3×10^8	298
	8.6^s	10^2 to 2.5×10^{10}	298
As_2S_3	8.1	10^3 to 10^6	–
AgCl	12.3	10^6	293
ADP	56.4 to 55.9^s	10^2 to 10^8	–
	16.4 to 13.7^p	10^2 to 10^8	–
BaF_2	7.33	2×10^6	–
$BaTiO_3$	1240 to 1100	10^2 to 10^8	298
$CaCO_3$	8.5^s	10^4	290 to 295
	8.0^p	10^4	290 to 295
CuCl	10.0	5×10^5	293
CaF_2	6.76	10^5	–
CsBr	6.51	2×10^6	298
CsI	5.65	10^6	298
CuBr	8.0	3×10^6	293
CdTe	11.0	1 to 10^5	5.5×10^{13} carriers $(cc)^{-1}$
$CaTiO_3$	140.0	1.5×10^6	294
GaAs	11.06 ± 0.14	–	–
Ge	16.6	9.37×10^9	9.0 ohm cm resistivity
KDP	44.5 to 44.3^s	10^2 to 10^8	–
	21.4 to 20.2^p	10^2 to 10^8	–
KCl	4.64	10^6	302.5
KRS-5	32.9 to 32.5	10^2 to 10^7	298
KRS-6	32.9 to 31.8	10^2 to 10^5	298

Table 7-7. Permittivity [7-2] (*Continued*)

Material	ϵ_r (Relative dielectric constant)	f (Hz)	T (K)
KBr	4.90	10^2 to 10^{10}	298
	4.97	10^2 to 10^{10}	360
KI	4.94	2×10^6	—
LiF	9.00	10^2 to 10^{10}	298
	9.11	10^2 to 10^{10}	353
$MgO \cdot 3.5\ Al_2O_3$	8.0 to 9.0	—	—
MgO	9.65	10^2 to 10^8	298
Muscovite	5.4	10^2 to 3×10^9	299
$NaNO_3$	6.85	2×10^5	292
NaCl	5.90	10^2 to 2.5×10^{10}	298
	6.35 to 5.97	10^2 to 2.5×10^{10}	358
NaF	6.0	2×10^6	292
$PbMoO_4$	26.8	4×10^8	—
$PbCl_2$	33.5	5×10^5	293
PbS	17.9	10^6	288
PbF_2	3.6	10^6	—
Se	6.0	10^2 to 10^{10}	298
Si	13.0	9.37×10^9	—
SrF_2	7.69	2×10^6	—
SiO_2 (crystal)	4.34^s	3×10^7	290 to 295
	4.27^p	3×10^7	290 to 295
SiO_2 (fused)	3.78	10^2 to 10^{10}	300
$SrTiO_3$	306.0	10^2 to 10^5	298
Se(As)	234 to 230.0	10^2 to 10^{10}	298
TlCl	31.9	2×10^6	—
TlBr	30.3	10^3 to 10^7	298
TiO_2	200 to 160	10^4 to 10^7	298

of glasses, plastics, and some compacts; values of specific heats, c_p, in terms of the ratio of heat capacity of the substance to that of water; and the thermal conductivity, k. For crystals which exhibit anisotropy, orientation of heat flow with respect to the c axis is noted. Values are given for the heat flow parallel and perpendicular to the c axis.

7.3.6. Thermal Expansion. The expansion is given in terms of coefficients A, B, and C in the equation shown in Table 7-9. Remarks of "parallel" and "perpendicular" specify expansion in those directions with respect to the c axis. The expansion can also be in the direction shown by the angular brackets.

Table 7-8. Thermal Properties

Materials					Thermal Conductivity Temperature (K)
Ammonium Dihydrogen Phosphate	ADP	—	—	1.7×10^{-3p}; 3.0×10^{-3s}	315; 313
Silver	Ag	0.0559	298	—	—
Silver Chloride	AgCl	0.0848; 0.0906	273; 323	2.6×10^{-3}; 2.71×10^{-3}	295; —
Silver Sulfide	Ag$_2$S	0.072	273	—	—
Aluminum	Al	0.218	298	—	—
Sapphire	Al$_2$O$_3$	0.180; 0.174	298; 273	60×10^{-3p}; 5.5×10^{-3s}	299; 296
Arsenic Trisulfide Glass	As$_2$S$_3$	—	—	4.0×10^{-4}	313
Gold	Au	0.0309	298	—	—
Boron	B	0.260	273	—	—
Barium Fluoride	BaF$_2$	—	—	28×10^{-3}; 17×10^{-3}; 1.6×10^{-3}; 3.2×10^{-3}	286; 311; 401; 273
Barium Titanate	BaTiO$_3$	0.01799; 0.03004; 0.04471; 0.05709; 0.06868; 0.07813; 0.02990	55; 75; 100; 125; 150; 175; 298	—	—
Bismuth	Bi	—	—	—	—
Carbon (diamond)	C	0.12000	298	34.0; 8.6; 6.59	76; 194; 273
Carbon (granular)	C	0.21600	298	—	—
Calcite	CaCO$_3$	0.20300; 0.21400	273; 373	1.32×10^{-2p}; 1.11×10^{-2s}	273; 273
Calcium Fluoride	CaF$_2$	0.20400; 0.21200	273; 373	9.32×10^{-2}; 3.60×10^{-2}; 2.47×10^{-2}; 2.32×10^{-2}; 1.91×10^{-2}	83; 200; 273; 298; 373
Calcium Titanate	CaTiO$_3$	—	—	—	—
Cadmium Fluoride	CdF$_2$	0.08820	273	—	—
Cadmium Sulfide	CdS	0.92200	323	3.8×10^{-2}	287
Cadmium Sulfide (pressed)	CdS	—	—	1.0×10^{-1}; 4.0×10^{-1}	10 & 100; 50
Cadmium Telluride	CdTe	0.01875	323	0.015	—
Cesium Bromide	CsBr	0.63000	293	2.3×10^{-3}; 2.2×10^{-3}; 2.6×10^{-3}	298; 318; 338

Material	Symbol					
Cesium Iodide	CsI	894.0	0.04800	293	2.7×10^{-3}	298
Copper	Cu	1356.0	0.00717×10^{-3}	2		
			0.038×10^{-3}	5		
			0.21×10^{-3}	10		
			1.8×10^{-3}	20		
			24.0×10^{-3}	50		
			61.0×10^{-3}	100		
			92.0×10^{-3}	300		
Copper Bromide	CuBr	777.0			0.125	300
Copper Chloride	CuCl	695.0			0.13	300
Copper Sulfide	CuS	—			0.105	300
Iron	Fe	1808.0	0.12900	298	0.14	293
Iron Oxide	Fe$_2$O$_3$	—	0.11000	298		
Gallium Arsenide	GaAs	1511.0	0.17000	298	0.085	329
Gallium Phosphide	GaP	>773.0			0.035	452
Gallium Antimonide	GaSb	993.0			0.026	327
Germanium	Ge	1209 to 1215	0.01828	273	0.037	447
Indium Arsenide	InAs	1215	0.074		0.026	353
			3.4	78 to 290		
			5.2	290 to 573		
			7.01	573 to 673		
Indium Phosphide	InP	1323 to 1343	0.0231	180	0.019	449
Indium Antimonide	InSb	796.0	0.248	300	0.015	327
Irtran 1		1528.0			0.031	695
Irtran 2		2103.0 (150 psi)			0.016	298
Irtran 3		1692.0			0.104	441
Irtran 4		1788.0			0.070	273
Irtran 5		3220.0			0.010	417
Irtran 6*		1363.0			0.0085	299
Potassium Bromide	KBr	1003.0	0.104	273	0.698×10^{-2}	319
			0.108	373		
Potassium Chloride	KCl	1049.0	0.162	273	1.15×10^{-2}	315
			0.168	373		
Potassium Dihydrogen Phosphate	KDP	525.6			1.56×10^{-2}	312
Potassium Iodide	KI	996.0	0.73	200	$2.9 \times 10^{-3}\,p$	312
			0.75	250	$3.2 \times 10^{-3}\,s$	319
			0.75	270	$5.0(\pm 3\%) \times 10^{-3}$	299
Thallium Bromide Iodide	KRS-5	687.5			1.3×10^{-3}	293

*Irtran 6 is no longer manufactured by the Eastman Kodak Co. Irtran® is a registered trademark of the Eastman Kodak Co.

Table 7-8. Thermal Properties (Continued)

Materials	Symbol	Melting Temperature (K)	Specific Heat	Specific Heat Temperature (K)	Thermal Conductivity (cal cm^{-1} sec^{-1} K)	Thermal Conductivity Temperature (K)
Thallium Bromide Chloride	KRS-6	696.5	0.0482	293	17.1×10^{-4}	329
Lithium Fluoride	LiF	1143.0	0.373	283	2.70×10^{-2}	314
Magnesium Oxide	MgO	3073.0	0.209	273	262×10^{-2}	10
					756×10^{-2}	30
					638×10^{-2}	100
					14.0×10^{-2}	300
					7.7×10^{-2}	500
	$MgO \cdot 3.5\ Al_2O_3$	2303.0	0.03	308	3.3×10^{-2}	308
			0.026	441		
			0.028	443		
Mica						
Sodium Chloride	Muscovite	1473 to 1573		293 to 373	0.0006 to 0.0014	—
	NaCl	1074.0	0.208	373	1.55×10^{-2}	289
Sodium Fluoride	NaF	1253.0	0.204	373	—	—
			0.217	273	0.124	83
			0.26	—	0.0252	273
					0.0220	298
Sodium Nitrate	$NaNO_3$	579.8	0.247	273	—	—
			0.270	373		
Lead	Pb	600.5	0.031	273	—	—
Lead Chloride	$PbCl_2$	774.0	0.0649	273	—	—
			0.0681	373		
Lead Fluoride	PbF_2	1128.0	—	—	—	—
Molybdate	$PbMoO_4$	1333 to 1343	0.100	288	—	—
Lead Sulfide	PbS	1387.0	0.0502	273	16×10^{-4}	—
			0.0511	373		
Lead Selenide	PbSe	1338.0	—	—	100×10^{-4}	—
Lead Telluride	PbTe	1190.0	—	—	120×10^{-4}	—
Polyethylene			0.3×10^{-3}	5	—	—
			2.3×10^{-3}	10		
			16.1×10^{-3}	20		
			78.8×10^{-3}	50		
			157×10^{-3}	100		
			566×10^{-3}	300		
Platinum	Pt	2046.5	0.0318	273	—	—
Pyrex**			0.006×10^{-3}	2	—	—
			0.09×10^{-3}	5		

Material	Formula	Melting Point	Value	Temp	Parallel	Perpend.	Temp
Selenium	Se	308 (amorphous)	1.0×10^{-3}	10	—	—	—
		490 (crystal)	6.5×10^{-3}	20	—	—	308
			0.068	85 to 291	3.1×10^{-3}	10^{-3}	
			0.072	276	2.6×10^{-3}	10^{-3}	
			0.077	293.5	0.4×10^{-3}	10^{-3}	
			0.085	302.5			
			0.127	305			
			0.131	311			
			0.95	291 to 311			
Silicon	Se(As)	~343					
Silicon Dioxide	Si	1693	0.177	298	3.3×10^{-4}		313
	SiO₂ (crystal)	2000	0.1657	273	0.39		314
			0.201	373	2.82×10^{-3}	10^{-3}	
					2.64×10^{-3}	10^{-3}	
					4.50×10^{-3}	10^{-3}	
Silicon Dioxide	SiO₂ (quartz)	>1743	0.188	285 to 373	Parallel / Perpend.		83
					0.117	0.586	195
					0.0467	0.0249	273
					0.0273	0.0163	323
					0.0224	0.0135	373
					0.0190	0.0118	423
					0.0168	0.0160	473
					0.0151	0.00967	523
					0.0136	0.00895	573
					0.0123	0.0084	623
					0.0113	0.0079	
Tin	Sn	504.85	0.0556	273			
Strontium Fluoride	SrF₂	1190.0					
Strontium Titanate	SrTiO₃	2353.0					
Tellurium	Te	722.8	0.0483	561 to 646	1.5×10^{-2}		
Teflon†			0.07×10^{-3}	2			
			0.57×10^{-3}	5			
			4.3×10^{-3}	10			
			18×10^{-3}	20			
			48×10^{-3}	50			
			92×10^{-3}	100			
			241×10^{-3}	300			
Titanium Dioxide	TiO₂	2093.0	0.17	293	3.0×10^{-2}		309 (parallel)
					3.3×10^{-2}		340 (parallel)
					2.1×10^{-2}		317 (perpend.)
					1.7×10^{-2}		340 (perpend.)
Thallium Bromide	TlBr	733.0	0.045	293	1.4×10^{-3}		316
Thallium Chloride	TlCl	703.0	0.0520	273	1.9×10^{-3}		311
Zinc	Zn	692.4	0.0939	273			

**Pyrex® is a registered trademark of Corning Glass Works.
†Teflon® is a registered trademark of the Dupont Corp.

OPTICAL MATERIALS

Table 7-9. Thermal Expansion [7-2]

$$\alpha = A \times 10^{-6} + B \times 10^{-8} \ T + C \times 10^{-11} \ T^2$$

Material	T (K)	A	B	C	Remarks
Al_2O_3	323	6.7	–	–	Parallel
	323	5.0	–	–	Perpendicular
As_2S_3	306-438	24.62	–	–	–
AgCl	298	30.01	–	–	–
	473	34.59	–	–	–
	623	52.09	–	–	–
	653	58.37	–	–	–
	673	63.19	–	–	–
	698	69.99	–	–	–
	293-333	30.00	–	–	–
ADP	297-407	39.3	–	–	–
		1.9	–	–	–
BaF_2	272-573	–	–	–	–
$BaTiO_3$	193-253	16.0	–	–	–
	283-343	19.0	–	–	–
	393-453	13.0	–	–	–
Borosilicate Crown Glass	295-771	9.0	–		–
$CaCO_3$	123-273	24.39	0.533	-30.7	Parallel
	123-273	-5.68	0.0333	-4.58	Perpendicular
	323	26.6	–	–	Parallel
	323	5.2	–	–	Parallel
	638	-3.8	–	–	Perpendicular
	348-673	24.71	3.775	-3.653	Parallel
CuCl	313-413	10.0	–	–	–
CaF_2	181-280	18.38	2.511	-21.10	–
	316-900	1.851	1.481	21.52	–
CsBr	293-323	47.9	–	–	–
	134-573	46.6	4.67	-1.78	–
CsI	298-323	50.0	–	–	–
CuBr	293-423	19.0	–	–	–
CdS	300-343	4.2	–	–	–
	323-773	3.5	–	–	Parallel
CdS Pressed	10	0	–	–	–
	50	-2.4	–	–	–
	110	0	–	–	–
	200	2.5	–	–	–
	300	4.2	–	–	–
CdTe	323	4.5	–	–	–
	873	5.9	–	–	–
GaAs	40	-0.5	–	–	–
	491	0.00	–	–	–
	78-290	3.64	–	–	–
	291-560	5.74	–	–	–
	560-680	7.44	–	–	–
CdF_2	293-393	27.0	–	–	–
GaSb	–	6.9	–	–	–
GaP	–	5.3	–	–	–
Ge	40	0.07	–	–	–

Table 7-9. Thermal Expansion [7-2] (*Continued*)

$$\alpha = A \times 10^{-6} + B \times 10^{-8} \, T + C \times 10^{-11} \, T^2$$

Material	T (K)	A	B	C	Remarks
Ge (*Continued*)	50	0.20	–	–	–
	60	0.39	–	–	–
	70	0.67	–	–	–
	80	1.05	–	–	–
	90	1.54	–	–	–
	100	2.20	–	–	–
	110	2.79	–	–	–
	120	3.25	–	–	–
	130	3.62	–	–	–
	140	3.91	–	–	–
	150	4.12	–	–	–
	160	4.29	–	–	–
	170	4.45	–	–	–
	180	4.58	–	–	–
	190	4.70	–	–	–
	200	4.82	–	–	–
	210	4.93	–	–	–
	220	5.03	–	–	–
	230	5.13	–	–	–
	240	5.23	–	–	–
	250	5.32	–	–	–
	260	5.42	–	–	–
	270	5.50	–	–	–
	280	5.59	–	–	–
	290	5.67	–	–	–
	300	5.75	–	–	–
InAs	–	5.3	–	–	–
InSb	10	–0.06	–	–	–
	30	–1.72	–	–	–
	50	–0.33	–	–	–
	70	0.89	–	–	–
	100	2.76	–	–	–
	160	4.08	–	–	–
	190	4.35	–	–	–
	220	4.58	–	–	–
	253	4.78	–	–	–
	270	4.89	–	–	–
	280	4.95	–	–	–
	300	5.04	–	–	–
InP	–	4.5	–	–	–
Irtran 1	298-573	11.0	–	–	–
Irtran 2	298-573	6.9	–	–	–
Irtran 3	298-573	20.0	–	–	–
Irtran 4	298-573	7.7	–	–	–
Irtran 5	298-573	12.0	–	–	–
Irtran 6*	298-573	5.9	–	–	–
KDP	123-293	21.6	–	–	–
KCl	293-333	36.0	–	–	–

*Irtran 6 is no longer manufactured by the Eastman Kodak Co. Irtran® is a registered trademark of the Eastman Kodak Co.

Table 7-9. Thermal Expansion [7-2] (*Continued*)

$$\alpha = A \times 10^{-6} + B \times 10^{-8} T + C \times 10^{-11} T^2$$

Material	T (K)	A	B	C	Remarks
KRS-5	223-293	61.0	–	–	–
	293-373	58.0	–	–	–
KRS-6	223	55.0	–	–	–
	233	56.0	–	–	–
	253	56.0	–	–	–
	273	55.0	–	–	–
	293	51.0	–	–	–
	313	48.0	–	–	–
	333	49.0	–	–	–
	353	51.0	–	–	–
	373	53.0	–	–	–
	393	56.0	–	–	–
	413	57.0	–	–	–
	433	58.0	–	–	–
	453	59.0	–	–	–
	473	59.0	–	–	–
KBr	113-573	27.6	4.1	–	–
	318-953	37.99	1.263	5.256	–
	293-333	43.0	–	–	–
KI	313	42.6	–	–	–
LiF	273-373	37.0	–	–	–
	123-273	31.95	5.049	–4.070	–
	320-1067	33.17	3.075	2.399	–
$MgOAl_2O_3$	313	5.9	–	–	–
MgO	323-988	10.98	–	–	–
	300	11.2	–	–	–
	481	12.3	–	–	–
	659	13.5	–	–	–
	825	14.6	–	–	–
	964	15.4	–	–	–
	1061	16.0	–	–	–
	293-1000	13.8	–	–	–
Muscovite	324	8.1	7.5	–	–
$NaNO_3$	323	12.0	–	–	–
	323	11.0	–	–	–
NaCl	223-473	44.0	–	–	–
NaF	Room temp.	36.0	–	–	–
PbTe	303	9.02	–	–	–
	313	12.08	–	–	–
	323	14.30	–	–	–
	333	15.57	–	–	–
	343	15.38	–	–	–
	353	16.42	–	–	–
	363	17.31	–	–	–
	373	17.70	–	–	–
	383	18.04	–	–	–
	393	18.33	–	–	–

Table 7-9. Thermal Expansion [7-2] (*Continued*)

$$\alpha = A \times 10^{-6} + B \times 10^{-8} \, T + C \times 10^{-11} \, T^2$$

Material	T (K)	A	B	C	Remarks
PbTe (*Continued*)	403	18.57	–	–	–
	413	18.78	–	–	–
	423	18.97	–	–	–
	433	19.42	–	–	–
	453	19.62	–	–	–
	473	19.74	–	–	–
	493	19.79	–	–	–
	513	19.80	–	–	–
	533	19.80	–	–	–
	553	19.80	–	–	–
	573	19.80	–	–	–
	593	19.80	–	–	–
	613	19.80	–	–	–
PbSe	303	7.65	–	–	–
	313	10.55	–	–	–
	323	12.92	–	–	–
	333	14.55	–	–	–
	343	15.63	–	–	–
	353	16.41	–	–	–
	363	16.97	–	–	–
	373	17.37	–	–	–
	383	17.66	–	–	–
	393	17.89	–	–	–
	403	18.09	–	–	–
	413	18.97	–	–	–
	423	18.43	–	–	–
	433	18.57	–	–	–
	433	18.57	–	–	–
	453	18.79	–	–	–
	473	18.94	–	–	–
	493	19.06	–	–	–
	513	19.16	–	–	–
	533	19.26	–	–	–
	553	19.34	–	–	–
	573	19.40	–	–	–
	593	19.46	–	–	–
	613	19.50	–	–	–
PbCl$_2$	293-393	31.0	–	–	–
Se	195-292	20.3	–	–	–
	195-273	42.7	–	–	–
	273-294	48.7	–	–	–
	293-373	22.9	–	–	–
	478	45.2	–	–	–
Si	323-373	2.5	–	–	<111>
	–	2.7	–	–	<110>
	373-473	3.1	–	–	<111>
	–	3.5	–	–	<110>
	473-573	3.9	–	–	<111>

Table 7-9. Thermal Expansion [7-2] (*Continued*)

$$\alpha = A \times 10^{-6} + B \times 10^{-8} \, T + C \times 10^{-11} \, T^2$$

Material	T (K)	A	B	C	Remarks
Si (*Continued*)	–	3.8	–	–	<110>
	673-773	4.3	–	–	<111>
	–	4.1	–	–	<110>
	773-873	4.7	–	–	<111>
	–	4.4	–	–	<110>
	873-973	5.0	–	–	<111>
	–	4.5	–	–	<110>
	298-1173	3.0024	0.1544	0.20576	<111>
SiO$_2$ (crystal)	173-310	7.067	2.11	–	Parallel
	283-607	–	–	–	Parallel
	273-633	7.067	1.6742	–	Parallel
	633-723	25.80	–	20.163	Parallel
	273-353	7.97	–	–	Parallel
	273-353	13.37	–	–	Perpendicular
SiO$_2$ (fused)	293-1173	0.5	–	–	–
SrTiO$_3$	–	9.4	–	–	–
Se(As)	–	34.0	–	–	–
Te	313	16.75	–	–	–
	293	–1.6	–	–	Parallel
	293	27.2	–	–	Perpendicular
	293-333	–1.7	–	–	Parallel
	293-333	27.0	–	–	Perpendicular
TlCl	293-333	53.0	–	–	–
TlBr	293-353	51.0	–	–	–
TiO$_2$	313	9.19	2.25	–	Parallel
	313	7.14	1.10	–	Perpendicular

7.3.7. Debye Temperatures.

Section 7.2.7 describes the application of the Debye temperature. Table 7-10 lists some and the methods used to obtain them. These are: c_{ij} = elastic constants; c_p = specific heat; *cal* = calorimetric; β = volume expansion; ρ = resistivity; T_m = melting temperature.

Table 7-10. Debye Temperature [7-3]

Material	θ_D (K)	Method
Ag	212, 220, 212, 220, 203	c_p, c_{ij}, cal, ρ
AgBr	140	c_p
AgCl	180	—
Al	385, 399, 396	c_p, c_{ij}, cal
As	275	—
Au	168-186, 180, 166, 186, 180, 175	c_p, c_{ij}, cal, β, ρ
B	1220	—
Be	940	—
Bi	120	—
BN	600	—
C diamond	2050, 1860, 1491, 2200	c_p, β, ρ
C graphite	760	—
Ca	230	—
CaF$_2$	470, 474	c_p, β
Cd hcp	280	—
Cd bcc	170	—
Cl	115	—
Cr	430	—
Cu	310-330, 310, 329, 313, 325, 333	cal, c_p, c_{ij}, cal, β, ρ
Ge	370, 360	—
Hg	100	—
In	140	—
InSb	208	c_{ij}
	200 ± 5	c_p
K	100	—
KBr	152-183, 180, 185, 171, 176, 162	cal, c_p
KCl	218-235, 230, 233, 229, 226, 203	—
KI	195, 162, 119	—
LiF	607-750, 680, 686, 1020, 845, 440	cal, c_p, k_T, T_m
Mg	330	—
MgO	800	—
MoS$_2$	290	—
Na	150	—
Na Cl	275-300, 280, 292, 294, 276, 235	cal, c_p, k_T, T_m
Ni	440, 435, 441	—
Pb	82-88, 85, 72, 86, 86	—
P+	225, 226, 220, 236	—
RbI	100-118, 115, 119, 109, 122, 122	—
Rh	350	—
Sb	140	—
SiO$_2$ (quartz)	255	—
Te	130	—
Ti	355	—
TiO$_2$ (rutile)	450	—
W	315	—
Zn	250	—
ZnS	260	—

7.3.8. Elastic Coefficients. Elastic coefficients can be thought of as the basis of the engineering moduli. They represent the stress-strain relationship along a particular direction in a crystal. Data are given in bar in Table 7-11.

Table 7-11. Elastic Coefficients [7-2]

Material	T (K)	c_{11} (Bar)	c_{12} (Bar)	c_{13} (Bar)	c_{33} (Bar)	c_{44} (Bar)
Al_2O_3	298	49.68	16.36	—	49.81	14.74
AgCl	—	6.01	3.62	—	—	0.625
ADP	—	617.00	0.72	1.94	3.28	0.85
BaF	—	9.01	4.03	—	—	2.49
$BaTiO_3$	298	8.18	2.98	1.95	6.76	18.30
$CaCO_3$	—	13.71	4.56	4.51	7.97	3.42
CaF_2	—	16.4 ± 0.1	5.3 ± 0.2	—	—	3.370 ± 0.01
CsBr	—	3.097	0.403	—	—	0.7500
CsI	—	2.46	0.67	—	—	0.624
CdS	—	8.432	5.212	4.638	9.397	1.489
CdTe	—	5.351	3.681	—	—	1.994
GaAs	—	1.192	0.5986	—	—	0.538
GaSb	—	8.849	4.037	—	—	4.325
GaP	—	14.7	—	—	—	—
Ge	—	1.29	4.83	—	—	6.71
InAs	—	8.329	4.526	—	—	3.959
InSb	300	6.472	3.625	—	—	3.071
InP	—	10.7	—	—	—	—
KDP	—	7.14	-0.49	1.29	5.62	1.27
KCl	—	3.98	0.62	—	—	—
KRS-5	—	3.31	1.32	—	—	0.597
KRS-6	—	3.85	1.49	—	—	0.737
KBr	—	3.45	0.54	—	—	0.508
KI	—	2.69	0.43	—	—	0.362
LiF	—	9.74	4.04	—	—	5.54
$MgAl_2O_3$	—	30.05	15.37	—	—	15.86
MgO	—	2.90	0.876	—	—	1.55
$NaNO_3$	—	8.67	1.63	1.60	3.74	2.13
NaCl	--	4.85	1.23	—	—	1.26
NaF	—	9.09	2.64	—	—	1.27
PbS	—	12.7	2.98	—	—	2.48
Si	—	1.67	0.65	—	—	0.80
SiO_2 (crystal)	—	8.675	0.687	1.13	10.68	5.786
$SrTiO_3$	—	31.56	10.27	—	—	12.15
Te	300	3.265	0.195	2.493	7.22	3.121
TlCl	—	4.01	1.53	—	—	0.760
TlBr	—	3.78	1.48	—	—	0.756
TiO_2	—	35.8	26.7	17.0	47.9	12.5
ZnS	—	9.45	5.70	—	—	4.36

7.3.9. Engineering Elastic Moduli. For practical use, the various elastic moduli listed in Table 7-12 are of value. They are described and defined in Section 7.2.8.

Table 7-12. Elastic Moduli* [7-2]

Material	Young's $(10^6$ psi)	Rigidity $(10^6$ psi)	Bulk $(10^6$ psi)	Rupture (psi)	Apparent Elastic Limit (psi)
Al_2O_3	50.00	21.50	0.30	–	–
As_2S_3	2.30	0.94	–	2.4×10^3	–
AgCl	0.02	1.03	6.39	–	3.8×10^3
	–	–	–	–	7.4×10^2
BaF_2	7.70	–	–	3.9×10^7	3.9×10^7
$BaTiO_3$	4.90	18.30	23.50	–	–
$CaCO_3$	10.50 parallel	–	18.80	–	–
	12.80 perpendicular	–	–	–	–
CaF_2	11.00	4.90	12.00	5.3×10^3	5.3×10^3
CsBr	2.30	–	–	23.9	12.2×10^2
CsI	0.769	–	–	–	8.1×10^2
CdTe	–	–	–	850.0	–
GaSb	9.19	6.28	8.19	–	–
Ge	14.90	9.73	11.30	–	–
InSb	6.21	4.45	6.28	–	–
Irtran 1	16.6×10^6 at 25°C	–	–	21,800.0 at 25°C	–
	16.6×10^6 at 500°C	–	–	10,000.0 at 500°C	–
Irtran 2	14.0×10^6 at 25°C	–	–	14,100.0 at 25°C	–
	10.6×10^6 at 250°C	–	–	13,500.0 at 250°C	–
Irtran 3	14.3×10^6 at 25°C	–	–	5,300.0 at 25°C	–
	14.0×10^6 at 500°C	–	–	9,000.0 at 500°C	–
Irtran 4	10.3×10^6 at 25°C	–	–	7,500.0 at 25°C	–
Irtran 5	48.2×10^6 at 25°C	–	–	19,200.0 at 25°C	–
	–	–	–	13,000.0 at 500°C	–
Irtran 6**	5.3×10^6 at 25°C	–	–	4,540.0 at 25°C	–
	4.5×10^6 at 100°C	–	–	5,880.0 at 100°C	–
KCl	4.30	0.906	2.52	6.4×10^2	3.3×10^2
KRS-5	2.30	0.840	2.87	1.81×10^4	3.8×10^4
KRS-6	3.00	1.230	3.31	–	3.05×10^3
KBr	3.90	0.737	2.18	4.8×10^2	1.6×10^2
KI	4.57	0.90	124.00	–	–
LiF	9.40	8.00	9.00	2.0×10^6	16.2×10^6
MgO	36.10	22.40	22.40	–	–
$NaNO_3$	–	–	3.80	–	–
NaCl	5.80	1.83	3.53	5.7×10^2	3.5×10^2
Si	1.9×10^7	1.16×10^7	1.48×10^7	–	–
SiO_2 (crystal)	11.10 perpendicular	5.28	–	–	–
	14.10 parallel	–	–	–	–
SiO_2 (quartz)	1.06×10^7	4.52	–	–	–
	$9.884 \pm 0.079 \times 10^7$	–	–	–	–
TlCl	4.60	1.10	3.42	–	–
TlBr	4.28	1.10	3.26	–	–

*To convert stresses and module from psi to dyne cm^{-2}, multiply the magnitude in psi by 6.90×10^4; to convert dyne cm^{-2} to psi, multiply the magnitude in dyne cm^{-2} by 1.45×10^{-5}.

**Irtran 6 is no longer manufactured by the Eastman Kodak Co. Irtran® is a registered trademark of the Eastman Kodak Co.

7.3.10. Hardness. Values of microhardness generally obtained by the Knoop test are tabulated in Table 7-13. The temperatures at which the measurement was taken as well as the orientation of the long direction of the diamond-shaped indenter and the load are given when available.

Table 7-13. Hardness of Optical Materials [7-2]

Material	Knoop $(kg\ mm^{-2})$	Temperature (K)	Load (g)
AgCl	9.5	—	200
AgTe	7.3	—	—
Al_2O_3	1370	—	1000
AlSb	400	—	—
As_2S_3	109	—	100
BaB_6	2900	—	120
BaF_2	82	—	500
C	8820	110	—
CaB_6	3150	—	120
CdS	55, 80	—	—
CdSe	90, 44, 66	—	—
CdTe	56	—	—
CeB_6	2350	—	120
Cr_2C	2160	—	120
Cr_2B_5	2150	—	120
CsBr	19.5	—	200
Cu	48, 17.5, 8	293, 773, 973	—
CuBr	21.2	—	—
CuTe	19.2	—	—
GaAs	721	—	—
GaSb	469	—	—
$GaSe_3$	316	—	—
$GaTe_3$	237	—	—
Ge	176, 83, 80, 24	873, 973, 1023, 1223	—
InAs	330	—	—
InSb	225	—	—
InP	430	—	—
In_2Te_3	180	—	—

Table 7-13. Hardness of Optical Materials [7-2] (*Continued*)

Material	Knoop (kg mm^{-2})	Temperature (K)	Load (g)
KBr	5.9, 7.0	—	200, 200
KCl	7.2, 9.3	—	200, 200
KRS-5	40.2, 39.8, 33.2	—	200, 500, 500
KRS-6	29.9, 38.5	—	500, 500
LaB$_6$	2500	—	120
LiF	102-113	—	600
MgO	692	—	600
MgO-3.5Al$_2$O$_3$	1140	—	1000
Mo$_2$B$_5$	2950	—	120
Mo$_2$C	1800	—	120
N$_6$B$_2$	2900	—	120
NaCl	15.2, 18.2	—	200, 200
NaNO$_3$	19.2	—	200
Si	1000, 500, 128	293, 773, 1273	—
SiO$_2$	461, 741	—	200, 500
SrTiO$_3$	595	—	—
TaB$_2$	2000	—	120
TaC	·1629	—	120
TaSi$_2$	1200	—	120
TiB$_2$	3400	—	120
TiC	~2600	—	120
TiN	2100	—	120
TiO$_2$	879	—	500
TlBr	11.9, 11.9	—	500, 500
TlCl	12.8, 12.8	—	500, 500
W$_2$B$_5$	2500	—	120
WC	1800	—	120
WSi$_2$	1430	—	120
ZnS	178	—	—
ZnSe	137	—	—
ZnTe	82	—	—
ZrB$_2$	1500	—	120
ZrC	2400	—	120
ZrN	930	—	120

7.3.11. Solubility, Molecular Weight, and Specific Gravity. Data are given in Table 7-14 of these properties.

Table 7-14. Solubility, Molecular Weight, and Specific Gravity [7-2]

Material	Solubility (g/100 g H_2O)	Molecular Weight	Specific Gravity
Al_2O_3	Insoluble	101.94	3.98 (3.95 to 4.10 for natural)
AgCl	8.9×10^{-5} at 283 K	143.34	5.589 at 273 K 5.56 at 293 K
As_2S_3	Insoluble	364.02	3.198
ADP	22.7 at 273 K	115.04	1.803 at 293 K
BaF_2	0.17	175.36	4.83 at 293 K
$BaTiO_3$	–	232.96	5.90 (single crystal)
$CaCO_3$	1.4×10^{-3} at 298 K 1.8×10^{-3} at 298 K	100.09	2.7102 at 293 K
CuCl	0.0062 at 293 K	99.00	3.53 at 293 K
CaF_2	0.0017 at 299 K Soluble in ammonia salt solutions	78.08	3.179 at 298 K
CdF_2	–	150.41	6.382 ± 0.006 at 293 K
CsBr	124.3 at 298 K Soluble in acid	212.83	4.44 at 293 K
CuBr	Insoluble	143.46	4.718 at 293 K
CsI	–	259.83	4.526
CdS	Insoluble	144.48	4.82 at 293 K
CdTe	Probably insoluble	240.02	5.854
$CaTiO_3$	–	135.98	4.10 at 293 K
GaAs	Insoluble	144.63	5.3161 ± 0.0002 at 298 K
GaSb	Insoluble	191.48	–
GaP	–	100.70	–
Ge	Insoluble in water; soluble in hot sulfuric acid and aqua regia; etched in CP-4	72.60	5.327 at 298 K
InAs	Insoluble	189.73	5.66
InSb	Insoluble	237.0	5.78
InP	–	145.80	4.8
<u>Irtran</u> 1	Insoluble	62.32	3.18
<u>Irtran</u> 2	Insoluble	97.45	4.09
<u>Irtran</u> 3	Insoluble	78.08	3.18
<u>Irtran</u> 4	Insoluble	144.34	5.27
<u>Irtran</u> 5	0.00062	40.32	3.58
<u>Irtran</u> 6*	Insoluble	240.02	5.85
KDP	–	136.09	2.338
KCl	34.7 at 293 K	74.55	1.984 at 293 K
KRS-5	0.05 at room temperature	–	7.371 at 289 K
KRS-6	0.32 at 293 K The solubility of a micro crystal is that of the more soluble component, in this case TICl.	–	7.192 at 289 K
KBr	53.48 at 273 K Slightly hygroscopic 102 at 373 K	119.01	2.75 at 298 K
KI	127.5 at 273 K	116.02	3.13
LiF	0.27 at 291 K	25.94	2.639 at 298 K
$MgOAl_2O_3$	Insoluble in water; not attacked by common acid NaOH; slightly etched by HF	356.74	3.61

*<u>Irtran</u> 6 is no longer manufactured by the Eastman Kodak Co. Irtran® is a registered trademark the Eastman Kodak Co.

Table 7-14. Solubility, Molecular Weight, and Specific Gravity [7-2] (*Continued*)

Material	Solubility (g/100 g H_2O)	Molecular Weight	Specific Gravity
MgO	Insoluble in water; soluble in acids and ammonia salts	40.32	3.567 at 298 K
Muscovite	Insoluble	–	2.8-2.9
$NaNO_3$	73 at 273 K 180 at 373 K	85.01	2.261
NaCl	35.7 at 273 K 39.12 at 373 K Soluble in glycerine; slightly soluble in alcohol and liquid ammonia; insoluble in hydrochloric acid	58.45	2.164 at 293 K
NaF	4.22 at 291 K	42.00	2.79 at 293 K 2.558 at 314 K
$PbMoO_4$	Insoluble	367.16	6.03/7.01 at 293 K
PbTe	Insoluble	334.82	8.16
PbSe	Insoluble	286.17	8.10 at 288 K
$PbCl_2$	0.673 at 273 K 0.99 at 283 K 3.34 at 373 K	278.12	5.85 at 293 K
PbS	Insoluble	239.28	7.5
PbF_2	0.064 at 293 K	245.21	8.24 at 293 K 7.763 ± .001 at 291 K
Se	Insoluble in water	–	4.82 4.26
SrF_2	0.011 at 273 K 0.012 at 27 K	125.63	4.24 at 293 K
SiO_2 (crystal)	Insoluble in water	60.06	2.648 at 298 K
SiO_2 (fused)	Insoluble in water; very slightly soluble in alkalis; soluble in hydrofluoric acid	60.06	2.202 at 293 K
$SrTiO_3$	Insoluble	183.53	5.122 at 293 K
SeAs	Insoluble	Not applicable	Not applicable
Te	Insoluble	–	6.24 at 293 K
TlCl	0.32 at 293 K	238.85	7.018 at 298 K
TlBr	0.05 at 298 K 0.25 at 341 K	284.31	7.453 at 298 K
TlO_2	Insoluble in water; soluble in acid	79.90	4.25 (4.18-5.13)

7.4. Detailed Values of Selected Properties

Refractive index and transmission play such important roles in the use of optical materials that their values are tabulated in more detail than the others. They are nevertheless still in summary form compared to data needed for final design. They usually represent measurements made on no more than a few samples and may be different from a specific sample.

7.4.1. Transmission Details. Figures 7-8 through 7-29 show the transmission (and reflection) of various materials grouped according to the generic type: glasses, plastics, refractory dielectric crystals, semiconducting crystals, dielectric crystals, and compacts.

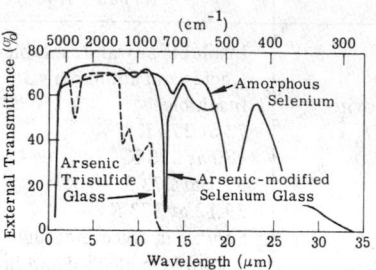

Fig. 7-8. The transmission of amorphous selenium, arsenic-modified selenium, and arsenic trisulfide. The properties of glass vary from batch to batch; these curves can be regarded as typical [7-4, 7-5, 7-6].

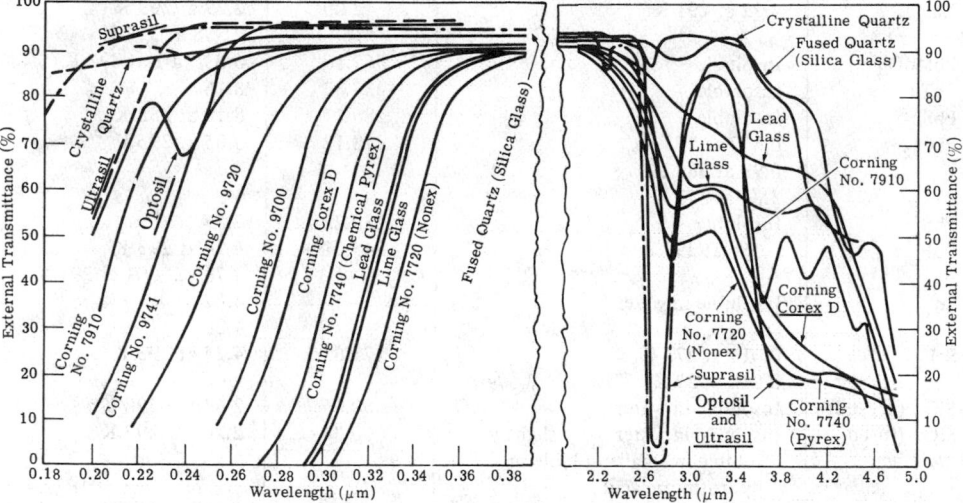

Fig. 7-9. The transmission of several samples of fused quartz and quartz glasses. Pyrex® and Corex® are registered trademarks of Corning Glass Works. Suprasil®, Optosil®, and Ultrasil® are registered trademarks of Heraeus-Amersil, Inc. [7-7].

Fig. 7-10. The transmission of several calcium aluminate glasses (2.0 mm thickness).

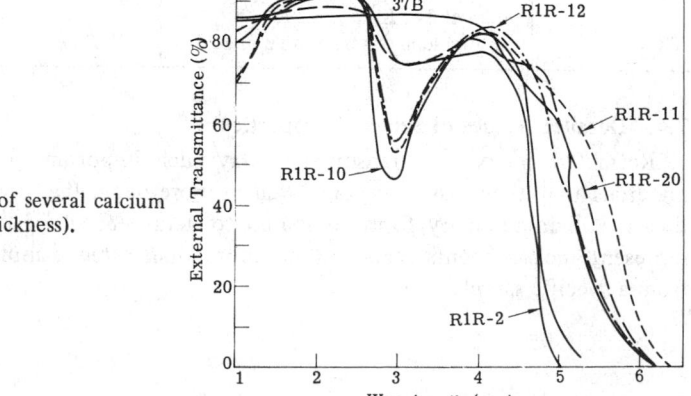

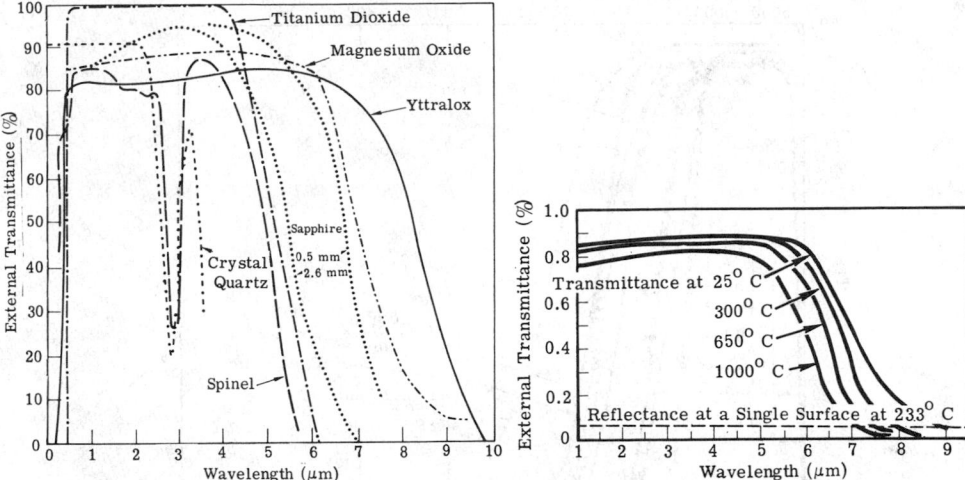

Fig. 7-11. The transmission of sapphire, spinel, titanium dioxide, crystal quartz (for the ordinary ray), yttralox, magnesium oxide, [7-8, 7-9, 7-10, 7-11, 7-12].

Fig. 7-12. The transmission of magnesium oxide at several temperatures (5.5 mm thickness) [7-12].

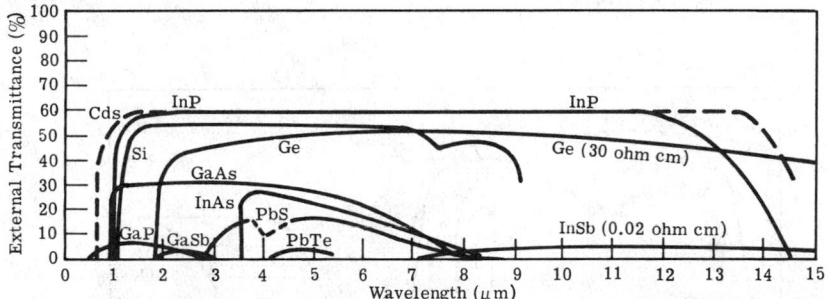

Fig. 7-13. The transmission of cadmium sulfide, indium phosphide, silicon, germanium, gallium arsenide, gallium phosphide, gallium antimonide, indium arsenide, indium antimonide, lead telluride, and lead sulfide.

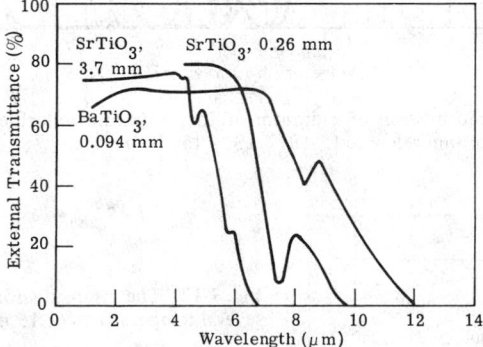

Fig. 7-14. The transmission of barium titanate and strontium titanate [7-13, 7-14].

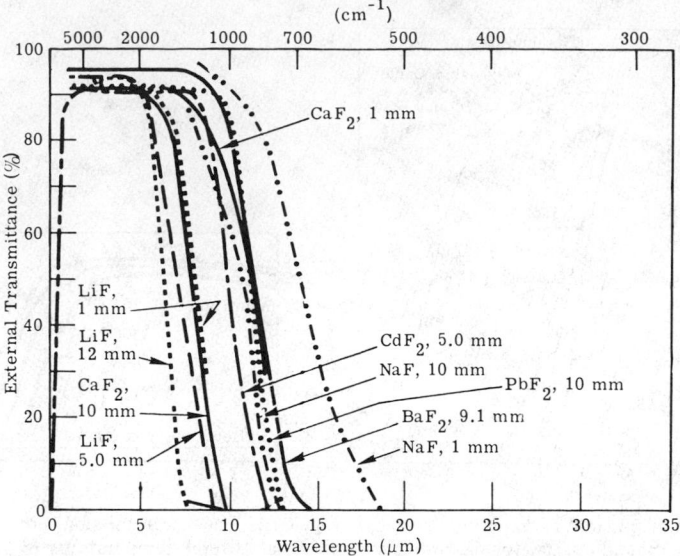

Fig. 7-15. The transmission of barium fluoride, cadmium fluoride, lithium fluoride, calcium fluoride, lead fluoride, and sodium fluoride [7-15, 7-16, 7-6, 7-17].

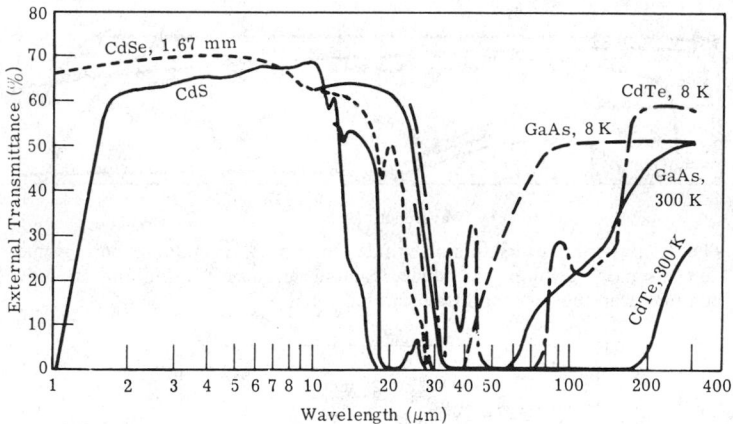

Fig. 7-16. The transmission of cadmium sulfide, cadmium telluride, gallium arsenide, and cadmium selenide [7-18, 7-19, 7-15, 7-20].

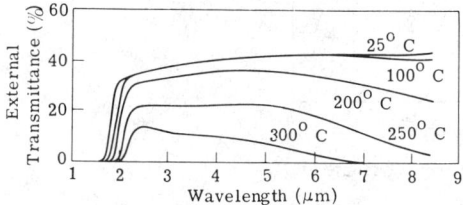

Fig. 7-17. The transmission of germanium for several temperatures (1.15 mm thickness) [7-15].

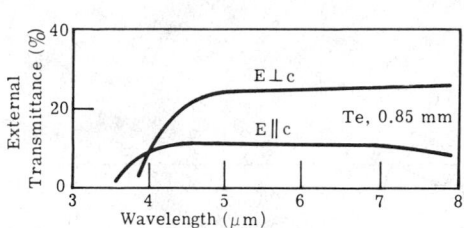

Fig. 7-18. The transmission of tellurium for two polarizations (0.85 mm thickness) [7-21].

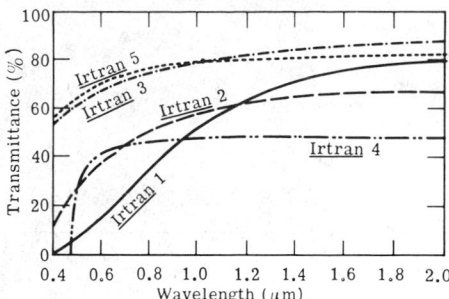

Fig. 7-19. The transmission of Irtran 1 through 5 (2.0 mm thickness). Irtran® is a registered trademark of the Eastman Kodak Co. [7-22].

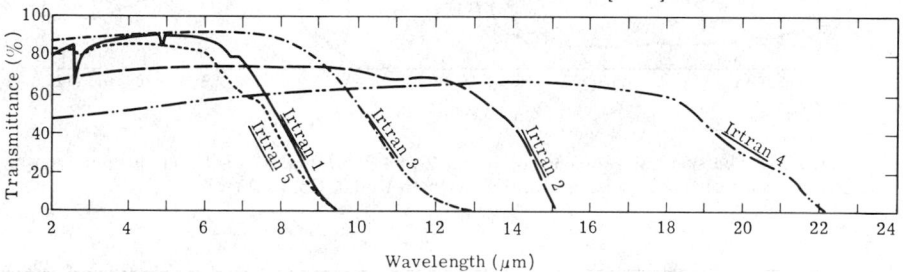

Fig. 7-20. The transmission of Irtran materials (2.0 mm thickness). Irtran® is a registered trademark of the Eastman Kodak Co. [7-22].

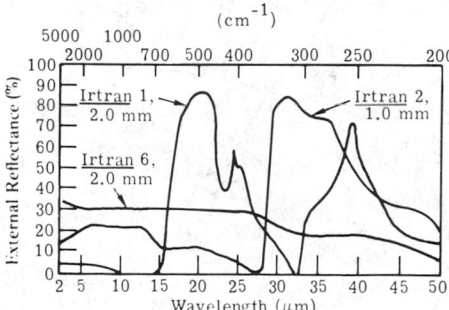

Fig. 7-21. The reflectance of Irtran 1, Irtran 2, and Irtran 6. Irtran® is a registered trademark of the Eastman Kodak Co. [7-6, 7-18].

Fig. 7-22. The spectral emissivity of Irtran 4 at 4.2, 77, and 195 K (0.37 mm thickness). Irtran® is a registered trademark of the Eastman Kodak Co. [7-23].

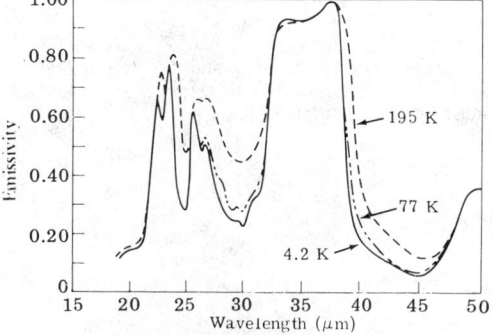

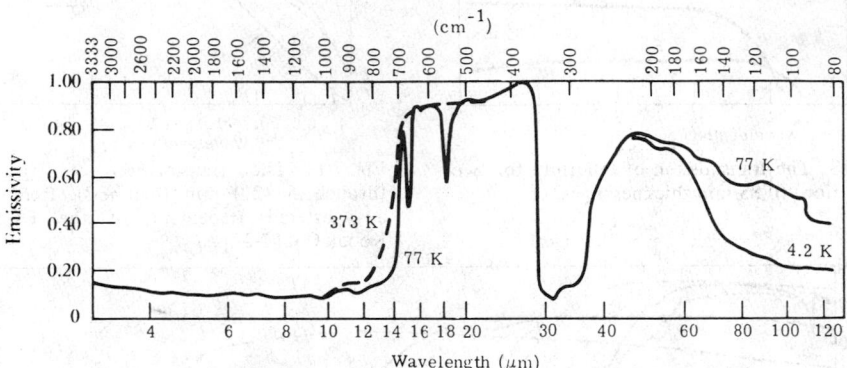

Fig. 7-23. The spectral emissivity of Irtran 2 at 4.2, 77, and 373 K (2.0 mm thickness). Irtran® is a registered trademark of the Eastman Kodak Co. [7-23].

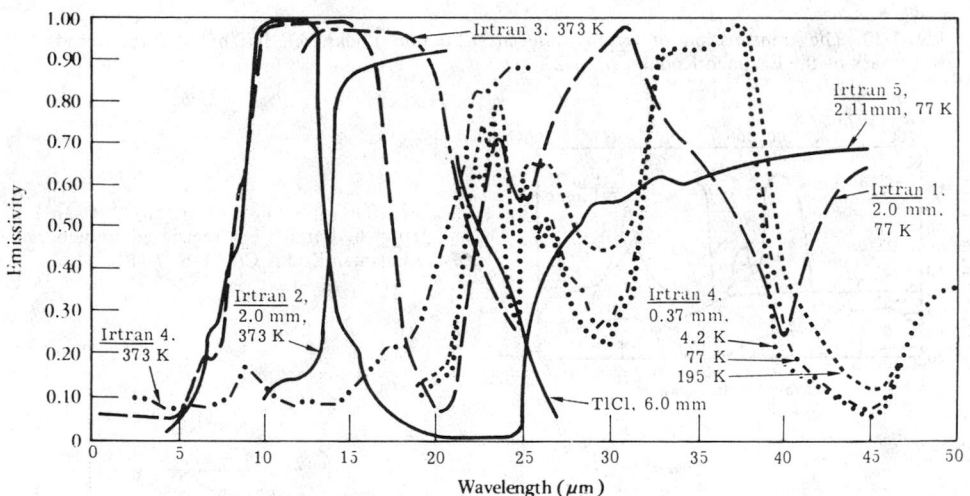

Fig. 7-24. The spectral emissivity of Irtran 1 at 77 K; Irtran 2 at 373 K; Irtran 3 at 373 K; Irtran 4 at 4.2, 77, 195, and 373 K; Irtran 5 at 77 K; and thallium chloride. Irtran® is a registered trademark of the Eastman Kodak Co. [7-23, 7-24].

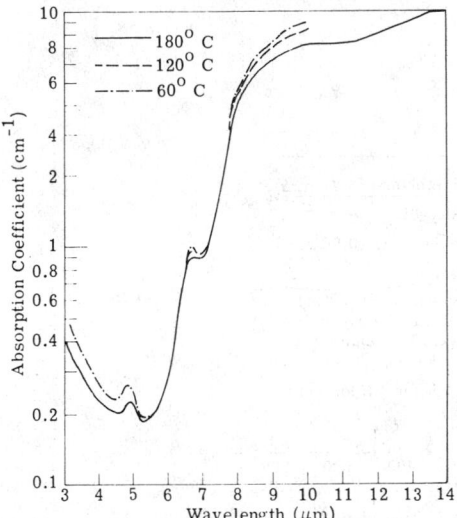

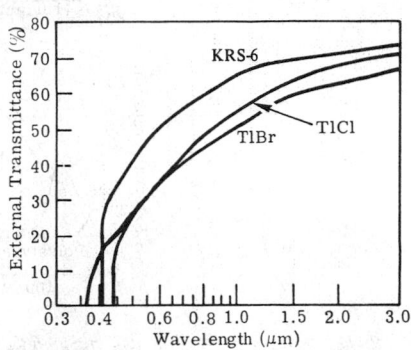

Fig. 7-25. The absorption coefficient of Irtran 1. Irtran® is a registered trademark of the Eastman Kodak Co. [7-24].

Fig. 7-26. The transmission of thallium bromide, thallium bromide-chlorine, and thallium chloride (1.65 mm all thickness). Thallium bromide can be ground a very small amount at a time without cracking or chipping. It bends like lead and is only slightly soluble in water [7-25].

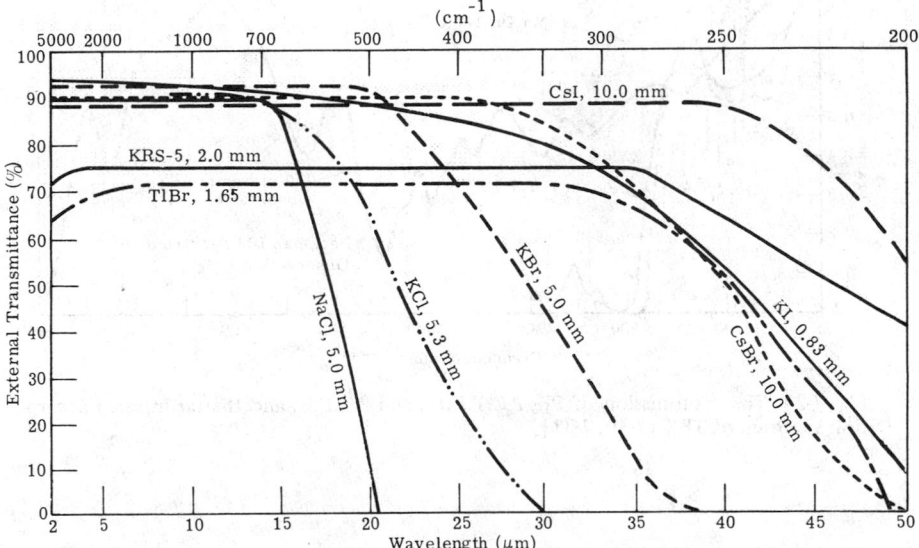

Fig. 7-27. The transmission of cesium iodide, potassium iodide, potassium bromide, thallium bromide, KRS-5, cesium bromide, sodium chloride, and potassium chloride [7-6, 7-26, 7-20, 7-27].

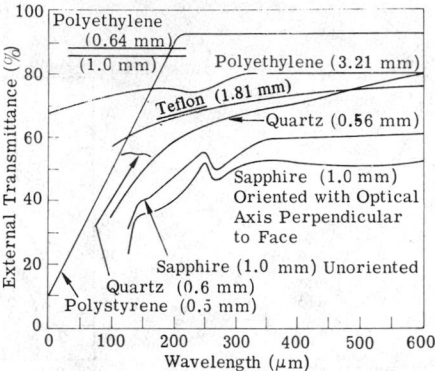

Fig. 7-28. The transmission of several long wave pass materials: quartz, sapphire, polystyrene, and polyethylene samples. Teflon® is a registered trademark of the Dupont Corp. [7-28, 7-29].

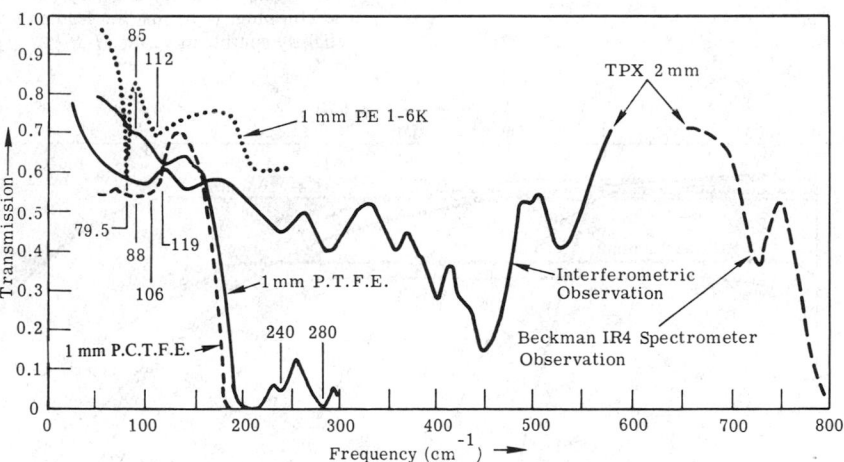

Fig. 7-29. The transmission of PE, P.C.T.F.E., and P.T.F.E. and the far infrared absorption spectrum of TPX [7-30, 7-31].

7.4.2. Refractive Index. Most data are given here in tabular form in Tables 7-15 through 7-52, with additional data given in Figures 7-30 through 7-37. The precision and accuracy change from author to author. Equations that are fit to the data are also given where they are available.

Table 7-15. Refractive Index of Arsenic Sulfur Glass at 25°C [7-32]

λ (μm)	n	λ (μm)	n	λ (μm)	n	λ (μm)	n
0.560	2.68689	0.960	2.48396	4.600	2.40878	8.600	2.39068
0.580	2.65934	0.980	2.48074	4.800	2.40802	8.800	2.38949
0.600	2.63646	1.000	2.47773	5.000	2.40725	9.000	2.38827
0.620	2.61708	1.200	2.45612	5.200	2.40649	9.200	2.38700
0.640	2.60043	1.400	2.44357	5.400	2.40571	9.400	2.38570
0.660	2.58594	1.600	2.43556	5.600	2.40493	9.600	2.38436
0.680	2.57323	1.800	2.43000	5.800	2.40414	9.800	2.38298
0.700	2.56198	2.000	2.42615	6.000	2.40333	10.000	2.38155
0.720	2.55195	2.200	2.42318	6.200	2.40250	10.200	2.38007
0.740	2.54297	2.400	2.42086	6.400	2.40166	10.400	2.37855
0.760	2.53488	2.600	2.41898	6.600	2.40979	10.600	2.37698
0.780	2.52756	2.800	2.41742	6.800	2.39991	10.800	2.37536
0.800	2.52090	3.000	2.41608	7.000	2.39899	11.000	2.37369
0.820	2.51483	3.200	2.41491	7.200	2.39306	11.200	2.37196
0.840	2.50928	3.400	2.41386	7.400	2.39709	11.400	3.37018
0.860	2.50418	3.600	2.41290	7.600	2.39610	11.600	2.36833
0.880	2.49949	3.800	2.41200	7.800	2.39503	11.800	2.36643
0.900	2.49515	4.000	2.41116	8.000	2.39403	12.000	2.36446
0.920	2.49114	4.200	2.41035	8.200	2.39294	–	–
0.940	2.48742	4.400	2.40956	8.400	2.39183	–	–

The dispersion equation for As_2S_3 is

$$n^2 - 1 = \sum_{i=1}^{i=5} \frac{K_i \lambda^2}{(\lambda^2 - \lambda_i^2)}$$

Table 7-16. The Constants for the Dispersion Equation for 25°C [7-32]

i	λ_i^2	K_i
1	0.0225	1.8983678
2	0.0625	1.9222979
3	0.1225	0.8765134
4	0.2025	0.1188704
5	750.0000	0.9569903

The properties of arsenic trisulfide glass vary with different batches. This difference is probably caused by various quantities of arsenic di- and pentasulfide as well as free sulfur.

Table 7-17. Refractive Index of Barium Fluoride at 25°C [7-33]

λ (μm)	n_{obs}	n_{cal}	Δn	λ (μm)	n_{obs}	n_{cal}	Δn
0.2652	1.51217	1.51216	+1	1.12866	1.46779	1.46778	+1
0.28035	1.50668	1.50669	−1	1.36728	1.46673	1.46671	+2
0.28936	1.50390	1.50390	0	1.52952	1.46613	1.46611	+2
0.296728	1.50186	1.50184	+2	1.681	1.46561	1.46561	0
0.30215	1.50044	1.50043	+1	1.7012	1.46554	1.46555	−1
0.3130	1.49782	1.49786	−4	1.97009	1.46472	1.46470	+2
0.32546	1.49521	1.49526	−5	2.1526	1.46410	1.46412	−2
0.334148	1.49363	1,49363	0	2.32542	1.46356	1.46356	0
0.340365	1.49257	1.49256	+1	2.5766	1.46262	1.46271	−9
0.31662	1.49158	1.49154	+4	2.6738	1.46234	1.46237	−3
0.361051	1.48939	1.48940	−1	3.2434	1.46018	1.46017	+1
0.366328	1.48869	1.48869	0	3.422	1.45940	1.45941	−1
0.404656	1.48438	1.48439	−1	5.138	1.45012	1.45014	−2
0.435835	1.48173	1.48174	−1	5.3034	1.44904	1.44905	−1
0.486133	0.47855	1.47856	−1	5.343	1.44878	1.44878	0
0.546074	1.47586	1.47586	0	5.549	1.44732	1.44736	−4
0.589262	1.47443	1.47443	0	6.238	1.44216	1.44217	−1
0.643847	1.47302	1.47301	+1	6.6331	1.43899	1.43890	+9
0.656279	1.47274	1.47273	+1	6.8559	1.43694	1.43696	−2
0.706519	1.47177	1.47176	+1	7.0442	1.43529	1.43526	+3
0.85211	1.46984	1.46981	+3	7.268	1.43314	1.43317	−3
0.89435	1.46942	1.46940	+2	9.724	1.40514	1.40511	+3
1.01398	1.46847	1.46846	+1	10.346	1.39636	1.39639	−3

The dispersion equation for barium fluoride is

$$n^2 - 1 = \sum \frac{(A_i \lambda^2)}{(\lambda^2 - \lambda_i^2)}$$

Table 7-18. The Constants of the Dispersion
Equation at 25°C [7-33]

$A_1 = 0.63356$	$\lambda_1^2 = 0.0033396$
$A_2 = 0.506762$	$\lambda_2^2 = 0.012030$
$A_3 = 3.8261$	$\lambda_3^2 = 2151.70$

Table 7-19. Temperature Coefficient of Refractive Index for
Barium Fluoride [7-33, 7-34]

λ (μm)	$-dn/dT$ $(10^{-6}\ {}^\circ C^{-1})$
0.4046563	15.05
0.4358342	15.00
0.4861327	15.15
0.5460740	15.20
0.589262	15.22
0.6562793	15.23
0.6678149	15.25
0.7065188	15.28
0.767858	15.45

Table 7-20. Refractive Index of Hexagonal Cadmium Sulfide [7-35]

λ (μm)	n_o		n_e		λ (μm)	n_o		n_e	
	Calc.	Obs.	Calc.	Obs.		Calc.	Obs.	Calc.	Obs.
0.5500	2.501	2.565	2.598	2.597	0.9000	2.349	2.353	2.358	2.358
0.5750	2.517	2.518	2.548	2.545	0.9500	2.341	2.340	2.350	2.352
0.6000	2.484	2.483	2.511	2.511	1.0000	2.336	2.338	2.343	2.341
0.6250	2.458	2.458	2.482	2.478	1.0500	2.330	2.332	2.338	2.338
0.6500	2.438	2.438	2.459	2.459	1.1000	2.326	2.325	2.334	2.333
0.6750	2.421	2.421	2.438	2.437	1.1500	2.321	2.322	2.328	2.330
0.7000	2.407	2.407	2.425	2.425	1.2000	2.319	2.317	2.324	2.322
0.7500	2.386	2.386	2.400	2.403	1.2500	2.317	2.316	2.321	2.320
0.8000	2.371	2.371	2.383	2.383	1.3000	2.314	2.315	2.319	2.319
0.8500	2.359	2.359	2.369	2.372	1.4000	2.310	2.311	2.314	2.314

The dispersion equation for cadmium sulfide is

$$n_o{}^2 = 5.235 + \frac{(1.819 \times 10^7)}{(\lambda^2 - 1.651 \times 10^7)}$$

$$n_e{}^2 = 5.239 + \frac{(2.076 \times 10^7)}{(\lambda^2 - 1.651 \times 10^7)}$$

Table 7-21. Refractive Index of Calcium Carbonate, Calcite*
[7-36, 7-37, 7-38, 7-39]

λ (μm)	n_o	n_e	λ (μm)	n_o	n_e	λ (μm)	n_o	n_e
0.198	–	1.57796	0.410	1.68014	1.49640	1.229	1.63926	1.47870
0.200	1.90284	1.57649	0.434	1.67552	1.49430	1.273	1.63849	–
0.204	1.88242	1.57081	0.441	1.67423	1.49373	1.307	1.63789	1.47831
0.208	1.80733	1.56640	0.508	1.66527	1.48956	1.320	1.63767	–
0.211	1.85692	1.56327	0.533	1.66277	1.48841	1.369	1.63681	–
0.214	1.84558	1.55976	0.560	1.66046	1.48736	1.396	1.63637	1.47789
0.219	1.83075	1.55496	0.589	1.65835	1.48640	1.422	1.63590	–
0.226	1.81309	1.54921	0.643	1.65504	1.48490	1.479	1.63490	–
0.231	1.80233	1.54541	0.656	1.65437	1.48459	1.497	1.63457	1.47744
0.242	1.78111	1.53782	0.670	1.65367	1.48426	1.541	1.63381	–
0.257	1.76038	1.53005	0.706	1.65207	1.48353	1.609	1.63261	–
0.263	1.75343	1.52736	0.768	1.64974	1.48259	1.615	–	1.47695
0.267	1.74864	1.52547	0.795	1.64886	1.48216	1.682	1.63127	–
0.274	1.74139	1.52261	0.801	1.64869	1.48216	1.749	–	1.47638
0.291	1.72774	1.51705	0.833	1.64772	1.48176	1.761	1.62974	–
0.303	1.71959	1.51365	0.867	1.64676	1.48137	1.849	1.62800	–
0.312	1.71425	1.51140	0.905	1.64578	1.48098	1.900	–	1.47573
0.330	1.70515	1.50746	0.946	1.64480	1.48060	1.946	1.62602	–
0.340	1.70078	1.50562	0.991	1.64380	1.48022	2.053	1.62372	–
0.346	1.69833	1.50450	1.042	1.64276	1.47985	2.100	–	1.47492
0.361	1.69317	1.50228	1.097	1.64167	1.47948	2.172	1.62099	–
0.394	1.68374	1.49810	1.159	1.64051	1.47910	3.324	–	1.47392

*Data for the ordinary ray at about 0.21 μm appear a little out of line.

Table 7-22. Temperature Coefficients of Refractive Index of
Calcium Carbonate, Calcite [7-40]

λ (μm)	dn/dT $(10^{-6}\ °C^{-1})$	dn/dT $(10^{-9}\ °C^{-1})$
0.211	2.150	–
0.231	1.397	2.198
0.298	0.604	1.641
0.361	0.360	1.449
0.441	0.325	1.318
0.467	0.319	–
0.480	0.305	1.287
0.508	0.287	1.234
0.589	0.240	1.213
0.643	0.208	1.185

Table 7-23. Refractive Index of Calcium Fluoride, Fluorite, and
Temperature Coefficient [7-41]

λ (μm)	n_{cal}	Synthetic Δn	Natural Δn	$-dn/dT$ $(10^{-6} \times °C^{-1})$
0.228803	1.47635	−2	+1	6.2
0.24827	1.46793	+3	+5	7.0
0.2537	1.46602	+9	+12	7.5
0.26520	1.46233	−1	0	8.1
0.28035	1.45828	−1	+1	8.4
0.296728	1.45167	−2	0	8.8
0.331148	1.44852	−1	+2	9.2
0.34662	1.44691	−3	0	9.4
0.365015	1.44190	−4	0	9.6
0.4016563	1.44151	−3	+1	9.8
0.4358312	1.43949	−3	0	10.0
0.4861327	1.43703	−4	0	10.2
0.546074	1.43494	−3	+1	10.4
0.589262	1.43381	−2	+2	10.4
0.643847	1.43268	−2	+3	10.4
0.6562793	1.43216	−2	+1	10.4
0.6678149	1.43226	−1	+1	10.5
0.7065188	1.43167	−2	+2	10.5
0.767858	1.43088	−2	+2	10.6
0.85212	1.43002	−1	+4	10.6

Table 7-23. Refractive Index of Calcium Fluoride, Fluorite, and
Temperature Coefficient, in $°C^{-1} \times 10^{-6}$ [7-41] (*Continued*)

λ (μm)	n_{cal}	Synthetic Δn	Natural Δn	$-dn/dT$
0.8944	1.42966	0	+1	10.6
1.91398	1.42879	-2	0	10.5
1.3622	1.42691	+1	+8	10.0
1.39506	1.42675	+1	+6	9.9
1.52952	1.42612	+4	+4	9.6
1.7012	1.42531	+2	+4	9.4
1.81307	1.42478	0	+9	9.1
1.97009	1.42401	+3	+3	8.9
2.1426	1.42306	-1	+1	8.7
2.32512	1.42212	+3	+4	8.5
2.4374	1.42147	0	+2	8.5
3.3026	1.41561	0	+3	8.2
3.422	1.41467	+2	+2	8.1
3.5070	1.41398	-1	+2	8.0
3.7067	1.41229	+2	+2	7.8
4.258	1.40713	+4	+4	7.5
5.01882	1.39873	+1	+5	7.3
5.3031	1.39320	+3	+3	7.2
6.0140	1.38539	+5	+5	7.0
6.238	1.38200	-6	0	7.0
6.63306	1.37565	0	+1	6.9
6.8559	1.37186	-8	+2	6.7
7.268	1.36113	+2	+7	6.5
7.4611	1.36070	+5	+6	6.4
8.662	1.33500	-4	+3	6.0
9.724	1.30756	+1	+5	5.6

The dispersion equation for calcium fluoride is

$$n^2 - 1 = \frac{A_i \lambda^2}{(\lambda^2 - \lambda_i^2)}$$

Table 7-24. The Constants of the Dispersion Equation at 25°C [7-41]

$\lambda_1 = 0.050263605$	$\lambda_1^2 = 0.002526430$	$A_1 = 0.5675888$
$\lambda_2 = 0.1003909$	$\lambda_2^2 = 0.01007833$	$A_2 = 0.4710914$
$\lambda_3 = 34.649040$	$\lambda_3^2 = 1200.5560$	$A_3 = 3.8484723$

Table 7-25. Refractive Index of Calcium Fluoride [7-42, 7-43]

λ (μm)	n	λ (μm)	n	λ (μm)	n	λ (μm)	n
0.19	1.50500	1.0140	1.42881	2.1608	1.42306	4.000	1.40963
0.20	1.49531	1.08304	1.42843	2.250	1.42258	4.1252.	1.40847
0.22	1.48119	1.1000	1.42834	2.3573	1.42198	4.2500	1.40722
0.24	1.47133	1.1786	1.42789	2.450	1.42143	4.4000	1.40568
0.26	1.46397	1.250	1.42752	2.5537	1.42080	4.6000	1.40357
0.28	1.45841	1.3756	1.42689	2.6519	1.42018	4.7146	1.40233
0.30	1.45400	1.4733	1.42642	2.700	1.41988	4.8000	1.40130
0.35	1.44658	1.5715	1.42596	2.750	1.41956	5.0000	1.39908
0.40	1.441857	1.650	1.42558	2.800	1.41923	5.3036	1.39522
0.48615	1.43704	1.7080	1.42502	2.880	1.41890	5.8932	1.38712
0.58758	1.43388	1.8400	1.42468	2.9466	1.41823	6.4825	1.37824
0.58932	1.43384	1.8688	1.42454	3.0500	1.41750	7.0718	1.36805
0.65630	1.43249	1.900	1.42439	3.0980	1.41714	7.6612	1.35675
0.68671	1.43200	1.9153	1.42431	3.2413	1.41610	8.2505	1.34440
0.72818	1.43143	1.9644	1.42407	3.4000	1.41487	8.8398	1.33075
0.76653	1.43093	2.0582	1.42360	3.5359	1.41367	9.4291	1.31605
0.88400	1.42980	2.0626	1.42357	3.8206	1.41119	–	–

Table 7-26. Temperature Coefficient of Refractive Index for
Calcium Fluoride [7-44]

λ (μm)	dn/dT (10^{-5} $°C^{-1}$)	T ($°C$)	λ (μm)	dn/dT (10^{-5} $°C^{-1}$)	T ($°C$)
0.185	-0.296	61.25	0.325	-0.948	61.25
0.186	-0.313	61.25	0.340	-0.964	61.25
0.193	-0.402	61.25	0.361	-0.979	61.25
0.197	-0.451	61.25	0.441	-1.028	61.25
0.198	-0.464	61.25	0.480	-1.035	61.25
0.200	-0.493	61.25	0.508	-1.056	61.25
0.204	-0.538	61.25	0.589	-1.111	60.5
0.208	-0.582	61.25	0.640	-1.113	59.2
0.211	-0.601	61.25	0.900	-1.031	59.9
0.214	-0.637	61.25	1.2	-1.040	59.9
0.219	-0.655	61.25	1.25	-1.029	60.0
0.224	-0.696	61.25	1.30	-1.018	60.2
0.231	-0.732	61.25	2.0	-0.932	60.2
0.257	-0.811	61.25	3.16	-0.881	59.4
0.274	-0.855	61.25	4.2	-0.831	59.6
0.288	-0.884	61.25	5.3	-0.821	59.0
0.298	-0.904	61.25	6.5	-0.787	56.8

Table 7-27. Refractivity of Cesium Bromide at 27°C [7-45]

λ (μm)	$(n-1) \times 10^5$									
	0.0	0.1	0.2	0.3	0.4	0.5	0.6	0.7	0.8	0.9
0	–	–	–	–	73,519	70,896	69,583	68,825	68,345	68,022
1	67,793	67,624	67,496	67,397	67,318	67,254	67,201	67,157	67,120	67,088
2	67,061	67,036	67,015	66,996	66,979	66,963	66,948	66,935	66,923	66,911
3	66,904	66,890	66,881	66,871	66,862	66,853	66,845	66,837	66,829	66,821
4	66,813	66,805	66,798	66,790	66,782	66,775	66,767	66,760	66,752	66,745
5	66,737	66,730	66,722	66,715	66,707	66,699	66,691	66,683	66,675	66,667
6	66,659	66,651	66,643	66,634	66,626	66,617	66,609	66,600	66,591	66,582
7	66,573	66,564	66,555	66,545	66,536	66,526	66,517	66,507	66,497	66,487
8	66,477	66,467	66,457	66,446	66,436	66,425	66,414	66,403	66,392	66,381
9	66,370	66,359	66,347	66,335	66,324	66,312	66,300	66,288	66,276	66,263
10	66,251	66,238	66,226	66,213	66,200	66,187	66,174	66,160	66,147	66,134
11	66,120	66,106	66,092	66,078	66,064	66,050	66,035	66,021	66,006	65,991
12	65,976	65,961	65,946	65,931	65,915	65,900	65,884	65,868	65,852	65,836
13	65,820	65,804	65,787	65,770	65,754	65,737	65,720	65,703	65,685	65,668
14	65,651	65,633	65,615	65,597	65,579	65,561	65,543	65,524	65,505	65,487

Table 7-27. Refractivity of Cesium Bromide at 27°C [7-45] (Continued)

λ (μm)	$(n-1) \times 10^5$									
	0.0	0.1	0.2	0.3	0.4	0.5	0.6	0.7	0.8	0.9
15	65,468	65,449	65,430	65,411	65,391	65,372	65,352	65,332	65,312	65,292
16	65,272	65,251	65,231	65,210	65,190	65,169	65,148	65,126	65,105	65,084
17	65,062	65,040	65,018	64,996	64,974	64,952	64,929	64,907	64,884	64,861
18	64,838	64,815	64,792	64,768	64,745	64,721	64,697	64,673	64,649	64,625
19	64,600	64,576	64,551	64,526	64,501	64,476	64,450	64,425	64,399	64,374
20	64,348	64,322	64,295	64,269	64,243	64,216	64,189	64,162	64,135	64,108
21	64,080	64,053	64,025	63,997	63,969	63,941	63,913	63,884	63,856	63,827
22	63,798	63,769	63,739	63,710	63,681	63,651	63,621	63,591	63,561	63,530
23	63,500	63,469	63,438	63,407	63,376	63,345	63,313	63,282	63,250	63,218
24	63,186	63,154	63,121	63,089	63,056	63,023	62,990	62,957	62,923	62,890
25	62,856	62,822	62,788	62,754	62,719	62,685	62,650	62,615	62,580	62,545
26	62,509	62,474	62,438	62,402	62,366	62,330	62,293	62,256	62,220	62,183
27	62,146	62,108	62,071	62,033	61,995	61,957	61,919	61,881	61,842	61,803
28	61,764	61,725	61,686	61,646	61,607	61,567	61,527	61,487	61,446	61,406
29	61,365	61,324	61,283	61,242	61,200	61,158	61,116	61,074	61,032	60,990
30	60,947	60,904	60,861	60,818	60,775	60,731	60,687	60,643	60,599	60,555
31	60,510	60,465	60,420	60,375	60,330	60,284	60,238	60,192	60,146	60,100
32	60,053	60,007	59,960	59,912	59,865	59,817	59,770	59,722	59,673	59,625
33	59,576	59,527	59,478	59,429	59,380	59,330	59,280	59,230	59,179	59,129
34	59,078	59,027	58,976	58,924	58,873	58,821	58,769	58,717	58,664	58,611
35	58,558	58,505	58,452	58,398	58,344	58,290	58,236	58,181	58,126	58,071
36	58,016	57,960	57,905	57,849	57,792	57,736	57,679	57,622	57,565	57,508
37	57,450	57,392	57,334	57,276	57,217	57,158	57,099	57,040	56,980	56,920
38	56,860	56,800	56,739	56,678	56,617	56,556	56,494	56,432	56,370	56,308
39	56,245	56,182	56,119	–	–	–	–	–	–	–

The dispersion equation for cesium bromide is

$$n^2 = 5.640752 - 0.000003338\lambda^2 + (0.0018612/\lambda^2)$$
$$+ (41,110.49/\lambda^2 - 14,390.4) + (0.0290764/\lambda^2 - 0.024964)$$

The average temperature coefficient of refractive index for two samples of different origin is given by Rodney and Spindler as 7.9×10^{-5} per °C.

Table 7-28. Refractivity of Cesium Iodide at 24°C [7-46]

λ (μm)	$(n-1) \times 10^5$	λ (μm)	$(n-1) \times 10^5$	λ (μm)	$(n-1) \times 10^5$	λ (μm)	$(n-1) \times 10^5$	λ (μm)	$(n-1) \times 10^5$
0.280	103 939	0.840	76 352	1.80	74 702	10.8	73 852	16.4	73 267
0.300	97 872	0.860	76 252	1.84	74 683	11.0	73 835	16.6	73 242
0.320	93 700	0.880	76 159	1.88	74 664	11.2	73 818	16.8	73 216
0.340	90 649	0.900	76 074	1.92	74 647	11.4	73 800	17.0	73 190
0.360	88 324	0.920	75 993	1.96	74 631	11.6	73 783	17.2	73 164
0.380	86 497	0.940	75 918	2.00	74 616	11.8	73 765	17.4	73 137
0.400	85 027	0.960	75 848	2.20	74 551	12.0	73 746	17.6	73 111
0.420	83 823	0.980	75 782	2.40	74 500	12.2	73 728	17.8	73 083
0.440	82 820	1.00	75 721	2.60	74 460	12.4	73 709	18.0	73 056
0.460	81 975	1.04	75 608	2.80	74 427	12.6	73 690	18.2	73 028
0.480	81 255	1.08	75 508	3.00	74 400	12.8	73 670	18.4	72 999
0.500	80 635	1.12	75 419	3.50	74 346	13.0	73 650	18.6	72 971
0.520	80 097	1.16	75 339	4.00	74 305	13.2	73 630	18.8	72 942
0.540	79 626	1.20	75 268	4.50	74 270	13.4	73 610	19.0	72 913
0.560	79 213	1.24	75 203	5.00	74 239	13.6	73 589	19.2	72 883
0.580	78 846	1.28	75 144	5.50	74 210	13.8	73 568	19.4	72 853
0.600	78 520	1.32	75 091	6.00	74 181	14.0	73 547	19.6	72 823
0.620	78 229	1.36	75 042	6.50	74 152	14.2	73 525	19.8	72 793
0.640	77 967	1.40	74 997	7.00	74 122	14.4	73 504	20.0	72 762
0.660	77 731	1.44	74 956	7.50	74 091	14.6	73 481	20.2	72 731
0.680	77 517	1.48	74 919	8.00	74 059	14.8	73 459	20.4	72 699
0.700	77 323	1.52	74 884	8.50	74 026	15.0	73 436	20.6	72 667
0.720	77 146	1.56	74 852	9.00	73 991	15.2	73 413	20.8	72 635
0.740	76 985	1.60	74 822	9.50	73 954	15.4	73 389	21.0	72 602
0.760	76 836	1.64	74 795	10.0	73 916	15.6	73 366	21.2	72 570
0.780	76 700	1.68	74 769	10.2	73 901	15.8	73 342	21.4	72 536
0.800	76 575	1.72	74 745	10.4	73 885	16.0	73 317	21.6	72 503
0.820	76 459	1.76	74 723	10.6	73 868	16.2	73 292	21.8	72 469
—	—	—	—	—	—	—	—	—	—

Table 7-28. Refractivity of Cesium Iodide at 24°C [7-46] *(Continued)*

λ (μm)	$(n-1) \times 10^5$	λ (μm)	$(n-1) \times 10^5$	λ (μm)	$(n-1) \times 10^5$	λ (μm)	$(n-1) \times 10^5$	λ (μm)	$(n-1) \times 10^5$
22.0	72 435	27.6	71 334	33.2	69 941	38.8	68 227	44.4	66 151
22.2	72 400	27.8	71 289	33.4	69 886	39.0	68 159	44.6	66 070
22.4	72 365	28.0	71 244	33.6	69 830	39.2	68 091	44.8	65 988
22.6	72 330	28.2	71 199	33.8	69 774	39.4	68 023	45.0	65 905
22.8	72 294	28.4	71 153	34.0	69 717	39.6	67 954	45.2	65 822
23.0	72 258	28.6	71 107	34.2	69 660	39.8	67 884	45.4	65 739
23.2	72 222	28.8	71 061	34.4	69 602	40.0	67 814	45.6	65 655
23.4	72 185	29.0	71 014	34.6	69 544	40.2	67 744	45.8	65 570
23.6	72 148	29.2	70 967	34.8	69 486	40.4	67 673	46.0	65 485
23.8	72 111	29.4	70 919	35.0	69 427	40.6	67 601	46.2	65 399
24.0	72 073	29.6	70 871	35.2	69 368	40.8	67 530	46.4	65 313
24.2	72 035	29.8	70 823	35.4	69 308	41.0	67 457	46.6	65 226
24.4	71 997	30.0	70 774	35.6	69 248	41.2	67 384	46.8	65 138
24.6	71 958	30.2	70 725	35.8	69 188	41.4	67 311	47.0	65 051
24.8	71 919	30.4	70 676	36.0	69 127	41.6	67 207	47.2	64 962
25.0	71 880	30.6	70 626	36.2	69 065	41.8	67 163	47.4	64 873
25.2	71 840	30.8	70 576	36.4	69 004	42.0	67 088	47.6	64 783
25.4	71 800	31.0	70 525	36.6	68 941	42.2	67 013	47.8	64 693
25.6	71 759	31.2	70 474	36.8	68 879	42.4	66 937	48.0	64 602
25.8	71 718	31.4	70 422	37.0	68 815	42.6	66 861	48.2	64 511
26.0	71 677	31.6	70 371	37.2	68 752	42.8	66 784	48.4	64 419
26.2	71 635	31.8	70 318	37.4	68 688	43.0	66 707	48.6	64 326
26.4	71 593	32.0	70 266	37.6	68 623	43.2	66 629	48.8	64 233
26.6	71 551	32.2	70 213	37.8	68 558	43.4	66 551	49.0	64 139
26.8	71 508	32.4	70 159	38.0	68 493	43.6	66 472	49.2	64 045
27.0	71 465	32.6	70 105	38.2	68 427	43.8	66 392	49.4	63 950
27.2	71 422	32.8	70 051	38.4	68 361	44.0	66 312	49.6	63 855
27.4	71 378	33.0	69 996	38.6	68 294	44.2	66 232	49.8	63 759
—	—	—	—	—	—	—	—	50.0	63 662

The dispersion equation for cesium iodide is

$$n^2 - 1 = \sum_i \frac{(K_i \lambda^2)}{(\lambda^2 - \lambda_i^2)}$$

Table 7-29. The Constants of the Dispersion
Equation for Cesium Iodide [7-46]

i	λ_i^2	K_i
1	0.00052701	0.34617251
2	0.02149156	1.0080886
3	0.032761	0.28551800
4	0.044944	0.39743178
5	25,921.0	3.3605359

Table 7-30. Refractive Index of Crystal Quartz*
[7-36, 7-37, 7-47, 7-48, 7-49, 7-50]

λ (μm)	n_o	n_e	λ (μm)	n_o	n_e
0.185	1.65751	1.68988	1.5414	1.52781	1.53630
0.198	1.65087	1.66394	1.6815	1.52583	1.53422
0.231	1.61395	1.62555	1.7614	1.52468	1.53301
0.340	1.56747	1.57737	1.9457	1.52184	1.53004
0.394	1.55846	1.56805	2.0531	1.52005	1.52823
0.434	1.55396	1.56339	2.30	1.51561	—
0.508	1.54822	1.55746	2.60	1.50986	—
0.5893	1.54424	1.55335	3.00	1.49953	—
0.768	1.53903	1.54794	3.50	1.48451	—
0.8325	1.53773	1.54661	4.00	1.46617	—
0.9914	1.53514	1.54392	4.20	1.4569	—
1.1592	1.53283	1.54152	5.00	1.417	—
1.3070	1.53090	1.53951	6.45	1.274	—
1.3958	1.52977	1.53832	7.0	1.167	—
1.4792	1.52865	1.53716	—	—	—

*The data are questionable at the extreme ends of the range (0.185, 5.00, 6.45 and 7.0 μm).

Table 7-31. Temperature Coefficients of Refractive Index for
Crystal Quartz [7-40]

λ (μm)	dn_o/dT (10^{-5} °C^{-1})	$dn_e dT$ (10^{-5} °C^{-1})	λ (μm)	dn_o/dT (10^{-5} °C^{-1})	$dn_e dT$ (10^{-5} °C^{-1})
0.202	+0.321	+0.267	0.298	-0.311	-0.415
0.206	0.253	0.198	0.313	-0.348	-0.450
0.210	0.193	0.143	0.325	-0.352	-0.469
0.214	0.124	0.083	0.340	-0.393	-0.501
0.219	0.074	0.027	0.361	-0.418	-0.521
0.224	0.017	-0.048	0.441	-0.475	-0.593
0.226	-0.008	-0.075	0.467	-0.485	-0.601
0.228	-0.027	-0.093	0.480	-0.499	-0.610
0.231	-0.052	-0.112	0.508	-0.514	-0.616
0.257	-0.186	-0.265	0.589	-0.539	-0.642
0.274	-0.235	-0.323	0.643	-0.549	-0.653
0.288	-0.279	-0.385	—	—	—

Table 7-32. Refractive Index
of Diamond [7-51]

λ (μm)	n
0.480	2.4368
0.486	2.4354
0.546	2.4235
0.589	2.4175
0.644	2.4114
0.656	2.4104

Table 7-33. Refractive Index of Fused Silica and Residuals at
20°C for Three Specimens [7-52]

Wavelength (μm)	Spectral Source	Computed Index	C-D-G.E.* Residual × 10⁶	Corning Residual × 10⁶	Dynasil Residual × 10⁶	General Electric Residual × 10⁶
0.213856	Zn	1.534307	−27	−29	−42	−31
0.214438	Cd	1.533722	− 2	−11	−21	−22
0.226747	Cd	1.522750	+70	+71	+68	+73
0.230209	Hg	1.520081	−21	−28	−31	−23
0.237833	Hg	1.544729	+ 1	+13	+23	+19
0.239938	Hg	1.513367	+ 3	+ 6	+ 2	+ 9
0.248372	Hg	1.508398	+ 2	+ 6	− 1	+ 7
0.265204	Hg	1.500029	−29	−32	−25	−13
0.269885	Hg	1.498047	+ 3	+ 7	− 4	+11
0.275278	Hg	1.495913	− 3	+ 2	+ 8	+12
0.280347	Hg	1.494039	+ 1	− 4	− 9	−11
0.289360	Hg	1.490990	+20	+18	+22	+20
0.296728	Hg	1.488734	−14	− 7	−12	− 4
0.302150	Hg	1.487194	− 4	− 9	− 2	+ 4
0.330259	Zn	1.480539	− 9	+ 1	+10	+ 3
0.334148	Hg	1.479763	− 3	− 8	− 1	+ 9
0.340365	Cd	1.478584	+ 6	+ 9	+ 2	− 8
0.346620	Cd	1.477468	+ 2	−17	−12	−14
0.361051	Cd	1.475129	+ 1	+ 3	− 9	− 8
0.365105	Hg	1.474539	−19	−11	−15	−21
0.404656	Hg	1.469618	+ 2	+ 1	− 1	+ 2
0.435835	Hg	1.466693	− 3	+ 5	+ 1	+ 3
0.467816	Cd	1.464292	+ 8	+ 5	+ 3	+ 6
0.486133	H	1.463126	+ 4	+ 6	+ 5	+ 7
0.508582	Cd	1.461863	+ 7	+ 4	+ 1	+ 5
0.546074	Hg	1.460078	+ 2	+ 4	+ 1	− 5
0.576959	Hg	1.458846	+ 4	+ 5	+ 3	+ 4
0.579065	Hg	1.458769	+ 1	+ 6	+ 6	+ 6
0.587561	He	1.458464	+ 6	+ 3	− 2	+ 1
0.589262	Na	1.458404	− 4	+ 6	+ 3	+ 7
0.643847	Cd	1.456704	+ 6	+ 9	+ 4	+ 7
0.656272	H	1.456367	+ 3	+ 7	+ 5	+ 7
0.667815	He	1.456067	+ 3	+ 8	+ 6	+ 3
0.706519	He	1.455445	+ 5	+10	+12	+ 7
0.852111	Cs	1.452465	+ 5	+ 8	+ 3	+ 5

*Residuals for arithmetical-mean table of values compiled from experimental data of Corning (C), Dynasil (D), and General Electric (G.E.)

Table 7-33. Refractive Index of Fused Silica and Residuals at
20°C for Three Specimens [7-52] *(Continued)*

Wavelength (μm)	Spectral Source	Computed Index	C-D-G.E.* Residual × 10⁶	Corning Residual × 10⁶	Dynasil Residual × 10⁶	General Electric Residual × 10⁶
0.894350	Cs	1.451835	+ 5	+11	+ 5	+10
1.01398	Hg	1.450242	+ 8	+ 6	+ 3	+ 6
1.08297	He	1.449405	- 5	+ 8	+ 1	+ 9
1.12866	Hg	1.448869	+ 1	+ 7	+ 8	+ 9
1.3622	Hg	1.446212	-12	- 6	-14	-12
1.39506	Hg	1.445836	+ 4	- 1	+ 4	- 3
1.4695	Cs	1.444975	- 5	+ 3	+ 9	+10
1.52952	Hg	1.444268	+ 2	+ 8	+ 6	0
1.6606	TCB†	1.442670	-20	-14	-19	-11
1.681	Poly‡	1.442414	+ 6	- 2	-10	+ 8
1.6932	Hg	1.442260	0	+ 7	- 6	+ 1
1.70913	Hg	1.442057	+ 3	0	+ 3	- 1
1.81307	Hg	1.440699	+21	- 7	- 7	+ 6
1.97009	Hg	1.438519	+ 1	+ 6	+12	+12
2.0581	He	1.437224	- 4	- 3	- 9	-11
2.1526	TCB	1.435769	-29	-22	-25	-24
2.32542	Hg	1.432928	-18	-10	- 3	- 6
2.4374	TCB	1.430954	-24	-23	-21	-14
3.2439	Poly	1.413118	+32	+21	+29	+25
3.2668	Poly	1.412505	+25	+20	+30	+25
3.3026	Poly	1.411535	+25	+32	+30	+28
3.422	Poly	1.408180	+20	+40	+42	+37
3.5070	Poly	1.405676	-16	-26	-20	-10
3.5564	TCB	1.401174	-24	-27	-29	-18
3.7067	TCB	1.399389	-19	-22	-14	- 9
Average of absolute values of residuals			10.5	11.9	12.2	11.7

*Residuals for arithmetical-mean table of values compiled from experimental data of Corning (C), Dynasil (D), and General Electric (G.E.)
† TCB = 1.2.4-Trichlorobenzene
‡ Poly = Polystyrene

The dispersion equation for fused silica is

$$n^2 - 1 = 0.6961663\lambda^2/[\lambda^2 - (0.0684043)^2] + 0.4079426\lambda^2/[\lambda^2 - (0.1162414)^2]$$
$$+ 0.8974794\lambda^2/[\lambda^2 - (9.896161)^2]$$

where λ is expressed in μm.

Table 7-34. Interspecimen Variation of the Refractive Index of Fused Silica at 20°C from Different Sources* [7-52]

λ (μm)	Corning Residual × 10⁶				Dynasil Residual × 10⁶				General Electric Residual × 10⁶			
	1	2	3	4	1	2	3	4	1	2	3	4
0.4047	11.0	26.0	24.0	20.0	4.0	27.0	21.0	21.0	11.0	12.0	-12.0	14.0
0.4861	16.0	27.0	16.0	21.0	5.0	29.0	19.0	21.0	7.0	15.0	- 7.0	12.0
0.5461	14.0	24.0	22.0	17.0	5.0	33.0	25.0	16.0	11.0	12.0	-13.0	12.0
0.5893	12.0	27.0	23.0	27.0	0.0	18.0	20.0	22.0	12.0	13.0	- 8.0	12.0
0.6563	12.0	23.0	19.0	19.0	1.0	27.0	22.0	23.0	13.0	10.0	- 7.0	14.0
0.7065	17.0	28.0	20.0	18.0	2.0	31.0	-23.0	23.0	13.0	15.0	-11.0	17.0
0.8944	11.0	26.0	22.0	19.0	3.0	31.0	19.0	25.0	11.0	12.0	- 8.0	13.0
1.014	16.0	21.0	20.0	20.0	6.0	30.0	26.0	22.0	14.0	13.0	-11.0	14.0
1.083	15.0	24.0	25.0	25.0	7.0	35.0	25.0	25.0	13.0	16.0	- 9.0	12.0
Av. Δn	13.5	24.9	20.0	21.6	3.5	27.4	20.4	20.5	11.3	12.9	- 9.6	13.2
r-value	67.78	67.78	67.86	67.80	67.80	67.80	67.85	67.85	67.88	67.77	67.82	67.84

*Numbered columns under each brand indicate individual specimens. The residuals are differences between measured values and those computed by the dispersion equation. Each Δn is an average for the 18 wavelengths that were used.

Table 7-35. Refractive Index and Extinction Coefficient of Gallium Antimonide [7-53]

λ (μm)	n	κ	λ (μm)	n	κ
1.49	—	9.70×10^{-2}	1.88	—	2.00×10^{-4}
1.51	—	9.45×10^{-2}			
1.53	—	9.06×10^{-2}	1.88	—	1.41×10^{-4}
1.55	—	8.67×10^{-2}	1.90	3.802	—
1.56	—	8.52×10^{-2}	1.91	—	1.41×10^{-4}
			1.94	—	1.18×10^{-4}
1.57	—	8.16×10^{-2}	1.97	—	1.10×10^{-4}
1.58	—	7.90×10^{-2}			
1.59	—	7.68×10^{-2}	2.00	—	1.08×10^{-4}
1.60	—	7.39×10^{-2}	2.00	—	9.87×10^{-3}
1.61	—	7.10×10^{-2}	2.00	3.789	—
			2.03	—	1.09×10^{-4}
1.62	—	6.80×10^{-2}	2.07	—	1.13×10^{-4}
1.63	—	6.49×10^{-2}			
1.64	—	6.14×10^{-2}	2.10	3.780	—
1.65	—	5.82×10^{-2}	2.20	3.764	—
1.66	—	5.47×10^{-2}	2.30	3.758	—
			2.40	3.755	—
1.68	—	5.10×10^{-2}	2.40	—	1.43×10^{-4}
1.69	—	4.70×10^{-2}	2.50	—	1.65×10^{-4}
1.70	—	4.06×10^{-2}	2.50	3.749	—
1.71	—	2.51×10^{-2}	2.80	—	2.65×10^{-4}
1.72	—	7.48×10^{-3}	3.00	—	3.65×10^{-4}
			3.00	3.898	—
1.73	—	6.68×10^{-3}			
1.73	—	2.14×10^{-3}	3.40	—	6.66×10^{-4}
1.74	—	3.05×10^{-3}	3.50	—	7.46×10^{-4}
1.75	—	1.83×10^{-3}	3.50	3.861	—
1.76	—	1.24×10^{-3}	3.70	—	9.25×10^{-4}
			4.00	—	1.26×10^{-3}
1.77	—	9.18×10^{-4}			
1.77	—	1.23×10^{-3}	4.00	3.833	—
1.80	—	5.40×10^{-4}	4.50	—	1.88×10^{-3}
1.80	—	5.51×10^{-4}	5.00	—	2.53×10^{-3}
1.80	3.820	—	5.00	3.824	—
			5.40	—	3.13×10^{-3}
1.82	—	3.56×10^{-4}			
1.82	—	3.55×10^{-4}	5.80	—	3.66×10^{-3}
1.84	—	1.96×10^{-4}	6.00	—	3.94×10^{-3}
1.85	—	2.52×10^{-4}	6.00	3.824	—

Table 7-35. Refractive Index and Extinction Coefficient of
Gallium Antimonide [7-53] *(Continued)*

λ (μm)	n	κ	λ (μm)	n	κ
6.20	–	4.22×10^{-3}	12.00	3.843	–
6.70	–	4.90×10^{-3}			
			12.40	–	1.21×10^{-2}
7.00	–	5.33×10^{-3}	12.80	–	1.26×10^{-2}
7.00	3.843	–	13.40	–	1.41×10^{-2}
7.40	–	5.90×10^{-3}	14.00	–	1.40×10^{-2}
8.00	–	6.68×10^{-3}	14.00	3.861	–
8.00	3.843	–			
			14.90	3.880	–
8.40	–	7.21×10^{-3}	–	–	–
9.00	–	7.99×10^{-3}	–	–	
9.00	3.843	–	–	–	–
9.50	–	8.63×10^{-3}	–	–	–
10.00	–	9.26×10^{-3}	–	–	–
			–	–	–
10.00	3.843	–	–	–	–
10.60	–	9.95×10^{-3}	–	–	–
11.10	–	1.06×10^{-2}	–	–	–
12.00	–	1.16×10^{-2}	–	–	–

Table 7-36. Refractive Index of Gallium Arsenide* [7-54, 7-55]

λ (μm)	n	λ (μm)	n
0.78 ± 0.01	3.34 ± 0.04	14.5 ± 0.05	2.82 ± 0.04
8.0 ± 0.05	3.34 ± 0.04	15.0 ± 0.05	2.73 ± 0.04
10.0 ± 0.05	3.135 ± 0.04	17.0 ± 0.05	2.59 ± 0.04
11.0 ± 0.05	3.045 ± 0.04	19.0 ± 0.05	2.41 ± 0.04
13.0 ± 0.05	2.97 ± 0.04	21.9 ± 0.1	2.12 ± 0.04
13.7 ± 0.05	2.895 ± 0.04	–	–

*The experimental data seem to be somewhat more scattered than the reported experimental errors indicate.

Table 7-37. Refractive Index of Germanium at 27°C
[7-56]

λ (μm)	n	λ (μm)	n
2.0581	4.1016	4.258	4.0216
2.1526	4.0919	4.866	4.0170
2.3126	4.0786	6.233	4.0094
2.4374	4.0708	8.66	4.0043
2.577	4.0609	9.72	4.0034
2.7144	4.0552	11.04	4.0026
2.998	4.0452	12.20	4.0023
3.3033	4.0369	13.02	4.0021
3.4188	4.0334	–	–

The resistivity of the germanium sample is about 50 ohm cm. The refractive indices of cast polycrystalline germanium differ from those of the single crystal by only a few parts in the fourth decimal place. The relative temperature coefficient of refractive index (dn/dT) is 4.0×10^{-4} °C^{-1} from 2.5 through 12 μm [7-98].

The dispersion equation for germanium is

$$n = A + BL + CL^2 + D\lambda^2 + E\lambda^4$$

where $A = 3.99931$
$B = 0.391707$
$C = 0.163492$
$D = -0.0000060$
$E = 0.000000053$
$L = (\lambda^2 - 0.028)^{-1}$ [7-57]

Note: Table 7-38 was deleted during revision of this handbook.

Table 7-39. Refractive Index* of Irtran Materials[†] [7-22]

Wavelength (μm)	Irtran 1	Irtran 2	Irtran 3	Irtran 4	Irtran 5
1.0000	1.3778	2.2907	1.4289	2.485	1.7227
1.2500	1.3763	2.2777	1.4275	2.466	1.7188
1.5000	1.3749	2.2706	1.4263	2.456	1.7156
1.7500	1.3735	2.2662	1.4251	2.450	1.7123
2.0000	1.3720	2.2631	1.4239	2.447	1.7089
2.2500	1.3702	2.2608	1.4226	2.444	1.7052
2.5000	1.3683	2.2589	1.4211	2.442	1.7012
2.7500	1.3663	2.2573	1.4196	2.441	1.6968
3.0000	1.3640	2.2558	1.4179	2.440	1.6920
3.2500	1.3614	2.2544	1.4161	2.438	1.6868
3.5000	1.3587	2.2531	1.4141	2.437	1.6811
3.7500	1.3558	2.2518	1.4120	2.436	1.6750
4.0000	1.3526	2.2504	1.4097	2.435	1.6684
4.2500	1.3492	2.2491	1.4072	2.434	1.6612
4.5000	1.3455	2.2477	1.4047	2.433	1.6536
4.7500	1.3416	2.2462	1.4019	2.433	1.6455
5.0000	1.3374	2.2447	1.3990	2.432	1.6368
5.2500	1.3329	2.2432	1.3959	2.431	1.6275
5.5000	1.3282	2.2416	1.3926	2.430	1.6177
5.7500	1.3232	2.2399	1.3892	2.429	1.6072
6.0000	1.3179	2.2381	1.3856	2.428	1.5962
6.2500	1.3122	2.2363	1.3818	2.426	1.5845
6.5000	1.3063	2.2344	1.3778	2.425	1.5721
6.7500	1.3000	2.2324	1.3737	2.424	1.5590
7.0000	1.2934	2.2304	2.3693	2.423	1.5452

*Irtrans 1-5 are hot-pressed samples of MgF_2, ZnS, CaF_2, ZnSe, and MgO, respectively. Index of refraction values were experimentally determined at selected wavelengths between 1 and 10 μm. Coefficients of interpolation formula were established and reduced by least squares methods, and the values computed. All values beyond 10 μm are extrapolated. Irtran® is a registered trademark of the Eastman Kodak Co.

†Irtran 6 is not longer manufactured by Eastman Kodak Company.

Table 7-39. Refractive Index* of Irtran Materials[†] [7-22] *(Continued)*

Wavelength (μm)	Irtran 1	Irtran 2	Irtran 3	Irtran 4	Irtran 5
7.2500	1.2865	2.2282	1.3648	2.422	1.5307
7.5000	1.2792	2.2260	1.3600	2.421	1.5154
7.7500	1.2715	2.2237	1.3550	2.419	1.4993
8.0000	1.2634	2.2213	1.3498	2.418	1.4824
8.2500	1.2549	2.2188	1.3445	2.417	1.4646
8.5000	1.2460	2.2162	1.3388	2.416	1.4460
8.7500	1.2367	2.2135	1.3330	2.415	1.4265
9.0000	1.2269	2.2107	1.3269	2.413	1.4060
9.2500	–	2.2078	1.3206	2.411	–
9.5000	–	2.2048	1.3141	2.410	–
9.7500	–	2.2018	1.3073	2.409	–
10.0000	–	2.1986	1.3002	2.407	–
11.0000	–	2.1846	1.2694	2.401	–
12.0000	–	2.1688	–	2.394	–
13.0000	–	2.1508	–	2.386	–
14.0000	–	–	–	2.378	–
15.0000	–	–	–	2.370	–
16.0000	–	–	–	2.361	–
17.0000	–	–	–	2.352	–
18.0000	–	–	–	2.343	–
19.0000	–	–	–	2.333	–
20.0000	–	–	–	2.323	–

*Irtrans 1-5 are hot-pressed samples of MgF_2, ZnS, CaF_2, ZnSe, and MgO, respectively. Index of refraction values were experimentally determined at selected wavelengths between 1 and 10 μm. Coefficients of interpolation formula were established and reduced by least squares methods, and the values computed. All values beyond 10 μm are extrapolated. Irtran® is a registered trademark of the Eastman Kodak Co.

† Irtran 6 is no longer manufactured by Eastman Kodak Company.

Table 7-40. Refractive Index of Lithium Fluoride [7-59, 7-60, 7-61, 7-62]

λ (μm)	n	λ (μm)	n	λ (μm)	n
0.1935	1.4450	0.366	1.40121	4.50	1.33875
0.1990	1.4413	0.391	1.39937	5.00	1.32661
0.2026	1.4390	0.4861	1.39480	5.50	1.31287
0.2063	1.4367	0.50	1.39430	6.00	1.29745
0.2100	1.4346	0.80	1.38896	6.91	1.260
0.2144	1.4319	1.00	1.38711	7.53	1.239
0.2194	1.4300	1.50	1.38320	8.05	1.215
0.2265	1.4268	2.00	1.37875	8.60	1.190
0.231	1.4244	2.50	1.37327	9.18	1.155
0.254	1.41792	3.00	1.36660	9.79	1.109
0.280	1.41188	3.50	1.35868	—	—
0.302	1.40818	4.00	1.34942	—	—

The dispersion equation for lithium fluoride is

$$n = A + BL + CL^2 + D\lambda^2 + E\lambda^4$$

where A = 1.38761
B = 0.001796
C = 0.000041
D = -0.0023045
E = -0.00000557
L = $(\lambda^2 - 0.028)^{-1}$ [7-57]

Table 7-41. Refractive Index of Magnesium
Fluoride at 19.5°C [7-63]

λ (μm)	n_e	n_o	$n_e - n_o$
0.21386	1.42897	1.41566	0.01331
0.22675	1.42251	1.40942	0.01309
0.24827	1.41615	1.40329	0.01286
0.25763	1.41382	1.40106	0.01276
0.27528	1.40967	1.39707	0.01260
0.29673	1.40592	1.39345	0.01247
0.31315	1.40364	1.39124	0.01240
0.33415	1.40116	1.38889	0.01227
0.36501	1.39834	1.38614	0.01220
0.40466	1.39567	1.38359	0.01208
0.43584	1.39048	1.38208	0.01200
0.46782	1.39276	1.38082	0.01194
0.47999	1.39232	1.38040	0.01192
0.50858	1.39142	1.37954	0.01188
0.54607	1.39043	1.37859	0.01184
0.58756	1.38955	1.37774	0.01181
0.64385	1.38858	1.37682	0.01176
0.66781	1.38823	1.37649	0.01174
0.79476	1.38679	1.37512	0.01167
1.0830	1.38465	1.37307	0.01158

The dispersion equation for magnesium fluoride is

$$n_o = 1.36957 + 0.0035821 \, (\lambda - 0.14925)^{-1}$$

$$n_e = 1.38100 + 0.0037415 \, (\lambda - 0.14947)^{-1}$$

[7-64]

Note that the wavelength must be given in angstroms in this equation.

Table 7-42. Temperature Coefficient of Refractive Index
for Magnesium Fluoride [7-64]

λ (μm)	dn_o/dT (10^{-5} °C^{-1})	dn_e/dT (10^{-5} °C^{-1})
0.4047	0.23	0.17
0.7065	0.19	0.10

Table 7-43. Refractive Index of Magnesium Oxide
at 23.3°C [7-65]

λ (μm)	n	λ (μm)	n
0.36117	1.77318	1.97009	1.70885
0.365015	1.77186	2.24929	1.70470
1.01398	1.72259	2.32542	1.70350
1.12866	1.72059	3.3033	1.08526
1.36728	1.71715	3.5078	1.68055
1.52952	1.71496	4.258	1.66039
1.6932	1.71281	4.138	1.63138
1.7092	1.71258	5.35	1.62404
1.81307	1.71108	—	—

The dispersion equation for magnesium oxide is

$$n^2 = 2.956362 - 0.01062387\lambda^2 - 0.0000204968\lambda^4 - \frac{0.02195770}{(\lambda^2 - 0.01428322)}$$

Table 7-44. Temperature Coefficients of Refractive Index
for Magnesium Oxide [7-65]

λ (μm)	dn/dT (10^{-6} °C^{-1})				
	20°C	25°C	30°C	35°C	40°C
7.679	13.6	13.7	13.8	13.9	14.0
7.065	14.1	14.2	14.3	14.4	14.5
6.678	14.4	14.5	14.6	14.7	14.8
6.563	14.5	14.6	14.7	14.8	14.9
5.893	15.3	15.4	15.5	15.6	15.7
5.461	15.9	16.0	16.1	16.2	16.3
4.861	16.9	17.0	17.1	17.2	17.3
4.358	18.0	18.1	18.2	18.3	18.4
4.047	18.9	19.0	19.1	19.2	19.3

Table 7-45. Refractive Index of Potassium Bromide at 22°C [7-66]

λ (μm)	n	λ (μm)	n	λ (μm)	n	λ (μm)	n
0.404656	1.589752	1.01398	1.54408	6.238	1.53288	17.40	1.50390
0.435835	1.581479	1.12866	1.54258	6.692	1.53225	18.16	1.50076
0.486133	1.571791	1.36728	1.54061	8.662	1.52903	19.01	1.49703
0.508582	1.568475	1.7012	1.53901	9.724	1.52695	19.91	1.49288
0.546074	1.563928	2.44	1.53733	11.035	1.52404	21.18	1.48655
0.587562	1.559965	2.73	1.53693	11.862	1.52200	21.83	1.48311
0.643847	1.555858	3.419	1.53612	14.29	1.51505	23.86	1.47140
0.706520	1.552447	4.258	1.53523	14.98	1.51280	25.14	1.46324

The dispersion equation for potassium bromide is

$$n^2 = 2.361323 - 0.00311497\lambda^2 - 0.000000058613\lambda^4$$

$$+ \frac{0.007676}{\lambda^2} + \frac{0.0156569}{\lambda^2} - 0.0324$$

The a⸱⸱⸱⸱⸱⸱⸱⸱⸱⸱⸱⸱⸱⸱⸱⸱⸱⸱rature coefficient of refractive index is given as 4.0×10^{-5}

Table 7-46. Refractive Index of Potassium Chloride [7-36, 7-62, 7-67]

λ (μm)	n	λ (μm)	n	λ (μm)	n
0.185109	1.82710	0.410185	1.50907	5.3039	1.470013
0.186220	1.81853	0.434066	1.50503	5.8932	1.468804
0.197760	1.73120	0.441587	1.50390	8.2505	1.462726
0.198990	1.72438	0.467832	1.50044	8.8398	1.460853
0.200090	1.71870	0.486149	1.49841	10.0184	1.45672
0.204470	1.69817	0.508606	1.49620	11.786	1.44919
0.208216	1.68308	0.53383	1.49410	12.965	1.44346
0.211078	1.67281	0.54610	1.49319	14.144	1.43722
0.21445	1.66188	0.56070	1.49218	15.912	1.42617
0.21946	1.64745	0.58931	1.49044	17.680	1.41403
0.22400	1.63612	0.58932	1.490443	18.2	1.409
0.23129	1.62043	0.62784	1.48847	18.8	1.401
0.242810	1.60047	0.64388	1.48777	19.7	1.398
0.250833	1.58979	0.656304	1.48727	20.4	1.389
0.257317	1.58125	0.67082	1.48669	21.1	1.379
0.263200	1.57483	0.76824	1.48377	22.2	1.374
0.267610	1.57044	0.78576	1.483282	23.1	1.363
0.274871	1.56386	0.88398	1.481422	24.1	1.352
0.281640	1.55836	0.98220	1.480084	24.9	1.336
0.291368	1.55140	1.1786	1.478311	25.7	1.317
0.308227	1.54136	1.7680	1.475890	26.7	1.300
0.312280	1.53926	2.3573	1.474751	27.2	1.275
0.340353	1.52726	2.9466	1.473834	28.2	1.254
0.358702	1.52115	3.5359	1.473049	28.8	1.226
0.394415	1.51219	4.7146	1.471122	—	—

The dispersion equations for potassium chloride (for the ultraviolet and visible, respectively) are

$$n^2 = a^2 + \left(\frac{M_1}{\lambda^2} - \lambda_1^2\right) + \left(\frac{M_2}{\lambda^2} - \lambda_2^2\right) - k\lambda^2 - h\lambda^4$$

$$n^2 = b^2 + \left(\frac{M_1}{\lambda^2} - \lambda_1^2\right) + \left(\frac{M_2}{\lambda^2} - \lambda_2^2\right) - \left(\frac{M_3}{\lambda_3^2} - \lambda^2\right)$$

where $a^2 = 2.174967$ $\quad\lambda^2 = 0.0255550$ $\quad b^2 = 3.866619$
$\quad M_1 = 0.008344206$ $\quad k = 0.000513495$ $\quad M_3 = 5,567.715$
$\quad \lambda_1^2 = 0.0119082$ $\quad h = 0.06167587$ $\quad \lambda_3^2 = 3,292.47$
$\quad M_2 = 0.00698382$

The temperature coefficient of refractive index for potassium chloride is

$$n_T + 1.490443 - (T - 15)\, 0.000034$$

Table 7-47. Refractive Index of Potassium Iodide [7-60, 7-68]

λ (μm)	n	λ (μm)	n	λ (μm)	n
0.248	2.0548	0.656	1.65809	10.02	1.6201
0.254	2.0105	0.707	1.6537	11.79	1.6172
0.265	1.9424	0.728	1.6520	12.97	1.6150
0.270	1.9221	0.768	1.6494	14.14	1.6127
0.280	1.8837	0.811	1.6471	15.91	1.6085
0.289	1.85746	0.842	1.6456	18.10	1.6030
0.297	1.83967	0.912	1.6427	19.0	1.5997
0.302	1.82769	1.014	1.6396	20.0	1.5964
0.313	1.80707	1.083	1.6381	21.0	1.5930
0.334	1.77664	1.18	1.6366	22.0	1.5895
0.366	1.74416	1.77	1.6313	23.0	1.5858
0.391	1.72671	2.36	1.6295	24.0	1.5819
0.405	1.71843	3.54	1.6275	25.0	1.5775
0.436	1.70350	4.13	1.6268	26.0	1.5729
0.486	1.68664	5.89	1.6252	27.0	1.5681
0.546	1.67310	7.66	1.6235	28.0	1.5629
0.588	1.66654	8.84	1.6218	29.0	1.5571
0.589	1.66643	–	–	–	–

The temperature coefficient of refractive index for 0.546 μm in the temperature region 38 to 90°C is -5.0×10^{-5} per °C.

Table 7-48. Refractive Index of Ruby Relative to Air at 22°C [7-70]

λ (μm)	n_o	n_e
0.4358	1.78115	1.77276
0.5461	1.77071	1.76258
0.5876	1.76822	1.76010
0.6678	1.76445	1.75611
0.7065	1.76302	1.75501

Table 7-49. Refractive Index for the Ordinary Ray of Synthetic
Sapphire [7-69, 7-71]

λ (μm)	n_{obs}	n_{cal}	$n_{obs} - n_{cal}$ $\times$ 10^5
0.26520	1.83360	1.83365	−5
0.28035	1.82427	1.82426	+1
0.28936	1.81949	1.81947	+2
0.29673	1.81595	1.81593	+2
0.30215	1.81351	1.81351	0
0.3130	1.80906	1.80909	−3
0.33415	1.80184	1.80181	+3
0.34662	1.79815	1.79820	−5
0.361051	1.79450	1.79451	−1
0.365015	1.79358	1.79358	0
0.39064	1.78826	1.78828	−2
0.404656	1.78582	1.78582	0
0.435834	1.78120	1.78120	0
0.546071	1.77078	1.77077	+1
0.576960	1.76884	1.76884	0
0.579066	1.76871	1.76871	0
0.64385	1.76547	1.76547	0
0.706519	1.76303	1.76303	0
0.85212	1.75885	1.75887	−2
0.89440	1.75796	1.75790	+6
1.01398	1.75547	1.75546	−1
1.12866	1.75339	1.75339	0
1.36728	1.74936	1.74935	+1
1.39506	1.74888	1.74888	0
1.52952	1.74660	1.74659	+1
1.6932	1.74368	1.74368	0
1.70913	1.74340	1.74339	+1
1.81307	1.74144	1.74144	0
1.9701	1.73833	1.73835	−2
2.1526	1.73444	1.73449	−5

Table 7-49. Refractive Index for the Ordinary Ray of Synthetic
Sapphire [7-69, 7-71] (Continued)

λ (μm)	n_{obs}	n_{cal}	$n_{obs} - n_{cal}$ $\times 10^5$
2.24929	1.73231	1.73232	−1
2.32542	1.73057	1.73055	+2
2.4374	1.72783	1.72784	−1
3.2439	1.70437	1.70433	+4
3.2668	1.70356	1.70354	+2
3.3026	1.70231	1.70231	0
3.3303	1.70140	1.70136	+4
3.422	1.69818	1.69814	+4
3.5070	1.69504	1.69501	+3
3.7067	1.68746	1.68743	+3
4.2553	1.66371	1.66363	+8
4.954	1.62665	1.62669	−4
5.1456	1.61514	1.61510	+4
5.349	1.60202	1.60204	−2
5.419	1.59735	1.59736	−1
5.577	1.58638	1.58642	−4

The constants of the dispersion equation at 24°C are

$$\lambda_1{}^2 = 0.00377588 \qquad A_1 = 1.023798$$

$$\lambda_2{}^2 = 0.0122544 \qquad A_2 = 1.058264$$

$$\lambda_3{}^2 = 321.3616 \qquad A_3 = 5.280792$$

The dispersion equation for sapphire is

$$n^2 - 1 = \sum_i (A_i \lambda^2)/(\lambda^2 - \lambda_i{}^2)$$

Temperature coefficients of index (dn/dT) were determined from the differences between the indices at 19°C and 24°C. The results indicate that the coefficient is positive and decreases from about $20 \times 10^{-6}/°C$ at the short wavelengths to about $10 \times 10^{-6}/°C$ near 4 μm. The average value of $13 \times 10^{-6}/°C$ for the visible region was determined from additional measurements made at 17°C, 24°C, and 31°C on a Wild precision spectrometer.

Table 7-50. Refractive Index of Trigonal Selenium [7-72]

λ (μm)	n_o	n_e
1.06	2.790 ± 0.008	3.608 ± 0.008
1.15	2.737 ± 0.008	3.573 ± 0.008
3.39	2.65 ± 0.01	3.46 ± 0.01
10.6	2.64 ± 0.01	3.41 ± 0.01

Trigonal selenium: for E perpendicular to the c axis, n is 2.78 from 9 to 23 μm at 27°C. For E parallel to it, n is 3.58 ± 0.0 from 9 to 23 μm [7-4].

Amorphous selenium: n is 2.44-2.46 at 2.5 μm [7-73]. n varies from 2.42 to 2.38 between 5 and 15 μm [7-74].

Liquid selenium: 2.44 from 30 to 152 μm [7-73].

Table 7-51. Refractive Index of Selenium Glass, Arsenic Modified at 27°C* [7-5]

λ (μm)	n Prism a	n Prism b	λ (μm)	n Prism a	n Prism b
1.0140	2.5774	2.5783	7.00	2.4778	2.4787
1.1286	2.5554	2.5565	7.50	–	2.4784
1.3622	2.5285	2.5294	8.10	2.4772	2.4778
1.5295	2.5173	2.5183	8.50	–	2.4775
1.7012	2.5089	2.5100	9.10	2.4765	2.4771
2.1526	2.4950	2.4973	9.50	–	2.476
3.00	2.4861	2.4882	10.00	2.4756	2.4767
3.4188	2.4841	2.4858	10.50	–	2.4759
4.00	2.4825	2.4835	11.00	2.4752	2.4758
4.50	–	2.4822	11.50	–	2.4753
5.00	2.4803	2.4811	12.00	–	2.4749
5.50	–	2.4804	13.00	–	2.4760 [sic]
6.00	2.4789	2.4798	13.50	–	2.4748
6.50	–	2.4792	14.00	–	2.4743

*The data for two different prisms illustrate some of the variability that can be expected.

Table 7-52. Refractive Index of Silicon at 26°C* [7-5]

λ (μm)	n	λ (μm)	n	λ (μm)	n	λ (μm)	n
1.3570	3.4975	2.1526	3.4476	4.00	3.4255	7.50	3.4186
1.3673	3.4962	2.3254	3.4430	4.258	3.4242	8.00	3.4184
1.3951	3.4929	2.4373	3.4408	4.50	3.4236	8.50	3.4182
1.5295	3.4795	2.7144	3.4358	5.00	3.4223	10.00	3.4179
1.6606	3.4696	3.00	3.4320	5.50	3.4213	10.50	3.4178
1.7092	3.4664	3.3033	3.4297	6.00	3.4202	11.04	3.4176
1.5131	3.4608	3.4188	3.4286	6.50	3.4195	–	–
1.9701	3.4537	3.50	3.4284	7.00	3.4189	–	–

*The purity of the silicon sample is not specified. The refractive index values of adequately pure (30 obm cm) cast polycrystal silicon should be very near those of single crystals. The temperature coefficient of refractive index is 1.5×10^{-4} per °C.

The dispersion equation for silicon is

$$n = A + BL + CL^2 + D\lambda^2 + E\lambda^4$$

where $A = 3.41696$ $B = 0.138497$ $C = 0.013924$ $D = -0.0000209$

$E = 0.000000148$ $L = (\lambda^2 - 0.028)^{-1}$

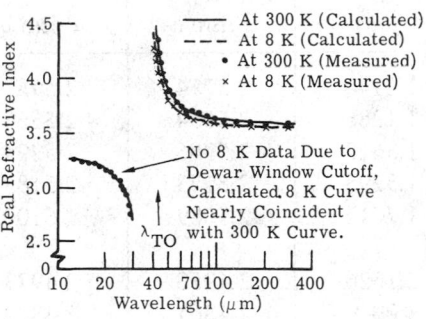

Fig. 7-31. The far infrared real refractive index of high resistivity cadmium telluride at 300 and 8 K [7-19].

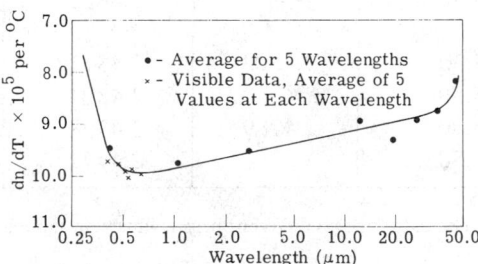

Fig. 7-32. The temperature coefficient of refractive index of cesium iodide [7-46].

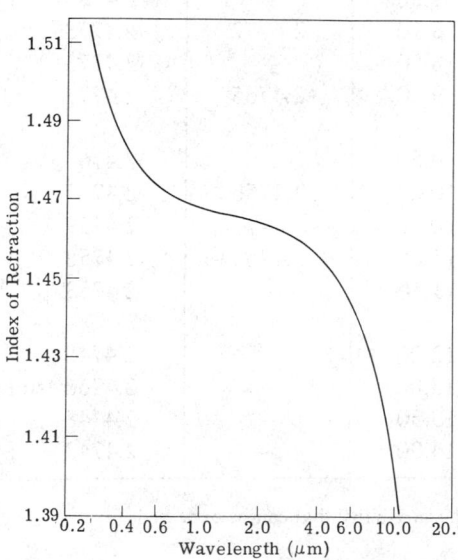

Fig. 7-30. The refractive index of barium fluoride [7-33].

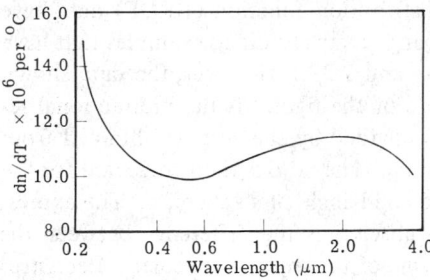

Fig. 7-33. The thermal coefficient of index of fused silica for a mean temperature of 25°C [7-52].

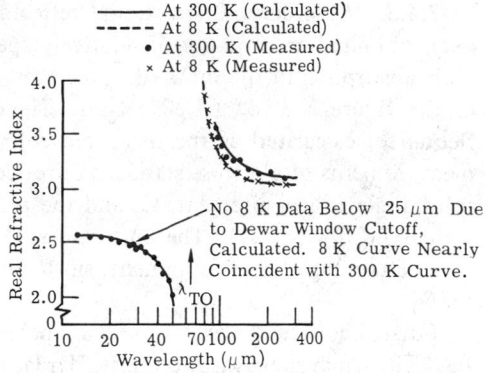

Fig. 7-34. The far infrared real refractive index of high resistivity gallium arsenide at 300 and 8 K [7-19].

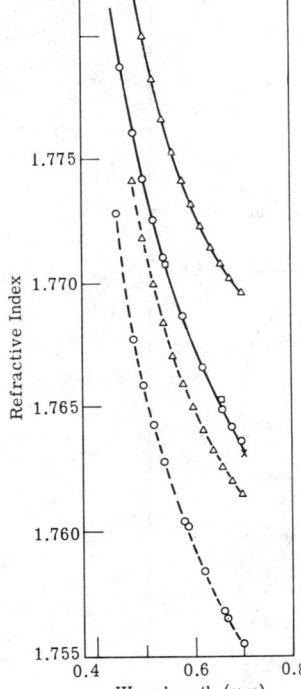

× = sapphire, Malitson

□ = sapphire, Dowell

△ = 0.01% Cr_2O_3, Houston

○ = 0.062% Cr_2O_3, NBS

□ = 0.077% Cr_2O_3, Dowell

○ = 0.11% Cr_2O_3, Mandarino

△ = 1.4% Cr_2O_3, Mandarino

—— Ordinary Ray

----- Extraordinary Ray

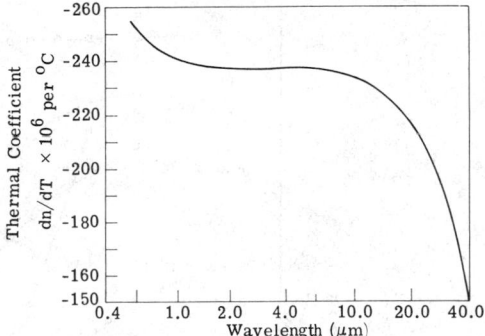

Fig. 7-35. The average thermal coefficient of refractive index of potassium chloride near room temperature [7-67].

Fig. 7-36. The comparison of the refractive index of sapphire and ruby as a function of wavelength. The solid and dashed lines indicate the ordinary and extraordinary rays, respectively [7-69].

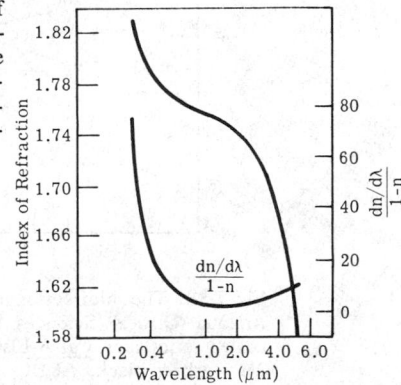

Fig. 7-37. The refractive index and relative dispersion of synthetic sapphire for the ordinary ray at 24°C [7-69, 7-71].

7.4.3. Absorbers. Bidirectional reflectance distribution function (BRDF) data have been obtained on a variety of relatively specular and relatively diffuse samples that have high absorption in the infrared. (See Figures 7-38 and 7-39.) However, the data shown in the figures are for a He-Ne laser. The ordinate of the figures is the bidirectional reflectance, calculated as the measured voltage, V_s, divided by the voltage obtained from measurements of a diffuse standard of reflectivity, ρ_0. The ratio is then corrected for the solid angle of measurement, Ω, and the projected solid-angle of a sphere, π. The expression is $(V_s/V_0)(\rho/\pi\Omega)$. The units are sr^{-1}. The abscissa is the difference between the sine of the angle of measurement, $\sin\theta$, and the sine of the angle of specular reflectance, $\sin\theta_0$.

Cat-A-Lac® * and Cat-A-List® * are paints manufactured by the Finch Co. in Torrance, CA. The sandpaper is a 280 C grit. Hardas is a black, anodized-aluminum coating. Carbon

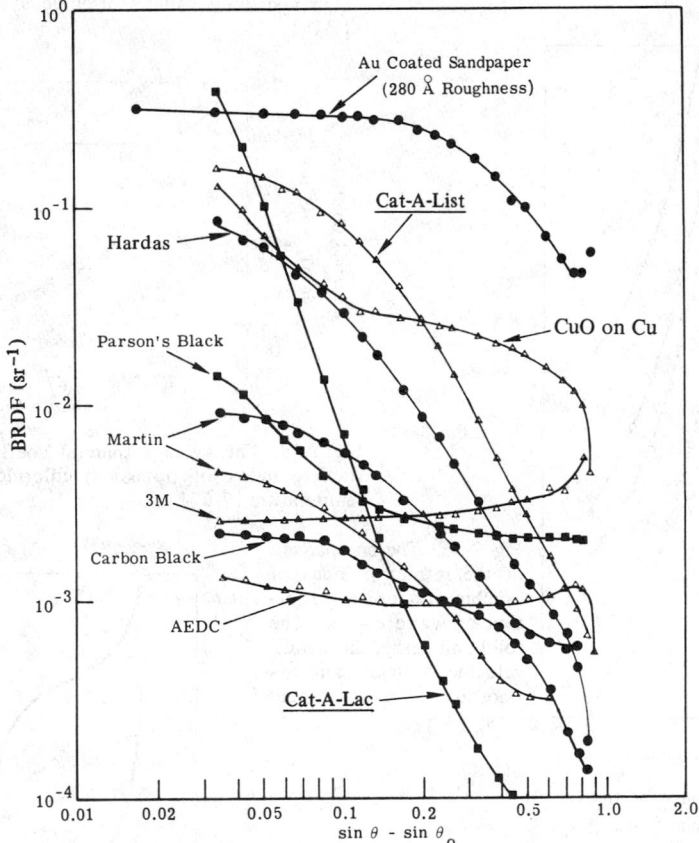

Fig. 7-38. The bidirectional reflectivity measured by the University of Arizona Optical Sciences Center with He-Ne laser radiation for: Au coated sandpaper, <u>Cat-A-List</u>, Hardas, CuO on Cu, Parson's black, Martin, 3M, carbon black, AEDC, and <u>Cat-A-Lac</u>. Cat-A-Lac® and Cat-A-List® are registered trademarks of Bostic-Finch, Inc. (AEDC = Arnold Engineering Development Co.)

---

*Cat-A-Lac® and Cat-A-List® are registered trademarks of Bostic-Finch, Inc.

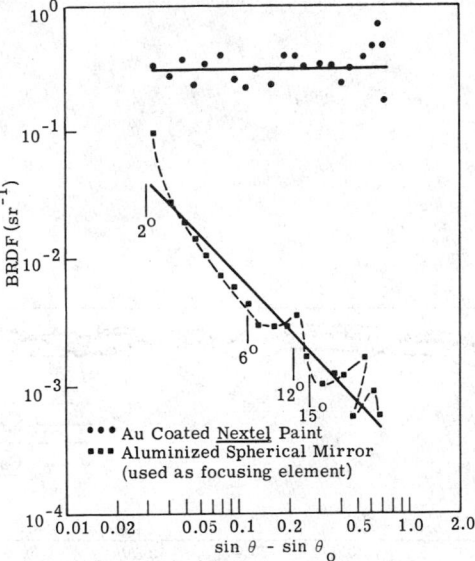

Fig. 7-39. The bidirectional reflectivity measured by the University of Arizona Optical Sciences Center with He-Ne laser radiation for: Au coated Nextel paint with an aluminated spherical mirror used as a focusing element. Nextel® is a registered trademark of the Minnesota Mining and Manufacturing Co.

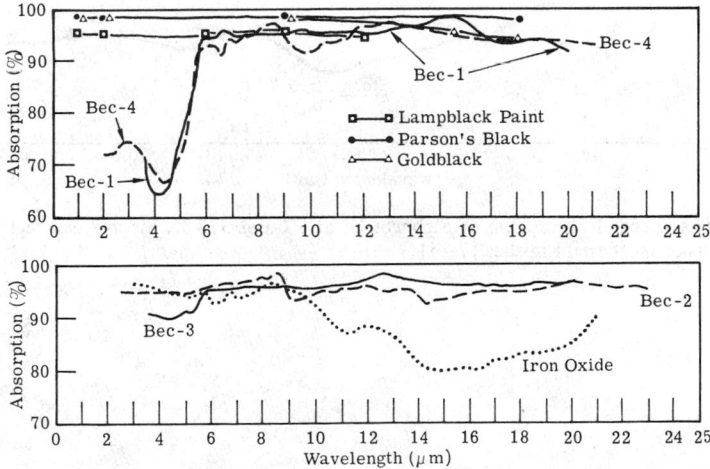

Fig. 7-40. The spectral absorption of Lampblack Paint, Parson's Black Gold-Black, and BEC-1,2,3, and 4. (BEC = Barnes Engineering Co.) [7-24].

black is soot from the flame of a candle. The two Martin samples were obtained from them at different times and may show the results of handling or batch-to-batch variations. Spectral absorption/emission data are shown in Figures 7-40, 7-41, and 7-42.

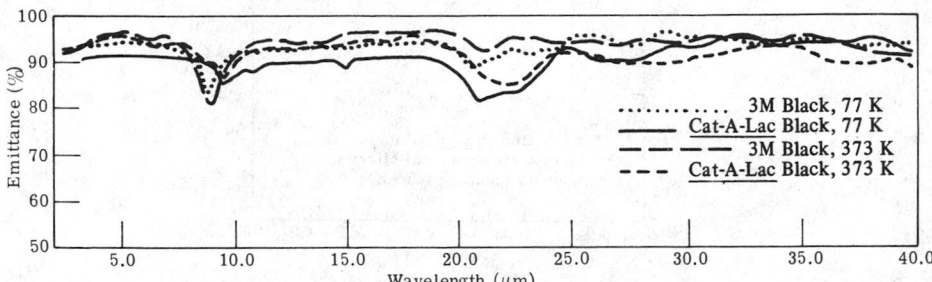

Fig. 7-41. The spectral absorption of 3M black and Cat-A-Lac black at 77 and 373 K. Cat-A-Lac®
is a registered trademark of Bostic-Finch, Inc. [7-23].

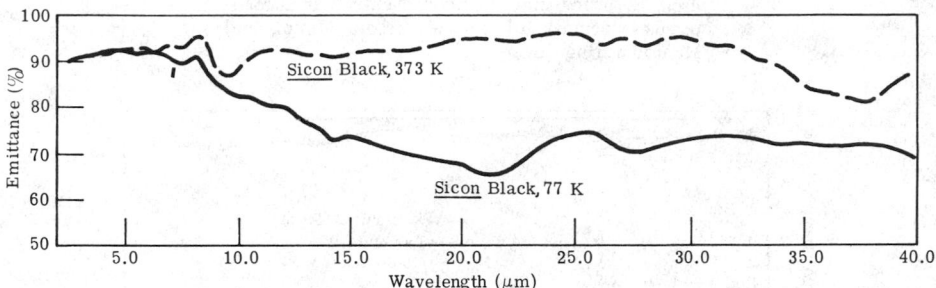

Fig. 7-42. The spectral absorption of Sicon black at 77 and 373 K. Sicon® is a registered trademark of Midland Industrial Finishes [7-23].

7.4.4. Reflectors. Although in the infrared most metal mirror surfaces are better than at shorter wavelengths, the requirements are often very stringent. Representative spectral curves are given in Figure 7-43. The reader should be aware that oxidation, dust, and tarnish from other chemical reactions can reduce the reflectivity of the specified surfaces below that reported here.

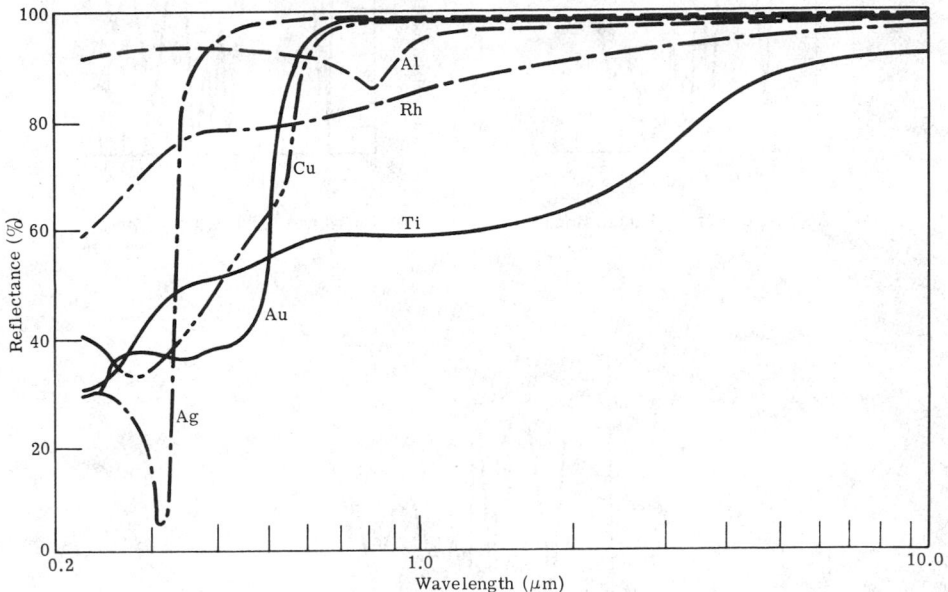

Fig. 7-43. The reflectance of various films of silver, gold, aluminum, copper, rhodium, and titanium [7-75].

7.4.5. The Optical Properties of Some Infrared Cements. Infrared optical elements are rarely cemented because few are available that hold well and are transparent. Some useful ones are listed here. In some cases the cement might also be used as an antireflection coating or matching coating. For this application, the refractive index is needed. These are given in Figure 7-44.

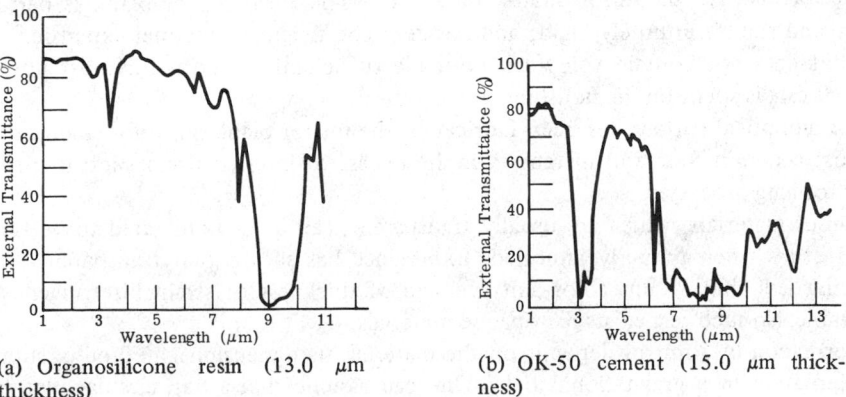

(a) Organosilicone resin (13.0 μm thickness)

(b) OK-50 cement (15.0 μm thickness)

Fig. 7-44. The transmission of selected infrared cements [7-76].

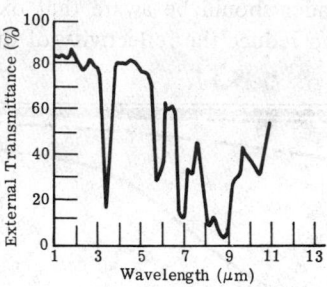

(c) Acrylic cement (14.0 μm thickness)

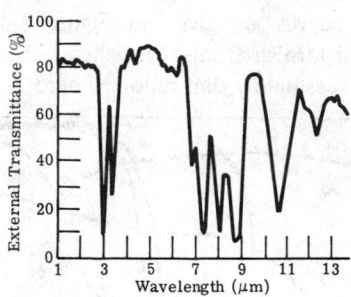

(d) Balsamine (15.0 μm thickness)

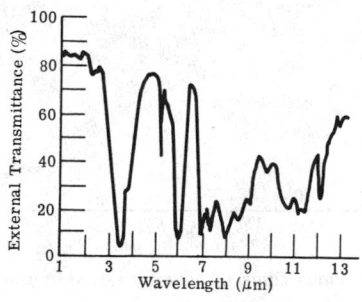

(e) Balsam (12.0 μm thickness)

Fig. 7-44. (Continued) The transmission of selected infrared cements [7-76].

7.4.6. Properties of Mirror Blank Materials.* The ideal mirror blank is perfectly smooth and stable, infinitely rigid, and has zero coefficient of thermal expansion. Any solid that does not contain voids can be made sufficiently smooth for most purposes if enough care is spent on the polishing process.

Once an optical surface has been fabricated, the mirror blank must not change shape. The most common source of instability is the release of internal stresses left by production or forming processes.

Vitreous materials, which are usually transparent, can be demonstrated to be free of internal stress when properly annealed. Experience has shown that transparent blanks containing less than 5 nm color shift per cm of thickness of strain birefringence are quite stable. No such test exists for opaque materials.

Deformation in a mirror depends on the material, its dimensions, method of support, and orientation in a gravitational field. One can assume that a flat, circular plate rests

*Original material provided by R. Noble, Institute of Optics, Tonanzintla, Mexico.

concentrically on a flat, annular ring. The combination is horizontal. The deformation is given by

$$\delta = K\left(\gamma, \frac{b}{r}\right) \left\{\left(\frac{r^2}{t}\right)^2 \frac{\rho}{E}\right\} \tag{7-39}$$

where δ = deformation at the edge
  r = radius of the plate
  t = thickness of the plate
  ρ = density of the plate
  E = Young's modulus Y
  K = a function of ν and b/r
  γ = Poisson's ratio of the plate
  b = inside radius of the support ring

Poisson's ratio, γ, enters into K as $(1 - \gamma^2)$. Fused silica has a ratio of 0.14, so that the factor $(1 - \gamma^2)$ is 0.98. Most metals have a ratio near 0.3, and thus a factor of 0.91. A change of material does not have a large effect on the term K. However, the ratio ρ/E may change quite markedly.

Table 7-53 gives some of the important mechanical strength properties. High strength-to-weight ratio is important when combined with low thermal expansion and high conductivity (Table 7-54). Even more accurate data on thermal expansion of low-expansion materials are given in Table 7-55 and temporal variations in Table 7-56. These data show the changes are about one part per 10^7 per year. They change about as much in a year as they do for each degree. Figure 7-45 displays the data on thermal expansion.

Table 7-53. Density and Young's Modulus Data for
Selected Mirror Blank Materials [7-2]

	Density, ρ (g cm^{-3})	Modulus, E (g cm^{-1} sec^{-2} 10^{-12})	ρ/E (sec^2 cm^{-2} 10^{-12})
Beryllium	1.82	2.8	0.65
Beryllia	3.03	—	—
Alumina	3.85	3.5	1.1
Cer-Vit*	2.50	0.92	2.7
Fused silica (7940)	2.20	0.73	3.0
ULE fused silica (7971)	2.21	0.68	3.2
Pyrex** (7740)	2.35	0.68	3.5
Aluminum	2.70	0.69	3.9
Magnesium	1.74	0.45	3.9

*Cer-Vit® is a registered trademark of Owens-Illinois.
**Pyrex® is a registered trademark of Corning Glass Works.

Table 7-54. Thermal Expansion* and Conductivity for
Selected Mirror Blank Materials [7-2]

Material	Expansion, α (cm cm^{-1} K^{-1} × 10^6)	Conductivity, k (cal cm^{-1} sec^{-1} K^{-1})	α/k (cm sec cal^{-1} × 10^6)
ULE fused silica	0.035	0.0031	11
Cer-Vit**	0.05	0.0040	13
Beryllia	9.5	0.42	23
Beryllium	12.0	0.38	32
Aluminum	24.0	0.53	45
Magnesium	26.0	0.38	68
Alumina	6.0	0.041	150
Fused silica (7940)	0.55	0.0033	170
Pyrex† (7740)	3.2	0.0027	1200

*All values in this table are positive, and for room temperature.
**Cer-Vit® is a registered trademark of Owens-Illinois.
†Pyrex® is a registered trademark of Corning Glass Works.

Table 7-55. Temperature of Samples During Dimensional Stability Measurements
and Thermal Expansion* of the Samples at Those Temperatures [7-77]

Material	Temperature (°C)	Expansion, α (cm cm^{-1} K^{-1} × 10^6)
Zerodur**	15.5	−0.040
ULE	26.5	−0.040
Cer-Vit†	15.5	−0.050
Superinvar	27.5	−0.200†††
Unispan LR-35 Invar††	27.0	+0.566
Homosil***	26.5	+0.495
7940	26.0	+0.472

*Note that some values are negative.
**Zerodur® is a registered trademark of Schott Optical Glass, Inc.
†Cer-Vit® is a registered trademark of Owens-Illinois.
††Unispan LR-35 Invar® is a registered trademark of Universal Cyclops Corp.
***Homosil® is a registered trademark of Heraeus-Amersil, Inc.
†††Superinvar has a negative expansion coefficient below 50°C.

Table 7-56. Average Length Change in Parts Per Billion at Constant Temperature

Material	Per Day	Per Second
Corning Code 7971 ULE Titanium Silicate	-0.18 ± 0.03	2.08×10^{-6}
Corning Code 7940 Fused Silica	-0.50 ± 0.03	5.8
Owens-Illinois Cer-Vit* C-101, Glass Ceramic	0.49 ± 0.03	5.7
Schott Zerodur** Glass Ceramic	0.0 ± 0.03	0.0
Heraeus-Amersil Homosil† Fused Silica	-0.20 ± 0.05	2.3
Universal Cyclops Unispan LR-35 Invar††	5.69 ± 0.03	6.6
Simonds Saw and Steel Superinvar	0.0 ± 0.03	0.0

*Cer-Vit® is a registered trademark of Owens-Illinois.
**Zerodur® is a registered trademark of Schott Optical Glass, Inc.
†Homosil® is a registered trademark of Heraeus-Amersil, Inc.
††Unispan LR-35 Invar® is a registered trademark of Universal Cyclops Corp.

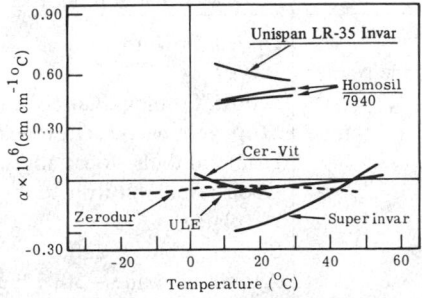

Fig. 7-45. Measured thermal expansion versus temperature for dimensional stability materials. Unispan LR-35 Invar® is a registered trademark of Universal Cyclops Corp. Homosil® is a registered trademark of Heraeus-Amersil, Inc. Cer-Vit® is a registered trademark of Owens-Illinois. Zerodur® is a registered trademark of Schott Optical Glass, Inc. [7-77].

7.5. Working Optical Materials *

Data useful for working with optical materials are presented in Table 7-57. These include grinding and polishing characteristics. The compilation uses the following descriptive terms:

(1) Rate of Grind Scale 1 to 10
      1 = fast, 10 = slow, BK7 = 5
(2) Rate of Polish Scale 1 to 10
      1 = fast, 10 = slow, BK7 = 5

---

*Compiled by Richard Sumner, Optician, University of Arizona, Optical Sciences Center.

Table 7-57. Data for Working Optical Materials

Name:	Cer-Vit*
Chemical Formula or Trade Name:	
Supplier:	Owens-Illinois, Toledo, Ohio
Form:	Cut blanks (special shapes and sizes on request)
Shaping:	Diamond tools—loose abrasives
Grinding Substrate:	Good results on iron—best results on glass
Grinding Abrasive-Coarse:	Silicon carbide = 80, 120, 220, 320 grit size
Grinding Abrasive-Fine:	Aluminum oxide = 30, 12, 3 μm size
Abrasive Carrier:	Water
Carrier Additives:	Detergent—about 10 ml per qt; glycerine—1 part to 3 parts water
Rate of Grind:	7—slower than BK-7
Rate of Polish:	7—slower than BK-7
Polishing Lap Material:	Medium-hard black optical pitch
Polishing Compound:	Cerium oxide, rouge-milled materials for final finishing
Automatic Feed:	Not required
Special Considerations:	Do all mechanical grinding and shaping before final figuring. Super-smooth ($<$ 10 Å) can be accomplished with finely-milled materials and a continuous slurry feed. Finish this operation by adding distilled water for the last polishing period greater than 4 h. Mohs hardness = 7.
Name:	Optical Crown
Chemical Formula or Trade Name:	BK-7
Supplier:	Schott, Corning, Bausch & Lomb, Burns
Form:	Strip, gob, blocks, cut blanks, and pressings
Shaping:	Diamond tools—loose abrasives
Grinding Substrate:	Good results on iron tools, best results on soft glass (plate)
Grinding Abrasive-Coarse:	Silicon carbide = 220, F
Grinding Abrasive-Fine:	Aluminum oxide = 30, 12, 3 μm
Abrasive Carrier:	Water
Carrier Additives:	Detergent and/or glycerine
Rate of Grind:	5 (standard glass)
Rate of Polish:	5 (standard glass)
Polishing Lap Material:	Medium-soft rosin pitch (Universal No. 850, Zobel)
Polishing Compound:	XOX by Grace, mixed 1 to 3 with water
Automatic Feed:	Desirable for critical work
Special Considerations:	Work to control all heat-producing operations. A clean slurry system in conjunction with block operations or annulus polishing machines will help to produce superior results.

*Cer-Vit® is a registered trademark of Owens-Illinois.

Table 7-57. Data for Working Optical Materials (*Continued*)

Name:	Optical Flint
Chemical Formula or Trade Name:	F4
Supplier:	Schott, Burns
Form:	Strip, gob, blocks, cut blanks, and pressings
Shaping:	Diamond tools—loose abrasives
Grinding Substrate:	Soft glass (plate)
Grinding Abrasive-Coarse:	Silicon carbide = 220, 320 grit size
Grinding Abrasive-Fine:	Aluminum oxide = 30, 12, 3 μm, 3 is optional
Abrasive Carrier:	Water
Carrier Additives:	Detergent and/or glycerine
Rate of Grind:	3—faster than BK-7
Rate of Polish:	8—slower than BK-7
Polishing Lap Material:	Soft rosin pitch (Universal No. 850 or soft Zobel)
Polishing Compound:	Milled Barnesite No. 85
Automatic Feed:	Desirable for critical work
Special Considerations:	Work to control all heat-producing operations. A slurry system is a help in this respect when used with a block or annulus polishing operation. Superior results can be obtained for hard work using a lap with small facets with deep cuts. This allows a large volume of polishing slurry that helps carry away heat generated during polishing. Sleeks can be avoided by polishing all bevels. For soft optical glasses that have poor resistance to water, Dow Corning silicone fluid No. 200 can be substituted.
Name:	Pyrex†
Chemical Formula or Trade Name:	Pyrex, E6, Duran 50
Supplier:	Corning, Burns, Schott, United Lens
Form:	Molded blanks, sheets, pressings (special shapes on request)
Shaping:	Diamond tools—loose abrasives
Grinding Substrate:	Iron
Grinding Abrasive-Coarse:	Silicon carbide = 80, 120, 220, 320 grit size
Grinding Abrasive-Fine:	Aluminum oxide = 40, 20, 9, 3 μm
Abrasive Carrier:	Water
Carrier Additives:	Detergent and/or glycerine, 1 to 3
Rate of Grind:	4—slow
Rate of Polish:	4—slow
Polishing Lap Material:	Black optical pitch
Polishing Compound:	Raw Cerox††—final finishing with milled Cerox or rouge
Automatic Feed:	Recommended for critical work
Special Considerations:	A constant feed system will help keep a constant temperature during working that will help produce superior work. Mohs hardness = 6.

†Pyrex® is a registered trademark of Corning Glass Works.
††Cerox® is a registered trademark of Babcock and Wilcox.

Table 7-57. Data for Working Optical Materials (*Continued*)

Name:	Vitreous Silica, Vycor**, ULE, Optical Quartz
Chemical Formula or Trade Name:	SiO_2
Supplier:	Corning, General Electric, Amercil
Form:	Blocks, cut blanks (special shapes on request)
Shaping:	Diamond tools—loose abrasives
Grinding Substrate:	Iron
Grinding Abrasive-Coarse:	Silicon carbide = 80, 120, 220, 320 grit size
Grinding Abrasive-Fine:	Aluminum oxide = 30, 12, 3 μm
Abrasive Carrier:	Water
Carrier Additives:	Detergent and/or glycerine
Rate of Grind:	Slow—3
Rate of Polish:	Slow—4
Polishing Lap Material:	Black optical pitch
Polishing Compound:	Raw Cerox—finish with milled Cerox, Barnesite, and rouge
Automatic Feed:	Required for supersmooth—not required for normal finishing
Special Considerations:	For surfaces of 10 Å or less, a slurry feed or milled Cerox for preliminary polishing is recommended; final finishing can be accomplished by cleaning the work and polishing lap and returning to polishing with a slurry feed of distilled water only. Good results can be expected in 16 to 24 hr—the lap slowly loses its polishing ability with time. Mohs hardness of 7.
Name:	Barium Fluoride
Chemical Formula or Trade Name:	BaF_2
Supplier:	Harshaw, Optovac
Form:	Boules—cut disks to 5 in.
Shaping:	Diamond tools—loose abrasives
Grinding Substrate:	Soft metal—lead
Grinding Abrasive-Coarse:	Aluminum oxide = 40, 30, 20 μm
Grinding Abrasive-Fine:	Aluminum oxide = 10, 5, 3 μm
Abrasive Carrier:	Dow Corning No. 200 silicone fluid
Carrier Additives:	None
Rate of Grind:	Fast—3
Rate of Polish:	Slow—6
Polishing Lap Material:	Soft optical pitch coated with thin layer of beeswax
Polishing Compound:	Linde A 3 μm
Automatic Feed:	Normally not required
Special Considerations:	Keep the pitch squares small in comparison to work size. To apply the beeswax, take a large piece of brown wrapping paper. Place on flat surface and rub with a block of beeswax. When the paper is coated, place the pitch lap to be coated face down on the paper and slide along the paper. The wax will be transferred to the pitch in a thin smooth layer. Turn the polisher as you slide it across the paper. This will improve the quality of

**Vycor® is a registered trademark of Corning Glass Works.

Table 7-57. Data for Working Optical Materials (*Continued*)

	the wax layer. Hardness: Mohs = 3+, Knoop = 65. Attacked by water; heat sensitive. Cleaves cubic system.
Name:	Calcite
Chemical Formula or Trade Name:	$CaCO_3$
Supplier:	Karl Lambrecht
Form:	Crystals and finished items
Shaping:	Cleaving and loose abrasives
Grinding Substrate:	Coarse work—iron, fine finishing on soft metal (tin, lead)
Grinding Abrasive-Coarse:	Aluminum oxide = 40, 20
Grinding Abrasive-Fine:	Aluminum oxide = 10, 5, 3 μm, 3 is optional
Abrasive Carrier:	Water
Carrier Additives:	Glycerine
Rate of Grind:	Fast—2
Rate of Polish:	Fast—3
Polishing Lap Material:	Soft optical pitch
Polishing Compound:	Aluminum oxide, 3 μm, Linde A
Automatic Feed:	Not required
Special Considerations:	Same as barium fluoride Temperature sensitive Hardness: Mohs = 3 Cleavage = distinct
Name:	Calcium Fluoride
Chemical Formula or Trade Name:	CaF_2
Supplier:	Harshaw
Form:	Cut disks—crystals
Shaping:	Diamond tools—loose abrasives
Grinding Substrate:	Soft glass, soft metal, iron
Grinding Abrasive-Coarse:	Aluminum oxide = 40, 30, 20
Grinding Abrasive-Fine:	Aluminum oxide = 10, 5, 3
Abrasive Carrier:	Water
Carrier Additives:	Soap and/or glycerine
Rate of Grind:	Fast—4
Rate of Polish:	Slow—7
Polishing Lap Material:	Optical pitch
Polishing Compound:	Linde A, 3 μm
Automatic Feed:	Not required
Special Considerations:	This material is thermally sensitive and can be broken with heat or cold. Also slow to polish. Good results can be obtained with polishing compound mixed 1 to 10 with water and used thin on the lap. Hardness: Mohs = 4+, Knoop = 120. Cleavage = distinct.

Table 7-57. Data for Working Optical Materials (*Continued*)

Name:	Germanium
Chemical Formula or Trade Name:	Ge
Supplier:	General Electric
Form:	Boules—blanks
Shaping:	Diamond tools—loose abrasives
Grinding Substrate:	Iron or hard glass
Grinding Abrasive-Coarse:	Silicon carbide = 220, 320
Grinding Abrasive-Fine:	Aluminum oxide = 30, 12, 3
Abrasive Carrier:	Water
Carrier Additives:	Detergent and/or glycerine
Rate of Grind:	Fast—4
Rate of Polish:	Fast—3
Polishing Lap Material:	Medium-soft optical pitch
Polishing Compound:	Aluminum oxide—Linde A, 3/10 μm, final finish Linde B 1/20 μm
Automatic Feed:	Not required but a help in block work
Special Considerations:	A relatively soft material that takes a good polish with 1/20 μm Linde B. Care should be taken to ensure that the polishing material does not become contaminated.
Name:	Lithium Iodate
Chemical Formula or Trade Name:	LIO_3
Supplier:	Isomet
Form:	Crystals
Shaping:	Loose abrasives—wet string sawing
Grinding Substrate:	Soft glass—soft metal
Grinding Abrasive-Coarse:	Aluminum oxide = 40, 30, 20
Grinding Abrasive-Fine:	Aluminum oxide = 10, 5, 3; care should be taken with 3 μm
Abrasive Carrier:	Dow Corning No. 200 fluid
Carrier Additives:	None
Rate of Grind:	Fast—3
Rate of Polish:	Fast—3
Polishing Lap Material:	Very soft pitch—Politex††† pad on optical flat
Polishing Compound:	Linde A—3/10 μm
Automatic Feed:	Not required
Special Considerations:	Very sensitive to temperature change. Highly soluble in water. Good cleavage.

†††Politex® is a registered trademark of Geoscience Instruments Corp.

Table 7-57. Data for Working Optical Materials (*Continued*)

Name:	Magnesium Fluoride
Chemical Formula or Trade Name:	MgF_2
Supplier:	Harshaw
Form:	Cut disks and cylinder 4 in. diameter, 4 in. long
Shaping:	Diamond tools and loose abrasives
Grinding Substrate:	Iron
Grinding Abrasive-Coarse:	Silicon carbide = 220, 320 grit sizes
Grinding Abrasive-Fine:	Aluminum oxide = 30, 12, 3 μm
Abrasive Carrier:	Water
Carrier Additives:	Soap and/or glycerine
Rate of Grind:	Slow—6
Rate of Polish:	Slow—6
Polishing Lap Material:	Medium-hard pitch or soft metal
Polishing Compound:	Pitch—Linde A. Soft metal—diamonds
Automatic Feed:	Not required
Special Considerations:	Very slightly soluble in water. Hardness: Mohs = 4+, Knoop = 415. No cleavage. A good, fast, but not the best optical finish may be had by using diamond paste with a soft lap metal or Politex. The best optical polish is obtained with a medium-hard optical pitch.
Name:	Potassium Bromide
Chemical Formula or Trade Name:	KBr
Supplier:	Harshaw
Form:	Cylinders 9½ in. diameter by 8 in. long, cut disks
Shaping:	String saw—loose abrasive—cleavage
Grinding Substrate:	Soft glass—soft metal—iron
Grinding Abrasive-Coarse:	Aluminum oxide = 40, 20
Grinding Abrasive-Fine:	Aluminum oxide = 12, 3; care should be taken with 3 μm
Abrasive Carrier:	Dow Corning No. 200 silicone fluid
Carrier Additives:	None
Rate of Grind:	Fast—2
Rate of Polish:	Fast—3
Polishing Lap Material:	Soft pitch—cover with thin layer of beeswax, Politex
Polishing Compound:	Linde A or B
Automatic Feed:	Not required
Special Considerations:	Temperature-sensitive. Water soluble. Hardness: Mohs = 2, Knoop = 7.0. Cleavage = fair to good. A fair polish and good figure can be obtained on soft pitch. A very good polish and a fair figure can be obtained on Politex.

Table 7-57. Data for Working Optical Materials (*Continued*)

Name:	Potassium Chloride
Chemical Formula or Trade Name:	KCl
Supplier:	Harshaw
Form:	Cylinders up to 9½ in. in diameter by 8 in. long, cut disks
Shaping:	String saw—loose abrasives
Grinding Substrate:	Soft glass—soft metal
Grinding Abrasive-Coarse:	Aluminum oxide = 40, 30, 20 μm
Grinding Abrasive-Fine:	Aluminum oxide = 12, 5, 3 μm
Abrasive Carrier:	Dow Corning No. 200 silicone fluid
Carrier Additives:	None
Rate of Grind:	Fast—3
Rate of Polish:	Polish—3
Polishing Lap Material:	Soft pitch—Politex
Polishing Compound:	Linde A
Automatic Feed:	Not required
Special Considerations:	Water soluble; Knoop hardness about 17, good cleavage, same as KBr.
Name:	Ruby and Sapphire
Chemical Formula or Trade Name:	Al_2O_3
Supplier:	Union Carbide
Form:	Whole boules, split boule, cut disks, rods (special shapes on request)
Shaping:	Diamond tools only
Grinding Substrate:	Coarse—aluminum, medium—iron, fine—tin
Grinding Abrasive-Coarse:	Diamond
Grinding Abrasive-Fine:	Diamond
Abrasive Carrier:	Oil—kerosene
Carrier Additives:	Kerosene—olive oil
Rate of Grind:	Slow—7
Rate of Polish:	Slow—10
Polishing Lap Material:	Iron—tin—phenolic
Polishing Compound:	Diamond—Linde A
Automatic Feed:	Not required
Special Considerations:	Ultimate finishing can be accomplished with Linde A on a tin or phenolic lap. Hardness—9.

Table 7-57. Data for Working Optical Materials (*Continued*)

Name:	Silicon
Chemical Formula or Trade Name:	Si
Supplier:	Exotic materials
Form:	Boules, cut disks
Shaping:	Diamond tools—loose abrasives
Grinding Substrate:	Iron—hard glass
Grinding Abrasive-Coarse:	Silicon carbide = 80, 220, 320 grit size
Grinding Abrasive-Fine:	Aluminum oxide = 30, 12, 3 μm
Abrasive Carrier:	Water
Carrier Additives:	Soap and/or glycerine
Rate of Grind:	Slow—7
Rate of Polish:	Slow—8
Polishing Lap Material:	Hard pitch
Polishing Compound:	Linde A
Automatic Feed:	Not required
Special Considerations:	Hardness—Mohs 7, Knoop 1150. For a fast polish, diamond paste on a pellon pad will produce quick results.
Name:	Sodium Chloride
Chemical Formula or Trade Name:	NaCl
Supplier:	Semi-Elements
Form:	Crystals—shaped disks
Shaping:	String saw—loose abrasive-cleavage
Grinding Substrate:	Soft glass—soft metal-iron
Grinding Abrasive-Coarse:	Aluminum oxide = 40, 20 μm
Grinding Abrasive-Fine:	Aluminum oxide = 12, 5 μm
Abrasive Carrier:	Dow Corning No. 200 silicone fluid, ethelene glycol
Carrier Additives:	None
Rate of Grind:	Fast—4
Rate of Polish:	Medium—5
Polishing Lap Material:	Pure soft pitch, Politex
Polishing Compound:	Linde A, 3/10 μm
Automatic Feed:	Not required
Special Considerations:	A fast scratch-free polish can be obtained with a Politex polishing pad. Work with a circular motion in the center, clean in ethyl alcohol. Temperature sensitive; water soluble; Knoop hardness = 18, cleavage = very good cubic.

Table 7-57. Data for Working Optical Materials (*Continued*)

Name:	Spinel
Chemical Formula or Trade Name:	$MgAl_2O_4$
Supplier:	Union Carbide
Form:	Boules—cut disks, and windows (special shapes on request)
Shaping:	Diamond tools
Grinding Substrate:	Iron
Grinding Abrasive-Coarse:	Diamond—silicon carbide
Grinding Abrasive-Fine:	Diamond—aluminum oxide
Abrasive Carrier:	Water
Carrier Additives:	Soap and/or glycerine
Rate of Grind:	Fast with diamonds—slow with aluminum oxide
Rate of Polish:	Slow—8
Polishing Lap Material:	Hard black optical pitch—soft metal
Polishing Compound:	Diamond—Linde A
Automatic Feed:	Not required
Special Considerations:	Wherever possible the use of oversize tools will speed up the work. In some cases good work can be done in blocks on phenolic laps. Hardness: Mohs = 8.
Name:	Thallium Bromide
Chemical Formula or Trade Name:	KRS-5, Ti(BrD)
Supplier:	Harshaw—cylinders 5 in. in diameter, 3 in. tall
Form:	Boules—shaped into disks
Shaping:	Wire saw—band saw loose abrasives
Grinding Substrate:	Soft glass—soft metal
Grinding Abrasive-Coarse:	Aluminum oxide = 40, 30, 20 μm
Grinding Abrasive-Fine:	Aluminum oxide = 12, 5, 3 μm
Abrasive Carrier:	Water
Carrier Additives:	Soap
Rate of Grind:	Fast—4
Rate of Polish:	Fast—4
Polishing Lap Material:	Very soft pitch—soft wood or Politex
Polishing Compound:	Aluminum oxide—Linde A
Automatic Feed:	No
Special Considerations:	Care should be taken to ensure that the grinding compounds do not embed in the material. Knoop Hardness = 40, poison, relatively unsoluble.

Table 7-57. Data for Working Optical Materials (*Continued*)

Name:	Yittrium Aluminum Garnet
Chemical Formula or Trade Name:	YAG
Supplier:	Union Carbide
Form:	Boules, cut disks, and doped rods (special shapes on request)
Shaping:	Diamond tools—loose abrasives
Grinding Substrate:	Iron
Grinding Abrasive-Coarse:	Diamond—silicon carbide—aluminum oxide
Grinding Abrasive-Fine:	Diamond—aluminum oxide
Abrasive Carrier:	Oil or water
Carrier Additives:	Detergent and/or glycerine
Rate of Grind:	Slow—3
Rate of Polish:	Slow—2
Polishing Lap Material:	Hard pitch—soft metal
Polishing Compound:	Diamond or aluminum oxide—(Linde A)
Automatic Feed:	Not required
Special Considerations:	Since this is a hard crystal, be sure the polish is good.
Name:	Aluminum metal
Chemical Formula or Trade Name:	Al
Supplier:	Kaiser—Alcoa
Form:	Cut to order
Shaping:	Lath, mill
Grinding Substrate:	Fixed aluminum oxide sanding paper—iron
Grinding Abrasive-Coarse:	Al_2O_3 fixed or loose
Grinding Abrasive-Fine:	Al_2O_3 fixed or loose
Abrasive Carrier:	Oil—WD40, Dow No. 200 silicone fluid
Carrier Additives:	None
Rate of Grind:	Slow
Rate of Polish:	Slow
Polishing Lap Material:	Soft pitch—Politex
Polishing Compound:	Al_2O_3
Automatic Feed:	
Special Considerations:	A simple wad of cotton soaked in WD40 or silicone fluid with Linde A will give a fine fast polish. An optical figure can be obtained on soft pitch with Linde A and silicone fluid.

Table 7-57. Data for Working Optical Materials (*Continued*)

Name:	Beryllium Oxide
Chemical Formula or Trade Name:	BeO
Supplier:	
Form:	Disks
Shaping:	Diamond tools
Grinding Substrate:	Iron
Grinding Abrasive-Coarse:	Diamonds
Grinding Abrasive-Fine:	Diamonds
Abrasive Carrier:	Oil—kerosene
Carrier Additives:	None
Rate of Grind:	Slow—2
Rate of Polish:	Slow—2
Polishing Lap Material:	Iron—soft metal
Polishing Compound:	Diamonds
Automatic Feed:	No
Special Considerations:	Material is toxic. Because of the material hardness, more weight can be applied during the polishing operation. Select a lap-to-work ratio that will facilitate the principle of equal work routine. This material comes sintered and is difficult to polish to optical tolerances.
Name:	Beryllium Copper
Chemical Formula or Trade Name:	BeCu
Supplier:	
Form:	Rods and shaped to order
Shaping:	Machine tools—lath and mill
Grinding Substrate:	Fixed abrasive—paper on iron
Grinding Abrasive-Coarse:	Al_2O_3
Grinding Abrasive-Fine:	Al_2O_3
Abrasive Carrier:	Water
Carrier Additives:	Detergent
Rate of Grind:	Fast
Rate of Polish:	Slow—6
Polishing Lap Material:	Medium soft pitch
Polishing Compound:	Linde A 3/10 μm Al_2O_3
Automatic Feed:	Not required
Special Considerations:	Material is toxic. Handle with care. Material should be annealed before final figuring.

Table 7-57. Data for Working Optical Materials (*Continued*)

Name:	Boron Carbide
Chemical Formula or Trade Name:	
Supplier:	Coors, Golden, Colorado
Form:	Cut disks
Shaping:	Diamond tools
Grinding Substrate:	Iron
Grinding Abrasive-Coarse:	Diamond
Grinding Abrasive-Fine:	Diamond
Abrasive Carrier:	Oil
Carrier Additives:	None
Rate of Grind:	Slow—10
Rate of Polish:	Slow—10
Polishing Lap Material:	Iron
Polishing Compound:	Diamond
Automatic Feed:	No
Special Considerations:	Be careful to select a tool size that allows the piece to be figured to the required tolerance.
Name:	Hardened Tool Steel
Chemical Formula or Trade Name:	
Supplier:	
Form:	As required
Shaping:	Fixed abrasive grinding
Grinding Substrate:	Fixed abrasive grinding
Grinding Abrasive-Coarse:	Silicon carbide
Grinding Abrasive-Fine:	Aluminum oxide
Abrasive Carrier:	Oil or water depending on steel
Carrier Additives:	Detergent
Rate of Grind:	Slow—8
Rate of Polish:	Medium—6
Polishing Lap Material:	Pitch or phenolic depending on shape and application
Polishing Compound:	Linde A—diamonds
Automatic Feed:	No
Special Considerations:	Good results have been accomplished with Linde A on phenolic. For optical figure use hard pitch.

Table 7-57. Data for Working Optical Materials (*Continued*)

Name:	360 Stainless
Chemical Formula or Trade Name:	
Supplier:	
Form:	As required
Shaping:	Machine tools and fixed abrasive machining
Grinding Substrate:	Iron
Grinding Abrasive-Coarse:	Aluminum oxide
Grinding Abrasive-Fine:	Aluminum oxide
Abrasive Carrier:	Water
Carrier Additives:	Soap
Rate of Grind:	Slow
Rate of Polish:	Slow
Polishing Lap Material:	Pitch—Politex
Polishing Compound:	Linde A
Automatic Feed:	Yes
Special Considerations:	The addition of soap and glycerine tends to help reduce sleeks and background scatter. Too great a pressure will generate heat and smear the metallic structure.
Name:	Anodized Aluminum
Chemical Formula or Trade Name:	Al_2O_3
Supplier:	
Form:	
Shaping:	Anodize after machined and fine ground
Grinding Substrate:	Iron
Grinding Abrasive-Coarse:	Al_2O_3 or diamond
Grinding Abrasive-Fine:	Al_2O_3 or diamond
Abrasive Carrier:	Water—oil
Carrier Additives:	None
Rate of Grind:	Slow—fast
Rate of Polish:	Slow—fast
Polishing Lap Material:	Pitch—pellon
Polishing Compound:	Linde A—diamond
Automatic Feed:	None
Special Considerations:	The anodized surface is very thin and care should be exercised not to polish through.

Table 7-57. Data for Working Optical Materials (*Continued*)

Name:	Cadmium Sulfide
Chemical Formula or Trade Name:	
Supplier:	Eastman Kodak Co.
Form:	Hot-pressed sheets
Shaping:	Loose abrasive
Grinding Substrate:	Glass—soft metal or iron
Grinding Abrasive-Coarse:	Al_2O_3 = 40, 20 μm
Grinding Abrasive-Fine:	Al_2O_3 = 12, 3 μm
Abrasive Carrier:	Water
Carrier Additives:	Soap and/or glycerine
Rate of Grind:	Fast—4
Rate of Polish:	Fast—4
Polishing Lap Material:	Soft pitch
Polishing Compound:	Linde A
Automatic Feed:	No
Special Considerations:	Material is toxic.
Name:	Electroless Nickel
Chemical Formula or Trade Name:	
Supplier:	
Form:	Coating applied to metal substrate
Shaping:	Loose abrasives
Grinding Substrate:	Pitch
Grinding Abrasive-Coarse:	Aluminum oxide = 30 μm
Grinding Abrasive-Fine:	Aluminum oxide = 12, 3 μm
Abrasive Carrier:	Water
Carrier Additives:	Soap and/or glycerine
Rate of Grind:	Slow—use only enough weight to hold work on the lap
Rate of Polish:	Medium
Polishing Lap Material:	Soft pitch covered with beeswax
Polishing Compound:	Linde A
Automatic Feed:	No
Special Considerations:	Care should be exercised to prevent comtamination. Excessive weight will deform the piece.

Table 7-57. Data for Working Optical Materials (*Continued*)

Name:	Glassy Carbon
Chemical Formula or Trade Name:	C
Supplier:	Coors
Form:	Cut disks
Shaping:	Loose abrasives
Grinding Substrate:	Iron—hard glass
Grinding Abrasive-Coarse:	Aluminum oxide = 40, 20 μm
Grinding Abrasive-Fine:	Aluminum oxide = 10, 3 μm
Abrasive Carrier:	Water
Carrier Additives:	Glycerine and/or soap
Rate of Grind:	Slow—7
Rate of Polish:	Slow—8
Polishing Lap Material:	Hard pitch
Polishing Compound:	Linde A
Automatic Feed:	
Special Considerations:	Material shows some tendency to be harder in one direction. The sample worked tended to develop a nonsymmetrical figure.
Name:	Irtran 6¶ (no longer available)
Chemical Formula or Trade Name:	
Supplier:	Eastman Kodak Co.
Form:	Cut and shaped disks
Shaping:	Loose abrasive
Grinding Substrate:	Iron or glass
Grinding Abrasive-Coarse:	Aluminum oxide = 40, 20 μm
Grinding Abrasive-Fine:	Aluminum oxide = 12, 3 μm
Abrasive Carrier:	Water
Carrier Additives:	Soap—glycerine
Rate of Grind:	Fast—3
Rate of Polish:	Slow—6
Polishing Lap Material:	Soft optical pitch
Polishing Compound:	Linde A
Automatic Feed:	No
Special Considerations:	Material tends to phase change—short rapid strokes help to control this condition.

¶ Irtran® is a registered trademark of the Eastman Kodak Co.

Table 7-57. Data for Working Optical Materials (*Continued*)

Name:	TI-20
Chemical Formula or Trade Name:	TI-20
Supplier:	Texas Instruments
Form:	Cast forms—cut disks
Shaping:	Diamond tools—loose abrasives
Grinding Substrate:	Soft glass—soft metals—iron
Grinding Abrasive-Coarse:	Aluminum oxide = 40, 30 μm
Grinding Abrasive-Fine:	Aluminum oxide = 12, 3 μm (careful with 3)
Abrasive Carrier:	Water
Carrier Additives:	Glycerine
Rate of Grind:	Fast—3
Rate of Polish:	Fast—3
Polishing Lap Material:	Soft pitch
Polishing Compound:	Linde A
Automatic Feed:	No
Special Considerations:	Work carefully, this material will chip very easily. Better finishes can be produced if the polishing mixture is allowed to work thin. Add only water for the last several wets.
Name:	Molybdenum
Chemical Formula or Trade Name:	
Supplier:	
Form:	
Shaping:	
Grinding Substrate:	
Grinding Abrasive-Coarse:	
Grinding Abrasive-Fine:	
Abrasive Carrier:	
Carrier Additives:	
Rate of Grind:	
Rate of Polish:	Slow—1
Polishing Lap Material:	Hard pitch
Polishing Compound:	Linde A
Automatic Feed:	Yes—for good finish
Special Considerations:	Hard pitch, wax covered with relatively high unit area pressures are necessary for a good polish, low scatter, and figure control. Seems to be some variance from piece to piece.

(3) Percent of Glycerine Used in Normal Grinding Solutions
  Ratio of 1 to 3, glycerine to water
  Ratio of 1 to 5, glycerine to water for polishing
(4) Hardness of Pitch

Graded Soft	Relative Hardness per Hilger-watts
Medium-soft	25
Medium	15
Medium-hard	10
Hard	7
Very Hard	3

(5) Additives to Polishing Materials
  Glycerine—reduces drying
  Evenflow—helps keep polishing agent in suspension
  Detergent—wets solution—Botler contact with polisher

7.6. Window Design—Plate Deformation [7-78]

The deformation for a circular plate, with three-point loading and zero deflection at the supports, located at $\theta = 0, 2\pi/3, 4\pi/3$, is given by

$$w(\rho) = \frac{pa^4}{64D} \left[\frac{5+\gamma}{1+\gamma} - \frac{2(3+\gamma)}{1+\gamma} \rho^2 + \frac{32\rho^4}{3+\gamma} \sum_{m=3,6,9} \left\{ \left[\frac{2}{m(m-1)(m+1)} \right. \right. \right.$$
$$\left. + \frac{2(3+\gamma)}{m(m-1)(1-\gamma)} \right] - \rho^m \cos m\theta \left[\frac{1}{m(m-1)} + \frac{2(1+\gamma)}{(1-\gamma)m^2(m-1)} - \frac{\rho^2}{m(m+1)} \right] \right\}$$

$$(7\text{-}40)$$

where w = window deformation h = window thickness
    p = pressure loading γ = Poisson's ratio
    a = window radius ρ = fractional radial distance
    $D = Eh^3[12(1-\gamma^2)]^{-1}$ θ = angle around plate
    E = Young's modulus

Any consistent set of units can be used. For a circular plate with a ring-support (i.e., at 6 or more points)

$$w(\rho) = \frac{pa^4}{64D}(1-\rho^2) \left[\left(\frac{5+\gamma}{1+\gamma} \right) - \rho^2 \right] \tag{7-41}$$

In addition, the radial stress is given by

$$\sigma_r = \frac{3pa^2}{8h^2}(3+\gamma)(1-\rho^2) \tag{7-42}$$

While the angular stress is

$$\sigma_\theta = \frac{3pa^2}{8h^2}[(3+\gamma)-(1+3\gamma)\rho^2] \tag{7-43}$$

For a rectangular plate with dimensions a and b, with a simple, continuous, edge support

$$w = \frac{4pa^4}{\pi^5 D} \sum_{m=1,3,5} \frac{1}{m^5} \left[1 - \frac{\alpha_m \tanh \alpha_m + 2}{2 \cosh \alpha_m} \cosh \beta_m \right.$$

$$\left. + \frac{\alpha_m\, 2yb^{-1}}{2 \cosh \alpha_m} \sinh \beta_m \right] \sin \gamma_m \qquad (7\text{-}44)$$

where $\alpha_m = m\pi b/2a$
$\beta_m = 2\alpha_m y/b$
$\gamma_m = m\pi x/a$

7.7. Interference Filters [7-97]

7.7.1. Introduction.
Filters may be classified by either their transmission characteristics or the physical phenomena upon which their action is based. A long-wave pass filter transmits all radiation with wavelengths greater than the specified value; a short-wave pass filter passes all radiation with wavelengths shorter than the specified value; a bandpass filter transmits only between two wavelengths.

Some of the physical phenomena that determine filter action are selective reflection and refraction, scattering, polarization, interference, and selective absorption.

7.7.2. Terminology.
The description of filters—and even curves of their transmission—has not been standardized. The terms given below have received some measure of acceptance, but for precise knowledge of characteristics nothing can substitute for a transmittance or reflectance curve. For the following definitions, refer to Figure 7-46.

Passband: The primary wavelength interval of transmission of a transmission filter, or reflection of a reflection filter.

Stopband: The primary region of rejection of a filter.

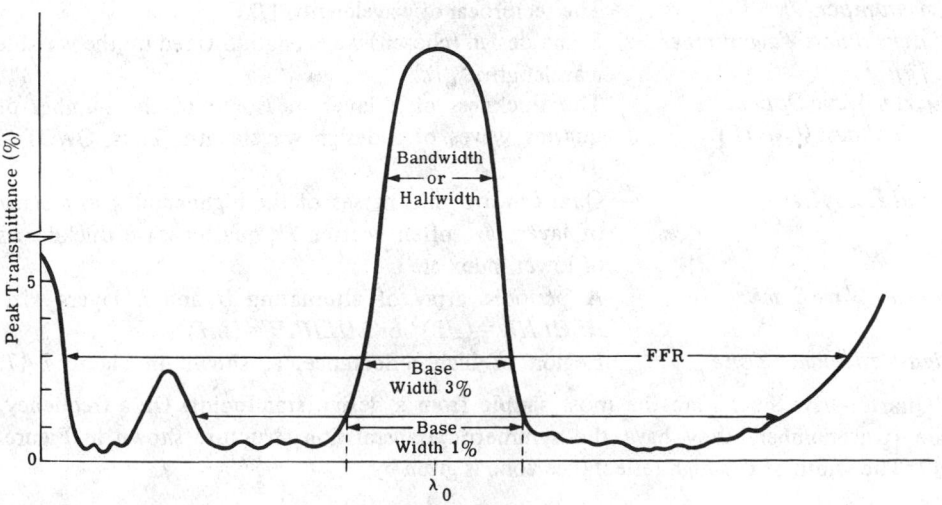

Fig. 7-46. Illustration of filter nomenclature.

Background Region: The region of low transmission or reflection of a transmission filter.

Center Wavelength (λ_0): The wavelength at the center of the passband. For interference filters, λ_0 is usually the mean of the long- and short-wave cutoffs.

Bandwidth or Half-width (HW): The full width of the passband at half maximum—often expressed as a percentage of λ_0.

Basewidth (BW): The width of the passband of 3% or 1% (or some other small percentage) of peak transmittance. It is more precise and useful to specify the 3% width, etc., as $(\Delta\lambda)_{0.03}$: thus HW becomes $(\delta\lambda)_{0.5}$.

Peak Transmittance (τ_0): The maximum transmittance in the passband. For interference filters this is often specified as a percent of the uncoated substrate.

Substrate Transmittance (τ_s): Transmittance of the uncoated substrate.

Free Filter Range (FFR): The wavelength interval over which the background is less than a specified amount except where the passband is.

Cutoff and Cuton Wavelengths (λ_c): The wavelengths of the limits of the passband, often at the 50%, 10%, or 5% transmission points.

Slope: The linear approximation to the cuton or cutoff slope, expressed as the ratio $(\lambda_{0.8}-\lambda_c)/\lambda_c$, where $\lambda_{0.8}$ is the wavelength of 80% transmittance.

Filter users should take care to understand the nomenclature given by individual manufacturers. In particular, they should note whether τ_0 is given in terms of the ratio of the filter transmission to the substrate transmission, whether λ_c is given as a 5%, 3%, or 1% cutoff, and whether FFR is specified for 1%, 0.1%, etc.

7.7.3. General Theory of Interference Filters

Phase Difference (δ): The phase difference in radians is $2\pi\tilde{\nu}nt_g$.

Wavenumber $(\tilde{\nu})$: The reciprocal of wavelength, $1/\lambda$.

Dimensionless Wavenumber $(\tilde{g})$: Some design (chosen) wavelength divided by the variable wavelength, λ_0/λ.

Quarter-Wave Optical Thickness (QWOT): The thickness of a layer measured in the number of quarter waves of a design wavelength. Thus, QWOT = $4nt_g$.

H and L Layers: Quarter-wave thicknesses of the higher index in a stack of layers are often written H; quarter-wave thicknesses of lower index are L.

Quarter-Wave Stack: A periodic array of alternating H and L layers—viz., $HLHLHL = (HL)^3$ or $LHLHLH = (LH)^3$.

High-Reflectance Zone: Region of high reflectance, as shown in Figure 7-47.

Quarter-wave stacks are the most simple from a design standpoint. On a frequency scale (wavenumber) they have the symmetric transmission structure shown in Figure 7-47. The width of the high reflectance zone is given by

$$\Delta\tilde{g} = \frac{4}{\pi} \text{ arc sin}\left(\frac{n_H - n_L}{n_H + n_L}\right) \tag{7-45}$$

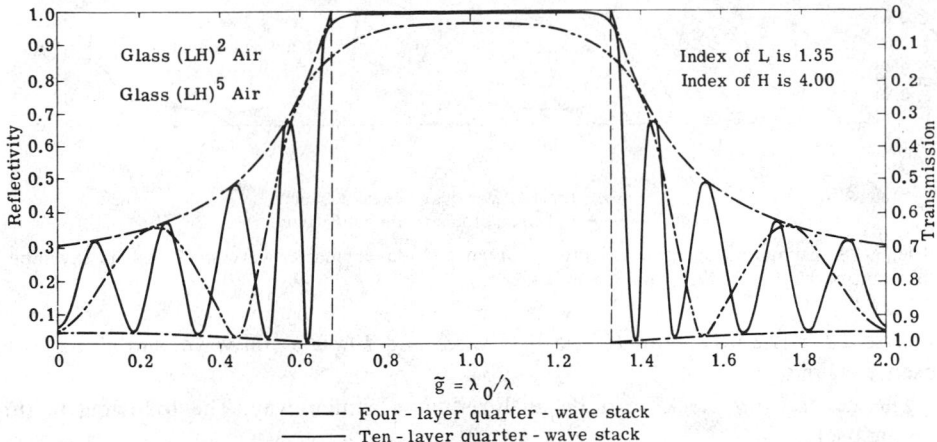

Fig. 7-47. Computed spectral reflectivity of a four-layer and a ten-layer quarter-wave stack, and envelope of maximum reflectivity.

where n_H = higher index of refraction
n_L = lower index of refraction
$\tilde{g}$ = λ_0/λ

The maximum reflectivity is given by

$$r_{max} = \frac{P + P^{-1} - 2}{P + P^{-1} + 2} \qquad (7\text{-}46)$$

where

$$P = \left[\frac{n_l}{n_{l-1}} \frac{n_{l-2}}{n_{l-3}} \cdots \frac{n_2}{n_1}\right]^2 \frac{n_0}{n_s} \qquad l \text{ even}$$

$$P = \left[\frac{n_l}{n_{l-1}} \cdots n_1\right]^2 \frac{1}{n_0 n_s} \qquad l \text{ odd} \qquad (7\text{-}47)$$

where l = number of layers
n_0 = refractive index of the medium
n_s = refractive index of the substrate

For $P \gg P^{-1}$,

$$r_{max} = \frac{P - 2}{P + 2} \qquad (7\text{-}48)$$

The theoretical curve given in Figure 7-47 is useful for beginning calculations. Figure 7-48 shows the transmission of a quarter-wave stack over a larger range of $\tilde{g}$.

Stacks with unequal optical thickness ratios can also be very useful. The 2.1 stack has the configuration

$$LLH\,LLH\,LLH \ldots = (LLH)^m$$

The first-order, high-reflectance zone occurs at $\tilde{g} = 1$, when LLH is $\lambda_0/2$. The second-order zone occurs when $\tilde{g} = 2$ and LLH is λ_0. There is no high reflectance when $\tilde{g} = 3$

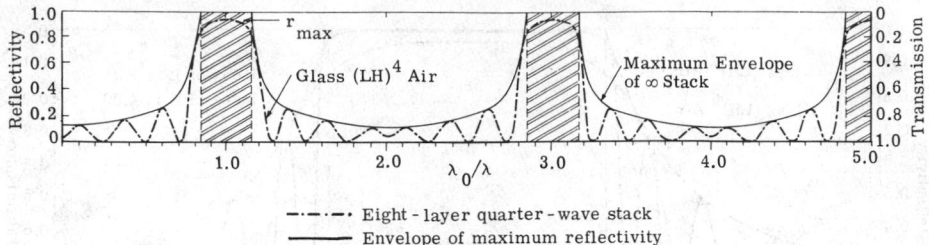

Fig. 7-48. Computed spectral reflectivity of an eight-layer quarter-wave stack and its envelope of maximum reflectivity.

because *LLH* is $3\lambda_0/2$; thus *H* is a half wave and *LL* is a full wave, and all *LLH*s are absentee layers.

The general *p:q* stack* can be analyzed in a similar way. The following features may be useful.

   (1) The high-reflectance zone may not be a center of symmetry.

   (2) The high-reflectance zone of a quarter-wave stack is wider than other *p:q* stacks.

   (3) The number of oscillations outside the high-reflectance zone increases as the number of layers is increased.

   (4) Other things being equal, the quarter-wave stack has the highest reflectivity.

Additional layers can be added to (1) increase reflection in the stop band, or (2) decrease reflection in the passband. The use of *H* layers at both ends of the basic period will increase the overall reflectivity:

$$HLHLHLH$$

To decrease reflection, one can replace the layer with an equivalent single layer and antireflection coatings designed for the desired wavelength. The stack can also be varied by computer techniques based on variational principles.

7.7.4. Long-Wave Pass Interference Filters. The design of a long-wave pass interference filter is based on Figure 7-47. Low-reflectance regions are regions of high transmission, and can be designed on that basis. The curves change with different ratios of n_H to n_L and different numbers of layers. The design proceeds with the choice of a useful substrate and a design λ_0. Some changes can then be made. The long-wave cutoff is determined by either the substrate absorption or the second-order maximum. (See Figure 7-48.) Commercially available long-wave pass filters are described in Section 7.13. They have the following properties:

   (1) The slope of the cuton increases with the number of layers.

   (2) The maximum reflectance increases with the number of layers.

   (3) The width of the reflectance zone increases as n_H/n_L increases.

   (4) A higher-order reflectance peak has a sharper cuton but a narrower transmission region ($\Delta\lambda$).

   (5) Angle shift is minimized by high values for n_H and n_L or by more high-index materials in the basic period.

---

*The term *p:q* is the ratio of *p*, the high index of some optical thickness (not a quarter-wave), to *q*, the low index of a different optical thickness (not a quarter-wave).

7.7.5. Short-Wave Pass Interference Filters. The comments applicable to long-wave pass filters (Section 7.7.4) also apply here. Short-wave pass filters are usually designed from quarter-wave stacks because these have the longest region of high transmission to the short-wave (high-frequency) side of the high-reflectance zone. Then an antireflection coat is applied to the stack in the transmission region. Some commercially available short-wave pass filters are described in Section 7.13.

7.7.6. Bandpass Interference Filters. Every filter is a bandpass filter. If the desired pass region is smaller than that obtained by a long-wave pass filter, a short-wave pass can be added; they can be deposited on opposite sides of the substrate. Narrower bandpasses are obtained by interference techniques, similar to that for the Fabry-Perot interferometer. The transmission is given by

$$\tau = \frac{\tau_1 \tau_2}{(1 - \sqrt{r_1 r_2})^2}\left[1 + \frac{4\sqrt{r_1 r_2}\sin\left(\frac{2\pi n t_g}{\lambda} - \frac{\epsilon_1 + \epsilon_2}{2}\right)}{(1 - \sqrt{r_1 r_2})^2}\right] \tag{7-49}$$

where τ_1, τ_2, r_1 and r_2 = transmittance and reflectance of the plate coatings (looking from the gap)

ϵ_1 and ϵ_2 = phase shift upon reflection

$n t_g$ = optical path of the gap with width t_g

The transmission is a sinusoidal function of $1/\lambda$ or $\tilde{\nu}$. The region between adjacent transmission peaks is the free spectral range. In filter language this is the free filter range, or *FFR*. This spacing $\tilde{\nu}_f$ is given by

$$\tilde{\nu}_f = \frac{1}{2t_g} \tag{7-50}$$

where t_g = width of gap. The narrowness of a line is given by the "Q" (also called the resolving power):

$$"Q" = \frac{\lambda_0}{(\Delta\lambda)_{0.5}} = \frac{\tilde{\nu}_0}{(\Delta\tilde{\nu})_{0.5}} = \frac{(r_1 r_2)^{1/4}}{1 - \sqrt{r_1 r_2}} m\pi \tag{7-51}$$

If r_1 and r_2 are large and if ϵ_1 and ϵ_2 are constant over $\Delta\lambda$, then the line has a Lorentz shape. Narrowband filters can have the following construction:

$$A \qquad\qquad B$$

$$HLH\ LL\ HLH$$

This can be thought of as two filters separated by a half-wave of low-index material. The curve for this filter is shown in Figure 7-49. These filters are usually combined with blocking filters to isolate the narrow band. Some commercially available bandpass filters are described in Section 7.13.

7.7.7. "Square-Band" Interference Filters. This type of filter—not really a square band—can be designed as a general p:q stack. The design is

$$(HL)^m\ LL\ (HL)^m\ LL\ \ldots\ LL\ (HL)^m$$

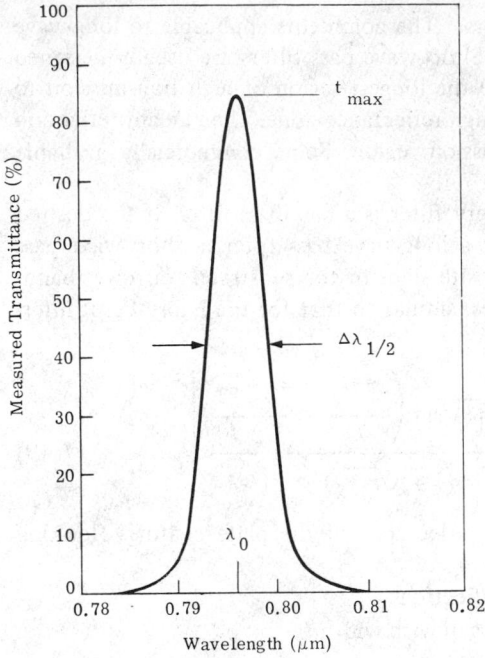

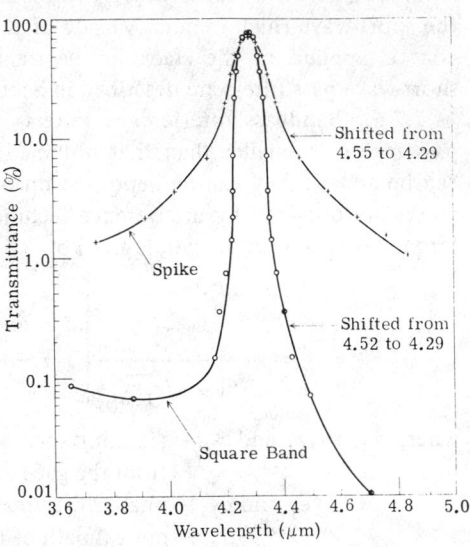

Fig. 7-49. Measured transmission of a narrow-band filter.

Fig. 7-50. Measured transmittance of two bandpass filters with nominal 2% halfwidth at 4.29 μm.

The filter is generally steeper and has a rippled top. The rejection is also better than that of a quarter-wave stack. A comparison of the "square" filter with the normal quarter-wave stack is given in Figure 7-50. These can also be thought of as multiple-wave filters.

7.7.8. The Filter Matrix. The relations between the electric and magnetic fields on two sides of the ith interface are given by

$$\begin{bmatrix} E \\ H \end{bmatrix} = M_i \begin{bmatrix} E' \\ H' \end{bmatrix} \tag{7-52}$$

Here M_i is the characteristic matrix of the ith surface. M_i is given by

$$M_i = \begin{bmatrix} \cos \delta_i & jn_{-i} \sin \delta_i \\ jn_i \sin \delta_i & \cos \delta_i \end{bmatrix} \tag{7-53}$$

The relation between

$$\begin{bmatrix} E_0 \\ H_0 \end{bmatrix} \quad \text{and} \quad \begin{bmatrix} E_m \\ H_m \end{bmatrix}$$

is

$$\begin{bmatrix} E_0 \\ H_0 \end{bmatrix} = \sum_{i=1}^{m} M_i \begin{bmatrix} E_m \\ H_m \end{bmatrix} = M \begin{bmatrix} E_m \\ H_m \end{bmatrix} = \begin{bmatrix} a_{11} & ja_{12} \\ ja_{21} & a_{22} \end{bmatrix} \begin{bmatrix} E_m \\ H_m \end{bmatrix} \tag{7-54}$$

The determinant $|M|$ is 1, so that the reflectivity, r, can be calculated from a knowledge of three elements.

For any periodic layer, the period can be reduced to a fictitious bilayer; if the period occurs m times, then

$$\begin{bmatrix} E_0 \\ H_0 \end{bmatrix} = (M_1 M_2)^m \begin{bmatrix} 1 \\ n_s \end{bmatrix} \tag{7-55}$$

For a symmetrical layer, the period can be replaced by a monolayer.

7.7.9. The Herpin Equivalent Layer. The thickness, t_g, of the layer can be written in terms of the phase δ_n and index n_n

$$
\begin{aligned}
\delta_n &= 360 n t_g / \lambda \quad \text{(degrees)} \\
&= 2\pi n t_g / \lambda \quad \text{(radians)} \\
&= n t_g / \lambda \quad \text{(wavelengths)} \\
&= \text{arc cos } a_{11}
\end{aligned}
$$

$$n_n = a_{21} [1 - (a_{11})^2]^{-1/2} \tag{7-56}$$

where $a_{21} = n \sin \delta$
$a_{11} = \cos\delta$

7.7.10. Analogies with Transmission-Line Theory [7-79]. The matching theorems involving calculation of line admittance, chracteristic admittance, reflection coefficient, etc., can be applied to optical multilayers by treating the refractive index as the admittance. Some useful equations are as follows.

For the nth element of an infinite lumped-constant line (s = series; sh = shunt):

$$i_n = Ae^{\gamma n}$$

$$\cosh \gamma = 1 + \frac{Z_s}{2Z_{sh}}$$

$$e^\gamma = 1 + \frac{Z_s}{2Z_{sh}} \pm \sqrt{\frac{Z_s}{Z_{sh}} + \frac{Z_s}{2Z_{sh}}^2}$$

$$Z_0 = \sqrt{Z_s Z_{sh} + (Z_s/2)^2}$$

$$\sinh \gamma = Z_0/Z_{sh} \tag{7-57}$$

For an infinite distributed-parameter line:

$$Z = R + j\omega L \qquad Y = G + j\omega C$$
$$Z_0 = \pm \sqrt{Z/Y} \qquad \gamma = \pm \sqrt{ZY} = Z_0 Y \qquad (7\text{-}58)$$

For a line of length l, terminated by Z_R:

$$R = \frac{Z_0 - Z_R}{Z_0 + Z_R} = \frac{Y_R - Y_0}{Y_R + Y_0}$$

$$Z(l) = Z_0 \frac{Z_0 \sinh \gamma l + Z_R \cosh \gamma l}{Z_0 \cos \gamma l + Z_R \sinh \gamma l}$$

$$Z_{cl} Z_{op} = (Z_0)^2 \qquad (7\text{-}59)$$

where Z_{cl} = short-circuit impedance
$\quad\ Z_{op}$ = open-circuit impedance

Impedance Matching. A section of lossless lumped line can be chosen to join generator and load for maximum power transfer if

$$(Z_0)^2 = Z_{in} Z_{term} \qquad (7\text{-}60)$$

and the length is a quarter wave.

7.7.11. Effects of Angle of Incidence. The center wavelength is a function of the angle of incidence, as shown in Figure 7-51. For non-normal incidence the center wavelength is shifted toward shorter wavelengths. An effective optical thickness can be used:

$$(nt_g)_{eff} = nt_g \cos \theta \qquad (7\text{-}61)$$

where θ is the angle of incidence. This technique can be used for a few layers, but since the angle of incidence for each layer is a function of the original angle and all the preceding layers, the technique is cumbersome. Substituted into the matrix formulation, however, the effective optical thickness is again useful.

An approximation for the amount of shift in the center wavelength is given by

$$\lambda_\theta = \lambda_0 \frac{[n^2 - \sin^2\theta]^{1/2}}{n} \qquad (7\text{-}62)$$

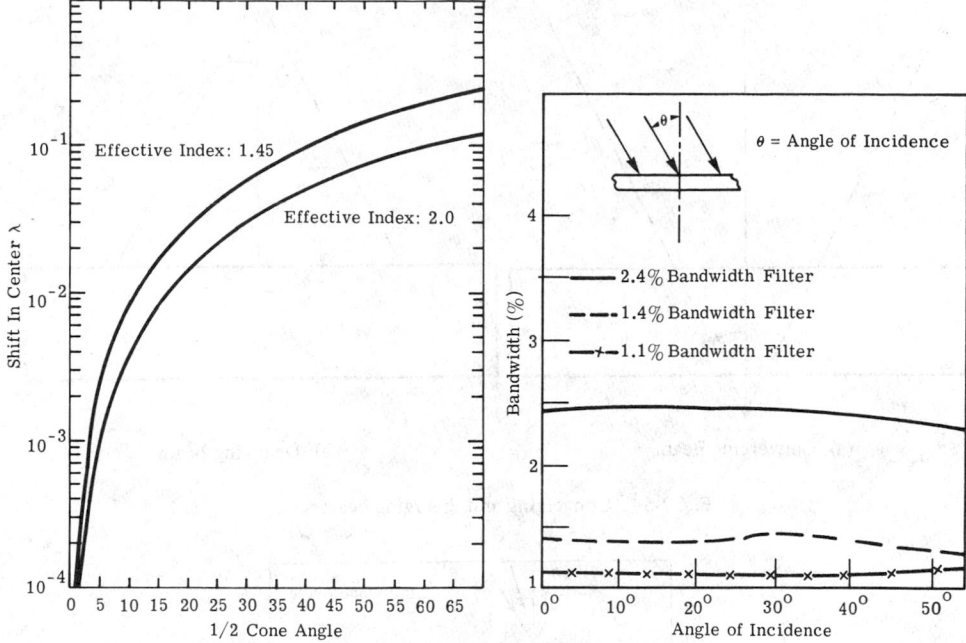

Fig. 7-51. Interference filters [7-80].

Fig. 7-52. Collimated radiation. Bandwidth variation of three narrow bandpass filters [7-81].

where λ_θ = central wavelength at angle of incidence θ

λ_0 = central wavelength at normal incidence

n = effective index of refraction of the filter

θ = angle of incidence

The variation in bandwidth is essentially negligible over a wide range of incident angles, as shown in Figure 7-52. For non-normal incidence, the filter also becomes polarizing; for simple layers, the standard equations for reflectance are useful, or they can be put into matrix form for iteration. Angle effects can be minimized by using higher indexes or more material of a higher index in a layer. The effective thickness is shown in Figure 7-53.

7.7.12. Effects of Converging and Diverging Beams.
Converging and diverging beams can be represented by symmetrical cones of radiation with their centroids assumed to be normal to the filter surface. (See Figure 7-54.)

The center wavelength is a function of the half-cone angle, as shown in Figure 7-55. As in the case of non-normal incidence, the center wavelength is shifted toward shorter wavelengths for

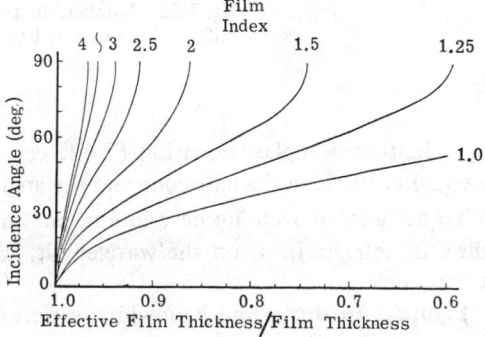

Fig. 7-53. Effective film thickness [7-82].

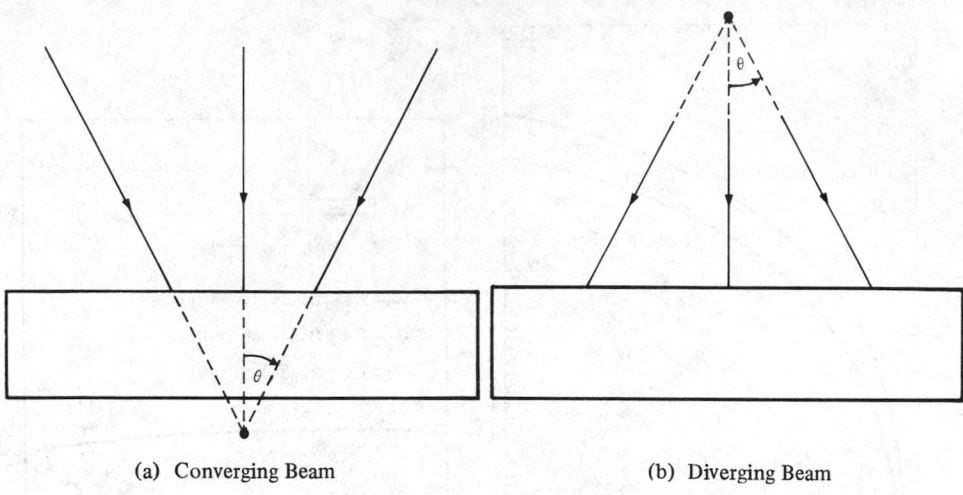

(a) Converging Beam   (b) Diverging Beam

Fig. 7-54. Converging and diverging beams.

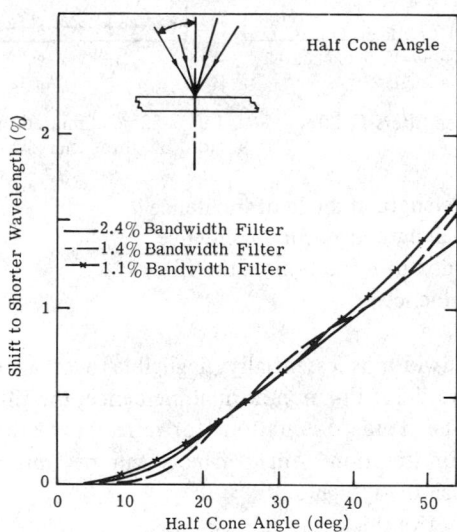

Fig. 7-55. Lambertian radiation. Wavelength
shift of three narrow bandpass filters [7-81].

larger half-cone angles. Equation (7-62) can be used as an approximation by assuming
an equality between the half-cone and the angle of incidence.

To get a more accurate description, one must include the dependence of the effective
index of refraction, n, on the wavelength, λ, and perform an integration over the cone
of half angle.

Figure 7-56 shows the bandwidth variation with half-cone angle for converging and
diverging beams (Lambertian radiation).

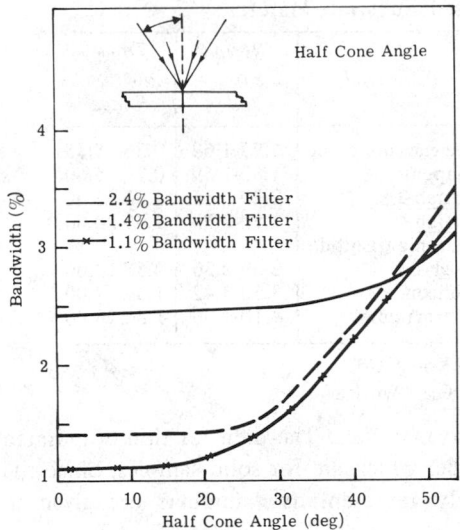

Fig. 7-56. Lambertian radiation. Bandwidth variation of three narrow bandpass filters [7-81].

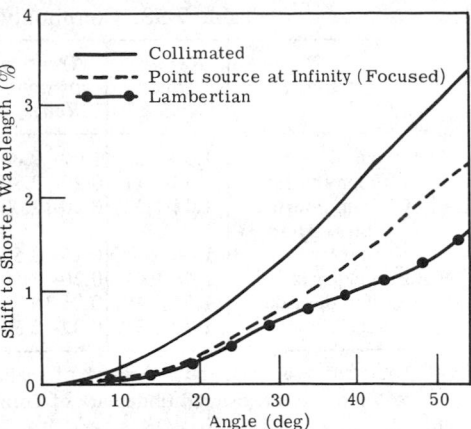

Fig. 7-57. Wavelength shift of 1.4% bandwidth filter [7-81].

Figure 7-57 compares the shift to shorter wavelengths for the non-normal angle of incidence beams (collimated) and the converging and diverging beams (Lambertian radiation) for a 1.4% bandwidth filter.

7.7.13. Effects of Temperature. Temperature variations change both the actual thickness and the indices of refraction of the layers in an interference filter. Hence, the wavelength shift is essentially a linear function of temperature because the optical path varies linearly with temperature. The equation is

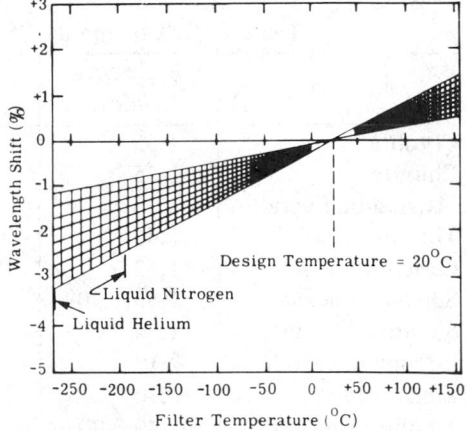

Fig. 7-58. Wavelength shift of interference filters with temperature variation [7-81].

$$\frac{\Delta n t_g}{\Delta T} = n \frac{\Delta t_g}{\Delta T} + t_g \frac{\Delta n}{\Delta T} \qquad (7\text{-}63)$$

where n = index of refraction
$\quad t_g$ = thickness
$\quad T$ = temperature

The shift is to longer wavelengths with an increase in temperature and to shorter wavelengths with a decrease in temperature, as shown in Figure 7-58.

7.7.14. Substrates and Films. Table 7-58 is a list of commonly used substrate materials. The physical data for these materials were given before in this Chapter. Sometimes it will be necessary to extrapolate the data to thinner samples.

Table 7-58. Commonly Used Substrate Materials

Material	Refractive Index	Transmission Range	Material	Refractive Index	Transmission Range
Irtran* 1	1.38-1.23	1.00- 9.00	Magnesium oxide	1.77-1.62	0.36- 5.35
Lithium fluoride	1.45-1.11	0.20- 9.80	Sapphire	1.83-1.59	0.27- 5.60
Calcium fluoride	1.44-1.32	0.20-12.00	Irtran 2	2.29-2.15	1.00-13.00
(also as Irtran 3)			Irtran 4	2.50-2.30	1.00-20.00
Vycor**	1.46	0.25- 3.50	Arsenic trisulfide		
Fused quartz	1.48-1.41	0.20- 4.50	glass	2.69-2.36	0.56-12.00
Barium fluoride	1.51-1.40	0.26-10.35	Silicon	3.50-3.42	1.36- 7.00
Glass	1.70-1.51	0.32- 2.50	Germanium	4.10-4.00	1.80-23.00

*Irtran® is a registered trademark of Eastman Kodak Co.
**Vycor® is a registered trademark of Corning Glass Works.

Table 7-59 is a list of commonly used film materials. The data for thin evaporated films are often different from those given earlier which are for solid samples. Since the values vary with deposition conditions, only representative numbers are given in Table 7-59.

Table 7-59. Commonly Used Film Materials [7-75]

Material	Refractive Index	Range of Transparency* from (nm)	to (μm)	Comments
Cryolite	1.35	<200	10	1
Chiolite	1.35	<200	10	1
Magnesium fluoride	1.38	230	5	2, 3
Thorium fluoride	1.45	<200	10	–
Cerium fluoride	1.62	300	>5	4
Silicon monoxide	1.45-1.90	350	8	5
Sodium chloride	1.54	180	>15	6
Zirconium dioxide	2.10	300	>7	2
Zinc sulfide	2.30	400	14	7
Titanium dioxide	2.40-2.90	400	>7	8
Cerium dioxide	2.30	400	5	2, 3
Silicon	3.50	900	8	–
Germanium	3.80-4.20	1400	>20	–
Lead telluride	5.10	3900	>20	–

1. Both materials are sodium-aluminum fluoride compounds, but differ in the ratio of Na to Al and have different crystal structure. Chiolite is preferable in the infrared, because it has less stress than cryolite.
2. These materials are hard and durable, especially when evaporated onto a hot substrate.
3. The long wavelength is limited by the fact that, when the optical thickness of the film is a quarter-wave at 5 μm, the film cracks because of the mechanical stress.
4. Other fluorides and oxides of rare earths have refractive indices in this range from 1.60 to 2.0
5. The refractive index of SiOx (called silicon monoxide) can vary from 1.45 to 1.90 depending upon the partial pressure of oxygen during the evaporation. Films with a refractive index of 1.75 and higher absorb at wave-lengths below 500 nm.
6. Sodium chloride is used in interference filters out to a wavelength of 20 μm. It has very little stress.
7. The refractive index of zinc sulfide is dispersive.
8. The refractive index of TiO_2 rises sharply in the blue spectral region.

*The range of transparency is for a film of quarter-wave optical thickness at this wavelength. These values are approximate and also depend quite markedly upon the conditions in the vacuum during the evaporation of the film.

7.8. Christiansen Filters [7-10, 7-83, 7-84]

These filters are made of small, closely packed particles of an infrared-transparent substance suspended in a liquid or a gas. The optical properties of the materials are so chosen that the indices of refraction of the particles and the suspending medium are the same at the wavelength that is to be transmitted. The $dn/d\lambda$ values of the liquid and the solid particles are chosen to be as widely different as possible. Thus, as the wavelength is progressively increased or decreased from the wavelength at which equality of the indices occurs, the difference in index between the particles and the suspending medium increases rapidly.

One form of Christiansen filter for the infrared is obtained by using quartz particles in air. Figure 7-59 shows the dispersion curve of ordinary quartz. The refractive index is unity at 7.4 μm. At this wavelength, therefore, quartz particles have the same refractive index as air, and high transmission occurs.

Table 7-60 lists other materials which, when suspended in air, can be used as Christiansen filters. Also shown are the wavelengths at which maximum transmission occurs. These are designated as Christiansen wavelengths.

Figure 7-60 shows the effect of quartz powder in a medium of pure CCl_4 and CS_2 as compared with quartz in air. Because values of dn/dT are relatively high, these filters are sensitive to temperature fluctuations. By the same token, in a controlled environment the center of the passband can be varied by changing the temperature.

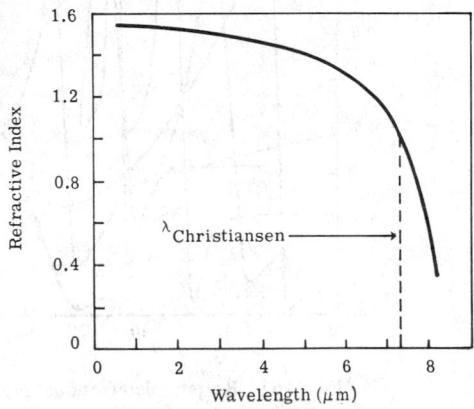

Fig. 7-59. Dispersion curve of quartz, showing Christiansen wavelength [7-10].

Table 7-60. Christiansen Wavelengths of Selected Materials

Crystal in Air	Christiansen Wavelength (μm)
Quartz	7.3
LiF	11.2
MgO	12.2
NaCl	32.0
NaBr	37.0
NaI	49.0
KBr	52.0
KI	64.0
RbI	73.0
TlI	90.0

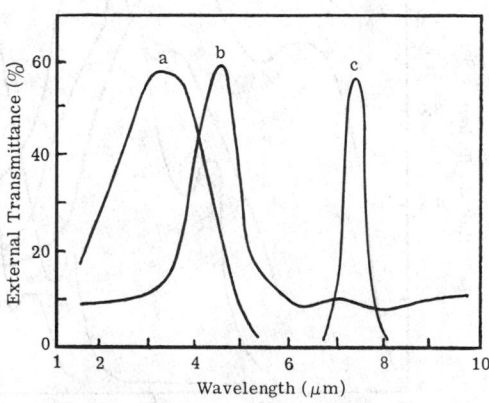

Fig. 7-60. Position of the Christiansen peak for quartz powder in liquids. a) Quartz in a 50% by volume mixture of CS_2 and CCl_4. b) Quartz in pure CCl_4. c) Quartz in air [7-86].

7.9. Selective Reflection Filters [7-85]

Selective reflection filters are made of crystalline materials that show selective reflection at certain wavelengths. These filters are useful to about 200 μm.

In practice, radiation from a suitable source is collimated and directed at the surface of a polished crystal whose residual ray occurs at the wavelength to be selectively reflected. After three or four successive reflections from similar crystal plates, only the residual ray is present with any appreciable intensity, the other wavelengths having been attenuated by a factor of several thousand. Figures 7-61 and 7-62 indicate the wavelengths of the residual rays of a number of materials.

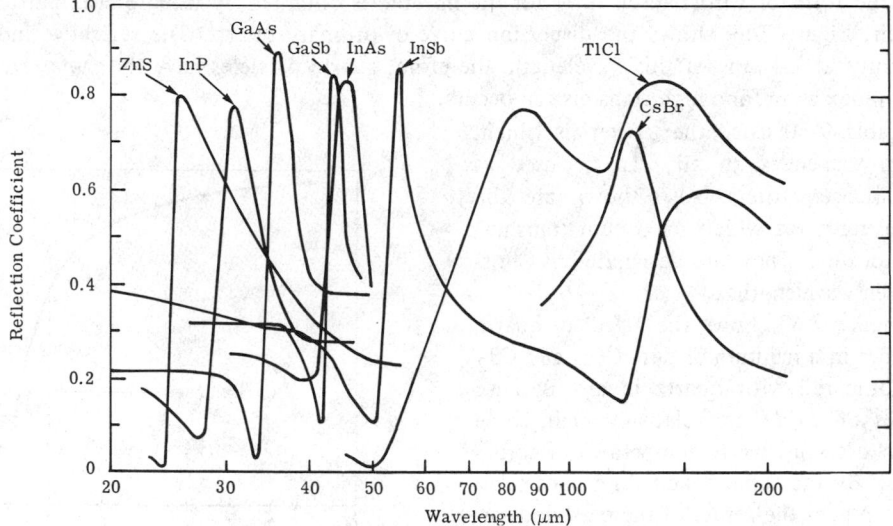

Fig. 7-61. Reststrahlen (residual ray) frequencies of alkali halide crystals.

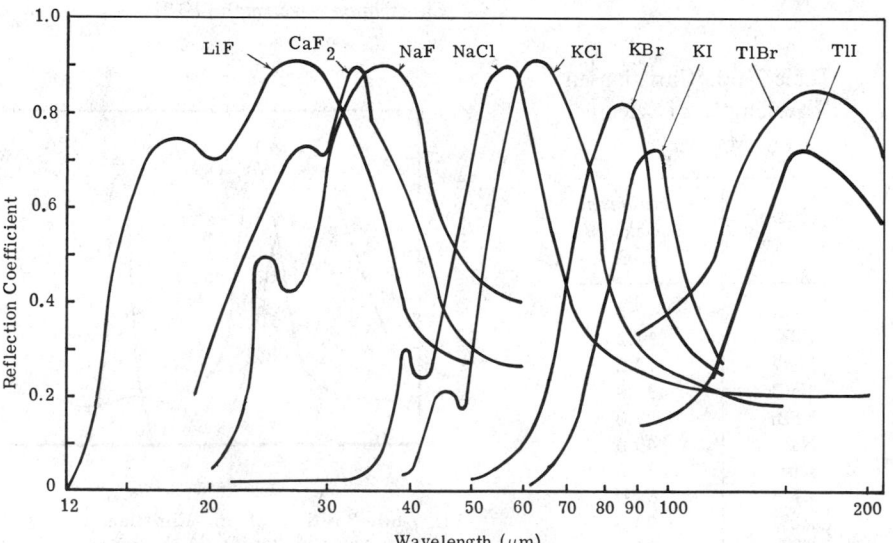

Fig. 7-62. Reststrahlen (residual ray) frequencies of polar crystals.

7.10. Selective Refraction Filters

Filtering by selective refraction depends upon the $dn/d\lambda$ of a lens material. Radiation of different wavelengths will be focused at different points along the optical axis if the lens is used in a wavelength region where it has high dispersion. This technique is particularly useful when the lens is used near an absorption band, because the refractive index will be considerably different on opposite sides of the band. This method of focal isolation or selective refraction is illustrated in Figure 7-63.

Two quartz lenses are usually employed in this method. The refractive index of quartz in the near infrared is about 1.5, and about 2.15 in the range 60 to 100 μm. (Quartz absorbs in the region around 9 μm.) In Figure 7-63, radiation in the far infrared is passed through the aperture, brought to a focus at aperture A', and transmitted. Visible and near-infrared radiation, being deviated less, impinges upon screen No. 2. Lens No. 2 focuses the desired radiation upon the detector. The two opaque discs, d_1 and d_2, obscure the paraxial zone of the lenses and prevent transmission of direct radiation.

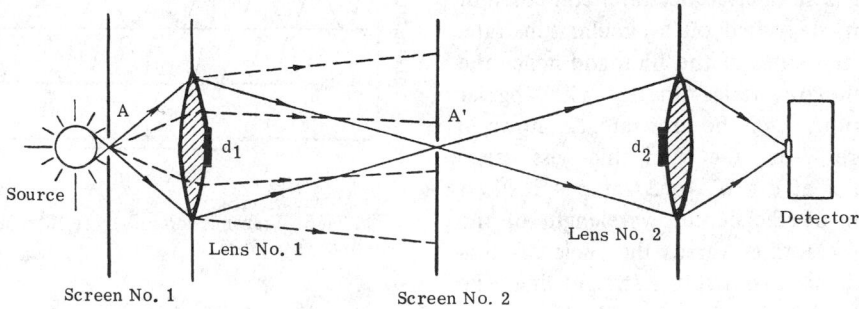

Fig. 7-63. Filtration by the method of focal isolation.

7.11. Polarization Interference Filters [7-85]

The polarization interference filter, sometimes called a Lyot-Ohman (or birefringent) filter, isolates a spectral band only a few angstroms wide. These filters are constructed of alternate plates of polarizers and birefringent crystals (e.g., quartz), as shown in Figure 7-64.

The birefringent crystal and quartz plates are cut with their optical axes parallel to the large faces. The axes of the polarizers are oriented at 45° to the quartz optical axes. Linearly polarized light, incident on the first quartz plate at 45°, would have its plane of polarization rotated by 90° if the plate were a half-wave plate (or if the

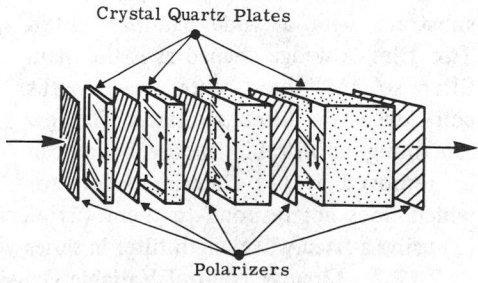

Fig. 7-64. Birefringent filter [7-85].

optical-path difference between the ordinary and extraordinary rays were any odd multiple of half-waves). If the plane of polarization is rotated at 90°, the radiation will not be transmitted by the second polarizer. Since the phase difference introduced between the ordinary and extraordinary rays depends on wavelength as well as on the thickness of the quartz plate, the same plate may be a 5/2-wave plate for one wavelength, a 7/2-wave plate for another wavelength, and 9/2-wave plate for still another wavelength. Each of these wavelengths will be blocked by the polarizer following the plates, whereas those wavelengths for which the difference in optical path between the two polarizations is an even number of half-waves will be completely transmitted.

Figure 7-65(a) illustrates the transmission of the first quartz plate and its polarizers. If each quartz plate is made twice as thick as the preceding one, it will have twice as many transmission maxima and minima in a given wavelength interval. The transmission curves for the second, third, and fourth plates are illustrated in Figure 7-65(b), (c), and (d). The transmission of the entire filter which is the result of all these transmission curves, is shown in Figure 7-65(e).

7.12. Circular Variable Filters (CVF)

7.12.1. Circular Variable Interference Filters.
One type of a circular variable filter is an interference filter composed of a film deposited on a circular substrate. The thickness of the film, and hence the wavelength, varies linearly with angular position, ϕ, on the substrate. As shown in Figure 7-66, the film thickness varies from d at $\phi = 0°$ to $2d$ at $\phi = 180°$. A graph of the center wavelength of the bandpass filter versus the angle of rotation is approximately a straight line. The bandwidth, in percent, is relatively constant over all wavelengths or rotation, as shown in Figure 7-67. By spinning the filter, one can get a rapid scan of its entire spectrum.

Other energy bands which may exist above and below the desired one can be eliminated by coating the back of the substrate with a wide bandpass filter. This film is wedge shaped as is the main filter so that their center wavelengths coincide as the filter is rotated. The few remaining undesirable energy bands can be removed by (1) choosing a detector

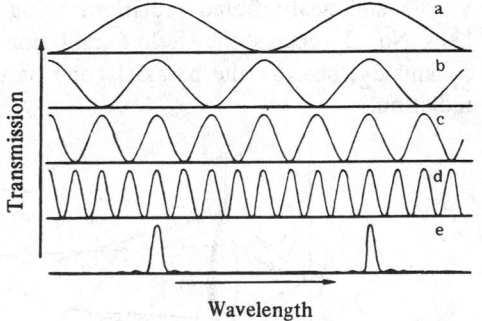

Fig. 7-65. Transmission of Lyot filter and its components [7-85].

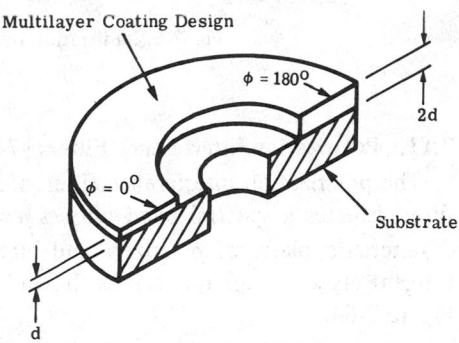

Fig. 7-66. Design principle of circular variable filters [7-87].

which does not respond to them, (2) placing absorbing materials in the viewing slit, or (3) using a fixed-wavelength filter in series with the circular variable filter.

7.12.2. Circular Neutral Variable Density Filters.
Another type of circular variable filter is the circular neutral variable density filter. This filter reduces the intensity of the transmitted beam, without changing its spectral content. The density of the filter coating on the substrate increases linearly with the angle of rotation; hence, the transmittance increases linearly from a minimum at a ϕ of $180°$ or $270°$ to a maximum at $360°$.

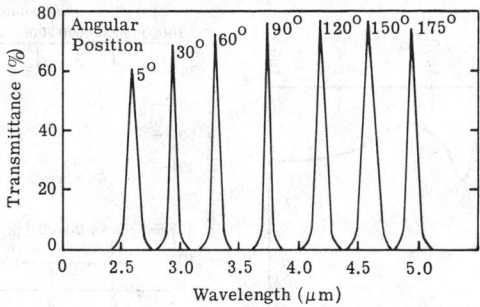

Fig. 7-67. Scan of a 2% bandpass CVF coated on a 180° segment for spectral coverage of 2.5 to 5.0 μm. The transmittance is equal to or less than 0.1% outside the bandpass [7-87].

Commercially available circular variable filters of this type are described in Section 7.13.

7.13. Commercially Available Filters*

The fact that customers desire a wide range of specifications for interference filters in the infrared region forces manufacturers to produce most of their filters by custom design and keep only a limited supply of stock filters. A representative sample is given here of stock filters available from various manufacturers. Individual manufacturers should be contacted for filters requiring different specifications.

Figure 7-68 through 7-73 show transmittance curves of interference filters available from Baird Atomic, Inc.: long-wave pass (Figures 7-68 and 7-69); wide-bandpass (Figure 7-70); narrow bandpass (Figures 7-71 and 7-72); and bandpass (Figure 7-73).

Figures 7-74 through 7-79 show transmittance curves of interference filters available from Eastman Kodak Company: short-wave pass (Figures 7-74 and 7-75); long-wave pass (Figures 7-76 and 7-77); and bandpass (Figures 7-78 and 7-79).

Optical Coating Laboratory, Inc. manufactures interference filters to exact specifications supplied by customers. They also maintain a stock of filters which fail to meet particular requirements in these specifications. These filters are listed in a catalogue in which all spectral specifications are within 2% and all in-band transmittance levels within 5% of the stated values.

Figure 7-80 shows the transmittance curve for a typical short-wave pass filter; Figure 7-81 for a long-wave pass filter; Figure 7-82 for a narrow-bandpass filter; and Figure 7-83 for a wide-bandpass filter.†

Their catalog lists circular variable filters with wavelength ranges from 1.2 to 2.3 μm to 8.6 to 16.3 μm, and bandwidth ranges from 0.9 to 9.2%. The substrates include quartz, glass, Vycor®, silicon, germanium, sapphire, and Irtran® 1.

---

*The material in this section was compiled and provided by Craig Mueller of The Environmental Research Institute of Michigan, Ann Arbor, Michigan.

†λ_c denotes cuton and cutoff wavelengths at 5% absolute transmittance. The average transmittance in the attenuation range is less than or equal to 1%.

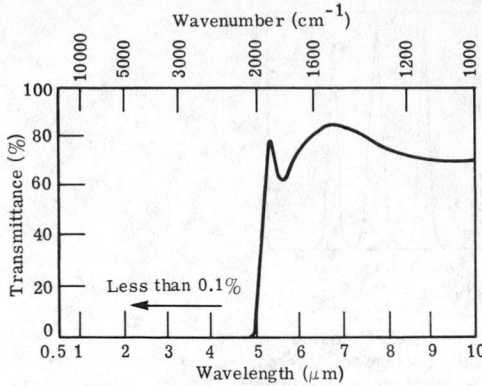

Fig. 7-68. Long wave pass filter. λ_c is available from 2 to 20 μm [7-88].

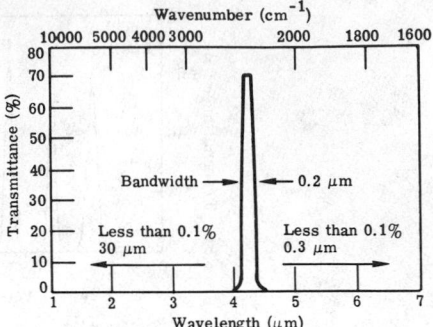

Fig. 7-71. Transmittance of a typical narrow-band filter. λ_0 is available from 2 to 20 μm [7-88].

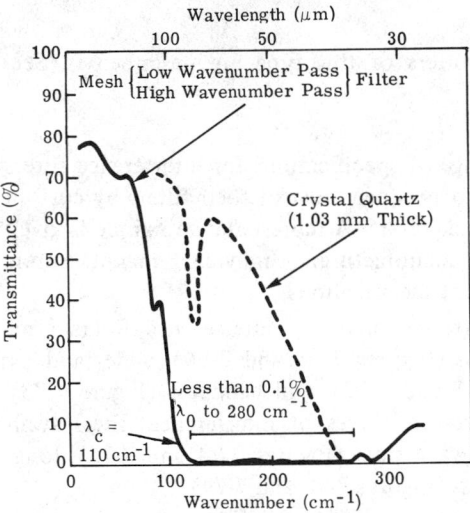

Fig. 7-69. Long-wave pass filter [7-88].

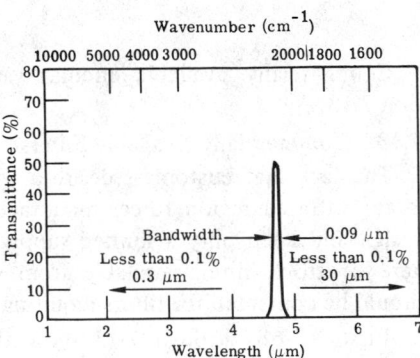

Fig. 7-72. Transmittance of a typical narrow-band filter. λ_0 is available from 2 to 15 μm [7-88].

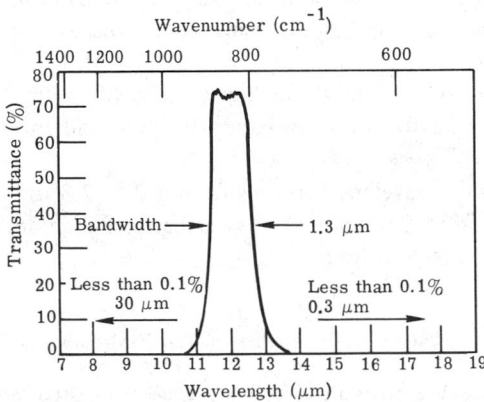

Fig. 7-70. Transmittance of a typical broad-band filter. λ_0 is available from 2 to 20 μm [7-88].

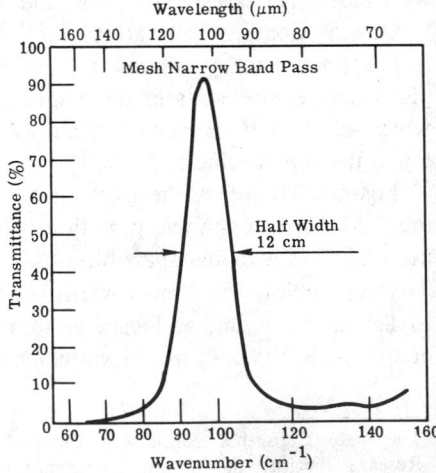

Fig. 7-73. Transmittance of a typical band-pass type 2 (MB2) filter at 96 cm^{-1}. λ_0 is available from 100 to 1000 μm [7-88].

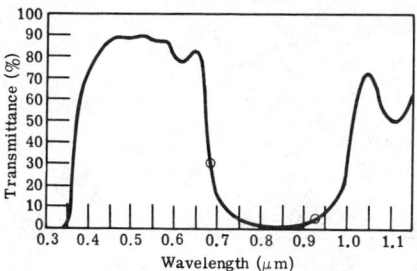

Fig. 7-74. Short-wave pass filter (plate glass substrate with Kodak No. 301 coating) [7-89].

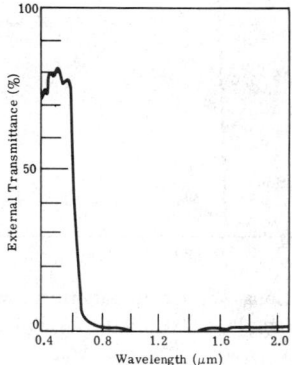

Fig. 7-75. Short-wave pass filter (heat-absorbing glass substrate with Kodak No. 301 coating) [7-89].

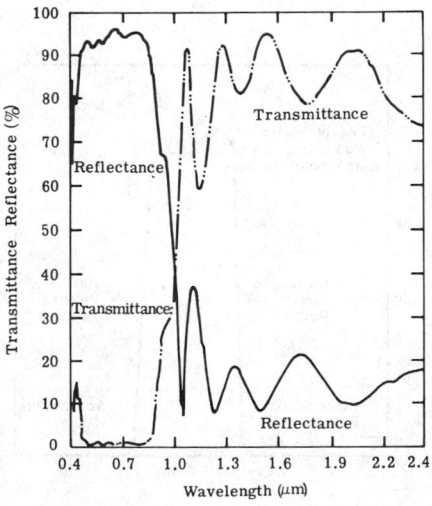

Fig. 7-76. Long-wave pass filter (Kodak cold mirror filter No. 310) [7-89].

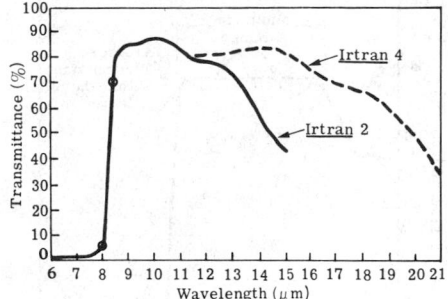

Fig. 7-77. Long-wave pass filters (Kodak infrared interference filters No. 480). Irtran® is a registered trademark of the Eastman Kodak Co. [7-89].

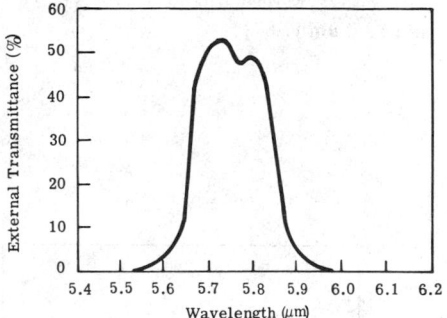

Fig. 7-78. Bandpass filter 5.75 μm (Kodak filter No. 530) [7-89].

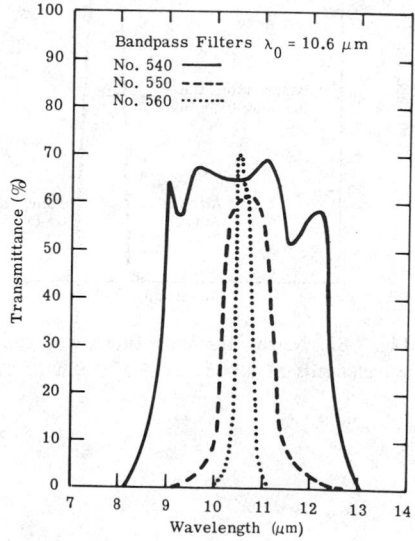

Fig. 7-79. Bandpass filters $\lambda|_o$ = 10.6 μm [7-89].

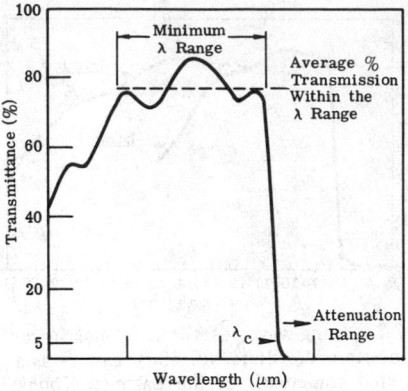

Fig. 7-80. Short wavelength pass filters with cutoff wavelengths of 1.429, 3.642, and 13.0 μm [7-87].

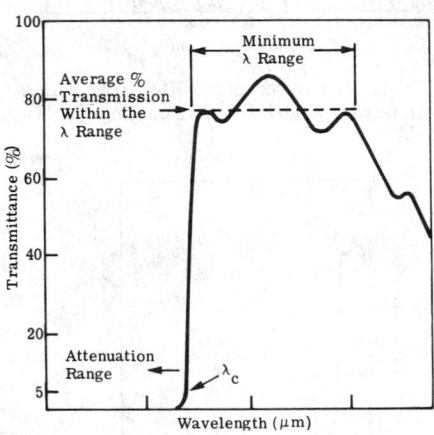

Fig. 7-81. Long wavelength pass filters with cuton wavelengths of 0.359 to 23.580 μm [7-87].

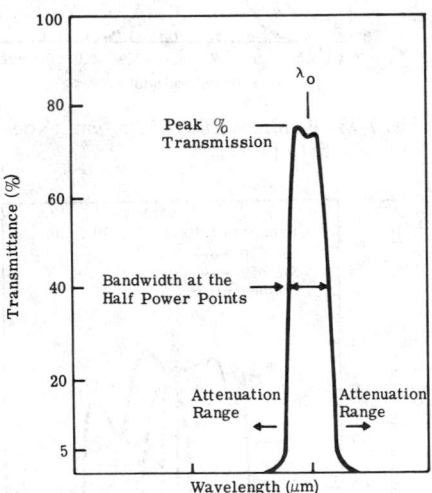

Fig. 7-82. Narrow bandpass filters with center wavelengths of 0.885 to 15.570 μm [7-87].

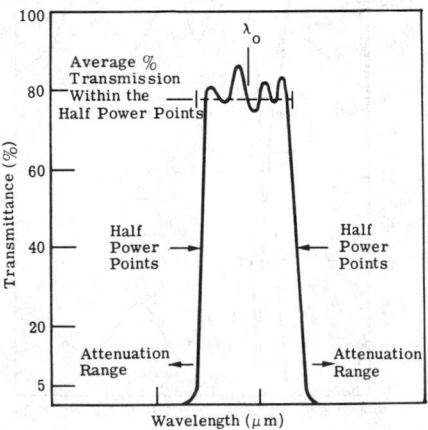

Fig. 7-83. Wide bandpass filters with center wavelengths of 0.381 to 19.888 μm [7-87].

The catalogue also lists circular neutral variable density filters with optical density ranges from 2.0 to 2.9 to 5.0 to 5.9. The relationship of optical density and transmittance is

$$\text{Optical Density} = \text{Log}_{10}\left(\frac{1}{\tau}\right) = -\text{Log}_{10}(\tau) \tag{7-64}$$

where τ is the transmittance.

Figure 7-84 shows the transmittance curve for an interference filter made by Omega Optical, Inc. This particular filter is called a "Raman Notch" filter.

Figures 7-85 and 7-86 show transmittance curves of interference filters available from Quantum Detector Technology, Inc.: short-wave pass (Figure 7-85); and long-wave pass (Figure 7-86).

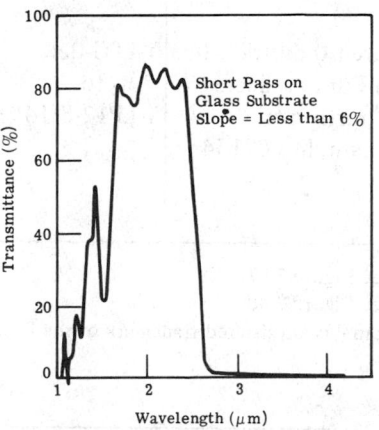

Fig. 7-85. Short-wave pass filter [7-90].

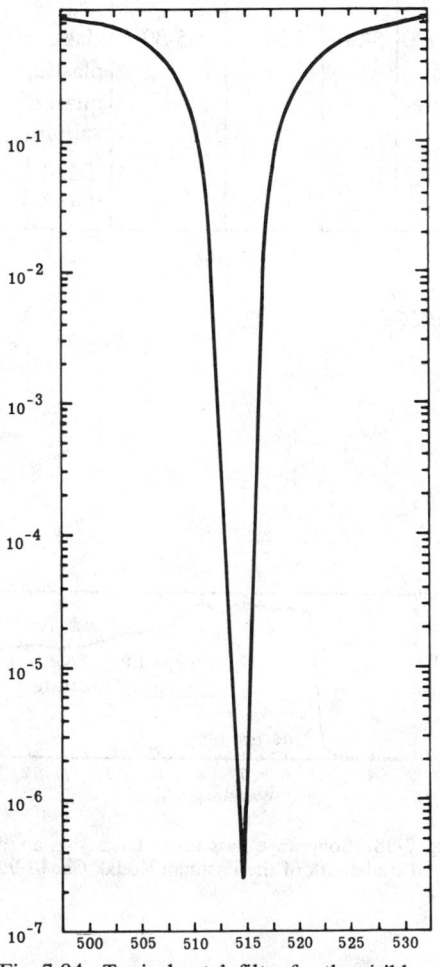

Fig. 7-84. Typical notch filter for the visible.

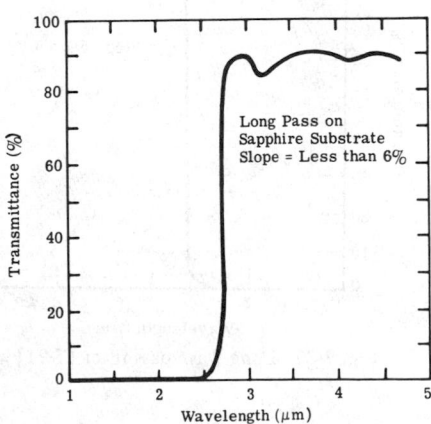

Fig. 7-86. Long-wave pass filter [7-90].

Table 7-61 gives specification of long-wave pass interference filters available from Corion Corporation and Infrared Industries, Inc. (See also Figures 7-87 and 7-88.)

Table 7-61. Commercially Available Long-Wave Pass Interference Filters

Manufacturer	Catalog No.	5% Cuton (μm)	Slope (%)	Transmission in Passband		Substrate Material
				Average (%)	Minimum (%)	
Corion Corporation 73 Jeffrey Ave. Holliston, MA 01746	ILP-1.5-1 to ILP-6.0-1*	1.5 to 6.0 in 0.5 increments	7-6	70	–	glass, sapphire, Irtran† 1, Irtran 2
Infrared Industries, Inc. Thin Film Products 84 Fourth Ave. Waltham, MA 02154	T-LPG-0.4 to T-LP12-8/14**	0.4 to 8.0 in 0.15-3.0 increments ± 2-3%	5-20	70-85	65-80	glass, plastic, quartz, sapphire, Irtran 1, Irtran 2

*See Figure 7-87.
**See Figure 7-88.
†Irtran® is a registered trademark of the Eastman Kodak Co.

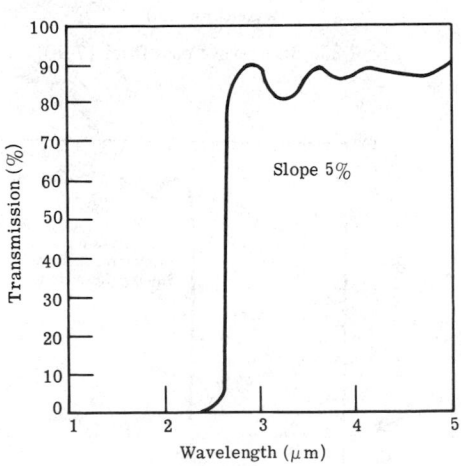

Fig. 7-87. Long-wave pass filter [7-91].

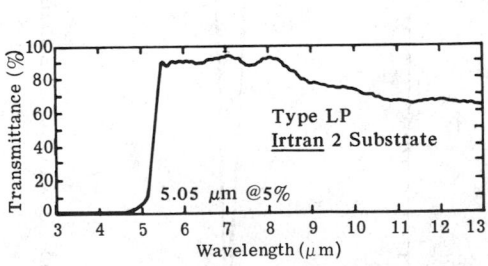

Fig. 7-88. Long-wave pass filter. Irtran® is a registered trademark of the Eastman Kodak Co. [7-92].

Table 7-62 gives specifications and short-wave pass interference filters available from Corion Corporation and Infrared Industries, Inc. and Table 7-63 gives specifications of bandpass interference filters available from Barnes Engineering Company, Corion Corporation, Infrared Industries, Inc., and Quantum Detector Technology, Inc. (See also Figure 7-89.)

Table 7-62. Commercially Available Short-Wave Pass Interference Filters

Manufacturer	Catalog No.	5% Cutoff (μm)	Slope (%)	Transmission in Passband		Substrate Material
				Average (%)	Minimum (%)	
Corion Corporation 73 Jeffrey Ave. Holliston, MA 01746	ISP-2.0-1 to ISP-6.0-1*	2.0 to 6.0 in 1.0 increments	5	70	—	glass, quartz
Infrared Industries, Inc. Thin Film Products 84 Fourth Ave. Waltham, MA 02154	T-SPG-2.0 to T-SPS-5.5**	2.0 to 5.5 ± 2% in 1.0 increments	5	75% to 1.5- 3.3 μm	70% to 1.5- 3.3 μm	glass, quartz, sapphire

*See Figure 7-89.
**Infrared Industries, Inc., can also manufacture short-wave pass filters with λ_c from 1.7 to 16.0 μm.

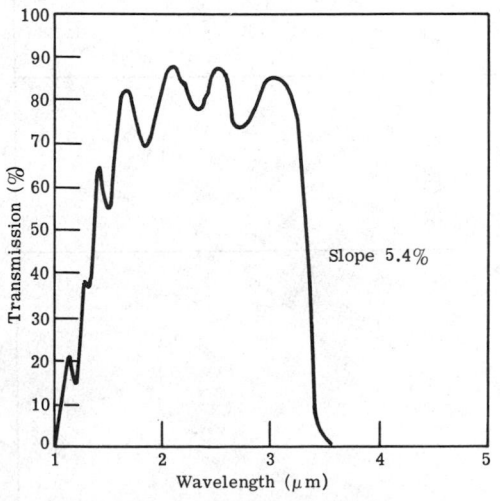

Slope 5.4%

Fig. 7-89. Short-wave pass filter [7-91].

Table 7-63. Commercially Available Bandpass Interference Filters

Manufacturer	Catalog No.	Center Wavelength Range or Center Wavelength, λ_0 (μm)	Tolerance $\lambda_0 \pm$ (μm)	Half-Power Bandwidth (% of λ_0) or (μm)	Minimum Transmittance τ (%)		Substrate Material
					Blocking to 3.2 μm	Complete	
Barnes Engineering Co. 30 Commerce Rd. Stamford, CT 06904	SS-IR-1	1.6-5.5	0.01-0.03	1 ± 0.5%	38-45	40-45	glass, quartz, sapphire
	SS-IR-1.5	1.6-5.5	0.01-0.034	1.5 ± 0.5%	45	45	glass, quartz, sapphire
	SS-IR-2.5	1.6-5.5	0.015-0.035	2.5 ± 0.5%	50	50	glass, quartz, sapphire
	SS-IR-5	1.6-5.5	0.015-0.050	5 ± 1%	50	50-60	glass, quartz, sapphire
	SS-IR-10	1.6-5.5	0.030-0.085	10 ± 2%	50	60-65	glass, quartz, sapphire
	—	5.501-7.5	+0.03, +0.04, -0.02, -0.02, +0.05, 2% of λ_0, 0.03, 3% of λ_0	1.0, 1.5, 2.5, 5.0, 10.0	40-65	—	—
	—	7.501-9.0	+0.04, +0.06, -0.03, -0.03, 2% of λ_0, 3% of λ_0	1.0, 1.5, 2.5, 5.0, 10.0	45-65	—	—
	—	9.01-14.0	+0.05, +0.07, -0.04, -0.03, 2% of λ_0, 3% of λ_0	1.0, 1.5, 2.5, 5.0, 10.0	40-65	—	—

Table 7-63. Commercially Available Bandpass Interference Filters (*Continued*)

Manufacturer	Catalog No.	Center Wavelength Range or Center Wavelength, λ_0 (μm)	Tolerance $\lambda_0 \pm$ (μm)	Half-Power Bandwidth (% of λ_0) or (μm)	Minimum Transmittance τ (%)		Substrate Material
					Blocking to 3.2 μm	Complete	
Corion Corporation 73 Jeffrey Ave. Holliston, MA 01746	SS-7500-1-2 to SS-10500-1-2	0.7500 to 1.05 in 0.01 μm increments	±0.02	0.1 μm	50	—	—
	SS-1.1-1 to SS-5.5-1	1.1 to 5.5 in 0.1 μm increments	±0.005-0.035	0.03-0.179 μm	40-50	—	optical glass, quartz, sapphire
Infrared Industries, Inc. Thin Film Products 84 Fourth Ave. Waltham, MA 02154	T-1.50 to T-11.0	1.50 to 11.0 in 0.1 μm increments	0.02 to 0.18	1.5-2.5 to 8-10	40-55	—	—
Quantum Detector Technology, Inc. 18 Terry Ave. P.O. Box 395 Burlington, MA 01803	—	1.5-10.0	—	1-1.5, 1-12, 1-20	—	—	—

7.14. Absorption Filters [7-10, 7-93, 7-94]

Many materials in a solid, liquid, or gaseous state can be used as selective absorption filters in various regions of the infrared spectrum. These filters have high transmission above or below a certain wavelength where high absorption produces a sharp cutoff.

Long-wave pass filters in the near infrared are normally made of plastic materials containing dyes, colored glass, or sublimated phthalocyanines upon glass. Other long-wave pass filters consist of glass coated with plastic dye solutions. Figure 7-90 shows the characteristics of these filters.

Figure 7-91 shows the variety of cuton wavelengths that are obtainable with various semiconductors. By proper doping, the location of the absorption limit can be moved, although this reduces the gradient. A similar effect also can be obtained by the use of mixed crystals (Figure 7-92) although no such filters are commercially available at present. An excellent work on mixed-crystal semiconductors is by R. H. Bube.

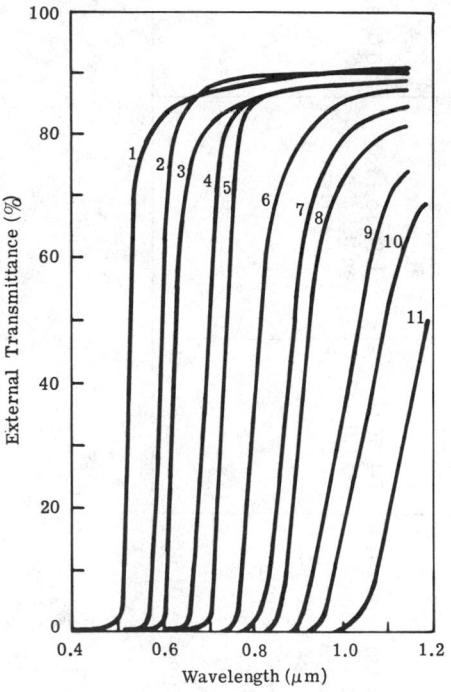

Fig. 7-90. Near infrared dyed-plastic filter characteristics [7-95].

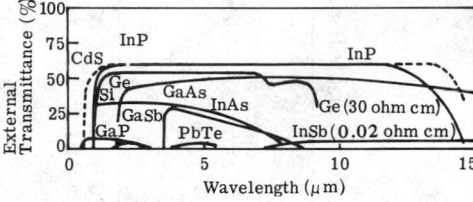

Fig. 7-91. Transmission of selected semiconductors [7-96].

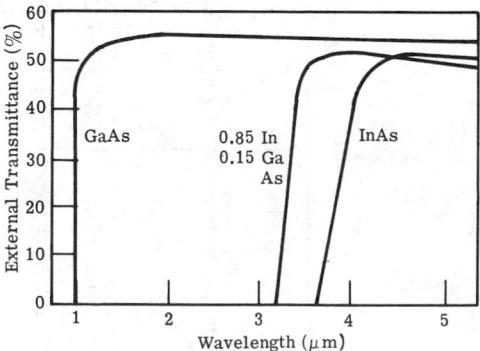

Fig. 7-92. Transmission of some mixed crystals.

7.15.   References and Bibliography

   7.15.1.   References

[7-1]   Max Born and Emil Wolf, *Principles of Optics: Electromagnetic Theory of Propagation, Interference and Diffraction of Light*, Pergamon Press, New York, NY, 1959, p. 90.

[7-2]   Adapted from: Stanley S. Ballard, Kathryn A. McCarthy, William L. Wolfe, "Optical Materials for Infrared Instrumentation," IRIA *State-of-the-Art Report*, The University of Michigan, Willow Run Laboratories, Ann Arbor, MI, Report No. 2389-11-S, Contract No. Nonr 1224 (12), January 1959, and Supplement, April 1961.

[7-3]   E. S. R. Gopal, *Specific Heats at Low Temperatures*, Plenum Press, New York, NY, 1966, p. 33.

[7-4]   R. S. Caldwell, "Optical Properties of Tellurium and Selenium," Purdue University, Department of Physics, West Lafayette, IN, January 1958.

[7-5]   C. D. Salzberg and J. J. Villa, "Infrared Refractive Indexes of Silicon Germanium and Modified Selenium Glass," *Journal of the Optical Society of America*, Vol. 47, No. 3, March 1957, p. 244.

[7-6]   D. E. McCarthy, "The Reflection and Transmission of Infrared Materials: I, Spectra From 2-50 Microns," *Applied Optics*, Vol. 2, No. 6, June 1963, pp. 591, 593.

[7-7]   General Electric Company, "Fused Quartz Catalog," Q-6, 1957; Heraeus-Amersil, Inc., Sayreville, NJ, "Optical Fused Quartz & Fused Silica," Bulletin EM-9227, 1977.

[7-8]   G. Calingaert, S. D. Heron, and R. Stair, *Journal of the Society of Automotive Engineers*, Vol. 39, 1936, p. 448.

[7-9]   M. D. Beals and L. Merker, *Materials in Design Engineering*, Reinhold Publishing, New York, NY, 1960.

[7-10]   R. P. Chasmar, F. E. Jones and R. A. Smith, *The Detection and Measurement of Infra-Red Radiation*, Oxford University Press, New York, NY, 1957.

[7-11]   General Electric Lamp Glass Department, Cleveland, OH, Data Sheets.

[7-12]   U. P. Oppenheim and A. Goldman, "Infrared Spectral Transmittance of MgO and BaF_2 Crystals between $27°$ and $1000°C$," *Journal of the Optical Society of America*, Vol. 54, No. 1, January 1964, p. 127.

[7-13]   C. Hilsum, "Infrared Transmission of Barium Titanate," *Journal of the Optical Society of America*, Vol. 45, 1955, p. 771.

[7-14]   S. B. Levin, et al., "Some Optical Properties of Strontium Titanate Crystals," *Journal of the Optical Society of America*, Vol. 45, 1955, pp. 737-739.

[7-15]   C. M. Phillippi and N. F. Beardsley, "Infrared Window Studies," Pt. I, Wright Air Development Center, Wright Patterson AFB, OH, Technical Note 55-194, June 1955, p. 16.

[7-16]   D. A. Jones, R. V. Jones, and R. W. H. Stevenson, "Infrared Transmission Limits of Single Crystals of Some Fluorides," *Proceedings of the Physical Society B-65*, A. C. Stickland (ed.), The Physical Society, London, 1952, p. 906.

[7-17]   S. S. Ballard, University of Florida, Gainesville, FL, Private Communication, 1955.

[7-18]   D. E. McCarthy, Beckman Instruments, Fullerton, CA, Private Communication, 1960.

[7-19]   C. J. Johnson, G. H. Sherman, and R. Weil, "Far Infrared Measurement of the Dielectric Properties of GaAs and CdTe at 300K and 8K," *Applied Optics*, Vol. 8, No. 8, August 1969, p. 1667.

[7-20] D. E. McCarthy, "The Reflection and Transmission of Infrared Materials: III, Spectra from 2μ to 50μ," *Applied Optics*, Vol. 4, No. 3, March 1965, p. 317.

[7-21] J. J. Loferski, "Infrared Optical Properties of Single Crystals of Tellurium," *Physical Review*, Vol. 93, 1954, p. 707.

[7-22] Eastman Kodak Co., Rochester, NY, "Condensed Data for Kodak Irtran Infrared Optical Materials," Publication No. U-71, Revised 5-75.

[7-23] D. L. Stierwalt, "Infrared Spectral Emittance Measurements of Optical Materials," *Applied Optics*, Vol. 5, No. 12, December 1966, pp. 1911-1915.

[7-24] D. L. Stierwalt, J. B. Bernstein, and D. D. Kirk, "Measurement of the Infrared Spectral Absorptance of Optical Materials," *Applied Optics*, Vol. 2, No. 11, November 1963, p. 1169.

[7-25] D. E. McCarthy, "Transmittance of Optical Materials from 0.17 μ to 3.0 μ," *Applied Optics*, Vol. 6, No. 11, November 1967, p. 1896.

[7-26] J. Strong, *Physical Review*, Vol. 38, 1931, p. 1818.

[7-27] E. K. Plyler, National Bureau of Standards, Washington, DC, Private Communication, 1955.

[7-28] D. Landers, et al., "Long Wavelength Infrared Components Measurements," Part II, Spencer Laboratory, Raytheon Corporation, Burlington, MA, August 1962.

[7-29] A. Hadni, E. Decamps, and C. Janot, "Spectrométrie Dans L'infrarouge Lointain (50 à 350 μ)," *Review d'Optique*, Vol. 38, 1959, p. 463.

[7-30] G. W. Chantry, et al., "TPX, a New Material for Optical Components in the Far Infra-Red Spectral Region," *Infrared Physics*, Vol. 9, 1969, p. 31.

[7-31] J. A. Alvarez, et al., "Far Infrared Measurements of Selected Optical Materials at $1.6°K$," *Infrared Physics*, Vol. 15, 1975, pp. 46-47.

[7-32] W. S. Rodney, I. H. Malitson, and T. A. King, "Refractive Index of Arsenic Trisulfide," *Journal of the Optical Society of America*, Vol. 48, No. 9, September 1958, p. 633.

[7-33] I. H. Malitson, "Refractive Properties of Barium Fluoride," *Journal of the Optical Society of America*, Vol. 54, No. 5, May 1964, p. 628.

[7-34] T. W. Houston, L. F. Johnson, P. Kisliuk, and D. J. Walsh, "Temperature Dependence of the Refractive Index of Optical Maser Crystals," *Journal of the Optical Society of America*, Vol. 53, No. 11, November 1963, p. 1286.

[7-35] S. J. Czyzak, W. M. Baker, R. C. Crane, and J. B. Howe, "Refractive Indexes and Single Synthetic Zinc Sulfide and Cadmium Sulfide Crystals," *Journal of the Optical Society of America*, Vol. 47, No. 3, March 1957, p. 240.

[7-36] F. F. Martens, *Annalen der Physik*, Vol. 6, 1901, pp. 602, 603, 619-626.

[7-37] J. W. Gifford, *Proceedings of the Physical Society*, Vol. 70, London, 1902, p. 329.

[7-38] A. Carvallo, *Comptes Rendus*, Vol. 126, 1896, p. 950.

[7-39] *J. Phys. Radium.*, Ser. 3, Vol. 9, 1900, p. 465.

[7-40] F. J. Micheli, *Annalen der Physik*, Vol. 4, No. 7, 1902, p. 772.

[7-41] I. H. Malitson, "A Redetermination of Some Optical Properties of Calcium Fluoride," *Applied Optics*, Vol. 2, No. 11, November 1963, p. 1103.

[7-42] F. Kohlrausch, *Praktische Physik*, B. G. Teubner, Stuttgart, West Germany, Band 3—Tafeln, 22 Aufl., 1968.

[7-43] W. W. Coblentz, *Journal of the Optical Society of America*, Vol. 4, 1920, p. 441.

[7-44] E. Liebreich, *Verhandlungen Deutsches Physik. Gesellschaft*, Vol. 13, 1911, p. 709.

[7-45] W. S. Rodney and R. J. Spindler, *Journal of Research*, National Bureau of Standards, Washington, DC, Vol. 51, 1953, p. 123.

[7-46] W. S. Rodney, "Optical Properties of Cesium Iodide," *Journal of the Optical Society of America*, Vol. 45, No. 11, November 1955, p. 987.

[7-47] H. Trommsdorff, *Zeitschrift Für Physik*, Vol. 2, 1901, p. 576.

[7-48] R. B. Sosman, *The Properties of Silica*, Chemical Catalog Company, Inc., New York, NY, 1927.

[7-49] A. Carvallo, *Comptes Rendus*, Vol. 126, 1898, p. 728.

[7-50] H. Rubens, *Wiedemann Annalen*, Vol. 54, 1895, p. 488.

[7-51] S. Rösch, "Die Optik des Fabulit, die Durbe des Brewsterwinkels und das Farbspielmoment," *Optica Acta*, Vol. 12, July 1965, p. 253.

[7-52] I. H. Malitson, "Interspecimen Comparison of the Refractive Index of Fused Silica," *Journal of the Optical Society of America*, Vol. 55, No. 1, October 1965, p. 1205.

[7-53] B. O. Seraphin and H. E. Bennett, Chapter 12 "Optical Constants," *Semiconductors and Semimetals*, R. K. Willardson and A. C. Beer (eds.), Academic Press, New York, NY, Vol. 3, 1967, pp. 525-526.

[7-54] L. C. Barcus, A. Perlmutter, and J. Callaway, "Effective Mass of Electrons in Gallium Arsenide," *Physical Review*, Vol. 111, No. 1, July 1958, p. 167.

[7-55] L. C. Barcus, Lowell Institute of Technology, Private Communication, 1968.

[7-56] C. D. Salzberg and J. J. Villa, "Index of Refraction of Germanium," *Journal of the Optical Society of America*, Vol. 48, No. 8, August 1958, p. 579.

[7-57] M. Herzberger and C. D. Salzberg, "Refractive Indices of Infrared Optical Materials and Color Correction of Infrared Lenses," *Journal of the Optical Society of America*, Vol. 52, No. 4, April 1962, p. 420.

[7-58] D. H. Rank and H. E. Bennett, "The Index of Refraction of Germanium Measured by an Interference Method," *Journal of the Optical Society of America*, Vol. 44, No. 1, January 1954, p. 13.

[7-59] Z. Gyulai, *Zeitschrift Für Physik*, Vol. 46, 1927, p. 84.

[7-60] H. Harting, *Sitzber, Deutsches Akad. Wiss. Berlin*, Vol. 4, 1948, pp. 1-25.

[7-61] L. W. Tilton and E. K. Pyler, *Journal of Research*, National Bureau of Standards, Washington, DC, Vol. 47, 1951, p. 25.

[7-62] H. W. Hohls, *Annalen der Physik*, Vol. 29, 1937, p. 433.

[7-63] I. H. Malitson and M. J. Dodge, National Bureau of Standards, Private Communication.

[7-64] A. Duncanson and R. W. H. Stevenson, "Some Properties of Magnesium Fluoride Crystallized from the Melt," *Proceedings of the Physical Society*, A. C. Stickland (ed.), The Physical Society, London, Vol. 72, 1958, p. 1001.

[7-65] R. E. Stephens and I. H. Malitson, *Journal of Research*, National Bureau of Standards, Washington, DC, Vol. 49, 1952, pp. 249-252.

[7-66] R. E. Stephens, E. K. Plyler, W. S. Rodney, and R. J. Spindler, "Refractive Index of Potassium Bromide for Infrared Radiant Energy," *Journal of the Optical Society of America*, Vol. 43, No. 2, February 1965, p. 110.

[7-67] F. Paschen, *Annalen der Physik*, Vol. 26, 1908, p. 120.

[7-68] K. Korth, *Zeitschrift Für Physik*, Vol. 84, 1933, p. 677.

[7-69] I. H. Malitson, "Refraction and Dispersion of Synthetic Sapphire," *Journal of the Optical Society of America*, Vol. 52, No. 12, December 1962, p. 1377.

[7-70] M. J. Dodge, I. H. Malitson, and A. I. Mahon, "A Special Method for Precise Refractive Index Measurement of Uniaxial Optical Media," *Applied Optics*, Vol. 8, No. 8, August 1969, p. 1703.

[7-71] E. Lowenstein, "Optical Properties of Sapphire in the Far Infrared," *Journal of the Optical Society of America*, Vol. 51, No. 1, January 1961, p. 108.

[7-72] L. Gampel and F. M. Johnson, "Index of Refraction of Single-Crystal Selenium," *Journal of the Optical Society of America*, Vol. 59, No. 1, January 1969, p. 72.

[7-73] E. W. Saker, "The Optical Constants of Liquid Selenium," *Proceedings of the Physical Society*, A. C. Stickland (ed.), The Physical Society, London, B65, 1952, p. 785.

[7-74] L. Henry, *Comptes Rendus*, Institute Geologic, Bucharest, Romania, 1953, pp. 237, 148.

[7-75] G. Hass and A. F. Turner, "Coatings for Infrared Optics," *Ergebnisse der Hochvakuumtechnik und der Physik dünner Schichten*, M. Auwarter (ed.), Wissenschaftliche Verlagsgesellschaft M.B.H., Stuttgart, 1957.

[7-76] M. S. Ishmuratova and L. V. Sergeyev, "Optical Cements Transparent in the Infrared," *Soviet Journal of Optical Technology*, Vol. 34, 1967, p. 801.

[7-77] John William Berthold, III, *Dimensional Stability of Low Expansivity Materials— Time Dependent Changes in Optical Contact Interfaces and Phase Shifts on Reflection From Multilayer Dielectrics*, A Dissertation submitted to the Faculty of the Committee on Optical Sciences, University of Arizona, Tucson, AZ, 1976.

[7-78] William P. Barnes, Itek Corporation, Private Communication.

[7-79] Bausch & Lomb Optical Company, Rochester, NY, "Near Infrared Transmission Filters," Progress Report No. 3, 1958.

[7-80] Omega Optical, Inc., Brattleboro, VT, "Data Sheets," 1976, p. 7.

[7-81] Martin L. Baker and Victor L. Yen, "Effects of the Variation of Angle of Incidence and Temperature on Infrared Filter Characteristics," *Applied Optics*, Vol. 6, No. 8, August 1967, pp. 1343-1351.

[7-82] C. F. Mooney and A. F. Turner, "Infrared Transmitting Interference Filters, Proceedings of the Conference on Infrared Optical Materials, Filters, and Films," Engineer Research and Development Laboratories, Fort Belvoir, VA, 1955.

[7-83] Astrionics Division, Aerojet-General Corporation, Azusa, CA, *Engineering Notebook*, 1961.

[7-84] E. D. McAlister, "The Christiansen Light Filter; Its Advantages and Limitations," Smithsonian Miscellaneous Collections, Washington, DC, Vol. 93, 1936, p. 7.

[7-85] J. Strong, *Concepts of Classical Optics*, W. H. Freeman and Company, San Francisco, CA, 1958.

[7-86] R. B. Barnes and T. W. Bonner, *Physical Review*, Vol. 49, 1936, p. 732.

[7-87] Optical Coating Laboratory, Inc., Technical Products Division, Santa Rosa, CA, "Stock Filter Catalog," Infrared Section, August 1975, pp. II-3 to IV-3.

[7-88] Baird Atomic, Inc., Bedford, MA, "Grubb Parsons Optical Filters and Thin Films," Catalog C-14, October 1974, p. 13-15.

[7-89] Eastman Kodak Company, Rochester, NY, "Special Filters from Kodak for Technical Applications," Kodak Pamphlet No. U-73, pp. 3-7.

[7-90] Quantum Detector Technology, Inc., Burlington, MA, "Quantum Detector Technology Optical Components," p. 11.

[7-91] Corion Corporation, Holliston, MA, "Thin Film Optical Filter Guide," pp. 15, 17.

[7-92] Infrared Industries, Inc., Thin Film Products Division, Waltham, MA, "Optical Filters Designers Guide and Catalog," 1973, p. 14.

[7-93] Infrared Industries, Inc., Waltham, MA, "Data Sheets," 1962.

[7-94] N. M. Mohler and J. R. Loofbourow, "Optical Filters," *American Journal of Physics*, Vol. 20, 1952, pp. 579-588.

[7-95] J. H. Shenk, et al., "Plastic Filters for the Visible and Near Infra-Red Regions," *Journal of the Optical Society of America*, Vol. 36, 1946, pp. 10, 569.

[7-96] W. L. Wolfe and S. S. Ballard, "Optical Materials, Films, and Filters for Infrared Instrumentation," *Proceedings of the Institute of Electrical and Electronics Engineers*, Vol. 47, 1959, p. 9.

[7-97] R. Hopkins, et al., *MIL 141 Military Standardization Handbook*, "Optical Design," Defense Supply Agency, 1962.

[7-98] H. W. Icenogle, B. C. Platt, and W. L. Wolfe, "Refractive Indexes and Temperature Coefficients of Germanium and Silicon," *Applied Optics*, Optical Society of American, Washington, DC, Vol. 15, No. 10, October 1976, pp. 2348-2351.

7.15.2. Bibliography

Aerojet-General Corporation, Astrionics Division, Azusa, CA, *Engineering Notebook*, 1961.

Alvarez, J. A., "Far Infrared Measurements of Selected Optical Materials at 1.6 K," *Infrared Physics*, Vol. 15, 1975, pp. 46-47.

Baird Atomic, Inc., Bedford, MA, "Grubb Parsons Optical Filters and Thin Films," Catalog C-14, October 1974, pp. 13-15.

Baker, Martin L., and Victor L. Yen, "Effects of the Variation of Angle of Incidence and Temperature on Infrared Filter Characteristics," *Applied Optics*, Vol. 6, No. 8, August 1967, pp. 1343-1351.

Ballard, S. S., Kathryn A. McCarthy, and William L. Wolfe, "Optical Materials for Infrared Instrumentation," IRIA *State-of-the-Art Report*, The University of Michigan, Willow Run Laboratories, Ann Arbor, MI, Report No. 2389-11-S, Contract No. Nonr 1224 (12), January 1959, and Supplement, April 1961.

Barcus, L. C., A. Perlmutter, and J. Callaway, "Effective Mass of Electrons in Gallium Arsenide," *Physical Review*, Vol. 111, No. 1, July 1958, p. 167.

Barnes, R. B., and T. W. Bonner, "The Christiansen Filter Effect in the Infrared," *Physical Review*, Vol. 49, 1936, p. 732.

Bausch & Lomb Optical Company, Rochester, NY, "Near Infrared Transmission Filters," Progress Report No. 3, 1958.

Beals, M. D., and L. Merker, *Materials in Design Engineering*, Reinhold Publishing Corporation, New York, NY, 1960.

Berthold, John William III, *Dimensional Stability of Low Expansivity Materials—Time Dependent Changes in Optical Contact Interfaces and Phase Shifts on Reflection From Multilayer Dielectrics*, A Dissertation submitted to the Faculty of the Committee on Optical Sciences, University of Arizona, Tucson, AZ, 1976.

Born, Max, and Emil Wolf, *Principles of Optics: Electromagnetic Theory of Propagation, Interference and Diffraction of Light*, Pergamon Press Inc., New York, NY, 1959, p. 90.

Börnstein, Landolt, *Zahlenwerte und Functionen*, Springer Verlag, Berlin, 1961, 6th Edition.

Briggs, H. B., "Optical Effects in Bulk Silicon Germanium," *Physical Review*, Vol. 77, 1950, p. 287.

Bube, R. H., *Photoconductivity of Solids*, John Wiley and Sons, Inc., New York, NY, 1960.

Burstein, E., J. J. Oberly, and E. K. Plyler, *Proceedings of Indian Academy of Science*, Vol. 38, 1948.

Caldwell, R. S., "Optical Properties of Tellurium and Selenium," Purdue University,

Department of Physics, West Lafayette, IN, January 1958.

Calingaert, G., S. D. Heron, R. Stair, *Journal of the Society of Automotive Engineers*, Vol. 39, 1936, p. 448.

Carvallo, A., *Comptes Rendus*, Vol. 126, 1896, p. 950.

Carvallo, A., *Comptes Rendus*, Vol. 126, 1898, p. 728.

Chantry, G. W., et al., "TPX, a New Material for Optical Components in the Far Infra-Red Spectral Region," *Infrared Physics*, Vol. 9, 1969, p. 31.

Coblentz, W. W., *Journal of the Optical Society of America*, Vol. 4, 1920, p. 441.

Corion Corporation, Holliston, MA, "Thin Film Optical Filter Guide," pp. 15, 17.

Corrucini, R. J., and J. J. Gniewek, "Specific Heats and Enthalpies of Solids at Low Temperatures," *National Bureau of Standards*, Boulder, CO, 1960.

Czyzak, S. J., W. M. Baker, R. C. Crane, and J. B. Howe, "Refractive Indexes and Single Synthetic Zinc Sulfide and Cadmium Sulfide Crystals," *Journal of the Optical Society of America*, Vol. 47, No. 3, March 1957, p. 240.

Dodge, M. J., I. H. Malitson, and A. I. Mahon, "A Special Method for Precise Refractive Index Measurement of Uniaxial Optical Media," *Applied Optics*, Vol. 8, No. 8, August 1969.

Duncanson, A., and R. W. H. Stevenson, "Some Properties of Magnesium Fluoride Crystallized from the Melt," *Proceedings of the Physical Society*, A. C. Stickland (ed.), The Physical Society, London, Vol. 72, 1958, p. 1001.

Eastman Kodak Co., Rochester, NY, "Condensed Data for Kodak Irtran Infrared Optical Materials," Publication No. U-71, Revised 5-75.

Eastman Kodak Company, Rochester, NY, "Special Filters from Kodak for Technical Applications," Kodak Pamphlet No. U-73, pp. 3-7.

Fuller, R. M., R. J. Bell, and D. G. Rothburn, "Transmittance of Cer-Vit Glass Ceramic in the Ultraviolet, Visible, Infrared, and Sub-millimeter Wavelength Regions," *Applied Optics*, Vol. 7, No. 6, June 1968.

Gampel, L., and F. M. Johnson, "Index of Refraction of Single-Crystal Selenium," *Journal of the Optical Society of America*, Vol. 59, No. 1, January 1969, p. 72.

General Electric Company, "Fused Quartz Catalog," Q-6, 1957.

Gifford, J. W., *Proceedings of the Physical Society*, A. C. Stickland (ed.), The Physical Society, London, Vol. 70, 1902, p. 329.

Gopol, E. S. R., *Specific Heats at Low Temperatures*, Plenum Press, New York, NY, 1966, p. 33.

Gray, Dwight C., Ed., *American Institute of Physics Handbook*, McGraw-Hill, New York, NY, 1963, 2nd Edition.

Gyulai, Z., *Zeitschrift Für Physik*, Vol. 46, 1927, p. 84.

Hadni, A., E. Decamps, and C. Janot, "Spectrométrie Dans L'infrarouge Lointan (50 á 350 μ)," *Review d'Optique*, Vol. 38, 1959, p. 463.

Harting, H., *Sitzber. Deutsch. Akad. Wiss. Berlin*, Vol. 4, 1948.

Hass, G., and A. F. Turner, "Coatings For Infrared Optics," *Ergebnisse der Hochvakuumtechnik und der Physik dünner Schichten*, M. Auwarter (ed.), Wissenschaftliche Verlagsgesellschaft M.B.H., Stuttgart, 1957.

Hausler, R. L., Thesis, Ohio State University, Columbus, Ohio, 1952.

Henry, L., *Comptes Rendus*, Institute Geologic, Bucharest, Romania, 1953, pp. 147, 238.

Heraeus-Amersil, Inc., Sayreville, NJ, "Optical Fused Quartz & Fused Silica," Bulletin EM-9227, 1977.

Herzberger, M., and C. D. Salzberg, "Refractive Indices of Infrared Optical Materials and Color Correction of Infrared Lenses," *Journal of the Optical Society of America*, Vol. 52, No. 4, April 1962, p. 420.

Hilsum, C., "Infrared Transmission of Barium Titanate," *Journal of the Optical Society of America*, Vol, 45, 1955, p. 771.

Hodgman, Charles D., Samuel M. Shelby, and Robert C. Weast (eds.), *Handbook of Chemistry and Physics*, Chemical Rubber Publishing Company, Cleveland, OH.

Hohls, H. W., *Annalen der Physik*, Vol. 29, 1937, p. 433.

Hopkins, R., et al., *MIL 141 Military Standardization Handbook*, "Optical Design," Defense Supply Agency, 1962.

Houston, T. W., L. F. Johnson, P. Kisliuk, and D. J. Walsh, "Temperature Dependence of the Refractive Index of Optical Maser Crystals," *Journal of the Optical Society of America*, Vol. 53, No. 11, November 1963, p. 1286.

Icenogle, H. W., B. C. Platt and W. L. Wolfe, "Refractive Indexes and Temperature Coefficients of Germanium and Silicon," *Applied Optics*, Optical Society of America, Washington, DC, Vol. 15, No. 10, October 1976, pp. 2348-2351.

Infrared Industries, Inc., Thin Film Products Division, Waltham, MA, "Optical Filters Designers Guide and Catalog," 1973, p. 14.

Ishmuratova, M. S., and L. V. Sergeyev, *Soviet Journal of Optical Technology*, Vol. 34, 1967, p. 801.

Johnson, C. J., G. H. Sherman, and R. Weil, "Far Infrared Measurement of the Dielectric Properties of GaAs and CdTe at 300 K and 8 K," *Applied Optics*, Vol. 8, No. 8, August 1969, p. 1667.

Johnson, V. J. (ed.), "A Compendium of the Properties of Materials at Low Temperatures," *National Bureau of Standards*, Boulder, CO, 1957.

Jones, D. A., R. V. Jones, and R. W. H. Stevenson, "Infrared Transmission Limits of Single Crystals of Some Fluorides," *Proceedings of the Physical Society*, A. C. Stickland (ed.), The Physical Society, London, B-65, 1952, p. 906.

J. Phys. Radium., Ser. 3, Vol. 9, 1900, p. 465.

Kohlrausch, F., *Praktische Physik*, B. G. Teubner, Stuttgart, West Germany, Band 3—Tafeln, 22 Aufl., 1968.

Korth, K., *Zeitschrift Für Physik*, Vol. 84, 1933, p. 677.

Landers, D., et al., "Long Wavelength Infrared Components Measurements," Part II, Spencer Laboratory, Raytheon Corporation, Burlington, MA, August 1962.

Levin, S. B., et al., "Some Optical Properties of Strontium Titanate Crystals," *Journal of the Optical Society of America*, Vol. 45, 1955, pp. 737-739.

Liebreich, E., *Verhandlungen Deutsches Physik Gesellschaft*, Vol. 13, 1911, p. 709.

Loferski, J. J., "Infrared Optical Properties of Single Crystals of Tellurium," *Physical Review*, Vol. 93, 1954, p. 707.

Lowenstein, E., "Optical Properties of Sapphire in the Far Infrared," *Journal of the Optical Society of America*, Vol. 51, No. 1, January 1961, p. 108.

Malitson, I. H., "Interspecimen Comparison of the Refractive Index of Fused Silica," *Journal of the Optical Society of America*, Vol. 55, No. 10, October 1965, p. 1205.

Malitson, I. H., "A Redetermination of Some Optical Properties of Calcium Fluoride," *Applied Optics*, Vol. 2, No. 11, November 1963, p. 1103.

Malitson, I. H., "Refraction and Dispersion of Synthetic Sapphire," *Journal of the Optical Society of America*, Vol. 52, No. 12, December 1962, p. 1377.

Malitson, I. H., "Refractive Properties of Barium Fluoride," *Journal of the Optical Society of America*. Vol. 54, No. 5, May 1964, p. 628.

Martens, F. F., *Annalen der Physik*, Vol. 6, 1901, pp. 602, 603, 619-626.

McAlister, E. D., "The Christiansen Light Filter; Its Advantages and Limitations," Smithsonian Miscellaneous Collections, Vol. 93, 1936, p. 7.

McCarthy, D. E., "The Reflection and Transmission of Infrared Materials: I, Spectra from 2-50 Microns," *Applied Optics*, Vol. 2, No. 6, June 1963, pp. 591-593.

McCarthy, D. E., "The Reflection and Transmission of Infrared Materials: III, Spectra from 2μ to 50μ," *Applied Optics*, Vol. 4, No. 3, March 1965, p. 317.

McCarthy, D. E., "Transmittance of Optical Materials From 0.17 μ to 30 μ," *Applied Optics*, Vol. 6, No. 11, November 1967, p. 1896.

Mentzel, A., *Zeitschrift Für Physik*, Vol. 88, 1934.

Micheli, F. J., *Annalen der Physik*, Vol. 4, No. 7, 1902, p. 772.

Mohler, N. M., and J. R. Loofbourow, "Optical Filters," *American Journal of Physics*, Vol. 20, 1952, pp. 579-588.

Mooney, C. F., and A. F. Turner, "Infrared Transmitting Interference Filters, Proceedings of the Conference on Infrared Optical Materials, Filters, and Films," Engineer Research and Development Laboratories, Fort Belvoir, VA, 1955.

Nysander, R. E., *Physical Review*, Vol. 28, 1909.

Oppenheim, U. P., and A. Goldman, "Infrared Spectral Transmittance of MgO and BaF_2 Crystals between 27° and 1000°C," *Journal of the Optical Society of America*, Vol. 54, No. 1, January 1964, p. 127.

Optical Coating Laboratory, Inc., Technical Products Division, Santa Rosa, CA, "Stock Filter Catalog; Infrared Section," August 1975, pp. II-3 to IV-3.

Packard, R. D., "A Bonding Material Useful in the 2-14 μm Spectral Range," *Applied Optics*, Vol. 8, No. 9, September 1969.

Paschen, F., *Annalen der Physik*, Vol. 26, 1908, p. 120.

Phillippi, C. M., and N. F. Beardsley, "Infrared Window Studies," Pt. I, Wright Air Development Center, Wright Patterson AFB, OH, Technical Note 55-194, June 1955, p. 16.

Quantum Detector Technology, Inc., Burlington, MA, "Quantum Detector Technology Optical Components," p. 11.

Rank, D. H., and H. E. Bennett, "The Index of Refraction of Germanium Measured by an Interference Method," *Journal of the Optical Society of America*, Vol. 44, No. 1, January 1954, p. 13.

Rodney, W. S., "Optical Properties of Cesium Iodide," *Journal of the Optical Society of America*, Vol. 45, No. 11, November 1955.

Rodney, W. S., and R. J. Spindler, National Bureau of Standards, *Journal of Research*, Vol. 51, 1953, p. 123.

Rodney, W. S., I. H. Malitson, and T. A. King, "Refractive Index of Arsenic Trisulfide," *Journal of the Optical Society of America*, Vol. 48, No. 9, September 1958, p. 633.

Rösch, S., "Die Optik des Fabulit, die Durbe des Brewsterwinkels und das Farbspielmoment," *Optica Acta*, Vol. 12, July 1965, p. 253.

Rubens, H., *Wiedemann Annalen*, Vol. 54, 1895, p. 488.

Saker, E. W., "The Optical Constants of Liquid Selenium," *Proceedings of the Physical Society*, A. C. Stickland (ed.), The Physical Society, London, B65, 1952, p. 785.

Salzberg, C. D., and J. J. Villa, "Index of Refraction of Germanium," *Journal of the Optical Society of America*, Vol. 48, No. 8, August 1958, p. 579.

Salzberg, C. D., and J. J. Villa, "Infrared Refractive Indexes of Silicon and Germanium and Modified Selenium Glass," *Journal of the Optical Society of America*, Vol. 47, No. 3, March 1957, p. 244.

Seraphin, B. O., and H. E Bennett, Chapter 12, "Optical Constants," *Semiconductors and Semimetals*, R. K. Willardson and A. C. Beer (eds.), Academic Press, New York, NY, Vol. 3, 1967, pp. 525-526.

Shenk, J. H., et al., "Plastic Filters for the Visible and Near Infra-Red Regions," *Journal of the Optical Society of America*, Vol. 36, 1946, pp. 569-575.

Simeral, W. G., Dissertation, University of Michigan, Ann Arbor, MI, June 1953.

Smith, R. A., R. P. Chasmar, and F. E. Jones, *The Detection and Measurement of Infra-Red Radiation*, Oxford University Press, New York, NY, 1957.

Sosman, R. B., *The Properties of Silica*, Chemical Catalog Company, Inc., New York, NY, 1927.

Stephens, R. E., and I. H. Malitson, *Journal of Research*, National Bureau of Standards, Washington, DC, Vol. 49, 1952, pp. 249-252.

Stephens, R. E., E. K. Plyler, W. S. Rodney, and R. J. Spindler, "Refractive Index of Potassium Bromide for Infrared Radiant Energy," *Journal of the Optical Society of America*, Vol. 43, No. 2, February 1953, p. 110.

Stierwalt, D. L., "Infrared Spectral Emittance Measurements of Optical Materials," *Applied Optics*, Vol. 5, No. 12, December 1966, pp. 1911-1915.

Stierwalt, D. L., J. B. Bernstein, and D. D. Kirk, "Measurement of the Infrared Spectral Absorptance of Optical Materials," *Applied Optics*, Vol. 2, No. 11, November 1963, p. 1169.

Strong, J., *Concepts of Classical Optics*, W. H. Freeman and Company, San Francisco, CA, 1958.

Strong, J., *Physical Review*, Vol. 38, 1931, p. 1818.

Tilton, L. W., and E. K. Plyler, *Journal of Research*, National Bureau of Standards, Washington, DC, Vol. 47, 1951, p. 25.

Trommsdorff, H., *Zeitschrift Für Physik*, Vol. 2, 1901, p. 576.

Willmott, J. C., *Nature*, Vol. 162, 1948.

Wolfe, W. L., and S. S. Ballard, "Optical Materials, Films, and Filters for Infrared Instrumentation," *Proceedings of the Institute of Electrical and Electronics Engineers*, Vol. 47, 1959, p. 9.

Chapter 8

OPTICAL DESIGN

Warren J. Smith
Infrared Industries, Inc.

CONTENTS

8. Optical Design

8.1. Introduction

This chapter deals with optical design, lenses, mirrors, and combinations of these elements. The media of propagation are always considered isotropic. Unless otherwise indicated, optical systems are assumed to be axially symmetrical, that is, to be composed of surfaces which are figures of rotation, whose axes of symmetry coincide with the optical axis.

In general, where both upper and lower case symbols are used for the same quantity, the upper case symbol represents the trigonometric (or "exact") quantity; the lower case symbol represents the corresponding paraxial, or first-order, value. Primed symbols refer to quantities after refraction (or reflection) by a surface or by a lens, or to quantities associated with the image. Subscripts are used to indicate the surface or element with which a symbol is associated, or to indicate a particular ray.

8.1.1. Symbols, Nomenclature and Units. Table 8-1 lists most of the symbols, nomenclature and units used in this chapter. Those symbols which appear as parameters in a particular calculation, however, have been omitted.

8.1.2. Definitions

Axis, Optical. The common axis of symmetry of an optical system; in an element, the line between the centers of curvature of the two (axially symmetric) surfaces.

Eye Relief. In a visual instrument (e.g., telescope or microscope), the distance from the last optical surface to the (usually external) exit pupil, thus the clearance or "relief" between the instrument and the eye.

Invariant, Optical. When two unrelated (i.e., with different axial intercepts) paraxial rays are traced through an optical system, their data are sufficient to completely define the system. If the ray height and ray slope data of the marginal and principal rays are identified by y, u, and y_p, u_p, the expression

$$\mathfrak{J} = n(y_p u - y u_p) \tag{8-1}$$

(where n is the index of refraction of the medium) is invariant across any surface or space of the optical system. At an object or image plane (where $y = 0$ and $y_p = h$), the invariant reduces to the Lagrange Invariant:

$$\mathfrak{J} = hnu = h'n'u' \tag{8-2}$$

where h and h' = object and image height, respectively

 nu and $n'u'$ = the ray slope-index products at the object and image, respectively

Magnification, Lateral or Linear. The ratio between the size of an image (measured perpendicular to the optical axis) and the size of the corresponding (conjugate) object.

Magnification, Longitudinal. The ratio between the length or depth (measured along the optical axis) of an image and the length of the corresponding object.

Table 8-1. Symbols, Nomenclature, and Units

Symbol	Nomenclature	Unit
A	Area	m^2
$A(y,z)$	The complex amplitude of the wavefront emerging from the optical system	$V\ m^{-1}$
AIM	Aerial image modulation	—
A_n	Coefficient of the nth order aspheric deformation term	$m^{(n-1)}$
B	The blur diameter	m
$B(u,v)$	Amplitude factor proportional to the square root of the flux density at point (u,v) in the pupil	—
BFL, bfl	Back focal length—the distance from the back vertex of the optical system to the back (or second) focal point (See Figure 8-2.)	m
C	Surface curvature, reciprocal of surface radius when subscripted for a particular surface, also total curvature of an element	m^{-1}
CC	Coma contribution (TOA)	m
C_e	Equivalent curvature	m^{-1}
$Coma_s$	Coma, sagittal	m
$Coma_t$	Coma, tangential	m
D	Diameter	m
D_e	Diameter for $(1/e^2)$ of optical beam	m
$D_{j,k}$	Distance along ray from surface j to surface k	m
DC	Distortion contribution (TOA)	m
d	Axial distance, especially between (thin) elements, or between principal points	m
E_v	Illuminance	$lm\ m^{-2}$
F, f, EFL, efl	Focal length (effective) or effective focal length—the distance from the second principal point to the back (or second) focal point. Also, the distance from the front (or first) focal point to the first principal point. (See Figure 8-2.)	m
FFL, ffl	Front focal length—the distance from the front vertex of optical system to the front focal point. (See Figure 8-2.)	m
$F/\#$	Relative aperture, speed. The ratio of focal length to entrance pupil diameter. If the object is at infinity, $(F/\#) = [2(NA)]^{-1}$	—
f, f_y, f_z	Spatial frequency in cycles per unit length in the direction indicated	m^{-1}
H, h	Image height	m
$H(\omega_y, \omega_z)$	Optical transfer function	—
$h(y,z)$	Point-spread function (impulse response)	—
I	Angle of incidence which is equal to (minus) the angle of reflection (See Figure 8-1.)	rad
$I(\omega_y, \omega_z)$	Spatial frequency spectrum of an image	—

Table 8-1. Symbols, Nomenclature, and Units (*Continued*)

Symbol	Nomenclature	Unit
I'	Angle of refraction, defined by Snell's Law	rad
$\mathcal{J}$	Optical invariant	rad m
L	Radiance	W m^{-2} sr^{-1}
L, l	Distance from surface to intersection of ray with axis, before refraction; L', l' same after refraction	m
LA	Spherical aberration	m
LA_m	Marginal spherical aberration (longitudinal)	m
LA_z	Zonal spherical aberration (longitudinal)	m
MTF	Modulation transfer function	m
M_j	Distance (vector) from vertex of surface j (perpendicular) to ray	m
M_x	The x component of M_j	m
m	Magnification, lateral	—
m_t	Magnification, longitudinal	—
NA	Numerical aperture, given by $n \sin U$ where n is the final index in an optical system and U is the slope angle of the axial ray at the image. If the object is at infinity, $(NA) = [2(F/\#)]^{-1}$	—
n	Index of refraction	—
$O(\omega_y, \omega_z)$	Spatial frequency spectrum of an object	—
OPD	Optical path difference	m
OSC	Offense against the sine condition	—
OTF	Optical transfer function	—
P_1, P_2	First and second principal points, respectively	—
p (subscript)	Used to denote a principal or chief ray	—
Q	Stop-shift ratio	—
R	Radius of curvature	m
r	Semi-aperture = $(y^2 + z^2)^{1/2}$	m
S	Nominal distance at which a system is focused	m
s	Distance, first principal point to object	m
s'	Distance, second principal point to image	m
TA	Transverse aberration	m
TAC	Transverse version of aberration contribution for astigmatism	m
TchA	Lateral chromatic aberration, chromatic difference of image size	m
TchC	Lateral chromatic aberration contribution	m
TLchC	Transverse version of aberration contribution for axial color	m
$T/\#$	The speed of a lens, taking its transmission into account $(T/\#) = (F/\#)$ (transmission)$^{-1/2}$	—
TOA	Third-order aberration	—
TPC	Transverse version of aberration contribution for Petzval curvature	m

Table 8-1. Symbols, Nomenclature, and Units (*Continued*)

Symbol	Nomenclature	Unit
TSC	Transverse version of aberration contribution for spherical aberration	m
t	Thickness; axial spacing between surfaces	m
U, u	Ray-slope angle, before refraction; U', u' same after refraction	rad
u, v	Spatial coordinates in exit pupil	m
V	Abbe number; reciprocal relative dispersion	—
v, v'	u/y and u'/y	m^{-1}
W_{ABC}	Numerical coefficient in the wave-aberration polynomial	—
W	Semi-diameter of a Gaussian beam (e^{-2} points); also, width of slit	m
x	Distance from first focal point to object	m
x'	Distance from second focal point to image	m
x_p	The longitudinal curvature (sag) of the image field: Petzval	m
x_s	The longitudinal curvature (sag) of the image field: sagittal	m
x_t	The longitudinal curvature (sag) of the image field: tangential	m
X, Y, Z	Direction cosines	—
x, y, z	Coordinate system; x is the optical axis; x and y define the meridional plane. The origin is at the vertex of the surface	m
Y, y	Height of intersection of ray with surface	m
β	Angular diameter of image blur spot	rad
$\Delta\phi$	Optical phase difference	m
δ, or Δ	The change in any quantity, such as Δn, δn and δs	—
δS	The longitudinal distance from the position of best focus	m
λ	Wavelength	μm
ν	Frequency of radiation	sec^{-1}
$\tilde{\nu}$	Wavenumber of radiation	cm^{-1} or wavenumber
ρ	Reflectivity	—
ρ, ϕ	Polar coordinates	m, rad
Φ	Radiant power or flux	W
$\phi(u, \mathrm{v})$	The wave-aberration function, equal to the OPD of the ray through point (u, v)	m
ϕ	Lens power, i.e., reciprocal focal length $\equiv 1/F$ or $1/f$	diopter, m^{-1}
$\psi(\omega_y, \omega_z)$	Phase transfer function	—
ω_y, ω_z	Radian spatial frequency in the y- and z-directions, i.e., $\omega_y = 2\pi f_y$ and $\omega_z = 2\pi f_z$	m^{-1}

Magnification, Angular. The ratio of the angular size of an image (produced by an afocal optical system) to the angular size of the corresponding object.

Magnification, Microscopic. The ratio of the angular size of an image to the angular size of the object, if the object were to be viewed at a conventional distance. For visual work, the conventional distance is 250 mm (10 in.).

Paraxial. Pertaining to an infinitesimal, thread-like region about the optical axis.

Plane of Incidence. The incident ray, the normal to the surface at the point of incidence, and the refracted (or reflected) ray all lie in the same plane of incidence. (See Figure 8-1.)

Planes, Principal. If each ray of a bundle, incident on an optical system parallel to the axis, is extended to meet the backward extension of the same ray after it has passed through the system, the locus of the intersections of all the rays is called a Principal Plane. The First Principal Plane is formed by rays from the right. The Second Principal Plane is formed by rays incident from the left. The Principal Planes are planes only in the paraxial region; at any finite distance from the axis they are figures of rotation, frequently approximating spherical surfaces.

Points, Cardinal. The Focal Points, Principal Points, and Nodal Points. (See Figure 8-2.)

Point, Focal. The point to which (paraxial) rays, parallel to the axis, converge, or appear to converge, after passing through the optical system. (See Figure 8-2.)

Point, Front (First) Focal. The Focal Point to which rays incident from the right are converged. (See Figure 8-2.)

Point, Back (Second) Focal. The Focal Point to which rays incident from the left are converged. (See Figure 8-2.)

Points, Principal. The intersection of the Principal Planes with the Optical Axis. (See Figure 8-2.)

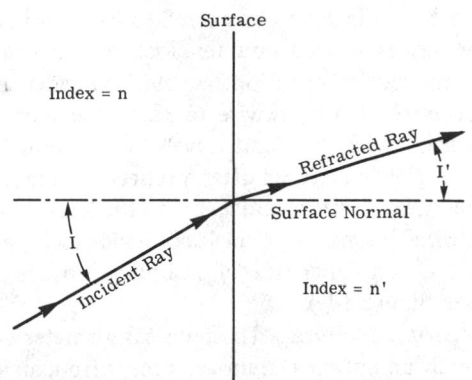

Fig. 8-1. Refraction at an optical surface. The Plane of incidence and refraction is the plane of the paper.

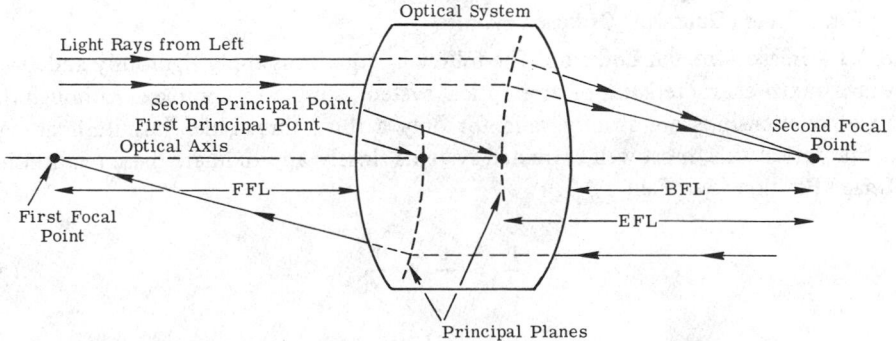

Fig. 8-2. The focal points and principal points of a generalized optical system.

Points, Nodal. Two axial points of an optical system, so located that an oblique ray directed toward the first appears to emerge from the second, parallel to its original direction. For systems in air, the nodal points coincide with the principal points.

Pupil, Entrance. The image of the aperture stop formed by the optical elements (if any) between the aperture stop and the object. The image of the aperture stop as "seen" from the object.

Pupil, Exit. The image of the aperture stop formed by the optical elements (if any) behind the aperture stop.

Ray, Chief. A ray directed toward the center of the entrance pupil of the optical system.

Ray, Principal. Strictly, a ray directed toward the first principal point, but commonly used to refer to the Chief Ray.

Ray, Marginal. The ray from the axial point on the object which intersects the rim of the aperture stop.

Sign Conventions. Light rays are assumed to progress from left to right. Radii and curvatures are positive if the center of curvature is to the right of the surface. Surfaces or elements have positive power if they converge light. Distances upward (or to the right) are positive, that is, points which lie above the axis (or to the right of an element, surface or another point) are considered to be a positive distance away. Slope angles are positive if the ray is rotated counter-clockwise to reach the axis. (This is the reverse of the usual geometrical convention.) Angles of incidence, refraction, and reflection are positive if the ray is rotated clockwise to reach the normal to the surface. The index of refraction is positive when the light travels in the normal left-to-right direction. When the light travels from right-to-left, as after a reflection, the index is taken as negative (as is the distance to the "next" surface, since it is to the left).

Snell's Law. The angles of incidence, I, and refraction, I', and the refractive indices, n and n', on either side of an optical surface are related by Snell's Law: $n \sin I = n' \sin I'$. (See Figure 8-1.)

Stop, Aperture. The physical diameter which limits the size of the cone of radiation which an optical system will accept from an axial point on the object. For off-axis points, the limiting aperture may be defined by more than one physical feature of the optical system.

Stop, Field. The physical diameter which limits the angular field of view of an optical system.

8.2. First Order (Gaussian) Optical Layout

8.2.1. Image Size and Location. The following equations apply rigorously and exactly to the paraxial characteristics of any optical system, simple or complex. Although these paraxial relationships are strictly valid for only a thin, thread-like, infinitesimal region near the optical axis, most well-corrected systems closely approximate these relationships.

Image Position (See Figure 8-3.):

$$\frac{1}{s'} = \frac{1}{s} + \frac{1}{f} \tag{8-3}$$

$$x' = -\frac{f^2}{x} \tag{8-4}$$

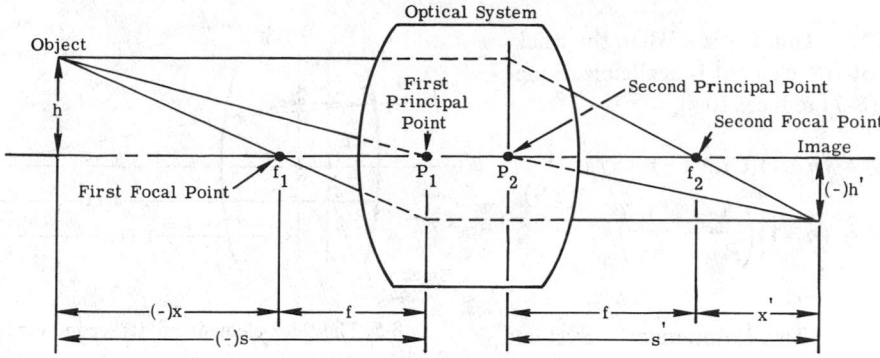

Fig. 8-3. Object and image relationships.

Image Size; Lateral Magnification (See Figure 8-3.):

$$m = \frac{h'}{h} = \frac{s'}{s} = \frac{f}{x} = -\frac{x'}{f} \tag{8-5}$$

Image Size; Longitudinal Magnification (See Figure 8-4.):

$$m_t = \frac{s_2' - s_1'}{s_2 - s_1} = \frac{s_1'}{s_1} \frac{s_2'}{s_2} = m_1 m_2 \approx m^2 \tag{8-6}$$

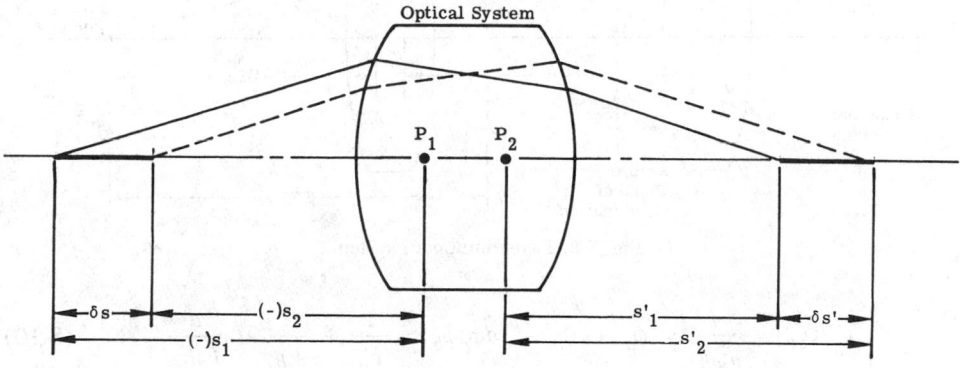

Fig. 8-4. Longitudinal magnification, m_t.

8.2.2. Thick Elements (See Figure 8-5.). The power, focal length, and back focal length of a single element in air are given by

$$\phi = \frac{1}{f} = (n-1)\left[C_1 - C_2 + \frac{tC_1C_2(n-1)}{n} \right]$$

$$= (n-1)\left[\frac{1}{R_1} - \frac{1}{R_2} + \frac{t(n-1)}{nR_1R_2} \right] \tag{8-7}$$

$$BFL = f\left[1 - \frac{tC_1(n-1)}{n} \right] = f\left[1 - \frac{t(n-1)}{nR_1} \right] \tag{8-8}$$

8.2.3. Thin Lenses.

When the thickness of the element is negligible, Equation (8-7) reduces to

$$\phi = (n-1)(C_1 - C_2)$$

$$\phi = (n-1)\left(\frac{1}{R_1} - \frac{1}{R_2}\right) \qquad (8-9)$$

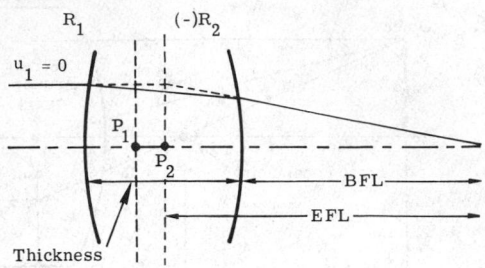

Fig. 8-5. The thick element and its second cardinal points.

8.2.4. Two-Component Systems

(See Figure 8-6.). When a system consists of two components, a and b, the following explicit expressions may be applied. Components a and b may be simple elements, mirrors, compound lenses, or complex systems in their own right. In Equations (8-10) through (8-16), the distances (d, BFL, FFL) are measured from the principal points of components a and b.

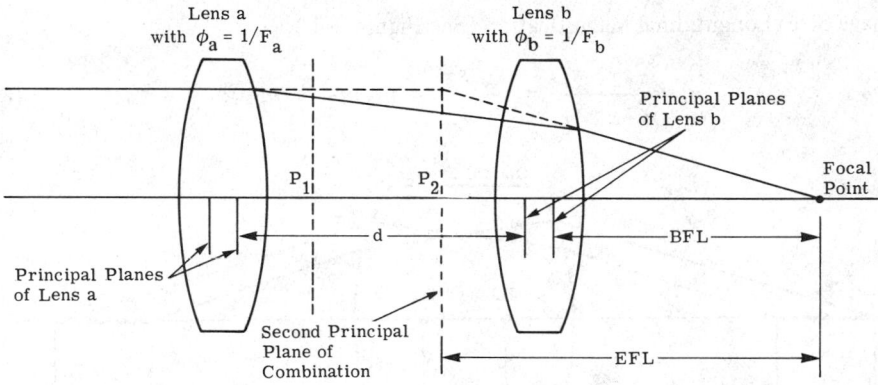

Fig. 8-6. Two-component system.

$$\phi_{ab} = \frac{1}{F_{ab}} = \phi_a + \phi_b - d\phi_a\phi_b = \frac{1}{F_a} + \frac{1}{F_b} - \frac{d}{F_a F_b} \qquad (8-10)$$

$$F_{ab} = \text{EFL}_{ab} = \frac{F_a F_b}{F_a + F_b - d} \qquad (8-11)$$

$$\text{BFL}_{ab} = F_b\left(\frac{F_a - d}{F_a + F_b - d}\right) = F_{ab}\left(\frac{F_a - d}{F_a}\right) \qquad (8-12)$$

$$-\text{FFL} = F_{ab}\left(\frac{F_b - d}{F_b}\right) \qquad (8-13)$$

The powers of the components which will produce a desired set of system characteristics can be determined from the following equations:

$$F_a = d\left(\frac{F_{ab}}{F_{ab} - BFL}\right) \tag{8-14}$$

$$F_b = -d\left(\frac{BFL}{F_{ab} - BFL - d}\right) \tag{8-15}$$

$$d = F_b\left(\frac{BFL}{F_b - BFL}\right) \tag{8-16}$$

$$= F_a + F_b - \frac{F_a F_b}{F_{ab}}$$

8.2.5. Paraxial Ray-Tracing Equations.
Paraxial Image Location, Single Surface (See Figure 8-7.)

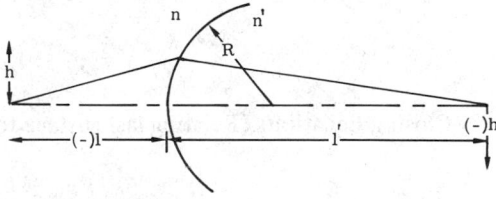

$$\frac{n'}{l'} = \frac{n}{l} + \frac{(n' - n)}{r} \tag{8-17}$$

$$= \frac{n}{l} + (n' - n)C$$

$$m = \frac{h'}{h} = \frac{nl'}{n'l} \tag{8-18}$$

Fig. 8-7. Image formation at a single surface by Equation (8-17).

Paraxial Ray-Tracing (See Figure 8-8.). The following equations are more convenient for tracing ray paths than Equations (8-17) and (8-18).

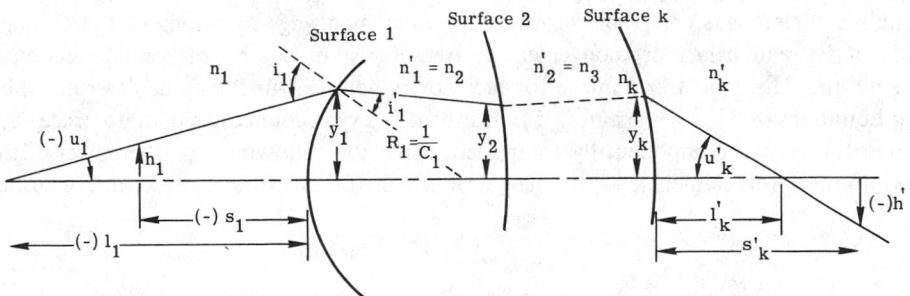

Fig. 8-8. Illustrating the nomenclature of the paraxial ray-tracing equations; Equations (8-19) through (8-28).

Opening Equations (Relating object to the first surface):

$$n_1 u_1 = \frac{n_1 y_1}{l_1} \tag{8-19}$$

or

$$n_1 u_1 = \frac{n_1 h_1}{(l_1 - s_1)} \tag{8-20}$$

Iterative Equations (Applied to each surface in turn, $j = 1, 2, ... k$):

$$n'_j u'_j = n_j u_j + (n'_j - n_j) y_j C_j \qquad (8\text{-}21)$$

$$y_{j+1} = y_j - \frac{t'_j n'_j u'_j}{n'_j} \qquad (8\text{-}22)$$

Alternate Iterative Equations:

$$i_j = y_j C_j - u_j \qquad (8\text{-}23)$$

$$i'_j = \frac{n_j i_j}{n'_j} \qquad (8\text{-}24)$$

$$u'_j = u_j + i_j - i'_j$$
$$= u_j + i_j \left(1 - \frac{n_j}{n'_j}\right) \qquad (8\text{-}25)$$

$$y_{j+1} = y_j - t'_j u'_j \qquad (8\text{-}26)$$

Closing Equations (Relating last surface to image):

$$l'_k = \frac{n'_k y_k}{n'_k u'_k} = \frac{y_k}{u'_k} \qquad (8\text{-}27)$$

$$h'_k = y_k - u'_k s'_k = u'_k (l'_k - s'_k) \qquad (8\text{-}28)$$

8.2.6. Multi-Element Systems (See Figure 8-9.). Although paraxial rays may be traced through complete systems, one surface at a time by using Equations (8-19) through (8-28), it is frequently more convenient to treat a system as a set of components separated by air. The object and image for each component (in turn) may be determined by using Equations (8-3) or (8-4) and (8-5). It is usually even more convenient to trace rays through the system component-by-component using the following: ϕ_j, the power of the jth component (or element); y_j, the height at which the ray strikes the principal planes

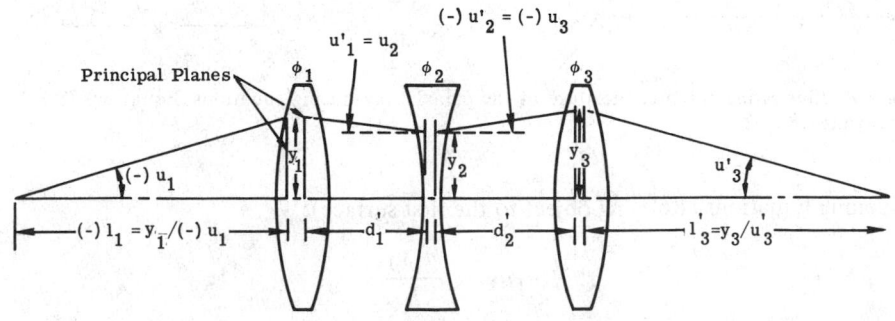

Fig. 8-9. Illustrating the ray-tracing nomenclature for use with component by component ray-tracing Equations (8-19) and (8-10).

of the jth component; and d'_j, the distance from the second principal plane of the jth component to the first principal plane of the $(j + 1)^{th}$ component. The ray slope after refraction by the jth component, u'_j, is determined from

$$u'_j = u_j + y_j \phi_j \qquad (8\text{-}29)$$

The ray height at the next component is given by

$$y_{j+1} = y_j - d'_j u'_j \qquad (8\text{-}30)$$

If the elements are thin, d is the space between them. If the ray from the axial intercept of the object has been traced, the magnification can be determined from the Lagrange Invariant (Section 8.1.2).

8.3. Exact Ray-Tracing

8.3.1. **The General, or Skew, Ray** (See Figure 8-10.) [8-1, 8-2, 8-3, 8-4]. A general ray is defined by its direction cosines $(X, Y, \text{and } Z)$ and by the coordinates $(x, y, \text{and } z)$ of its intersection with a surface of the optical system. The subscript notation for this section is illustrated in Figure 8-11.

Spherical Surfaces. Opening (At the initial reference surface):

$$C(x^2 + y^2 + z^2) - 2x = 0 \qquad (8\text{-}31)$$

$$X^2 + Y^2 + Z^2 = 1.0 \qquad (8\text{-}32)$$

Intersection of ray with next surface:

$$e = tX - (xX + yY + zZ) \qquad (8\text{-}33)$$

$$M_{1x} = x + eX - t \qquad (8\text{-}34)$$

$$M_1^2 = x^2 + y^2 + z^2$$
$$- e^2 + t^2 - 2tx \qquad (8\text{-}35)$$

$$\cos I_1 = E_1 =$$
$$[X^2 - C_1(C_1 M_1^2 - 2M_{1x})]^{1/2} \qquad (8\text{-}36)$$

$$D_{0,1} = e + \left(\frac{C_1 M_1^2 - 2M_{1x}}{X + E_1} \right) \qquad (8\text{-}37)$$

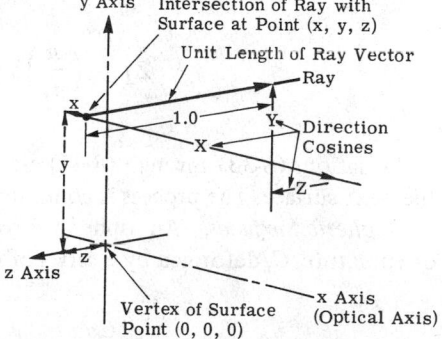

Fig. 8-10. Illustrating the symbols used in the general ray-tracing equations of Section 8.3.3. The spatial coordinates of the intersection point of the ray with the surface are x, y, and z. The ray direction cosines are X, Y, and Z [8-8].

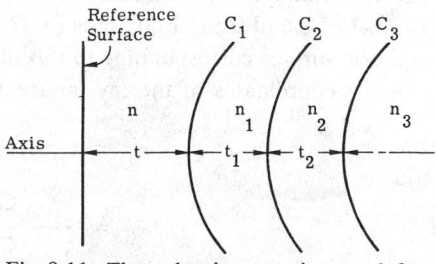

Fig. 8-11. The subscript notation used for the constructional parameters in Section 8.3.1 [8-8].

$$x_1 = x + D_{0,1}X - t \tag{8-38}$$

$$y_1 = y + D_{0,1}Y \tag{8-39}$$

$$z_1 = z + D_{0,1}Z \tag{8-40}$$

Direction Cosines of Ray after Refraction:

$$\cos I_1' = E_1' = \left[1 - (1 - E_1^2)\left(\frac{n}{n_1}\right)^2\right]^{1/2} \tag{8-41}$$

$$g_1 = E_1' - \frac{n}{n_1}E_1 \tag{8-42}$$

$$X_1 = \left(\frac{n}{n_1}\right)X - g_1C_1x_1 + g_1 \tag{8-43}$$

$$Y_1 = \left(\frac{n}{n_1}\right)Y - g_1C_1y_1 \tag{8-44}$$

$$Z_1 = \left(\frac{n}{n_1}\right)Z - g_1C_1z_1 \tag{8-45}$$

Equations (8-33) through (8-45) are repeated with the subscripts advanced by one for the next surface. The process is continued until the final (image) surface is reached.

Aspheric Surfaces. An aspheric surface of revolution may be represented as a sphere of curvature, C, deformed by a series of terms in even powers of the semi-diameter, r.

$$x = \frac{Cr^2}{1 + \sqrt{1 - C^2r^2}} + A_2r^2 + A_4r^4 + \cdots + A_jr^j \tag{8-46}$$

where $r^2 = y^2 + z^2$
    j = an even integer

Intersection of Ray with Aspheric. The sphere of curvature, C, is presumed to be a fair approximation to the aspheric. The intersection of the ray with the sphere at (x_0, y_0, z_0) is found using Equations (8-33) through (8-40). The actual x coordinate of the aspheric surface corresponding to this distance from the axis is found by substituting the y and z coordinates of the ray intersection with the sphere into Equation (8-46) to get

$$r_0^2 = y_0^2 + z_0^2 \tag{8-47}$$

and

$$\tilde{x}_0 = \frac{Cr_0^2}{1 + \sqrt{1 - C^2r_0^2}} + A_2r_0^2 + \cdots \tag{8-48}$$

Thus, a measure of the approximation error is the difference $(\tilde{x} - x)$ between the true sag of the aspheric and the approximation. Then one computes

$$l_0 = (1 - C^2 r_0^2)^{1/2} \tag{8-49}$$

$$m_0 = -y_0 [C + l_0(2A_2 + 4A_4 r_0^2 + \cdots + jA_j r_0^{(j-2)})] \tag{8-50}$$

$$n_0 = -z_0 [C + l_0(2A_2 + 4A_4 r_0^2 + \cdots + jA_j r_0^{(j-2)})] \tag{8-51}$$

An improved approximation to the intersection of the ray with the aspheric (Figure 8-12) can be obtained from

$$G_0 = \frac{l_0(\tilde{x}_0 - x_0)}{(Xl_0 + Ym_0 + Zn_0)} \tag{8-52}$$

$$x_1 = G_0 X + x_0 \tag{8-53}$$

$$y_1 = G_0 Y + y_0 \tag{8-54}$$

$$z_1 = G_0 Z + z_0 \tag{8-55}$$

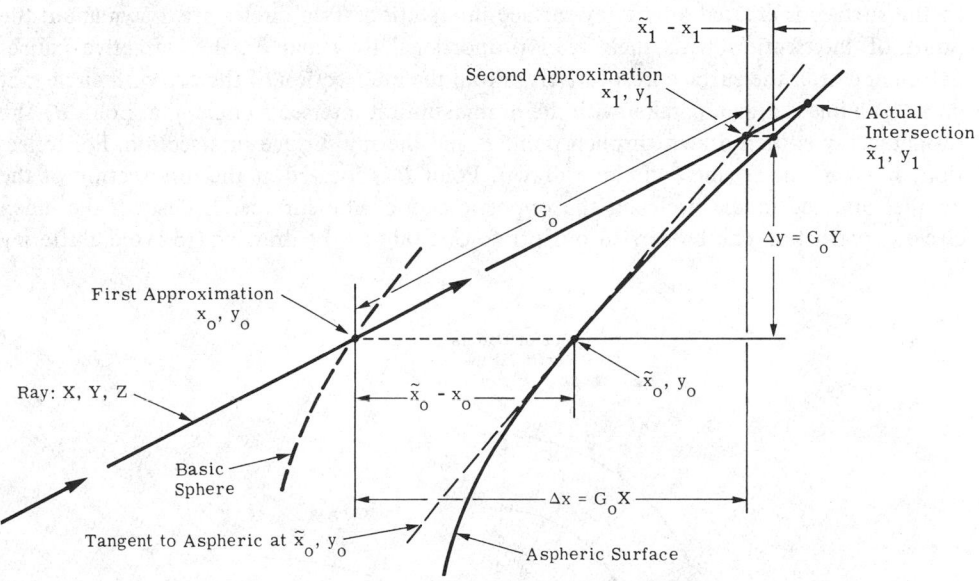

Fig. 8-12. Determination of ray intersection with an aspheric surface. The intersection of the ray with aspheric surface is found by a convergent series of approximations. Shown above are the relationships in finding the first approximation after the intersection with the basic sphere has been found.

and the new error is $(\tilde{x}_1 - x_1)$. This approximation process is repeated, from Equation (8-47) to Equation (8-55), with the subscripts advanced by one at each iteration, until the error $(\tilde{x}_k - x_k)$ after the kth iteration is negligible. Refraction at the aspheric surface is then

$$P^2 = l_k^2 + m_k^2 + n_k^2 \tag{8-56}$$

$$P \cos I = F = Xl_k + Ym_k + Zn_k \tag{8-57}$$

$$P \cos I' = F' = \sqrt{P^2 \left(1 - \left(\frac{n}{n_1}\right)^2\right) + \left(\frac{n}{n_1}\right)^2 F^2} \tag{8-58}$$

$$g = \frac{\left(F' - \left(\frac{n}{n_1}\right)F\right)}{P^2} \tag{8-59}$$

$$X_1 = X\left(\frac{n}{n_1}\right) + gl_k \tag{8-60}$$

$$Y_1 = Y\left(\frac{n}{n_1}\right) + gm_k \tag{8-61}$$

$$Z_1 = Z\left(\frac{n}{n_1}\right) + gn_k \tag{8-62}$$

8.3.2. Graphical Ray-Tracing (See Figure 8-13.). Meridional rays can be traced using only a scale, straight edge, and compass. The ray is drawn to the surface, and the normal to the surface is erected at the ray-surface intersection. Two circles are drawn about the point of intersection with their radii proportional to n and n', the refractive indices before and after the surface, respectively. From the intersection of the ray with circle n at point A, a line is drawn parallel with the normal until it intersects circle n' at point B. The refracted ray is then drawn through point B and the ray-surface intersection. For reflection, $n' = -n$, and a single circle is drawn. Point B is located at the intersection of the parallel and the index circle on the opposite side of the surface. If desired, the index circle construction can be carried out off to one side of the drawing (to avoid cluttering

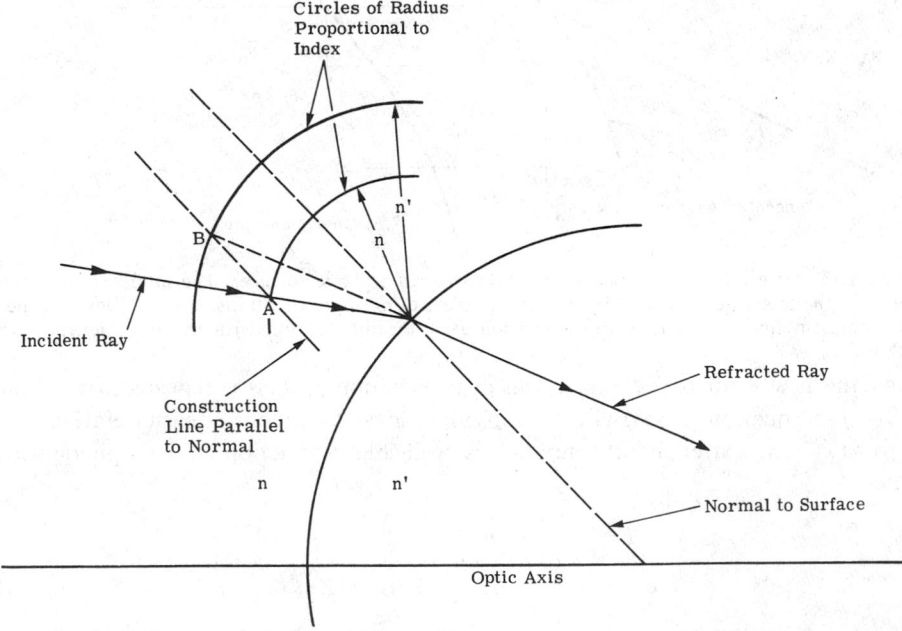

Fig. 8-13. Graphical ray-tracing. Starting with the construction of circles (with radii proportional to the indices on either side of the surface) about the point of intersection of the ray and surface and the development of the refracted ray.

the diagram) and the angles transferred to the drawing. An alternative is to measure the angle of incidence and compute the angle of refraction using Snell's Law (Section 8.1.2). The accuracy of this technique is poor and the process is laborious. Thus, it is rarely used except for crude condenser-type design. It is usually preferable to use a computer and draw the rays from the computed data.

8.4. Aberrations

8.4.1. The Wave-Aberration Polynomial [8-5]. If the image aberrations of an optical system are expressed as an optical path difference (OPD) it can be shown (by reasons of symmetry) that the OPD can be expressed as a series expansion of the following form:

$$\text{OPD}(h, \rho, \phi) = \sum_{l,m,n} W_{2l+n, 2m+n, n} h^{2l+n} \rho^{2m+n} (\cos \phi)^n \qquad (8\text{-}63)$$

where $W_{2l+n, 2m+n, n}$ = wave-aberration numerical coefficient

$\quad h$ = image height

$\quad \rho, \phi$ = polar coordinates of ray intersection with the system entrance pupil

$\quad l, m, n$ = running indices, all positive integers, $0, 1, 2, 3, \ldots$

The term $W_{2l+n,0,0}$ is zero by definition; the term $W_{020}\rho^2$ is equivalent to a simple longitudinal shift of the image (or reference) plane; the term $W_{111} h \rho \cos\phi$ is equivalent to a vertical shift of the reference point (i.e., a change in image-height). These latter two terms are called the first-order terms and become zero when the reference point is chosen at the paraxial focus. The "order" is given by $[(2l + n) + (2m + n) - 1]$.

The next five terms in the series are the third-order, or Seidel, aberrations:

Spherical Aberration	$W_{040}\rho^4$
Coma	$W_{131}\rho^3 \cos\phi$
Astigmatism and Petzval	$W_{220}h^2\rho^2 + W_{222}h^2\rho^2\cos^2\phi$
Distortion	$W_{311}h^3\rho \cos\phi$

The fifth-order aberrations are

Spherical Aberration	$W_{060}\rho^6$
Linear Coma	$W_{151}h\rho^5 \cos\phi$
Elliptical Coma	$W_{331}h^3\rho^3 \cos\phi + W_{333}h^3\rho^3 \cos^3\phi$
Oblique Spherical	$W_{240}h^2\rho^4 + W_{242}h^2\rho^4 \cos\phi$
Astigmatism and Petzval	$W_{420}h^4\rho^2 + W_{422}h^4\rho^2 \cos^2\phi$
Distortion	$W_{511}h^5\rho \cos\phi$

There are 2 first-order terms, 5 third-order terms, 9 fifth-order terms, 14 seventh-order terms, 20 ninth-order terms, and $[2 + 3 \cdots + 0.5 (N + 3)] N^{\text{th}}$-order terms.

The wave aberration polynomial Equation (8-63), represents the departure of the actual wavefront from a perfect spherical reference surface which passes through the axial intercept of the exit pupil and is centered on the ideal (or reference) image point. Thus, TA_y and TA_z, the y and z components of the transverse ray aberration, can be expressed in terms of the wave aberration as

$$TA_y = -\left(\frac{l}{n}\right)\frac{\partial OPD}{\partial y} \tag{8-64}$$

$$TA_z = -\left(\frac{l}{n}\right)\frac{\partial OPD}{\partial z} \tag{8-65}$$

where　　l = distance from the exit pupil to the image point
　　　　　n = index of the final medium
　　OPD = optical path difference (Equation (8-63))

8.4.2. Ray-Aberration Polynomial. The result of the operations indicated in Equations (8-64) and (8-65) is a pair of polynomial expressions for the ray aberrations:

$$
\begin{aligned}
TA_y = -\left(\frac{l}{n}\right)\Big[& W_{020}\cdot 2\rho\cos\phi + W_{111}\cdot h + W_{040}\cdot 4\rho^3\cos\phi \\
& + W_{131}\cdot h\rho^2(2 + \cos 2\phi) + (W_{220} + W_{222})\cdot 2h^2\rho\cos\phi \\
& + W_{311}\cdot h^3 + W_{060}\cdot 6\rho^5\cos\phi + W_{151}\cdot h\rho^4(3 + 2\cos 2\phi) \\
& + W_{331}\cdot h^3\rho^2(2 + \cos 2\phi) + W_{333}\cdot\frac{3}{2}h^3\rho^2(1 + \cos 2\phi) \\
& + W_{240}\cdot 4h^2\rho^3\cos\phi + W_{242}\cdot h^2\rho^3\cos\phi(3 + \cos 2\phi) \\
& + (W_{420} + W_{422})\cdot 2h^4\rho\cos\phi + W_{511}\cdot h^5 \\
& + (\text{7th and higher-order terms})\Big]
\end{aligned}
\tag{8-66}
$$

$$
\begin{aligned}
TA_z = -\left(\frac{l}{n}\right)\Big[& W_{020}\cdot 2\rho\sin\phi + W_{040}\cdot 4\rho^3\sin\phi + W_{131}\cdot h\rho^2\sin 2\phi \\
& + W_{220}\cdot 2h^2\rho\sin\phi + W_{060}\cdot 6\rho^5\sin\phi + W_{151}\cdot 2h\rho^4\sin 2\phi \\
& + W_{331}\cdot h^3\rho^2\sin 2\phi + W_{240}\cdot 4h^2\rho^3\sin\phi + W_{242}\cdot h^2\rho^3\sin\phi(1 + \cos 2\phi) \\
& + W_{420}\cdot 2h^4\rho\sin\phi + (\text{7th and higher-order terms})\Big]
\end{aligned}
\tag{8-67}
$$

In these expressions, the sum of the exponents of the aperture and field terms, ρ and h, indicates the order of the aberration represented by that term. Thus, the first order terms contain either ρ or h; the third-order aberration terms have ρ^3, $\rho^2 h$, ρh^2, or h^3; the fifth-order terms are in ρ^5, $\rho^4 h$, $\rho^3 h^2$, $\rho^2 h^3$, ρh^4, or h^5; and so forth.

8.4.3. Aberration Descriptions. Optical aberrations are faults or defects of the image. They are described in terms of the amount by which a geometrically-traced ray misses a desired location in the image formed by the optical system. Ordinarily, the desired location for a ray in the image is that indicated by the first-order laws of image formation, as set forth in Section 8.2.

Spherical aberration can be defined as the longitudinal variation of focus with aperture. (See Figure 8-14.) Longitudinal spherical aberration is the distance from the paraxial focus to the axial intersection of the ray. Lateral (or transverse) spherical aberration is the vertical distance from the axis to the intersection of the ray with the paraxial image plane.

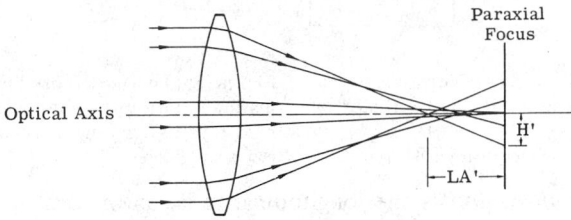

Fig. 8-14. A simple converging lens with undercorrected spherical aberration. The rays further from the axis are brought to a focus nearer the lens.

Coma is the variation of magnification (i.e., image size) with aperture. (See Figure 8-15.) Tangential coma is the vertical distance from the chief ray to the intersection of the upper and lower rim rays. The appearance of a comatic point image is simulated in Figure 8-28.

Field curvature describes the amount by which the off-axis image departs longitudinally from the surface (usually flat) in which it should be located. Curvature of field may differ for rays in different meridians. The focus of a fan of rays in the meridional plane is called the tangential focus; the focus of rays lying in a plane normal to the meridional plane is called the sagittal focus. The distance from the sagittal to the tangential focus is the astigmatism, and the longitudinal distance from the paraxial image plane to the focii are the tangential and sagittal field curvatures, x_t and x_s, respectively. The *Petzval curvature* is the basic field curvature of an optical system. The Petzval surface lies three times as far from the tangential focal surface as from the sagittal focal surface, as indicated in Figure 8-16. Field curvatures may be described as inward, undercorrected, and negative (as shown in Figure 8-16) or as backward, overcorrected, and positive.

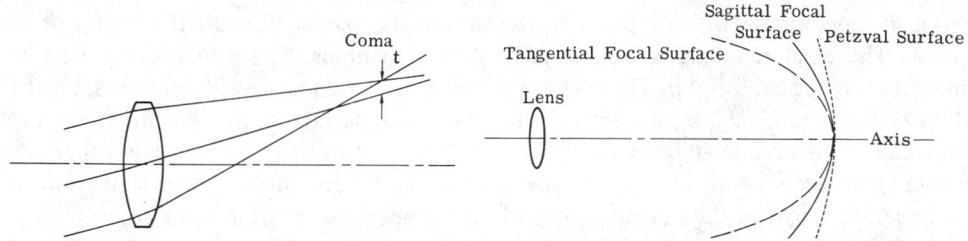

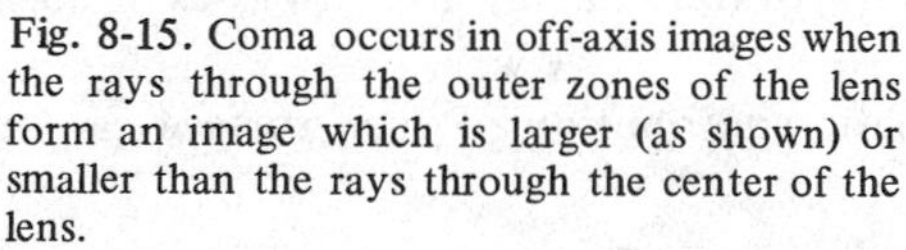

Fig. 8-15. Coma occurs in off-axis images when the rays through the outer zones of the lens form an image which is larger (as shown) or smaller than the rays through the center of the lens.

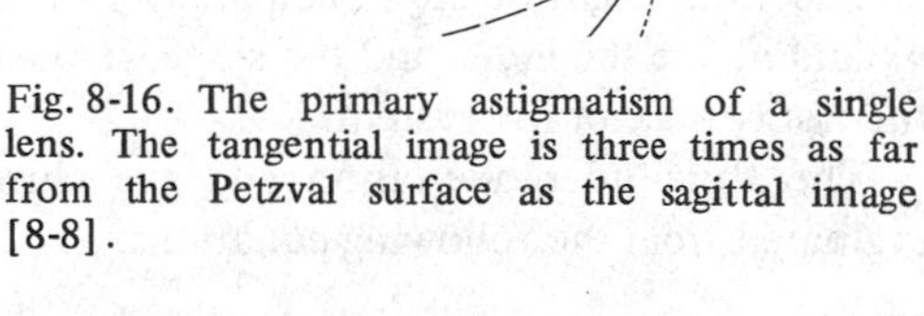

Fig. 8-16. The primary astigmatism of a single lens. The tangential image is three times as far from the Petzval surface as the sagittal image [8-8].

Distortion is the amount by which an image is closer to or further from the axis than its position as given by first-order optics. The linear amount of simple distortion varies with h^3. Thus, nonradial straight lines are imaged as curved lines (as shown in Figure 8-17).

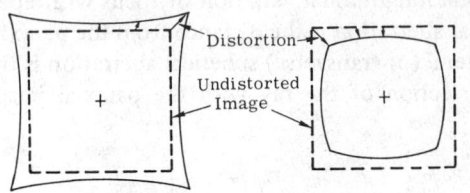

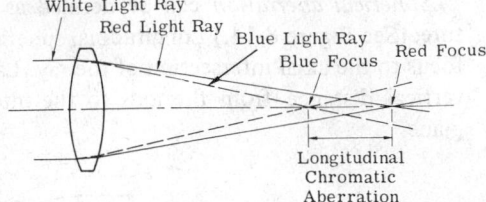

(a) Positive, or Pin-   (b) Negative, or
cushion Distortion   Barrel Distortion

Fig. 8-17. Distortion [8-8].

Fig. 8-18. The undercorrected, longitudinal, chromatic aberration of a simple lens. This is due to the blue rays undergoing a greater refraction than the red rays [8-8].

Axial chromatic aberration is the longitudinal variation of focal position with wavelength. The longitudinal chromatic aberration is the distance from the long wavelength focus to the short wavelength focus. (See Figure 8-18.)

Lateral chromatic aberration is the variation of image size with wavelength. It is the vertical distance from the off-axis image of a point in long wavelength light to the corresponding image point in short wavelength light. It is also known as chromatic difference of magnification (CDM).

Monochromatic aberrations vary with wavelength. Chromatic variation of spherical aberration, or sphero-chromatism, is the most commonly encountered. Usually spherical aberration for the shorter wavelength is more overcorrected (or less undercorrected) than for the longer wavelength. Thus, a typical system might have the spherical aberration overcorrected in short wavelength light, corrected in the center of the bandpass, and undercorrected in long wavelength light. Chromatic variations of the other aberrations are less frequently encountered and are usually less serious.

8.4.4. Third-Order Aberrations. The third-order portion of the aberration polynomials (Equations (8-63), (8-66), and (8-67)), may be computed from the data of two paraxial rays traced through the optical system. The equations given here indicate directly the amount of transverse aberration at the image. Two paraxial rays are traced through the system. The axial or aperture ray is traced from the axial intercept of the object and passes through the rim of the system's pupil. The *principal*, *chief*, or *field ray* is traced from an appropriate off-axis point in the object and passes through the center of the pupil. The axial ray data are indicated by plain symbols. The principal ray data are indicated by the subscript p. The rays are traced using the equations of Section 8.2.5. The data i, y, u, and i_p, y_p, u_p are thus available for each surface of the system. The optical invariant, $\mathcal{J}$, is evaluated from the data of the two rays at the first (or any other convenient) surface: $\mathcal{J} = n(y_p u - y u_p)$. The final image height can be determined from the intersection height of the principal ray with the image plane, or from $h = \mathcal{J}/(n'_k u'_k)$, where n'_k and u'_k are the index and the slope, respectively, of the axial ray after it passes through the last surface of the system.

The third-order aberration and the chromatic contribution of each surface can be evaluated from the following equations:

Spherical	$TSC = Bi^2 h + Wy^4$	(8-68)
Sag Coma	$CC = Bii_p h + Wy^3 y_p$	(8-69)
Astigmatism	$TAC = Bi_p^2 h + Wy^2 y_p^2$	(8-70)

Petzval $\quad\quad\quad$ TPC $= \dfrac{(n - n')\, \mathfrak{I}\, hC_e}{2nn'}$ $\hfill$ (8-71)

Distortion $\quad\quad\quad$ DC $= h\left[B_p ii_p + \dfrac{1}{2}(u_p'^2 - u_p^2)\right]$

$\hspace{5cm} + \, Wyy_p^3$ $\hfill$ (8-72)

Axial Chromatic $\quad$ TLchC $= \dfrac{yi\left(\Delta n - \dfrac{\Delta n' n}{n'}\right)}{u_k'}$ $\hfill$ (8-73)

Lateral Chromatic $\quad$ TchC $= \dfrac{yi_p\left(\Delta n - \dfrac{\Delta n' n}{n'}\right)}{u_k'}$ $\hfill$ (8-74)

where

$$B = \frac{n(n' - n)y(u' - i)}{2n'\mathfrak{I}} \hspace{3cm} (8\text{-}75)$$

$$B_p = \frac{n(n' - n)y_p(u_p' - i_p)}{2n'\mathfrak{I}} \hspace{3cm} (8\text{-}76)$$

Δn = dispersion of the medium

$\quad$ = $n_{short} - n_{long}$

W = $4Kh(n - n')/\mathfrak{I}$

K = $A_4 - \dfrac{1}{4}A_2(4A_2^2 + 6CA_2 + 3C^2)$

C_e = $C + 2A_2$

The surface equation is in the form of Equation (8-46). Equivalent curvature, C_e, is used in the paraxial ray-tracing of any aspheric surfaces. The longitudinal values of the spherical aberration, astigmatism, Petzval curvature, and axial chromatic contributions may be obtained by dividing the transverse contributions (TSC, TAC, TPC, or TLchC) by u_k', the final slope of the axial ray.

The third-order aberration at the image of the complete optical system is the sum of the contributions from each of the surfaces.

8.4.5. **Stop-Shift Equations.** When the stop (or pupil) of an optical system is shifted, the changes produced in the third-order aberrations can be computed from the following equations, in which the starred (*) terms are the aberration contributions after the stop is shifted and the unstarred terms are the contributions before the shift:

$$\text{TSC}^* = \text{TSC} \hspace{4cm} (8\text{-}77)$$

$$\text{CC}^* = \text{CC} + \text{TSC} \cdot Q \hspace{3cm} (8\text{-}78)$$

$$\text{TAC}^* = \text{TAC} + \text{CC} \cdot 2Q + \text{TSC} \cdot Q^2 \hspace{1.5cm} (8\text{-}79)$$

$$\text{TPC*} = \text{TPC} \tag{8-80}$$

$$\text{DC*} = \text{DC} + (\text{TPC} + 3\text{TAC}) \cdot Q + \text{CC} \cdot 3Q^2 + \text{TSC} \cdot Q^3 \tag{8-81}$$

$$\text{TLchC*} = \text{TLchC} \tag{8-82}$$

$$\text{TchC*} = \text{TchC} + \text{TLchC} \cdot Q \tag{8-83}$$

Q represents the amount of the pupil shift:

$$Q = \frac{(y_p^* - y_p)}{y} \tag{8-84}$$

where y_p^* = ray height of the principal ray after the stop is shifted
$\quad\quad y_p$ = original principal ray height, before the stop shift
$\quad\quad y$ = height of the axial ray

Q is an invariant and thus may be evaluated at any convenient surface of the system; a further consequence of the invariance of Q is that the starred and unstarred terms of Equations (8-77) through (8-83) may represent the contributions of an entire optical system or any part of it (e.g., a single surface or element).

8.4.6. Thin-Lens Aberrations. If the elements of an optical system can be regarded as thin lenses (i.e., of zero thickness) surrounded by air, the following equations give the third-order and chromatic aberration contributions of each element, assuming that the stop (pupil) is in contact with the element:

$$\text{TSC} = \frac{-y^4(G_1 C^3 - G_2 C^2 C_1 + G_3 C^2 v + G_4 CC_1^2 - G_5 CC_1 v + G_6 Cv^2)}{u_k'}$$

$$= \frac{-y^4(G_1 C^3 + G_2 C^2 C_2 - G_3 C^2 v' + G_4 CC_2^2 - G_5 CC_2 v' + G_6 Cv'2)}{u_k'} \tag{8-85}$$

$$\text{CC} = -hy^2\left(\frac{1}{4} G_5 CC_1 - G_7 Cv - G_8 C^2\right)$$

$$= -hy^2\left(\frac{1}{4} G_5 CC_2 - G_7 Cv' + G_8 C^2\right) \tag{8-86}$$

$$\text{TAC} = -\frac{1}{2} h^2 \phi u_k' \tag{8-87}$$

$$\text{TPC} = \frac{-\frac{1}{2} h^2 \phi u_k'}{n} = \frac{\text{TAC}}{n} \tag{8-88}$$

$$\text{DC} = 0 \tag{8-89}$$

$$\text{TLchC} = \frac{-y^2 \phi}{V u_k'} \tag{8-90}$$

$$\text{TchC} = 0 \tag{8-91}$$

Paraxial rays are traced through the system using Equations (8-29) and (8-30); a principal and an axial ray (as defined in Section 8.4.4) are traced. Equations (8-77) through (8-83) are used with $Q = y_p/y$ to determine the contributions of each element for the actual stop position. The contributions from the elements are summed to determine the aberrations of the whole system. The aspheric terms of Equations (8-68) to (8-81) can be added to the thin-lens aberrations obtained from Equations (8-77) to (8-83) when the thin lenses have aspheric surfaces.

In Equations (8-85) through (8-91), C represents the total curvature of the element, and C_1 and C_2 are the curvatures of the left and right surfaces, respectively, so that $C = C_1 - C_2 = \phi/(n - 1)$. The symbol v is the reciprocal object-distance (for the element), and $v = u/y$. Similarly, $v' = u'/y$. The reciprocal relative dispersion of the lens material, V, is conventionally equal to $(n_D - 1)/(n_F - n_C)$ in the visible. In the infrared, the reciprocal relative dispersion is $(n_M - 1)/(n_S - n_L)$, where the subscripts M, S, and L identify the refractive index at middle, short, and long wavelengths, respectively.

G_1 through G_8 are functions of the index, as follows [8-6] :

$$G_1 = \frac{n^2(n-1)}{2} \qquad G_5 = \frac{2(n+1)(n-1)}{n}$$

$$G_2 = \frac{(2n+1)(n-1)}{2} \qquad G_6 = \frac{(3n+2)(n-1)}{2n}$$

$$G_3 = \frac{(3n+1)(n-1)}{2} \qquad G_7 = \frac{(2n+1)(n-1)}{2n}$$

$$G_4 = \frac{(n+2)(n-1)}{2n} \qquad G_8 = \frac{n(n-1)}{2} \qquad (8\text{-}92)$$

8.4.7. Interpretation of Third-Order Aberration Contributions. To the extent that third-order aberrations approximate the complete aberration polynomial, aberration values computed from equations of the preceding three sections (8.4.4, 8.4.5, and 8.4.6) will correspond to the aberration values as determined by actual ray-tracing. Thus ΣTSC (the sum of the transverse, third-order, spherical aberration contributions) will approximate the intersection height (in the paraxial image plane) of a trigonometrically traced axial ray (i.e., the transverse spherical aberration). Similarly, 3ΣCC will approximate the tangential coma and ΣDC the linear distortion, with 100ΣDC$/h$ the percentage distortion. The longitudinal curvature (or sag) of the field is approximated by

$$x_s \approx \frac{(\Sigma\text{TAC} + \Sigma\text{TPC})}{u'_k}$$

and

$$x_t \approx \frac{(3\Sigma\text{TAC} + \Sigma\text{TPC})}{u'_k} \qquad (8\text{-}93)$$

for the sagittal and tangential fields, respectively. The chromatic aberration equations are

$$\frac{\Sigma\text{TLchC}}{u'_k} = \text{LchC} \approx \ell_F - \ell_C$$

and

$$\Sigma TchC \approx h_F - h_C \tag{8-94}$$

where $(\ell_F - \ell_C)$ = paraxial longitudinal axial-chromatic aberration
$(h_F - h_C)$ = paraxial lateral chromatic aberration

8.5. Depth of Field and Focus

8.5.1. Photographic Depth of Focus. Based upon the concept that some arbitrarily-selected level of blur (of diameter B) caused by lack of focus of the optical system can be tolerated, the depth of focus of the optical system is calculated to be $\pm B/2(NA)$—assuming purely geometrical optics and no aberrations. The corresponding depth of field at the object ranges from S_{near} to S_{far}:

$$S_{far} = \frac{fS(D-B)}{(fD+SB)} \tag{8-95}$$

and

$$S_{near} = \frac{fS(D+B)}{(fD-SB)} \tag{8-96}$$

where S = nominal distance at which system is focused
D = diameter of entrance pupil
B = "tolerable" blur diameter in image

The hyperfocal distance is the distance at which the optical system must be focused so that S_{far} is infinitely large (S_{near} is one-half the hyperfocal distance):

$$S_{hyp} = -f\frac{D}{B} \tag{8-97}$$

8.5.2. Physical Depth of Focus. Actually, there is no sharp demarcation between being in-focus and out-of-focus; the image worsens gradually as the amount of defocus is increased. The wavefront aberration due to defocusing is given by

$$OPD = \frac{1}{2}(\delta S)n \sin^2 U_m \tag{8-98}$$

where δS = longitudinal distance from position of best focus
n = index of the medium
U_m = slope angle of the marginal ray at the image

Thus, a depth-of-focus tolerance corresponding to the Rayleigh quarter-wave criterion is given by

$$\delta S = \pm \frac{\lambda}{2n \sin^2 U_m} \tag{8-99}$$

Note that the same equation applies to both depth of field or focus if U_m is taken as the marginal ray slope at the object or image, respectively.

8.6. Vignetting and Baffling

8.6.1. Vignetting. Vignetting is the blocking of an oblique bundle of light rays by the limiting diameters of the optical system. Figure 8-19 shows an example of vignetting. To eliminate vignetting, one must make the lens elements toward the ends of the optical

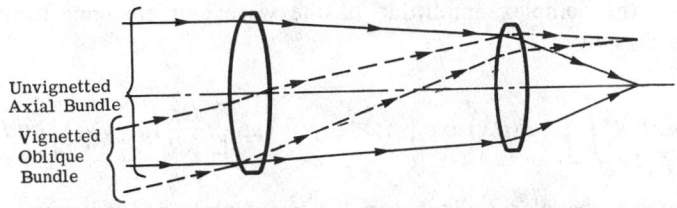

Fig. 8-19. Vignetting. The passage of the full diameter of the oblique ray bundle (dashed lines) is prevented by the lower edge of the front lens and the upper edge of the near lens.

system large enough to pass all the rays of the oblique bundle. A long system is likely to require large-diameter elements to prevent vignetting; conversely, a short system can be made quite compact.

8.6.2. Baffles. Baffles are opaque diaphragms which prevent the propagation of light through the system by reflection or by scattering from the mechanical (or nonoptical) elements. Figure 8-20 shows a system of baffles designed to prevent stray light from being directly reflected onto the detector from the walls of its mount.

8.6.3. Glare Stop. A glare stop is an opaque diaphragm located at the image of the optical system aperture stop. The diameter of the glare stop is exactly the same size as the image of the aperture stop (which is a pupil of the system); thus, it will pass all the rays passing directly through the system from the aperture stop, but it will intercept those scattered from walls, etc., as shown in Figure 8-21.

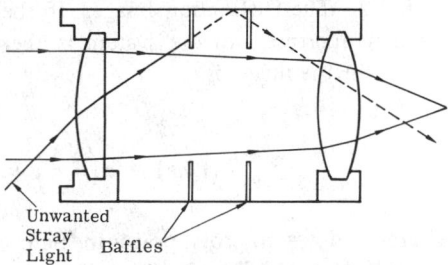

Fig. 8-20. Baffles.

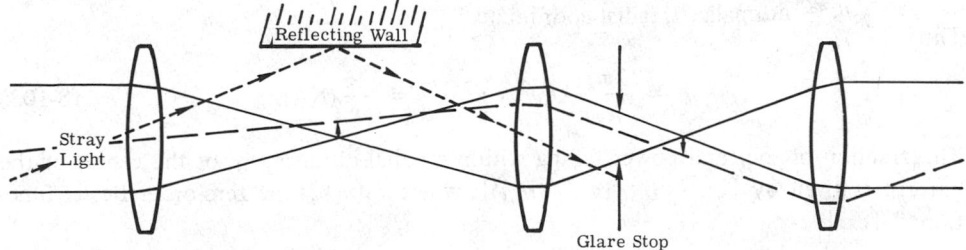

Fig. 8-21. Glare stop. An internal glare stop prevents the passage of unwanted stray light reflected from outside the field of view, but permits passage of all the useful light.

8.7. Measures of Optical Performance

8.7.1. The Diffraction Integral. The exact point-spread function, $h(y,z)$, of an image (as contrasted to the approximate geometrical function derived from a ray-traced spot diagram) is given by

$$h(y,z) = |A(y,z)|^2 \qquad (8\text{-}100)$$

where $A(y,z)$ is the complex amplitude of the wavefront emerging from the optical systems, and is given by

$$A(y,z) = \int\int_{-\infty}^{\infty} B(u,v)\exp\left[-i\,\Delta\phi(u,v)\right]\exp\left[i\,\frac{2\pi}{\lambda R}(uy + vz)\right] du\,dv \qquad (8\text{-}101)$$

where y,z and u,v = spatial coordinates in the image plane and exit pupil, respectively, normal to the principal (or chief) ray

$B(u,v)$ = an amplitude factor proportional to the square root of the flux density (i.e., transmission) at point (u,v) in the pupil; $B(u,v) = 0$ outside the pupil

$\Delta\phi(u,v)$ = wave-aberration phase function difference, which equals the $(2\pi/\lambda)$ OPD of the ray through point (u,v)

R = reference sphere radius

i = the imaginary $\sqrt{-1}$

The terms $B(u,v)$ and $\exp(-i\Delta\phi(u,v))$ are often combined and referred to as the pupil function of the system. Equation (8-101) is sometimes written to include a focussing term, $\exp(i\pi x(u^2 + v^2)/\lambda R^2)$, where x is the coordinate along the chief ray.

8.7.2. The Diffraction Image. If the transmission of the system is uniform over a circular aperture and the system is aberration free, the illuminance or irradiance distribution in the image is

$$E_v(y,z) = \pi\left(\frac{\text{NA}}{\lambda}\right)^2 \Phi_t\left[\frac{2J_1(m)}{m}\right]^2 = E_0\left[\frac{2J_1(m)}{m}\right]^2 \qquad (8\text{-}102)$$

where Φ_t = the total power in the point image
$J_1(m)$ = first-order Bessel function*
E_0 = peak illuminance
NA = $n'\sin U'$, the numerical aperture
m = normalized, radial coordinate

Thus,

$$m = \frac{2\pi}{\lambda}\text{NA}(y^2 + z^2)^{1/2} = \frac{2\pi}{\lambda}(\text{NA})r \qquad (8\text{-}103)$$

The fraction of the total power falling within a radial distance, r_0, of the center of the pattern is given by $[1 - J_0^2(m_0) - J_1^2(m_0)]$, where $J_0(m)$ is the zero-order Bessel function** [8-7].

---

*
$$J_1(m) = \frac{m}{2} - \frac{\left(\frac{m}{2}\right)^3}{1^2 2} + \frac{\left(\frac{m}{2}\right)^5}{1^2 2^2 3} - \cdots$$

**
$$J_0(m) = 1 - \left(\frac{m}{2}\right)^2 + \frac{\left(\frac{m}{2}\right)^4}{1^2 2^2} - \frac{\left(\frac{m}{2}\right)^6}{1^2 2^2 3^2} + \cdots$$

Equation (8-102) is plotted in Figure 8-22. The pattern consists of a circular patch of light (the Airy disc) surrounded by rings of rapidly decreasing intensity. Table 8-2 indicates the size and distribution of energy in the pattern for both a circular aperture and a slit aperture. When the aperture is rectangular and uniformly illuminated,

$$E_v(y,z) = E_0 \left\{ \frac{\sin\left[\dfrac{2\pi(NA)y}{\lambda}\right]}{\dfrac{2\pi(NA)y}{\lambda}} \right\}^2 \left\{ \frac{\sin\left[\dfrac{2\pi(NA)z}{\lambda}\right]}{\dfrac{2\pi(NA)z}{\lambda}} \right\}^2 \qquad (8\text{-}104)$$

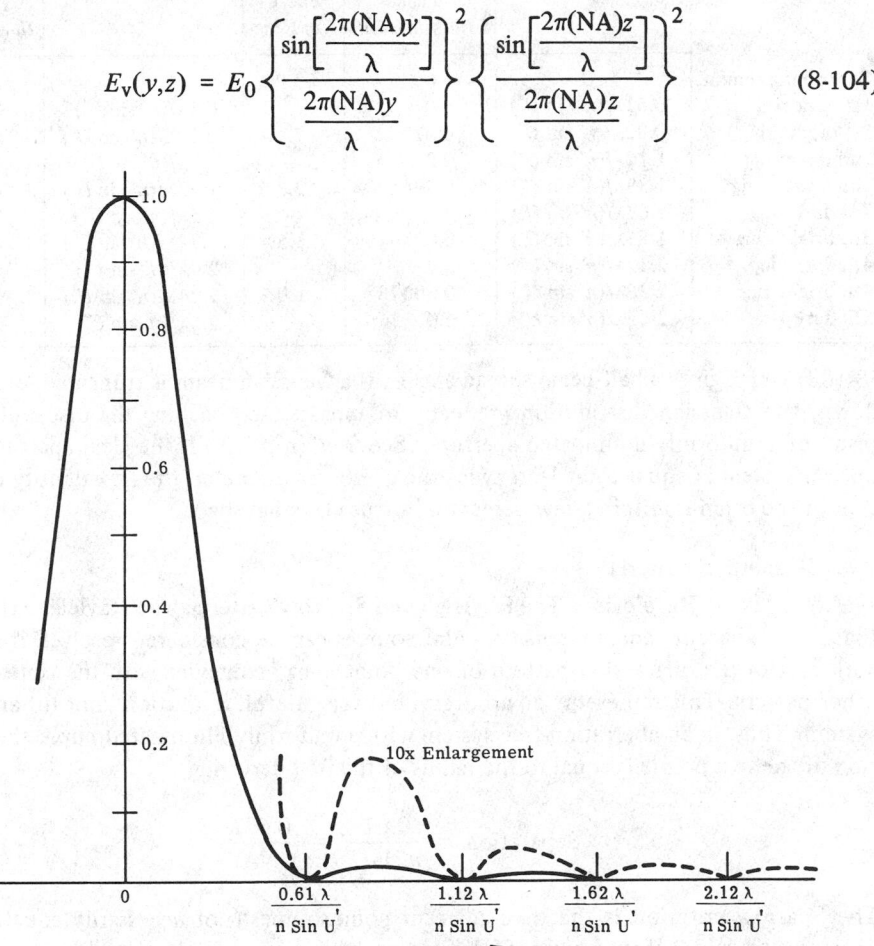

Fig. 8-22. The distribution of illumination in the diffraction pattern [8-8].

8.7.3. Gaussian (Laser) Beams.

When the amplitude factor, $B(u,v)$, in Equation (8-101) is such that the beam cross-section has a Gaussian flux-density distribution,

$$E(r) = E_0 \, e^{-2(r/w)^2} \qquad (8\text{-}105)$$

then the diffraction pattern in an unaberrated image also has a Gaussian distribution. The size of a Gaussian beam or diffraction pattern is usually given in terms of the semi-diameter, w, at which the radiation falls to e^{-2}, or about 0.135 of its central value. At large distances, the angular spread of a Gaussian beam is $4\lambda/\pi D$ between e^{-2} points, where D is the e^{-2} beam diameter at the optical system. (Compare this with Equation

Table 8-2. Tabulation of the Size of and Distribution of Energy
in the Diffraction Pattern at the Focus of a Perfect Lens

Ring (or band)	Circular Aperture			Slit Aperture	
	z	Peak Illumination	Energy in Ring	z	Peak Illumination
Central maximum	0	1.0	83.9%	0	1.0
1st dark ring	$0.61\lambda/(n' \sin U')$	0.0	–	$0.5\lambda/(n' \sin U')$	0.0
1st bright ring	$0.82\lambda/(n' \sin U')$	0.017	7.1%	$0.72\lambda/(n' \sin U')$	0.047
2nd dark ring	$1.12\lambda/(n' \sin U')$	0.0	–	$1.0\lambda/(n' \sin U')$	0.0
2nd bright ring	$1.33\lambda/(n' \sin U')$	0.0041	2.8%	$1.23\lambda/(n' \sin U')$	0.017
3rd dark ring	$1.62\lambda/(n' \sin U')$	0.0	–	$1.5\lambda/(n' \sin U')$	0.0
3rd bright ring	$1.85\lambda/(n' \sin U')$	0.0016	1.5%	$1.74\lambda/(n' \sin U')$	0.0083
4th dark ring	$2.12\lambda/(n' \sin U')$	0.0	–	$2.0\lambda/(n' \sin U')$	0.0
4th bright ring	$2.36\lambda/(n' \sin U')$	0.00078	1.0%	$2.24\lambda/(n' \sin U')$	0.0050
5th dark ring	$2.62\lambda/(n' \sin U')$	0.0	–	$2.5\lambda/(n' \sin U')$	0.0

(8-108) which gives a half-beam spread angle.) If a Gaussian beam is truncated, or stopped down, the Gaussian distribution gradually disappears, approaching the distribution as a result of a uniformly illuminated aperture. (See Section 8.7.2.) If the clear aperture of the optical system is equal to at least twice the e^{-2} beam diameter, the flux density distribution of the beam is within a few percent of a true Gaussian shape.

8.8. Resolution Criteria

8.8.1. Point Resolution: The Rayleigh and Sparrow Criteria. The Rayleigh criterion is that two, adjacent, equal-intensity, point sources can be considered resolved if the first dark ring of the diffraction pattern of one point image coincides with the center of the other pattern. This represents an arbitrary, but very useful, resolution limit for an optical system. Thus, in an aberration-free system with a uniformly-illuminated pupil, the separation of the two points is equal to the radius of the first dark ring:

$$\text{Separation} = \frac{0.61\lambda}{n'\sin U'} = \frac{0.61\lambda}{\text{NA}} \qquad (8\text{-}106)$$

The Sparrow criterion is that two adjacent point-sources (not necessarily equal) can be considered resolved if the combined diffraction pattern has no minimum between the two point-images. This occurs when

$$\text{Separation} = \frac{0.5\lambda}{n'\sin U'} = \frac{0.5\lambda}{\text{NA}} \qquad (8\text{-}107)$$

where U' is the angle to the axis of the marginal rays of the image-forming cone. The separations can be converted to object separations either by using the object-space numerical aperture, NA, in the equations, or by dividing the image separation as given above by the system magnification. In object space, the angular separation of distant object-points (for axially symmetric systems) is given by

$$\text{Rayleigh Angular Resolution} = \frac{1.22\lambda}{D} \text{ rad} \qquad (8\text{-}108)$$

$$\text{Sparrow Angular Resolution} = \frac{\lambda}{D} \text{ rad} \qquad (8\text{-}109)$$

where D is the effective aperture (entrance pupil diameter) of the optical system.

Since these resolution criteria are based on an aberration- and defect-free optical system, they are frequently used as a standard for excellence of design and construction, as well as an indication of the limiting performance for a given size system.

8.8.2. The Aerial Image Modulation (AIM) Curve. The AIM curve is a plot of the minimum image modulation required to produce a response in a responsive element as a function of spatial frequency. AIM curves are commonly used for such detectors as photographic film, image tubes, and the human eye. A typical AIM curve rises with frequency, indicating that a higher modulation in the image is necessary to produce a response at higher spatial frequencies. For example, if the AIM curve and the MTF curve for a film and camera-lens combination are plotted on the same graph, the frequency at which the two curves intersect is the limiting frequency, or resolution of the combined system. More complicated measures than this are used for "system performance" as opposed to resolution.

8.9. Image Quality Criteria

8.9.1. The Rayleigh Quarter-Wave Limit. The Rayleigh limit for image quality states that if the wavefront aberration, OPD, varies no more than one quarter-wavelength over the aperture of an optical system, the image will be sensibly perfect. When the wavefront is relatively smooth and free of high-order ripples, this is the reliable criterion.

The amounts of certain aberrations which correspond to a maximum wavefront deformation of one quarter-wave are as follows [8-6]:

Out of Focus
$$\Delta \ell = \frac{\pm\lambda}{2n \sin^2 U_m} \qquad (8\text{-}110)$$

Spherical Aberration (third-order)
$$LA_m = \frac{\pm 4\lambda}{n \sin^2 U_m} \qquad (8\text{-}111)$$

Zonal Spherical Aberration
$$LA_z = \frac{\pm 6\lambda}{n \sin^2 U_m} \qquad (8\text{-}112)$$

Axial Chromatic Aberration
$$LchA = \frac{\pm\lambda}{n \sin^2 U_m} \qquad (8\text{-}113)$$

Sagittal Coma
$$Coma_s = \frac{\pm\lambda}{2n \sin U_m} \qquad (8\text{-}114)$$

These values assume that the reference point is chosen to minimize the OPD.

The effect of a wavefront deformation on the diffraction pattern (Section 8.7.2) is to shift some radiation from the central disc into the rings. In a perfect system, 84% of the energy is in the central disc and 16% is in the rings. A quarter-wave of defocussing produces a pattern with 68% in the central disc and 32% in the rings. This is a detectable change. In practice, however, the change is difficult to measure, and a system with less than a quarter-wave of aberration is an excellent one for most applications.

Because the Rayleigh quarter-wave limit assumes a smooth wavefront, it is less reliable when the wavefront has large, high spatial-frequency components or abrupt deformations. The rms (root-mean-square) OPD is a somewhat more widely applicable measure of the quality of a system. A rms OPD of between one-fourteenth and one-twentieth of a wave is approximately the equivalent of the classical quarter-wave Rayleigh limit.

8.9.2. Strehl Definition [8-7]. The Strehl definition is the ratio between the illuminance at the peak of the diffraction pattern of an aberrated point-image and the illuminance at the center of an aberration-free image. A Strehl ratio of 0.8 is equal to the Rayleigh quarter-wave limit and has a much broader applicability.

The Strehl definition of a system can be evaluated by calculating the normalized illuminance at the center of the diffraction pattern:

$$\text{Strehl Definition} = \left[\frac{1}{A}\iint e^{i\Delta\phi}dA\right]^2 \tag{8-115}$$

where the integration is over the pupil area, A, and $\Delta\phi$ is the optical phase difference of the wavefront (Section 8.7.1). The Strehl resolution is equal to the ratio of the integral of the three-dimensional MTF for the system divided by the integral of the MTF for an unaberrated system.

8.10. Transfer Functions

The performance of any linear, shift-invariant system (see Goodman, esp. Chapter 6) can be described by its impulse function, or the ratio of the output spectrum to that of the corresponding input spectrum. The impulse (or transfer) function of an optical system is generally complex, is a function of spatial frequencies in two dimensions, and is not limited by causality (in contrast to an electrical system which depends upon time and cannot provide an output before there is an input).

8.10.1. Optical Transfer Function (OTF).* The point-spread function, $h(y,z)$, is defined as the response of an optical system to a point source of light (a two-dimensional delta function). The response of the system can be written directly as the convolution of the spatial distribution and the point-spread function. It can also be written in terms of the spectra of these quantities:

$$I(\omega_y,\omega_z) = H(\omega_y,\omega_z)\,O(\omega_y,\omega_z) \tag{8-116}$$

where $I(\omega_y,\omega_z)$ = spatial frequency spectrum of the image
$O(\omega_y,\omega_z)$ = spatial frequency spectrum of the object
$H(\omega_y,\omega_z)$ = transfer function of the optical system
ω_y = radian spatial frequency in the y direction
ω_z = radian spatial frequency in the z direction

The transfer function can be written as the Fourier transform of the point spread function:

$$H(\omega_y,\omega_z) = \int_{-\infty}^{\infty} h(y,z)\,e^{-i(y\omega_y + z\omega_z)}\,dy\,dz$$

$$= \int_{-\infty}^{\infty} h(y,z)\,e^{-2\pi i(y\,f_y + z\,f_z)}\,dy\,dz \tag{8-117}$$

*Sections 8.10.1 and 8.10.2 contributed by William L. Wolfe, The University of Arizona, Tucson, AZ.

8.10.2. Modulation and Phase Transfer Functions. In general, the optical transfer function is complex. It can be written in Cartesian or polar form as follows:

$$H(\omega_y, \omega_z) = Re\{H\} + i\,Im\{H\}$$

$$= |H|\,e^{i\psi}$$

(8-118)

where $\psi = \arctan[Im\{H\}/Re\{H\}]$
$Re\{H\}$ = real part of the complex OTF
$Im\{H\}$ = imaginary part of the complex OTF
$|H|$ = absolute magnitude of the OTF

Many authors use the following abbreviations:

$$H(\omega_y, \omega_z) = \text{OTF (optical transfer function)}$$

$$|H(\omega_y, \omega_z)| = \text{MTF (modulation transfer function)}$$

$$\psi(\omega_y, \omega_z) = \text{PTF (phase transfer function)}$$

(8-119)

Modulation is a measure of the relation between the dimmest and brightest portions of the scene and the average level. It is one measure of what is commonly called *contrast*. Modulation of radiance is defined as follows:

$$\text{Modulation} = \frac{L_{max} - L_{min}}{L_{max} + L_{min}}$$

(8-120)

The modulation transfer is the ratio of the modulation in the image to that in the object:

$$\text{MT} = \frac{\left(\dfrac{L_{max} - L_{min}}{L_{max} + L_{min}}\right) \text{image}}{\left(\dfrac{L_{max} - L_{min}}{L_{max} + L_{min}}\right) \text{object}}$$

(8-121)

In the visible spectral region, the scene is usually illuminated by the sun, so that the modulation results only from reflectivity differences:

$$\text{Modulation} = \frac{\rho_{max} - \rho_{min}}{\rho_{max} + \rho_{min}}$$

(8-122)

The maximum value of modulation in this case is "one" and the minimum zero; it is never negative. In the infrared, the modulation can be caused by differences in emissivity, ϵ, and in temperature, T, so that it is given by

$$\text{Modulation} = \frac{L_{max} - L_{min}}{L_{max} + L_{min}}$$

(8-123)

and since $L = \epsilon L^{BB}(T)$, where L^{BB} is the radiance from a blackbody,

$$\text{Modulation} = \frac{\epsilon_1 L^{BB}(T_1) - \epsilon_2 L^{BB}(T_2)}{\epsilon_1 L^{BB}(T_1) + \epsilon_2 L^{BB}(T_2)} = \frac{\epsilon_1 - \epsilon_2 \dfrac{L^{BB}(T)}{L^{BB}(T_1)}}{\epsilon_1 + \epsilon_2 \dfrac{L^{BB}(T_2)}{L^{BB}(T_1)}}$$

(8-124)

$$\text{Modulation} = \frac{1 - \dfrac{\epsilon_2(e^{c_2/\lambda T_2} - 1)^{-1}}{\epsilon_1(e^{c_2/\lambda T_1} - 1)^{-1}}}{1 + \dfrac{\epsilon_2(e^{c_2/\lambda T_2} - 1)^{-1}}{\epsilon_1(e^{c_2/\lambda T_1} - 1)^{-1}}} \qquad (8\text{-}125)$$

The modulation can be any value between 0 to +1. The expression can be written

$$\text{Modulation} = \frac{1 - a}{1 + a} \qquad (8\text{-}126)$$

when

$$a = \frac{\epsilon_2(e^{c_2/\lambda T_2} - 1)^{-1}}{\epsilon_1(e^{c_2/\lambda T_1} - 1)^{-1}}$$

$$= \frac{\epsilon_2(e^{c_2/\lambda T_1} - 1)}{\epsilon_1(e^{c_2/\lambda T_2} - 1)} \qquad (8\text{-}127)$$

Figure 8-23 is a curve of modulation as a function of a.

The modulation transfer of an infrared·system is sometimes described in terms of the minimum variations in temperature it can sense. This description is based on the assumption that changes in flux sensed by the system are due only to changes in temperature from one part of the infrared source to the other. This minimum resolvable temperature (MRT) is proportional to the MTF.

The MTF is a more complete expression of the resolution performance than Rayleigh, Strehl, or other single measures, but it is still incomplete when presented as only a single curve. The MTF may be different in one direction of the field of view than it is in the other. This is especially true for scanning devices in which the MTF is one value in the direction of scan, but quite a different value perpendicular to the scanning direction. The MTF can be different for different field angles; this is especially true for wide-angle systems. It can also be different for different object and image positions and different focal positions. It will almost always vary with wavelength, because the diffraction limit is dependent upon wavelength. To a much smaller degree, the MTF depends upon the location of baffle and glare stops in the scanning optics as well as the general level of background·radiation.

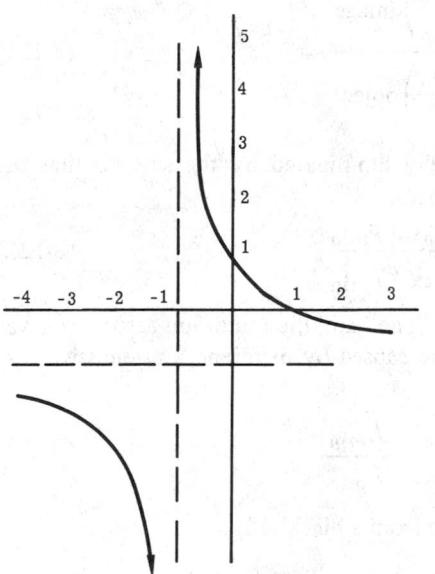

Fig. 8-23. Graph of $(1 - a)(1 + a)^{-1}$. The asymptotes are $a = -1$ and $(1 - a)(1 + a)^{-1} = -1$.

The geometrical MTF can be calculated from ray-trace data; both this and diffraction MTF calculations are available as subroutines on some lens-design computer programs.* The usual procedure is to calculate the point-spread function (sometimes with the source point at a variety of different field angles) and then to calculate its Fourier transform. Some programs also include a diffraction contribution as part of the MTF calculation.

8.10.3. Specific Modulation Transfer Functions [8-10 through 8-13]. An optical system is incapable of transmitting spatial frequencies higher than f_c, which can be determined from

$$f_c = \frac{2NA}{\lambda} = \frac{1}{\lambda(F/\#)} \tag{8-128}$$

For many applications, it is convenient to express the limit as an angular frequency at the object by $f_c = D/\lambda$ cycles per radian, where D is the effective clear aperture of the system.

The MTF of an optical system without aberrations and with a uniformly transmitting circular aperture is given by

$$MTF(f_y) = \frac{2}{\pi}\left[\arccos\left(\frac{\lambda f_y}{2(NA)}\right) - \frac{\lambda f_y}{2(NA)} \sin\left(\arccos\left(\frac{\lambda f_y}{2(NA)}\right)\right)\right] \tag{8-129}$$

Equation (8-129) is plotted in Figure 8-24. For a slit or rectangular aperture, the MTF is

$$MTF(f_y) = 1 - \frac{f_y}{f_c} = 1 - \frac{f_y\lambda}{2NA} \tag{8-130}$$

When the center of the pupil is obscured, as in a Cassegrain mirror system, the MTF of an aberration-free system is reduced at low frequencies and increased slightly at high ones. This is shown in Figure 8-25.

Figure 8-26 shows the effect of various amounts of defocus. The amount of the focus shift is given as a function of sin U, so that the graph can be applied to systems of any aperture. When the defocussing is relatively large, to the order of $4\lambda/n \sin^2 U$ or more, the diffraction effects can be neglected and the MTF is well approximated by

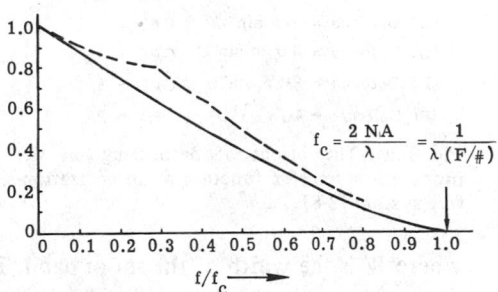

Fig. 8-24. The modulation transfer function of an aberration-free system.   Note that the frequency is given in terms of the limiting resolution frequency, f_c. This curve is based on diffraction effects.   The dashed line is the MTF for a square-wave target; the solid line is for a sine-wave target [8-8].

$$MTF(f_y) = \frac{2J_1(\pi B f_y)}{\pi B f_y} = \frac{J_1(2\pi\, \delta d\, NA f_y)}{\pi\, \delta d\, NA f_y} \tag{8-131}$$

---

*Computer program packages are available from Scientific Calculation Inc., Rochester, NY, and Genesee Computer Center, Inc., Rochester, NY.

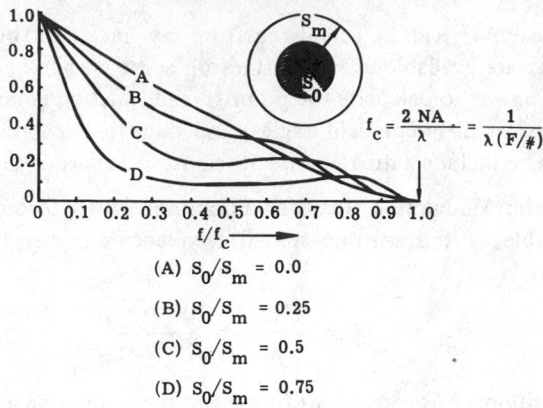

(A) $S_0/S_m = 0.0$

(B) $S_0/S_m = 0.25$

(C) $S_0/S_m = 0.5$

(D) $S_0/S_m = 0.75$

Fig. 8-25. The effect of a central obscuration on the modulation transfer function of an aberration-free system [8-8].

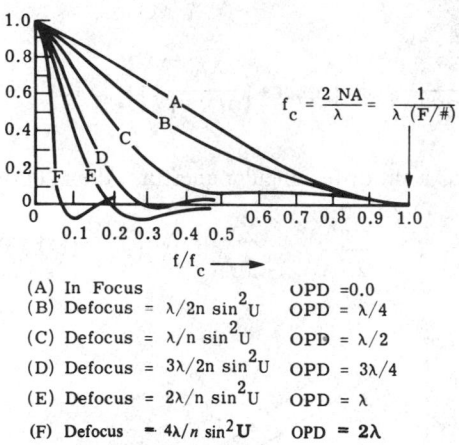

(A) In Focus   OPD $= 0.0$
(B) Defocus $= \lambda/2n \sin^2 U$ OPD $= \lambda/4$
(C) Defocus $= \lambda/n \sin^2 U$ OPD $= \lambda/2$
(D) Defocus $= 3\lambda/2n \sin^2 U$ OPD $= 3\lambda/4$
(E) Defocus $= 2\lambda/n \sin^2 U$ OPD $= \lambda$
(F) Defocus $= 4\lambda/n \sin^2 U$ OPD $= 2\lambda$

Fig. 8-26. The effect of defocusing on the modulation transfer function of an aberration-free system [8-8].

where $J_1(\ )$ = first-order Bessel function (Section 8.7.2)

B = diameter of blur spot produced by defocussing

δd = longitudinal defocussing

NA = $n \sin U$ = numerical aperture

f_y = spatial frequency in cycles per unit length

A system whose image is a uniformly illuminated slit, or band, of light has an MTF given by

$$\text{MTF}(f_y) = \frac{\sin(\pi W f_y)}{\pi W f_y}$$

$$= \text{sinc}(W f_y) \qquad (8\text{-}132)$$

where W is the width of the slit or band. The equivalent cut-off frequency at which MTF drops to zero is equal to $1/W$. This is applicable in the study of blurs caused by image motion.

8.10.4. Square Waves and Sine Waves [8-9]. The OTF and the MTF apply (by definition) to the imagery of an object or target whose radiance can be described by a sine function. A convenient target for the testing of optical systems is the square-wave target, which is a pattern of alternate bright and dark bars of equal width. When the MTF (i.e., the sine-wave response) is known, the modulation transfer of a square wave, $S(f_y)$, can be calculated:

$$S(f_y) = \frac{4}{\pi}\left[M(f_y) - \frac{M(3f_y)}{3} + \frac{M(5f_y)}{5} - \frac{M(7f_y)}{7} + \cdots \right] \qquad (8\text{-}133)$$

Where $M(f_y)$ is the MTF. Conversely, when $S(f_y)$ is known,

$$M(f_y) = \frac{\pi}{4}\left[S(f_y) + \frac{S(3f_y)}{3} - \frac{S(5f_y)}{5} + \frac{S(7f_y)}{7} - \cdots\right] \qquad (8\text{-}134)$$

In general, the modulation transfer factor is higher for a square-wave target than for a sine-wave target. For example, the factor for a perfect optical system at frequencies between 0.25 f_c and 0.5 f_c is about 0.1 greater for a square-wave target than for a sinusoidal target. (See Figure 8-24.)

The most common form of bar (i.e., square-wave) target is the USAF 1951 target, which consists of only three bright bars on an extended dark background (or the reverse) for each frequency. If the frequency of the target is taken as the reciprocal of the center-line spacing of the bars, the modulation transfer factor is higher than that indicated by the sine-wave MTF for this frequency. This is because the frequency content of a three-bar pattern is heavily concentrated in frequencies lower than the basic frequency of the target, i.e., a spectral breakdown (Fourier analysis) of a three-bar target shows lots of power at frequencies less than 1/(bar spacing).

8.10.5. Pupil Convolution. The computation of the OTF can be carried out by an autoconvolution of the system pupil function (Section 8.7.1). For aberration-free systems, the MTF can be readily computed regardless of the shape of the aperture since the MTF is simply the normalized area common to the pupil and the pupil is displaced laterally by an amount proportional to the frequency. The displacement corresponding to f_c, the cut-off frequency, is of course equal to the maximum dimension of the pupil in the direction of the spatial frequency (i.e., the displacement beyond which there is no common area). Note that Equation (8-129) in Section 8.10.3 can be derived using this principle; it is simply the area common to a circle (the aperture) and that same circle displaced, normalized by dividing by the area of the circle, and with f_c corresponding to a displacement equal to the circle diameter.

8.11. Ray-Intercept Plots and Spot Diagrams

8.11.1. Ray-Intercept Plot (H-tan U curve). To prepare a ray-intercept plot, one traces a fan of rays (either meridional or sagittal) from an object point through the optical system, and the coordinates of the ray intersection with the image surface, H, are plotted against the position of the ray in the aperture, which is often represented by the slope of the ray, tan U, at the image. The spread of radiation in the image can thus be read directly from the plot, and an estimate of image blur size can easily be made. When the plot coordinates are H and tan U, the effects of refocussing on the size of the image blur are readily evaluated by rotating the tan U axis of the plot.

Figure 8-27 illustrates ray-intercept plots for several common aberrations. Ray-intercept plots are often (incorrectly) called *rim-ray curves*.

8.11.2. Spot Diagrams and Spread Functions. If the aperture of an optical system is divided into a large number of equal, small areas, and if a ray from a selected point is traced through the center of each small area, then a plot of the intersection points (spots) of the rays with the image surface is an approximate representation of the (geometrical) irradiance distribution at the image. (See Figure 8-28.) The more rays traced, the better the approximation. Such a representation is called a *spot diagram*. Assuming the geometrical spot diagram to be a reasonable representation of the actual image irradiance

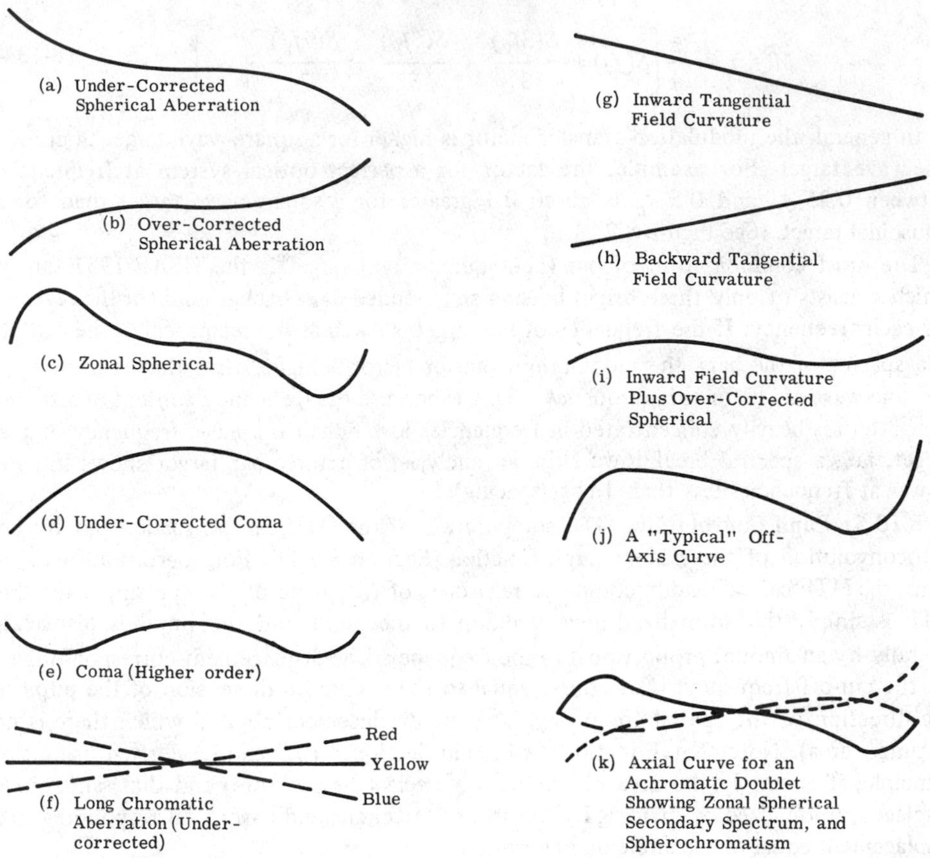

(a) Under-Corrected
    Spherical Aberration

(g) Inward Tangential
    Field Curvature

(b) Over-Corrected
    Spherical Aberration

(h) Backward Tangential
    Field Curvature

(c) Zonal Spherical

(i) Inward Field Curvature
    Plus Over-Corrected
    Spherical

(d) Under-Corrected Coma

(j) A "Typical" Off-
    Axis Curve

(e) Coma (Higher order)

Red
Yellow
Blue

(f) Long Chromatic
    Aberration (Under-
    corrected)

(k) Axial Curve for an
    Achromatic Doublet
    Showing Zonal Spherical
    Secondary Spectrum, and
    Spherochromatism

Fig. 8-27. Ray-intercept curves for various aberrations. The ordinate for each curve is H, the height at which the ray intersects the (paraxial) image plane; the abscissa is tan U', the final slope of the ray with respect to the optical axis [8-8].

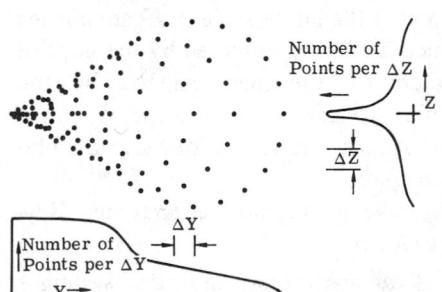

Number of
Points per ΔZ

Z

ΔZ

ΔY

Number of
Points per ΔY

Y—

Fig. 8-28. Spot diagram (for a system with pure coma) and the two line-spread functions (below and on the right) obtained by counting the number of points between parallel lines separated by a small distance, ΔY or ΔZ [8-8].

distribution, several other representations can be derived from it. The radial energy-distribution is obtained by arbitrarily selecting a center point and plotting the percentage of the energy (i.e., the number of spots) encircled within a radius, R, as a function of R. If the irradiance (i.e., the spot density) is represented as a function of the y and z coordinates of the image plane, this is the point-spread function of the system. It can be compared to the measured value or transformed to get a geometrical approximation to MTF.

8.12. Relationship Between Surface Imperfections and Image Quality

The effect of a manufacturing defect on the image quality of an optical system can be estimated by converting it into a wavefront deformation or OPD. For example, the irregularity of figure (i.e., the departure of a surface from its ideal geometrical configuration) is usually measured in interference fringes, each of which represents a departure from the nominal surface of one half-wavelength of visible light. This can be converted to a wavefront deformation:

$$OPD = \frac{1}{2}Fr(n' - n) \text{ wavelengths}$$

where Fr = number of fringes of irregularity (or asphericity in the case of a spherical surface)

$(n' - n)$ = change in index across the surface

When the OPD is summed for the entire system, its effect on the image can be estimated using Figure 8-26 as a guide, since OPD indicates the reduction in MTF caused by a low-order distortion of the wavefront. The defect can also be evaluated as an rms defect and related to the peak-to-peak measure, as indicated in Section 8.9.1.

8.13. References and Bibliography

8.13.1. References

[8-1]   Donald P. Feder, "Optical Calculations with Automatic Computing Machinery," *Journal of the Optical Society of America*, Optical Society of America, Washington, DC, Vol. 41, No. 4, April 1951, pp. 630-635.

[8-2]   Lucian Montagnino, "Ray Tracing in Inhomogeneous Media," *Journal of the Optical Society of America*, Optical Society of America, Washington, DC, Vol. 58, No. 12, December 1968, pp. 1667-1668.

[8-3]   G. H. Spencer and M. V. R. K. Murty, "Generalized Ray-Tracing Procedure," *Journal of the Optical Society of America*, Optical Society of America, Washington, DC, Vol. 52, No. 6, June 1962, pp. 672-678.

[8-4]   *Handbook of Optical Design*, U.S. Government Printing Office, Washington, DC, MIL-HDBK-141, 1962.

[8-5]   Harold H. Hopkins, *Wave Theory of Aberrations*, Oxford University, London, 1950, University Microfilms, Ann Arbor, MI, No. OP17185.

[8-6]   A. E. Conrady, *Applied Optics and Optical Design*, Dover, New York, NY, Two Volumes, 1957 and 1960.

[8-7]   Max Born and Emil Wolf, *Principles of Optics*, Macmillan, New York, NY, 1964.

[8-8]   Warren Smith, *Modern Optical Engineering: the Design of Optical Systems*, McGraw-Hill, New York, NY, 1966, pp. 55-57, 70, 254, 301, 319, 320, 322.

[8-9]   John W. Coltman, "The Specification of Imaging Properties by Response to a Sine Wave Input," *Journal of the Optical Society of America*, Optical Society of America, Washington, DC, Vol. 44, No. 6, June 1954, pp. 468-471.

[8-10] Richard Barakat, "Numerical Results Concerning the Transfer Functions and Total Illuminance for Optimum Balanced Fifth-Order Spherical Aberration," *Journal of the Optical Society of America*, Optical Society of America, Washington, DC, Vol. 54, No. 1, January 1964, pp. 38-44.

[8-11]   Richard Barakat and Agnes Houston, "Diffraction Effects of Coma," *Journal of the Optical Society of America*, Optical Society of America, Washington, DC, Vol. 54, No. 9, September 1964, pp. 1084-1088.

[8-12]   Richard Barakat and Agnes Houston, "The Effect of a Sinusoidal Wavefront on the Transfer Function of a Circular Aperture," *Applied Optics*, Optical Society of America, Washington, DC, Vol. 5, No. 11, November 1966, pp. 1850-1852.

[8-13]   Edward L. O'Neill, "Transfer Function of an Annual Aperture," *Journal of the Optical Society of America*, Optical Society of America, Washington, DC, Vol. 46, No. 4, April 1956, pp. 285-288.

8.13.2. Bibliography

Barakat, Richard, "Numerical Results Concerning the Transfer Functions and Total Illuminance for Optimum Balanced Fifth-Order Spherical Aberration," *Journal of the Optical Society of America*, Optical Society of America, Washington, DC, Vol. 54, No. 1, January 1964, pp. 38-44.

Barakat, Richard, and Agnes Houston, "Diffraction Effects of Coma," *Journal of the Optical Society of America*, Optical Society of America, Washington, DC, Vol. 54, No. 9, September 1964, pp. 1084-1088.

Barakat, Richard, and Agnes Houston, "The Effect of a Sinusoidal Wavefront on the Transfer Function of a Circular Aperture," *Applied Optics*, Optical Society of America, Washington, DC, Vol. 5, No. 11, November 1966, pp. 1850-1852.

Born, Max, and Emil Wolf, *Principles of Optics*, Macmillan, New York, NY, 1964.

Buchdahl, H. A., *Optical Aberration Coefficients*, Dover, New York, NY, 1968.

Coltman, John W., "The Specification of Imaging Properties by Response to a Sine Wave Input," *Journal of the Optical Society of America*, Optical Society of America, Washington, DC, Vol. 44, No. 6, June 1954, pp. 468-471.

Conrady, A. E., *Applied Optics and Optical Design*, Dover, New York, NY, Two Volumes, 1957 and 1960.

Feder, Donald P., "Optical Calculations with Automatic Computing Machinery," *Journal of the Optical Society of America*, Optical Society of America, Washington, DC, Vol. 41, No. 4, April 1951, pp. 630-635.

Goodman, J. W., *Introduction to Fourier Optics*, McGraw-Hill, New York, NY, 1968.

Handbook of Optical Design, U.S. Government Printing Office, Washington, DC, MIL-HDBK-141, 1962.

Hardy, Arthur C., and Fred H. Perrin, *The Principles of Optics*, McGraw-Hill, New York, NY, 1932.

Hertzberger, Max, *Modern Geometrical Optics*, Interscience, New York, NY, 1958.

Hopkins, Harold H., *Wave Theory of Aberrations*, Oxford University, London, 1950, University Microfilms, Ann Arbor, MI, No. OP17185.

Kingslake, Rudolph, *Applied Optics and Optical Engineering*, Academic Press, New York, NY, Five Volumes, 1965 to 1969.

Linfoot, E. H., *Fourier Methods in Optical Design*, Focal Press, New York, NY, 1964.

Martin, Louis C., *Technical Optics*, Pitman, London, Two Volumes, 1960.

Montagnino, Lucian, "Ray Tracing in Inhomogeneous Media," *Journal of the Optical Society of America*, Optical Society of America, Washington, DC, Vol. 58, No. 12, December 1968, pp. 1667-1668.

O'Neill, Edward L., *Introduction to Statistical Optics*, Addison-Wesley, Reading, MA, 1963.

O'Neill, Edward L., "Transfer Function of an Annular Aperture," *Journal of the Optical Society of America*, Optical Society of America, Washington, DC, Vol. 46, No. 4, April 1956, pp. 285-288.

Perrin, Fred H., "Methods of Appraising Photographic Systems," *Journal of the Society of Motion Picture and Television Engineers*, Society of Motion Picture and Television Engineers, New York, NY, Vol. 69, March 1960, pp. 151-156, April 1960, pp. 239-249.

Smith, Warren, *Modern Optical Engineering: the Design of Optical Systems*, McGraw-Hill, New York, NY, 1966, pp. 55-57, 70, 254, 301, 319, 320, 322.

Southall, James P. C., *Mirrors, Prisms and Lenses*, Dover, New York, NY, 1964.

Spencer, G. H., and M. R. V. K. Murty, "Generalized Ray-Tracing Procedure," *Journal of the Optical Society of America*, Optical Society of America, Washington, DC, Vol. 52, No. 6, June 1962, pp. 672-678.

Stephens, Robert E., and Loyd Sutton, "Diffraction Images of a Point in the Focal Plane and Several Out-of-Focus Planes," *Journal of the Optical Society of America*, Optical Society of America, Washington, DC, Vol. 58, No. 7, July 1968, pp. 1001-1002.

Chapter 9

OPTICAL ELEMENTS -
LENSES AND MIRRORS

Warren J. Smith
Infrared Industries, Inc.

CONTENTS

9. Optical Elements—Lenses and Mirrors

9.1. Introduction

This chapter deals with the following: telescopes and projection systems; elements, such as relay optics, light pipes, and fiber optics; lenses, mirrors, and combinations of these.

9.1.1. Symbols, Nomenclature and Units. The symbols, nomenclature and units used in this chapter are listed in Tables 8-1 (in Chapter 8) and 9-1 (below).

9.1.2. Performance Estimation. A large portion of this chapter presents equations and graphs used to estimate the performance of a number of objective systems. In most instances, the angular diameter, β, of the blur spot is given (that is, the diameter, in radians, of the best-focussed smallest spot containing all of the geometrically traced rays in the image). Diffraction effects are not included, and a separate computation of the diffraction blur should be made for them. (See Chapter 8, Section 8.7.2, and Figure 9-33.) The values of β are reliable for modest apertures and fields; but they must be viewed with caution for large apertures or fields, since in most cases the expressions for β are based on either third-order aberration theory or simplified empirical expressions.

To convert between β and the linear values of the aberrations causing the image blur, one may use the following relationships:

Spherical Aberration: $\quad \beta = (\text{longitudinal spherical aberration}) \left(\dfrac{U}{2F} \right)$

$$\beta = \frac{(\text{transverse spherical aberration})}{2F}$$

Coma: $\qquad\qquad\qquad \beta = \left(\dfrac{\text{coma}}{F} \right)$

Astigmatism: $\qquad\quad \beta = \dfrac{1}{2} \text{astigmatic focus difference} \left(\dfrac{2U}{F} \right)$

Field Curvature: $\qquad \beta = (\text{defocus}) \left(\dfrac{2U}{F} \right)$

where U = slope angle between marginal ray and axis at the image
$\quad\ F$ = focal length (for infinitely distant objects), or distance from second principal point to image (for finite conjugates)

When several aberrations are present, a conservative estimate of the resultant blur is given by summing the blurs resulting from the individual aberrations. The angular blur, β, is the angle subtended by the linear image blur-diameter B, from the second nodal point of the system.

9.2. Afocal Systems

Afocal systems are without focal length. The most common example of an afocal system is the ordinary telescope. Such a system forms its image of an infinitely distant

Table 9-1. Symbols, Nomenclature and Units

Symbols	Nomenclature	Units
A	Area	m^2
B	Blur diameter	m
C	Surface curvature, reciprocal of surface radius	m^{-1}
CC	Coma contribution (TOA)	m
$Coma_s, Coma_t$	Coma, sagittal or tangential	m
D	Diameter	m
d	Axial distance, spacing between elements	m
DC	Distortion contribution (TOA)	m
D_i	Exit pupil diameter	m
D_0	Entrance pupil diameter	m
d_f	Diameter of individual fiber	m
E_v	Illuminance	$lm\ m^{-2}$
F	Focal length or effective focal length—the distance from the second principal point to the back (or second) focal point; the distance from the front (or first) focal point to the first principal point	m
$F/\#$	Relative aperture, speed. The ratio of focal length to entrance pupil diameter. If the object is at infinity, $(F/\#) = [2(NA)]^{-1}$	—
f_y, f_z	Spatial frequency in cycles per unit length in the direction $\lfloor$indicated	m^{-1}
H, h	Image height	m
I	Angle of incidence, which is equal to minus the angle of reflection	rad
I'	Angle of refraction, defined by Snell's Law	m rad
$\mathcal{I}$	Optical invariant	rad
L	Radiance	$W\ m^{-2}\ sr^{-1}$
LA_m	Marginal spherical aberration, longitudinal	m
LA_z	Zonal spherical aberration, longitudinal	m
$L_{s,i}$	Distance from surface to intersection of ray with axis, before refraction	m
$L'_{s,i}$	Distance from surface to intersection of ray with axis, after refraction	m
MTF	Modulation transfer function	—
m	Magnification, linear	—
m_a	Magnification, angular	—
NA	Numerical aperture, given by $n \sin U$ where n is the final index in an optical system and U is the slope angle of the axial ray at the image. If the object is at infinity, $(NA) = [2(F/\#)]^{-1}$	—
n	Index of refraction	—
OTF	Optical transfer function	—
R	Radius of curvature	m

Table 9-1. Symbols, Nomenclature and Units (*Continued*)

Symbols	Nomenclature	Units
r	Semi-aperture $= (y^2 + z^2)^{1/2}$	m
S	Distance from first principal point to object	m
S'	Distance from second principal point to image	m
S_d	Diameter or linear dimension of a source or detector	m
S_1, S_2	Object (1) image (2) distances	m
TA	Transverse aberration	m
TAC	Transverse version of aberration contribution for astigmatism (TOA)	m
TchA	Lateral chromatic aberration, chromatic difference of image size	m
TchC	Lateral chromatic aberration contribution	m
TLchC	Transverse version of aberration contribution for axial color	m
TOA	Third-order aberration	—
TPC	Transverse version of aberration contribution for Petzval curvature (TOA)	m
TSC	Transverse version of aberration contribution for spherical aberration (TOA)	m
t	Thickness; axial spacing between surfaces	m
U, u	Ray-slope angle, before refraction; U', u' same after refraction	rad
V	Abbe ν-number; reciprocal relative dispersion	—
X, Y, Z	Ray direction cosines	—
x, y, z	Coordinate system; x is the optical axis; x and y define the meridional plane. The origin is at the vertex of the surface	m
Y, y	Height of intersection of ray with surface	m
α	Half-field angles	rad
β	Angular diameter of image blur spot	rad
θ	Angular field of view (half angle)	rad
λ	Wavelength	μm
τ	Transmission	—
Φ	Radiant flux	W
ϕ	Lens power; reciprocal focal length $\equiv 1/F$	Diopter, m^{-1}
Ω	Solid angle	sr

Subscripts

i	Image, image space, or system nearest the image	—
o	Object, object space, or objective system	—
p	Principal or chief ray	—

object at infinity. Thus, the usual definition of focal length is meaningless.

The angular magnification of an afocal system is the ratio of the angular size of the image to the angular size of the object. An afocal device also forms a finite image of any object at a finite distance. For an afocal system, linear magnification, which is the ratio of image size to object size, is the inverse of angular magnification; longitudinal magnification for finite conjugates is the square of the linear magnification. Since the pupils of an optical system are conjugates of each other, the magnifications, the pupil diameters, and the fields of view for any afocal system are related by

$$m = \frac{1}{m_a} = \frac{D_i}{D_o} = \frac{\tan \alpha_o}{\tan \alpha_i} \tag{9-1}$$

where m and m_a = linear and angular magnifications, respectively

D_i and D_o = exit and entrance pupil diameters, respectively

α_o and α_i = half-field angles in object and image space, respectively (See Figure 9-1.)

In any afocal system such as a telescope, angular magnification is given by

$$m_a = -\frac{F_o}{F_i} \tag{9-2}$$

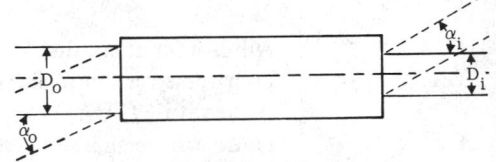

Fig. 9-1. Generalized afocal system.

where F_o = effective focal length of the objective system (that member of the telescope system near the object)

F_i = focal length of the member near the image (usually referred to as the eyepiece)

A positive magnification indicates an erect image. When a telescope consists of three major components (i.e., objective, erector, and eyepiece), the erector may be considered a part of the objective member or part of the image member. Either approach results in an angular magnification for the system given by

$$m_a = -\left(\frac{F_o}{F_i}\right)\left(\frac{S_2}{S_1}\right) \tag{9-3}$$

where (S_2/S_1) is the magnification of the erector component. (See Figure 9-4.) The objective member forms an image of an infinitely distant object image at its second focal point. For the final image of the afocal system to be focused at infinity, this internal image must lie at the first focal point of the eyepiece. Thus, for a simple telescope, as shown in Figures 9-2 and 9-3, the spacing, d, between the objective and eyepiece equals the algebraic sum of their focal lengths:

$$d = F_o + F_i \tag{9-4}$$

9.2.1. Astronomical Telescope. The astronomical telescope is an afocal system in which both members have positive focal lengths. There is a real, internal image, and the final image is inverted (and reversed). Figure 9-2 shows simplified schematic diagrams of refracting and pure reflecting astronomical-type telescopes. Actually, the normal telescope for visual astronomy has a reflecting objective and a refracting eyepiece. Refracting members, either objective or eyepiece, are usually of compound construction to correct aberrations. For most telescopes, the aperture stop and the entrance pupil are located at the objective to minimize its size and cost. Thus, the exit pupil is the image of the objective formed by the eyepiece.

9.2.2. Galilean Telescope. The Galilean telescope has a positive objective and a simple negative eyepiece. There is no real internal image, and a reticle or cross hair cannot be introduced into the system. The final image is erect. The Galilean telescope is compact, since the negative eyepiece causes the length, as indicated by Equation (9-4), to be equal to the difference between the absolute values of the members' focal lengths. The aperture stop is usually the pupil of the viewer's eye, which is also the exit pupil. Schematic drawings of refracting and reflecting Galileans are shown in Figure 9-3.

9.2.3. Terrestrial Telescope. The terrestrial telescope is essentially an astronomical telescope with an erecting system inserted between eyepiece and objective so that the final image is erect. A refracting telescope with a lens erector is shown in Figure 9-4. The image may also be erected by a system of prisms or plane mirrors. An all-reflector analogue of Figure 9-4 is possible, but uncommon, since the beam obscuration caused by the three mirrors markedly reduces the utility of such systems when a significant field of view is required.

9.3. Projection Systems

9.3.1. Irradiation Limits. The fraction of the radiant power from a source which a projection system can utilize is limited in two ways. The most apparent limitation is imposed by the size of the exit pupil. An optical system that directs the radiation from a source into a solid angle, Ω, has a maximum possible flux-output given by

$$\Phi = \tau L A \Omega \qquad (9\text{-}5)$$

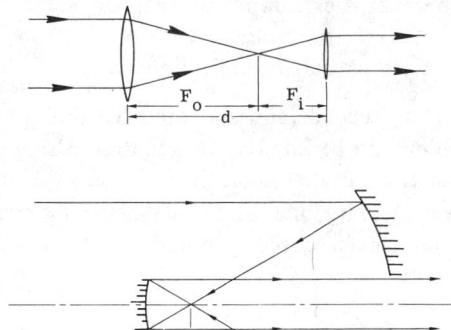

Fig. 9-2. The astronomical telescope.

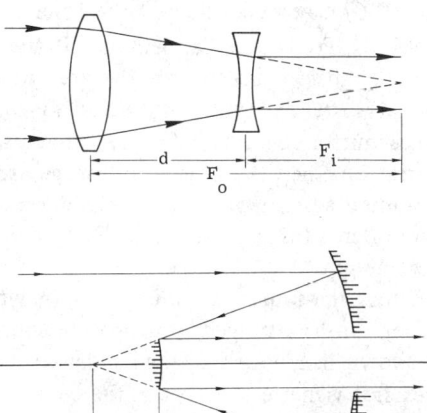

Fig. 9-3. The Galilean telescope.

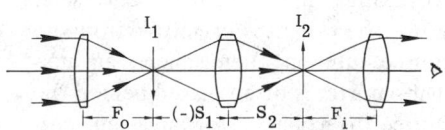

Fig. 9-4. The lens-erecting telescope [after 9-1].

where A = exit pupil area
    τ = transmission
    L = radiance of the source

The second (and less obvious) limitation is on the maximum size of the exit pupil which can be filled with radiation. Using the optical invariant, the Abbe sine condition, and the fact that in air the numerical aperture ($NA = n \sin U$) of an optical system cannot exceed unity, one can show that the maximum exit pupil diameter, D, which can be filled from a given source is limited to

$$D \leqslant \frac{S_d}{\theta} \tag{9-6}$$

where S_d = diameter (or linear dimension) of the source of radiation
    θ = angular half field to be irradiated from pupil

Equation (9-6) implies the use of an optical system with the extremely large numerical aperture of 1.0 (equivalent to an operating $F/0.5$ cone). This size exit pupil is very difficult to fill effectively and efficiently because of aberrations. Thus, because it corresponds to $NA = 0.5$ ($F/1.0$), a pupil diameter half as large as that indicated by Equation (9-6) may be considered a reasonable practical limit for most systems.

9.3.2. Projection Condensers. If the radiance of the source is sufficiently uniform, it may be imaged directly on the area to be irradiated. For small sources, a compound optical system provides additional magnification, as indicated in Figure 9-5. An ellipsoidal mirror (Section 9.9.4) is often used for this purpose since it is free of spherical aberration when the object and image are located at the geometrical foci. Note that the combined system of condenser and projection lens can be considered a system of two components (as in Chapter 8, Section 8.2.4) which together produce a short, negative, focal length.

When the source is nonuniform, or when very uniform irradiation is required, a condenser can be arranged to image the source in the pupil of the projection lens or mirror. As shown in Figure 9-6, the condenser member must be large enough so that its diameter does not vignette the rays at the edge of the field. The image of the source in the pupil then irradiates the desired area. If the source is Lambertian, the uniformity of irradiation is limited only by the geometry of the situation (e.g., the cosine-fourth distribution of irradiation).

9.3.3. Concentric Source Reflectors. Since the condenser collects, at most, the radiant flux emitted from the source into one hemisphere, an auxiliary mirror is often placed behind the source to collect the radiation into the other hemisphere and redirect it into the primary system. A spherical

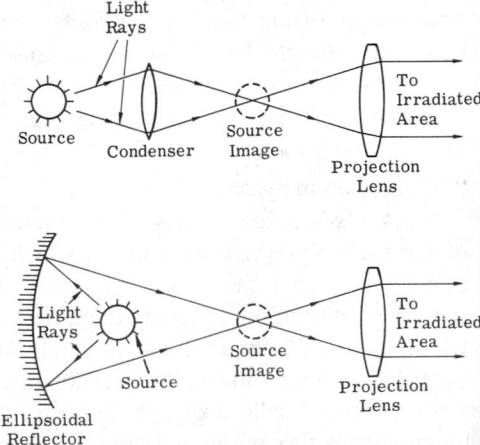

Fig. 9-5. Relay condensers.

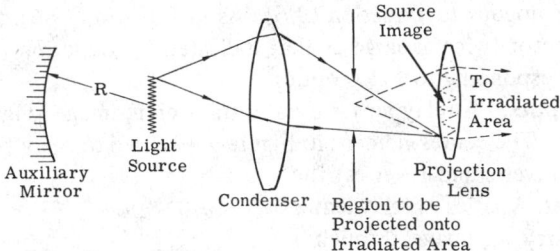

Fig. 9-6. Condenser for uniform illumination. Solid lines show imagery of a point on the source into the pupil of the projection lens. Dashed lines indicate imagery of an axial point in the region to be projected [9-1].

reflector, with its center of curvature at the source, will image the source back on itself. Depending on the source construction, the reflector may nearly double the average radiance (by filling in the empty areas of the source), or it may merely increase the temperature of the source.

9.4. Field and Relay Optics

9.4.1. Field Lenses. A field lens is placed at or near an image plane for the purpose of changing the location of the pupil without significantly modifying the magnification or image location of the system. The field lens ordinarily images the pupil onto subsequent elements of the system to prevent the diverging beams which form the images in the outer portions of the field from missing the subsequent elements of the system.

In systems with detectors, a field lens is often utilized to increase the field of view without increasing the size of the detector. Figure 9-7 shows a simple radiometer system in which the field lens images the exit pupil of the objective system (which may be refracting, as shown, or reflecting) onto the surface of the detector. This use of a field lens permits the use of small detectors, provides uniform irradiation over the detector surface, and makes detector irradiation independent of the position of the image point within the field of view. (See Chapter 20.)

9.4.2. Detector Size Limitations. The detector field lens of Figure 9-7 is exactly analogous to a projection condenser (Section 9.3.2, Figure 9-6) and is subject to similar limitations. For *any* optical system associated with a detector, the limit on the smallness of the detector is given by

$$S_d \geqslant \left(\frac{D\theta}{n}\right) \tag{9-7}$$

where S_d = diameter (or linear dimension) of the detector

D = the corresponding diameter or linear dimension of the entrance pupil of the optical system

θ = half-angle of field of view, or angular half field

n = refractive index of the medium in which the detector is immersed

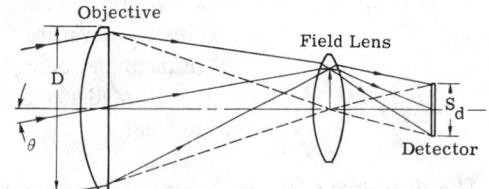

Fig. 9-7. Radiometer with field lens to increase the field of view with a small detector [9-1].

This equation is analogous to Equation (9-6). As in Equation (9-6), this limit is difficult to achieve. A detector twice as large as that indicated by Equation (9-7) is a reasonable practical limit, corresponding to $F/1.0$ optics.

9.4.3. Relay Optics. A relay lens is used to transfer an image longitudinally from one location to another. The center element in Figure 9-4 is used to relay the image from I_1 to I_2, and it erects (inverts and reverses) the image in the process. A series of relay lenses can be used to carry an image through a long path of restricted diameter.

Figure 9-8 shows a train of alternating field and relay lenses, illustrating both the transference of an image along the axis by the relay lenses, as well as the action of the field lenses in imaging (and thus relaying) the pupil from one relay lens to the next. This is the basic system used in periscopes and endoscopes.

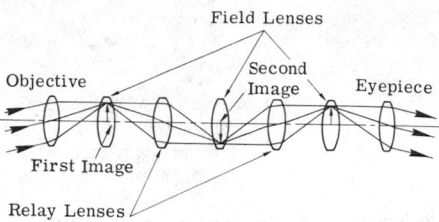

Fig. 9-8. The optical system of a periscope. This system consists of alternating field lenses and relay lenses, which transmit the image through a long bore of limited diameter [after 9-1].

9.5. Immersion Lenses and Aplanats

9.5.1. Aplanatic Surfaces. There are three aplanatic object positions for which the spherical aberration and coma of a simple spherical surface are equal to zero:

(1) The usually trivial case of object and image at the surface,

$$L_{s,i} = L'_{s,i} = 0 \quad \text{(See Figure 9-9(a).)}$$

(2) The object and image at the center of curvature of the surface,

$$L_{s,i} = L'_{s,i} = R \quad \text{(See Figure 9-9(b).)}$$

(3) The aplanatic case where

$$L_{s,i} = \frac{R(n' + n)}{n} = \frac{n'}{n} L'_{s,i}$$

$$n \sin U' = n' \sin U$$

$$U = I'; \quad U' = I \quad \text{(See Figure 9-9(c).)} \tag{9-8}$$

where $L_{s,i}, L'_{s,i}$ = the distances from the surface to the axial intercept of the object and image ray, before and after refraction, respectively

n, n' = the refractive indices before and after the surface, respectively, which has a radius of curvature R

U, U' = the ray-slope angles before and after refraction, respectively

I, I' = the angles of incidence and refraction, respectively, measured from the surface normal

The third-order spherical aberration, coma, and astigmatism of a single, air-glass surface are plotted in Figure 9-10 as a function of the position of the object. The three aplanatic conditions are indicated at A, B and C.

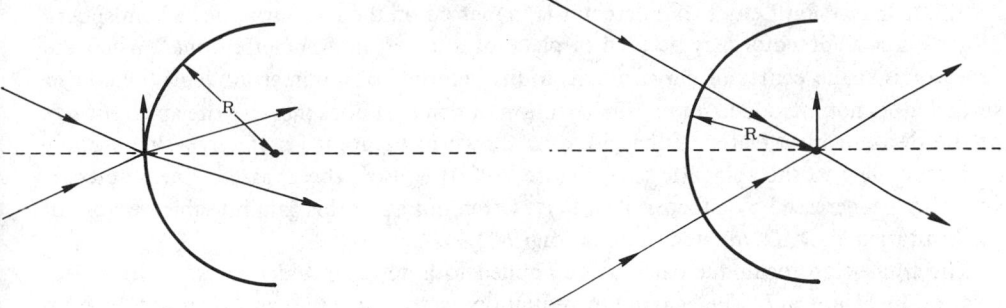

(a) Object and image lie at the surface.

(b) The object and image both lie at the center of curvature of the surface.

(c) The object and image positions indicated produce a useful magnification, equal to the ratio of the indices (n/n').

Fig. 9-9. Aplanatic surfaces. Surfaces for which both spherical aberration and coma are zero.

$$L_{s,i} = \frac{R(n'+n)}{n}$$

$$L'_{s,i} = \frac{nL_{s,i}}{n'}$$

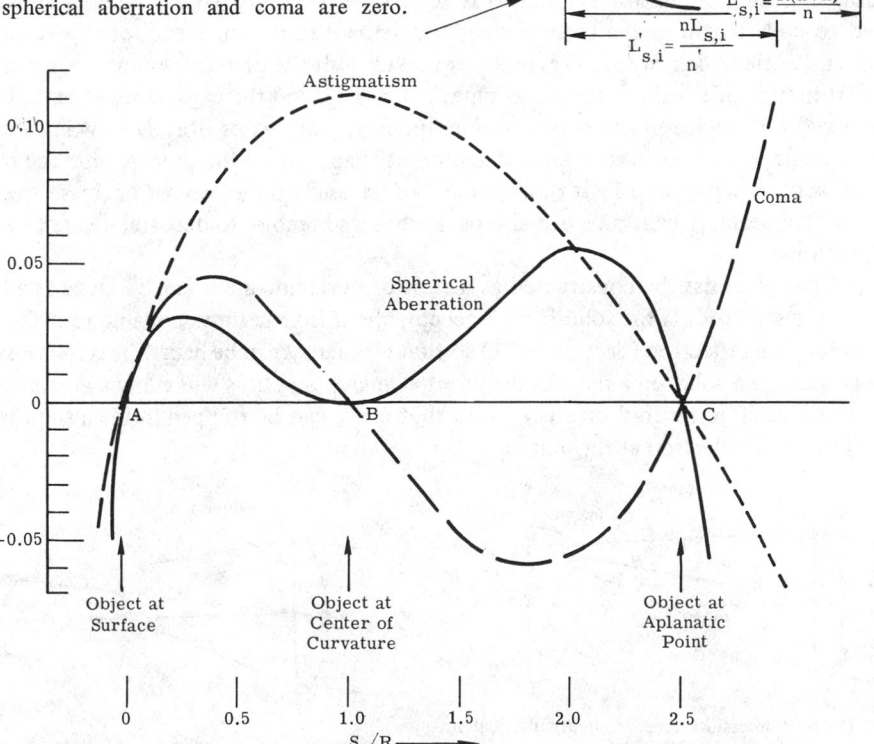

Fig. 9-10. Aplanatic surface. The aberrations are plotted versus S_1/R, the ratio of object distance to surface radius.

9.5.2. Immersion Lenses. If a detector is immersed on the rear surface of a hemispherical lens, a small detector may be used in place of a larger, non-immersed one. When the detector is at the center of curvature, as in the hemispherical immersion lens, the convex surface does not introduce spherical aberration or coma. It does increase the apparent size of the detector by a factor of the index, as shown in Figure 9-11. If a hyperhemisphere (corresponding to the aplanatic case, Figure 9-9(c)) is used, the apparent linear detector size may be increased by a factor of $(n'/n)^2$. Often not all of this gain is usable because of the limitation $S_d \geqslant D\theta/n$ cited in Equation (9-7).

The immersion technique can also be applied to a detector which is used with a field lens, as in Figure 9-7. The maximum reduction factor on detector size resulting from simple immersion is equal to the index. Any additional reduction in connection with an immersion lens (e.g., as in the hyperhemisphere) should be regarded as caused by a field-lens effect.

9.6. Light Pipes and Cone Channel Condensers

Light pipes conduct radiation from one end to the other by multiple reflections from their polished, interior walls. In many respects, they function as large-scale optical fibers, as discussed in Section 9.7. Figure 9-12 illustrates the collecting action of a reflecting, tapered, light pipe in redirecting rays which would otherwise miss the detector. Figure 9-13 indicates a convenient method of tracing ray paths through a pipe, using the standared unfolding technique which allows the reflected rays to be drawn as straight lines through prism and mirror systems. The light pipe of Figure 9-12 is tapered and can be used to collect light at the large end and condense it to the small end, or vice versa. Note that the angle which ray A (Figure 9-13) makes with the pipe axis-of-symmetry is an inverse function of the diameter of the pipe; at the large end the angle is small, and the angle increases with each reflection as the pipe dimension decreases. Ray B, which enters at a larger angle, is actually turned around and comes back out of the pipe. A pipe can be used to expand or contract a light beam, or it can be used to increase or decrease the divergence of a beam. A light pipe can also be used as a scrambler to decorrelate light beam relationships.

A light pipe may be constructed as a cone or pyramid, and it can be made as a hollow tube or a solid rod. If the solid form is acceptable, it is sometimes possible to make use of total internal reflection (Section 9.7.1) so that efficiency can be high. The refraction at the entry face of a solid pipe reduces the internal angles, and thus will allow a greater acceptance angle before reversal or a ray. Note that a ray can be trapped inside a solid pipe by total internal reflection at the exit face.

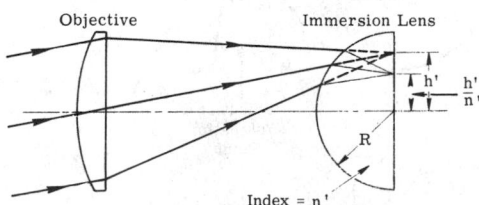

Fig. 9-11. Immersion lens. A hemispherical immersion lens concentric with the focus of an optical system reduces the linear size of the image by a factor of its index [9-1].

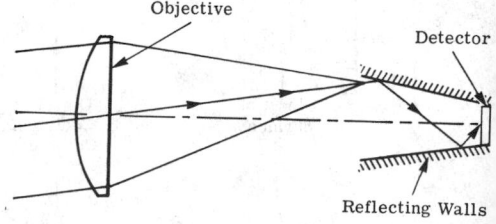

Fig. 9-12. The action of a reflecting light pipe in increasing the field of view of a radiometer [9-1].

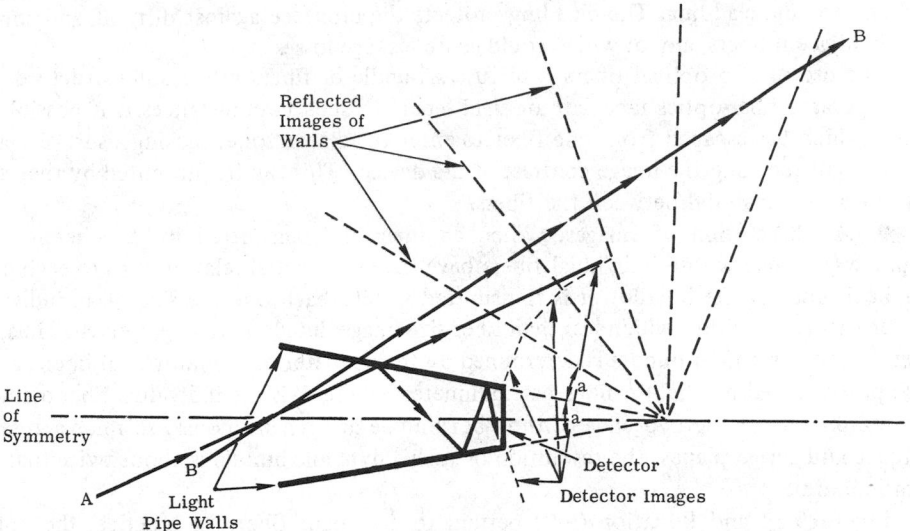

Fig. 9-13. Ray-tracing through a light pipe by means of an unfolded diagram [9-1].

9.7. Fiber Optics

9.7.1. Total Internal Reflection. When light is incident on a surface separating two media of indices, n and n', the light experiences total internal reflection if the angle of incidence exceeds the critical angle, $I_c = \arcsin(n'/n)$, and if $n > n'$.

9.7.2. Optical Fibers. A long polished cylinder of glass will transmit light without loss by leakage through the cylinder walls provided that all surface reflections occur at angles exceeding the critical angle, as shown in Figure 9-14. If the cylinder has length l, then the total path of the meridional ray (i.e., a ray in the plane of the drawing) is given by $l/\cos U'$, and the number of reflections which occur are $(l/d_f)\tan U'$, where d_f is the cylindrical fiber diameter.

9.7.3. Numerical Aperture and Cladding. The numerical aperture of a fiber is

$$\text{NA} = n_0 \sin U = (n_1{}^2 - n_2{}^2)^{1/2}. \tag{9-9}$$

(See Figure 9-14.) Skew rays make greater angles of incidence with the cylindrical surface of the fiber and therefore have somewhat greater numerical apertures than indicated by Equation (9-9). They also travel longer paths and are reflected more often than meridional rays.

Because of the very large number of reflections necessary to transmit a ray a useful distance through a small diameter fiber, optical fibers are often "clad" with a material of lower index, so that the total internal reflection occurs at the interface between the

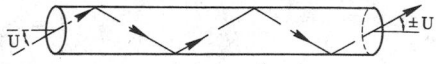

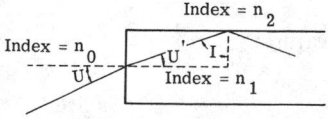

Fig. 9-14. Light is transmitted through a long polished cylinder by means of total internal reflection [9-1].

fiber core and cladding. The cladding protects the interface against dirt, oil, and contact with adjacent fibers, any of which could cause leakage losses.

One use of clad optical fibers is to fuse a bundle of fibers into a solid structure (e.g., for use as a fiber-optics faceplate or field lens). In such circumstances it is possible for light which has escaped from one fiber to enter an adjacent one, causing a sort of "cross-talk," and reducing the image contrast of the device. This can be prevented by the use of an absorbing material between the fibers.

9.7.4. Resolution of Images. When an image is transmitted by a coherent fiber bundle (i.e., one whose individual fibers have the same spatial relationships to each other at both ends of the bundle), it is transmitted in bits, each one the size of an individual fiber diameter, within which less than fiber-size image details are not preserved. Thus, the resolution for a fiber bundle is determined by the diameter of the individual fibers, and it is approximated by $(2d_f)^{-1}$ lines per millimeter, where d_f is the individual fiber diameter in millimeters. If the two ends of the fiber bundle are synchronously scanned across the object and image planes, the resolution of such a dynamic bundle is about twice that of a static bundle.

Figure 9-14 and Equation (9-9) pertain to a straight fiber. In practice, the optical fibers are usually curved. As a result, the numerical aperture is reduced because the reduced angle of incidence in the curved area allows leakage. Another bending limit is determined by the mechanical (tensile) strength of the fibers.

9.8. Refracting Objectives

9.8.1. The Refracting Singlet Objective. A singlet with spherical surfaces which is shaped or bent to minimize spherical aberration has the following image characteristics with the object at infinity:

Shape Factor for Minimum Spherical Aberration:

$$K = \frac{C_1}{C_1 - C_2} = C_1 F(n-1) = \frac{n(2n+1)}{2(n+2)} \qquad (9\text{-}10)$$

Spherical Aberration:

$$\beta = \frac{n(4n-1)}{128(n+2)(n-1)^2(F/\#)^3} \qquad (9\text{-}11)$$

Sagittal Coma:

$$\beta = \frac{\theta}{16(n+2)(F/\#)^2} \qquad (9\text{-}12)$$

Astigmatism (at best focus):

$$\beta = \frac{\theta}{2(F/\#)} \qquad (9\text{-}13)$$

Field Curvature:

$$\text{Sagittal} \quad \beta = \frac{\theta^2}{(F/\#)} \frac{(n+1)}{2n} \qquad (9\text{-}14)$$

$$\text{Tangential} \quad \beta = \frac{\theta^2}{(F/\#)} \frac{(3n+1)}{2n} \qquad (9\text{-}15)$$

Chromatic Aberration:

$$\beta = \frac{1}{2V(F/\#)} \tag{9-16}$$

For singlet shapes other than the minimum spherical form, the spherical aberration can be estimated from Figure 9-15, where r is the lens semi-aperture.

If a singlet objective is made with an aspheric surface, its shape factor can be chosen to eliminate coma, and the aspheric surface deformation can be chosen to eliminate spherical aberration. The astigmatism, field curvature, and chromatic aberration are given by Equations (9-13), (9-14), (9-15), and (9-16) above.

9.8.2. The Doublet Refracting Objective. A simple doublet, composed of two positive elements of equal power, will have much less spherical aberration than the singlet of Section 9.8.1. The amount of reduction depends upon the index of the lens

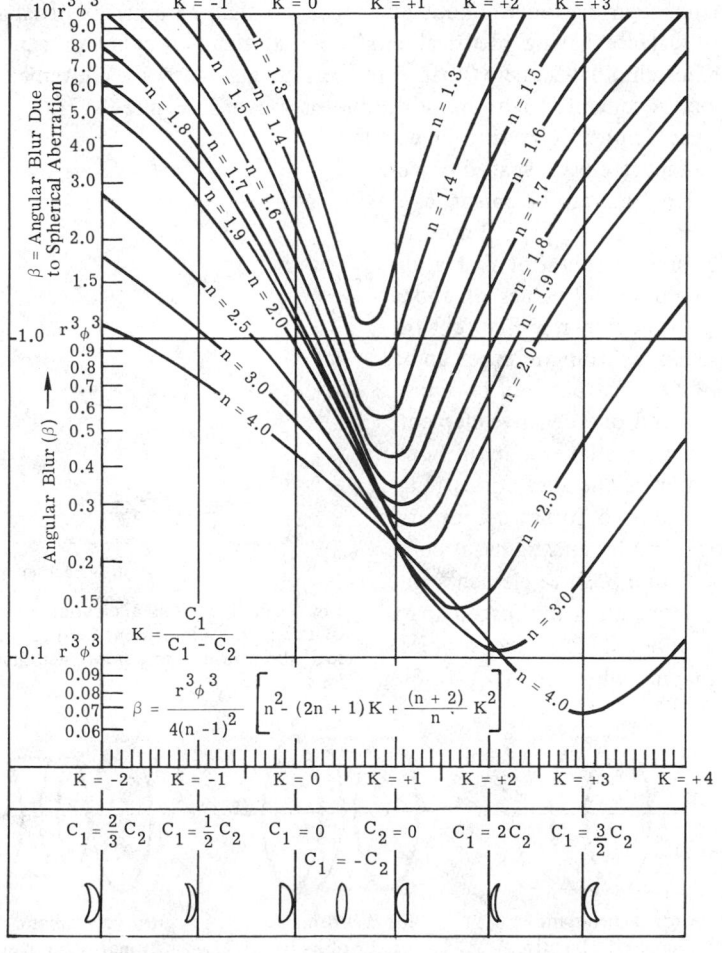

Fig. 9-15. Angular blur of simple lens as function of shape for various indices of refraction. ϕ is the lens power, and r is the semi-aperture.

material: at $n = 1.5$ the spherical aberration is reduced by a factor of 5; at $n = 2.5$ the spherical aberration is eliminated. These reductions assume that each element is shaped to minimize spherical aberration. Figure 9-16 indicates these relationships for compact singlets, doublets and triplets of this type.

9.8.3. The Achromatic Doublet Objective. If one element of a doublet is made negative in power while the other is increased in positive power to maintain the sum of the powers at the desired value, shapes can be found for the elements so that both primary spherical aberration and coma are eliminated. High-order aberrations remain for large-aperture systems. Depending on the spectral region and the availability of materials, the primary chromatic aberration may be eliminated (if the positive element is made from low-dispersion material and the negative from high-dispersion material). In the visible spectral region, crown and flint optical glasses are used; in the near infrared, calcium fluoride and glass can be combined. Silicon and germanium form a suitable pair for longer wavelengths.

Smaller diameter achromatic doublets (those less than 75 mm) are often designed with their inner surfaces having identical radii so that the two elements can be cemented together. This eliminates these surface reflections and allows a convenient, monolithic construction. A cemented achromatic doublet is shown in Figure 9-17(a).

9.8.4. The Triplet Refracting Objective. The compact, closely spaced objectives of Section 9.8.2 can be constructed in triplet form. This simple, all-positive type can be free of spherical aberration for index-of-refraction values of about 1.75 or higher, as shown in Figure 9-16. The compound, achromatic type can be advantageously constructed of two positive elements and one negative element. This construction allows a more complete correction of the aberrations. Figure 9-17 shows two triplet achromats.

The Cooke triplet anastigmat is composed of two outer positive elements and a central negative element, spaced apart from each other. This is the simplest construction in which it is possible

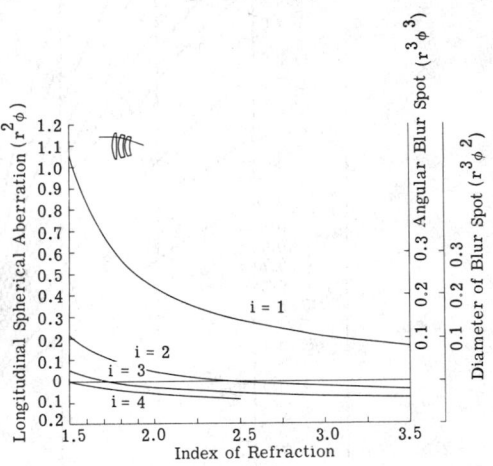

Fig. 9-16. Thin lens aberration of i thin elements of index n, each element bent for minimum spherical aberration. Lens pupil semidiameter = r and lens power = θ

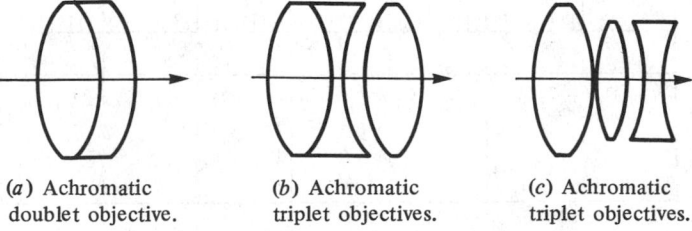

(a) Achromatic doublet objective.

(b) Achromatic triplet objectives.

(c) Achromatic triplet objectives.

Fig. 9-17. Achromatic objectives.

to correct all the primary aberrations. Photographic Cooke triplets are designed with speed and total-field combinations such as $f/5.6$ and $55°$ or $f/3$ and $40°$. In general, speed can be increased at the expense of field coverage. The use of high-index materials such as silicon or germanium allows combinations of extremes of speed and field, especially when resolution requirements are not rigorous. The Cooke triplet and several other more complex forms of photographic objectives are illustrated in Figure 9-18.

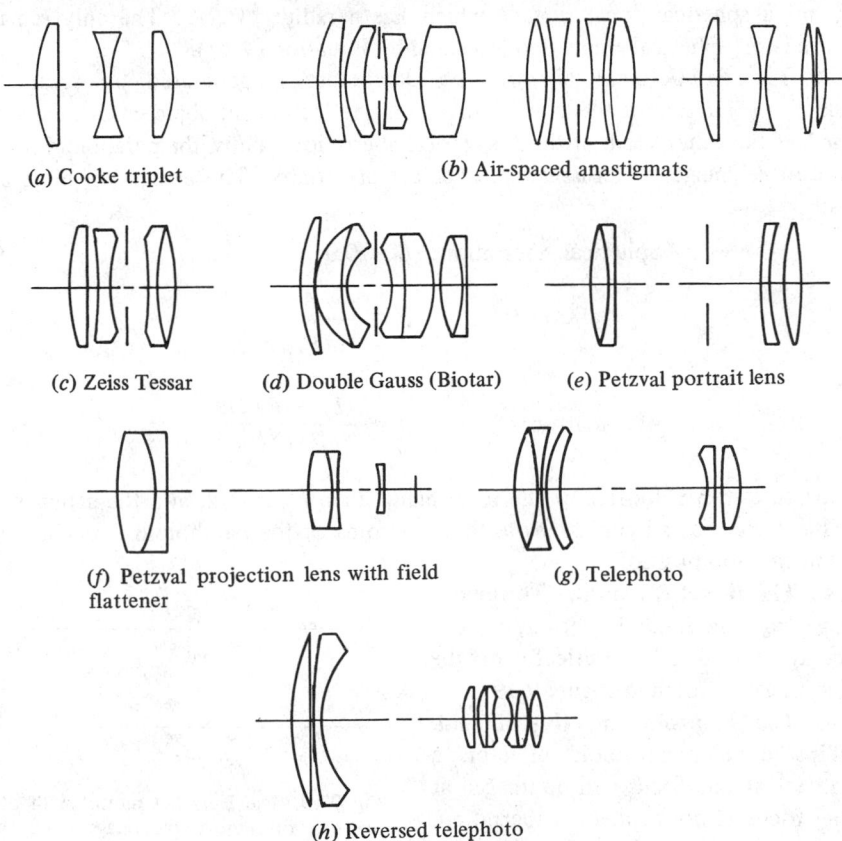

(a) Cooke triplet   (b) Air-spaced anastigmats

(c) Zeiss Tessar   (d) Double Gauss (Biotar)   (e) Petzval portrait lens

(f) Petzval projection lens with field flattener   (g) Telephoto

(h) Reversed telephoto

Fig. 9-18. Compound photographic objectives

9.9. Reflecting Objectives: Single Mirrors

9.9.1. The Spherical Mirror.

The simple, reflecting sphere is possibly the most useful infrared objective because it is cheap, achromatic, and (at low apertures) produces an excellent image. The image blurs are as follows:

$$\text{Spherical Aberration:} \quad \beta = \frac{1}{128(F/\#)^3} \tag{9-17}$$

$$\text{Sagittal Coma:} \quad \beta = \frac{(L_p - R)\theta}{16R(F/\#)^2} \tag{9-18}$$

$$\text{Astigmatism (at best focus):} \quad \beta = \frac{(L_p - R)^2 \theta^2}{2R^2(F/\#)} \tag{9-19}$$

where L_p is the distance from mirror surface to aperture stop.

When the aperture stop is placed at the center of curvature, $L_p = R$; and, as indicated by Equations (9-18) and (9-19), both coma and astigmatism are eliminated. The image is formed on a spherical focal surface which has a radius of $R/2$. The only remaining aberration is the spherical aberration indicated by Equation (9-17).

9.9.2. The Paraboloidal Mirror. The paraboloid is generated by rotating the parabola $x = y^2/2R$ about the x axis. With conic reflectors, an object at one conic focus is imaged at the other focus without spherical aberration. Thus, the paraboloid forms an aberration-free image of an *axial point* object at infinity. The aberration blurs are as follows:

$$\text{Spherical Aberration:} \quad \beta = 0.0 \tag{9-20}$$

$$\text{Sagittal Coma:} \quad \beta = \frac{\theta}{16(F/\#)^2} \tag{9-21}$$

$$\text{Astigmatism:} \quad \beta = \frac{(L_p - R/2)\theta^2}{R\,(F/\#)} \tag{9-22}$$

If the aperture stop is located at the focal plane, then $L_p = R/2$, and the astigmatism is zero. This is true for all conics. Note that the coma of the paraboloid is unchanged by changes in the stop position.

9.9.3. The Herschel Mount. To remove the image from the incoming beam, one can shift the aperture stop of a reflector off the optical axis, as indicated in Figure 9-19.

9.9.4. The Ellipsoid and Hyperboloid. With ellipsoid and hyperboloid reflectors, a point object at one focus will be imaged at the other focus without spherical aberration.

Off-axis images are afflicted with coma. Both object and image are real in the ellipsoid,

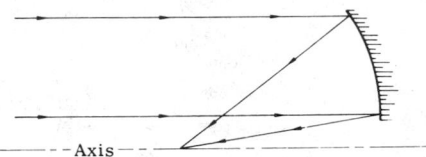

Fig. 9-19. The Herschel mount with off-axis aperture to remove the image from the incoming beam.

as shown in Figure 9-20(a); one or the other is virtual in the hyperboloid, as shown in Figure 9-20(b). Both are free of astigmatism if the aperture stop is at one of the foci.

9.10. Reflecting Objectives: Two-Mirror Systems

9.10.1. Newtonian Configuration. A diagonal plane mirror displaces the image of a reflecting objective to a location outside the incoming beam, making the image readily accessible—as shown in Figure 9-21(a).

9.10.2. Newtonian Complement. A large diagonal mirror placed before the reflecting objective isolates the image from the incoming beam by means of an axial hole, as shown in Figure 9-21(b).

9.10.3. Folded Configuration. A plane secondary mirror can be used to shift the image formed by a primary mirror through a central hole in the primary to an accessible

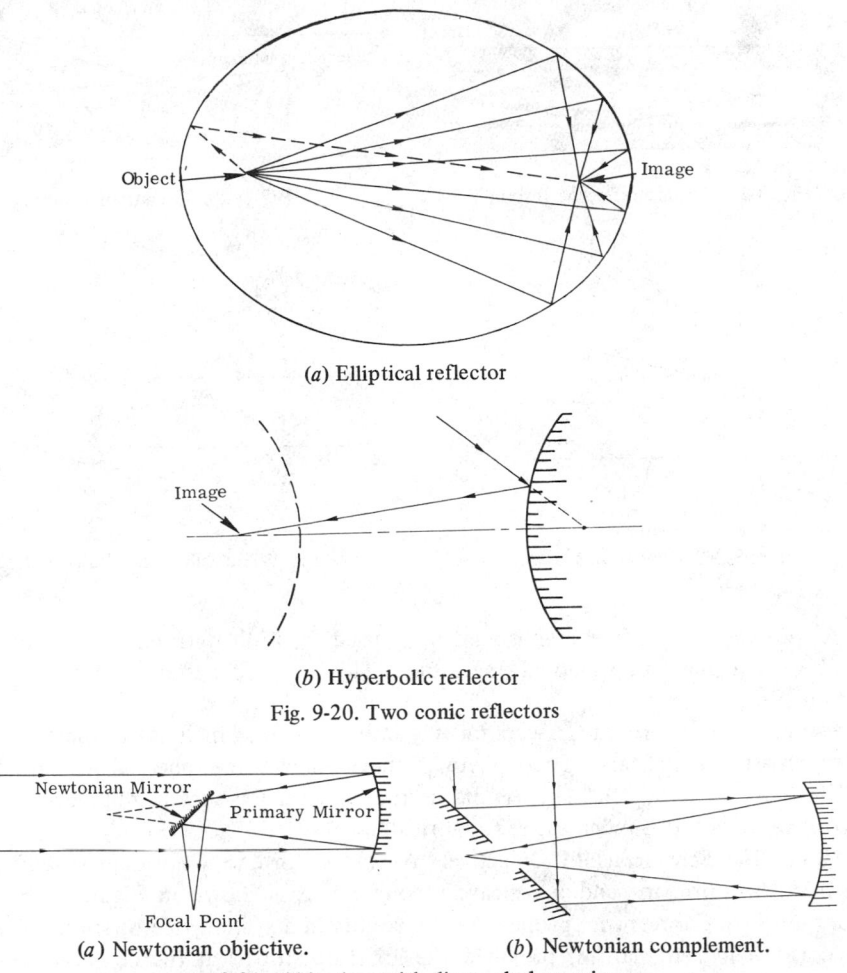

(a) Elliptical reflector

(b) Hyperbolic reflector

Fig. 9-20. Two conic reflectors

(a) Newtonian objective.

(b) Newtonian complement.

Fig. 9-21. Objectives with diagonal plane mirrors.

location behind it, as indicated in Figure 9-22. Although this is a very simple system, its drawback is that at least 50% of the beam diameter is obscured by the folding mirror.

9.10.4. Gregorian System. The Gregorian objective consists of two concave mirrors arranged so that the secondary mirror relays the real image from the primary mirror through a hole in the primary, as shown in Figure 9-23. The classical Gregorian consists of a paraboloidal primary and an ellipsoidal secondary so that the final image is free of spherical aberration. This arrangement is afflicted with coma. If general aspherics are used, the system can be designed to be free simultaneously of both coma and spherical aberration. A drawback of the Gregorian is its length, which is large compared to the Cassegrain configuration.

9.10.5. Cassegrain System. A concave primary and a convex secondary, combined as shown in Figure 9-24, comprise a short system with an accessible image. The classical Cassegrain consists of a paraboloidal primary mirror and a hyperboloidal secondary to achieve an image free of spherical aberration. As with the classical Gregorian, the image

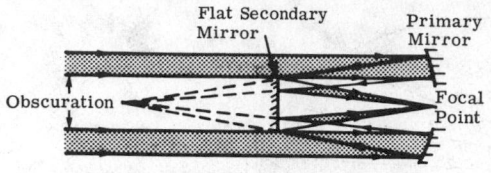

Fig. 9-22. Folded reflector system.

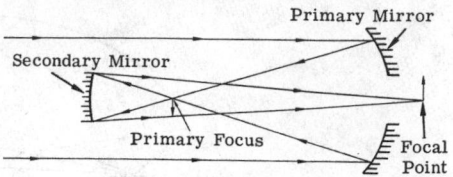

Fig. 9-23. Gregorian system.

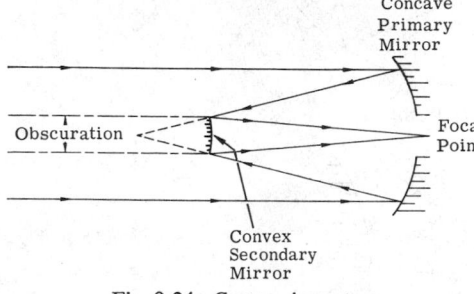

Fig. 9-24. Cassegrain system.

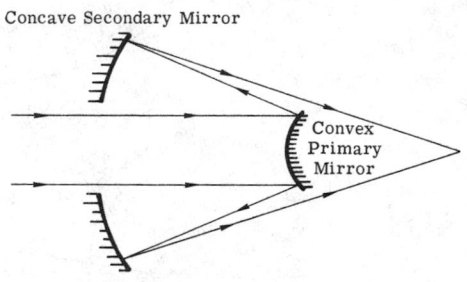

Fig. 9-25. Schwarzschild objective.

here is quite comatic. If general aspherics are used for both mirrors, the system can be free of both coma and spherical aberration. This type of system is named Ritchey-Chrétien [9-2].

Another popular form of Cassegrain is the Dall-Kirkham, which has a spherical secondary and an aspheric primary. The aspheric primary removes the spherical aberration. The Dall-Kirkham has about the same coma as the classical Cassegrain, but because of the spherical secondary it is much easier to fabricate.

9.10.6. The Schwarzschild Objective. A sort of inverse Cassegrain system results from a convex primary and a concave secondary, as is shown in Figure 9-25. If the mirrors are made concentric, proper spacing results in a system free of spherical aberration, coma, and astigmatism, provided the aperture stop is at the common center of curvature. The great advantage of this system is that it utilizes inexpensive spherical surfaces; the drawback is that the concave secondary mirror is much larger than the entrance pupil. The prescription for the concentric case is $R_1 = 1.236F$, $R_2 = 3.236F$, and $d = 2F$.

9.10.7 Generalized Two-Reflector Systems [9-3]. The following equations are derived specifically for a generalized two-mirror system with the stop at the primary mirror, as shown in Figure 9-26:

$$C_1 = \frac{(B-F)}{2DF} \tag{9-23}$$

$$C_2 = \frac{(B+D-F)}{2DB} \tag{9-24}$$

$$\Sigma TSC = \frac{Y^3 \left[F(B-F)^3 + 64D^3F^4K_1 + B(F-D-B)(F+D-B)^2 - 64B^4D^3K_2 \right]}{8D^3F^3} \tag{9-25}$$

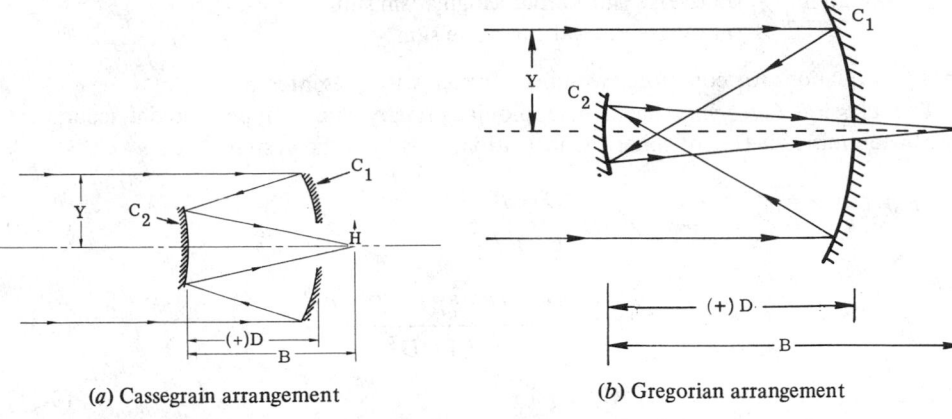

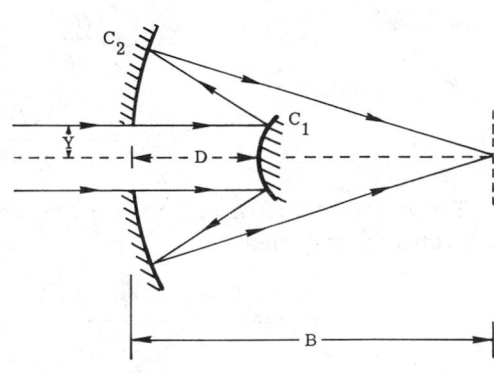

(a) Cassegrain arrangement

(b) Gregorian arrangement

Fig. 9-26. Schematics of generalized mirror objective illustrating the symbols for the system parameters as used in the Section 9.10.7

(c) Schwarzschild arrangement

$$\Sigma CC = \frac{HY^2[2F(B-F)^2 + (F-D-B)(F+D-B)(D-F-B) - 64B^3D^3K_2]}{8D^2F^3} \tag{9-26}$$

$$\Sigma TAC = \frac{H^2Y[4BF(B-F) + (F-D-B)(D-F-B)^2 - 64B^3D^3K_2]}{8BDF^3} \tag{9-27}$$

$$\Sigma TPC = \frac{H^2Y[DF - (B-F)^2]}{2BDF^2} \tag{9-28}$$

where C_1 and C_2 = curvatures of the primary and secondary mirrors

B = distance from mirror 2 to image (i.e., the back focal length)

F = system focal length

D = spacing (use positive sign)

K_1 and K_2 = fourth-order aspheric deformation coefficients for primary and secondary mirrors, respectively

H = image height

ΣTSC = transverse third-order spherical aberration sum

ΣCC = third-order coma sum

ΣTAC = transverse third-order astigmatism sum

ΣTPC = transverse Petzval curvature sum

Specific equations for common, two-mirror systems are presented below.

The classical *Cassegrain* has a paraboloidal primary and a hyperboloidal secondary. Since each mirror is free of spherical aberration, so is the total system.

$$K_1 = \frac{(F-B)^3}{64D^3F^3} \tag{9-29}$$

$$K_2 = \frac{(F-D-B)(F+D-B)^2}{64B^3D^3} \tag{9-30}$$

$$\Sigma TSC = 0.0 \tag{9-31}$$

$$\Sigma CC = \frac{HY^2}{4F^2} \tag{9-32}$$

$$\Sigma TAC = \frac{H^2Y(D-F)}{2BF^2} \tag{9-33}$$

The *Ritchey-Chrétien* aspheric surfaces are chosen so that both spherical aberration and coma are corrected.

$$K_1 = \frac{[2BD^2 - (B-F)^3]}{64D^3F^3} \tag{9-34}$$

$$K_2 = \frac{[2F(B-F)^2 + (F-D-B)(F+D-B)(D-F-B)]}{64B^3D^3} \tag{9-35}$$

$$\Sigma TSC = 0.0 \tag{9-36}$$

$$\Sigma CC = 0.0 \tag{9-37}$$

$$\Sigma TAC = \frac{H^2Y(D-2F)}{4BF^2} \tag{9-38}$$

The *Dall-Kirkham* has an aspheric primary and a spherical secondary for ease of fabrication.

$$K_1 = \frac{[F(F-B)^3 - B(F-D-B)(F+D-B)^2]}{64D^3F^4} \tag{9-39}$$

$$K_2 = 0.0 \tag{9-40}$$

$$\Sigma TSC = 0.0 \tag{9-41}$$

$$\Sigma CC = \frac{HY^2[2F(B-F)^2 + (F-D-B)(F+D-B)(D-F-B)]}{8D^2F^3} \tag{9-42}$$

$$\Sigma TAC = \frac{H^2 Y[4BF(B-F) + (F-D-B)(D-F-B)^2]}{8DBF^3} \tag{9-43}$$

Spherical Primary, Aspheric Secondary

$$K_1 = 0.0 \tag{9-44}$$

$$K_2 = \frac{[F(B-F)^3 + B(F-D-B)(F+D-B)^2]}{64B^4 D^3} \tag{9-45}$$

$$\Sigma TSC = 0.0 \tag{9-46}$$

$$\Sigma CC = \frac{HY^2[2BD^2 - (B-F)^3]}{8BD^2 F^2} \tag{9-47}$$

$$\Sigma TAC = \frac{H^2 Y[(F-B)^3 + 4BD(D-F)]}{8B^2 DF^2} \tag{9-48}$$

9.10.8. Clamshell System. Two concave mirrors facing each other, and spaced apart by their common focal length, form a convenient relay device, as shown in Figure 9-27. Both mirrors may be paraboloidal, or all of the aspheric correction can be put on one mirror—the other remaining spherical to reduce cost.

9.11. Reflecting Objectives: Three-Mirror System

9.11.1. The Generalized Three-Mirror System [9-4]. In most two-mirror systems only spherical aberration and coma can be corrected; astigmatism remains. Another mirror can be used, however, and all three of these aberrations can be corrected. Unfortunately, the complexity of the resulting system is such that the mirrors can get in each other's way;* as a result, three-mirror systems are seldom designed. (Note that by adding a refracting corrector plate to a two-mirror system, one can correct the three aberrations in any of several useful configurations.)

9.11.2. The Clamshell-Schmidt. The clamshell relay of Section 9.10.8 can be combined with a concave primary to make an interesting system, as shown in Figure 9-28. If mirror C is made aspheric, it can correct the spherical aberration of the two spherical mirrors, A and B. Note that mirror C is imaged by A and B onto the front focal plane of mirror A, where it can be considered a virtual Schmidt corrector—although it is not at the ideal center of curvature location for such a corrector (which is not too important in view of the small field of view necessitated by the configuration).

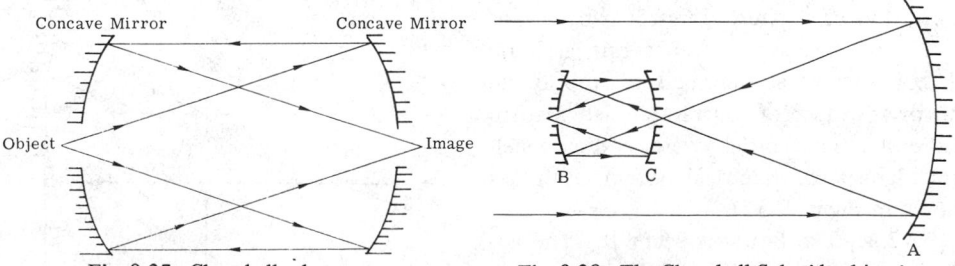

Fig. 9-27. Clamshell relay system.   Fig. 9-28. The Clamshell-Schmidt objective.

---

*Sections of spherical mirrors, eccentric systems, and folding flats are useful for special applications.

9.11.3. The Schiefspiegler, or Oblique Telescope [9-5]. To achieve a reflecting objective without a central obstruction (which greatly reduces image contrast), amateur telescope-makers have utilized two- and three-mirror configurations in which the mirrors are used obliquely. Although there are exceptions, the surfaces tend to be more or less toroidal. In many instances, the mirrors are thin spheres, and warped into toroidal shapes by straps and harnesses.

9.12. Catadioptric Objectives

9.12.1. The Mangin Mirror. The spherical aberration of a spherical mirror can be corrected by a refracting first-surface, as shown in Figure 9-29. This is accomplished at the penalty of a very heavy chromatic aberration. The aberration blurs are as follows:

Residual Spherical Aberrations:
$$\beta = \frac{0.00025}{(F/\#)^4} \quad (9\text{-}49)$$

Chromatic Aberration:
$$\beta = \frac{1}{6(F/\#)V} \quad (9\text{-}50)$$

Sagittal Coma:
$$\beta = \frac{\theta}{32(F/\#)^2} \quad (9\text{-}51)$$

Astigmatism:
$$\beta = \frac{\theta^2}{2(F/\#)} \quad (9\text{-}52)$$

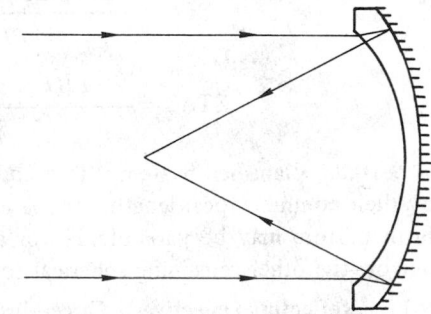

Fig. 9-29. The Mangin mirror objective [9-1].

9.12.2. The Bouwers-Maksutov System. The spherical aberration of a spherical mirror also can be corrected by a thick meniscus lens, as shown in Figure 9-30. In one version, the meniscus is shaped to eliminate chromatic aberration. In another, all radii of the system are concentric and the stop is put at the common center. Thus, the coma and astigmatism are zero, and the image lies on a curved surface which has a radius equal to the focal length and which is concentric with rest of the system.

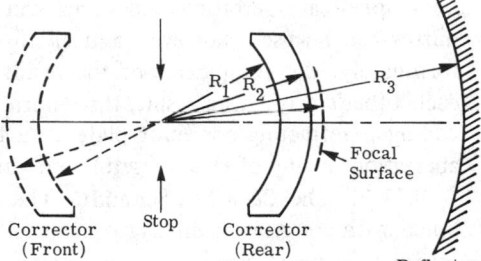

Figure 9-30. The Bouwers concentric meniscus corrector [9-13].

9.12.3. The Schmidt System [9-6]. Another way of correcting the spherical aberration of a spherical mirror is by means of a refracting aspheric. If a thin aspheric plate is located at the aperture stop at the mirror's center of curvature, astigmatism and coma are essentially zero, and an excellent high-speed, wide-field system results (as shown in Figure 9-31).

9.12.4. The Bouwers-Schmidt. The advantages of both the Schmidt and the meniscus correctors can be combined, as shown in Figure 9-32.

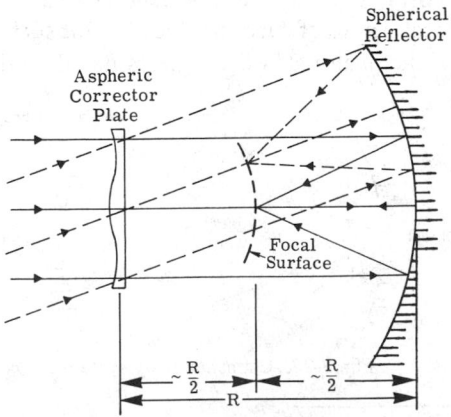

Fig. 9-31. The Schmidt objective.

9.12.5. The Ross Corrector [9-4, 9-7]. The coma inherent in a paraboloid reflector can be corrected by placing a simple positive lens between the mirror and its focus. The field curvature is also reduced. Since the corrector is close to the focal plane, its effect on the spherical and chromatic aberrations is modest, and a significant increase in the field coverage is obtained. An improved version uses an achromatic doublet.

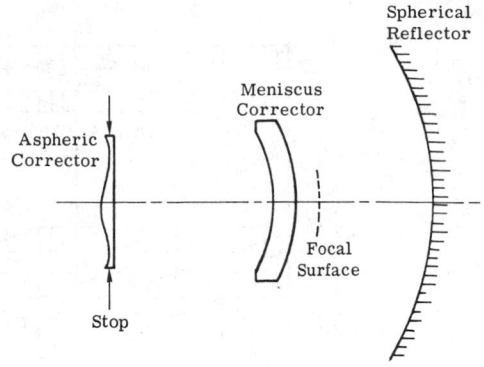

Fig. 9-32. The Bouwers-Schmidt corrected concentric objective [9-1].

The Ross corrector is only one of a large family of refracting correctors located near the focal plane of a mirror system. Some are used as correctors (i.e., to improve the image quality of a system). Others may be used as conversion devices (e.g., to enable the primary mirror of a Ritchey-Chrétien to be used without its secondary).

9.13. Blur Spot Charts

The charts in Figures 9-33 through 9-38 may be used to make a rapid estimate of the limiting size of the blur spot in an optical system.

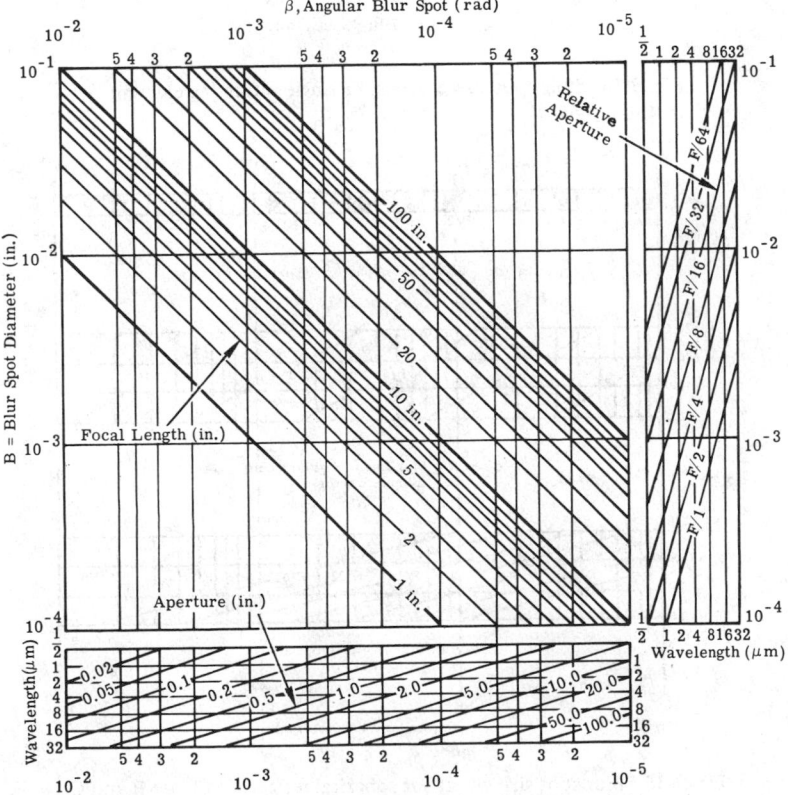

Fig. 9-33. Blur spot size charts for diffraction-limited systems [9-1].

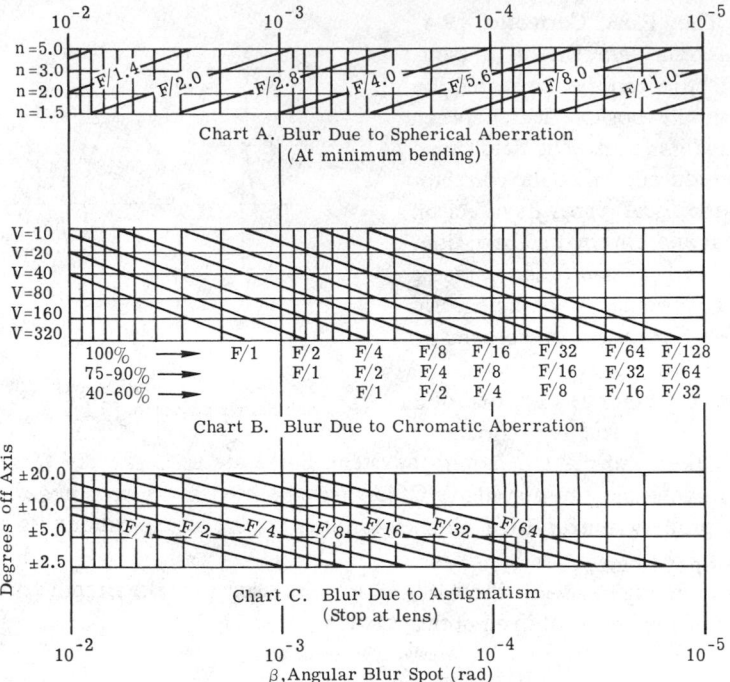

Fig. 9-34. Blur spot size charts for a single refracting element
(a simple lens) [9-1].

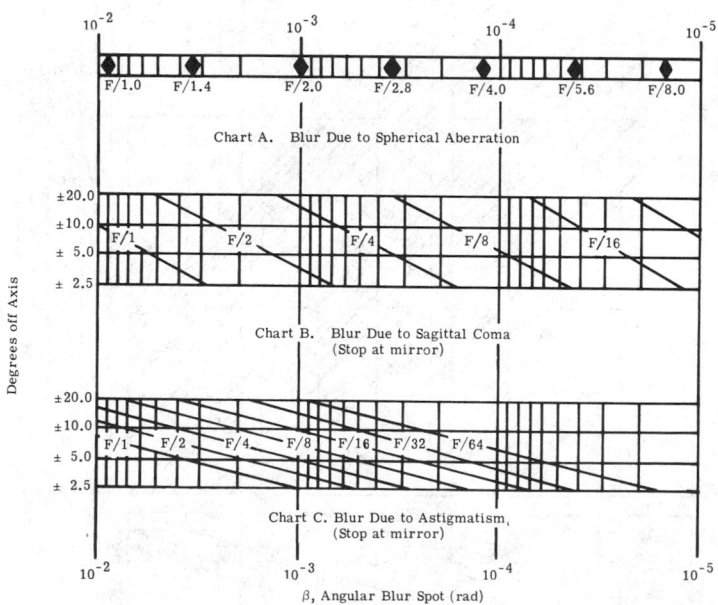

Fig. 9-35. Blur spot size charts for spherical reflector. Charts B and C
also apply to a paraboloidal reflector [9-1].

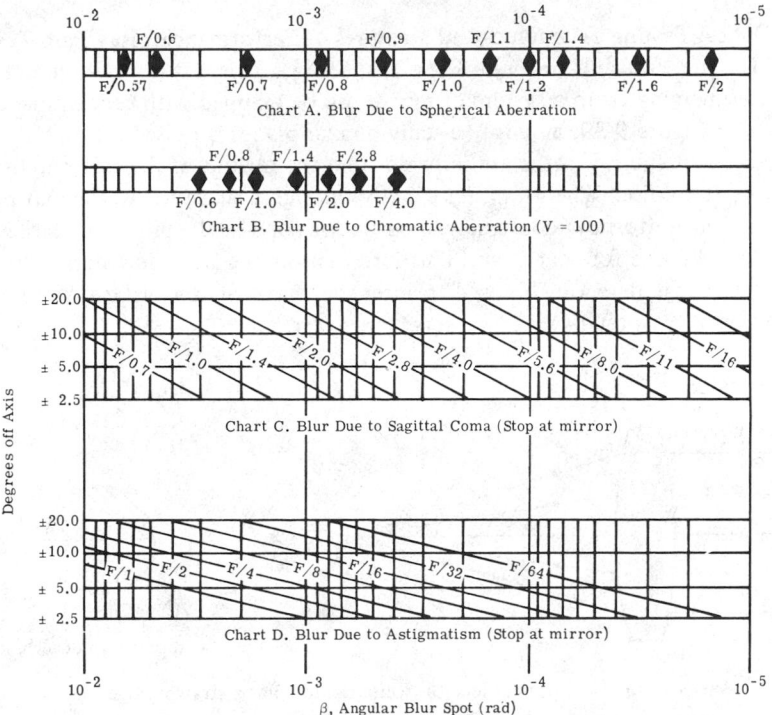

Fig. 9-36. Blur spot size charts for Mangin mirrors [9-1].

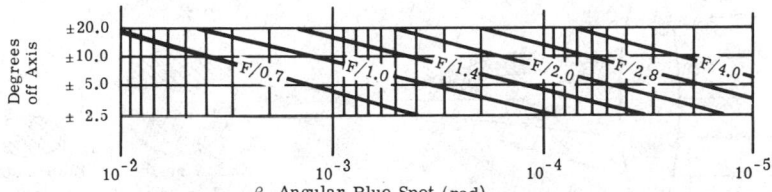

Fig. 9-37. Blur spot size chart for Schmidt objectives.

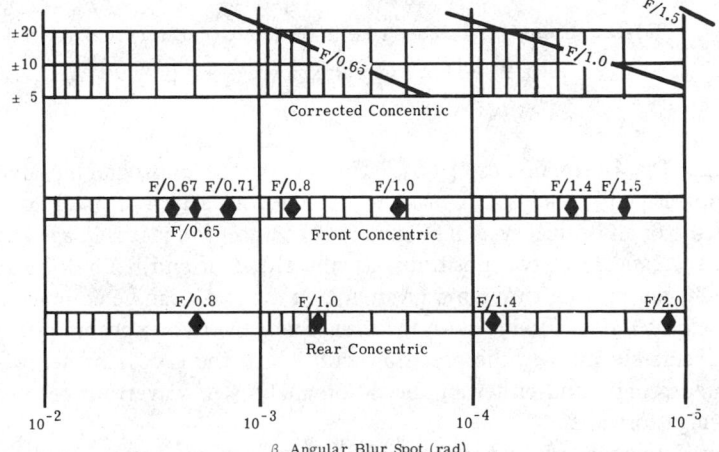

Fig. 9-38. Blur spot size charts for concentric meniscus corrector systems [9-1].

9.14. Optical Testing Techniques and Measures of Performance: Ray Path Tests

9.14.1. The Foucault or Knife Edge Test [9-8]. This test is widely used because the wavefront emerging from a system or mirror can be mapped with very simple equipment. As shown in Figure 9-39, an edge (usually a razor blade) is passed through the focus of a point image and the ray paths are inferred from the regions of the exit pupil successively blocked by the knife. When it passes exactly through a sharp focus, the exit pupil can be seen to darken uniformly; conversely, a particular zone of the pupil will darken when the knife cuts the focus for that zone. This information can be used to map the wavefront normals. The test data can be used to infer the shape of the surface of a mirror under construction, or as a relatively crude acceptance test.

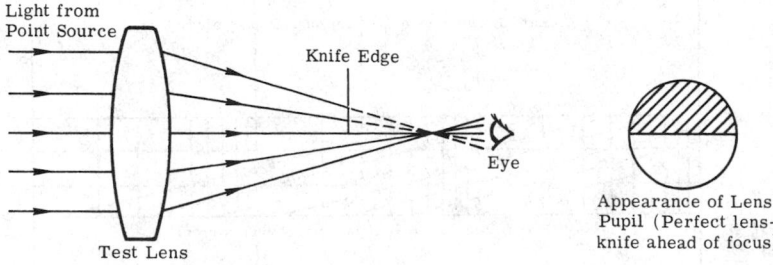

(*a*) On a perfect lens the knife shadow has a straight edge.

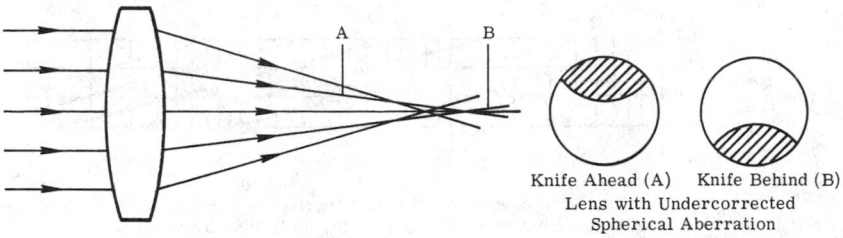

(*b*) The shadow has a curved edge in the presence of spherical aberration.

Fig. 9-39. The Foucault knife edge test [9-8, 9-1].

9.14.2. The Hartman Test [9-8]. This test is the photographic analogue of a ray-trace (spot diagram) analysis. A mask with a known pattern of small holes is placed over the aperture of an optical system (Figure 9-40). A point object is imaged on photographic material successively in two positions, usually ahead of and behind the focus. The path of the light ray passing through a given hole in the mask can be computed by connecting the two exposed dots produced on the film. Since the two exposure positions of the film can be accurately known, the precision with which the ray paths near the focus can be determined is high, and either an aberration analysis or wavefront reconstruction is relatively straightforward.

An optical bench can be used to perform tests analogous to the Hartman test. A mask with one or more small holes placed in front of a system defines a ray, and the traveling

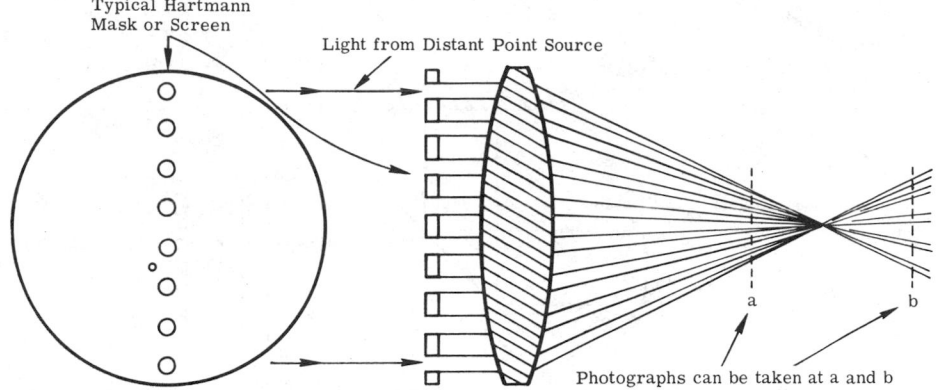

Fig. 9-40. The Hartmann test [9-8].

microscope of the lens bench is used to measure the position of the intersection of the ray with a selected focal plane.

9.14.3. The Ronchi Test. This test can be considered as a variation of the Foucault knife edge test. Instead of a single edge, a grating of relatively coarse structure is placed in the beam near the focus. Thus, the multiplicity of edges in the grating (which is often either a photographic or evaporated-metal pattern of alternating clear and opaque bands) produces shadow contours over the entire exit pupil of the system without the necessity of shifting the knife to scan across the aperture.

9.15. Optical Testing Techniques and Measures of Performance: Interferometric Tests

9.15.1. The Twyman-Green Interferometer [9-8, 9-9]. As shown in Figure 9-41(a), the Twyman-Green interferometer utilizes a collimated, monochromatic point source and a beam splitter to create two coherent wavefronts which are recombined by the beam splitter. The two wavefronts produce interference patterns by which any differences in the wavefronts can be measured. If a plate to be tested is inserted into one beam, the changes it produces in the interference pattern (as seen in the exit pupil of the system) can be interpreted and the optical characteristics of the test piece can be inferred. The Twyman-Green can be adapted to test prisms or lenses, as shown in Figure 9-41(b) and (c).

9.15.2. The Fizeau Interferometer [9-8]. The Fizeau Interferometer, shown in Figure 9-42, is extremely useful in testing the accuracy of polished surfaces, either flat or spherical, and it is widely used in optical shops. The standard configuration is used to test an unknown surface against a master flat or sphere. If one surface of a nominally plane parallel plate is tested against the other surface of the plate, the parallelism of the piece can be measured.

9.15.3. The Scatterplate Interferometer. This interferometer is based on a scatterplate (similar to a piece of ground glass) which has symmetry about its center. As indicated in Figure 9-43, when a ray of the illuminating cone is focussed on the center of the optical system under test, it is, at a given point on the plate (A of the figure), partly scattered and partly transmitted without scattering. The scattered beam fills the aperture of the test system, and the unscattered ray passes through the center of the system aperture without being affected. Both the scattered beam and the unscattered ray are

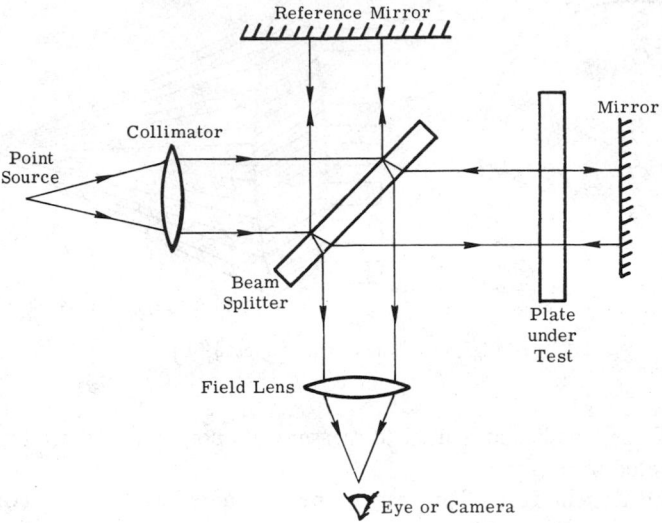

(a) Set-up to test plano plate.

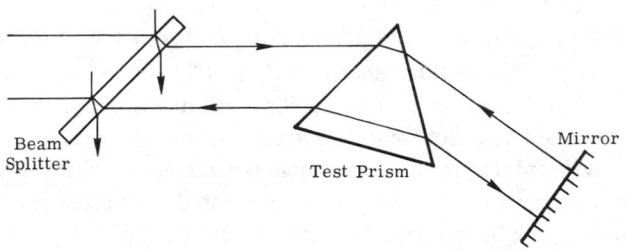

(b) Set-up to test deviating prism.

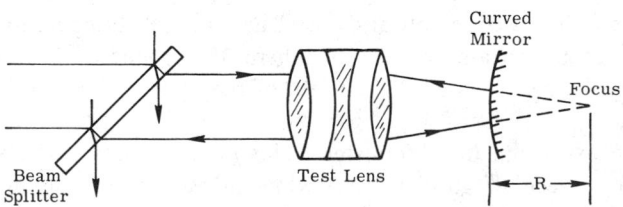

(c) Set-up to test objective lens.

Fig. 9-41. Twyman-Green interferometer.

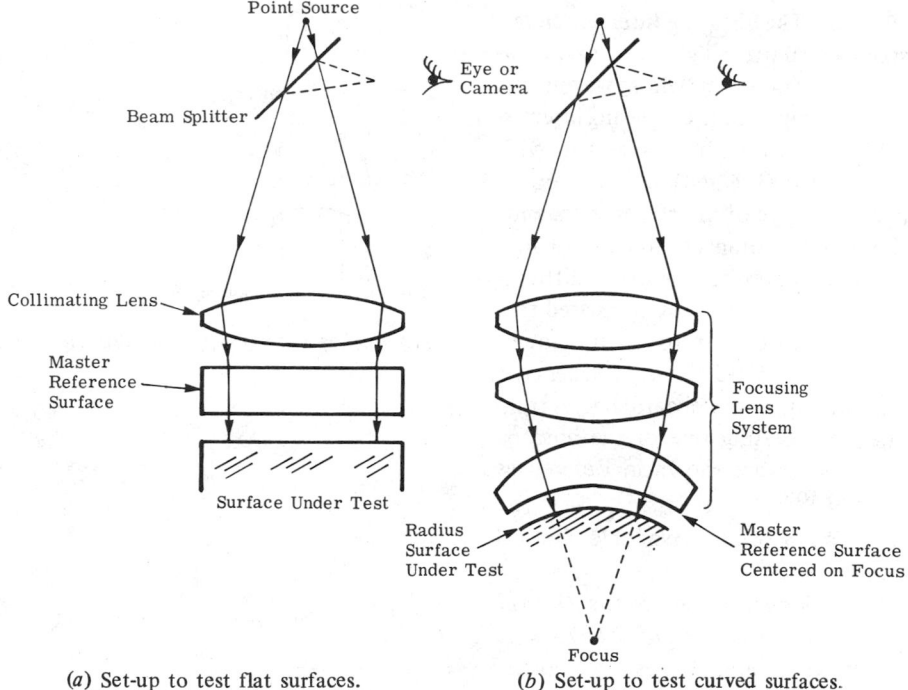

(*a*) Set-up to test flat surfaces.   (*b*) Set-up to test curved surfaces.

Fig. 9-42. Fizeau interferometer.

imaged back on the plate at the same point (B in the figure). Part of the scattered beam passes through un-scattered, and part of the unscattered ray will be scattered. The result is two wavefronts, one of which has passed through the full aperture and the other of which has passed through only the central portion of the aperture. They interfere and produce interference fringes. The scatterplate interferometer is simple and convenient, although

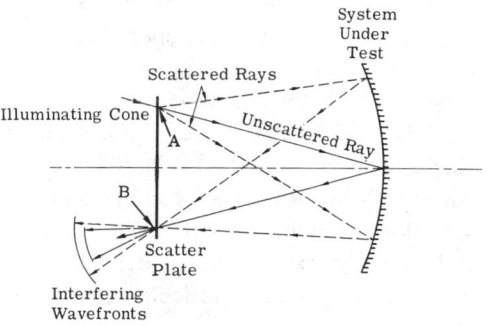

Fig. 9-43. The scatterplate interferometer.

optically inefficient, and can be applied to devices much larger than it is. A different arrangement is shown in Reference [9-8].

9.15.4. Laser Unequal Path Interferometer (LUPI) [9-10]. The extreme coherence of the laser has made possible the use of interferometers with greatly differing pathlengths. The Twyman-Green and Fizeau types discussed above have two beams with nearly equal paths and can operate using the relatively broad spectral lines of ordinary arc lamps. With a laser source, this equal-path limitation is eased, and considerably broader utility and greater convenience result. The schematic of Figure 9-44 shows one of the many possible ways of utilizing this advantage of laser coherence. Note that a very large system can be tested with a relatively small LUPI.

9.15.5. The Shearing Interferometer.

Instead of splitting light into two completely separate beams whose wavefronts are then compared, the shearing interferometer compares the wavefront with itself, shifted (sheared) laterally or radially. This type of interferometer tends to be simple, compact, and undemanding, but interpretation is often difficult since the wavefront is not compared to a known standard, but to an unknown: itself. Figure 9-45 shows a schematic device in which the shearing is accomplished by rotating one of the mirrors. Other schemes are shown in References [9-8] and [9-9].

9.16. Commercially Available Optical Systems*

Manufacturers listed in the *Optical Industry & Systems Directory 1976* were asked what infrared lenses, and optical systems they have commercially available. Almost all of these are "off the shelf." Other optical systems can be designed and manufactured to special order by most of these companies and by others listed in the Directory.

Table 9-2 contains descriptions of single element lenses. Table 9-3 and Figures 9-46 through 9-52 contain descriptions of multi-element lenses. Table 9-4 and Figures 9-53 through 9-56 include catadioptric and reflective (catoptric) systems.

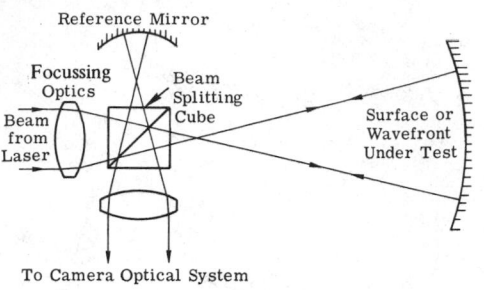

Fig. 9-44. Laser unequal path interferometer (LUPI).

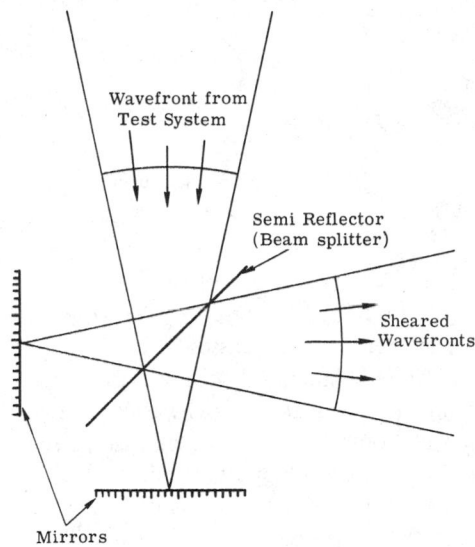

Fig. 9-45. Shearing interferometer.

---

*Contributed by Craig Mueller, Environmental Research Institute of Michigan, Ann Arbor, MI.

Table 9-2. Single-Element Lenses

Manufacturer	Model	Type	Wavelength Range, λ (μm)	Design Wavelength, λ (μm)	Diameter of Clear Aperture, D (mm)	Effective Focal Length, efl (mm)	Back Focal Length* (mm)	Focal Ratio f/#	Circle of Confusion (C. of C.) (mm)	Distance of C. of C. from Vertex, bfl (mm)	Lens Material
Barnes Engineering Co. 30 Commerce Rd. Stamford, CT 06904	SPBL-164 to SFBL-193	Biconvex	0.7- 1.3	1.0	12.5-100.0	25.00- 600.0	22.19- 595.57	f/2-f/6	0.010-0.800	—	SiO_2
	SASL-194 to SASL-217	Meniscus	2.0- 5.0	3.5	12.5-100.0	12.50- 400.0	9.76- 396.70	f/1-f/4	0.007-0.500	—	As_2S_3
	SASL-218 to SASL-241	Meniscus	4.0- 8.0	6.0	12.5-100.0	12.50- 400.0	10.60- 396.60	f/1-f/4	0.007-0.500	—	As_2S_3
	SASL-242 to SASL-265	Meniscus	6.0-10.0	8.0	12.5-100.0	17.50- 400.0	10.55- 396.55	f/1-f/4	0.007-0.500	—	As_2S_3
	SBFL-266 to SBFL-290	Biconvex	0.7- 3.1	1.9	12.5- 75.0	25.00- 450.0	22.25- 444.75	f/2-f/6	0.010-0.600	—	BaF_2
	SBFL-291 to SBFL-297	Biconvex	—	6.0	25.0- 50.0	25.00- 500.0	21.36- 498.53	f/1-f/10	0.010-0.415	—	BaF_2
	SBFL-298 to SBFL-301	Biconvex	—	10.6	25.0- 37.5	37.50- 150.0	34.08- 147.68	f/1.5-f/6	0.020-0.250	—	BaF_2
	SCFL-302 to SCFL-331	Biconvex	0.7- 2.3	1.5	12.5-100.0	25.00- 600.0	22.20- 595.45	f/2-f/6	0.010-0.800	—	CaF_2
	SKRL-332 to SKRL-343	Meniscus	8.0-15.0	11.5	12.5- 25.0	12.50- 100.0	10.55- 98.05	f/1-f/4	0.010-0.060	—	KRS-5 Thallium Bromide
	SLSI-344 to SLSI-367	Meniscus	2.0-11.0	8.0	12.5-100.0	12.50- 400.0	10.87- 397.50	f/1-f/4	0.005-0.480	—	Si
	SSGL-368 to SSGL-391	Meniscus	6.0-15.0	10.6	12.5-100.0	12.50- 400.0	10.87- 397.52	f/1-f/4	0.010-0.470	—	Ge
Eastman Kodak Co. Rochester, NY 14650	IR-100	Meniscus all-spherical	1.5-10.0	—	25.4	25.4	22.8 / 23.2	f/1.0	0.71 / 0.71	21.3* / 21.4	ZnS**
	IR-200	Meniscus all-spherical	1.5-10.0	—	50.8	50.8	45.7 / 48.6	f/1.0	1.52 / 1.40	42.1 / 45.0	ZnS**
	IR-101	Meniscus concave aspheric	3.0-10.0	—	25.4	25.4	21.8 / 22.9	f/1.0	0 / 0.28	21.8 / 22.1	ZnS**
	IR-201	Meniscus concave aspheric	4.26-10.0	—	50.8	50.8	42.6 / 44.5	f/1.0	0.025 / 0.025	42.6 / 44.5	ZnS**

*The first number is for the short wavelength end of the range; the second is for the long wavelength end.

**Other materials available upon request.

Table 9-2. Single-Element Lenses (*Continued*)

Manufacturer	Model	Type	Wavelength Range, λ (μm)	Design Wavelength, λ (μm)	Diameter of Clear Aperture, D (mm)	Effective Focal Length, efl (mm)	Back Focal Length* (mm)	Focal Ratio f/#	Circle of Confusion (C. of C.) (mm)	Distance of C. of C. from Vertex, bfl (mm)	Lens Material
Servo Corp. of America 111 New South Rd. Hicksville, NY 11802	IR-301	Meniscus concave aspheric	4.26-10.0	—	76.2	76.2	69.1 72.2	f/1.0	0.025 0.025	69.1 72.2	ZnS**
	1022.002	Convex-Convex	0.70- 1.5	1.1	—	101.6	93.55	f/4	0.280	—	Fused Quartz
	1022.003	Convex-Convex	0.70- 1.5	1.1	—	50.8	44.53	f/2	0.710	—	Fused Quartz
	1022.004	Convex-Convex	0.70- 1.5	1.1	—	101.6	89.48	f/2	1.400	—	Fused Quartz
	1022.008	Convex-Convex	0.70- 2.0	1.4	—	50.8	44.40	f/2	0.760	—	CaF$_2$
	1022.009	Convex-Convex	0.70- 2.0	1.4	—	101.6	88.67	f/2	1.520	—	CaF$_2$
	1022.010	Convex-Convex	0.70- 2.0	1.4	—	101.6	93.09	f/4	0.280	—	CaF$_2$
	1022.005	Convex-Convex	0.70- 3.0	1.8	—	50.8	44.68	f/2	0.710	—	BaF$_2$
	1022.006	Convex-Convex	0.70- 3.0	1.8	—	101.6	94.31	f/4	0.280	—	BaF$_2$
	1022.007	Convex-Convex	0.70- 3.0	1.8	—	101.6	89.81	f/2	1.400	—	BaF$_2$
	1024.044	Meniscus	1.00- 2.0	1.5	—	50.8	47.07	f/2	0.560	—	MgO
	1024.045	Meniscus	1.00- 2.0	1.5	—	101.6	98.48	f/4	0.080	—	MgO
	1024.024	Meniscus	2.00- 5.0	3.5	—	19.1	16.43	f/1	0.690	—	Servofrax®†
	1024.026	Meniscus	2.00- 5.0	3.5	—	38.1	35.00	f/3	0.130	—	Servofrax
	1024.012	Meniscus	2.00- 5.0	3.5	—	50.8	48.39	f/4	0.060	—	Servofrax
	1024.013	Meniscus	2.00- 5.0	3.5	—	50.8	48.39	f/2	0.230	—	Servofrax
	1360.563	Meniscus	2.00- 5.0	3.5	—	91.2	87.99	f/1.2	1.300	—	Servofrax
	1024.014	Meniscus	2.00- 5.0	3.5	—	101.6	91.06	f/4	0.130	—	Servofrax
	1024.019	Meniscus	2.00- 5.0	3.5	—	101.6	91.06	f/2	0.430	—	Servofrax
	1024.025	Meniscus	2.00- 5.0	3.5	—	139.7	129.39	f/1	4.830	—	Servofrax
	1024.015	Meniscus	2.00- 5.0	3.5	—	263.2	194.84	f/4	0.250	—	Servofrax
	1024.020	Meniscus	2.00- 5.0	3.5	—	263.2	194.84	f/2	0.940	—	Servofrax
	1360.567	Meniscus	2.00- 5.0	3.5	—	228.6	223.42	f/3	0.760	—	Servofrax
	1024.018	Meniscus	2.00- 5.0	3.5	—	363.0	352.78	f/1.9	1.680	—	Servofrax
	1024.022	Meniscus	2.00- 5.0	3.5	—	910.6	904.09	f/4.5	1.040	—	Servofrax
	1024.023	Meniscus	2.00- 5.0	3.5	—	910.6	909.09	f/3.6	1.420	—	Servofrax
	1024.038	Meniscus	2.00-11.0	6.5	—	50.8	48.92	f/2	0.150	—	Silicon
	1024.039	Meniscus	2.00-11.0	6.5	—	101.6	98.45	f/2	0.300	—	Silicon
	1024.040	Meniscus	2.00-11.0	6.5	—	101.6	99.03	f/4	0.130	—	Silicon
	1024.035	Meniscus	6.00-10.0	8.0	—	50.8	48.67	f/2	0.230	—	Servofrax

*The first number is for the short wavelength end of the range; the second is for the long wavelength end.
**Other materials available upon request.

Table 9-2. Single-Element Lenses (Continued)

Manufacturer	Model	Type	Wavelength Range, λ (μm)	Design Wavelength, λ (μm)	Diameter of Clear Aperture, D (mm)	Effective Focal Length, efl (mm)	Back Focal Length (mm)	Focal Ratio f/#	Circle of Confusion (C. of C.) (mm)	Distance of C. of C. from Vertex, bfl (mm)	Lens Material
	1024.036	Meniscus	6.00-10.0	8.0	—	101.6	97.94	f/2	0.480	—	Servofrax
	1024.037	Meniscus	6.00-10.0	8.0	—	101.6	98.70	f/4	0.150	—	Servofrax
	1024.041	Meniscus	6.00-16.0	11.0	—	50.8	48.92	f/2	0.080	—	Ge
	1024.042	Meniscus	6.00-16.0	11.0	—	101.6	98.55	f/2	0.160	—	Ge
	1024.043	Meniscus	6.00-16.0	11.0	—	101.6	99.09	f/4	0.030	—	Ge
	1020.001	Planoconvex	—	5.0	—	10.16	—	f/2	—	—	Servofrax
	1020.002	Planoconvex	—	5.0	—	76.2	—	f/.8	—	—	Servofrax
	1020.003	Planoconvex	—	5.0	—	50.8	—	f/2	—	—	Servofrax
	1020.004	Planoconvex	—	5.0	—	38.1	—	f/.75	—	—	Servofrax
	1020.005	Planoconvex	—	5.0	—	19.05	—	f/.8	—	—	Servofrax
	1020.006	Planoconvex	—	4.0	—	14.48	—	f/.65	—	—	Servofrax
	1022.031	Equiconvex	—	4.0	—	13.97	—	f/.63	—	—	Servofrax
	1360.569	Equiconvex	—	6.0	—	50.80	—	f/1	—	—	Servofrax
	1360.570	Equiconvex	—	6.0	—	101.60	—	f/1.3	—	—	Servofrax
Unique Optical Co. P.O. Box 585 Farmingdale, NY 11735	-050	Meniscus 1.1" Dia.	—	10.6	—	25.40- 762.00	—	—	—	—	Ge
	-030	Meniscus 1.1" Dia.	—	10.6	—	25.40- 762.00	—	—	—	—	ZnSe
	-010	Meniscus 1.1" Dia.	—	10.6	—	25.40- 762.00	—	—	—	—	GaAs
	-050W	Negative Meniscus	—	10.6	12.5- 50.0	12.50- 200.00	13.59- 201.53	f/1-f/4	—	—	Ge
	-060W	Negative Biconcave	—	10.6	250.0- 50.0	50.00- 400.00	51.12- 401.66	f/2-f/8	—	—	BaF2
	-040W	Negative Biconcave	—	1.5	25.0	50.00- 400.00	51.43- 402.82	f/2-f/8	—	—	CaF2
	-030	Meniscus	—	10.6	12.5- 50.0	12.50-1000.0	11.00- 998.34	f/1-f/20	—	—	ZnSe
	-050	Meniscus	—	10.6	12.5-100.0	12.50-2000.0	10.87-1996.15	f/1-f/20	—	—	Ge
	-053	Meniscus	—	3.5	12.5-2000.0	12.50-2000.0	11.37-1997.12	f/1-f/20	—	—	Ge
	-010	Meniscus	—	10.6	12.5- 50.0	12.50-1000.0	11.10- 997.59	f/1-f/20	—	—	GaAs
	-176	Meniscus	—	11.5	12.5- 50.0	12.50-1000.0	10.55- 998.17	f/1-f/20	—	—	KRS-5 Thallium Bromoiodide

Table 9-2. Single-Element Lenses (Continued)

Manufacturer	Model	Type	Wavelength Range, λ (μm)	Design Wavelength, λ (μm)	Diameter of Clear Aperture, D (mm)	Effective Focal Length, efl (mm)	Back Focal Length (mm)	Focal Ratio f/#	Circle of Confusion (C. of C.) (mm)	Distance of C. of C. from Vertex, bfl (mm)	Lens Material
	-090	Meniscus	—	8.0	12.5-100.0	12.50-2000.0	10.87-1992.61	f/1-f/20	—	—	Si
	-094	Meniscus	—	4.3	12.5-100.0	12.50-2000.0	11.20-1994.86	f/1-f/20	—	—	Si
	-060	Biconvex	—	10.6	12.5- 50.0	12.50-1000.0	10.40- 998.47	f/1-f/20	—	—	BaF$_2$
	-060	Biconvex	—	6.0	12.5- 50.0	12.50-1000.0	9.58- 998.47	f/1-f/20	—	—	BaF$_2$
	-060	Biconvex	—	1.9	12.5- 50.0	12.50-1000.0	9.50- 996.90	f/1-f/20	—	—	BaF$_2$
	-198	Meniscus	—	8.0	12.5-100.0	12.50-2000.0	16.55-1993.35	f/1-f/20	—	—	As$_2$S$_3$
	-196	Meniscus	—	6.0	12.5-100.0	12.50-2000.0	10.60-1993.34	f/1-f/20	—	—	As$_2$S$_3$
	-190	Meniscus	—	3.5	12.5-100.0	12.50-2000.0	9.76-1993.34	f/1-f/20	—	—	As$_2$S$_3$
	-040	Biconvex	—	1.5	12.5- 75.0	12.50-1500.0	10.38-1497.07	f/1-f/20	—	—	CaF$_2$
	-000	Biconvex	—	1.0	12.5- 75.0	12.50-1500.0	10.61-1496.32	f/1-f/20	—	—	SiO$_2$

Table 9-3. Multi-Element Lens Systems

Manufacturer	Model	Wavelength Range (μm)	Design Wavelength (μm)	Diameter of Clear Aperture (mm)	Effective Focal Length (mm)	Back Focal Length (mm)	Focal Ratio, f/#	Field of View (deg)	Resolution (mrad)	Lens Material	Number of Elements	Figure Number
Barnes Engineering Co.	SROS-392	1.0- 5.0	3.0	10.0	10.5	3.15	f/1.04	2	2.3	BaF$_2$-BaF$_2$-LiF$_2$	3	9-47
30 Commerce Rd.	SROS-393	1.5- 5.5	3.5	12.5	12.5	9.15	f/1	2	0.16	Ge-Ge	2	9-48
Stamford, CT 06904	SROS-394	1.5- 5.5	3.5	12.5	25.0	18.50	f/2	2	DL*	Ge-Ge	2	9-48
[9-10]	SROS-395	1.5- 5.5	3.5	12.5	31.3	28.70	f/2.5	2	DL	Ge-Ge	2	9-48
	SROS-396	1.5- 5.5	3.5	12.5	37.5	36.78	f/3	2	DL	Ge-Ge	2	9-48
	SROS-397	1.5- 5.5	3.5	12.5	50.0	47.12	f/4	2	DL	Ge-Ge	2	9-48
	SROS-398	8.0-14.0	10.6	12.5	12.5	9.27	f/1	2	DL	Ge-Ge	2	9-48
	SROS-399	8.0-14.0	10.6	12.5	25.0	18.82	f/2	2	DL	Ge-Ge	2	9-48
	SROS-400	8.0-14.0	10.6	12.5	31.3	28.95	f/2.5	2	DL	Ge-Ge	2	9-48

*Diffraction limited.

Table 9-3. Multi-Element Lens Systems (Continued)

Manufacturer	Model	Wavelength Range (μm)	Design Wavelength (μm)	Diameter of Clear Aperture (mm)	Effective Focal Length (mm)	Back Focal Length (mm)	Focal Ratio, f/#	Field of View (deg)	Resolution (mrad)	Lens Material	Number of Elements	Figure Number
	SROS-401	8.0-14.0	10.6	12.5	37.5	37.10	f/3	2	DL*	Ge-Ge	2	9-48
	SROS-402	8.0-14.0	10.6	12.5	50.0	47.35	f/4	2	DL	Ge-Ge	2	9-48
	SROS-403	8.0-14.0	10.6	18.7	18.7	15.59	f/1	2	DL	Ge-Ge	2	9-48
	SROS-404	3.0- 7.0	5.0	18.7	18.7	15.50	f/1	6	2.24	Ge-Ge	2	9-48
	SROS-405	10.0-14.0	12.0	32.5	109.0	113.00	f/3.3	12	0.91	Ge-Ge-Ge	3	9-47
	SROS-406	3.5- 5.5	4.5	50.0	100.4	76.13	f/2	2	0.25	As₂S₃-CaF₂	3	9-48
	SROS-407	1.5- 3.5	2.5	81.5	103.8	79.35	f/1.3	4	0.6	As₂S₃-As₂S₃-CaF₂	3	9-46
	SROS-408	8.0-14.0	10.6	100.0	120.0	104.95	f/1.2	6	2.5	Ge-Ge	2	9-48
	SROS-409	8.0-14.0	10.6	100.0	300.0	226.10	f/3	2	DL	Ge-Ge	2	9-48
	SROS-410	3.5- 5.5	4.5	100.0	120.0	105.20	f/1.2	6	2.51	Si-Ge	3	9-48
	SROS-411	6.0-14.0	10.6	37.0	25.0	14.70	f/0.7	7	DL	Ge-Ge	3	9-47
Servo Corp. of America 111 New South Rd. Hicksville, NY 11802 [9-11]	1025.029	1.2- 4.0	2.5	63.5	127.0	—	f/2	6	0.40-0.80	BaF₂-37A-Glass	3	9-47
	1025.024	2.0- 5.0	3.5	42.42	50.8	—	f/1.2	6	2.00-5.72	Servofrax-CaF₂	2	9-48
	1025.025	2.0- 5.0	3.5	84.84	101.6	—	f/1.2	6	2.00-5.72	Servofrax-CaF₂	2	9-48
	1025.027	2.0- 5.0	3.5	101.6	203.2	—	f/2.0	6	0.38-0.90	Servofrax-CaF₂	2	9-48
	1025.050	3.5- 5.5	4.5	100.58	120.65	—	f/1.2	6	0.55-2.52	Si-Ge	2	9-48
	1025.033	6.0-16.0	11.0	34.04	101.6	—	f/3	9	DL	Ge-Ge	2	9-49
	1025.034	6.0-16.0	11.0	37.34	55.88	—	f/1.5	—	DL	Ge-Ge-Ge-Ge	4	9-49
	1025.110	3.8- 5.5	4.7	114.30	114.30	—	f/1	2.3	0.3-0.7	Si-Ge-Si	3	9-47
	1025.060	2.0- 4.0	—	34.80	226.26	116.84	f/6.5	25	0.3	Si-Ge-Si	3	9-47
Unique Optical Co. P.O. Box 585 Farmingdale, NY 11735 [9-12]	BL.4101-5	1.0- 5.0	3.0	12.5	10.50	3.15	f/1.04	2	2.3	BaF₂-BaF₂-LiF₂	3	9-47
	G.5101-5	1.5- 5.5	3.5	12.5	12.50	9.15	f/1.0	2	0.61	Ge-Ge	2	9-48
	G.5201-5	1.5- 5.5	3.5	12.5	25.00	18.50	f/2.0	2	DL	Ge-Ge	2	9-48
	G.5251-5	1.5- 5.5	3.5	12.5	31.30	28.70	f/2.5	2	DL	Ge-Ge	2	9-48
	G.5301-5	1.5- 5.5	3.5	12.5	37.50	36.78	f/3.0	2	DL	Ge-Ge	2	9-48
	G.5401-5	1.5- 5.5	3.5	12.5	50.00	47.12	f/4.0	2	DL	Ge-Ge	2	9-48
	G.5102-5	2.5- 5.5	3.5	12.5	13.86	9.85	f/1.0	2	0.40	Ge-Ge-Ge	3	9-50
	G.5108-14	8.0-14.0	10.6	12.5	12.50	9.27	f/1.0	2	DL	Ge-Ge	2	9-48
	G.5208-14	8.0-14.0	10.6	12.5	25.00	18.82	f/2.0	2	DL	Ge-Ge	2	9-48

*Diffraction limited

Table 9-3. Multi-Element Lens Systems (Continued)

Manufacturer	Model	Wavelength Range (μm)	Design Wavelength (μm)	Diameter of Clear Aperture (mm)	Effective Focal Length (mm)	Back Focal Length (mm)	Focal Ratio, f/#	Field of View (deg)	Resolution (mrad)	Lens Material	Number of Elements	Figure Number
	G.5258-14	8.0-14.0	10.6	12.5	31.30	28.95	f/2.5	2	DL*	Ge-Ge	2	9-48
	G.5308-14	8.0-14.0	10.6	12.5	37.50	37.10	f/3.0	2	DL	Ge-Ge	2	9-48
	G.5408-14	8.0-14.0	10.6	12.5	50.00	47.35	f/4.0	2	DL	Ge-Ge	2	9-48
	G.7103-7	3.0- 7.0	5.0	18.7	18.70	15.50	f/1.0	6	2.24	Ge-Ge	2	9-48
	SG.7103-5	3.5- 5.5	5.0	18.7	22.42	16.33	f/1.0	6	4.4	Si-Ge	2	9-48
	G.7108-14	8.0-14.0	10.6	18.7	18.70	15.59	f/1.0	2	DL	Ge-Ge	2	9-48
	G.25108-14	8.0-14.0	10.6	25.0	25.00	20.00	f/1.0	4	1.5	Ge-Ge	2	9-48
	AC.30102-5	2.0- 5.0	3.0	30.0	312.00	301.83	f/10.0	11	0.83	As$_2$S$_3$-CaF$_2$	2	9-51
	AC.30352-5	2.0- 5.0	3.5	30.5	109.00	98.81	f/3.5	12	2.3	As$_2$S$_3$-CaF$_2$	2	9-51
	G.323310-14	10.0-14.0	12.0	32.5	109.00	113.00	f/3.3	12	0.91	Ge-Ge-Ge	3	9-50
	G.33338-14	8.0-14.0	10.6	33.0	105.70	81.75	f/3.3	12	3.8	Ge-Ge	2	9-48
	G.37706-14	6.0-14.0	10.6	37.0	25.00	14.70	f/0.7	7	DL	Ge-Ge-Ge	3	9-50
	C.39293-4	3.5- 4.0	3.8	39.37	121.92	113.07	f/2.9	2	1.3	CaF$_2$-CaF$_2$	2	9-52
	SG.48183-5	3.5- 5.0	4.5	48.77	87.78	75.00	f/1.8	2	DL	Si-Ge	2	9-48
	AC.50203-5	3.5- 5.5	4.5	50.00	100.40	76.13	f/2.0	2	0.25	As$_2$S$_3$-CaF$_2$	2	9-51
	SG.75122-5	2.5- 5.5	2.7	75.00	91.44	72.19	f/1.2	6	2.5	Si-Ge	2	9-48
	C.75331-4	0.9- 4.0	2.5	75.00	250.00	229.60	f/3.3	3.4	0.55	CaF$_2$-CaF$_2$	2	9-52
	AC.81131-3	1.5- 3.5	2.5	81.50	103.80	79.35	f/1.3	4	0.6	As$_2$S$_3$-As$_2$S$_3$	3	9-46
	SG.100123-5	3.5- 5.5	4.5	100.0	120.00	105.20	f/1.2	6	2.51	Si-Ge	2	9-48
	G.100128-14	8.0-14.0	10.6	100.0	120.00	104.95	f/1.2	6	2.5	Ge-Ge	2	9-48
	G.100308-14	8.0-14.0	10.6	100.0	300.00	226.10	f/3.0	2	DL	Ge-Ge	2	9-48
Zoomar, Inc. 55 Sea Cliff Ave. Glen Cove, L.I., NY 11542	IR50	2-5 or 8-14	—	—	50.00	34.50	f/0.8	20	—	All Ge	—	—
	IR80	2-5 or 8-14	—	—	80.00	34.00	f/0.8	12.83	—	All Ge	—	—
	IR100	2-5 or 8-14	—	—	100.00	64.00	f/0.7	10	—	All Ge	—	—
	IR HS 100	2-5 or 8-14	—	—	100.00	61.60	f/0.7	10	—	All Ge	—	—
	IR140	2-5 or 8-14	—	—	140.00	101.00	f/0.7	7.83	—	All Ge	—	—
	IR270	2-5 or 8-19	—	—	270.00	182.40	f/1.35	3.83	—	All Ge	—	—
	IR2X30	2-5 or 8-14	—	—	30- 60	33.00	f/0.8	33-16.5	—	All Ge	—	—
	IR2X50	2-5 or 8-14	—	—	50-100	66.00	f/1	20-10	—	All Ge	—	—

*Diffraction limited.

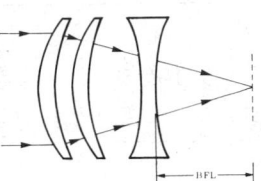

Fig. 9-46. System
configuration [9-11].

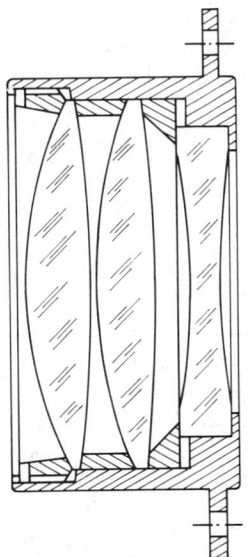

Fig. 9-47. Refractive
optical system [9-12].

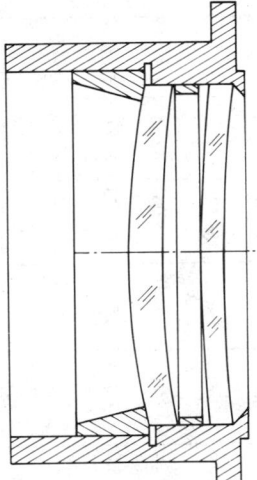

Fig. 9-48. Refractive
optical system [9-12].

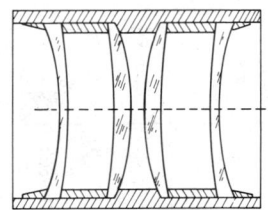

Fig. 9-49. Refractive
optical system [9-12].

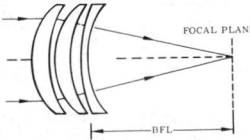

Fig. 9-50. Lens system
[9-13].

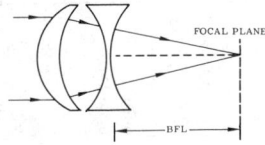

Fig. 9-51. Lens system
[9-13].

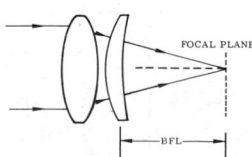

Fig. 9-52. Lens system
[9-13].

Table 9-4. Catadioptric and Reflective Systems

Manufacturer	Model	Optical System	Wavelength Range (μm)	Design Wavelength (μm)	Diameter of Clear Aperture (mm)	Effective Focal Length (mm)	Back Focal Length (mm)	Focal Ratio, f/#	Primary Diameter (mm)	Field of View (deg)	Resolution (mrad)	Type	Figure Number
Barnes Engineering Co. 30 Commerce Rd. Stamford, CT 06904	SRXS-412	–	16.0- 22.0	19.0	75.0	225.00	–	f/3	–	8	0.55	KRS-5 Mirror-Mirror	9-53
Nye Optical Co. 8781 Troy St. Spring Valley, CA 92077	Flectan-I Lens	–	0.6-500.0	–	–	200.00	–	f/2.8	–	12	60-25 line pair mm^{-1}	Reflective	–
Servo Corp. of America 111 New South Rd. Hicksville, NY 11802	1025.013	–	0.4- 2.0	1.2	–	102.87	–	f/1.6	–	0-6	0.88-2.61	Quartz corrector	9-54
	1025.032	–	1.0- 4.0	2.5	–	243.84	–	f/1.6	–	0-4	1.5	Servofrax corrector	9-55
	1025.031	–	0.3- 0.6	0.45	–	256.54	–	f/3.3	–	0-6	0.15	Quartz corrector	9-56
Unique Optical Co. P.O. Box 585 Farmingdale, NY 11735	T75R316-22	Bouwers-Maksutov	16.0- 22.0	19.0	75.0	225.00	42.35	f/3	–	8	0.55	KRS-5 corrector	9-53
	T151.635-5	Bouwers Maksutov	3.5- 5.0	4.0	150.0	254.32	109.66	f/1.6	–	2.5	0.65	As$_2$S$_3$ corrector	9-53
	R6334.53-0	–	0.5- 30.0	–	63.0	2200.00	22.80	f/34.0	–	0.06	DL*	Mirror-Mirror	–
Zoomar, Inc. 55 Sea Cliff Ave. Glen Cove, L.I., NY 11542	IR CAT. 250	–	2-5 or 8-14	–	–	250.0	–	f/1	265	4	–	Ge corrector	–
	IR CAT. 2X125	–	2-5 or 8-14	–	–	125-250	–	f/1	265	8-4	–	Ge corrector	–

*Diffraction Limited

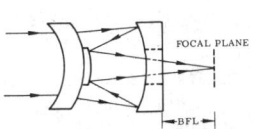

Fig. 9-53. Catadioptric (Bowers-Maksutov) optical system [9-11].

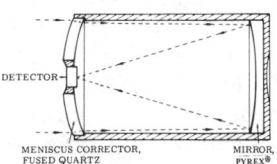

Fig. 9-54. Catadioptric optical system. Pyrex is a registered trademark of Dow Corning [9-12].

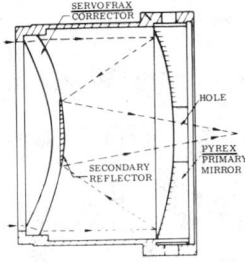

Fig. 9-55. Catadioptric optical system [9-12].

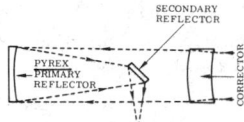

Fig. 9-56. Catadioptric optical system [9-12].

9.17. References and Bibliography

9.17.1. References

[9-1] Warren J. Smith, *Modern Optical Engineering: The Design of Optical Systems*, McGraw-Hill, New York, NY, 1966, pp. 55, 70, 205, 213, 220, 223-236, 254, 301, 319, 320, 322, 395-397, 402-404, 439.

[9-2] H. Chrétien, *Revue d'Optique*, Paris, France, Vol. 1, 1922, pp. 13-22 and 49-64 and G. W. Ritchey, and H. Chrétien, *Comptes Rendus*, Institute Geologic, Bucharest, Romania, Vol. 185, 1927, p. 266.

[9-3] S. C. B. Gascoigne, "Some Recent Advances in the Optics of Large Telescopes," *Quarterly Journal of the Royal Astronomical Society*, Royal Astronomical Society, Oxford, England, Vol. 9, 1968.

[9-4] J. G. Baker, "On Improving the Effectiveness of Large Telescopes," *IEEE Transactions on Aerospace and Electronic Systems*, The Institute of Electrical and Electronics Engineers, New York, NY, March 1969, pp. 261-272.

[9-5] R. A. Buchroeder, "Application of Aspherics for Weight Reduction in Selected Data," Design Examples of Tilted Component Telescopes, Optical Sciences Center, University of Arizona, Tucson, AZ, Technical Report 68, 1971.

[9-6] B. Schmidt, *Mitthandlungen Hamburg Sternwalte Bergedorf*, Vol. 7, 1932, p. 15.

[9-7] F. E. Ross, "Lens Systems for Correcting Coma of Mirrors," *Mount Wilson Observatory Yearbook*, Vol. 35, 1936, p. 191, and *Astrophysical Journal*, American Astronomical Society, University of Chicago, Chicago, IL, Vol. 81, 1935, p. 156.

[9-8]   J. L. Strong, *Concepts of Classical Optics*, W. H. Freeman and Company, San Francisco, CA, 1958, pp. 295-298, 354, 355, 382, 227-230, 393-399, 383-384.

[9-9]   W. H. Steel, *Interferometry*, Cambridge University Press, New York, NY, 1967, pp. 162-165, 168-182.

[9-10]   J. B. Houston, et al., "A Laser Unequal Path Interferometer for the Optical Shop," *Applied Optics*, Optical Society of America, Washington, DC, Vol. 6, 1967, pp. 1237-1242.

[9-11]   Barnes Engineering Staff, "Standard Optical Components," Barnes Engineering Company, Stamford, CT, Bulletin 6-050, 1976.

[9-12]   Servo Tech Data, Infrared/Electro-Optical Division, Servo Corporation of America, New York, NY, 1976, pp. 3, 4.

[9-13]   "Unique Optical Company," Unique Optical Company, Farmingdale, NY, December 1975, pp. 10-11.

9.17.2. Bibliography

Amon, M., and S. Rosin, "Mangin Mirror Systems," *Applied Optics*, Optical Society of America, Washington, DC, Vol. 7, 1968, p. 214.

Amon, M., and S. Rosin, "Color-Corrected Mangin Mirror," *Applied Optics*, Optical Society of America, Washington, DC, Vol. 6, 1967, p. 963.

Baker, J. G. "On Improving the Effectiveness of Large Telescopes," *IEEE Transactions on Aerospace and Electronic Systems*, The Institute of Electrical and Electronics Engineers, New York, NY, March 1969, pp. 261-272.

Barnes Engineering Staff "Standard Optical Components," Barnes Engineering Company, Stamford, CT, Bulletin 6-050, 1976.

Born, Max, and Emil Wolf, *Principles of Optics*, Macmillan, New York, NY, 1964.

Bouwers, A., *Achievements in Optics*, Elsevier, New York, NY, 1946.

Buchroeder, R. A., "Application of Aspherics for Weight Reduction in Selected Data," Design Examples of Tilted Component Telescopes, Optical Sciences Center, University of Arizona, Tucson, AZ, Technical Report 68, 1971.

Chrétien, H., *Revue d' Optique*, Paris France, Vol. 1, 1922, pp. 13-22.

Coltman, John W., "The Specification of Imaging Properties by Response to a Sine Wave Input," *Journal of the Optical Society of America*, Optical Society of America, Washington, DC, Vol. 44, 1954, p. 234.

Conrady, A. E., *Applied Optics and Optical Design*, Dover Publishing, New York, NY, Two Volumes, 1957 and 1960.

Cox, A., *A System of Optical Design*, Focal Press, New York, NY, 1964.

Gascoigne, S. C. B., "Some Recent Advances in the Optics of Larger Telescopes," *Quarterly Journal of Royal Astronomical Society*, Royal Astronomical Society, Oxford, England, Vol. 9, 1968.

Handbook of Optical Design, U.S. Government Printing Office, Washington, DC, MIL-HDBK-141, 1962.

Hardy, A., and F. Perrin, *The Principles of Optics*, McGraw-Hill, New York, NY, 1932.

Hertzberger, M., *Modern Geometrical Optics*, Interscience, New York, NY, 1958.

Houston, J. B., et al, "A Laser Unequal Path Interferometer for the Optical Shop," *Applied Optics*, Optical Society of America, Washington, DC, Vol. 6, 1967, pp. 1237-1242.

Kelly, D. H., "Spatial Frequency, Bandwidth and Resolution," *Applied Optics*, Optical Society of America, Washington, DC, Vol. 4, 1965, pp. 435-438.

Kingslake, Rudolph, *Applied Optics and Optical Engineering*, Academic Press, New York, NY, Five Volumes, 1965 to 1969.

Lauroesch, T. J., and I. C. Wing, "Bouwers Concentric Systems for Materials of High Refractive Index," *Journal of the Optical Society of America*, Optical Society of America, Washington, DC, Vol. 49, 1959, pp. 410-411.

Linfoot, E. H., *Recent Advances in Optics*, Clarendon Press, London, England, 1955.

Maksutov, D. D., "New Catadioptric Meniscus Systems," *Journal of the Optical Society of America*, Optical Society of America, Washington, DC, Vol. 34, 1944, pp. 270-284.

Martin, Louis C., *Technical Optics*, Pitman, London, England, Two Volumes, 1960.

Perrin, F. H., "Methods of Appraising Photographic Systems," *Journal of the Society of Motion Picture and Television Engineers*, New York, NY, Vol. 69, 1960, pp. 151-156.

Potter, R. J., "Transmission Properties of Optical Fibers," *Journal of the Optical Society of America*, Optical Society of America, Washington, DC, Vol. 51, 1961, pp. 1079-1089.

Ritchey, G. W., and H. Chrétien, *Compes Rendus*, Bucharest, Romania, Vol. 185, 1927, p. 266.

Ross, F. E., "Lens Systems for Correcting Coma of Mirrors," *Mount Wilson Observatory Yearbook*, Vol. 35, 1936, p. 191, and *Astrophysical Journal*, American Astronomical Society, University of Chicago Press, Chicago, IL, Vol. 81, 1935, p. 156.

Schmidt, B., *Mitthandlungen Hamburg Sternwalte Bergedorf*, Vol. 7, 1932.

Servo Tech Data, Infrared/Electro-Optical Division, Servo Corporation of America, New York, NY, 1976.

Smith, W., *Modern Optical Engineering: the Design of Optical Systems*, McGraw-Hill, New York, NY, 1966, pp. 55, 70, 205, 213, 220, 233-236, 254, 301, 319, 320, 322, 395-397, 402-404, 439.

Southall, J. C., *Mirrors, Prisms, and Lenses*, Dover Publishing, New York, NY, 1964.

Steel, W. H., *Interferometry*, Cambridge University Press, New York, NY, 1967.

Strong, J. L., *Concepts of Classical Optics*, W. H. Freeman and Company, San Francisco, CA, 1958, pp. 227-230, 295-298, 354, 355, 382, 383-384, 393-399.

"Unique Optical Company," Unique Optical Company, Farmingdale, NY, December 1975, pp. 10-11.

Williamson, D. E., "Cone-Channel Condenser Optics," *Journal of the Optical Society of America*, Optical Society of America, Washington, DC, Vol. 42, 1952, pp. 712-715.

Chapter 10

OPTICAL-MECHANICAL SCANNING
TECHNIQUES AND DEVICES

William L. Wolfe
University of Arizona

CONTENTS

10. Optical-Mechanical Scanning Techniques and Devices

10.1. Introduction

Point and array detectors are used to make an image or sample a field of view. Accordingly, optical-mechanical methods are often used to scan or move the instantaneous field of view (IFOV) of the detector over the total field of view. Scanning can be done with single or point detectors and arrays in either image space or object space. With arrays, the rotation or other deformation of the projection of the IFOV on the scene surface is more important than with single detectors.

10.1.1. Symbols, Nomenclature, and Units. Table 10-1 lists the symbols, nomenclature, and units used in this chapter.

Table 10-1. Symbols, Nomenclature and Units

Symbols	Nomenclature	Units
A	Area	m^2
A_T	Target area	m^2
B	Bandwidth	Hz
D	Entrance-pupil diameter	m
D_s	Distance from sensor	m
D_1, D_2	Entrance-pupil diameter of lens 1 and lens 2	m
d	Thickness of scanning plate	m
d_E	Distance to Earth's surface	m
d_i	Designators for detectors in an array $i = 1, 2, 3, \ldots$	–
F	Focus distance	m
fr	Fractional part of scan ratio	–
f_r	Rotational frequency	rad sec^{-1}
h	Altitude or perpendicular range of the scanner from the target	m
int	Integer part of scan ratio	–
k_1, k_2, k_s	Light wave vectors	–
l	Length of side of scanning cube	m
l_t, l_s	Linear measure of instantaneous field of view at the target: tangential and perpendicular to line of sight, respectively	m
m	Number of detectors in an array	–
N	Number of resolution elements in a frame	–
N_r	Number of resolution elements in a scan line	–
N_l	Number of scan lines in a frame	–
n	Number of sides of scanner of element; number of faces in a polygon scanner; scan ratio of prism frequencies	–

Table 10-1.   Symbols, Nomenclature and Units (*Continued*)

Symbols	Nomenclature	Units
R	Range to target	m
R_E	Radius of the earth	m
r	Radius of circumscribed circle of polygon scanner	m
r_o	Overlap ratio	—
T_f	Frame time	sec
t	Time	sec
t_c	Time to scan a circle	sec
t_d	Dwell time	sec
$t_{d,\min}$	Minimum dwell time	sec
t_{dd}	Dead time	sec
t_l	Line time	sec
v	Scan velocity, or vehicle velocity	rad sec^{-1} or m sec^{-1}
v, w	Length of prism sides	m
α, β	Offset angles from some line of sight	rad
ΔF	Change in focal length	m
ΔF_x	Change for x direction	m
ΔF_y	Change for y direction	m
δ_i	Prism deviation magnitude = $\sqrt{\delta_x^2 + \delta_y^2}$ for prism i	rad
δ_x	Prism deviation magnitude in the x-direction	rad
δ_y	Prism deviation magnitude in the y-direction	rad
η_s	Scan efficiency	—
θ, ϕ	Linear angles of the instantaneous field of view (Section 10.2)	rad
θ_t, ϕ_t	Linear angles of the total field of view	rad
κ	Ratio of prism deviation magnitudes	—
ξ	Rotation angle of a scanner element	rad
ψ	Phase relation	—
$\omega_1, \omega_2, \omega_s$	Light frequencies	rad sec^{-1}

10.2.   Scanning Parameters and Signal-to-Noise Ratio

In a general scanning-system, the IFOV has angular dimensions θ and ϕ and points in the direction (α, β). (See Figure 10-1.) The total field angle is θ_t in one direction and ϕ_t in the other direction. The perpendicular distance from the scanner to the object at the nadir point is usually designated h and the distance to any particular resolution element is R. The linear dimensions of a resolution element are given by l_s and l_t.

The signal-to-noise ratio (SNR) of a scanning infrared system can be written in the simplified form

$$\text{SNR} = D^* \Phi_d A_d^{-1/2} B^{-1/2} \tag{10-1}$$

where D^* = specific detectivity (See Chapter 11, "Detectors.")

$\quad\quad \Phi_d$ = power or flux on the detector

$\quad\quad A_d$ = effective area of the detector

$\quad\quad B$ = noise bandwidth

The noise bandwidth is directly related to the information bandwidth and inversely proportional to the dwell time on a single resolution element. The electrical information-bandwidth of an infrared system should be proportional to the reciprocal of the shortest dwell time on any resolution element in the total field of view. If scanning velocity is constant over the total field and only the field is covered (i.e., there is no overlap and no dead time in retrace or at the edges), then the dwell time, t_d, for a frame is

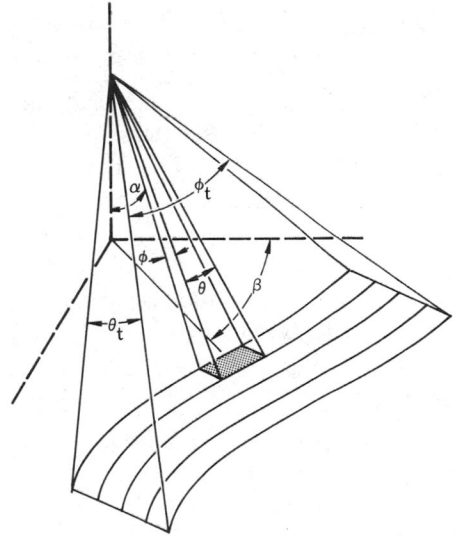

Fig. 10-1. A general scan pattern.

$$t_d = \frac{T_f}{N} \tag{10-2}$$

where N = number of resolution elements

$\quad\quad T_f$ = frame time

10.2.1. Dead Time Relationships. If there is time, T_f', during the frame during which no useful information is generated, then the dwell time is

$$t_d = \frac{T_f - T_f'}{N} \tag{10-3}$$

The scan efficiency, η_s, can be defined as the ratio of the dwell time in a perfect scanner (T_f/N) to the dwell time of the actual scanner:

$$\eta_s = \frac{\left(\dfrac{T_f - T_f'}{N}\right)}{\left(\dfrac{T_f}{N}\right)} = \frac{t_d}{t_{d,\,min}} = \frac{T_f - T_f'}{T_f} \tag{10-4}$$

The scan efficiency can be less than 100% as a result of three different causes: the scan velocity is not constant; there is dead time at the end of a line, at the end of a frame or in the unrecorded retrace; or there is overlap of scan lines.

10.2.2. Variable Scan Velocity. If scanning velocity is not constant, then the bandwidth must be determined by the maximum scan velocity. The scan efficiency will be degraded by the degree to which the scan velocity is not constant:

$$\eta = \frac{\bar{v}}{v_{max}} = \frac{N t_{d,\,min}}{T_f} \qquad (10\text{-}5)$$

where v_{max} = the maximum scan velocity
$\bar{v}$ = the average scan velocity

This assumes that overlap and dead time characteristics are ignored.

10.2.3. Overlap Relationships. If scan lines overlap, the scanning is not 100% efficient; but the loss in SNR is not related simply to the scan efficiency. If the line-overlap ratio is designated r_o, then the frame time for a frame with N_l lines is given by

$$T_f = N_l T_l (1 + r_o) \qquad (10\text{-}6)$$

If there are N_r resolution elements (i.e., IFOV) in a line, then the dwell-time for constant scan velocity is given by

$$t_d = T_l N_l^{-1} (1 + r_o)^{-1} \qquad (10\text{-}7)$$

Therefore, the scan efficiency is given by

$$\eta_s = \frac{N_l}{N_r (1 + r_o)} \qquad (10\text{-}8)$$

Overlap provides a gain in SNR according to the general square-root relationship. (See Chapter 19, "Imaging Systems".) This applies if $r_o > 0.50$. Otherwise the display or image output can appear banded.

10.3. Scan Patterns and Characteristics

10.3.1. Raster Scans. Most scan patterns used for forming an image are rectilinear raster patterns that are sequential or interlaced. Figure 10-2 shows a typical, single-line, contiguous, two-directional scan. The cross-hatching is meant to show that each successive line advances by an average of only one-half of a line-width at the scan-line edges so that the entire raster pattern has 100% overlap or a scan efficiency of 50% at most. Sometimes the retrace scan is not active (i.e. providing a signal) and is done very quickly in comparison to the forward (active) scan.

In another way to create a rectilinear raster pattern, sometimes called a *stepped raster*, the lines are scanned in a short time when compared to the time required for the generation of a full frame. At the end of each frame, during the dead time, the scanning spot moves down one line and retraces (to the left). Here, dead time accounts for the lack of a 100% duty cycle. This is shown in Figure 10-3.

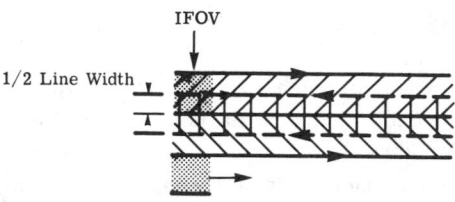

Fig. 10-2. Lines of a two-direction active scan.

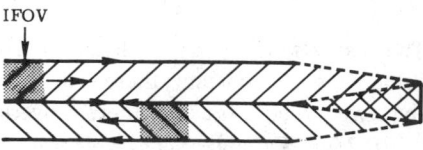

Fig. 10-3. Lines of a dead-time turn around scan.

Raster scans with detector arrays can be of the interlace type as shown in Figure 10-4 or they can be moved an entire array length (a step-scan) as shown in Figure 10-5. The arrays illustrated contain three elements, d_1, d_2, and d_3; the Roman numerals indicate the first, second, and third positions of the array. Basically, the two arrays differ in the spacing of the detectors (or their fields of view) and the relative positioning of the lines. A gradual error-buildup in the interlace scan will cause a maximum change in every third line, whereas the step scan will cause a bigger error in the contiguity between scans 6 and 7 than in the others.

These multi-element scans can also be done with staggered elements as shown in Figure 10-6. They can be considered as either a six-element array or a three-element array in two positions. Detectors d_1, d_3 and d_5 in Figure 10-6 have an electronic phase-delay so that they appear as if they were in positions d_1', d_2', d_3'. This allows additional separation between the detector elements.

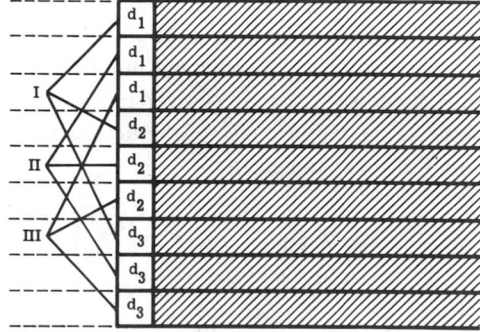

Fig. 10-4. An interlace scan, showing three positions.

The scans described above are usually described as parallel scans because the detectors scan in parallel. The mechanical scan rate is reduced in direct proportion to the number of detectors (excluding dead time); the signal-to-noise ratio of the system is increased in proportion to the square root of the number of detectors. A frame of N lines takes a time, T_f, given by

$$T_f = N(t_l + t_{dd}) \tag{10-9}$$

where t_l = the length of time to scan a line
t_{dd} = the dead time for each line

If, for a fixed frame time, m detectors are used and there is the same percentage of dead time in a scan line, then both the line time and dead time are increased by m.

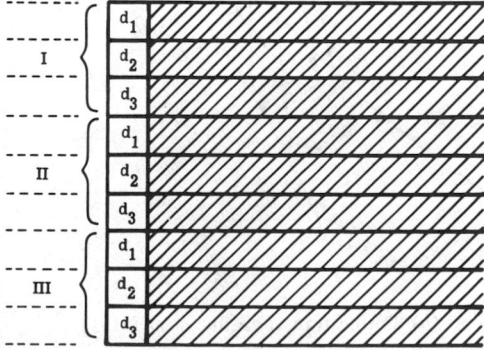

Fig. 10-5. A stepped array in which the whole array is moved an array length between scans.

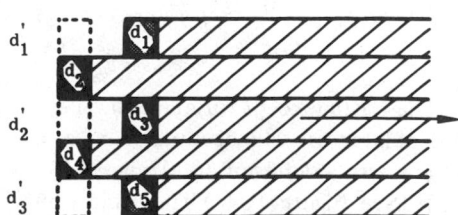

Fig. 10-6. A two-row staggered five-element array with contiguous scan lines.

$$\frac{mT_f}{N} = t_l + t_{dd}$$

$$\frac{T_f}{N} = \frac{t_l}{m} + \frac{t_{dd}}{m} \tag{10-10}$$

Serial scanning makes use of an array of detectors in what is basically the single-detector scanning mode, as shown in Figure 10-7(a). Compared to single detector scanning, serial-scanning has the same scan rate, decreased scan efficiency, and a SNR that is increased approximately by the square root of the number of detectors.

10.3.2. Transverse Line Scan. In many imaging situations, a simple line-scan, transverse to the direction of travel of a platform, is generated as shown in Figure 10-7(b). The geometry forces the projected line-scan to have the shape of a bow tie (Figure 10-7(b)) and successive scans to overlap. The dimensions of the image of the detector on a flat plane are given by

(a) A five-element array used in serial scanning, with dead-time bidirectional scanning.

$$l_t = \frac{h \tan \theta}{\cos(\alpha + \theta)\cos\alpha} \tag{10-11}$$

$$l_s = 2h \sec\alpha \tan\frac{\phi}{2} \tag{10-12}$$

For small values of θ (usually the case)

$$l_t = h \tan\theta \sec^2\alpha = h\theta\sec^2\alpha$$
$$= h\theta(1 + \tan^2\alpha) \tag{10-13}$$

$$l_s = h \sec\alpha \tag{10-14}$$

At the nadir, where most contiguity calculations are made, $\alpha = 0$ and

$$l_s = 2h \tan\frac{\phi}{2} \simeq h\phi \tag{10-15}$$

$$l_t = 2h \tan\frac{\theta}{2} \simeq h\theta \tag{10-16}$$

The Earth (except for hills and valleys and oblateness) can be thought of as a sphere with a radius of 4000 miles (6437 km). (See Figure 10-8.) The projection of the detector image on the sphere can be taken as the projection on a plane

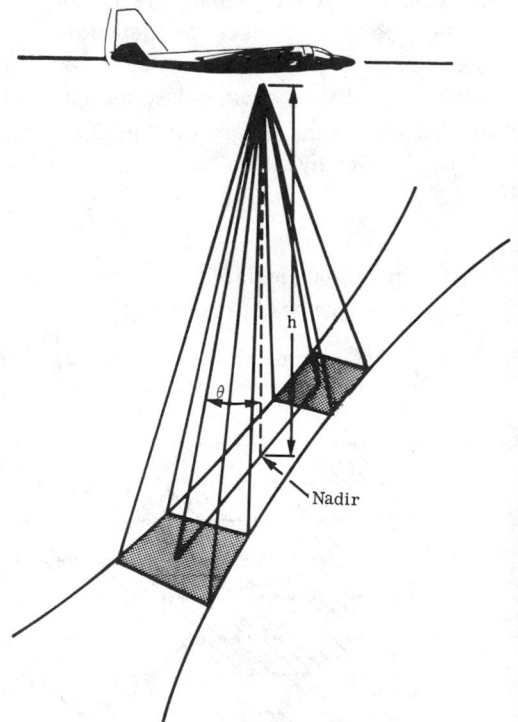

(b) Simple transverse line scan [10-1].

Fig. 10-7. Serial scanning using the single-detector scanning mode.

tangent to any point on the sphere rather than on the curved surface of the sphere. The total distance in the spherical case is

$$D_s + d_E = (R_E + h)\cos\alpha \pm (R_E + h)[\cos^2\alpha - h(2R_E + h)(R_E + h)^{-2}]^{1/2} \quad (10\text{-}17)$$

where D_s = distance from the sensor to the tangent plane
d_E = distance from the tangent plane to the Earth's surface
R_E = radius of the Earth
h = distance from the Earth's surface to the sensor

The angle α can be replaced with the angle between the distance $D_s + d_E$ and the tangent plane. It is

$$\alpha' = \frac{\pi}{2} + \alpha - \arcsin\left[\frac{D_s + d_E}{R_E}\sin\alpha\right] \quad (10\text{-}18)$$

10.3.3. Stationary Circular Pattern. This pattern is formed when a single detector or an array scans a small annulus at a constant rate as shown in Figure 10-9. With an array, all detectors on a radius do not take the same dwell time. If they are of uniform size, the central detectors will have longer dwell times—inversely proportional to their radial position. If the field of view is circular so that the pattern matches the periphery, then the scan efficiency (Section 10.2.3) is given by

$$\eta_s = \frac{1}{4}\frac{m\theta}{\theta_t} \quad (10\text{-}19)$$

where m is the number of elements.

10.3.4. Translated Circle Scan. The Palmer (or translated-circle) scan is generated by the translation of a pure circular scan. For contiguous scan, the translation amounts to

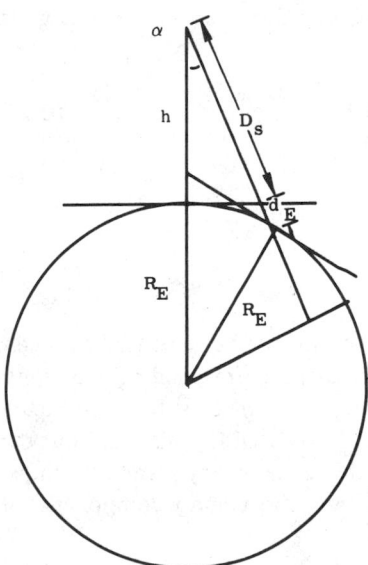

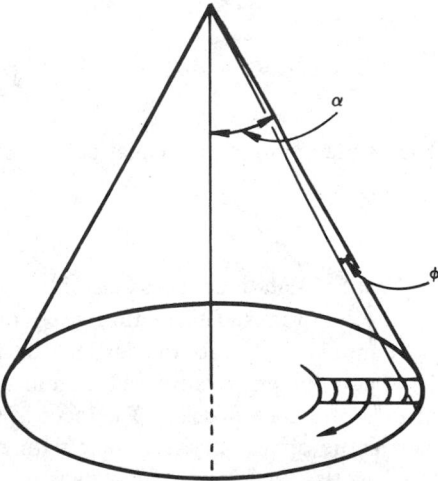

Fig. 10-9. Geometry of a circular scan with an array of m elements each subtending an IFOV of ϕ and covering a circular field of $\theta_t = 2\alpha$.

Fig. 10-8. Geometry for scanning on the Earth's sphere.

the width of one resolution element during the time it takes to make one full revolution. The pattern is shown in Figure 10-10. If the rotational frequency is f_r, then the time it takes to scan a circle is $1/f_r$. If the resolution angle is θ, then the linear velocity, v, for a contiguous scan can be approximated by $l_s\theta = v/f_r$. A more detailed treatment yields

$$l_s = 2h \sec \alpha \tan \frac{\theta}{2}$$

$$l_t = h[\tan \alpha - \tan(\alpha - \theta)] \tag{10-20}$$

This is the same formula as for the straight-line scanning technique. The difference is that the resolution element in the contiguous scan formula is constant for a given altitude and a given α; whereas in the straight-line scanning technique the resolution element changes with α. The rate of advance of the vehicle for contiguous scan is

$$v = l_t f_r \tag{10-21}$$

$$f_r = \frac{v \cos \alpha \cot \dfrac{\theta}{2}}{2h} \tag{10-22}$$

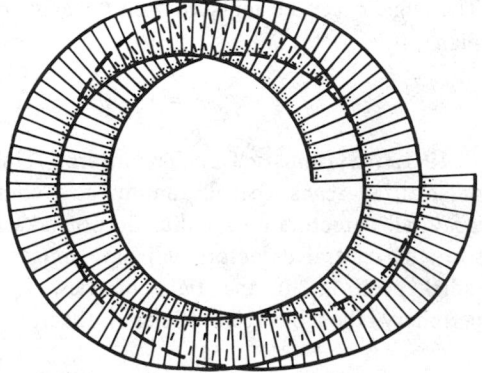

Fig. 10-10. Palmer scan pattern for contiguous scan.

The total lateral field of view, θ_t, is 2α. Contiguous scanning can be accomplished at the exact forward direction by moving the vehicle at a constant horizontal velocity equal to one resolution element on the ground. The time, t_c, to scan the circle is given by

$$t_c = 2 \frac{h}{f_r} \sec \alpha \tan \frac{\phi}{2} \tag{10-23}$$

The overlap ratio in one semicircular scan is

$$\frac{h \tan \theta}{\cos(\alpha + \theta) \cos \alpha} \tag{10-24}$$

10.3.5. Rotating Prisms Scan. Many different scan patterns can be generated by a pair of rotating prisms. This section gives some of the patterns and corresponding equations.

A simple single prism of angle A deviates a ray through an angle δ as shown in Figure 10-11. As the prism is rotated around the optical axis, the deviated ray will trace out the surface of a cone in space. The locus of intersection of that cone on any plane perpendicular to the optical axis is a circle with z as the optical axis. The x and y components of the deviation angle can be written as

$$\delta_x = \delta \cos \omega_r t$$

$$\delta_y = \delta \sin \omega_r t \tag{10-25}$$

where ω_r = the angular rotation rate of the prism, rad sec^{-1}
　　　δ = the angle of deviation

Two prisms can be used in series to generate more complicated patterns. The two prisms can have equal or different prism angles, A (A_1, A_2); this results in equal or different ray deviation angles, δ (δ, $\kappa\delta$). The prisms can be rotated at the same or at different rates and in the same or opposite directions, $\pm\omega_r$ (ω_r, $\pm n\omega_r$). They can start the rotation with their apexes in the same position or at any angular offset, ϕ. If the prisms are aligned with their apexes in the same direction, $\phi = 0$. If the apexes are in opposite direction, $\phi = 180°$.

The simplest case of *two-prism* rotation is for $\delta_1 = \delta_2 = \delta$, $\omega_1 = \omega_2 = \omega_r$ and $\phi = 0$. Then

$$\delta_x = 2\,\delta\,\cos\,\omega_r t$$

$$\delta_y = 2\,\delta\,\sin\,\omega_r t \qquad (10\text{-}26)$$

The two prisms act as one, generating a scan with twice the amplitude in Equation (10-25) above.

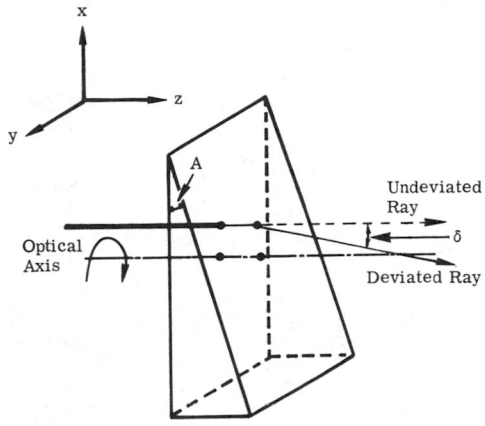

Fig. 10-11. A simple single prism of angle A deviating a ray through angle δ.

If one prism is displaced in rotational orientation with respect to the other by a fixed angle, $\phi/2$, and they continue with equal co-rotation, then

$$\delta_x = 2\,\delta\,\cos\left(\omega_r t - \frac{\phi}{2}\right)\cos\left(\frac{\phi}{2}\right) \qquad (10\text{-}27)$$

$$\delta_y = 2\,\delta\,\sin\left(\omega_r - \frac{\phi}{2}\right)\cos\left(\frac{\phi}{2}\right) \qquad (10\text{-}28)$$

The resultant deviation is proportional to the cosine of the half-angle of rotation displacement (Figure 10-12). The figure shows the pattern scanned out by a single ray through the two prisms on some plane perpendicular to the optical axis of the system. The first row shows equal prism co-rotation, counter-rotation, co-rotation with a phase shift and counter-rotation with a phase shift, respectively. The first column shows co-rotation of equal-angle prisms rotating at four different frequency ratios. The bottom two figures in this column shown precession resulting from a nonintegral frequency ratio. The second column is identical to the first except that counter-rotation is illustrated. Illustration 10-12(g) shows the effect of a one percent error in frequency equality when one tries to obtain a line scan like Figure 10-12(b). Figure 10-12(h) shows the result for one percent error in prism-angle equality. (The line is one percent different in length but does not "feather.") Figure 10-12(k) illustrates a possible horizon sensor scan obtained by making one prism larger than the other by a factor of three and using a co-rotation ratio of 10. Figure 10-12(l) shows the same for counter-rotation (probably a better horizon-sensor pattern). Figure 10-12(o) shows a spiral pattern with almost equal rotation rates of the two prisms.

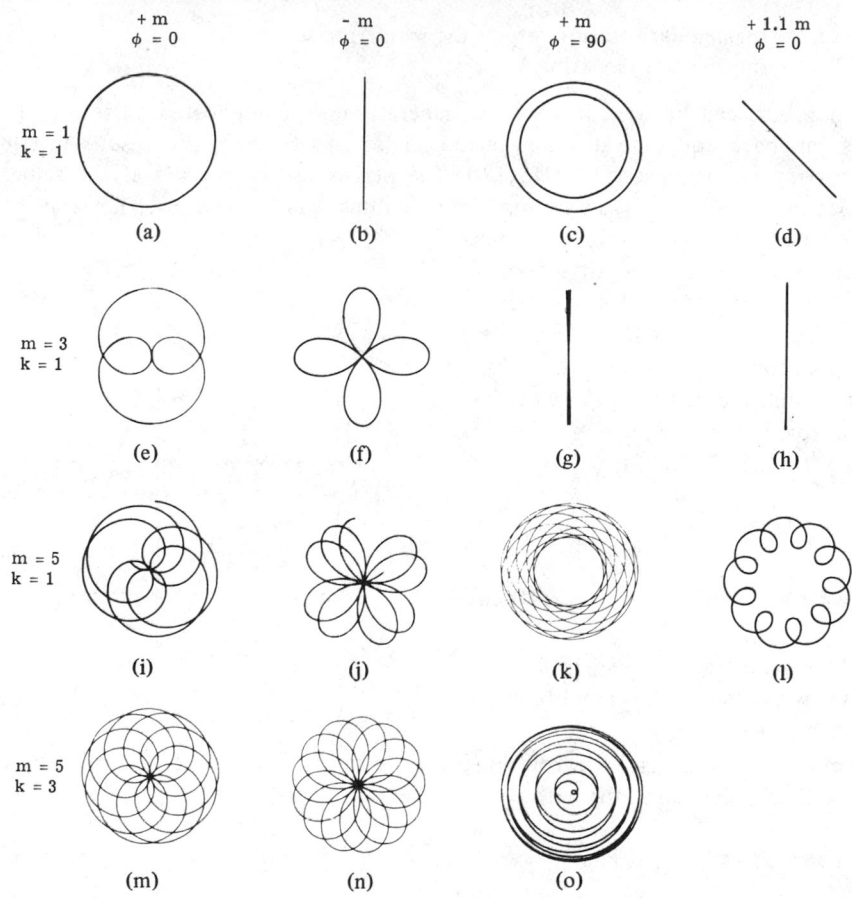

Fig. 10-12. Rotating prism scan patterns. The ratio of the rotational frequencies of the two prisms is m; the ratio of the prism angles is k; the phase relation at the time they start is ϕ. A negative value of m means the prisms counter-rotate; a zero value for ϕ means the prism apexes are oriented in the same direction.

If the two prisms counter-rotate but start in phase, then the two frequencies are ω and $-\omega$, and $\phi = 0$ (Figure 10-12).

$$\delta_x = 2k\delta\cos\omega t \tag{10-29}$$

$$\delta_y = 0 \tag{10-30}$$

If the deviations of the prisms are different, then each of the cases above is evaluated as indicated:

In-Phase Co-Rotation (Figure 10-12)

$$\delta_x = \delta\cos\omega t + k\delta\cos\omega t \tag{10-31}$$

$$\delta_y = \delta\sin\omega_r t + k\delta\cos\omega_r t \tag{10-32}$$

Phase-Shifted Co-Rotation

$$\delta_x = \delta_1\cos\omega_r t + \delta_2\cos(\omega_r t - \phi) \tag{10-33}$$

$$\delta_y = \delta_1\sin\omega_r t + \delta_2\sin(\omega_r t - \phi) \tag{10-34}$$

For two prisms that have the same deviation, δ, and start with shift ϕ, a more general expression is

$$\delta_x = (k+1)\delta \cos \omega_r + \cos(n\omega_r + \phi) \qquad (10\text{-}35)$$

$$\delta_y = (k+1)\delta \sin \omega_r t + \sin(n\omega_r t + \phi) \qquad (10\text{-}36)$$

where n = rotation-rate ratio
        ω_r = rotation rate of the reference prism

If the prisms co-rotate at the frequency ratio

$$n = int + fr \qquad (10\text{-}37)$$

where *int* = the integer portion of n
      fr = the fractional portion of n

then a spiral will be generated as shown. If n is an integer, i.e. *fr* = 0, then the pattern changes as shown in figure. There are $n + 1$ circular loops in the corresponding pattern. If the prisms counter-rotate, a rosette will be generated in which there are $n + 1$ petals with the scan crossing through the center to the other side of the circular envelope. If n is an integer, there is an integral number of petals.

The pattern is repeated consistently. If the prisms do not start in phase, the pattern is rotated by half the phase-shift. If n is non-integral, then the pattern precesses by an amount $(360°/n)fr$. Patterns which might be useful for horizon scanning kinds of applications are also shown in Figure 10-12. When the apex ratio is not one, unscanned space is left in the middle of the field. If an array is used in the focal plane, its orientation in space will remain fixed regardless of the type of pattern or its position in the pattern.

10.3.6. Cycloidal Patterns. The two main cycloidal patterns are hypocycloidal and epicycloidal. They are generated by superimposing circular motions of different angular radii, frequencies, and centers of rotation. A circular pattern with angular radius $\theta_t/2$ is shown in Figure 10-13. Another, with angular radius $\theta/2$ is shown offset a distance corresponding to angle α. These are cones in space of angles θ_t and θ respectively. In a manner similar to the composite circular scans, ω_1 and ω_2 can be of the same or opposite sign, and the two radii can be equal, unequal, and equal or unequal to the offset. The cycloidal patterns are part of the more general class of superimposed circular rotations considered in the previous section.

10.4. Scan Techniques

Many different techniques and implementations are used to cause an IFOV to scan or

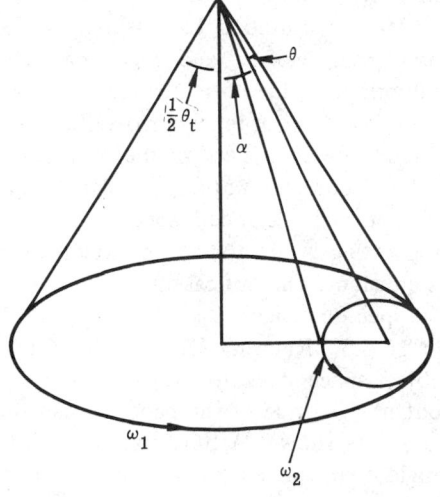

Fig. 10-13. Geometry of cycloidal scans.

sample a total field of view. Important parameters include scan rate, scan rate constancy, scan efficiency, optical efficiency and vignetting.

10.4.1. Nipkow Scanner. One of the earliest implementations was the disc scanner used in very early TV systems. An image is formed on a portion of the Nipkow scanning disc (Figure 10-14). Since all the radiation must be collected by the lens that focuses onto

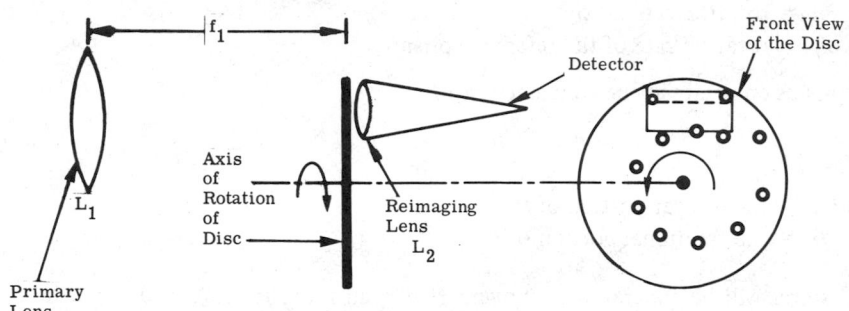

Fig. 10-14. Nipkow scanning system.

the detector, the lens must be of a diameter that is equal to size of the field plus $2l\tan(\theta_t/2)$, where θ_t is the full-field angle (with diameter d_1) and l is the distance from the primary image to the remaining lens. The lens can be used to form an image of the primary lens, L_1, on the detector. Thus, $f_1 + l$ is the object distance for the second lens, L_2, (with focal length f_2 and diameter d_2). The magnification is

$$\left[\frac{f_1}{f_2} + \frac{l}{f_2} - 1\right]^{-1} \tag{10-38}$$

The ratio of l/f_2 is small with respect to f_1/f_2, or $(F/\#)_1 d_1/[(F/\#)_2 d_2]$. If the optical speed of the two lenses is the same, the magnification is approximately proportional to the ratio of the lens diameters. Values of 0.01 to 0.001 are typically required.

10.4.2. Rotating Lens Wheels. A form of image space scanning which provides constant scan velocity and high scan efficiency is a wheel of lenses [10-2]. In this embodiment, the corrector for the Schmidt objective is incorporated on the surfaces of the relay lenses or relay mirrors. Each of the spokes of the wheel (e.g. 20) covers 1° and is tilted 1° with respect to the preceeding spoke for a 20-spoke configuration. Therefore only 1° correction at a specific angle is necessary behind the spherical mirror. The optical efficiency is degraded somewhat by the wide field required in the folded system. Vignetting is minimal and occurs at the beginning and end of each line. The scan velocity is constant. The optical efficiencies are about 70% to 80%; the exact values depend upon the specific designs.

10.4.3. Rotating Mirror. Figure 10-15 shows a mirror, M, which can be rotated in object space at a scan rate, $\dot{\alpha}$, around the axis shown, to move the collimated beam in and out of the plane of the paper. It can also be rotated around the orthogonal axis in a scan rate $\dot{\beta}$ as shown. A third rotation can be used with the rate shown as $\dot{\gamma}$. Of course, the angle of inclination of the mirror need not be 45° as shown. If, for instance, the mirror is inclined at an angle of ϵ to the vertical and rotated about either the $\dot{\alpha}$ or the $\dot{\gamma}$ axis, then a circle of half-angle ϵ will be generated around either axis. Usually a scan around the $\dot{\beta}$ axis is used for v/h compensation in a moving vehicle; the $\dot{\gamma}$ scan is usually used to generate the across-track scan of a stripmapper. For such an application, the rotating mirror scanning technique provides a constant scan rate, but only looks at the ground a little less than 50% of the time. It has 100% scan efficiency for scanning the inside of a cylinder. The field of view also rotates at twice the rate of the rotation of the mirror.

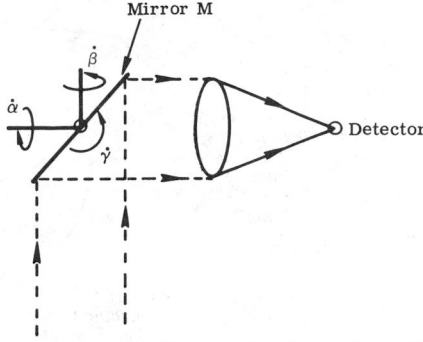

Fig. 10-15. Simple rotating flat mirror M with rotation around any of three axes at rates of $\dot{\alpha}$, $\dot{\beta}$, $\dot{\gamma}$.

10.4.4. The Axe-Blade or Knife-Edge Scanning Systems [10-3]. This type of optical-mechanical scanning system was pioneered by Haller-Raymond and Brown (now HRB-Singer) for airborne mappers. Since the single, flat, scanning mirror has low scan-efficiency, a second mirror is added to scan the ground while the first is scanned upward. (See Figure 10-16.) Even when the rotational axis is aligned to prevent

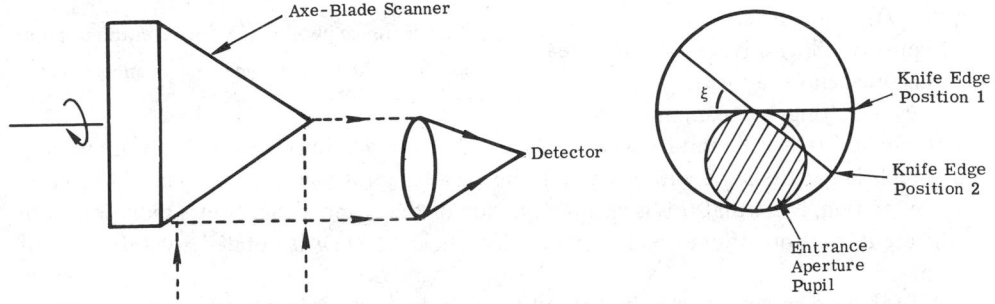

Fig. 10-16. Axe-blade or knife-edge-scanning system.

high-speed asymmetric rotations, as shown, some manipulation is necessary to prevent the detector from seeing both sides of the field of view, i.e., from recording radiation reflected in part by both facets. In Figure 10-16, the left-side view shows how radiation from opposite sides of the field of view can be incident on the detector at the same time. The right side shows this in more detail. In position 1, the mirror is such that the telescope looks straight down. When it has rotated through an angle ξ, a very small portion of the field is reflected in from segment of the knife blade which the rest comes from a field centered on the direction ξ from the vertical. Figure 10-17 shows vignetting when the pupil fills the scan mirror.

For the system shown in Figure 10-18(a), the vignetting ratio $1/\pi(2\xi - 1/2 \sin 2\xi)$, with reference to the area of the circular entrance pupil. The area of the entrance pupil is one-half the area of the scan mirror as shown. It is possible to trim the edges of the scan mirror. Figure 10-18(b) shows the entrance pupil in a more realistic situation. Here the area of the pupil is smaller than in Figure 10-18(a) and does

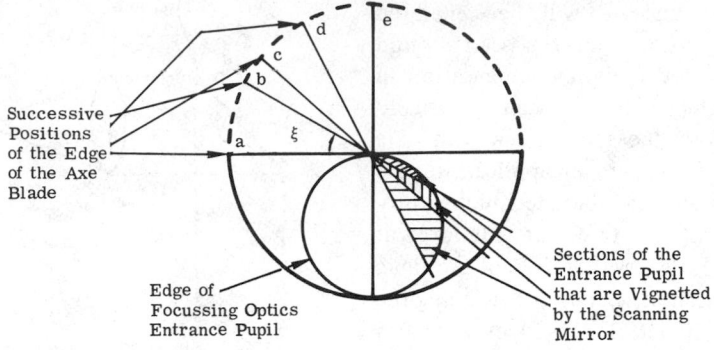

Fig. 10-17. Vignetting in an axe-blade scanner. Five positions of the scan mirror are shown (*a* through *e*). In position *a*, the scan field is straight down and no vignetting occurs. In position *e* the orientation is horizontal and 50% vignetting occurs.

not meet the edges of the scan mirror. If the projected diameter of the scan mirror is $D + d$ and the separation as shown is s, then the area ratio is $2(1/2 - s/(D + d))^2$. For the situation illustrated in Figure 10-18(a), s is zero and the area ratio is one-half (as above).

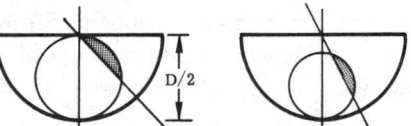

(*a*) Full entrance pupil.   (*b*) Partial entrance pupil.

Fig. 10-18. Details of vignetting on an axe-blade scanning mirror.

The vignetting by scanning is somewhat smaller. It is zero until $\cos \theta$ equals $1 - 2s/(D + d)$. Since scan mirrors are usually not much larger than the entrance pupil, the vignetting ratio above can be used to a good approximation. Note that this is also the portion of the scanner that brings radiation from a different portion of the field of view. The field of view is rotated by this type of scanner.

A four-sided prism has also been used (in principal, an *n*-sided wedge can be used) to obtain scan angles of $360°/n$, with some loss in the field of view, to prevent the multiple-look problem. All of the scanners mentioned rotate the field of view.

10.4.5. Polygon Scanners with Exterior Mirrors. Mirrors constructed in a many-sided polygon, as shown in Figure 10-19, will also scan a line pattern. In a sense, the mirror is a many-sided blade with no taper of the blade faces. The angular field of view is 360° divided by the number of sides. Each face of the polygon must be $\sqrt{2}$ times the diameter of the aperture, but only as wide as the aperture. The distance from the center to an apex is given by

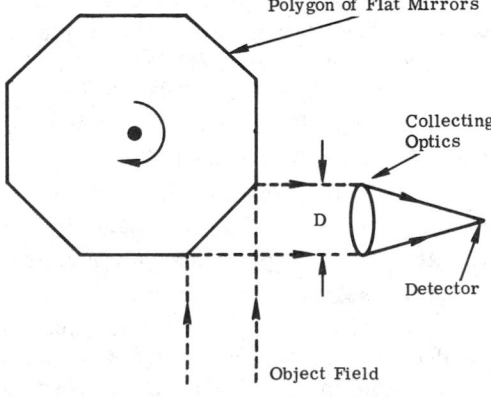

Fig. 10-19. An exterior-mirror polygon scanner.

$$\frac{\sqrt{2}D \cos\left(\dfrac{180°}{n}\right)}{\sin \dfrac{360°}{n}} \tag{10-39}$$

where D = the entrance pupil diameter
n = the number of sides

Vignetting in the polygon scanner is generated as the scanner rotates, as shown in Figure 10-19. The beam moves down as $r \tan \xi$, in which r is the radius of the polygon scanner (circumscribed circle) and ξ is the angle through which it is rotated. If the entrance pupil is circular, the vignetting related to the circular entrance pupil is

$$\pi^{-1}\left[\cos^{-1}(1 - \tan \xi) - \frac{1}{2}(1 - \tan \xi)(2 \tan \xi - 1)^{1/2}\right] \tag{10-40}$$

If the entrance pupil is square the vignetting is $1 - [1/2 \cdot (\tan \xi)]$. If the entrance pupil does not "fill," i.e., meet the edges for a circle, then the vignetting starts when $D/2 \cdot (\tan \xi)$ is equal to the separation and the above expression can be used as a reasonable approximation. Two fields of view are sometimes sensed by the detector, but one is usually of the instrument housing.

10.4.6. Polygon Scanners with Interior Mirrors. In Figure 10-20, a folding mirror is used to redirect the incident-collimated beam to the top of the rotating polygon, which reflects it back down to another folding flat, which in turn directs it back out the other side. One of the two folding flat mirrors can be just the size of the entrance aperture, but the other must be large enough to accommodate the scan field. The vignetting and double-fielding conditions are not easy to analyze or describe for the inside-mirror

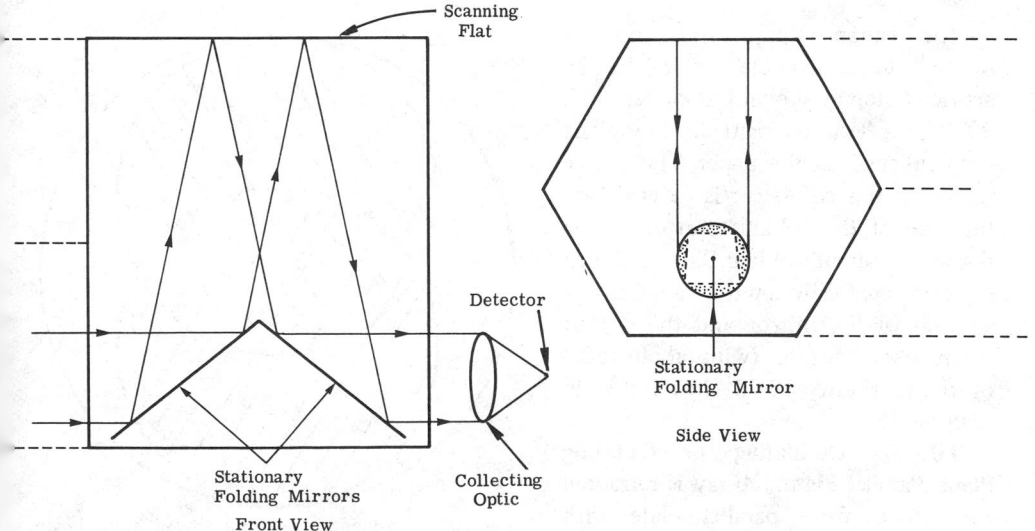

Fig. 10-20. Inside polygon scanner with inside mirrors.

polygon. Figure 10-21 shows one type of geometry where the polygon is rotated an angle ξ from the "on-axis" position. The beam of radiation which reaches the detector is from 2ξ. The first folding mirror, shown in dashed lines, directs the radiation up and from an angle 2ξ to the left. The required size of the mirror is shown. The second folding flat is shown in dashed lines.

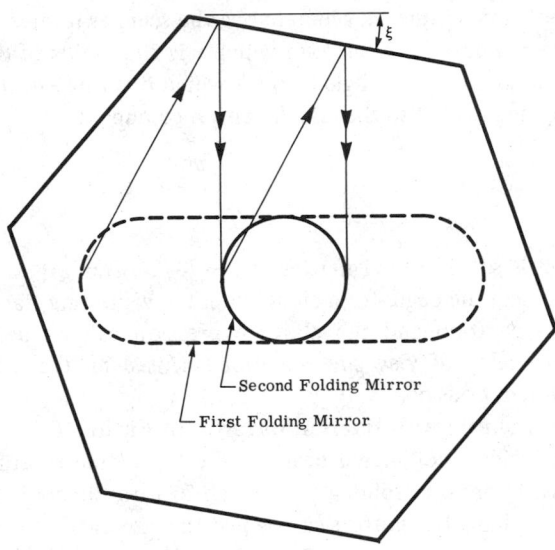

Fig. 10-21. Geometry of the inside-polygon scanner. Single arrows show the incoming rays. Double arrows show the rays that are reflected from the scanning mirrors.

The entrance pupil need not be limited to a diameter of D by an artificial stop as shown but can scan all 90° from left to right in the plane perpendicular to the paper. The degree to which this can occur is governed by the size of the folding mirror. In the figure, a pair of folding flats is shown superimposed. The input beam uses one portion of the mirror and the output beam uses another (with about 50% overlap). Figure 10-22 shows this in plan view.

10.4.7. Oscillating or Rotating Plane Parallel Plate. A ray is refracted through a plane parallel plate with displacement but no deviation as shown in Figure 10-23. If the plate is tilted at

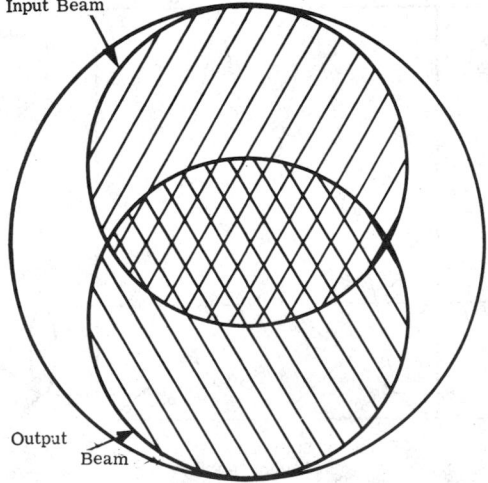

Fig. 10-22. An entrance pupil whose input beam uses one portion of a mirror and output beam uses another portion of a mirror.

an angle ξ, and rotated about the optical axis, then a circle will be generated. If the beam is collimated, it will only be displaced. A similar action occurs if the plate is oscillated or rotated about the axis perpendicular to the paper, but then the scan is vertical. These motions do little good unless coupled to an imaging system. This is one of few scanning schemes that performs linear displacement of a beam rather than angular deviation. The change in focal length, ΔF, caused by the insertion of a plate with plane and parallel sides is given by

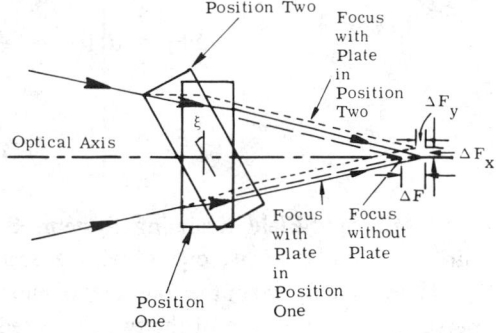

Fig. 10-23. Geometrical representation of change of focus due to insertion and rotation of a plane parallel plate.

$$\Delta F = d \left[1 - \frac{\cos \theta_i}{\sqrt{n^2 - \sin^2 \theta_i}} \right] \qquad (10\text{-}41)$$

where d = the plate thickness
θ = the angle of incidence on the front face
n = the refractive index of the plate

The focal shift is a function of the angle of incidence. Figure 10-23 shows how this changes with angle (from 0 to 30°). These values stop when the angle is sufficiently large that $n \sin \theta = 1$, the angle of total internal reflection.

The change in focus is a function of wavelength because of dependence of the refractive index on wavelength.

$$\frac{\partial}{\partial \lambda} \left[\frac{\Delta F}{d} \right] = \frac{1}{d} \frac{\partial (\Delta F)}{\partial \lambda} = \frac{n \cos \theta_i}{(n - \sin^2 \theta_i)^{3/2}} \frac{dn}{d\lambda} \qquad (10\text{-}42)$$

The position of focus moves as the plate is rotated through an angle α. The angle of incidence changes as the rotation angle. Two rays at equal angles from the axis of the beam have angles with incidence of $\theta_i + \xi$ and $\theta_i - \xi$. If the cone angle is 2θ, then the location at which these rays cross (their focus) is given by

$$\Delta F_y = \frac{1}{2} (\Delta F_1 + \Delta F_2) \tan \theta \qquad (10\text{-}43)$$

$$\Delta F_x = \frac{1}{2} (\Delta F_1 + \Delta F_2) \qquad (10\text{-}44)$$

where

$$\Delta F_1 = d\left[1 - \frac{\cos(\theta + \xi)}{\sqrt{n^2 - \sin^2(\theta + \xi)}}\right] \qquad (10\text{-}45)$$

$$\Delta F_2 = d\left[1 - \frac{\cos(\theta - \xi)}{\sqrt{n^2 - \sin^2(\theta - \xi)}}\right] \qquad (10\text{-}46)$$

10.4.8. Split-Field Scanning System. By using the rotating-reflecting cube and a pair of folding mirrors, one obtains a scanning system of constant scan rate (Figure 10-24). In general, two of the four sides of cube (each with area A) are used at any time. The aperture stop is the sum of the two projected areas (as long as no other mirrors vignette).

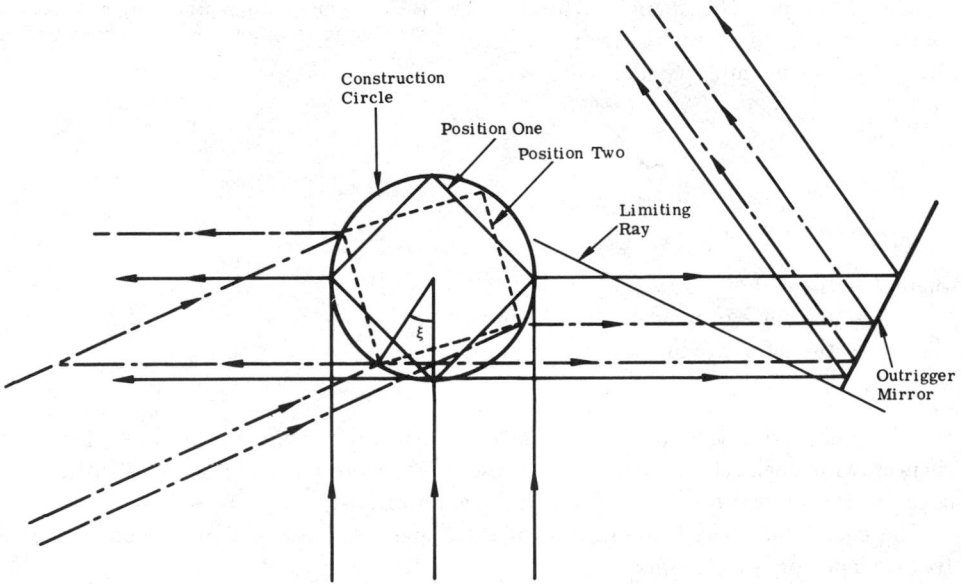

Fig. 10-24. Geometry of a rotating reflecting cube. The two positions show: on-axis (straight down); and about 70° off-axis. This gives a 140° field.

As the mirror rotates through an angle ξ from position 1 to position 2, one side has a projected area of $A \cos \xi$ and the second side has a projected area of $A \cos((\pi/2) - \xi) = A \sin \xi$. The total area is $A (\cos \xi + \sin \xi)$. When the cube is in position 1, its total aperture is $\sqrt{2} A$. These are the maximum and minimum values of the aperture. The orientation of the outrigger mirror is such that it must go all the way down to the central position or $1/2 \cdot \sqrt{2} \, l$, where l is the length of a side of the cube. When the cube rotates 20° from the vertical position, rays 70° from the vertical will enter the system, etc. The outrigger must be far enough out to avoid blocking the incoming beam.

Figure 10-24 shows the distances and geometries required by different angular positions. Each side of the cube is used for one fourth of the rotation so that if there were no outrigger obscuration the scan efficiency would be 100%. This configuration can be used with other polygon-mirror scanners—triangles, pentagons and octagons. Figure 10-25 shows how a baffle might be used to eliminate stray radiation.

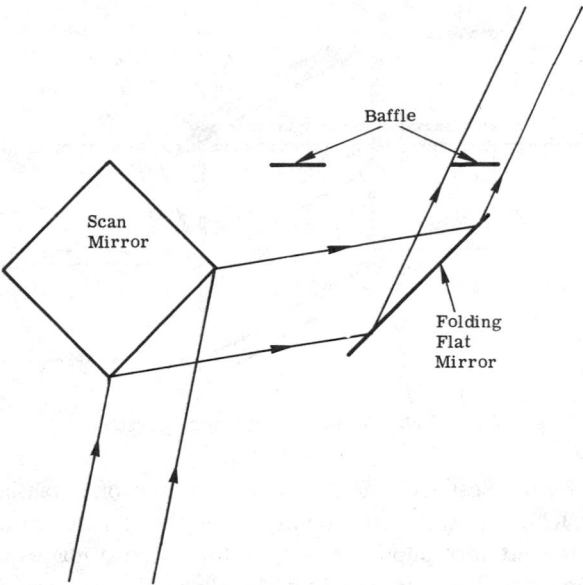

Fig. 10-25. Appropriate baffles keep stray light from the beam.

10.4.9. Fresnel Scanning Prisms and Mirrors. Rotating prisms are heavy for a relatively large aperture. A partial solution to this is to make Fresnel devices out of them (as shown in Figure 10-26) though a loss in resolution results because of the smaller dimensions. The Rayleigh diffraction limit would become

$$\theta\phi = \frac{\lambda}{l_x} \frac{\lambda}{l_y} \tag{10-47}$$

where l_x, l_y are the x and y dimensions of each facet. The flux distributions would be as shown in Figure 10-27. For narrow radiation bandwidths the steps can be made an integral number of wavelengths. To a first approximation the steps will then not be seen.

 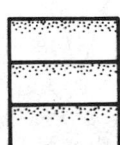 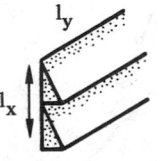 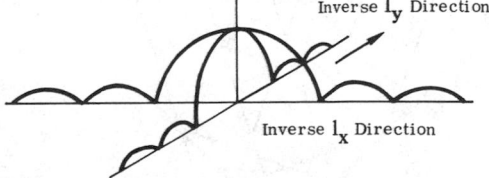

Fig. 10-26. Fresnel scanning prisms.

Fig. 10-27. Energy distribution in the field of rectangular Fresnel prism.

An almost equivalent way to accomplish this light-weighting is to use the mirror version of these prisms shown in Figure 10-28.

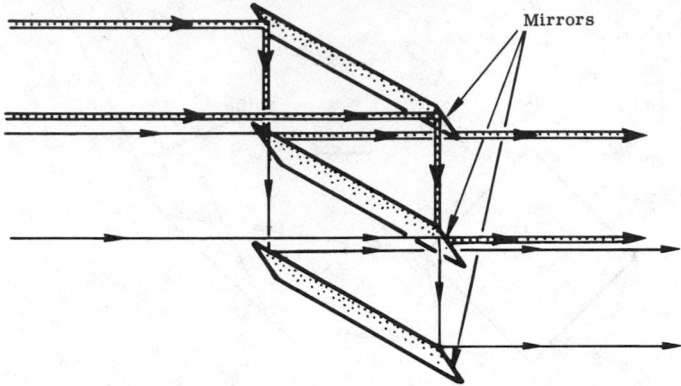

Fig. 10-28. Fresnel mirror scanning system.

10.4.10. Roof-Mirror Scanner. The scanning operation of a translating roof is illustrated in Figure 10-29. Part (a) of the figure shows the relative motion by illustrating three positions of the entrance pupil, resulting in three virtual images at A, B, and C. In reality, the roof moves as shown in part (b). This shows that for a single position of the entrance pupil the images are at A, B, and C, with resulting light cones emanating to the three exit pupils as indicated. Motion of the image points is twice that of the motion of the roofs. These same elements mounted on a rotating wheel, as shown in Figure 10-30, also provide translation and rotation to the axis.

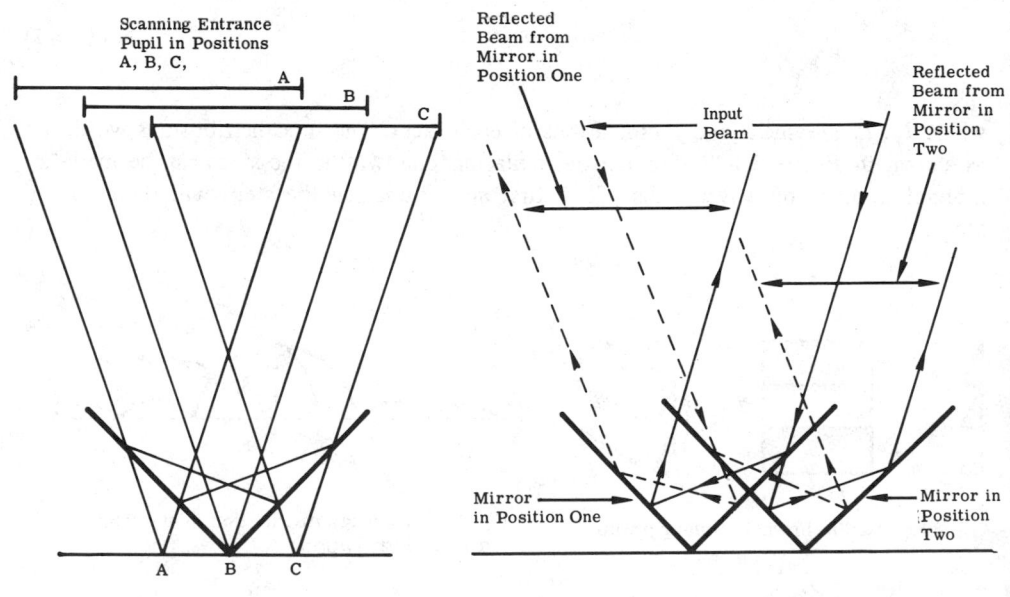

(a) Three entrance pupil positions.   (b) Actual roof movement.

Fig. 10-29. Scanning operation of a translating roof.

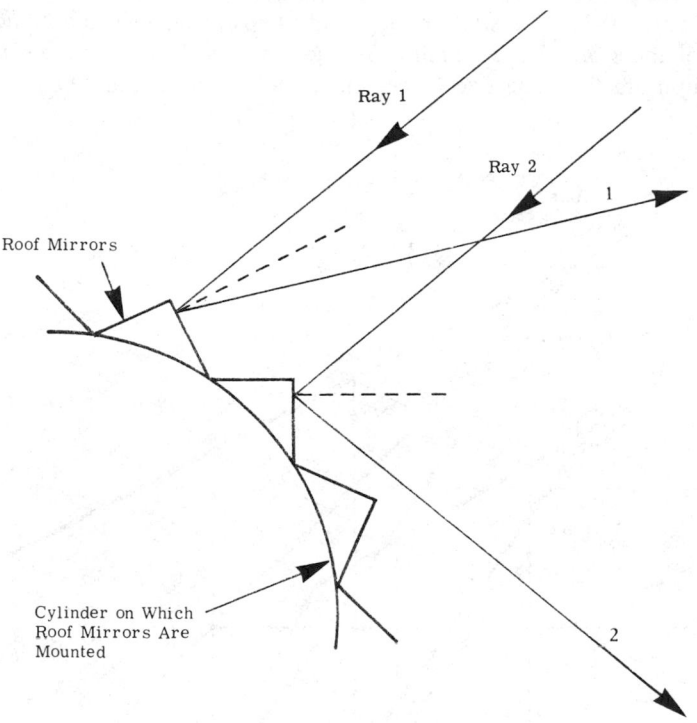

Fig. 10-30. Rotation of rays with circular tooth scanner. An extreme situation is shown. Rays 1 and 2 enter parallel and reflect off different roof mirrors. They are not parallel as they exit. In practice, the angular deviation would be smaller and occur only over the space occupied by one roof mirror.

10.4.11. Window and Mirror Vignetting. Generally, when a mirror scanner is used, the window of the system and the mirror must be larger than the diameter of the optical-system entrance aperture. For simple 90° folding, the mirror must be $\sqrt{2}D$, where D is the diameter of the entrance pupil. If $1/2\,\theta_t$ is the half angle of the field of view, then the length of the mirror of this configuration must be $\sqrt{2}D\,[\cos(\theta/8) - \sin(\theta/8)]^{-1}$. The center of the mirror must be set back from the window a distance of $-D[1 + \tan(\theta/8)]/[1 - \tan(\theta/8)]$ and the window must be of a height of

$$\left[D \tan \frac{\theta}{4} \; \frac{\tan\left(\dfrac{\theta}{8}\right) - 1}{\tan\left(\dfrac{\theta}{8}\right) + 1} + \frac{1}{2} \; \frac{1 - \tan\left(\dfrac{\theta}{8}\right)}{1 + \tan\left(\dfrac{\theta}{8}\right)} \right]$$

For other examples and geometries the calculations are similar.

The window can be reduced in size by rotating the mirror about a different axis. For example, see Figure 10-31. The window need only be large enough to accomodate the projected area of the beam. The scan mirror is larger than the beam, but the window and the entrance pupil are the same size (except that the windows should be larger by the projected area).

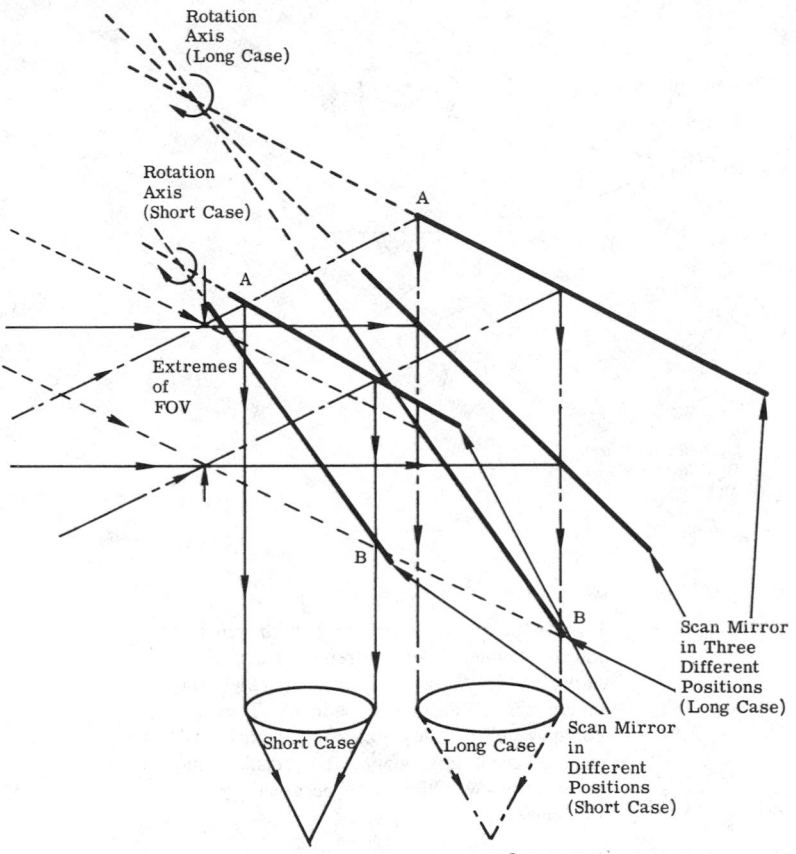

Fig. 10-31. Fixed window scanning system for a 90° full field. The window and 45° mirror for the central beam are shown. The beam-to-entrance aperture is also shown. The intersection points of the incoming beams with the folded beam determines the rotation axis. Intersection points A and B then determine the size of the scan mirror.

In framing systems, Forward Looking Infrared (FLIR) systems for instance, the problem of the beam moving across the mirror or window is a two-dimensional one. The first mirror and window arrangement can be the same as shown in Figure 10-15. Since a second scanning motion orthogonal to the first is required, the same process may be used.

10.4.12. Stator-Rotor Scanners. Relatively small mirrors like those in horizon sensors, can be mounted on the rotor of an electric motor. The currents in the stator can then be used to aim the mirror in virtually any direction.

With these orthogonal drives, almost any desired motion that is a combination of sinusoids can be generated. These include simple circles, circles with sinusoidal dither on the circumference, and ellipses of various eccentricities with sawtooth dither on the periphery. Since these have been used mostly in horizon sensors, the most frequently occurring patterns have been dithers on circles. One example is a mirror that is aimed at the nominal horizon for a given satellite altitude with appropriate currents through the coils. Superimposed upon this is a smaller current that causes the system to nod up and down an appropriate amount above and below the nominal horizon. This is one-dimensional scan. A similar (or different) motion could be generated in another direction. Scanners of this type are best used in horizon sensors and small tracking heads. One example is the OGO Horizon Sensor, shown schematically in Figure 10-32.

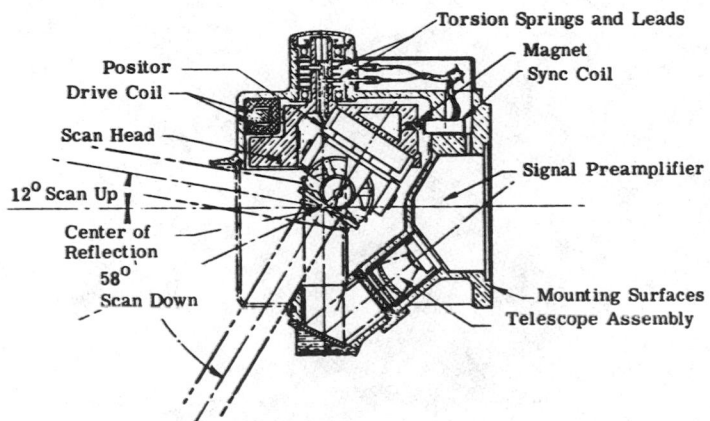

Fig. 10-32. OGO Horizon Sensor [10-4].

10.4.13. Tuning Forks. Tuning forks can be made in many sizes and shapes. Generally they are smaller than about 20 cm in their largest dimension. They are designed mechanically to have a specific resonant frequency appropriate for the design application. Schematics of several typical varieties are shown in Figures 10-33 and 10-34. The mirrors are usually attached at the ends of the tines of the tuning fork which are driven by a small vibrator at the resonant frequency. The mirrors can be mounted at any reasonable angle and thereby scan a beam through translation. More often, they are used as light choppers or episcotisters. Another form is the torsional tuning fork which twists at its resonant torsional frequency.

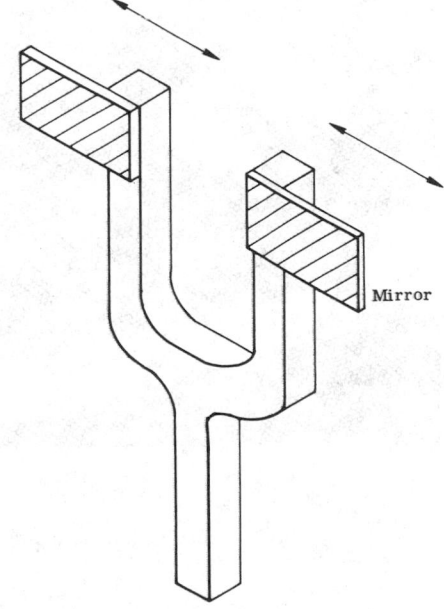

Fig. 10-33. Schematic of a tuning fork.

These devices can be quite rugged (withstanding the shock of a drop to the laboratory floor) and quite reliable, with a lifetime of 10^9 cycles. Typical frequencies are 10 to 10^4 Hz with amplitudes of about 2 cm [10-5].

(*a*) Bulova type L51c, taut band chopper.

Fig. 10-34. Several typical tuning fork types [10-5].

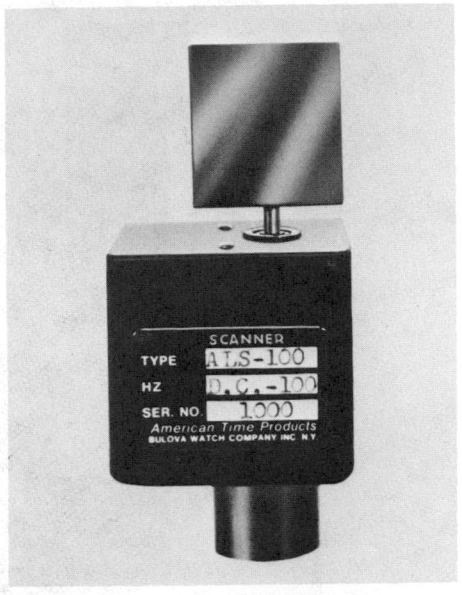

(c) Bulova type ALS, galvanometer movement scanner.

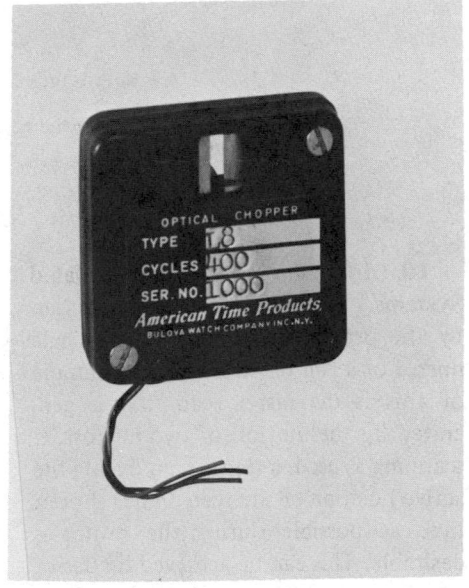

(b) Bulova type L40c, tuning fork chopper.

(d) Bulova type L8c, miniature tuning fork chopper.

Fig. 10-34 (Continued). Several typical tuning fork types [10-5].

(*e*) Bulova type L-45 torsion rod scanner.

Fig. 10-34 (Continued). Several typical tuning fork types [10-5].

10.4.14. Cam-Driven and Related Systems. Many systems generate a scan by the programmed motion of a single mirror or a pair of mirrors. One example of this is the raster scan that is generated by the motion of two mirrors. If scanning is used, a linear scan during the active position of the scan and as short a time as possible during the return is desirable. This can be achieved by use of camdrives as shown in Figure 10-35. As the cam rotates (because it is affixed firmly to the shaft at a), the coupling pushes the end of the mirror further until it reaches the end of its excursion (approximately as shown in the

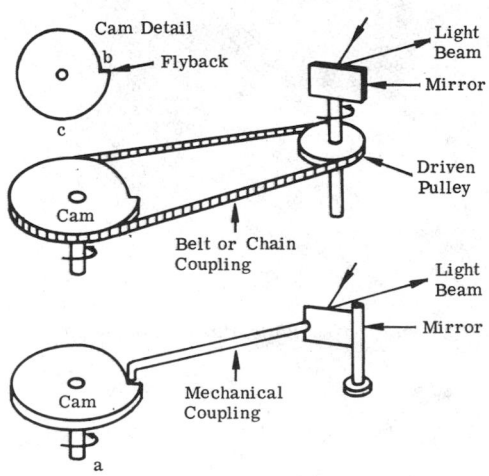

Fig. 10-35. Cam-driven scanners.

diagram). It then returns rapidly by sliding down the portion of the cam labeled b. The side view of the cam is shown in Figure 10-35. The portion labeled c has a radius r proportional to the angle $(r - k\theta)$. The portion b is shaped such that a minimum time for flyback is accomplished without causing chatter, overshoot, and intolerable wear. Cams are convenient for drives of this type but are limited in accuracy and lifetime because the mechanical coupling is accomplished through friction.

10.4.15. Galvanometer Mirrors. The guidance of small beams can be accomplished with galvanometer mirrors in much the same way that galvanometers have been used in electrical measurements. A mirror is mounted on a thin wire which has a coil of many turns as shown in Figure 10-34. The wire is held in place in a constant magnetic field produced by either permanent magnets or an electromagnet. Information of this kind is found in most elementary physics or control system texts.

10.4.16. Solid State or Electro-Optical Scanners. One approach to the generation of line scan or more complicated patterns is to impose sound waves on a transparent plate through which radiation travels (Figure 10-36). The maxima of the sound-wave amplitudes are regions of increased density and therefore increased refractive index. The sound pattern of traveling waves can be thought of as moving sinusoidal grating [10-6].

The conditions governing this scan follow directly from the laws of conservation of energy and conservation of momentum. An incident optical wave of frequency ω_1 and wave vector k_1 is scattered by the sound wave (of ω_s, k_s) into a new wave (ω_2, k_2). Some of the original wave is unscattered. Conservation of energy requires

$$\omega_2 - \omega_1 = \omega_s \qquad (10\text{-}48)$$

Typical values are

$$\omega_s < 10^6$$

$$\omega_1 < 10^{13}$$

The frequency shift is trivial for infrared applications. This relationship can be represented as a vector triangle in $\mathbf{k}$ space (Figure 10-37). Conservation of momentum requires

$$k_2 - k_1 - k_s = 0 \qquad (10\text{-}49)$$

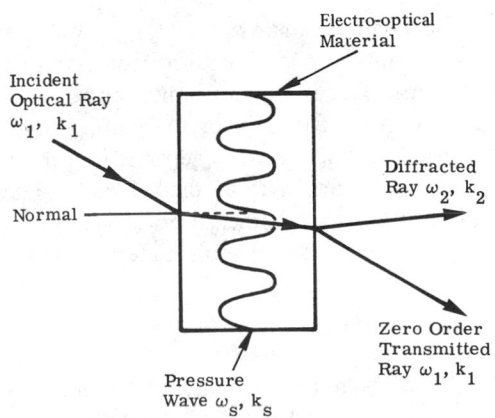

Fig. 10-36. Electro-optical acoustical scanning.

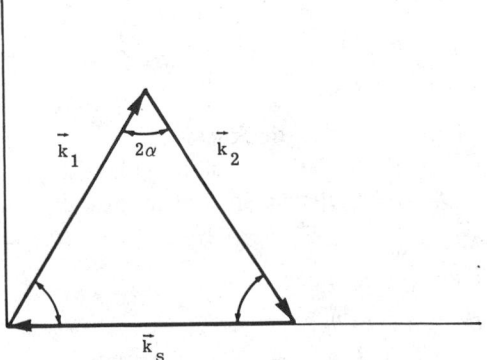

Fig. 10-37. Vector triangle in k-space.

The magnitudes of vectors $\mathbf{k}_1$ and $\mathbf{k}_2$ are almost equal. Therefore the vector triangle is isosceles, so that $k_s = 2k_2 \sin \alpha$ and therefore the deviation angle α that can be generated is

$$\alpha = \sin^{-1}\left(\frac{k_s}{2k_2}\right) = \sin^{-1}\left(\frac{\lambda_2}{2\lambda_s}\right) \tag{10-50}$$

For small angles it is a simple proportionality so that

$$\alpha \doteq \frac{1}{2}\left(\frac{\lambda_2}{\lambda_s}\right) = \frac{1}{2}\frac{\nu_s c}{\nu_2 v} \tag{10-51}$$

$$\alpha \doteq \frac{1}{2}\frac{\nu_s}{\nu_2} \times 10^6 \tag{10-52}$$

Typical values can be $\nu_s = 10^6$, $\nu_2 = 3 \times 10^{13}$ (10 μm), so that α is about 1/60 rad ($\approx 1°$). For an optical system with a resolution of 0.1 mrad, this represents only 170 resolution elements. This is a major limitation to scanning of this type. A second limitation has to do with the efficiency of the grating that is produced, and this in turn relates to the change in refractive index generated by the acoustical pressure wave [10-6].

10.4.17. Scanning by the Use of Nonlinear Materials. A plate with plane and parallel sides made of a nonlinear optical material can be used as a nonmoving scanning device if the refractive index of some materials can be written as

$$n = n_o - n_o^3 r_{63} E_z \tag{10-53}$$

where n_o = refractive index of the ordinary ray
r_{63} = the element in the sixth row, third column of the r_{ij} tensor where r_{ij} is defined by $\Delta(1/n^2)i = r_{ij} E_j$
E_z = z component of the electric field

A ray incident at an angle θ_i (Figure 10-38) will emerge parallel but displaced by

$$l_d = d\left\{\tan \theta_i - \left(1 - \frac{\sin^2 \theta_i}{n^2}\right)^{-1/2}\right\} \tag{10-54}$$

where d = plate thickness
n = relative refractive index
l_d = lateral displacement of the beam

Thus l_d is given by

$$l_d = d\{\tan \theta_i - [1 - \sin^2\theta_i(n_o - n_o^3 r_{63} E_z)^{-2}]^{-1/2}\} \tag{10-55}$$

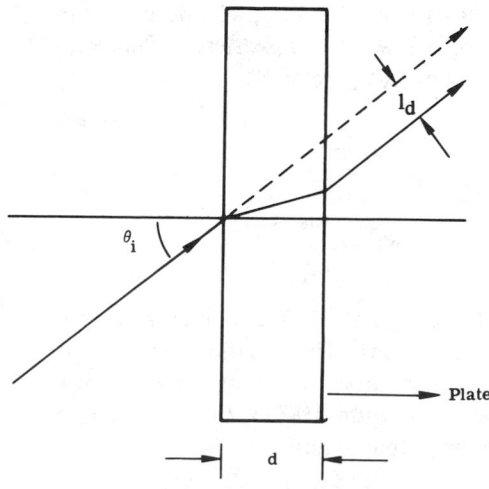

Fig. 10-38. A scanner using nonlinear materials.

A prism made of such a material can be used to obtain angular scans. A ray is shown in Figure 10-39 normal to the front face of a right-angle prism. The deviation angle, δ, is given by

$$\delta = \sin^{-1}[n \sin A] - A$$

$$= \sin^{-1}\{(n_o - n_o^3 r_{63} E_z) \sin A\} - A \qquad (10\text{-}56)$$

The change in deviation with respect to a change in the applied electric field is given by

$$\frac{d\delta}{dE_z} = n_o^3 r_{63}[1 - (n_o - n_o^3 r_{63} E_z)^2]^{-1/2} \qquad (10\text{-}57)$$

Values for n_o and r_{63} for selected materials can be found in Chapter 23.

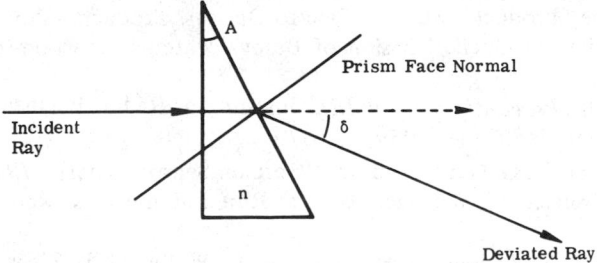

Fig. 10-39. A prism scanner using nonlinear materials.

10.4.18. Limitations on Scan Rate. Several factors are influential in limiting the rapidity of the rotation or oscillation of a scanning element. These include the forces generated, the strength of the materials, size, weight, lever arm, windage, and friction heating. For a polygonal scanner the maximum rate $\dot{\omega}$ is given by

$$\dot{\omega} = \frac{1}{2\pi r_o} \sqrt{\frac{8UTS}{\rho(3 + \eta)}} \qquad (10\text{-}58)$$

where r_o = the distance from the center to an edge
$\quad UTS$ = the ultimate tensile strength of the (solid) mirror
$\quad \eta$ = Poisson's ratio for the mirror material
$\quad \rho$ = its volumetric density

The maximum would probably be reached before this because of mirror deformation.

10.5. References and Bibliography

10.5.1. References

[10-1]   M. R. Holter, S. Nudelman, G. H. Suits, W. L. Wolfe, G. J. Zissis, *Fundamentals of Infrared Technology,* The Macmillan Co., New York, NY, 1963.

[10-2]   W. L. Wolfe and B. C. Platt, "Proceedings of the International Commission on Optics," Santa Monica, CA, 1974, National Academy of Science, Washington, DC.

[10-3]   William L. Wolfe, University of Arizona, Tucson, AZ, Personal Files.

[10-4]   John Duncan, William Wolfe, George Oppel, James Burn, "Horizon Sensor Report," IRIA State of the Art Report, University of Michigan, Willow Run Laboratories, Report No. 2389-80-T, April 1965.

[10-5]   American Time Products Staff, "Electro-Optical Products—Short Form Catalog," American Time Products, Division of Bulova Watch Co., Woodside, NY, September 1975.

[10-6]   Amnon Yariv, *Nonlinear Optics,* John Wiley & Sons, New York, NY, 1967.

10.5.2. Bibliography

American Time Products Staff, "Electro-Optical Products—Short Form Catalog," American Time Products, Division of Bulova Watch Co., Woodside, NY, September 1975.

Ballard, S.S. (ed.), *Proceedings of the IRE,* Institute of Radio Engineers, New York, NY, Vol 47, No. 9, September 1959, p. 1679

Duncan, J., W. Wolfe, G. Oppel, J. Burn, "Horizon Sensor Report," *IRIA State of the Art Report,* University of Michigan, Willow Run Laboratories, Report No. 2389-80-T, April 1965.

Holter, M. R., S. Nudelman, G. H. Suits, W. L. Wolfe, G. J. Zissis, *Fundamentals of Infrared Technology,* The Macmillan Co., New York, NY, 1963.

Lloyd, J. M., *Thermal Imaging Systems,* Plenum Press, New York, NY, 1975.

NASA, "Scanners and Images for Earth Resources Applications," Sponsored by Goddard Space Flight Center on behalf of the National Aeronautics and Space Administration, Cocoa Beach, Florida, December 11-15, 1972, National Aeronautics and Space Administration, Washington, DC, 1973.

Yariv, Amnon, *Nonlinear Optics,* John Wiley & Sons, New York, NY, 1967.

Chapter 11

DETECTORS

Thomas Limperis
Argo Sciences, Inc.

Joseph Mudar
Environmental Research Institute of Michigan

CONTENTS

11. Detectors

11.1. Introduction

The purpose of this chapter is to provide the critical data, formulae, and written text for evaluation of infrared detectors for specific applications. The chapter is limited to single-element detectors and detector arrays that are sensitive to the 0.7 to 1000 μm spectral region. A detector may be defined as:

> ...a device that provides an electrical output that is a useful measure of the radiation incident on the device. It is intended to include not only the responsive element, but also the physical mounting of the responsive element, as well as any other elements—such as windows, area-limited apertures, Dewar flasks, internal reflectors, etc.—that form an integral part of the detector as it is received from the manufacturer [11-1].

11.1.1. Symbols, Nomenclature and Units. Table 11-1 lists the symbols, nomenclature and units used in this chapter.

Table 11-1. Symbols, Nomenclature and Units

Symbols	Nomenclature	Units
A_c	Cross-sectional area	cm^2
A_d	Area of the detector	cm^2
A_e	Effective area of detector	cm^2
a	Absorption coefficient	cm^{-1}
B	Magnetic field	G
b	Ratio of electron to hole mobility	—
C	Electrical capacitance	F
$\mathcal{C}$	Heat capacitance	$J\,K^{-1}$
C_e	Equivalent capacitance	F
c_1	First radiation constant	$W\,cm^{-2}\,\mu m^4$
c_2	Second radiation constant	$\mu m\,K$
c	Velocity of light	$m\,sec^{-1}$
D	Detectivity	W^{-1}
D^*	Detectivity, specific (normalized with regard to detector area and electrical bandwidth)	$cm\,Hz^{1/2}\,W^{-1}$
$D^*(\lambda)$	Spectral, specific (normalized) detectivity	$cm\,Hz^{1/2}\,W^{-1}$
D^{**}	Detectivity normalized with regard to detector area, electrical bandwidth, and effective, weighted angular field of view	$cm\,Hz^{1/2}\,W^{-1}\,sr^{1/2}$
DQE	Detective quantum efficiency	—
D_e	Electron diffusion constant	$cm^2\,sec^{-1}$
D_h	Hole diffusion constant	$cm^2\,sec^{-1}$

Table 11-1. Symbols, Nomenclature and Units

Symbols	Nomenclature	Units
d	Thickness of responsive element	cm
E	Irradiance	W cm^{-2}
$\mathcal{E}$	Electric field	V cm^{-1}
E_g	Photon energy	J
E_i	Impurity activation energy of photoconductor	J
E_k	Kinetic energy of the freed electron	J
E_q	Photon flux density, or photon irradiance	photons cm^{-2} sec^{-1}
$E_{q,B}$	Photon flux density for background radiation	photons cm^{-2} sec^{-1}
$E_{q,s}$	Photon flux density for signal radiation	photons cm^{-2} sec^{-1}
e	Charge on an electron	C
FOV	Detector geometric field of view	sr or cone-angle degrees
f	Electrical frequency	Hz
f_c	Chopping frequency	Hz
f_o	Modulation frequency	Hz
$\mathcal{G}$	Thermal conductance	W K^{-1}
$\mathcal{G}_e$	Effective thermal conductance	W K^{-1}
$\mathcal{G}_0$	Combined effective and radiative thermal conductance	W K^{-1}
G_{gen}	Generation rate of free charge carriers	sec^{-1}
G_p	Photoconductive gain	—
G_{sh}	Effective shunt conductance	mho
g	Gain	—
h	Planck's constant	eV sec
I	dc current	A
I_d	dc diffusion current	A
I_s	dc signal current	A
I_{sa}	dc saturation current	A
I_{sc}	dc short-circuit current	A
i, i_s, i_n	ac rms generalized, rms signal, or rms noise current	A
$i(t), i_s(t), i_n(t)$	Instantaneous generalized, signal, or noise current	A
K	Constant	—
k	Boltzmann's constant	J K^{-1}
L	Diffusion length	cm
L_d	Effective diffusion length	cm
L_e	Electron diffusion length	cm
L_h	Hole diffusion length	cm
l	Length; also electrode separation	cm
M	Free charge carrier multiplication factor	—
N	Total number of free charges	—

Table 11-1. Symbols, Nomenclature and Units

Symbols	Nomenclature	Units
NEE	Noise equivalent irradiance	$W\ cm^{-2}$
NEP	Noise equivalent power	W
N_s	Signal photon rate	sec^{-1}
N_λ	Average photon rate per unit wavelength and per unit area	$sec^{-1}\ \mu m^{-1}\ cm^{-2}$
n	Density of free electrons	cm^{-3}
$\mathcal{P}$	Pyroelectric coefficient	$A\ cm^{-2}$
P_{ab}	Thermoelectric power	$V\ K^{-1}$
p	Density of free holes	cm^{-3}
R	Resistance	ohms
$\mathcal{R}$	Responsivity	$V\ W^{-1}$ or $A\ W^{-1}$
$\mathcal{R}_{bb}$	Blackbody responsivity	$V\ W^{-1}$ or $A\ W^{-1}$
$\mathcal{R}_{ref}(\lambda)$	Relative spectral responsivity of the reference	$V\ W^{-1}$ or $A\ W^{-1}$
R_d	Detector resistance	ohms
R_{dyn}	Dynamic resistance	ohms
R_e	Equivalent input resistance of the detector-preamplifier circuit	ohms
R_L	Load resistor	ohms
R_m	Resistance of the current meter	ohms
RQE	Responsive quantum efficiency	—
SNR	Signal to noise ratio	—
s	Surface recombination velocity	$m\ sec^{-1}$
T	Temperature	K
T_B	Background temperature	K
T_d	Detector temperature	K
T_0	Sink temperature	K
t	Time	sec
V	dc voltage	V
V_B	dc bias voltage	V
V_{bd}	dc breakdown voltage	V
V_0	dc open-circuit voltage	V
V_p	Peltier voltage	V
v, v_s, v_n	ac rms generalized, rms signal, or rms noise voltage	V V
v_c	ac calibration signal voltage, rms	V
v_n^*	Root-power-spectrum	$V\ Hz^{-1/2}$
v_0	ac open-circuit voltage	V
$\nu(t), \nu_s(t), \nu_n(t)$	Instantaneous generalized, signal, or noise voltage	V
v_T	Thermal noise voltage, rms	V
W_h	Heat generated in a detector due to $I^2 R_d$ heating	J
w	Width of the detector	cm

Table 11-1. Symbols, Nomenclature and Units

Symbols	Nomenclature	Units
Z	Impedance	ohms
$\mathcal{Z}$	Thermal impedance	$K\ W^{-1}$
$\tilde{Z}$	Complex impedance	ohms
α	Temperature coefficient of resistance	K^{-1}
β	Efficiency factor for a photodiode	—
γ	Coherence factor	—
ΔT_d	Temperature change of the detector	K
δ	Phase angle	rad
ϵ	Emissivity	—
η	Quantum efficiency	—
λ	Optical wavelength	μm
λ_c	Cutoff wavelength	μm
λ_p	Peak wavelength	μm
λ_s	Signal wavelength	μm
μ	Carrier mobility	$cm^2\ sec^{-1}\ V^{-1}$
μ_e	Electron mobility	$cm^2\ sec^{-1}\ V^{-1}$
μ_h	Hole mobility	$cm^2\ sec^{-1}\ V^{-1}$
ν	Optical frequency	sec^{-1}
π_{ab}	Peltier coefficient	V
ρ	Surface reflectance	—
σ	Stefan-Boltzmann constant	$W\ m^{-2}\ K^{-4}$
σ_c	Capture cross-section	cm^2
σ_e	Electrical conductivity	$mho\ cm^{-1}$
τ	Time constant	sec
τ_c	Average, free charge carrier lifetime	sec
τ_e	Electrical time constant	sec
τ_{el}	Electron lifetime	sec
τ_h	Hole lifetime	sec
τ_T	Thermal time constant	sec
Φ	Flux, or radiant power	W
$\Phi(t)$	Instantaneous radiant power	W
Φ_B	Background radiant power	W
$\Phi_{q,\lambda}(\lambda)$	Photon flux per unit wavelength	$photons\ sec^{-1}\ \mu m^{-1}$
$\Phi_{q,\lambda,B}(\lambda)$	Photon flux per unit wavelength from the background	$photons\ sec^{-1}\ \mu m^{-1}$
$\Phi_{q,\lambda,s}(\lambda)$	Photon flux per unit wavelength from the signal	$photons\ sec^{-1}\ \mu m^{-1}$
Φ_s	rms signal radiant power	W
$\Phi_\lambda(\lambda)$	Spectral radiant power or flux	$W\ \mu m^{-1}$
ϕ	Surface work function of a material	$J\ C^{-1}$
χ_i	Phase shift between input flux and output voltage	rad
Ω	Solid angle (field of view)	sr
Ω_e	Effective, weighted detector solid angle	sr
ω	Angular frequency	$rad\ sec^{-1}$

11.1.2. Symbols and Descriptions for Detector Parameters. Tables 11-2 and 11-3 present the currently acceptable symbols and preferred units for the important detector parameters and noise equations. This nomenclature was assembled from the archival literature, government reports, and the standards report prepared by Jones, et al. [11-1].

11.1.3. Responsive Elements. The responsive element is a radiation transducer. It changes the incoming radiation into electrical power which can be amplified by the accompanying electronics.

The methods of transduction can be separated into two groups: thermal detectors and photon detectors. The responsive element of thermal detectors is sensitive to changes in temperature brought about by changes in incident radiation. The responsive element of photon detectors is sensitive to changes in the number or mobility of free charge-carriers, i.e. electrons and/or holes, that are brought about by changes in the number of incident, infrared photons. Thermal detectors employ transduction processes including the bolometric, thermovoltaic, thermopneumatic, and pyroelectric effects. Photon detectors employ transduction processes including the photovoltaic, photoconductive, photoelectromagnetic, and the photoemissive effects. Each process of transduction is described in the following text.

11.1.4. Descriptions of the Processes of Transduction.

Bolometric Process: Changes in the temperature of the responsive element, induced by the incident, infrared radiation, cause changes in the electrical conductivity, monitored electrically.

Photoconductive Process: A change in the number of incident photons on a semiconductor causes a change in the average number of free charge-carriers in the material. The electrical conductivity of the semiconductor is directly proportional to the average number of free charge-carriers in the material. Therefore, the change in electrical conductivity is directly proportional to the change in the number of photons incident on the semiconductor.

Photoelectromagnetic Process: Photons absorbed at or near the surface of a semiconductor generate free charge-carriers which diffuse into the bulk and are separated enroute by a magnetic field. This charge separation produces an output voltage which is directly proportional to the number of incident photons.

Photovoltaic Process: A change in the number of photons incident on a semiconductor p-n junction causes a change in the voltage generated by the junction.

Pyroelectric Process: The incident, infrared radiation increases the temperature of the crystalline responsive element. This temperature change alters the dipole moment which produces an observable, external, electric field.

Thermopneumatic Process: The radiation incident on a gas in a chamber increases the temperature (and therefore the pressure) of the gas, causing the chamber to expand, and thus moving a mirror attached to an exterior wall. This movement can be detected optically.

Thermovoltaic Process: The temperature of a junction of dissimilar metals is varied by changes in the level of incident radiation absorbed at the junction and thus causes the voltage generated by the junction (due to the Seebeck effect) to fluctuate.

11.1.5. Windows. Windows are used to isolate the ambient environment from the special environment which is often required around the responsive element. In cooled detectors, the responsive element is kept in a vacuum. The window affects the spectral distribution of photons incident upon the responsive element.

DETECTORS

Table 11-2. Detector Parameters

Name, Symbol, and Preferred Units	Definition
Responsive area, A_d (cm^2)	For responsive elements made of thin films or single crystals, the responsive area is usually the geometric area. For detectors using integrating chambers, the responsive area is the entrance aperture. An effective area, A_e, can be defined by integrating a normalized responsivity over the responsive area.
Impedance, Z_d (ohms)	The slope of the instantaneous voltage-instantaneous current curve at bias voltage V_B.
Resistance, R_d (ohms)	The ratio of the dc voltage across the detector to the dc current through it.
Background temperature, T_B (K)	The temperature of a uniform blackbody completely filling the detector field of view that would give the observed total flux on the detector.
Detector solid angle, Ω (sr)	The solid angle (field of view) from which the detector receives radiation.
Effective, weighted detector solid angle, Ω_e (sr)	The solid angle (field of view), weighted by a cosine function, from which the detector receives radiation.

Table 11-2. Detector Parameters (*Continued*)

Equation Definition	Functional Relationship
A_d = area of detector (geometric)	
$A_e = \displaystyle\int\int_{A_d} \frac{\mathcal{R}(x, y)dxdy}{\mathcal{R}_{max}}$	
where $\mathcal{R}_{max}$ = maximum value of $\mathcal{R}(x, y)$ $\mathcal{R}$ = responsivity x, y = coordinates in plane of responsive area	
$Z_d = \dfrac{d\nu(t)}{di(t)}\bigg\|_{V_B}$	Z_d is a function of the bias voltage, the inter-electrode capacitance, and the level of irradiance.
$R_d = V/I$	R_d is a function of the detector temperature and the level of irradiance.

$$\Omega_e = \int\int_{A_d}\left[\int_0^{\pi/2}\int_0^{2\pi}\frac{\cos\theta\,\sin\theta\,\mathcal{R}(x, y, \phi, \theta)}{A_e\,\mathcal{R}_{max}(0, 0)}\,d\phi\,d\theta\right]dx\,dy$$

where $\mathcal{R}_{max}(0, 0)$ = the maximum value of $\mathcal{R}(x, y, 0, 0)$; measured with a small field of view, $d\Omega$

ϕ and θ = spherical coordinates, with ϕ being the azimuthal angle

$\cos\theta$ = the weighting function

The z-axis is normal to the plane of the responsive element. If the respon-

Table 11-2. Detector Parameters (*Continued*)

Name, Symbol, and Preferred Units	Definition
Instantaneous signal voltage, $v_s(t)$ or current, $i_s(t)$ (V) or (A), respectively	That component of the electrical output voltage (or current) which is coherent with $\Phi_s(t)$, the instantaneous value of the input signal radiant power. $\Phi_s(t)$ can be monochromatic or have a blackbody character.
rms amplitude of the fundamental signal voltage or current component, v_s or i_s (V) or (A), respectively	The rms amplitude of the fundamental signal component determined by taking the square root of the time average of the square of the first time varying component in the series, i.e., the fundamental.
rms noise voltage, v_n, or current, i_n (V) or (A)	That component of the electrical output voltage (or current) which is incoherent with the signal radiant power. This value is determined with the signal power removed.
Spectral responsivity, $\mathcal{R}(\lambda)$ (V W^{-1}) or (A W^{-1}),	The ratio of the rms signal voltage (or current) to the rms value of the monochromatic incident signal power, referred to an infinite load impedance and to the terminals of the detector.
Blackbody responsivity, $\mathcal{R}_{bb}$ (V W^{-1}) or (A W^{-1})	Same as spectral responsivity except that the incident signal power is from a blackbody.

Table 11-2. Detector Parameters (*Continued*)

Equation Definition	Functional Relationship
sivity is not a function of ϕ the element is said to have circular symmetry and	

$$\Omega = \pi \sin^2 (\Theta/2)$$

where Θ is the total cone angle.

If the incident radiant power, $\phi_s(t)$, is periodic in time:	The signal voltage is a function of electrical frequency, f. For a single-time-constant detector,

$$\phi_s(t) = \Phi_0 + \Phi_1 \cos (2\pi ft + \delta_1)$$

$$+ \Phi_2 \cos (2 \cdot 2\pi ft + \delta_2) + \ldots$$

then

$$v_s = \frac{v_{s,max}}{(1 + 4\pi^2 f^2 \tau^2)^{1/2}}$$

$$\nu_s(t) = V_0 + V_1 \cos (2\pi ft + \psi_1)$$

$$+ V_2 \cos (2 \cdot 2\pi ft + \psi_2) + \ldots$$

$v_s = (2)^{-1/2} V_1$	v_s is related to f, and bias voltage.

If the dc gain of the associated electronics is zero,	v_n is related to the detector area, Δf, f, and in some cases to Ω and T_B.

$$v_n = (\overline{\nu_n^2(t)})^{1/2}$$

$\mathcal{R}(\lambda) = \dfrac{v_s}{\Phi_{s,\lambda} \Delta\lambda}$	Responsivity is a function of λ, f, T, and bias voltage or current.

$\mathcal{R}_{bb} = \dfrac{v_s}{\Phi_{s,bb}}$	Responsivity is a function of f, T, and bias voltage or current.

Table 11-2. Detector Parameters (*Continued*)

Name, Symbol, and Preferred Units	Definition
Time constant, τ (sec)	A measure of the detector's speed of response. The alternative equations for τ (next column) become identical if the noise has a flat power spectrum (See Section 11.4.3.) and if the responsivity varies with frequency according to the relation $$\mathcal{R}(f) = \frac{\mathcal{R}(0)}{(1 + 4\pi^2 f^2 \tau^2)^{1/2}}$$

Table 11-2. Detector Parameters (*Continued*)

Equation Definition	Functional Relationship

(a) The decay time constant is given as
$$\tau = 1/(2\pi f_c)$$
where f_c is that chopping frequency at which the responsivity has fallen to $2^{-1/2}$ of its maximum value. | — |

(b) The rise time constant is the time required for the signal voltage (or current) to rise to $1 - 1/e$ or 0.63 times its asymptotic value. It is measured by the lightpulse method: exposing the detector to a square-wave pulse of radiation | — |

(c) Responsive time constant
$$\tau_r = \frac{\mathcal{R}^2_{max}}{4 \int_0^\infty [\mathcal{R}(f)]^2 df}$$

(d) Detective time constant
$$\tau_d = \frac{(D^*_{max})^2}{4 \int_0^\infty [D^*(f)]^2 df}$$

(e) Empirical responsive time constant

$$\tau_{rs} = \frac{1}{2\pi} \left\{ \frac{[\mathcal{R}(f_1)]^2 - [\mathcal{R}(f_2)]^2}{[f_2 \mathcal{R}(f_2)]^2 - [f_1 \mathcal{R}(f_1)]^2} \right\}^{1/2}$$

f_1 and f_2 must be specified.

(f) Empirical detective time constant

$$\tau_{ds} = \frac{1}{2\pi} \left\{ \frac{[D^*(f_1)]^2 - [D^*(f_2)]^2}{[f_2 D^*(f_2)]^2 - [f_1 D^*(f_1)]^2} \right\}^{1/2}$$

f_1 and f_2 must be specified.

Table 11-2. Detector Parameters (*Continued*)

Name, Symbol, and Preferred Units	Definition
Spectral noise equivalent power, $(NEP)_\lambda$ (W)	That value of monochromatic incident rms signal power of wavelength λ required to produce an rms signal-to-rms-noise ratio of unity. The chopping frequency, the electrical bandwidth used in the measurement, and the detector area should be specified.
Blackbody noise equivalent power, NEP_{bb} (W)	That value of incident rms signal power (with a blackbody spectral character) required to produce an rms signal-to-rms-noise ratio of unity. The blackbody temperature must be specified along with the detector area, the electrical bandwidth used in the measurement, and the chopping frequency.
Spectral detectivity $D(\lambda)$ (W^{-1})	The reciprocal of spectral noise equivalent power. The chopping frequency, the electrical bandwidth used in the measurement, and the detector sensitive area should be specified.
Blackbody detectivity, D_{bb} (W^{-1})	The reciprocal of the blackbody noise equivalent power. The blackbody temperature should be specified, along with the electrical bandwidth used in the measurement, the detector area, and the chopping frequency.
Spectral D-star $D^*(\lambda, f_c)$ $(cm\ Hz^{1/2}\ W^{-1})$	A normalization of spectral detectivity to take into account the area and electrical bandwidth dependence. The chopping frequency (f_c) used in the measurement is specified by inserting it in the parentheses as indicated in the last column. For detectors limited by the fluctuation in arrival rate of background photons, Ω and T_B must be specified.
Blackbody D-star, $D^*(T_{bb}, f_c)$ $(cm\ Hz^{1/2}W^{-1})$	A normalization of blackbody detectivity to take into account the detector area and the electrical bandwidth. The chopping frequency (f_c) and the blackbody temperature (T_{bb}) are specified in the parentheses as indicated. For detectors that are background noise limited, Ω and T_B must also be specified.

Table 11-2. Detector Parameters (*Continued*)

Equation Definition	Functional Relationship
$(NEP)_\lambda = \Phi_{s,\lambda} \Delta\lambda \left(\dfrac{v_n}{v_s}\right)$ $= \dfrac{v_n}{\mathscr{R}_\lambda}$	Depends upon λ, A, f, Δf, and in some cases Ω and T_B.
$NEP_{bb} = \Phi_{s,bb} \left(\dfrac{v_n}{v_s}\right)$ $= \dfrac{v_n}{\mathscr{R}_{bb}}$	Depends upon blackbody temperature, A, f, Δf, and in some cases Ω and T_B.
$D(\lambda) = 1/(NEP)_\lambda$	Depends upon λ, A, f, Δf, and in some cases Ω and T_B.
$D_{bb} = 1/NEP_{bb}$	Depends upon blackbody temperature, A, f, Δf, and in some cases Ω and T_B.
$D^*(\lambda, f_c) = \sqrt{A_d \Delta f}\, D(\lambda)$	For background-noise-limited detectors, $D^*(\lambda, f_c)$ depends upon Ω and T_B.
$D^*(T_{bb}, f_c) = \sqrt{A_d \Delta f}\, D_{bb}$	For background-noise-limited detectors, $D^*(T_{bb}, f_c)$ depends upon Ω and T_B.

Table 11-2. Detector Parameters (*Continued*)

Name, Symbol, and Preferred Units	Definition
Maximized D-star, $D^*(\lambda_p, f_c)$ (cm Hz$^{1/2}$W^{-1})	A quantity obtained when the wavelength is λ_p and the chopping frequency used yields a maximum rms signal-to-rms-noise ratio.
Spectral D-double star, $D^{**}(\lambda, f_c)$ (cm Hz$^{1/2}$sr$^{1/2}$W^{-1})	A normalization of $D^*(\lambda, f_c)$ to account for the detector effective weighted field of view Ω_e (Note: if $\Omega_e = \pi$, $D^{**} = D^*$.)
Peak wavelength, λ_p or λ_{max} (μm)	The wavelength at which spectral detectivity is a maximum.
Cutoff wavelength, λ_c (μm)	The wavelength at which $D^*(\lambda, f_c)$ has degraded to one-half its peak value.
Responsive quantum efficiency, RQE	The ratio of the number of countable output events, N_o, to the number of incident photons, N_p.
Detective quantum efficiency, DQE	The square of the ratio of measured detectivity to the theoretical limit of detectivity. Both detectivities must be for the same set of conditions
D^*f^*	The product of the maximum D-star and f^*, the highest frequency at which $D^*(f)$ has decreased to $2^{-1/2}$ of its maximum value.

Table 11-2. Detector Parameters (*Continued*)

Equation Definition	Functional Relationship
$D^*(\lambda_p, f_c) = \dfrac{\sqrt{A_d \Delta f}}{(\text{NEP})_{\lambda_p}}$	Same as for $D^*(\lambda, f)$.
$D^{**}(\lambda, f_c) = (\Omega/\pi)^{1/2} D^*(\lambda, f_c)$	—
—	Depends upon cell temperature and detector material used.
—	Depends upon cell temperature and detector material used.
$\text{RQE} = N_o/N_p$	Depends upon bias voltage, time constant, and cell geometry.
$\text{DQE} = \left[\dfrac{D(\lambda) \quad \text{measured}}{D(\lambda) \quad \begin{array}{l}\text{theoretical}\\ \text{limit}\end{array}}\right]^2$	—
D^*f^*, where $D^*(f^*) = 2^{-1/2}D^*_{\text{max}}$	—

Table 11-3. Detector Noises [11-2]

Type of Noise	Physical Mechanism
Johnson (also called Nyquist or thermal)	At thermal equilibrium the random motion of charge carriers in a resistive element generates a random electrical voltage across the element. As the temperature of the resistor is increased, the mean kinetic energy of the carriers increases, yielding an increased electrical noise voltage.
Temperature	The fluctuations in temperature of the sensitive element, due to either radiative exchange with the background or conductive exchange with the heat sink, produce a fluctuation in signal voltage. For thermal detectors, the detector is said to be at its theoretical limit if the temperature noise is due to radiative exchange with the background.
Modulation (or $1/f$)	The mechanism is not well understood. As its name implies, it is characterized by a $1/f^n$ noise power spectrum, where n varies from 0.8 to 2.
Generation-recombination, G-R	Statistical fluctuations in the rate of generation and in the rate of recombination of charge carriers in the sensitive element result in an electrical noise. These fluctuations can be caused by charge-carrier-phonon interactions or by the random arrival rate of photons from the background. If the background photons are the prime contributors to the fluctuation in G-R rates, then the noise is often called photon, radiation, or background noise.
Shot	Noise caused by the discreteness of electronic charge. The current, I, flowing through the responsive element, is the result of current pulses produced by the individual electrons and/or holes.

Table 11-3. Detector Noises [11-2] *(Continued)*

Detectors Concerned	Equation for v_n
All detectors	$$v_n = (4kT_dR_d\Delta f)^{1/2} \qquad \text{(See p. 11-45.)}$$
All detectors but especially those made of thin films	For thermal detectors, $$\overline{\Delta T^2} = \frac{4kT_d^2\mathcal{G}\Delta f}{\mathcal{G}^2 + 4\pi^2f^2C^2} \qquad \text{(See p. 11-28.)}$$ The relation between $\overline{\Delta T^2}$ and v_n should be determined for each detector.
All detectors	$$v_n \propto R_dI\left(\frac{\Delta f}{A_dd}\right)^{1/2}\left(\frac{1}{f}\right)^n \qquad \text{(See p. 11-44.)}$$
All photon detectors	$$v_n = R_dI\left(\frac{2\tau\Delta f}{N(1 + 4\pi^2f^2\tau^2)}\right)^{1/2} \qquad \text{(See p. 11-44.)}$$ For photovoltaic detectors, the value of v_n is smaller by a factor of $\sqrt{2}$ since only fluctuations in the generation rate of free charge carriers contribute to the noise. Fluctuations in the rate of recombination of free charge carriers do not affect the detector output voltage.
Photovoltaic detectors and thin-film detectors	$$v_n = R_d(2eI\Delta f)^{1/2} \qquad \text{(See p. 11-45.)}$$ where e is the charge of an electron.

11.1.6. Apertures. Apertures are used to restrict the field of view of the responsive element. This is often done in cooled detectors that are photon-noise limited to cut down on the extraneous, background photons and thus reduce noise. (See Section 11.4.)

11.1.7. Dewar Flask. Dewar flasks are used to house the coolant needed to reduce the operating temperature of the responsive element and thus improve detectivity.

11.2. Theoretical Descriptions of Thermal Detectors

As indicated earlier, thermal detectors rely on one of four basic processes to accomplish infrared radiation detection. The four processes are:

(1) The bolometric effect.
(2) The thermovoltaic effect.
(3) The thermopneumatic effect.
(4) The pyroelectric effect.

The elementary theory of each process is given below.

11.2.1. Bolometers. The bolometric effect is a change in the electrical resistance of the responsive element due to temperature changes produced by absorbed, incident, infrared radiation. Figures 11-1 and 11-2 show two electronic circuit configurations that use this effect.

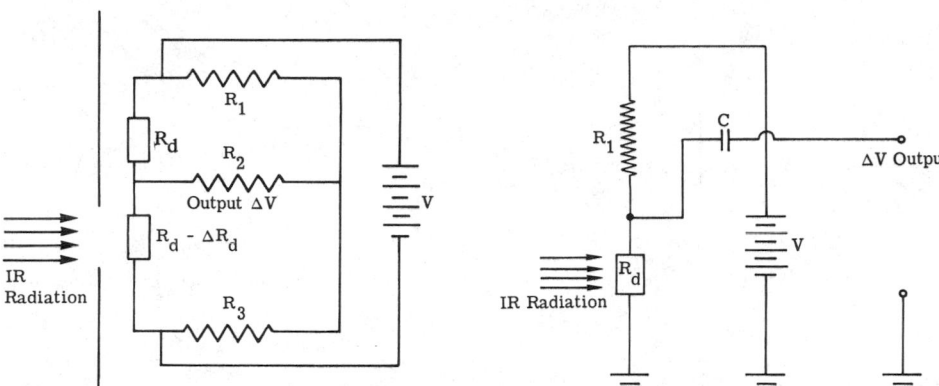

Fig. 11-1. Bolometer detector circuit with bridge configuration for dc operation.

Fig. 11-2. Bolometer detector circuit.

When the bridge circuit is used (Figure 11-1), the two detectors are placed close to each other with one shielded from any incident radiation in excess of the ambient levels. The bridge is balanced when no excess radiation is on the exposed detector. Incident, infrared radiation will then cause a rise in the temperature of the exposed detector, thereby causing a drop in its resistance. This electrically unbalances the bridge, causing a current to flow through R_2. In the ac circuit of Figure 11-2, only changes in voltage across the bolometer pass through the coupling capacitor to the electronics.

The change in electrical resistance resulting from the increased temperature of the bolometer depends upon the temperature coefficient of resistance, α, which is

$$\alpha = \frac{1}{R_d} \frac{dR_d}{dT_d} \tag{11-1}$$

where R_d = the resistance of the detector

T_d = the temperature of the detector

The signal equations for the circuits in Figures 11-1 and 11-2 are, respectively,

$$\nu_s = \Delta v = \frac{I(\Delta R_d)R_2}{2R_2 + R_1 + R_3} \tag{11-2}$$

$$\nu_s = \Delta v = \frac{R_1 V \Delta R_d}{(R_d + R_1)^2} \tag{11-3}$$

$$\Delta R_d = \frac{dR_d}{dT_d}\Delta T_d \tag{11-4}$$

where I = the steady state current through the bolometers in the bridge circuit

$$\Delta R_d = \frac{dR_d}{dT_d}\Delta T_d$$

ΔT_d = the time-variation of T_d

Δv = the resulting change in voltage

ν_s = the ac signal voltage

V = the dc bias voltage

R_1, R_2, R_3 are identified in Figures 11-1 and 11-2 R_d and T_d are as in Equation (11-1) above

Responsivity.   The responsivity, $\Re$, is defined as

$$\Re = \frac{\Delta v}{\Delta \Phi} \tag{11-5}$$

where Δv is the open-circuit output voltage appearing across the load resistor for an incremental increase in the infrared radiation power input, $\Delta \Phi$. The increase in bolometer temperature caused by $\Delta \Phi$ is ΔT_d, which is expressed in the following differential equation:

$$\mathcal{C}\frac{d\Delta T_d}{dt} + \mathcal{G}_0 \Delta T_d = W_h + \Delta \Phi \tag{11-6}$$

where $\mathcal{C}$ = the heat capacitance of the bolometer element, J K^{-1}

$\mathcal{G}_0 \Delta T_d$ = the conductive and radiative heat flow for the element

W_h = the thermal power generated in the bolometer due to $I^2 R_d$ heating

In the steady-state condition,

$$\mathcal{G}_0 \Delta T_d = W_h = I^2 R_d \tag{11-7}$$

From Equations (11-6) and (11-7), one can write the following equation when ΔT_d is small:

$$\mathcal{C}\frac{d\Delta T_d}{dt} + \mathcal{G}\Delta T_d = \frac{dW_h}{dT}\Delta T_d + \Delta \Phi \tag{11-8}$$

where $\mathcal{G}$ is the thermal conductance defined for small temperature changes, in units of $W\,K^{-1}$. The rate of change of W_h with T depends on the electronic circuit arrangement. For the circuit shown in Figure 11-2,

$$\frac{dW_h}{dT} = \alpha W_h \left(\frac{R_1 - R_d}{R_1 + R_d} \right) \tag{11-9}$$

$$\frac{dW_h}{dT} = \alpha (\Delta T_d) \, \mathcal{G}_0 \left(\frac{R_1 - R_d}{R_1 + R_d} \right) \tag{11-10}$$

Equation (11-8) can now be rewritten

$$\mathcal{C} \frac{d\Delta T_d}{dT} + \mathcal{G}_e \Delta T_d = \Delta \Phi \tag{11-11}$$

where $\mathcal{G}_e$ is the effective thermal conductance, given as

$$\mathcal{G}_e = \mathcal{G} - \alpha \mathcal{G}_0 (\Delta T_d) \left(\frac{R_1 - R_d}{R_1 + R_d} \right) \tag{11-12}$$

If $\mathcal{G}_e < 0$, then Equation (11-11) has an exponentially increasing solution when $\Delta \Phi = 0$. The bolometer is unstable in this condition and will burn out. For stable operation, the requirement is

$$\mathcal{G} > \mathcal{G}_0 \alpha (\Delta T_d) \tag{11-13}$$

where R_1 is chosen such that $R_1 \gg R_d$ for maximizing the signal voltage, Δv. The solution of Equation (11-11) is given below for a sinusoidally varying input radiation function ($\Delta \Phi = \Delta \Phi \cos \omega t$)

$$\Delta T_d = \frac{\epsilon \Delta \Phi_0}{\mathcal{G}_e (1 + \omega^2 \tau^2)^{1/2}} \tag{11-14}$$

where $\quad \tau = \mathcal{C}/\mathcal{G}_e$
$\qquad \epsilon = $ the emissivity of the bolometer
$\qquad \Delta \Phi_0 = $ the periodic function with angular frequency ω and a peak amplitude $\Delta \Phi_{max}$

The thermal response to an arbitrary, periodic, radiation impact can be determined by expressing the arbitrary, periodic function in terms of its Fourier-series components and applying the superposition principle. From Equation (11-1) it can be shown that

$$\Delta R_d = \Delta T_d R_d \alpha \tag{11-15}$$

Therefore,

$$\Delta R_d = \frac{R_d \alpha \epsilon \Delta \Phi_0}{\mathcal{G}_e (1 + \omega^2 \tau^2)^{1/2}} \tag{11-16}$$

The responsivity, $\mathcal{R}$, of the bolometer in the circuit shown in Figure 11-2 is obtained by combining Equations (11-16) and (11-3) to obtain the following expression:

$$\mathcal{R} = \left[\frac{R_1}{R_1 + R_d}\right] \frac{I\epsilon R_d \alpha}{\mathcal{G}_e(1 + \omega^2\tau^2)^{1/2}} \tag{11-17}$$

For the bridge circuit shown in Figure 11-1, the responsivity becomes

$$\mathcal{R} = \frac{1}{2}\epsilon I R_d \alpha \frac{1}{\mathcal{G}_e} \tag{11-18}$$

Several numerical examples given in Reference [11-2] are presented below.

Case 1: An Ideal, Metal Bolometer with Predominantly Conductive Cooling. The temperature coefficient of resistance is given as $\alpha = (1/R_d)\,(dR_d/dT_d)$. For a metal, the resistance over a wide temperature range is approximately proportional to the temperature so $\alpha \approx 1/T_d$. If one assumes that $R_1 \gg R_d$ and $\mathcal{G} \approx \mathcal{G}_0$, then Equation (11-12) becomes

$$\mathcal{G}_e = \mathcal{G}\left[1 - \left(\frac{\Delta T_d}{T_d}\right)\right] \tag{11-19}$$

If $T_0 = 300$ K, $T_d = 450$ K, $R_d = 50$ ohms, and $\mathcal{G} = \mathcal{G}_0 = 10^{-4}$ W K^{-1}, then $\mathcal{G}_e = 6.7 \times 10^{-5}$. Using the assumptions above, and assuming $\omega^2\tau^2 \ll 1$, one can simplify Equation (11-17):

$$\mathcal{R} = \frac{IR_d\epsilon}{T_d(6.7 \times 10^{-5})} \tag{11-20}$$

If one sets $\epsilon = 1$ and solves for I in Equation (11-7), the responsivity, $\mathcal{R}$, becomes 30 V W^{-1}.

Case 2: Metal Bolometer, Predominantly Conductive Cooling. Assume $\mathcal{G} = \mathcal{G}_0 = 10^{-4}$ W K^{-1}, $\alpha = $ constant, $\epsilon = 1$, $\omega^2\tau^2 \ll 1$, $T_0 = 300$ K, $R_d = 50$ ohms, $R_1 \gg R_d$, and $T_d = 375$ K.

From Equation (11-12),

$$\mathcal{G}_e = \mathcal{G}\,[1 - \alpha(\Delta T)] \tag{11-21}$$

To prevent thermal instability and detector burnout, $(\Delta T_d) < 1/\alpha$. If $\Delta T_d = 1/2\alpha$, then $\mathcal{G}_e = (1/2)\,(\mathcal{G})$. The responsivity equation is

$$\mathcal{R} = \frac{I\epsilon R_d \alpha}{\mathcal{G}_e} \tag{11-22}$$

I is computed using Equation (11-7). Substitution into Equation (11-22) gives $\mathcal{R} = 82$ V W^{-1}.

Case 3: Semiconducting Bolometer. Assume that $\alpha = -10T_0/T^2$, $\mathcal{G} = \mathcal{G}_0 = 10^{-4}$ W K^{-1}, $R_1 \gg R_d$, $\epsilon = 1$, $\omega^2\tau^2 \ll 1$, $T_d = 315$ K, $T_0 = 300$ K, and $R_d = 10^6$ ohms.

$$\mathcal{G}_e = \mathcal{G} - \alpha\mathcal{G}_0(\Delta T_d)$$

$$\mathcal{G}_e = \mathcal{G}[1 - \alpha(\Delta T_d)] = 5.5 \times 10^{-5} \tag{11-23}$$

If one substitutes this value for $\mathcal{G}_e$, as well as the value for I derived from Equation (11-7) into the responsivity equation, then

$$\mathcal{R} = \frac{IR_d\alpha}{\mathcal{G}_e} \tag{11-24}$$

$$\mathcal{R} = 21,000, \text{V W}^{-1}$$

Noise. The noise voltage from commercially available thermistor bolometers is composed of $1/f$ noise (also called current, excess, or modulation noise) and Johnson noise. Current noise is expressed as

$$v_n \propto IR_d\left(\frac{\Delta f}{A_d}\right)^{1/2}\left(\frac{1}{f}\right)^{1/2} \tag{11-25}$$

and Johnson noise is expressed as

$$v_n = (4\,kT_dR_d\Delta f)^{1/2} \tag{11-26}$$

For bias current values high enough to give maximum detector performance (optimum signal-to-noise ratio), $1/f$ noise predominates throughout most of the useful part of the frequency spectrum to which the detector is responsive. If the bias current is reduced sufficiently, then the $1/f$ noise is reduced and the Johnson noise predominates. In this case, the spectrum of the detector noise is flat, depending only upon the resistance and temperature of the responsive element. (See the noise spectrum for bolometers in the data enumeration section of this chapter.)

Some bolometers have been specially designed and built to operate at low temperatures to increase detector sensitivity and decrease the time constant. Significant improvements in the detectivity, D^*, and time constant have been observed. These cooled bolometers have not done well commercially because of their increased complexity and cost, caused by the need for cryogenic apparatus. Photon detectors are more attractive by comparison.

11.2.2. Thermocouples and Thermopiles. A junction of two dissimilar materials will, when heated, produce a voltage across the two open leads. This is the thermovoltaic effect. Such a junction is called a thermocouple. When more than one thermocouple is combined in a single responsive element, it is termed a thermopile.

Figure 11-3 contains a schematic of a thermocouple made of two dissimilar materials, A and B, connected with an electrical conductor, C. The junction, J_1, is attached to the responsive element that is irradiated with infrared radiation. Upon absorbing the infrared radiation, the temperature of the responsive element increases from T_d to $(T_d + \Delta T_d)$, which causes heating at J_1. If one assumes that the temperature at J_1 is also $T_d + \Delta T_d$, then the open-circuit thermoelectric electromotive force (emf) established in the circuit is

$$V_0 = P_{ab}\Delta T_d \tag{11-27}$$

where P_{ab} is a characteristic of the two materials and is known as the thermo-electric power. When radiation is incident on the responsive element, thus heating it, a current will flow in the circuit. This current will flow through the junction and tend to cool it by the Peltier effect. The cooling, ΔW_h, is given by

$$\Delta W_h = - \pi_{ab} I \qquad (11\text{-}28)$$

where π_{ab} is known as the Peltier coefficient. The quantity π_{ab} is related to P_{ab} by

$$\pi_{ab} = T_d P_{ab} \qquad (11\text{-}29)$$

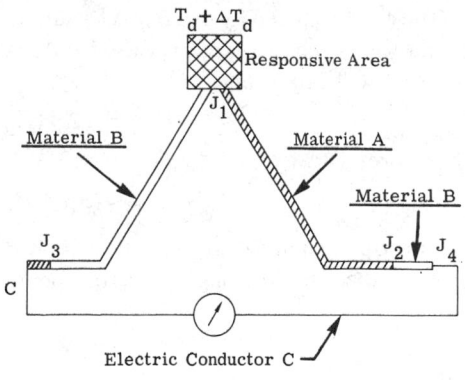

Fig. 11-3. Thermocouple schematic.

As soon as current flows, the value of ΔT_d is changed by the Peltier cooling of the junction, J_1. If the cold junction, J_2, is kept at a constant temperature, then, using Equations (11-28) and (11-29), one can show that the hot junction will be cooled at a rate $I P_{ab} \Delta T_d$. If $\mathcal{Z}$ is the thermal impedance of the hot junction plus the responsive element, then

$$\Delta(\Delta T_d) = I P_{ab} \mathcal{Z} T_d \qquad (11\text{-}30)$$

This cooling induces an emf, V_p, in the circuit, where

$$V_p = P_{ab} \Delta(\Delta T_d) = - I P_{ab}^2 \mathcal{Z} T_d \qquad (11\text{-}31)$$

The total emf, V_t, caused by the increased temperature at J_1 resulting from the incident infrared radiation and the Peltier effect is

$$V_t = V_0 + V_p = V_0 - I P_{ab}^2 \mathcal{Z} T_d \qquad (11\text{-}32)$$

The current, I, is related to V_t by

$$I = \frac{V_t}{(R_d + R_m)} \qquad (11\text{-}33)$$

where R_d = detector resistance
R_m = meter resistance

Therefore,

$$V_0 = I(R_d + R_m + P_{ab}^2 \mathcal{Z} T_d) \qquad (11\text{-}34)$$

This shows that, in effect, Peltier cooling increases the electrical resistance of the circuit by the dynamic resistance of the thermocouple,

$$R_{dyn} = P_{ab}^2 \mathcal{Z} T_d \qquad (11\text{-}35)$$

When there is constant, external radiant power, Φ, feeding into the responsive element, J_1, the equation for the balance of the heat flow becomes

$$\frac{\Delta T_d}{\mathcal{Z}} = \Phi - P_{ab} I T_d \qquad (11\text{-}36)$$

The emf produced by Φ is $P_{ab}\Delta T_d$, and the current in the circuit is $P_{ab}\Delta T_d(R_d + R_m)^{-1}$. If one solves the latter expression for ΔT_d, Equation (11-35) for T_d, and substitutes into Equation (11-36), then

$$I = P_{ab} \mathfrak{Z} \Phi (R_d + R_m + R_{dyn})^{-1} \qquad (11\text{-}37)$$

Therefore,

$$\Delta T_d = \mathfrak{Z} \Phi (R_d + R_m)(R_d + R_m + R_{dyn})^{-1} \qquad (11\text{-}38)$$

The open circuit emf, V_0, and the level of ΔT_d for the open circuit case, ΔT_0, can be computed by omitting the Peltier heating term:

$$\Delta T_0 = \mathfrak{Z} \Phi \qquad (11\text{-}39)$$

and

$$V_0 = P_{ab}\Delta T_0 = P_{ab} \mathfrak{Z}\Phi \qquad (11\text{-}40)$$

Therefore, Equation (11-37) can be rewritten in terms of the open circuit voltage, V_0:

$$I = \frac{V_0}{R_d + R_m + R_{dyn}} \qquad (11\text{-}41)$$

and ΔT_d can be written in terms of ΔT_0:

$$\Delta T_d = \Delta T_0 \frac{(R_d + R_m)}{(R_d + R_m + R_{dyn})} \qquad (11\text{-}42)$$

The responsivity, $\mathcal{R}$, of a thermovoltaic detector for constant input power, Φ, is then

$$\mathcal{R} = \frac{V_0}{\Phi} = \epsilon P_{ab} \mathfrak{Z} \qquad (11\text{-}43)$$

where ϵ is the fraction of incident infrared power that is absorbed. To obtain a high responsivity, one needs to select materials with a high value of thermoelectric power and high thermal resistance.

The responsivity of a thermovoltaic detector to an alternating input power is given as

$$\mathcal{R} = \epsilon P_{ab} \mathfrak{Z}(1 + \omega^2 \tau^2)^{-1/2} \qquad (11\text{-}44)$$

where τ = time constant, $\tau = \mathfrak{Z}\mathcal{C}$
$\mathcal{C}$ = the thermal capacitance of the responsive element
ω = the angular frequency of the alternating input power, Φ

A derivation of this equation is given in Reference [11-2].

The time constant, τ, of evaporated thermopiles ranges from 4 to 50 msec, depending upon the type and thickness of the radiation-absorbing material used on the thermopile surface. This absorbing material increases the thermal capacitance of the responsive element. If the responsive element is enclosed in a sealed housing, the thermal conductive paths will affect the time constant. A high, effective, thermal conductance leads to a decreased time constant.

According to manufacturers' data, Johnson noise is the predominant noise mechanism in currently produced thermocouples and thermopiles.

11.2.3. Thermopneumatics. In this detection process, an infrared-radiation-absorbing element is placed in a chamber filled with gas. A window in one of the chamber walls allows incident, infrared radiation to irradiate the absorbing element. When an increased flux of infrared radiation strikes the element, it heats and, by conduction, heats the gas in the chamber. The increase in temperature of the gas results in increased pressure in the chamber, which distorts a thin, flexible mirror mounted on one of the chamber walls. The degree of distortion is sensed by a separate optical system consisting of a light source and a detector.

Golay [11-3] has described the theory of operation for this type of detector.

11.2.4. Pyroelectrics. When the responsive element absorbs an incrementally increased amount of infrared radiation, its temperature rises, changing its surface charge. With the appropriate external circuit, this change in surface charge leads to a signal voltage. The change in temperature, ΔT_d, of the responsive element is related to its thermal capacitance, $\mathcal{C}$, and its thermal conductance, $\mathcal{G}$, by the following equation:

$$\mathcal{C}\,\frac{d\Delta T_d}{dt} + \mathcal{G}\,\Delta T_d = \Phi \tag{11-45}$$

where Φ is the incident, incremental infrared radiation. The solution of this equation for a periodic, incident, infrared radiation power, Φ, is

$$\Delta T_d = \epsilon\,\Phi\,\mathcal{G}^{-1}\left[1 + \omega^2\left(\frac{\mathcal{C}}{\mathcal{G}}\right)^2\right]^{-1/2} \tag{11-46}$$

where ϵ = emissivity
  ω = angular frequency of Φ

This analysis assumes that the radiation is absorbed uniformly throughout the sample. Putley [11-4] has derived the responsivity for a pyroelectric detector used in the circuit shown in Figure 11-4:

$$v_s = I_p|Z| = I_p R_e (1 + \omega^2 \tau_e^2)^{-1/2} \tag{11-47}$$

where $I_p = \omega \mathcal{P} A_d \Delta T_d$
  $\mathcal{P}$ = the pyroelectric coefficient
  Z = impedance
  R_e = equivalent input resistance of the detector-preamplifier circuit
  $\tau_e = R_e C_e$
  C_e = equivalent capacitance

Therefore, the voltage, v_s, can be expressed as

$$v_s = \omega\,\mathcal{P} A_d \Delta T_d(\omega) R_e (1 + \omega^2 \tau_e^2)^{-1/2} \tag{11-48}$$

where A_d is the sensitive area of the responsive crystal. Substituting the expression for ΔT_d from Equation (11-46) into Equation (11-48), one gets

$$v_s = \omega \mathcal{P} A_d \epsilon \Phi R_e\,\mathcal{G}^{-1}\left[1 + \omega^2\left(\frac{\mathcal{C}}{\mathcal{G}}\right)^2\right]^{-1/2} [1 + \omega^2 \tau_e^2]^{-1/2} \tag{11-49}$$

The expression for the responsivity, $\mathcal{R}$, is then

$$\mathcal{R} = \frac{v_s}{\Phi} = \frac{\omega \mathcal{P} A_d \epsilon R_e}{\mathcal{G}} [1 + \omega^2 \tau_T^2]^{-1/2} [1 + \omega^2 \tau_e^2]^{-1/2} \qquad (11\text{-}50)$$

where $\tau_T = \mathcal{C}/\mathcal{G}$ is the thermal time constant.

The relationship between the angular frequency, ω, of the incoming infrared radiation, the incremental temperature rise, ΔT_d, of the pyroelectrical crystal, and the responsivity, $\mathcal{R}$, of the detector is shown in Figure 11-5. The responsivity is 0 at $\omega = 0$ and increases as ω increases until the angular frequency reaches the value $\omega = 1/\tau_T$. In the range $1/\tau_T \leqslant \omega \leqslant 1/\tau_e$, the value of responsivity is a constant. For values of ω larger than τ_e, the responsivity is inversely proportional to ω [11-4].

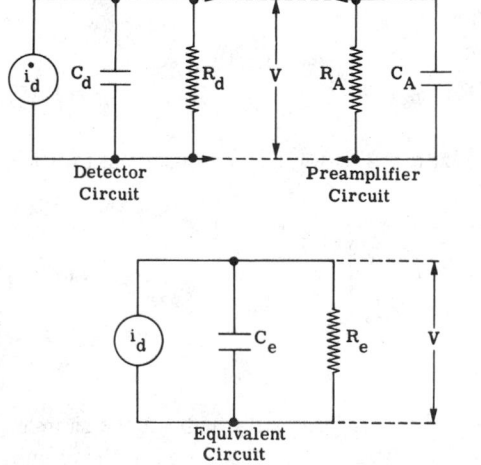

Fig. 11-4. Equivalent circuit for pyroelectric detector and amplifier input [11-4].

Fig. 11-5. Log-log plots of ΔT and $\mathcal{R}$ versus ω for a pyroelectric detector [11-4].

11.2.5. Theoretical Limit of Performance for Thermal Detectors.

The radiation power, Φ, incident upon the responsive element of a thermal detector and the power emitted by the responsive element consist of streams of photons. The rate of arrival and the rate of emission of these photons fluctuate randomly; they are not correlated spatially or temporally to any significant degree. These random arrival and emission rates lead to random fluctuations in the responsive element's temperature, which in turn produces a random output voltage. Smith, et al. [11-2] have shown that fluctuations in arrival and emission rates of photons on a thermal detector lead to a mean square fluctuation in radiation power, $\overline{\Delta\Phi^2}$:

$$\overline{\Delta\Phi^2} = 4kT_d^2 \mathcal{G} \Delta f \qquad (11\text{-}51)$$

where k = Boltzmann's constant
   Δf = the electrical frequency bandwidth
   T_d = the temperature of the responsive element
   $\mathcal{G}$ = the thermal conductance between the responsive element and its surroundings

For the case of a detector responsive element with an area, A_d, and constant emissivity, ϵ, connected to its surroundings by radiation alone,

$$\mathcal{G} = 4\sigma\epsilon A_d T_d^3 \qquad (11\text{-}52)$$

where σ is the Stefan-Boltzmann constant.

Substituting for $\mathcal{G}$ into Equation (11-51), one gets

$$\overline{\Delta\Phi^2} = 16A_d k\sigma\epsilon T_d^5 \Delta f \qquad (11\text{-}53)$$

For the case of a thermal detector at 300 K with a 1 mm^2 area and $\Delta f = 1$ Hz, the value of $[(\overline{\Delta\Phi^2})]^{1/2}$ is 5.5×10^{-12} W. This means that a thermal detector (with a 1 mm^2 area and 300 K temperature), limited only by the fluctuation in power flowing to and from the responsive element (neglecting all other noise sources), will have a minimum detectable power of

$$\text{NEP} = [(\overline{\Delta\Phi^2})]^{1/2} = 5.5 \times 10^{-12} \qquad (11\text{-}54)$$

The theoretical limit for D* in this case is 1.8×10^{10} cm Hz$^{1/2}$W^{-1}.

In the case of bolometers (excluding $1/f$ noise), the minimum detectable power can be expressed as the sum of the mean square noise resulting from temperature fluctuations and Johnson noise:

$$(\text{NEP})^2 = \overline{\Delta\Phi^2} + \mathcal{R}^{-2}v_n^2 \qquad (11\text{-}55)$$

where $\overline{\Delta\Phi^2}$ = that shown in Equation (11-53)

$\mathcal{R}$ = the responsivity

v_n = rms Johnson noise voltage

Using Equation (11-52), one obtains

$$(\text{NEP})^2 = 4kT_d^2\mathcal{G}\Delta f + \mathcal{R}^{-2}4kT_d R_d \Delta f \qquad (11\text{-}56)$$

This equation can be rearranged and Equations (11-7), (11-12), and (11-17) can be combined with Equation (11-56) to produce the following expression. (It is assumed that $\omega^2\tau^2 \ll 1$.)

$$(\text{NEP})^2 = 4kT_d^2\Delta f \left[\mathcal{G} + \frac{[\mathcal{G} - \alpha\mathcal{G}_0(\Delta T_d)]^2}{T_d\mathcal{G}_0(\Delta T_d)\epsilon^2\alpha^2} \right] \qquad (11\text{-}57)$$

If $\epsilon \approx 1$ and $\alpha(\Delta T_d) \ll 1$, then the mean square fluctuation, $(\text{NEP})^2$ reduces to

$$(\text{NEP})^2 = 4kT_d\Delta f \left[\frac{\mathcal{G}^2}{\mathcal{G}_0(\Delta T_d)\alpha^2} \right] \qquad (11\text{-}58)$$

This expression shows that for a bolometer, $(\text{NEP})^2$ is not dependent upon the detector resistance, R_d. It also points out that a bolometer can be optimized by choosing materials with a higher α and a low thermal conductance, $\mathcal{G}$. Since the thermal time constant is expressed as $\tau_T = \mathcal{C}/\mathcal{G}$, reducing the value of $\mathcal{G}$ leads to a longer time constant unless the thermal capacitance, $\mathcal{C}$. is reduced to the same extent.

For the case of the thermopile, the minimum detectable power, NEP, can be obtained using Equations (11-56) and (11-44).

$$(\text{NEP})^2 = 4kT_d^2\Delta f \left[\mathcal{G} + \frac{R_d(1 + \omega^2 \tau_T^2)\mathcal{G}^2}{\epsilon^2 P_{ab}^2 T_d} \right] \tag{11-59}$$

and, since $\tau_T = \mathcal{C}/\mathcal{G}$,

$$(\text{NEP})^2 = 4kT_d^2\Delta f \left[\mathcal{G} + \frac{R_d(\mathcal{G}^2 + \omega^2\mathcal{C}^2)}{\epsilon^2 P_{ab}^2 T_d} \right] \tag{11-60}$$

In this equation the thermal conductance, $\mathcal{G}$, and the thermal capacitance, $\mathcal{C}$, are made up of contributions from the following: the responsive element, $\mathcal{G}_R$ and $\mathcal{C}_R$; the gas used to fill the chamber containing the responsive element, $\mathcal{G}_g$ and $\mathcal{C}_g$; and the electrical leads from the responsive element to the connector pins, $\mathcal{G}_c$ and $\mathcal{C}_c$. Thus

$$\mathcal{G} = \mathcal{G}_R + \mathcal{G}_g + \mathcal{G}_c \tag{11-61}$$

$$\mathcal{C} = \mathcal{C}_R + \mathcal{C}_g + \mathcal{C}_c \tag{11-62}$$

If the responsive element of a thermopile detector is located in a housing which has been evacuated, then the contributions resulting from the ambient gas are zero (i.e., $\mathcal{G}_g = 0$ and $\mathcal{C}_g = 0$). This reduces the NEP somewhat; but since $\tau_T = \mathcal{C}/\mathcal{G}$, a decrease in $\mathcal{G}$ also increases the thermal time constant. The most fundamental method of increasing the performance of a thermopile is to use materials with higher thermoelectric powers, P_{ab}.

The ultimate limit of a pyroelectric detector is given by Equation (11-53). The other source of noise contributed by the responsive element is Johnson noise. Therefore, Equation (11-55) applies for the pyroelectric detector as well in cases where the responsivity, $\mathcal{R}$, is given by Equation (11-50). Assuming that $\omega^2 \tau_e^2 \ll 1$, $\omega^2 \tau_e^2 > 1$ and that $\epsilon \approx 1$, gives the following expression for the mean square noise:

$$(\text{NEP})^2 = 4kT_d^2 \mathcal{G}\Delta f + 4kT_d R_e \Delta f \mathcal{R}^{-2} \tag{11-63}$$

Therefore,

$$(\text{NEP})^2 = 4kT_d\Delta f \left[T_d\, \mathcal{G} + \frac{\mathcal{G}^2 \tau_T^2}{\mathcal{P}^2 A_d^2 R_e} \right] \tag{11-64}$$

where $\mathcal{P}$ is the pyroelectric coefficient and the other symbols are as previously defined. Since $\tau_T = \mathcal{C}/\mathcal{G}$, Equation (11-64) can be rewritten as

$$(\text{NEP})^2 = 4kT_d\Delta f \left[T_d\, \mathcal{G} + \frac{\mathcal{C}^2}{\mathcal{P}^2 A_d^2 R_e} \right] \tag{11-65}$$

This means that the mean square noise can be reduced by decreasing the area of the detector, A_d, and by choosing a material with a high value of pyroelectric coefficient, $\mathcal{P}$, and a low thermal capacitance, $\mathcal{C}$. ($\mathcal{C}$ and $\mathcal{G}$ are both area, A_d, dependent terms as analyzed by Putley [11-4]. Therefore, the area dependency cancels in the second term in brackets on the right side of Equation (11-65) and remains only in the first term. This analysis ignores noise contributions from other parts of the infrared system in which these detectors are used, such as in post-detector electronics.)

In 1946, R. Havens developed an empirical relationship between the signal power generated by a thermal detector and the change in incident optical power [11-5]:

$$\frac{V_s^2}{R_d} = N \, d\Phi \, \frac{\Delta T}{T} \tag{11-65a}$$

where V_s = signal voltage due to the change in optical power

R_d = resistance of the detector element

$d\Phi$ = optical power change incident on the detector

ΔT = temperature change due to the change in incident optical power $d\Phi$

T = temperature of detector, K

N = empirically determined constant that is a function of thermal detection type and ranged from 1-100

He further postulated that the theoretical minimum detectable optical energy would be approximately the same for all thermal detectors operating at room temperature and would be:

$$d\Phi \tau = 3 \times 10^{-12} \text{ J} \tag{11-65b}$$

where τ is the duration of the pulse in seconds and the detector size is 1 mm square.

Equating $d\Phi$ in this case with NEP we have:

$$\text{NEP} = \frac{3 \times 10^{-9}}{\tau} \text{ W} \tag{11-65c}$$

for τ measured in milliseconds. Hudson [11-6] has converted this limit to its equivalent D* value and obtained the value

$$\text{D}^*_{\text{Havens Limit}} = 5.38 \times 10^8 \, (\tau)^{1/2} \tag{11-65d}$$

where τ is in milliseconds.

Contemporary room-temperature detectors, pp. 11-63 — 11-65, exhibit performance characteristics that are close to (or slightly exceed) this empirically devised, and somewhat arbitrary, limit.

11.3. Theoretical Description of Photon Detectors

In photon detectors, incident infrared photons are absorbed producing free charge carriers which change an electrical characteristic of the responsive element. This process is carried out without any significant temperature change in the responsive element. Theoretical descriptions of the four most commonly used processes are briefly given below.

11.3.1. Photoconductive Effect.
In this type of photon detector, incident infrared photons are absorbed, producing free charge carriers which change the electrical conductivity of the responsive element. This change in conductivity is detected in the associated electronic circuit (Figure 11-6). If the conductivity of the responsive element (Figure 11-7) increases because of absorbed infrared photons, then the resistance, R_d, of the responsive element will decrease since

$$R_d = \frac{l}{\sigma_e A_c} \tag{11-66}$$

where l = the length
A_c = the cross-sectional area = wd
σ_e = the electrical conductivity of the responsive element.

This change in resistance produces a signal voltage, which is fed to a preamplifier. The signal voltage can be expressed as

$$v_s = I\Delta R_d + R_d\Delta I \qquad (11\text{-}67)$$

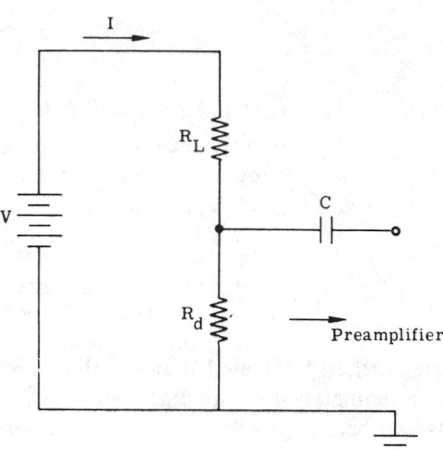

Fig. 11-6. Photoconductor detector circuit.

where I is the current through the circuit. If the circuit is operated in a constant current condition (i.e., $\Delta I=0$ and $R_L \gg R_d$), then this equation reduces to

$$v_s = I\Delta R_d \frac{R_L}{(R_L + R_d)}$$

$$\approx I\Delta R_d \qquad (11\text{-}68)$$

and further:

$$v_s = \frac{V\Delta R_d}{(R_L + R_d)} \qquad (11\text{-}69)$$

$$v_s = \frac{V_B R_d}{(R_L + R_d)} \frac{\Delta N}{N} \qquad (11\text{-}70)$$

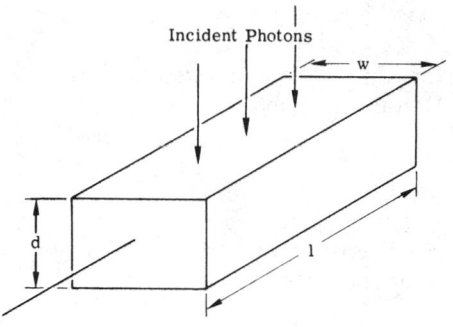

Fig. 11-7. Photoconductor geometry.

where $R_d = l/\sigma_e wd$
σ_e = the electrical conductivity; $\sigma_e = ne\mu$
l = length of detector
w = width of detector
d = depth of detector
n = the density of free charge carriers per unit volume
N = the total number of free charges in the responsive element when there are no incident, infrared signal photons; $N = nlwd$
ΔN = the increase in total number of free charges caused by the incident infrared signal photons
V_B = bias voltage

According to Petritz [11-7], the detector small-signal properties are governed by the following two equations:

$$\frac{d}{dt}\Delta N = A_d\eta E_q - \frac{\Delta N}{\tau_c} \qquad (11\text{-}71)$$

$$\Delta N = N(t) - N \qquad (11\text{-}72)$$

where η = the efficiency in converting incident, infrared photons into free charge carriers

A_d = the detector area, wl

E_q = incident photon flux density, photons cm^{-2} sec^{-1}

τ_c = the average, free charge-carrier lifetime

The solution for a sinusoidal signal input rate, $E_{q,s}$, at a modulation frequency, $f_c = \omega_c/2\pi$, where $E_q = E_{q,o} + E_{q,s} \cos \omega_c t$, is

$$|\Delta N(f)| = \frac{A_d\eta E_{q,s}\tau_c}{(1 + \omega^2\tau_c^2)^{1/2}} \qquad (11\text{-}73)$$

Therefore,

$$\frac{|\Delta N(f)|}{N} = \frac{\Delta N}{N} = \frac{A_d\eta E_{q,s}\tau_c}{N(1 + \omega^2\tau_c^2)^{1/2}} \qquad (11\text{-}74)$$

The signal voltage expression for a photoconductor is obtained by substituting Equation (11-74) into (11-70):

$$v_s = \frac{V_B R_d}{(R_L + R_d)} \frac{A_d\eta E_{q,s}\tau_c}{N(1 + \omega^2\tau_c^2)^{1/2}} \qquad (11\text{-}75)$$

Equation (11-75) is an expression for the magnitude of the detector output voltage for an input signal flux that has a sinusoidal waveform. The superposition theorem for linear systems permits use of Equation (11-75) for other more complex waveforms, and for modulation at any frequency. One can also write this in terms of the spectra of the quantities involved:

$$\mathcal{F}\{v_s\} = \frac{V R_d}{R_L + R_d} \frac{A_d\eta\tau_c}{N(1 + i\omega\tau_c)} \mathcal{F}\{\Phi\} \qquad (11\text{-}76)$$

where $\mathcal{F}\{v_s\}$ = Fourier transform of the voltage change, or the voltage spectrum

$\mathcal{F}\{\Phi\}$ = modulation spectrum of the signal flux

The complex voltage spectrum can be separated into its modulus and phase:

$$\mathcal{F}\{v_s\} = |\mathcal{F}\{v_s\}| + i\, arg\,(\mathcal{F}\{v_s\}) \qquad (11\text{-}77)$$

$$|v_s| = \frac{V R_d}{R_L + R_d} \frac{A_d\eta\tau_c}{N\sqrt{1 + \omega^2\tau_c^2}} \qquad (11\text{-}78)$$

$$arg\,(\mathcal{F}\{v_s\}) = arc\,\tan\,(-\omega\tau_c) + arg\,(\mathcal{F}\{\Phi\}) \qquad (11\text{-}79)$$

The modulus of the transfer function relating the output signal voltage spectrum to the input flux spectrum is given by

$$\left| \frac{\mathcal{F}\{v_s\}}{\mathcal{F}\{\Phi\}} \right| = \frac{VR_d}{R_L + R_d} \frac{A_d \eta \tau_c}{N\sqrt{1 + \omega^2 \tau_c^2}} \tag{11-80}$$

The phase shift is given by arc tan $(-\omega\tau_c)$.

The noise encountered in photoconductors and other photon detectors is described in Section 11.3.5. In that section, expressions for the signal-to-noise ratio will be examined using Equation (11-75) as the signal voltage for photoconductors.

11.3.2. Photovoltaic Effect. In the photovoltaic process, a $p - n$ junction is formed in or on a semiconductor. Infrared photons, that are absorbed at or near the junction, are separated by the junction producing an external electrical voltage. The magnitude of this voltage is related to the number of incident, infrared photons.

A simplified, energy-band picture of the photovoltaic process is shown in Figure 11-8. Part (a) of the figure shows the energy-band arrangement for an unirradiated $p - n$ junction.

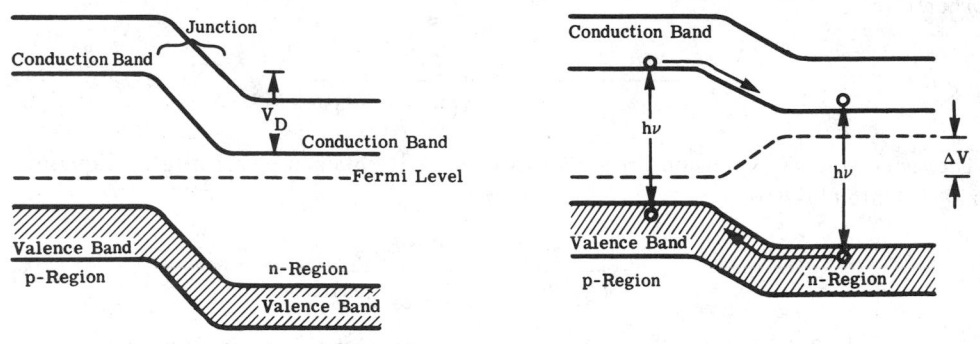

(a) Unilluminated p-n junction.   (b) Illuminated p-n junction.

Fig., 11-8. Energy band model of A p–n junction.

Incident, infrared photons with energy, $h\nu$, greater than the energy-band gap will create electron-hole pairs in the semiconductor at or near the junction region. As shown in part (b), if the photons are absorbed in the p-region, the free electron will move down along the conduction band, c-band, to the n-region. This means that the Fermi levels in the n- and p-regions will be displaced because of the presence of the free electron and hole respectively. This shift in the Fermi level produces a voltage, ΔV, which can be observed with an external electronic circuit. When the electrical characteristics of the responsive element are observed with an external electronic circuit, the current-voltage relationship is that shown in Figure 11-9, where curve (a) is the unilluminated case and (b) is the illuminated case. In the unilluminated case, no short-circuit current, I_{sc}, will be observed if the diode is externally shorted. Also, no open-circuit voltage, V_0, will be observed. In practice, however, some background photons with sufficient energy to cause band-to-band transitions will be incident upon the responsive element, producing free electron-hole pairs. Thus the unilluminated I-V curve really lies between curves (a) and (b). In curve (b), the open-circuit voltage, V_0, and the short-circuit current, I_{sc}, are shown. The current through the junction is given by

$$I = I_{sa}\left[\exp\left(\frac{eV}{\beta kT_d}\right) - 1\right] \quad (11\text{-}81)$$

where β = an efficiency factor for photo-diodes ($\beta = 1$ for an ideal diode and $\beta = 2$ to 3 for the real case)

I_{sa} = the saturation current

I_{sa} is given by

$$I_{sa} = e\left(\frac{D_h p}{L_h} + \frac{D_e n}{L_e}\right) \quad (11\text{-}82)$$

where D_h and D_e = the hole- and electron-diffusion constants, respectively, cm^2 sec^{-1}

L_h and L_e = the hole- and electron-diffusion lengths, respectively, cm

These are defined by the equations

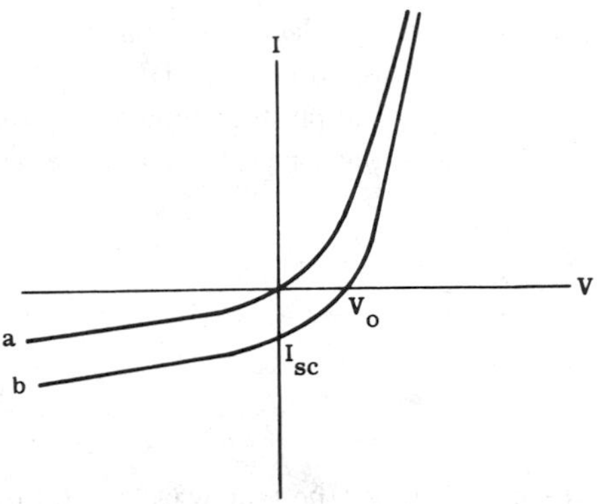

a = unilluminated condition
b = illuminated condition

Fig. 11-9. Current voltage characteristics of a photodiode.

$$L_h = \sqrt{D_h \tau_h} = \sqrt{\left(\frac{kT}{e}\right)\mu_h \tau_h} \quad (11\text{-}83)$$

$$D_h = \frac{kT}{e}\mu_h$$

$$L_e = \sqrt{D_e \tau_e} = \sqrt{\left(\frac{kT}{e}\right)\mu_e \tau_e}$$

$$D_e = \frac{kT}{e}\mu_e \quad (11\text{-}84)$$

where μ_h and μ_e = the hole and electron mobilities, respectively

τ_h and τ_e = the hole and electron lifetimes, respectively

However, Equation (11-81) is not complete. As pointed out earlier in this section, when the photodiode is operated in the short-circuit condition (i.e., $V_B = 0$), the current is not zero as curve (a) in Figure 11-9 implies. The existence of background photons produce some free charges that lead to a short-circuit current, I_{sc}. Also, in practice there is a leakage current in photodiodes which can be represented by the term $G_{sh}V$, where G_{sh} is the effective shunt conductance. Equation (11-81) can now be completed [11-7]:

$$I = I_{sa}\left[\exp\left(\frac{eV}{\beta kT_d}\right) - 1\right] - I_{sc} + G_{sh}V \quad (11\text{-}85)$$

Since I_{sc} is due to background radiation, it can be expressed as

$$I_{sc} = e\eta E_{q,B} A_d \tag{11-86}$$

where e = the electronic charge
η = the quantum efficiency
$E_{q,B}$ = the photon flux density from the background
A_d = the sensitive area of the responsive element

$E_{q,B}$ is given by

$$E_{q,B} = \int_0^{\lambda_c} E_{q,\lambda,B}(\lambda)d\lambda \tag{11-87}$$

where λ_c = the long wavelength limit of the detector's spectral response
$E_{q,\lambda,B}(\lambda)$ = the spectral photon flux density from the background

Pruett and Petritz [11-7] have shown that for the small signal case, the signal current, I_s, can be derived from Equation (11-85). The expression for I_s becomes

$$I_s = - e\eta A_d E_{q,s} \tag{11-88}$$

where $E_{q,s}$ is the signal photon flux density.

The best operating point (producing a maximum signal-to-noise ratio) is at or near $V_B = 0$, which necessitates a back-bias configuration, shown in Figure 11-10. The back-bias also increases the width of the depletion region, thereby reducing the time constant, because the wider depletion region ensures a higher probability that signal photons will be absorbed in the vicinity of the depletion region. This cuts down the time it takes free charge carriers to move from the site of absorption to the depletion region.

If the back-bias voltage, V_B, is raised to a large enough value, the free electrons and holes moving in the field will be accelerated and, thus, acquire sufficient energy to produce additional free charge-carriers on collision with the lattice. As V_B is increased further, the multiplication of free charge-carriers increases. The limit to this is a breakdown condition where the electric field produced by the battery is causing free charge-carriers to be generated. The multiplication factor, M, defined as the average number of electron-hole pairs produced from a single initiating electron-hole pair, is related to the bias voltage, V_B, and the breakdown voltage, V_{bd}, as shown below:

$$M^{-1} \propto \left[1 - \left(\frac{V_B}{V_{bd}}\right)^m\right] \tag{11-89}$$

The values of the empirical constant, m, have been observed to be between 1.4 and 4. Avalanche photodiodes made from silicon with M values up to 10^6 have been reported by Haitz, et al. The signal current in an avalanche diode can be expressed as

$$I_s = - e\eta A_d E_{q,s} M \qquad (11\text{-}90)$$

However, the mean-square noise current, as calculated by McIntyre [11-8] is given by

$$i_n{}^2 = 2eIM^3 \Delta f \qquad (11\text{-}91)$$

where I is the current from Equation (11-85). The signal-to-noise current ratio is then proportional to $1/\sqrt{M}$. This means that although the signal current rises with M, the noise rises faster. Because of this, avalanche photodiodes are very useful in situations where amplifier noise predominates. In these cases, an increase in noise caused by M would not significantly increase the system noise but would substantially increase the signal level.

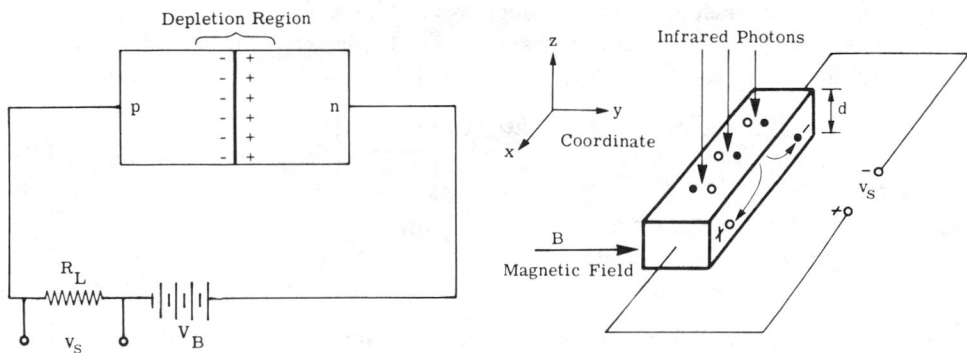

Fig. 11-10. Back bias in photovoltaic detector.   Fig. 11-11. Schematic of the PEM effect.

11.3.3. Photoelectromagnetic (PEM) Effect.

The photoelectromagnetic effect is not often used in modern infrared photon detectors. In this effect, incident, infrared photons are absorbed at or near the surface of a semiconductor, producing free electron-hole pairs. These pairs diffuse from the surface and down into the crystal. The presence of a magnetic field, B, directed along the y axis causes the electrons to separate from the holes as they diffuse away from the surface. (See Figure 11-11.) This separation of charge produces an electrical voltage across the terminals. If the external circuit is shorted, a current will continue to flow as long as infrared photons are arriving at the surface. If the exterior circuit is left open, as shown in the figure, a voltage will appear across the open terminals and remain there as long as the surface is irradiated. Moss, et al. [11-9] have derived the open-circuit voltage, V_0, and the short-circuit current, I_{sc}, per unit length for photo electromagnetic detectors. These expressions are given below:

$$I_{sc} = \frac{e E_{q,s} \theta L}{(1 + \gamma)} \qquad (11\text{-}92)$$

$$V_0 = \frac{e E_{qs} \theta L l}{(1 + \gamma)\sigma d} \qquad (11\text{-}93)$$

where E_{qs} = photon flux density for signal radiation, photons $\sec^{-1}$ cm^{-2}
     L = ambipolar diffusion length, either L_e or L_h, whichever is the majority carrier
     γ = $\tau s/L$
     l = the electrode separation
     σ = electrical conductivity of the crystal
     d = thickness of the responsive element
     e = electronic charge
     τ = time constant
     s = surface recombination velocity
     θ = $\theta_e + \theta_h$
     θ_e = $B \mu_e$, Hall angle for electrons
     θ_h = $B \mu_h$, Hall angle for holes

These expressions are valid under the following conditions:
(1) The magnetic field, B, is small enough that $\mu^2 B^2 \ll 1$.
(2) The specimen is thick enough that $d/2L_d \approx 1$, where L_d is the effective diffusion length:

$$L_d = \left[\frac{bD_h \tau (n + p)}{b\,n + p + (bp + n)\theta_h \theta_e} \right]^{1/2} \tag{11-94}$$

where b = μ_e/μ_h = ratio of electron to hole mobility
     n = electron volume density
     p = hole volume density

(3) The crystal has high enough conductivity to assume charge neutrality throughout the responsive element.
(4) The surface recombination velocity is small enough to allow a large fraction of free charge-carriers to diffuse from the generation site near the surface down into the crystal bulk [11-9].

11.3.4. Photoemissive Effect. In the photoemissive effect, an incident photon is absorbed by the sensitive surface. It gives up all its energy, hc/λ, to a free electron at or near the surface of the sensitive material. Thus, the kinetic energy of the electron is increased by an amount equal to the photon energy. Before the electron can escape from the surface, it may give up part of its kinetic energy to atoms through collisions. The amount of energy lost in this manner varies considerably. If the electron still has enough kinetic energy by the time it arrives at the surface, it can escape from the surface. The kinetic energy of the escaped electron can be expressed as follows:

$$E_k = \frac{hc}{\lambda} - e\phi \tag{11-95}$$

where E_k = the kinetic energy of the freed electron
   hc/λ = the energy of the absorbed photon
     ϕ = the surface work function of the material
     e = the charge of the electron

If such a sensitive surface is placed in an evacuated chamber along with an anode and attached to an exterior circuit, as shown in Figure 11-12, the electrons freed from the cathode surface by absorbed photons will be attracted to the anode. A current will then flow in the circuit through R_L as long as photons of sufficient energy arrive at the sensitive surface.

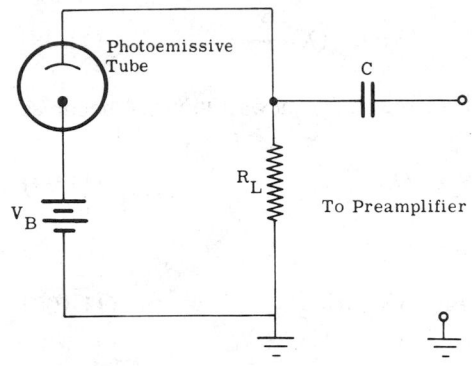

Fig. 11-12. Photoemissive bias circuit.

The time constant is determined by the spread in transit times of the electrons between the cathode and the anode. This can be as small as 10^{-9} sec.

In addition to the current induced by infrared photons, a small current flows even when the cell is in the dark. This dark current sets the limit to the minimum detectable radiation. It is mostly due to thermionic emission, and its level depends upon cathode temperature, sensitive area, and the work function.

The lowest work functions that have been achieved to date are approximately 1 eV. This means that the photoemissive detectors are sensitive in the very short wavelength region of the infrared spectrum. Spectral responses as long as 1.5 μm have been achieved.

A considerable gain in responsivity can be produced by adding a small amount of inert gas into the detector chamber. Electrons emitted from the cathode may then be accelerated by the field toward the anode. Before reaching the anode, an electron may collide with a gas atom and ionize it. With this technique, the number of electrons that reach the anode could be 100 times the number of electrons photoemitted from the cathode. The presence of positive ions in the cell lengthens the transit time for the electrons, which increases the dispersion and thus increases the time constant.

Electrons are emitted from a material surface when it is bombarded with high-velocity electrons. This process of secondary emission has been used to develop a photocell having high internal amplification. Such cells are called photomultipliers. Photoemitted electrons are focused onto another electrode where each electron produces a number of secondary electrons. These, in turn, are focused onto a third electrode, and the process is repeated several times. The photomultiplier has advantages over both the simpler photocells. The current can be multiplied by about 10^6, compared with a gain of about 10^2 for the gas-filled photocell. Since no positive ions are involved in the photomultiplier photocells, the time constants are shorter than for the gas photocell, but longer than for the simple vacuum photocell.

The responsivity, $\mathfrak{R}$, of the photoemissive detector can be derived as follows. The effective spectral, signal, photon flux-density on the sensitive cathode is given by

$$\eta(\lambda)\Phi_{s,\lambda}(\lambda)\left(\frac{\lambda}{hc}\right), \qquad (11\text{-}96)$$

where $\Phi_{s,\lambda}(\lambda)$ = the spectral signal power
$\eta(\lambda)$ = the quantum efficiency

If one assumes that all the electrons will be collected at the anode, the photocurrent, i_s, is expressed as

$$i_s(\lambda) = e\eta(\lambda)\Phi_{s,\lambda}(\lambda)\frac{\lambda}{hc} \tag{11-97}$$

The signal voltage, v_s, will be the change in voltage across the load resistor, R_L. (See Figure 11-12.)

$$v_s(\lambda) = i_s(\lambda)R_L = R_L e\eta(\lambda)\Phi_{s,\lambda}(\lambda)\frac{\lambda}{hc} \tag{11-98}$$

Since the responsivity, $\Re$, is defined as the signal voltage per signal watt input, $\Re$ becomes

$$\Re = \frac{v_s(\lambda)}{\Phi_{s,\lambda}(\lambda)} = R_L e\eta(\lambda)\frac{\lambda}{hc} \tag{11-99}$$

The total current is given as

$$i_s = \frac{e}{hc}\int_0^\infty \eta(\lambda)\Phi_{s,\lambda}(\lambda)\lambda d\lambda \tag{11-100}$$

The responsivity is then given by

$$\Re = R_L\frac{\dfrac{e}{hc}\displaystyle\int_0^\infty \eta(\lambda)\Phi_{s,\lambda}(\lambda)\lambda d\lambda}{\displaystyle\int_0^\infty \Phi_{s,\lambda}(\lambda)d\lambda} \tag{11-101}$$

For photomultipliers, the responsivity equation should be multiplied by the gain. The limitation to sensitivity will be set by fluctuations in the dark current manifested by changes in voltage across the load, R_L.

11.3.5. Theoretical Limit of Performance of Photon Detectors. In general, the limit of performance of infrared, photon detectors is set by the fluctuation in arrival rate of background photons. This fluctuation, called photon noise, appears as a random voltage at the output terminals of the detector. The total noise from the detector is the rms sum of the inherent noise generated within the detector plus the photon noise.

This section deals with the various noise mechanisms encountered in the use of infrared detectors, including photon noise, lattice G-R noise, $1/f$ noise, Johnson noise, and shot noise.

Photon Noise. The mean square fluctuations in the number of photons arriving at a surface in a small wavelength interval, $\Delta\lambda$, is given by

$$\overline{(\Delta\Phi_{q,\lambda}(\lambda)\Delta\lambda)^2} = \Phi_{q,\lambda}(\lambda)\Delta\lambda[1 + \gamma(e^x - 1)^{-1}] \tag{11-102}$$

where $\Phi_{q,\lambda}(\lambda)$ = the photon flux per unit wavelength
   γ = a coherence factor
   $x = c_2/\lambda T, c_2 = hc/k$

The quantity $\Phi_{q,\lambda}(\lambda)\Delta\lambda$ can be expressed as the sum of a signal flux and a background flux. The background photon flux is assumed to be generated by a uniform temperature, hemispheric blackbody surrounding the detector. The mean square fluctuation in total photon flux on the detector is

$$\overline{(\Phi_{q,\lambda}(\lambda)\Delta\lambda)^2} = \overline{(\Phi_{q,\lambda,s}(d)\Delta\lambda)^2} + \overline{(\Phi_{q,\lambda,B}(\lambda)\Delta\lambda)^2} \qquad (11\text{-}103)$$

where $\Phi_{q,\lambda,s}(\lambda)\Delta\lambda$ = photon flux from the signal
$\Phi_{q,\lambda,B}(\lambda)\Delta\lambda$ = photon flux from the background

The total photon flux incident on the detector is given by

$$\Phi_q = \int_0^\infty (\Phi_{q,\lambda,s}(\lambda) + \Phi_{q,\lambda,B}(\lambda))d\lambda \qquad (11\text{-}104)$$

A monochromatic signal-flux generates free charge-carriers at a rate G_{gen} given by

$$(G_{gen})_s = \eta(\lambda)\frac{\lambda}{hc}\Phi_{s,\lambda}(\lambda)\Delta\lambda \qquad (11\text{-}105)$$

The variance (noise squared) is thus

$$\overline{(\Delta G_{gen})_s^2} = 2\eta(\lambda)E_{q,\lambda,s}(\lambda)\Delta\lambda A_d\Delta f \qquad (11\text{-}106)$$

The noise from the background is given by

$$\overline{(\Delta G_{gen})_B^2} = 2\eta(\lambda)E_{q,\lambda,B}(\lambda)\Delta\lambda A_d\Delta f \qquad (11\text{-}107)$$

The total mean square noise from the two sources is given by the sum of these two generation rates:

$$\overline{(\Delta G_{gen})_t^2} = 2\eta(\lambda)A_d\Delta f(E_{q,\lambda,s}(\lambda) + E_{q,\lambda,B}(\lambda))\Delta\lambda \qquad (11\text{-}108)$$

The NEP is the flux (power) divided by the signal-to-noise ratio:

$$\text{NEP} = \Phi_s/\text{SNR} \qquad (11\text{-}109)$$

$$\text{NEP} = \Phi_s[\overline{(\Delta G_{gen})_t^2}]^{1/2}$$

$$\times (G_{gen})_s^{-1} = [2\eta(\lambda)A_d\,\Delta f(E_{q,\lambda,s}(\lambda) + E_{q,\lambda,B}(\lambda))\Delta\lambda]^{1/2}\left[\eta(\lambda)\frac{\lambda}{hc}\right]^{-1}$$
$$(11\text{-}110)$$

The specific detectivity, D^*, is given by

$$D^*(\lambda) = \frac{\sqrt{A_d\Delta f}}{\text{NEP}} \qquad (11\text{-}111)$$

Therefore

$$D^*(\lambda) = \eta(\lambda)\frac{\lambda}{hc}[2\eta(\lambda)(E_{q,\lambda,s}(\lambda) + E_{q,\lambda,B}(\lambda))(\Delta\lambda)]^{-1/2} \qquad (11\text{-}112)$$

If the assumption of monochromatic signal and noise is relaxed,

$$D^* = \frac{\displaystyle\int_0^\infty \eta(\lambda)\frac{\lambda}{hc}\Phi_{s,\lambda}(\lambda)d\lambda}{\left[\displaystyle\int_0^\infty \Phi_{s,\lambda}(\lambda)d\lambda\right]\left[\displaystyle\int_0^\infty 2\eta(\lambda)(E_{q,\lambda,s}(\lambda) + E_{q,\lambda,B}(\lambda))d\lambda\right]^{1/2}} \qquad (11\text{-}113)$$

Wolfe [11-10] has defined a D_n^*, the response of detectors to photons rather than power, as

$$D^*_n = \eta(\lambda) \left[\int_0^\infty 2\eta(\lambda)(E_{q,\lambda,s}(\lambda) + E_{q,\lambda,B}(\lambda))d\lambda \right]^{-1/2} \qquad (11\text{-}114)$$

Several cases can be conveniently evaluated: noise due to signal alone, noise due to background alone, coherent detection and incoherent detection.

If the background can be neglected (which is almost never the case) and the signal is monochromatic, then

$$D^*(\lambda) = \eta(\lambda)\frac{\lambda}{hc} [2\eta(\lambda)E_{q,\lambda,s}\Delta\lambda]^{-1/2} = \left(\frac{\eta(\lambda)}{2}\right)^{1/2} \frac{\lambda}{hc} \left[E_{q,\lambda,s}\Delta\lambda \right]^{-1/2} \qquad (11\text{-}115)$$

$$D_n^* = \left(\frac{\eta(\lambda)}{2}\right)^{1/2} \left[E_{q,\lambda,s}\Delta\lambda \right]^{-1/2} = [2\eta(\lambda)E_{q,\lambda,s}\Delta\lambda]^{-1/2} \qquad (11\text{-}116)$$

The noise is contained in the (assumed) narrow band of the signal and can be evaluated (See Chapter 1.) as

$$E_{q,\lambda,s} \propto \frac{e^x - 1 + \gamma}{(e^x - 1)^2} \qquad (11\text{-}117)$$

where $x = c_2/\lambda T$, and $c_2 = hc/k$.

If the noise in the signal can be neglected (almost always the case) and the signal is monochromatic then

$$D^*(\lambda_s) = \frac{\lambda_s}{hc} \eta(\lambda_s)[2\eta(\lambda_s)E_{q,\lambda,B}\Delta\lambda]^{-1/2} \qquad (11\text{-}118)$$

While the signal carriers are excited only in a narrow spectral band, the background contribution is generally over a wide spectral band, so that

$$D^*(\lambda_s) = \frac{\lambda_s}{hc} \eta(\lambda_s) \left[2 \int_0^\infty \eta(\lambda)E_{q,\lambda,B}(\lambda)d\lambda \right]^{-1/2} \qquad (11\text{-}119)$$

For many detectors $\eta(\lambda)$ is essentially constant, in which case

$$D^*(\lambda_s) = \frac{\lambda_s \sqrt{\eta}}{hc} \left[2 \int_0^{\lambda_c} E_{q,\lambda,B}(\lambda)d\lambda \right]^{-1/2} \qquad (11\text{-}120)$$

where λ_c = the cutoff wavelength of the detector

$$D_n^* = \sqrt{\eta} \left[2 \int_0^{\lambda_c} E_{q,\lambda,B}(\lambda)d\lambda \right]^{-1/2} \qquad (11\text{-}121)$$

The effective background photon flux is also often referred to as Q_B [11-6], where:

$$Q_B \equiv \int_0^{\lambda_c} E_{q,\lambda,B}(\lambda)d\lambda \qquad (11\text{-}122)$$

and

$$D^*(\lambda_c) = \frac{\lambda_c}{\sqrt{2}\,hc} \left[\frac{\eta}{Q_B}\right]^{1/2} \tag{11-123}$$

If λ_c is expressed in μm, c in cm sec^{-1}, h in W sec^2, and $\eta = 1$, then:

$$D^*(\lambda_c) = 3.56 \times 10^{18} \times \lambda_c \times \left[\frac{1}{Q_B}\right]^{1/2} \tag{11-124}$$

For a photoconductor, the recombination noise introduces a factor of $\sqrt{2}$ into the denominator of Equation (11-124) which, therefore, reduces the value of the BLIP D^*. Therefore,

$$D_{pc}^*(\lambda_c) = 2.52 \times 10^{18} \,\lambda_c \left[\frac{1}{Q_B}\right]^{1/2} \tag{11-125}$$

The background flux can be reduced by the use of a cold spectral filter and by a cold aperture stop that limits the field of view. The $D^*(\lambda)$ value for the filter spectral band is given by

$$D^*(\lambda_s) = \frac{\lambda_s \sqrt{\eta}}{hc} \left[2 \int_{\lambda_1}^{\lambda_2} E_{q,\lambda,B}(\lambda)d\lambda\right]^{-1/2} \tag{11-126}$$

$$D_n^* = \sqrt{\eta}\left[2 \int_{\lambda_1}^{\lambda_2} E_{q,\lambda,B}(\lambda)d\lambda\right]^{-1/2} \tag{11-127}$$

The reduction of the background flux by a cooled aperture is a geometrical consideration involving the ratio of the square root of the projected solid angle of view to that of a hemisphere. For a conical field of view, $D^*(\lambda)$ can be multiplied by $(\sin\Theta/2)^{-1}$, where Θ is the total cone angle.

Figure 11-13 is an example of the above problem, done for $\gamma = 1$ and $\gamma = 0$ respectively. If $\lambda_c = 10.7\mu$m, the background is a 300 K blackbody, and the spectral filter characteristics are those shown in Figure 11-14, then the theoretical limit in D^* for this case is

$$D^*(\lambda_c) = \frac{\lambda_c}{2hc\,\sqrt{\pi c}} \left[\int_0^{10.5\mu m} \frac{e^{c_2/\lambda T_1}}{\lambda^4(e^{c_2/\lambda T_1} - 1)^2}\,d\lambda\right.$$

$$\left. + \int_{10.5\mu m}^{10.7\mu m} \frac{e^{c_2/\lambda T_2}}{\lambda^4(e^{c_2/\lambda T_2} - 1)^2}\,d\lambda\right]^{-1/2} [\sin\pi/40]^{-1} \tag{11-128}$$

The factor $(\sin \pi/40)$ is due to the cooled cone, which has the effect of increasing $D^*(\lambda_c)$ by $(\sin \Theta/2)^{-1}$, with Θ equal to $\pi/20$ in the example. One can assume that, since the absorption of the filter is 100% outside the 10.5 to 10.7 μm region, the filter has an emittance of unity in that region. Also, $T_1 = 77$ K and $T_2 = 300$ K. This also assumes that $\eta(\lambda) = 1$ for $0 < \lambda < \lambda_c$. It is reasonable to assume further that the first integral within the brackets has an effective value of zero because of the small value of T_1. D^* can then be written

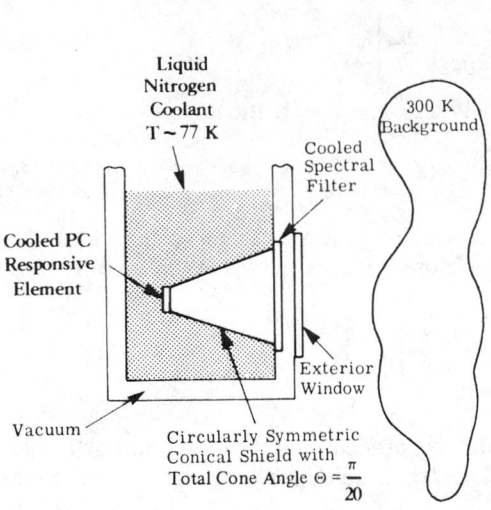

Fig. 11-13. Schematic of spectrally filtered detector with a cold stop.

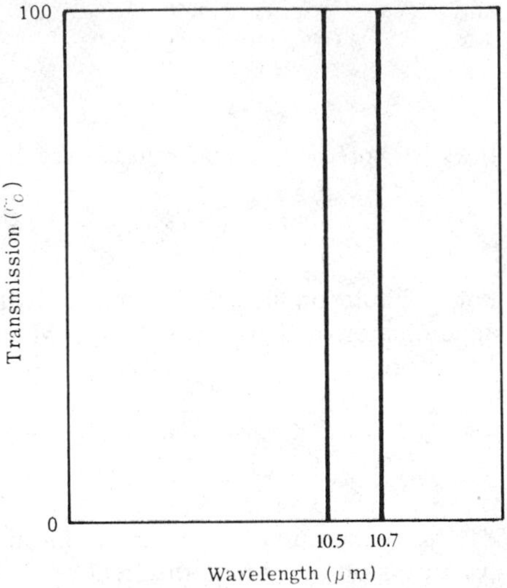

Fig. 11-14. Transmission characteristics of an ideal 10.5 to 10.7 μm spectral filter.

$$D^*(\lambda_c) = \frac{\lambda_c}{2hc\sqrt{\pi c}} \left[\int_{10.5}^{10.7} \frac{e^{c_2/\lambda T_2}}{\lambda^4 (e^{c_2/\lambda T_2} - 1)^2} \, d\lambda \right]^{-1/2} (\sin \Theta/2)^{-1} \quad (11-129)$$

For $\gamma = 0$, one has

$$D^*(\lambda_c) = \frac{\lambda_c}{2hc\sqrt{\pi c}} \left[\int_{10.5}^{10.7} \frac{1}{\lambda^4 (e^{c_2/\lambda T} - 1)} \, d\lambda \right]^{-1/2} (\sin \Theta/2)^{-1} \quad (11-130)$$

Using the mks system of units, one can evaluate the integral where

h = Planck's constant   = 6.627×10^{-34} W sec^2 or J Hz^{-1}
c = velocity of light   = 3×10^8 m sec^{-1}
k = Boltzmann's constant = 1.38×10^{-23} W sec K^{-1} or J K^{-1}
λ = optical wavelength in meters

Therefore, the theoretical limit in D^* is given by

$$D^*(\lambda_c) = 1.3 \times 10^{11} (\sin \Theta/2)^{-1}, \qquad \text{cm Hz}^{1/2} \text{ W}^{-1} \quad (11-131)$$

$$D^*(\lambda_c) = 1.7 \times 10^{12}, \qquad \text{cm Hz}^{1/2} \text{ W}^{-1} \quad (11-132)$$

This is about a factor of 50 times better than the $D^*(\lambda_c)$ value obtained without a cooled spectral filter and a cold aperture stop.

Figure 11-15 gives the $D^*(\lambda_c)$ value versus long wavelength threshold, λ_c, at different background temperatures, assuming the detector looks into a hemisphere. Figure 11-16 gives the effect on the theoretical limit of D^* caused by the reduction in the solid angle through which the detector views the background.

Van Vliet [11-11] has shown that, at equilibrium, a photoconductor has a total noise power that can be no less than twice the value of the photon noise power alone. This means that for photoconductors, the values of $D^*(\lambda_c)$ given in Figure 11-15 should be divided by the square root of two [11-11].

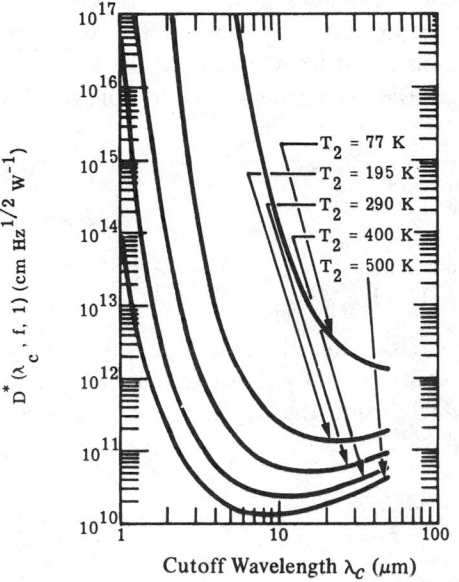

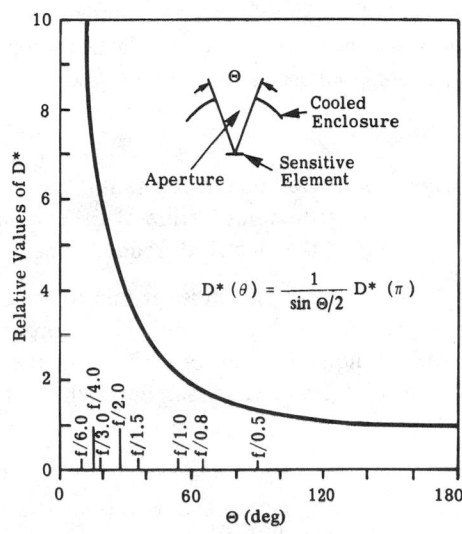

Fig. 11-15. Photon noise limited photovoltaic, D^*, at peak wavelength, λ, for various background temperatures [11-12].

Fig. 11-16. Improvement factor in D^* versus cone angle Θ.

Lattice G-R Noise. This noise is a random generation and recombination of free charge-carriers resulting from interactions with the vibrating atoms of the crystal lattice. The noise-voltage spectrum caused by this process has been derived by Van Vliet and Kruse, et al. [11-12] and [11-11].

$$v_n = R_d I \left[\frac{4\tau\Delta f}{\overline{N}(1 + 4\pi^2 f^2 \tau^2)} \right]^{1/2} \qquad (11-133)$$

where $\overline{N}$ = the average total number of free charge-carriers in the responsive element
τ = the carrier lifetime [11-11]

Current, 1/f, or Modulation Noise. The physical mechanisms causing this noise are not understood. If has been observed in non-ohmic contacts, crystal surfaces, and in some cases the bulk of the crystal itself. The noise can be minimized with proper fabrication procedures. A general empirical expression for the noise voltage is

$$v_n \propto \left[\frac{I^2}{fA_d} \Delta f \right]^{1/2} R_d \qquad (11\text{-}134)$$

where I = the dc current through the responsive element
A_d = the detector area
f = the electrical frequency

Johnson Noise. This noise is caused by the random motion of free charge-carriers in the responsive element, which is the result of collisions between the free charge-carriers and the lattice atoms and exists regardless of the presence of a bias current. Since the degree of motion of the lattice sites will depend upon the lattice temperature, the degree of Johnson noise depends upon lattice temperature. The expression derived by Johnson for this noise voltage is

$$v_n = \sqrt{4kTR_d\Delta f} \qquad (11\text{-}135)$$

where k = Boltzmann constant
T = the temperature of the responsive element
Δf = the electrical frequency bandwidth [11-13]

Shot Noise. This noise is due to the discreteness of free electrons and holes as they pass across p - n junctions. They appear as minute current pulses which show up as a random noise current or voltage in the exterior circuit. The expression for the shot-noise spectrum given by van der Ziel [11-14] is

$$v_n = R_d(2eI\Delta f)^{1/2} \qquad (11\text{-}136)$$

where I = the dc current across the p - n junction
e = the charge of an electron

The total noise in a photoconductive detector is the sum of the noises (excluding shot noise) discussed in the above sections.

$$(v_n)_{\text{total}} = \sqrt{(v_n)^2_{ph} + (v_n)^2_J + (v_n)^2_{1/f} + (v_n)^2_{gr}} \qquad (11\text{-}137)$$

Since a photoconductive detector does not have a p - n junction, shot noise is not present. The frequency spectrum of the sum of all these noises is given in Figure 11-17.

11.4 Detector Characterization
This section contains an enumeration and description of the important parameters associated with infrared detectors. The figures-of-merit that are used for comparing detectors are also discussed.

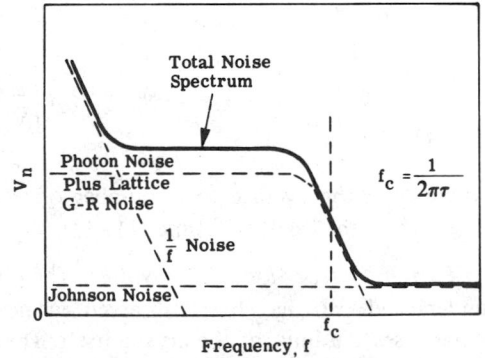

Fig. 11-17. Frequency spectrum for the total noise spectrum in a photoconductor.

This is followed by a description of the currently accepted procedures for testing detectors.

11.4.1. Detector Parameters. The following parameters are important when one measures the properties of infrared detectors or interprets data measured by others:

(1) The infrared radiation incident on the detector.
(2) The electrical output of the detector (which consists of the signal and noise voltages mixed).
(3) Geometrical properties of the detector.
(4) The detector as a circuit element.
(5) The detector temperature.
(6) The bias.

Incident Infrared Radiation. The radiation incident on a detector is characterized by the distribution of the radiant power with respect to wavelength, modulation frequency, position on the responsive element, and the direction of arrival. In the laboratory, the source of infrared radiation is usually a blackbody operated at 500 K. The accuracy of the measurements will be determined by the accuracy of the temperature setting and the uniformity of temperature in the source cavity. (See Chapter 2.)

The axis of the conical blackbody cavity should be slightly off-axis so that the apex is not visible, normal to the sensitive surface of the detector. The radiant power incident upon the detector is measured in watts. The sensitive surface of the detector is also irradiated by optical power from the surroundings; this is called ambient radiation. The signal radiation is usually modulated periodically in time. This modulation is usually achieved by rotating a multibladed wheel at constant angular frequency, ω, between the radiation source and the detector. This device, called a chopper, produces a fundamental frequency, f. The instantaneous power, $\Phi(t)$, incident on the detector can be expressed as the sum of Fourier components:

$$\Phi(t) = \Phi_0 + \Phi_1 \cos(2\pi f t + \delta_1) + \Phi_2 \cos(2 \cdot 2\pi f t + \delta_2) + \ldots \quad (11\text{-}138)$$

Each Fourier component has amplitude, Φ_k, and phase, δ_k. Each component represents a sinusoidally modulated signal, with frequency f, $2f$, $3f$, etc. The fundamental component is the sinusoidal component of frequency, f. The rms value is

$$\Phi_s = 2^{-1/2}\Phi_1 \quad (11\text{-}139)$$

The radiation can have a 500 K blackbody spectral distribution, or it can be optically filtered in various ways to produce a specific character. (See Section 11.4.3.)

In measuring detector characteristics, one uses the rms value of the sinusoidal fundamental component to compute the detector figures-of-merit.

The Electrical Output of the Detector. There are two additive components of the electrical voltage from a detector. First is the signal voltage, v_s, (or signal current, i_s), and second is the noise voltage, v_n (or the noise current, i_n). Over a sufficiently long time period, the two voltages (or currents) can be distinguished because the signal voltage, v_s, is coherent with the signal radiation power, Φ_s, while the noise is not. If the signal power is a periodic function of time, the signal voltage is as well.

$$v_s(t) = V_0 + V_1 \cos(2\pi f t + \psi_1) + V_2 \cos(4\pi f t + \psi_2) + \ldots \quad (11\text{-}140)$$

In general, the magnitude of the signal is a function of the bias voltage, V_B, applied to the detector, the modulation frequency, f, the wavelength, λ, the incident irradiance, E, and the detector area, A_d. In functional notation, this can be written

$$v_s = v_s(V_B, f, \lambda, E, A_d) \tag{11-141}$$

Over a wide range of input powers, the signal is a linear function of the incident radiation power. Since only those detectors that have this property will be considered, v_s/EA_d is a constant for fixed V_B, f, and λ. In many detectors, the signal's dependence on the modulation frequency can be separated from its dependence on the wavelength of the incident signal radiation power. Detectors are said to have the factorability property in those wavelength regions in which separation is possible. Only detectors having this factorability property are considered here. Under these conditions the detector signal is given by

$$v_s = v_s(V_B, E, A_d)v(\lambda)v(f) \tag{11-142}$$

where $v(\lambda)$ and $v(f)$ are normalized functions with values ranging from 0 to 1.

The functional dependence of the signal on the applied bias may also be separable. There are useful detectors which do not have this property, however, therefore caution must be exercised in making this assumption.

The dependence of the detector signal on the applied bias is measured at a specified modulation frequency with specified incident radiation. These data are usually reported as a graph labeled "determination of optimum bias."

The dependence of a detector signal upon the frequency at which the incident radiation is modulated, is a measure of its temporal response. The measurements are usually made with constant radiation signal power and bias value. The results can be reported in a plot of relative signal voltage versus modulation frequency. Because such a plot represents the frequency response of the detector signal, a time constant, called the responsive time constant, can be determined. The peak detective frequency is the modulation frequency that maximizes the signal-to-noise ratio.

The signal dependence of a detector upon the wavelength of the radiation signal power is a measure of its ability to respond to the radiation of different wavelengths. These spectral responsivity measurements are made at a constant bias value and modulation frequency. Such data are reported on a plot of relative signal, normalized to constant radiation signal power at each wavelength, versus wavelength. Such a plot is usually labeled "relative spectral response."

The second additive component of the electrical voltage from a detector is the noise voltage. The detector noise is the electrical voltage or current output from the detector that is not coherent with the signal radiation power. The rms voltage (or current) of the electrical noise is defined as the square root of the time average of the square of the difference between the instantaneous voltage (or current) and the time average voltage (or current).

$$v_n = [(v_n(t) - \overline{v_n(t)})^2]^{1/2} \tag{11-143}$$

$$i_n = [(i_n(t) - \overline{i_n(t)})^2]^{1/2} \tag{11-144}$$

In practice, the average values of $v(t)$ and $i(t)$ are usually zero because ac-coupled amplifiers are usually used. Then v_n and i_n become

$$v_n = [\overline{(v_n(t))^2}]^{1/2} \qquad (11\text{-}145)$$

$$i_n = [\overline{(i_n(t))^2}]^{1/2} \qquad (11\text{-}146)$$

The detector noise, which is random, is a function of the bias voltage, modulation frequency, and area of the detector. With certain assumptions, it can be shown that the detector signal-to-noise ratio varies directly as the square root of the detector area when the modulation frequency and bias are constant. The noise voltage is measured with the detector shielded from the radiation signal source. The detector noise, as reported, is referred to the output terminals of the detector and normalized to a 1 Hz effective noise bandpass.

The detector noise may be reported as a family of curves representing the different noise values obtained at several bias values plotted against frequency and labeled noise spectrum.

Geometrical Properties of the Detector. Usually there is no ambiguity in selecting the responsive planes of an infrared detector. If the responsive element itself is in the form of a thin, flat layer (such as evaporated lead salt photoconductors), the responsive plane is the plane of the sensitive lead salt film. In the case of curved photoemissive detectors, the adopted responsive plane is the plane that contains the straight edges of the photocathode. Positions on the adopted responsive plane are defined by a rectangular cartesian coordinate system. The responsivity, $\mathcal{R}(x, y)$, of a small spot on the plane is measured with a small spot of radiation. The effective area of the adopted responsive plane is defined as

$$A_e = \int_{A_d} \int \frac{\mathcal{R}(x, y)dxdy}{\mathcal{R}_{max}} \qquad (11\text{-}147)$$

The detector solid angle, Ω, is the solid angle from which the detector receives radiation from the outside. For flat detector responsive elements, the effective solid angle is the solid angle weighted at each angle by the cosine of the angle of incidence. This is called the effective weighted solid angle and is designated as Ω_e. For detectors with circular symmetry (i.e., the detector responsivity is independent of the azimuthal angle, ϕ_a), the total cone angle, Θ, may be used to define Ω_e by

$$\Omega_e = \pi \sin^2 (\Theta/2) \qquad (11\text{-}148)$$

The projected solid angle of a detector that has a hemispherical field of view is π sr.

The Detector as a Circuit Element. The properties of a detector as a circuit element will usually depend on the electrical frequency and on the amount of ambient radiant power, Φ, and on the dc current, I_B, through the detector. The impedance of the detector can be written as a complex impedance

$$\widetilde{Z}_d = R_d + iX_d \qquad (11\text{-}149)$$

where R_d is the resistance of the detector, given by the ratio of the dc voltage to the dc current, i.e.

$$R_d = \frac{V}{I} \tag{11-150}$$

where $\quad i = \sqrt{-1}$

$\quad X_d$ = the detector reactance

When a detector is tested, it is connected to an amplifier and in some cases to bias sources. The load impedance, $\widetilde{Z}_L$, is the impedance of the external circuit as seen from the terminals of the detector. Usually the load impedance is almost purely resistive.

The Detector Temperature. Detectors that operate without refrigerators have responsive elements with temperatures equal to or higher than the ambient temperature. Refrigerated detectors have designated temperatures equal to the temperature of the coolant. The detector signal and noise voltages and the resistance are greatly influenced by the operating temperature of the detector.

The Bias. Many detectors require an external bias of some kind. In the case of a photoconductive detector, the bias is an applied electrical voltage. In general, the signal and noise characteristics of a detector are measured over the entire useful range of bias values. Since both the detector signal and noise are functions of the bias and modulation frequency, the optimum bias value reported is the value that maximizes the signal-to-noise ratio at a stated modulation frequency.

11.4.2. Detector Figures-of-Merit. Using the parameters described in Section 11.5.1, one can compare different detectors to determine the one best suited for an application. The first coarse comparisons are done with the following figures-of-merit.

Responsivity. The responsivity of an infrared detector can be defined, in general, as the ratio of the output electrical signal to the incident radiation power. However, it is usually rigorously defined only for periodically modulated radiation. For this case, the responsivity is defined as the ratio of the rms value of the fundamental component of the electrical output signal of the detector to the rms value of the fundamental component of the input radiation power. The units of responsivity are volts per watt or amperes per watt. Two cases are of particular importance: the spectral responsivity, defined by the equation

$$\mathcal{R}(\lambda) = \frac{V_s}{\Phi_{s,\lambda} \Delta\lambda} \tag{11-151}$$

and the blackbody responsivity, defined by the equation

$$\mathcal{R}_{bb} = \frac{V_s}{\Phi_s bb} \tag{11-152}$$

where the incident radiant power is the integral over all wavelengths of the spectral power distribution from a blackbody. The responsivity is usually a function of the bias voltage, V_B, the operating electrical frequency, f, and the optical wavelength, λ.

Noise Equivalent Power (NEP). The NEP is that level of rms, incident, infrared signal radiation required to produce a signal-to-noise ratio of unity.

$$\text{NEP} = \frac{\Phi_s}{\left[\dfrac{v_s}{v_n}\right]} \tag{11-153}$$

It can be written in terms of the responsivity:

$$NEP = \frac{v_n(V_B, f)}{\mathscr{R}(V_B, f, \lambda)} \qquad (11\text{-}154)$$

Specific Detectivity (D*). This figure-of-merit is defined in cm $Hz^{1/2}$ W^{-1} as

$$D^*(V_B, f, \lambda) = \frac{\mathscr{R}(V_B, f, \lambda)}{v_n(V_B, f)} (A_d \Delta f)^{1/2} \qquad (11\text{-}155)$$

where A_d = the detector area
Δf = the electrical bandwidth of the post-detector electronics

D^* can be written more fundamentally as

$$D^*(V_B, f, \lambda) = (A_d \Delta f)^{1/2} \frac{v_s(V_B, f, \lambda)/v_n(V_B, f)}{\Phi_{s,\lambda} \Delta \lambda} \qquad (11\text{-}156)$$

Responsive Time Constant. The signal voltage versus electrical frequency curve of many detectors can be analytically described by the equation

$$v_s(f) = v_s(0) [1 + \omega^2 \tau^2]^{-1/2} \qquad (11\text{-}157)$$

where ω = $2\pi f$
τ = the time constant

If this equation is true for a detector, then the factor τ is the responsive time constant of that detector. If the above relationship does not hold true, then there is no common agreement upon the definition of the responsive time constant, and any use of this figure must be accompanied by an explicit definition.

Detectivity-Frequency Product (D*f*). Another figure-of-merit for performance of an infrared detector suggested by Williams [11-15] and Borrello [11-16] is the product D^*f^* where D^* is the maximum specific detectivity as defined previously and f^* is the upper frequency at which D^* has decreased to 0.707 of its maximum value. This figure-of-merit allows the performance potential of detectors to be compared for high-frequency, low-background applications. Borrello has derived the maximum value of D^*f^*:

$$(D^*f^*)_{max} = \frac{1}{2\pi} \left(\frac{\sigma_c}{4E_g h}\right)^{1/2} \qquad (11\text{-}158)$$

where σ_c = the photon capture cross-section
E_g = the energy of a photon
h = Planck's constant

Curves of maximum D^* versus frequency for some typical detectors are shown in Figure 11-18. The experimental results [11-16] give a D^*f^* value of about 10^{17}, whereas calculations based on capture cross-sections predict values of 10^{18} and 10^{19}.

11.4.3. Detector Performance Tests*

Test Equipment. The equipment used to measure important detector parameters

---

*Sections 11.4.3 and 11.4.4 essentially reproduce material prepared by W.L. Eisenman, J.D. Merrian, and R.F. Potter of the Electro Materials Sciences Center, Naval Electronics Laboratory Center, Corona, CA (now the Naval Ocean Systems Center, San Diego, CA) and originally published in the 1965 *Handbook of Military Infrared Technology.*

consists of electronic equipment, black-body source, variable frequency source, monochromatic source, and reference detector. For the most part, these instruments can be obtained from various commercial suppliers and can be assembled to meet specific measurement requirements. Equipment costs for a modest detector test set-up may approach $100,000, and a large facility may easily require five times this amount.

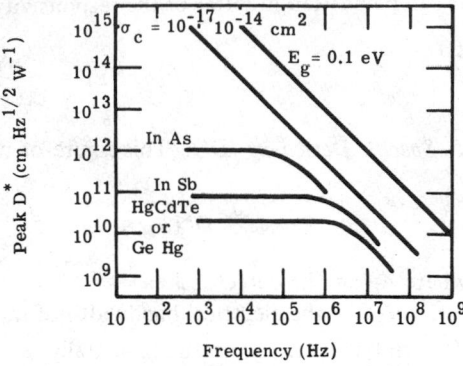

Fig. 11-18. Maximum D* versus frequency: calculations for two σ_c values compared with measured detectors [11-16].

The electronics generally used to measure detector signal and noise characteristics consist of a preamplifier, amplifier, and a spectrum analyzer. Auxiliary equipment consists of an oscillator and calibrated attenuator used for electrical calibration, and a variable voltage supply suitable for providing detector bias. Figure 11-19 is a block diagram of a typical arrangement of equipment suitable for signal and noise measurements in the 1 kHz to 2 MHz frequency range. If a variety of detectors is to be measured, it will be convenient to have several similar arrangements of instruments to cover the necessarily wide frequency range. A number of preamplifiers will also be necessary to cover a wide range of detector impedances [11-17]. When possible, the preamplifier should be placed with the detector in a shielded enclosure to minimize input capacity. Apparatus suitable for measuring signal and noise of thermal type detectors over a frequency range of 1 to 100 Hz are shown in Figure 11-20 [11-18].

The primary requirement for measuring detector responsivity is a stable, modulated source of infrared radiation with known spectral characteristics. A blackbody simulator is a convenient means of meeting these requirements. If fixed-frequency modulators are used,

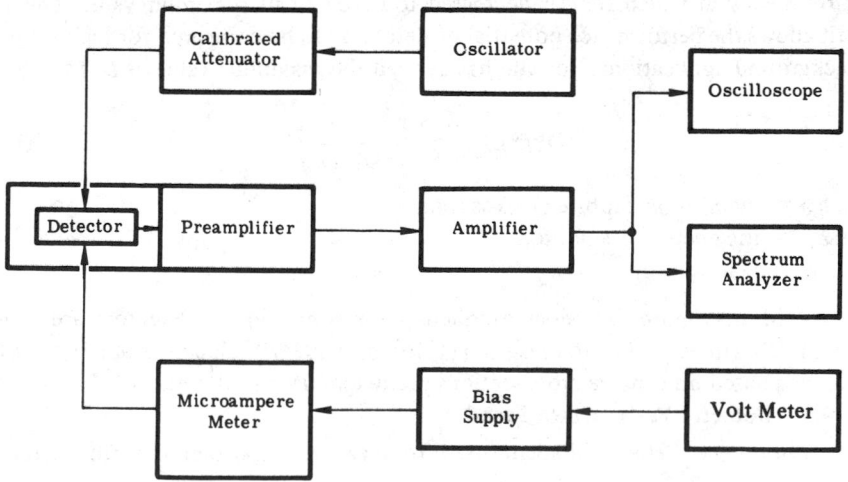

Fig. 11-19. Block diagram for detector signal and noise measurements with a range of 1.0 kHz to 1.0 MHz.

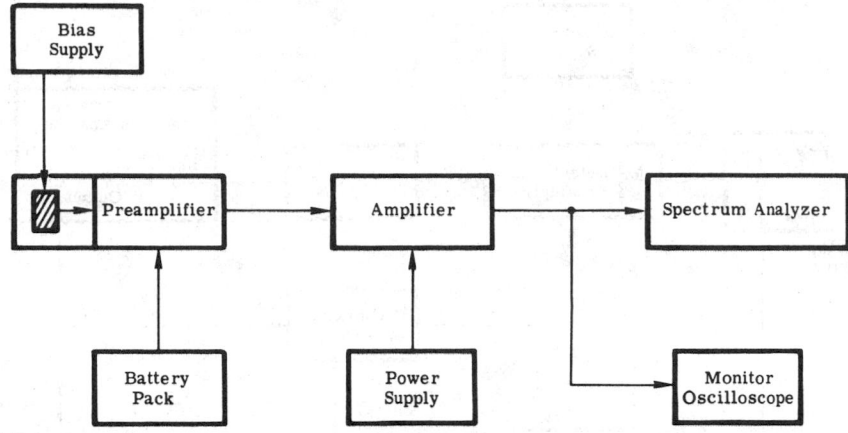

Fig. 11-20. Block diagram for detector signal and noise measurements with a range of 1.0 to 100.0 Hz.

it will be convenient to have at least three available, operating at frequencies of approximately 10, 100, and 1000 Hz. The source must produce an accurately known irradiance over the responsive plane of the detector, and the irradiance must be uniform over the sensitive area of the detector. The spectral irradiance used is the rms value of the fundamental component of the modulation frequency. In determining this value, one must take into account the radiation from the modulator. To compare measured data easily, operators of many detector laboratories operate their blackbody sources at a temperature of 500 K. Irradiance values in the order of microwatts per centimeter squared are appropriate for many types of detectors.

A variable frequency source is required to measure the dependence of detector signal on modulation frequency. Basically, any stable source having a suitable spectral output and equipped with a variable speed modulator may be used. The irradiance, E, produced by the source must be uniform over the sensitive surface of the detector. At audio and subaudio frequencies, satisfactory mechanical modulators are easily fabricated [11-19]. However, above 50 kHz mechanical modulators become awkward, blade diameters become large, aperture sizes are reduced to small dimensions, and the blades, must be driven at quite dangerous speeds.

Light emitting diodes (LEDs) can also serve as variable frequency sources for frequencies into the megahertz region. The apparatus is quite simple. One instrument arrangement is shown in Figure 11-21. The system utilizes the beat frequency oscillator (BFO) output voltage available on several models of audio frequency wave analyzers. This BFO voltage has a frequency that is always equal to the input frequency of the analyzer. Thus, when the input frequency of the analyzer is varied over the frequency range of the equipment, the frequency of the BFO voltage automatically follows. The BFO voltage from the analyzer is amplified and applied to a light emitting diode. The modulated radiation from the diode is focused on the detector being measured. The electrical signal from the detector is amplified and then applied to the input of the analyzer.

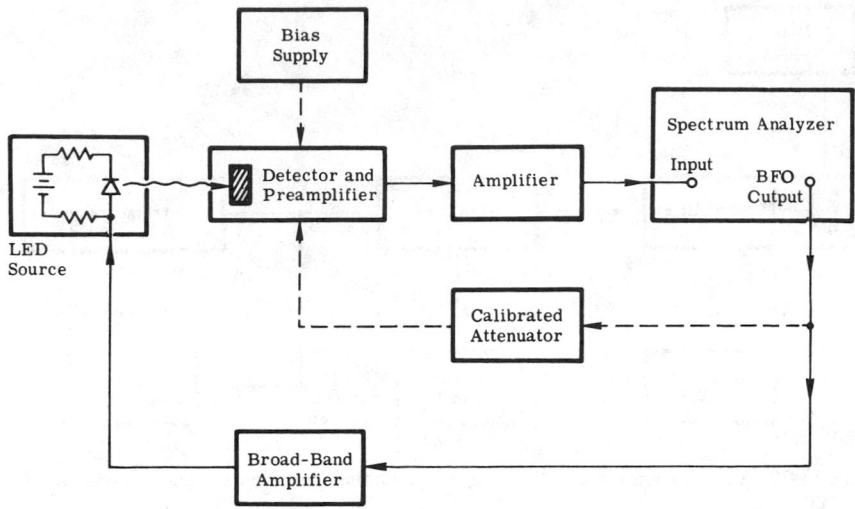

Fig. 11-21. Block diagram for a variable frequency source using LEDs with a range up to 1.5 MHz.

This arrangement has two distinct advantages in addition to its simplicity. The measurement of detector frequency response is quite rapid, since only one control on the wave analyzer needs be adjusted, and the narrow bandpass of the wave analyzer produces a large detector signal-to-noise ratio. Thus, a relatively small amount of input radiation is necessary.

The major disadvantage of this method is the narrow spectral distribution of the infrared radiation emitted from the LED. Gallium arsenide and indium arsenide diodes emitting at 0.8 and 3.2 μm, respectively, are commercially available. Diodes of the ternary alloys are available which provide emission at longer wavelengths.

The monochromatic, infrared-radiation source consists of a stable, broad spectral-band source of infrared radiation, a modulator, a monochromator or other spectral filter, and an optical system which directs the monochromatic radiation to the detectors. Tungsten filament lamps, Nernst glowers, and glow bars are commonly used sources depending upon the wavelength region of interest. (See Chapter 2.) Since the radiant output of both the glower and glow bar will be affected by air currents, these sources should be provided with a suitable housing or chimney. The modulator may be a fixed frequency device and is normally placed at the entrance slit of the monochromator, which should be capable of providing a wavelength band of radiation not wider than about 1/25th of the center wavelength. Because scattered radiation may be a problem, particularly at the longer wavelengths, a double monochromator is preferable for this application. However, satisfactory performance may be obtained from a single monochromator by fitting the instrument with suitable rejection filters [11-19]. The detector being measured and the reference detector are alternately illuminated by an optical system placed at the exit slit of the monochromator. A typical arrangement is shown in Figure 11-22.

The spectral response of a detector is obtained by comparing the signal from the detector to the signal from a reference detector as a function of the wavelength of the incident radiation. Detectors utilizing a cavity as the radiation receiver have been used as

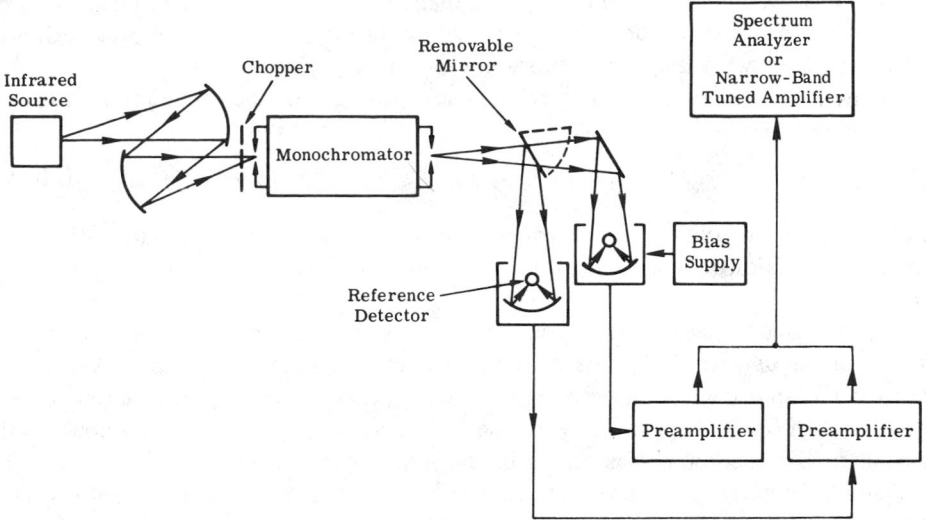

Fig. 11-22. Block diagram of instrument arrangement for measuring spectral response.

reference standards [11-20, 11-21, 11-22] but they are difficult to obtain and hard to use because of their low sensitivity and slow speed of response. A radiation thermocouple is a convenient detector for use as a reference standard, provided its spectral sensitivity has been determined (by comparison to a cavity-type detector). The relative spectral sensitivity of a typical radiation thermocouple compared to such a cavity is shown in Figure 11-23. The decline in sensitivity is relatively smooth and there is no difficulty in correcting the data. Since the sensitivity of a thermal detector may not be uniform across the responsive plane [11-23], the optical system used at the exit slit of the monochromator should be adjusted to flood completely the thermocouple receiver. The monochromator should also have a reasonably uniform exit pupil.

Test Procedures. Test procedures may be divided into two groups. In the first group are the measurements that are necessary to determine the detector responsivity, and in the second are the measurements that yield the root power spectrum of the noise. The arrangement of the electrical equipment is the same for both groups. Figure 11-19 shows such an arrangement for use on common photoconductive detectors that require a single bias voltage, V_B.

The determination of detector responsivity involves three separate measurements. In each of these measurements the incident signal radiation must be normal to the detector's

Fig. 11-23. Relative spectral response of a typical radiation thermocouple as compared to a special black conical cavity reference detector.

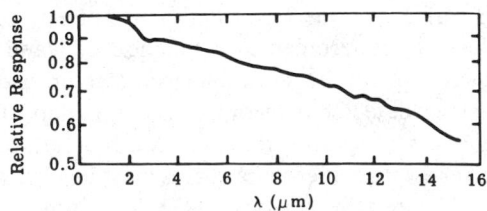

responsive plane, and the amount of signal radiation must be confined to a range in which the output signal from the detector is proportional to the incident radiant power. Confirmation of this linearity may be necessary in some cases.

The measurement of the responsivity involves the use of the factorability property (See Equation (11-141).):

$$v_s(V_B, \Phi_s, A_d, \lambda, f) = v_s(V_B, \Phi_s, A_d)v(\lambda)v(f) \tag{11-159}$$

The blackbody and the variable frequency sources must be equipped with a filter that limits the radiation to a wavelength band within which the factorability property holds [11-22].

The first step in measuring responsivity is to establish the range of bias values to be used with the detector being measured. Experience with similar detectors will usually indicate the approximate range of bias values. The range will normally cover at least one decade of bias voltage or current. The highest value of bias, normally known as the manufacturer's maximum bias, is explicitly stated by the manufacturer. Considerable care should be exercised if measurements are to be made at biases greater than this value. The detector noise should be carefully monitored and the bias should be increased in small steps. Operating some types of detectors in a region of high bias is very risky and experience must be relied upon.

The blackbody source equipped with a modulator of frequency, f, is then used to irradiate the detector. The center frequency of the spectrum analyzer is set at f and the signal generator is set to zero. The reading, v_s, of the output meter is noted. Then the radiation is adjusted to a value which gives the same reading, v_s, on the meter.

The open-circuit detector signal voltage, v_s, is the voltage across the calibrating resistor. These measurements are then repeated until the complete range of bias values has been covered.

The radiant power, Φ_s, incident on the detector is obtained from the known irradiance, E, upon multiplication by the detector area, A_d:

$$\Phi_s = A_d E \tag{11-160}$$

The corresponding detector responsivity is given by

$$\mathscr{R}_{bb}(V_B, f) = \frac{v_s(V_B, f)}{\Phi_s} \tag{11-161}$$

Next, with the spectrum analyzer set at frequency f and the bias applied, the detector is irradiated by the monochromatic source. The center wavelength of the monochromator is varied over the wavelength range of interest, and the relative signal voltage, v_s, of the detector is recorded as a function of wavelength. The detector under measurement is then replaced by the reference detector, and the relative signal voltage, v_{ref}, of the reference detector is recorded as a function of wavelength. (Some detectors may exhibit changes in spectral response as a function of bias. If these changes are significant, then several spectral response curves must be obtained for different bias values.)

The relative response, $v(\lambda)$, as a function of the wavelength is then calculated by

$$v(\lambda) = \frac{v_s \mathcal{R}_{ref}(\lambda)}{v_{ref}} \tag{11-162}$$

where $\mathcal{R}_{ref}(\lambda)$ is the relative spectral responsivity of the reference detector.

The detector is irradiated with the variable frequency source. As the modulation frequency of the source is varied, the center frequency of the spectrum analyzer is continuously adjusted to the modulation frequency. The detector signal voltage, v_s, read on the meter is recorded as a function of frequency. The source is then removed, and the signal generator with a fixed attenuator setting is varied over the same range of frequencies. As the frequency of the signal generator is varied, the center frequency of the spectrum analyzer is continuously adjusted. The voltage, v_c, read on the meter is recorded as a function of frequency. The relative response, $v(f)$, as a function of modulation frequency, is then computed by

$$v(f) = \frac{v_s(f)v_c(f_0)}{v_c(f)v_s(f_0)} \tag{11-163}$$

where f_0 is such that $v(f_0) = 1$.

The frequency response of some detectors may vary as a function of applied bias. If the change in frequency response is significant, then several frequency response curves must be measured over the entire range of bias values.

The measurement of the noise characteristics of a radiation detector requires good judgment and experience to ensure that the noise recorded is only the noise generated in the detector, load resistor, and the amplifier. Constant attention is required to prevent external sources of noise from influencing the results. It may be convenient to place a wideband oscilloscope in the electronic system ahead of the bandpass filter. The appearance of the noise trace on the oscilloscope is helpful in determining the presence of any extraneous noise.

In particular, the bias supply must not contribute appreciable noise. The bias source can be checked for internal noise by substituting a wire-wound resistor in place of the detector in the input circuit. The resistance of the wire-wound resistor should be approximately equal to the detector resistance. Bias is then applied to the circuit and the noise noted on the output meter. The noise generated in the wire-wound resistor should be independent of the current flowing through the resistor.

Bias is then applied to the detector. All radiation sources except ambient background are removed. With the signal generator producing zero signal, the rms noise voltage indicated by the output meter is recorded as a function of frequency over the entire frequency range of interest. The voltage read is then denoted v_0.

The detector and load resistor are replaced by a wire-wound resistor having approximately the same resistance as the parallel combination of the detector and load resistor. The temperature of this wire-wound resistor is maintained such that the Johnson noise generated in the resistor is small compared to the noise generated in the amplifier. The rms noise voltage indicated by the output meter is again recorded as a function of frequency. This voltage is denoted v_a.

The signal generator is then adjusted to produce a calibration signal, v_c, across the calibrating resistor. This calibration signal is made approximately 100 times larger than the detector noise, the spectrum analyzer is tuned to the frequency of the calibration signal, and the voltage indicated on the output meter recorded. This procedure is repeated over the entire frequency range of interest. The system gain, $g(f)$ is thus determined as a function of frequency.

The root-power-spectrum, v_n^*, referred both to the terminals of the detector and to an infinite load impedance and corrected for amplifier noise, is calculated in units of rms volts per root Hertz from the following formula:

$$v_n^*(f, V_B) = \frac{\left[\dfrac{v_0^2(f,V_B) - v_a^2(f)}{g^2(f)} - v_T^2\left(\dfrac{R_d}{R_L}\right)^2\right]^{1/2}}{(\Delta f)^{1/2}} \tag{11-164}$$

where R_d is the resistance of the detector. The thermal noise voltage, v_T, generated by the load resistor, R_L, in the noise bandwidth, Δf, is given by

$$v_T^2 = 4kTR_L\Delta f \tag{11-165}$$

Note that Δf, which is the effective noise bandwidth of the measurement equipment, is defined by

$$\Delta f = \int_0^\infty \frac{g^2(f)}{g_m^2}\,df \tag{11-166}$$

where $g(f)$ = the gain of the system as a function of frequency, f
$\quad g_m$ = the maximum value of the gain

The frequency, f_m, that corresponds to g_m is defined as the center frequency of the passband.

11.4.4. Performance Calculations. A limited set of conditions must be selected for the tests, but if these are properly chosen it is possible to predict the detector response under a variety of other operating conditions.

Responsivity in functional form is written

$$\mathcal{R} = \mathcal{R}(V_B, f, \lambda) \tag{11-167}$$

The parameters V_B, f, and λ enter into $\mathcal{R}$ only through the signal, since E and A are independent of them. Therefore, we have

$$v_s = v_s(V_B, f, \lambda) \tag{11-168}$$

Because the parameters V_B, f, and λ are factorable (separable) in any usable detector, one may write

$$v_s = v_s(V_B)v(f)v(\lambda) \tag{11-169}$$

The parameters $v(V_B)$, $v(f)$, and $v(\lambda)$ can be measured separately and reported graphically in the charts as follows:

$v_s(V_B)$ in a chart for the determination of optimum bias;

$v(f)$ in a frequency response chart;

$v(\lambda)$ in a spectral response chart.

The subscript r will be used to designate the reported value and the subscript 1 to designate a desired value. The responsivity at bias, V_{B1}, modulation frequency, f_1, and radiation wavelength, λ_1, is given by the relation

$$\mathcal{R}(V_{B1}, f_1, \lambda_1) = \frac{\mathcal{R}(V_{Br}, f_r, \lambda_r)v_s(V_{B1})v(f_1)v(\lambda_1)}{v_s(V_{Br})v(f_r)v(\lambda_r)} \tag{11-170}$$

Since the responsivity is measured with a blackbody, usually at 500 K, the following relation is used:

$$\mathcal{R}(V_{Br}, f_r, \lambda_{\max}) = \frac{\mathcal{R}_{bb}(V_{Br}, f_r, T)}{\gamma_p} \tag{11-171}$$

where $\lambda_{\max}$ is the infrared wavelength at which $v(\lambda)$ is a maximum. The symbol γ_p is given by the expression

$$\gamma_p = \frac{\displaystyle\int_0^\infty v(\lambda)\Phi_{s,\lambda}(T)d\lambda}{\displaystyle\int_0^\infty \Phi_{s,\lambda}(T)d\lambda} \tag{11-172}$$

where $\Phi_{s,\lambda}(T)$ is the spectral radiant power from a blackbody at temperature T and wavelength λ. Also, since $v(\lambda)$ is normalized to the peak wavelength, one can evaluate at $\lambda_r = \lambda_{\max}$ and $v(\lambda_{\max}) = 1$. Then

$$\mathcal{R}(V_{B1}, f_1, \lambda_1) = \frac{\mathcal{R}_{bb}(V_{Br}, f_r, T)}{\gamma_p} \frac{v_s(V_{B1})}{v_s(V_{Br})} \frac{v(f_1)}{v(f_r)} \frac{v(\lambda_1)}{1} \tag{11-173}$$

where $\mathcal{R}_{bb}(V_{Br}, f_r, T)$ and γ_p (sometimes reported as the ratio $\mathcal{R}(\lambda_{\max})/\mathcal{R}_{bb})$ are reported parameters.

Unfortunately, noise is not factorable into independent functions of the parameters V_B and f. (The internal noise does not depend upon λ.) Thus noise values desired at other than the measured values must be interpolated or estimated by the theoretical formulae given in Table 11-3.

The set of values of V_B, f, and λ that give the maximum or "peak" D^*, i.e., $D^*_{\max}$, is designated $(V_{Bp}, f_p,$ and $\lambda_p)$. At bias value, V_{Bp}, the signal-to-noise ratio is maximum. At chopping frequency, f_p, the signal-to-noise ratio is maximum. And at λ_p the signal voltage is maximum. (The noise voltage is independent of λ.) Given $D^*_{\max}$, one can find D^* at other values of the set V_B, f, and λ with the following equation:

$$D^*(V_{B1}, f_1, \lambda_1) = D^*_{\max}(V_{Bp}, f_p, \lambda_p) \frac{v_n(V_{Bp}, f_p)}{v_n(V_{B1}, f_1)} \frac{v_s(V_{B1})}{v_s(V_{Bp})} \frac{v(f_1)}{v(f_p)} \frac{v(\lambda_1)}{v(\lambda_p)} \tag{11-174}$$

A sample calculation of $\mathcal{R}$, D^*, and NEP will be performed below as an example for a detector operated at conditions different from those given in standard reported data. For this example, a photoconductive lead selenium (PbSe) detector operated at liquid nitrogen temperature will be used. Its characteristics are listed in Table 11-4 and shown in Figures 11-24 and 11-25.

Table 11-4. Detector Characteristics Performance Parameters, and Test Conditions for a PbSe Detector [11-24]

Test Results

$\mathcal{R}$ (500 K, 90 Hz)	3.2×10^4 V W^{-1}
NEE (500 K, 90 Hz) ($\Delta f = 1$ Hz)	2.9×10^{-9} W cm^{-2}
NEP (500 K, 90 Hz) ($\Delta f = 1$ Hz)	1.1×10^{-10} W
D^* (500 K, 90 Hz)	1.8×10^9 cm Hz$^{1/2}$ W^{-1}
Peak wavelength, λ_p or λ_{max}	4.5 μm
Peak modulation frequency	7.0×10^2 Hz
$D^*_{m,m}$†	1.2×10^{10} cm Hz$^{1/2}$ W^{-1}
Effective time constant	1.2×10^2 μsec
$\mathcal{R}(\lambda_{max})/\mathcal{R}_{bb}$	4.8

Conditions of Measurement

Blackbody temperature	500 K
Blackbody flux density	7.7 μW cm^{-2}, rms
Chopping frequency	90 Hz
Noise bandwidth	5 Hz
Cell temperature	78 K
Cell current	50 μA
Load resistance	1.0×10^6 ohms
Transformer	–
Relative humidity	31%
Responsive plane from window	2.5 cm
Ambient temperature	24°C
Ambient radiation on detector	297 K (only)

Cell Description

Type	PbSe (chemical)
Area	3.9×10^{-2} cm^2
Dark resistance	1.0×10^6 ohms
Field of view	180°
Window material	Sapphire

†D^* evaluated at maximum wavelength and maximum chopping frequency.

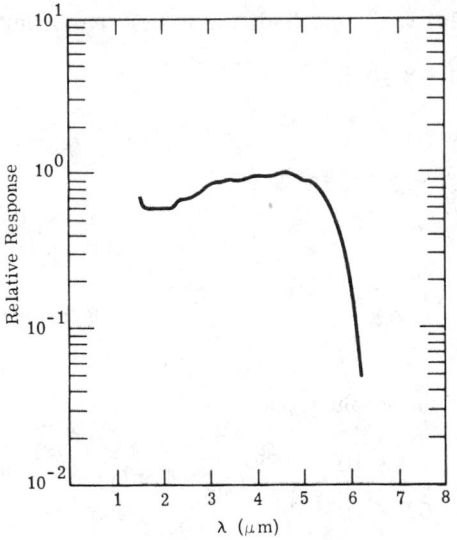

Fig. 11-24. Spectral response [11-24].

Suppose one wishes to find the responsivity and signal-to-noise ratio at the following operating point:

$$f_1 = 10^3 \text{ Hz}$$
$$I_{B1} = 75 \,\mu\text{A}$$
$$\lambda_1 = 6.0 \,\mu\text{m} \qquad (11\text{-}175)$$

From the frequency response plot (Figure 11-25(a)), the relative response, $v(f)$, at 90 and 10^3 Hz is

$$v(90 \text{ Hz}) = 1.0$$
$$v(10^3 \text{ Hz}) = 0.78 \qquad (11\text{-}176)$$

From the determination of an optimum bias plot (Figure 11-25 (b)), the values of $v(I_B)$ at bias currents of 50 and 75 μA are

$$v(50 \,\mu\text{A}) = 9.0 \times 10^{-3} \text{ V}$$
$$v(75 \,\mu\text{A}) = 1.3 \times 10^{-2} \text{ V} \qquad (11\text{-}177)$$

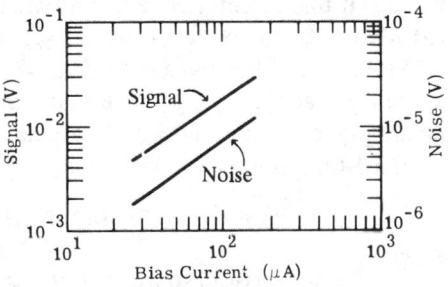

(a) Frequency response.

(b) Determination of optimum bias.

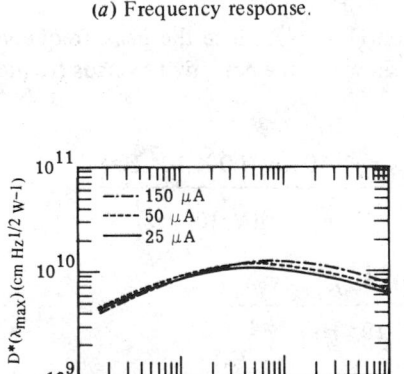

(c) Detectivity versus frequency.

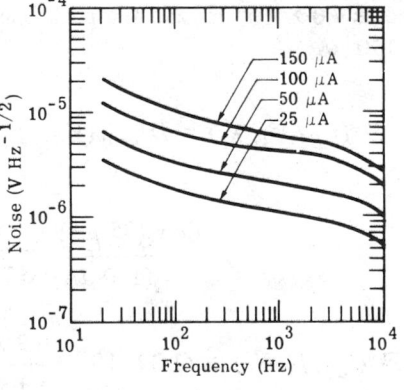

(d) Noise spectrum.

Fig. 11-25. Photoconductive lead selenium (PbSe) detector operated at liquid nitrogen temperature [11-25].

From the spectral response plot (Figure 11-24) $v(\lambda)$, one finds the values at 6.0 μm:

$$v(6.0\ \mu m) = 1.5 \times 10^{-1} \tag{11-178}$$

From the test results in Table 11-4 one finds

$$\mathcal{R}_{bb}(50\ \mu A, 90\ Hz, 500\ K) = 3.2 \times 10^4\ V\ W^{-1}$$

and

$$\frac{1}{\gamma_p} = \frac{\mathcal{R}(\lambda_{max})}{\mathcal{R}_{bb}} = 4.8 \tag{11-179}$$

Using Equation (11-173) and substituting in the above values, one has

$$\mathcal{R}(75\ \mu A, 10^3\ Hz, 6.0\ \mu m) = 3.2 \times 10^4 \frac{1.3 \times 10^{-2}}{9.0 \times 10^{-3}} \frac{0.78}{1.0} (1.5 \times 10^{-1}) 4.8$$

$$= 2.6 \times 10^4,\ V\ W^{-1} \tag{11-180}$$

Therefore, if this detector were of unit area and the radiation source had a flux density of 1.0 μW cm^{-2}, the expected signal level would be 2.6 $\times$ 10^4 μV from the detector at $I_B = 75$ μA, $f = 10^3$ Hz, and $\lambda = 6.0$ μm.

To find D^* at f_1, V_{B1}, λ_1, one must find the noise level at f_1, V_{B1}. From the noise spectrum plot of Figure 11-25(d), interpolate at $f = 10^3$ to find the noise level for $I_B = 75$ μA. For a 1-Hz system bandwidth, this gives

$$v_n(75\ \mu A, 10^3\ Hz) = 3.1 \times 10^{-6},\ V$$

$$v_n(150\ \mu A, 7 \times 10^2\ Hz) = 6.2 \times 10^{-6},\ V \tag{11-181}$$

D^* at V_{B1}, f_1, λ_1 can be found with Equation (11-174). Since the peak frequency is reported as 7.0 $\times$ 10^2 Hz and the peak bias is shown in the detectivity versus frequency plot at 150 μA,

$$D^*(V_{B1}, f_1, \lambda_1) = D^*_{max}(V_{B1}, f_p, \lambda_p) \frac{v_n(150\ \mu A, 7.0 \times 10^2\ Hz)}{v_n(75\ \mu A, 10^3\ Hz)}$$

$$\frac{v_s(75\ \mu A)}{v_s(150\ \mu A)} \frac{v(10^3\ Hz)}{v(7.0 \times 10^2\ Hz)} \frac{v(\lambda_1)}{1} \tag{11-182}$$

$$D^*(I_{B1}, f_1, \lambda_1) = (1.2 \times 10^{10}) \frac{6.2 \times 10^{-6}}{3.1 \times 10^{-6}} \frac{1.3 \times 10^{-2}}{2.9 \times 10^{-2}} \frac{0.78}{0.83} (1.5 \times 10^{-1}) \tag{11-183}$$

$$D^*(I_{B1}, f_1, \lambda_1) = 1.5 \times 10^9,\ Hz^{1/2}\ cm\ W^{-1} \tag{11-184}$$

If the detector has a 1 cm^2 area, a post-detector electronic system with unit bandwidth, and an incident radiation flux density of 1.0×10^{-6} W cm^{-2}, the expected signal-to-noise ratio at I_{B1}, f_1, λ_1, would be 1.5×10^3.

The noise equivalent power at I_{B1}, f_1, λ_1, is

$$\text{NEP} = \frac{A_d^{1/2}}{D^*(I_{B1}, f_1, \lambda_1)} = \frac{(3.9 \times 10^{-2})^{1/2}}{1.5 \times 10^9} = 1.3 \times 10^{-10}, \text{W} \qquad (11\text{-}185)$$

Some precautions and reservations should be kept in mind when extrapolating the performance data to low-background operations.

Figure 11-16 shows the theoretical improvement in D^* that can be achieved by reducing the field of view of the detector and thus reducing the background radiation flux. A reduction in background radiation can also occur through the use of cooled filters, or in space applications where the background temperatures can be very low. In general, the improvement in D^* is accompanied by a decrease in some other performance parameter.

Since the resistance of the detector is an inverse function of the background radiation, operation at low backgrounds results in an increased resistance and can cause the time constant of the system to be RC-limited. Thus, one price that the system designer should be prepared to pay is a decrease in frequency response.

Reduction of the background noise will improve the D^* only if the background noise is the dominant noise mechanism. Thus, successful operation at low backgrounds can involve the reduction of thermal generation-recombination noise by operating at lower temperatures. This can increase the complexity and power consumption of an infrared system.

There are a number of complex and, in general, nonlinear phenomena that can arise from low-background operations, including high-field carrier sweep-out effects [11-26, 11-27, 11-28, 11-29] and nonlinearities due to bias and memory effects (the previous irradiation history of the detector) [11-30, 11-31].

These nonlinear phenomena make precise calibration measurements a very complex and difficult undertaking.

11.5. Performance of Commercial Thermal Detectors

The following figures present data on the performance one might expect from thermal detectors that can be purchased today, including:

(1) Thermocouples (Figure 11-26).
(2) Thermopiles (Figure 11-27).
(3) Thermistors (Figure 11-28).
(4) Golay cells (Figure 11-29).
(5) Pyroelectrics (Figure 11-30).

Detailed information on commercially available devices can be obtained from manufacturer's data, the *Optical Industry and Systems Directory*, the Naval Electronics Laboratory Center, and the IRIA Library.

Thermocouple

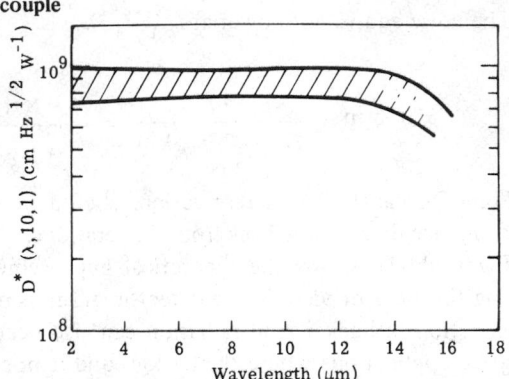

T_d = 300 K

A_d = 0.01 x 0.1 to 0.03 x 3 cm^2

R_d = 5 to 15 ohms

τ = 10 to 20 msec

FOV = 2π sr

$\mathscr{R}_{\lambda_p}$ ~ 5 V W^{-1}

$\mathscr{R}_{\lambda_p}/\mathscr{R}_{bb}$ ≈ 2

Note

(1) Noise from the detector is Johnson noise only and is at relatively low levels. The low R_d makes the use of geoformer-type transformers to couple the detector to the preamplifier desirable if compatible low-impedance preamplifiers are not available.

(2) Spectral response and time constant are greatly determined by the type and amount of black used on the responsive plane surface. The spectral response is also influenced by the spectral characteristics of the window material. Black absorbers are available that extend the spectral response to beyond 40 μm.

Fig. 11-26. Thermocouple performance data. Spectral response of detector D* (λ, 10, 1).

Thermopile

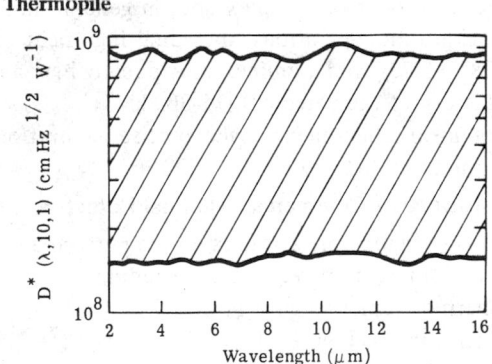

T_d = 300 K

A_d = 10^{-2} to 1 cm^2

R_d = 3 to 50 ohms

τ = 4 to 30 msec

FOV = 2π sr

$\mathscr{R}_{bb}$ = 1 to 10 V W^{-1}

Note

(1) Noise from the detector is Johnson noise and is at moderately low levels. Although no transformer is necessary, the preamplifier should be designed to ensure a detector-limited noise performance in the system.

(2) These detectors are often mounted in TO-5 cans.

(3) Spectral response and time constant are determined by type and amount of black absorbing material in responsive plane and choice of window material.

Fig. 11-27. Thermopile performance data. Spectral response of detector D* (λ, 10, 1).

Thermistor

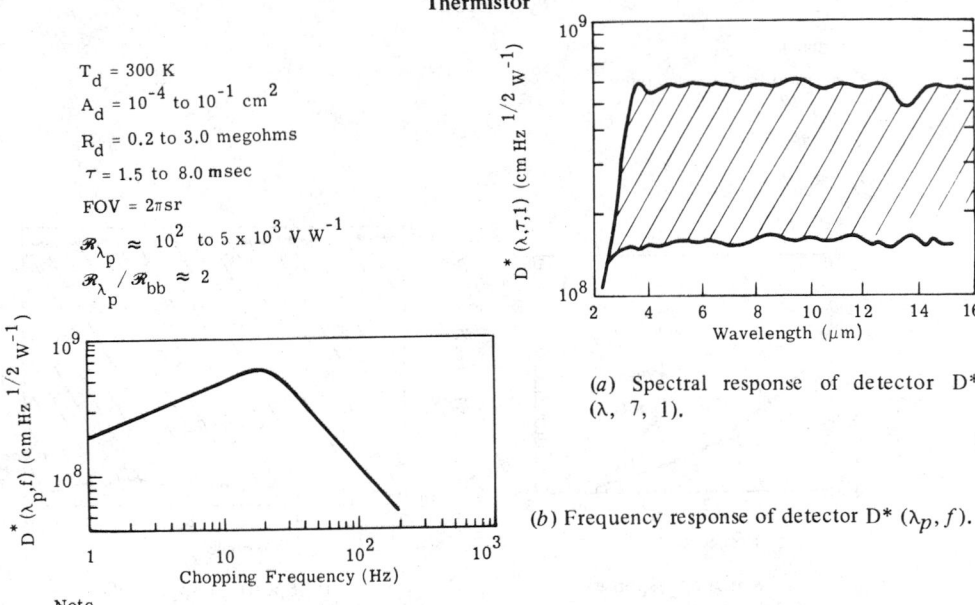

$T_d = 300$ K

$A_d = 10^{-4}$ to 10^{-1} cm^2

$R_d = 0.2$ to 3.0 megohms

$\tau = 1.5$ to 8.0 msec

FOV $= 2\pi$ sr

$\mathscr{R}_{\lambda_p} \approx 10^2$ to 5×10^3 V W^{-1}

$\mathscr{R}_{\lambda_p} / \mathscr{R}_{bb} \approx 2$

(a) Spectral response of detector D* $(\lambda, 7, 1)$.

(b) Frequency response of detector D* (λ_p, f).

Note

Thermistors with immersion lenses are available which provide increased effective D*
values. Also, tradeoffs can be obtained between D* and τ (i.e. D* $= 8 \times 10^9 \tau^{1/2}$) when τ is
in seconds.

Fig. 11-28. Thermistor performance data.

Golay Cell

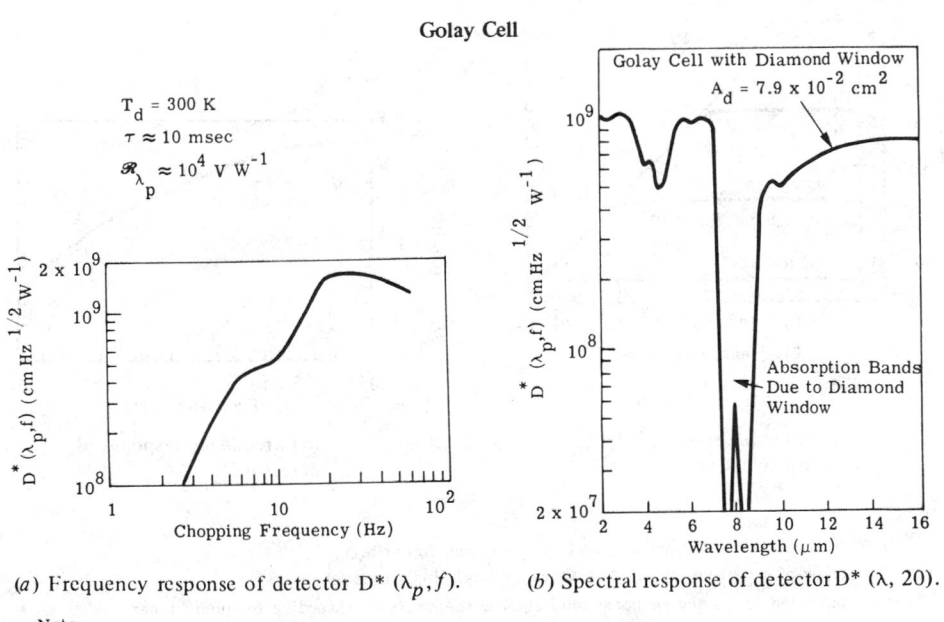

$T_d = 300$ K

$\tau \approx 10$ msec

$\mathscr{R}_{\lambda_p} \approx 10^4$ V W^{-1}

Golay Cell with Diamond Window
$A_d = 7.9 \times 10^{-2}$ cm^2

Absorption Bands
Due to Diamond
Window

(a) Frequency response of detector D* (λ_p, f).   (b) Spectral response of detector D* $(\lambda, 20)$.

Note

(1) This device uses a pneumatic circuit coupled to a photoemissive detector.

(2) Window materials such as diamond and quartz are available.

Fig. 11-29. Golay cell performance data.

Pyroelectric

T_d = 300 K

A_d = 1 mm to 1 cm diameter

$R_d \approx 10^{12}$ ohms

FOV = 2π sr

$\mathscr{R}_{\lambda_p} / \mathscr{R}_{bb}$ = 1 to 2

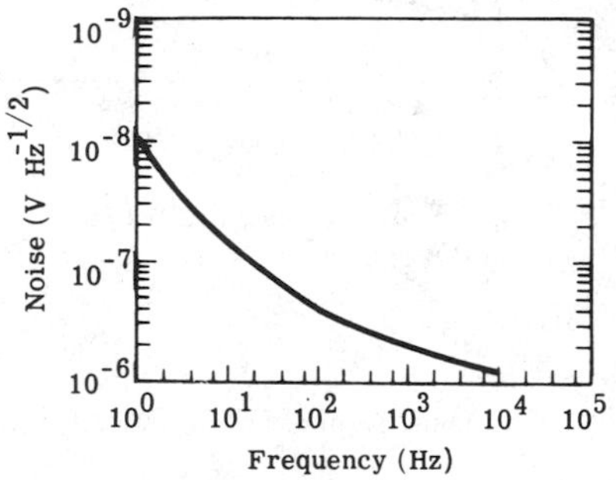

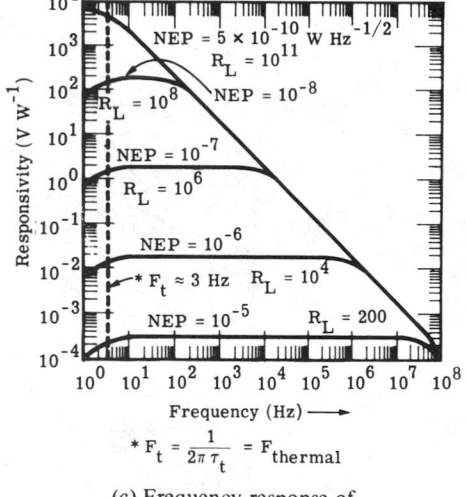

| (a) Frequency response of detector noise. | (b) Spectral response of detector D* (λ, 100, 1). |

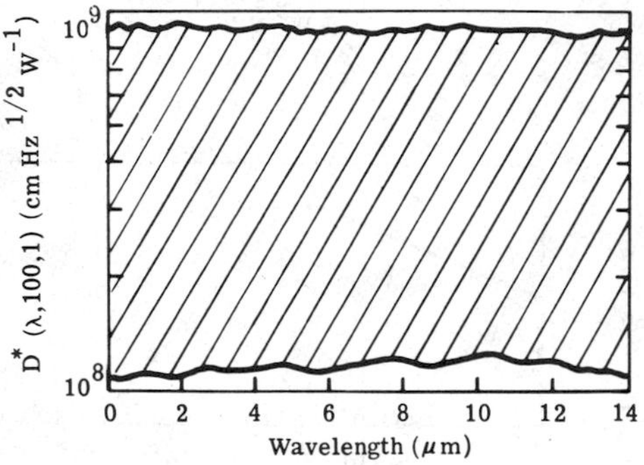

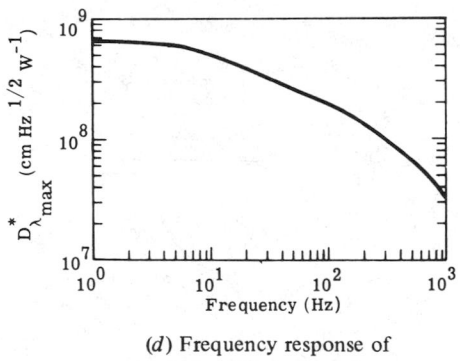

$$* F_t = \frac{1}{2\pi \tau_t} = F_{thermal}$$

| (c) Frequency response of detector signal. | (d) Frequency response of detector $D^*_{\lambda_{max}}$ |

Note

(1) Detector materials include KDP, PVE, $LiTaO_3$ and $SrBaNb_2O_6$.

(2) Some types of sensitive materials give exceedingly high microphonics.

(3) Most published frequency response data show a responsivity extending to much higher frequencies than can be obtained with other uncooled thermal detectors. In general, a single τ cannot be determined from these data.

Fig. 11-30. Pyroelectric performance data.

11.6. Performance of Commercial Photon Detectors

This section contains performance data on commercially available photon detectors, followed by a discussion of some unique properties of the ternary alloy detectors. Detectors covered in this section are presented in increasing wavelengths, as listed in Table 11-5 and shown in Figures 11-31 through 11-51.

Spectral D^* (λ) curves for a number of commercially available quantum detectors are shown in Figure 11-52. The theoretical limit curves for photoconductive and photovoltaic operations with a 300 K background and a 180° field of view are also included.

More detailed information is available from the commercial suppliers and the IRIA Center, as well as from the Naval Electronics Laboratory Center.

A list of suppliers can be found in the *Optical Industry and Systems Directory*.

Table 11-5. Commercially Available Photon Detectors

Detector Material	*Temperature* (K)	*Figure*
Si	300	11-31
Si	300	11-32
PbS	300, 193, 77	11-33, -34, -35
PbSe	300, 145 to 250, 77	11-36, -37, -38
InAs	300, 195, 77	11-39, -40, -41
InSb	300, 77	11-42, -43
Ge:Au	77	11-44
Ge:Hg	5	11-45
Ge:Cd	5	11-46
Ge:Cu	5	11-47
Ge:Zn	5	11-48
Si:As and Si:Ga	20	11-49
$Hg_{1-x}Cd_xTe$	77	11-50
$Pb_{1-x}Sn_xTe$	77	11-51

Si Avalanche Diode

T_d = Ambient operation (300 K)

A_d = 5 x 10^{-4} to 10^{-1} cm^2

Z_d = 50 to 2000 ohms

τ = 10 nsec fall time

$\mathscr{R}_{\lambda_p} \approx 10^5$ V W^{-1}

Gain = 75 to 500

Avalanche voltage = 140 to 3000 V

Dark current = 0.5 to 1.5 μA

$D^*(\lambda_p)$ = 5.6 × 10^{13} cm $Hz^{1/2} W^{-1}$

$D^*(\lambda = 1.06 \mu m)$ = 2.2 10^{13} cm $Hz^{1/2}$ W^{-1}

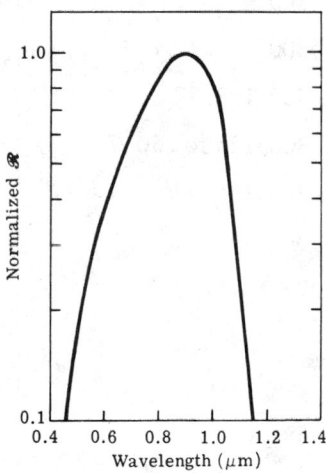

Normalized $\mathscr{R}$ / Wavelength (μm)

<u>Note</u>

By driving the detector into an avalanche condition, a percentage increase in signal voltage and a greater percentage increase in noise voltage will result. The advantage in using this type of detector comes when the system noise is higher than the standard silicon detector noise.

Fig. 11-31. Si avalanche diode detector performance data at 300 K. Spectral response of detector normalized.

Si

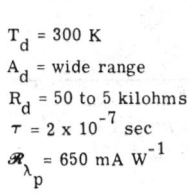

$T_d = 300$ K
A_d = wide range
$R_d = 50$ to 5 kilohms
$\tau = 2 \times 10^{-7}$ sec
$\mathcal{R}_{\lambda_p} = 650$ mA W^{-1}

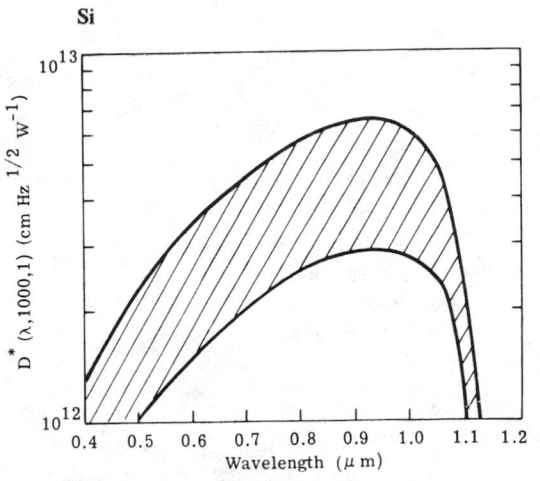

(a) Spectral response of detector D* (λ, 1000, 1).

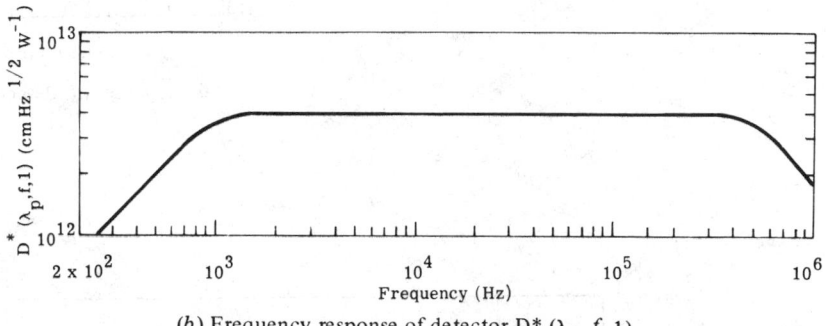

(b) Frequency response of detector D* (λ_p, f, 1).

Note

(1) Small detectors have been made in arrays on 0.6 mm centers or less, and large detectors several centimeters in diameter or more have been made.

(2) $\mathcal{R}$ depends upon bias voltage.

(3) Capacitance can vary from 2 to 800 pF, depending on bias.

Fig. 11-32. Si detector performance data at 300 K.

PbS

$T_d = 300$ K

$A_d = 4 \times 10^{-6}$ to 16 cm^2

$R_d = 4$ megohms

$\tau = 100$ to 500 μ sec

FOV = 180° or 2π sr

$\mathcal{R}_{\lambda_p} \approx 10^6$ V W^{-1}

$\mathcal{R}_{\lambda_p}/\mathcal{R}_{bb} \approx 95$

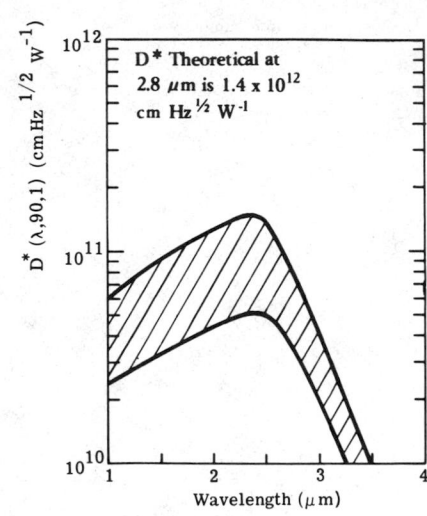

(a) Spectral response of detector D* (λ, 90, 1).

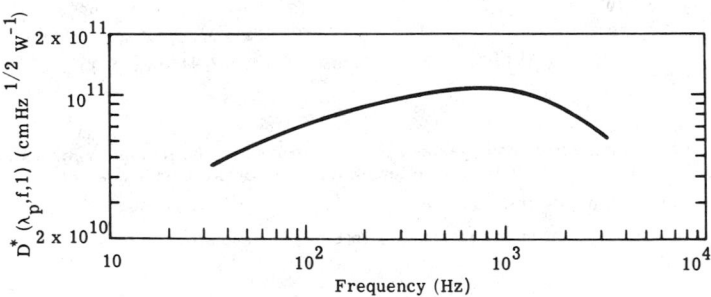

(b) Frequency response of detector D* (λ_p, f, 1).

Note

Resistance depends upon the l/w ratio of a responsive element. The values stated above are for a square detector.

Fig. 11-33. PbS detector performance data at 300 K.

PbS

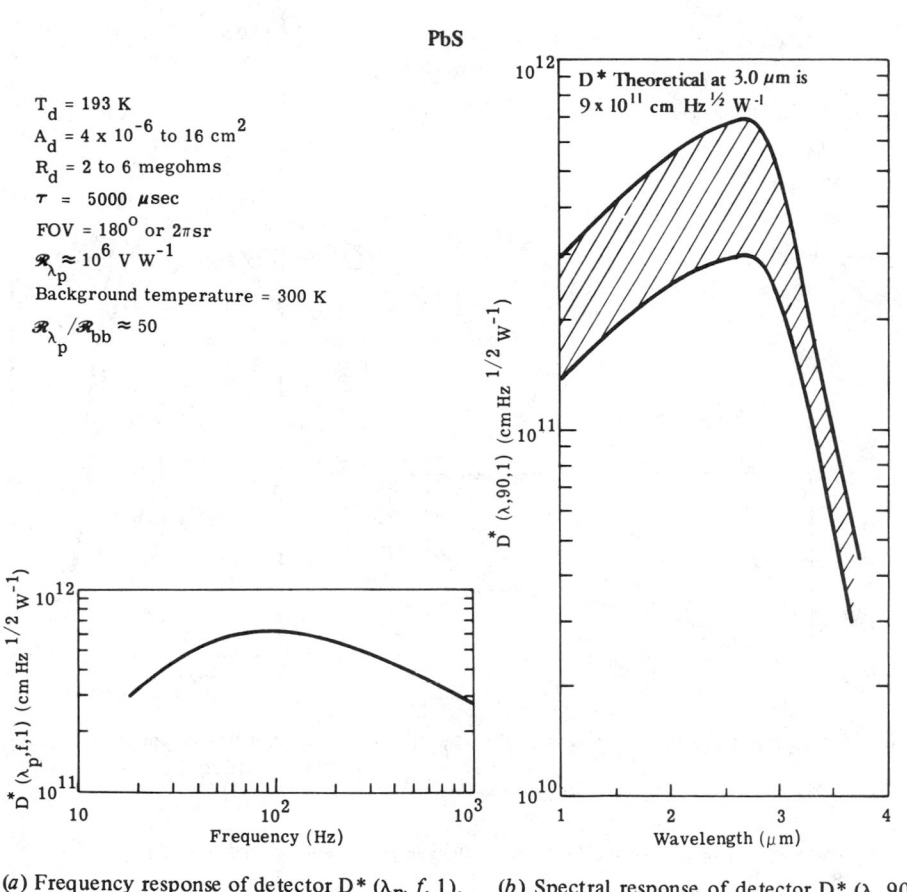

T_d = 193 K

A_d = 4 x 10^{-6} to 16 cm^2

R_d = 2 to 6 megohms

τ = 5000 μsec

FOV = 180° or 2πsr

$\mathscr{R}_{\lambda_p} \approx 10^6$ V W^{-1}

Background temperature = 300 K

$\mathscr{R}_{\lambda_p} / \mathscr{R}_{bb} \approx 50$

D* Theoretical at 3.0 μm is 9×10^{11} cm $Hz^{1/2}$ W^{-1}

(a) Frequency response of detector D* $(\lambda_p, f, 1)$.   (b) Spectral response of detector D* $(\lambda, 90, 1)$.

Note

    Resistance depends upon the l/w ratio of the responsive element. The values stated above are for square detectors.

Fig. 11-34. PbS detector performance data at 193 K.

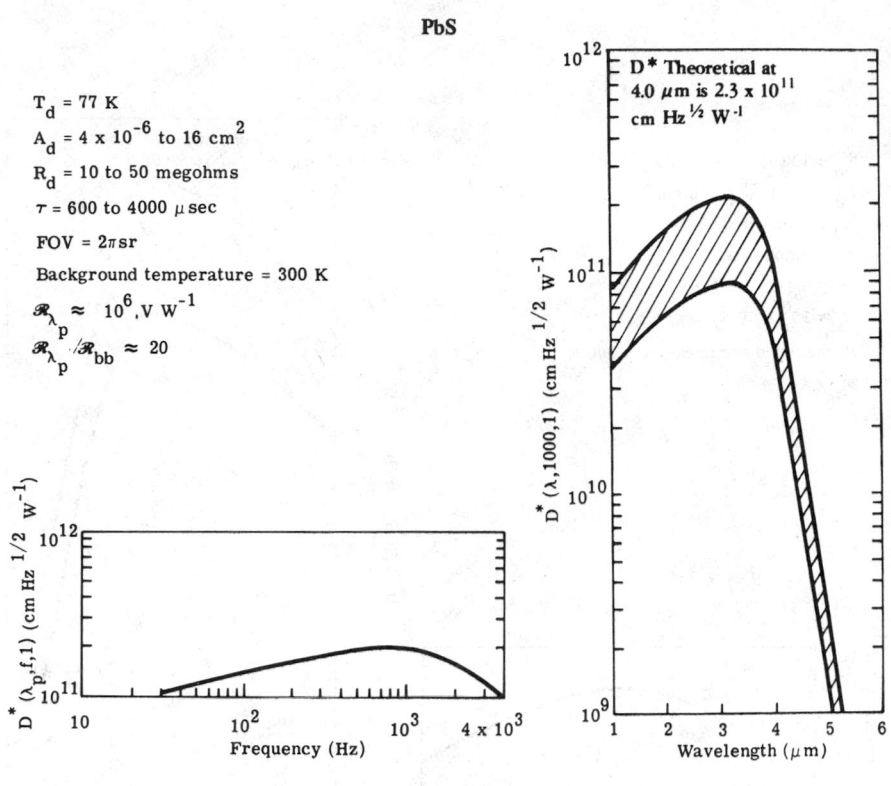

PbS

T_d = 77 K

A_d = 4 x 10^{-6} to 16 cm^2

R_d = 10 to 50 megohms

τ = 600 to 4000 μ sec

FOV = 2π sr

Background temperature = 300 K

$\mathcal{R}_{\lambda_p} \approx 10^6$, V W^{-1}

$\mathcal{R}_{\lambda_p}/\mathcal{R}_{bb} \approx 20$

D* Theoretical at 4.0 μm is 2.3 x 10^{11} cm Hz$^{1/2}$ W^{-1}

(a) Frequency response of detector D* (λ_p, f, 1).

(b) Spectral response of detector D* (λ, 1000, 1).

Note

Resistance depends upon the l/w ratio of the responsive element. The range stated above is for $l/w = 1$.

Fig. 11-35. PbS detector performance data at 77 K.

$T_d = 300$ K

$A_d = \sim 10^{-4}$ to 1 cm^2

$R_d = 0.3$ to 20.0 megohms

$\tau \leq 3$ μsec

FOV $= 2\pi$ sr

Background temperature $= 300$ K

$\mathscr{R}_{\lambda_p} \approx 10^3$ V W^{-1}

$\mathscr{R}_{\lambda_p} / \mathscr{R}_{bb} \approx 9$

PbSe

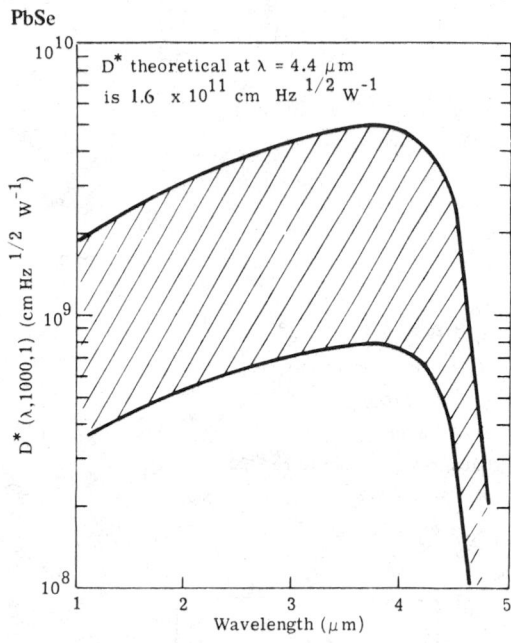

(a) Spectral response of detector D* (λ, 1000, 1).

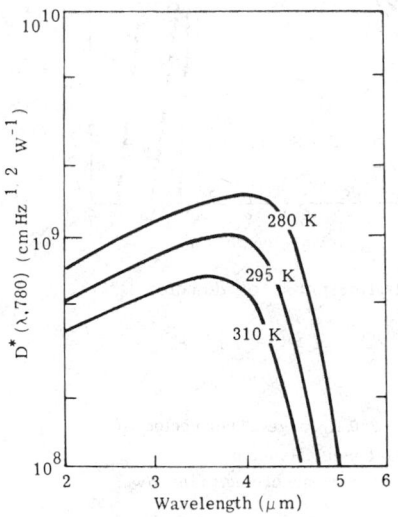

(b) Spectral response of detector D* (λ, 780). SBRC data.

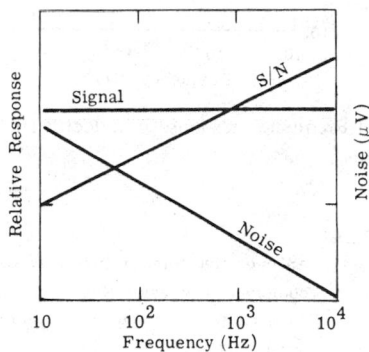

(c) Frequency response of detector signal. SBRC data.

Note

(1) Resistance depends upon the l/w ratio. The data given are for square detectors ($l/w = 1$)

(2) Responsivity varies roughly as the inverse of the square root of the detector area. The value stated above is also for approximately 0.1 to 1.0 cm^2

Fig. 11-36. PbSe detector performance data at 300 K.

PbSe

T_d = 145 to 250 K

A_d = 25 x 10^{-6} to 1 cm^2

R_d = 2 to 100 megohms

τ = 10 to 1000 μsec

FOV = 2π sr

Background temperature = 300 K

$\mathscr{R}_{\lambda_p}$ ≈ 10^5 V W^{-1}

$\mathscr{R}_{\lambda_p}/\mathscr{R}_{bb}$ ≈ 5

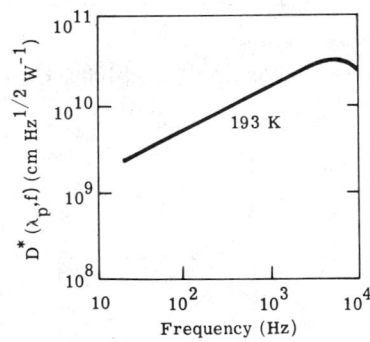

(a) Frequency response of detector D* (λ_p, f).

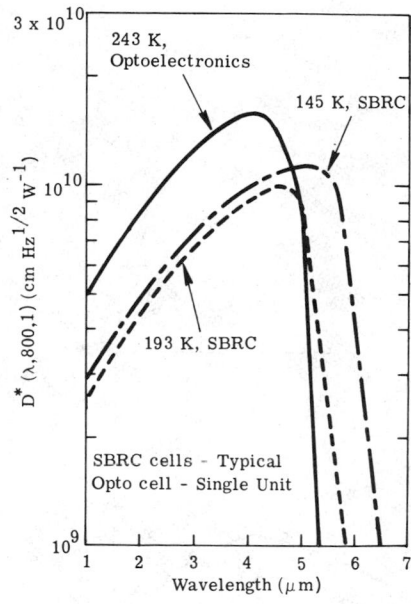

(b) Spectral response of detector D* (λ, 800, 1).

<u>Note</u>

  (1) An operating temperature may be selected in the 145 to 250 K range. Thermoelectric coolers can be used. Known manufacturers are SBRC and Optoelectronics.

  (2) This range is for square configurations. The resistance will vary according to the l/w ratio.

Fig. 11-37. PbSe detector performance data at 145 to 250 K.

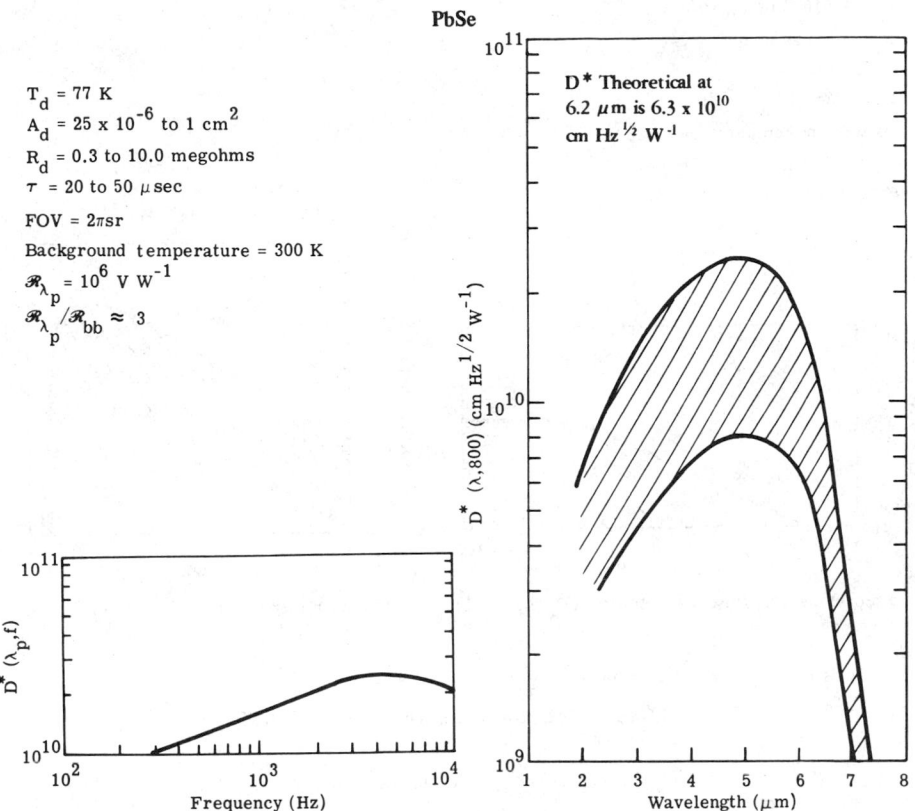

(a) Frequency response of detector D* (λ_p, f). (b) Spectral response of detector D* (λ, 800).

Note

This resistance range is for square geometries. For other geometries, R_d is directly related to l/w.

Fig. 11-38. PbSe detector performance data at 77 K.

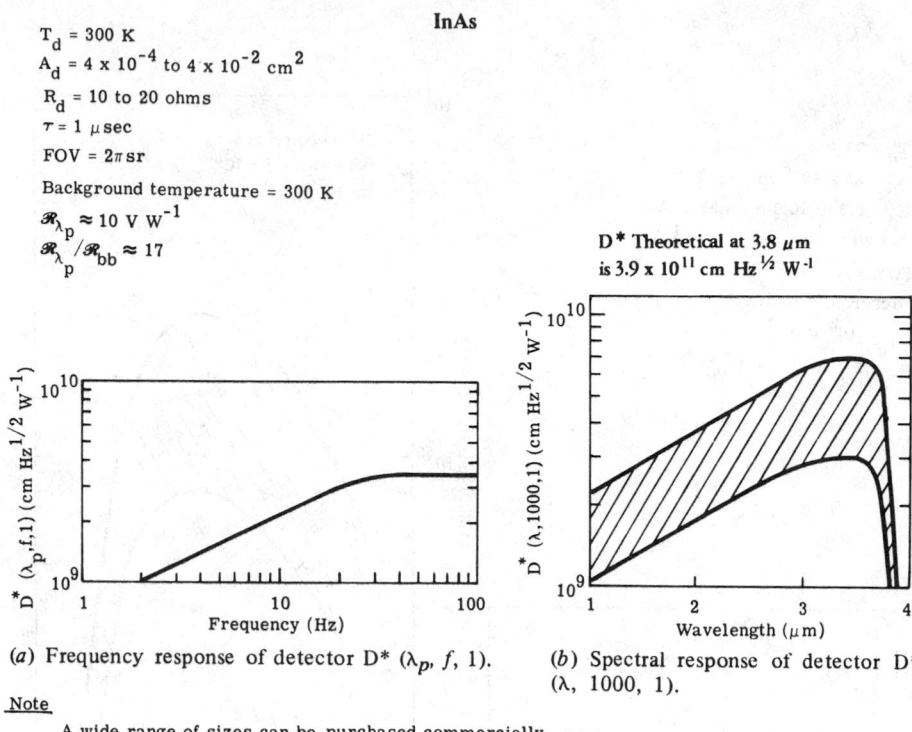

InAs

T_d = 300 K
A_d = 4 x 10^{-4} to 4 x 10^{-2} cm^2
R_d = 10 to 20 ohms
τ = 1 μ sec
FOV = 2π sr
Background temperature = 300 K
$\mathscr{R}_{\lambda_p}$ ≈ 10 V W^{-1}
$\mathscr{R}_{\lambda_p}/\mathscr{R}_{bb}$ ≈ 17

D* Theoretical at 3.8 μm
is 3.9 x 10^{11} cm Hz$^{1/2}$ W^{-1}

(a) Frequency response of detector D* (λ_p, f, 1).

(b) Spectral response of detector D* (λ, 1000, 1).

Note

A wide range of sizes can be purchased commercially.

Fig. 11-39. InAs detector performance data at 300 K.

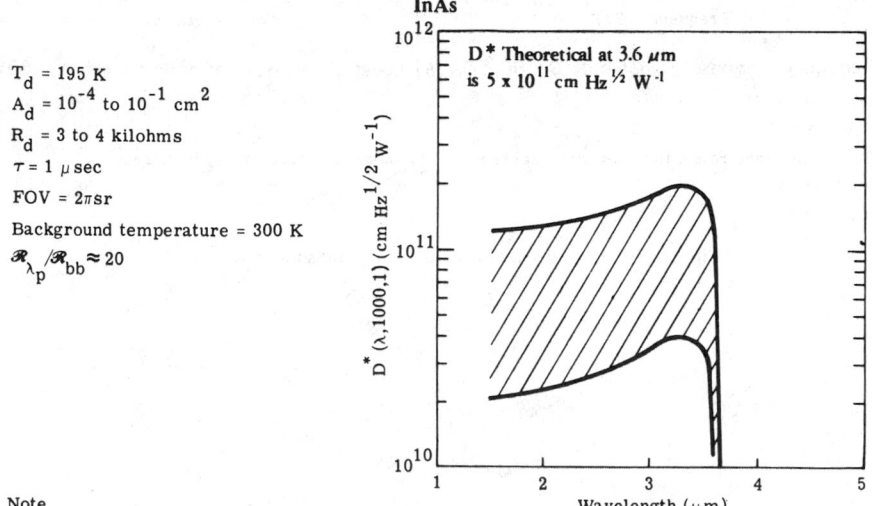

InAs

D* Theoretical at 3.6 μm
is 5 x 10^{11} cm Hz$^{1/2}$ W^{-1}

T_d = 195 K
A_d = 10^{-4} to 10^{-1} cm^2
R_d = 3 to 4 kilohms
τ = 1 μ sec
FOV = 2π sr
Background temperature = 300 K
$\mathscr{R}_{\lambda_p}/\mathscr{R}_{bb}$ ≈ 20

Note

This detector, mounted in thermoelectric coolers, is available from Barnes and SBRC.

Fig. 11-40. InAs detector performance data at 195 K.
Spectral response of detector D* (λ, 1000, 1).

InAs

T_d = 77 K

A_d = 2 x 10^{-3} to 10^{-1} cm^2

R_d = 100 kilohms

$\tau \approx 1 \, \mu sec$

FOV = $2\pi sr$

$\mathscr{R}_{\lambda_p}$ = 1 A W^{-1}

$\mathscr{R}_{\lambda_p}/\mathscr{R}_{bb} \approx 61$

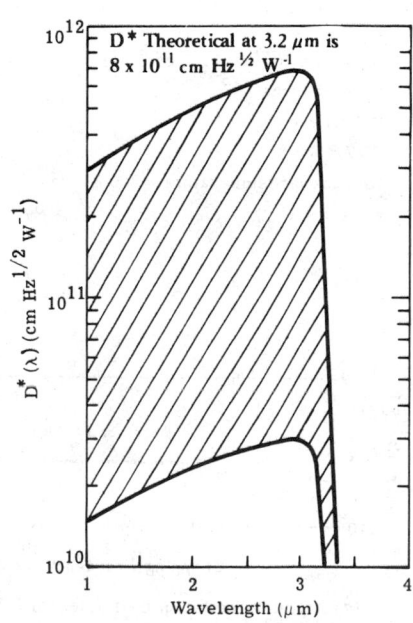

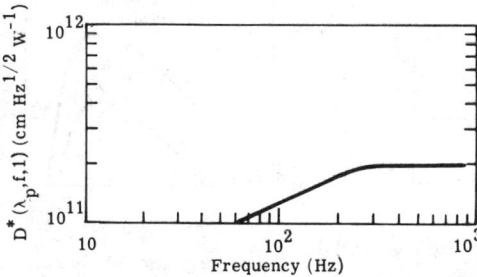

(a) Frequency response of detector D* (λ_p, f, 1). (b) Spectral response of detector D* (λ).

Note

(1) Check with vendors for availability of areas outside the indicated range.

(2) Resistance is area dependent.

(3) D* can increase when the FOV is reduced with cold shields or cold filters.

Fig. 11-41. InAs detector performance data at 77 K.

InSb PC

T_d = 300 K

A_d = 25 x 10^{-4} to 10^{-1} cm^2

$R_d \approx$ 15 ohms

$\tau <$ 1 μ sec

FOV = 2πsr

Background temperature = 300 K

$\mathscr{R}_{\lambda_p}$ = 5 V W^{-1}

D* Theoretical at 7 μm is 5 x 10^{10} cm $Hz^{1/2}$ W^{-1}

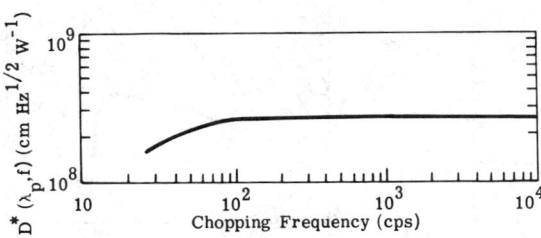

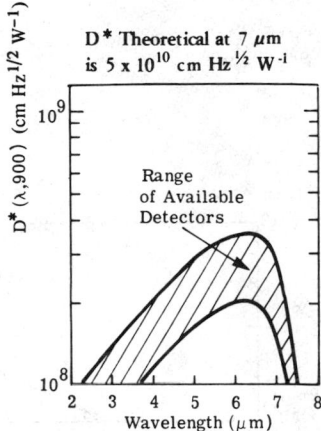

(a) Frequency response of detector D* (λ_p, f).

(b) Spectral response of detector D* $(\lambda, 900)$.

Note

It operates in the photoelectromagnetic or photoconductive mode.

Fig. 11-42. InSb PC detector performance data at 300 K.

InSb PV

T_d = 77 K

A_d = 10^{-4} to 10^{-1} cm^2

R_d = 1 to 100 megohms

τ < 1 μsec

FOV = 60°

Background temperature = 300 K

$\mathcal{R}_\lambda \approx 10^5$ V W^{-1}

$\mathcal{R}_\lambda^p / \mathcal{R}_{bb}^p \approx 5.0$

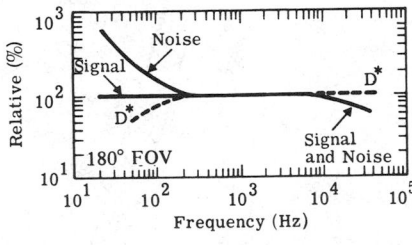

(a) Frequency response of detector signal and noise.

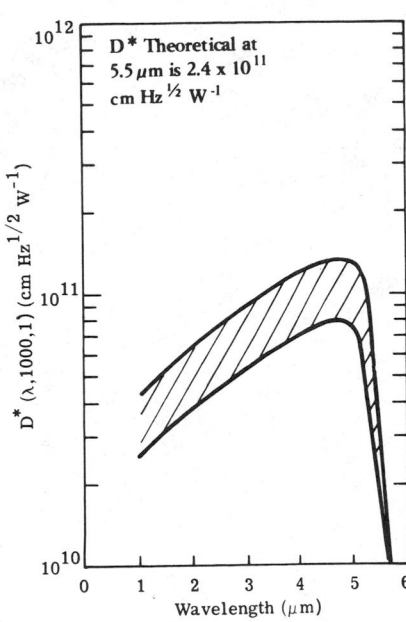

(b) Spectral response of detector D* (λ, 1000, 1).

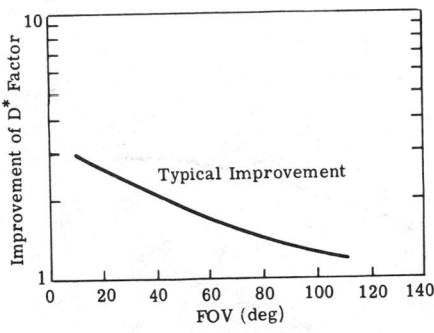

(c) Improvement in D* versus FOV for typical detectors.

Note

(1) Resistance is area and bias voltage dependent.

(2) D* will increase when the FOV is restricted with cold shields or when the spectral extent is restricted with cold spectral filters.

Fig. 11-43. InSb PV detector performance data at 77 K.

T_d = 77 K

A_d = 2 x 10^{-3} to 2 x 10^{-1} cm^2

R_d ≤ 500 kilohms

τ = 100 nsec

FOV = 2π sr

$\mathscr{R}_{\lambda_p}$ ≈ 10^4 v w^{-1}

$\mathscr{R}_{\lambda_p}/\mathscr{R}_{bb}$ ≈ 2.5

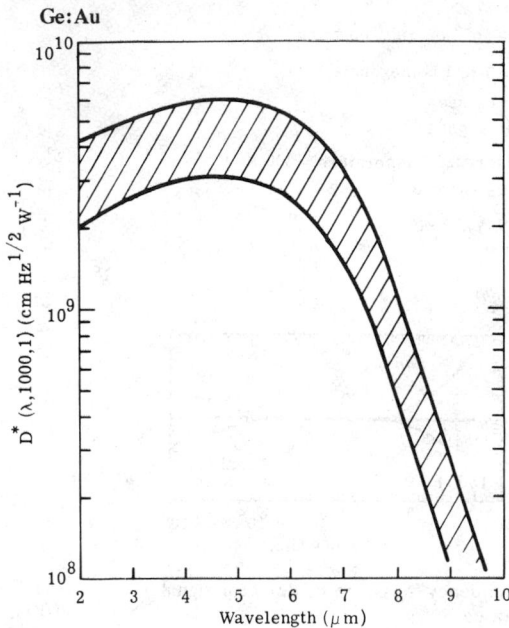

(a) Spectral response of detector D* (λ, 1000, 1).

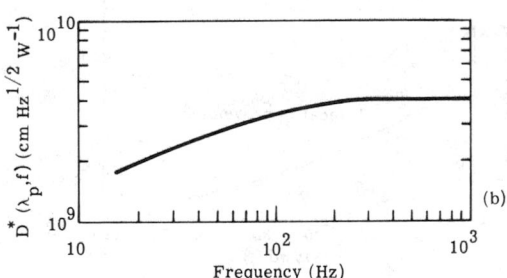

(b) Frequency response of detector D* (λ_p, f).

Note

(1) The detector is generally used with a liquid N_2 cryostat.

(2) SBRC offers a high-speed Ge:Au material with resistances greater than 1 megohm.

(3) Two nanoseconds can be obtained when using a special high-speed dewar.

Fig. 11-44. Ge:Au detector performance data at 77 K.

Ge:Hg

T_d = 5 K

$A_d = 10^{-1}$ to 10^{-4} cm^2

R_d < 500 kilohms per square

τ < 100 nsec

FOV = 60°

Background temperature = 300 K

$\mathscr{R}_{\lambda_p} \approx 10^6$ V W^{-1}

$\mathscr{R}_{\lambda_p} / \mathscr{R}_{bb} \approx 1.7$

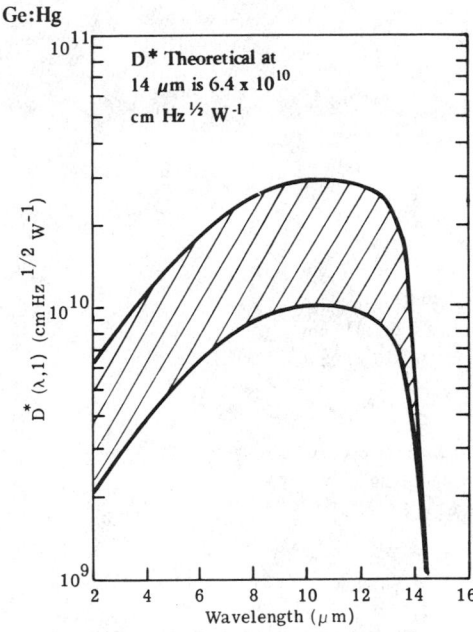

D* Theoretical at 14 μm is 6.4 x 10^{10} cm Hz $^{1/2}$ W^{-1}

(a) Spectral response of detector D* (λ, 1).

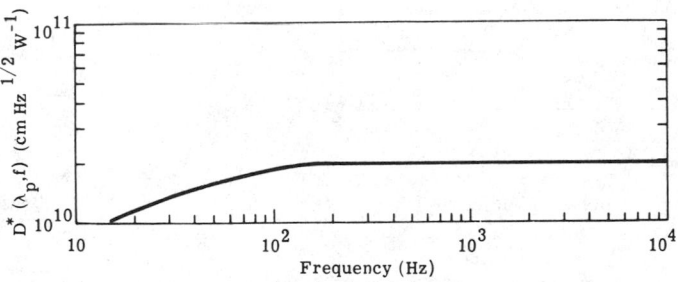

(b) Frequency response of detector D* (λ_p, f).

Note

A τ of 2 nsec can be achieved.

Fig. 11-45. Ge:Hg detector performance data at 5 K.

Ge:Cd

T_d = 5 to 26 K

A_d = 10^{-4} to 10^{-1} cm^2

R_d = 20 to 200 kilohms

τ < 1 μ sec

FOV = 60^0

Background temperature = 300 K

$\mathscr{R}_{\lambda_p}$ ≈ 0.5 to 3.0 A W^{-1}

$\mathscr{R}_{\lambda_p}/\mathscr{R}_{bb}$ ≈ 2.8

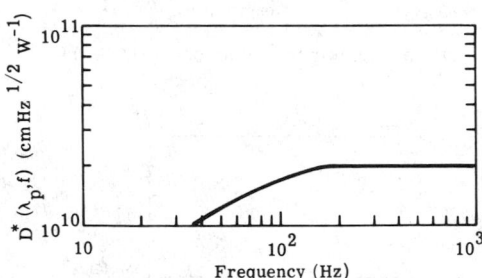

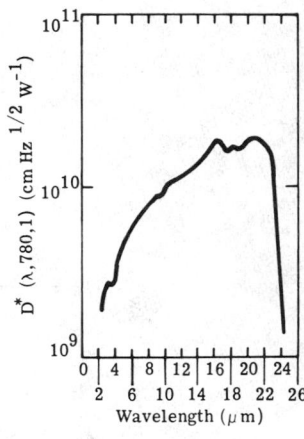

(a) Frequency response of detector D* (λ_p, f).

(b) Spectral response of detector D* (λ, 780, 1).

Fig. 11-46. Ge:Cd detector performance data at 5 to 26 K.

Ge:Cu

T_d = 5 K
A_d = 10^{-1} to 10^{-4} cm^2
R_d < 500 kilohms per square
τ < 100 nsec
FOV = 60°
Background temperature = 300 K
$\mathcal{R}_{\lambda_p}$ = 10^4 to 10^6 V W^{-1}
$\mathcal{R}_{\lambda_p} / \mathcal{R}_{bb} \approx 2$

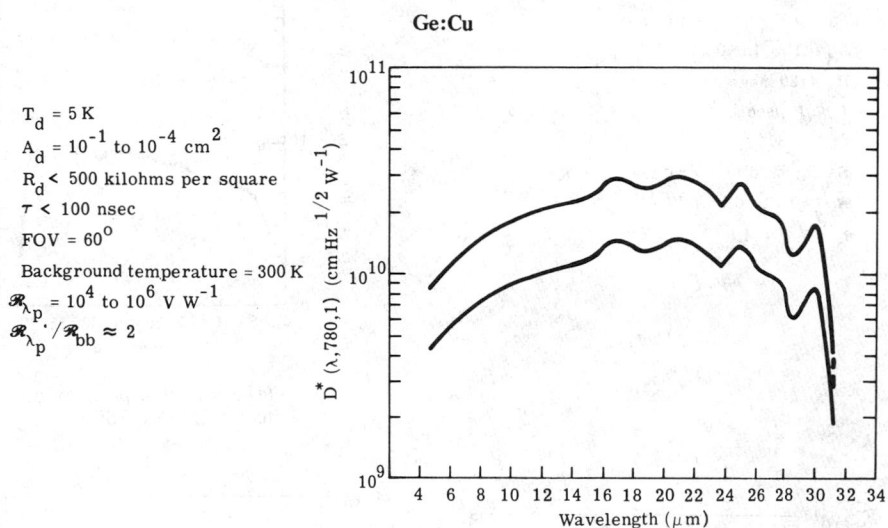

(a) Spectral response of detector D* (λ, 780, 1).

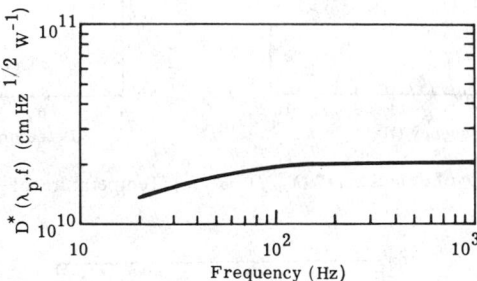

(b) Frequency response of detector D* (λ_p, f).

Note

(1)   Cold spectral filters or shields will increase R_d and D*

(2)   A τ of less than 1 nsec can be achieved.

Fig. 11-47.   Ge:Cu detector performance data at 5 K.

Ge:Zn

$T_d = 5$ K
$A_d = 10^{-4}$ to 10^{-1} cm^2
$R_d < 20$ megohms
$\tau < 1$ μsec
FOV = 60°
Background temperature = 300 K
$\mathscr{R}_{\lambda_p} = 0.1$ to 0.5 A W^{-1}
$\mathscr{R}_{\lambda_p}/\mathscr{R}_{bb} \approx 1.8$

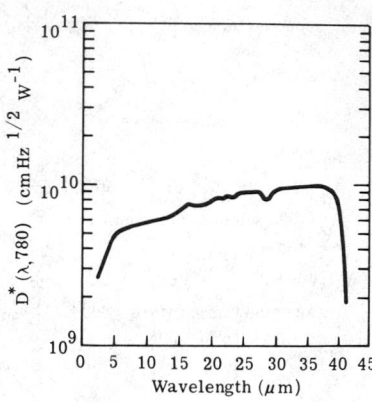

(a) Spectral response of detector D^* $(\lambda_p, 780)$.

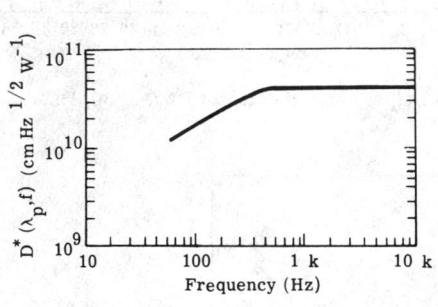

(b) Frequency response of detector D^* (λ_p, f).

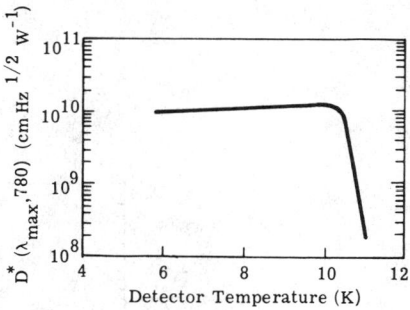

(c) Temperature of detector D^* $(\lambda_p, 780)$.

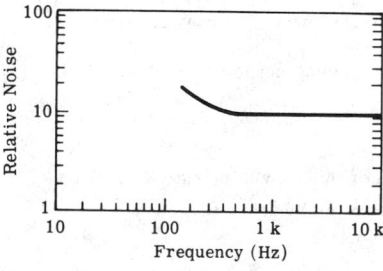

(d) Frequency response of detector noise.

Fig. 11-48. Ge:Zn detector performance data at 5 K.

Si:As and Si:Ga

$T_d < 20$ K

FOV $= 60^\circ$

Background temperature $= 300$ K

$\mathscr{R}_{\lambda_p}/\mathscr{R}_{bb} = \boxed{\text{Si:As} = 3.2}\ \boxed{\text{Si:Ga} = 2.2}$

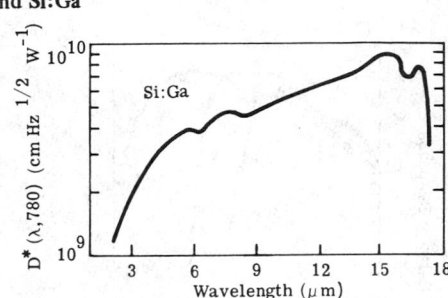

(a) Spectral response of Si:Ga detector D* (λ, 780).

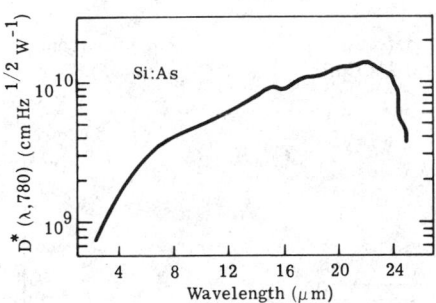

(b) Spectral response of Si:As detector D* (λ, 780).

Fig. 11-49. Si:As and Si:Ga detector performance data at less than 20 K.

$Hg_{1-x}Cd_xTe$

T_d = 77 K

A_d = 4 x 10^{-6} to 4 x 10^{-2} cm^2

R_d = 20 to 600 ohms

τ = 100 to 800 nsec

FOV = 10^0 to 130^0

Background temperature = 300 K

$\mathscr{R}_\lambda$ = 10^4 V W^{-1}

$\mathscr{R}_{\lambda_p}/\mathscr{R}_{bb} \approx 2$

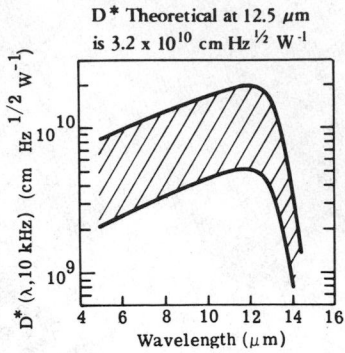

D* Theoretical at 12.5 μm is 3.2 x 10^{10} cm Hz $^{1/2}$ W^{-1}

(a) Spectral response of detector D* (λ, 10 kHz).

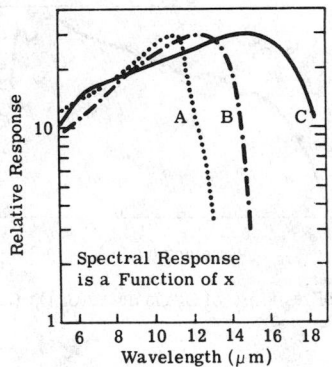

(b) Spectral response of detector responsivity. Curve A:x = 0.21; Curve B:x = 0.20; Curve C:x = 0.17.

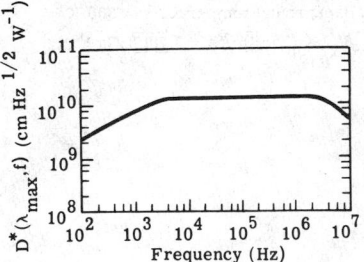

(c) Frequency response of detector D* (λ_p, f).

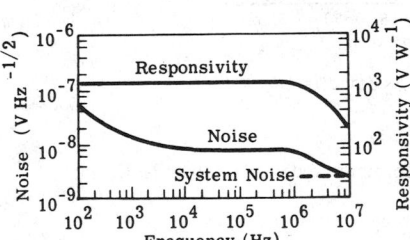

(d) Frequency response of detector noise.

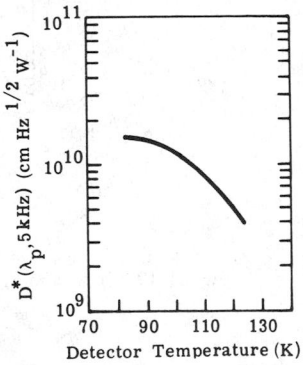

(e) Temperature of detector D* (λ_p, 5 kHz).

Note

    Spectral response is determined by the alloy composition.

Fig. 11-50. $Hg_{1-x}Cd_xTe$ detector performance data at 77 K.

$$Pb_{1-x}Sn_xTe$$

T_d = 77 K

A_d = 2.5 x 10^{-5} to 10^{-2} cm^2

R_d = Dynamic impedance > 100 ohms

$\tau < 1\ \mu$sec

FOV = 60°

Background temperature = 300 K

$\mathscr{R}_{\lambda_p}$ = 2.0 to 4.0 A W^{-1}

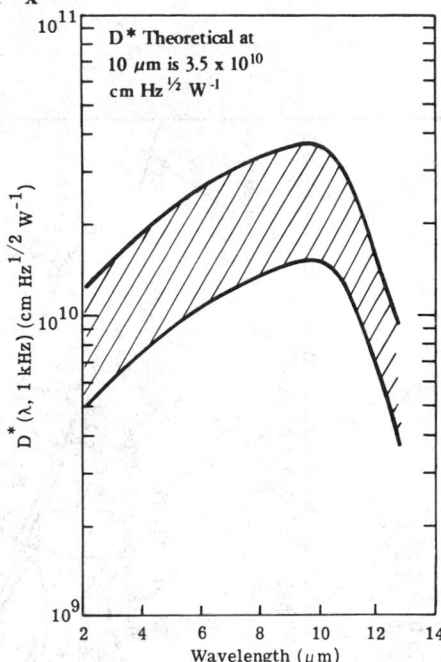

D* Theoretical at 10 μm is 3.5 x 10^{10} cm Hz$^{1/2}$ W^{-1}

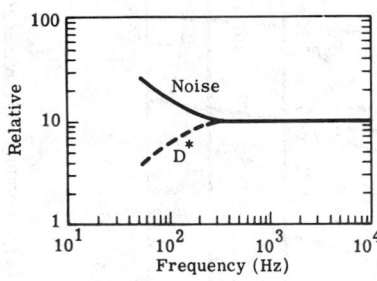

(a) Frequency response of detector D* and noise.

(b) Spectral response of detector D* (λ, 1 kHz).

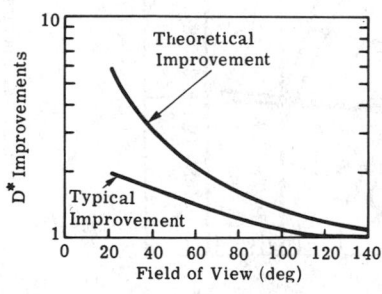

(c) Dependence of detectivity improvement on FOV.

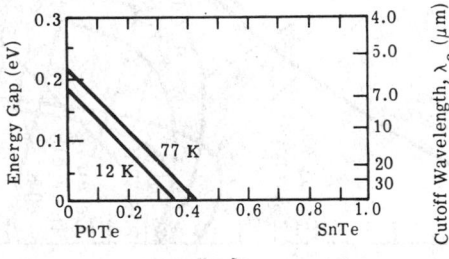

(d) Energy gap of $Pb_{1-x}Sn_xTe$ as a function of x, the mole fraction of Sn Te [11-32].

Note

Spectral response is determined by the alloy composition.

Fig. 11-51. $Pb_{1-x}Sn_xTe$ detector performance data at 77 K.

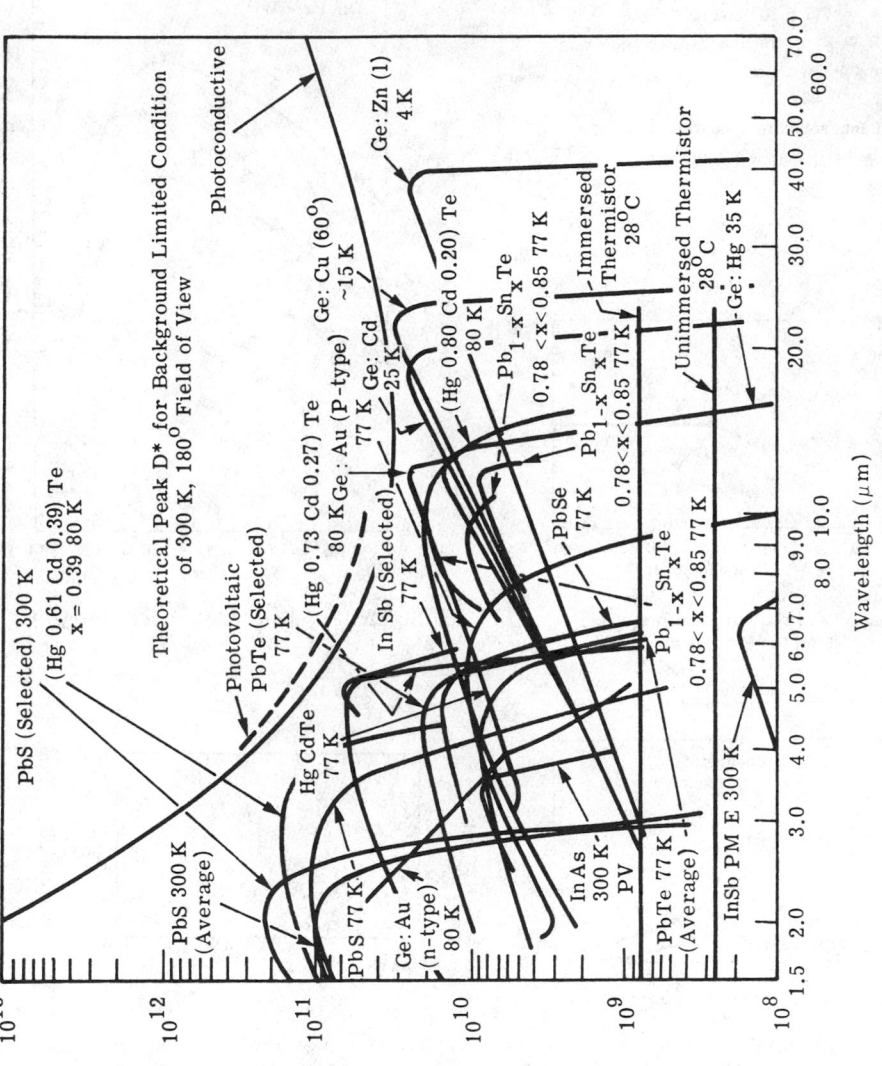

Fig. 11-52. Spectral D* for a number of commercially available detectors.

11.6.1. Extrinsic Silicon Characteristics*. The development of extrinsic silicon (Si:X) detectors has been spurred in the past few years by the possibility of using integrated-circuit technology to achieve monolithic detector and preamp arrays. Since the development of high performance Si:X detectors is relatively recent, a brief review of their properties is presented in this section. For a comprehensive treatment of the photoconductive properties of silicon (and germanium) the papers of E. Burstein, et al. [11-33] and E. H. Putley [11-34] are recommended.

In the intrinsic photoconductivity process, photons with energy greater than the band-gap are absorbed resulting in electronic transitions from the valence to the conduction band (Figure 11-53). In the intrinsic absorption process, both free electrons and free holes are generated. In extrinsic or impurity photoconductivity, photons with energy greater than the ionization level of the impurity are absorbed, resulting in the generation of one free carrier and an ionized impurity site. The third photoconductive process in a semiconductor is that of free-carrier absorption. Free electrons can be excited to higher energy levels by absorption of radiation, resulting in changes in their mobility, and hence conductivity.

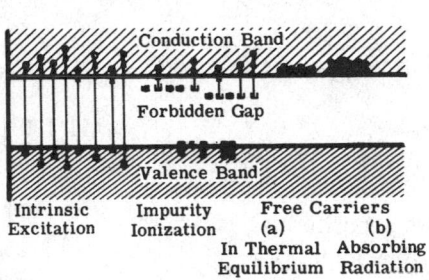 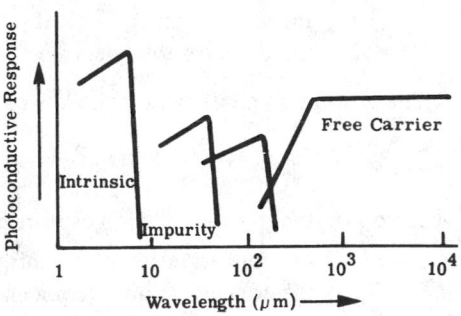

Fig. 11-53. Three photoconductive processes [11-34].

The impurity activation energy of a photoconductor is related to response cut-off wavelength, λ_c, by

$$E_i(eV) \approx \frac{hc}{\lambda_c} \approx \frac{1.24}{\lambda_c}, \text{ for } \lambda_c \text{ in } \mu m \qquad (11\text{-}186)$$

where h = Planck's constant
  c = the speed of light

For the 3 to 5 μm and 8 to 14 μm spectral bands, the appropriate activation energies are approximately 0.25 eV and 0.09 eV.

The extrinsic-detector spectral responsivity at wavelength λ is given by

$$\mathcal{R}(\lambda) = \frac{e\eta\lambda}{hc}G_p \qquad (11\text{-}187)$$

---

*This section was written by Andrew J. Steckl, Rensselaer Polytechnic Institute, Troy, NY.

where e = the electronic charge
        η = the quantum efficiency
        G_p = the photoconductive gain

In turn,

$$G_p = \frac{\mu \tau \mathcal{E}}{l} \tag{11-188}$$

and

$$\eta = \frac{(1-\rho)(1-e^{-ad})}{1-\rho e^{-ad}} \tag{11-189}$$

where μ = the carrier mobility
        τ = excess carrier lifetime
        $\mathcal{E}$ = the applied electric field
        l = the interelectrode spacing
        ρ = the surface reflectance of Si(≈ 0.31)
        a = the absorption coefficient
        d = the detector thickness

The absorption coefficient is the key to the performance of the detector

$$a = \sigma_A(N_i - N_c) \tag{11-190}$$

where σ_A = the photoionization capture cross-section
        N_i = the concentration of the infrared active impurities
        N_c = the concentration of compensated residual impurities

The lifetime of the photogenerated carriers is given by

$$\tau = \frac{1}{B_r N_c} \tag{11-191}$$

where B_r is the recombination coefficient. Increases in N_c result in a degradation of the responsivity through the combined effect of lower absorption coefficient and shorter lifetime. It therefore becomes apparent that a stringent control is required over the residual impurity concentration in the detector material.

The peak wavelength detectivity in the BLIP limit is given by the relation

$$D^*(\lambda_p) = \frac{\lambda_p}{2hc}\sqrt{\frac{\eta}{E_{q,B}}} \tag{11-192}$$

where $E_{q,B}$ is the background photon flux density.

A number of impurities whose peak spectral response fall in or near the 3 to 5 μm or 8 to 14 μm bands are listed along with their important parameters, in Table 11-6. For the 8 to 14 μm region, the extrinsic silicon detectors that have received most attention, [11-35] to [11-37], are Si:Ga and Si:Al; while, for the 3 to 5 μm band, Si:In has received

Table 11-6. Properties of Extrinsic 3 to 5 and 8 to 14 μm Silicon Detectors

Dopant	$(\Delta E_i)_{OPT}$ (eV)	$(\Delta E_i)_{THERM}$ (eV)	λ_p (μm)	$\sigma_A(\lambda_p)$ (10^{-15} cm^2)	(μm)	$m^*/m^{\dagger\dagger}$	$g^{\dagger\dagger\dagger}$	$B_r^{\#}$ (cm^3 sec^{-1})	$B_r^{\#}$ (cm^3 sec^{-1})
Si:Al	0.0685	0.067 ± 0.003	15.0	0.85	18.4 (27 K)	0.59	4	—	1.8×10^{-6} (30 to 40 K; 600 V cm^{-1})
Si:Bi	0.0706	0.069 ± 0.002	17.5	0.70	18.7 (29 K)	0.08	2	1.4×10^{-5} (1 to 4 K)	1.0×10^{-5} (30 to 40 K; 300 V cm^{-1})
Si:Ga	0.0723	0.074 ± 0.002	15.0	0.56	17.8 (27 K)	0.59	4	1.5×10^{-6} (30 K)	1.0×10^{-4} (30 to 40 K; 570 V cm^{-1})
Si:Mg	0.107	0.108; 0.044	11.5	1.7	12.1 (5 K)	1.08	2	—	—
Si:In	0.155	0.153 ± 0.003	5.0	0.033	7.4 (78 K)	0.59	6	2×10^{-6} (70 to 120 K)	8.3×10^{-6} (60 to 80 K; 1000 V cm^{-1})
Si:S	0.187	0.170 ± 0.010	5.5	0.2	6.8 (78 K)	1.08	2	—	0.55×10^{-6} (75 to 90 K; 600 V cm^{-1})
Si:Tl	—	0.230 ± 0.010	3.5	—	4.3 (78 K)	0.59	4	—	—

† Cut-off wavelength defined as the wavelength at which the spectral response has dropped to 1/2 of its peak value.
†† Effective mass ratio of charge carriers to the free carriers.
††† Degeneracy factor of impurity ground state.
# Recombination coefficient (Equation (11-191)).

recent attention, [11-35], [11-38]. Considerable work has also been performed on shallower impurities such as Si:As for longer wavelength ($\lambda_p \approx 23$ μm) operation [11-39].

The relative spectral responses of typical Si:Ga and Si:In detectors are shown in Figures 11-54 and 11-55. Detectivity of Si:Ga detectors as a function of temperature is shown in Figure 11-56. That of Si:In detectors is shown in Figure 11-57.

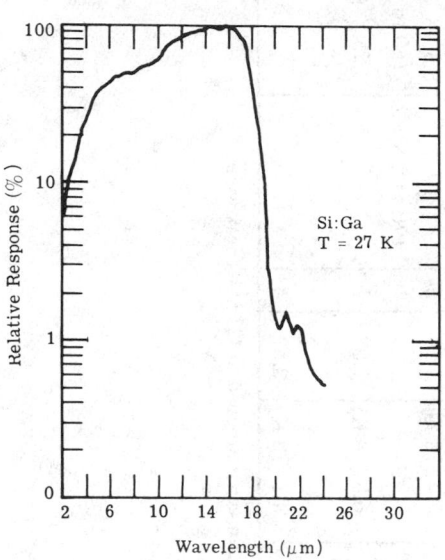

Fig. 11-54. Spectral response of Si:Ga [11-35].

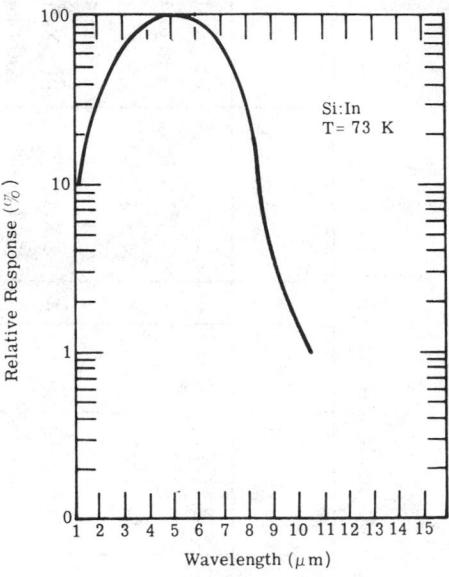

Fig. 11-55. Spectral response of Si:In [11-35].

11.6.2. Ternary Alloy Detectors. Ternary alloy detectors provide an added dimension in infrared detector characteristics. They permit some of the detector parameters to be tailored by varying the alloy constituents. The two currently popular alloy systems are $Hg_{(1-x)}Cd_xTe$ (Mercury-Cadmium-Telluride) and $Pb_{(1-x)}Sn_xTe$ (Lead-Tin-Telluride). The discussion in this section of the detector parameter variations as functions of constituent variation will, in general, be limited to $Hg_{(1-x)}Cd_xTe$ since more data are currently available on that system.

The spectral response of a semiconductor material is determined primarily by the one or more energy band gaps that exist between various energy levels within the material. In $Hg_{(1-x)}Cd_xTe$ the energy gap is determined by the ratios of HgTe to CdTe in the detector material. Curves showing energy gap versus composition, x, and cutoff wavelength versus composition are presented in Figures 11-58 and 11-59, respectively. Thus, by specifying a particular x value in the $Hg_{(1-x)}Cd_xTe$, a designer could, in principle, reduce background noise by obtaining a detector that would respond only out to the signal wavelength. In general, as the energy gap is reduced to obtain longer wavelength capability, the cooling requirements are increased. Figure 11-60 shows the maximum operating temperature possible for background-limited operation as a function of the detector cutoff wavelength or energy gap. The maximum possible D^* (λ) will depend on the energy gap and the

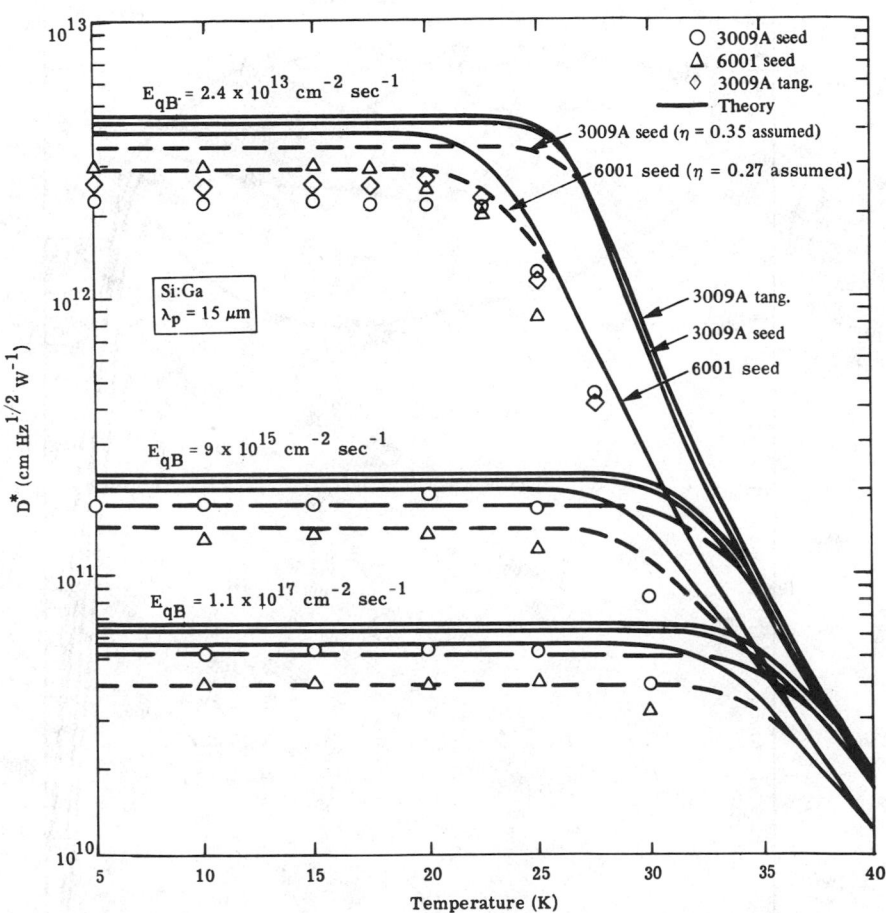

Fig. 11-56. D* versus temperature for various background photon flux densities. The detector material is Si:Ga. Several different samples, labeled according to method of growth, are given: 3009A seed, 6001 seed, 3009A tang. [11-37].

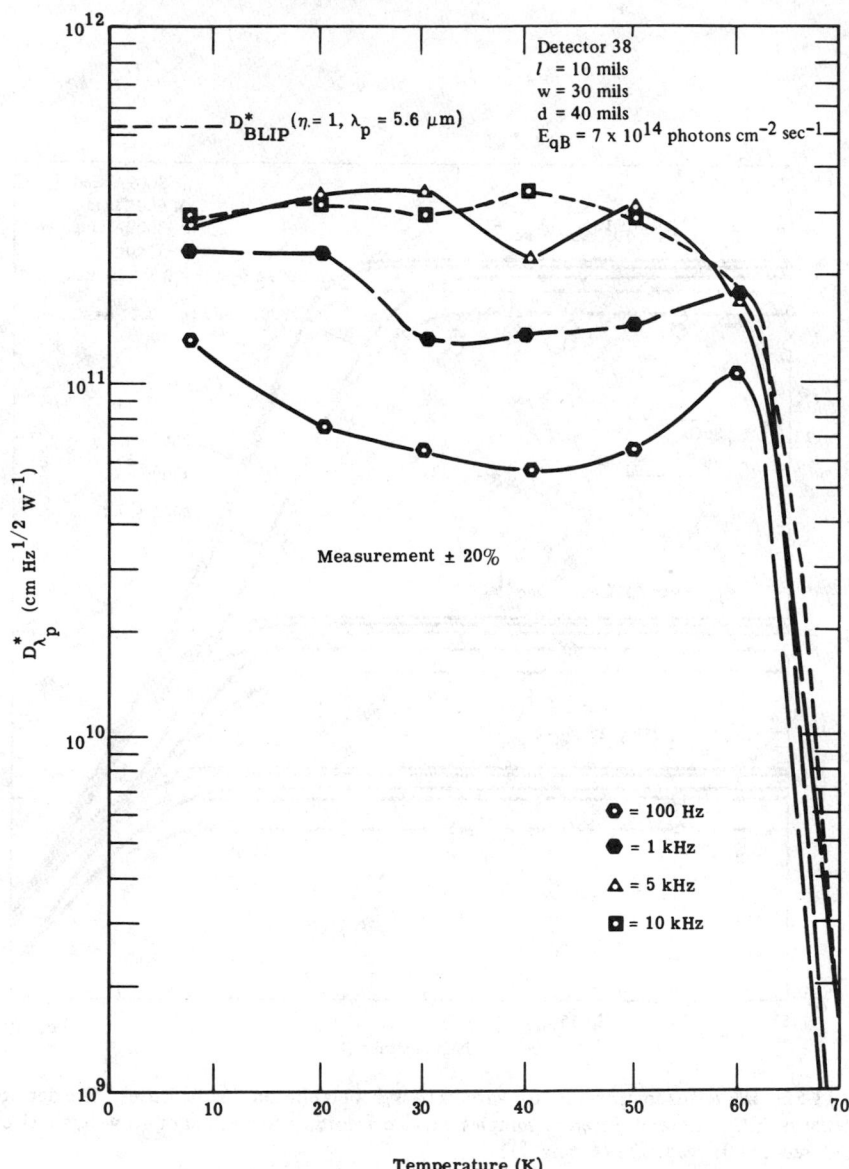

Fig. 11-57. $D_{\lambda_p}^*$ as a function of temperature [11-38].

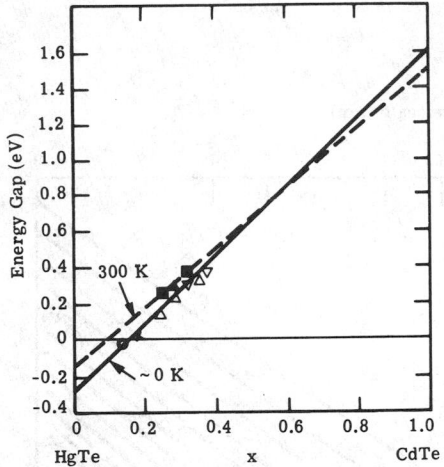

o Interband Magnetoreflection at 77 K
x Interband Magnetoreflection at 4 K
■ Optical Absorption at 300 K
△▲ Photovoltaic Studies at 77 and 300 K
▽ Photoluminescence at 12 K

Fig. 11-58. Energy gap versus composition. The solid line represents the dependence at ~ 0 K and the dashed line is dependence at 300 K [11-40].

Fig. 11-59. Cut-off wavelength versus alloy composition for (Hg, Cd) Te at 77 K [11-41].

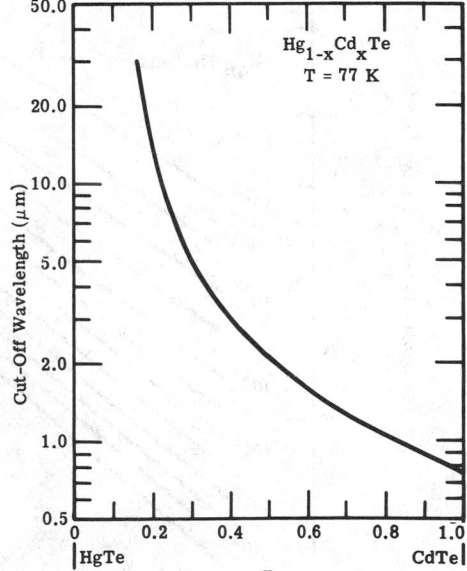

effective background photon flux incident on the detector. Families of curves of theoretically achievable D^* (λ) values as functions of the cutoff wavelength, λ_c, for various background photon flux values are shown in Figure 11-61. The case where the background flux level can be expressed in terms of an effective temperature rather than photon flux is covered in Figure 11-62.

If a particular cutoff wavelength, λ_c, is required, but signal-to-noise requirements can be decreased without detracting from overall system performance, the designer may be able to simplify the system by reducing the cooling capability. The curves shown in Figures 11-63 and 11-64 demonstrate how the detector's D^* (λ) and responsivity, $\mathcal{R}$, vary as a function of temperature for typical, 14 μm cutoff, HgCdTe detectors.

Similar curves can be generated for $Pb_{(1-x)}Sn_x Te$ detectors, or for any other type of ternary detector.

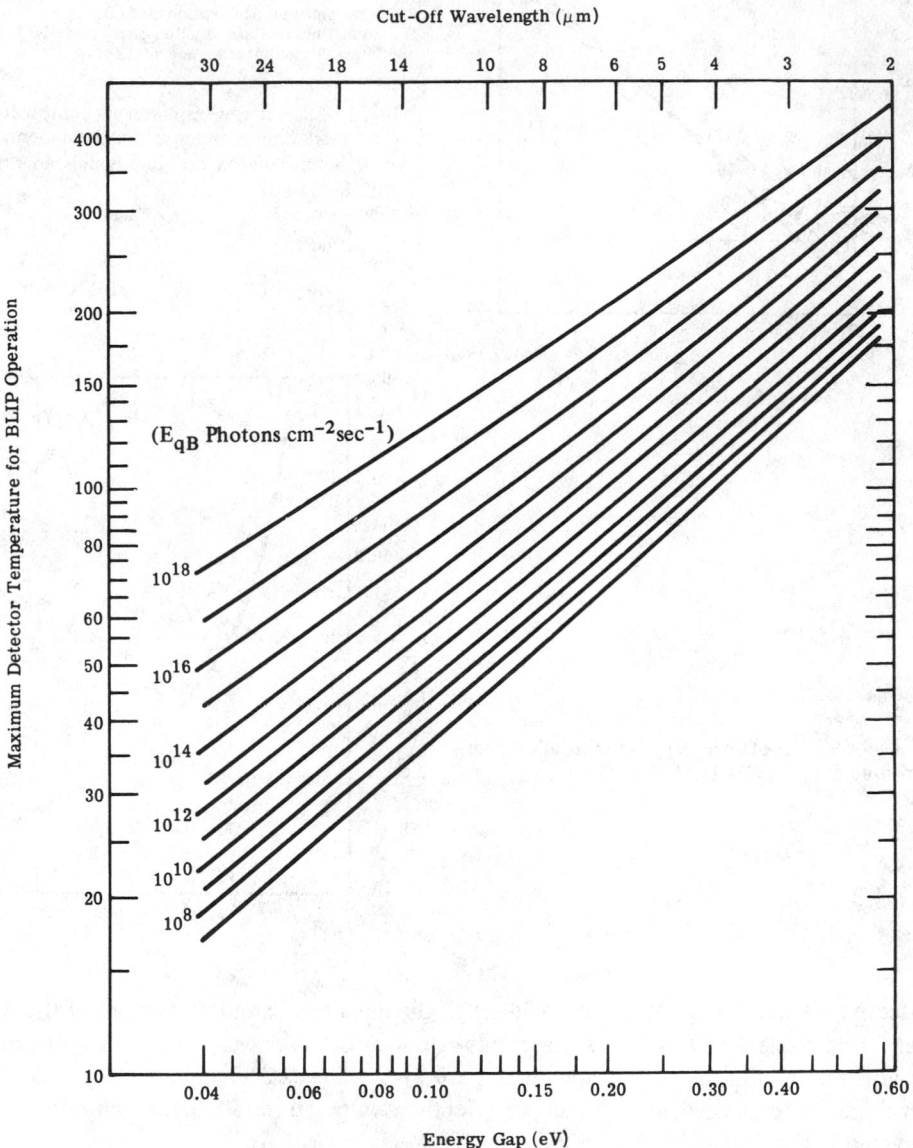

Fig. 11-60. Detector temperature required to achieve background limited performance for an intrinsic photoconductor with a given energy gap operating into a background with a flux shown as a parameter [11-42].

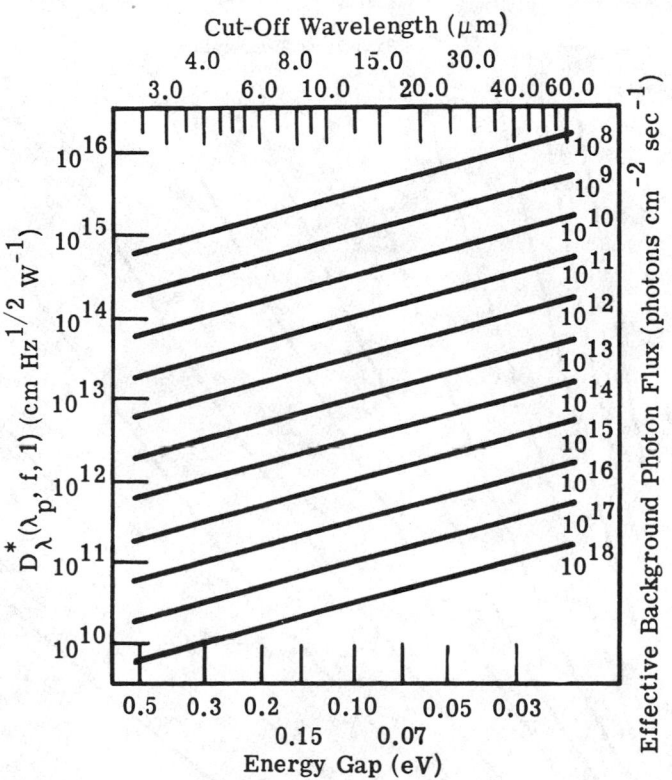

Fig. 11-61. D* versus cut-off wavelength for the ideal photoconductor with the effective background flux as a parameter [11-42].

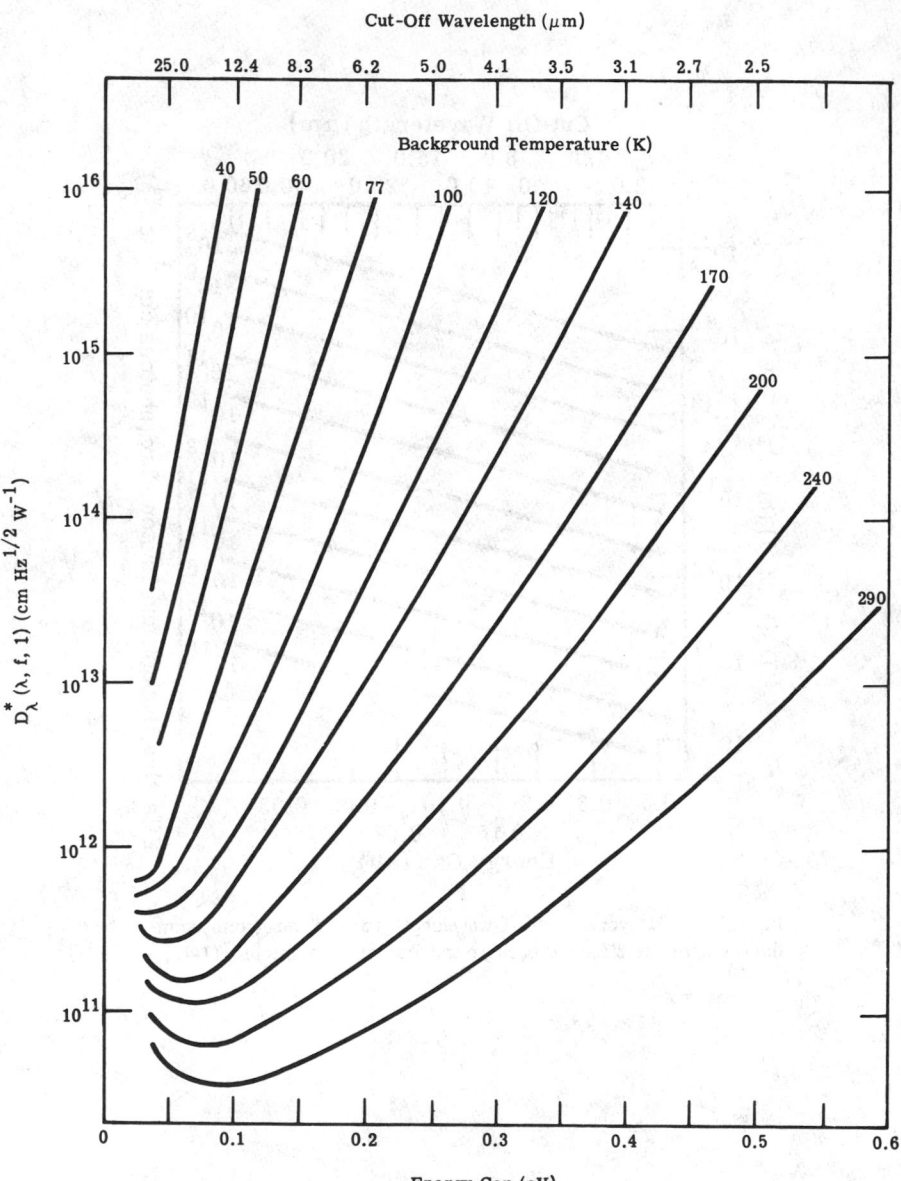

Fig. 11-62. $D^*_{\lambda_p}$ as a function of bandgap for the ideal photoconductor with the background temperature as a parameter [11-42].

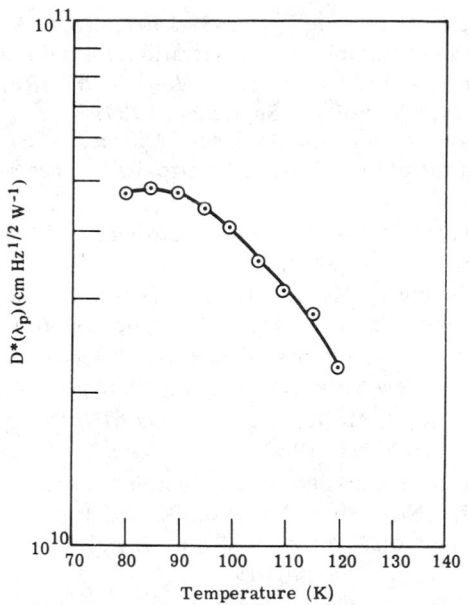

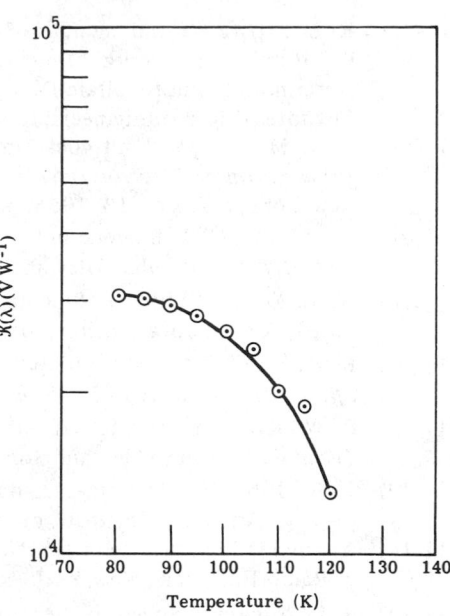

Fig. 11-63. $D^*_{\lambda_p}$ as a function of detector temperature for a typical detector [11-43].

Fig. 11-64. Responsivity as a function of temperature [11-43].

11.7. References and Bibliography

11.7.1. References

[11-1]   R. Clark Jones, D. Goodwin and G. Pullan, "Standard Procedures for Testing Infrared Detectors and for Describing Their Performance," AD No. 257597, Office of Defense and Development Research and Engineering, Washington, DC, September 1960, pp. 1-45.

[11-2]   R. A. Smith, F. E. Jones, and R. P. Chasmar, *The Detection and Measurement of Infrared Radiation,* Clarendon Press, Oxford, England 1968.

[11-3]   M. J. E. Golay, "Theoretical Considerations in Heat and Infra-Red Detection, with Particular Reference to the Pneumatic Detector," p. 347, and "A Pneumatic Infra-Red Detector," p. 357, *Review of Scientific Instruments,* American Institute of Physics, New York, NY, Vol. 18, No. 5, May 1947. Also M. J. E. Golay, "The Theoretical and Practical Sensitivity of the Pneumatic Infra-Red Detector," *Review of Scientific Instruments,* American Institute of Physics, New York, NY, Vol. 20, 1949, p. 816.

[11-4]   E. H. Putley, "The Pyroelectric Detector," *Semiconductors and Semimetals,* R. K. Willardson and A. C. Beer (eds.), Academic Press, New York, NY. Vol. 5, 1970, pp. 259-285.

[11-5]   R. Havens, "Theoretical Comparison of Heat Detectors," *Journal of the Optical Society of America,* Optical Society of America, Washington, DC, Vol. 36, 1946, p. 355.

[11-6]   R.D. Hudson, *Infrared System Engineering,* John Wiley and Sons, New York, NY, 1969, p. 357.

[11-7]   R. L. Petritz, "Fundamentals of Infrared Detection," pp. 1459-1467; also, G. R. Pruett and R. L. Petritz, "Detectivity and Preamplifier Considerations for Indium Antiomonide Photovoltaic Detectors," pp. 1524-1529; *Proceedings of the IRE,* Institute of Radio Engineers, New York, NY, Vol. 47, September 1959.

[11-8]   R. J. McIntyre, "Multiplication Noise in Uniform Avalanche Diodes," *IEEE Transactions on Electron Devices,* Institute of Electrical and Electronics Engineers, New York, NY, Vol. 13, 1966, p. 164.

[11-9]   T. S. Moss, G. J. Burrell, and B. Ellis, *Semiconductor Opto-Electronics,* Halsted Press Division of John Wiley and Sons, New York, NY, 1973.

[11-10]   W. L. Wolfe, "Photon Number D* Figure of Merit," *Applied Optics,* Optical Society of America, Washington, DC, Vol. 12, No. 3, March 1973, pp. 619-621.

[11-11]   K. M. Van Vliet, "Noise in Semiconductors and Photoconductors," *Proceedings of the IRE,* Institute of Radio Engineers, New York, NY, Vol. 46, 1958, p. 1004.

[11-12]   P. W. Kruse, L. D. McGlauchlin, and R. B. McQuistan, *Elements of Infrared Technology,* John Wiley and Sons, New York, NY, 1962.

[11-13]   J. B. Johnson, "Thermal Agitation of Electricity in Conductors," *Physical Review,* American Institute of Physics, New York, NY, Vol. 32, 1928, p. 97.

[11-14]   A. van der Ziel, "Noise in Junction Transistors," *Proceedings of the IRE,* Institute of Radio Engineers, New York, NY, Vol. 46, 1958, p. 1019.

[11-15]   R. L. Williams, "Speed and Sensitivity Limitations of Extrinsic Photoconductors," *Infrared Physics,* Pergamon Publishing, Elmsford, NY, Vol. 9, 1969, pp. 37-40.

[11-16]   S. R. Borrello, "Detection Uncertainty," *Infrared Physics,* Pergamon Publishing, Elmsford, NY, Vol. 12, 1972, pp. 267-270.

[11-17]   J. A. Jamieson, "Preamplifiers for Nonimage-Forming Infrared Systems," *Proceedings of the IRE,* Institute of Radio Engineers, New York, NY, Vol. 47, 1959, p. 1522.

[11-18]   P. C. Caringella and W. L. Eisenman, "System for Low-Frequencies Noise Measurements," *Review of Scientific Instruments,* American Institute of Physics, New York, NY, Vol. 33, 1962, p. 654.

[11-19]   R. F. Potter, J. M. Pernett, and A. B. Naugle, "The Measurement and Interpretation of Photodetector Parameters," *Proceedings of the IRE,* Institute of Radio Engineers, New York, NY, Vol. 47, 1959, p. 1503.

[11-20]   A. E. Martin, *Infrared Instrumentation and Techniques,* Elsevier, New York, NY, 1966.

[11-21]   W. L. Eisenman, R. L. Bates, and J. D. Merriam, "Black Radiation Detector," *Journal of the Optical Society of America,* Optical Society of America, Washington, DC, Vol. 53, 1963, p. 729.

[11-22]   W. L. Eisenman and R. L. Bates, "Improved Black Radiation Detector, *Journal of the Optical Society of America,*" Optical Society of America, Washington, DC, Vol. 54, 1964, p. 1280.

[11-23]   R. Stair, W. E. Schneider, W. R. Walters, and J. K. Jackson, "Some Factors Affecting the Sensitivity and Spectral Response of Thermoelectric (Radiometric) Detectors," *Applied Optics,* Optical Society of America, Washington, DC, Vol. 4, 1965, p. 703.

[11-24]   Naval Ocean Systems Center, San Diego, CA, *Properties of Photoconductive Detectors,* NOLC Report 564 (a continuing series begun 30 June 1952).

[11-25]   Infrared Industries, Santa Barbara, CA, Data Sheet No. 698-A, Cell D52-9P, PbSe. February 1961.

[11-26] R. B. Emmons, Sylvania Electronics System, Western Division, Mountain View, CA, Private Communication, 1970.

[11-27] A. F. Milton, Institute for Defense Analyses, Arlington, VA, Private Communication, 1970.

[11-28] M. M. Blouke, C. B. Burgett, and R. L. Williams, "Sensitivity Limits for Extrinsic and Intrinsic Infrared Detectors," *Infrared Physics*, Pergamon Publishing, Elmsford, NY. Vol. 13, No. 1, January 1973, pp. 61-72.

[11-29] M. A. Kinch, S. R. Borrello, "0.1 eV HgCdTe Photodetectors," *Infrared Physics*, Pergamon Publishing, Elmsford, NY, Vol. 15, No. 2, May 1975, pp. 111-124.

[11-30] N. Sclar, G. J. Hoover, W. C. Milo, and R. L. Pierce, Rockwell International, Anaheim, CA, Private Communication, 1974.

[11-31] H. Macurda and R. Baxter, Philco-Ford Corporation, Aeronutronic Division, Newport Beach, CA, Private Communication, 1974.

[11-32] I. Melngailis and T. Harman, "Single-Crystal Lead-Tin Chalcogenides," *Semiconductors and Semimetals, Infrared Detectors*, R. K. Willardson and A. C. Beer (eds.). Vol. 5, 1970, p. 113.

[11-33] E. Burstein, G. Pines and N. Sclar, "Optical and Photoconductive Properties of Silicon and Germanium," *Photoconductive Conference*, R. G. Breckenridge et al. (eds.), John Wiley and Sons, New York, NY, 1956, pp. 353-413.

[11-34] E. H. Putley, "Far Infrared Photoconductivity," *Physics Status Solidi* (B, Basic Research), Vol. 6, 1964, p. 571.

[11-35] N. Sclar, Rockwell International, Anaheim, CA, Private Communication, 1975.

[11-36] R. A. Soref, "Extrinsic IR Photoconductivity of Si Doped with B, Al, Ga, P, As or Sb," *Journal of Applied Physics*, American Institute of Physics, New York, NY, Vol. 38, December 1967, p. 5201.

[11-37] M. Pines, D. Murphy, D. Alexander, R. Baron and M. Young, "Characteristics of Gallium Doped Silicon Infrared Detectors," *Technical Digest of the International Electron Devices Meeting*, Washington, DC, December 1975, Institute of Electrical and Electronics Engineers, New York, NY, p. 502.

[11-38] M. Pines and R. Baron, "Characteristics of Indium Doped Silicon Infrared Detectors," *Proceedings 1974 National Electron Devices Meeting*, Washington, DC, December 1974, Institute of Electrical and Electronics Engineers, New York, NY, p. 446.

[11-39] N. Sclar, Rockwell International, Anaheim, CA, Private Communication, 1972.

[11-40] D. Long, J. L. Schmit, "Mercury-Cadmium Telluride and Closely Related Alloys," *Semiconductors and Semimetals*, Academic Press, New York, NY, Vol. 5, 1970, pp. 175-255.

[11-41] H. Halpert, B. L. Musicant, M. B. Reine, R. P. Murasako, J. W. Reynolds, and J. J. Schlickman, Honeywell Radiation Center (HRC), Lexington, MA, Private Communication, 1971.

[11-42] S. Borrello, G. Roberts, B. Breazeale, G. Pruett, Texas Instruments, Dallas, TX, Private Communication, 1971.

[11-43] C. B. Burgett, Texas Instruments, Dallas, TX, Private Communication, 1971.

11.7.2. Bibliography

Andrews, D. H., R. M. Milton, and W. DeSorbo, "A Fast Superconducting Bolometer," *Journal of the Optical Society of America*, Optical Society of America, Washington, DC, Vol. 36, No. 9, 1946, pp. 518-524.

Bell, R. L. and W. E. Spicer, "3-5 Compound Photocathodes: A New Family of Photoemitters with Greatly Improved Performance," *Proceedings of the IEEE*, Institute of Electrical and Electronics Engineers, New York, NY, Vol. 58, 1970, p. 1788.

Biard, J. R., and W. E. Spicer, "A Model of the Avalanche Photodiode," *IEEE Transactions on Electron Devices*, Institute of Electrical and Electronics Engineers, New York, NY, Vol. 14, 1967, pp. 233-238.

Blouke, M. M., C. B. Burgett, R. L. Williams, "Sensitivity Limits for Extrinsic and Intrinsic Infrared Detectors," *Infrared Physics*, Pergamon Publishing, Elmsford, NY, Vol. 13, No. 1, January 1973, pp. 61-72.

Borrello, S. R. "Detection Uncertainty," *Infrared Physics*, Pergamon Publishing, Elmsford, NY, Vol. 12, 1972, pp. 267-270.

Boyle, W. S. and K. F. Rodgers, Jr., "Performance Characteristics of a New Low-Temperature Bolometer," *Journal of the Optical Society of America*, Optical Society of America, Washington, DC, Vol. 49, No. 1, 1959, pp. 66-69.

Burstein, E., G. Pines and N. Sclar, "Optical and Photoconductive Properties of Silicon and Germanium," *Photoconductive Conference*, R. G. Breckenridge et al. (eds.), John Wiley & Sons, New York, NY, 1956, pp. 353-413.

Coringella, P. C., and W. L. Eisenman, "System for Low-Frequencies Noise Measurements," *Review of Scientific Instruments*, American Institute of Physics, New York, NY, Vol. 33, 1962, p. 654.

Eisenman, W. L., and R. L. Bates, "Improved Black Radiation Detector," *Journal of the Optical Society of America*, Optical Society of America, Washington, DC. Vol. 54, 1964, p. 1280.

Eisenman, W. L., R. L. Bates, and J. D. Merriam, "Black Radiation Detector," *Journal of the Optical Society of America*, Optical Society of America, Washington, DC, Vol. 53, 1963, p. 729.

Emmons, R. B., and G. Lucovsky, "The Frequency Response of Avalanching Photodiodes," *IEEE Transactions on Electron Devices*, Institute of Electrical and Electronics Engineers, New York, NY, Vol. 13, 1966, p. 297.

Golay, M. J. E., "Theoretical Considerations in Heat and Infra-Red Detection, with Particular Reference to the Pneumatic Detector," p. 347, and "A Pneumatic Infra-Red Detector," p. 357; *Review of Scientific Instruments*, American Institute of Physics, New York, NY, Vol. 18, No. 5, May 1947. Also, M. J. E. Golay, "The Theoretical and Practical Sensitivity of the Pneumatic Infra-Red Detector," *Review of Scientific Instruments*, American Institute of Physics, New York, NY, Vol. 20, 1949, p. 816.

Haitz, R. H., A. Goetzberger, R. M. Scarlett, and W. Schockley, *Journal of Applied Physics*, American Institute of Physics, New York, NY, Vol. 34, 1963, p. 1581.

Havens, R., "Theoretical Comparison of Heat Detectors," *Journal of the Optical Society of America*, Optical Society of America, Washington, DC, Vol. 36, 1946, p. 355.

Holter, M., S. Nudelman, G. Suits, W. Wolfe, and G. Zissis, *Fundamentals of Infrared Technology*, The Macmillan Co., New York, NY, 1962.

Hudson, R. D., *Infrared System Engineering*, John Wiley and Sons, New York, NY, 1969, p. 357.

Infrared Industries, Santa Barbara, CA, Data Sheet No. 698-A, Cell D52-9P, PbSe, February 1961.

Jamieson, J. A., "Preamplifiers for Nonimage-Forming Infrared Systems," *Proceedings of the IRE*, Institute of Radio Engineers, New York, NY, Vol. 47, 1959, p. 1522.

Johnson, J. B., "Thermal Agitation of Electricity in Conductors," *Physical Review,* American Institute of Physics, New York, NY, Vol. 32, 1928, p. 97.

Johnson, K. M. "High-Speed Photodiodes Signal Enhancement at Avalanche Breakdown Voltage," *IEEE Transactions on Electron Devices,* Institute of Electrical and Electronics Engineers, New York, NY, Vol. 12, 1965, p. 55.

Jones, R. Clark, D. Goodwin and G. Pullan, "Standard Procedures for Testing Infrared Detectors and for Describing Their Performance," Office of Defense and Development Research and Engineering, Washington, DC, AD No. 257597, September 1960, pp. 1-45.

Kinch, M. A., S. R. Borrello, "0.1 eV HgCdTe Photodetectors," *Infrared Physics,* Pergamon Publishing, Elmsford, NY, Vol. 15, No. 2, May 1975, pp. 111-124.

Kruse, P. W., L. D. McGlauchlin, and R. B. McQuistan, *Elements of Infrared Technology,* John Wiley and Sons, New York, NY, 1962.

Long, D., "Generation—Recombination Noise Limited Detectivities of Impurity and Intrinsic Photoconductive 8-14 μ Infrared Detectors," *Infrared Physics,* Pergamon Press, Ltd, Oxford, England, Vol. 7, 1967, pp. 121-128.

Long, D., J. L. Schmit, "Mercury-Cadmium Telluride and Closely Related Alloys," *Semiconductors and Semimetals,* Academic Press, New York, NY, Vol. 5, 1970, pp. 175-255.

Low, F. J., "Low-Temperature Germanium Bolometer," *Journal of the Optical Society of America,* Optical Society of America, Washington, DC, Vol. 51, No. 11, 1961, pp. 1300-1304.

Low, F. J. and A. R. Hoffman, "The Detectivity of Cryogenic Bolometers," *Applied Optics,* Optical Society of America, Washington, DC, Vol. 2, No. 6, 1963, pp. 649-650.

Martin, A. E., *Infrared Instrumentation and Techniques,* Elsevier, New York, NY, 1966.

McIntyre, R. J., "Multiplication Noise in Uniform Avalanche Diodes," *IEEE Transactions on Electron Devices,* Institute of Electrical and Electronics Engineers, New York, NY, Vol. 13, 1966, p. 164.

Melngailis, I., and T. Harman, "Single-Crystal Lead-Tin Chalcogenides," *Semiconductors and Semimetals, Infrared Detectors,* R. K. Willardson and A. C. Beer (eds.), Vol. 5, 1970, p. 113.

Miller, S. L., "Ionization Rates for Holes and Electrons in Silicon," *Physical Review,* American Institute of Physics, New York, NY, Vol. 105, 1957, pp. 1246-1249.

Morton, G. A., "Infrared Photoemission," *Proceedings of the IRE,* Institute of Radio Engineers, New York, NY, Vol. 47, 1959, p. 1467.

Moss, T. S., G. J. Burrell, and B. Ellis, *Semiconductor Opto-Electronics,* Halsted Press Division of John Wiley & Sons, New York, NY, 1973.

Naval Ocean Systems Center, San Diego, CA, *Properties of Photoconductive Detectors,* NOLC Report 564 (a continuing series begun 30 June 1952).

North American Rockwell, Anaheim, CA Detector 632-1, Data Sheet 976, 9 March, 1972.

Optical Publishing Co., The Pittsfield, MA, *Optical Industry and Systems Directory,* 1977, 23rd edition.

Petritz, R. L., "Fundamentals of Infrared Detection," pp. 1459-1467; also G. R. Pruett and R. L. Petritz, "Detectivity and Preamplifier Considerations for Indium Antimonide Photovoltaic Detectors," pp. 1524-1529; *Proceedings of the IRE,* Institute of Radio Engineers, New York, NY, Vol. 47, September 1959.

Pines, M., and R. Baron, "Characteristics of Indium Doped Silicon Infrared Detectors," *Proceedings of the 1974 National Electron Devices Meeting,* Washington, DC, December 1974, Institute of Electrical and Electronics Engineers, New York, NY, p. 446.

Pines, M., D. Murphy, D. Alexander, R. Baron and M. Young, "Characteristics of Gallium Doped Silicon Infrared Detectors," *Technical Digest of the IEEE*, The Electron Devices Meeting, Washington, DC, December 1975, Institute of Electrical and Electronics Engineers, New York, NY, p. 502.

Potter, R. F., J. M. Pernett, and A. B. Naugle, "The Measurement and Interpretation of Photodetector Parameters," *Proceedings of the IRE*, Institute of Radio Engineers, New York, NY, Vol. 47, 1959, p. 1503.

Putley, E. H., "Far Infrared Photoconductivity," *Physica Status Solidi* (B, Basic Research), Vol. 6, 1964, p. 571.

Putley, E. H., "The Pyroelectric Detector," *Semiconductors and Semimetals*, R. K. Willardson, A. C. Beer (eds.), Academic Press, New York, NY, Vol. 5, 1970, pp. 259-285.

Scheer, J. J. and J. Van Loar, "GaAs-Cs: A New Type of Photoemitter," *Solid State Communications*, Pergamon Press, Elmsford, NY, Vol. 3, 1965, p. 189.

Smith, R. A., F. E. James, and R. P. Chasmar, *The Detection and Measurement of Infrared Radiation*, Clarendon Press, Oxford, England, 1968.

Sonnenberg, H., "Low-Work-Function Surfaces for Negative-Electron-Affinity Photoemitters," *Applied Physics Letters*, American Institute of Physics, New York, NY, Vol. 14, 1969, p. 289.

Soref, R. A., "Extrinsic IR Photoconductivity of Si Doped with B, Al, Ga, P, As or Sb," *Journal of Applied Physics*, American Institute of Physics, New York, NY, Vol. 38, December 1967, p. 520.

Stair, R., W. E. Schneider, W. R. Walters, and J. K. Jackson, "Some Factors Affecting the Sensitivity and Spectral Response of Thermoelectric (Radiometric) Detectors," *Applied Optics*, Optical Society of America, Washington, DC, Vol. 4, 1965, p. 703.

Tolman, R. C., *The Principles of Statistical Mechanics*, Clarendon Press, Oxford, England 1938, pp. 94, 95, 97, 145, 163.

Uebbing, J. J., and R. L. Bell, "Improved Photoemitters Using AsAs InGaAs," *Proceedings of the IEEE*, Institute of Electrical and Electronics Engineers, New York, NY, Vol. 56, 1968, p. 1624.

Van Vliet, K. M., "Noise in Semiconductors and Photoconductors," *Proceedings of the IRE*, Institute of Radio Engineers, New York, NY, Vol. 46, 1958, p. 1004.

van der Ziel, A., "Noise in Junction Transistors," *Proceedings of the IRE*, Institute of Radio Engineers, New York, NY, Vol. 46, 1958, p. 1019.

Williams, R. L., "Speed and Sensitivity Limitations of Extrinsic Photoconductors," *Infrared Physics*, Pergamon Publishing, Elmsford, NY, Vol. 9, 1969, pp. 37-40.

Wolfe, W. L., "Photon Number D^* Figure of Merit," *Applied Optics*, Optical Society of America, Washington, DC, Vol. 12, No. 3, March 1973, pp. 619-621.

Zworykin, V. K., and E. G. Ramberg, *Photoelectricity and Its Applications*, John Wiley and Sons, New York, NY, 1949.

Chapter 12

CHARGE-COUPLED DEVICES

Andrew J. Steckl
Rensselaer Polytechnic Institute

CONTENTS

12. Charge-Coupled Devices

12.1. Introduction

The invention of charge-coupled devices (CCDs) [12-1] has brought a new perspective to the development of infrared optical imaging and detection systems. It impacts both the need for ever increasing detectivity and spatial resolution, and the wider use and deployment of such systems through lower unit cost, component miniaturization, and relaxed operating requirements.

The charge-coupling concept is based upon the formation and transfer of charge packets in a repetitive metal-insulator-semiconductor (MIS) structure. When charge packets are photogenerated in the substrate and then electronically manipulated to a single output, the CCD operates as the solid-state, self-scanned equivalent of a vidicon tube. When the charge packets are introduced electronically, the CCD operates as a sequential shift-register in which the time delay can be electronically varied by changing the clock frequency. The simplicity of the basic CCD combined with the fact that it can be operated as an imager and/or shift register has resulted in a remarkably versatile device.

12.1.1. Symbols, Nomenclature and Units. The symbols, nomenclature and units used in this chapter are listed in Table 12-1.

Table 12-1. Symbols, Nomenclature and Units

Symbols	Nomenclature	Units
A, A_0	Ephemeral constants	—
C_D	Total detector capacitance	F
C_I	Insulator capacitance per unit area	$F\ m^{-2}$
C_{gs}	CCD input gate-to-source capacitance	F
D	Diffusion constant	$cm^2\ sec^{-1}$
D_{eff}	Effective diffusion constant incorporating drift and diffusion mechanisms	$cm^2\ sec^{-1}$
E	Electric field	$V\ m^{-1}$
E_I	Electric field amplitude in the insulator	$V\ m^{-1}$
E_s	Electric field amplitude in the semiconductor at the interface	$V\ m^{-1}$
E_m	Minimum value of the fringing field	$V\ m^{-1}$
f_c	Clock frequency	Hz
f_s	Signal frequency	Hz
G	Gain	—
G_D	Total detector conductance	mho
g_m	CCD transconductance	mho
i_D	Total detector current	A
i_d	Drain current	A

Table 12-1. Symbols, Nomenclature and Units (*Continued*)

Symbols	Nomenclature	Units
i_1	Current flowing into detector	A
i_2	Current injected into CCD from detector	A
J_B	Background photoelectron current density	A m^{-2}
J_D	Diffusion current density	A m^{-2}
J_G	Depletion region current density	A m^{-2}
J_s	Surface state current density	A m^{-2}
k	Boltzmann constant	J K^{-1}
L	Gate width	m
L_D	Diffusion length	m
N_A	Substrate acceptor doping density	m^{-3}
N_{max}	Maximum number of stored charges per unit area	m^{-2}
N_{ss}	Number of surface states per unit area	m^{-2}
n	Number of output stages, number of cells	—
n	Surface normal	—
n_i	Intrinsic carrier concentration	m^{-3}
Q	Charge packet	C
Q_B	Background photon flux density	sec^{-1}m^{-2}
Q_D	Depletion region charge per unit area	C m^{-2}
Q_{FT}	Value of charge packet in final stage	C
Q_{IT}	Value of charge packet in initial stage	C
Q_{inv}	Surface charge density in the inversion region	C m^{-2}
Q_s	Value of charge packet at equilibrium	C
q	Electronic Charge, 1.6×10^{-19}	C
S_0	Surface recombination velocity	m sec^{-1}
S_1	Surface	m^2
s	Laplace variable	rad sec^{-1}
t_i	Time, special case indicated by i	sec
t_D	Time delay	sec
V_{FB}	Flatband voltage	V
V_T	Threshold voltage	V
V_a	Applied voltage across semiconductor	V
V_G	Gate voltage	V
V_I	Voltage across insulator	V
V_{ref}	Reference voltage	V
v_{th}	Average thermal electron velocity	m sec^{-1}
x	Distance	m
x_I	Insulator thickness	m
x_d	Depletion layer thickness	m
α	$= (qn_i/2\tau_n C_I)(2\epsilon_s/qN_A)^{1/2}$	V$^{1/2}$ sec^{-1}
β	$= C_I^{-1}(2q\epsilon_s N_A)^2$	V$^{1/2}$
γ	$= C_I^{-1}(J_G + J_S + J_D)$	V sec^{-1}
Δf	Bandwidth	Hz

Table 12-1. Symbols, Nomenclature and Units (*Continued*)

Symbols	Nomenclature	Units
$\Delta\phi$	$= \|\phi_{sf} - \phi_{so}\|$	V
$\Delta\psi$	Phase shift	rad
ϵ_I	Insulator dielectric constant	F m^{-1}
ϵ_s	Semiconductor substrate dielectric constant	F m^{-1}
ϵ_o	Permittivity of vacuum	F m^{-1}
ϵ_r	Relative dielectric constant	—
ϵ_x	Charge transfer inefficiency = $(1 - \eta_x)$	—
η	Quantum efficiency	—
η_{INJ}	Injection efficiency	—
η_x	Charge transfer efficiency	—
$\eta_{x,T}$	Total charge transfer efficiency	—
μ	Mobility	m^2 V^{-1} sec^{-1}
μ_n	Electron mobility	m sec^{-1} V^{-1}
ρ	Volume density of charge inside surface S	C m^{-3}
$\rho_I(x)$	Charge area density inside insulator	C m^{-2}
σ_{ss}	Surface state capture cross-section	m^2
τ_D	Thermal diffusion time constant	sec
τ_c	Charge-up time	sec
τ_e	Trap emission time constant	sec
τ_f	Final stage time constant	sec
τ_n	Minority carrier (electron) lifetime	sec
ϕ_{ms}	Work function	V
ϕ_i	ith phase electrode and waveform	—
ϕ_s	Potential of insulator-semiconductor interface with respect to the substrate contact (potential well)	V
ϕ_{sf}	Final value of surface potential	V
ϕ_{so}	Initial surface potential	V
ω	Radian frequency	rad sec^{-1}

12.2. Fundamentals of Charge-Coupled Devices

12.2.1. Device Physics. Charge-coupled devices operate by storing information in the form of packets of charge in an array of closely-spaced capacitors. The information in the array can be electronically manipulated by pulsing capacitors sequentially such that the charge packet is transferred from one to the next and so on.

Charge Storage. The basic unit of most CCDs is the MIS capacitor. Figure 12-1 illustrates the storage process for a representative MIS structure and a p-type semiconductor substrate. By applying a positive voltage, V_G, between the top metal electrode and the substrate contact, one repels the majority carriers (holes) from the insulator-semiconductor interface, creating a depletion region in the vicinity of the interface. This effect can be represented in an energy-band diagram, as shown in Figure 12-2. The net voltage drop across the MIS capacitor is

$$V_a = V_G - V_{FB} \qquad (12\text{-}1)$$

where V_{FB} is the flat-band voltage generated by the difference, ϕ_{ms}, between the work function of the metal, M, and electron affinity of the semiconductor, S, and by the presence of fixed charge in the insulator layer:

$$V_{FB} = \phi_{ms} - \frac{1}{C_I x_I} \int_0^{x_I} x\, \rho_I(x)\, dx \quad (12\text{-}2)$$

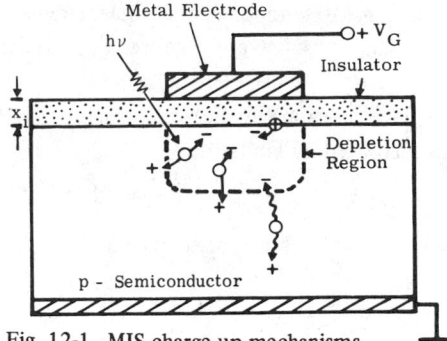

Fig. 12-1. MIS charge-up mechanisms.

where C_I = insulator capacitance per unit area

x_I = insulator thickness

x = distance from metal electrode into insulator

$\rho_I(x)$ = charge density in the insulator

In the M-SiO$_2$-Si system, V_{FB} is typically no more than a few volts.

The effective applied voltage is divided between a voltage drop across in the insulator, V_I, and a voltage drop across the semiconductor

$$V_a = V_I + \phi_s \qquad (12\text{-}3)$$

where ϕ_s is the potential of the insulator-semiconductor interface with respect to the substrate contact. Deep in the semiconductor, the screening action of the mobile carriers shields the presence of the applied electric field resulting in an energy band of constant energy. Toward the interface, however, the applied voltage bends the energy bands such that empty states of lower energy are now available for minority carrier electrons, creating, in effect, a potential "well" of depth ϕ_s. Initially, the well is empty resulting in a large depletion region width and a relatively large surface potential, ϕ_{so}. (See Figure 12-2(a).) As minority carriers are generated, they accumulate at the interface creating an inversion region. (See Figure 12-2(b).) Many more field lines now

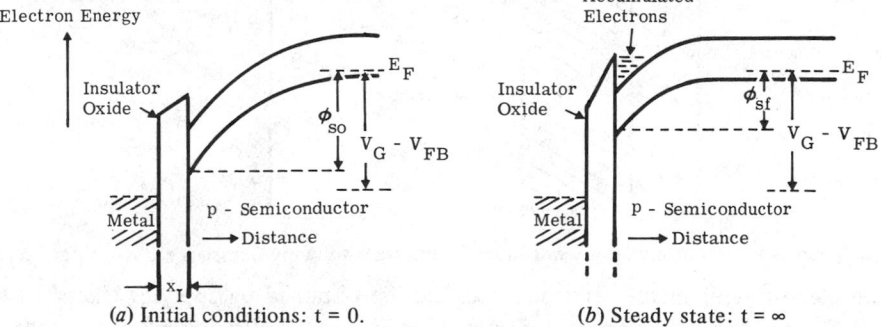

(a) Initial conditions: t = 0.   (b) Steady state: t = ∞.

Fig. 12-2. MIS energy band diagram biased into inversion.

terminate at the interface resulting in a smaller surface potential, ϕ_{sf}, and a narrower depletion region, x_d, given by

$$x_d(t) = \left(\frac{2\epsilon_s\, \phi_s(t)}{q\, N_A} \right)^{\frac{1}{2}} \qquad (12\text{-}4)$$

where ϵ_s = semiconductor dielectric constant

N_A = substrate acceptor doping density

q = electronic charge

The relationship between the electric field and the amount of charge at the interface is determined by Gauss' Law:

$$\int_S \epsilon \, \mathbf{E} \cdot \mathbf{n} \, dS = \int_V \rho \, dV \qquad (12\text{-}5)$$

where V = volume of surface S

ϵ = dielectric constant

$\mathbf{E}$ = electric field

$\mathbf{n}$ = unit vector normal to dS

ρ = charge density enclosed within S

By applying Gauss' Law to the surface S_1 enclosing the charge accumulated at the insulator-semiconductor interface (as shown in Figure 12-3(a)), one can reduce Equation (12-5) to

$$\epsilon_I E_I = Q_{\text{inv}} + \epsilon_s E_s \qquad (12\text{-}6)$$

where ϵ_I = insulator dielectric constant

E_I = magnitude of the insulator electric field at the interface, $|\mathbf{E}_I|$

E_s = magnitude of the semiconductor electric field at the interface, $|\mathbf{E}_s|$

Q_{inv} = charge density in the inversion region

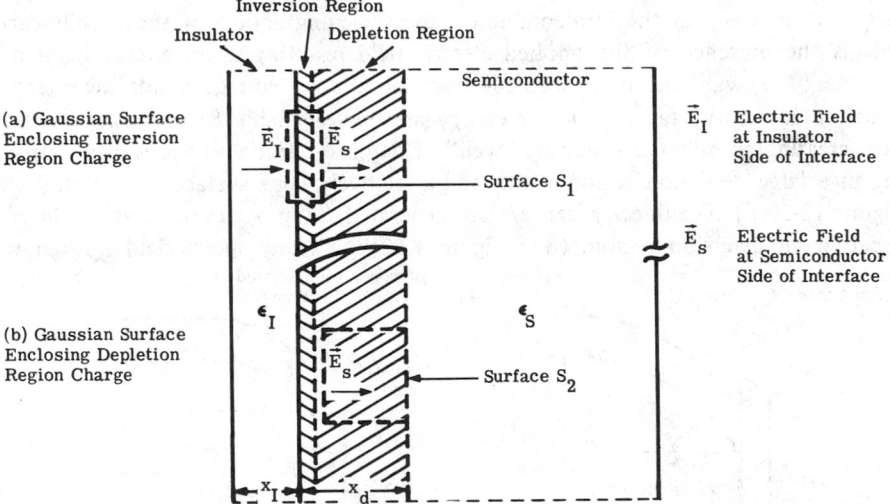

Fig. 12-3. Cross-section of insulator-semiconductor interface showing Gaussian surfaces S_1 and S_2.

The electric field in the semiconductor can be obtained by applying Gauss' Law to surface S_2 enclosing the depletion region. (See Figure 12-3(b).) Since at the substrate edge of the depletion region the electric field is zero, E_s is given by

$$E_s = \frac{Q_D}{\epsilon_s} \qquad (12\text{-}7)$$

where

$$Q_D = q \, N_A \, x_d \qquad (12\text{-}8)$$

is the depletion region charge per unit area. The electric field in the insulator layer is proportional to the voltage drop across it:

$$E_I = \frac{V_I}{x_I}$$

(12-9)

The electric field and voltage in the MIS structure are shown as a function of distance in Figure 12-4. By introducing Equations (12-4), and (12-6) through (12-9) into Equation (12-3), one can relate the semiconductor surface potential to the applied voltage:

$$V_a = C_I^{-1} Q_{inv} + \beta \phi_s^{1/2} + \phi_s \quad (12\text{-}10)$$

where $C_I^{-1} = x_I/\epsilon_I$

$$\beta = C_I^{-1} (2 q \epsilon_s N_A)^{1/2}$$

Minority carriers are continuously produced by thermal generation and by the absorption of incident background radiation. As the carriers reach the insulator-semiconductor interface, they reduce the surface potential, in effect filling up the potential well. The charge-up time, τ_c, is the length of time required for the surface potential to reach its steady state level, ϕ_{sf}, (Figure 12-2(b)) and for the potential well to reach equilibrium. One can store and preserve information in the

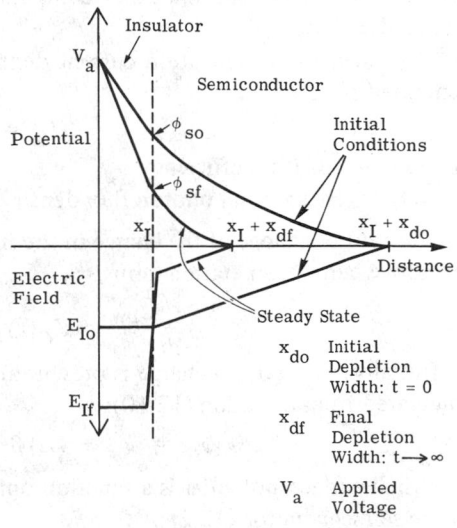

Fig. 12-4. Voltage and electric field versus distance in insulator-semiconductor.

form of a proportional amount of charge in the MIS capacitor only for a length of time considerably shorter than τ_c.

The thermal or dark current consists of a number of components, each associated with a particular generation process. The depletion region dark-current density, J_G, is associated with generation-recombination, $g\text{-}r$, centers within the region:

$$J_G(t) = \frac{q n_i x_d(t)}{2\tau_n}$$

(12-11)

where n_i = intrinsic carrier concentration
     τ_n = minority carrier lifetime

The depletion region dark current is, therefore, a function of time depending on the instantaneous value of the depletion width (Equation (12-4)).

The surface-state current density, J_s, is associated with the $g\text{-}r$ process via interface states:

$$J_s = (\tfrac{1}{2}) q n_i S_0 = (\tfrac{1}{2}) q n_i \sigma_{ss} v_{th} N_{ss}$$

(12-12)

where $S_0 = \sigma_{ss} v_{th} N_{ss}$ = surface recombination velocity
     σ_{ss} = surface state capture cross-section
     v_{th} = average electron thermal velocity
     N_{ss} = surface states per unit area

The diffusion current density, J_D, is associated with the diffusion of minority carriers from the substrate into the depletion region. Only the carriers generated within a diffusion length, L_D, of the depletion region are actually collected.

$$J_D = \frac{q\, n_i^2\, L_D}{N_A\, \tau_n} \qquad (12\text{-}13)$$

The intrinsic current resulting from band-to-band transitions in the depletion region is very much smaller than the other components of thermal current and is, therefore, not included.

The background-generation current density is the result of absorption of background-generated photons

$$J_B = \eta\, q\, Q_B \qquad (12\text{-}14)$$

where η = quantum efficiency

    Q_B = background photon flux density

The rate of change of the charge in the inversion layer is given by the summation of all the above current generation terms:

$$\frac{d\, Q_{inv}}{dt} = J_G(t) + J_s + J_D + J_B \qquad (12\text{-}15)$$

Initially, at $t = 0$, no charge is present at the interface and the surface potential can be calculated from Equation (12-10):

$$\phi_{so} = V_a + (\tfrac{1}{2})\beta^2\, [1 - (1 + 4V_a)^{\frac{1}{2}}\beta^{-1}] \qquad (12\text{-}16)$$

The final surface potential is a function only of the impurity and intrinsic concentrations in the semiconductor [12-2]:

$$\phi_{sf} = \left(\frac{2kT}{q}\right) \ln\left(\frac{N_A}{n_i}\right) \qquad (12\text{-}17)$$

where k is Boltzmann's constant. Therefore, the total charge that accumulates in the inversion region can be obtained from Equations (12-10) and (12-17):

$$Q_{inv} = C_I \left(V_a - \beta \sqrt{\phi_{sf}} - \phi_{sf}\right) \qquad (12\text{-}18)$$

Generally, $V_a \gg \phi_{sf}$, and the maximum number of charges that can be stored is related to the insulator capacitance:

$$N_{max} = \frac{Q_{inv}}{q} \cong \frac{C_I V_a}{q} \qquad (12\text{-}19)$$

For a typical case where $x_I = 1500$ Å and $V_a = 10$ V, the insulator capacitance is approximately 2×10^{-8} F cm^{-2} and $N_{max} \approx 1 \times 10^{12}$ electrons cm^{-2}.

The time-dependent solution for τ_c can be obtained after combining Equations (12-10) and (12-15):

$$\tau_c = \frac{2}{\alpha}\left(\phi_{so}^{\frac{1}{2}} - \phi_{sf}^{\frac{1}{2}}\right) + \frac{2}{\alpha^2}\left(\frac{\beta\alpha}{2} - \gamma\right)\ln\left[\frac{\gamma + \alpha\,\phi_{so}^{\frac{1}{2}}}{\gamma + \alpha\,\phi_{sf}^{\frac{1}{2}}}\right] \qquad (12\text{-}20)$$

where

$$\alpha = \frac{q\, n_i}{2\, \tau_n\, C_I}\sqrt{\frac{2\epsilon_s}{q\, N_A}} \qquad (12\text{-}21)$$

$$\gamma = (J_B + J_s + J_D) \, C_I^{-1} \tag{12-22}$$

Equation (12-20) is fairly general. Once the characteristics (e.g., lifetime, doping density, background flux) are known, simplifying approximations can be made. For example, in the case of high-quality silicon, the major current contribution is the depletion-region current [12-3]:

$$J_G \gg (J_B + J_s + J_D) \tag{12-23}$$

Therefore, $\gamma \to 0$. In addition, $\phi_{so} \gg \phi_{sf}$. In this case, τ_c can be approximated by

$$\tau_c \cong \frac{2}{\alpha} \, \phi_{so}^{1/2} + \left(\frac{\beta}{2\alpha}\right) \ln \frac{\phi_{so}}{\phi_{sf}} \tag{12-24}$$

If one takes into consideration the material parameters for silicon [12-2], then typical values are

$$
\begin{aligned}
\tau_n &= 100 \ \mu\text{sec} \\
n_i &= 1.5 \times 10^{10} \, \text{cm}^{-3} \\
N_A &= 5 \times 10^{14} \, \text{cm}^{-3} \\
C_I &= 2 \times 10^{-8} \ \text{F cm}^{-2} \\
V_a &= 10 \ \text{V}
\end{aligned} \tag{12-25}
$$

resulting in

$$
\begin{aligned}
\alpha &= 0.1 \ \text{V}^{1/2} \, \text{sec}^{-1} \\
\beta &\cong 0.75 \ \text{V}^{1/2} \\
\phi_{so} &\cong 10 \ \text{V} \\
\phi_{sf} &\cong 0.5 \ \text{V}
\end{aligned} \tag{12-26}
$$

These values lead to a charge-up time of

$$\tau_c = 74 \ \text{sec} \tag{12-27}$$

In practice, charge-up times for silicon can vary from 0.1 to 100 sec.

Charge Transfer. The charge-transfer process, in which a packet of charge is controllably manipulated from one metal electrode (gate) to the next, is at the basis of the operation of the CCD. The charge transfer process is initiated by a change in gate voltages which shifts the potential well from one gate to the next in each cell. The operation of a three-phase CCD (i.e., with a unit cell consisting of three gates) is shown in Figure 12-5. A three-phase CCD is essentially a linear array of closely-spaced MIS capacitors with every third one connected to the same clock voltage. At time $t = t_1$, the charge packet is stored in the potential well underneath the ϕ_1 electrode because of the higher positive value of the voltage on ϕ_1 than on ϕ_2 and ϕ_3. At time $t = t_2$, the same voltage is present on ϕ_1 and ϕ_2. However, the potential well is initially deeper under ϕ_2 than under ϕ_1 since the latter is partially collapsed as a result of its stored charge. (See Figure 12-4.) This initiates the charge-transfer process from the ϕ_1 to the ϕ_2 electrode. As the voltage on the ϕ_1 electrode decreases, $t_2 < t < t_4$, the charge continues to spill into the region under the ϕ_2 electrode. At $t = t_4$, the charge packet is isolated under the ϕ_2 electrode, and thus a shift of one-third of a cell has been effected. As the phase clock waveforms continue to change periodically with time, the packet will be shifted from ϕ_2 to ϕ_3 and so on through the device.

There are three basic mechanisms involved in the charge transfer process: (1) self-induced drift, (2) thermal diffusion, (3) drift as a result of fringing fields. The self-induced drift field is produced by the electrostatic repulsion of the carriers while fringing fields between adjacent electrodes produce electrical fields with components in the direction of charge transfer. The approach to be followed here is that of Reference [12-4] whereby an approximate closed-form solution is developed that accounts for all three mechanisms.

Whereas more accurate numerical [12-5] and closed-form expressions [12-6] do exist, the method used here will allow for a reasonable estimation of the charge-transfer time and efficiency of various semiconductors on the basis of their physical parameters.

The transport of charge packet Q is described by the continuity equation [12-7]:

$$\frac{\partial Q}{\partial t} = \frac{\partial}{\partial x}\left(D_{eff}\,\frac{\partial Q}{\partial x}\right) \qquad (12\text{-}28)$$

D_{eff} is an effective diffusion constant which incorporates both thermal diffusion and drift:

$$D_{eff} = \mu\,\Delta\phi\left(\delta + \frac{Q}{Q_s}\right) \qquad (12\text{-}29)$$

$$\Delta\phi = |\phi_{sf} - \phi_{so}| \qquad (12\text{-}30)$$

$$\delta = \frac{D/\mu}{\Delta\phi} \qquad (12\text{-}31)$$

where $\Delta\phi$ = the difference between the surface potentials in equilibrium and in deep depletion, respectively

Q_s = the equilibrium value of the charge packet

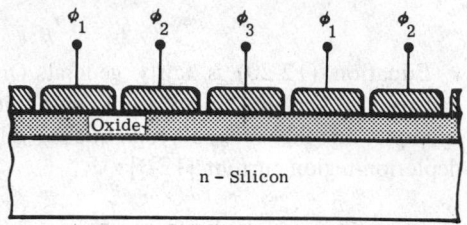

(a) Three-phase CCD cross-section.

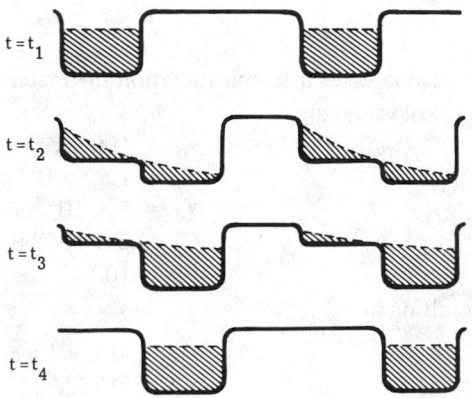

(b) Surface potential profile at sequential times $t = t_1, t_2, t_3, t_4$, showing charge transfer from ϕ_1 to ϕ_2 electrodes.

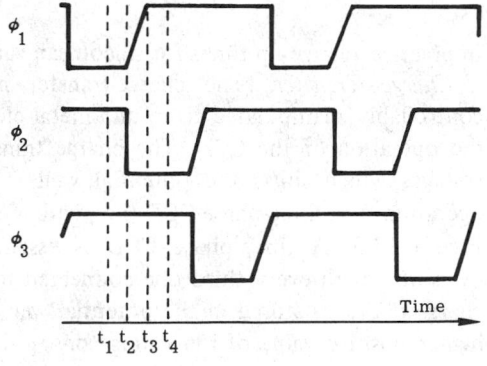

(c) Phase wave-forms.

Fig. 12-5. Charge transfer in a three-phase p-channel CCD.

Initially, the contribution of the self-induced drift current dominates the transfer of initial transfer charge, Q_{IT}:

$$\frac{\partial Q_{IT}}{\partial t} = \mu \frac{\Delta\phi}{2Q_s} \frac{\partial^2 [Q_{IT}^2]}{\partial x^2}$$ (12-32)

Based on numerical results, the spatial and temporal dependences of the charge as it decays under the influence of the self-induced field have been shown to be independent [12-8]. One can, therefore, use a separation of variables technique to solve Equation (12-32):

$$Q_{IT} = \frac{Q_s}{\left(1 + \dfrac{t}{t_0}\right)}$$ (12-33)

in which

$$t_0 = \frac{4L^2}{\pi^2 \mu \Delta\phi}$$ (12-34)

where L is the gate width.

During the final stage of the charge-transfer process, thermal diffusion and fringing field currents are the dominant contributions. Daimon, et al. [12-9] show that these processes result in an exponential decay of the form

$$Q_{FT} = C_1 e^{-t/\tau_f}$$ (12-35)

where τ_f = the final stage transfer time-constant
Q_{FT} = the charge during the final transfer process

The final transfer time constant, τ_f, is approximated [12-9] by

$$\frac{1}{\tau_f} \approx \frac{4}{\tau_D} + \frac{\mu^2 E_{min}^2}{4D}$$ (12-36)

$$\tau_D = \frac{4L^2}{\pi^2 D}$$ (12-37)

The fringing field process appears in the first term of Equation (12-36) through a fourfold decrease in the relative importance of the thermal diffusion time constant, τ_D. The second term of Equation (12-36) is entirely due to the fringing field and involves the minimum value of that field [12-10]:

$$E_{min} = \frac{2\pi}{3} \frac{\epsilon_s}{\epsilon_0} \frac{x_i (V_G - V_T)}{L^2} \left[\frac{\dfrac{5x_d}{L}}{\dfrac{5x_d}{L} + 1}\right]^4$$ (12-38)

where ϵ_0 is the permittivity of vacuum.

The constant in Equation (12-35) can now be obtained by matching the amplitudes and slopes of the expressions given by Equations (12-33) and (12-35) at time $t = t_{min}$:

$$Q_{IT} = Q_{FT}$$ (12-39)

$$\frac{dQ_{IT}}{dt} = \frac{dQ_{FT}}{dt}$$ (12-40)

Solving Equations (12-39) and (12-40), one finds t_m to be

$$t_m = \tau_f - t_0 \tag{12-41}$$

As a convenient yardstick, t_4 is defined as the time required for the transfer of 99.99% of the charge:

$$\frac{Q_{FT}}{Q_s} = 10^{-4} = \frac{C_1 e^{-t_4/\tau_f}}{Q_s} \tag{12-42}$$

$$t_4 = \tau_f \ln \left[\frac{10^4 \, e^{t_m/\tau_f}}{1 + \dfrac{t_m}{t_0}} \right] \tag{12-43}$$

If E_{min} is small enough that the first term in Equation (12-36) is dominant,

$$E_{min} \ll \frac{2\pi D}{\mu L} \tag{12-44}$$

Then the equation for t_4 can be further simplified to

$$t_4 \cong \frac{L^2}{10\mu} \frac{q}{kT} \ln \left[10^5 \frac{kT/q}{\Delta\phi} \right] \tag{12-45}$$

Time t_4 is proportional to the square of the gate length, so L should be small. It is also inversely proportional to the minority carrier mobility. As shown in Section 12.3.2, it is due to this factor that certain narrow-bandgap semiconductors which have a small effective mass and a large mobility (e.g., InSb) could have very high transfer efficiency.

Figure 12-6 shows t_4 versus gate length, L, based on Equation (12-45) for an n-channel CCD with silicon substrate having the same parameters as in the example used in this section. For a typical gate length of 10 μm, t_4 of 7.8×10^{-8} sec is obtained, indicating that a transfer efficiency of $\eta_x = 0.9999$ should be obtainable in this case for frequencies up to 5 to 6 MHz.

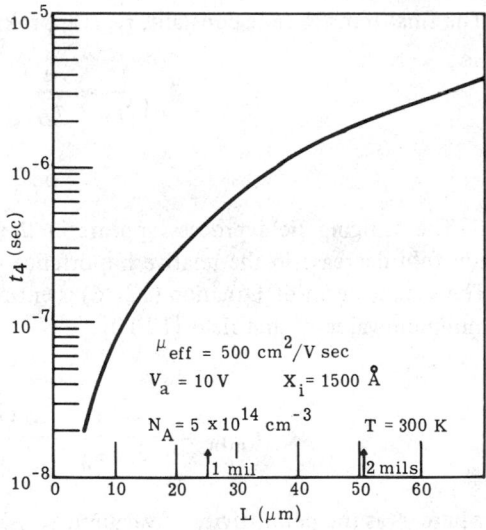

Fig. 12-6. Time (t_4) required to achieve $\eta = 0.9999$, versus gate length for n-channel CCD.

12.2.2. Device Implementation. The basic CCD shift-register consists of three sections (Figure 12-7): (1) an *input section* containing a diffused p-n junction which serves as a source of minority carriers, and one or more input gates which control the charge flow into the first potential well; (2) the *transfer section*, consisting of a series of electrodes which, through the application of periodic potentials result in the shift-

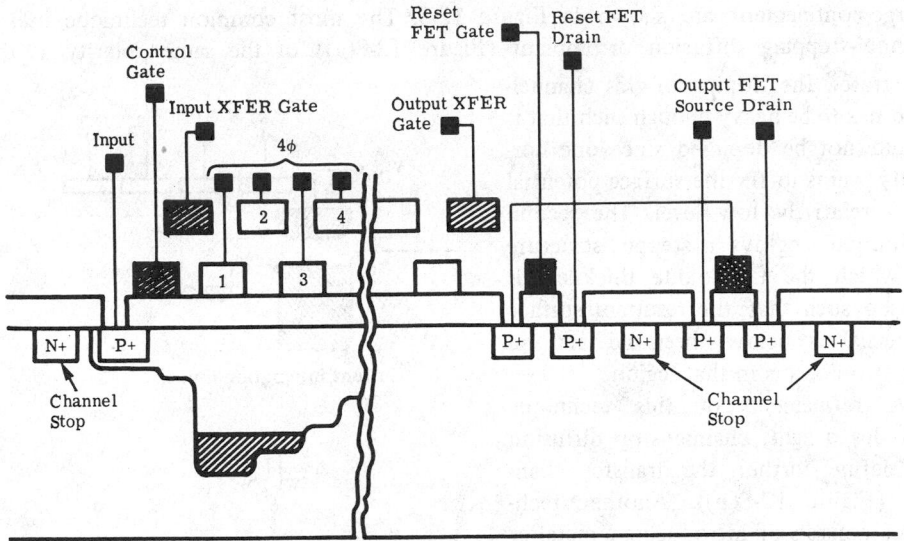

Fig. 12-7. Structure of two-level, overlapping gate CCD.

register-like transfer of charge; (3) the *output section*, consisting of an output gate, the output FET, and the reset FET.

Cell Architecture. The first CCD shift-register, built by Tompsett [12-11] and his colleagues at Bell Laboratories in 1970, consisted of only 8 cells, but was sufficient to show that charge could be transferred by moving potential wells. This first CCD had a three-phase clocking system consisting of coplanar metal electrodes. (As discussed in Section 12.2.1, at least three consecutive electrodes are the minimum number necessary to provide directionality of charge density for this electrode structure.) Since the gap between electrodes could not exceed a few micrometers for efficient coupling between potential wells, gate-to-gate shorts presented a serious yield problem. Furthermore, the gaps exposed the oxide, and stray charges which accumulated there altered the potential profile—resulting in an unstable operation. To passivate the oxide in the inter-electrode region and provide a completely sealed channel, Kim and Snow [12-12] used a continuous layer of polycrystalline silicon. The electrodes are formed by selectively doping periodic regions, with the undoped, highly-resistive region providing the isolation. This structure results in somewhat larger minimum cell dimensions since the process required for the very heavy doping of the polycrystalline silicon cannot be controlled as well as the metal etching definition used in the previous structure. If the doping of the electrodes is not sufficiently high, the resulting sheet resistance tends to limit the high frequency range of the device. A further refinement of this approach [12-13] uses coplanar polysilicon electrodes which are edge-overlapped and thus result in a sealed channel and zero effective gap. Four-phase CCD structures have been fabricated using two metal levels or a combination of metal and polysilicon.

To control fully both the charge-storage and charge-transfer processes, one must confine the carriers in the lateral direction. The edges of the potential well, which define the transfer channel, should be under accumulation to prevent additional dark-current minority carriers from entering the channel. The three basic techniques for lateral charge-confinement are shown in Figure 12-8. The most common technique uses a channel-stopping diffusion or implant (Figure 12-8(a)) of the same polarity as the substrate. The doping of this channel-stop has to be heavy enough such that it would not be depleted since one normally wants to fix the surface potential to a relatively low level. The second technique employs a stepped-structure in which the field-oxide thickness is chosen such that the resultant surface potential is below threshold and no depletion occurs in that region.

A refinement of this technique includes a light, channel-stop diffusion to define further the transfer channel (Figure 12-8(b)). Another technique consists of introducing a metal or

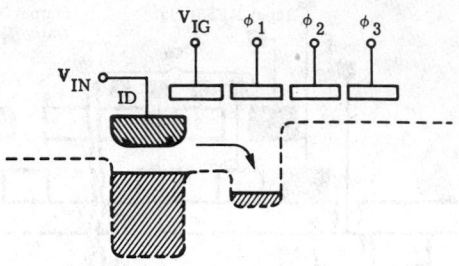

(a) Current integration input.

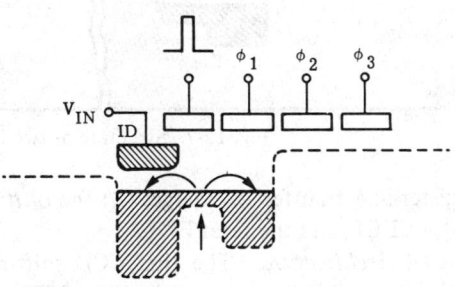

(b) Strobed input.

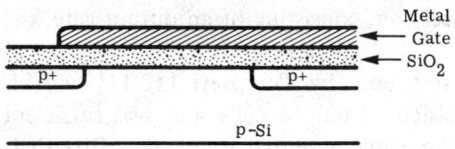

(a) Channel stop diffusion: p- and n-channel CCDs.

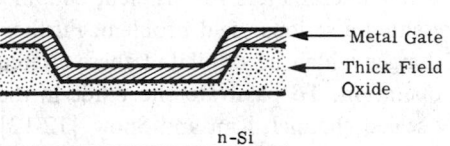

(b) Field oxide: p-channel CCD on low substrate.

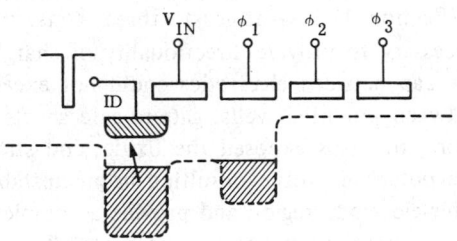

(c) "Fill and spill" input.

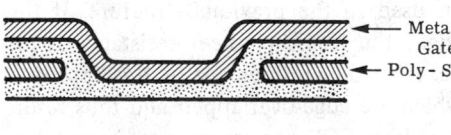

(c) Field shield of polysilicon: p- and n-channel CCDs.

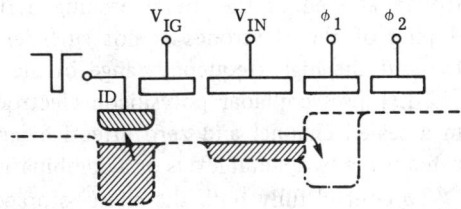

(d) Improved "fill and spill" input.

Fig. 12-8. Lateral confinement of charge transfer channel.

Fig. 12-9. Introduction of electrical input into the CCD.

polysilicon field shield which surrounds the transfer channel and is biased under accumulation (Figure 12-8(c)). However, the field shield requires additional processing and more complicated design for implementation.

Input/Output Structures and Techniques. Signal charge can be introduced into the CCD either optically or electrically. Optical input (which is achieved by the collection of photogenerated carriers) is used in visible CCD imagers (Section 12.2.4) and monolithic IRCCD imagers (Section 12.3.2).

At the heart of most input/output techniques is a diffusion which forms a p-n junction with the substrate (Figures 12-9 and 12-10). In the input circuit, the diffusion serves as the source of minority carriers, while at the output it serves as collector of minority carriers. In the basic input technique (Figure 12-9(a)), the input circuit consists of the input diffusion and an input gate. With the electrical signal applied to the input diffusion, as shown in Figure 12-9(a), and the input gate biased at a dc potential higher than the MOS threshold, current will flow into the well formed under the first transfer electrode [12-11]. Alternatively, the diffusion can be held at a constant level and the signal applied to the input gate. Since the amount of charge injected is a function of the clock period, this input technique is frequency dependent, an unsuitable characteristic for most applications.

An alternative input technique is shown in Figure 12-9(b). In this method, the input signal is applied to the diffusion and the input gate is held nominally open, isolating the diffusion from the transfer section. When the input gate is temporarily but strongly pulsed on or "strobed," charge flows from the diffusion into the well under the first transfer electrode until it is filled to the potential of the diffusion as set by the input signal. The input gate is then turned off, isolating the charge packet during the transfer to the next electrode. However, there is still some nonlinearity between the input signal voltage and the corresponding charge packet resulting from the inversion-charge-dependent surface potential under the collecting electrode.

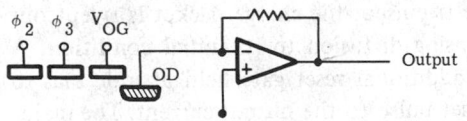

(a) Output circuit using an external preamplifier connected to the output diode.

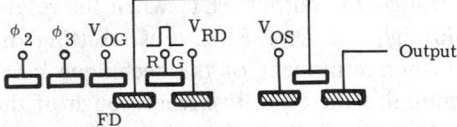

(b) Floating diffusion (FD) output circuit using on-chip sense and reset FETs.

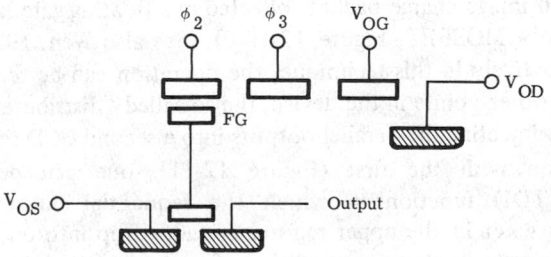

(c) Floating gate (FG) output circuit.

Fig. 12-10. CCD output circuits.

If the roles of the input diffusion and input gate are reversed (Figures 12-9(c) and (d)) an even more linear input can be obtained. This is the so-called "fill-and-spill" or potential equilibration method [12-14, 12-15]. In its simplest implementation (Figure 12-9(c)), the input signal is applied to the input gate with the input diffusion normally reverse biased. When the diffusion is set into forward bias, charge flow completely fills the collecting well. When the pulse is off, the diffusion is once again reverse biased, thereby collecting some but not all of the charge initially introduced. Because of the potential difference between the input gate and first transfer electrodes, a certain amount of charge is trapped under the latter, forming the charge packet. The size of the packet is proportional precisely to that potential difference (and therefore to the input signal). A further improvement of this technique [12-16] uses two adjacent input electrodes with the input signal being applied between the two (Figure 12-9(d)). In this case, the potential equilibration forces are at roughly the same values as the surface potential, with the collecting electrode varying only slowly with time (i.e., with the input signal) rather than being pulsed as in the first fill-and-spill technique. These two factors result in an input which displays both good linearity and low noise, critical factors in the operation of the majority of IRCCDs. (See Sections 12.3.2 and 12.3.3.)

In the output circuit, a diffusion similar to the one used in the input circuit forms a p-n junction with the substrate. When reverse biased, this diode acts as a sink for minority carriers in its vicinity. The basic output circuit consists of the output diffusion and an output gate which isolates the last transfer electrode from the diffusion (Figure 12-10(a)). The output signal in this case has the form of a spike in the output current whenever a charge packet arrives at the output diffusion. An external preamplifier is generally required with this method to provide useful signal levels.

A second output technique uses a simple on-chip preamplifier consisting of an output MOSFET and a reset MOSFET. The output-sensing diffusion, which is connected to the gate of the output FET (Figure 12-10(b)), now also serves as the source of the reset FET. The charge packet collected by the sensing diffusion first modulates the current flowing through the output FET. When the reset gate is pulsed, the charge packet is swept out through the reset FET, thus resetting the sensing diffusion to its initial condition. A further refinement of this technique uses an additional reset gate held at a dc bias to minimize the capacitive feedthrough of the reset pulse to the output current. The major noise contribution of this type of output circuit is due to the periodic resetting of the sensing diffusion [12-17]. (See Section 12.2.3.)

To avoid the addition of reset noise to the output, one can have the charge packet sensed through an image charge packet collected in a floating gate laid across the channel and connected to a MOSFET, Figure 12-10(c). (See also Wen, 1974.) Since the charge is read nondestructively in this technique, the operation can be repeated with the same charge packet at other points in the device, the so-called "distributed floating gate amplifier" [12-18]. By injecting the parallel outputs into a second CCD register which operates in synchronization with the first (Figure 12-11), one can perform a time-delay-and-integration (TDI) function in which the sequential outputs corresponding to any one charge packet in the upper register are added up into one packet in the lower register. The TDI output, therefore, will be a factor of n times bigger than the original packet for n output stages. The noise associated with the charge packet adds incoherently (i.e., in rms fashion), and the TDI register noise will increase only by $\sqrt{n}$.

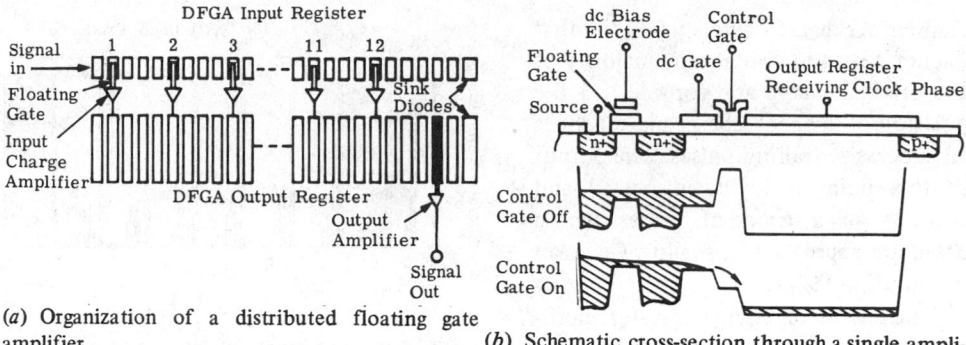

(a) Organization of a distributed floating gate amplifier.

(b) Schematic cross-section through a single amplifier stage.

Fig. 12-11. A time-delay-and-integration (TDI) function [12-19].

12.2.3. Device Characteristics.

This section explores the characteristics and perform-ance of CCDs. First, the effect of charge transfer inefficiency (CTI), which was intro-duced in Section 12.2.1., on practical devices is discussed. Second the noise mechanisms of a CCD are considered and evaluated. In general, these two properties define the input/output characteristics of a CCD since they combine to determine the signal-to-noise ratio (SNR) of the device. Since this discussion is concerned with IR applications in which one usually operates at cryogenic temperatures, the operation of CCDs at low temperatures is also discussed.

Transfer Efficiency. The charge transfer efficiency, η_x, is the ratio of the charge trans-ferred from one CCD *cell* to the next, divided by the charge originally present in the first cell. Therefore

$$\eta_x + \epsilon_x = 1 \qquad (12\text{-}46)$$

where ϵ_x is the transfer loss factor or the charge transfer inefficiency. As discussed in Section 12.2.1, η_x and ϵ_x (for small transfer times) are functions of time and therefore should be considered in light of the CCD clock frequency. The input/output, or total, transfer efficiency, $\eta_{x,T}$, of the CCD, is a function of the number of cells, n, in the device:

$$\eta_{x,T} = \eta_x^n = (1 - \epsilon_x)^n$$
$$\approx 1 - n\epsilon_x \qquad (12\text{-}47)$$

when $\epsilon_x \ll 1$. If a single charge packet is injected into a CCD shift-register, then $\eta_{x,T}$ represents the fraction of charge which appears at the output after the correct delay. The remaining fraction of the charge, $\epsilon_{x,T}$, will appear in one or more trailing packets whose relative value will depend on the $n\epsilon_x$ product. It is clear that for a device with a large number of cells, the transfer efficiency must be very high to obtain an undistorted output signal. If, for example, an $\eta_{x,T} = 0.9$ is required for a 100-cell shift-register, then ϵ_x must be smaller than 10^{-3}, equivalent to a transfer efficiency of almost 0.999. Figure 12-12 shows a computer simulation [12-20] of the spreading of the individual charge packet into a number of trailing packets as a function of the $n\epsilon_x$ product. The left-hand side of each case shows the first charge packet arriving after the correct delay followed by trailing packets. For $n\epsilon_x \gtrsim 0.6$, the first packet ceases to contain the majority of the original charge and for $n\epsilon_x \gtrsim 1$ the position of the peak charge packet shifts to successive trailing pulses. The output at this point is highly dispersed and consists of a train of pulses whose envelope approaches a shallow Gaussian distribution [12-3].

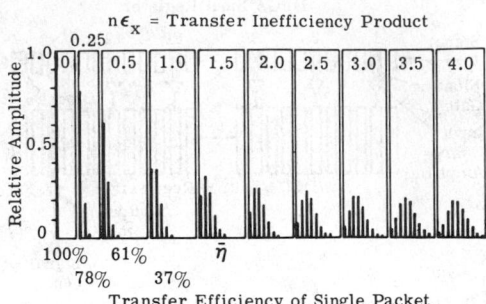

Fig. 12-12. Appearance of output signals upon injection of a single charge packet and transfer through charge transfer devices for several different values of total transfer inefficiency products $n\epsilon$ (linear model) [12-20].

The effect of charge transfer inefficiency on a sinusoidal input signal [12-24] of amplitude A_{in} is characterized by a gain $G = A_{out}/A_{in}$ and a phase shift $\Delta\psi$:

$$G = \exp\left\{-n\epsilon_x \left[1 - \cos\left(2\pi \frac{f_s}{f_c}\right)\right]\right\} \tag{12-48}$$

$$\Delta\psi = -n\epsilon_x \left\{2\pi \frac{f_s}{f_c} - \sin 2\pi\left(\frac{f_s}{f_c}\right)\right\} \tag{12-49}$$

where f_s = signal frequency
$\quad f_c$ = clock frequency
$\quad n$ = number of cells in CCD shift-register

Since the CCD is a time-sampled device with the sampling frequency given by the clock frequency, the Nyquist frequency (beyond which no useful information is transmitted) is equal to $0.5 f_c$.

In Figure 12-13, the gain and the phase shift of a CCD shift-register are plotted versus signal frequency and as a function of the $n\epsilon_x$ product. As a rough guideline, it can be seen that $n\epsilon_x \gtrsim 0.1$ is generally acceptable since it results in a gain factor between 0.8 and 1.0 and a phase shift between 0 and 20°.

Typically, transfer efficiency *per gate* for *surface*-channel CCDs runs between 0.999 and 0.9999 per transfer, with values as high 0.99999 having been reported [12-21]. At relatively low frequencies ($f_c \gtrsim 1$ MHz), the major transfer-loss mechanism is due to interface-state trapping. (See Section 12.2.1.) This effect can be minimized by transferring a constant amount of charge (or "fat-zero") in each potential well of the CCD. The fat-zero charge fills most of the interface states, thus minimizing the interaction with the signal charge. Typical improvement in transfer efficiency with fat-zero level is shown in Figure 12-14 for 128-cell, p-channel, two-phase CCDs [12-22] with channel widths of 0.5, 1.0 and 5.0 mils. In the absence of any fat-zero, the transfer inefficiency is roughly the same for all three devices. As the percentage of the full well represented by the fat-zero is increased, the inefficiency initially decreases and eventually reaches a different lower limit for each device. This lower limit in ϵ_x is the so-called "edge effect" related to the extent to which the interface states along the edges of the transfer electrodes can be neutralized [12-23].

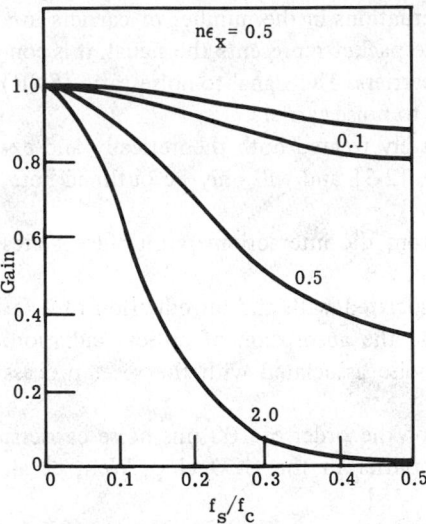

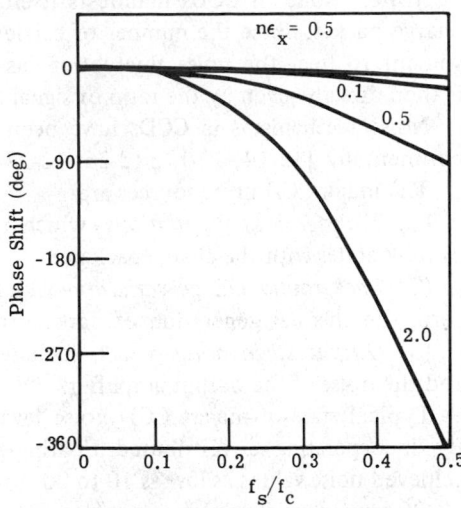

(a) Gain versus signal frequency (normalized to clock frequency).

(b) Phase shift versus signal frequency.

Fig. 12-13. Effect of transfer inefficiency on the propagation of sinusoids through a CCD shift-register [12-24].

At frequencies above approximately 1 MHz, increasing transfer loss resulting from the other mechanisms discussed in Section 12.2.1 begins to dominate. Typical transfer inefficiency versus frequency characteristics [12-22] are shown in Figure 12-15 for 500-cell, n-channel CCDs. In a *bulk* channel CCD, the charge is in contact with bulk traps which have a relatively low density, resulting in transfer losses in the range 10^{-4} to 10^{-5}. The high frequency limit of operation of bulk channel CCDs is theoretically very high—being ultimately limited by the drift velocity of the carriers. To date, operation as high as 235 MHz has been reported [12-25].

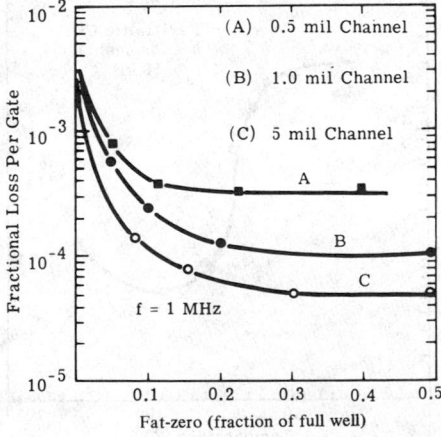

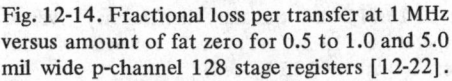

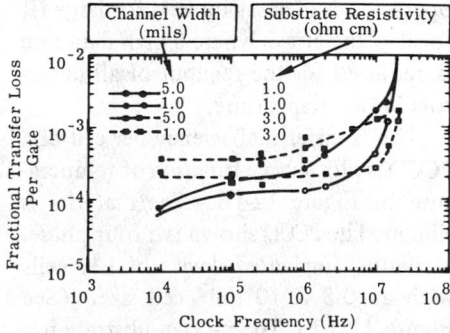

Fig. 12-14. Fractional loss per transfer at 1 MHz versus amount of fat zero for 0.5 to 1.0 and 5.0 mil wide p-channel 128 stage registers [12-22].

Fig. 12-15. Transfer loss versus clock frequency for 500 stage n-channel registers operating with 30% fat zero [12-22].

Noise. Noise in CCDs manifests itself as fluctuations in the number of carriers in a charge packet. Since the number of carriers in the packet represents the signal, it is convenient to treat the noise fluctuations as noise carriers. The signal-to-noise ratio (SNR) is then directly given by the ratio of signal carriers to noise carriers.

Noise mechanisms in CCDs have been extensively treated both theoretically and experimentally [12-14, 12-17, 12-26, 12-27, 12-28, 12-3] and will only be outlined here.

The major CCD noise sources are:

(1) *Transfer loss fluctuations*—which result from the interaction of interface states or bulk states with the charge packet.

(2) *Background charge variation*—which is associated with the introduction of a fat zero, the thermal generation of dark current, and the absorption of optical radiation.

(3) *Output stage noise*—which includes the noise associated with the reset process and the noise of the output amplifier.

Typical state-of-the-art CCD noise levels are of the order of 100 rms noise carriers. By incorporating a distributed floating-gate amplifier in the CCD chip, Wen, et al. achieved noise values as low as 10 to 20 carriers [12-18].

Of particular importance in infrared applications is the so-called "fixed pattern" noise which arises from cell-to-cell nonuniformity within the device. Nonuniformities in dark current, responsivity, bias charge, etc., in the CCD represent a serious noise limitation since the IR signal is typically 1% or less of the total incident radiation.

Low-Temperature Operation. The low-temperature operation of silicon CCDs is a critical factor in the utilization of CCDs in infrared systems [12-29]. The operating temperature of infrared charge-coupled devices (IRCCDs) is a function of a number of factors: the type of CCD, the wavelength region of interest, the IR detector material, the sensitivity requirements, etc. These parameters are covered in detail in Section 12.3. In this section, the effect of temperature on transfer efficiency is discussed.

Both bulk and surface channel CCDs are of interest. The epitaxial bulk (peristaltic) channel CCD (PCCD) [12-25, 12-30] with its high frequency of operation is of great interest for large IR focal plane arrays where a high data rate is required for the readout of all detectors in one frame time.

The transfer inefficiency per cell of a PCCD is plotted as a function of temperature in Figure 12-16. The material is silicon. The PCCD shown is a four-phase, n-channel implanted device of 130 cells with an 0.8×10 mil^2 cell size. (See Figure 12-17.) The p-type substrate has a doping of 5×10^{14} cm^{-3}, and the 5 μm thick epitaxial layer is doped at 1×10^{15} cm^{-3}. The driving gate system consists

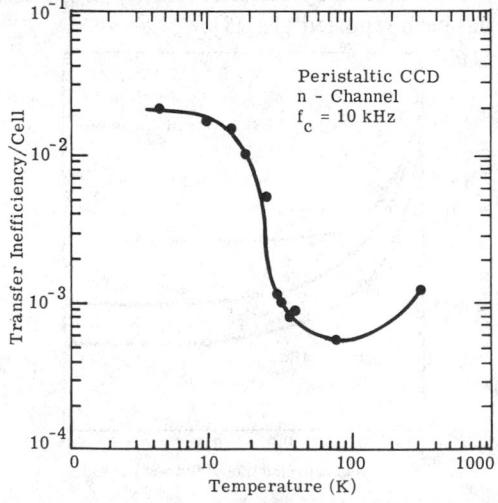

Fig. 12-16. PCCD transfer inefficiency versus temperature [12-29]. The cell consists of four gates, ϵ_x(gate) ≈ 0.25 ϵ_x(cell).

of a two-level overlapped Al structure with a silox (silicon oxide) intergate isolation layer.

At 300 K, the transfer inefficiency *per cell* of this particular PCCD was $\epsilon_x = 1.2 \times 10^{-3}$, with $\epsilon_x = 3 \times 10^{-4}$ having been observed under optimum conditions. At liquid nitrogen temperature, 77 K, the transfer inefficiency decreased to $\epsilon_x = 5.5 \times 10^{-4}$. For temperatures between 77 and 30 K, the pattern is reversed with the transfer inefficiency now slowly increasing with decreasing temperature. At 30 K, the transfer inefficiency is 10^{-3}, which is approximately the same as at room temperature. Below 30 K, the transfer inefficiency increases abruptly, an order of magnitude over a range of ten degrees, to 10^{-2} at 18 K. Finally, below 18 K, ϵ_x continues to increase but much more gradually and it approaches what appears to be a saturation level of 2×10^{-2} at 4.8 K.

The major factor contributing to the transfer inefficiency of a PCCD is associated with the bulk traps. To measure the trap-emission time-constant, one introduces a first charge packet into the CCD to fill all the traps. A second charge packet is introduced after a variable time interval, Δt. The output corresponding to the second packet will be a function of the number of traps empty at time Δt and thus capable of capturing charge: $A = A_0 e^{-\Delta t/\tau_e}$. The output signals for various time intervals are shown in Figure 12-18 (left-hand axis) for the PCCD operated at 80 K and clocked at 10 kHz. The trap emission time constants are obtained by plotting $\ln(A_0/A)$ versus Δt on the same graph (right-hand axis). A value for A_0 of 55 mV is obtained by extrapolating the curve to the origin. A time constant with an apparent value of 0.26 msec is calculated from the data. At room temperature, a time constant with a value of 0.55 msec was found for the same device [12-25].

The transfer inefficiency *per cell* of a surface channel (SCCD) is shown in Figure 12-19 as a function of temperature. The material is silicon. This particular CCD is a 100-cell, four-phase p-channel device with 1×2 mil^2 cells. The gate structure is the same as for the PCCD described above, namely two-layer overlapping Al gates. The substrate doping is $\sim 1 \times 10^{15}$ cm^{-3}. Between 300 and 30 K, the general behavior is very similar to that of the PCCD with an apparent minimum of $\epsilon_x = 4\text{-}5 \times 10^{-4}$ at 77 K. However, below 30 K the SCCD does not exhibit the same transition region shown by the PCCD at the onset of freeze-out. Rather, only a gradual increase in transfer efficiency is observed at 5 K where $\epsilon_x = 1.4 \times 10^{-3}$.

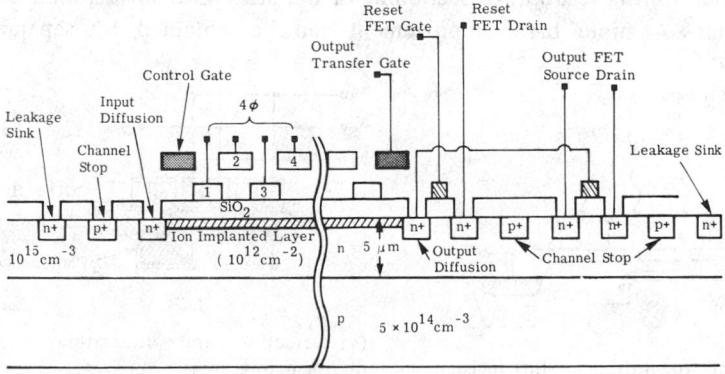

Fig. 12-17. Cross-section of PCCD in direction of charge propagation [12-29, 12-31].

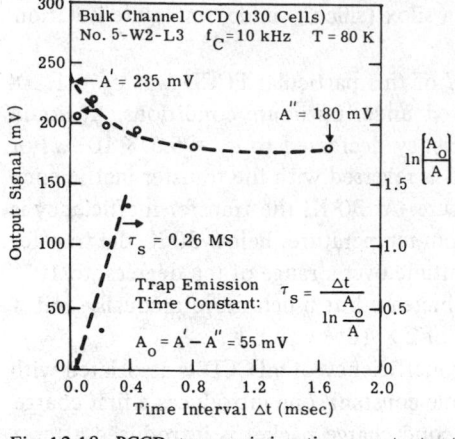

Fig. 12-18. PCCD trap emission time constants at 80 K [12-29].

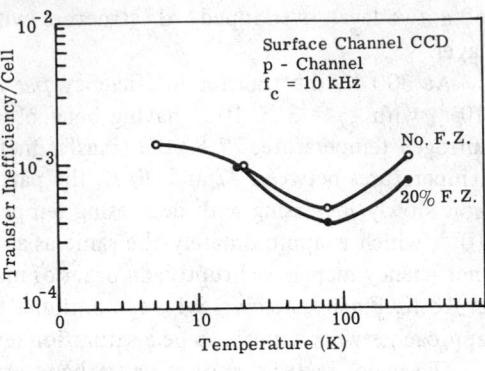

Fig. 12-19. SCCD transfer inefficiency versus temperature. The cell consists of four gates, $\epsilon_x(\text{gate}) \approx 0.25\, \epsilon_x(\text{cell})$ [12-29].

12.2.4. Applications Overview. Section 12.3 covers infrared imaging CCDs. Memories, delay lines, and signal processing can also be accomplished efficiently by these solid-state devices. They are described here briefly. The reader should consult the references for CCD details.

Visible Image Sensing. Visible CCD imagers take the form of either linear or area arrays. Linear arrays can be used to obtain a one-dimensional picture (which is quite useful in many process control applications) or they can be scanned to obtain a two-dimensional image. Various linear array configurations have been implemented [12-32, 12-33].

In Figure 12-20 different organizations of linear imagers are shown. In the simplest approach, both the accumulation of photogenerated carriers and their readout takes place in the same transfer channel. First, the electrodes are held at constant potentials to generate individual storage potential wells for an integration time. Then the usual periodic potentials are applied to clock out the stored charge. The readout time has to be considerably shorter than the integration time to prevent "smearing" the information. Other ways to minimize smearing include mechanical or electronic chopping to prevent carrier generation during readout, or sectioning of the array into smaller units each with its own output. A more basic improvement can be obtained by separating the

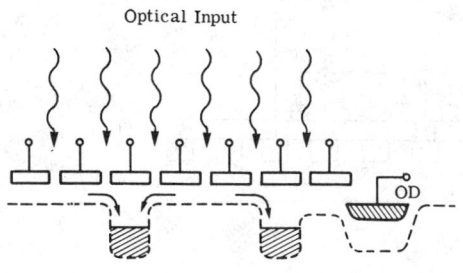

(a) Detection and integration in shift register.

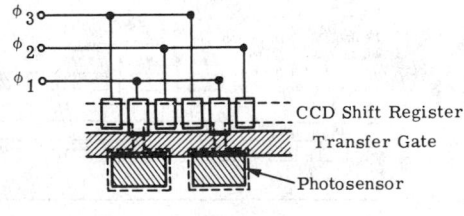

(b) Detection and integration in separate photoemissive areas.

Fig. 12-20. CCD line image sensors.

photosensitive area from the shift register proper (Figure 12-20(b)) through a common transfer gate. In this configuration, the charge is first integrated in the photosensitive area. Then, the transfer gate is pulsed to allow the charge to enter the shift register; then the detectors are isolated once again. As another line of information is being generated, the previous line is being readout of the shift register—resulting in an almost continuous integration.

CCD line imagers with as many as 1600 elements have been reported [12-34]. This high resolution line imager had an electrode structure consisting of three levels of polysilicon and a cell length of 16 μm.

CCD two-dimensional or area imagers are usually organized along an interline transfer approach or a frame transfer approach [12-35]. The interline transfer approach essentially extends the line imager with separate photosensitive sites to a two-dimensional array consisting of such line sensors in parallel. The outputs of the line sensors are fed into a single output register. In the frame transfer system, the imager consists of two equal halves: a photosensitive array and a temporary storage array. The image is first detected in the photosensitive region and then transferred into the opaque storage region. During the detection and integration of the next image in the photosensitive array, the previous image is transferred out line-by-line from the storage array. Numerous large CCD area imagers have been fabricated. For example, a fully TV-compatible 320 × 512 CCD array was reported by Rodgers in 1975 [12-36].

Memories. CCD memories are projected to fill a gap in access time between cores and RAMs (random access memories) on the one hand, and discs and drums on the other [12-3, 12-37, 12-38]. (See Figure 12-21.) The implementation of CCD memories generally takes the form of recirculating digital devices because of the finite transfer inefficiency (which results in gradual signal erosion) and the finite storage time which is due to dark current generation. Special regenerator circuits are periodically required within the CCD memory to maintain appropriate signal level and, thus, low error-rate. (See also [12-39].)

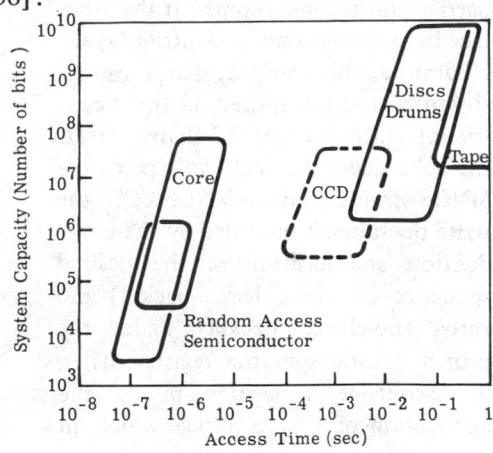

Fig. 12-21. Memory hierarchies [12-37].

The two basic CCD memory designs employ either a serpentine or serial-parallel-serial (SPS) organization, as shown in Figure 12-22. In the single-loop serpentine organization, Figure 12-22(a), all the information circulates through the entire array. Refresh amplifier stages are located between individual shift registers. The worst-case access time for this organization is the time required to completely recirculate the information in the entire array. An alternative serpentine organization is shown in Figure 12-22(b). In this organization, multiple loops, each consisting of two shift-registers and two refresh stages, are addressed by a multiplexer. In this fashion, loop-level random access is achieved, and the worst-case access time is reduced to the time it takes to recirculate only two of the shift registers of the unit.

The SPS unit (Figure 12-22(c)) consists of input and output serial shift-registers with the main body consisting of a multichannel parallel shift-register. Information is introduced into the input shift-register at the normal data rate. As the data is transferred into the parallel shift-register, it is clocked through the main body at a lower frequency corresponding to the number of parallel channels.

The output shift register reads out the information at the original data rate. The SPS organization has three features which makes it attractive: (1) only one regeneration stage per unit; (2) low power dissipation as a result of the lower clock rate in the main body; (3) relatively short input-to-output transfer path.

Recirculating CCD memories are at present organized around 4K cell blocks ([12-40, 12-41, 12-42, 12-43, 12-44]), with total chip storage ranging up to 64 kilobits [12-44]. For example, the memory chip reported by Mohsen, et al. [12-44] consists of 16 blocks of four kilobits. The device has a maximum data rate of 10 MHz, an associated mean access time of 200 μsec, and average power dissipation of 10 μW per bit. The cell size is 0.4 mil^2 and the total chip size is 218 $\times$ 235 mil^2.

In addition to recirculating memories, nonvolatile CCD memories have been implemented using the storage capacity of the metal-nitride-oxide-semiconductor (MNOS) structure [12-45, 12-46, 12-47, 12-48]. When the oxide layer of the MNOS structure is only 10 or 20 Å thick, carriers under the influence of a strong field can tunnel through this barrier and become trapped at the interface between the oxide and nitride layers. If desired, this trapped charge can be stored almost indefinitely at the dielectric interface. The trapped charge affects the subsequent threshold voltage of the MNOS structure. In an MNOS CCD, the write operation is preceded by the introduction and location of the desired sequence of ones (charge packet) and zeros (no-charge packet) under the proper locations in the register. Then the sequence is written in by the application of a large voltage which, in turn, results in the trapping process discussed above. The readout process is initiated by filling the device with a string of ones. Then the gate voltages are lowered so that charge from the lower potential wells (which contain a stored one) is spilled into the substrate. In this fashion, a complementary sequence to that stored in the memory is formed and can be then transferred out. To erase the memory, one can pulse the MNOS storage sites into accumulation.

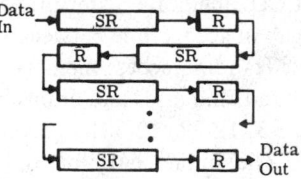

(a) Serpentine, single loop.

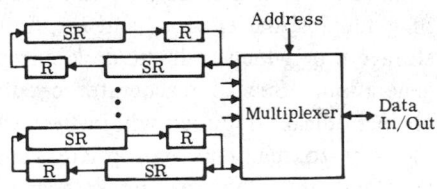

(b) Serpentine, multi-loop.

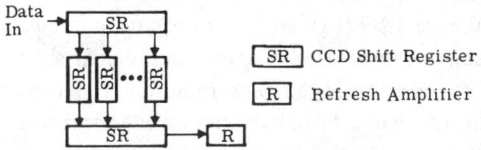

(c) Series-parallel-series.

Fig. 12-22. Arrangements for basic memory elements [12-37].

Signal Processing. Signal processing functions can be implemented with CCDs in either analog or digital form. It is, however, in analog signal processing that CCDs promise to make the more significant contribution [12-49, 12-50, 12-51, 12-52]. The CCD operates as an analog sampled-data or pulse amplitude-modulation system. Furthermore, the CCD is governed by a master clock, thus incorporating one of the major advantages of digital systems.

The basic building-block for the CCD implementation of any signal processing function is the delay line. The delay, t_D, for a device is

$$t_D = \frac{n}{f_c} \qquad (12\text{-}50)$$

where n = the number of cells

f_c = the clock frequency

Since the CCD samples the signal once per clock cycle, the input signal can have a bandwidth given by

$$\Delta f \leqslant \frac{f_c}{2} \qquad (12\text{-}51)$$

It can be seen by comparing Equations (12-50) and (12-51) that a longer delay can be obtained only at the expense of a smaller bandwidth and vice-versa. The (delay) time-bandwidth product is

$$t_D \Delta f = \frac{n}{2} \qquad (12\text{-}52)$$

The product is independent of frequency. The number of cells—and, therefore, the time-bandwidth product for the CCD delay line—is limited by the transfer efficiency which can be practically obtained. As discussed in Section 12.2.3 a nonzero transfer inefficiency has the effect of reducing the gain of the CCD monotonically with frequency. For $ne_x \approx 0.1$, the gain at $f_s = f_c/2$ is approximately 0.8, thus still allowing for the maximum bandwidth with less than 3 dB signal degradation. (See Figure 12-13, Section 12.2.3.)

A conventional state-of-art transfer inefficiency of $\epsilon_x = 2 \times 10^{-4}$ per cell would allow, for a delay line with $n = 500$, a value of $t_D \Delta f = 250$.

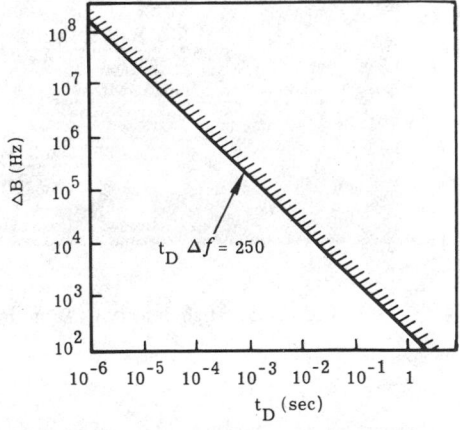

Fig. 12-23. Time-bandwidth regime for typical CCD delay time.

The usable regime for $t_D \Delta f = 250$ is shown in Figure 12-23, where a practical limit for t_D of one second is chosen and an upper limit for Δf of 200 MHz (for a peristaltic CCD) is chosen [12-31]. It should be noted that when the CCD delay line is used in conjunction with an IR detector array and is on or near the focal plane, improvement in characteristics are expected as a result of the lower temperature. Transfer efficiency generally improves at low temperature (Section 12.2.3, [12-29]) resulting in a smaller ne_x product and, therefore, allowing for a device with more cells. In addition, dark current decreases basically exponentially with temperature thus extending the maximum time delay limit.

A second basic signal processing function performed with CCDs is time division multiplexing. In this operation, several parallel signal channels are simultaneously fed into an input-tapped CCD delay line (Figure 12-24) and are read out in serial form. The design of the input tap to the CCD is similar to the conventional input circuits discussed in Section 12.2.2.

Among their other applications, CCD multiplexers are very useful in simplifying the operation of large-scale IR(LSIR) focal plane arrays where signals from 10^2 to 10^4 detectors have to be conveniently and effectively readout. (See Sections 12.3.1 and 12.3.3.) A prototype of a high-density CCD multiplexer [12-53] designed for such applications is shown

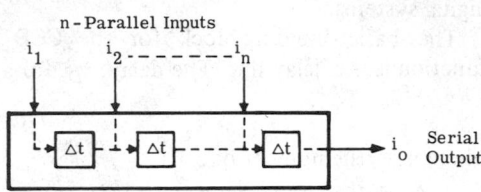

Fig. 12-24. CCD input-tapped shift-register: time division multiplexing.

in Figure 12-25. The device has a total of 10 taps, 5 contiguous taps on each side. The $\phi 1$ set of electrodes actually consists of two interdigitated subsets which allows for the in-phase loading of the multiplexer from both sides. The tap separation, equal to the cell length, is 25 μm.

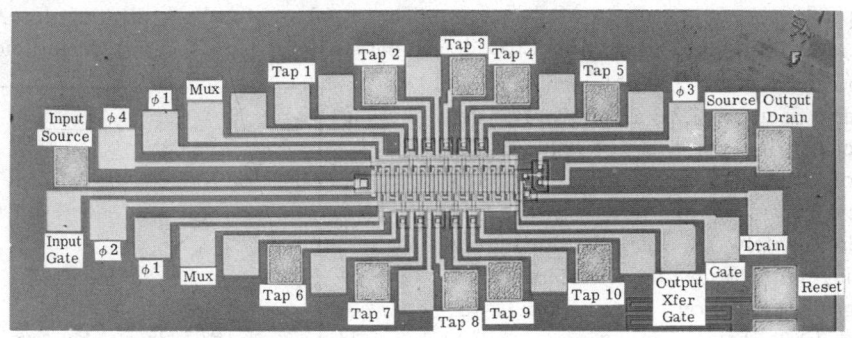

Fig. 12-25. High density CCD multiplexer with contiguous input taps.

which allows for the in-phase loading of the multiplexer from both sides. The tap separation, equal to the cell length, is 25 μm.

Another critical function for LSIR that can also be implemented with an input-tapped CCD shift-register is time-delay-and-integration (TDI) [12-54]. In TDI operation, signals generated by a moving scene-segment in successive detectors are sequentially delayed and added in the CCD. (See Figure 12-26 and Sections 12.3.1 and 12.3.3.) Since the total detector noise is given by a root mean summation while the total signal is given by a simple sum, an increase in signal-to-noise ratio (SNR) equal to the square root of the number of detectors is obtained. The significance of the CCD in this operation is not in

the simple increase in SNR (this technique has previously been implemented
with passive elements), but rather in the
fact that the operation can be economically (e.g., volume, weight, and cost)
performed, and thus extended to larger
arrays with a compounded increase in
SNR.

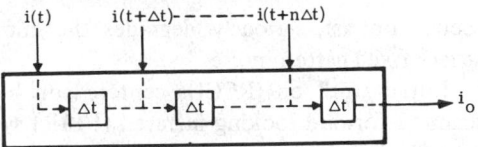

Fig. 12-26. CCD input tapped shift-register: time-delay-and-integration.

Finally, it should be noted that much more sophisticated signal processing functions can be accomplished with CCD transversal and recursive filters. See References [12-49, 12-55] and [12-3].

12.3. Infrared Charge-Coupled Devices (IRCCDs)
12.3.1. Infrared Applications

CCDs on the Focal Plane. A variety of IRCCDs, explored or proposed, generally fall into two main categories: monolithic and hybrid devices. The monolithic IRCCD category consists of devices in which both the photodetection and charge generation as well as the charge transfer are achieved in a structure built around a one-material system. Monolithic devices generally use the basic CCD concept and structure, in conjunction with an IR sensitive substrate consisting of either a narrow bandgap semiconductor or an extrinsic semiconductor having an appropriate impurity energy level.

The hybrid IRCCD category consists of devices in which the major functions occur in distinct but integrable components and materials. Photodetection and charge generation in a hybrid device can take place in any one of various suitable IR photodetectors. Charge transfer takes place in a standard silicon CCD. The role of the CCD in this case is that of a signal processor performing appropriate functions, such as multiplexing, delay-and-add operation, or amplification. The nature of the coupling between the two components subdivides the hybrid IRCCD category into two subclasses: direct and indirect injection devices. In the direct injection technique, the photodetector is coupled directly to an input tap of the CCD without any signal conditioning. In the indirect injection approach, a buffer stage is interposed between the two components to improve the signal characteristics prior to the shift-register stage.

Requirements and Operations. A major difference between the operation of CCDs in the visible and in the infrared is the large background flux rate, Q_B, in the infrared. This results in increasingly shorter background chargeup or saturation time with increasing wavelength of operation (from about 1 sec for the 2.0 to 2.5 μm window to about 10 μsec for the 8 to 14 μm region). In addition, internally generated charge is simultaneously integrated, further reducing the saturation time. (See Section 12.3.2.) Thus, one limitation of practical IRCCD devices is the length of exposure time.

Another important consideration, especially for IR thermal imaging applications, is the small contrast between the photon flux generated by an incremental change in background temperature and the total photon flux. For 0.1 K change (typical for IR imaging) of the average 300 K background, the contrast is on the order of 1.0% in the 2.0 to 2.5 μm window and less than 0.1% in the 8 to 14 μm region. The presence of such low contrast points to another practical limitation: the uniformity of IR material properties. Nonuniformity in the responsivity of IR detectors, of the same order or higher than the scene contrast, seriously degrades the minimum resolvable temperature and results in severe fixed pattern noise.

Initial work on IRCCDs centered on hybrid approaches for the focal plane of the scanned forward looking infrared (FLIR) systems, where it was quickly realized [12-54, 12-56] that neither of the above problems represented insurmountable difficulties. The serially-scanned FLIR, for example, uses a linear array of detectors raster-scanned across the scene, the scan direction being parallel to the array. Since each detector in turn scans the entire field of view, the dwell time per resolution element is quite short (Figure 12-27(a)). The amount of charge generated during this dwell time in even an 8 to 14 μm IR detector by a 300 K blackbody is well within the charge handling capacity of a typical CCD shift-register. By processing the outputs of the detectors through a time-delay-and-integration (TDI) operation, one can sum the components of the signals corresponding to the same resolution element linearly.

With the CCD, TDI operation is achieved by clocking the device at a rate corresponding to the image velocity across the detector array (Figure 12-28). Because the noise contribution of each detector is independent and thus uncorrelated, the total noise is obtained by an rms summation. For an array of m detectors, the TDI operation could, therefore, result in a potential maximum improvement of $\sqrt{m}$ (in the detector-noise-limited case) in the SNR of the entire array over that of an individual detector. The upper limit of m and, therefore, of the maximum achievable SNR

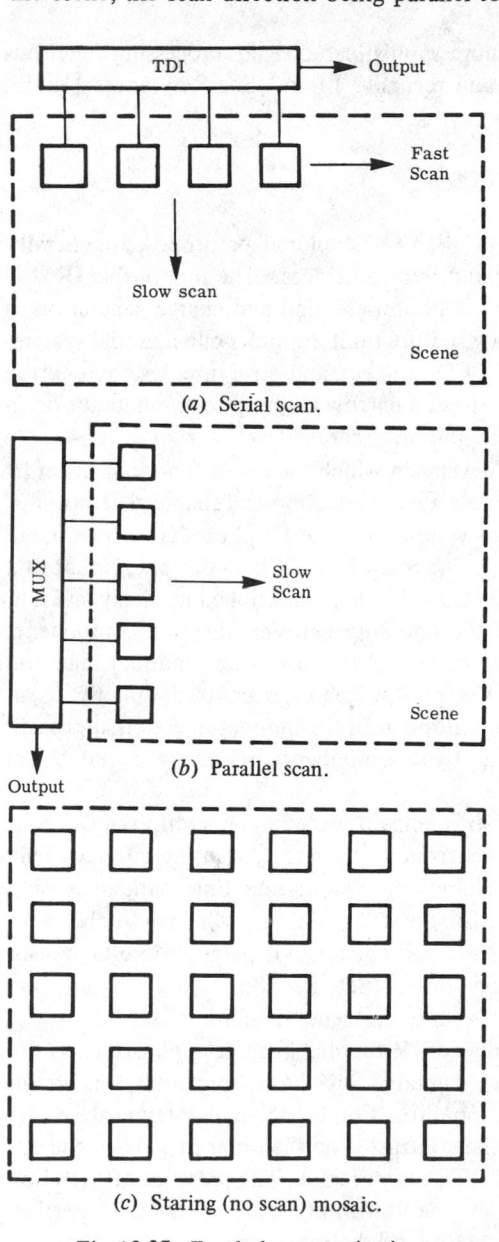

(a) Serial scan.

(b) Parallel scan.

(c) Staring (no scan) mosaic.

Fig. 12-27. Focal plane organizations.

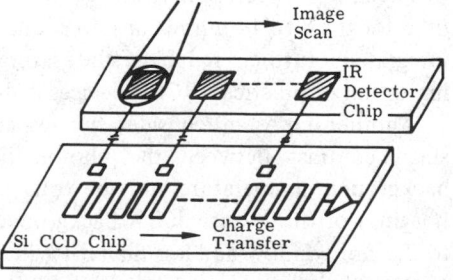

Fig. 12-28. Time-delay-and-integration (TDI) operation of serial scan IRCCD.

improvement is set by the total detector-array signal versus the CCD charge capacity. The operation of the entire array appears at the CCD output essentially as one detector. This results in reduced fixed-pattern noise, lowered detector-response uniformity requirements, and built-in redundancy. These features, together with the potential $\sqrt{m}$ improvement in SNR, help circumvent the problems posed by the high background and low contrast present in thermal imaging systems at IR wavelengths.

In the parallel scan FLIR, a vertical linear array of detectors is slowly scanned across the scene, the scan direction being perpendicular to the array (Figure 12-27(b)). In this case, an input-tapped CCD shift-register is used as a multiplexer which reads out the information line by line. The serial-parallel FLIR consists of a two-dimensional array which is scanned across the scene. TDI is performed in each channel along the scan direction and all channels are read out by a multiplexer. The "staring" two-dimensional array fully covering the entire scene field of view (Figure 12-27(c)) requires no scanning and is the eventual goal of IRCCD focal planes.

12.3.2. Monolithic IRCCDs. The category of monolithic IRCCDs includes both intrinsic and extrinsic photodetection. In turn, the extrinsic IRCCD can be operated in both inversion and accumulation modes. The Schottky-barrier IRCCD uses yet another photodetection mechanism, namely internal photoemission. The common feature of all these devices is that they are constructed around a one-material system such that both the sensor and the transfer section are fabricated on the same substrate.

Extrinsic-Silicon (Si:X) IRCCDs. By monolithically integrating extrinsic-silicon (Si:X) IR detectors with silicon CCD circuits, one can use integrated circuit (IC) technology in the development of sophisticated IR sensors. Because of the maturity of IC technology, the emphasis in Si:X IRCCD work is on the development of large-scale arrays using such sophisticated tools as computer-aided design and automated processing and testing. At the same time, by the use of batch processing one can expect that the eventual unit cost of extrinsic silicon detectors will be substantially reduced. (See Chapter 11 for Si:X IR detector characteristics.)

The main drawback of the Si:X IRCCD is the low temperature of operation required by Si:X detectors. For the 8 to 14 μm region operation, a temperature between 15 and 30 K is generally required, while for the 3 to 5 μm region the temperature range is typically between 40 and 65 K. This represents a serious problem in certain applications and higher cooling system cost in most applications. The Si:X IRCCD also inherits the relatively low quantum efficiency ($\sim$5 to 30%) associated with the long impurity absorption-length in extrinsic detectors. This effect, in turn, can result in increased crosstalk due to repeated reflection of radiation within a detector array.

Extrinsic Silicon IRCCD *Operation.* Si:X IRCCD development is now centered on Si:Ga for 8 to 14 μm applications and Si:In for 3 to 5 μm applications [12-57]. Depending on the background photon flux, the temperature range of operation for Si:Ga detectors is between 15 and 30 K while for Si:In the range is between 40 and 60 K. As discussed in "Low Temperature Operation," Section 12.2.3, CCDs operate with very good characteristics at these low temperatures.

Since extrinsic detectors are by their very nature majority-carrier devices, and CCDs use minority-carrier transport, the normal CCD structure is not adequate for a Si:X IRCCD. A number of devices have been conceived and are being studied which could resolve this incompatibility. One solution is to grow an epitaxial layer of opposite type to the substrate and then fabricate the CCD on this layer. In this fashion, when photo-generated majority carriers from the substrate are injected into the epilayer, they represent minority carriers and can be handled in the usual fashion. Another solution is to invert the photocurrent polarity before introducing it into the CCD proper. This buffer circuit can be designed and processed simultaneously with the CCD, thus retaining the total integrated-circuit aspect of the Si:X IRCCD. The special impurities used in Si:X detectors (e.g., Ga, Al, and In) require special processing techniques when combined with the usual Si IC impurities (e.g., P, As, and B). Furthermore, the special device configurations, such as the growth of an epilayer and subsequent CCD fabrication, represent additional processing-compatibility requirements.

Considerable Si:X IRCCD research is now in progress, with relatively little published yet. A Si:Ga IRCCD has been reported [12-60] which uses inversion-mode aluminum-polysilicon gate CCD shift-registers connected to linear arrays of 8, 16, and 32 detectors. When operated with an 800 K blackbody source, the Si:Ga IRCCD had a SNR between 10^3 and 10^4 for signal frequencies between 10 and 10^3 Hz and for clock frequencies of 10 kHz and 100 kHz. A 2×32 array with an element area of 4×4 mils has also been reported [12-57]. The response uniformity of the array is reported to be limited by the variation of the Ga concentration in the substrate to a value of approximately ~15%.

The majority-carrier or accumulation-mode CCD has also been investigated [12-60, 12-4, 12-59] for use as an Si:X IRCCD. Accumulation-mode operation of an MIS structure results when the proper polarity gate-voltage induces, in the vicinity of the insulator-semiconductor interface, the same majority carrier-type as is in the bulk of the semiconductor. Figure 12-29, shows the energy band diagram in an MIS structure, biased into accumulation at a sub-freeze-out temperature. The typical transfer-efficiency of accumulation-mode CCDs has been found to be of the order of 0.9 at best. Thus, they are at least an order of magnitude less efficient than inversion-mode devices operated at the same temperature.

An experimental Si:Ga CCD test chip [12-53] is shown in Figure 12-30. The chip includes both inversion-mode and accumulation-mode devices. Detectors with an area of 8×8 mils^2 are connected by direct injection to input-tapped CCD shift-registers for TDI operation.

Intrinsic IRCCDs. The intrinsic IRCCD is the infrared equivalent of the original Si IRCCD where the substrate now consists of a narrow-bandgap semiconductor. The operation and structure of these devices is, therefore, also based on the generation of an inversion region at the insulator-semiconductor interface where photogenerated carriers are collected. The bandgap of the semiconductor substrate determines the main optical absorption and, therefore, the useful atmospheric IR window. These narrow-bandgap semiconductors are usually found among the binary and ternary III-V, II-VI, and IV-VI compounds. (See [12-61, 12-62].)

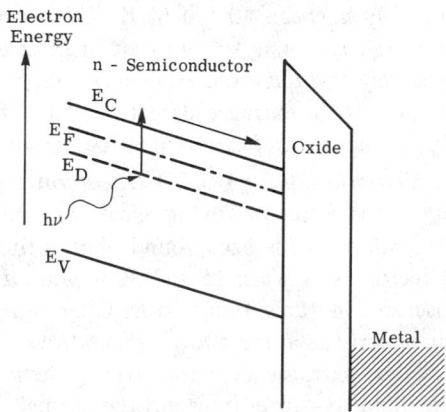

Fig. 12-29. Energy band diagram for MIS structure biased into accumulation at sub-freeze-out temperature.

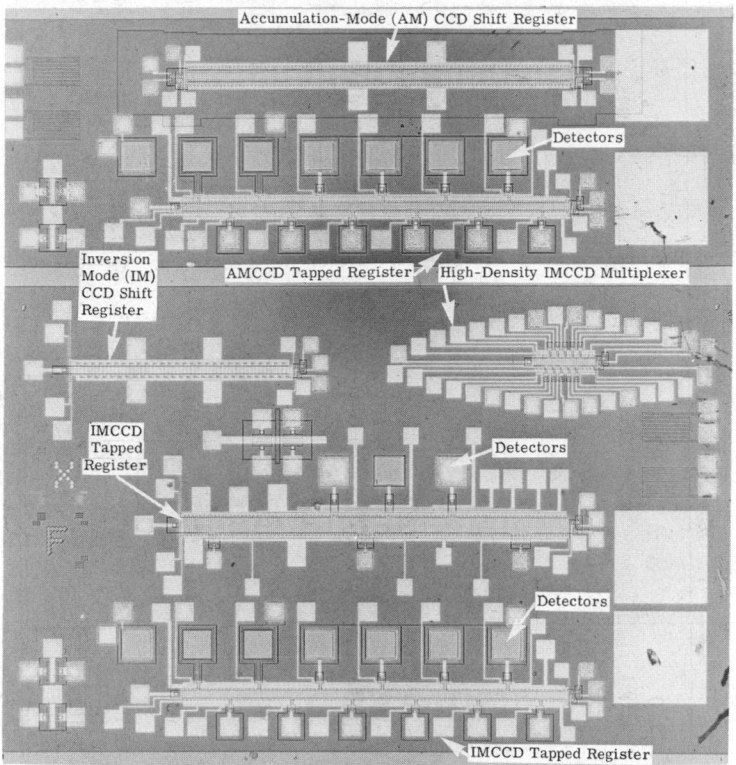

Fig. 12-30. Experimental Si:Ga CCD test chip.

Narrow-Bandgap Semiconductor Materials. Three representative IR materials are InAs ($E_g \approx 0.4$ eV, $\lambda_c \approx 3.0$ μm), InSb ($E_g \approx 0.23$ eV, $\lambda_c \approx 5.4$ μm) and $Hg_{0.8}Cd_{0.2}$ Te ($E_g \approx 0.09$ eV, $\lambda_c \approx 14$ μm). The charge-up time and potential transfer efficiency are calculated here for these materials. In Table 12-2, the appropriate material parameters and the results of the calculations are summarized. For a field of view (FOV) = 30° against the 300 K background, the incident photon flux density rate increases from $Q_B \approx 2 \times 10^{11}$ cm^{-2} sec^{-1} for InAs, to $\sim 6 \times 10^{16}$ cm^{-2} sec^{-1} for HgCdTe, while the contrast for $\Delta T_B = 0.1$ K decreases from $\Delta Q_B/Q_B \simeq 0.7\%$ for InAs to $\simeq 0.05\%$ for (Hg,Cd) Te. The very low contrast poses serious demands on the material uniformity required to obtain BLIP-limited performance. Under these circumstances, InAs has the longest charge-up time (almost 2 sec at 77 K and 20 msec at 150 K). The InSb thermal charge-up time increases two orders of magnitude from p-type ($\tau_{th} \sim 10$ msec) to n-type ($\tau_{th} \sim 2$ sec). However, since the background charge-up time is the dominant term in both cases, the total (combined) charge-up time changes by only 60%, from 3.6 to 5.8 msec. For $Hg_{0.8}Cd_{0.2}$Te, the combined charge-up time is 13 μsec when SiO_2 is the insulator, and 250 μsec when TiO_2 is the insulator. The charge-transfer time required for 99.99% transfer efficiency, t_4, has been calculated assuming a 10 μm gate width. For InAs, t_4 is equal to 320 nsec at 77 K and 220 nsec at 150 K. As might be expected, p-type InSb has the shortest t_4, only 1.1 nsec due to its high minority-carrier mobility, while n-type InSb has a t_4 of 29 nsec. The influence of the insulator on the value of t_4 is shown to be minimal in the case of Hg CdTe, where t_4 of 130 nsec has been calculated.

Table 12-2. Material Properties and Charge-Up Time for Selected Metal-Insulator-IR Semiconductors [12-4]

Semiconductor		InAs		InSb		Hg$_{0.8}$Cd$_{0.2}$Te	
Substrate Doping		n	n	p	n	SiO$_2$	TiO$_2$
Substrate Temperature	K	77	150	77		77	
Band gap energy, E_g	eV	0.41	0.39	0.23		0.09	
λ_c	μm	3.0	3.2	5.4		14	
Spectral band (λ_1 to λ_2)	μm	2.0 to 2.5		3.5 to 4.2		8 to 14	
Q_B(a) (300 K, λ_1 to λ_2)	Photons cm^{-2} sec^{-1}	1.8×10^{11}	1.8×10^{11}	1.6×10^{14}		5.9×10^{16}	
ΔQ_B(b)$/Q_B$		0.7%		0.3%		0.05%	
Semiconductor relative dielectric constant(c)		14.5		17.0		18.0	
Minority carrier lifetime, τ	sec	7.0×10^{-8}(d)	8.0×10^{-9}(d)	2.0×10^{-10}(e)	5.0×10^{-8}(e)	10^{-7}(c)	
Intrinsic carrier conc., n_i	cm^{-3}	6.8×10^{2}(f)	2.5×10^{10}(f)	6.4×10^{8}(f)	6.4×10^{8}(f)	2×10^{13}(c)	
Impurity conc., N	cm^{-3}	6.0×10^{16}	6.0×10^{16}	6.0×10^{15}	6.0×10^{15}	10^{15}	
Inversion layer mobility, μ_{eff}	cm^2 V^{-1} sec^{-1}	350.0(g)	280.0(g)	3.0×10^{4}(e)	3.0×10^{3}(e)	600(h)	
Initial surface potential, ϕ_{so} (i)	V	0.8(j)	0.8(j)	2.5(j)	2.5(j)	3.7(j)	4.9(k)
Final surface potential, ϕ_{sf}	V	0.42	0.38	0.21	0.21	0.05	0.05
Max. inversion region charge density, Q_{inv}	C cm^{-2}	5.5×10^{-8}	6.3×10^{-8}	1.4×10^{-7}	1.4×10^{-7}	1.6×10^{-7}	3.3×10^{-6}
Thermal charge up time, τ_{th}(l)	sec	2.7×10^{6}	1.8×10^{-2}	1×10^{-2}	2.1	5.2×10^{-5}	9×10^{-4}
Background charge up time, τ_B (m)	sec	1.7	1.9	5.9×10^{-3}	5.9×10^{-3}	1.8×10^{-5}	3.6×10^{-4}
Combined charge up time, τ_C	sec	1.7	1.8×10^{-2}	3.6×10^{-3}	5.8×10^{-3}	1.3×10^{-5}	2.5×10^{-4}
Charge transfer time, t_4(n)	sec	3.2×10^{-7}	2.2×10^{-7}	1.1×10^{-9}	2.9×10^{-8}	1.3×10^{-7}	1.3×10^{-7}

(a) FOV = 30°; (b) $\Delta T = 0.1$ K; (c) for comprehensive list of materials, parameters and references see Moss et al.; (d) see Mikhailova et al.; (e) see Hilsum and Rose-Innes; (f) calculated; (g) see Dixon; (h) see Tasch et al.; (i) $V_G - V_{FB} = 5$ V; (j) SiO$_2$, 10^3 Å, $K_I = 3.9$; (k) TiO$_2$, 10^3 Å, $K_I = 75$; (l) Nominal, $S_0 = 200$ cm sec^{-1}; (m) $\eta = 1$; (n) $L = 10$ μm, $\epsilon_x = 10^{-4}$.

Both the charge-up and charge transfer times are dependent not only on material properties but also on the device structure. Therefore, the numbers presented above and in Table 12-2 do not necessarily indicate ultimate performance, but rather some potential values to be expected for monolithic IRCCDs fabricated on these narrow-bandgap semiconductors.

Intrinsic IRCCD *Operation.* Most intrinsic IRCCD work has centered around InSb, possibly the best developed narrow-bandgap semiconductor. InSb MIS arrays organized in both CCDs and CIDs (Charge Injection Devices) have been reported [12-63, 12-64]. In the CID approach, the signal charge is confined to its cell during sensing (cf [12-65]) while the addressing is performed with X-Y coincident voltage electrodes. The interface-state density for InSb MIS devices is reported to be in the 10^{11} to 10^{12} cm^{-2} eV^{-1} range [12-63, 12-64]. At 77 K, a thermal charge-up time of around 0.5 sec due predominantly to bulk generation has been reported [12-63]. Thom, et al. [12-63] reported on the fabrication and operation of a four-bit InSb CCD with a four-phase overlapping gate-structure. This 50 μm gate device exhibited a transfer efficiency of η_x = 0.9 at a clock frequency of 5 kHz.

InSb CID investigation has been reported by Kim, et al. [12-64, 12-66]. Two-dimensional arrays with as many as 32 × 32 elements have been fabricated and mosaic operation has been demonstrated [12-67]. The unit cell dimensions for this array are 4 × 4 mils on 5 mil centers. A linear 32-element unit, operated at 77 K and clocked at 1 kHz, was reported [12-66] to be BLIP at a photon flux rate of $Q_B \geqslant 5 \times 10^{-12}$ cm^{-2} sec^{-1}. For a smaller Q_B, the linear array was shot-noise limited with approximately 200 noise-equivalent-carriers.

MIS devices using other narrow bandgap semiconductors have been reported for PbS [12-68], PbTe and Pb$_{1-x}$Sn$_x$Te [12-69].

In the near IR, the investigation of Ge CCDs has been reported [12-70]. The Ge bandgap of ~0.67 eV extends the near IR response to approximately 1.8 μm. At 200 K, the 10-cell device had a reported [12-70] transfer efficiency of η_x = 0.975 with a 20% fat-zero.

Schottky-Barrier IRCCD (*or* SBIRCCD). Another monolithic IRCCD approach is based on internal photoemission from metal-semiconductor Schottky-barrier arrays on a silicon substrate [12-71, 12-72]. The photoemission process is shown in Figure 12-31. Photons with energy $h\nu < E_g$ are absorbed in the metal resulting in the excitation of hot-carriers. Carriers with energy larger than the contact barrier, and with sufficient momentum in that direction, are emitted into the semiconductor. With radiation incident on the Schottky barrier through the semiconductor, the photogeneration process depends only on the absorption in the metal and emission over the barrier, resulting in uniform responsivity. The quantum efficiency of the Schottky-barrier detector decreases with decreasing photon energy and is quite small at the IR wavelengths of interest. For example, for a Au-p-Si diode, a quantum efficiency of 1% at 3 μm was reported [12-73]. However, the combined characteristic

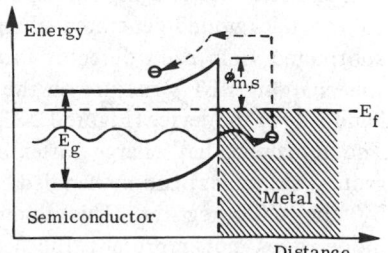

Fig. 12-31. Band diagram of Schottky-Barrier detector.

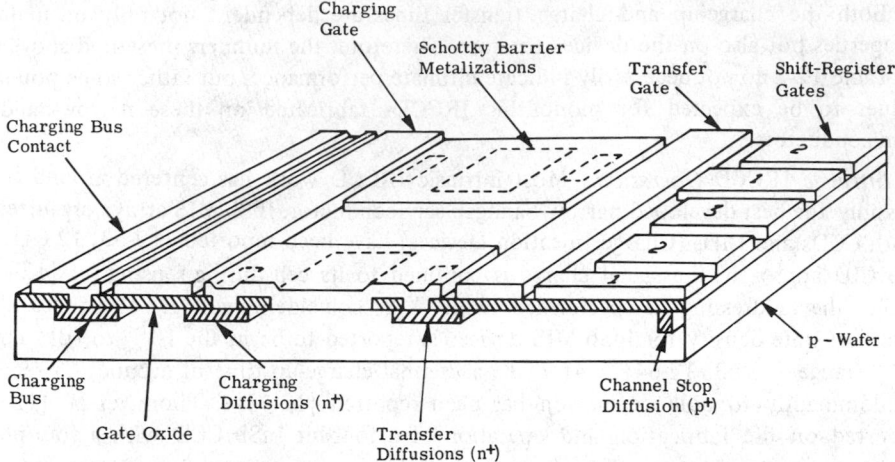

Fig. 12-32. Layout of SBIRCCD [12-72].

of high uniformity and low-responsivity can be potentially useful in large staring arrays where the longer integration time increases the detectivity.

The operation of a 64-cell Schottky-barrier linear-array coupled to a CCD shift-register has been reported [12-71, 12-72]. The layout of the device is shown in Figure 12-32. The transfer gate, which overlaps each individual transfer diffusion, couples the detectors to the CCD. Figure 12-33(a) shows a cross-section of an SBIRCCD. The transfer gate is turned on during the on-time of the phase gate, thus simultaneously injecting the signal from all the detectors into the shift register (Figure 12-33(b)). The IRCCD can also be operated in the skimming mode, in which a fixed amount of charge roughly corresponding to the background-generated charge is subtracted from each detector through the charging FET structure on the left-hand side of the device (Figure 12-33(c)). The remaining net charge (after background skimming) can be handled easily by the CCD register. This skimming mode does not represent full-fledged background subtraction where the background charge subtracted in each

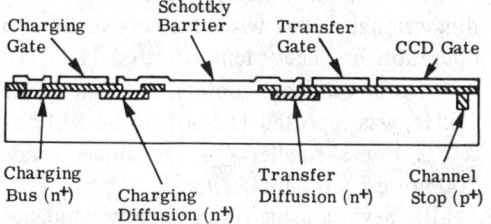

(a) Cross-section of device.

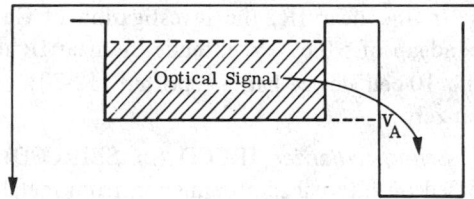

(b) Potential profile during normal (vidicon) operation.

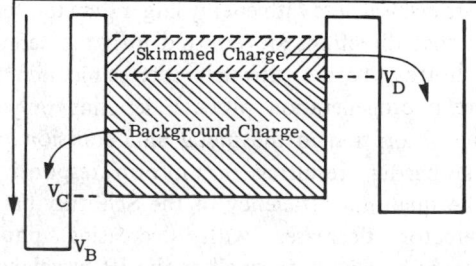

(c) Potential profile during background skimming mode.

Fig. 12-33. Electron potential profile of SBIRCCD during operation [12-72].

detector is that generated in the individual detector. However since the responsivity uniformity for large Schottky-barrier detector arrays will apparently be very high [12-72], background skimming could be sufficient.

12.3.3. Hybrid IRCCDs. Since the hybrid IRCCD involves basically the coupling of two fairly well developed technologies, it initially received more attention than monolithic IRCCDs. The detector materials investigated in various IRCCD configurations have included InSb [12-75, 12-76] for 3 to 5 μm operation, PbSnTe [12-58, 12-79] and HgCdTe [12-77, 12-78] for 8 to 14 μm operation.

The two basic methods for introducing IR signals from the detector array into the tapped CCD shift-register are the direct coupled input and the buffer interface input. (See Figure 12-34.) In the direct coupled input, the detector is connected either to the input diffusion (direct injection) or to the input gate (gate modulation) of the CCD tap. In the second method, the IR signal is first processed through a buffer stage before being introduced into the CCD.

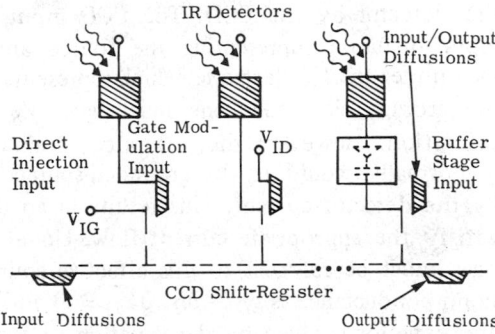

Fig.12-34. Hybrid IRCCD input coupling schemes

Direct Injection. In the direct injection IRCCD, the photogenerated charge is directly introduced into the CCD shift-register [12-54, 12-56]. Since this is in effect a dc coupled system, only IR detectors which exhibit relatively small dc currents (e.g., photovoltaic, extrinsic detectors) can be coupled to the CCD due to the latter's limited charge handling capacity.

Figure 12-35 illustrates the basic injection concept for a hybrid IRCCD consisting of a (Pb, Sn)Te photodiode and an n-channel CCD. The (Pb, Sn)Te/PbTe heterostructure [12-80] is particularly attractive for a hybrid IRCCD array since integration can be achieved in a relatively simple sandwich structure with full use of the detector active area thus avoiding interconnections. As shown in Figure 12-35, the IR diode is connected in parallel to a silicon diode which serves as an input tap to the CCD. The first MOS gate, V_G, serves

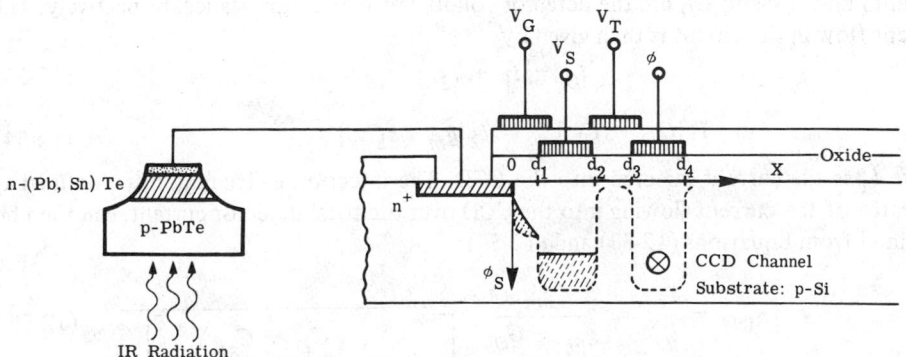

Fig. 12-35. Hybrid IRCCD: direct injection photodiode/CCD interface [12-79].

to reverse bias both diodes. While the charge accumulates under the storage gate, V_S, it is isolated from the CCD channel by the transfer gate, V_T. After one read time, t_R, V_T is biased into inversion and the accumulated charge is transferred into the CCD channel.

Injection Efficiency Model. A critical parameter of the direct injection IRCCD is the injection efficiency, defined as the ratio of the charge introduced into the CCD to the total charge generated by the detector [12-74, 12-79]. To evaluate the direct injection efficiency, one must first consider the equivalent load presented to the detector by the CCD. The CCD input stage is effectively a MOSFET with the input diffusion representing the source and the potential, ϕ_s, of the inverted surface under the V_S electrode, which represents the drain. To accumulate the charge under the storage electrode, one must have $V_S > V_G$, thus driving the MOSFET into saturation. However, the saturated drain current is not free to take the value it normally would in the grounded-source MOSFET configuration since it is driven by the detector current. This results in an increase in the source potential required to satisfy the appropriate current flow. Under these conditions, the gate and drain voltages must be referred to this effective source potential. In the saturation region, the drain conductance is $g_D = \partial i_D/\partial V_D \cong 0$ and the small signal input conductance seen by the detector is given by the variation of the drain current with changes in the gate-to-source voltage, V_{GS}, or the transconductance, g_m. However, as charge accumulates under V_S, the surface potential decreases to the point where the drain-to-source potential is lower than the gate-to-source potential, thus forcing the MOSFET out of the saturation regime. As the drain-to-source potential decreases further, the drain current will decrease accordingly, resulting in a potentially very useful self-limiting action. The CCD input capacitance is the parallel combination

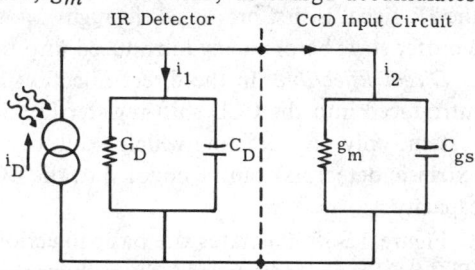

Fig. 12-36. Direct injection equivalent circuit model [12-79].

of the source diode capacitance, channel capacitance and gate-to-source capacitance with the latter being the dominating factor [12-81].

The equivalent circuit for the detector and CCD input circuit can thus be simply shown as in Figure 12-36, where i_D is the detector current (signal + background + dark current) and G_D and C_D are the detector conductance and capacitance, respectively. The current flow in the circuit is then given by

$$i_D = i_1 + i_2 \tag{12-53}$$

$$i_1 [G_D + sC_D]^{-1} = i_2 [g_m + sC_{gs}]^{-1} \tag{12-54}$$

where i_2 is the current injected into the CCD. The injection efficiency, η_{INJ}, defined as the ratio of the current flowing into the CCD over the total detector current, can then be obtained from Equations (12-53) and (12-54):

$$\eta_{INJ} = \frac{i_2}{i_D} = \frac{g_m}{g_m + G_D} \frac{\mathcal{P}}{\left[1 + \omega^2 \left(\dfrac{C_D + C_{gs}}{G_D + g_m} \right)^2 \right]} \tag{12-55}$$

where $\mathcal{P}$ is given by

$$\mathcal{P} = \sqrt{(1 + \omega^2)\, \frac{2C_{gs}^2(1 + G_D/g_m) + H/g_m^2}{(G_D + g_m)^2}} \tag{12-56}$$

and $H = C_{gs}^2 G_D^2 + g_m^2 C_D^2 + [(C_{gs} + C_D)^2\, C_{gs}^2\, \omega^2]$.

For reasonably good photodiode characteristics, the injection efficiency is well approximated in the submegahertz range by

$$\eta_{INJ} = \frac{\eta_{INJ}(0)}{\left[1 + \omega^2 \left(\dfrac{C_D + C_{gs}}{g_m + G_D}\right)^2\right]}, \quad \omega \leqslant 1 \text{ MHz} \tag{12-57}$$

where $\eta_{INJ}(0) = g_m/(g_m + G_D)$.

The transconductance in the saturation region is given by [12-2]

$$g_m = \left(\frac{2\,Z\,\mu_n C_I}{L}\, i_D\right)^{\frac{1}{2}} \tag{12-58}$$

where Z = width of the channel
$\quad L$ = length
$\quad \mu_n$ = electron mobility at the temperature of operation
$\quad C_I$ = insulator capacitance per unit area
$\quad i_D$ = drain current, or i_2

Therefore, since g_m is a function of the injected current, the injection efficiency depends both statically on g_m, as in Equation (12-55), as well as dynamically, as in Equation (12-58).

A good example is the direct injection of current from a (Pb, Sn)Te/PbTe diode sensitive over the entire 8 to 12 μm region. The background photon flux density rate is roughly 1×10^{17} photons cm^{-2} sec^{-1} which, for an optical area of 130 $\times$ 130 μm, results in a background current of ~2 μA. Typical detector parameters [12-80] at a reverse bias of 0.1 V are G_D = 200 μmho, C_D = 20 pF and a total detector current of i_D = 20 μA.

The CCD input parameters can be used within a certain range as design parameters for the optimization of the injection efficiency. Both the transconductance, g_m, and the capacitance, C_{gs}, are also a function of the dimensions of the input tap circuit. The transconductance is directly proportional to the aspect ratio of input channel, while the capacitance is a function of the overlap area between the source diffusion and the input gate as well as of the insulator thickness. Since the photodiode capacitance, C_D, is generally much larger than C_{gs}, the input-tap channel width can be increased to improve the transconductance without greatly affecting the total capacitance. This can be best accomplished in a "razorback" type CCD [12-82] where the dimensions of the input tap are not entirely dictated by the CCD cell size. In particular, the aspect ratio, Z/L, of the input tap channel can be varied considerably. The baseline parameters for the CCD input tap at 77 K are taken to be

$$g_m = 500 \ \mu mho$$
$$C_{gs} = 1 \ pF$$

$$Z = 2.54 \times 10^{-2} cm$$
$$L = 2.54 \times 10^{-3} cm$$
$$Z/L = 10$$
$$\mu_n(77 \ K) = 4 \times 10^4 \ cm^2 \ V^{-1} sec^{-1} \ (\text{See Reference [12-83].})$$
$$C_I = 2.3 \times 10^{-8} F \ cm^{-2}$$
$$x_I = 1500 \ \text{Å}$$

(12-59)

The injection efficiency versus frequency is plotted in Figure 12-37 for different values of the CCD input circuit parameters generated by varying the input-tap channel width. In this manner, both the aspect-ratio and the total area of the input channel are varied, thus affecting simultaneously the transconductance and the input capacitance. If the IRCCD is part of a serially-scanned focal plane operating at a TV compatible rate, the injection efficiency at, say, 6

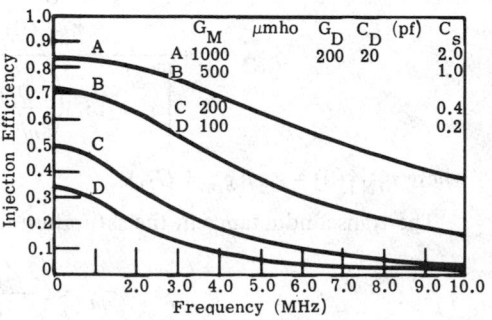

Fig. 12-37. Hybrid IRCCD: direct injection [12-79].

MHz is $\eta_{INJ} \approx 0.57$ for curve A and $\eta_{INJ} \cong 0.32$ for curve B. It therefore becomes advantageous to have a serial-parallel system where one can lower the frequency by the number of channels and thus approach as much as possible the maximum injection efficiency.

In the case of BLIP-limited detectors where the background current is the dominant detector current, the transconductance and thus the injection efficiency is related to the background photon flux [12-75]. In the 3 to 5 μm region, the background current is low enough to result in the subthreshold or weak inversion operation of the input MOSFET where the transconductance is given by [12-75]

$$g_m = \frac{q \, i_D}{kT}$$

(12-60)

where $i_D = \eta q A_D Q_B$
η = quantum efficiency
A_D = detector area
Q_B = background flux density rate
k = Boltzmann constant
T = temperature

Direct Injection IRCCD *Performance.* In the 3 to 5 μm region, direct-injection hybrid IRCCDs using InSb detectors have been reported. Milton and Hess [12-75] have discussed the operation of IRCCD and IRCID focal plane arrays employing InSb. French et al. [12-76], also working with InSb detectors, used a CCD shift-register on-the-focal plane multiplexer followed by demultiplexing and TDI in a CCD chip off the focal plane. (See Figure 12-38.) The InSb array had 32 elements, and both the array and the CCD chip were operated at a temperature of 77 K. The CCD multiplexer was clocked at 163 kHz while the TDI register was operated at 2105 Hz. A larger dwell time of 0.95 msec was possible due to a background subtraction circuit. Figure 12-39 shows the results obtained by French, et al. at a background of 3.3×10^{13} photons cm^{-2} sec^{-1}. A maximum $D^*(\lambda) \approx 8 \times 10^{12}$ cm Hz$^{1/2}$ W^{-1} for the TDI output of the array is shown at a signal frequency of around 30 Hz. The appropriate BLIP condition in this case is $\sqrt{32}$ times $D^*(\lambda)$(BLIP) of an individual detector or $\sim 10^{13}$ cm Hz$^{1/2}$W^{-1}. The performance of the IRCCD is very good, coming within 20% of the BLIP limit.

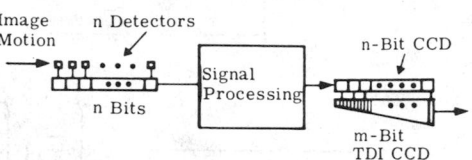

Fig. 12-38. Off-focal-plane TDI using CCDs [12-76].

Buffer Interface. The use of a buffer stage is sometimes required between the IR detector array and the CCD shift-register. The buffer stage generally performs signal conditioning: amplification, ac coupling, etc. The main advantage of a buffer stage is in the versatility inherent in the type of preprocessing it provides. There are serious drawbacks however: increased electrical power requirements, increased thermal power dissipation, increased device complexity, and, inevitably, increased cost.

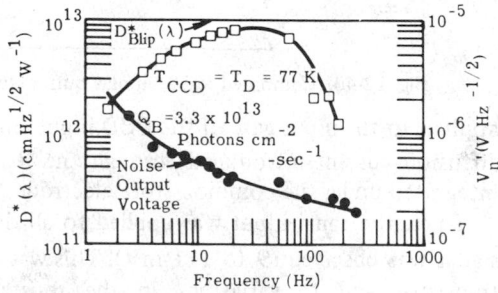

Fig. 12-39. D* and output noise voltage of InSb hybrid IRCCD as a function of frequency [12-76].

To date, published work on buffered-input hybrid IRCCDs has dealt with photoconductive (Hg, Cd) Te detectors operating in the 8 to 14 μm region, [12-77, 12-78]. Photoconductive (Hg,Cd)Te detectors typically have an impedance of 10 to 100 ohms [12-84]. By comparison, the CCD input impedance in the gate-modulation mode is for practical purposes infinite. If the signal is introduced through an input diffusion, the input impedance will be frequency dependent (See Section 12.3.3.) varying typically between 10^4 and 10^5 ohms over the 0.5 to 5 MHz range. It is, therefore, advantageous to include a transimpedance amplifier in the buffer stage. The amplifier and the buffer stage as a whole must contribute less noise than the detector, provide proper bandwidth and uniform gain in all channels and, if integrated with the CCD chip, typically dissipate no more than a few milliwatts per channel.

Buffer Input IRCCD *Performance.* Grant, et al. [12-78] reported the development of an integrated bipolar buffer/CCD chip for use with (Hg, Cd) Te detectors in a serial scan/TDI configuration. Each buffer channel consisted of a MOSFET bias resistor for the detector, a bipolar amplifier operating in the grounded base configuration, and a MOSFET load resistor for the amplifier. The CCD was a surface channel, tapered register for maintaining a constant charge-fill percentage. The buffer stage is shown schematically in Figure 12-40. The signal voltage is first amplified and then

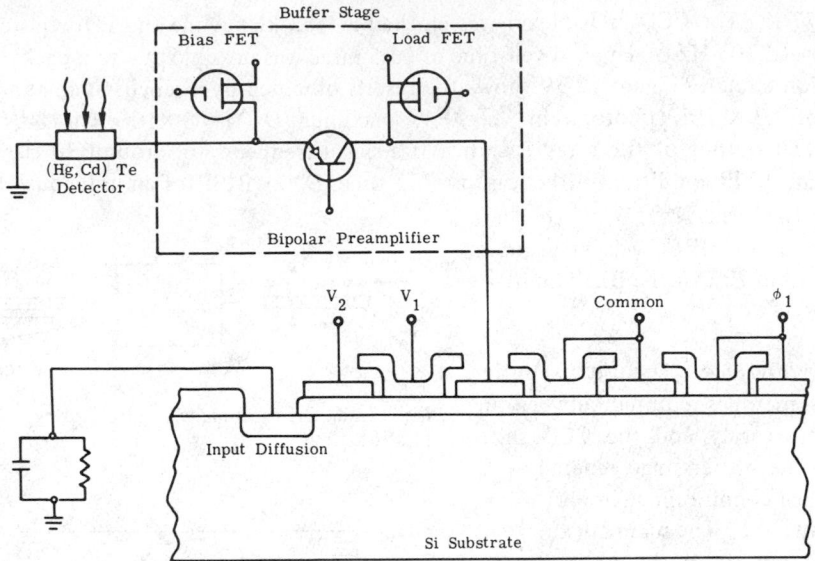

Fig. 12-40. Combined MOS/bipolar buffer stage for (Hg, Cd) Te hybrid IRCCD [12-78].

applied to the input gate of the CCD input circuit. The signal voltage modulates the input diffusion current introduced through the RC integrating network. The signal current is integrated under the common store electrode.

When the same input was applied to all the buffer stages, a large variation in output signal was observed (9 to 111 mV). This was attributed to offset variations in the bipolar transistor and to variations in the resistance of the MOSFET load resistor. The undistorted (<3% harmonic content) signal dynamic range of the buffer stage/CCD was measured at 77 K with a 10 kHz input signal and 1 MHz clock, and was reported to be in excess in 65 dB.

A CCD tapped shift-register used as a multiplexer for a (Hg,Cd)Te array in a parallel scan configuration has been reported by Emmons, et al. [12-77]. The shift register was a buried (ion implanted) n-channel device with forty cells, each connected to a floating diffusion input. (See Figure 12-41.) Each channel of the integrated buffer stage included a single-stage MOSFET preamplifier, an ac coupling capacitor, and a clamping MOSFET. (See Figure 12-42.) The preamplified signal was ac coupled to the input gate of the CCD input circuit through the series coupling capacitor. The input circuit was operated in the potential equilibration or "fill-and-spill" mode [12-85] with respect to the floating diffusion. The potential of the floating diffusion was a function of two factors: (a) the signal voltage, $V_{G1} = V_{in}$, on gate 1 minus its respective threshold, and (b) the reference voltage, $V_{G2} = V_{ref}$, applied to gate 2 minus its respective threshold. For two closely spaced electrodes the threshold voltage is expected to be practically identical such that the potential of the floating diffusion will be equal only to the difference, $\phi_{FD} = V_{in} - V_{ref}$ and thus be insensitive to threshold variations. In the variation on the "fill-and-spill" technique used by Emmons et al. [12-77], charge is introduced onto the floating diffusion by first pulsing the input diode into forward bias. Then gate 2 is pulsed to the reference voltage, thus isolating a packet of charge equal to the product of the floating diffusion potential and its capacitance, $Q_{in} = \phi_{FD} C_{FD}$. The process is terminated as the charge

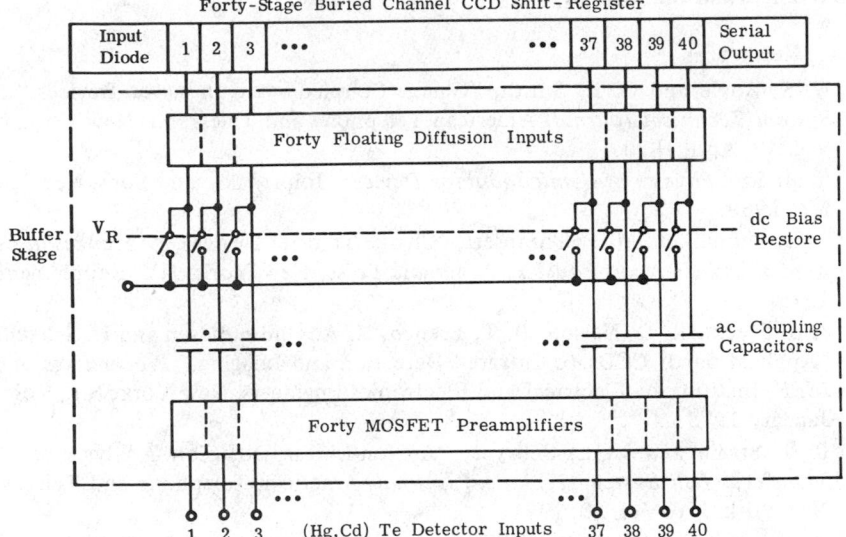

Fig. 12-41. Block diagram for buffered input hybrid IRCCD [12-77].

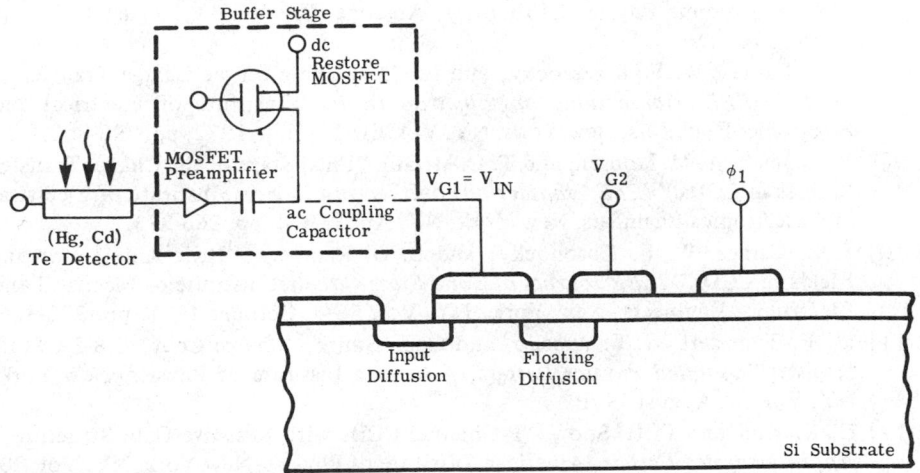

Fig. 12-42. An ac coupled MOSFET buffer stage for (Hg, Cd) Te hybrid IRCCD [12-77].

packet is transferred into the CCD channel and the input diffusion is returned to reverse bias, thus extracting the excess charge. The charge transfer efficiency of the CCD multiplexer was reported to be 0.99996 at 4 MHz resulting in a very low interchannel cross-talk, –44 dB in the worst case. At an input signal frequency of 100 kHz the buffer interface/CCD has a wide band (over the Nyquist bandwidth of 50 kHz) input noise of 220 μV. The linear signal dynamic-range was reported to be over 50 dB, while the dc rejection was better than 75 dB. The power dissipation of the MOSFET preamplifiers was reported to be only 25 μW per channel for a total power load of 1 mW.

12.4. References and Bibliography

12.4.1. References

[12-1] W. S. Boyle and G. E. Smith, "Charge Coupled Semiconductor Devices," *Bell System Technical Journal*, American Telephone and Telegraph, New York, NY, Vol. 49, April 1970.

[12-2] S. M. Sze, *Physics of Semiconductor Devices*, John Wiley and Sons, New York, NY, 1969.

[12-3] C. H. Sequin and M. F. Tompsett, "Charge Transfer Devices," *Advances in Electronics and Electron Physics*, Academic Press, New York, NY, Supplement 8, 1975.

[12-4] A. J. Steckl, R. D. Nelson, B. T. French, R. A. Gudmundsen and D. Schechter, "Application of CCDs to Infrared Detection and Imaging," *Proceedings of the IEEE*, Institute of Electrical and Electronics Engineers, New York, NY, Vol. 63, January 1975.

[12-5] R. J. Strain and N. L. Schryer, "A Non-Linear Diffusion Analysis of CCD Transfer," *Bell System Technical Journal*, American Telephone and Telegraph, New York, NY, Vol. 50, 1971.

[12-6] H. S. Lee and L. G. Heller, "Charge Control Method of CCD Transfer Analysis," *IEEE Transactions on Electron Devices*, Institute of Electrical and Electronics Engineers, New York, NY, Vol. ED-19, 1972, p. 1270.

[12-7] C. K. Kim and M. Lenzlinger, "Charge Transfer in CCDs," *Journal of Applied Physics*, Argonne Physics Laboratory, Argonne, IL, Vol. 42, August 1971, pp. 3586-3594.

[12-8] J. E. Carnes, W. F. Kosonocky, and E. G. Ramberg, "Free Charge Transfer in CCDs," *IEEE Transactions on Electron Devices*, Institute of Electrical and Electronics Engineers, New York, NY, Vol. ED-19, June 1972, pp. 798-808.

[12-9] Y. Daimon, A. M. Mohsen, and T. C. McGill, "Final Stage of the Charge Transfer Process in CCDs," *IEEE Transactions on Electron Devices*, Institute of Electrical and Electronics Engineers, New York, NY, April 1974, pp. 266-272.

[12-10] J. E. Carnes, W. F. Kosonocky, and E. G. Ramberg, "Drift Aiding Fringing Fields in CCDs," *IEEE Journal of Solid State Circuits*, Institute of Electrical and Electronics Engineers, New York, NY, Vol. SC-6, October 1971, pp. 322-326.

[12-11] M. F. Tompsett, G. F. Amelio, and G. E. Smith, "Charge Coupled 8-Bit Shift Register," *Applied Physics Letters*, American Institute of Physics, New York, NY, Vol. 17, August 1970.

[12-12] C. K. Kim, and E. H. Snow, "P-Channel CCDs with Resistive Gate Structure," *Applied Physics Letters*, American Institute of Physics, New York, NY, Vol. 20, 1972, p. 514.

[12-13] W. J. Bertram, A. M. Mohsen, F. J. Morris, D. A. Sealer, C. H. Sequin, and M. F. Tompsett, "A Three-Level Metallization Three-Phase CCD," *IEEE Transactions on Electron Devices*, Institute of Electrical and Electronics Engineers, New York, NY, Vol. ED-21, December 1974, p. 758.

[12-14] J. E. Carnes, W. F. Kosonocky, and P. A. Levine, "Measurements of Noise in CCDs," *RCA Review*, RCA Research and Engineering, RCA, Princeton, NJ, Vol. 34, December 1973, p. 553.

[12-15] M. F. Tompsett and E. J. Zimany, "Use of CCDs for Delaying Analog Signals," *IEEE Journal of Solid State Circuits*, Institute of Electrical and Electronics Engineers, New York, NY, Vol. SC-8, 1973, p. 151.

[12-16] C. H. Sequin, and A. M. Mohsen, "Linearity of Electrical Charge Injection into CCDs," *IEEE Journal of Solid State Circuits*, Institute of Electrical and Electronics Engineers, New York, NY, Vol. SC-10, 1975.

[12-17] J. E. Carnes, and W. F. Kosonocky, "Noise Sources in CCDs," *RCA Review*, RCA Research and Engineering, RCA, Princeton, NJ, Vol. 33, June 1972, p. 327.

[12-18] D. D. Wen, J. M. Early, C. K. Kim, and G. F. Amelio, "A Distributed Floating Gate Amplifier in CCDs," *Digest of Technical Papers*, ISSC Conference, Philadelphia, PA, Institute of Electrical and Electronics Engineers, New York, NY, February 1975, p. 24.

[12-19] G. F. Amelio, "The Impact of Large CCD Image Sensing Area Arrays," *Proceedings of the 1974 CCD Applications International Conference*, Edinburgh, Scotland, October 1974, University of Edinburgh, E118, 9JZ, Great Britain.

[12-20] M. F. Tompsett, "Charge Transfer Devices," *Journal of Vacuum Science Technology*, American Vacuum Society, American Institute of Physics, New York, NY, Vol. 9, 1972, p. 1166.

[12-21] D. A. Sealer, C. H. Sequin, and M. F. Tompsett, "High Resolution Charge Coupled Image Sensors," *1974 IEEE Intercon*, Institute of Electrical and Electronics Engineers, New York, NY, Session 2, March 1974.

[12-22] W. F. Kosonocky and J. E. Carnes, "Two-Phase CCDs with Overlapping Polysilicon and Aluminum Gates," *RCA Review*, RCA Research and Engineering, RCA, Princeton, NJ, Vol. 34, March 1973, p. 164.

[12-23] M. F. Tompsett, "The Quantitative Effects of Interface States on the Performance of CCDs," *IEEE Transactions on Electron Devices*, Institute of Electrical and Electronics Engineers, New York, NY, Vol. ED-20, January 1973, p. 45.

[12-24] D. F. Barbe, "Imaging Devices Using the Charge-Coupled Concept," *Proceedings of the IEEE*, Institute of the Electrical and Electronics Engineers, New York, NY, Vol. 63, January 1975, pp. 38-67.

[12-25] Y. T. Chan, B. T. French, and P. E. Green, "Extremely High Speed CCD Analog Delay Line," *Proceedings of the 1975 CCD Applications International Conference*, San Diego, CA, October 1975, Naval Oceans System Center, San Diego, CA, p. 389.

[12-26] D. F. Barbe, "Noise and Distortion Considerations in CCDs," *Electronics Letters*, Institute of Electrical Engineers, Stevenage, Herts, England, Vol. 8, April 1972, p. 209.

[12-27] A. M. Mohsen, M. F. Tompsett, and C. M. Sequin, "Noise Measurements in CCDs," *IEEE Transactions on Electron Devices*, Institute of Electrical and Electronics Engineers, New York, NY, Vol. ED-22, May 1975, p. 209.

[12-28] R. W. Brodersen, and S. P. Emmons, "Noise in Buried Channel CCDs," *IEEE Journal of Solid-State Circuits*, Institute of Electrical and Electronics Engineers, New York, NY, Vol. SC-11, February 1976, p. 147.

[12-29] A. J. Steckl, "Low Temperature Silicon CCD Operation," *Proceedings of the 1975 CCD Applications International Conference*, San Diego, CA, October 1975, Naval Oceans System Center, San Diego, CA, p. 383.

[12-30] L. J. M. Esser, "Peristaltic CCD," *Electronics Letters*, Institution of Electrical Engineers, Stevenage, Herts, England, Vol. 8, December 1972, p. 620.

[12-31] Y. T. Chan, Rockwell International, Anaheim, CA, Personal Communication, 1976.

[12-32] C. K. Kim, and R. H. Dyck, "Low Light Level Imaging with Buried Channel CCDs," *Proceedings of the IEEE*, Institute of Electrical and Electronics Engineers, New York, NY, Vol. 61, 1973, p. 1146.

[12-33] M. F. Tompsett, W. J. Bertram, D. A. Sealer, and C. H. Sequin, "Charge Coupling Improves its Image, Challenging Video Camera Tubes," *Electronics*, Vol. 46, No. 2, 1973, p. 162.

[12-34] C. H. Sequin, D. A. Sealer, W. J. Bertram, R. R. Buckley, F. J. Morris, T. A. Shankoff and M. F. Tompsett, "Charge Coupled Image Sensing Devices Using Three Levels of Polysilicon," *Digest of Technical Papers*, International Solid State Circuits Conference, Philadelphia, PA, 1974, Institute of Electrical and Electronics Engineers, New York, NY, 1974.

[12-35] D. F. Barbe and M. H. White, "A Tradeoff Analysis for CCD Area Imagers: Front-Side Illumination Interline Transfer vs. Back-Side Illuminated Frame Transfer," *Proceedings of the 1973 CCD Applications Conference*, Naval Oceans System Center, San Diego, CA, September 1973.

[12-36] R. L. Rodgers, "A 512 × 320 Element Silicon Imaging Device," *Digest of Technical Papers*, International Solid State Circuits Conference, Philadelphia, PA, 1975, Institute of Electrical and Electronics Engineers, New York, NY, 1975.

[12-37] J. M. Chambers, D. J. Sauer, and W. F. Kosonocky, "CCDs as Drum and Disc Equivalents," Western Electronics Show and Convention, Los Angeles, CA, 1974. See Roger Melen and Dennis Buss (eds.), *Charge-Coupled Devices: Technology and Applications*, IEEE Press, New York, NY, 1977.

[12-38] H. A. R. Wegener, "Appraisal of the Charge Transfer Technologies for Peripheral Memory Applications," *Proceedings of the 1973 CCD Applications International Conference*, San Diego, CA, 1973, Naval Oceans System Center, San Diego, CA, p. 43.

[12-39] W. E. Tchon, and J. S. T. Huang, "256 Bit Repeater Chained 2-Phase CCD Digital Shift Register," *Proceedings of the 1973 CCD Applications International Conference*, San Diego, CA, 1973, Naval Oceans Systems Center, San Diego, CA, p. 73.

[12-40] S. Chou, "Design of a 16 384-Bit Serial Charge Coupled Memory Device," *IEEE Transactions on Electron Devices*, Institute of Electrical and Electronics Engineers, New York, NY, Vol. ED-23, February 1976, p. 78.

[12-41] W. E. Tchon, B. R. Elmer, A. J. Benboer, S. Negishi, K. Hirabayashi, I. Nojima, and S. Kohama, "4096-Bit Serial Decoded Multiphase Serial-Parallel-Serial CCD Memory," *IEEE Transactions on Electron Devices*, Institute of Electrical and Electronics Engineers, New York, NY, Vol. ED-23, February 1976, p. 93.

[12-42] S. D. Rosenbaum, C. H. Chan, J. T. Caves, S. C. Poon, and R. W. Wallace, "A 16 384-Bit High Density CCD Memory," *IEEE Transactions on Electron Devices*, Institute of Electrical and Electronics Engineers, New York, NY, Vol. ED-23, February 1976, p. 101.

[12-43] A. M. Mohsen, M. F. Tompsett, E. N. Fuls, and E. J. Zimany, "A 16-K Bit Block Addressed Charge Coupled Memory Device," *IEEE Transactions on Electron Devices*, Institute of Electrical and Electronics Engineers, New York, NY, Vol. ED-23, February 1976, p. 108.

[12-44] A. M. Mohsen, R. W. Bower, E. M. Wilder, and D. M. Erb, "A 64-K Bit Block Addressed Charge-Coupled Memory," *IEEE Transactions on Electron Devices*, Institute of Electrical and Electronics Engineers, New York, NY, Vol. ED-23, February 1976, p. 117.

[12-45] C. A. T. Salama, "Two-Phase MNOS CCD," *Electronics Letters*, Institution of Electrical Engineers, Stevenage, Herts, England, Vol. 8, 1972, p. 21.

[12-46] Y. T. Chan, B. T. French, and R. A. Gudmundsen, "Charge Coupled Memory Device," *Applied Physics Letters*, American Institute of Physics, New York, NY, Vol. 22, 1973, p. 650.

[12-47] M. H. White, D. R. Lampe, and J. L. Fagan, "CCD and MNOS Devices for Programmable Analog Signal Processing and Digital Nonvolatile Memory," *Digest of Technical Papers*, International Electron Device Meeting, Washington, DC, 1973, Institute of Electrical and Electronics Engineers, New York, NY, p. 130.

[12-48] M. H. White, D. R. Lampe, J. L. Fagan, and D. A. Barth, "A Nonvolatile Charge-Addressed Memory Cell," *Digest of Technical Papers*, International Electron Device Meeting, Washington, DC, 1974, Institute of Electrical and Electronics Engineers, New York, NY, p. 115.

[12-49] D. D. Buss, and W. H. Bailey, "Applications of Charge Transfer Devices to Communication," *Proceedings of the 1973 CCD Applications International Conference*, San Diego, CA, October 1973, Naval Oceans System Center, San Diego, CA, p. 83.

[12-50] M. H. White, and D. R. Lampe, "CCD Analog Signal Processing," *Proceedings of the 1975 CCD Applications International Conference*, San Diego, CA, October 1975, Naval Oceans System Center, San Diego, CA, p. 189.

[12-51] H. J. Whitehouse, "Signal Processing with Charge Transfer Devices," *IEEE Transactions on Electron Devices*, Institute of Electrical and Electronics Engineers, New York, NY, Vol. ED-23, February 1976, p. 132.

[12-52] D. F. Barbe, and W. D. Baker, "Signal Processing Using the Charge Coupled Concept," *Microelectronics Journal*, Plenum Publishing, New York, NY, Vol. 7, December 1975, p. 36.

[12-53] R. A. Aguilera, A. J. Steckl and N. Sclar, "Low-Cost Arrays for the Detection of Infrared (LADIR)," Air Force Avionics Laboratory, Wright-Patterson Air Force Base, OH, Report No. AFAL-TR-76-13.

[12-54] A. J. Steckl, and T. Koehler, "Theoretical Analysis of Directly Coupled 8-12 μm Hybrid IRCCD Serial Scanning," *Proceedings of the 1973 CCD Applications International Conference*, San Diego, CA, September-October 1973, Naval Oceans System Center, San Diego, CA, pp. 247-258.

[12-55] W. H. Bailey, W. Eversole, J. Holmes, W. Arens, W. Hoover, J. McGhee, and R. Ridings, "CCD Applications to Synthetic Aperture Radar," *Proceedings of the 1975 CCD Applications International Conference*, San Diego, CA, October 1975, Naval Oceans System Center, San Diego, CA, p. 301.

[12-56] D. M. Erb, and K. Nummedal, "Buried Channel CCD for Infrared Applications," *Proceedings of the 1973 CCD Applications Conference*, San Diego, CA, September 1973, Naval Oceans System Center, San Diego, CA, pp. 157-167.

[12-57] K. Nummedal, J. C. Fraser, S. C. Su, R. Baron, and R. M. Finnila, "Extrinsic Silicon Monolithic Focal Plan Array Technology and Applications," *Proceedings 1975 CCD Applications International Conference*, San Diego, CA, October 1975, Naval Oceans System Center, San Diego, CA, p. 79.

[12-58] J. M. Tracy, A. M. Andrews, C. C. Wang, M. Ewbank, and J. T. Longo, "Hybrid Infrared Focal Plane Construction and Operation," Device Research Conference, Ithaca, NY, June 1977, *IEEE Transactions on Electron Devices*, Institute of Electrical and Electronics Engineers, New York, NY, Vol. ED-24, September 1977, p. 1194.

[12-59] R. D. Nelson, "Accumulation-Mode CCD," *Applied Physics Letters*, American Institute of Physics, New York, NY, Vol. 35, November 1974, pp. 568-570.

[12-60] J. C. Fraser, D. H. Alexander, R. M. Finnila, and S. C. Su, "An Extrinsic SiCCD for Detecting Infrared Radiation," *Digest of Technical Papers*, International Electron Device Meeting, Washington, DC, December 1973, Institute of Electrical and Electronics Engineers, New York, NY, pp. 442-445.

[12-61] P. W. Kruse, L. D. McGlauchlin, and R. B. McQuistan, *Elements of Infrared Technology*, John Wiley and Sons, New York, NY, 1962.

[12-62] R. K. Willardson, and A. C. Beer (eds.), *Semiconductors and Semimetals*, Academic Press, New York, NY, Vol. 5, 1970.

[12-63] R. D. Thom, R. E. Eck, J. D. Phillips, and J. B. Scorso, "InSb CCDs and other MIS Devices for Infrared Applications," *Proceedings 1975 CCD Applications Conference*, San Diego, CA, October 1975, Naval Oceans System Center, San Diego, CA, p. 31.

[12-64] J. C. Kim, "InSb MIS Technology and CID Devices," *Proceedings 1975 CCD Applications Conference*, San Diego, CA, October 1975, Naval Oceans System Center, San Diego, CA, p. 1.

[12-65] H. K. Burk and G. J. Michon, "Charge-Injection Imaging: Operating Techniques and Performance Characteristics," *Proceedings of the IEEE*, Institute of Electrical and Electronics Engineers, New York, NY, Vol. ED-23, February 1976, p. 189.

[12-66] J. C. Kim, J. M. Hooker, and P. E. Howard, "Noise Measurements in InSb CID Linear Arrays and Operation of Two Dimensional InSb CID Arrays," Device Research Conference, Salt Lake City, UT, June 22, 1976, *IEEE Transactions on Electron Devices*, Institute of Electrical and Electronics Engineers, New York, NY, Vol. ED-23, No. 11, November 1976, p. 1257.

[12-67] J. C. Kim, General Electric, Syracuse, NY, Private Communication, 1976.

[12-68] F. J. Leonberger, A. L. McWhorter, T. C. Harman, and C. Hurwitz, "PbS MIS Devices for Charge Coupled Infrared Imaging Applications," *IEEE Transactions on Electron Devices*, Institute of Electrical and Electronics Engineers, New York, NY, Vol. ED-21, 1974, p. 738.

[12-69] T. F. Tao, J. R. Ellis, L. Kost, and A. Doshier, "Feasibility Study of PbTe and $Pb_{0.76}Sn_{0.24}Te$ Infrared Charge Coupled Imager," *Proceedings of the CCD Applications Conference*, San Diego, CA, September-October 1973, Naval Oceans System Center, San Diego, CA, pp. 259-268.

[12-70] D. K. Schroeder, "A Two-Phase Germanium CCD," *Applied Physics Letters*, American Institute of Physics, New York, NY, Vol. 25, 1974, p. 747.

[12-71] E. S. Kohn, S. A. Roosild, F. D. Sheperd, and A. C. Yang, "Infrared Imaging with Monolithic, CCD Addressed Schottky Barrier Detector Arrays: Theoretical and Experimental Arrays," *Proceedings 1975 CCD Applications International Conference*, San Diego, CA, October 1975, Naval Oceans System Center, San Diego, CA, p. 59.

[12-72] E. S. Kohn, "A Charge Coupled Infrared Imaging Array with Schottky Barrier Detectors," *Proceedings of the IEEE*, Institute of Electrical and Electronics Engineers, New York, NY, Vol. ED-23, February 1976, p. 207.

[12-73] J. Cohen, J. Vilms, and R. Archer, "Investigation of Schottky Barriers for Optical Detection and Cathodic Emission," Air Force Cambridge Research Laboratory, L. G. Hanscomb Field, Bedford, MA, Report No. AFCRL-68-0651 and AFCRL-69-0287.

[12-74] A. J. Steckl, "Infrared CCDs," *Infrared Physics*, Pergamon Press, Elmsford, NY, Vol. 16, January 1976, pp. 65-73.

[12-75] A. F. Milton and M. R. Hess, "Series-Parallel Scan IRCID Focal Plane Array Concept," *Proceedings 1975 CCD Applications International Conference*, San Diego, CA, October 1975, Naval Oceans System Center, San Diego, CA, p. 71.

[12-76] D. E. French, J. A. Sekula, and J. M. Hartman, "Off-Focal-Plane Time Delay and Integration Using CCDs," *Digest of Technical Papers*, International Electron Device Meeting, Washington, DC, December 1974, Institute of Electrical and Electronics Engineers, New York, NY, pp. 437-441.

[12-77] S. P. Emmons, T. F. Cheek, J. T. Hall, P. W. Van Atta, and R. Balcerak, "A CCD Multiplexer with Forty AC Coupled Inputs," *Proceedings 1975 CCD Applications Conference*, San Diego, CA, October 1975, Naval Oceans System Center, San Diego, CA, pp. 43-52.

[12-78] W. Grant, R. Balcerak, P. Van Atta, and J. T. Hall, "Integrated CCD-Bipolar Structure for Focal Plane Processing of IR Signals," *Proceedings 1975 CCD Applications Conference*, San Diego, CA, October 1975, Naval Oceans System Center, San Diego, CA, pp. 53-58.

[12-79] A. J. Steckl, "Injection Efficiency in Hybrid IRCCDs," *Proceedings 1975 CCD Applications Conference*, San Diego, CA, October 1975, Naval Oceans System Center, San Diego, CA, pp. 85-91.

[12-80] A. M. Andrews, et al., "Backside-Illuminated $Pb_{1-x}Sn_xTe$ Heterojunction Photodiode," *Applied Physics Letters*, American Institute of Physics, New York, NY, Vol. 26, No. 8, 15 April 1975, p. 438.

[12-81] P. Richman, *MOS Field-Effect Transistors and Integrated Circuits*, John Wiley and Sons, New York, NY, 1973.

[12-82] J. Shott and R. Melen, "The Razorback CCD," *Digest*, 1975 ISSC Conference, Philadelphia, PA, Institute of Electrical and Electronics Engineers, New York, NY, February 1975.

[12-83] G. W. Ludwig and R. L. Watters, "Drift and Conductivity Mobility in Silicon," *Physical Review*, American Physical Society, New York, NY, Vol. 101, 15 March 1956, p. 1699.

[12-84] D. L. Long and J. L. Schmit, "Mercury-Cadmium Telluride and Closely Related Alloys," *Semiconductors and Semimetals*, R. K. Willardson and A. C. Beer (eds.), Academic Press, New York, NY, Vol. 5, 1970.

[12-85] S. P. Emmons, and D. D. Buss, "Techniques for Introducing a Low Noise Fat Zero in CCDs," Device Research Conference, Boulder, CO, June 1973, *IEEE Transactions on Electron Devices*, Institute of Electrical and Electronics Engineers, New York, NY, Vol. ED-20, December 1973, p. 1172.

12.4.2. Bibliography

Amelio, G. F., "The Impact of Large CCD Image Sensing Area Arrays," *Proceedings of the 1974 CCD Applications International Conference*, Edinburgh, Scotland, October 1974, University of Edinburgh, E118, 9JZ, Great Britain.

Andrews, A. M., et al., "Backside-Illuminated $Pb_{1-x}Sn_xTe$ Heterojunction Photodiode," *Applied Physics Letters*, American Institute of Physics, New York, NY, Vol. 26, No. 8, 15 April 1975, p. 438.

Aguilera, R. A., A. J. Steckl, and N. Sclar, "Low-Cost Array for the Detection of Infrared (LADIR)," Air Force Avionics Laboratory, Wright-Patterson Air Force Base, OH, Report No. AFAL-TR-76-13.

Bailey, W. H., D. D. Buss, L. R. Hite, and M. W. Whatley, "Radar Video Processing Using the CCD Chirp Z Transform," *Proceedings of the 1975 CCD Applications International Conference*, San Diego, CA, October 1975, Naval Oceans System Center, San Diego, CA.

Bailey, W. H., W. Eversole, J. Holmes, W. Arens, W. Hoover, J. McGhee, and R. Ridings, "CCD Applications to Synthetic Aperture Radar," *Proceedings of the 1975 CCD Applications International Conference*, San Diego, CA, October 1975, Naval Oceans System Center, San Diego, CA, p. 301.

Barbe, D. F., "Imaging Devices Using the Charge-Coupled Concept," *Proceedings of the IEEE*, Institute of Electrical and Electronics Engineers, New York, NY, Vol. 63, January 1975.

Barbe, D. F., "Noise and Distortion Considerations, in CCDs," *Electronics Letters*,

Institution of Electrical Engineers, Stevenage, Herts, England, Vol. 8, April 1972.

Barbe, D. F., and W. D. Baker, "Signal Processing Using the Charge Coupled Concept," *Microelectronics Journal*, Plenum Publishing, New York, NY, Vol. 7, December 1975, p. 36.

Barbe, D. F., and M. H. White, "A Tradeoff Analysis for CCD Area Imagers: Front-Side Illumination Interline Transfer vs. Back-Side Illuminated Frame Transfer," *Proceedings of the 1973 CCD Applications Conference*, Naval Oceans System Center, San Diego, CA, September 1973.

Berglund, C. N., and K. K. Thornber, "Incomplete Transfer in Charge-Transfer Devices," *IEEE Journal of Solid-State Circuits*, Institute of Electrical and Electronics Engineers, New York, NY, Vol. SC-8, April 1973.

Bertram, W. J., A. M. Mohsen, F. J. Morris, D. A. Sealer, C. H. Sequin, and M. F. Tompsett, "A Three-Level Metallization Three-Phase CCD," *IEEE Transactions on Electron Devices*, Institute of Electrical and Electronics Engineers, New York, NY, Vol. ED-21, December 1974, p. 758.

Boyle, W. S., and G. E. Smith, "Charge Coupled Semiconductor Devices," *Bell System Technical Journal*, American Telephone and Telegraph, New York, NY, Vol. 49, April 1970.

Brodersen, R. W., and S. P. Emmons, "Noise in Buried Channel CCDs," *IEEE Journal of Solid-State Circuits*, Institute of Electrical and Electronics Engineers, New York, NY, Vol. SC-11, February 1976.

Burke, H. K., and G. J. Michon, "Charge-Injection Imaging: Operating Techniques and Performance Characteristics," *Proceedings of the IEEE*, Institute of Electrical and Electronics Engineers, New York, NY, Vol. ED-23, February 1976.

Burstein, E., G. Pines, and N. Sclar, "Optical and Photoconductive Properties of Silicon and Germanium," *Photoconductive Conference*, R. G. Breckenridge, et al. (eds.), John Wiley and Sons, New York, NY, 1956.

Buss, D. D., and W. H. Bailey, "Applications of Charge Transfer Devices to Communication," *Proceedings of the 1973 CCD Applications International Conference*, San Diego, CA, October 1973, Naval Oceans System Center, San Diego, CA, p. 83.

Carnes, J. E., and W. F. Kosonocky, "Noise Sources in CCDs," *RCA Review*, RCA Research Engineering, RCA, Princeton, NJ, Vol. 33, June 1972, p. 327.

Carnes, J.E., W. F. Kosonocky, and P. A. Levine, "Measurements of Noise in CCDs," *RCA Review*, RCA Research and Engineering, RCA, Princeton, NJ, Vol. 34, December 1973, p. 553.

Carnes, J. E., W. F. Kosonocky, and E. G. Ramberg, "Drift Aiding Fringing Fields in CCDs," *IEEE Journal of Solid-State Circuits*, Institute of Electrical and Electronics Engineers, New York, NY, Vol. SC-6, October 1971, pp. 322-326.

Carnes, J. E., W. F. Kosonocky, and E. G. Ramberg, "Free Charge Transfer in CCDs," *IEEE Transactions on Electron Devices*, Institute of Electrical and Electronics Engineers, New York, NY, Vol. ED-19, June 1972, pp. 798-808.

Chambers, J. M., D. J. Sauer, and W. F. Kosonocky, "CCDs as Drum and Disc Equivalents," Western Electronics Show and Convention, Los Angeles, CA, 1974. See Roger Melen and Dennis Buss (eds.), *Charge-Coupled Devices: Technology and Applications*, IEEE Press, New York, NY, 1977.

Chan, Y. T., B. T. French, and P. E. Green, "Extremely High Speed CCD Analog Delay Line," *Proceedings of the 1975 CCD Applications International Conference*, San Diego, CA, October 1975, Naval Oceans System Center, San Diego, CA, p. 389.

Chan, Y. T., B. T. French, and R. A. Gudmundsen, "Charge Coupled Memory Device," *Applied Physics Letters*, American Institute of Physics, New York, NY, Vol. 22, 1973, p. 650.

Chou, S., "Design of a 16 384-Bit Serial Charge Coupled Memory Device," *IEEE*

Transactions on Electron Devices, Institute of Electrical and Electronics Engineers, New York, NY, Vol. ED-23, February 1976, p. 78.

Cohen, J., J. Vilms, and R. Archer, "Investigation of Schottky Barriers for Optical Detection and Cathodic Emission," Air Force Cambridge Research Laboratory, L. G. Hanscomb Field, Bedford, MA, Report No. AFCRL-68-0651 and AFCRL-69-0287.

Daimon, Y., A. M. Mohsen, and T. C. McGill, "Final Stage of the Charge Transfer Process in CCDs," *IEEE Transactions on Electron Devices,* Institute of Electrical and Electronics Engineers, New York, NY, Vol. ED-21, April 1974, pp. 266-272.

Dixon, J. R., "Anomalous Electrical Properties of P-type Indium Arsenide," *Journal of Applied Physics,* Argonne Physics Laboratory, Argonne, IL, Vol. 30, 1959.

Emmons, S. P., and D. D. Buss, "Techniques for Introducing a Low Noise Fat Zero in CCDs," Device Research Conference, Boulder, CO, June 1973, *IEEE Transactions on Electron Devices,* Institute of Electrical and Electronics Engineers, New York, NY, Vol. ED-20, December 1973, p. 1172.

Emmons, S. P., T. F. Cheek, J. T. Hall, P. W. Van Atta, and R. Balcerak, "A CCD Multiplexer with Forty AC Coupled Inputs," *Proceedings of the 1975 CCD Applications Conference,* San Diego, CA, October 1975, Naval Oceans System Center, San Diego, CA, pp. 43-52.

Engeler, W. E., J. J. Tiemann, and R. D. Baertsch, "Surface Charge Transport in a Multi-element Charge Transfer Structure," *Journal of Applied Physics,* Argonne Physics Laboratory, Argonne, IL, Vol. 43, 1972, p. 2277.

Erb, D. M., and K. Nummedal, "Buried Channel CCDs for Infrared Applications," *Proceedings of the 1973 CCD Applications Conference,* San Diego, CA, September 1973, Naval Oceans System Center, San Diego, CA, pp. 157-167.

Esser, L. J. M., "Peristaltic CCD," *Electronics Letters,* Institution of Electrical Engineers, Stevenage, Herts, England, Vol. 8, December 1972.

Fraser, J. C., D. H. Alexander, R. M. Finnila, and S. C. Su, "An Extrinsic SiCCD for Detecting Infrared Radiation," *Digest of Technical Papers,* International Electron Device Meeting, Washington, DC, December 1973, Institute of Electrical and Electronics Engineers, New York, NY, pp. 442-445.

French, D. W., J. A. Sekula, and J. M. Hartman, "Off-Focal-Plane Time Delay and Integration Using CCDs," *Digest of Technical Papers,* International Electron Device Meeting, Washington, DC, December 1974, Institute of Electrical and Electronics Engineers, New York, NY, pp. 437-441.

Grant, W., R. Balcerak, P. Van Atta, and J. T. Hall, "Integrated CCD-Bipolar Structure for Focal Plane Processing of IR Signals," *Proceedings of the 1975 CCD Applications Conference,* San Diego, CA, October 1975, Naval Oceans System Center, San Diego, CA, pp. 53-58.

Hilsum, C., and A. C. Rose-Innes, *Semiconducting III-V Compounds,* Pergamon Press, New York, NY, 1961.

Hudson, R. D., *Infrared System Engineering,* John Wiley and Sons, New York, NY, 1969.

Institute of Electrical and Electronics Engineers, New York, NY, *IEEE Journal of Solid State Circuits,* Special Issue on Charge Transfer Devices, Vol. SC-8, April 1973.

Institute of Electrical and Electronics Engineers, New York, NY, "Infrared Technology for Remote Sensing," Special Issue *Proceedings of the IEEE,* Vol. 63, January 1975.

Joyce, W. B., and W. J. Bertram, "Linearized Dispersion Relation and Green's Function for Discrete Charge Transfer Devices with Incomplete Transfer," *Bell System Technical Journal,* American Telephone and Telegraph, New York, NY, Vol. 50, 1971, p. 1741.

Kim, C. K., and R. H. Dyck, "Low Light Level Imaging with Buried Channel CCDs," *Proceedings of the IEEE*, Institute of Electrical and Electronics Engineers, New York, NY, Vol. 61, 1973, p. 1146.

Kim, C. K., and M. Lenzlinger, "Charge Transfer and on CCDs," *Journal of Applied Physics*, Argonne Physics Laboratories, Argonne, IL, Vol. 42, August 1971, pp. 3586-3594.

Kim, C. K., and E. H. Snow, "P-Channel CCDs with Resistive Gate Structure," *Applied Physics Letters*, American Institute of Physics, New York, NY, Vol. 20, 1972, p. 514.

Kim, J. C., "InSb MIS Structures for Infrared Imaging Devices," *Digest of Technical Papers*, International Electron Device Meeting, Washington, DC, December 1973, Institute of Electrical and Electronics Engineers, New York, NY.

Kim, J. C., "InSb MIS Technology and CID Devices," *Proceedings 1975 CCD Applications Conference*, San Diego, CA, October 1975, Naval Oceans System Center, San Diego, CA, p. 1.

Kim, J. C., J. M. Hooker, and P. E. Howard, "Noise Measurements in InSb CID Linear Arrays and Operation of Two Dimensional InSb CID Arrays," Device Research Conference, Salt Lake City, UT, June 22, 1976, *IEEE Transactions on Electron Devices*, Institute of Electrical and Electronics Engineers, New York, NY, Vol. ED-23, No. 11, November 1976, p. 1257.

Kohn, E. S., "A Charge Coupled Infrared Imaging Array with Schottky Barrier Detectors," *Proceedings of the IEEE*, Institute of Electrical and Electronics Engineers, New York, NY, Vol. ED-23, February 1976, p. 207.

Kohn, E. S., "An Infrared Sensitive Charge Coupled Imager," Device Research Conference, Santa Barbara, CA, June 1974.

Kohn, E. S., S. A. Roosild, F. D. Sheperd and A. C. Yang, "Infrared Imaging with Monolithic, CCD Addressed Schottky Barrier Detector Arrays: Theoretical and Experimental Arrays," *Proceedings 1975 CCD Applications International Conference*, San Diego, CA, October 1975, Naval Oceans System Center, San Diego, CA, p. 59.

Kosonocky, W. F., and J. E. Carnes, "Two-Phase CCDs with Overlapping Polysilicon and Aluminum Gates," *RCA Review*, RCA Research and Engineering, RCA, Princeton, NJ, Vol. 34, March 1973, p. 164.

Kruse, P. W., L. D. McGlauchlin, and R. B. McQuistan, *Elements of Infrared Technology*, John Wiley and Sons, New York, NY, 1962.

Lee, H. S., and L. G. Heller, "Charge Control Method of CCD Transfer Analysis," *IEEE Transactions on Electron Devices*, Institute of Electrical and Electronics Engineers, New York, NY, Vol. ED-19, 1972, p. 1270.

Leonberger, F. J., A. L. McWhorter, T. C. Harman, and C. E. Hurwitz, "PbS MIS Devices for Charge Coupled Infrared Imaging Applications," *IEEE Transactions on Electron Devices*, Institute of Electrical and Electronics Engineers, New York, NY, Vol. ED-21, 1974, p. 738.

Leonberger, F. J., A. L. McWhorter, T. C. Harman, and C. E. Hurwitz, "PbS MIS Devices for Charge Coupled Imaging Applications," Device Research Conference, Santa Barbara, CA, June 1974.

Leonberger, F. J., A. L. McWhorter, T. C. Harman, and C. E. Hurwitz, "PbS MIS Devices for Charge Coupled Infrared Imaging Applications," *IEEE Transactions, on Electron Devices*, Institute of Electrical and Electronics Engineers, New York, NY, Vol. ED-21, 1974, p. 738.

Levinstein, H., "Extrinsic Detectors," *Applied Optics*, Optical Society of America, Washington, DC, Vol. 4, June 1965.

Long, D., *Infrared Physics*, Pergamon Press, Elmsford, NY, Vol. 7, 1967.

Long, D. L., and J. L. Schmit, "Mercury-Cadmium Telluride and Closely Related Alloys," *Semiconductors and Semimetals*, R. K. Willardson and A. C. Beer (eds.), Academic Press, New York, NY, Vol. 5, 1970.

Ludwig, G. W. and R. L. Watters, "Drift and Conductivity Mobility in Silicon," *Physical Review*, American Physical Society, New York, NY, Vol. 101, 15 March 1956, p. 1699.

Michon, G. J., and H. K. Burke, "Charge Injection Imaging," *Digest of Technical Papers*, IEEE International Solid State Circuits Conference, Philadelphia, PA, Institute of Electrical and Electronics Engineers, New York, NY, February 1973.

Mikhailova, M. P., D. N. Nasledov, and S. V. Slobodchikov, "Temperature Dependence of Carrier Lifetimes in InAs," *Sov. Phys. Solid State*, Vol. 5, February 1974.

Milton, A. F. and M. R. Hess, "Series-Parallel Scan IRCID Focal Plane Array Concept," *Proceedings 1975 CCD Applications International Conference*, San Diego, CA, October 1975, Naval Oceans System Center, San Diego, CA, p. 71.

Mohsen, A. M., and M. F. Tompsett, "The Effects of Bulk Traps on the Performance of Bulk Channel CCDs," *IEEE Transactions on Electron Devices*, Institute of Electrical and Electronics Engineers, New York, NY, Vol. ED-21, November 1974.

Mohsen, A. M., M. F. Tompsett, and C. M. Sequin, "Noise Measurements in CCDs," *IEEE Transactions on Electron Devices*, Institute of Electrical and Electronics Engineers, New York, NY, Vol. ED-22, May 1975, p. 209.

Mohsen, A. M., R. W. Bower, E. M. Wilder, and D. M. Erb, "A 64-K Bit Block Addressed Charge-Coupled Memory," *IEEE Transactions on Electron Devices*, Institute of Electrical and Electronics Engineers, New York, NY, Vol. ED-23, February 1976, p. 117.

Mohsen, A. M., M. F. Tompsett, E. N. Fuls, and E. J. Zimany, "A 16-K Bit Block Addressed Charge Coupled Memory Device," *IEEE Transactions on Electron Devices*, Institute of Electrical and Electronics Engineers, New York, NY, Vol. ED-23, February 1976, p. 108.

Moss, T. S., G. J. Burrel, and B. Ellis, *Semiconductor Optoelectronics*, John Wiley and Sons, New York, NY, 1973.

Nelson, R. D., "Accumulation-Mode CCD," *Applied Physics Letters*, American Institute of Physics, New York, NY, Vol. 35, November 1974, pp. 568-570.

Nummedal, K., J. C. Fraser, S. C. Su, R. Baron, and R. M. Finnila, "Extrinsic Silicon Monolithic Focal Plane Array Technology and Application," *Proceedings 1975 CCD Applications International Conference*, San Diego, CA, October 1975, Naval Oceans System Center, San Diego, CA, p. 79.

Ovum Ltd., London, England, *International Abstracts on CCDs*, 1970-1974.

Pines, M., and R. Baron, "Characteristics of Indium Doped Silicon Infrared Detectors," *Proceedings of the 1974 International Electron Devices Meeting*, Washington, DC, December 1974, Institute of Electrical and Electronics Engineers, New York, NY.

Pines, M., D. Murphy, D. Alexander, R. Baron, M. Young, "Characteristics of Gallium Doped Silicon Infrared Detectors," *Proceedings of the 1975 International Electron Devices Meeting*, Washington, DC, December 1975, Institute of Electrical and Electronics Engineers, New York, NY.

Proceedings of the 1973 CCD Applications Conference, Naval Electronics Laboratory Center, San Diego, CA.

Proceedings of the 1975 CCD Applications Conference, University of Edinburgh, E118, 9JZ, Great Britain.

Putley, E. H., "Far Infrared Photoconductivity," *Physica Status Solidi*, Academic Press, New York, NY, Vol. 6, 1964.

Richman, P., *MOS Field-Effect Transistors and Integrated Circuits,* John Wiley and Sons, New York, NY, 1973.

Roberts, J. B. G., R. Eames, D. V. McCaughan, and R. F. Simons, "A Processor For Pulse-Doppler Radar," *IEEE Transactions on Electron Devices,* Institute of Electrical and Electronics Engineers, New York, NY, Vol. ED-23, February 1976.

Rodgers, R. L., "A 512 X 320 Element Silicon Imaging Device," *Digest of Technical Papers,* International Solid State Circuits Conference, Philadelphia, PA, 1975, Institute of Electrical and Electronics Engineers, New York, NY, 1975.

Rosenbaum, S. D., C. H. Chan, J. T. Caves, S. C. Poon, and R. W. Wallace, "A 16384-Bit High Density CCD Memory," *IEEE Transactions on Electron Devices,* Institute of Electrical and Electronics Engineers, New York, NY, Vol. ED-23, February 1976, p. 101.

Salama, C. A. T., "Two-Phase MNOS CCD," *Electronics Letters,* Institute of Electrical Engineers, Stevenage, Herts, England, Vol. 8, 1972, p 21.

Schroder, D. K., "A Two-Phase Germanium CCD," *Applied Physics Letters,* American Institute of Physics, New York, NY, Vol. 25, 1974, p. 747.

Sealer, D. A., C. H. Sequin, and M. F. Tompsett, "High Resolution Charge Coupled Image Sensors," *1974 IEEE Intercon,* Institute of Electrical and Electronics Engineers, New York, NY, Session 2, March 1974.

Sequin, C. H., and A. M. Mohsen, "Linearity of Electrical Charge Injection into CCDs," *IEEE Journal of Solid State Circuits,* Institute of Electrical and Electronics Engineers, New York, NY, Vol. SC-10, 1975.

Sequin, C. H., D. A. Sealer, W. J. Bertram, R. R. Buckley, F. J. Morris, T. A. Shankoff and M. F. Tompsett, "Charge Coupled Image Sensing Devices Using Three Levels of Polysilicon," *Digest of Technical Papers,* International Solid State Circuits Conference, Philadelphia, PA, 1974, Institute of Electrical and Electronics Engineers, New York, NY, 1974.

Sequin, C. H., and M. F. Tompsett, "Charge Transfer Devices," *Advances in Electronics and Electron Physics,* Academic Press, New York, NY, Supplement 8, 1975.

Shepard, F. D., and A. C. Yang, "Silicon Schottky Retinas for Infrared Imaging," *Digest of Technical Papers,* International Electron Device Meeting, Washington, DC, December 1973, Institute of Electrical and Electronics Engineers, New York, NY, pp. 310-313.

Shott, J., and R. Melen, "The Razorback CCD," *Digest,* 1975 ISSC Conference, Philadelphia, PA, Institute of Electrical and Electronics Engineers, New York, NY, February 1975..

Soref, R. A., "Extrinsic IR Photoconductivity of S_i Doped with B, Al, Ga, P, As or Sb," *Journal of Applied Physics,* Argonne Physics Laboratory, Argonne, IL, Vol. 38, December 1967.

Steckl, A. J., "Infrared CCDs," *Infrared Physics,* Pergamon Press, Elmsford, NY, Vol. 16, January 1976, pp. 65-73.

Steckl, A. J., "Injection Efficiency in Hybrid IR CCDs." *Proceedings of the 1975 CCD Applications Conference,* San Diego, CA, October 1975, Naval Oceans System Center, San Diego, CA, pp. 85-91.

Steckl, A. J., "Low Temperature Silicon CCD Operation," *Proceedings of the 1975 CCD Applications International Conference,* San Diego, CA, October 1975, Naval Oceans Systems Center, San Diego, CA.

Steckl, A. J., and T. Koehler, "Theoretical Analysis of Directly Coupled 8-12 μm Hybrid IRCCD Serial Scanning," *Proceedings of the 1973 CCD Applications International Conference,* San Diego, CA, September 1973, Naval Oceans System Center, San Diego, CA. pp. 247-258.

Steckl, A. J., R. D. Nelson, B. T. French, R. A. Gudmundsen, and D. Schechter, "Application of CCDs to Infrared Detection and Imaging," *Proceedings of the IEEE*, Institute of Electrical and Electronics Engineers, New York, NY, Vol. 63, January 1975.

Strain, R. J., and N. L. Schryer, "A Non-Linear Diffusion Analysis of CCD Transfer,"*Bell System Technical Journal*, American Telephone and Telegraph, New York, NY, Vol. 50, 1971.

Sze, S. M., *Physics of Semiconductor Devices*, John Wiley and Sons, New York, NY, 1969.

Tao, T. F., J. R. Ellis, L. Kost, and A. Doshier, "Feasibility Study of PbTe and $Pb_{0.76}Sn_{0.24}Te$ Infrared Charge Coupled Imager," *Proceedings of the CCD Applications Conference*, San Diego, CA, September 1973, Naval Oceans System Center, San Diego, CA, pp. 259-268.

Tasch, A. F., R. A. Chapman, and B. H. Breazeale, "Field-Effect Measurements of the HgCdTe Surface," *Journal of Applied Physics*, Argonne Physics Laboratories, Argonne, IL, Vol. 41, September 1970.

Tchon, W. E., and J. S. T. Huang, "256 Bit Repeater Chained 2-Phase CCD Digital Shift Register," *Proceedings of the 1973 CCD Applications International Conference*, San Diego, CA, 1973, Naval Oceans System Center, San Diego, CA, p. 73.

Tchon, W. E., B. R. Elmer, A. J. Denboer, S. Negishi, K. Hirabayaski, I. Nojima, and S. Kohyama, "4096-Bit Serial Decoded Multiphase Serial-Parallel-Serial CCD Memory," *IEEE Transactions on Electron Devices*, Institute of Electrical and Electronics Engineering, New York, NY, Vol. ED-23, February 1976, p. 93.

Thom, R. D., R. E. Eck, J. D. Phillips, and J. B. Scorso, "InSb CCDs snd other MIS Devices for Infrared Applications," *Proceedings of the 1975 CCD Applications Conference*, San Diego, CA, October 1975, Naval Oceans System Center, San Diego, CA, p. 31.

Tompsett, M. F., "Charge Transfer Devices," *Journal of Vacuum Science Technology*, American Vacuum Society, American Institute of Physics, New York, NY, Vol. 9, 1972, p. 1166.

Tompsett, M. F., "The Quantitative Effects of Interface States on the Performance of CCDs," *IEEE Transaction on Electron Devices*, Institute of Electrical and Electronics Engineers, New York, NY, Vol. ED-20, January 1973, p. 45.

Tompsett, M. F., and E. J. Zimany, "Use of CCDs for Delaying Analog Signals," *IEEE Journal of Solid State Circuits*, Institute of Electrical and Electronics Engineers, New York, NY, Vol. SC-8, 1973, p. 151.

Tompsett, M. F., G. F. Amelio, and G. E. Smith, "Charge Coupled 8-Bit Shift Register," *Applied Physics Letters*, American Institute of Physics, New York, NY, Vol. 17, August 1970.

Tompsett, M. F., W. J. Bertram, D. A. Sealer, and C. H. Sequin, "Charge Coupling Improves Its Image, Challenging Video Camera Tubes," *Electronics*, Vol. 46, No. 2, 1973, p. 162.

Tracy, J. M., A. M. Andrews, C. C. Wang, M. Ewbank, and J. T. Longo, "Hybrid Infrared Focal Plane Construction and Operation," Device Research Conference, Ithaca, NY, June 1977, *IEEE Transactions on Electron Devices*, Institute of Electrical and Electronics Engineers, New York, NY, Vol. ED-24, September 1977, p. 1194.

Wegener, H. H. R., "Appraisal of Charge Transfer Technologies for Peripheral Memory Applications," *Proceedings of the 1973 CCD Applications International Conference*, San Diego, CA, 1973, Naval Oceans System Center, San Diego, CA, p. 43.

Wen, D. D., "Design and Operation of a Floating Gate Amplifier," *IEEE Journal of Solid State Circuits*, Institute of Electrical and Electronics Engineers, New York, NY, Vol.SC-9, 1974.

Wen, D. D., J. M. Early, C. K. Kim, and G. F. Amelio, "A Distributed Floating Gate Amplifier in CCDs," *Digest of Technical Papers,* ISSC Conference, Philadelphia, PA, Institute of Electrical and Electronics Engineers, New York, NY, February 1975, p. 24.

White, M. H., and D. R. Lampe "CCD Analog Signal Processing," *Proceedings of the 1975 CCD Applications International Conference,* San Diego, CA, October 1975, Naval Oceans System Center, San Diego, CA, p. 189.

White, M. H., D. R. Lampe, and J. L. Fagan, "CCD and MNOS Devices for Programmable Analog Signal Processing and Digital Nonvolatile Memory," *Digest of Technical Papers,* International Electron Device Meeting, Washington, DC, 1973, Institute of Electrical and Electronics Engineers, New York, NY, p. 130.

White, M. H., D. R. Lampe, J. L. Fagan, and D. A. Barth, "A Nonvolatile Charge-Addressed Memory Cell," *Digest of Technical Papers,* International Electron Device Meeting, Washington, DC, 1974, Institute of Electrical and Electronics Engineers, New York, NY, p. 115.

Whitehouse, H. J. "Signal Processing with Charge Transfer Devices," *IEEE Transactions on Electron Devices,* Institute of Electrical and Electronics Engineers, New York, NY, Vol. ED-23, February 1976, p. 132.

Willardson, R. K., and A. C. Beer (eds.), *Semiconductors and Semimetals,* Academic Press, New York, NY, Vol. 5, 1970.

Chapter 13

IMAGING TUBES

James A. Hall
Westinghouse Advanced Technology Laboratory

CONTENTS

13. Imaging Tubes

13.1. Introduction

13.1.1. Symbols, Nomenclature, and Units.
Table 13-1 lists the symbols, nomenclature and units used in this chapter.

Table 13-1. Symbols, Nomenclature and Units

Symbols	Nomenclature	Units
A	Area	m^2
C	Contrast	—
C_I	Contrast in image at sensor	—
CTF	Contrast transfer function, relative response to bar chart inputs as a function of spatial frequency	—
CCD	Charge-coupled device	—
C_F	Feedback capacitance in an operational amplifier	F
C_G	First amplifier input capacitance	F
C_o^*	Active gate-to-channel capacitance in a MOS field-effect transistor	F
C_R	Vidicon retina shunt capacitance at signal terminal	F
C_r	Vidicon retina storage capacitance	F
C_S	Shunt capacitance at vidicon signal electrode terminal, including wiring and preamplifier input capacitance	F
C_w	Wiring capacitance	F
c	Material density	$kg\ m^{-3}$
c_P	Specific heat	$J\ kg^{-1}\ K^{-1}$
c'	Volume specific heat	$J\ m^{-3}\ K^{-1}$
$D(f)$	MTF of a pyroelectric retina	—
d	Retina thickness	m
E	Electric field	$V\ m^{-1}$
E_B	Irradiance in image due to background	$W\ m^{-2}$
E_S	Irradiance in image from scene	$W\ m^{-2}$
E_T	Irradiance in image due to target	$W\ m^{-2}$
$F/\#$	Focal ratio or F number of an optical system	—
f	Signal frequency; also, spatial frequency	Hz; cycles mm^{-1}
f_h	Horizontal spatial frequency	Cycles mm^{-1}
G	Voltage gain of an amplifier	—
G_o	Voltage gain at midband or low signal frequencies	—
g_m	Transconductance	mho

Table 13-1. Symbols, Nomenclature and Units (*Continued*)

Symbols	Nomenclature	Units
g'_m	Transconductance of a field effect transistor under saturation conditions (high drain-to-source voltage)	mho
H^2	Space-charge smoothing factor	—
I_B	Electron beam current	A
I_d	Dark current; current flowing through the target of a visible-light-sensitive vidicon when irradiance is zero	A
I_n	Noise current, always rms	A
$I_n(\omega)$	Noise current spectral density	$A\,Hz^{1/2}$
I_p	Pedestal current, in a pyroelectric vidicon	A
I_S	Signal current, usually peak-to-peak	A
K	Relative dielectric constant	—
k	Boltzmann constant	—
l_c	Channel length in a field effect transistor	m
M	Exitance (of scene or object being viewed)	$W\,m^{-2}$
MTF	Modulation transfer function, the magnitude of the optical transfer function	—
m	Linear magnification of an optical system	—
P	Pyroelectric coefficient	$C\,m^{-2}\,K^{-1}$
PEV	Pyroelectric vidicon	—
P-I-N	p-type-intrinsic-n-type, a specific solid state diode structure	—
Q_B	Background photon exitance	$photons\,sec^{-1}\,cm^{-2}$
q	Charge of the electron	C
q_n	rms noise charge per Nyquist sample, number of electrons	—
$\Re$	Responsivity	$V\,W^{-1}$
R_B	Effective resistance of the scanning electron beam in a vidicon	ohms
R_{eff}	Effective load resistance for cascode output	ohms
R_F	Feedback resistance, in an operational amplifier	ohms
R_L	Vidicon load resistor	ohms
R_P	Parallel resistance of R_F and R_L in an operational preamplifier	ohms
SNR_D	Display signal-to-noise ratio	—
T	Temperature	K
T_B	Effective electron temperature of the scanning electron beam in a vidicon	K
TGS	Triglycine sulfate, a pyroelectric material	—
T_R	Retina temperature	K
t_e	Integration time of the eye	sec

Table 13-1. Symbols, Nomenclature and Units (*Continued*)

Symbols	Nomenclature	Units
t_f	Frame time	sec
t_h	Active line scan time, in a television scanning system	sec
V_{en}	Envelope noise voltage: apparent peak-to-peak noise envelope amplitude when the video signal is viewed on an oscilloscope (to be deprecated)	V
V_{GS}	Operating gate-to-source voltage	V
V_N	Noise voltage, usually rms	V
V_o	Output voltage	V
V_{PO}	Pinch-off voltage in a junction field effect transistor	V
V_R	Retina electrode bias voltage	V
V_T	Threshold gate-to-source voltage	V
V_S	Signal voltage, usually peak-to-peak	V
Z_T	Operational amplifier transfer impedance, volts out/ampere in	ohms
α	Aspect ratio of a rectangular image, usually width-to-height	–
α_c	Signal reading factor in the chopped mode	–
$\alpha(I_p, t_f, f)$	Signal-reading factor	–
Δf_v	Video amplifier frequency bandwidth	Hz
ΔT_m	Change in temperature from one frame to the next	K
δ	Secondary electron emission ratio	–
ϵ	Generic for emissivity	–
ϵ_n	rms noise voltage spectral density of an amplifier	$V\,Hz^{-1/2}$
ϵ_o	Permittivity of free space	$F\,m^{-1}$
λ	Radiation wavelength	μm
μ_n^*	Effective carrier mobility in the surface channel	$m^2\,V^{-1}\,sec^{-1}$
ρ	Resistivity of a material; also, sensor material density	$ohm\,m; kg\,m^{-3}$
σ	Stefan Boltzmann constant	$W\,m^{-2}\,K^{-4}$
τ	Time constant	sec
τ_A	Internal amplifier time constant	sec
τ_a	Transmittance of the atmosphere	–
τ_f	Transmittance of the face plate	–
τ_m	Electron transmission fraction of a mesh electrode	–
τ_o	Transmission of optics	–
Φ	Power (flux)	W
ω	Signal frequency	$rad\,sec^{-1}$

13.1.2. Definitions. The term Plumbicon® is a trademark of N. V. Philips of Holland for a camera tube like a vidicon with a lead oxide or lead oxisulfide retina. In this chapter, the term *vidicon* is used for a television camera tube in which a single multilayer structure, the *retina*, absorbs the image irradiance, converts it to a charge pattern, integrates the charge information for a frame time, and is the signal output electrode when the charge image is neutralized by a scanning electron beam. The retina of a vidicon is the radiation-sensing electrode structure in a vidicon, often called a target electrode in the literature. The *frame time* is defined as the time to scan a complete television image once.

13.2. Television Sensing Systems

Television sensing systems using electron-beam scanning have long shown promise of providing compact, economical, infrared imaging systems. Several limitations have become apparent, however. These include the requirement of a uniformity of the sensing layer of greater than 99%, or a means of compensating for the nonuniformities, as well as improvement in the maximum signal-current capacity. If the latter is not accomplished, the background radiation can cause the imaging tube to saturate unless the tube and optics are cooled and the spectral response is limited to the region under 3 μm.

Two approaches have been tried for improving these tubes: an attempt to improve broadcast television sensor techniques; and an attempt to change the sensor concept so that the background can be cancelled or ignored.

13.3. Vidicons—Common Characteristics

A vidicon, shown in Figure 13-1, was originally an RCA trademark for a simple television camera tube in which a thin noncrystalline layer of high-resistivity photo-conductive material, antimony trisulphide, served as both the photon sensor and the charge-storage element which was scanned by a beam of low-velocity electrons to develop a video signal. The video signal output was taken from the transparent backing electrode of the photoconductive layer.

Today, the term vidicon has become generic and is used to identify any television camera tube in which the same thin-layer structure serves as radiation sensor, signal

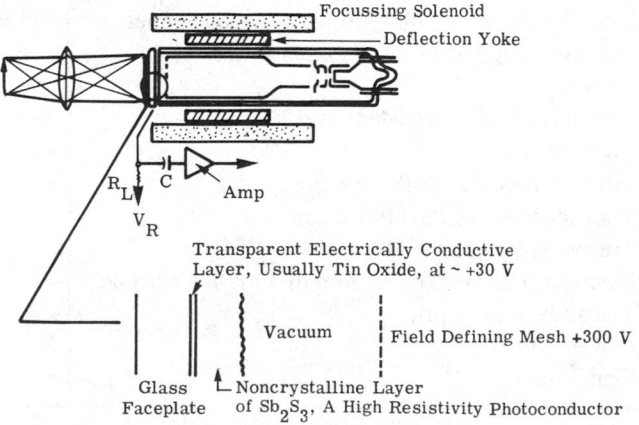

Fig. 13-1. Original vidicon. In the original vidicon the irradiance image varied the resistivity of a high resistivity photoconductor to form a charge image which was erased by a scanning electron beam to generate a video signal.

output electrode, and the image-charge storage member which is scanned by a beam of low-velocity electrons. This is true whether the structure is crystalline or amorphous, whether the radiation is absorbed in either direct excitation of charge carrier pairs by photon absorption or through an induced change in the temperature of the layer, and whether the temperature change alters the resistivity of the layer structure or produces a pyroelectric potential pattern across it. The principle, infrared-sensitive, television camera tubes which have been used in infrared television camera systems have all been vidicons.

13.3.1. Electron-Beam Reading Characteristics. All vidicons share certain common mechanisms. The first is the interaction of the scanning electron beam and the thin layer retina or target* in producing a video signal. As shown in Figure 13-2, the retina can be modeled as a set of capacitors, each corresponding to a small area of the surface. One side of each capacitor is tied together by the transparent conductive layer, to form a common signal electrode, and connected through the input load resistor of the video amplifier circuit to a retina bias voltage, V_R, some volts or tens of volts positive with respect to the thermionic cathode of the scanning beam electron gun from which the scanning electrons originate. The electron beam is scanned over the free side of the retina, making momentary contact with each small elemental area or group of such areas in turn, and beam electrons are deposited on the free surface, charging it more negatively toward cathode potential. If no image irradiation is falling on the sensing sur-

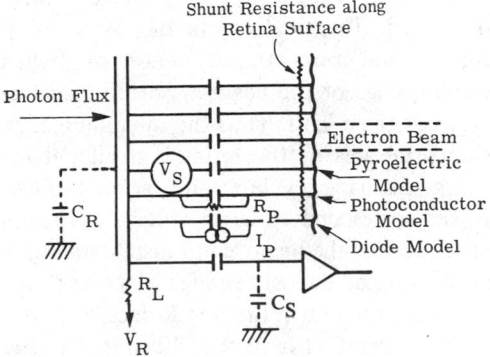

Fig. 13-2. Vidicon sensing layer (retina). A vidicon sensing layer modeled as an array of elemental capacitors, each corresponding to a small area of the layer. Radiation in the image acts to charge the free surface more positively. Beam electrons are deposited on the free surface to erase this positive charge, constituting a video signal current.

face, the charging action stops when the surface is slightly more negative than the cathode. Then the beam electrons can no longer reach the sensing surface, but are accelerated instead to the mesh or back into the electron gun. The retina bias potential then appears across every elemental capacitor.

The action of the incident photon flux of an image on the retina is to discharge the capacitors, most rapidly where the image irradiance is largest, so that the free or scanned surface becomes more positive than the reference cathode potential of the electron gun. When the scanning electron beam returns to a given elemental area after a scanning frame time, electrons are deposited on that area to charge the surface back toward cathode potential. These electrons constitute a displacement current as the elemental capacitor is recharged, and this current flows through the parallel combination of the load resistor and the shunt capacitance from the retina electrode and video amplifier input to ground, causing a small negative signal voltage at the video amplifier input. By scanning the electron beam in a raster pattern over the retina area, one can generate a video signal for

---

*In this chapter, "retina" will be used for this member of the vidicon. "Target" will be reserved for a radiating part of the scene unless there is no possibility of confusion.

each elemental image area on the retina in turn. The amount of charge deposited by the beam on each retina area (and hence the magnitude of the video signal from that area) is a function of the exposure of that area in the preceding frame time.

In all vidicons, the action of the electron beam is to erase the charge from each elemental retina area in turn as it returns that area essentially to cathode potential. The maximum signal current which can be developed from each such area is therefore limited first by the maximum current which can be supplied by the electron beam (on the order of 2 μA). As suggested in Figure 13-3, the beam cross-sectional area is often the appropriate elemental area for this model, and the beam usually sweeps across this elemental area in a small fraction of a microsecond. The first electrons in the leading edge of the beam find a positively charged surface at a given point and many of them are deposited. The electrons in the center of the beam therefore approach a less positive surface and proportionally fewer of them are deposited. Finally, the electrons in the trailing edge of the beam approach a surface at or below cathode potential, and few or none of them land. Thus the maximum signal current which can be developed from any elemental area of the retina is smaller than the total current in the electron beam. The charge-discharge cycle at the retina surface is illustrated in Figure 13-3. Larger photon fluxes correspond to larger voltage excursions of the scanned surface. But the recharging operation by the beam is not instantaneous, and a signal current limit is reached when the beam cannot deposit enough electrons on each elemental area as it scans across each retina area to return the free surface of that area to gun cathode potential.

The operation is further illustrated in Figure 13-4. A stream of electrons approaches normally a planar sensing surface, whose potential is initially negative with respect to the electron gun cathode. The electrons stop before reaching the surface and are accelerated back to the field mesh electrode. Hence, no current flows to the surface. As the retina surface is made more positive, the most energetic beam electrons can reach the surface and many are accepted, while some are scattered elastically by the fields around the surface atoms and return to the field mesh. Except for this scattered fraction, which is somewhat energy dependent, the curve of current collected by the surface versus surface potential is at first a measure of the energy spread in the electron beam, increasing approximately exponentially because of the roughly Maxwellian energy distribution among the beam electrons. The current to the surface actually relates only to the electron

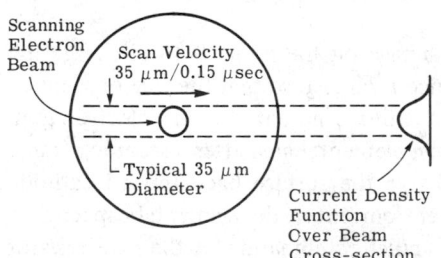

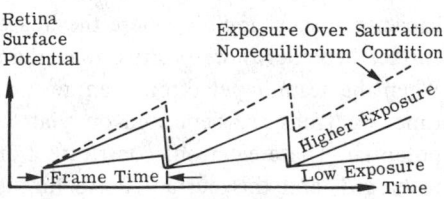

(a) The electron beam has an approximately Gaussian current density distribution with an effective diameter larger than the raster line spacing.

(b) The beam recharges the target surface nearly to the gun cathode potential for exposures below saturation.

Fig. 13-3. Electron beam action.

velocity components perpendicular to the retina surface, and the process of forming the electrons into a beam broadens that velocity spread significantly above that produced by the hot thermionic cathode.

As the retina surface is made still more positive, the energy of the electrons reaching the surface increases above that required for the onset of secondary electron emission. Further increases in surface potential excite electrons within the retina to higher energies, increasing the probability of emission, until as many electrons leave the surface as reach it and the retina charging current falls to zero. Any further positive excursion now causes a net removal of electrons from the surface and acts to charge it positively toward mesh potential. To avoid this unstable operating mode and to assure that the retina remains cathode potential stabilized, one normally operates the retina surface only to the left of potential "A" in Figure 13-4. Although potential A can be made fairly high (on the order of a few tens of volts) by forming the surface for a low secondary-emission coefficient, this "cross-over" effect limits maximum signal currents.

In fact, since the electron beam deposits electrons on the surface of a highly resistive structure, the potential of any elemental area of retina surface changes rapidly as the beam is scanned across it. At large signal levels, the surface potential may start at point A, but moves rapidly along the curve to the left. At point A, the signal current may be a major fraction of the beam current, but the fraction rapidly becomes smaller as the charging cycle continues until the signal current deposited becomes very small indeed. The scanning electron beam has often been modeled as a resistor in series with a commutating switch connecting each elemental retina area in turn to ground. But this

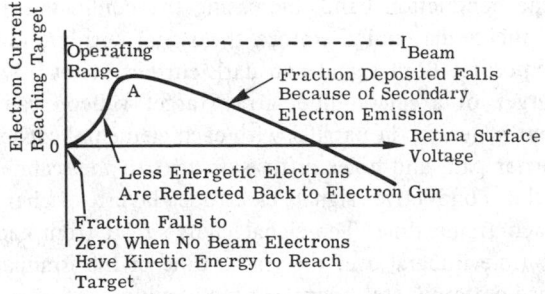

(a) Fraction is dependent on instantaneous surface potential. Read and erase action moves surface potential along curve toward zero.

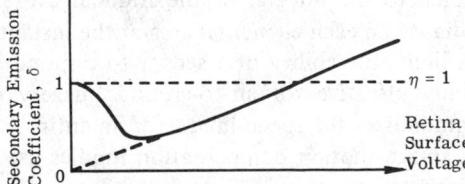

(b) Secondary emission curve appears as mirror image of beam acceptance curve, but actually falls to 0, 0.

Fig. 13-4. Fraction of beam electrons deposited on retina.

model predicts a charging current which varies linearly with surface voltage above ground. In fact, the variation is exponential for small voltage excursions. If the scene radiation is blocked, the electron beam may require several or many scan cycles to completely charge the retina surface to its dark reference potential, and a significant time may also be required at low signal levels to establish a new equilibrium when scene content changes. This effect, known as beam lag, appears as a low-amplitude, persistent, after-image, and significantly affects the operation of pyroelectric vidicons where the signal voltage is developed by changes in scene content. The broad axial-velocity spread of the beam electrons prevents reading the full signal amplitude from the retina in such a non-equilibrium operating mode.

The maximum signal amplitude is also set by the quotient obtained from the product of the effective charge-storage capacitance and maximum surface voltage excursion divided by the frame time. Many television cameras are operated with interlaced scanning, but with electron beam cross-sections large enough so that the retina is essentially completely recharged in each interlaced field, twice in a frame time. Then the proper divisor is the field time.

13.3.2. Full-Frame Integration: A Second Common Characteristic of the Vidicon Family. Full-frame integration is implied in the capacitor mosaic model of the vidicon retina of Figure 13-2, which shows several ways in which the image radiation can act to change the retina surface potential.

The sensing layer in the traditional Sb_2S_3 vidicon behaves like a photoconductor with no junctions or blocking contacts. The model is a photosensitive resistance in parallel with each elemental capacitor. Absorbed photons from the image excite carriers across the bandgap into the conduction band, increasing the conductance. The signal charge transferred through the retina, and therefore the signal current, vary as the 0.7 power of the integrated exposure. Both signal and dark current increase with applied voltage.

A Plumbicon target or a silicon-diode-array target vidicon can be modeled by a photon-driven current generator in parallel with each elemental capacitor. Each absorbed photon creates a carrier pair, and holes diffuse or drift to the scanned side of the target, and electrons to the conductive signal electrode layer, discharging the elemental capacitors during each frame time. The signal charge read from each area is therefore proportional to the time integral over the frame time of the irradiance times that area, whether the irradiance pattern is stationary or time varying.

For temperature-sensitive retinas, like those of the pyroelectric vidicon or the thermicon, the change in temperature of the retina in a frame time, and hence the signal voltage, is also proportional to the integral of the irradiance over the frame time, not to the instantaneous irradiance on each elemental area at the instant it is scanned.

Full-frame integration limits the ability of a sensor to respond to moving scenes. For example, a vidicon is fully effective for air-to-ground imagery only if optical image motion compensation immobilizes the scene image for an entire frame time to a fraction of a resolution element. Image motion compensation implies use of intermittent exposure, then intermittent readout, to provide a uniform exposure of all parts of the sensitive surface.

13.3.3. Amplifier Noise. All vidicon cameras made to date have the property that amplifier noise is dominant in determining system signal-to-noise ratio so that a vidicon is never completely background limited. Scanning beam current, or retina storage capacitance and the allowable surface voltage-swing, act to limit signal current to approximately 2 μA for 30 frame-per-second scanning; typical video-amplifier input-noise-currents for a 4.5 MHz bandwidth are about 2 nA. Noise in a 2 μA signal in 4.5 MHz is about 1.7 $\times 10^{-9}$ A, about equal to amplifier noise. Thus, achieving background-limited performance would require either an increase in maximum signal handling capability or a decrease in amplifier noise. Because video preamplifier performance so often determines vidicon performance at low image irradiances or low image contrasts, the design of video preamplifiers is treated separately in Section 13.4.

13.3.4. Nonuniformity and Shading. Even if a vidicon is exposed uniformly over the entire image area, the video signal is generally not constant over the image. Fine-textured variations are often caused by variations in point-to-point response, variations in the retina thickness, by the "graininess" of polycrystalline or noncrystalline retina layers, or by the "electron shadows" cast by the field mesh. These effects are often called "coherent noise," since they repeat in successive scanning frames and have a similar effect on image visibility as true time varying noise. A better name is "disturbance." Larger area nonuniformities may appear as shading, or darker or lighter areas in the image. These are usually caused by local variations in photo sensor properties.

Most vidicons show darker shading toward the corners of the image. These may be caused by off-axis vignetting in the lens, and by non-normal beam landing. After the scanning beam in the vidicon has been deflected, fields are provided to curve the electron paths in reverse so that all electrons approach the retina surface normally. Beam normalization is seldom complete, however, and the electrons scanning the corners of the image normally have some transverse motion whose energy is subtracted from that associated with motion parallel to the axis. The electrons therefore cannot charge the surface in the corners to cathode potential, but rather to an equilibrium potential a few volts more positive. Thus the bias across the retina is larger in the center, smaller toward the edge. This variation should have little effect for the Plumbicon (described in Section 13.5.2), but shading is significant for a photoconductive vidicon, reaching 10 to 30% of the center signal amplitude.

The P-I-N structure of the Plumbicon, where signal and dark current are both saturated at the normal operating target bias, should be essentially immune to beam-landing shading, but may have objectionable granularity for low-contrast imaging. Diode-array targets should also be relatively satisfactory for infrared imaging, although carrier collection efficiency may vary with bias to cause some second-order shading effects.

13.4. Video Preamplifiers for Vidicon Cameras

A vidicon has a signal output electrode which can be electrically represented as a current generator shunted by a capacitance to ground. Signal-current levels range from a few nanoamperes to about one microampere. The passband extends from low frequencies to several megahertz. The noise produced by the camera-tube load circuit and by the first video preamplifier is of the same order as the signal at low image irradiance values or low contrasts. Thus, amplifier noise is a major factor limiting vidicon television camera performance, and video preamplifier design has been extensively studied.

13.4.1. Characteristics of the Signal. As shown in Figure 13-5, the scanning beam of a vidicon is blanked during the horizontal and vertical retrace intervals so that no beam electrons reach the target to erase the stored charged image. The signal current from the camera tube varies during each active scan line but falls to zero during the beam blanking interval. The time scale of the final television signal in Figure 13-6 is for U.S. broadcast television, and must be modified for other television applications. To reproduce the average value of a television picture, one could use an amplifier whose passband extended below the frame frequency of the television system, but there are important advantages instead in limiting the low-frequency response to the scanning-line frequency and clamping the output signal to a reference voltage before beginning each scanning line. As shown in Figure 13-6, a clamp placed after one or more high-gain, low-noise, ac-coupled amplifier stages, and keyed on during the blanking interval, fixes the zero level so that synchronizing and display blanking signals can be added independently of picture content. As shown, the inserted display blanking level is usually "blacker" than the zero signal-current level to ensure that retrace lines will not be visible in remote display devices. The difference between the display blanking and zero signal-current levels is known as the pedestal.

Use of a line frequency, zero-level, keyed clamp has another advantage, since excess low-frequency noise from input preamplifier stages, and power supply or other drift phenomena with frequencies below f_h, are effectively removed.

The maximum video frequency is that needed to reproduce the finest detail required in the direction of the scanning lines. In the television field, as shown in Figure 13-7, it is standard practice to express the spatial frequency of image structure in terms of "television lines" or half cycles of video information per pattern height, even though the

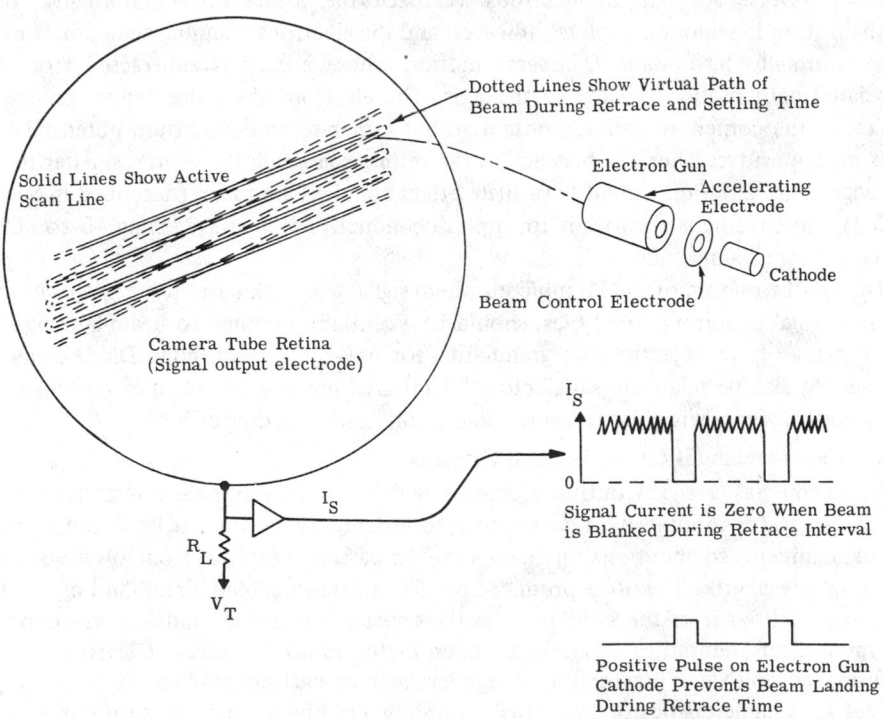

Fig. 13-5. Electron beam blanked during retrace. The stored charge is read only during active scan times. Zero signal current during retrace provides a convenient reference level.

spatial frequency is measured along the scan lines, i.e., in the direction of picture width. The maximum video frequency (bandwidth) required is

$$\Delta f_v = \frac{1}{2} \alpha f_h t_h^{-1}, \text{Hz} \qquad (13\text{-}1)$$

where α = the ratio of width to height

t_h = the active line scan time, sec

f_h = horizontal spatial frequency, half cycles per pattern height

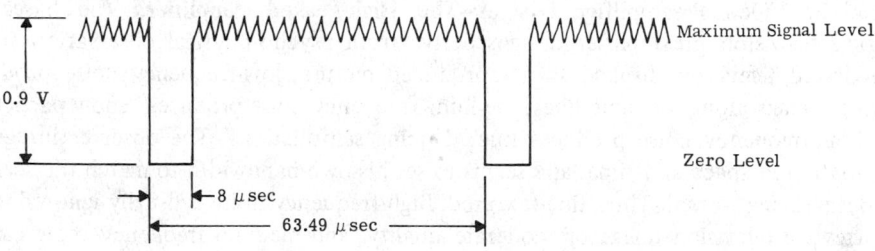

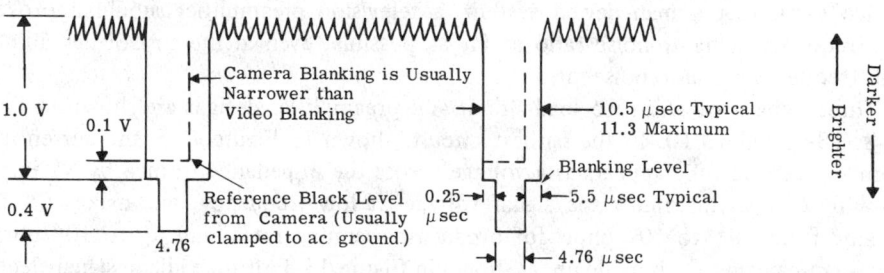

Fig. 13-6. Television signal. It includes added blanking and sync pulses referenced to the black zero current level of the video signal from the sensor.

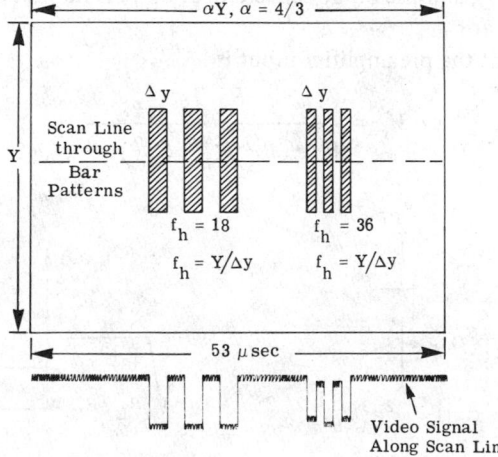

Fig. 13-7. Spatial frequency response. It is measured horizontally along a scanning line but is always referred to the pattern height, $f_h = Y/\Delta y$.

For the 4 × 3 aspect ratio scanning pattern of U.S. commercial television, with t_h = 53 × 10^{-6} sec, the required maximum video frequency is

$$\Delta f_v = \frac{4}{6} \frac{f_h}{53 \times 10^{-6}} \text{ , Hz} \tag{13-2}$$

For a "350 TV line" system, Δf_v is 4.41 MHz, approximately the video signal bandwidth allocated for U.S. broadcast television.

13.4.2. Video Preamplifier Designs–The High-Peaked Amplifier. The bases for existing television preamplifier designs derive from psycho-physical observations for a man-viewed television display. In a reproduced picture, low-frequency noise produces moving streaks along scanning lines, medium frequency noise produces "snow particles," and high-frequency noise produces fine, dancing scintillations. The observer integrates information in space and time, and seems to set his own bandwidth to match the scale of the detail being viewed. Thus, fine-textured, high-frequency noise is usually ignored when one views a television image of moderate quality, and medium frequency snow can be tolerated by moving away from the display and visualizing only gross image features, but noise becomes subjectively more objectionable as the frequency decreases. Thus if a choice exists, for a man-viewed system, a television preamplifier should improve the low-frequency signal-to-noise ratio as far as possible, even at the cost of degrading the high frequency signal-to-noise ratio.

Three typical camera-tube load-circuit and preamplifier designs are shown in Figures 13-8, 13-9, and 13-10. In the earliest circuit, shown in Figure 13-8, the current signal from the camera tube appears as a voltage across the impedance formed by R_L in parallel with C_S. To maximize the signal voltage at low frequency, one makes R_L large, usually from 10^4 to 10^6 ohms for broadcast scanning rates, and up to 10^8 ohms in slow-scan cameras. C_S is made up, as shown in Figure 13-8, of the vidicon signal-electrode capacitance to its surroundings, C_R, ranging from 5 to 30 pF depending on tube size and structural details, the capacitance of interconnections, C_W (which can be kept to 2 or 3 pF by careful design), and the input capacitance, C_G, of the first amplifier stage, shown as a field-effect transistor.

The signal voltage at the preamplifier input is

$$V_s = \frac{R_L I_S}{(1 + \omega^2 R_L^2 C_S^2)^{1/2}} \tag{13-3}$$

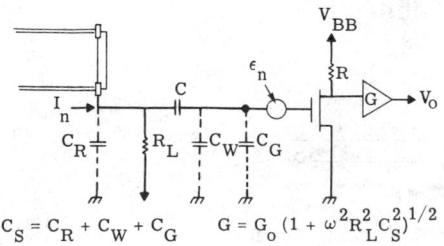

$$C_S = C_R + C_W + C_G \qquad G = G_o (1 + \omega^2 R_L^2 C_S^2)^{1/2}$$

(a) High peaking is used in later stages to provide flat signal frequency response.

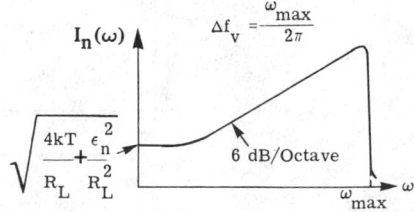

(b) First stage transistor noise is high peaked but has low visibility to the human observer.

Fig. 13-8. Classic TV camera analog preamplifier.

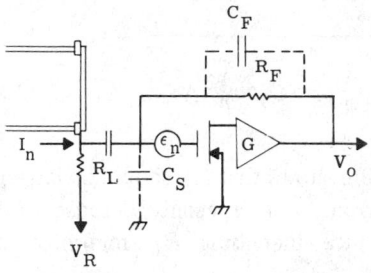

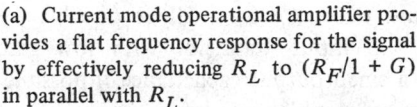

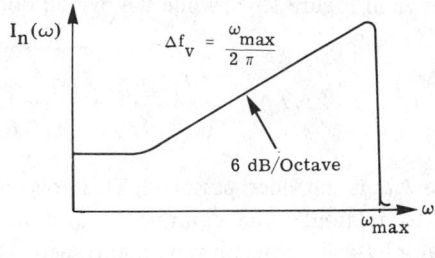

(a) Current mode operational amplifier provides a flat frequency response for the signal by effectively reducing R_L to $(R_F/1 + G)$ in parallel with R_L.

(b) First stage transistor noise is high peaked by falling impedance from signal electrode to ground caused by C_S. The zero-frequency value is $\sqrt{(4kT/R_P)(\epsilon_n^2/R_P^2)}$ where $1/R_P = 1/R_L + (1/R_F)$.

Fig. 13-9. Current mode operational amplifier.

To provide a signal gain independent of frequency through the video passband, one can design a later stage of the preamplifier for a rising gain versus frequency characteristic so that the amplifier gain is

$$G = G_o \sqrt{1 + \omega^2 R_L^2 C_S^2} \qquad (13\text{-}4)$$

The noise sources are the Johnson-Nyquist noise of R_L and the noise of the first and following stages of the preamplifier. For a field-effect-transistor input-stage, transistor noise can be represented by a white-noise voltage-generator internal to the transistor in series with the gate*. Factors determining this noise voltage are discussed below.

Since the camera tube acts like a current source, noise contributions are described as equivalent noise currents at the amplifier input terminal. The noise power spectral density is proportional to

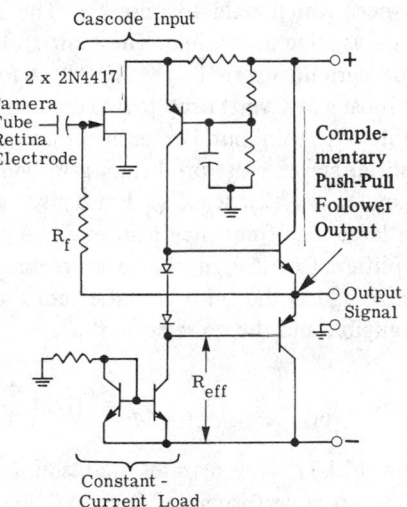

Fig. 13-10. Complementary op-amp with JFET cascode input stage (parallel devices are not required).

$$\frac{d(I_n^2)}{df} = \frac{4kT}{R_L} + \frac{\epsilon_n^2}{R_L^2}(1 + 4\pi^2 f^2 R_L^2 C_S^2) \qquad (13\text{-}5)$$

---

*The white-noise model is accurate only for junction field effect transistors in systems where a scanning line frequency keyed-clamp removes excess low-frequency (1/f) noise. Both MOSFETs and JFETs exhibit a white-noise spectrum above f_c, the (1/f) corner. For JFETs, f_c is approximately 10^4 Hz, well below the scanning line frequency of broadcast television. For slow-scan applications, a more complex analysis is required to include the (1/f) noise contribution.

as shown in Figure 13-8, while the overall noise current is

$$I_n = \sqrt{\left(\frac{4kT}{R_L} + \frac{\epsilon_n^2}{R_L^2}\right)\Delta f_v + \frac{4}{3}\,\pi^2\,C_S^2\,\epsilon_n^2\,\Delta f_v^3} \qquad (13\text{-}6)$$

where Δf_v is the video passband. This noise expression shows that R_L should be large not only to minimize the equivalent input noise current, but especially because of the psycho-physical observations mentioned above since increasing R_L minimizes the low-frequency noise contributions.

To minimize I_n, an equipment designer should make C_S as small as possible by minimizing interconnect capacity, by choosing an input transistor with low gate capacitance, and by choosing a camera tube with a small output capacitance if possible. The upper video-frequency limits should be as low as possible, using slower scan rates if systems requirements allow. The input transistor should be selected for low input noise, with the qualification that low-noise field effect transistors often have large gate capacitances which tend to raise C_S. The trade-off is discussed below. In addition, R_L should be as large as feasible. The limit to large R_L is, first, that the camera tube signal electrode current, up to 1 or 2×10^{-6} A for broadcast scan rates, must pass through R_L with at most a few volts drop to avoid changing the tube operating point with signal level. This limits R_L to about 10^6 ohms in broadcast type equipment. Second, the following preamplifier stages must produce a gain which rises at 6 dB per octave above the corner frequency, $f_c = 1/(2\pi R_L C_S)$. For a given C_S, a large R_L means a low corner frequency, hence a large high-frequency gain must be provided without objectionable phase shifts in the amplifier. One design choice is to make R_L large enough so that the first Johnson noise term equals the third amplifier noise term. The second term in Equation (13-6) is then negligible and the noise current is

$$I_n = \sqrt{2\left(\frac{4}{3}\,\pi^2\,C_S^2\,\epsilon_n^2\,\Delta f_v^3\right)} \qquad (13\text{-}7)$$

Equation (13-7) demonstrates that shunt capacitance and ϵ_n at the amplifier input determine noise performance. Once Δf_v is set by system considerations and the vidicon choice has established C_R, the task of the designer is to choose an input transistor and circuit to minimize both C_G and ϵ_n. Often these requirements are mutually contradictory. The trade-offs are discussed in the next subsection.

13.4.3. Choice of First-Stage Transistor. Discrete-component junction field-effect transistors (JFETs) suitable for television preamplifiers include the 2N4391, 2N4392, and 2N4393 n-channel JFETs intended for switching applications, and the microwave transistor 2N4417. Most so called "low noise" JFETs have high gate capacitances, and are less suitable. Table 13-2 includes data measured at Westinghouse on the 2N4392, and 2N4417 and on the "low noise amplifier" JFET 2N4139.

In noise models for field-effect transistors, the input noise is related to channel resistance, and hence to transconductance as

$$\epsilon_n = \sqrt{\frac{4kT}{g_m}\,A} \qquad (13\text{-}8)$$

Table 13-2. Westinghouse Data on Selected JFETs

JFETs	ϵ_n at Midband (nV Hz$^{-1/2}$)	C_g at $V_{GS} = 0$ (pF)	g_m eff (mmho)
2N4392	1.2	18	10
2N4139	1.8	20	5
2N4417	1.7	4	5
2 × 2N4417*	1.2	8	10

*Parallel connection

The factor A is a function of operating conditions, transistor design, and processing which can in theory be as low as 2/3. This expression applies for midband noise for both JFETs and MOSFETs. High transconductance is therefore required for low noise.

The transconductance for MOSFETs in the common-source mode under saturation conditions is related to input capacitance as

$$g'_m = \frac{C_o^* \mu_n^*}{l_c^2} (V_{GS} - V_T), \text{ for MOSFETs} \tag{13-9}$$

where C_o^* = the active gate-to-channel capacitance
μ_n^* = effective carrier mobility at the semiconductor surface
l_c = channel length between source and drain diffusions
V_{GS} = operating gate-to-source voltage
V_T = threshold gate-to-source voltage

The transconductance for a JFET with both gates active is more complex, but similar:

$$g'_m = \frac{K\mu_n C_{A,0}}{(V_{PO} + \psi)l_c^2} \left[1 - \left(\frac{V_{GS} + \psi}{V_{PO} + \psi} \right)^{1/2} \right] \approx \frac{K\mu_n C_{A,0}}{2(V_{PO} + \psi)^2 l_c^2} (V_{GS} - V_{PO}) \tag{13-10}$$

for $V_{GS} < \frac{1}{2} V_{PO}$.

where K = a constant depending on doping profiles and levels
$C_{A,0}$ = active capacitance from gates to channel at $V_{GS} = 0$
μ_n = effective carrier mobility in the channel
V_{PO} = pinch-off gate-to-source voltage
ψ = built-in voltage of the gate junction

Any capacitances from gate-to-substrate, or from gate-to-drain or -source outside the channel region, or gate lead capacitances, are not active capacitances but parasitics and must be minimized.

Combining Equations (13-8) and (13-9) or (13-10), one has

$$\epsilon_n = \frac{const}{\sqrt{g_m}} = \frac{const}{\sqrt{C_{A,0}}} \tag{13-11}$$

for a given channel length and a given voltage above threshold or above pinch-off. The Table 13-2 data follow the first equality of Equation (13-11), but not the second, primarily because gate capacitance values for many JFETs include extensive stray capacitance from the back gate-to-substrate. The 2N4417 has an exceptional combination of low capacitance and low noise, probably because this unit has little stray back-gate-to-substrate capacitance.

In designing a field-effect-transistor structure for preamplifier use, one should strive to minimize

$$I_n = \sqrt{\frac{8}{3} \ \pi^2 \ (C_R + C_W + C_{GA})^2 \ \Delta f_v^3 / C_{GA}} \tag{13-12}$$

where Equation (13-12) is (13-7) and (13-11) combined. This means that one must minimize $(C_R + C_W + C_{GA})^2 / C_{GA}$ which requires that $C_{GA} = C_R + C_W$, where C_{GA} is the active gate capacitance and any stray gate-capacitance is lumped with C_W.

For selection of stock, discrete JFETs (since manufacturers do not separate the active gate capacitance), one must balance low-noise voltage, ϵ_n, against C_G, the total transistor input-capacitance, to minimize

$$I_n = \sqrt{\frac{8}{3} \ \pi^2 \ (C_R + C_W + C_G)^2 \ \epsilon_n^2 \ \Delta f_v^3} \tag{13-13}$$

For a specific example, assume a 1 in. vidicon with C_R = 10 pF when the tube is mounted in its coil assembly, and C_W = 3 pF. The transistor choice should minimize (13 + C_G) ϵ_n. The choices, from Table 13-2, are the 2N4392 and the 2N4417, or two 2N4417 units in parallel. For the conditions given, the best choice would be two 2N4417 transistors in parallel as an input stage.

The remaining parameter determining amplifier noise is the video bandwidth. If the application permits, a narrower bandwidth will reduce the noise current, but it will also reduce resolving power in the along-scan direction unless the scanning frequencies are reduced proportionally. Table 13-3 lists the noise current versus the video frequency bandwidth for the components in the example above, and also shows q_n, the amplifier noise, expressed as an rms charge-fluctuation in number of electrons per Nyquist sample referred to the amplifier input. The exposure during a frame time must be sufficient to produce about this many carriers at the scanned surface of the target per Nyquist sample

Table 13-3. Preamplifier Noise Current Versus Video Bandwidth. The conditions are:
$C_S = C_R + C_W + C_G = (10 + 3 + 8)$ pF = 21 pF for paralleled 2N4417 FETs;
$\epsilon_n = 1.2 \times 10^{-9}$ V Hz$^{-1/2}$; R_L is large enough so that

$$I_n = \sqrt{\frac{8}{3} \pi^2 C_S^2 \epsilon_n^2 \Delta f_v^3} \text{ and } q_n = \frac{1}{e} \sqrt{\frac{2}{3} \pi^2 C_S^2 \epsilon_n^2 \Delta f_v}$$

Δf_v	10^4	10^5	10^6	4.5×10^6	10^7	Hz
I_n	1.29×10^{-13}	4.09×10^{-12}	1.29×10^{-10}	1.23×10^{-9}	4.09×10^{-9}	A
q_n	40.3	128	403	854	1280	Electrons

to achieve a signal equal to amplifier noise. However, the concept is useful principally for large-area image details which are completely resolved. For point images, the signal charge is spread over several Nyquist samples and the signal charge in each sample must equal the noise fluctuation.

The example shows that for U.S. broadcast scanning standards in a well-designed system, I_n is between 1 and 2 nA, and q_n is of the order of 1000 electrons per Nyquist sample.

13.4.4. Video Preamplifier Designs—The Current-Mode Input Operational Amplifier. The current-mode input operational amplifier shown in Figure 13-9 reduces the effective low-frequency, vidicon load-impedance from R_L to R_L in parallel with $R_F/(G+1)$. Thus in gain calculations, the input capacitive-susceptance to ground can usually be neglected, and the amplifier will provide an approximately flat response over the video frequency spectrum without need for a high-peaked post amplifier. Surprisingly, the input transistor noise is still high-frequency peaked by the action of shunt capacitance at the input, and the equation for the equivalent, input, noise-power spectral density has essentially the same form as Equation (13-5) for the classic uncompensated preamplifier with a following high-peaked stage. The overall noise performance of the current-mode input operational preamplifier is usually slightly poorer than for the classic voltage amplifier of Section 13.4.2 because of the Johnson-Nyquist noise contribution from R_F.

In Figure 13-9, the resistor labelled R_L now simply provides the bias current for the vidicon and is made large, typically 10^6 ohms for broadcast television applications. The inverting operational amplifier has a typical low-frequency open-loop gain, G_o, of several hundred, and acts to maintain an approximate virtual ground at the gate of the first transistor. For stability, phase shifts within the feedback loop must be kept small, and the amplifier is often designed with a single, uncompensated, high-gain stage as shown in Figure 13-10. Open-loop amplifier gain, G, varies as $G = G_o/(1 + j\omega\tau_A)$, falling 6 dB per octave above the corner-frequency, f_c, through the video passband. The closed-loop gain of the amplifier is expressed as a transfer impedance $Z_T = V_o/I_S$. For the circuit of Figure 13-9, the transfer impedance can be calculated based on a two-pole analysis:

$$Z_T = \frac{-R_F}{1 + j\omega R_F \left(C_F + \dfrac{C_S + C_F}{G_o} + \dfrac{\tau_A}{R_P G_o} \right) - \dfrac{\omega^2 R_F (C_S + C_F)\tau_A}{G_o}} \qquad (13\text{-}14)$$

where $1/R_p = 1/R_L + 1/R_F$, and the term $R_F/R_p\, A_o \ll 1$ has been neglected in the denominator.

This is of the form

$$Z_T = \frac{-R_F}{1 + jb\omega - a\omega^2} \qquad (13\text{-}15)$$

where $\quad a = \dfrac{R_F(C_S + C_F)\tau_A}{G_o}$

$$b = R_F \left(C_F + \frac{C_S + C_F}{G_o} + \frac{\tau_A}{R_P G_o} \right)$$

The magnitude of the transfer impedance is

$$|Z_T| = \frac{R_F}{\sqrt{(1 - a\omega^2)^2 + b^2\omega^2}} = \frac{R_F}{\sqrt{1 + (b^2 - 2a)\omega^2 + a^2\omega^4}} \tag{13-16}$$

For a flat or monotonically-decreasing transfer impedance versus frequency characteristic, the coefficient of ω^2 should not be negative, requiring $b^2 \geqslant 2a$, with equality preferred. This usually requires adding a bandwidth-limiting feedback capacitor, C_F, to the typical 0.3 pF stray shunt-capacitance of the feedback resistor, R_F. The following typical example is based on Figures 13-9 and 13-10, with values chosen for a 4.0×10^6 Hz nominal video bandwidth.

$$R_F = 3 \times 10^4 \text{ ohms} \qquad\qquad G_o = 152$$
$$C_s = 2.1 \times 10^{-11} \text{F} \qquad\qquad \tau_A = 1.06 \times 10^{-6} \text{ sec}$$
$$C_F = 1.8 \times 10^{-12} \text{F} \qquad\qquad R_L = 1 \times 10^6 \text{ ohms}$$

Hence $R_p = 2.91 \times 10^4$ ohms, $g_m = 10$ mmho, and $R_{eff} = 51.2 \times 10^3$ ohms.
Then

$$a = 1.42 \times 10^{-15} \text{ sec}^2$$

$$b = 5.75 \times 10^{-8} \text{ sec}$$

and

$$|Z_T| = \frac{R_F}{\sqrt{1 + (b^2 - 2a)\omega^2 + a^2\omega^4}} = \frac{3 \times 10^4}{\sqrt{1 + 1.83 \times 10^{-14}f^2 + 3.15 \times 10^{-27}f^4}} \tag{13-17}$$

This equation is plotted in Figure 13-11.

The noise performance is calculated by representing the noise of the first field effect transistor by a voltage generator in series with the gate. Since it is the gate itself which is held at a virtual ground, the model predicts an equal but opposite noise voltage at the amplifier output at low frequencies where the amplifier input is essentially open circuited, i.e., $R_L \gg R_F$. At higher frequencies, as the shunt susceptance of C_S becomes significant, the transistor noise voltage at the output becomes approximately $(1 + \omega^2 R_F^2 C_S^2)^{1/2}\epsilon_n$ and the equivalent rms input noise current is approximately

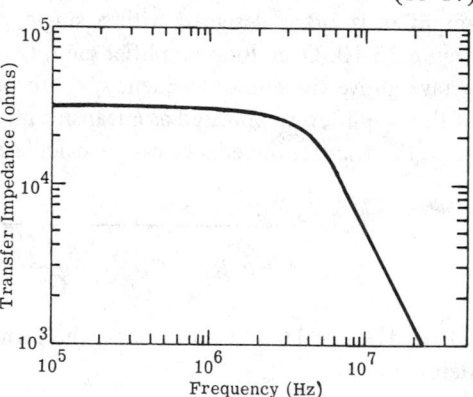

Fig. 13-11. Transfer impedance versus frequency for the current mode input operational amplifier.

$$I_n = \sqrt{\left(\frac{4kT}{R_P} + \frac{\epsilon_n^2}{R_P^2}\right)\Delta f_v + \frac{4}{3}\pi^2 C_S^2 \epsilon_n^2 \Delta f_v^3} \tag{13-18}$$

as in Equation (13-6) for the classical amplifier, except R_L is now replaced by $R_P = R_L R_F/(R_L + R_F)$. Because R_F is usually limited to 5×10^4 ohms, the first term (i.e., the Johnson noise) contribution cannot usually be made negligible with this circuit.

The operational amplifier is normally used where insensitivity to wiring and component capacitance variations outweighs this Johnson noise contribution. Again, the shunt capacitance at the preamplifier input and the first transistor noise voltage set noise performance and must be minimized.

13.4.5. Video Preamplifier Designs—Inductance-Peaked Vidicon Load Circuit. Larger vidicons have larger output electrode capacitances, and inductive peaking of the load circuit can increase the available signal voltage at higher frequencies before the input stage noise is added. Shunt peaking was used in some early work at RCA, and a successful series-peaking embodiment, shown in Figure 13-12, was disclosed by Percival [13-1] and described by James [13-2]. The Percival/James pi-network splits the shunt capacitance into its components and places a "Percival coil" between C_R and $C_W + C_G$ to resonate with the series combination at about 85% of the upper video frequency limit.

In James's analysis for the circuit of Figure 13-12,

$$C_S = C_R + (C_W + C_G)$$

$$a = \frac{C_W + C_G}{C_S} \qquad (13\text{-}19)$$

$$b = \frac{\Delta f_v}{f_r}$$

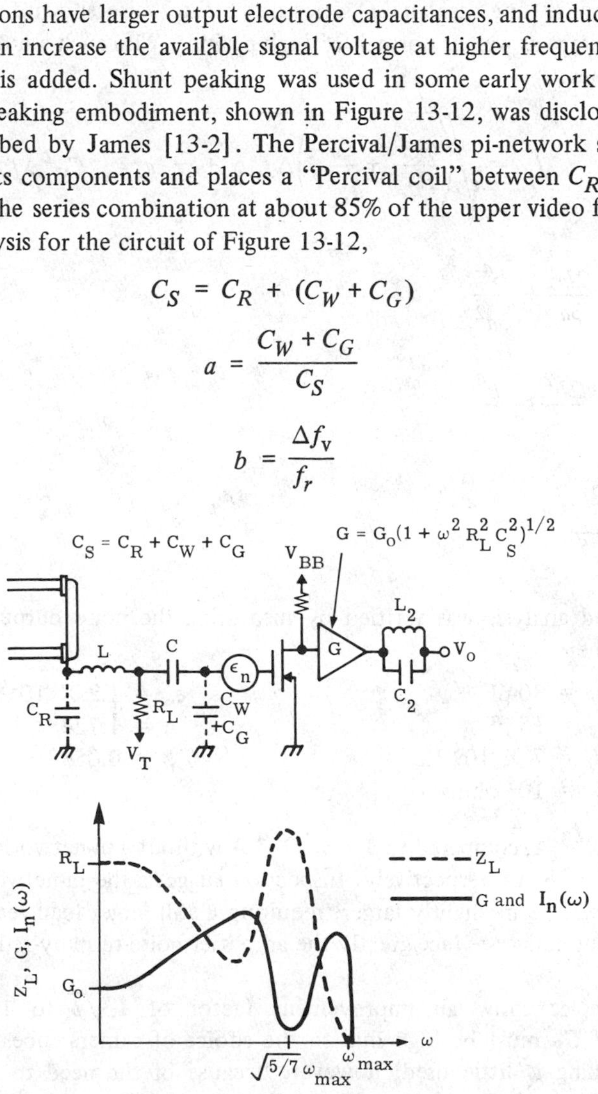

Fig. 13-12. Series peaking. It increases signal power at preamplifier input for ω near ω_{max}. Notch filter flattens the signal frequency response and reduces integrated noise current by as much as 6 dB [13-1, 13-2].

where Δf_v = video bandwidth

f_r = resonant frequency for L and $C_W + C_G$ and C_T in series

$$f_r = \frac{1}{2\pi\sqrt{LC(1-a)a}}$$

Optimum results are reported when R_L is large and $b = \sqrt{7/5}$. The impedance of the vidicon load circuit varies approximately as shown in Figure 13-12. To produce a flat overall response, one uses an LC series notch filter, also shown in Figure 13-12, in a later amplifier stage as well as normal RC high peaking. With such a notch filter, the noise equation becomes

$$I_n = \sqrt{\left(\frac{4kT}{R_L} + \frac{\epsilon_n{}^2}{R_L{}^2}\right)\Delta f_v \, \alpha + \frac{4}{3}\,\pi^2 \epsilon_n{}^2 C_S{}^2 \Delta f_v{}^3 3\beta} \qquad (13\text{-}20)$$

where $\alpha = 1 - \dfrac{2b^2}{3a} + \dfrac{b^4}{5a^2}$

$\beta = \dfrac{1}{3} - \dfrac{2b^2}{5} + \dfrac{b^4}{7}$

and $a = 1/3$

$b = \sqrt{7/5}$ $(13\text{-}21)$

This circuit and analysis was verified by measuring the noise output of a JFET input preamplifier when

$$\begin{aligned}
C_R &= 30 \text{ pF} & \epsilon_n &= 1.2 \times 10^{-9} \text{ V Hz}^{-1/2}\\
C_W + C_G &= 15 \text{ pF} & \alpha &= 1.728\\
\Delta f_v &= 7 \times 10^6 \text{ Hz} & \beta &= 0.0533\\
R_L &= 10^5 \text{ ohms}
\end{aligned}$$

and $I_n = 2 \times 10^{-9}$ A, compared to 3.7×10^{-9} A without a pi-network. Calculated values were 2.01 and 3.78 nA respectively. Inspection suggests the pi-network actually makes the Johnson noise term slightly larger, requiring a still larger load resistor for optimum performance, but acts to reduce greatly the amplifier noise term by reducing the effective capacitance.

Other examples show an improvement factor of $1/\sqrt{2}$ to $1/2$, a worthwhile improvement if C_S must be high due to the choice of camera tubes. For conventional television, peaking is little used, however, because of the need to individually adjust both the peaking and notch filter networks in case of component replacement, and because the noise reduction is in the higher video frequencies where the human observer is remarkably noise tolerant. A more fruitful application of this technique is in imaging systems with automatic output, where, for example, a thresholding circuit passes signal locations to a computer memory and the human observer is no longer a link in an analog system. There, the human-factors design criteria no longer apply, and use of series or shunt peaking should give a significant performance improvement.

13.5. Vidicons of Interest

13.5.1. Silicon Diode Array Vidicon.
A silicon diode array can provide a photon-sensing and charge-storage retina using low resistivity material with a 1.1 eV bandgap

which responds to radiation in the near infrared. The silicon diode array retina is formed from a single-crystal wafer of approximately 10 ohm cm, n-type silicon. As shown in Figure 13-13, a mosaic of p^+ islands is formed on one face, usually by boron diffusion through an oxide mask or by epitaxial growth. The resulting diodes are back-biased by scanning them with a low-velocity electron beam. The n-type wafer is biased 5 or 10 V positive with respect to the electron gun cathode so that in the dark each diode becomes back-biased by approximately this bias voltage, V_R. The photon flux in the image is focussed on the reverse face of the silicon retina, which is usually only 10 to 15 μm thick over the image area. Light is absorbed in the n-type silicon, thereby generating carrier pairs; many of the photo-generated holes diffuse to a nearby depletion region and are swept across the

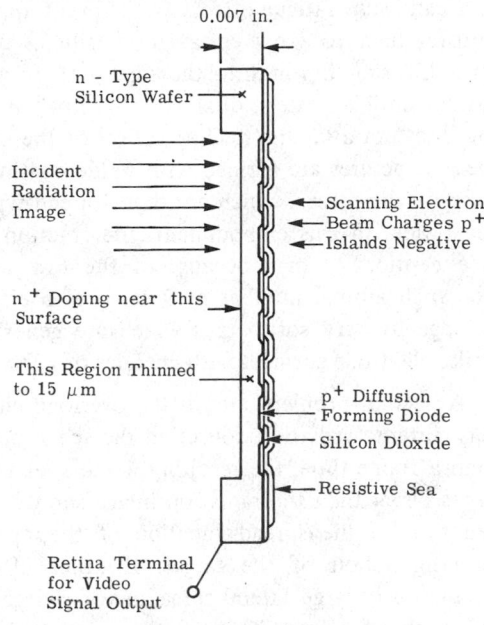

Fig. 13-13. Silicon-diode array retina. The retina stores the charge image on the depletion layer capacitance of each back-biased diode.

junction and collected on the p^+ island, partially discharging its reverse bias. During a frame time, the positive charge on each island is very nearly proportional to the time integral of the photon flux density in that part of the image.

The silicon-diode-array retina vidicon as described above would have two undesirable characteristics. First, in a diode array fabricated by depositing p^+ diffusions through holes in an oxide mask, the oxide area would be larger than that of the exposed p^+ islands, and only a small fraction of the beam electrons would be deposited on the p^+ diffusions to generate a video signal. Many of the rest of the electrons would be deposited on the oxide, charging the oxide surface negatively until the resulting negative potential above the target surface was sufficient to repel all beam electrons and vidicon operation would stop. The oxide film must be left in place both to passivate the silicon surface over each depletion region and to prevent beam electrons from landing on the regions between the diffusions. The first practical target design used a "resistive sea," as shown in Figure 13-13, a thin layer of high resistivity material like Sb_2S_3 deposited by vacuum evaporation over the whole target and making contact with the p^+ diffusions. The resistive sea provides adequate lateral conductivity over the oxide surface so that electrons deposited anywhere are conducted to a nearby diode within a frame time. The whole target scanned surface is brought approximately to gun-cathode potential after reading. Thus, stray oxide charging is eliminated, and additionally all of the scanning beam is able to provide signal electrons since the target has 100% area efficiency.

The resistive sea approach is still in use, but is not ideal since it acts to degrade the modulation transfer function (MTF) by providing lateral charge conduction, and since it introduces an effective resistance in series with each diode which contributes to lag and can cause fatigue effects. A different approach is to etch holes in the oxide mask as before, then to grow epitaxial p^+ silicon up through the holes instead of depositing and diffusing boron into the silicon to form p^+ islands. The epitaxial growth is continued until a pattern of silicon "bumps" almost covers the oxide, so that oxide charging does not alter the reading action of the scanning beam. In another approach, silicon mesa structures are formed with oxide coating in the valleys. Boron is diffused into the tops of the mesas, which are then topped with plated gold caps which essentially mask the oxide. In this embodiment, the fraction of beam electrons deposited on the target is exceptionally high because of the low secondary-emission ratio of the gold caps. The high atomic number material is also effective in shielding the silicon surface from damage by very soft x-rays which are generated within the tube when beam electrons strike electrode surfaces with energies of a few hundred electron volts.

A second problem area is the overload characteristic of the target. If the image of a very intense radiation source in the scene discharges the diodes in a time much shorter than a frame time, the resulting white spot in the reproduced image often appears many times larger than the radiation image should. This blooming effect is, of course, partly a function of the spread function of the optical system together with the peak signal clipping action of the target. If other effects are removed, however, one finds an anomalously large lateral spreading of charge carriers in the target itself, explainable by the Webster effect. That is, within a single frame time, each diode which has been completely discharged acts as a source of holes to propagate the hole-spreading from its edge, so that carrier diffusion need take place only across the relatively narrow spaces between p^+ islands.

Approaches to reducing blooming include making the target thin and treating the first surface for a controlled, nonzero surface recombination-velocity to provide a nearby sink for laterally diffusing electrons. Obviously this requires a tradeoff of some responsivity. A second approach is to include a p^+ linear diffusion after every second row of diodes and to bias this diffusion to form a sink for excess holes. This effectively inhibits blooming, but produces an effectively coarser mosaic structure which may interfere with televising fine detail. Variations in the width of the diffusion can also cause shading in the reproduced picture. A third approach to reducing blooming is the Westinghouse mesa, deep-etched, metal-capped target. This target provides a significant reduction in blooming because the junctions are up in the mesas, and excess holes must diffuse down around the bottom of the adjoining valleys to reach the next diodes.

Performance of typical silicon vidicons is described in Figures 13-14 through 13-19.

Spectral Response. The spectral response shown in Figure 13-14 is broader than for most camera tubes and extends into the near infrared beyond 1.1 μm with response of 60 to 70 mA W^{-1} at 1.06 μm. Response falls at wavelengths below about 500 nm because a

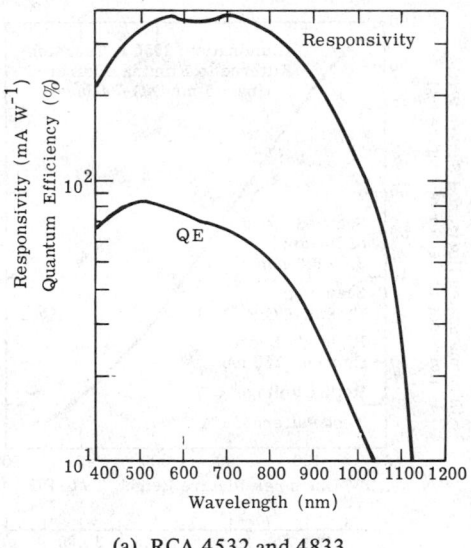

(a) RCA 4532 and 4833.

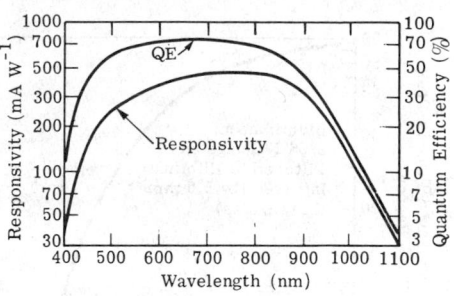

(b) Amperex S1200.

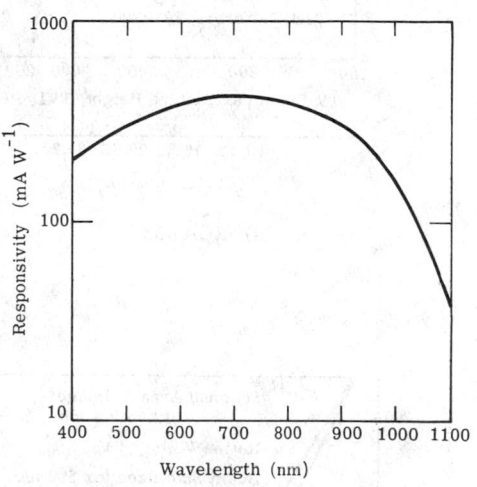

(c) Westinghouse WX-32834.

Fig. 13-14. Spectral response and quantum efficiency.

larger fraction of shorter wavelength photons are reflected from the first silicon surface and because the photons entering the silicon are strongly absorbed, generating carrier pairs near the first surface. Holes generated near the surface have a stronger probability of being captured by localized surface states and recombining. To minimize such recombination, one customarily enriches the surface by depositing additional donor atoms to make the surface n^+, hence building in a field to repel the photogenerated holes away from the surface and toward the diodes. However, precise control of this doping profile is difficult to obtain, and a layer at the surface with slightly lower donor density will act to accelerate some holes toward the surface and to neutralize them. The spectral response curves of Figure 13-14 demonstrate that a target with an antireflection coating on the first surface can achieve quantum efficiencies of 70% or greater over a significant spectral interval, with peak quantum efficiency about 80%.

The falling response in the near infrared, beyond 850 nm, is due to the increasing transparency of the silicon near the band edge, which is at about 1.13 μm. Near the band edge many of the photons pass through the target without being absorbed. Infrared response can, of course, be increased by use of a thicker silicon target, but at the cost of decreased response in the visible and of a decrease in the modulation transfer function (MTF).

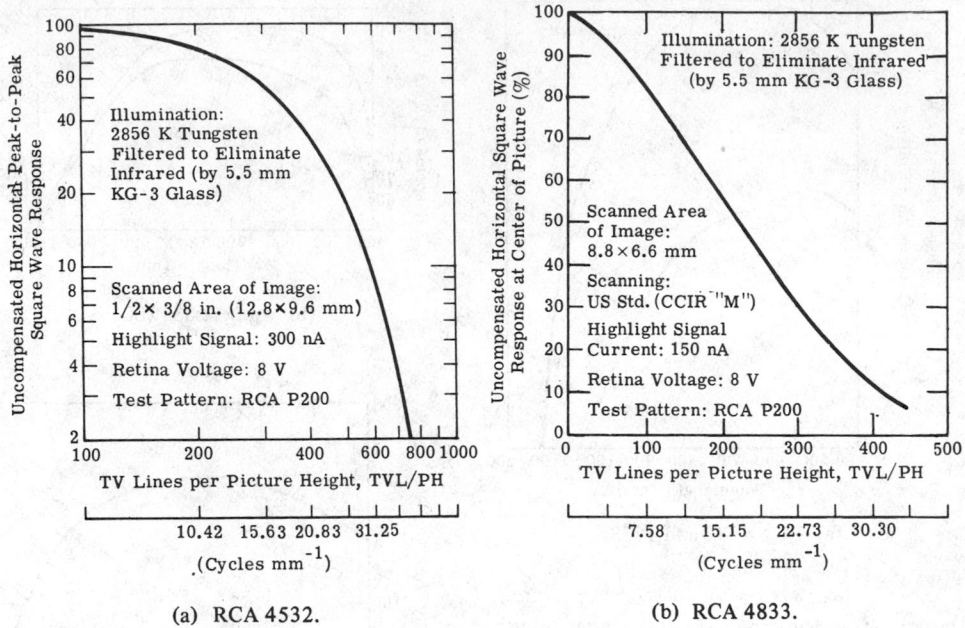

(a) RCA 4532.

(b) RCA 4833.

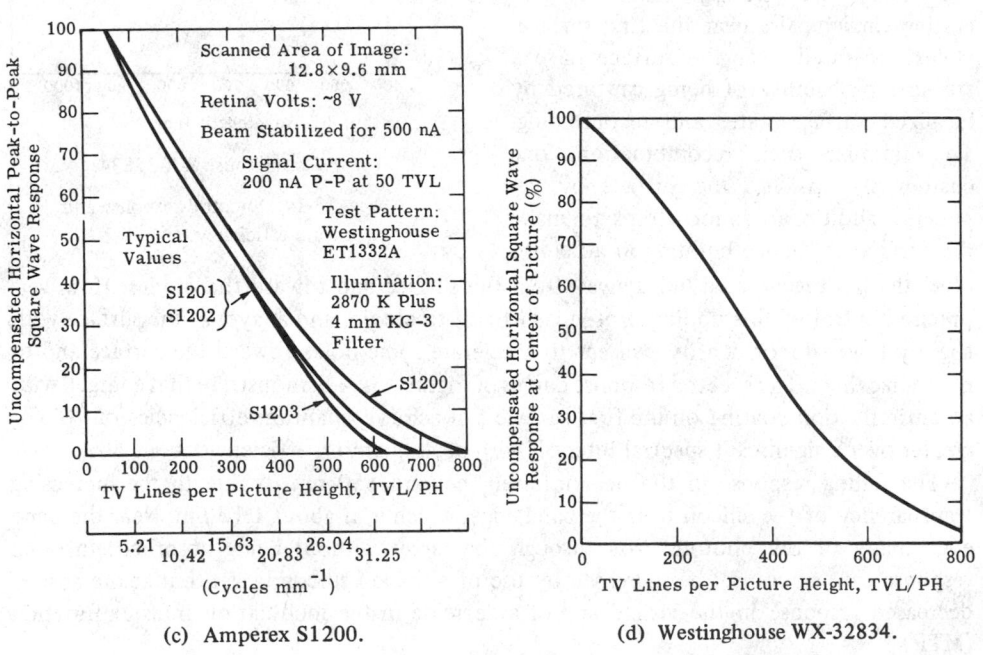

(c) Amperex S1200.

(d) Westinghouse WX-32834.

Fig. 13-15. Square wave response measurements of typical silicon vidicons.

Resolving Power. The square-wave response shown in Figure 13-15 is a measure of the resolving power of a silicon vidicon. The MTF may be calculated from these data. (See Figure 13-16.) For 2856 K tungsten illumination modified by a KG-3 filter to remove most radiation at wavelengths beyond 720 nm, the horizontal square-wave response falls to 50% at about 16 cycles mm^{-1}. This value is nearly the same for tubes with both 11 and 16 mm diameter retina.

Crowell and Labuda [13-3] have analyzed silicon-diode array retina resolving power as determined by the lateral diffusion of holes in the bulk n^{+} silicon before they reach a depletion region and are collected. They found close agreement with measured results (Figure 13-16), indicating that carrier diffusion rather than the diameter of the scanning electron beam determined the MTF in their experiment. They postulated a 15 μm thick structure with 5 μm deep depletion layer, a first-surface recombination velocity of 10^{3} cm sec^{-1}, and a diffusion length of 47.5 μm. Their analysis and experimental results agree that resolving power can be increased by thinning the target from 15 to 10 μm, but at the cost of reduced response in the near infrared.

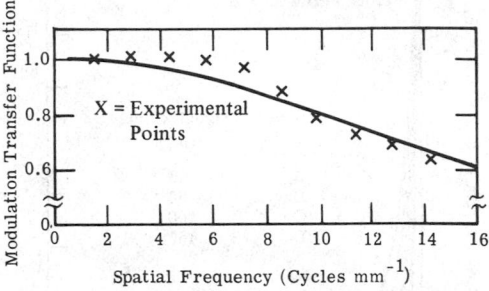

Fig. 13-16. Measured and calculated values of MTF of a diode-array camera tube as a function of spatial frequency. The optical wavelength is 0.55 μm [13-3].

Signal Transfer Characteristic. Since essentially all photogenerated holes diffuse to and are collected by the diodes, the light-to-signal transfer characteristic is essentially linear, as shown in Figure 13-17. This linearity facilitates use of silicon vidicons for photometry or radiometry or for certain signal processing algorithms. The response to 2856 K tungsten illumination is approximately 80 mA W^{-1}.

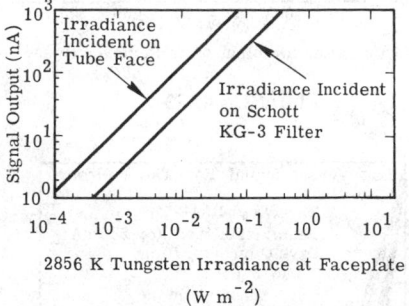

Fig. 13-17. Light-to-signal transfer characteristic. All silicon-diode array vidicons have essentially linear light-to-signal transfer characteristics. Those shown are for an RCA 4532 with a 12.8 × 9.8 mm scanned area; 30°C faceplate temperature; and an 8 V retina voltage.

Image Lag. The silicon-diode array target provides essentially no trapping centers for photo-excited carriers, and therefore shows no "photoconductive" lag. It does show some image persistence due to the dynamic impedance of the scanning electron beam and the target storage capacitance, like all other vidicons. The mechanism is described in Section 13.3.1. The effective charge-storage capacitance of a typical 0.95 × 1.27 μm active target area is about 2.4×10^{-9} F when the diodes are reverse biased to about 10 V. At high signal levels, about 10% of the image charge is left after one complete reading of the target when operated at a 10 V target bias. The fraction is larger at lower signal levels.

Lag data normally published as in Figure 13-18 are somewhat misleading. Although the manufacturer draws a smooth curve of persistent image signal versus time, only the points at the beginning of each field after the illumination has been interrupted are unambiguous measurements of lag. The data are taken by interrupting the image illumination either during the vertical retrace interval just before the start of the first scanning

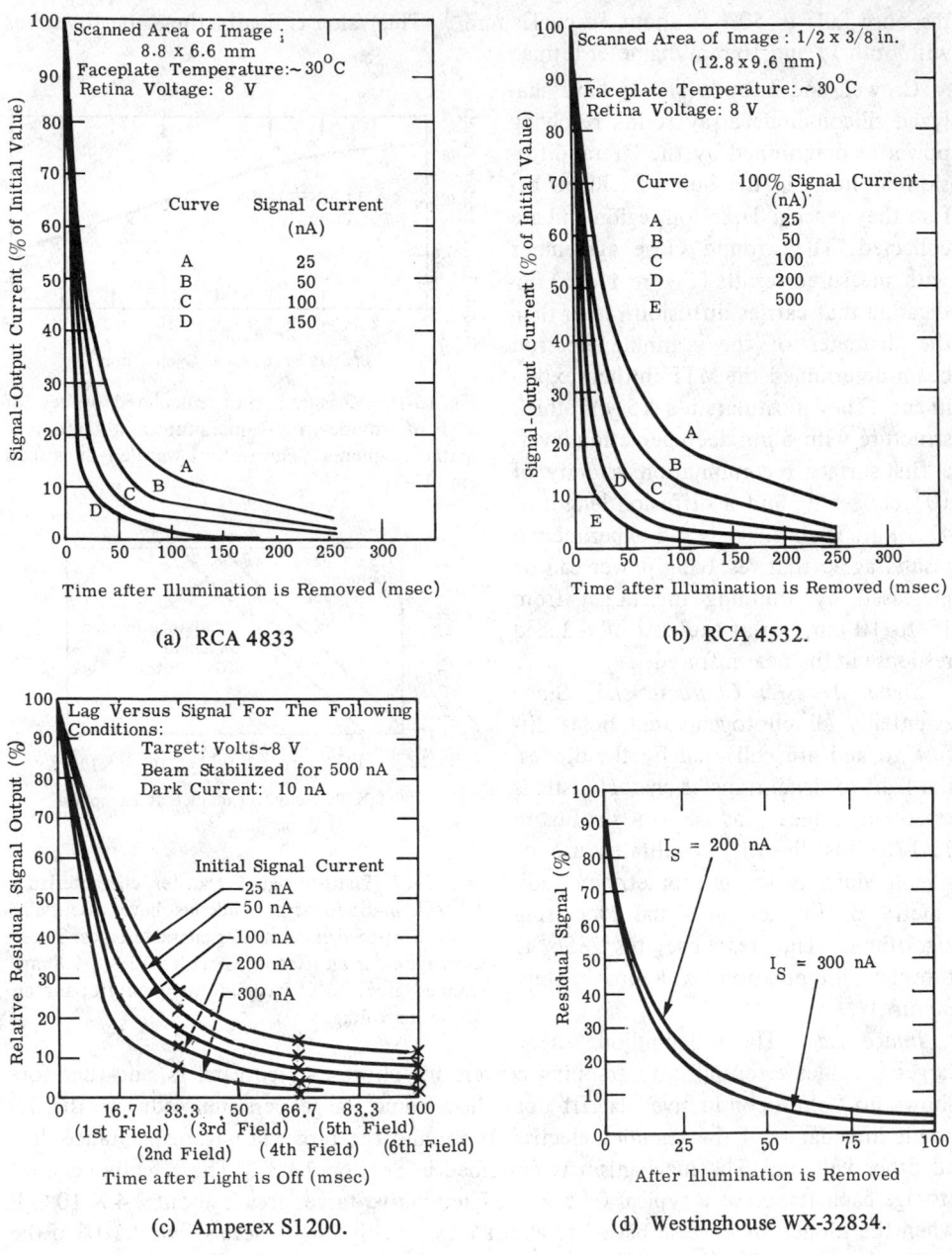

(a) RCA 4833

(b) RCA 4532.

(c) Amperex S1200.

(d) Westinghouse WX-32834.

Fig. 13-18. Image persistence. The combination of beam resistance and retina storage capacitance causes larger image persistence and lower signal levels.

field or during the scan when the beam is about to scan a line near the center of the image. The first point scanned after the interruption was exposed for a full frame time after its previous scan and gives full signal amplitude. Each point after the first has slightly less exposure, 63.5 μsec per scan line, and the signal in the initial or "zeroth" field, 0 to 16.7 msec, should fall nearly linearly with time, that is nearly linearly with scanning line number, simply because the exposure is shorter, not because of any "lag" mechanism. The "first" field signal, 16.7 to 33 msec, should also fall nearly linearly with time and scan-line number, but the relation between the "zeroth" and "first" field signals is undetermined because the beam diameter is larger than the interline spacing; and scanning the even numbered lines also removes a substantial charge from the areas associated with the odd numbered lines. The signal versus scan-line relation in each field is not quite linear because the effective beam impedance increases as the signal decreases, hence a larger fraction of the charge is left for smaller charge.

Lag is most meaningfully specified for a 60 field, 30 frame per second, interlaced system, as the ratio of the signal after 33.3 msec, at the beginning of the second frame, to the signal at time zero, and similarly for the signal at 66.7 msec, 100 msec, etc. In each case, the entire target has been read one or more times. The "X" data points on the curves of Figure 13-18(c) are these values. But the curve between the full frame points, if accurately drawn from real data, also contains information. Between any two points 33.3 msec apart, the target has been read once, and the ratio of the two signal currents is a valid measure of lag for an image signal current corresponding to the earlier point. This can be seen by plotting the curves of Figure 13-18(b) for the 4532 as a single curve of signal current versus time, shifting the abscissas to match the ordinates. The resulting

curve of Figure 13-19 is general. For any initial signal level within the range of the curve, the signal after one full frame is found 33.3 msec to the right. This curve demonstrates how the effective beam impedance increases as the signal and target surface potential fall. It takes one frame time to halve the signal current from 200 to 100 nA. It takes three frame times to go from 5 to 2.5 nA. Obviously, this is not exponential decay.

This curve also describes the lag behavior of the smaller type 4833 silicon-

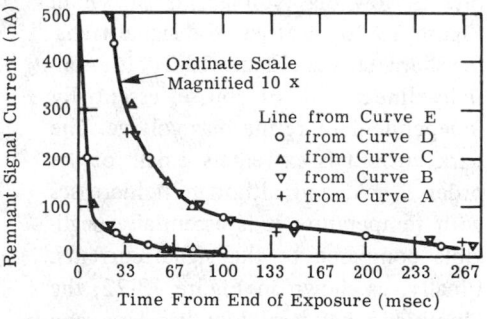

Fig. 13-19. Single curve of signal current versus time for RCA 4532. The data were taken from Figure 13-18(b).

diode array vidicon if the signal currents are doubled, that is, scaled by the ratio of the image areas.

Note that present television sensor practice is to specify the "3rd-field" lag as a single number descriptor. As defined in data sheets, "3rd-field" lag is that measured 50 msec after illumination has been interrupted, when the target has been scanned 1 1/2 times. This number is difficult to interpret, and is only useful to compare tubes with similar characteristics.

Operating Temperatures. Typically, specified operating temperatures range from -54 to $+70°C$. However, the maximum operating temperature for a silicon-diode array vidicon is set by the level of dark current which can be tolerated. Typical dark currents at a faceplate temperature of 25°C are 3 to 6 mA, doubling for every 8 to 10°C rise in faceplate temperature. Since maximum allowable signal current decreases from its low-temperature value as dark current increases, operation at 65°C, for example, could produce dark current of 190 nA, reducing the maximum current capability from a typical 750 to 560 nA, a significant change.

Minimum operating temperatures are set in resistive-sea tubes when the resistance of the overlayer becomes too high, resulting in excessive image lag and possible surface charging effects. There appears to be no similar limit for epitaxial bump and deep-etched metal-capped target tubes.

13.5.2. The Plumbicon® * Camera Tube.

A typical Plumbicon retina is made from polycrystalline red tetragonal lead oxide with a bandgap of about 1.9 eV, corresponding to an absorption limit at 650 nm. The layer is formed by evaporation through a gas at low pressure onto a glass faceplate precoated with transparent, electrically-conducting tin oxide. The resulting camera tube, which is operated very much like a traditional Sb_2S_3 vidicon, has signal and dark currents which saturate at low bias voltage, as shown in Figure 13-20. Further, the signal transfer characteristic (Figure 13-21) is essentially linear and, of course, essentially independent of retina bias voltage. The dark current is extremely small, of the order 2 nA, and although it increases with temperature it is essentially negligible compared to the signal currents. Finally, as shown in Figure 13-22, the Plumbicon has very low lag, less than 5% residual signal in the fourth field after the image irradiance is shuttered off.

These characteristics make the Plumbicon nearly ideal for three-tube color television cameras, and this tube type dominates the entertainment broadcasting industry.

The properties of the Plumbicon can be explained by a model of the target as a surface composed of P-I-N diodes as

*Trademark of N. V. Philips, Holland.

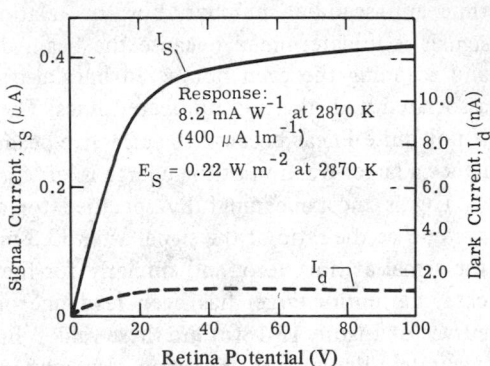

Fig. 13-20. Signal and dark currents of typical Plumbicon versus retina potential with fixed level of target illumination.

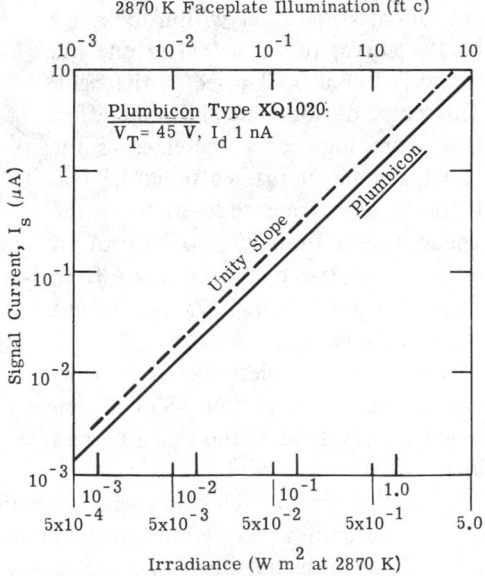

Fig. 13-21. Plumbicon versus vidicon sensitivity (signal current versus faceplate irradiance). The scanned area of this Plumbicon is 40% larger than that of the 1 in. vidicon. A line with unity slope is shown for comparison.

suggested in Figure 13-23. In this model, it is postulated that the layer near the first lead-oxide surface behaves like an n-type semiconductor because of diffusion of tin from the signal electrode layer into the individual lead-oxide crystallites. The crystallites near the electron-beam scanned surface are made p-type by the action of a gas introduced into the envelope following layer deposition. The balance of the layer behaves like an intrinsic semiconductor, not because of great crystalline or stoichiometric purity, but rather because the Fermi level is fixed near mid-gap by surface states. This model is based on the microstructure of the lead oxide layer. The crystallites are platelets, typically 1 μm $\times$ 1 μm $\times$ 0.1 μm, and Philips workers postulate, as suggested in Figure 13-24 (where E_F = Fermi level), that surface states on the flat crystallite faces act as compensating acceptors to balance the donor centers in the bulk of each crystallite, causing it to act as a nearly intrinsic semiconductor with high effective resistivity. In such a structure (Figure 13-23), most of the applied target bias-voltage appears as a field across a high resistivity, nearly intrinsic region occupying most of the layer thickness. Photogenerated

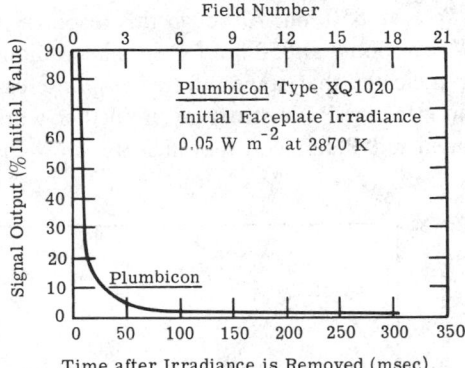

Fig. 13-22. Lag characteristic of XQ1020 Plumbicon.

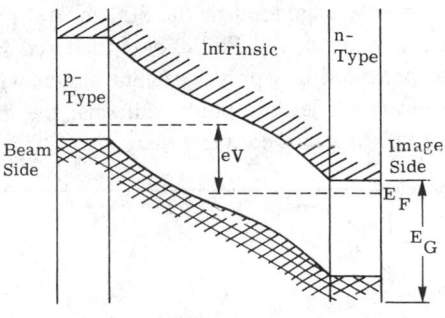

Fig. 13-23. Simplified energy-band picture of a reversed-biased P-I-N diode.

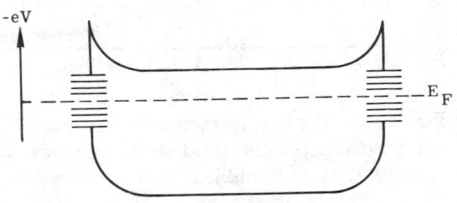

(a) Single crystal of n-type lead oxide.

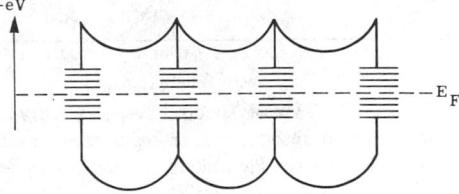

(c) Polycrystalline PbO with crystalline thickness less than two times barrier width.

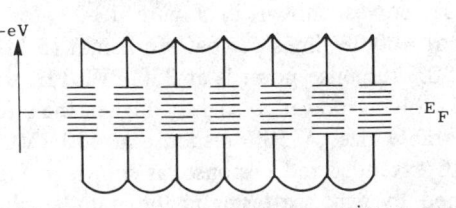

(b) Polycrystalline PbO (large crystallites).

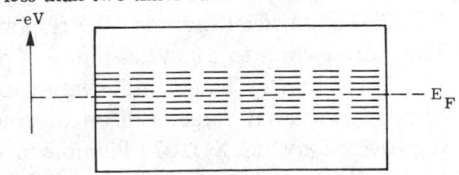

(d) Limiting case of small-crystalline PbO.

Fig. 13-24. Energy-band picture.

carriers are swept quickly to the surfaces, where blocking contacts of some nature prevent injection of carriers from either the signal electrode or the electron beam. The necessary high lateral resistance in the assumed p-type surface layer is apparently provided by the structure, whose oriented platelets must be somewhat insulated from one another. This model would predict the characteristics which are actually observed.

Characteristics of typical Plumbicon tubes are given in Figures 13-25 through 13-28 for the XQ1070 and XQ1073 series of 1 in. diameter tubes. Table 13-4 shows a comparison of the characteristics of these tubes.

Spectral Response. The spectral response of the standard lead-oxide layer shown in Figure 13-25 has a threshold at approximately 675 nm in the visible, and has in fact insufficient red response for fully satisfying performance in the red channel of a three- or four-tube color camera. To augment red response without altering the other desirable Plumbicon characteristics, Philips engineers have formed an extended red response Plumbicon by introducing enough sulphur into the lead oxide layer to extend the threshold beyond 850 nm. The spectral response curve for the XQ1073 Plumbicon shown in Figure 13-26 has typical response of 3 mA W^{-1} at 850 nm. Although this response is miniscule compared to the 300 to 400 mA W^{-1} response at 850 nm for the silicon diode array vidicon, the Philips-extended red layer indicates that some infrared response can be provided in a polycrystalline target with an effective P-I-N structure. If further work is done on lead sulphide vidicons, the extended red Plumbicon may help to show the way to blocking contacts there also.

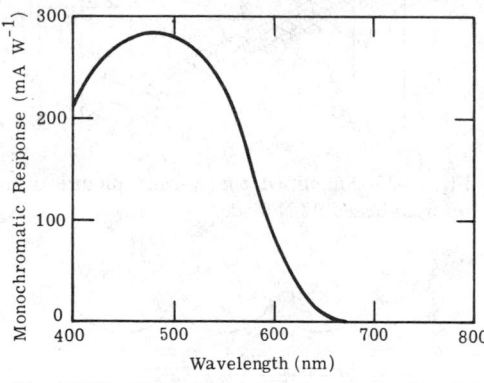

Fig. 13-25. Typical spectral response curve for special high resolution lead oxide retina as used in XQ1070 series Plumbicon camera tubes.

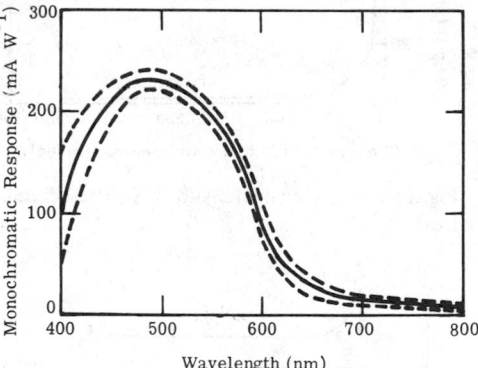

Fig. 13-26. Typical spectral response curve for extended red response lead oxide retina as used in XQ1073 series Plumbicon camera tubes.

Resolving Power. Square-wave response curves shown in Figure 13-27 for the XQ1070 Plumbicon indicate 40% response at 400 TV lines per pattern height (5 MHz). This corresponds to 21 cycles mm^{-1}. The 50% response point is at 350 TVL/PH, about 18 cycles mm^{-1}. Thus, this measure of resolving power is slightly higher than for a silicon-diode array target vidicon of comparable size. A 50% response at 400 TVL/PH is provided by the XQ1073 Plumbicon with extended red response, as shown in Figure 13-28. Resolving power is largely determined by light scattering in the granular photosensitive layer.

Table 13-4. A Comparison of Characteristics for 1 in. and 2/3 in.
Plumbicon Camera Tubes

Characteristic	1 in. Plumbicon Camera Tube		2/3 in. Plumbicon Camera Tube
	XQ1070	XQ1073	XQ1423
Picture area (mm)	12.8 × 9.6	12.8 × 9.6	8.8 × 6.6
Bulb diameter (mm)	25.4	25.4	18.0
Length (mm)	163.0	163.0	103.0
Response (2854) (mA W^{-1})	8	7	8
Spectral response			
maximum (nm)	480	500	475
cutoff (nm)	650	850 to 950	850 to 950
CTF			
400 TVL	40%	50%	25%
E_{G4} (V)	960	960	500
I_d (nA)	⩽3	⩽3	⩽3
Lag, typical, 60 msec	1.5%*	2%*	2%*
C_R (pF)	3 to 5	3 to 5	1.5 to 3.0

*Measured through a Calflex B1/K1 infrared blocking filter which has 50% transmission at 700 nm. Condition of measurement is 200 nA signal current, with the beam set to handle 400 nA, for the XQ1070 and XQ1073, and 150 nA current and 300 nA beam setting for XQ1423.

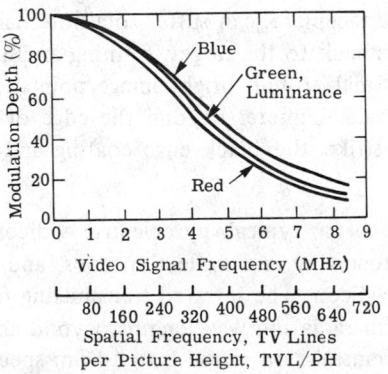

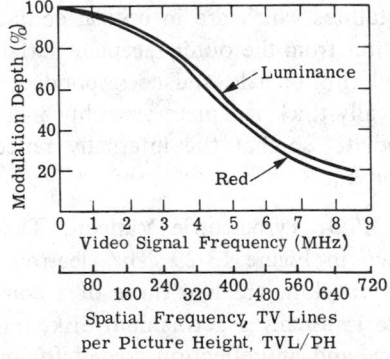

Fig. 13-27. Square wave response curve for XQ1070 series Plumbicon camera tubes. The test conditions are E_{c2} = 300 V, E_{c3} = 600 V, and E_{c4} = 900 V, measured with Philips AT1116 focus-deflection coil assembly.

Fig. 13-28. Square wave response curve for XQ1073 series Plumbicon camera tubes. The test conditions are E_{c2} = 300 V, E_{c3} = 600 V, and E_{c4} = 900 V, measured with Philips AT1116 focus-deflection coil assembly.

Lag. Remanent signal versus time after removal of image irradiance is shown in Figure 13-22 for the XQ1070 Plumbicon. Remanent signal was about 7% one frame after the image irradiance was removed, for an initial signal of 100 μA. This compares with about 25% for a type 4532 silicon-diode array vidicon, a significant reduction probably indicating the effective charge-storage capacitance of the Plumbicon target is only about 1/5 that of the silicon-diode array target. This lower capacitance and the higher target bias voltage permit a significantly larger signal voltage excursion at the scanned target surface, better matched to the velocity spread of the scanning beam electrons.

Operating Temperature. Philips list –30 to +50°C for either operation or storage, with the proviso that during long term storage, temperature should not exceed 30°C. Plumbicon tubes have been operated from 77 K for long time image integration, up to several hours at 60°C. The upper-limit is set by outgassing from the photosensitive layer.

Mechanical Characteristics. The XQ1070 series of Plumbicon camera tubes is mechanically interchangeable with 1 in. diameter vidicons with separate mesh construction and has the same base pin connections. The focus and deflection coil assembly must produce magnetic fields which accommodate the specified G_3 and G_4 electrode voltages. Amperex recommends the Philips AT1116 deflection assembly for use with the XQ1070 or XQ1073 series of Plumbicon camera tubes.

Also available is the XQ1423 series of Plumbicon camera tubes using 2/3 in. diameter bulbs, intended for compact color television camera equipment. These tubes are mechanically interchangeable with 2/3 in. diameter vidicons with separate mesh construction and have the same base pin connections. Amperex recommends the Philips AT1105 or KV12 deflection and focus coil assemblies for use with these tubes.

Anti-halation Button. All three series of Plumbicon camera tubes described in this section are equipped with anti-halation buttons. These are glass discs bonded to the outer surface of the borosilicate glass faceplate to increase substantially the optical thickness of the faceplate assembly and to provide a light-absorbing black coating around the periphery of the assembly. Light reflected randomly from the interior of target crystallites which are in optical contact with the faceplate, can suffer total internal reflection from the outer faceplate surface and pass back to the target, forming an illuminated ring or halo and corresponding spurious signal around bright image points. The optically-thick faceplate assembly increases the halo diameter beyond the edge of the faceplate, so that the internally reflected rays strike the black edge coating and are absorbed.

13.5.3. Pyroelectric Vidicon. The structure of a typical pyroelectric vidicon is shown in Figure 13-29. The electron gun, the focussing and deflection coils, and the tube envelope are like those of a conventional vidicon. The infrared transmitting faceplate is usually a germanium disk, transparent for radiation wavelengths beyond about 2 μm, and antireflection coated for maximum transmission in the 8 to 14 μm spectral band. The radiation-sensing retina is a single crystal disk of an insulating pyroelectric material like triglycine sulphate (TGS) with a semitransparent electrically-conducting signal electrode formed by evaporation on the face next to the faceplate. For a 25 mm tube, the active image area is 16 to 18 mm in diameter, and the retina is approximately 30 μm thick.

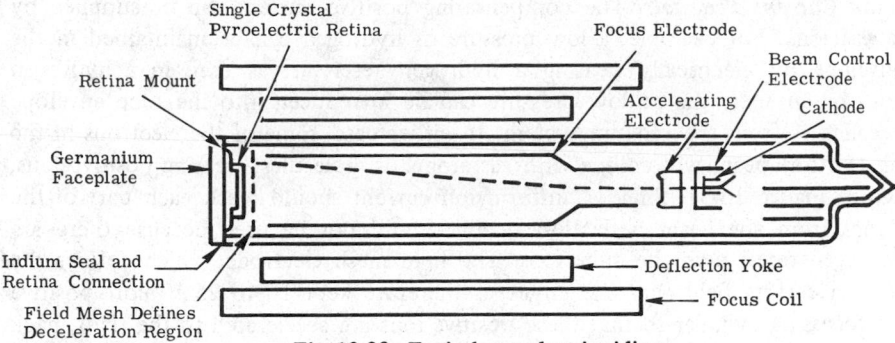

Fig. 13-29. Typical pyroelectric vidicon.

To televise the thermal image of a stationary scene with a pyroelectric vidicon, one must modulate the radiation pattern, because the pyroelectric material senses only changes in its temperature. In a typical camera, shown schematically in Figure 13-30, a chopper-wheel interrupts the image irradiance in synchronization with the electron beam scanning pattern so that in alternate frames the retina views first the scene image, then the chopper blade. If a scene area is hotter than the chopper, the retina warms during scene exposure, producing a corresponding positive charge pattern. It cools during

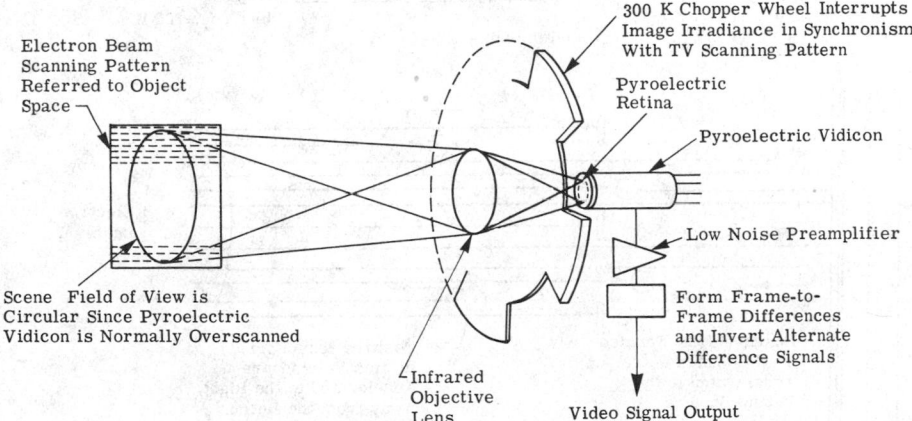

Fig. 13-30. Television camera using pyroelectric vidicon. Scene irradiance must be modulated with a chopper, shown here, or by panning the camera or nutating the lens.

chopper exposure, producing a negative charge pattern. This presents a problem, since normally a low-energy scanning electron beam reads a charge pattern from the retina surface by depositing electrons to charge the surface approximately to the electron gun cathode potential. If a positive charge pattern were read in one frame, the negative charge pattern in the following frame would charge the surface below cathode potential and no signal would be produced. For continuing operation, a compensating positive current must flow to the target surface to balance the negative current of signal electrons. Three general schemes are in use to provide this positive current, which is known as the pedestal.

Gas Ion Current Pedestal. The compensating positive current can be supplied by positive gas ions. For example, a low pressure of hydrogen can be maintained in the tube envelope by electrically heating a hydrogen reservoir, as used in a hydrogen thyratron, or an inert gas at low pressure can be introduced into the tube envelope before sealing it from the vacuum system. In either case, some of the electrons in the scanning electron beam will collide with gas atoms or molecules, creating positive ions, as shown in Figure 13-31. Since a uniform ion current should reach each part of the retina, ions from the beam deflection region should not be used because there are more ions generated near the tube axis. The field mesh electrode, which defines the electron decelerating field near the target, is therefore kept 10 to 25 V more positive than the focussing cylinder so that these positive ions are accelerated to the right, back into the electron gun. The positive ions used for pedestal current generation are created by the scanning beam in the space between the field mesh and the retina, and the electron beam scans across the retina together with a nearly constant ion beam generated in this space. As suggested in Figure 13-31, the ion beam current is proportional to the scanning beam current and to the residual gas pressure. The ion generating geometry should be invariant across the tube if the mesh is parallel to the target surface and if the electron beam path is everywhere normal to the surface. Hence the ion pedestal current should be uniform across the target.

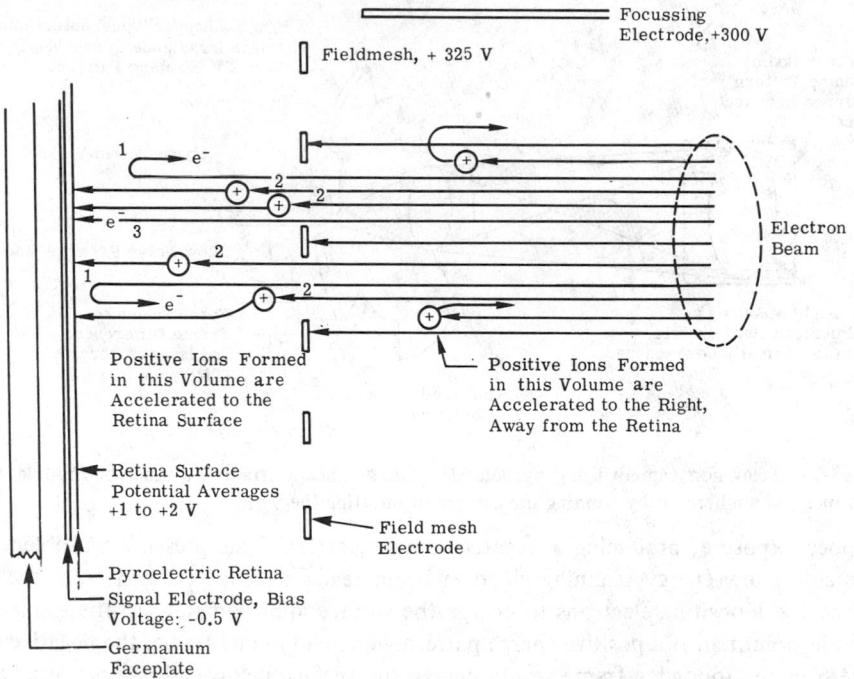

Fig. 13-31. Ionization of residual gas atoms. In the electron deceleration space between the mesh and the target, a positive ion current is provided which is equal to the average electron current reaching the target. Slower electrons (1) fail to reach the target. Faster electrons (3) reach more positive points of the target surface. Many electrons (2) impact gas atoms to create positive ions for pedestal current.

Assume the retina surface is initially at or just below cathode potential, and no electrons land from the electron beam, but the ion beam steadily deposits positive ions, charging the surface positively. As illustrated in Figure 13-32, more electrons can reach the surface as the surface becomes more positive. Eventually the potential is positive enough so that the electron current landing on the retina equals the ion current reaching the retina. Further, since the ion stream is generated by the electron beam, the effect of the two beams at the target is approximately simultaneous so that the target potential is unchanged by either. Thus there is no signal current, that is, no net current flow from the signal electrode through the load resistor, once this equilibrium has been established. The ion current, the pedestal, can be as high as 100 nA, but is usually 30 to 50 nA.

The retina is now exposed alternately to a warmer scene image and a cooler chopper blade. For example, the chopper motion can be synchronized with the vertical scanning motion of the electron beam so that at any point on the retina the irradiance transition occurs just before (or during or just after) the electron beam scans that point. Thus any retina point is heated for a full frame and the resulting charge is read. That point is then cooled for a full frame and the negative charge is available for reading. With interlaced electronic scan, the scanning action is more complicated, as described below.

The detailed reading cycle is illustrated in Figure 13-32. During the time between scans essentially no external current reaches any elemental area, but the change in temperature produces a surface charge pattern on the scanned surface. If the charge is positive, the electron beam lands additional electrons to restore the surface potential approximately to the average pedestal value. If the charge is negative, the beam lands fewer electrons, and the ion beam restores the surface potential approximately to the average pedestal value. In either case, the ion beam current is essentially constant, while the electron current to each retina area varies above and below the pedestal value as a function of the retina temperature excusion in the preceding frame time. These variations in electron landing current are capacitively coupled through the retina to the signal electrode and appear as a varying ac voltage across the load resistor at the input of the preamplifier.

This type of operation could be called the pedestal-current stabilized mode, as contrasted with the cathode-potential stabilized mode employed in the conventional vidicon. The equilibrium target surface potential is that at which the average electron current deposited equals the ion current. As illustrated in Figure 13-32, the ion beam current is constant and essentially independent of target surface voltage, for a given electron beam current and gas pressure. It is shown by a horizontal line at I_{ion}. The equilibrium target surface potential is the intersection of this line with the beam acceptance curve.

As drawn in Figure 13-32, the slope of the beam acceptance curve, the rate at which electron current changes with retina surface voltage, is small at the indicated equilibrium point. Hence a substan-

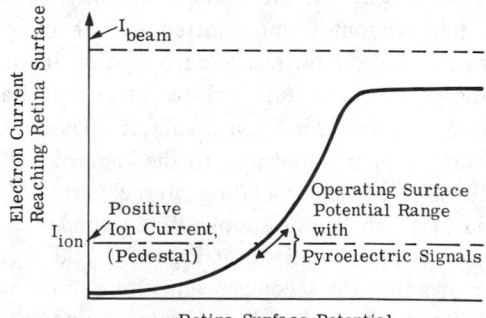

Fig. 13-32. Beam landing characteristics. Constant positive ion current charges target surface positively. The process continues until the average electron current equals the ion current.

tial surface voltage swing is required to change the landing current significantly. However, this voltage swing must be accomplished by charging the retina capacitance with the electron beam current which lands. Hence the reading action has an effective time constant, equal to the product of the effective beam impedance R_B and the retina storage capacitance of the scanned area, C_r. The beam impedance is the reciprocal of the slope of the beam acceptance curve, which is approximated in the low current tail by a Maxwellian distribution. Thus,

$$R_B = \frac{kT_B}{q} \frac{1}{I_p} \qquad (13\text{-}22)$$

where I_p is the pedestal current to the retina at the operating point. T_B is associated with the distribution of beam velocities normal to the target surface. For conventional, negative-grid electron guns T_B is far higher than the 1200 K temperature of the thermionic cathode, because the paraxial velocity distribution is broadened when electrons from an area on the cathode are concentrated into the crossover. Workers have measured T_B as high as 3700 K, and a T_B of 2400 K is usual. If the time constant, $R_B C_r$, is of the order of or longer than a frame time, the video signal will be small because only a small fraction of the pyroelectric charge will be read in each frame time. To make the beam impedance smaller, one should increase the pedestal current by, for example, increasing the electron beam current to increase the gas ion current. Unfortunately, the effective beam temperature usually increases when the beam current increases, and pedestal current in ion current stabilized tubes is usually set for a compromise value of about 30 nA, well down on the beam acceptance curve. For a typical storage capacitance of 5×10^{-9} F (9.2×10^{-9} F when referred to the scan pattern size) $I_p = 30$ nA, and $T_B = 2400$ K

$$\tau = \frac{kT_B}{q} \frac{C_r}{I_p} = 6.1 \times 10^{-2}, \text{sec} \qquad (13\text{-}23)$$

Thus, as usually operated with the circular retina just filling the height of a 4 × 3 aspect ratio scanning pattern, a typical 30 nA pedestal current gives a readout time constant of 61 msec, and the beam reads only about 32% of the available signal charge in the single readout frame allotted by the chopper action. Nonuniformities in the gas ion pedestal also have restricted operation to the 30 to 50 nA range. Methods for decreasing the beam temperature and the retina capacitance are described below.

Secondary Emission Pedestal. The positive pedestal current to the scanned surface of an insulating pyroelectric surface can also be supplied by secondary electron emission. Figure 13-33 shows that the secondary emission ratio for a typical triglycine sulphate retina increases to a broad maximum of approximately 1.6:1 which is essentially constant for primary energies between 60 and at least 80 eV. In the secondary

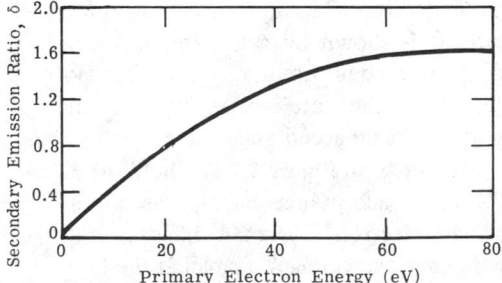

Fig. 13-33. Secondary emission characteristic of triglycine sulphate [13-4].

emission pedestal model (SEPM) of operation, pioneered in the United States by workers at Philips Laboratories [13-4] and at Amperex [13-5], the voltage between the electron-gun thermionic cathode and the retina is increased to approximately 70 V during the horizontal retrace interval while the beam is scanned in reverse from the right-hand to the left-hand edge of the image. (The directions are referred to object space or to the displayed image.) The arrangement is illustrated in Figure 13-34. The retina backplate potential is left constant at -10 V with respect to ground, while the scanned surface averages close to ground. The resulting field inhibits depoling of the retina. The gun cathode is grounded during the reading part of each scanning line period, but pulsed to -70 V during the retrace time. The 70 V electrons cause a net secondary electron current flow from the surface of

$$I_S = I_B \tau_m \ (\delta - 1) \qquad (13\text{-}24)$$

where I_S = secondary electron current leaving retina surface
I_B = electron beam current
τ_m = transmission or open area fraction of the field mesh electrode
= 0.65
δ = secondary emission ratio of retina surface
= 1.6 for triglycine sulphate at 70 eV

Since the secondary emission ratio, δ, varies only very slowly with retina surface potential for 70 eV primary electrons, this mechanism provides a pedestal current which is essentially independent of the pyroelectrical signal. In the equilibrium state, the electronic charge removed by the electron beam during the retrace time, area b in Figure 13-34(d), must equal the electronic charge deposited by the beam during the reading time, area a in Figure 13-34, both averaged over many frame times. For U.S. broadcast scanning standards the line scan period is 63.5 μsec, the reading time is approximately 53 μsec, and the blanking time approximately 10.5 μsec. The retrace time is about half of this, since part of the blanking period is provided for scan settling after retrace. Thus, the net current due to secondary emission during retrace must be about ten times larger than the average signal current, which is defined as the pedestal current. The secondary emission current is adjusted by setting the beam current during retrace by pulsing the beam control electrode negative with the cathode so that the control-electrode-to-cathode-potential is less negative during retrace than it was during normal operation. To produce a pedestal of 100 nA, one must have a net secondary emission current of about 1000 nA, and Equation (13-24) shows that the beam current must then be about 2600 nA. This beam current can be provided by a vidicon gun, especially since the cathode to accelerating electrode voltage is increased by 70 V during the retrace time. However, the calculation demonstrates that providing pedestal currents much larger than 100 nA would require a modified electron gun design.

In the secondary-electron-pedestal mode of operation, the detailed signal generating process at the retina differs from that described earlier in this section for gas ion pedestal operation. Positive current flows to the retina surface and out through the load resistor only during the retrace interval, while negative signal current flows to the sensor

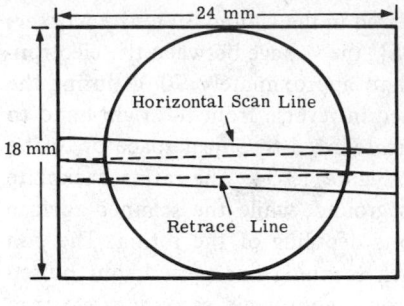

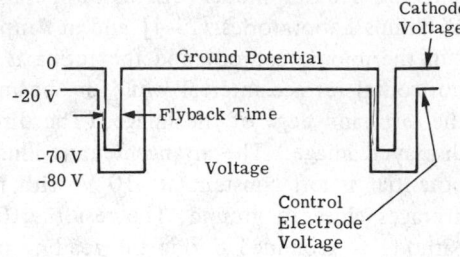

(a) Scanning pattern.

(b) Cathode and control electrode waveforms.

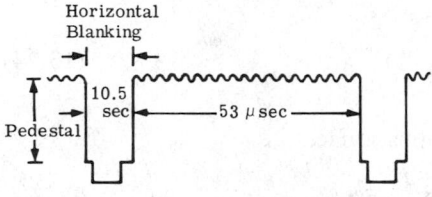

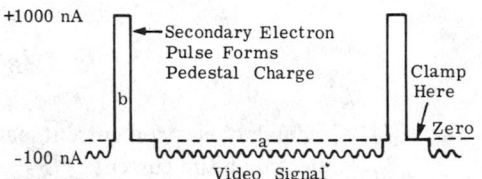

(c) Composite video signal.

(d) Signal electrode current.

Fig. 13-34. Secondary emission pedestal mode. The higher energy beam during retrace charges the retina surface positive by secondary electron mission. The average signal charge read from the retina area, a, equals the average pedestal charge area, b [13-6].

surface and out through the load resistor during the reading time. Thus, a current flows in the load resistor during every reading time, whether the target has sensed a changing thermal image or not. The secondary electron current is constant, essentially independent of retina surface potential, and is set by the beam current during the retrace interval.

The electron current landing during the reading time is determined by the instantaneous retina surface potential as the electron beam scans across each elemental area. As suggested in Figure 13-35, if the surface potential is initially negative no reading electrons land, but electrons are removed from the surface during each retrace time, charging the surface more positively, and increasing the rate of deposition of reading electrons until equilibrium is reached. Actually, if the potential of surface element is considered in detail, the equilibrium is a complex and dynamic one. During retrace, the elemental surface potential increases by

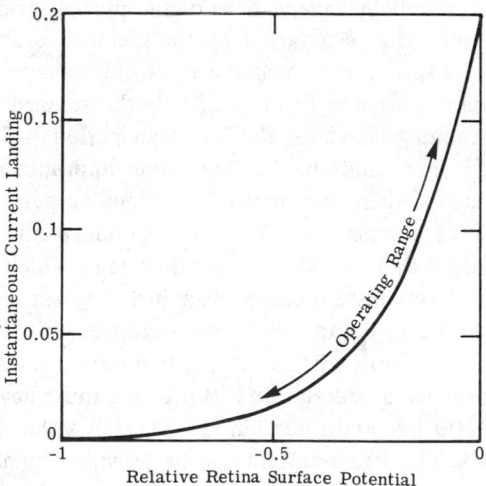

Fig. 13-35. Typical beam landing characteristic for a dielectric retina near cathode potential. For a 100 nA pedestal current the retrace charge moves the potential about 0.3 V more positive. The reading action slides the potential down the curve so that $1/C \int I(V) dt = 0.3$ V, where the integral is over the dwell time.

approximately 0.3 V. If there is no change in retina temperature, during the next reading cycle the potential decreases by approximately 0.3 V, back toward the electron gun cathode potential, for this is cathode potential stabilized (CPS) operation (just as in the conventional vidicon). But the exact operating range, and in particular the fraction of charge left on the retina, is a function of the shape of the beam acceptance curve. At equilibrium, the voltage excursion is 0.3 V for a typical retina at 100 nA pedestal current. The charge-fraction remaining can be reduced by use of a lower capacitance retina.

A pyroelectric vidicon used in a camera with a scan synchronized chopper, viewing a scene which is warmer than the chopper blade, operates as follows. An elemental sensor area is exposed to the warmer scene just after the electron beam has read a signal from that area and returned the surface essentially to gun cathode potential. The area is charged approximately 0.3 V positive by secondary emission action during electron beam retrace—probably a total swing from the effect of several retrace scans since the higher voltage electron beam during retrace will not be as well focussed as during the reading cycle. But, in addition, the retina temperature will increase during the frame time, and an additional positive pyroelectric charge will appear, so that the electron-beam current landing on the retina will be larger than 100 nA when the beam next reads that area. As before, the surface will be recharged toward thermionic cathode potential by the reading action of the electron beam. The shutter now closes, and since by assumption the shutter is cooler than the scene, the retina temperature falls during the next frame time. The secondary emission current generated during the retrace periods again charges the surface about 0.3 V positive, but the negative pyroelectric charge now decreases the surface potential, and fewer reading electrons are required to charge the surface again toward cathode potential. The reading current is smaller than 100 nA in this frame.

The variation of the reading current above and below the pedestal value, capacitively coupled through the retina, constitutes the video information. The resulting electron current is larger where a scene area is warmer, smaller where it is cooler. But the signal is always superposed on the 100 nA pedestal current.

As described, the pyroelectric vidicon, operated with a chopper, would produce a signal of alternating polarity in successive frame times. For normal U.S. broadcast scanning standards, the picture information is a 15 Hz signal superposed on a 30 Hz pedestal. To convert the signal to drive a display, one could in theory invert the signal from every other frame, but the large pedestal signal would then become a 15 Hz signal and would produce a very objectionable flicker. Helmick and Woodworth [13-7] suggest instead subtracting each frame of information from the preceding frame, then inverting every other difference signal. This method has the important result of cancelling the pedestal signal, which is unchanging from frame to frame, so that pedestal nonuniformities do not appear in the signal.

The secondary emission pedestal mode of operation of a pyroelectric vidicon has two important advantages. First, the tube may be processed for a better vacuum so that thermionic cathode life can be several thousand hours, like that of a conventional vidicon. Second, the pedestal current may be set high enough to produce reasonably lag free operation. A minor technical problem with secondary emission pedestal operation is that the retrace scanning pattern must be linear. Conventional television circuits seek to linearize the scanning pattern during the reading period, and simply ensure that the retrace cycle is completed and any transients damped out before the start of the next reading period. One should not be surprised if a conventional television camera requires some scanning circuit modification to provide linear retrace scanning action to obtain satisfactory SEPM operation, when used with a pyroelectric vidicon.

Conduction Pedestal and Ramp Pedestal. The pedestal current can also be provided through the sensing surface by conduction, if the pyroelectric crystal has appropriate resistivity in the order of 10^{10} to 10^{12} ohm cm so that a 30 to 100 V positive bias on the backing electrode results in a 100 nA average current through the scanned area of the retina. However, the resistivity must be high to avoid bleeding off the pyroelectric signal in a frame time. That is

$$\tau = RC = \rho \frac{l}{A} K_r \epsilon_o \frac{A}{l} = 8.85 \times 10^{-12} \rho K_r \qquad (13\text{-}25)$$

must be much larger than 1/30 sec. A typical K_r is 72. This requires $\rho \gg 10^8$ ohm $m = 10^{10}$ ohm cm. Thus a choice of $\rho = 5 \times 10^{11}$ to 10^{12} ohm cm is appropriate. Unfortunately, pyroelectric materials like triglycine sulphate (TGS) and triglycine fluoroberyllate (TGFB) are insulating crystals with resistivities far above 10^{12} ohm cm. Of materials currently being considered, only lead germanate ($Pb_5Ge_3O_{11}$) and doped lead zirconate titanate (PBZT) have resistivities low enough to be considered, but the former is fragile and the latter has a high dielectric coefficient.

Pedestal current through the retina can also be provided as a displacement current. The capacitance through a typical TGS retina is large enough so that a slowly increasing backing electrode voltage can provide a useful pedestal current for 10 or 20 sec of operation. After this time, the video signal is interrupted for several frames, the electron gun voltage is changed to bombard the sensing surface with 70 V electrons to charge the surface more positively (toward collector mesh potential), the beam is cut off, and the back plate voltage is then decreased to a negative potential. Normal reading-mode voltages are restored to the electron gun, and the retina voltage ramp cycle is repeated.

This mode of operation has been used to produce a pedestal current of 30 to 50 nA, at the cost of a flicker-like interruption for repriming every 10 or 20 sec. The limitation, beside operating time, is point-to-point capacitance variations due to retina thickness nonuniformities which produce pedestal nonuniformities.

For both the conduction and ramp pedestal modes, the irradiance pattern in the image must be modulated to produce a pyroelectric signal, either by panning the camera or by chopping the radiation. The pyroelectric signal is superposed on the pedestal signal current and pedestal nonuniformities will appear in the signal unless a pedestal subtraction scheme is employed.

Image Modulation—Chopping versus Panning. To televise a stationary scene, one must modulate the image irradiance pattern reaching the pyroelectric vidicon either with a chopper or by panning the camera. Both methods have advantages. Most literature written before 1975 reports better low-contrast imaging indicated by smaller minimum-resolvable-temperature differences in the panning mode than in the chopping mode, despite analysis which predicts better thermal MTF when chopping. Further, the early data of Watton and coworkers described below show signal amplitude in the panning mode roughly 10 times larger than that in the chopping mode at equal test pattern spatial frequencies. Analysis indicates both these results are probably due to readout

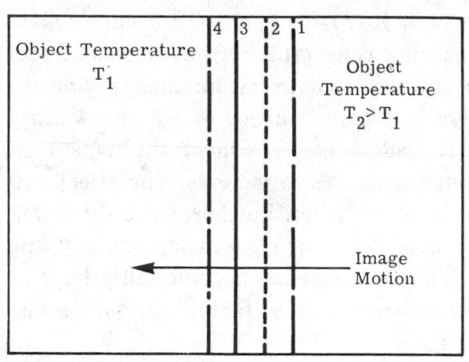
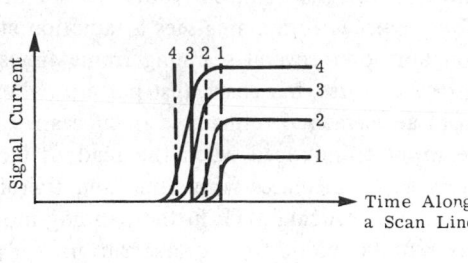

(a) Panning makes image (two half-fields each at a uniform temperature) move to the left at the sensor. Vertical lines show position of edge in successive frames.

(b) Subsequent scans show greater phase shift, lower MTF, but larger signal amplitude.

Fig. 13-36. Panning mode.

lag effects, complicated by the thermal time constant of the retina. Watton's data were taken with pedestal currents of only 1 or 2 nA, at which the effective beam resistance, $R_B = 2.5 \times 10^8$ ohms, and the readout time constant, $\tau = 2.3$ sec, were far larger than a frame time. Thus, in a 25 or 30 frame per second system, only a very small fraction of the charge generated on the retina surface was read out in a frame time.

In a panned system, the charge not read in one frame is added to the charge generated in the next frame, increasing the voltage swing and therefore the signal charge read out in each frame. Consider a simple image, shown in Figure 13-36—a straight vertical edge between two different uniform irradiance regions. When the edge passes across an elemental retina area during a frame time the area warms, and a positive charge pattern appears on the scanned surface. Some of this charge is read in the following scan cycle, but the retina continues to warm in subsequent frames. The rate of change of retina temperature depends on the radiation balance between the retina and the surroundings, including the scene, by radiation through the lens. With $\Delta T = T_s - T$, i.e., the scene temperature minus the retina temperature,

$$\frac{d(\Delta T)}{dt} = \frac{\tau_a \tau_o}{8 \, (F/\#)^2} \; \frac{4\sigma \, \epsilon T^3}{c' \, d} \; \Delta T \qquad (13\text{-}26)$$

where τ_a = transmission factor of atmosphere
  τ_o = transmission factor of optical system
  σ = Stefan-Boltzmann constant
  ϵ = emissivity
  $F/\#$ = focal ratio of optical system
  c' = volume specific heat of retina material
  d = retina thickness

Using typical values $\tau_a\tau_o$ = 0.8, ϵ = 0.7, T = 290 K, $F/\#$ = 1, c' = 2.5 J cm^{-3} °C^{-1}, and d = 3 $\times$ 10^{-3} cm, one finds that the thermal time constant is between 1.5 and 2 sec. Thus, when a retina area sees a radiation step, the rate of temperature change is nearly constant over several scanning frame times. For a large-area image or for the warmer bar of a coarse, bar-chart, test pattern, when the readout factor is small, the signal read from an elemental retina area in successive frames continues to increase. The effect will be more pronounced when the readout factor is even smaller. Further, since the developed signal increases with time, and therefore with position, readout lag has a strong effect on reducing MTF in the panning mode. That is, signal can become fairly large in the panning mode for a coarse pattern for a small beam readout factor, although signal will be small at higher spatial frequencies.

In contrast, in the chopping mode, the retina heats for one frame time, then cools for one frame time. The signal charge does not add from frame to frame, and in fact the charge not read from the retina during the scene frame subtracts from the opposite charge developed during the chopper frame. Thus, pyroelectric vidicons for use in the chopping mode must have a high beam-reading factor (in vidicon terms, have low beam lag) to provide high response. This appears to be the basic reason that the early gas-pedestal tubes gave low signal in the chopping mode. At a 1 or 2 nA pedestal value, the reading factor, i.e., the fraction of signal read in a frame time, was only a few percent. Thus, in the panning mode, the integrated charge image could increase the voltage developed at the retina surface to at least 10 times the single frame value, in effect increasing the average signal current and therefore increasing the readout factor in the later frames after a transition. In the chopping mode, no such increase was possible. Recent developments should change this picture. The use of the secondary emission pedestal mode, and of image difference processing to cancel pedestal nonuniformities, and the development of low beam-temperature electron guns, all permit use of large pedestal currents with low beam-temperatures for beam discharge factors as high as 40%, which should make signal currents nearly equal in both the panning and chopping modes.

Pyroelectric vidicons are usually operated in conventional broadcast type television equipment with interlaced scanning. The cycle of operations is illustrated in Figure 13-37 with a scan-synchronized chopper. (See also Figure 13-30.) The chopper exposes the retina to the scene for one frame time, then to the chopper blade for one frame time, with the edge of the chopper blade moving from the top to the bottom of the picture in one field time, approximately in synchronism with the scanning electron beam. When the trailing chopper edge starts from the top, the electron beam scans areas of the retina which have viewed the blade for 1/30 sec. In the next field, the scanned retina areas have viewed the scene for 1/60 sec, and in the third field the scanned areas have viewed the scene for 1/30 sec. As shown above, the rate of temperature change is essentially the same for each field, and, if the two scans were independent, the second field in each frame would have twice the signal amplitude of the first. But in fact the electron beam in any vidicon camera tube has an approximately Gaussian cross-section with an effective width at half amplitude of about 46 μm. When the beam scans each line of one field, about half the signal is derived from the charge geometrically associated with that line and one quarter from the charge associated with each adjoining line in the other field. Thus, in each field the beam erases the stored charge pattern fairly uniformly although not completely, and, to a reasonable approximation for signal amplitude and lag modeling, the retina can be considered as being scanned completely 60 times per second. To this approximation, the exposure history matters only for the preceding field time, and the frame synchronized chopper should give equal amplitude signals in both fields of a frame. Note that calculated signal amplitude is not reduced, however, since both the exposure and readout times have been halved in this model.

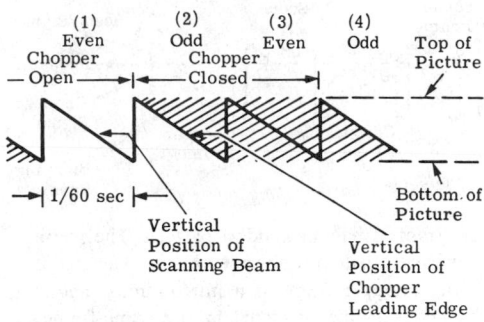

Fig. 13-37. Chopping mode. The shutter moves from top to bottom together with the electron beam. Because the beam reads the entire retina in each field, only the exposure in the last 1/60 sec determines the first-order signal, and the signal in scan (1) equals that in scan (2); that in scan (3) equals that in scan (4).

Because of the better results observed with early gas-pedestal pyroelectric vidicons and because conventional television cameras could be used without modification, many workers through 1975 concentrated on use of camera panning for image modulation. This panning mode has limitations, however, which were recognized and set forth by Petito and Garn [13-8], among others. First, the panning mode produces a signal only when there are scene temperature variations in the direction of motion. This leads to unsymmetrical response, so that, for example, horizontal panning gives very little signal from the irradiance transition at the horizon. Since the direction of temperature gradients in the field of view is not known in most cases, a possible solution is circular panning or nutating, rather like the orbiting scheme once used to prevent image fatigue on image orthicon camera tubes.

Since panning sensors respond to temperature or emissivity gradients, all panning cameras are difference detectors. This effect, which helps image visibility, is probably partly responsible for the small minimum-resolvable-temperature values reported in the panning case.

A second problem relates to the lag effects in the pyroelectric vidicon. If the scene contains, for example, a large object whose temperature is much above the average, the corresponding area of the pyroelectric retina will reach a high temperature. This area of the pyroelectric retina will then take many frame times to return to an equilibrium temperature, and during the entire recovery time it will produce a dark area in the reproduced image. Thus, in the panning mode, any hot object is followed by a dark or black trail in the image which can and often does mask other scene details. This effect is compounded by beam readout lag, since the repeated negative signal in successive frames shifts the target operating-point negative and several more frames may be required to restore electronic equilibrium.

A final drawback to the panning mode is observer fatigue. Continuous motion in the entire displayed image requires the observer to make a conscious effort to examine the fine detail. With circular panning, the picture can be immobilized by use of counter deflection in the display, but only at the cost of some reduction in effective field of view. Further, any hot object will then result in a counter-rotating black trail in the otherwise stationary displayed picture.

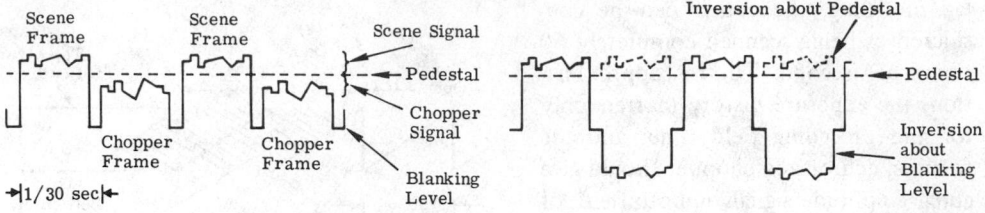

(a) Frame frequency video pattern. The scene is warmer than the chopper blade. The signal in the chopper frame is a mirror image about the pedestal line of signal in the scene frame. A secondary emission pedestal mode is illustrated.

(b) Inversion of alternate frame signals about blanking level would leave a large 15 Hz flicker. Inversion about pedestal would match signals from successive frames, but is difficult to implement because of pedestal nonuniformities.

Fig. 13-38. Video signal characteristics with a frame synchronized chopper.

Frame-to-Frame Image Difference Processing (IDP). When a pyroelectric vidicon is used with a frame-synchronized chopper, as shown in Figure 13-30, the image information is the signal above pedestal (Figure 13-38(a)) when the retina has viewed the scene for a frame time, that is, in every other frame. Conversely, in the alternate frames when the retina has viewed the chopper blade for a frame time, the image information is the signal below pedestal. Since the information-bearing part of the signal changes sign in successive frames, the video signal for the display could be provided by inverting the signal in alternate frames. Unfortunately, the pedestal even for the gas ion pedestal mode is, in practice, larger than the signal, and for the secondary emission pedestal mode it is far higher. If video signal polarity is simply inverted in alternate frames, the pedestal signal produces a high amplitude 15 Hz flicker as shown in Figure 13-38(b). The flicker could be eliminated by subtracting the pedestal from the video signal in each frame before processing (the dotted line in Figure 13-38(b)). The inversion would then ideally contain only the information-bearing signal. In practice, however, pedestal currents in pyroelectric vidicons are always somewhat nonuniform, and simple subtraction of an average pedestal value from the signal leaves pedestal nonuniformity components added to the signal. The alternate-frame inversion technique then results in flicker, due to pedestal nonuniformities, which is much smaller but often objectionable.

Up to this point, all signal processing methods suggested could be performed with modest circuit modification to a standard closed circuit vidicon camera to subtract pedestal and invert video polarity in alternate frames. In 1975, Helmick and Woodworth [13-7] and other workers examined imagery produced with panning cameras and with chopping mode cameras, and decided that neither was adequate. More radical signal processing seemed necessary. The image difference processing idea used with a chopping mode camera was therefore revived and built into a more complicated camera which produced the best television imagery seen to date from a pyroelectric vidicon.

Figure 13-39 shows in schematic form a pyroelectric vidicon camera system with image-difference processing. The key addition is a video disc recorder, or other memory, used to store each frame of video information for one frame time. In this method, the real-time video signal for each frame is subtracted from the stored video signal from the preceding frame. If the current frame is that taken after the pyroelectric target views the chopper, the resulting difference signal has correct polarity and is transmitted as

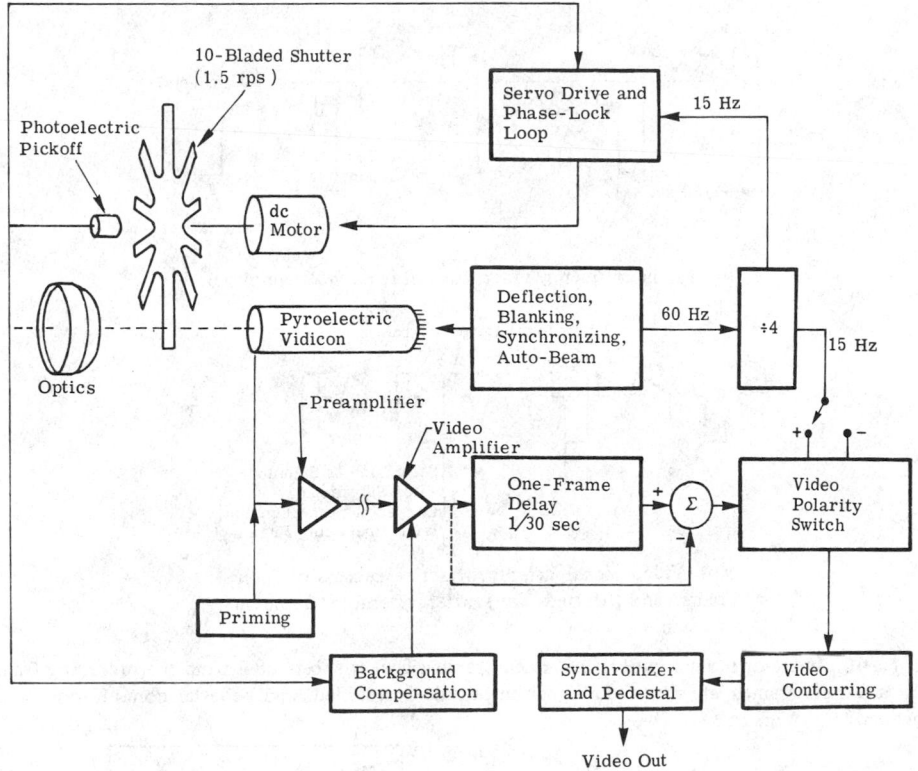

Fig. 13-39. Pyroelectric camera system with image-difference processing [13-7].

formed. If the current frame is that taken after the pyroelectric target views the scene image, the resulting difference signal has negative polarity and is inverted before transmission. These manipulations are illustrated in Figure 13-40.

Image-difference processing used with a chopping-mode camera has significant advantages. First, the pedestal in each frame is subtracted from the pedestal in the preceding frame. Thus pedestal nonuniformities cancel, greatly improving the quality of the reproduced picture. Second, the signal amplitude in each frame is essentially doubled since it is now the difference of the heating signal and the cooling signal. Minimum resolvable temperature is lowered as a result. Third, this method produces symmetrical resolving power, and actually improves thermal MTF, as explained below. Lastly, the effect of a negative after-image following hot objects is essentially eliminated, and the entire image is stationary and essentially free of flicker. The cost of IDP is substantial. Nevertheless, the superior image quality and smaller minimum-resolvable-temperature differences achieved with IDP make this an approach to consider.

Present Performance of Pyroelectric Vidicons: Modulation Transfer Function (MTF). Figure 13-41 gives the calculated thermal MTF of a uniform triglycine sulphate retina 18 mm diameter and 25 μm thick [13-9]. The upper solid curves are calculated for panning in the direction of irradiance variation of a sinusoidal test pattern. The lower dotted curves are for chopping mode. The low-frequency thermal response in the

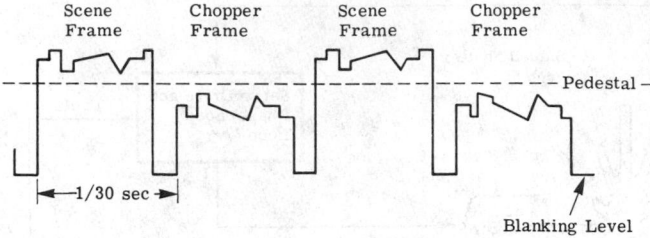

(a) Frame frequency video pattern from video amplifier.

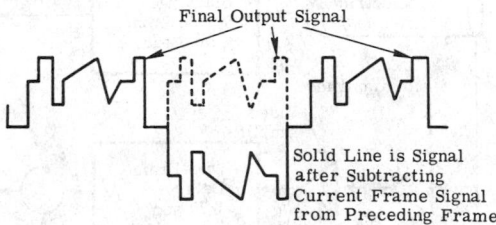

(b) Video signal after subtracting successive frame pairs, and (dotted line) after alternate frame inversion.

Fig. 13-40. Image-difference processing. Each frame signal is substracted from the preceding frame, then alternate frames are inverted. Signal amplitude is doubled, and pedestal nonuniformities are cancelled.

chopping mode is approximately half that in the panning mode. Therefore the chopped mode curves approach 50% rather than unity at low frequencies.

There is an additional effect known as signal mixing which degrades response in the panning mode, shown by the 10 mm sec^{-1} solid curve of Figure 13-41. If panning speed is too fast, and the image moves a whole cycle in a field time (interlaced scanning is assumed), the response will fall to zero since each small sensor area will average the input irradiance pattern to a neutral level. Thus at 10 mm sec^{-1}, response is zero at 5 cycles mm^{-1} for a 25 frame, 50 field per second system no matter how good the tube may be. Not apparent from these curves is the fact that response in the panning mode also falls to zero as the image spatial frequency approaches zero.

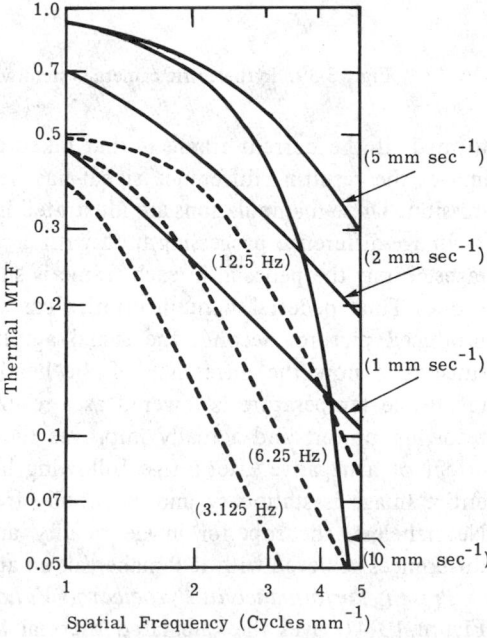

Fig. 13-41. Thermal modulation transfer functions. Solid curves are for panning mode at four panning speeds. Dotted lines are for chopping mode at three chopping frequencies [13-9].

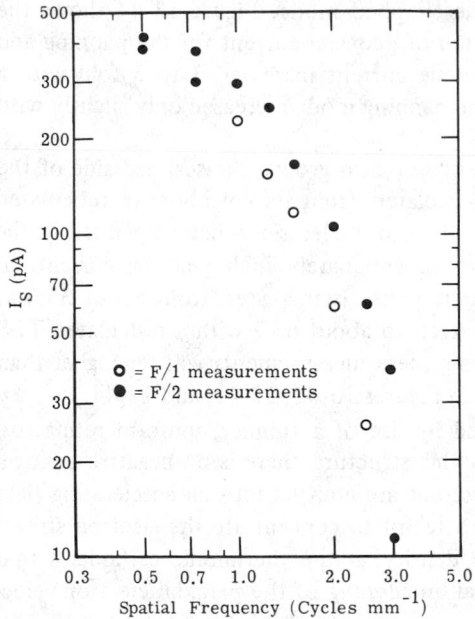

Fig. 13-42. Measured video signal-panning mode [13-9].

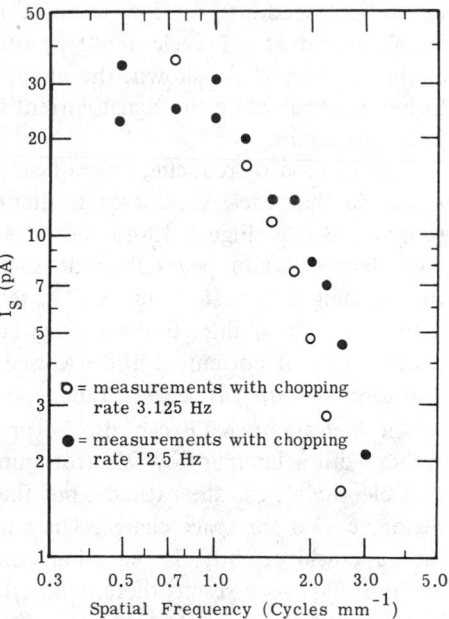

Fig. 13-43. Measured video signal-chopping mode [13-9].

In the chopped mode, MTF is highest for the highest chopping frequency, 12.5 Hz, because there is less time available for lateral heat flow. This curve, taken with a 25 frame per second system, describes characteristics where the sensor views the scene and the chopper blade in alternate frames.

As shown, for this retina the lateral heat spread or signal mixing reduces calculated thermal MTF to a small value at 3 to 5 cycles mm^{-1} in the image. Thus, tubes using these retinas are expected to have maximum useful resolutions of 108 to 180 TV lines per picture height.

Experimental measurements of video

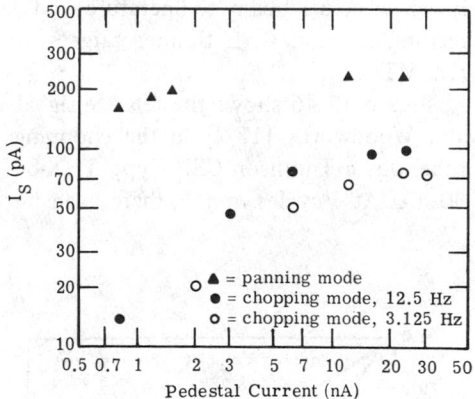

Fig. 13-44. Signal variation with pedestal current. Spatial frequency is 1.1 cycle mm^{-1} [13-9].

signal as a function of test-pattern spatial frequency are shown in Figures 13-42 and 13-43 for the panning and chopping modes respectively. These data were taken for a gas ion current pedestal tube with an 18 mm, 25 μm thick retina, using a thermal bar-chart test pattern. It is apparent that MTF is far below that predicted by thermal spread calculations, and that the signal amplitude in the chopping mode is only about 10% that in the panning mode. The pedestal current was set between 1 and 2 nA during these measurements for good pedestal uniformity, which produced a very small beam readout factor and accounts for low response in the chopped mode. Figure 13-44 shows the signal current at 1.1 cycle mm^{-1} as a function of pedestal current for the panning and chopping modes. As shown, the chopping-mode current increased 3 to 5 fold with a higher pedestal while the signal current in the panning mode increased only slightly with pedestal current.

One method of reducing lateral heat conduction is to groove the scanned side of the retina so that each small area is thermally isolated from its neighbors (a reticulated retina) [13-10]. Figure 13-45 shows results for a pyroelectric vidicon operated in the secondary emission pedestal mode to assure an adequately high pedestal current. In the panning mode at 2 mm sec^{-1}, relative response is increased from about 35% at 3 lp mm^{-1} for a thin, uniform-layer TGS target, to about 63% with a reticulated TGS target. Pedestal current of 100 nA used during these measurements was far higher than that used with the gas pedestal tube reported in Figures 13-42, 13-43, and 13-44.

An increase in MTF can also be provided by use of a thinner, uniform retina, together with a laminar-flow electron gun. In this structure there is no negative control-grid electrode near the cathode, but the electrons are emitted into an accelerating field region, except for space charge. There is no attempt to concentrate the electron stream into a crossover, but the emission current density at the thermionic cathode is that at the retina. As a result, there is no artificial broadening of the paraxial electron velocity distribution, and effective beam temperatures of 1500 to 1800 K have been measured. In addition, the laminar flow gun permits use of a larger pedestal current since the beam temperature does not increase significantly with beam current. Together the lower effective beam temperature and larger pedestal current maintain a high signal reading fraction with thinner targets, reducing lateral heat spread and improving system MTF.

Figure 13-46 shows the relative signal versus spatial frequency measured by Helmick and Woodworth [13-7] in the chopping mode using image difference processing. The tube was a Thomson-CSF Type THX-840 which uses a gas ion pedestal set for about 30 nA. At 3 cycles mm^{-1}, there is an uncorrected 40% signal which is nearly 50% if the

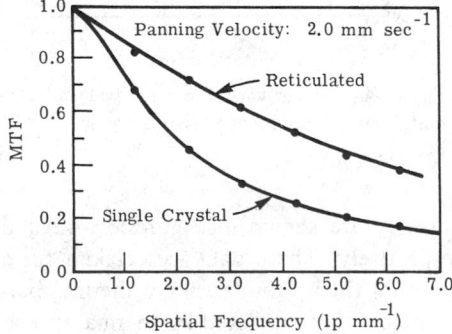

Fig. 13-45. MTF versus spatial frequency for reticulated TGS retina [13-10].

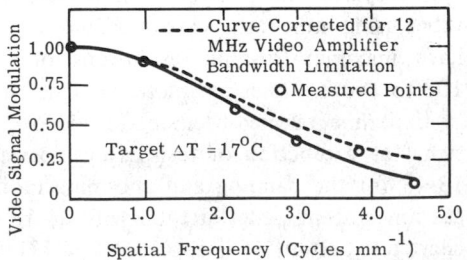

Fig. 13-46. Signal modulation versus spatial frequency. A 15 Hz chopper and IDP are used [13-7].

MTF-limiting effects of a 1 MHz video amplifier are factored out. Part of their excellent results is due to their use of image difference processing, and part may be explained by their excellent optics which had an optical MTF greater than 50% at 25 cycles mm^{-1}, hence close to 100% at 3 to 5 cycles mm^{-1}. The good results are also probably due to use of the chopping mode, which limits the time available for lateral heat spread.

Responsivity. When the temperature of a pyroelectric crystalline retina is changed, the resulting change in spontaneous volume polarization causes an electrical surface charge to appear on the polar faces of the crystal. This pyroelectric effect is described by the pyroelectric coefficient, P, in units of C m^{-2} K^{-1}. If the pyroelectric target is operated in a vidicon, the resulting signal current is determined by the rate at which this charge is read. Thus, the available signal current is $I_S = P(A/t)\,[T_m(t) - T_m(t - t_f)]$, and if all net radiation incident on the target is absorbed, then, in the chopping mode where the target is exposed to scene radiation for a frame time,

$$I_S = P\frac{A}{t_f}\Delta T_m = P\frac{A}{t_f}\left(\frac{EAt_f}{A\rho dc_P}\right) = \frac{P}{\rho dc_P}(AE)$$

(13-27)

$$I_S = \frac{P}{\rho d\, c_P}\,\Phi$$

where P = pyroelectric coefficient, C m^{-2} K^{-1}
  d = retina thickness, m
  ρ = sensor material density, kg m^{-3}
  c_P = specific heat, J kg^{-1} K^{-1}
  A = effective scanned area including overscanned portion, m^2
  t_f = frame time of scanning system, sec
  E = net irradiance at target, W m^{-2}
  Φ = net power incident on effective scanned area at target, W
  ΔT_m = change in retina temperature from one frame to the next, K

The responsivity of $\mathfrak{R} = I_S/\Phi = P/(\rho dc_P)$ may be expressed in A W^{-1} as for nonimaging detectors.

Equation (13-27) actually applies only to a pyroelectric vidicon in the chopping mode and when the image irradiance is uniform over the active target area. It can be made more general by including the MTF of the optical system and the tube and camera electronics, and by expressing the radiation signal in terms of scene parameters. The MTFs are discussed in more detail in the following section.

One may assume that the scene includes a multiple-bar-type test chart whose exitance varies sinusoidally in the cross-bar direction. If the average scene temperature and the average retina temperature are approximately equal, the irradiance pattern in the image at the sensor is given by

$$E(y) = \frac{\tau_a\tau_o\tau_f\,4\sigma T^3\Delta T\cos(2\pi fy)}{4(F/\#)^2\,(1+m)^2}\,O(f,x,y)$$

(13-28)

where $\tau_a\tau_o\tau_f$ = respectively, the fractions of a blackbody emission passed by the atmosphere, the optics, and the sensor face plate. Let $\tau_a\tau_o\tau_f = \tau$.

ΔT = the peak temperature variation from the mean temperature of the sinusoidal test pattern (modulation amplitude)

f = spatial frequency in, e.g., cycles per millimeter in the image at the pyroelectric retina

T = average scene temperature (and pyroelectric retina temperature)

$F/\#$ = focal ratio of the optical system

m = linear magnification of the optical system

$0(f,x,y)$ = modulation transfer function of the optical system, expressed to indicate an expected variation of MTF at off-axis points x, y in the image, but used hereafter as $0(f)$

Combining Equations (13-27) and (13-28), one gets

$$I_S = \frac{PA}{\rho d c_P} \frac{\tau \, 4\sigma T^3 \, \Delta T \cos{(2\pi fy)}}{4(F/\#)^2 \, (1+m)^2} \, 0(f) \, D(f) \, \alpha(I_P, t_f, f) \qquad (13\text{-}29)$$

where $D(f)$ is the thermal MTF of the pyroelectric retina, and α, the signal reading factor, is defined later.

The inclusion of the MTF as part of the responsivity equation seems strange at first. For conventional vidicons, one defines a responsivity by irradiating the whole image uniformly, or, as a compromise, by using a bar chart test pattern so coarse that the MTF of all components is unity. This approach does not work at all for the pyroelectric vidicon which senses only changing images. For the chopping mode, the signal current is found to be a function of chopping frequency and of pedestal current, as well as of test pattern spatial frequency. For the panning mode, it is a function of panning speed, electronic scanning frequencies, and pedestal current, as well as test pattern spatial frequency. Further, for the panning mode, the response at zero spatial frequency is zero, so that the normal practice of normalizing MTF to unity at zero spatial frequency must be discarded. Further, in making comparisons between sensor response in the chopping and panning mode, Watton and coworkers have described pyroelectric vidicon performance by Equation (13-29), but used a definition of $D(n)$, the retina thermal MTF, which yields MTF = 2 for low spatial frequencies in the panning mode [13-9].

Equation (13-29) also contains the factor $\alpha(I_p, t_f, f)$, the fraction of the available signal which is actually read by the electron beam. As explained above, Equation (13-23), the electron-beam reading process has a time constant, $R_B C_r$, where R_B, the effective beam resistance, is determined by the pseudo-Maxwellian distribution of scanning beam electron velocities approaching the retina and varies as a function of retina surface potential. The readout time constant is approximately

$$\tau = R_B C_T = \frac{kT_B}{q} \frac{C_r}{I_P} \qquad (13\text{-}30)$$

where C_r = effective target storage capacitance of the entire scanned area including overscanning

T_B = effective beam temperature for measured distribution of electron velocity components normal to the target surface

I_P = average pedestal current at the target

The fraction, α, of the available pyroelectric charge on the target which is read out as signal current in a frame time has been shown by Watton [13-11] for the chopped mode to be

$$\alpha_c = \frac{1}{2}\frac{t_f}{\tau} = \frac{1}{2}\frac{q}{kT_B}\frac{t_f}{C_r}I_P \tag{13-31}$$

The fraction of available charge read is also a function of the spatial frequency of the image at the target, although for uniform pyroelectric target structures the electron beam MTF is usually much higher than the pyroelectric target thermal MTF and the spatial frequency dependence of the reading fraction may be disregarded.

Equations (13-27), (13-28), and (13-29) indicate that to increase pyroelectric response, one should select a material with a high pyroelectric coefficient, low volume-specific-heat, and low thermal conductivity, and should minimize thickness to minimize both the retina heat capacity and the lateral heat flow. However, to maximize α, the charge storage capacitance should also be small, setting a lower limit on retina thickness, and making a low dielectric coefficient desirable. Figure 13-47 shows that both pyroelectric coefficient, P, and relative dielectric constant, K, for TGS are strong functions of material temperature. TGS retinas in Philips or Amperex tubes are normally operated at about 40°C, while British or French cameras have provided target temperatures of 30 to 35°C.

Figure 13-48 shows recent comparative data on the relative dielectric constants of TGS, TGFB, and DTGFB as a function of retina temperature. Figure 13-49 shows comparable data on the responsivity, which is proportional to the pyroelectric coefficient. Finally, Figure 13-50 shows the ratio of responsivity to dielectric coefficient as a figure of merit. Both DTGFB, shown, and TGFB have figures of merit significantly higher than TGS and should permit imaging with smaller scene temperature differences than is possible with TGS retinas.

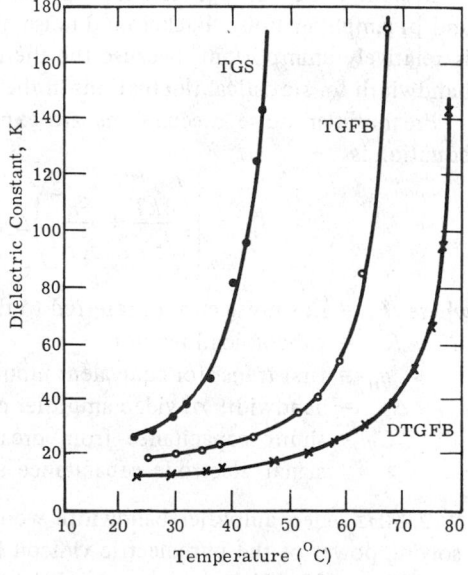

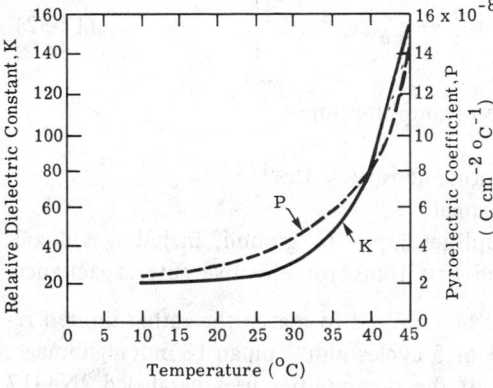

Fig. 13-47. Pyroelectric coefficient, P, and relative dielectric constant, K, versus temperature for TGS. P is measured in the poled state [13-12].

Fig. 13-48. Dielectric constants of TGS, TGFB, and DTGFB as a function of temperature [13-12].

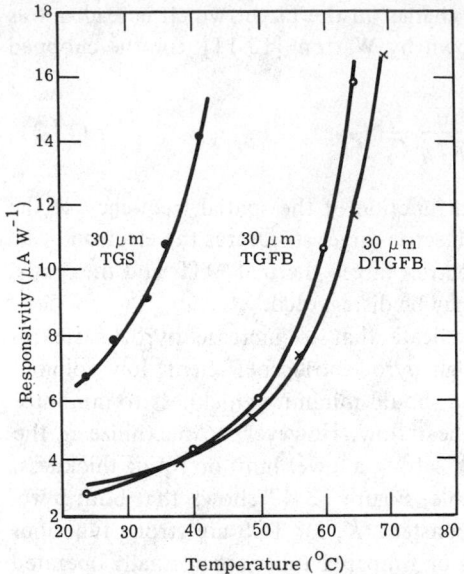

Fig. 13-49. Responsivities of TGS, TGFB, and DTGFB as a function of temperature [13-12].

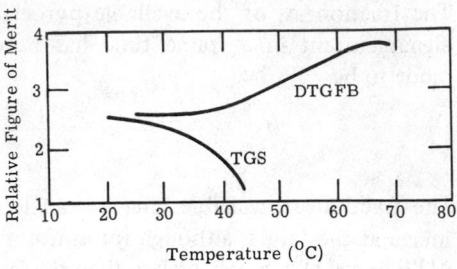

Fig. 13-50. Relative figure of merit for TGS and DTGFB [13-12].

Noise. While pedestal nonuniformities, sometimes called fixed pattern noise, have traditionally limited performance of pyroelectric vidicons, improved target fabrication techniques, use of a secondary emission pedestal, and image-difference processing which cancels pedestal nonuniformities should leave system noise as the principal factor limiting the observer's performance. The relationship between signal-to-noise ratio and observer performance has been extensively studied by Anderson, Rosell and Wilson, and others. For an overview, see Chapter 18.

System noise for pyroelectric vidicon cameras includes pedestal current shot noise and preamplifier noise. Background noise, the limiting mechanism for single-cell scanners, is relatively unimportant, because the thermal time constant of the target sets a narrow bandwidth for statistical fluctuations in the background flux.

Preamplifier noise mechanisms are explained in Section 13.4. The applicable noise equation is

$$I_n = \left[\left(\frac{4kT}{R_L} + \frac{\epsilon_n^2}{R_L^2}\right)\Delta f_v + \frac{4}{3}\pi^2 C_S^2 \epsilon_n^2 \Delta f_v^3\right]^{1/2} \tag{13-32}$$

where I_n = rms noise current referred to the preamplifier input

R_L = vidicon load resistor

ϵ_n = first transistor equivalent input noise density, V Hz$^{-1/2}$

Δf_v = bandwidth of video amplifier channel

C_S = shunt capacitance from preamplifier input to ground, including vidicon signal electrode capacitance and first transistor effective gate capacitance

A 2 MHz video amplifier bandwidth would be required to match the rather limited resolving power of the pyroelectric vidicon (4 or 5 cycles mm^{-1} on an 18 mm high image, or 144 to 180 TV lines per raster height). If the preamplifier uses paralleled 2N4417 transistors as the input stage, $\epsilon_n = 1.2 \times 10^{-9}$ V Hz$^{-1/2}$, and, with $R_L = 2 \times 10^6$ and $C_S = 14$ pF, $I_n = 0.214$ nA. Thus a reasonable state-of-the-art preamplifier should have an equivalent input noise current of about 200 pA. To reduce this figure, one may further narrow the bandwidth and increase the load resistance with a tradeoff of system resolving power.

Shot noise is associated with the pedestal current, the relatively large average electron current which is deposited on the target even in the absence of signal. In general, for a secondary emission pedestal, both the secondary emission statistics and the beam shot noise contribute. According to Conklin, et al. [13-13],

$$I_n = \left[2q\, I_P\, \Delta f \right]^{1/2} \left[\frac{H^2 + \delta}{\delta - 1} \right]^{1/2} = 0.42 \text{ nA} \qquad (13-33)$$

where H^2 = 0.2, the space-charge smoothing factor in the electron beam
δ = 1.65, the secondary emission ratio
I_P = 100 nA, the pedestal current
Δf = 2 MHz

Thus, for the example chosen, pedestal shot noise is larger than preamplifier noise and their total is 0.47 nA. To reduce the noise, one could reduce the pedestal current and decrease the bandwidth.

To complete the example, one can calculate the expected signal current for a 1°C peak-to-peak, scene-temperature excursion from Equation (13-27) as about I_S = 0.75 nA, assuming an $F/1$ lens and an 18 mm diameter, 30 μm thick, TGS target operated at 40°C for a pyroelectric coefficient of 8 $\times$ 10^{-8} C cm^{-2} K^{-1}. Thus NEΔT at the scene should be of the order of 0.5 K, for a one or two cycle mm^{-1} bar-chart image.

To estimate the minimum resolvable temperature, one must use, for example, Rosell's model to compute display signal-to-noise ratio, SNR$_D$. (See Chapter 18.) Approximately

$$\text{SNR}_D = \left(2t_e\, \Delta f_v\, \frac{a}{A} \right)^{1/2} \text{SNR}'_V \qquad (13-34)$$

where t_e = 0.1 sec, integration time of the eye
Δf_v = 2 $\times$ 10^6 Hz, video bandwidth
a = area of one bar in the pattern
A = scanned area
SNR$'_V$ = measured signal-to-noise ratio based on average rather than peak signal, approximately 0.64 times SNR$_V$ in normal terms of PP signal/rms noise

For a 2 cycle mm^{-1} bar-chart image on an 18 mm high television raster, assuming a 10:1 bar length-to-width ratio,

$$\text{SNR}_D = 15.3 \text{ SNR}_V \qquad (13-35)$$

For detection of a bar pattern, SNR$_D$ must be at least 3; hence threshold SNR$_V$ = 0.2, and MRΔT should be about 0.1 K.

Performance Levels Summary, Minimum Resolvable Temperature Difference. Measured and calculated performance data are described in the preceding sections for experimental pyroelectric vidicons in experimental cameras. Several recent advances in the tube or the system have significantly improved performance, and older data are therefore obsolete. However, these advances have not all been tested together and the latest data, listed here, are still tentative. The recent advances include secondary emission pedestal which permits a larger pedestal current for a larger signal current reading fraction, the laminar flow electron gun for reduced electron beam temperature and a larger reading fraction, use of triglycine fluoroberyllate retinas which have a lower dielectric constant, and image difference processing which doubles available signal in the chopping mode and removes most retina nonuniformity effects (but at the cost of increased equipment complexity with a full frame analog memory). In addition, examination of persistence effects and human-factor considerations now establish the chopping mode as preferred, while many previous observations were made in the panning mode.

When one compares overall performance, minimum resolvable temperature difference data are important since they include the effects of response, MTF, noise, and sensor nonuniformities. Figures 13-51 and 13-52 show MRΔT data as reported by Watton using 1 in. pyroelectric vidicons made by English Electric Valve Company. These tubes use the gas-ion pedestal mode with a pedestal current of only 2 to 3 nA. As shown, with F/1 optics, the MRΔT at the scene for a 1 lp mm^{-1} pattern was about 0.7°C for the panning mode and 2.8°C for the chopping mode. These data were taken in 1973 at the Royal Radar Establishment (RRE). Figure 13-53 shows MRΔT of 0.12°C, for the panning mode at 2.75 cycles mm^{-1}. These data were reported by Conklin and Singer [13-13] in December, 1975, and taken with a laminar flow electron gun (to reduce beam temperature) and a thin, 13 μm thick TGS target. Helmick and Woodworth [13-7] report an MRΔT at 1°C for a 7.5 cycle mm^{-1} pattern using image difference processing with a Thomson-CSF THX 840 pyroelectric vidicon using a gas ion pedestal

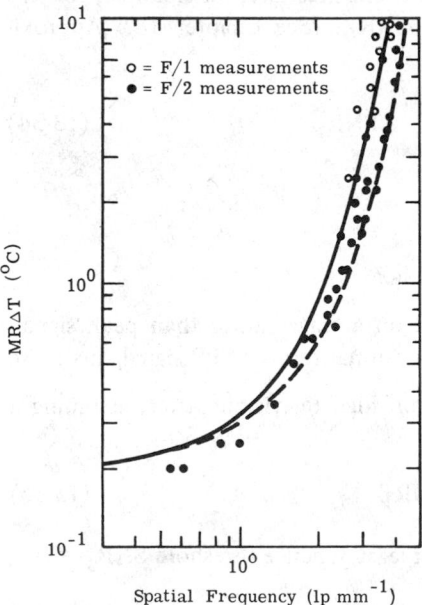

Fig. 13-51. Minimum resolvable temperature difference—panning mode. The lines represent theoretical values; circles and dots represent experimental values [13-9].

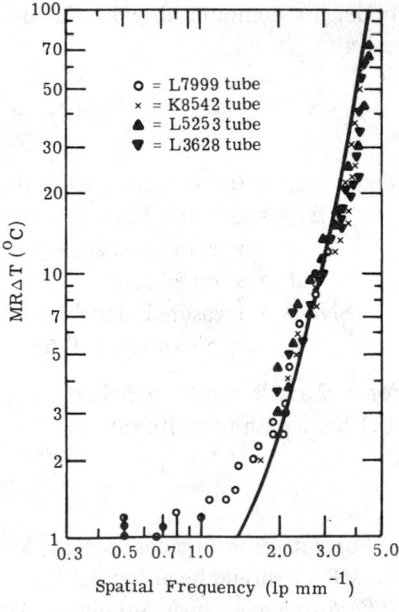

Fig. 13-52. Minimum resolvable temperature difference—chopping mode. The line represents theoretical values; data points represent experimental values [13-9].

at about 30 nA in the chopping mode. Helmick and Woodworth give no other MRΔT data, but the contrast transfer function curve in this paper indicates the signal at 2 cycles mm^{-1} should be at least 10 times higher than at 7.5 cycles mm^{-1}, hence the MRΔT should be about 0.1°C. All the MRΔT data listed here were taken with $F/1$ optical systems. Both these results compare well with the 0.1 K estimate of the preceding section. (Note that the Helmick and Woodworth data were taken with a wedge pattern rather than a burst pattern, and the extrapolated MRΔT results given here are not as accurate as directly measured data.)

Finally, the DTGFB curve in Figure 13-54 shows an MRΔT of 0.07°C at 2.75 cycles mm^{-1} (100 TVL/PH) [13-12]. Significantly, MRΔT at 250 TVL/PH is more than halved, from 1° to less than 0.5°, by the change of target material from TGS to TGFB. These data were taken in the

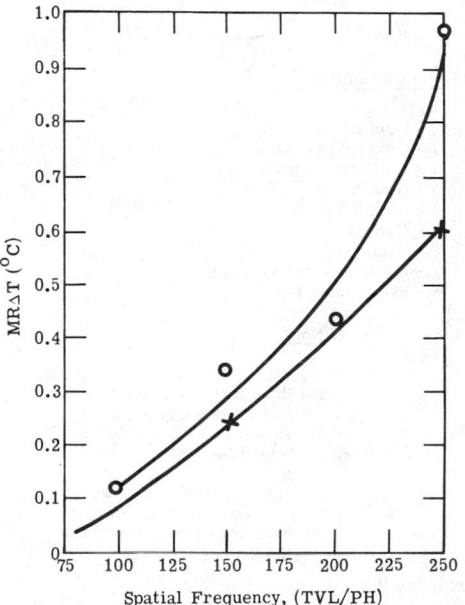

Fig. 13-53. MRΔT versus spatial frequency. This graph shows measurements for a pyroelectric vidicon with a laminar flow electron gun and a 13 μm thick TGS target, operated in the SEP mode [13-13].

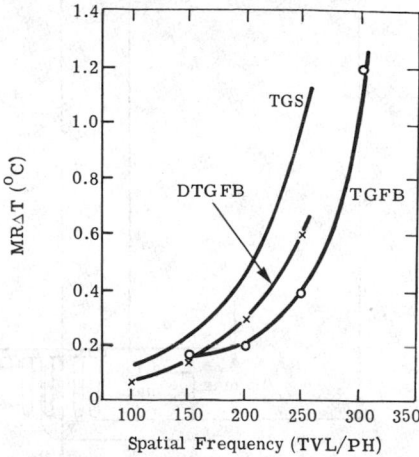

Fig. 13-54. MRΔT for TGS, TGFB, and DTGFB PEVs measured with I_p equaling 100 nA; germanium F/1 optics optimized for 8 to 14 μm; panning mode at 0.3 cm sec^{-1} (6 sec per target diameter); and a 4.0 MHz bandwidth.

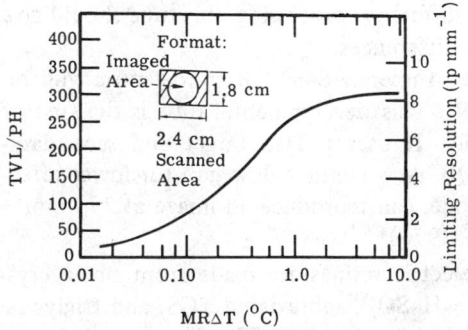

Fig. 13-55. Typical limiting resolution–MRΔT characteristics.

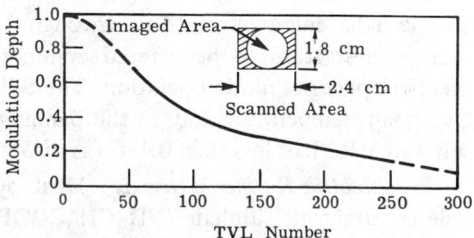

Fig. 13-56. Typical contrast transfer characteristics [13-14].

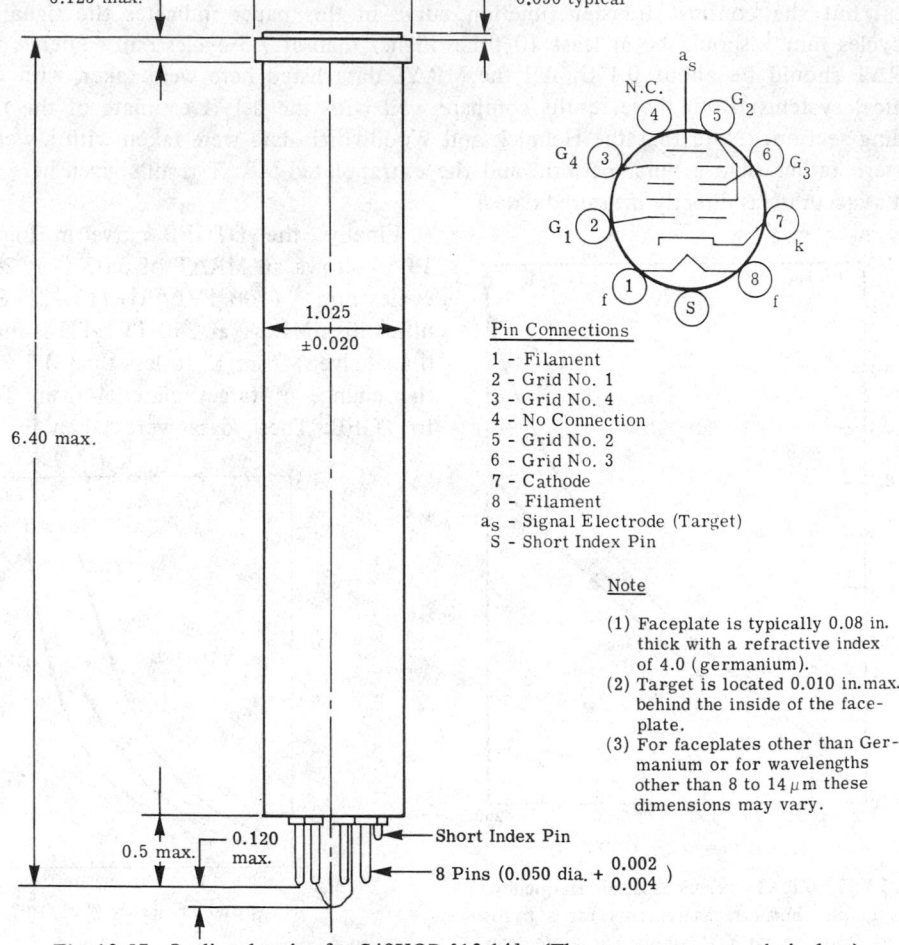

Fig. 13-57. Outline drawing for S49XQB [13-14]. (The measurements are in inches.)

panning mode, with a pedestal current of 100 nA, in an experimental tube using a laminar flow electron gun and a thin DTGFB target. It represents the state-of-the-art as of late 1976. With a synchronized chopper and image difference processing, this tube should give useful TV imagery with 1°C scene temperature differences.

This section concludes with data on the Amperex S49XQB pyroelectric vidicon (Table 13-5 and Figures 13-55 through 13-59). This developmental tube is the first of U.S. manufacture to be offered commercially. It uses a TGS target and secondary-electron pedestal mode operation. The S49XQB has a laminar flow gun for lower effective beam temperature and, in the panning mode, can reproduce an image at 2 lp mm^{-1} with an MRΔT of less than 0.1°C (as of March, 1976)*.

Pyroelectric Retina Materials. Most pyroelectric retinas are made from single crystals of triglycine sulphate $(NH_2CH_2COOH)_3 \cdot H_2SO_4$, abbreviated TGS, and triglycine fluoroberyllate, $(CH_2NH_2COOH)_3 \cdot H_2BeF_4$, abbreviated TGFB. These crystals can

---

*The S49XQB has been replaced by the S58XQ as of July 1977.

be grown readily from a solution formed
by reacting an aqueous glycine solution
with the proper amount of sulphuric or
beryllic acid, then evaporating the solvent, or by cooling to effect supersatura-
tion. Both materials form monoclinic
crystals which are ferroelectrics, i.e., their
polarization along the monoclinic b axis
as a function of an applied electric field
shows a substantial hysteresis loop, (Figure 13-59(a)), and the material remains
polarized when the applied electric field
is returned to zero. Actually these materials are inherently polarized because of
the asymmetry in the basic cell of their
crystalline structures, but the effect is not

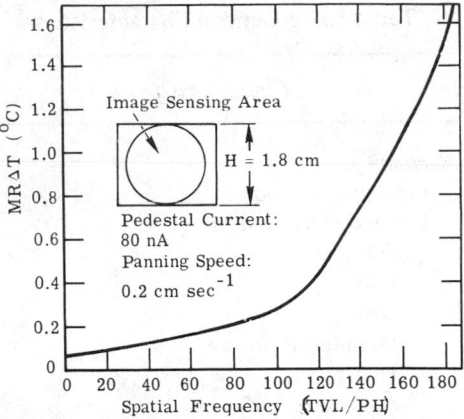

Fig. 13-58. Minimum resolvable temperature difference as a function of spatial frequency for a
S49XQB.

externally observed in the static case because the resulting unbalanced charges on faces
perpendicular to the polar axis are soon neutralized by compensating charges conducted
through the volume or across the surface of the crystal. The spontaneous polarization is
strongly temperature dependent, however, and, when the crystal temperature is changed,
a charge appears on these faces which can be measured with applied electrodes or sensed
by a scanning electron beam. The hysteresis loop shows that the structure may be polarized either way, and, before use in a camera tube, the single crystal is poled by applying
an electric field across the target to orient all polarization vectors the same way. The

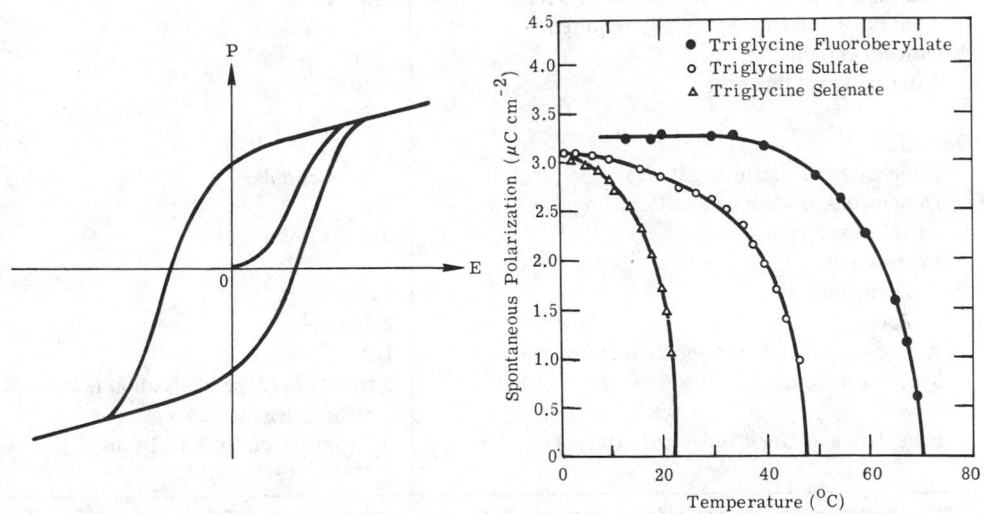

(a) Ferroelectric hysteresis loop [13-15].

(b) Spontaneous polarization versus temperature
for previously poled crystals of TGS, TGFB, and
TGSe [13-15, 13-16].

Fig. 13-59. Polarization as a function of electric field and temperature.

Table 13-5. General Characteristics of a Pyroelectric Vidicon (S49XQB)* [13-14]

Characteristic	Pyroelectric Vidicon (S49XQB)
Mechanical	
Focusing method	Magnetic
Deflection method	Magnetic
Dimensions	
Bulb	T8
Base	EIA No. E8-11**
Mounting positions	Any
Weight	2.5 oz
Accessories	
Socket	Cynch No. 8VT or equivalent
Focussing and deflection	
Coil assembly	Amperex Type AT1116 or equivalent
Electrical	
Heating	Indirectly, ac or dc
Heater voltage	6.3 V ± 10%
Heater current	1/2 W cathode, 100 mA max.
Capacitance	
Target to all other electrodes	3 to 5 pF
Grid no. 1 voltage for cutoff at grid no. 2	
voltage = 300 V	-10 V
Blanking voltage, peak to peak on grid no. 1	100 V min.
Grid no. 1 current at normally required	
beam current	3 mA
Internal target heater	TBD†
Optical	
Dimensions of useful target area	1.8 cm circular
Dimensions of scanned area	
(4:3 aspect ratio)	2.4 × 1.8 cm
Imaging size	
Diameter	1.8 cm
Area	2.54 cm^2
Average gamma of transfer characteristics	1.0
Spectral response	2 to 40 μm (dependent upon faceplate transmission)
Faceplate and target reflectivity (max.)	0.3 (optimized for 8 to 14 μm)

*The Amperex pyroelectric vidicon, type S49XQB, is an infrared sensitive, TV camera pick-up tube, in a 1 in. diameter "vidicon style" envelope, using thinned triglycine sulfate (TGS) as the target material. The faceplate is normally AR coated and can be selected to provide the desired spectral response from 2 to 400 μm. The tube operates at room temperature. No chopping or panning mechanism is included.

**Electronic Industries Association, Washington, DC.

†To be determined.

Table 13-5. General Characteristics of a Pyroelectric Vidicon
(S49XQB)* [13-14] *(Continued)*

Characteristic	Pyroelectric Vidicon (S49XQB)
Absolute Maximum Ratings	
Grid no. 4 voltage	500 V
Grid no. 3 voltage	300 V
Grid no. 4 to grid no. 3 voltage difference	350 V
Grid no. 2 voltage	350 V
Grid no. 1 voltage (with respect to cathode)	
Positive	10 to 30 V
Negative	80 V
Cathode to heater voltage	
Peak to peak positive	125 V
Peak to peak negative	-120 V max.
Cathode current	3 mA
Heater warm-up	1 minute, min.
Target voltage	-50 V
Faceplate temperature (operation)	+40°C
Internal target heater	TBD†
Typical Operating Conditions and Performance	
Operating Conditions:	
Cathode voltage	0 V, forward scan; -70 V flyback
Grid no. 1 voltage	+10 V, forward scan; -50 V flyback
Grid no. 2 voltage	300 V
Grid no. 3 voltage	300 V
Grid no. 4 voltage	440 V
Target voltage	-10 V
Pedestal current	100 nA
Faceplate temperature	30 to 40°C
Internal target heater	TBD†
Performance: [Panning mode, F/1 lens, with 10°C contrast in scene temperature. (300 K blackbody source and 100 nA peak pedestal current.)]	
Responsivity (panning mode)	5 μA W^{-1} (22 μA W^{-1} cm^2)
Sensitivity (peak)	3 nA °C^{-1} in scene
Minimum resolvable temperature	0.3 °C at 150 TVL
Resolution, limiting	300 TVL
Modulation, depth at 150 TVL	30%
Lag (persistence of 20 nA signal)	
Residual signal after 50 msec	25%
Uniformity of responsivity, (center to edge)	50%
Uniformity of pedestal current (center to edge)	10%
Life	1000 to 3000 h

ferroelectric behavior occurs only below a transition temperature, T_C, called the Curie point, in analogy to the similar point in ferromagnetic materials.

Figure 13-59(b) shows the spontaneous or remanent polarization as a function of temperature for TGS, TGFB and TGSe, triglycine selenide. A high slope of the polarization versus temperature curve is needed to sense an infrared image.

As indicated in earlier sections, for maximum signal-reading-efficiency the storage capacitance of the pyroelectric target should be low so that a given signal charge produces as large a surface potential excursion as possible. Figure 13-47 shows the relative dielectric coefficient of TGS as a function of temperature. As shown, both the pyroelectric coefficient and the dielectric coefficient increase with temperature below the Curie temperature, and the best tradeoff of a large response and reasonably small capacitance is found at about 40°C.

A number of other materials have been or are being studied for use as pyroelectric targets. Table 13-6 lists the pyroelectric coefficients, specific heats, and other properties for some of these materials. The actual choice of materials is influenced strongly by their ruggedness, immunity to processing at elevated temperatures and stability in a vacuum environment as well as by specific thermal and electric properties. At present TGFB, triglycine fluoroberyllate seems the most promising candidate material, but work continues on ceramics and plastic films in the hope that more ideal properties can be provided.

Table 13-6. Properties of Several Candidate Materials for Pyroelectric Targets

Material	d**	$P/c'd^\dagger$	$\epsilon^{\dagger\dagger}$	Modes†††	Vacuum Compatibility	Handling	$K,^\# 10^{-3}$ $cm^2 sec^{-1}$	Curie Temperature
TGS*	3×10^{-3}	10.6	80	I, S, R	poor	good	3.0	49
TGFB	3×10^{-3}	4.0	15	I, S, R	poor	poor	3.0	70
LiTaO$_3$	3×10^{-3}	1.4	45	I, S, R	good	good	12.0	618
Pb$_5$Ge$_3$O$_{11}$	3×10^{-3}	1.4	40	I, S, R, C	good	poor	>3.0	178
Doped Pb$_5$Ge$_3$O$_{11}$	3×10^{-3}	4.0	100	I, S, R	good	good	>3.0	80
PBZT	3×10^{-3}	4.0	300	I, S, R, C	good	good	>3.0	200
PVF$_2$	6×10^{-4}	2.5	15	I, S, R	good	good	0.5	(80)

*TGS has a pyroelectric coefficient, P, of 8×10^{-8} C cm^{-2} K^{-1} and a volume specific heat, c', of 2.50 J cm^{-3} K^{-1}.

**d = assumed retina thickness.

†For high performance p/c'd should be large, ϵ and K small.

††ϵ = relative dielectric constant.

†††Modes: I = gas ion pedestal
           S = secondary emission pedestal
           R = ramp pedestal
           C = conduction pedestal

#K = thermal diffusivity.

13.6. Tubes for Special Uses and Historical Interest

13.6.1. The Thermicon, A Bolometric Vidicon.

The temperature sensitive dark current of the photoconductive vidicon described in Section 13.3 rose as the retina voltage was increased to a value which was of the same order as the photon excited signal current. The Westinghouse thermicon, shown in Figure 13-60, was the most successful of a series of devices to exploit this effect to sense an infrared image. The retina was isolated from the massive heat-conducting faceplate by either of two structures shown in Figure 13-61(a) and (b). In the "smoke" supported retina, a thin "smoke" of Sb_2S_3 was deposited on a cooled infrared-transmitting faceplate by evaporation through argon gas at low pressure in an evacuated chamber. The evaporated molecules are thought to undergo many collisions with argon atoms and with each other in the space between the electrically-heated evaporation boat and the substrate. During these collisions, the Sb_2S_3 molecules apparently link together to form fiber-like particles before being deposited. The resulting layer has a density only a few percent of that of the solid material. The structures of such smokes have been examined with an electron microscope, and appear consistent with this description. Obviously the thermal conductivity of such a layer in vacuum would be very low.

A thin layer of gold black was formed over the insulating layer by a similar evaporation technique. The layer was made thick enough to absorb most

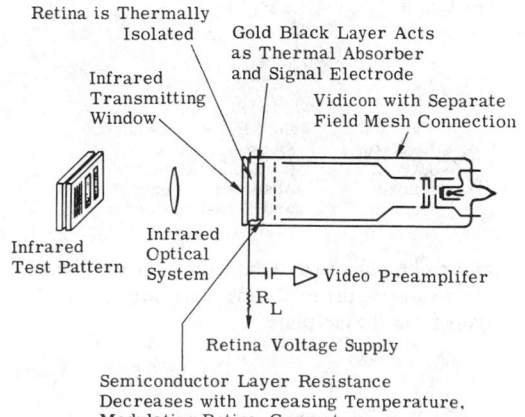

Fig. 13-60. Schematic cross-section of Westinghouse thermicon.

infrared radiation incident through the faceplate, and to be electrically conductive to serve as the signal electrode. The gold black extended over an evaporated, metal, contact ring around the edge of the faceplate. Over the gold black was deposited another smoke layer of a material like Sb_2S_3, As_2S_3, or GeTe. This three-layer retina on its faceplate was then made part of a vidicon tube. The gold black layer was biased 20 to 40 V positive with respect to the electron gun cathode, and the free surface of the final layer was scanned with electrons to bias it approximately uniformly. When an infrared image was focused on the gold black, a corresponding pattern of temperature variation resulted from absorption in the gold black, and the final layer was warmed by conduction. This effected a corresponding conductivity change, which modulated the dark current to produce a stored image on the scanned layer which was read by the scanning beam to produce a video signal. This "smoke supported retina" was used primarily in tubes operated near ambient temperature.

An alternate retina structure is also shown in Figure 13-61. A thin edge-supported nitrocellulose film was stretched like a drum head across a metal ring and provided mechanical support for the absorbing and temperature sensing layers. Gold-black smoke was evaporated onto the support film as before to constitute both an infrared absorber and signal electrode. A temperature-sensing smoke layer of Sb_2S_3 or a related com-

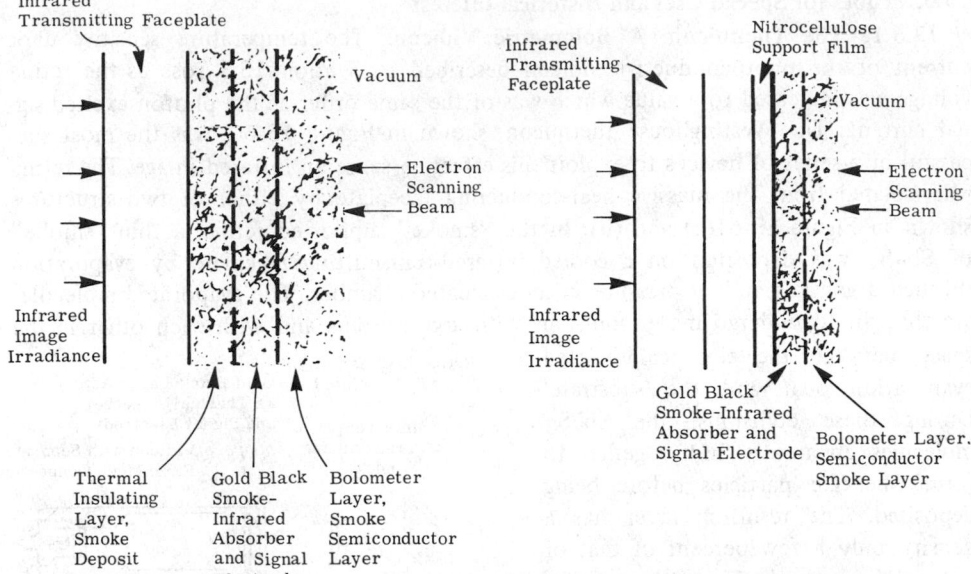

(a) Smoke-supported thermicon used low density film to isolate thermally the temperature sensitive layers from the faceplate.

(b) The nitrocellulose supported retina was used principally with cooled tubes. Thin nitrocellulose had low heat capacity and low lateral heat conductance.

Fig. 13-61. Two types of thermicon retinas.

pound, was then deposited over the gold black. The retina assembly was mounted within a vidicon envelope behind an infrared transmitting faceplate. An infrared image formed on the gold black layer was absorbed, causing a corresponding pattern of temperature variations. The temperature variations produced conductivity variations in the final layer, and a corresponding charge image which is sensed and erased by the scanning electron beam. This second retina embodiment was used primarily on structures which were cooled below ambient temperature to enhance sensitivity. Note that in both cases the infrared photon energy was too small to excite electrons across the 1.6 to 1.9 eV bandgap in the semiconductor, but that the free carriers were excited thermally.

Observed performance of thermicon tubes is indicated in Figures 13-62 through 13-64. Thermicons were made in 1 in. vidicon envelopes and in 2 in. diameter envelopes, using either magnetic or electrostatic focussing, and with a return-beam electron-multiplier output as an option in some of the larger tubes. The data are from a 2 in. diameter tube with a 40 mm scan diagonal. Table 13-7 shows a summary of characteristics taken by several investigators. The gold-black heat-absorbing retina is otherwise panchromatic. Figure 13-63 is the signal transfer characteristic, relating signal-to-rms-noise ratio to faceplate irradiance. The response is linear over the range measured, with NEE about 2×10^{-4} W cm^{-2}. Figure 13-64 shows the thermicon resolving power expressed in terms of the square-wave relative response as a function of spatial frequency. Limiting resolution was 4 or 5 lp mm^{-1}, 200 to 240 TV lines per pattern height, but was limited by the obscuring effect of retina nonuniformities to lower values in many tubes.

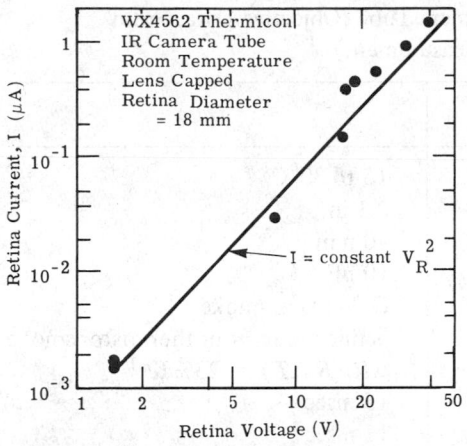

Fig. 13-62. Retina current-voltage characteristic.

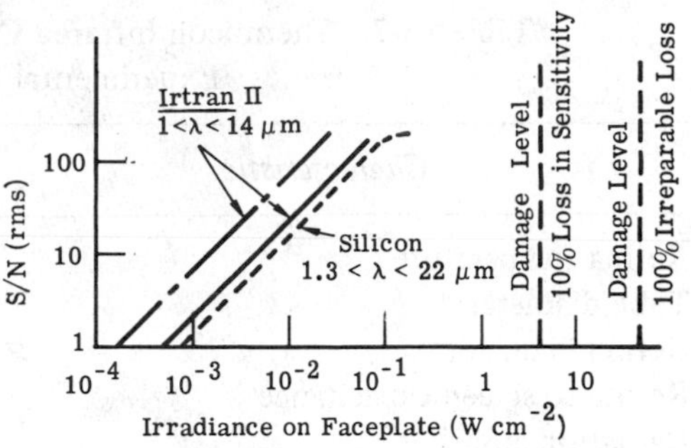

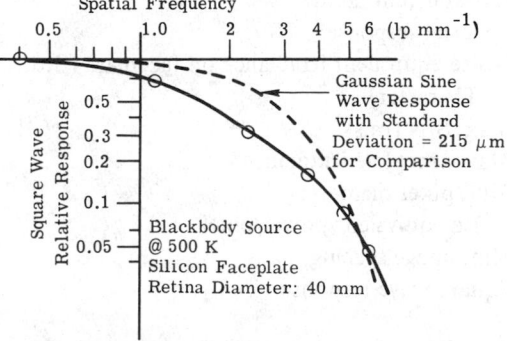

Fig. 13-63. Characteristics of thermicon infrared camera tubes.

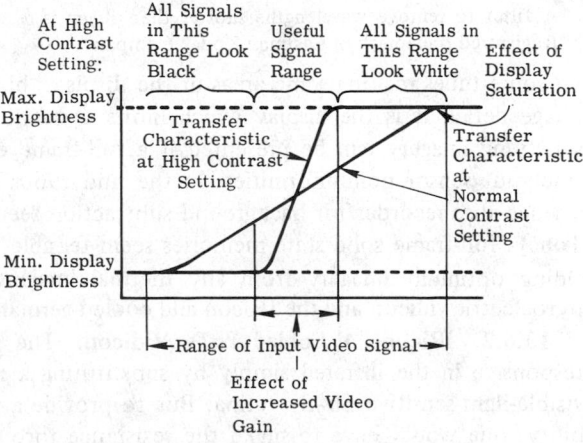

Fig. 13-64. Square wave response of thermicon infrared camera tube.

Nonuniformities were the principal limitation on overall thermicon performance. Although the thermicon is a temperature sensor, and although the retina operates at room temperature, the scanning process reads the patterns of local conductance through the retina as a measure of local retina temperature. Small nonuniformities in thickness or density of the bolometric layer, or in bias across the layer because of electron optical nonuniformities, can alter the background-signal current-density from point to point. Contrast in a thermal radiation image is very low, and to provide a reproduced image with sufficient contrast to satisfy the observer's requirements, one advances the video gain or contrast control and adjusts the background or brightness control as suggested in Figure 13-65. Thus a very small range of signal currents through the thermicon retina causes the display brightness to vary from black to maximum display brightness. The effects of retina current nonuniformities are thus greatly accentuated,

Fig. 13-65. Use of high contrast to enhance part of image.

Table 13-7.   Thermicon Infrared Camera Tube (Objective Data, From Experimental Measurement)

Characteristic	Data
Retina temperature	15 to 30°C
Tube diameter	2.1 in.
Retina diameter	40 mm
Retina to screen capacitance	10 pF
Radiation absorber	Gold black smoke
Sensitive layer	Semiconductor or thermistor smoke
Thermistor response	$\Delta R/(R \Delta T) = 23\% \, K^{-1}$
Signal response time constant	15 msec
Thermal time constant	11 msec
Electrical time constant	11 msec
Retina voltage	15 to 35 V
Dark current factor	$6.3 \times 10^{-4} \, \mu A \, V^{-2}$
Spectral response	Irtran®** 2 or 4, etc.
Noise equivalent irradiance on faceplate (Std. TV raster)	$1.6 \times 10^{-4} \, W \, cm^{-2}$
Max. S/N (rms)	100
Max. faceplate irradiance*	$4 \, W \, cm^{-2}$
Min. pixel diameter	$180 \, \mu m = 7 \, mil$
Noise equivalent power/pixel	$10^{-7} \, W/pixel$
Min. image spacing	$5 \, mm^{-1}$
Square wave response	50% @ 2 lp mm⁻¹ $\quad$ 50% @ $2 \, lp \, mm^{-1}$
	10% @ $5 \, lp \, mm^{-1}$
	5% @ $6 \, lp \, mm^{-1}$
NET (noise equivalent scene temperature difference for F/# = 1 and 300 K scene)	1.2°C

*A filter to remove wavelengths shorter than 8 μm is recommended to prevent radiation damage.
**Registered trademark of Eastman Kodak Company.

on some tubes making some areas of the display "blacker than black" and masking any image detail. It is the display which limits dynamic range under these conditions, and satisfying imagery can be presented if a full-frame electrical memory is added to subtract out sensor nonuniformities. In the mid 1960s (when this work was done), use of a video disc recorder for background subtraction seemed feasible only in the laboratory. Today, full-frame solid state memories seem feasible, and indeed may be useful for providing optimum imagery from any thermal imaging television camera, including the pyroelectric vidicon and the IRicon and cooled-germanium retina vidicon.

13.6.2. IRicon, A Cooled PbTe Vidicon. The vidicon could in theory be made responsive in the infrared simply by substituting a narrower bandgap material for the visible-light-sensitive Sb_2S_3 retina. But to provide a full frame-time image-storage capability, one would have to make the resistance through the scanned area of the retina on the order of 10^8 ohms at retina operating temperature with background radiation incident through the lens as well as from the various housing and tube electrode and envelope surfaces. This corresponds to a resistivity of the order of 10^{12} ohm cm, a value difficult to realize even with wider bandgap materials appropriate for the visible. In the continuing research at RCA Laboratories, a series of materials were tried, starting with lead sulphide, then lead selenide, and finally porous, amorphous, lead telluride formed by evaporation through an inert gas at low pressure. The high resistivity is thus provided by low carrier mobility, aided by the porous structure which increases the effective thickness of the layer for electrical conduction.

Two typical lead telluride IRicons, made by RCA, are shown in Figures 13-66 and 13-67. The first, designated C23105, provides signal current directly in the retina

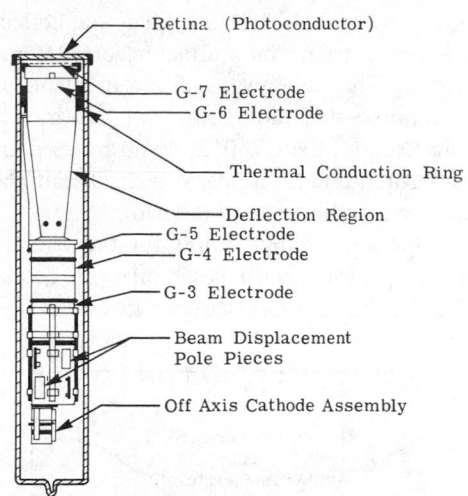

Fig. 13-66. Direct-readout electrostatic IRicon with brazed electrode structure and off-axis gun (RCA developmental type C23105) [13-4].

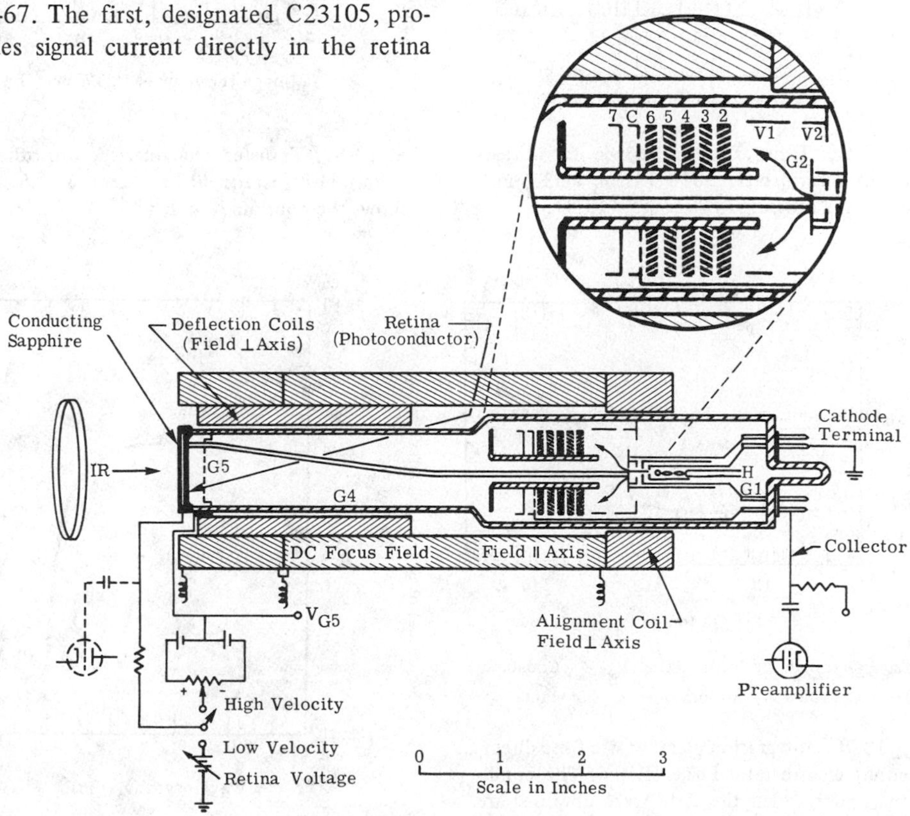

Fig. 13-67. IRicon. The magnetic focus and deflection type with an inverted multiplier (RCA development type C74-12) [13-4].

lead, and uses electrostatic focus and deflection with an off-axis electron gun so that direct radiation from the warmer electrodes near the thermionic cathode cannot reach the retina. This tube was 3.5 cm in diameter by about 17 cm long and was operated in a stainless-steel liquid-nitrogen dewar, with conduction cooling of interior electrodes. The second, the C74012, could be used either with signal current taken from the retina or with internal secondary electron multiplication to amplify the signal current information on the returning electron beam.

Operating temperatures for the sensitive surface of either tube were below 100 K, easily provided with liquid-nitrogen cooling of the tube and lens housing. Performance for the 3.5 cm electrostatic IRicon is described in Figures 13-68 through 13-71, using

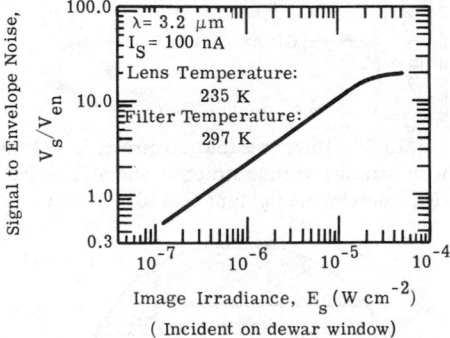

Fig. 13-68. Transfer characteristic for direct-readout electrostatic PbTe IRicon at 3.2 μm, follows the relation $I_S \propto E^{0.69}$ [13-4].

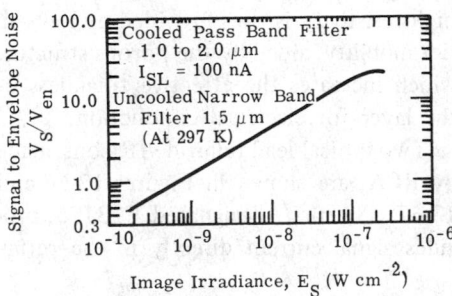

Fig. 13-69. Transfer characteristic for direct-readout electrostatic PbTe IRicon at 1.5 μm, follows the relation $I_S \propto E^{0.58}$.

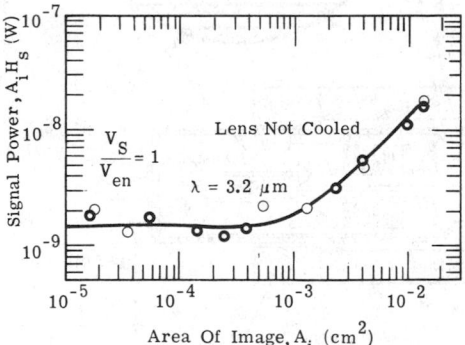

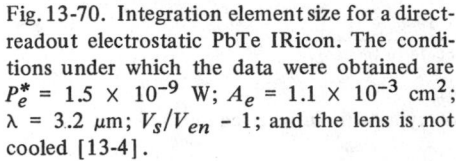

Fig. 13-70. Integration element size for a direct-readout electrostatic PbTe IRicon. The conditions under which the data were obtained are $P_e^* = 1.5 \times 10^{-9}$ W; $A_e = 1.1 \times 10^{-3}$ cm^2; $\lambda = 3.2$ μm; $V_s/V_{en} - 1$; and the lens is not cooled [13-4].

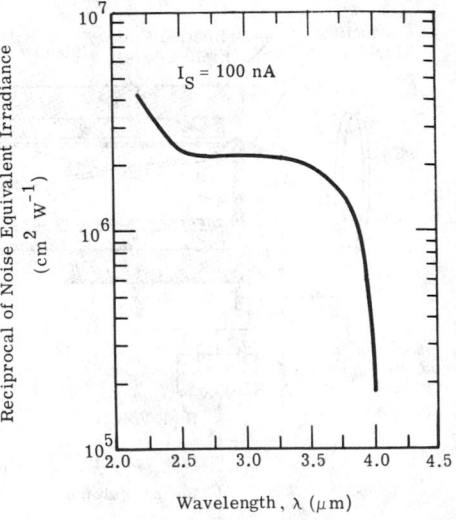

Fig. 13-71. Spectral response for a PbTe direct-readout camera tube [13-4].

methods of test once standard for IR camera tubes. These results may be compared with the tubes of Sections 13.6.1 and 13.6.3, but not with those of Section 13.5, since these test methods have now been discarded because they conflicted with methods standard in the industry for visible spectrum television, giving numbers for signal-to-noise ratio which are significantly poorer.

Figures 13-68 and 13-69 show that the lead telluride IRicon followed the relation $I_S \propto E^{0.6}$ to $E^{0.7}$, much like the Sb_2S_3 vidicon. Figure 13-71 shows that response extended to nearly 4 μm. Maximum resolving power reached nearly 500 TV lines per pattern height and scene temperature differences less than 1°C could be detected using $F/1.2$ optics. As with most infrared sensitive vidicons, image nonuniformities rather than noise appeared to limit performance.

13.6.3. Doped Germanium Retina Cooled Vidicon. A copper-doped germanium retina vidicon was developed by the General Electric Company in the late 1950s, and small quantities of tubes and television cameras were manufactured and sold until 1975. The tube, designated Z-7880, used a thinned single-crystal germanium wafer about 25 μm thick as a uniform layer photoconductive retina. There was no diode structure. For response to about 4.3 μm, the germanium was doped with copper, then counterdoped with arsenic to eliminate long wavelength response which would otherwise have extended to beyond 30 μm. The resulting extrinsic photoconductor has a quantum efficiency of about 10^{-3}, but the large photon flux from a 300 K background still causes a large background signal from the vidicon. The tube, lens, and housing must be cooled to 77 K for operation. Quantitative performance data reported by GE includes a noise equivalent temperature difference of 0.2°C with cooled $F/2$ optics and a cooled narrowband optical filter, and a limiting resolving power of 300 TV lines per pattern height for large temperature contrasts. Characteristic curves are shown in Figures 13-72 through 13-75. The Z-7880 used return-beam scan with electron-multiplier output like an image orthicon and had bent electrostatic electron optics so that no radiation from the warm gun parts near the thermionic cathode could reach the cooled retina. The scanned area was 15 × 20 mm, and the tube envelope measured 2.75 in. in diameter × 11 in. in length and included a built-in liquid-nitrogen container to cool the retina, the electron-gun parts, and the periphery of the faceplate. Figure 13-72 shows the relative spectral

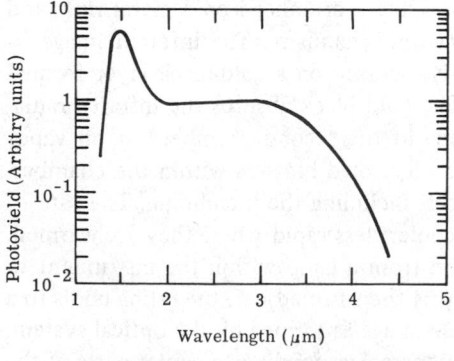

Fig. 13-72. Z-7880 spectral response (no integral filter).

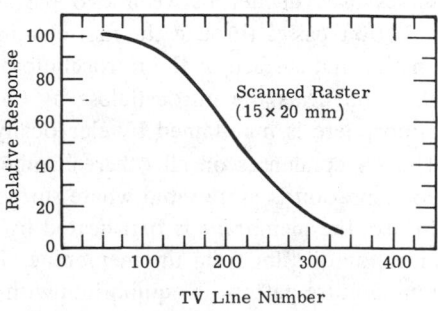

Fig. 13-73. Z-7880 square wave response.

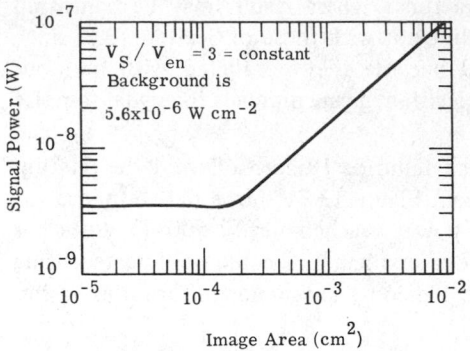

Fig. 13-74. Z-7880 integration element size.　　Fig. 13-75. Transfer characteristic [13-17].

response, Figure 13-73 the square wave response, Figure 13-74 the integration element size, and Figure 13-75 the signal transfer characteristic, which was linear. The test methods do not correspond to methods used with camera tubes for the visible, and they have not been used in any discussions of currently active camera tubes in Section 13.5.

13.6.4. Nonelectronic Infrared Imaging Devices. Conversion of infrared images to visible images does not require an electron tube at all, let alone a television system. Descriptions are given here of three nonelectronic, directly-viewed, infrared-image-conversion devices: the evaporograph, the edgegraph, and the mesoscope.

Evaporograph. A schematic diagram of a simplified evaporograph is shown in Figure 13-76(a). The evaporograph principle was first described in 1846 by Sir William Herschel, who used it to demonstrate the existence of absorption bands in the infrared portion of the sun's spectrum. The principle was adopted by M. Czerny in Germany in the 1920s to build an instrument for recording infrared spectra. More recently, Baird Atomic designed a useful evaporographic instrument for viewing thermal infrared images and for recording them photographically, and marketed several versions of that instrument from about 1950 to about 1968.

The heart of the instrument is the evaporographic cell. (See Figure 13-76(b).) A thin, 5 μm thick, nitrocellulose membrane is mounted like a drumhead on a metal ring, and serves as a divider between two partially-evacuated chambers. The infrared image irradiation passes through the first chamber and is focussed on a gold-black layer formed on the first surface of the nitrocellulose film. The gold black absorbs the infrared radiation and heats the nitrocellulose by conduction. In the second chamber, an oil vapor atmosphere is maintained by electrically-heated oil-soaked blotters within the chamber. The oil condenses on all other chamber surfaces, including the membrane. The rate of condensation is more rapid where surfaces are cooler, less rapid where they are warmed. In use, the membrane is first heated by radiation from a lamp within the instrument to clear the oil film from the membrane. The lamp is then turned off, the retina cools to a temperature pattern in equilibrium with the scene image as formed by the optical system, and an oil film begins to form, faster in the regions corresponding to cooler parts of the scene. The oil film is observed using a white viewing light, which produces interference

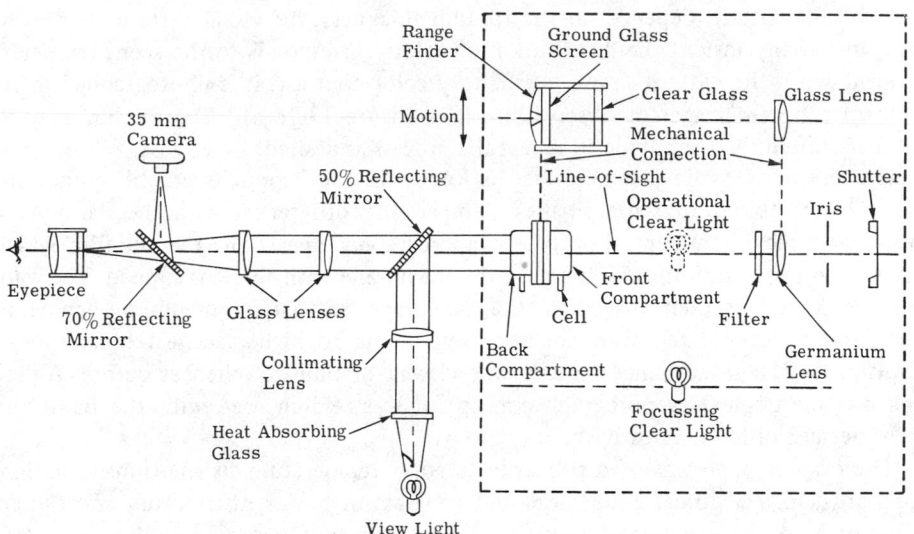

(a) Simplified diagram of an evaporograph.

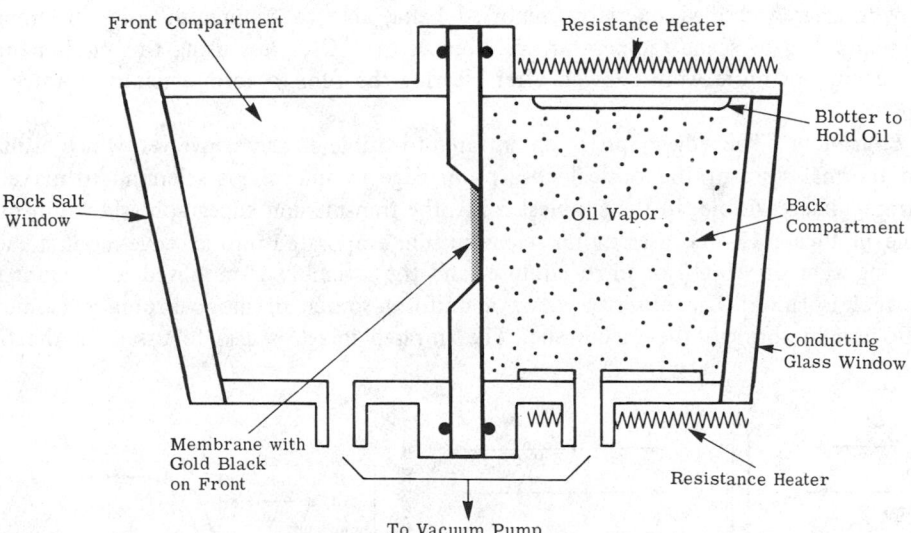

(b) Image conversion cell of the evaporograph.

Fig. 13-76. An evaporograph.

patterns by multiple reflections very much like those of an oil film on water. Since the color of the pattern depends on the oil-film thickness, the visual pattern changes with time, but at any instant the distribution of colors corresponds to the scene temperature pattern. When the pattern shows satisfactory color contrast, it is photographed in color to form a stored image for observation. (See Figure 13-76(a).) The oil film may then be cleared from the membrane to repeat the process as desired.

The operating mode just described is known as the "condensograph" mode, but it was the one found to sense smallest temperature differences with the Baird Atomic evaporographic instrument. For true evaporographic operation, the oil film was first allowed to reach uniform thickness on the membrane with the lens capped. The warmer infrared scene was then imaged on the membrane, so that the membrane temperature rose, and oil evaporated from areas corresponding to higher scene temperature. The resulting optical interference pattern was viewed or photographed as before. Although this was the original evaporograph concept, it was seldom used with the Baird instrument because of lower sensitivity.

The evaporograph was used primarily to study temperature distributions in stationary apparatus, like electrical connectors and contacts in power distribution and switching equipment, or to examine the surface temperatures of insulated boilers, etc. It was not suited for quantitative radiometry, but could be used to show whether one scene object was warmer than a neighboring object. The need for clearing the sensing film, then taking pictures as the image changed, makes this device unsuitable for viewing moving images. Skilled operators reported being able to distinguish color differences corresponding to scene temperature differences of 1°C or less using this equipment if 30 seconds or more were allowed after clearing the film to form a relatively thick oil film.

Edgegraph. The edgegraph is an infrared-to-visible image converter which utilizes the thermal shift of the optical absorption edge in amorphous selenium to make an infrared image visible. In the simplest form, the transmission edgegraph (shown in principle in Figure 13-77), uses a thin selenium film evaporated into an edge-mounted supporting film of Al_2O_3 or nitrocellulose, and the assembly is mounted in vacuum for thermal isolation. The observer views a uniform source of monochromatic (sodium) yellow light through the membrane. The infrared image is also focussed on the film

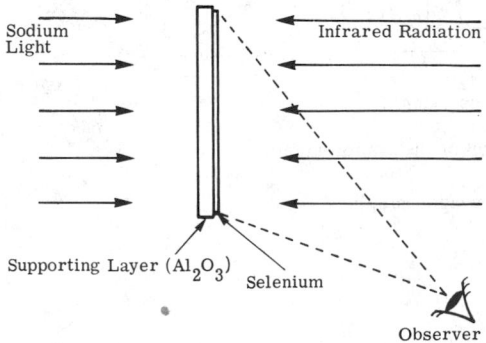

Fig. 13-77. Transmission edgegraph.

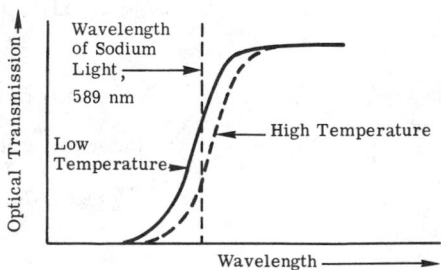

Fig. 13-78. The absorption edge in amorphous selenium. Absorption at $\lambda = 589$ nm increases at higher temperatures.

assembly to create a corresponding temperature pattern. The sodium "D" line wavelength falls in the middle of the selenium absorption edge at room temperature, as shown in Figure 13-78. Since the transmission at the D line falls as temperature increases, the visible image is the negative of the infrared image.

The transmission edgegraph was investigated before 1955 by the Services Electronics Research Laboratories in England. They reported detection of an 8°C scene temperature difference using $F/0.5$ optics. A principal limiting mechanism was low contrast in the converted visible image.

To provide a high-contrast converted image, one must find a mechanism which gives an essentially dark image when viewing a uniform background. To enhance image contrast, researchers at The University of Michigan's Engineering Research Institute developed the interference edgegraph in 1954 to provide high-contrast imagery while retaining the edgegraph principle.

The four-layer image-converter membrane, shown in Figure 13-79, is formed on a thin edge-supported aluminum-oxide support film. A gold-black infrared-absorbing layer is formed on one face by evaporation through an inert gas at low pressure. A thin gold mirror layer is formed by high vacuum evaporation on the other face of the support film, followed by a 3/4 wave thickness of amorphous selenium. The membrane is viewed at essentially normal incidence by specularly reflected sodium yellow light. The infrared image is focussed on a gold-black absorber on the reverse side of the membrane. Where the membrane is cooler, transmission through the selenium is higher, and the first and second surface reflections cancel in part to give low image brightness.

Where the layer is warmer, corresponding to higher infrared image irradiance, the second surface reflection is reduced by absorption in the selenium and the image brightness increases.

With the interference edgegraph, Vis of Michigan reported minimum resolvable temperature differences of about 10% those observed with the transmission edgegraph. Combining Vis's results with the SERL results, one should find that the interference edgegraph permits the detection of approximately 2° to 3° temperature differences with more realistic $F/0.7$ to $F/1.0$ infrared optics.

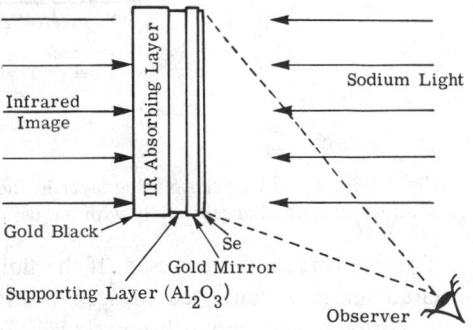

Fig. 13-79. The interference edgegraph.

The advantages of the edgegraph appeared to be continuous operation, fast response (on the order of milliseconds) to follow moving images, and easy adaptation to a sealed vacuum environment with no need for vacuum pumps. The problems were the need for reasonably precise film-temperature-control to keep the absorption edge at the D line wavelength.

Work on this approach was apparently concluded in 1959.

The Mesoscope, A Liquid Crystal Infrared Viewer. Certain cholesteric liquid crystals reflect light in a narrow spectral region by scattering within the liquid crystal layer. Further, for certain of these mesomorphic materials, the wavelength of maximum reflection varies strongly with film temperature. This effect was used by Westinghouse in the mesoscope, a nonelectronic infrared image converter, developed in part for the U.S. Army Night Vision Laboratory.

As shown schematically in Figure 13-80, a thin, edge-mounted, plastic support-layer is coated with a gold-black absorbing layer on one surface, and with a mesomorphic liquid-crystal film on the other. For thermal isolation, the retina is enclosed in a cylindrical vacuum envelope, and the infrared image is focussed on the gold-black layer through a sodium chloride window at one end of the cylinder, while the liquid-crystal layer is viewed through a glass window at the other.

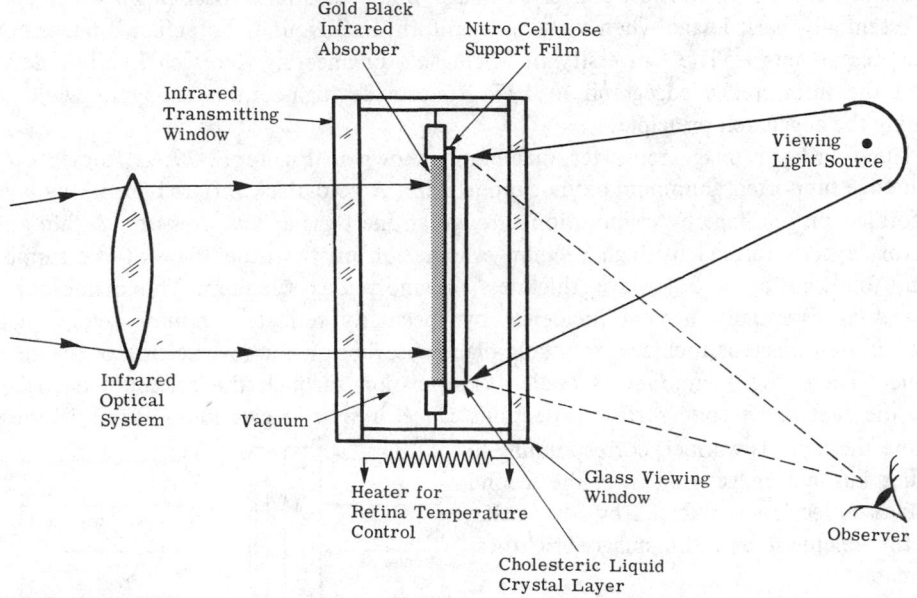

Fig. 13-80. Liquid crystal sensing layer in the mesoscope reflects light in a 200 Å band, whose peak wavelength changes rapidly with temperature.

Two viewing modes are used. If the liquid crystal is illuminated with white light, the infrared image is converted to a color-shift image, somewhat like the evaporograph. If illumination is with monochromatic light the shifting wavelength of the reflectance peak with temperature modulates the amount of the reflected light. In either case, the wavelength of maximum reflectance must be tuned to the desired value by adjusting the operating temperature of the layer. Since resistance to lateral heat flow is necessarily high for reasonable thermal MTF, the retina temperature is determined by the radiation balance to its surroundings. The optimum temperature can be preselected by the material parameters, and these are usually chosen for a temperature a few degrees above ambient. This retina temperature is achieved by electrically warming the metal envelope of the vacuum cell, or in some cases by a local heat lamp, filtered to remove visible light and aimed through the viewing window.

Mesomorphic layers can be made extremely sensitive, with spectral shifts as high as 1000 Å K^{-1}. Such extreme sensitivity is required for observing fractional-degree differences in the scene, but proved a handicap in the mesoscope, since the working temperature range of the liquid-crystal layer was then only a few tenths of a degree. Further, the local retina temperature depended on radiation from the scene, and, if retina temperature and scene temperature differed significantly, was even affected by the normal off-axis vignetting of the optical system. For example, if the average retina temperature was set higher than the scene temperature, the temperature at the center of the retina would be lower than at the edges because of the normal $\cos^4\theta$ image-irradiance fall-off of the lens. This sort of problem seems unavoidable for any area imaging device where signal is a sharp function of retina temperature, and where a low focal ratio, and hence wide field-of-view optical system, is required for sensitivity.

Westinghouse made mesoscope devices for studying far-field laser patterns and for thermal infrared viewing in the early 1960s, with typical image diameter of 6 cm. The highest temperature-sensitivity liquid-crystals were physically unstable, however, and the general liquid crystal activity was sold. It has recently been reported that the French Atomic Energy Commission's Nuclear Research Center has revived the concept. Data are not yet available to compare their work with the earlier work in the U.S., both at Westinghouse and in the Far Infrared Technical Area at the Night Vision Laboratories.

13.7. References and Bibliography

13.7.1. References

[13-1] W. S. Percival, British Patent No. 528179, 1939, British Patent Office, London, England.

[13-2] I. J. P. James, "Fluctuation Noise in Television Camera Head Amplifiers," *IEEE Proceedings*, Institution of Electrical Engineers, London, England, No. 20, Part III A, 1952.

[13-3] M. H. Crowell and E. F. Labuda, "The Silicon-Diode-Array Camera Tube," *Photoelectronic Imaging Devices*, L. M. Biberman and Sol Nudelman (eds.), Plenum Press, New York, NY, 1971.

[13-4] S. V. Forgue, L. D. Miller, and J. O. Schroeder, "The IRicon: A Passive Infrared Camera Tube," *IEEE Transactions on Electron Devices*, Institute of Electrical and Electronics Engineers, New York, NY, Vol. ED-22, No. 10, October 1975, pp. 904–910.

[13-5] R. Kurczewski and R. S. Levitt, "Pyroelectric Vidicon Camera System, Modern Utilization of Infrared Technology," *Proceedings, Society of Photo-Optical Instrumentation Engineers*, Society of Photo-Optical Instrumentation Engineers, Palos Verdes Estates, CA, Vol. 62, 1976.

[13-6] United States Broadcast Scanning Standards, Electronic Industries Association, Washington, DC, Standard No. EIA-RS-170, 1976.

[13-7] C. N. Helmick, Jr. and W. H. Woodworth, "Improved Performance from Pyroelectric Vidicons by Image-Difference Processing," *Ferroelectrics*, Gordon and Breach, London, Vol. 10, 1976.

[13-8] F. Petito and L. Garn, "A Critical Systems Assessment of the Pyroelectric Vidicon," *Technical Digest*, International Electron Devices Meeting, Institute of Electrical and Electronics Engineers, New York, NY, 1975, pp. 74–77.

[13-9] R. Watton, C. Smith, B. Harper, and W. M. Wreathall, "Performance of the Pyroelectric Vidicon for Thermal Imaging in the 8-14 Micron Band," *IEEE Transactions on Electron Devices*, Institute of Electrical and Electronics Engineers, New York, NY, Vol. ED-21, No. 8, August 1974, pp. 462–469.

[13-10] B. Singer and J. Lalak, "Pyroelectric Vidicon with Improved Resolution," *IEEE Symposium on Applications of Ferroelectrics*, Institute of Electrical and Electronics Engineers, New York, NY, June 1975.

[13-11] R. M. Logan and R. Watton, "Analysis of Cathode Potential Stabilization of the Pyroelectric Vidicon," *Infrared Physics*, Pergamon Publishing, Elmsford, NY, Vol. 12, 1972, p. 17.

[13-12] E. Stupp, T. Conklin, and B. Singer, Philips Laboratories, Briar-Cliff Manor, NY, Private Communication, November, 1976.

[13-13] T. Conklin, B. Singer, M. H. Crowell, and R. Kurczewski, "Theory and Performance of Pyroelectric Vidicon with an Electronically Generated Pedestal Current for the Cathode Potential Stabilized Mode," *Technical Digest*, IEEE International Electron Devices Meeting, Institute of Electrical and Electronics Engineers, New York, NY, T-ED-74, November 1974, pp. 451–454.

[13-14] Amperex Electronic Corporation Staff, "Pyroelectric Vidicon S49XQB," Development Sample Data, Electro-Optical Devices Division, Amperex Electronic Corporation (a North American Philips Company), Slatersville, RI, January, 1976. See also, "Development Type S58XQ Pyroelectric Vidicon (Previously S49XQB)," July 1977.

[13-15] Franco Jona and G. Shirane, *Ferroelectric Crystals*, Pergamon Publishing, New York, NY, 1962.

[13-16] S. Hoshino, T. Mitsui, F. Jona, and R. Pepinsky, "Dielectric and Thermal Study of Tri-Glycine Sulphate and Tri-Glycine Fluoberyllate," *Physical Review*, American Institute of Physics, New York, NY, Vol. 107, No. 5, 1957, p. 1255.

[13-17] P. J. Daly, U. S. Army Engineering Research and Development Laboratories, Ft. Belvoir, VA, Private Communication, April, 1966.

13.7.2. Bibliography

Amperex Electronic Corporation Staff, "Pyroelectric Vidicon S49XQB," Development Sample Data, Amperex Electronic Corporation, Electro-Optical Devices Division (a North American Philips Company), Slatersville, RI, January 1976. See also, "Development Type S58XQ Pyroelectric Vidicon (Previously S49XQB)," July 1977.

Anderson, R. W., "Thermal Mapping in Color Using the Evaporagraph," *Photomethods for Industry*, Ziff-Davis Publishing, New York, NY, Vol. 6, No. 10, October, 1963.

Biberman, L. M. (ed.), *Perception of Displayed Information*, Plenum Press, New York, NY, 1973.

Blumenfeld, S. M., G. W. Ellis, R. W. Redington, and R. H. Wilson, "The Epicon Camera Tube: An Epitaxial Diode Array Target Vidicon," *IEEE Transactions on Electron Devices*, Institute of Electrical and Electronic Engineers, New York, NY, Vol. ED-18, No. 11, November 1971.

Conklin T., and B. Singer, "High Performance Pyroelectric Vidicon," *Technical Digest*, IEEE International Electron Devices Meeting, Institute of Electrical and Electronics Engineers, New York, NY, 1965, pp. 66–69.

Conklin, T., B. Singer, M. H. Crowell, and R. Kurczewski, "Theory and Performance of a Pyroelectric Vidicon with an Electrically Generated Pedestal Current for the Cathode Potential Stabilized Mode," *Technical Digest*, IEEE International Electron Devices Meeting, Institute of Electrical and Electronics Engineers, New York, NY, Vol. T-ED 74, November 1974, pp. 451–454.

Crowell, M. H., and E. F. Labuda, "The Silicon-Diode-Array Camera Tube," *Photoelectronic Imaging Devices*, L. M. Biberman and Sol Nudelman (eds.), Plenum Press, New York, NY, 1971.

Dimmock, J. O., "Capabilities and Limitations of Infrared Vidicons vs Infrared Scanning Systems," *Lincoln Laboratory Technical Note 1971-79*, Lincoln Laboratory, Lexington, MA, December, 1971.

Forgue, S. V., L. D. Miller, and J. O. Schroeder, "The IRicon: A Passive Infrared Camera Tube," *IEEE Transactions on Electron Devices*, Institute of Electrical and Electronics Engineers, New York, NY, Vol. ED-22, No. 10, October 1975, pp. 904–910.

Hall, J. A., "Problem of Infrared Television-Camera Tubes vs Infrared Scanners," *Applied Optics,* Optical Society of America, Washington, DC, Vol. 10, No. 4, April 1971, pp. 838–844.

Hansen, J. R., J. L. Fergason and A. Okaya, "Display of Infrared Laser Patterns by a Liquid Crystal Viewer," *Applied Optics,* Optical Society of America, Washington, DC, Vol. 3, No. 8, August 1964, pp. 987–988.

Helmick, C. N., Jr., and W. H. Woodworth, "Improved Performance from Pyroelectric Vidicons by Image Difference Processing," *Ferroelectrics,* Gordon and Breach, London, England, Vol. 10, 1976.

Hoshino, S., T. Mitsui, F. Jona, and R. Pepinsky, "Dielectric and Thermal Study of Tri-Glycine Sulphate and Tri-Glycine Fluoberyllate," *Physical Review,* American Institute of Physics, New York, NY, Vol. 107, No. 5, 1957, p. 1255.

James, I. J. P., "Fluctuation Noise in Television Camera Head Amplifiers," *IEE Proceedings,* Institution of Electrical Engineers, London, England, No. 20, Part III A, 1952.

Jona, Franco, and G. Shirane, *Ferroelectric Crystals,* Pergamon Publishing, New York, NY, 1962.

Kurczewski, R., and R. S. Levitt, "Pyroelectric Vidicon Camera System, Modern Utilization of Infrared Technology," *Proceedings, Society of Photo-Optical Instrumentation Engineers,* Society of Photo-Optical Instrumentation Engineers, Palos Verdes Estates, CA, Vol. 62, 1976.

Logan, R. M. and R. Watton, "Analysis of Cathode Potential Stabilization of the Pyroelectric Vidicon," *Infrared Physics,* Pergamon Publishing, Elmsford, NY, Vol. 12, 1972, p. 17.

McDaniel, G. W., and D. Z. Robinson, "Thermal Imaging by Means of the Evaporograph," *Applied Optics,* Optical Society of America, Washington, DC, Vol. 1, May 1962.

Percival, W. S., British Patent No. 528179, 1939, British Patent Office, London, England.

Petito, F., and L. Garn, "A Critical Systems Assessment of the Pyroelectric Vidicon," *Technical Digest,* International Electron Devices Meeting, Institute of Electrical and Electronics Engineers, New York, NY, 1975, pp. 74–77.

Singer, B., "Pyroelectric Vidicons, the IR Imaging Dark Horse," *Electro-Optical Systems Design,* Milton S. Kiver Publications, Chicago, IL, Vol. 7, No. 7, July 1975.

Singer, B., and J. Lalak, "Pyroelectric Vidicon with Improved Resolution," *IEEE Symposium on Applications of Ferroelectrics,* Institute of Electrical and Electronics Engineers, New York, NY, June 1975.

Vis, Vincent, "The Interference Edgegraph (A New Thermal Imaging Device)," The University of Michigan Engineering Research Institute, Ann Arbor, MI, Report No. 2144-31-R, January 1955.

Watton R., "Pyroelectric Materials: Operation and Performance in Thermal Imaging Camera Tubes and Detector Arrays," *Ferroelectrics,* Gordon and Breach, London, Vol. 10, 1976.

Watton, R., C. Smith, B. Harper and W. M. Wreathall, "Performance of the Pyroelectric Vidicon for Thermal Imaging in the 8-14 Micron Band," *Transactions on Electron Devices* Institute of Electrical and Electronics Engineers, New York, NY, Vol. ED-21, No. 8, August 1974, pp. 462–469.

Weed and Merkel, Pennsylvania Power and Light, *Transmission and Distribution,* Cleworth Publications, Cos Cob, CT, 1967.

Chapter 14

PHOTOGRAPHIC FILM

Gwynn H. Suits
Environmental Research Institute of Michigan

CONTENTS

14. Photographic Film

14.1 Introduction

Most characteristics of infrared films and plates are like those of visible-light films and plates. Infrared emulsion responds to near infrared radiation as well as to visible and ultraviolet radiation. The parameters characterizing infrared photographic materials are most often based upon photometric rather than radiometric units.

14.1.1. Symbols, Nomenclature and Units.

The symbols, nomenclature and units used in this chapter are listed in Table 14-1.

Table 14-1. Symbols, Nomenclature and Units

Symbols	Nomenclature	Units
C	Proportionality constant	–
CI	Contrast index; slope of a straight line drawn between two points on the H-D curve that usually represent the highest and the lowest useful densities in a continuous-tone, black-and-white negative.	–
D	Photographic density*; $= \log_{10} 0 = \log_{10}(\tau_D^{-1})$	–
E_v	Illuminance; photometric irradiance, the incident luminous flux per unit area	lm
$E_{v,\lambda}(\lambda)$	Spectral illuminance; $= \partial E_v / \partial \lambda$	lm nm^{-1}
E_e	Irradiance; the incident radiant flux per unit area, $\partial\phi/\partial A$	W m^{-2}
$E_{e,\lambda}(\lambda)$	Spectral irradiance; $= \partial E_e/\partial\lambda$	W m^{-2} nm^{-1}
H	Exposure; the time integral of E_v or E_e; $= \int_0^t E_e(t)\, dt$	J m^{-2}
H_v	Photometric exposure; $= K_m \int_0^t \int_{\lambda_1}^{\lambda_2} V(\lambda) E_{e,\lambda}(\lambda)\, d\lambda\, dt$	m cd sec lm sec m^{-2}
H_e	Radiometric exposure; $= \int_0^t \int_{\lambda_1}^{\lambda_2} E_{e,\lambda}(\lambda)\, d\lambda\, dt$	J m^{-2}
K_m	Maximum luminous efficacy; $\doteq 683$	lm W^{-1}
0	Opacity*; $= \tau_D^{-1}$	–
S	Sensitivity	cm^2 erg^{-1}
$S(\lambda)$	Spectral sensitivity; $= [H_e(\lambda)]^{-1}$	cm^2 erg^{-1}
t	Time	sec
$V(\lambda)$	Relative spectral luminous efficiency	–
$\Delta\lambda$	Spectral bandwidth	nm, or μm
γ	Slope of the straight-line portion of the H-D curve	–
λ	Wavelength	μm, or nm
$\bar{\lambda}$	Center wavelength	μm, or nm
ρ	Reflectance	–
$\bar{\rho}$	Reflectance, average	–
τ	Transmittance	–
$\bar{\tau}$	Transmittance, average	–
τ_D	Transmittance, diffuse; the fraction of incident radiation	–
$\tau(\lambda)$	Transmittance, spectral	–

*All data in this chapter are based upon diffuse densities (diffuse transmittance).

Table 14-1. Symbols, Nomenclature and Units (*Continued*)

Symbols	Nomenclature	Units
Subscripts		
e	Radiometric quantities	–
v	Photometric quantities	–
λ	The operation of partial differentiation with respect to wavelength	–

14.1.2. General Remarks. Infrared films can be handled like similar conventional materials, but extra precautions must be observed to insure that they will not be fogged by stray, infrared radiation. Loading and unloading of cameras, magazines, or cassettes should be done in total darkness. Certain woods, hard rubbers, plastics, and black papers are not opaque to near infrared radiation; thus, accidental fogging of infrared film can occur when an improper protective covering is used for the plates. Metal foil on black paper is a dependable covering. Since most safe-lights transmit infrared freely, they must be used with a filter specifically recommended. Kodak®* safe-light filter, Wratten® number 7 (green), is recommended for use with black-and-white infrared materials, except Kodak high-speed infrared films.

Additional problems arise when infrared film is used in equipment designed primarily for visible-light photography. The lens focal positions indicated by barrel markings or range finders are not accurate with infrared film. In a single-lens reflex (SLR) camera, the visible image on the screen in not a reliable indicator of focus, either. The longer wavelength rays are refracted less by lenses so that the effective focal length of the lens is about 0.5% longer than the visible-light focal length. In addition, aberration corrections made to optimize the sharpness of the image in the visible portion of the spectrum are not generally optimum for the infrared portion. Exposure meters designed for visible light use do not respond to near-infrared radiation. Hence, these meter readings are not useful as exposure indicators unless the ratio of visible to infrared radiation is known.

14.2 Available Infrared Films and Plates

Infrared films are available in roll form (for still and motion picture cameras) and in sheet form, as are spectroscopic plates and films with spectral sensitivity in the near-infrared region. As of January 1976, their availability was the subject of a brief survey of major film manufacturers [14-1]. The results indicate that Eastman® Kodak Company is the only manufacturer and supplier of stock products in the U.S.A., but that large, special orders can be met by other manufacturers (Table 14-2). Table 14-3 lists brief descriptions of Eastman Kodak materials.

---

*The following terms used in this chapter are trademarks of the Eastman Kodak Company: Aerochrome, Aerographic, D-17, D-19, D-76, DK-50, Eastman, Ektachrome, Estar, Estar-Ah, Kodak, Versamat, Wratten.

Table 14-2. Availability of IR Films and Plates [14-1]

Companies	Availability
Agfa-Gevaert, Inc. Teterboro, NJ	Currently make none, though have capability for large orders.
E.I. Dupont DeNemours & Co., Inc. Wilmington, DE	Currently make none, though have capability for large orders.
Eastman Kodak Co. Rochester, NY	Currently only manufacturer and supplier of both color and black-and-white infrared films and plates in the U.S.A.
Fuji Photo Film U.S.A. Inc. New York, NY	Manufacture infrared sensitive materials in Japan and do not normally export. In large and special orders it can be brought into the U.S.A.
GAF Corp. Binghamton, NY	Currently make none, though have capability for large orders.
Ilford, Inc. Paramus, NJ	Currently make none, though have capability for large orders.
Konishirokun Corp. Englewood Cliff, NJ	Produce black-and-white infrared materials under trade-name Sakura®. Not normally exported from Japan, but can be obtained in large orders.
Minnesota Mining & Mfg. Co. (3M Company) St. Paul, MN	Currently make none, though have capability for large orders.

Table 14-3. Infrared Films and Plates

Kodak Name & Type	Characteristics	Remarks
Infrared Aerographic film – 2424 [14-2, 14-6]	Available in 70 mm and other aerial film sizes. Negative, black-and-white, response to 900 nm with peak about 760 to 880.	Refrigerated shipment and storage below 4°C (Extended storage, –18°C to –23°C). This requirement exists, to some extent, for all IR films.
Infrared Aerographic film – 2481 [14-2, 14-3]	Available in 35 and 16 mm rolls. Negative, high speed black-and-white, response to 900 nm, as for 2424.	Also called high speed infrared film in shorter rolls.
Kodak high-speed infrared film – 4143 [14-2, 14-3]	Available in 4 × 5 sheets. Negative, high speed, moderately high contrast, black-and-white response to 900 nm.	Estar thick base.
Infrared Aerographic film – SO-289* [14-5]	Was available as aerial film made to special order. Very fine grain, very high resolution, negative, black-and-white. Contrast decreases somewhat with increasing wavelength.	Special order, nonstock. Specifications subject to change.* This film was discontinued in 1977.
Aerochrome infrared – 2443 [14-2, 14-7]	Available in 70 mm and other aerial film sizes. Three-layer, false-color reversal with moderate resolution.	Estar base.
Aerochrome infrared – 3443 [14-2, 14-7, 14-8]	Was available in 70 mm and other aerial film sizes. Emulsion as above for 2443.	Estar thin base. Discontinued in 1976.
Ektachrome infrared [14-3]	Available in 35 and 16 mm rolls. Emulsion as above for Aerochrome.	
High-definition Aerochrome infrared – SO-127* [14-7, 14-10]	Available as aerial film on special order. False-color reversal with extremely fine grain and very high resolution, but low speed. Response to 900 nm.	Special order, Estar base.
High-definition Aerochrome infrared – SO-130* [14-7, 14-10]	Emulsion as for SO-127 above.	Special order, Estar thin base.
High-definition Aerochrome infrared – SO-131* [14-7, 14-10]	Emulsion as for SO-127 above.	Special order. Estar ultra thin base.
Spectroscopic films and plates [14-4]:		
Type I-N	Moderately high speed, high contrast, coarse granularity, high resolution, N-type spectral response to 900 nm, peak at around 800 nm.	Type I-Z plates must be hypersensitized just before exposure. Refrigeration particularly important.
Type I-Z	High contrast, high resolution, moderately coarse granularity, response to 1.15 μm, peak near 1.08 μm.	
Type IV-N	High contrast, less sensitive, very high resolution, fine granularity, N-type spectral response.	

*SO (special order) aerial films are usually either newly developed products or products not manufactured on any regular production schedule. SO films (1) are subject to specification changes, (2) may show more batch-to-batch variability, (3) are subject to minimum-order requirements, (4) may require longer time for delivery, and (5) may cost more than regular name-grade products [14-2].

14.3. Hypersensitizing

The speeds of some infrared films and plates can be increased by hypersensitizing them. For best results, this should be done just before exposure. Kodak spectroscopic films and plates with M and N sensitizings may be hypersensitized a few days before use. However, type Z plates must be hypersensitized just before use.

Films or plates are hypersensitized by immersion in a weak solution of ammonia: 4 parts of 28% ammonia (the strongest available commercially) with 100 parts of water. (Kodak spectroscopic plate and film, type I-N can be hypersensitized by using plain water instead of the ammonia solution.) *Concentrated ammonia is a dangerous irritant and should be handled cautiously.* The diluted ammonia solution should be used at a temperature of about 5°C (40°F) and never higher than 15°C (55°F). The films or plates should be bathed for about 3 minutes and then bathed for 2 to 3 minutes in methyl or ethyl alcohol and dried as rapidly as possible in a stream of cool, dust-free air from a blower or fan.

Care must be exercised in hypersensitizing to prevent streaking. Particular care must be taken to keep dust from spotting the emulsion surface during drying.

14.4. Definition of Density and Exposure

The exposure and development of photographic film produces an image consisting of areas having different transmittances, depending on the number and size of the silver grains present. Specular transmittance may be defined as the ratio of the amount of the undeviated light passing through the plate or film to that of the incident collimated light on the back. Diffuse transmittance is the ratio of the intensity of the undeviated and scattered light together to that of the incident collimated light. The opacity, 0, is defined as the reciprocal of the transmittance. The density, D, is defined as $\log_{10} 0$; it can be either specular or diffuse density. The data in this chapter represent diffuse density values.

The exposure, H, may be expressed as either the time integral of the illuminance, E_v, or the time integral of irradiance, E_e, so that

$$H_v = K_m \int_0^t \int_0^\infty V(\lambda) E_e(\lambda) \, d\lambda \, dt \qquad (14\text{-}1)$$

and

$$H_e = \int_0^t \int_0^\infty E_e(\lambda) \, d\lambda \, dt \qquad (14\text{-}2)$$

where $V(\lambda) =$ the relative, spectral, luminous efficiency for the CIE-standard photometric observer

$K_m =$ the maximum, luminous efficacy

When exposure is expressed as a photometric quantity in the photographic literature, the units of H_v are customarily meter-candle-seconds. When exposure is expressed as a radiometric quantity, the units of H_e should be joule per square meter.

14.5. Sensitometric Characteristics

The curves presenting density, D, as a function of $\log_{10} H$ are known as the H and D curves (after Hurter and Driffield), or simply the characteristic curves. (See Figure 14-1.) The standard source of radiation for obtaining these curves is an artificial source made to provide a spectral distribution irradiance close to that provided by a 6000 K blackbody in the restricted spectral interval of interest. This source is intended to simulate average daylight, i.e., sunlight plus skylight. From the illuminance from this source, one may derive an unique irradiance in the infrared spectral region. Although more direct use of H_e for characteristic curves of infrared films makes good sense, H_v has been used in the literature and is a valid indicator of the magnitude of H_e. When a filter is specified along with the characteristic curves, the effects of the filter losses are implicit in the characteristic curves, i.e., the filter is considered to be an integral part of the film.

The slope of the straight-line portion (gamma) of the H and D curve is one parameter used to select appropriate development times. Another parameter which may be used instead of gamma is the contrast index, CI, which also is a slope derived from the D vs $\log_{10} H_v$ plots. CI is defined [14-9] as the slope of a line between the two points on the H and D curve which represent the highest and lowest useful densities for a continuous-tone, black-and-white negative. The CI is used more widely in pictorial photography than in scientific and industrial work.

The characteristic curves are average properties. (See Figures 14-2 through 14-8.) In addition, these characteristics hold only for the specified conditions of exposure and processing. Changes in light quality or in processing will yield a different set of characteristic curves. For example, Figure 14-4(a) shows a different CI curve than Figure 14-4(b) for the same film. The response of photographic materials to parameter variations tends to be nonlinear with almost every parameter. Apparatus used for making photometric or radiometric measurements must be calibrated for photographic response. Extreme care must be exercised to obtain reproducible processing conditions. When possible, calibrating exposures should be made adjacent to the areas to be measured. A wealth of information concerning the use and problems of photographic materials for photometry may be found in the literature of astronomy.

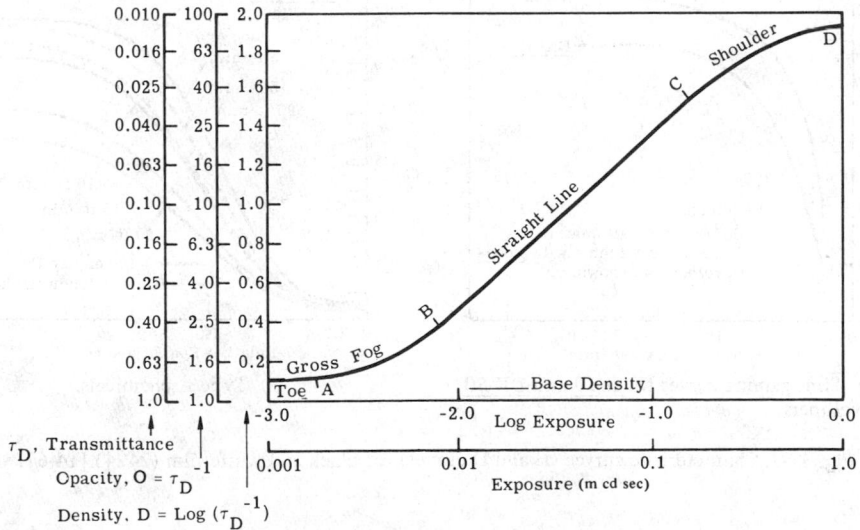

Fig. 14-1. Characteristic curve representative of a negative photographic material [14-4].

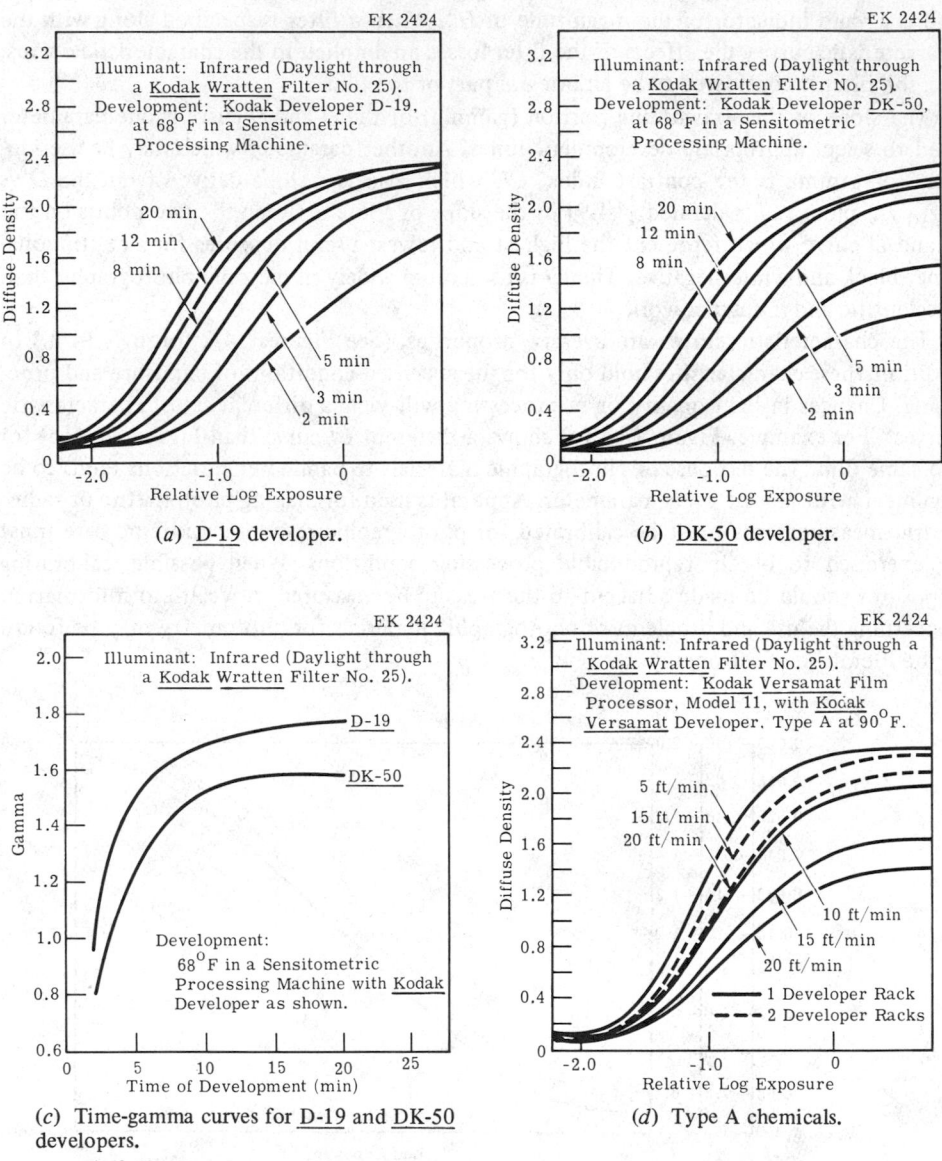

(a) D-19 developer.

(b) DK-50 developer.

(c) Time-gamma curves for D-19 and DK-50 developers.

(d) Type A chemicals.

Fig. 14-2. Characteristic curves (H and D curves) for black and white film (2424) [14-6].

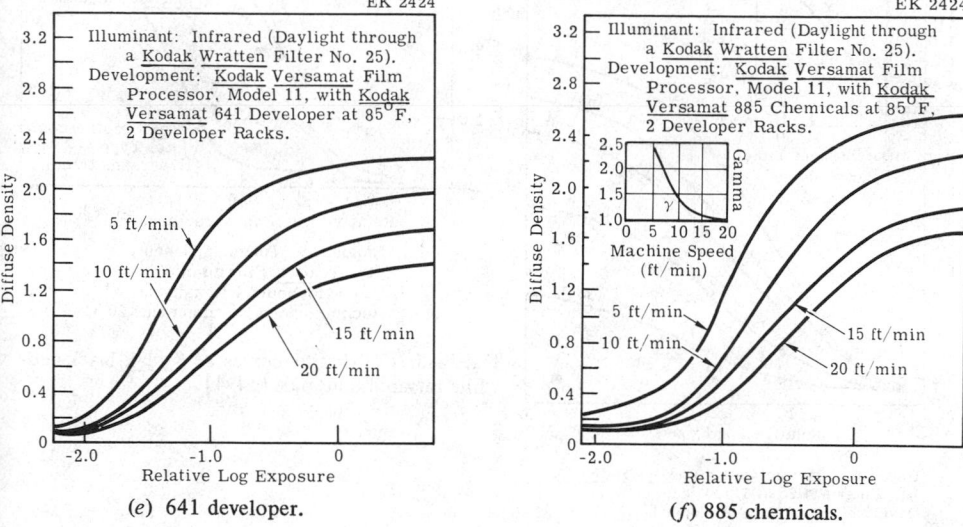

(e) 641 developer.   (f) 885 chemicals.

Fig. 14-2 (Continued). Characteristic curves (H and D curves) for black and white film (2414) [14-6].

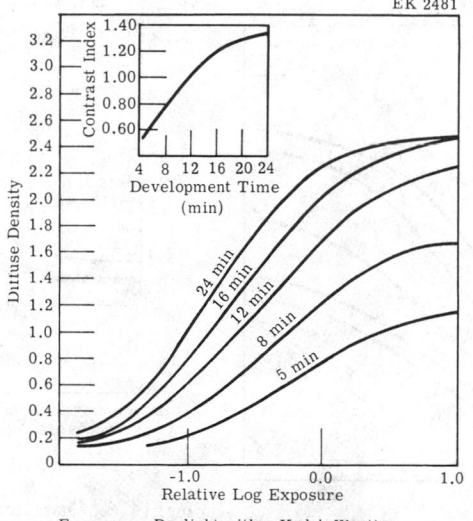

Fig. 14-3. Characteristic curves; Kodak high speed infrared film 2481 (Estar base) and Kodak high speed infrared film [14-3].

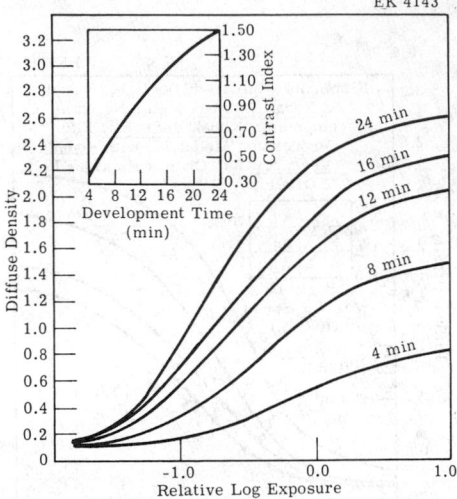

Exposure: Daylight with a Kodak Wratten
Filter No. 25 at 1/100 sec.
Development: Kodak Developer D-76
at 20°C (Continuous agitation) for
4, 8, 12, 16, and 24 min.

(a) Kodak high speed infrared film 4143 (Estar thick base) [14-3].

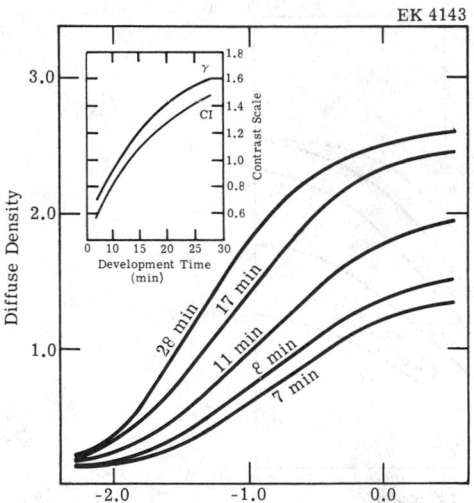

Tungsten Exposure: 1 sec
with Wratten Filter
No. 25 + 1.0 Neutral Density
integrated over a range
of 520 to 900 nm.

(b) Kodak high speed infrared film 4143 (Estar thick base) [14-4].

Fig. 14-4. Characteristic curves.

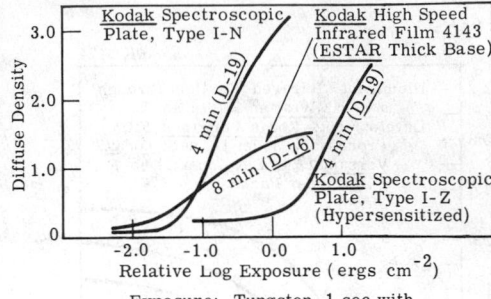

Exposure: Tungsten, 1 sec with
Wratten Filter No. 25
+1.0 Neutral Density
Recommended Development: 20°C

Fig. 14-5. Composite curves of Kodak black-and-white infrared materials [14-4].

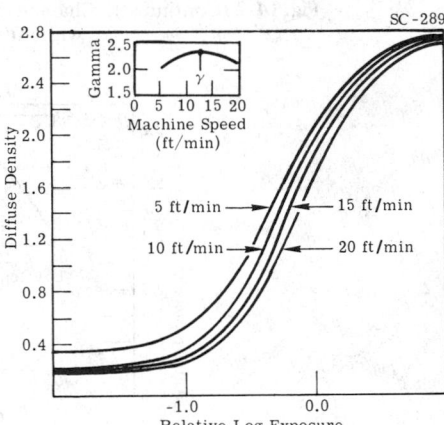

Exposure: Daylight through a Kodak
Wratten Filter No. 25 at 1/100 second.

Development: Kodak Versamat Film
Processor, Model 11, with Kodak
Versamat 885 Chemicals at 85°F
(29.4°C), 2 Developer Racks.

Fig. 14-6. Characteristic curves for SO-289 (885 chemicals). These data represent product tested under the conditions of exposure and processing specified. They are representative of production coatings and, therefore, do not apply directly to a particular box or roll of photographic material. These data do not represent standards or specifications which must be met by Eastman Kodak Company. The Company reserves the right to change and improve product characteristics at any time [14-5].

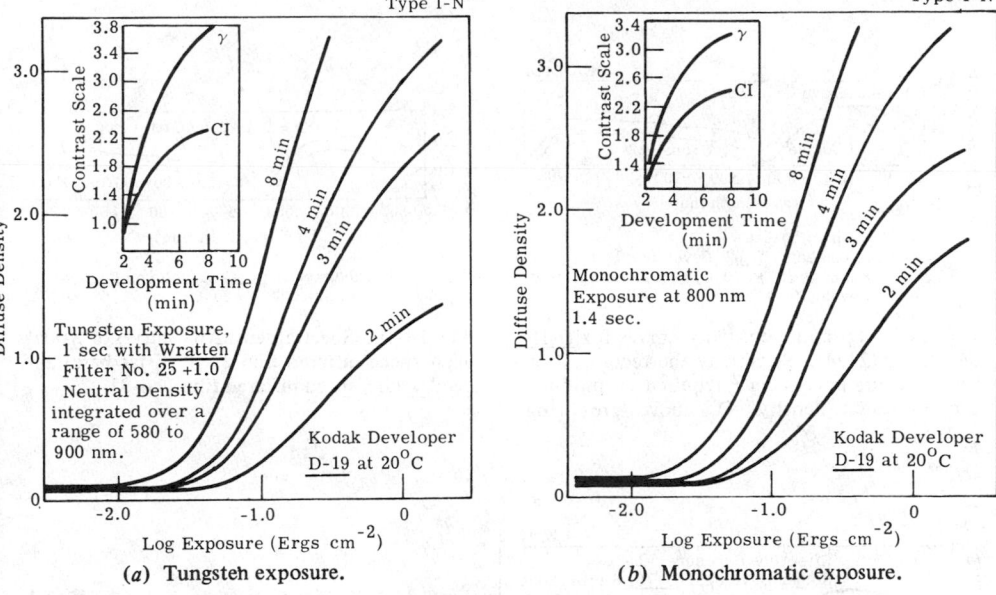

(a) Tungsten exposure.

(b) Monochromatic exposure.

Fig. 14-7. <u>Kodak</u> spectroscopic plate, type I-N [14-4].

14.6.   Spectral Sensitivity and Filter Transmittances

The sensitivity, $S(\lambda)$, of a photographic film to monochromatic light is defined as the reciprocal of the monochromatic exposure, $H_e(\lambda)$ (most often in erg cm^{-2}), required to produce a stated density above the fog level. The common logarithm of the sensitivity as a function of wavelength is given in Figures 14-9 through 14-13. Notice that in Figure 14-10, the sensitivity curve for $D = 0.3$ is not exactly a vertical translation of the curve for $D = 1.0$. This indicates that the gamma is a function of wavelength and development time. This is shown in Figure 14-14 in terms of the CI for type I-N plates. Transmission curves for useful filters are found in Figure 14-15.

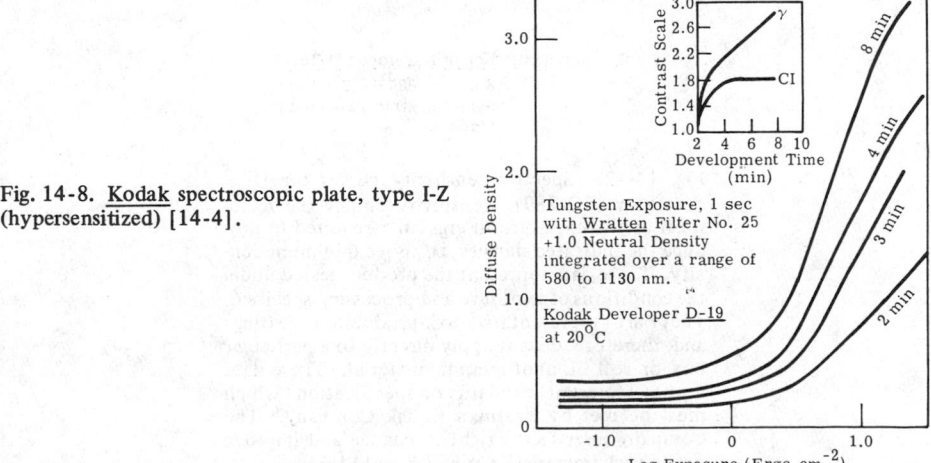

Fig. 14-8. <u>Kodak</u> spectroscopic plate, type I-Z (hypersensitized) [14-4].

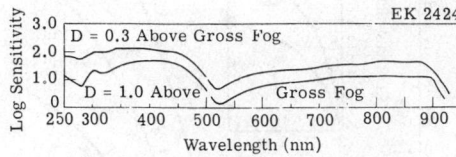

Exposure: 1.4 sec
Development: Kodak Developer D-19,
    8 min at 68°F (20°C) in a Sensitometric
    Developing Machine.

Fig. 14-9. Spectral sensitivity curves for D-19 developer (2424). Sensitivity the reciprocal of the exposure in ergs cm^{-2} required to produce the indicated density, D, above gross fog [14-6].

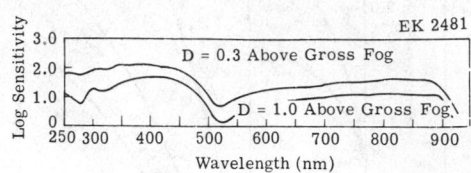

Development: Kodak Developer D-17,
    10 min at 68°F (20°C).

Fig. 14-10. Spectral sensitivity curves of Kodak high speed infrared film 2481 (Estar base) and Kodak high speed infrared film [14-3].

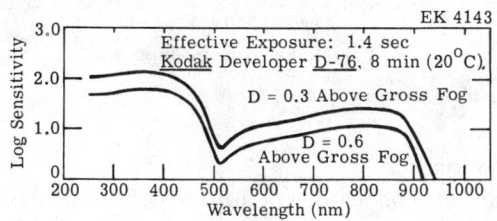

Fig. 14-11. Spectral sensitivity curves of Kodak high speed infrared film 4143 (Estar thick base). These curves are essentially the same as the curves for Kodak high speed infrared film 2481 and Kodak high speed infrared film [14-4].

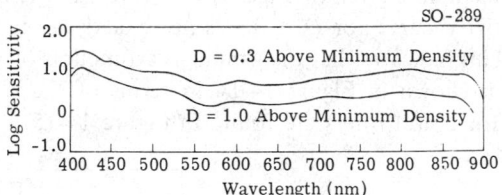

Development: Kodak Developer D-19,
    8 min at 68°F (20°C) in a
    Sensitometric Processing
    Machine.

Fig. 14-12. Spectral sensitivity curves for D-19 developer (SO-289). Sensitivity equals the reciprocal of the exposure in ergs cm^{-2} required to produce the indicated density, D, above minimum density. These data represent the product tested under the conditions of exposure and processing specified. They are representative of production coatings and, therefore, do not apply directly to a particular box or roll of photographic material. These data do not represent standards or specifications which must be met by Eastman Kodak Company. The Company reserves the right to change and improve product characteristics at any time [14-5].

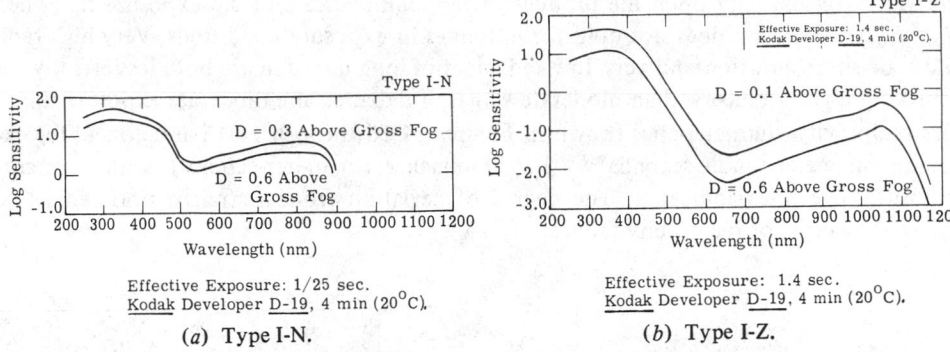

Effective Exposure: 1/25 sec.
Kodak Developer D-19, 4 min (20°C).

Effective Exposure: 1.4 sec.
Kodak Developer D-19, 4 min (20°C).

(a) Type I-N.

(b) Type I-Z.

Fig. 14-13. Spectral sensitivity of Kodak spectroscopic plates [14-4].

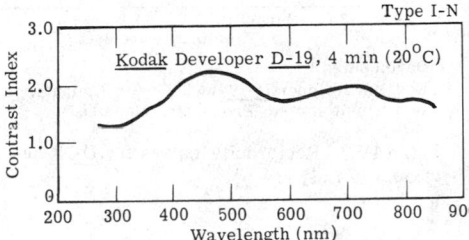

Fig. 14-14. Contrast index versus wavelength. This curve relates contrast index and wavelength of the exposing radiation and was generated from the same D-log exposure curve and was used to plot the spectral sensitivity curve for this material [14-4].

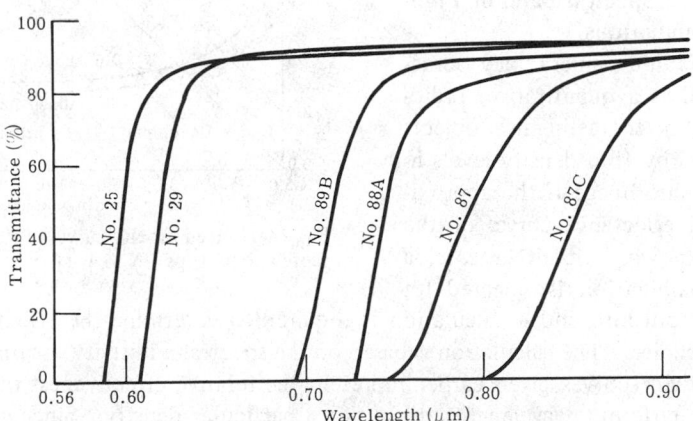

Fig. 14-15. Transmittance of Kodak Wratten filters used for infrared photography [14-2].

14.7.   Reciprocity Characteristics

The reciprocity law states that, with all other parameters fixed, the final density on the film depends only upon the product of the illuminance and the exposure time, i.e., $H_v = E_v \Delta t$. This law does not hold for extremes in exposure conditions. Very high radiances of short duration and very low radiances of long duration are both less effective in achieving a given density than moderate values of radiance and moderate exposure times. The reciprocity characteristics shown in Figures 14-16 through 14-18 are plots of log exposure (in meter-candle-seconds) vs. log illuminance (in meter-candles), with constant-exposure-time lines indicated. The degree of deviation from a straight horizontal line shows the degree of reciprocity failure.

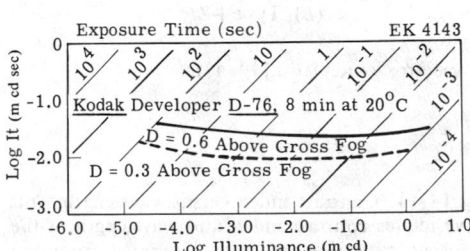

Fig. 14-16. Reciprocity curves of Kodak high speed infrared film 4143 (Estar thick base) [14-4].

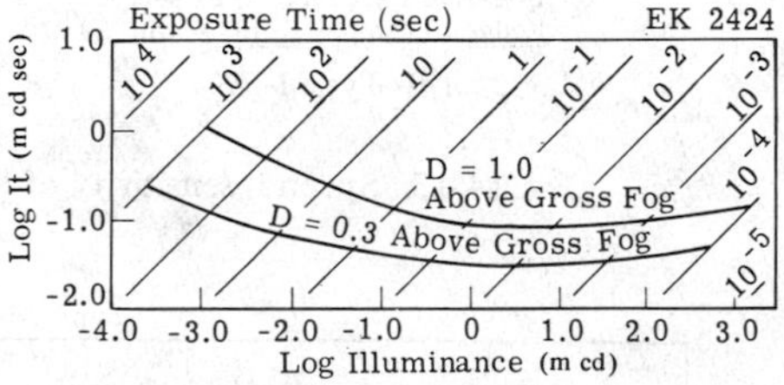

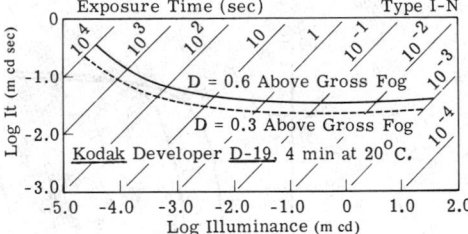

Development:
Kodak Developer D-19 at 68°F for 8 minutes in a Sensitometric Processing Machine.

Fig. 14-17. Reciprocity curves for D-19 developer (2424) [14-6].

14.8.   Effective Spectral Band of Film-Filter Combinations

A photographic camera may sometimes be used as a quantitative, radiometric device or to distinguish objects on the ground by film density levels in the photographic image if the approximate spectral-reflectance curves of the objects are known.   In either case, a film-filter combination is selected for

Fig. 14-18. Reciprocity curves of Kodak spectroscopic plate, type I-N [14-4].

the particular purpose and a calculation is required to determine the effective spectral band of that choice. The calculation is based on the spectral-additivity assumption. The spectral-sensitivity curves given earlier represent the relative effectiveness of monochromatic flux at different wavelengths in causing a particular density. Since more flux at some wavelengths is required to produce a certain density than flux at other wavelengths, one assumes that the process is merely less efficient for the former wavelengths. When levels of flux in two different wavelength bands are incident on the film at the same time, it is assumed that the total exposure will be determined by the efficiency-weighted sum of the exposures in each wavelength band.

The accuracy of the additivity assumption is best when the gamma of the sensitometric curve of the film is the same for the bands. It becomes worse as the variation in gamma becomes greater. Though film may respond in a variety of ways, the spectral-additivity assumption has been found accurate for practical purposes. Over relatively wide spectral bands (when the additivity assumption is likely to result in reduced accuracy) the inherent inaccuracy of broad-band radiometric measurements is also increased; in narrow spectral bands, both the validity of the additivity assumption and the radiometric accuracy will improve.

The relation used to calculate the effective exposure, H, or the efficiency-weighted sum of the exposures for each wavelength is

$$H = C \int_0^\infty \int_0^t \tau(\lambda)S(\lambda)\rho(\lambda)E_\lambda(\lambda)\,d\lambda\,dt \qquad (14\text{-}3)$$

where C = a proportionality constant
$\tau(\lambda)$ = the filter transmittance
$S(\lambda)$ = the relative spectral sensitivity of the film
$\rho(\lambda)$ = the spectral reflectance of the object to be photographed
$E_\lambda(\lambda)$ = the spectral irradiance caused by the source of illumination

If the value of $\rho(\lambda)$ is not known, one assumes that it is suffciently constant over the band of operation to be replaced by its average value over that band, $\bar{\rho}$. Thus

$$H = C\bar{\rho} \int_0^\infty \int_0^t \tau(\lambda)S(\lambda)E_\lambda(\lambda)\,d\lambda\,dt \qquad (14\text{-}4)$$

The effective spectral bandwidth of operation, $\Delta\lambda$, can be found by normalizing the remaining spectrally varying quantity, i.e., $[\tau(\lambda)S(\lambda)E_\lambda(\lambda)]$, to its peak value. (For a general discussion of normalization procedure see Chapter 20.)

$$\Delta\lambda = \frac{\left[\int_0^\infty \tau(\lambda)S(\lambda)E_\lambda(\lambda)\,d\lambda\right]}{[\tau(\lambda)S(\lambda)E_\lambda(\lambda)]_{\text{peak}}} \qquad (14\text{-}5)$$

The center wavelength of the band is then found from

$$\bar{\lambda} = \frac{\left[\int_0^\infty \tau(\lambda)S(\lambda)E_\lambda(\lambda)\lambda\,d\lambda\right]}{\Delta\lambda[\tau(\lambda)S(\lambda)E_\lambda(\lambda)]_{\text{peak}}} \qquad (14\text{-}6)$$

The photographic exposure, H, should then be proportional to the average reflectance of the object in the effective band from $\bar{\lambda} - \Delta\lambda/2$ to $\bar{\lambda} + \Delta\lambda/2$.

If maximum density separation of two objects with known spectral reflectances is desired, the value of H will depend upon $\rho(\lambda)$ for the objects, and $\tau(\lambda)$ for the filter transmittance. Best separation is obtained when the relationship $[H\,(\text{object 1})/H(\text{object 2})]$ is farthest from a one-to-one ratio. Equation (14-1) is used repeatedly with different filters to compare the effective pairs of exposures for the two objects.

The spectral irradiance, $E_\lambda(\lambda)$, is often caused by sunlight. Although the exact spectral distribution of $E_\lambda(\lambda)$ may not be known for a given photographic measurement, usually it will not greatly differ from day-to-day. Therefore, any reasonable curve of $E_\lambda(\lambda)$ for a clear day may be used. Further, if the spectral band of operation is narrow, the value of $E_\lambda(\lambda)$ may not change significantly with wavelength in that range. In the latter case, normalization may proceed as if E_λ were constant. The result is identical to Equation (14-5) in that

$$\Delta\lambda = \frac{\overline{E}_\lambda \int_0^\infty \tau(\lambda)S(\lambda)\,d\lambda}{\{[\tau(\lambda)S(\lambda)]_{\text{peak}}\overline{E}_\lambda\}} \tag{14-7}$$

where $\overline{E}_\lambda$ is the average E_λ in the band $\Delta\lambda$. The value of $\overline{E}_\lambda$ cancels; it does not affect the value of $\Delta\lambda$. The band center is found as before in Equation (14-6).

14.9. Color Infrared Film

Color infrared film is made with a three-layer emulsion very much like normal color film. The spectral sensitivity of each layer is controlled by an appropriate dye sensitization. Because the primary photosensitive material consists of silver halide crystals, each layer will respond not only to exposure in its dye-sensitized spectral range, but also to exposure in the short-wavelength, spectral range.

In normal color film, the top layer is sensitized to the blue spectral band. Below the top layer, a yellow (minus blue) filter is included to avoid exposure of the lower layers by the blue light. In the color infrared film, the layers are sensitized to a green band, a red band, and a near-infrared band. (See Figure 14-19.) A yellow filter may not be built into these layers for some films so the user of such a film must supply a suitable yellow filter. The Wratten 12 filter is commonly used for this purpose.

In normal color film, chemical processing introduces suitable dye colorants in such a way that, for example, where only blue light produced an exposure on the film, the processed image is made to appear blue. The color rendition of the various colors is based upon the tristimulus theory of color vision [14-22]. In color infrared film, the colorants are assigned to the layers in a simple color translation scheme, i.e., where only short-wavelength infrared radiation exposed the film, the processed image is made to appear red; where the exposure was due only to radiation in the red band, the image is made to appear green; and where the exposure was due only to radiation in the green band, the image is made to appear blue. Exposure by radiation in the blue band is avoided by the yellow or minus-blue filter. Table 14-4 lists schematically, by color band, the relationship between the spectral

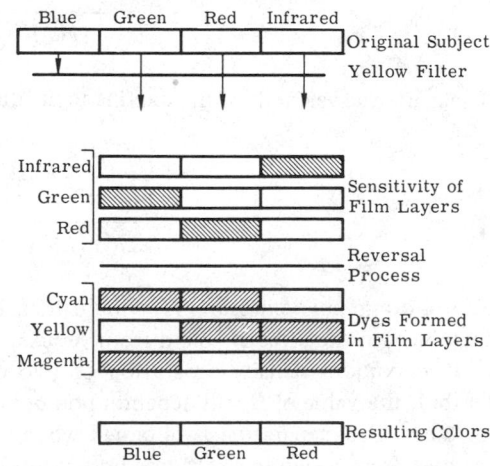

Fig. 14-19. Color formation with Kodak color-infrared film [14-3].

Table 14-4. Schematic Relationship Between Relative Exposures and
Image Color of Color Infrared Film

Relative Exposure			Image Color
Green	Red	IR	
0	0	0	Black
0	0	1	Red
0	1	0	Green
0	1	1	Yellow
1	0	0	Blue
1	0	1	Magenta
1	1	0	Cyan
1	1	1	White

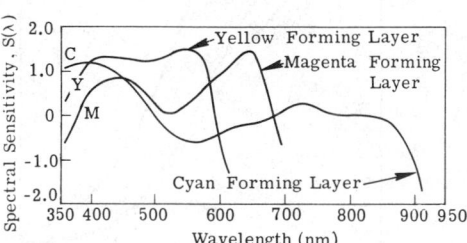

Fig. 14-20. Spectral sensitivity curves of Kodak Ektachrome infrared film [14-3].

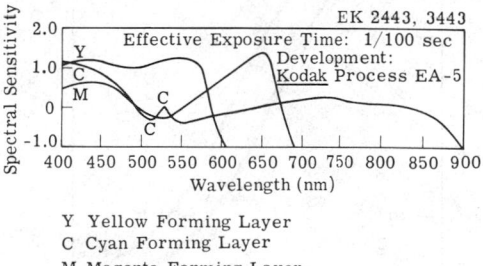

Y Yellow Forming Layer
C Cyan Forming Layer
M Magenta Forming Layer

Fig. 14-21. Spectral sensitivity curves of Kodak Aerochrome infrared film 2443 (Estar base) and Kodak Aerochrome infrared film 3443 (Estar thin base) [14-7].

distribution of the exposure and the color of the processed, infrared color image.

Figure 14-20 shows the spectral sensitivity of the Ektachrome, trilayer, infrared film. The curves are those for an equivalent neutral density of 1.0 above minimum density with EA-5 processing. Figure 14-21 shows the spectral sensitivity of Kodak Aerochrome® infrared films 2443 and 3443. The spectral sensitivities of new, experimental products SO-127, SO-130, and SO-131 are given in Figure 14-22, while the characteristic curves are in Figure 14-23.

The detection of very small changes in the reflectance properties of distant objects in the green, red and near-infrared spectral bands can be enhanced. The gamma of the three layers may be made greater than the gamma of normal color film. The greater gamma results in greater density changes and, hence, color-shifts in the layers (and therefore in the image) in response to small changes in the spectral distribution of the spectral radiance of the scene.

These large values of gamma make proper exposure difficult. Noticeable color changes in the processed image occur when the exposure of these films is changed by one camera stop (a factor of 2 in exposure). Small changes in the spectral irradiance of the illuminant

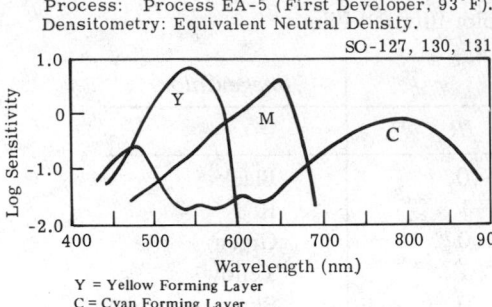

Density: 1.0
Process: Process EA-5 (First Developer, 93°F).
Densitometry: Equivalent Neutral Density.

Y = Yellow Forming Layer
C = Cyan Forming Layer
M = Magenta Forming Layer

Fig. 14-22. Spectral sensitivity curves of Kodak high definition Aerochrome infrared films: (Estar base) SO-127, (Estar ultra-thin base) SO-130, and (Estar thin base) SO-131. These data represent product tested under the conditions of exposure and processing specified. They are representative of production coatings and, therefore, do not apply directly to a particular box or roll of photographic material. These data do not represent standards or specifications which must be met by Eastman Kodak Company. The Company reserves the right to change and improve product characteristics at any time [14-10].

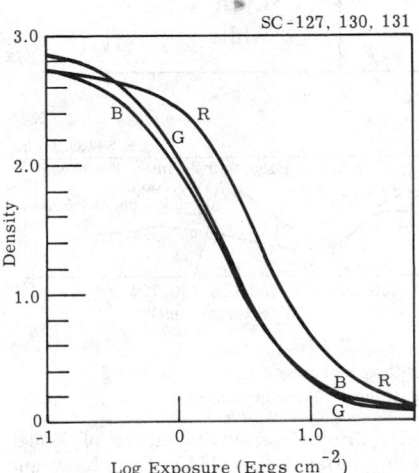

Exposure: 1/10 sec Daylight (500-900 nm).

Process: Process EA-5
           (First Developer, 93°F).
Densitometry: Status A .

(a) Status A density

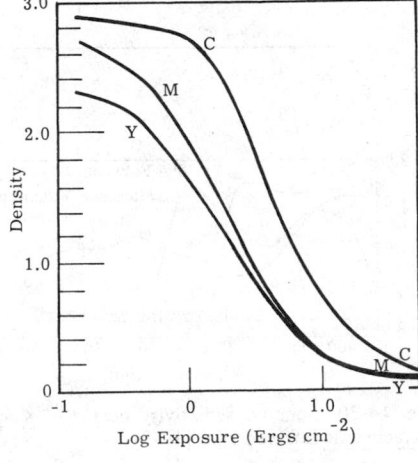

Exposure: 1/10 sec Daylight
            (500-900 nm)
Process: Process EA-5
            (First Developer, 93°F).
Densitometry: Equivalent Neutral
               Density.

(b) Equivalent neutral density

Fig. 14-23. Characteristic curves of Kodak high definition Aerochrome infrared films: (Estar base) SO-127, (Estar ultra thin base) SO-130, and (Estar thin base) SO-131. These data represent product tested under the conditions of exposure and processing specified. They are representative of production coatings and, therefore, do not apply directly to a particular box or roll of photographic material. These data do not represent standards or specifications which must be met by Eastman Kodak Company. The Company reserves the right to change and improve product characteristics at any time [14-10].

are also detectable by the film. The sun's angle and the sky's conditions may alter the balance between irradiance in the near-infrared band and the visible spectral range. Such alteration will not be evident to the eye or to a normal photographic light meter. By limiting the use of color infrared film to clear days and high-sun-elevation conditions, one may be fairly confident that the spectral balance between visible and infrared bands will be consistent.

14.10. Densitometry

Photographic film generally responds to exposure in a nonlinear fashion. The quantitative result of exposure may be measured by the transmittance of the resulting processed transparency. The commonly derived expression for this transmittance is the density, $D = \log_{10}(1/\tau)$. The density of a processed transparency depends in part upon the method which is used to measure transmittance. Consequently, different densitometers may not produce identical density readings on the same transparency. However, any densitometer which produces self-consistent density readings is entirely suitable for use in radiometric photography. The film must be calibrated by exposure to a sensitometric step-wedge which provides the relationship between exposure and final density readings. As long as the functional relationship between exposure and density is single-valued, a direct relationship can be established between output signal, input exposure, or target radiance for unknown targets.

14.10.1. Black and White Film.

The densitometer and the photographic camera with an associated filter and single-emulsion film can be combined to make a useful and economical radiometric system provided that certain features of the camera and film are taken into account. These features are those which alter the single-valued relationships between density and scene radiance.

The primary, camera-related variations are off-axis vignetting (or fall-off), flare, and in the case of cameras with focal plane shutteres, shutter speed variations across the film plane. Because of these variations, exposure varies on the film itself. A photograph of an infinite, perfectly uniform, diffuse (i.e., Lambertian) target will result in nonuniform density across the processed film. Such variations may be taken into account by making one of the exposures of a photographic sequence an exposure of a reasonably uniform, diffuse scene. A sheet of matte white typing paper, back-illuminated by daylight, and placed directly over the camera lens or filter will produce the effect of a uniform Lambertian illuminator or diffuse scene in the photographic spectral range. The relative variation in exposure as a function of film position may be found using the calibrated, step-wedge, exposure data along with the photograph of the Lambertian scene.

Primary, film-related variations are the result of the limiting spatial resolution of the film in combination with the camera and the chemical processing. High-contrast edges tend to develop preferentially. As a result, regions appear (near the edges) with higher densities than would be predicted by a density step-wedge calibration curve. Also, the density of detail near the limit of resolution approaches the density corresponding to the average, scene radiance of that detail and the surrounding area, rather than to the radiance of the detail. This is a result of the size of the grains and the scattering of light in the film causing a reduction in the contrast modulation transfer function. A bar-pattern resolution of 50 line-pairs mm^{-1} marks the point at which the averaging is nearly complete. Thus, the centers of regions no smaller than 0.5 to 1.0 mm should be used to obtain an accurate density-to-scene-radiance correspondence for a particular object.

To obtain the average scene-radiance of a fairly large area, one must measure the scene radiance of each part separately by using a densitometer on the photograph. The use of a large densitometer spot will usually lead to an incorrect scene-radiance average unless the variation in density within the spot is much less than 0.1. The transmittance measurement using the large spot size is equivalent to an average of film transmittance which is nonlinearly related to scene-radiance. The most accurate method of achieving the average scene-radiance of objects which contain much high-contrast texture, such as forests and agricultural fields, is to take the photograph in such a way that the texture falls below the resolving power of the camera and film system. A small-format camera, hand-held and set at small aperture and with a long exposure time, can produce ideal images of textured areas suitable for radiometric photography.

14.10.2. Color Film. All of the features of single-layer film relating to a radiometric densitometry apply also to trilayer color film. Unfortunately, there are additional peculiarities of trilayer films which make the use of such film for radiometric purposes not only difficult, but also less accurate.

In the color film layers, density is produced by the deposition of dye colorants (instead of grains of metallic silver) in the finished image. Dye colorants are chosen which will produce visual color from the white light used for the image illuminant by means of the tristimulus, color-subtractive system. Generally, the spectral absorption coefficient of the dye colorant in one layer is intended to control only one visual band—red, green, or blue—in accordance with the dye concentration within that layer. However, the spectral absorption coefficients are significantly nonzero in the other visual bands which are to be controlled by the

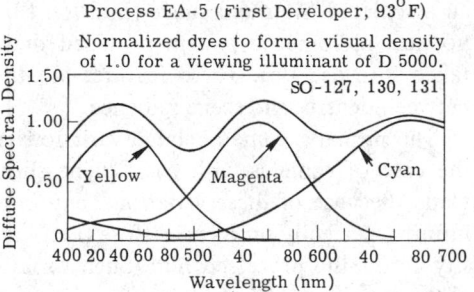

Fig. 14-25. Spectral dye density curves of <u>Kodak</u> high definition <u>Aerochrome</u> infrared films: (<u>Estar</u> base) SO-127, (<u>Estar</u> ultra-thin base) SO-130, and (<u>Estar</u> thin base) SO-131. These data represent product tested under the condition of exposure and processing specified. They are representative of production coatings and, therefore, do not apply directly to a particular box or roll of photographic material. These data do not represent standards or specifications which must be met by Eastman Kodak Company. The Company reserves the right to change and improve product characteristics at any time [14-10].

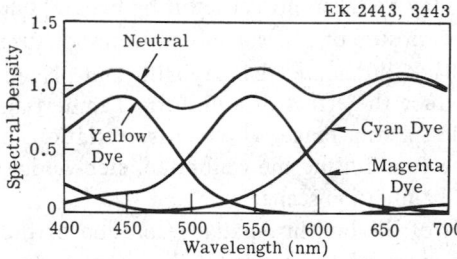

Fig. 14-24. Spectral dye density curves of <u>Kodak</u> <u>Aerochrome</u> infrared film 2443 (<u>Estar</u> base) and <u>Kodak</u> <u>Aerochrome</u> infrared film 3443 (<u>Estar</u> thin base) [14-7].

dye colorants as shown in Figure 14-24, for Kodak Aerochrome infrared films 2443 and 3443, and in Figure 14-25 for the experimental, special order films.

Densitometry of trilayer film requires the use of color-separation filters for transmittance measurements. A red filter, for instance, is used over the densitometer light source to obtain the density in the visual tristimulus, red band of the image. Because the film layers are stacked together, the products of the transmittances of all three layers must be measured at the same time. The spectral overlap of the colorant, spectral absorption coefficients permits the exposures in all three layers to control, to some extent, the density in one visual tristimulus band. The image densities taken with color-separation filters are called *integral densities*.

The spectral absorption coefficients of the dye colorants may be used to compute the densities one would have found for the isolated layers. These computed densities are called *analytic densities* and are computed by using appropriate, linear combinations of the three integral densities. If there were no further difficulties, trilayer film could be used for radiometric purposes in the same way as single-layer film.

However, interimage effects appear in trilayer film beyond those of the overlapping, spectral absorption coefficients. There is a fundamental asymmetry to the film processing. Processing chemicals must enter the emulsion stack by diffusion from only one side. Exhausted chemicals and by-products must leave by the same route. The speed of chemical reactions depends upon the concentration of the constituents and, for diffusion to occur, there must be a concentration gradient through the trilayer emulsion. Therefore, the extent of the reaction in one layer will depend upon its location and the extent of the reaction in adjacent layers. Image processing will differ depending upon the spectral balance and magnitude of the exposure. The consequences are that sensitometric curves derived from a grey, stepwedge source will not apply accurately to nongrey sources. The internal chemical processing is not always the same.

14.11. Modulation-Transfer Curves

Information on the spatial properties of photographic materials is available in the form of modulation-transfer curves. Some of these curves appear in Figures 14-26 through 14-31.

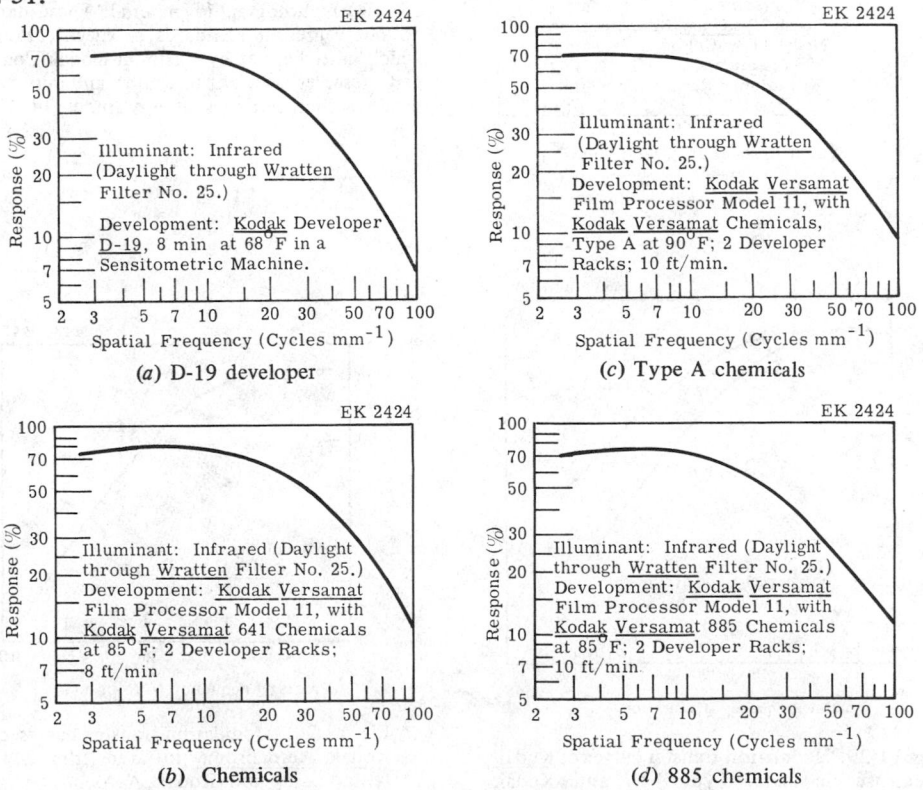

Fig. 14-26. Modulation-transfer function curves (2424) [14-6].

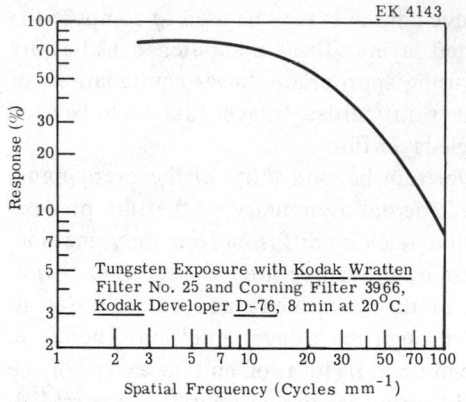

Fig. 14-27. Modulation-transfer curves of Kodak high speed infrared film 4143 (Estar thick base) [14-4].

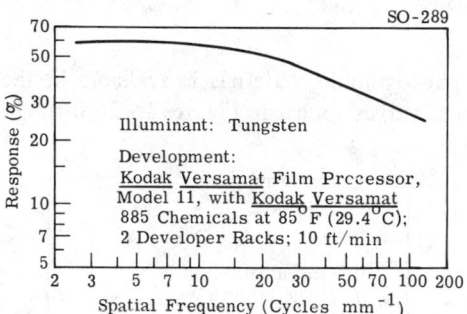

Fig. 14-28. Modulation-transfer curve for 885 chemicals (SO-289). These data represent product tested under the conditions of exposure and processing specified. They are representative of production coatings and, therefore, do not apply directly to a particular box or roll of photographic material. These data do not represent standards or specifications which must be met by Eastman Kodak Company reserves the right to change and improve product characteristics at any time [14-4].

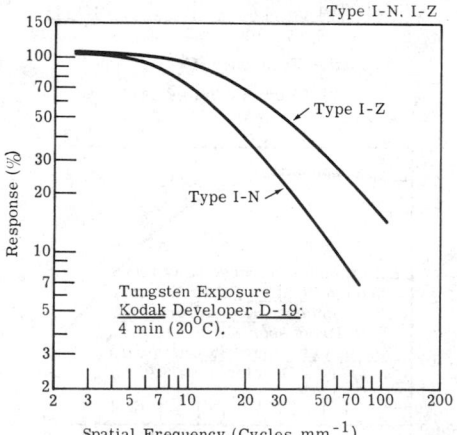

Fig. 14-29. Modulation-transfer curves of Kodak spectroscopic plate, Type I-N, and Kodak spectroscopic plate, Type I-Z [14-4].

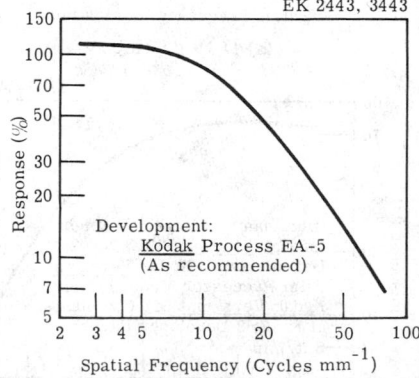

Fig. 14-30. Modulation-transfer curves of Kodak Aerochrome infrared film 2443 (Estar base) and Kodak Aerochrome infrared film 3443 (Estar thin base) [14-7].

Fig. 14-31. Modulation-transfer curve of Kodak high definition Aerochrome infrared films: (Estar base) SO-127 (Estar ultra-thin base) SO-130 and (Estar thin base) SO-131. These data represent product tested under the conditions of exposure and processing specified. They are representative of production coatings and, therefore, do not apply directly to a particular box or roll of photographic material. These data do not represent standards or specifications which must be met by Eastman Kodak Company. The Company reserves the right to change and improve product characteristics at any time [14-10].

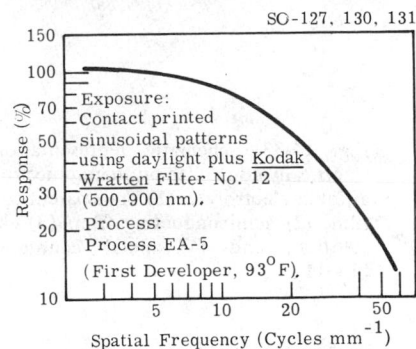

14.12. Infrared Semiconductor Photography*

Common photographic films have a long-wavelength sensitivity limit of about 1.0 μm. An extension of photographic techniques to longer wavelengths involves two basic problems: the development of processes photographically sensitive to the longer wavelengths and the development of techniques to limit fogging of the sensitive surface by background radiation. The more success one has in solving the first problem the more severe the second problem becomes. The severity of the fogging problem is graphically illustrated in Figure 14-32 where curves of fogging time of 300 K versus long-wavelength cutoff are presented for various sensitivities. For sensitivities of 1 cm² erg⁻¹, the fogging time is on the order of seconds for a long-wavelength cutoff of about 2.5 μm. The use of cooled enclosures reduces the fogging problem, but increases system complexity.

Recently, work has been done on the development of near-infrared photographic systems which use the photo-sensitive properties of semiconductors and can be activated only during the exposure time [14-11, 14-12]. Basically, the process consists of placing a large-area semiconducting layer in contact with an electrolyte that can be electrically biased either positive, negative, or zero. (See Figure 14-33.) Irradiation of the semiconductor material with a spatially varying, infrared image results in the generation of a corresponding spatial variation of chemical activity. This chemical activity can be: the deposition of metallic atoms out of the electrolyte solution onto the semiconducting layer; the dissolving of the semiconducting material into the electrolyte solution; or the formation of an oxide layer on the semiconductor. Electrically biased chemical reactions are initiated

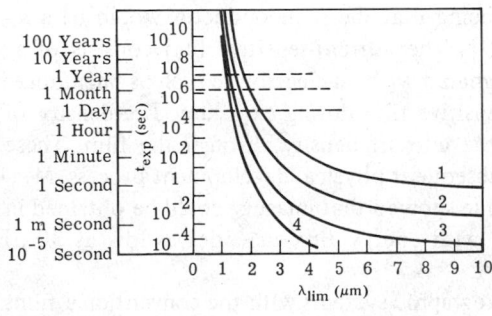

Fig. 14-32. Dependance of the fogging time of a photographic material τ_0 on the photosensitivity limit λ_{lim} in the case of storage at 300 K. Sensitivity (cm² erg⁻¹): (1) 10^{-4}, (2) 10^{-2}, (3) 1, and (4) 10^2 [14-11].

---

*Section 14.12 was prepared by Joseph Mudar, Environmental Research Institute of Michigan, Ann Arbor, MI.

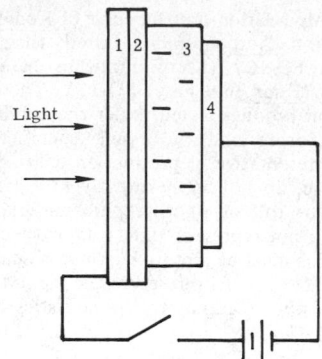

Fig. 14-33. Schematic representation of an electrically controlled photographic process with a liquid electrolyte. (1) Transparent conducting film, (2) semiconducting film, (3) electrolyte solution, and (4) metal counter-electrode [14-11].

by the presence of an electric field across the electrolyte. Unbiased reactions are controlled by the presence or absence of the electrolyte itself. These reactions can generate a visible image on the semiconductor surface. The electrically unbiased mode of operation had been termed *contact-sensitized semiconductor photography* and the biased mode termed *electrically controlled semiconductor photography* [14-11]. Typical irradiance values are 10^{-4} to 10^{-5} W cm^{-2} and typical exposure times are 1 to 5 sec [14-11]. (The time required is a function of the amount of the irradiation.) Semiconductor materials used successfully as of 1973 in this imaging process are silicon [14-11, 14-12, 14-13, 14-14], germanium, lead sulfide [14-11, 14-12, 14-15, 14-16, 14-17], lead selenide [14-11, 14-12, 14-15], gallium arsenide [14-11, 14-12, 14-18, 14-19], indium arsenide [14-11, 14-12], indium antimonide [14-20], gallium selenide [14-21], and indium selenide [14-21].

Imagery has been obtained with energy density levels as low as 20 μJ cm^{-2} using silicon as the semiconductor layer [14-11, 14-12]. Measured spatial resolution of 20 line-pairs mm^{-1} have been reported [14-14]; theoretical analyses indicate a limiting resolution, limited by the transverse diffusion of minority carriers, of approximately 300 line-pairs mm^{-1} [14-19].

A visible, permanent image can be produced by exposure alone; however, a reduced exposure time can produce a latent image that can be made visible by a physical development process. The images developed on the semiconductor wafers are permanent (i.e., not short-term transients).

Formation of an image in a current-sensitive film in series with the semiconductor element was also investigated, the advantage being that the semiconductor would be a reusable element [14-11, 14-12, 14-13, 14-19]. The current-sensitive film consists of a gelatine or polyethylene-type of layer impregnated with an electrolyte such as phenidone. Latent-image centers form in this current-sensitive film during exposure. The density of the latent-image centers is a function of the current density through the film. These latent-image centers are made visible by a subsequent physical development process. Measurements using high-resistivity gallium arsenide showed that imagery could be obtained in an electrolyte impregnated cellophane film with energy flux densities as low as 30 μJ cm^{-2} [14-19].

A comparison of the semiconductor photographic systems with the conventional films shows that semiconductor systems require total energy flux densities greater by factors of about 10^2.

14.13. References and Bibliography

14.13.1. References

[14-1] David B. Eisendrath, Photographic Consultant, Brooklyn, NY, Private Communication, January and April 1976.

[14-2] "Applied Infrared Photography," Eastman Kodak Co., Rochester, NY, Kodak Publication No. M-28, 1970.

[14-3] "Kodak Infrared Films," Eastman Kodak Co., Rochester, NY, Kodak Publication No. N-17, 1971 and 1974, pp. 3, 10, 11, 12, and 14.

[14-4] "Kodak Plates and Films for Scientific Photography," Eastman Kodak Co., Rochester, NY, Kodak Publication No. P-315, 1973, pp. 2, 16d, 17d, 20d, 25d, 31d, 32d, and 33d.

[14-5] "Kodak Multi-Spectral Infrared Aerial Film (Estar-Ah® Base) SO-289," Eastman Kodak Co., Rochester, NY, Aerial Data, Kodak Publication No. M-127, July 1975, p. 7.

[14-6] "Kodak Infrared Aerographic Film 2424 (Estar Base)," Eastman Kodak Co., Rochester, NY, Aerial Data, Kodak Publication No. M-58, January 1973, pp. 8-10.

[14-7] "Kodak Data for Aerial Photography," Eastman Kodak Co., Rochester, NY, Kodak Publication No. M-29, 4th. Edition, May 1976, pp. 54-62.

[14-8] J. Berezansky, Jr., ed., Eastman Kodak Co., Rochester, NY, Private Communication, April 1976.

[14-9] C.J. Niederpruem, C.N. Nelson, J.A.C. Yule, "Contrast Index," *Photographic Science and Engineering,* Society of Photographic Scientists and Engineers, Washington, DC, Vol. 10, 1966, pp. 35-41.

[14-10] J. Berezansky, Jr., ed., Eastman Kodak Co., Rochester, NY, Private Communication; data on SO-127, -130, -131 film, MI-24283, TI-785, February-March 1974, pp. 3-7.

[14-11] L.G. Paritskii and S.M. Ryvkin, "Use of Semiconductors in Long-Wavelength Photography," *Soviet Physics-Semiconductors,* American Institute of Physics, New York, NY, Vol. 4, No. 4, October 1970, pp. 645-650.

[14-12] L.G. Paritskii and S.M. Ryvkin, "Some Principles of Long Wavelength, Semiconductor Photography," *Zhurnal Nauchoy i Prikladnoy Fotografiya i Kimematografiya,* Akademiya Nauk SSSR, Moscow Tzentr, USSR, May-June 1970, pp. 184-191.

[14-13] Zh. G. Dokholyan, L.G. Paritskii and S.M. Ryvkin, "Photographic Process Based on the Photoelectric Action of a Surface-Barrier Junction in Silicon," *Soviet Physics-Semiconductors,* American Institute of Physics, New York, NY, Vol. 4, No. 8, February 1971, pp. 1377-1378.

[14-14] D.N. Goryachev, L.G. Paritskii, and S.M. Ryvkin, "Photographic Process Based on Oxidation-Reduction Reactions on the Surface of Silicon," *Soviet Physics-Semiconductors,* American Institute of Physics, New York, NY, Vol. 4, No. 8, February 1971, pp. 1356-1357.

[14-15] D.N. Goryachev, L.G. Paritskii, and S.M. Ryvkin, "Photographic Process Based on Oxidation-Reduction Reactions on Thin Films of Lead Sulfide and Selenide," *Soviet Physics-Semiconductors,* American Institute of Physics, New York, NY, Vol. 4, No. 8, February 1971, pp. 1354-1355.

[14-16] L.V. Belyakov, L.G. Paritskii, S.M. Ryvkin, and V.B. Yarzhembitskii, "Use of Evaporated Lead Sulfide Film in Electrolytic Photography," *Soviet Physics-Semiconductors,* American Institute of Physics, New York, NY, Vol. 5, No. 7, January 1972, pp. 1283-1284.

[14-17] D.N. Goryachev, L.G. Paritskii, and S.M. Ryvkin, "Formation of Photographic Images on Thin Lead Sulfide Films by Internal Electrolysis," *Soviet Physics-Semiconductors,* American Institute of Physics, New York, NY, Vol. 6, December 1972, pp. 1003-1004.

[14-18] G.B. Gorlin, L.G. Paritskii, S.M. Ryvkin and A.A. Bagdanavichus, "Possible Use of the Semiconductor-Dielectric Electrophotographic System in Long-Wavelength Semiconductor Photography," *Soviet Physics-Semiconductors,* American Institute of Physics, New York, NY, Vol. 6, No. 2, August 1972, p. 371.

[14-19] E.I. Ivanova, B.V. Novogrudskii and L.G. Paritskii, "Semiconductor Photographic System Based on High-Resistivity Gallium Arsenide," *Soviet Physics-Semiconductors,* American Institute of Physics, New York, NY, Vol. 6, No. 7, March 1973, pp. 1585-1587.

[14-20] L.G. Paritskii, S.M. Ryvkin, and V.B. Yarzhembitskii, "Formation of Photographic Images at the Interface Between a Semiconductor and an Active Gas Medium," *Soviet Physics-Semiconductors,* American Institute of Physics, New York, NY, Vol. 6, No. 7, January 1973, pp. 1224-1225.

[14-21] G.B. Abdullaev, M.Kh. Alieva, D.N. Goryachev, F.N. Kaziev, L.G. Paritskii and S.M. Ryvkin, "Formation of Photographic Images on Thin Films of Gallium and Indium Selenides," *Soviet Physics-Semiconductors,* American Institute of Physics, New York, NY, Vol. 6, No. 6, December 1972, pp. 1024-1025.

[14-22] *The Science of Color,* Optical Society of America, Washington, DC, 1973, 3rd. Edition.

14.13.2 Bibliography

Abdullaev, G.B. M.Kh. Alieva, D.N. Goryachev, F.N. Kaziev, L.G. Paritskii, and S.M. Ryvkin, "Formation of Photographic Image on Thin Films of Gallium and Indium Selenides," *Soviet Physics-Semiconductors,* American Institute of Physics, New York, NY, Vol. 6, No. 6, December 1972.

Belyakov, L.V., L.G. Paritskii, S.M. Ryvkin, and V.B. Yarzhembitskii, "Use of Evaporated Lead Sulfide Films in Electrolytic Photography," *Soviet Physics-Semiconductors,* American Institute of Physics, New York, NY, Vol. 5, No. 7, January 1972.

Clark, W., *Photography by Infrared,* John Wiley and Sons, New York, NY, 1946.

Dokholyan, Zh.G., L.G. Paritskii, and S.M. Ryvkin, "Photographic Process Based on the Photoelectric Action of a Surface-Barrier Junction in Silicon," *Soviet Physics-Semiconductors,* American Institute of Physics, New York, NY, Vol. 4, No. 8, February 1971.

Gorlin, G.B., L.G. Paritskii, S.M. Ryvkin, and A.A. Bagdanavichus, "Possible Use of the Semiconductor-Dielectric Electrophotographic System in Long-Wavelength Semiconductor Photography," *Soviet Physics-Semiconductors,* American Institute of Physics, New York, NY, Vol. 6, August 1972.

Goryachev, D.N., L.G. Paritskii, and S.M. Ryvkin, "Formation of Photographic Images on Thin Lead Sulfide Films by Internal Electrolysis," *Soviet Physics-Semiconductors,* American Institute of Physics, New York, NY, Vol. 6, No. 6, December 1972.

Goryachev, D.N., L.G. Paritskii, and S.M. Ryvkin, "Photographic Process Based on Oxidation-Reduction Reactions on the Surface of Silicon," *Soviet Physics-Semiconductors,* American Institute of Physics, New York, NY, Vol. 4, No. 8, February 1971.

Goryachev, D.N., L.G. Paritskii, and S.M. Ryvkin, "Photographic Process Based on Oxidation-Reduction Reactions on Thin Films of Lead Sulfide and Selenide," *Soviet Physics-Semiconductors,* American Institute of Physics, New York, NY, Vol. 4, No. 8, February 1971.

Grinberg, A.A., L.G. Paritskii, and L.V. Udod, "Enhancement of the Resolving Power of *p-n* Junction Semiconductor Photographic System in a Magnetic Field," *Soviet Physics-Semiconductors,* American Institute of Physics, New York, NY, Vol. 6, No. 3, September 1972.

Hardy, A.C. and F.F. Perrin, *The Principles of Optics,* McGraw-Hill, New York, 1932.

Ivanova, E.I., B.V. Novogrudskii, and L.G. Paritskii, "Semiconductor Photographic System Based on High-Resistivity Gallium Arsenide," *Soviet Physics-Semiconductors,* American Institute of Physics, New York, NY, Vol. 6, No. 7, March 1973.

Kodak "Applied Infrared Photography," Eastman Kodak Co., Rochester, NY, Kodak Publication No. M-28, 1970.

"Kodak Data for Aerial Photography," Eastman Kodak Co., Rochester, NY, Kodak Publication No. M-29, 4th Edition, May 1976, pp. 54-62.

Kodak data on SO-127, -130, -131 film, MI-24283, TI-785, February-March 1974, pp. 3-7.

"Kodak Infrared Aerographic Film 2424 (Estar Base)," Eastman Kodak Co., Rochester, NY, Aerial Data, Kodak Publication No. M-58, January 1973.

"Kodak Infrared Films," Eastman Kodak Co., Rochester, NY., Kodak Publication No. N-17, 1971 and 1974.

"Kodak Multi-Spectral Infrared Aerial Film (Estar-Ah Base) SO-289," Eastman Kodak Co., Rochester, NY, Aerial Data, Kodak Publication No. M-127, June 1975.

"Kodak Plates and Films for Scientific Photography," Eastman Kodak Co., Rochester, NY, Kodak Publication No. P-315, 1973.

Niederpruem, C.J., C.N. Nelson, and J.A. Yule, J.A.C., "Contrast Index," *Photographic Science and Engineering,* Society of Photographic Scientists and Engineers, Washington, DC, Vol. 10, 1966.

Paritskii, L.G. and S.M. Ryvkin, "Some Principles of Long Wavelength, Semiconductor Photography," *Zhurnal Nauchoy i Prikladnoy Fotografiya i Kinematorgrafiya,* Akademiya Nauk SSSR, Moscow Tzentr, USSR, May-June 1970.

Paritskii, L.G., S.M. Ryvkin, "Use of Semiconductors in Long-Wavelength Photography," *Soviet Physics-Semiconductors,* American Institute of Physics, New York, NY, Vol. 4, October 1970.

Paritskii, L.G., S.M. Ryvkin, and V.B. Yarzhembitskii, "Formation of Photographic Images at the Interface Between a Semiconductor and an Active Gas Medium," *Soviet Physics-Semiconductors,* American Institute of Physics, New York, NY, Vol. 6, No. 7, January 1973.

Ross, F.E., *The Physics of the Developed Photographic Image,* Van Nostrand, New York, NY, 1924.

Science of Color, The, Optical Society of America, Washington, DC, 1963, 3rd. edition.

Chapter 15

COOLING SYSTEMS

Martin Donabedian
The Aerospace Corporation

CONTENTS

15. Cooling Systems

15.1. Introduction

This chapter treats cooling systems, often called "coolers" by workers in infrared technology. Such systems are designed to attain and maintain desired temperatures in electro-optical systems. Most often they are of greatest importance for the detecting elements.

15.1.1. Symbols, Nomenclature, Units. Table 15-1 lists the symbols, nomenclature, and units used in this chapter.

Table 15-1. Symbols, Nomenclature, Units

Symbols	Nomenclature	Units
A	Surface area	in.^{2}, m^2
a	Albedo	–
B	Bottle weight	lb
b	Nusselt, first exponent	–
C	Nusselt, second exponent	–
c_p, c_v	Specific heats	J kg^{-1} K^{-1}
COP	Coefficient of performance; the produced cooling power (output power), per unit of power supplied (input power)	W W^{-1}
D	Diameter, a dimension of the flow path or configuration	cm or in.
E	Irradiance or incident heat flux density: solar, albedo or earth emission	W m^{-2}, or Btu h^{-1} ft^{-2}
F_a	Radiation interchange configuration factor	–
F_e	Radiation interchange emissivity factor	–
F_{ER}	Radiation interchange factor for earth emission to a flat plate	–
FS	Factor of safety	–
h	Altitude above planet	nmi
h_c	Convective heat transfer coefficient	W cm^{-2} K^{-1}
I	Current	A
k	Thermal conductivity	W cm^{-1} K^{-1}
K	Thermal conductance	W K^{-1}
l	Insulation or gap thickness	cm or in.
M_o	Mach number	–
P	Pressure	psia

Table 15-1. Symbols, Nomenclature, Units (*Continued*)

Symbols	Nomenclature	Units
P_{ab}	Peltier coefficient	W A^{-1}
p	Contact pressure	psi
Q, q	Thermal energy	J or Btu
$\dot{q}$	Heat transfer rate	W or Btu h^{-1}
R	Radius of planet or earth, also electrical resistance	nmi, ohms
r	Recovery factor for fluid flow	—
S	Entropy	J kg^{-1} K^{-1} or Btu kg^{-1} K^{-1}
$\mathcal{S}$	Seebeck coefficient for a single material	V K^{-1}
S_O	Solar irradiance	W cm^{-2} or Btu h^{-1} ft^2
T	Temperature	K or °R
t	Time	h or sec
V	Volume	in.^{3}
v	Velocity	m sec^{-1} or ft sec^{-1}
W_c	Weight of cryogen	lb or g
W_o	Mechanical work	J or ft lb
Z	Figure of merit for thermoelectric coolers	K^{-1}
α	Absorptance	—
α_s	Solar absorptivity	—
β	Angle between orbital plane and the planet-earth line	degrees
γ	Ratio of specific heats; also angle between the surface normal and the local vertical	—
ϵ	Emissivity	—
θ	Angle between the surface normal and the sun line	degrees
θ_S	Angle between the local vertical and the earth-sun line	degrees
μ	Viscosity	kg m^{-1} sec^{-1} or lb ft^{-1} h^{-1}
ρ	Reflectance; or electrical resistivity; or density	Ohm cm kg m^{-3} or lb ft^{-3}
σ	Stefan-Boltzmann constant	W cm^{-2} K^{-4} or Btu h^{-1} ft^{-2} °R^{-4}
—	Refrigeration, refrigeration performance (see COP above)	—
—	Specific Power = [COP]$^{-1}$; = (Power)$_{in}$/(Power)$_{out}$	W W^{-1}
—	Specific Weight = Weight of Refrigeration, Cooling System, or Refrigerant per unit of cooling capacity or cooling power	lb W^{-1} or kg W^{-1}

Table 15-1. Symbols, Nomenclature, Units (*Continued*)

Symbols	Nomenclature	Units
Subscripts		
a	Albedo, or hot boundary	—
c	Convective, cold boundary, cone, or cryogen	—
d	Detector	—
ER	Earth emission	—
eff	Apparent or effective value when used in conjunction with k or ϵ	—
f	Film	—
H	Hemispherical	—
h	Hot boundary	—
max	Maximum	—
p	Patch	—
r	Recovery	—
s	Surface or solar	—

15.1.2. Abbreviations

Organizations

AESC	Aerojet ElectroSystems Co., Azusa, CA
AFFDL	Air Force Flight Dynamics Laboratory, Wright-Patterson Air Force Base, OH
ARPA	Defense Advanced Research Projects Agency, Washington, DC
BREL	Boeing Radiation Effects Laboratory, Seattle, WA
GSFC	Goddard Space Flight Center, Greenbelt, MD
ITT	International Telephone and Telegraph, Optical Division, Ft. Wayne, IN
IITRI	Illinois Institute of Technology Research Institute, Chicago, IL
JPL	Jet Propulsion Laboratory, Pasadena, CA
MVE	Minnesota Valley Engineering, Inc., New Prague, MN
SAMSO	Space and Missile Systems Organization, USAF, Los Angeles, CA
SBRC	Santa Barbara Research Center, Goleta, CA

Other Terms

ATS	Applications Technology Satellite
DCA	Detector capsule assembly
DAM	Double aluminized Mylar
ELMS	Earth Limb Measurement Satellite Program
ERTS	Earth Resources Technology Satellite
ESH	Equivalent sun hours
G-M	Gifford-McMahon
HEAO	High Energy Astronomical Observatory
HTTA	Hydrogen thermal test article
IMP	Integrated Materials Program
J-T	Joule-Thomson
LM	Lunar Module

MLI	Multilayer insulation
MOL	Manned Orbiting Laboratory
OAO	Orbiting Astronomical Observatory
OSO	Orbiting Solar Observatory
OTTA	Oxygen thermal test article
SAM	Single aluminized Mylar®*
VISSR	Visible IR Spin-Scan Radiometer
VM	Vuilleumier

15.2. Summary of Cooling-System Characteristics and Selection Guidelines

The primary factors which determine the type of cooling system or cryogenic cooler best suited for a specific application are: the cooling capacity; operating temperature; operating or mission time (and maintenance requirements); environmental conditions; the weight, volume and power availability; the configuration and dimensions of the device to be cooled; development time and/or cost considerations; cool-down time; and compatibility with sensor and vehicle. The various types of coolers discussed in this chapter can be segregated into four fundamental categories:

(1) Open-cycle, expendable systems which use: stored cryogens in either the subcritical or supercritical liquid state; solid cryogens; or stored, high-pressure gas with a Joule-Thomson (J-T) expansion.

(2) Passive radiators which cool systems to cryogenic temperatures by radiation to the low-temperature, deep-space environment.

(3) Closed-cycle, mechanical-refrigerator systems which provide cooling at low temperatures and reject heat at high temperatures.

(4) Thermoelectric coolers which use the Peltier cooling effect.

Table 15-2 summarizes the primary characteristics of various cooler types, such as the operating temperature range, the power input, and the coefficient of performance (COP). COP is defined as the ratio of the produced cooling power to the input or supplied power.

Cryogenic fluids stored as liquids in equilibrium with their vapors (subcritical storage) can provide a convenient, constant-temperature, control system. Liquids are available which provide temperatures ranging from 4.2 to 240.0 K. The primary limitations of this approach are the complex tank designs required to minimize boil-off, the direct relation of weight and volume requirements to elapsed time, and phase separation in a weightless environment.

Fluids can be stored at pressures above their critical pressures (supercritical storage) as homogeneous fluids, thus eliminating the phase-separation problems encountered during weightless conditions in space. The high pressures of supercritical storage require heavier containers than subcritical storage.

Stored, solid cryogens provide a reliable refrigeration system for low power heat sources for one to three years or longer depending on weight and volume limitations. A solid cryogen is used in conjunction with an insulated container, an evaporation path to space, and a conduction path from the coolant to the device being cooled. Advantages over the use of cryogenic liquids include a higher heat-content per mass and volume. Examples of temperatures are 10 K (using hydrogen), 60 K (using methane), and 125 K (using carbon dioxide). Limitations include restrictions on detector mounting, specialized filling procedures, and temperature-control requirements. Solid-cryogen designs have operated successfully in space [15-1, 15-2].

---

*Trademark of the DuPont Corporation.

Table 15-2. Cryogenic Cooler Characteristics and Selection Guidelines

Cooler Concept	Cooler Type	Temperature Range (K)	Cooling Capacity (W)	Power Input
Open-cycle expendable systems	Liquid storage (subcritical)	4.2 to 77.0	Unlimited	None
	Single-phase storage (supercritical)	5.2 to 126.0	Unlimited	Minimal
	Gas storage with Joule-Thomson (J-T) expansion	4.2 to 87.4	Up to 20	None
	Solid cryogen storage	8.3 to 150.0	0* to 1.0	None
Mechanical refrigerators	Vuilleumier (VM)	10 to 77	0* to 15	–
	Stirling	10 to 77	0* to 15	–
	Gifford-McMahon (G-M)	10 to 77	0* to 15	–
	Turbine and rotary-recip. Brayton cycle systems	4.2 to 77.0	10 to 100	–
	Closed-cycle J-T	$77^\dagger$	0.5 to 5.0	–
Thermoelectric	Single stage	230 to 300	0.10 to 100.0	–
	Multiple stage, cascaded	145 to 230	0.01 to 1.0	–
Radiation	Small radiator	80 to 100	0* to 0.10	None††
	Large, external radiators	100 to 200	1 to 10	None††

Table 15-2. Cryogenic Cooler Characteristics and Selection Guidelines *(Continued)*

COP**	Advantages or Features	Disadvantages or Limitations
–	Extensive experience, reliable, constant temperature sink over a wide range	Limited operating time, complex dewar design logistics, phase-separation problems in space
–	Homogeneous fluid provides increased design flexibility, eliminates two-phase problems	Higher pressure and temperature, increased dewar weight, may require internal heater
–	Remote cooling, simple, can be used intermittantly	Weight penalty for high pressure gas storage and high consumption rates, high gas purity required
–	Higher heat content per pound, lower storage density, reliable, ideal for long term space operation	Limitations of detector mounting, complex dewar design, specialized filling procedures
3×10^{-4} to 2×10^{-2}	Can be powered directly by heat, low pressures can provide potentially longer life than Stirling or G-M	Limited operational experience, moderate power input
10^{-3} to 5×10^{-2}	Minimum power input, compact, extensive operational experience in ground and airborne applications	Limited life capability
2×10^{-4} to 10^{-2}	Fully developed for airborne operations, split cycle permits remote cooling	High power input required, limited life capability
10^{-4} to 5×10^{-3}	Gas supported bearings have potential for longest life, best suited for large cooling capacities	Highest power input required, substantial development effort still required
3×10^{-3} to 10^{-2}	Eliminates logistics associated with open-cycle system	High input power, limited life compressors
10^{-1} to 5×10^{-1}	Lightweight, compact, high reliability, low cost	Limited minimum temperature attainable
2×10^{-4} to 3×10^{-2}	Lightweight, compact, high reliability, low cost	High power input required at low temperature
	Completely passive, reliable, long life capabilities, low cost	Orientation may be limited, parasitic heat leak and area become prohibitive as temperature decreases and cooling load increases

*"0" means on the order of milliwatts
**COP = Coefficient of Performance
† With nitrogen or air as the working fluid.
†† Except for temperature control

A simple, inexpensive approach to cooling is a J-T cooler in which the expansion of a high-pressure gas (1000 to 6000 psia) through a J-T expansion valve results in cooling of the gas and provides a source of liquid coolant at the point to be cooled. The use of helium, hydrogen, argon, and nitrogen enable units to provide cooling from approximately 4.2 to 77 K (although certain of these gasses require precooling) at capacities of 0.50 to 10.0 W. The primary limitation is the large mass associated with storage of high-pressure gas (approximately 4 lb of tank weight per pound of gas for N_2 and 15 for H_2). One advantage of this approach over cryogenic storage is the ability to provide intermittent operation over a long period of time.

The simplest, potentially most reliable method of providing cryogenic cooling in space employs a radiator which uses the low-temperature sink of space. This system is passive, requires little or no power, and is highly reliable for extended periods. The radiator must be shielded from radiation from sunlight, the parent spacecraft, and, in the case of near earth-orbits, from thermal emission and reflected sunlight from the earth and its atmosphere. The primary limitations are a rapid increase in required radiator size for obtaining lower temperatures and a parasitic heat leak into the radiator. Radiators have been designed to provide cooling to between 85 K for milliwatt-level loads and 135 K for loads up to 5 W [15-3, 15-4].

Stirling-cycle (mechanical-refrigerator) systems developed and fabricated for ground vehicles, aircraft and spacecraft use are the most efficient, in terms of power requirements, down to about 10 K. Systems using the Vuilleumier (VM) cycle have an inherent capability for longer life than Stirling-cycle units, and have the potential advantage of being powered directly by a heat source. A VM unit was operated successfully in space [15-5]. Two high-capacity, three-stage, VM units developed for long-life operation in space have been undergoing life testing since 1975 [15-6, 15-7].

Gifford-McMahon/Solvay, separable-component systems have been fully developed for aircraft, and currently possess the longest, maintenance-free, operating life of all systems (about 3000 h). However, low efficiency requires significantly more power and weight for these systems than for comparable VM or Stirling systems. (See Table 15-12.)

Turbomachinery or rotary-reciprocating machinery using Brayton or Claude cycles are usable for temperatures from 4 to 40 or 50 K. The low efficiency associated with turbomachinery systems restricts their application to the larger capacities. Both of these systems may have potential for extremely long life because of gas-supported bearings. Components have been developed and fabricated [15-8, 15-9]. However, little overall, system-performance data or operating experience are available.

No mechanical-refrigerator cycle yet developed can operate without maintenance for more than a few thousand hours.

Thermoelectric coolers can be used to provide cryogenic temperatures for low-power heat sources. By use of the Peltier cooling-effect, thermoelectric coolers provide a simple, lightweight, reliable method of cooling. Their primary limitations are low efficiency and the maximum operating-temperature difference between the hot and cold junctions. Production units are available in capacities ranging from 10 mW at 180 K to 30 W at 250 K (both based on a heat-sink of 300 K). Prototype systems operate at a minimum temperature of 170 K. Experimental systems have operated as low as 145 K under no-load conditions [15-10].

15.3. Cryogenic Data

15.3.1. Operating-Temperature Range of Selected Coolants. Temperature ranges of expendable coolants which use the heat of sublimation or vaporization are illustrated in Figure 15-1. The limits of the temperature ranges for each coolant are based on a minimum, as defined by the solid phase at 0.1 mm Hg pressure, and on a maximum, defined approximately by the critical point. Cooling can be provided from about 2 K to over 300 K by selection of the proper coolant under the proper environmental conditions. The thermodynamic properties of selected materials are given in Table 15-3.

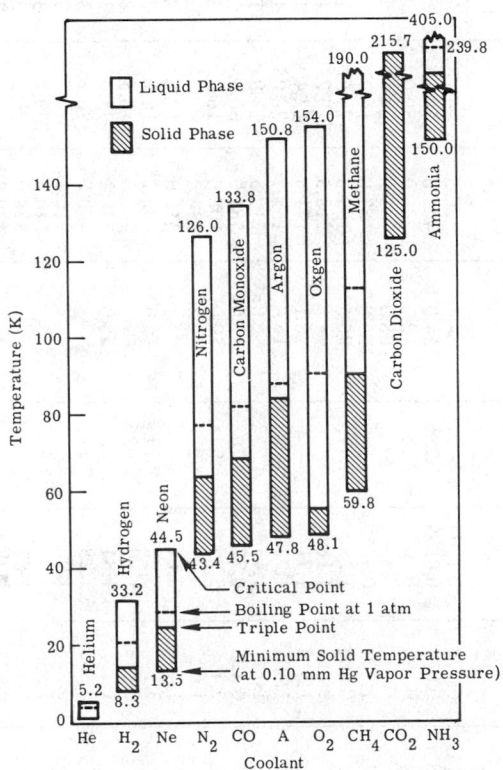

Fig. 15-1. Operating temperature range of selected expendable coolants.

15.4. Open-Cycle, Expendable Cooling Systems

Open-cycle or expendable refrigeration systems include those using high-pressure gas combined with a J-T expansion valve, cryogenic liquids in either the subcritical or supercritical state, and, for special applications, cryogenic solids. The advantages of these systems are simplicity, reliability, relative economy, and negligible power requirements. In most cases, the technology is developed. The basic disadvantages of cryogenic storage systems are their limited life due to heat leakage and the rapid increase in weight and volume for extended durations of use. Although high-pressure-gas storage systems with J-T valves can overcome the long-storage limitation, the penalties associated with the storage of high-pressure gas generally make system-weights prohibitive as operating time increases.

Table 15-3. Properties Of Miscellaneous Liquids, Gases, Refrigerants and Fuels [15-11]

Substance	Formula	Boiling Point at 1 atm Pressure (K)	Melting Point at 1 atm Pressure (K)	Liquid Density at bp ($g\,ml^{-1}$)	Gas Density at 273 K & 1 atm ($g\,l^{-1}$)	Vapor Density at bp ($g\,l^{-1}$)	Vapor Pressure Solid at mp (mm Hg)	Heat of Vapor at bp ($J\,g^{-1}$)	Heat of Fusion at mp ($J\,g^{-1}$)	Critical Temperatures (K)	Critical Pressure (atm)	Critical Volume* ($l\,g^{-1}$)
Helium	He^3	3.2	~1.0†	—	—	—	—	—	—	—	—	—
Helium	He^4	4.2	~2.0††	0.125	0.1735	17.0	—	20.5	4.183	5.2	2.26	0.0144
Hydrogen	H_2	20.39	13.98	0.071	0.0899	1.286	54.0	448.0	58.15	33.19	12.98	0.03321
Deuterium	D_2**	23.6	18.7	0.173	0.167	2.58	12.8	286.0	50.0	38.3	16.2	0.0142
Tritium	T_2**	25.1	21.6	—	—	—	188.0	—	—	43.7	20.8	0.0089
Neon	Ne	27.2	24.47	1.2	0.901	9.5	323.0	87.0	16.72	44.5	26.8	0.002
Nitrogen	N_2	77.37	63.4	0.808	1.250	4.415	96.5	199.0	25.52	126.1	33.5	0.00321
Carbon monoxide	CO	81.6	68.0	0.812	1.186	—	—	213.5	29.27	138.8	35.0	0.0032
Fluorine	F_2	85.24	53.6	1.513	1.71	5.03	0.1	171.5	13.4	144.8	55.0	—
Argon	A	87.4	83.6	1.391	1.78	4.75	516.0	162.7	28.05	150.8	48.0	0.0019
Oxygen	O_2	90.1	54.9	1.14	1.43	—	2.0	212.5	13.8	154.1	50.1	0.0023
Methane	CH_4	111.7	90.7	0.425	0.72	1.76	71.0	581.0	60.25	190.5	45.8	0.008
Krypton	Kr	120.3	116.0	2.4	3.75	8.33	550.0	108.0	16.3	209.3	54.5	—
Freon-14	CF_4	145.14	89.5	1.62	—	7.2	—	134.8	—	227.5	37.0	—
Ozone	O	161.3	80.5	1.46	2.14	—	—	316.0	13.8	261.1	54.6	0.00306
Xenon	Xe	165.3	150.5	3.1	5.93	9.77	615.0	96.25	119.1	290.0	58.0	0.00086
Ethylene	C_2H_4	169.3	104.0	0.578	1.19	2.08	—	481.0	148.5	282.8	50.9	0.0045
Nitrous oxide	N_2O	183.6	183.0	1.23	1.84	—	858.0	250.2	95.0	399.7	71.7	0.0022
Ethane	C_2H_6	184.8	90.0	0.562	1.28	0.32	—	490.0	96.25	305.0	48.8	0.6048
Acetylene	C_2H_2	189.1	191.2	0.623	1.09	—	—	916.0	—	309.0	62.0	—
Freon-13	$CCLF_3$	192.0	91.6	1.505	—	7.9	—	146.4	179.9	302.0	38.9	—
Carbon dioxide	CO_2	194.6	215.7	1.51	1.87	—	—	574.0	71.6	304.5	73.0	0.0022
Propylene	C_3H_6	226.1	77.5	0.604	1.78	—	—	439.5	—	365.0	45.0	—
Freon-22	$CHCLF_2$	232.5	113.0	1.414	21.3	4.65	—	235.0	35.15	369.0	48.7	0.0042
Ammonia	NH_3	239.8	195.0	0.683	0.77	0.898	45.0	1363.0	—	405.0	111.2	—
Freon-12	CCL_2F_2	243.1	118.0	1.488	17.7	6.25	—	167.2	—	384.0	39.6	—
Methyl chloride	CH_3Cl	249.4	—	0.993	5.93	2.56	—	427.0	—	—	—	—
Sulfur dioxide	SO_2	263.1	198.0	0.80	4.49	3.2	—	388.0	—	430.0	77.7	0.002
Isobutane	C_4H_{10}	272.5	—	—	2.53	—	—	—	—	426.0	36.0	—
Freon-11	CCL_3F_2	296.8	162.7	1.48	2.47	5.93	—	237.3	—	471.0	43.1	—
Propane	C_3H_8	230.8	85.9	0.595	1.92	2.08	—	342.0	79.9	370.0	42.0	—

*The critical volume is the volume at the critical temperature and critical pressure.

**Isotopes of hydrogen.

Liquid cryogens have been used for cooling in military aircraft. The logistics and costs associated with liquid cryogens, however, can be serious drawbacks. In recent years, many military systems have changed to high-pressure-gas, J-T systems and mechanical-refrigerator systems. Closed-cycle J-T systems have also been used.

15.4.1. Open-Cycle, J-T, Gas-Storage, Cooling Systems. In open-cycle, J-T gas-storage systems, high-pressure gas (1000 to 6000 psia) combined with a J-T cooler expansion valve produces the necessary cooling (Figure 15-2). The J-T effect involves the ratio of temperature change to pressure change of an actual gas in the process of throttling or expansion (during a constant enthalpy process) without doing work or transfering heat. Under normal pressure and temperature conditions, a perfect gas provides no cooling effect or temperature change for a throttling process. However, in actual gases under conditions of high pressure and/or low temperatures, molecular forces cause a change in internal energy when the gas expands. The change in internal energy during the expansion process results in cooling of the gas. The cooled, expanded gas is passed back over the incoming gas to cool it. This results in regenerative cooling. The process continues until liquid begins to form at the orifice to produce a bath of liquid at the cooling temperature of the gas. For

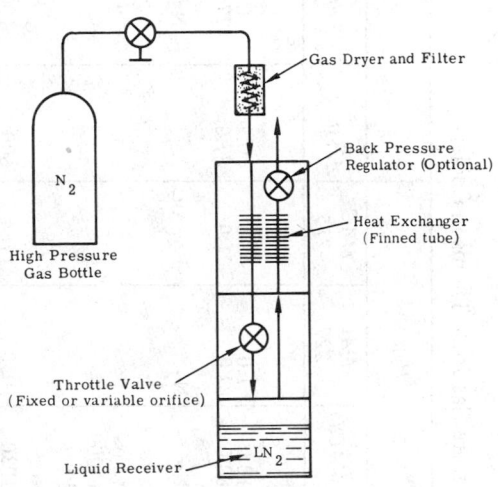

Fig. 15-2. Schematic representation of a J-T system.

certain gases, this effect occurs only below a specific inversion-temperature. Helium (40 K), hydrogen (204 K) and neon (250 K), for example, require precooling to the indicated temperature before the J-T expansion-cooling effect occurs. Most other gases, such as nitrogen, argon and air, have inversion points well above room temperature and no special precooling is required.

The J-T cooler consists of a finned tube in the form of a coil, an orifice and orifice cap, and an outer shield or coil. The finned tube is made of very small-inside-diameter tubing to provide the large ratio of surface-area to volume necessary for effective heat exchange. For a fixed-orifice cryostat, the flow will vary with pressure. Thus there is only one pressure that will provide just the desired refrigeration, as illustrated in Figure 15-3. Traditional J-T systems have suffered because of the inefficient matching of the pressure to the desired refrigeration. In recent years, self-regulating (or variable orifice-size) cryostats have increased the capability for correct matching, or for compensation for changes in heat-load or the gradual tendency of clogging.

Representative J-T Coolers. Table 15-4 includes data on representative J-T coolers. The use of helium, hydrogen, argon, and nitrogen gas enables the units to provide cooling from approximately 4.2 to 87.4 K at capacities of 0.50 to 10.0 W. A typical J-T cryostat is shown in Figure 15-4. J-T units have been fabricated for general-purpose commercial use, military missile-system applications (e.g., Sidewinder and Falcon), and for use in space. Air Products Model AC-2 was used in the Mariner program to provide 30 minutes of cooling at approximately 23 K [15-12].

Table 15-4. Representative High-Pressure Gas Joule-Thomson Coolers

Manufacturer	Air Products					Hymatic			SBRC			
Model	AC-1	AC-2	AC-2L	AC-3L	VF2020	MAC8	MAC 227	MAC 215	42902-1-F	42902-2-F	9174	9186
Nominal cooling capacity load (W) primary temp. (K)	7.0 80.0	4.0 23.0	6.0 22.0	4.0 22.0	0.20 77.0	10.0 80.0	1.0 82.0	2.0 21.0	0.20 87.0	0.20 77.0	0.25 77.0	0.19 77.0
Working fluid primary	N_2	N_2,H_2	N_2,LN_2	He,H_2,LN_2	N_2	N_2	N_2	H_2, Air	A	N_2	N_2	N_2
Gas consumption rate at nominal capacity (1 min^{-1}) 2000 psia	22.6	26.0(N_2) and 22.6(H_2)	27.2(H_2) and 0.008(LN_2)	36.0(He) and 24.0(H_2) and 0.008(LN_2)	—	14.0 (4000 psia)	—	—	1.2	0.8	14.0	9.5
Minimum operating pressure (psi)	1160	1160	1160	—	—	1030	1030	1764	600	1000	1000	900
Cool-down time	5 min	10 min	10 min	40 min	20 sec (2000 psi)	30 sec	—	3 min	20 sec (2000 psi)	35 sec (2000 psi)	2 min (2000 psi)	1.5 min (2000 psi)
Orifice characteristics type and diameter (in.)	Fixed	Fixed	Fixed	Fixed	Variable 0.326	Fixed 0.285	Variable	Fixed	Variable 0.21 max.	Variable 0.21 max	Fixed 0.326	Fixed 0.326
Overall dimensions (in.) length diameter	0.5 3.5	3.625 7.75	17.0 6.0	23.0 6.0	2.55 0.875	1.7 0.3	1.8 —	2.5 1.0	2.75 0.25	2.75 0.25	2.5 0.1875	2.5 0.1875
Primary Use	Lab.	Lab.	Lab.	Lab.	Airborne	Missile	Missile	—	Missile	Missile	Lab.	Lab.
Remarks	—	—	Incl. LN_2 precooler	Incl. LN_2 precooler	Cooling rate variable from 0.20 to 2.0 W	—	A, air, Freon 13 or Freon 14 are acceptable	—	Demand flow	Demand flow	Open-loop	Closed-loop
Reference	[15-12]	[15-12]	[15-12]	[15-12]	[15-12]	[15-14]	[15-14]	[15-14]	[15-15]	[15-15]	[15-15]	[15-15]

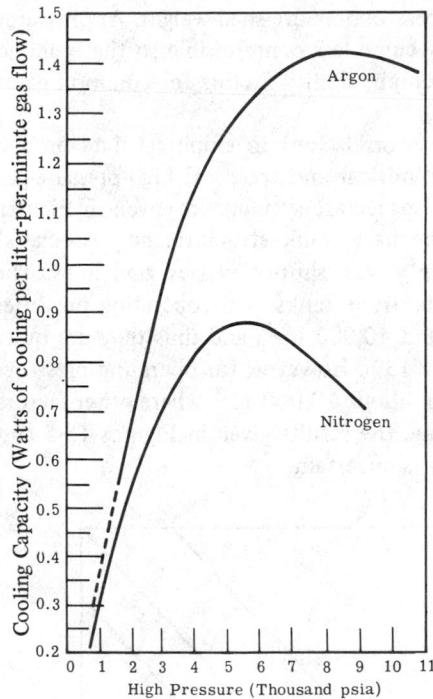

Fig. 15-3. Cooling capacity versus inlet pressure for an ideal J-T cooler (300 K gas temperature). Gas flow is standard liters per minute @ 1 atm and 273 K. The liquid is at 1 atm pressure [15-13].

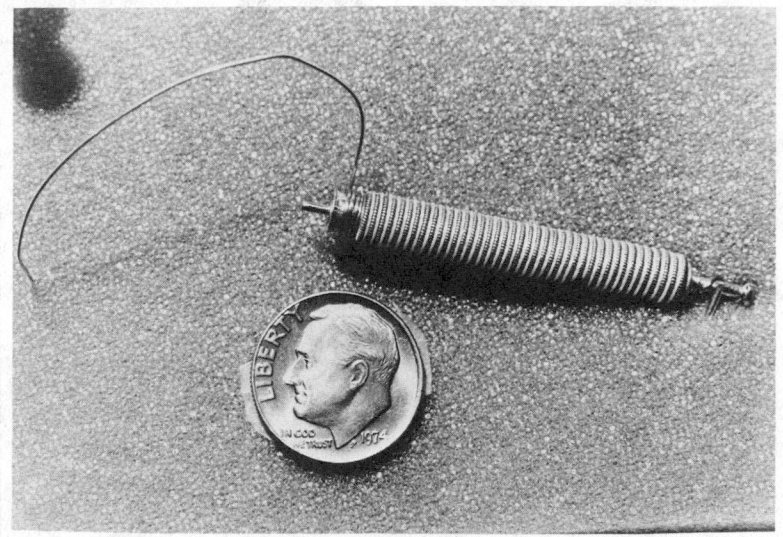

Fig. 15-4. Variable flow (Variflow®*) J-T cryostat [15-12].

Registered trademark of Air Products and Chemicals, Inc., Allentown, PA.

Ambient-Temperature, High-Pressure-Gas Storage. The design and optimization of high-pressure storage vessels include minimization of container volume (by the use of elevated storage pressures) without incurring excessive pressure-shell weight. At pressures above about one thousand atmospheres, gases become less compressible so that volume savings at high pressures are diminished. Gas compressibility factors for common gases can be found in the work by Kunkle (ed.).

Curves fitted (on the basis of a least-squares correlation) to empirical data on the weight, volume, and operating pressures of cylindrical and spherical high-pressure-gas bottles designed and produced for aircraft and spacecraft systems are given in Figures 15-5 and 15-6. These weights include only the basic tank structure, not associated hardware such as pressure relief and regulating valves, shut-off valves, and lines. The resulting equations are based primarily on data from tanks with operating pressures above 3000 psia and volumes between 10,000 and 40,000 in.^{3}, and thus they are quite good in these regions, accurate to within about ±15%. However, for operating pressures below about 3000 psia and for volumes less than about 10,000 in.^{3} where other factors (such as minimum design gauge) may be limiting, the results given in Figures 15-5 and 15-6 are not necessarily satisfactory and accuracy is uncertain.

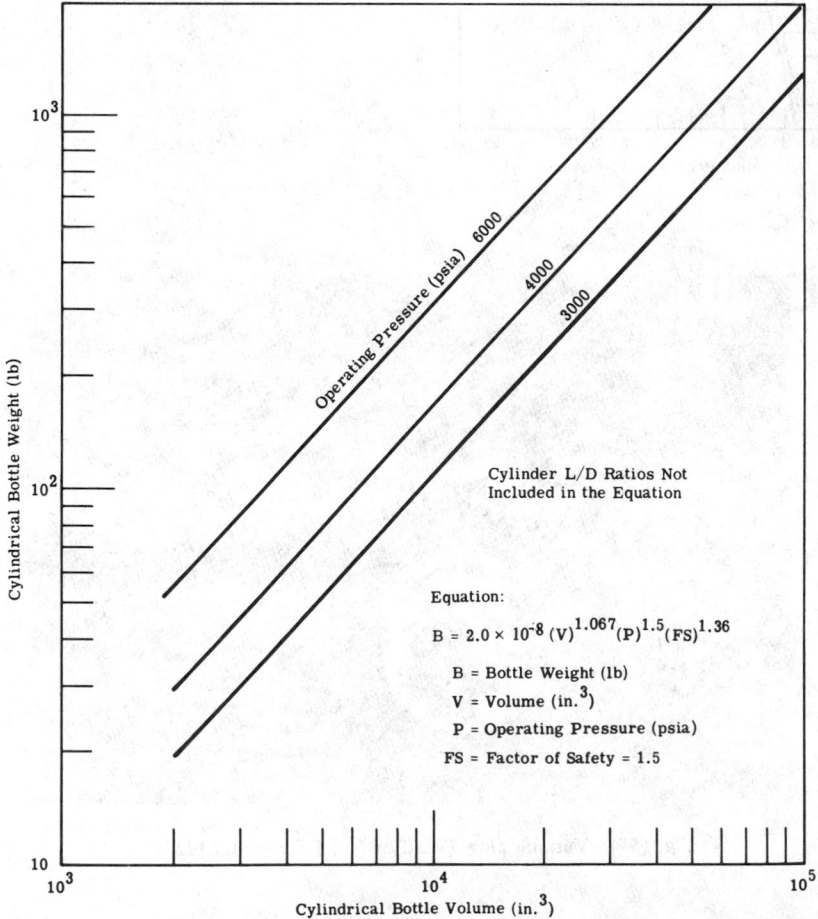

Cylinder L/D Ratios Not Included in the Equation

Equation:

$$B = 2.0 \times 10^{-8} (V)^{1.067} (P)^{1.5} (FS)^{1.36}$$

B = Bottle Weight (lb)
V = Volume (in.^{3})
P = Operating Pressure (psia)
FS = Factor of Safety = 1.5

Fig. 15-5. **Weight of high-pressure, cylindrical tanks based on correlation of production hardware.**

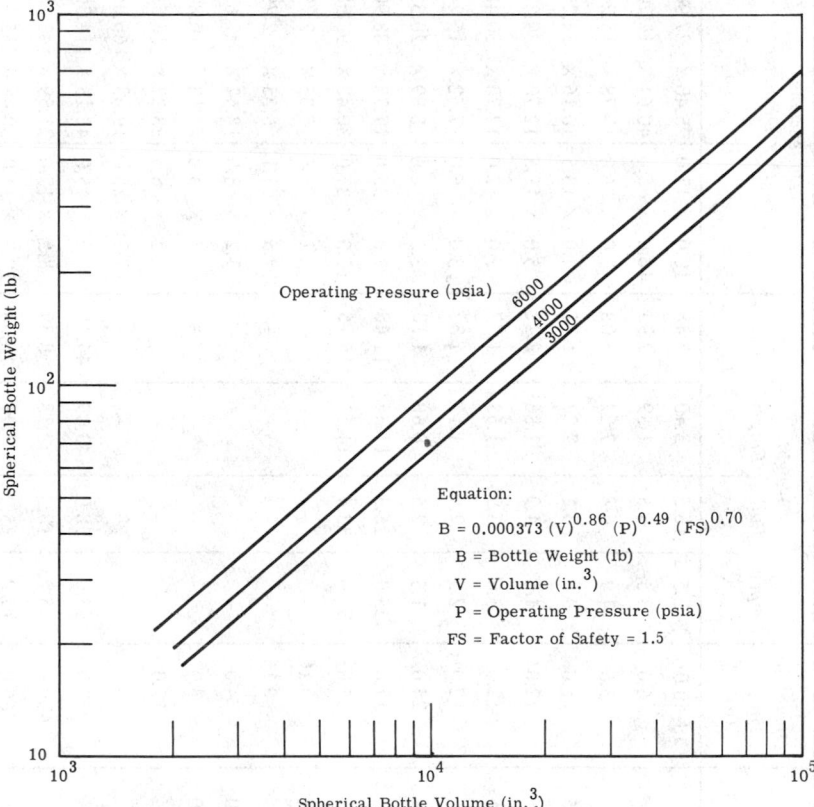

Fig. 15-6. Weight of high-pressure-gas, spherical tanks based on correlation of production hardware data.

15.4.2. Liquid-Cryogen-Storage Cooling Systems. Cryogenic fluids may be stored as liquids in equilibrium with their vapors (subcritical) or, at higher pressures and temperatures, as supercritical, homogeneous fluids. In laboratory or airborne operations, the fluid is usually stored in the liquid, two-phase form because of simplicity and weight advantages of the storage dewar. Because of gravity, the liquid is constantly exposed to the supply port of the dewar.

In space-system storage of cryogenic liquids, the absence of gravity or acceleration from orientation-forces prevents the use of standard two-phase systems (unless special capillary retention devices are utilized) since random orientation of the liquid phase during weightless conditions prevents continual communication between the liquid phase and the supply port. Space-system storage of cryogenic liquids is currently accomplished by pressurizing the cryogen to the supercitical pressure or single-phase state. The absence of gravity or acceleration does not affect the delivery of fluid, since the supply port is in direct communication with a relatively homogeneous fluid at all times.

Temperatures available from use of baths of liquefied gasses, shown in Figure 15-1 and Table 15-3, range from approximately 2 to 300 K. Varying the pressure allows variation of the temperature of a liquefied gas to provide cooling from the triple-point to the critical-point.

Two basic types of storage systems are direct-contact (or integral) and liquid-feed. The integral system is basically a detector built into a cryogenic dewar in which the cryogenic

Table 15-5. Representative Commercial Dewars For Cryogenic Liquids

Manufacturer	Model	Type Liquid Feed (L) or Integral (I)*	Coolant	Capacity (l)	Operating Time (h)	Weight Empty (lb)	Full (lb)	Empty (kg)	Full (kg)	Dimensions Length × Diameter (in.)	(cm)
Aieresearch Mfg., Los Angeles, CA	134338	L	N$_2$	5.0	30–50.0	7.5	16.5	3.40	7.48	10.0 × 11.0	25.40 × 27.94
	134642	L	N$_2$	1.5	6.0	3.5	6.5	1.59	2.95	16.0 × 5.0	40.64 × 12.70
	134548	L	N$_2$	1.0	3.5	4.9	6.8	2.22	3.08	7.0 × 6.0	17.78 × 15.24
Cryogenic Associates, Indianapolis, IN	IR 12	I	He	0.35	5.0	3.0	3.1	2.27	1.41	4.0 × 13.5	10.16 × 34.29
	IR 15	I	He	1.35	12.0	4.1	4.9	1.86	2.22	5.0 × 15.5	12.70 × 39.37
	IR 43	I	He	1.15	12.0	3.6	3.9	1.63	1.77	5.0 × 12.0	12.70 × 30.48
	IR 13	I	N$_2$	2.9	12.0	2.5	2.9	1.13	1.32	3.0 × 6.0	7.62 × 15.24
	IR 14	I	N$_2$	2.5	24.0	1.5	2.5	0.68	1.13	3.5 × 8.0	8.89 × 20.32
	CT 53	I	He	2.5	17.0	25.0	25.3	11.34	11.48	7.0 × 35.0	17.78 × 88.90
	CT 14	L	Ne	9.5	65.0	32.0	33.4	14.51	15.15	12.0 × 37.0	30.48 × 93.98
SBRC, Goleta, CA	520 A	I	N$_2$	–	0.16	–	–	–	–	2.1 × 1.1	5.33 × 2.79
	520 AS	I	N$_2$	–	0.16	–	–	–	–	2.5 × 1.1	6.35 × 2.79
	40742	I	N$_2$	0.10	4.0	–	–	–	–	7.25 × 3.0	18.42 × 7.62
	9144-1	I	He	0.40	8.0	–	–	–	–	9.5 × 4.25	24.13 × 10.80
	9145-1	I	He	1.0	12.0	–	–	–	–	9.2 × 6.5	23.37 × 16.51
MVE, New Prague, MN	HLDT-3	L	N$_2$,He	4.0	–	–	–	–	–	7.75 × 26.0	19.69 × 66.04
	DELTA-10	L	He	10.0	250.0	95.0	98.0	43.09	44.45	17.0 × 45.0	43.18 × 114.30
	HEMI-3	L	N$_2$	5.0	430.0	5.25	18.5	2.38	8.39	8.5 × 17.0	21.59 × 43.18
	A-200	Flask (L)	N$_2$	10.0	360.0	17.0	35.0	7.71	15.88	12.0 × 14.0	30.48 × 35.56

*Detector is in direct contact with coolant.

liquid is stored. In the liquid-feed cooler, coolant is fed to the detector from a storage tank in a remote location through transfer lines. Commercially available coolers of both types are summarized in Table 15-5.

The integral system consists of a detector in direct thermal contact with a supply of liquid coolant. In the examples shown in Figure 15-7, the detector is integrally mounted in a dewar which serves as detector mount and liquid container. A limitation of the direct-contact cooler can be its operating attitude. The dewar must be maintained in an essentially vertical position to keep the coolant in direct physical contact with the detector. For airborne and tracking-instrument applications where the detector is moved through 360 degrees of arc, thermal contact between the coolant and the detector is maintained regardless of the dewar attitude by using copper conducting plates which remain in contact with the coolant (Figure 15-7(b)). Typical, open-mouthed, laboratory, glass dewars are shown in Figure 15-7(c).

The liquid-feed system consists of an insulated, liquid-storage container, transfer lines, a cooling head, and the necessary controls. The transfer mechanism is either gravity or gas pressure. The gas pressure to force the liquid from the storage container to the cooling head originates from the natural pressure buildup resulting from thermal leakage into the storage container or from the residual pressure of the filling operation. This concept is illustrated in Figure 15-8.

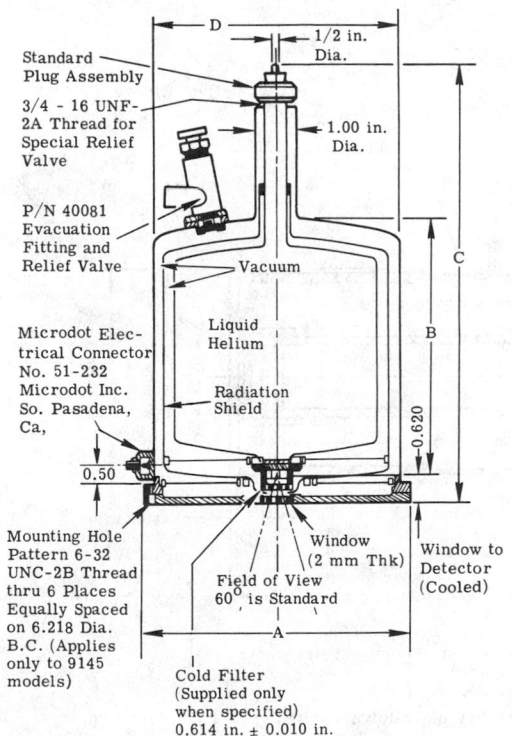

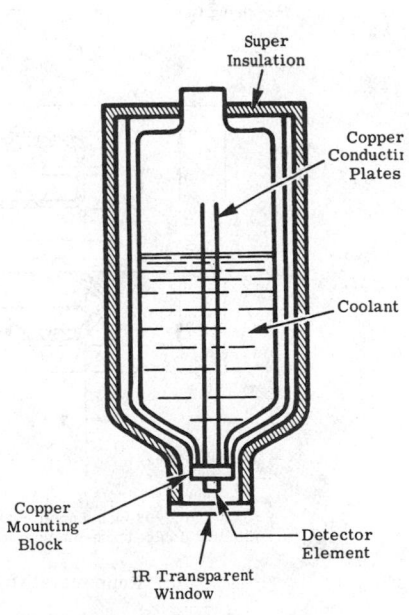

(a) Typical Metal 8 h helium dewar, SRBC part no. 9145-1; A = 6.5 in., B = 5.4 in., C = 9.2 in. D = 6 in. [15-15].

(b) Direct-contact, variable attitude dewar.

Fig. 15-7. Direct contact coolers.

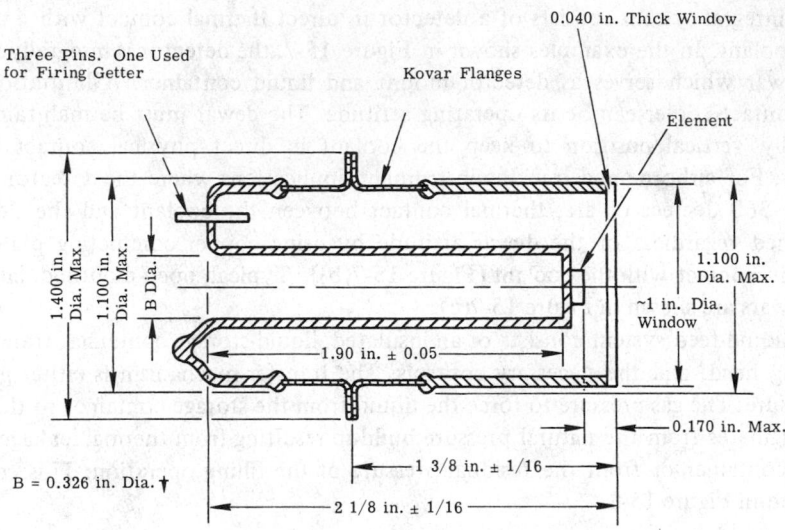

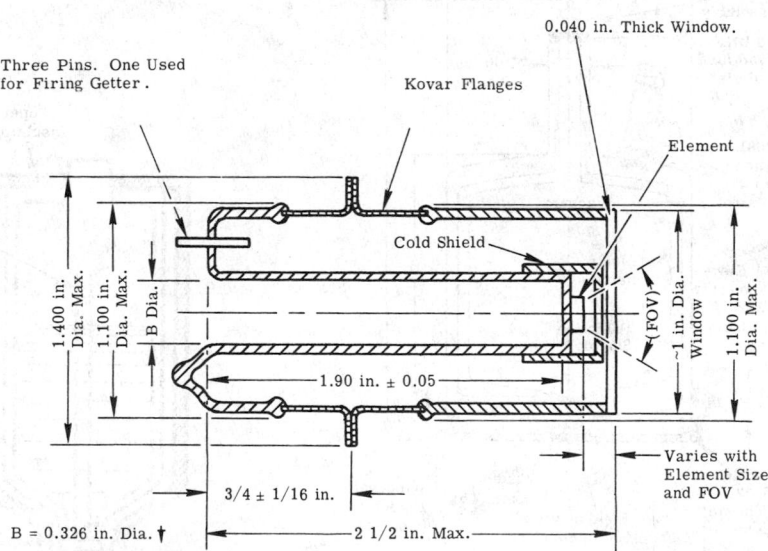

† 0.204 and 0.208 inch diameter available for small elements when more efficient Joule-Thomsom cooling is required.

(c) Representative laboratory glass dewars, SBRC [15-15].

Fig. 15-7 (Continued). Direct contact coolers.

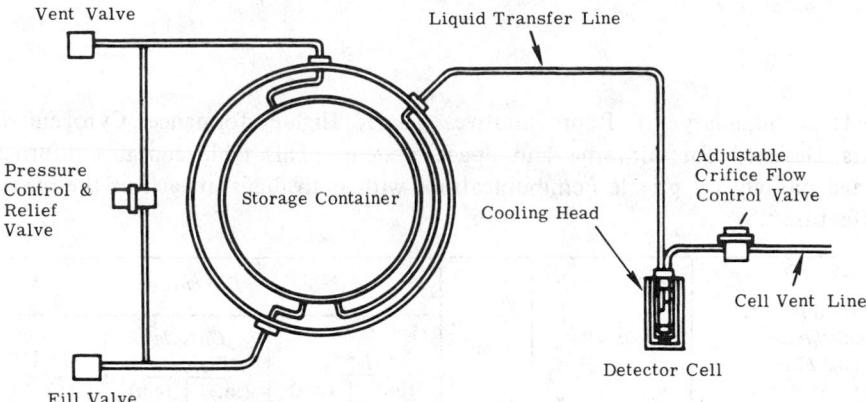

Fig. 15-8.   Liquid-feed cooler system.

Characteristics of a variety of large, high-performance cryogenic dewars (tanks) designed for aircraft or space vehicles (plus a few large, commercial dewars) are summarized in Table 15-6. In general, these dewars utilize vacuum, multilayered-insulation (MLI) blankets and, in some cases, rigid radiation-shields cooled by the vented vapor, to achieve low values of heat-leaks necessary for extended operation. The technology used in the design of liquid oxygen and hydrogen storage systems for the Gemini and Apollo programs has been extended, so that the storage of fluids such as hydrogen and helium for periods of a year or longer in space is now feasible. (See items 9, 16 and 21 in Table 15-6.)

Transfer Lines. Flexible, vacuum, super-insulated lines are available in a variety of sizes and strengths to provide for the efficient, low-heat-loss transfer of cryogenic liquids. Data on representative, flexible cryogenic-lines and the associated heat-loss factors for liquid-helium transfer (at room temperature) are provided in Table 15-7.

15.4.3.   Solid-Cryogen-Storage Cooling Systems

Description and Operating Characteristics. A cooling system based on the sublimation of a solid coolant into the high vacuum of space avoids several problems associated with either subcritical or supercritical storage of liquids. This system consists of a solidified substance, an insulated container, a vent-gas path to space, and a conduction path from the coolant to the device cooled. The operating temperature depends upon the choice of coolant and the back-pressure of the vent gas maintained in the system. The system's operating time depends upon the amount of coolant and the heat-load. The cooling capacity is governed by the heat of sublimation, which is equal to the sum of the heat of vaporization and the heat of fusion.

The advantages of a solid versus liquid system include independence from operating attitude, a higher cooling capacity, a higher density storage, and a lower temperature solid phase, permitting a gain in sensititivity in certain infrared detector systems.

Cryogen Design Properties. The operation of a cryogenic-solid system is based on the interrelation of the pressure and temperature of a solid in equilibrium with its vapor. Addition of heat sublimes the solid coolant, increasing the vapor pressure, and thus causing a temperature increase. Pressure and temperature are maintained at constant levels by venting the vapor to space at the appropriate pressure level. Normally, the coolers are designed without valves and the vent-gas ducting is designed to maintain a given pressure.

Table 15-6. Summary of Representative, Large, High-Performance Cyrogenic-Fluid Dewars Designed for Airborne and Space Systems (This table contains information obtained entirely by private communications with individuals at each of the identified manufacturers.*)

Item No., Manufacturer and Fluid	Program	Press. (psia)	Dimensions					
			L**		Outside Diameter		Vol.	
			(in.)	(cm)	(in.)	(cm)	(ft³)	(m³)
Oxygen/Nitrogen								
1 Airesearch (O$_2$)	MOL	880††	–	–	34.3	87.1	12.1	0.34
2 Airesearch (O$_2$)	Gemini	1000††	–	–	22.9	58.2	2.5	0.07
3 Beech (O$_2$)	Apollo	900††	–	–	26.5	67.3	4.8	0.14
4 Airesearch (N$_2$)	707 Galley	55-0	43.2	109.7	23.6	59.9	5.5	0.16
5 Airesearch (N$_2$)	727 Galley	55-80	23.9	60.7	16.9	42.9	1.5	0.04
6 Airesearch (N$_2$)	747 Galley	55-80	54.5	138.4	20.5	52.1	7.1	0.20
7 Essex (O$_2$)	C-5, C-141	300	–	–	24.0	61.0	2.7	0.08
8 Airesearch (O$_2$)	Development	200	–	–	25.7	65.3	2.6	0.07
9 Beech (O$_2$)	OTTA (Test)	150	–	–	109.0	276.0	227.0	6.42
10 Beech (O$_2$)	Shuttle	950††	–	–	36.9	93.7	11.2	0.32
Hydrogen								
11 Beech	Shuttle	285††	–	–	45.6	115.8	21.3	0.60
12 Airesearch	MOL	280††	–	–	42.6	108.2	19.1	0.54
13 Airesearch	Gemini	250	–	–	28.9	73.4	5.4	0.15
14 Beech	Apollo	245	–	–	31.8	80.8	6.8	0.19
15 Airesearch	Development	100	–	–	31.1	79.0	5.7	0.16
16 Beech	HTTA (Test)	50	262.0	615.5	110.0	279.4	800.0	22.65
Helium								
17 Airesearch	Apollo LM	1300††	–	–	33.0	83.8	5.9	0.17
18 Beech	Apollo AGE	14.7	36.0	91.4	75.0	190.5	44.1	1.25
19 Minn. Valley Engr.	Commercial	15.2	54.0	137.2	78.0	198.1	65.2	1.85
20 Cryogenic Engr.	Commercial	15.2	44.0	111.8	72.0	182.9	36.2	1.03
21 Airesearch	NASA HEAO	14.7	93.5	237.5	72.0	182.9	128.0	3.63
22 Beech	ELMS	90.0††	84.4	214.4	45.4	115.3	45.0	1.27
Miscellaneous								
23 Beech (N$_2$O)	Airborne Test	2200††	–	–	34.0	86.4	7.2	0.20
24 Beech (CO)	Airborne Test	2200††	–	–	38.8	98.6	7.2	0.20

*Airesearch Mfg. Co., Torrance, CA; Beach Aircraft Corp., Boulder, CO; Cryogenic Engineering Co., Div. of Cryogenic Technology, Inc., Denver, CO; Essex Cryogenics, Inc., St. Louis, MO; Minnesota Valley Engineering, Inc., New Prague, MN.

**Tanks with no entries are spherical in shape.

Table 15-6. (*Continued*)

Weight				Materials† (Pressure Shell/ Outer Shell)	Heat Leakage at Room Temp.		
Dry Tankage		Fluid			Total (Q)	Q/A	Q/A
(lb)	(kg)	(lb)	(kg)		(Btu h^{-1})	(Btu h^{-1} ft^{-2})	(Btu h^{-1} m^{-2})
167	76	715	324	Inconel 718/Al 2219	24.75	0.98	10.55
68	31	180	82	Inconel/Titanium	15.20	1.33	14.32
91	41	330	150	Inconel 718/Inconel 750	27.20	2.04	21.96
174	79	250	113	304 Stainless	–	–	–
60	27	71	32	304 Stainless	–	–	–
206	93	370	168	304 Stainless	–	–	–
75	34	188	85	85 Stainless	–	–	–
98	44	184	84	2169 Stainless	11.40	1.70	18.30
2000	~907	15,730	7,135	Aluminum	13.10	0.066	0.71
214	97	809	367	Inconel 718/Al 2219	21.0***	0.71	7.64
199	90	96	44	Al 2219/Al 2219	6.6	0.318	3.42
126	57	84	38	Al 2219/Al 2219	6.75	0.19	2.05
47	21	23	10	Titanium	5.4	0.36	3.88
80	36	29	13	Titanium	5.0	0.291	3.13
53	24	24	11	2169 Stainless	2.98	0.192	2.07
4,700	2,132	3,520	1,597	Aluminum	17.1	0.034	0.37
115	52	48	22	Titanium	8.0	0.336	3.62
354	161	185	84	6061 Aluminum	1.1	–	
1,750	794	276	125	Stainless	1.17	~0.10	~1.08
780	354	153	69	Stainless	0.79	–	–
1,342	609	950	431	2219 Aluminum	0.77	0.005	0.054
324	147	351	159	Al 2219/Al 6061	Variable	–	–
523	237	443	201	Al 2219/Al 6061	75.0	3.5	37.67
723	328	375	170	Al 2219/Al 6061	132.0	4.1	44.13

†Single entry indicates same material for both inner and outer shell.
††Supercritical storage.
***@ 150°F.

Table 15-7. Typical Heat Loss For Vacuum Insulated (10^{-5} torr*) Flexible Cryogenic Lines** [15-16]

Nom. Hose Size		Inner Hose ID		Outer Hose ID		Heat Loss (Btu h^{-1})		Typical Minimum Bend Diameter	
(in.)	(cm)	(in.)	(cm)	(in.)	(cm)	per foot of length	per meter of length	(in.)	(cm)
1/4	0.635	0.125	0.3175	2.25	5.715	0.30	0.98	5.0	12.70
1/2	0.127	0.50	1.27	2.85	7.239	0.40	1.31	5 1/2	13.97
3/4	1.905	0.750	1.905	3.50	8.89	0.50	1.64	6.0	15.24
1.0	2.54	1.00	2.54	4.10	10.414	0.56	1.84	7 1/4	18.415
1 1/2	3.81	1.50	3.81	4.95	12.573	0.66	2.17	11.00	27.94
2.0	5.08	2.00	5.08	5.35	13.589	0.73	2.40	15.00	38.10
2 1/2	6.35	2.50	6.35	6.50	16.51	0.84	2.76	19.00	48.26
3.0	7.62	3.00	7.62	6.50	16.51	0.91	2.99	23.00	58.42
3 1/2	8.89	3.50	8.89	7.60	19.304	1.04	3.41	27.00	68.58
4.0	10.16	4.00	10.16	8.25	20.955	1.13	3.71	30.00	76.20
5.0	12.70	5.00	12.70	9.60	24.384	1.30	4.27	37.00	93.98
6.0	15.24	6.00	15.24	10.60	26.924	1.45	4.76	44.00	111.76

*1 torr = 1 mm Hg
**Heat loss based on LHe transfer in room temperature environment.

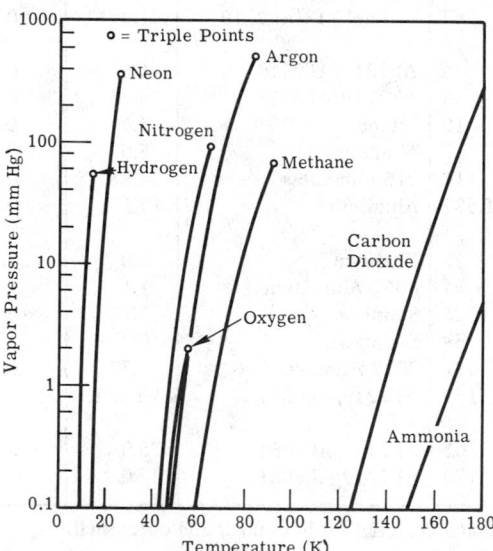

Fig. 15-9. Operating-temperature range versus vapor pressure for various cryogenic solids.

Figure 15-1 shows the nominal operating-temperature ranges available from solid cryogens. Figure 15-9 gives more precise values of temperature as a function of the vapor pressure maintained over the coolant. The vapor-pressure range shown varies from 0.10 mm Hg (about the minimum that can be maintained in practice) to that found at the triple-point of the specific coolant (the maximum temperature at which the coolant can be maintained in the solid state). Solid coolants are available from a minimum of 8.3 K (hydrogen) to 215.7 K (carbon dioxide) with the only gaps existing between 24.5 and 43.4 K, and between 90.7 and 125.0 K. Table 15-8 summarizes the more precise temperature ranges, together with solid densities and cooling capacities (heat of sublimation). If the required detector temperature is substantially higher than the triple-point-temperature of the solid cryogen being used, the sensible heat of the effluent vapor can be used to further advantage.

Table 15-8. Selected Solid-Cryogen Properties

Symbol	Cryogen Name	Heat of Sublimation		Density of Solid at Melting Point		Operating Temperature Range (K)	
		(Btu lb^{-1})	(J g^{-1})	(lb ft^{-3})	(kg m^{-3})	0.10 mm Hg	Triple Point*
NH$_3$	Ammonia	739.0	1718.9	51.3	821.8	150.0	195.4
CO$_2$	Carbon dioxide	246.6	573.6	97.5	1562.0	125.0	215.7
CH$_4$	Methane	244.5	568.7	31.1	498.2	59.8	90.4
O$_2$	Oxygen	97.5	226.8	81.3	1302.4	48.1	54.9
A	Argon	79.8	185.6	107.0	1714.0	47.8	83.6
CO	Carbon monoxide	126.0	293.1	58.0	929.2	45.5	68.0
N$_2$	Nitrogen	96.6	224.7	63.8	1022.1	43.4	63.4
Ne	Neon	45.5	105.8	89.8	1438.6	13.5	24.5
H$_2$	Hydrogen	218.5	508.2	5.02	80.4	8.3	13.98

*Corresponds to the highest temperature at which the solid phase can exist.

Satellite System Designs. A number of solid-cryogen coolers have been developed which use nitrogen, argon, carbon dioxide, neon, methane, oxygen, and ammonia. A few of these coolers (Table 15-9) represent both single-stage and two-stage units wherein a secondary coolant (such as CO$_2$ or ammonia) with a higher operating temperature and higher cooling capacity is used to shield a lower-temperature, primary coolant (Figure 15-10). In this way, for a given operating life of the cooler, weight and volume can generally be minimized.

Two systems have been flown. A one-year-life CO$_2$ cooler was developed for the Navy by Lockheed [15-2] and launched by the Air Force aboard SAMSO STP72-1 on 20 October 1972 (Figure 15-11). A larger, two-stage, one-year-life ammonia-methane cooler was developed by Lockheed for NASA as part of the Limb Radiance Inversion Experiment [15-1] and launched in June 1975 aboard the Nimbus-F spacecraft (Figure 15-12). Cooler efficiency normally increases with size since the volume/surface-area ratio increases. Thus, current technology can extend cooler operating-life beyond a year and is limited only by the weight and volume available. Solid cryogens can meet long-life operating requirements which may be difficult to achieve with, for example, mechanical refrigerators. Also, the need for power and waste-heat rejection associated with mechanical refrigerators is eliminated and can be traded off against the larger fixed-weight of the solid-cryogen systems.

Special Design Problems. The design of a practical solid-cooler involves integration of a number of solutions to special technical problems including: minimizing the parasitic heat-leak into the cryogen caused by penetrations of the device to be cooled (including vent gas lines and support members); special filling and solidification techniques; and maintaining necessary thermal contact as the cryogen sublimes.

Table 15-9. Representative Solid-Cryogenic Coolers

Manufacturer/ Developer	Aerojet/General (AESC)	Lockheed (LMSC)	Aerojet/General (AESC)	Ball Brothers Research Corp.	Aerojet/General (AESC)	Lockheed (LMSC)	Lockheed (LMSC)
Nature of program Primary use	Experimental IR detector cooling	Experimental IR detector cooling	Design study O$_2$ storage	Experimental IR detector cooling	Experimental IR telescope	Orbital experiment Gamma ray experiment	Nimbus-F Limb radiance measurement
Experimental operating temperature (K)	58	50	12 to 54	76	78	130	65
Cryogen operating pressure (mm Hg)	29.4	0.06 - 0.15	0.01 - 2.0	189	–	0.10	0.10
Design operating life (days)	41.6	1 year	250	1 year	210	1 year	1 year
Primary heat load (mW)	750	25	None	14.3†	100	–	100
Total parasitic heat leak (mW)	310	102.6	200	134.7	–	240	355
Total system weight (lb)	38	34.1	–	47.4	95	51.3	56.5
Refrigerant	Nitrogen	Argon/CO$_2$	Oxygen	Argon/CO$_2$	Methane	CO$_2$	Methane/NH$_3$
Refrigerant weight (lb)	27	Argon 13.7 CO$_2$ 8.8	270	Argon 9.02 CO$_2$ 11.62		33	28.8
Configuration	Cylinder	Cylinder	Sphere	Cylinder	Cylinder	Cylinder	Cylinder
Cooler dimensions (in.)	L=D=10.5	L=D=9.25	Dia.=24	D=13,L=22	D=18,L=22	D=14.5,L=16	D=14,L=26
Standby time* (days)	4.5	17 (max.)	250**	–			20
Sponsor of original program	AFFDL	NASA	AFAMRL	In-house	NASA	NAVY	NASA
Current status of program	Completed	Completed	Completed	Testing	Completed	Launched 20 Oct. '72	Launched July 1975
Reference	[15-17]	[15-18]	[15-19]	[15-20]	Private Communication	[15-2]	[15-1]

*Standby time is time required for refrigerant to begin to melt after vacuum source is disconnected.

**Estimated time for oxygen to melt and boil-off to begin.

†Includes IR radiation through optical window.

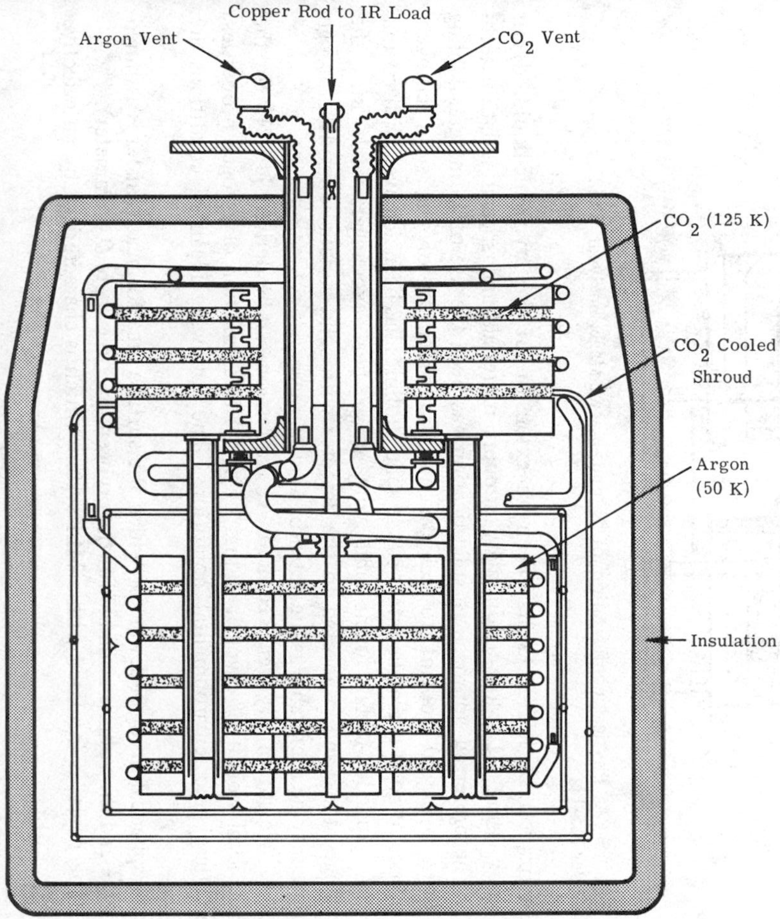

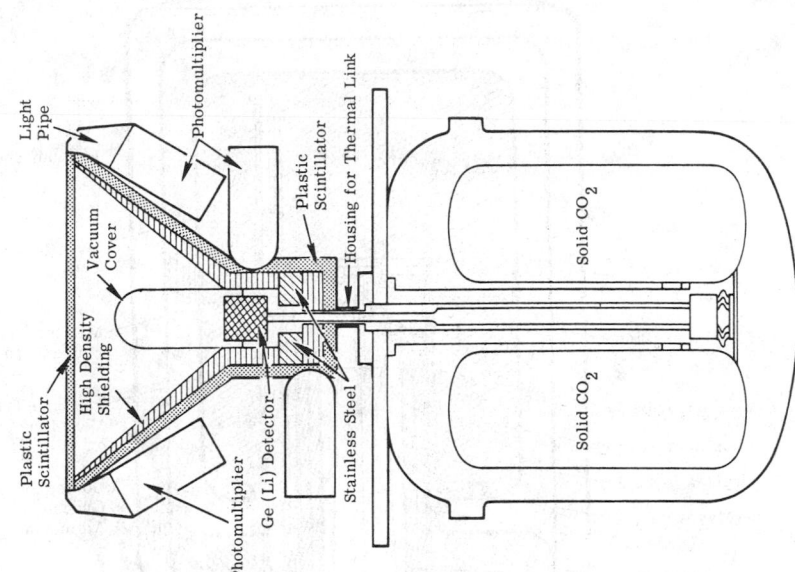

Fig. 15-11. Orbital, gamma-ray spectrometer, solid CO₂ cooler [15-2].

Fig. 15-10. Schematic of solid-argon, solid carbon-dioxide infrared detector cooler [15-18].

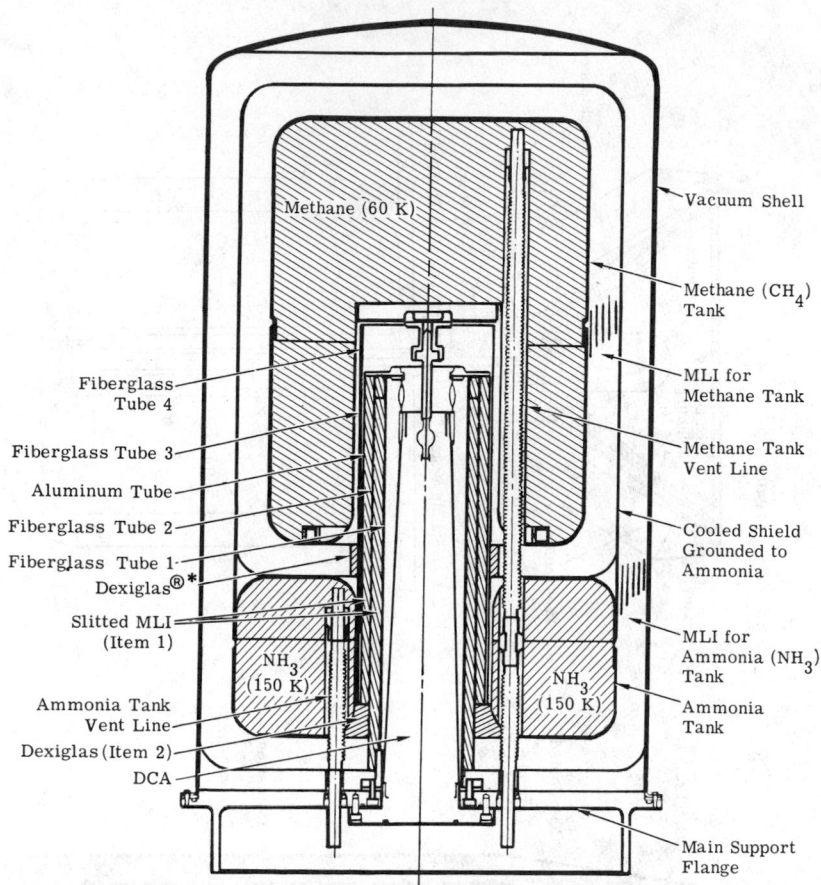

Fig. 15-12. Physical configuration of Nimbus-F, two-stage, one-year, solid cryogen cooler [15-1].

Minimization of heat-leaks has been an extension of the cryogenic-liquid, dewar technology using MLI, low-conductivity support materials and special isolation techniques. The preferred approach to solidification has been the use of an integral cooling loop using a liquid coolant to achieve maximum density of the solid cryogen. Maintenance of good thermal contact as the solid sublimes has been achieved with various types of expanded-metal foams and wire mesh heat exchangers built into the solid-cryogen tank and connected to the detector with solid conduction-rods, although heat pipes may be preferable as the detector load increases.

15.4.4. Weight of Cryogen Required. An expression can be derived [15-21] for the required, ideal weight of cryogen (liquid or solid) as a function of cryogen properties, mission time, and environmental conditions. The following idealized conditions are assumed:

(1) A spherical cryogen-dewar for which the surface area for heat transfer is assumed to be πD^2 and the volume available for the cryogen, $\pi D^3/6$ (where D = diameter).

(2) The dewar is insulated with a MLI system which is characterized by an effective emissivity, ϵ_{eff}.

---

*Trademark of Dexter Paper Co. Dexiglas is a glass-fiber-mat spacer material use for MLI.

(3) The only heat-leak into the cryogen is through the MLI.

(4) The heat absorbed by the cryogen is the product of the heat-leak and total elapsed time, t.

The weight of cryogen required under these assumptions is given by

$$W_c = \frac{36\pi}{\rho^2} \left[\frac{\sigma \epsilon_{\text{eff}} \left(T_s^4 - T_c^4 \right) t}{h_c} \right]^3 \tag{15-1}$$

where W_c = weight of cryogen required, lb, or g
ρ = density of cryogen, lb ft^{-3}, or g cm^{-3}
σ = Stefan-Boltzmann constant, Btu h^{-1} ft^{-2} °R^{-4}, or W cm^{-2} K^{-4}
ϵ_{eff} = effective emissivity, dimensionless
T_s = absolute temperature of the outer shell of vessel, °R, or K
T_c = absolute temperature of the cryogen and inner shell of the MLI, °R, or K
t = total elapsed time in units consistent with the Stephan-Boltzmann constant, h, or sec
h_c = latent heat of cryogen (heat of vaporization for a liquid or heat sublimation for a solid), Btu lb^{-1}, or J g^{-1}

The constant, 36π in Equation (15-1), results from the spherical geometry assumed and is not dependent on the units used. Data for various cryogens are provided in Figures 15-1 and 15-9 and Tables 15-3 and 15-8. Dry weights of cryogenic-liquid dewars are provided in Tables 15-5 and 15-6. Representative data on solid-cryogen, vessel weights are provided in Table 15-9. Heat transfer and effective emissivity data for MLI systems are provided in Section 15.8.2.

15.5. Mechanical-Refrigerator Cooling Systems

15.5.1. Ideal Thermodynamic Cycles. A mechanical refrigerator is a device that absorbs heat at one temperature, T_c, and rejects it at some higher temperature, T_a, as shown in Figures 15-13 and 15-14. To perform this task, expenditure of work is required. The work for a mechanically driven refrigerator, W_o, and the thermal energy for a heat driven refrigerator, q_h, can be expressed as

$$W_o \geqslant q_c \left[\frac{T_a - T_c}{T_c} \right] \tag{15-2}$$

and

$$q_h \geqslant q_c \left[\frac{T_h}{T_c} \times \frac{(T_a - T_c)}{(T_h - T_a)} \right] \tag{15-3}$$

The input power is at a minimum when all the processes of the thermodynamic cycle are reversible (i.e., $S = 0$). Although reversible cycles can be achieved theoretically and are used as a standard to compare performance, practical refrigeration cycles are irreversible. The coefficient of performance (COP) for a refrigerator is customarily expressed as

$$\text{COP} = \frac{\text{Refrigeration Produced}}{\text{Power Supplied}} \tag{15-4}$$

The COP serves as a basis of comparing mechanically-driven refrigerators or heat-powered refrigerators if the heat energy is supplied by electrical resistance heaters. However, if the

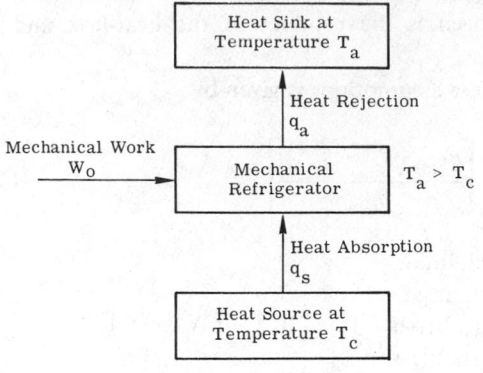

Fig. 15-13. Mechanically-powered, refrigerator operation.

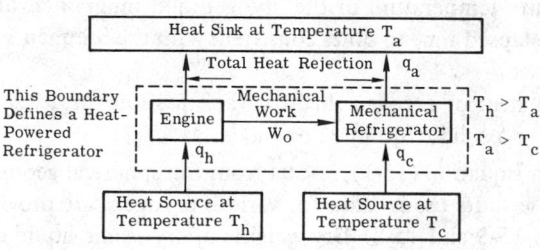

Fig. 15-14. Heat-powered, refrigeration operation.

source of power for a heat-powered refrigerator is not electrical, i.e., chemical, radioisotope, nuclear, etc., then the COP values should be examined in light of the cost and availability of the power source.

The Carnot cycle is used as a standard of comparison for heat-engine cycles because its efficiency is the maximum for a given temperature limit. In a similar manner, the reversed-Carnot cycle is used as a standard of comparison for refrigeration because for given temperature limits its COP is the maximum. Figure 15-15 shows a T-S diagram for the reversed-Carnot cycle. The heat absorbed at temperature T_c during process 1-2 is represented by area 1-2-b-a-1; the heat rejected at T_a during process 3-4 is represented by area 3-4-a-b-3; and the net work input is represented by the difference between these two areas, which is area 1-2-3-4-1.

The compression, cooling and expansion, and heating processes are accomplished isothermally. The heat transfer processes during these phases are effected over negligibly small temperature differences, resulting in no overall increase in entropy. The fluid is cooled and heated between these temperatures by isentropic expansion and compression, respectively.

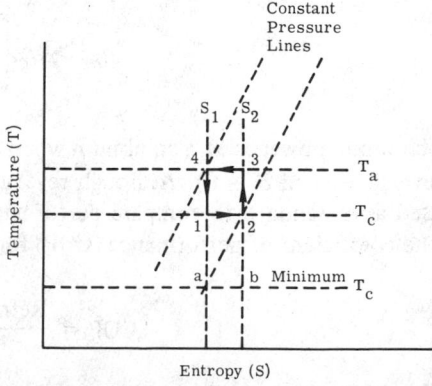

Fig. 15-15. The reversed-Carnot, refrigeration cycle.

15.5.2. Practical, Mechanical-Refrigerator Cooling Systems. Discussion of practical, mechanical refrigerators is divided into five basic categories: Stirling, VM, Gifford-McMahon/Solvay, reversed-Brayton/Claude, and J-T closed-cycle system. (Open-cycle, J-T systems are covered in Section 15.4.1.)

Stirling-Cycle Systems. Stirling-cycle refrigerators possess several major advantages for military refrigerator systems, such as low power-consumption and small size and weight. Many units have been fabricated for use with ground and/or airborne infrared detector systems. Production units available have characteristically a limited life (about 2000 h between maintenance periods) and suffer somewhat from vibration of the cold head. A two-stage unit designed by Philips Laboratories for space application (ID No. 89, Table 15-10) which has undergone substantial space qualification testing (e.g., vibration, acoustic, and shock loading equivalent to booster environments) is, as of January 1976, undergoing one-year-life testing at the Applied Physics Laboratory of Johns Hopkins University [15-22]. Stirling refrigerators have traditionally been constructed as a single unit; however, substantial development has occurred since 1972. Hughes Aircraft Company has fabricated modified Stirling-cycle units wherein the cold head is separated from the compressor by a single flexible line (Figure 15-16). This increases flexibility and substantially reduces the vibration environment of the cold head.

Representative, Stirling refrigerators are listed in Table 15-10. The specific power (i.e., $(COP)^{-1}$) and specific weight of some units are plotted as a function of temperature in Figure 15-17. The refrigerator identification number corresponds to that shown in Table 15-10. Additional details of these units can be found in Reference [15-5] or the manufacturer's literature cited in Table 15-10.

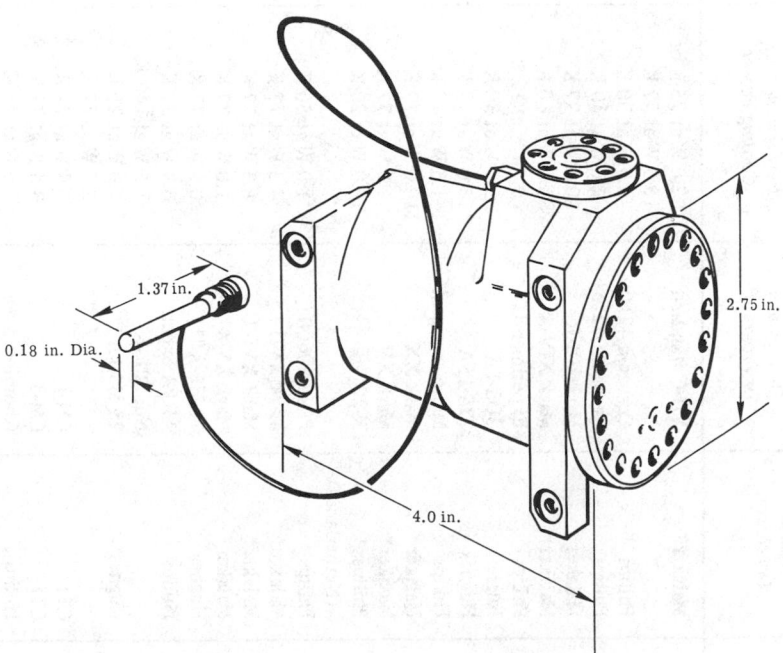

Fig. 15-16. Modular (split-component), modified-Stirling refrigerator.

Table 15-10. Identification Of Representative, Stirling-Cycle Refrigerators

Ident. No.	Mfg. or Developer	Model, Description or Program	Refrigeration Performance, or Cooling Power	Status	Power		Weight		Reference
					Input (W)	Specific Power (W W^{-1})	Total (lb)	Specific Wt. (lb W^{-1})	
29	Malaker*	VII-C standard	2.0 W at 25 K or	Production*	480.0	240.0	15.5	7.75	[15-23]
			15.0 W at 77 K		295.0	26.4	–	1.03	
36	Philips	Cryogem 42100	2.0 W at 30 K	Production	350.0	175.0	12.0	6.00	[15-24]
37	Philips	Cryogem 42151	2.0 W at 30 K	Production	550.0	275.0	25.0	12.5	[15-24]
30	Malaker*	Mark VII-R	60.0 W at 77 K	Production*	1220.0	20.5	40.0	0.67	[15-23]
32	Malaker*	Mark XIV-A	1.5 W at 60 K	Production*	120.0	80.0	5.5	3.7	[15-24]
34	Philips	Experimental	0.50 W at 12 K	Prototype	700.0	1400.0	35.0	70.0	[15-24]
24	Hughes	3OS-1A	1.0 W at 30 K	Prototype	620.0	620.0	16.0	16.0	[15-25]
27	Hughes	77S-15A	15.0 W at 77 K	Prototype	500.0	33.5	10.0	0.67	[15-25]
38	Philips	Micro cryogem	1.5 W at 77 K	Production	90.0	60.0	3.0	2.0	[15-24]
31	Malaker*	Mark XX	110.0 W at 77 K	Production*	1990.0	18.1	65.0	0.59	[15-23]
33	Malaker*	Mark XV	1.0 W at 77 K	Production*	29.5	29.5	5.0	5.0	[15-23]
39	Philips (Netherlands)	X-20	10.0 W at 20 K	Prototype	1750.0	175.0	112.0	11.2	[15-26]
52	Philips	P/N 460600	1.0 W at 50 K	Prototype	120.0	120.0	4.0	4.0	[15-24]
73	Malaker*	Mark XVII-1	4.3 W at 77 K	Production*	280.0	65.3	13.0	3.0	[15-23]
74	Malaker*	Mark XVI-3	8.3 W at 77 K	Production*	208.0	25.0	10.0	1.2	[15-23]
75	Malaker*	Mark XV-4	1.0 W at 77 K	Production*	29.5	29.5	5.0	5.0	[15-23]
76	Hughes	77 MS-A**	0.5 W at 77 K	Pre-production	80.0	160.0	4.0	8.0	[15-27]
89	Philips	ARPA exper. No. 301	0.3 W at 77 K	Flight Unit	30.0	100.0	15.0	–	[15-22]
77	Hughes	25 MS-1A**	1.5 W at 140 K	Testing	30.0	20.0	15.0	1.0	[15-27]
			1.0 W at 25 K	Pre-production	850.0	850.0	29.0	29.0	
			2.5 W at 77 K		–	340.0	–	11.6	
78	CTI	CM-1	0.9 W at 77 K	Prototype	90.0	100.0	3.25	3.6	[15-28]
79	CTI	CM-3	3.5 W at 77 K	Development	180.0	51.0	6.5	1.85	[15-28]
81	Philips	Rhombic drive	1.0 W at 75 K	Development	42.0	42.0	7.7	7.7	[15-29]
82	Hughes	77 MS-2B**	2.0 W at 77 K	Pre-production	200.0	100.0	14.5	7.25	[15-25]

*Malaker units are no longer available, but were retained in this chart for weight and power projections.

**Modified, Stirling, split-component units.

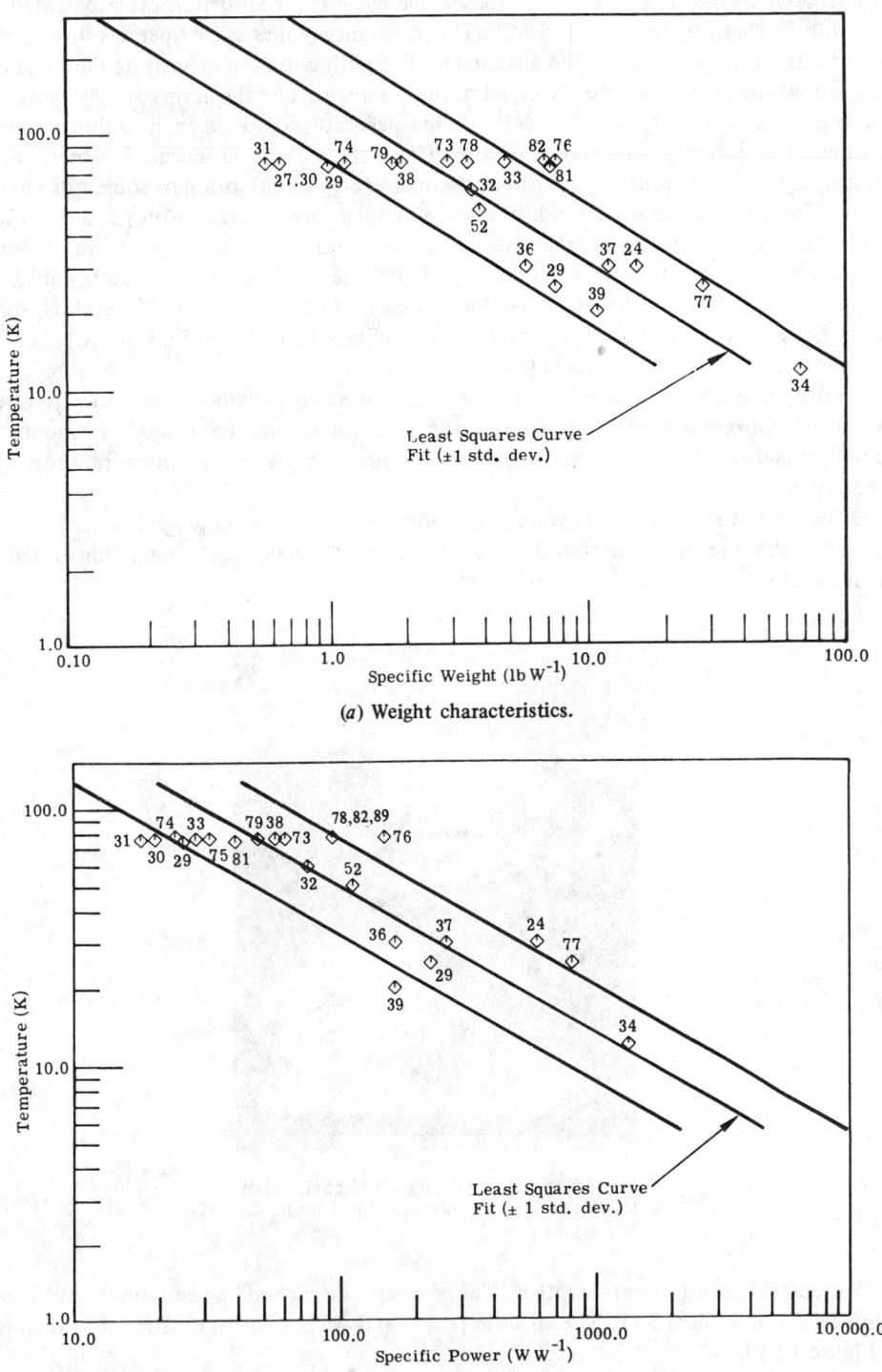

(a) Weight characteristics.

(b) Power characteristics.

Fig. 15-17. Stirling-cycle-refrigerator characteristics (Listed by ID. no. See Table 15-10.).

VM-Cycle Systems. The VM thermodynamic cycle is a heat-driven cycle patented by Rudolph Vuilleumier in 1918 [15-62]. This constant-volume cycle operates through the use of displacers which have the advantage of requiring minimum seals as the pressures throughout the system are nearly equal at any moment. The displacers simply move the gas from one section to another without the need for compressing it within a closed volume. Two primary advantages of the VM cycle are the absence of a mechanical compressor and its ability to use direct thermal energy as the primary source of energy input. The latter allows the flexibility to employ solar, nuclear, radioisotope, or electrical resistance heating, or combustible fuels or other chemical reactions to provide thermal energy directly to the hot-cylinder head. Although electrical resistance heaters imbedded in the head of the hot cylinder have been used in most of the VM refrigerators, some units have used a radioisotope heat source, Plutonium-238 [15-25, 15-30], and a propane combustion heater [15-31].

Ideally, no work is required to drive the crank mechanism which causes the displacers to move approximately 90 degrees out-of-phase, but in practice a modest amount of power supplied by an electric motor is required to overcome pressure drop and mechanical friction.

Different VM refrigerators have been developed for mobile ground applications and for use with airborne and spaceborne systems. A small, single-stage unit is illustrated in Figure 15-18.

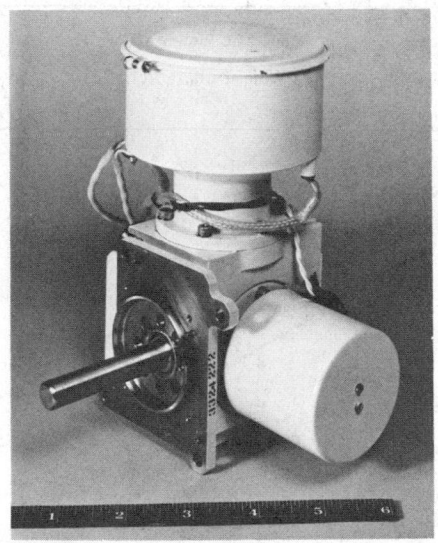

Fig. 15-18. A single-stage, VM cooler (1 W @ 77 K; Hughes Aircraft Co., Model 77 MVM-1-B) [15-25].

Characteristics of representative VM coolers developed since about 1965 are summarized in Table 15-11. The specific power and weight characteristics are presented in Figure 15-19.

Gifford-McMahon/Solvay-Cycle Systems. The basic refrigeration cycle used in this type of system was originally conceived by Ernest Solvay in 1886 as a derivative of the Stirling cycle. K. W. Taconis, W. E. Gifford [15-63], and H. O. McMahon have made a

Table 15-11. Identification Of Representative Vuilleumier-Cycle Refrigerators

Identification Number	Manufacturer or Developer	Model, Description or Program	Refrigeration Performance, or Cooling Power	Status	POWER		WEIGHT		Reference
					Input (W)	Specific Power (W W⁻¹)	Total (lb)	Specific Wt. (lb W⁻¹)	
22	Hughes	AFFDL development (AFLIR) 25 VM-2A	2.0 W at 25 K 3.0 W at 75 K	Prototype	750 –	375 250	19.5 –	9.8 6.5	[15-25]
26	Hughes	AFFDL development for aircraft IR scanner	1.5 W at 75 K	Accumulated 1100 h testing at FDL	200	133	5.75	3.84	[15-32]
40	Hughes	Army Night Vision Lab development	0.6 W at 77 K	Prototype	105	175	6.5	10.8	[15-5]
41	Hughes	X447550 (early model for ID #42)	0.50 W at 30 K 6.0 W at 75 K	Inactive	480 –	960 80	9.6 –	19.2 1.6	[15-30]
42	Hughes	SAMSO experiment 002 for SESP-712	0.15 W at 15 K 3.5 W at 55 K	Experimental flight model (completed 1 flight)	540 –	– 154	60.0 –	– 17.4	[15-5]
53	Philips	Early development model	1.0 W at 77 K	Prototype	120	120	15.0	15.0	[15-34]
71	Philips	Army Night Vision Lab development	0.5 W at 77 K	Completed 1000 h testing	90	180	10.3	20.6	[15-5]
80	Hughes	77 MVM-1B	1.0 W at 77 K	Prototype	260	260	3.0	3.0	[15-25]
83	Hughes	77 VM-2B	2.0 W at 77 K	Prototype	340	170	8.5	4.25	[15-25]
84	Hughes	77 VM-2C	2.0 W at 77 K	Exploratory	*	–	9.0	4.5	[15-25]
85	Hughes	77 VM-1C	1.0 W at 85 K	Exploratory	**	–	8.0	4.0	[15-25]
87	Hughes	15 VM-A	0.50 W at 15 K 3.0 W at 66 K	Prototype	400	800 133	19.5 –	39.0 6.5	[15-25]
88	Airesearch	GSFC	5.0 W at 75 K	Prototype	300	60	–	–	[15-35]

*Powered by Pu238 and 30 W.
**Powered by propane heater and 35 W.

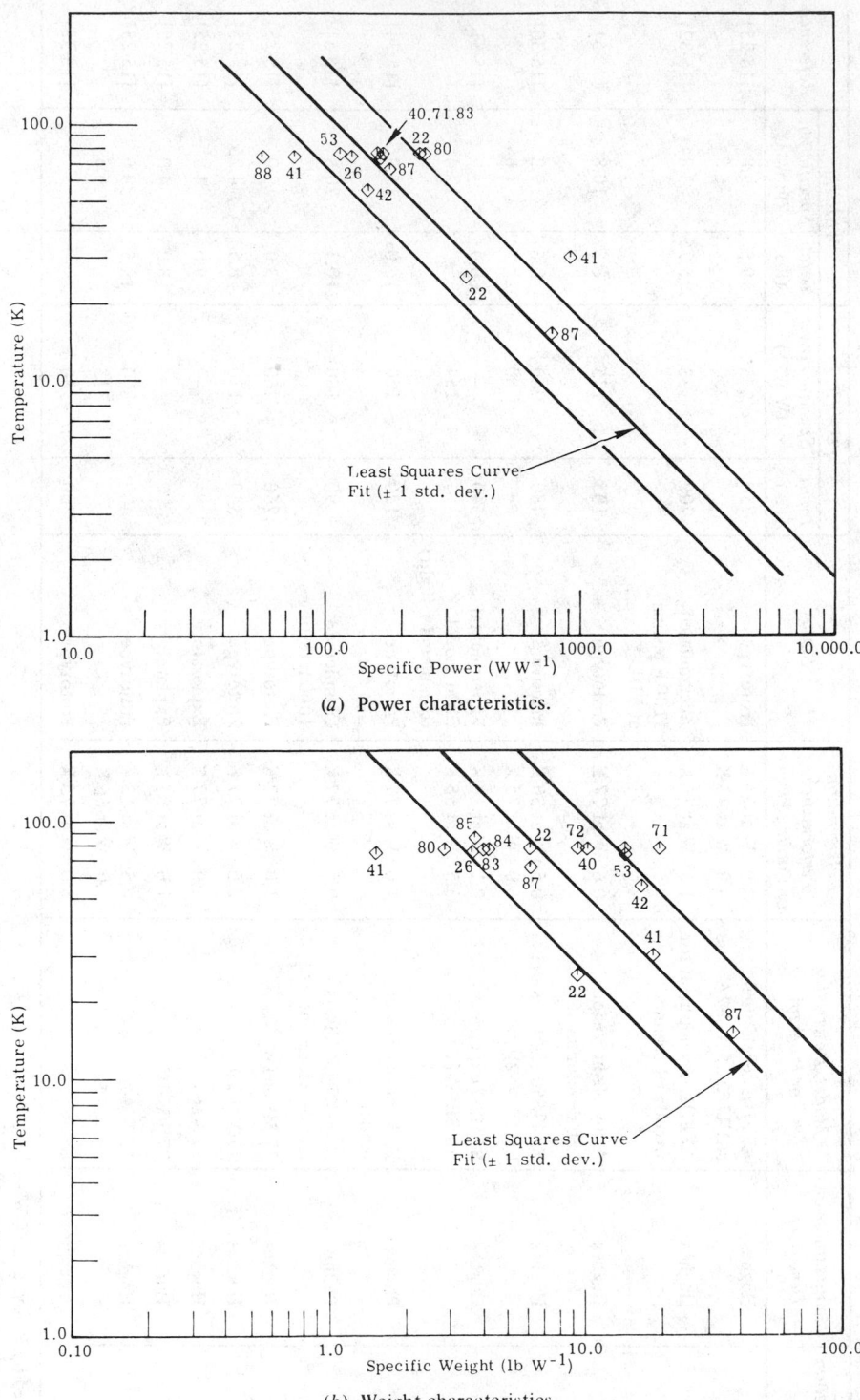

(a) Power characteristics.

(b) Weight characteristics.

Fig. 15-19. VM-cycle-refrigerator characteristics (Listed by ID. no. See Table 15-11.).

number of modifications. Manufactured systems are usually marketed using various names as Gifford-McMahon (G-M) and Solvay, with or without the adjective "modified." They are basically the same cycle, but with different expander modifications.

By separating the expander from the compressor, one obtains a system with a simple, lightweight, compact cooling-unit, which can be more easily integrated with the load, and a compressor which can be located separately. The compressor is then connected to the expander with long, flexible lines which carry the high- and low-pressure, working fluid.

There are many varieties of these systems on the market and nearly all use hermetically-sealed compressors. System variations are primarily confined to methods of operating the expander unit and various design, material, and manufacturing techniques to produce more reliable, long-life, low-cost systems. Table 15-12 presents a representative listing of available G-M refrigerators. The specific power and weight of these units are illustrated in Figure 15-20.

Brayton-Cycle Systems. The term Brayton cycle is used here to denote a class of thermodynamic cycles which are all variants of the reversed-Brayton, thermodynamic cycle (which includes the Claude cycle). A reversed-Brayton-cycle refrigerator contains a compressor, an after-cooler, a heat-exchanger and an expansion turbine as shown in Figure 15-21. The compressor, operating at ambient temperature, compresses the cycle gas, and heat is rejected to ambient temperature. The high-pressure gas is then passed through a regenerative heat exchanger and is expanded across a turbine where the energy is extracted. The gas is then directed through the refrigeration load. The Claude cycle is similar except that an additional heat-exchanger and J-T valve are added at the cold end to achieve a further reduction in the temperature.

As the operating temperature of the Brayton refrigerator is lowered, point 5 (Figure 15-21) will enter the two-phase region of the working fluid and the fluid will leave the expander as a two-phase mixture. For refrigeration at temperatures within the two-phase region of the working fluid, the expansion process is performed through a throttle valve (as in the J-T cycle) rather than in an expansion engine. The Claude cycle (Figure 15-22) is effectively a J-T cycle in which the effective sink temperature is lowered by a Brayton cycle refrigerator.

Refrigerators operating on the reversed-Brayton cycle have been used for years in large, ground-based installations. In recent years, efforts have been directed toward developing gas-bearing-supported turbomachinery [15-8] and rotary-reciprocating machinery [15-9] suitable for use in small, cryogenic refrigerators. Turborefrigerators employing gas-bearing turbomachinery have the potential for high reliability and long, maintenance-free life. Lubrication of the bearings with the cycle working fluid excludes lubricants as a source of contamination and fouling in the low-temperature regions of the cycle. The absence of continuously rubbing parts eliminates the failure by wear that is typical of nonlubricated or dry-lubricated machinery. Hence, with gas-bearing turbomachinery there is the expectation for long life, probably limited by the number of start-stop cycles rather than by hours of operation.

The performance of miniature turbomachinery refrigerators is a strong function of their capacity. This is primarily because the working fluid flow-rates in small refrigerators are substantially lower than the normal range appropriate to turbomachinery. As capacity and cycle flow are increased, tubomachinery design requirements move in a more favorable direction, and the relative size, weight, and performance of the refrigerators are substantially improved. Power and weight estimates for turbomachinery refrigerators are presented in Figure 15-23.

Table 15-12. Identification Of Representative Gifford-McMahon/Solvay-Cycle Refrigerators

Ident. No.	Mfg. or Developer	Model, Description, or Program	Operating Temperature Range (K)	Typical Refrigeration Performance, or Cooling Power	Status	Power		Weight		Remarks	Reference
						Input (W)	Input/Refrig. (W W^{-1})	Total (lb)	Total/Refrig. (lb W^{-1})		
11	Cryomech, Inc.	Model CB02	7.5 to 25	1.0 W at 9.5 K	Production	3000	3000	200.0	200.0	G-M	[15-24]
12	Cryomech, Inc.	Model CB04	9 – 30	4.0 W at 13 K	Disconted.	3000	750	200.0	50.0	G-M	[15-26]
6	Cryogenic Technology	Model 1023	10 – 28	2.0 W at 13 K	Production	6100	3050	458.0	229.0	G-M	[15-36]
7	Cryogenic Technology	Model 350	15 – 28	2.0 W at 18.5 K 15.0 W at 77 K	Production	2100	1050	197.0	98.5	G-M	[15-35]
8	Cryogenic Technology	Model 0120/PC-30	19 – 28	1.0 W at 26 K	Production	650	650	25.5	25.5	G-M	[15-24]
89	Cryomech Inc.	Model GB03	8 – 25	5.0 W at 20 K	Production	3000	60	200.0	40.0	G-M	[15-24]
10	Cryogenic Technology	SP-77	77	0.60 W at 77 K	Production	135	225	6.3	10.5	G-M	[15-38]
14	Cryomech, Inc.	Model AL05	23 – 89	10.0 W at 30 K	Production	3000	300	200.0	20.0	G-M	[15-24]
2	Air Products	Model I-1	30 – 150	20.0 W at 77 K	Production	1700	85	140.0	7.0	Solvay	[15-24]
45	Air Products	Military Application	77	1.5 W at 77 K	Production	340	226	11.0	7.3	Solvay	[15-12]
54	Air Products	CS-1003	50 – 300	1.0 W at 77 K	Production	400	400	63.0	63.0	Solvay	[15-37]
46	Kinergetics	SRS-07	50 – 77	1.0 W at 38 K 2.0 W at 50 K	Production	400	400 200	12.4	12.4 6.2	Solvay	[15-39]
50	Air Products	CS-102	30 – 200	17.0 W at 77 K	Production	1735	102	150.0	8.75	Solvay	[15-12]
51	Air Products	CS-202	12 – 300	1.0 W at 17 K	Production	1735	1735	150.0	150.0	Solvay	[15-12]
61	Air Products	MS-1003	30 – 77	1.0 W at 77 K	Production	368	368	14.5	14.5	Dev. for airborne appl.	[15-12]
66	Cryogenic Technology	0120	19 – 30	1.0 W at 26 K	Production	800	800	25.0	25.0	Designed for airborne use (MIL-E5400)	[15-36]
82	Air Products	DE-1003	44 – 77	1.0 W at 76 K	Development	500	500	135.0	13.5	Light Weight Dry Compr.	[15-12]

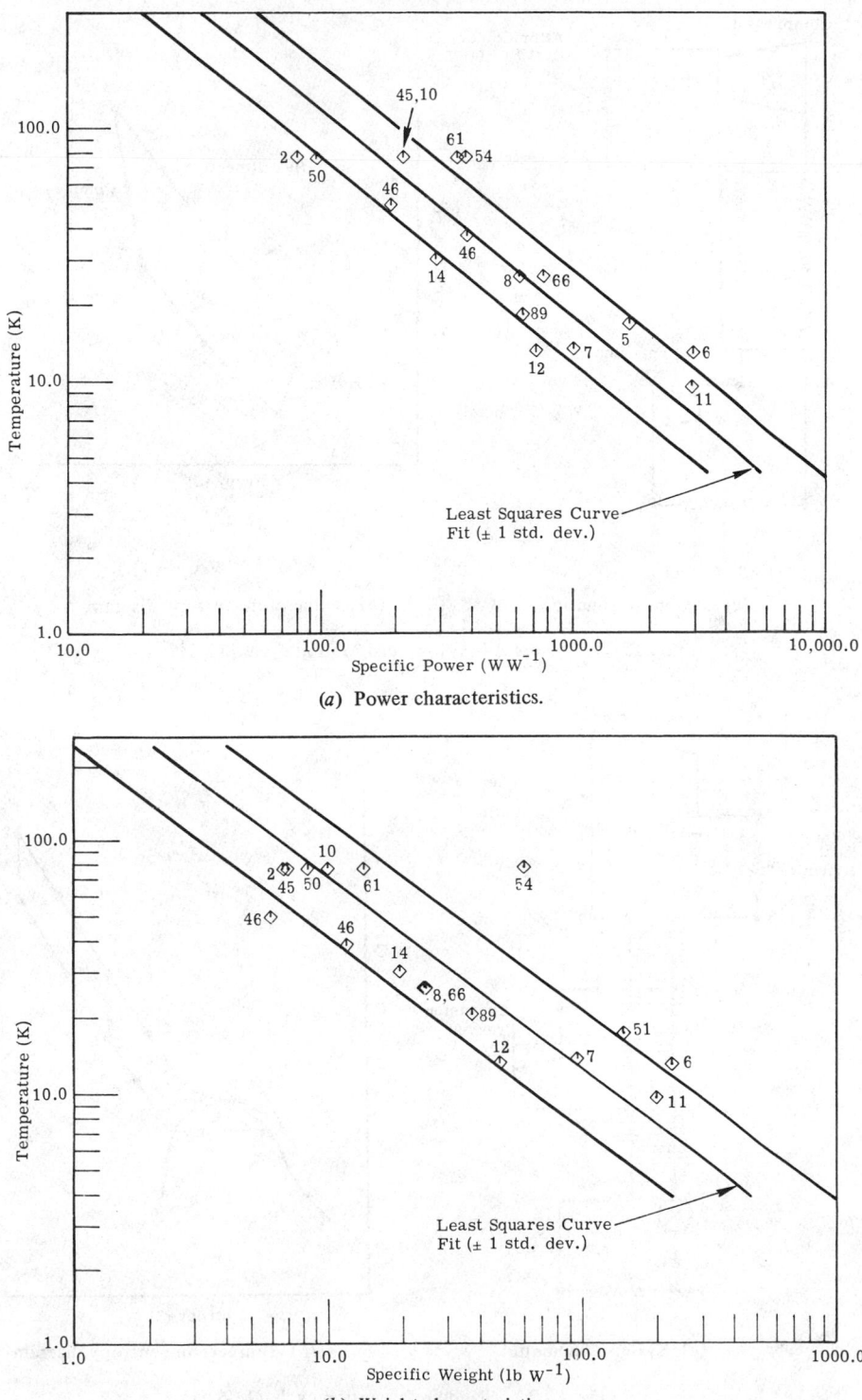

(a) Power characteristics.

(b) Weight characteristics.

Fig. 15-20. Gifford-McMahon-cycle-refrigerator characteristics. (Listed by ID. no. See Table 15-12).

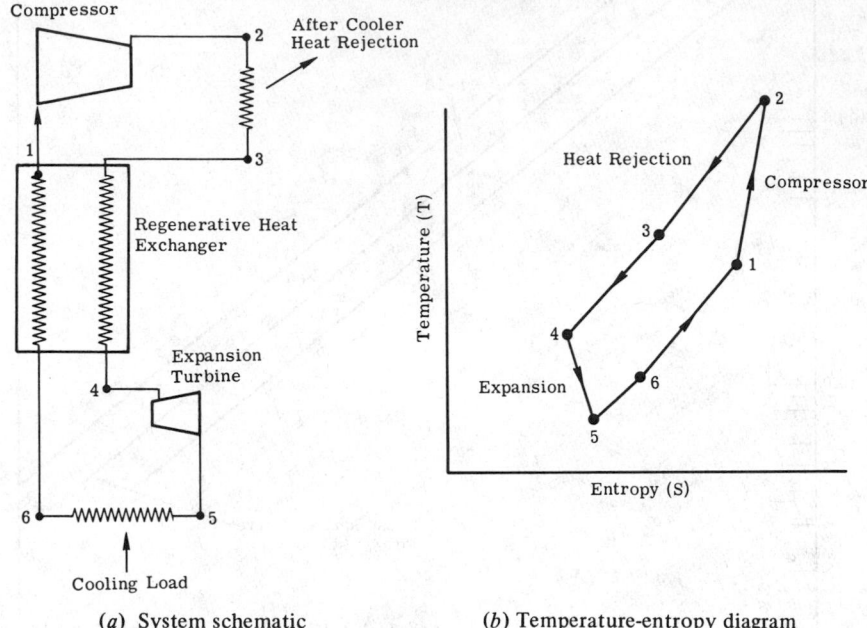

(a) System schematic   (b) Temperature-entropy diagram

Fig. 15-21. Reversed-Brayton, refrigeration cycle [15-5].

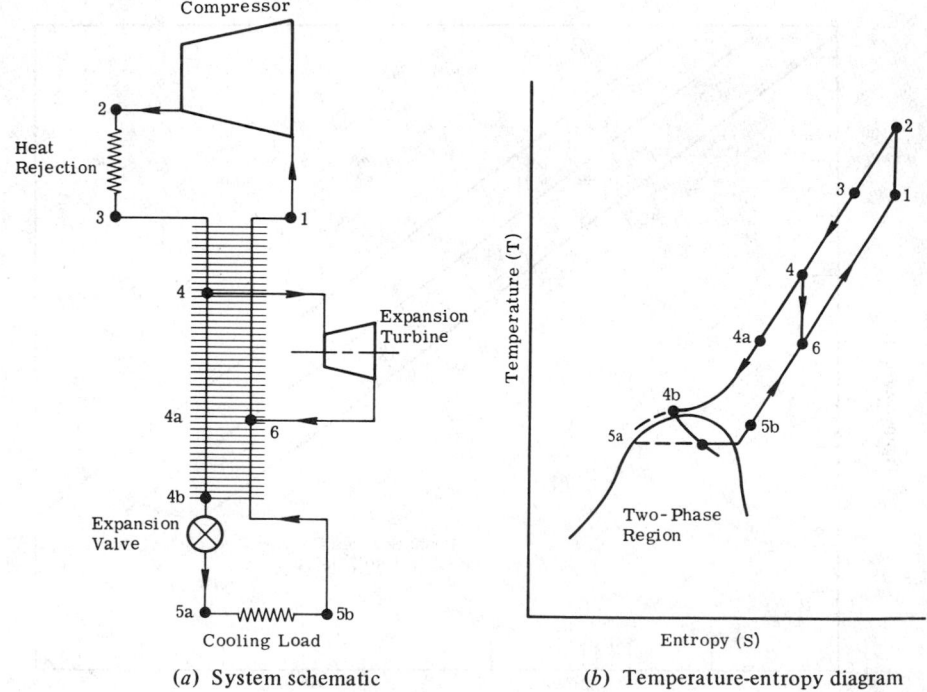

(a) System schematic   (b) Temperature-entropy diagram

Fig. 15-22. Claude refrigeration cycle [15-5].

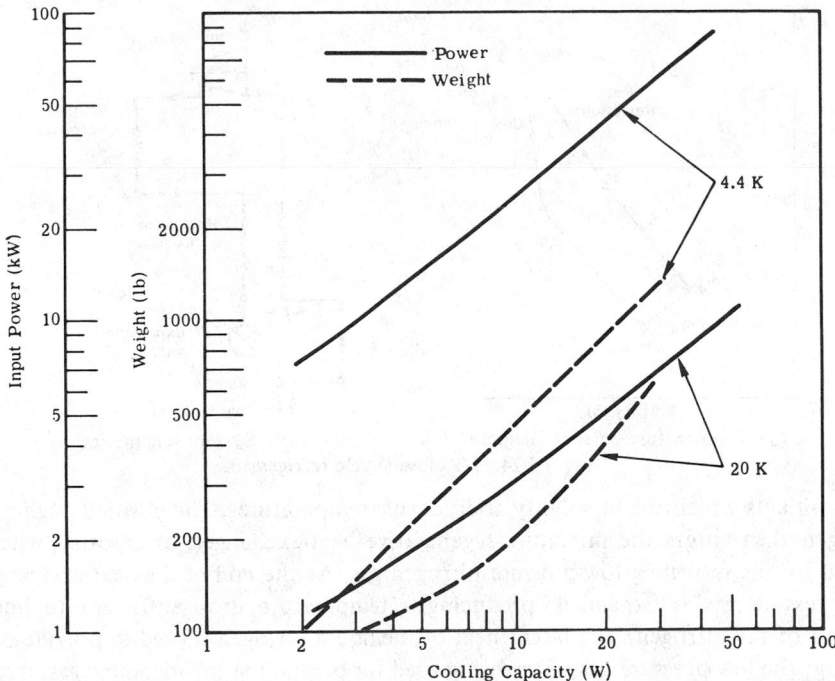

Fig. 15-23. Power and weight estimates for 4.4 and 20 K turbomachinery refrigerators [15-27].

Miniature refrigerators using reciprocating machinery in the Brayton cycle have recently been developed for space applications [15-9]. For reliability, the pistons in these machines are rotated as well as reciprocated. This permits the use of ports (to control gas flow) and clearance seals (to limit leakage). Electromagnetic actuators drive the pistons. The few moving parts are completely supported on gas bearings. There are no rubbing or sliding surfaces as in conventional reciprocating equipment. The refrigeration machinery required to execute the cycle is contained in two separate units: a compressor assembly and an expander package. Power conditioning equipment is required to convert the basic source of electrical power to voltages of the proper frequency, amplitude, and phase for operating the refrigerator.

The ADL development program [15-9] has demonstrated the feasibility of this concept, with a test unit which produced 2 W at 77 K; however, testing experience has not been sufficient to verify operating reliability.

J-T Closed-Cycle Systems. A practical J-T refrigerator cycle is shown in Figure 15-24. It is essentially identical to the reversed-Brayton cycle except for one fundamental difference. The expansion process, 4 to 5, is accomplished by expansion through a throttling valve rather than through a turbine. In the J-T system, point 5 is in the two-phase region, and the heat of vaporization of the coolant is used to absorb heat from the cooling load in process 5 to 6. In a typical application, the working fluid (gaseous nitrogen) is compressed to approximately 250 psia in a multistage, oil-lubricated, reciprocating compressor. The heat of compression is removed by ram air or by a fan mounted on the compressor assembly. After compression, the gaseous nitrogen passes through an absorber-filter component which removes oil vapor and other trace

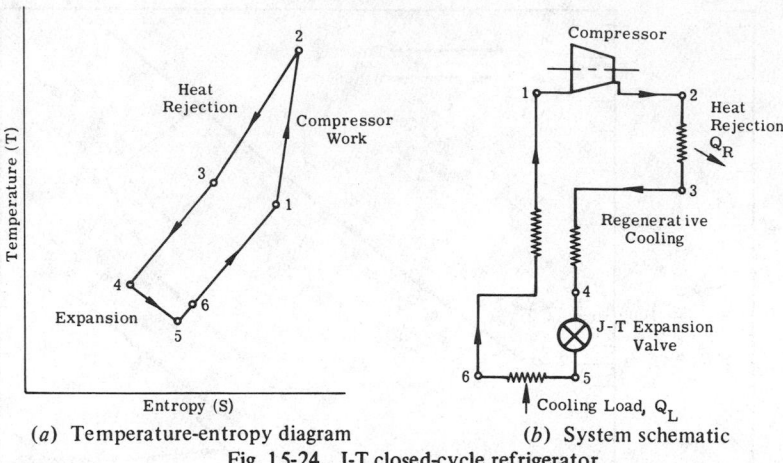

(a) Temperature-entropy diagram   (b) System schematic

Fig. 15-24. J-T closed-cycle refrigerator

contaminants which might solidify at cryogenic temperatures. The purified, high-pressure nitrogen then enters the miniature regenerative heat-exchanger, or cryostat, where it is cooled by the returning low-pressure nitrogen gas. At the end of the heat-exchanger, the high-pressure gas is expanded, producing a temperature drop sufficient to liquefy a portion of the nitrogen. The latent heat of the liquid nitrogen is used to provide the spot cooling; the low-pressure gas, after being used for precooling the incoming gas, returns to the first stage of the compressor. A small gas reservoir (accumulator) is connected to the low-pressure return line to adjust the gas volume to compensate for the increased density of that portion of the working fluid that is liquefied in the course of normal operation. Representative J-T units are presented in Table 15-13. Specific power and weight characteristics are presented in Figure 15-25.

15.6. Thermoelectric Cooling Systems

When two dissimilar metals are connected in series with a source of electromotive force (emf), one junction will be cooled while the other will be heated. The generation or absorption of heat depends on the direction of the current, while the rate of heat pumping is a function of the current and material properties. This basic phenomenon was first observed by Jean C. A. Peltier in approximately 1834 [15-40].

The outstanding features of thermoelectric coolers are simplicity and reliability. There are no moving parts. Other advantages include the following:

(1) The heating and cooling function can be easily interchanged by reversing the polarity of the direct current.

(2) Noise is absent during operation.

(3) Operation is independent of orientation or gravity.

(4) Weight and volume are small.

The primary limitation of thermoelectric coolers is the maximum temperature-difference (or temperature-lift) attainable on a practical basis: about 150°C based on available materials, which results in a minimum attainable temperature in the order of 150 to 170 K. The chief disadvantage of these devices is the relatively low COP, especially where large temperature-differences are required.

15.6.1. Description and Design Characteristics. A typical, single-stage, thermoelectric cooler consists of a p- and n-type semiconductor connected together by a metallic

Table 15-13. Identification Of Joule-Thomson, Closed-Cycle Refrigerators

Identification Number	Manufacturer or Developer	Model, Description or Program	Operating Temperature Range (K)	Typical Refrigeration Performance, or Cooling Power	Status	Power		Weight		Remarks	Ref.
						Input (W)	Input/Refrig. (W W^{-1})	Total (lb)	Tot./Refrig. (lb W^{-1})		
15	Garrett Airesearch	133386	77	Three cooling pts. 0.75 W ea. at 77 K	Production	450	200.0	25.0	11.1	Utilizes ram air cooling	[15-24]
16	Garrett Airesearch	133488	77	5.0 W at 77 K	Production	650	130.0	22.5	4.5	Includes fan power	[15-24]
17	Garrett Airesearch	144406	77	3.0 W at 77 K	Production	450	150.0	19.5	6.5	Utilizes ram air cooling	[15-24]
18	Garrett Airesearch	80034	77	Two cooling pts. 1.0 W ea. at 77K	Production	460	230.0	22.5	11.25	Includes fan power	[15-24]
19	Garrett Airesearch	800398	77	Two cooling pts. 1.0 W ea. at 77 K	Production	650	325.0	23.0	11.50	Includes fan power	[15-24]
20	Garrett Airesearch	800656	77	2.5 W at 77 K	Production	530	211.0	20.0	8.0	Includes fan power	[15-24]
55	Air Products	J-80-1000	77	2.0 W at 77 K	Production	600	300.0	18.0	9.0	—	[15-21]
56	Air Products	J-30-3500	23 and 77	2.0 W at 77 K 0.35 W at 23 K	Production	1050	300.0 1350.0	—	—	Two-stage unit	[15-21]

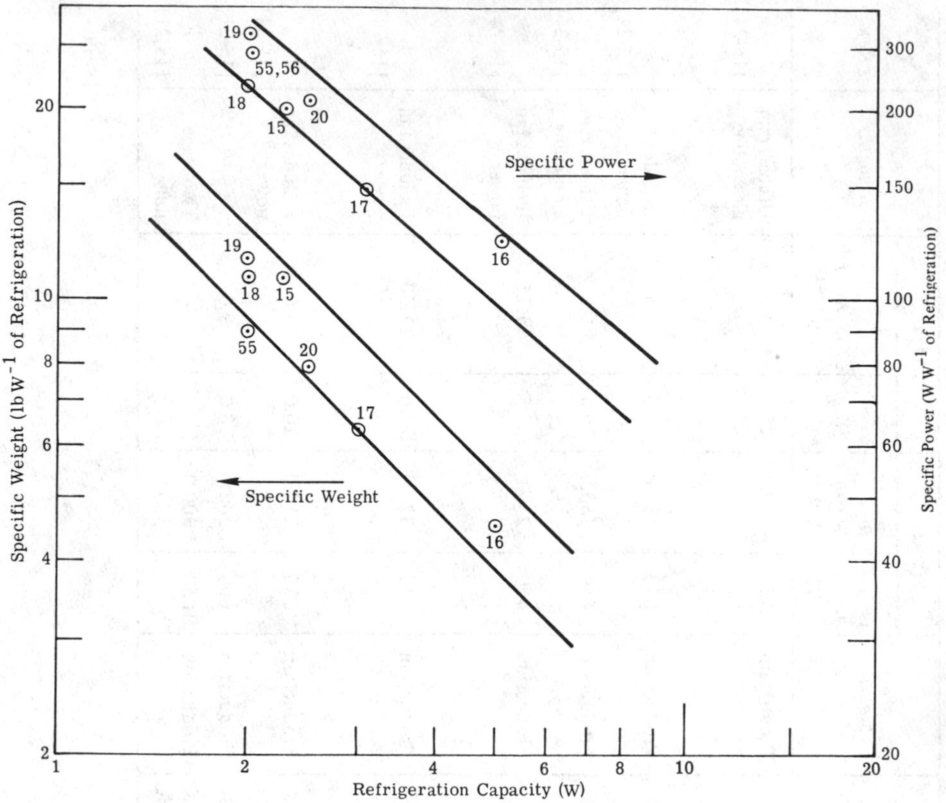

Fig. 15-25. J-T closed-cycle refrigerator: specific power and weight versus capacity. All units are for airborne use at 77 K and are listed by ID. in Table 15-13.

conductor (Figure 15-26). When a voltage is applied (normally in the range of 0.10 to 5.0 V for typical IR cooling units), the flow of current in the direction shown produces a temperature difference between two junctions, absorbing heat at one end and releasing it at the other. The heat removed from the cold junction is the difference between the Peltier cooling effect and the sum of the Joule heat generated by the current and the heat conducted from the hot to the cold junction. The result is the net cooling capacity of the couple.

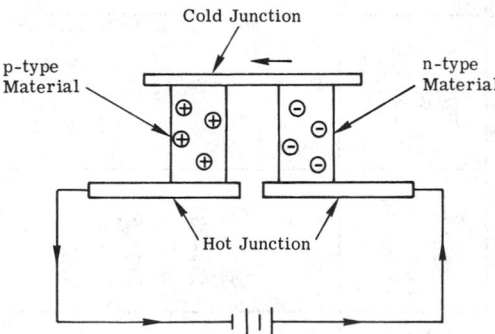

Fig. 15-26. Peltier thermoelectric couple.

Thermoelectric couples of similar design can be arranged in parallel to increase the heat-pumping capacity. When identical couples are placed in parallel and supplied with equal currents, they pump as many times more heat at the same temperature difference and with the same COP as there are couples. Thermoelectric couples can also be connected in series (cascaded) for one of two purposes: to provide a temperature difference greater than that attainable from a single couple; or to achieve a higher COP for a given heat-pumping rate and overall temperature difference. Manufactured units are generally available with up to four stages in which temperature differences of 80 to 120°C are achieved between the cooling-load temperature and the available heat-sink temperature. Theoretical and actual performance data for various operating conditions are given in the following paragraphs.

15.6.2. Theoretical Performance Data. A brief review of theoretical-performance parameters is presented here only to the extent necessary to assess the capabilities of thermoelectric coolers. For detailed discussions of performance characteristics and derivations of basic equations, consult the bibliography for this chapter.

When a steady state is established at the cold junction of a couple, the net heat absorbed at the cold junction less the Joule heat, and the heat conducted from the hot to the cold end is given by

$$Q_i = \mathcal{S} \, T_c I = \left(\frac{1}{2} \, I^2 R \right) - \mathcal{K} \, (T_h - T_c) \tag{15-5}$$

where Q_i = net heat absorbed at the cold junction (Peltier effect), W
$\mathcal{S}$ = Seebeck coefficient for a couple, V K^{-1}
I = current, A
R = electrical resistance, ohms
$\mathcal{K}$ = thermal conductance, W K^{-1}
T_c = temperature of cold junction, K
T_h = temperature of hot junction, K

The center terms (those between the two equal signs) of Equation (15-5) are derived from the fact that the Peltier cooling effect is given by

$$Q_i = P_{ab} I \tag{15-6}$$

and that

$$P_{ab} = \mathcal{S} \, T_c \tag{15-7}$$

where P_{ab} = Peltier coefficient between materials a and b, W A^{-1}
T_c = cold junction temperature, K

Thus

$$Q_i = \mathcal{S} \, T_c I \tag{15-8}$$

The pertinent performance parameters of thermoelectric coolers most useful for evaluation purposes are: the figure of merit, Z, which is a function of the material properties; the temperature difference, ΔT, or lift attainable between a single-stage couple; and the COP.

The figure of merit characterizes a material for its cooling potential and is given for a single material by

$$Z = \frac{\mathcal{S}^2}{\rho k} \tag{15-9}$$

where $\mathcal{S}$ = Seebeck coefficient for a single material, V K^{-1}
ρ = electrical resistivity, ohm cm
k = thermal conductivity, W cm^{-1} K^{-1}
Z = figure of merit, K^{-1}

15.6.3. Material Considerations. The figure of merit involves the parameters of thermal conductivity, electrical resistivity, and the Seebeck coefficient, and is the primary factor for evaluating materials for thermoelectric coolers. The properties which determine the Z value are interrelated and, in general, dependent on the electron density and the relative emf difference between two materials. The value of Z is normally maximized with electron densities associated with materials in the semiconductor class. The materials exhibit a high Seebeck coefficient and electrical resistivity.

The average Z at room temperature for the combined n- and p-type commercial materials is 2.5 to 3.0×10^{-3} K^{-1}. However, the value of Z decreases significantly with temperature and has a dramatic effect on reducing the COP of the multi-stage coolers which are required to achieve low temperatures. Typical ranges of Z versus temperature for commercially available materials are presented in Figure 15-27. The COP of a single-stage cooler versus Z and cold junction temperature is presented in Figure 15-28.

The COP of multi-stage, cascaded coolers requires computerized numerical approaches for accurate design. The maximum COP of multi-stage coolers at 170 and 193 K (with a 300 K sink-temperature) are illustrated in Figure 15-29 (based on the Z-values of Figure 15-17).

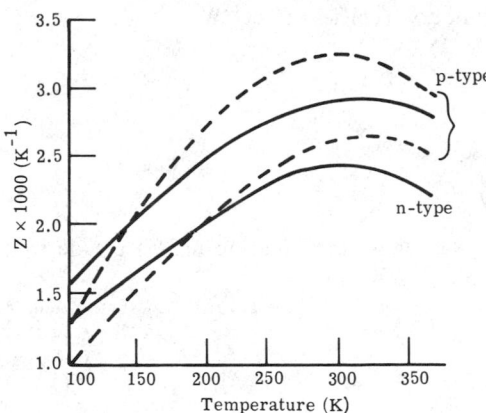

Fig. 15-27. Figure of merit versus temperature for p- and n-type material [15-10].

$$COP_{MAX} = \left(\frac{T_C}{T_H - T_C} \right) \frac{\sqrt{1 + Z\overline{T}} - T_H/T_C}{\sqrt{1 + Z\overline{T}} + 1}$$

T_H = Hot Junction Temperature (K)
T_C = Cold Junction Temperature (K)
Z = Figure of Merit (K^{-1})

$$\overline{T} = \frac{T_C + T_H}{2}$$

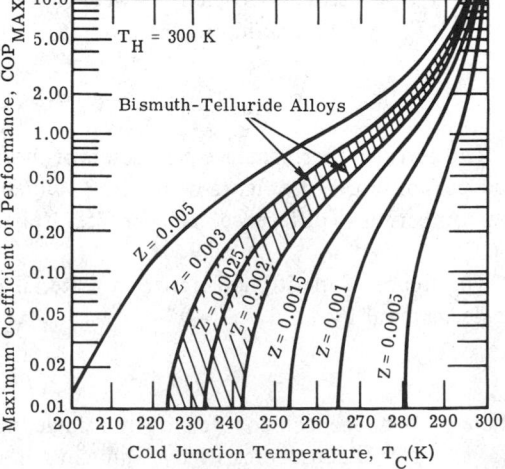

Fig. 15-28. Maximum COP versus cold-junction temperature and figure of merit for a single-stage, thermoelectric cooler [15-41].

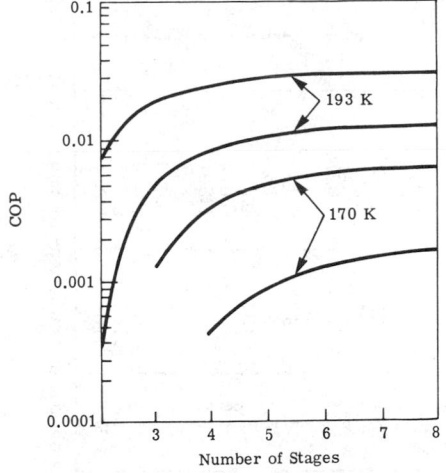

Fig. 15-29. Calculated, maximum coefficients of performance: focus of optimized designs for 300 K heat-sink operation using material properties shown in Figure 16-27 [15-10].

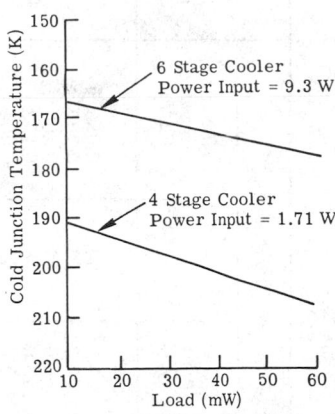

Fig. 15-30. Performance test data on low-power, low-temperature coolers [15-10].

15.6.4. Available Capabilities.

Characteristics of typical production units designed for cooling of IR detectors or other electronic components are presented in Table 15-14. A number of low-temperature, multiple-stage, cascaded coolers are currently under development by Borg-Warner Thermoelectrics. Figure 15-30 shows the performance of a four-stage, 193 K (nominal) cooler. A six-stage, 170 K cooler is shown in Figure 15-31. The cooler dimensions are approximately 0.40 × 0.67 in. at the base and 0.75 in. high. An eight-stage cooler has cooled a 0.20 × 0.04 in. cold plate to 145 K in a 325 K environment using 40 W of power [15-10]. The estimated, total effective load was 10 mW, with a COP of about 0.00025.

Fig. 15-31. Six-stage, low-power, 170 K cooler [15-10].

Table 15-14. Characteristics Of Typical, Production, Thermoelectric Coolers

Manufacturer/Developer and Model Number	System Design Limitations*			System Performance Characteristics*				Number of Stages
	ΔT max. at No Load (°C or K)	Cold Junction Temp. at ΔT max. (K)	Maximum Cooling Load (Qmax.) at $\Delta T=0.0$	Nominal Cooling Load	Power Input (W)	ΔT (°C or K)	Coefficient of Performance (COP)	
Borg-Warner [15-41]								
837	70	230	1.2 W	150 mW at 240 K	1.75	60	0.085	1
920	70	230	19.0 W	5.5 W at 250 K	27.20	50	0.202	1
950	66	234	30.0 W	10 W at 255 K	50.0	45	0.200	1
970	70	230	130.0 W	30 W at 252 K	175.0	48	0.172	1
447	90	210	200.0 mW	15 mW at 215 K	3.8	85	0.004	2
638	90	210	800.0 mW	200 mW at 220 K	6.0	–	0.033	2
623	117	183	140.0 mW	20 mW at 200 K	7.5	100	0.0027	3
724	100	200	900.0 mW	–	–	–	–	3
493	95	205	90.0 mW	10 mW at 230 K	1.1	70	0.009	3
670	119	181	200.0 mW	10 mW at 185 K	5.4	115	0.0019	4
605	126	174	–	10 mW at 178 K	6.0	122	0.0016	4
Cambridge Thermionic Corporation [15-42]								
800-3955	77	223	–	1 W at 240 K	29.4	60	0.034	2
800-3956	84	216	–	100 mW at 220 K	28.0	80	0.0036	3
800-1004	90	210	4.3 W	1.0 W at 236 K	29.6	70	0.034	2
800-1005	85	215	5.7 W	1.0 W at 228 K	24.5	72	0.041	2
800-1006	105	195	1.6 W	100 mW at 200 K	13.5	100	0.0074	3
Marlow Industries, Inc. [15-43]								
MI1020	67	233	1.0 W	0.2 W at 253 K	1.6	47	0.15	1
MI2060	97	203	1.1 W	0.2 W at 223 K	4.4	87	0.05	2
MI3026	111	189	0.7 W	0.15 W at 203 K	3.5	107	0.043	3
MI4012	115	185	116.0 mW	20 mW at 186 K	6.6	105	0.014	4
MI6020	127	173	70.0 mW	20 mW at 180 K	10.2	–	0.0019	6
SBRC [15-15]								
TE100-1	–	–	–	50 mW at 250 K	1.0	50	0.050	1
TE100-2	–	–	–	50 mW at 220 K	5.0	80	0.010	2
TE100-3	–	–	–	50 mW at 196 K	16.0	104	0.003	3
TE100-4	–	–	–	80 mW at 196 K	7.0	104	0.011	4

*Based on hot junction temperature of 300 K (27°C) and a vacuum of 10^{-4} to 10^{-6} mm Hg except for the 5 Cambridge units whose performance data is specified for still air at atmospheric pressure.

The chief limitation on achieving more-efficient, low-temperature, thermoelectric coolers is the lack of high-quality, low-temperature materials.

15.7. Radiant-Cooling Systems

15.7.1. Concept.

In spaceborne applications, temperatures of about 80 to 200 K can be achieved by use of a suitably-designed, radiant cooler radiating into space. The low effective sink-temperature of deep space provides an ideal environment for passive radiant-cooling of infrared detectors and related devices to the temperatures indicated. This approach involves no moving parts, provides inherently long life, and requires no power.

The effective temperature of deep space is approximately 4 K. One or more detectors mounted to a suitably sized cold-plate of high emissivity can radiate to this sink. The high vacuum of orbital altitudes minimizes the effect of convective heating. The cold-plate must be shielded (with a cone, for example) against heat from direct sunlight, and, in the case of near earth-orbits, against the heat inputs from thermal emission and re-flected sunlight from the earth and its atmosphere (Figures 15-32 and 15-33). Further-more, the cold-plate must be thermally shielded from the parent spacecraft. These con-siderations usually result in a passive-cooler design which is tailored to a particular spacecraft system.

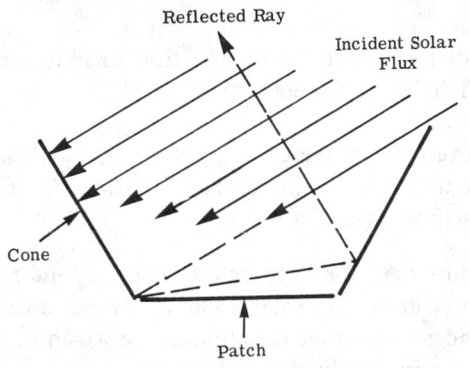

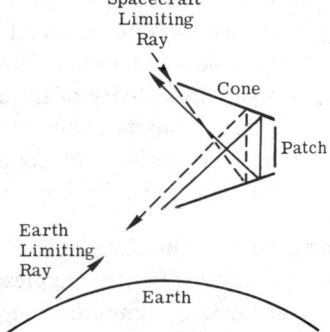

Fig. 15-32. Solar-flux reflec-tion by cooler cone [15-4].

Fig. 15-33. Earth and spacecraft flux reflected by cooler cone [15-4].

The type of orbit (e.g., near-polar, equatorial), orbit altitude, orientation of the space-craft relative to the earth or sun, and the location of the radiator, all significantly influence radiator design. Ideally, the radiant-cooler patch (i.e., the detector-mounting surface) is large enough so that thermal inputs (e.g., Joule heat, lead conduction, and radiative input through the optics) produced by the attachment of detectors or other components are small compared to the total power radiated by the patch at its equilibrium temperature. This permits flexibility in the optical and electrical design.

15.7.2. Thermal-Balance Equations.

For preliminary radiator-sizing, thermal-balance equations for both the patch and the cone are needed to determine the equilibrium (steady-state) temperature of the patch. Adding the conductive and detector thermal-inputs to the radiative input, one gets the thermal-balance equation for a black (i.e., $\epsilon = 1.0$), isothermal patch of an ideal cooler mounted on a spacecraft. (See Figure 15-33.)

$$\sigma A_p T_p^4 = \sigma A_p F_a \epsilon_{cp} T_c^4 + K_{cp}(T_c - T_p) + Q_d \qquad (15\text{-}10)$$

where A_p = external area of patch, cm^2
T_p = absolute temperature of patch, K
T_c = absolute temperature of the cone, K
F_a = cone-to-patch configuration view-factor (Section 15.8.2, Table 15-18)
ϵ_{cp} = effective cone-to-patch emissivity
K_{cp} = thermal conductance from cone to patch, W K^{-1}
Q_d = thermal inputs from detector, W
σ = Stefan-Boltzmann radiation constant, W cm^{-2} K^{-4}

Equation (15-10) assumes no direct environmental-radiative inputs to the patch, but only to the cone and its supporting structure. The thermal-balance equation for the cone can be obtained by equating the radiative output of the cone to the radiative and conductive inputs. In general, the heat transferred from the cone to the patch is small compared to that emitted and absorbed by the cone, and therefore can be neglected for preliminary design. The resulting equation is given by

$$\sigma \epsilon_c A_c T_c{}^4 = A_c \alpha_s (E_s + E_a) + A_c E_{ER} \epsilon_c + Q_{ac} \qquad (15\text{-}11)$$

where ϵ_c = emissivity of the cone
A_c = effective area of the cone, cm^2
E_s, E_a, E_{ER} = irradiance due to direct solar radiation, earth-reflection solar radiation (albedo), and earth infrared emission, respectively, W cm^{-2}
α_s = solar absorptivity of the cone
Q_{ac} = thermal inputs from the spacecraft to the cone which normally includes both radiative and conductive mechanisms of heat transfer, W. (See Section 15.8.2 for basic equations and data.)

Data for computing the incident radiative-inputs to the cone walls (i.e., E_s, E_a and E_{ER}) with the appropriate view-factors are presented in Section 15.8. Optical properties, such as α_s and ϵ_c, needed to compute absorbed and emitted radiation are provided in Section 15.8.4. Information helpful in determining conductive heat transfer and the conductance, K_{cp}, used in Equation (15-10) is given in Section 15.8.2.

15.7.3. Spacecraft, Radiant-Cooling Programs and Characteristics. Since about 1965, radiation coolers designed to maintain an IR device aboard a spacecraft at temperatures from 80 to 200 K have been developed. A listing of the various instruments, programs, orbital data and basic radiator characteristics is presented in Table 15-15.

Typical low-temperature coolers have involved a rectangular or conical, second-stage, high-reflective surface. This surface shields the cold patch from the spacecraft and earth, and reflects shallow-angle, solar input away from the cooler before it reaches the patch. (See Figures 15-32 and 15-33.) A schematic of the VISSR cooler, which was designed for synchronous, geostationary orbit and is representative of a conical design, is presented in Figure 15-34. A two-stage International Telephone and Telegraph (ITT) cooler representing a rectangular design for low earth-orbits is illustrated in Figure 15-35.

Large radiators (Figure 15-36) which provide a cooling capacity of 1 to 5 W, represent a major impact on spacecraft design. For example, deployable sun and earth shields are used to minimize parasitic-heat loading.

The net cooling capacity versus temperature for a number of selected designs illustrated in Figure 15-37 indicate the range of typical performance attainable in practice.

Table 15-15. Characteristics Of Selected Radiant Coolers

Instrument of Experiment	Sponsor/ Mfg.	Flight or Program	Launch Date	Orbit	Altitude (nmi)	Temp (K)	Net Cooling Capacity, Q(mW)	Cold Patch Area, A (cm²)	No. of Stages	Field of View (Deg.)	Overall Dimensions (cm)	Weight (lb)	Q/A (W ft⁻²)	Q/A (mW cm⁻²)
High Resolution Infrared Radiometer (HRIR)	NASA/ ITT	NIMBUS I, II, III	1964-1969	Near polar, sun synchronous	~590	195	1.7	9.8	1	± 30	8 × 6 × 10	1.0	0.16	0.17
Filter Wedge Spectrometer (FWS)	NASA/ ITT	NIMBUS IV	1970	Near polar, sun synchronous	~590	175	5.8	8.9	1	± 48 horiz.	21 × 12 × 13	2.0	0.605	0.65
Very High Resolution Radiometer (VHRR)	NASA/ RCA	ITOS	1972	Circular, polar	–	97	1.4	85.0	3	–	20 × 18 × 15	3.3	0.0153	0.01
High Resolution Radiometer (WHR)	USAF/ SAMSO/ ADL	DMSP	1972-1974	Near polar, sun synchronous	450	100	10.0	95.0	2	101	30 × 16 × 12	3.5	0.098	0.10
Surface Composition Mapping Radiometer (SCMR)	NASA/ ITT	NIMBUS V	1972	Near polar, sun synchronous	500	110	4.7	23.0	2	104 horiz. 76 vert.	17 × 25 × 13	9.0	0.189	0.20
Visible IR Spin-Scan Radiometer (VISSR)	NASA/ SBRC	SMS-1	1974	Synchronous, geostationary	–	80	2.0	45.0	2	130 circ.	Dia. = 45 L = 22	8.0	0.041	0.04
High Resolution Infrared Radiation Sounder (HIRS)	NASA/ ITT	NIMBUS F(VI)	1975	Near polar, sun synchronous	~620	120	8.0	23.0	2	+ 52 horiz. + 50 vert. - 28	18 × 13 × 25	9.0	0.32	0.35
Multiple Spectral Scanner (MSS)	NASA/ SBRC	ERTS-1	1972	Near polar, sun synchronous	~495	95	1.0	64.0	2	100 horiz. 72 vert.	56 × 56 × 26	16.0	0.0145	0.0156
RM 20A	USAF/ SAMSO (LMSC)	STP-72-2	–	Near polar (i = 98°)	400	100*	1.0W	12 ft²	1	–	Dia. = 46 in. (Radiator only)	–	0.0825	0.088
RM 20B	USAF/ SAMSO (NAR)	STP-72-2	–	Near polar (i = 98°)	400	135*	5.0W	14 ft²	1	–	65 × 31 in. (Radiator only)	35.0**	0.358	0.38

*Radiator temperature

**Does not include detector and transport heat pipes

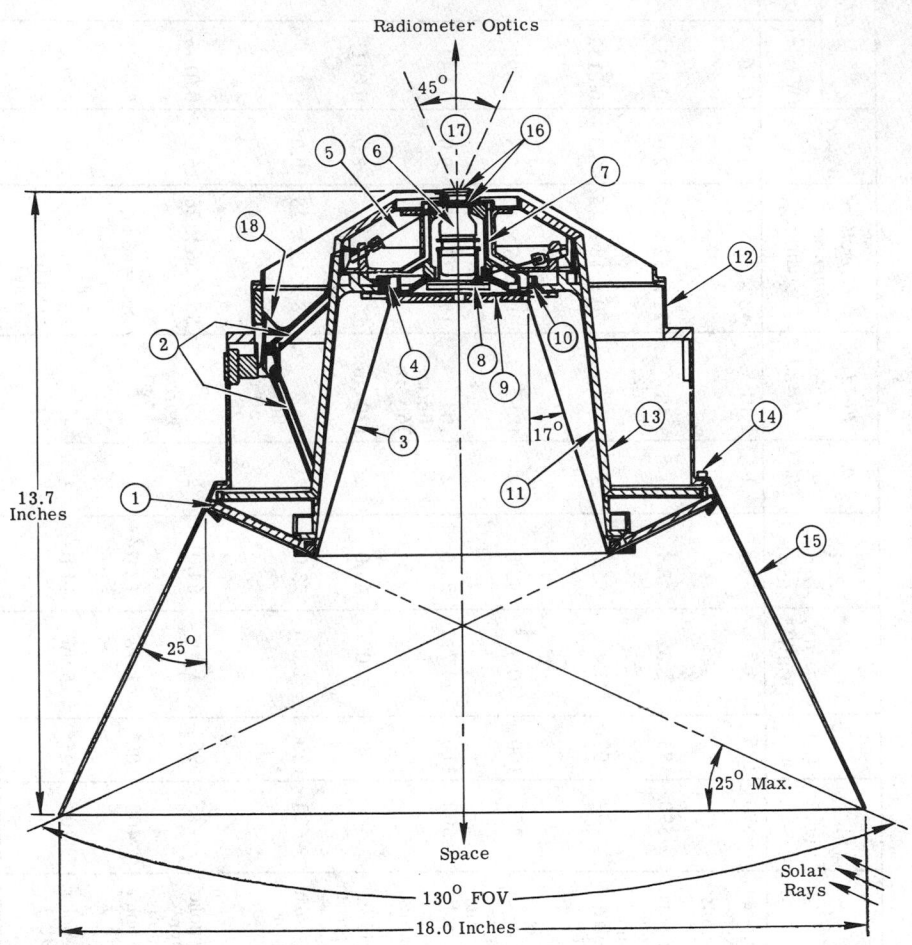

1. Intermediate Stage Radiator
2. Intermediate Stage Support Bands
3. Intermediate Stage Shield
4. Cold Stage Heater
5. Cold Stage Support Bands
6. Detector/Dewar Assembly
7. Cold Stage Assembly
8. Cold Stage Damper
9. Cold Stage Radiator

10. Intermediate Stage Heater
11. Intermediate Stage Assembly
12. Ambient Housing
13. Superinsulation
14. Shield Dampers
15. Sun Shield
16. Warm and Cold Filters
17. Detector FOV
18. Cable Assembly

Figure 15-34. Schematic of SBRC, VISSR, radiant cooler [15-44].

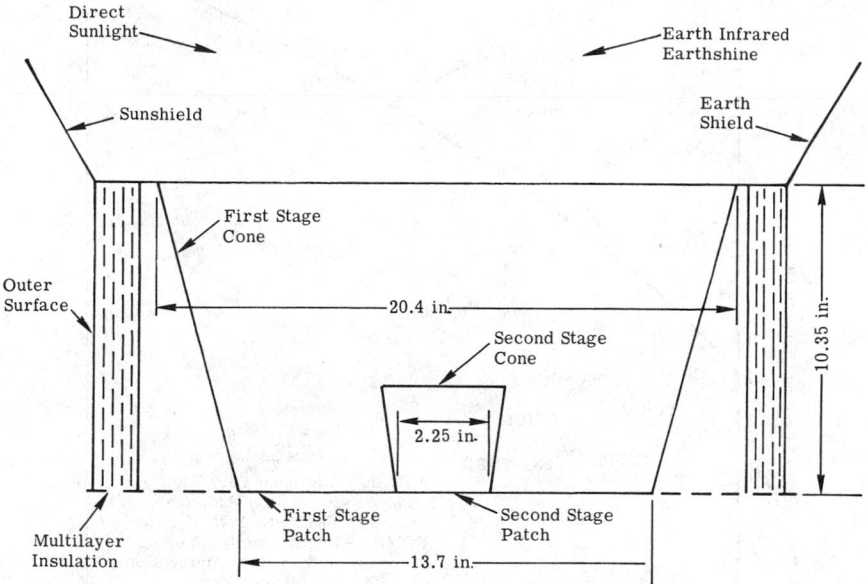

Fig. 15-35. A generalized, two-stage, rectangular, 77 K, radiant cooler for low-earth altitude [15-45].

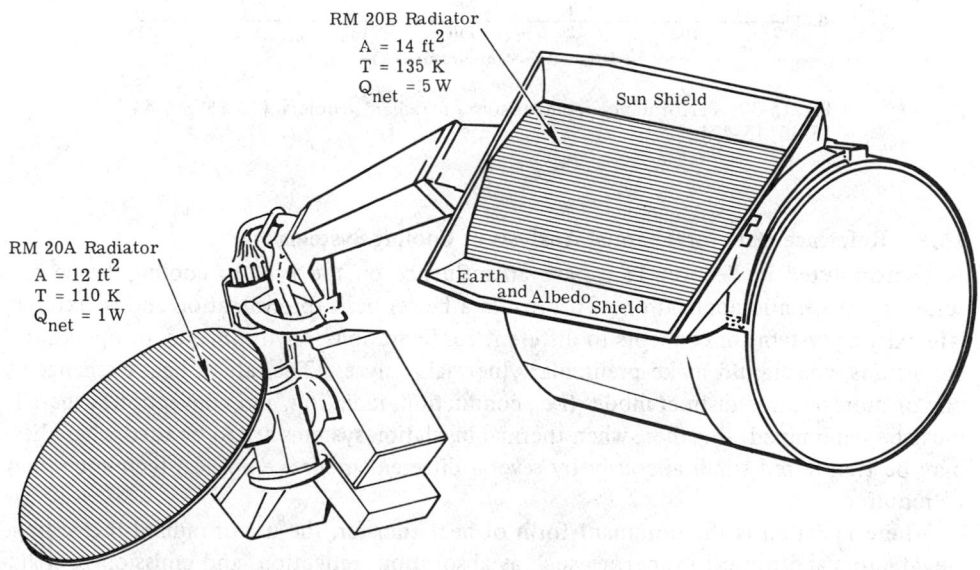

Fig. 15-36. RM 20A and RM 20B radiator concept [15-3].

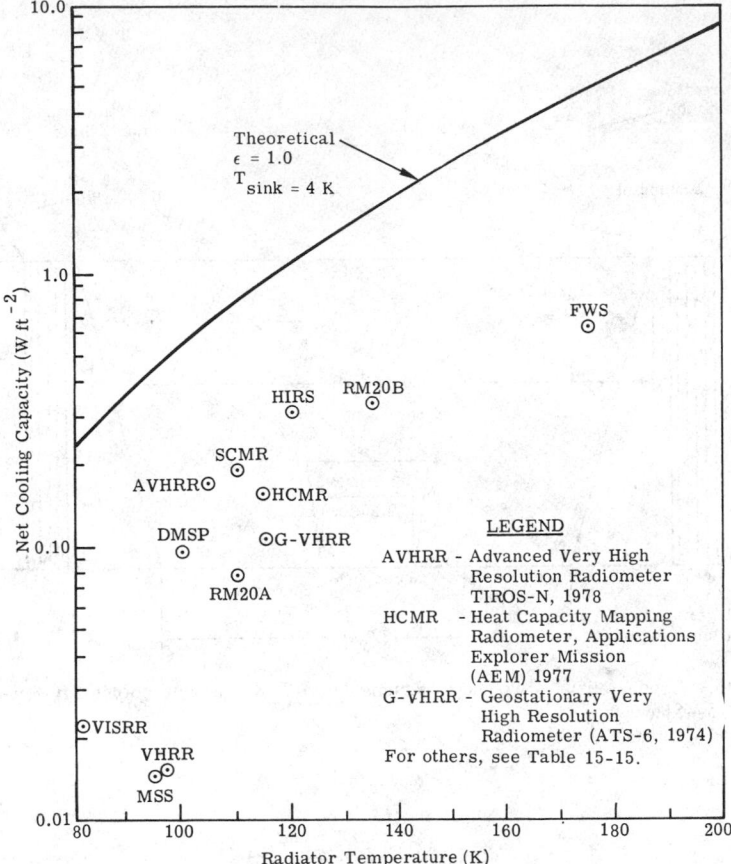

Fig. 15-37. Performance comparison of radiant coolers [15-15, 15-44, 15-46, 15-47].

15.8. Reference Data for Thermal Analysis of Cooling Systems

Factors listed in Section 15.2 have an influence on the type of cooling device best suited for a specific application. To provide a better basis for evaluation and to extrapolate existing systems or concepts to different configurations, environments, or operational conditions, one should make preliminary thermal analyses. The rate of heat exchange by one or more of three distinct modes (i.e., conduction, radiation, and convection) generally must be determined. Further, when thermal insulation systems are to be used where heat may be transferred simultaneously by several different modes, empirical techniques may be required.

Where radiation is the dominant form of heat transfer, the use of radiant-interchange view-factors and optical properties such as absorption, reflection, and emission of materials and coatings are required for proper evaluation. For a convenient reference for this type of analysis, data are provided in the following paragraphs on the space thermal-environment, heat-transfer theory including performance of thermal-insulation systems, and optical properties of useful thermal-control materials and coatings. The basic texts and references cited at the end of this chapter provide additional background.

15.8.1. Space Thermal-Environment. A spacecraft orbiting the earth is irradiated from three external sources: solar radiation; solar radiation reflected off the earth (albedo); and radiation emitted from the earth's surface.

Solar Radiation. Variations of solar irradiance at a near earth-orbit with changes in earth-sun distance during the year are presented in Table 15-16. In orbit, a satellite in the shadow of a planet is occulted from the sun. This shadowing and the time in the sunlight vary according to β, the angle between the satellite's orbital plane and the sun-planet plane (the ecliptic). (See Figure 15-38.) Because of the precession of the orbit and the rotation of the earth about the sun, the angle β varies continuously (except for a sun-synchronous orbit). The maximum and minimum average solar-heat-load over the orbit are determined when β is a maximum and when $\beta = 0$ degrees, respectively.

Table 15-16. Variation in Solar Irradiance With Earth-Sun
Distance [15-48]

Date	W ft^{-2}	Btu h^{-1} ft^{-2}	mW cm^{-2}
January 1	130.01	443.6	139.9
February 1	129.5	441.9	139.3
March 1	128.1	437.0	137.8
April 1	125.9	429.4	135.5
May 1	123.8	422.4	133.2
June 1	122.3	417.2	131.6
July 1	121.6	415.0	130.9
August 1	122.0	416.4	131.3
September 1	123.4	421.1	132.9
October 1	125.4	427.9	135.0
November 1	127.7	435.6	137.4
December 1	129.3	441.2	139.2

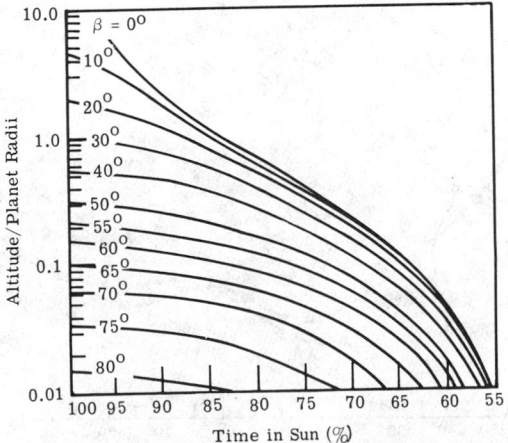

Fig. 15-38. Percentage of time in the sun versus altitude with orbit inclination β as parameter for vehicle in circular orbit.

Consequently, the orbital average, solar-heat flux-density incident at a point on the surface of a spatially-oriented satellite is:

$$E_s = S_O(\cos \theta)(\% \text{ time in sun}) \qquad (15\text{-}12)$$

where θ = the angle between the surface-normal and the sun-line
S_O = the solar irradiance, W cm^{-2}

However, when albedo and earth emission are considered, the instantaneous-peak heat-load will occur at minimum β angle (for a low circular orbit) at a local noon.

Earth Albedo. The incident irradiance resulting from albedo for a surface in orbit about the earth is

$$E_a = S_O a F_a \qquad (15\text{-}13)$$

where E_a = incident albedo-irradiance, Btu h^{-1} ft^{-2}, or W cm^{-2}
S_O = solar irradiance, Btu h^{-1} ft^{-2}, or W cm^{-2}
a = albedo (the fraction of solar radiation reflected from earth)
F_a = geometrical view-factor between the surface and the earth

Although orbital average values can be used in most instances, the value of albedo varies significantly on the surface of the earth, depending on the type of terrain, season, latitude, and cloud cover. (See Chapter 3.) A representation of the variation of albedo with latitude is presented in Figure 15-39. Where long, thermal time-constants exist for the surface in question, a mean-orbital seasonal-average value of 0.30 ± 0.02 is recommended from measurements reported in Reference [15-49].

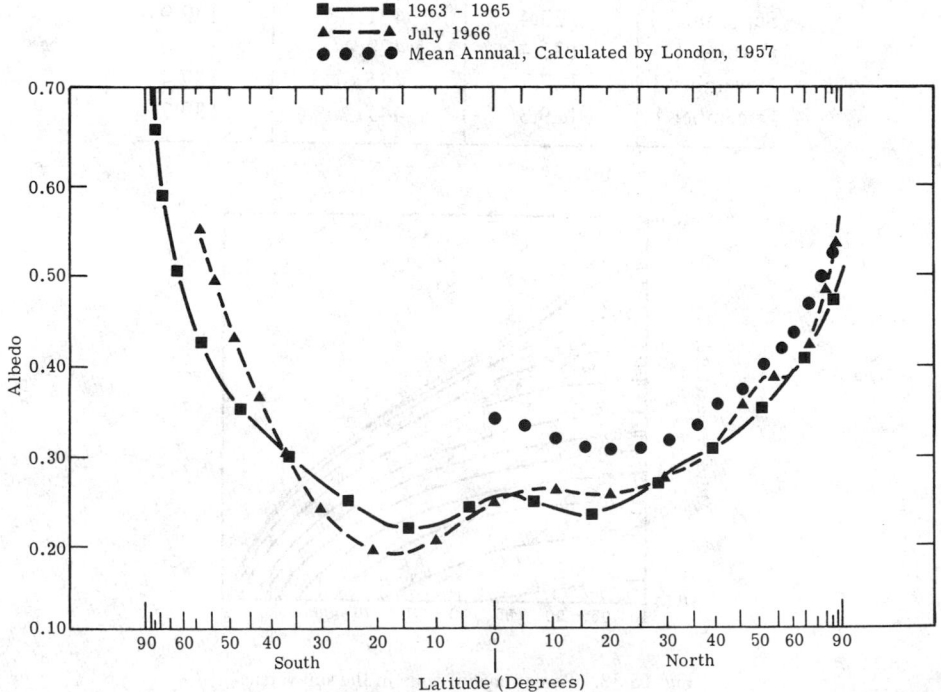

Fig. 15-39. Variation in earth albedo versus latitude [15-49].

The geometrical view-factor, F_a, varies with surface geometry, altitude, orientation to the earth's surface, and the relative position of the sun. Tabulated values of F_a for flat plates, cylinders, and spheres for most geometries can be found in Ballinger and Christensen and Reference [15-50]. Values of F_a for a flat plate are presented in Figure 15-40.

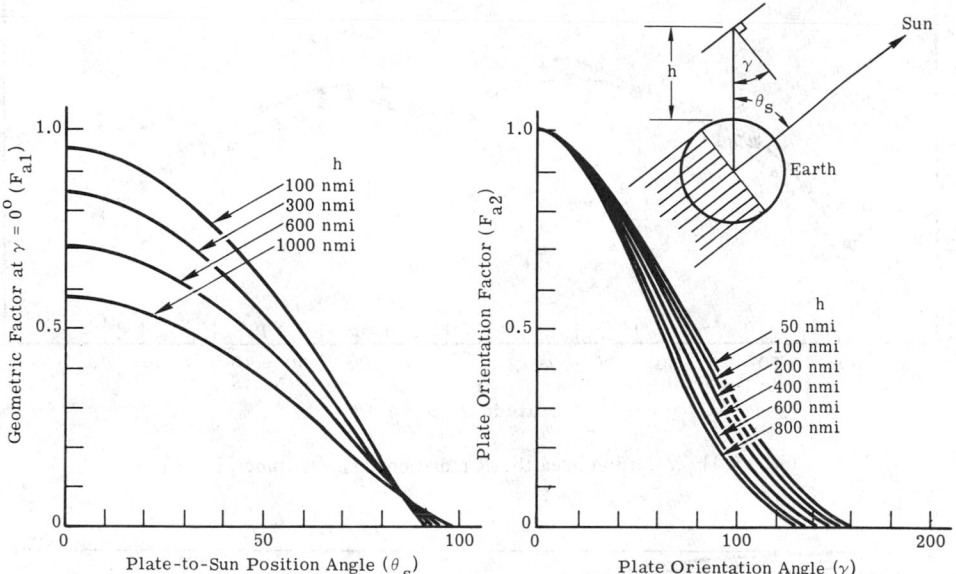

(a) Geometric factor (F_{a1}) from earth to a plate parallel to the earth's surface

(b) Plate orientation correction factor (F_{a2}) versus altitude (γ is the true angle between plate normal and earth plate line, θ_s is true angle between earth plate line and earth sun line, h is the plate distance)

Fig. 15-40. Geometric factors for earth albedo to a flat plate ($F_a = F_{a1} \times F_{a2}$).

Earth IR Emission. The earth infrared emission incident onto a surface is

$$E_{ER} = M_{ER} F_{ER} \qquad (15\text{-}14)$$

where E_{ER} = earth infrared-irradiance (heat flux incident), Btu h^{-1} ft^{-2}, or W cm^{-2}
M_{ER} = earth emission, Btu h^{-1} ft^{-2}, or W cm^{-2}
F_{ER} = geometric view-factor

The earth emission-rate also varies with season and with latitude, as shown in Figure 15-41; however, in most instances the earth can be considered to be a uniform, diffuse emitter with a mean-annual average of 75.5 Btu h^{-1} ft^{-2} or 23.8 mW cm^{-2} [15-49]. View-factors, F_{ER}, for a flat plate as a function of altitude and angle γ (angle between the normal to the plate and the local vertical) are presented in Figure 15-42.

Effective Sink-Temperature. The concept of effective sink temperature, T_S, conveniently describes the overall space thermal-radiation environment for the purpose of thermal analysis. This is the temperature which the entire environment, as seen by a surface, would have to assume to exchange the same amount of energy with the vehicle as the individual sources and sinks. It can be calculated from the environmental heat-fluxes (previously defined) as follows:

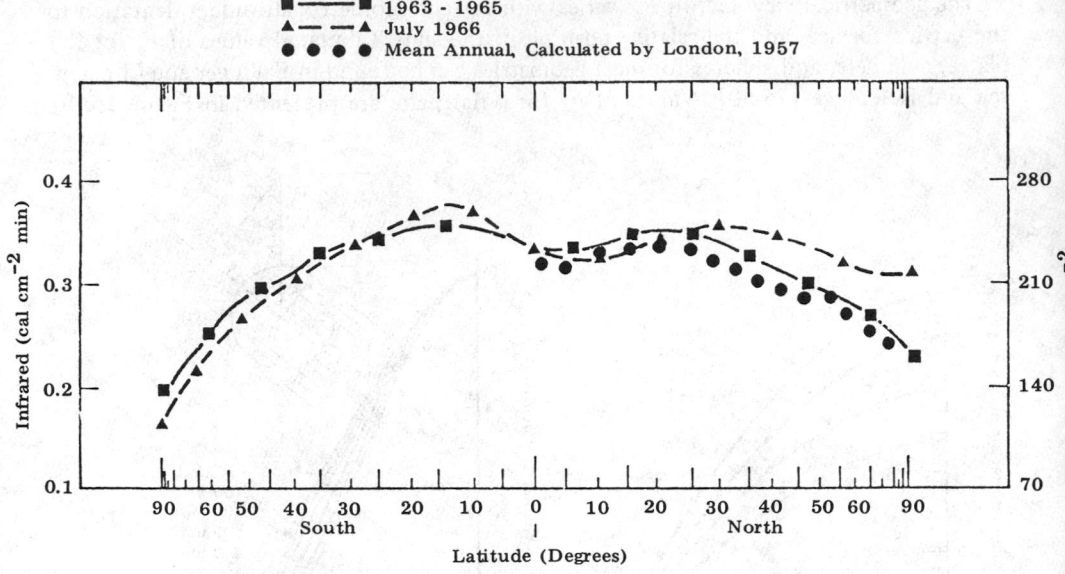

Fig. 15-41. Variation in earth, IR radiation versus latitude [15-49].

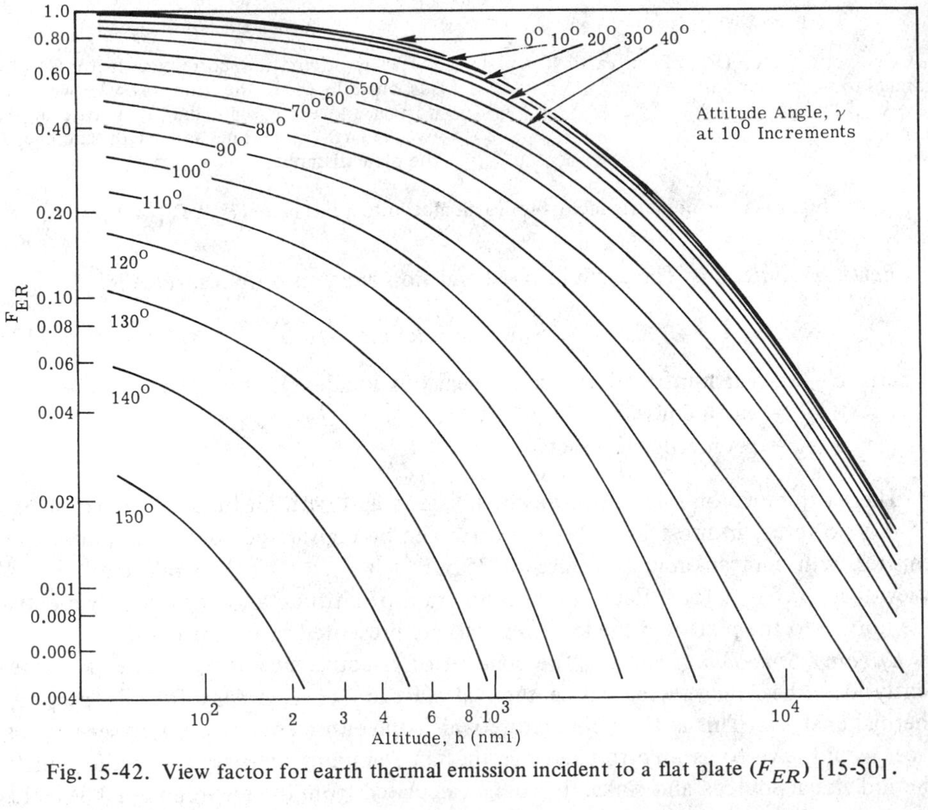

Fig. 15-42. View factor for earth thermal emission incident to a flat plate (F_{ER}) [15-50].

$$T_S = \left[\frac{1}{\sigma \epsilon} \left(\alpha_s E_s + \alpha_s E_a + \epsilon E_{ER} \right) \right]^{1/4} \qquad (15\text{-}15)$$

where
σ = Stefan-Boltzmann constant
α_s = solar absorptance
ϵ = total infrared emissivity
E_s = incident irradiance due to direct solar, Btu h^{-1} ft^{-2} or W cm^{-2}
E_a = incident irradiance due to earth reflected solar (albedo) Btu h^{-1} ft^{-2}, or W cm^{-2}
E_{ER} = incident irradiance due to earth infrared emission, Btu h^{-1} ft^{-2} or W cm^{-2}
T_S = effective sink-temperature

15.8.2. Heat-Transfer Theory

Conduction. The basic equation for conduction is

$$\dot{q} = \frac{kA}{l} (T_1 - T_2) \qquad (15\text{-}16)$$

where
$\dot{q}$ = heat flow through the material, W
k = thermal conductivity, W cm^{-1} K^{-1}
A = area, cm^2
T_1, T_2 = boundary temperatures, K
l = thickness of the material, cm

In the case of conduction across joints or interfaces, the value of k/l in the above equation is replaced by a joint conductance (discussed below). In the case of heat transfer through thermal insulations, the value of k in Equation (15-16) is replaced by an apparent or effective conductivity, k_{eff}, across the inner and outer boundary of the insulation.

Thermal conductivities of selected metals and solids at low temperatures are presented in Figures 15-43 and 15-44. Extensive thermal conductivity data can be found in Touloukian, et al., and NBS Circular 556.

At the interface between two materials held together in unbonded contact, there is a resistance to heat flow due to imperfect contact. Heat transfer through an interface takes place by the combined mechanisms of conduction across entrapped interstitial fluid and radiation across interstitial gaps. The resulting overall conductance of the joint is therefore a function of the materials in contact (i.e., their conductivity, surface finish, flatness, and hardness), the contact pressure, the mean temperature, the nature of the interstitial fluid (liquid, gas, vacuum), and the presence of oxide films or interface shim materials. The interface contact conductance has been measured for various conditions and varies widely. Figure 15-45 shows examples of measured conductance data as a function of contact pressure, p, for the interface conditions identified in Table 15-17. Other data can be found in Fontenot and Cunnington. Theoretical expressions for thermal contact resistance in a vacuum have been derived in Clausing and Chao.

Interface contact conductance can also be estimated analytically (for nonvacuum conditions) from the properties of the joining surfaces and the interstitial fluid. These procedures are given in Reference [15-51].

Radiation. The transfer of heat between two surfaces by the process of radiation (Chapter 1) is given by

$$\dot{q} = \sigma A F_e F_a (T_1^4 - T_2^4) \qquad (15\text{-}17)$$

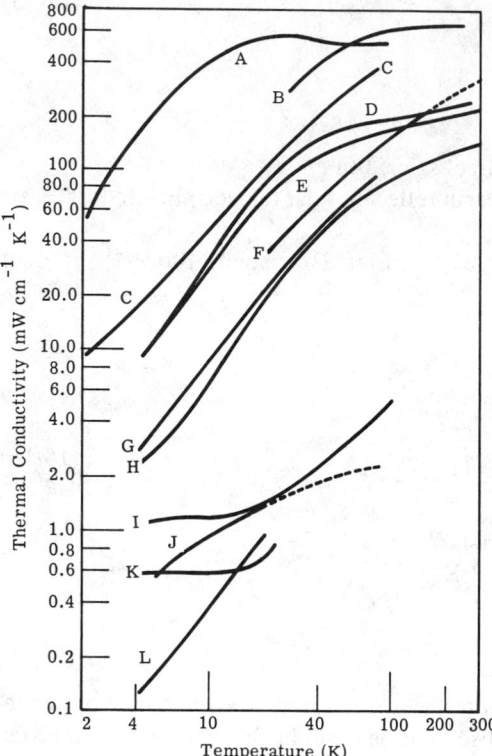

Fig. 15-43. Low-temperature, thermal conductivities of some solids with relatively low conductivities. A: 50-50 lead-tin solder; B: steel SAE 1020; C: berylium copper; D: constantan; E: monel; F: silicon bronze; G: inconel; H: type 347 stainless steel; I: fused quartz; J: polytetrafluorochylene (Teflon); K: polymethacrylate; L: nylon [15-46].

Fig. 15-44. Low-temperature, thermal conductivities of some metals with relatively high conductivities. A: silver, 99.999% pure; B: high purity copper; C: coalesced copper; D: copper; E: aluminum, single crystal; F: free-machining tellurium copper; G: aluminum, 1100; H: aluminum, 6063-T5; I: copper, phosphodeoxidized; J: aluminum, 2024-T4; K: free-machining leaded brass [15-46].

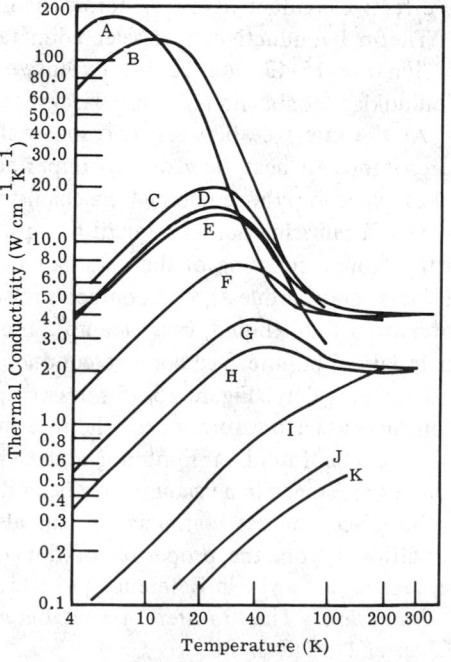

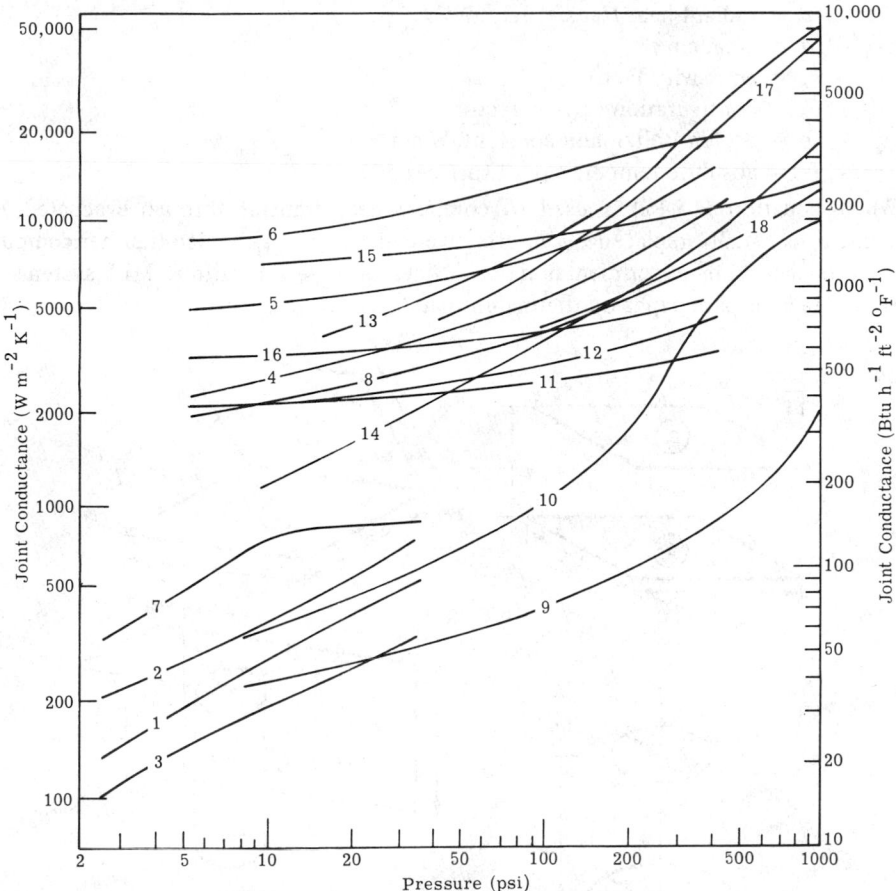

Fig. 15-45. Thermal-interface, conductance data [15-52].

Table 15-17. Interface Conditions For Conductance Data in Figure 15-45 [15-52]

Curve	Material Pair	rms Surface Finish (μ in.)	Gap Material	Mean Contact Temp. (°F)
1	Aluminum (2024-T3)	48-65	Vacuum (10^{-4} mm Hg)	110
2	Aluminum (2024-T3)	8-18	Vacuum (10^{-4} mm Hg)	110
3	Aluminum (2024-T3)	6-8 (not flat)	Vacuum (10^{-4} mm Hg)	110
4	Aluminum (75S-T6)	120	Air	200
5	Aluminum (75S-T6)	65	Air	200
6	Aluminum (75S-T6)	10	Air	200
7	Aluminum (2024-T3)	6-8 (not flat)	Lead foil (0.008 in.)	110
8	Aluminum (75S-T6)	120	Brass foil (0.001 in.)	200
9	Stainless (304)	42-60	Vacuum (10^{-4} mm Hg)	85
10	Stainless (304)	10-15	Vacuum (10^{-4} mm Hg)	85
11	Stainless (416)	100	Air	200
12	Stainless (416)	100	Brass foil (0.001 in.)	200
13	Magnesium (AZ-31B)	50-60 (oxidized)	Vacuum (10^{-4} mm Hg)	85
14	Magnesium (AZ-31B)	8-16 (oxidized)	Vacuum (10^{-4} mm Hg)	85
15	Copper (OFHC)	7-9	Vacuum (10^{-4} mm Hg)	115
16	Stainless/aluminum	30/65	Air	200
17	Iron/aluminum	–	Air	80
18	Tungsten/graphite	–	Air	270

where $\dot{q}$ = radiant heat transfer rate, W
           A = area, cm^2
           F_e = emissivity factor
           F_a = configuration or view-factor
           σ = Stefan-Boltzmann constant, W cm^{-2} K^{-4}
   T_1, T_2 = absolute temperature of surfaces of 1 and 2, K

When Equation (15-17) is used to compute heat transfer through evacuated MLI systems, F_e is usually replaced by an effective emissivity, ϵ_{eff}, which is either computed or determined from experimental data. Values of ϵ_{eff} for various MLI systems are presented in later paragraphs on thermal-insulation systems.

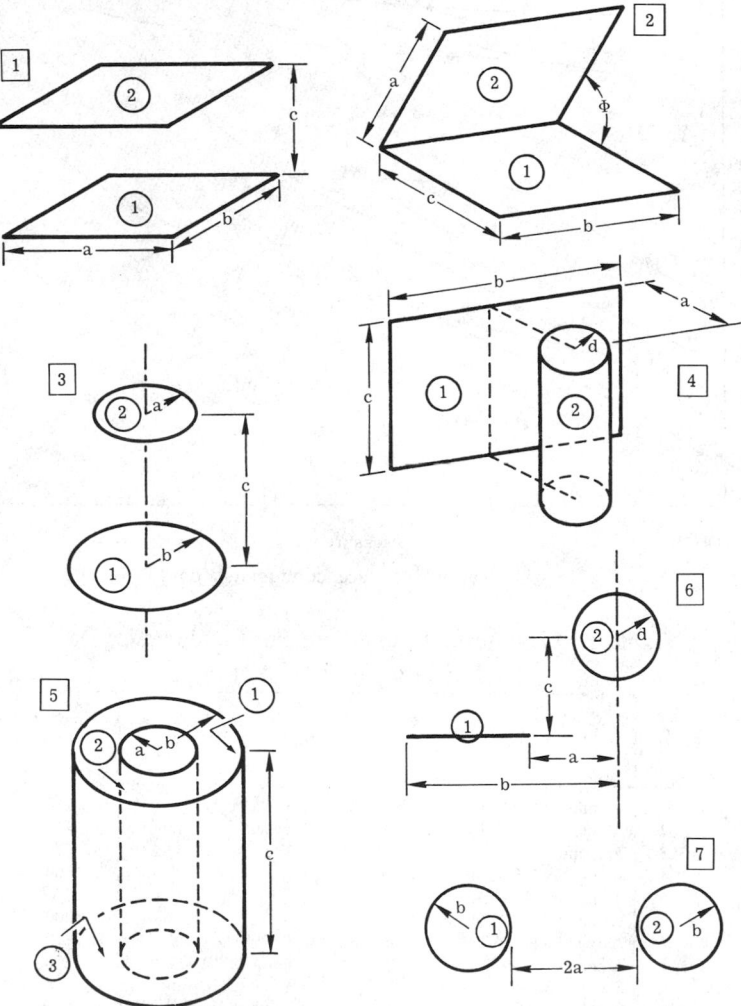

Fig. 15-46. Schematic representations of configurations 1-7. Circled numbers correspond to subscripts for equations given in Table 15-18 [15-55].

Various catalogs of radiation-interchange, configuration, and emissivity-factors have been prepared by numerous authors. Schematic representations of 15 configurations are presented in Figures 15-46 and 15-47. The configuration factor, F_a, and emissivity factor,

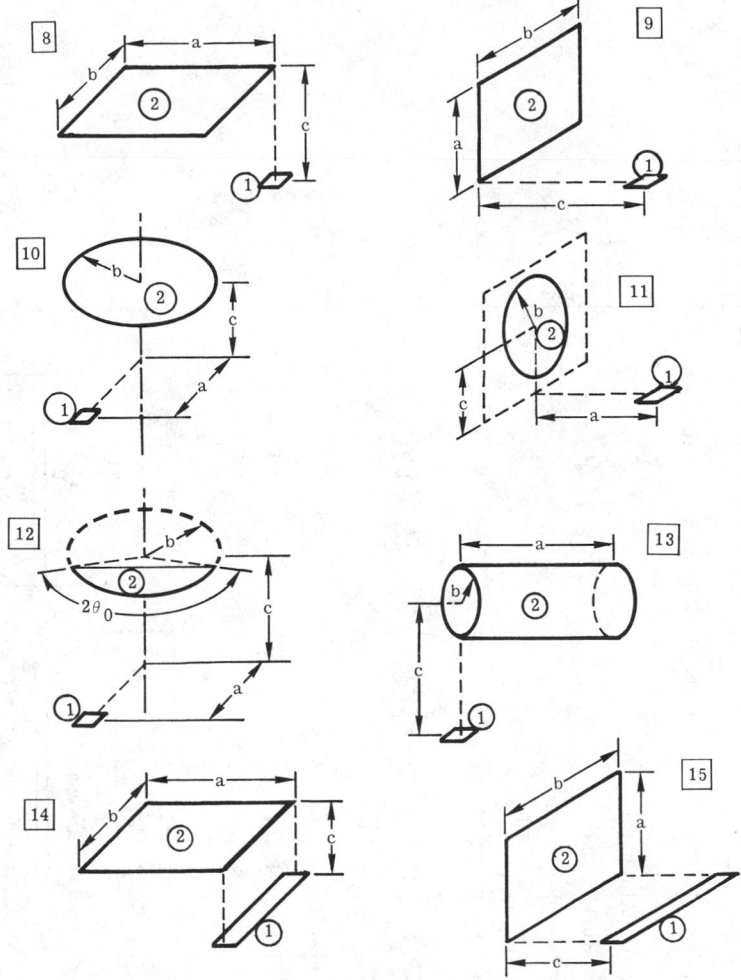

Fig. 15-47. Schematic representations of configurations 8-15 [15-55].

F_e, for these and others are summarized in Table 15-18, or in Figure 15-48 either in the form of an algebraic expression or a graphical representation. Additional configurations can be found in References [15-53] and [15-54]. Values of emissivity for the specific materials and coatings can be found in Section 15.8.4. Where the specific desired configuration has not been previously evaluated, the view-factors must be derived from the basic geometrical considerations. A number of ray-tracing (Monte Carlo) computer programs are available. (See Davis and Drake.) A generalized, thermal-radiation, analyzer system (TRASYS) developed by Martin Marietta Co. is also available. (See Jensen and Gobel.)

Convection. The transfer of heat from a fluid flowing over a surface of a hotter or colder body is by the process known as convection. When circulation is produced because of differences in density resulting from temperature changes, the convection is termed "free." When the circulation is made positive by some mechanical means such as a pump or fan, or a relative motion of the object or fluid, the convection is termed "forced."

Table 15-18. Radiation Interchange Configuration and Emissivity Factors [15-55, 15-56]

No.	Configuration Description (See Figures 15-46 & 15-47.)	Configuration Factor (F_a)	Emissivity Factor (F_e)
1	Parallel rectangles	See Figure 15-48 (a)	$\epsilon_1\epsilon_2$
2	Rectangles with a common side	See Figure 15-48 (c) for $\phi = 30°$ and $60°$ and Figure 15-48 (d) for $\phi = 90°$ and $120°$	$\epsilon_1\epsilon_2$
3	Parallel discs	See Figure 15-48 (b)	$\epsilon_1\epsilon_2$
4	Rectangle to a cylinder	See Figure 15-48 (e)	$\epsilon_1\epsilon_2$
5	Concentric cylinders (finite length)	See Reference [15-56]	$\epsilon_1\epsilon_2$
6	Cylinder and rectangle offset	$F_{A_1-A_2} = \dfrac{1}{Z-Y}\left(\tan^{-1}\dfrac{Z}{X} - \tan^{-1}\dfrac{Y}{X}\right)$, $X = c/d,\ Y = a/d,\ Z = b/d$	$\epsilon_1\epsilon_2$
7	Parallel cylinders	$F_{A_1-A_2} = \dfrac{2}{\pi}\left(\sqrt{X^2-1} - X + \dfrac{\pi}{2} - \cos^{-1}\dfrac{1}{X}\right)$, $X = 1 + (a/b)$	$\epsilon_1\epsilon_2$
8	Element dA_1 parallel to area A_2	See Figure 15-48 (f)	$\epsilon_1\epsilon_2$
9	Element dA_1 normal to area A_2	See Figure 15-48 (g)	$\epsilon_1\epsilon_2$
10	Element dA_1 parallel to disc	See Figure 15-48 (h)	$\epsilon_1\epsilon_2$
11	Element dA_1 normal to disc	See Figure 15-48 (i)	$\epsilon_1\epsilon_2$
12	Element dA_1 parallel to segment of a disc	$F_{dA_1-A_2} = \dfrac{\theta_0}{2\pi} + \dfrac{1 - X^2 - Y^2}{\pi Z}\tan^{-1}\left[\dfrac{Z\tan(\theta_0/2)}{\sqrt{1+X^2+Y^2-2X}}\right]$ $+ \dfrac{X-\cos\theta_0}{\pi\sqrt{(X-\cos\theta_0)^2+Y^2}}\tan^{-1}\left[\dfrac{\sin\theta_0}{\sqrt{(X-\cos\theta_0)^2+Y^2}}\right]$, $X = a/b,\ Y = c/b,\ Z = [(1 + X^2 + Y^2)^2 - 4X^2]^{1/2}$	$\epsilon_1\epsilon_2$

Table 15-18. Radiation Interchange Configuration and Emissivity Factors [15-55, 15-56] (Continued)

No.	Configuration Description (See Figures 15-46 & 15-47.)	Configuration Factor (F_a)	Emissivity Factor (F_e)
13	Element dA_1 parallel to cylinder	$$F_{dA_1-A_2} = \frac{1}{\pi Y}\tan^{-1}\left(\frac{X}{\sqrt{Y^2-1}}\right) + \frac{X}{\pi}\left\{\frac{A-2Y}{Y\sqrt{AB}}\tan^{-1}\left[\sqrt{\frac{A(Y-1)}{B(Y+1)}}\right]\right.$$ $$\left. - \frac{1}{Y}\tan^{-1}\left(\sqrt{\frac{Y-1}{Y+1}}\right)\right\}, \quad X = a/b, \; Y = c/b, \; A = (1+Y)^2$$ $$+ X^2, \quad B = (1-Y)^2 - X^2$$	$\epsilon_1\epsilon_2$
14	Element dA_1 normal to area A_2	$$F_{dA_1-A_2} = \frac{1}{\pi X}\left[\sqrt{1+X^2}\,\tan^{-1}\left(\frac{Y}{\sqrt{1+X^2}}\right) - \tan^{-1} Y\right.$$ $$\left. + \frac{XY}{\sqrt{1+Y^2}}\tan^{-1}\left(\frac{X}{\sqrt{1+Y^2}}\right)\right], \quad X = b/c, \; Y = a/c$$	$\epsilon_1\epsilon_2$
15	Element dA_1 adjacent to A_2	$$F_{dA_1-A_2} = \frac{1}{\pi}\left\{\tan^{-1}\frac{1}{Y} + \frac{Y}{2}\ln\left[\frac{Y^2(Z+1)}{(Y^2+1)Z}\right] - \frac{Y}{\sqrt{Z}}\tan^{-1}\left(\frac{1}{\sqrt{Z}}\right)\right\},$$ $$X = a/b, \; Y = c/b, \; Z = X^2 + Y^2$$	$\epsilon_1\epsilon_2$
16	A_1 small and completely enclosed by A_2	1.0	ϵ_1
17	Infinite parallel planes	1.0	$\dfrac{1}{(1/\epsilon_1 + 1/\epsilon_2 - 1)}$
18	Concentric spheres or infinite cylinders	1.0	$1/\left[1/\epsilon_1 + \dfrac{A_1}{A_2}\,(1/\epsilon_2 - 1)\right]$

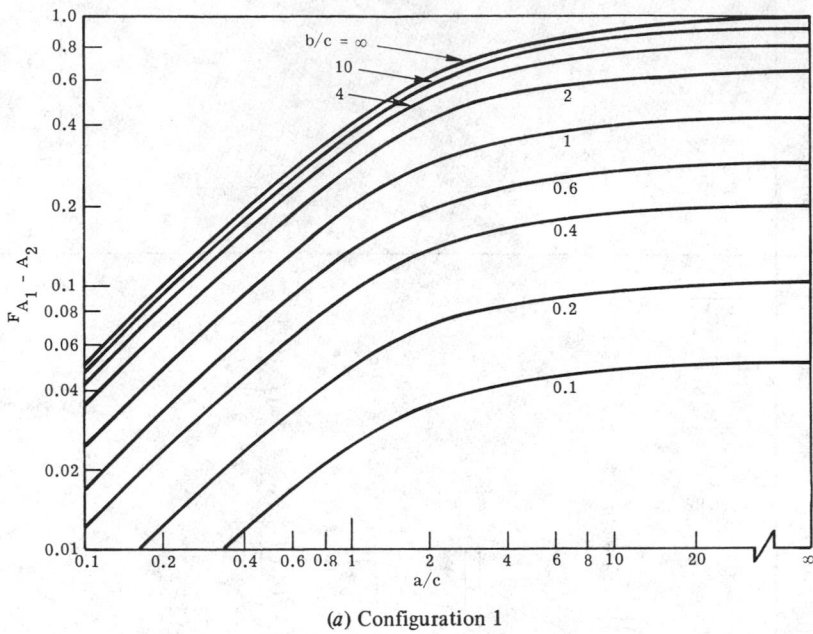

(a) Configuration 1

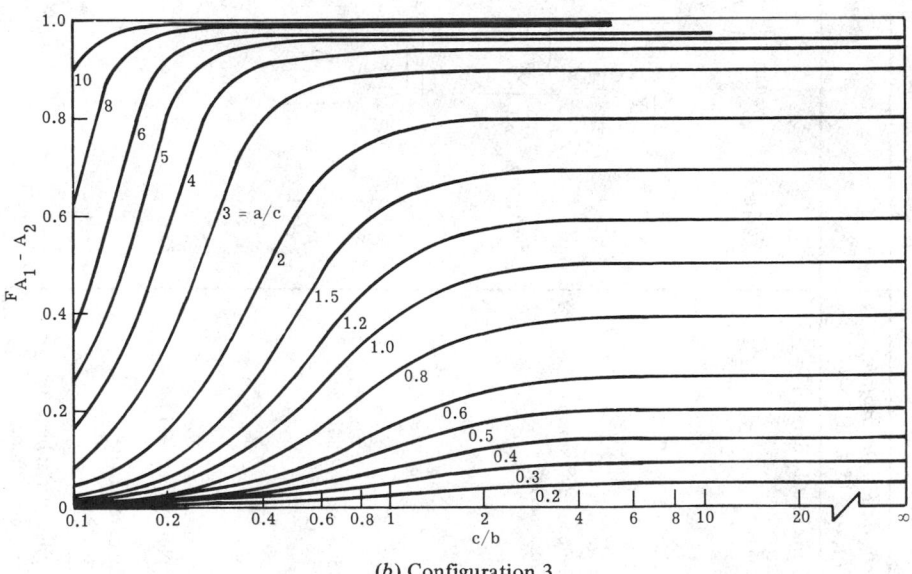

(b) Configuration 3

Fig. 15-48. View factors [15-55].

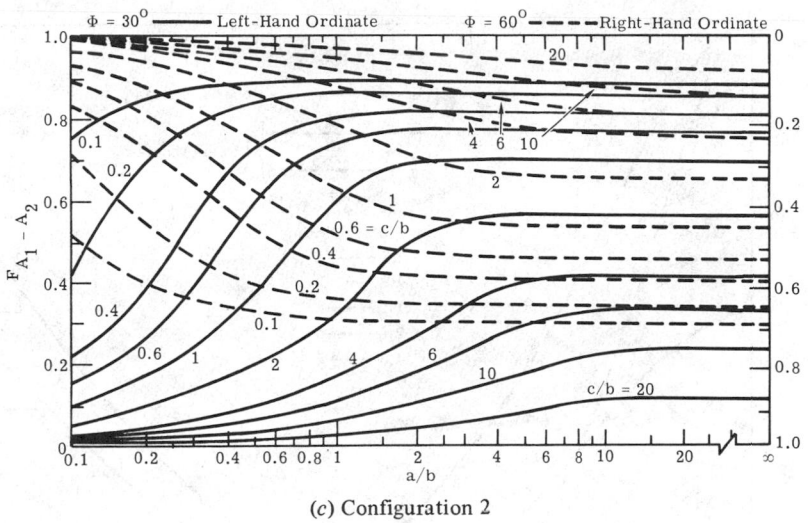

(c) Configuration 2

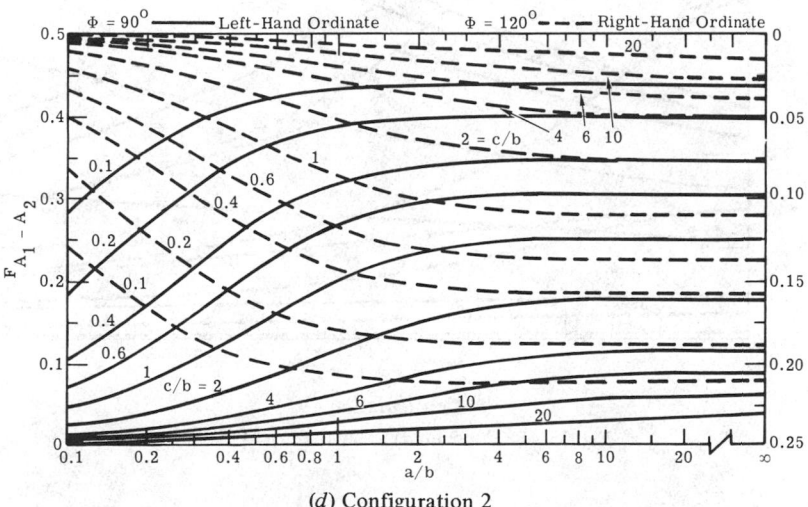

(d) Configuration 2

Fig. 15-48 (Continued). View factors [15-55].

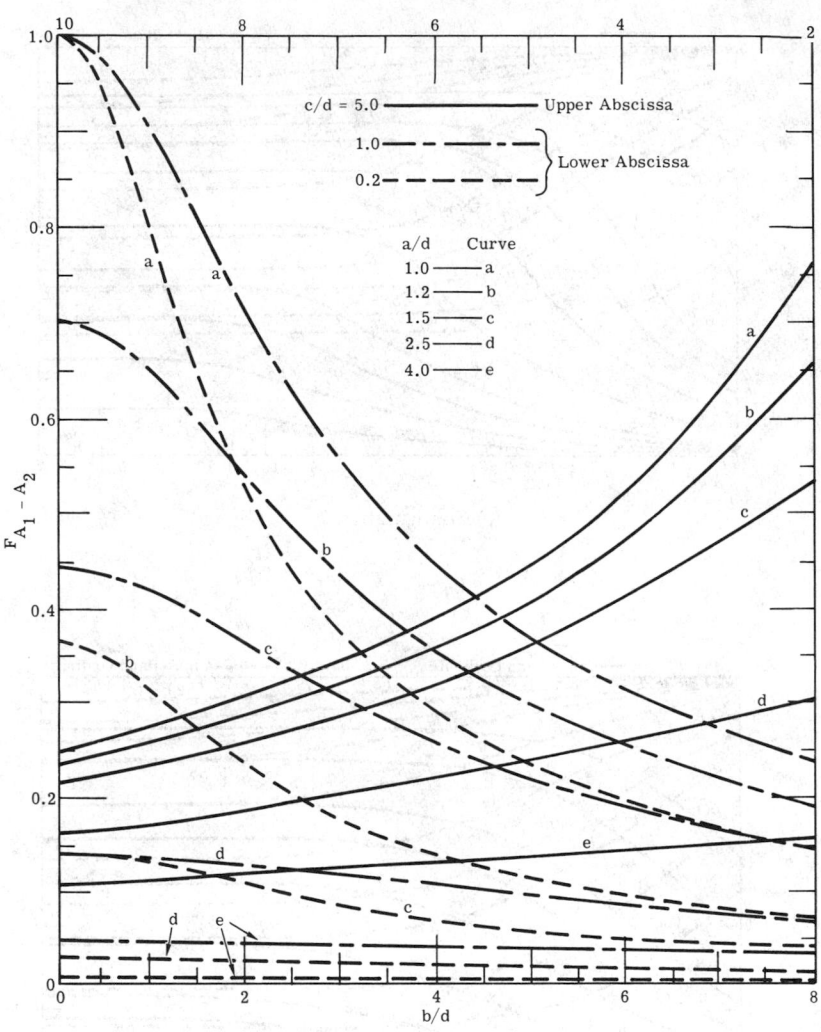

(e) Configuration 4

Fig. 15-48 (Continued). View factors [15-55].

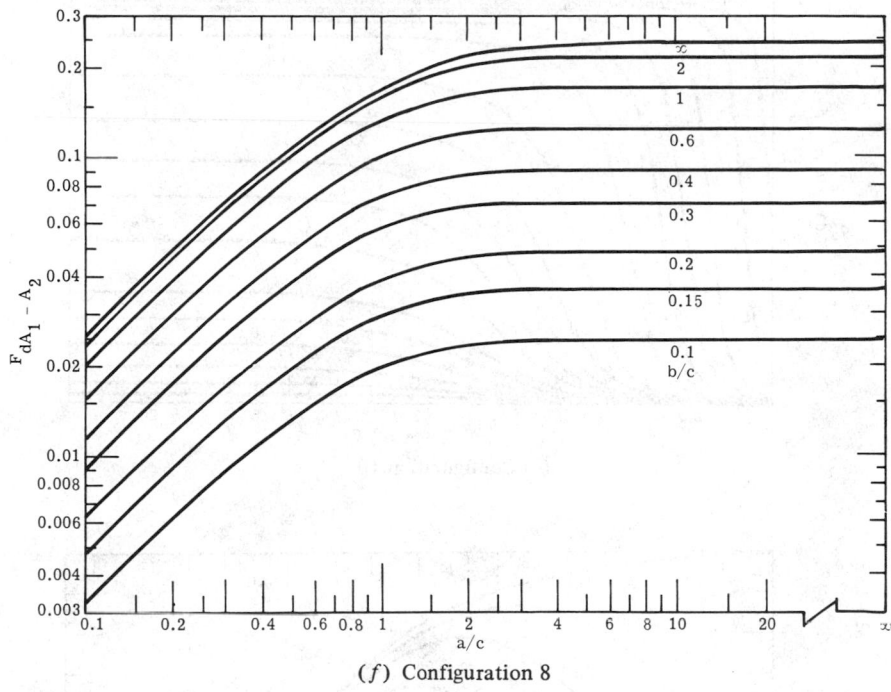

(f) Configuration 8

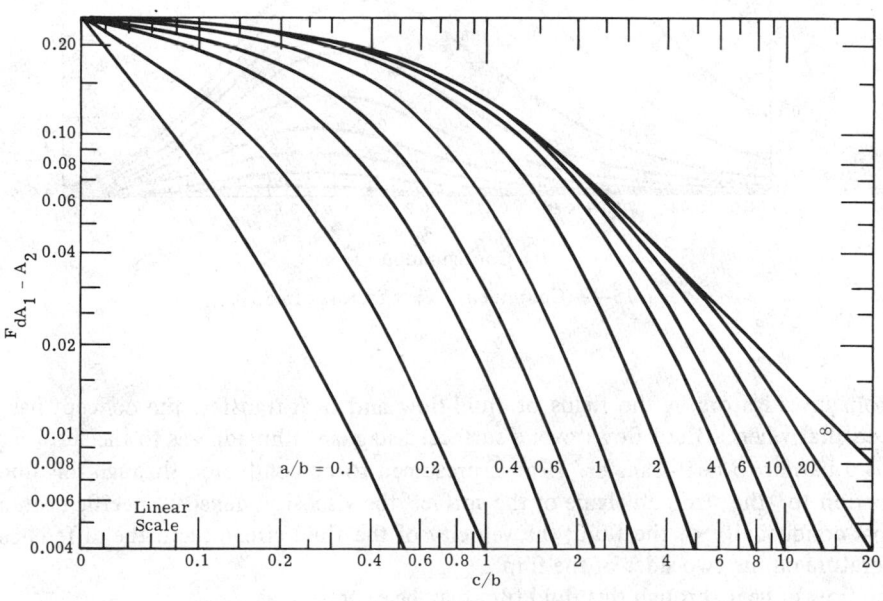

(g) Configuration 9

Fig. 15-48 (Continued). View factors [15-55].

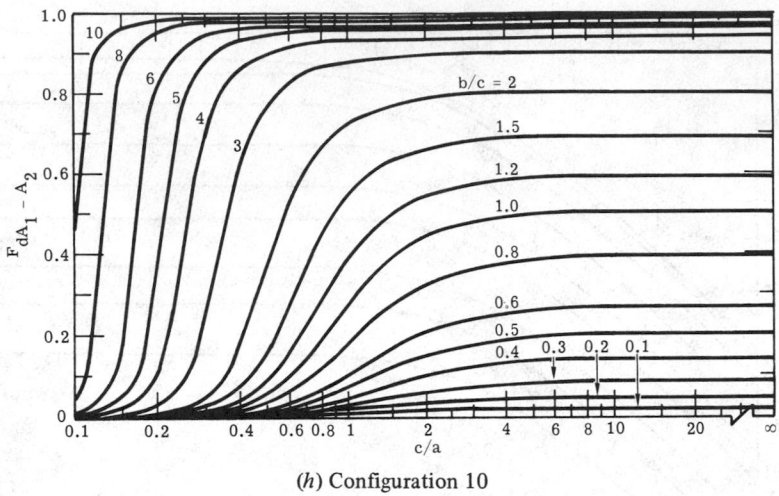

(h) Configuration 10

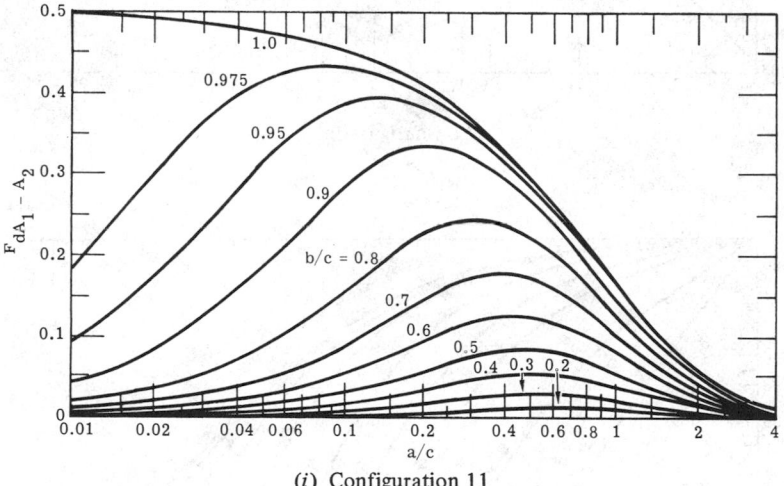

(i) Configuration 11

Fig. 15-48 (Continued). View factors [15-55].

From investigators in the fields of fluid-flow and heat-transfer, the concept has developed that when a fluid flows over a surface, a stagnant film adheres to the surface and
acts as a barrier to heat-transfer. Heat is presumed to be conducted through this film in
proportion to: the size and shape of the surface; the viscosity, density, specific heat, and
thermal conductivity of the fluid; the velocity of the fluid stream; and the difference in
temperature on the two sides of the film.

The flow of heat through this fluid film may be expressed as

$$\dot{q}_c = h_c A(T_f - T_s) \qquad (15\text{-}18)$$

where $\dot{q}_c$ = rate of convective heat-flow, Btu h^{-1} or W
h_c = surface-film, heat-transfer coefficient, Btu h^{-1} ft^{-2} °F^{-1}, or W cm^{-2} K^{-1}

A = area of the surface, ft^2 or cm^2

T_s, T_f = temperature of the surface and film, respectively, °F or K

Equation (15-18) may be applied for the purpose of establishing the rate of convective heat-flow, $\dot{q}_c$, whenever T_s and T_f are known. Where the value of $\dot{q}_c$ is known, calculation of h_c makes it possible to determine the temperature difference, $T_f - T_s$, and further to determine T_f or T_s when one or the other is known.

For surfaces in unbounded convection (such as plates, tubes, and bodies of revolution) which are immersed in a large body of fluid, it is customary to define T_f as the temperature of the fluid far away from the surface, often identified as T_∞. For bounded convection (as for fluids flowing in tubes, in channels, and across tubes in bundles), T_f is usually taken as the enthalpy-mixed-mean temperature, customarily identified as T_m.

For convective processes involving high-velocity gas-flows (as in aircraft applications) a more meaningful expression of Equation (15-18) is to replace T_f with T_r, commonly called the adiabetic wall-temperature or the recovery temperature, which is the equilibrium temperature which the surface would obtain in the absence of radiation exchange between the surroundings and the surface. Generally, the adiabatic wall-temperature is reported in terms of a dimensionless recovery factor, r, as follows:

$$T_r = T_f + r\,\frac{V^2}{2c_p} = T_f\left(1 + \frac{\gamma - 1}{2}\,M_o^2\right) \qquad (15\text{-}19)$$

where V = the free-stream velocity

c_p = the specific heat of the fluid

γ = the ratio of specific heats (i.e., c_p/c_v)

M_o = the free-steam Mach number

The value of r for gases normally lies between 0.80 and 1.0, while γ ranges from 1.1 to 1.7.

Three fundamental approaches may be employed to determine the convective heat-transfer coefficient, h_c: mathematical analyses of fluid flow; application of principles of dimensional analysis together with numerical constants derived from experimental data; and representation of experimental data by purely empirical formulas. For practical purposes, the second approach has been found to be the best in most instances. A generalized equation used in various forms is presented below:

$$\left(\frac{h_c D}{k}\right) = C\left(\frac{DV\rho}{\mu}\right)^b \left(\frac{c_p\mu}{k}\right)^c \qquad (15\text{-}20)$$

Equation (15-20) is known as Nusselt's expression and the three groups of symbols within the brackets are known as follows (all dimensionless)

$$\frac{h_c D}{k} = \text{Nusselt number}$$

$$\frac{DV\rho}{\mu} = \text{Reynolds number}$$

$$\frac{c_p\mu}{k} = \text{Prandtl number}$$

This basic relationship can be applied to a large number of conditions and a range of values for the constant, C. The exponents b and c have been established by many investigators for specific geometries and, in some cases, for certain fluids. For specific applications, equations and data can be found in numerous heat transfer texts listed in the bibliography of this chapter and other handbooks such as Reference [15-55].

15.8.3. Thermal-Insulation Systems

Mechanisms of Heat Transfer. Heat can flow through an insulation system by the simultaneous action of several different mechanisms:

(1) solid-conduction through the materials making up the insulation and conduction between individual components of the insulation across areas of contact

(2) gas-conduction in void spaces contained within the insulation material

(3) radiation across these void spaces and through the components of the insulation.

Because these heat-transfer mechanisms operate simultaneously and interact with each other, the thermal conductivity of an insulation is not strictly definable, analytically, in terms of variables such as temperature, density, or physical properties of the component materials. It is therefore useful to refer to either an apparent thermal-conductivity, k_{eff}, which is derived experimentally during steady-state heat-transfer using the basic conduction-equation, or an effective emissivity, ϵ_{eff}, which can be determined experimentally by substituting ϵ_{eff} for F_e in Equation (15-17), or computed using Equation (15-21).

The low thermal-conductivity of evacuated insulation systems can largely be attributed to removal of gas from the void spaces within the insulation. The degree of vacuum necessary to achieve the desired effectiveness can be established by considering the mechanism by which the heat flows. The gas-conduction can be divided into two regions: the region ranging from atmospheric pressure down to a few torrs (1 torr = 1 millimeter of mercury), in which gas-conduction is independent of pressure, and the region at pressures below a few torrs, in which gas-conduction depends on pressure. The transition from one type of gas-conduction region to the other depends upon the dimensions of the system with respect to the mean-free-path of the gas molecules. The effect of gas pressure on conductivity can be characterized by the curves in Figure 15-49. The effective conductivity begins to decrease sharply between 1 and 10 torr until about 10^{-4} to 10^{-5} torr, where the heat conducted by the gas is only a small portion of the residual heat-transfer. A finite value of effective thermal-conductivity remains at lower pressures due to heat transfer by solid conduction and radiation between the elements of the insulation.

Insulation Types and Performance. Various types of insulation systems in use are listed in Table 15-19 with their characteristic properties. Illustrations of the basic concepts and their advantages and disadvantages are listed in Table 15-20. At present, one of the more popular concepts for cryogenic systems involves MLI using, for example, thin Mylar sheets (0.25 mil) with vacuum-deposited aluminum ($\sim 2.5 \times 10^{-6}$ cm) on one or both sides of Mylar®*. Spacing is maintained either by crinkling or wrinkling each sheet, or by using separation sheets of fiberglass, nylon, dacron, silk, or similar material.

In theory, for highly evacuated MLI systems (i.e., with gas pressures of 10^{-5} torr or less), the theoretical effective emissivity, ϵ_{eff}, for a blanket of N isolated surfaces, or layers of emissivity of ϵ_1 and ϵ_2 on opposite sides, is computed as

$$\epsilon_{eff} = \frac{1}{\dfrac{1}{\epsilon_1} + \dfrac{1}{\epsilon_2} - 1}\left(\frac{1}{N+1}\right) \tag{15-21}$$

---

*Trademark of the Dupont Corporation.

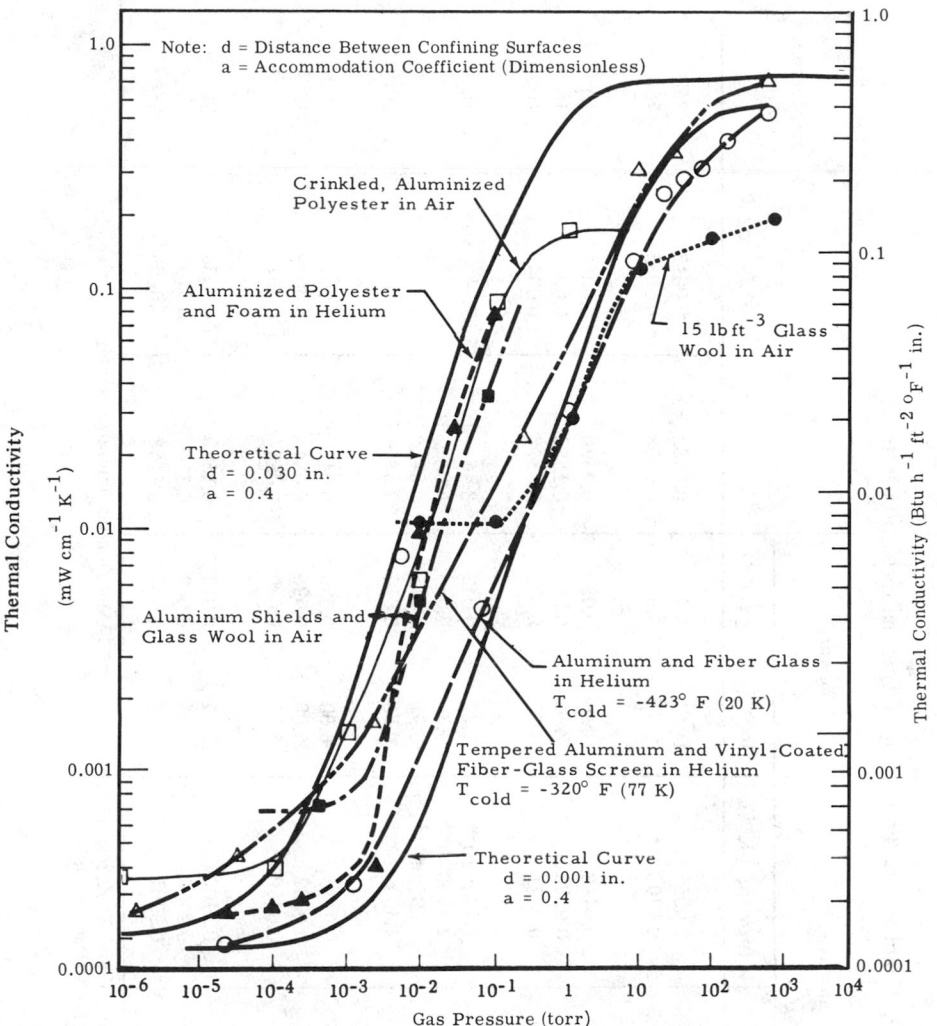

Fig. 15-49. Effect of gas pressure on thermal conductivity [15-57].

This is based on no significant contact between the blankets and no significant heat-leakage through the blanket edges. In real installations, actual ϵ_{eff} can be significantly larger because residual pressure may exist between layers, individual layers may touch each other to varying degrees depending on the compression of the blanket, or the emissivity of the aluminized sheets may increase due to damage, contamination, etc. All these factors may contribute to making the practical effectiveness of the blanket substantially less than theory would indicate. Experimental and theoretical data for Mylar sheets, aluminized on both sides, with various combinations of spacer materials are shown in Figure 15-50. The use of ϵ_{eff} has increased in recent years in place of the traditional effective thermal conductivity, k_{eff}, since ϵ_{eff} is more general and easier to use, and because its theoretical value can be easily computed by Equation (15-21). Data for ϵ_{eff} are normally presented independent of temperature, although a significant change in the hot-boundary temperature will change the true ϵ_{eff} (or k_{eff}). In this case, if the coefficient is

Table 15-19. Characteristic Properties Of Thermal Insulation Materials

Insulation Type	Thermal Conductivity* k_{eff}		Density ρ		$\rho\,k_{eff}$		Gas Pressure (torr)
	(Btu-in. h⁻¹ ft⁻² °F⁻¹)	(mW cm⁻¹ K⁻¹)	(lb ft⁻³)	(g cm⁻³)			
Fibrous							
Fiberglass-ordered	0.0039	0.0056	14.98	0.24	0.0582	0.0013	760
Fiberglass-random	0.118	0.1702	3.12	0.05	0.0368	0.0085	760
Microspheres							
Al coated	0.0311	0.0448	4.99	0.08	0.2320	0.0036	10^{-6}
Al hemispherical	0.020	0.0288	4.99	0.08	0.1552	0.0023	10^{-6}
Uncoated	0.0465	0.0671	4.37	0.07	0.0878	0.0047	10^{-6}
Multilayer (50 layers/in.)							
**DAM/silk net	3.00×10^{-4}	4.33×10^{-4}	2.82	0.045	8.46×10^{-4}	19.49×10^{-6}	10^{-6}
DAM/nylon net	2.04×10^{-4}	2.94×10^{-4}	3.36	0.054	6.85×10^{-4}	15.88×10^{-6}	10^{-6}
NRC-2 †SAM crinkled	3.12×10^{-4}	4.50×10^{-4}	0.91	0.015	2.84×10^{-4}	6.75×10^{-6}	10^{-6}
††Superfloc	3.00×10^{-4}	4.33×10^{-4}	0.86	0.014	2.58×10^{-4}	6.06×10^{-6}	10^{-6}
Powder							
Perlite	0.0080	0.0115	6.00	0.096	0.048	0.0011	760
Santocel A (Monsanto)	0.0140	0.0202	6.00	0.096	0.084	0.0019	10^{-6}
Silica aerogel	0.0111	0.0160	5.00	0.080	0.0554	0.0013	10^{-3}
Preformed							
Expanded polystyrene	0.1665	0.2401	0.94	0.015	0.1560	0.0036	760
Polyurethane foam	0.1734	0.2500	3.06	0.049	0.530	0.0123	760
Resin bonded fiberglass	0.222	0.3201	1.87	0.030	0.4158	0.0096	760

*Approximate hot and cold boundary temperatures = 300 K and 77 K, respectively.   **DAM = Double-aluminized Mylar

†SAM = Single-aluminized Mylar   ††Superfloc = DAM/Dacron Tufts

Table 15-20. Thermal Insulation Concepts

	Foams	Powders and Fibers	MLI	Microspheres	Honeycomb
Advantages	Fabrication Low cost Self supporting	Fabrication	Lowest thermal conductivity Lowest density	Flexible contour Predictable	Self supporting load bearing
Disadvantages	High thermal conductivity Moderate density	High thermal conductivity Evacuation High density	Poor predict-ability No compressive strength High shell weight	High thermal conductivity High density	High thermal conductivity High density

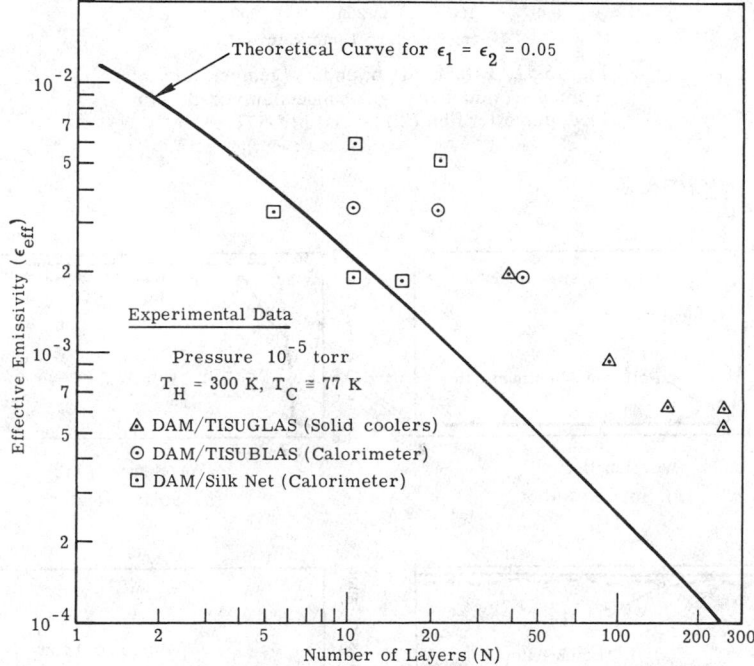

Fig. 15-50. Effective emissivity of double-aluminized Mylar (DAM) MLI.

a measured value, it should be modified if used under a set of boundary temperatures other than the measured condition. An example of the effect of boundary temperature on the k_{eff} of a blanket of crinkled, single-aluminized Mylar is presented in Figure 15-51.

15.8.4. Optical Properties of Thermal-Control Materials and Coatings. In analyzing the interchange of radiant energy and/or the computation of equilibrium surface-temperatures in space, one must know the emission, reflection, and solar-absorption characteristics of the surfaces involved. The fundamental relationships between these properties are covered in Chapter 1. The characteristics of selected materials and coatings are presented below.

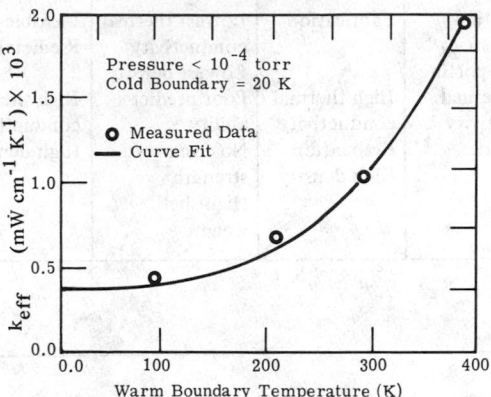

Fig. 15-51. Effect of boundary temperature on thermal conductivity of single-aluminized, crinkled, polyester film (20 layers) [15-57].

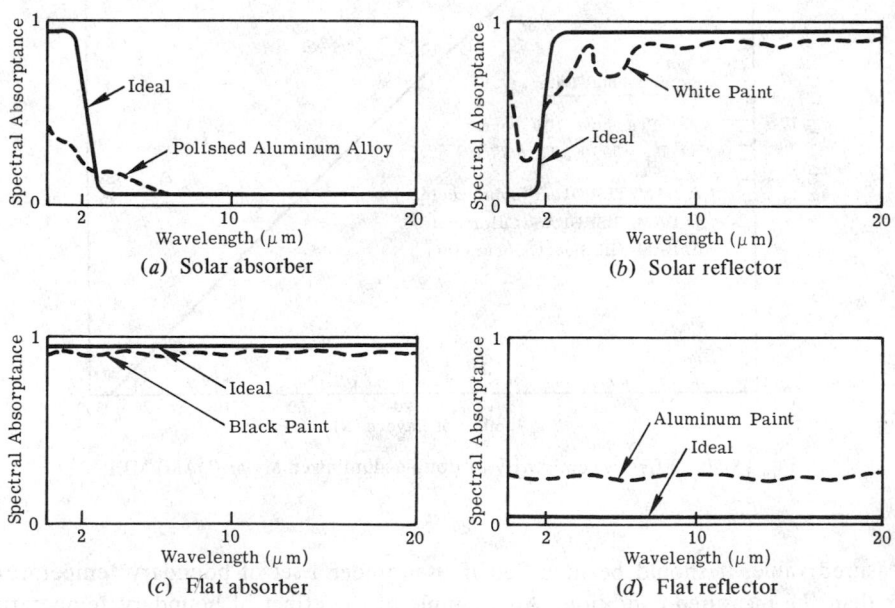

Fig. 15-52. Ideal representation for four basic, thermal-control surfaces [15-58].

Ideally, thermal-control surfaces can be divided into four basic classes: solar absorbers; solar reflectors; flat absorbers; and flat reflectors, as illustrated in Figure 15-52. The solar absorbers are primarily polished metals which exhibit a high solar-absorptance, α_s, and a relatively low emissivity, ϵ, in the longer IR wavelengths. Flat absorbers, such as black paint, generally exhibit a high absorptance over both the solar and IR wavelengths. Aluminum paints are flat reflectors which exhibit moderate and relatively flat reflectance, ρ, over the wavelength range through the IR. Specifically developed white paints (Table 15-21) and second surface mirrors, which are also referred to as optical solar reflectors, OSRs, typify solar reflectors with low α_s and high ϵ in the IR region.

Table 15-21. White Paints Suitable For Spacecraft Thermal Control

Designation	Characteristic Formulation	Manufacturer or Developer	Typical Spacecraft Applications	Approximate Thermal Properties		Reference
				(α_s)	(ϵ_H)*	
Thermatrol (DC 92-007)	Titanium dioxide (TiO_2) pigment/ silicone binder	Lockheed/Dow Corning	Lunar Orbiter, Skynet	0.19	0.82	[15-58]
S-13	Zinc oxide (ZnO)/ RTV 602 binder	Illinois Institute of Technology Research Institute (IITRI)	ATS-I, OSO I	0.21	0.88	[15-58]
S-13G	Zinc oxide/ potassium silicate with RTV 602 binder	Illinois Institute of Technology Research Institute (IITRI)	Lunar Orbiter, II, III, IV, Mariner V	0.19	0.88	[15-59]
Z-93	Zinc oxide/ potassium silicate binder	Illinois Institute of Technology Research Institute (IITRI)	OSO III, Mariner IV, Apollo SM Radiators	0.18	0.88	[15-58]
PV-100	Titanium dioxide/ silicone alkyd binder	Vita-Var/GE	Nimbus, ERTS-A	0.22	0.82	[15-58]
Kemacryl	Titanium dioxide/ acrylic binder	Sherwin-Williams	–	0.24	0.86	[15-60]
Skyspar	Titanium dioxide/ epoxy binder	Andrew Brown Co.	–	0.22	0.91	[15-60]

*Hemispherical emissivity at room temperature.

A typical OSR consists of a film of vapor-deposited metal (usually aluminum, silver, or gold) on a rigid, silica face-sheet, or on flexible, transparent films such as Teflon®*, Kapton®** or Mylar. Although the transparent film OSRs degrade somewhat in the space environment (i.e., α_s increases with time), they are in general considerably more stable than most white paints. A variation of the conventional OSR is a specially anodized aluminum, wherein a transparent, aluminum-oxide film is chemically formed on a bright, polished-aluminum substrate. These surfaces also tend to be somewhat more stable than white paints. Typical degradations of white paints, anodized aluminum surfaces, and OSRs when exposed to the space environment are presented in Figures 15-53, 15-54, and 15-55, respectively.

The α_s and ϵ for a number of materials, including paints, coatings, and OSRs, are summarized in Table 15-22.

*Trademark of the Dupont Corporation.
**Trademark of the Dupont Corporation.

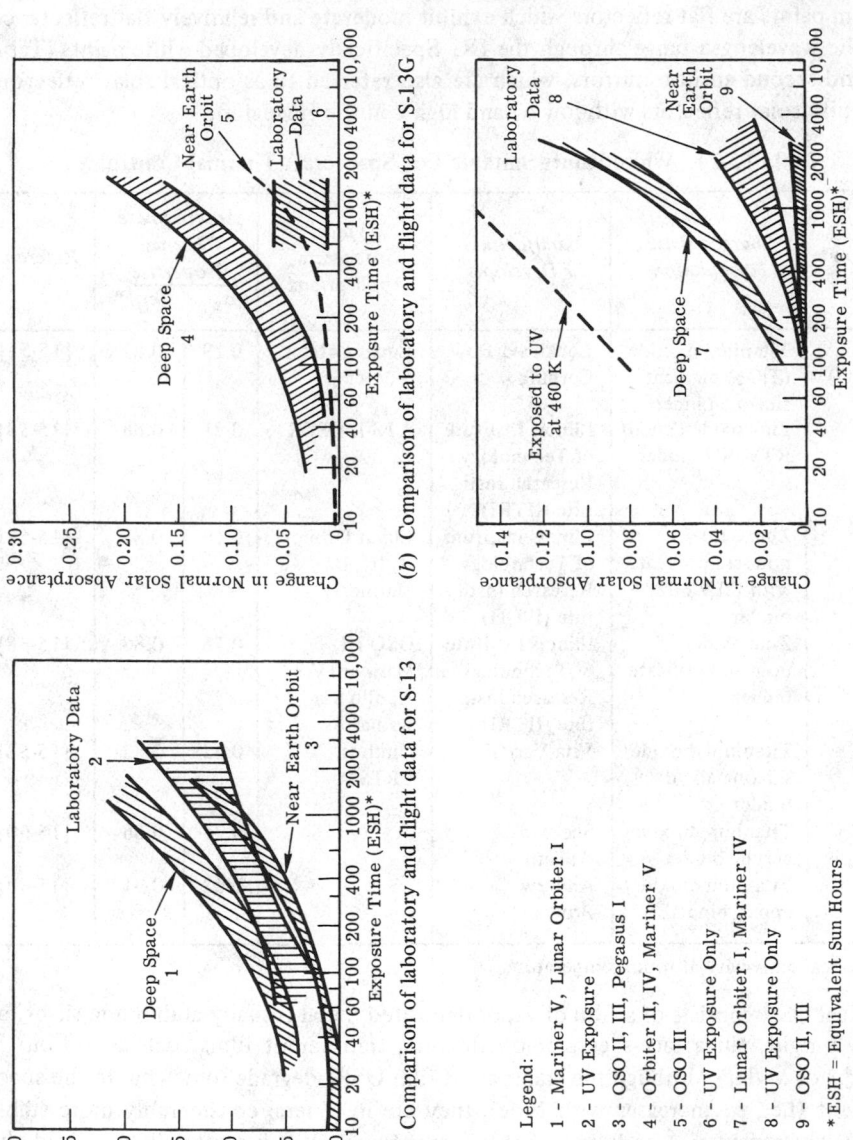

(a) Comparison of laboratory and flight data for S-13

(b) Comparison of laboratory and flight data for S-13G

(c) Comparison of laboratory and flight data for Z-93

Legend:

1 – Mariner V, Lunar Orbiter I
2 – UV Exposure
3 – OSO II, III, Pegasus I
4 – Orbiter II, IV, Mariner V
5 – OSO III
6 – UV Exposure Only
7 – Lunar Orbiter I, Mariner IV
8 – UV Exposure Only
9 – OSO II, III

*ESH = Equivalent Sun Hours

Fig. 15-53. Change in normal solar absorptance of three white paints [15-59].

Table 15-22. Solar Absorptance and Infrared Emissivity of Selected Materials and Coatings

	α_s^*	ϵ_H^{**}	α_s/ϵ_H
Metals			
Aluminum 6061 (polished)	0.19	0.042	4.5
Aluminum 6061 (unpolished)	0.37	0.042	8.8
Aluminum foil, adhesive backed	0.12	0.04	3.0
Aluminum, anodized (clear)	0.16	0.76	0.21
Titanium (6 A14V)	0.57	0.18	3.2
Beryllium (extruded)	0.70	0.17	4.1
Beryllium (polished)	0.49	0.09	5.5
Stainless steel (polished)	0.40	0.05	8.0
Magnesium (AZ 31B)	0.27	0.10	2.7
Silver	0.04	0.02	2.0
Chrome	0.24	0.08	3.0
Gold (vacuum deposited)	0.24	0.03	8.0
Gold (plated)	0.30	0.03	10.0
Nickel	0.45	0.18	2.5
White Paints (Solar reflectors)			
Thermatrol (DC-92-007), Dow Corning	0.19	0.82	0.23
S-13 (IITRI)	0.21	0.88	0.24
S-13G (IITRI)	0.19	0.88	0.22
Z-93 (IITRI)	0.18	0.88	0.20
PV-100 (GE)	0.22	0.82	0.27
Kemacryl	0.24	0.86	0.28
Skyspar	0.22	0.91	0.24
Velvet White (3M Co.)	0.24	0.85	0.28
Black Paints (Flat absorbers)			
Chemglase Z-306 (Hughson Chem. Co.)	0.95	0.88	1.08
Black Velvet 407 (3M)	0.95	0.92	1.03
Black Kemacryl (Sherwin-Williams)	0.93	0.88	1.06
Black Silicone 517-B-2 (W.P. Fuller)	0.89	0.88	1.01
Cat-a-Lac (Finch Paint and Chemical)	0.85	0.90	0.94
Parson's Black	0.98	0.92	1.06
Aluminum Paints (Flat reflectors)			
GE D4D	0.29	0.27	1.03
Rinshed Mason	0.26	0.22	1.18
Transparent Polymers, Rear Surface Coated			
Aluminized Mylar†			
0.25 mil thickness	0.14	0.36	0.39
1.0 mil thickness	0.16	0.54	0.30
2.0 mil thickness	0.17	0.70	0.24
5.0 mil thickness	0.18	0.75	0.24
Aluminized Kapton††			
1.0 mil thickness	0.36	0.54	0.67
3.0 mil thickness	0.44	0.78	0.38
5.0 mil thickness	0.53	0.80	0.66
Aluminized Teflon (FEP)***			
1.0 mil thickness	0.15	0.60	0.25
2.0 mil thickness	0.15	0.66	0.23
5.0 mil thickness	0.15	0.78	0.19
Silverized Teflon (FEP)***			
1.0 mil thickness	0.06	0.52	0.12
5.0 mil thickness	0.09	0.80	0.11
Second Surface Metallized Glass Mirrors			
Silver-Fused Silica (8 mil)	0.07	0.80	0.09
Silver-Microsheet (6 mil)	0.07	0.82	0.085
Aluminum-Fused Silica (8 mil)	0.10	0.80	0.125

*Solar absorptance, undegraded
**Infrared emissivity (hemispherical) at room temperature
† Dupont's trade name for polyethylene terephthalate
†† Dupont's trade name for polymide
***Fluorinated ethylene propylene

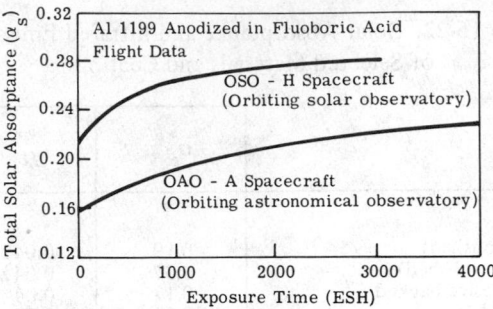

Fig. 15-54. Solar-absorptance degradation of anodized, aluminum coatings [15-61].

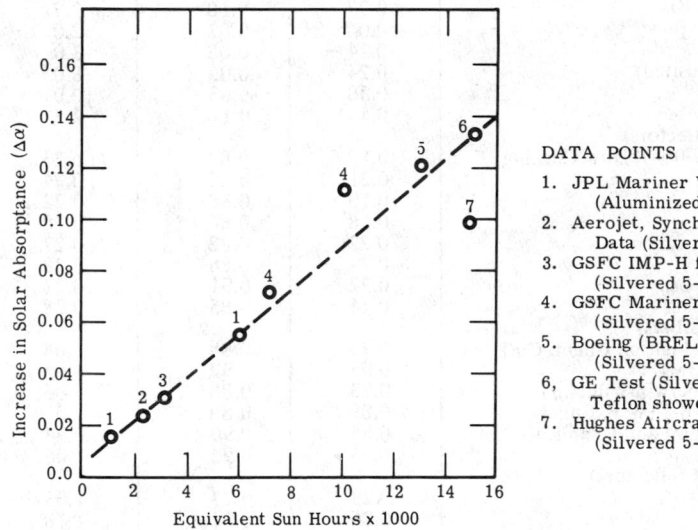

DATA POINTS

1. JPL Mariner V Flight
    (Aluminized 1-mil Teflon)
2. Aerojet, Synchronous Orbit
    Data (Silvered 2-mil Teflon)
3. GSFC IMP-H flt. exp.
    (Silvered 5-mil Teflon)
4. GSFC Mariner X
    (Silvered 5-mil Teflon)
5. Boeing (BREL) Helios Test
    (Silvered 5-mil Teflon)
6, GE Test (Silver and aluminum
    Teflon showed identical results)
7. Hughes Aircraft Lab Tests
    (Silvered 5-mil Teflon)

Fig. 15-55. Increase in α_s of silvered and aluminized Teflon films exposed to the combined UV and particle bombardment associated with deep space environment.

15.9. References and Bibliography
15.9.1   References

[15-1]   "Nimbus F Two-Stage Solid Cryogen Cooler," Final Report, Report No. LMSC-D358720, Lockheed Missiles and Space Company, Inc., Sunnyvale, CA, April 15, 1974.

[15-2]   G. H. Nakano, et al., "A Satellite Borne High-Resolution GE(Li) Gamma Ray Spectrometer System," presented at Institute of Electrical and Electronics Engineers Nuclear Science Symposium, San Francisco, CA, 14-16 November 1973.

[15-3]   J. P. Wright and W. R. Pence, "Development of A Cryogenic Heat Pipe Radiator for a Detector Cooling System," American Society of Mechanical Engineers, paper presented at the Society of American Engineers-American Society of Mechanical Engineers-American Institute of Aeronautics and Astronautics Inter-Society Conference, San Diego, CA, 16-19 July 1973. American Society of Mechanical Engineers, New York, NY, 73-ENAs-47, 1973.

[15-4]   M. J. Donohoe, et al., "Radiant Coolers-Theory, Flight Histories, Design Comparisons and Future Applications," American Institute of Aeronautics and Astronautics Paper 75-184, 20 January 1975.

[15-5]   M. Donabedian, "Survey of Cryogenic Cooling Techniques," Space and Missile Systems Organization-Technical Report 73-74, available from National Technical Information Service, AD 755 780, October 1972.

[15-6]   D. Lehrfeld and G. Pitcher, "An Oil Lubricated Triple Expansion VM Cooler for Long Duration Space Missions," Air Force Flight Dynamics Laboratory-Technical Report 73-149, Vol. I, Philips Laboratory, Briarcliff Manor, NY, December 1973.

[15-7]   "Long Life, High Capacity Vuilleumier Refrigerator for Space Applications," Air Force Flight Dynamics Laboratory-Technical Report 75-108, Hughes Aircraft Company, Culver City, CA, September, 1975.

[15-8]   D. B. Colyer, et al., "Design and Development of Cryogenic Turbo Refrigerator Systems," Air Force Flight Dynamics Laboratory-Technical Report 72-154, General Electric R&D Center, Schenectady, NY, April 1973.

[15-9]   R. W. Breckinridge, Jr., et al., "Development of a Rotary-Reciprocating Refrigerator for Space Applications," Air Force Flight Dynamics Laboratory-Technical Report 75-77, Arthur D. Little, Inc., Cambridge, MA, August 1975.

[15-10] R. J. Buist, "Low Power, Low Temperature Thermoelectric Coolers," presented at Electro-Optical Systems Design Conference–1974 West, San Francisco, CA, 5 November 1974.

[15-11] "Cryogenic Data" Cryenco, A Division of Cryogenic Technology, Inc., Denver, CO, March 1965.

[15-12] "Specification, Cryogenic Refrigeration Systems," Brochure, Air Products and Chemicals, Inc., Advanced Products Dept., Allentown, PA, 1975.

[15-13] J. S. Buller, "Miniature Self-Regulating Rapid Cooling Joule-Thomson Cryostat," Santa Barbara Research Center, Goleta, CA, May 1969, p. 12.

[15-14] "Miniature Cryogenic Systems," The Hymatic Engineering Company, Worcestershire, England, July 1975.

[15-15] "The Infrared Brochure," Santa Barbara Research Center, Goleta, CA, 1973, pp. 25,30.

[15-16] Minnesota Valley Engineering, Catalog, New Prague, MN, June 1970.

[15-17] U. E. Gross et al., "Solid Cryogen Cooler Design Studies and Development of an Experimental Cooler," Air Force Flight Dynamics Laboratory-Technical Report 68-1, Aerojet-General Corp., Azusa, CA, March 1968.

[15-18] R. P. Caren and R. M. Coston, "Design and Construction of an Engineering Model Solid Cryogen Refrigerator for Infrared Detector Cooling at 50K," National Aeronautics and Space Administration-Contractor Report-988, Lockheed Missiles and Space Co., January 1968.

[15-19] J. A. Ahern and T. W. Lawson, Jr., "Cryogenic Solid Oxygen Storage and Sublimation Investigation," Aerospace (USAF) Medical Research Laboratory-Technical Report-68-105, Aerojet-General Corp., Azusa, CA, December 1968.

[15-20] R. N. Herring, et al., "Solid Cryogen Cooler Development," Ball Brothers Research Corp., Boulder, Colo., presented at the National Aeronautics and Space Administration Cryogenic Workshop, Marshall Space Flight Center, Huntsville, AL, 29-30 March 1973.

[15-21] R. Breckenridge, "A Solid Cryogen Cooler for an X-Ray Spectrometer," presented at the National Aeronautics and Space Administration Cryogenic Workshop, Marshall Space Flight Center, Huntsville, AL, 29-30 March 1973.

[15-22] C. Balas, Jr., and C. A. Wingate, Jr., "An Efficient Long-Life Cryogenic Cooling System for Spacecraft Applications," Technical Note-1218, presented at the

International Astronautical Federation, 26th Congress, Lisbon, Portugal, 8 September 1975.

[15-23] Malaker Corporation Brochure, Malaker Corp., High Bridge, NJ. revised 1 December 1970.

[15-24] J. G. Daunt, et al., "Miniature Cryogenic Refrigerators," AD 697 972, Stevens Institute of Technology, NJ, August 1969.

[15-25] H. Yoshimoto, Hughes Aircraft Company, Personal Communication, 7 July 1975.

[15-26] H. L. Jensen, et al., "Investigation of External Refrigeration Systems for Long Term Cryogenic Storage," Report No. LMSC-A903162, Lockheed Missiles and Space Co., 28 May 1970.

[15-27] A. L. Jokl and R. B. Flemming, "Miniature Turbomachinery Cryogenic Refrigeration Systems," paper presented at the Intersociety Energy Conversion Conference, Las Vegas, NV, 20 September 1970, American Society of Mechanical Engineers, New York, NY, 1970.

[15-28] New Products Brochure, Cryogenic Technology, Inc., Waltham, MA, August 1975.

[15-29] C. Balas, Jr., "An Acoustically Quiet, Low Power Minimum Vibration Stirling Cycle Refrigerator, Air Force Flight Dynamics Laboratory-Technical Report-73-149, Philips Labs., Briarcliff Manor, NY, December 1973.

[15-30] B. Miller, "New Sensor Cooling Unit Developed," *Aviation Week and Space Technology,* 20 November 1972, pp. 76-78.

[15-31] R. J. Buist, "Low Power, Low Temperature Thermoelectric Coolers," presented at Electro-Optical Systems Design Conference—1974 West, San Francisco, CA, 5 November 1974.

[15-32] "Vuilleumier Cycle Cryogenic Refrigeration Systems for Infrared Scanner Applications," Air Force Flight Dynamics Laboratory-Technical Report-71-18, Hughes Aircraft Company, August 1971.

[15-33] F. N. Magee, et al., "Vuilleumier Cycle Cryogenic Refrigerator Development," AD 841 543, Hughes Aircraft Company, Culver City, CA, August 1968.

[15-34] G. K. Pitcher and F. K. du Pre, *Advances in Cryogenic Engineering,* Vol. 15, 1969, pp. 447-451.

[15-35] C. W. Browning, W. S. Miller and V. L. Potter, "75 K Vuilleumier Refrigerator, Final Report for Task II Analytical and Test Program," Report 72-8687, Airesearch Manufacturing Co., Los Angeles, CA, November 1972.

[15-36] "Design and Performance Data for Cryodyne Refrigerators," Bulletin 7-75-001, Technical Documents Service-5058, Cryogenic Technology, Inc., Waltham, MA, July 1975.

[15-37] "Cryocooler Specifications," Bulletin 7-75-013, Cryogenic Technology, Inc., Waltham, MA, July 1975.

[15-38] "Cryogenic Detector Cooler, Displex Model C3-1003 Data Sheet," Air Products and Chemicals, Advanced Products Dept., Allentown, PA, July 1970.

[15-39] "Cryogenic Refrigerator System for Infrared Detector Arrays," Specifications SC02-0670, CS01-0470, Submarine Systems Corp., Chatsworth, CA, 1972.

[15-40] P. H. Egli, *Thermoelectricity,* John Wiley and Sons, New York, NY, 1958.

[15-41] "Standard Thermoelectric Devices Specification Chart," Borg-Warner Corp., Des Planes, IL, 9 May 1975.

[15-42] G. S. Bird, *The Cambion Thermoelectric Handbook,* Cambion Thermionic Corp., Cambridge, MA, 1971.

[15-43] "Thermoelectric Cooler Specifications," Marlow Industries, Inc., Garland, TX, January 1977.

[15-44] J. S. Buller, et al., "Mercury Cadmium Telluride Space Radiation Cooler for

Synchronous Orbiting Satellites," Santa Barbara Research Center, Goleta, CA, May 1974, p. 8.

[15-45] K. L. DeBrosse, et al., "A Day-Night High Resolution Infrared Radiometer Employing Two-Stage Radiant Cooling," Part I, National Aeronautics and Space Administration-Contractor Report-94600, 11 December 1967.

[15-46] R. B. Scott, *Cryogenic Engineering,* D. Van Nostrand Co., Inc., Princeton, NJ, 1959.

[15-47] R. V. Annable, "Radiant Cooling," *Applied Optics,* Vol. 9, No. 1, January 1970, pp. 185-193.

[15-48] "Solar Electromagnetic Radiation," National Aeronautics and Space Administration-Special Publication-8005, revised May 1971.

[15-49] "Earth Albedo and Emitted Radiation," National Aeronautics and Space Administration-Special Publication-8067, July 1971.

[15-50] "Radiator Design for Space Vehicles," Airesearch Manufacturing Div., Garret Corp., Los Angeles, CA, March 1972.

[15-51] T. N. Veziloglu, *Correlation of Thermal Contact Conductance Experimental Results,* Prog. Of Astronautics and Aeronautics, 20, Academic Press, NY 1967.

[15-52] P. J. Schneider, "Conduction," Section 3, *Handbook of Heat Transfer,* W. M. Rohsenow and J. P. Hartrett (eds.), McGraw-Hill Book Co., 1973.

[15-53] A. J. Buschman, et al., "Configuration Factors for Exchange of Radiant Energy Between Axisymmetrical Sections of Cylinders, Cones, Hemispheres, and Their Bases," National Aeronautics and Space Administration-Technical Notes D-944, 1961.

[15-54] C. J. Sotos, et al., "Radiant Interchange View Factors and Limits of Visibility for Differential Cylindrical Surfaces with Parallel Generating Lines," National Aeronautics and Space Administration-Technical Note-D-2556, 1964.

[15-55] F. R. G. Eckert, "Radiation," *Handbook of Heat Transfer,* Part A, Section 15, W. M. Rohsenow and J. P. Hartnett (eds.) McGraw-Hill Book Co., 1973.

[15-56] H. C. Hottel, "Radiation Heat Transmission," *Heat Transmission,* by W. H. McAdams, Chapter 3, 3rd Edition, McGraw-Hill Book Co., 1954.

[15-57] P. E. Glaser, et al., "Thermal Insulation Systems, A Survey," National Aeronautics and Space Administration-Special Publication-5027, 1967.

[15-58] N. J. Broadway, *Radiation Effects Handbook,* Section 2, "Thermal Control Coatings," National Aeronautics and Space Administration-Contractor Report-1986, prepared by Battelle Memorial Institute, Columbus, OH, June 1971.

[15-59] Y. S. Touloukian, et al., *Thermophysical Properties of Matter,* "Thermal Radiative Properties," Thermophysical Properties Research Center, Purdue University, Plenum Publishing Co., NY, 1972.

[15-60] J. B. Rittenhouse, et al., *Space Materials Handbook,* Supplement 1, National Aeronautics and Space Administration-Special Publication-3025, 1966.

[15-61] H. Fine, "An Insight into the Features of OAO Thermal Design," American Society of Mechanical Engineers Paper 73 ENAS-46, 20 August 1973.

[15-62] Rudolph Vuilleumier, "Method and Apparatus for Inducing Heat Changes," U.S. Patent No. 1,275,507, 13 August 1918.

[15-63] W. E. Gifford and H. O. McMahon, "A Low Temperature Heat Pump," Proceedings of the Tenth International Conference of Refrigeration, Vol. 1, Copenhagen, Denmark, August 1959, American Society of Heating, Refrigerating, and Air Conditioning Engineers, New York, NY., U.S. Patent Nos. 2,966,035 and 3,119,237.

15.9.2. Bibliography

Ahern, J. A., and T. W. Lawson, Jr., "Cryogenic Solid Oxygen Storage and Sublimation Investigation," Aerospace (USAF) Medical Research Laboratory-Technical Report-68-105, prepared by Aerojet-General Corporation, Azusa, CA, December 1968.

Air Products and Chemicals Inc., "Cryogenic Detector Cooler, Displex Model C3-1003 Data Sheet," Advanced Products Dept., Allentown, PA, July 1970.

Air Products and Chemicals, Inc., "Specification, Cryogenic Refrigeration System," Advanced Products Dept., Allentown, PA, 1975.

Annable, R. V., "Radiant Cooling," *Applied Optics,* Vol. 9, No. 1, January 1970, pp. 105-193.

Balas, C., Jr., "An Acoustically Quiet, Low Power Minimum Vibration Stirling Cycle Refrigerator," Air Force Flight Dynamics Laboratory-Technical Report-73-149, Phillips Labs., Briarcliff, Manor, NY, December 1973.

Balas, C., Jr. and C. A. Wingate, Jr., "An Efficient Long-Life Cryogenic Cooling System for Spacecraft Applications," Technical Note-No. 1218, Phillips Labs., presented at the International Astronautical Federation, 26th Congress, Lisbon, Portugal, 8 September 1975.

Ballinger, J. C., and E. H. Christensen, "Thermal Environment of Space," Supplement B, Engineering Research Report-Aerodynamics Note-016, Convair-Astronautics Division, General Dynamics Corp., San Diego, CA.

Bird, G. S., *The Cambion Thermoelectric Handbook,* Cambion Thermionic Corp., Cambridge, MA, 1971.

Borg-Warner Thermoelectrics, Standard Thermoelectric Devices Specification Chart, Borg-Warner Corp., Des Plaines, IL, 9 May 1975.

Breckingridge, R. W., Jr., et al., "Development of a Rotary-Reciprocating Refrigerator for Space Applications," Air Force Flight Dynamics Laboratory-Technical Report-75-77, Arthur D. Little, Inc., Cambridge, MA, August 1975.

Broadway, N. J., *Radiation Effects Handbook,* Section 2, "Thermal Control Coatings," National Aeronautics and Space Administration-Contractor Report-1786, prepared by Battelle Memorial Institute, Columbus, OH, June 1971.

Browning, C. W., W. S. Miller and V. L. Potter, "75 K Vuilleumier Refrigerator Final Report for Task II Analytical and Test Program," Report 72-8687, Airesearch Manufacturing Co., Los Angeles, CA, November 1972.

Buist, R. J., "Low Power, Low Temperature Thermoelectric Coolers," presented at Electro-Optical Systems Design Conference-1974 West, San Francisco, CA, 5 November 1974.

Buller, J. S., "Miniature Self-Regulating Rapid Cooling Joule-Thomson Cryostat," Santa Barbara Research Center, Goleta, CA, May 1969.

Buller, J. S., et al., "Mercury Cadmium Telluride Space Radiation Cooler for Synchronous Orbiting Satellites," Santa Barbara Research Center, Goleta, CA, May 1974.

Buschman, A. J., et al., "Configuration Factors for Exchange of Radiant Energy between Axisymmetrical Sections of Cylinders, Cones, Hemispheres, and Their Bases," National Aeronautics and Space Administration-Technical Note-D-944, 1961.

Caren, R. P., and R. M. Coston, "Design and Construction of an Engineering Model Solid Cryogen Refrigerator for Infrared Detector Cooling at 50K," National Aeronautics and Space Administration-Contractor Report-988, Lockheed Missiles and Space Co., January 1968.

Clausing, A. M., and B. T. Chao, "Thermal Contact Resistance in a Vacuum Environment," *J. of Heat Transfer,* May 1965, pp. 243-251.

Colyer, D. B. et al., "Design and Development of Cryogenic Turbo Refrigerator Systems," Air Force Flight Dynamics Laboratory-Technical Report-72-154, General Electric R & D Center, Schenectady, NY, April 1973.

Cryenco, "Cryogenic Data," A Division of Cryogenic Technology, Inc., Denver, CO, March 1965.

Cryogenic Technology Staff, "Cryocooler Specifications," Bulletin 7-75-013, Cryogenic Technology, Inc., Waltham, MA, July 1975.

Cryogenic Technology Staff, "Design and Performance Data for Cryodyne Refrigerators," Bulletin 7-75-001, Technical Document Service-5058, Cryogenic Technology, Inc., Waltham, MA, July 1975.

Cryogenic Technology Staff, New Products Brochure, Cryogenic Technology, Inc., Waltham, MA, August 1975.

Cryomech Staff, "Cryorefrigerator Data Sheets," Cryomech, Inc., Jonesville, NY, 1975.

Cunnington, G. R., "Thermal Conductance of Filler Aluminum and Magnesium Joints in a Vacuum Environment," American Society of Mechanical Engineers, paper presented at American Society of Mechanical Engineers Meeting, NY, 29 November 1964.

Daunt, J. G., et al., "Miniature Cryogenic Refrigerators," AD 697 972, Stevens Institute of Technology, August 1969.

Davis, D. A., and R. L. Drake, "Monte Carlo Direct View Factor Program and Generalized Radiative Heat Transfer Program," Document No. AS-2110. The Boeing Co., Seattle, WA, 1966.

DeBrosse, K. L., et al., "A Day-Night High Resolution Infrared Radiometer Employing Two-State Radiant Cooling," Part I, National Aeronautics and Space Administration-Contractor Report-94600, 11 December 1967.

Donabedian, M., "Survey of Cryogenic Cooling Techniques," SAMSO-Technical Report-73-34, AD 755 780, October 1972.

Donohoe, M. J., et al., "Radiant Coolers-Theory, Flight Histories, Design Comparisons and Future Applications," American Institute of Aeronautics and Astronautics Paper 75-184, 20 January 1975.

Eckert, E. R. G., "Radiation," *Handbook of Heat Transfer*, Part A, Section 15, W. M. Rohsenow and J. P. Hartnett (eds.), McGraw-Hill Book Co., 1973.

Eckert, E. R. G., and R. M. Drake, Jr., *Heat and Mass Transfer*, McGraw-Hill Book Co., NY, 1959.

Egli, P. H., *Thermoelectricity*, John Wiley and Sons, NY, 1958.

Faires, V. M., *Applied Thermodynamics*, MacMillan and Co., NY, 1947.

Fine, H., "An Insight into the Features of OAO Thermal Design," American Society of Mechanical Engineers Paper 73-ENAs-46, 20 August 1973.

Fontenot, J. E., "Thermal Conductance of Contacts and Joints," Report No. D5-12206, AD 479 008, The Boeing Co., Aerospace Division, 1964.

Garret Corp., "Radiator Design for Space Vehicles," Airesearch Manufacturing Division, Los Angeles, CA, March 1972.

Gifford, W. E., and H. O. McMahon, "A Low Temperature Heat Pump," Proceedings of the Tenth International Conference of Refrigeration, Vol. 1, Copenhagen, Denmark, August 1959, American Society of Heating, Refrigerating, and Air Conditioning Engineers, New York, NY, U.S. Patent Nos. 2,966,035 and 3,119,237.

Glaser, P. E., et al., "Thermal Insulation System, A Survey," National Aeronautics and Space Administration-Special Publication-5027, 1967.

Green, W., *Thermoelectric Handbook*, Westinghouse Electric Corporation, 1962.

Gross, U. E., et al., "Solid Cryogen Cooler Design Studies and Development of an Experimental Cooler," Air Force Flight Dynamics Laboratory-Technical Report-68-1, Aerojet-General Corp, Azusa, CA, March 1968.

Herring, R. N., et al., "Solid Cryogen Cooler Development," Ball Brothers Research Corp., Boulder, CO, presented at the National Aeronautics and Space Administration Cryogenic Workshop, Marshall Space Flight Center, Huntsville, AL, 29-30 March 1973.

Hughes Aircraft Co., "Long Life, High Capacity Vuilleumier Refrigerator for Space Applications," Air Force Flight Dynamics Laboratory-Technical Report-75-108, Hughes Aircraft Co., Culver City, CA, September 1975.

Hughes Aircraft Company, "Vuilleumier Cycle Cryogenic Refrigeration Systems for Infrared Scanner Applications," Air Force Flight Dynamics Laboratory-Technical Report-71-18, August 1971.

Hymatic Engineering Staff, "Miniature Cryogenic Systems," The Hymatic Engineering Co., Worcestershire, England, July 1975.

Jensen, C. L., and R. G. Gobel, "Thermal Radiation Analysis System (TRASYS) User's Manual," MCR-73-105, Martin Marietta, May 1973.

Jensen, H. L., et al., "Investigation of External Refrigeration Systems for Long Term Cryogenic Storage," Lockheed Missiles and Space Company Report A903162, Lockheed Missiles and Space Co., 28 May 1970.

Jokl, A. L., and R. B. Flemming, "Miniature Turbomachinery Cryogenic Refrigeration Systems," Paper presented at the Intersociety Energy Conversion Conference, Las Vegas, NV, 20 September 1970, American Society of Mechanical Engineers, New York, NY.

Jones, J. B., and G. A. Hawkins, Engineering Thermodynamics, John Wiley and Sons, NY, 1960.

Kreith, F., Principles of Heat Transfer, International Textbook, Scranton, PA, 1958.

Kunkle, J. S., (ed.), Compressed Gas Handbook, National Aeronautics and Space Administration-Special Publication-3045, National Aeronautics and Space Administration, Washington, DC, 1960.

Lehrfeld, D., and G. Pitcher, "An Oil Lubricated Triple Expansion VM Cooler for Long Duration Space Missions," Air Force Flight Dynamics Laboratory-Technical Report-73-149, Vol. 1, Phillips Lab., Briarcliff Manor, NY, December 1973.

Lockheed Missiles and Space Company, "Nimbus F Two-Stage Solid Cryogen Cooler," Lockheed Missiles and Space Co., Inc., Sunnyvale, CA, Final Report-O358720, 15 April 1974.

London, J., "A Study of the Atmospheric Heat Balance," Final Report Contract Air Force 19(122)-165, AD 117 227, Air Force Cambridge Research Center-Technical Report-57-287, Dept. of Meteorology and Oceanography, New York University, July 1957.

Magee, F. N., et al., "Vuilleumier Cycle Cryogenic Refrigerator Development," AD 841 543, Hughes Aircraft Co., Culver City, CA, August 1968.

Marlow Industries Staff, "Thermoelectric Cooler Specifications," Marlow Industries, Inc., Garland, TX, January 1977.

McAdams, W., (ed.), Heat Transmission, 3rd Edition, McGraw-Hill, NY, 1954.

Miller, B., "New Sensor Cooling Unit Developed," Aviation Week and Space Technology, 20 November 1972, pp. 76-78.

Minnesota Valley Engineering Catalog, New Prague, MN, June 1970.

Nakano, G. H., et al., "A Satellite Borne High-Resolution GE(Li) Gamma Ray Spectrometer System," presented at Institute of Electrical and Electronics Engineers Nuclear Science Symposium, San Francisco, CA, 14-16 November 1973.

National Aeronautics and Space Administration, "Earth Albedo and Emitted Radiation," Special Publication-8067, July 1971.

National Aeronautics and Space Administration, "Solar Electromagnetic Radiation," Special Publication-8005, Revised May 1971.

National Bureau of Standards, "A Compendium of the Properties of Materials at Low Temperature," Wright Air Development Division-Technical Report-60-56, Part I, National Bureau of Standards, Washington, DC, October 1960.

National Bureau of Standards, "Thermal Conductivity of Metals and Alloys at Low Temperatures, A Review of the Literature," U.S. Dept. of Commerce, National Bureau of Standards, NBS Circular 556, 1954.

Pitcher, G. K., and F. K. du Pre, Advances in Cryogenic Engineering, Vol. 15, 1969, pp. 447-451.

Rittenhouse, J. B., et al., Space Materials Handbook, Supplement No. 1, National Aeronautics and Space Administration-Special Publication-3025, 1966.

Rosi, F. D., et al., "Materials for Thermoelectric Refrigeration," J. Physical Chemistry, Vol. 10, 1959, pp. 191-200.

Santa Barbara Research Center, Goleta, CA, "The Infrared Brochure," 1973.

Schneider, P. J., "Conduction," Section 3, Handbook of Heat Transfer, W. M. Rohsenow and J. P. Hartrett (eds.) McGraw-Hill Book Co., 1973.

Scott, R. B., Cryogenic Engineering, D. Van Nostrand Co., Inc., Princeton, NJ, 1959.

Nuclear Systems, Inc., "Solid State Cooling-Thermoelectric Module Selection Guide," Nuclear Systems, Inc., Garland, TX, Bulletin No. 70-001, 1970.

Siegel, R., and J. R. Howell, Thermal Radiation Heat Transfer, McGraw-Hill Book Co., New York, NY, 1972.

Sotos, C. J., et al., "Radiant Interchange View Factors and Limits of Visibility for Differential Cylindrical Surfaces with Parallel Generating Lines," National Aeronautics and Space Administration-Technical Note-D-2556, 1964.

Stoecker, W. F., Refrigeration and Air Conditioning, McGraw-Hill Book Co., New York, 1958.

Submarine Systems Staff, "Cryogenic Refrigerator System for Infrared Detector Arrays," Submarine Systems Corp., Chatsworth, CA, Specifications SC02-0670, CS01-0470, 1972.

Tang, K. T., "Peltier (Thermoelectric) Cooling of Infrared Detectors," The Garrett Corp., Los Angeles, CA, Report GR-16-R, June 1960.

Touloukian, Y. S., et al., Thermophysical Properties of Matter, Thermal Conductivity, Vol. 2, Non-Metallic Solids, Vol. 3, Metallic Elements and Alloys, Thermophysical Properties Research Center (TPRC), Purdue University, Plenum Publication, New York, NY, 1970.

Touloukian, Y. S., et al., Thermophysical Properties of Matter, "Thermal Radiative Properties," Thermophysical Properties Research Center, Purdue University, Plenum Publication, New York, NY, 1972.

Veziloglu, T. N., Correlation of Thermal Contact Conductance Experimental Results, Progress of Astronautics and Aeronautics, 20, Academic Press, NY, 1967.

Weibelt, J. A., Engineering Radiation Heat Transfer, Holt, Rinehart and Winston, New York, NY, 1966.

Wright, J. P., and W. R. Pence, "Development of a Cryogenic Heat Pipe Radiator for a Detector Cooling System," paper presented at the Society of Automotive Engineers-American Society of Mechanical Engineers-American Institute of Aeronautics and Astronautics Intersociety Conference, San Diego, CA, 16-19 July 1973. American Society of Mechanical Engineers, New York, NY, 73-ENAs-47.

DETECTOR-ASSOCIATED ELECTRONICS

William W. Sloan
Santa Barbara Research Center

CONTENTS

16. Detector-Associated Electronics

16.1. Introduction

Preamplifiers are required for virtually all types of detectors, either to amplify the detector's output signal, transform the detector's output impedance, bias the detector, extend the detector's bandwidth by modifying shunt capacitance effects, or perform some combination of these functions. Pulse-shaping and filtering are usually done in the signal-processor section or in a postamplifier section. The primary objective in preamplifier design is to design an amplifier which performs the desired function(s) mentioned above with minimum degradation of detector performance. The output of the preamplifier should be capable of driving the postamplifier stage and any interconnecting cable.

The primary requirements of such an amplifier are low noise, high gain, low output impedance, large dynamic range, good linearity, and relative freedom from microphonics. It must be compact, since it is usually mounted near the detector, and carefully shielded to eliminate unwanted signals from stray fields.

This chapter describes noise mechanisms in preamplifier components and analyzes how these noise sources can degrade detector performance. Several different types of preamplifier designs and configurations are discussed.

16.1.1. Symbols, Nomenclature and Units.
Table 16-1 lists the symbols, nomenclature, and units used in this chapter.

Table 16-1. Symbols, Nomenclature and Units

Symbols	Nomenclature	Units
A	Gain of amplifier	V V^{-1} or A A^{-1}
A_d	Area of detector surface	cm^2
B, B_1 and B_2	Dimensionless constants	—
BPT	Bipolar transistor	—
C	Capacitance	F
D^*	Specific detectivity	cm Hz$^{1/2}$ W^{-1}
D_s^*	D^* of the system consisting of the IR detector, preamplifier, and bias or load resistor	cm Hz$^{1/2}$ W^{-1}
e	Charge on an electron (1.59×10^{-19} coulombs)	C
F	Noise factor	—
FET	Field-effect transistor	—
f	Frequency	Hz
g_m	Low-frequency transconductance	ohms^{-1} = mhos
h_{fe}	Small-signal current-gain of a transistor	A^{-1}
I	Current	A
$\bar{I}$	Root-mean-square, rms, noise current	A
JFET	Junction field-effect transistor	—

Table 16-1. Symbols, Nomenclature and Units (*Continued*)

Symbols	Nomenclature	Units
j	Imaginary operator, equal to $\sqrt{-1}$	—
k	Boltzmann constant	$J\,K^{-1}$
MOSFET	Metal-oxide field-effect transistor (also called MOST)	—
NF	Noise figure	$V\,V^{-1}$
R	Resistance	ohms
$\mathcal{R}_V$	Voltage responsivity of the IR detectors	$V\,W^{-1}$
$\mathcal{R}_{VS}$	Voltage responsivity of the detector-preamplifier system	$V\,W^{-1}$
s	Complex-domain Laplace operator	sec^{-1}
T	Temperature	K
t	Time	sec
V	Voltage	V
$\overline{V}$	rms noise voltage	V
Z	Impedance	ohms
Δf	Noise bandwidth	Hz
$\mathcal{E}$	Differential input voltage (error voltage)	V
τ	Time constant	sec
τ_0	Time constant of single pole	sec
ω	Angular frequency	$rad\,sec^{-1}$

16.1.2. General Theory. The D^* for a detector (Chapter 11) can be defined by

$$D^* = \frac{\mathcal{R}_V A_d^{1/2} \Delta f^{1/2}}{\overline{V}_{Det}} \tag{16-1}$$

where $\mathcal{R}_V$ = voltage responsivity of the detector, $V\,W^{-1}$
A_d = area of the detector, cm^2
$\overline{V}_{Det}$ = rms noise voltage of the detector, V
Δf = electrical frequency noise bandwidth, Hz

The noise bandwidth can be defined as a theoretical bandwidth, having step function turn-on and turn-off characteristics which contain the same noise power as the system bandwidth. The effective D^* of the system with the detector, preamplifier, and bias (or load) resistor combined can be defined by

$$D_S^* = \frac{\mathcal{R}_{VS} A_d^{1/2} \Delta f^{1/2}}{\overline{V}_{Sys}} \tag{16-2}$$

where $\mathcal{R}_{VS}$ = voltage responsivity of the detector-preamplifier, $V\,W^{-1}$
$\overline{V}_{Sys}$ = rms noise voltage of detector-preamplifier system, V

The voltage responsivity of the detector and the noise voltage of the detector-preamplifier system are usually dependent on frequency.

An optimum system signal-to-noise ratio for an arbitrary input power can be obtained for a given type of detector having a predetermined size and operating condition (i.e., temperature, background, spectral response, bias, etc.) by optimizing the ratio $\mathfrak{R}_{VS}/\overline{V}_{Sys}$ over the desired bandwidth for the system.

The frequency-dependence of the voltage responsivity of the detector-preamplifier, $\mathfrak{R}_{VS}$, is a function of the detector, the preamplifier, the detector bias network, and shunt capacitance across the detector. The voltage responsivity of the detector, $\mathfrak{R}_V$, can be modified or shaped by using operational amplifiers with various input and band-pass characteristics. Sections 16.4 and 16.5 discuss operational amplifiers and their characteristics.

The noise sources which determine the magnitude and frequency-response characteristics of $\overline{V}_{Sys}$ are discussed in Sections 16.2 and 16.3.

16.2. Noise

Detector-preamplifier noise is described by the term $\overline{V}_{Sys}$. This noise voltage can be divided into two types: those internally generated and those externally induced. If these noise components are not correlated, then the noise power, which is proportional to the noise-voltage squared, can be represented by the following equation:

$$\overline{V}_{Sys}^{2} = \overline{V}_{E}^{2} + \overline{V}_{I}^{2} \qquad (16\text{-}3)$$

where $\overline{V}_E$ = externally induced rms noise voltage, V

$\overline{V}_I$ = internally generated rms noise voltage, V

The externally induced rms noise voltage, $\overline{V}_E$, can be caused by external electromagnetic fields, radiation (including photon noise), or mechanical vibration (i.e., microphonics). The magnitude and the frequency characteristics of externally induced noise are functions of the detector-preamplifier system environment. A noiseless, external environment is assumed in this chapter, $\overline{V}_E = 0$.

$$\overline{V}_{Sys}^{2} = \overline{V}_{I}^{2} \qquad (16\text{-}4)$$

The internally generated noise-voltage-squared, $\overline{V}_I^{2}$ can be divided into a detector-generated, noise-voltage-squared, $\overline{V}_{Det}^{2}$, and a preamplifier and bias-circuit noise-voltage-squared, $\overline{V}_A^{2}$. The interaction of the preamplifier and bias-circuit noise with the detector can result in a correlation between these noise components. This noise correlation can usually be neglected for detectors having a relatively flat responsivity and a constant impedance as a function of bias changes.

Therefore, the total detector-preamplifier, mean-square noise voltage, which is proportional to noise power, can be represented by the following equation if noise correlation is neglected:

$$\overline{V}_{Sys}^{2} = \overline{V}_{Det}^{2} + \overline{V}_{A}^{2} \qquad (16\text{-}5)$$

The remainder of Section 16.2 discusses the noise components of $\overline{V}_A^{2}$.

16.2.1. Noise Sources and Characteristics

Thermal Noise [16-1, 16-2, 16-3]. The movement of free charges caused by thermal energy generates an electrical noise in all resistors. This is a random noise, which therefore contains a constant power distribution with frequency. This power distribution is often referred to as a white, power spectrum, and the noise is referred to as white noise.

Nyquist has shown theoretically [16-3] and Johnson has proved experimentally [16-2] that the value of the mean-square thermal-noise voltage, $\overline{V_T}^2$, produced in a resistor is given by

$$\overline{V_T}^2 = 4kTR\Delta f \qquad (16\text{-}6)$$

where k = Boltzmann's constant, 1.38×10^{-23} J K^{-1}
       T = temperature, K
       R = resistance, ohms
       Δf = noise-equivalent power bandwidth, Hz

Johnson noise will occur in a feedback or load resistor, in the base spreading resistance of a bipolar transistor, in the channel resistor of a field-effect transistor (FET) and in the contact resistance of a diode or transistor. The noise-equivalent circuits of these devices will aid in determining the total noise resulting from a summation of a circuit containing these noise components.

Shot Noise [16-4, 16-5, 16-6, 16-7]. Root-mean-square shot-noise current, $\overline{I_S}$, can be attributed to the random and independent crossing of junctions by individual current carriers.

The basic equation for mean-square shot noise $\overline{I_S}^2$, in a junction at frequencies below the cutoff frequency for the device is

$$\overline{I_S}^2 = 2eI\Delta f \qquad (16\text{-}7)$$

where $\overline{I_S}$ = rms shot-noise current in the junction, A
       e = charge on an electron, 1.60×10^{-19}, C
       I = current, A

Surface Noise [16-4, 16-5, 16-6, 16-7]. Surface noise is caused by slow and fast states at the surface of a semiconductor. Slow states act mainly as traps for the majority carriers and fast states act as recombination centers for minority carriers. The fluctuating occupancy of slow states modulates the conductivity and the capture cross-section of the recombination centers. The fluctuating current of minority carriers which disappear at the surface causes current flow through the junction(s), thereby modulating the resistance of the junction(s).

Forward-biased junctions and devices which depend upon conduction close to the surface, such as metal-oxide semiconductor field-effect transistors (MOSFETs or simply MOSTs) are prone to surface noise. This noise can be reduced by proper surface treatment.

Surface noise exhibits a noise power spectrum which varies as the reciprocal of frequency. It is referred to as *semiconductor noise, flicker noise, 1/f noise,* or *excess noise.*

Mean-square surface-noise current for a diode junction, $\overline{I_{Sur}}^2$, can be approximated by the integral

$$\overline{I_{Sur}}^2 \approx \int_{f_1}^{f_2} \frac{\{K_{Sur}I_B^2\}df}{f} \qquad (16\text{-}8)$$

$$\overline{I_{Sur}}^2 \approx K_{Sur}I_B^2 \ln\left(\frac{f_2}{f_1}\right) \qquad (16\text{-}9)$$

where f_1 = equivalent-power lower cutoff frequency, Hz
         f_2 = equivalent-power upper cutoff frequency, Hz
         I_B = dc bias current, A
         K_{Sur} = a constant dependent on surface conditions

Leakage Noise [16-4, 16-5, 16-6, 16-7]. Most semiconductor and film resistors have leakage noise caused by a conductance on the surface of the device. This noise is highly dependent upon the surface condition, temperature, bias voltage, and physical layout of the device. Leakage noise can be divided into three types, depending on the noise-power spectral density: a $1/f$ frequency component, a white-noise component, and random-burst noise component.

The random-burst noise component is frequently referred to as *popcorn noise*. Its mechanism is not completely understood, but it can be eliminated almost entirely by proper manufacturing process. Popcorn noise is characterized by temporary shifts in the noise baseline or average value of the noise.

The white-noise component has been accredited to a channel in the surface layer. The conductance of this channel is voltage- and temperature-dependent. An expression which approximates the mean-square value of this white-noise component is

$$\overline{V}_{Lek}{}^2 \approx K_V V_B{}^2 \qquad (16\text{-}10)$$

where $\overline{V}_{Lek}$ = rms white-noise voltage component of leakage noise, V
      V_B = bias voltage across the channel, V
      K_V = a constant, dependent upon surface condition and temperature

The $1/f$ component is caused by spontaneous changes in the conductance of the leakage film, and is highly dependent upon bias voltage and surface condition. An expression which approximates the mean square of this component is

$$\overline{I}_{Lek}{}^2 \approx K_{Lek} I_{Lek}{}^2 \ln\!\left(\frac{f_2}{f_1}\right) \qquad (16\text{-}11)$$

where $\overline{I}_{Lek}$ = rms leakage $1/f$ noise current, A
      I_{Lek} = leakage current, A
      K_{Lek} = a constant, dependent upon surface condition for leakage $1/f$ noise
      f_1 = lower cutoff frequency for the equivalent-power bandwidth, Hz
      f_2 = upper cutoff frequency for the equivalent-power bandwidth, Hz

16.2.2. Noise Factor *(F)*. Noise factor is usually defined as the ratio of the total noise power in a load resistance to the noise power in the load resistance resulting from the source. This ratio can be expressed as

$$F = \frac{\text{total noise power in load}}{\text{noise power in load due to source}} \qquad (16\text{-}12)$$

In a detector-preamplifier, the noise power from the detector source is a function of several system parameters (e.g., the temperature of the detector, the detector material type, the bias conditions, and the background). A more meaningful representation of noise factor for comparing preamplifiers can be obtained by substituting the thermal-noise power caused in the load by an equivalent source resistance for the noise power caused by the detector source. With this substitution, the noise factor expression becomes

$$F = \frac{\text{total noise power in load}}{\text{thermal noise power in load due to an equivalent source resistance}} \quad (16\text{-}13)$$

This expression is used most frequently for detector-preamplifier systems. The noise figure, NF, is defined as a decibel expression of noise factor F, or $NF = 10 \log_{10} F$.

Therefore,
$$NF = 10 \log_{10} \frac{\bar{V}^2}{\bar{V}_T{}^2} = 10 \log_{10} \frac{\bar{I}^2}{\bar{I}_T{}^2} \quad (16\text{-}14)$$

$$NF = 10 \log_{10} \left(1 + \frac{\bar{V}_A{}^2}{\bar{V}_T{}^2} \right) \quad (16\text{-}15)$$

$$NF = 10 \log_{10} \left(1 + \frac{\bar{I}_A{}^2}{\bar{I}_T{}^2} \right) \quad (16\text{-}16)$$

where $\bar{V}$ = total root-mean-square noise voltage at load, V

$\bar{V}_T$ = root-mean-square noise voltage at load caused by thermal noise of an equivalent source resistance, V

$\bar{V}_A$ = root-mean-square noise voltage at load caused by preamplifier and bias current, V

$\bar{I}$ = total root-mean-square noise current in load, A

$\bar{I}_T$ = root-mean-square noise current in load caused by thermal noise of an equivalent source resistance, A

$\bar{I}_A$ = root-mean-square noise current in load caused by preamplifier and bias current, A

16.2.3. Noise Models

Resistor Noise Models [16-1] The noise power in a resistor can be modeled in one of two ways: as a mean-square noise-voltage source in series with a noiseless resistor (Figure 16-1(a)), or as a mean-square noise-current source in parallel with a noiseless resistor (Figure 16-1(b)).

For a resistor, R, having only thermal noise,

$$\bar{V}^2 = \bar{V}_T{}^2 = 4kTR\Delta f \quad (16\text{-}17)$$

for the series model and

$$\bar{I}^2 = \bar{I}_T{}^2 = \frac{4kT\Delta f}{R} \quad (16\text{-}18)$$

for the parallel model.

For semiconductor resistors that have excess noise the following voltage (or current) must be summed with either Equation (16-17) or (16-18), as appropriate. Expressions are

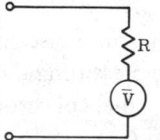
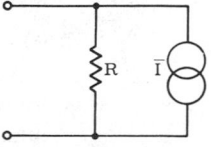

(a) Voltage source in series with a noiseless resistor.

(b) Current source in parallel with a noiseless resistor.

Fig. 16-1. Resistor noise models. $\bar{I}$ equals root mean square (rms) noise current, in amps; R equals resistance, in ohms; and $\bar{V}$ equals rms noise voltage, in volts.

$$\bar{I}_{Sur}^{2} = K_{Sur} I \ln \frac{f_2}{f_1} \tag{16-19}$$

$$\bar{V}_{Sur}^{2} = R^2 K_{Sur} \bar{I} \ln \frac{f_2}{f_1} \tag{16-20}$$

where R is a measured equivalent-noise resistance.

Caution should be exercised in using these models. The multiple, equivalent-noise voltage or current sources sum as power because of the incoherent nature of noise, and not as voltage or current sources directly. These noise components can be expressed as

$$\bar{V}^2 = \bar{V}_1^{\;2} + \bar{V}_2^{\;2} + \;.... \; \bar{V}_n^{\;2} \tag{16-21}$$

$$\bar{I}^{\;2} = \bar{I}_1^{\;2} + \bar{I}_2^{\;2} + \;.... \; \bar{I}_n^{\;2} \tag{16-22}$$

where $\bar{V}_1, \bar{V}_2, \;.... \; V_n$ = incoherent-equivalent-noise voltage sources, V

$\bar{I}_1, \bar{I}_2, \;.... \; \bar{I}_n$ = incoherent-equivalent-noise current sources, A

$\bar{V}^2$ = mean-square total voltage noise, V

$\bar{I}^{\;2}$ = mean-square total current noise, A

Diode Noise Model [16-4, 16-5, 16-6, 16-7]. The equivalent circuit for the low-frequency noise components in a diode can be represented by the model shown in Figure 16-2.

Bipolar Transistor Model [16-4, 16-5, 16-6, 16-7, 16-8]. The equivalent circuit for the low-frequency noise components in a bipolar transistor can be represented as shown in Figure 16-3. The expressions for the square of the currents $\bar{I}_{1s}^{\;2}$ and $\bar{I}_{2s}^{\;2}$ are

$$\bar{I}_{1s}^{\;2} = 2e(I_e + I_{ee})\Delta f + 2eI_{ee}\Delta f \tag{16-23}$$

$$\bar{I}_{2s}^{\;2} = 2eI_{Lek}\Delta f \tag{16-24}$$

where I_{Lek} = dc collector leakage current, A

$\bar{I}_{2s}$ = equivalent-rms-noise current source representing the shot noise in the base-emitter junction, A

$\bar{I}_{2s}$ = equivalent-rms-noise current source representing the shot noise in the collector-base junction, A

I_e = emitter current in a bipolar transistor, A

I_{ee} = base current in a bipolar transistor, A

These currents are correlated for many bipolar transistors; the percent of correlation must be determined experimentally.

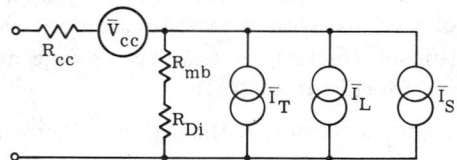

Fig. 16-2. Diode equivalent noise circuit for low-frequencies. R_{Di} equals dc forward resistance of diode at a given current, in ohms; R_{mb} equals modulation part of dc resistance of the diode, in ohms; R_{cc} equals contact resistance of diode, in ohms; $\bar{V}_{cc}$ equals equivalent-rms thermal noise voltage source of the contact, in volts; $\bar{I}_S$ equals equivalent-rms shot noise current source of the junction, in amps; $\bar{I}_L$ equals equivalent-rms surface and leakage noise current source of the junction, in amps; and $\bar{I}_T$ equals equivalent-rms thermal noise current source of the junction, in amps [16-7].

Typical noise curves for the Motorola semiconductor 2N3811A transistor are shown in Figures 16-4 and 16-5. The noise of this transistor is predominately $1/f$ at frequencies below 20 Hz (Figure 16-4), caused by the surface and leakage noise currents $\bar{I}_{2L}$ and $\bar{I}_{1L}$ respectively. The $1/f$ component of the noise current, $\bar{I}_{1L}$, can be approximated from Figure 16-5(b).

For a collector current, $\bar{I}_C$, of 100 μA, a frequency of 10 Hz, and a generator resistance, R_g, of 100 ohms, the noise figure is 8 dB. Therefore, based on the noise figure definition (Section 16.2.2.) and T = 300 K, the total noise voltage, $\bar{V}$, of the amplifier circuit under low-impedance conditions is as follows:

$$\bar{V} \,(10 \text{ Hz}) \approx 3.24 \times 10^{-9} \text{ V} \tag{16-25}$$

where $\bar{V}$ (10 Hz) is the total noise consisting of the rms sum of an amplifier noise component, $\bar{V}_A$, and a Johnson noise component, $\bar{V}_T$, of the generator resistor (100 ohms).

The Johnson noise, $\bar{V}_T$, is 1.29×10^{-9} V (from Section 16.2.1)

$$\bar{V}_T = 1.29 \times 10^{-9} \text{ V} \tag{16-26}$$

and therefore the vector difference is

$$\bar{V}_A \,(10 \text{ Hz}) = 2.97 \times 10^{-9} \text{ V} \tag{16-27}$$

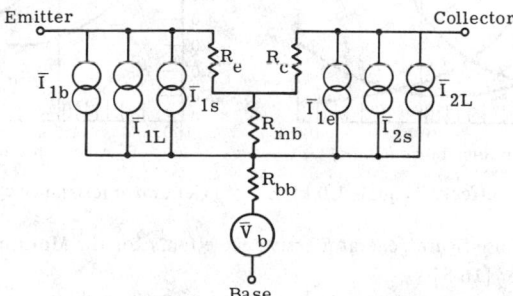

Fig. 16-3. Bipolar transistor equivalent noise circuit for low-frequency noise. $\bar{I}_{1e}$ equals equivalent-rms noise current source due to the collector current, in amps; $\bar{I}_{1b}$ equals equivalent-rms noise current source due to conductance of the base-emitter junction, in amps; $\bar{I}_{1L}$ equals equivalent-rms noise current source due to the shot noise on the base-emitter junction, in amps; $\bar{I}_{2L}$ equals equivalent-rms noise current source due to the surface and leakage current across the collector-base junction, in amps; $\bar{I}_{2s}$ equals equivalent-rms noise current source due to the shot noise in the collector-base junction, in amps; $\bar{V}_b$ equals equivalent-rms noise voltage source due to the thermal or Johnson noise of the base spreading resistance, R_{bb}, in volts; R_{bb} equals base spreading resistance, in ohms; R_{mb} equals modulated part of the dc resistance of the diode junction (base-emitter junction), in ohms; R_e equals dynamic resistance of the base-emitter junction, in ohms; and R_c equals dynamic resistance of the collector-base junction, in ohms [16-7].

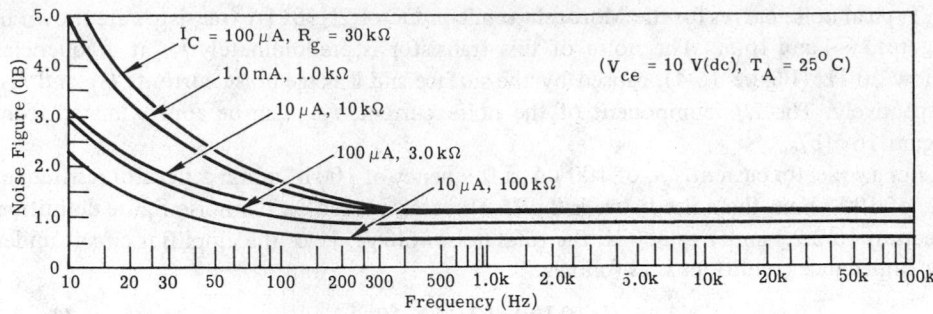

Fig. 16-4. Spot noise-figure frequency-effects for the Motorola semiconductor 2N3811A transistor. I_C equals collector current, in amps; R_g equals generator resistance, in ohms; T_A equals ambient temperature; and V_{ce} equals collector-emitter bias voltage, in volts [16-8].

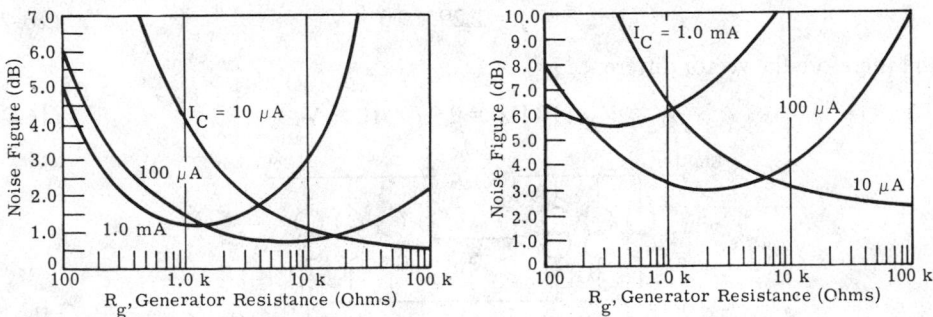

(a) Generator resistance effects; f equals 1.0 kHz.   (b) Generator resistance effects; f equals 10 kHz.

Fig. 16-5. Spot noise-figure generator resistance effects for the Motorola semiconductor 2N3811A transistor [16-8].

where $\overline{V}_A$ (10 Hz) is the amplifier noise component at 10 Hz ($NF = 8.0$ dB from Figure 16-5(b)) from $\sqrt{\overline{V}^2 \,(10 \text{ Hz}) - \overline{V}_T{}^2}$.

If a similar calculation is performed for 1 kHz, the following results will be obtained, based on a noise figure $F = 6.0$ dB from Figure 16-5(a):

$$\overline{V} \,(1 \text{ kHz}) = 2.57 \times 10^{-9} \text{ V} \qquad (16\text{-}28)$$

$$\overline{V}_A \,(1 \text{ kHz}) = 2.22 \times 10^{-9} \text{ V} \qquad (16\text{-}29)$$

where $\overline{V}_A$ (1 kHz) is the amplifier noise component at 1 kHz.

If one assumes a correlation of 0, the difference between voltages $\overline{V}_A$ (10 Hz) and the nonfrequency (white noise) component of $\overline{V}_A$, which is approximated by $\overline{V}_A$ (1 kHz), is the $1/f$ component of the surface and leakage noise at 10 Hz. Therefore

$$\overline{V}_{2L} \,(10 \text{ Hz}) = 1.69 \times 10^{-9} \text{ V} \qquad (16\text{-}30)$$

The shot noise of the base-emitter junction can be calculated as follows:

$$\bar{I}_{1S} = \sqrt{2e(I_e - I_{ee})\Delta f + 2eI_{ee}\Delta f}$$

$$= 5.64 \times 10^{-12} \text{ A} \tag{16-31}$$

A base resistance of 260 ohms can be calculated from a typical silicon junction characteristic of 26 mohms A if $I_C = 100 \ \mu\text{A}$.

Therefore, the shot noise voltages in the base-emitter junction is

$$\bar{V}_{1S} = \bar{I}_{1S} \times (R_e + R_{bb} + R_{mb}) = 1.47 \times 10^{-9} \text{ V} \tag{16-32}$$

The excess noise between $\bar{V}_{1S}$ and $\bar{V}_A$ (1 kHz) is due to the noise of the base spreading resistance, R_{bb}, and excess surface and leakage noise. A similar calculation can be performed for a high generator resistance (20 kohms). The results of this calculation are

$$\bar{V} (10 \text{ Hz}) = 32.7 \times 10^{-9} \text{ V} \ (NF = 5.1 \text{ dB}) \tag{16-33}$$

$$\bar{V}_T (20 \text{ kohms}) = 18.2 \times 10^{-9} \text{ V} \tag{16-34}$$

$$\bar{V}_A (10 \text{ Hz}) = 27.2 \times 10^{-9} \text{ V} \tag{16-35}$$

$$\bar{V} (1 \text{ kHz}) = 20.4 \times 10^{-9} \text{ V} \ (NF = 1.0 \text{ dB}) \tag{16-36}$$

$$\bar{V}_A (1 \text{ kHz}) = 9.26 \times 10^{-9} \text{ V} \tag{16-37}$$

$$\bar{V}_{2L} (10 \text{ Hz}) = 17.29 \times 10^{-9} \text{ V} \tag{16-38}$$

The noise voltage, $\bar{V}_{2L}$, is dominated by the surface and leakage current $1/f$ components of the collector-base junction, I_{2L}. Therefore

$$\bar{I}_{2L} \approx \frac{\bar{V}_{2L}}{20 \text{ kohms}} \approx 0.865 \times 10^{-12} \text{ A} \tag{16-39}$$

where $\bar{I}_{2L} = 1/f$ component of the surface and leakage current, I_{2L}. The contribution of this current to the low-impedance (100 ohms) calculation for $\bar{V}_{2L}$ is 86.5×10^{-12} V. If there is no correlation, the value for $\bar{V}_{1L}$ (the $1/f$ noise voltage component due to base-emitter noise current) for a 100 ohm generator, resulting from base-emitter leakage, is

$$\bar{V}_{1L} (10 \text{ Hz}) = 7.5 \times 10^{-10} \text{ V} \tag{16-40}$$

Therefore, if one assumes the base resistance is 260 ohms, the $1/f$ component of the leakage current can be calculated as follows:

$$\bar{I}_{1L} (10 \text{ Hz}) = 2.9 \times 10^{-12} \text{ A} \tag{16-41}$$

where

$$\bar{I}_{1L} = \frac{\bar{V}_{1L}}{R_e + R_{mb} + R_{bb}} \tag{16-42}$$

The noise sources $\bar{I}_{1L}$ and $\bar{I}_{2L}$ are usually partially correlated and must be determined experimentally for each device.

Junction Field-Effect Transistor (JFET) [16-9, 16-10]. The equivalent circuit for the low-frequency noise in a junction field-effect transistor can be represented as shown in Figure 16-6. The general descriptions and/or the defining equations for the current and voltage noise sources for the JFET are

$$\bar{I}_1 = \left[2eI_g\Delta f + \bar{I}_{1L}{}^2\right]^{1/2} \tag{16-43}$$

$$\bar{I}_2 = \left[4kT\Delta f g_m B_2 + \bar{I}_{2L}{}^2\right]^{1/2} \tag{16-44}$$

$$\bar{V}_c = \left[\frac{B_1 4kT\Delta f}{g_m}\right]^{1/2} \tag{16-45}$$

$$\bar{V}_d = \left[4kTR_{dc}\Delta f\right]^{1/2} \tag{16-46}$$

$$\bar{V}_s = \left[4kTR_{sc}\Delta f\right]^{1/2} \tag{16-47}$$

where the symbols are defined, in part, in Figure 16-6 and Equations (16-43) through (16-47) and

$\bar{I}_{1L}$ = equivalent noise current source due to the surface noise power which modulates the gate, A

$\bar{I}_{2L}$ = equivalent noise current source due to the surface noise power between source and drain, A

B_1, B_2 = dimensionless numbers which depend upon the particular device and the bias conditions

A typical noise versus frequency curve for a Siliconix 2N5564-66 JFET is shown in Figure 16-7. This curve is for a shorted gate transistor operating in the common-source configuration. A typical value for the common-source forward transconductance, for this device operating at a 5 mA bias, is 10 mmho. Therefore $R_{ds} \approx 1/g_m = 100$ ohms.

Fig. 16-6. JFET equivalent-noise circuit. C_{gs} equals gate-to-gate source capacitance, in farads; C_{gd} equals gate-to-drain capacitance, in farads; g_m equals low-frequency transconductance, in ohms; $\bar{V}_1$ equals voltage between gate and source of bulk material; R_{ds} equals drain-to-source resistance, in ohms; R_{sc} equals source-contact resistance, in ohms; R_{dc} equals drain-contact resistance, in ohms; $\bar{I}_1$ equals noise due to shot and $1/f$ noise of gate current; $\bar{I}_2$ equals noise due to thermal and $1/f$ noise of the drain current; $\bar{V}_c$ equals channel thermal noise; $\bar{V}_d$ equals thermal noise due to drain-contact resistance; and $\bar{V}_s$ equals thermal noise due to source-contact resistance.

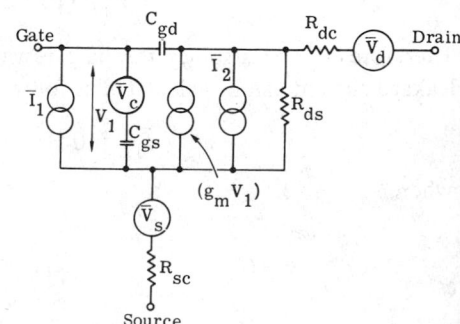

By assuming that $R_{dc} \ll R_{ds}$ and $R_{sc} \ll R_{ds}$, one can make the following approximation:

$$\bar{V}_1 \simeq \bar{V}_c = \sqrt{\frac{B_1 4kT\Delta f}{g_m}} \qquad (16\text{-}48)$$

Therefore,

$$g_m \bar{V}_1 = \sqrt{B_1 4kT\Delta f g_m} \qquad (16\text{-}49)$$

The current generator, $\bar{I}_1{}^2$, can be neglected because of its small value.

The total noise which can be represented by the noise current at the drain of the JFET is

$$\bar{I}_2 = g_m{}^2 \bar{V}_1{}^2 + \bar{I}_2{}^2 = |(B_1 + B_2)| 4kTg_m \Delta f + \bar{I}_{2L}{}^2 \qquad (16\text{-}50)$$

where $|(B_1 + B_2)|$ denotes the absolute value of $B_1 + B_2$.

$$\bar{V} = \sqrt{\bar{I}_T{}^2 R_{ds}{}^2} \qquad (16\text{-}51)$$

$$\bar{V} = \sqrt{\frac{|(B_1 + B_2)| 4kT\Delta f}{g_m} + I_{2L}{}^2 R_{ds}{}^2} \qquad (16\text{-}52)$$

The minimum absolute value for $B_1 + B_2$ is one. For this minimum value, and for $g_m = 10$ mmho, the noise resulting from channel effect is 1.27×10^{-9} V. In Figure 16-7, where $\bar{V}_n = 1.4 \times 10^{-9}$ V at 10 kHz, the source-drain surface noise current is*

$$\bar{I}_{2L} = 5.9 \times 10^{-12} \text{ A} \qquad (16\text{-}53)$$

Typical calculated values for surface and leakage noise obtained from the leakage current specification are

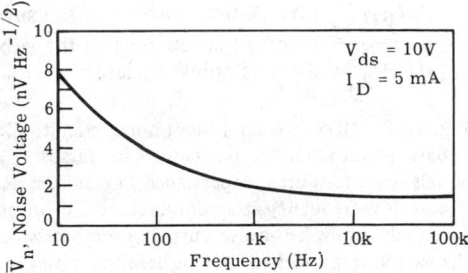

Fig. 16-7. Typical curve for Siliconix 2N5564-66 JFET. V_{ds} equals drain-to-source bias voltage; I_d equals drain current; and V_n equals equivalent input noise voltage [16-11].

$$\bar{I}_{2L} = 4 \times 10^{-15} \text{ A} \qquad (16\text{-}54)$$

Since the value obtained by assuming that all excess channel noise is due to surface and leakage is more than three orders of magnitude greater than the value obtained by assuming Johnson noise characteristics for the leakage current, part of the excess channel noise must be due to channel effects, or R_d and R_s must be only slightly less than R_{ds}.

---

*This assumes that all excess noise is due to the surface and leakage noise.

The $1/f$ component of the surface and leakage noise current, $\bar{I}_{2L}$, at 10 Hz is

$$\bar{I}_{2L} = \sqrt{(\bar{V}_n^2 g_m^2)\Big|_{10\text{ kHz}} - (\bar{V}_n^2 g_m^2)\Big|_{10\text{ Hz}}} \qquad (16\text{-}55)$$

$$= 7.88 \times 10^{-11} \text{ A} \qquad (16\text{-}56)$$

Metal-Oxide Field-Effect Transistor [16-9, 16-12, 16-13, 16-14, 16-15]. Figure 16-8 represents the model for a metal-oxide or insulated-gate field-effect transistor, MOSFET and IGFET. The general descriptions, and/or defining equations for the current and voltage noise sources of the MOSFET, are

$$\bar{I}_2 = \sqrt{4kT\Delta f g_m B_2 + \bar{I}_{2L}^{\,2}} \qquad (16\text{-}57)$$

$$\bar{I}_{\text{sub }1} = \sqrt{2eI_{BD}\Delta f} \qquad (16\text{-}58)$$

$$\bar{I}_{\text{sub }2} = \sqrt{2eI_{BS}\Delta f} \qquad (16\text{-}59)$$

$$\bar{V}_s = \sqrt{4kT\Delta f R_{sc}} \qquad (16\text{-}60)$$

$$\bar{V}_d = \sqrt{4kT\Delta f R_{dc}} \qquad (16\text{-}61)$$

$$\bar{V}_c = \sqrt{\frac{B_1 4kT\Delta f}{g_m}} \qquad (16\text{-}62)$$

where the symbols are defined, in part, in Figure 16-8 and

$\bar{I}_{2L}$ = equivalent-noise current source which is due to the surface state and leakage current between the source and drain (Part of I_2), A

I_{BD} = reverse-bias current of the substrate-to-drain junction, A

I_{BS} = reverse-bias current of the substrate-to-source junction, A

B_1, B_2 = dimensionless variables

Fig. 16-8. MOSFET equivalent-noise circuit. C_{gd} equals gate-to-drain capacitance, in farads; C_{gs} equals gate-to-source capacitance, in farads; g_m equals low-frequency transconductance, in ohms; $\bar{I}_1$ equals equivalent-noise current source between the gate and source in a FET or base-to-emitter in a bipolar transistor, in amps; $\bar{I}_{\text{sub }1}$ equals substrate modulation-noise current coupled to the drain, in amps; $\bar{I}_2$ equals equivalent-noise current source between the drain-to-source in FETs or collector-to-base of BPTs, in amps; $\bar{I}_{\text{sub }2}$ equals substrate modulation-noise current coupled to the gate, in amps; R_{dc} equals drain-contact resistance, in ohms; R_{ds} equals drain-to-source resistance, in ohms; R_{sc} equals source-contact resistance, in ohms; V_1 equals voltage between gate and source of the bulk material for FET, in volts; $\bar{V}_d$ equals equivalent-rms thermal noise due to drain-contact resistance, in volts; and $\bar{V}_s$ equals equivalent-rms thermal noise due to source-contact resistance, in volts.

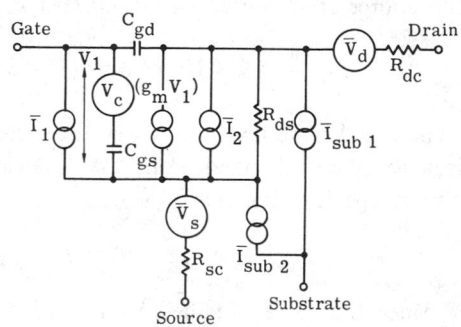

A typical curve of noise voltage versus frequency for Siliconix M119 P Channel MOSFET is shown in Figure 16-9.

The noise at high frequencies approaches the channel noise, or

$$\overline{V}_n(100 \text{ kHz}) =$$

$$\sqrt{\frac{|(B_1 + B_2)|\Delta f 4kT}{g_m}} \quad (16\text{-}63)$$

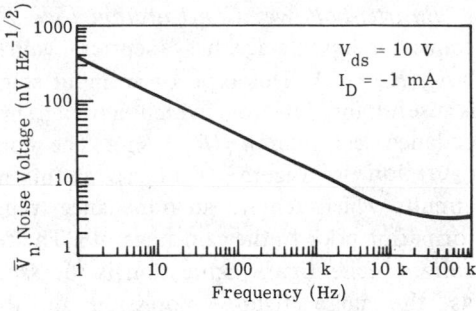

Fig. 16-9. Typical short-circuit equivalent input noise voltage versus frequency for the Siliconix M119·P channel MOSFET. V_{ds} equals drain-to-source bias voltage; I_D equals drain current; and $\overline{V}_n$ equals equivalent input noise voltage.

The $1/f$ component of the surface and leakage noise current, $\overline{I}_{2L}$, at 1 Hz is $\overline{I}_{2L} = V_n g_m$, where $\overline{I}_{2L} = 3.80 \times 10^{-10}$ A (assuming $g_m = 1000 \ \mu$mhos).

16.3 Active Circuits

16.3.1. Bipolar Transistors.
Bipolar transistors are used in three basic configurations for the detector-preamplifier systems. Examples of these configurations are discussed in this section.

Common-Emitter Configurations (See Figure 16-10). The common-emitter, bipolar-transistor input stage of an amplifier has both current gain ($\simeq$ h$_{fe}$) and voltage gain ($\simeq R_C/(R_S + R_e)$) where h_{fe} is small-signal amplification factor for the transistor and R_e is the emitter resistance. This type of input stage is useful for detectors which have an impedance greater than the emitter resistance of the transistor, and less than h_{fe} ($R_S + R_e$). The input noise of the bipolar transistor, Q_1, and the detector signal and noise will be amplified by the gain, A_1, of the input stage. The resultant noise figure expression is

$$NF = 10 \log_{10}\left(1 + \frac{\overline{V}_{A1}^2}{\overline{V}_T^2} + \frac{\overline{V}_{A2}^2}{A_1^2 \overline{V}_T^2}\right)$$

$$(16\text{-}64)$$

where $\overline{V}_{A1}$ = rms voltage noise of the transistor, V

$\overline{V}_T$ = rms thermal-noise voltage of equivalent source resistor, V

$\overline{V}_{A2}$ = rms voltage noise of gain stage (A_2), V

A_1 = voltage gain of transistor (Q_1), V

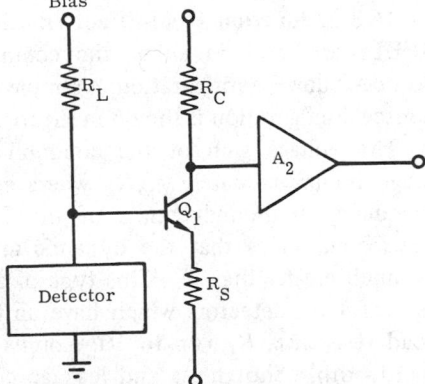

Fig. 16-10. Common-emitter bipolar transistor amplifier. R_L equals load resistance, in ohms; R_C equals collector resistance, in ohms; A_2 equals voltage gain of the gain stage; Q_1 equals bipolar transistor; and R_S equals source resistance, in ohms

Emitter-Follower Configuration (See Figure 16-11). The emitter-follower, bipolar-transistor input stage has a current gain, $\approx h_{fe}$, and a voltage gain of less than one, $\approx R_S(R_S + R_e)$. This type of an input stage is useful for detectors which have an impedance less than $h_{fe}(R_s + R_e)$. The configuration in Figure 16-11 is useful in circuits which require an impedance transformation prior to the gain stage, A_2. The resultant noise figure expression is the same as the noise figure expression of the common-emitter stage when the voltage gain of the emitter-follower stage ($A_1 < 1$) is used.

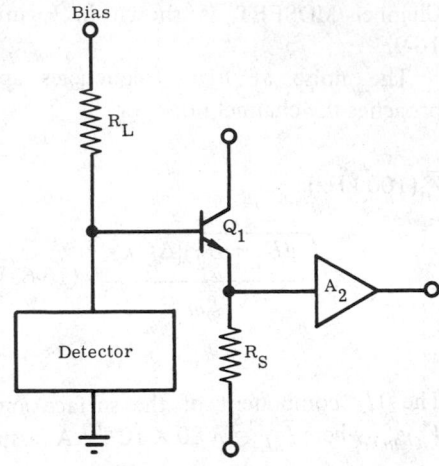

Fig. 16-11. Emitter-follower bipolar transistor. R_L equals load resistance, in ohms; Q_1 equals bipolar transistor; A_2 equals voltage gain of the gain stage; and R_S equals source resistance, in ohms.

Common-Base Configurations (See Figure 16-12.). The common-base, bipolar-transistor input stage has a voltage gain greater than unity, $\approx R_C/(R_{Det} + R_e)$, and a current gain slightly less than unity, $\approx h_{fe}(1 + h_{fe})$. This type of input stage is useful for low-impedance, photoconductive detectors. The noise figure expression is the same as for the common-emitter configuration.

The mean-square noise of the bipolar transistor stage, $\overline{V}_{A1}^2$, can be reduced for any given transistor by using parallel devices. If the parallel devices are operating at the same bias conditions, the net reduction in noise that can be achieved is $\overline{V}_{A1}^2/N$, assuming the noises are not correlated.

In the above expression, N is the number of parallel devices. The input-capacitance noise currents, and the bias current required for the first stage will increase proportionally.

16.3.2. Junction Field-Effect Transistors (JFETs).
JFETs are used in either the common-source or source-follower configuration. A sample of a common-source configuration is shown in Figure 16-13.

The voltage gain of the common-source, JFET stage is approximately $g_m R_c$ where g_m is the low-frequency transconductance of the FET. This expression assumes that the dynamic drain resistance is much greater than R_c. This type of an input stage is useful for detectors which have an impedance or load resistance, R_L, up to 10^{10} ohms with reduced bandwidth. Shot noise and leakage currents across the reverse-biased gate junction begin to limit the effectiveness of this circuit at an input impedance greater than 10^{10} ohms unless a cooled JFET is used. Shot noise and leakage current across the reverse-biased gate junction will be reduced by an

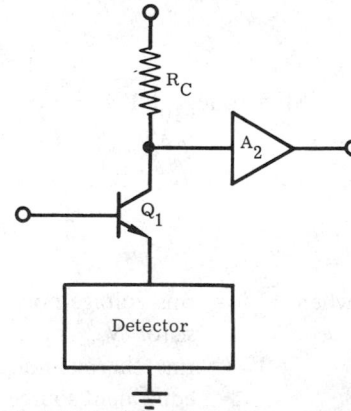

Fig. 16-12. Common-base stage bipolar transistor. R_C equals collector resistance, in ohms; A_2 equals voltage gain of the gain stage; and Q_1 equals bipolar transistor.

order of magnitude at 77 K, which will extend the input impedance range to greater than 10^{11} ohms.

An example of a source-follower JFET configuration is shown in Figure 16-14. The voltage gain of this stage is approximately equal to $R_f/(R_f + (1/g_m))$, which is less than unity. The source-follower JFET input stage is useful for detectors which have an input resistance of up to 10^{10} ohms and has a higher bandwidth than the common-source configuration. The primary application for this type of circuit occurs when an impedance transform is desired between the JFET stage

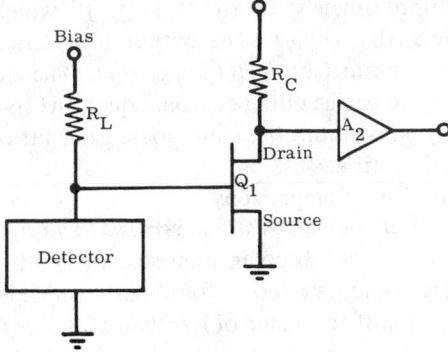

Fig. 16-13. Common-source JFET. R_L equals load resistance, in ohms; R_C equals collector resistance, in ohms; A_2 equals voltage-gain of the gain stage; and Q_1 equals JFET.

and the gain stage, A_2. The output impedance of the source-follower stage is approximately $R_f/(1 + g_m R_f)$.

The noise figure for the JFET's configurations can be obtained from Equation (16-64) by substituting the appropriate gains into the expression where $A_1 = R_f/(R_f + (1/g_m))$.

JFETs can be used in applications which require the input stage to operate at cryogenic temperatures for optimum performance. Silicon JFETs remain operational at 30 K and germanium JFETs have been operated at 4 K.

16.3.3. Metal-Oxide Semiconductor Field-Effect Transistors (MOSFETs).

MOSFETs can be used in either the common-source or source-follower configuration. An example of a common-source configuration is shown in Figure 16-15. The voltage gain of the common-source MOSFET state is approximately $g_m R_c$, where g_m is the low-frequency transconductance of the FET. The MOSFET input stage is useful for detectors which have an impedance or load resistance, R_L, of up to 10^{11} ohms.

MOSFETs are most commonly used in the source-follower configuration. A sample configuration is shown in Figure 16-16. The voltage gain of the source follower, A_1, is

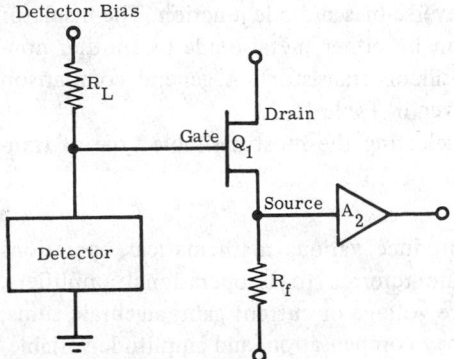

Fig. 16-14. Source-follower JFET. R_L equals load resistance, in ohms; R_f equals follower resistance, in ohms; Q_1 equals JFET; and A_2 equals voltage gain of the gain stage.

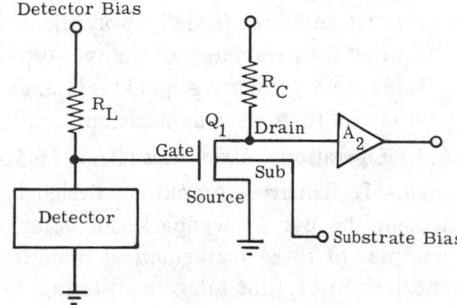

Fig. 16-15. Common-source MOSFET. R_L equals load resistance, in ohms; R_C equals collector resistance, in ohms; A_2 equals voltage gain of the gain stage; and Q_1 equals MOSFET.

approximately $R_f/(R_f + (1/g_m))$, which is less than unity. The output impedance is approximately $R_f/(1 + g_mR_f)$. The noise figure can be obtained from Equation (16-64) by substituting the appropriate gains into the expression.

16.3.4. Comparisons

Bipolar Transistors versus Field-Effect Transistors. Bipolar transistor preamplifiers are usually preferred for circuits which have an input resistance of less than 500 ohms because a lower noise figure can be obtained. Parallel bipolar transistors frequently used tc optimize the noise figure in preamplifier operating from inputs of less than 100 ohms.

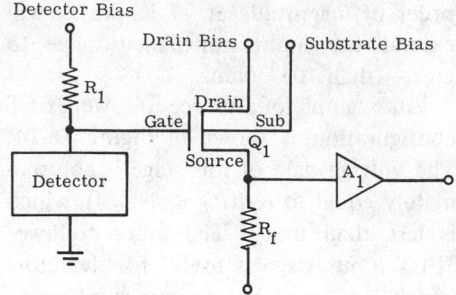

Fig. 16-16. Source-follower MOSFET. R_L equals load resistance, in ohms; R_f equals follower resistance, in ohms; A_1 equals voltage gain of amplifier; and Q_1 equals MOSFET.

For an input resistance between 500 ohms and 10 kohms, either bipolar transistors or JFETs could yield optimum noise figures. The choice of device is usually determined by system constraints (e.g., the gain required, the bandpass, the available voltages and currents, the common-ground rejection, or the power supply rejection), or simply by the circuit designer's preference.

FETs will usually yield a lower noise figure than bipolar transistors for preamplifiers operating from an input resistance greater than 10 kohms. Hybrid preamplifier configurations using a FET first stage with bipolar postamplifier stages are good designs for low-noise preamplifiers.

Bipolar transistors have an operating temperature range of 218 to 398 K. FETs have been operated at temperatures below 4 K. They usually are not recommended for use above 373 K.

JFETs versus IGFETs or MOSFETs. The basic difference between a JFET and a MOSFET or IGFET is that the JFET uses a reverse-biased diode junction to isolate the gate from the bulk material and the source or drain, whereas the MOSFET and IGFET use an insulator to isolate the gate. The source and drain in the MOST and IFGET are, in turn, isolated from the bulk material by a reverse-biased diode junction. The insulator in a MOST is metal oxide; in an IGFET it can be either metal oxide or another non-conductive material (usually polysilicon in a silicon transistor). A general comparison of some of the parameters of the two types is given in Table 16-2.

Table 16-3 presents a quick reference for selecting the most desirable type of transistor circuit for a given application.

16.4. Operational Amplifiers (OA) [16-16]

16.4.1. Theory. Amplifiers designed to produce various mathematical operations through the use of feedback characteristics are referred to as operational amplifiers. Examples of these mathematical operations are voltage or current gain, algebraic sums, time derivatives, time integrals, filtering, frequency compensation, and amplitude-variable-gain control (log suppression).

The most versatile operational amplifiers are the differential input amplifiers. These amplifiers can be used in either the current-mode (inverting) configuration or the voltage-mode (noninverting) configurations. A fundamental OA circuit used in the voltage-mode

Table 16-2. Transistor Circuits and Applications

Applications	JFET	MOST	IGFET
Gate noise	Shot noise across isolation diode to gate.	Does not have shot noise at gate.	Does not have shot noise at gate.
Channel noise	Buried channel results in low $1/f$ noise.	Surface channel results in $1/f$ noise. Can be reduced by ion inplantation to bury channel.	Surface channel results in $1/f$ noise. Can be reduced by ion inplantation to bury channel.
Substrate noise	The gate is the substrate. Some channel modulation of noise is usually observable.	Up to 80% of substrate noise has been observed modulating the channel.	Up to 80% of substrate noise has been observed modulating the channel.
Transconductance	Rate of decrease in g_m with a reduction in temperature and bias current is greater at the low extreme than in MOSTs.	Rate of decrease in g_m with a reduction in temperature and bias current is less at low extremes than in JFETs.	–
Gate capacitance	Varies with bias.	Remains relatively constant.	Remains relatively constant.
Gate leakage current	Has leakage currents across the reverse-biased diode junctions which are usually an order of magnitude greater than the leakage of the insulation in a MOST or IGFET.	Leakage through the insulation is lower by an order of magnitude than the leakage currents in a JFET.	Leakage through the insulation is lower by an order of magnitude than the leakage currents in a JFET.
Temperature effects	Has a g_m of less than 10% of its room temperature value at 30 K.	Maintains greater than 50% of its room temperature g_m at 4 K.	–
Element density	Obtainable density of devices per chip size is less than for MOSTs and IGFETs.	Obtainable density of devices per chip size is greater than for a comparable JFET.	Obtainable density of devices per chip size is greater than for a comparable JFET.
Sensitive to change by static charge buildup	–	Sensitive to damage by static charge buildup.	Sensitive to damage by static charge buildup.

Table 16-3. Types of Transistor Circuits

Transistor Type	Amplifier Configuration	Input Z Range (ohms)	State of Art Input Noise Voltage (rms nV) -55 to + 75° C		Operating Temperature Range (K)	Comments
			1 kHz	100 kHz		
Bipolars	Common-Emitter	20×10^3	0.5	0.25	220 - 398	—
	Emitter-Follower	20×10^3	0.7	0.35	220 - 398	—
	Common-Base	10^3	0.5	0.25	220 - 398	—
JFET	Common-Source	10^{10}	2.0	1.5	30 - 398	Silicon units
	Source-Follower	10^{10}	2.0	1.5	30 - 398	—
MOSTs and IGFETs	Common-Source	10^{12}	12.0	4.0	4 - 398	Input Z range for cooled device
	Source-Follower	10^{12}	12.0	4.0	4 - 398	

configuration is shown in Figure 16-17. The following expressions can be developed from this circuit model:

$$V_o = A_o \xi \tag{16-65}$$

$$\xi = V_{in} - \frac{V_o Z_{ref}}{Z_{fb} + Z_{ref}} \tag{16-66}$$

where A_o = open-loop gain of the OA, V V^{-1} or A A^{-1}
   ξ = differential input voltage (error voltage), V
   V_{in} = input signal voltage, V
   V_o = voltage output from the OA. V_o is in phase with the plus (+) input terminal of the OA and 180° out of phase with the minus (−) input terminal, V
   Z_{ref} = reference impedance, ohms
   Z_{fb} = feedback impedance, ohms

Combining these expressions yields the following equation for a closed-loop, OA:

$$A_c = \frac{V_o}{V_{in}} = \frac{Z_{ref} + Z_{fb}}{Z_{ref} + \dfrac{(Z_{fb} + Z_{ref})}{A_o}} \tag{16-67}$$

where A_c is the closed-loop gain of the OA. This expression is frequently written as

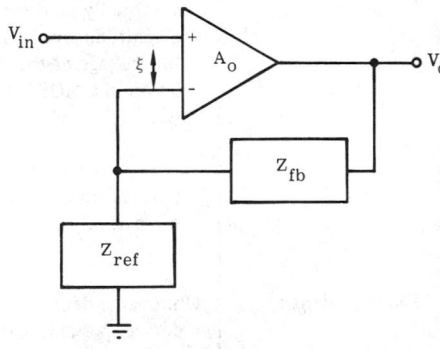

Fig. 16-17. Voltage-mode operational amplifier circuit. V_{in} equals input signal voltage, in volts; ξ equals differential input voltage (error voltage); A_o equals open-loop gain of the operational amplifier, in volts/volt or amps/amp; V_o equals output signal voltage, in volts; Z_{fb} equals feedback impedance, in ohms; and Z_{ref} equals reference impedance, in ohms.

$$A_c = \frac{V_o}{V_{in}} = \frac{\frac{1}{\beta}}{1 + \frac{1}{A_o \beta}} \tag{16-68}$$

where

$$\beta = \frac{1}{1 + \frac{Z_{fb}}{Z_{ref}}} \tag{16-69}$$

This circuit configuration is primarily used when high input and low output impedance are desired while retaining some voltage gain without a signal-phase inversion.

Figure 16-18 shows a fundamental circuit of an OA used in the current-mode or inverting configuration. The following expressions can be developed from this circuit model:

$$V_o = A_o \xi \tag{16-70}$$

$$\frac{V_{in} - \xi}{Z_S} = - \frac{V_o + \xi}{Z_{fb}} \tag{16-71}$$

where Z_S is the source impedance.
Combining these expressions yields the following equation for the closed-loop gain, A_c, of the current-mode OA.

$$A_c = \frac{V_o}{V_{in}} = \frac{-Z_{fb}}{Z_S} \frac{A_o Z_S}{(1 + A_o)Z_S + Z_{fb}}$$

$$\tag{16-72}$$

This expression is frequently written as

$$A_c = \frac{1 - \frac{1}{\beta}}{1 + \frac{1}{A_o \beta}} \tag{16-73}$$

where

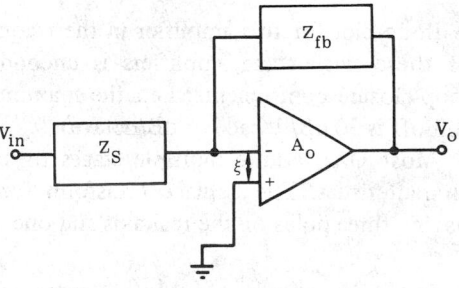

Fig. 16-18. Current-mode operational amplifier circuit. V_{in} equals input signal voltage, in volts; Z_S equals source impedance, in ohms; Z_{fb} equals feedback impedance, in ohms; A_o equals open-loop gain of the operational amplifier, in volts/ volt or amps/amp; V_o equals voltage output, in volts; and ξ equals differential input voltage (error voltage).

$$\beta = \frac{1}{1 + \frac{Z_{fb}}{Z_S}} \tag{16-74}$$

This circuit configuration is primarily used for high gain when a nominal input impedance, low output impedance, and overall signal-phase inversion are desired.

16.4.2. Characteristics. The open-loop gain, A_o, of the OA can contain one or more poles and possible zeroes, and can be a function of frequency. A single-gain-stage amplifier will usually contain a single pole at frequencies below the cut-off frequency of the transistor. The Laplace transformation of the open-loop gain for an amplifier having a single pole on the real axis can be represented by

$$A_o(s) = \frac{A_{ol}}{1 + \tau_o s} \tag{16-75}$$

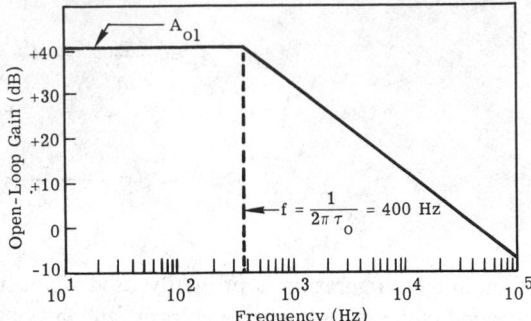

Fig. 16-19. Open-loop gain of the operational amplifier versus frequency for an amplifier having a single pole on the real axis. A_{ol} equals open-loop gain at low frequency, volts/volt or amp/amps.

where A_{ol} = the low-frequency open-loop gain, V V^{-1} or A A^{-1}
τ_o = the time constant of the single pole, sec
s = the Laplace transform variable, sec^{-1}

A Bode plot for this amplifier in the frequency domain is shown in Figure 16-19. Either of these single-stage amplifiers is unconditionally stable for a system using resistive, loop-closure components; i.e., the maximum phase shift is 90° and the maximum gain-roll-off is 20 dB/decade (6 dB/octave).

Most OAs require multiple stages to achieve open-loop gain and desired functional characteristics. The Laplace transform function of the open-loop gain for an amplifier having three poles on the real axis and one pole at zero can be represented by

$$A_o(s) = \frac{A_o(1 + \tau_a s)}{(1 + \tau_1 s)(1 + \tau_2 s)(1 + \tau_3 s)} \tag{16-76}$$

where τ_1, τ_2, τ_3 = the time constants of the poles, sec
τ_a = the time constant of the zero, sec

A Bode plot for this amplifier in the frequency domain is shown in Figure 16-20. This amplifier will tend to be unstable for gains of less than 82 dB when used as an OA with resistive, loop-closure components. It may, therefore, be desirable to relocate the poles and zero in the frequency spectrum by adding compensation. The location of a pole or zero is the frequency at which the magnitude of the imaginary part equals the real part. The effect of a pole is to decrease the roll-off slope by 20 dB/decade and the effect of a zero is to increase the slope by 20 dB/decade. The open-loop gain in the frequency domain for a compensated OA could be represented by the compensated OA curve in Figure 16-20.

The β term in the closed-loop (Equation (16-68)) for an OA can be complex if the dominant loop-closure components are reactive. This will frequently be the case for bandpass amplifiers, active filters, time-integral and time-derivative amplifiers, frequency-compensation amplifiers, etc.

A transfer function and stability analysis should be performed to ensure a stable loop closure. See Reference [16-17] for a discussion of these techniques.

16.5. Detector-Preamplifier Circuit

The two basic types of OAs used for detector-preamplifiers are voltage-mode (non-inverting) and current-mode (inverting).

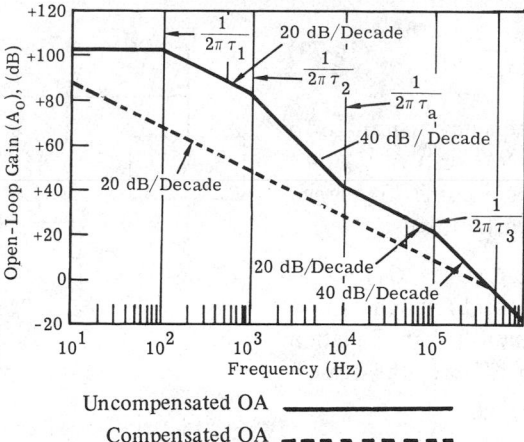

Fig. 16-20. Open-loop gain versus frequency for an operational amplifier having three poles on the real axis and one pole at zero. Poles: τ_1 equals 1.6×10^{-3} seconds; τ_2 equals 1.6×10^{-4} seconds; and τ_3 equals 1.6×10^{-6} seconds. Zero: τ_a equals 1.6×10^{-5} seconds.

16.5.1. Voltage Mode. The circuit for a voltage-mode preamplifier is shown in Figure 16-21. From Equation (16-69)

$$\beta = \frac{1}{1 + \dfrac{Z_{fb}}{Z_{ref}}} = \frac{R_2(1 + j\omega R_1 C_1)}{R_2(1 + j\omega R_1 C_1) + R_1(1 + j\omega R_2 C_2)} \tag{16-77}$$

where C_2 and C_1 = capacitance across the reference and feedback resistances R_2 and R_1, respectively

C_3 = the input capacitance, the OA plus stray capacitance

Therefore, from Equation (16-68),

$$A_c = \frac{V_o}{V_{in}} = \frac{\dfrac{1}{\beta}}{1 + \dfrac{1}{A_o \beta}} = \frac{1 + j\omega R_p(C_1 + C_2)}{\dfrac{R_p}{R_1}(1 + j\omega R_1 C_1) + \dfrac{1}{A_o}[1 + j\omega R_p(C_1 + C_2)]} \tag{16-78}$$

where $R_p = \dfrac{R_1 R_2}{R_1 + R_2}$

j = imaginary operator = $\sqrt{-1}$

ω = frequency, rad sec^{-1}

A Bode plot for an OA containing the above poles and zero of Equation (16-78) is shown in Figure 16-22. For this plot, the open-loop gain, A_o, of the OA can be represented by the following expression:

$$A_o = \frac{A_{ol}}{1 + j\omega\tau_o} \qquad (16\text{-}79)$$

where A_{ol} = low-frequency open-loop gain, V V^{-1} or A A^{-1}

τ_o = single-pole corner-frequency time constant, sec

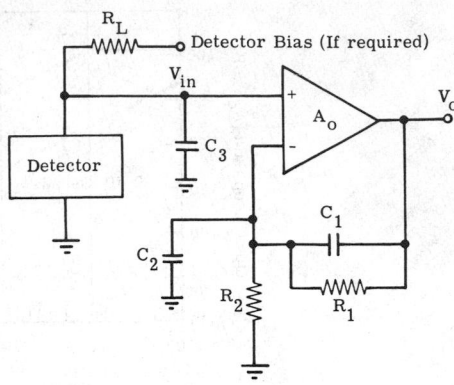

For the conditions $\omega R_p(C_1 + C_2) \ll 1$, and $\omega R_1 C_1 \ll 1$ (case 1) which can be mathematically accomplished in Equation (16-78) by setting ω to zero, the closed-loop gain is

$$A_{c1} = \frac{R_1 + R_2}{R_2 + \dfrac{1}{A_o}(R_1 + R_2)} \qquad (16\text{-}80)$$

For the conditions $\omega R_p(C_1 + C_2) \ll 1$, and $\omega R_1 C_1 \gg 1$ (case 2), obtained by setting $1 + j\omega R_p(C_1 + C_2) = 1$ and $1 + j\omega R_1 C_1 + j\omega R_1 C_1$, the second closed-loop is

Fig. 16-21. Voltage-mode operational amplifier circuit R_L equals load resistance, in ohms; V_{in} equals input signal to OA; C_3 equals input capacitance of OA plus stray capacitance; C_2 equals capacitance across the reference resistance R_2; A_o equals open-loop gain of the operational amplifier; V_o equals output signal from OA; C_1 equals capacitance across the feedback resistance R_1; R_2 equals reference resistance; and R_1 equals feedback resistance, in ohms.

$$A_{c2} = \frac{1}{j\omega R_p C_1 + \dfrac{1}{A_o}} \qquad (16\text{-}81)$$

For $\omega R_p(C_1 + C_2) \gg 1$, $\omega R_1 C_1 \gg 1$ (case 3), the closed-loop gain is

$$A_{c3} = \frac{C_1 + C_2}{C_1 + \dfrac{1}{A_o}(C_1 + C_2)} \qquad (16\text{-}82)$$

When the condition $\dfrac{A_{ol}}{j\omega\tau_o} < \dfrac{C_1 + C_2}{C_1}$

occurs (case 4), the open- and closed-loop response will be the same, and can be approximated by

$$A_{c4} = \frac{A_{ol}}{j\omega\tau_o} \qquad (16\text{-}83)$$

The input voltage was defined, for the derivation of the voltage-mode equations, as the signal which occurs at the input of the OA. Therefore, the expression for the closed-loop gain, Equation (16-78), is independent of detector parameters, load resistance, R_L, and the input and stray capacitance across the detector, C_3. However, the frequency response of the detector-preamplifier combination will be influenced by these parameters.

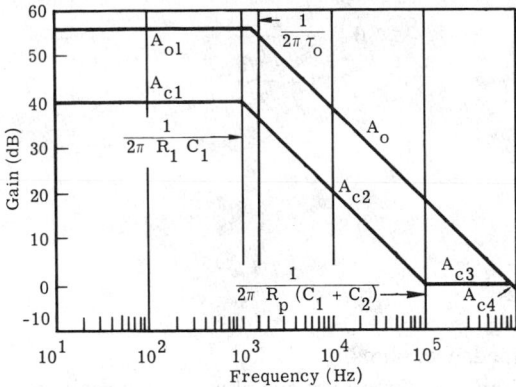

Fig. 16-22. Voltage-mode amplifier Bode plot for the circuit in Figure 16-21. A_{ol} equals 56 dB and τ_o equals 1.06×10^{-4} seconds.

A Bode plot for the closed-loop gain of a voltage-mode OA is shown in Figure 16-22 where

$$A_o = 56 \text{ dB} \qquad\qquad C_2 = 7.5 \times 10^{-12} \text{ F}$$
$$\tau_o = 1.06 \times 10^{-4} \text{ sec} \qquad R_L = 10^3 \text{ ohms}$$
$$R_1 = 10^5 \text{ ohms} \qquad\qquad R_{Det} = 10^6 \text{ ohms}$$
$$C_1 = 1.6 \times 10^{-9} \text{ F} \qquad\quad C_3 = 12.5 \times 10^{-12} \text{ F}$$
$$R_2 = 10^3 \text{ ohms}$$

The time constant for the detector-preamplifier combination (assuming the detector time constant is not charge-carrier lifetime limited) is

$$\frac{R_L R_{Det} C_3}{R_L + R_{Det}} = 1.136 \times 10^{-6} \text{ sec} \qquad (16\text{-}85)$$

This corresponds to a reduction in response by 3 dB at 1.4×10^5 Hz.

16.5.2. Current Mode. The circuit for a current-mode preamplifier is shown in Figure 16-23.

From Equation (16-74)

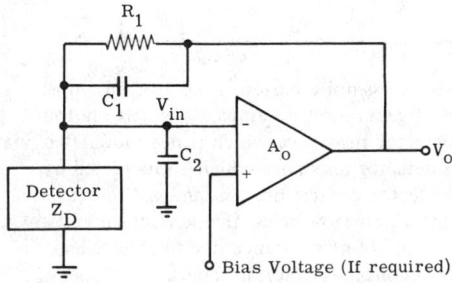

Fig. 16-23. Current-mode operational amplifier circuit. R_1 equals feedback resistance, in ohms; C_1 equals shunt capacitance across feedback resistance; C_2 equals stray capacitance at summing node plus input capacitance of preamplifier; A_o equals open-loop gain of the operational amplifier, in volts/volt or amps/amp; V_o equals output signal of preamplifier; Z_D equals detector impedance, in ohms; and V_{in} equals input signal to OA.

$$\beta = \frac{1}{1 + \dfrac{Z_{fb}}{Z_S}} \qquad (16\text{-}86)$$

$$\text{where } Z_{fb} = \frac{R_1}{1 + j\omega R_1 C_1} \qquad (16\text{-}87)$$

$$Z_S = \frac{Z_D}{1 + j\omega Z_D C_2} \qquad (16\text{-}88)$$

Z_D = detector impedance, ohms
C_2 = stray capacitance at summing node plus input capacitance of preamplifier, F
R_1 = feedback resistor, ohms
C_1 = shunt capacitance across feedback resistor, F

The β expression and the closed-loop gain equation for the current mode amplifier is a direct function of the detector's impedance. A circuit model for a detector is shown in Figure 16-24. For detectors which have a contact resistance, R_{cc}, much smaller than the detector resistance, $R_d + R_{mb}$, Figure 16-24 can be reduced to the simplified model shown in Figure 16-25, where $C_{Det} = C_{Di} + C_{shunt}$

$$R_{Det} = \frac{R_{shunt}(R_d + R_{mb})}{R_{shunt} + R_d + R_{mb}} \qquad (16\text{-}89)$$

$\bar{I}_{Det}$ = equivalent-noise current generator
for detector, A

I_p = photon- or radiation-current generator, A

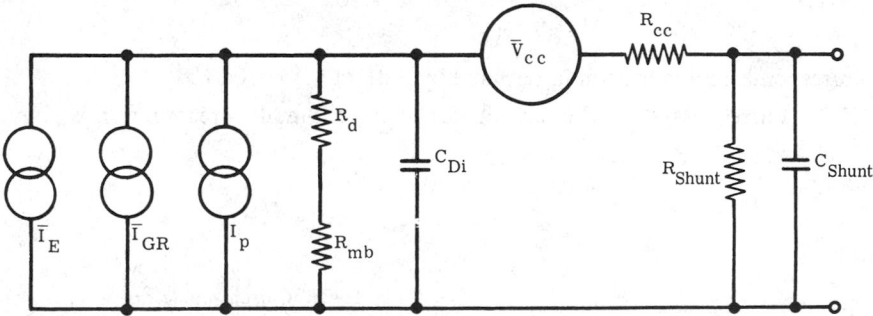

Fig. 16-24. Circuit model of a detector $\bar{I}_E$ equals excess-noise current generator, in amps; $\bar{I}_{GR}$ equals generation-recombination noise current generation, in amps; I_p equals photon or radiation-current (signal circuit); R_d equals detector resistance which is not modulated by current flow in detector, in ohms; R_{mb} equals detector resistance which is modulated by current flow in detector, in ohms; C_{Di} equals detector capacitance, in farads; $\bar{V}_{cc}$ equals equivalent-noise voltage generator representing the thermal noise of the contact, in volts; R_{cc} equals contact resistance, in ohms; R_{shunt} equals shunt resistance due to surface leakage, in ohms; and C_{shunt} equals shunt capacitance at contact surface, in farads.

Therefore,

$$Z_D = \frac{R_{Det}}{1 + j\omega R_{Det} C_{Det}} \qquad (16\text{-}90)$$

The closed-loop gain expression from Equation (16-72) for the current-mode amplifier, expressed in terms of the detector impedance and the feedback characteristics, is

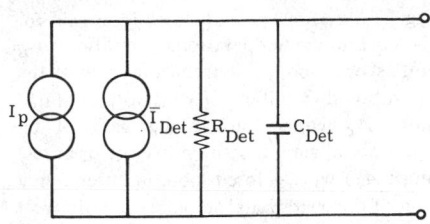

Fig. 16-25. Simplified detector model. I_p equals signal current due to optical signal incident on the detector, in amps; $\bar{I}_{Det}$ equals detector noise, in amps; R_{Det} equals detector resistance, in ohms; and C_{Det} equals equivalent detector capacitance, in farads.

$$A_c = \frac{V_o}{V_{in}} = -\frac{Z_{fb}}{Z_D}\left(\frac{A_o Z_D}{(1 + A_o)Z_D + Z_{fb}}\right)$$

$$(16\text{-}91)$$

where $Z_{fb} = \dfrac{R_1}{1 + j\omega R_1 C_1}$

Assuming that $A_o \gg 1$ (i.e., $1 + A_o \approx A_o$) one can express this equation as

$$A_c = \frac{-V_o}{V_{in}} = \frac{-R_1[1 + j\omega R_{Det}(C_{Det} + C_2)]}{R_{Det}(1 + j\omega R_1 C_1) + \dfrac{1}{A_o}[R_1\{1 + j\omega R_{Det}(C_{Det} + C_2)\}]} \qquad (16\text{-}92)$$

For this expression

$$V_{in} = (I_p + \bar{I}_{Det})Z_{S1} \qquad (16\text{-}93)$$

where $Z_{S1} = \dfrac{R_{Det}}{1 + j\omega R_{Det}(C_{Det} + C_2)}$

In its transimpedance form, the closed-loop gain expression is

$$A_{cz} = \frac{V_o}{(I_p + \bar{I}_{Det})} = -\frac{R_1}{(1 + j\omega R_1 C_1) - \dfrac{1}{A_o}\left[\dfrac{R_1}{R_{Det}}\{1 + j\omega R_{Det}(C_{Det} + C_2)\}\right]} \qquad (16\text{-}94)$$

where A_{cz} is the transimpedance form of the closed-loop gain.

A Bode plot for an OA containing the poles and zero in Equation (16-92) is shown in Figure 16-26. The open-loop gain for the OA can be expressed as

$$A_o = \frac{A_{ol}}{1 + \tau_o s} \qquad (16\text{-}95)$$

where s = the Laplace operator, $\sec^{-1}$

A_{ol} = the low-frequency open-loop gain of the amplifier, $V\,V^{-1}$, or $A\,A^{-1}$

τ_o = the corner frequency of the amplifier response, sec

In Case 1, A_c approaches the open-loop gain, A_o, before $R_1 C_1$ can have any effect. At this point, the closed-loop expression approaches the complex form of

Fig. 16-26. Bode plot of closed-loop gain for the current-mode operational amplifier. A_{ol} equals open-loop, low-frequency gain of the operational amplifier, in volts/volt or amps/amp; A_c equals closed-loop gain of an operational amplifier, in volts/volt or amps/amp; A_{cl} equals closed-loop, low-frequency gain of the operational amplifier, in volts/volt or amps/amp; and A_o equals open-loop gain of an operational amplifier in volts/volt or amps/amp.

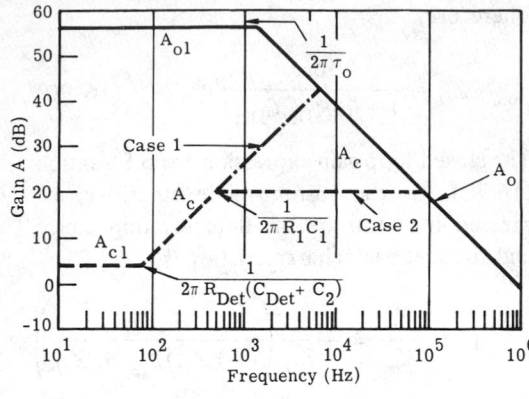

$$A_c = \frac{s}{\dfrac{1}{R_1(C_{Det} + C_2)} + \dfrac{\tau_o}{A_{ol}}} \qquad (16\text{-}96)$$

In Case 2, the closed-loop-gain expression, A_c, remains basically a single zero expression. When $\omega R_e(C_e + C_2) \gg 1$ and $\omega R_1 C_1 \ll 1$,

$$A_c = R_1(C_{Det} + C_2)s \qquad (16\text{-}97)$$

When $\omega R_{Det}(C_{Det} + C_2) \gg 1$ and $\omega R_1 C_1 \gg 1$,

$$A_c = \frac{C_{Det} + C_2}{C_1} \qquad (16\text{-}98)$$

For the condition

$$\frac{A_{ol}}{\tau_o s} \ll \frac{C_1}{C_{Det} + C_2}$$

then

$$A_c = \frac{A_{ol}}{\tau_o s} = A_o \qquad (16\text{-}99)$$

A plot of the frequency response for Cases 1 and 2 is shown in Figure 16-27.

A Bode plot of the closed-loop gain for a current-mode OA is shown in Figure 16-26, where

$$\tau_0 = 1.06 \times 10^{-4} \text{ sec}$$
$$R_1 = 10 \text{ ohms}$$
$$C_1 = 3 \times 10^{-12} \text{ F (Case 2)}$$
$$C_1 < 2.2 \times 10^{-13} \text{ F (Case 1)}$$
$$R_{Det} = 1.6 \times 10^8 \text{ ohms}$$
$$C_{Det} = 2.5 \times 10^{-12} \text{ F}$$
$$C_2 = 10^{-11} \text{ F}$$

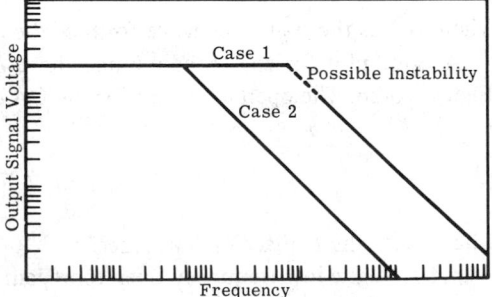

Fig. 16-27. Bode plots of the frequency response for the current-mode amplifier shown in Figure 16-23.

16.6. Synchronous Detection

Radiometric instruments frequently use choppers to interrupt the radiation periodically in the optical path. Choppers permit operation in the high-frequency region where detector and amplifier characteristics are better. The frequency of the amplitude-modulated signal produced by the chopper is determined by the chopping rate. If the chopper blade is shaped to produce sine wave modulation, the detector output will be

$$V_s = \frac{1}{2} V_{so}(t) \sin \omega t \qquad (16\text{-}100)$$

where $V_{so}(t)$ = the peak-to-peak signal amplitude, V
ω = angular frequency produced by chopper, rad sec^{-1}

A synchronous detector is essentially a narrowband detection system in which the target signal is beat with a reference signal of the same frequency producing a dc output. A block diagram of a synchronous detector is shown in Figure 16-28. The output of the target detector is an intelligence signal which is amplified and then multiplied by a reference signal. The reference signal is generated by a magnetic or photoelectric device synchronized by the chopper. Therefore, the frequency and phase of the reference signal and the intelligence signal are the same. The output of the reference detector is then

$$V_r(t) = V_{ro} \sin \omega t \qquad (16\text{-}101)$$

The product of the two signals appearing at the output of the multiplier is therefore

$$V_m(t) = B V_{so}(t) V_{ro} \sin^2 \omega t \qquad (16\text{-}102)$$

or

$$V_m(t) = B V_{so}(t) V_{ro} (\cos 2\omega t - 1) \qquad (16\text{-}103)$$

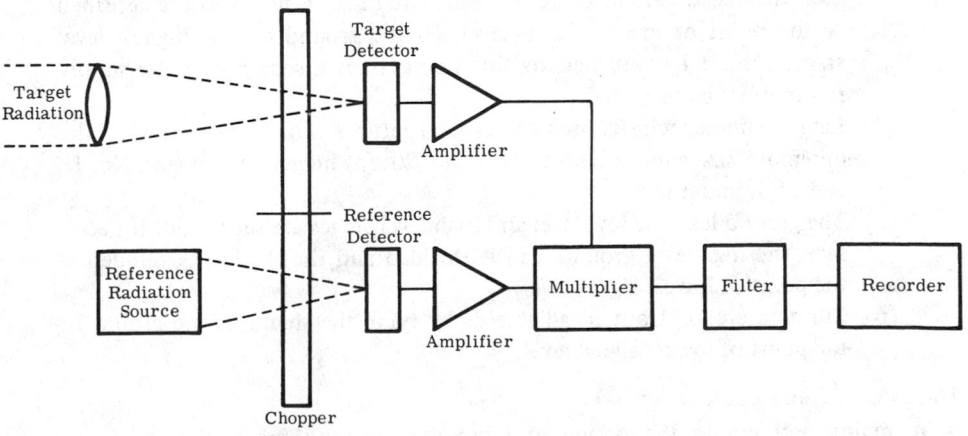

Fig. 16-28. Block diagram of synchronous detector.

The passband of the filter is made much less than 2ω to eliminate the unwanted frequency component. The output of the filter is then

$$V(t) = \frac{BV_{ro}}{2} V_{so}(t) \tag{16-104}$$

Thus the original intelligence has been recovered.

There is marked similarity between synchronous detection and cross correlation. The improvement in the signal-to-noise ratio obtained by synchronous rectification depends upon the form and frequency of the target signal, the form of the reference signal, and the power spectrum of the noise entering the correlator. If one assumes that the radiation and the reference signals vary sinusoidally with time and that the system suffers from band-limited white noise, a figure of merit can be derived by determining the ratio of the signal-to-noise ratio at the output of the filter to the signal-to-noise ratio at the input to the preamplifier.

16.7. Power Supply Considerations

Because of the very low-level signals amplified by preamplifiers, the noise level introduced through the power supplies should be much less than the noise introduced at the input. In most preamplifiers, the noise or ripple voltage on the B+ supply should be less than 200 μV.

16.8. Grounding Considerations

Improper grounding of preamplifiers and associated circuits can cause self-sustained oscillations, gain distortions, and numerous other undesirable effects. The following grounding techniques are desirable:

(1) The amplifier circuit is grounded to the chassis at the point of lowest signal level.

(2) There are no ground loops. The grounding of shielded leads at both ends is avoided. The power supply return is tied to the amplifier at the point of lowest signal level.

(3) A ground bus for every amplifier stage having a gain of 100 or more is used instead of a single ground bus. Grounds of all stages are returned to the point of lowest signal level. (If the ground for the highest level stages cannot be returned to this point, then a separate power supply return must be provided.)

(4) Large diameter wire is used for ground returns. For low-level stages, the minimum size should be No. 22 or No. 20; for higher level stages, No. 18 is the minimum size.

(5) The ground leads of low-level and high-level stages are separated. If necessary, the high-level ground lead is shielded and the shield is grounded at the point of lowest signal level.

(6) Currents are kept out of all shields by tying the shield to the ground at the point of lowest signal level.

16.9. Low-Noise Cable

A major problem in the design of low-noise preamplifiers is the spurious audiofrequency noise or microphonic noise generated in coaxial cables resulting from shock, excitation, and vibration. Tribo-electricity (i.e., the electric charges separated by friction between bodies), a major contributor of microphonic noise, is generated in cables under conditions of vibration. To reduce microphonic noise caused by tribo-electricity, one can shield the signal-carrying conductor from all tribo-electric fields or prevent the generation of such fields by special low-noise cable construction [16-18].

16.10 References and Bibliography

16.10.1 References

[16-1] J. F. Pierce, *Transistor Circuit Theory and Design,* C. E. Merrill Books, Columbus, OH, 1963

[16-2] J. B. Johnson, "Thermal Agitation of Electricity in Conductors," *Physical Review,* American Physical Society, Menasha, WI, Vol. 32, No 97, 1928, pp. 97-109.

[16-3] H. Nyquist, "Thermal Agitation Electric Charge in Conductors," *Physical Review,* American Physical Society, Menasha, WI, Vol. 32, No. 110, 1928, p. 110.

[16-4] A. van der Ziel and A. G. T. Becking, "Theory of Junction Diode and Junction Transistor Noise," *Proceedings of the IRE,* J. D. Ryder (ed.), The Institute of Radio Engineers, New York, NY, Vol. 46, No. 1, 1958, pp. 589-594.

[16-5] A. van der Ziel, "Noise Aspects of Low Frequency Solid-State Circuits," *Solid-State/Design,* Horizon House, Brookline, MA, March 1962, pp. 39-44.

[16-6] Edward G. Nielsen, "Behavior of Noise Figure in Junction Transistors," *Proceedings of the IRE,* D. G. Funk (ed.), The Institute of Radio Engineers, New York, NY, 1957, pp. 957-963.

[16-7] A. van der Ziel, "Noise in Junction Transistors," *Proceedings of the IRE,* J. D. Ryder (ed.), The Institute of Radio Engineers, New York, NY, 1958, pp. 1019-1038.

[16-8] Richard Brubaker, "Semiconductor Noise Figure Considerations," Motorola Semiconductor Products, Phoenix, AZ, Note AN-421, August 1968.

[16-9] Rudolph D. Kasser, "Noise Factor Contours for Field-Effect Transistors at Moderately High Frequencies," *IEEE Transactions on Electron Devices,* John A. Copeland (ed.), Institute of Electrical and Electronics Engineers, New York, NY, Vol. ED-19, No. 2, February 1972, p. 165.

[16-10] T. L. Chiu and H. N. Ghosh, *Solid-State Electronics,* Pergamon Press, Oxford, England, Vol. 14, 1971, pp. 1307-1317.

[16-11] Siliconix Staff, *Siliconix Field Effect Transistors Data Book,* Siliconix Semiconductor Devices, Siliconix, Santa Clara, CA, 1976, pp. 4-24.

[16-12] W. Milton Gosney, "Subthreshold Drain Leakage Currents in MOS Field-Effect Transistors," *IEEE Transactions on Electron Devices,* John A. Copeland (ed.), Institute of Electrical and Electronics Engineers, New York, NY, Vol. ED-19, No. 2, February 1972, pp. 213-219.

[16-13] F. M. Klaassen, "On the Substrate Current Noise in MOS Transistors Beyond Pinchoff," *Proceedings of the IEEE,* Joseph E. Rowe (ed.), Institute of Electrical and Electronics Engineers, New York, NY, Vol. 59, No. 2, February 1971, pp. 331-332.

[16-14] Hughes Technical Staff, "The FET as a Switch," Hughes Aircraft Company, MOS Division, Newport Beach, CA. No. 68-04, 1968.

[16-15] Hughes Technical Staff, "MOSFET Operation in Basic Circuitry," Hughes Aircraft Company, MOS Division, Newport Beach, CA., No. 68-03, 1968.

[16-16] G. E. Tobey, et al., *Operational Amplifiers: Design and Application*, McGraw-Hill, New York, NY, 1971.

[16-17] Texas Instruments Staff, *Transistor Circuit Design*, McGraw-Hill, New York, NY, 1963.

[16-18] Z. Neumark, "Tribo-Electricity and Microphonic Noise in Infrared Systems," *Infrared Physics*, Pergamon Press, Oxford, England, Vol. 4, 1964, pp. 67-71.

16.10.2 Bibliography

Bell, D. A., *Electrical Noise*, D. Van Nostrand, New York, NY, 1960.

Brubaker, Richard, "Semiconductor Noise Figure Considerations," Motorola Semiconductor Products, Phoenix, AZ, Note AN-421, August 1968.

Chiu, T. L., and H. N. Ghosh, *Solid-State Electronics*, Pergamon Press, Oxford, England, Vol. 14, 1971, pp. 1307-1317.

Electronics Instruments Digest Staff, "Grounding Low-Level Instrumentation Systems." *Electronic Instruments Digest*, January 1966.

Gardner, Murray F., and John L. Barnes, *Transients on Linear Systems*, John Wiley and Sons, New York, NY, 1958.

Gosney, W. Milton, "Subthreshold Drain Leakage Currents in MOS Field-Effect Transistors," *IEEE Transactions on Electron Devices*, John A. Copeland (ed.), Institute of Electrical and Electronics Engineers, New York, NY, Vol. ED-19, No. 2, February 1972, pp. 213-219.

Hudson, Richard D., Jr., *Infrared System Engineering*, John Wiley and Sons, New York, NY, 1969.

Hughes Aircraft Staff, "The FET as a Switch," Hughes Aircraft Company, MOS Division Newport Beach, CA, No. 68-04, 1968.

Hughes Technical Staff, "MOSFET Operation in Basic Circuitry," Hughes Aircraft Company, MOS Division, Newport Beach, CA, No. 68-03, 1968.

Johnson, J. B., "Thermal Agitation of Electricity in Conductors," *Physical Review*, American Physical Society, Menasha, WI, Vol. 32, No. 97, 1928, pp. 97-109.

Kasser, Rudolph, "Noise Factor Contours for Field-Effect Transistors at Moderately High Frequencies," *IEEE Transactions on Electron Devices*, John A. Copeland (ed.), Institute of Electrical and Electronics Engineers, New York, NY, Vol. ED-19, No. 2, February 1972, p. 165.

Klaassen, F. M., "On the Substrate Current Noise in MOS Transistors Beyond Pinchoff," *Proceedings of the IEEE*, Joseph E. Rowe (ed.), Institute of Electrical and Electronics Engineers, New York, NY, Vol. 59, No. 2, February 1971, pp. 331-332.

Neumark, Z., "Tribo-Electricity and Microphonic Noise in Infrared Systems," *Infrared Physics*, Pergamon Press, Oxford, England, Vol. 4, 1964, pp. 67-71.

Nielsen, Edward G., "Behavior of Noise Figure in Junction Transistors," *Proceedings of the IRE*, D. G. Funk (ed.), The Institute of Radio Engineers, New York, NY, 1957, pp. 957-963.

Nyquist, H., "Thermal Agitation Electric Charge in Conductors," *Physical Review*, American Physical Society, Menasha, WI, Vol. 32, No. 110, 1928, p. 110.

Pettit, Joseph M., and M. McWhorter, *Electronic Amplifier Circuits: Theory and Design,* McGraw-Hill, New York, NY, 1961.

Pierce, J. F., *Transistor Circuit Theory and Design,* C. E. Merrill Books, Columbus, OH, 1963.

Riddle, Robert L., and M. P. Ristenbatt, *Transistor Physics and Circuits,* Prentice Hall, Englewood, NJ, 1960, First Edition.

Siliconix Staff, *Siliconix Field Effect Transistors Data Book,* Siliconix Semiconductor Devices, Siliconix, Santa Clara, CA, 1976.

Texas Instruments Staff, *Transistor Circuit Design,* McGraw Hill, New York, NY, 1963.

Tobey, G. E., et al., *Operational Amplifiers: Design and Application,* New York, NY, 1971.

van der Ziel, A., "Noise Aspects of Low-Frequency Solid-State Circuits," *Solid-State/ Design,* Horizon House, Brookline, MA, March 1962, pp. 39-44.

van der Ziel, A., "Noise in Junction Transistors," *Proceedings of IRE,* J. D. Ryder (ed.), The Institute of Radio Engineers, New York, NY, 1958, pp. 1019-1038.

van der Ziel, A., and A. G. T. Becking, "Theory of Junction Diode and Junction Transistor Noise," *Proceedings of IRE,* J. D. Ryder (ed.), The Institute of Radio Engineers, New York, NY, Vol. 46, No. 1, 1958, pp. 589-594.

Chapter 17

RETICLE AND IMAGE ANALYSES

Richard Legault
Environmental Research Institute of Michigan

CONTENTS

17. Reticle and Image Analyses

17.1. Introduction

This chapter introduces some analytic tools useful in the design and analysis of electro-optical systems. Emphasis is placed on the spatio-temporal analysis of reticles, sampled data imagery, optimal spectral filters and the signal processing of multispectral data. The intent is two-fold: to introduce the analytic techniques and to provide the reader with some examples of their applications. The designer or analyst of electro-optical systems will not find models treated in this chapter. He must then rely on other data: tables and computer-assisted computation.

17.1.1. Symbols and Definitions.
Table 17-1 lists the symbols and definitions used in this chapter.

17.2. Fourier Analysis

Fourier methods provide the analyst with methods for determining the frequency content of a signal.

17.2.1. Fourier Series—One Dimensional.
If $v(t)$ is a periodic function of period T, i.e., $v(t) = v(t + T)$, then $v(t)$ may be written in exponential form

$$v(t) = \sum_{-\infty}^{\infty} c_n \, e^{i2\pi nt/T} \tag{17-1}$$

where $c_n = (1/T) \int_0^T v(t) \, e^{-i2\pi nt/T} dt$,

or in trigonometric form

$$v(t) = \frac{1}{2} a_0 + \sum_1^{\infty} \left\{ a_n \cos \frac{2\pi nt}{T} + b_n \sin \frac{2\pi nt}{T} \right\} \tag{17-2}$$

where $a_n = 2/T \int_{-T/2}^{T/2} v(t) \cos (2\pi nt/T) dt$

$b_n = 2/T \int_{-T/2}^{T/2} v(t) \sin (2\pi nt/T) dt$

The term $2\pi/T$ may be written as ω_0 in units of radians per unit time.

17.2.2. Fourier Integral—One Dimensional.
If the integrals $|v(t)|$ or $v^2(t)$ exist over the integration path, then

$$v(t) = \int_{-\infty}^{\infty} V(f) \, e^{2\pi itf} \, df \tag{17-3}$$

with $v(t)$ and $V(f)$ called transform pairs. $V(f)$ may be obtained from $v(t)$ by the relation

$$V(f) = \int_{-\infty}^{\infty} v(t) \, e^{-2\pi itf} \, dt \tag{17-4}$$

There are a number of excellent compilations of Fourier series expansions and Fourier transform pairs. References [17-1] and [17-2] contain two such important compilations.

Table 17-1. Symbols and Definitions

Symbol	Definition
A	Parallelogram defined by $\mathbf{a}_1$ and $\mathbf{a}_2$; region defined by the aperture
$A(\mathbf{k})$	Fourier transform of $a(\mathbf{x})$; and optical response
$A_m(\mathbf{k})$	Fourier transform of $a_m(\mathbf{x})$
$a, b, c,$	Coefficients
$a(\mathbf{x})$	Aperture transmission pattern function
$a_m(\mathbf{x})$	Coefficients
$a(\rho)$	Optimal scanning aperture
$B(\mathbf{k})$	Fourier transform of display response
$b(\mathbf{x})$	Display response
$b(\lambda)$	Average background signal
$b_\omega(\lambda)$	Random background signal
$\langle b^2(\lambda) \rangle$	Averaged, squared, background signal; variance of background signal
$C(i/j)$	Cost
c	Radius of nutation
$D(\mathbf{k})$	Detector spatial frequency response
$D(\mu)$	Decision function
$d(\mathbf{x})$	Response of detector over its surface
$d(\mu, \mu_o)$	Distance metric, in units of length
f	Temporal frequency
$f(\mu)$	Multivariant density
$f(\lambda)$	Spectral filter function or response
G	Fourier transform of g
g	Function
H	Hadamard matrix
I	Intensity; also, identity matrix
$\mathbf{k}$	Spatial frequency, $\mathbf{k} = (k_1, k_2)$, in units of cycles per unit length
k_ρ, ϕ	Polar coordinate transforms
m	Running index; 0,1,2,3 . . .
N	Noise: detector/preamplifier, sensor
n	Running index; 0,1,2,3 . . .
$O(\mathbf{k})$	Fourier transform of $o(\mathbf{x})$
$o(\mathbf{x})$	Display output spatial signal
p	Number of spoke pairs
$p(j/\mu)$	Conditional probability
R	Matrix, (r_{ij})
$R(\mathbf{k})$	Fourier transform of $r(\mathbf{x})$
r_{ij}	Matrix element
$r(\mathbf{x})$	Reticle transmission coefficient
$S(\mathbf{k})$	Fourier transform of $s(\mathbf{x})$
$s(\mathbf{x})$	Average scene radiation distribution
$s(\lambda)$	Spectral target signal

Table 17-1. Symbols and Definitions (*Continued*)

Symbol	Definition
$s_\omega(\mathbf{x})$	Random scene radiation distribution
T	Period; as superscript, transpose of matrix; as subscript, target
$T(\mathbf{k})$	Fourier transform of target $t(\mathbf{x})$
t	Time
$t(\mathbf{x})$	Target function
V, k, λ	Integers
$v(t)$	Voltage signal output from sensor detector
$v_\omega(t)$	Random voltage signal
$W(\mathbf{k})$	Wiener spectrum
x, y	Coordinates
β	Parallelogram defined by $\mathbf{b}_1$ and $\mathbf{b}_2$
β_m	Modulation coefficients
δ	Phase
$\delta(\mathbf{x} - \mathbf{x}_o)$	Delta function
(θ_{ij}^B)	Background variance covariance matrix
(θ_B^{ij})	Inverse of background variance covariance matrix
(θ_{ij}^T)	Target variance covariance matrix
(θ_T^{ij})	Inverse of target variance covariance matrix
λ	Wavelength
μ_i	Total radiation in i^{th} spectral region
ρ_{12}	Correlation coefficient
ρ, θ	Polar coordinates
σ	Bar width in translating reticle
$\Phi_\lambda(\lambda)$	Spectral radiant power
$\phi(\mathbf{x})$	Second-order correlation statistic
Ω	A set of real numbers
ω	Average domain
ω_0	Angular frequency, $\omega_0 = 2\pi/T$
-1	As superscript, inverse of matrix
$*$	As superscript, complex conjugate

17.2.3. Fourier Series—Two Dimensional. The definition of two-dimensional periodicity and the two-dimensional counterpart of the one-dimensional reciprocal, $1/T$, are less familiar. Let $\mathbf{a}_1 = (x_1, x_2)$ and $\mathbf{a}_2 = (y_1, y_2)$ be any two noncolinear vectors in the plane, and n_1, n_2 any two positive or negative integers, then

$$\mathbf{a}_n = n_1 \mathbf{a}_1 + n_2 \mathbf{a}_2 \tag{17-5}$$

The vectors (points) defined by $\mathbf{a}_n$ in Equation (17-5) form a lattice in the plane x_1', x_2', depicted in Figure 17-1. The definition of periodicity is given by

$$s(\mathbf{x}) = s(\mathbf{x} + \mathbf{a}_n) \tag{17-6}$$

for all $\mathbf{n} = (n_1, n_2)$ integers, and can be pictured by referring to the figure. The vectors (points) $\mathbf{a}_n$ break the plane up into parallelograms. A periodic pattern is one which repeats itself from one parallelogram to the next. The pattern in one parallelogram equals the pattern in another.

The vectors defining the reciprocal lattice are

$$\mathbf{b}_n = n_1 \mathbf{b}_1 + n_2 \mathbf{b}_2, \qquad n_1, n_2 \text{ integer} \tag{17-7}$$

The generating vectors $\mathbf{b}_1, \mathbf{b}_2$ are derived from $\mathbf{a}_1, \mathbf{a}_2$ by use of the notion of dot or inner product:

$$\mathbf{a}_1 \cdot \mathbf{a}_2 = (x_1, x_2) \cdot (y_1, y_2) = x_1 y_1 + x_2 y_2 \tag{17-8}$$

This results in four linear equations in four unknowns.

$$\begin{array}{ll} \mathbf{b}_1 \cdot \mathbf{a}_2 = 0 & \mathbf{b}_1 \cdot \mathbf{a}_1 = 1 \\ \mathbf{b}_2 \cdot \mathbf{a}_1 = 0 & \mathbf{b}_2 \cdot \mathbf{a}_2 = 1 \end{array} \tag{17-9}$$

The first two equations specify that the spatial and reciprocal lattice generating vectors must be perpendicular, as illustrated in Figure 17-2.

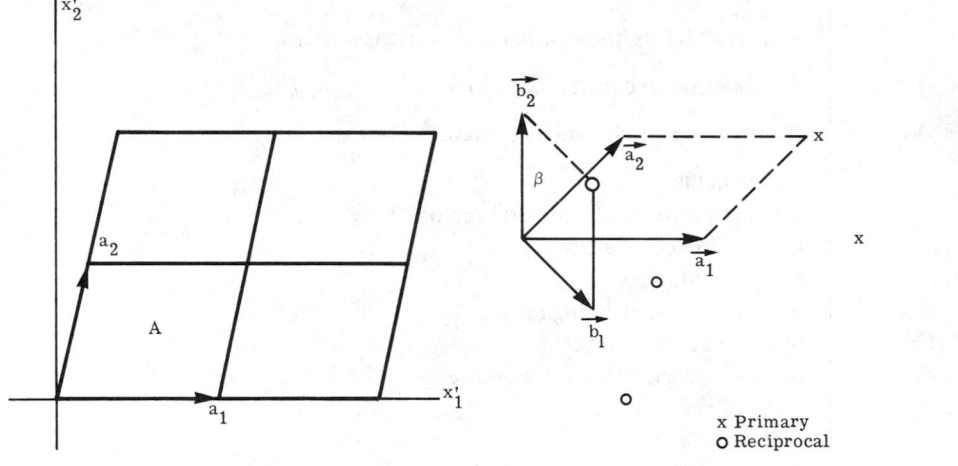

Fig. 17-1. Spatial periodic lattice, a_n.   Fig. 17-2. Primary and reciprocal lattice.

The two-dimensional Fourier series expansion of a spatial pattern is given by

$$s(\mathbf{x}) = \sum_n s_n \, e^{2\pi i \, \mathbf{x} \cdot \mathbf{b}_n} \tag{17-10}$$

where $s_n = \int_A s(\mathbf{x}) \, e^{-2\pi i \mathbf{x} \cdot \mathbf{b}_n} \, d\mathbf{x}$

   A = the parallelogram defined by the vectors $\mathbf{a}_1$ and $\mathbf{a}_2$

17.2.4. Fourier Transform—Two Dimensional. If the integral $\int_{-\infty}^{\infty} s^2(\mathbf{x}) \, d\mathbf{x} < \infty$ exists, then

$$s(\mathbf{x}) = \int_{-\infty}^{\infty} S(\mathbf{k}) \, e^{2\pi i \mathbf{k} \cdot \mathbf{x}} \, d\mathbf{k} \tag{17-11}$$

where $\mathbf{k} = (k_1, k_2)$
          $s(\mathbf{x})$ and $S(\mathbf{k})$ = Fourier transform pairs.

If $s(\mathbf{x})$ is known, then $S(\mathbf{k})$ can be found from the relation

$$S(\mathbf{k}) = \int_{-\infty}^{\infty} s(\mathbf{x}) \, e^{-2\pi i \mathbf{k} \cdot \mathbf{x}} \, d\mathbf{x} \qquad (17\text{-}12)$$

A spatial frequency representation $S(\mathbf{k})$ gives the amplitude of the sinusoidal wave with spatial frequency k_1 in the x_1 direction and spatial frequency k_2 in the x_2 direction. Figure 17-3 depicts an approximation to such a wave with $k = k_1 = k_2$. Clearly, negative spatial frequencies have a physical interpretation.

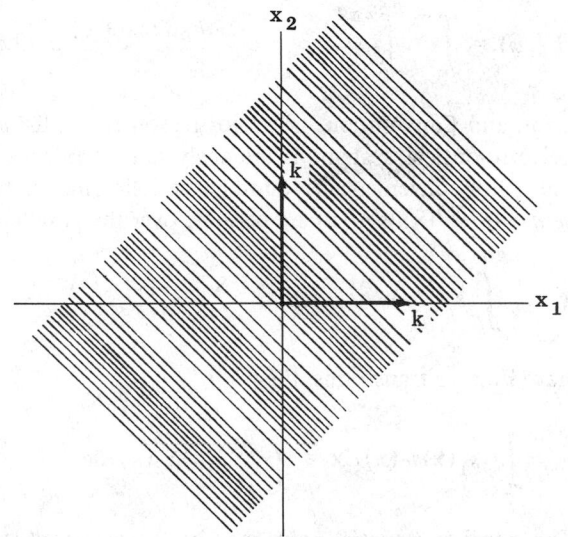

Fig. 17-3. Bar approximation to $\cos 2\pi(k_x x + k_y y)$.

17.2.5. Properties of Fourier Transform Pairs. Certain relationships between Fourier transform pairs will be required in the further development.

Scale changes: $s(ax_1, bx_2) \longleftrightarrow \dfrac{1}{|a||b|} S\left(\dfrac{k_1}{a}, \dfrac{k_2}{b}\right)$

Translation: $s(\mathbf{x} + \mathbf{x}_0) \longleftrightarrow S(\mathbf{k}) \, e^{-2\pi i \mathbf{k} \cdot \mathbf{x}_0}$

Conjugates: $s^*(\mathbf{x}) \longleftrightarrow S^*(-\mathbf{k})$, $s(\mathbf{x})$ complex

$s^*(\mathbf{x}) = s(\mathbf{x}) \longleftrightarrow S(\mathbf{k})$, $s(\mathbf{x})$ real

$S^*(\mathbf{k}) = S(-\mathbf{k})$

where $s^*(\mathbf{x})$ is the complex conjugate of $s(\mathbf{x})$.

17.2.6. Fourier Transforms—Polar Coordinates. Many times it is convenient to perform the analysis in polar coordinates, particularly when spatial patterns are constant radially or angularly. The usual coordinate transforms are:

$$
\begin{array}{ll}
\text{for } \mathbf{x} = (x_1, x_2), & \text{for } \mathbf{k} = (k_1, k_2), \\
x_1 = \rho \cos \theta & k_1 = k_\rho \cos \phi \\
x_2 = \rho \sin \theta & k_2 = k_\rho \sin \phi
\end{array}
\tag{17-13}
$$

The Fourier transform pair relationship is

$$
s(\rho, \theta) = \int_0^\infty \int_0^{2\pi} S(k_\rho, \phi) e^{2\pi i k_\rho \rho \cos(\theta - \phi)} k_\rho \, d\phi \, dk_\rho
$$

$$
S(k_\rho, \phi) = \int_0^\infty \int_0^{2\pi} s(\rho, \theta) e^{-2\pi i k_\rho \rho \cos(\theta - \phi)} \rho \, d\theta \, d\rho
\tag{17-14}
$$

17.2.7. Correlation and Convolution. A major reason for utilizing Fourier methods in the analysis of electro-optical systems is the analytic convenience for some types of systems. The reason for this convenience is found in the following relationships.

Parseval's Theorem. If $s^{(1)}(\mathbf{x})$, $s^{(2)}(\mathbf{x})$ are periodic over the parallelograms A, then

$$
\int_A s^{(1)}(\mathbf{x}) s^{(2)}(\mathbf{x}) \, d\mathbf{x} = \sum_{n=-\infty}^\infty s_n^{*(1)} s_n^{(2)}
\tag{17-15}
$$

If $s_1(\mathbf{x})$ and $s_2(\mathbf{x})$ have Fourier transforms, then

$$
\int_{-\infty}^\infty s_1(\mathbf{x}) s_2(\mathbf{x}) \, d\mathbf{x} = \int_{-\infty}^\infty S_1^*(\mathbf{k}) S_2(\mathbf{k}) \, d\mathbf{k}
\tag{17-16}
$$

Convolution. The signal at a spatial point may be the weighted sum of the signals at surrounding points. Then the output spatial signal, $o(\mathbf{x})$, is written as a convolution integral

$$
o(\mathbf{x}) = \int_{-\infty}^\infty g(\mathbf{x} - \mathbf{y}) s(\mathbf{y}) \, d\mathbf{y}
\tag{17-17}
$$

The Fourier transform of $o(\mathbf{x})$ takes a simple form

$$
O(\mathbf{k}) = G(\mathbf{k}) S(\mathbf{k})
\tag{17-18}
$$

Correlation. A similar result is found for the correlation integral

$$
o(\mathbf{x}) = \int_{-\infty}^\infty g(\mathbf{x} + \mathbf{y}) s(\mathbf{y}) \, d\mathbf{y}
\tag{17-19}
$$

The Fourier transform of $o(\mathbf{x})$ takes the form

$$
O(\mathbf{k}) = G(\mathbf{k}) S^*(\mathbf{k})
\tag{17-20}
$$

17.2.8. Wiener Spectrum. Analysis may require only the statistics of the output signal. Further, one may find that background scenes do not have a Fourier transform in the usual sense. Each value of ω identifies a sample $s_\omega(y)$ from some two-dimensional process. The second-order correlation statistic can be defined as

$$\underset{\omega}{\text{Average }} s_\omega(y)\, s_\omega(x + y) \qquad (17\text{-}21)$$

It is usually assumed that the process is stationary. Then the average in expression (17-21) depends only on the displacement, x i.e.,

$$\underset{\omega}{\text{Average }} s_\omega(y)\, s_\omega(x + y) = \phi(x) \qquad (17\text{-}22)$$

If the statistics of a single realization of the random process represent the process (ergodicity), then the average $\phi(x)$ can be written as

$$\phi(x) = \int_{-\infty}^{\infty} s(y)\, s(x + y)\, dy \qquad (17\text{-}23)$$

The Fourier transform of $\phi(x)$ is called the Wiener spectrum of the scene and is defined by

$$W(k) = \int_{-\infty}^{\infty} \phi(x)\, e^{-2\pi i k \cdot x}\, dx \qquad (17\text{-}24)$$

R. C. Jones postulated a model for the Wiener spectrum defined by

$$W(k) = \frac{B}{(k_0^2 + k \cdot k)^q} \qquad (17\text{-}25)$$

where k_0 = constant
$\quad q$ = a measure of the fuzziness of the disc edges
$\quad B$ = total scene radiance

One should realize that the Wiener spectrum is the frequency representation of the autocorrelation of the scene. It is the distribution of the total power over the spatial frequencies and is a second moment statistic. It is a representation of the noise power in the scene, but only in specialized cases can it be used as an estimator of false alarms which usually are signals exceeding a high threshold.

17.3. The Scanning Aperture

Initially, the aperture transmission pattern, $a(x_1, x_2)$, is centered on the origin of the scene coordinate system. Suppose at time, t, the aperture is moved, but not rotated, and centered on the scene point $[x_1(t), x_2(t)]$. The output signal is given

$$g[x(t)] = \int_{-\infty}^{\infty} a(x)\, s[x + x(t)]\, dx \qquad (17\text{-}26)$$

where $a(x)$ = aperture transmission pattern
$\quad s(x)$ = radiant signal from the scene point x

If the aperture is translated so that the center of the pattern is pointed in turn at *every* scene point **x**, the function $g[\mathbf{x}(t)]$ may be considered as a filtered version of the original imaged scene $s(\mathbf{x})$.

The Wiener spectrum, $W_g(\mathbf{k})$, of $g(\mathbf{x})$ is given by

$$W_g(k_1, k_2) = |A(k_1, k_2)|^2 \, W_S(k_1, k_2) \tag{17-27}$$

where $A(k_1, k_2)$ = Fourier transform of $a(\mathbf{x})$
       $W_S(k_1, k_2)$ = Wiener spectrum of the scene

The filtering of the scene by the reticle system is obvious and the Fourier transform of the aperture transmission pattern is seen to be descriptive of scanning aperture performance.

It is of considerable interest that this formulation of the scanning aperture model permits an optimization of the scanning aperture. The criterion for optimization is the maximization of the ratio of the instantaneous target signal squared to the mean squared background signal, that is, maximization of

$$\frac{\left| \int_{-\infty}^{\infty} A^*(\mathbf{k}) \, T(\mathbf{k}) \, d\mathbf{k} \right|^2}{\int_{-\infty}^{\infty} |A(\mathbf{k})|^2 \, W_B(\mathbf{k}) \, d\mathbf{k}} \tag{17-28}$$

where $A(\mathbf{k})$ = aperture Fourier transform
       $T(\mathbf{k})$ = target Fourier transform
       $W_B(\mathbf{k})$ = Wiener spectrum of background

Using the Schwarz inequality, one finds the aperture, $A(\mathbf{k})$, which maximizes the ratio (17-28). The Schwarz inequality is given by

$$\left| \int g(\mathbf{x}) \, s(\mathbf{x}) \right|^2 \leqslant \int g^2(\mathbf{x}) \, d\mathbf{x} \int s^2(\mathbf{x}) \, d\mathbf{x} \tag{17-29}$$

The upper bound is obtained when $g(\mathbf{x}) = s(\mathbf{x})$. Ratio (17-28) is now written as

$$\frac{\left| \int_{-\infty}^{\infty} A^*(\mathbf{k}) \, W_B^{1/2}(\mathbf{k}) \, \dfrac{T(\mathbf{k})}{W_B^{1/2}} \, d\mathbf{k} \right|^2}{\int_{-\infty}^{\infty} |A(\mathbf{k})|^2 \, W_B(\mathbf{k}) \, d\mathbf{k}} \leqslant \frac{\int |A(\mathbf{k})|^2 \, W_B(\mathbf{k}) \, d\mathbf{k} \int \dfrac{T^2(\mathbf{k})}{W_B(\mathbf{k})} \, d\mathbf{k}}{\int_{-\infty}^{\infty} |A(\mathbf{k})|^2 \, W_B(\mathbf{k}) \, d\mathbf{k}}$$

$$\tag{17-30}$$

$$= \int_{-\infty}^{\infty} \frac{T^2(\mathbf{k})}{W_B(\mathbf{k})} \, d\mathbf{k}$$

Fortunately, the upper bound is independent of $A(\mathbf{k})$ and is obtained when

$$A(\mathbf{k}) = \frac{T^*(\mathbf{k})}{W_B(\mathbf{k})} \tag{17-31}$$

The optimal scanning aperture specified in Equation (17-31) is the two-dimensional counterpart of the matched filter of electronics. This result supports the feeling that a scanning aperture should essentially be matched to the target shape but modified by the spatial characteristics of the background.

Consider an example proposed by Jones. The target is assumed to be a Gaussian pulse.

$$t(x_1, x_2) = a e^{-(x_1^2 + x_2^2)/2b} \tag{17-32}$$

where a = peak radiance of Gaussian pulse
       b = second moment of radiance density of Gaussian pulse

The Wiener spectrum of the background is given by

$$W_B(k_1, k_2) = \frac{b}{k_1^2 + k_2^2} \tag{17-33}$$

The optimal scanning aperture is given by

$$a(\rho) = \left(1 - \frac{\rho^2}{2b}\right) e^{-\rho^2/2b} \tag{17-34}$$

Unfortunately, for some targets the time scale of events is so short that it is not possible to scan the required field of view with a small instantaneous field of view within the time required and keep within the response time of available detectors.

There are, in general, two solutions to the dilemma. The straightforward one is to scan many detectors, each scanning some smaller part of the total field of view, the number selected to permit covering the total field of view in the required time. Another, and more commonly employed solution, is the use of a reticle.

17.4. Reticle Systems

A reticle system is essentially a mask or pattern placed in the image plane of an optical system. The transmission of this mask varies spatially. Usually the mask transmits certain portions of the imaged scene and completely blocks other portions of the scene. The radiation from the transmitted portions is focussed upon a detector. The detector output is assumed proportional to the total incident radiation. The reticle mask may be moved in the image plane, the imaged scene may be moved over a fixed reticle mask, or both. Reference [17-3] contains a fairly complete exposition of the current uses of reticle systems.

The reticle mask in scene coordinates is specified by a real valued function $r(\mathbf{x}, t)$. The function $r(\mathbf{x}, t)$ specifies the transmission coefficient for the intensity of an image scene-point $\mathbf{x}$ at the time t. The radiation distribution of the image scene in scene coordinates is represented by a positive real valued function $s(\mathbf{x})$. Since the transmitted fluxes are integrated, the output $v(t)$ from a reticle system is

$$v(t) = \int_{-\infty}^{\infty} r(\mathbf{x}, t) s(\mathbf{x}) \, d\mathbf{x} \tag{17-35}$$

Consequently, the scene $s(\mathbf{x})$ is encoded into a temporal signal, $v(t)$, by a reticle system. Equation (17-35) is a general model of a reticle system.

The correlation properties of the output signal give

Average $v_\omega(t) v_\omega(t + \tau)$
  ω

$$= \underset{\omega}{\text{Average}} \int_{-\infty}^{\infty} r(\mathbf{x}, t) s_\omega(\mathbf{x}) \, dx \int_{-\infty}^{\infty} r(\mathbf{x}', t + \tau) s_\omega(\mathbf{x}') \, dx'$$

$$(17\text{-}36)$$

$$= \int_{-\infty}^{\infty} \underset{\omega}{\text{Average}} \; s_\omega(\mathbf{x}) s(\mathbf{x} + \mathbf{x}'') \int_{-\infty}^{\infty} r(\mathbf{x}, t) r(\mathbf{x} + \mathbf{x}'', t + \tau) \, dx \, dx''$$

If the process is stationary, Average $\underset{\omega}{} s_\omega(\mathbf{x}) s(\mathbf{x} + \mathbf{x}'')$ is independent of $\mathbf{x}$, and its transform pair is the Wiener spectrum $W(\mathbf{k})$. Using the Parseval relation (Equation (17-16)), one obtains

$$\underset{\omega}{\text{Average}} \; v_\omega(t) v_\omega(t + \tau) = \int_{-\infty}^{\infty} W(\mathbf{k}) R(-\mathbf{k}, t) R(\mathbf{k}, t + \tau) \, dk \qquad (17\text{-}37)$$

The correlation and ultimately the power spectrum of $v(t)$ are weighted averages of the scene's Wiener spectrum. The weights are derived from the reticle patterns.

Reticles are commonly used in guidance systems. The reticle and its motion encodes the coordinates of the object to be located. Generally, this object is assumed to be a point source. It is necessary to determine how an object's coordinates modulate the signal from the reticle system. The models considered are too general for this purpose.

17.4.1. Analysis of Reticle Modulation. A stylized reticle system is illustrated in Figures 17-4 and 17-5.

The general reticle system considered consists of a transparent aperture in the image plane with a reticle pattern moving across the aperture. The area, A, and shape of the aperture are independent of time. Take a coordinate system fixed in the aperture A. Since the reticle moves across the aperture, the reticle pattern transmission will be time-dependent in aperture coordinates. As the aperture is scanned, the scene will be time-dependent in aperture coordinates.

Then

$$v(t) = \int_A r(\mathbf{x}, t) s(\mathbf{x}, t) \, dx \qquad (17\text{-}38)$$

where $\mathbf{x}$ = a point in A
    $r(\mathbf{x}, t)$ = reticle transmission
    $s(\mathbf{x}, t)$ = scene radiation distribution in the aperture limited plane
    dx = an elemental area in A
    $v(t)$ = voltage output from the detector

Here, $\mathbf{x}$ may be any set of two coordinates specifying a point in the plane; for example, $\mathbf{x} = (x_1, x_2)$ if Cartesian coordinates are used, or $\mathbf{x} = (\rho, \phi)$ if polar coordinates are used. The integral on the right-hand side of Equation (17-38) is independent of the coordinate

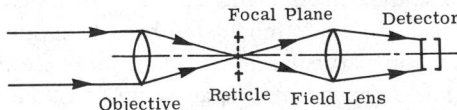

Fig. 17-4. Conceptual reticle system.

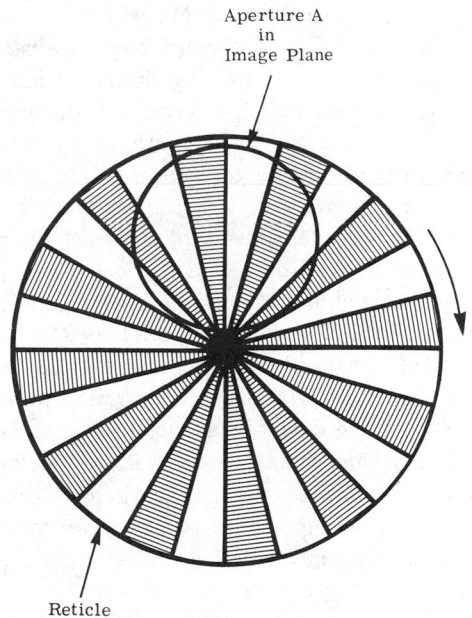

choice. The coordinate system most convenient for calculation is dictated by the geometry of the aperture. If A is rectangular, Cartesian coordinates make calculation simplest; if A is circular, polar coordinates are the choice.

A reticle pattern is finite. The reticle seen through the aperture is a moving pattern. This pattern is repeated at regular temporal intervals. The assumption of periodicity stated in Equation (17-39) below does not restrict the class to realizable reticle systems.

Fig. 17-5. Reticle, aperture, and motion.

$$r(\mathbf{x}, t) = r\left(\mathbf{x}, t + \frac{2\pi}{\omega_0}\right) \quad \text{for all } \mathbf{x} \tag{17-39}$$

The fundamental frequency, ω_0, should not be confused with any rotational or translational frequency; they may or may not correspond. Generally, there is a simple relation between the two frequencies. The reticle function defined in aperture coordinates has a Fourier series representation in time, with fundamental frequency ω_0.

$$r(\mathbf{x}, t) = \sum_{m=-\infty}^{\infty} a_m(\mathbf{x})\, e^{im\omega_0 t}$$

$$a_m(\mathbf{x}) = \frac{\omega_0}{2\pi} \int_0^{2\pi/\omega_0} r(\mathbf{x}, t)\, e^{-im\omega_0 t}\, dt \tag{17-40}$$

The substitution of Equation (17-40) in Equation (17-38) yields

$$v(t) = \sum_{m=-\infty}^{\infty} \beta_m(t)\, e^{im\omega_0 t}$$

$$\beta_m(t) = \int_A a_m(\mathbf{x})\, s(\mathbf{x}, t)\, d\mathbf{x} \tag{17-41}$$

At this level, Equations (17-40) and (17-41) are revealing. Equation (17-41) is generally not a Fourier series even though Equation (17-40) always is. The coefficients in the summation of Equation (17-41) are time dependent. Each component $e^{im\omega_0 t}$ of Equation (17-41) may be considered as a carrier. Each carrier is modulated by a temporal function, $\beta_m(t)$, reflecting the scene information, in particular an object's coordinates. When the aperture is not scanned, the scene is time-independent. The analysis of tracking systems generally assumes $s(\mathbf{x}, t)$ is not time dependent, i.e., the scene does not change much before the target is located. The reticle pattern is reflected in the coefficients $a_m(\mathbf{x})$ and in the interaction of scene and reticle pattern by $\beta_m(t)$.

The time dependence of $v(t)$ has two origins; reticle and scan motion. A portion of the temporal dependence, in particular ω_0, arising from reticle motion, is contained in the exponential terms, and all time dependence arising from aperture scan (and inherent scene time dependence) is contained in $\beta_m(t)$. However, $\beta_m(t)$ is affected by $a_m(\mathbf{x})$, and $a_m(\mathbf{x})$ is generally not independent of reticle motion. Inspection of Equation (17-40) indicates the relation between reticle motion and $a_m(\mathbf{x})$. Reticle motion determines both ω_0 and the time dependence of $r(\mathbf{x}, t)$. Generally, reticle pattern and reticle motion are inextricably interdependent in producing time dependence in $v(t)$. From Equation (17-41), one sees that a similar remark can be made about scene pattern and scan motion.

In some important cases, $a_m(\mathbf{x})$ is independent of reticle motion. For example, if one has a uniformly rotating reticle, then ω_0 is its rotational frequency, and $a_m(\mathbf{x})$ is independent of ω_0.

In summary, the reticle and its motion are completely specified by ω_0 and the set $a_m(\mathbf{x})$. The voltage output is completely speficied by ω_0 and the set $\beta_m(t)$. The harmonics of ω_0 are the carrier frequencies and $\beta_m(t)$, the modulation placed upon the mth harmonic by the interaction of scene and reticle patterns. To define a reticle system, one must calculate ω_0 and $a_m(\mathbf{x})$. To specify $v(t)$ for a given scene, one must calculate $\beta_m(t)$. In most of the cases, the convergence of Equation (17-41) is sufficiently rapid to insure the practicality of computing $v(t)$ numerically.

The basic method of reticle system analysis as represented in Equations (17-40) and (17-41) is to determine a fundamental frequency, ω_0, for the system. The interaction of the system with the scene is seen as a modulation of the carrier frequencies. Target location is based on the relation between the target location in aperture-fixed coordinates and the modulation, $\beta_m(t)$. Discrimination between targets and backgrounds is based on modulation differences. References [17-4] and [17-5] consider the modulation from various sources for a number of reticle systems.

17.4.2. Aperture Effects. The modulation coefficients, $\beta_m(t)$, of Equation (17-41) indicate integration over an aperture area, A. The integral expression may be written with infinite limits and the introduction of an aperture function $a(\mathbf{x})$.

$$\beta_m(t) = \int_A a_m(\mathbf{x}) s(\mathbf{x}, t) \, d\mathbf{x} = \int_{-\infty}^{\infty} a_m(\mathbf{x}) a(\mathbf{x}) s(\mathbf{x}, t) \, d\mathbf{x} \qquad (17\text{-}42)$$

where $a(\mathbf{x}) = 1$, with $\mathbf{x}$ contained in A
$\phantom{where a(\mathbf{x})} = 0$, elsewhere

The calculation of modulation characteristics is usually facilitated by the use of Equation (17-41). Introduction of the aperture function provides some insight, and, occasionally, computational ease. Using the Parseval relation and the convolution theorem, one may write

$$\beta_m(t) = \int_{-\infty}^{\infty} A_m^*(\mathbf{k}) \, S'(\mathbf{k}) \, d\mathbf{k} \tag{17-43}$$

where $A_m^*(\mathbf{k})$ = the conjugate of the Fourier transform of $a_m(\mathbf{x})$

$S'(\mathbf{k}) = \int_{-\infty}^{\infty} S(\mathbf{k}') \, O(\mathbf{k} - \mathbf{k}') \, d\mathbf{k}'$

The aperture smears the scene in the spatial frequency.

The transform of a rectangle with sides of length a and b is

$$A(\mathbf{k}) = A(k_1, k_2) = \frac{\sin \pi a k_1 \, \sin \pi b k_2}{\pi^2 k_1 k_2} \tag{17-44}$$

The transform of a circular aperture of radius a is

$$A(\mathbf{k}) = A(k_1, k_2) = \frac{a J_1 \left(2\pi a \sqrt{k_1^2 + k_2^2}\right)}{\sqrt{k_1^2 + k_2^2}} \tag{17-45}$$

where J_1 is a Bessel function of the first order. The most common aperture is circular.

17.4.3. Reticle Motion. While many types of reticle motion are mathematically possible, convenient mechanical implementation tends to drive the designer to translation, rotation, and nutation of the reticle.

The simplest translational motion is along one axis of the aperture-fixed coordinates which are used to express the reticle pattern. (See Figure 17-6.) If the translational motion is in the negative x_2 direction,

$$r(x_1, x_2) = r(x_1, x_2 + st) \tag{17-46}$$

where s is the velocity. Thus, if the point $(x_1, x_2) = \mathbf{x}$ in the aperture is seen at time t, the point $(x_1, x_2 + st)$ is seen on the reticle.

The next form of motion to be considered is the rotating reticle. The center of rotation may not coincide with the aperture center, as is shown in Figure 17-7.

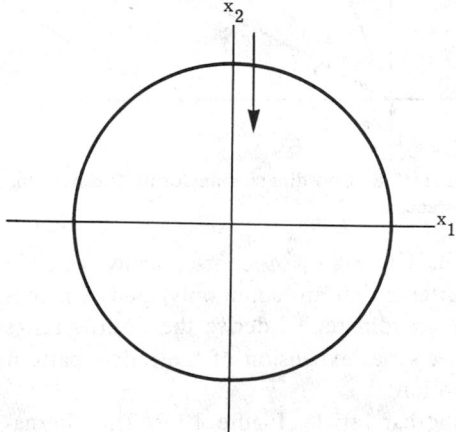

Fig. 17-6. Aperture coordinates.

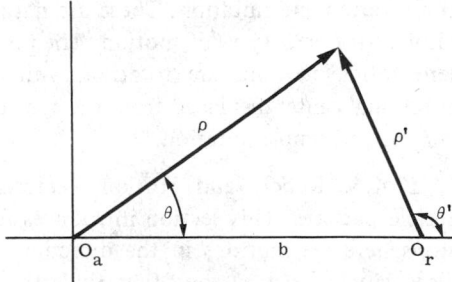

Fig. 17-7. Coordinate transforms for rotation. O_a equals aperture center and O_r equals center of rotation, usually taken as the center of the reticle pattern.

The reticle pattern is usually obtained in rotational-center coordinates (ρ', θ'), but the relation can be transformed to aperture centered coordinates, (ρ, θ), by

$$r(\rho, \theta, t) = r'(\rho'(\rho, \theta), \theta'(\rho, \theta), t) \tag{17-47}$$

From the law of cosines, and the fact that three complex vectors, which form a triangle, sum to zero, one obtains

$$\rho' = \rho'(\rho, \theta) = (\rho^2 + b^2 - 2\rho b \cos \theta)^{1/2}$$

$$e^{-i\theta'} = e^{-i\theta'(\rho,\theta)} = (\rho e^{-i\theta} - b)(\rho^2 + b^2 - 2\rho b \cos \theta)^{1/2} \tag{17-48}$$

A nutating system centers the reticle in the aperture and nutates the scene, as shown in Figure 17-8. Reticle motion is the same as for an aperture-centered reticle (Figure 17-7). The scene-center, O_S, is rotated around the aperture-center, O_a. The radius of scene nutation is c. Generally, scene nutation is at a uniform rate.

$$\theta(t) = \Omega_0 t \tag{17-49}$$

where $\Omega_0 = 2\pi/T$
     T = period

The temporal dependence of the scene $s(\rho, t)$ becomes

$$s(\rho, t) = s(\rho, \theta, t) = s[\rho'(\rho, \theta) \theta'(\rho, \theta)t] = s[\rho'(\rho, \theta - \Omega_0 t) \theta'(\rho, \theta - \Omega_0 t)] \tag{17-50}$$

Analysis of the modulation of point sources requires the relations

$$\rho = [c^2 + \rho'^2 - 2c\rho \cos (\theta' - \Omega_0 t)]^{1/2}$$

$$e^{-i\theta} = e^{-i\Omega_0 t}(\rho'e^{-i(\theta'-\Omega_0 t)} - c)[\rho'^2 + c^2 - 2\rho'c \cos (\theta' - \Omega_0 t)]^{-1/2} \tag{17-51}$$

This permits writing the intensity of a scene point (ρ', θ') in aperture coordinates. (For a description of a nutating-scene system, see Chapter 22.) The reader should note that Cartesian coordinates were chosen for translation, while polar coordinates were chosen for rotation and nutation. These are natural choices for each type of motion. The prevalent reticle motions are rotation, with the rotational center displaced from the aperture center, and simple nutation.

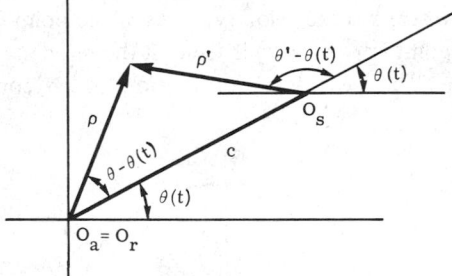

Fig. 17-8. Coordinate transforms for nutating scene.

17.4.4. Reticle and Motion Representation.

There are, of course, many possible reticle patterns. This section introduces four patterns that are commonly used. The motions here are motions in the direction of one coordinate. To derive the Fourier series expansion of $r(x, t)$, one first finds the Fourier series expansion of the reticle pattern with respect to the coordinate involved in the motion.

The first pattern considered is the translating bar reticle, Figure 17-9. The alternately opaque and transparent bars of dimension σ are translated in the negative x_2 direction. There is no pattern variation in the x_1 direction.

$$r'(x_1', x_2') = r'(0, x_2') \qquad (17\text{-}52)$$

The pattern periodicity in x_2' is now exploited.

$$r'(x_1', x_2') = \sum_{k=\infty}^{\infty} a_m e^{im\pi x_2'/\sigma}$$

$$\qquad (17\text{-}53)$$

$$a_m = \frac{1}{2\sigma} \int_0^{2\sigma} r_c'(x_1', x_2') e^{-im\pi x_2'/\sigma} dx_2'$$

Evaluation of a_m requires the introduction of an arbitrary phase δ, illustrated in Figure 17-10. The evaluation of a_m shows that $a_m = 0$ for even m, and one finds that

$$a_0 = \frac{1}{2}$$

$$\qquad (17\text{-}54)$$

$$a_{(2k+1)} = \frac{e^{-i(2k+1)\pi\delta}}{i\pi(2k+1)}, \qquad k = \pm 1, \pm 2, \ldots$$

$$= 0, \text{ otherwise.}$$

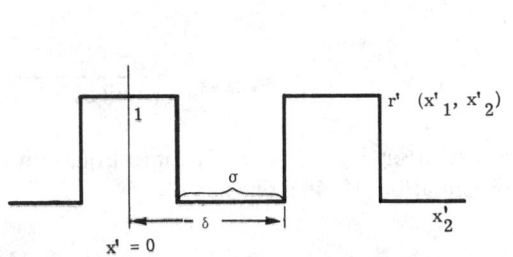

Fig. 17-9. Translating bar reticle.   Fig. 17-10. Bar reticle phase.

Substituting Equation (17-54) in Equation (17-53) and using Equation (17-45), one gets

$$r(x_1, x_2 + st) = \frac{1}{2} + \frac{1}{i\pi} \sum_{k=-\infty}^{\infty} \frac{e^{-i(2k+1)\pi\delta}}{2k+1} e^{i(2k+1)\pi x_2/\sigma} e^{i(2k+1)\pi st/\sigma} \qquad (17\text{-}55)$$

From Equations (17-40) and (17-41)

$$\beta_0(t) = \frac{1}{2} \int_A s(\mathbf{x}, t) d\mathbf{x}$$

$$\beta_{2k+1}(t) = \frac{1}{i\pi} \int_A \frac{e^{-i(2k+1)\pi\delta}}{2k+1} e^{i(2k+1)\pi x_2/\sigma} s(\mathbf{x}, t) d\mathbf{x} \qquad (17\text{-}56)$$

The second pattern, the radial, uniformly-rotating reticle known as the wagonwheel or episcotister, is commonly employed. On occasion, the basic pattern is modified but not in important ways. Figure 17-11 illustrates the pattern. The reticle's radial property simplifies the reticle description

$$r'(\rho', \theta', t) = r(0, \theta', t) \tag{17-57}$$

If there are p transparent and opaque spoke pairs, the basic period of θ' is $2\pi/p$. The one dimensional Fourier series expansion of $r(0, \theta', t)$ follows the same analytic pattern as the bar reticle with the same definition of the arbitrary phase δ.

$$r'(0, \theta') = \sum_{k=-\infty}^{\infty} a_m \, e^{im\theta'p}$$

$$\tag{17-58}$$

$$a_m = \frac{p}{2\pi} \int_0^{2\pi/p} r'(0, \theta') e^{-im\theta'p} \, d\theta'$$

After the phase shift of $e^{im\pi\delta}$, $r'(0, \theta') = 1$, $(0 \leq \theta' \leq \pi/p)$, and 0 for the rest of the period. Thus, one finds $a_m = 0$ for even m and one has

$$a_0 = \frac{1}{2}$$

$$\tag{17-59}$$

$$a_{(2k+1)} = \frac{e^{-i(2k+1)\pi\delta}}{i\pi(2k+1)}, \qquad k = \pm 1, \pm 2, \ldots.$$

Substituting Equation (17-58) into Equation (17-57) and using the rotational definition of Equation (17-49), one has

$$r(\rho', \omega_0 t - \theta') = \frac{1}{2} + \frac{1}{i\pi} \sum_{k=\infty}^{\infty} \frac{e^{-i(2k+1)\pi\delta}}{2k+1} \, e^{-i(2k+1)p\theta'} \, e^{i(2k+1)p\omega_0 t} \tag{17-60}$$

Then from Equations (17-40), (17-41), and the definition of $e^{-i\theta'}$ from Equation (17-48), one gets

$$\beta_0(t) = \frac{1}{2} \int_0^{2\pi} \int_0^a \rho s(\rho, \theta, t) \, d\rho \, d\theta$$

$$\tag{17-61}$$

$$\beta_{2k+1}(t) = \frac{e^{-i(2k+1)\pi\delta}}{i\pi(2k+1)} \int_0^{2\pi} \int_0^a \frac{\rho(\dot{\rho}e^{-i\theta} - b)^{(2k+1)p}}{(\rho^2 + b^2 - 2b\rho \cos\theta)^{(2k+1)p/2}} \, s(\rho, \theta, t) \, d\rho \, d\theta$$

The carrier frequency is $p\omega_0$ and one sees at once that the modulation will be AM.

The third pattern considered is the sun-burst or rising-sun reticle. This reticle adds a phasing sector to the episcotister (Figure 17-12) to obtain angular-positional information. The phasing sector has a transmission of $1/2$. The one-dimensional pulse train is shown in Figure 17-13. The period of this reticle is 2π and one has

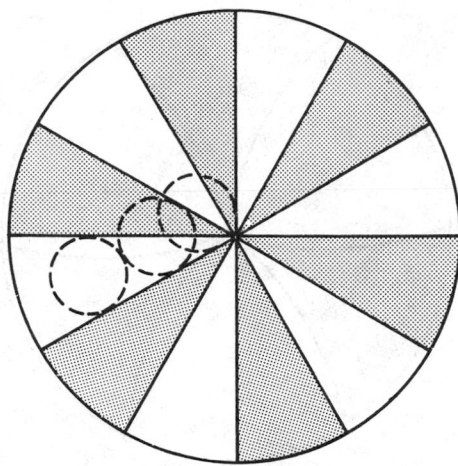

Fig. 17-11. Episcotister pattern.

Fig. 17-12. Episcotister with phasing sector.

$$r(\rho', \omega_0 t - \theta') = \sum a_m e^{-im\theta'} e^{im\omega_0 t}$$

$$a_0 = \frac{1}{2}$$

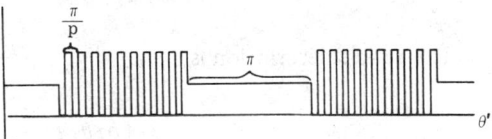

Fig. 17-13. Annular pattern of sectioned episcotister.

$$a_m = -\frac{e^{-im\pi\delta}}{im2\pi} \left\{ \sum_{k=0}^{p-1} (e^{-im\pi(2k+1)/2p} - e^{-im\pi k}) + \frac{1}{2} (e^{-2im\pi} - e^{-im\pi}) \right\}$$

$$(17-62)$$

The last term, $(1/2) (e^{-2im\pi} - e^{-im\pi})$, is 0 if m is even, and 1 if m is odd. Intuition suggests considerable modulation at $2p\omega_0$ and its harmonics $2np\omega_0$. Examination of Equation (17-62) shows $a_m = 0$ for n even, and a_m a maximum at $n = 1$, which is $e^{-2p\pi\delta}/i2\pi$. One finds that

$$\beta_m(t) = \frac{1}{2} \int_0^{2\pi} \int_0^a \rho s(\rho, \theta, t) \, d\rho \, d\theta$$

$$(17-63)$$

$$\beta_m(t) = \frac{e^{-im\pi\delta}}{im2\pi} \int_0^{2\pi} \int_0^a \frac{\rho(\rho e^{-i\theta} - b)^m}{(\rho^2 + b^2 - 2\rho b \cos \theta)^{m/2}} s(\rho, \theta, t) \, d\rho \, d\theta$$

Again, this is an AM modulation.

The last reticle pattern considered is more general than the preceding three. It is presented both as a potential reticle system and as a method for approximating more complex patterns. The basic motion is a reticle pattern which is a sum of concentric ring reticles. Figure 17-14 illustrates such a reticle pattern.

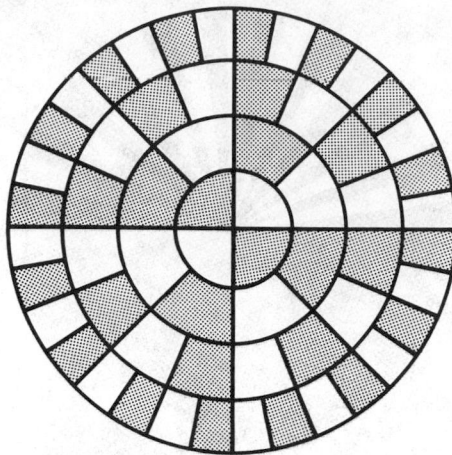

Fig. 17-14. Rotating concentric annular ring reticle.

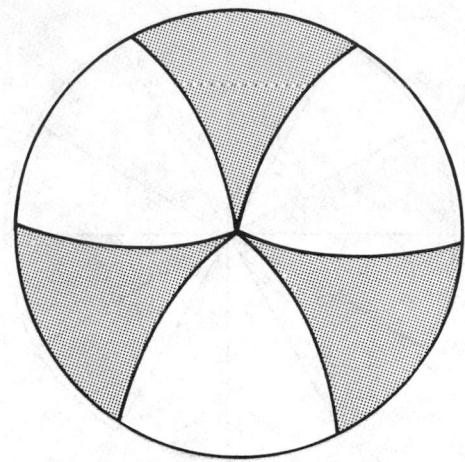

Fig. 17-15. Complex reticle pattern which is approximately annular.

The reticle description is

$$r(\rho, \theta, t) = \sum_{n=1}^{N} r_n(\rho, \theta, t) \tag{17-64}$$

where $r_n(\rho, \theta, t)$ has p_n equal transparent and opaque spokes in the ring $\rho_{n-1} < \rho < \rho_n$. It is assumed that the reticle is uniformly rotated with the center of the aperture as the center of rotation. The values of δ_n may differ, but they are equal for Figure 17-14. The expression for $v_n(t)$ is

$$v_n(t) = \int_0^{2\pi} \int_{\rho_{n-1}}^{\rho_n} \rho\, r_n(\rho, \theta, t)\, s(\rho, \theta, t)\, d\rho\, d\theta \tag{17-65}$$

$$v_n(t) = \beta_0^{(n)}(t) + \sum_{k=-\infty}^{\infty} \beta_{2k+1}^{(n)}\, e^{i(2k+1)p_n\omega_0 t}$$

where $\beta_0^{(n)}(t) = (1/2)\int_0^{2\pi}\int_{\rho_{n-1}}^{\rho_n} \rho s(\rho, \theta, t)\, d\rho\, d\theta$

$\beta_{(2k+1)}^{(n)} = \left[e^{-i(2k+1)\pi\delta_n} / i\pi(2k+1) \right] \int_0^{2\pi}\int_{\rho_{n-1}}^{\rho_n} \rho\, e^{-i(2k+1)p_n\theta}\, s(\rho, \theta, t)\, d\rho\, d\theta$

Then $v(t)$ is the sum of $v_n(t)$:

$$v(t) = \sum_{1}^{N} v_n(t) \tag{17-66}$$

The analysis of a pattern such as shown in Figure 17-15 would be tedious. Clearly, if one makes N large enough one can approximate the pattern's signal modulation $v(t)$ as closely as one chooses.

17.4.5. Modulation of Point Sources. The delta function representation for a point source scene, $I\delta(\mathbf{x} - \mathbf{x}_0)$, is a convenient total for the analysis of modulation characteristics of reticle systems:

$$I \int_A a_m(\mathbf{x})\, \delta(\mathbf{x} - \mathbf{x}_0)\, dx = I\, a_m(\mathbf{x}_0), \quad \text{for } \mathbf{x}_0 = A$$

$$= 0, \text{ otherwise}$$

(17-67)

where I is the target intensity. The delta function representation for a point source in polar coordinates is $I\,\delta(\rho - \rho_T,\, \theta - \theta_T)$. For the episcotister, one has the modulation

$$v(t) = \frac{I(\rho_T)}{2} + I(\rho_T) \sum_{k=-\infty}^{\infty} \frac{e^{-i(2k+1)\pi\delta}}{i\pi(2k+1)}\, \frac{\rho_T(\rho_T e^{-i\theta_T} - b)^{(2k+1)p}}{(\rho_T^2 + b^2 - 2b\rho_T \cos\theta_T)}\, e^{i(2k+1)p\omega_0 t}$$

(17-68)

This is an amplitude modulation of the carrier $\rho\omega_0$. The modulation of I occurs as in Figure 17-11 because the target blur-disc is not completely modulated by the reticle.

A point source in a nutated scene has the following representation: $s\rho(\rho', \theta')s(\rho', \theta') = S(\rho(\rho', \theta') - \rho_T',\, \theta(\rho', \theta') - \theta_T')$. The nutated scene with the centered reticle has, with $b = 0$ in Equation (17-61),

$$\beta_0(t) = \frac{I}{2}\left(c^2 + \rho_T'^2 - 2c\rho_T' \cos(\theta_T - \Omega_0 t)\right)^{1/2}$$

$$\beta_{2k+1}(t) = \frac{I e^{-i(2k+1)\pi\delta}}{i\pi(2k+1)}\, e^{-\Omega_0 t}\left(\rho_T' e^{-i(\theta_T - \Omega_0 t)} - c\right)^{(2k+1)p}$$

(17-69)

The modulation clearly depends on the relative size of the error, ρ_T', and the radius of nutation, c. For $\rho_T' = c$, the modulation is AM. For small values of the ratio, ρ_T'/c, one may approximate Equation (17-51) by

$$\rho \sim c$$

$$\theta \sim \frac{\rho_T'}{c} \sin(\theta' - \Omega_0 t) + \Omega_0 t$$

(17-70)

Using the Approximation (17-70) instead of Equation (17-51) in Equation (17-61), with $b = 0$, one has, for a stationary reticle, i.e., $\omega_0 = 0$, the system of Chapter 22:

$$\beta_0(t) = \frac{Ic}{2}$$

$$\beta_{2k+1}(t) = \frac{I e^{-i(2k+1)\pi\delta}}{i\pi(2k+1)}\, e^{-i(2k+1)p\left(\frac{\rho_T'}{c}\sin(\theta_T - \Omega_0 t) + \Omega_0 t\right)}$$

(17-71)

$$v(t) = \frac{Ic}{2} + \frac{I}{i\pi}\sum_{k=-\infty}^{\infty} \frac{e^{-i(2k+1)\pi\delta}}{2k+1}\, e^{i(2k+1)p\left(\frac{\rho_T'}{c}\sin(\theta_T - \Omega_0 t) + \Omega_0 t\right)}$$

The modulation is clearly FM. The carrier frequency is $\rho\Omega_0$ where ρ is the number of spokes and Ω_0, the nutation frequency, becomes the modulation frequency. The target's radial position, ρ'_T, is proportional to the amplitude (modulation index) of the modulating frequency, Ω_0. The amplitude is $p/c\ \rho'_T$. The phase of the modulation is determined by θ_T.

Another FM reticle system can be constructed from the translating bar-reticle and rotation of the scene with the center of the scene's rotation, the aperture center. Then one has

$$s(\rho, \theta, t) = s(\rho, \theta + \Omega_0 t) \tag{17-72}$$

Figure 17-16 represents such a system. Changing to polar coordinates in Equation (17-56) and introducing the delta function representation for the scene, one gets

$$s(\rho, \theta - \Omega_0 t) = \delta(\rho - \rho_T, \theta - \theta_T - \Omega_0 t) \tag{17-73}$$

The modulation of the point source is, by substitution into Equation (17-56),

$$\beta_0(t) = \frac{I}{2}$$

$$\beta_{2k+1}(t) = \frac{I}{i\pi} \frac{e^{-i(2k+1)\pi\delta}}{2k+1}\ e^{i(2k+1)\pi\rho_T \sin(\theta_T - \Omega_0 t)}$$

then

$$v(t) = \frac{I}{2} + \frac{I}{i\pi} \sum_{k=-\infty}^{\infty} \frac{e^{-i(2k+1)\pi\delta}}{2k+1}\ e^{i(2k+1)\pi \left\{ \frac{\rho_T}{\sigma} \sin(\theta_T - \Omega_0 t) + st/\sigma \right\}} \tag{17-74}$$

where $\pi s/\sigma$ = carrier frequency

$\quad\quad\quad \Omega_0 t$ = modulation frequency

$\quad\quad\quad \pi\rho_T/\sigma$ = amplitude of modulation (carries information about ρ_T)

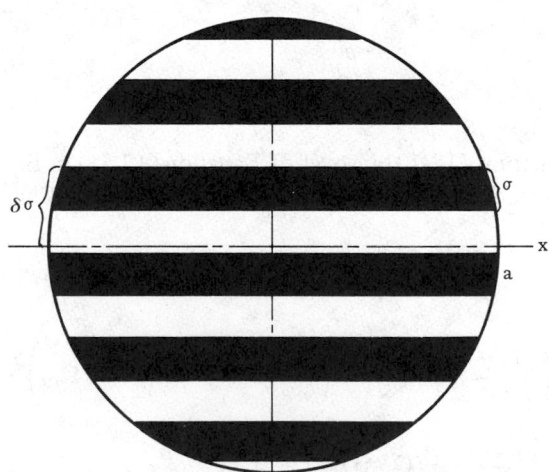

Fig. 17-16.   Translating bar-reticle circular aperture and rotating scene at aperture center.

ρ_T = radial error

θ_T = angular error

The phase of the modulated signal carries the information about θ_T.

17.4.6. Reticles Coding Spatial Frequencies in Temporal Frequencies. The analytic tools developed so far are discussed towards the analysis of the modulation characteristics of reticle system. This section covers some reticle system synthesis results.

Doubly Periodic Reticles. This method of reticle synthesis was proposed by D. Montgomery [17-4]. These reticles are fairly simple to design. The pattern of openings in the reticle mask is assumed to have two-dimensional periodicity. (For the definition of two-dimensional periodicity, see Section 17.2.3.) The aperture is considered arbitrarily large. This reticle system may be regarded as a device which codes linear combinations of certain spatial frequencies of the scene into the time frequencies of the voltage output.

Let the infinite reticle mask have a periodicity defined by a_1 and a_2. The reticle pattern may be represented by a series, as in Section 17.2.3.

$$r(\mathbf{x}) = \sum_n r_n \, e^{2\pi i \, \mathbf{x} \cdot \mathbf{b}_n} \tag{17-75}$$

The output voltage, $v(t)$, when the reticle is translated, may be represented by

$$v(t) = \int_{-\infty}^{\infty} s(\mathbf{x}) \, r \, [\mathbf{x} - \mathbf{x}(t)] \, d\mathbf{x} \tag{17-76}$$

Substituting Equation (17-75) into Equation (17-76) and using the fact that $r(\mathbf{x})$ is real, and consequently that the complex conjugate of r_n, r_n^*, is r_{-n}, one has

$$v(t) = \sum_n r_n^* \, S(\mathbf{b}_n) \, e^{2\pi i \mathbf{b}_n \cdot \mathbf{x}(t)} \tag{17-77}$$

where $S(\mathbf{b}_n)$ is the Fourier transform of $s(\mathbf{x})$ evaluated at the lattice point $\mathbf{b}_n$. For the signal to be periodic with period T, $\mathbf{x}(t)$ must equal $t/T \, a_k$ and a_k must be associated with only one primary lattice point. Using the $\mathbf{x}(t)$ defined above, and the definition of the primary and reciprocal lattice, one has

$$e^{2\pi i \, \mathbf{b}_n \cdot \mathbf{x}(t)} = e^{2\pi i \, \mathbf{k} \cdot \mathbf{n} \, t/T} \tag{17-78}$$

where $\mathbf{k} = (k_1, k_2)$

$\mathbf{n} = (n_1, n_2)$

Using the periodicity of $v(t)$, one may write the series expansion for $v(t)$ as

$$v(t) = \sum_m V_m \, e^{2\pi i m \, t/T} \tag{17-79}$$

Equating Equations (17-79) and (17-77) after substituting Equation (17-78) in Equation (17-77), one obtains

$$\sum_m V_m \, e^{2\pi i m t/T} = \sum_n r_n^* \, S(\mathbf{b}_n) \, e^{2\pi i \, \mathbf{n} \cdot \mathbf{k} t/T} \tag{17-80}$$

Equating coefficients, one has the mth harmonic of $v(t)$ expressed as

$$V_m = \sum_{\mathbf{n} \cdot \mathbf{k} = m} r_n^* \, S(\mathbf{b}_n) \tag{17-81}$$

Equation (17-81) says that the mth harmonic of the output is a linear combination of the spatial frequencies along a line $(\mathbf{n} \cdot \mathbf{k} = m)$ perpendicular to the scan direction, $\mathbf{a}_k$. Since $\mathbf{k}$ has been chosen so that k_1 and k_2 are relatively prime, there will be at least one lattice point on every line $\mathbf{n} \cdot \mathbf{k} = m$. If only a finite number of spatial frequencies are required to distinguish between targets and their backgrounds, then one sets the coefficients r_n to zero on the remaining reciprocal lattice points. Figure 17-17 represents the discrimination set in reciprocal space. The coefficients r_n are set equal to zero for all lattice points not in S. Choose a scan direction $\mathbf{a}_k$ so that for each pair of lattice points in S, $\mathbf{a}_k$ is not perpendicular to a line joining them.

$$(\mathbf{n} - \mathbf{m}) \cdot \mathbf{k} \neq 0, \qquad \text{for all } \mathbf{n}, \mathbf{m} \text{ in } S \tag{17-82}$$

Then, since each harmonic v_m of the output voltage is the sum of spatial frequencies perpendicular to the scan direction, the sum of the right-hand side of Equation (17-81) will contain only one term.

$$V_m = r_n^* \, S(\mathbf{b}_n) \tag{17-83}$$

where $\mathbf{n} \cdot \mathbf{k} = m$. Discrimination is achieved by analyzing the output signal, $v(t)$, for the harmonics, m/T. Some combination of harmonic amplitudes will indicate a target while other, hopefully distinct combinations, simply a background.

Consider a reticle that will separate a point source from a straight edge. A point source will have equal spatial frequency content in all directions. The spatial frequency content of a straight edge is maximum in the direction perpendicular to the edge and zero in the direction parallel to the edge. Our discrimination set contains only four, perpendicular, reciprocal lattice points. (See Figure 17-18.) Let the temporal period of the scan be $2T$ seconds. Then the spatial frequency pair, $\mathbf{b}_1, -\mathbf{b}_1$, is encoded in temporal frequency $1/2T$, since $m = \mathbf{n} \cdot \mathbf{k} = (1, 0)(1, 2) = 1$; and the spatial frequency pair, $\mathbf{b}_2, -\mathbf{b}_2$, is encoded in temporal frequency $1/T$, since $m = \mathbf{n} \cdot \mathbf{k} = (0, 1)(1, 2) = 2$. This includes conjugate pairs and requires $r_n^* = r_{-n}$, real values for $r(x)$. The reticle mask $r(\mathbf{x})$ is given by

$$r(\mathbf{x}) = \frac{1}{2} e^{2\pi i x_1} + \frac{1}{2} e^{2\pi i x_2} \tag{17-84}$$

The two sinusoidal patterns could be approximated by the superposition of two bar patterns. Figure 17-19 shows the resulting reticle pattern and scan direction. The dark areas have zero transmission, the lightly shaded areas have transmission $1/2$, and the unshaded areas have transmission 1. Suppose a point source is scanned by the reticle, making a path $\mathbf{a}_k$. The output signal (a), depicted in Figure 17-20, is the sum of the signals (b) and (c). If the point is shifted, the relative phase of (b) and (c) will change, but the output signal will remain the same. Discrimination of the symmetric source from the straight edge is achieved by comparing the amplitude of frequency $1/T$ with the

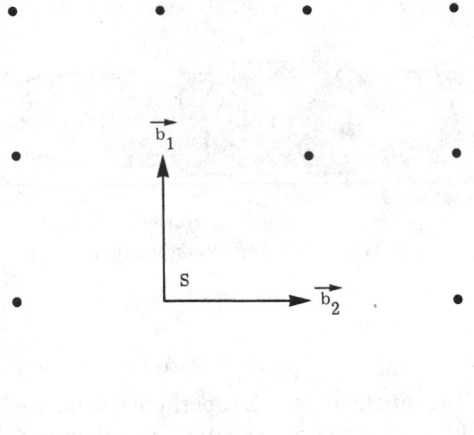

Fig. 17-17. Discrimination set in reciprocal space.

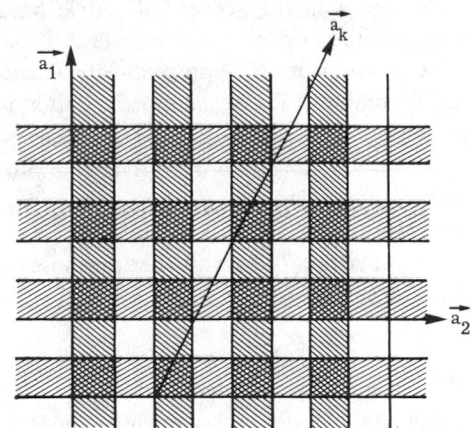

Fig. 17-19. Bar approximation to reticle pattern Equation (17-85).

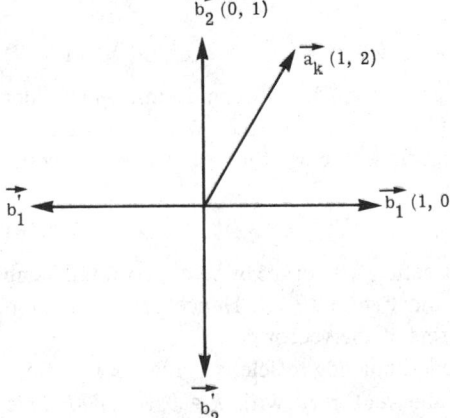

Fig. 17-18. Discrimination set and scan direction.

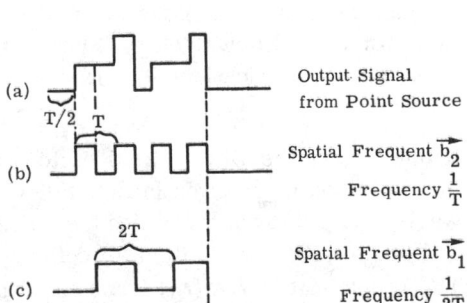

Fig. 17-20. Output from point source.

amplitude of frequency $1/2T$. A symmetric source exists if the two amplitudes are equal. Such an experimental reticle has been constructed and works as predicted [17-5].

17.4.7. Coded Imaging Reticles. The output signal $v(t)$ of the reticles considered so far cannot be used to recreate an image of the scene from $v(t)$. There do exist reticle codes which permit recovery of the image. The simplest and most familiar is the Nipkow scanner, the first TV camera. Such a system produced N raster lines with M resolution elements in each line as illustrated in Figure 17-21. The reticle contains N transparent apertures. The first aperture scans the image and produces a signal $v_1(t)$ representing the first line. The second aperture scans the second line until finally the frame is complete, with the scanning of the Nth aperture producing $v_N(t)$. The reticle was implemented on a rotating disc. In order to analyze such reticle codes an analytic model is required.

So far both the scene and reticle have been treated as continuous entities. A discrete model is more appropriate for coded reticle analysis. The scene representation is presented in Figure 17-22. The intensity s_{ij} represents the signal from the ith resolution element from the jth line. The representation of the scene is in column vector. The superscript T denotes the transpose of the row vector.

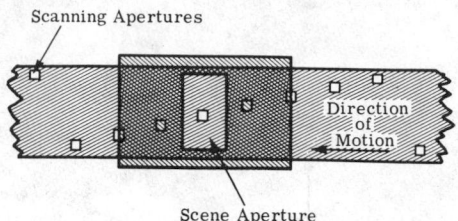

Scanning Apertures

Direction of Motion

Scene Aperture

Fig. 17-21. Idealized Nipkow reticle.

$$S = (S_{11}, \ldots, S_{1M}, \ldots, S_{i1}, \ldots, S_{ij}, \ldots, S_{iM}, \ldots, S_{N1}, \ldots, S_{NM})^T \quad (17\text{-}85)$$

The reticle pattern is depicted in Figure 17-23. The reticle pattern is superimposed on the scene. As one moves the reticle to the left, one has new block of the reticle superimposed on the scene. The kth reticle block superimposes on the scene the reticle code given by

$$\mathbf{r}_k = (r_{1k}, \ldots, r_{1,k+M-1}, r_{2k}, \ldots, r_{2,k+M-1}, \ldots, r_{N,k}, \ldots, r_{N,k+M-1}) \quad (17\text{-}86)$$

The voltage, v_k, observed when the kth reticle block is superimposed on the scene is simply the inner or dot product.

$$v_k = \mathbf{s}^T \cdot \mathbf{r}_k, \qquad k = 1, \ldots, \ell - M + 1 \quad (17\text{-}87)$$

The number of v_ks produced by a reticle of length ℓ is $\ell - M + 1$. The definition of inner product Equation (17-7) shows that Equation (17-87) is a set of $\ell - M + 1$ linear equations for the unknown NM s_{ij}. For a unique solution to exist there must be NM equations. Thus, the reticle length, ℓ, is

$$\ell = M(N + 1) - 1 \quad (17\text{-}88)$$

If one is given the $M(N + 1) - 1$ vectors, $\mathbf{r}_k$, the reticle pattern can be constructed using the definition of $\mathbf{r}_k$ given in Equation (17-86) and Figure 17-23. Henceforth, attention will be directed towards defining the reticle in terms of the vector $\mathbf{r}_k$.

The most compact method for defining the coded imaging reticle is in matrix notation. The reticle matrix $R = (r_{ij})$ has as its kth row the vector $\mathbf{r}_k$ with $k = 1, \ldots, NM$. One should understand at least the rules for matrix multiplication, addition, identity matrix, and the definition of a matrice's inverse. Reference [17-6] (or any elementary text on linear algebra) contains this information. The output vector from a coded reticle in matrix notation is

$$\mathbf{v} = \mathbf{s}R \quad (17\text{-}89)$$

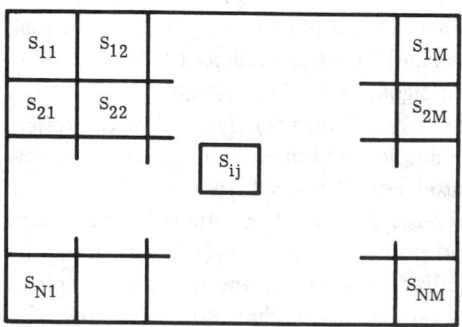

Fig. 17-22. Discrete scene representation.

$$
\begin{matrix}
r_{11} & r_{12} & \bullet & r_{1M} & \bullet & \bullet & r_{1\ell} \\
r_{21} & r_{22} & \bullet & \bullet & \bullet & \bullet & \bullet \\
\bullet & \bullet & \bullet & \bullet & \bullet & \bullet & \bullet \\
r_{N1} & \bullet & \bullet & r_{NM} & \bullet & \bullet & r_{N\ell}
\end{matrix}
$$

Fig. 17-23. Physical reticle pattern.

The encoded signal, $\mathbf{v}$, must be decoded to recover the scene, $\mathbf{s}$. The decoding transform is the inverse of the matrix, R, denoted by R^{-1}. Then, operating on $\mathbf{v}$ with R^{-1}, one finds the scene is recovered:

$$\mathbf{v}R^{-1} = \mathbf{s}RR^{-1} = \mathbf{s}I = \mathbf{s} \tag{17-90}$$

where I is the identity.

There are important restrictions on the matrices, R, which can be implemented as reticles. The rows of R, $\mathbf{r}_k$, are representations of *overlapping* reticle blocks. (See Equation (17-86) and Figure 17-23.) The key restriction is overlapping. This leads to a restriction on the values (r_{ij}) of the matrix, R.

$$r_{ij} = r_{i-1,j+1} \tag{17-91}$$

A matrix, R, with the property shown in Equation (17-91) is called a *circulant*. A circulant matrix is represented in Figure 17-24. The circulant requirement for R is a direct and inescapable consequence of the reticle implementation. The circulant has the diagonal elements from the lower left to upper right equal. For example, inspection shows the matrix R associated with the Nipkow scanner has ones in the diagonal $r_{k,NM-k+1}$. (See Figure 17-25.)* The mathematical development of coded reticles may be found in Reference [17-7].

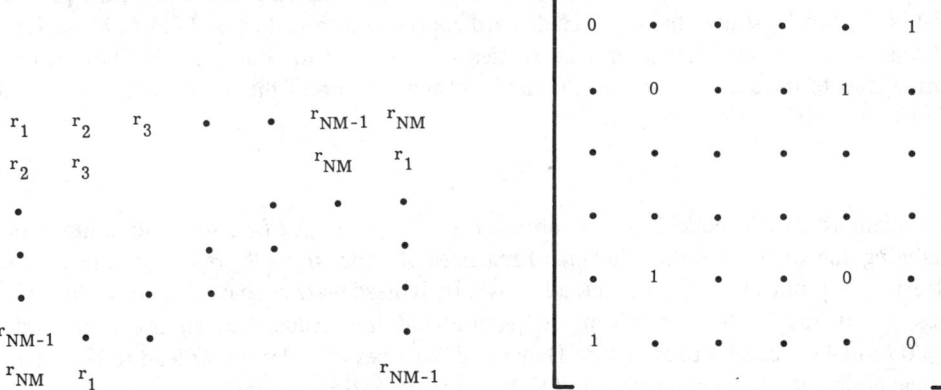

Fig. 17-24. Circulant reticle matrix. Fig. 17-25. Reticle matrix for Nipkow.

A further restriction on the coded reticle is that the transmission, r_{ij}, is such that $0 \leqslant r_{ij} \leqslant 1$. In fact, the easiest reticles to construct are those which set $r_{ij} = 0$ or 1, i.e., either opaque or transparent. Letting $V = NM$, one has a class of widely studied matrices, the (V, k, λ) configuration. (See Chapter 8 of Reference [17-8].) These configurations have been studied mathematically [17-8] and have found application in statistical design of experiments [17-9]. The matrices of zeroes and ones associated with (V, k, λ) have the following important properties:

---

*The critical reader will note that while every circulant matrix can be implemented in a reticle, not every matrix associated with a reticle is a strict circulant. The circulant restriction must hold for submatrices of R. The less restrictive condition on R has, to this author's knowledge, not been studied. On the other hand, the literature on circulants is rich and the electro-optical engineer may draw on a large mathematical literature. Coded reticles which are not full circulants do not appear in the literature.

(a) R is of order V $0 < \lambda < k < V - 1$
(b) $r_i \cdot r_i = k$ $i = 1, \ldots, V$,   where r_i is the ith row of R
(c) $r_i \cdot r_j = \lambda$ $i = j$ (17-92)
(d) $RR^T = R^R R = (k - \lambda)I + \lambda J,$ where J is the $V \times V$ matrix of ones

(e) $R^{-1} = \dfrac{1}{(k - \lambda)} R - \dfrac{\lambda}{k} J$

Relation (17-90) says that each row of R contains k ones. the Nipkow scanner has $k = 1$, $\lambda = 0$, and $R^{-1} = R$ which is easily checked. Of great importance is the fact that the matrix, R, associated with a (V, k, λ) configuration has an inverse, R^{-1}.

Not all combinations of integers (V, k, λ) form a (V, k, λ) configuration. If there exists a (V, k, λ) configuration, the $V, k,$ and λ must satisfy

   (a) $\lambda = k(k - 1)/V - 1.$
   (b)   If V is even, then $(k - \lambda)$ is a square.
   (c)   If V is odd, then the equation

$$x^2 = (k - \lambda)y^2 + (-1)^{(V-1)/2} \lambda z^2$$ (17-93)

has a solution in integers x, y, z not equalling zero.

The coded reticles used in the current literature are called Hadamard codes [17-10, 17-11]. It can be shown that such Hadamard codes are a special case of (V, k, λ) configurations. A Hadamard matrix has as entries only ones and minus ones. They can be normalized to having only ones in the first row and column of the matrix. The Hadamard of order n matrix satisfies the relation

$$HH^T = nI$$ (17-94)

A Hadamard reticle code matrix is obtained by dropping the first row and column and changing the ones to zeros. Hadamard matrices of order $n = 2^m$ are easily constructed. (See [17-12] and [17-13].) In fact, all known Hadamard matrices have order divisible by 4, i.e., $n \equiv 0 \bmod 4$. An unproven conjecture has *all* Hadamard matrices of order $n \equiv 0 \bmod 4$. It can be proven that Hadamard matrices of order $n = 4t$ lead to Hadamard reticle codes which are equivalent to (V, k, λ) configurations where

$$V = 4t - 1 \qquad RR^T = tI + (t - 1)J$$
$$k = 2t - 1 \qquad R^{-1} = \frac{1}{t} R - \frac{t - 1}{2t - 1} J \qquad (17\text{-}95)$$
$$\lambda = t - 1$$

The relation between Hadamard codes and maximal length sequences (m-sequences) is discussed in Reference [17-12].

So far, the discussion has not introduced circulant (V, k, λ) configurations. Such circulants do exist. (See Chapter 9 of Reference [17-8].) Circulant (V, k, λ) configurations are derived from perfect different set [17-14] or cyclic projective planes [17-15]. If $4t - 1$ is a prime, then a circulant Hadamard reticle code can be found by the method of quadratic residues. A detailed discussion of methods for constructing such circulant reticle codes lies outside the scope of this chapter. References [17-14] and [17-15] plus some computational skills enable the reader to construct circulant coded reticles. The reader should note that the Hadamard configurations are a subset of the more general (V, k, λ) configurations that the reticle designer has at his disposal.

Attention must now be paid to decoding the encoded scene **v**. The quantity **v** may be recorded digitally and the inverse, R^{-1}, implemented by a general or special purpose computer. From Equations (17-92) and (17-95), one observes that R^{-1} is also a circulant. Since R^{-1} is a circulant, the decoding operation can be implemented by shift registers. The encoded image, **v**, must be stored. The price and size of digital storage and shift registers today warrants considering the construction of special purpose decoders.

An early method of decoding, [17-7], was to make an encoded photographic image of **v** and then to decode the image with a decoding reticle. The natural decoding reticle is R^T. This decoding reticle produced a signal pedestal, $\lambda \mathbf{S} J$, Equations (17-92) and (17-95), which had to be reduced by a bias or dc correction.

One should ask what advantages can be gained from a coded reticle as compared with a single small detector scanned in the object or image plane. The answer that can be given is equivocal. The coded reticle increases the dwell time on the scene element by a factor k, the number of ones in the reticle. If an imaging system's limiting noise is photon noise from the housing, or preamplifier noise, then there is a gain in signal-to-noise of $\sqrt{k}$. Again, one would maximize k. The coded reticle shows no improvement for photon noise arising from the scene. If the system is detector noise limited, and the f-numbers of the scanned element and coded reticle system are equal, then the detector area must increase by a factor, V. In this case, the signal-to-noise gain is $\sqrt{k/V}$ which is always less than one (a loss). These results are summarized in Table 17-2.

Table 17-2   S/N Gains From Coded Reticle

Gain in S/N	Noise Source
1	Scene photon noise
$\sqrt{k}$	Preamplifier, or housing photon noise
$\sqrt{k/V}$	Detector noise

The factors considered so far lead to maximizing k. The configuration which maximizes k sets $k = N - 1$. This trivial circulant makes the transparent portions of the Nipkow reticle opaque, the remainder of the reticle transparent in Figure 17-21. This would be a practical solution except that the modulation amplitude is very small compared to the average signal. Electronic circuits do not handle such a signal well. A compromise solution is to set $k \approx 1/2\, V$, a reticle code implemented by a Hadamard configuration. The designer must make his own compromise.

17.5. Sampled Data Images

An image produced by a sampled-data, imaging, electro-optical sensor has some peculiar problems. The sampling process can introduce a form of image distortion called *aliasing*. The choice of a sampling lattice, the point spread response of the optical system, and the detector areal response will affect the amount of aliasing distortion in the image.

The display of sampled data imagery cannot affect the aliasing distortion of the imagery, but it can introduce additional distortions. A discussion of two-dimensional, sampled-data imagery appears in Reference [17-16]. The methods of analysis introduced in References [17-17] and [17-18] are the most appropriate for the analysis of sampled imagery. These methods are introduced by considering first the familiar one-dimensional sampling.

The samples of $v(t)$ are equally spaced T units of time apart. The sampled values are $v(mT)$. A reconstruction (interpolation) function is used to estimate the signal $v(t)$

$$\hat{v}(t) = \sum_{m=-\infty}^{\infty} v(mT)g(t - mT) \tag{17-96}$$

The frequency representation of the estimated signal is revealing.

$$\hat{V}(f) = \frac{G(f)}{T} \sum_{m=-\infty}^{\infty} V\left(f + \frac{m}{T}\right) \tag{17-97}$$

This relation shows that at every frequency f, $\hat{V}(f)$ is the sum of the aliased frequencies, $f + m/T$. If $V(f)$ is nonzero only on an interval $|f| \leqslant 1/2T$, and one sets $G(f) = T$ for $|f| \leqslant 1/2T$, then

$$\hat{V}(f) = V(f), \quad |f| \leqslant \frac{1}{2T}$$
$$= 0, \text{ elsewhere} \tag{17-98}$$

Thus, if $V(f)$ is band limited, with bandwidth $B = 1/2T$, then $V(f)$ can be exactly reproduced from the sampled values by passing the values through a bandpass filter of bandwidth $1/2T$. The reconstruction function is

$$g(t) = \frac{T \sin (\pi t/T)}{\pi t} \tag{17-99}$$

The radiation distribution at the focal plane is given by $s'(x) = \int s(\mu) a(x - \mu) d\mu$, where $a(x)$ is the aperture point-spread-function. Either elements in the detector array or electronic sampling integrate the radiation from small areas in the focal plane. The areal response of each element in the array is $d(n_1 a_1 + n_2 a_2 - x)$. The center of a detector element in the array is specified by a primary lattice point, $n_1 a_1 + n_2 a_2$. The signal from each detector drives a light source at the display. The light source may be a defocused or shaped electron beam exciting a phosphor or LED. The distribution of light at the display surface corresponding to the sampled point is $b(n_1 a_1 + n_2 a_2 - x)$. An analysis of the sampled data system is given here.

Distribution of Radiation at the Focal Plane:

$$s'(\mathbf{x}) = \int s(\mu) \, a(\mathbf{x} - \mu) \, d\mu$$

Signal from Detector at Sample Point $n_1 \mathbf{a}_1 + n_2 \mathbf{a}_2$:

$$v(n_1 \mathbf{a}_1 + n_2 \mathbf{a}_2) = \int s'(\mathbf{x}) \, d(n_1 \mathbf{a}_1 + n_2 \mathbf{a}_2 - \mathbf{x}) \, dx$$

Distribution of Light at the Display: (17-100)

$$v(n_1 \mathbf{a}_1 + n_2 \mathbf{a}_2) \, b(\mathbf{x} - n_1 \mathbf{a}_1 - n_2 \mathbf{a}_2)$$

Displayed Image:

$$o(\mathbf{x}) = \sum_n v(n_1 \mathbf{a}_1 + n_2 \mathbf{a}_2) \, b(\mathbf{x} - n_1 \mathbf{a}_1 - n_2 \mathbf{a}_2)$$

The Fourier transform of the displayed image $o(\mathbf{x})$ is

$$O(\mathbf{k}) = \frac{B(\mathbf{k})}{A} \sum_n A(\mathbf{k} + \mathbf{b}_n) \, D(\mathbf{k} + \mathbf{b}_n) \, S(\mathbf{k} + \mathbf{b}_n) \qquad (17\text{-}101)$$

where $\qquad \Sigma_n = \Sigma_{n_1=-\infty}^{\infty} \, \Sigma_{n_2=-\infty}^{\infty}$

$D(\mathbf{k}), S(\mathbf{k}), A(\mathbf{k})$ = the Fourier transforms of $\mathbf{d}(x)$, $\mathbf{s}(x)$, and $\mathbf{a}(x)$, respectively

$B(\mathbf{k})$ = the reconstruction function

$D(\mathbf{k})A(\mathbf{k})$ = a prefilter on the scene

Clearly, the pattern specified by $\Sigma_n A(\mathbf{k} + \mathbf{b}_n) \, D(\mathbf{k} + \mathbf{b}_n) \, S(\mathbf{k} + \mathbf{b}_n)$ is periodic and repeats on parallelograms, β, with vertices $\pm 1/2 \, (\mathbf{b}_1 + \mathbf{b}_2)$, $\pm 1/2 \, (\mathbf{b}_1 - \mathbf{b}_2)$. (See Figure 17-26.) Thus, in a sense, the displayed information, $O(\mathbf{k})$, conveys only that information about the scene, $S(\mathbf{k})$, contained in the parallelogram, β, defined by $\mathbf{b}_1$ and $\mathbf{b}_2$. The parallelogram defines the frequency reproduction capability of the sampling lattice defined by $\mathbf{a}_1$ and $\mathbf{a}_2$ in the same sense that $1/2T$ is the frequency reproduction limit (Nyquist limit) for a one-dimensional sampling lattice of spacing T. The amplitude of a spatial frequency, $\mathbf{k}_0$, at the display is the weighted sum of the amplitude of the aliased spatial frequencies present in the scene. The weights are provided by the prefilter $A(\mathbf{k}) \, D(\mathbf{k})$.

The requirement for $O(\mathbf{k})$, the Fourier transform of the displayed scene, to be identical to the transform of the real scene, $S(\mathbf{k})$, is now apparent. If the spatial frequency representation of the scene is nonzero only on the parallelogram formed by $\mathbf{b}_1$ and $\mathbf{b}_2$, then $S(\mathbf{k} + \mathbf{b}_n)$ equals zero for $n = \pm 1, \pm 2, \ldots$, and is nonzero for $\mathbf{k}$ contained in β. Then Equation (17-101) can be written as

$$O(\mathbf{k}) = \frac{B(\mathbf{k})}{A} A(\mathbf{k}) \, D(\mathbf{k}) \, S(\mathbf{k}) \qquad (17\text{-}102)$$

If one lets the spatial frequency distribution of luminance at the display, $B(\mathbf{k})$, equal $A/A(\mathbf{k})D(\mathbf{k})$ over the spatial frequency parallelogram, β, then $O(\mathbf{k})$ equals $S(\mathbf{k})$ and one has perfect reproduction.

If the scene's spatial frequency representation, $S(\mathbf{k})$, is nonzero over a bounded region in the frequency plane, one can call the scene spatial-frequency-limited. For a spatial-frequency-limited scene, one can find a pair of vectors $\mathbf{b}_1$ and $\mathbf{b}_2$ such that the associated parallelogram covers the spectrum of the scene, $S(\mathbf{k})$. The reciprocal lattice specifies the sampling lattice $\mathbf{a}_1$ and $\mathbf{a}_2$ and, in principle, perfect reproduction can be achieved. Spatial-frequency amplitude-distributions of real scenes are decreasing functions of frequency. One can find a spatial-frequency limit above which there is little power.

If the spatial frequency distribution of the scene lies outside the parallelogram, β, defined by $\mathbf{b}_1$ and $\mathbf{b}_2$, then perfect reproduction of the scene is not possible with the sampling lattice $\mathbf{a}_1$ and $\mathbf{a}_2$. It is the summation Σ_n which makes the output display from a sampled-data image system an aliased image. The aliased image is a distorted version of the original scene. It is not simply a bandpassed image with the bandpass determined by the reciprocal vectors $\mathbf{b}_1$ and $\mathbf{b}_2$. For a discussion of electronic sampling, see [17-16].

17.5.1. Image Distortion by Aliasing. Two types of aliasing effects appear. The first is a result of periodic patterns in the scene, and the second is a result of aliased, continuous frequency spectra. If a scene containing periodic components is sampled, Moiré patterns appear in the displayed image.

The more complex case is aliasing from continuous spectra. The discussion is simplified by assuming that the aperture response and the detector-area-weighting optical-frequency response are unity. The display frequency response is unity over β. Then Equation (17-101) becomes

$$O(\mathbf{k}) = \frac{B(\mathbf{k})}{A} \sum_n S(\mathbf{k} + \mathbf{b}_n)$$

$$A(\mathbf{k}) = D(\mathbf{k}) = 1 \quad \text{for all } \mathbf{k}$$

$$(17\text{-}103)$$

The displayed scene $o(\mathbf{x})$ is the Fourier transform of $O(\mathbf{k})$ given by

$$o(\mathbf{x}) = \frac{1}{A} \int_{-\infty}^{\infty} e^{2\pi i \mathbf{x} \cdot \mathbf{k}} B(\mathbf{k}) \sum_n S(\mathbf{k} + \mathbf{b}_n) \, d\mathbf{k} \qquad (17\text{-}104)$$

Interchanging the summation and integration, one obtains

$$o(\mathbf{x}) = \frac{1}{A} s_\beta(\mathbf{x}) + \frac{1}{A} \sum_{n \neq 0} e^{-2\pi i \mathbf{x} \cdot \mathbf{b}_n} \int_{\beta + \mathbf{b}_n} S(\mathbf{k}) \, e^{2\pi i \mathbf{k} \cdot \mathbf{x}} \, d\mathbf{k} \qquad (17\text{-}105)$$

where $B(\mathbf{k}) = 1$ for k contained in β, and is zero otherwise. The image, $s_\beta(x)$, is the one which would have been obtained if the original scene image, $s(x)$, had been bandpassed. It is a "smeared" version of the original picture. The bandpass is determined by the sampling lattice. The remaining term is a weighted sum of bandpassed images where the bandpasses are given by $\beta + \mathbf{b}_n$ (Figure 17-26). The display image is then given as a bandpassed image plus the weighted sum of images with higher spatial frequency bandpasses.

$$o(\mathbf{x}) = \frac{1}{A} s_\beta(\mathbf{x}) + \frac{1}{A} \sum_{n \neq 0} e^{-2\pi i \mathbf{x} \cdot \mathbf{b}_n} s_{\beta + \mathbf{b}_n}(\mathbf{x}) \qquad (17\text{-}106)$$

The high spatial-frequency bandpass images $s_{\beta+\mathbf{b}_n}(\mathbf{x}) + s^*_{\beta+\mathbf{b}_n}(\mathbf{x})$ are essentially outlines of the edges of the original scene. One might consider the images $s_{\beta+\mathbf{b}_n}(\mathbf{x}) + s^*_{\beta+\mathbf{b}_n}(\mathbf{x})$ as a set of pictures of edges in the scene. The weightings $e^{-2\pi i \mathbf{x} \cdot \mathbf{b}_n}$ are periodic over the sampling lattice and play an interesting role. The expression $e^{-2\pi i \mathbf{x} \cdot \mathbf{b}_n} s_{\beta+\mathbf{b}_n}(\mathbf{x}) + e^{2\pi i \mathbf{x} \cdot \mathbf{b}_n} s_{\beta+\mathbf{b}_n}(\mathbf{x})$ is the high spatial-frequency bandpass image, frequency-shifted down into the bandpass β.

17.5.2. Best Sampling Lattices. The customary use of a square sampling lattice is a product of our one-dimensional experience. One can select any noncollinear sampling lattice vectors $\mathbf{a}_1$ and $\mathbf{a}_2$. The

Fig. 17-26. Bandpasses for displayed image.

only requirement is that the reciprocal lattice $\mathbf{b}_1$ and $\mathbf{b}_2$ provide a parallelogram which covers the spatial frequency distribution of the scene.

Atmospheric and optical spatial frequency responses are usually symmetric. There is no directional preference for reproducing spatial frequencies. One should consider the spatial frequency distribution of a natural scene to be approximated by a circle in the reciprocal (frequency) plane.

The pattern in the frequency plane which minimizes the area not covered by circles of radius $1/2d$ is formed by placing the centers of the circles at the vertices of an equilateral triangle of side $1/d$ (Figure 17-27(a)). The reciprocal lattice shown in Figure 17-27(b) has $\mathbf{b}_1$ and $\mathbf{b}_2$ $60°$ apart and both vectors are of length $1/d$. The sampling lattice (Figure 17-28) shows that $\mathbf{a}_1$ and $\mathbf{a}_2$ are $120°$ apart and of length $2d/\sqrt{3}$. The detector array which covers the focal plane consists of rectangular hexagons whose sides are of length $2d/3$.

If one compares a square lattice and a $60°$ rhombic lattice with the same limiting

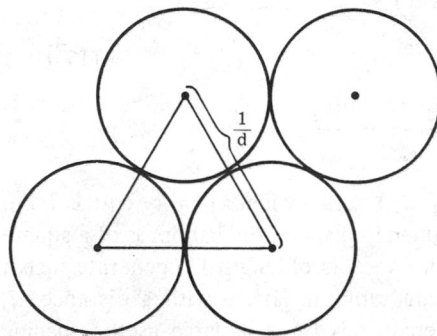

(a) Close packing of circles in the plane ($60°$ rhombic lattice).

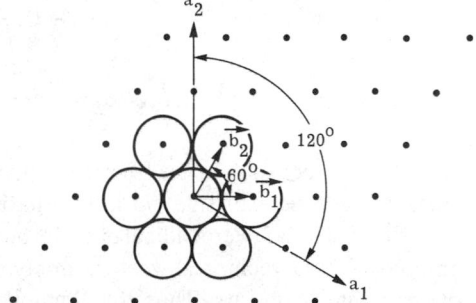

(b) Spectrum repetition for circular spectrum (reciprocal lattice of Figure 17-27(a)).

Fig. 17-27. Lattices.

spatial frequency response (both reciprocal lattices cover the circular spatial frequency spectrum), the 60° rhombic primary lattice is coarser (in fact, the hexagon detectors are 27% larger), and therefore, more advantageous, since the larger detectors collect more radiation. The 60° rhombic lattice requires 13% fewer detectors to cover the same area in the focal plane covered by the square lattice.

17.5.3. Prefiltering of the Scene by Aperture and Detector Response. One sees from Equation (17-101) that the detector and optical response can be used to prefilter the scene's spatial-frequency distribution. As Equation (17-101) shows, aliasing occurs before the display. Thus there is no display response, $B(\mathbf{k})$, which can eliminate aliasing.

It is apparent that the response of the optical system, $A(\mathbf{k})$, should be matched to the sampling lattice. If one has coarse sampling of the scene, one does not want a high resolution optical system. One wants an optical response, $A(\mathbf{k})$, that passes *only*

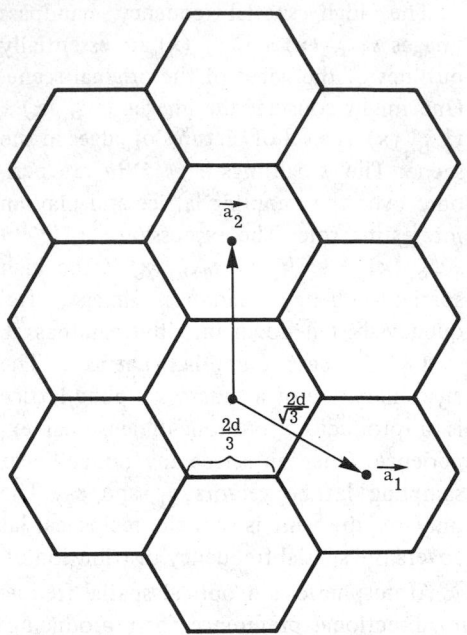

Fig. 17-28. Best detector pattern for circular spectrum.

the limiting frequencies, β, implied by the sampling lattice.

The detector spatial frequency response, $D(\mathbf{k})$, should also be selected so $D(\mathbf{k})$ has a flat response. If $A(\mathbf{k})$ bandpasses the image, then $D(\mathbf{k})$ need not, but if $A(\mathbf{k})$ does not bandpass the scene, $D(\mathbf{k})$ should. The area responsivity of the detector is generally not controllable, but the shape and size of the detector are. A square detector of side $2d$ is represented by

$$d(x_1, x_2) = 1 \begin{cases} -d \leqslant x_1 \leqslant d \\ -d \leqslant x_2 \leqslant d \end{cases}$$

$$= 0, \text{ elsewhere} \qquad (17\text{-}107)$$

$$A(k_1, k_2) = \frac{\sin 2\pi d_{k_1} \sin 2\pi d_{k_2}}{\pi^2 k_1 k_2}$$

The first zero contour in the frequency plane (k_1, k_2) occurs when k_1 or k_2 equals $1/2d$. Consider the square of side $1/d$ in the spatial-frequency plane as the bandpass of a square detector of side $2d$. Perpendicular reciprocal-lattice vectors of length $1/d$ generate such a bandpass. These reciprocal vectors imply a square sampling-lattice with a distance, d, between sample points. Thus the "matched" detector is twice as large as the spacing between sampling centers and the edge of one detector should coincide with the center of the adjacent detector. This construction cannot be realized in a mosaic, but can be realized in a scanning system by overscanning. The usual mosaic has adjacent detectors nearly touching. The detector area is $1/2\beta$ and the detector bandpass is 2β. The aliased frequencies then lie in the band β to 2β, with 2β as the parallelogram which has all sides doubled in length.

17.5.4. Requirements for the Display Light Distribution. The specification of the spatial frequency distribution of a display light source can be easily obtained from Equation (17-101). If one has prefiltered the image, then $O(\mathbf{k}) = S_\beta(\mathbf{k})$ if the display element specification is

$$B(\mathbf{k}) = \frac{1}{D(\mathbf{k})A(\mathbf{k})}, \qquad \text{for } \mathbf{k} \text{ contained in } \beta$$
$$= 0, \text{elsewhere} \qquad\qquad (17\text{-}108)$$

Generally, $b(\mathbf{x})$ for each discrete light source is an oscillating spatial distribution of light. Suppose one has a square sampling lattice with spacing d between the sample points. Then β is a square with $k_1 \leqslant |1/2d|$ and $k_2 \leqslant |1/2d|$. Let $A(\mathbf{k})$ and $D(\mathbf{k})$ equal 1. Then $B(\mathbf{k}) = 1$ for $k_1 \leqslant |1/2d|$ and $k_2 \leqslant |1/2d|$. The distribution of light $b(\mathbf{x})$ equals

$$\frac{\sin(x_1 \pi/d)\sin(x_2 \pi/d)}{4\pi^2 x_1 x_2} \qquad (17\text{-}109)$$

The first zero of the light distribution occurs when $|x_1|$ and $|x_2|$ equals d. If the light distribution is centered on a sample point of the display, then the light is zero at adjacent sampling centers. The first zeros of the light intensity distribution from a display sample point fall on the centers of the adjacent sample points. This is the motive for defocusing or shaping of the spot on a cathode ray tube display so that the spot covers half of the adjacent scan line.

17.6. Spectral Discrimination

Let $\Phi_\lambda(\lambda)$ be the spectral radiant power of the IR signal in an interval, $d\lambda$, centered at wavelength λ. The total power available from the IR source is $\int_0^\infty \Phi_\lambda(\lambda)\,d\lambda$. This source description $\Phi_\lambda(\lambda)$ is known as the spectral distribution or density, often approximated by a finite sequence of numbers $\mu_1, \ldots, \mu_N$ where μ_i specifies the total radiation in the ith spectral region. There is not *a* spectral distribution of an object. An object's "spectral distribution" is in reality a collection of spectral distributions and the statistics associated with this collection. Consider a discrete spectral distribution $\mu_1, \ldots, \mu_N$; the mathematical model describing the variations in the observed spectra is a multivariate probability distribution. If the spectral model is continuous, then an observed spectrum, $\Phi_{\lambda,\omega}(\lambda)$, is a realization from a random process. See References [17-19] and [17-20] for a development of the theory of random processes as applied to the problem of signal detection.

17.6.1. Spectral Filtering. Infrared devices discriminate spectrally because of the detector/filter spectral response. This response is characterized by a filter function, $f(\lambda)$. The detection of a target signal, $s(\lambda)$, is attempted in the presence of a random background signal, $b_\omega(\lambda)$ which is related to $\Phi_{\lambda,\omega}$ for the background. The detector and preamplifier introduce noise, N, which is not affected by the choice of filter.

17.6.2. Filters Maximizing Average Contrast. The statistics used to represent the background and equipment noise processes depend upon the application. If one wishes to increase the average contrast between the average background signal and the target signal, the statistic of interest is the average, over ω, of $b_\omega(\lambda)$ and the average equipment noise. In this treatment,

$b_\omega(\lambda)$ = the average background variations in radiant power as a function of wavelength
   $t(\lambda)$ = the average target radiant power
   $s(\lambda)$ = average $(t(\lambda) - b_\omega(\lambda))$ radiant power as a function of wavelength
      N = the rms (average) equipment noise
   $f(\lambda)$ = the filter function, $0 \leqslant f(\lambda) \leqslant 1$,   all λ

The optimal filter should maximize the target signal, $t(\lambda)$, relative to the noise signal. The criterion for an optical filter is that filter, $f(\lambda)$, which maximizes the ratio

$$\frac{\displaystyle\int_0^\infty f(\lambda) s(\lambda)\, d\lambda}{\displaystyle\int_0^\infty f(\lambda)\, b(\lambda)\, d\lambda + N} \tag{17-110}$$

The quantity $s(\lambda)$ may assume both positive and negative values over the spectral region of interest. Since $s(\lambda)$ is the average difference between the average target and average background spectrum, when $s(\lambda)$ changes sign, there is a reversal in target-to-background contrast. Inspection of ratio (17-110) is enough to determine that spectral regions of positive and negative contrast should not be combined. One may optimize for positive or negative contrast, whichever gives the better result. The rule for optimizing the spectral response consists of

(a)   Let s^+ be the set of wavelengths where $s(\lambda) > 0$.
(b)   Let s^- be the set of wavelengths where $s(\lambda) < 0$.
(c)   Compute the optimum filters $f^+(\lambda), f^-(\lambda)$ for both s^+ and s^-.
(d)   Choose the filter which maximizes ratio (17-110).

Subsequent discussion will assume $s(\lambda) \geqslant 0$. In computing the optimum filter for negative contrast, use $-s(\lambda)$ for λ in s^-.

For an optimum contrast filter, one concept used here may be unfamiliar: that of integration over point sets. If Ω is any set of real numbers, one defines $\int_\Omega g(x)\, dx$ to be the area under that part of the curve $y = g(x)$, for which x is in the set Ω. For example, if one denotes by Ω the set of all positive λ, one can write

$$\int_0^\infty f(\lambda)\, s(\lambda)\, d\lambda = \int_\Omega f(\lambda)\, s(\lambda)\, d\lambda \tag{17-111}$$

It can be concluded that an optimal filter can be sought among filters whose filter functions take on only the values 0 and 1, i.e., filters consisting of only transparent and opaque spectral regions [17-21]. The spectral regions where $f(\lambda) = 1$ must be found.

For the background limited case, $N \cong 0$, no optimal $f(\lambda)$ exists. A "good" $f(\lambda)$ is obtained by letting $f(\lambda) = 1$ in the smallest practical region where $s(\lambda)/b(\lambda)$ is large. The quantity $\gamma(c)$ given below (Equation (17-112)) provides an effective means for calculating the set of wavelengths where $f(\lambda) = 1$.

$$\gamma(c_n) = \frac{\displaystyle\int_{\Omega_{c_n}} s(\lambda)\,d\lambda}{\displaystyle\int_{\Omega_{c_n}} b(\lambda)\,d\lambda + N} \qquad (17\text{-}112)$$

Ω_{c_n} is the set of λ's where $s(\lambda)/b(\lambda) \geqslant c_n$.

One computes iteratively as follows: $c_1 = 0$, $c_2 = \gamma(c_1)$, $c_3 = \gamma(c_2)$, ..., $c_i = \gamma(c_{i-1})$, ..., $c_n = \gamma(c_{n-1})$.

The set Ω_{c_n} can be computed numerically or graphically. This scheme converges rapidly to the optimal set of wavelength and is amenable to machine or hand computation. The quantity $f(\lambda)$ is set equal to 1 on the optimal set and to zero otherwise.

17.6.3. Filters Maximizing rms Contrast. Suppose one wants a spectral filter which maximizes the signal-to-noise ratio in the presence of rms variations in the background and equipment noise. If there is no correlation of the background signal between wavelengths, one can measure the variation in the background by the variance of the output signal:

$$\int_0^\infty f^2(\lambda) \langle b^2(\lambda) \rangle \, d\lambda \qquad (17\text{-}113)$$

where $\langle b^2(\lambda) \rangle = \text{average}_\omega (b_\omega(\lambda) - \bar{b}(\lambda))^2$

$\bar{b}(\lambda) = \text{average}_\omega b_\omega(\lambda)$

The rms noise signal is

$$\sqrt{\int_0^\infty f^2(\lambda) \langle b^2(\lambda) \rangle \, d\lambda + N^2} \qquad (17\text{-}114)$$

where N is the detector noise. The signal-to-noise ratio one wants to optimize with respect to the filter is given by

$$\frac{\displaystyle\int_0^\infty f(\lambda)\,s(\lambda)\,d\lambda}{\sqrt{\displaystyle\int_0^\infty f^2(\lambda) \langle b^2(\lambda) \rangle \, d\lambda + N^2}} \qquad (17\text{-}115)$$

The optimum filter for this application is *not* either spectrally opaque or transparent, as it was in the optimal average contrast case above.

One may optimize the square of the discrete version of ratio (17-115)

$$\frac{\left(\displaystyle\sum_{i=1}^n f_i s_i\right)^2}{\displaystyle\sum_{i=1}^n f_i^2 b_i^2 + N^2} \qquad (17\text{-}116)$$

where s_i = the signal in the ith spectral band

$b_i{}^2$ = the variance of the background in the ith spectral band

f_i = the value of the spectral filter in the ith spectral band

The optimal filter problem is then to find the coefficients $f_1, \ldots, f_n$ which maximize ratio (17-116).

The set of coefficients $f_1 = r\, s_i/b_i$ maximizes the ratio

$$\frac{\left(\sum f_i s_i\right)^2}{\sum f_i^2 b_i}$$

The computing algorithm for the optimal spectral response is as follows:

(a) Rank the ratios s_i/b_i in descending order. Let s'_{k+1}/b'_{k+1} be the maximum of ratios S_i/b_i after deleting the k largest ratios. $s'_1/b'_1 \geqslant s'_2/b'_2 \ldots \geqslant s'_n/S'_n$. Keep track of the spectral interval associated with s'_k/b'_k

(b) Starting with $k = 1$, compute

$$\frac{\displaystyle\sum_1^k s'_1}{\displaystyle\sum_1^k b'_i + N^2}$$

iteratively until one reaches the first k_0 with

$$\frac{\displaystyle\sum_1^{k_0} s'_i}{\displaystyle\sum_1^{k_0} b'_1 + N^2} \geqslant \frac{s'_{k_0} + 1}{b'_{k_0} + 1} \tag{17-117}$$

(c) Set

$$r = \frac{\displaystyle\sum_1^{k_0} s'_i}{\displaystyle\sum_1^{k_0} b'_i + N^2}$$

On those spectral intervals corresponding to $i \leqslant k_0$, set $f_i = 1$. On the remaining intervals, set $f_i = r\, s'_i/b'_i$, and one has the optimal filter. There is a rather remarkable similarity between the average and rms optimal contrast filters. Since the mechanisms are remarkably similar one should expect this similarity in the optimal filters.

17.6.4. Optimal Spectral Response (Photon Noise). The limiting noise for a well designed IR sensor is photon noise. If $\Sigma_{i=1}^{N}(t_i - b_i)$ is the signal, where i denotes the ith spectral band (discrete version), then the photon noise associated with the signal is $[\Sigma_{i=1}^{N}(t_i + b_i)]^{1/2}$. (See Chapter 11.) The signal-to-noise ratio to be optimized by selection of an optimal spectral filter is

$$\frac{\displaystyle\sum_{i=1}^{n} f_i(t_i - b_i)}{\left[\displaystyle\sum_{i=1}^{n} f_i(t_i + b_i) + N^2\right]^{1/2}} \tag{17-118}$$

It can be shown that the optimal spectral filter $\mathbf{f} = (f_1, \ldots, f_n)$ has components equal to 0 or 1. It can also be shown that if $\mathbf{f_0}$ is the optimal filter, then

$$\text{Property 1} \quad \begin{cases} f_{i_o} = 1 \\ f_{k_o} = 0 \end{cases}$$

Then

$$\frac{f_{i_o}(t_{i_o} - b_{i_o})}{f_{i_o}(t_{i_o} + b_{i_o})} \geqslant \frac{f_{k_o}(t_{k_o} - b_{k_o})}{f_{k_o}(t_{k_o} + b_{k_o})} \tag{17-119}$$

for any i_o, k_o with Property 1. The algorithm for computing the optimal spectral response is given below.

(a) As before, order the ratios, keeping track of the spectral bands involved.

$$\frac{t_1' - b_1'}{t_1' + b_1'} \geqslant \frac{t_2' - b_2'}{t_2' + b_2'} \geqslant \cdots \geqslant \frac{t_n' - b_n'}{t_n' + b_n'}$$

(b) Compute successively starting with $k = 1$.

$$s(k) = \frac{\displaystyle\sum_{i=1}^{k} t_i' - b_i'}{\left(\displaystyle\sum_{i=1}^{k} t_i' + b_i' + N^2\right)^{1/2}} \quad k = 1, \ldots, n \tag{17-120}$$

(c) Let k_o be the first index such that

$$s(1) < s_2 < \cdots < s(k_o) > s(k_o + 1)$$

(d) Set $f_i = 1$ on those spectral intervals i such that

$$\frac{t_i' - b_i'}{t_i' + b_i'} \geqslant \frac{t_{k_o}' - b_{k_o}'}{t_{k_o}' + b_{k_o}'}$$

It should be noted that all the optimal filters get broader with increasing equipment noise, N.

17.7. Multispectral Discrimination

The statistics of target and background spectral signatures can be used to design an "optimal" signal processor. The methods used to design such an optimal system are those of statistical decision theory. These methods, when applied to a problem such as spectral discrimination, are called *pattern recognition techniques*. These design methods are developed in abstract mathematical terms with little or no reference to the equipment intended to implement the discrimination process. This mathematical system definition is then translated into equipment specifications.

As an illustration, consider a discrete representation of the spectral distribution, limited to two spectral samples. (Everything will apply to more than two spectral samples and to continuous spectral distributions.) Let μ_1 be the radiant power observed in the first spectral region, μ_2 the radiant power observed in the second, with $\mu = (\mu_1, \mu_2)$.

The random variation in the observed spectra forms a scatter diagram as in Figure 17-29. Seek a multivariate density $f(\mu)$ which defines the probability of an observation occurring in any region A by

$$\text{Probability [observed spectrum lies in } A] \ = \ \int_A f(\mu)\, d\mu \qquad (17\text{-}121)$$

Since the observed spectrum from an object varies randomly, one must abandon the notion of *an* object spectrum and consider regions or sets of spectra when deciding what object is present. This decision rule is a partition of the set of all possible observed spectrum into disjoint regions. If the observed spectrum lies in Region A, the object present is A. Figure 17-30 illustrates such a decision partition. The decision rule specifies a partition in sets. The probability distribution specifies the probability of these sets. These two notions provide a basis for characterizing and evaluating proposed spectral discrimination schemes. More important, they provide a basis for the design of discrimination schemes.

The output from a spectral filter has the form

$$f_1\mu_1 + f_2\mu_2 = y \qquad (17\text{-}122)$$

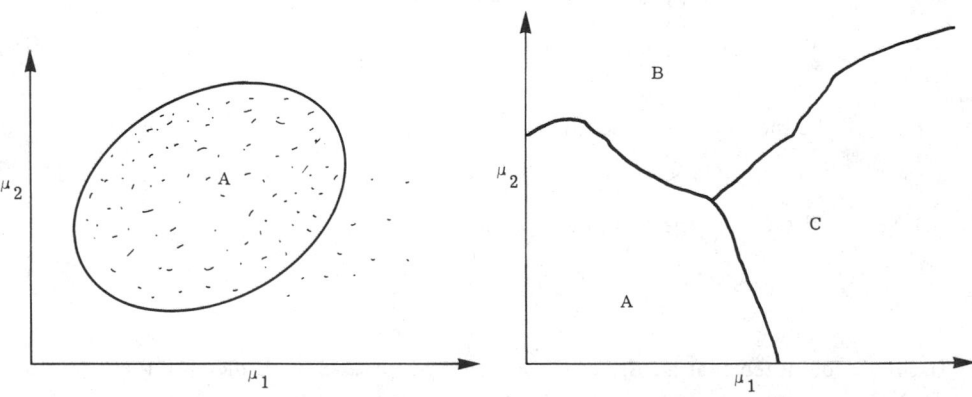

Fig. 17-29. Distribution of observed spectra.   Fig. 17-30. Decision partition.

where f_1 and f_2 are the percent transmission of the filter for μ_1 and μ_2, $0 \leqslant f_1, f_2 \leqslant 1$. One may say there is a target if y exceeds a threshold T. The target and noise regions are shown in Figure 17-31, which illustrates an important characteristic of any spectral filter. The discrimination regions are separated by lines, whose slopes are always negative. Suppose that both the target and noise spectrum distributions have concentrations of, say, 99 and 44/100% of the observations, that are oriented positively (Figure 17-32). Clearly, no line with negative slope will separate target from background as well as will a line with positive slope.

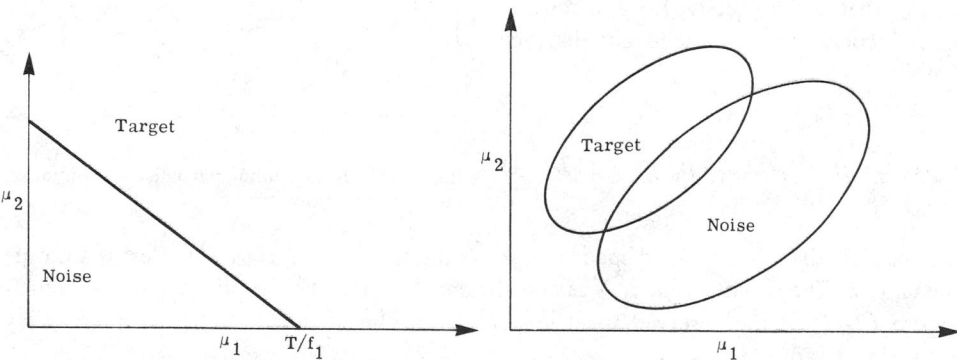

Fig. 17-31. Regions specified by spectral filter.   Fig. 17-32. Observed spectra with positive correlation.

This example indicates the importance of the orientation of the concentration of the spectral distribution. The correlation ρ_{12}, Equation (17-123), between two spectral bands is such a measure.

$$\rho_{12} = \text{Cor}\,(\mu_1, \mu_2) = \frac{\text{Ave}\,(\mu_1 - \bar{\mu}_1)(\mu_2 - \bar{\mu}_2)}{\sqrt{\text{Ave}\,(\mu_1 - \bar{\mu}_1)^2\;\text{Ave}\,(\mu_2 - \bar{\mu}_2)^2}} \qquad (17\text{-}123)$$

where $\rho_{12} = \mu_1$ and μ_2. If, on the average, $\mu_1 > \bar{\mu}_1$, this implies that μ_2 is greater than its average, $\bar{\mu}_2$, and conversely, the concentration is at least approximately positively oriented. If both target and noise distributions have positive correlations, as in Figure 17-32, any spectral filter, even an optimum, does an inadequate job of discrimination. If the correlations of both background and target are negative (Figure 17-33) then a spectral filter will work well. Thus, the correlation statistic has value in evaluating a discrimination scheme.

The two schemes considered are similar in that the boundary between decision regions is a straight line. If the distribution of target and background have concentrations similar to Figure 17-34 (the "kidney" concentration), then no line will adequately partition the spectra observations.

The spectral matching problem is

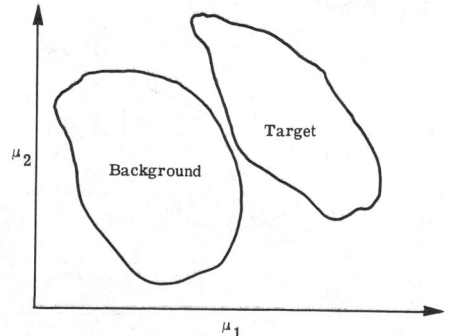

Fig. 17-33. Observed spectra with negative correlation.

illustrative of the approach to the design of spectral discrimination in a statistical context. Suppose one has a spectrum $(\mu_1, \ldots, \mu_n) = \mu$ that defines a target A. One observes a spectrum $(\mu_1{}^0, \ldots, \mu_n{}^0) = \mu^0$. One must define the meaning of "close" to determine if the observed spectrum is close enough to μ to justify stating that the object is A. As a starting point, consider the Euclidean distance given by

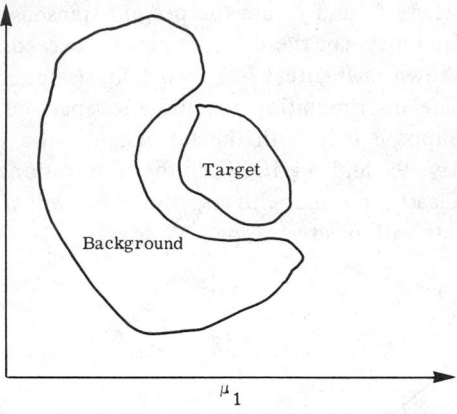

$$d(\mu, \mu_o) = \sqrt{\Sigma(\mu_i - \mu_i{}^0)^2} \quad (17\text{-}124)$$

Fig. 17-34. Background partially surrounding target distributions.

One may say that the observed spectrum, μ^0, is the target, A, if $d(\mu, \mu^0)$ is less than some constant, c. Then, with $d(\mu, \mu^0)$ as the distance metric, the decision region is circular (Figure 17-35). If the observed spectrum, μ^0, lies in the circle with radius c, one may say that target A is present.

If one spectral channel, number one, is noisier than the other, the concentration of the distribution might appear as in Figure 17-36. Consequently, the probability of a deviation from μ^0 as large or larger than c_1 in the first channel might be the same as the probability of a deviation in the second channel as large or larger than c_2. The dispersion in a channel may be measured by its variance. The larger the variance, the less weight one wants to give that coordinate of the distance. If one weighs each coordinate with the reciprocal of its variance, the new distance metric is then

$$d^{(1)}(\mu, \mu^0) = \sqrt{\sum \frac{\left(\mu_i - \mu_i{}^0\right)^2}{\sigma^2_{\mu_i}}} \quad (17\text{-}125)$$

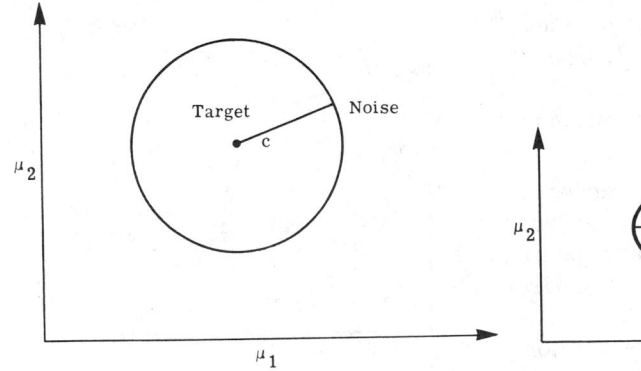

Fig. 17-35. Constant distance decision rule (circular).

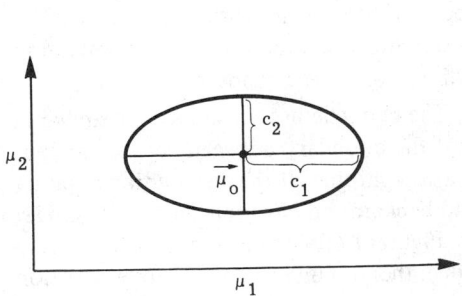

Fig. 17-36. Elliptical distribution.

When one considers the region $d^{(1)}(\mu, \mu^0)^2 \leqslant c$, one has an ellipse with axes of length $\overline{c\sigma_{\mu_1}}$ and $\overline{c\sigma_{\mu_2}}$. (See Figure 17-36.) If the observed value μ^0 lies in the ellipse, it is "close" to μ and one has a target.

The concentrations may have orientations, i.e., the axes of the ellipse may be tilted. An indication of this orientation is given by the correlation ρ_{ij}, Equation (17-123). Consider the array of numbers $[\rho_{ij}]$, where $[\rho^{ij}]$ are the elements of the reciprocal matrix. A new distance is then defined by

$$d^{(2)}(\mu, \mu^0) = \sum_i \sum_j (\mu_i - \mu_i^0)\rho^{ij}(\mu_j - \mu_j^0) \qquad (17\text{-}126)$$

This is a distance in a coordinate system with oblique axis. For example, suppose $\rho_{12} = \rho_{21} = 1/2$ and $\sigma_1 = \sigma_2 = 1$,

$$[\rho_{ij}] = \begin{pmatrix} 1 & 1/2 \\ 1/2 & 1 \end{pmatrix}$$

then $\qquad\qquad\qquad\qquad\qquad\qquad\qquad\qquad\qquad\qquad\qquad\qquad\qquad (17\text{-}127)$

$$[\rho^{ij}] = \begin{pmatrix} 4/3 & -2/3 \\ -2/3 & 4/3 \end{pmatrix}$$

Equation (17-126) becomes $4/3 \, [(\mu_1 - \mu_1^0)^2 - (\mu_1 - \mu_1^0)(\mu_2 - \mu_2^0) + (\mu_2 - \mu_2^0)^2]$. The major axis of the ellipse is in the direction $(1, 1)$. (See Figure 17-37.)

The definition of "close" given here is to be used in a spectral matching system that is essentially a characteristic of the target's spectral distribution and its variation. The distance defined in Equation (17-126) is specified by the first and second moment statistics of the target's probability distribution.

17.7.1. Optimal Statistical Spectral Discrimination.
No attempt has been made so far to optimize the performance of a spectral discrimination scheme. In

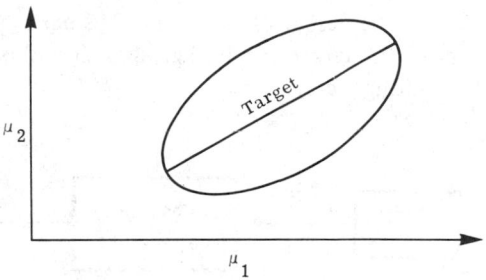

Fig. 17-37. Oriented ellipse of concentration.

fact, discussion of the spectral matching problem has ignored the statistics of the background. It can be shown that no spectral discrimination scheme based solely upon first and second moment statistics of the target and background can work well when the "kidney" distribution (Figure 17-34) holds for *either* distribution.

A spectral discrimination scheme is considered optimal if the probability of a false alarm for the optimal system is smaller than the false alarm probability for *any* other system. Since any spectral discrimination scheme can be represented as a partition of the observable spectrum space, the design problem reduces formally to the selection of a target detection region, A, with the following properties:

(a) The probability of an observed spectrum from a target falling in A is P.

(b) Among all detection regions, R, the region A should have the smallest probability of an observed background spectrum falling in A, i.e., Probability [μ is in A, given background spectrum] $\leqslant$ Probability [μ is in R, given background spectrum].

The solution to this problem is well known in mathematical statistics. Let A be the set of observed spectrum $(\mu_1, \ldots, \mu_n)$ that satisfy the inequality:

$$\frac{f_T(\mu_1, \ldots, \mu_n)}{f_B(\mu_1, \ldots, \mu_n)} \geqslant c \qquad (17\text{-}128)$$

where $f_T(\mu_1, \ldots, \mu_n)$ and $f_B(\mu_1, \ldots, \mu_n)$ = probability densities of the signal, given the target and background, respectively

c = constant, which is the solution to Equation (17-129)

$$\text{Probability} \left\{ \frac{f_T(\mu)}{f_B(\mu)} \geqslant c \text{ given a target present} \right\} = P \qquad (17\text{-}129)$$

The ratio in Equation (17-129) is known as the *likelihood ratio*. As Birdsall and Peterson [17-22] point out, this ratio is central in the theory of signal detectability. The Region A defined by Equations (17-128) and (17-129) satisfies conditions (a) and (b).

The implementation of the spectral discrimination scheme specified by Equation (17-128) can be seen in Figure 17-38. The schemes discussed before, with the exception of the spectral filters, all require some computational manipulation of the detected signals. This computation is generally done electronically. Spectral filter schemes use the filter and detector as a computer, an attractive feature of their implementation. One can see that the design of an optimal system requires knowledge of the statistical properties of both the target and background. It is this knowledge rather than the theory which is the limiting factor.

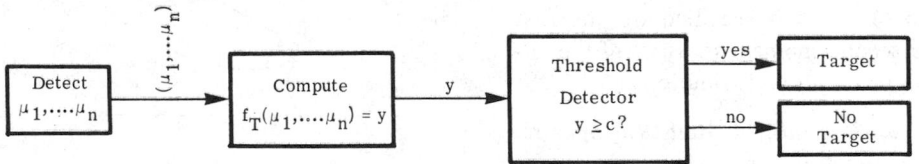

Fig. 17-38. Block diagram of likelihood ratio scheme.

More may be required of a spectral discrimination scheme than the simple indication of the presence or absence of a target. In many instances one wants to decide which of many objects is present on the basis of the observed spectrum, μ. The likelihood ratio test discussed above partitions the sample space μ into *two* regions. A *classification scheme* partitions the space into many regions. One may represent this multiple partition by a function, $D(\mu)$, defined as

$$D(\mu) = i, \qquad i = 1, \ldots, n \qquad (17\text{-}130)$$

if μ lies in the region specifying the ith target. The likelihood ratio test assumes $n = 2$, and essentially finds an optimal partition or decision function, $D(\mu)$.

To construct an optimal decision function $D(\mu)$, one needs additional data. These include some a priori estimates of the probability, p_i, that an object will occur in the scene. Let p_i be that a priori probability that an object, i, will be present in a scene. Generally, there is some cost associated with classifying the jth object as being the ith object. Let this cost be $C(i/j)$. Given a spectral observation, μ, the conditional probability of the j population is given by the Bayes theorem as

$$p(j|\mu) = \frac{p_j f_j(\mu)}{\displaystyle\sum_{i=1}^{n} p_i f_i(\mu)} \tag{17-131}$$

where $f_i(\mu)$ = the probability density of μ, given that the observation comes from the ith population. Classification of an observation, μ_0, as coming from the ith object incurs an average cost:

$$\text{Average cost } (\mu_0 \text{ comes from } i) = \sum_{\substack{j=1 \\ j \neq i}}^{n} \frac{p_j f_j(\mu)}{\displaystyle\sum_{i=1}^{n} p_i f_i(\mu)} C(i|j) \tag{17-132}$$

One chooses i so that the cost, Equation (17-132), is minimized. The decision function, $D(\mu)$, which minimizes the cost of a misclassification is given by

$$D(\mu) = i$$

when

$$\sum_{\substack{j=1 \\ j \neq i}}^{n} p_j f_j(\mu) C(i|j) \leqslant \sum_{\substack{j=1 \\ j \neq k}}^{n} p_j f_j(\mu) C(k|j) \tag{17-133}$$

for all $k = 1, \ldots, n$

$D(\mu)$ defines a partition of the space into n regions.

One now considers a situation closely related to the binary decision space of the likelihood ratio model. Suppose the target is object 1, and the objects in the background are $2, \ldots, n$. The costs of calling a target any background object are equal, and the costs of calling any background object the target are equal. The costs of calling one background object another background object are zero. The cost situation is summarized as

$$C(i|1) = a \qquad\qquad i = 2, \ldots, n$$

$$C(1|j) = b \qquad\qquad i = 2, \ldots, n \tag{17-134}$$

$$C(i|j) = 0 \qquad\qquad i,j = 2, \ldots, n$$

From Equation (17-133), one obtains the following specification for the target region:

$$D(\mu) = 1$$

when

$$b \sum_{j=2}^{n} p_j f_j(\mu) \leqslant a p_1 f_1(\mu) \tag{17-135}$$

Equivalently, one says a target is present when Equation (17-136) holds.

$$\frac{f_1(\mu)}{\displaystyle\sum_{j=2}^{n} p_j f_j(\mu)} \geqslant \frac{b}{ap_1} \tag{17-136}$$

This is the likelihood ratio where the background is composed of many objects. In most practical cases, the background does, in fact, consist of a number of distinct objects. A more detailed discussion of this theory is found in Reference [17-23].

Suppose one partially specifies the target and background distributions to derive the optimal system. Suppose both $f_T(\mu_1, \ldots, \mu_n)$ and $f_B(\mu_1, \ldots, \mu_n)$ are Gaussian with means $(\mu_1^T, \ldots, \mu_n^T)$, $(\mu_1^B, \ldots, \mu_n^B)$ and variance covariance matrices (θ_{ij}^B), (θ_{ij}^T).

Ignoring constants, one finds that the densities are

$$f_T(\mu) = \exp\left[-\sum_i \sum_j \theta_T^{ij}(\mu_j - \mu_j^T)(\mu_j - \mu_j^T)\right]$$

$$f_B(\mu) = \exp\left[-\sum_i \sum_j \theta_B^{ij}(\mu_i - \mu_i^B)(\mu_j - \mu_j^B)\right] \tag{17-137}$$

where θ_B^{ij} and θ_T^{ij} = elements of the inverse of the background and target variance covariance matrices (θ_{ij}^B), (θ_{ij}^T). Using the likelihood ratio (Equation (17-128)) and taking the log of the ratio, one obtains, for the target detection region specification,

$$\sum_i \sum_j \theta_B^{ij}(\mu_i - \mu_i^B)(\mu_j - \mu_j^B) - \theta_T^{ij}(\mu_j - \mu_i^B)(\mu_j - \mu_j^T) \geqslant c \tag{17-138}$$

If the outputs as computed have exceeded the threshold c, the spectral discrimination system will indicate a target detection. Figure 17-39 illustrates the type of partition that might arise in a likelihood ratio scheme. The computing circuitry to implement Equation (17-138) requires adders, subtractors, and multipliers. The complexity of this circuitry depends upon the constants θ_B^{ij}, θ_T^{ij}, μ_i^B, μ_i^T, i.e., upon the complexity of the statistics of both background and target.

While individual background objects may be said to be Gaussian, the background consists of multiple objects. The probability that a given resolution element is an object is the proportion of the scene occupied by the object. Let p_i be this proportion. The discrimination scheme developed generally

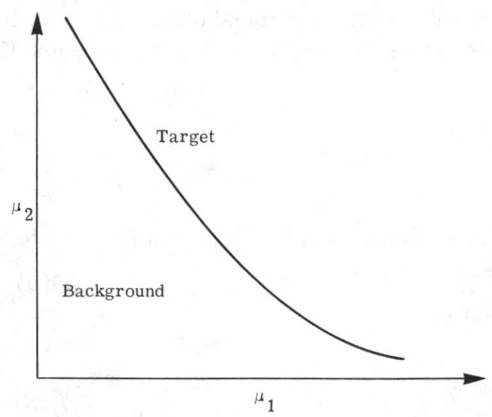

Fig. 17-39. Likelihood ratio discrimination regions.

in Equation (17-136) specifies the spectral processor. The mathematical specification of the spectral processing hardware is given by

$$
\text{A target is present if} \quad \left\{ \frac{|\theta_1^{ij}| \exp\left[-\sum_i \sum_j \left(\mu_i - \mu_i^{T}\right) \theta_1^{ij}\left(\mu_j - \mu_j^{T}\right)\right]}{\sum_{k=2}^{n} p_k |\theta_k^{ij}| \exp\left[-\sum_i \sum_j \left(\mu_{ik} - \mu_{ik}^{B}\right) \theta_k^{ij}\left(\mu_{jk} - \mu_{jk}^{B}\right)\right]} \right\} \geq c \quad (17\text{-}139)
$$

The parameters μ_i^{T}, μ_{jk}^{B}, θ_k^{ij} specify the target and background statistics. The resolution element is said to be a target when the spectrum $(\mu_1, \ldots, \mu_n)$ satisfies Equation (17-139).

17.8. References and Bibliography

17.8.1. References

[17-1] F. Oberhettinger, *Fourier Expansions: A Collection of Formulas,* Academic Press, New York, NY, 1973.

[17-2] I. S. Gradshteyn and I. M. Ryzhik, *Tables of Integrals Series and Products,* Academic Press, Washington, DC, 1965, 4th Edition.

[17-3] L. M. Biberman, *Recticles in Electrooptical Devices,* Pergamon Press, Elmsford, NY, 1966.

[17-4] J. P. Ulrich, W. D. Montgomery, and J. L. Alward, "Analysis of Reticle System," Background Analysis Center, Willow Run Laboratories, University of Michigan, Ann Arbor, MI, Report No. 6054-2-T, October 1965.

[17-5] S. Sternberg, J. Ulrich, and R. Hamilton, "Analysis and Testing of a Special Rotating-Translating Reticle," Willow Run Laboratories, University of Michigan, Ann Arbor, MI, Report No. 7102-1-T, June 1967.

[17-6] G. Birkhoff and S. MacLane, *A Survey of Modern Algebra,* Macmillan Company, New York, NY, 1941.

[17-7] S. Sternberg, et al., "Designing Reticles for Imaging Systems by Use of Linear Algebra," Transactions of the Twelfth Conference of Army Mathematicians, U. S. Army Research Office, Durham, NC, ARO-D Report 67-1, February 1976.

[17-8] H. J. Ryser, "Combinatorial Mathematics," Carus Mathematical Monographs, No. 14 (Mathematical Association of America), John Wiley and Sons, New York, NY, 1963.

[17-9] R. C. Bose, "On the Construction of Balanced Incomplete Block Designs," *Annals of Eugenics,* Cambridge University Press, New York, NY, Vol. 9, 1939, pp. 353-399.

[17-10] R. D. Swift, R. B. Wattson, J. A. Decker, Jr., R. Paganetti, and M. Harwit, "Hadamard Transform Imaging and Imaging Spectrometer," *Applied Optics,* Optical Society of America, Washington, DC, Vol. 15, No. 6, June 1976, pp. 1595-1609.

[17-11] E. D. Nelson and M. L. Fredman, "Hadamard Spectroscopy," *Journal of the Optical Society of America,* Optical Society of America, Washington, DC, Vol. 60, No. 12, December 1970, p. 1665.

[17-12] W. W. Peterson, *Error Correcting Codes,* MIT Press, Cambridge, MA, and John Wiley and Sons, New York, NY, 1961, p. 106.

[17-13] F. J. McWilliams and N. J. A. Sloane, "Pseudo-Random Sequences and Arrays," *Proceedings of the IEEE,* Institute of Electrical and Electronics Engineers, New York, NY, Vol. 64, No. 12, December 1976, pp. 79-80, 1715-1729.

[17-14] M. Hall, "A Survey of Different Sets," *Proceedings of the American Mathematical Society,* American Mathematical Society, Providence, RI, Vol. 7, 1956, pp. 975-986.

[17-15] J. Singer, "A Theorem Infinite Projective Geometry and Some Applications to Number Theory," *Transactions,* American Mathematical Society, Providence, RI, Vol. 43, 1938, pp. 377-385.

[17-16] L. Biberman, *Perception of Displayed Information,* Plenum Press, New York, NY, 1973.

[17-17] D. Petersen and D. Middleton, "Sampling and Reconstruction of Wave Number Limited Functions in N-Dimensional Euclidean Spaces," *Information and Control,* Academic Press, New York, NY, Vol. 5, No. 4, December 1962, pp. 279-323.

[17-18] W. D. Montgomery, "Some Consequences of Sampling in FLIR Systems," Institute for Defense Analyses, Arlington, VA, Research Paper P-543, September 1969.

[17-19] W. D. Davenport and W. L. Root, *Random Signals and Noise,* McGraw-Hill, New York, NY, 1958.

[17-20] Y. W. Lee, *Statistical Theory of Communications,* John Wiley and Sons, New York, NY, 1960.

[17-21] R. Legault and J. Riordan, "Optimal Spectral Filters," *Applied Optics,* Optical Society of America, Washington, DC, Vol. 2, June 1964, p. 735.

[17-22] T. G. Birdsall and W. W. Petersen, "The Theory of Signal I Detectability," The University of Michigan, Ann Arbor, MI, Technical Report No. 13 EDG, June 1953.

[17-23] D. Blackwell and M. A. Girshick, *Theory of Games and Statistical Decisions,* John Wiley and Sons, New York, NY, 1954.

17.8.2. Bibliography

Biberman, L., *Perception of Displayed Information,* Plenum Press, New York, NY, 1973.

Biberman, L., *Reticles in Electrooptical Devices,* Pergamon Press, Elmsford, NY, 1966.

Birdsall, T. G., and W. W. Petersen, "The Theory of Signal I Detectability," The University of Michigan, Ann Arbor, MI, Technical Report No. 13 EDG, June 1953.

Birkhoff, G. and S. MacLane, *A Survey of Modern Algebra,* Macmillan Company, New York, NY, 1941.

Blackwell, D., and M. A. Girshick, *Theory of Games and Statistical Decisions,* John Wiley and Sons, New York, NY, 1954.

Bose, R. C., "On the Construction of Balanced Incomplete Block Designs," *Annals of Eugenics,* Cambridge University Press, New York, NY, Vol. 9, 1939, pp. 353-399.

Davenport, W. D., and W. L. Root, *Random Signals and Noise,* McGraw-Hill, New York, NY, 1958.

Gradshteyn, I. S., and I. M. Ryzhik, *Tables of Integrals Series and Products,* Academic Press, Washington, DC, 1965, 4th Edition.

Hall, M., "A Survey of Different Sets," *Proceedings of the American Mathematical Society,* American Mathematical Society, Providence, RI, Vol. 7, 1956, pp. 975-986.

Lee, Y. W., *Statistical Theory of Communications,* John Wiley and Sons, New York, NY, 1960.

Legault, R. R., and J. Riordan, "Optimal Spectral Filters," *Applied Optics,* Optical Society of America, Washington, DC, Vol. 2, June 1964, p. 735.

McWilliams, F. J. and N. J. A. Sloane, "Pseudo-Random Sequences and Arrays," *Proceedings of the IEEE*, Institute of Electrical and Electronics Engineers, New York, NY, Vol. 64, No. 12, December 1976, pp. 1715-1729.

Montgomery, W. D., "Some Consequences of Sampling in FLIR Systems," Institute for Defense Analyses, Arlington, VA, Research Paper P-543, September 1969.

Nelson, E. D. and M. L. Fredman, "Hadamard Spectroscopy," *Journal of the Optical Society of America*, Optical Society of America, Washington, D.C., Vol. 60, No. 12, December 1970, p. 1665.

Oberhettinger, F., *Fourier Expansions: A Collection of Formulas*, Academic Press, New York, NY, 1973.

Petersen, D. and D. Middleton, "Sampling and Reconstruction of Wave Number Limited Functions in N-Dimensional Euclidean Spaces," *Information and Control,* Academic Press, New York, NY, Vol. 15, No. 4, December 1962, pp. 279-323.

Peterson, W. W., *Error Correcting Codes*, MIT Press, Cambridge, MA, and John Wiley and Sons, New York, NY, 1961.

Ryser, H. J., "Combinatorial Mathematics," Carus Mathematical Monographs, No. 14 (Mathematical Association of America), John Wiley and Sons, New York, NY, 1963.

Singer, J., "A Theorem Infinite Projective Geometry and Some Applications to Number Theory," *Transactions,* American Mathematical Society, Providence, RI, Vol. 43, 1938, pp. 377-385.

Sternberg, S., et al., "Designing Reticles for Imaging Systems by Use of Linear Algebra," Transactions of the Twelfth Conference of Army Mathematicians, U.S. Army Research Office, Durham, NC, ARO-D Rept. 67-1, February 1976.

Sternberg, S., J. Ulrich, and R. Hamilton, "Analysis and Testing of a Special Rotating-Translating Reticle," Willow Run Laboratories, The University of Michigan, Ann Arbor, MI, Rept. No. 7102-1-T, June 1967.

Swift, R. D., R. B. Wattson, J. A. Decker, Jr., R. Paganetti and M. Harwit, "Hadamard Transform Imaging and Imaging Spectrometer," *Applied Optics*, Optical Society of America, Washington, DC, Vol. 15, No. 6, June 1976, pp. 1595-1609.

Ulrich, J. P., W. D. Montgomery, and J. L. Alward, "Analysis of Reticle System," Background Analysis Center, Willow Run Laboratories, The University of Michigan, Ann Arbor, MI, Rept. No. 6054-2-T, October 1965.

Chapter 18

DISPLAYS

Lucien M. Biberman
Institute For Defense Analyses

CONTENTS

18. Displays

18.1. Introduction

The choice of specifications for a display terminal of a sensor system is not different in principle from the choice of specifications for detectors, amplifiers, or data processors. The designer must first establish the needs or requirements for the overall system output and insure that the display can provide gain and bandwidth sufficient to fill the data throughput requirements. The designer must then insure that the inevitable introduction of noise in the subsystem does not degrade system utility below the design requirements. Finally, the critical problem of coupling the sensor system to the observer must be treated with greater than usual care since this last interface can and often does degrade real systems far more than most other related major hardware design decisions. Once the designer has considered these primary issues he can proceed to examine the available devices that may fill the matrix of parameters established by overall system requirements. However, considerable attention must be given to the special problems caused by the introduction of a human into the system as a critical data processing element.

This chapter first reviews the factors that govern the transfer of information across the display-observer interface in Section 18.2 and then examines the technologies for display implementation in Section 18.3. Finally, standards and calibrations are discussed in Sections 18.4 and 18.5.

18.1.1 Symbols, Nomenclature and Units. The symbols, nomenclature and units used in this chapter are summarized in Table 18-1.

18.2. Display Performance Requirements

18.2.1. General. Display performance requirements are primarily determined by the characteristics of a human operator and the visual task to which he has been assigned. The environment seriously impacts the observer's abilities to do even quite simple tasks.

It is clear to most who stop to think about it that the information in a report contained in a sheet of microfiche is degraded in the process of projection. Yet, in spite of these losses an observer can only obtain an understanding of the microfiche content by using such a projector or "reader", even though the microfiche contains more signal and less noise than its poorly displayed projection on the granular screen. The microfiche size is a bad mismatch to the eye. This situation is akin to the mismatch in the display of information in many electro-optical systems. Though signal may well be there, it may also be nearly useless because of its form, not because of its content. The error is often made in providing excellent displays that are nearly useless only because of their inadequate size. Conversely, too large displays require too much time to scan visually. Tradeoffs in choice of display size are very important since display size may govern strongly the overall utility of an otherwise good sensor system.

Table 18-1. Symbols, Nomenclature and Units

Symbols	Nomenclature	Units
A	Total display area	m^2
a	Image area on display	m^2
B	Brightness	ft L
D_V/D_H	Display-distance to display-height ratio	—
SNR_D	Displayed image signal-to-noise ratio	—
SNR_{DT}	Threshold (50% probability of detection) displayed-image signal-to-noise ratio	—
SNR_P	Perceived image signal-to-noise ratio	—
SNR_V	Video signal-to-noise ratio	—
t_e	Time constant of eye	sec
α	Width-to-height aspect ratio of the display	—
Δf	Video bandwidth	Hz
$\Delta x, \Delta y$	Rectangular image dimensions	m
E_v	Illumination	ft c

18.2.2. The Display and the Observer [18-1]. The detail discrimination threshold of the eye, i.e., visual acuity, has been investigated exhaustively. Various types of acuity such as minimum detectable, minimum separable, vernier and stereo have been defined. Minimum separable visual acuity applies in the case of shape recognition in which, generally, closely-spaced image details must be discerned. It is known to vary as a function of adaptation level, image brightness, contrast, exposure time, image motion, vibration, spectral characteristics, angular position of the target relative to the line of sight, etc. Visual acuity is usually defined in terms of arbitrary regular test patterns with generally sharp edges, although some studies have been conducted with sine wave patterns.

Discrimination of imagery detail differs from visual acuity measurements in that it requires detection of discontinuities characterized by diffuse edges and irregular brightness distributions. The published acuity data are statistics representing specified performance levels (usually 50% detection probability). Thus they provide information in a probabilistic rather than in a deterministic sense. Therefore, in any specific instance, visual performance may fall far short, or exceed, predictions based on published data. In general, standard visual acuity data are modified by field factors to obtain realistic operator performance estimates under operational conditions. Unmodified data can be used to establish average expected limits of performance under ideal conditions.

A minimum contrast threshold visual acuity curve is plotted in Figure 18-1. These data are for a sine wave test pattern with an average brightness of 100 ft L and viewed at 25 in. This curve neglects image motion, exposure time, wavelength and vibration effects. The visual acuity curve sets the lower limit on useful system contrast. To be visually discernible, an image detail must exceed the threshold modulation of Figure 18-1. The maximum usable resolution of a sensor-display system (for a specified viewing distance) is indicated by the point at which the modulation transfer function (MTF) of the system crosses the corresponding visual acuity modulation threshold.

18.2.3. Brightness, Contrast and Gray Scale. The display contrast must be sufficient to provide a clearly visible image when the ambient light is as high as 10,000 ft c. From a display design standpoint, sunlight shining directly on the display, e.g. cathode ray tube (CRT) phosphor, represents the most severe lighting condition. In this case, to be clearly visible, the maximum brightness of the image must be many times greater than the brightness of the ambient light reflected back from the phosphor.

The display-contrast ratio as a function of ambient illumination and display (CRT) highlight brightness is shown in Figure 18-2. A 10% neutral density filter (considered to be approximately optimum) placed in front of the display and a 70% reflectance display phosphor are assumed.

The contrast ratio required by the observer's eye for any given number of successive gray shades when the observer

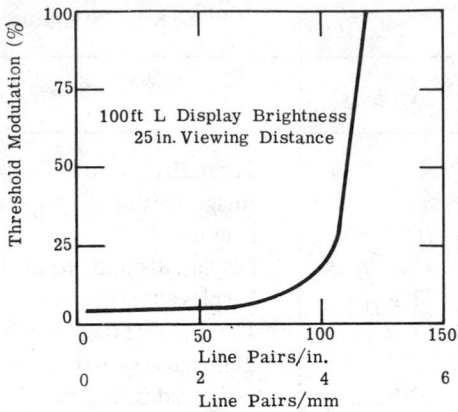

Fig. 18-1. Visual acuity threshold modulation [18-2].

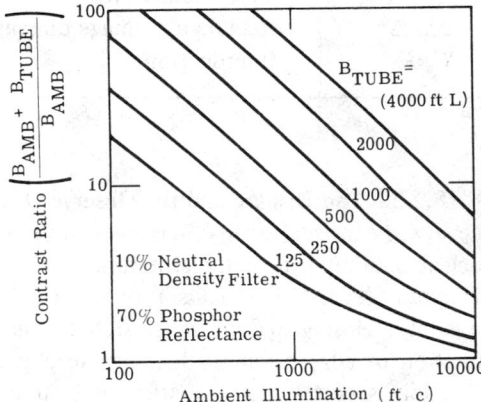

Fig. 18-2. Contrast ratio as a function of ambient illumination and display CRT highlight brightness [18-1].

is scanning the display is shown in Figure 18-3. The number of gray shades seen by the eye with a given contrast ratio depends markedly on the image size as well as the brightness. Three different target sizes are plotted: 3, 8 and 60 minutes of arc. These curves are based on data from Blackwell and include the "field factors" that transform laboratory threshold data into a form more appropriate for operational use.

Using the curves in Figures 18-2 and 18-3, one can calculate the brightness required for a given number of successive gray shades under a variety of conditions.

The maximum brightness that can be obtained in high-resolution sensor displays is generally about 500 to 2000 ft L. If 2000 ft L of ambient light fall on a CRT with a highlight brightness of 500 ft L, the maximum image-contrast ratio is about 4.5 to 1 when a 10% neutral density filter is used on the display. If gray shades generated on the display subtended 8 minutes of arc at the operator's eye, he could discriminate about 5 shades of gray. Under the same conditions, about 7 shades of gray could be seen on a 1000 ft L display. To facilitate target recognition by an operator, at least 7 shades of gray are desired.

Environment impacts strongly upon the display parameter choices. Again, a reference to homilies points out it is more difficult to read a newspaper when riding a bus on a cobblestone street than on a smooth highway. The result of such relative vibratory motion between observer and display can be reduced materially through the use of a much larger display at a much greater viewing distance. The same amplitude of vibration now is small compared to the display imagery while the overall angular resolution of the observer is more or less unaffected. When the level of vibration becomes very se-

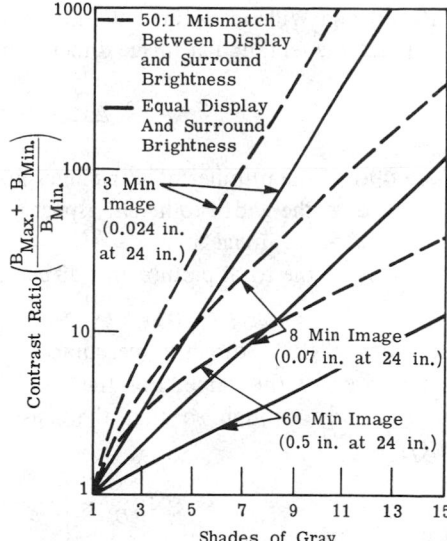

Fig. 18-3. Contrast ratio versus shades of gray [18-1].

vere or when acceleration effects are large, additional important physiological effects cause further degradations of observer performance. Many and varied trials have been made and many simulations proposed and/or carried out. The most definitive work, to date, however, was carried out by Rosell and Willson* who showed, in a series of flight tests with carefully controlled image presentation to parallel observers, the increase required in signal-to-noise at a display for an observer to maintain a given level of performance as g-loading and/or turbulence increased. This material, abridged from first results, is presented in the next section after a short review of basic theory.

18.2.4. The Concept of Signal-to-Noise Ratio in a Displayed Image and the Effects of Turbulent Flight Conditions upon Signal-to-Noise Thresholds. The concept of signal-to-noise ratio in a displayed image is not widely understood and is one that is basic, important, and necessary in the design and/or specification of a display system. This concept, most completely developed by Rosell, is treated in depth in Chapters 4, 5, and 6 in Reference [18-3].

Historical Background. The earliest psychophysical experiments performed by Rosell and Willson employed simple rectangular images on a uniform background. These images were electronically generated, mixed with additive white noise, and displayed on a television monitor. The same amount of noise was added to both the rectangular image and its background. The purpose of the experiments was to determine the probability that an observer will detect a displayed image as a function of the image's signal-to-noise ratio (SNR). These experiments proved easy to perform and, over the years, reruns of the experiment to establish equipment calibration have produced highly consistent results.

---

*Performed under the USAF 698DF program in an attempt to establish the signal-to-noise requirements for an electro-optical imaging system [18-4, 18-5, 18-6]. Earlier tests and experiments by many psychophysicists are reviewed in Chapter 2 of [18-3].

Rosell and Willson defined the image size in terms of the dimension of a single scan line. Thus, the rectangular image dimensions are

$$\Delta x \Delta y = (490)^2 \ \alpha \left(\frac{a}{A}\right) \qquad (18\text{-}1)$$

where 490 = the number of active lines in a conventional 525-line television display
  α = the width-to-height aspect ratio of the total effective picture on the CRT
  a = the image area
  A = the total picture area displayed

The video SNR is directly measurable. The video SNR and the image SNR, though related, are not the same. For the special case where the noise is white and where the image is large enough so that the sensor MTFs can be ignored, the displayed image, SNR_D, is

$$SNR_D = \left[2\Delta f t_e \left(\frac{a}{A}\right)\right]^{1/2} SNR_V \qquad (18\text{-}2)$$

where Δf = the video bandwidth, Hz
  t_e = the time constant of the eye, sec
  SNR_V = the video signal-to-noise ratio as measured in the video channel

The subscript D on SNR_D differentiates it from SNR_P which is the signal-to-noise ratio perceived by the observer. For displayed images that are bright enough and neither too large or too small, SNR_D will be very nearly equal to SNR_P.

Figure 18-4 shows the probability of detection corrected for chance as a function of the video SNR. The images ranged from a square 4 × 4 scan lines in size to a long rectangle subtending 4 × 180 lines. The display-distance to display-height ratio, (D_V/D_H), was fixed at 3.5. A large SNR_V is required to detect the small square while the long thin rectangles are readily detected with a very small SNR_V.

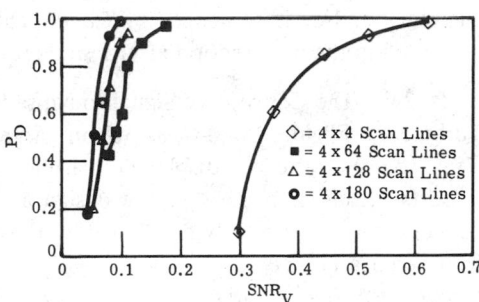

Fig. 18-4. Probability of detection versus signal-to-noise (P_D versus SNR_V). There are 30 frames per second with D_V/D_H equaling 3.5 [18-7].

By the use of Figure 18-4 and Equations (18-1) and (18-2), one can calculate SNR_D and plot the probability of detection as a function SNR_D as shown in Figure 18-5. Observe that the four curves now become a single curve. The angular subtense of the rectangles relative to the observer's eye varied from 0.13° × 0.13° to 0.13° × 6.2°.

Figure 18-6 shows the result of an experiment using squares. The SNR_D needed to detect the 4 × 4 and 8 × 8 line squares is the same within the experimental error. However, a small increase is needed to detect the 16 × 16 line square which subtends 0.5° at the observer's eye and the increase has been found to be statistically significant. Substantial increases are needed to detect the 32 × 32 and 64 × 64 line squares at edges.

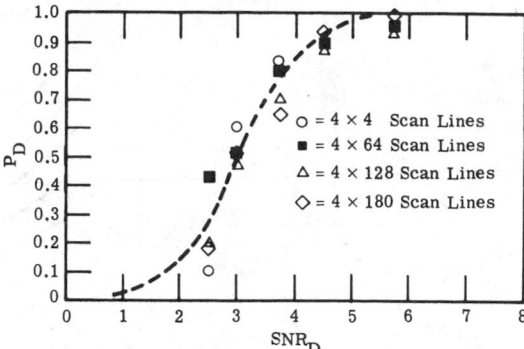

Fig. 18-5. Corrected probability of detection versus SNR_D required for rectangular images. The televised images are at 30 frames per second and 525 scan lines with D_V/D_H equaling 3.5 [18-7].

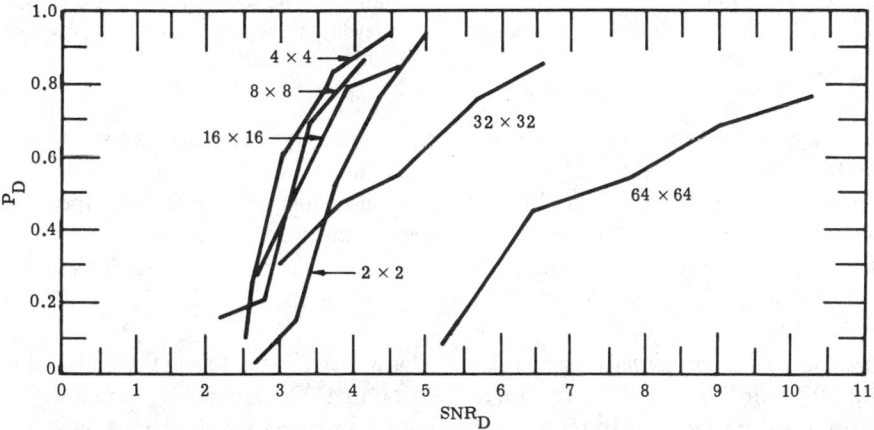

Fig. 18-6. Probability of detection versus SNR_D for squares of various size at a display distance of height ratio of 3.5 [18-7].

The value of SNR_D is constant over a range of sizes and grows rapidly for sizes that are either much larger or smaller. (See threshold SNR_D, plotted in Figure 18-7.) The smaller-size images are smeared or blurred by the MTF of the eye and the reduced contrast of that image necessitates a larger SNR_D. For large images, the limitations of the eye as a spatial integrator come into play again requiring compensation in the form of SNR_D. These effects are discussed more completely in [18-8, 18-9, 18-10, 18-11, 18-12].

The angular subtense of a 2×2 line square at the observer's eye as a function of viewing distance for various display diagonal dimensions is shown in Figure 18-8.

Values of the *threshold* SNR_D which, by definition, correspond to a 50% probability of detection, are plotted in Figure 18-7. Note the increase in SNR_{DT} needed for the 2×2 line square. This subtends about 1 mrad at the eye. Thus, the ability of the observer to detect images of size smaller than about 1 mrad will be limited by the MTF of the

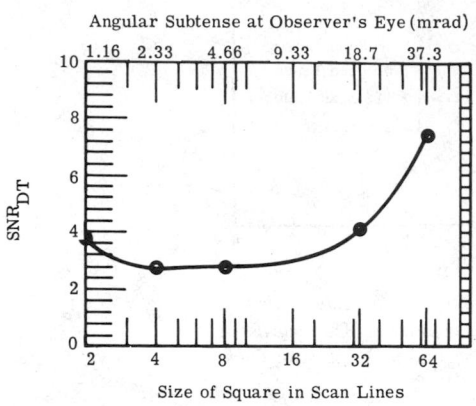

Fig. 18-7. Threshold signal-to-noise ratio versus size of square in scan lines with D_V/D_H equaling 3.5 [18-7].

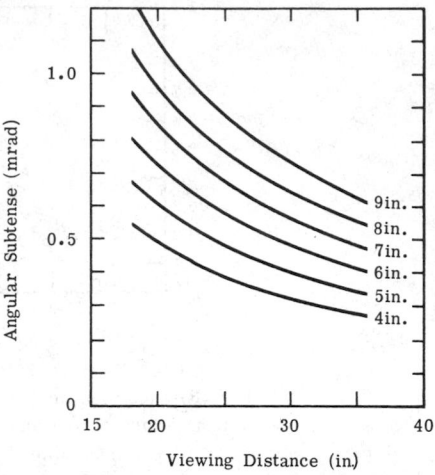

Fig. 18-8. Angular subtense of a 2 × 2 square at the observer's eye as a function of viewing distance for various display diagonal dimensions [18-7].

observer's eye. The values of SNR_{DT} required for various square sizes are summarized in Table 18-2 for the listed squares at a viewing distance to display height ratio, D_V/D_H, of 3.5. If the viewing distance is doubled using the same display, the SNR_{DT} applying to a 4 × 4 line square becomes that for a 2 × 2 square as illustrated in Table 18-2.

It is clear the factor of 2 in D_V/D_H corresponds to a factor of 2 in viewing angle. Thus the 4 × 4 square at $D_V/D_H = 7$ appears to the eye as identical to a 2 × 2 at $D_V/D_H = 3.5$. Thus the SNR_{DT} is 3.7 for both.

Analysis of Experimental Results. Four flights were flown during the first half of the Spring 1975 Air Force 698DF program and a total of 1,928 data points were taken for all conditions. Figure 18-9 illustrates the placement of observers and displays.

In Table 18-3 a summary of the data points and conditions for the four flights is shown for the four observers.

Table 18-2. Threshold SNR_{DT} Versus Square Size for Two Ratios of Display Distance to Display Height [18-7]

Image Size (Scan lines)	SNR_{DT} $D_V/D_H = 3.5$	SNR_{DT} $D_V/D_H = 7.0$
2 × 2	3.7	–
4 × 4	2.8	3.7
8 × 8	2.8	2.8
16 × 16	3.3	2.8
32 × 32	4.0	3.3
64 × 64	7.5	4.0

Table 18-3. The Number of Data Points Taken Each Flight for Various Conditions for Squares [18-7]

Flight No.	Condition			
	Straight & Level	Rough Air	2 1/2 g Turns	1 1/2 g Turns
1	400	40	32	20
2	324	28	48	–
3	60	400	56	–
4	60	400	60	–

In Figures 18-10 and 18-11 the data from the four flights and the laboratory test are plotted. Straight and level data from flight no. 1 have been averaged with the laboratory data and are represented by the solid line in the figure. The data from the two turbulent conditions have also been averaged and are represented by the dashed line. Figure 18-10 shows the data versus image size, while Figure 18-11 shows their spatial frequency in lines per picture height.

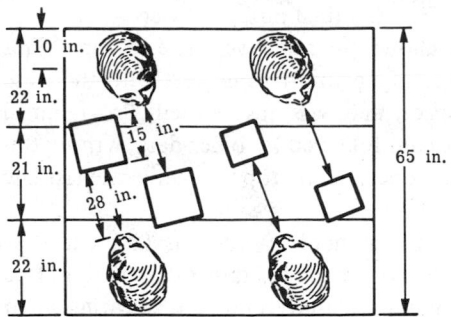

Fig. 18-9. Layout of observer's compartment [18-7].

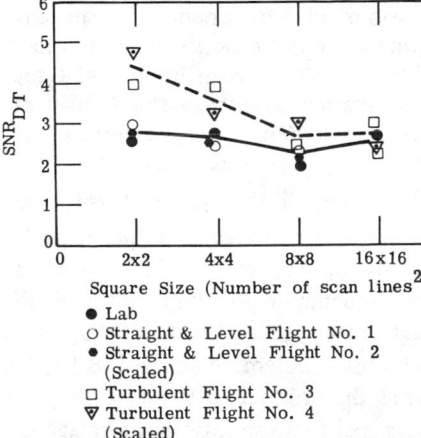

Fig. 18-10. Threshold SNR_{DT} as a function of square size for the four flights and lab comparison data [18-7].

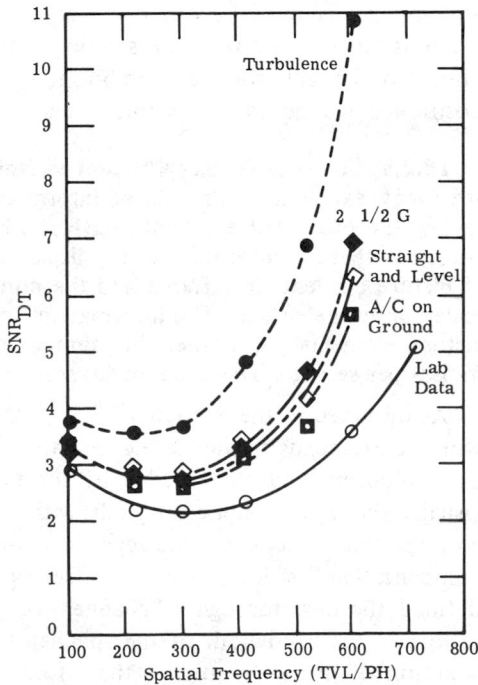

Fig. 18-11. Composite average threshold SNR_{DT} for bar pattern recognition [18-7].

18.2.5. Display Storage and Other Requirements. The storage, or phosphor persistence for short-time storage requirements of a display system or display tube vary from less than 1/60 of a second for sixty field per second refreshed TV-type displays to as much as a few minutes for a "rolling map" display. Such storage characteristics may be a preset property of the phosphor, a programmed storage characteristic of a display subsystem, or may be entirely under the control of the operator who chooses to "freeze" the image.

Storage may be designed into the system through a number of elections: the direct-view storage tube (DVST), the scan converter tube of various forms, or straightforward memory devices with suitable data access such as tape, disk, or the new MOS or related solid-state devices.

Straightforward memory, such as in simple, direct-view storage tubes, serve the simple storage problem fairly well, costing the designer a moderate amount of deterioration in both resolution and dynamic range over that achieved either with a simple CRT or more advanced storage techniques that feed simple CRTs as the final output device.

The use of the more sophisticated memory allows for selective write features. This makes possible the continuous writing into a memory bank of some sort with the fresh data being written into the oldest memory space that was just erased. Programmed reading-beam routines then read out the newest data followed by older data, writing this upon a CRT in a manner so that the newest data appear as the top line while the last line on bottom disappears.

Most sensor displays will have requirements for symbology such as target marker symbols, artificial horizon and steering symbology, or cursors for target designation. It is obviously desirable for these symbols to create minimum interference and smearing when positioned and moved across storage images.

18.2.6 Forms of Imagery, Signal-to-Noise Ratio and Information Transfer.

There are many excellent treatments of information content of an image, such as the good review (an older, but excellent, text) by Linfoot. In our brief treatment, one can only point out that the information in a display is determined by the product of the number of picture elements in a frame and the number of discernible steps in brightness (grey scale) of each element. The information flow rate is thus the product of the number of picture elements per frame, the number of levels per element, and the number of frames per second. This product governs the bandwidth requirements of the system.

As discussed in the Section 18.2.4 on the signal-to-noise ratio in the displayed image which corresponds to the "levels" mentioned above, the signal-to-noise ratio produced by an equipment, and that needed by the eye in an image, vary with spatial frequency. Usually the signal-to-noise ratio falls with increasing frequency in an equipment while the corresponding needs of the eye increase, though often some form of "aperture compensation," is introduced to offset or diminish this problem. Shannon [18-13] has defined the flow through a "channel" of a system as the product of the signal-to-noise ratio and the bandwidth of the channel. Since the signal-to-noise ratio in an image is a function of spatial frequency, the information flow is the integral of signal-to-noise at each frequency over the spatial frequency of the display. The corresponding evaluation of the electrical signal involves the temporal equivalent of the spatial frequencies mentioned above.

In halftone or pictorial representations, the number of levels or grey scales per picture element tends to be a number such as 6, 8, 10 or possibly 12 levels or "shades of grey," though the latter two are quite exceptional. For many forms of information transfer, only two levels are required—on (or white) and off (or black). It is because of this that the problem of symbolic data display becomes highly simplified compared to a picture of a scene with features represented by small variations in level from picture element to picture element.

In digital systems, such elements are often referred to in terms of pixels (or picture elements) and levels (or shades of grey). Generally, the analog systems are poorly described by the pixel concept since the very number of separable elements or resolved elements is a direct consequence of the signal-to-noise ratio dependence upon frequency. If signal-to-noise is low, variation between adjacent elements may not exceed in level the random variation in level, and thus they will not be separable, or, in more common terms, not "resolvable." It is because of these effects that both the bandwidth, and dynamic range (and thus signal-to-noise ratio) must be much greater in a pictorial display than in an alphanumeric device.

18.3. Display Technologies

This section first summarizes, then compares some of the more completely developed display technologies before briefly examining them individually. A bibliography (Section 18.6.2) is supplied to make possible a more thorough review for the specific interests of the reader.

CRT display technology is so highly developed that we merely tabulate its strengths and weaknesses, and concentrate on the less-well-known versions such as the Digisplay,®* and the Multi-Mode Tonotron,®** and single and double-ended scan converters and storage tubes.

Table 18-4 is a summary of CRT effectiveness factors according to Reingold [18-8]. It is clear there are but two serious objections to the CRT for all its many good features. The CRT is a big, empty (vacuum) bottle that can and often does takes a lot of valuable space for the display area it offers. Secondly, in some specific modes of beam addressing at high speeds, it can require significant amounts of power in the beam deflection circuits. Its technology, on the other hand, is very well developed and the associated driving circuits so completely debugged that Anderson [18-14] was moved to write an excellent paper asking the question, 'Why consider anything else?'

Why else indeed? For small or otherwise limited demands both the light-emitting diodes and liquid crystal displays offer cheaper, more compact displays. As size increases, complexity increases at a severe exponential rate. The problem shifts from the display to the problem of interconnections and to the driver circuits that now become a tour de force in microelectronics.

A comparison is in order. The conventional home television set has three driving circuits, one to brighten the beam to the level required for each picture element and one each to deflect the beam in the X and Y coordinates. The first of these three drive circuits is a video driver, while the X-driver operates in the upper audio region and the Y-driver is usually about a sixty cycle circuit. As an interesting comparison, note that the largest non-CRT video-compatible display in operation prior to this publication was the Hughes 100×100 element liquid crystal display.

Figures 18-12 and 18-13 show the battery of drive circuits for each axis and a detail of the circuits associated with each crystalline element. Thus there are two 100 element banks of inputs (X and Y inputs), but within the crystalline array there is an individual transistor capacitor and crystal cell for each picture element.

Thus there are two controls, each with an output of 100 leads, and an internal matrix of 10,000 active transistor circuits. In about 1972, the imperfections of such circuits gave rise to picture imperfections such as shown in Figure 18-14.

As of the summer of 1975, production of that same device resulted in a defect-free

---

*Trademark of the Northrup Corporation.
**Trademark of the Hughes Aircraft Company.

Table 18-4. Summary of Cathode Ray Tubes (CRTs)
Display-Effectiveness Factors [18-8]

Factor	Required	Achieved
Brightness	Average—min. of 50 ft L	Yes (far exceeded)
	Max.—3,000 ft L	Yes (far exceeded)
Contrast	Viewable in shade for stationary displays	Yes
	Viewable in direct sunlight	Only with direct view storage tubes (DVSTs)
Half-Tones	Radar displays—two-tone acceptable	Yes
	Television—5 or more required	Yes
Resolution	Min. size commensurate with eye acuity	Yes
	Size constant with brightness and position	No, but adequate for most purposes
Flicker	None present	Yes, for most applications
Distortion	Size constant with brightness and position	See "Resolution" above
Accuracy	Position linear with input voltage	System rather than device limited
Blemishes	Radar displays—min. loss of resolution elements	0.04 to 0.04% max. blemished area
	Television—indiscernible loss of picture detail	0.005 to 0.01% max. lost resolution elements
Volume-to-area	Overall volume small for desired viewing area	Poor; display device volume and shape may dictate equipment volume and shape
Power consumption	Negligible fraction of total equipment power	Yes, for TV No, for random access

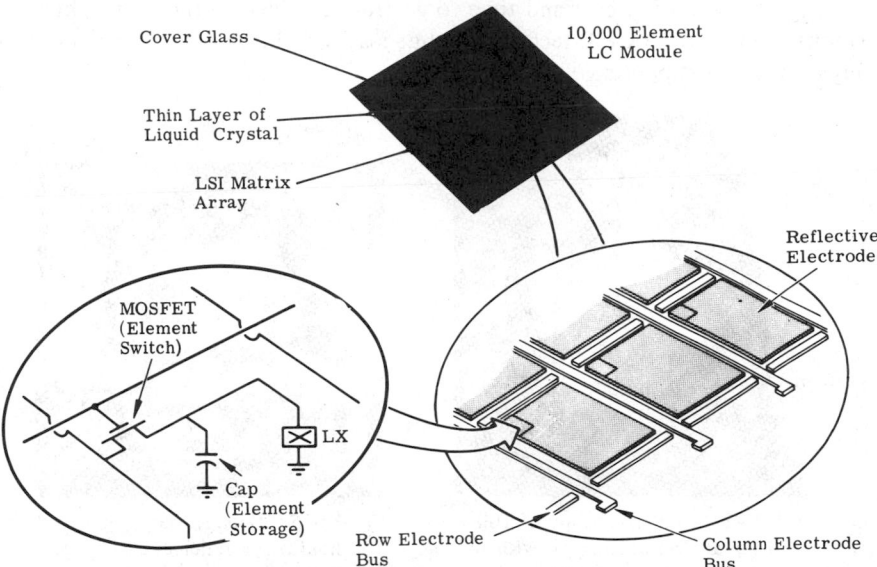

Fig. 18-12. Liquid crystal display module construction [18-15].

Fig. 18-13. Early version of the 100 × 100 display. The liquid crystal elements and directly associated electronics form a 25 mm × 25 mm square. Leads to external driving circuits radiate to two hundred peripheral contact pads [18-15].

image, Figure 18-15. The cost and time to go from the idea to the device, however, are enormous compared to CRT technology, but may well be much cheaper for individual displays if use ever supports volume production.

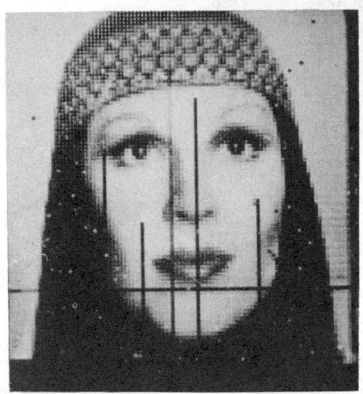

Fig. 18-14. Enlarged copy of 100 × 100 liquid crystal imagery with picture imperfections [18-15].

Fig. 18-15. Enlarged copy of 100 × 100 liquid crystal defect-free imagery [18-15].

On a smaller scale, it is clear the alphanumeric displays for pocket calculators and cash registers have driven the production costs of drivers and displays to the point that in 1975 these devices were offered at prices at which "a five dollar bill would supply display array, driver circuits and change."

In the case of the 100 × 100 liquid crystal array, this has yet to happen, although the liquid crystal cells are but "printed circuits" of silicon and sapphire with less than a drop of liquid crystalline material per display. The development has been an expensive multiyear effort leading to a clear, highly developed new technology.

The result, however, presents about 25% of the information displayed on a domestic television set. Four such liquid crystals could and perhaps will be interconnected to equal home television quality. On the other hand, the development of a new CRT display, including the design of a new cathode ray tube, is a matter of a few weeks for an electronic designer, machinist, and glass blower.

Thus, this example of a comparison is meant to show that while the new flat panel display technologies may well replace the CRT for some purposes, the costs are low for new limited developments using CRTs and low for flat panels in very high volume production.

18.3.1. Summary of Conventional CRT Display Techniques [18-1]. Although a complete treatise of all available CRTs is beyond the scope of this chapter, several types of CRTs are of particular interest for sensor displays. These include high-brightness CRTs used with scan converters, miniature (1 in. diameter) CRTs, which due to their small size can be mounted on the pilot's helmet, and long-persistence CRTs.

Short-Persistence CRTs. A short-persistence phosphor CRT with high brightness is ideally suited for television imagery. If a short-persistence CRT is to be used for a low-frame-rate sensor display, a scan converter must be used for intermediate storage of the sensor imagery which is read out electrically at TV rates and displayed on the CRT.

At present, there are a number of CRTs that can provide brightness levels of 800 to 2000 ft L at resolutions of 150 TV_{50} lines per inch or greater. (See Figure 18-25 and Table 18-15 and accompanying text on "Sensor Resolution" for a definition of TV_{50} and other terms used for sensor resolution.)

Long-Persistence CRTs. For many years, CRTs with both P-7 and P-14 long-persistence phosphors have been used for the display of air-to-air and air-to-ground radar and IR displays. There are several disadvantages in using tubes of this type:

(1) The brightness is low.
(2) Persistence is fixed and nonlinear.
(3) Selective erasure is not possible.

The low brightness makes it impossible, for example, to use a display of this type in an aircraft cockpit environment without ambient light shielding by means of a visor or hood. Even then the operator is forced to adjust quickly from exterior ambient light levels as high as 10,000 ft L to the low brightness of the CRT long-persistence phosphor.

Where ambient light can be controlled and cost is an overriding consideration, a long-persistence CRT can provide an acceptable display for low-resolution sensors. Since both P-7 and P-14 phosphors have short- and long-persistence components of different colors, it is possible to filter out the long component and present TV-frame-rate imagery. (See Tables 18-5 and 18-6.) For even longer persistence, storage techniques like a direct view storage tube (DVST) are indicated.

Table 18-5. Typical Characteristics of a Fast
Erase Storage Tube [18-1]

Characteristics	Fast Erase Storage Tube
Resolution (Shrinking raster)	100 stored lines/in.
Brightness	800 ft L (at maximum resolution, brighter if smearing or resolution compromised)
Writing speed	100×10^3 in. sec^{-1}
Storage time	10 sec
Shades of gray	6

Fast-erase and conventional DVSTs are fade-erased continuously (analogous to phosphor decay) across the whole display area. Fade erasure results in a loss of resolution and gray shades, and scan-to-scan interference. Symbology presented on fast-erase and conventional DVSTs causes smearing when moved across the stored image. How a DVST would be used for the display of sensors listed earlier is shown in Figure 18-16.

Available Phosphors for CRTs. The many and varied phosphor screens available for CRTs give rise to families of subtechnologies for color and storage effects in display tubes. The depth of penetration achieved by variable beam velocities can yield multicolor displays with proper beam controls.

Table 18-6.   Typical Characteristics of the Multi-Mode
Tonotron® (MMT) [18-1]

Characteristics	Multi-Mode Tonotron®
Resolution (Shrinking raster)	
Stored writing gun	120 lines/in.
TV gun	150 lines/in.
Symbology gun	100 lines/in.
Erase gun	70 lines /in.
Brightness	1000 ft L
Writing speed	60×10^3 in. sec^{-1}
Erase speed	10×10^3 in. sec^{-1}
Storage time	2 min
Shades of gray	7
Deflection	Electrostatic (3 focused guns)

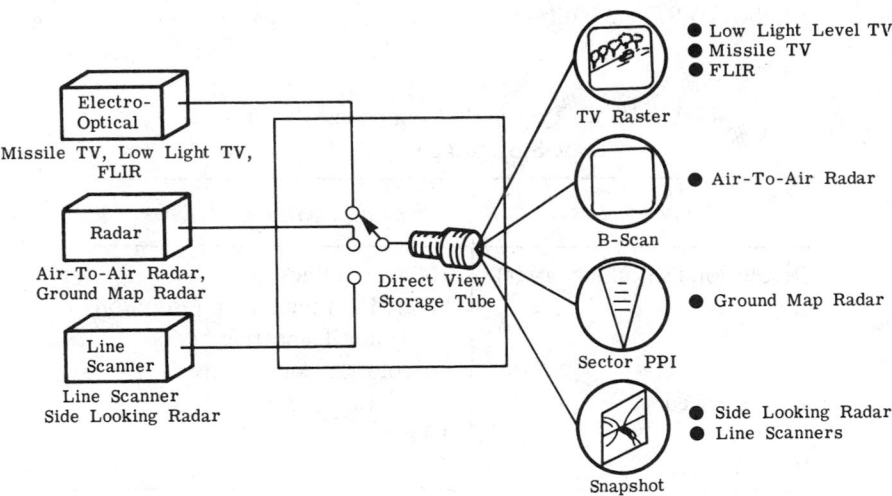

Fig. 18-16. Multi-sensor display system using a direct view storage tube.

The factors governing penetration and luminous efficiency of such phosphors are covered in References [18-17 through 18-28].

Table 18-7 shows the results of Kingsley and Ludwig's [18-16] measurements of the cathodoluminescence efficiency of a variety of phosphors.

The composition of the more common phosphors and their "P" designators are shown in Table 18-8.

A comparison of the more common display technologies is shown in Tables 18-9 through 18-13.

Table 18-7. CR Efficiency of Various Phosphors [18-16]

Phosphor	Efficiency (%)	Phosphor	Efficiency (%)
Zn_2SiO_4:Mn	4.7	Y_2O_3:Eu	6.5
ZnS:Cu	12.4	Gd_2O_3:Eu	9.6–11.7
(Zn, Cd)S:Ag	18.7	Y_2O_2S:Eu	13.1
(Zn, Cd)S:Cu	9.7	La_2O_2S:Eu	10.6
$Zn_3(PO_4)_2$:Mn	3.1	Gd_2O_2S:Eu	10.2
$CaWO_4$:Pb	3.4	YOCl:Eu	12.9
$MgWO_4$	2.9	La_2O_2S:Tb	11.8
Zn_2SiO_4:Mn	6.8	—	—

Table 18-8. Composition of the More Common Phosphors
and Their P Designators

P Designator	Material	P Designator	Material
P1	Zinc silicate:manganese	P12	Zinc Magnesium fluoride:
P2	Zinc sulfide:copper		manganese
P3	Zinc beryllium silicate:	P13	Magnesium silicate:manganese
	manganese	P14	Similar to P7
P4	Zinc sulfide:silver and	P15	Zinc oxide
	zinc cadmium sulfide	P16	Calcium magnesium silicate:
	silver		cerium
P5	Calcium tungstate	P17	Zinc oxide and zinc cadmium
P6	Similar to P4		sulfide:copper
P7	Zinc sulfide:silver and	P18	Calcium magnesium silicate:
	zinc cadmium sulfide:		titanium and calcium
	copper		beryllium silicate:manganese
P8	No information	P19	Potassium magnesium fluoride:
P9	Calcium pyrophosphate		manganese
P10	Potassium chloride (Dark	P20	Zinc cadmium sulfide:silver
	trace—nonluminescent—	P21	Magnesium fluoride:manganese
	called a scotophor)		
P11	Zinc sulfide:silver		

Table 18-8. Composition of the More Common Phosphors
and Their P Designators (*Continued*)

P Designator	Material	P Designator	Material
P22	Zinc sulfide:silver, zinc silicate:manganese, zinc phosphate manganese	P31	Zinc sulfide:copper
	Zinc sulfide:silver zinc cadmium sulfide:silver, zinc cadmium sulfide: silver	P32	Calcium magnesium silicate: titanium zinc cadmium sulfide:copper
	Zinc sulfide:silver, zinc cadmium sulfide:silver, yttrium oxysulfide:euro pium	P33	Magnesium fluoride:mang-anese
		P34	Zinc sulfide:lead:copper
	Zinc sulfide:silver, zinc cadmium sulfide:silver, yttrium oxysulfide:euro-pium	P35	Zinc sulfide selenide:silver
		P36	Zinc cadmium sulfide:silver nickel
	Zinc sulfide:silver, zinc cadmium sulfide:copper, yttrium oxide:europium	P37	Zinc sulfide:silver:nickel
		P38	Zinc magnesium fluoride: manganese
		P39	Zinc silicate:manganese: arsenic
	Zinc sulfide:silver zinc cadmium sulfide:copper, yttrium oxysulfide:euro-pium	P40	Zinc sulfide:silver zinc cadmium sulfide:copper
		P41	Zinc magnesium fluoride manganese calcium magnesium silicate: cerium
P23	Similar to P4	P42	Zinc sulfide:copper, zinc silicate:manganese:arsenic
P24	Zinc oxide	P43	Gadolinium oxysulfide: terbium
P25	Calcium silicate:lead: manganese	P44	Lanthanum oxysulfide: terbium
P26	Same as P19	P45	Yttrium oxysulfide:terbium
P27	Zinc phosphate:manganese	P46	Yttrium aluminate:cerium
P28	Zinc cadmium sulfide:copper	P47	Yttrium silicate:cerium
P29	Similar to P2 and P25	P48	70:30 mix P46-P47
P30	This phosphor is no longer available		

Table 18-9. Technologies Comparison [18-8, 18-9, 18-29]

Device	Comments
Cathode ray tube (CRT)	A mature technology of high reliability in wide-spread use for black and white or color. Associated circuits and hardware readily available. Requires moderately high voltages for bright displays. Requires substantial depth behind display surfaces.
Video-driven image reproducers of pictorial or symbolic material	Basically a television-like reproducer of imagery which may be halftone or *only* black *or* white. Image usually refreshed at thirty frame, sixty field rate.
Extruded beam signal generators for alphanumeric symbols	Symbols limited to those built into the beam-shaping structure. Any shapes possible if built into the tube or formed by alternate super position of existing characters. Refresh necessary.
Stroke-driven symbol generation	Basic tube driven in direct strokes by program of beam addressing, beam blanking or brightening, and beam moving instructions. Repetitive programming necessary to maintain brightness of display.
Storage tubes	Large variety of methods for achieving storage of the functions discussed. Storage achieved by electrical charge stored on dielectrics within the tube in a manner to spatially modulate electron flow to the screen.
Digitally addressed flat-panel CRT	This quite recently developed nonconventional CRT is about 2 in. thick. It employs an area-cathode and dynode aperture plates. A multiplicity of electron beams is formed by the plates, one for each resolution element. A particular beam is selected by applying proper voltages to each plate in a binary selection scheme. The beam passing through the final plate impinges on a phosphor screen as in a conventional CRT. Resolution up to 80 lines per inch has been achieved. Viewing areas up to 7 × 7 in. are available.
Plasma panels	Transparent panel often containing a large array of discharge electrodes usually in a common gas cavity. Both ac and dc versions exist. The plasma display technology is being developed in many sizes and for many applications. For large graphic displays it is the only technology seriously challenging the cathode ray tube.

Table 18-9. Technologies Comparison [18-8, 18-9, 18-29] (Continued)

Device	Comments
dc plasma panels	In one type of dc structure two sets of parallel electrodes oppose each other in the gas with one set directed orthogonally to the other. An aperture plate, placed between the electrode sets, confines the discharges. Appropriate addressing voltages on two intersecting electrodes cause an electrical breakdown and emission of light at the intersection; an addressing voltage on only one electrode is too small to ignite the discharge. In this matrix arrangement, the gas discharge cell, which is sufficiently nonlinear for the purpose, functions as a two input "and" circuit. This device is usually operated one row (or column) at a time with signals on the opposing electrodes determining which cells in the row (or column) will be on. Since the duty cycle in these devices becomes smaller as the device becomes larger, the peak currents limit the array size.
ac plasma panels	In the ac plasma display, as in the dc panel, two sets of electrodes oppose each other across a discharge gap and are directed orthogonally to one another. However, at each discharge site, the dielectric surfaces that isolate the electrodes from the gas define two capacitances which are in series with the discharge. An alternating voltage applied across the two electrode sets is too small by itself to ignite discharges. However, a pair of write voltages applied across two selected electrodes will ignite a pulsed discharge that extinguishes as ions and electrons flow to the dielectric surfaces and charge the capacitors. This charge augments the applied voltage on the next half cycle to ignite a second discharge, which then charges the capacitors in preparation for the third discharge. This sequence of pulsed discharges which characterizes the "on" state of a cell, terminates when an erase signal on the two intersecting electrodes produces a controlled discharge that reduces the charge on the series capacitance below the minimum required for ignition.
Electroluminescent (El) panels	Electroluminescent displays consist of an El powder or evaporated film between two electrodes, one of which is transparent. El displays can be made in many colors. However most displays are single-color, usually green or orange because of the higher efficiency achieved with copperactivated and manganese-activated materials. When a potential is applied across the El material, visible light is emitted. The potential may be ac or dc depending upon the specific structure, but

Table 18-9. Technologies Comparison [18-8, 18-9, 18-29] *(Continued)*

Device	Comments
Electroluminescent (El) panels (Continued)	El displays usually operate in the ac mode. The resolution or pattern is defined by the electrodes. Luminance is typically 5 to 30 ft L, although luminance in the thousands of ft L has been achieved and demonstrated in 1974.
Liquid crystals	A thin clear layer of a cholesteric material placed between transparent electrically conducting covers when excited by an electric field becomes turbulent and scatters ambient light in a manner that yields an apparent brightness related to the applied field. When an aggregate of such cells forms a two-dimensional array, a digitally-addressed display results. These displays can be small, light and relatively inexpensive. The driving circuitry for large arrays is the costly part, not the liquid crystalline materials.
Light emitting diodes (LEDs)	LED displays are now a mature technology for small scale displays such as in pocket calculators, small area indicators and related applications. Their utility for larger area or ambient brightness applications depends upon improved luminous efficiency, lower power dissipation in driving circuits, and costs to challenge other technologies.
Projection displays	Both high luminosity CRTs and light valves of the oil film type are available and useful. The projection CRTs fill the need for heads-up display and small screen systems. The oil film systems fill the need for small-to-large theater screen displays in both black and white or color. A typical device is the Eidophor. Projection CRT displays are finding increasing application in displays for tactical systems in sizes from 3 to 6 ft on a side. Current tubes, with typical F/0.9 optics, can develop 200 to 300 lm output after accounting for optical surface losses. Resolutions of 1,000 TV lines have been achieved on 5 in. projection CRTs. New longer-life and more efficient phospors are necessary to expand the application of projection CRTs.
	Oil-film light valves are currently being considered for a wide variety of command and control display applications of the fixed-site type. Devices can typically provide 525 line TV images with light outputs of 5,000 lm. They are in general large, complex, and expensive systems. Small sealed-off light-valves are available, but light-output and resolution are limited by light-source and cooling requirements.

Table 18-10.   Functions to be Filled by Various Technologies

Display Function	Display Device	*Comments*
Numeric readout	Plasma	Available (orange-red); moderate voltage-low current.
	LED	Available (red or green); low voltage-high current
	EL, liquid crystal	Available; not used as much as plasma or LED; Industry working on improved devices.
	Nixie tubes	Becoming obsolete and being replaced by other technologies.
Message board display (up to several hundred characters, and up to 8 in. dimension)	Plasma	Available.
	CRT	Available.
	Digital address flat CRT	Feasibility demonstrated; beginning to have an impact in the field.
Small alphanum- eric vector-graphic (up to 4 in.)	CRT	Available.
	Plasma	Available.
	Digital address flat CRT	Available.
	LED, EL, liquid crystal	Technologically feasible; further engineering needed.
Medium-Sized alphanumeric- vector-graphic (5 to 10 in.)	CRT	Available.
	Plasma	Available.
	Digital address flat CRT	Feasible up to 9 in. and should be available in that size.
	EL, liquid crystal	Technologically feasible; further engineering needed.
Large-sized alpha- numeric vector- graphic (greater than 10 in.)	CRT	Available.
	Plasma	17 × 17 in. demonstrated; sizes up to 4 × 4 ft under development.
	Projection	Available.
Video Picture	CRT	Available.
	Plasma	Demonstrated in laboratory.
	Liquid crystal	Demonstrated in laboratory.
Color	CRT	Full spectral range.
	Plasma, LED	3 colors possible.
Tactical display (rigged, viewable in high ambient)	CRT	Requires filters or shielding; use of narrow line width phosphor a potentially good solution.
	Liquid crystal	Excellent potential; development and engineer- ing needed.
	EL	Good potential; requires development and engineering.

Table 18-10. Functions to be Filled by Various Technologies (*Continued*)

Display Function	Display Device	Comments
Map overlays	CRT	Possible with ported tubes; expensive; not a desirable approach; limited in size.
	Projection	Feasible; development and engineering needed.
	Plasma	Available up to 8-½ in.; feasible for larger displays.
	LED	Technology feasible; cost effectiveness controversial.

Table 18-11. Plasma Panels [18-8]

Plasma Panels (ac)

Features	Current Status	Limitations
Flat format; inherent storage; low voltage compared to CRT (several hundred volts); direct compatibility with digital computer; matrix-addressable; panel is transparent, hence can add fixed format information; switching speed is in the micro-second region. Selective write and erase.	Red-orange emission; 512 X 512 element panels (8.5 X 8.5 in.) with 60 element/in. resolution are available off-the-shelf; average brightness of 50 ft. Lamberts has been measured; life in excess of 10,000 hours has been demonstrated. Compatibility with military environments demonstrated. Three color capability demonstrated at reduced brightness in 17 X 17 in. panels; large panels (2 X 4-1/4 in) at a resolution of 10 element/in. have been fabricated using capillary tubes. Several levels of intensity on same panel shown possible. No full MIL spec. displays yet.	Need for hundred volt ac waveforms in addressing complicates integration of addressing circuitry. Limited gray scale capability.

Plasma Panels (dc)

Features	Current Status	Limitations
Flat format; low voltage compared with CRT; direct digital interface. Gray-scale capability. Not suitable for group viewing.	Characteristic neon emission. 33 lines/in. resolution, 256 character panels available. Reduced frame size TV images have been presented.	Brightness dependent on number of characters. Data must be formatted for presentation; limited resolution capability.

Table 18-12. Light-Emitting Diodes [18-8]

Features	Current Status	Limitations
Solid state; IC compatible; long life (many years); rugged, lightweight, compact; compatible with digital drive voltage levels (5 V). Modulatable at megahertz rates.	Red, yellow, and green lamps available (GaAsP, GaAlAs, GaP), numeric indicators available with character size from 0.1 to 0.5 in. (with or without addressing circuitry); individual diodes have yielded 300 ft L at 5 mA dc. Limited MIL spec. qualification tests have been performed.	Discrete diode emitting area is 0.01 to 0.025 in. on a side; integral lenses to enlarge apparent area decrease viewing angle; energy conversion efficiency is not the same for all colors; however, luminous efficiency is; limited multicolor capability; fabrication of large areas is difficult and costly; lack of monolithic techniques for the addressing circuitry and LED.

Table 18-13. Digital Address Flat CRT [18-8]

Features	Current Status	Limitations
Digital addressing; storage mode operation possible; fixed linearity; capability for alphanumeric/graphics/TV raster type formats. Device also capable of multiple as well as single beam scan. Beam selection scheme reduces number of address circuits. Color capability.	512 X 512 element, 80 lines per in. switching plates with an active area of 7 X 7 in. demonstrated. Low resolution (10 lines per in.), small devices (3 X 5 in. active area) have been made. 512 character displays at 40 dots per in. have been fabricated. 512 X 512 element displays with a 9 X 9 in. viewing area have been made.	It is premature at this time to predict what long term limitations may evolve. Some vibration and environmental data have been gathered and appear favorable. Active areas larger than 9 in. on a side are limited by envelope problems and cathode problems.

18.3.2. The Digitally-Addressed Flat Panel Display CRT. In the past several years, the Northrup Corporation has combined the microchannel plate amplifier with the older, conventional CRT principles to achieve a relatively thin CRT of high sustained brightness. This tube has been assigned the Digisplay® trademark.

Lester Jeffries [18-30] described the Digisplay® as follows*:

> The display, known as the DIGISPLAY, utilizes an areal electron source followed by a series of thin, apertured control plates, which are aligned and act collectively to generate a group of scanning electron beams. The position of the group of beams is determined by the digital addressing signals applied to decoding electrodes that are deposited on control plates.
>
> The display envelope consists of two standard rectangular CRT-type faceplates, sealed to the center glass mounting plate. The area cathode is attached to one side of the mounting plate, and the fuzed stack of six control plates is attached to the opposite side of the mounting plate. The leads for addressing the display are deposited on the mounting plate and come out through the solder glass seal between the front faceplace and the mounting plate. The external dimensions of the tube are 4 in. × 5 in. × 1.5 in. The phosphor connection is brought through the front of the faceplate in order to eliminate voltage breakdown problems between the high voltage on the phosphor and the relatively low voltage on all other parts of the tube.

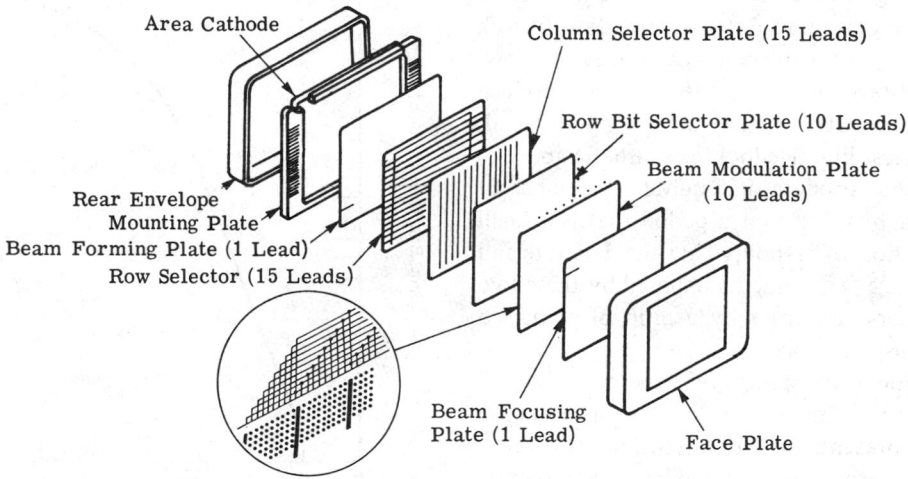

Fig. 18-17. Exploded view of 150 × 150 element digisplay [18-30].

Figure 18-17 is an exploded view of the display tube. The display has 150 × 150 resolution elements over a 2.7 inch by 2.7 inch format. The specific functions of the six apertured plates that form the heart of the device are as follows:

First Plate
Forms electrons from cathode into 150 × 150 beams.

---

*This material has been lightly edited.

Second Plate
Reduces the number of beams to a group 10 beams high by 150 beams wide.

Third Plate
Reduces the number of beams to a 10 X 10 beam group.

Fourth Plate
Reduces the number of beams to 10 co-linear beams.

Fifth Plate
Modulates the 10 beams individually.

Sixth Plate
Controls and adjusts the focus of the electron beams en route to 'the phosphor faceplate.

The six plates are fuzed together prior to tube assembly to maintain permanent hole alignment and to increase their ruggedness.

The tube utilizes a type P-44 phosphor target. A plot of spot brightness as a function of phosphor voltage is given in Figure 18-18. The maximum brightness measured was 6,000 foot-lamberts at a phosphor potential of 19 kilovolts. Total input power to the tube was approximately 10 watts at maximum brightness.

18.3.3. Electrical and Visual Output Storage Tubes for CRTs.* Often, a single display is needed to receive data from a variety of input sources. Some, like low-light-level TV or FLIR sensors, produce data at rates of many megabits per second. Others, like side-looking synthetic aperture radars, produce an equivalent number of data bits, but over a period that is a small fraction of an hour. Thus the data rate and storage requirements imposed by these two sensors can differ by a ratio of perhaps a thousand to one.

Electrical storage tubes permit economical buffering of low-data-rate information for presentation at conventional TV rates. In alternate system modes, other sensors which operate at TV rates may then bypass the storage tube buffer and the video data directly drive a conventional CRT.

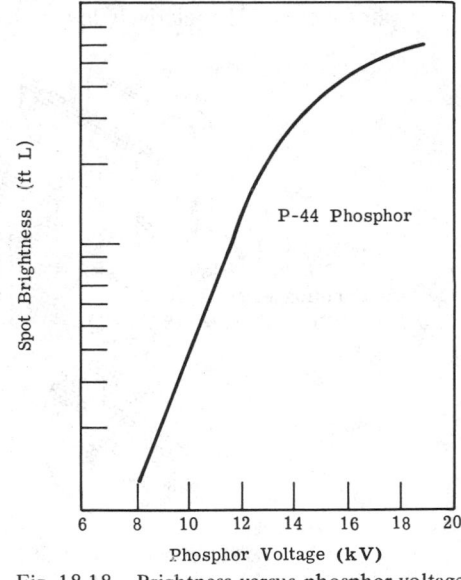

Fig. 18-18.   Brightness versus phosphor voltage [18-30]

Electrical storage tubes are sometimes used to store the more complex data in displays and, by video mixing at the monitor, these data may then be combined with information such as a nonstored alphanumeric overlay. An economical, high-quality presentation with desirable man-machine interactive properties results.

As digital, solid-state memory costs decrease and as more information is in a digital

---

*Abridged from [18-31]

form, the digital scan-conversion memory systems are becoming more attractive. When the latter are used, the analog information is converted to digital format by an A/D converter. A digital memory is used for storage following which the digital information is converted back to analog form for display. These techniques have been used primarily in the lower information density displays and where real-time interaction between the operator and machine is critical and cost is secondary.

For a large class of high-density, high-resolution systems, the economy and performance of analog scan converters and analog DVSTs still give them a considerable lead and a firm market for a multitude of applications.

In some CR storage tubes, storage takes place directly on the viewing surface. The cathodochromic or dark trace storage CRT uses a viewing surface whose optical properties are changed following electron beam excitation, and this surface is viewed either by transmitted or reflected light.

The substrate in these 12 to 16 in. diameter tubes is green backlighted for viewing. Writing speed is on the order of 10,000 in. $\sec^{-1}$ and erasure is accomplished by an internal heater which raises the substrate temperature for 5 to 10 sec. A high resolution display with good gray scale rendition may be presented.

In other devices, such as those used in some computer-graphics terminals, storage takes place at the light emitting phosphor [18-36]. The display is bistable, e.g., no halftones. A writing electron beam establishes a high resolution electron charge pattern on the phosphor which then modulates a lower velocity electron beam that is continually flooding the entire viewing surface. Luminance is adequate for a controlled office or laboratory environment. Writing speed is about 5000 in. $\sec^{-1}$ in these displays, which can present about 10 characters per inch.

When storage with higher light output and gray scale capability is required, another class of DVST is available [18-37] in which storage is accomplished near the viewing surface.

These tubes utilize a low-voltage electrostatic charge on an insulator supported by a metal mesh, located several tenths of an inch away from the light-emitting phosphor or viewing screen which is at a high voltage. An electrostatic charge, representing the image to be sorted, is deposited on the insulator by a focused high-velocity electron beam. This charge, located at the individual mesh webs, then modulates a lower-velocity electron beam which is flooding the entire mesh surface. The more positively-charged mesh elements allow the approaching low-velocity flood electrons to pass through those mesh openings. The electrons are then accelerated to strike the viewing phosphor at high energy. Less positively charged elemental areas restrict the quantity of electrons, resulting in lower light output. This design results in a transfer characteristic that permits a moderate gray scale capability, depending on the local charge pattern.

The integrating characteristics of the storage tube surface are especially important in bringing signals or a picture up out of noise. For this function, successive lines or frames of information can be cummulatively added to each other. The signal is spatially repetitive while the random noise is not; thus the signal-to-noise builds rapidly and a clearer picture is obtained. To erase, or return the charge image to the unwritten, or more negative, state, a flood of electrons is caused to land at a low velocity. This is accomplished by the metal mesh driving positive; the insulator follows, and the flood electrons are able to land at low velocity, charging the insulator in a negative direction until landing stops. When the metal mesh is returned to its normal potential, the more negative charge creates a black or erased condition.

One specialized tube, the projection Tonotron ® DVST is designed for projection applications and provides a 4-in. diameter image with a luminance of 10,000 ft L.

The high luminance in the output of this tube results primarily from continual phosphor excitation, rather than the usual repetitive, but short, high peak-loading of more common projection CRTs. Because of this, operating life of this projection storage tube is not adversely affected by operating at such high output liminance levels which tend to cause short lifetimes in conventional CRTs.

A further advancement in DVSTs allows a more versatile usage. An example is the Multi-Mode Tonotron ® [18-31] which utilizes a storage surface that enables writing and erasing by two different mechanisms. This is achieved with multiple electron guns operating at various beam energies.

The high-performance, economical, analog scan-converter may be used as a link in a display system. Electrical-output scan-converter storage tubes [18-32] are used in display terminals as buffer-storage elements and for video storage, scan conversion, and integration. A display of one format is often converted to television scan rates because of the availability of such display monitors in many packaging configurations and the ease with which the converted TV signal may be transmitted to one or many display stations.

In double-ended storage tubes, the memory target is at the tube center with writing and reading guns on either end. This arrangement permits simultaneous writing and reading. Several types of memory targets are available: (1) a transmission grid-modulation type similar to the DVSTs described previously. With it, the read beam electrons are modulated as they scan and pass through the central mesh. Since both the write and read beams strike the signal output mesh, RF intensity modulation of the read beam, and consequent demodulation at the tube signal output electrode, are used to separate the currents coming from the write and read beams. (2) The EBIC, or electron bombardment induced conductivity type, which uses a very high-energy writing-beam to discharge or write on a thin, continuous insulator with a metallized backing. (3) The membrane scan-converter target [18-33] utilizes a thin membrane target so that a charge deposited on the write-side transfers to the read-side. The writing electron-beam energy is low; thus, writing electrons do not get into the read-output, and special signal-separation techniques are not necessary. Limiting resolution is 2000 TV lines per target diameter and writing speeds are up to 8 μsec per target diameter.

In double-ended scan converters, the read collector is used for the output electrode. The transfer characteristic is shown in Reference [18-33]. Selective erasure is readily achieved by adjusting the beam energy of one of the guns, permitting displays similar to those on the Multi-Mode Tonotron ® as well as "passing scene" and gradual fade types of display. These tubes are generally used in a nondestructive readout mode where the reading action does not remove written information.

For some applications it is desirable that the reading process also remove the written charge; this is called "destructive read."

Another destructive, erase, double-ended scan converter uses a silicon diode array target [18-34]. This is similar to a target used in a sensitive TV camera tube and the target exhibits electron gain.

Writing speeds of about 1.3 nsec per target diameter are achieved and permit applicability to transient recording of signals up to 1 GHz in bandwidth. The target can retain information on a signal for as long as 100 msec. Readout is an orthogonal TV scan with the transient intersections for each scan line recorded and processed. Alternately, the trace can be read out and displayed on a TV monitor.

Another variation of scan converter technology is the single-ended scan converter. A single-ended scan converter is a tube having only one electron gun which is then time-shared between accepting information for writing and the presentation of it during reading. Most use a solid target with an insulating island. Its written charge image controls the proportion of read beam allowed to land on the adjacent conducting surface which then constitutes the signal current. Complete 2000 TV line limiting resolution scan conversion memory units using these tubes were readily available for 3000 to 4000 dollars in 1974 [18-38].

A very high-resolution storage tube [18-39], of which only a few have been built, is based upon RCA's 4½ in. return beam vidicon. The target is similar to other silicon-target storage tubes except that 7000 silicon-dioxide stripes are contained in the 2 in. X 2 in. target.

The limiting resolution of this storage tube is about four times that of the more popular silicon storage tubes providing about sixteen times the information capacity. Storage of 50 to 100 pages of printed text is feasible.

Resolution, shading, gray-scale, write-time, erase-time, repeatability, linearity, and storage time are the tradeoff parameters in selection of a video memory.

18.3.4 Plasma Panels.

Introduction. Gas discharge tube displays in the form of banks of neon or argon indicator lamps have long been used. The main problem of their large scale use in displays has been associated with the fabrication, wiring, and driving of such lamps. Plasma panels are a logical outgrowth to achieve the functions of large arrays of lamps without the conventional difficulties. Plasma panels can be divided into two basic forms: the ac type, typified by the Owens-Illinois "Digivue" ®* panel; and the dc type, represented by the Burroughs' "Self Scan" ®** panel.

The two technologies are first briefly described and compared, then each is examined in a small amount of detail. H. G. Slottow [18-35] compares the two technologies as follows†:

> The term plasma display was first used to describe a gas discharge device in which the discharges are insulated from the exciting electrodes by dielectric surfaces. Although these surfaces prevent the development of continuous discharges, this structure can support a sequence of pulsed AC discharges in response to an alternating voltage. In recent years the term plasma display has also been used to describe devices in which the electrodes are immersed in the gas. This kind of structure can support continuous or pulsed discharges, but in most applications the discharges are unidirectional. To distinguish between the two kinds of structure we refer to the first class as AC plasma displays and to the second class as DC plasma displays. Because they are simpler conceptually, we begin our discussion with the DC plasma displays.

---

*Registered trademark of the Owens-Illinois.
**Registered trademark of the Burroughs Corporation.
 †This material has been lightly edited.

DC Plasma Displays

In one DC structure two sets of parallel electrodes oppose each other in the gas with one set directed orthogonally to the other. An aperture plate, placed between the electrode sets, confines the discharges. Appropriate addressing voltages on two intersecting electrodes cause an electrical breakdown and emission of light at the intersection; an addressing voltage on only one electrode is too small to ignite the discharge. In this matrix arrangement, the gas discharge cell, which is sufficiently nonlinear for the purpose, functions as a two input "and" circuit. This device is usually operated one row (or column) at a time with signals on the opposing electrodes determining which cells in the row (or column) will be on. Since the duty cycle is these devices becomes smaller as the device becomes larger, the peak currents limit the array size.

Except for a difference in geometry, segmented gas discharge displays are also operated and connected in the same way. Corresponding segments are connected together to define a "row", while the segment electrode that opposes all the segments of a character define a "column".

Information storage can be added to this structure in two ways. One technique depends on exciting the panel with voltage pulses that exceed the normal ignition voltage but are too narrow to allow new discharges to mature. A cell that is already "on", however, retains sufficient ionization products during the interval between pulses to insure a discharge at the next pulse. A second technique requires a current limiting resistance at each cell in series with the discharge. The combination of resistance and gas cell provides a bistable luminous element in which a voltage that is too small to ignite a discharge is more than large enough to maintain one. In both cases, the memory is associated with the volume properties of the discharge.

Although the matrix structure reduces the required number of connections to the panel, it would be desirable economically to reduce the number further. An important nonstorage DC discharge device, the Self Scan® Panel, accomplishes this objective at the cost of increased panel complexity. In this device, a set of scanning discharges, only slightly visible to the viewer, is provided by a multiphase (usually three phase) driver. The ignition of a scanning discharge requires both adequate voltage and the volume priming from the preceding discharge. A separate set of electrodes transfers the scanning discharges to the front of the panel according to the information content. These devices do not have internal memory.

Because the intensity of these nonstorage panels can be controlled by current or pulse width, provision of gray scale is not difficult, and their application to new kinds of television displays is being widely studied. Multi-color, important for television, as well as for some information display applications, can be provided by phosphors which are excited by the ultraviolet components of gas discharges.

AC Plasma Displays

In the AC plasma display, as in the DC panel, two sets of electrodes oppose each other across a discharge gap and are directed orthogonally to one another. However, at each discharge site, the dielectric surfaces that isolate the electrodes from the gas define two capacitances which are in series with discharge. An alternating voltage applied across the two electrode sets is too small by itself to ignite discharges. However, a pair of write voltages applied across two selected electrodes will ignite a pulsed discharge that extinguishes as ions and electrons flow to the dielectric surfaces and charge the capacitors. This charge augments the applied voltage on the next half cycle to ignite a second discharge, which then charges the capacitors in preparation for the third discharge. This sequence of pulsed discharges which characterizes the "on" state of a cell, terminates when an erase signal on the two intersecting electrodes produces a controlled discharge that reduces the charge on the series capacitance below the minimum required for ignition.

In the Digivue ® version of the AC plasma panel no aperture plate is required in the gas. The discharge is confined instead by the electric field and by the pressure of the gas. This panel is self-registering with discharges occurring at the electrode intersection. The simplicity of the structure has made possible the development of commercial devices with over 250,000 discharge sites at a density of 3,600 per square inch. Panels with over one million cells at a density of 6,889 per square inch have been made experimentally.

The basic structure of the AC panel has also been realized in segmented form for numeric displays but in this case the device is excited beyond the bistable range and the surface memory is used to inhibit spurious effects of half select signals.

This AC plasma panel device is also a matrix array, but here each discharge site functions not only as a two-input "and" circuit, but also as a bistable storage element. The memory depends, in this case, on the storage of electric charge on dielectric surfaces. Volume memory effects, however, are present and these are now being exploited in extensions of the technique. One departure provides additional electrodes spaced between the usual display electrodes. The response to three phase drive signals is a shifting image. The condition for a new discharge at a site is the application of the appropriate voltage and the presence of an adjacent discharge. Programmed charge to single phase causes the image to remain stationary. The benefit is reduced circuitry at a cost of some increase in panel complexity and less of random access. Volume effects are also exploited in a dual line panel in which electrode pairs are interconnected to perform a second level of logic in a random access panel, and also in the development of light pen techniques.

The AC plasma display is also being studied for application to television and it, too, can provide multicolor through excitation of phosphors.

The plasma display technology is being developed in many sizes and for many applications. For large graphic displays it is the only technology seriously challenging the cathode ray tube.

18.3.5. Liquid Crystals. For many years, a class of cholesteric organic material has generated research interest because of the crystal-like properties of the liquid materials. The chemistry and properties of liquid crystals are treated extensively and thoroughly by Creagh [18-40]. They have been applied to a number of problems with useful, but not dramatic, results.

In 1968 and 1970, however, Heilmeier, Zanoni and Barton, [18-29] and [18-41], showed in considerable detail that, by electrically exciting the materials, a controlled amount of optical scattering could be produced that made the liquid crystals seem brighter and darker than surrounding material. From this early work, a variety of technologies have emerged, from wristwatch and pocket computer displays, to TV compatible displays and projection systems.

The work of Kobayashi [18-42] is singled out to illustrate just one class of data on but one class of liquid crystals operated in but one mode. The amount of data needed to make a choice depends upon temperature, power efficiency, color, dynamic range needed, and form of crystalline material and form of excitation. While Kobayashi considered dynamic scattering that is a result of the turbulence caused by passing a current through the liquid crystals, a related paper by D. Jones and S. Lu [18-43], considered field-effect liquid crystalline devices which employed twisted nematic crystals. They observed the resultant changes in polarization.

Although this work was an advance, the slow response made liquid crystals suitable only for low-information-rate displays. Lipton and Koda [18-44] reviewed this problem at the 1973 SID meeting and indicated that if the technology was to be useful for such high-rate application as TV or other dynamic imaging applications, the problems of cross-talk between elements and the slow response time of the elements must be overcome. For an excellent current review, see [18-45].

Two years later Lipton, Meyer and Massetti [18-46] again addressed SID. The amount of progress can be judged from their paper and the two year span from the 1973 report.

> Live television has been demonstrated on a liquid crystal display. The display uses a transistor switch at each matrix point in order to solve the traditional liquid crystal problems of crosstalk between elements, slow response time and limited multiplexing capability.

> The use of a transistor at each matrix point of the display represents the most general approach towards solving these problems. When combined with a suitable storage capacitor, such a transistor-capacitor combination can act as a sample and hold circuit to allow rapid addressing of a selected element. This approach has previously been shown capable of allowing TV rate operation of liquid crystal display. This paper reports on a further development and refinement of these concepts whereby a ten-thousand element transistor array is fabricated as a 1″ × 1″ monolithic array using silicon-on-sapphire (SOS) technology. The particular advantages of the SOS approach are discussed as are the general fabrication procedures and the operating characteristics of a working display.

Results

SOS arrays have been processed and completed displays fabricated. Live television has been demonstrated with characteristics as follows:

    Contrast Ratio: 18:1 (full on to full off)
    Response Time: 100 msec risetime to 90%
                    160 msec decay to 10%

                    Under special conditions and with reduced grey scale, decay times of 40 msec have been obtained.

High Ambient Operation: No fading apparent under 10,000 ft L illumination. Contrast ratio same as under 100 ft L.

    Shades of Grey: Approximately 6

It is clear that there are picture defects in the sample reported at the SID meeting in March 1975. By August 1975 the picture from a defect-free circuit and crystal assembly typified later production controls. (See Figure 18-15.) The LCD approach taken by Ernstoff has been particularly successful [18-48, 18-49, 18-50] and is being vigorously pursued by Hughes Aircraft Co. [18-51].

18.3.6. Light Emitting Diodes. The application of electroluminescent materials in the form of light-emitting diodes (LEDs) is thoroughly reviewed by Nuese, Kressel and Ladany (1972).

Loebner (1973) has reviewed the properties of electroluminescent materials. That article particularly treats the colors available from various materials rather thoroughly.

18.3.7. The Relative Merits of LEDs and LCDs. The strong potential of the liquid crystal display (LCD) and light-emitting diode (LED) technologies for display purposes needs a careful comparison when making a choice for display design. Such an analysis has been prepared by L. A. Goodman [18-47] in 1975. His paper is reproduced here in strongly abridged form.

Visual Appearance

Since LEDs emit radiation, they are more visible in dim ambient light and less visible in bright ambient illumination ... the luminescence intensity is proportional to the current passing through the diode. The visibility of an LED for a given power or current input is dependent primarily upon two factors, the external quantum efficiency of the radiation process (number of emitted photons per current flow through the diode), and the relative sensitivity of the human eye to the wavelength of the emitted radiation (the luminous efficiency in lumens/watt). The luminous power efficiency of the LED (the ratio of the light output in lumens to the input power in watts) is proportional to the product of the external quantum efficiency and the luminous efficiency. Table 2-28 in Chapter 2 lists performance data for $GaAs_{1-x}P_x$ diodes grown either by the vapor-phase epitaxy or liquid-phase techniques.

$GaAs_{0.6}P_{0.4}$ diodes are more efficient at high currents than they are at low currents because of the occurrence of nonradiative recombination that is less important at high current levels than it is at low current levels. Due to this fact, and the integration capability of the eye, diodes operated in a pulsed or strobed mode require less average current than diodes operated in a nonpulsed manner to produce the same time-averaged luminous output. Whether operated in a pulsed or nonpulsed mode, the upper limit on the light output is set by the luminous power efficiency and the maximum current that can be passed through the diode.

One of the important factors limiting the efficiency of LEDs is the mismatch in index of refraction between air (n=1) and the luminescent material (n≅3.6). As a result of this mismatch, the critical angle for total internal reflection is quite small (~16°) so that almost all of the randomly emitted junction radiation is reflected back into the semiconductor. This is particularly harmful with the common LEDs made from $GaAs_{0.6}P_{0.4}$ material because the energy of the radiation is close to the bandgap of the material and the totally internally reflected radiation is strongly absorbed by the semiconductor.

When a diode is encapsulated in a plastic lens, the total internal reflection is reduced because of the lens' index of refraction (n≅1.8 for epoxy) and the shape of the lens. A simple hemispherical dome lens improves the external efficiency by a factor of approximately 2 to 3 [18-52].

The plastic encapsulation can also be used to shape the angular distribution of the emitted radiation, as shown in Figure 18-19.

When GaP substrate is used instead of the standard GaAs substrate, it is possible to improve the external quantum efficiency, not only by utilization of plastic encapsulation, but also by making the undersurface of the substrate optically reflecting [18-53]. Plastic encapsulation and usage of a GaP substrate improves the efficiencies of $GaAs_{1-x}P_x$ LEDs above those listed in Table 2-28.

Because LCDs are passive devices which modulate the passage of light, they are at least as legible in bright ambient illumination as they are in dim ambient lighting. Furthermore, their legibility is strongly influenced by the spatial distribution of the lighting.

Typical transmitted light curves are shown in Figure 18-20 for a dynamic scattering device. As the data demonstrate, high contrast ratios can be obtained with a collimated source. The curves also indicate that the contrast ratio, as a function of voltage, is not monotonically increasing for all angles. Indeed, for some angles, the inverse is true, for backlit dynamic scattering displays, the

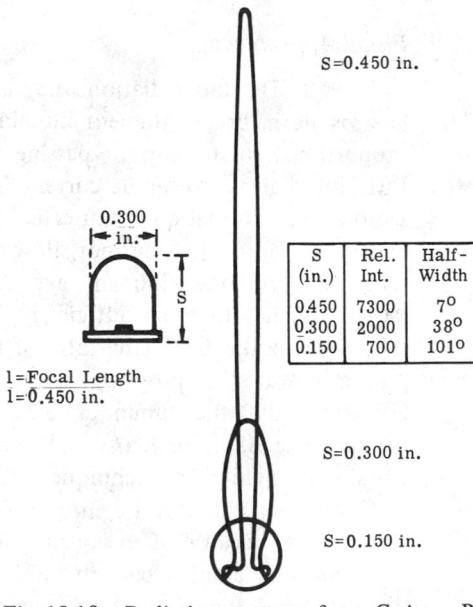

S (in.)	Rel. Int.	Half-Width
0.450	7300	7°
0.300	2000	38°
0.150	700	101°

l=Focal Length
l=0.450 in.

Fig. 18-19. Radiation patterns for a $GaAs_{1 \mp x}P_x$ diode encapsulated with a plastic lens [18-47].

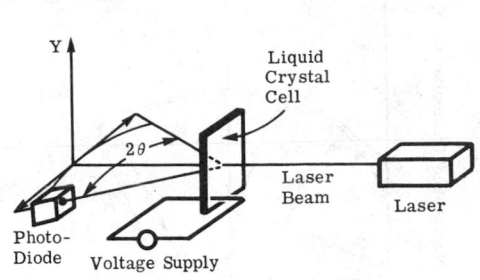

(a) Schematic illustration of the measuring apparatus for transmissive dynamic scattering cells.

(b) Scattering light intensity as a function of viewing angle and voltage.

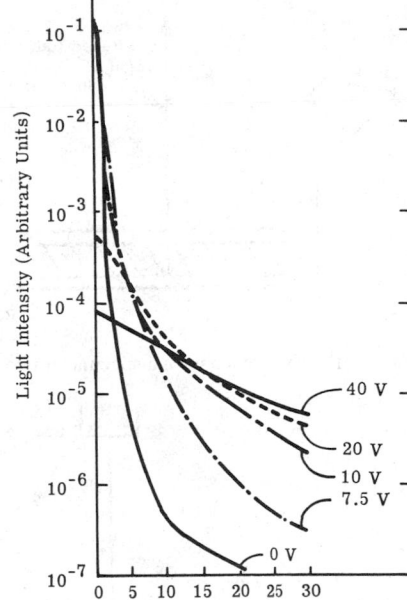

Fig. 18-20. Typical transmitted light curves [18-47].

strong angular dependence of the contrast ratio curves is not a serious problem, because it is relatively easy to use a quasi-collimated source so situated that the contrast ratio for normal viewing angles increases with voltage.

However, the proper orientation of the illuminating source, the display, and the observer is not as easy to arrange with reflective dynamic scattering displays in ambient lighting. Because dynamic scattering is a forward-scattering phenomenon, it is common to use a specular mirror back electrode to reflect the scattered light back to the observer when the illumination source is on the same side of the display as the observer. (See Figure 18-21.) The observer should be at an appropriate position so that only the reflected, scattered light is within his field of view and none of the specularly reflected, unscattered light is observable by him.

Although the contrast ratio versus voltage curves for twisted nematic displays are strongly dependent upon the viewing angle as shown in Figure 18-22, the glare problem in reflective applications is minimal. Because the twisted nematic effect does not induce the scattering of light, a diffuse reflector can be used behind the display instead of a specular reflector. Consequently, the observer on the same side of the display as the illumination can see the voltage-induced change in the light transmission of the device without the presence of reflective glare from the back reflector. It should be noted from the curves in Figure 18-22 that, with a sufficiently high voltage,

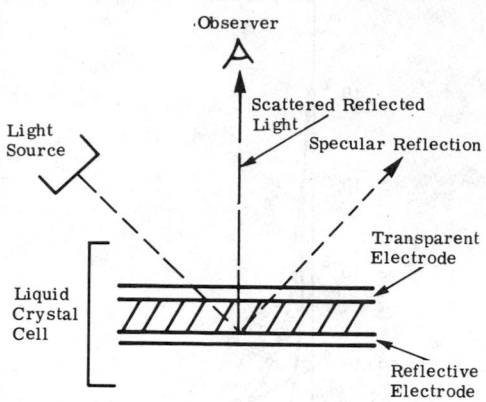

(a) Reflective device with a specular mirror.

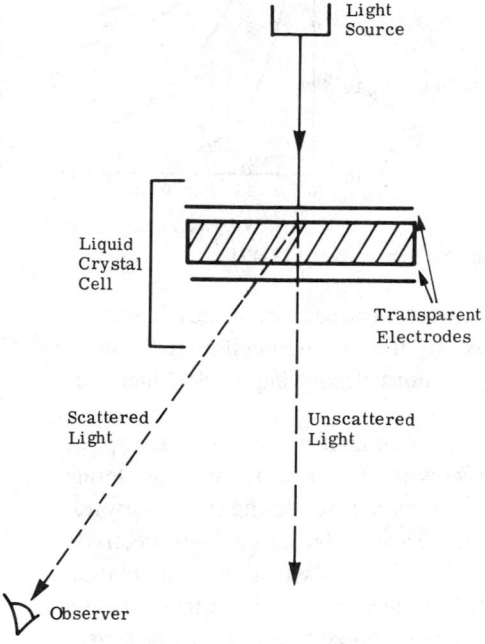

(b) Transmissive Device

Fig. 18-21. Schematic illustrations of the viewing conditions for dynamic scattering cells [18-47].

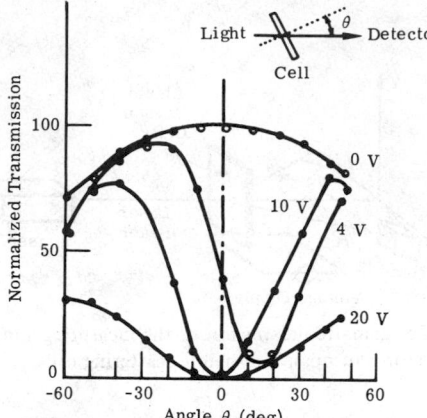

Fig. 18-22. Light transmission of a twisted nematic cell with crossed polarizers as a function of turning angle for various values of applied field at 5 kHz (cell thickness is 30 μm) [18-47].

the transmitted light curves are fairly isotropic with viewing angle and that the viewing cone can be as much as ±50°.

There is no intrinsic wavelength dependence in the visible for either of the effects; however, color can be introduced into both effects by external means. With dynamic scattering, either dichroic mirrors or colored filters can be used if so desired. For twisted nematic displays, colored polarizers can be utilized instead of neutral polarizers.

Power Dissipation

The power dissipation in an LED is the product of the voltage drop across the diode and the current passing through it.

It is not possible to give a precise value for the power dissipation of LEDs because of the wide range of possible uses, but typical power densities range between $0.1 - 10$ W/cm^2 for 2 V DC across the diode. For example, a typical power dissipation for a segment in a calculator display with an integral lens is $0.6 - 0.7$ mW; whereas, the power dissipation of a segment in an LED watch display without a lens is about 3 mW [18-54].

With LCDs, the power dissipated by the liquid crystal display itself is very small. For devices using the dynamic scattering effect, typically

$0.1 - 1$ mW/cm^2 is used at a drive voltage of 15 volts and at 30 Hz. The power dissipation is due to the fact that, for satisfactory operation, dynamic scattering devices require that the resistivity of the fluid be in the 10^8 to 10^9 ohm-cm range. The per-segment power dissipation of an eight-digit, seven-segment calculator using a dynamic scattering display is about 0.02 mW.

By the nature of the effect, twisted nematic displays do not need any resistive current flow in the material when the voltage is applied. Therefore, the material resistivity can be as high as obtainable by chemical purification and is usually above $10^{10} - 10^{11}$ ohm-cm. Hence, at the normal operating frequency of 30 Hz, the power dissipation is entirely capacitive. With an applied voltage of 5 volts, the capacitive power dissipation is about 5μW/cm^2.

Response Times

The rise and decay times for LEDs are primarily controlled by electronic processes and are very fast. Typically, the response times are in the 10-1000 nsec range.

The same is not true for LCDs. The transient response of the electro-optic effects are determined by the motion of the bulk fluid; consequently, the speeds are in the msec to sec regime. The rise is proportional to the fluid viscosity, the square of the thickness, and approximately inversely proportional to the difference between the square of the applied voltage and the threshold voltage squared. The decay time is proportional to the viscosity and the square of the liquid crystal layer thickness. Because of the thickness dependence of the speed, it is highly desirable to keep the fluid thickness as small as technically feasible. In practice, typical thickness values are on the order of 10-20 μm.

At 20°C with an applied voltage of 10 V_{rms} and for a 12 μm thick layer, rise times on the order of 10 msec have been observed with twisted nematic devices and 200 to 300 msec for dynamic scattering displays. For the same value of applied voltage, the much lower threshold voltage for the twisted nematic effect as compared to the dynamic scattering mode results in a significantly faster rise time. This difference is not true for the decay times where, for the same thickness and temperature given above, the decay times for both effects are on the order of 100 to 500 msec.

It is difficult to give exact numbers for the response times not only because of the expected variations from one material to another in the threshold voltage and other material dependent parameters, but also because the response times are angle-dependent and can vary by as much as a factor of three depending upon the angular relationship between the display, the light source, and the observer [18-55, 18-56].

Temperature Dependence

The device parameters are all temperature dependent, but not enough to significantly degrade device performance.

The temperature dependencies of the response times are not noticeable because the variation in transient response is not very large with temperature

[18-57] and, more importantly, the response times are so short that the temperature dependence is not detectable by the eye.

At constant current, the light output of a $GaAs_{0.6}P_{0.4}$ LED decreases at a rate of approximately 1%/°C between 0°C and 50°C [18-58], which is unimportant in most display applications.

The last relevant parameter is the forward voltage temperature coefficient of the current-voltage characteristic. The change with temperature is about minus 2 mV/°C, but this is also unimportant because LEDs are usually driven a constant current source.

Unlike LEDs, whose operation is basically unaffected by ±25°C excursions about 20°C, many of the electro-optic properties of the LCDs are distinctly affected by the same temperature variations. Indeed, until about 5 years ago, there were very few materials which exhibited the mesophase at 20°C. However, today there are many liquid crystal materials which possess a mesophase range of at least 70°C with the low end of the range at or below 0°C.

Of more significance is the fact that the viscosity, conductivity, and response times are exponentially dependent on temperature with an activation energy of 0.5 to 1.0 eV. In dynamic scattering devices, this means that both the power dissipation and the cutoff frequency are strongly temperature dependent. Furthermore, the speed of operation of devices must be carefully considered not only at 20°C but also at lower temperatures where the speed of response becomes somewhat slow [18-59, 18-60].

The threshold voltage for dynamic scattering is quite constant with temperature, but for the twisted nematic effect, the threshold voltage decreases at a rate of about 0.5%/°C as long as the operating temperature is not too close (within 10°C) to the critical temperature separating the mesophase and the isotropic phase.

Circuit Compatibility

LEDs are high-current, low-voltage devices which are very compatible with discrete and integrated bipolar transistors and discrete MOS transistors. However, they are not readily driven by MOS integrated circuits (ICs), particularly CMOS ICs, due to the current required by the LEDs. It is easy to matrix address LED displays.

The electrical properties of LCDs and LEDs are very different. They (LCDs) are low-to-medium voltage devices with low current requirements. Consequently, they can be easily driven by CMOS ICs. In order to obtain long operating life for LCD displays, it is necessary to excite them with bipolar waveforms with the dc component being less than 0.2 V. The relatively slow speed of LCDs limits the rate at which information in an LCD can be changed to less than a few times per second.

LCDs do not possess a strongly asymmetric current-voltage characteristic like LEDs. Rather, in the useful range of operation, they can be modeled as leaky capacitors. For dynamic scattering devices, the dielectric relaxation time is about 1 msec, whereas, for field effect devices, the dielectric relaxation time can be in excess of 100 msec.

The electro-optic transfer function of an LCD changes rapidly, for applied voltages greater than the threshold voltage and is symmetric about zero volts. The nonlinearity in the electro-optic transfer function does permit some multiplexing of LCDs, but to nowhere the degree that is feasible with LEDs. There are two prime factors which affect the multiplying capability of LCDs. One condition is that the applied voltage signals to the matrix coincide so that only certain specified matrix elements turn on and no others. In addition, it has been established that, unlike LEDs, both field effect [18-61] and dynamic scattering [18-62] devices respond to the driving signal in a root-mean-square fashion. This statement is true as long as the response times are much longer than the duration of the scanning pulse, which is typically a few msec or less. Because of the coincidence requirement, the rms value of the scanned waveform on a selected element is less than three times the threshold voltage and decreases rapidly as the number of the digits to be multiplexed and increases [18-61], 18-62]. Under these driving conditions, the contrast ratio is very viewing angle dependent, as indicated by the data in Figures 18-20 and 18-22.

Multiplexing of both dynamic scattering and field effect devices is further complicated by the temperature dependence of some of their properties. In the field effect case, the threshold voltage varies at a rate of about $0.5\%/°C$, which can be significant since it is desirable to maintain a small constant voltage difference between the threshold voltage and the bias level across the non-selected elements. The threshold voltage for the dynamic scattering effect is not as temperature sensitive as for the twisted nematic effect, but the cutoff frequency varies quite rapidly with temperature. Since the dynamic scattering effect only occurs for driving signals whose frequencies are less than the cutoff frequency, the temperature dependence of the cutoff frequency must be taken into account in selecting the rate at which the multi-element display is scanned.

The so-called two-frequency approaches can be used to multiplex both dynamic scattering [18-63, 18-64] and twisted nematic devices [18-65], however, they require larger voltage amplitudes and the dissipation of far more power than the single frequency method described above.

For displays using dynamic scattering or the twisted nematic effect, anywhere from 4 to 7 digits have been reported as the maximum addressing capability [18-61, 18-66].

Packaging

The rapid development of LEDs has been greatly assisted by the utilization of integrated circuit processing techniques. The most common LED technology today uses vapor-phase epitaxity of n-type $GaAs_{1-x}P_x$ on either GaAs or GaP substrates. Zn is diffused into the n-type layer to form the p-n junction. This technology is reproducible and provides relatively high quality devices at high yields. Furthermore, the surface of the device through which the Zn diffusion is done is usually passivated with a double layer of SiO_2 and Si_3N_4 which helps to provide long-term device stability. Liquid phase epitaxy can also be used for the creation of the diodes, but it is primarily restricted to the manufacturing of red GaP diodes doped with Zn and O.

Given a specific technology for making the doped p-n junction, a number of approaches can be used to package the LEDs. This variety is prompted by the desire to maximize the display's contrast ratio and minimize the power consumption and cost of the device.

The packaging approaches for segmented numeric displays can be divided into three basic categories, which are the monolithic, hybrid or silver, and light pipe or stretched segment structures. Figure 18-23 is a schematic illustration of all three types.

In the monolithic approach, the whole seven-segment numeric consists of a single chip of $GaAs_{1-x}P_x$. The Zn diffusions are done so as as to define each of the segments. Since this approach uses a lot of the luminescent material, it is usually restricted to small (0.1″ high) numerics with lenses to provide magnification.

In hybrid or silver displays, each segment consists of a discrete bar of the luminescent

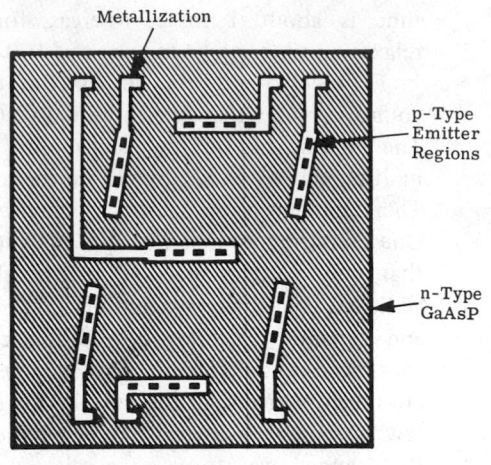

(a) Monolithic structure in which the entire device is a single chip of $GaAs_{1-x}P_x$.

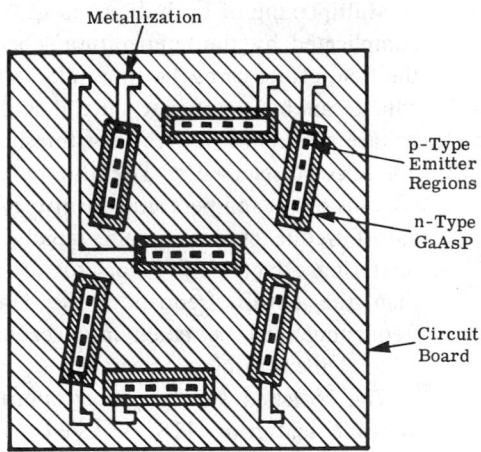

(b) Bar segment device in which a bar of $GaAs_{1-x}P_x$, containing a series of diffused regions, is used to define each segment of the numeric.

Fig. 18-23. Schematic drawings of different types of LED Numeric Packages [18-47].

material with the diffused p-type regions. The bars are mounted in the seven-segment layout on some low cost substrate such as a circuit board. This approach is utilized in displays ranging in numeric height from 0.1″ to 0.3″.

There are different variations of the light pipe method [18-52]. One of these is illustrated in Figures 18-23(c) and 18-23(d) however, they all involve the same concept. A single light-emitting chip is placed at the bottom of a cavity whose top surface is in the shape of a rectangular segment. By means of the appropriate passive optics, the radiation emitted by the chip irradiates the top surface of the cavity so that it appears to an observer that the whole segment is uniformly emitting. This technique is favored for large character heights ($\geqslant 0.3''$) because it requires the least raw semiconductor material of the three methods.

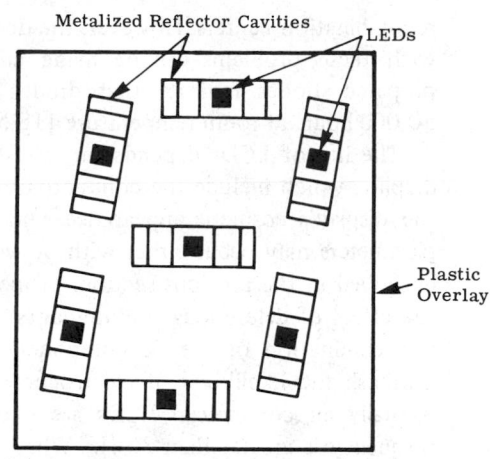

(c) Light pipe structure, top view.

LCD packaging technology is particularly flexible with respect to size variations. The basic LCD is a sandwich structure with the liquid crystal material enclosed between two glass plates that are partially covered with conductive coating. The conductive coating is easily patterned so that the character height, font, and layout can be readily changed. Commercial liquid crystal displays vary in size from several tenths of an inch in height to more than ten inches on a side.*

Moisture, oxygen, and ultraviolet (UV) light can seriously degrade many liquid crystal materials, thereby impairing device performance.

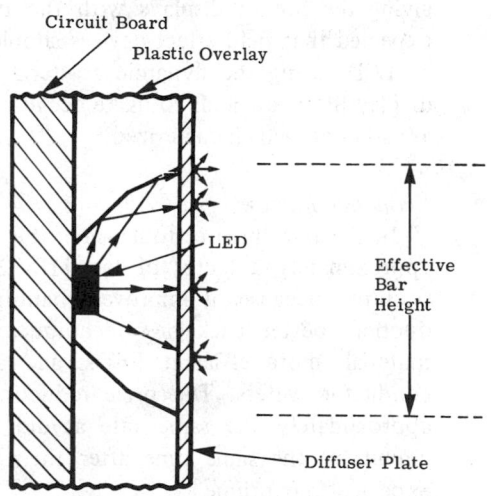

(d) Cross-section of a light pipe structure.

Fig. 18-23 (Continued). Schematic drawings of different types of LED Numeric Packages [18-47].

Appropriate sealing techniques must be used to diminish the effect of moisture and other gases on the properties of the liquid crystal material [18-68].

Reliability

The radiative efficiency of LEDs can decrease with time for several reasons including; surface leakage, diffusion of contaminating impurities such as Cu into the p-n junction region, and the formation of intrinsic nonradiative

---

*For a good review of larger, flat panel displays, see [18-67].

recombination centers. However, the decrease in LED performance associated with these problems can be made fairly small. Test results for properly prepared diodes indicate that diodes can operate continuously for over 50,000 hours at room temperature [18-57].

The life of LCDs depends on the stability of different properties of the display, which include the conductivity, the temperature range of operation, the display's cosmetic appearance, and the response times. Changes in these parameters may occur either with or without the application of a voltage. As described in the previous section, proper packaging is necessary to minimize the effect of deleterious contaminants on the fluid's chemical stability. Also, the elimination of the dc components of the excitation is important to diminish the likelihood of electrochemically-induced failure. Consequently, virtually all commercial LCDs are excited by an ac signal whose driving frequency is greater than 30 Hz. When the proper precautions are taken, the operating life for LCDs using dynamic scattering is cited as being greater than 15,000 hours at 20°C [18-68]. Adequate data is not present in the literature giving the life of displays with the twisted nematic effect; however, it is expected that field effect devices should have operating lives at least as long as LCDs using the dynamic scattering mode. An exact prediction of the display life is difficult to make because many of the degradation mechanisms are still not well characterized.

Economic Factors

In the last three or four years, the cost of LED displays has dropped by approximately a factor of ten [18-52]. This sharp decrease in price has occurred because of improved manufacturing methods, high volume production, better packaging techniques which allow the use of less LED material, more efficient LEDs, and a decrease in the cost of the semiconductor wafers. The price reduction for LED devices has occurred at approximately the same rate as did the price decrease for Si integrated circuits at the same time after their introduction [18-69]. This trend is expected to continue.

LCDs have also decreased in price in the last year to the extent that a typical 3½ digit watch display with 0.2″ high numerics sells for about $5.00 in production quantities. Today, approximately one-third of the cost of an LCD is for the raw materials with the biggest fraction of this being due to the transparent conductive coating, and to the polarizers used in field effect displays. LCD technology is relatively new and unsophisticated compared to LED technology and the selling price of LCDs should continue downward as LCD manufacturers acquire more experience.

A few extra processing steps, whose cost scales with size, are needed to manufacture twisted nematic displays than are necessary for dynamic scattering devices. Also, the polarizers' cost is not negligible. For many applications which only use displays up to 2 or 3 inches in lateral dimension, the extra cost of the additional steps is unimportant in the choice between twisted nematic and dynamic scattering displays. However, for displays on the order of 5-10 inches or more on a side, the monetary factors tend to out-weigh technical considerations and dynamic scattering displays are preferred.

The direct expense of the display, be it an LED or an LCD, is only one part of the total system cost which is the final economic factor that the user must consider. For example, the cost of the driving circuit is also important. LEDs can be easily multiplexed and require about 2 volts for excitation; whereas, LCDs are not easily multiplexed and most need more than 3 or 4 volts excitation. This advantage for LEDs is offset by the fact that they require much more current than LCDs and, as a result, extra bipolar transistors are normally needed to interface the LED with the MOS driving circuit while LCDs are directly MOS compatible.

Table 18-14. Summary of LED and LCD Characteristics [18-47]

| Category | Comments | |
	LEDs	LCDs
Visual appearance	Medium to wide viewing angle. Visible in dim ambient illumination but not as visible in bright ambient. All colors available except blue.	Medium viewing angle. Viewability insensitive to intensity of ambient illumination. All colors available.
Power dissipation	0.1 to 10 W cm^{-2} at 2 V	5 to 1.0 μW cm^{-2} at 3 to 15 V
Response times	10 to 1,000 nsec	10 to 500 msec
Temperature dependence	Unimportant. Operating range −40 to 100°C	The temperature dependence of the operating parameters can be significant. Operating range about 0 to 70°C
Circuit compatibility	High-current, low-voltage devices. Bipolar transistors usually required. Unipolar waveforms adequate for excitation. Easily multiplexed (>20 lines).	Low-current, low-to-medium voltage devices. CMOS IC compatible Bipolar waveforms necessary. Multiplexing is limited (~4 to 8 lines).
Packaging	Semiconductor processing techniques. Different structures are used to maximize light output with a minimum of LED material.	Glass and organic fluid technology. Need for hermiticity. Flexible with respect to size variations.
Reliability	>50,000 hours	>10 to 20,000 hours
Economic	Prices have dropped sharply in last few years. Well along learning curve. Cost is area-sensitive.	New relatively immature technology. Relatively low cost raw materials.

SUMMARY

A short description of the operating principles of LEDs and LCDs has been presented. The properties of both types of devices have been listed in each of eight different categories for characterizing displays. A summary of these properties is presented in Table 18-14.

LEDs have numerous positive features as compared to LCDs. These include having; (1) good viewability in low ambient light levels; (2) lower cost at present for small display; (3) good multiplexing properties; (4) a wide range of operating temperature of operation; (5) high speed; and (6) proven reliability.

On the other hand, LCDs possess the following advantages: (1) enhanced visibility in high ambient lighting; (2) much lower power consumption; (3) direct compatibility with MOS integrated circuitry; (4) good size and format flexibility; and (5) lower cost for larger displays. Since neither device is ideal, the predominance of either technology will depend upon which of the above criteria are more important in the specific application.

18.3.8. The Single Gun Color Display Projector. The need for a ground-based display with characteristics suitable for color images after extensive and sophisticated data processing may be met by an interesting, innovative approach described by William E. Good [18-70]. His paper, in an extensively edited and abridged form, is reproduced here.

The single gun color TV light-valve projector was introduced several years ago [18-71]. It uses a separate xenon light source, a fluid control layer and a projection lens. Optically it is much like a slide or movie projector. Miniature grooves are created on the deformable surface of the control layer by the electrostatic forces from the charge deposited by the scanning electron beam which is modulated by video information. These groove patterns are made visible by use of a "dark field" or schlieren optical system consisting of a set of input slots and output bars. The resulting television picture is imaged on the screen by the projection lens.

Figure 18-24 shows the xenon lamp, sealed light valve, and schlieren-projection lens. The cross-sections of the light body, color filters, and input slots and output bars are shown below the light valve. Green light is passed through the horizontal slots and is controlled by modulating the width of the raster lines themselves. This is done by means of a high frequency carrier applied to the vertical deflection plates and modulated by the green video signal. Magenta (red and blue) light is passed through the vertical slots and is modulated by diffraction gratings created at right angles to the raster lines by velocity modulating the electron spot in the horizontal direction. This is done by applying a 16 MHz (12 MHz for blue) signal to the horizontal deflection plates and modulating it with the red video signal. The grooves created have the proper spacing to diffract the red portion of the spectrum through the output slots while the blue portion is blocked. For the 12 MHz carrier the blue light is passed and the red is blocked. Thus, three simultaneous and superimposed primary color pictures are written with the same electron beam and projected to the screen as a completely registered full color picture.

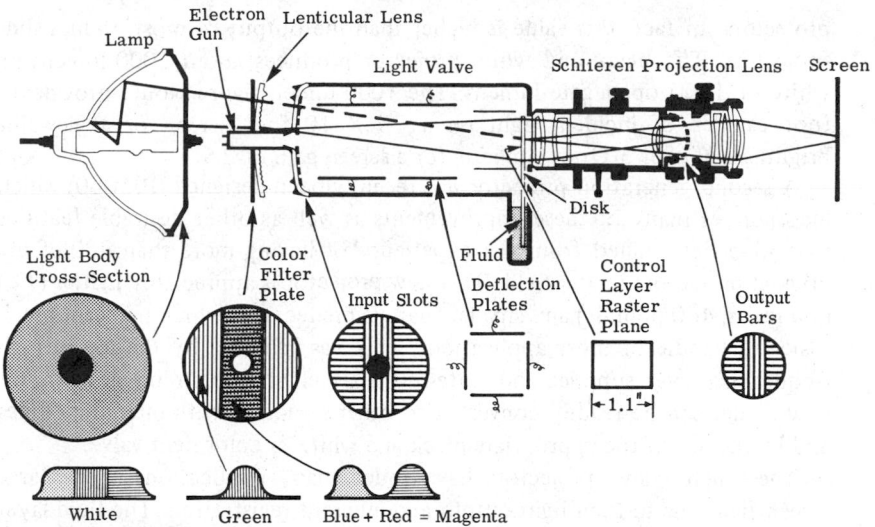

Fig. 18-24. Schematic diagram of the G.E. single-gun color TV light valve assembly [18-70].

The resolution or sharpness of the projected image depends on the quality of the projection optics, the diffraction limit of the output slots, and the definition of the actual patterns created on the fluid surface itself. Recently T. True and W. Bates [18-72] have been able to improve this latter situation by developing modulation techniques that consider the filtering action of the output bars and slots, in the frequency plane, to the sidebands that exist in the first and second order optical spectra due to the video modulation. By a combination of the above technique and the use of higher performance optics, the horizontal resolution of the green image of the single-gun color unit has been increased to over 900 TV lines. The horizontal resolution of the red and blue pictures has been increased to over 500 TV lines, being restricted primarily by the diffraction limit of the output slots. The combined horizontal resolution of a white picture tends to be more of an average and may be as high as 800 TV lines due to the luminance contribution of the green picture. These improvements are particularly helpful in reproducing the output of high quality television sources such as an R.G.B. color TV camera, a computer generated image, or computer generated graphics and alpha-numerics. The quality of an off-the-air broadcast is somewhat better but in this case the bandwidth has already been limited to 270 or 280 TV lines by the encoding and decoding of the NTSC color signal.

The light output of the color projector has more than doubled since the original projector was introduced due to increased lamp power (650 W vs. 500 W), the use of low reflection optical surfaces and optimized fluid writing conditions. Recently, experiments have shown that light output can be increased several fold by further increase in lamp power. Current light output for the single gun color unit is typically around 300 lumens on white or equivalent to 800 open gate lumens as measured for slide and movie projectors. In fact, this value is higher than the output of most 35 mm slide projectors. The black and white projector produces about 1000 lumens on white or 1800 open gate lumens. The 1000 lumen figure would provide 13 foot candles of incident light on a 7½ X 10 foot screen or a viewable brightness of over 30 foot lamberts for a screen gain of 2.5.

A second generation projector has recently been designed (PJ5000) which incorporates many of these improvements as well as other desirable features that were determined from the experience of having more than 300 of the original projectors in the field. This new projector is appreciably lighter (125 pounds vs. 460 pounds) and smaller than its predecessor which makes it much easier to handle in most applications. This was achieved by the use of high frequency power supplies and state-of-the-art electronic circuit design. This new model can be readily converted to other standards with plug-in modules and by the use of the appropriate black and white or color light valve.

These light valve projectors have found many applications in the large screen field due to their light output and inherent registration. The fluid layer has proved to be extremely stable under motion platform accelerations so that the light-valve projector is serving in the flight simulation field at a number of locations both with camera and probe pick-up as well as with computer generated images. Units are fed from a variety of signal sources such as computer generated graphics, TV camera pickup and video tape recorders. The screens are typically 5 to 10 feet wide and usually used in the rear-projection mode. A "wide screen" color TV demonstration showed the feasibility of using this projection system in a mode which is analagous to cinemascope or panavision in the movies. A standard cinemascope ana-morphic lens was added to the projector which increased the picture width by a factor of two. The projector was driven from a video tape which had been recorded from a "squeezed" format cinemascope movie film. No changes were made in the NTSC signal or the original raster format. The resulting 4 foot by 11 foot picture appeared quite acceptable in spite of the limited bandwidth of the NTSC encoded video signal.

18.4. Display Specification and Calibration

Several techniques exist for measuring and specifying display performance, most specify "resolution." These can result in widely different resolution numbers for the same device. The cascading of several devices in series such as a scan converter tube, video amplifier and CRT creates additional complexities in specifying or predicting a total system resolution especially when the resolution of each individual device is specified differently. Therefore, a standard for comparing and combining the respective resolution of several devices would be useful. This resolution standard may be arbitrarily selected but should be meaningful in application to sensor displays and should be capable of convenient and consistent measurement.

There are no "official" standards or procedures for defining or measuring display device performance. Although the EIA undertook a study to set up a new MIL Specification on display and display storage tubes, it stopped after outlining definitions and methods for the storage and reproduction of the electrical signals. It did not address the topic of the optical outputs of such tubes stating that the task was too difficult. (See Ehrlich, et al.) That EIA report is an otherwise useful guide and should be in the hands of every display designer.

The most frequently used specifications for display device resolution are: shrinking raster, limiting television response, and spatial frequency response or MTF. In general all but the last do not allow one to predict operator performance when viewing a display. Nevertheless, the following definitions are common throughout the industry worldwide, and thus are restated below [18-1].

Shrinking Raster Resolution

Shrinking raster resolution is determined by writing a raster of equally spaced lines on the display and reducing or "shrinking" the raster line spacing until the lines are just on the verge of blending together to form an indistinguishable blur. An experienced observer normally determines this flat field condition at about 2 to 5 percent peak-to-peak light intensity variation. Since the energy distribution in a CRT spot is very nearly gaussian, the flat field response factor occurs at a line spacing of approximately 2σ where σ is the spot radius at the 60 percent amplitude of the spot intensity distribution.

Television Resolution (TV Limiting Response)

A television wedge pattern measures spot size by determining the point at which the lines of the wedge are just detectable. The number of TV lines per unit distance is then the number of black and white lines at the point of limiting resolution. The wedge pattern is equivalent to a square wave modulation function, and, therefore, the TV resolution is often referred to as the limiting square wave response. (One needs to be careful to remember that, in television parlance, one cycle of the square wave produces a black interval and a white interval and is considered as *two* television lines.) Assuming a gaussian spot distribution, the limiting square wave response occurs at a television line space of 1.18σ. Thus, there are approximately 1.7 times as many limiting television lines per unit distance as shrinking raster lines for a display with the same spot size.

Modulation Transfer Function (MTF)

O. H. Schade in his sine wave response technique analyzes the display resolution by the use of a sine wave test signal, rather than the square wave signals employed in a TV test pattern or the photographic bar patterns commonly employed in the optical field. The sine wave response test produces a curve of response called the modulation transfer function. This is shown in Figure 18-25. When several devices are cascaded such as a scan converter and CRT, the MTFs of the individual devices are multiplied together to provide the total system MTF. This capability for computing the system MTF from individual device MTFs is a major advantage of using the MTF resolution measurement. Another advantage of the MTF technique is the graphic capability it provides for the determination of the visual acuity limit of a given display system. The MTF response can be related to the other resolution measurements (shrinking raster and television) if a gaussian spot shape is assumed. For example, if a sine wave test signal were set on the display at a half cycle spacing corresponding to the shrinking raster resolution line spacing, the resultant observable modulation on the display would be approximately 29 percent. Table 18-15 can be used to convert from one resolution measurement to another.

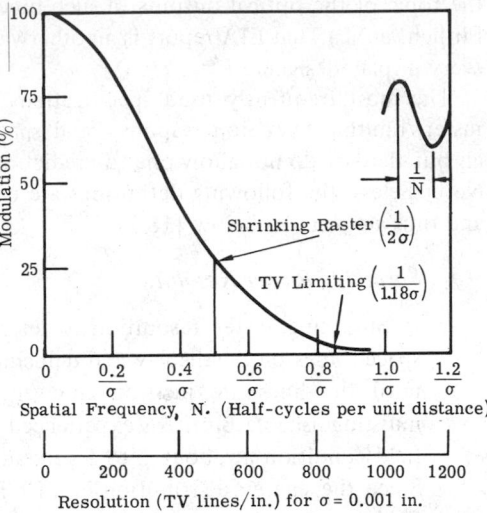

Fig. 18-25 Relative modulation transfer function [18-1].

Sensor Resolution. Sensor resolution is usually defined by the 3 dB response (1.67σ) which is equivalent to TV_{50} display resolution and is 1.2 times the shrinking raster solution. (See Figure 18-25 and Table 18-15.)

It is clear that the definitions and measured quantities defined above *seem* quantitative and useful. In fact these are the "specifications" that are used by the more serious responsible designers and manufacturers. As a matter of fact, Harry L. Snyder [18-3] of Virginia Polytechnic Institute (VPI) has shown that characteristics discussed above are

Table 18-15. Conversion Table for Various Measures of Display Resolution [18-1]

To Convert From To →		TV Limiting	10% MTF	TV_{50}	Shrinking Raster	50% Amplitude	50% MTF	Optical	Equivalent Passband
TV Limiting	1.18σ	—	0.80	0.71	0.59	0.50	0.44	0.42	0.33
10% MTF	1.47σ	1.25	—	0.88	0.74	0.62	0.55	0.52	0.42
TV_{50}(3 dB)	1.67σ	1.4	1.14	—	0.84	0.71	0.63	0.59	0.47
Shrinking raster	2.00σ	1.7	1.36	1.2	—	0.85	0.75	0.71	0.56
50% Amplitude	2.35σ	2.0	1.6	1.4	1.17	—	0.88	0.83	0.66
50% MTF	2.67σ	2.26	1.8	1.6	1.33	1.14	—	0.94	0.75
Optical (1/e)	2.83σ	2.4	1.9	1.7	1.4	1.2	1.06	—	0.80
Equivalent Passband (N_e)	3.54σ	3.0	2.4	2.1	1.77	1.5	1.33	1.25	—

not time-invariant, and Schade [18-3] has said that the properties based upon beam half-power points are only valid if the beam cross-section is Gaussian and independent of beam position on the display face, and these criteria have yet to be found to apply in available hardware!

As a first rough cut the designer may well consider tubes on these factors. But resolution is often stated in terms of spot size (or beam diameter) for some minimal value of beam current, while maximum brightness is specified for a much higher beam current, usually without reference to either beam size or distribution.

One needs to choose a relatively large beam diameter to avoid the black unlighted stripes, between active lines, that cause serious masking and other forms of interference to display observation. One also needs to ensure such beam size is constant over all the display area, or degrades to an *acceptable* degree off axis.

Actually one would like to have a plot of signal-to-noise ratio on the face of the display for a given signal-to-noise input to the display over the entire face of the tube for various signal-to-noise ratios and for various spatial frequencies.

Rosell [18-6] conducted his experiments at spatial frequencies sufficiently low that the MTF of the display did not interfere with his experiments. Snyder [18-3], on the other hand, had great difficulty in maintaining calibrations over even relatively short experiments even after long warm up times.

This handbook cannot list tables of specifications for displays since meaningful specifications have yet to be accepted by the manufacturers and yet to be used by the commercial display designers.

Detailed methods for measuring spot size as a display criterion are well documented. See References [18-74 through 18-78].

Standards for alphanumerics and other symbols are harder to establish, because of the human factors studies required. Shurtleff, at The MITRE Corporation, [18-73] has begun such studies. He developed capital letters and numbers (Figure 18-26) to serve as a reference standard for evaluating the quality of symbols shown on display equipment. Test procedures and criteria were developed to ensure that candidates provide identification performance equivalent to that obtained by Shurtleff which was designed specifically to minimize intersymbol confusion.

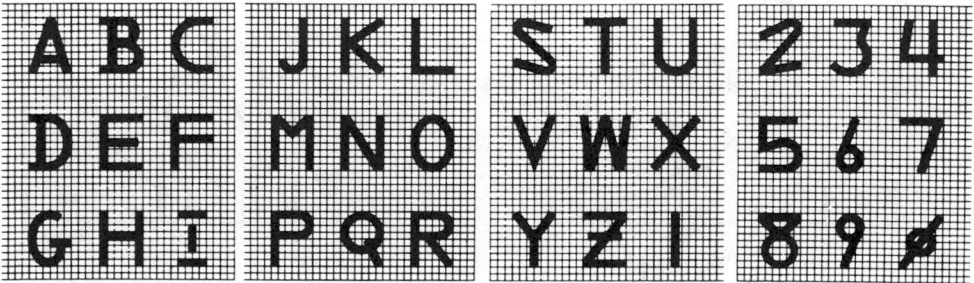

Fig. 18-26. Lincoln/Mitre font (L/M) [18-73].

18.5. Display Tradeoffs [18-1]

Some of the factors influencing the selection of the multi-sensor display system can be considered quantitatively but many factors are as yet subjective judgments. Let us compare two sizes (5 inches and 7 inches) of the Multi-Mode Tonotron® (MMT) to a membrane scan converter tube (SCT) and two sizes of CRTs (8 inches and 11 inches) for the real time display of a high resolution line scan (strip map) sensor. For this sensor, image storage is required.

The Modulation Transfer Function (MTF) of the four display systems and the scan converter tube alone are plotted in Figure 18-27. The 7-inch MMT and 8-inch CRT and SCT have identical MTF curves.

The line scan sensor response alone and the combined sensor and display system response for the four displays are shown in Figure 18-28.

In Figure 18-29 the visual acuity of the eye for each of the four image sizes in addition to the response of the display systems is shown. The intersection of the appropriate visual acuity curve and display curve indicates the maximum usable resolution and is marked with a small circle. Any signal to the left of the intersection has enough display response to be visible to the operators. Conversely, signals to the right of the intersection have insufficient

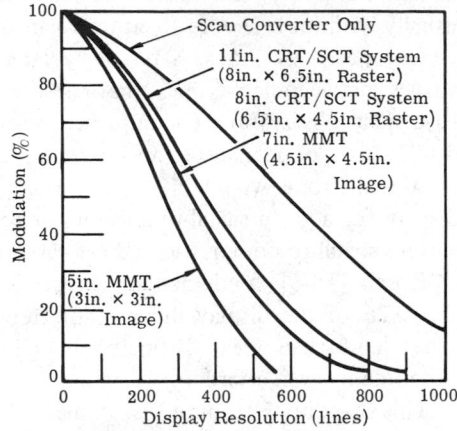

Fig. 18-27.   Display response [18-1].

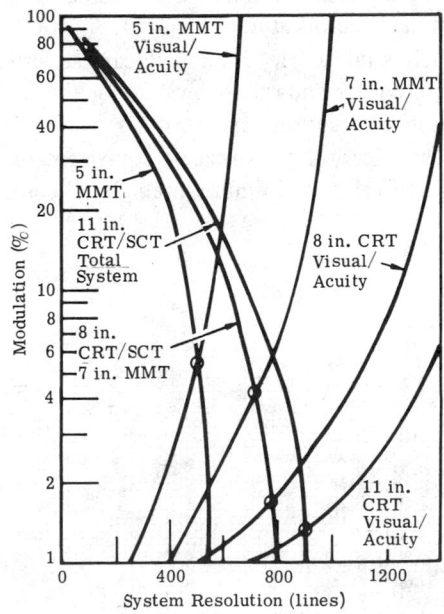

Fig. 18-29.   Display system response and visual acuity for four display systems [18-1].

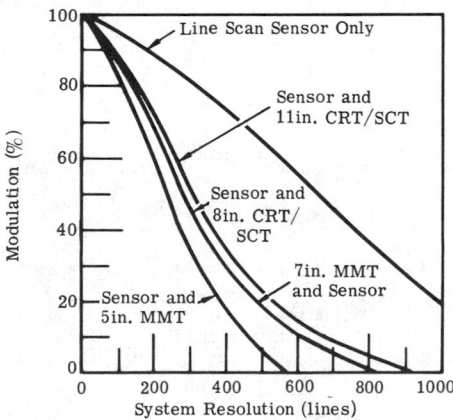

Fig. 18-28.   Overall sensor plus display systems response [18-1].

response to be visible. The maximum usable system resolution for each display approach is shown in Table 18-16.

Table 18-16. Maximum Usable Display System Resolution [18-1]

	Image Size (in.)	Maximum Usable Resolution (TV Lines Limiting)
5 in. MMT	3.0 × 3.0	505
7 in. MMT	4.5 × 4.5	725
Scan Converter plus 8 in. CRT	6.5 × 4.5	795
Scan Converter plus 11 in. CRT	8.0 × 6.5	885

It can be seen from Table 18-16 that the scan converter plus 11-inch CRT has the best system response. While a significant part of maximum usable resolution results from the total number of resolution elements in the display, the size of the display image is also very important. For example, assume that the number of resolution elements of the 5-inch MMT could be increased so that its total system resolution would be the same as that obtained with the scan converter plus 11-inch CRT. Under these conditions, it can be seen from Figure 18-29 that the maximum usable resolution would increase from 505 to only 560 TV lines even though the display system resolution is increased from 550 to 900 TV lines. The small increase in resolution is due to the MMT display size (3 inches by 3 inches) not being large enough for the operator to see more resolution.

Summary of Display Performance. A summary of the overall performance of the five candidate display systems for each primary sensor mode is listed in Table 18-17. Additional tradeoff factors are also included. There is no single display that is optimum or even best in all tactical aircraft systems. Some general statements can be made to summarize the types of system applications in which each of these displays are best suited.

Long Persistence Phosphor CRT appears most applicable where:
— No ambient light falls on the display
— Air-to-air radar or air-to-air IR are the only sensors
— The operator has little time pressure and no competing tasks
— Cost is an overriding consideration
Fast Erase Storage Tube appears most applicable where:
— Radar ground mapping is of poor quality
— Non-stored symbology is not required simultaneously with a stored image
— Operator task does not require target or pattern recognition (image fade, smear, and gray scale not important)
— TV image brightness is more important than resolution or gray scale
— Continuous strip map modes and FLIR modes are not required
Multi-Mode Tonotron® appears most applicable where:
— Radar ground mapping is of good quality
— Simultaneous stored image and non-stored symbology are required
— Operator can not continuously watch display

Table 18-17. Summary of Display Performance [18-1]

Display of Sensors	Long-Persistence Phosphor CRT	Fast-Erase Storage Tube	Multi-Mode Tonotron®	EBIC and Transmission Grid Scan-Converter and CRT	Membrane Scan Converter and CRT
Air-to-air radar	Fair	Good	Excellent	Good	Excellent
Forward ground map radar	Poor	Fair	Excellent	Fair	Excellent
High-resolution strip map Radar (single image)	N.A.	Poor	Excellent	Good	Excellent
High-resolution strip map radar (continuous passing scene strip map)	N.A.	N.A.	Fair	Poor	Excellent
Air-to-air infrared	Fair	Good	Good	Good	Good
Forward ground map infrared	Good	Very poor	Good	Excellent	Excellent
Missile guidance television					
Resolution	Good	Good	Good	Excellent	Excellent
Brightness	Good Filter	Good	Poor	Excellent	Excellent
Gray scale	Good Req'd.	Fair	Fair	Excellent	Excellent
Low light television					
Resolution	Excellent	Good	Excellent	Excellent	Excellent
Gray scale	Excellent	Fair	Excellent	Excellent	Excellent
Additional factors:					
Nonstored symbology	Separate gun available	Complex time-sharing circuits required for video writing gun	Separate gun available—symbols move without smearing	Complex timing circuits required to form symbols by selective raster brightening—no symbol smearing	Complex timing circuits required to form symbols by selective raster brightening—no symbol smearing
Erasure and smearing	Symbol smears Image fades	Symbol smears Image fades: changing image smears	Image selectively erased before writing	Image fades: changing image smears	Image selectively erased before writing
Display size available	4 in., 6 in., usable	4 in., 6 in. usable	4 in., 8 in., or 10 in. usable	Wide variety	Wide variety
Development risk	Low	Low	Low	Medium	Medium
Growth capability	Poor	Good	Excellent	Good	Excellent
Production cost	Low	Medium	Medium	Medium+	Medium+
Weight and volume	Medium	Medium	Medium	Medium+	Medium+
Reliability	Good	Good	Good	Good	Good

— Operator task requires target or pattern recognition (image fade, smear, and gray scale critical)
— TV and FLIR not used in high ambient light
— Continuous strip map modes are not required

EBIC *and Transmission Grid Scan-Converter and* CRT appear most applicable where:
— Forward looking radar ground mapping is of poor quality
— Operator tasks in forward looking modes do not require target or pattern recognition (image fade and smear not critical)
— More than one display of same imagery is required
— TV image brightness and resolution are important
— High quality FLIR is required
— Continuous strip map "passing scene" modes are not required
— Large display or non-standard display format is required

Membrane Scan Converter and CRT appear most applicable where:
— Forward looking radar is of good quality
— Operator tasks in ground map modes require pattern or target recognition
— Continuous strip map "passing scene" mode is required
— High quality FLIR is required
— TV image brightness and resolution are important
— More than one display of same sensor imagery is required
— Large display size or non-standard display format is required

18.6. References and Bibliography

18.6.1. References

[18-1] Edited and abridged from G. K. Slocum, lectures given at the University of Tel Aviv, 1974, derived from "Airborne Sensor Display Requirements and Approaches," Report TM-888, Hughes Aerospace Group, Display Systems Dept., September 1967. This material is also available in Opto-Electronics seminar sponsored by AGARD, 9/16-9/20/74, Paris, France and Balkesjφ, Norway-published as "Lecture Notes" by AGARD.

[18-2] A. S. Patel, "Spatial Resolution by the Human Visual System. The Effect of Mean Retinal Illuminance," *Journal of the Optical Society of America*, Vol. 56, No. 5, May 1966, pp. 689-694.

[18-3] L. M. Biberman, (ed), *Perception of Displayed Information* Plenum Press, 1973; Chapters 3 (H. L. Snyder), 4 (Schitzler), 5 (F. A. Rosell and R. H. Willson), and 6 (Otto Schade, Sr.).

[18-4] F. A. Rosell and R. H. Willson, "Performance Synthesis: Electro-Optical Sensors," Report No. AFAL-TR-72-279, Westinghouse Defense and Electronic Systems, August 1972.

[18-5] F. A. Rosell and R. H. Willson, "Performance Synthesis: Electro-Optical Sensors," Report No. AFAL-TR-73-260, Westinghouse Defense and Electronic Systems, August 1973.

[18-6] F. A. Rosell and R. H. Willson, "Performance Synthesis: Electro-Optical Sensors," Report No. AFAL-TR-74-104, Westinghouse Defense and Electronic Systems, April 1974.

[18-7] R. H. Willson, F. A. Rosell, and Heru Ra Walmsley, "Effects of Vibration and G-Loading on Airborne E-O Sensor Augmented Observers," Report No. AFAL-TR-75-172, Westinghouse Defense and Electronic Systems, April 1976.

[18-8] Irving Reingold, "Display Devices: A Perspective on Status and Availability," *Proceedings of the Society for Information Display*, Society for Information Display, Montvale, NJ, Vol. 15, No. 2, 1974, pp. 63-73.

[18-9] Solomon Sherr, *Fundamentals of Display System Design*, John Wiley and Sons, New York, NY, 1970.

[18-10] L. M. Biberman, and Sol Nudelman, *Photoelectronic Imaging Devices,* Vols. 1 and 2, Plenum Press, New York, NY, 1971.

[18-11] J. E. Grimaldi, "A Computer Driven Photochromic Image Display," Dicomed Corporation, Minneapolis, MN, 1972.

[18-12] B. Kazan, and M. Knoll, *Electronic Image Storage,* Academic Press, New York, NY, 1968.

[18-13] Shannon, Claude, and Warren Weaver, *Mathematical Theory of Communication*, University of Illinois Press, Urbana, IL, 1959.

[18-14] L. K. Anderson, "The Cathode Ray Tube Display," *Journal of Vacuum Science and Technology*, American Institute of Physics, New York, NY, Vol. 10, No. 5, 1973, p. 67.

[18-15] M. N. Ernstoff, Hughes Aircraft Company, Aerospace Division, Culver City, CA, Private Communication.

[18-16] J. D. Kingsley and G. W. Ludwig, *Journal of the Electrochemical Society*, Electrochemical Society, Manchester, NH, Vol. 117, 1970, p. 353.

[18-17] G. Garlick, *Luminescence of Inorganic Solids*, P. Goldberg (ed.), Academic Press, New York, NY, 1966.

[18-18] S. G. Tomlin, *Proceedings of the Physics Society*, London, England, Vol. 82, 1963.

[18-19] H. A. Bethe, *Annalen der Physik*, Leipzig, East Germany, Vol. 5, 1930.

[18-20] L. V. Spencer, "Theory of Electron Penetration," *Physical Review*, American Physical Society, New York, NY, Vol. 98, 1955, p. 1597-1615.

[18-21] J. H. Jacob, "Penetration and Energy Deposition of Electrons in Thick Targets," *Journal of Applied Physics*, Argonne National Laboratory, Argonne, IL, Vol. 45, No. 1, January 1974, pp. 467-475.

[18-22] C. Feldman, *Physical Review*, American Physical Society, New York, NY, Vol. 117, 1960, p. 455.

[18-23] W. Ehrenberg and D. L. N. King, *Proceedings of the Phys. Society*, London, England, Vol. 81, 1963, p. 751.

[18-24] P. H. Hoff and T. E. Everhart, *10th Symposium on Electron, Ion and Laser Beam Technology*, L. Marton (ed.), San Francisco Press, 1969; *Journal of Applied Physics*, Argonne National Laboratory, Argonne, IL, Vol. 42, 1971.

[18-25] J. D. Kingsley and J. S. Prener, "Voltage Dependence of Cathode-Ray Efficiency of Phosphors: Phosphor Particles with Nonluminescent Coatings," *Journal of Applied Physics*, Argonne National Laboratory, Argonne, IL, Vol. 43, No. 7, 1972, pp. 3073-3079.

[18-26] P. H. Dowling and J. R. Sewell, *Journal of the Electrochemical Society*, Electrochemical Society, Manchester, NH, Vol. 100, 1953, p. 22.

[18-27] V. D. Meyer, *Journal of Applied Physics*, Argonne National Laboratory, Argonne, IL, Vol. 41, 1970.

[18-28] A. Bril, and H. A. Klasens, Philips Research Reports, No. 7, 1952.

[18-29] G. H. Heilmeier, L. A. Zanoni and L. A. Barton, "Dynamic Scattering: A New Electrooptic Effect in Certain Classes of Nematic Liquid Crystals," *Proceedings of the Institute of Electronics Engineers*, Vol. 56, No. 7, July 1968, pp. 1162-1171.

[18-30] Lester A. Jeffries, "Digitally-Addressed High Brightness Flat Panel Display," *SID Digest*, Society for Information Display, Montvale, NJ, 1973.

[18-31] Phillip Damon, *Fifth Technical Session Proceedings*, Society for Information Display, Montvale, NJ, 1965.

[18-32] J. W. Sandberg, "Operation of Scan Converter Tubes," Hughes Aircraft Company, Carlsbad, CA, 1974.

[18-33] L. S. Yaggy and N. J. Koda, *Eighth Technical Session Proceedings*, Society for Information Display, Montvale, NJ, 1967.

[18-34] R. Hayes, R. G. Culter, and K. W. Hawken, *Electronics*, 30 August 1973.

[18-35] H. G. Slottow, *SID Digest*, Society for Information Display, Montvale, NJ, 1975, p. 138.

[18-36] C. Curtin et al., *1973 Symposium Digest of Technical Papers*, Society for Information Display, Montvale, NJ.

[18-37] J. W. Sandberg, "Operation of Direct View Storage Tubes," Hughes Aircraft Company, Carlsbad, CA, 1972.

[18-38] Hughes Technical Staff, "Scan Conversion Memory Model 639," Hughes Aircraft Company, Carlsbad, CA, 1972.

[18-39] Michael J. Cantella, *Optical Society of American Technical Digest on Optical Displays*, Session WB3, 1975.

[18-40] L. Creagh, "Nematic Liquid Crystal Materials for Displays," *Proceedings of the Institute of Electrical and Electronics Engineers*, Vol. 61, No. 7, July 1973, pp. 814-822.

[18-41] G. H. Heilmeier, L. A. Zanoni, and L. A. Barton, "Further Studies of the Dynamic Scattering Mode in Nematic Liquid Crystals," *IEEE Transactions on Electron Devices*, Vol. ED-17, No. 1, January 1970, p. 22-26.

[18-42] S. Kobayashi, Conference Record of 1970 Institute of Electrical and Electronics Engineers Conference on Display Devices, New York, NY, 2-3 December 1970, p. 135.

[18-43] D. Jones and S. Lu, *SID Digest*, Society for Information Display, Montvale, NJ, 1972.

[18-44] Lewis T. Lipton and N. John Koda, "Matrix-Address Liquid Crystal Panel Display," *SID 1973 Symposium Digest of Technical Papers*, Society for Information Display, Montvale, NJ, pp. 46-47.

[18-45] A. R. Kmetz and F. K. VonWillisen, *Non-emissive Electro-optic Displays*, Plenum Press, 1976.

[18-46] Lewis T. Lipton, et al., "A Liquid Crystal Television Display using a Silicon-On-Sapphire Switching Array," *SID 1975 International Symposium Digest of Technical Papers*, Society for Information Display, Montvale, NJ, pp. 78-79.

[18-47] L. A. Goodman, "The Relative Merits of LEDs and LCDs," *Proceedings of the Society for Information Display*, Society for Information Display, Montvale, NJ, Vol. 16, No. 1, First Quarter 1975.

[18-48] M. N. Ernstoff, et al., "Liquid Crystal Pictorial Display," *IEEE 1973 International Electron Devices Meeting Conference Record*, pp. 548-551.

[18-49] M. N. Ernstoff, "Liquid Crystal Pictorial Display," *AIAA 1975 Digital Avionics Systems Conference Reprints*, paper 602.

[18-50] M. N. Ernstoff, "Liquid Crystal Pictorial Display," SID 1975 Technical Meeting, Society for Information Display, Montvale, NJ.

[18-51] Craig P. Stephens and Lewis T. Lipton, "A Multichip MOS Video Rate Liquid Crystal Display," *SID 1976 Symposium Digest of Technical Papers*, Society for Information Display, Montvale, NJ, pp. 44-45, VG-3177B & VG-3970.

[18-52] C. J. Neuse, J. J. Tietjen, J. J. Gannon, and H. F. Gossenberger, "Optimization of Electroluminescent Efficiencies for Vapor-Grown $GaAs_{1-x}P_x$ Diodes," *Journal of the Electrochemical Society*, Electrochemical Society, Manchester, NH, Vol. 116, 1968.

[18-53] R. H. Haitz, "Trends in LED Display Technology," *Proceedings of the 24th Electronics Components Conference*, sponsored by the Institute of Electrical and Electronics Engineers, Washington, D.C., May 13-15, 1974, p. 2.

[18-54] D. A. Laws and R. R. Ady, "Should You use LCD or LED Displays?" *Electronic Design*, Vol. 23, 1974, p. 88.

[18-55] L. Cosentino, "On the Transient Scattering of Light by Pulsed Liquid Crystal Cells," *IEEE Transactions on Electron Devices*, Vol. ED-18, 1971, p. 1172.

[18-56] G. Baur, "Angular Dependence of Transmitted Light in Deformed Twisted Nematic Cells," *Conference Record of 1974 Conference on Display Devices and Systems*, sponsored by the Institute of Electrical and Electronics Engineers and the Society for Information Display, 1974, p. 139.

[18-57] A. A. Bergh and P. J. Dean, "Light-Emitting Diodes," *Proceedings of the Institute of Electrical and Electronics Engineers*, Vol. 60, 1972, p. 182.

[18-58] National Semiconductor data sheet for NSN 71 numeric display.

[18-59] L. Creagh, A. Kmetz, and R. Reynolds, "Performance Characteristics of Nematic Liquid Crystal Display Devices," *IEEE Transactions on Electron Devices*, Vol. ED-18, 1971, p. 672.

[18-60] A. Sussman, "Electro-Optic Liquid Crystal Devices: Principle and Application," *IEEE Transactions on Parts, Hybrids, and Packaging*, Vol. PHP-8, Institute of Electrical and Electronics Engineers, 1972, p. 28.

[18-61] A. R. Kmetz, "Liquid Crystal Displays in Perspective," *IEEE Transactions on Electron Devices*, Vol. ED-20, 1973, p. 954.

[18-62] P. M. Alt and P. Pleshko, "Scanning Limitations of Liquid Crystal Displays," *IEEE Transactions on Electron Devices*, Vol. ED-21, 1974, p. 146.

[18-63] C. R. Stein and R. A. Kashnow, "A Two Frequency Coincidence Addressing Scheme for Nematic Liquid Crystal Displays," *Applied Physics Letters*, American Institute of Physics, New York, NY, Vol. 19, 1971, p. 343.

[18-64] P. J. Wild and J. Nehring, "An Improved Matrix Addressed Liquid Crystal Display," *Applied Physics Letters*, American Institute of Physics, New York, NY, Vol. 19, 1971, p. 335.

[18-65] H. K. Bucher, R. T. Klingbiel, and J. P. VanMeter, "Frequency-Addressed Liquid Crystal Field Effect," *Applied Physics Letters*, American Institute of Physics, New York, NY, Vol. 25, 1974, p. 186.

[18-66] H. Takata, O. Kogure, and K. Murase, "Matrix-Addressed Liquid Crystal Display," *IEEE Transactions on Electron Devices*, Vol. ED-20, 1973, p. 990.

[18-67] *Proceedings of the SID*, Special Issue on Flat-Panel and Large Screen Displays, Society for Information Display, Montvale, NJ, Vol. 17, No. 1, 1976.

[18-68] K. Nakada, T. Ishibashi, K. Toriyama, "A Design of Multiplexing Liquid Crystal Display for Calculators," *Conference Record of 1974 Conference on Display Devices and Systems*, sponsored by the Institute of Electrical and Electronics Engineers and the Society for Information Display, 1974, p. 139.

[18-69] M. George Craford and W. O. Groves, "Vapor Phase Epitaxial Materials for LED Applications," *Proceedings of the Institute of Electrical and Electronics Engineers*, Vol. 61, 1973, p. 863.

[18-70] W. E. Good, "Recent Advances in the Single-Gun Color Television Light-Valve Projector," Meeting of Society for Photo-Optical Instrumentation Engineers, Anaheim, CA, March 17-18, 1975.

[18-71] T. T. True, "Color Television Light Valve Projection System," Institute of Electrical and Electronics Engineers International Convention Session 26, 1973.

[18-72] T. T. True and W. C. Bates, General Electric, Private Communication, 1976.

[18-73] D. Shurtleff, The MITRE Corp., Bedford, MA, Private Communication.

[18-74] J. E. Bryden, "Some Notes on Measuring Performance of Phosphors Used in CRT Displays," *Proceedings of 7th National Symposium on Information Display*, Society for Information Display, Montvale, NJ.

[18-75] John M. Constantine, "Two-Slit Spot Analyzer," *Proceedings of 7th National Symposium on Information Display*, Society for Information Display, Montvale, NJ.

[18-76] R. J. Doyle, F. P. Heiman, and M. Kerman, "Modulation Transfer Function of Electrical Output Cathode Ray Storage Tubes," *SID Journal*, Society for Information Display, Montvale, NJ, Vol. 1, No. 4, pp. 20-22.

[18-77] Edward M. Sawtelle and George W. Gonyou, "Dynamic CRT Spot Measurement Techniques," *Proceedings of the National Symposium of the Society for Information Display*, Society for Information Display, Montvale, NJ, May 1968.

[18-78] Laurence E. White, *Measuring Spot Size in High-Resolution Cathode-Ray Tubes*, Sutton Publishing Company, 1959.

18.6.2. Bibliography

Alt, P. M., and P. Pleshko, "Scanning Limitations of Liquid Crystal Displays," *IEEE Transactions on Electron Devices*, Vol. ED-21, 1974.

Anderson, L. K., "The Cathode Ray Tube Display," *Journal of Vacuum Science and Technology*, American Institute of Physics, New York, NY, Vol. 10, No. 5, 1973.

Andrews, Gene, *Tekscope*, Vol. 6, No. 2, Tektronix, Inc., Beaverton, OR, 1974.

Baur, G., "Angular Dependence of Transmitted Light In Deformed Twisted Nematic Cells," *Conference Record of 1974 Conference on Display Devices and System*, sponsored by the Institute of Electrical and Electronics Engineers and the Society for Information Display, 1974.

Bergh, A. A., and P. J. Dean, "Light-Emitting Diodes," *Proceedings of the Institute of Electrical and Electronics Engineers*, Vol. 60, 1972.

Bethe, H. A., *Annalen der Physik*, Leipzig, East Germany, Vol. 5, 1930.

Biberman, L. M., (ed.), *Perception of Displayed Information*, Plenum Press, 1973; Chapters 3 (H. L. Snyder), 4 (Schitzler), 5 (F. A. Rosell and R. H. Willson), and 6 (Otto Schade, Sr.).

Biberman, L. M., and Sol Nudelman, *Photoelectronic Imaging Devices*, Vols. 1 and 2, Plenum Press, New York, N.Y. 1971.

Bitzer, D. L., and D. Alpert, "Advances in Computer-based Education," *Science*, 20 March 1970.

Bitzer, D. L., and H. G. Slottow, "Principles and Application of the Plasma Display Panel," *Proceedings of the O.A.R. Research Applications Conference*, Office of Aerospace Research, Arlington, VA, March 1968; also in Proceedings of the 1968 Micro-electronics Symposium, Institute of Electrical and Electronics Engineers, St. Louis, MO, 1968.

Bitzer, D. L., and H. G. Slottow, "The Plasma Display Panel—A Digitally Addressable Display with Inherent Memory," *Proceedings of the Fall Joint Computer Conference*, San Francisco, CA, November 1966.

Blackwell, H. R., "Contrast Thresholds of the Human Eye," *Journal of the Optical Society of America*, Vol. 36, No. 11, 1946.

Blackwell, H. R., "Specification of Interior Illumination Levels," *Illum. Engineering,* June 1959.

Bril, A., and H. A. Klasens, Philips Research Reports, No. 7, 1952.

Bryden, J. E., "Some Notes on Measuring Performance of Phosphors Used in CRT Displays," *Proceedings of 7th National Symposium on Information Display,* Society for Information Display, Montvale, NJ.

Bucher, H. K., R. T. Klingbiel, and J. P. VanMeter, "Frequency-Addressed Liquid Crystal Field Effect," *Applied Physics Letters,* American Institute of Physics, New York, NY, Vol. 25, 1974.

Cantella, Michael J., *Optical Society of America Technical Digest on Optical Displays,* Session WB3, 1975.

Constantine, John M., "Two-Slit Spot Analyzer," *Proceedings of 7th National Symposium of Information Display,* Society for Information Display, Montvale, NJ.

Cosentino, L., "On the Transient Scattering of Light by Pulsed Liquid Crystal Cells," *IEEE Transactions on Electron Devices,* Vol. ED-18, 1971.

Craford, M. George, and D. L. Keume, "LED Technology," *Solid State Technology,* January 1974.

Craford, M. George, and W. O. Groves, "Vapor Phase Epitaxial Materials for LED Applications," *Proceedings of the Institute of Electrical and Electronics Engineers,* Vol. 61, 1973.

Creagh, L., "Nematic Liquid Crystal Materials for Displays," *Proceedings of the Institute of Electrical and Electronics Engineers,* Vol. 61, No. 7, July 1973, pp. 814-822.

Creagh, L., A. Kmetz, and R. Reynolds, "Performance Characteristics of Nematic Liquid Crystal Display Devices," *IEEE Transactions on Electron Devices,* Vol. ED-18, 1971.

Curtin, et al., *1973 Symposium Digest of Technical Papers,* Society for Information Display, Montvale, NJ, 1973.

Damon, Phillip, *Fifth Technical Session Proceedings,* Society for Information Display, Montvale, NJ, 1965.

Dowling, P. H., and J. R. Sewell, *Journal of the Electrochemical Society,* Electrochemical Society, Manchester, NH, Vol. 100, 22, 1953.

Doyle, R. J., F. P. Heiman, and M. Kerman, "Modulation Transfer Function of Electrical Output Cathode Ray Storage Tubes," *SID Journal,* Society for Information Display, Montvale, NJ, Vol. 1, No. 4, pp. 20-22.

Ehrenberg, W., and D. L. N. King, *Proceedings of the Phys. Society,* London, England, Vol. 81, 1963, p. 751.

Ehrlich, Nathan, et al., "Review of MIL-E-1 Test Methods for Cathode Ray Charge Storage Tubes," Navy Contract N00039-69-C-1580, EIA, Washington, D.C., March 1972.

Ernstoff, M. N., et al., "Liquid Crystal Pictorial Display," *IEEE 1973 International Electron Devices Meeting Conference Record,* pp. 548-551.

Ernstoff, M. N., "Liquid Crystal Pictorial Display," *AIAA 1975 Digital Avionics Systems Conference Reprints,* paper 602.

Ernstoff, M. N., "Liquid Crystal Pictorial Display," *SID 1975 Technical Meeting,* Society for Information Display, Montvale, NJ.

Feldman, C., *Physical Review,* Vol. 117, 1960, p. 455.

Garlick, G., *Luminescence of Inorganic Solids,* P. Goldberg (ed.), Academic Press, Inc., New York, NY, 1966.

Glenn, W. E., "Principles of Simultaneous Color Projection Using Fluid Deformation," *SMPTE,* Vol. 79, September 1970.

Good, William E., "Recent Advances in the Single-Gun Color Television Light-Valve Projector," Meeting of the Society for Photo-Optical Instrumentation Engineers, Anaheim, CA, March 17-18, 1975.

Goodman, L. A., "Liquid Crystal Displays—Electro-Optical Effects and Addressing Techniques," *RCA Review,* Vol. 35, 1974. (The March, September, and December issues of *RCA Review,* Vol. 35, 1974, contain a series of papers on the physics, chemistry, and application of liquid crystals.)

Goodman, L. A., "The Relative Merits of LEDs and LCDs," *Proceedings of the Society for Information Display,* Society for Information Display Montvale, NJ, Vol. 16, No. 1, First Quarter, 1975.

Gregory, R., M. S. Bishop, and Raoul Weil, "Electron Beam Addressed Plasma Display Panel," *Proceedings of the Institute of Electrical and Electronics Engineers,* May 1969.

Grimaldi, J. E., "A Computer Driven Photochromic Image Display," Dicomed Corporation, Minneapolis, MN, 1972.

Haitz, R. H., "Trends in LED Display Technology," *Proceedings of the 24th Electronics Components Conference,* sponsored by the Institute of Electrical and Electronics Engineers and Electronics Industries Association, Washington, DC, May 13-15, 1974.

Hayes, R., R. G. Culter, and K. W. Hawken, *Electronics,* 30 August 1973.

Heilmeier, G. H., L. A. Zanoni, and L. A. Barton, "Dynamic Scattering: A New Electro-optic Effect in Certain Classes of Nematic Liquid Crystals," *Proceedings of the Institute of Electrical and Electronics Engineers,* Vol. 56, No. 7, July 1968, pp. 1162-1171.

Heilmeier, G. H., L. A. Zanoni, and L. A. Barton, "Further Studies of the Dynamic Scattering Mode in Nematic Liquid Crystals," *IEEE Transactions on Electron Devices,* Vol. ED-17, No. 1, January 1970, pp. 22-26.

Hoehn, H. J., and R. A. Martel, "A 60 Line per Inch Plasma Display Panel," 1970 Institute of Electrical and Electronics Engineers Conference on Display Devices, New York, NY, December 1970.

Hoff, P. H., and T. E. Everhart, *10th Symposium on Electron, Ion and Laser Beam Technology,* L. Marton (ed.), San Francisco Press, 1969; *Journal of Applied Physics,* Argonne National Laboratory, Argonne, IL, Vol. 42, 1971.

Holz, G. E., "The Primed Gas Discharge Cell—A Cost and Capability Improvement for Gas Discharge Matrix Displays," *Proceedings of the Society for Information Display,* Society for Information Display, Montvale, NJ, Vol. 13, First Quarter 1972.

Hughes Technical Staff, "Scan Conversion Memory Model 639," Hughes Aircraft Company, Carlsbad, CA, 1972.

Jackson, R. N., and K. E. Johnson, "Gas Discharge Displays: a Critical Review," *Advances in Electronics and Electron Physics,* Vol. 35, Academic Press, Inc., New York, NY, 1974.

Jacob, J. H., "Penetration and Energy Deposition of Electrons in Thick Targets," *Journal of Applied Physics,* Argonne National Laboratory, Argonne, IL, Vol. 45, No. 1, January 1974, pp. 467-475.

Jeffries, Lester A., "Digitally-Addressed High Brightness Flat Panel Display," *SID Digest,* Society for Information Display, Montvale, NJ. 1974.

Johnson, R. L., "The Application of the Plasma Display Technique to Computer Memory Systems," Report No. R-461, Coordinated Science Laboratory, University of Illinois, Urbana, IL, April 1970.

Johnson, W. E., and L. J. Schmersal, "A Quarter Million Element AC Plasma Display with Memory," *Proceedings of the Society for Information Display,* Vol. 13, First Quarter 1972.

Jones, D., L. T. Creagh, and S. Lu, "Dynamic Scattering in a Room-Temperature Nematic Liquid Crystal," *Applied Physics Letters*, Vol. 16, No. 2, January 1970, p. 61.

Jones, D., and S. Lu, *SID Digest*, Society for Information Display, Montvale, NJ, 1972.

Kazan, B., and M. Knoll, *Electronic Image Storage*, Academic Press, New York, NY, 1968.

Kingsley, J. D., and G. W. Ludwig, *Journal of Electrochemical Society*, Electrochemical Society, Manchester, NH, Vol. 117, 1970, p. 353.

Kingsley, J. D., and J. S. Prener, "Voltage Dependence of Cathode-Ray Efficiency of Phosphors: Phosphor Particles with Nonluminescent Coatings," *Journal of Applied Physics*, Argonne National Laboratory, Argonne, IL, Vol. 43, No. 7, 1972, pp. 3073-3079.

Kmetz, A. R., "Liquid Crystal Displays in Perspective," *IEEE Transactions on Electron Devices*, Vol. ED-20, 1973.

Kmetz, A. R., and F. K. VonWillisen, *Non-emmissive Electro-optic Displays*, Plenum Press, 1976.

Kobayashi, S., Conference Record of 1970 Institute of Electrical and Electronics Engineers Conference on Display Devices, New York, NY, 2-3 December 1970.

Kobayashi, S., and F. Takeuchi, "Multicolor Field-Effect Display Devices with Twisted Nematic Liquid Crystals," *Proceedings of the Society for Information Display*, Vol. 14, 1973.

Laws, D. A., and R. R. Ady, "Should You Use LCD or LED Displays?", *Electronics Design*, Vol. 23, 1974.

Letchner, B. J., *Pertinent Concepts in Computer Graphics*, M. Faiman and J. Nievergeht, (eds.), University of Illinois Press, 1969.

Linfoot, E. H., *Fourier Methods in Optical Image Evaluation*, Focal Press, London and New York, NY.

Lipton, Lewis T., et al., "A Liquid Crystal Television Display using a Silicon-on-Sapphire Switching Array," *SID 1975 International Symposium Digest of Technical Papers*, Society for Information Display, Montvale, NJ, pp. 78-79.

Lipton, Lewis T., and N. John Koda, "Matrix-Address Liquid Crystal Panel Display," *SID 1973 Symposium Digest of Technical Papers*, Society for Information Display, Montvale, NJ, pp. 46-47.

Lipton, L. T., M. A. Meyer, H. G. Dill, and D. O. Massetti, "SOS Liquid Crystal TV Display," paper presented at 1974 International Electron Devices Meeting, Washington, DC, 9 December 1974.

Loebner, E. E., "The Future of Electroluminescent Solids in Display Applications," *Proceedings of the IEEE*, Vol. 61, No. 7, July 1973, pp. 837-861.

Meyer, V. D., "Absorption of Electron Kinetic Energy by Inorganic Phosphors," *Journal of Applied Physics*, Argonne National Laboratory, Argonne, IL, Vol. 41, 1970, p. 4059.

Meyer, M. A., L. T. Lipton, G. H. Hershman, and P. G. Hilton, "Processing of a Monolithic SOS Array for Liquid Crystal Display Applications," paper presented at 1975 Electrochemical Society Meeting, Toronto, Canada, May 1975.

Nakada, K., Ishibashi, K. Toriyama, "A Design of Multiplexing Liquid Crystal Display for Calculators," *Conference Record of 1974 Conference on Display Devices and Systems*, sponsored by Institute of Electrical and Electronics Engineers and Society for Information Display, 1974.

National Semiconductor data sheet for NSN71 numeric display.

Nuese, C. J., H. Kressel, and I. Ladnay, "Light-Emitting Diodes and Semiconductor Material for Displays," *Journal of Vacuum Science Technology*, American Institute of Physics, New York, NY, Vol. 10, 1973.

Nuese, C. J., H. Kressel, and I. Ladnay, "The Future for LEDs," *IEEE Spectrum*, May 1972, pp. 28-38.

Nuese, C. J., J. J. Tietjen, J. J. Gannon, and H. F. Gossenberger, "Optimization of Electroluminescent Efficiencies for Vapor-Grown $GaAs_{1-x}P_x$ Diodes," *Journal of the Electrochemical Society,* Electrochemical Society, Manchester, NH, Vol. 116, 1968.

Optical Society of America, "Displays: Technology of Operator Performance and Design," *Technical Digest, Optical Society of American Topical Meeting,* March 19-20, 1975.

Patel, A. S., "Spatial Resolution by the Human Visual System. The Effect of Mean Retinal Illuminance," *Journal Optical Society of America,* Vol. 56, No. 5, May 1966.

Petty, W. D., "Multiple States and Variable Intensity in the Plasma Display Panel," Report R-497, Coordinated Science Laboratory, University of Illinois, Urbana, IL, November 1970.

Reingold, Irving, "Display Devices: A Perspective on Status and Availability," *Proceedings of the Society for Information Display,* Society for Information Display, Montvale, NJ, Vol. 15, No. 2, 1974.

Rosell, F. A., and R. H. Willson, "Performance Synthesis: Electro-Optical Sensors," Report No. AFAL-TR-72-279, Westinghouse Defense and Electronic Systems, August 1972.

Rosell, F. A., and R. H. Willson, "Performance Synthesis: Electro-Optical Sensors," Report No. AFAL-TR-73-260, Westinghouse Defense and Electronic Systems, August 1973.

Rosell, F. A., and R. H. Willson, "Performance Synthesis: Electro-Optical Sensors," Report No. AFAL-TR-74-104, Westinghouse Defense and Electronic Systems, August 1974.

Sandberg, J. W., "Operation of Direct View Storage Tubes," Hughes Aircraft Company, Carlsbad, CA, 1972.

Sandberg, J. W., "Operation of Scan Converter Tubes," Hughes Aircraft Company, Carlsbad, CA, 1974.

Sawtelle, Edward M. and George W. Gonyou, "Dynamic CRT Spot Measurement Techniques," *Proceedings of the National Symposium of the Society for Informantion Display,* Society for Information Display, Montvale, NJ, May 1968.

Schadt, M., and W. Helfrich, "Voltage Dependent Optical Activity of a Twisted Nematic Liquid Crystal," *Applied Physics Letter,* American Institute of Physics, New York, NY, Vol. 18, 1971.

Schiekel, M. F., and K. Fahrenschon, "Deformation of Nematic Liquid Crystals with Vertical Orientation in Electric Fields," *Applied Physics Letters,* American Institute of Physics, New York, NY, Vol. 19, No. 10, 15 November 1971, pp. 391-393.

Schott, D. J., "The Plasma Display as a Computer Input Device," *Digest of Papers—The IDEA Symposium,* Society for Information Display, New York, NY, May 1970.

Shade, O. H., "A Method of Measuring the Optical Sine-Wave Spatial Spectrum of Television Image Display Devices," *Society of Motion Picture and Television Engineers,* September 1958.

Shannon, Claude and Warren Weaver, *Mathematical Theory of Communication,* University of Illinois Press, Urbana, IL, 1959.

Sherr, Solomon, *Fundamentals of Display System Design,* John Wiley, New York, NY, 1970.

Slocum, G. K., Lectures given at the University of Tel Aviv, 1974, derived from "Airborne Sensor Display Requirements and Approaches, G. K. Slocum, Hughes Aerospace Group, Display Systems Department, Report TM-888, September 1967.

Also available in Opto-Electronics seminar sponsored by AGARD, 9/16-9/20/74, Paris, France and Balkesjø, Norway—published as "Lecture Notes" by AGARD.

Slottow, H. G., *SID Digest*, Society for Information Display, Montvale, NJ, 1975, p. 138.

Slottow, H. G., "The Plasma Display Panel—Principles and Prospects," *Proceedings of 1970 Institute for Electrical and Electronics Engineers Conference on Display Devices*, 1970.

Sobel, A., Review Lecture, Third International Conference on Gas Discharges, Institution of Electrical Engineers, London, September 1974.

Society for Information Display, *Proceedings of the SID,* Special Issue on Flat-Panel and Large Screen Displays, Vol. 17, No. 1, 1976.

Society for Information Display, The IDEA Symposium, New York, NY, May 1970. Institute of Electrical and Electronics Engineers International Electron Device Meeting, October 1970. Institute of Electrical and Electronics Engineers Conference on Electron Devices, December 1970.

Spencer, L. V., "Theory of Electron Penetration," *Physical Review*, Vol. 98, No. 6, 15 June 1955, p. 1597-1615.

Stephens, Craig P., and Lewis T. Lipton, "A Multichip MOS Video Rate Liquid Crystal Display," *SID 1976 Symposium Digest of Technical Papers,* Society for Information Display, Montvale, NJ, pp. 44-45. VG-3177B & VG-3970.

Stein, C.R., and R. A. Kashnow, "A Two Frequency Coincidence Addressing Scheme for Nematic Liquid Crystal Displays," *Applied Physics Letters*, American Institute of Physics, New York, NY, Vol. 19, 1971.

Stredde, E., "The Development of a Multi-Color Plasma Display Panel," Report R-370, Coordinated Science Laboratory, University of Illinois, November 1967.

Sussman, A., "Electro-Optic Liquid Crystal devices: Principles and Applications," *IEEE Transactions on Parts, Hybrids; and Packaging,* Vol. PHP-8, 1972.

Takata, H., O. Kogure, and K. Murase, "Matrix-Addressed Liquid Crystal Display," *IEEE Transactions on Electron Devices,* Vol. ED-20, 1973.

Tomlin, S. G., *Proceedings of the Physics Society,* London, England, Vol. 82, 1963.

Trogden, "Transformer Coupled Driving Circuits for Plasma Panel Displays," *Digest of Papers—The IDEA Symposium,* Society for Information Display, New York, NY, May 1970.

True, T. T., "Color Television Light Valve Projection Systems," *IEEE International Convention Session 26,* 1973.

University of Illinois, Coordinated Science Laboratory Progress Report for September 1968-June 1969, University of Illinois, Urbana, IL, August 1969.

van Raalte, J. A., "Survey of Developmental Light Valve Systems," *IEEE International Convention Session 26,* 1973.

Weston, G. F., "Cold Cathode Glow Discharge Tubes," Iliffe, London, 1968.

White, Laurence E., *Measuring Spot Size in High-Resolution Cathode-Ray Tubes,* Sutton Publishing Company, 1959.

Wild, P. J., and J. Nehring, "An Improved Matrix Addressed Liquid Crystal Display," *Applied Physics Letters,* American Institute of Physics, New York, NY, Vol. 19, 1971.

Willson, R. H., F. A. Rosell, and Heru Ra Walmsley, "Effects of Vibration and G-Loading on Airborne E-O Sensor Augmented Observers," Report No. AFAL-TR-75-172, Westhouse Defense and Electronic Systems, April 1976.

Yaggy, L. S., and N. J. Koda, *Eighth Technical Session Proceedings,* Society for Information Display, Montvale, NJ. 1967.

Chapter 19

IMAGING SYSTEMS

William L. Wolfe
University of Arizona

CONTENTS

19. Imaging Systems

19.1. Introduction

Infrared imaging systems can be used for medical diagnosis, nondestructive testing of airplane parts, night vision, real-time aircraft reconnaissance, imaging extraterrestrial objects, and weather mapping. The image can be generated by the use of a pyroelectric vidicon or other area sensor. (See Chapter 13.) In such a case, the main system calculations pertain to the tube and the optical system used to form an image on the tube. More common techniques use single-element detectors or arrays (Chapters 11 and 12) and an optical-mechanical scanning mechanism (Chapter 10).

19.1.1. Symbols, Nomenclature, and Units. The symbols, nomenclature and units used in this chapter are listed in Table 19-1.

Table 19-1. Symbols, Nomenclature and Units

Symbols	Nomenclature	Units
A_d	Detector area	m^2
D	Entrance pupil diameter	m
D^*	Normalized detectivity	$W^{-1}\ Hz^{1/2}\ cm$
E	Irradiance	$W\ m^{-2}$
EIFOV	Effective instantaneous field of view	rad
FAR	False alarm rate	sec^{-1}
$F/\#$	Focal ratio, optical speed or "F number"	−
I	Intensity	$W\ sr^{-1}$
IFOV	Instantaneous field of view	rad
L	Radiance	$W\ m^{-2}\ sr^{-1}$
M	Exitance	$W\ m^{-2}$
MRT	Minimum resolvable temperature (difference)	K
MTF	Modulation transfer function	−
NEFD	Noise equivalent flux density	$W\ m^{-2}$
NEP	Noise equivalent power	$W,\ W\ Hz^{-1/2}$
NETD	Noise equivalent temperature difference	K
n	Refractive index	−
OTF	Optical transfer function	−
PTF	Phase transfer function	−
Q_B	Photon irradiance spectrum	m^{-2}
q	(As a subscript, photon)	−
r	Distance	m
SiTF	Signal transfer function	$cm\ K^{-1}$
T	Temperature (as a subscript, target)	K
t_f	Frame time	sec
Δf	Temporal bandwidth	Hz
Φ	Radiant flux	W

19.1.2. Definitions of Terms. The performance and descriptions of performance of an infrared imaging device are complicated. The terms used to describe these quantities can also be complex. They are defined in Table 19-2.

Table 19-2. Definitions of Terms

Term	Definition
NEP (Specific NEP)	Noise equivalent power: the power or flux difference on the entrance pupil of the receiver that generates an rms difference in signal equal to the rms value of the noise. The effective noise bandwidth needs to be specified. An alternate definition is the power that gives a ratio of rms signal to rms noise of one in a one-hertz bandwidth; the latter will be called specific NEP in this chapter.
NETD	Noise equivalent temperature difference: the temperature difference between two blackbody sources that both fill the field of view and give an rms signal difference equal to the rms value of the noise.
IFOV	Instantaneous field of view: the angular subtense of a linear dimension of the detector at the entrance pupil of the infrared optical system. It can also be the geometric image of the detector element on the object plane and the solid angle of the detector projection.
EIFOV	Effective instantaneous field of view: the reciprocal of the spatial frequency at which the modulation transfer is 50%.
Frame time	The time required to cover every point in a field of view.
Frame rate	Reciprocal of the frame time.
Interlace	One plus the number of times per frame that a gapped detector array is stepped to obtain contiguous, complete coverage.
Field rate	Product of the frame time and the interlace.
Overscan	The number of times each point in a stationary scene is covered per frame.
Impulse response (Point spread function)	The ratio of system output to a point-source input, usually plotted as a function of the spatial variables in the image plane.
Line spread function	The system response to a line source, normalized to the line source output.
SiTF	Signal transfer function: the photopic luminance output of the system divided by the temperature difference of the scene for specified gain, brightness, gamma, and dc restoration conditions.
OTF	Optical transfer function: the complex Fourier transform of the point spread function. The function must be

Table 19-2. Definitions of Terms (*Continued*)

Term	Definition
	specified for at least two directions, perpendicular and parallel to the scan direction.
MTF	Modulation transfer function: the amplitude of the optical transfer function.
PTF	Phase transfer function: the argument of the optical transfer function.
MRT	The minimum resolvable temperature is the smallest element-to-element temperature difference between two blackbody elements in a standard periodic test pattern for which the pattern is observable. Unlimited viewing time is allowed; the system is operated in a noise-limited, linear mode; and the MRT is specified as a function of the fundamental frequency of the test pattern.
Pixel	A picture element or resolution element. For a system with switching, it is a single exposure of the detector or the spatial coverage of the detector during one dwell time, sampling time, or integration time.
Reselm	See pixel.

19.2. Sensitivity Analysis

Several expressions describe the ability of an infrared system to sense a radiant source or a change in radiant power. These include signal-to-noise ratio (SNR), noise equivalent power (NEP), minimum resolvable temperature (MRT), etc. They can be calculated for different conditions: whether a point or extended source is viewed; whether the limit of detectivity is within the detector, is photon noise, or is background clutter; and whether or not the system is diffraction limited.

19.2.1. Signal-to-Noise Ratio.

An infrared imaging device records a signal that is proportional or almost proportional to the power on the detector for a thermal detector, and to the number of photons per unit time for a photon detector. In the absence of a signal, S, the system is said to record only noise, N. A dc system will record the average background radiance plus the fluctuations around that average. An ac system will record only the variations. Thus as an imaging device scans a scene element-by-element, a dc system will record the average value plus the changes, and the ac system only the changes. The output for a dc system with and without a signal will be, respectively, $S+N$ and N. The ratio of these two readings is $(S+N)/N$, which is equal to the signal-to-noise ratio only if $S/N \gg 1$. An ac system records the signal-to-noise ratio directly because it records only the changes. If the target is larger than the instantaneous field of view (IFOV) the signal will correspond to the radiance of the target. If the target is smaller than the IFOV, the signal will correspond to the target radiance in a portion of the field of view, but the system output will also be related to the level of background radiation.

19.2.2. Extended Source, Detector-Noise Limited, Detector Is Field Stop. The SNR is given by

$$\text{SNR} = \int_0^{\Omega} (A_d)^{1/2} \cos\theta \, d\Omega \int_0^{\infty} (\Delta f)^{-1/2} \tau_a \tau_o L_\lambda D^* n^{-2} \, d\lambda \qquad (19\text{-}1)$$

where A_d = detector sensitive area

θ = angle the optical axis makes with the detector surface

$d\Omega$ = differential solid angle formed by a portion of the entrance pupil at the image surface

Ω = full solid angle of the entrance pupil at the image surface

τ_a = transmission of the atmosphere

τ_o = transmission of the optics

L_λ = spectral radiance of the source

D^* = specific detectivity $\equiv D^*(\lambda)$

n = refractive index of the material which surrounds the detector

λ = wavelength

When the projected solid angle can be approximated by the ratio of the area of the entrance pupil divided by its distance to the image surface, and that distance is the focal length, SNR is given by

$$\text{SNR} = \frac{\pi \alpha D}{4(F/\#)(\Delta f)^{1/2}} \int_0^{\infty} D^* \tau_a \tau_o L_\lambda \, n^{-2} \, d\lambda \qquad (19\text{-}2)$$

where D = entrance pupil diameter

α = the length of a (square) detector divided by the focal length

$F/\#$ = focal ratio or "F number"

Δf = temporal bandwidth

For a detector which has a constant quantum efficiency that drops abruptly to zero at the cutoff wavelength, this can be rewritten as

$$\text{SNR} = \frac{\pi \alpha D D_{\max}^*}{4(F/\#)(\Delta f)^{1/2}} \int_0^{\infty} \tau_a \tau_o L_\lambda \, n^{-2} \lambda \lambda_c^{-1} \, d\lambda$$

$$\qquad\qquad\qquad\qquad\qquad\qquad\qquad\qquad\qquad (19\text{-}3)$$

$$= \frac{\pi \alpha D D_{\max}^* c h}{4(F/\#)(\Delta f)^{1/2}} \int_0^{\infty} \tau_a \tau_o L_{q,\lambda} \, n^{-2} \lambda_c^{-1} \, d\lambda$$

where $D_{\max}^*$ = maximum value of the specific detectivity

$L_{q,\lambda}$ = spectral photon sterance (radiance)

λ_c = cutoff wavelength (or maximum wavelength)

19.2.3. Extended Source, Detector-Noise Limited, Detector Is Field Stop, Diffraction-Limited Optics. When the optics are limited by diffraction, the energy is distributed according to a bessinc function $J_1(x)/x$, where $J_1(x)$ is the Bessel function of the first kind. A good rule of thumb seems to be to make the detector length about equal to twice the radius of the first zero of the bessinc. In that case, only about 84% of the irradiance falls on the detector and the IFOV is given approximately by $2.44\,\lambda_c/D$. Then

$$\text{SNR} = \frac{\pi\,(0.84)\,(2.44\,\lambda_c)}{4\,(\Delta f)^{1/2}\,(F/\#)} \int_0^\infty \tau_a \tau_o D^* L_\lambda\, n^{-2}\, d\lambda \qquad (19\text{-}4)$$

A more detailed calculation requires calculating the value of $\int D^*_{\max}(x,y)\, E(x,y)\, dA_d$ over the surface area of the detector, A_d, at any one sample time. If the system is not sampled, a one-dimensional convolution in the scan direction must be taken.

19.2.4. Extended Source, Photon-Noise Limited, Detector Is Field Stop. This situation is similar to that of 19.2.3 except that D^* is given by

$$D^* = D^*_{\text{BLIP}} = \left(\frac{\eta}{2Q_B}\right)^{1/2} \frac{\lambda}{hc} \qquad (19\text{-}5)$$

where h = Planck's constant
  Q_B = Wiener, variance or power spectrum of the background photon flux per unit area (usually a constant)
  c = speed of light
  η = quantum efficiency

The quantity $2^{-1/2}$ is valid for photoconductors without sweepout. Then

$$\text{SNR} = \frac{\pi\,\alpha\,D}{4(\Delta f)^{1/2}\,(F/\#)\,2^{1/2}\,c\,h\,Q_B^{1/2}} \int_0^\infty \tau_a \tau_o L_\lambda\, n^{-2}\, d\lambda \qquad (19\text{-}6)$$

The quantity Q_B is the power spectrum of the variation in the number of photons emitted, or incident, per unit area. As Kruse [19-1] shows, this power or Wiener spectrum, is given by

$$Q_B = 2\,\bar{n}\left|\int_0^\infty \{f(t)\exp(j\omega t)\}\,dt\right|^2 \qquad (19\text{-}7)$$

If the function $f(t)$ corresponds to the emission of a photon, then

$$f(t) = \delta(t - t_o) \qquad \text{(Dirac Delta Function)} \qquad (19\text{-}8)$$

and

$$Q_B = 2\,\bar{n} \qquad (19\text{-}9)$$

Photons are emitted as independent, random events. The power spectrum of the variation in photons is twice the average rate. This quantity can be integrated over the spectral region of interest so that

$$Q_B = 2 \int_{\Delta\nu} 2\pi\nu^2 c^{-2} (e^x - 1)^{-1} \, d\nu = 2 E_q \tag{19-10}$$

The units of Q_B are number of photons squared per unit area per second squared per hertz, and it is numerically equal to twice the photon irradiance.

The irradiance on the detector from the background is the combination of radiation from the scene and from different parts of the optical system, modified by the transmission of the appropriate elements. Figure 19-1 shows a representative optical system. The irradiance on the detector is that from the background within the field of view of the system, plus the emission from the atmosphere, plus that from the focusing mirror, plus that from the folding flat. The irradiance on the detector from the flat can be calculated from the radiance of the flat, L, by

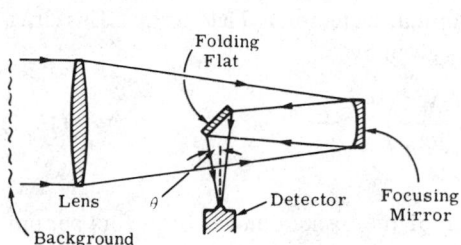

$$E_{ff} = \int_\Omega L \cos\theta \, d\Omega \tag{19-11}$$

where E_{ff} = the detector irradiance from the folding flat

θ = angle between the detector normal and a line that connects the detector to any arbitrary point on the flat.

Fig. 19-1. Background radiation geometry.

If the mirror radiates isotropically, the radiance is constant and the irradiance is given by

$$E_{ff} = L \int \cos\theta \, d\Omega \tag{19-12}$$

The term on the far right is the total projected solid angle, which can be written as Ω'. Then, if the mirror has a uniform temperature and an emissivity ϵ,

$$E_{ff} = \Omega' \int \epsilon_{ff}(\lambda) \, c_1 \, \pi^{-1} \lambda^{-5} (e^x - 1)^{-1} \, d\lambda \tag{19-13}$$

where Ω' = projected solid angle
c_1 = first radiation constant (See Chapter 1.)
λ = wavelength
$x = c_2/\lambda T$
c_2 = second radiation constant
T = temperature of the folding flat
$\epsilon_{ff}(\lambda)$ = spectral emissivity of the folding flat

The focusing mirror subtends the same solid angle as the folding flat so that it contributes an irradiance given by

$$E_{fm} = \Omega' \int \epsilon_{fm}(\lambda) \, \rho_{ff}(\lambda) \, c_1 \, \pi^{-1} \lambda^{-5} (e^x - 1)^{-1} \, d\lambda \tag{19-14}$$

where T = temperature of the focusing mirror

$\epsilon_{fm}(\lambda)$ = spectral emissivity of the focusing mirror

$\rho_{ff}(\lambda)$ = spectral reflectivity of the folding flat

The contribution of the lens is

$$E_l = \Omega' \tau_{ff} \int \epsilon_l(\lambda)\, \rho_{fm}(\lambda)\, \rho_{ff}(\lambda)\, L_\lambda^{BB}(T_l)\, d\lambda \tag{19-15}$$

where $\epsilon_l(\lambda)$ = spectral emissivity of the lens

ρ_{fm} = reflectivity of the focusing mirror

L_λ^{BB} = blackbody spectral radiance of the lens

T_l = (uniform) temperature of the lens

τ_{ff} = transmission factor due to the obscuration, by the folding flat, in the beam between the lens and the focusing mirror

A good design will incorporate a cold field-stop in the detector dewar such that the folding flat and the cold aperture subtend the same angle (or the cold aperture subtends a slightly smaller angle, for design margin). In that case all irradiance terms have been considered. The total irradiance is

$$E_{tot} = E_{ff} + E_{fm} + E_l + E_{Bk} \tag{19-16}$$

These should all be calculated on the basis of photon flux density so that

$$Q_B = Q_{B,ff} + Q_{B,fm} + Q_{B,l} + Q_{B,Bk} \tag{19-17}$$

Some simplifying approximations can be made for calculations that will have inaccuracies of about 10%. The main approximations are that the radiance of each element is Lambertian and that the projected solid-angle subtended is given by $\pi(2F/\#)^{-2}$. Furthermore, if all the elements in the system have the same emissivity, then for l_e elements

$$Q_B = \frac{\epsilon M_q}{4(F/\#)^2}(1-\rho)^{l_e} \tag{19-18}$$

where $M_q = \dfrac{c_1}{hc}\lambda^{-4}(e^x - 1)^{-1}$. The signal-to-noise ratio can therefore be written

$$\frac{S}{N} = \left(\frac{\eta}{\Delta f}\right)^{1/2}\left(\frac{0.84 \times 1.22}{2}\left(\frac{\pi}{2}\right)^{1/2}\right)\frac{\int \tau_a \tau_o\, L_{q,\lambda}\,(T_S) n\, d\lambda}{\left[\int L_{q,\lambda}\,(T_B)\, d\lambda\right]^{1/2}} \tag{19-19}$$

19.2.5. Small Source, Detector-Noise Limited. If the source is smaller than the IFOV, then the radiation transfer is based on the source's radiant intensity, I. The power on the aperture of area A_c at a distance d is given by

$$\Phi_a = \frac{\tau_a I A_c}{d^2} \tag{19-20}$$

If power is lost only in optical transmission and obscuration, then the power on the detector is

$$\Phi_d = \frac{\tau_o \tau_a I A_c}{d^2} \tag{19-21}$$

The signal-to-noise ratio is given by

$$\frac{S}{N} = \frac{D^* \Phi_d}{\sqrt{A_d \Delta f}} = \frac{\pi D^2}{4 A_d^{1/2} \Delta f^{1/2} d^2} \int_0^\infty \tau_o \tau_a D_\lambda^* I_\lambda \, d\lambda \tag{19-22}$$

The linear IFOV can be approximated as $\sqrt{A_d}/F$ so that

$$\frac{S}{N} = \frac{\pi D}{4(F/\#)d^2 \alpha \Delta f^{1/2}} \int_0^\infty \tau_o \tau_a D_\lambda^* I_\lambda \, d\lambda \tag{19-23}$$

The D_λ^* for a detector that has a constant quantum efficiency can be written as

$$D_\lambda^* = \frac{\lambda}{\lambda_c} D_{\max}^* \tag{19-24}$$

Therefore the SNR can be written as

$$\frac{S}{N} = \left(\frac{\pi}{4}\right)\left(\frac{D^2}{d^2 F \alpha}\right) \int \tau_o \tau_a D_\lambda^* I_\lambda \, d\lambda$$

$$= \left(\frac{\pi}{4}\right)\left(\frac{D^2 D_{\max}^*}{d^2 F \alpha}\right)\left(\lambda_{\max} F\alpha\right)^{-1}\left(\frac{hc}{\lambda_{\max}}\right)\int \tau_o \tau_a I_{q,\lambda} \, d\lambda \tag{19-25}$$

19.2.6. Small Source, Detector-Noise Limited, Diffraction-Limited Optics. If the optics are sufficiently good that they can be said to be diffraction-limited, and the conditions of Section 19.2.5 apply, then the SNR is given by

$$\frac{S}{N} = \frac{0.84\pi}{2\sqrt{2}\,2.44}\left(\frac{D}{d}\right)^2 \lambda_{\max}^{-1}\left(\frac{\eta}{\Delta f}\right)^{1/2} \frac{\int \tau_o \tau_a I_{q,\lambda} \, d\lambda}{\left[\int M_{q,\lambda} \, d\lambda\right]^{1/2}} \tag{19-26}$$

19.2.7. Background Effects (Clutter). Sometimes the imaging system has sufficiently good system SNR, but the target is hidden in the variations of the background. Some simple criteria can be developed to determine whether a given target can be discerned. For an extended source, target irradiance at the entrance pupil should exceed background irradiance variations. As an extreme measure one may require that the target irradiance from an IFOV, plus the remaining background irradiance from that IFOV, must be greater than the maximum background irradiance from an IFOV minus the minimum background irradiance from another IFOV. This can be written in terms of the power emitted by the target and the background (because the geometrical relations are the same, and the target and background may be assumed to be isotropic radiators):

$$L_T A_T d^{-2} - L_{B_{min}} A_T d^{-2} + L_{B_{min}} \Omega > L_{B_{max}} \Omega - L_{B_{min}} \Omega$$

$$L_T > L_{B_{min}} + \left(L_{B_{max}}\right) d^2 A_T^{-1} - 2 L_{B_{min}} \Omega d^2 A_T^{-1}$$

$$L_T > L_{B_{min}} + \left(L_{B_{max}} - 2 L_{B_{min}}\right) \Omega / \Omega_T$$

where L_T = target radiance

A_T = target area

Ω_T = target solid angle

$L_{B_{max}}$ = maximum background radiance

$L_{B_{min}}$ = minimum background radiance

Ω = solid angle IFOV

d = target-imaging system distance

The background usually has some statistical distribution of radiance variations so that rather than exceeding the maximum background variation, the target radiance need only exceed some fraction of this variation. In this case, some percentage of the time targets will be missed. This is closely allied to the warning problem. (See Chapter 21.)

19.2.8. Simplified Analysis and Examples. The analysis of an imaging infrared system can proceed from the basic equation for SNR, which in this case is the ratio of the rms signal to the rms noise. The SNR of a bare detector is given by

$$\frac{S}{N} = \frac{D^* \Phi_d}{\sqrt{A_d \Delta f}} \tag{19-27}$$

The SNR of the system is determined by relating the power on the detector to the source power.

$$\Phi_d = \frac{\tau_a \tau_o L A_d \cos\theta_d A_o \cos\theta_o}{n^2 F^2} \tag{19-28}$$

where L = radiance (any place in the system)

θ_d = the angle the detector makes with the optical axis

A_o = the area of the optics

θ_o = the angle the optics make with the optical axis

F = the focal length of the optical system

n = refractive index of the medium

τ_a = atmospheric transmission

τ_o = optics transmission (including obscurations)

If one assumes that the system works in air ($n = 1$ everywhere), that all solid angles are just an area divided by the square of a length, and that the image is at the focal distance, then

$$\frac{S}{N} = \frac{A_d A_o}{\sqrt{A_d \, \Delta f} F^2} \int_{\Delta \lambda} \tau_a(\lambda) \, \tau_o(\lambda) \, D^*(\lambda) \, L_\lambda(\lambda) \, d\lambda \qquad (19\text{-}29)$$

The integral is introduced because both D^* and L_λ are wavelength dependent and their product must be integrated over the spectral bandwidth of the system. For a detector with a quantum efficiency that does not vary with wavelength in the region $\Delta \lambda$, one can write

$$D^* = \frac{\lambda}{\lambda_{max}} D^*_{max} \qquad (19\text{-}30)$$

Therefore

$$\frac{S}{N} = \frac{D^*_{max} \sqrt{A_d} \, A_o}{\lambda_{max} \sqrt{\Delta f} \, F^2 \, c \, h} \int_{\Delta \lambda} \tau_a(\lambda) \, \tau_o(\lambda) \, L_{q,\lambda}(\lambda) \, d\lambda \qquad (19\text{-}31)$$

where λ_{max} = wavelength at which D^* is a maximum
$\quad\quad D^*_{max}$ = the maximum D^* value
$\quad\quad\quad c$ = speed of light
$\quad\quad\quad h$ = Planck's constant
$\quad L_{q,\lambda}(\lambda)$ = spectral photon sterance (radiance)

The factor of $c \, h$ enters because

$$L_{q,\lambda} = \frac{\lambda L_\lambda}{ch} \qquad (19\text{-}32)$$

The SNR as developed so far assumes implicitly that there is some change in $L_{q,\lambda}$ that causes a change in the output. Some systems sense the dc value of the radiance and additive noise, but they are rare. Since most imaging systems are ac-coupled, a change in radiance is the better assumption. Therefore, from here on, $dL_{q,\lambda}$ will be written

$$dL_{q,\lambda} = \frac{\partial L_{q,\lambda}}{\partial T} \, dT + \frac{\partial L_{q,\lambda}}{\partial \epsilon} \, d\epsilon \qquad (19\text{-}33)$$

The change in $L_{q,\lambda}$ can be the result of a change in temperature or of a change in emissivity as the imaging device moves from one pixel to the next. Since $L_{q,\lambda}$ can be written as

$$L_{q,\lambda} = \epsilon \, c_1 \, \lambda^{-4} \, (e^x - 1)^{-1} \, (ch)^{-1} \qquad (19\text{-}34)$$

then

$$dL_{q,\lambda} = L_{q,\lambda} x (e^x - 1)^{-1} \, dT + L_{q,\lambda} \, d\epsilon \qquad (19\text{-}35)$$

A familiar figure of merit in the field is the noise equivalent temperature difference, NETD. It is commonly defined as the difference in temperature that gives a SNR of one. Its reciprocal, the SNR for a one-degree temperature difference, SNR_{Temp}, is given by

$$SNR_{Temp} = (NETD)^{-1} = \frac{\partial (S/N)}{\partial T} = \frac{D^*_{max}}{\lambda_{max}} \frac{A_d^{1/2}}{\Delta f^{1/2}} \frac{A_o}{F^2 c \, h} \int_{\Delta \lambda} \tau_a \tau_o \frac{\partial L_{q,\lambda}}{\partial T} \, d\lambda \qquad (19\text{-}36)$$

The noise equivalent emissivity difference (NEϵD) and its reciprocal, SNR_ϵ, are given by

$$SNR_\epsilon = (NE\epsilon D)^{-1} = \frac{\partial (S/N)}{\partial \epsilon} = \frac{D^*_{max} A_d^{1/2} A_o}{\lambda_{max} \Delta f^{1/2} F^2 c h} \int \tau_a \tau_o \frac{\partial L_{q,\lambda}}{\partial \epsilon} d\lambda \quad (19\text{-}37)$$

The total SNR is given by

$$SNR_L = (NELD)^{-1} \equiv SNR_{Temp} \, dT + SNR_\epsilon \, d\epsilon \quad (19\text{-}38)$$

It cannot be made specific for either quantity in combination. The change in SNR for a unit change in radiance is the S/N given above, but not normalized. Division by $\int_{\Delta\lambda} L_{q,\lambda} \, d\lambda$ will accomplish this. The bandwidth can be written as

$$\Delta f \geqslant \frac{1}{2t_d} = \frac{\theta_f \phi_f}{2 \, t_f \, \theta^2} = \frac{N}{2t_f} \quad (19\text{-}39)$$

This assumes that there are no overlaps, dead times, retrace times, turn-around times, etc. (See Chapter 10.). It also assumes that a scanning efficiency of $\eta_s = 1$; however, since it never is,

$$\Delta f \geqslant \frac{1}{2t_d} = \frac{\theta_f \phi_f}{2 \, t_f \, \theta^2 \, \eta_s} = \frac{N}{2 \, \eta_s t_f} \quad (19\text{-}40)$$

This assumes that only one detector is used. If there are m detectors, the dwell time can be increased by the factor m, so that

$$\Delta f \geqslant \frac{1}{2t_d} = \frac{\theta_f \phi_f}{2 \, t_f \, \theta^2 \, \eta_s \, m} = \frac{N}{2 \, \eta_s t_f m} \quad (19\text{-}41)$$

For the purposes of simplicity it is assumed that τ_a and τ_o are both constant over the spectral band of interest. This is a good first approximation for most systems of interest. The remaining manipulation involves the quantity

$$\frac{A_d^{1/2} A_o}{\Delta f^{1/2} F^2} = \frac{\pi \alpha D}{4(F/\#) (\Delta f)^{1/2}} \quad (19\text{-}42)$$

where α = the linear IFOV angle, where the IFOV is assumed square
     D = optics diameter
   $F/\#$ = focal ratio of the optics

The bandwidth is calculated from the scanning dynamics. The sampling theorem states that $\Delta f \geqslant 1/(2t_d)$ where t_d is the dwell time on a pixel. The frame is θ_f by ϕ_f radians. Thus there are N pixels where

$$N = \frac{\theta_f \phi_f}{\theta^2} \quad (19\text{-}43)$$

The dwell time, t_d, is given in terms of the frame time as

$$t_d = \frac{t_f}{N} = \frac{t_f \, \theta^2}{\theta_f \, \phi_f} \tag{19-44}$$

Therefore

$$\text{SNR}_{\text{Temp}} = \left(\frac{D^*_{\max}}{\lambda_{\max}}\right)\left(\frac{\theta D \, \tau_o}{F/\#}\right)\left(\frac{2\eta_s \, t_f m}{N}\right)^{1/2}\left(\frac{\pi \, \tau_a}{4 \, c \, h}\right)\int_{\Delta\lambda} \frac{\partial L_{q,\lambda}}{\partial T} \, d\lambda \tag{19-45}$$

$$= \left(\frac{D^*_{\max}}{\lambda_{\max}}\right)\left(\frac{\theta^2 D \, \tau_o}{F/\#}\right)\left(\frac{2\eta_s \, t_f m}{\theta_f \phi_f}\right)^{1/2}\left(\frac{\pi \, \tau_a}{4 \, c \, h}\right)\int_{\Delta\lambda} \frac{\partial L_{q,\lambda}}{\partial T} \, d\lambda$$

These two forms explicitly show the effect of the different parameters on the system. The term on the far right is a function of the target. The next term to the left consists of constants (except for τ_a which will be carried through the calculations and is difficult to evaluate).

The next term to the left deals with scanning; it involves the specified field of view, θ_f and ϕ_f, the specified frame time, t_f, the designed scan efficiency, η_s, and the chosen number of detectors, m, all in a square-root relationship to SNR_{Temp}. The next term involves the size of the optics, the $F/\#$, their transmission, and the solid-angle field of view. Finally the detector parameters are in the equation.

Figure 19-2 is a plot of

$$\pi \int_{\Delta\lambda} \frac{\partial L_{q,\lambda}}{\partial T} \, d\lambda$$

for various values of temperature. If all the items in the field have the same emissivity, it is sufficient to evaluate this term as the only one generating target contrast. The relative effects of temperature and emissivity changes can be evaluated directly. For monochromatic radiation

$$\frac{dL_{q,\lambda}}{L_{q,\lambda}} = \frac{1}{L_{q,\lambda}} \frac{\partial L_{q,\lambda}}{\partial T} dT = \frac{1}{L_{q,\lambda}} \frac{\partial L_{q,\lambda}}{d\epsilon} d\epsilon \tag{19-46}$$

$$= xe^x(e^x - 1)^{-1} \frac{dT}{T} + \frac{d\epsilon}{\epsilon}$$

At the wavelength that maximizes the contrast (for a given temperature), $xe^x(e^x - 1)$ ≈ 6. For broadband radiation $L_q^{BB} = \sigma_q T^3$ (Chapter 1) and

$$\frac{dL_{q,\lambda}}{L_{q,\lambda}} = 3 \frac{dT}{T} + \frac{d\epsilon}{\epsilon} \tag{19-47}$$

For $T \approx 300 \text{ K}$ and $\epsilon \approx 1$,

$$\frac{\partial L_{q,\lambda}}{L_{q,\lambda}} \approx 0.01 \, dT + d\epsilon \tag{19-48}$$

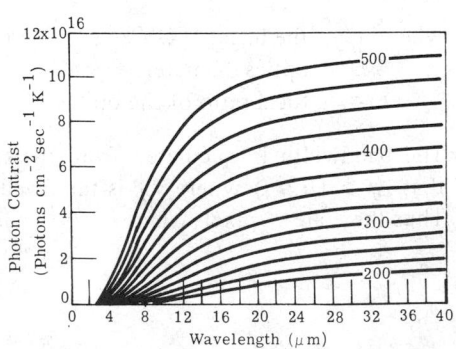

Fig. 19-2. Photon contrast plot.

Therefore, 1% emissivity changes are about the equivalent to 1 K temperature changes. This does not take into account the radiation from the background that is reflected into the sensor. For a full hemisphere at temperature T_B, this is $(1 - \epsilon) L_{q,\lambda} (T_B)$. If an element has 0.01 less emissivity, it will have 0.01 more reflectivity. The relative change incorporating this effect is

$$\frac{dL_{q,\lambda}}{L_{q,\lambda}} = \frac{1}{L_{q,\lambda}} \frac{\partial L_{q,\lambda}}{\partial T} dT + \frac{1}{L_{q,\lambda}} \frac{\partial L_{q,\lambda}}{\partial \epsilon} d\epsilon - \frac{L_{q,\lambda}(T_B)}{L_{q,\lambda}} d\epsilon$$

$$\approx 3 \frac{dT}{T} + \frac{d\epsilon}{\epsilon} \left(1 - e^{x - x_B} \right)$$

(19-49)

The last step is possible because, for most conditions, x is between 3 and 7. If $d\epsilon$ is positive, then pixel 2 radiates more than pixel 1. If $x = x_B$, at any λ the temperatures of the pixel and the background are the same; there is no emissivity contribution. If $x > x_B$, the background reduces the emissivity contribution so that it subtracts from the temperature effect. But $x > x_B$ at a given wavelength means $T < T_B$; the background is hotter than the scene. Conversely, the usual situation is $x < x_B$, by values varying from 0 (for a complete low overcast) to 2.5 (for a very dry, clear, cold night). This latter condition would give

$$\frac{dL_{q,\lambda}}{L_{q,\lambda}} = 3 \frac{dT}{T} + 0.3 \frac{d\epsilon}{\epsilon}$$

(19-50)

For representative conditions of $T = 300$ K and $\epsilon = 1$,

$$\frac{dL_{q,\lambda}}{L_{q,\lambda}} = 0.03 \, dT + 0.3 \, d\epsilon$$

(19-51)

Emissivity then has only about ten times the effect.

The next term involving the constant, is evaluated as follows:

$$\frac{\pi \tau_a}{4 c h} = \frac{3.14 \, \tau_a}{4 x 3 x 10^{10} \times 6.6 x 10^{-34}}$$

$$= \frac{3.14 \, \tau_a}{4 x 19.8 \times 10^{-24}}$$

(19-52)

$$= 4 \times 10^{-22} \, \tau_a$$

Probably as much time is spent on designing scanning systems and scanning optics as on any other part of infrared system design. Scan designs are intimately related to the requirements, to the size of detectors and the number of detectors one can get in an array, and to the wide-angle performance of optical systems. These are all subject to detailed design and changes in technology, but some rules of thumb, and even some data, can be given.

Scanning efficiency is the ratio of the minimum dwell time of the scanner under consideration to the dwell time of an ideal scanner, e.g., the frame time divided by the total number of pixels. Thus

$$\eta_s = t_{d,\min}\frac{N}{t_f} \qquad (19\text{-}53)$$

If the scanner spends any time in retrace or in turnaround, the efficiency is decreased. If the scan velocity is not constant, the efficiency is decreased (because, with no dead time, $t_f/N = t_{\text{ave}}$ and $t_{\min} < t_{\text{ave}}$). Various schemes for accomplishing different geometries of scan with different field of view geometries and efficiencies are described in Chapter 10.

The choice of more than one detector is usually dictated by the requirement for high data-rate. If one scans a typical TV frame of 525 lines with a 4/3 aspect ratio in 1/30 sec, the dwell time is given by

$$t_d = 525^2\,(0.75)\,(1/30)$$
$$\approx 1.5 \times 10^{-6}\ \text{sec} \qquad (19\text{-}54)$$

Although this is about the dwell time for modern detectors, it also requires a very wide bandpass, thereby reducing the SNR. The other extreme would be to use one detector for every pixel; in this case 2.75×10^5 detectors would be required. Detector channels presently cost from 10 to 100 dollars. But even at 1 dollar per channel such a system is no bargain.

Cost is not the only hindrance to the mosaic approach. A chopper is needed to retain the usual ac system, and 50% of the energy is lost in chopping. Perhaps the largest technical difficulty in mosaic systems is the uniformity of response that is required—from 99 to 99.9% (sometimes even a little better). The sheer lack of space in a focal plane is another hindrance to mosaic systems. It appears that the only way to get contiguous coverage is to have (infrared) transparent conductors on the front face, with leads. It would seem that a full-blown, high-resolution mosaic is only feasible when scanned by some sort of technique that does not require leads, e.g., electron guns or CCDs.

The next step seems to be to scan the mosaic just a little—say one line—in one direction. This might mean 25 to 50 μm length scans. The array would alternate with detectors and leads. This is still a difficult manufacturing job, requiring many detectors. For the example chosen, 525 detectors could be used in a linear array and the dwell time would be 8×10^{-4} sec. Problems would still remain with line-to-line uniformity. One solution is to determine the spectral bandpass by a single filter for all detectors. This fixes the longwave cutoff, but it still cannot adjust for individual shapes. Each detector can have its own amplifier with individually adjustable gain. However, these can be expensive, and represent a tremendous challenge for solid state circuit manufacturers.

Some systems have been developed that use the serial scan, in which every detector samples every part of the field of view. It might at first appear that no gain in SNR accrues, but in fact the signals from the detectors are delayed and added. The SNR is proportional to $m^{1/2}$ (the square root of the number of detectors), as with parallel scanning, but a little more dead time and faster scan rates are required.

The series-parallel or hybrid scanner uses rows and columns of detectors. It has some of the advantages and disadvantages of both. The outputs from the detectors in a row are delayed and added, but the system does not have to scan as rapidly because there are several rows.

There are two types of scanning systems: object scanning and image scanning. In the first, the image of the detector (or array) sweeps over the field of view because the optical axis moves. Such a system generally requires large scanning members, but optics that perform only over a narrow field of view (that of the detector or array). The second forms an image of the entire optics and that image is scanned. Small scanning-optics can be used, but they must produce good imagery over the total field of view. Figure 19-4 (Section 19.5.1) shows the approximate performance of a variety of telescope systems. It can be seen that for resolutions much better than 0.1 mrad, over a field of view greater than 150 mrad, it is necessary to scan in object space.

19.3. Temporal Frequency Response Effects

The low-frequency cutoff affects the droop of the system and the high-frequency cutoff is a determining factor in resolution dc-restoration. Boost and other effects also influence the performance of an imaging system.

If the low-frequency cutoff of the system is too high, the picture can display a fading or "droop" toward the end of a line or of a frame. This result can be obtained by determining the band-limited output of an impulse or of a step. If the high-frequency response of the system is too low, the sharp edges of picture elements will appear fuzzy or ill-defined. Both of these effects can be determined by the analysis of the typical response of a system.

The relatively simple system described by the following transfer function, is shown in Figure 19-3.

$$\frac{s\tau_l}{(1 + s\tau_l)(1 + s\tau_r)} \qquad (19\text{-}55)$$

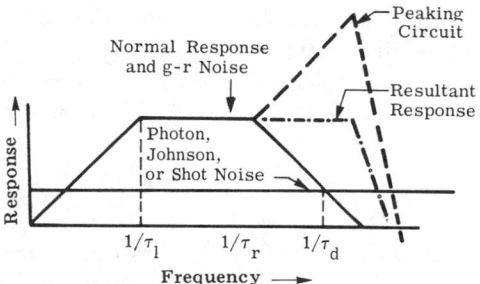

where s = Laplace frequency variable
τ_l = low frequency time constant
τ_r = responsive time constant

This is equivalent to

$$\frac{\tau_l}{\tau_r - \tau_l}\left[\frac{1}{1 + s\tau_r} - \frac{1}{1 + s\tau_l}\right] \qquad (19\text{-}56)$$

Fig. 19-3. Plot of the asymptotes of the voltage transfer function.

The weighting function equivalent to this transfer function is

$$\frac{\tau_l}{\tau_r - \tau_l}\left[\frac{1}{\tau_r}e^{-t/\tau_r} - \frac{1}{\tau_l}e^{-t/\tau_l}\right] \qquad (19\text{-}57)$$

The response of this system to a unit step is given by

$$\frac{\tau_l}{(1 + s\tau_l)(1 + s\tau_r)} = \frac{\tau_r - \tau_l - 1}{(\tau_r - \tau_l)(1 + s\tau_l)} + \frac{1}{(\tau_r - \tau_l)(1 + s\tau_r)} \qquad (19\text{-}58)$$

This response in the time domain is

$$\frac{1}{\tau_r - \tau_l}\left[(\tau_r - \tau_l - 1)e^{-t/\tau_l} + e^{-t/\tau_r}\right] \qquad (19\text{-}59)$$

19.3.1. Direct Current Restoration Techniques. Most imaging systems do not have dc response; they are ac-coupled through a capacitor. Accordingly, there is no average level of scene radiance recorded on the image unless some method is used to restore a dc level to the system. This can be done arbitrarily by the insertion of a dc voltage level after several stages of signal amplification. Usually, in such a case the level is adjustable; the adjustment is chosen by the operator. All radiometric information is lost. A second technique is to force the detector to view the radiation from an internal standard. The black level of the picture can be chosen by setting to any desired reference temperature, or the equivalent blackbody temperature of the black level can be found by trial and error with these adjustments. A third method is to set the reference level from scene information. There are several variations of this. One can obtain the average of the last several lines, or the average value of the last frame, or even the average value of a designated portion of the scene from the previous frame. Each of these techniques is a form of automatic gain control (AGC). An averaging circuit is used. The differences are mainly in the use of the initiation and completion pulses for the averaging. The dc restoration is quite similar to the brightness control in commercial television. Reference to books on television engineering should be helpful.

19.3.2. High-Frequency Boost. Most detectors have a 6 dB/octave rolloff of the frequency response of their signal and of their generation-recombination, g-r, noise. Some system gain can be obtained by boosting the gain of the amplifier between the cutoff frequency of the response and the frequency at which the g-r noise is equal to either Johnson, shot, or photon noise. This is shown schematically in Figure 19-3. This procedure improves the high-frequency response of the system. The peaking circuit for this application should have a +6 dB/octave slope from $1/\tau_r$ to $1/\tau_d$ and a sharp cutoff slope (12 to 18 dB/octave) at $1/\tau_d$. The overall transfer function would then be

$$\frac{(1 + s\tau_l)\,(1 + s\tau_r)}{(1 + s\tau_r)\,(1 + s\tau_d)^m} \tag{19-60}$$

Other similar filter designs can be used. A resonant, or Tchebychef, or Butterworth filter could be used. A representative filter gives a response $(1 + s\tau_l)\,(1 + s\tau_d)^{-m}$. For $m = 2$, the response to an impulse is

$$\left(\frac{1}{\tau_l} - \frac{1}{\tau_d}\right) t e^{-t/\tau_d} - e^{-t/\tau_d} \tag{19-61}$$

For $m = 3$, it is

$$\left(\frac{1}{\tau_l} - \frac{1}{\tau_d}\right) e^{-t/\tau_d} + t e^{-t/\tau_d} \tag{19-62}$$

19.4. Specifications for Thermal Imaging Systems [19-2]

19.4.1. Specimen Specifications. The following example can be used, where applicable, to purchase an imaging device or to describe one. It can be used for framing devices, strip-mappers and for real-time or reconnaissance imagery.

Primary Performance Parameters
    Total field of view (TFOV)
    Pointing range (of the center of the TFOV)
    Signal transfer function (as a function of λ and T)

MRT (as a function of angular spatial frequency)
- at center of frame
- at four corners and four sides of frame
- at 0.707 points along diagonals
- at 0.707 points along azimuth and elevation

Dynamic range
Reference level range
Frame rate
V/H capability
Information update rate
Size, type, luminance, viewing distance and apparent magnifications of display

Secondary Performance Parameters

Static uniformity (as a function of scene temperature)
Moving patterns
Geometric distortion
Image stability (for static and dynamic scenes)
System MTF*
Blemishes
Optical noises
Scan pattern
Display spectrum
Display aspect ratio

Other Characteristics

Size
Weight
Form factor
Power consumption (amount, type, regulation, etc.)
Time between attendance
MTBF (for shelf life and use life)
Level of maintenance
Test features
Environment
Electromagnetic compatibility
Indicators, switches, dials, knobs
Warmup time
Standby mode
Detector quality
Focal plane operational life

Special Specifications for CRT, *Diode and Gas*

Arrays, DVST, scan converters and similar displays
Size of display and virtual image size
Dynamic range and background level
Light reflections from display face
MTF of the display

---

*This specification differs from that of Reference [19-2] by relegating MTF to the secondary parameters on the basis that MRT versus spatial frequency includes MTF implicitly.

        Video response and format
        Luminance variations
        Phosphor protection
        Dimensional stability of raster
        Positional
        Line jitter
        Line pairing
        Video input
        Protection against ac phase-reversal
    Check List for Other Features
        Self tests
        Panel lighting
        Electrical power requirements
        Nonoperating period
        Environmental features
            thermal
            pressure
            humidity
            vibration
            shock
            acoustic levels
            sand and dust
            explosive conditions
            salt
            rain
            portability
        Human interfacing
        Electrical hazards
        Conformal coatings

19.4.2. Discussion of Specifications. There is considerable evidence that observers perform better (i.e., make fewer mistakes or detect more targets) when the area, MTFA, is higher. The MTFA is the area between the MTF curve of the system and the aerial image modulation (AIM) curve (the minimum modulation the eye needs at a given spatial resolution just to discern a difference):

$$\text{MTFA} = \int_0^\infty [\text{MTF}(f_x) - \text{AIM}(f_x)] \, df_x \qquad (19\text{-}63)$$

(The AIM curve is also called the DMF, or demand modulation function curve.) There seems to be no one "correct" measure, and almost all of the measures depend upon the overall system MTF. Thus, it seems that the MRT curve, which may be viewed as the system MTF times the minimum detectable temperature difference at zero spatial frequency, is the best measure of infrared image performance to date. The MRT should be measured under a variety of conditions. The SNR equations show that MRT is a function of the average scene radiance. Most optical systems have more aberrations at the edges of the field than at the center. Many scanning systems do not have constant-velocity scans. All the detectors in arrays are not identical. This means that scenes with different spectral distributions may yield different performance evaluations, as may operation at different ambient temperature, look-angles, or even background spatial distributions.

19.4.3. Four-Bar Targets. The standard test for resolution for visible-spectrum systems has for years used the three-bar target of alternating black and white bars of different width. More recently, Biberman [19-2] has proposed a four-bar chart with an aspect ratio of 7 to 1. In theory, of course, this is not a sine-wave target, but a square-wave target. One obtains the square-wave response and must correct. (See Chapter 8.) It also is a truncated square wave. If the bars have a width, w, and a period $2w$, then the spectrum can be written

$$\text{Bar}\,(\omega) = \left[\left(\text{sinc}\,\omega\,\frac{w}{2}\right)*\text{comb}\,\omega w\right] \times \left(\text{sinc}\,m\omega\,\frac{w}{2}\right) \tag{19-64}$$

This is a broad sinc function with its first zero at π, combined with a series of narrow sinc functions with first zeros at π/m from their centers, but repeated at frequencies of $\omega = 1/w$. The number of lines in the bar chart, m, determines how well these sinc functions approach delta functions, and therefore the quality of the target.

19.5. Optical System Considerations

Before a system is given to an optical designer (and after it comes back) several things must be considered. These include layout, narcissus, vignetting, etc.

19.5.1. Optical System Layout. Chapters 8 and 9 provide much information on the design and performance of optical systems. It is sometimes convenient to have some rules of thumb from which to start the design. Figure 19-4 provides one such approximate set of data. The figure is derived from the approximate design equations and tables at the end of Chapter 9. The curves show the diameter of the blur spot of the optical system as a function of the half angle of the field of view. The horizontal curves labelled "$F/1$, $F/2$, and $F/4$ Spherical, Mirror" are the spherical aberrations of a simple mirror of any figure. The horizontal curves labeled "$F/1$ and $F/1.5$ Rear Bouwers" describe the total aberration for a Bouwers with the corrector plate curved toward the mirror (and similarly for a Front Bouwers with the corrector curved in the opposite direction).

There are four lines with a slope of one that represent the coma of simple mirrors. The astigmatism is shown by four curves with a slope of two. All these simple mirrors are assumed to have the stop at the mirror. Finally, there are four lines in the lower right part of the figure that represent the performance of Schmidt systems of four different focal ratios. The figure should be viewed with considerable skepticism for the region at about 0.3 rad off-axis and beyond. Higher-order aberrations can be important then.

The use of the chart can be illustrated by an example. Assume a system needs to provide 0.1 mrad imagery about 6° off axis. An $F/1$ mirror has two orders-of-magnitude too much spherical aberration (at 0.1 rad off-axis). An $F/4$ spherical mirror would be a little "out of spec" based on spherical aberration alone, but the coma, when added, would bring the blur to $0.12 + 0.38 = 0.4$ mrad. The astigmatism would be over 0.1! A Schmidt system between $F/1.4$ and $F/2$ would be all right. A rear Bouwers with a focal ratio of about $F/1.3$ or a front Bouwers of about $F/1.1$ could also be used. After very preliminary optical considerations of this sort, one should go to Chapter 9 and thence to an appropriate computer program.

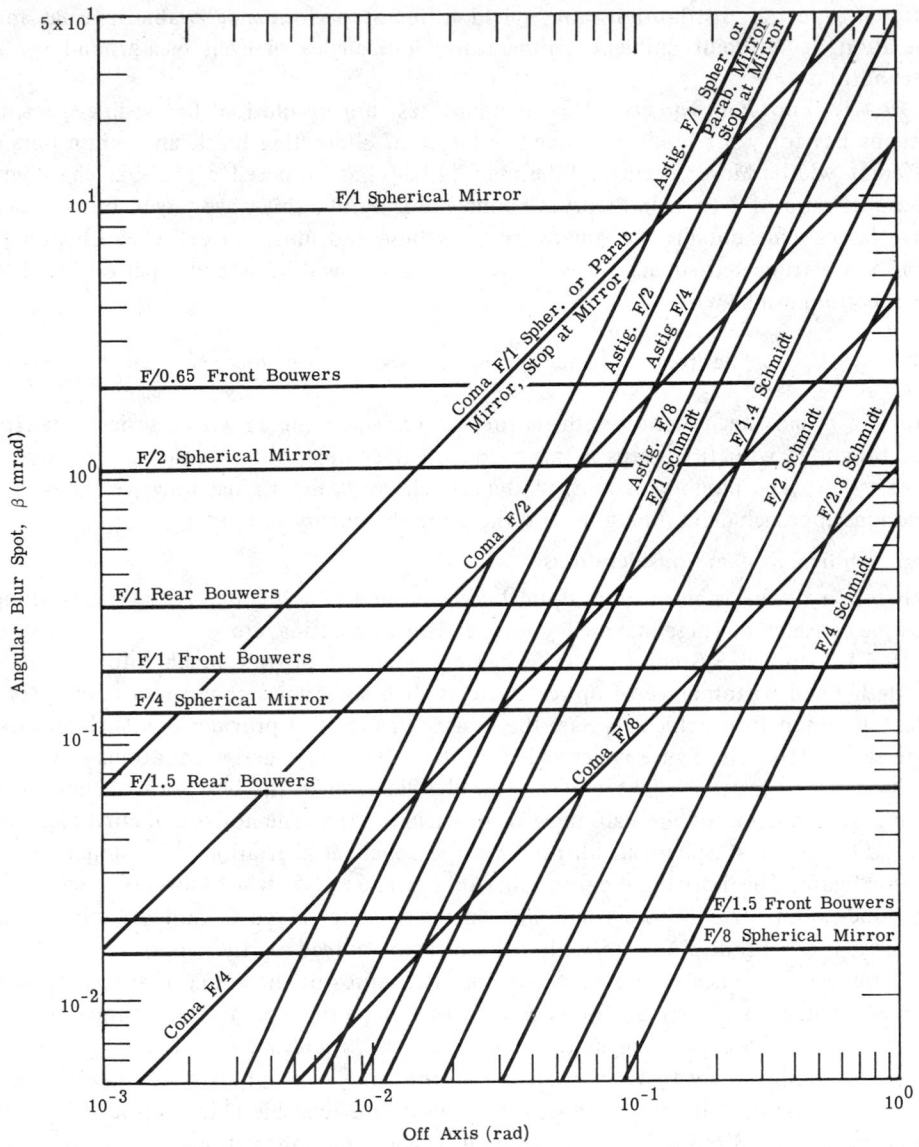

Fig. 19-4. Approximate set of data for design of optical system layout.

19.5.2. Narcissus. The irradiance on the detector can be calculated by the magnification. The irradiance on the detector from the detector is

$$E_d = \rho M_d \left(\frac{2d}{R} + 1\right)^2 \tag{19-65}$$

where M_d = the radiant exitance of the detector

d = the distance from the detector to the lens

R = the radius of curvature of the lens

ρ = the reflectivity of the lens surface

This is the constant input to a detector from the detector itself as compared to that from an ambient background. This is a good, rather than a bad effect because most systems are ac-coupled. The detector sees a constant field, and the reflected amount is negligible compared to the emission from the lens and the transmission of radiation through the lens. To indicate this, one can calculate the emission from the lens. If one assumes a very low absorption coefficient of 10^{-4} cm^{-1}, then the absorptivity (equal to the emissivity) is given by

$$\epsilon = 1 - e^{-\alpha x} \approx 1 - 1 + \alpha x = \alpha x \qquad (19\text{-}66)$$

So, for a 2 cm thick lens, the emissivity is about 2×10^{-4}. Therefore the irradiance on the detector is given by

$$E = \frac{\phi}{A_d} = \int L d\Omega \approx \frac{\alpha x \, c_1 \, \lambda^{-5} \, (e^x - 1)^{-1} \, A_l}{2 \pi \, d^2} \qquad (19\text{-}67)$$

The real difficulties with narcissus occur when the reflection of the detector moves in and out of the detector's field of view. In theory, any reflection whether focused on the detector or not, can cause a cold spike which is a sudden transient narcissus. Such a calculation need only be done for reflections involving a moving member of the optical system.

19.5.3. Dynamic Noises (Scanning Noise). These are often the most difficult to analyze and to measure. Because they are not amenable to general analysis, some dynamic or scanning noise causes and some of the ways in which noise or variation in response can occur are explained below.

19.5.4. Spurious Detector Reflections. Most detectors have evaporated gold contact pads or other reflecting areas including leads that can be flat or cylindrical. Target signals can reflect from these areas to optics or other surfaces and reflect back to the same detector or neighboring ones in an array. This can generate scanning noise if scan elements are involved; it can be the source of crosstalk; and it can be the cause of asymmetrical scanning characteristics.

19.5.5. Polarization Properties. Almost every scanner is somewhat sensitive to some degree of polarization of input radiation. In some cases, parts of the scanner polarize the input radiation, which may or may not be polarized. Other parts of the scanner can then act as analyzers.

19.5.6. Vignetting. As shown in Chapter 10, some scanners can have an entrance pupil which is a function of the position of the scanning member. This causes a regular and periodic variation in the sensor responsivity. Such an effect is dynamic vignetting.

19.5.7. Scan Reflections Variations. In some scanners, even with a constant entrance pupil, various portions of the overall field are seen by scanning members that are not participating in the active part of the scan. These members can reflect different levels of radiation to the detector from different parts of the background. This changes both the apparent signal level and that of the noise. One important example is the noise from a Nipkow scanning wheel when it is not uniform in temperature, emissivity or both.

19.5.8. Detector Nonlinearities. Many detectors are extrinsic crystals or polycrystalline films. They can operate differently under different levels of illumination or of bias current. The current for a constant voltage can change if the detector resistance changes. The detector resistance can change with the amount of background radiation on the detector. All of these factors contribute to a nonlinear (in the sense of non-constant parameters) response. The response and the noise levels (and character of the noise) can change with irradiation.

19.5.9. Baffle Designs. The use of baffles varies in importance depending upon the application. There are presumably two main methods for the design of these baffles. They both involve analysis of results, after basic guidelines are used, to eliminate any direct radiation on the detector and any single-reflection radiation. The two techniques both use computer methods of tracing rays to find the amount of off-axis rejection: GUERAP (General Unwanted Radiation Analysis Program) is a Monte Carlo ray-tracing technique [19-3, 19-4]; Arizona Paraxial Analysis of Radiation Transfer (APART) [19-5] uses the y-$\bar{y}$ diagram to image surfaces and their BRDF data to calculate transfers. Both programs use lots of computer core and are too complicated for handbook presentation. Some representative results for simple designs are given here as a guideline.

Two types of Cassegrainians are shown. Their properties are: 10 cm aperture, $1°$ field of view; baffles were 0.01 Lambertian reflectors; mirrors were 0.01 BRDF at $1°$ from specular, with a θ^{-1} falloff; and an obscuration ratio of 0.33. The first system has its aperture stop located at the primary mirror. Figure 19-5 is a drawing of an $F/8$ system of this type. Both an $F/8$ and $F/5$ system were analyzed, and the off-axis point source transmittance (PST) characteristics of the two focal ratios are plotted in Figure 19-6. The $F/8$ system had lower PST values because of its greater length (i.e.,

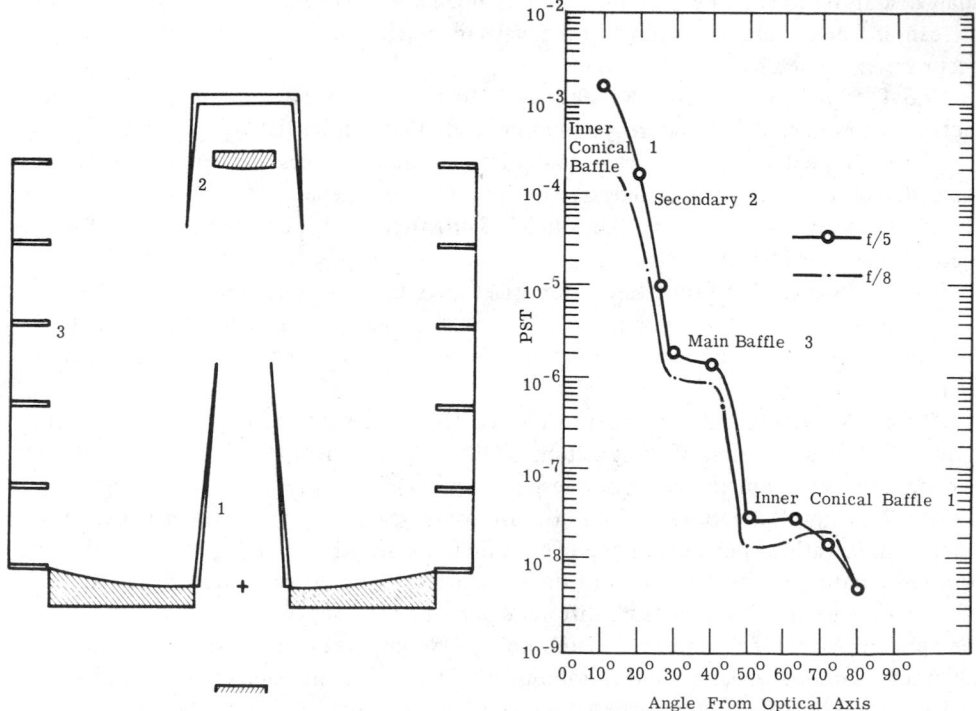

Fig. 19-5. F/8 Cassegrain with the stop at the primary mirror.

Fig. 19-6. F/5 and F/8 Cassegrains with the stop at the primary mirror.

smaller solid angles involved). Those objects that were the primary source of stray radiation for each angle are also indicated on the figure. For off-axis angles between 30° and
40°, those sections of the main tube near the primary mirror, as seen in reflection
through the secondary mirror, contributed significantly to the stray radiation in the
image plane.

In a second type of system, the stop is at the secondary mirror. This means a larger
primary, but smaller secondary mirror. Figure 19-7 shows the effect of the stop shift on
the PST in the region of 30 to 40° off-axis. Figure 19-8 shows the PSTs for a Gregorian
and four-mirror system.

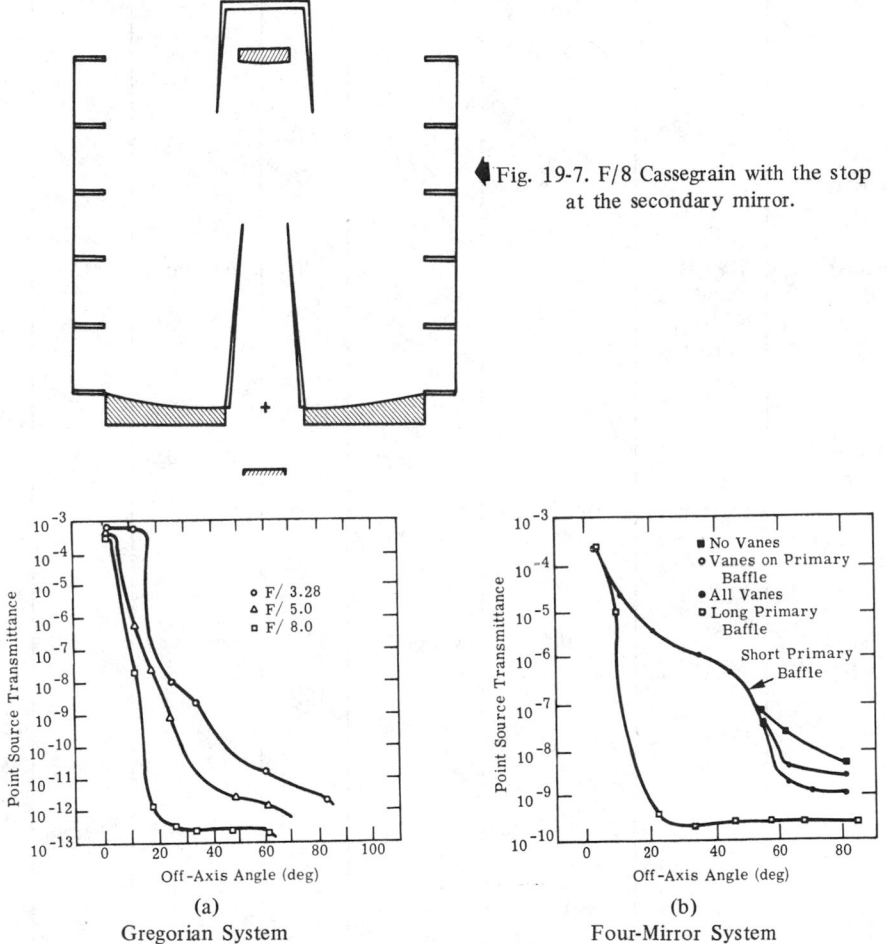

Fig. 19-7. F/8 Cassegrain with the stop
at the secondary mirror.

(a)
Gregorian System

(b)
Four-Mirror System

Fig. 19-8. PSTs for Gregorian and Four-Mirror Systems

19.6. Existing Imaging Instruments

Many companies have developed their own instruments for various applications to
military programs, and many of these programs are sponsoring equipment development.
Development of those instruments used for aerial-survey, remote-sensing, and multispectral scanning, as well as medical and industrial thermography, is fairly mature. Instruments illustrative of the state-of-the-art are listed in Tables 19-3 to 19-6.

Table 19-3. Commercially Available Infrared Scanners

Manufacturer	Method of Scanning	Lines/sec Rate of Scan (rpm)	Size of Resolution Element (mrad)	Swath Width (deg)	Type of Collecting Optics	Focal Length (cm)	F/no.
Texas Instrument* RS-310	4-sided prism	200	1.5	90°	Reflective	15	2.1
TRW Hawker† Saddeley			1.5	120°			
Daedalus* DEI-100	Double-faced 45° mirror	120	2.5	120°	Newtonian	15	1.2
Daedalus† DS-1200		80	2.5	77°	Newtonian	15	1.2
HRB Singer* Reconofax IV	45° Mirror	105	2 or 3	120° or 140°	Newtonian	15	1.8
Reconofax VI*	Double-faced 45° mirror	370	2 or 3	120° or 140°	Newtonian	15	1.8

*Data obtained from the Infrared Information and Analysis Center, Environmental Research Institute of Michigan, Ann Arbor, MI.
†Data obtained from manufacturer literature.

Table 19-3. Commercially Available Infrared Scanners (*Continued*)

Effective Collector Area (cm^2)	V/H, Maximum	Spectral Bands of Each Channel (μm)	Type of Detector	NEΔT (K)	Method of Recording	Weight (lbs)	Power
40.5	0.3		5 to 13 PM Silicon		Film	96.5	dc: 28 V 9 A
			InSb	0.48	Magnetic tape is optional		ac: 110 V 400 Hz 8 A
			GeHg HgCdTe	0.20			
	0.75	7.5 to 14.0	HgCdTe	0.5	Film	25	dc: 28 V 6 A
45.0	0.3	0.35 to 1.0	PM	0.3	Magnetic tape and film are optional	65	dc: 28 V 5 A
		1.0 to 5.5	InSb	0.2			
		0.35 to 1.0	PM				
40.7	0.2	0.35 to 1.0	PM	0.3	Magnetic tape and film are optional	85	dc: 28 V 5 A
		1.0 to 5.5	InSb	0.2			
		8.0 to 14.0	PbSnTe				
47.6	0.37 to 0.56		GeHg InSb	0.3 0.5	Film	55 scanner, plus 20 for closed-cycle cyrostat	ac: 3 phase, 115 V, 400 Hz, 4 A surge, 1.5 A steady
							ac: 3 phase, 115 V, 400 Hz, 5 A surge, 3 A steady (cooler)
21.0	0.75 to 1.1		GeHg InSb	0.5 0.7		55 scanner plus 20 for closed-cycle cyrostat	Same as Reconofax IV

IMAGING SYSTEMS

Table 19-3. Commercially Available Infrared Scanners (*Continued*)

Manufacturer	Method of Scanning	Lines/sec Rate of Scan (rpm)	Size of Resolution Element (mrad)	Swath Width (deg)	Type of Collecting Optics	Focal Length (cm)	F/no.
Reconofax XI*	Double-faced 45° mirror	133	1.5	120°		25	2.0
HRB Singer† Reconofax XIIA Model 13-21	4-sided pyramidal mirror	800	1 and 2 mirror (in same drawer)	120°	Reflective	10	1.5
Bendix T/M* LN-3	45° mirror	100	2.5	120°	Dall-Kirkham	30	4.0

*Data obtained from the Infrared Information and Analysis Center, Environmental Research Institute of Michigan, Ann Arbor, MI.
†Data obtained from manufacturer literature.

Table 19-3. Commercially Available Infrared Scanners (*Continued*)

Effective Collector Area (cm^2)	V/H, Maximum	Spectral Bands of Each Channel (μm)	Type of Detector	NEΔT (K)	Method of Recording	Weight (lbs)	Power
48.0	0.20	8.0 to 14.0	InSb GeHg	0.32 0.2	Film	109	dc: 28 V, 8 A
35.0	1.6 rad sec^{-1}	8.0 to 14.0	HgCdTe	0.3 at 1 mirror film 0.2 at 2 mirror	Film	135, including cooler and stabilized mount	28 V dc at 4.3 A; 115 V ac, 400 Hz, 3-phase; 400 W
45.5	0.25	0.38 to 0.6 0.2 to 0.6 1.0 to 5.5 8.0 to 14.0	5 to 11 PM 5 to 13 PM InSb HgCdTe	0.3 0.2	Film Magnetic tape optional	55	dc: 28 V, 5 A

Table 19-4. Medical Thermographs

Manufacturer's Name	Method of Scanning	Frame Time (sec)	Lines/Frame	Size of Resolution Element (mrad)	FOV	Type of Optics
AGA 680 M	Horizontal and vertically rotating Ge prism	1/16 (real time)	90	–	25 × 25	Refractive Reflective
Inframetrics	2 mirrors, wobbling orthogonally	1/30 (real time)	200	2.0	156 × 20 horizontal	Reflective, doublet, and refractive
Spectrotherm	2 mirrors, 6-sided rotating prism, rocking in sync.	2.0	528 vertical 600 elements/horizontal line (2% point of MTF)	1.3	30 × 30	Reflective and refractive
Unico	Rotating, hexagonal, front-surface mirror, oscillating in vertical direction	4.3	540	0.75	33 × 33	Reflective and refractive

Table 19-5. Spacecraft Scanners

Manufacturer	ITT Aerospace Optical Division	ITT Aerospace Optical Division	Santa Barbara Research Center	Santa Barbara Research Center	Santa Barbara Research Center
Scanner model	High-resolution infrared radiometer	Surface-composition mapping radiometer	Medium resolution infrared radiometer	Spin scan cloud camera	Multi-spectral scanner
Satellite	Nimbus I, Nimbus II, Nimbus III	Nimbus I	Nimbus II, Nimbus III	ATS I, ATS II	ERTS A (LANDSAT I)

Table 19-4. Medical Thermographs (*Continued*)

Effective Optical Collective Area (cm²)	Spectral Bands (μm)	Type of Detector	NETD (K)	Method of Recording	Weight (kg)	Power Requirement	Price
4 in. diameter	3.5 to 5.5	InSb and HgCdTe	0.2	CRT plus Polaroid 195, 70 mm, 35 mm, Video magnetic tape	50	110 V ac	$45,000
1.27	8 to 12	HgCdTe	0.2	CTR plus Polaroid 195, plus video magnetic tape	–	110 V ac, 20 W; or 12 V dc	$25,000
2.5 in. diameter	8 to 12	HgCdTe	0.2	CRT (dual mode-red or video) Video tape, Polaroid or 70 mm	125	100 V ac, 200 W	
103	8 to 12	HgCdTe	0.07	CRT, Polaroid or 70 mm	180	110 V ac, 300 W	$36,000

Table 19-5. Spacecraft Scanners (*Continued*)

Santa Barbara Research Center	Santa Barbara Research Center	Santa Barbara Research Center	Santa Barbara Research Center	RCA	Honeywell
Multi-spectral scanner	Temperature humidity IR radiometer	ITOS scanning radiometer	Visible IR spin scan radiometer	Very high resolution radiometer	S-192, Multispectral scanner
ERTS B (LANDSAT II)	Nimbus IV, K, R	NOAA 1,-,5 ITOS 1	SMS 1,2 GOES 1,2,3	NOAA 2,3,4,5	SKYLAB A

Table 19-5. Spacecraft Scanners (*Continued*)

Manufacturer	ITT Aerospace Optical Division	ITT Aerospace Optical Division	Santa Barbara Research Center	Santa Barbara Research Center	Santa Barbara Research Center
Method of scanning	Single 45° mirror	Single 45° mirror	Single 45° mirror	Spinning spacecraft plus tilting mirror	Pivoting mirror
Scan lines per second	0.8 0.8 0.8	10	0.131	1.66 1.66	13.6 scans/ sec 6 lines/scan
Swathwidth (deg)	180 180 180	90	160	15.0 17.3	11.5
Orbit	Sun synchronous	Sun synchronous	Sun synchronous	Earth synchronous	Sun synchronous (500 nmi)
Resolution IFOV (mrad)	7.9 7.9 7.2	0.6	50	0.1 0.1	0.086
Type of collecting optics	Modified cassegrain	Afocal Dall-Kirkham	Cassegrain for each band	Paraboloidal Newtonian Wynne-Rosin	Ritchey Chrétien
Focal length and F/no.	102 mm, F/1	166 mm, F/0.8	107 mm, F/2.44	254 mm, F/2 380 mm, F/3	825 mm, F/3.6
Diameter of aperture (mm)	102	203	43.7	127 127	229
Effective aperture area (mm^2)	5000	10,795	1160	–	32,900
Calibration in flight	Blackbody plus outerspace	Blackbody plus outerspace	Blackbody, space, sun	–	Lamp, sun

Table 19-5. Spacecraft Scanners (*Continued*)

Santa Barbara Research Center	Santa Barbara Research Center	Santa Barbara Research Center	Santa Barbara Research Center	RCA	Honeywell
Pivoting mirror	Single 45° mirror	Single 45° mirror	Spinning spacecraft and tilting mirror	45° mirror	Image plane control
13.6 scans/ sec 6 lines/scan (refl. bands) 2 lines/scan (thermal bands)	0.8	0.8	1.6 scans/ sec 6 lines/scan (visible) 1 line/scan (IR)	6.67	100
11.5	180	180	18	180	9
Sun synchronous (500 nmi)	Sun synchronous (600 nmi)	Sun synchronous (790 nmi)	Geostationary	Sun synchronous (790 nmi)	435 km
0.086 (reflective bands) 0.26 (thermal band)	7(10.5 to 12.5 µm) 21(6.7 µm)	2.8(0.52 to 0.72 µm) 5.3(10.5 to 12.5 µm)	0.021(0.55 to 0.75 µm) 0.25(10.5 to 12.6 µm)	0.6	0.182
Ritchey Chrétien	Cassegrain	Cassegrain	Ritchey Chrétien	Dall-Kirkham	Reflective Schmidt
825 mm, F/3.6	430 mm, F/3.38	430 mm, F/3.38	2913 mm, F/7.2	113 mm F/0.89	635 mm
229	125	127	406	127	–
32,900	10,170	11,000	109,000	11,400	73,000
Lamp, sun	Blackbody, space	Blackbody, space, lamps	Blackbody, space, sun	Blackbody, space, sun	–

Table 19-5. Spacecraft Scanners (*Continued*)

Manufacturer	ITT Aerospace Optical Division	ITT Aerospace Optical Division	Santa Barbara Research Center	Santa Barbara Research Center	Santa Barbara Research Center
Method of recording (ground)	Photo-facsimile film, digital or analog	Photo-facsimile film, digital or analog	Analog, output chart digital tape photo display	Photo-facsimile	Computer compatible magnetic tape
Method of spectral separation	Filters	Filters plus dichroics	Filters	Filters	Filters
Size				10 × 11 × 17 in.	15 × 24 × 35 in.
Weight (kg)	8.45(15.6 lb) 8.45(15.6 lb) 8.67(19.1 lb)	22.7(50 lb)	6.9(15.2 lb) 6.9(15.2 lb)	9.0(20 lb) 10.4(23 lb)	51(112 lb)
Power (W)	4 4 4	15.0	7.5	24 maximum 3.08 average 22.5 maximum 10.0 average	42 maximum
Source of specifications	IRIA	IRIA	IRIA	IRIA	Manufacturer literature
Detector	PbSe	HgCdTe	Bolometer	Photo-multipliers	Photo-multiplier S-1 photo-diode
Detector cooling	Radiation cooled (200)	Radiation cooled (~ 100 K)	No	No	No
Spectral bands (μm)	3.4 to 4.2 3.4 to 4.2 3.4 to 4.2	8.3 to 9.3 10.2 to 11.2 0.8 to 1.1	0.2 to 4.0 5 to 30 6.4 to 6.9 10 to 11 14 to 16; 0.2 to 0.4 6.5 to 7.0 10 to 11 15.4 to 15.5 20 to 23	0.47 to 0.61 0.38 to 0.48 0.48 to 0.56 0.55 to 0.63	0.5 to 0.6 0.6 to 0.7 0.7 to 0.8 0.8 to 1.0

Table 19-5. Spacecraft Scanners (*Continued*)

Santa Barbara Research Center	Santa Barbara Research Center	Santa Barbara Research Center	Santa Barbara Research Center	RCA	Honeywell
Computer compatible magnetic tape	Magnetic tape	Magnetic tape and facsimile	Magnetic tape	Magnetic tape and facsimile	Computer compatible magnetic tape
Filters	Dichroic	Dichroic	Filters	Dichroic	Spectrometer
15 × 24 × 35 in.	7 × 7.5 × 15.6 in.	6.4 × 8.4 × 15.9 in.	26 × 26 × 60 in.	19 × 12 × 10 in.	19.3 ft^3
51(112 lb)	9.03(19.9 lb)	8.30(18.0 lb)	64.5(142 lb)	9.5 (20.9 lb)	159 (350 lb)
42 maximum	7.5	6.5	22 average 32 maximum	6.5	266
Manufacturer literature	Manufacturer literature	Manufacturer literature	Manufacturer literature	Manufacturer literature	NASA literature
Photo-multiplier S-1 photo-diode, HgCdTe	Bolometer	Bolometer S-1 photo-diode	S-1 photo-diode(8) HgCdTe(2)	Hg:Cd:Te/ S-1 photo-diode	HgCdTe
Yes(HgCdTe)	No	No	Yes	Radiation cooled (~100 K)	Closed-cycle cryostat (77 K)
0.5 to 0.6 0.6 to 0.7 0.7 to 0.8 0.8 to 1.1 10.5 to 12.6	6.5 to 7.0 10.5 to 12.5	0.55 to 0.73 10.5 to 12.5	0.55 to 0.73 10.5 to 12.5	0.6 to 0.7 10.5 to 12.5	0.41-0.46 0.46-0.51 0.52-0.556 0.565-0.609 0.620-0.667 0.680-0.762 0.783-0.88 0.98-1.09 1.09-1.19 1.20-1.30 1.55-1.75 2.10-2.35 10.2-12.5

Table 19-6. Airborne Multispectral Scanners

Manufacturer Model	University of Michigan Experimental Multispectral Scanner M-7	Texas Instruments RS-14	Bendix Aerospace Systems Division Modular Multispectral Scanner (M^{2}S)
V/H maximum	0.12 or 0.20	0.2	0.25
Method of scanning	Single 45° mirror	4-sided mirror	Single 45° mirror
Scan lines/sec	60 or 100		100
Swathwidth (deg)	90	80	100
Resolution IFOV (mrad)	2 maximum	1 or 3	2.5
Type of collecting optics	Dall-Kirkham plus concentric lens system	Dall-Kirkham	Dall-Kirkham
Focal length and F/no.	Dall-Kirkham— 508 mm, F/4.0 Lens system— 76 mm, F/1.0	354 mm, F/3.7	400 mm, F/4
Diameter of aperture (mm)	Dall-Kirkham—217 Lens system—76	—	100
Effective aperture area (mm^2)	Dall-Kirkham—8107 Lens system—4560	3470	6500
Roll compensation	Signal ± 10°	Signal ± 8°	Signal ± 10°
Calibration in flight	Blackbodies, Q-I lamps, sky	Blackbodies, Q-I lamp	Blackbodies, Q-I lamp sunlight
Method of recording	Magnetic tape	CRT with film	Magnetic tape
Method of spectral separation	Beam divider, dichroic beam splitter, filters and prism	Dichroic	Diffraction grating Dichroic
Total weight (less recorder)	225 kg (550 lbs)	120 kg (264 lbs)	118 kg (261 lbs)

Table 19-6. Airborne Multispectral Scanners (*Continued*)

Bendix Aerospace Systems Division Multispectral Data System	Daedalus Enterprises Daedalus Spectrometer Module DS-1050	Actron Industries, Inc. HMS-564X
0.18	0.2	0.1
Single 45° mirror	Single 45° mirror	Conical
9.2 to 90	80	50
80	77	51.3
2.0	2.5	2
Dall-Kirkham	Paraboloidal Newtonian	Cassegrain
1142 mm, F/5.0	152 mm, F/2	—
228	127	150
38100	4560	—
Signal ± 8°	Signal ± 10°	None, stabilized mount optional
Blackbody, Q-H lamp, sky ref.	Q-H lamp	Lamp
14 track digital magnetic tape	7 track magnetic tape	Magnetic tape
Diffraction grating	Prism	Prism beam splitter
1273 kg (2800 lbs)	54.3 kg (120 lbs)	79 kg (175 lbs)

Table 19-6. Airborne Multispectral Scanners (*Continued*)

Manufacturer Model	University of Michigan Experimental Multispectral Scanner M-7	Texas Instruments RS-14	Bendix Aerospace Systems Division Modular Multispectral Scanner (M^{2}S)
Power	882 W 28 V dc 1250 W 400 Hz 115 V 594 W 60 Hz 115 V	154 W 28 V dc 876 V A 115 V ac	7 A 115 V 400 Hz 15 A 28 V dc
Source of specification	University of Michigan	IRIA, manufacturer literature	Manufacturer literature
Spectral bands (μm)	0.32 to 0.38 0.40 to 0.44 0.44 to 0.46 0.46 to 0.48 0.48 to 0.50 0.50 to 0.52 0.52 to 0.55 0.55 to 0.58 0.58 to 0.62 0.62 to 0.66 0.66 to 0.72 0.72 to 0.82 0.82 to 0.96 1.0 to 1.4 1.5 to 1.8 2.0 to 2.6 8.0 to 13.5 Any 12 bands can be recorded	0.3 to 0.55 0.7 to 0.9 1.0 to 1.5 1.5 to 1.8 2.0 to 2.5 3.0 to 5.5 8.0 to 14.0 Only two of the bands can be used at one time.	0.38 to 0.44 0.44 to 0.49 0.49 to 0.54 0.54 to 0.58 0.58 to 0.62 0.62 to 0.66 0.66 to 0.70 0.70 to 0.74 0.76 to 0.86 0.97 to 1.05 8.0 to 12.0

Table 19-6. Airborne Multispectral Scanners (*Continued*)

Bendix Aerospace Systems Division Multispectral Data System	*Daedalus Enterprises Daedalus Spectrometer Module DS-1050*	*Actron Industries, Inc. HMS-564X*
1155 W 28 V dc 10.2 kV A 115 V	132 W 28 V dc	280 W 28 V dc
IRIA, manufacturer	Manufacturer literature	Manufacturer
0.34 to 0.40	0.38 to 0.42	0.5 to 0.6
0.40 to 0.44	0.42 to 0.45	0.6 to 0.7
0.46 to 0.50	0.45 to 0.50	0.7 to 0.8
0.53 to 0.57	0.50 to 0.55	0.8 to 1.1
0.57 to 0.63	0.55 to 0.60	10.4 to 12.6
0.64 to 0.68	0.60 to 0.65	
0.71 to 0.75	0.65 to 0.70	
0.77 to 0.81	0.70 to 0.80	
0.82 to 0.87	0.80 to 0.90	
0.97 to 1.06	0.90 to 1.10	
1.06 to 1.095		
1.13 to 1.17		
1.18 to 1.3		
1.52 to 1.73		
2.1 to 2.4		
3.54 to 4.0		
4.5 to 4.75		
6.0 to 7.0		
8.3 to 8.8		
8.8 to 9.3		
9.3 to 9.8		
10.1 to 11.0		
11.0 to 12.0		
12.0 to 13.0		

19.7. References and Bibliography

19.7.1. References

[19-1]   P. W. Kruse, R. McGlauchlin, R. B. McQuistan, *Elements of Infrared Technology*, John Wiley and Sons, Inc., New York, NY, 1962, p. 58.

[19-2]   L. M. Biberman (ed.), "A Guide for the Preparation for Real-time Thermal Imaging Systems," Institute for Defense Analyses, Arlington, VA, Paper P-676, 1971.

[19-3]   Honeywell Staff, "GUERAP: General Unwanted Energy Rejection, Analysis Program User's Manual," Honeywell Aerospace Division, St. Petersburg, FL, SAMSO TR No. 72-314, Vol. 1, 5 October 1972, Revised 5 January 1973.

[19-4]   Perkin Elmer Staff, "GUERAP II Users Guide," Perkin Elmer Corp., Norwalk, CT, SAMSO TR 73-309, February 1974.

[19-5]   R. P. Breault, *Users Manual for University of Arizona APART Program* (Analysis Program—Arizona Radiation Trace) Optical Sciences Center, University of Arizona, Tucson, AZ, August 1975.

[19-6]   F. O. Bartell, R. P. Breault, A. G. DeBell, B. Fannin, A. Greynolds, D. S. Goodman, J. Gunderson, J. Harvey, S. Lange, R. V. Shack, A. F. Turner, W. L. Wolfe and R. D. Wooden, "A Study Leading to Improvements in Radiation Focusing and Control in Infrared Sensors," University of Arizona, Tucson, AZ, AMMRC CTR 76-42, December 1976.

19.7.2. Bibliography

Bartell, F. O., R. P. Breault, A. G. DeBell, B. Fannin, A. Greynolds, D. S. Goodman, J. Gunderson, J. Harvey, S. Lange, R. V. Shack, A. F. Turner, W. L. Wolfe, and R. D. Wooden, "A Study Leading to Improvements in Radiation Focusing and Control in Infrared Sensors," University of Arizona, Tucson, AZ, AMMRC CTR 76-42, December 1976.

Biberman, L. M. (ed.), "A Guide for the Preparation for Real-time Thermal Imaging Systems," Institute for Defense Analyses, Arlington, VA, Paper P-676, 1971.

Breault, R. P., *User's Manual for University of Arizona APART Program* (Analysis Program—Arizona Radiation Trace), Optical Sciences Center, University of Arizona, Tucson, AZ, August 1975.

Honeywell Staff, "GUERAP: General Unwanted Energy Rejection, Analysis Program User's Manual," Honeywell Aerospace Division, St. Petersburg, FL, SAMSO TR No. 72-314, Vol. 1, 5 October 1972, Revised 5 January 1973.

Kruse, P. W., R. McGlauchlin, R. B. McQuistan, *Elements of Infrared Technology*, John Wiley and Sons, Inc., New York, NY, 1972, p. 58.

Lloyd, J. M., *Infrared Imaging Systems Devices*, Plenum Press, New York, NY, 1974.

NASA Working Group Report, "Advanced Scanners and Imaging Systems for Earth Observations," NASA, Prepared by Goddard Space Flight Center, SP-335, 1973.

Perkin Elmer Staff, "GUERAP II Users Guide," Perkin Elmer Corp., Norwalk, CT, SAMSO TR 73-309, February 1974.

Chapter 20

RADIOMETRY

George J. Zissis
Environmental Research Institute of Michigan

CONTENTS

20. Radiometry

20.1. Introduction

20.1.1. Symbols, Nomenclature and Units.
The symbols and units of quantities used in this chapter are those defined and used earlier in the handbook, especially in Chapters 1-6. For more extensive discussion of them see those chapters. See also Table 20-1.

Table 20-1. Symbols, Nomenclature, Units

Symbols	Nomenclature	Units
A, a	Area	m^2
A_p	Angle of prism	rad
$A(\phi)$	Airy function	—
B	Prism base length	m
C	Distance of mirror motion	m
c	Speed of light in vacuum	$m\ sec^{-1}$
D	Detectivity; also diameter	W^{-1} or m
D_g	Angle of deviation for a grating	rad
D_p	Angle of deviation for a prism	rad
D_E, D_L, D_Φ	Detectivity in terms of the subscripted radiometric quantity	[radiometric unit]$^{-1}$
d	Grating constant; also, distance	m
E	Irradiance	$W\ m^{-2}$
E_0	Reference irradiance	$W\ m^{-2}$
E_λ	Spectral irradiance	$W\ m^{-2}\ \mu m^{-1}$
F	Focal length	m
f	Temporal frequency	sec^{-1}
h	Depth of Michelson Echelon Steps	m
I	Intensity	$W\ sr^{-1}$
$^\circ I$	Spectral intensity normalization constant	$W\ sr^{-1}\ \mu m^{-1}$
I_λ	Spectral intensity	$W\ sr^{-1}\ \mu m^{-1}$
i	Integer running index	—
i_λ	Relative spectral intensity	—
i_1, i_2	Angles of entry and exit for a prism	rad
j	$\sqrt{-1}$	—
k	An integer	—

Table 20-1. Symbols, Nomenclature, Units (*Continued*)

Symbols	Nomenclature	Units
L	Radiance	$W\ m^{-2}\ sr^{-1}$
$°L$	Spectral radiance normalization constant	$W\ m^{-2}\ sr^{-1}\ \mu m^{-1}$
L_λ	Spectral radiance	$W\ m^{-2}\ sr^{-1}\ \mu m^{-1}$
L_λ^{BB}	Spectral radiance from a blackbody	$W\ m^{-2}\ sr^{-1}\ \mu m^{-1}$
l_λ	Relative spectral radiance	—
m	Integer running index	—
N	Number of lines or slits, or steps	—
n	Index of refraction	—
P	Order number at the center of the Fabry-Perot circular fringe pattern	—
p	Order number, integer	—
$\mathcal{R}$	Responsivity	$V\ W^{-1}$
R_A	Amplitude reflectance	—
$\mathcal{R}_E, \mathcal{R}_L, \mathcal{R}_\Phi$	Responsivity in terms of the subscripted radiometric quantity	$V\ [\text{Radiometric Unit}]^{-1}$
RP	Resolving Power	—
$\mathcal{R}_0, \mathcal{R}_{0,E}, \mathcal{R}_{0,L},$ $°\mathcal{R}_E, °\mathcal{R}_L$	Responsivity normalization constants	$V\ [\text{Radiometric Unit}]^{-1}$
r	Relative responsivity; also, radius of curvature	$-;\ m$
r_1, r_2	Internal angles of reflection in a prism	rad
S	Distance	m
s_p	Length of prism face	m
s	Width	m
T	Temperature	$°C, K$
Th	Throughput	$sr\ m^2$
T_A	Amplitude transmittance	—
T_m	Time period of Michelson mirror	sec
t	Time	sec
V	Output signal	V
V_n	rms noise fluctuation of output signal	V
W	Width of prism entrance beam	m
w	Width of one slit	m
X, Y, Z	Cartesian coordinates	—
x	Linear distance, also path difference	m
$\langle \Delta\lambda \rangle$	Spectral range in wavelength	μm
$\langle \Delta\tilde{\nu} \rangle$	Spectral range in wavenumber	cm^{-1} or wavenumber
δ	Phase increment	—
ϵ	Emissivity	—
θ	Angle of a ray, usually from the normal to a surface of an optical element	rad

Table 20-1. Symbols, Nomenclature, Units (*Continued*)

Symbols	Nomenclature	Units
θ_o	Half-angle field of view	rad
λ	Radiation wavelength	μm
ν	Radiation frequency	$\sec^{-1}$, Hz
$\tilde{\nu}$	Wavenumber, $\tilde{\nu} = 1/\lambda$	cm^{-1} or wavenumber
ρ	Reflectivity	—
$\tau(\lambda)$	Spectral transmittance	—
Φ	Radiant flux or power	W
Φ_λ	Spectral radiant flux or power	$W \mu m^{-1}$
ϕ	Angle of incidence; Also, phase	rad
ϕ_d	Angle of deviation	rad
ψ	Angle between rays	—
Ω	Field of view	sr
ω	Angular radiation frequency, $\omega = 2\pi c/\lambda$	rad $\sec^{-1}$

20.1.2. Definitions. *Radiometry* is the measurement of some radiometric quantity such as E or L. *Measurement* is the process by which a quantitative comparison is made between two or more states of a physical observable such as length, temperature, or spectral radiance. The performance of an instrument may be defined as a measure of its usefulness to make unambiguous quantitative comparison; thus, *calibration* is the special measurement process by which one determines all parameters significantly affecting an instrument's performance. If subsequent measurements are referred to standard values, such as those maintained by the U.S. National Bureau of Standards, then the calibration process is called *absolute calibration*. Calibrations are performed to make the results of measurement as independent as possible of the particular measuring instrument employed.

Radiometric measurements include the determination of the intensity, I, or the radiance, L, of some distant source which is not accessible for close-in measurement. Thus, one defines a radiometer as an instrument designed to allow measurement of some radiometric quantity such as L, or irradiance, E. The output of a radiometric instrument (after calibration has been performed) is determined by the incident radiation which is being measured. The purpose of calibration is to determine, within experimental error, the quantitative relationship between instrumental outputs and the incident radiation that is being measured. The instrument's receiving aperture is selected to be the reference position for incident radiation, since incident radiation would affect any other instrument similarly placed. Determination of the significance of all possible instrumental outputs requires a complete determination of the instrument's three performance characteristics, i.e., responsivity, $\mathfrak{R}$, detectivity, D, and reference radiation. Finally, estimates of the accuracy (precision and systematic error) of the results of calibration measurements must be obtained to provide the basis for similar estimates of the accuracy of subsequent measurement results. Calculation of the desired radiometric quantity at some point remote to the instrument by use of the values determined at the radiometer aperture requires more information. For example, the determination of the radiant intensity of a star requires, in addition to measurement of the irradiance at the observatory radiometer aperture, knowledge of the effects of the intervening atmosphere and a value of the distance to the star.

For the most part, the challenge of radiometry stems from the desire to state completely the significance of all possible instrument outputs. Radiometers measure the magnitude of radiometric quantities within some region of the wavelength-time-space-polarization domains created by the dependency of the radiation upon these parameters. Broadly, a radiometer is any apparatus for quantifying some property of incident radiation.

Names of radiometric devices are usually created by combinations of the prefix and suffix words (or modifications of those words), shown in Table 20-2, to correspond with the design purposes of these instruments. Additional prefixes specify the spectral region of the band, e.g., infrared radiometer, ultraviolet radiometer, microwave radiometer. Photometer is used instead of photoradiometer, but with the additional implication that the relative spectral response is the same as that of the visibility curve. Certain modifying terms are used with these names. "Double-pass" refers to the use of an optical component twice by causing the radiation to pass through it twice in the same direction. For example, a double-pass prism spectrometer disperses the radiation by sending radiation through a prism first forward and reflected back through once, then reflected from another mirror so as to enter forward and back through a second time. A "double monochromator" is a monochromator composed of two monochromators used in series. The output of the first forms the input to the second.

Table 20-2. Instrumentation Nomenclature

Prefix (Connotation)	Suffix	Output Form
Radio, i.e. pertaining to general electromagnetic radiation	meter	Numerical, but not recorded
Photo, i.e. visible or photographic spectral region	graph	Recorded graphically
Spectro, i.e., spectrally dispersive	scope	As seen directly by the eye

If one seeks the variation of the amount of radiation as a function of any of the spatial, spectral, or temporal parameters, then an instrument is used which responds to a selectable narrow range of values of the parameters and the total range of interest is scanned. The narrow range is specified by statement of the instrumental resolution. The specifications are not independent. One cannot, for example, arbitrarily reduce $\Delta\lambda$ and Δt together. The limits are determined by the signal-to-noise ratio attainable for the specific experiment.

20.2. Radiometers

A radiometer is a radiation-measuring instrument having substantially equal spectral response to radiation having a relatively wide band of wavelengths. Radiometers measure the difference between source radiation incident on the radiometer detector and a radiation reference level. All radiometers and radiometric measuring instruments consist of at least three essential components (Figure 20-1):

(1) Optics which collect the radiation through an aperture of area A, disperse or spectrally filter the radiation, and focus it on a field stop of area a.

(2) Detector (s) which transduce the radiation which comes from the field stop to a signal, usually an electrical voltage, of magnitude V.

(3) Electronics/Display-Recorder which process (amplify) the signal and record it.

The three major radiometric performance characteristics are defined as

(1) The responsivity, $\Re$—the output per input of incident radiation.

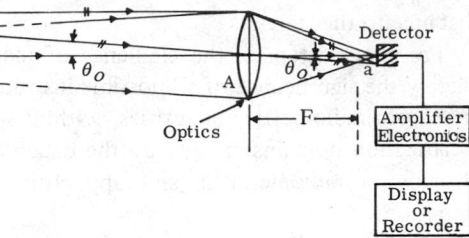

Fig. 20-1. Basic radiometer. The collecting optics form the (circular) aperture stop of area A; the detector element forms a circular field stop of area a; θ_O is the half-angle field of view measured in radians; $\Omega = 2\pi(1-\cos\theta_O) \simeq \pi\theta_O^2 \simeq a/F^2$ is the field of view (the solid angle) in steradians, of the corresponding conical field; and F is the focal length of the radiometer optics.

(2) The detectivity, D—the reciprocal of the noise equivalent incident (input) radiation, or the responsivity divided by the noise voltage, i.e., the rms noise fluctuation of the output.

(3) The reference radiation—the level of incident radiation corresponding to a zero reading on the output scale of the instrument.

20.2.1. Responsivity of a Radiometer. Of the three defined characteristics, responsivity is the most useful since it is always involved in transforming an output reading into the corresponding value of incident radiation, or input. Thus an exact determination is valuable. In contrast, it is often sufficient to know the rough order of the detectivity if the measurements to be made are so far above the instrument noise level that noise considerations are not involved. Similarly, a rough measure of the reference radiation is often sufficient if a truly negligible amount of radiation is represented by an output reading of zero. These last two characteristics are easily overlooked, however, in those rather rare instances when they may be highly significant factors in the interpretation of an instrument output. For this reason, they should always be checked.

The responsivity of a detector is defined as the change in output signal, ΔV, divided by the change in incident flux, $\Delta\Phi$, on the detector. For a radiometer as a complete system, it is useful to extend the meaning of responsivity to include either the change in output signal, divided by the change in irradiance, ΔE, at the entrance pupil of the radiometer, or the change in output signal divided by the change in radiance, ΔL. Thus, three kinds of instrument responsivity can be considered for a radiometer.

$$\Re_\Phi = \frac{\Delta V}{\Delta\Phi} \qquad (20\text{-}1)$$

$$\Re_E = \frac{\Delta V}{\Delta E} \qquad (20\text{-}2)$$

$$\mathcal{R}_L = \frac{\Delta V}{\Delta L} \tag{20-3}$$

Similarly, since the detectivity of a detector is defined as: $D = \mathcal{R}/V_n$, where V_n is the rms noise fluctuation of the detector output, the radiometer detectivity is

$$D_\Phi = \frac{\mathcal{R}_\Phi}{V_n} \tag{20-4}$$

$$D_E = \frac{\mathcal{R}_E}{V_n} \tag{20-5}$$

$$D_L = \frac{\mathcal{R}_L}{V_n} \tag{20-6}$$

The output voltage of a simple radiometer as a function of E at the radiometer aperture is illustrated in Figure 20-2. The simplest radiometer is an isolated detector element which therefore is simultaneously the aperture stop, field stop, entrance pupil, exit pupil, entrance window and exit window. The output signal value is indicated directly by an electric current meter. Ordinary photographic exposure meters are of this kind. The equation for the function is

$$V = \mathcal{R}_E (E - E_0) + V_n \qquad (20\text{-}7)$$

where $\mathcal{R}_E \equiv \dfrac{\Delta V}{\Delta E}$

$E_0 \equiv$ reference irradiance

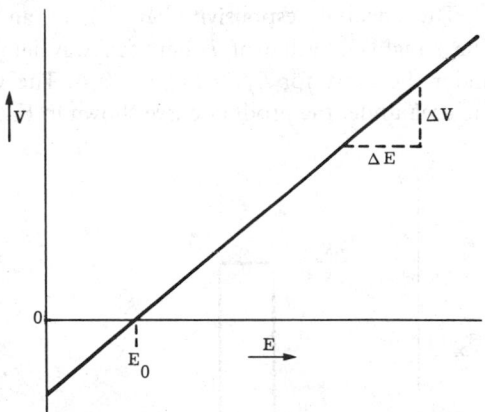

The reference irradiance, E_0, can be determined by use of an external calibration source which uniformly irradiates the instrument, with adjustment of the source until the radiometer output reads zero. The value of E at the radiometer aperture due to this adjustment is calculated to establish the value of E_0, provided that the noise voltage is negligibly small. Frequently, the values of E to be measured are quite large so that

Fig. 20-2. Radiometer response to input irradiance.

$$E \gg E_0 \tag{20-8}$$

$$V \gg V_n \tag{20-9}$$

Then the relation may be approximated well enough by

$$V = \mathcal{R}_E E \tag{20-10}$$

This relation can be derived for $\mathcal{R}_L$ under similar circumstances with the use of uniform radiance sources. Calibration is discussed in some detail in Section 20.2.4.

Since radiance and irradiance are functions of wavelength, location, direction, polarization, and time, the radiometer responsivity must also be a function of these parameters.

The spatial dependency defines the field of view (FOV); the spectral dependency, the spectral response; and the dependency upon time, the fluctuation-frequency response or temporal-frequency bandwidth. Radiometer designers aim at making these dependencies mutually *independent factors* so that the responsivity is the product of several relative responsivity factors and a responsivity normalization constant, $\mathcal{R}_0$. Thus

$$\mathcal{R} = \mathcal{R}_0 \cdot r_1(\lambda) \cdot r_2(t) \cdot r_3(x,y) \cdot r_4(\theta,\phi) \ldots \tag{20-11}$$

In the idealized case of a "perfect radiometer," the responsivity is zero outside the geometrical bounds established by a solid angle, Ω, and is constant within these bounds. Similarly,

$$r_1(\lambda) = 1 \text{ for } \lambda_1 \leq \lambda \leq \lambda_2 \tag{20-12}$$

$$r_1(\lambda) = 0 \text{ outside of } \Delta\lambda = \lambda_2 - \lambda_1 \tag{20-13}$$

and, with f for the temporal frequency,

$$r_2(f) = 1 \text{ for } f_1 \leq f \leq f_2 \tag{20-14}$$

$$r_2(f) = 0 \text{ outside of } \Delta f = f_2 - f_1 \tag{20-15}$$

The spectral responsivity, $\mathcal{R}_L(\lambda)$, of an ideal radiometer is shown in Figure 20-3. The sought-for value of L between wavelength λ_1 and λ_2 is equal to the shaded area under the curve for L_λ in Figure 20-4. The voltage output of the radiometer is equal to the area under the product curve shown in Figure 20-5. In equation form, this is

$$V = \int_0^\infty \mathcal{R}_L(\lambda) L_\lambda \, d\lambda \tag{20-16}$$

Since $\mathcal{R}_L(\lambda) = \mathcal{R}_{0,L}$ (which is constant between λ_1 and λ_2 and is otherwise zero), then

$$V = \int_{\lambda_1}^{\lambda_2} \mathcal{R}_L(\lambda) L_\lambda \, d\lambda = \mathcal{R}_{0,L} \int_{\lambda_1}^{\lambda_2} L_\lambda \, d\lambda \tag{20-17}$$

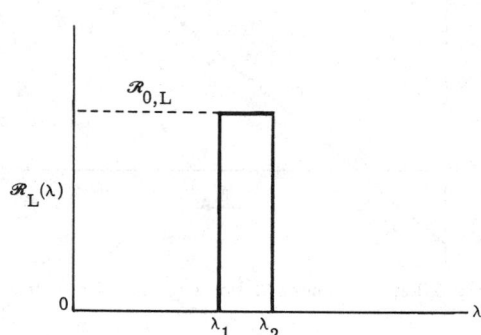

Fig. 20-3. Ideal radiometer spectral responsivity.

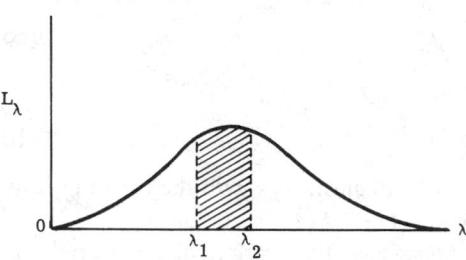

Fig. 20-4. A smoothly varying spectral radiance.

The relationships between $\mathcal{R}_\Phi$, $\mathcal{R}_E$, and $\mathcal{R}_L$ are obtainable from the relation between Φ, E, and L. Thus

$$\mathcal{R}_\Phi \equiv \frac{\Delta V}{\Delta \Phi} = \mathcal{R}_E A^{-1} = \mathcal{R}_L \, [Th]^{-1} \tag{20-18}$$

where $Th \equiv \int \cos\theta \, d\Omega \, dA$. ("*Th*" is often called the instrumental "through-put.") The $\cos \theta$ factor arises from the definition of $L \equiv \partial^2 \Phi / \cos \theta \; \partial A \partial \Omega$. If in the example L is defined as

$$\int_{\lambda_1}^{\lambda_2} L_\lambda \, d\lambda = L \qquad (20\text{-}19)$$

then

$$V = \mathcal{R}_{0,L} \, L \qquad (20\text{-}20)$$

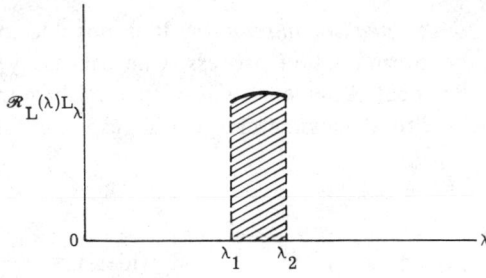

Fig. 20-5. The product of spectral radiance and spectral responsivity.

Most radiometers are not ideal in spectral responsivity. Yet a direct relation between voltage output and the value of radiance between two wavelengths is required. With real radiometers, $\mathcal{R}_L(\lambda)$ is measured as a property of the radiometer and used in the process (discussed in Section 20.2.2) of normalization.

20.2.2. Normalization. In Figure 20-6 a radiometer spectral responsivity is shown as triangular in shape. In Figure 20-7 the unknown spectral radiance, L, of the source is shown. The value of L_λ between λ_1 and λ_2 changes only slowly. The value of V, which the actual radiometer would exhibit, is equal to the area under the product curve in Figure 20-8. The product curve is not quite triangular in shape because L_λ is not quite constant. The equivalent, ideal, radiometer spectral responsivity which has the same maximum responsivity as the actual radiometer is shown as a dotted rectangle in Figure 20-6. It is that rectangle of height $\mathcal{R}_{0,L}$ which has the same area as the triangle. That ideal radiometer would produce a product curve as shown by the dotted line in Figure 20-8. The area under the dotted product curve is equal to the value of V which an ideal radiometer would provide. Notice that these two product curve areas are approximately equal and would be exactly equal if L_λ were truly constant.

The determination of the equivalent ideal radiometer having the same maximum spectral responsivity as the real radiometer is called *normalizing the radiometer response*

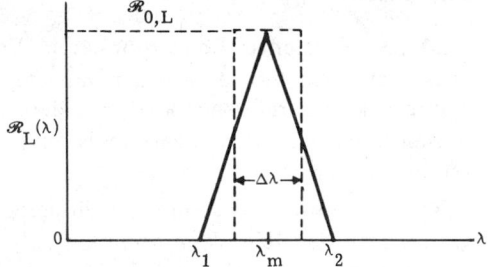

Fig. 20-6. Triangular spectral responsivity.

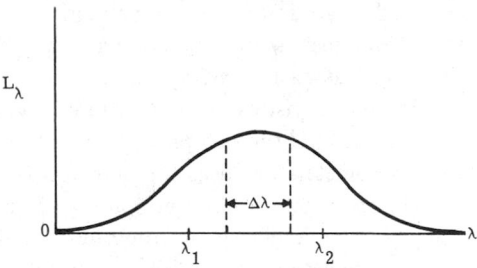

Fig. 20-7. Smoothly varying spectral radiance.

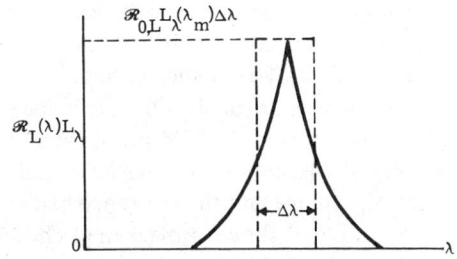

Fig. 20-8. Production of triangular spectral responsivity and spectral radiance.

to the peak or maximum. It is possible to utilize other equivalent ideal radiometers in the normalization process. One can easily determine the peak spectral responsivity of the ideal radiometer whose spectral bandwidth is limited by the wavelengths for which the actual radiometer spectral response has fallen to one half its maximum value, i.e.,

$$\mathcal{R}_L \text{ (ideal)} = \frac{\int_0^\infty \mathcal{R}_L(\lambda)\, d\lambda}{(\Delta\lambda \text{ for half response})} \tag{20-21}$$

The values would then be stated as that radiance within the half-value spectral responsivity points.

Normalization to specify the equivalent ideal radiometer responsivity for the other parameters such as field of view and fluctuation frequency bandpass follows analogous reasoning. Likewise, symmetry of the responsivity for these other parameters permits good accuracy for the linear variation of radiance with these other parameters for the same reasons.

20.2.3. Reference Radiation Level. Either absolute or relative infrared radiation levels may be obtained with a radiometer. Absolute measurements of radiation levels require a known reference level. The detector itself or a radiation chopping system can be used to provide the reference radiation level. For absolute measurements, the chopping system is preferred.

Radiometers that measure the difference in radiation from two neighboring spatial positions provide relative information only as no stationary reference level exists. Radiometers used to compare any element of a large area to the average radiation associated with the entire area can supply an absolute measurement, providing the average is known and used as a reference level. Similarly, in the time domain, the power level due to radiation at one instant may be compared to that of a previous instant or to an average associated with all past measurements.

The characteristics of three different types of absolute radiometers, including the form of reference level they employ, are listed in Table 20-3.

A radiometer that uses the detector as a radiation reference level is often referred to as a dc radiometer. This terminology is applicable because the instrument measures a change in the dc electrical properties of a thermoelectric or bolometric infrared detector. These radiometers are subject to drift because the reference level is determined by the temperature of the detector; however, drift can often be tolerated.

Chopped or ac radiometers provide an ac output from the detector for signal processing. They are particularly suitable for absolute radiation measurements, not having the drift problems associated with dc radiometers. The electrical output of this type radiometer is proportional to the difference between radiation falling upon the detector from the source within its field of view and that of a blackened chopper blade or a controlled reference blackbody. In the latter case, a chopper mirror alternately directs radiation from the source and the reference blackbody onto the detector.

Figure 20-9 shows a blackened chopper where the detector alternately sees the source image and the blackened chopper. The temperature and emissivity of the blackened chopper determine the reference radiation level. At wavelengths out to about 1 μm, a

Table 20-3. Principal Characteristics of Different Type
Nonselective Radiometers [20-1]

Type	Detector	Detector Response Speed (Time constant)	System Speed	Reference Radiation Level	Radiation Measurements	Electrical Signal
dc	Thermo-pile	2 sec	2 sec	Emissivity and temperature of thermopile	Difference of source radiation and thermopile radiation	Electromotive force from compensated dc thermopile
ac Blackened Chopper	Therm-istor	1 msec	25 msec	Emissivity and temperature of blackened chopper	Difference of source radiation and blackened chopper radiation	ac signal from compensated thermistor bolometer bridge
ac Chopper Mirror	Therm-istor	1 msec	Adjustable, 16 msec to 1.6 sec	Temperature of blackbody reference (emissivity = 1.0); temperature controlled or monitored within 0.2°C	Difference of source radiation and reference blackbody radiation; null method may be used with temperature-controlled blackbody	ac signal from compensated thermistor bolometer bridge; null detection method when using controlled blackbody

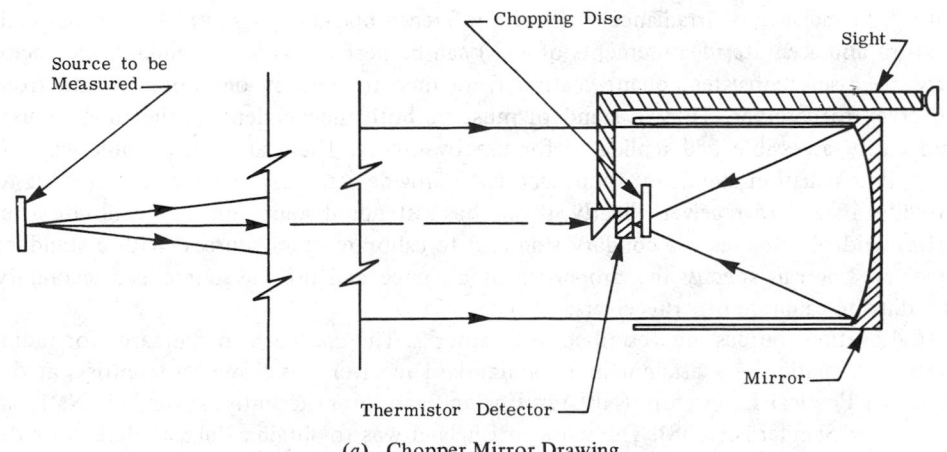

(a) Chopper Mirror Drawing

Fig. 20-9. Chopper mirror drawing and block diagram.

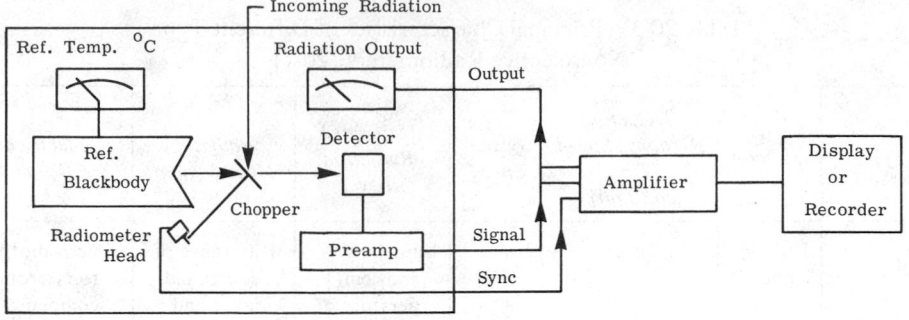

(b) Chopper Block Diagram

Fig. 20-9 (Continued). Chopper mirror drawing and block diagram.

very stable reference level is obtained by ensuring that the chopper has a uniformly coated black surface. To about 3 μm, chopper temperature is not usually a critical factor because at ambient temperatures there is little radiation in this region from a blackbody or a graybody. At wavelengths longer than about 3 μm, however, the effects of variations in chopper temperature and emissivity become a serious consideration.

Emissivity effects can be minimized by using a highly polished chopper blade. However, incident radiation reflected from the surrounding area to the detector by the polished chopper must be controlled. This is done in the radiometer shown in Figure 20-9 where reflected incident radiation is controlled by a reference blackbody and a chopper mirror.

A small reference blackbody is placed in such a position that, by specular reflection from the chopper mirror, radiation from the reference blackbody and the source image are alternately directed onto the detector. This not only provides a stable source of reference radiation but also an adjustable one when temperature controls are provided for changing the reference blackbody temperature. In addition, if the reference blackbody source is adjustable, its radiance can be matched to the source radiance and the detector used as a sensitive null or quantitative error detector.

20.2.4. Calibration of Radiometers. In this discussion, it is assumed that a standard source of radiance or irradiance, such as a reference blackbody, is available so that calibration and accurate measurements of $\Re(\lambda)$ can be performed. Such standards are necessary to assure consistent quantification from time to time by one observer and from observer to observer. Thus, a standard must be both independent of the time of use, and easily accessible and replicated for use by others. There are two possible ways of providing radiation standardization. One is to provide a standard source, the other is to provide a standard receiver. Clearly, if one has a standard source one can calibrate a receiver and use that as a secondary standard to calibrate other sources. With a standard receiver, one can specify the properties of a source and use the source as a secondary standard to calibrate other receivers.

Calibration sources are described in Chapter 2. The approach to standards for radiometric calibrations by attainment of a standard receiver was taken by scientists at the National Physical Laboratory (NPL) in England and, more recently, at the U.S. National Bureau of Standards (NBS). The work in England was to obtain a "black" detector with extremely stable response characteristics and to use it with electrical heating introduced to balance the heat due to incoming radiation. The resulting configuration, often called an absolute radiometer, or electrically calibrated radiometer (ECR) has been used by NPL with Moll thermopiles as transfer or working standards. The original Guild drift radiometer [20-2] was extended by Gilham [20-3, 20-4, and 20-5] to higher levels of precision.

The NBS work along these lines (from 1968 to 1971) has been described by Geist. The status following the attainment of an electrically calibrated pyroelectric radiometer (ECPR), [20-6 and 20-7], was described by Karl Kessler, Chief of the NBS Optical Physics Division, at the June 1976 meeting of the Council for Optical Measurements (CORM) as follows:

"With these devices (ECPR's), NBS now has the capability of 1% accuracy in the measurement of radiant power from 0.1 to 100 mW over the wavelength range from 250 nm to 10 micrometers. The ECPR has recently been used to run an intercomparison between itself, the C-series calorimeter used for laser measurements at NBS Boulder Laboratories, the electrically calibrated thermopile radiometer at the West German national standards lab (PTB), and the electrically calibrated solar radiometer at the World Radiation Center in Davos, Switzerland. In all cases the virtually independent measurement communities represented by these installations were found to be in very good agreement with each other. . . . ECPR's are now commercially available . . ." [20-8]

Further details have been given in References [20-38, 20-39].

Calibrations are categorized into two groups: those in which the radiometer field stop is partially illuminated by the calibration source; and those in which the stop is totally filled with uniform calibration radiation. The first may be achieved by a *distant point (small) source* (Figure 20-10), or by a collimator with suitable magnification and a small source (Figure 20-11). The latter can be realized with a so-called *extended source*, either *distant* (Figure 20-12) or *close* (at the entrance aperture in Figure 20-13), by a *collimator* (with different magnifications) or by proper placement of a point source very close to the entrance aperture (the so-called "Jones method" of calibration shown in Figure 20-14).

A good general rule is always to carry out a calibration under conditions which reproduce as completely as possible the situation expected when measurements based on that calibration are made. On the other hand, for maximum versatility and as a check on consistency, it is important to conduct additional calibrations covering as wide a range of variation of as many parameters as possible. Thus the calibration should be in

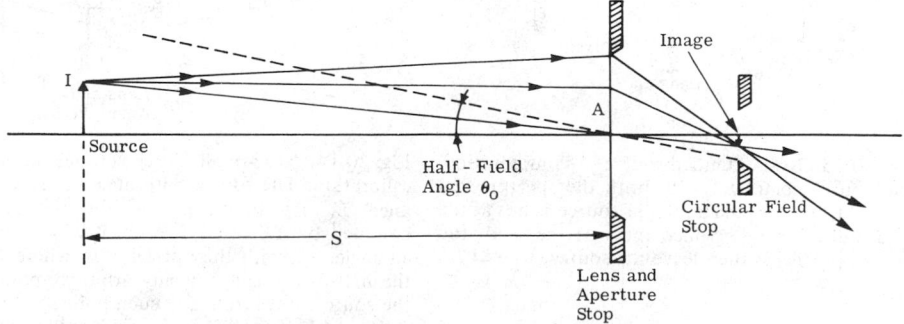

Fig. 20-10. Distant-small-source configuration. The source image is entirely within the field stop and $\mathfrak{R}_E \equiv V/E = VS^2/I$.

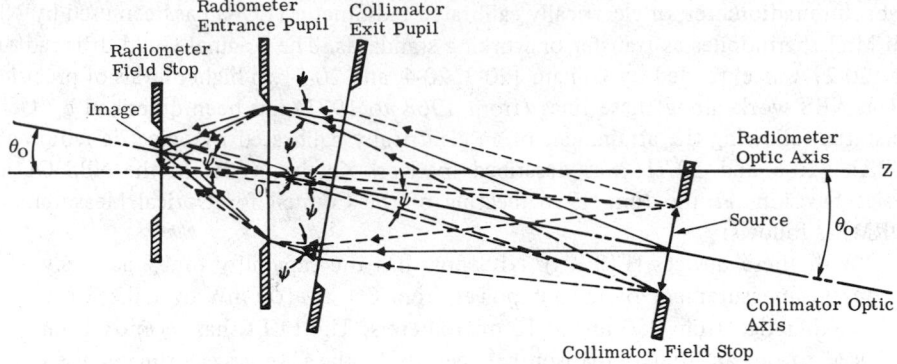

Fig. 20-11. Geometry for mapping the field with a collimator. 0 is the center of the radiometer entrance pupil; θ_0 is the angle between the optic axes, which intersect at 0; and ψ is the angle between the extreme rays in the collimated beam, and the angle subtended at 0 by the image of the source $\psi \ll \theta_0$ (the relative dimensions are exaggerated to emphasize the angles).

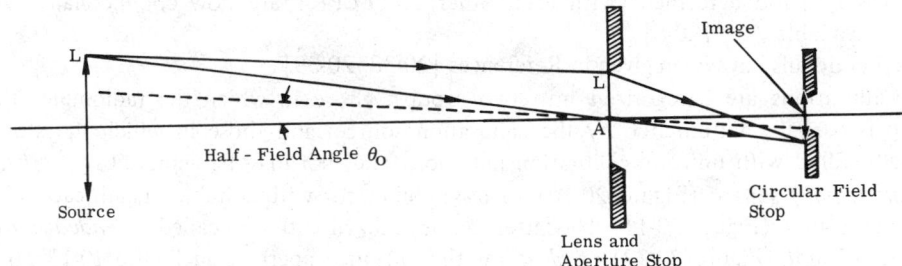

Fig. 20-12. Distant-extended-source configuration. The source image completely fills the circular field stop and $\mathscr{R}_L \equiv V/L$.

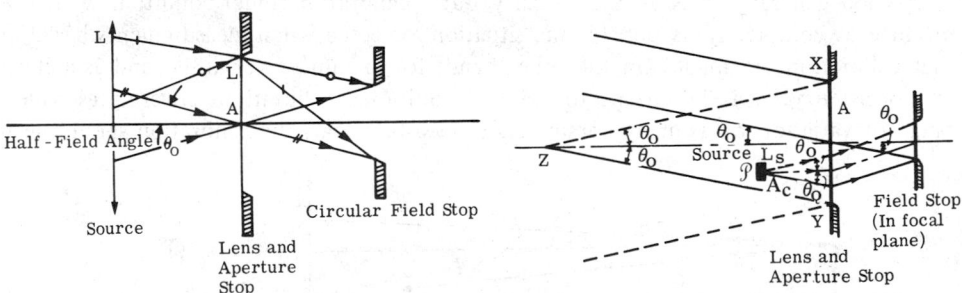

Fig. 20-13. Near-extended-source configuration. The source completely fills both the aperture and the field, but the image of the source is not at the field stop. The rays which meet at the field stop are not parallel as they leave the source, $\mathscr{R}_L \equiv V/L$.

Fig. 20-14. Near-small-source (Jones method) calibration. The source with area A_c, and radiance L_s is completely within the region bounded by XZ and YZ, each of which makes an angle, θ_0, with the optical axis, where θ_0 is the half-field angle. $\mathscr{P}$ is any arbitrary point of the source. Rays from any such point, $\mathscr{P}$, within a cone of half-angle θ_0 will uniformly irradiate the circular field stop as shown.

terms of E if measurements of E are to be made with the radiometer. Using the distant point source method one would obtain, for an ideal radiometer,

$$\text{Distant Point Source: } \frac{\Delta V S^2}{\Delta I} = \frac{\Delta V}{\Delta E} = \Re_E, \quad \text{V W}^{-1} \text{ cm}^2 \qquad (20\text{-}22)$$

where S is the distance from the source to the radiometer aperture area, A. Here atmospheric attenuation has been neglected, i.e., $\tau = 1$ is assumed, an asssumption not always valid. (See Figure 20-15.)

Similarly, with a distant extended source of radiance L,

$$\text{Extended Source: } \frac{\Delta V}{\Delta L} \left[\frac{Th}{A}\right]^{-1} =$$

$$\Re_E = \Re_L \left[\int \cos \theta \, d\Omega\right], \text{V W}^{-1}\text{cm}^2$$

$$(20\text{-}23)$$

where Th is the throughput of the radiometer, given by

$$Th = \int_A \int_\Omega \cos \theta \, dA d\Omega \qquad (20\text{-}24)$$

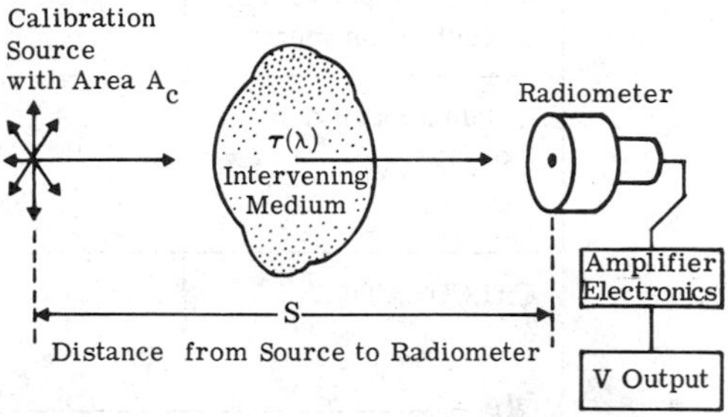

Fig. 20-15. Calibration of radiometers.

Here, besides accounting for the intervening atmosphere, one must determine Ω, the angular field of view of an ideal radiometer. The extended source gives a radiance calibration directly. When the extended source is placed at the radiometer aperture, the adjustment for atmospheric attenuation is minimized.

A near small calibration source of area, A_c, and radiance, L_c, can be used to establish $\Re_L$ by using the relation

$$\Re_L = \frac{(A)(\Delta V)}{(A_c)(\Delta L)} \qquad (20\text{-}25)$$

Calibration is summarized and compared to measurement in Table 20-4.

Table 20-4. Calibration-Measurement Taxonomy

	Calibration	Measurement	
		Case I	Case II
Distant Small (i.e. Point Source)	GIVEN		
	1. $\tau(\lambda)$, transmittance of intervening medium	1. $\tau(\lambda)$	1. $\tau(\lambda)$
	2. S, distance, source to radiometer aperture	2. S	2. S
	3. $r_E(\lambda)$, relative spectral irradiance-responsivity	3. $r_E(\lambda)$ and $°\Re_E$ so that $\Re_E(\lambda) = °\Re_E r_E(\lambda)$	3. $r_E(\lambda)$ and $°\Re_E$ so that $\Re_E(\lambda) = °\Re_E r_E(\lambda)$
	4. $\epsilon_c(\lambda)$, spectral-emissivity of calibration source, c		4. $i_\lambda(\lambda)$, relative spectral radiant intensity of source to be measured.

Table 20-4. Calibration-Measurement Taxonomy (*Continued*)

	Calibration	Measurement Case I	Measurement Case II
	5. A_c, Area of calibration source; A_c < Projected IFOV 6. T_c, Temperature of calibration source		
	Obtain a *measure* of output: V_c	V	V
	CALCULATE $$^\circ\!R_E = \dfrac{V_c}{A_c S^{-2} \int_{\text{All }\lambda} r_E(\lambda)\tau(\lambda)\epsilon_c(\lambda)L_\lambda^{BB}(\lambda, T_c)d\lambda}$$ where $L_\lambda^{BB}(\lambda, T_c)$ is the Planck function for T_c	1. $E = V/^\circ\!R_E$ (at Radiometric Aperture) and 2. $I' = ES^2$ (apparent radiant intensity)	$^\circ I = \dfrac{V}{^\circ\!R_E S^{-2} \int_{\text{All }\lambda} r_E(\lambda)\tau(\lambda)i_\lambda(\lambda)d\lambda}$ and $I_\lambda(\lambda) = ^\circ I\, i_\lambda(\lambda)$
Distant Extended Source	GIVEN 1. $\tau(\lambda)$, transmittance of intervening medium 2. S, distance, source to radiometer aperture 3. $r_L(\lambda)$, relative spectral radiance-responsivity 4. $\epsilon_c(\lambda)$, spectral emissivity of calibration source 5. A_c, Area of calibration source; $A_c \geqslant$ Projected IFOV 6. T_c, Temperature of calibration source	1. $\tau(\lambda)$ 2. S 3. $r_L(\lambda)$ and $^\circ\!R_L$ so that $R_L(\lambda) = ^\circ\!R_L r_L(\lambda)$	1. $\tau(\lambda)$ 2. S 3. $r_L(\lambda)$ and $^\circ\!R_L$; $R_L(\lambda) = ^\circ\!R_L r_L(\lambda)$ 4. $l_\lambda(\lambda)$, relative spectral radiance of source to be measured

Table 20-4. Calibration-Measurement Taxonomy (*Continued*)

	Measurement	
Calibration	*Case I*	*Case II*
Obtain a *measure* of output: V_c	V	V
CALCULATE $$°\mathcal{R}_L = \frac{V_c}{\int_{All\ \lambda} r_L(\lambda)\tau(\lambda)\epsilon_c(\lambda)L_\lambda^{BB}(\lambda,T_c)d\lambda}$$	$$L' = \frac{V}{°\mathcal{R}_L}$$ (apparent radiance)	$$°L = \frac{V}{°\mathcal{R}_L \int_{All\ \lambda} r_L(\lambda)\tau(\lambda)l_\lambda(\lambda)d\lambda}$$ and $L_\lambda(\lambda) = °L\, l_\lambda(\lambda)$

20.2.5. Commercially Available Radiometers. Commercially available radiometric instruments can be grouped into general-purpose radiometers and specialized instruments. The latter are designed to perform one particular function as, for example, laser power meters. The results of a brief Infrared Information and Analysis (IRIA) Center survey* are tabulated in Tables 20-5 and 20-6.

20.3. Spectroradiometers

Spectroradiometers are instruments designed to allow determination of the wavelength distribution of radiation. Separation of the radiation into its spectral components, or dispersion, is accomplished by the use of an optical element possessing a known functional dependence on wavelength. Prisms and diffraction gratings are among the optical elements which can be used to produce spectral dispersion.

The essential components of any prism or grating spectroradiometer are the same as those of any radiometer with the addition of the following:

(1) An entrance slit, which usually acts as the field stop of the collecting optics.

(2) A collimator, which may be a lens or a mirror, with the entrance slit at its focus.

(3) A dispersing element, either a prism or grating.

(4) A focusing element, which produces an image of the entrance slit from the parallel beam at each wavelength so that these images are dispersed linearly to form the familiar spectrum.

(5) One or more exit slits to select the radiation in any desired region of the spectrum and allow it to pass on to the detector.

A spectroradiometer can be said to be a radiometer with a monochromator located between the collecting optics and the detector. Figure 20-16 shows a schematic for any simple prism or grating spectroradiometer.

Radiometers which rapidly sequence through a set of narrow filters also can be called

*Conducted by Craig Mueller, Environmental Research Institute of Michigan (ERIM), Ann Arbor, MI.

Table 20-5. Radiometers

Manufacturer	Model	Spectral Range (μm)	Temperature Range (K)	Detector Type	Optics	Field of View (mrad)	Range of Focus (cm)	Focal Length (cm)	Effective Focal Ratio (F/#)	Remarks
Barnes Engineering 30 Commerce Rd. Stamford, CT 06904	12-550	1-20	323-1273 Close range, 273-503 Long range	Thermistor, bolometers InAs, InSb, HgCdTe, GeAu, Si, PbS, PbSnTe, PbSe	Cassegrain	2.5-17.5	91-∞	21.59	—	Circular variable filters
	12-660	1.8-26	—	Thermistor bolometer (others available)	—	35-350, 70	—	—	F/0.2	—
	12-880 spectro-radiometer	2-20 or 0.5-14 or 1.2-35	313-5273	—	—	35, 350, 105, 140	—	—	—	4 positions of filter wheel
EG&G Electro-Optics Division 35 Congress St. Salem, MA 01970	95	9.9-12.5	233-343 ±0.1°	GeHg	—	7.0	—	—	—	Single channel
	450	0.4-1.1	—	Si	—	140	—	—	—	—
	550	0.4-1.1	—	Si	—	140	—	—	—	—
	580	0.35-1.15	—	580-23A S-1	—	244	—	—	—	8 filter positions
Electro Optical Industries, Inc. P.O. Box 3770 Santa Barbara, CA, 93105	454	2.0-2.6*	—	PbS**	Cassegrain	1.5	152.4-∞	—	F/2.5	—
	456	2.0-2.6*	—	PbS**	Cassegrain	1.5	152.4-∞	—	F/2.5	—
	458	2.0-2.6*	—	PbS**	Cassegrain	1.5	152.4-∞	—	F/2.5	—
	401	1.8-2.8	—	—	Cassegrain	5.0	—	—	F/2.5	—
	467A	2.0-4.5	—	—	Cassegrain	5.0	—	—	—	—
	464	1.0-20	—	Thermistor bolometer	Cassegrain	1.5	304.8-∞	—	F/2.5	—
	481D	0.2-50	—	—	—	316	—	—	—	—
	4103	0.5-6.5	323-873	InSb	—	—	—	—	—	—
	4105	6-15 3-5	—	HgCdTe InSb	Cassegrain	1.5	152.4-∞	—	F/2.5	—

Manufacturer	Model	Spectral range (μm)	Range	Detector	Optics	Field of view (Instantaneous)	Range (distance)	Remarks
Gamma Scientific, Inc. 3777 Ruffin Rd. San Diego, CA 92123	470	8-14	323-873	HgCdTe	Cassegrain	Instantaneous: 0.5, 1.0, 2.0, 4.0, 6.0	—	Airborne, 4 filter positions
	470A	2.5-5.1	—	InSb	Cassegrain	3, 6, 12	—	Airborne, 4 filter positions
	4107	8-14	—	HgCdTe	—	3, 6, 12	91.44 - ∞	mK accuracy
	A425D	3-5	253-293	—	—	13, 1-350	—	—
	700	0.45-0.95	—	—	—	—	—	—
	721	—	—	—	—	—	—	—
	2020	0.3-0.75 extended range available	—	—	—	—	—	—
International Light, Inc. Dexter Industrial Green Newburyport, MA 01950	IL-500	0.22-1.1	—	—	—	—	—	—
	IL-510	0.4-0.7	—	—	—	—	—	—
	IL-700	0.3-1.1	—	—	—	—	—	—
	IL-710	0.4-0.7	—	—	—	—	—	—
Irtronics, Inc. 57 Commerce Rd., Stamford, CT 06902	2140	1.9-2.9	333-1088	PbS	Quartz lens	21	45 - ∞	3 models, each covering different temperature ranges
	2144	0.9-1.1	755-2530	Si	Refractive	21	30.5 - ∞	4 models, each covering different temperature ranges
	2040	1.9-2.6	310-2089	PbS	Refractive	21	—	5 models, each covering different temperature ranges

*1-20 available.
**Others available.

Table 20-5. Radiometers (Continued)

Manufacturer	Model	Spectral Range (μm)	Temperature Range (K)	Detector Type	Optics	Field of View (mrad)	Range of Focus (cm)	Focal Length (cm)	Effective Focal Ratio (F/#)	Remarks
Laser Precision Corp. 5 W. Whitesboro St. Yorkville, NY 13495	Pyrometer RS-3900	UV-far IR	288-308	—	—	—	—	—	—	—
	RK-3400	—	—	—	—	—	—	—	—	—
	RK-3440	0.3-500 optional	—	—	—	—	—	—	—	—
	RK-3442	—	—	—	—	—	—	—	—	—
Macam Photometrics, Ltd. 559 Lanark Rd. Edinburgh (031) 441-6375	1000	0.22-1.1	—	—	—	—	—	—	—	—
	2000	0.22-1.1, 2.0	—	—	—	—	—	—	—	—
	3000	0.22-1.1, 2.0	—	—	—	—	—	—	—	—
Molectron 177 N.Wolfe Rd. Sunnyvale, CA 94086	PR 100	—	—	—	—	—	—	—	—	—
	PR 200	0.2-500	253-358	Lithium tantalate pyroelectric crystal	Cassegrain	316	—	5.6	—	—
Optical Associates, Inc. 2906 Lafayette St, Santa Clara, CA 95050	203	0.35-1.1	—	Si	—	—	—	—	—	—
	201	0.4-0.7	—		—	—	—	—	—	—
	202	0.4-0.7	—	Si	—	—	—	—	—	—
Optronic Laboratories, Inc. 7676 Fenton St. Silver Spring, MD 20910	Radiometer/ photometer 730A	0.2-1.1	—	Si	—	—	—	—	—	2 filters

Manufacturer	Model										Notes
Oriel Corp. of America 15 Market St. Stamford, CT 06902	IR Golay detector	0.4-1000	—	Golay	—	—	—	—	—	—	—
	7000	0.2-1.1	—		—	—	—	—	—	—	Photomultiplier
	7010	0.2-1.1	—		—	—	—	—	—	—	Photodiode
Photo Research 3000 W. Hollywood Way Burbank, CA 91505	PR-1000 Spotmeters	0.45-1.0	—		—	—	—	—	—	—	4 filter positions
	UBA	0.45-1.0	—		—	4.36	6.35 · ∞	—	F/2.8	—	6 filter positions
	UBD	0.35-0.72, 0.82	—		—	17.44	—	—	—	—	—
Sensors, Inc. P.O. Box 1383 Ann Arbor, MI 48106	1061	0.6-40	213-773	Thermopile	Newtonian	332	0 · ∞	—	—	—	—
	1150H	0.6-40	763-1273	Thermopile	Newtonian	105	0 · ∞	—	—	—	—
	1150L	0.6-40	273-773	Thermopile	Newtonian	105	0 · ∞	—	—	—	—
United Detector Technology, Inc. 2644 30th St. Santa Monica, CA 90405	11A Power meters	—	—		—	—	—	—	—	—	—
	21A	0.25-1.15	—		—	—	—	—	—	—	—
	80X	—	—		—	—	—	—	—	—	—
	10A	3.2-11.0	—		—	—	—	—	—	—	—

Table 20-6. Specialized Radiometers

Manufacturer	Model	Spectral Range (μm)	Temperature Range (K)	Detector Type	Optics	Field of View (mrad)	Range of Focus (cm)	Focal Length (cm)	Effective Focal Ratio (F/#)	Remarks
Barnes Engineering 30 Commerce Rd. Stamford, CT 06904	Thermometers									Temperature sensitivity
	IT-4A	4.95-17.05	233-573	–	–	35, 350, short focus	–	–	–	1% of temp. range
	IT-4B	8.4-12.6	233-573	–	–	35, 350, short focus	–	–	–	1% of temp. range
	IT-4C	3.3-3.5	303-673	–	–	35, 350, short focus	–	–	–	1% of temp. range
	IT-4D	7.4-8.4	303-573	–	–	35, 350, short focus	–	–	–	1% of temp. range
	IT-4E	5.0-5.2	303-1973	–	–	35, 350, short focus	–	–	–	1% of temp. range
	IT-7A	0.7-1.0	873-3273	–	–	12.2, 6.3	–	–	–	1% of temp. range
	IT-7B	2.0-2.6	373-1773	–	–	12.2, 6.3	–	–	–	1% of temp. range
	IT-7C	3.3-3.5	323-673	–	–	12.2, 6.3	–	–	–	1% of temp. range
	IT-7D	5.0-6.2	323-1973	–	–	12.2, 6.3	–	–	–	1% of temp. range
	IT-7E	7.4-8.4	303-573	–	–	12.2, 6.3	–	–	–	1% of temp. range
	Portable									
	PRT-10	6.5-20	263-333	–	–	489	Close range	–	–	±0.2° @ 293 K
	PRT-10L	6.5-20	263-333	–	–	48.9	Long range	–	–	±0.2° @ 293 K
	PRT-11	6.5-20	273-473	–	–	489	Close range	–	–	+0.4° @ 373 K
	PRT-11L	6.5-20	273-473	–	–	48.9	Long range	–	–	+0.4° @ 373 K
EG&G Electro-Optics Division 35 Congress St. Salem, MA 01970	Laser radiometer									
	581	0.35-1.1	–	Si photodiode	–	15.6	–	–	–	–
Electro Optical Industries, Inc. P.O. Box 3770 Santa Barbara, CA 93105	Thermometers									
	4101	8-14	243-353	Bolometer	–	6.9, 69, 690	30.48-∞	–	–	–
	4104	9.5-11.5	243-353		–	35	30.48-∞	–	F/4	–
	Laser power									–

Manufacturer / Instrument									
430 Laser pulse meters	0.3-25	—	—	—	—	—	—	—	—
431	0.4-1.1	—	—	—	—	—	—	—	—
432	0.4-1.1	—	—	—	—	—	—	—	—
Laser radiometer 468A	—	—	—	—	—	—	—	—	—
The Epply Laboratory, Inc. 12 Sheffield Ave. Newport, RI 02840 — Black and white pyranometer	—	—	—	—	—	—	—	—	Measurement of solar radiation
Ultraviolet radiometer (photometer)	—	—	—	—	—	—	—	—	Measurement of solar radiation
Normal incidence pyrheliometer	—	—	—	—	—	—	—	—	Measurement of solar radiation
Precision infrared radiometer (pyrgeometer)	—	—	—	—	—	—	—	—	Measurement of solar radiation
Gamma Scientific, Inc. 3777 Ruffin Rd. San Diego, CA 92123 — Telephotometers 2009	—	—	—	—	—	121.92 · ∞	20	—	8 filter positions
IC 2000	—	—	—	—	—	121.92 · ∞	—	—	—
Infrared Industries, Inc. P.O. Box 989 Santa Barbara, CA 93102 — TD-1	—	311-1366	Indium antimonide	F/1.6 mirror system	—	—	—	—	—
International Light Dexter Industrial Green Newburyport, MA 01950 — IL 150	0.4-1.0	—	—	—	—	—	—	—	—
IL 155	0.35-0.9	—	—	—	—	—	—	—	—
IL 335	0.22-1.1	—	Photodiode	—	—	—	—	—	—
IL 463 Laser power meter	0.3-1.1	—	—	—	—	—	—	—	Multichannel
IL 550	0.4-0.94	—	—	—	—	—	—	—	—

Table 20-6. Specialized Radiometers (Continued)

Manufacturer	Model	Spectral Range (μm)	Temperature Range (K)	Detector Type	Optics	Field of View (mrad)	Range of Focus (cm)	Focal Length (cm)	Effective Focal Ratio (F/#)	Remarks
Ircon, Inc. 7555 N. Linden Ave. Skokie, IL 60076	**Modline®**									
	2000 series	0.7-0.97	773-3273	—	—	—	—	—	—	Accuracy equal to the greater of 1% of full scale or 5.6°
	3400 series	3.36-3.50	323-1073	—	—	—	—	—	—	
	4000 series	8-14	273-773	—	—	—	—	—	—	
	6000 series	2.0-2.6	353-3273	—	—	—	—	—	—	
	7000 series	4.8-5.2	323-1573	—	—	—	—	—	—	
	8000 series	7.4-8.4	293-673	—	—	—	43.18 - ∞	—	—	Equal to the greater of 1% of temp. or 11.1°
	300 series	2.0-2.6	353-1973	—	—	—	17.78 - ∞	—	—	
	700 series	4.8-5.6	308-2473	—	—	—	55.58 - ∞	—	—	
	CH-34 series	3.29-3.57	303-673	—	—	—	17.78 - ∞	—	—	2.8-5.6°
	1100 series	0.2-1.0	823-2273	—	—	—	50.8 - ∞	—	—	
Irtronics, Inc. 57 Commerce Rd. Stamford, CT 06902	2060 or 2160	4.5-5.0	366-1033 or 338-1088	PbSe	Servofax lens, focussing	42 or 21	30 - ∞	—	—	Used for glass processing monitoring
	2080	3.1-3.7	322-700	PbSe	Quartz lens, focussing	42	30 - ∞	—	—	2 models, used for plastic processing monitoring
	2090 or 2180	3.45 ±3%	366-700 or 338-700	PbSe	Quartz lens, focussing	42 or 21	30 - ∞	—	—	Monitoring in the C-H absorption band for thin film processing
Laser Precision Corp. 5 W. Whitesboro St. Yorkville, NY 13495	Pyro E meter RK 3200	0.3-500								—
	RK 3230	(optional)								—
	RK 3232	—	—							—
	E meters									
	AK-2930/N	—	—							—
	AK-2940/N	—	—							—

Manufacturer	Model					
Mikron Instrument Co., Inc. P.O. Box 211 Ridgewood, NJ 07451	Power meter AK-2940/16932	298-2773			15.24 - ∞	
	5	1-20	313-1973		17-35	76.2 - 228.6
	10	1-20	273-2273		13.1-26.2	2.54 - ∞
	22	1-20	2730-2273			2.54 - ∞
	10L	1-20	273-773		174	2.54 - 12.7
	22L	1-20	273-773		174	2.54 - 12.7
	15	7-20	253-473		35	2.54 - ∞
	21	7.9-8.5*	253-503			2.54 - ∞
	25	7-20	253-333		698	0 - ∞
	44	8-14	253-673		3.5, 5.8	304.8 - 6096
	55	1-20	273-2273		13.1-6.5	12.7 - ∞
	56	1-20	273-2273		13.1-6.5	12.7 - ∞
	57	0.5-1.1	1023-2023		2.9, 13.1	25.4 - ∞
	66	—	273-2273		35, 707	2.54 - ∞
Milletron 63 E. Sandford Blvd. Mt. Vernon, NY 10550	Thermometers KT 4	0.6-40	173-673	Metal film bolometer (Heimann)	10.5, 5.2 (optional) Cassegrain	20 - ∞
	KT 13/1	4-40	223-473			
	KT 13/2	0.6-5.5	323-1273			
	KT 14	2-20	273-1073			
	KT 16	2-20	273-673, 1073			
	KT 24	7.8	293-333			
	KT 30	2-2.7	273-2273			
	KT 31	6.8	273-573			
	KT 32	8.0	273-573			
	KT 33	—	573-1873			
	KT 36	—	273-773			
	KT 41	—	298-313			
	Pyrometer	0.81	923-2473		21-140	
	Pyrometer	0.95	923-2473		10.5-698	
Molectron Corp. 177 N. Wolfe Rd. Sunnyvale, CA 94086	Laser photometer LP-20	—				

*Depends on material.

Table 20-6. Specialized Radiometers (Continued)

Manufacturer	Model		Spectral Range (μm)	Temperature Range (K)	Detector Type	Optics	Field of View (mrad)	Range of Focus (cm)	Focal Length (cm)	Effective Focal Ratio (F/#)	Remarks
Raytek. Inc. 1277 Terra Bella Ave. Mountain View, CA 94043	**Raynger®**										
	R-38		8-14	293-703	Thermistor	—	70	15.24 - ∞	—	—	—
	R-380		8-14	273-1073	Thermistor	—	70	15.24 - ∞	—	—	—
	R-45		2.1-2.3	623-1973	PbS	—	17	15.24 - ∞	—	—	—
	R-450		2.1-2.3	623-1973	PbS	—	17	15.24 - ∞	—	—	—
	R-150		0.6-1.0	1273-1973	Si	—	7	91.44 - ∞	—	—	—
	MR-150		2.1-2.3	873-1973	PbS	—	7	91.44 - ∞	—	—	—
	LC-814	A	8-14	273-423		—	264	30.48 - ∞	—	—	—
		B	8-14	323-473		—	264	30.48 - ∞	—	—	—
		C	8-14	333-573		—	264	30.48 - ∞	—	—	—
		D	8-14	373-773		—	264	30.48 - ∞	—	—	—
	S/L										
	300 A		2.1-2.3	523-773		—	7	45.72 - ∞	—	—	—
	300 B		2.1-2.3	623-973		—	7	45.72 - ∞	—	—	—
	300 C		2.1-2.3	723-1273		—	7	45.72 - ∞	—	—	—
	300 D		2.1-2.3	873-1773		—	7	45.72 - ∞	—	—	—
	400 A		0.65-1.0	1073-1573		—	—	—	—	—	—
	400 B		0.65-1.0	1273-1773		—	—	—	—	—	—
	400 C		0.65-1.0	1473-1973		—	—	—	—	—	—
	Circuit ryder										
	R38E A		IR	293-393		—	133		—	—	—
	R38E B		IR	313-473		—	133		—	—	—
	R38E C		IR	373-583		—	133		—	—	—
	R38E D		IR	533-803		—	133		—	—	—
	R38E E		IR	333-533		—	133		—	—	—
	R38E F		IR	393-703		—	133		—	—	—
	T-1000		IR	292-773		—	50	1.27 - 5.08	—	—	—
Sensors, Inc. P.O. Box 1383 Ann Arbor, MI 48106	Portable thermometers										
	1205		0.6-40	273-773		—	279	0 - ∞	—	—	—
	1205 N		0.6-40	273-773		—	105	0 - ∞	—	—	—

Company	Model					
The Te Co.	Portable 906	—	223-323	—	34.9	152.4 - ∞
	Airborne 907	—	223-323	—	34.9	152.4 - ∞
United Detector Technology, Inc. 2644 30th St. Santa Monica, CA 90465	Laser power meter 20A	3.0-11.5	—	—	—	—
William Wahl Corp. 12908 Panama St. Los Angeles, CA 90066	Thermometers					
	HSA-1E	1.8-20	273-593	—	175	2.54 - ∞
	HSA-1G	4.8-20	273-593	—	175	2.54 - ∞
	HSA-4E	1.8-20	273-873	—	175	2.54 - ∞
	HSA-4G	4.8-20	273-873	—	175	2.54 - ∞
	HSA-6E	1.8-20	373-1273	—	175	2.54 - ∞
	HSA-8E	—	273-593	—		2.54 - ∞
Williamson Corp. 1152 Main St. Concord, MA 01742	Portable thermometers					
	600	2.3-3.3	323-1373		62.5	7.62 - ∞
	1200	2.3-3.3	323-1423		6.25	7.62 - ∞
	1206/1236	—	323-1273	Quartz lens	5.6-33.3	7.62 - ∞
	1211SD	3.43	573-1473	PbS	16.7	7.62 - ∞
	1220		323-1023			
	Thermometers					
	HL-4	0.5-25	248-373	—	384	7.62 - ∞
	4000 series					
	4100	1.0	753-2173	—	27.8	7.62 - ∞
	4200	2.8-3.3	298-1273	—	20.1	7.62 - ∞
	4300	3.43	313-1073	—	34.7	7.62 - ∞
	4400	5.1	573-1923	—	5.6-33.3	7.62 - ∞
	4500	7.95	313-1673	—	41.7	7.62 - ∞
	4800		323-648	—	41.7	7.62 - ∞
	3300	1.0	973-1673	—	16.7	7.62 - ∞
	2000	2.8-3.3	298-1923	—	27.8	15.24 - ∞
	2200	2.8-3.3	298-1373	—	27.8	15.24 - ∞
	2400	—	698-1923	—	6.9	15.24 - ∞

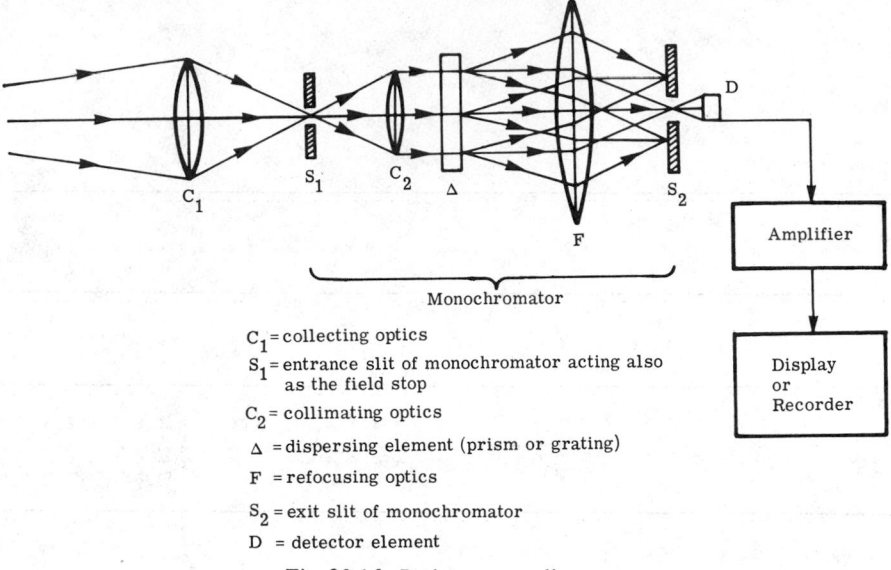

C₁ = collecting optics
S₁ = entrance slit of monochromator acting also
     as the field stop
C₂ = collimating optics
Δ = dispersing element (prism or grating)
F = refocusing optics
S₂ = exit slit of monochromator
D = detector element

Fig. 20-16. Basic spectroradiometer.

spectroradiometers. The filter is often sequenced by the rotation of a filter wheel. If the speed of rotation is sufficiently high, the dwell time for any one filter can approach the response time of the unfiltered radiometer. In such cases, the response time should be determined under dynamic rather than static conditions.

20.3.1. Prisms. The wavelength dependence of the index of refraction is used in prism spectrometers. Such an optical element disperses parallel rays or collimated radiation into different angles from the prism according to wavelength. Distortion of the image of the entrance slit is minimized by the use of plane wave illumination. Even with plane wave illumination, the image of the slit is curved because not all of the rays from the entrance slit can traverse the prism in its principal plane. The prism is shown in the position of minimum angular deviation of the incoming rays in Figure 20-17. At minimum angular deviation, maximum power can pass through the prism.

For a prism adjusted to the position of minimum deviation,

$$r_1 = r_2 = \frac{A_p}{2} \tag{20-26}$$

and

$$i_1 = i_2 = \frac{D_p + A_p}{2} \tag{20-27}$$

where D_p = angle of deviation
          A_p = angle of the prism
          r_1 and r_2 = internal angles of refraction
          i_1 and i_2 = angles of entry and exit

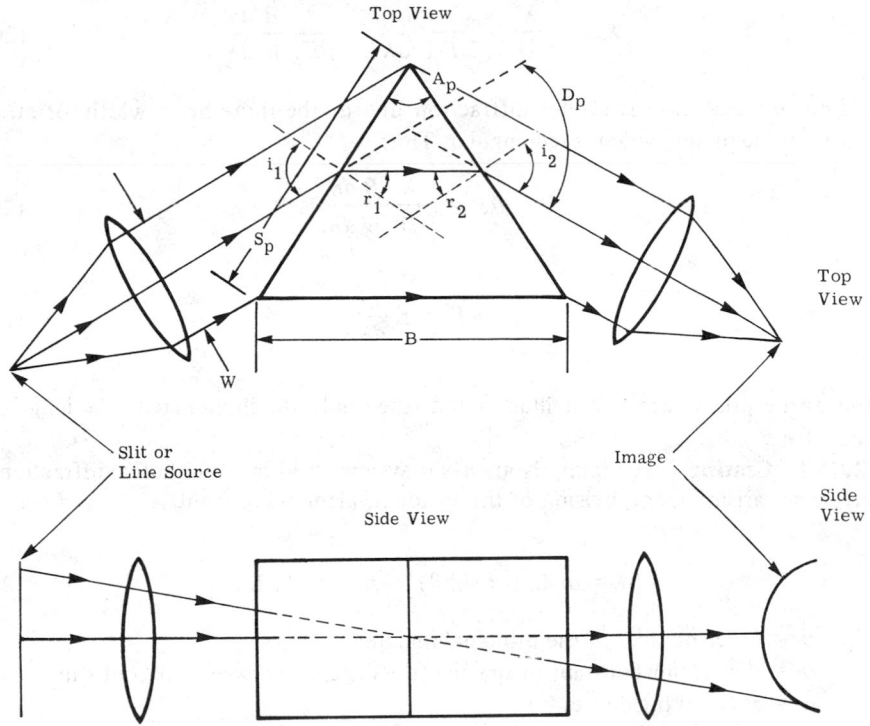

Fig. 20-17. Elementary prism spectrometer schematic. W is the width of the entrance beam; S_p is the length of the prism face; and B is the prism base length.

The angle of deviation, D_p, varies with wavelength. The resulting angular dispersion is defined as $dD_p/d\lambda$, while the linear dispersion is $dx/d\lambda = F \, dD_p/d\lambda$, where F is the focal length of the camera or imaging lens and x is the distance across the image plane. It can be shown [20-9] that

$$\frac{dD_p}{d\lambda} = \frac{B}{W}\frac{dn}{d\lambda} = \frac{dD_p}{dn}\frac{dn}{d\lambda} \qquad (20\text{-}28)$$

where B = base length of the prism
$\quad\;\; W$ = width of the illumination beam
$\quad\;\; n$ = index of refraction

while

$$\frac{dx}{d\lambda} = F\frac{B}{W}\frac{dn}{d\lambda} \qquad (20\text{-}29)$$

One may define the resolving power, RP, of an instrument as the smallest resolvable wavelength difference, according to the Rayleigh criterion (Chapter 8), divided into the average wavelength in that spectral region. Thus,

$$RP \equiv \frac{\bar{\lambda}}{\Delta\lambda} = \frac{\lambda}{dD_p}\frac{dD_p}{d\lambda} = \frac{\lambda}{dD_p}\frac{B}{W}\frac{dn}{d\lambda} \qquad (20\text{-}30)$$

The limiting resolution is set by diffraction due to the finite beam width, or effective aperture of the prism, which is rectangular. Thus,

$$RP = \frac{\lambda}{\lambda/W}\frac{B}{W}\frac{dn}{d\lambda} \qquad (20\text{-}31)$$

so that

$$RP = B\frac{dn}{d\lambda} \qquad (20\text{-}32)$$

If the entire prism face is not illuminated, then only the illuminated base length must be used for B.

20.3.2. Gratings. A grating is an n-slit system used in Fraunhofer diffraction with interference arising from division of the incident, plane wave front.

$$p\lambda = d(\sin\phi + \sin\theta) \qquad p = 0, 1, 2, \dots \qquad (20\text{-}33)$$

where p = order number of the principal maxima
d = the grating constant or spacing (the distance between adjacent slits)
ϕ = angle of incidence
θ = angle of diffraction
w = width of any one slit

The most common case is $\phi = 0$, so that

$$p\lambda = d\sin\theta \qquad (20\text{-}34)$$

and the irradiance distribution is

$$E = E_0 \left[\frac{\sin\frac{\pi w \sin\theta}{\lambda}}{\frac{\pi w \sin\theta}{\lambda}}\right]^2 \left[\frac{\sin\frac{N\pi d \sin\theta}{\lambda}}{\sin\frac{\pi d \sin\theta}{\lambda}}\right]^2 \qquad (20\text{-}35)$$

where N is the number of lines or slits. This equation is more often written as

$$E = E_0 \frac{\sin^2 \beta}{\beta^2}\frac{\sin^2 N\gamma}{\sin^2 \gamma} \qquad (20\text{-}36)$$

which can be considered to be

$$E = (\text{constant}) \times (\text{single-slit diffraction function}) \times (N\text{-slit interference function}) \qquad (20\text{-}37)$$

These considerations are for unblazed gratings. For a diffraction grating, the angular dispersion is given (for angle ϕ constant) by

$$\frac{dD_g}{d\lambda} \text{ or } \frac{d\theta}{d\lambda} = \frac{p}{d \cos \theta} \qquad (20\text{-}38)$$

The resolving power is given by

$$RP_g = pN \qquad (20\text{-}39)$$

where N is the total number of slits.

Several examples of specialized prisms and spectrometers are given below.

20.3.3. Prism Configurations and Instruments [20-9]. One of the oldest spectrometer configurations is shown in Figure 20-18. Reflective interactions are used in Figure 20-19, 20-20 and 20-21. Dispersion without deviation is realized in Figures 20-22 and 20-23, while half-prisms are used in Figure 20-24.

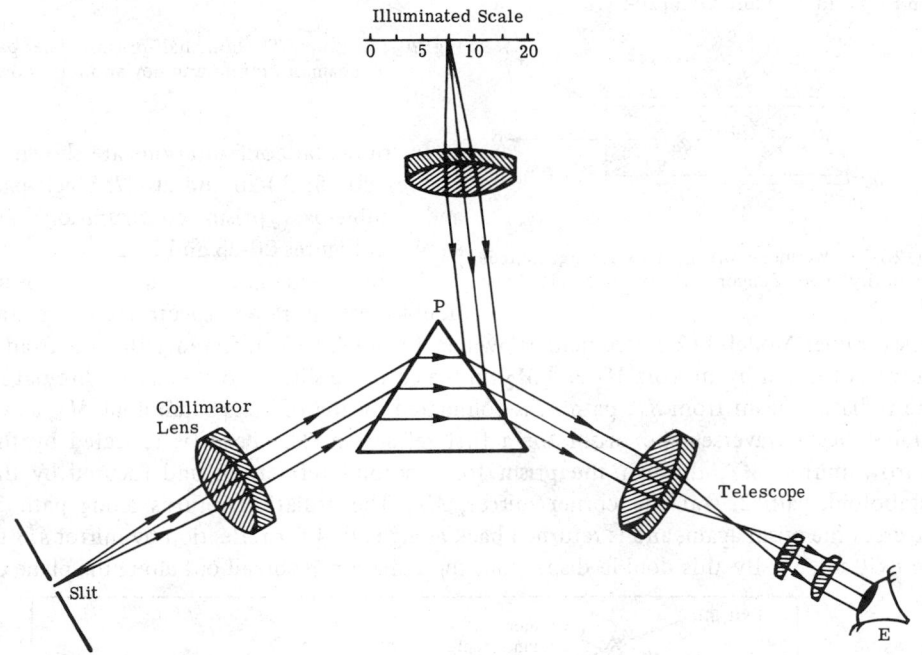

Fig. 20-18. Bunsen-Kirchhoff spectrometer. An illuminated scale is reflected from the prism face into the telescope.

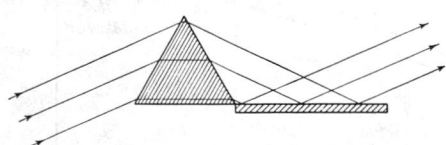

Fig. 20-19. Wadsworth constant-deviation, prism-mirror arrangement. The beam enters the prism at minimum deviation and emerges displaced but not deviated from its original direction [20-9].

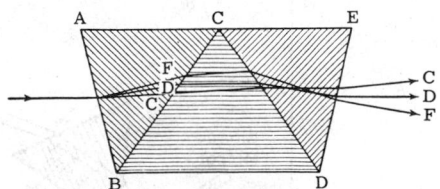

Fig. 20-20. Amici prism. The central ray, D, enters and leaves parallel to the base. The C and F rays are deviated and dispersed [20-9].

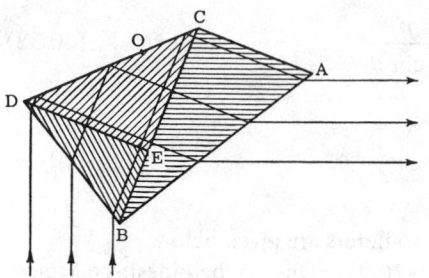

Fig. 20-21. Pellin-Broca prism. This prism is equivalent to two 30° prisms, *ABC* and *BED*, and one 45° prism, *DEC*, but is made in one place. The beam shown, entering at minimum deviation, emerges at 90° deviation to its entrance direction [20-9].

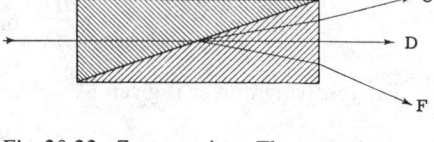

Fig. 20-22. Zenger prism. The central ray, *D*, is undeviated. The *C* and *F* rays are deviated and dispersed [20-9].

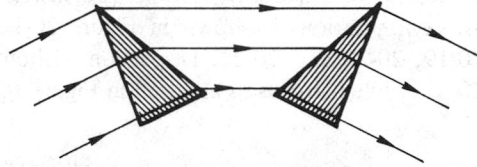

Fig. 20-24. Young-Thollon half-prisms. The passage of a beam at minimum deviation is shown [20-9].

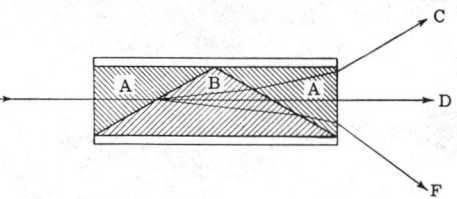

Fig. 20-23. Wernicke prism. This arrangement is essentially two Zenger prisms, back to back [20-9].

Instrumental configurations are shown in Figures 20-25, 20-26 and 20-27. Single-pass and double-pass prism configurations are shown in Figures 20-28 and 20-29.

Another example of a single beam double-pass infrared spectrometer is the Perkin-Elmer Model 112 instrument shown in Figure 20-30. Infrared radiation from a source is focused by mirrors M_1 and M_2 on the entrance slit, S_1, of the monochromator. The radiation beam from S_1, path 1, is collimated by the off-axis paraboloid, M_3, and a parallel beam traverses the prism for a first refraction. The beam is reflected by the Littrow mirror, M_4, through the prism for a second refraction, and focused by the paraboloid, path 2, at the corner mirror, M_6. The radiation returns along path 3, traverses the prism again, and is returned back along path 4 for reflection by mirror M_7 to the exit slit, S_2. By this double dispersion, the radiation is spread out along the plane of

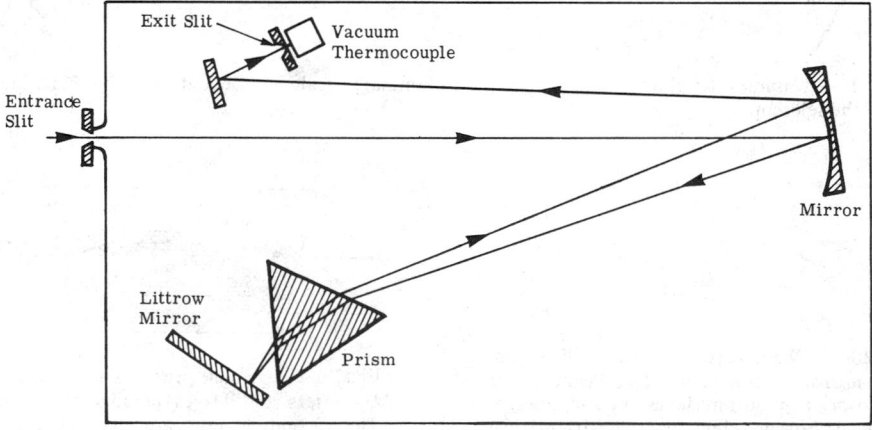

Fig. 20-25. Infrared spectrograph of the Littrow-type mount with a rock salt prism [20-9].

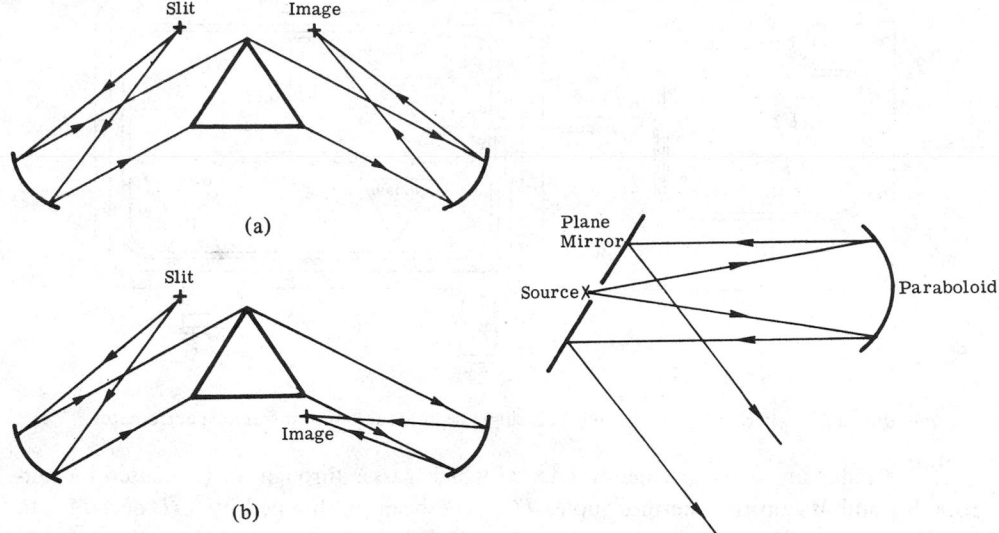

Fig. 20-26. Mirror spectrometer with two choices of the location of the image. Arrangement (b) leads to smaller aberrations than arrangement (a) and is used in the Czerny-Turner mount [20-9].

Fig. 20-27. Pfund mirror. The use of a plane mirror to avoid astigmatism in the use of a paraboloidal mirror [20-9].

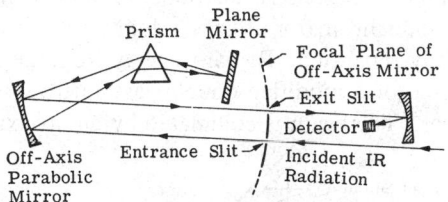

Fig. 20-28. Single-pass monochromator [20-10].

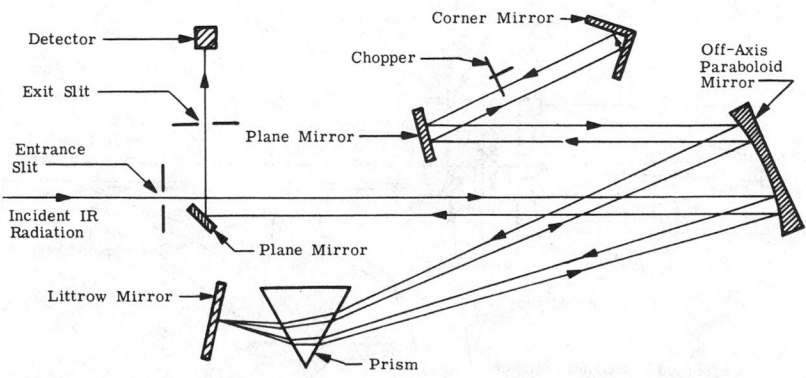

Fig. 20-29. Perkin-Elmer Model 99 double-pass monochromator [20-11].

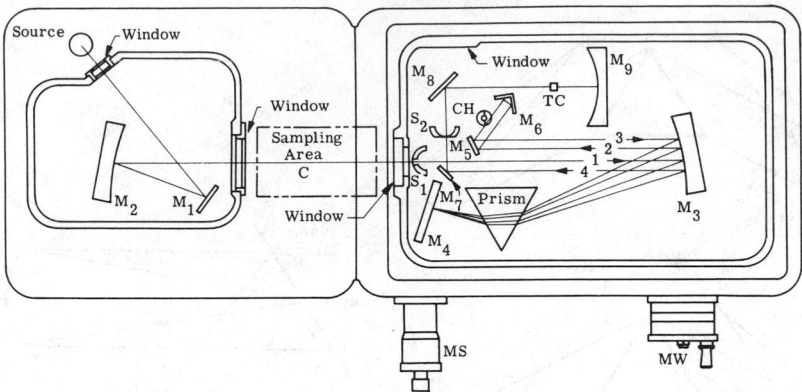

Fig. 20-30. Perkin-Elmer Model 112 single-beam double-pass infrared spectrometer.

S_2. The radiation of the frequency interval which passes through S_2 is focused by mirrors M_8 and M_9 on the thermocouple, *TC*. The beam is chopped by *CH*, near M_6, to produce an ac voltage (at the thermocouple) which is proportional to the radiant power or intensity of the beam. This voltage is amplified and recorded by an electronic potentiometer. Motor driven rotation of Littrow mirror M_4 causes the infrared spectrum to pass across exit slit S_2, permitting measurement of the radiant intensity of successive frequencies.

Prisms have been used in rapid-scan instruments as by Perkin-Elmer. (See Figure 20-31.) The rapid-scan monochromator utilizes a double-pass system where the first and second passes are physically separated. Radiant energy from the source is imaged by the collecting optics, composed of a modified double-pass Littrow system, on the entrance slit of the monochromator. The beam is collimated by an off-axis paraboloid mirror on-

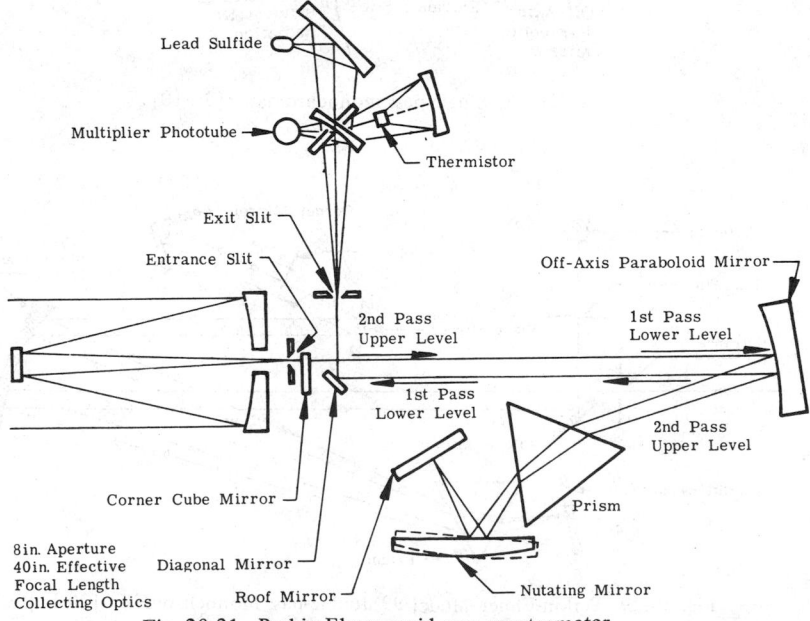

Fig. 20-31. Perkin-Elmer rapid-scan spectrometer.

to the prism. The energy beam is then refracted to the nutating mirror, reflected to the roof mirror (where the vertical, light-motion component is eliminated), and then back through the prism system.   The returned beam is brought to a focus by the parabolidal mirror at the cube corner mirror, where it is displaced up and across, and back to the paraboloid mirror. It then passes through the monochromator for the second time.   The second pass is 1 in. higher than the first pass, and the second pass only is intercepted by the diagonal mirror and focused on the exit slit.

   20.3.4.   Grating Configurations and Instruments [20-12]. Gratings can be used either in transmission or reflection. A more interesting variation comes from their use in plane or concave reflection form. The latter was treated most completely by Rowland, who achieved a useful combination of focussing and grating action. He showed that the radius of curvature of the grating surface is the diameter of a circle (called the Rowland circle). Any source placed on the circle will be imaged on the circle, with dispersion, if the rulings are made so that d is constant on the secant to the grating-blank (spherical) surface. The astigmatism acts so that a point source on a Rowland circle is imaged as a vertical line perpendicular to the plane of the circle. Rowland invented and constructed the first concave-grating mounting, illustrated in Figure 20-32.

   If dispersion is sufficiently large, one may find overlapping of the lines from one order with members of the spectra belonging to a neighboring order. Errors and imperfections in the ruling of gratings can produce spurious images which are called "ghosts." Also, the grooves in a grating can be shaped so as to send more radiation along a preferred direction corresponding to an order other than the zero order. Such gratings are said to be blazed in that order.

(1) *Paschen-Runge,* illustrated in Figure 20-33.

   In this arrangement, one or more fixed slits are placed to give an angle of incidence suitable for the uses of the instrument. The spectra are focused along the Rowland circle $P\ P'$, and photographic plates, or other detectors, are placed along a large portion of this circle.

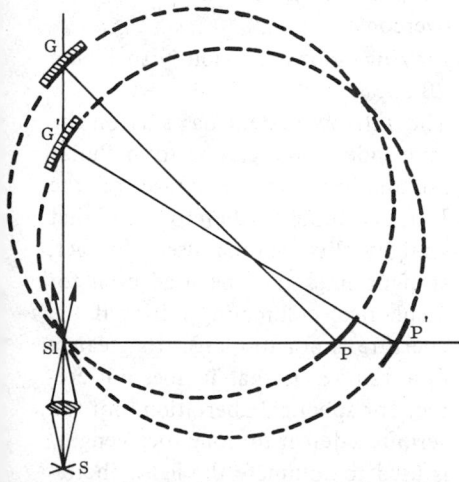

Fig. 20-32. Rowland mounting of the concave grating. The grating plate-holder bar which slides on the two perpendicular ways is shown in two positions, GP and $G'P'$. The slit, Sl, and source, S, remain fixed [20-9].

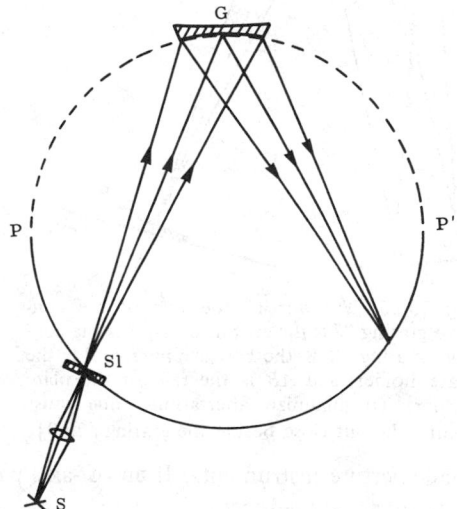

Fig. 20-33. Paschen-Runge mounting of the concave grating. Sl is the slit; G is the grating; and S is the light source [20-9].

(2) *Eagle*, shown in Figure 20-34.

This is similar to the Littrow prism spectrograph. The slit and plate holder are mounted close together on one end of a rigid bar with the concave grating mounted on the other end.

(3) *Wadsworth*, shown in Figure 20-35.

The Rowland circle is not used in this mounting in which the grating receives parallel light.

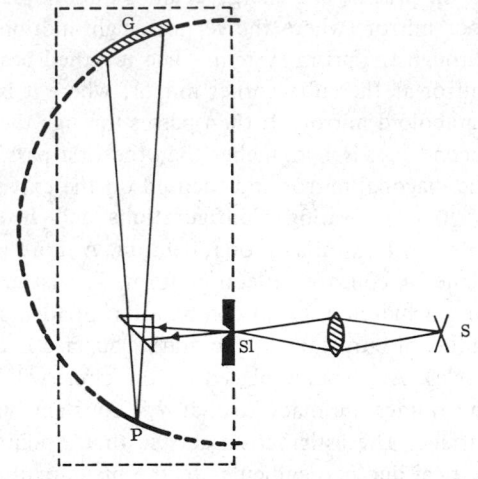

Fig. 20-34. Eagle mounting on the concave grating. *Sl* is the slit; *G* is the grating; *S* is the light source; and *P* is the plate holder [20-9].

(4) *Ebert-Fastie*, shown in Figure 20-36.

The Ebert-Fastie features a single, spherical, collimating mirror and a grating placed symmetrically between the two slits. The major advantage of the Ebert system is the fact that it is self-correcting for spherical aberration. With the use of curved slits, astigmatism is almost completely overcome.

(5) *Littrow*, shown previously in Figure 20-25.

The Littrow system has slits on the same side of the grating to minimize astigmatism. An advantage of the Littrow mount, therefore, is that straight slits can be used. In fact, straight lines may be used even for a spherical collimating mirror if the aperture is not too large. Its greatest disadvantage is that it does not correct for spherical aberration, not too serious a defect for long focal-length/

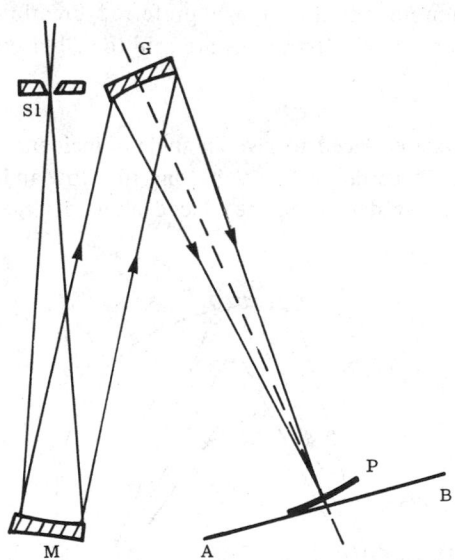

Fig. 20-35. Wadsworth mounting of the concave grating. *Sl* is the entrance slit; *G* is the concave grating; *M* is the concave mirror; *P* is the plate holder; and *AB* is the rail for the plate holder. To minimize aberrations, one must locate the slit close beside the grating [20-9].

small-aperture instruments. If an off-axis parabola is used to collimate the light, aberrations are greatly reduced.

Gratings may also be used in single-pass and double-pass configurations. Figure 20-37 shows an on-axis, Pfund-type grating instrument. Incident infrared radiation, focused by

a collimating lens on the entrance slit
and modulated by a chopper, passes
through the central aperture of plane
mirror M_1. Reflected by the paraboloidal mirror, P_1, it emerges as a parallel
beam of radiation, which is reflected by
mirror M_1 to the grating. The grating
is accurately located on a turntable,
which may be rotated to scan the spectrum. From the grating, the diffracted
beam, reflected by mirror M_2, is focused
by a second paraboloid, P_2, through the

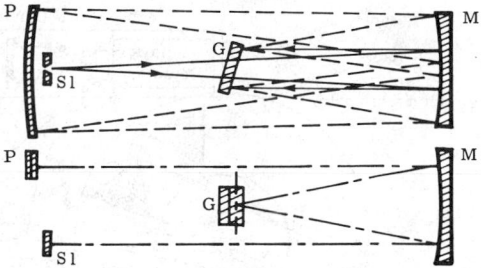

Fig. 20-36. Ebert mounting of the plane grating
designed by Fastie. Sl is the entrance slit; G is
the grating; M is the concave mirror; and P is
the photographic plate. The horizontal section is
at the top and the vertical section is at the bottom
[20-9].

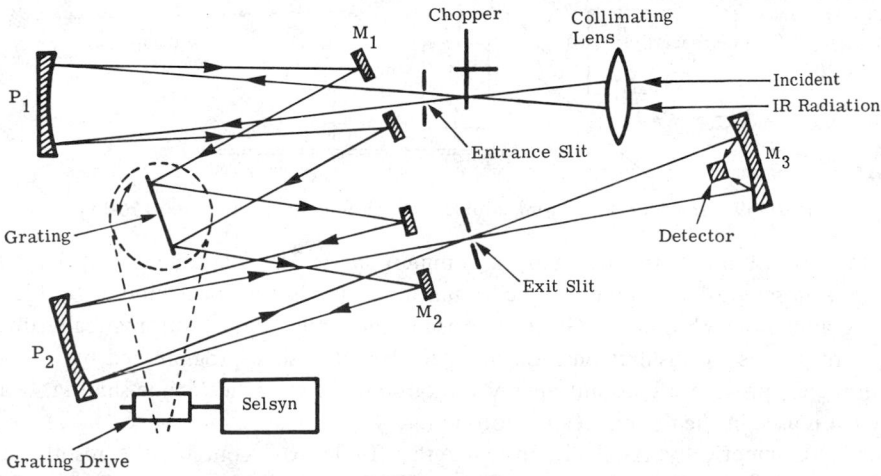

Fig. 20-37. On-axis Pfund grating spectrograph [20-11].

central aperture of mirror M_2 to the
exit slit. The emerging beam is then
focused by the ellipsoidal mirror, M_3,
on the detector.

An off-axis, double-pass instrument
is shown in Figure 20-38.

**20.3.5. Prism-Grating Combina-
tions.** Combinations of prisms and
gratings are not uncommon. A prism-
grating, double-monochromator spectrometer by Unicam Instruments, Ltd.
is shown in Figure 20-39. The prism

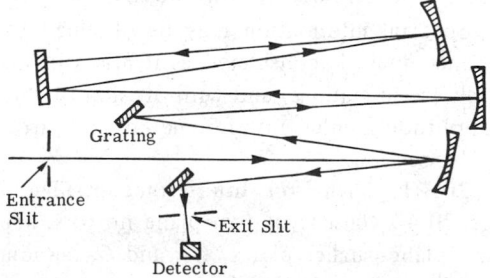

Fig. 20-38. Off-axis, double-pass grating spectrograph [20-12].

monochromator has four interchangeable prisms, and the grating monochromator has
two interchangeable gratings. The two monochromators, ganged by cams which are

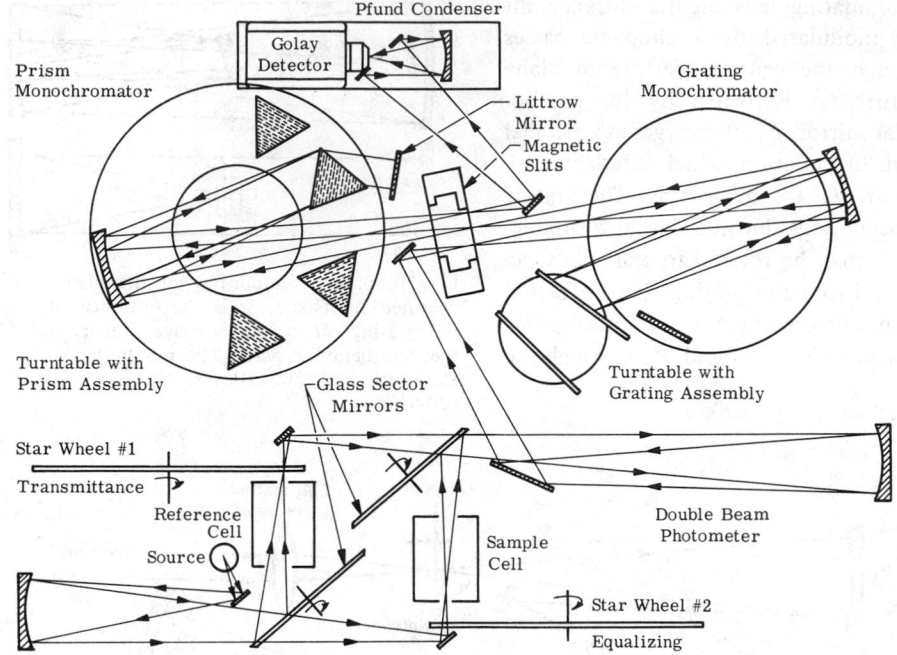

Fig. 20-39. Unicam prism-grating double monochromator spectrometer [20-11].

linear in wave number, are driven by a common shaft. The instrument may be used either as a prism-grating double monochromator, or as a prism spectrometer by blanking the grating monochromator. Gratings, prisms, and cams may be automatically interchanged by means of pushbuttons. Magnetically operated slits, programmed by a taped potentiometer, provide a constant energy background. A star-wheel, time-sharing, beam attenuator is used in the double-beam photometer.

20.3.6. Commerically Available Instruments. Table 20-7 contains the results of a brief Infrared Information and Analysis Center (IRIA) survey* of commercially-available, prism and grating spectroradiometers.

20.4. Interferometers: Two-Beam

Spectral information may be obtained by use of interferometers. Classically, interference may be considered as it arises in instruments with division of the wave front itself (as in gratings) and with division of the amplitude associated with the wave front. Amplitude-dividing interferometers are particularly useful instruments and are stressed in this section.

20.4.1. Michelson Interferometer. The Michelson interferometer, shown in Figure 20-40 consists of two plane mirrors, M_1 and M_2, one of which is adjustable, and two plane-parallel plates, G_1 and G_2. Radiation from an extended source is incident at 45° on G_1, (partially silvered on the rear surface) and is divided into reflected (path A) and transmitted beams (path B) of equal strength. The radiation reflected from mirror M_1 passes through plate G_1 a third time before it reaches the eye or the

---

*Conducted by Craig Mueller, Environmental Research Institute of Michigan (ERIM), Ann Arbor, MI.

detector. The radiation reflected from mirror M_2 passes back through G_2 a second time, is reflected from the surface of plate G_1, and onto the detector. The two beams have a phase difference governed by the difference in the two paths.

Compensating plate G_2 compensates for the phase change caused by passage of radiation through the plate in path A. Its use is not essential for producing fringes in monochromatic light but is indispensable when white light is used. The radiation from every point on the extended source interferes with itself according to the distance between mirrors, or according to the difference in length of arms A and B. Constructive interference will occur when

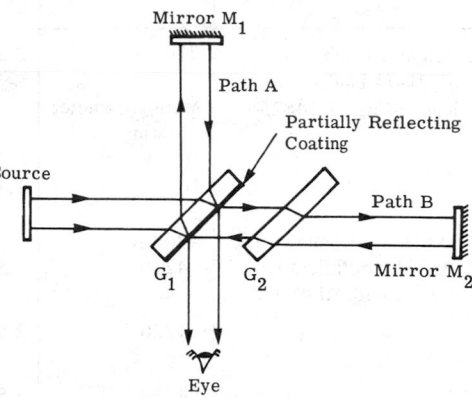

Fig. 20-40. Michelson interferometer.

$$2x \cos \theta = p\lambda \qquad (20\text{-}40)$$

where x = path difference
 θ = angle to a source element imaged by M_1 or M_2
 λ = wavelength
 p = the order number, an integer = 0, 1, 2, . . .

The adjustment of M_1 perpendicular to M_2 gives circular fringes, while a slight deviation from this condition leads to straight-line fringes. The shape or profile of the fringes is given by

$$E = E_0 \left(1 + \cos \frac{4\pi x}{\lambda}\right) \qquad (20\text{-}41)$$

or

$$E = 2E_0 \cos^2 \left(\frac{2\pi x}{\lambda}\right) \qquad (20\text{-}42)$$

where E is the irradiance at the detector and x, λ, are as defined for Equation (20-40). Since $E_0 \equiv$ maximum value of E, one finds that

$$\frac{2\pi x}{\lambda} = p\pi \qquad (20\text{-}43)$$

The minimum is $E = 0$ where

$$\left(p + \frac{1}{2}\right)\pi = \frac{2\pi x}{\lambda} \qquad (20\text{-}44)$$

Michelson used the "visibility" of fringes to determine small wavelength differences, e.g., the sodium doublet separation. He defined this quantity as follows:

$$\text{Vis.} = \frac{E_{max} - E_{min}}{E_{max} + E_{min}} \qquad (20\text{-}45)$$

Table 20-7. Spectroradiometers

Manufacturer	Model	Spectral Range (μm)	Monochromator Type F/#	Operation
Bausch and Lomb 61-21-04 Linden Ave. Rochester, NY 14625	710	0.2-1.0	–	DB single grating
	Monochrometer grating			
	IR #1	0.7-1.6	Grating	–
	IR #2	1.4-3.2	Grating	–
Beckman Instruments, Inc. 25511 Southfield Rd. Southfield, MI 48075	4200 series			
	4210	2.5-16.6	Filter-grating (single)	–
	4220	2.5-40.0	Filter-grating (single)	–
	4230	2.5-16.6	Filter-grating (single)	–
	4240	2.5-40.0	Filter-grating (single)	–
	4250	2.5-50.0	Filter-grating (single)	–
	4260	2.5-50.0	–	–
	Acculab			
	1 and 2	2.5-16.6	Rotating wedge filter; grating 100 ℓ mm^{-1}	DB optical null
	3 and 4	2.5-16.6	Rotating wedge filter; grating 100 ℓ mm^{-1}	DB optical null
	5 and 6	2.5-26.6 2.5-40	Rotating wedge filter; grating 100 and 40 ℓ mm^{-1}	DB optical null
	M-IV	0.19-3.0	Filter/grating blazer @ 1.2 μm for NIR (single)	DB and SB ratio recording
	M-VII	0.19-3.0	Filter/grating blazed @ 1.2 μm for NIR (double)	DB and SB ratio recording
Block Engineering, Inc. 19 Blackstone Street, Cambridge, MA 02139	E-8 H	0.225-0.8925	Diffraction F/2 grating 600 & 1200 ℓ mm^{-1}	Ebert

Table 20-7. Spectroradiometers (*Continued*)

Resolution	Scan Speed or Scan Time	Total Accuracy	Wavelength Reproducibility	Remarks
–	–	–	–	Wavelength accuracy ± 1.0 nm Dispersion
–	–	–	–	12.8 nm mm^{-1}
–	–	–	–	25.0 nm mm^{-1}
3 cm^{-1} @ 3000 cm^{-1} 1.5 cm^{-1} @ 900 cm^{-1}	750-5 cm^{-1} min^{-1}	± 1.0%	–	–
3 cm^{-1} @ 3000 cm^{-1} 1.5 cm^{-1} @ 900 cm^{-1}	750-5 cm^{-1} min^{-1}	± 1.0%	–	–
1.4 cm^{-1} @ 3000 cm^{-1} 0.5 cm^{-1} @ 900 cm^{-1}	1000-2 cm^{-1} min^{-1}	± 1.0% ± 1.0%	–	–
1.4 cm^{-1} @ 3000 cm^{-1} 0.5 cm^{-1} @ 900 cm^{-1}	1000-2 cm^{-1} min^{-1}	± 1.0%	–	–
1.4 cm^{-1} @ 3000 cm^{-1} 0.5 cm^{-1} @ 900 cm^{-1}	1000-2 cm^{-1} min^{-1}	± 1.0%	–	–
0.7 cm^{-1} @ 2400 cm^{-1} 0.35 cm^{-1}@ 1000 cm^{-1}	1000-2 cm^{-1} min^{-1}	± 1.0%	–	–
10 cm^{-1} @ 3000 cm^{-1} 5 cm^{-1} @ 1000 cm^{-1}	2.5-22 min	2%	1%	–
5 cm^{-1} @ 3000 cm^{-1} 3 cm^{-1} @ 1000 cm^{-1}	3.2-28 min	2%	1%	–
5 cm^{-1} @ 3000 cm^{-1} 3 cm^{-1} @ 1000 cm^{-1}	3.2-28 min	2%	1%	–
0.2-1.2 nm	1/64 – 16 nm sec^{-1}	–	–	–
0.05-0.3 nm	1/64 – 16 nm sec^{-1}	–	–	–
0.0015	10 sec^{-1}	–	–	–

Table 20-7. Spectroradiometers (*Continued*)

Manufacturer	Model	Spectral Range (μm)	Monochromator Type F/#	Operation
Cary Instrument (Varian Subsidiary) 611 Hansen Way Palo Alto, CA 94303	14	0.186-2.65	Prism/grating F/8 (double)	DB water cooled
	17	0.186-2.65	Prism/grating F/8 (double)	–
	17I	0.275-3.0	Prism/grating (double)	–
	17H	0.186-2.65	Prism/grating (double)	–
EG & G Electro-Optics Div. 35 Congress St. Salem, MA 01970	580/585	0.2-3.2	Grating	–
	585 IR	0.2-3.2	Grating	–
Electro Optical Industries P.O. Box 3770 Santa Barbara, CA 93105	474S	0.3-1.15	Holographic F/3.5 grating	–
	474P	0.6-40	1/4 Ebert F/3 type grating	–
	474MOS	0.17-1.15	Holographic F/3.5 grating	–
	474LS	0.5-3.2	1/4 Ebert F/3.5 type grating	–
Gamma Scientific, Inc. 3777 Ruffin Rd. San Diego, CA 92123	2900SR	0.35-0.75	Grating	–
	2400SR	0.35-0.75	Grating	–
	2900TR	0.35-0.75	Diffraction grating	–
	2400TR	0.35-0.75 (0.35-1.1, optional)	Diffraction grating	–

Table 20-7. Spectroradiometers (*Continued*)

Resolution	Scan Speed or Scan Time	Total Accuracy	Wavelength Reproducibility	Remarks
–	0.05-50 nm sec^{-1} (8 speeds)	–	0.05 nm in UV-Vis 0.25 nm in NIR	Wavelength accuracy ± 0.4 nm
0.1 nm UV-Vis 0.3 nm NIR	0.005-10 nm sec^{-1}	–	0.005 nm in UV-Vis 0.025 nm in NIR	Wavelength accuracy ± 0.4 nm
1-2 cm^{-1}	0.005-10 nm sec^{-1}	–	0.005 nm in UV-Vis 0.025 nm in NIR	Wavelength accuracy ± 0.4 nm
0.15 nm UV-Vis 0.4 nm NIR	0.005-10 nm sec^{-1}	–	0.005 nm in UV-Vis 0.025 nm in NIR	Wavelength accuracy ± 0.4 nm
–	–	–	–	–
–	–	–	–	–
–	–	–	–	–
–	–	–	–	Range of focus 180 cm
–	–	–	–	–
–	–	–	–	–
–	–	–	–	Range of focus 180 cm
–	–	–	± 0.5 nm	Wavelength accuracy ± 2 nm
–	–	–	± 0.5 nm	Wavelength accuracy ± 2 nm
–	–	–	–	Wavelength accuracy ± 2 nm
–	–	–	–	Wavelength accuracy ± 2 nm

Table 20-7. Spectroradiometers (*Continued*)

Manufacturer	Model	Spectral Range (μm)	Monochromator Type F/#	Operation
GCA/McPherson Instrument Div. 530 Main St. Acton, MA 01720	EU-72i	0.185-1.0	Czerny-Turner 350 mm grating (double)	Programmed Alternating cell
	EU-707	0.185-1.0 (optional)	Czerny-Turner 350 mm grating (double)	DB
	EU-701	0.2-0.7 (0.185-1.0, optional)	Czerny-Turner 350 mm grating	SB
International Light, Inc. Dexter Industrial Green Newburyport, MA 01950	IL700 760 780	0.24-0.81	Grating $F/4.7$ off-axis Ebert	–
Jarrel-Ash Div. of Fisher Science Co. 590 Lincoln St. Waltham, MA 02154	Monochromator 82-415	0.3-10	Ebert $F/3.5$ grating	–
J–Y Optical Systems Division of ISA 173 Essex Ave. Metuchen, NJ 08840	Ramanor® H.G. 2 Monochromator	0.31-0.83	Holographic $F/8$ gratings (double)	–
	H-20V-IR Monochromator	0.4-1.6	Diffraction $F/4.2$ grating	–
	H-20 IR Monochromator	0.8-3.2	Diffraction $F/4.2$ grating	–
	HR-1000	0-1.5 (0-50 optional)	Diffraction $F/6.8$ grating	–
Optronic Laboratories, Inc. 7676 Fenton St. Silver Spring, MD 20910	740A	0.3-1.07	Grating $F/4$ (single)	–
	740IR	0.4-12	Grating $F/4$	Si & PbS
Oriel Corp. of America 15 Market St. Stamford, CT 06902	7240	0.19-24	Grating $F/3.7$	–

Table 20-7. Spectroradiometers (*Continued*)

Resolution	Scan Speed or Scan Time	Total Accuracy	Wavelength Reproducibility	Remarks
0.1 nm	1-200 nm min^{-1}	± 0.01 nm	± 0.01 nm	–
0.1 nm	1-200 nm min^{-1}	± 0.01 nm	± 0.01 nm	–
0.1 nm	1-120 nm min^{-1}	± 0.1 nm	± 0.01 nm	–
–	–	–	–	–
0.3 nm @ 313.1 nm	–	–	–	Dispersion 3.3 nm mm^{-1}
0.5 cm^{-1} @ 19,436 cm^{-1}	1-10 cm^{-1} min^{-1}	–	–	–
2.5 nm	–	–	–	Dispersion and focal length 8 nm mm^{-1} 20.0 cm
5 nm	–	–	–	Dispersion and focal length 16 nm mm^{-1} 20.0 cm
0.8 nm	–	–	–	0.78-0.26 nm mm^{-1} 100.0 cm
–	–	± 3-5%	–	–
–	–	–	–	–
0.25 nm mm^{-1}	–	–	–	Dispersion and focal length 3.2-6.4 nm mm^{-1} 12.5 cm

Table 20-7. Spectroradiometers (*Continued*)

Manufacturer	Model	Spectral Range (μm)	Monochromator Type F/#	Operation
Perkin-Elmer Main Ave. Norwalk, CT 06856	180	2.5-55.5	Ebert $F/6$ 5 plane gratings	DB, SB electrical null; ratio recording
		19-312.5	Ebert $F/6$ 5 plane gratings	DB, SB electrical null; ratio recording
	281	2.5-16.6	Filter $F/5$ grating	DB optical null
	283	2.5-50	Filter $F/5$ grating	DB optical null
	580	2.5-55.5	4 gratings $F/5.7$ Littrow	DB ratio recording
	67 series 167	2.5-16.6	Filter $F/5$ single grating	DB optical null
	267	2.5-16.6	Filter $F/5$ single grating	DB optical null
	467	2.5-40	Filter $F/5$ double grating	DB optical null
	567	2.5-50	Filter $F/5$ double grating	DB optical null
	700 series 710B	2.5-16.6	Filter $F/6$ grating Littow (single)	DB optical null
	727B	2.5-16.6	Filter $F/6$ grating Littow (single)	DB optical null
	735B	2.5-24	Filter $F/6$ grating Littow (single)	DB optical null
	197	2.5-16.6	Filter $F/5$ grating	DB optical null
	297	2.5-16.6	Filter $F/5$ grating	DB optical null
	Computer dispersive spectro- photometers			

Table 20-7. Spectroradiometers (*Continued*)

Resolution	Scan Speed or Scan Time	Total Accuracy	Wavelength Reproducibility	Remarks
0.2 cm^{-1} @ 1030 cm^{-1} 0.4 cm^{-1} @ 3000 cm^{-1}	0.2 cm^{-1} min^{-1} to 1 min per grating range	0.1-0.25 cm^{-1}	0.1 cm^{-1}	Spectral range cm^{-1} 4000-180
0.75 cm^{-1} @ 200 cm^{-1}	0.25 cm^{-1} min^{-1} to 1 min per grating range	0.25 cm^{-1}	0.1 cm^{-1}	525-32
3 cm^{-1} @ 4000-200 cm^{-1} 115 cm^{-1} @ 2000-600 cm^{-1}	3 min-10 h	–	1.5-0.8 cm^{-1}	4000-600
3 cm^{-1} @ 4000-2000 cm^{-1} 1.5 cm^{-1} @ 2000-200 cm^{-1}	3 min-10 h	–	1.5-0.8 cm^{-1}	4000-200
0.4-1.0 cm^{-1}	0.5-10 cm min^{-1}	–	0.25 cm^{-1} -0.5 cm^{-1}	4000-180
4 cm^{-1} @ 4000-2000 cm^{-1} 2 cm^{-1} @ 2000-600 cm^{-1}	5-48 min	–	2-1 cm^{-1}	4000-600
4 cm^{-1} @ 4000-2000 cm^{-1} 2 cm^{-1} @ 2000-600 cm^{-1}	5-192 min	–	2-1 cm^{-1}	4000-600
4 cm^{-1} @ 4000-2000 cm^{-1}	6-200 min	–	2-1 cm^{-1}	4000-250
4 cm^{-1} @ 4000-2000 cm^{-1} 2 cm^{-1} @ 2000-600 cm^{-1}	6-224 min	–	2-1 cm^{-1}	4000-200
8 cm^{-1} @ 4000-2000 cm^{-1} 4 cm^{-1} @ 2000 cm^{-1} end of range	2 2/3-8 min	–	4-2 cm^{-1}	4000-600
8 cm^{-1} @ 4000-2000 cm^{-1} 4 cm^{-1} @ 2000 cm^{-1} end of range	2 2/3-8 min	–	4-2 cm^{-1}	4000-600
8 cm^{-1} @ 4000-2000 cm^{-1} 4 cm^{-1} @ 200 cm^{-1} end of range	3 2/3-11 min	–	4-2 cm^{-1}	4000-600
1.5-4 cm^{-1} @ 1100 cm^{-1}	3-10 min	–	3-1.5 cm^{-1}	4000-600
3-4 cm^{-1} @ 1100 cm^{-1}	3-50 min	–	2-1 cm^{-1}	4000-600

Table 20-7. Spectroradiometers (*Continued*)

Manufacturer	Model	Spectral Range (μm)	Monochromator Type	F/#	Operation
Ray Lee Instruments, Inc 7103 Milford Industrial Rd. Pikesville, MD 21208	12-150	–	Ebert grating	F/5	–
Schoeffel Instrument Corp. 24 Bocker St. Westwood, NJ 07675	High sensitivity spectro- radiometer Monochromator	0.18-1.1	Grating		–
	GM 100 Monochromator	0.185-1.4	Off-axis Ebert grating	F/4.7	–
	GM 250	0.18-3.0	Off-axis Ebert grating	F/3.6	–
	Quartz prism spectro- radiometer Monochromator	0.2-1.1	Quartz prism		–
	QPM 30S	0.185-2.5	Quartz prism	F/7.6	–
Princeton Applied Research (OMA) P.O. Box 2565 Princeton, NJ 08540	1205E 1205KE 1205A	– 0.9-1.1 –	– – –		– Cooled vidicons –

If mirror M_1 is moved at a constant speed, so that dx/dt is constant, in a sawtooth pattern, repeated over the distance interval of

$$-\frac{C_m}{4} \leqslant x \leqslant \frac{C_m}{4} \tag{20-46}$$

in a time interval of T_m, then

$$x(t) = \frac{C_m t}{2T_m} \tag{20-47}$$

and

$$E = E_0 \left(1 + \cos \frac{2\pi C_m t \tilde{\nu}}{T_m} \right) \tag{20-48}$$

Thus the modulation frequency, f, depends upon the wave number of the radiation so that

$$f = \frac{\tilde{\nu} C_m}{T_m} \tag{20-49}$$

and

$$E = E_0 \left(1 + \cos 2\pi f t \right) \tag{20-50}$$

Table 20-7. Spectroradiometers (*Continued*)

Resolution	Scan Speed or Scan Time	Total Accuracy	Wavelength Reproducibility	Remarks
1.0 nm	–	–	–	–
–	–	–	–	–
–	–	–	–	Dispersion and focal length 8.5 nm mm^{-1} 10.0 cm
–	–	–	–	Dispersion and focal length 3.3 nm mm^{-1} 25.0 cm
–	–	–	–	–
–	–	–	–	Focal length 10.0 cm
–	–	–	–	–
–	–	–	–	–
–	–	–	–	–

This result illustrates the basis of scanning-Michelson interferometry for spectroscopic instrumentation.

A pioneer in practical implementation of this concept was M. Block. An early configuration of the Block Engineering interferometer is shown in Figure 20-41. Incoming infrared radiation is received by the interferometer, and a fringe pattern is produced. When one of the mirrors in the interferometer is moved back and forth at a slow, constant velocity, the motion is manifested as an alternate brightening and darkening of the central fringe.

An infrared detector placed at the central fringe converts these cyclic changes into an ac signal. If the mirror velocity is kept constant at a predetermined value, the frequency

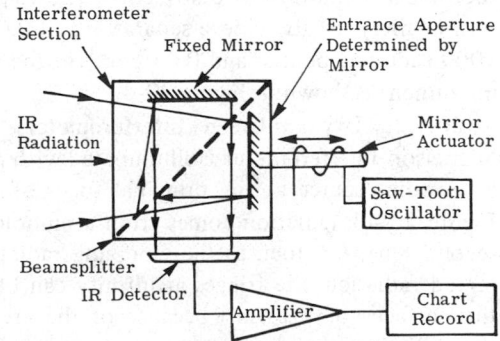

Fig. 20-41. Block Associates interferometer spectrometer.

of the ac signal from the detector is directly related to the wavelength of incident radiation, assuming that the incident radiation is at one given wavelength (monochromatic).

If another wavelength twice as long as the first (half the frequency) should be substituted as the incident radiation source, the ac output signal from the detector would be at one-half the frequency of the first. The amplitudes of the two signals would remain the same if the maximum brightness of the two sources were the same.

If incident radiation containing any wavelengths were introduced into the system, the output of the detector would consist of a superposition of all the ac signals which correspond to all wavelengths in the source.

The output of the interferometer system is tape recorded and played back through an audio wave analyzer to recover the infrared spectrum.

The scan drive must be linear and constant and the effects of source-intensity variations must be negligible. Either of these can cause apparent spectral peaks in the output.

The Block Engineering interferometer-spectrometer has greater throughput than a conventional spectrometer, because the interferometer has a large entrance aperture determined by the mirror size. This enables the instrument to accept more radiant flux from the source than prism or grating instruments in which the entrance aperture is limited by narrow slits.

High-sensitivity gain is due to the instrument's examining each wavelength throughout the entire time period of each scan, i.e., the multiplex advantage. In a conventional instrument, each wavelength is examined for only a very short part of the scan time ($1/n$th the scan time if n is the number of resolution elements); the interferometer achieves a gain which is $\sqrt{n}$ for the same scan time. For typical instruments this can be a factor of 50. Furthermore, this gain is realized even when one examines point sources where the throughput gain is not large.

The general field of Fourier Transform Spectroscopy (FTS) for the infrared spectral region has been reviewed in considerable depth by Bates [20-13]. Discussions of Fellgett's (the multiplex) advantage and the throughput gain are referenced in his treatment of slow-scanning (including step-scanned) and fast-scanning Michelson interferometers. The work by Janine and Pierre Connes was reported in a now classic paper which treats the step-scanned interferometer, an instrument they used so successfully, primarily in planetary spectroscopy [20-14]. In 1976, a far-infrared (50 to 100 μm) Michelson-type Fourier transform spectrometer was constructed and reported to have a theoretical resolution of 0.007 cm^{-1} [20-15]. The instrument was used with an absorption (White) cell to achieve separations of 0.05 cm^{-1} in the 200 cm^{-1} region using about 1000 sampling points, and 0.04 cm^{-1} in the 100 cm^{-1} region. The optical layout of the instrument is shown in Figure 20-42.

20.4.2. Twyman-Green Interferometer. In the Twyman-Green interferometer, a Michelson interferometer is illuminated with parallel monochromatic radiation, produced by a point source at the principal focus of a well-corrected lens [20-16]. As shown in Figure 20-43, radiation comes from a pinhole, P, at the principal focus of a lens, L_1. A second lens, L_2, focuses the emerging radiation onto the detector. By the use of collimated radiation, the fringes at infinity can be seen at finite distance (such as on one of the surfaces of a prism) because of the greatly increased depth-of-focus. In addition, the radiation is made to traverse the optical part under test (a prism in Figure 20-43), and the result of the test is explicit. The use of interferometers for tests of optics is treated in more detail in Chapter 9.

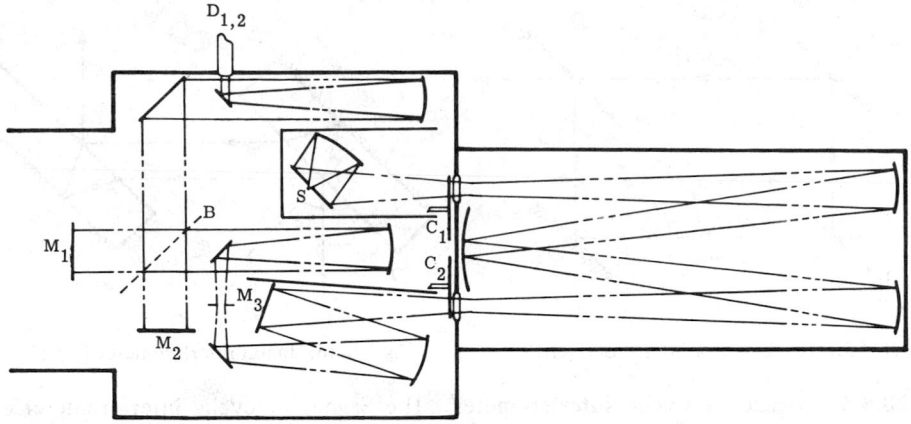

Fig. 20-42. Optical layout of the instrument. M_1 and M_2 are the Michelson mirrors; M_3 may be replaced by a diffraction grating; C_1 and C_2 are choppers; S is the source; $D_{1,2}$ are light pipes to the detectors; and B is the beam splitter [20-15].

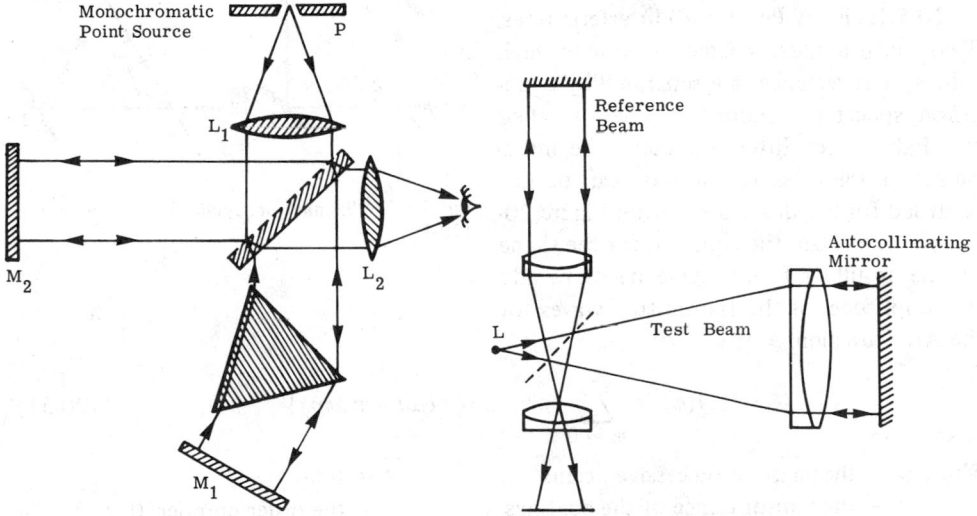

Fig. 20-43. Twyman-Green interferometer [20-16].   Fig. 20-44. Williams interferometer [20-18].

Modifications of the Twyman-Green have been described by Françon [20-17], e.g., the Martin-Watt-Weinstein interferometer, and by Steel [20-18], e.g., the Williams interferometer, shown in Figure 20-44.

20.4.3. Mach-Zehnder Interferometer. Another two-beam amplitude-division interferometer is the Mach-Zehnder interferometer, shown in Figure 20-45. It has two beam-splitters, or semitransparent reflectors, L_1 and L_2, and two mirrors, M_1 and M_2. Derivative from this configuration is the Jamin interferometer (Figure 20-46) which uses two plane-parallel plates, L_1 and L_2, with reflective coatings, A_1 and A_2.

Other interesting variations of the Mach-Zehnder are described by Steel [20-18]. These include the use of diffraction gratings as the beam-splitters and the use of arms of different cross-sections as realized by Johnson and Scholes.

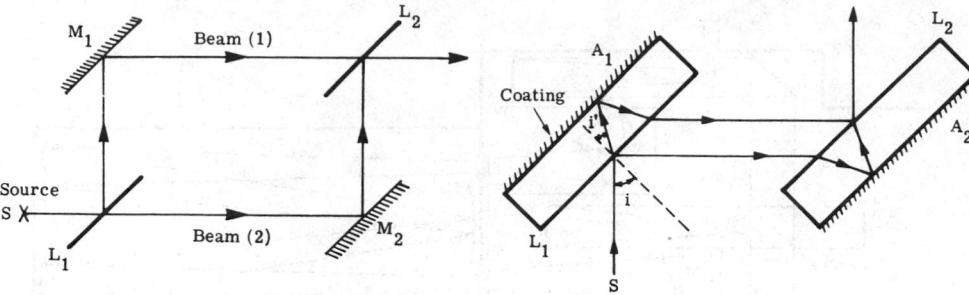

Fig. 20-45. Mach-Zehnder interferometer [20-17].   Fig. 20-46. Jamin interferometer [20-17].

20.4.4. Sagnac or Cyclic Interferometer. The Sagnac or cyclic interferometer can be realized by either an even- or odd-number of reflections in each beam. Two forms are shown in Figure 20-47.

20.5. Interferometers: Multiple-Beam

20.5.1. Fabry-Perot (F-P) Interferometer. Two optical flats, coated to obtain high values of reflectance, are separated by a precision spacer (or etalon) in what is called the Fabry-Perot Interferometer. The materials and thickness of the flats can be disregarded for this discussion. From Figure 20-48, one can obtain the equation for the shape of the resulting fringes. One needs to add the amplitudes of the transmitted waves for the Airy function:

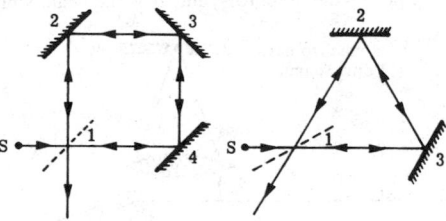

Fig. 20-47. Sagnac or cyclic interferometer [20-18].

$$A(\phi) = \sum_{m=0}^{\infty} \tau \rho^m \exp\left\{ j\left(\omega t - m2\pi p\right)\right\} \qquad (20\text{-}51)$$

where ϕ = the phase of successive beams t = time
   τ = the transmittance of the coatings p = the order number, 0, 1, 2, . . .
   ρ = the reflectivity of the coatings m = the summation running index,
   $\omega = 2\pi\nu = 2\pi c/\lambda$ 0, 1, 2, . . .

The time dependence and τ can be removed from the summation, so that the equation may be rewritten as

$$A(\phi) = \tau e^{j\omega t}\left\{ \frac{1}{1 - \rho\, e^{-j2\pi p}}\right\} \qquad (20\text{-}52)$$

since $(a - bx)^{-1} = (1/a)(1 + bx/a + b^2x^2/a^2 + ...)$, for $b^2x^2 < a^2$. The resulting irradiance is

$$E = A^*A = \tau^2\left\{ \frac{1}{(1 - \rho e^{-j2\pi p})(1 - \rho e^{j2\pi p})}\right\}$$

$$= \tau^2\left\{ \frac{1}{1 + \rho^2 - 2\,\rho\cos(2\pi p)}\right\} \qquad (20\text{-}53)$$

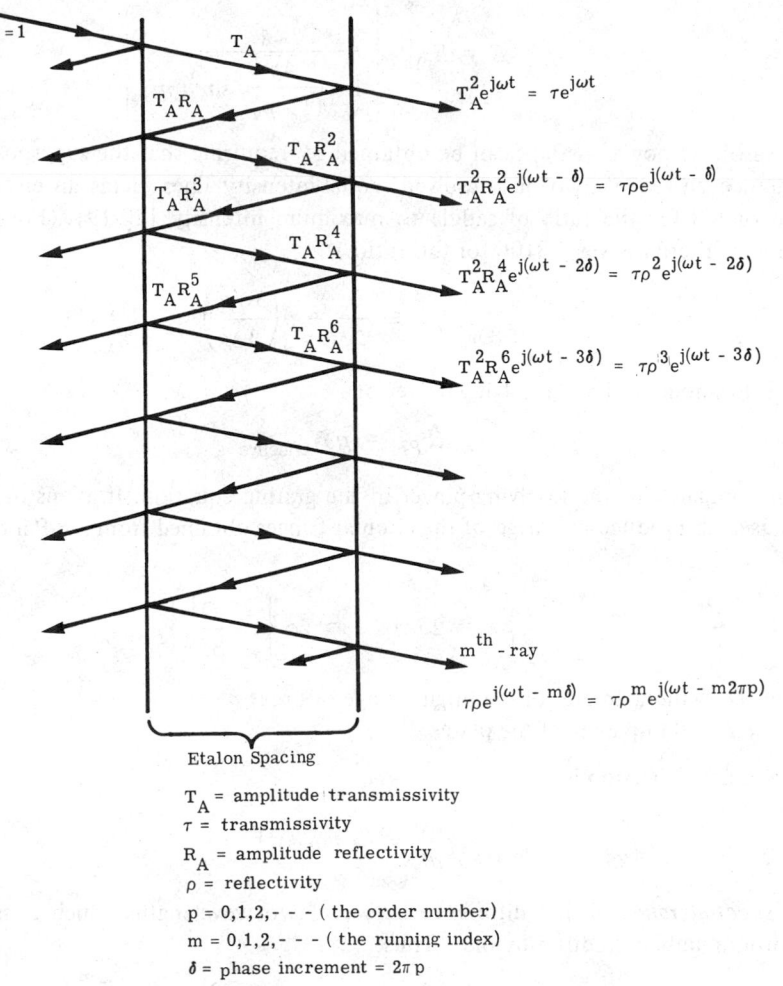

Fig. 20-48. Fabry-Perot interferometer schematic.

$$\cos(2\pi p) = 1 - 2\sin^2(\pi p) \qquad (20\text{-}54)$$

$$E = \frac{\dfrac{\tau^2}{(1-\rho)^2}}{1 + \dfrac{4\rho}{(1-\rho)^2}\sin^2(\pi p)} \qquad (20\text{-}55)$$

$$E_{max} = \frac{\tau^2}{(1-\rho)^2} \quad \text{at } p = \text{integer} \qquad (20\text{-}56)$$

$$E_{min} = \frac{\tau^2}{(1+\rho)^2} \quad \text{at } p = \text{integer} + 1/2 \qquad (20\text{-}57)$$

$$E = E_{\max} \left[\cfrac{1}{1 + \cfrac{4\rho}{(1-\rho)^2} \sin^2(\pi p)} \right] \qquad (20\text{-}58)$$

The resolving power, RP_{PF}, can be obtained by requiring that the superposition of the irradiance curves of two, just-resolved, equal-intensity lines yields an envelope with a value of 0.8 for the ratio of saddle to maximum intensity [20-19]. (For gratings, the Rayleigh criterion gives 0.8106 for this ratio.)

$$RP_{PF} \equiv \frac{\lambda}{\Delta\lambda} \doteq p\left(\frac{3\sqrt{\rho}}{1-\rho}\right) \qquad (20\text{-}59)$$

This interferometer has a resolving power of

$$RP_{PF} = p\,N_{\text{effective}} \qquad (20\text{-}60)$$

when compared to the resolving power in the grating equation. If a lens of focal length F is used to produce an image of the circular fringes obtained from a F-P interferometer, then

$$p\lambda = 2d \cos\phi = 2d\left\{1 - \frac{D^2}{8F^2}\right\} \qquad (20\text{-}61)$$

where D = the diameter of the bright fringe of order p
$\quad\quad d$ = the spacing of the plates

The angular dispersion is

$$\frac{d\phi}{d\lambda} = (\lambda \tan\phi)^{-1} \qquad (20\text{-}62)$$

The *spectral range* is that difference between two wavelengths which is sufficient for the order numbers to differ by one. Thus

$$p_1\lambda_1 = p_2\lambda_2 = (p_1 + 1)\lambda_2 \qquad (20\text{-}63)$$

$$|\lambda_1 - \lambda_2| = \lambda_{2/p_1} \qquad (20\text{-}64)$$

$$\Delta\lambda = \frac{\lambda}{p} = \frac{\lambda^2}{2d} \qquad (20\text{-}65)$$

Often this is shown as

$$\langle\Delta\lambda\rangle = \frac{\lambda^2}{2d} \qquad (20\text{-}66)$$

or, in wavenumber

$$\langle\Delta\tilde{\nu}\rangle = \frac{1}{2d} \qquad (20\text{-}67)$$

Thus, one can predict the overlapping of orders.

At the center of the pattern (Figure 20-49), assume an order number of P which is an integer plus a fraction smaller than one. The first bright fringe or ring has order number p_0;

center		1st ring		2nd ring		3rd ring		----		$(k+1)$ ring		----
P	$>$	p_0	$>$	p_1	$>$	p_2	$>$ ---- $>$			p_k	$>$ ----	

$$p_k = 2d\tilde{\nu}\left\{1 - \frac{D_k^2}{8F^2}\right\}\tilde{\nu}2d - \frac{\tilde{\nu}d}{4F^2}D_k^2 \qquad (20\text{-}68)$$

At the center, since $\phi = 0$ (or $D = 0$)

$$P = p_0 + \epsilon = \tilde{\nu}2d = p_k + k + \epsilon \qquad (20\text{-}69)$$

$$\epsilon = P - p_k - k = \tilde{\nu}2d - p_k - k \qquad (20\text{-}70)$$

Subtracting p_k from p_i, one gets

$$p_i - p_k = \frac{\tilde{\nu}d}{F^2}\left\{D_k^2 - D_i^2\right\} = (p_0 - i) - (p - k) = k - i \qquad (20\text{-}71)$$

Thus, the difference between the squares of the ring's diameters is

$$\Delta D^2 \equiv \frac{D_k^2 - D_i^2}{k - i} = \frac{4F^2}{\tilde{\nu}d} \qquad (20\text{-}72)$$

and

$$\epsilon = \frac{D_k^2}{\Delta D^2} - k \qquad (20\text{-}73)$$

For the first ring

$$\epsilon = \frac{D_0^2}{\Delta D^2} \qquad (20\text{-}74)$$

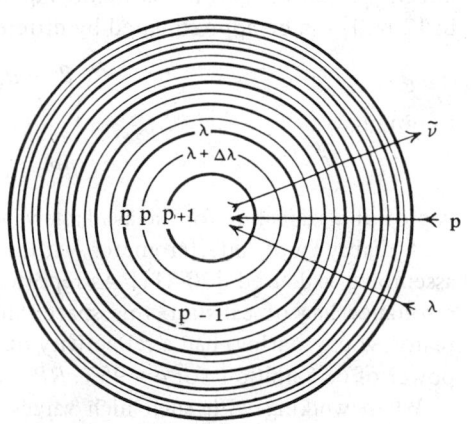

Fig. 20-49. Interference system of two close lines [20-19].

and thus

$$D_k^2 = D_0^2 + k\,\Delta D^2 \qquad (20\text{-}75)$$

D_0^2 and ΔD^2 are treated as constants to be determined such that the sum of the squares of the deviations is minimized. Then these values are used to calculate ϵ:

$$\epsilon = \frac{\overline{D_0^2}}{\overline{\Delta D^2}} \qquad (20\text{-}76)$$

The method of least squares is applied as follows. One measures the n-innermost rings; the results are

$$D_0, D_1, D_2, ----, D_k, ---, D_{n-1} \tag{20-77}$$

$$\overline{\Delta D^2} = \frac{6}{n(n^2-1)} \sum_{i=1}^{i \geqslant \frac{n}{2}} [n+1-2i] \left[D_{n-1}^2 - D_{i-1}^2 \right] \tag{20-78}$$

$$\overline{D_0^2} = \overline{D_k^2} - k \overline{\Delta D^2} \tag{20-79}$$

where k is the average of $0, 1, 2, ---, n-1$. The process of absolute wavelength measurement starts with determination of d from use of a standard wavelength, λ_s or $\tilde{\nu}_s$.

$$(p_s + \epsilon_s) = \tilde{\nu}_s 2d \tag{20-80}$$

To find p_x, one starts with approximate values of the unknown wave number $\tilde{\nu}_x$ and the now known value of d. If the values of $\tilde{\nu}_x$ are very poor, one must start with *small* values of d.

$$(p_x + \epsilon_x) = \tilde{\nu}_x 2d \tag{20-81}$$

This also helps avoid the confusion of overlapping orders, a problem which can be reduced in impact by various nomographic methods. One method, rigorously presented in [20-20] can be approximated by differentiation of

$$P = p_0 + \epsilon = \tilde{\nu} 2d \tag{20-82}$$

to obtain

$$dP = d\epsilon = (2d) \, d\tilde{\nu} \tag{20-83}$$

A plot of d against $d\epsilon$ for the available values of $d\tilde{\nu}$ will give a useful nomograph.

A Fabry-Perot interferometer with a value of $d = 1.0197$ m has been successfully assembled and used [20-21]. As reported by Meissner and Kaufman, for the 422.6 nm resonance line of calcium, this spacer thickness corresponds to $p = 4.82 \times 10^6$. With plates whose coatings had a reflectivity of about 81%, this interferometer yields a resolving power of 68.5 million; for $\rho = 95\%$, $RP = 10^7$.

When working with such high values of resolving power, investigators often turn to the spherical Fabry-Perot interferometer, a more efficient configuration for high-resolution work, especially with weak sources. One design, shown in Figure 20-50, consists of two spherical mirrors whose separation is equal to the radius of curvature r, so that the paraxial foci coincide and the instrument is an afocal system. One half of the surface of the mirrors is semi-reflecting and the other half is fully reflecting. Any incident ray gives rise to an infinite number of outgoing rays which are coincident, and not only parallel, as with the plane Fabry-Perot instrument. Neglecting aberrations, their path difference is $4r$,

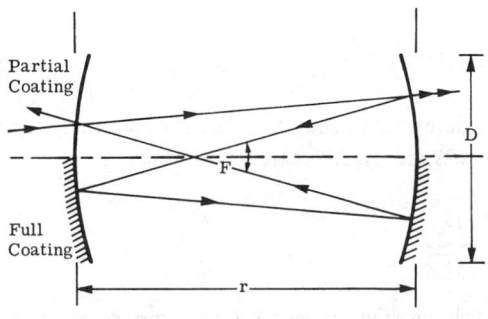

Fig. 20-50. Spherical Fabry-Perot interferometer [20-22].

a constant. This requires that both mirrors be stopped down with circular diaphragms of diameter D. It can be shown that D increases with r, and that the light-gathering power is proportional to r (and thus to the theoretical resolving power RP, instead of being inversely proportional to RP). For practical values of r (a few centimeters), D always remains quite small (a few millimeters).

The use of two Fabry-Perot interferometers in series was introduced to solve the problem of overlapping orders encountered in studies of complex spectra [20-23]. The resulting "compound interferometer" has a resolving power greater than that of either interferometer alone. This concept, whose history was traced and reported by Mack, et al. [20-24], was extended by Mack and his coworkers into the PEPSIOS spectrometer [20-24, 20-25, and 20-26]. The pilot model PEPSIOS spectrometer used three etalons and pressure scanning (e.g., variation of the ambient pressure). A resolving power exceeding 4×10^5 was achieved by this pilot model.

Comparison of the single Fabry-Perot, the PEPSIOS, and the SISAM* instruments was made by Meaburn [20-27] for use in studies of visible nebular lines. Döbele and Massig have compared the use of a single Fabry-Perot spectrometer to the use of a grating spectrometer to measure spectral line shifts much smaller than the line width [20-28]. They concluded that the Fabry-Perot instrument was superior, experimentally showing a lower limit to the measurable shift of 1/1000 of the line width.

A single-etalon Fabry-Perot interferometer with blocking filters has been combined with an SEC vidicon for use in observational astronomy [20-29]. The configuration, dubbed the SPIFI (spectroscopic photoelectric imaging Fabry-Perot interferometer), is shown in Figure 20-51. The interferometer was step-scanned by means of piezo-electric elements used to change the spacing of the plates.

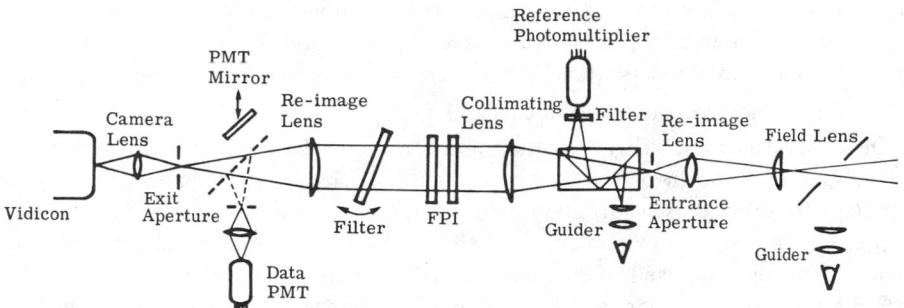

Fig. 20-51. Layout of the optical system of the SPIFI indicating the positions of all components [20-29].

20.5.2. Michelson Echelon. This instrument consists of a stack of optically flat plates "wrung" together so as to produce a stair-step structure of steps s-wide and h-deep (Figure 20-52). In transmission, as realized by Michelson, for near-normal angles of incidence and small diffraction angles

$$p\lambda = (n-1)h + s\phi_d \qquad (20\text{-}84)$$

---

*The SISAM uses two identically-blazed reflection-gratings to replace the two totally-reflecting mirrors of a conventional Michelson interferometer.

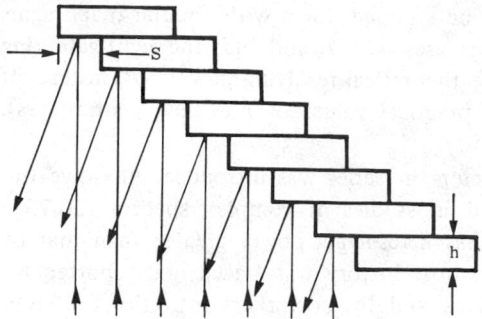

Fig. 20-52. Reflection Michelson Echelon.

Fig. 20-53. Lummer-Gehrcke plate interferometer.

where p = order number
n = index of refraction
ϕ_d = angle of deviation

In reflection, as produced by Williams [20-18]

$$p\lambda = (2h + s\phi_d) \qquad (20\text{-}85)$$

The dispersion is

$$\frac{d\phi_d}{d\lambda} = \frac{p}{s} = \frac{2h}{\lambda s} \qquad (20\text{-}86)$$

The spectral range is

$$\langle \Delta\tilde{\nu} \rangle = \frac{1}{2h} \qquad (20\text{-}87)$$

and

$$RP_{ME} \equiv \frac{\bar{\lambda}}{\Delta\lambda} = pN \simeq \frac{2hN}{\lambda} \qquad (20\text{-}88)$$

where N is the number of steps (often from 25 to 40).

20.5.3. The Lummer-Gehrcke Plate [20-18]. The Lummer-Gehrcke plate (Figure 20-53) utilizes the interference between successive reflections in a thin flat plate. Radiation can be introduced to the plate by a total reflection prism from lens L_1. It then undergoes multiple internal reflections very near the critical angle of total reflection. The beams emerging at a grazing angle are brought to interference by a second lens, L_2. High reflectance and resolving power are thus obtained with unsilvered surfaces.

20.5.4. Commercially Available Interferometers. Table 20-9 (pages 20-60 and 20-61) presents some commercially available interferometers as found by a brief IRIA survey.*

20.6. Other Spectroscopic Instruments and Techniques

Other spectroscopic instrumentation and techniques have emerged in relatively recent times. One of these involves the use of coding masks as first demonstrated by Golay [20-30]. He showed that infrared spectra produced by use of a dispersing system (e.g., grating or prism spectrometer) could be treated by use of modulating patterns or masks placed at the entrance and/or exit apertures. The resulting modulated beam would allow obtaining spectra at a far greater rate or with increased signal-to-noise ratios. Others have followed this direction to achieve multiplexing by use of binary optical encoding masks. As reported by Swift, et al. [20-31], the development of Hadamard transform spectrometers was achieved in the late 1960's. The combination of two-dimensional spatial encoding with one-dimensional spectral encoding has been realized by Swift and his coworkers in their Hadamard Transform Imaging Spectrometer. The design specifications are shown in Table 20-8. Since the subject of such encoding techniques is treated in Chapter 17, they are not discussed further here.

The advent of laser technology has made possible another set of techniques. For example, it has become possible to carry out what is called *tunable laser spectroscopy* [20-32] or, more broadly, "laser spectroscopy." The review by Letokhov [20-33] points

*Conducted by Craig Mueller, Environmental Research Institute of Michigan, (ERIM), Ann Arbor, MI.

Table 20-8. Imaging Spectrometer Design
Specifications [20-31]

Wavelength range:	2-5 μm and 8-14 μm	
Spectral resolving power:	~60	
Spectral channels:	63	
Field of view:	8.88° X 9.45°	
Spatial resolution:	0.005 rad	
Spatial resolution elements:	1023 (31 X 33)	
Detectors:	(a) PbSe (2-5 μm only)	
	(b) Pyroelectric	
	PbSe	**Pyroelectric**
Scanning drive speed:	720 rpm	20 rpm
Spatial frame time:	25 msec	1 sec
Spatial/spectral frame time:	2.6 sec	95 sec

out that laser spectroscopy brings a new level of sensitivity and resolution, spectrally, spatially, and temporally. Spectral analyses on very small volumes, on the order of the wavelength cubed, are also within reach. Spectral resolving powers of up to 10^{11} are attainable.

Pine has reported work with a cW difference-frequency spectrometer which operates in the 2.2 to 4.2 μm region by mixing radiation from an argon-ion laser with that from a tunable dye laser in the nonlinear optical crystal, $LiNbO_3$ [20-34]. His drift-compensation and scan-calibration techniques allowed high resolution spectra of methane to be produced with a precision and reproducibility of 5×10^{-4} cm^{-1}.

Work at NBS has applied saturated absorption in neutral molecules to the wavelength stabilization of gas lasers, leading to systems exceeding the stability and reproducibility of the ^{86}Kr standard by several orders of magnitude [20-35]. Thus, absolute wavelength determinations can be expected to be made with precision exceeding that of the present internationally-adopted standard of length.

Laser resonance fluorescence has been applied to make absolute measurements of sodium-vapor density. The use of a cW dye laser makes it possible to make measurements from 10^2 atoms cm^{-3} at $-28°$C to 10^{11} atoms cm^{-3} at $144°$C [20-36].

Infrared heterodyne spectroscopy can be accomplished by mixing infrared radiation from the source being measured and radiation from a much stronger local oscillator (laser). The signal is detected at the difference frequency. When the local oscillator power is sufficiently large, its shot noise dominates all other detector noise, and the minimum detectable power at this difference frequency can be considered to be on the order of $h\nu \, \Delta\nu$, for an assumed detector quantum efficiency of one. Abbas and his colleagues have analyzed the effect of factors degrading performance from this ideal and conclude that a practical astronomical heterodyne spectrometer will differ only by about a factor of 30 [20-37]. This technique will provide a powerful radiometric tool, especially for astronomical observations such as line profile determinations in stellar sources.

Table 20-9. Some Commercially Available Interferometers

Manufacturer	Model	Spectral Range (μm)	Type	F/#	Operation	Resolution	Scan Speed or Scan Time	Total Accuracy	Wavelength Reproducibility	Remarks
Block Engineering Inc. 19 Blackstone St. Cambridge, MA 02139	Spectrometer 600	1.3-5.0	Michelson	–	–	16-4 cm⁻¹	64-16 scans sec⁻¹	–	–	–
	296C	5-15	Michelson	F/22	Cryogenic	4-0.5 cm⁻¹	1.26 cm sec⁻¹**	–	–	FOV 7.2°
	496C	5-15	Michelson	F/22	Cryogenic	0.1 cm⁻¹	1.26 cm sec⁻¹**	–	–	FOV 7.2°
	I-8	0.86-5.3	Michelson	–	–	20 cm⁻¹	10 scans sec⁻¹	–	–	FOV 7.2°
	197/RS	1.3-5.5	Michelson	–	–	20 cm⁻¹	8-100 scans sec⁻¹	–	–	FOV 7.2°
	197A	1.3-20	Michelson	–	–	0.5-32 cm⁻¹	1-60 spectra sec⁻¹	–	–	FOV 240-20 mrad
	6.0	5.5-15	Michelson	–	Fourier transform	8 cm⁻¹	0.8 sec⁻¹	–	–	Wavelength accuracy ± 0.05 cm⁻¹
	620	2-15	Michelson	–	Fourier transform	0.5 cm⁻¹ 32 cm⁻¹	1 spectra sec⁻¹ 764 spectra sec⁻¹	–	–	FOV 175 mrad
	E-41	2-13	–	–	Wedge filter	1.35% @ 5000 cm⁻¹ and 1666 cm⁻¹ 1.15% @ 2500 cm⁻¹ and 833 cm⁻¹	5 sec⁻¹	–	–	–
Burleigh Instruments, Inc. P.O. Box 388 East Rochester, NY 14445	RC-10	–	Fabry-Perot	–	–	–	NA†	–	NA	–
	RC-40	–	Fabry-Perot	–	–	–	NA	–	NA	–
	RC-4C IR	–	Fabry-Perot	–	–	–	NA	–	NA	–
	RC-50-A	–	Fabry-Perot	–	–	–	NA	–	NA	–
	RC-50-IR	–	Fabry-Perot	–	–	–	NA	–	NA	–
Digilab 237 Putnam Ave. Cambridge, MA 02139	FTS-10	2.5-22.2*	Michelson	–	–	4 cm⁻¹ - 2 cm⁻¹	0.5 sec	–	–	–
	FTS-15	2-25*	Michelson	–	Fourier transform	8-0.25 cm⁻¹	0.3 cm sec⁻¹	–	–	–

The Ealing Co. 2225 Mass. Ave. Cambridge, MA 02140	25K-9218	—	—	F/2	—	—	NA	—	—	NA	—
Gamma Scientific, Inc. 377 Ruffin Rd. San Diego, CA 92123	Scanning 3100	0.4-0.7	—	—	Interference wedge filter	—	305-5 min	—	—	—	Wavelength accuracy ±2.5 nm
ISCO P.O. Box 5347 Lincoln, NB 68505	SR portable	0.38-1.35	—	—	Interference wedge	—	—	—	7-10%	—	FOV 180°
United Detector Technology, Inc. 2644 30th St. Santa Monica, CA 90405	1100A	0.38-0.7	—	—	Variable filter	—	—	—	—	—	—
	1100B	0.38-1.1	—	—	Variable filter	—	—	—	—	—	—

*Optional
**5 cm sec^{-1} available
†NA = not applicable

20.7. References and Bibliography

20.7.1. References

[20-1] "Barnes Engineering Techniques," *Techniques*, Spring Issue, Barnes Engineering Company, Stamford, CT, 1956.

[20-2] J. Guild, "Investigations in Absolute Radiometry," *Proceedings of the Royal Society of London,* London, England, A, Vol. 161, 1937, p. 1.

[20-3] E. J. Gillham, "The Measure of Optical Radiation," *Research Applied in Industry,* Butterworth Scientific, London, England, Vol. 12, 1959, pp. 404-411.

[20-4] E. J. Gillham, "Radiometric Standards and Measurements," *Notes, Applied Science,* National Physics Laboratory, London, England, Vol. 23, 1961.

[20-5] E. J. Gillham, "Recent Investigations in Absolute Radiometry," *Proceedings of the Royal Society of London,* London, England, A, Vol. 269, 1962, pp. 240-276.

[20-6] R. J. Phelan, Jr., and A. R. Cook, "Electrically Calibrated Pyroelectric Optical-Radiation Detector," *Applied Optics,* Optical Society of America, Washington, DC, Vol. 12, 1973, p. 2494.

[20-7] J. Geist and W. R. Blevin, "Chopper Stabilized Null Radiometer Based Upon an Electrically Calibrated Pyroelectric Detector," *Applied Optics,* Optical Society of America, Washington, DC, Vol. 12, 1973, p. 2532.

[20-8] E. S. Steeb, "Minutes of the CORM-NBS Spring Meeting," General Electric Company, Nela Park, Cleveland, OH, June 1976.

[20-9] R. A. Sawyer, *Experimental Spectroscopy,* esp. Chapter 4, Dover Press, New York, NY, 1963, pp. 79, 78, 76, 77, 75, 291, 87, 155, 152, 158, 137, 191a, Third Edition.

[20-10] A. Walsh, "Multiple Monochromators. II. Application of a Double Monochromator to Infrared Spectroscopy," *Journal of the Optical Society of America,* Optical Society of America, Washington, DC, Vol. 42, 1952, p. 95.

[20-11] H. L. Hackforth, *Infrared Radiation,* McGraw Hill, New York, NY, 1960, pp. 209, 211, 214.

[20-12] A. H. Nielsen, "Recent Advances in IR Spectroscopy," Office of Ordnance Research, Durham, NC, Tech. Memo 53-2, December, 1953.

[20-13] J. B. Bates, "Fourier Transform Infrared Spectroscopy," *Science,* American Association for the Advancement of Science, Washington, DC, Vol. 191, January 1976, p. 31.

[20-14] Janine and Pierre Connes, "Near-Infrared Planetary Spectra by Fourier Spectroscopy. I. Instruments and Results," *Journal of the Optical Society of America,* Optical Society of America, Washington, DC, Vol. 56, July 1966, p. 896.

[20-15] J. Kachmarsky, C. Belorgeot, A. Pluchino, and K. D. Möller, "Far-infrared high-resolution Fourier transform spectrometer: Applications to H_2O, NH_3, NO_2 Lines," *Applied Optics,* Optical Society of America, Washington, DC, Vol. 15, March 1976, p. 708.

[20-16] J. Strong, *Concepts of Classical Optics,* Freeman Press, San Francisco, 1958, Appendix B, p. 382.

[20-17] M. Françon, *Optical Interferometry,* Academic Press, New York, NY, 1966, p. 202, 98, 99.

[20-18] W. H. Steel, *Interferometry,* Cambridge University Press, New York, NY, 1967, pp. 88, 166, 176-177, 94, 201.

[20-19] K. W. Meissner, "Interference Spectroscopy, Part I," *Journal of the Optical Society of America,* Optical Society of America, Washington, DC, Vol. 31, June 1941, pp. 405-413.

[20-20] G. V. Deverall, K. W. Meissner, and G. J. Zissis, "Selection of Optical Spacers in Perot-Fabry Interferometer," *Journal of the Optical Society of America,* Optical Society of America, Washington, DC, Vol. 43, August 1953, p. 673.

[20-21] K. W. Meissner, and V. Kaufman, "Calcium Atomic Beam Source and Interference Beyond Two-Meter Retardation," *Journal of the Optical Society of America,* Optical Society of America, Washington, D.C. Vol. 49, October 1959, p. 942.

[20-22] P. Jacquinot, "New Developments in Interference Spectroscopy," *Reports on Progress in Physics,* American Institute of Physics, New York, NY, and The Institute of Physics and the Physical Society, London, England, Vol. 23, 1960, pp. 267-312.

[20-23] K. W. Meissner, "Interference Spectroscopy, Part II," *Journal of the Optical Society of America,* Optical Society of America, Washington, DC, Vol. 32, April 1942, p. 185.

[20-24] J. E. Mack, D. P. McNutt, F. L. Roesler, and R. Chabbal, "The PEPSIOS Purely Interferometric High-Resolution Scanning Spectrometer, I. The Pilot Model," *Applied Optics,* Optical Society of America, Washington, DC, Vol. 2, September 1963, p. 873.

[20-25] D. P. McNutt, "Pepsios Purely Interferometric High-Resolution Scanning Spectrometer, II. Theory of Space Ratios," *Journal of the Optical Society of America,* Optical Society of America, Washington, DC, Vol. 55, March 1965, p. 288.

[20-26] J. O. Stoner, Jr., "PEPSIOS Purely Interferometric High-Resolution Scanning Spectrometer. III. Calculation of Interferometer Characteristics by a Method of Optical Transients," *Journal of the Optical Society of America,* Optical Society of America, Washington, DC, Vol. 56, March 1966, p. 370.

[20-27] J. Meaburn, "Nebular Fabry-Perot, Pepsios, and Sisam Monochromators," *Applied Optics,* Optical Society of America, Washington, DC, Vol. 12, February 1973, p. 279.

[20-28] H. F. Döbele and J. H. Massig, "Application of a Fabry-Perot Spectrometer to the Measurement of Spectral Line Shifts Much Smaller than Line Width," *Applied Optics,* Optical Society of America, Washington, DC, Vol. 15, January 1976, p. 79.

[20-29] W. H. Smith, J. Born, W. D. Cochran, and J. Gelfand, "Spectroscopic Photoelectric Imaging Fabry-Perot Interferometer: Its Development and Preliminary Observational Results," *Applied Optics,* Optical Society of America, Washington, DC, Vol. 15, March 1976, p. 717.

[20-30] M. J. E. Golay, "Multi-Slit Spectrometry," *Journal of the Optical Society of America,* Optical Society of America, Washington, DC, Vol. 39, 1949, p. 437.

[20-31] R. D. Swift, R. B. Wattson, J. Decker, Jr., R. Paganetti and M. Harwit, "Hadamard Transform Imager and Imaging Spectrometer," *Applied Optics,* Optical Society of America, Washington, DC, Vol. 15, June 1976, p. 1595.

[20-32] M. R. Querry, "Tunable Laser Spectroscopy," *Methods of Experimental Physics,* Vol. 13, *Spectroscopy,* Part B, D. Williams (ed.), Academic Press, New York, NY, August 1976.

[20-33] V. S. Letokhov, "Nonlinear High Resolution Laser Spectroscopy," *Science,* American Association for the Advancement of Science, Washington, DC, Vol. 190, October 1975, p. 344.

[20-34] A. S. Pine, "Doppler-Limited Molecular Spectroscopy by Difference-Frequency Mixing," *Journal of the Optical Society of America,* Optical Society of America, Washington, DC, Vol. 64, 1974, p. 1683.

[20-35] H. P. Layer, R. D. Deslattes, and W. G. Schweitzer, Jr., "Laser Wavelength Comparison by High Resolution Interferometry," *Applied Optics*, Optical Society of America, Washington, DC, Vol. 15, March 1976, p. 734.

[20-36] W. M. Fairbanks, T. W. Hänsch, and A. L. Shawlow, "Absolute Measurement of Very Low Sodium-Vapor Densities Using Laser Resonance Fluorescence," *Journal of the Optical Society of America*, Optical Society of America, Washington, DC, Vol. 65, 1975, p. 199.

[20-37] M. M. Abbas, M. J. Mumma, T. Kostiuk, and D. Buhl, *Applied Optics*, Optical Society of America, Washington, DC, Vol. 15, February 1976, p. 427.

[20-38] E. F. Zalewski, "A Modern Approach to Accurate Radiometry," 12th Informal Conference on Photochemistry, National Bureau of Standards, Washington, DC, Paper A1, June 1976.

[20-39] E. F. Zalewski, and M. A. Lind, "Improving the Accuracy of Radiant Power Measurements Based on Photodetector Instrumentation," *Proceeding of the Spring 1976 Symposium*, Bureau of Radiological Health, Rockville, MD, 1976.

20.7.2. Bibliography

Abbas, M. M., J. J. Mumma, T. Kostiuk and D. Buhl, "Senstivity Limits of an Infrared Heterodyne Spectrometer for Astrophysical Applications," *Applied Optics*, Optical Society of America, Washington, DC, Vol. 15, February 1976, p. 427.

"Barnes Engineering Techniques," *Techniques*, Spring Issue, Barnes Engineering Company, Stamford, CT, 1956.

Bates, J. B., "Fourier Transform Infrared Spectroscopy," *Science*, American Association for the Advancement of Science, Washington, DC, Vol. 191, January 1976, p. 31.

Born, M., and E. Wolf, *Principles of Optics*, Pergamon Press, 1966.

Connes, Janine and Pierre, "Near-Infrared Planetary Spectra by Fourier Spectroscopy. I. Instruments and Results," *Journal of the Optical Society of America*, Optical Society of America, Washington, DC, Vol. 56, July 1966, p. 896.

Deverall, G. V., K. W. Meissner, and G. J. Zissis, "Selection of Optimal Spacers in Perot-Fabry Interferometer," *Journal of the Optical Society of America*, Optical Society of America, Washington, DC, Vol. 43, August 1953, p. 673.

Döbele, H. F., and J. H. Massig, "Application of a Fabry-Perot Spectrometer to the Measurement of Spectral Line Shifts Much Smaller than Line Width," *Applied Optics*, Optical Society of America, Washington, DC, Vol. 15, January 1976, p. 79.

Fairbanks, W. M., T. W. Hänsch, and A. L. Shawlow, "Absolute Measurement of Very Low Sodium-Vapor Densities Using Laser Resonance Fluorescence," *Journal of the Optical Society of America*, Optical Society of America, Washington, DC, Vol. 65, 1975, p. 199.

Françon, M., *Optical Interferometry*, Academic Press, New York, NY, 1966, pp. 202, 98, 99.

Geist, J. and W. R. Blevin, "Chopper Stabilized Null Radiometer Based Upon an Electrically Calibrated Pyroelectric Detector," *Applied Optics*, Optical Society of America, Washington, DC, Vol. 12, 1973, p. 2532.

Gillham, E. J., "The Measure of Optical Radiation," *Research Applied in Industry*, Butterworth Scientific, London, England, Vol. 12, 1959, p. 404-411.

Gillham, E. J., "Radiometric Standards and Measurements," *Notes, Applied Science*, National Physics Laboratory, London, England, Vol 23, 1961.

Gillham, E. J., "Recent Investigations in Absolute Radiometry," *Proceedings of the Royal Society of London*, London England, A, Vol. 269, 1962, p. 240-276.

Golay, M. J. E., "Multi-Slit Spectrometry," *Journal of the Optical Society of America*, Optical Society of America, Washington, DC, Vol. 39, 1949, p. 437.

Guild, J., "Investigations in Absolute Radiometry," *Proceedings of the Royal Society of London,* London, England, A, Vol. 161, 1937, p. 1.

Hackforth, H. L., *Infrared Radiation,* McGraw Hill, New York, NY, 1060, pp. 209, 211, 214.

Holter, M., S. Nudelman, G. Suits, W. Wolfe, and G. Zissis, *Fundamentals of Infrared Technology,* Macmillan, New York, NY, 1962 (out of print).

Jacquinot, P., "New Developments in Interference Spectroscopy," *Reports on Progress in Physics,* American Institute of Physics, New York, NY, and The Institute of Physics and the Physical Society, London, England, Vol. 23, 1960, p. 267-312.

Jenkins, F. A., and H. E. White, *Fundamentals of Optics,* McGraw Hill, New York, NY, 1957, Third Edition.

Kachmarsky, J., C. Belorgeot, A. Pluchino, and K. D. Möller, "Far-infrared high-resolution Fourier transform spectrometer: Applications to H_2O, NH_3, NO_2 lines," *Applied Optics,* Optical Society of America, Washington, DC, Vol. 15, March 1976, p. 708.

Layer, H. P., R. D. Deslattes, and W. G. Schweitzer, Jr., "Laser Wavelength Comparison by High Resolution Interferometry," *Applied Optics,* Optical Society of America, Washington, DC, Vol. 15, March 1976, p. 734.

Letokhov, V. S., "Nonlinear High Resolution Laser Spectroscopy," *Science,* American Association for the Advancement of Science, Washington, DC, Vol. 190, October 1975, p. 344.

Mack, J. E., D. P. McNutt, F. L. Roesler, and R. Chabbal, "The PEPSIOS Purely Interferometric High-Resolution Scanning Spectrometer, I. The Pilot Model," *Applied Optics,* Optical Society of America, Washington, DC, Vol. 2, September 1963, p. 873.

McNutt, D. P., "Pepsios Purely Interferometric High-Resolution Scanning Spectrometer, II. Theory of Space Ratios," *Journal of the Optical Society of America,* Optical Society of America, Washington, DC, Vol. 55, March 1965, p. 288.

Meaburn, J., "Nebular Fabry-Perot, Pepsios, and Sisam Monochromators," *Applied Optics,* Optical Society of America, Washington, DC, Vol. 12, February 1973, p. 279.

Meissner, K. W., "Interference Spectroscopy, Part I," *Journal of the Optical Society of America,* Optical Society of America, Washington, DC, Vol. 31, June 1941, p. 405-413.

Meissner, K. W., "Interference Spectroscopy, Part II," *Journal of the Optical Society of America,* Optical Society of America, Washington, DC, Vol. 32, April 1942, p. 185.

Meissner, K. W., and V. Kaufman, "Calcium Atomic Beam Source and Interference Beyond Two-Meter Retardation," *Journal of the Optical Society of America,* Optical Society of America, Washington, DC, Vol. 49, October 1959, p. 942.

Nicodemus, F. N., and G. J. Zissis, "Methods of Radiometric Calibration," University of Michigan, Ann Arbor, MI, WRL Report 4613-20-R, AD 289 375, October 1962.

Nielson, A. H., "Recent Advances in IR Spectroscopy," Office of Ordnance Research, Durham, NC, Technical Memo 53-2, December 1953.

Phelan, R. J., Jr., and A. R. Cook, "Electrically Calibrated Pyroelectric Optical-Radiation Detector," *Applied Optics,* Optical Society of America, Washington, DC, Vol. 12, 1973, p. 2494.

Pine, A. S. "Doppler-Limited Molecular Spectroscopy by Difference-Frequency Mixing," *Journal of the Optical Society of America,* Optical Society of America, Washington, DC, Vol. 64, 1964, p. 1683.

Querry, M. R., "Tunable Laser Spectroscopy," *Methods of Experimental Physics,* Vol. 13, *Spectroscopy,* Part B, D. Williams (ed.), Academic Press, New York, NY, August 1976.

Sawyer, R. A., *Experimental Spectroscopy,* esp. Chapter 4, Dover Press, New York, NY, 1963, Third Edition.

Smith, W. H., J. Born, W. D. Cochran, and J. Gelfand, "Spectroscopic Photoelectric Imaging Fabry-Perot Interferometer: Its Development and Preliminary Observational Results," *Applied Optics*, Optical Society of America, Washington, DC, Vol. 15, March 1976, p. 717.

Steeb, E. S., "Minutes of the CORM-NBS Spring Meeting," General Electric Company, Nela Park, MD, June 1976.

Steel, W. H., *Interferometry*, Cambridge University Press, New York, NY, 1967, pp. 88, 166, 176-177, 94, 201.

Stoner, J. O., Jr., "PEPSIOS Purely Interferometric High-Resolution Scanning Spectrometer. III. Calculation of Interferometer Characteristics by a Method of Optical Transients," *Journal of the Optical Society of America*, Optical Society of America, Washington, DC. Vol. 56, March 1966, p. 370.

Strong, J., *Concepts of Classical Optics*, Freeman Press, San Francisco, 1958, Appendix B, p. 382.

Swift, R. D., R. B. Wattson, J. Decker, Jr., R. Paganetti and M. Harwit, "Hadamard Transform Imager and Imaging Spectrometer," *Applied Optics*, Optical Society of America, Washington, DC, Vol. 15, June 1976, p. 1595.

Walsh, A., "Multiple Monochromators. II. Application of a Double Monochrometer to Infrared Spectroscopy," *Journal of the Optical Society of America*, Optical Society of America, Washington, DC, Vol. 42, 1952, p. 95.

Zalewski, E. F., "A Modern Approach to Accurate Radiometry," 12th Informal Conference on Photochemistry, National Bureau of Standards, Washington, DC, Paper A1, June 1976.

Zalewski, E. F., and M. A. Lind, "Improving the Accuracy of Radiant Power Measurements Based on Photodetector Instrumentation," *Proceeding of the Spring 1976 Symposium*, Bureau of Radiological Health, Rockville, MD, 1976.

Chapter 21

WARNING SYSTEMS

John A. Jamieson
Jamieson Consultants, Inc.

CONTENTS

21. Warning Systems

21.1. Introduction

"Warning systems" comprise many systems for search and surveillance; the sensor subsystem often has characteristics in common with imaging and radiometric sensors and tracking, but not with guidance sensors. The distinguishing characteristic is that the end product is a statistical decision [21-1, 21-2] about the existence or state of a target phenomenon. Criteria of merit are detection probability and false-alarm rate. Processing may often distort the shape of the signal away from a linear representation of observables to facilitate a decision [21-3, 21-4]. Processing always includes at least one nonlinear step.

21.1.1. Symbols, Nomenclature and Units.
The symbols, nomenclature and units used in this chapter are listed in Table 21-1.

21.1.2. Definition.
An *infrared warning system* is an instrument which yields automatic decisions about the state of a phenomenon based on data from observations by an infrared sensor and on predetermined decision algorithms.

Examples of infrared-warning-system applications include the following: (1) detection and characterization of ICBM or SLBM missile launchings [21-5], (2) detection of poisonous gases [21-5], (3) detection of nuclear detonations, (4) warning to military aircraft of ground fire, anti-aircraft missiles, or fighter pursuit, (5) terrain avoidance, (6) hazards from mines, (7) detection of hazardous levels of air pollution or of damage to the atmosphere by alien gases [21-6], (8) intrusion detection and burglar alarms, and (9) warning of a fire in a fuel tank [21-7].

21.2. General Theory

Because of stochastic (random or unpredictable) processes in the sensor or incident radiation, a warning system in good order will sometimes make erroneous decisions:

(1) False alarms (Type-I errors) are alarms called when the phenomenon is not in fact at an alarm state.

(2) Missed alarms (Type-II errors) are failures of the system to recognize alarm states of the target phenomenon. Probability of detection is one minus the probability of a missed alarm.

The signal-to-noise ratio before making an automatic decision may be maximized by making the entrance pupil of the sensor large, the integration time long, and the sensor as well matched to the contrast between the target phenomenon and its absence as economics and design constraints allow [21-8]. Maximizing the signal-to-noise ratio allows total errors to be minimized. A decision rule may then be chosen to approach the minimum level of total errors and select the proportions of Type-I and Type-II errors to minimize their consequences.

Table 21-1. Symbols, Nomenclature and Units

Symbols	Nomenclature	Units
A_d	Area of a detector	cm^2
A_0	Area of entrance aperture	cm^2
A_r	Area of receiver optics	cm^2
A_t	Area of transmitter optics	cm^2
$B = B_1, B_2$	Region of measurement of an autocorrelation function	km^2, nmi^2
C	Cost of a decision	Varies
C_s	Shunt capacitance	F
C_0, C_1	Covariance matrices for two Gaussian populations	—
D_0	Diameter of an entrance pupil	cm
$D(\gamma|V)$	Decision rule (to make decision γ given data V)	—
D^*	Specific detectivity of a detector	$W^{-1}\ cm\ Hz^{1/2}$
d	Diameter of a detector	cm
E_t	Energy of a laser pulse	J
e	Charge of the electron	C
F	A factor describing increased noise in a detector due to internal gain	—
$F(V|S)$	Conditional probability of observing V given n-sample signal S	—
$F_n(V|S)$	Conditional probability of observing n-sample sequence V given n-sample signal S	—
$F/\#$	Optical speed	—
f	Frequency	Hz
f_n	A Doppler frequency shift	Hz
g_m	Transconductance (of a FET)	—
H	A spatial transfer function	—
h	Planck's constant	$W\ sec^2$
$h(t)$	Filter time response	—
I	Intensity of a target	$W\ sr^{-1}$
I_0	Modified Bessel function of the first kind	—
$<i_n{}^2>$	Variance of a noise current	A^2
K	A decision threshold	Varies (usually V)
K_B	Bhattacharyya Mismatch Coefficient	(Bhattacharyya distance)
k	Boltzmann's constant	$W\ sec\ deg^{-1}$
L	Radiance of a source	$W\ cm^{-2}\ sr^{-1}$
L_G	Optical gain of a laser transmitter	—
l, m, n	Numbers	—
M	Internal gain, e.g., of a photomultiplier	—
m_1, m_0	Arrays of means of samples of a Gaussian population	—
m_1, m_2	Spatial frequency	km^{-1}, cycles nmi^{-1}
N	Set of noise samples	—
NEP	Noise equivalent power	W

Table 21-1. Symbols, Nomenclature and Units (*Continued*)

Symbols	Nomenclature	Units
$<n_i^2>$	Variance of noise in ith cell of signal space	—
P	Peak amplitude of a sinusoid	—
P_0	A priori probability	—
R	Slant range	km, nmi
$\mathcal{R}$	Responsivity	V W^{-1}
R_L	Resistance of a load resistor	ohms·
$R(\xi)$	Two-dimensional autocorrelation function	(W cm^{-2} sr^{-1})^2
S	Set of signal samples	—
S_i	Signal in ith cell of signal space	—
$S(t, \epsilon, \rho, \theta)$	A signal as a function of time, phase, relative amplitude, and other parameters	—
T	Temperature	K
T_s	Target duration	sec
t	Time variable	sec
V	Set of observed voltage samples	V
V_T	Set of voltages above a threshold	V
V_{dc}	Voltage at zero frequency	V
V_{nc}	Noise voltage due to clutter	V
V_{nl}	Noise voltage due to load resistor	V
V_{ngr}	Noise voltage due to generation and recombination of carriers	V
V_{npa}	Noise voltage due to the preamplifier	V
$<V_n^2>$	Variance of a noise variable	V^2
V_s	Signal voltage	V
v	A speed	cm sec^{-1}
W_{xx}	Wiener spectrum	(W cm^{-2} sr^{-1})^2 (cycle nmi^{-1})$^{-2}$
w	Probability density for a noise	—
$\mathbf{x} = x_1, x_2$	A vector distance	cm
x, y	Coordinates	Varies
y	Response variable, signal volts plus noise	V
y_i	Sensor response to ith cell of signal space	V or A
α	An angular coordinate; also, probability of false alarm	rad; —
β	An angular coordinate; also, probability of missed alarm	rad; —
γ	A decision variable; also, heterodyne efficiency	Varies; —
Δf	A bandwidth	Hz
ϵ	Phase of a signal	rad
η	Responsive quantum efficiency	—
η_e	A time filtering factor	—
θ	A signal parameter; also, an angle	—; rad

Table 21-1. Symbols, Nomenclature and Units (*Continued*)

Symbols	Nomenclature	Units
Λ_n	Likelihood ratio (from n samples)	—
λ	Wavelength	m
ν	Infrared frequency	Hz
$\xi = \xi_1, \xi_2$	An autocorrelation lag	cm
ρ	rms voltage signal-to-noise ratio	—
σ	Scattering cross-section of a target	cm^2
τ	A time coordinate	sec
τ_a	Transmittance of the atmosphere	—
τ_0	Transmittance of an optical system	—
Φ_{LO}	Local-oscillator power	W
Φ_{min}	Minimum transmitted radiant flux (power)	W
Φ_r	Received radiant flux (power)	W
Φ_λ	Spectral radiant power	$W\,\mu m^{-1}$
ψ	Variance of the envelope of a noise	V^2 or A^2
Ω	Angular field of view	sr
ω	Angular frequency	$rad\,sec^{-1}$

An extensive theory of this design process has been developed for communication systems [21-3] and for radar systems [21-9, 21-10, 21-11, 21-12]. It has been applied less thoroughly to infrared detection systems. The theory is drawn from statistical decision theory [21-1, 21-2]. It is based on models of the form of the system, on assumptions about the statistics of signal and noise, and on the relevant criteria of excellence. Results include prescriptions for the optimum processing structure of the system (which are usually nonlinear and relatively complex) as well as relationships between error rates and the quality of the input data.

This theory is useful as a general guide to design and estimate performance limits for comparison with realized performance, particularly to indicate whether worthwhile payoffs may be available from further design refinements. Very often the models cannot be applied fully to infrared systems as a result of various factors: (1) the statistics of the noise are nonstationary beyond the useful capability of the theory; (2) data on signal and noise distributions are incomplete; (3) phenomena not usually modeled as noise (such as telemetry dropouts in a later part of the system) may dominate residual errors; or (4) the prescription for the optimum system may be excessively complex so that simplifications must be adopted for reasonable economy and availability.

In particular, the theory will prescribe processing of all available information before application of a single irreversible decision threshold. In many infrared systems the amounts of raw data are very great. Considerations of telemetry and memory costs and of system organization frequently make it desirable to divide the processing into two steps. First, a signal processing step which shapes raw data and compares them against first thresholds; and second, a data-processing step which further tests data which pass these first primitive tests.

21.2.1. Sensor Matching. If the four-dimensional object-space (in angle, wavelength, and time) is divided into cells numbered in some convenient order, 1, 2, ... i, ... N, then the signal response to the ith cell expressed in photons is

$$S_i = \int_{\alpha}^{\alpha+\Delta\alpha} \int_{\beta}^{\beta+\Delta\beta} \int_{\lambda}^{\lambda+\Delta\lambda} \int_{t}^{t+\Delta t} S(x, y, \lambda, t)\, dx\, dy\, d\lambda\, d\tau \qquad (21\text{-}1)$$

and the noise associated with observation of the ith cell is additive (but not necessarily Gaussian, stationary, or isotropic), with a variance, $<n_i^2>$ and a signal-to-noise ratio of

$$\rho = \frac{\sum_i S_i}{\left[\sum_i <n_i^2> \right]^{1/2}} \qquad (21\text{-}2)$$

This is a fundamental variable in the analysis of the operating characteristic. Then ρ may be maximized [21-8] and an optimum operating characteristic designed by choosing a sensor response, y_i, given by

$$y_i = \frac{S_i}{\sqrt{<n_i^2>}} \qquad (21\text{-}3)$$

This sensor response yields a signal-to-noise ratio given by

$$\rho = \left[\sum_i \frac{S_i^2}{<n_i^2>} \right]^{1/2} \qquad (21\text{-}4)$$

This algorithm implies the following:

(1) The degrees of freedom of the sensor subsystem in optical throughput, spectral resolution, and time should be made sufficient so that, when fluctuations are negligible, the envelope of the target phenomenon can be reconstructed exactly from the number of independent observations.

(2) The response at each degree of freedom should be weighted by

$$\frac{S_i}{\sqrt{<n_i^2>}} \qquad (21\text{-}5)$$

(3) Each decision by the sensor should be made on the basis of all relevant degrees of freedom derived in (1), added with the weights given in (2).

21.2.2. Formulation of Signal Processing in Terms of Statistical Decision Theory. Detection of a warning-state symptom in the data from the infrared sensor is considered as a statistical decision. This chapter summarizes binary tests of the hypothesis H_1, that any one of a class of warning states is present, against the null hypothesis, that the sensor output is only a random fluctuation (noise). This summary also considers tests based on finite amounts of data. It is an adaptation of analysis developed for radar and communications systems. The theory can be extended to multiple alternative hypotheses and arbitrary amounts of data [21-3].

It is assumed that one detector channel of the sensor yields a response to incident infrared power represented by n successive samples

$$V = (V_1, V_2, ..., V_n) \tag{21-6}$$

which in general contains noise

$$N = (N_1, N_2, ..., N_n) \tag{21-7}$$

and signal derived from the incident radiation

$$S = (S_1, S_2, ..., S_n) \tag{21-8}$$

These samples may be considered values of the voltage at successive instants in time spaced closely enough that the relevant waveforms can be reproduced exactly. In the case of heterodyne systems, this spacing may need to be finer than an optical frequency cycle; for incoherent and passive systems a fraction of a detection time constant will usually be sufficient. (Infrared photon detectors give a response proportional to the square of the electric field of the incident radiation averaged over a response time many orders of magnitude longer than one optical frequency cycle. For analysis of some phenomena such as photon shot noise and heterodyne responses, one will find it useful to investigate this process, but the squaring and averaging are unavoidable parts of processing the signal of passive infrared systems).

Many infrared systems utilize arrays of detectors to sample an image of the object space in parallel (or successively) by scanning or various kinds of modulation. The total set of data used to make a warning decision is usually acquired by several detectors on several successive observations of the image surface. In a missile warning system, for example, the whole data gathering process may take hundreds of seconds and consist of observations by tens of detectors, each making several observations on the phenomena for dwell times on the order of milliseconds separated by seconds. The combination of data from such multiple observation is modeled in the next section by multivariate analysis.

Each of the quantities S, N, and V may be considered a vector with n components that is represented by a point in an abstract signal space. Signal vectors occur with *a priori* probability $P_0 = p(S)dS$ and noises with probability density $w(N)$. Only the combination variable, V, is observable. Its conditional probability density, given the signal S, is $F_n(V|S)$. Designing the signal processing amounts to choosing a decision rule, $D(\gamma|V)$, for using the raw data, V, to make a decision, γ, about the state of the warning phenomenon. For a binary decision where γ can have either the value γ_0, corresponding to acceptance of hypothesis H_o, or γ_1, corresponding to H_1, four "costs," C, of correct and incorrect decision may be identified. (See Table 21-2.) The probabilities of error are as follows:

False Alarm: $$\alpha = \int F_n(V|0)D(\gamma_1 |V) \, dV \tag{21-9}$$

Table 21-2. The "Costs," C, of Correct and Incorrect Binary Decisions

Decision	γ_o	γ_1
True state of warning — Hypothesis accepted	H_o Warning states absent	H_1 A warning state present
Warning states absent	$C_{1-\alpha}$	C_α
A warning state present	C_β	$C_{1-\beta}$

Missed Alarm: $\qquad \beta = \int <F_n(V|S)> D(\gamma_o|V)dV \qquad$ (21-10)

The integrals are over the signal space. The pointed brackets denote an average of F_n over the possible signal states.

The expected cost associated with a decision rule is

$$<C> = P_0\beta(C_\beta - C_{1-\beta}) + (1 - P_0)\alpha(C_\alpha - C_{1-\alpha}) + P_0C_{1-\beta} + (1 - P_0)C_{1-\alpha} \quad (21\text{-}11)$$

Decisions which are optimum in the sense of minimizing expected costs may be specified in terms of a likelihood ratio:

$$\Lambda_n = \frac{P_0 <F_n(V|S)>_s}{(1 - P_0)F_n(V|O)} \qquad (21\text{-}12)$$

where P_0 and $1 - P_0$ are the *a priori* probabilities that a warning state is present or absent. The conditional probability, P_0, of obtaining a particular data set, V, is to be averaged over all the signal states, S, corresponding to warning states.

The decision rule is usually specified as a comparison of the logarithm of Λ_n to a threshold value $\log K$:

$$D(\gamma_o|V) = 1 \qquad \text{when } \log \Lambda_n < \log K \qquad (21\text{-}13)$$

$$D(\gamma_1|V) = 1 \qquad \text{when } \log \Lambda_n \geq \log K \qquad (21\text{-}14)$$

$$K = \left(\frac{C_\alpha - C_{1-\alpha}}{C_\beta - C_{1-\beta}}\right) > 0 \qquad (21\text{-}15)$$

These rules amount to partitioning the space of observables, V, by a surface $\Lambda_n(V) = K$.

Equations (21-9), (21-13), and (21-14) specify the structure of the signal processing. The structure is optimum if the appropriate criteria are used in choosing the threshold K. If K is chosen to minimize expected cost (Equation (21-11)), the rules are known to statisticians as Bayes rules.

A particularly useful class of such optimum decision rules are known as Neyman Pearson rules, those for which the false alarm probability, α, or the probability of detection, $1 - \beta$, is fixed and the threshold K adjusted to minimize β or α. For these rules the design prescription is that the processor should calculate the likelihood ratio and compare it to a threshold which is a function of the signal and the allowable false alarm probability. The thresholds are biased by the logarithm of the ratio of *a priori* probabilities P_0 and $1 - P_0$. Since the form of the likelihood ratio is complicated in general, and since the greatest interest is in optimizing for small signals, the analysis may proceed by expanding log Λ_n in a Taylor series in S and using just the first two terms [21-3]. It is then found that an important distinction arises between coherent and incoherent signals. In general, the signal corresponding to a warning state may be written

$$S(t, \epsilon, \rho, \theta) \tag{21-16}$$

where t = the dependence on time
   ϵ = the phase of the optical frequency signal
   ρ = the ratio of the signal power to rms noise power of the sensor
   θ = a general description of other distinguishing characteristics

If the optical-frequency phase is meaningful and fixed, then the signal is coherent. This may be true for an active infrared warning system whose target is a specular reflector such as a single flat plate or a corner reflector. In this case, the average of terms in S arising from log $\langle F_n(V|S)\rangle$ do not vanish. The data processing will be dominated by a linear cross-correlation of the known signal form with the received data. This case can be implemented by a heterodyne receiver followed by a linear filter, an envelope detector, and a threshold; this comprises an exceptionally high-performance system depending asymptotically on the ratio of signal and noise amplitudes.

Usually the signal corresponding to a warning state consists of a superposition of many signal components of comparable amplitude, and random and variable phases. This case corresponds to an active sensor with a rough target with several scattering centers, or to a passive sensor receiving samples of a thermal noise signal. In these cases, the term in S will vanish when averaged over phase ϵ, so that the structure of the resulting incoherent receiver will be dominated by terms in S^2 and higher powers. The detection performance of a warning system is often shown by graphs of probability of detection $(1 - \beta)$ (found from Equation (21-10)) versus signal-to-noise ratio with probability of false alarm (found from Equation (21-9)) as a parameter.

The fundamental analyses of detection probability, false-alarm rates, threshold value, and signal-to-noise ratio have been made by Marcum, in 1947 [21-10], and Swerling, in 1954 [21-11], for radar systems assuming Gaussian noise added to a sinusoidal signal at radio frequency, the sum being passed through a filter with bandwidth much less than its center frequency. The resulting envelope of noise has a distribution

$$dP = w(N)dN = \left(\frac{N}{\psi}\right) \exp\left(\frac{-N^2}{2\psi}\right) dN \tag{21-17}$$

where ψ is the variance of the distribution, and the envelope of signal plus noise is

$$F(V|S) = \frac{N}{\psi} \exp\left[\left(\frac{-(N^2 + P^2)}{2\psi}\right)\right] I_0\left(\frac{NP}{\psi}\right) \tag{21-18}$$

where I_0 is the modified Bessel function of the first kind. It is assumed the envelope values are extracted by a square-law rectifier and low-pass filter before they are tested with a threshold. Marcum assumed that the power scattered back to the radar receiver is constant. Swerling [21-11] extended the analysis to targets whose scattering cross-section is itself a random variable, because real targets usually consist of a number of scattering centers of comparable magnitudes which give independent returns which, in turn, are superposed at the receiver. The number, strength, and radio-frequency phase of these components vary as the geometry of target and radar change from scan to scan and sometimes from pulse to pulse. Swerling [21-12] has further extended these results by analyzing a more comprehensive set of distribution for scattering cross-sections.

These results may be used for active, infrared, warning systems [21-13], and they may be used with care to estimate threshold settings and performance for passive warning systems. The cases which are usually more useful are for single dwell times of observation [21-4]. These results are plotted in Figures 21-1 through 21-3. (The axes have been re-labelled from the originals in terms of rms signal-to-noise ratios at the output of the detector, rather than range ratios, because of the strong-effect of atmospheric absorption and scattering on many infrared systems.) Table 21-3 gives values of probability of detection for a useful range of rms signal-to-noise ratios corresponding to Figure 21-1. Figure 21-2 shows values of threshold voltage which should be set to achieve specified false alarm rates. The right-hand scale shows threshold voltage after the diode in units of rms noise which is called bias number. The left-hand scale shows signal-to-noise ratio in dB at the input to the diode. The bias number is independent of the diode characteristic.

Table 21-4 lists several cases of signal and noise distributions relevant to infrared warning systems. In all cases, the consequences of photon quantization is more likely to significantly affect the statistical distribution at infrared wavelengths than at radio wavelengths. In particular, the minimum noise power for an active system (laser radar) is limited to

$$P_n \geqslant \frac{h\nu\Delta f}{\eta} \qquad (21\text{-}19)$$

where η = quantum efficiency
Δf = effective optical-frequency band-pass

The correlation time for infrared radiation is of the order of $\Delta t \sim 1/4\pi\Delta f$. In some particular cases this limit may be set by molecular absorption or emission processes at the source or in the intervening atmosphere. For long wavelengths or hot sources where

$$\frac{h\nu}{kT} \lesssim 5 \qquad (21\text{-}20)$$

photon correlation or "bunching" may be observed as a consequence of the basic, underlying, Bose-Einstein statistics.

At wavelengths of the order of 1 μm or less, and for low backgrounds, the number of

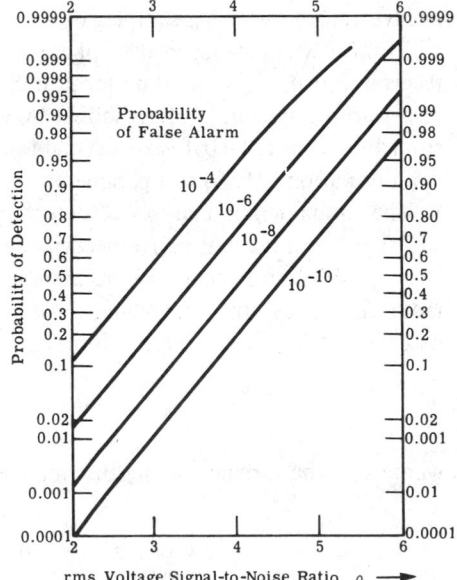

Fig. 21-1. Probability of detection vs rms voltage signal-to-noise ratio ρ (envelope detection of constant sinusoid in Gaussian noise) [21-4, 21-10, 21-11].

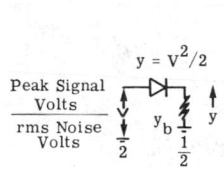

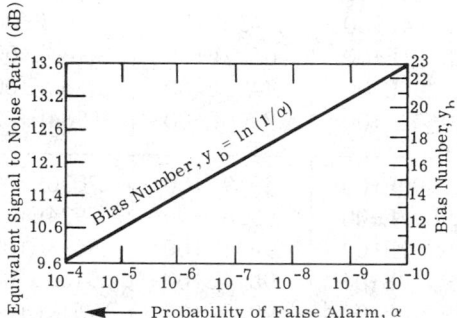

Fig. 21-2. Threshold numbers for expected false alarm rates from the envelope of narrow band noise [21-9, 21-10, 21-11].

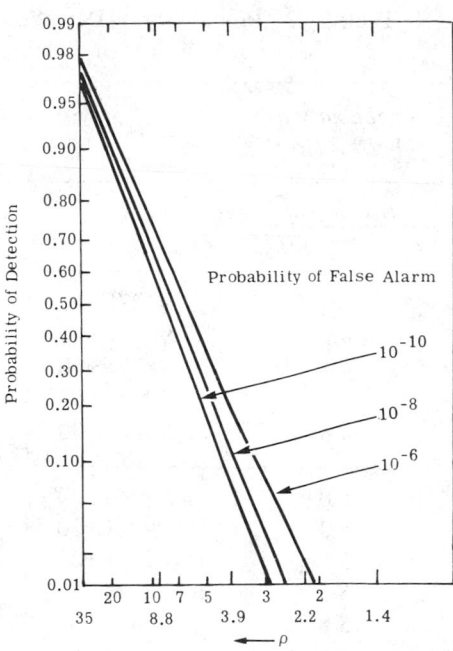

Fig. 21-3. Probabilities of detection versus ρ for incoherent phenomena (envelope detection of a target behaving like a number of independent targets of comparable strength) [21-4, 21-10, 21-11].

Table 21-3. Probability of Detection vs rms Voltage Signal-to-Noise Ratio for Four Values of Probability of False Alarm*

Probability of False Alarm	10^{-4}	10^{-6}	10^{-8}	10^{-10}
rms Signal / rms Noise Voltage	Detection Probability			
2.0	.1127E+00	.1488E-01	.1423E-02	.8378E-04
2.2	.1749E+00	.2911E-01	.3437E-02	.2992E-03
2.4	.2554E+00	.5312E-01	.7705E-02	.8469E-03
2.6	.3523E+00	.9076E-01	.1609E-01	.2154E-02
2.8	.4600E+00	.1454E+00	.3132E-01	.5050E-02
3.0	.5709E+00	.2188E+00	.5695E-01	.1099E-01
3.2	.6766E+00	.3100E+00	.9681E-01	.2227E-01
3.4	.7697E+00	.4148E+00	.1542E+00	.4208E-01
3.6	.8456E+00	.5262E+00	.2305E+00	.7421E-01
3.8	.9029E+00	.6356E+00	.3243E+00	.1224E+00

*This table is calculated from Equations (21-17) and (21-18). It assumes a sinusoidal signal in Gaussian noise, tested by a square law rectifier and threshold, following a narrow band electronic filter.

Table 21-3. Probability of Detection vs rms Voltage Signal-to-Noise Ratio for
Four Values of Probability of False Alarm* *(Continued)*

Probability of False Alarm	10^{-4}	10^{-6}	10^{-8}	10^{-10}
$\dfrac{rms\ Signal}{rms\ Noise\ Voltage}$	Detection Probability			
4.0	.9427E+00	.7350E+00	.4310E+00	.1892E+00
4.2	.9684E+00	.8184E+00	.5430E+00	.2747E+00
4.4	.9837E+00	.8831E+00	.6517E+00	.3760E+00
4.6	.9921E+00	.9295E+00	.7493E+00	.4866E+00
4.8	.9964E+00	.9602E+00	.8300E+00	.5984E+00
5.0	.9984E+00	.9790E+00	.8919E+00	.7026E+00
5.2	.9992E+00	.9897E+00	.9356E+00	.7924E+00
5.4	.9995E+00	.9952E+00	.9641E+00	.8638E+00
5.6	–	.9979E+00	.9814E+00	.9163E+00
5.8	–	.9991E+00	.9910E+00	.9519E+00
6.0	–	.9996E+00	.9959E+00	.9742E+00

*This table is calculated from Equations (21-17) and (21-18). It assumes a sinusoidal signal in Gaussian noise, tested by a square law rectifier and threshold, following a narrow band electronic filter.

Table 21-4. Signal and Noise Statistics for Infrared Warning Systems

Application	Basic Noise Statistics	Signal Statistics	Suggested Performance Estimate and References
Active systems; Specular target	Gaussian	$P \sin (2\pi v t + \epsilon)$ (P and ϵ unknown but fixed)	Figure 21-8 [21-10, 21-13]
Active system; Rough target	Gaussian	$\sum_i P_i \sin (2\pi v t + \epsilon_i)$ (P_i and ϵ_i random variables)	Figure 21-9 [21-22, 21-12, 21-13]
Active system; Photon counting	Poisson/Gaussian	Poisson	[21-14]
Passive system limited by photon noise or preamplifier noise or Nyquist noise	Gaussian	C for $t_1 < t < t_1 + \tau_d$ (C constant for one dwell-time τ_d; t, unknown)	Figure 21-1 [21-4, 21-10, 21-11]
Passive system limited by background clutter noise	Poisson, Rayleigh or mixed	same	–

photons per observation may be small enough that the noise distribution is Poisson. If, in addition, the signal is from several scatterers of comparable magnitude, then the signal-plus-noise statistics may also be Poisson [21-14]. At wavelengths of the order of 3 to 5 μm, the number of background photons is often of the order of 10^6 per observation, and the number of target photons of the order of 10^4. Even if photon noise dominates, the distribution may be adequately approximated as Gaussian. (These orders of magnitude occur in passive systems because the more intense target is not resolved.) At longer wavelengths such as 8 to 14 μm, the photon counts may be less either because of short integration times (in systems with high background radiance), or because the external background photon radiance is low (e.g., in space) and the optics are cryogenically cooled. However, background photons usually exceed 10^4 per observation.

When active systems which employ heterodyne processing are used against targets dominated by a single, steady, specular reflector, the warning signal will be a sinusoid—so the conditions of Marcum hold. For targets such as a tumbling flat plate, the signals will scintillate so that conditions similar to Swerling's 1970 condition may hold. If the target has a number of scattering centers, then the earlier Swerling conditions may be a good model.

For passive systems, the target is usually a thermal radiator. In the infrared, the number of photons is usually large so that target-photon statistics may be considered Gaussian. In fact, however, the variance of signal-plus-noise is usually dominated by background photons, or by amplifier or resistor dissipation noise, so that the hypothesis test is between one Gaussian distribution with one mean (warning phenomenon absent) and a second distribution with a slightly larger mean. The sensor may scan so that it dwells for a short time on each resolution element; it may stare for a longer time; a field may be modulated spatially by a reticle or spectrally by an interferometer; etc. In any case, the optical-frequency radiation is passed through a square-law detector, which destroys optical phase information, and then through a filter approximately matched to the expected audio-frequency modulation [21-4]. The hypothesis test is then applied to a single pulse of the envelope at the output of the filter. A thorough analysis of this problem has apparently not been published, but Marcum's analysis for single pulses may give useful guidance.

Passive scanning systems are often limited by noise arising from spatial inhomogeneities of background radiance. This noise may exhibit very strong local correlations (e.g., if a sensor scans over an array of clouds structured by a trade wind). There is some evidence [21-16] that spatial clutter may tend to a Poisson distribution in the near infrared (where local, strong solar-scattering centers predominate) and to a Gaussian (or Rayleigh) distribution at longer wavelengths. For very fine spatial resolution in the middle infrared an apparently uniform distribution (over a range of radiance) has been observed.

21.2.3. Multivariate Data Processing. The sensor of an infrared warning system may frequently make several observations of a warning phenomenon. The probability of detection and the false alarm probability based on decisions which include all of the observations combined, may be quite acceptable, even though the performance after signal processing from each "primitive" observation is quite poor. Examples include: detection of rockets or aircraft against cluttered backgrounds where the primitive observations are very uncertain because of the variability of clutter statistics (Section 21.3.2); automatic detection of military vehicles in real time using cluttered data from infrared imaging systems; detection of a temperature change in a space vehicle by multispectral observations (see the example in Section 21.3.2); and detection of obstruction by wires in a terrain avoidance system (Section 21.3.5).

If the probability of a missed threshold exceedance at the ith opportunity is β_i and the probability of a false alarm is α_i, and if there are N statistically-independent opportunities to observe the target phenomenon, then a decision rule which is often used is

$$D(\gamma_1 | V_T) = 1 \qquad V_T \geqslant m \qquad (21\text{-}21)$$

$$D(\gamma_0 | V_T) = 1 \qquad V_T < m \qquad (21\text{-}22)$$

where V_T is the number of threshold exceedances observed from the signal processing.
Probability of detection is

$$1 - \beta = \sum_{l=m}^{n} \left[1 - {_nC_l} \prod_{i=1}^{n} \beta_i \right] \qquad (21\text{-}23)$$

where ${_nC_l} = n! / [(n-l)!l!]$ is the number of ways of missing exactly $(n-l)$ exceedances in n tries.
Probability of false alarm is

$$\alpha = \sum_{l=m}^{n} {_nC_l} \prod_{i=1}^{n} \alpha_i \qquad (21\text{-}24)$$

The probabilities α_i and β_i usually vary systematically with index i for many reasons. For example: the intensity of the signal from an aircraft will vary with its range and aspect; the signal from a missile will vary from boost to sustainer phases; and the aspect of a scattering target will vary.

In general, the chances of successive, primitive threshold-crossings will not be independent. For example, a target may be observed in two spectral regions so that one may estimate a change in its temperature. In addition, a persistent target may be observed to make a track in object space. The primitive decision-threshold then may be lowered locally for successive looks along that track. Examples of these mechanisms are described in Sections 21.3.2 and 21.3.5.

If the statistics for exceeding a primitive threshold are not independent, then an accurate calculation will take account of the correlations according to the circumstances of the case. For multidimensional, Gaussian statistics with nondiagonal covariance matrices, the Bhattacharyya distance [21-17] can be used to plot the probability of detection. This distance is defined as

$$K_B = \left\{ (m_1 - m_0)^T \left[\frac{C_1 + C_0}{2} \right]^{-1} (m_1 - m_0) + 4 \, ln \left[\frac{det \left(\dfrac{C_1 + C_0}{2} \right)}{\sqrt{det(C_1)det(C_0)}} \right] \right\}^{1/2} \qquad (21\text{-}25)$$

where m_1, m_0 = the array of means corresponding to the presence and absence of the warning phenomena, respectively

C_1, C_0 = the corresponding covariance matrices

T = transpose of the matrices

$- 1$ = inverse of the matrices

det = the determinant of the matrices

Detection-probability curves for this case, given by Evans [21-18], are shown in Figure 21-4.

In some cases, pattern-recognition techniques have been used. For example, Geokezas [21-19] has developed a real-time automatic classifier which warns an operator about data (in the output of FLIR systems) which correspond to military vehicles in the images. These techniques utilize strong spatial correlations in the raw data. Primitive observation may include positive and negative

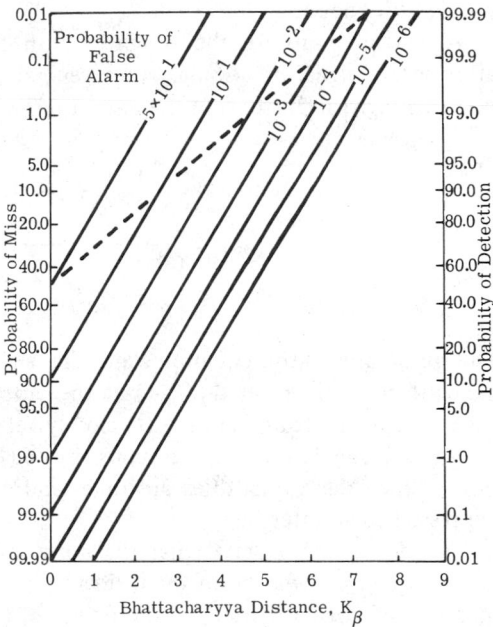

Fig. 21-4. Probabilities of detection for warning phenomena with correlated Gaussian multivariate distributions [21-18].

crossings of a threshold, as well as intervals (i.e., the number of resolution elements separating a positive from the succeeding negative crossing). Combinations of interval lengths in nearby subfields are used to estimate features of a particular image, such as its length-to-width ratio, the straightness of its sides, and its area. A decision vector is then constructed whose components are numerical measures of the values of these features. The decision to cue the operator is then made by partitioning feature-space.

21.3. Types of Warning Systems

21.3.1. Classification of Systems. In this chapter warning systems are classified by the type of sensor used. The first dichotomy is between active sensors, which both transmit and receive radiation, and passive sensors, which only receive radiation. The second is between coherent sensors, which utilize spatial- or temporal-coherence properties of the amplitude of incident radiation for detection, [21-20, 21-21], and incoherent sensors, which detect incident radiation as the square of the absolute magnitude of the amplitude. These considerations lead to greatly different analyses.

21.3.2. Warning Systems with Passive Sensors Detecting Radiation Incoherently. Most warning systems are of this class. Two important subclasses are sensors for targets against warm (usually room temperature) backgrounds, and space-surveillance sensors with low background-photon rates. Equipment of the first subclass usually has to contend with background clutter as an important noise and tends to be limited by one of various kinds of noise due to the background; equipment of the second subclass tends both to have cryogenically-cooled optics so that all background rates are low and to be limited by preamplifier noise.

A starting point for the design of both kinds is an equation for signal-to-noise ratio after sensor signal-processing. In a general formalism, the ratio, ρ, of signal-voltage to noise-voltage (or current) at the output of linear processing of voltages from one detector is

$$\rho = \frac{\int\int \Phi_\lambda(\lambda, t)\tau_0(\lambda)\,\eta_e(t)\,\Re(\lambda, t)\,d\lambda\,dt}{\left[V_{ngr}^2 + V_{nl}^2 + V_{npa}^2 + V_{nc}^2\right]^{1/2}} \tag{21-26}$$

The numerator of this equation states that signal voltage is proportional to the product of incident spectral power $\Phi_\lambda(\lambda, t)$ at the entrance aperture, transmittance of the optics $\tau_0(\lambda)$, a time filtering factor $\eta_e(t)$ closely related to scanning and filter efficiencies (e.g., Chapters 7 and 10), and the responsivity, $\Re(\lambda, t)$, of the detector-preamplifier combination. Since these quantities are time- and wavelength-dependent, the product must be expressed as an integral.

The denominator states that the noise voltage may have components resulting from (1) the quantized nature of the incident photon stream and of the generation and recombination of charge carriers in the detector, (2) thermodynamic fluctuations in the load resistor, (3) various mechanisms in the preamplifier, and (4) various clutter mechanisms in the background. This formalism assumes that the detector is a photon detector rather than a thermal detector. (See Chapter 11.) Military warning systems usually use photon detectors because of their greater detectivity and speed of response; for a warning system using thermal detectors, the expression could be generalized following analyses given in Chapter 11.

Equation (21-26) assumes that the sensor is linear in incident radiant power and in resulting signal voltage to the point where ρ is calculated, that all of the power represented by $\Phi_\lambda(\lambda, t)\,\tau_0(\lambda)$ is impressed on the detector in question, and that the term of $\Re(\lambda, t)$ allows for any effects of not impressing the power uniformly over the detector's receiving surface. Linearity in response to radiance is critical if clutter noise is important. It is more likely to be satisfied if the detector is ac-coupled so that the mean of any background is not an issue, if the dynamic range of incident radiation is not too great, and if ρ is calculated prior to any analog-to-digital conversion or other nonlinear processing. The effect of nonlinearities is to create cross-products between signal and noise, and between noise terms. Since these cross-products always increase the noise variance, the signal-to-noise ratio and consequent performance potential are maximized if all spectral, spatial, and temporal linear-filtering are performed before any nonlinear processing. If performance is the most important consideration, the linear part of the system should be matched to target contrast, as indicated in Section 21.2.1.

To write Equation (21-26) in a useful form, one must expand the term $\Phi_\lambda(\lambda, t)$ and all terms of the denominator, which is best done for each of the system subclasses.

Space Warning Systems. A system which senses targets against the low background level of space has a relatively low incident-flux of background photons. If the optics are also cooled to a sufficiently low temperature ($\sim$20 to 70 K depending on the wavelength band) and the detector cavity and interference filter are further cooled ($\sim$2 to 15 K), then the internal flux of photons is also relatively low. Under these conditions, only the terms V_{nl} and V_{npa} are important in the denominator of Equation (21-26). Furthermore, if the sensor and target phenomena are both outside almost all of the atmosphere, then the term Φ_λ usually does not require an atmospheric-transmittance factor.

As an example, consider a hypothetical application where a warning system in one satellite is to sense remotely the radiant signal from another body in space, such as an asteroid, a satellite which has failed, or a piece of space debris. In such a case, the signal-to-noise equation, Equation (21-26), might become

$$\rho = \frac{\iiint \left[\dfrac{A_0 L\,(\lambda,t,\mathbf{x})}{R^2} \right] \tau_0(\lambda)\eta_e(t)\,R\,(\lambda,\,t)\,d\lambda\,dt\,dx}{\left[\left(\dfrac{K}{\omega} + \dfrac{8kT}{3g_m} + \dfrac{4kT_S}{g_m^2\,R_S} \right) + \dfrac{4kT_L\,R_L}{1 + \omega^2 R^{2L} C_{sh}^2} \right]^{1/2} [\Delta f]^{1/2}} \tag{21-27}$$

where
$\qquad A_0$ = area of the entrance aperture
$\qquad L(\lambda, t, \mathbf{x})$ = radiance of the object to be examined
$\qquad R$ = slant range
$\qquad K$ = constant

The terms in the denominator represent (1) "inverse frequency noise," (2) channel thermodynamic noise, (3) source resistance thermodynamic noise—all in the first-stage, field-effect preamplifier which has transconductance g_m, and (4) thermodynamic noise in a load resistor shunted by capacitance C_{sh}. It is assumed that the dynamic impedance of the detector is very much greater than that of the load resistance. This expression may be manipulated into various forms depending on design assumptions. It will be found for this case that the D^* of the detector is an artificial concept. The detector detectivity tends to depend upon the area of the detector rather than the square root of the area, but is strongly influenced by electronic effects, especially preamplifier and stray capacitance. The signal-to-noise ratio tends to depend on the area of the entrance aperture. The term $\tau_0(\lambda)$ is often dominated spectrally by the interference filters and in magnitude by aperture blocking, since optics are usually reflective and folded.

If the phenomenon to be warned about involves the temperature of part of the target body (e.g., as a symptom that a solar-array drive is failed), then the spectral band may be divided into two parts defined by two, separately-filtered detector arrays. For example, a linear array of detectors filtered in one band, λ_{11} to λ_{12}, may be placed in the focal-surface normal to a direction of scan over the object. A second array of detectors, filtered at λ_{21} to λ_{22}, follows the first (Figure 21-5(a)).

The two spectral bands (Figure 21-5(b)) are chosen to have different ratios according to whether the temperature is high or low. The resulting partition function is shown in Figure 21-5(c). In that diagram the two axes represent responses of the two-detector arrays (with suppressed zeros and different scales). Shaded area A represents locations of signal and noise in the absence of the warning symptom, shaded area B in its presence. Lines X, Y, and Z represent three alternative decision algorithms which might be chosen relative to whether the importance of Type-I errors is more or less than that of Type-II errors.

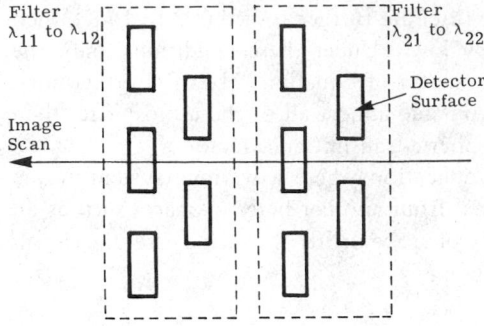

(a) Arrangement of Hypothetical Focal Space

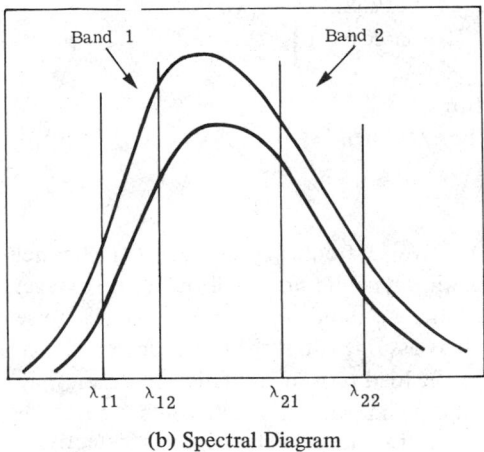

(b) Spectral Diagram

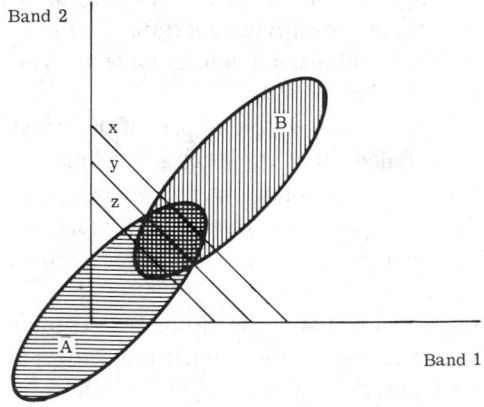

(c) Partition Diagram

Fig. 21-5. System to warn of a temperature change in a space object.

In practice, numerous, more subtle considerations enter the design: discrimination against celestial bodies, acquisition of the correct object, fluctuations in the signal from the target, the desirability of a sequence of observations to avoid possible errors due to noise artifacts, precise alignment of two arrays and the scan vector, provision of adequate overlap between detector loci to assure full image detector, etc. But to a first approximation, the noise sources might be considered additive, Gaussian, and stationary; and the signals might be considered to be changes in the two mean levels of minimum separation of the probability density functions. In such a case, the two errors can be integrated straightforwardly from known bivariate, Gaussian distributions.

Background Limited Systems. A system in the atmosphere is usually limited by generation-recombination noise. In Equation (21-26), therefore, the term V_{ngr} dominates all other noise terms. Examples are: airborne systems to detect mines; gunflash detectors; and intrusion detectors. In such cases, the signal-to-noise ratio of an observation can be written in terms of specific detectivity D^* as

$$\rho = \frac{\left[\iint \Phi_\lambda(\lambda, t)\tau_0(\lambda)\, \eta_e(t)\, D^*(\lambda)\, d\lambda\, dt \right]}{(A_d \Delta f)^{1/2}}$$

(21-28)

where A_d is the area of the detector [21-4]. In this case, the signal-to-noise ratio is proportional to the diameter rather than to the area of the entrance aperture. Since the background photon rate is large, the benefit from cooled optics is small, and refractive optics can be used to achieve a wide field-of-view.

For example, consider a hypothetical gunflash detector for which the target is distant, unresolved, and brief, and for which the time of occurrence is unknown. Suppose the sensor has been designed with an array of m detectors covering a total field of view Ω, each detector having a specific detectivity essentially constant over the (modulation) frequencies in the signal response. Then the signal voltage at the detectors will be

$$V_s(t) = \frac{A_0}{R^2} \int I_\lambda(\lambda, t)\tau_a(\lambda)\tau_0(\lambda) \, \Re(\lambda)d\lambda \qquad (21\text{-}29)$$

To optimize performance, one must filter this signal and the additive, Gaussian, white noise for maximum peak signal-to-rms-noise by making the electronic response [21-4]

$$h(t) = V_s(-t) \qquad (21\text{-}30)$$

or a practical approximation to this form. The resulting peak signal-to-noise ratio will be approximately proportional to

$$\hat{\rho} \propto \frac{\pi}{4} \frac{D_0}{F/\#} \frac{1}{R^2} \left[\frac{mt_s}{\Omega} \right]^{1/2} \int \tau_a \tau_0 D^* I_\lambda d\lambda \qquad (21\text{-}31)$$

where $F/\#$ = optical speed

$\quad t_s$ = an equivalent target duration

The filtered voltages are applied to voltage thresholds which, when exceeded, indicate a warning. If there are m detectors, each with a threshold, sharing the field with a processing-circuit, then an indication of the alarm direction is obtained. To estimate the performance required of the sensor, suppose that the conditions are close to Marcum's (i.e., additive Gaussian noise, sinusoidal signal, envelope detection). If the sensor had a response time of 1 msec and if a false alarm rate on the order of 1 per 14-hours per detector were allowable (i.e., 1 per 5 minutes for 165 detectors), then in Figure 21-1 the curve with false alarm rate 10^{-8} could be used. At a probability of detection of 0.99, a rms signal-to-noise ratio of ~ 5.7 would be required; at a probability of detection of 0.8, a peak signal-to-noise ratio of about 6.7 would suffice. The difference between these two performances is nearly a factor of 2 in entrance-pupil area.

For one to design the electronic filter precisely, the signal, as a function of time, must be constant and known. In practice, however, the signal may be variable since it can arise from various weapons at various stages. Since the spectral signature may vary with weather conditions, particularly humidity, $\tau_0(\lambda)$ cannot be chosen always to maximize $V_s(t)$. Such problems are typically handled by designing for a difficult, but not worst, possible case and assuming that matching is not critical under better conditions.

Clutter Limited Systems. A background of clouds or terrain, which scatters solar radiation, exhibits a pattern of irradiance varying from point to point. A structured background emitting radiation in the long-wavelength infrared is similarly variable. A sensor scanning over such a structured background generates a response which is irrelevant to the target phenomenon and which normally interferes with target sensing. This response must be treated stochastically in the sense that it cannot be predicted even though it may contain a high degree of internal order. For many systems, such as surveillance systems for missile detection and warning, artillery spotting, and anti-aircraft missile detection, this kind of noise source may often be limiting.

Clutter noise is a severe limitation because it is linear in background average radiance and because it usually integrates with time in the same way that the target integrates (though, if the target is moving relative to the background, a tracking sensor may gain an intensity-pattern-coherence advantage). The dynamic range of a clutter process is usually much greater than that of individual scenes. For example, a study of terrain correlation for navigational guidance recently found a dynamic range of 12 bits for its correlation, with 4 bits an adequate range for correlation of an individual scene. The dynamic range of solar scatter from clouds at 2 to 3 μm may vary four orders of magnitude over scattering angles of 360 to $0°$. (See Chapter 3.)

Statistical description of clutter is difficult because the process is neither ergodic nor stationary. As normally encountered, it is a phenomenon which is positive-definite. In bands of major interest, the standard deviation is of the same order of magnitude as the mean. In warning systems the greatest concern is with the tails of the probability-density function since they have the greatest influence on probability of detection and false alarm rate. Since background clutter may arise from scattering or emission from modest numbers of features, there is no assurance that it will be Gaussian. Empirical data are relatively sparse or specialized.

The principal method of enhancing target-to-clutter contrast is by shaping and scanning instantaneous fields of view to take advantage of any spatial correlations. In particular, the contrast of an unresolved target brighter than the background (such as a rocket), or dimmer (e.g. an aircraft), can be enhanced by decreasing the instantaneous field of view. Such a process tends to decrease the term V_{nc}^2 in Equation (21-26). However, if the reduction is carried beyond the diffraction limit of the optics, the terms V_{nl}^2 and V_{npa}^2 remain constant while τ_0, V_{ngr}^2, and V_{nc}^2 are reduced.

An analogy has been drawn between spatial filtering of background clutter and electronics filtering of electrical noise [21-4]. A two-dimensional autocorrelation function, $R(\xi)$, is defined in the usual way between the irradiances at arbitrary separations, $\xi = \xi_1$, ξ_2; and a Wiener spectrum is defined as the two-dimensional Fourier transform of $R(\xi)$. As a practical matter, measurements are made over ensembles of finite scenes to find

$$R'(\xi) = \lim_{B \to \infty} \frac{1}{4B_1 B_2} \int_B L_1(\mathbf{x}) L_2(\mathbf{x} + \xi) d\mathbf{x} \tag{21-32}$$

The corresponding Wiener functions are found to be steep, power-law functions in the region of interest (Figure 21-6). For a sensor scanning linearly at constant speed v, a power spectrum (for the clutter averaged over the scene used to measure the Wiener spectrum) can be found:

$$W'(f) \propto \int W_{xx}\left(\frac{f}{v,m_2}\right) \left|H\left(\frac{f}{v,m_2}\right)\right|^2 dm_2 \tag{21-33}$$

The product of this power spectrum with the absolute square of an electronic transfer-function can be integrated to find noise variance.

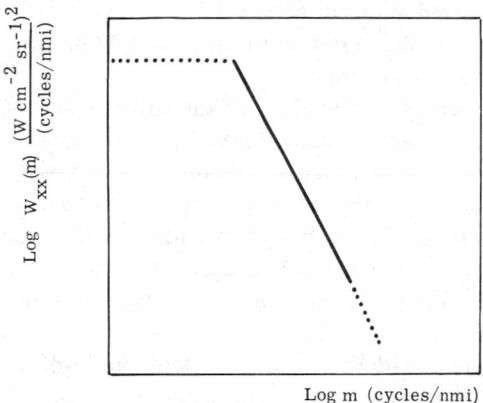

Fig. 21-6. Typical background Wiener
spectrum [21-23].

However, since the process is not assured to be Gaussian, a second-moment description is not sufficient, particularly for describing rare events; and since the process is not stationary, even the second-order description is empirical. The analysis can be used to give general guidance about forms and sizes of field-stops, especially about how small the field-stops should be and, therefore, how large an array of detectors is required to reduce V_{nc}^{2} to be comparable with other noise terms when scanning most scenes. A series of field tests is usually necessary to characterize performance. To assure adequate performance, one must make the data-processing part of the system sufficiently powerful to compensate for uncertain performances in signal processing.

The following is a hypothetical system to detect high-flying aircraft from space. A 3 to 5 μm sensor scans a raster pattern over an area of several hundred square miles. At night, the aircraft exhaust plume and the hot engine parts at a color temperature of about 600 K radiate a signal satisfactorily larger than the signal from an instantaneous detector field of view of 10^{3} m^{2} filled with a background at a temperature of about 270 K. During the day, however, solar-scatter from clouds provides a clutter response with peaks larger than those of the signal. Furthermore, a designer usually does not have adequate data on statistical descriptions of sufficient samples of background scenes to design a threshold system. He also finds that the variability of the scenes is so great that there is no simple, fixed design which would give both adequate probability of detection and a sufficiently low false alarm rate when background clutter is severe. (He is further hampered by intermittent observation of the target signal resulting from clouds between sensor and aircraft.)

Fortunately, the aircraft will take about an hour to fly through the field of view, and a plausible aircraft track has characteristics which are improbable in the random component of the background process. Therefore, the sensor makes a series of observations at 30-second intervals, peak-holding each for 30 seconds. Each observation is tested against a threshold which is set adaptively in response to the average level of the background. Observations which exceed the threshold are stored in an associated memory structured analogously to the field of view and time of observation. A real-time computer tests the stored exceedances every five minutes to associate them into plausible tracks. The decision rule demands that a series of exceedances which are accepted as a plausible track must meet the following requirements:

(1) They must be ordered in one direction.

(2) The total length of the series must not exceed the distance an aircraft could plausibly fly since the start of the track.

(3) The track must eventually extend across the entire field of view.

(4) The track must eventually contain at least 30 exceedances.

The resulting decision rule is a highly structured, multivariate algorithm which can tolerate a wide range of sensor signal-to-noise ratios at each individual observation and yield good system performance even though individual observations are of poor quality. Even if the probability of detection is 0.4 (due to observation) and the false-alarm rate at each observation is 10^{-2}, the probability of track detection is on the order of 99.95 and the false-alarm rate is on the order of 10^{-30} per channel.

21.3.3. Warning Systems with Passive Sensors Detecting Radiation Coherently. Passive sensors can be built that employ heterodyne detection [21-24]. Figure 21-7 shows a schematic view illustrating the principle. Incident radiation from the object surface is combined with locally generated laser radiation. The detector responds to the square of the incident electric field which will contain cross-products between components from the object surface and the local oscillator field, as well as many others [21-25]. For incident components at frequencies, ω_i, close to the local oscillator laser frequency, ω_1, a difference frequency, ω_{if}, is developed

$$\omega_{if} = \omega_i - \omega_1 \qquad (21\text{-}34)$$

to which the detector can respond. Such difference frequencies can be on the order of hundreds of megahertz. (See Section 21.3.5.) The limited frequency response of the detector causes it to average the faster components at infrared frequencies ($\sim 10^{12}$ to 10^{14} Hz) yielding a dc response. The conversion gain for intermediate frequencies, i.f., in the detector bandwidth is very great. If the laser local-oscillator amplitude is large enough, then noise from preamplifiers and load resistors is small compared to the shot noise on the local oscillator. Noise equivalent powers, NEPs, approaching

$$\text{NEP} = \frac{2h\nu\Delta f}{\eta} \qquad (21\text{-}35)$$

can be attained using photoconductive detectors in which Δf is the i.f. bandwidth.

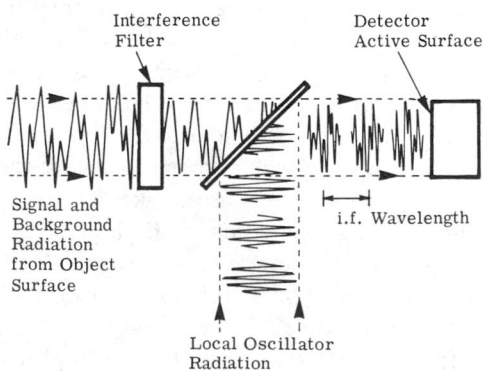

Fig. 21-7. Basic principle of heterodyne receiver.

At first sight these low NEPs are very attractive. However, a heterodyne detector is intrinsically diffraction-limited at the detector so that the maximum throughput of each detector is

$$(A_0\Omega)_{max} = \lambda^2 \tag{21-36}$$

Also, the spectral response is limited to the intermediate-frequency bandwidth of the detector, which is on the order of the reciprocal of its time constant. One detector with one laser local-oscillator can have access only to a very limited region of signal space. Siegman [21-26] puts it that

> optical heterodyne reception . . . is very insensitive for detecting incoherent thermal radiation emitted by any purely thermal source. The optical heterodyne cannot detect less than about one photon per resolving time in a single spatial mode. Thermal sources at reasonable temperatures emit much less than this . . . at optical frequencies.

Passive optical heterodynes are useful where very high spectral resolution is required (e.g., for detecting particular vibrational-rotational lines). Thus, they may be useful to warn against poisonous gases and to detect pollutants [21-24].

Passive heterodyne systems can also be useful where the target is a laser. For example, Sepp [21-27] describes a system to monitor accurately the orientation of geosynchronous satellites by observing a laser beacon on the ground. He points out that heterodyne reception would be far superior for this application at 10.6 μm; however, he is obliged to reject it because its technical complexity is too great for these satellites.

21.3.4. Active Systems. An infrared lidar (infrared radar) irradiates a target with energy from an infrared laser and collects a small part of the scattered energy with receiver optics. A laser-augmented imaging system scans a narrow beam of energy over a scene and displays the amplified, scattered returns as a map of the distributed scattering. The optical radar equation for incoherent detection is

$$\rho^2 \propto \frac{\eta}{h\nu} \frac{A_r A_t}{R^4} \left(\frac{E_t L_G \sigma}{\lambda^2} \right) \tag{21-37}$$

where A_r and A_t = transmitting and receiving aperture areas
η = detector quantum efficiency
R = slant range
E_t = energy of the transmitted pulse
L_G = angular gain of the transmitter
σ = scattering cross-section of the target

This equation is useful provided that the number of photons received in the integration time of the receiver is large. In assessing the feasibility of an optical radar, one should first calculate the photons received. Scattering cross-sections of typical military targets are frequently as small as 1 to 0.1% of geometrical areas, and the spherical spreading factors $(A_t A_r / R^4)$ are frequently very large; consequently, $(E_t / h\nu)$, the number of photons transmitted per pulse, and the transmitter directivity must be large if signal-to-noise ratios greater than 1 are to be achieved.

Unlike passive systems, these systems use narrow infrared bandwidths determined by the laser transmitters. In some cases, bandwidths are comparable to a single vibrational-rotational line selected by a highly tuned cavity. Optical radars with argon and YAG lasers and photomultipliers or silicon photodiode detectors have been used extensively at 1 μm. The greatest interest in the infrared is in systems at 10.6 μm using powerful CO_2 lasers, photovoltaic or photoconductive detectors (such as HgCdTe, PbSnTe, and Ge:Cu, and extrinsic dopings of silicon) [21-28]. The longer wavelength offers better weather penetration (especially through haze and light fog), easier alignment for a given size of optics, and a greater number of photons for a given transmitted energy. Much longer wavelengths, such as 496 μm (using the methyl fluoride laser, [21-29]), and point-contact detectors, may become of increasing interest.

Scanning of active optical systems presents difficulties because of the high angular resolution, high speeds required to cover useful fields in times dictated by system applications, and relatively cumbersome feed equipment. Mechanical techniques employing multifaceted mirrors, rotated or vibrated rapidly, are currently in use. Electro-optical scanning techniques are in development [21-36], as are phased array methods. For applications in the atmosphere, further difficulties arise because of variations in the refractive index of air. (See Chapter 6.) If very large powers are used, as may be necessary for airborne equipment looking out over long ranges, then the radiant beam will deposit sufficient energy to alter the local optical properties of the atmosphere and generate effects generically known as blooming. (See Chapter 6.)

Because the transmitted signal is determined by the warning system, a designer may choose to detect returns directly or by heterodyne coherent-detection [21-25]. These two cases will be discussed separately in the next two sections. However, a brief comparision is given first.

The quantum limit to the sensitivity of an optical detection system is

$$\text{NEP} = \frac{h\nu\Delta f}{\eta} \tag{21-38}$$

where Δf is the system's optical frequency bandwidth. When heterodyne detection is employed, power from a local oscillator is combined coherently and in the same direction and polarization with the return signal. In this case

$$\Phi_{min} = \rho^2 \Delta f \left\{ \left(\frac{h\nu}{\eta}\right)F + \left(\frac{h\nu}{e\eta}\right)^2 \frac{1}{\Phi_{LO}M^2}\left(\frac{i_{nL}^2}{2} + \frac{i_{na}^2}{2} + \frac{i_{ngr}^2}{2} M^2 F\right)\right\} \tag{21-39}$$

where M = internal gain factor

Δf = intermediate frequency bandwidth

F = factor allowing for increased noise due to internal gain

$i_{nL}^2, i_{na}^2, i_{ngr}^2$ = mean-square noise currents per unit bandwidth due to detector loads, preamplifiers and generation-recombination noise

If the local oscillator power, Φ_{LO}, is increased sufficiently, the second term may be made negligible compared to the first. (Physically, the only significant noise term at the i.f. amplifier is photon shot-noise of the signal beating with the local oscillator). In this case,

$$\text{NEP} = \frac{h\nu}{\eta}\,\Delta f\,F \tag{21-40}$$

For photoconductive or photovoltaic detectors [21-25] used at 10 μm, $F = 2$; although for ideal back-biased photodiodes with very low leakage, $F = 1$. For direct detection, it is possible that photon shot noise on the signal is the dominant noise. In that case [21-28]

$$\text{NEP} = \frac{4h\nu}{\eta} \, \Delta f \, F \tag{21-41}$$

However, for weak returns, sensitivity is more usually limited by amplifier or load-resistor noise, in which case

$$\Phi_{min} = \frac{h\nu}{e\eta} \, \rho \sqrt{\Delta f} \, \sqrt{2(i_{nL}^{\,2} + i_{na}^{\,2})} \tag{21-42}$$

An infrared, heterodyne system can frequently be several orders of magnitude more sensitive than a direct-detection system. However, to achieve this advantage the local oscillator must be aligned with the return signals to less than the diffraction angle, λ/d, at the detector; the maximum throughput must be λ^2; and the detection bandwidth (including uncorrected Doppler shifts) must be accommodated by the detector.

Goodman [21-13] points out that the performance of heterodyne systems is much less reliable for low false alarm rates and for rough targets (containing several scattering centers) than for specular targets (as would be expected from predictions of the general theory for coherent and incoherent systems). He also notes that energy-detection systems are superior to heterodyne systems where the expected number of background photons per primitive observation is small. The latter condition is important at short wavelengths in the near infrared. Figures 21-8 and 21-9 show probability of detection versus signal-to-noise ratio where the noise is entirely due to photoelectron generation for false alarm

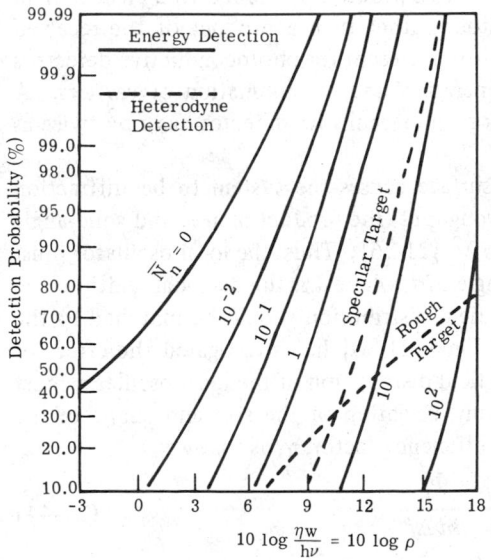

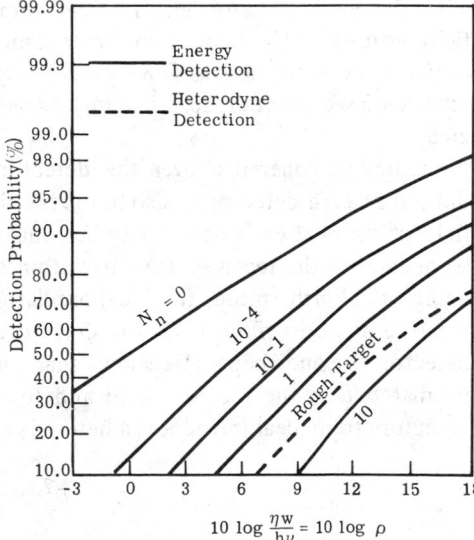

Fig. 21-8. Probabilities of detection for active systems with a specular or well-resolved rough target (false alarm probability 10^{-6}) [21-10, 21-13].

Fig. 21-9. Probabilities of detection for active systems with rough targets (false alarm probability 10^{-6}) [21-10, 21-13].

probabilities of 10^{-6}. Curves for energy detection with average number of background photons as a parameter are shown solid. Curves for heterodyne detection are shown dashed. (The heterodyne curves are adapted from Marcum [21-10] and are the same as in Figure 21-1.)

Active systems may experience severe target fluctuations as a result of glint (i.e., the difference between near spectral reflection in particular narrow solid angles and diffuse scattering at most angles) and scattering absorption and refraction in the narrow spectral bands utilized.

21.3.5. Active System with Coherent Detection. In a coherent heterodyne system, the received energy originates from a laser, is scattered by a target, and is applied to a detector in spatial coherence with and in the same polarization as a second signal from a local oscillator, which may be the same laser or one phase-locked to it or to the received signal [21-25].

For a sufficiently intense, local oscillator, the voltage-squared, signal-to-noise ratio of the envelope at i.f. is

$$\rho^2 = \frac{\eta \Phi_r}{2h\nu\Delta f} \tag{21-43}$$

where η = detector quantum efficiency
Φ_r = received power
Δf = intermediate frequency bandwidth

It will be noted that ρ^2 is proportional to the received power, unlike sensors with incoherent processing in which ρ is proportional to Φ_r. This behavior is typical of a coherent receiver for which the preferred structure is a cross-product of the received electric field with the local-oscillator, electric-field amplitude rather than a product of the received field with itself. The factor 2 in the denominator is present for photoconductive detectors because the noise derives from both the generation and recombination of carriers. A reverse-biased photodiode, without leakage, or a photoemissive detector has a ρ^2 twice as great.

Achieving coherence over the detector surface causes the system to be diffraction limited at each detector; it also limits the throughput (the product of area and solid-angle field of view) at each detector to be equal to λ^2 [21-26]. Thus, the local oscillator must be aligned to the received wave to within angle λ/d, where d is the diagonal width of the detector. Furthermore, the local-oscillator field distribution should be matched to the received field distribution at the detector. Fink [21-30] has investigated the effect of detector size and shape. He shows that the field distribution of the local-oscillator must be matched to the received field and that any deviation of the size and shape of the detectors from ideal introduces a heterodyne efficiency factor, γ, as follows:

$$\rho^2 = \frac{\gamma\eta}{2} \frac{\Phi_r}{h\nu\Delta f} \tag{21-44}$$

Many authors incorporate the heterodyne efficiency within η. The effect of nonoptimum size is shown in Figure 21-10 for one case. Degnan and Klein have discussed the effects of central obscuration, local-oscillator-field distribution, and detector size.

If the target has a radial velocity with respect to the axis of the system, the scattered radiation will be received at a frequency Doppler-shifted from the transmitted frequency.

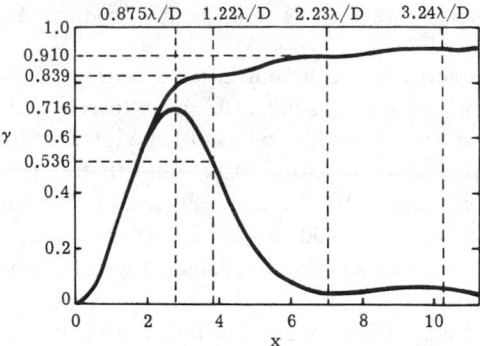

Fig. 21-10. Effect of nonoptimum detector size on heterodyne efficiency [21-29].

If the Doppler shift is not known, the intermediate frequency will be uncertain. Relative velocities of military targets often produce large Doppler shifts, particularly in space applications. For example, a satellite traveling at 5 nmi sec^{-1} with respect to the earth and using a 10.6 μm laser to sense at $45°$ would see a Doppler shift of

$$f_n = \frac{v}{\lambda} \cos \theta \sim 600 \text{ MHz} \qquad (21\text{-}45)$$

If the same sensor must scan a range of angles around its relative velocity, the intermediate frequencies may easily range over more than 10^9 Hz. However, the available 10.6 μm detectors are limited to bandwidths of 1 to 5×10^9 Hz [21-32]. A large i.f. bandwidth may be required, resulting in a poor signal-to-noise ratio; alternatively, a frequency-scanning local-oscillator may be used.

It is difficult to arrange a stable, tunable local-oscillator in the infrared. One solution to this problem was proposed by Teich and analyzed by Teich and Yen [21-33]. The laser transmitter is arranged to transmit two signals differing in frequency by amount, $\Delta \nu$, which is fixed, known, and much smaller than either signal ν_1 or $\nu_1 + \Delta \nu$. The return signals and a local-oscillator signal in the same polarization are applied to an infrared detector. The return signals are first filtered by a bandpass filter with a width set by a maximum Doppler frequency and then applied to a square-law device. The squaring results in several cross-products of signal and noise, including terms at $\pm \Delta \nu$. The signals are then applied to a bandpass filter with a width Δf, which is narrower than the maximum Doppler width, f_n, but wide enough to encompass the radar-information bandwidth. The output (voltage squared) signal-to-noise ratio, ρ^2, of this second filter is

$$\frac{f_n}{4\Delta f} \left[\frac{\rho_1^4}{1 + 2\rho_1^2} \right] \leqslant \rho_2^2 \leqslant \frac{f_n}{2\Delta f} \left[\frac{\rho_1^4}{1 + 2\rho_1^2} \right] \qquad (21\text{-}46)$$

where $\rho_1^2 = \eta \Phi_r [h\nu f_n]^{-1}$, so that the procedure can improve the signal-to-noise ratio or reduce it to within about a factor of 6 of the ideal heterodyne result in the absence of the unknown Doppler. However, this refinement introduces an additional substantial complexity.

As an example of an active and coherent infrared warning system, consider a terrain-avoidance system for a helicopter, similar to the experimental system described by Buczek [21-34]. This system warns the pilots of helicopters or low-flying aircraft about obstacles, especially wires, in their paths. It is found that a light, compact unit with a 2-in. aperture and transmitting $\sim$ 10 W can detect, locate, and measure relative velocity (by sensing Doppler) to ranges of 1000 to 4000 ft. At these short ranges, attenuation by fog, haze, and rain is not excessive. Wires are detected with a probability of 0.98 and at a false alarm rate of 10^{-6}.

A block schematic-diagram of the system is shown in Figure 21-11. A carbon-dioxide laser is associated with a saturable absorber and a grating (or prisms) to form a stable-frequency, high-prf (pulse repetition frequency) transmitter. The Doppler frequency-shift for 100 mph ground speed is about 10 MHz. The local oscillator is offset from the transmitter by this amount, as well as by a chosen intermediate frequency, and approximately stabilized to about 2 MHz by sensing ground-return Doppler. A transmitted pulse-width chosen to match this bandwidth provides adequate range resolution. A wide field ahead of the vehicle is sampled by mechanically scanning the common transmit/receive telescope in an open mesh. The scan pattern, particularly the spacing between scanned lines, is

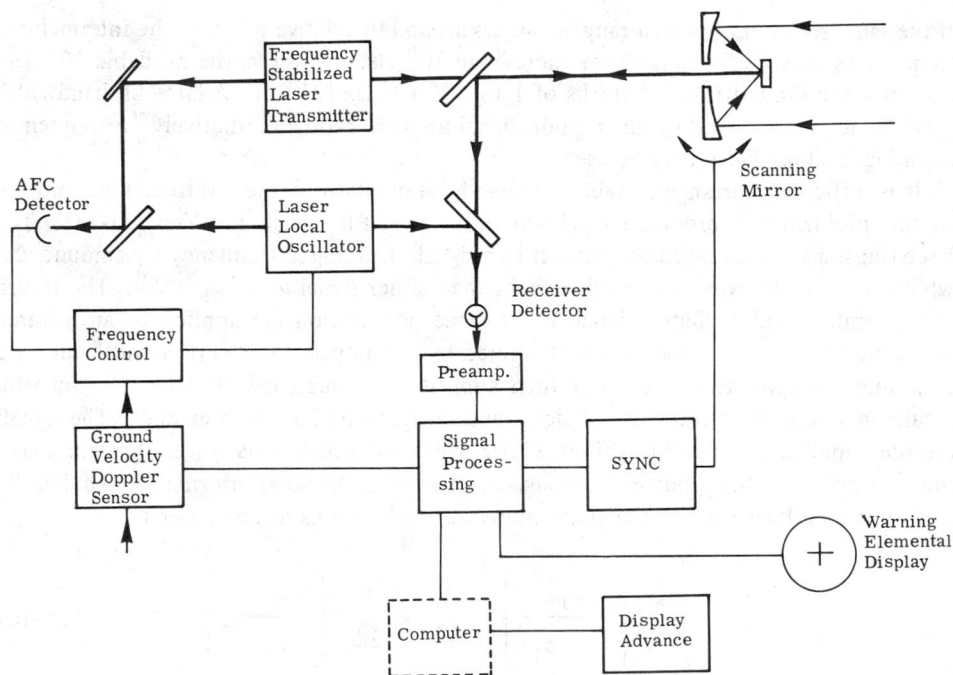

Fig. 21-11. Schematic diagram of an active coherent terrain avoidance system [21-34].

chosen as a best compromise between the high transmitter-power for detection on an intercept versus the high prf required to assure multiple intercepts. (In this design, the "cost" of high probability of detection and low false alarm rate is power-supply energy.)

Scattered returns are heterodyned with the local oscillator on a HgCdTe detector, which is cooled by liquid nitrogen and has a quantum efficiency of 0.4. Detector signals are processed and displayed on A-scope and low-resolution, imaging displays. Range-gating is used to enhance the imaging. Avoidance decisions are made by a pilot (or observer) watching the display.

21.3.6. Active System with Direct Detection. Active systems with direct detection are similar to passive systems, except that the signal energy is known *a priori* to be concentrated in the narrow spectral band of the laser transmitter—for pulsed systems there may be an advantage in knowing the prf. A range-gate can frequently be set up to ignore inputs which are received too soon or too late relative to the transmitted pulse. Unlike heterodyne systems, the spectral band in this system must be defined in the receiver by an interference filter or by some other form of spectral selection which must be made very narrow and stable with temperature.

If the dominant noise is generation-recombination noise, so that the detector may be described by a specific detectivity D^*, then

$$\rho = \frac{\Phi_s D^*}{\sqrt{A_d}} \sqrt{\Delta t} \qquad (21\text{-}47)$$

More generally, if noise variance is proportional to bandwidth $\Delta f \propto 1/\Delta t$, then

$$\rho = \frac{\Phi_s}{\Phi_n} \sqrt{\Delta t} \qquad (21\text{-}48)$$

where Φ_s = received power
$\quad\ \Phi_n$ = noise power per unit bandwidth
$\quad\ \Delta t$ = integration time

For a pulsed system, no advantage is gained for integrating beyond the duration of the pulse-length (plus the uncertainty of the range-gate); therefore, the signal-to-noise ratio is improved by shortening the pulse and increasing the maximum power.

Direct-detection systems are simpler than heterodyne systems, do not require such precise alignment, and are not ordinarily subject to difficulties arising from Doppler shifts. Furthermore, the voltage signal-to-noise ratio is proportional to returned power.

An example of a direct-detection laser-radar is given by Hoge [21-35]. A ruby-laser lidar augments a microwave radar, used for geodesy, by tracking and measuring the range to satellites. This lidar uses separate transmitter and receiver optics mounted through holes in the antenna dish of the associated microwave radar. The transmitter emits one-joule pulses, 30-nsec long, at a rate of 1 pps, into a beam with 0.5-mrad divergence. The receiver uses the existing 20-cm-aperture boresight-telescope of the radar. A 30% beam splitter diverts part of the scattered returns through a 1-nm bandpass interference filter to a photomultiplier tube.

In operation, the radar approximates the location of the satellite based on its ephemeris. The laser transmitter is fired in response to a timing pulse from the radar processing. A photodiode monitoring the scattered return from the back of the objective lens starts a counter at the exact time of the transmit pulse. The receiver is range-gated to 50 μsec around the nominal range for acquisition. The decision algorithm requires six consecutive returns within the range-gate to indicate acquisition.

After acquisition, the system switches to an automatic mode to perform its principal function of tracking and measuring range to the satellite to calibrate a network of radars.

21.4. References and Bibliography

21.4.1. References

[21-1] A. Wald, *Statistical Decision Functions,* John Wiley & Sons, New York, NY, 1950.

[21-2] H. Chernoff and L. E. Moses, *Elementary Decision Theory,* John Wiley & Sons, New York, NY, 1959.

[21-3] D. Middleton, *Statistical Communication Theory,* McGraw Hill, New York, NY, 1960.

[21-4] J. A. Jamieson, R. H. McFee, G. N. Plass, R. H. Grube, R. G. Richards, *Infrared Physics and Engineering,* McGraw Hill, New York, NY, 1963.

[21-5] R. D. Hudson and J. W. Hudson, "The Military Applications of Remote Sensing by Infrared," *Proceedings of the IEEE,* Vol. 63, No. 1, January 1975, p. 104.

[21-6] A. J. Grobecker, "Progress Report on the Climactic Impact Assessment Program," *Proceedings of the Third Conference on Climatic Impact Assessment Program,* DOT-TSC-OST-74-15, United States Department of Transportation.

[21-7] J. Humphreys, "Optimum Utilization of Lead Sulfide Infrared Detectors under Diverse Operating Conditions," *Applied Optics,* Vol. 4, 1965, p. 665.

[21-8] J. A. Jamieson, "Passive Infrared Sensors Limitations on Performance," *Applied Optics,* Vol. 15, April 1976, p. 891.

[21-9] P. M. Woodward, *Probability and Information Theory and Applications to Radar,* McGraw Hill, New York, NY, 1955.

[21-10] J. I. Marcum, "A Statistical Theory of Target Detection by Pulsed Radar," *IRE Transactions on Information Theory,* Vol. IT-6, April 1960, p. 59. (Reprinted from RAND Memorandum RM-754, December 1947.)

[21-11] P. Swerling, "Probability of Detection for Fluctuating Targets," *IRE Transactions on Information Theory,* IT-6, April 1960, p. 269. (Reprinted from RAND Memorandum RM-1217, March 1954.)

[21-12] P. Swerling, "Recent Developments in Target Models for Radar Detection Analysis," *AGARD Technical Symposium on Advanced Radar Systems,* Istanbul, Turkey, 25-29 May 1970. (Available from Technology Service Corporation, Santa Monica, CA.)

[21-13] J. W. Goodman, "Comparative Performance of Optical-Radar Detection Techniques," *IEEE Transactions on Aerospace and Electronic Systems,* Vol. AES-2, No. 5, September 1966, p. 526.

[21-14] B. Reiffen and H. Sherman, "An Optimum Demodulation for Poisson Processes: Photon Source Detectors," *Proceedings of the IEEE,* Vol. 51, No. 10, October 1963, p. 1316.

[21-15] R. Clow, E. Hanson, and F. McNolty, "Detection Probabilities for Fluctuating Infrared Targets," *Applied Optics,* Vol. 14, No. 1, January 1975, p. 61.

[21-16] Y. Itakura, S. Tsutsumi, and T. Takagi, "Statistical Properties of the Background Noise for the Atmospheric Windows in the Immediate Infrared Region, *Infrared Physics,* Vol. 14, No. 1, February 1974, p. 17.

[21-17] H. Bhattacharyya, "On a Measure of Divergence between Two Statistical Populations Defined by their Probability Distributions," *Bulletin Calcutta Mathematical Society*, Vol. 35, 1943, pp. 99-109.

[21-18] R. Evans, Aerojet ElectroSystems Company, Azusa, CA, private communication.

[21-19] M. Geokezas, "FLIR Image Target Analysis," *Sixth Annual Automatic Pattern Recognition Symposium*, University of Maryland, College Park, MD, 2-3 June 1976. (Available from Honeywell Systems and Research Center, Minneapolis, MN.)

[21-20] M. J. Beran, and G. B. Parrent, *Theory of Partial Coherence*, Prentice-Hall, New York, NY, 1964.

[21-21] M. Born and E. Wolf, *Principles of Optics*, Chapter 10, Pergamon Press, London, 1959.

[21-22] G. W. Autio, and J. F. Heintz, Private Communication. September 1970.

[21-23] N. G. Kulgein, Lockheed Missiles and Space Company, Sunnyvale, CA, private communication.

[21-24] R. T. Menzies, "Use of CO and CO_2 Lasers to Detect Pollutants in the Atmosphere," *Applied Optics*, Vol. 10, No. 7, July 1971, p. 1532.

[21-25] M. C. Teich, "Infrared Heterodyne Detection," *Proceedings of the IEEE*, Vol. 56, No. 1, January 1968.

[21-26] A. E. Siegman, "The Antenna Properties of Optical Heterodyne Receivers," *Proceedings of the IEEE*, Vol. 54, No. 10, October 1966.

[21-27] G. Sepp, "Earth Laser Beacon Sensor for Earth Oriented Geosynchronous Satellites," *Applied Optics*, Vol. 14, No. 7, July 1975, p. 1719.

[21-28] H. Melchior, M. B. Fisher, and F. R. Arams, "Photodetectors for Optical Communication Systems," *Proceedings of the IEEE*, Vol. 58, No. 10, October 1970, p. 1466.

[21-29] F. Brown, "High Power Optically Pumped Far-Infrared Lasers," *Infrared Physics*, Vol. 16, March 1976, p. 171.

[21-30] D. Fink, "Coherent Detection Signal to Noise," *Applied Optics*, Vol. 14, No. 3, March 1965, p. 689.

[21-31] S. C. Cohn, "Heterodyne Detection: Phase Front Alignment, Beam Spot Size, and Detector Uniformity," *Applied Optics*, Vol. 14, No. 8, August 1975, p. 1953.

[21-32] R. Heinisch, Systems Research Center, Honeywell, Minneapolis, MN, private communication.

[21-33] M. C. Teich, and R. Y. Yen, "Three Frequency Non-Linear Heterodyne Detection," *Applied Optics*, Vol. 14, No. 3, March 1975, pp. 666, 680.

[21-34] C. J. Buczek, "Ten-Micron Laser Radar for Low Level Flight," *Proceedings of the Tri-Service Radar Symposium*, Air Force Academy, 10-12 July, 1973.

[21-35] F. E. Hoge, "Integrated Laser/Radar Satellite Ranging and Tracking System," *Applied Optics*, Vol. 13, No. 10, October 1974, p. 2352.

[21-36] M. King and L. Schlom, "Rapid Scanning Laser Receiver," *Applied Optics*, Vol. 14, No. 7, July 1975.

21.4.2. Bibliography

Beran, M. J. and G. B. Parrent, *Theory of Partial Coherence*, Prentice-Hall, New York, NY, 1964.

Bhattacharyya, H., "On a Measure of Divergence between Two Statistical Populations Defined by their Probability Distributions," *Bulletin Calcutta Mathematical Society*, Vol. 35, 1943, pp. 99-109.

Born, M. and E. Wolf, *Principles of Optics,* Chapter 10, Pergamon Press, London, 1959.

Brown, F., "High Power Optically Pumped Far-Infrared Lasers," *Infrared Physics,* March 1976.

Buczek, C. J., "Ten-Micron Laser Radar for Low Level Flight," *Proceedings of the Tri-Service Radar Symposium,* Air Force Academy, 10-12 July, 1973.

Chernoff, H., and L. E. Moses, *Elementary Decision Theory,* John Wiley & Sons, New York, NY, 1959.

Clow, R., E. Hanson, and F. McNolty, "Detection Probabilities for Fluctuating Infrared Targets," *Applied Optics,* Vol. 14, No. 1, January 1965, p. 61.

Cohn, S. C., "Heterodyne Detection: Phase Front Alignment, Beam Spot Size, and Detector Uniformity," *Applied Optics,* Vol. 14, No. 8, August 1975.

Degnan, J. J., and B. J. Klein, "Optical Antenna Gain 2: Receiving Antennas," *Applied Optics,* Vol. 13, No. 10, October 1974.

Fink, D., "Coherent Detection Signal to Noise," *Applied Optics,* Vol. 14, No. 3, March 1965.

Geokezas, M., "FLIR Image Target Analysis," *Sixth Annual Automatic Pattern Recognition Symposium,* University of Maryland, College Park, MD, 2-3 June 1976. (Available from Honeywell Systems and Research Center, Minneapolis, MN.)

Goodman, J. W., "Comparative Performance of Optical-Radar Detection Techniques," *IEEE Transactions on Aerospace and Electronics Systems,* Vol. AES-2, No. 5, September 1966.

Grobecker, A. J., "Progress Report on the Climactic Impact Assessment Program," *Proceedings of the Third Conference on Climatic Impact Assessment Program,* DOT-TSC-OST-74-15, United States Department of Transportation.

Hoge, F. E., "Integrated Laser/Radar Satellite Ranging and Tracking System," *Applied Optics,* Vol. 13, No. 10, October 1974.

Hudson, R. D., and J. W. Hudson, "The Military Applications of Remote Sensing by Infrared," *Proceedings of the IEEE,* Vol. 63, No. 1, January 1975.

Humphreys, J., "Optimum Utilization of Lead Sulfide Infrared Detectors under Diverse Operating Conditions," *Applied Optics,* Vol. 4, 1965.

Itakura, Y,, S. Tsutsumi and T. Takagi, "Statistical Properties of the Background Noise for the Atmospheric Windows in the Immediate Infrared Region," *Infrared Physics,* Vol. 14, No. 1, February 1974.

Jamieson, J. A., "Passive Infrared Sensors Limitations on Performance," *Applied Optics,* Vol. 15, April 1976.

Jamieson, J. A., R. H. McFee, G. N. Plass, R. H. Grube, R. G. Richards, *Infrared Physics and Engineering,* McGraw Hill, New York, NY, 1963.

King, M., and L. Schlom, "Rapid Scanning Laser Receiver," *Applied Optics,* Vol. 14, No. 7, July 1975.

Marcum, J. I., "A Statistical Theory of Target Detection by Pulsed Radar," *IRE Transactions on Information Theory,* Vol. IT-6, April 1960, p. 59. (Reprinted from RAND Memorandum RM-754, December 1947.)

Melchior, H., M. B. Fisher, and F. R. Arms, "Photodetectors for Optical Communication Systems," *Proceedings of the IEEE,* Vol. 58, No. 10, October 1970.

Menzies, R. T., "Use of CO and CO_2 Lasers to Detect Pollutants in the Atmosphere," *Applied Optics,* Vol. 10, No. 7, July 1971.

Middleton, D., *Statistical Communication Theory*, McGraw Hill, New York, NY, 1960.

Reiffen, B., and H. Sherman, "An Optimum Demodulation for Poisson Processes: Photon Source Detectors," *Proceedings of the IEEE*, Vol. 51, No. 10, October 1963.

Sepp, G., "Earth Laser Beacon Sensor for Earth Oriented Geosynchronous Satellites," *Applied Optics*, Vol. 14, No. 7, July 1975.

Siegman, A. E., "The Antenna Properties of Optical Heterodyne Receivers," *Proceedings of the IEEE*, Vol. 54, No. 10, October 1966.

Swerling, P., "Probability of Detection for Fluctuating Targets," *IRE Transactions Information Theory*, IT-6, April 1960, p. 269. (Reprinted from RAND Memorandum RM-1217, March 1954.)

Swerling, P., "Recent Developments in Target Models for Radar Detection Analysis," *AGARD Technical Symposium on Advanced Radar Systems*, Istanbul, Turkey, 25-29 May 1970. (Available from Technology Service Corporation, Santa Monica, CA.)

Teich, M. C., "Infrared Heterodyne Detection," *Proceedings of Institute of Electrical and Electronics Engineers*, Vol. 56, No. 1, January 1968.

Teich, M. C., "Infrared Heterodyne Detection," *Proceedings of the IEEE*, Vol. 56, No. 1, January 1968.

Wald, A., *Statistical Decision Functions*, John Wiley & Sons, New York, NY, 1950.

Woodward, P. M., *Probability and Information Theory and Applications to Radar*, McGraw Hill, New York, NY, 1955.

Chapter 22

TRACKING SYSTEMS

Gordon Gerson *
Arthur K. Rue **
Hughes Aircraft Company

CONTENTS

*Sections 22.1, 22.5 through 22.7, and 22.9.
** Sections 22.2 through 22.4.

22. Tracking Systems

22.1. Introduction to Optical Trackers

Trackers are systems designed to follow the position of a selected remote object, the target, by responding to electromagnetic radiation from it.

Optical tracking systems contain essential components which may be mechanized in many different forms. Most trackers contain the following: a sensor which collects the target radiation and converts it to one or more electrical signals; tracker electronics which process the sensor output signals and produce tracking error signals, i.e., measurements of the target position relative to the sensor optical axis; a movable sensor pointing assembly which permits isolation from disturbances to the tracking system and allows the sensor to follow target motion; and a servo and stabilization system to control the pointing assembly positon. These components and their configuration for a typical tracker are shown schematically in Figure 22-1. Active trackers also contain a source of radiation which illuminates the target and is reflected back to the sensor. Semi-active trackers respond to reflected radiation from a source that is removed from the tracker.

There are numerous military applications for optical trackers. One of the most common is in the guidance systems of guided weapons, i.e., missiles and bombs. Passive intermediate infrared (2 to 5 μm) trackers guide some air-to-air and surface-to-air missiles to their targets. The visible and long wave infrared (8 to 12 μm) portions of the spectrum are used in passive air-to-surface missiles and guided bombs. Battlefield mis-

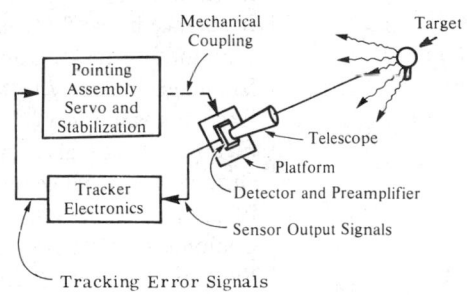

Fig. 22-1. Typical optical tracking system.

siles with command guidance systems contain an infrared beacon which is passively tracked by the launcher so that it can generate steering commands which are transmitted to the missile. Semi-active pulsed-laser trackers are applicable to all of the above weapon categories.

Passive infrared trackers are used in some aircraft fire control systems for initial target acquisition and to serve as a pointing reference to which infrared missiles can be slaved. Horizon sensors and star trackers are used for spacecraft attitude control. Optical communication systems employ semi-active trackers to maintain alignment between transmitters and receivers. Many other applications are described in Hudson [22-1].

22.1.1. Symbols, Nomenclature, and Units. The symbols, nomenclature and units used in this chapter are listed in Table 22-1.

22.1.2. Simplified Tracker Model. Tracking systems are most commonly designed as Type-1 servos, which, by definition, contain one integration in the open-loop transfer function. Insight into the dynamic performance of tracking systems can be gained from a simplified, single-axis model of a Type-1 tracking system as shown in Figure 22-2. The input signal, σ, is the angle from an arbitrary inertial-reference axis to the target-

Table 22-1. Symbols, Nomenclature, Units

Symbols	Nomenclature	Units
A	Gain	—
A_d	Detector area	cm^2
A_o	Collecting aperture area	cm^2
A_r, A_e, A_d	Moments of inertia of body A about r, e, d axes	in. lb sec^2, m kg sec^2
A_T	Area of total track field	sr
B_n, B_e, B_k	Body B moments of inertia about n, e, k axes	in. lb sec^2, m kg sec^2
c	Arbitrary constant	—
D	Drive displacement	in.^{3} rad^{-1}, m^3 rad^{-1}
D_t	Maximum target dimension	m
D_o	Collecting aperture diameter of optics	cm
D^*	Specific detectivity	cm Hz$^{1/2}$ W^{-1}
DV	Digitized video signal	V
E	Irradiance	W cm^{-2}
E_{Tj}	Irradiance in band j from interference	W cm^{-2}
ETP	Equilibrium tracking point	—
E_{Ij}	Irradiance in band j from target, T	W cm^{-2}
$E\{x\}$	Expected value of x	Units of x
e	Electron charge; also error	C, as appropriate
e_g	Output axis signal generator output	V
F	Effective focal length	cm
F_c	Coulomb friction torque	in. lb, m kg
FOV	Field of view angular diameter	rad
$F/\#$	Focal ratio	—
f	Frequency	Hz
f_c	Servo open-loop unity-gain frequency	Hz
f_N	Nutation frequency	Hz
f_P	Rosette petal frequency	Hz
f_R	Reticle rotation frequency	Hz
H	Angular momentum	in. lb sec, m kg sec
$H(\omega)$	Filter transfer function	—
$h(x, y)$	Relative detector response	—
I	dc current	A
I_B	dc detector current from the background	A
I_D	Detector dark current	A
i	ac current	A
i, j	Integer summation indices	—
i_n	Noise current	A
i_T	Integrating gyro torquing current	A

Table 22-1. Symbols, Nomenclature, Units (*Continued*)

Symbols	Nomenclature	Units
$\bar{i}_{\text{tgt}}$	Peak target current	A
J	Inertia	in. lb sec^2, m kg sec^2
K_T	Integrating gyro torquing scale factor	rad sec^{-1} A^{-1}
K_I	Current feedback and gain compensation	V V^{-1}
K_L	Leakage constant	in.^{5} lb^{-1} sec^{-1}, m^5 kg^{-1} sec^{-1}
K_P	Pressure feedback and gain compensation	in.^{3} V^{-1} sec^{-1}, m^3 V^{-1} sec^{-1}
K_M	Motor torque constant	in. lb A^{-1}, m kg A^{-1}
K_v	Rate or integrating gyro rate scale factor; also, motor back emf constant	V rad^{-1} sec
k	Arbitrary constant	—
k_B	Boltzmann constant	J K^{-1}
L	Motor inductance	H
L_B	Background radiance	W m^{-2} sr^{-1}
LOS	Line of sight	—
M, N	Integer variables	—
NA	Numerical aperture	—
NEFD	Noise equivalent flux density	W cm^{-2}
$n(t)$	Noise signal	As appropriate
P_D	Probability of detection	—
Q_L	Laser source energy	J
R	Motor resistance; also range	ohms; m
$\Re$	Detector responsivity	A W^{-1}
R_D	Angular radius of detector array	rad
R_L	Load resistance	ohms
R_N	Nutation radius	rad
$R(\tau)$	Autocorrelation function	As appropriate
r, θ	Polar coordinates	m, rad
s	Laplace transform argument	rad sec^{-1} (complex)
SNR	Signal to noise ratio	—
$S(\omega)$	Fourier transform of signal $s(t)$	—
T	Absolute temperature	K
T_D	Drive torque	in. lb, m kg
T_d	Torque disturbance	in. lb, m kg
T_o	Optics transmission	—
T_{go}	Time to go	sec
t	Time	sec
u	Time interval	sec
V	dc voltage; also velocity	V; m sec^{-1}
Var $\{x\}$	Variance of x	Units of x^2

Table 22-1. Symbols, Nomenclature, Units (*Continued*)

Symbols	Nomenclature	Units
V_e	Total entrapped volume	in.^{3}, m^3
V_I	Input voltage	V
V_ϵ	Error signal	V
v	Signal output; also, velocity	V; m sec^{-1}
v_T	Threshold signal output	V
$W(\omega)$	Power spectral density (of the background	—
w	Rosette petal width at widest point	—
x, y	Rectangular coordinates	m
β	FM modulation index; also, fluid bulk modulus	—; lb in.$^{-2}$, kg m^{-2}
Δf	Electrical bandwidth	Hz
Δf_v	Video bandwidth	Hz
Δf_s	Servo noise equivalent bandwidth	Hz
δ	Resolution element size	rad
ϵ	Tracking error; also, emissivity	rad; —
$\dot\epsilon, \dot\eta$	Outer and inner gimbal angular rates	rad sec^{-1}
θ	Target subtense; angle between target surface normal and target-to-tracker LOS	rad
$\dot\theta_c$	Angular rate of actuator housing	rad sec^{-1}
$\dot\theta_o$	Angular output rate	rad sec^{-1}
Λ	Likelihood ratio	—
λ	Wavelength	m
ξ	Rate gyro damping factor	—
ρ	Target difference reflectance	—
σ	Target LOS angle (inertial reference)	rad
$\sigma_x{}^2$	Variance of x	Units of x^2
τ	Transmission factor; also, time constant, time interval	—; sec
$\phi(t)$	Nutation phase angle	—
Ψ	Fluence	J m^{-2}
ψ	Tracker pointing angle	rad
ω	Angular diameter; also, detector instantaneous solid angle field of view; also, electrical frequency	rad; sr; rad sec^{-1}
$\omega_{A_r}, \omega_{A_e}, \omega_{A_d}$	Inner gimbal angular rates about r, e, d axis with respect to inertial space	rad sec^{-1}

Table 22-1. Symbols, Nomenclature, Units (*Continued*)

Symbols	Nomenclature	Units
$\omega_{B_n}, \omega_{B_e}, \omega_{B_k}$	Outer gimbal angular rates about n, e, k axis with respect to inertial space	rad sec^{-1}
ω_c	Open-loop unity gain frequency	rad sec^{-1}
ω_{IA}	Input axis angular rate	rad sec^{-1}
ω_n	Rate gyro natural frequency	rad sec^{-1}
$\dot{\omega}_{OA}$	Case angular acceleration about output axis	rad sec^{-2}
$\omega_{pi}, \omega_{pj}, \omega_{pk}$	Platform angular rates with respect to inertial space about i, j, k axes	rad sec^{-1}
$\partial Q/\partial P$	Valve flow-pressure gain	in.^{5} lb^{-1} sec^{-1}, m^5 kg^{-1} sec^{-1}
$\cdot$	Super dot for time derivative	—
$*$	Superscript for complex conjugate	—
$\hat{}$	Super meaning estimate	—
$'$	Prime (or apostrophe) for derivative	—

line-of-sight, and ψ is the sensor pointing-angle measured from the same reference. The angle-error, ϵ, is $\sigma - \psi$. The tracker error-characteristic represents the operation of the sensor and electronics in measuring the target position relative to the optical axis. Integration in the feedback path represents pointing assembly motion.

A critical factor in determining system performance is the *error characteristic*: the tracking error signal versus the actual error. The error characteristic shown in Figure 22-2 is typical of many trackers. The tracking system operation is linear for $|\epsilon| < \theta_1$. For $\theta_1 \leqslant |\epsilon| \leqslant \theta_2$, the tracker develops its maximum slew-rate, which is $K\theta_1$ rad sec^{-1}. The tracking field of view is $|\epsilon| \leqslant \theta_2$; the tracker will acquire a stationary target that is displaced from the sensor axis by θ_2 or less. If the angle error of the tracked target ever exceeds θ_2, then the tracker will break lock and the target will escape.

Over the linear region, the error characteristic is just a gain factor K. Then the transfer function from σ to ϵ (Figure 22-2) is

$$\frac{\epsilon(s)}{\sigma(s)} = \frac{s}{s+K} \qquad (22\text{-}1)$$

where $|\epsilon| \leqslant \theta_1$. The linear system-error responses (from rest) to step, ramp, and paraboloid inputs are given in Table 22-2. A Type-1 tracker can track a constant input with zero steady-state error, and can track, with constant steady-state error, a line of sight rate of up to $K\theta_1$ rad sec^{-1}. With a constant line-of-sight acceleration input, the error magnitude will increase as shown in the table until it reaches θ_1, at which time the

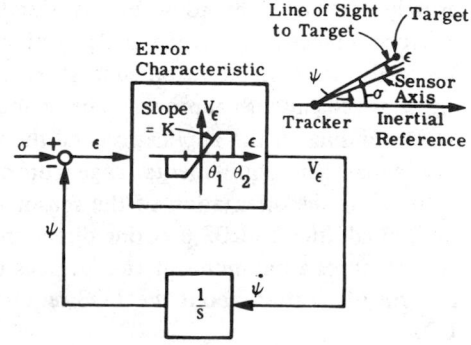

Fig. 22-2. Simplified single-axis Type-1 tracking system model.

Table 22-2. Error Responses of Linear Type-1 System
to Step, Ramp, and Paraboloid Inputs

$\sigma(t), t \geqslant 0$	$\epsilon(t)$	$\lim_{t \to \infty} \epsilon(t)$
Step $\Delta\sigma$	$\Delta\sigma\, e^{-Kt}$	0
Ramp $\dot{\sigma} t$	$\dfrac{\dot{\sigma}}{K} (1 - e^{-Kt})$	$\dfrac{\dot{\sigma}}{K}$
Paraboloid $\dfrac{\ddot{\sigma}}{2} t^2$	$\ddot{\sigma} \left[\dfrac{t}{K} - \dfrac{1}{K^2} (1 - e^{-Kt}) \right]$	∞

linear model no longer applies. Then the output rate $\dot{\psi}$ is limited to $\pm K\theta_1$ rad sec^{-1}, the error magnitude increases faster than shown in the table, and the tracker breaks lock. Thus, the error characteristic is very important in determining tracking system performance.

The remaining sections of this chapter present design and performance information on various kinds of optical tracking systems. Section 22.2 contains an introduction to gimbal systems and describes pointing assembly stabilization techniques. Section 22.3 describes tracking loop design. Section 22.4 describes performance analysis methods. Sections 22.5 through 22.8 describe different design approaches toward implementing the tracker error-characteristic, and Section 22.9 discusses some special problems often encountered in tracking system design.

22.2. Sensor Stabilization

22.2.1. Introduction. Tracking systems that operate from nonstationary platforms are subjected to angular jitter that results from platform linear and angular motions. These line-of-sight (LOS) disturbances are direct inputs to the tracking system and are usually attenuated by some means. For those situations in which the magnitude of the platform induced angular jitter is excessive, stabilization subsystems are generally utilized. Typically, these stabilization subsystems are implemented by using some type of gimbaling arrangement with appropriate drive and inertial sensing components.

Depending on the application and the type of tracking sensor being used, the stabilization subsystem requirements range from simply stabilizing the LOS pointing direction, to controlling the orientation of the sensor field of view with regard to rotation about the LOS in addition to LOS pointing direction stabilization. LOS pointing direction stabilization requires a minimum of two degrees of angular freedom orthogonal to the LOS; and control of rotation about the LOS requires a third degree of angular freedom about the LOS.

22.2.2. Basic Gimbaling Concepts. Many gimbal arrangements provide the desired pointing capability. The sensor is gimbaled in the most straightforward of these. In some applications, however, it may be desirable to use gimbaled optics to direct the LOS with the sensor mounted off the gimbal. Schematic representations of some gimbaled-optics concepts are given in Figure 22-3, where the outermost gimbal is taken to be azimuth and the innermost gimbal is elevation. For purposes of illustration, all but the third configuration are shown in the gimbal-lock position (i.e., the LOS is coincident with the outer gimbal axis).

A coelostat configuration, employing a pair of pointing mirrors, is shown in part (a) of the figure. For this configuration, the mirror on the azimuthal axis, AZ, is fixed to the azimuth gimbal and the elevation mirror rotates relative to the azimuth gimbal about the elevation axis, EL. Part (b) shows a coeloscope. In this configuration, the sensor is mounted on the outer gimbal rather than on the gimbal base. The LOS is directed in elevation by a pointing-mirror which rotates about the elevation axis. A heliostat or half-angle mirror configuration is shown in part (c). This configuration employs a single pointing mirror supported by an outer-azimuth, inner-elevation gimbal system. The angular motion of the LOS corresponds directly to the motion of the mirror in azimuth. The angular motion of the LOS in elevation is twice that of the mirror minus the sensor or platform rotation about the elevation axis. To provide LOS stabilization for the heliostat, one must derive a measure of the LOS motion since there is no mechanical element intrinsic to the basic mechanism that rotates such that its motion corresponds with the LOS. If the assembly is made to retain the required alignment, an appropriate reference can be established by utilizing an additional, inner-gimbal, mechanical element that is coupled to the mirror elevation-axis by means of a two-to-one drive linkage. Part (d) of the figure shows a gimbal-mounted telescope which requires a series of relay mirrors. In this configuration, angular misalignments in the optical relay-train are reduced by the telescope magnification. Performance may be improved by gimbaling the telescope assembly and relaying the unexpanded optical beam to the telescope, rather than relaying the beam after it has been expanded. For a telescope, the output-beam angular-pointing deviations are due to input-beam misalignments and to both rotation and relative translation of the optical elements. Figure 22-4 is a schematic of a reflective Cassegrainian telescope. The focal lengths of the primary and secondary mirrors are designated F_P and F_S, respectively, and the telescope magnification is expressed as F_P/F_S. The other symbols used in the figure are defined as follows:

$\Delta\theta_O$ = angular rotation of output beam
$\Delta\theta_I$ = angular rotation of input beam
$\Delta\theta_P$ = angular rotation of primary mirror
$\Delta\theta_S$ = angular rotation of secondary mirror
ΔX_P = translation of primary mirror
ΔX_S = translation of secondary mirror

If the telescope assembly is rigid, and it is rotated through angle $\Delta\theta_A$, $\Delta\theta_P = \Delta\theta_S = \Delta\theta_A$, and $\Delta X_S - \Delta X_P = \Delta\theta_A (F_S - F_P)$. In this case, it follows from the relationship given in Figure 22-4 that

$$\Delta\theta_O = \Delta\theta_I \left(\frac{F_S}{F_P}\right) + \Delta\theta_A \left(1 - \frac{F_S}{F_P}\right) \tag{22-2}$$

Furthermore, if both the input beam and the assembly are rotated by angle $\Delta\theta_T$, $\Delta\theta_O$ is equal to $\Delta\theta_T$.

The sources and characterization of relay misalignments in gimbaled optical assemblies are discussed in some detail in References [22-2 and 22-3].

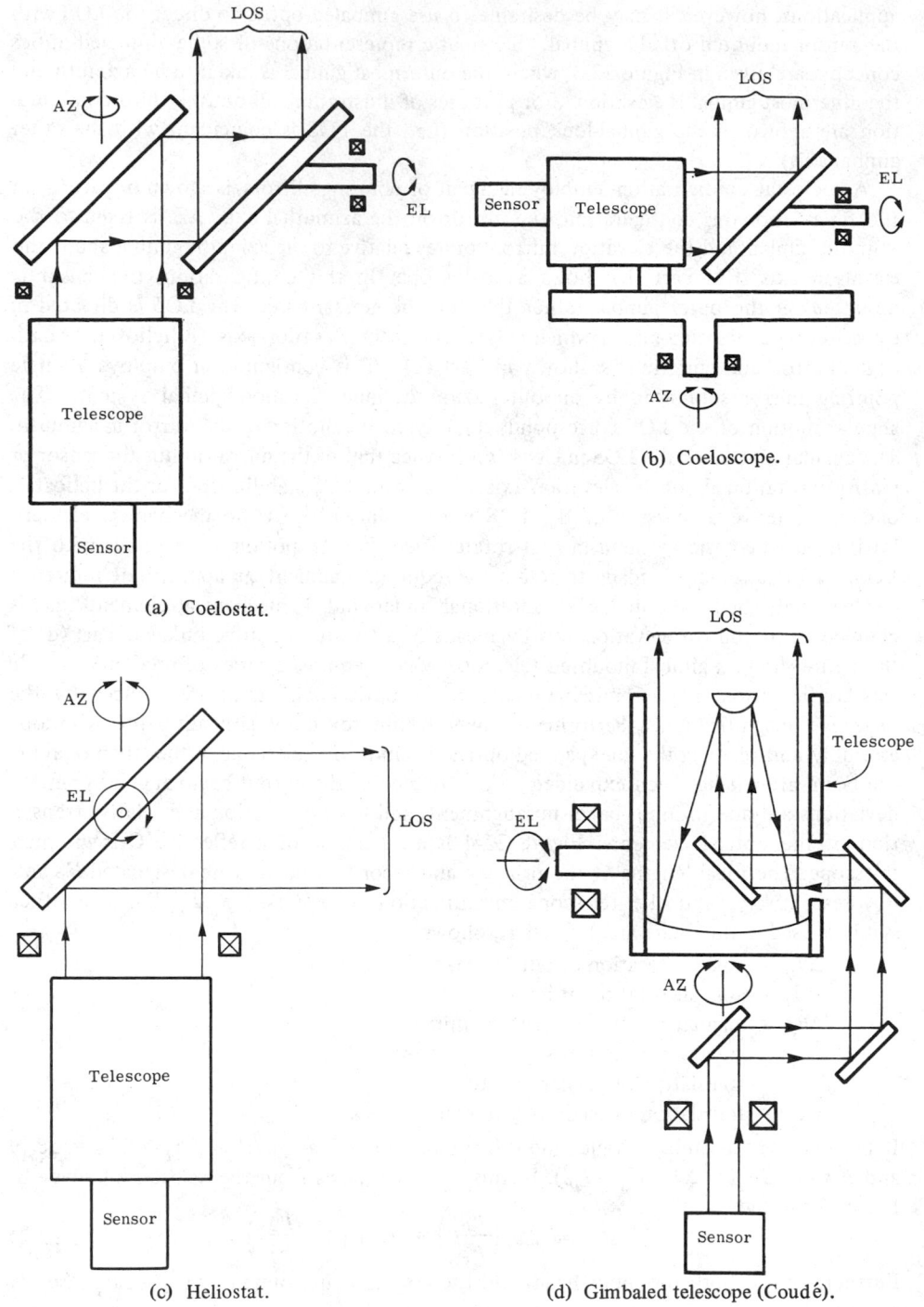

(a) Coelostat.

(b) Coeloscope.

(c) Heliostat.

(d) Gimbaled telescope (Coudé).

Fig. 22-3. Basic gimbaled-optics concepts.

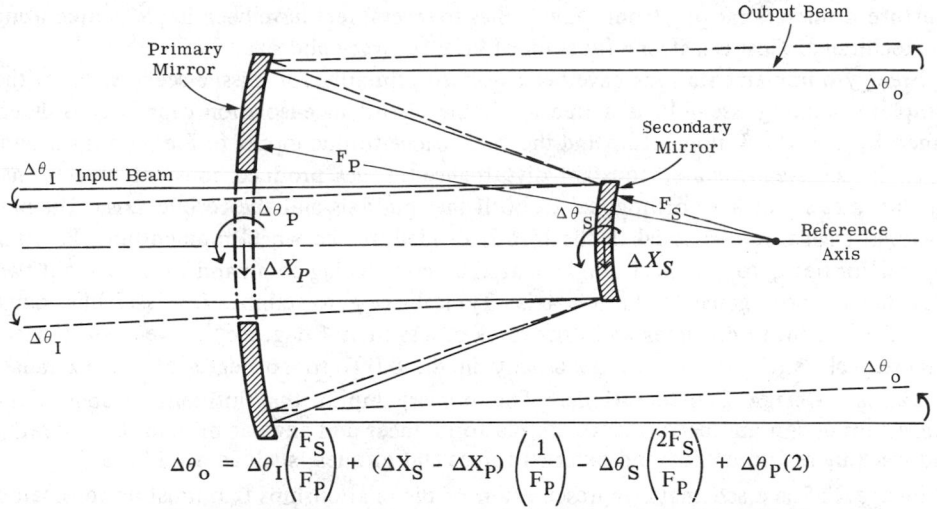

$$\Delta\theta_{o} = \Delta\theta_{I}\left(\frac{F_S}{F_P}\right) + (\Delta X_S - \Delta X_P)\left(\frac{1}{F_P}\right) - \Delta\theta_S\left(\frac{2F_S}{F_P}\right) + \Delta\theta_P \quad (2)$$

Fig. 22-4. Reflective telescope angular pointing relationships [22-4].

For optics configurations employing relay elements, sensor field of view (image) rotation may occur as the pointing assembly is moved through its angular travel. If the pointing configurations are implemented as shown in Figure 22-3, image rotation occurs for both azimuth and elevation angular-motion with the coelostat and gimbaled telescope. For the coeloscope, image rotation occurs with elevation angular-motion; and for the heliostat it occurs with azimuth angular-motion. If these types of gimbaling arrangements are used, the image can be derotated by rotating the sensor or by utilizing optical derotation elements such as a K-prism.

22.2.3. Stabilization Implementation. Implementation of the stabilization control function can be accomplished with: (1) gimbal-angle-control loops to slave the sensor-pointing assembly to follow a master stabilization-reference (such as an attitude-reference unit located elsewhere on the platform); (2) the use of inertial-sensing components mounted directly on a mechanical element that defines the pointing-assembly LOS to close local-control loops; (3) implementation of a free-gyro pointing assembly which has a momentum wheel that is an integral part of the inner gimbal assembly and that is oriented with its spin axis parallel to the LOS.

The first of these techniques is usually limited to applications that require only moderate accuracy. The errors are associated with gimbal and reference angle readout, coordinate transformations, gimbal misalignments [22-3], relative bending that may occur between the master reference mounting base and the sensor mounting base, disturbance inputs to the pointing assembly itself, and servo-control-loop error-rejection limitations. Accuracy values in the 1 to 10 mrad range are usually obtained with this technique.

The potential stabilization-performance of systems implemented with on-gimbal inertial-sensors and local-control-loops is much greater than that of slaved systems. In this case, the primary sources of stabilization errors are inertial-sensor imperfections and limitations, along with the residual servo-errors arising from pointing-assembly disturbance inputs. Accuracy values in the 1 to 500 μrad range are obtained using this technique. Very accurate systems are usually more complex than the less accurate ones and require the use of precision sensing and drive components. Stabilization systems for sensors with aperture diameters ranging from a few inches to several feet have been implemented using this technique. Cost is a strong function of both accuracy and size.

Free-gyro implementations have been utilized primarily for missile seekers. Since the complete pointing-assembly is a free gyro, the disturbance-isolation capability is determined by the wheel momentum and the disturbance-torque inputs to the pointing assembly. In the steady state, constant disturbance-torques produce constant precessional angular-rates about axes orthogonal to both the spin axis and the torque axis. The precessional angular-rate is equal to the torque divided by the wheel momentum. Pointing control for free-gyro systems is implemented by introducing command torques about two axes that are orthogonal to the spin axis. Typical free-gyro stabilization assemblies range from 3 to 15 in. in diameter and drift rates of less than 4 deg. sec^{-1} have been realized. Precessional "stiffness" values are usually in the 0.007 to 16 deg. sec^{-1}/in. oz range.

22.2.4. Design Considerations. The primary inputs that influence stabilization-subsystem design and performance are platform linear and angular motion, LOS slewing and tracking requirements, and external forces and torques, such as wind loads.

Figure 22-5 is a schematic representation of the relationships that must be considered in a typical design. For example, the forces required to accelerate the suspended mass due

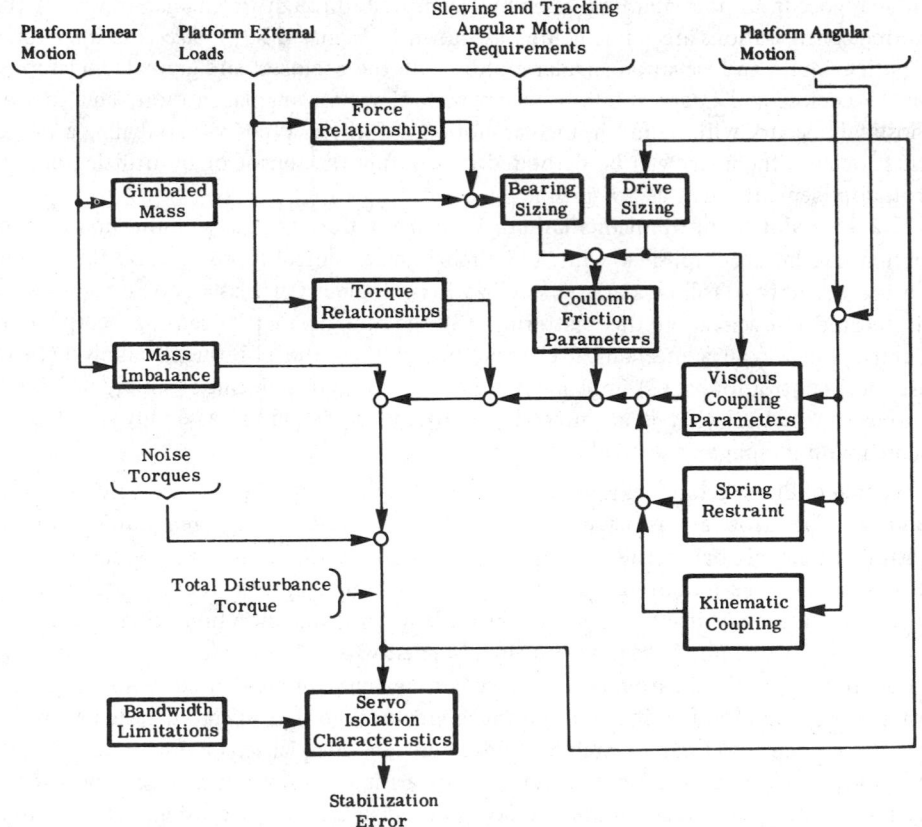

Fig. 22-5. Stabilization system design relationships [22-4].

to platform linear motion together with the other significant external forces, determine the size of the gimbal bearings and their preload, which, in turn, determine the values of bearing coulomb-friction torque. Also, the linear-motion and the values of the imbalance for the suspended mass result in torque-disturbance terms. The requirements for slewing and tracking angular-motion, together with the platform angular-motion, give rise to disturbance torques because of kinematic coupling, viscous damping, and spring restraint. In addition to bearing-friction, coulomb-friction torque terms also occur because of actuators. As indicated in Figure 22-5, drive-actuator coulomb-friction and viscous coupling are both related to the full-scale torque capability required and the type of actuators used.

The total torque-requirements are determined by all the dynamic torques, and by any torque demands that exist due to noise inputs such as high-frequency angle or rate noise from the control-loop sensors. Although these high-frequency noise inputs may not be important from a system-accuracy viewpoint, they may give rise to torque terms of significant magnitude and must be considered to avoid system torque saturation. Once the torque inputs have been identified and evaluated, the stabilization errors can be found by modifying the transfer function (from torque or angular acceleration to system error) to reflect the action of stabilization control-loops.

Torque is the fundamental driving parameter for all rotational systems, and the basic dynamics relating torque to angular position is a double integration with the input scaled to reflect the system inertia. The output of the first of the two integrators is the system angular-rate. As noted in Figure 22-5, these basic dynamics are usually modified due to the existence of viscous coupling and spring restraint. If viscous coupling or spring restraint occurs between the controlled load and the mounting base, stabilization-disturbance coupling-terms exist along with the modified dynamics.

The characteristics of simplified angular-rate and angular-position loops are summarized in Table 22-3. In the table, the disturbance inputs of interest for each loop are identified, and the appropriate steady-state errors are listed for each case.

In the case of the Type-0 and Type-1 angular-position servos, the closed-loop control-loop dynamics become lightly damped if the open-loop unity-gain frequency is much greater than load natural-frequency or the frequency associated with the viscous-coupling time-constant. When loops of this sort are actually implemented, they are usually stabilized by use of a lead-network similar to the one used for the Type-2 loop. Furthermore, integral networks can be utilized to augment the compensation for any of these loops, thus further reducing the errors due to disturbance inputs that occur at low frequencies.

When the stabilization performance for a particular system is evaluated, the use of the steady-state error values is usually not adequate since the control-loop disturbance-rejection gain-values usually vary with frequency and the disturbance inputs undoubtedly exist over some finite frequency-spectrum. In this case, the complete disturbance coupling transfer function is required. The open-loop unity-gain frequency relationships for the various loops all show that if the open-loop gain is increased to reduce disturbance coupling, the unity-gain frequency also increases. In general, the disturbance rejection that will be realized with the stabilization control-loops is limited because of bandwidth constraints that occur due to structural, noise, inertial sensor, or drive restrictions. Typical open-loop bandwidth values for stabilization systems are in the 5 to 50 Hz range.

Table 22-3. Basic Servo Isolation Characteristics

Servo Description	Open-Loop Transfer Function	Open-Loop Unity-Gain Frequency	Servo Type	Input
Angular-rate loop with viscous coupling between load and mounting base. Viscous coupling time constant τ.	$\dfrac{K_p}{\tau s + 1}$	$\omega_c = \dfrac{(K_p^2 - 1)^{1/2}}{\tau}$	0	Angular rate command, $\dot{\theta}_c$
				Angular acceleration command, $\ddot{\theta}_c$
Angular-rate loop without viscous coupling.	$\dfrac{K_v}{s}$	$\omega_c = K_v$	1	Angular rate command, $\dot{\theta}_c$
				Angular acceleration command, $\ddot{\theta}_c$
Angular-position loop with spring restraint and viscous coupling between load and mounting base. Spring restraint and load inertia yield a natural frequency of ω_n and the viscous coupling yields a damping ratio of δ.	$\dfrac{K_p}{\dfrac{s^2}{\omega_n^2} + \dfrac{2\delta}{\omega_n} s + 1}$	$\omega_c = \omega_n(1 - 2\delta^2 + A)^{1/2}$ where $A = 2\left[\delta^4 - \delta^2 + \left(\dfrac{K_p}{2}\right)^2\right]^{1/2}$	0	Angular position command, θ_c
				Angular rate command, $\dot{\theta}_c$
				Angular acceleration command, $\ddot{\theta}_c$

Table 22-3. Basic Servo Isolation Characteristics (*Continued*)

Steady-State Following Error	Input	Steady-State Inertial Error	Input	Steady-State Inertia Error (J = load inertia)
$\dfrac{\dot{\theta}_c}{K_p + 1}$ rad sec^{-1}	Mounting base angular rate, $\dot{\theta}_B$	$\dfrac{\dot{\theta}_B}{K_p + 1}$ rad sec^{-1}	Torque disturbances, T_d	$\dfrac{T_d}{J}\dfrac{\tau}{K_p + 1}$ rad sec^{-1}
Unbounded angular-rate error	Mounting base angular acceleration, $\ddot{\theta}_B$	Unbounded angular-rate error		
Zero steady-state angular-rate error	Mounting base angular rate, $\dot{\theta}_B$	No coupling	Torque disturbances, T_d	$\dfrac{T_d}{J}\dfrac{1}{K_v}$ rad sec^{-1}
$\dfrac{\ddot{\theta}_c}{K_v}$ rad sec^{-1}	Mounting base angular acceleration, $\ddot{\theta}_B$	No coupling		
$\dfrac{\theta_c}{K_p + 1}$ rad	Mounting base angular position, θ_B	$\dfrac{\theta_B}{K_p + 1}$ rad		
Unbounded angular-position error	Mounting base angular rate, $\dot{\theta}_B$	Unbounded angular-position error	Torque disturbances, T_d	$\dfrac{T_d}{J}\dfrac{1}{\omega_n^2(K_p + 1)}$ rad
Unbounded angular-position error	Mounting base angular acceleration, $\ddot{\theta}_B$	Unbounded angular-position error		

Table 22-3. Basic Servo Isolation Characteristics (*Continued*)

Servo Description	Open-Loop Transfer Function	Open-Loop Unity-Gain Frequency	Servo Type	Input
Angular-position loop with viscous coupling between load and mounting base. Viscous coupling time constant τ.	$\dfrac{K_v}{s(\tau s + 1)}$	$\omega_c = \dfrac{1}{\tau}\left(B - \dfrac{1}{2}\right)^{1/2}$ where $B = \left(K_v^2\tau^2 + \dfrac{1}{4}\right)^{1/2}$	1	Angular position command, θ_c
				Angular rate command $\dot{\theta}_c$
				Angular acceleration command, $\ddot{\theta}_c$
Angular-position loop without spring restraint or viscous coupling.	$\dfrac{K_a\left(\dfrac{s}{\omega_z} + 1\right)^*}{s^2\left(\dfrac{s}{N\omega_z} + 1\right)}$	$\omega_c = K_a^{1/2}N^{1/4}$ when $\omega_z = K_a^{1/2}N^{-1/4}$	2	Angular position command, θ_c
				Angular rate command, $\dot{\theta}_c$
				Angular acceleration command, $\ddot{\theta}_c$

*Type-2 loops are stabilized with a lead network. In this transfer function, the lead zero frequency is designated ω_z. Type-2 loop compensation design is discussed further in Section 22.3.2.

22.2.5. Kinematic Coupling. Kinematic disturbance-torques are due to the dynamic relationships between the gimbal-assembly inertia parameters and system linear and angular motion terms. In the general case, the kinematic torques applied to a body are discussed in Reference [22-5] and are given by

$$\mathbf{L} = \frac{d}{dt}(\mathbf{H})_s + (\mathbf{V}_P \times \Sigma M_i \mathbf{V}_{C_i}) \tag{22-3}$$

where $\mathbf{L}$ = vector sum of kinematic torques applied to body
        $\mathbf{H}$ = angular momentum of body

Table 22-3. Basic Servo Isolation Characteristics (*Continued*)

Steady-State Following Error	Input	Steady-State Inertial Error	Input	Steady-State Inertia Error (J = load inertia)
Zero steady-state angular-position error	Mounting base angular position, θ_B	Zero steady-state angular-position error		
$\dfrac{\dot\theta_c}{K_v}$ rad	Mounting base angular rate, $\dot\theta_B$	$\dfrac{\dot\theta_B}{K_v}$ rad	Torque disturbances, T_d	$\dfrac{T_d}{J}\dfrac{\tau}{K_v}$ rad
Unbounded angular-position error	Mounting base angular acceleration, $\ddot\theta_B$	Unbounded angular-position error		
Zero steady-state angular-position error	Mounting base angular position θ_B	No coupling		
Zero steady-state angular-position error	Mounting base angular rate, $\dot\theta_B$	No coupling	Torque disturbances, T_d	$\dfrac{T_d}{J}\dfrac{1}{K_a}$ rad
$\dfrac{\ddot\theta_c}{K_a}$ rad	Mounting base angular acceleration, $\ddot\theta_B$	No coupling		

$\mathbf{V}_P$ = vector linear velocity, relative to inertial space, of the origin about which the angular momentum is computed

M_i = individual mass elements of the body

$\mathbf{V}_{C_i}$ = vector linear velocity, relative to inertial space, of the center of mass for mass element M_i

(In Equation (22-3), the use of the subscript, s, denotes the derivative with respect to inertial space or a selected Newtonian frame of reference.)

The kinematic torques consist of gyroscopic terms which are functions of the product of angular rates, and conventional inertia torques which are functions of the various angular-acceleration terms.

A good illustration of kinematic coupling is given by a two degree-of-freedom gimbal-assembly. A sketch of a typical two-axis gimbal is given in Figure 22-6, where i, j, and k axes are defined in relation to the platform, the n, e, and k axes are defined in relation to the outer gimbal and the r, e, and d axes are defined in relation to the inner gimbal. Rotation of the outer and inner gimbals occurs about the k and e axes, respectively, and the respective gimbal edges are designated η and ϵ.

If each gimbal is balanced and if the gimbaled elements, bodies A and B, are suspended about principal axes, the torque relationships are considerably simplified. In this case, the various parameters of interest for the gimbal assembly shown in Figure 22-6 are identified as follows:

$\dot{\omega}_{P_i}, \dot{\omega}_{P_j}, \dot{\omega}_{P_k}$ = platform angular acceleration in relation to inertial space about the i, j, and k axes

$\omega_{P_i}, \omega_{P_j}, \omega_{P_k}$ = platform angular rates in relation to inertial space about the i, j, and k axes

$\omega_{B_n}, \omega_{B_e}, \omega_{B_k}$ = outer gimbal angular rates in relation to inertial space about the n, e, and k axes

$\omega_{A_r}, \omega_{A_e}, \omega_{A_d}$ = inner gimbal angular rates in relation to inertial space about the r, e, and d axes

$\dot{\eta}, \dot{\epsilon}$ = outer- and inner-gimbal angular-rates, respectively

A_r, A_e, A_d = body A moments-of-inertia about r, e, and d axes

B_n, B_e, B_k = body B moments-of-inertia about n, e, and k axes

T_{EL}, T_{AZ} = elevation and azimuth drive torques

The kinematic coupling relationships for this type of gimbal assembly are considered in detail in References [22-2 and 22-6]. If the torque relationships derived in these references are modified to reflect the simplifying assumptions previously mentioned, basic torque relationships of interest can be obtained. These simplified relationships are expressed in block-diagram form in Figure 22-7. To stabilize the gimbal-assembly LOS (r-axis), the angular-rates orthogonal to the r-axis (ω_{A_e} and ω_{A_d}) must be equal to zero. The relationships shown in Figure 22-7 have been derived in such a way that $\dot{\omega}_{A_e}$ and $\dot{\omega}_{A_d}$ are given explicitly and, hence, the appropriate torque relationships are obtained directly.

It can be seen from this figure that most of the disturbance-torque terms for the two-axis assembly are minimized when the inner-gimbal angle is equal to zero degrees. If an additional degree of gimbal freedom is incorporated so that the inner-gimbal angle is maintained near null, a considerable amount of intrinsic disturbance-isolation

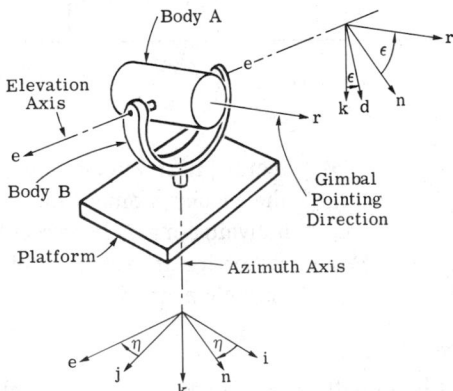

Fig. 22-6. Two-axis gimbal assembly [22-4].

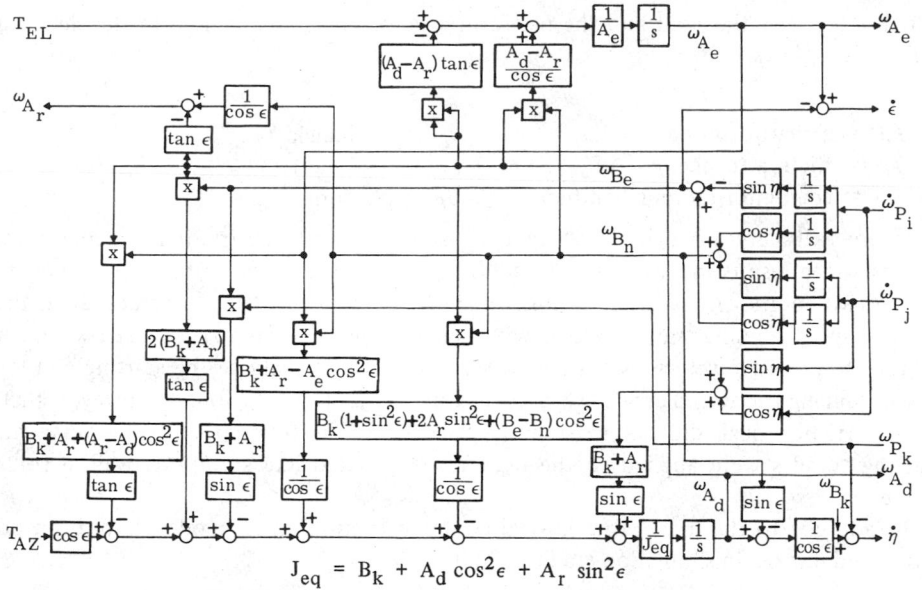

$$J_{eq} = B_k + A_d \cos^2\epsilon + A_r \sin^2\epsilon$$

Fig. 22-7. Two-axis gimbal torque relationships [22-4].

can be realized. Azimuth-elevation-azimuth or elevation-azimuth-elevation three-axis gimbal assemblies are representative of those types that can be implemented in this way.

LOS relationships for coelostats and coeloscopes are characterized by kinematics that
correspond to those for a two-axis gimbal. However, since the mirror for a heliostat must
be driven to compensate for platform motion about the half-angle axis, the LOS kinematic
relationships for this type of pointing assembly are somewhat more involved. These
kinematic torque-relationships are considered in Reference [22-4].

22.2.6. Coulomb Friction. Classical coulomb friction is a constant friction torque
(or force) that acts to oppose the torques
(or forces) inducing relative motion between
two bodies. The relationship between torque
and relative velocity is usually represented
as shown in Figure 22-8. This representation
is misleading in that when the relative velocity
between the two bodies is zero, the coulomb-
friction torque can take any value between
$-F_c$ and F_c. The value that exists in this case
is determined by the torques applied.

The angular motion of the mounting platform is usually specified rather than the
torques that induce the motion. In this case,
classical coulomb friction can be accurately
represented as shown in Figure 22-9 where
the platform angular acceleration, actuator
torque and driven load-inertia define the
value of coulomb friction when the relative

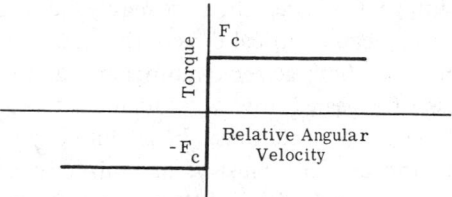

Fig. 22-8. Coulomb friction torque-relative
angular velocity characteristic.

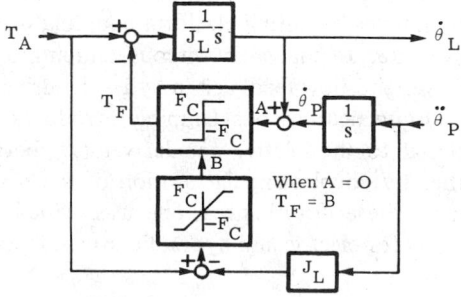

Fig. 22-9. Coulomb friction model.

velocity between the load and the platform is zero. The symbols used in the figure are defined as follows:

T_A = actuator torque θ_L = load angle

T_F = friction torque J_L = load inertia

F_C = coulomb friction magnitude θ_P = platform angle

There is usually a roughness component associated with the friction in actual mechanisms which results in variations of friction values as a function of the relative position of the two bodies. In addition to this roughness component, friction values larger than the low-speed running friction values have generally been accepted as realistic when zero or very low relative velocities exist. In a series of tests to evaluate the starting and low-speed running-friction characteristics of bearings [22-7] it was found, however, that a starting-friction peak did not exist. Furthermore, it was found that when very small starting torques were applied to the bearings they acted like springs as well as rolling elements.

In addition to the model for classical coulomb friction, other models have been proposed and utilized in simulation studies [22-8, 22-9].

22.2.7. Actuators. Depending on the precision and power required, direct-drive or geared actuators can be used for stabilization gimbal drives. In general, direct drives are utilized for applications that require the most precision. The factors that usually lead to this selection are the backlash, drive compliance, friction, and platform motion coupling that exist with geared drives. If high power is required for rapid acquisition over large angles, however, a geared drive may be the only feasible implementation. In these cases, coarse-fine gimbaling may be used to advantage if the precision gimbal is driven with a high-torque, limited-travel, direct drive and the coarse gimbal is driven with a high-power geared drive. For low-power applications with accuracy requirements that are consistent with geared drives, their use may yield smaller and lighter drive packages.

Typical, electrical direct-drive actuators are permanent magnet and moving-coil torque motors. Both ac servo-motors and dc motors with various types of field windings can be used for geared drives. Hydraulic actuators can also be utilized in both direct and geared-drive implementations. In satellite applications, for which drive reaction torques acting on the platform must be minimized, control-moment gyros or reaction-wheel drives are often utilized. Typical characteristics for commonly used stabilization actuators are summarized in Table 22-4.

The ideal drive would be a directly-coupled torquer with no viscous coupling; it minimizes mounting-platform coupling disturbances. No such actuator is available, however, due to equivalent viscous-damping terms that occur because of back-emf in electric drives and flow-feedback in hydraulic drives.

Equivalent viscous-damping introduces torques on the actuator rotor that are proportional to the relative angular velocity between the actuator stator and the rotor, and it thereby couples angular motion of the platform into the stabilization system. To minimize these disturbances, one uses wide-bandwidth current or pressure minor feedback-loops for electric and hydraulic drives, respectively.

Table 22-4. Representative Stabilization System Actuators

Actuator Type	Drive Power Capability and Example Values*	Available Rotational Travel	Typical Drive Implementation	Breakaway Torque (Percent of full scale torque)	Comments
Comutated dc torque motors	Low to moderate (100 in. oz at 300 rpm)	Continuous	Direct	1 to 4%	Probably the most commonly used precision stabilization system actuator. Lower breakaway torque achieved with electronic comutation.
Brushless dc torque motors	Low to moderate (100 in. oz at 300 rpm)	Limited ($\simeq \pm30°$)	Direct	< 1%	Directly wired rotor. Motors with very low breakaway torque values are usually larger and weigh more for the same peak torque output.
Moving coil dc torquers	Low to moderate (100 in. oz at 300 rpm)	Very limited ($\simeq \pm2°$)	Direct	< 1%	Similar to a loudspeaker. Two linear actuators are usually used to form a couple. Peak torque output is usually limited by coil heating.
ac servo motors	Low to moderate (1.5 in. oz at 5000 rpm)	Continuous	Geared	2 to 5%	Form low-cost compact drive system when the inaccuracies associated with geared drives can be tolerated.
Conventional dc motors	Low to high (40 in. oz at 7000 rpm)	Continuous	Geared	3 to 10%	Usually used to drive the coarse gimbal in coarse-fine implementations. Provides high slewing rates and accelerations.
Vane-type hydraulic torquers	Moderate to high (100 in. lb at 60 rpm)	Limited ($\approx \pm30°$)	Direct	0.5 to 1.5%	The hydraulic equivalent of a brushless dc torque motor with much greater torque available for the same size and weight.
Precision piston-type hydraulic torquers	Moderate to high (100 in. lb at 60 rpm)	Very limited ($\approx \pm2°$)	Direct	0.2 to 3%	The hydraulic equivalent of a moving coil dc torquer. Torque to size and weight ratios similar to those for the vane type.
Conventional hydraulic motors	Moderate to very high (500 in. lb at 2500 rpm)	Continuous	Geared	10 to 15%	Used for coarse gimbal drives. May be the only feasible actuator when high rates and accelerations are required for large loads.

*At least a ten-to-one range of power value sizes is available for each type of actuator.

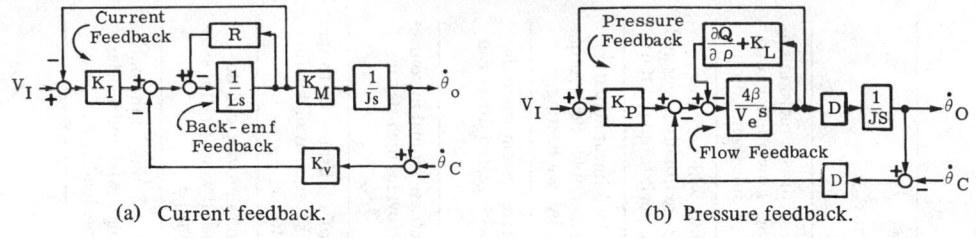

(a) Current feedback.   (b) Pressure feedback.

Fig. 22-10. Current and pressure minor loops [22-4].

Figure 22-10 shows simplified block diagrams of typical directly coupled electric and hydraulic drives, respectively. The symbols used in the figure are defined as follows:

$$V_I = \text{input voltage}$$
$$\dot\theta_o = \text{angular rate of output in relation to inertial space}$$
$$\dot\theta_C = \text{angular rate of actuator housing in relation to inertial space}$$
$$J = \text{inertia}$$
$$K_I = \text{current feedback gain and compensation}$$
$$R = \text{motor resistance}$$
$$L = \text{motor inductance}$$
$$K_M = \text{motor torque constant}$$
$$K_v = \text{motor back-emf constant}$$
$$K_P = \text{pressure feedback gain and compensation}$$
$$\beta = \text{fluid bulk modulus}$$
$$V_e = \text{total entrapped volume}$$
$$\partial Q/\partial P = \text{valve flow-pressure gain}$$
$$K_L = \text{leakage constant}$$
$$D = \text{drive displacement}$$

If they are properly implemented, the current or pressure feedback loops yield low-frequency drive characteristics with small equivalent viscous-damping terms and high-frequency characteristics that are consistent with closing wide-bandwidth stabilization loops.

In addition to eliminating various problems associated with geared drives (such as backlash and drive compliance), directly coupled drives also eliminate an additional disturbance-coupling mechanism. If a geared drive is employed and if platform angular motion occurs, the actuator rotor must accelerate to allow the load to remain stationary in relation to inertial space.

The block diagrams of Figure 22-11 show this coupling relationship for geared drives. The symbols used in the figure are defined as follows:

$$J_M = \text{motor inertia}$$
$$J_L = \text{load inertia}$$
$$N = \text{gear ratio}$$
$$T_D = \text{drive torque}$$
$$T_R = \text{reflected load torque}$$
$$\theta_M = \text{motor angle in relation to inertial space}$$
$$\theta_{MG} = \text{motor angle in relation to gear train case}$$
$$\theta_L = \text{load angle in relation to inertial space}$$
$$\theta_P = \text{platform angle in relation to inertial space}$$

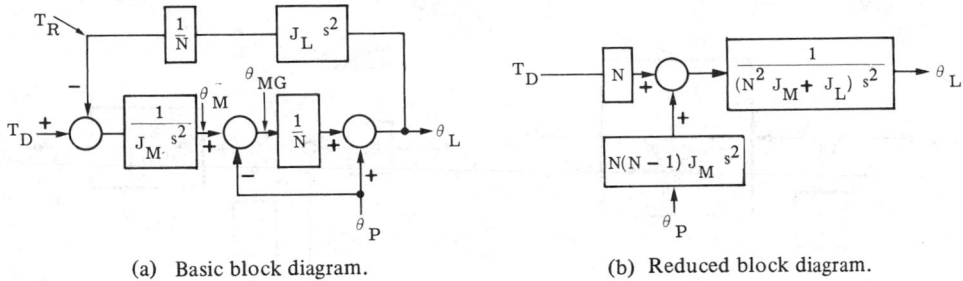

(a) Basic block diagram.   (b) Reduced block diagram.

Fig. 22-11. Geared drive relationships [22-4].

It is evident from Figure 22-11(b) that significant platform-motion coupling can occur if geared drives are used. In the case of a heliostat, it is possible to use this property to advantage by gearing or linking an inertia to the mirror. In this way, the torque required to accelerate the mirror about the half-angle axis (to compensate for platform angular-motion) is provided by the linkage assembly, rather than by the drive actuator.

22.2.8. Inertial Sensing Components. Gyros and angular accelerometers are used for stabilization subsystem inertial references. Although gyros have been developed that utilize various physical phenomena and techniques, the large majority of gyros are momentum-wheel types [22-10]. As noted previously (Section 22.2.3), the complete sensor gimbal-assembly can be a free or a two degree-of-freedom gyro. Two degree-of-freedom gyros also exist as basic sensing components. In this case, a momentum wheel is supported on a limited-travel suspension and the complete mechanism is mounted on the gimbal assembly which is to be stabilized. The gimbal control-loops are then driven so as to null the momentum-wheel suspension angles. LOS positioning is accomplished by torquing the gyro, thus causing momentum wheel precession about an axis orthogonal to both the torquing and momentum axes.

An elementary single-degree-of-freedom gyro is shown schematically in Figure 22-12. Angular motion about the input axis produces a torque about the output axis. If the output axis is spring restrained, the steady-state output-axis deflection-angle, which is measured with a signal generator, is proportional to the angular rate about the input axis. Use of viscous-damping restraint rather than spring restraint yields an integrating gyro since the steady-state angle about the output axis is proportional to the time integral of the angular rate about the input axis. In addition to the use of viscous damping, a torque generator is utilized in integrating gyros to provide a means for precessing the gyro about the output axis. Block diagrams for rate and integrating gyros are given in Figures 22-13(a) and (b), respectively, where

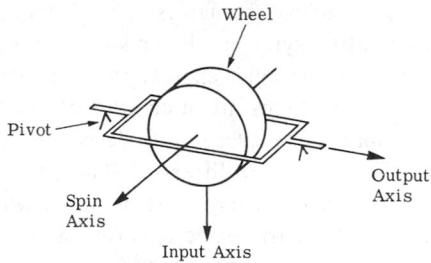

Fig. 22-12. Elementary gyro.

ω_{IA} = input-axis angular rate
e_g = output-axis signal-generator output
$\dot\omega_{OA}$ = case angular-acceleration about the output axis

H = gyro momentum
J = gyro output-axis inertia
ω_n = rate-gyro natural frequency
ζ = rate-gyro damping factor

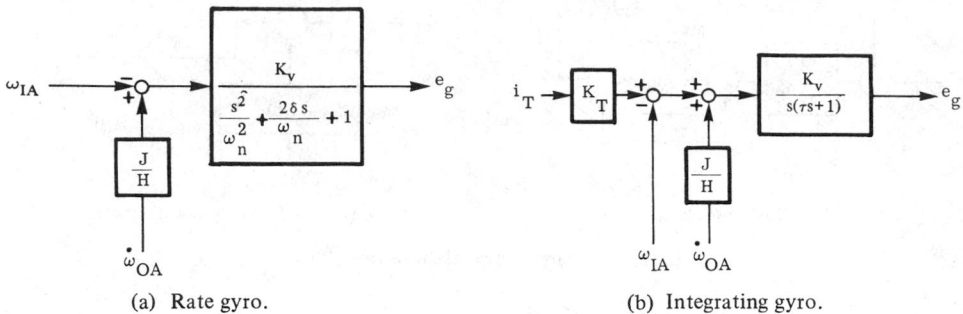

(a) Rate gyro.   (b) Integrating gyro.

Fig. 22-13. Gyro block diagrams.

τ = integrating-gyro time constant
K_v = gyro rate scale factor
K_T = integrating-gyro torquing scale-factor
i_T = integrating-gyro torquing current

As shown in the figures, angular acceleration of the case about the output axis causes an undesirable signal-generator output [22-10]. Other undesirable outputs usually specified for rate and integrating gyros include drift terms that are characterized as bias, random, and those that are acceleration dependent. Since various gyro parameters are temperature dependent, it is often necessary to provide a controlled-temperature environment to obtain the required performance.

In general, the accuracy characteristics of integrating gyros are superior to those of rate gyros. If accurate angular-rate measurements are required, integrating gyros can be caged by using the gyro torquer to null the signal-generator output. In this case, gyro-torquer current provides the measure of input-axis angular rate.

The gimbal actuators for a stabilization control-loop implemented with uncaged integrating gyros are driven so as to null the gyro signal-generator outputs, and the stabilization subsystem is angular-rate commanded by torquing the gyros. If rate-gyros or caged integrating-gyros are used, the measured angular rate is compared with the desired or commanded angular rate to derive gimbal-actuator command-signals.

The laser gyro differs fundamentally from a conventional gyro. Its operation is based on optical frequency shift obtained when an angular rate is applied about an axis normal to the geometric plane of a ring laser. The optical frequencies of clockwise- and counter-clockwise-propagating laser beams differ in proportion to the applied angular-rate. This frequency difference is detected and used as an angular-rate measurement signal. As the applied rate approaches zero, however, the modes approach the same optical frequency and tend to lock on a common frequency, thus producing zero output for small angular rates (a "deadband"). Various techniques are used to bias the device so that the deadband region is excluded from the normal operating range. Outside the lock-in region, the laser gyro functions as an accurate, highly-linear sensor with a wide dynamic range.

Conventional angular accelerometers consist basically of an inertial mass that is viscously damped and either elastically or torque generator restrained. The input axis is coincident with the output axis and, in the steady-state, the angle of rotation is proportional to the applied angular acceleration. Implementation of a somewhat different angular accelerometer consists of a fluid rotor contained in a ring and force-balance servoed sensor [22-11].

Pertinent characteristics for representative rate-gyros, integrating-gyros, and an angular accelerometer are summarized in Table 22-5.

22.2.9. Structural Compliance. Structural compliance of the gimbal assembly comes into play in a number of ways. Structural modes that occur due to compliance in the directly-driven load are usually fundamental in limiting the stabilization bandwidth that can be realized. Dynamic coupling due to gimbal kinematics, actuator viscous coupling, actuator reaction torques and gimbal-angle transducer outputs may introduce structural modes that are derived from compliance in the supporting structure. In general, the dynamics that arise due to structural compliance are dependent on both the modes that exist and the gimbaling configuration utilized [22-4].

For systems that must operate in a severe linear-motion environment, anisoelasticity may be of particular importance. That is, if the gimbal assembly does not exhibit identical compliance properties in all directions, linear accelerations produce unbalanced torques. These torques are proportional to the square of the linear acceleration [22-10]. Generally, anisoelastic errors are of greatest concern for free gyro stabilization gimbals and for gyroscopic instruments.

22.3. Tracking Loop Design

22.3.1. Deterministic Inputs. Tracking errors result from the residual stabilization errors, tracking-sensor angle-noise inputs, and relative translation between the target and the tracking system. A fly-by situation, in which the relative velocity between the target and the tracking system remains constant, is not a trivial tracking problem [22-12]. This situation is depicted in Figure 22-14 where

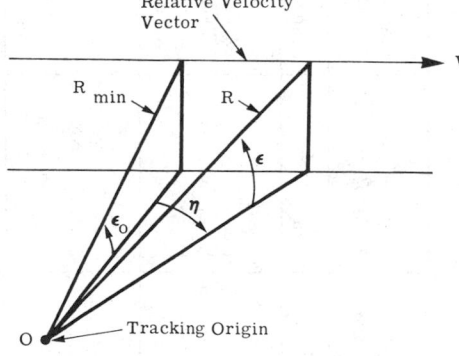

Fig. 22-14. Fly-by tracking situation.

V = relative velocity (between target and tracker)

R = range to target

R_{min} = minimum range to the target (i.e., range at crossover)

ϵ = target elevation angle

ϵ_o = target elevation angle at crossover

η = target azimuth angle

t = time in seconds (t = 0 at crossover)

The azimuth and elevation angles are given by the relationships

$$\eta = \tan^{-1}\left[\frac{Vt}{R_{min}\cos\epsilon_o}\right] \tag{22-4}$$

and

$$\epsilon = \tan^{-1}\left[\frac{\sin\epsilon_o\cos\eta}{\cos\epsilon_o}\right] \tag{22-5}$$

Table 22-5. Inertial Sensor Summary

Parameter \ Type	Rate-Gyro Spring Restrained	Rate-Gyro Spring Restrained	Integrating-Gyro Floated	Integrating-Gyro Floated-Heated	Integrating-Gyro Floated-Gas Bearing	Force-Balance Angular Accelerometer
Full scale input	2 rad sec^{-1}	5 rad sec^{-1}	2 rad sec^{-1}	1 rad sec^{-1}	3 rad sec^{-1}	50 rad sec^{-2}
Output accuracy:						
Non-G sensitive*	240 deg h^{-1} (braced)	1000 deg h^{-1} (braced)	12.0 deg h^{-1}	5.0 deg h^{-1}	< 1.0 deg h^{-1}	0.05 rad sec^{-2} (braced)
Random/friction			5.0 deg h^{-1}	3.0 deg h^{-1}	< 0.1 deg h^{-1}	
Anisolastic			0.3 deg h^{-1} g^{-2}	0.3 deg h^{-1} g^{-2}	0.1 deg h^{-1} g^{-2}	—
G sensitive			18.0 deg h^{-1} g^{-1}	10.0 deg h^{-1} g^{-1}	< 1.0 deg h^{-1} g^{-1}	0.2 rad sec^{-2} g^{-1}
Torquer accuracy:						
Set point	—	—	1%	0.02% up to 0.5 rad sec^{-1} (braced)	< 0.1%	—
Linearity	—	—	1%		< 0.05%	—
Temperature sensitivity	—	—	2% over temperature	—	—	
Response characteristics:						
Time constant	—	—	1.5 msec	1.5 msec	0.6 msec	—
Natural frequency	60 Hz	60 Hz	—	—	—	30 Hz
Damping	0.4 to 1.2, 0.7 nominal	0.35 to 1.05, 0.7 nominal	—	—	—	0.55 to 0.75, 0.6 nominal
Operating temperature range	−40 to 160°F	−65 to 240°F	−65 to 240°F	−40 to 160°F	20 to 100°F	−40 to 200°F
Miscellaneous considerations	Temperature compensated	Temperature compensated	Nonheated	For precision torquer, temperature controlled at 180°F	Temperature controlled at 130°F and expensive	Nonheated
Size and weight	1 in. diameter by 2.4 in. long; 5 oz	1 in. diameter by 2.4 in. long; 5 oz	1 in. diameter by 2.4 in. long; 4.5 oz	13/16 in. diameter by 2.4 in. long; 5 oz	2.32 in. diameter by 4.7 in. long; 1.65 lb	2.60 in. by 1.10 in. by 1.39 in.; 3 oz

*In this table, the symbol G indicates linear acceleration sensitivity (gravity).

These relationships can be differentiated with respect to time twice to obtain angular rates and accelerations.

$$\dot{\eta} = \left[\frac{V}{R_{min} \cos \epsilon_o} \right] \cos^2 \eta \qquad (22\text{-}6)$$

$$\ddot{\eta} = - \left[\frac{V}{R_{min} \cos \epsilon_o} \right]^2 \cos^2 \eta \sin 2 \eta \qquad (22\text{-}7)$$

$$\dot{\epsilon} = - \left[\frac{V}{R_{min} \cos \epsilon_o} \right] \left[\frac{\sin 2 \epsilon_o \sin 2 \eta \cos \eta}{4 (1 - \sin^2 \epsilon_o \sin^2 \eta)} \right] \qquad (22\text{-}8)$$

$$\ddot{\epsilon} = \left[\frac{V}{R_{min} \cos \epsilon_o} \right]^2 \left[\frac{\sin 2 \epsilon_o \cos^3 \eta [(2 - \cos^2 \eta)(1 - \sin^2 \epsilon_o \sin^2 \eta) - 2 \cos^2 \eta]}{2(1 - \sin^2 \epsilon_o \sin^2 \eta)^2} \right] \qquad (22\text{-}9)$$

The maximum values for azimuth-angle time-derivatives are easily determined and are found to occur at azimuth angles of $0°$ and $\pm 30°$ for the angular rate and acceleration terms, respectively. Using these values, one finds that

$$|\dot{\eta}_{max}| = \frac{V}{R_{min} \cos \epsilon_o} \qquad (22\text{-}10)$$

and

$$|\ddot{\eta}_{max}| = \left[\frac{V}{R_{min} \cos \epsilon_o} \right]^2 \left[\frac{3 \sqrt{3}}{8} \right] \qquad (22\text{-}11)$$

The azimuth angular rates and accelerations are normalized to their maximum values and plotted versus the azimuth angle in Figure 22-15. In addition to these plots, a plot of $V t [R_{min} \cos \epsilon_o]^{-1}$ versus η is also included in the figure to provide a convenient means for obtaining time relationships. Maximum values for the elevation angle rates and accelerations are somewhat more tedious to obtain and vary as a function ϵ_o. Peak values and the azimuth angles at which they occur are summarized for selected ϵ_o values as seen in Table 22-6.

Elevation angles along with normalized plots of the elevation angular rate and acceleration values for the selected ϵ_o cases are depicted versus azimuth angle in Figures 22-16, 22-17 and 22-18, respectively.

From examination of the plots for the fly-by situation it is apparent that significant LOS angular dynamics can occur even if neither the target nor the tracking platform are maneuvering. Although the tracking dynamics that may exist for a particular application do not correspond exactly with the fly-by case, these LOS dynamics are useful in establishing a first-cut at the tracking requirements for many situations. For example, the tracking dynamics associated with a missile on an intercept course can be related to the fly-by situation if a target miss-distance is specified and related to R_{min}.

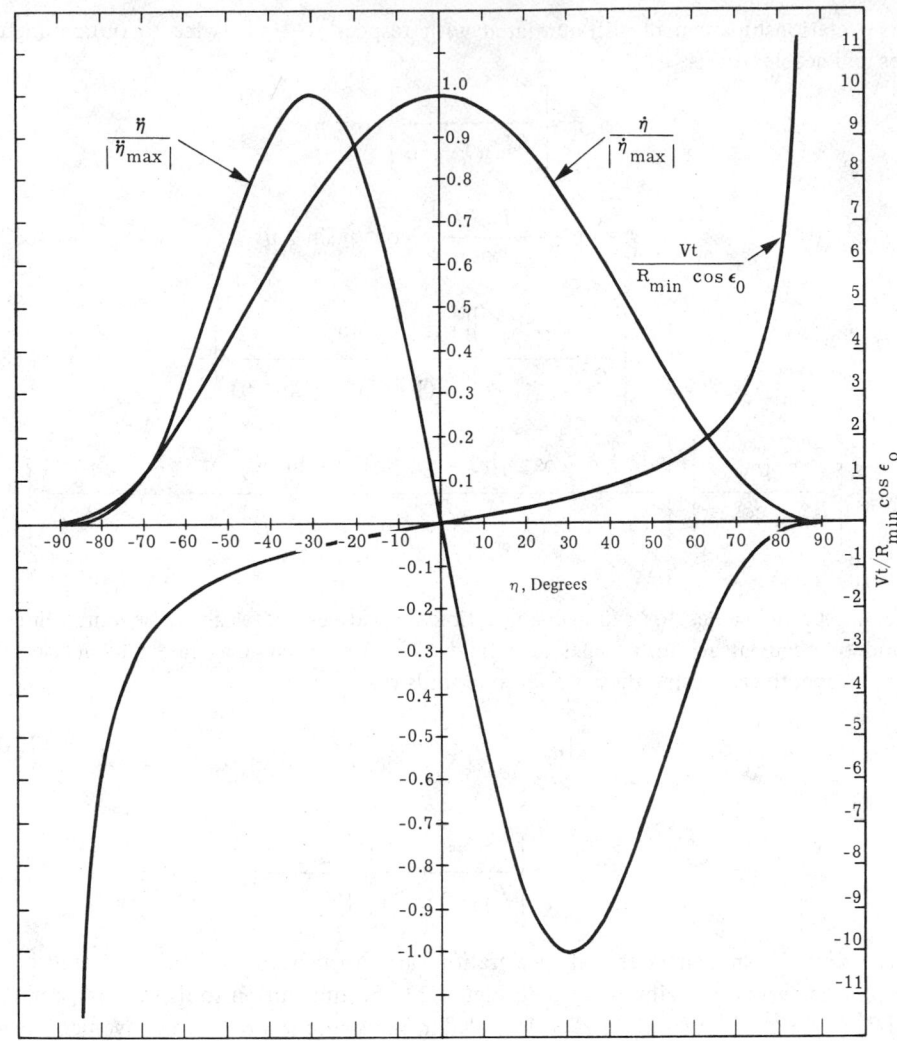

Fig. 22-15. Fly-by azimuth tracking parameters.

Table 22-6. Maximum Fly-by Elevation Angular Rates and Accelerations

ϵ_o	$\eta_{\dot\epsilon_{max}}$	$\dfrac{|\dot\epsilon_{max}|}{[V/R_{min}\cos\epsilon_o]}$	$\eta_{\ddot\epsilon_{max}}$	$\dfrac{|\ddot\epsilon_{max}|}{[V/R_{min}\cos\epsilon_o]^2}$
15°	± 35.9°	0.0985	0°	0.2500
30°	± 37.9°	0.1829	0°	0.4330
45°	± 41.5°	0.2381	0°	0.5000
60°	± 47.4°	0.2460	0°	0.4330
75°	± 57.7°	0.1810	0°	0.2500

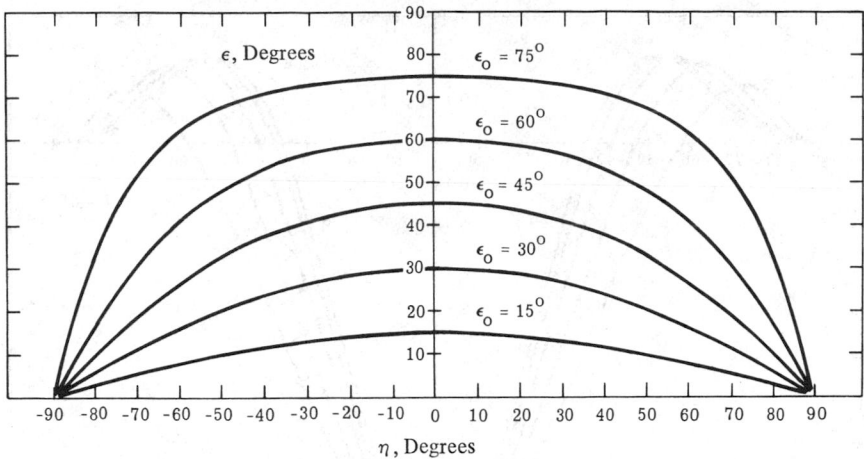

Fig. 22-16. Fly-by elevation angle relationships.

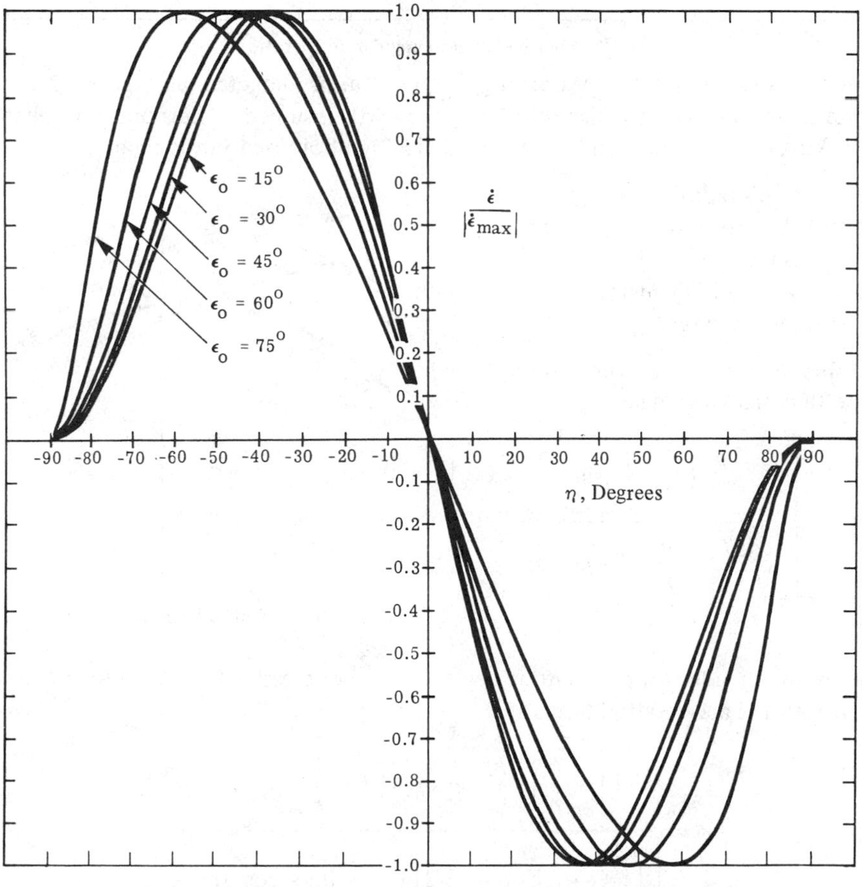

Fig. 22-17. Fly-by elevation angular rates.

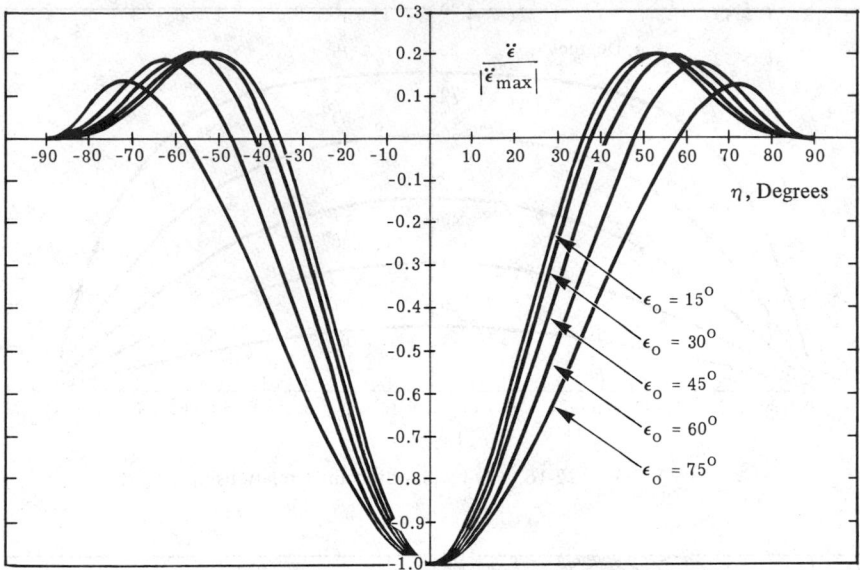

Fig. 22-18. Fly-by elevation angular accelerations.

Another example, satellite tracking from earth, is depicted in Figure 22-19 for a satellite in an idealized circular orbit with the earth assumed to be a uniform sphere and the tracking origin in the plane of the orbit. The symbols used in the figure are

r = earth's radius
h = orbit altitude
ρ = orbit angle
ϕ = tracking LOS angle
R = range to target

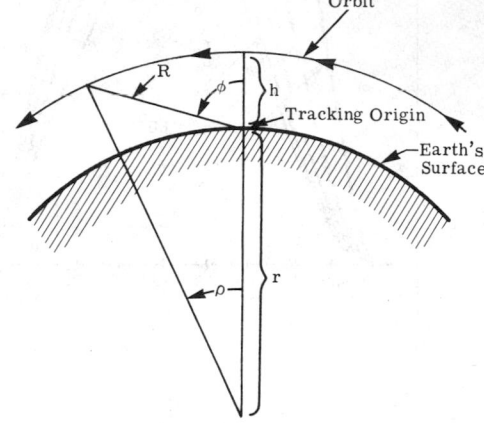

From this geometry, the expression obtained for the LOS tracking angle is

$$\phi = \sin^{-1}\left[\frac{\left(1 + \dfrac{h}{r}\right)\sin \rho}{\sqrt{2\left(1 + \dfrac{h}{r}\right)(1 - \cos \rho) + \dfrac{h^2}{r^2}}}\right] \quad (22\text{-}12)$$

Fig. 22-19. Idealized circular orbit relationships.

If the earth's angular-rotation rate is neglected, the expressions obtained for the tracking angular rate and acceleration terms are

$$\dot{\phi} = \frac{\dot{\rho}\left(1 + \dfrac{h}{r}\right)\left[\dfrac{h^2}{r^2}\cos \rho - \left(1 + \dfrac{h}{r}\right)(1 - \cos \rho)^2\right]}{\left[\left(1 + \dfrac{h}{r}\right)\cos \rho - 1\right]\left[2\left(1 + \dfrac{h}{r}\right)(1 - \cos \rho) + \dfrac{h^2}{r^2}\right]} \quad (22\text{-}13)$$

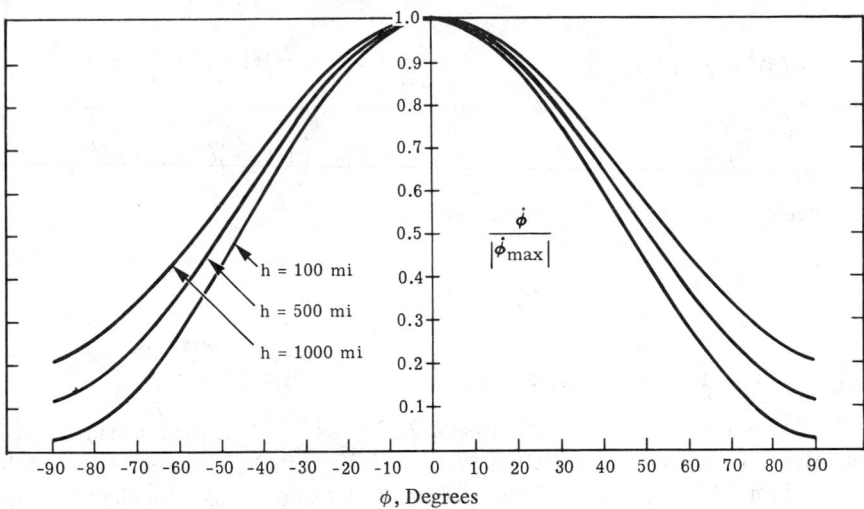

Fig. 22-20. Orbital angular rates.

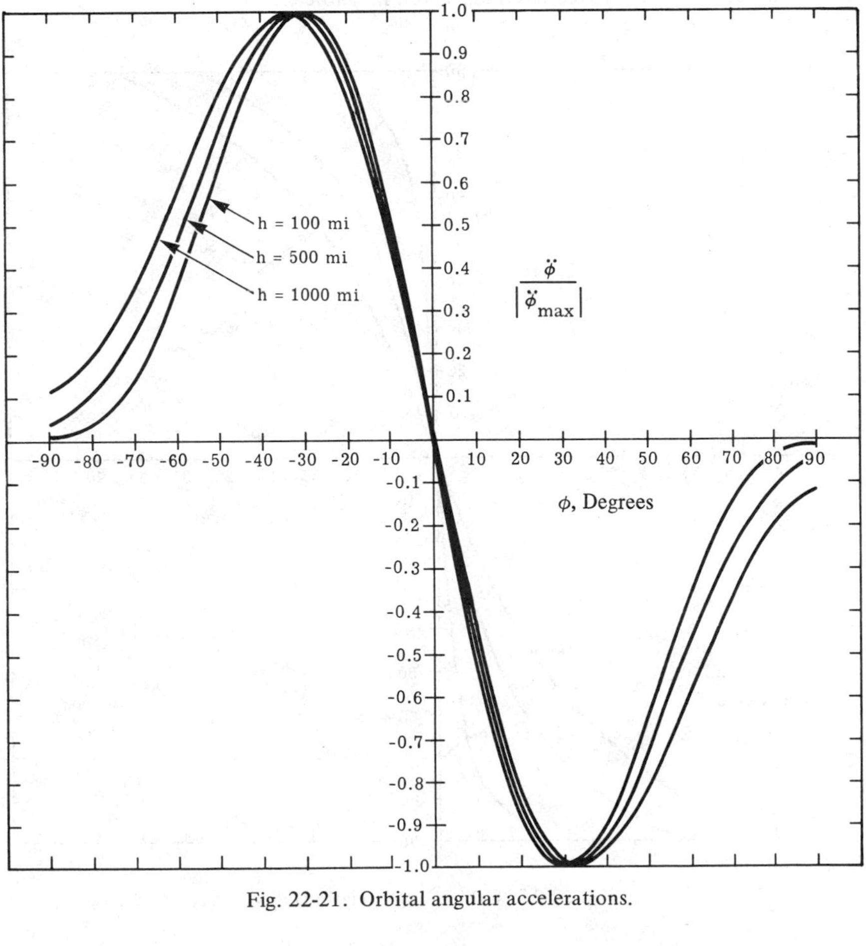

Fig. 22-21. Orbital angular accelerations.

and

$$\ddot{\phi} = \cfrac{-(\dot{\rho})^2 \sin \rho \left(1 + \cfrac{h}{r}\right)}{\left[\left(1 + \cfrac{h}{r}\right)\cos \rho - 1\right]} + (\dot{\phi})^2 \sin \rho \cfrac{\left[\cfrac{h^2}{r^2} + 2\left(1 + \cfrac{h}{r}\right)(1 - 2\cos \rho) + 2\right]}{\left[\cfrac{h^2}{r^2}\cos \rho - \left(1 + \cfrac{h}{r}\right)(1 - \cos \rho)^2\right]} \qquad (22\text{-}14)$$

where $\dot{\rho} = \left(\cfrac{1}{1 + \cfrac{h}{r}}\right)\sqrt{\cfrac{\mathcal{G}}{r + h}}$

$\mathcal{G}$ = earth's gravitational acceleration

With 3962 miles for the earth's radius and 32.2 ft sec^{-2} for $\mathcal{G}$, the normalized tracking angular rates and accelerations plotted in Figures 22-20 and 22-21, respectively (p. 22-31), are obtained for 100, 500 and 1000 mile orbits. A plot of ϕ versus ρ for the three cases is given in Figure 22-22. The peak angular rate and acceleration values obtained along with the LOS angles at which they occurred are given in Table 22-7.

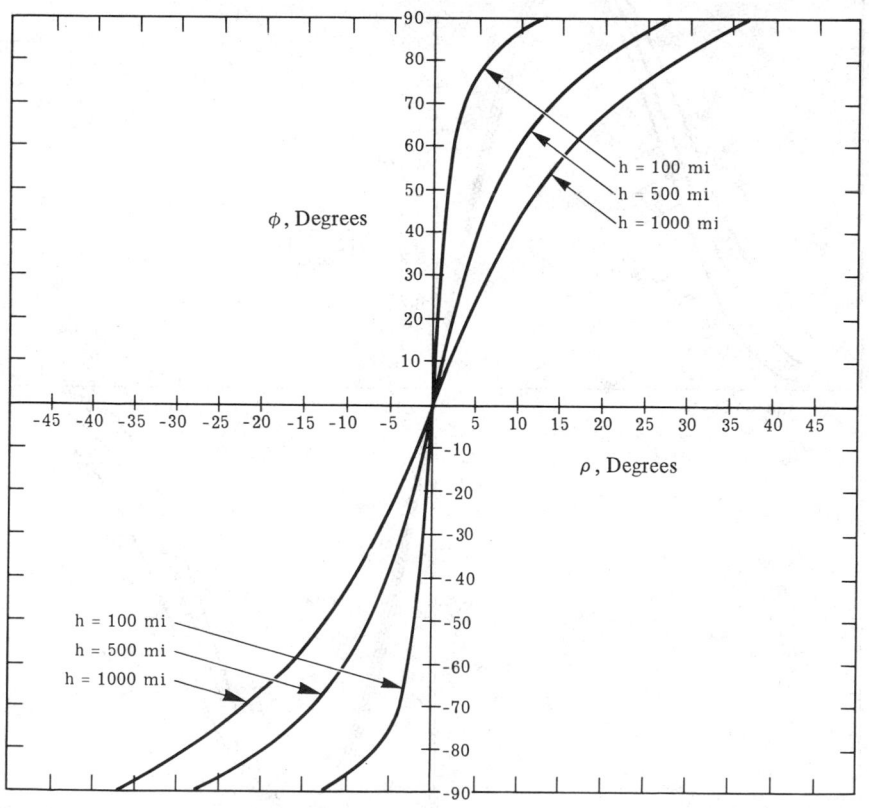

Fig. 22-22. Orbital angle relationships.

Table 22-7. Orbital Angular Rate and Acceleration Values

h (mi)	$\lvert \dot{\phi}_{max} \rvert$ (rad sec^{-1})	ϕ (at $\dot{\phi}_{max}$) (deg)	$\lvert \ddot{\phi}_{max} \rvert$ (rad sec^{-2})	ϕ (at $\ddot{\phi}_{max}$) (deg)
100	4.855 (10^{-2})	0	1.492 (10^{-3})	± 29.65
500	9.264 (10^{-3})	0	4.955 (10^{-5})	± 31.79
1000	4.392 (10^{-3})	0	1.005 (10^{-5})	± 34.07

The orbital velocity is given by $\dot{\rho}$ times the quantity $(r + h)$. If the velocity and the orbit altitude are related to $V R_{min}^{-1}$ for the fly-by situation, the peak angular-rate values given by the fly-by and the circular orbit relationships will match. In this case, the peak angular-acceleration values given by the fly-by relationships are all somewhat larger than those obtained from the orbit relationships (i.e., 3%, 12% and 24% higher for the 100, 500 and 1000 mile cases, respectively). For the fly-by case, the angular acceleration peak occurs at tracking angles of ±30° which is reasonably close to the values obtained for all three circular orbit cases.

22.3.2. Servo Aspects. Although higher-order control-loops have been proposed and implemented, typical angle tracking loops are either Type-1 or Type-2 servos. Type-1 loops, either with or without integral compensation to augment the basic low-frequency performance, are the most common.

In the steady state, Type-1 loops yield a constant following-error for constant LOS angular-rate inputs. Similarly, Type-2 loops yield constant following-errors for constant LOS angular-acceleration inputs. If the track-loop bandwidth is relatively high with respect to the variation frequencies for nonconstant inputs and integral networks are not utilized, preliminary estimates of the tracking performance can be obtained by assuming that the instantaneous following-error will correspond with the steady-state value obtained by using the instantaneous value of the input. For example, given a simple azimuth fly-by tracking situation and a track-loop open-loop bandwidth that is relatively high compared to a frequency which is on the order of $V R_{min}^{-1}$ Hz, the following error-time response for a Type-1 track will approximately correspond with the changing azimuth LOS angular-rate input divided by the loop-velocity constant. Likewise, the following-error response for a Type-2 loop will approximately correspond with the response obtained by dividing the changing LOS angular-acceleration input by the track-loop acceleration constant.

The velocity constant for continuous Type-1 loops is given by the open-loop unity-gain frequency in rad sec^{-1}. For Type-2 loops, the acceleration constant is proportional to the square of the open-loop unity-gain frequency. In this case, the loop is stabilized with a lead network that is typically such that the open-loop unity-gain frequency corresponds with the geometrical mean of the lead-network zero- and pole-frequencies. If this is done, the closed-loop poles consist of a quadratic and a single-order term all located at the open-loop unity-gain frequency. The loop acceleration-constant and the damping ratio for the quadratic term are given by

$$K_a = \omega_c^2 K^{-1/2} \tag{22-15}$$

and

$$\zeta = \frac{K^{1/2} - 1}{2} \tag{22-16}$$

where ω_c = the open-loop unity-gain frequency, rad sec^{-1}
K = the ratio of the lead-network pole and zero frequencies

Depending on the type of tracking sensor employed, the tracking-error information may be modulated in amplitude or frequency, or contained in a sampled data signal. If the error signals are AM or FM, phase information, as well as the modulated signal, is usually required to demodulate the error signals and recover the tracking-error information. For these types of error sensors, the bandwidth limitations imposed by the modulation are usually determined by noise considerations. That is, the bandwidth is typically limited to frequencies considerably lower than the modulation frequency to provide filtering to attenuate residual carrier-ripple and demodulation-noise terms.

Control system stability and transient response considerations do impose fundamental bandwidth limitations for sampled data tracking loops. For a simple Type-1 sampled-data loop (i.e., a single integrator with gain, a sampler and a zero-order hold), the loop is unstable if the open-loop unity-gain frequency for just the integrator with the selected loop-gain is greater than the sampling frequency divided by π. To achieve suitable transient response characteristics, one usually limits the open-loop bandwidth of sampled-data loops to frequencies less than the sampling frequency divided by 2π. In many cases, the sampled data aspects of the control loop problem are further complicated by the fact that the sampling rate is not periodic. This situation occurs when the measurement is obtained at a time that is a function of the tracking error at that time. If this is the case, a rigorous analysis is usually impossible and the loop design is determined either by simulation or by assuming the worst-case data-rate.

Since, in general, tracking loops include either the closed stabilization loop-dynamics, or the gimbaling and drive dynamics when a stabilization subsystem is not utilized, these dynamics must be taken into account in the track-loop design. If the tracking sensor carrier-frequency or sample data-rate is sufficiently high, the track-loop bandwidth may be limited by stability and transient response considerations that arise due to these dynamics, rather than tracking sensor related bandwidth limitations. For example, the bandwidth of an ideal Type-1 track-loop with a minor rate loop is typically limited to open-loop bandwidth values that are less than half the bandwidth of the closed minor loop.

In the case of gated video trackers with a tracking gate that is smaller than the tracking sensor field of view, bandwidth limitations imposed by the stabilization loop or drive dynamics can be eliminated, to some extent. This is accomplished if the basic track-loop positions the gate within the sensor field of view and the pointing assembly is commanded to locate the sensor LOS so that the gate is in the center of the field of view.

22.3.3. Tracking Sensor Angle Noise. Tracking sensor angle-noise occurs because received-signal fluctuations, and additive noise introduced within the tracking sensor and processing electronics, are converted to erroneous angle information. The mechanisms for this conversion are dependent on the type of tracking sensor and signal processing techniques utilized.

Rigorous characterization and analysis of the effect of angle noise can be tedious and time consuming. In many analyses, however, a rms angle-noise is calculated using a video bandwidth associated with the sensor information processing, and it is assumed that the spectrum of the angle noise is flat over this bandwidth. The resulting rms tracking jitter is then given by

$$\theta_j = \left(\frac{\Delta f_s}{\Delta f_v}\right)^{1/2} \theta_{AN} \tag{22-17}$$

where $\quad \theta_j$ = angular tracking jitter
θ_{AN} = angular noise input
Δf_v = video bandwidth
Δf_s = servo noise-equivalent-bandwidth (closed loop)

If the tracking loop is a sampled-data servo, the situation is more complex due, among other things, to the occurrence of frequency folding. A rigorous analysis of noise for sampled-data loops involves the use of pulse spectral-density functions and Z-transforms of the appropriate system transfer-functions [22-13]. However, if the input angle-noise is uncorrelated at the moments of sampling and the loop is a simple Type-1 servo, the rms angular-jitter at the instants of sampling is given by

$$\theta_{j_i} = \left(\frac{f_c}{\dfrac{f}{\pi} - f_c}\right)^{1/2} \theta_{AN} \tag{22-18}$$

where θ_{j_i} = angular jitter at the sampling instants
f_c = servo open-loop unity gain frequency
f = sampling rate

And the rms value of the jitter for the continuous time function is

$$\theta_{j_c} = \left(1 - \frac{2\pi f_c}{3f}\right)^{1/2} \theta_{j_i} \tag{22-19}$$

Although these relationships are only exact for the simple Type-1 case, they are relatively good approximations for more complex loops.

22.3.4. Sensor Angle Error Characteristics. A generalized angle error-characteristic for tracking sensors is linear over some region about the null point, and then flattens out and ultimately becomes zero for increasing tracking errors. For a linear or proportional tracking control-loop, the system design is predicated on operation in the linear region after the target has been acquired. This requirement gives rise to a fundamental trade-off involved in tracking system design. That is, both the extent of the linear region and the characteristics of the sensor angle-noise are a function of sensor parameters, such as the angle subtended by the detector, the optical blur size, or the sensor field of view.

Unfortunately, the sensor angle-noise usually increases as the linear region is extended. Thus, the designer, in most cases, is driven toward minimizing the linear region. In dynamic tracking situations, however, the following-errors which are to be kept within the linear region are a function of the tracking loop bandwidth. Again, the angle noise tends to increase if the data rate is increased to realize higher bandwidth tracking loops. The designer's task is then to determine a suitable compromise with regard to these system parameters.

System considerations of importance with regard to the sensor angle error-characteristics include scale factor stability and initial acquisition. If excessive variations in the angle error scale factor occur due for example, to target image-size changes or received-signal variations, the track-loop design may be restricted to a simple Type-1 servo, since Type-1 servos with integral network or higher-order servos are either marginally or conditionally stable. A similar problem exists if the saturated portion of the error characteristic is utilized for initial acquisition. In this case, it may be necessary to *acquire* with a simple Type-1 loop and then switch to a more complex configuration once the acquisition error is reduced to an acceptable value. It should also be noted that use of the saturated portion for acquisition imposes an intrinsic limit on the system acquisition angular slew-rate.

A special class of tracking error sensors exist that do not have any linear region. For these types of sensors, the error information is "bang-bang" and yields only sense and not magnitude. In this case, the track-loop causes the system to limit-cycle about the null point and the magnitude of the limit-cycle is determined by the maximum drive signal provided. Steady-state limit-cycling can be eliminated by introducing a dead zone about the null point. If this technique is used, the system will still exhibit some sort of non-symmetrical limit-cycle when tracking LOS angular-rates. As was the case for the linear system with a saturated error characteristic, the angular slew rate is limited for these types of systems unless some method and rationale for modifying the maximum drive signal is provided.

22.3.5. Recursive Filtering and Estimation. Various types of recursive filters have been proposed and implemented for angle tracking systems. In general, recursive filters yield time-varying gain-factors that are a function of the tracking-geometry model selected and the various statistics germane to the tracking problem. Filter formulation is usually based on minimizing the mean-squared error in some sense. A rigorous formulation of the Kalman filter is the most sophisticated and accurate recursive filter. However, since it is also the most complex and costly to implement, use of other simplified alternatives is often of interest [22-14].

The fundamental output of the recursive filter is in the form of state estimates for the system. In fire control systems, for example, where the tracking system serves to provide measures of the target-LOS angular-rates, these estimates can be used directly. The filter state estimates can also be combined in some fashion and used to aid the tracking system. In this case, commands to the tracking system are derived from the filter state-estimates and the filter serves to improve the tracking performance. The application of Kalman filters to angle-tracking systems has been investigated extensively [22-15 and 22-16] , and digital implementations of the derived filter structures have been incorporated in operational tracking systems.

22.4. Tracking Loop Performance

22.4.1. Deterministic Analysis. Basic techniques for frequency and transient-response analysis are appropriate for determining system-stability and performance characteristics when the system is linear, and deterministic inputs or disturbances are specified. Frequency response analysis characterizes the system input-output relationships in terms of sinusoids, and transient response analysis yields input-output time responses. If the inputs can be expressed in terms of standard transforms, closed-form expressions for the output transforms can be obtained directly. For those cases, such as fly-by LOS dynamics, where the input cannot be expressed in terms of closed-form transforms, the performance is often evaluated by use of power-series expansions in system dynamic error-coefficients and the appropriate input time-derivatives [22-17] .

Virtually all physical systems contain nonlinear elements such as saturation, coulomb friction, hysteresis, backlash, threshold and quantization. These nonlinear elements may cause system limit cycling or large signal-stability problems. Performance analysis techniques that are appropriate with nonlinear elements include describing-function analysis, phase plane (or space) analysis, and time-response simulation analysis. In many cases, linear analysis is adequate for small signal conditions and nonlinear analysis is only required to determine large signal or acquisition responses.

22.4.2. Statistical Analysis. For linear, time-invariant systems, the response to stationary random processes is usually obtained by utilizing the power spectral density function for the input variable and the appropriate system transfer function to derive the mean square response [22-18] . The input power-spectral-density function is the Fourier transform of the input autocorrelation function and represents the statistical characteristics of the random input in the frequency domain.

Although statistical input-output relationships have been derived for various nonlinear processes, these cases are rather specialized. Statistical approximations with some generality are given in Landau and Leondes. The use of a time-response simulation is usually the most practical method for tracking loop analysis when system nonlinearities come in to play. In this case, the inputs are represented by random time series with the appropriate statistics, and a series of Monte Carlo runs are made.

The signal-to-noise ratio required for acquisition, and the probability of loss of track are tracking system parameters that are determined by statistical analysis. The signal-to-noise ratio for acquisition is related to detection false-alarm rates, and probability of loss of track is related to the tracking angle noise statistics. Both parameters are dependent on the acquisition or tracking-loop response characteristics and the temporal statistics of the appropriate noise parameters. In many cases, the noise pulse rate (resulting from band-limited white-noise) and a reference time-period are assumed, in order to derive threshold and signal-to-noise ratios corresponding to achieving the required probability-of-detection for acquisition, or the probability that the angle noise is less than the output value of the full-scale angle error, which is assumed to produce loss of track.

22.4.3. Computer Simulation Techniques. Both conventional analog and digital techniques are suitable for simulations of system-dynamic time responses. For digital simulations, two basically different programming approaches can be considered. The first is to solve the differential equations in a way similar to that used in an analog computer in which the system differential equations are formulated in terms of interconnected integrators. If this approach is taken, a numerical integration routine and some type of error evaluation criteria are needed for each integrator. For wide-band systems, the computation intervals may be very small, and hence many computations will be necessary per real-time interval. A number of simulation programs implemented by utilizing this technique are commercially available.

The second approach is to solve the differential equations in terms of piecewise-continuous time response expressions with constant forcing functions and initial condition inputs. If this approach is followed, the intervals between calculations still have to be consistent with the system control loop and input signal bandwidths as well as with any system nonlinear functions. However, for sampled data control loops with internal minor loops that can be represented in closed-loop form, this approach is particularly attractive. In most cases, the time response calculations can be minimized to one calculation per sample interval. A disadvantage of this approach is that rather minor modifications of the system dynamics may require changes in the time-response expressions used, rather than just a change in the gain of one integrator.

22.5. Reticle Trackers

The basic arrangement of a reticle tracker is illustrated in Figure 22-23. The telescope collects electromagnetic radiation and focuses an image of the field of view upon the reticle which contains a spatial pattern of varying optical transmission. Relative motion between the image and the reticle modulates the total amount of radiation transmitted through the reticle. This radiation is collected and deposited on a detector which produces an electrical signal proportional to the amount of incident radiant power. The electronics amplify the detector signal and demodulate it to recover an error signal which represents the location of the target within the field of view. The error signal is fed back to point the telescope so that the magnitude of the error is reduced. The zero-error target position is usually the telescope field of view center.

Reticles normally serve two functions. In trackers, the primary function is to modulate the detector signal with information indicating the target position. Reticles also perform spatial filtering, as described in Chapter 17, suppressing signals from large background objects relative to the signals from small target sources. Reticle trackers are inherently point-target trackers; they function best when tracking sources which subtend angles that are small relative to the field of view, and primarily are used in such applications.

Tracker electronics usually contain low-noise amplification to bring the detector signal up to a practical working level, bandpass filtering to retain target position modulation and minimize random noise, AGC (automatic gain control) to maintain a constant signal level, and demodulation to extract the error signal. Systems utilizing FM for target position encoding can be designed with limiters and no AGC; but the use of AGC has the advantage of permitting strong target signals, when they occur, to suppress background signals and thereby provide improved tracking performance.

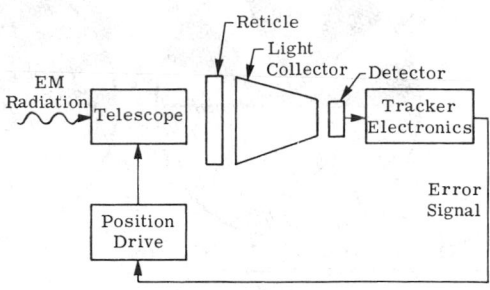

Fig. 22-23. Basic reticle tracker.

There are many variations possible in designing a reticle tracker. The relative motion between the image and the reticle can be provided by reticle motion, by telescope motion, or both. Multiple reticles can be used, and there is a virtually unlimited number of possible reticle patterns. Descriptions of various approaches are contained in References [22-1] and [22-19]. The remainder of this section will treat only two common types of reticle trackers: rotating-reticle, and stationary-reticle trackers. An extensive discussion of reticles and their applications may be found in Biberman [22-20].

22.5.1. Rotating Reticles. Systems with rotating reticles are common among infrared trackers, due to the simplicity of the concept and relative ease of mechanization. The basic reticle consists of transparent and opaque regions which are at least as large as the image of a point target over most of the reticle; it is impossible to maintain this desirable size relationship near the reticle center. Reticle rotation produces an ac signal as the target image is chopped by the reticle. To be useful in a tracker, the reticle pattern must be asymmetric so that the target position is unambiguously encoded as detector signal-modulation. The minimum information required for tracking is the target polar angle, i.e., the direction of the target from the center of the tracker field of view. In addition, it is desirable to have a signal indicating the target radius.

Examples of rotating reticles are shown in Figure 22-24. The prototype of rotating reticles is the "rising sun." It contains a target-sensing sector and a phasing sector, each sector usually occupying half of the reticle. The target-sensing sector contains N pairs of transparent and opaque spokes which chop the target image as the reticle rotates, producing N pulses per half rotation-period. The phasing sector introduces asymmetry into the signal modulation, encoding the target polar angle as the phase of the pulse-burst envelope. The transmission of the phasing sector is 0.5, matching the average transmission of the target-sensing sector, thereby minimizing the signal modulation caused by background objects of moderate size. The rising sun reticle generates no target radius information, except possibly near the center; here the target image overlaps more than one transparent spoke, and the signal amplitude is reduced as the image approaches the center.

Although the target position is encoded as the phase of the pulse-burst envelope, the rising sun type of reticle is usually referred to as an AM reticle. This misnomer is due to the fact that AM signal processing is used to recover the target-position signal. A predetection filter transmits the carrier and first pair of sidebands, which combine to form an AM signal with modulation index $M = 4/\pi$. Then an AM detector is used to obtain the pulse-burst envelope, followed by a postdetection filter.

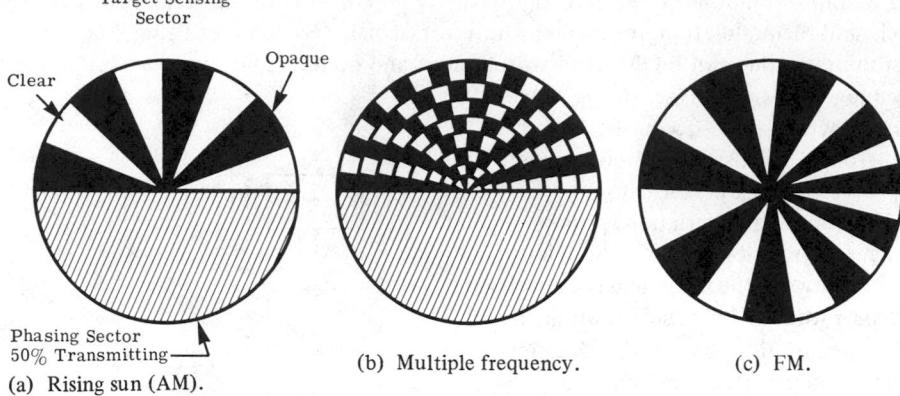

Target Sensing
Sector

Opaque

Clear

Phasing Sector
50% Transmitting

(a) Rising sun (AM).

(b) Multiple frequency.

(c) FM.

Fig. 22-24. Rotating reticle examples.

The multiple frequency reticle [22-1], provides both polar-angle and quantized, radial, target-position information. It is an extension of the rising sun reticle, obtained by partitioning the target-sensing sector into several rings, each with a different number of elements. This pattern gives a variable number of pulses in the pulse burst, depending upon the target radius. The radial information can be extracted by using a set of bandpass filters to determine which ring contains a target. This approach provides improved resolution among multiple targets if their radial positions correspond to different reticle rings.

The FM reticle has N-pairs of spokes, the widths of which vary around the reticle. As the reticle rotates, an FM signal is generated. The carrier frequency is N times the reticle rotation frequency, and the modulation index is determined by the amount of variation in the widths of the reticle spokes. The phase of the modulation corresponds to the target polar angle; as with the rising sun reticle, radial information is not generated.

Typical tracker electronics for rising sun and FM reticles is shown in Figure 22-25. The first bandpass filter, H_1, transmits the carrier frequency and enough sidebands to preserve the target position information. AGC is used to maintain a nearly constant signal level despite a usually large dynamic range of target irradiance. An AM demodulator is used with the rising sun reticle, and an FM demodulator with the FM reticle. The postdetection bandpass filter, H_2, is centered about the modulation frequency, which is the reticle spin-frequency. The bandwidth should be minimized to reduce output noise, but should be wide enough to respond to tracking transients. The phase detector measures the phase of the signal modulation with respect to the reticle phase reference, and generates the tracking error signal.

In the case of the multiple frequency reticle, the tracker electronics are functionally equivalent to parallel AM channels. There is one channel for each reticle ring, with the predetection bandpass filter tuned to the carrier frequency of that ring. Since a single error signal is required, the parallel channels must be combined. This may be done at any point between the output of the predetection filter and that of the phase detector.

The power signal-to-noise ratios at the output of the postdetection filter for AM and FM systems are derived in Reference [22-21], and asymptotic expressions for small and large signals are given in Table 22-8.

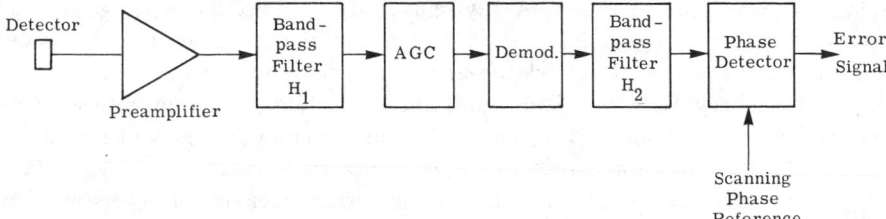

Fig. 22-25. Rotating reticle tracker electronics.

Table 22-8. Asymptotic Expressions for Output
Signal/Noise Power Ratio [22-21]

Modulation	Small Signal	Large Signal
AM	$\left[\dfrac{M}{2\pi}\right]^2 \dfrac{\gamma^2}{\pi^2 f_R b_2}$	$\left[\dfrac{M}{2\pi}\right]^2 \dfrac{\gamma}{b^2}$
FM	$\dfrac{\gamma^2}{\pi^4 \beta f_R b_2}$	$\left[\dfrac{\beta}{\pi}\right]^2 \dfrac{\gamma}{b_2}$

$\gamma = [D^*\Phi_o]^2/A_d$

Φ_o = radiant power in target image

D^* = detector specific detectivity

A_d = detector area

M = AM modulation index $(= 4/\pi)$

β = FM modulation index

f_R = reticle rotation frequency

b_2 = H_2 noise equivalent bandwidth

These expressions apply as long as the target image is small compared to the width of the reticle spokes at the target radius. As noted earlier, the signal-to-noise ratio vanishes for both AM and FM rotating reticle systems as the target radius approaches zero.

It is seen from Table 22-8 that FM systems, compared to AM systems, have one more parameter which determines system sensitivity—the FM modulation index, β. The value of β is determined by the reticle pattern. FM systems with β less than $\pi^2/4$ have better small-signal performance than comparable AM systems. Decreasing β degrades the large-signal performance, but frequently this is an acceptable tradeoff. Threshold extension techniques for FM systems [22-22] can be used to improve the small signal performance.

A characteristic which often is important is the equilibrium tracking point (ETP) when multiple targets are present in the field of view. An ETP is a point in a stationary scene such that zero error-signal results when the tracker points at it. A particular tracker and scene combination may produce more than one ETP. Precise results on the ETP characteristics of rotating reticle trackers require analysis of each particular case, but some generalizations have been observed. AM systems tend to average the signals generated by multiple targets, producing an ETP at the centroid of the several sources; FM systems have a capture effect and tend to track the strongest target in the field of view.

22.5.2. Stationary Reticles. Two significant limitations of rotating-reticle trackers are loss of carrier signal for zero tracking-error and, generally, the lack of target radius information. The zero-error carrier loss causes increased tracking noise with on-axis targets and is especially troublesome for high precision tracking. Trackers with only target polar-angle error signals are "bang-bang" nonlinear trackers, and exhibit relatively large transient errors in the presence of target motion or other disturbances. Both of these limitations are overcome by stationary-reticle trackers.*

A prototype stationary reticle, commonly designated the wagon wheel reticle, is illustrated in Figure 22-26. The reticle consists of N pairs of spokes, and has a radius corresponding to a field angle R_R. Nutation scanning is used, in which the image of a point source is moved around a circle of fixed radius R_N at the nutation frequency f_N. The center of the nutation circle corresponds to the target position in the field of view; the nutation circle of an on-axis target is concentric with the reticle.

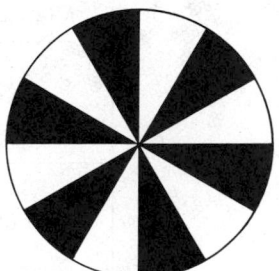

Stationary reticle systems incorporate a wide variety of reticle patterns and tracking electronics which are designed for specialized applications. However, the tracking of point targets by such systems is usually very similar to the wagon wheel system, which has the virtue of analytical amenability. The present discussion therefore concentrates on wagon wheel reticles, but also is applicable (in a qualitative sense) to other reticle patterns.

Fig. 22-26. Wagon wheel reticle.

The target image generates a pulse for each pair of spokes it crosses, with equal pulse-widths and interpulse periods for an on-axis target. As the target position moves off-axis, the pulse-widths and periods vary over the nutation cycle, producing a recoverable modulation which accurately indicates the target radius and polar angle. For tracking errors that are small relative to the nutation radius, the modulation is pure FM, i.e., the detector signal is a carrier of frequency Nf_N which is frequency modulated by sinusoidal-modulating signal of frequency f_N with amplitude proportional to the tracking error. The modulation phase, with respect to the nutation reference, is determined by the target polar angle.

As the error increases up to $R_R - R_N$, at which value the nutation circle reaches the edge of the reticle at some time during the nutation cycle, the FM is fairly well behaved. Up to this point the frequency deviation increases with increasing error, with the modulation waveform exhibiting an increasing third-harmonic component. The instantaneous frequency over the nutation cycle typically behaves as shown in Figure 22-27. If the tracking error normalized to R_N is designated ϵ, then the FM modulation index β is given by[†]

---

*Although usually accomplished by a stationary reticle and rotating optics, the same relative motion between the image and the reticle can be achieved by suitable reticle motion.

[†]The standard definition of β is used to derive this expression, i.e., $\beta = (f_{max} - f_{min})/2f_N$, even through the modulating waveform is not exactly sinusoidal.

$$\beta = \pi/2 \ \{\cot^{-1} \ [(1 - \epsilon) \tan (\pi/N)]$$
$$- \cot^{-1} \ [(1 + \epsilon) \tan (\pi/N)]\}$$
$$(22\text{-}20)$$

For large N this is approximately $\beta \simeq N\epsilon/(1 - \epsilon^2)$.

Figure 22-28 contains curves of maximum β, which occur for $\epsilon = (R_R - R_N)/R_N = \epsilon_1$, plotted versus R_R/R_N, with the number of spoke pairs, N, as a parameter. The bandwidth of the modulated signal is approximately $2\beta f_N$ [22-23], so the maximum β determines the bandwidth required of the H_1 filter.

In a stationary reticle tracker, the choice of system parameters involves a number of compromises. Large β improves the signal-to-noise ratio of the demodulated signal, but also corresponds to a large minimum trackable signal (the threshold effect of FM systems). A large number of reticle spokes is often desired for rejection of large background objects, but N is limited by the resolution of the sensor optics and the largest size of target image, as well as by reticle fabrication technology. The extent of the linear tracking field is proportional to $(R_R/R_N) - 1$.

The stationary reticle with nutation scanning can be designed to track targets with radial errors, ϵ, in excess of ϵ_1, the error corresponding to maximum β. As seen in Figure 22-29, a signal is produced for errors up to $(R_R + R_N)/R_N = \epsilon_2$. However, for target radii in the interval $\epsilon_1 < \epsilon < \epsilon_2$, the target image is off of the reticle during part of the nutation cycle, and the signal modulation departs significantly from FM. The demodulator output typically drops to small values in this region, and may even change sign, leading to unstable operation. Efforts to extend the lin-

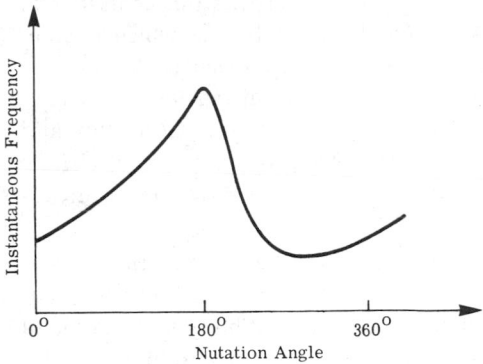

Fig. 22-27. Instantaneous frequency for wagon wheel stationary reticle.

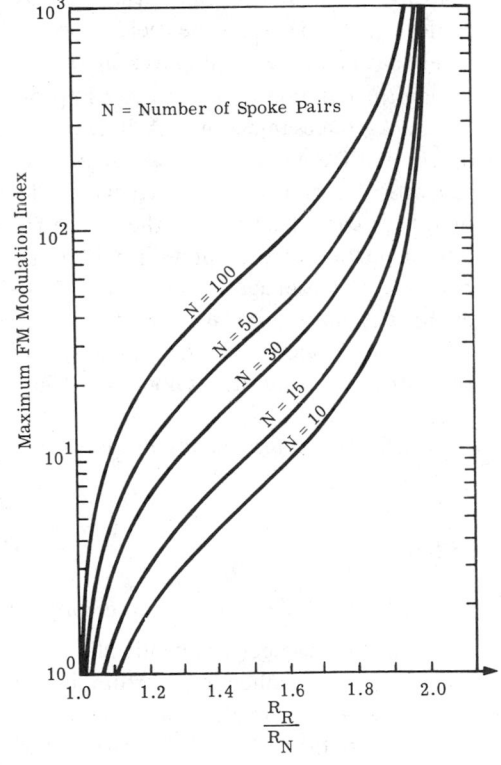

Fig. 22-28. Maximum FM modulation index for wagon wheel reticle.

ear tracking field beyond ϵ_1 by special demodulation techniques have been unsuccessful. The best performance that seems to be attainable is to produce a demodulated error signal that is approximately constant in magnitude over $\epsilon_1 < \epsilon < \epsilon_2$, and equal to the error signal corresponding to ϵ_1, as shown in Figure 22-30. To do this generally requires modification of the central region of the reticle pattern so that the target image produces strong pulses as it crosses this central area. A demodulation technique which is compatible with this approach is slope detection [22-23]. Careful and coordinated design of the reticle pattern, AGC circuit, and demodulator can produce a tracker characteristic similar to that shown in Figure 22-30.

Extension of the tracker field as described above incurs a performance penalty in tracker susceptibility to background sources. If such an extension is not incorporated, an optical field stop can be used to block incoming rays from sources that are removed from the optical axis by angles exceeding ϵ_1. The field stop ensures that no image point leaves the reticle during the nutation cycle, thereby improving background rejection. A background object with even moderate gradients will generate large interfering signals if the image (or part of it) crosses the reticle for only a portion of the nutation period. By contrast, if the image remains on the reticle for the entire nutation period, the reticle is much more effective as a spatial filter in suppressing signals due to distributed sources.

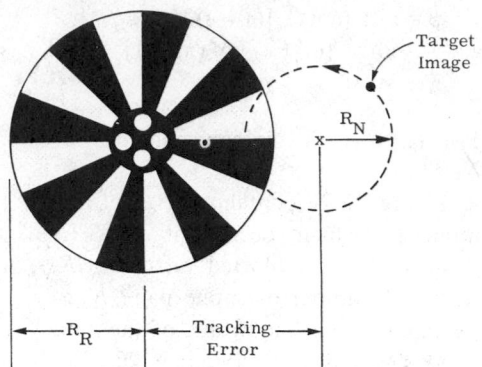

Fig. 22-29. Extended tracking field.

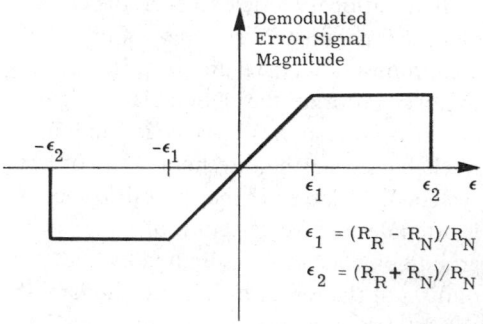

$$\epsilon_1 = (R_R - R_N)/R_N$$
$$\epsilon_2 = (R_R + R_N)/R_N$$

Fig. 22-30. Idealized extended tracker characteristic.

A general expression for noise equivalent flux density (NEFD) is given in Reference [22-1]:

$$\text{NEFD} = \frac{2\sqrt{\omega \Delta f}}{\pi D_o (\text{NA}) D^* \tau_o} \qquad (22\text{-}21)$$

where ω = the detector instantaneous field of view, sr

Δf = the electrical bandwidth
D_o = the collecting aperture diameter
NA = the numerical aperture of the optics
D^* = the specific detectivity of the detector
τ_o = the transmission factor of the optics

This expression corresponds to a unity signal-to-noise ratio prior to demodulation, i.e., at the output of filter H_1. NEFD also can be expressed as a function of parameters which are more directly related to tracker performance. In stationary reticle trackers, ω is related to the angular diameter of the tracking field of view (FOV) by

$$\omega \simeq \pi \left(\frac{\text{FOV}}{a}\right)^2 \tag{22-22}$$

where $a = \begin{cases} 2(1 - R_N/R_R), & \text{with field stop*} \\ 2(1 + R_N/R_R), & \text{without field stop} \end{cases}$

As indicated earlier, Δf is approximately $2\beta f_N$. An additional transmission factor for the reticle, τ_R, must be included in the NEFD expression. The value of τ_R is the average reticle transmittance around the image nutation circle, which is 0.5 for the wagon wheel reticle over $\epsilon < \epsilon_1$. A field lens or condensing light pipe often is placed between the reticle and the detector to permit the use of a smaller, hence less noisy, detector [22-1]. In such a case, NA is replaced by $(\text{NA})_f$, the field-lens (or condenser) numerical aperture. Combining the above and making the appropriate substitutions, one gets

$$\text{NEFD} = \frac{\text{FOV}\sqrt{\dfrac{8\beta f_N}{\pi}}}{aD_o(\text{NA})_f D^* \tau_o \tau_R} \tag{22-23}$$

The wagon wheel reticle serves well as a model for describing stationary reticle trackers, but it has a limitation which is overcome with a different pattern. Examination of the wagon wheel pattern discloses a variation in spatial filtering performance at different radial positions of the target image on the reticle. This limitation does not occur in the involute pattern, which is shown in Figure 22-31. Recalling the geometric definition of the involute of a circle—the curve described by the end of a string that is unwound from the circumference of a circle—one can see that the spoke widths and the spacing between spokes is constant over the pattern. This leads to the uniform spatial-filter performance mentioned above. This in turn allows a maximum number of reticle spokes to be used, which produces a large FM modulation index.

Fig. 22-31. Involute reticle.

The rectangular coordinates of an involute of a circle are given by

$$x = r_b(\cos\phi + \phi\sin\phi) \tag{22-24}$$

$$y = r_b(\sin\phi - \phi\cos\phi) \tag{22-25}$$

In polar coordinates (r, θ), this is

$$r^2 = r_b^2(1 + \phi^2) \tag{22-26}$$

$$\theta = \phi - \tan^{-1}\phi \tag{22-27}$$

---

*If the image is stationary and nutation is due to reticle motion, then $\omega = \pi(\text{FOV})^2/4$, or $a = 2$.

An involute reticle normally is used with a field stop so that the spatial filtering uniformity inherent in the pattern is actually realized.

22.6. Scanning Trackers

22.6.1. General Considerations. Scanning trackers incorporate one or more detectors which have instantaneous fields of view that are small fractions of the total field of view. Together, the detector fields scan the total field repeatedly, thereby transforming the scanned scene into a set of detector signals. Reference signals are also generated, derived from the scan motion, which represent the instantaneous position of each detector field of view within the total field. The signal processor identifies the target signal in the detector outputs, usually by a thresholding process, and samples the reference signals at that time to determine the target position. These position signals are then used to point the sensor so that the target is centered within the total field.

Scanning trackers have numerous performance advantages over reticle trackers. The instantaneous field of view of each detector is smaller than that of reticle trackers, giving smaller background signals. On the other hand, scanning trackers do not lend themselves to optical spatial filtering, as afforded by reticles, to suppress large background objects in favor of small targets. The burden of background rejection for scanning trackers is transferred to the signal processor.

With relatively small instantaneous fields of view, scanning trackers preserve more of the scene information in the detector signals than do reticle trackers. This permits resolution of multiple targets and selection of the desired target based upon observable criteria. For extended targets, i.e., targets which are larger than a detector instantaneous field of view, the signal processor can be designed to track a particular point on the target, such as the centroid, an edge, or some other identifiable point.

Scanning trackers offer flexibility to the system designer in cost versus performance tradeoff analyses. For a given optical design, background signals and detector noise can be reduced by using a larger number of smaller detectors. Target detectability is thereby improved so long as each detector instantaneous field of view exceeds the size of the target image. In some systems, tracking resolution is also improved. The costs of this improvement are associated with additional preamplifiers and processing electronics.

Scanning trackers naturally lend themselves to the use of digital signal processing, since the signal processors for this class of systems usually contain a substantial amount of logic functions.

Tracker Electronics. The tracker electronics for scanning trackers generally are equivalent to the block diagram shown in Figure 22-32. Each detector drives a target detection circuit, the output of which is a pulsed logic signal. The target detection circuit is designed to produce a pulse when a target is within the instantaneous field of view of the detector. Motion of the scanning mechanism is sensed by some means, and used to generate reference signals which identify the location of the instantaneous field of view of each detector throughout the scan. When a target detection circuit produces a pulse, the tracker logic samples the reference signals for the corresponding detector. These signals give the target position and may be used to point the sensor to keep the target centered in the field of view.

Tracker electronics do not necessarily correspond physically to the block diagram of

Figure 22-32. For example, some systems may use only one target-detection circuit which is time-shared among the several detector channels. Another alternative is to implement the target-detection circuits digitally. In these systems, the amplifier outputs are sampled, converted to a digital format, and filtered and thresholded by means of digital arithmetic operations.

Basic design principles apply to all scanning tracker electronics no matter how they are mechanized. The following material is organized in a way corresponding to Figure 22-32, but applies as well to different structures.

Target Detection Circuit. The essential elements of the target detection circuit are shown in Figure 22-33. The output z is a logic signal which is 1 only when the threshold detector input-signal, v, exceeds the threshold, v_T. The filter is designed to maximize the ratio of the target-signal to interfering noise plus background. Amplification is necessary to bring the relatively small target-signal well above the noise associated with the subsequent circuits. Automatic gain control, while not essential, is usually required to achieve reliable tracking over the wide range of target intensities and background conditions commonly encountered.

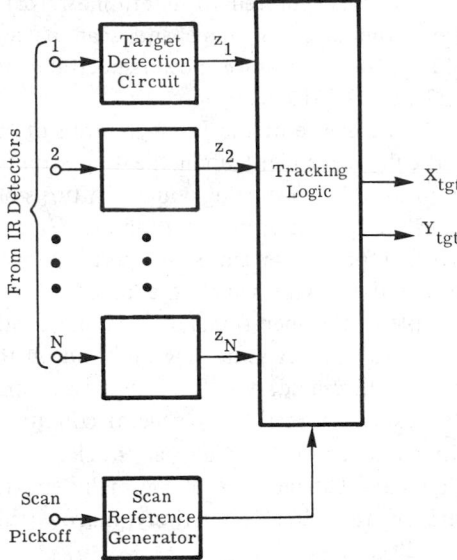

Fig. 22-32. Generalized tracker electronics for a scanning tracker.

The filter should be designed so that the ratio of the peak target-signal to the rms noise at the filter output is maximized. This is a standard problem with a well known solution: the matched filter [22-24].

To design a matched filter, one must have the following conditions at the filter input: a target signal with a known Fourier transform $S(\omega)$; and a random background plus noise with a zero mean, stationarity, and a power spectral density $W(\omega)$. Then the transfer function of the matched filter is

$$H(\omega) = k \, \frac{S^*(\omega)}{W(\omega)} \, e^{-j\omega t_O} \qquad (22\text{-}28)$$

where k = arbitrary, real constant

     $*$ = complex conjugate

     t_O = time delay equal to the duration of the target-signal at the filter input

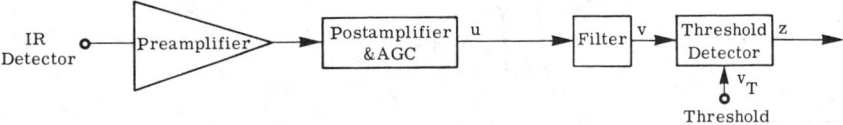

Fig. 22-33. Target detection circuit.

The data required to determine $H(\omega)$, i.e., $S(\omega)$, t_o, and $W(\omega)$ can be determined from the detector shape and scan pattern, the optical point-spread-function (or the MTF), detector-noise power-spectral-density, and the background Wiener spectrum [22-25, 22-26].

Performance of the filter, in terms of maximizing the signal-to-noise ratio, often is not critically dependent upon the fine structure of $H(\omega)$. Thus the matched filter can be used as a design starting point; if it turns out to have a complicated realization, as it often does, simplified approximations to $H(\omega)$ can be considered. This is a reasonable engineering procedure since the performance penalty is usually modest, and because the $W(\omega)$ data, with which the matched filter is determined, often contain only a small sample of the operational background conditions.

Threshold. A remaining question on the target detection circuit relates to the choice of the threshold, v_T. This requires some results on threshold crossings by stochastic processes. Under fairly general conditions, the random background plus noise at the threshold point is a Gaussian stochastic process. Electronic (thermal and shot) noise is Gaussian. The radiance of thermally emitting backgrounds tends to be Gaussian [22-27] and, if the filter is relatively narrowband, the entire interference process is Gaussian [22-28].

For a stationary, zero-mean, Gaussian process $v(t)$ with autocorrelation function $R(\tau)$, the expected number of distinct times per second that $v(t)$ exceeds a positive threshold v_T, denoted by λ, is [22-28, 22-29],*

$$\lambda = \frac{1}{2\pi} \sqrt{-\frac{R''(O)}{R(O)}} \; e^{-0.5(v_T/\sigma_v)^2} \tag{22-29}$$

where $\sigma_v^2 = E\{v(t)^2\} = R(O)$. A restriction for this result to hold is that the process be *regular*, which means $R'(O)$ exists and equals zero. If the process is *singular,* then λ is not defined, but there is an expression for the probability of one threshold crossing in a small time interval of length u [22-29], i.e.,

$$p_1(u) = \frac{1}{2\pi} \sqrt{-\frac{2R'(O+)u}{R(O)}} \; e^{-0.5(v_T/\sigma_v)^2} \tag{22-30}$$

In terms of the power spectral density $S_v(\omega)$ of the process $v(t)$, which is the Fourier transform of $R(\tau)$, λ is given by

$$\lambda = \frac{1}{2\pi} \sqrt{\frac{\displaystyle\int_{-\infty}^{\infty} \omega^2 S_v(\omega)\,d\omega}{\displaystyle\int_{-\infty}^{\infty} S_v(\omega)\,d\omega}} \; e^{-0.5(v_T/\sigma_v)^2} \tag{22-31}$$

---

*Note that Papoulis gives the rate of threshold up and down crossings, which is twice that given here.

The exponential coefficient is a measure of the $S_v(\omega)$ bandwidth. As an example, consider the ideal bandpass process with

$$S_v(\omega) = \begin{cases} S, & \omega_a < |\omega| < \omega_b \\ O, & \text{otherwise} \end{cases}$$

(22-32)

For this process,

$$\lambda \cong \frac{1}{2\pi} \sqrt{\frac{\omega_b^3 - \omega_a^3}{3(\omega_b - \omega_a)}} \; e^{-0.5(v_T/\sigma_v)^2}$$

(22-33)

If $\omega_a \ll \omega_b = 2\pi\Delta f$, then

$$\lambda \cong 0.58 \; \Delta f \; e^{-0.5(v_T/\sigma_v)^2}$$

(22-34)

Note $\sigma_v^2 = \int_{-\infty}^{\infty} S_v(\omega) \, d\omega$

$$S_v(\omega) = |H(\omega)|^2 \, W(\omega)$$

Thus given $W(\omega)$ and $H(\omega)$, one can determine λ as a function of the threshold v_T, and can choose v_T to give an acceptably-low rate of threshold crossings due to background plus noise.

The target detection performance is characterized by the probability of the target signal exceeding the threshold during one scan period. Let $\hat{v}$ denote the peak value of $v(t)$ due to a target only. At the time corresponding to $\hat{v}$, the total signal, $v(t)$, is normally distributed with mean $\hat{v}$ and variance σ_v^2. Then the probability of target detection is given by

$$P_D = \frac{1}{\sqrt{2\pi\sigma_v^2}} \int_{v_T}^{\infty} e^{-0.5[(v - \hat{v})/\sigma_v]^2}$$

(22-35)

or

$$P_D = 0.5 + \text{erf} \; [(\hat{v} - v_T)/\sigma_v]$$

(22-36)

where $\text{erf} \, (x) = \frac{1}{\sqrt{2\pi}} \int_0^x e^{-0.5y^2} \, dy$

is a tabulated quantity.

Signal Averaging. A variation on the target-detection circuit structure is the incorporation of signal-averaging before thresholding. This consists of averaging the signal in each detector channel over several scan periods before thresholding. Since averaging and filtering are linear operations, the averaging can be done either before or after the filter. A weighting factor K is used, with $0 < K < 1$, to diminish the weighting of old information. The structure of an ideal signal averager is shown in Figure 22-34, where the delay T is the scan period. The output is given by

$$y(t) = \sum_{n=0}^{\infty} K^n v(t - nT) \quad (22\text{-}37)$$

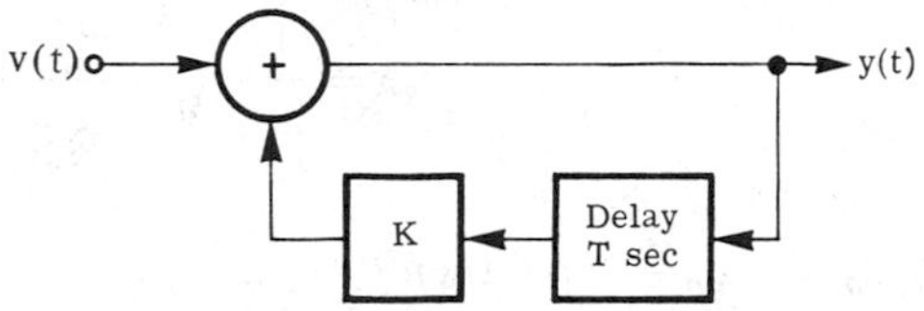

This technique improves the signal-to-noise ratio if the target position in the field of view is fixed during the averaging time.

Fig. 22-34. Signal averager.

Continuous delay lines with delays of many milliseconds, i.e., typical scan periods, are not presently available. Furthermore, scan-frequency fluctuations would degrade performance of signal averaging with a fixed delay. To overcome both objections, one can use signal averaging which is usually discrete in time, and often also discrete in amplitude. That is, the input signal is sampled with a clock that is synchronized to the scan, and the samples are then averaged, either in analog (continuous) or digital (discrete) form. As long as the sampling frequency is large enough (in practice at least five times the signal bandwidth), discrete time operation suffers no information loss.

Output Angle Noise. In the presence of noise, there will be an error of e seconds between the occurrence of the target pulse at the threshold point and the time of the threshold crossing. Since target position decoding is by means of pulse position demodulation, the error becomes angle noise at the tracker output. If the scanning velocity is μ rad sec^{-1}, then the equivalent angle noise is μe radian. Averaging or low-pass filtering can be used to reduce the output angle noise at the cost of reduced tracking bandwidth.

The exact statistics of e and their relation to signal-to-noise ratio, threshold setting, and other relevant quantities represents an unsolved mathematical problem. A recent, comprehensive treatment of the problem is given in Reference [22-30], which provides the following estimate* for the variance of e under small error conditions and white Gaussian noise:

$$\sigma_e^2 \cong \frac{2\pi N_o}{\displaystyle\int_{-\infty}^{\infty} \omega^2 |S(\omega)|^2 \, d\omega} \quad (22\text{-}38)$$

where $S(\omega)$ = the Fourier transform of the target signal at the preamp output
$2N_o \delta(t)$ = the noise autocorrelation function
$\delta(t)$ = Dirac delta function

---

*This expression is derived from that in Reference [22-30] by elementary manipulations and the use of Parseval's theorem on Fourier transforms.

Additional results are found in the cited work.

Tracking Logic. The primary function of the tracking logic is to produce the target position signals. This is done in principle by sampling the scan reference signals when triggered by the target detection circuit(s). The exact logic used for target position decoding depends upon the type of scanning used. In multiple detector systems, the instantaneous field of view of each detector has a fixed displacement from some reference point. The tracking logic must account for this displacement by identifying which detector channel contains the target pulse, and by adding the detector displacement to the sampled reference signals.

In addition to generating target position signals, the tracking logic can perform numerous other functions which may be important in specific applications. Acquisition of a very weak stationary target in a uniform background can be facilitated by integration within the tracking logic. In such a case, the interfering signal is not periodic at the scan frequency, as it would be for a background with structure. Lowering the threshold increases the rate of noise-induced threshold crossings, but they will occur randomly in time. By contrast, the threshold crossings due to a stationary target will occur repetitively and will correspond to the same position in the field of view. By correlating the threshold crossings over a number of scan periods, the noise crossings can be rejected and the target crossings identified. The penalty is a reduced tracking bandwidth, but this is often acceptable since the weakest targets occur at the longer ranges, for which acquisition speed is not critical and the stationary target assumption tends to hold. This smoothing procedure often is found in pulsed radar systems and is variously called binary integration, post-detection integration, or coincidence detection. Considerable performance improvement is realized using coincidence detection. Design curves may be found in References [22-31 and 22-32].

Another useful tracking logic function is called "window-gating." Once the target has been acquired, the effective field of view can be reduced to a small region around the target, thereby preventing strong interfering sources outside of this region from capturing the tracker. Window-gating is mechanized by generating a logic signal for each detector which is 1 when the detector is scanning within the window centered on the estimated target position. This logic signal gates the target detection signal, z, before sampling of the reference signals. Generation of the logic signal is a straight-forward process, and is based upon the instantaneous position of the detector field of view (which is available since it is required for the tracking function).

There is a tradeoff between how small the tracking window can be and the trackable line-of-sight motion. For type-1 trackers which track a constant line-of-sight rate, $\dot{\sigma}$, with a constant error, window-gating imposes a limit on the trackable line of sight acceleration, $\ddot{\sigma}$. Additional logic can be incorporated to enlarge the window as the target maneuvers, or as the target signal-level gets large.

22.6.2. Rosette Trackers. The basic rosette tracker is a single-detector scanner with a rosette scan pattern which contains a number of loops or petals emanating from a common center. The scan pattern represents the path followed by the projection, into object space, of the center of the detector instantaneous field of view. Mechanization of a rosette scan is easily achieved by means of two counter-rotating optical elements, each of which deflects incoming rays by the same angle. (See Chapter 10.) The deflection elements can be optical wedges (prisms), tilted mirrors, or off-centered lenses. Ease of mechanization makes rosette scanners attractive candidates for tracking systems.

Rosette Patterns. Rosettes are members of a three-parameter family of curves, defined as a function of time by

$$x(t) = \frac{\rho}{2}(\cos 2\pi f_1 t + \cos 2\pi f_2 t) \tag{22-39}$$

$$y(t) = \frac{\rho}{2}(\sin 2\pi f_1 t - \sin 2\pi f_2 t) \tag{22-40}$$

In polar coordinates, the rosette equations are

$$r(t) = \rho \cos \pi(f_1 + f_2)t \tag{22-41}$$

$$\theta(t) = \pi(f_1 - f_2)t \tag{22-42}$$

The rotating elements spin at frequencies f_1 and f_2, the values of which determine the nature of the pattern. The parameter ρ is a scale factor which defines the angle subtended by the pattern radius. The frequencies are both taken as positive, with f_1 greater than f_2.

If f_2/f_1 is a rational number, then the pattern repeats exactly, i.e., is closed. In practice, the two rotating elements are usually geared together, producing a closed pattern, so this discussion concentrates on this subclass. Examples of closed rosettes are illustrated in Figure 22-35. Important rosette properties for tracker applications include the number of petals, the petal width, and the amount of overlap between adjacent petals. These all depend upon f_1 and f_2.

The petal frequency, f_p, is $f_1 + f_2$ petals per second. If f_2/f_1 is rational, then f_1 and f_2 have a greatest common divisor f, such that $N_1 = f_1/f$ and $N_2 = f_2/f$ are both positive integers. Moreover, N_1 and N_2 are the smallest integers satisfying

$$\frac{N_2}{N_1} = \frac{f_2}{f_1} \tag{22-43}$$

The rosette period, T, is $1/f$, also given by N_i/f_i, where $i = 1, 2$. The number of petals in the pattern is $N = N_1 + N_2$. The size of the instantaneous field of view should be minimized to minimize interfering background signals and detector noise, but should be large enough to provide full coverage of the scanned region. The required instantaneous field of view for full scan-coverage depends upon petal width and the amount of overlap between adjacent petals.

The petal width, w, is the width of a petal at its widest point. This corresponds to the point t_w on the first petal at which $dy/dt = 0$. The time, t_w, is obtained from

$$\frac{dy}{dt} = \pi\rho(f_1 \cos 2\pi f_1 t - f_2 \cos 2\pi f_2 t) \tag{22-44}$$

The condition for a solution of $dy/dt = 0$ is

$$\cos \alpha - a \cos a\alpha = 0 \tag{22-45}$$

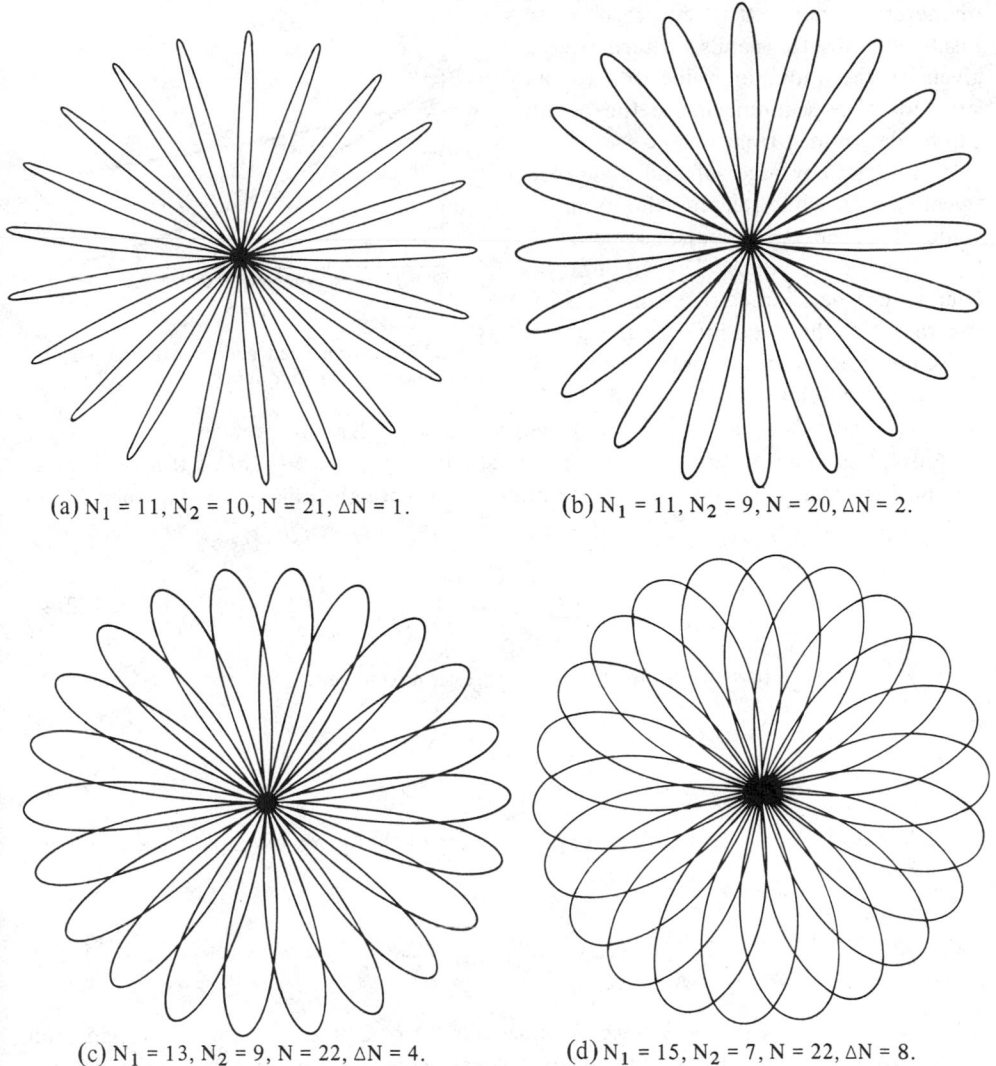

(a) $N_1 = 11, N_2 = 10, N = 21, \Delta N = 1.$

(b) $N_1 = 11, N_2 = 9, N = 20, \Delta N = 2.$

(c) $N_1 = 13, N_2 = 9, N = 22, \Delta N = 4.$

(d) $N_1 = 15, N_2 = 7, N = 22, \Delta N = 8.$

Fig. 22-35. Rosette patterns.

where $\alpha = 2\pi f_1 t_w$

$a = f_2/f_1$, and $0 \leqslant a \leqslant 1$

Then

$$w = 2y(t_w) = \rho(\sin \alpha - \sin a\alpha) \qquad (22\text{-}46)$$

Figure 22-36 contains plots of α and w/ρ versus $a = f_2/f_1$. Note a is also given by N_2/N_1. The system designer can use w as a preliminary estimate of the instantaneous field of view diameter, ω, required to ensure no gaps inside the rosette petals. Alternatively, given ω, the minimum value of a to ensure complete scanning inside the petals can be determined from $w \leqslant \omega$.

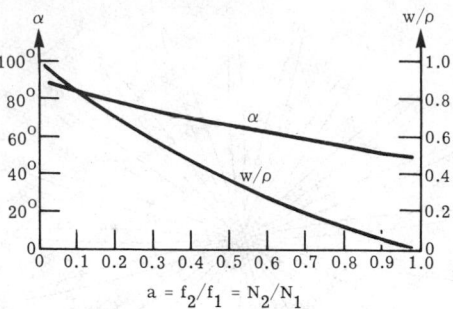

Fig. 22-36. Normalized petal width and phase of widest point versus frequency ratio.

Complete coverage of the field between petals, while keeping the minimal value of ω, generally requires some overlap of adjacent petals. The amount of overlap can be characterized in terms of the radius of the petal intersection point r_i. An expression for r_i is necessary to see how it is related to N_1 and N_2, and to indicate how these parameters can be chosen to give full scan-coverage.

Along the circle of radius r_i, each petal subtends an angle $\phi = 2\pi/N$. If t_i is the point on the first petal corresponding to the intersection with the adjacent petal, then $\theta(t_i) = \phi/2$, or

$$\pi(f_1 - f_2)t_i = \frac{\pi}{N} \tag{22-47}$$

Solving for t_i and substituting into the $r(t)$ expression, one gets

$$r_i = \rho \cos \frac{\pi}{\Delta N} \tag{22-48}$$

for $\Delta N = N_1 - N_2 \geqslant 2$. A refined expression for ω that gives no gaps inside the petals is

$$\omega = \frac{2\pi\rho}{N} \cos \frac{\pi}{\Delta N} \tag{22-49}$$

This *only* applies for $\Delta N \geqslant 3$ since the cases $\Delta N = 1$ or 2 produce no petal overlap at all.

At $\Delta N = 3$, r_i has the value 0.5ρ. As ΔN increases to larger values, r_i increases from 0.5ρ to a limiting value of ρ. The corresponding limits on ω are $\pi\rho/N$ and $2\pi\rho/N$, respectively.

With $\omega = 2\pi r_i/N$, the minimum value for no interior gaps, there will be cusps in the scanned field between petal tips. The depth of these cusps is generally negligible for $\Delta N = 4$, and diminishes further as ΔN increases. The effect of the cusps is further reduced when the optical resolution is optimized and becomes comparable to ω.

Other patterns for which the frequencies are opposite in sign and the two rotations are of different magnitude, and in which precession is permitted, are given in Chapter 10.

Sensitivity. The basic sensitivity equation, given before, is

$$\text{NEFD} = \frac{\sqrt{A_d \Delta f}}{A_o T_o D^*} \tag{22-50}$$

Assuming a circular field stop with angular diameter ω, a circular detector, and a condenser lens between the two, one finds that the detector area is

$$A_d = \frac{\pi}{4}\left[\omega F \frac{(F/\#)_c}{F/\#}\right]^2 \tag{22-51}$$

where F and $F/\#$ = the effective focal length and focal ratio of the optical system, respectively

$(F/\#)_c$ = the focal ratio of the condenser lens

If the interfering signal is white noise, the electrical bandwidth should be about

$$\Delta f = \frac{1}{2\tau} \tag{22-52}$$

where τ is the target signal pulse width. The value of τ can be approximated by

$$\tau \simeq \frac{\omega}{v} \tag{22-53}$$

where v is the scan velocity in rad $\sec^{-1}$. In rosette scan patterns, v is not constant. An expression for v is obtained by differentiating $x(t)$ and $y(t)$, setting

$$v(t) = \sqrt{\dot{x}(t)^2 + \dot{y}(t)^2} \tag{22-54}$$

and simplifying. The result is

$$v(t) = \pi f_1 \rho \sqrt{1 + a^2 - 2a \cos(1-a)2\pi f_1 t} \tag{22-55}$$

The extreme values of v are

$$v_{max} = \pi\rho(f_1 + f_2) \tag{22-56}$$

which occurs at the pattern center, and

$$v_{min} = \pi\rho(f_1 - f_2) \tag{22-57}$$

which occurs at the petal tips. The bandwidth is chosen on the basis of v_{max} to avoid undue attenuation of pulses from targets in the center of the scan pattern.* Then

$$\Delta f = \frac{\pi\rho(f_1 + f_2)}{2\omega} \tag{22-58}$$

and

$$A_d \Delta f = \frac{[\pi F((F/\#)_c/F/\#)]^2 \, \omega\rho(f_1 + f_2)}{8} \tag{22-59}$$

---

*It might be possible to exploit the greater frequency of central target pulses, compared to peripheral target pulses, to reduce Δf by utilizing multiple pulse integration.

Using the previously obtained expression for ω, one has

$$\omega = \frac{2\pi\rho}{N} \cos \frac{\pi}{\Delta N} \tag{22-60}$$

which gives

$$A_d \Delta f = \frac{\pi}{4N} \left[\pi F((F/\#)_c/F/\#)\rho\right]^2 (f_1 + f_2) \cos \frac{\pi}{\Delta N} \tag{22-61}$$

Assuming the collecting aperture is circular with diameter D_o, and substituting $F/\# = F/D$, one has

$$\text{NEFD} = \frac{2(F/\#)_c \rho}{D_o T_o D^*} \sqrt{\frac{\pi (f_1 + f_2) \cos\left(\dfrac{\pi}{\Delta N}\right)}{N}} \tag{22-62}$$

The frequency at which the rosette pattern repeats, f_R, is

$$f_R = \frac{f_1 + f_2}{N} \tag{22-63}$$

Then NEFD may be expressed as

$$\text{NEFD} = \frac{\text{FOV}(F/\#)_c}{D_o T_o D^*} \sqrt{\pi f_R \cos\left(\frac{\pi}{\Delta N}\right)} \tag{22-64}$$

where FOV = 2ρ is the diameter of the field of view.

Since NEFD increases only by $\sqrt{2}$ as ΔN varies upward from 3, sensitivity is essentially independent of the rosette parameters N_1 and N_2.

Perhaps the single most important parameter is N, the number of petals. The choice of N determines how small ω can be and still provide full scan-coverage. The value of ω, in turn, determines the amplitude of interfering background signals. Tracker performance is only weakly dependent upon ΔN, so long as $\Delta N \geq 4$. However, for a given f_R, the values of f_1 and f_2 and the attendant mechanical design will generally indicate some preference for ΔN.

22.6.3. Crossed Array Trackers.
The basic crossed-detector array is illustrated in Figure 22-37. It consists of four rectangular detectors arranged in a cross. The detector array is located in the telescope image plane. Conical, or nutation scanning is utilized, in which the image of each point in the field of view traces out a circle of angular radius R_N, called the nutation radius. Usually, R_N is a little smaller than the radius of the

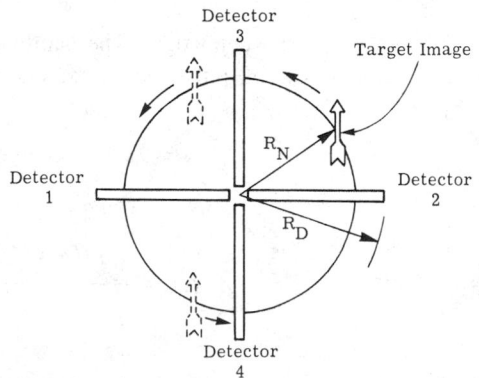

Fig. 22-37. Crossed detector array.

detector array, R_D. The center of the circle corresponds to the location of the source point with respect to the optical axis. As shown in Figure 22-37, an extended (nonpoint) image does not rotate about its own axis; its spatial orientation is maintained constant over the scan cycle. Each detector produces a separate signal proportional to the IR energy instantaneously incident upon the sensitive area.

Nutation scanning is achieved by a linear displacement of the image by an angular displacement, R_N, between the optical and mechanical axes of the telescope, combined with rotation of the telescope about its mechanical axis. This may be implemented in refractive systems by off-center lens mounting, or by adding an optical wedge. In reflective systems, nutation may be implemented by placing the secondary mirror at an angle to the optical axis.

The image of a point source on the optical axis will trace a nutation circle of radius R_N centered on the array. A point source with arbitrary angular coordinates (X, Y) with respect to the optical axis will produce the same size image-nutation circle, but the center will be at the angular coordinates of the source, as shown in Figure 22-38. The corresponding linear dimensions in the detector or image plane are obtained by multiplying these angular dimensions (expressed in radians) by the effective focal length. As the source position changes, only the position of the image nutation circle center changes; the radius and phasing of the nutation circle remain fixed.

Angle Decoding. The nutation phase angle $\phi(t)$ is given by

$$\phi(t) = 2\pi f_N t \tag{22-65}$$

where f_N is the nutation scanning frequency. With $\phi = 0$ defined along the positive x-axis, a point source with angular coordinates (X, Y) will produce image coordinates (x, y) given by

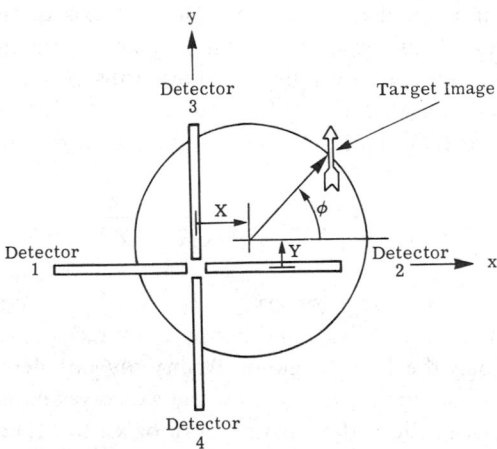

Fig. 22-38. Target image nutation circle.

$$x(t) = X + R_N \cos \phi(t) \tag{22-66}$$

$$y(t) = Y + R_N \sin \phi(t) \tag{22-67}$$

Each detector produces an output pulse when it is crossed by the image. Thus, detectors 1 and/or 2 produce a pulse when $y(T) = 0$ and $|x(T)| < R_D$; the $y(T) = 0$ condition is equivalent to

$$\sin \phi(T) = - \frac{Y}{R_N} \tag{22-68}$$

Similarly, detectors 3 and/or 4 produce a pulse when

$$\cos \phi(T) = - \frac{X}{R_N} \tag{22-69}$$

so long as $|y(T)| < R_D$ is satisfied.

Signals for tracking are now easily generated. It is straightforward to generate reference signals

$$U_x(t) = - \cos \phi(t) \tag{22-70}$$

$$U_y(t) = - \sin \phi(t) \tag{22-71}$$

derived from motion of the scanning mechanisms. Reference $U_x(t)$ is sampled upon the occurrence of a pulse on detector 3 or detector 4, giving the x-error signal

$$v_x = \frac{X}{R_N} \tag{22-72}$$

Similarly, sampling $U_y(t)$ upon the occurrence of a pulse on detector 1 or detector 2 gives the y-error signal

$$v_y = \frac{Y}{R_N} \tag{22-73}$$

This logic generates a linear tracker characteristic in each axis as long as the error in that axis does not exceed R_N. Thus the linear tracking region is a square of side $2 R_N$. Note that linear tracking does not require pulses on both arms of a detector axis; pulses on detector 1 are sufficient to generate v_y without the need for pulses on detector 2.

It is clear from Figure 22-38 that for $|Y| < R_N$, target pulses will be generated in only one detector if

$$R_N < |X| < R_D + \sqrt{R_N^2 - Y^2} \tag{22-74}$$

where R_D is the radius of the detector array. By symmetry, X and Y may be interchanged in the above statement. This condition is easily recognized and used to extend the tracking field beyond the linear region. When only one detector-axis produces a target pulse, the error signal due to the pulse on that axis is generated as described above. However, the error signal in the orthogonal axis can be set to ±1, corresponding to angle errors of $\pm R_N$, with the polarity based upon which detector contains the target pulse.

For example, if $|Y| < R_N$ and $R_N < X < R_D + \sqrt{R_N^2 - Y^2}$, then the target pulses will occur only in detector 2. The error signal v_y is generated, as before, by sampling $U_y = - \sin \phi(t)$ with the target pulses, and v_x is set equal to +1. Similarly, if target pulses occur only on detector 1, this must mean

$$-\left[R_D + \sqrt{R_N^2 - Y^2}\right] < X < -R_N \qquad (22\text{-}75)$$

Hence, v_x is set to -1.

The result of the above logic is illustrated in Figure 22-39, which shows the crossed-array tracking field. Linear tracking in both axes is obtained over a square of side $2R_N$. Semilinear tracking is obtained in the four outer segments shown in the figure. The area of the total track field is

$$A_T = R_N^2(2\pi - 4) + 8 R_N R_D \qquad (22\text{-}76)$$

For $R_D/R_N = 1.2$, a typical ratio, $A_T = 11.9 \, R_N^2$. This corresponds to an equivalent, circular, track field with radius $R_c = 1.95 \, R_N$.

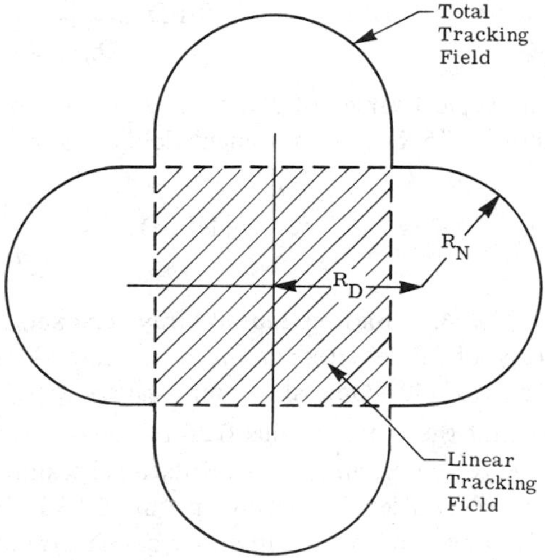

Sensitivity. If the electrical filter is designed for white noise, then, in the basic sensitivity equation

$$\text{NEFD} = \frac{\sqrt{A_d \Delta f}}{A_o T_o D^*} \qquad (22\text{-}77)$$

the bandwidth Δf is about

Fig. 22-39. Crossed array tracking field.

$$\Delta f = \frac{1}{2\tau} \qquad (22\text{-}78)$$

where τ is the target signal pulse-width. The value of τ depends upon the target position coordinate, orthogonal to the detector axis. The electrical bandwidth is based upon the minimum τ,

$$\tau_{\min} = \frac{\alpha}{2\pi R_N f_N} \qquad (22\text{-}79)$$

where α is the angular width of the detector. Then

$$\Delta f = \frac{\pi R_N f_N}{\alpha} \qquad (22\text{-}80)$$

As mentioned earlier, multiple detector systems allow the use of additional detectors to improve performance at the penalty of higher costs. In the crossed array, this means dividing each of the four basic detectors into M separate, smaller detectors, each with its own preamplifier and processing electronics. Neglecting the necessary spaces between detectors, the area of each detector is

$$A_d = \alpha \frac{R_D}{M} F^2 \tag{22-81}$$

where F is the equivalent focal length. Then, with a circular collecting aperture of diameter $D_o = F(F/\#)^{-1}$

$$NEFD = \frac{4 (F/\#)}{D_o T_o D^*} \sqrt{\frac{R_N R_D f_N}{M\pi}} \tag{22-82}$$

For typical values of $R_D = 1.2 R_N$, and an equivalent, circular, tracking-field radius of about $1.95 R_N$, the equivalent field of view diameter is $3.9 R_N$, and

$$NEFD \cong \frac{1.1 \; FOV(F/\#)}{D_o T_o D^*} \sqrt{\frac{f_N}{M\pi}} \tag{22-83}$$

22.6.4. Rotating Linear-Array Trackers. Rotating linear-array trackers contain an array of IR detectors arranged in a line along a radius of the field of view, as shown in Figure 22-40. Optical scanning causes the field of view image to rotate about its center, so that the instantaneous field of view of each detector covers an annulus. Rotational scanning can be realized by a Dove prism [22-33, 22-34], or a roof mirror. In either case, the image rotates at twice the rate of the scan element.

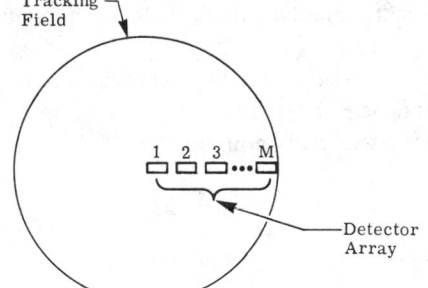

Fig. 22-40. Rotating linear array.

Target-position signals naturally occur in polar coordinates for this class of trackers. The target radius corresponds to the detector channel containing the target signal, and the polar angle corresponds to the scan phase at which the target signal occurs. Point targets therefore produce position signals which are discrete in radius, and can be either discrete or continuous in polar angle, depending upon the mechanization. Conversion to a Cartesian coordinate system is easily accomplished if necessary.

Sensitivity. Again, the basic sensitivity equation is

$$NEFD = \frac{\sqrt{A_d \Delta f}}{A_o T_o D^*} \tag{22-84}$$

The detector area is

$$A_d = \alpha \frac{FOV}{2M} aF^2 \tag{22-85}$$

where α = detector instantaneous field of view width, rad
    FOV = total field of view angular-diameter, rad
    M = number of detectors (all assumed of equal length)
    F = effective focal length

$$a = \text{detector geometric efficiency} = \frac{\text{detector length}}{\text{FOV}/(2\,M)}$$

The bandwidth, assuming a design for white noise interference, is

$$\Delta f = \frac{1}{2\tau} \tag{22-86}$$

where τ is the target dwell-time. If one assumes the optical resolution is smaller than the detector width,

$$\tau = \frac{\alpha}{2\pi r f_s} \tag{22-87}$$

where r = the target-position angular-radius
$\quad f_s$ = the scan frequency

The value of r is taken as the midpoint of each detector, i.e.,

$$r_n = \frac{\text{FOV}}{2\,M}\,(n - 0.5) \tag{22-88}$$

where n is the detector number ($n = 1, 2, \ldots, M$) starting with the one closest to the center. Then

$$\Delta f = \frac{(\pi\,\text{FOV}\,f_s)\,(n - 0.5)}{2M\alpha} \tag{22-89}$$

and

$$A_d\Delta f = \left(\frac{\text{FOV}\,F}{2M}\right)^2 \alpha\pi f_s(n - 0.5) \tag{22-90}$$

Therefore, the sensitivity of channel n is

$$\text{NEFD}_n = \frac{2\,\text{FOV}\,(F/\#)}{D_o T_o D^* M}\sqrt{\frac{af_s(n - 0.5)}{\pi}} \tag{22-91}$$

where a circular collecting aperture of diameter D_o is assumed, $F/\# = F/D_o$, and $n = 1$, $2, \ldots, M$. It is seen that the bandwidth for a channel increases at larger detector radii, causing a larger NEFD.

Centered Target Tracking. This class of trackers has a characteristic that is common to all trackers employing rotational scanning (e.g., rotating reticles). As the target position approaches the center of rotation, the target-signal pulse decreases in amplitude and increases in duration. This degrades discrimination between the target signal and the interfering signals due to uniform backgrounds and electronic noise, which increase at lower frequencies. Thus, if the sensor field contains the center of rotation, as it does in the present case, tracking necessarily degrades in the vicinity of the scan center. This degradation becomes a loss of signal-to-noise ratio as a function of target position.

It is assumed that both the centerline and the inner edge of detector number one intersect the center of rotation. This assumption results in minimum tracking degradation. The target is assumed to be a point target, i.e., the angle subtended by the target is much smaller than the extent of the optical point-spread-function, $h(x, y)$; the latter is assumed to be symmetric, but this is only for convenience and is not essential to the analysis.

The peak target signal occurs when the center of the target image crosses the detector centerline. In terms of detector current, the peak target signal is

$$\bar{i}_{tgt} = E_{tgt} A_o T_o \Re \int_{D1} h(x, y) \, dx dy \tag{22-92}$$

where E = the received target irradiance
$\Re$ = the detector responsivity, A W^{-1}

and the integration domain $D1$ is the sensitive area of detector one. It is assumed that $h(x,y)$ is normalized such that it integrates to one over the whole plane. If the image center (i.e., target position) is at a radius r, and if the detector width and length are $2w$ and l, respectively, then the integral is

$$\int_{D1} h(x, y) \, dx \, dy = \int_{-w}^{w} \int_{0}^{l} h(x - r, y) \, dx \, dy \tag{22-93}$$

$I(r)$ denotes this integral. The minimum target signal usually will be displaced from the peak by one-half of a scan period, in which case it is given by

$$i_{tgt} = E_{tgt} A_o T_o \Re I(-r) \tag{22-94}$$

The target signal peak-to-peak amplitude is then

$$i_{tgt} = E_{tgt} A_o T_o \Re [I(r) - I(-r)] \tag{22-95}$$

As the target radius, r, becomes comparable to the diameter of $h(x,y)$, or less, the target signal pulse-width becomes a large fraction of a scan period. Accurate generation of the polar angle signal requires estimating the scan phase corresponding to the peak target signal. This is readily accomplished by determining the midpoint of the scan interval for which the detector signal is above a threshold.

The rms detector noise current is

$$i_{ND} = \frac{\sqrt{A_d \Delta f}}{D^*} \Re \tag{22-96}$$

A uniform background of radiance L_B will produce a dc detector current, I_B, given by

$$I_B = L_B A_o T_o \left(\frac{A_d}{F^2}\right) \Re \tag{22-97}$$

The rms shot noise due to I_B is

$$i_{NB} = \sqrt{2 e I_B \Delta f} \tag{22-98}$$

where e is the electron charge. Thus

$$i_{NB} = \sqrt{2eL_B A_o T_o\left(\frac{A_d}{F^2}\right) \Re \Delta f} \qquad (22\text{-}99)$$

The expressions for i_{tgt}, i_{ND}, and i_{NB} can be evaluated to determine the signal-to-noise ratio as a function of r, and, in turn, the accuracy in identifying the scan phase of the signal peak.

If the signal-to-noise ratio is too small for the required accuracy, steps can be taken to gain some improvement. If the length of detector 1 substantially exceeds the diameter of $h(x, y)$, then it can be reduced, giving a smaller A_d with no loss of i_{tgt}. However, to maintain the same FOV, the remaining M-1 detectors must be made longer, or else another detector added.

A second modification is to reduce the filter bandwidth, Δf. This could be done by the use of a bandpass filter centered at the scan frequency. In choosing the bandwidth, there is a tradeoff between signal-to-noise ratio and response speed, as well as the practical problem of maintaining the scan frequency within the pass band. An equivalent approach is the use of signal averaging as described in Section 22.6.1.

Another way to improve central tracking is to improve the optical resolution by making $h(x,y)$ and the detector width smaller. In practice, usually a combination of several such modifications are made to achieve a balanced overall design.

22.7. Imaging Trackers

Trackers which are designed using imaging sensors are called imaging trackers. An imaging sensor is one using single or multiple detectors which produce video signals by means of a linear raster-scan* of the target scene. Examples of imaging sensors are TV cameras—vidicons, image orthicons, etc.,—as well as linearly-scanned detector arrays, two-dimensional electronically-scanned arrays, and radars which organize the received data into an image format. Imaging sensors generally preserve more scene information than nonimaging sensors of the same resolution, spectral band, and speed. Thus, imaging trackers can discriminate between objects by various criteria, satisfying performance requirements beyond the capabilities of most nonimaging trackers.

Basically, there are two types of imaging trackers: gated-video trackers (GVTs), and correlation trackers. GVTs define some portion of the received scene to be the target image, and generate tracking error-signals from the target information so identified. Typically, the target-definition function is mechanized by means of electronic filters, gates, and thresholding operations. The gate is electronically movable, and serves to prevent background clutter or other spurious sources from interfering with target tracking. Numerous algorithms are available for generating tracking error-signals, some of which are described below.

Correlation trackers compute a cross-correlation function between the scene from the sensor and a stored reference scene. The tracking point is the location of the best match of the two scenes, e.g., the peak of the correlation function. Thus, while GVTs

---

*Although nonlinear scanning can also produce an image of the target scene, such sensors normally are not referred to as imaging.

may be used to track specific objects, or targets contained within the scene, correlation trackers generally utilize most or all of the scene information, and may be used to track targets that are too small or that have insufficient contrast for a GVT. Correlation trackers also may be used to track a changing target scene in which the scene dynamics exceed GVT capabilities. Such conditions arise in some homing missile applications.

In most imaging trackers, the values of some internal tracker variables are determined by the scene content over an extended area. For example, a GVT threshold level used to discriminate between target and background can be taken as the midpoint between the maximum and minimum values of the gated video.

The straightforward approach toward determining the values of such variables is to store the gated video, determine the variable values, then make a second pass through the video to perform the tracking operation. But such video storage is too costly for many, if not most, tracker applications. An alternative approach is to choose the frame rate sufficiently large so that the target scene changes very little from frame-to-frame.* Then the internal tracker variables can be determined on the basis of scene content of previous frames, and the video can be processed as it is generated by the sensor. This basic approach underlies the specific tracker techniques described below.

22.7.1. Gated-Video Trackers (GVTs). A typical GVT system is illustrated in Figure 22-41. The sensor generates one or more video signals, mapping the target scene radiance into video amplitude. Position within the field of view is mapped into time within the frame according to the scan pattern utilized. The video signal is gated to prevent scene information outside of the gated region from interfering with target tracking. The gated video is processed to produce tracking error-signals $(\epsilon_{tx}, \epsilon_{ty})$ which represent the target tracking-point position with respect to the gate center (x_G, y_G). New error-signals are generated for each scan frame, and they are used to reposition the gate for the next frame. The gate is repositioned to drive $(\epsilon_{tx}, \epsilon_{ty})$ to zero, i.e., to keep the gate centered on the target tracking-point.

The outer loop develops a sensor-position error-signal which repositions the sensor to keep it aimed at the target tracking point. The sensor-position error-signal $(\epsilon_{sx}, \epsilon_{sy})$ is formed by

$$(\epsilon_{sx}, \epsilon_{sy}) = (x_G, y_G) + (\epsilon_{tx}, \epsilon_{ty}) \qquad (22\text{-}100)$$

In some cases, if (x_G, y_G) is determined with sufficient precision, and if the inner loop keeps $(\epsilon_{tx}, \epsilon_{ty})$ small, then the gate position alone can be used as the sensor-position error.

The cost and power requirements of the sensor drive components (gimbals, torquers, power amplifiers, etc.) generally increase rapidly with increasing outer-loop bandwidth. Therefore, the outer-loop bandwidth is often made only large enough to satisfy slewing and steady-state tracking dynamics; the inner-loop bandwidth, gate size, and sensor field of view are then made large enough to accommodate tracking transients.

Gate Geometry. The gate geometry of a typical GVT is illustrated in Figure 22-42. A Cartesian coordinate system is shown with origin at the center of the sensor field of view. A rectangular gate is shown, aligned with the field of view, as is usually the case. The gate center (x_G, y_G) is determined by the tracker inner-loop as described above.

---

*Small frame-to-frame scene changes are also necessary to provide smooth target tracking.

Figure 22-43 is an expanded sketch of the gate superimposed upon a target-image cartoon. The tracking gate can be either fixed in size or adaptive. A fixed gate is simpler to mechanize, and can be used if the angular target size is fixed, or nearly so. A fixed gate also can be used for a large range of target sizes, if the target position computer selects a corner of the target as the tracking point. If a wide range of target sizes must be tracked, and if the desired tracking point is at a target interior point, then an adaptive gate is required. The size of an adaptive gate is made slightly larger than the target so as to include the target silhouette within the gate, but to exclude nontarget background clutter.

Numerous adaptive-gate-sizing algorithms can be used. A direct approach is to determine the location of the target edges, often done anyway in the video processor, and to size the gate to be slightly larger. Another technique is to measure the target area, A_t, and aspect ratio, ρ, and to choose the corresponding gate dimensions to match ρ and to be a little larger than A_t. Gate shapes which closely match those of arbitrary targets also can be generated if the system requirements justify the added complexity.

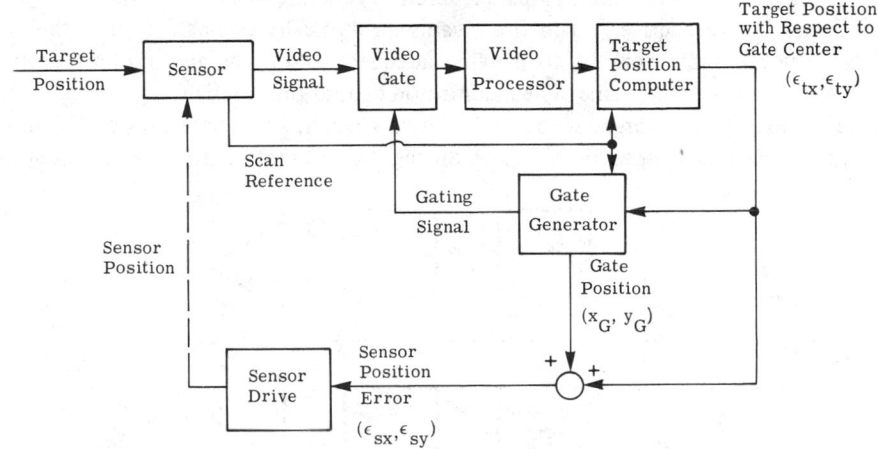

Fig. 22-41. Typical gated-video tracker system.

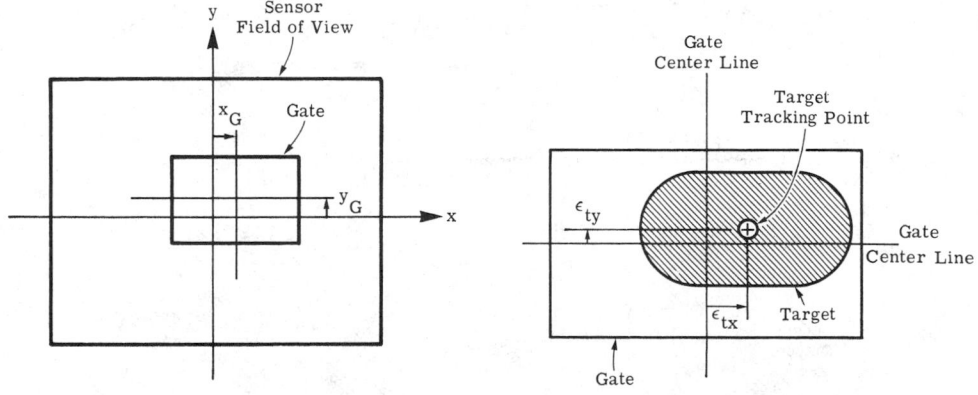

Fig. 22-42. Typical gate geometry.   Fig. 22-43. Typical tracking geometry.

Video Processors. The video processor operates upon the video signal and produces a suitable input signal for the target position computer (TPC). An electronic filter is often included as part of the video processor. To avoid spurious transients due to gating of the video signal, one can place the filter before the video gate. Three classes of video processors are described below: target edge detectors, one-bit digitizers, and multiple-bit digitizers. The video processor (VP) design is one of the most critical factors determining tracker performance. In most cases, the VP identifies the target within the scene, and this must be done under widely varying combinations of targets, backgrounds, and spurious sources. The goal in VP design is to arrive at an acceptable compromise of the many conflicting and often vaguely specified operational conditions.

An edge-detector video-processor is usually used in a fixed-gate target-corner tracker. It can also be used with a system that tracks the target center. Figure 22-44 gives an example of this type of VP. Its function is to identify the leading and/or trailing edge of the target on each video line. The VP contains a differentiator and two processing channels, one for rising edges, and one for falling. The peak detectors measure the extreme values of $v'(t)$, which are low-pass filtered to generate positive and negative thresholds, V_+ and V_-, respectively. The filter gains are typically one-half; typical time constants are several frame periods, to provide adequate smoothing and minimal droop of V_+ and V_- during a single frame. The assumption behind this configuration is that frame-to-frame changes in the target scene are relatively small, allowing values of V_+ and V_- based upon previous frames to be used during the current frame. The differentiator

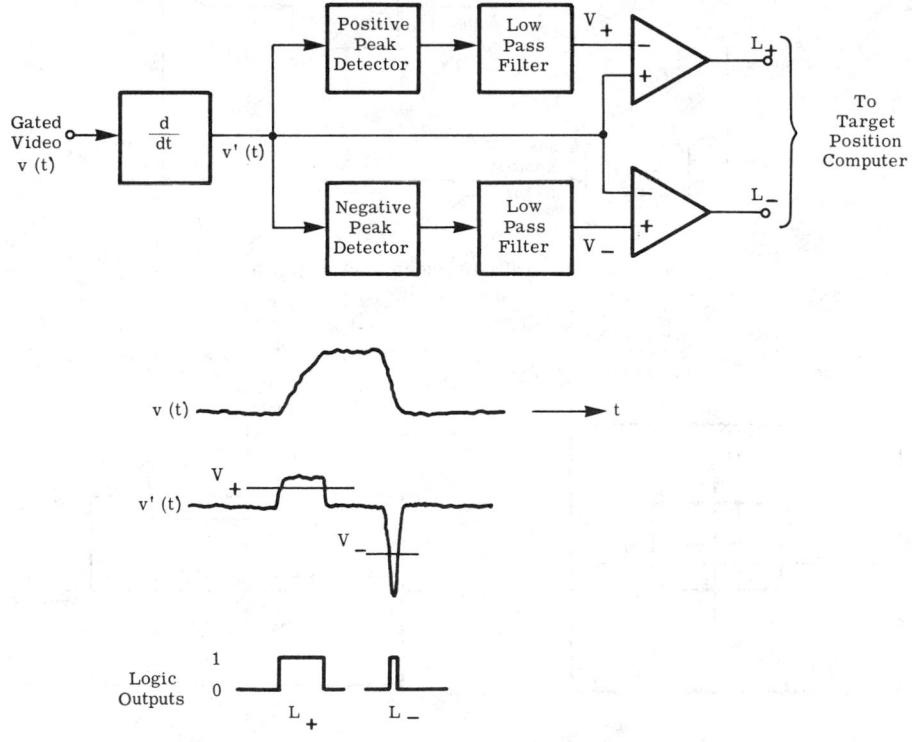

Fig. 22-44. Edge-detector video-processor.

output is thresholded by V_+ and V_- as shown in the figure, generating logical outputs L_+ and L_-, respectively. Output L_+ is one when $v'(t)$ exceeds V_+, and L_- is one when $v'(t)$ is less than V_-. The logic outputs go to the target position computer, where they are used to determine the tracking error-signals.

A different type of video processor, a one-bit digitizer VP, is required for use with a target position computer designed to operate with a bilevel input signal. This type of VP essentially determines whether each raster point within the gate is to be treated as being part of the target or part of the background. The VP output is called "digitized video" (DV); the points at which $DV(t) = 1$ or 0 are treated by the TPC as target or background, respectively.

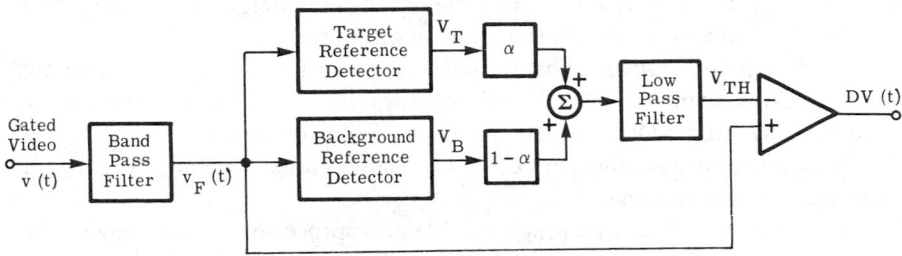

Fig. 22-45. One-bit digitizer video processor.

A one-bit digitizer VP is shown in Figure 22-45. The band-pass filter limits noise and enhances any predictable target characteristics. The filtered video signal $v_F(t)$ is processed in two parallel-channels to generate a target reference level V_T, and a background reference level V_B. These are then weighted as shown in the figure (with $0 < \alpha < 1$), summed, and filtered to generate a threshold level V_{TH} at a point between V_B and V_T determined by α. The filtered video is thresholded by V_{TH}, producing $DV(t)$ according to the rule

$$DV(t) = \begin{Bmatrix} 1 \\ 0 \end{Bmatrix} \text{ if } v_F(t) \begin{Bmatrix} \geq \\ < \end{Bmatrix} V_{TH} \qquad (22\text{-}101)$$

There are many different ways to generate the reference levels V_T and V_B. Perhaps the simplest is to use a positive-peak detector for V_T and a negative-peak detector for V_B. A limitation of this simple approach is that it allows the VP to be captured by a small, intense hot-spot, which may be undesirable in some applications. This can be prevented by a low-pass filter in the front end of the V_T circuit. A variation is to use a separate background gate G_B, such as shown in Figure 22-46, and to generate V_B as the average (or minimum) of $v_F(t)$ over G_B. In the figure, G_B is a narrow frame around the target gate, G_T. This

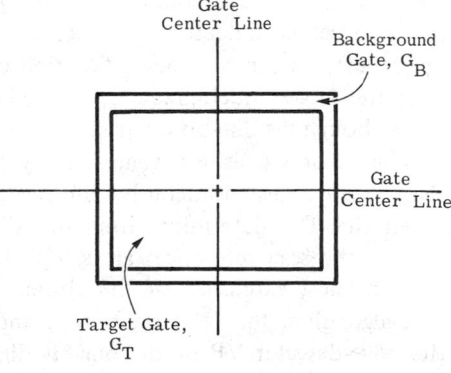

Fig. 22-46. Gate geometry variation.

technique has the advantage of spatially separating the scene data used for V_B from that used for V_T, thereby improving performance under some instances of cluttered backgrounds.

Multiple-bit digitizers trade increased complexity for potentially improved performance with severely cluttered target-scenes. One type of multiple-bit digitizer is a simple extension of the one-bit case; it generates several $DV(t)$ signals using different values of α. Comparison of the $DV(t)$, or comparison of the TPC outputs using the different $DV(t)$ signals, can provide improved performance.

Another type of multiple-bit digitizer is a conventional analog-to-digital converter. In this case, $v_F(t)$ is sampled and the samples are converted to digital form. This type of VP can be used if the TPC is a programmable digital computer, in which case the target/background discrimination function is performed within the TPC.

Target Position Algorithms. The tracking error signals $(\epsilon_{tx}, \epsilon_{ty})$ are generated in the target-position computer (TPC), and represent the angle between the gate center and the tracking point. Many algorithms are available for generating $(\epsilon_{tx}, \epsilon_{ty})$. Four of the more common algorithms are described below: corner tracking, dual edge tracking, area balance, and centroid.

The TPC often is realized by a programmable microprocessor, having many similarities to a general-purpose digital computer. In such cases it is convenient to perform certain preliminary processing steps to validate the VP output before applying the target position algorithm. For example, the output signals of an edge-detector VP corresponding to the left and right target edges should be of opposite polarities. If the VP produces a scan line with a sequence of ones in the order L_+, L_+, L_-, then the TPC must decide whether the first or second L_+ pulse represents the target leading edge. The false pulse could be due to noise or due to a spurious gradient in the scene. If it is assumed that the TPC has been tracking the target successfully on prior frames, then the L_+ pulse can be chosen which conforms more closely with the previously determined target width (for the line in question). This requires retaining in the TPC memory the necessary data from previous frames, but ordinarily this would not significantly increase the TPC requirements. Similarly, other validation criteria can be incorporated to overcome tracking problems which arise in specific applications.

The *corner-tracking* algorithm is one of the simplest and is suitable for trackers using fixed-gate size. One of the target corners is selected as the tracking point, and the error signals $(\epsilon_{tx}, \epsilon_{ty})$ are computed as the position of the selected corner with respect to the gate center. Selection of which corner to track can be fixed by the program (e.g., upper left), or can be made a function of the video-signal characteristics (e.g., the corner with the largest video-gradient). The edge-detector VP naturally goes with corner tracking, although the one-bit digitizer also can be used.

The error signals are generated by means of horizontal and vertical scan-reference signals, which instantaneously indicate the raster point corresponding to the video signal. When the TPC determines from the VP output that the scan point is at the selected corner, the scan references are sampled and held. The held values become ϵ_{tx} and ϵ_{ty}.

A modest extension of the corner tracker is the *dual-edge tracking* algorithm. In this algorithm, the left, right, top, and bottom target edges are located, with either the edge-detector VP or the one-bit digitizer. The tracking-point coordinates are taken as the midpoints between opposite edges. An advantage of this algorithm is the capability to track a target interior point. However, if the target size varies, an adaptive gate may be necessary. Furthermore, the minimum tracking range occurs where the target fills the field of view, but the corner tracker can operate as long as the target image remains in focus, which normally is much closer.

Each error-signal component in the dual-edge algorithm is determined by only two data points. This makes the error signals strongly dependent upon spatial quantization, scene changes, and video noise. Smoother tracking is obtained by some form of integration over the target signal.

The *area-balance* algorithm uses the one-bit digitizer VP and seeks to equalize the target-image area within the gate on opposite sides of the gate centerlines, thus utilizing integration. The error-signal equations for the area-balance algorithm are

$$\epsilon_{tx} = C_x^{-1} \int_{y_{GB}}^{y_{GT}} dy \left[\int_{x_{GL}}^{x_G} DV(x, y)\, dx - \int_{x_G}^{x_{GR}} DV(x, y)\, dx \right] \quad (22\text{-}102)$$

$$\epsilon_{ty} = C_y^{-1} \int_{x_{GL}}^{x_{GR}} dx \left[\int_{y_{GB}}^{y_G} DV(x, y)\, dy - \int_{y_G}^{y_{GT}} DV(x, y)\, dy \right] \quad (22\text{-}103)$$

where x_{GL} and x_{GR} = the x-coordinates of the left and right gate edges, respectively
y_{GB} and y_{GT} = the bottom and top gate edges
C_x and C_y = normalization constants

The digitized video-signal DV is here considered to be a function of position in the image plane, with DV = 1 corresponding to target points, and DV = 0 corresponding to nontarget points. C_x and C_y are chosen to maintain a constant proportionality between the actual tracking errors and the tracking error signals.

While the area-balance (AB) algorithm incorporates integration, thereby providing improvement over edge-trackers, the AB algorithm does have some limitations of its own. The major limitation is dependence of the tracker-error characteristic upon target size and shape.

As can be seen from the AB algorithm, as the error along an axis increases, the error signal stops increasing when the target leaves a gate-half. Thus the linear tracking-field in each axis—the error angle over which the error characteristic is linear—is, at most, equal to the target subtense in that axis. Furthermore, if a portion of the target moves outside the gate, the linear tracking-field is reduced further.

The tracking-loop dynamics depend strongly upon tracker gain (error signal/error), so it is desirable to design the target-position algorithm for constant gain independent of target size or shape. In general, there is no simple expression in the AB algorithm for C_x or C_y that maintains constant gain. For example, consider a convex target function $S(x, y)$ that is equal to 1 at target points, and zero otherwise. The origin is chosen to coincide with the area-balance null-position, i.e.,

$$\int_{-\infty}^{\infty} \int_{-\infty}^{0} dx\, S(x, y)\, dy = \int_{-\infty}^{\infty} \int_{0}^{\infty} dx\, S(x, y)\, dy \quad (22\text{-}104)$$

A similar condition holds with interchange of the x and y integration limits. Assume that the entire target is contained within the gate, and define

$$Y(x) = \int_{-\infty}^{\infty} S(x, y)\, dy \qquad (22\text{-}105)$$

which is the target height at horizontal position x. If there is a horizontal tracking error ϵ, the area-balance error-signal is

$$\epsilon_{tx} = C_x^{-1} \left[\int_{-\infty}^{\epsilon} Y(x)\, dx - \int_{\epsilon}^{\infty} Y(x)\, dx \right] \qquad (22\text{-}106)$$

or

$$\epsilon_{tx} = C_x^{-1}\, 2g(\epsilon) \qquad (22\text{-}107)$$

where $g(\epsilon)$ is the indefinite integral of $Y(x)$. Constant tracking-gain cannot be maintained unless $g(\epsilon)$ is of the form $C\epsilon$ with C known or computable. This corresponds to constant $Y(x)$, a rectangular target, which does not hold in general. Hence area-balance cannot give constant gain.

An approximation can be made by assuming the target to be rectangular, forming some estimate, $\hat{h}$, for target height, and setting $C_x = 2\hat{h}$. A similar procedure is used to generate ϵ_{ty}. The approximation error depends upon the actual shape of the target, and how $\hat{h}$ is formed. This approximation has been used successfully in practice.

A target-position algorithm, which is relatively simple to compute and inherently provides constant gain and does not limit the linear tracking-field, is the *centroid* algorithm, defined by

$$\epsilon_{tx} = A_t^{-1} \int \int x S(x, y)\, dy\, dx \qquad (22\text{-}108)$$

$$\epsilon_{ty} = A_t^{-1} \int \int y S(x, y)\, dy\, dx \qquad (22\text{-}109)$$

where A_t is the target area, given by

$$A_t = \int \int S(x, y)\, dy\, dx \qquad (22\text{-}110)$$

Like area-balance, the centroid algorithm has the advantages of integration. It can be applied to a one-bit digitized video-signal, or to a video signal in which amplitude variations have been retained. As shown below, the centroid algorithm is, in fact, optimum if the target radiation spatial-distribution is a paraboloid and if the noise is additive, white, and Gaussian.

Optimality of Centroid Algorithm. The maximum likelihood estimate approach is used to demonstrate optimality of the centroid algorithm. Let the received data (i.e., video signal) be $r(x, y)$, where (x, y) denotes angular position in the field of view. There are two hypotheses regarding the presence of a target, namely

$$H0: r(x, y) = n(x, y) \qquad\qquad \text{noise only} \qquad (22\text{-}111)$$

$$H1: r(x, y) = s(x - x_o, y - y_o) + n(x, y) \quad \text{target present} \qquad (22\text{-}112)$$

where (x_o, y_o) = the unknown target location
$n(x, y)$ = white Gaussian noise with mean square value N_o

A standard result in detection theory gives the likelihood ratio*

$$\Lambda = \frac{Pr\{r(\cdot, \cdot) \mid H1\}}{Pr\{r(\cdot, \cdot) \mid H0\}} = \exp\left\{ \frac{1}{N_o} \int\int r(x, y)s(x - x_o, y - y_o)\, dx\, dy \right.$$

$$\left. - \frac{1}{2N_o} \int\int s^2(x - x_o, y - y_o)\, dx\, dy \right\} \qquad (22\text{-}113)$$

Considering Λ as a function of (x_o, y_o), one finds that the maximum likelihood estimate of target position is the position $(\hat{x}_o, \hat{y}_o)$ which maximizes Λ. The second term in the exponential of Equation (22-113) gives the target-signal strength, and is independent of (x_o, y_o) provided that the entire signal is contained within the integration limits. Therefore it suffices to maximize

$$\int\int r(x, y)s(x - x_o, y - y_o)\, dx\, dy = C(x_o, y_o) \qquad (22\text{-}114)$$

The function $C(x_o, y_o)$ is a correlation function between the received data, r, and the target signal distribution, s.

To compute $C(x_o, y_o)$, one must know or assume the function s. The paraboloid is a reasonable function to use for s for general design and analysis purposes:

$$s_p(x - x_o, y - y_o) = \mu - \xi(x - x_o)^2 - 2\eta(x - x_o)(y - y_o) - \zeta(y - y_o)^2$$
$$(22\text{-}115)$$

The equal-amplitude contours of s_p are ellipses centered about the target center (x_o, y_o). The parameters μ, ξ, η, ζ determine the amplitude, orientation, eccentricity, and dispersion of the contours. Only the form of s_p is assumed; the parameters can have arbitrary values.

The best estimate of the target location is found by determining the point $(\hat{x}_o, \hat{y}_o)$ at which the partial derivatives of $C(x_o, y_o)$ vanish. The resulting equations, using s_p, are

---

*This form of the likelihood ratio holds in two dimensions, provided that the target is nonrandom. See Wong [22-35].

$$\xi \hat{x}_o - \eta \hat{y}_o = \frac{\xi \int\int x \, r(x, y) \, dx \, dy - \eta \int\int y \, r(x, y) \, dx \, dy}{\int\int r(x, y) \, dx \, dy} \qquad (22\text{-}116)$$

$$\eta \hat{x}_o - \zeta \hat{y}_o = \frac{\eta \int\int x \, r(x, y) \, dx \, dy - \zeta \int\int y \, r(x, y) \, dx \, dy}{\int\int r(x, y) \, dx \, dy} \qquad (22\text{-}117)$$

The x centroid of $r(x, y)$ is

$$x_c = \frac{\int\int x \, r(x, y) \, dx \, dy}{\int\int r(x, y) \, dx \, dy} \qquad (22\text{-}118)$$

With the corresponding expression for y_c, one can reduce the above equations to

$$\xi (\hat{x}_o - x_c) = \eta (\hat{y} - y_c) \qquad (22\text{-}119)$$

$$\eta (\hat{x}_o - x_c) = \zeta (\hat{y}_o - y_c) \qquad (22\text{-}120)$$

The only way these equations can be satisfied for arbitrary ξ, η, and ζ is by

$$\hat{x}_o = x_c \qquad (22\text{-}121)$$

$$\hat{y}_o = y_c \qquad (22\text{-}122)$$

i.e., the best estimates in the maximum-likelihood-sense are the centroids.

Centroid Tracker Error Analysis. The centroid algorithm for estimating target location from the received signal $r(x, y)$ is

$$x_c = \frac{\int\int x \, r(x, y) \, dx \, dy}{\int\int r(x, y) \, dx \, dy} \qquad (22\text{-}123)$$

$$y_c = \frac{\int\int y \, r(x, y) \, dx \, dy}{\int\int r(x, y) \, dx \, dy} \qquad (22\text{-}124)$$

Since r contains noise, x_c and y_c are random variables. The calculation of the rms error of x_c due to noise and background is a perturbation of the zero-noise, zero-background case. The results strictly apply only to the high signal-to-noise ratio and high signal-to-background ratio cases. The corresponding error of y_c is obtained by symmetry.

The tracking error is calculated by a generalization of a technique for approximating the mean and variance of a function of a random variable [22-29]. If Z is a random variable, and $f(Z)$ is a real-valued function that is smooth in an interval about $\bar{Z}$ (the mean value of Z), then $f(Z)$ can be expanded in a Taylor series about $\bar{Z}$, and integrals for the moments of $f(Z)$ can be integrated term-by-term. This gives

$$E\{f(Z)\} = f(\bar{Z}) + \ddot{f}(\bar{Z})\frac{\sigma_Z{}^2}{2} + \ldots \tag{22-125}$$

and

$$E\{f^2(Z)\} = f^2(\bar{Z}) + [\dot{f}^2(\bar{Z}) + f(\bar{Z})\ddot{f}(\bar{Z})]\,\sigma_Z{}^2 + \ldots \tag{22-126}$$

where $\sigma_Z{}^2$ is the variance of Z. The variance of $f(Z)$ is then approximated by

$$\text{Var}\,\{f(Z)\} \simeq \dot{f}^2(\bar{Z})\,\sigma_Z{}^2 \tag{22-127}$$

The approximation is good if σ_Z is small and $f(Z)$ is smooth around $\bar{Z}$. If $f(Z)$ varies linearly, the approximation is exact.

For the centroid tracker, the random variable $r(x, y)$ corresponds to Z, and the expression for x_c corresponds to the function $f(Z)$. If $r(x, y)$ is given by the sum of a target function $r_t(x, y)$ plus a random function $h(x, y)$, then, formally expanding x_c about r_t, one gets

$$x_c(r_t + h) = x_c(r_t) + \nabla x_c(r_t)h + \ldots \tag{22-128}$$

where the differential term can be defined [22-36] as

$$\nabla x_c(r_t)h = \frac{\partial}{\partial\lambda}\,[x_c(r_t + \lambda h)]_{\lambda=0} \tag{22-129}$$

This gives

$$\nabla x_c(r_t)h = \frac{\displaystyle\int\int[x - x_c(r_t)]\,h(x, y)\,dx\,dy}{\displaystyle\int\int r_t(x, y)\,dx\,dy} \tag{22-130}$$

By analogy with the earlier expression for Var $\{f(Z)\}$, the variance of x_c is approximated by

$$\sigma_{x_c}^2 \simeq \nabla x_c(r_t)E\,\{hh^T\}\,\nabla x_c(r_t)^T \tag{22-131}$$

where T denotes transpose. From the expression for $\nabla x_c(r_t)h$,

$$\sigma_{x_c}{}^2 = \frac{\int \int [x - x_c(r_t)] \int \int [x' - x_c(r_t)] E\{h(x, y)h(x', y')\} \, dx \, dy \, dx' \, dy'}{\left[\int \int r_t(x, y) \, dx \, dy \right]^2} \qquad (22\text{-}132)$$

Again, $h(x, y)$ is a small deviation from the target-only function $r_t(x, y)$. The deviation may be due to noise, or background, or both. When both are present,

$$E\{h(x, y)h(x', y')\} = R_b(x - x', y - y')\tilde{I}_t(x, y)\tilde{I}_t(x', y') + R_n(x - x', y - y') \qquad (22\text{-}133)$$

where R_b and R_n = the background and noise covariance functions, respectively

$\tilde{I}_t$ = the complement of the target-indicator function I_t

$$I_t(x, y) = \begin{cases} 1 \text{ if } (x, y) \text{ is a point on the target} \\ 0 \text{ otherwise} \end{cases} \qquad (22\text{-}134)$$

It is implied above that the background and noise are wide-sense stationary, i.e., that the target can mask a portion of the background, but that noise occurs over the entire field of view.

For results which will suggest how $\sigma_{x_c}{}^2$ may depend upon tracker parameters for the general case, the target is assumed to be of uniform amplitude A, rectangular in shape, and centered within and parallel to a rectangular tracking gate. The width and height of the target are T_x and T_y, respectively, and the corresponding tracking-gate dimensions are G_x and G_y. The background covariance-function is assumed to be of the form

$$R_b(x, y) = \sigma_b^2 \exp(-\alpha|x| - \beta|y|) \qquad (22\text{-}135)$$

and the noise is assumed to be white, i.e.,

$$R_n(x, y) = N_o \delta(x)\delta(y) \qquad (22\text{-}136)$$

The integration domain for $\sigma_{x_c}{}^2$ is the interior of the tracking gate. Then the denominator of $\sigma_{x_c}{}^2$ is

$$\text{Den} = (A T_x T_y)^2 \qquad (22\text{-}137)$$

The noise term contribution to the numerator of $\sigma_{x_c}{}^2$ is, with $x_c(r_t) = 0$,

$$\text{Num}_n = \frac{N_o G_x^3 G_y}{12} \qquad (22\text{-}138)$$

The background term contribution is given by

$$\text{Num}_b = \sigma_b{}^2 \int \int x \int \int x' \exp\left(-\alpha|x-x'| - \beta|y-y'|\right) \tilde{I}_t(x, y)\tilde{I}_t(x', y')\, dx\, dy\, dx'\, dy' \tag{22-139}$$

Evaluation gives

$$\sigma_{x_c}{}^2 = \frac{1}{(A\,T_x T_y)^2}\left[\frac{N_o G_x{}^3 G_y}{12} + \frac{\sigma_b{}^2 S}{4\alpha^4 \beta^2}\right] \tag{22-140}$$

The shape factor, S, is given by

$$S = \left[\frac{(\alpha G_x)^3}{12} + \left(1 + \frac{\alpha G_x}{2}\right)\exp\left(-\alpha G_x/2\right)(\alpha G_x \cosh \alpha G_x/2 - 2\sinh \alpha G_x/2)\right]$$

$$\cdot\, [\beta G_y - 2\exp\left(-\beta G_y/2\right)\sinh \beta G_y/2]$$

$$-2\left[\frac{(\alpha T_x)^3}{12} + \left(1 + \frac{\alpha G_x}{2}\right)\exp\left(-\alpha G_x/2\right)(\alpha T_x \cosh \alpha T_x/2 - 2\sinh \alpha G_x/2)\right]$$

$$\cdot\, [\beta T_y - 2\exp\left(-\beta G_y/2\right)\sinh \beta T_y/2]$$

$$+\left[\frac{(\alpha T_x)^3}{12} + \left(1 + \frac{\alpha T_x}{2}\right)\exp\left(-\alpha T_x/2\right)(\alpha T_x \cosh \alpha T_x/2 - 2\sinh \alpha T_x/2)\right]$$

$$\cdot\, [\beta T_y - 2\exp\left(-\beta T_y/2\right)\sinh \beta T_y/2] \tag{22-141}$$

The noise variance $\sigma_n{}^2$ is related to N_o by

$$\sigma_n{}^2 = 4B_x B_y N_o \tag{22-142}$$

where B_x and B_y are the horizontal and vertical bandwidths determined by the modulation transfer function.
Then

$$\sigma_{x_c} = \frac{1}{4\sqrt{3}}\left(\frac{A_g}{A_t}\right)^{1/2}\frac{1}{\text{SNR}}\left[\frac{G_x{}^2}{B_x T_x B_y T_y} + \frac{12(\text{BNR})^2 S}{\alpha^4 \beta^2 A_t A_g}\right]^{1/2} \tag{22-143}$$

where
$$\text{signal-to-noise ratio} = \text{SNR} = A/\sigma_n$$
$$\text{background-to-noise ratio} = \text{BNR} = \sigma_b/\sigma_n$$
$$\text{target area} = A_t = T_x T_y$$
$$\text{gate area} = A_g = G_x G_y$$

An idealized, special case results from the following conditions: no background, or BNR = 0; $G_x = T_x$, $G_y = T_y$; $B_x = B_y = (2\delta)^{-1}$, where δ is the resolution cell diameter. These conditions give the useful result

$$\sigma_{x_c} = \frac{\delta}{\sqrt{12}\ \text{SNR}} \tag{22-144}$$

Another special case is: a small (with respect to the gate) target for which $B_x T_x \simeq 1$ and $B_y T_y \simeq 1$; large background correlation lengths, giving $\alpha G_x \ll 1$ and $\beta G_y \ll 1$;

$$S \simeq \frac{(\alpha G_x)^3 (\beta G_y)^2}{12} \tag{22-145}$$

These conditions give

$$\sigma_{x_c} \simeq \frac{1}{4\sqrt{3}} \left(\frac{A_g}{A_t}\right)^{1/2} \frac{1}{\text{SNR}} \left[G_x^2 + \left(\frac{A_g}{A_t}\right) \frac{G_x}{\alpha} (\text{BNR})^2 \right]^{1/2} \tag{22-146}$$

In the case of negligible internal-noise (compared to the background)

$$\sigma_{x_c} \simeq \frac{1}{4\sqrt{3}} \left(\frac{A_g}{A_t}\right) \sqrt{\frac{G_x}{\alpha}} \frac{1}{\text{SBR}} \tag{22-147}$$

where SBR is the signal-to-background ratio A/σ_b. Note that the rms tracking-error is here proportional to A_g/A_t. This might have been expected since $A_t \ll A_g$ implies a larger probability of background induced errors than $A_t \gtrsim A_g$.

Precision Tracking. Systems which are to achieve high-precision pointing and tracking accuracies require special design considerations. In Section 22.2, the basic approach of minimizing pointing-jitter due to local noise-sources (other than those due to the tracking sensor) by means of LOS stabilization is discussed in some detail. There it is indicated that multigimbal (more than two) systems may be required to sufficiently isolate the LOS from base motions. Since greater isolation can be achieved with larger stabilization-loop bandwidths, the mechanical design should eliminate structural resonances close to the desired open-loop crossover-frequencies. It is also indicated that torquing methods must have low friction and virtually no backlash. In Section 22.3.3, the effect of tracking-loop bandwidth on tracking-jitter is considered. Potentially better performance can be achieved over that resulting from a conventional tracking loop (i.e., one with fixed parameters) if the target dynamics are taken into account in the control-loop design via a Kalman filter. Ultimately, however, the tracking accuracies required may be limited by the noise associated with the tracking sensor itself.

Often in infrared systems which involve high-precision pointing and tracking, the accuracies required are much smaller than the target angular-subtense. In these systems, the sensor resolution is usually smaller than the target subtense, providing extended-target imagery. A common approach to generating a two-axis tracking-error signal for each scan cycle, or frame, is to compute the horizontal and vertical displacements of the image centroid from the sensor zero-reference (e.g., the sensor LOS).

For a rectangular target of uniform radiance, and an ideal tracking-gate which limits the FOV to the target angular-subtense, the rms tracking-noise in each axis is $\delta/(\sqrt{12}\ \text{SNR})$, where δ is the resolution element size, and SNR is the peak-signal to rms-noise ratio. The derivation of this result is outlined earlier in this section. One assumes the noise is uncorrelated from element to element. In principle, for relatively small values of SNR, tracking accuracies to small fractions of δ can be achieved. However, in general, the tracking gate is larger than the target, and rms tracking-noise increases proportionally. To counter this effect, one often incorporates thresholding to eliminate the noise resulting from resolution elements outside of the target angular-subtense. The ideal level of tracking noise can be obtained through thresholding if SNR > 10. For lower values of SNR, the interrelationship between threshold setting and gate size becomes more complex; the effect is shown in Figure 22-47 for two types of thresholding:

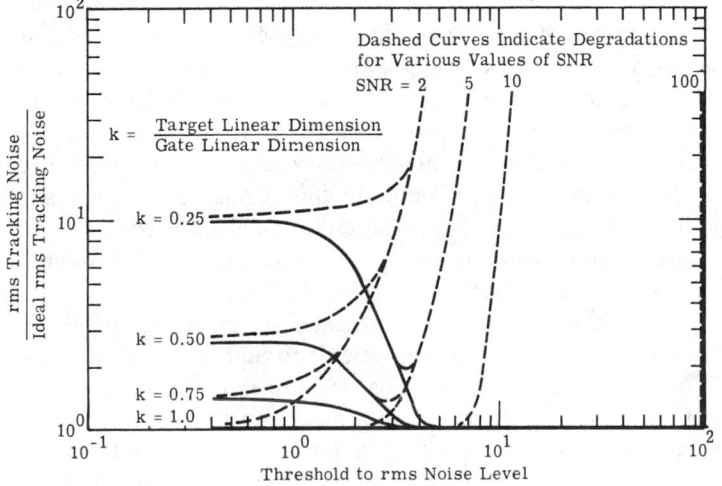

(a) Effects of Type-I thresholding.

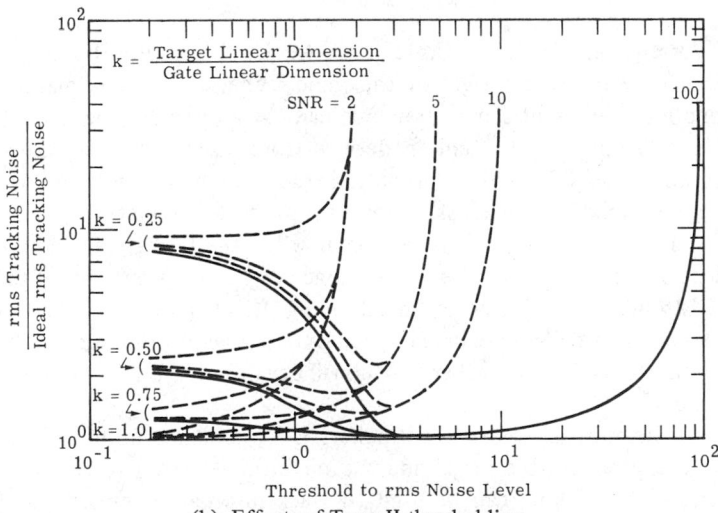

(b) Effects of Type-II thresholding.

Fig. 22-47. Tracking noise for different gate sizes and threshold settings.

Type I—The signal above threshold maintains its value. The signal below threshold is set to zero.

Type II—The threshold level is subtracted from the signal above threshold. The signal below threshold is set to zero.

For either case, if the SNR is low (e.g., 5 or less), there is an optimal threshold for each gate size. Better performance can be achieved with Type-I thresholding. However, Type-II thresholding may be more amenable to ad hoc adaptive approaches for automatic gate- and threshold-selection because changes due to SNR variations are more gradual, which implies less likelihood of loss of lock.

A higher degree of tracking accuracy can be achieved with imaging sensors than with point-source sensors. This follows from the improved clutter rejection that is possible with proper threshold and gate settings. However, special consideration must be given to an imaging sensor designed for providing good imagery to a viewer. For example, in such a sensor, if alternate fields are interlaced, then an image with detail may cause tracking jitter if the error signals are based upon each field. To achieve subresolution element tracking, one must either remove the interlace or average the error signals over a complete frame, thereby halving the data rate. Similarly, the tolerances of the scanning mechanism may be sufficient for imaging but not for subresolution element tracking.

There are other approaches to extended source tracking other than centroid tracking, such as area tracking or edge tracking. In general, these approaches can be mechanized quite simply if the imagery is digitized to one bit, i.e., truncated to zeros and ones. For high SNR, the performance would be comparable to that of a centroid tracker. However, for low SNR this may not be the case, particularly if the imagery detail is lost in the digitized signal. A correlation tracker would make most use of imagery detail by comparing the current scene to a stored reference (an earlier frame of the sensed image) and determining the displacement between the two. A correlation tracker is the most complex to mechanize but provides the best tracking performance in a low SNR condition. A properly designed correlation tracker would remove the requirements for optimizing threshold levels and gate sizes.

22.7.2. Correlation Trackers. Correlation trackers measure the relative displacement between two different images of the same scene. One of the images is called the reference, and may represent a measurement made at an earlier time. The other image is called the received image, and usually is derived from a sensor viewing the "live" scene.

There are many cases for which correlation trackers are inherently better than GVTs. For example, a correlation tracker is preferred if the target to be tracked cannot be identified readily and automatically. This can occur when the target is in the midst of other objects (clutter) which are too close to the target to be gated out, and if the clutter signals are too strong to be rejected by thresholding. In such a case, if the target location within the scene is initially designated by an external acquisition aid (such as a human operator), then a correlation tracker can continue alone to track the target during the tracking mission.

A useful application of correlation trackers is in guidance units of homing missiles. As the missile approaches the target and the target image size exceeds the tracker field of view, correlation tracking can be used to track internal target-detail and maintain a stable, terminal tracking-point. This overcomes the blind-range limitation of gated-video trackers. Correlation trackers can also be used to slave the missile guidance-unit to the pointing direction of a target acquisition system containing another sensor.

A correlation tracker is the natural choice for identifying one object within a cluster of similar objects, where the target is specified by its relative position. For example, automatic acquisition of a particular star by a spaceborn tracker is readily performed by a correlation tracker. A similar application is automatic map-matching for periodic update of airborne, inertial, navigation systems. In this case, reference ground-maps corresponding to known positions over the flight path are stored in tracker memory. The tracker indicates when the sensor image is in registration with a stored map, at which time the inertial system is updated.

Correlation Tracker Algorithm. The stored reference-image is $s(x, y)$ defined over a region S. The received image $r(x, y)$, displaced from the reference and containing noise, is given by

$$r(x, y) = As(x - x_o, y - y_o) + n(x, y) \qquad (22\text{-}148)$$

where A = an unknown constant
(x_o, y_o) = the horizontal and vertical translations* between s and r
n = the receiver noise

The image, r, is defined over a region, R, which usually is larger than S and contains it. The tracking loop is designed to shift the sensor axis to bring the received signal $r(x, y)$ into registration with the stored reference $s(x, y)$, i.e., to drive (x_o, y_o) to zero. Thus, it is necessary to form error signals, ϵ_{tx}, ϵ_{ty}, that are good estimates of the tracking errors x_o, y_o.

The cross-correlation function between s and r is

$$C(x, y) = \frac{\displaystyle\int\int s(u, v)\, r(u + x, v + y)\, du\, dv}{\left\{\displaystyle\int\int [s(u, v)]^2\, du\, dv \int\int [r(u, v)]^2\, du\, dv\right\}^{1/2}} \qquad (22\text{-}149)$$

where the integrations are over the region S. If the noise term is neglected, application of the Schwartz inequality [22-37] shows that $C(x, y)$ has a global maximum of unity at (x_o, y_o). The operation of correlation trackers is based upon this fact. Different correlation-tracker designs represent different methods of estimating the location of the $C(x, y)$ maximum.

The most obvious design-approach is the direct computation of $C(x, y)$, over a range of x, y values large enough to contain the error x_o, y_o, and taking the coordinates of the $C(x, y)$ maximum as the error signals. Ordinarily, this is very cumbersome and imposes a large computational requirement. However, practical trackers have been built using this approach. Another approach is to compute the two-dimensional Fourier transforms of $r(x, y)$, and $s(x, y)$, multiply the transforms, and compute the inverse

---

*A more general model could include other differences between s and r, such as rotation, magnification difference, and image distortions.

transform of the product, which is $C(x, y)$. This also imposes a large computational requirement, but is practical in some cases.

The correlation function $C(x, y)$ typically has a main peak near x_o, y_o, and several, smaller, local maxima at other points. If the tracking error is contained within the main peak, then $\partial C/\partial x$ and $\partial C/\partial y$ can be used as error signals in the x and y axes, respectively. The partial derivatives can be obtained much more readily than the maximum of $C(x, y)$. The s and r signals obtained in practice satisfy mathematical smoothness requirements. Then

$$C_x(x, y) = \frac{1}{\text{Den}} \int \int s_x(u - x, v - y) r(u, v) \, du \, dv \qquad (22\text{-}150)$$

$$C_y(x, y) = \frac{1}{\text{Den}} \int \int s_y(u - x, v - y) r(u, v) \, du \, dv \qquad (22\text{-}151)$$

where subscripts = partial derivatives
       Den = the denominator in the $C(x, y)$ expression

The error signals are then

$$\epsilon_{tx} = KC_x(0, 0) \qquad (22\text{-}152)$$

and

$$\epsilon_{ty} = KC_y(0, 0) \qquad (22\text{-}153)$$

where K is a suitable constant. This represents a significant reduction in computation compared to determining $C(x, y)$ at many points. The C_x and C_y expressions are of comparable complexity as C, but they need only be evaluated at $(0, 0)$. However, the tracking error must be contained within the main peak for this method to work.

The partial-derivative correlation-tracking concept has been developed and studied in detail [22-38], with emphasis on the common, practical case of images which are sampled in both directions. The approach taken is to determine optimal weighting-functions corresponding to s_x and s_y in the C_x and C_y expressions. The optimality criterion is minimization of the mean-square value of the random components of ϵ_{tx} and ϵ_{ty}. Values are determined for K such that, for small tracking errors, the expected values of $(\epsilon_{tx}, \epsilon_{ty})$ are (x_o, y_o). This is done by taking the first two terms of the Taylor series expansions of C_x and C_y about $(0, 0)$ and solving for the point at which C_x and $C_y = 0$. This approach overcomes the limitation of tracking accuracy to ± one-half the sampling interval for direct, peak-locating correlation-trackers.

The algorithm developed in Reference [22-38] is

$$\epsilon_{tx} = \frac{a}{2} + \frac{a}{AS_x} \sum_{i=1}^{M} \sum_{j=1}^{M} W_x(i, j) r(i, j) \qquad (22\text{-}154)$$

$$\epsilon_{ty} = \frac{b}{2} + \frac{b}{AS_y} \sum_{i=1}^{M} \sum_{j=1}^{M} W_y(i, j) r(i, j) \qquad (22\text{-}155)$$

The optimum weighting-functions are*

$$W_x(i, j) = -[s(i + 1, j) - s(i, j)]$$ (22-156)

$$W_y(i, j) = -[s(i, j + 1) - s(i, j)]$$ (22-157)

and

$$S_x = \sum_{i=1}^{M} \sum_{j=1}^{M} [W_x(i, j)]^2$$ (22-158)

$$S_y = \sum_{i=1}^{M} \sum_{j=1}^{M} [W_y(i, j)]^2$$ (22-159)

The parameters a and b are the sample intervals in the x and y directions, respectively, and A can be estimated by

$$\hat{A} = \frac{\sum_{i=1}^{M} \sum_{j=1}^{M} [r(i, j) - \bar{n}]}{\sum_{i=1}^{M} \sum_{j=1}^{M} s(i, j)}$$ (22-160)

where $\bar{n}$ is the mean noise-level (usually zero). This summation region corresponds to the area of the stored image.

Since the received image $r(i, j)$ contains noise, the error signals $(\epsilon_{tx}, \epsilon_{ty})$ will contain noise components. If the noise is white, the mean-square values of these noise components are given by

$$\sigma_x^2 = \frac{a^2}{S_x} \left(\frac{1}{\text{SNR}}\right)^2$$ (22-161)

$$\sigma_y^2 = \frac{b^2}{S_y} \left(\frac{1}{\text{SNR}}\right)^2$$ (22-162)

where SNR is the signal-to-noise ratio of the received image, i.e.,

$$\text{SNR} = \frac{A}{\sigma_n}$$ (22-163)

and σ_n is the rms value of the noise component of $r(x, y)$.

---

*It is assumed in Reference [22-38] that the stored reference is normalized so that

$$\sum_{i=1}^{M} \sum_{j=1}^{M} s(i, j) = 1$$

Examination of the algorithm shows that the weighting functions W_x and W_y are proportional to the directional differences of the stored reference, and that the algorithm converges to the partial-derivative method for continuous systems as the sample intervals go to zero. The variables S_x and S_y are, in a sense, measures of the amount of tracking information in the stored reference; hence the output-noise variances are inversely proportional to S_x and S_y.

Quantization and Sampling Effects. Advances in the technology of integrated circuits have led to widespread use of digital techniques in signal processing applications. Although digital processing has many advantages, these techniques necessarily introduce errors. One error-source is the loss of information caused by the representation of continuous signals by a finite set of discrete values (amplitude quantization). Quantization error is reduced as the number of quantization levels is increased. Another error source is the loss of information due to sampling of signals which may originate as continuous time-functions. Although the Nyquist sampling theorem states that no information is lost if the sampling frequency is at least twice the signal bandwidth, practical constraints usually require sampling frequencies appreciably greater than the Nyquist rate to avoid large errors. Since equipment cost increases with increasing quantization precision and sampling frequency, a cost versus performance trade off often is called for in choosing these parameters.

Projections of equipment cost for different system parameters are elusive and difficult to obtain, especially for new designs. They are also very specialized to each design and to each manufacturer, and therefore must be addressed separately for each particular case.

Something can be said, however, about the performance effects of quantization and sampling. Performance degradation due to quantization and sampling has been studied for a one-dimensional product correlator [22-39]. Although the present concern is for two-dimensional correlators, the one-dimensional results are indicative of the performance expected in two dimensions.

Performance loss due to quantization and sampling is given as a degradation factor D, defined as the output signal-to-noise ratio (average signal/rms noise) of an analog correlator divided by that for a digital correlator. The digital-correlator parameters are:

Q_s = number of quantized values of the stored image, $s(x, y)$.
Q_r = number of quantized values of the received image, $r(x, y)$.
K = sampling frequency/B.
B = bandwidth of s and r.

Thus, if samples of $s(x, y)$ are stored as n-bit binary numbers, $Q_s = 2^n$.

Calculated curves of D versus K are given in Figure 22-48 for differ nt combinations of Q_r and Q_s. The quantization levels used are optimized to give minimum D. The normalized, sampling frequency $K = 2$ represents the Nyquist rate. It is seen that the worst case is $D = \pi/2$ for $Q_r = Q_s = 2$ (one-bit quantization) and $K = 2$, and that D drops rapidly as K increases to about 4 or more. The figure also shows that D is not far above unity for two-bit quantizers ($Q_r = Q_s = 4$) and $K \geqslant 6$.

Sequential Similarity Detection Algorithms (SSDA). If the initial error is not confined to the correlation-function main-peak, the partial-derivative method may not work, at least for initial target acquisition. In such cases, numerical search for the correlation-function global maximum can be used. However, as noted earlier, this imposes a large computational burden. An alternative class of algorithms has been suggested which is much simpler to implement and which appears to locate efficiently the correlation-function global maximum [22-40]. The efficiency of these algorithms makes them attractive, even with tracking errors within the main peak.

SSDAs are directed toward simplifying computational requirements, and therefore were introduced in terms of sampled, discrete-valued signals. Thus the stored reference image $x(i, j)$ is an $M \times M$ array of digital picture-elements, each of which has one of K grey levels, $0, 1, 2, ..., K-1$. The received image is similarly defined on an $L \times L$ array, with $L > M$:

$$0 \leqslant r(i, j) \leqslant K - 1 \qquad (22\text{-}164)$$

$$0 \leqslant i, j \leqslant L \qquad (22\text{-}165)$$

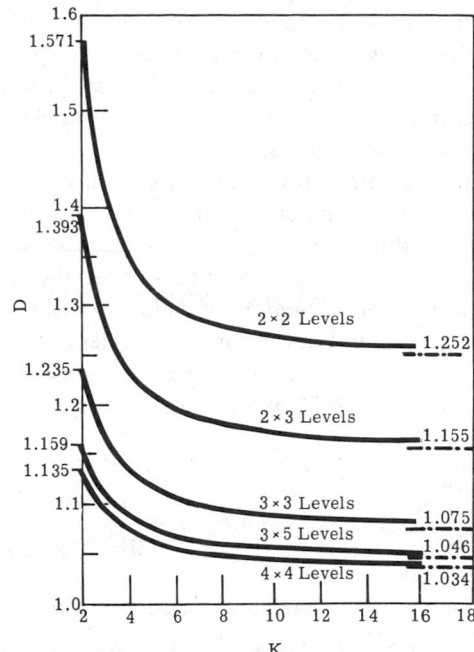

Fig. 22-48. Degradation factor versus normalized sampling rate. Correlators with various numbers of quantized levels are compared [22-39].

Assume that a subarray of the received image $r(i, j)$ contains the stored reference image s. An image-registration error-function $E(m, n)$ can be defined as

$$E(m, n) = \sum_{i=1}^{M} \sum_{j=1}^{M} |r(m+i, n+j) - s(i, j)| \qquad (22\text{-}166)$$

If the coordinates (m, n) correspond to the translations (x_o, y_o) between r and s, then $E(m, n) = 0$; otherwise $E(m, n) > 0$. In the ideal, noise-free case, it is only necessary to search for the point at which $E = 0$. In a realistic case, some small, positive threshold is used instead of 0, or else the minimum value of E is taken as the image registration point.

The computational requirements are reduced by replacing multiplications with additions, subtractions, and absolute value functions. A further computational reduction is achieved by monitoring the value of the partial sum of $E(m, n)$ as each new term is added. Then, if the partial sum grows rapidly, the point (m, n) can be rejected as the image match-point before computing the full sum of M^2 terms. Most nonregistration points will be rejected quickly, resulting in significant computational savings.

These factors are combined to form the following algorithm. A point (m, n) is selected, and $E(m, n)$ is accumulated. As each new term is added, the pth partial sum $E_p(m, n)$ is compared to a threshold T_p. If $E_p(m, n) < T_p$ and $p < M^2$, then $E_{p+1}(m, n)$ is computed; otherwise the computation terminates for (m, n), and the value of p is stored for the point (m, n). When all points (m, n) have been processed, the point having the largest value of p is taken as the image registration point.

It remains to specify the threshold sequence T_p. The sequence should be chosen to eliminate points of misregistration quickly, but to discriminate effectively at points near registration. The simplest case is T_p = constant. Better performance has been observed for monotonically increasing sequences, some of which are given in [22-40].

Many variations of the basic procedure are possible, resulting in an entire class of algorithms. For example, the ordering of the points (m, n), at which the registration is tested, may be specified so that the registration point can be found without testing all possible $(L - M + 1)^2$ points. Another variation is to design the threshold sequence so that it adapts to the data and accelerates identification of the registration point.

An SSDA variation, known as mean absolute difference (MAD) correlation, was studied [22-41], and compared to a product-correlation algorithm. The MAD correlation-function is

$$D(m, n) = \frac{1}{M^2} \sum_{i=1}^{M} \sum_{j=1}^{M} |r(m + i, n + j) - s(i, j)| \qquad (22\text{-}167)$$

The product-correlation function, to which it is compared, is

$$C(m, n) = \frac{1}{M^2} \sum_{i=1}^{M} \sum_{j=1}^{M} r(m + i, n + j)s(i, j) \qquad (22\text{-}168)$$

In the MAD algorithm, $D(m, n)$ is computed over a given search area, and the point at which $D(m, n)$ is a minimum is taken as the registration point. In the product algorithm, $C(m, n)$ is computed over the search area, and registration is taken to occur at the maximum of C.

Simplifying assumptions permit analytical expressions for the probability of correlation, P_c, the probability that the indicating extremum occurs at the true registration-point. Although the simplifications may appear to be overly restrictive, the results compare favorably to experimental data. Figure 22-49 contains analytical results for P_c as a function of normalized search-area A_s, and normalized correlation-area, A_c, for signal-to-noise ratios of 5 and 10. The normalized search-area A_s is defined as the number of independent points at which the correlation functions $D(m, n)$ and C are computed, and is

$$A_s = \min \left\{ (L - M + 1)^2, \frac{(L - M + 1)^2}{(\pi l)^2} \right\} \qquad (22\text{-}169)$$

where l is the image correlation-length in units of sample intervals. The normalized correlation-area is the number of independent data samples in each value of $D(m, n)$ and C, and is

$$A_c = \min \left\{ M^2, \frac{M^2}{(\pi l)^2} \right\} \qquad (22\text{-}170)$$

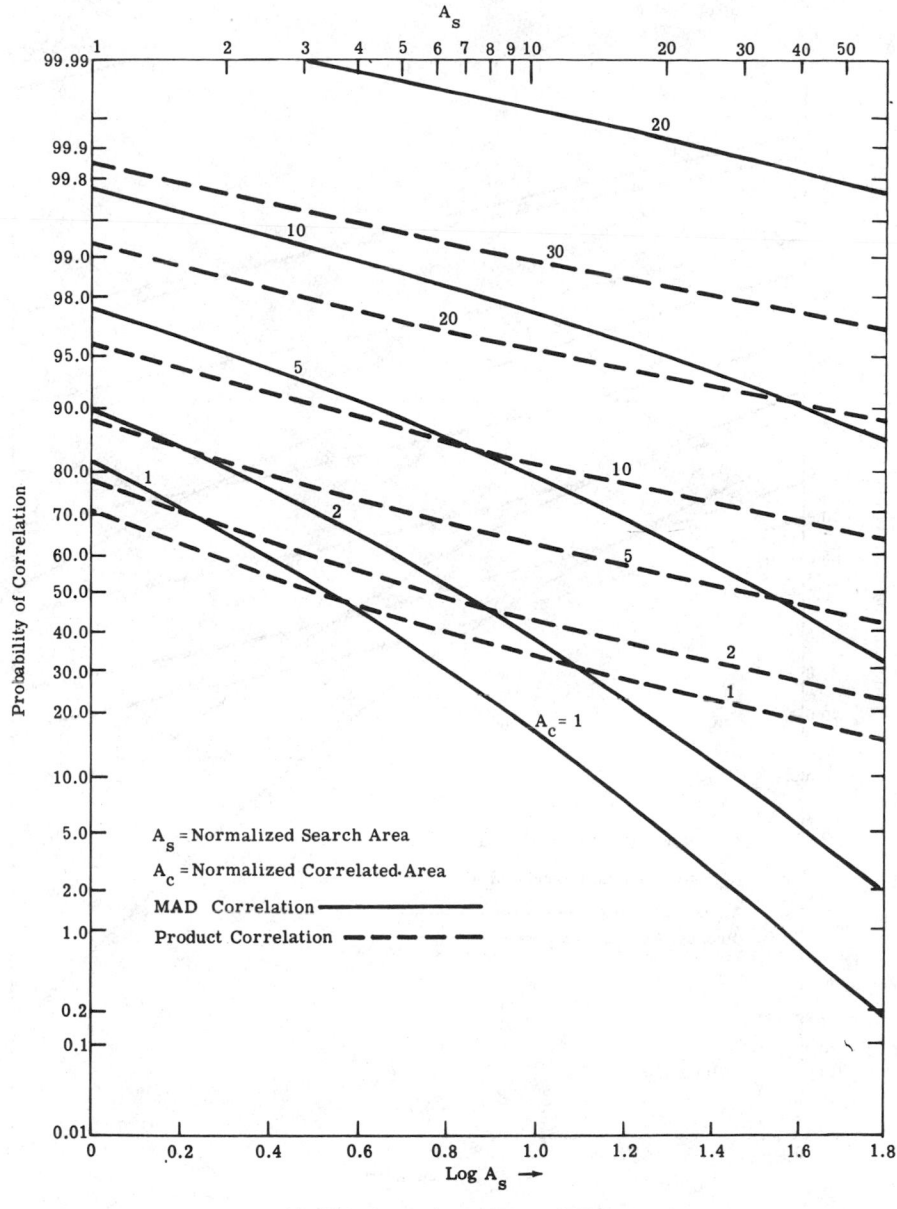

(a) Signal-to-noise ratio equals five.

Fig. 22-49. Probability of correlation for mean absolute difference and product algorithms [22-41].

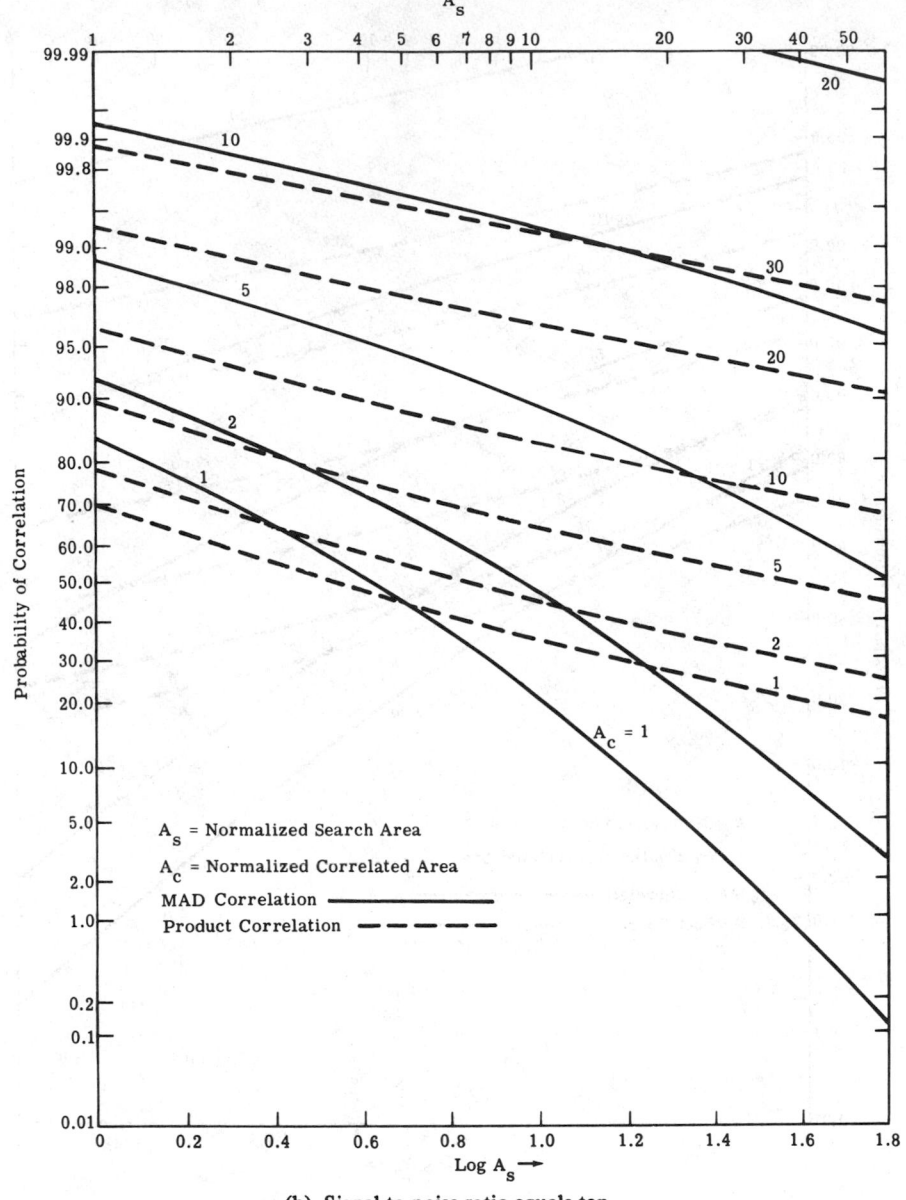

(b) Signal-to-noise ratio equals ten.

Fig. 22-49 (Continued). Probability of correlation for mean absolute difference and product algorithms [22-41].

The signal-to-noise ratio is defined as the ratio of the mean-square value of the scene data to the mean-square noise. It is seen that, over large intervals of A_s and A_c, the MAD algorithm performs better than the product algorithm. The MAD algorithm has the further advantage of much greater computational efficiency.

22.8. Pulsed Laser Trackers*

22.8.1. General Considerations. Pulsed laser tracking-systems detect and track narrow pulses of radiant energy. The pulses may come directly from a laser beacon, or may be reflected from an illuminated target. The tracker computes estimates of position error (from each detected pulse) to recenter the receiver. In reflected-beam trackers, the laser source may be included as part of the tracker system or positioned at some remote site.

Figure 22-50 shows a simplified diagram of the optics and detection system. The angular subtense of the laser beam radiated from the target must be consistent with the target-to-laser distance and the target size. The received spot-size should not exceed the diameter of the optical blur circle of the tracking receiver. Figure 22-51 shows how the received spot produces a position error estimate. The four-element array is connected in a quadrant-detector configuration. This scheme, similar to monopulse radar detection, provides a nearly-linear error-signal versus position-offset. A small spot can be optically defocused to a fixed diameter, but when the spot becomes large, the defocused spot-size begins to grow as a function of received beam width, degrading tracker performance.

The output of each detector is proportional to the image power incident upon its sensitive area. The error signals for the elevation and azimuth axes are

$$\epsilon_{\text{EL}} = \frac{(I_A + I_B) - (I_C + I_D)}{I_A + I_B + I_C + I_D} \tag{22-171}$$

and

$$\epsilon_{\text{AZ}} = \frac{(I_B + I_D) - (I_A + I_C)}{I_A + I_B + I_C + I_D} \tag{22-172}$$

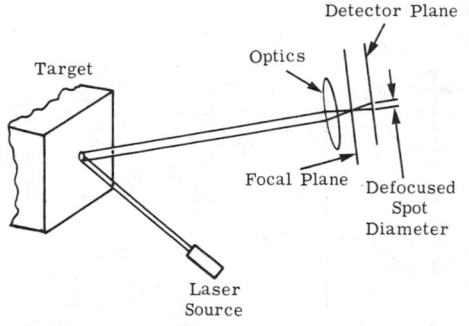

Fig. 22-50. Optics and detection system.

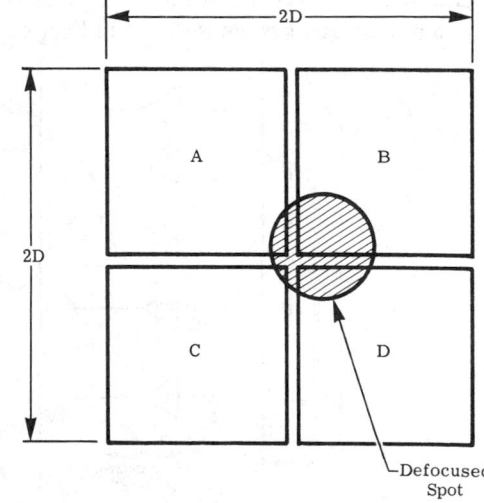

Fig. 22-51. Quadrant detector array.

*This section was prepared by Lawrence C. Caplan, Hughes Aircraft Co., Canoga Park, CA.

where I_x for $x = A, B, C, D,$ is the peak signal current of detector x. Normalization to ΣI_x makes the error signals independent of pulse-to-pulse energy variations in the received laser beam. The curves of idealized error-signal versus position are given in Figure 22-52 for uniform, round and square spots. Perfect linearity occurs with a square spot, since the area is linear with position along one dimension. However, the linearity degrades slightly for a round spot. When the position error exceeds the defocused-spot radius r, the error signal saturates. As the error exceeds $D - r$, the spot goes off the outer edge of the detectors, and the error signal tends toward zero, at an error $D + r$. These three regions are identified on Figure 22-52. The curves also show that an increase in the defocused-spot size, r, will result in a change of slope for the linear region.

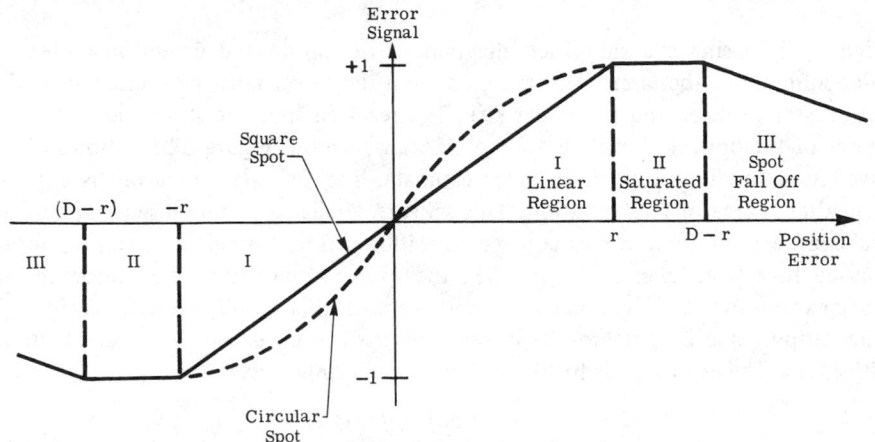

Fig. 22-52. Idealized error-signal versus position error.

22.8.2. Tracker Electronics. The signal processing for a pulsed-laser system must operate over a wide dynamic range. In general, the dynamic range of a linear amplification chain is not sufficient and some form of automatic gain control or logarithmic amplification must be included. A block diagram showing the signal processing functions in a pulsed laser system is shown in Figure 22-53.

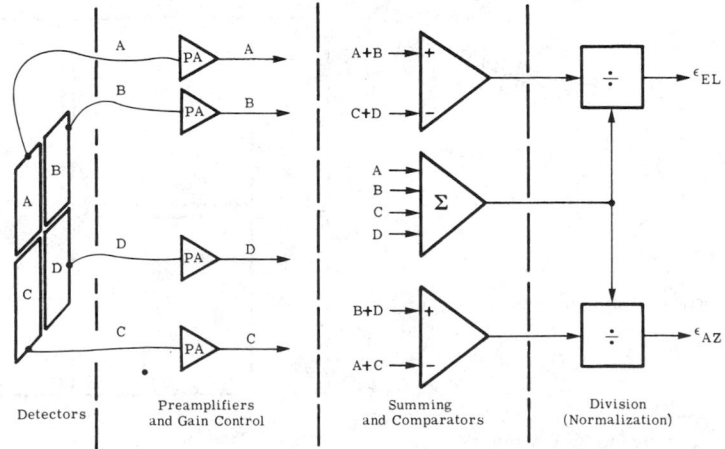

Fig. 22-53. Signal processing functions in a pulsed laser system.

22.8.3. Sensitivity and Noise Effects. There are two primary limitations in detecting low-energy laser-pulses. First is the shot-noise current from each quadrant due to background, dark currents, and thermal noise. The shot-noise current is

$$\bar{i}_{n_1} = \sqrt{2e(I_B + I_D)\Delta f} \tag{22-173}$$

where e = the electron charge

I_B and I_D = background and dark currents, respectively

Δf = the detector electrical bandwidth

The thermal-noise current in the resistive part of the detector load circuitry is

$$\bar{i}_{n_2} = \sqrt{\frac{4k_B T\Delta f}{R_L}} \tag{22-174}$$

where R_L = the load resistance

k_B = Boltzmann's constant

T = the absolute temperature of R_L

The theoretical noise-current competing with the signal-current at the preamplifier input is

$$\bar{i}_n = \sqrt{\bar{i}_{n_1}^2 + \bar{i}_{n_2}^2} \tag{22-175}$$

The signal current at the quadrant-detector output is

$$i_{s_q} = \delta \left(\frac{\Psi_R}{\Delta t}\right) A_o \Re \tag{22-176}$$

where Ψ_R = the received laser-beam energy-density or "fluence," J m^{-2}

Δt = the pulse width

A_o = the collecting aperture area

$\Re$ = the detector responsivity

δ = the fraction of the spot area on the detector

The energy density at the dome, Ψ_R, is found from

$$\Psi_R = \frac{Q_L \tau_{LT} \tau_{TR} \, \rho \cos \theta}{\pi R_{TR}^2} \qquad (22\text{-}177)$$

where Q_L = the laser-source energy
$\quad\quad \tau_{LT}$ = the laser-to-target transmission
$\quad\quad \tau_{TR}$ = the target-to-receiver transmission
$\quad\quad \rho$ = the target diffuse-reflectance
$\quad\quad \theta$ = the angle between the target-surface normal
$\quad\quad\quad$ and the tracker line-of-sight
$\quad\; R_{TR}$ = the target-to-receiver range

The signal-to-noise ratio at the input to the quadrant preamplifier is

$$\left(\frac{S}{N}\right) = \frac{i_{s_q}}{\bar{i}_n} \qquad (22\text{-}178)$$

To evaluate the quality of the error signals ϵ_{EL} and ϵ_{AZ} in the presence of noise, one can assume a Gaussian noise contribution i_{n_3} from the preamplifier. The $\bar{i}_n$ equation is modified to reflect the noise at the quadrant preamplifier output. The noise current becomes

$$i_{n_q} = \sqrt{\bar{i}_{n_1}^{\;2} + \bar{i}_{n_2}^{\;2} + \bar{i}_{n_3}^{\;2}} \qquad (22\text{-}179)$$

An expression for the probability density of a single-axis error-signal can be obtained for the case when each quadrant has a Gaussian noise-source i_{n_q} with zero-mean and variance σ_n^2. The actual error-signal for an axis is

$$\epsilon = \frac{S_D + N_1}{S_S + N_2} = \frac{U - V}{U + V} \qquad (22\text{-}180)$$

where S_D = the pairwise difference-signal for a given axis
$\quad\;\; S_S$ = the sum-signal of all quadrants
$\quad\;\; N_1$ = the rms noise associated with the difference signal
$\quad\;\; N_2$ = the rms noise of the sum-signal

N_1 and N_2 are not independent since they are different combinations of the same four quadrant noises. The Gaussian random variables U and V represent detector signals taken pairwise such that

$$U = \bar{u} + N_u \qquad (22\text{-}181)$$

$$V = \bar{v} + N_v \qquad (22\text{-}182)$$

and

$$\bar{u} - \bar{v} = S_D \qquad (22\text{-}183)$$

$$\bar{u} + \bar{v} = S_S \qquad (22\text{-}184)$$

The resulting expression for the probability density of the error signal ϵ has been found to be

$$f(\epsilon) = \frac{1}{\pi\,(1 + \epsilon^2)}\,e^{(-1/2\sigma_p^2)(\bar{u}^2 + \bar{v}^2)}\left[1 + \sqrt{2\pi}\,Be^{B^2/2}\int_0^B \frac{1}{\sqrt{2\pi}}\,e^{-1/2\alpha^2}\,d\alpha\right] \quad (22\text{-}185)$$

where $B = \dfrac{1}{\sqrt{2\sigma_p^2}}\left(\dfrac{(1 + \epsilon)\bar{u} + (1 - \epsilon)\bar{v}}{\sqrt{1 + \epsilon^2}}\right)$

σ_p^2 = the variance of both U and V and equal to $2\sigma_n^2$

A family of curves can be drawn for a given value of signal-to-noise ratio $(S_S/2\sigma_p^2)$. Each curve in the family represents a probability density about the expected value of the error, S_D/S_S, where $0 \leqslant S_D/S_S \leqslant 1$. Figure 22-54 shows two such curves when $S_S/2\sigma_p^2$ is 11.4 and S_D/S_S is 0.2 and 1.0. The important property of these systems which is evident from the curves in Figure 22-54 is the increasing spread of the density function for increasing errors within the linear tracking range. The error commands become more noisy with increasing error.

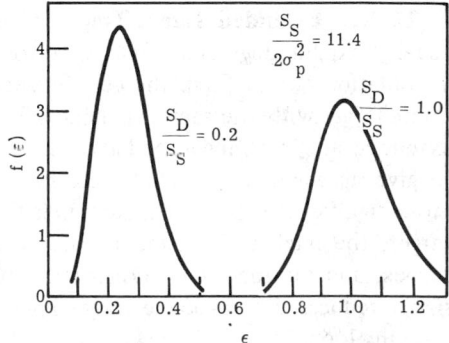

Fig. 22-54. Probability densities for normalized error signal.

22.8.4. Other Design Considerations.

The possibility of laser energy returns from the surroundings of the target exists for some conditions. These include large spot-size to target-size ratio, jitter of the laser illuminator, and target motion. The techniques used to avoid tracking an object other than the target include receiver time-gating and amplitude-discrimination when differences in reflectivities exist between the different illuminated objects. Not only must one consider the reflectivity variation between objects, but also the differences in reflectivity within the target surface, if the laser spot-size is small.

Another concern is the effect of atmospheric turbulence upon the laser beam. This manifests itself by introducing scintillation-noise upon the received pulse amplitude. Reflectivity variations and scintillation both introduce a requirement for increased pulse-to-pulse dynamic-range of the tracker.

A final item of importance is the need for matched gains of the four-quadrant processing channels. If the gains become mismatched, the equilibrium tracking-point will correspond to an offset of the laser spot from the center of the quadrant detectors. Depending on which processing algorithm is used, the error may become a function of error angle and signal-to-noise ratio.

22.9. Special Tracking Requirements

Many applications impose special performance requirements upon the tracker. In addition to tracking targets which subtend relatively small angles, one must, in some cases, also track large or extended targets, i.e., infrared sources which subtend angles which are large fractions of the field of view, or even exceed the field of view. For example, this requirement applies to some IR homing missiles with targets which may appear as extended source near the end of flight. Another possible missile requirement is to generate estimates of trajectory variables, such as the target range, closing velocity, or time-to-go to target intercept, for use in the autopilot. If certain conditions are satisifed, the tracker will contain the information necessary to generate these estimates.

A requirement that almost always occurs is the need to track targets in the presence of interfering radiation sources. Interfering sources may arise from natural backgrounds, sun glints, extraneous friendly sources, and hostile countermeasures. In such cases, it is necessary to determine if (and how) the detectable characteristics of the interfering sources differ from those of the targets. Often the targets and interfering sources have different spectral signatures, permitting discrimination on the basis of the received radiation spectral content.

22.9.1. Extended Target Tracking. Trackers designed on the basis of point targets usually exhibit degraded tracking performance against extended targets. Two types of degradation occur. First, the peak target-signal generally will be lower than that from a point target with the same irradiance. This may not be serious since targets only become extended at short ranges, and at these ranges the received irradiance often is large enough to give an adequate signal-to-noise ratio even with an extended image. The second and more significant type of degradation is that of the tracker characteristic. For an extended target, the tracker characteristic may contain multiple equilibrium tracking-points, i.e., zeroes, and the maximum error-signal may be reduced. Reduction of maximum error-signal reduces the trackable target line-of-sight rate in the common class of first-order tracking loops.

Figure 22-55 is a simplified, single-axis, first-order, tracking-loop block diagram. The line-of-sight angle to the target, with respect to an inertial reference, is σ. The corresponding tracker pointing-angle is ψ, and the tracking error is $\epsilon = \sigma - \psi$. The tracker slew-rate, $\dot{\psi}$, is equal to the error signal, V_ϵ. The maximum trackable $\dot{\sigma}$ is equal to the maximum value of V_ϵ.

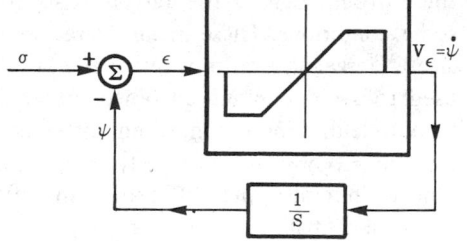

Fig. 22-55. Simplified tracking loop.

Tracker performance, both static and dynamic, can be affected adversely by an extended target image. These effects are not inevitable, but do almost always occur in trackers designed solely for point targets. For example, anti-jet-aircraft missile-trackers, originally designed only for tail attacks, have been considered for use in all aspect missiles. Whereas the target IR signature from the tail is the small tailpipe, at side and forward aspects the signature is primarily due to the relatively long exhaust plume. The increased size of the source causes tracker-characteristic degradation. A typical case is illustrated in Figure 22-56 for a stationary-reticle tracker.

As long as the target image is small enough so that radiation does not pass through two reticle sectors simultaneously, the tracker-characteristic extreme-values will be constant at $\pm V_o$. As the image size grows, the tracker-characteristic will degrade, as shown in the figure. Let V_+ and V_- denote the positive and negative extreme-values of the tracker-characteristic due to a side-aspect plume at some given range. Then, if a target line-of-sight rate $\dot{\sigma}_o$ can be tracked for small targets, the side-aspect tracking-rate for this case is re-

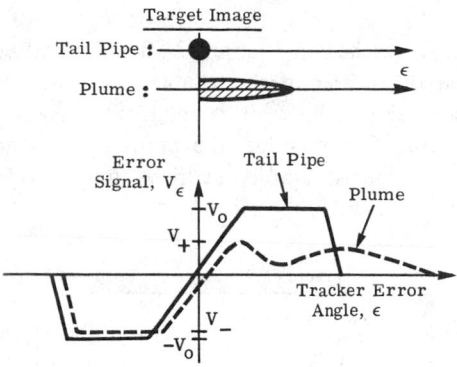

Fig. 22-56. Typical degradation of stationary reticle trackers against extended target.

duced to $(V_+/V_o)\,\dot{\sigma}_o$ for $\dot{\sigma}$ in the negative, $-\epsilon$, direction, and $(V_-/V_o)\,\dot{\sigma}$ for $\dot{\sigma}$ in the positive, ϵ, direction.

A "fix" is possible which restores part of the tracking-rate capability. Assume some indication exists that the missile has entered the terminal flight-phase in which the target image is extended, e.g., preamplifier signal amplitude. Then a constant bias-signal, V_B, can be added to V_ϵ which will shift the tracker operating point from V_ϵ to 0 to $V_\epsilon = -V_B$, as shown in Figure 22-57. This would result in an increased tracking-rate capability for $\dot{\sigma}$ in the $-\epsilon$ direction, at the price of decreased capability for $\dot{\sigma}$ in the $+\epsilon$ direction. The point in the loop at which the bias is inserted corresponds to the tracker line-of-sight rate; hence, this fix is called rate-bias.

The bias signal V_B actually is a two-dimensional vector in the tracker field of view, and some means must be provided for determining its direction. In the example considered here, the infrared sensor is gimbaled, and the gimbal angle is a very good approximation to the desired direction of V_B.

For scanning or imaging trackers, the problem of tracking an extended target reduces to that of determining, from the detector signals, the time when the scan pattern crosses the desired tracking-point. It is then only necessary to sample the scanning reference-signals at this time to track the desired point on the target.

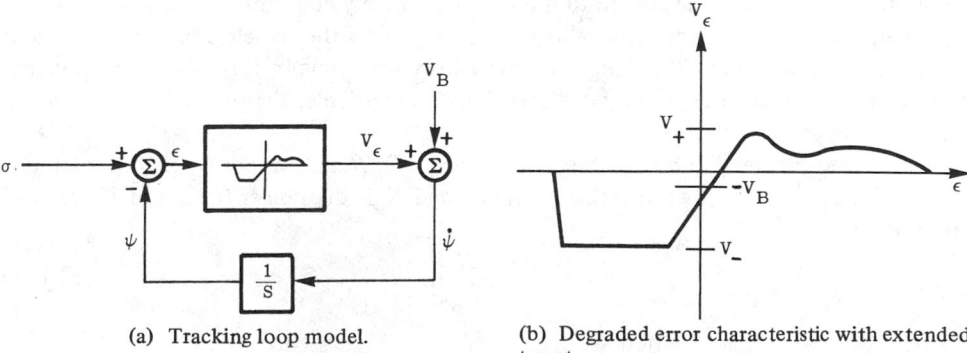

(a) Tracking loop model.

(b) Degraded error characteristic with extended target.

Fig. 22-57. Illustration of rate bias.

For example, the detector signals of an imaging sensor viewing an exhaust plume are sketched in Figure 22-58. The desired tracking-point for a missile would ordinarily be the center of the plume forward-edge, as shown. As the image grows and subtends several scan-lines, it is relatively easy to identify, within each scan line crossing the plume forward-edge, the point corresponding to this crossing. The average of the reference signals, sampled at these times, gives an error signal which vanishes near the desired tracking-point.

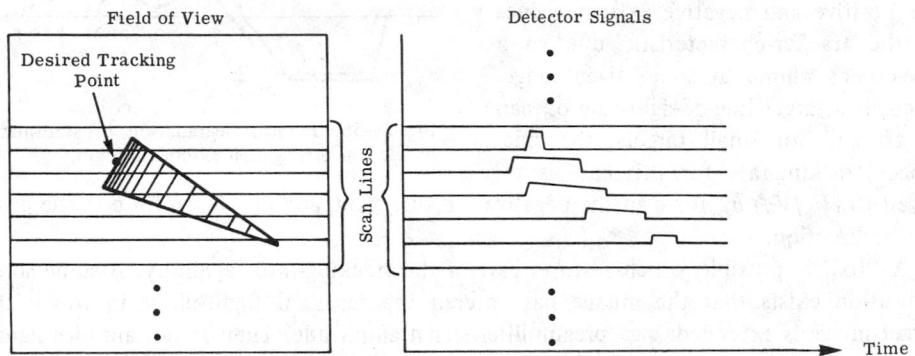

Fig. 22-58. Imaging sensor viewing exhaust plume.

In some applications, the infrared image is less predictable than that of the jet-engine exhaust plume, but the tracking requirement is to establish stable tracking of any point within the image. In these cases, as the image starts to fill the field of view, an expanding gate can be centered within the target and processed to search for internal contrast.

22.9.2. Estimating Trajectory Parameters. In the case of a homing vehicle closing on a target, estimates of the trajectory parameters may be necessary for guidance purposes. Optimum guidance laws often require estimates of time-to-go, T_{go}, until intercept. In some cases, target range, R, and/or closing velocity, V, are needed. The most direct, and accurate method for estimating these parameters is with an active ranging system, such as a pulsed radar or a pulsed laser. Measurements of R, in the form of two-way propagation delay, can be differenced to estimate V, and T_{go} can be estimated from R/V, under the assumption that V is relatively constant.

An active ranging-system is not always available, but if the vehicle contains an imaging or scanning tracker, then T_{go} can be estimated from the time history of the target image size. If a priori estimates of target dimensions are available, then R and V can also be estimated by passive systems.

The angle θ subtended by the target is given by D_t/R rad, where D_t is the maximum target dimension normal to the line of sight, and R is the range. If D_t and V are constant, then

$$\dot{\theta} = \frac{D_t V}{R^2} \tag{22-186}$$

where $\quad \dot{\theta} = d\theta/dt, \quad T_{go} = \dfrac{R}{V} = \dfrac{\theta}{\dot{\theta}}$

The angle θ is measured directly by the tracker, and $\dot{\theta}$ can be obtained by differencing successive $\theta(t)$ measurements. However, this procedure is sensitive to noise in the $\theta(t)$ data, and more accurate results can be obtained by an alternative procedure which utilizes the known (or assumed) structure of the problem. The expression for $\theta(t)$ can be put in the form

$$\theta(t) = \frac{D}{R_o - Vt} \qquad (22\text{-}187)$$

In this expression, the time origin, $t = 0$, is selected at any convenient point, and R_o is the corresponding target range. Only two independent parameters are contained in $\theta(t)$, since D, R_o, and V all may be multiplied by a constant without changing $\theta(t)$. Thus

$$\theta(t) = \frac{1}{b_1 - b_2 t} \qquad (22\text{-}188)$$

where $b_1 = R_o/D$
$\quad\;\; b_2 = V/D$

Estimates $\hat{b}_1$ and $\hat{b}_2$ of b_1 and b_2 can be generated from samples of $\theta(t)$. These then give the smoothed estimates

$$\hat{\theta}(t) = \frac{1}{\hat{b}_1 - \hat{b}_2 t} \qquad (22\text{-}189)$$

$$\dot{\hat{\theta}}(t) = \frac{\hat{b}_2}{(\hat{b}_1 - \hat{b}_2 t)^2} \qquad (22\text{-}190)$$

and

$$\hat{T}_{go}(t) = \frac{\hat{\theta}(t)}{\dot{\hat{\theta}}(t)} = \frac{\hat{b}_1}{\hat{b}_2} - t \qquad (22\text{-}191)$$

This can be reduced to a standard estimation problem, the solution to which is available [22-42] by minimizing the mean-squared-error between samples of $\theta^{-1}(t)$ and $(\hat{b}_1 - \hat{b}_2 t)$.

22.9.3. Spectral Discrimination. An important tracker performance parameter is the minimum level of irradiance from a target which can be acquired and tracked. This is often determined by interfering signals due to extraneous sources in the field of view, such as reflected sunlight, backgrounds, or countermeasures. If the spectral distributions of the target and interfering sources are significantly different, then spectral discrimination techniques can be used to improve performance. Other techniques commonly used to suppress interference include spatial filtering, electrical filtering, and nonlinear signal-processing. These techniques often are used in combination with spectral discrimination.

One of many criteria in choosing the sensor spectral band is maximizing the ratio of target-to-interference signals. For example, if the targets are 800 K blackbodies (often viewed against a nominal 300 K background) then allowing for the spectral transmission of the atmosphere, one can make a reasonable choice for the spectral band as 3.5 to 4.2 μm.

Suppose the tracker views reflected sunlight by background objects, and that irradiance in the selected spectral band due to the received sunlight is larger than the required, minimum, target irradiance. Since the sun radiates approximately as a 5900 K blackbody, the spectral distribution of sunlight will be easily distinguishable from that of 800 K targets. The discrimination could be based upon the ratio of the signal in a selected band to that in a second band. Choices for the second band could be 2.0 to 2.4 μm or 1.5 to 1.7 μm.

Most approaches to spectral discrimination are based upon the same property; the ratio* of received irradiance in two (or more) spectral bands due to a target is measurably different from that due to interference. Thus, if a second spectral band can be chosen to satisfy this property, spectral discrimination can be considered as a candidate for interference rejection. The secondary band can be distinct from the primary band, or it can contain the primary plus additional spectral intervals. Inclusion of one within the other may result from filter-fabrication constraints. The selection of the second band should also include consideration of mechanization costs. A large separation (e.g., more than a factor of 2) between minimum and maximum wavelengths imposes stringent demands upon the optics, especially refractive elements.

Carrier Suppression. There are at least four different approaches toward spectral discrimination. The simplest approach, called carrier suppression, is applicable primarily to reticle systems, and involves replacing the clear and opaque reticle sectors with optical filters with spectral transmission bands B_1 and B_2 for the two-color case. The respective transmission factors, τ_1 and τ_2, are chosen to satisfy

$$E_{I1}\tau_1 = E_{I2}\tau_2 \tag{22-192}$$

where E_{Ij} is the interference irradiance in band j. If this condition is satisfied, the interference source produces no modulation of the detector signal, only an increase of the dc level and the accompanying shot-noise. The latter two effects are normally attenuated to negligible levels by electrical filtering.

The amplitude of the detector-signal modulation due to the target is

$$S_T = c(E_{T1}\tau_1 - E_{T2}\tau_2) \tag{22-193}$$

where E_{Tj} is the target irradiance in band j and c is a constant representing the system parameters which are identical for both bands B_1 and B_2. A similar expression holds for modulation due to the interference source, S_I. A reticle improvement-factor, F, can be defined as

$$F = \frac{S_T/S_I}{E_{T1}/E_{I1}} \tag{22-194}$$

The numerator is the target-to-interference modulation-ratio using the two-color reticle, and the denominator is the same ratio using a clear/opaque reticle. From the expressions for S_T and S_I

$$F = \frac{1 - (E_{T2}\tau_2/E_{T1}\tau_1)}{1 - (E_{I2}\tau_2/E_{I1}\tau_1)} \tag{22-195}$$

---

*The ratio is the simplest measure; all that is really required is that some property of the irradiance due to a target is measurably different from that due to interference in the two (or more) bands.

The degree to which $E_{I1}\tau_1$ equals $E_{I2}\tau_2$, and E_{T1}/E_{T2} differs from E_{I1}/E_{I2}, determines how large F can be. Success of the approach depends upon the predictability and constancy of E_{I1}/E_{I2}.

Performance of the two-color reticle requires particularly close attention to two items. First, the filter spectral-response depends upon the angle of the incident rays; hence, precise spectral control is difficult to maintain with fast optical systems. Variations in filter temperature also cause spectral shifts; this is avoided in cooled detector systems which have the reticle located inside the dewar. Second, the borders between reticle sectors are critical areas which are difficult to fabricate. The lack of a sharp, well-defined boundary can produce anomalous signals and cause performance degradation.

Coded Pulse Filter. A second approach to spectral discrimination uses optical filters and electronic signal-processing to distinguish between target and interference pulses. This approach is primarily applicable to imaging or scanning sensors, although limited discrimination performance also is achievable with some reticle systems. The target and interference are assumed to be point sources which generate detector pulses in a scanning or reticle sensor. A set of optical filters covers the instantaneous field of view so that sources with different radiation temperatures produce distinguishable pulse waveforms from the detector.

Figure 22-59 illustrates the use of a coded, two-color, pulse filter with an imaging sensor. The filter is located in the image plane, and the transmitted optical radiation is collected and brought to the detector. The filter length in the scan-direction for the configuration in this example is about three optical-resolution diameters. There are three elements in the filter: the central element transmits spectral band, B1, which is chosen to maximize the target-to-(noise-plus-background) ratio. The outer-element spectral-band, B2, and transmission-factor are chosen so that these elements transmit more interference radiation than the center element, but less target radiation. This produces different pulse waveforms for targets and interference sources, as shown in Figure 22-60. These waveforms are easily distinguished electronically. As long as the circuits are not saturated by large interference pulses, the discrimination performance is independent of the irradiances from the target and interference.

If the second band can be chosen so that it does not contain appreciable target radiation, then target resolution is unaffected by the lengthening of the sensitive area in the image plane. If the second band does contain target radiation, then target resolution is degraded, as shown in Figure 22-60.

Design and fabrication of the filter involve similar considerations as for the carrier-suppression

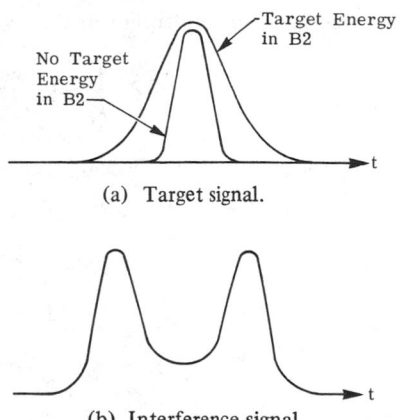

(a) Target signal.

(b) Interference signal.

Fig. 22-60. Detector pulse waveforms with a coded pulse filter.

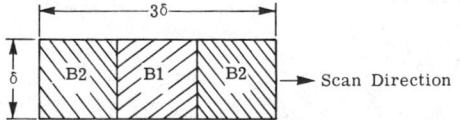

Fig. 22-59. Coded pulse filter for an imaging sensor.

reticle: spectral accuracy versus optical speed and temperature variations, and edge effects at the boundaries between filter elements. The difficulties associated with the edges are lessened by choosing one band to include the other. Then one filter is used for all three elements and the central element has an additional bandpass filter. It may be necessary to use neutral density filters on the outer elements to maintain the desired target pulse-waveform.

Two-Color Cancellation. This approach uses two detectors responding to different spectral bands. Band B1 contains both the target and interference, and band B2 is chosen to contain only interference. Both detector signals are derived from the same image, so that the interference signals in the two bands tend to have identical waveforms and phases. Interference rejection is accomplished by a closed loop which adjusts the gain of the second channel to match the amplitude of the interference component in the first. By subtracting the output of the B2 channel from that of B1, one suppresses the interference and preserves the target signal for tracking purposes. Two-color cancellation is applicable to reticle, scanning, and imaging systems.

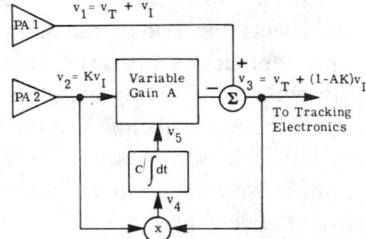

Fig. 22-61. Signal processing for two-color cancellation.

A block diagram of one signal-processing method is shown in Figure 22-61. The two preamplifier signals are assumed to be of the form

$$v_1(t) = v_T(t) + v_I(t) \tag{22-196}$$

$$v_2(t) = Kv_I(t) \tag{22-197}$$

where $v_T(t)$ = the target signal
$v_I(t)$ = the interference signal in the B1 channel
K = an unknown constant

The gain A of the B2 channel is variable and controlled by the integrator output-voltage, v_5. The closed-loop operation adjusts A so that v_5 is constant. When this occurs, the integral of the multiplier output, v_4, over a scan period T is zero, corresponding to

$$C \int_0^T Kv_I(t) [v_T(t) + (1-AK)v_I(t)]\ dt = 0 \tag{22-198}$$

This leads to the following expression for the coefficient of the residual interference-signal in the output, v_3:

$$1 - AK = \frac{\displaystyle\int_0^T v_I(t)v_T(t)\ dt}{\displaystyle\int_0^T v_I(t)^2\ dt} \tag{22-199}$$

Thus the loop serves to decorrelate the output $v_3(t)$ and the interference signal $v_I(t)$. If the target and interference signals are orthogonal, i.e., if

$$\int_0^T v_I(t)v_T(t)\,dt = 0 \qquad (22\text{-}200)$$

then the interference is completely suppressed.

The block diagram of Figure 22-61 needs to be augmented by additional circuits. The preamplifiers should have limited bandwidths for noise suppression, and some means is needed for sensing the absence of a v_2 signal, in which case A should be set equal to zero to avoid adding noise to the output. Proper loop-operation is impossible if the interference signal causes saturation in either channel. If the input signals have a large dynamic range, then AGC may be necessary. The B2 channel AGC can be incorporated with the variable gain element, A.

An important loop-parameter is the integrator gain, C. The loop transient-response speed increases with increasing C, but large C may also produce scan-frequency ripple in v_5, which degrades the interference suppression. Determination of C requires evaluation of loop performance using signal waveforms $v_T(t)$ and $v_I(t)$, derived from the particular sensor used.

The two-color cancellation technique will work well for multiple interference sources if they are all at the same effective temperature. Interference sources having different temperatures produce different values of K, so simultaneous cancellation of all sources is impossible. Only partial suppression is possible in this case.

Two-Color Ratio Discrimination. Two detectors are used in this approach, which is applicable to imaging and scanning sensors with small instantaneous fields of view. The detectors are sensitive to different, relatively narrow, spectral bands, and they develop signals due to identical scanning of the scene image. The ratio of the two signals is compared to a threshold at each point along the scan pattern. For greybodies the signal ratio varies monotonically with temperature. Thus the thresholding process identifies those points in the scene with apparent greybody temperatures greater than a discrimination temperature T_D, which is determined by the threshold level. The tracking logic can then reject signals from interference sources with apparent temperatures greater than T_D, and continue tracking the apparently cooler target.

The two signals ideally represent identical and simultaneous image scanning over the two spectral bands. This can be achieved by using a dichroic optical-wedge or a dichroic beam-splitter to separate spatially the scene image into the two spectral bands, as shown schematically in Figure 22-62. Alternatively, the two signals can be generated in a simpler way by giving up scan point-simultaneity. This concept consists of the two detectors, each covered by its own optical filter, placed side-by-side in a line along the scan direction. This results in a time delay between one signal and the corresponding value of the other. The time delay can be removed electronically by delaying the first signal to regain simultaneity prior to the ratio processing. The advantages in this can be lower cost, less volume, and less weight.

If the scene is composed of greybody sources, the spectral radiance, L_λ, at each point depends upon the apparent temperature at that point according to Planck's law

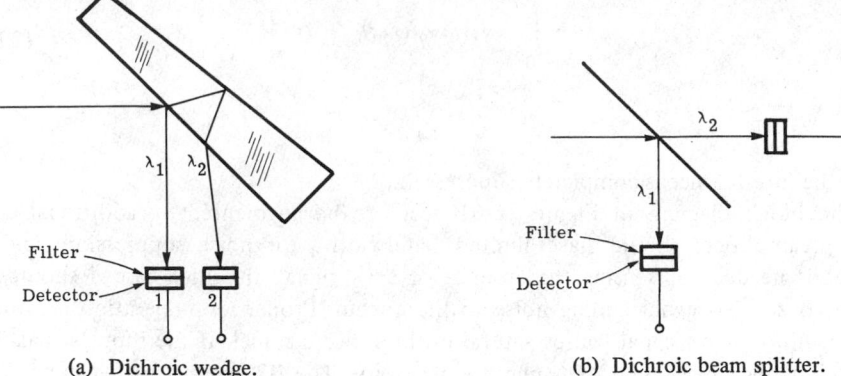

(a) Dichroic wedge.   (b) Dichroic beam splitter.

Fig. 22-62. Signal generation for two-color ratio discrimination.

$$L_\lambda = \frac{\epsilon\, c_1}{\pi\, \lambda^5\, [e^{c_2/\lambda T} - 1]}, \qquad \text{W cm}^{-2}\ \text{sr}^{-1}\ \mu\text{m}^{-1} \tag{22-201}$$

where ϵ = source emissivity
$\quad c_1$ = first radiation constant
$\quad c_2$ = second radiation constant
$\quad \lambda$ = wavelength
$\quad T$ = absolute temperature of source

If one assumes that the spectral bands are the same width, and sufficiently narrow so that L_λ is approximately constant over each band* then the ratio of the two signals at a point in the scene is

$$\frac{v_2}{v_1} = \frac{L_{\lambda_2}}{L_{\lambda_1}} \tag{22-202}$$

Substituting the L_λ expression, one gets

$$\frac{v_2}{v_1} = \frac{\lambda_1^5 \left[e^{c_2/\lambda_1 T} - 1\right]}{\lambda_2^5 \left[e^{c_2/\lambda_2 T} - 1\right]} \tag{22-203}$$

As an example, Figure 22-63 contains plots of v_2/v_1 versus T for $\lambda_2 = \lambda_1/2$ and

*This assumption is not essential to the concept, and is introduced only to simplify the discussion.

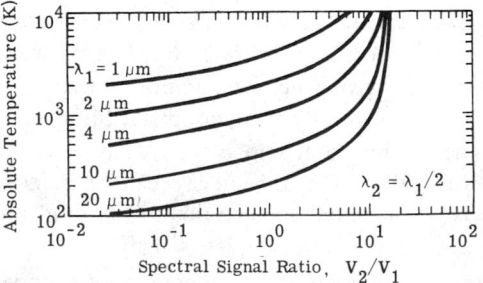

Fig. 22-63. Spectral signal ratio for greybodies.

selected values of λ_1. Once λ_1 and λ_2 are fixed, a desired discrimination temperature, T_D, is realized by setting the signal-ratio threshold at the value of v_2/v_1 corresponding to T_D.

Spectral discrimination is based upon the *received* target and interference spectral irradiances, which are proportional to the products of source intensities and atmospheric transmission. Spectral transmission of the atmosphere may produce significant changes in a received spectral distribution with respect to that of the corresponding source. Caution should be exercised in applying these results on the basis of target and interference-source temperatures or emission spectra, alone.

One possible signal-processing scheme for two-color ratio-discrimination is given in Figure 22-64. AGC is shown since the signal dynamic-range is usually sufficiently large to require gain control. The two variable-gain elements should have identical gain versus control-voltage characteristics. Control for the AGC is derived from the primary channel signal, which is also the input to the tracking circuits.

The effective comparator input, $u(t)$, is

$$u(t) = Gv_2(t) - GAv_1(t) \tag{22-204}$$

The comparator output, $c(t)$, is a logic signal which is one if $u(t)$ is positive, and zero if $u(t)$ is negative. This is equivalent to

$$c(t) = \begin{Bmatrix} 1 \\ 0 \end{Bmatrix} \quad \text{if} \quad \frac{v_2}{v_1} \begin{Bmatrix} > \\ < \end{Bmatrix} A \ . \tag{22-205}$$

Thus the discrimination temperature, T_D, is determined by the value of A.*

The comparator output, $c(t)$, is in the one state when the image scan-point covers an apparent temperature greater than T_D, and such points are to be ignored by the tracking

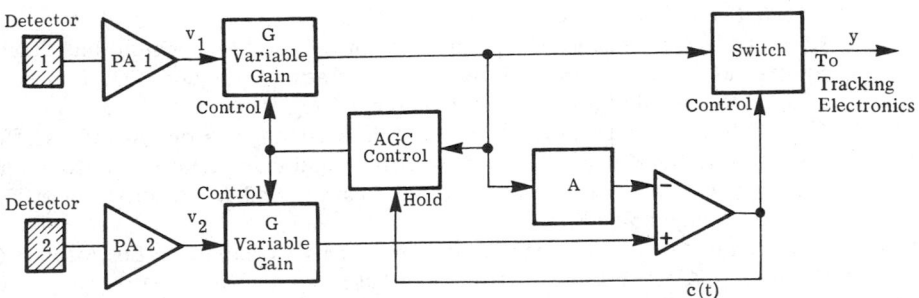

Fig. 22-64. Two-color ratio discrimination processor.

---

*In a practical system, it may be preferable to scale signal-levels and distribute gains differently, but the principle is as described.

circuits. This can be accomplished by the switch in series with the output signal, $y(t)$, which mechanizes

$$y(t) = \begin{Bmatrix} Gv_1(t) \\ 0 \end{Bmatrix} \quad \text{if} \quad c(t) = \begin{Bmatrix} 0 \\ 1 \end{Bmatrix} \qquad (22\text{-}206)$$

To prevent interference sources from capturing the AGC (and thereby suppressing target signals down to very low levels), one sets the AGC control circuit to prevent it from responding to the signal input when $c(t) = 1$ occurs.

22.10 References and Bibliography

22.10.1 References

[22-1] R. D. Hudson, *Infrared System Engineering*, Wiley-Interscience, New York, NY, 1969, pp. 249, 420, 202.

[22-2] A. K. Rue, "Stabilization of Precision Electro-Optical Pointing and Tracking Systems," *IEEE Transactions on Aerospace and Electronic Systems*, Institute of Electrical and Electronics Engineers, New York, NY, Vol. AES-5, No. 5, September 1969, pp. 805-819.

[22-3] A. K. Rue, "Calibration of Precision Gimbaled Pointing Systems," *IEEE Transactions on Aerospace and Electronic Systems*, Institute of Electrical and Electronics Engineers, New York, NY, Vol. AES-6, No. 5, September 1970, pp. 697-706.

[22-4] A. K. Rue, "Precision Stabilization Systems," *IEEE Transactions on Aerospace and Electronic Systems*, Institute of Electrical and Electronics Engineers, New York, NY, Vol. AES-10, No. 1, January 1974, pp. 34-42.

[22-5] W. T. Thomson, *Introduction to Space Dynamics*, John Wiley and Sons, New York, NY, 1963, pp. 230-232.

[22-6] A. K. Rue, "Correction to 'Stabilization of Precision Electro-Optical Pointing and Tracking Systems,'" *IEEE Transactions on Aerospace and Electronic Systems*, Institute of Electrical and Electronics Engineers, New York, NY, Vol. AES-6, No. 6, November 1970, pp. 855-857.

[22-7] R. C. Culver, "Evaluation of Concepts for a Laser Acquisition and Tracking Subsystem," Ball Brothers Research Corporation, Boulder, CO, Report No. AFAL-TR-73-180, AFAL/TEL/405 B, June 1973.

[22-8] N. A. Osborne, and D. L. Rittenhouse, "The Modeling of Friction and Its Effects on Fine Pointing Control," AIAA Mechanics and Control of Flight Conference, Anaheim, CA, American Institute of Aeronautics and Astronautics, New York, NY, AIAA Paper No. 74-875, August 1974.

[22-9] P. R. Dahl, "A Solid Friction Model," Aerospace Corporation, Los Angeles, CA, Report No. AFO 4695-67-C-D158, May 1968.

[22-10] P. H. Savet, *Gyroscopes: Theory and Design*, McGraw-Hill, New York, NY, 1961, pp. 54-75, 228-230.

[22-11] H. D. Morris, "Angular Acceleration Measurements for Geokinetic Stability," AIAA Guidance, Control and Flight Mechanics Conference, Hofstra University, Hempstead, NY, American Institute of Aeronautics and Astronautics, New York, NY, AIAA Paper No. 71-909, 1971.

[22-12] A. S. Locke, *Guidance*, D. Van Nostrand, Princeton, NJ, 1955, pp. 246-250.

[22-13] J. T. Tou, *Digital and Sampled Data Control Systems*, McGraw-Hill, New York, NY, 1959, pp. 534-549.

[22-14] R. A. Singer and K. W. Behnke, "Real-Time Tracking Filter Evaluation and Selection for Tactical Applications," *IEEE Transactions on Aerospace and Electronic Systems*, Institute of Electrical and Electronics Engineers, New York, NY, Vol. AES-7, No. 1, January 1971, pp. 100-110.

[22-15] J. M. Fitts, "Aided Tracking as Applied to High Accuracy Pointing Systems," *IEEE Transactions on Aerospace and Electronic Systems*, Institute of Electrical and Electronics Engineers, New York, NY, Vol. AES-9, No. 3, May 1973, pp. 350-368.

[22-16] J. B. Pearson, and E. B. Stear, "Kalman Filter Application in Airborne Radar Tracking," *IEEE Transactions in Aerospace and Electronic Systems*, Institute of Electrical and Electronics Engineers, New York, NY, Vol. AES-10, No. 3, May 1974, pp. 319-329.

[22-17] J. G. Truxal, *Control Engineer's Handbook*, McGraw-Hill, New York, NY, 1958, pp. 2-36.

[22-18] S. H. Crandall and W. D. Mark, *Random Vibration in Mechanical Systems*, Academic Press, New York, NY, 1963, pp. 71-73.

[22-19] D. W. Fisher, R. F. Leftwich, H. W. Yates, "Survey of Infrared Trackers," *Applied Optics*, Optical Society of America, Washington, DC, Vol. 5, No. 4, April 1966, pp. 507-515.

[22-20] L. M. Biberman, *Reticles in Electro-Optical Devices*, Pergamon Press, London, 1966.

[22-21] Robert O'B. Carpenter, "Comparison of AM and FM Reticle Systems," *Applied Optics*, Optical Society of America, Washington, DC, Vol. 2, No. 3, 1963, pp. 229-236.

[22-22] P. F. Panter, *Modulation, Noise, and Spectral Analysis*, McGraw-Hill, New York, NY, 1965.

[22-23] M. Schwartz, *Information Transmission Modulation and Noise*, McGraw-Hill, New York, NY, 1970, Second Edition.

[22-24] Ralph J. Schwarz, and Bernard Friedland, *Linear Systems*, McGraw-Hill, New York, NY, 1965.

[22-25] Hans Samuelsson, "Infrared Systems: Expressions for Signal and Background Induced Noise with Space Filters," and "Accuracy of Angle Measurement," *IEEE Transactions on Aerospace and Electronic Systems*, Institute of Electrical and Electronics Engineers, New York, NY, Vol. AES-7, No. 1, January 1971, pp. 27-33, 34-41.

[22-26] William Earl Major, "Optimum Threshold Detection for an Infrared Nutating Detection System," Electrical Engineer Thesis, Naval Postgraduate School, Monterey, CA, March 1973.

[22-27] Y. Itakura, A. Tsutsumi, and T. Takagi, "Statistical Properties of the Background Noise for the Atmospheric Windows in the Intermediate Infrared Region," *Infrared Physics*, Pergamon Publishing, Elmsford, NY, Vol. 14, 1974, pp. 17-29.

[22-28] S. O. Rice, "Mathematical Analysis of Random Noise," *Selected Papers on Noise and Stochastic Processes*, Nelson Wax (ed.), Dover, New York, NY, 1965.

[22-29] Athanasios Papoulis, *Probability, Random Variables, and Stochastic Processes*, McGraw-Hill, New York, NY, 1965.

[22-30] R. S. Orr and R. D. Yates, "On the Estimation of the Arrival Time of Pulse Signals in Gaussian Noise," Lincoln Laboratory, Massachusetts Institute of Technology, Cambridge, MA, Report No. DOT/TSC-RA-3-8(6), 26 August 1974.

[22-31] Raymond A. Berkowitz, *Modern Radar*, John A. Wiley and Sons, New York, NY, 1965.

[22-32] J. V. DiFranco, and W. L. Rubin, *Radar Detection*, Prentice-Hall, Englewood Cliffs, NJ, 1968.

[22-33] W. J. Smith, *Modern Optical Engineering*, McGraw-Hill, New York, NY, 1966.

[22-34] M. Born and E. Wolf, *Principles of Optics*, Pergamon Press, Elmsford, NY, 1965, Third Edition.

[22-35] E. Wong, "A Likelihood Ratio Formula for Two-Dimensional Random Fields," *IEEE Transactions on Information Theory*, Institute of Electrical and Electronics Engineers, New York, NY, Vol. IT-20, No. 4, July 1974, pp. 418-422.

[22-36] L. A. Liusternik and V. J. Sobolev, *Elements of Functional Analysis*, Frederick Ungar, New York, NY, 1961.

[22-37] D. V. Widder, *Advanced Calculus*, Prentice-Hall, Englewood Cliffs, NJ, 1961, Second Edition.

[22-38] J. M. Fitts, "Correlation Tracking Via Optimum Weighting Functions," Electro-Optical Division, Hughes Aircraft Company, Culver City, CA, Report No. P73-240, April 1973.

[22-39] F. K. Bowers and R. J. Klinger, "Quantization Noise of Correlation Spectrometers," *Astronomy and Astrophysics*, Springer-Verlag, New York, NY, Supplement 15, 1974, pp. 373-380.

[22-40] D. I. Barnea and H. F. Silverman, "A Class of Algorithms for Fast Digital Image Registration," *IEEE Transactions on Computers*, Institute of Electrical and Electronics Engineers, New York, NY, Vol. C-21, No. 2, 1972, pp. 179-186.

[22-41] M. W. Johnson, "Analytical Development and Test Results of Acquisition Probability for Terrain Correlation Devices Used in Navigational Systems," AIAA Tenth Aerospace Sciences Meeting, San Diego, CA, 17-19 January 1972, American Institute of Aeronautics and Astronautics, New York, NY, AIAA Paper No. 72-122.

[22-42] Ralph Deutsch, *Estimation Theory*, Prentice-Hall, Englewood Cliffs, NJ, 1965.

22.10.2 Bibliography

Allen, Donn A., and Marvin S. Stone, "The Effect of Scintillating Target Intensity upon the Performance of a Pulse-Position Modulation Optical Tracker," *Applied Optics*, Optical Society of America, Washington, DC, Vol. 5, No. 4, April 1966, pp. 525-532.

Arcese, A., P. H. Mengert, and E. W. Trombini, "Image Detection Through Bipolar Correlation," *IEEE Transactions on Information Theory*, Institute of Electrical and Electronics Engineers, New York, NY, Vol. IT-16, No. 5, September 1970, pp. 534-541.

Barnard, Thomas W., and Carroll R. Fencil, "Digital Laser Ranging and Tracking Using a Compound Axis Servomechanism," *Applied Optics*, Optical Society of America, Washington, DC, Vol. 5, No. 4, April 1966, pp. 497-505.

Barnea, D. I., and H. F. Silverman, "A Class of Algorithms for Fast Digital Image Registration," *IEEE Transactions on Computers*, Institute of Electrical and Electronics Engineers, New York, NY, Vol. C-21, No. 2, February 1972, pp. 179-186.

Berkowitz, Raymond A., *Modern Radar*, John Wiley and Sons, New York, NY, 1965.

Bershad, N. J., P. Merryman, and J. Sklansky, "Adaptive Trackers Based on Continuous Learning Theory," *IEEE Transactions on Aerospace and Electronics Systems*, Institute of Electrical and Electronics Engineers, New York, NY, Vol. AES-10, No. 2, March 1974, pp. 245-254.

Biberman, L. M., *Reticles in Electro-Optical Devices*, Pergamon Press, London, 1966.

Born, M., and E. Wolf, *Principles of Optics*, Pergamon Press, Elmsford, NY, 1965, Third Edition.

Bowers, F. K., and R. J. Klinger, "Quantization Noise of Correlation Spectrometers," *Astronomy and Astrophysics*, Springer-Verlag, New York, NY, Supplement 15, 1974, pp. 373-380.

Bowers, F. K., D. A. Whyte, T. L. Landecker, and R. J. Klinger, "A Digital Correlation Spectrometer Employing Multiple-Level Quantization," *Proceedings of the IEEE*, Institute of Electrical and Electronics Engineers, New York, NY, Vol. 61, No. 9, September 1973, pp. 1339-1343.

Carpenter, Robert O'B., "Comparison of AM and FM Reticle Systems," *Applied Optics*, Optical Society of America, Washington, DC, Vol. 2, No. 3, March 1963, pp. 229-236.

Cooke, C. R., "Automatic Laser Tracking and Ranging System," *Applied Optics*, Optical Society of America, Washington, DC, Vol. 11, No. 2, February 1972, pp. 277-284.

Crandall, S. H. and W. D. Mark, *Random Vibration in Mechanical Systems*, Academic Press, New York, NY, 1963, pp. 71-73.

Culver, R. C., "Evaluation of Concepts for a Laser Acquisition and Tracking Subsystem," Ball Brothers Research Corporation, Boulder, CO, Report No. AFAL-TR-73-180, AFAL/TEL/405B, June 1973.

Dahl, P. R., "A Solid Friction Model," Aerospace Corporation, Los Angeles, CA, Report No. AFO 4695-67-C-D158, May 1968.

De Cou, Anthony B., "Interferometric Star Tracking," *Applied Optics*, Optical Society of America, Washington, DC, Vol. 13, No. 2, February 1974, p. 414.

Delashmit, W. H., Jr., and R. F. Webber, "Analytical and Experimental Results for Maximizing the Pull-in Range of Product Correlators," *IEEE Transactions on Aerospace and Electronics Systems*, Institute of Electrical and Electronics Engineers, New York, NY, Vol. AES-10, No. 2, March 1974, pp. 216-222.

Deutsch, Ralph, *Estimation Theory*, Prentice-Hall, Englewood Cliffs, NJ, 1965.

DiFranco, J. V., and W. L. Rubin, *Radar Detection*, Prentice-Hall, Englewood Cliffs, NJ, 1968.

Fisher, D. W., R. F. Leftwich, and H. W. Yates, "Survey of Infrared Trackers," *Applied Optics*, Optical Society of America, Washington, DC, Vol. 5, No. 4, April 1966, p. 507-515.

Fitts, J. M., "Aided Tracking as Applied to High Accuracy Pointing Systems," *IEEE Transactions on Aerospace and Electronic Systems*, Institute of Electrical and Electronics Engineers, New York, NY, Vol. AES-9, No. 3, May 1973, pp. 350-368.

Fitts, J. M., "Correlation Tracking via Optimum Weighting Functions," Electro-Optical Division, Hughes Aircraft Co., Culver City, CA, Report No. P73-240, April 1973.

Fried, D. L., "Statistics of Laser Beam Fade Induced by Pointing Jitter," *Applied Optics*, Optical Society of America, Washington, DC, Vol. 12, No. 2, February 1973, p. 422.

Gagliardi, Robert M., and Sherman Karp, *Optical Communications*, Wiley-Interscience, New York, NY, 1976.

Hansen, J. P., and S. Madhu, "Angle Scintillations in the Laser Return From a Retroreflector," *Applied Optics*, Optical Society of America, Washington, DC, Vol. 11, No. 2, February 1972, pp. 233-238.

Hayes, J. N., P. B. Ulrich, and A. H. Aitken, "Effects of the Atmosphere on the Propagation of 10.6μ Laser Beams," *Applied Optics*, Optical Society of America, Washington, DC, Vol. 11, No. 2, February 1972, pp. 257-260.

Hertel, Richard J., "Using an Image-Dissector Tube for Tracking a Point Source," *Laser Focus*, Advanced Technology, Newton, MA, August 1976.

Hoge, F. E., "Integrated Laser/Radar Satellite Ranging and Tracking System," *Applied Optics*, Optical Society of America, Washington, DC, Vol. 13, No. 10, October 1974, pp. 2352-2358.

Hudson, R. D., *Infrared System Engineering*, Wiley-Interscience, New York, NY, 1969, pp. 249, 420, 202.

Itakura, Y., A. Tsutsumi, T. Takagi, "Statistical Properties of the Background Noise for the Atmospheric Windows in the Intermediate Infrared Region, *Infrared Physics*, Pergamon Publishing, Elmsford, NY, Vol. 14, 1974, pp. 17-29.

Johnson, M. W., "Analytical Development and Test Results of Acquisition Probability for Terrain Correlation Devices Used in Navigational Systems," AIAA Tenth Aerospace Sciences Meeting, San Diego, CA, 17-19 January 1972, American Institute of Aeronautics and Astronautics, New York, NY, AIAA Paper No. 72-122.

Landau, M., and C. T. Leondes, "Volterra Series Synthesis of Nonliner Stochastic Tracking Systems," *IEEE Transactions on Aerospace and Electronic Systems*, Institute of Electrical and Electronics Engineers, New York, NY, Vol. AES-11, No. 2, March 1975, pp. 245-265.

Liusternik, L. A. and V. J. Sobolev, *Elements of Functional Analysis*, Frederick Ungar Publishing, New York, NY, 1961.

Locke, A. S., *Guidance*, D. Van Nostrand, Princeton, NJ, 1955, pp. 246-250.

Lucy, R. F., C. J. Peters, E. J. McGann, and K. T. Lang, "Precision Laser Automatic Tracking System," *Applied Optics*, Optical Society of America, Washington, DC, Vol. 5, No. 4, April 1966, pp. 517-524.

Maher, W. E., "Laser Beam Propagation Model With a Hydrodynamic Treatment of the Transmission Medium," *Applied Optics*, Optical Society of America, Washington, DC, Vol. 11, No. 2, February 1972, pp. 249-256.

Major, William Earl, "Optimum Threshold Detection for an Infrared Nutating Detection System," Electrical Engineer Thesis, Naval Postgraduate School, Monterey, CA, March 1973.

Marloth, Rudolf, "Design Trade-Offs for Scanning Infrared Sensors," *Optical Engineering*, Society of Photo-Optical Instrumentation Engineers, Palos Verdes Estates, CA, Vol. 14, No. 1, January-February 1975, pp. 15-20.

Marquis, D. C., "Optical Tracking; A Brief Survey of the Field," *Applied Optics*, Optical Society of America, Washington, DC, Vol. 5, No. 4, April 1966, pp. 481-487.

Morris, H. D., "Angular Acceleration Measurements for Geokinetic Stability," AIAA Guidance, Control and Flight Mechanics Conference, Hofstra University, Hempstead, NY, American Institute of Aeronautics and Astronautics, New York, NY, AIAA Paper No. 71-909, 1971.

Neuweg, L. G., "Performance Evaluation of Scanning IR Optical Systems," *Optical Engineering*, Society of Photo-Optical Instrumentation Engineers, Palos Verdes Estates, CA, Vol. 14, No. 1, January-February 1975, pp. 6-14.

Olitsky, M., "Theoretical Analysis of SMAC Signal Processing," United States Naval Avionics Facility, Indianapolis, IN, Report No. TR-1420, 30 December 1969.

Orr, R. S., and R. D. Yates, "On the Estimation of the Arrival Time of Pulse Signals in Gaussian Noise," Lincoln Laboratory, Massachusetts Institute of Technology, Cambridge, MA, Report No. DOT/TSC-RA-3-8(6), 26 August 1974.

Osborne, N. A., and D. L. Rittenhouse, "The Modeling of Friction and Its Effects on Fine Pointing Control," AIAA Mechanics and Control of Flight Conference, Anaheim, CA, American Institute of Aeronautics and Astronautics, New York, NY, AIAA Paper No. 74-875, August 1974.

Panter, P. F., *Modulation, Noise, and Spectral Analysis*, McGraw-Hill, New York, NY, 1965.

Papoulis, Athanasios, *Probability, Random Variables, and Stochastic Processes*, McGraw-Hill, New York, NY, 1965.

Pearson, John B., "Basic Studies in Airborne Radar Tracking Systems," Ph.D. Dissertation, University of California, Los Angeles, CA, University Microfilms, Ann Arbor, MI, Publication No. 7103838, 1970.

Pearson, J. B., and E. B. Stear, "Kalman Filter Applications In Airborne Radar Tracking," *IEEE Transactions on Aerospace and Electronic Systems*, Institute of Electrical and Electronics Engineers, New York, NY, Vol. AES-10, No. 3, May 1974, pp. 319-329.

Pratt, W. K., "Correlation Techniques of Image Registration," *IEEE Transactions on Aerospace and Electronic Systems*, Institute of Electrical and Electronics Engineers, New York, NY, Vol. AES-10, No. 3, May 1974, pp. 353-358.

Rice, S. O., "Mathematical Analysis of Random Noise," *Selected Papers on Noise and Stochastic Processes*, Nelson Wax (ed.), Dover, New York, NY, 1965.

Robinson, A. H., "Multidimensional Fourier Transforms and Image Processing with Finite Scanning Apertures," *Applied Optics*, Optical Society of America, Washington, DC, Vol. 12, No. 10, October 1973, p. 2344.

Roe, G. M., and G. M. White, "Probability Density Functions for Correlators with Noisy Reference Signals," *IRE Transactions on Information Theory*, Institute of Radio Engineers, New York, NY, Vol. 7-8, No. 1, January 1961, pp. 13-18.

Rue, A. K., "Calibration of Precision Gimbaled Pointing Systems," *IEEE Transactions on Aerospace and Electronic Systems*, Institute of Electrical and Electronics Engineers, New York, NY, Vol. AES-6, No. 5, September 1970, pp. 607-706.

Rue, A. K., "Correction to 'Stabilization of Precision Electro-Optical Pointing and Tracking System,' " *IEEE Transactions on Aerospace and Electronic Systems*, Institute of Electrical and Electronics Engineers, New York, NY, Vol. AES-6, No. 6, November 1970, pp. 855-857.

Rue, A. K., "Precision Stabilization Systems," *IEEE Transactions on Aerospace and Electronic Systems*, Institute of Electrical and Electronics Engineers, New York, NY, Vol. AES-10, No. 1, January 1974, pp. 34-42.

Rue, A. K., "Stabilization of Precision Electro-Optical Pointing and Tracking Systems," *IEEE Transactions on Aerospace and Electronic Systems*, Institute of Electrical and Electronics Engineers, New York, NY, Vol. AES-5, No. 5, September 1969, pp. 805-819.

Saleh, Bahaa E. A., "Estimation of the Location of an Optical Object with Photodetectors Limited by Quantum Noise," *Applied Optics*, Optical Society of America, Washington, DC, Vol. 13, No. 8, August 1974, pp. 1824-1827.

Samuelsson, Hans, "Infrared Systems: Expression for Signal and Background Induced Noise with Space Filters," and "Accuracy of Angle Measurement," *IEEE Transactions on Aerospace and Electronic Systems*, Institute of Electrical and Electronics Engineers, New York, NY, Vol. AES-7, No. 1, January 1971, pp. 27-33, 34-41.

Savet, P. H., *Gyroscopes: Theory and Design*, McGraw-Hill, New York, NY, 1961.

Schuck, Walter H., "Vidicon Star Tracker," *Applied Optics*, Optical Society of America, Washington, DC, Vol. 5, No. 4, April 1966, pp. 489-496.

Schwartz, M., *Information Transmission Modulation and Noise*, McGraw-Hill, New York, NY, 1970, Second Edition.

Schwartz, M., and L. Shaw, *Signal Processing: Discrete Spectral Analysis, Detection and Estimation*, McGraw-Hill, New York, NY, 1975.

Schwarz, Ralph J., Bernard Friedland, *Linear Systems* McGraw-Hill, New York, NY, 1965.

Sher, Laurence, "Tracking Systems Requirements for Atmospheric Steering Compensation," *Applied Optics*, Optical Society of America, Washington, DC, Vol. 14, No. 11, November 1975, pp. 2750-2755.

Singer, R. A., and K. W. Behnke, "Real-Time Tracking Filter Evaluation and Selection for Tactical Applications," *IEEE Transactions on Aerospace and Electronic Systems*, Institute of Electrical and Electronics Engineers, New York, NY, Vol. AES-7, No. 1, January 1971, pp. 100-110.

Smith, W. J., *Modern Optical Engineering*, McGraw-Hill, New York, NY, 1966.

Spiro, Irving J., "The Optimization of an Optical Missile Guidance Tracker," *Applied Optics*, Optical Society of America, Washington, DC, Vol. 8, No. 7, July 1969, pp. 1365-1371.

Staron, M., "Control Optimization of a Laser Automatic Tracking System: Influence of the Space-Time Returns of the Echoes," *Applied Optics*, Optical Society of America, Washington, DC, Vol. 11, No. 2, February 1972, pp. 285-290.

Thomson, W. T., *Introduction to Space Dynamics*, John Wiley and Sons, New York, NY, 1963, pp. 230-232.

Tou, J. T., *Digital and Sampled Data Control Systems*, McGraw-Hill, New York, NY, 1959, pp. 534-549.

Truxal, J. G., *Control Engineer's Handbook*, McGraw-Hill, New York, NY, 1958, pp. 2-36.

Webb, W. E. and J. T. Marino, Jr., "Threshold Detection in an On-Off Binary Communications Channel with Atmospheric Scintillation," *Applied Optics*, Optical Society of America, Washington, DC, Vol. 14, No. 6, June 1975, pp. 1413-1417.

Widder, D. V., *Advanced Calculus*, Prentice-Hall, Englewood Cliffs, NJ, 1961, Second Edition.

Wong, E., "A Likelihood Ratio Formula for Two-Dimensional Random Fields," *IEEE Transactions on Information Theory*, Institute of Electrical and Electronics Engineers, New York, NY, Vol. IT-20, No. 4, July 1974, pp. 418-422.

Yap, B. K., "Effects of Object and Detector Sizes on the Spatial Frequencies of a One-Dimensionally-Scanned Optical System," *Applied Optics*, Optical Society of America, Washington, DC, Vol. 14, No. 3, March 1975, pp. 567-568.

Chapter 23

RANGING, COMMUNICATIONS, AND SIMULATION SYSTEMS

Claude Cooke*
GTE Sylvania, Inc.

John Cernius**
Aerojet ElectroSystems, Co.

Anthony J. LaRocca***
Environmental Research Institute of Michigan

CONTENTS

---

*Section 23.2
**Section 23.3
***Section 23.4

23. Ranging, Communications, and Simulation Systems

23.1. Introduction

This chapter treats systems designed for optical ranging and for infrared communications. A third, smaller section briefly reviews techniques and systems for simulation of field-tests of optical systems using selected targets and backgrounds. Sections 23.2 and 23.3 contain separate summaries of the symbols and nomenclature used in them. See Tables 23-1 and 23-3, respectively.

Table 23-1. Symbols, Nomenclature and Units

Symbols	Nomenclature	Units
A_E	Retroreflector area	m^2
A_{TAR}	Area of target	m^2
B	Bandwidth	Hz
B_{OPT}	Optical bandpass	Å or μm
C_I	Variance of log intensity	
C_l	Variance of normalized amplitude modified by aperture averaging	—
c	Speed of propagation of radiation in vacuum	m sec^{-1}
D	Diameter of receiver	m
D_L	Diameter of laser beam	m
D_R	Diameter of corner reflector	m
D_T	Diameter of transmitter	m
e	Electronic charge	C
F	Excess noise (due to multiplication, M)	—
h	Planck's constant	J sec
I	Intensity	W sr^{-1}
I_b	Photocurrent due to background radiation	A
I_d	Dark current	A
I_{LO}	Photocurrent due to local oscillator	A
I_N	Noise photocurrent	A
I_S	Signal photocurrent	A
I_T	Threshold value of photocurrent	A
k	Boltzmann constant	J K^{-1}
k_{Mie}	Mie scattering coefficient	—
k_{mol}	Molecular absorption coefficient	—
k_{RAY}	Rayleigh scattering coefficient	—
k_α	Atmospheric attenuation coefficient	—
$L_\lambda(\lambda)$	Spectral radiance	W m^{-2}sr$^{-1}\mu$m^{-1}, and W m^{-2}sr^{-1}Å^{-1}
M	Electron multiplication factor	—

Table 23-1. Symbols, Nomenclature and Units (*Continued*)

Symbols	Nomenclature	Units
m	Modulation index	—
N_p	Number of received photons per second	$\sec^{-1}$
n	Number of samples or events	—
n_m	Number of modulation wavelengths	—
$n(r)$	Index of refraction along propagation path	—
$\bar{n}$	Average number of events	—
P_D	Detection probability	—
$P(I)$	Cumulative probability for intensity	—
Q	Energy	J
q	Attenuation parameter	—
R	Range	m
R_{eq}	Equivalent resistance	ohm
r	Distance along path	m
$(S/N)_V$	Voltage signal-to-noise ratio	—
$(S/N)_\Phi$	Power signal-to-noise ratio	—
T	Temperature	K
t_p	Pulse duration, pulse width	sec
V	Visibility	km
v_T	Transverse target speed	$m\ \sec^{-1}$
$\bar{v}$	Average speed of radiation in a medium	$m\ \sec^{-1}$
α	Angle between velocity vector and line-of-sight	rad
Δt	Time between transmission and reception	sec
$\Delta\phi$	Phase difference	—
δR	Single-pulse range-error	m
η	Quantum efficiency	—
θ_R	Receiver field of view angle	rad
θ_T	Transmitter beamwidth full-angle	rad
λ	Radiation wavelength	μm, Å
ν	Optical radiation frequency	Hz
ρ	Reflectivity of target	—
σ_{TAR}	Target cross-section	$m^2\ sr^{-1}$
τ_A	Transmittance of the atmosphere	—
τ_r	Transmission of receiver	—
τ_T	Transmission of transmitter	—
τ_1, τ_2	Path transmissions	—
Φ_B	Radiant power from background	W
Φ_R	Radiant power on receiver photodetector	W
Φ_S	Radiant power-signal	W
Φ_T	Radiant power transmitted	W
Ω_T	Transmitter solid angle	sr
Ω_{TAR}	Solid angle of return beam from target	sr

23.2. Optical Ranging Systems

Modern, active, optical-ranging systems measure distance by use of the knowledge of the propagation speed of electromagnetic radiation. Some also rely on the precise knowledge of optical wavelength. A general expression for the range to the target, assuming measurement is made of distance from a single location (i.e. assuming a reciprocal or two-way path for the radiation), is

$$R = \frac{c}{2} \frac{\Delta t}{\langle n(r) \rangle} \tag{23-1}$$

where R = range to the target

c = speed of propagation of electromagnetic radiation in a vacuum

Δt = time between transmission and reception

$\langle n(r) \rangle$ = index of refraction averaged over the path.

Corrections for refractive effects can be made in many ways, depending upon the required accuracy [23-1].

23.2.1. Types of Systems.

There are three basic types of systems used for distance measurements from a single location: interferometric systems; CW ranging systems; and pulsed systems.*

Interferometric Systems. In these systems, an optical beam with a precisely controlled and accurately known wavelength is projected to a target. The return-signal beam is aligned with a reference beam. The optical phase relationship between the two beams causes either reinforcement or cancellation. By counting the number of "interference fringes" from a reference point to the point of interest, the distance traversed can be determined to accuracies comparable to the optical wavelength. Practical instruments have been developed using this technique for precise machine tooling, meteorology, geodesy, and seismology [23-1]. In some cases, multiple wavelengths are used to desensitize the system to atmospheric turbulence. For example, a commercially available laser interferometer has a usable range of approximately 60 m and an uncertainty of approximately 0.01 mm (10 μm). Interferometric systems have a limited useful range in free air (typically less than 100 m).

CW *Ranging System.* Unlike the interferometric system, this system does not use the optical wavelength as a standard of measurement. Instead, the optical beam is modulated with a sinusoidal carrier and the return signal is compared with the modulation signal. Polarization modulation is also used, but the net result is a modulation of the magnitude of the received signal.

Analysis of the modulation scheme is given later in Section 23.2.5. Early surveying instruments used modulated light sources and were limited in range. Current systems use CW laser sources and have substantially improved range. Typical performance of a commercially available system is a maximum range, R, of up to 50 miles, and uncertainty of approximately $\pm$ 1 mm plus $10^{-6} R$ [23-1]. Although less accurate than the interferometric system, this type of instrumentation has widespread use in the surveying industry.

Pulsed Ranging Systems. Here, the light source is pulsed (with pulse width duration usually 5 to 500 nsec) and the time of flight of the pulse is measured to determine the

---

*Techniques using the location of an image are generally too crude for serious consideration.

range. Compared to CW systems, larger ranges are possible, unambiguous readings can be obtained with a single pulse, and false targeting and backscatter can be suppressed by range gating. This type of system is used extensively for ranging on military, noncooperative targets where medium accuracy (1 to 10 m) is required. When used with cooperative targets this type of system has achieved accuracies between 2 cm and 0.5 m pulsewidth [26-2, 26-3, and 26-4]. Except for very sophisticated short-pulse (less than 1 nsec pulsewidth) systems, the CW system is superior in accuracy.

23.2.2. Received Power–Geometrical Factors. The fraction of the transmitted power returned to the receiver photodetector depends upon the divergence of the transmitted beam, the attenuation of the propagation medium, the effective cross-section of the target, the attenuation of the optical system, and the receiver collection-area. The general equation for received radiant power at the detector is

$$\Phi_R = \frac{\Phi_T}{\Omega_T} \tau_1 \frac{A_{TAR}}{R^2} \tau_2 \frac{\frac{\pi D^2}{4}}{\Omega_{TAR} R^2} \tau_T \tau_r \rho \tag{23-2}$$

where Φ_T = transmitted power
$\quad \Phi_R$ = power incident on the photodetector
$\quad \Omega_T$ = transmitter-beam solid-angle ($\pi \theta_T^2/4$ for the transmitter beamwidth, θ_T)
$\quad \tau_1, \tau_2$ = one-way path transmission
$\quad A_{TAR}$ = target area (A_{TAR}/R^2 = solid-angle subtended by the target)
$\quad R$ = one-way range
$\quad \Omega_{TAR}$ = solid angle of the return beam
$\quad D$ = receiver diameter, such that $\pi D^2/4R^2$ = solid-angle subtended by the receiver
$\quad \tau_T$ = transmitter optical transmission
$\quad \tau_r$ = receiver optical transmission
$\quad \rho$ = target reflectivity

To characterize the target, a target cross-section, σ_{TAR}, is introduced:

$$\sigma_{TAR} = \frac{\rho A_{TAR}}{\Omega_{TAR}} \tag{23-3}$$

Then

$$\Phi_R = \frac{\pi \Phi_T D^2}{4 \Omega_T R^4} \tau_1 \tau_2 \tau_T \tau_r \sigma_{TAR} \tag{23-4}$$

Targets vary substantially in σ_{TAR} as discussed below.

Diffusely Reflecting Targets. In the simplest case, a flat, perfectly diffuse (Lambertian) target, reflecting at normal incidence, has a cross-section of

$$\sigma_{TAR} = \frac{\rho\, A_{TAR}}{\pi} \tag{23-5}$$

This assumes that A_{TAR} is the target area illuminated and observed by the ranging system. If the target area is smaller than the transmitter beam-size and the receiver field of view, it should be used for A_{TAR}. If the target area is larger than either the transmitter beam or the receiver field of view, then

$$A_{TAR} = \frac{\pi}{4}\, R^2\, \theta^2 \tag{23-6}$$

where $\theta = \theta_T$, if $\theta_T < \theta_R$; or $\theta = \theta_R$, if $\theta_R < \theta_T$.

This is a reasonable characterization of noncooperative targets such as terrain, vehicles with flat paint, and foliage. There is substantial variation in ρ with wavelength. (See Chapter 3.) Typically, the cross-section is approximately 10% of the target area.

Retrodirective Targets. This type of target is usually quite small, but has a large effective target cross-section due to its ability to return the incident radiation in a small solid angle. The small size also provides the potential for very high-precision ranging accuracy. The most common types of retrodirective targets are corner reflectors and reflective sheeting.

The *corner reflector* is known synonymously as rectroreflector, rectrodirective reflector, retro, cube corner, and reflex prism. It consists of three, mutually-perpendicular, reflecting surfaces that have the property of returning the incident light back to the source, independent of the orientation angle. The area available for retroreflection, A_E, changes with the incident angle. Figure 23-1 shows the normalized area for four retroreflector types as a function of incidence angle [23-5].

Due to diffraction effects, the return solid-angle also depends on the area available for retroreflection and the optical wavelength, for a single retroreflector. Arrays of retroreflectors increase the area proportionately, but do not decrease the return-angle divergence. As the angle of incidence changes, the reflecting area changes from a circle (or hexagon) at normal incidence to a narrow slit at the extreme angles.

The return solid angle of a uniformly illuminated circular aperture is

$$\Omega_{cir} = (1.22)^2\, \frac{\pi\lambda^2}{D_R^{\,2}} = \frac{3.7\,\lambda^2}{A_E} \tag{23-7}$$

where λ = wavelength

D_R = corner reflector aperture diameter

A_E = corner reflector aperture area $(\pi D_R^{\,2}/4)$

The return solid angle of a uniformly illuminated rectangular aperture is

$$\Omega_{rect} = \frac{4.0\,\lambda^2}{A_E} \tag{23-8}$$

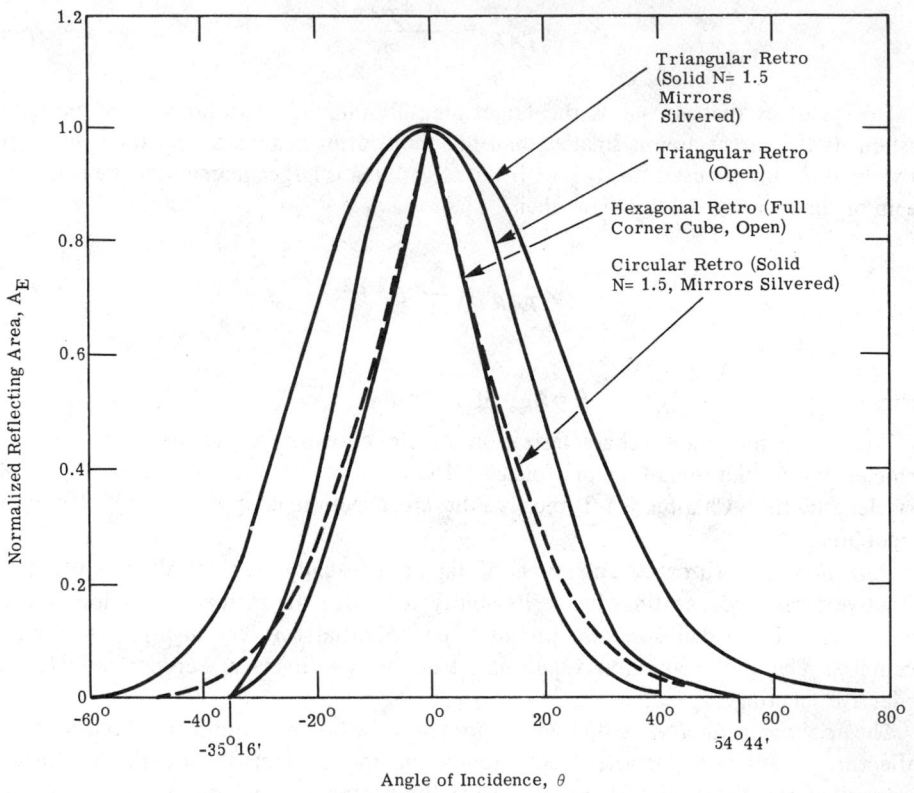

Fig. 23-1. Effective area versus angle of incidence [23-5].

Thus a reasonable approximation for the solid angle of the return beam from a corner reflector illuminated by a plane wave at various angles of incidence is

$$\Omega_{TAR} = \frac{4\lambda^2}{A_E} \qquad (23\text{-}9)$$

where A_E is the area used for retroreflection, which is not necessarily the retroreflector area. As a result, the cross-section is given by

$$\sigma_{TAR} = \frac{\rho A_E{}^2}{4\lambda^2} \qquad (23\text{-}10)$$

Figure 23-2 shows the cross-section for various corner reflector sizes. There are practical limitations to the cross-section that can be achieved with corner reflectors. These are fabrication accuracy and dimensional stability in the environment, limitation in physical size, and atmospheric coherence area. Practical fabrication limits are typically 10^8 m^2 sr^{-1}; practical environmental stability limits are typically 10^7 m^2 sr^{-1}; atmospheric turbulence limits the useful cross-section to between 10^5 m^2 sr^{-1} and 10^6 m^2 sr^{-1} depending on the atmospheric conditions and the wavelength.

One must choose materials to accommodate a selected wavelength. Glass is a popular medium for the visible and near IR. Either open-faced, first-surface mirrors or IR-transmitting materials are used at infrared wavelengths.

To provide retroreflection over large angles of incidence, one can array individual retroreflectors at different orientations. Figure 23-3 is an example of a hemispherical array of 14 individual elements with an average cross-section of 10^6 m^2 sr^{-1} at 1 μm over a hemisphere.

If there is relative motion between the target and the ranging system, there may be an additional limitation on achievable retroreflector cross-section. The retroreflector returns the beam back to the source, but during the return transmit time, the receiver may move outside the return beam with a displacement, Δx, of

$$\Delta x = \frac{2v_T R}{c} \qquad (23\text{-}11)$$

For one to assume that the receiver is within the return beam, the return divergence must satisfy the requirement

$$\frac{2.44\lambda R}{D_R} = \theta R \geqslant 2\Delta x = \frac{4v_T}{c} R$$

or

$$D_R \leqslant \frac{0.61\lambda c}{v_T} \qquad (23\text{-}12)$$

where D_R = corner-reflector diameter (circular retro assumed)
   v_T = transverse speed
   c = speed of radiation in vacuum
   λ = wavelength

This only becomes a consideration at velocities encountered in orbit (>1000 m sec^{-1}) where there is no limitation on the retroreflector cross-section because of atmospheric turbulence. Note that the range does come into the relationship.

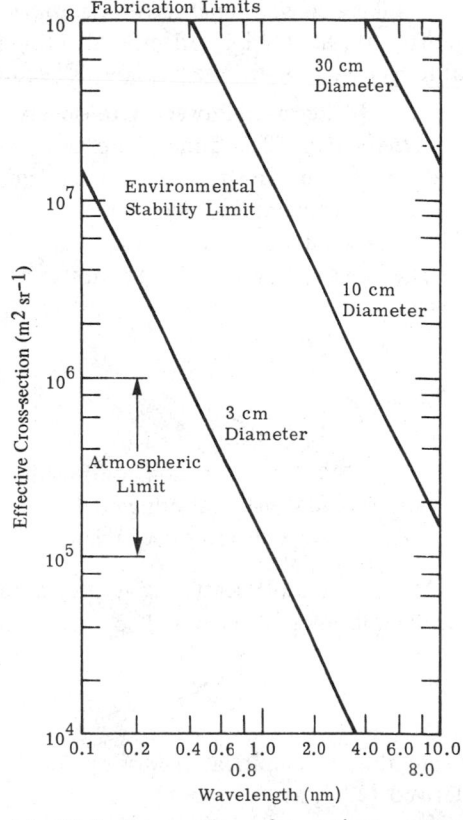

Fig. 23-2. Cross-section for various corner reflector sizes.

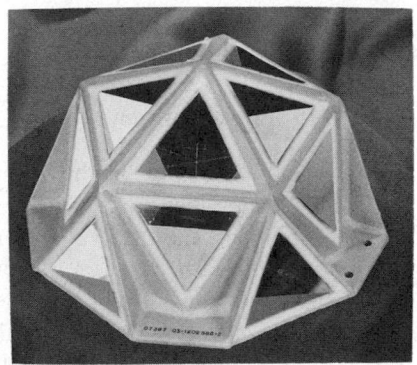

Fig. 23-3. Hemispherical retro-array.

Reflective sheeting consists of an array of small glass beads, reflectively coated on the rear-halves of the spheres. The return divergence is due to the inherent properties of the glass beads, so increasing the size does not reduce the return divergence. Reflective sheeting comes in paint and sheet form, and has a cross-section between 100 and 1500 times that of diffuse white paint.

Unlike a conventional retroreflector, which may use first-surface mirrors and thus be useful into the far IR, reflective sheeting is only useful in the visible and near IR because at longer wavelengths the glass beads become opaque.

23.2.3. Received Power–Atmospheric Effects

Attenuation. The atmosphere is a significant source of signal loss due to scattering and absorption. The relationships are quite complicated and dependent on wavelength, pressure, temperature, moisture content, particle size, distribution, and quantity. Chapters 4 and 5 provide details.

The transmission loss due to atmospheric attenuation for a round-trip path [23-6] is given by

$$\tau_A = \exp[2k_\alpha R] \tag{23-13}$$

where $k_\alpha = k_{mol} + k_{Ray} + k_{Mie}$
k_{mol} = ozone absorption coefficient
k_{Ray} = Rayleigh scattering coefficient
k_{Mie} = Mie scattering coefficient

Mie (or aerosol) scattering is the more significant term for most wavelengths in the visible and infrared regions. The coefficient is a function of wavelength [23-7] given by

$$k_{Mie}(\lambda) \sim \lambda^{-q} \tag{23-14}$$

From this, an empirical expression for τ_A in terms of wavelength and visibility may be derived [23-7] as

$$\tau_A = \exp\left[\frac{-7.82}{V}\left(\frac{\lambda}{0.55}\right)^{-q}R\right] \tag{23-15}$$

where V = visibility, km
R = range, km
λ = wavelength, μm

The values of q depend on the visibility and the size and distribution of the scattering particles. Some values are as follows: 1.6 for exceptionally good visibility, and 1.3 for average conditions, and $0.585 V^{1/3}$ for visibilities less than 6 km [23-7].

The expression for τ_A given in Equation (23-13) is for a horizontal path at sea level. For slanted paths and higher altitudes, the transmission increases. An estimate can be made of this factor [23-5] based on the assumption that as visibility changes, the particle distribution profile, as defined by Elterman [23-8], remains the same, with only a linear shift on the number-of-particle axis.

The results show that the attenuation coefficient is dependent on visibility and target altitude, and is independent of either slant-range or elevation angle. Figure 23-4 shows a series of transmission curves for a two-way path at 1.06 μm for various target elevations and sea-level visibility conditions. As shown, system performance and range are strongly affected by visibility conditions.

When tracking at high elevation angles (e.g., satellite tracking), the attenuation is significantly less. Figure 23-5 shows the one-way transmission from sea level through the entire atmosphere at 1.06 and 0.53 μm [23-9].

Scintillation Margin. Although not identified as a signal loss, signal margin must be provided to compensate for fluctuations in the magnitude of the return signal. Atmospheric turbulence causes fluctuations in the radiation level of a beam propagating

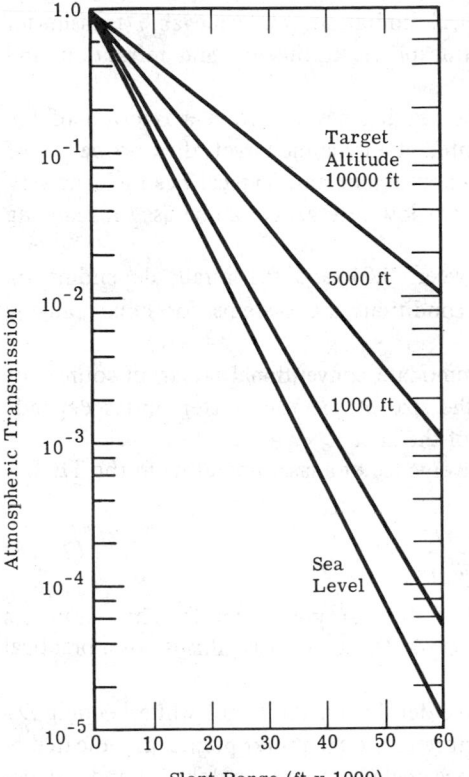

Fig. 23-4. Atmospheric transmission versus range. $\lambda = 1$ μm, visibility 30000 ft.

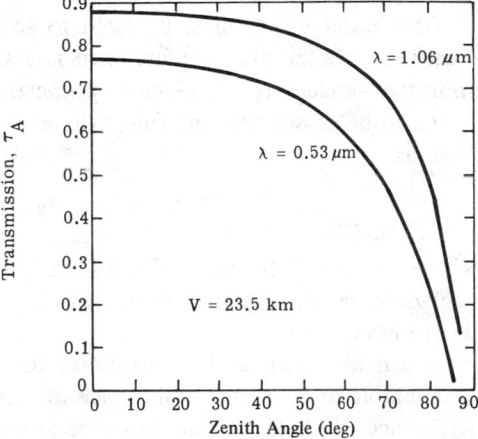

Fig. 23-5. One-way atmospheric transmission for paths through the atmosphere [23-9].

through the atmosphere. (See Chapter 6.) The turbulence consists of fluctuations in the refractive index of the air, which are caused by temperature fluctuations and gradients. A log-normal distribution is a well-accepted model for the radiation fluctuations [23-10, 23-11, 23-12, 23-13]. The strength of the turbulence is usually defined by the variance of this distribution; from this, a required margin to allow for scintillation-caused losses may be derived.

The cumulative probability expression for a log-normal distribution is

$$P(I) = \frac{1}{2} \left(1 + \mathrm{erf} \left\{ \frac{\ln I + \frac{1}{2} C_I(o)}{[2C_I(o)]^{1/2}} \right\} \right) \qquad (23\text{-}16)$$

where I = intensity

   $C_l(o)$ = variance of the log intensity

Receiver aperture averaging [23-14] effectively reduces the variance and is accounted for by

$$C_I(o) = 4C_l(o) \tag{23-17}$$

where $C_l(o)$ is the variance of a normalized log amplitude modified by aperture averaging.

Transmitter Beamwidth. There are practical limitations on achievable transmitter beamwidths due to pointing requirements, atmospheric turbulence and refraction, and transmitter optics size.

There is a basic requirement to point the ranging device within a fraction of the beamwidth. If the ranging system must be pointed at a dynamic target, then properties of the pointing system, such as dynamic lag, become important. Typical beamwidths vary between 0.5 and 50 mrad. Beamwidths with very low divergence can be used if scanning is allowed [23-3].

Atmospheric turbulence sets a limit between 0.01 and 0.1 mrad, depending on conditions. Examples of the best possible conditions are systems for lunar ranging [23-15].

Transmitter optics must be sized to accommodate conventional radiation sources of specified physical size. For use with lasers, the size of the transmitter optics depends upon the wavelength and the mode properties of the laser.

It can be shown that the full-angle beam-divergence of a laser operating in the TEM_{00} mode is

$$\theta_T = \frac{4\lambda}{\pi D_L} \tag{23-18}$$

where D_L is the diameter of the laser beam at the $1/e^2$ power points. This assumes a negligible transmitter limiting-aperture, D_T, or $D_T \gg D_L$. Obviously, this is not a practical design in most cases.

It can be shown that to maximize the flux density on the target while keeping D_T comparable to D_L, one should place the aperture at the 8% power-points, as indicated in Reference [23-16]. This increases the divergence angle by approximately 1.4 and reduces the flux density by approximately 1/2 due to increased aperture diffraction.

To achieve the atmospheric limit of divergence (typically 0.01 mrad) at 10.6 μm, an aperture satisfying the above criteria would be approximately 2 m in diameter. This would be an extremely large optical system. By comparison, at the second harmonic frequency of Nd:YAG wavelength of 0.53 μm, the required aperture would be 0.1 m, a much more practical design.

Multimode lasers have many modes sharing the transmitter aperture; thus, considerably larger optics are required to achieve divergence angles comparable to TEM_{00}. For example, a Nd:YAG laser used for rangefinder applications has a typical divergence of 1 mrad in a 0.25 in. aperture. This divergence is reduced by magnifying the beam diameter.

23.2.4. Detection Systems. The two basic types of detection systems are briefly described below.

Incoherent Detection. With this type of system, the radiation returned from the target is collected and converted into electrical energy by the photodetector. No coherence of the optical signal is required. The ranging information is contained in the intensity envelope of the optical signal. The system must be designed so that competitive signals such as background, leakage currents, and amplifier noise do not mask the ranging signal.

Coherent Detection. With this type of system, the coherence property of the optical signal is used to allow nonlinear mixing (in the photodetector) with a coherent local-oscillator to develop a beat-frequency which is then processed. The local-oscillator is a strong signal (usually derived from the transmitter) which raises the beat-frequency signal above the competitive noise sources (usually detector noise). This configuration is analogous to the microwave radar system using heterodyne detection.

Under proper conditions both systems can approach the theoretical, detection quantum-noise limit:

$$N_p = \frac{\Phi_R}{h\nu} \tag{23-19}$$

where N_p = number of photons sec^{-1} received
  h = Planck-constant (6.626 × 10^{-34} J sec)
  ν = frequency
  Φ_R = received radiant power

Incoherent Detection Systems. Figure 23-6 shows the basic incoherent-detection ranging-system which consists simply of a modulated optical-radiation source (usually a laser), beam-forming transmitter optics, receiver optics, a narrow-band background-rejection filter, and a photodetector. A general expression for the SNR in a direct-detection system under strong signal conditions [23-17] is

$$\left(\frac{S}{N}\right)_\Phi = \frac{(0.5m^2)I_s{}^2M^2R_{eq}}{2eBFM^2R_{eq}(I_s + I_b + I_d) + 4kTB} \tag{23-20}$$

where m = modulation index of the source (m=1 for a pulsed system)
  I_s = photo-current due to signal
  M = electron multiplication factor
  R_{eq} = equivalent resistance of detector load
  e = electronic charge
  B = detection bandwidth
  I_b = photocurrent due to background radiation
  I_d = dark current
  k = Boltzmann's constant
  T = absolute temperature (including amplifier noise figure)
  F = excess noise factor due to the multiplication, M

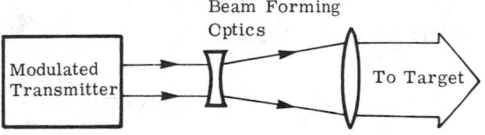

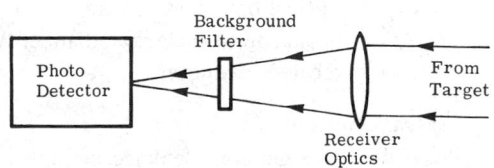

Fig. 23-6. Basic incoherent detector system.

The photocurrents are related to incident radiant power by the equations

$$I_s = \frac{\eta e}{h\nu} \Phi_R \tag{23-21}$$

and

$$I_b = \frac{\eta e}{h\nu} \Phi_B \tag{23-22}$$

where η = quantum efficiency of the detector
e = electron charge
$h\nu$ = photon energy
Φ_R = received radiant power incident on the photodetector
Φ_B = radiant power generated by the ambient background

Using Equations (23-21) and (23-22), one finds that the SNR in terms of radiant power is given by

$$\left(\frac{S}{N}\right)_\Phi = \frac{\frac{1}{2} m^2 \left(\frac{\eta e}{h\nu}\right)^2 \Phi_R^{\,2}}{2e \frac{\eta e}{h\nu} BF \left(\Phi_R + \Phi_B + \frac{h\nu I_d}{\eta e}\right) + \left(\frac{4kTB}{M^2 R_{eq}}\right)} \tag{23-23}$$

An alternative expression for the SNR is in terms of electrical voltages, with the relation

$$\left(\frac{S}{N}\right)_V \propto \left[\left(\frac{S}{N}\right)\right]_\Phi^{1/2} \tag{23-24}$$

Regardless of which of the above equations are used, the SNR formulation is useful only for large numbers of generated photo-electrons. For such large numbers, the dominant noise source (i.e., the largest term in the denominator) is determined, the SNR required for a given system performance is specified, and the required Φ_R derived.

Noise sources include (1) ambient backgrounds, (2) dark current, (3) thermal noise, and (4) quantum noise.

(1) Ambient Backgrounds
The governing equation is

$$\Phi_B = \left(\frac{\pi D^2}{4}\right) \Omega_R B_{OPT} L_\lambda(\lambda) \tag{23-25}$$

where Ω_R = receiver field of view (= $\pi\theta R^2/4$, for θ_R = receiver full-angle field of view)
B_{OPT} = optical bandpass of filter
$L_\lambda(\lambda)$ = background spectral radiance, W m^{-2} sr^{-1} Å^{-1}
D = receiver diameter

(2) Dark Current
Photodetectors generate leakage or dark current in the absence or presence of signal. This contributes to the overall noise depending on the relative magnitude compared to

the other sources. The magnitude depends on many parameters such as bias voltage, photosurface area, and temperature. (See Chapter 11.)

(3) Thermal Noise

Whether a passive or active load, the noise contributed by the amplifier system following the photodetector can be represented by the thermal-noise from a resistor. (This term may dominate in wideband receivers required for detection of narrow pulses.) A photomultiplier, which is useful only in the visible and near IR, can provide large relatively noise-free gain ($F \cong 1$) to suppress this noise source. Avalanche photodetectors have gain to suppress the thermal noise, but introduce avalanche noise in the process.

(4) Quantum Noise

As mentioned earlier, the quantum noise limit can be achieved under certain circumstances. This occurs when Φ_R in the denominator dominates; then

$$\left(\frac{S}{N}\right)_\Phi = \frac{\frac{1}{2}m^2\frac{\eta e}{h\nu}\Phi_R}{2eB} \qquad (23\text{-}26)$$

However this model only applies under strong signal conditions.

Coherent Detection Systems. Figure 23-7 is a block diagram of a basic heterodyne-type coherent detection system. It differs from the incoherent detection system in that a coherent local oscillator (either at the same or a different frequency) is inter-ferometrically aligned on the photodetector. Nonlinear mixing takes place, generating sum and difference components, the latter of which is in the RF or baseband detection range.

The power SNR for an "Ideal" optical heterodyne with single sideband (SSB) modu-lation [23-18, 23-19] is

$$\left(\frac{S}{N}\right)_\Phi = \frac{2I_{LO}I_s M^2 R_{eq}}{2eI_{LO}BM^2 R_{eq}} = \frac{I_s}{eB} = \frac{\Phi_s}{\dfrac{h\nu}{\eta}}B$$

$$(23\text{-}27)$$

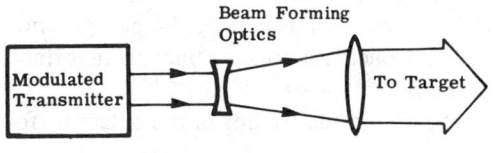

where I_{LO} = the photocurrent due to
the local oscillator (LO)
laser

Φ_s = the signal radiant power

The other terms are defined as in Equations (23-20, 23-21, and 23-22).

It is assumed that the local oscillator generates enough "coherent" noise to dominate all other terms. The 3 dB improvement over incoherent detection is due to the preservation of phase information in the detection process.

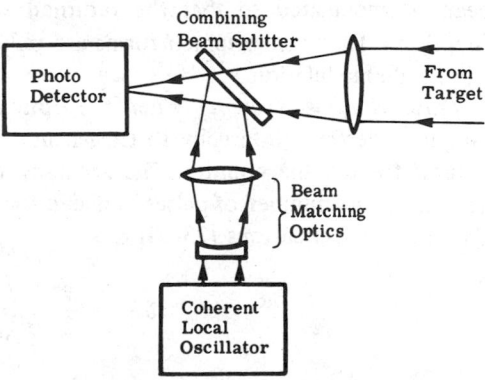

Fig. 23-7. Basic coherent detector system.

Although the coherent detection system shows a theoretical advantage, there are several very practical limitations. To date, only the 10.6 μm laser system has been widely used, so most of the comments apply to this system type. The limitations are listed below.

(1) Coherence. Careful control of the wavelength and operating mode is required. Diffraction gratings, low-expansion material cavities, and electronic control of frequency are commonly used.

(2) Doppler sensitivity. This may be an advantage if velocity measurement is important.

$$\Delta f = \frac{2v_T}{\lambda} \cos \alpha \qquad (23\text{-}28)$$

where Δf = doppler shift in frequency of the carrier
$\quad v_T$ = target velocity
$\quad \lambda$ = laser wavelength
$\quad \alpha$ = angle between the velocity vector and the line-of-sight

The sensitivity is $\cong$ 200 kHz (m sec^{-1})$^{-1}$. Since systems have been implemented with frequency stabilities better than 1 kHz, measurement accuracies (short-term) of about 1 cm sec^{-1} are possible. However, this may be a disadvantage for high velocity targets, e.g., a ground-based system which ranges on satellites in low orbit. Velocities approaching 10^4 m sec^{-1} are possible, causing shifts as large as 2 GHz. Either the detector must have the response or the local-oscillator the tuning range to accommodate the doppler.

(3) Receiver alignment. Since the receiver local-oscillator must be interferometrically aligned with the incoming beams, large optical receivers must meet stringent pointing tolerances. For example, a 0.5 m receiver must be pointed to within approximately $0.5(\lambda/D) \cong 10$ μrad. An incoherent detection receiver is limited only by the field of view, determined usually by the background.

(4) Coherence area. Atmospheric effects distort the coherent wavefront limiting the aperture.

(5) Cryogenic cooling of the detector (for 10.6 μm operation).

23.2.5. Modulation and Detection Methods. In optical ranging systems, the optical beam is modulated so that the returned signal can be compared to the modulation function. The most common forms are pulse modulation and CW multitone, sinusoidal amplitude modulation.

Pulse Ranging System. When the optical pulse is transmitted toward the target, a sample is detected internally to trigger an interval counter. Detection of the return signal is used to stop the counter. The accuracy depends on the pulse duration, the S/N of detection, the number of pulses available for averaging, and errors in the interval counter. The simplified equations [23-20] are

$$\delta R = \frac{\bar{v}}{2} \; t_p \left(\frac{S}{N}\right)_\Phi^{-1/2} \qquad (23\text{-}29)$$

$$\overline{\delta R} = \delta R \, (n)^{-1/2} \qquad (23\text{-}30)$$

where δR = the single-pulse range-error
  $\bar{v}$ = the average speed of radiation in the refractive medium
  t_p = the pulse rise time
  $(S/N)_\Phi$ = the power signal to noise ratio
  $\overline{\delta R}$ = the range error after averaging
  n = the number of samples

Other error sources include clock errors, biases, and atmospheric propagation delays.

CW *Ranging System.* The radiation level of the CW optical carrier is modulated and the modulation phase is compared to the detected return signal.

If the target range exceeds the modulation wavelength, there is a target ambiguity. By sequentially or simultaneously modulating with multiple tones, the ambiguity can be removed and high accuracy can be achieved. For the CW system, the simplified equation is

$$R = \lambda \left[\frac{\Delta\phi}{4\pi} + n_m \right] \qquad (23\text{-}31)$$

where R = target range
  λ = modulating wavelength
  $\Delta\phi$ = phase difference between the return signal and the modulation signal
  n_m = number of modulation wavelengths

23.2.6. Detection of Pulses in Noise. The most widely used military, laser, ranging-systems use direct pulse-detection receivers. The detection of a pulse in the presence of noise is a statistical problem discussed below.

Pulse Detection in White Noise. For systems that are amplifier-noise or background limited, the detection statistics follow the model given by Rice [23-21] of a matched filter followed by a threshold detector. The false alarm rate ($\overline{FAR}$) depends on the rms noise current, I_N, and the threshold current, I_T,

$$\overline{FAR} = \frac{1}{2\sqrt{3t_p}} \exp\left[-\frac{I_T^2}{2I_N^2} \right] \qquad (23\text{-}32)$$

where t_p is the pulse duration or "width."

The detection probability can be shown to be approximately

$$P_D \simeq \frac{1}{2}\left(1 + \text{erf}\left\{ \frac{I_S - I_T}{\sqrt{2}\, I_N} \right\} \right) \qquad (23\text{-}33)$$

where I_S = the instantaneous peak signal current
  erf { } = the conventional error function

Equations (23-32) and (23-33) can be used in designing a system. An acceptable pulse width, t_p(FAR), is set and the corresponding, achievable detection probability, P_D, is obtained by setting the peak signal-to-rms-noise value.

Pulse Detection in Quantum Noise. Near the quantum noise limit, the arrival of both signal and noise photons follows Poisson statistics where

$$P(n = k) = \frac{e^{-\bar{n}}(\bar{n})^k}{k!} \tag{23-34}$$

where $\bar{n}$ is the average number of events for the detection period. Then the probability that threshold n_T is exceeded is

$$P(n \geqslant n_T) = \sum_{k=n_T}^{\infty} \frac{e^{-\bar{n}}(\bar{n})^k}{k!} \tag{23-35}$$

The number of photoelectrons is determined when the signal and noise are present and a threshold is set to give an acceptable detection probability. Then the average number of photoelectrons is determined for noise alone and the probability of a false alarm is determined.

Useful systems can be designed using a large number of statistically correlated returns, even though on a single-event basis the detection probability is very low [23-15, 23-22].

23.2.7. Laser Sources. Lasers are the most suitable optical source for ranging because of:

(1) Ease of collimation with small optics.
(2) Ease of modulation.
(3) Narrow spectral bandwidth allowing background suppression.
(4) Coherence (if required for coherent detection systems).

Only a few laser sources have been found to have practical application in ranging systems, to date. These are summarized in Table 23-2.

23.2.8. Laser Detectors. As ranging systems developed around the available lasers, so also did matching detectors. The two important parameters are sensitivity and speed. The data below are basically empirical and treat the detector/preamplifier system as a "black box."

The commonly used detectors are the photomultiplier, silicon PIN-junction photo-diodes, silicon and germanium avalanche photodiodes, photoconductors, and photovoltaic detectors. Their characteristics are listed in Chapter 11 (with the exception of the photomultiplier).

The photomultiplier resembles most closely the ideal photodetector. Photons incident on a photosensitive surface produce photoelectrons which are accelerated into dynode targets producing secondary electrons. Several stages produce significant current gain (to suppress thermal noise of the following amplifiers) in a basically noiseless process. There are many photosurfaces available, depending on the wavelength used. All are limited to the visible and near IR. The most popular are the S-20, for the doubled Nd:YAG ($\sim$10% quantum efficiency), Ruby (3%), and the S-1 for Nd:YAG (0.015%). The rise-time depends upon the dynode configuration. Crossed-field dynodes are available with carefully-controlled transit-time dispersion. Rise times approach 0.1 nsec. Most conventional electrostatic tubes have rise times of 1 to 20 nsec. Photomultipliers require high voltage supplies (over 1 kV), but are rugged, simple, and suitable for field use. Under most daylight conditions, the systems are background limited. However, with low backgrounds, single photoelectron detection is possible.

23.2.9. Ranging Accuracy. Several effects contribute to the degradation of accuracy in a pulse ranging system:

(1) Pulse rise time and duration.
(2) Receiver bandwidth.
(3) Range counter accuracy.
(4) Pulse spreading by the target.
(5) Refractive index changes along path.
(6) Signal fluctuation.
(7) Statistical pulse distortion.
(8) S/N.

Most tactical ranging systems require accuracies of a few meters over ranges of 20 km or less. Thus, many of the sources of degradation are negligible, but instrumentation laser ranges using corner reflector targets may require accuracy to a fraction of a meter. Then some form of normalization electronics must be used to trigger the counter on a constant point of the pulse (usually the half-power point), independent of the signal strength.

Table 23-2. Common Laser Sources for Ranging Systems

Name	Type	Operating Wavelength (μm)	Pulse or CW	Pulse Width or PRF*	Energy per Pulse or Peak Power	Average Power
HeNe	Gas	0.6328	CW	–	–	1 to 100 mW
		1.1523	CW	–	–	1 to 20 mW
		3.3913	CW	–	–	1 to 10 mW
HeCd	Gas	0.325 to 0.44	CW	–	–	1 to 10 mW
Argon	Gas	0.4880	CW	–	–	0.1 to 5 W
		0.5145	CW	–	–	0.1 to 5 W
CO_2	Gas	10.6	CW	–	–	1 to 1000 W
CO_2	Gas	10.6	Pulsed	300 nsec/50 to 500 Hz	Greater than 1 kW	–
		10.6	Pulsed	1 to 10 μsec/ 0.1 to 10 kHz	Greater than 10 kW	–
Nd:YAG	Crystal	1.06	CW	–	–	0.1 to 1000 W
Nd:YAG	Crystal	1.06	Pulsed	5 to 100 nsec 1 to 10,000/ sec	0.1 to 1000 mJ/ pulse	–
X 2 Nd:YAG	Crystal	0.53	Pulsed	5 to 200 nsec 1 to 10,000/ sec	0.1 to 500 mJ/ pulse	–
Ruby	Crystal	0.6943	Pulsed	2 nsec to 50 nsec 1 to 60/min	0.1 to 50 J/pulse	–
GaAs	Solid state	0.9	Pulsed	2 to 100 nsec 1 to 10,000/ sec	1 W to 1 kW, peak	–

*Pulse repetition frequency.

23.2.10. Practical Applications

Example 1: Tactical Laser Range Finder

One of the most important uses for laser rangers has been to measure target distance for weapons direction. Two of the most commonly used lasers are ruby (0.6943 μm) and

Nd:YAG (1.06 μm). In recent years Nd:YAG has displaced ruby because of much better efficiency (energy out/energy in), covertness, better atmospheric transmission, and wider temperature operating range.

For lightweight systems, a single Nd:YAG laser rod oscillator is usually used, producing a 150 mJ pulse with 7 to 10 nsec pulsewidth, and repetition rates up to 10 per sec. Higher energy per pulse can be achieved with oscillator-amplifier combination.

The laser operates in multimode and beam expansion optics are used to reduce divergence. A typical minimum divergence is 0.1 mrad, limited by pointing accuracies; this is achieved with transmitter optics less than 10 cm in diameter.

Receiver-optics size depends primarily on weight, size constraints, and cost. Typical diameters vary from 5 to 25 cm. Either a PIN silicon photodiode or a silicon avalanche photodiode is most commonly used.

The PIN photodiode is rugged, simple, and has high quantum efficiency (over 50%) compared to photomultipliers (less than 1%). The junction is back biased, acting as a current source with good linearity over 7 orders of magnitude of signal strength. The junction capabilities are sufficiently low that rise times of under 10 nsec are achievable. A typical sensitivity of a detector-preamp combination is 2×10^{-12} W Hz$^{-1/2}$.

Avalanche photodiodes have internal gain and thus improved sensitivity, but the detector size is much smaller (typically 10^{-3} cm^2). This limits the receiver optics size and the available field of view. Another limitation is the gain sensitivity to temperature, which requires bias compensation. Because photodiode amplifiers are inherently noisy, sensitivity to background is low. This allows simple background filtering and relatively wide fields of view.

Typical range accuracies of 3 to 10 m satisfy tactical requirements. Thus, simple range-counters and ordinary threshold-detection are usually acceptable. To determine a practical maximum range, assume the following parameters (Refer to Equation (23-4).):

$$\Phi_T = \frac{150 \times 10^{-3} \text{ J}}{7 \times 10^{-9} \text{ sec}} = 2.1 \times 10^7 \text{ W} \tag{23-36}$$

$$\Omega_T = \frac{\pi (10^{-4})^2 \text{ rad}^2}{4} = 7.9 \times 10^{-9} \text{ rad}^2 \tag{23-37}$$

$$\frac{\pi}{4} D^2 = \frac{\pi}{4} (0.25)^2 = 4.9 \times 10^{-2} \text{ m}^2 \tag{23-38}$$

$$\sigma_{TAR} = \frac{\rho A_{TAR}}{\Omega_{TAR}} = \frac{0.1 \times 1 \text{m}^2}{\pi \text{ sr}} = 3.18 \times 10^{-2} \text{ m}^2 \text{ sr}^{-1} \tag{23-39}$$

(Note: this assumes a low reflectance, 1 m^2 diffuse target)

$$\tau_1 \tau_2 = 0.25 \tag{23-40}$$

If one assumes a 50 MHz detection bandwidth,

$$\Phi_R \geqslant \left(2 \times 10^{-12} \text{ W}/\sqrt{\text{Hz}}\right)\left(\sqrt{5 \times 10^7 \text{ Hz}}\right) = 1.4 \times 10^{-8} \text{ W} \qquad (23\text{-}41)$$

Using Figure 23-4 and assuming a 10 dB S/N, one can calculate the following maximum ranges:

$$
\begin{array}{lll}
\text{Sea level} & R_{\max} \cong 10,000 \text{ m} & \\
\text{5000 ft altitude} & R_{\max} \cong 15,000 \text{ m} & (23\text{-}42) \\
\text{10,000 ft altitude} & R_{\max} \cong 18,000 \text{ m} &
\end{array}
$$

There is a substantial difference in maximum range due to atmospheric attenuation. As the visibility decreases, the maximum range falls off dramatically. With no atmospheric attenuation, a range of approximately 52,000 m is achievable.

Example 2: Instrumentation Laser Ranger

This type of ranger is unique in that the target is cooperative (a retroreflector of some kind) and the accuracy requirements are more stringent (less than 1 m).

One can expect that because a retroreflector is used, substantial maximum range improvement can be achieved. Also, the susceptibility to atmospheric attenuation will increase substantially. Finally, the high-accuracy requirement dictates pulse normalization and signal returns well above minimum detectable thresholds. To allow comparison with the tactical system, a 0.1 mrad beamwidth is chosen. (This again depends on the pointing method available.) A practical combination of instrumentation for a ranging and tracking system (i.e., one which provides complete space position in 3 coordinates) uses a 3 mrad beam to accomodate dynamic targets [23-2].

Assume values of the four parameters to be the same as in Example 1:

$$\Phi_T = 2.1 \times 10^7 \text{ W} \qquad (23\text{-}43)$$

$$\Omega_T = 7.9 \times 10^{-9} \text{ rad}^2 \qquad (23\text{-}44)$$

$$\frac{\pi D^2}{4} = 4.9 \times 10^{-2} \text{ m}^2 \qquad (23\text{-}45)$$

$$\tau_1 \tau_2 = 0.25 \qquad (23\text{-}46)$$

$$\sigma_{TAR} = \frac{\rho A_E^2}{4\lambda^2} = 10^6 \text{ m}^2 \text{ sr}^{-1} \text{ (typical of retroreflector in} \qquad (23\text{-}47)$$
$$\text{Figure 23-3)}$$

$$\Phi_R \geqslant 1.4 \times 10^{-7} \ (S/N = 10) \qquad (23\text{-}48)$$

From Figure 23-4, again, the following ranges are found:

$$
\begin{array}{lll}
\text{Sea level} & R_{\max} \cong 30,000 \text{ m} & \\
\text{5000 ft altitude} & R_{\max} \cong 46,000 \text{ m} & (23\text{-}49) \\
\text{10,000 ft altitude} & R_{\max} \cong 65,000 \text{ m} &
\end{array}
$$

With no atmospheric attenuation, a range of approximately 3900 km is achieved.

Example 3: Satellite/Lunar Ranging System

This type of system is characterized by extremely high precision requirements (1 to 10 cm), cooperative retroreflective targets, and computer pointing of the transmitter beam and receiver. A practical system used for these purposes is described here and in more detail in the Reference [23-23]. Most of the same parameters are used to facilitate comparison to the previous examples, with the following exceptions:

(1) The laser operates at 0.53 μm, allowing the use of a quantum-noise-limited photomultiplier. In addition, the laser pulse-width is extremely short, allowing high precision.

(2) A 50-element retroreflector array is used, designed to accommodate point-ahead limitations (Equation (23-12)).

(3) The path is at high elevation angles through the entire atmosphere.

Assume

$$\Phi_T = \frac{150 \times 10^{-3}\,\text{J}}{200 \times 10^{-12}\,\text{sec}} = 7.5 \times 10^8\,\text{W} \tag{23-50}$$

and, as in Example 1,

$$\Omega_T = 7.9 \times 10^9\,\text{rad}^2 \tag{23-51}$$

$$\frac{\pi}{4}\,D^2 = 4.9 \times 10^{-2}\,\text{m}^2 \tag{23-52}$$

$$\tau_1\tau_2 = 0.25 \tag{23-53}$$

Then

$$\sigma_{TAR} = \frac{\rho A_E^2}{4\lambda^2} \quad \text{(from Equation (23-10))} \tag{23-54}$$

Assume an array of fifty elements, each with a 1 cm diameter retroreflector ($\rho = 0.5$).

$$50\,\sigma_{TAR} = \frac{50 \times 0.5 \times (7.9 \times 10^{-5})^2}{4 \times (5.3 \times 10^{-7})^2} = 1.38 \times 10^5\,\text{m}^2\,\text{sr}^{-1} \tag{23-55}$$

Assume a negligible background, an S-20 photomultiplier with a quantum efficiency of 0.1, and a requirement of 10 photoelectrons. Also assume the Figure 23-5 values for total atmospheric transmission. Computing on an energy basis, one finds the required energy is

$$Q_R = \frac{hc}{\eta\lambda} = 3.74 \times 10^{-18}\,\text{J} \tag{23-56}$$

At the zenith, $R_{max} = 8.47 \times 10^6$ m. At a 20 degree elevation angle, $R_{max} = 6.43 \times 10^6$ m.

For comparison, a system suitable for lunar ranging would have the following properties:

$$\Phi_T = \frac{250 \times 10^{-3}\,\text{J}}{200 \times 10^{-12}\,\text{sec}} = 1.25 \times 10^9\,\text{W} \tag{23-57}$$

$$\Omega_T = \frac{\pi}{4} (10^{-5})^2 = 7.85 \times 10^{-11} \text{ sr} \qquad (23\text{-}58)$$

(This is the practical atmospheric limit and sets much more stringent pointing accuracy requirements.)

$$\frac{\pi D^2}{4} = \frac{\pi}{4} (0.75)^2 = 4.4 \times 10^{-1} \text{ m}^2 \qquad (23\text{-}59)$$

$$\sigma_{TAR} \cong 10^6 \text{ m}^2 \text{ sr}^{-1} \text{ (lunar retroreflector at Apollo} \qquad (23\text{-}60)$$
$$\text{15 site)}$$

Using the above, one finds that the average return per pulse is 0.003 photoelectrons, which gives an extremely low probability of detection. But by pulsing at a high rate (typically 5 pps), and coherently processing the return after a several-minute observation period, the orbit of the moon can be determined to an accuracy of a few centimeters [23-15].

23.3. Infrared Communications Systems

23.3.1. Introduction. Infrared communications may be conveniently subdivided into incoherent and coherent communications. Simply stated, the concept of coherence is the property of an IR wave to be in-phase with itself after some time interval (temporal coherence), and in-phase at time-points in space (spatial coherence) [23-24]. The temporal and spatial coherences are related* to the frequency bandwidth of an IR radiation source by

$$\Delta t = \frac{1}{\Delta \nu} \text{ and } \Delta x = c\Delta t = \frac{c}{\Delta \nu} \qquad (23\text{-}61)$$

where Δt = temporal coherence
Δx = spatial coherence
c = speed of radiation propagation in vacuum
$\Delta \nu$ = frequency bandwidth

Most natural radiation sources such as blackbodies, radiant plasma sources, and even light emitting diodes (LEDs) have relatively wide frequency bands (LEDs have $\Delta \lambda \approx 40$ nm or $\Delta \nu \approx 1.2 \times 10^{13}$ Hz at 1 μm wavelength. The corresponding coherence time is 8.3

---

*A more basic definition of coherence is expressed by the complex degree-of-coherence function, $\gamma_{12}(j, \tau)$ [23-25]. In this case, j designates complex representation and τ designates time. It is assumed that two displaced-in-space radiation sources generate electric fields $E_1(j)$ and $E_2(j)$ with complex conjugates $E^*_1(j)$ and $E^*_2(j)$. Then the complex degree-of-coherence at a point P in space is given by

$$\gamma_{12}(j, \tau) = \frac{\langle E_1(j, t + \tau) E_2^*(j, t) \rangle}{\sqrt{\langle |E_1(j)|^2 \rangle \langle |E_2(j)|^2 \rangle}}$$

where brackets $\langle \rangle$ indicate an average. If $|\gamma_{12}(j, \tau)| = 1$, the coherence limit is designated; if $|\gamma_{12}(j, \tau)| = 0$ it designates the incoherent limit; values within the interval $0 < |\gamma_{12}(j, \tau)| < 1$ designate partial coherence.

$\times$ 10^{-14} sec [23-26]). These sources are considered *incoherent*, and communication systems using them are called *incoherent communication systems*. However, lasers have considerably narrower spectral bands. GaAs lasers have a $\Delta\lambda$ of about 1.0 nm, with many lasers having spectral bandwidths of $\Delta\lambda \leqslant 0.01$ nm. A practical 10.6 μm laser may have a line-width, $\Delta\nu$ as narrow as 100 kHz with a corresponding $\Delta\lambda$ of 3.7×10^{-5} nm. Thus, the systems employing lasers as sources of the carrier power are referred to as *coherent communications*.

Another distinct difference between incoherent IR sources and lasers is the structure of the output radiation. The incoherent IR sources are usually Lambertian radiators. (See Chapter 2.) Lasers produce a diffraction-limited output beam. In other words, the phase of the laser output beam is planar. This results in a diffraction-limited radiation pattern which is very narrow and is susceptible to the atmospheric effects caused by fluctuations in the atmospheric index of refraction. (See Chapter 6.)

The symbols, nomenclature and units used in this portion of Chapter 23 are listed in Table 23-3.

Table 23-3. Symbols, Nomenclature and Units

Symbols	*Nomenclature*	*Units*
A	Area	m^2
B	Bandwidth	Hz
c	Speed of radiation propagation in vacuum	$m\ sec^{-1}$
$c(t)$	Carrier	Various
$D*$	Specific detectivity	$m\ Hz^{1/2}W^{-1}$
E	Electric field magnitude	$V\ m^{-1}$
$\mathbf{e}$	Electric field vector	$V\ m^{-1}$
F	Free space loss	—
f	Frequency, usually not of light or optical radiation	Hz
f_c	Carrier frequency	Hz
f_m	Set of discrete frequencies, m = 0,1,2, . . .	Hz
f_{mx}	Highest frequency (maximum) of information signal	Hz
G	Gain	—
H_o	Height of acoustic beam	m
h	Planck's constant	J sec
I	Intensity	$W\ sr^{-1}$
I_d	Dark current	A
J_n	n^{th}-order Bessel function of the first kind	—
j	$\sqrt{-1}$	—
k	Magnitude of propagation vector, $|\mathbf{k}| = 2\pi/\lambda$; also, Boltzmann's constant	m^{-1}; $J\ K^{-1}$
k_n	Factor accounting for the difference in actual noise distribution from the Gaussian white noise	—
k_y	Factor accounting for nonuniform responsivity of the detector in frequency	—

Table 23-3. Symbols, Nomenclature and Units (*Continued*)

Symbols	Nomenclature	Units
L	Length	m
L_D	Detector dimension	m
L_c	Cavity length; length	m
M	Set of discrete messages; also, radiant exitance	$-$; W m^{-2}
M_p	Polarization modulation matrix	$-$
m	Estimate of message; also, pulsewidth modulation index	$-$
m_a	Amplitude modulation index	$-$
m_f	Frequency modulation index	$-$
m_k	kth message	$-$
m_{ln}	Matrix elements	$-$
N	Number of locked modes	$-$
NEFD	Noise equivalent flux density	W m^{-2}
n	Index of refraction	$-$
$n(t)$	Noise	Various
P_a	Acoustic power	W
P_e	Probability of error	$-$
P_m	Polarization matrix	$-$
p	Photo-elastic constant; also, probability density function	$-$
q	Charge of an electron	C
R	Quadratic electro-optic coefficient; also, range	m^2 V^{-2}; m
R_L	Load resistor	ohms
$R(f)$	Responsivity	A W^{-1}
$R(f)/R(f_o)$	Normalized responsivity of the detector as a function of frequency	$-$
r	Linear electro-optic coefficient	m V^{-1}
$r(t)$	Incoming signal at receiver	$-$
$r_k(t)$	Received modulated carrier	$-$
r	Range vector	m
r_1, r_2	Radial coordinates	m
S/D	Signal-to-distortion ratio	$-$
S/N	Signal-to-noise ratio (SNR)	$-$
$\bar{S}$	Amplitude of strain	$-$
$s(t)$	Signal	Various
$s_k(t)$	Transmitted modulated carrier	Various
T	Temperature; also, period	K; sec
t	Time	sec
V	Voltage	V
$V_k(t)$	Received demodulated signal	V
$V_n(t)$	Received demodulated noise	V
$V_r(t)$	Receiver front-end demodulated noise	V

Table 23-3. Symbols, Nomenclature and Units (*Continued*)

Symbols	Nomenclature	Units
V_t	Threshold voltage	V
v	Acoustic velocity	$m\ sec^{-1}$
W_o	Width of waist of acoustic beam	m
$X(f)$	Frequency distribution of $\Phi_\lambda(\lambda)$	Various
x_1, x_2, x_3	Coordinate axes	Various
α	Off-axis angle	rad
Γ	Phase retardation	rad
$\Gamma_{12}(j,\tau)$	Complex coherence factor	—
Δf_n	Noise equivalent bandwidth for Gaussian white noise	Hz
Δn	Increment of index of refraction	—
Δt	Temporal coherence	sec
Δx	Spatial coherence	m
$\Delta\lambda$	Bandwidth in wavelength	m
$\Delta\nu$	Radiation frequency bandwidth	Hz
$\Delta\nu_c$	Laser line width	Hz
η	Efficiency; quantum efficiency	—
θ	Phase reference	rad
θ_B	Bragg angle	rad
θ_1, θ_2	Angular coordinates	rad
λ	Wavelength	μm, Å, nm, m (Various)
ν	Radiation frequency, especially for optical frequencies	Hz
ν_c	Laser carrier frequency	Hz
ρ	Density of a material	Various
σ	$1/2e^2$ radius of laser beam	m
τ	Pulsewidth	sec
Φ	Radiant power	W
ϕ	Phase	rad
ω	Radian frequency, $\omega = 2\pi\nu$	rad Hz
ω_c	Carrier radian frequency	rad Hz

Subscripts

a	Acoustic	—
d	Detector	—
i	Running index; 0,1,2, . . .	—
j	Running index; 0,1,2, . . .	—
k	Running index; 0,1,2, . . .	—
l	Running index; 0,1,2, . . .	—
n	Running index; 0,1,2, . . .	—
r	Receiver	—
T	Transmitter	

23.3.2. The Communications Process. In the communications process, there are two major considerations: (1) the question of how efficiently and accurately a message, $m_k(t)$, is transmitted to the receiving terminus of a communications link; and (2) the components which make up the communication system hardware. Norbert Wiener [23-27] and Claude E. Shannon [23-28, 23-29] laid the foundation for the modern communications theory. References [23-30] through [23-35] provide an excellent review of the subject.

The functional elements of an IR communication system are shown in Figure 23-8. As shown, at any one time the system is required to transmit a message, m_k, one of a set of M discrete messages, $\{m_i\}$, where $i = 0, 1, 2 \ldots, k, \ldots, M-1$. The communication system is most efficient if the occurrence of the message m_k is random. Otherwise, superfluous or redundant information is transmitted, making inefficient use of the channel capacity. The function of the communication system, therefore, is to transmit the message in an efficient manner, such that when the receiver produces an estimate of the message, $\hat{m}_k$, the difference $|\hat{m}_k - m_k|$ is small, whereas any other difference $|\hat{m}_k - m_i|$ is large. This abstract definition of what a communication system should do is useful because it allows one to evaluate quantitatively and compare various communication systems when the signal is distorted and corrupted by noise.

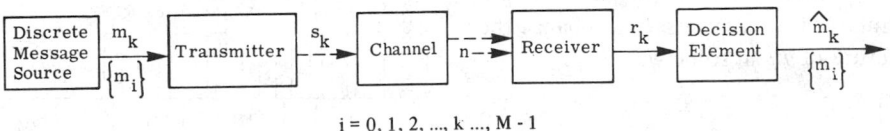

$$i = 0, 1, 2, ..., k ..., M - 1$$

Fig. 23-8. Key functional elements of IR communication system.

The modulation processes encompass techniques by which the baseband information is imprinted on the carrier. The baseband information may consist of the set of discrete messages discussed above, or it may be the output of a microphone or a television camera. In the receiver, the carrier is demodulated to extract the baseband signal such that the receiver output is the best estimate of the original baseband input.

The transferral of the baseband into the carrier may be subdivided into three distinct categories, as illustrated in Figure 23-9.

In the continuous-wave (CW) case, a parameter of the carrier is continuously changed in proportion to the magnitude of the input baseband signal. Thus, as shown in Figure 23-9(a), the electric field amplitude (V m^{-1}), or the intensity (W sr^{-1}) of an IR carrier is continuously changed as the baseband signal amplitude changes.

Pulse modulation is illustrated in Figure 23-9(b). In this example, the baseband signal is sampled in a discrete manner. The amplitudes of each sample are then transmitted. To reproduce the baseband signal in the receiver, one must sample it at a minimum rate of at least twice the highest frequency of the baseband signal [23-36]. As shown in the figure, the sampled value of the baseband signal may be expressed as pulse amplitude, pulse width, or pulse position.

The pulse code case is illustrated in Figure 23-9(c). In this case, the amplitude of each sample is converted into a digital word, normally by use of a binary system. Thus, if the desired resolution of the amplitude is at least 2%, then an 8-bit analog-to-digital (AD) converter can be used to encode the amplitude of each sample. The converted baseband signal then consists of a train of two distinct voltages, V_1 corresponding to a zero, and V_2 corresponding to a one. The techniques used to modulate a parameter of the carrier in each of the three methods are discussed in subsequent sections.

There are three key functional elements of a communication system, shown in greater detail in Figure 23-10. These are the transmitter, the channel, and the receiver.

The transmitter consists of the IR carrier source, a carrier modulation device, and the optical elements which properly confine the modulated carrier in the transmitted beam-shape. Thus, within the transmitter the message, m_k, is used to modulate the carrier, $c(t)$, and produce a corresponding signal waveform, $s_k(t)$. The signal processing block represents the processing which may be applied to the message prior to its application to the modulator if it is necessary to change the structure of m_k in some way.

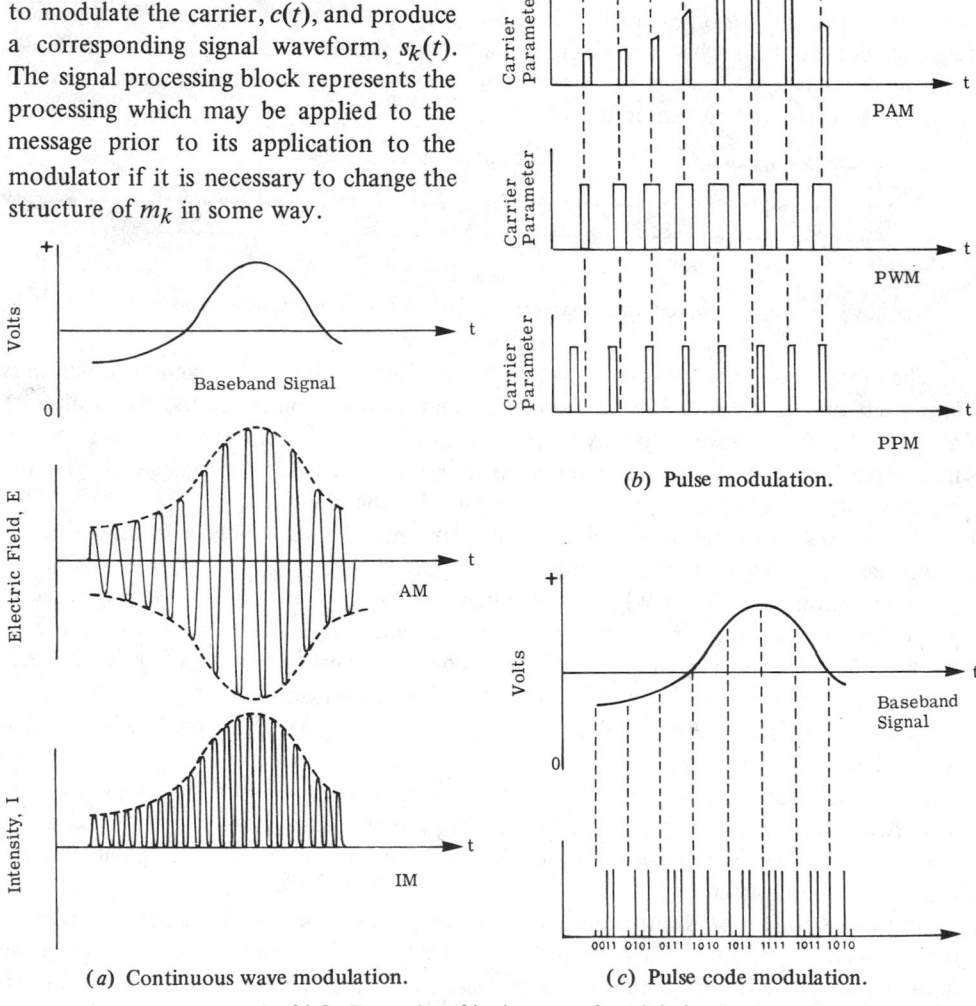

(a) Continuous wave modulation.

(b) Pulse modulation.

(c) Pulse code modulation.

Fig. 23-9. Examples of basic types of modulation.

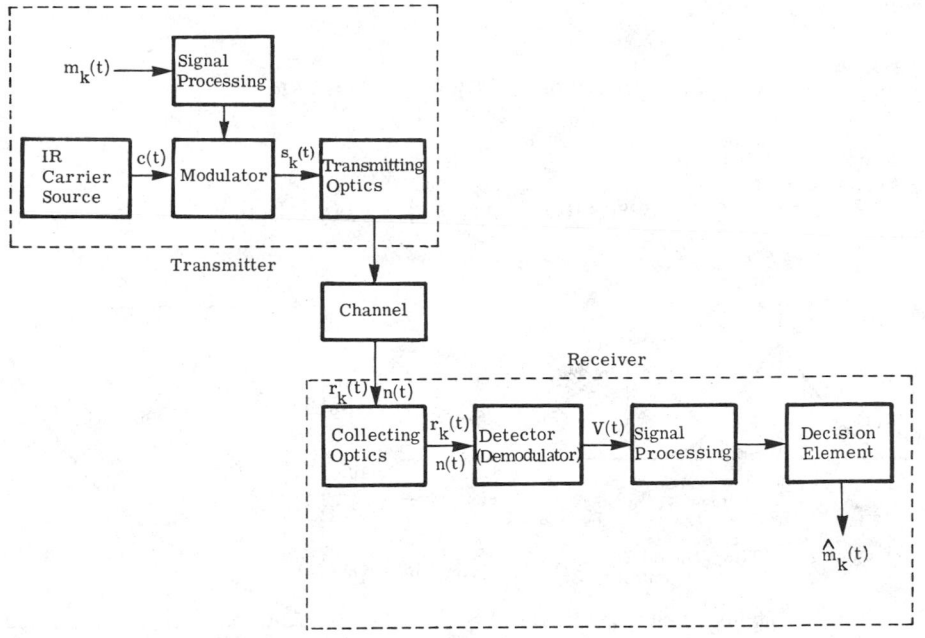

Fig. 23-10. Block diagram of IR communication system.

The signal is attenuated in the channel. The channel may be the traversed distance between a ground station and a satellite, a path of a few kilometers through the atmosphere between two stations, or a finite length of dielectric waveguide, again connecting two stations.

At the receiver, the incoming signal, $r_k(t)$, is combined with the incoming noise, $n(t)$. The receiver consists of the front-end optics which selectively enhance the signal, $r_k(t)$, as a function of wavelength and the direction of arrival, and a detector(s) which converts the received signal and noise into a corresponding electrical signal, $V(t) = V_k(t) + V_n(t) + V_r(t)$. The term $V_r(t)$ represents the additional noise contributed to the total system noise by the detector and the front-end electronics of the receiver. The output of the detector, after the required amplification and filtering, is applied to a decision element which compares the received signal, $V(t)$, with the replicas of all of the M message alternatives which are stored in the receiver. The decision element then produces the estimated message, $\hat{m}_k$.

As an example, in the case of a binary coded system, a decision of whether the signal represents a zero or a one must be made. Then the string of zeros and ones comprising the message, m_k, are further evaluated [23-37]. The statistics associated with deciding on one or zero are illustrated in Figure 23-11. Two probability density functions are shown. One corresponds to the presence of noise only. (In this case, the signal-level of zero volts corresponds to the digital zero.) The second corresponds to the presence of noise plus signal voltage, V_2, where V_2 corresponds to a digital one. The quantity V_t represents the threshold voltage, where, if $V \geqslant V_t$, the decision is made that a one has been transmitted and, if $V < V_t$, a decision is made that a zero has been transmitted. In this example, the probability of error, P_e, is given by

$$\text{Zero-digit error: } P_e(0) = \int_{V_t}^{\infty} p(n)\, dV$$

(23-62)

$$\text{One-digit error: } P_e(1) = \int_{0}^{V_t} p(n + V_2)\, dV$$

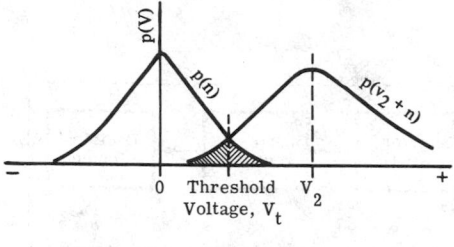

Zero Volts = Digital Zero
V_2 Volts = Digital One

Fig. 23-11. Probability density functions at $\langle V \rangle = 0$, and $\langle V \rangle = V_2$.

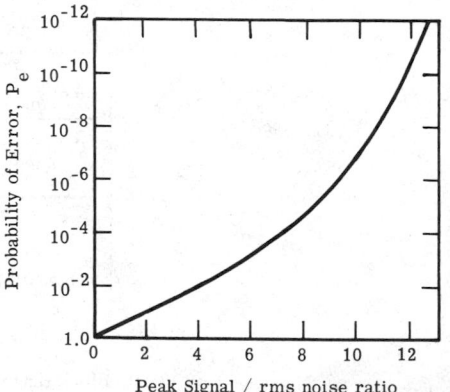

Peak Signal / rms noise ratio

Fig. 23-12. Error probability in binary transmission.

In Figure 23-12, the error probability in the transmission of a binary code is shown as a function of peak signal-to-rms-noise ratio. If the ratio is 7.4, the error probability, P_e, is found to be 10^{-4}. This means that, on the average, one digit in 10^4 transmitted will be judged incorrectly. The error curves of Figure 23-12 are based on the noise having a Gaussian distribution. In the case of IR communication systems, care must be taken to properly identify the noise probability function. For example, at signal levels of a few photons per second, the noise statistics follow a Poisson distribution, assuming all other sources of noise within the receiver are negligible [23-38]. Practical IR communication systems, operating in a dynamic environment, must contend with the corrupting signals due to background gradients, atmospheric index-of-refraction fluctuations, shot noise, and from still other noise mechanisms. All these noises have distinct probability-density-functions whose effects on the system depend greatly on the mechanization of the receiver, the modulation and demodulation techniques employed, and whether it is an incoherent or coherent communications system.

In the next decision element, the string of binary ones and zeros representing a message, m_k, are further evaluated with a corresponding reduction in the error associated with the estimate, $\hat{m}_k$.

In the case of speech transmission, the message alternatives and the decision element are in the human brain. Thus, for speech transmission (and transmission of other analog data), the overall, communication-link performance may be summarized as a function of the received signal-to-noise (S/N) and signal-to-distortion (S/D) ratios. The distortion components result from nonlinearities in the transmission system. For example, the speech baseband consists of certain frequency amplitudes of a given phase relationship. If the frequency content is band-limited by the use of an electrical filter with nonuniform amplitude response, together with the nonlinear phase-relationships, it will cause distortion of the input baseband. Thus, both the noise and the distortion in the communication system must be carefully controlled.

23.3.3. Incoherent Communications Systems.

Incoherent communications systems that do not use LEDs are limited in the amount of data which can be transmitted. Typical frequency bandwidths are under 10 kHz. Their use is limited to local, short transmission-ranges and applications where the security of the communications link is paramount. Thus, incoherent communications systems of this type find application in ship-to-ship communications and in area installations such as air fields. LED systems (although incoherent) are capable of handling high data-rates but are limited to short ranges.

Incandescent-Source (Low Data-Rate) Systems. Two types of IR sources are used to produce the radiant power that serves as a carrier of the low data-rate, incoherent communication-system. These are incandescent solid and hot plasma sources. Their properties are described in Chapters 1 and 2. Typical incandescent solid sources are summarized in Table 23-4. Because of the high thermal inertia of the incandescent sources, their time constants are on the order of several seconds. Accordingly, they are operated in a continuous mode with the modulation of the radiant power accomplished externally.

In Figure 23-13, a typical transmitter configuration with an incandescent source is shown. The radiant source is centered at the focal point of the optics which are designed to produce the required radiation pattern. To obtain enough radiant power from these sources, one must operate them at temperatures sufficiently high to radiate considerably in the visible. Thus, appropriate spectral filters must be used to confine the radiation to the IR. Candidate materials are silicon, germanium-coated quartz, or Irtran®*. The external modulation is implemented mechanically. Mechanical modulators have very narrow information-bandwidths, are usually

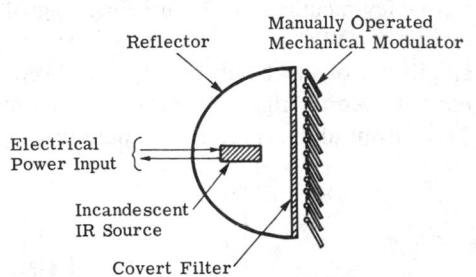

Fig. 23-13. Typical transmitter configuration of an incandescent source, incoherent communication system.

Table 23-4. Typical Incandescent Solid IR Radiation Sources

IR Source	Comment
Silicon carbide	T_{max} = 2700°F, operates in air
Super kanthal	T_{max} = 3100°F, operates in air
Rare earth oxides	$T_{max} \approx$ 5000°F, requires external heaters to initiate operation
Tungsten	$T_{max} \approx$ 4500°F, requires operation in vacuum
Graphite	$T_{max} \approx$ 5000°F, requires nonoxidizing atmosphere

*Irtran is a registered trademark of the Eastman Kodak Company.

manually operated, and employ a coding technique, such as Morse code. For example, in Figure 23-13, the radiant beam is modulated by the opening and closing of the louvers.

Heated plasma sources include xenon, krypton, cesium, or rubidium lamps. (See Chapter 2.) The hot-plasma IR sources are basically line radiators; however, as the pressure and the current density increase, the radiation becomes a continuum and its spectrum approaches that of a gray body, related to the temperature of the plasma. The time-constant of hot-plasma sources is on the order of a few microseconds. They can therefore be operated in the continuous mode with an external modulator or can employ internal modulation by varying the current through source. The modulation techniques which are practical with these sources are pulse-frequency-modulation, pulse-position-modulation, or pulse-width-modulation. Intensity modulation or pulsed intensity-modulation is seldom used, since hot-plasma sources are highly nonlinear and appear electrically as constant voltage loads, making the technique difficult to implement. Figure 23-14 shows a typical emission spectrum of a cesium lamp.

A pulse-frequency-modulated xenon-source IR communication transmitter is shown in Figure 23-15. The operation of the xenon tube requires three different types of power sources. To initiate the discharge path between the electrodes, one must apply a starting high-voltage spike. Depending on the spacing of the electrodes and the internal pressure of the gas, the starting voltage varies between 5 and 30 kV. Once the gas becomes ionized and an arc develops, it is sustained by the "keep alive" current supply operating at a low voltage level (15 to 30 V). The pulse modulation of the source is accomplished by discharging the current pulses through the lamp. In the case of the pulse-frequency-modulation, the baseband signal controls the frequency of a voltage-controlled oscillator (VCO); this frequency may vary between 5 and 10 kHz, depending on the amplitude of the baseband input-voltage. The output of the VCO on a cycle-by-cycle basis is used to trigger the discharge of a pulse-forming network through the xenon lamp. The output of the transmitter, therefore, is a pulse-frequency-modulated IR carrier.

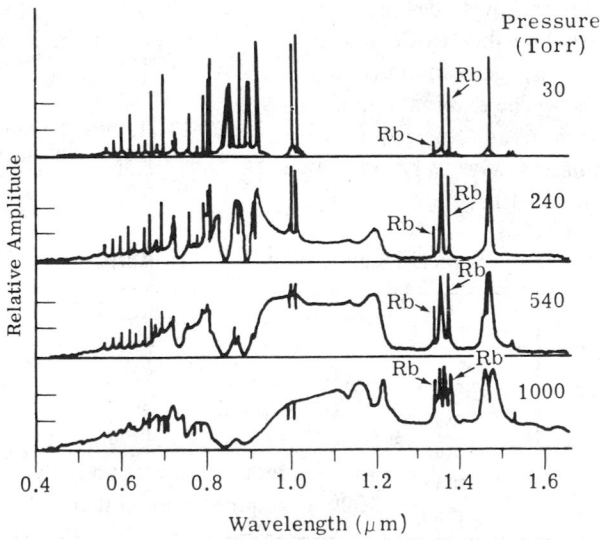

Fig. 23-14. Spectral radiation from cesium lamp [23-39].

LED *(High Data Rate) System.* Several LEDs have been developed in recent years. The spontaneous emission of the diode is the result of energy released when the electrons and holes recombine in the junction of a forward-biased semiconductor diode. The radiation is proportional to current through the diode and increases as the current increases until it is dissipation-limited by the junction temperature.

Gallium Arsenide (GaAs) LEDs are most developed in the infrared. The characteristics of the GaAs diodes are summarized in Table 23-5. At cryogenic temperatures, the diodes generate more average power with improved efficiency.

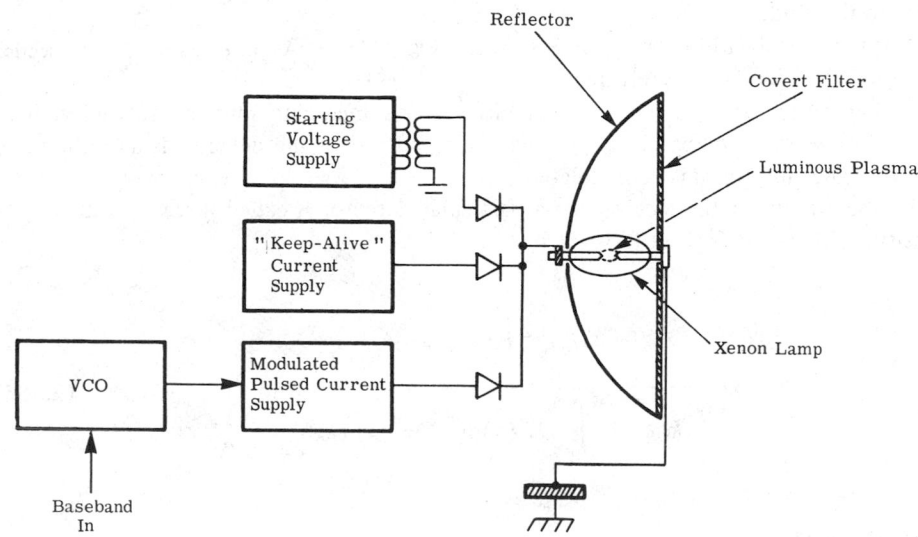

Fig. 23-15. Pulse frequency-modulated xenon-source IR communication transmitter.

Table 23-5. Typical Characteristics of GaAs Light Emitting Diode

Peak output power	—	1 to 40 mW
Average output power	Room temperature	3 mW
	Cryogenic temperature	5 mW
Power efficiency	Room temperature	4%
	Cryogenic temperature	20%
Max. duty factor	—	100%
Rise time	—	300 nsec
Spectral width	—	40 nm

The cross-section of a typical GaAs assembly is illustrated in Figure 23-16. To increase the power output, one usually arranges the diodes in an array assembly. The radiative recombination-time of GaAs is very short, approximately 1 nsec [23-40]. Thus, GaAs diodes have been modulated at rates up to 100 MHz [23-40]. The modulation technique consists of modulating the input current through the diode.

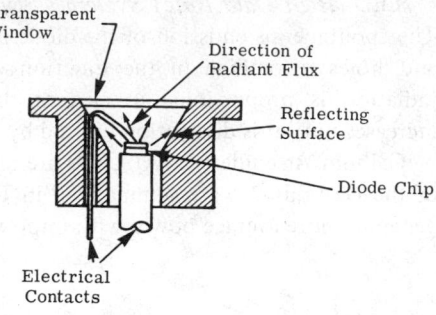

Fig. 23-16. GaAs LED assembly cross-section [23-26].

A typical receiver block-diagram is shown in Figure 23-17. The key elements are the collecting optics and the detector which converts incoming pulse-modulated radiant power into corresponding electrical pulses. The output of the detector is amplified and then the baseband signal is demodulated.

The sensitivity of the receiver for a particular detector is called noise equivalent flux density (NEFD) and is given [23-42] by

$$
\text{NEFD} = \frac{(A_d \Delta f_n)^{1/2} \int_0^\infty \Phi_\lambda(\lambda) d\lambda}{k_y k_n A_o \int_0^\infty D_{f_0}^*(\lambda) \Phi_\lambda(\lambda) \tau_o(\lambda) d\lambda} \, , \text{W m}^{-2} \tag{23-63}
$$

where A_d = area of the detector

Δf_n = noise equivalent bandwidth of the receiver, assuming Gaussian white noise

$\Phi_\lambda(\lambda)$ = spectral radiant power illuminating the receiving aperture

A_o = collecting aperture area

$D_{f_0}^*(\lambda)$ = specific spectral detectivity at frequency, f_o

$\tau_o(\lambda)$ = spectral transmissivity of the optics

k_n = factor accounting for the difference in actual noise distribution from the Gaussian white noise

k_y = factor accounting for nonuniform responsivity of the detector in frequency given by

$$
k_y = \sqrt{\langle V^2 \rangle} = \left\{ \frac{1}{(t_2 - t_1)} \int_{t_1}^{t_2} \left[\int_{-\infty}^{\infty} X(f)[R(f)/R(f_0)] \exp(j2\pi ft) df \right]^2 dt \right\}^{1/2} \tag{23-64}
$$

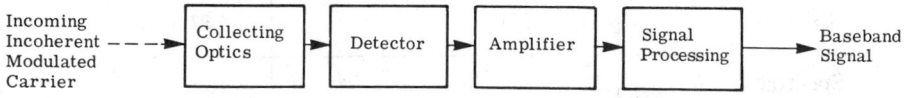

Fig. 23-17. Incoherent receiver block diagram.

where $X(f)$ = the frequency distribution of $\Phi_\lambda(\lambda)$
$R(f)/R(f_o)$ = normalized responsivity of the detector as a function of frequency

If, over the spectral range of the optics, the D* (the specific detectivity, Chapter 11) and $\tau_o(\lambda)$ are reasonably constant, the time constant of the detector is small compared to f_{mx}^{-1} (where f_{mx} is the highest frequency of the information signal), and the noise is Gaussian white noise, then the expression for noise equivalent irradiance input (NEFD) reduces to

$$\text{NEFD} = \frac{(A_d \Delta f_n)^{1/2}}{A_o D^*} \tag{23-65}$$

To ensure that the baseband signal is not lost, one must be sure that the frequency response of the detector and the amplification stages are sufficient to transmit significant frequency components of the received signal. Below, the frequency distributions of the pulse-frequency-modulated, pulse-width-modulated, and pulse-amplitude-modulated waveforms are shown [23-43, 23-44, 23-45].

Pulse Frequency Modulation:

$$f(t) = \frac{V \omega_c \tau}{2\pi} + \frac{V \omega_c \tau}{\pi} \sum_{k=1}^{\infty} \frac{\sin(k \omega_c \tau/2)}{k \omega_c \tau/2} \left\{ J_0(km_f) \cos(k\omega_c t) \right.$$

$$+ \sum_{n=1}^{\infty} J_n(km_f) \{ \cos[(k \omega_c + n \omega_m)t + n\phi]$$

$$\left. + (-1)^n \cos[(k \omega_c - n \omega_m)t - n\phi] \} \right\} \tag{23-66}$$

Pulse Width Modulation (Symmetrical Double-Edge Modulation):

$$f(t) = \frac{V \omega_c \tau}{2\pi} + \frac{V \omega_c m \tau}{2\pi} \sin(\omega_m t + \phi)$$

$$+ \frac{V \omega_c \tau}{2\pi} \sum_{k=1}^{\infty} \left[2J_0\left(k \omega_c m \frac{\tau}{2}\right) \frac{\sin(k \omega_c \tau/2)}{k \omega_c \tau/2} \cos(k \omega_c t) \right.$$

$$+ 4 J_1\left(k \omega_c m \frac{\tau}{2}\right) \frac{\cos(k \omega_c \tau/2)}{k \omega_c \tau/2} \cos(k \omega_c t) \sin(\omega_m t + \phi)$$

$$\left. + 4 J_2\left(k \omega_c m \frac{\tau}{2}\right) \frac{\sin(k \omega_c \tau/2)}{k \omega_c \tau/2} \cos(k \omega_c t) \cos(2\omega_m t + 2\phi) + \dots \right] \tag{23-67}$$

Pulse Amplitude Modulation (Exact Scanning):

$$f(t) = \frac{V\omega_c \tau}{2\pi} \left[1 + m_a \cos(\omega_m t + \phi) + 2\sum_{k=1}^{\infty} \frac{\sin(k\,\omega_c\,\tau/2)}{k\,\omega_c\,\tau/2} \left\{ \cos(k\,\omega_c\,t) \right. \right.$$

$$\left. \left. + \frac{m_a}{2}\cos\left[(k\,\omega_c + \omega_m)t + \phi\right] + \frac{m_a}{2}\cos\left[(k\,\omega_c - \omega_m)t - \phi\right] \right\} \right] \tag{23-68}$$

where V = peak pulse amplitude
ω_c = $(2\,\pi f_c)$; radiation carrier frequency
τ = rectangular pulse width
k = integer: $1,2,3,\ldots,\infty$
J_k = k^{th} order Bessel function of the first kind
m_f = $\Delta f_c/f_m$, frequency index of modulation given by the ratio of peak carrier frequency deviation to modulating frequency
n = integer: $1,2,3,\ldots,\infty$
ϕ = reference phase angle
m = $\Delta\tau/\tau$, pulse width modulation index
m_a = amplitude modulation index

The expressions are quite complex and represent modulation by a single frequency component. However, if the frequency is selected to represent the maximum frequency of the baseband signal, useful conclusions of the bandwidth characteristics of the receiver can be made. As an example, in the case of pulse-frequency-modulation, there is no sideband accompanying the zero or dc component of the pulse spectrum. Hence, employment of a low-pass filter for demodulation would not yield any modulating frequency component. The sidebands of the modulating frequency are clustered around the fundamental and the harmonics of the unmodulated carrier pulse-repetition frequency. Therefore, the fundamental and the harmonics can each be looked upon as frequency-modulated carriers with index of modulation km_f. Also, as the harmonic order of the pulse-repetition frequency increases, the amplitudes of the carrier and sidebands decrease as a function of $(\sin x)/x$, where $x = k\omega_c\,\tau/2$. Thus, the best signal would be obtained by the employment of a filter which would encompass the fundamental and its sidebands. The output of the filter would be applied to a frequency discriminator which would convert the input signal into a replica of the transmitted baseband signal.

On the other hand, in the case of pulse-width modulation, the second term of the frequency structure contains the modulating frequency component by itself. Thus, filtering the received signal with a low-pass filter encompassing the $[(V\omega_c m\tau)/2\pi]$ sin $(\omega_m t + \phi)$ term, would recover the baseband signal from the pulse-width-modulated pulse train.

If the baseband signal consists of several frequencies, f_m, each frequency modulating a pulse train, then for each harmonic of the pulse repetition frequency and its sidebands, the frequency structure is given [23-46] by

$$f(t) = V \prod_{m=1}^{N} \sum_{n_m=-\infty}^{\infty} J_{n_m}(m_{f_m}) \exp(jn_m\omega_m t) \exp(j\omega_c t) \tag{23-69}$$

Assuming the modulating baseband consists of only frequencies f_1 and f_2, one can reduce the equation to the following components:

(1) Carrier:

$$J_o(m_{f_1})\, J_o(m_{f_2})\; V \exp (j\, \omega_c\, t)$$

(2) Sidebands due to ω_1:

$$J_n(m_{f_1})\, J_o(m_{f_2})\; V \exp \left[j\, (\omega_c \pm n\omega_1)\, t\right]$$

(3) Sidebands due to ω_2:

$$J_m (m_{f_2})\, J_o(m_{f_1})\; V \exp \left[j(\omega_c \pm m\omega_2)\, t\right]$$

(4) Beat frequencies at $\omega_c \pm n\, \omega_1 \pm m\, \omega_2$:

$$J_n (m_{f_1})\, J_m (m_{f_2})\; V \exp \left[j\, (\omega_c \pm n\, \omega_1 \pm m\, \omega_2)\, t\right]$$

Figure 23-18 shows the significant bandwidth (normalized) versus the modulation index, m_f.

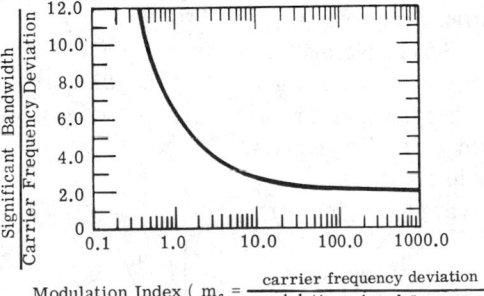

$$\text{Modulation Index} \left(m_f = \frac{\text{carrier frequency deviation}}{\text{modulating signal frequency}} \right)$$

Fig. 23-18. Significant bandwidth (normalized) versus modulation index, m_f. The bandwidth is occupied by significant sidebands. [23-47].

23.3.4. Coherent IR Communications Systems.
The invention of the laser made coherent communications possible [23-48, 23-49]. To date, thousands of lasing transitions have been discovered employing different types of materials, [23-50]. Four commonly used lasers are listed in Table 23-6.

Table 23-6. Typical Coherent Communication Carrier Laser Sources

Laser	Wavelength (μm)	Type	Operation
Nd:YAG	1.06	Glass	CW/pulse
HeNe	1.15, 3.39	Arc discharge	CW/pulse
CO_2	9.6, 10.6	Glow discharge	CW
GaAs	~0.9	Injection semiconductor	CW/pulse

Figure 23-19 shows the basic configuration of communication lasers. For communication purposes, the important parameters of the laser are the power output, axial and transverse mode control, efficiency, output beam-diameter, and polarization of the carrier. The degree to which these parameters need to be controlled depends on the type of modulation and reception technique employed. In almost every case, it is desirable to control the transverse modes such that the radiation pattern is TEM_{00}. In the case of heterodyne detection, the laser should operate in a single axial mode. However, if the detection process is direct, multimode axial operation is acceptable. Many of the modulation, and duplexing techniques depend on the fact that the laser is linearly polarized.

Noise characteristics of the laser carrier are important with continuous-wave modulation techniques. For heterodyne detection, the frequency purity of the carrier and of local oscillators are of paramount importance. Several noise mechanisms operate in a laser. The basic quantum-mechanical limitations to the spectral purity of the laser output are, respectively: (1) the laser carrier-envelope amplitude and frequency fluctuation spectrum, (2) the laser carrier-envelope intensity fluctuation spectrum, and (3) the laser carrier frequency, ν_c fluctuation spectrum [23-51]. Normally, quantum-mechanical noise effects are negligible compared to other sources in a practical laser. Based on Reference [23-51], the CO_2 laser linewidth is $\Delta\nu_c = 1.5 \times 10^{-3}$ Hz, which is negligible, but for GaAs LEDs it is $\Delta\nu_c = 1.3 \times 10^6$ Hz, which is excessive. The other sources of noise are due to the fluctuation in the effective cavity length of the laser. This is caused by the fluctuation in the refractive index of the materials in the laser cavity and mechanical vibration. As an example, the axial-mode resonance condition is

$$\nu_c = \frac{mc}{2L_c} \qquad (23\text{-}70)$$

where m = an integer
$\quad c$ = the speed of radiation propagation in vacuum
$\quad L_c$ = the cavity length

The change in frequency is given by

$$\Delta\nu_c = \left(\frac{\partial\nu_c}{\partial L_c}\right)\Delta L_c = \nu_c\left(\frac{\Delta L_c}{L_c}\right) \qquad (23\text{-}71)$$

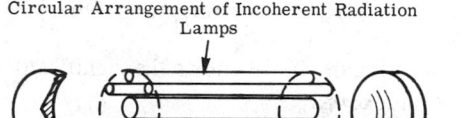

Circular Arrangement of Incoherent Radiation Lamps

Mirror   Doped Glass   Mirror

(a) Basic elements of a glass laser.

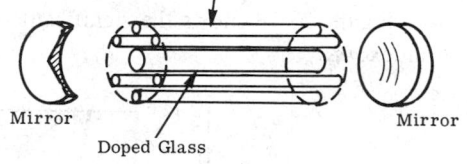

Tube Filled with Gas

Brewster Window   Glow Discharge

Mirror   Electrodes   Brewster Window   Mirror

(b) Basic Elements of a gas laser.

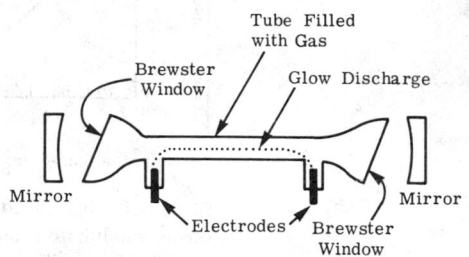

Current Flow   Electrode

n-Type Material

Laser Beam   Junction

p-Type Material

(c) Basic elements of injection semi-conductor (or injection) laser.

Fig. 23-19. Basic configuration of communication lasers.

Thus, at 1.06 μm and a cavity length of 1 m, a change in length of 2.5×10^{-8} m (one micro-inch) will produce a frequency shift of 7 MHz.

A source of energy input into the lasing material is associated with each laser. Typical sources appropriate for communications lasers are listed in Table 23-7.

Table 23-7. Typical Pump Sources for Communication Lasers

Type	Energy Coupling Causing Population Inversion
Incoherent radiation source	Radiant input
Electrical current	Glow discharge
Electrical current	Ion acceleration
Electrical current	Collisions in the active region of semiconductor

The laser control functions include the search for, and lock-on to, the correct laser line upon the start of the lasing action, and the tuning of the laser cavity to match its resonant length with the maximum of the laser gain-curve.

The modulation of the carrier laser can be either internal or external to the laser cavity. Techniques are discussed in the next section. The function of the signal processing is to adapt the baseband signal into the format most suitable for the modulation of the laser carrier. Thus, it can encompass a large number of techniques employed in RF and microwave communications.

As an example, assume the modulated carrier bandwidth is 2 GHz, and several TV channels are being transmitted using pulse modulation. Each TV baseband signal occupies 5 MHz bandwidth, therefore requiring 10 MHz sampling frequency. To fill the carrier bandwidth, each TV channel would be either time-division, or frequency-division multiplexed, and then the resulting signal used to modulate the laser carrier. In the case of heterodyne detection system, the signal processing may consist of shifting the baseband frequency distribution to above the frequency noise components of the laser carrier. If, for example, the TV signal consists of 5 MHz extending from 20 Hz to 5 MHz, then, in the signal processing, the baseband signal is transposed to cover 1 to 6 MHz prior to its application to the modulation of the carrier.

The optics perform several functions, the most important of which are listed below:

(1) Transmitting aperture matching.
(2) Duplexing.
(3) Spectral filtering.
(4) Signal tracking and pointing of the beam.
(5) Focusing received signal on the detector.
(6) Local oscillator injection (for heterodyne receivers).

The matching of the transmitting aperture entails adapting the laser-output beam-diameter to a diameter required for optimum illumination of the radiating aperture. A typical aperture-matching arrangement is shown in Figure 23-20. The flux density distribution of the laser-output beam, as a function of the radial distance, r_1, from the axis of symmetry of the beam, is usually Gaussian, so that

$$M = \frac{\Phi_l}{2\pi\sigma^2}\left[\exp\left(\frac{-r_1^2}{2\sigma^2}\right)\right]^2 , \text{W m}^{-2} \qquad (23\text{-}72)$$

where Φ_l = the laser output power
2σ = the $1/e^2$ radius

The output beam of the laser is normally diffraction-limited. Assuming TEM_{00} and single-axial-mode operation, one can compute the far-field pattern, $I(\text{W sr}^{-1})$, by using the geometry of Figure 23-21, and solving the integral [23-52, 23-53].

$$E(\alpha) = \frac{k}{R}\sqrt{\frac{\Phi_T}{2\pi\sigma^2}}\int_{r_b}^{r_a}\exp\left(\frac{-r_1^2}{2\sigma^2}\right)J_0(k\alpha r_1)r_1 dr_1 \qquad (23\text{-}73)$$

where $M = |E(\alpha)|^2$
$I = R^2 |E(\alpha)|^2$
R = Range
$k = 2\pi/\lambda$, λ = wavelength
Φ_T = laser power illuminating the radiating aperture in a beam from r_b to r_a
α = off-axis angle, shown in Figure 23-21
$J_0(k\alpha r_1)$ = zero-order Bessel function of the first kind

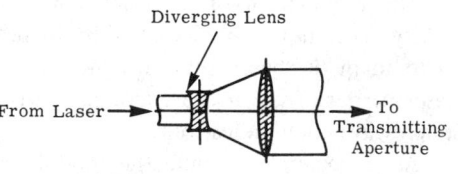

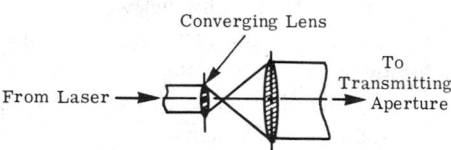

Fig. 23-20. Aperture matching arrangement.

The results are summarized in Table 23-8. The far-field patterns for a uniform illumination of the aperture are also included in Table 23-8. Using these expressions, one can readily compute the gain of the transmitting and receiving apertures defined by

$$G = \frac{\text{Far-Field Intensity of the Radiating Aperture}}{\text{Intensity of an Isotropic Radiator}}$$

$$= \frac{4\pi I(\alpha)}{\Phi_T} \qquad (23\text{-}74)$$

The received power is then given by

$$\Phi_r = \Phi_T G_T G_r F \qquad (23\text{-}75)$$

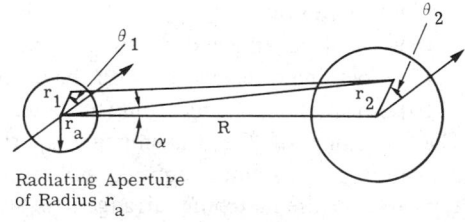

Fig. 23-21. Fraunhofer diffraction geometry, for axially symmetric configuration.

where F = the free-space loss given by $(\lambda/4\pi R)^2$

G_T and G_r = the transmitting and receiving aperture gains, respectively

This manner of computing the received power is widely used by communication workers. By adding a factor to account for losses due to the atmosphere, one completes the equation. This equation reduces to the more familiar form

$$\Phi_r = \frac{I_T A_r}{R^2} \tag{23-76}$$

where A_r is the area of the receiving aperture. By expansion, Equation (23-75) becomes

$$\Phi_r = \Phi_T \left(\frac{4\pi I_T(\alpha)}{\Phi_T}\right)\left(\frac{\lambda^2}{(4\pi R)^2}\right)\left(\frac{4\pi I_r(\alpha)}{\Phi}\right) = \left(\frac{I_T(\alpha)}{R^2}\right)\left(\frac{\lambda^2 I_r(\alpha)}{\Phi}\right) \tag{23-77}$$

Table 23-8. Far Field Intensity Distribution of a Diffraction Limited Laser Beam

Aperture Intensity Illumination	Aperture Configuration	Far Field Intensity Distribution, I (W sr^{-1})
Gaussian	Circular ($\sigma \ll r_a$)	$I(\alpha) = 8\pi\left(\frac{\sigma}{\lambda R}\right)^2 \Phi_T \exp\left[-8\left(\frac{\pi\sigma\alpha}{\lambda}\right)^2\right]$
Gaussian	Circular	$I(\alpha) = \pi\left(\frac{1}{\lambda R\sigma}\right)^2 \Phi_T \left[\int_0^{r_a} \exp(-r_1^2/2\sigma^2) J_0\,(\alpha r_1 k)\, r_1 dr_1\right]^2$
Gaussian	Circular with blockage of radius r_b in the center	$I(\alpha) = \pi\left(\frac{1}{\lambda R\sigma}\right)^2 \Phi_T \left[\int_{r_b}^{r_a} \exp(-r_1^2/2\sigma^2) J_0\,(\alpha r_1 k)\, r_1 dr_1\right]^2$
Uniform	Circular	$I(\alpha) = \frac{\pi r_a^2 \Phi_T}{\lambda^2} \left[\frac{2J_1\left(\frac{2\pi\alpha r_a}{\lambda}\right)}{\frac{2\pi\alpha r_a}{\lambda}}\right]^2$
Uniform	Rectangular ($a \times b$)	$I(x_2, y_2) = \frac{\Phi_T}{\lambda^2} \frac{\sin\left(\frac{\pi a x_2}{\lambda R}\right)}{\frac{\pi a x_2}{\lambda R}} \frac{\sin\left(\frac{\pi b y_2}{\lambda R}\right)}{\frac{\pi b y_2}{\lambda R}}$

where $J_0(\ )$ and $J_1(\ )$ designate zero- and first-order Bessel functions of the first kind.

Using the expression $I(\alpha)$ (from Table 23-8), for a uniformly illuminated aperture for the on-axis case ($\alpha = 0$), one has

$$\Phi_r = \left(\frac{I_T(\alpha)}{R^2}\right)\left(\frac{\lambda^2}{\Phi}\right)\left(\frac{\pi\, d_r^2\, \Phi}{4\lambda^2}\right) = \left(\frac{I_T(\alpha)\, A_r}{R^2}\right) \tag{23-78}$$

In a transceiver, if the carrier is linearly polarized, duplexing is fairly easy to implement. The function of the duplexer is to couple the transmitter laser output to the radiating aperture and, in turn, to couple the received power from the aperture to the detector. One way to do this is illustrated in Figure 23-22. The transmitted power entering the wire grid polarizer is horizontally polarized. Since polarization is perpendicular to the direction of the wires, the grid is transparent to the electromagnetic radiation and the power is coupled through to the radiating aperture [23-54]. The incoming received power is vertically polarized. In this case, the grid acts as a reflector, directing the received power to the detector. (To ensure complete isolation, one should be sure that the transmitted and received power are operating at two different transition lines.) The polarizer is placed at $45°$ to the optical axis.

For the heterodyne receiver, employment of a narrow-band spectral filter is not critical since, as will be shown later, a properly designed heterodyne receiver operates at the quantum limit $h\nu/\eta$ in W Hz^{-1} where η is the detector quantum efficiency, at most practical background irradiance levels. However, in the case of direct detection, incorporation of a narrow-band spectral filter to reduce the background irradiance is required for improved system sensitivity.

The diffraction-limited beamwidth of the transmitting and receiving apertures is very narrow at IR frequencies. For example, if one uses the formula for a centrally blocked, Gaussian-illuminated aperture, the three dB beamwidth is calculated. For an aperture diameter of 17.78 cm and a central blockage diameter of 2.54 cm, the radiation pattern of the aperture can be calculated to be as plotted in Figure 23-23; the 3 dB width is 70 μrad or 14.6 arc sec. If one assumes there is no *a priori* information on the relative directions between the two stations, and that there is a uniform probability that beam is pointed in any direction, the probability that both beams are in each other's FOV is 1 X

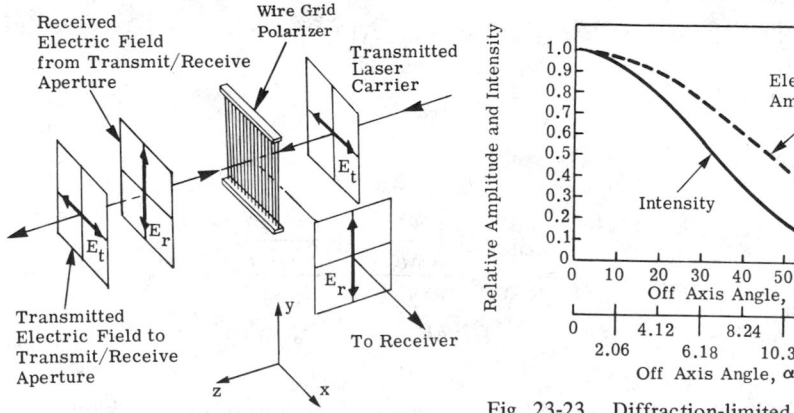

Fig. 23-22. Typical IR duplexer arrangement.

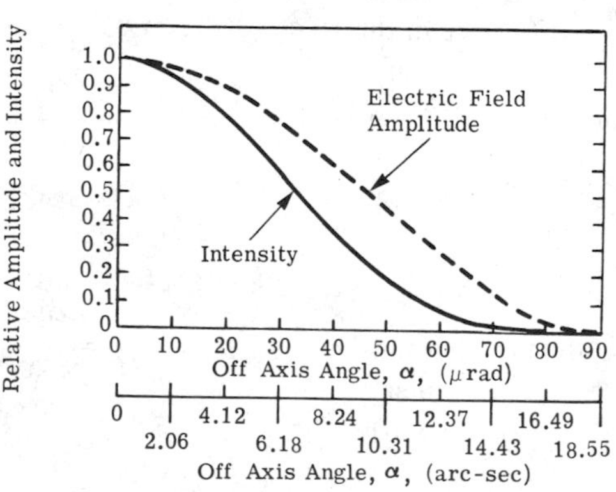

Fig. 23-23. Diffraction-limited far-field radiation pattern.

10^{-19}. Fortunately, *a priori* information is available on the relative location of the two stations. Thus the pointing direction for each station can be determined with, however, an uncertainty angle which may exceed the beamwidth of each aperture. The major sources of direction uncertainty are the angular motion of the earth surface, the pointing tolerances of the equipment, the fluctuations in the angle of arrival of the phase front due to the atmosphere, and, initially, the errors associated with the attitude and the location of each transmitting station. Thus, provisions must be made for the two stations to acquire each other and then continuously track each other's signals such that the pointing direction is maintained. Means to adjust the beam direction must be provided within the optics. An example of such optics is shown in Figure 23-24. The two mirrors are mounted on piezo-electric supports. The axes of rotation of the respective mirrors are orthogonal. The mirror rotation is induced by the application of appropriate voltages to the mirror mounts. With this technique, fine adjustments in the beam direction can be made. Larger scale adjustments are usually made with a set of precision-gimbals mounting a mirror which is external to the transmit/receive optics.

If a heterodyne receiver is used, provisions must be made to combine the local oscillator signal with the received signal. For proper operation, the two wavefronts must have their phase-front and polarization vectors aligned. The receiver optics are discussed under coherent infrared receivers.

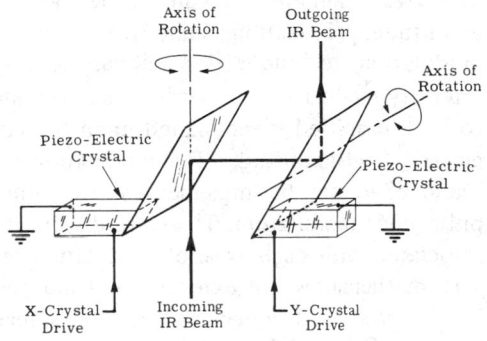

Fig. 23-24. Beam pointing mirror arrangement.

The Carrier Modulation. In principle, the modulation techniques which have been developed for the modulation of radio frequency waves can also be used at infrared wavelengths. Further, because of the short wavelengths in the IR and the corresponding small size of the hardware, modulation techniques, impractical in the case of radio waves, are feasible at the IR laser wavelength. Thus, whereas in radio communications the information bandwidth is almost always up-converted by the heterodyne or homodyne process and then down-converted in the receiver, in the case of laser communication, direct detection of the information modulated on the carrier is quite feasible. At the same time, the practical implementation of many of the potential laser carrier-modulation techniques is still to be demonstrated. The basic constraint is the development of various devices needed to implement the modulation techniques.

The propagating-carrier electric-field equation is

$$e = EP_m \exp[-j(\mathbf{k \cdot r} - \omega_c t - \theta)] \tag{23-79}$$

where E = peak electric field amplitude

P_m = polarization matrix. As examples, typical polarization matrices are given below:

Vertical Polarization, $P_m = \begin{bmatrix} 1 \\ 0 \end{bmatrix}$

$$\text{Horizontal Polarization}, P_m = \begin{bmatrix} 0 \\ 1 \end{bmatrix}$$

$$\text{Polarization } 45° \text{ to Horizontal}, P_m = \frac{1}{\sqrt{2}} \begin{bmatrix} 1 \\ 1 \end{bmatrix}$$

$$\text{Right Circular Polarization}, P_m = \frac{1}{\sqrt{2}} \begin{bmatrix} -j \\ 1 \end{bmatrix}$$

$\mathbf{e}$ = complex rotating vector

$\mathbf{k}$ = propagation vector; its magnitude given by $|2\pi/\lambda|$, where λ is the wavelength

$\mathbf{r}$ = range vector

ω_c = radian frequency given by $2\pi/T_c = 2\pi\nu_c$, where T_c is the period of the carrier and ν_c is its frequency

t = time

θ = phase-reference

The five parameters of the carrier which can be modulated are the electric field amplitude, polarization, phase, frequency, and direction of the light beam; each specific modulation technique is summarized in mathematical form in Table 23-9. The intensity-modulation format has also been included. In the intensity case, the information to be transmitted is modulated upon the laser carrier in proportion to the square of the electric field amplitude. The modulation of the laser carrier parameters, summarized in Table 23-9, may be implemented as a form of analog modulation, pulse modulation, or pulse code modulation. These are discussed in Section 23.2.2. The energy in the carrier associated with each type of modulation technique is redistributed into the side bands. The mathematics are extensive and are covered well in the literature [23-55, 23-56]. However, at IR frequencies the total spread of the frequency rarely exceeds a spectral range of 0.3 to 0.5 nm. Over this extent, the optics of a laser transmitter and a laser receiver can be considered to have constant frequency and phase response.

There are several devices and techniques which can be used for modulation of the coherent carrier. The action itself may take place internally in the laser cavity or externally to the cavity. The majority of practical modulation techniques are based on the availability of the materials whose index of refraction is a function of direction and, further, materials whose index of refraction can be changed by application of a perturbation such as caused by an electrical field or mechanical pressure. The combination of materials, devices, and arrangements present an almost limitless way of implementing the modulation of the laser carrier. This is an active field with advances and new discoveries being made continuously.

The parameters of a modulation technique which should be considered to assess its usefulness are the bandwidth, the depth of modulation, the linearity of modulation, the modulation efficiency both in terms of the loss of carrier power and the modulator driving power input, the stability over the environmental range, the lifetime, and the cost.

During the modulation of the carrier with a particular technique to achieve a given form of modulation, other secondary forms of modulation may simultaneously take place. As an example, frequency modulation of the carrier also causes a certain amount of amplitude modulation. Intensity modulation will also cause amplitude modulation and vice versa. In fact, both of these modulations always occur together and, whether

Table 23-9. Modulation Formats

Name	Parameter Affected	Modulating Expression	Propagating Carrier Expression
Amplitude modulation	Peak electric field amplitude, E	$[1 + m_a s(t)]$ where m_a is amplitude modulation index, $s(t)$ is the baseband signal, and $m_a s(t) \leq 1$.	$E[1 + m_a s(t)]\, P_m \exp[-j(\mathbf{k} \cdot \mathbf{r} - \omega_c t + \theta)]$
Intensity modulation	Peak intensity amplitude $I = c_1 E^2$, where c_1 is constant	$[1 + m_i s(t)]$ where m_i is intensity modulation index, $s(t)$ is the baseband signal, and $m_i s(t) \leq 1$.	$E^2[1 + m_i s(t)]\, P_m^2 \exp[-2j(\mathbf{k} \cdot \mathbf{r} - \omega_c t + \theta)]$
Polarization modulation	Polarization vector, P_m	Polarization modulation matrix, $$M_p = \begin{bmatrix} \exp j\dfrac{\Gamma}{2} & 0 \\ 0 & \exp\left(-j\dfrac{\Gamma}{2}\right) \end{bmatrix}$$ where $\Gamma = m_p s(t)$ is the phase shift, m_p is polarization modulation index, and $s(t)$ is the baseband signal.	$EM_p\, P_m\, exp[-j(\mathbf{k} \cdot \mathbf{r} - \omega_c t + \theta)]$
Frequency modulation	Carrier radian frequency, ω_c	$m_f \int_0^t s(\tau)\, d\tau$ where m_f is frequency modulation index and $s(\tau)$ is the baseband signal.	$EP_m \exp\left[-j\left(\mathbf{k} \cdot \mathbf{r} - \omega_c t - m_f \int_0^t s(\tau)\, d\tau + \theta\right)\right]$
Phase modulation	Carrier phase, θ	$m_p s(t)$ where m_p is the phase modulation index and $s(t)$ is the baseband signal.	$EP_m \exp[-j(\mathbf{k} \cdot \mathbf{r} - \omega_c t + \theta + m_p s(t))]$
Propagation vector modulation	Propagation vector, $$\mathbf{k} = \begin{bmatrix} i\, k_x \\ j\, k_y \\ k\, k_z \end{bmatrix}$$	Propagation vector rotation matrix, $$M_k = \begin{bmatrix} m_{11} & m_{12} & m_{13} \\ m_{21} & m_{22} & m_{23} \\ m_{31} & m_{32} & m_{33} \end{bmatrix}$$ where $|M_k| = \pm 1$ and m_{tn} = rotation matrix components whose magnitude is a function of baseband signal, $s(t)$.	$EP_m \exp[-j(\mathbf{k}\, M_k \cdot \mathbf{r} - \omega_c t + \theta)]$

they are called amplitude or intensity modulation depends on which parameter is linearly related to the modulating signal. Therefore, the design of the receiver must render it impervious to these secondary modulations. As an example, in the case of frequency modulation in the receiver after detection, an amplitude limiter would be included to remove the amplitude fluctuation prior to the application of the received signal to a frequency discriminator.

Electro-optic modulators employ birefringent crystals, polarizers, and other optical devices designed to operate on the polarization of a propagating electric wave. Birefringent crystals are materials which display an anisotropic optical property; in other words, their indices of refraction differ as a function of direction. The electro-optic effect refers to that property of a material which changes its index of refraction as a function of the magnitude, E, of an applied electric field. Thus, an isotropic transparent substance becomes birefringent when placed in an electric field. When the change in index of refraction is linearly dependent on the applied electric field, it is known as Pockels effect [23-57]. Quadratic dependence of induced birefringence or change in birefringence on the applied electric field is known as the Kerr effect [23-58].

In general, the index of refraction can be expressed by an indicatrix as

$$\frac{x_1^2}{n_1^2} + \frac{x_2^2}{n_2^2} + \frac{x_3^2}{n_3^2} = 1 \tag{23-80}$$

where $x_{1,2,3}$ = coordinate axes

$n_{1,2,3}$ = the principal refractive indices [23-59]

Application of an electric field will cause the index of refraction ellipsoid to rotate with respect to the crystallographic axes and change its shape. A typical effect on a crystal is illustrated in Figure 23-25.

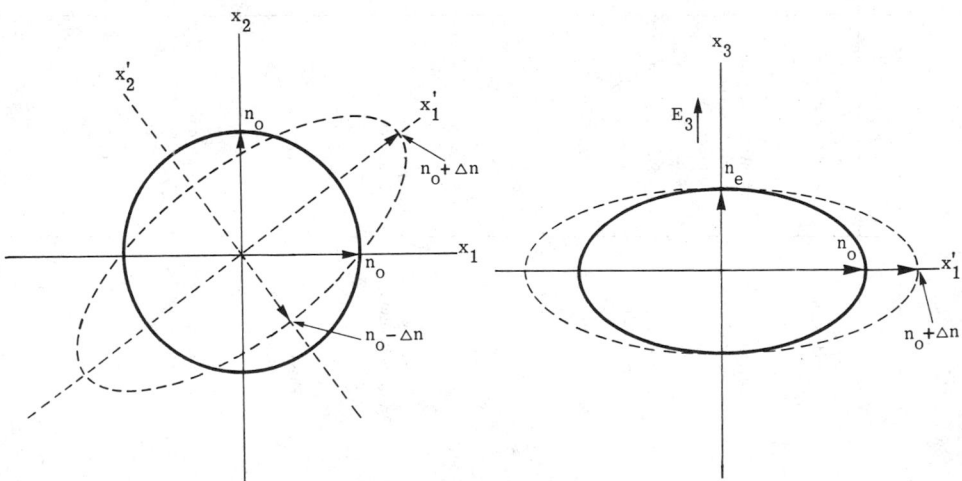

Fig. 23-25. Indicatrix of an electro-optical material before (solid line) and after (dashed line) an electric field is applied.

Along a crystal axis, the index of refraction can be expressed by

$$\frac{1}{(n + \Delta n)^2} = \frac{1}{n^2} + rE + RE^2 + \ldots \tag{23-81}$$

where r = the linear electro-optic coefficients
$\quad R$ = the quadratic electro-optic coefficients
$\quad n$ = the index of refraction without an electric field

Assuming a crystal in which the linear effect is the dominant effect (and neglecting R and higher-order terms), one can solve Equation (23-81) for Δn, assuming $n \gg \Delta n$. To a close approximation,

$$\Delta n = -\frac{n^3 rE}{2} \tag{23-82}$$

In many crystals, the tensor elements, r, in general, reduce to simple coefficients per crystal axis. Applying an electric field parallel to x_1 axis, one finds that the changes in the index of refraction along x_1 and x_2 axes are given by

$$\Delta n_1 = -\frac{n_1^3 r_{11} E_1}{2}$$

$$\Delta n_2 = -\frac{n_2^3 r_{21} E_1}{2} \tag{23-83}$$

If one assumes a wave with propagation vector, $\mathbf{k}$, parallel to the x_3 axis and with the electric field of the wave parallel to x_1 axis, after traversing a length $x_3 = L$, the change in phase due to electro-optic effect is given by

$$-\frac{2\pi L}{\lambda_o} [n_1 - (n_1 + \Delta n_1)] = -\frac{\pi n_1^3 r_{11} E_1 L}{\lambda_o} \tag{23-84}$$

The electric field propagating in the direction of x_3 axis is given by

$$e = E_1 \exp\left\{-j\left(kx_3 - \omega_c t - \frac{2\pi n_1 L}{\lambda_o} - \frac{\pi n_1^3 r_{11} E_1 L}{\lambda_o}\right)\right\} \tag{23-85}$$

where $x_3 > L$
$\quad k = 2\pi/\lambda_o$, and λ_o is the wavelength in free space

The characteristics of typical electro-optic materials in the IR are tabulated in Table 23-10.

Several configurations of the electro-optic modulators are possible. A frequency modulator is illustrated in Figure 23-26. The electro-optic crystal is placed within the cavity of the laser. The two faces of the crystal are cut at the Brewster angle (a surface cut at Brewster angle will reflect only the component of a polarized wave which is parallel to the surface). Thus, only polarized lasing action will be sustained. If the placement of the crystal is such that the polarization of the laser electric field coincides with one axis of the crystal and the externally-applied electric field is also in alignment with

Table 23-10. Characteristics of Typical Electro-Optical Materials [23-60]

| Material Crystal | Electro-Optic Coefficients | | λ (μm) | Refractive Index | |
	$r_{i3}(10^{-12}\text{mV}^{-1})$	$r_{ij}(10^{-12}\text{mV}^{-1})$		n_i	λ (μm)
LiNbO$_3$	$r_c = 16$	$r_{22} = 3.1$	3.39	$n_3 = 2.05$	4.0
LiTaO$_3$	$r_{33} = 27$	$r_{51} = 15$	3.39	$n_3 = 2.03$	4.0
GaAs	–	$r_{41} = 1.2$	1.0 to 3.0	$n_0 = 3.42$	1.25
	–	$r_{41} = 1.0$ to 1.2	4.0 to 12.0	$n_0 = 3.30$	<5.0
ZnTe	–	$r_{41} = 1.4$	10.6	$n_0 = 2.76$	1.24
ZnSe	–	$r_{41} = 2.2$	10.6	$n_0 = 2.3$	10.6
ZnS	–	$r_{41} = 1.4$	3.39	$n_0 = 2.26$	2.4
CdTe	–	$r_{41} = 6.8$	3.39	$n_0 = 2.82$	1.3
		$r_{41} = 6.8$	10.6	$n_0 = 2.60$	10.6
CdSe	$r_{13} = 1.8$	$r_{33} = 4.3$	3.39	$n_3 = 2.54$	1.15
				$n_1 = n_2 = 2.52$	1.15
CdS	$r_c = 5.5$	–	10.6	$n_3 = 2.3$	10.0

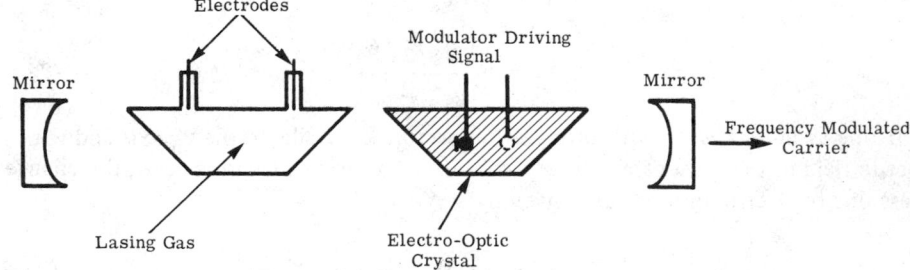

Fig. 23-26. Frequency modulation of laser carrier.

that axis, then as the external field is varied, the frequency of the lasing action will also change correspondingly. If the baseband signal is $f(t)$, the carrier frequency, at a particular point in space, will be given by

$$\nu = \nu_c - \frac{n_1 L}{\lambda_o} - \frac{n_1^3 r_{11} L E_m\,[f(t)]}{2\lambda_o} \tag{23-86}$$

An arrangement for intensity modulation is shown in Figure 23-27. The modulating components are placed outside the laser cavity and consist of an electro-optic crystal and a polarizer. When electric field, E_m is applied, the electro-optic crystal produces a phase retardation, Γ, given by $\Gamma = \Delta n_1 - \Delta n_2$. The polarization of the incoming electric field of amplitude, E_c, is set at $45°$ to the x_1 and x_2 axes of the crystal. Without voltage applied to the crystal, the output is orthogonal to the polarizer and the carrier is attenuated. Upon application of the electric field, E_m, the polarization is rotated, and intensity is coupled out of the modulator.

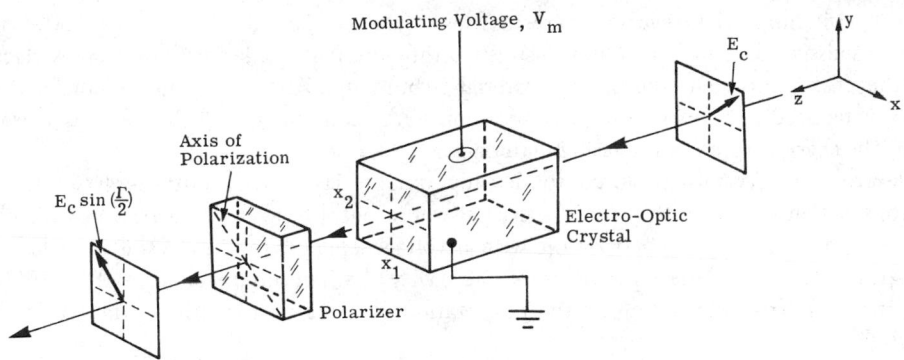

Fig. 23-27. Intensity modulation of laser carrier.

The polarization of the input laser beam may be represented by

$$E_c = \frac{E}{\sqrt{2}} \begin{bmatrix} 1 \\ 1 \end{bmatrix}$$ (23-87)

The modulator with a total phase retardation of Γ may be expressed in the matrix form

$$M' = \begin{bmatrix} \exp(j\,\Gamma/2) & 0 \\ 0 & \exp(-j\,\Gamma/2) \end{bmatrix}$$ (23-88)

and the polarizer by

$$M_p = \frac{1}{2} \begin{bmatrix} 1 & -1 \\ -1 & 1 \end{bmatrix}$$ (23-89)

The modulator output electric field is then given by

$$E'_c = M'M_pE_c = \frac{E}{\sqrt{2}} \, j \sin\frac{\Gamma}{2} \begin{bmatrix} 1 \\ -1 \end{bmatrix}$$ (23-90)

which represents an electric field polarized at $-45°$. The output intensity is given by

$$I_o = |E_c|^2 \sin^2\left(\frac{\Gamma}{2}\right)$$ (23-91)

Intracavity pulse modulation can be generated using electro-optic materials. The three techniques are known as mode-locking, cavity-dumping, and Q-switching [23-61].

Mode-locking results in a single transverse, multi-axial-mode laser by introduction of an intracavity loss or a phase modulator which is operated at a frequency equal to the axial mode separation frequency given by $\Delta\nu = c/2L$. The effect is to couple the axial modes in an AM mode. For a larger number, N, of locked modes, the output is a pulse train, with period $T = 2L/c$, pulse width T/N, and a peak-power N times the average power.

In cavity dumping, the transmission of the output mirror is varied between 0 and 100%. The 100% transmission is switched for a time interval $\geqslant 2L/c$, during which an output pulse is transmitted. During the zero mirror-transmission period, the energy stored in the laser is built up.

In Q-switching, the reverse operation takes place. The mirror is either 100% or low-transmissive. During the fully-transmissive time interval, no lasing action takes place and the energy stored in the lasing material is built up. When the transmission of the mirror is reduced to a low number, lasing action starts, coupling out of the laser (as a laser pulse) the energy stored in the lasing material.

Electro-optic crystals are used to develop variable laser-cavity output-mirror transmissions. As an example, the use of a crystal whose output is zero polarization with no voltage applied and 90° polarization with a voltage applied to the crystal (and has a wire-grid polarizer on the end-mirror of the cavity), will result in either a 0 or 100% transmissive mirror, depending on the polarization of the E-vector striking the wire-grid polarizer.

The engineering implementation of a practical electro-optic modulator is not a simple matter. Pulse widths, as small as $\approx$ 100 picoseconds, are generated at rates of a GHz or more. The capacitances of the crystals, the delays associated with line lengths, and proper impedance-matching to control reflections become important considerations in the design.

A class of materials known as *magneto-optic* exists which can be useful for modulation of the laser carrier. As the name implies, the index of refraction is a function of the magnetic field strength within the materials. These materials have not achieved the popularity that the electro-optic crystals enjoy.

Another class of materials used for the modulation of laser carrier is known as *acousto-optic* materials. These materials exhibit changes in the index of refraction when subjected to mechanical strain. An acousto-optic modulator is shown in Figure 23-28. The modulating signal is used to drive an acoustic transducer which then transmits strain waves along the modulator material. At the other end, these waves are dissipated in an acoustic absorber. Thus, a sinusoidal index of refraction structure exists within the modulator corresponding to the strain-waves propagating in the material. This index of refraction structure can be considered to be a diffraction grating which interacts with the incident laser beam. The intensity and frequency of the diffracted beams are proportional to the strain waves in the material. The waves, in turn, are proportional to the modulating signal. The most efficient acousto-optic modulation results when the propagation vector of the incident laser carrier and the acoustic wavefronts are oriented at the Bragg angle, θ_B. The acousto-optic modulator operating in the Bragg scattering limit is shown in Figure 23-29. The Bragg angle is given by

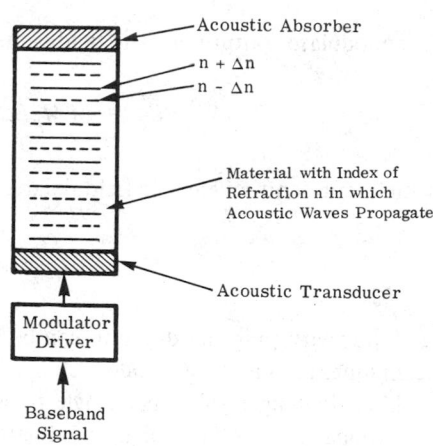

Fig. 23-28. Acousto-optic modulator.

$$\sin \theta_B = \frac{k_a}{2k_c} = \frac{\lambda_c}{2\lambda_a n} \qquad (23\text{-}92)$$

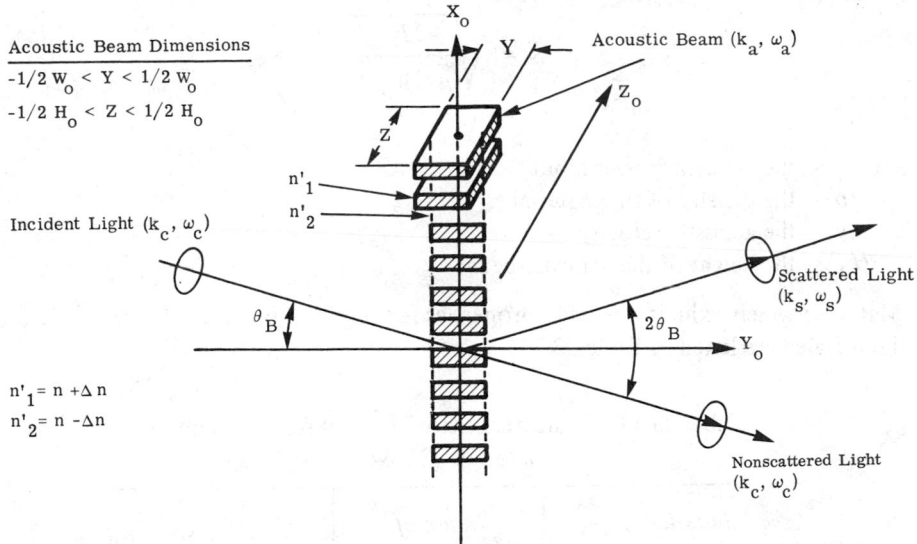

Fig. 23-29. Acousto-optic modulator operating at Bragg angle. The n'_1 and n'_2 regions are stratified for clarity. In reality, n'_1 and n'_2 are maxima/minima peaks of sinusoidal distribution.

Using this technique, one can pulse-modulate the incident laser carrier by periodically deflecting the beam. The energy and momentum considerations governing the scattering require

$$\omega_s = \omega_c \pm \omega_a \qquad (23\text{-}93)$$

and

$$k_s = k_c \pm k_a \qquad (23\text{-}94)$$

Thus, the deflected beam has a frequency transposed by ω_a and the direction of the deflected beam is at $2\theta_B$ with respect to the incident beam. In Table 23-9, this type of modulation was referred as propagation vector modulation.

The efficiency, η, being the ratio of the deflected power to incident laser power, is given [23-62, 23-63] by

$$\eta = \frac{k_c^2 \, W_o^2 \, (\Delta n)^2}{4 \cos^2 \theta_B} \qquad (23\text{-}95)$$

where Δn = the change in the index of refraction
$\quad W_o$ = the width of the waist of the acoustic beam

The change in index of refraction is given by

$$\Delta n = \frac{1}{2} n^3 \, p \, \bar{S} \qquad (23\text{-}96)$$

where p = the photo-elastic constant
$\quad \bar{S}$ = the amplitude of the strain given by

$$\bar{S} = \sqrt{\frac{2P_a}{\rho v^3 W_o H_o}}$$ (23-97)

where P_a = the acoustic power input
ρ = the density of the material
v = the acoustic velocity
H_o = the height of the acoustic beam

Materials which exhibit desirable properties for acousto-optic modulation applications in the infrared are listed in Table 23-11.

Table 23-11. Characteristics of Typical Acousto-Optic
Materials [23-64]

Material (Approximate Range of IR Transmission)	Index of Refraction, $n(10.6\ \mu m)$	Acoustic Velocity, $v \times 10^5$ cm sec^{-1}
Ge (2 to 20 μm)	4.0	5.5
CdS (0.5 to 11 μm)	2.22	4.32
GaAs (1 to 11 μm)	3.10	5.3
Si (1.5 to 10 μm)	3.42	9.85
Te (5 to 20 μm)	4.8	2.2
As$_2$S$_3$ Glass (0.6 to 11 μm)	2.38	2.6

Pump-power modulation is another technique which can be used to modulate the intensity of the laser carrier. It is especially effective with the injection semiconductor lasers. Modulating the laser-driving current in either continuous mode or pulse mode results in a corresponding modulation of the laser output. Changing the laser cavity length will produce frequency modulation of the carrier. The cavity of the laser is arranged with one window having a low inertia configuration. The window is mounted on a piezo-electric crystal mount. Upon application of a voltage on the crystal, the cavity length, L_c, changes with a corresponding change in carrier frequency, ν_c. For small changes in length, the change in frequency is given by

$$\Delta \nu_c = \nu_c \left(\frac{\Delta L_c}{L_c} \right)$$ (23-98)

Table 23-12. The Infrared Detectors for IR Communications* [23-65]

Material	Maximum Temperature for Background Limited Operation (K)	Test Temperature (K)	Long Wavelength Cutoff, 50% (μm)	Peak Wavelength, λ_m (μm)	Quantum Efficiency, η	Resistance (ohms)	$D_p^*(\lambda_p)$ (cm Hz$^{1/2}$ W^{-1})	Approximate Response Time (sec)
InAs	110	195.0	3.6	3.3	0.3 to 0.8	10^3 to 10^4	3×10^{11}	5×10^{-7}
InSb	60	77.0	5.6	5.3			6×10^{10} to -1×10^{11}	5×10^{-6}
Ge:Au	60	77.0	9.0	6.0	0.2 to 0.3	4×10^5	3×10^9 to 10^{10}	3×10^{-8}
Ge:Au(Sb)		77.0	9.0	6.0		10^6	6×10^9	1.6×10^{-9}
Ge:Hg	35	4.2	14.0	11.0	0.2 to 0.6	1 to 4×10^4	7×10^9 to -4×10^{10}	3×10^{-8} to 10^{-9}
		27.0		10.5	0.62	1.2×10^5	4×10^{10}	
Ge:Hg(Sb)	35	4.2	14.0	11.0		5×10^5	1.8×10^{10}	3×10^{-10} to 2×10^{-9}
								3×10^{-10} to 3×10^{-9}
Ge:Cu	17	4.2	27.0	23.0	0.2 to 0.6	2×10^4	2 to 4×10^{10}	3×10^{-9} to 10^{-8}
		20.0						4×10^{-9} to 1.3×10^{-7}
Ge:Cu(Sb)	17	4.2	27.0	23.0		2×10^5	2×10^{10}	$<2.2 \times 10^{-9}$
Hg$_{1-x}$Cd$_x$Te (x = 0.2)		77.0	14.0	12.0	0.05 to 0.3	60 to 400	10^{10}	$<10^{-6}$
						20 to 200	6×10^{10}	$<4 \times 10^{-6}$
Pb$_{1-x}$Sn$_x$Te (x = 0.17 to 0.2)		77.0	11.0	10.0		42	3×10^9	1.5×10^{-8}
		4.2	15.0	14.0		52	1.7×10^{10}	1.2×10^{-6}

*See also Chapter 11.

There are other phenomena which can be used for modulation purposes. However, their state of development or practicality of implementation makes it unlikely that they will become operational in the near future.

Coherent Infrared Receivers. Both direct and photomixing detection techniques are employed in coherent communication receivers using detectors such as those listed in Table 23-12. The direct receiver mechanization does not differ much from the mechanization of an incoherent receiver. The basic detection mechanism is the conversion of the incoming intensity envelope into a corresponding voltage signal within the limits of the transfer function of the detector. Because of the relatively low energy of an infrared photon, detection devices such as the photomultiplier or dynamic cross-field multiplier tubes are not available in the infrared, except in the case of GaAs and Nd: YAG lasers which are sufficiently close to the visible spectrum. The infrared detectors with time-constants sufficiently short to handle large data rates are summarized in Table 23-12. The sensitivity of a direct detection receiver is calculated by the expressions provided in Section 23.3.3.

One of the advantages of direct detection receivers is their relative insensitivity to the multiple effects of carrier modulation (e.g., simultaneous modulation of intensity, frequency, and polarization), the coherence distortion as a result of the atmospheric effects, and the doppler shift in the carrier frequency. The chief disadvantage of a direct receiver over that employing the photomixing technique is the power sensitivity of a practical, direct-detection receiver.

In a typical direct-detection receiver, the received radiant power is collected by the receiving aperture and focused upon a detector. Since the spectral bandwidth of a coherent carrier is very narrow, a spectral filter can be employed to reduce the background irradiance into the receiver. The output of the detector is amplified, appropriately filtered, and further processed if necessary to extract the baseband information out of the received signal. The frequency distribution of pulse-modulated signals, summarized in Section 23.3.3 is also applicable to coherent systems. Since the direct detection receivers may be used to receive continuous-wave intensity-modulated signals, the frequency distributions of amplitude-modulated and frequency-modulated signals are as follows:

Amplitude Modulation

$$f(t) = \cos \omega_c t + m_a \sum_{n=1}^{M} |C_n| \{ \cos[(\omega_c + n\omega)t + \theta_n] + \cos[(\omega_c - n\omega)t - \theta_n] \} \tag{23-99}$$

where $C_n = \dfrac{1}{T_c} \displaystyle\int_{-T_c/2}^{T_c/2} f(t) \exp(-jn\omega t)\, dt$, $n = 0, \pm 1, \pm 2, \pm 3, \ldots$

Frequency Modulation

$$f(t) = E_c \left[\prod_{m=1}^{N} \sum_{n_m = -\infty}^{\infty} J_{n_m}(m_f) \exp(j n_m \omega_m t) \right] \exp(j \omega_c t) \tag{23-100}$$

where m = the frequency component of the modulating signal

n = the harmonic of m frequency fundamental

$J_{n_m}(\ )$ = the Bessel function of the first kind of order n

With the help of these expressions, filter characteristics for the proper handling of the received signal can be determined. In the case of direct modulation of the laser carrier, the detected signal is reduced to the frequency components comprising the baseband signal. Often the baseband signal is modulated on an electrical carrier prior to the modulation of the laser carrier. In that case, the detected signal must be processed by a second demodulator and then low-pass-filtered to extract the baseband signal. As an example, in the case of frequency-modulated intensity envelope of the laser carrier, the detected signal must be passed through a frequency discriminator to extract the baseband information.

The photomixing receivers employ a local IR oscillator which is mixed with the received carrier prior to the detection process. Two basic techniques are possible: the heterodyne and the homodyne processes. The homodyne process requires a phase-locked local oscillator which, within today's technology, is not as practical at IR frequencies. The block diagram of a heterodyne receiver is shown in Figure 23-30. The received carrier power is combined with a local oscillator frequency which is offset at a fixed frequency amount, $\Delta\nu$, from the modulated carrier frequency. The detected signal is thus down-translated in frequency, with the corresponding intermediate frequency given by f_{if} $= \nu_c - \nu_o$. Whatever the modulated format of the laser carrier is, the resulting sidebands will arrange themselves on either side of the f_{if}. Typical values for f_{if} may be 30 or 60 MHz or higher, depending on the baseband frequency content. The combination of the carrier and local oscillator signals may be accomplished in several ways [23-66]. As shown in Figure 23-31, the received carrier is coupled through a 99% reflective beamsplitter. One percent of the local oscillator power is added to the combined signal path. Since the received carrier power is very low, this preserves most of the carrier power. The combined wavetrain is then focused on the detector. The interaction between the received carrier and the local oscillator is an interference effect. Thus, the electric field amplitudes either add or subtract, the net modulation envelope being detected by the IR detector. Therefore, to obtain the highest detection efficiency, one must properly

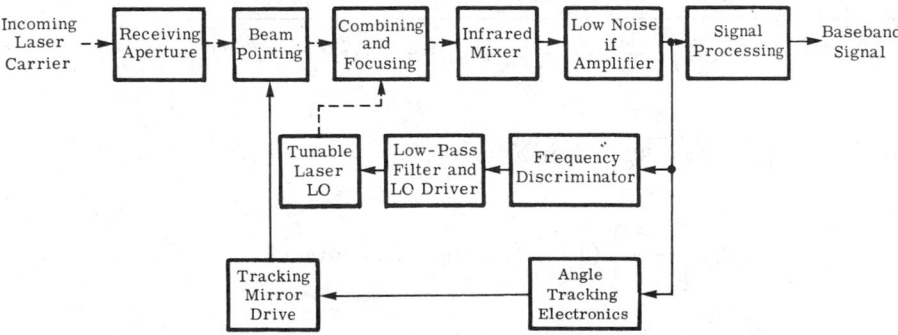

Fig. 23-30. Heterodyne receiver block diagram.

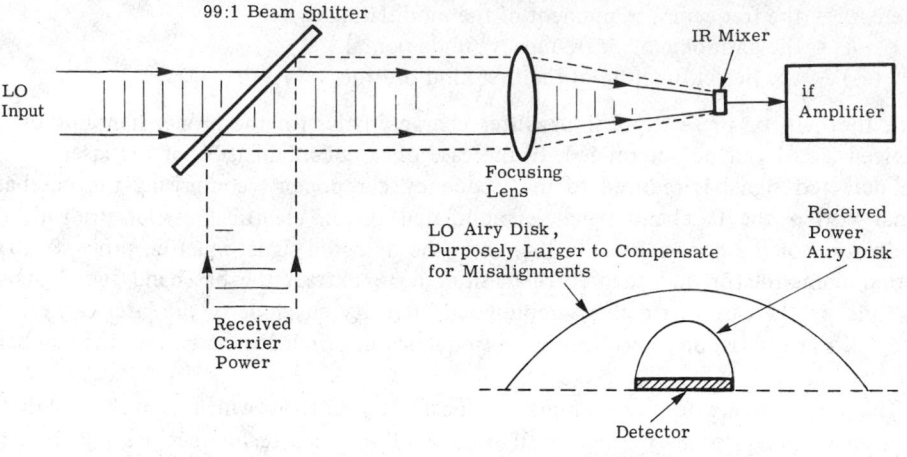

Fig. 23-31. Received and local oscillator (LO) spatial combination optics.

align the polarization and propagation wave vectors. In other words, the phase fronts of the received carrier and local oscillator must remain parallel over the sensitive area of the detector. If one assumes a square detector of area L_d^2, the ratio of signal power from the mixer (when there is no relative phase tilt) to the signal power (when there is a tilt of α radians), is given [23-19] by

$$\frac{\Phi(\alpha)}{\Phi(o)} = \left[\frac{\sin\left(\frac{\pi}{\lambda} L_d \sin\alpha\right)}{\frac{\pi}{\lambda} L_d \sin\alpha} \right]^2 \tag{23-101}$$

The variation in the detected power, as a result of polarization-vector misalignment, follows the cosine of the angle between the two polarization vectors.

The signal-to-noise ratio out of the if-filter of the heterodyne receiver is [23-67]

$$\left(\frac{S}{N}\right)_{if} = \frac{\left[\frac{G\eta q}{h\nu_c}\right]^2 \Phi_o \Phi_c R_L}{G^2 q\, B_{if}\left\{\frac{\eta q}{h\nu_c}\left[\Phi_c + \Phi_o + \Phi_B\right] + I_D\right\} R_L + 2kTB_{if}} \tag{23-102}$$

where $i_{if} = \dfrac{G\eta q}{h\nu_c}\sqrt{2\Phi_o \Phi_c}$, Mean if current

$n_s = 2G_q^2 B_{if}\left[\dfrac{\eta q}{h\nu_c}\right](\Phi_c + \Phi_o + \Phi_B)R_L$, Shot noise

$n_d = 2G^2 q B_{if} I_D R_L$, Dark current noise

$n_{th} = 4kTB_{if}$, Thermal noise

where G = detector current gain
η = quantum efficiency
q = charge of an electron
h = Planck's constant
ν_c = laser carrier frequency
Φ_o = local oscillator power
Φ_c = received carrier power
R_L = receiver load resistance
B_{if} = if amplifier noise equivalent bandwidth
Φ_B = background power
I_D = detector dark current
k = Boltzmann's constant
T = absolute temperature

If the local oscillator power, Φ_o, is large, then the signal, background, dark-current, shot and thermal noise effects become negligible compared to noise due to the local oscillator power. The signal-to-noise expression then reduces to

$$\left(\frac{S}{N}\right)_{if} = \frac{\eta\,\Phi_c}{h\,\nu_c\,B_{if}} \tag{23-103}$$

At a value of $(S/N)_{if}$ equal to one, the spectral noise density reduces to

$$\Phi_n = \frac{h\nu_c}{\eta}, \text{W Hz}^{-1} \tag{23-104}$$

Thus, in the case of a 10.6 μm receiver, for an efficiency $\eta = 0.5$,

$$\Phi_n = 3.7 \times 10^{-20}, \text{W Hz}^{-1} \tag{23-105}$$

To date, practical laser heterodyne receivers have been limited to CO_2 lasers operating at 10.6 μm.

23.4. Simulation

23.4.1. Introduction. Simulation can be used in the design, analysis, testing, and evaluation of systems. Simulation can vary in complexity and sophistication from simple portions of inputs or outputs as well as components of systems, to complete substitutions for an entire system, and its inputs and outputs. This section presents examples to illustrate the diversity found in this field.

Decisions at critical points in the evolution from the design to the procurement phase of electro-optical systems require a knowledge of specified and attainable system performance. Performance testing with credible results at the procurement phase can be an expensive task, especially when the system is designed for field use in a dynamic configuration. In these cases, early, preliminary simulation of field results in the laboratory can be extremely valuable. The value of simulation, however, is directly related to the credibility of the results it produces.

Feasible simulation in this context may take three forms:

(1) Mathematical—in which all system and environment characteristics are given mathematical analogs and performance is computed.

(2) Component test—in which the optical portion of the system, physical analogs of the optical system, or the optical system components are performance-tested, and the rest of the system is provided target-and-background electrical signals representing the radiation as degraded by optical system performance.

(3) Simulation—in which the system is provided actual radiation from simulated targets and backgrounds.

Each of these has unique and desirable features, but none produce wholly incontestable results. The first and last forms are relatively easily defined; the second can be considered a repository for what does not reasonably fit in (1) or (3).

23.4.2. Mathematical Simulation. In this type of simulation, the signal (voltage) output of a system is computed entirely from the specified or derived characteristics of the system via a mathematical analog of the interaction with the radiation input. The radiation-input analog can be derived either from a set of results of experimentally-measured physical entities, or from calculations using parameters of physical quantities such as reaction rates, temperatures, and pressures.

A schematic of the elements for the mathematical simulation of a system field-test is shown in Figure 23-32 where the target might be, for example, an aircraft observed through the real atmosphere. Given the necessary inputs (not all of which are explicitly shown here), the simulated sensor output (voltage) can be calculated to within the accuracy determined by the uncertainty of the input parameter values and the approximations used.

23.4.3. Component Test. Any system test not describable under either category (1) or (3) can be classed in this category. The examples given here, as simulations, actually contain some of the characteristics of either or both of the other categories.

CRT *Presentation of Target Data**. A particularly useful technique for simulating reticle-trackers of point or small-area targets uses a cathode ray tube (CRT) to present points or areas as targets within the instantaneous field of view (IFOV) of a tracker. A replica of the reticle used in the tracker is placed on the CRT face and any relative motion (such as seeker spin) between the target image and the reticle is controlled by imparting the motion electronically to the target image on the CRT. A phototransistor is substituted for the remainder of the tracker, i.e., the detector. No attempt is made to simulate the spectral characteristics of the tracker. The tracker amplifiers, detectors, and the tracking-loop servo-mechanisms are represented by transfer functions, created by operational amplifiers and related networks. The actual electronic circuits from a tracker may be used.

The tracker-simulator can be used on a stand-alone basis, or as part of a larger simulation loop. This specialized analog computer can be used in conjunction with either an analog or digital computer for simulating problems such as a missile-target encounter.

---

*This section on CRT presentation of target data was prepared by Mr. Yuji Morita, Environmental Research Institute of Michigan (ERIM), Ann Arbor, MI.

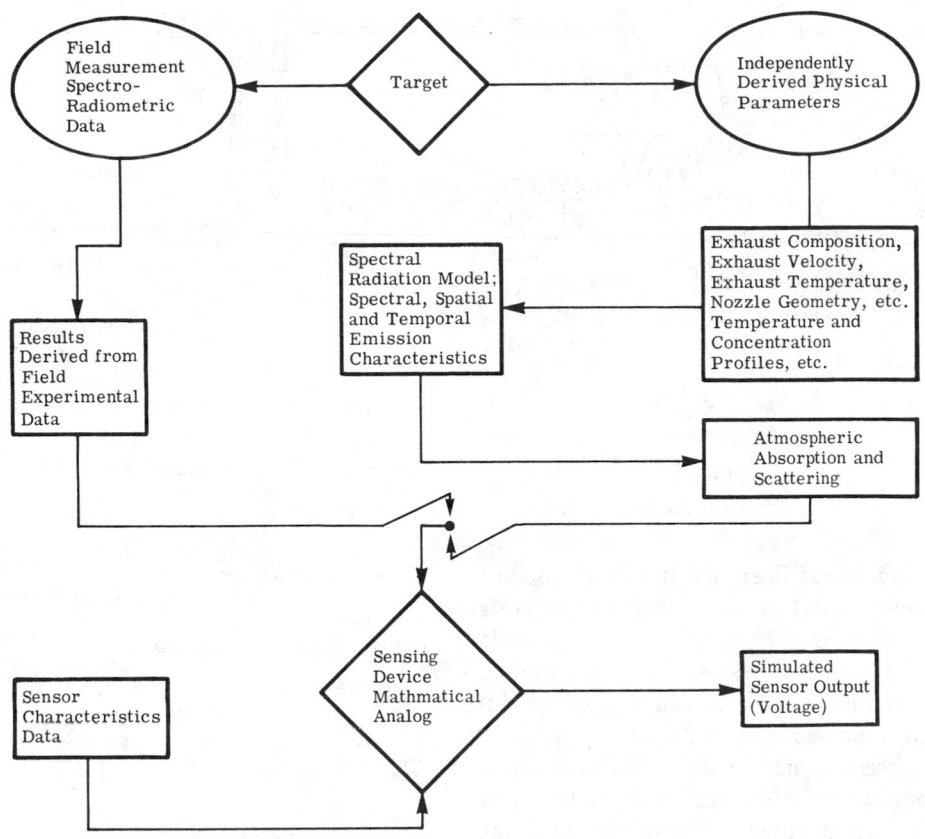

Fig. 23-32. Mathematical simulation of field test of a system.

The simulator is useful in the design-development of a tracker, particularly in determining the effects of nonlinearities which are caused by elements such as reticles, limiters, and automatic gain control (AGC) circuits. Reticles can be easily replaced and circuits modified. This type of simulator has been used for other purposes as well, such as studying the effects on a seeker of tracking extended targets (e.g., engine exhaust-plumes), determining the effect of two radiating sources on tracking action, and assessing the susceptibility of a seeker to countermeasures. Such studies can be conducted in slow, real, or fast time, depending on the simulator design. A salient feature in conducting studies such as those listed is the ability to monitor a large number of points within the simulated tracker unit, points which often are not available on a real tracker.

Space Simulation Facility (ARO) [23-68]. Space simulation chambers are used at the Arnold Engineering Development Center, Air Force Systems Command, to test IR sensors in a space environment. The Sensor Test Chamber (7 V) is used to provide a low-pressure and low-temperature environment for operational testing of IR sensors. A schematic of the chamber is shown in Figure 23-33, showing the pumping system, the liquid-nitrogen

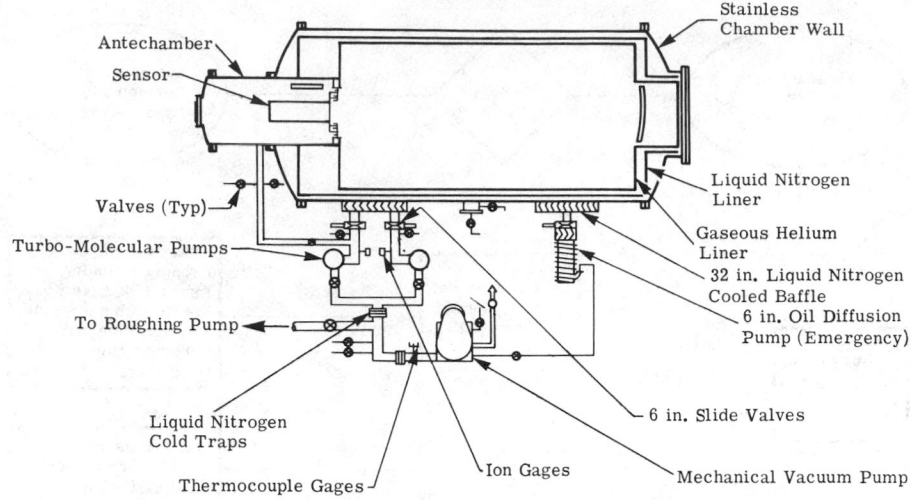

Fig. 23-33. Sensor test chamber (7 V) [23-68].

(LN$_2$) cooled liner, and the 20 K gaseous-helium (GHe) liner. Target simulation is provided by a combination of a movable radiation-source assembly and a rotating, collimating mirror. A plan view of the system is shown in Figure 23-34.

Target Signature Facility (McDonnell-Douglas) [23-69]. In addition to a test chamber designed to perform functions similar to those performed by the one described above, McDonnell-Douglas Astronautics Company has a facility to measure passive optical signatures of space objects to substantiate predicted target characteris-

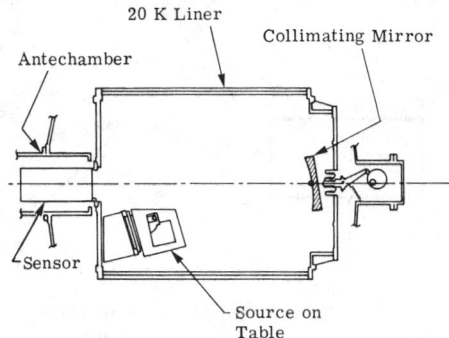

Fig. 23-34. Plan view of a 7 V target simulation system [23-68].

tics for conditions simulating exoatmospheric flight. The main vacuum chamber is a 39 ft diameter sphere with an internal shroud, cooled with LN$_2$ to prevent radiation from the warm chamber walls. (See Figure 23-35.) A horizontal, "solar" beam, derived from un-filtered xenon arc lamps, illuminates the targets, which are mounted to allow various aspects with respect to the simulated earth disc and the solar source. KRS-5 windows are used for viewing ports into the vacuum chamber, allowing spectral measurements over the long-wavelength IR range. A schematic of the system is shown together with a sketch of the scene to be simulated in Figure 23-35.

23.4.4. Field-Test Simulation Facilities for Viewing, Seeking and Tracking Devices [23-70]

Terrain Models. A terrain model is a nearly-real, scaled-down version of actual terrain containing scaled-down targets. It is used as the radiative input to the system being tested. One limit to realism is the obvious absence of real vegetation and other background materials. Terrain models thus far have been built for use in the visible spectral region

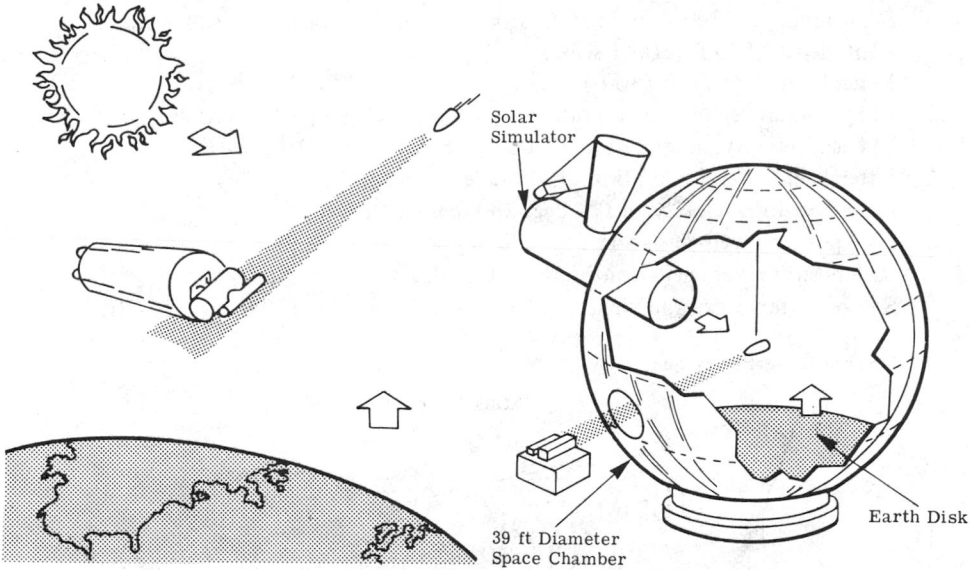

Fig. 23-35. Target signature measurements in space chamber [23-69].

with an emphasis on spatial detail and fidelity with respect to geographical features. Most of the sensings, furthermore, are not highly spectrally discriminating so that spectral purity of the scenery is not an essential characteristic. What is important, however, is that contrasts be retained so that system responses to radiation gradients can be fully evaluated.

The sensor is often mounted with ancillary optics so that it views the scene (the terrain model) as if it were viewing a real scene, retaining relative motion between it and the terrain. The model can be illuminated with artificial light or can be moved into sunlight and illuminated naturally. The whole platform on which the model rests can be tilted to change the angle viewed, or the angle of illumination.

Examples of this type of facility are located at Martin-Marietta, Orlando, Florida, and at Redstone Arsenal's Advanced Simulation Center. The latter is called the Electro-Optical Simulation System (EOSS).

The EOSS, shown in Figure 23-36, provides controlled environments for the testing of a wide variety of ultra-violet, visible, and near IR sensor systems. The sensors are hybrid-computer controlled in 6 degrees of freedom, while targets are viewed under controlled illumination levels (10^{-4} to 10^3 ft c) in an indoor simulation chamber, and under ambient conditions on an outdoor test range. Three-dimensional target simulation is provided by a 32 × 32 ft terrain model/transporter which features a variety of topographical and man-made target complexes at 600:1 and 300:1 scales. The model has removable sections, and can accommodate fixed and moving targets appropriately scaled.

A gimballed mount, which holds the sensor under test, is capable of simulating pitch, roll, and yaw movements to the sensor airframe and is attached to a transport which moves both vertically and laterally. A specially designed autofocusing lens system provides infinity focusing of targets from 5 to 160 ft. The terrain model is moved toward the flight table to provide the sixth degree of freedom. Optical guidance simulation flight parameters are

Slant range—31,800 ft indoor, 63,600 ft or horizon outdoor (300:1 scale)
Altitude—10,500 ft (300:1 scale)
Lateral range—9750 ft (300:1 scale)
Longitudinal velocity, acceleration—4500 ft sec^{-1}, 93 g (300:1 scale)
Vertical velocity, acceleration—1800 ft sec^{-1}, 56 g (300:1 scale)
Lateral velocity, acceleration—1200 ft sec^{-1}, 93 g (300:1 scale)
Pitch position, velocity—±120 deg, 200 deg sec^{-1}
Yaw position, velocity—±45 deg, 200 deg sec^{-1}
Roll position, velocity—continuous, 2000 deg sec^{-1}
Sensor systems size and weight—14 in dia × 36 in long, 150 lb   [23-70]

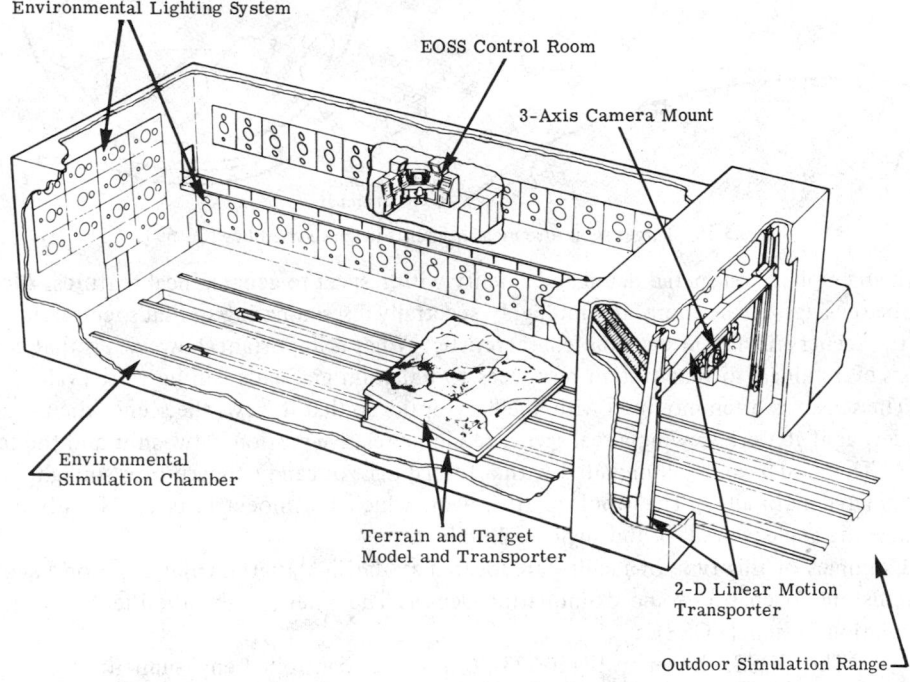

Fig. 23-36. Electro-optical simulation system (EOSS) [23-70].

A photograph of a terrain model, under conditions simulating a height of 13,000 ft is shown in Figure 23-37.

One obvious limitation of the terrain model is its lack of simple terrain changability, although this is not serious since several different types of terrain can be incorporated on a single platform if necessary. A more serious limitation is its restriction to the shorter wavelength regions. Attempts have been made to utilize terrain models in the thermal IR and some proposals for doing this have been offered [22-71]. Briefly, the change from visible to thermal status amounts to incorporating heating elements into the terrain model and heating them differentially.

Thermal Analogs [23-71]. A second method, applicable in the longer wavelength, thermal IR regions, utilizes two-dimensional radiative analogs to real targets. The method

Fig. 23-37. Terrain Model. Simulated (scaled) height is 13,000 ft [23-70].

involves several steps, starting with obtaining a thermogram of the target. (See Chapter 2.) The densities in the thermogram, which are related to the radiation levels in the target, can be digitized and transformed into a half-tone pattern etched on to an anodized aluminum plate. The result is similar to that shown in Figure 23-38. The plate can be held

magnetically in place, along with other possible "targets," on a large metal plate. This provides a thermally uniform background, being heated from the rear with sheet-heating elements. The temperature of the half-tone target is uniform, but the correct radiative contrast is achieved with emissivity changes incurred by the shades of grey etched into the target plate.

Fig. 23-38. Computer generated half-tone image [23-71].

Very fine detail can be achieved with this method, but the representation is two-dimensional instead of three-dimensional, and the scene is static with a uniform background incapable of creating clutter problems, unless the simulation is made quite complicated.

Redstone Arsenal IR Simulation System*. The third method utilizes artificial sources radiating through simple geometrical apertures to simulate the radiation from targets in a dynamic situation with all functions of the evaluated system incorporated into a closed loop. One such system, called the IR Simulation System (IRSS), exists at the Redstone Advanced Simulation Center. The following description is edited and abridged from Reference [23-70].

The IRSS is a simulation tool for the design, development and evaluation of infrared sensor systems applicable to surface-to-air and air-to-air missiles. Sensors in the 0.3 to 0.7 and 1 to 5 micron bands are hybrid-computer controlled in six degrees of freedom during the target engagement sequence. A gimballed flight table provides pitch, roll and yaw movements to the sensor airframe. A target generator simulates a variety of target/background combinations which include tailpipes, plumes, flares, and fuselages in single or multiple displays against clear sky, dark clouds, overcast sky, and sunlit cloud backgrounds. These are then displayed in azimuth, elevation, range, and aspect by the target projection system through a mirror/lens network, a display arm, and a display mirror. Simulation capability ranges from open loop component testing to closed loop total system simulation . . .

Functionally, the IRSS is divisible into five major entities: the guidance unit mount, the display subsystem, the target projection subsystem, the operator controls and servo control electronics subsystem, and the command and data management system (CDMS). The general relationship between these elements is illustrated in Figure 23-39 . . .

The Target Generation System consists of an assembly of equipment and components, which provide for generation of simulated aircraft targets, backgrounds, and countermeasures. The purpose of this assembly is to present to the guidance unit under test suitable radiation sources to simulate the physical, radiometric, and dynamic characteristics of targets, backgrounds, and countermeasures. These characteristics are designed to be manually or automatically controlled. Local instrumentation provides manual control, while automatic programmed control is provided through either local computers or those from the hybrid computer facility . . .

A source array suitable for simulation of a wide variety of tactical aircraft is provided. Target size, shape, radiant intensity, and relative motion are simulated as shown in Tables 23-13 and 23-14.

The guidance unit mount is a three-axis, hydraulically powered, servo-controlled table on which the seeker is mounted. It will produce, on command, all of the rotational motions that a missile would experience during an actual flight. See Table 23-15.

The functions of the display and projection subsystem are: (1) to provide the sensor with a window into target space at all times when the sensor is track-

---

*There are other simulation systems in various laboratories similar to the one at Redstone Arsenal. This one was chosen to describe because of the availability of a complete description. Examples of others in existence are those at the Rome Air Development Center and Wright-Patterson Air Force Base.

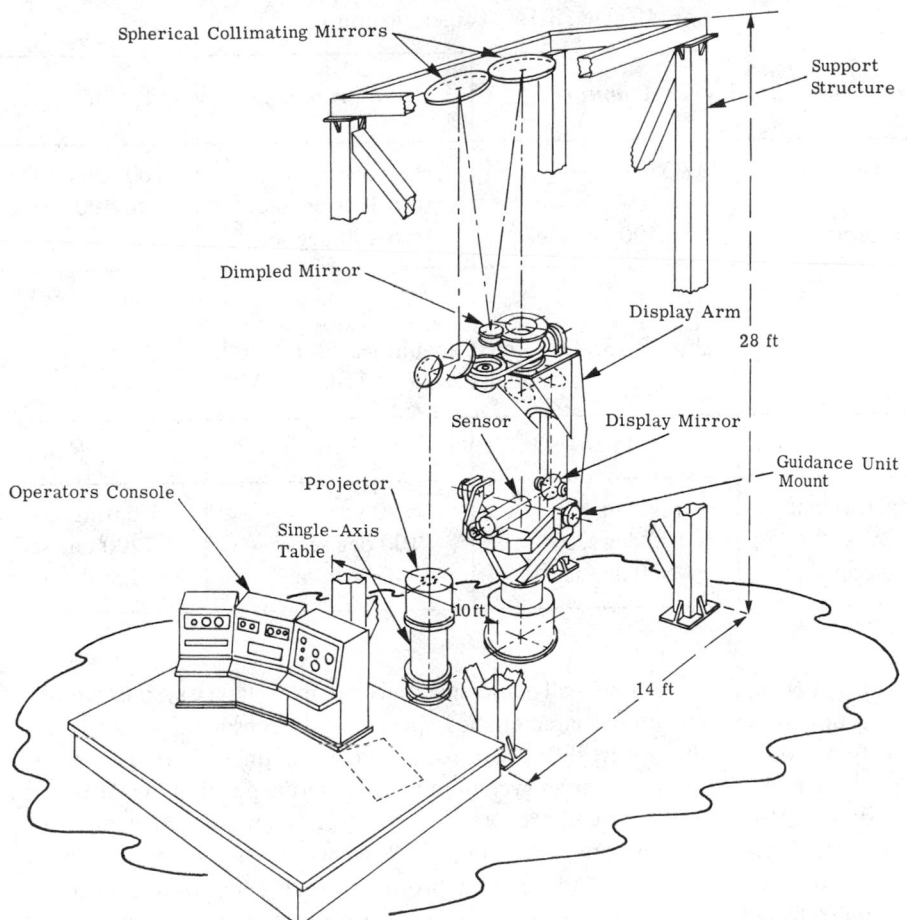

Fig. 23-39. Infrared simulation system (IRSS) [23-72].

Table 23-13. Target/Background Elements [23-70]

Element	Quan.	Shape	Size	Spectral Bandwidth	Apparent Target Radiance
Tailpipe/ flare	2	Circular	0.15 to 1 m dia.	1.0 to 5.0 μm	1.3×10^{-4} to 0.13 W cm^{-2} sr^{-1}
Plume	1	Triangular	1×1 to 1×5 m	3.0 to 5.0 μm	4×10^{-5} to 0.02 W cm^{-2} sr^{-1}
Fuselage	1	Rectangular	1×1 to 3×20 m	0.3 to 0.7 μm	10^{-6} to 10^{-3} W cm^{-2} sr^{-1}
Back- ground	1	—	—	—	10^{-3} W cm^{-2} sr^{-1} (max)

Table 23-14. Target Motion [23-70]

Variation in:	Azimuth	Elevation	Range
Displacement	± 90 deg	± 30 deg	160 to 16,000 ft
Velocity	0 to 100 deg sec^{-1}	0 to 100 deg sec^{-1}	0 to 4900 ft sec^{-1}
Acceleration	0 to 400 deg sec^{-2}	0 to 400 deg sec^{-2}	—

Table 23-15. Dynamic Capabilities with 25 lb Load
10 in. Dia. × 25 in. Long [23-70]

	Pitch	Yaw	Roll
Displacement	± 80 deg	± 90 deg	Continuous
Velocity	100 deg sec^{-1}	100 deg sec^{-1}	7200 deg sec^{-1}
Acceleration	7800 deg sec^{-2}	4400 deg sec^{-2}	12,300 deg sec^{-2}

ing, (2) to generate a spatially and spectrally complex target system whose geometry and radiation characteristics appear to the sensor substantially as they would in the real world, to full scale, and in real time, (3) to display to the sensor the generated target system in its true inertial position in real time, and (4) to cause the radiation from each target or target element to fill the sensor aperture fully and uniformly as long as the sensor is tracking.

An assembly of eight independent projectors (one being a dual-purpose one), focused at infinity, projects as many as seven scene elements plus two spectral backgrounds into the spherical collimater, which forms a composite, in-register image of the complex scene on a special dimpled spherical mirror. The dimpling expands the solid angle of radiation from each projector to ensure filling the sensor aperture from each scene element. The spherical collimator then forms a virtual image at infinity of this composite scene. The sensor in the guidance unit mount observes this scene through the window held before it by the display arm and mirror. The target projection subsystem is mounted on a single-axis table which removes the display scene rotation caused by azimuth motion of the display arm. Fine positioning of the targets within the display window is done by small 2-axis gimballed mirrors (i.e., directional mirrors) in the projection subsystem. Each projector provides control of spectral radiance, size, shape, and aspect for a single target by using servo-driven transparencies, irises, and spectral and neutral density filters.

The operator control and servo control electronics subsystem, as the name implies, contains all of the systems operator controls and displays, power distribution, servo control, projector lamp control, analog computation, and mode control electronics.

The Command and Data Management Subsystem (Figure 23-40) provides system control, timing, and command and performance data distribution and management during all testing. It acts as a digital interface between a hybrid computer and the IRSS during closed loop operation. It acts as a multichannel, high-speed function generator during open loop operation, provides for data recording and data reduction, and has the capability of exercising all of the system's servos and providing a 'go/no-go' decision on servo performance in the self-test mode [23-70].

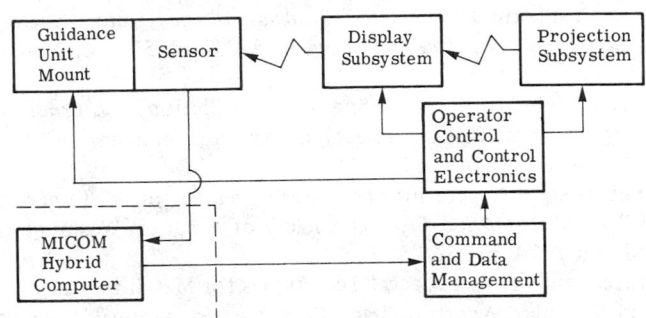

Fig. 23-40. Principal subsystems in the IRSS [23-70].

23.5. References and Bibliography

23.5.1. References

[23-1] F. T. Arecchi, et al. (eds.), *Laser Handbook,* North Holland Publishing, Amsterdam, The Netherlands, 1972, pp. 1479-1485, 1469-1476.

[23-2] C. R. Cooke, "Automatic Laser Tracking and Ranging System," *Applied Optics,* Optical Society of America, Washington, DC, Vol. 11, No. 2, February 1972, p. 277.

[23-3] T. Flom and D. Coombes, "Multiple Target Tracking and Target Altitude Determination with a Scanning Laser," *Journal of Navigation,* London, England, Winter, 1974-75.

[23-4] P. Wilson, et al., "Short Pulse Satellite Ranging System," EO/IL '77 Conference, October 1977, Anaheim, CA, sponsored by Electro Optic System Design.

[23-5] Defense Electronic Products, *RCA Electro Optics Handbook,* Aerospace Systems Division, Burlington, MA, 1968.

[23-6] S. Valley (ed.), *Handbook of Geophysics,* Macmillan, New York, NY, 1960, pp. 13-14.

[23-7] P. W. Kruse et al., *Elements of Infrared Technology: Generation, Transmission and Detection,* John Wiley and Sons, New York, NY, 1962, p. 189-191.

[23-8] L. Elterman, "A Model of a Clear Standard Atmosphere for Attenuation in the Visible Region and Infrared Windows," Air Force Cambridge Research Laboratories, Bedford, MA, Report 63-675, AFCRL, AD 422 014, 1963.

[23-9] R. A. McClatchey, R. W. Fenn, F. E. Volz, J. S. Garing, and J. E. A. Selby, "Optical Properties of the Atmosphere (Revised)," Air Force Cambridge Research Laboratories, L. G. Hanscom Field, Bedford, MA, AFCRL-71-0279, AFCRL-ERP-354, 10 May 1971.

[23-10] V. I. Tatarsky, *Wave Propagation in a Turbulent Medium,* McGraw-Hill, New York, NY, 1961.

[23-11] *Radiofizika* (Radiophysics and Quantum Electronics) Plenum Publishing, New York, NY. "On Strong Amplitude Fluctuations of a Wave Propagating in a Medium with Weak Random Inhomogeneities," Vol. 10, No. 1, 1967.

[23-12] D. L. Fried, G. E. Meyers, and M. P. Keister, "Measurements of Laser-Beam Scintillation in the Atmosphere," *Journal of the Optical Society of America,* Optical Society of America, Washington, DC, Vol. 57, No. 6, June 1967, p. 787.

[23-13] G. R. Ochs and R. S. Lawrence, "Saturation of Laser-Beam under Conditions of Strong Atmospheric Turbulence," *Journal of the Optical Society of America,* Optical Society of America, Washington, DC, Vol. 59, No. 2, February 1969, p. 226.

[23-14] D. L. Fried, "Aperture Averaging of Scintillation," *Journal of the Optical Society of America,* Optical Society of America, Washington, DC, Vol. 57, No. 2, February 1967, p. 169.

[23-15] E. C. Silverberg, "Operation and Performance of a Lunar Laser Ranging Station," *Applied Optics,* Optical Society of America, Washington, DC, Vol. 13, No. 3, March 1974, pp. 565-575.

[23-16] W. N. Peters and A. M. Ledger, "Techniques for Matching Laser TEM_{oo} Mode to Obscured Circular Aperture," *Applied Optics,* Optical Society of America, Washington, DC, Vol. 9, No. 6, June 1970, pp. 1435-1442.

[23-17] J. R. Kerr, "Microwave-Bandwidth Optical Receiver Systems," *Proceedings of the IEEE,* Institute of Electrical and Electronics Engineers, New York, NY, Vol. 55, No. 10, October 1967, pp. 1686-1700.

[23-18] B. M. Oliver, "Signal-to-Noise Ratios in Photoelectric Mixing," *Proceedings of the IRE,* Institute of Radio Engineers, New York, NY, Vol. 49, December 1961, pp. 1960-61.

[23-19] M. Ross, *Laser Receivers,* John Wiley and Sons, New York, NY, 1966.

[23-20] Merrill I. Skolnik, *Introduction to Radar Systems,* McGraw-Hill, New York, NY, 1962.

[23-21] S. O. Rice, "Mathematical Analysis of Random Noise," *Bell System Technical Journal,* American Telephone and Telegraph Company, New York, NY, Vol. 23, No. 3, p. 282; *ibid,* Vol. 24, No. 1, July 1944; *ibid,* January 1945, p. 46.

[23-22] R. I. Abbot, P. J. Shelus, J. D. Mulholland, and E. C. Silverberg, "Laser Observations of the Moon: Identification and Construction of Normal Points for 1969-1971," *The Astronomical Journal,* (American Astronomical Society) American Institute of Physics, New York, NY, Vol. 78, No. 8, October 1973, pp. 784-793.

[23-23] C. G. Lehr, M. R. Pearlman, J. L. Scott, and J. Wohn, "Laser Satellite Ranging," *Laser Applications in the Geosciences,* J. Gauger and F. Hall (eds.), Western Periodicals, CA, 1970, pp. 111-130.

[23-24] E. Hecht, and A. Zajac, *Optics,* Addison-Wesley Publishing, Reading, MA, 1974.

[23-25] M. Born, and E. Wolf, *Principles of Optics,* Pergamon Press, New York, NY, 1959.

[23-26] RCA Staff, "Solid State Infrared-Emitting Diodes, Injection Lasers, Silicon Diodes," RCA, Harrison, NJ, Brochure NO. OPT-100B, 1972; See also, "Solid State IR Emitters and Injection Lasers," Brochure No. OPT-113C, July 1977.

[23-27] N. Wiener, *The Extrapolation, Interpolation, and Smoothing of Stationary Time Series with Engineering Applications,* John Wiley and Sons, New York, NY, 1949.

[23-28] C. E. Shannon and W. Weaver, *The Mathematical Theory of Communication*, University of Illinois Press, Urbana, IL. 1949.

[23-29] C. E. Shannon, "Communication in the Presence of Noise," *Proceedings of the IRE*, Institute of Radio Engineers, New York, NY, Vol. 37, No. 1, January 1949, pp. 10-21.

[23-30] V. A. Kotel'nikov, *The Theory of Optimum Noise Immunity*, McGraw-Hill, New York, NY, 1959.

[23-31] H. S. Black, *Modulation Theory*, Van Nostrand, Princeton, NJ, 1953.

[23-32] W. B. Davenport, Jr., and W. L. Root, *An Introduction to the Theory of Random Signals and Noise*, McGraw-Hill, New York, NY, 1958.

[23-33] Y. W. Lee, *Statistical Theory of Communication*, John Wiley and Sons, New York, NY, 1960.

[23-34] C. W. Helstrom, *Statistical Theory of Signal Detection*, Pergamon Press, New York, NY, 1960.

[23-35] J. M. Wozencraft and I. M. Jacobs, *Principles of Communication Engineering*, John Wiley and Sons, New York, NY, 1967.

[23-36] H. Nyquist, "Certain Factors Affecting Telegraph Speed," *Bell System Technical Journal*, American Telephone and Telegraph Company, New York, NY, Vol. 3, No. 1, April 1924, pp. 324-346.

[23-37] M. Schwartz, W. R. Bennett, and S. Stein, *Communication Systems and Techniques*, McGraw-Hill, New York, NY, 1966.

[23-38] A. Papoulis, *Probability, Random Variables, and Stochastic Processes*. McGraw-Hill, New York, NY, 1965.

[23-39] I. Liberman, "Incoherent Optical Sources," *Handbook of Lasers with Selected Data on Optical Technology*, CRC Press, Cleveland, OH, 1971, pp. 13-38.

[23-40] C. H. Gooch, *Gallium Arsenide Lasers*, Wiley-Interscience, John Wiley and Sons, New York, NY, 1969.

[23-41] H. A. Elion, "Optical Communications in the Atmosphere," *Electronic, Electro-Optic, and Infrared Countermeasure*, Hamilton Burr Publishing, Santa Clara CA, Vol. 1, No. 4, August 1975, p. 43.

[23-42] J. A. Jamieson, R. H. McFee, G. N. Plass, R. H. Grube, and R. C. Richards, *Infrared Physics and Engineering*, McGraw-Hill, New York, NY, 1963.

[23-43] E. Fitch, "The Spectrum of Modulated Pulses," *Journal of the Institution of Electrical Engineers*, Institution of Electrical Engineers, London, England, Vol. 94, Part 3A, 1947, pp. 556-564.

[23-44] V. N. Rao, "Some Aspects of the Spectrum of Modulated Pulses," *Journal of the Indian Institute of Science*, Indian Institute of Science, Bangalore-560012, India, Vol. 35, Sec. B. July 1953, pp. 125-136.

[23-45] S. H. Moss, "Frequency Analysis of Modulated Pulses," *Philosophical Magazine and Journal of Science*, Taylor and Francis Ltd., London, England, Series 7, Vol. 39, September 1948, pp. 663-691.

[23-46] P. F. Panter, *Modulation, Noise, and Spectral Analysis*, McGraw-Hill, New York, NY, 1965.

[23-47] C. E. Tibbs and G. G. Johnstone, *Frequency Modulation Engineering*, John Wiley and Sons, New York, NY, 1956.

[23-48] A. L. Schawlow and C. H. Townes, "Infrared and Optical Masers," *Physical Review*, American Physical Society, New York, NY, Vol. 112, December 1958, pp. 1940-1949.

[23-49] T. H. Maiman, "Stimulated Optical Radiation in Ruby," *Nature*, Macmillan Journals Ltd., Washington, DC, Vol. 187, No. 4736, August 1960, pp. 493-494.

[23-50] R. J. Pressley (ed.) "Coherent Optical Sources" (by various authors) *Handbook of Lasers with Selected Data on Optical Technology*, CRC Press, Cleveland, OH, 1971, pp. 183-420.

[23-51] A. Yariv and W. M. Caton, "Frequency, Intensity, and Field Fluctuations in Laser Oscillators," *IEEE Journal of Quantum Electronics*, Institute of Electrical and Electronics Engineers, New York, NY, Vol. QE-10, No. 6, June 1974, pp. 509-515.

[23-52] R. Kingslake, (ed.), *Applied Optics and Optical Engineering*, Vol. I, Academic Press, Washington, DC, 1965.

[23-53] M. Françon, *Diffraction Coherence in Optics*, Pergamon Press, Elmsford, NY, 1966.

[23-54] W. K. Pursley, *The Transmission of Electro-Magnetic Waves through Wire Diffraction Gratings*, Doctoral Dissertation Series Publication No. 21, 348, University Microfilms, Ann Arbor, MI, 1956.

[23-55] M. Schwartz, *Information Transmission, Modulation and Noise*, McGraw-Hill, New York, NY, 1959.

[23-56] A. Hund, *Frequency Modulation*, McGraw-Hill, New York, NY, 1942.

[23-57] F. Pockels, *Lehrbuch der Kristalloptic*, Tübner, Leipzig, Germany, 1906.

[23-58] P. A. Franken and J. F. Ward, "Optical Harmonics and Non-Linear Phenomena," *Review of Modern Physics*, American Institute of Physics, New York, NY, Vol. 35, 1963, pp. 23-29.

[23-59] I. P. Kaminow and E. H. Turner, "Electro-Optic Light Modulators," *Proceedings of the IEEE*, Institute of Electrical and Electronics Engineers, New York, NY, Vol. 54, No. 10, October 1966, pp. 1374-1390.

[23-60] I. P. Kaminow and E. H. Turner, "Linear Electro-optic Materials," *Handbook of Lasers with Selected Data on Optical Technology*, CRC Press, Cleveland, OH, 1971, pp. 447-459.

[23-61] J. E. Geusic, W. B. Bridges, and J. I. Pankove, "Coherent Optical Sources for Communications," *Proceedings of the IEEE*, Institute of Electrical and Electronics Engineers, New York, NY, Vol. 58, No. 8, October 1970, pp. 1419-1439.

[23-62] E. I. Gordon, "A Review of Acousto-optical Deflection and Modulation Devices," *Proceedings of the IEEE*, Institute of Electrical and Electronics Engineers, New York, NY, Vol. 54, No. 10, October 1966, pp. 1391-1401.

[23-63] D. Maydan, "Acousto-optical Pulse Modulators, *IEEE Journal of Quantum Electronics*, Institute of Electrical and Electronics Engineers, New York, NY, Vol. QE-6, No. 1, January 1970, pp. 15-24.

[23-64] R. W. Dixon, "The Photoelastic Properties of Selected Materials and Their Relevance for Applications to Acoustic Light Modulators and Scanners," *Journal of Applied Physics*, American Institute of Physics, New York, NY, Vol. 38, No. 13, December 1967, pp. 5149-5153.

[23-65] H. Melchior, M. B. Fisher, and F. R. Arams, "Photodetectors for Optical Communication Systems," *Proceedings of the IEEE*, Institute of Electrical and Electronics Engineers, New York, NY, Vol. 58, No. 10, October 1970, pp. 1466-1486.

[23-66] A. E. Siegman, "The Antenna Properties of Optical Heterodyne Receivers," *Proceedings of the IEEE*, Institute of Electrical and Electronics Engineers, New York, NY, Vol. 54, No. 10, October 1966, pp. 1350-1356.

[23-67] W. K. Pratt, *Laser Communication Systems*, John Wiley and Sons, New York, NY, 1969, p. 188.

[23-68] F. A. Arnold and F. W. Nelms, "Arnold Engineering Development Center Long Wavelength Infrared Test Facilities," *Optical Engineering,* Society of Photo-Optical Instrumentation Engineers, Palos Verdes Estates, CA, Vol. 15, No. 6, November/December 1976, pp. 549-553.

[23-69] R. M. Hartman, McDonnell Douglas Astronautics Company, Huntington Beach, CA, Private Communication, 1976.

[23-70] U.S. Army Missile Research and Development Command Staff, "Advanced Simulation Center," Advanced Simulation Center Brochure, Redstone Arsenal, DRDMI—TD, AL, 1976, pp. 5, 7-10, 12, 13, 21.

[23-71] V. Bly, Night Vision Laboratories, Ft. Belvoir, VA, "Simulation Techniques in the Far Infrared," Paper delivered at the Meeting of the Association of Old Crows, Elgin AFB, FL, 1973.

[23-72] R. J. Baessler and H. Popper, "Infrared Simulation System (IRSS) Phase I," Army Missile Command, Redstone Arsenal, AL, Final Report 71SD484, Contract DAAH01-71-C-0511, July 1971.

23.5.2. Bibliography

Abbot, R. I., P. J. Shelus, J. D. Mulholland, and E. C. Silverberg, "Laser Observations of the Moon; Identification and Construction of Normal Points for 1969-1971," *The Astronomical Journal,* (American Astronomical Society) American Institute of Physics, New York, NY, Vol. 78, No. 8, October 1973, pp. 784-793.

Arecchi, F. T., et al. (eds.), *Laser Handbook,* North Holland Publishing Company, Amsterdam, The Netherlands, 1972, pp. 1469-1476, 1479-1485.

Arnold, F. A., and F. W. Nelms, "Arnold Engineering Development Center, Long Wavelength Infrared Test Facilities," *Optical Engineering,* Society of Photo-Optical Instrumentation Engineers, Palos Verdes Estates, CA, Vol. 15, No. 6, November/ December 1976, pp. 549-553.

Baessler, R. J., and H. Popper, "Infrared Simulation System (IRSS) Phase I," Army Missile Command, Redstone Arsenal, AL, Final Report 71SD484, Contract DAAH01 71-C-0511, July 1971.

Black, H. S., *Modulation Theory,* Van Nostrand, Princeton, NJ, 1953.

Born, M., and E. Wolf, *Principles of Optics,* Pergamon Press, New York, NY, 1959

Cooke, C. R., "Automatic Laser Tracking and Ranging System," *Applied Optics,* Optical Society of America, Washington, DC, Vol. 11, No. 2, February 1972, p. 277.

Davenport, W. B., Jr., and W. L. Root, *An Introduction to the Theory of Random Signals and Noise,* McGraw-Hill, New York, NY, 1958.

Defense Electronic Products, *RCA Electro Optics Handbook,* Aerospace Systems Division, Burlington, MA, 1968.

Dixon, R. W., "The Photoelastic Properties of Selected Materials and Their Relevance for Applications to Acoustic Light Modulators and Scanners," *Journal of Applied Physics,* American Institute of Physics, New York, NY, Vol. 38, No. 13, December 1967, pp. 5149-5153.

Elion, H. A., "Optical Communications in the Atmosphere," *Electronic, Electro-optic, and Infrared Countermeasure,* Hamilton Burr Publishing, Santa Clara, CA, Vol. 1, No. 4, August 1975, p. 43.

Elterman, L., "A Model of a Clear Standard Atmosphere for Attenuation in the Visible Region and Infrared Windows," Air Force Cambridge Research Laboratories, Bedford, MA, Report 63-675, AFCRL, AD 422 014, 1963.

Fitch, E., "The Spectrum of Modulated Pulses," *Journal of the Institution of Electrical Engineers,* Institution of Electrical Engineers, London, England, Vol. 94, Part 3A, 1947, pp. 556-564.

Flom, T., and D. Coombes, "Multiple Target Tracking and Target Altitude Determination with a Scanning Laser," *Journal of Navigation*, Royal Institute of Navigation, London, England, Winter, 1974-75.

Françon, M., *Diffraction Coherence in Optics*, Pergamon Press, Elmsford, NY, 1966.

Franken, P. A., and J. F. Ward, "Optical Harmonics and Non-Linear Phenomena," *Review of Modern Physics*, American Institute of Physics, New York, NY, Vol. 35, 1963, pp. 23-29.

Fried, D. L., "Aperture Averaging of Scintillation," *Journal of the Optical Society of America*, Optical Society of America, Washington, DC, Vol. 57, No. 2, February 1967, p. 169.

Fried, D. L., G. E. Meyers, and M. P. Keister, "Measurements of Laser-Beam Scintillation in the Atmosphere," *Journal of the Optical Society of America*, Optical Society of America, Washington, DC, Vol. 57, No. 6, June 1967, p. 757.

Geusic, J. E., W. B. Bridges, and J. I. Pankove, "Coherent Optical Sources for Communications," *Proceedings of the IEEE*, Institute of Electrical and Electronics Engineers, New York, NY, Vol. 58, No. 8, October 1970, pp. 1419-1439.

Gooch, C. H., *Gallium Arsenide Lasers*, Wiley-Interscience, John Wiley and Sons, New York, NY, 1969.

Gordon, E. I., "A Review of Acousto-optical Deflection and Modulation Devices," *Proceedings of the IEEE*, Institute of Electrical and Electronics Engineers, New York, NY, Vol. 54, No. 10, October 1966, pp. 1391-1401.

Hecht, E., and A. Zajac, *Optics*, Addison-Wesley Publishing Company, Reading, MA, 1974.

Helstrom, C. W., *Statistical Theory of Signal Detection*, Pergamon Press, New York, NY, 1960.

Hund, A., *Frequency Modulation*, McGraw-Hill, New York, NY, 1942.

Jamieson, J. A., R. H. McFee, G. N. Plass, R. H. Grube, and R. C. Richards, *Infrared Physics and Engineering*, McGraw-Hill, New York, NY, 1963.

Kaminow, I. P., and E. H. Turner, "Linear Electro-optic Materials," *Handbook of Lasers with Selected Data on Optical Technology*, CRC Press, Cleveland, OH, 1971, pp. 447-459.

Kaminow, I. P., and E. H. Turner, "Electro-Optic Light Modulators," *Proceedings of the IEEE*, Institute of Electrical and Electronics Engineers, New York, NY, Vol. 54, No. 10, October 1966, pp. 1374-1390.

Kerr, J. R., "Microwave-Bandwidth Optical Receiver System," *Proceedings of the IEEE*, Institute of Electrical and Electronics Engineers, New York, NY, Vol. 55, No. 10, October 1967, pp. 1686-1700.

Kingslake, R., (ed.), *Applied Optics and Optical Engineering*, Vol. I, Academic Press, Washington, DC, 1965.

Kotel'nikov, V. A., *The Theory of Optimum Noise Immunity*, McGraw-Hill, New York, NY, 1965.

Kruse, P. W., et al., *Elements of Infrared Technology: Generation, Transmission and Detection*, John Wiley and Sons, New York, NY, 1962, pp. 189-191.

Lee, Y. W., *Statistical Theory of Communication*, John Wiley and Sons, New York, NY, 1960.

Lehr, C. G., M. R. Pearlman, J. L. Scott, and J. Wohn, "Laser Satellite Ranging," *Laser Applications in the Geosciences*, J. Gauger and F. Hall (eds.), Western Periodicals, CA, 1970, pp. 111-130.

Liberman, I., "Incoherent Optical Sources," *Handbook of Lasers with Selected Data on Optical Technology*, CRC Press, Cleveland, OH, 1971, pp. 13-38.

Maiman, T. H., "Stimulated Optical Radiation in Ruby," *Nature*, Macmillan Journals Ltd., Washington, DC, Vol. 187, No. 4736, August 1960, pp. 493-494.

Maydan, D., "Acousto-optical Pulse Modulators," *IEEE Journal of Quantum Electronics*, Institute of Electrical and Electronics Engineers, New York, NY, Vol. QE-6, No. 1, January 1970, pp. 15-24.

McClatchey, R. A., R. W. Fenn, F. E. Volz, J. S. Garing, and J. E. A. Selby, "Optical Properties of the Atmosphere (Revised)," Air Force Cambridge Research Laboratories, L. G. Hanscom Field, Bedford, MA, AFCRL-71-0279, AFCRL-ERP-354, 10 May 1971.

Melchior, H., M. B. Fisher, and F. R. Arams, "Photodetectors for Optical Communication Systems," *Proceedings of the IEEE*, Institute of Electrical and Electronics Engineers, New York, NY, Vol. 58, No. 10, October 1970, pp. 1466-1486.

Moss, S. H., "Frequency Analysis of Modulated Pulses," *Philosophical Magazine and Journal of Science*, Taylor and Francis Ltd., London England, Series 7, Vol. 39, September 1948, pp. 663-691.

Nyquist, H., "Certain Factors Affecting Telegraph Speed," *Bell System Technical Journal*, American Telephone and Telegraph Company, New York, NY, Vol. 3, No. 1, April 1924, pp. 324-346.

Ochs, G. R., and R. S. Lawrence, "Saturation of Laser-Beam under Conditions of Strong Atmospheric Turbulence," *Journal of the Optical Society of America*, Optical Society of America, Washington, DC, Vol. 59, No. 2, February 1969, p. 226.

Oliver, B. M., "Signal-to-Noise Ratios in Photoelectric Mixing," *Proceedings of the IRE*, Institute of Radio Engineers, New York, NY, Vol. 49, December 1961, pp. 1960-1961.

Panter, P. F., *Modulation, Noise, and Spectral Analysis*, McGraw-Hill, New York, NY, 1965.

Papoulis, A., *Probability, Random Variables, and Stochastic Processes*, McGraw-Hill, New York, NY, 1965.

Peters, W. N., and A. M. Ledger, "Techniques for Matching Laser TEM_{00} Mode to Obscured Circular Aperture," *Applied Optics*, Optical Society of America, Washington, DC, Vol. 9, No. 6, June 1970, pp. 1435-1442.

Pockels, F., *Lehrbuch der Kristalloptic*, Tübner, Leipzig, Germany, 1906.

Pratt, W. K., *Laser Communication Systems*, John Wiley and Sons, New York, NY, 1969, p. 188.

Pressley, R. J., (ed.) "Coherent Optical Sources" (by various authors) *Handbook of Lasers with Selected Data on Optical Technology*, CRC Press, Cleveland, OH, 1971, pp. 183-420.

Pursley, W. K., *The Transmission of Electro-Magnetic Waves through Wire Diffraction Gratings*, Doctoral Dissertation Series Publication No. 21, 348 University Microfilms, Ann Arbor, MI, 1956.

Radiofizika, (Radiophysics and Quantum Electronics) Plenum Publishing, New York, NY, "On Strong Amplitude Fluctuations of a Wave Propagating in a Medium with Weak Random Inhomogeneities," Vol. 10, No. 1, 1967.

Rao, V. N., "Some Aspects of the Spectrum of Modulated Pulses," *Journal of the Indian Institute of Science*, Indian Institute of Science, Bangalore-560012, India, Vol. 35, Sec. B, July 1953, pp. 125-136.

RCA Staff, "Solid State Infrared-Emitting Diodes, Injection Lasers, Silicon Diodes," RCA, Harrison, NJ, Brochure No. OPT-100B, 1972; See also, "Solid State IR Emitters and Injection Lasers," Brochure No. OPT-113C, July 1977.

Rice, S. O., "Mathematical Analysis of Random Noise," Bell System Technical Journal, American Telephone and Telegraph Company, New York, NY, Vol. 23, No. 3, p. 282; *ibid*, Vol, 24, No. 1, July 1944; *ibid*, January 1945, p. 46.

Ross, M., *Laser Receivers*, John Wiley and Sons, New York, NY, 1966.

Schawlow, A. L., and C. H. Townes, "Infrared and Optical Masers," *Physical Review*, American Physical Society, New York, NY, Vol. 112, December 1958, pp. 1940-1949.

Schwartz, M., *Information Transmission, Modulation and Noise*, McGraw-Hill, New York, NY, 1959.

Schwartz, M., W. R. Bennett, and S. Stein, *Communication Systems and Techniques*, McGraw-Hill, New York, NY, 1966.

Shannon, C. E., "Communication in the Presence of Noise," *Proceedings of the IRE*, Institute of Radio Engineers, New York, NY, Vol. 37, No. 1, January 1949, pp. 10-21.

Shannon, C. E., and W. Weaver, *The Mathematical Theory of Communication*, University of Illinois Press, Urbana, IL, 1949.

Siegman, A. E., "The Antenna Properties of Optical Heterodyne Receivers," *Proceedings of the IEEE*, Institute of Electrical and Electronics Engineers, New York, NY, Vol. 54, No. 10, October 1966, pp. 1350-1356.

Silverberg, E. C., "Operation and Performance of a Lunar Laser Ranging Station," *Applied Optics*, Optical Society of America, Washington, DC, Vol. 13, No. 3, March 1974, pp. 565-575.

Skolnik, Merrill, I., *Introduction to Radar Systems*, McGraw-Hill, New York, NY, 1962.

Tatarsky, V. I., *Wave Propagation in a Turbulent Medium*, McGraw-Hill, New York, NY, 1961.

Tibbs, C. E., and G. G. Johnstone, *Frequency Modulation Engineering*, John Wiley and Sons, New York, NY, 1956.

U.S. Army Missile Research and Development Command Staff, "Advanced Simulation Center," Advanced Simulation Center Brochure, Redstone Arsenal, DRDMI—TD, AL, 1976, pp. 5, 7-10, 12, 13, 21.

Valley, S. (ed.), *Handbook of Geophysics*, Macmillan, New York, NY, 1960, pp. 13-14.

Wiener, N., *The Extrapolation, Interpolation, and Smoothing of Stationary Time Series with Engineering Applications*, John Wiley and Sons, New York, NY, 1949.

Wilson, P., et al, "Short Pulse Satellite Ranging System," EO/IL '77 Conference, October 1977, Anaheim, CA, sponsored by Electro Optical System Design.

Wozencraft, J. M., and I. M. Jacobs, *Principles of Communication Engineering*, John Wiley and Sons, New York, NY, 1967.

Yariv, A., and W. M. Caton, "Frequency, Intensity, and Field Fluctuations in Laser Oscillators," *IEEE Journal of Quantum Electronics*, Institute of Electrical and Electronics Engineers, New York, NY, Vol. QE-10, No. 6, June 1974, pp. 509-515.

Chapter 24

AERODYNAMIC INFLUENCES ON
INFRARED SYSTEM DESIGN

Lawrence D. Lorah
Private Consultant

Eugene Rubin
Sanders Associates, Inc.

CONTENTS

24. Aerodynamic Influences on Infrared System Design

24.1. Introduction

This chapter discusses the following major problems encountered by supersonic missiles and aircraft: the influence of high-speed flight on the window covering the infrared equipment; the optical influence exerted by the aerodynamic flow field around the vehicle; and the background radiation produced by the heated air surrounding the vehicle.

The discussions in Section 24.2 and 24.3 are restricted to supersonic flight over the Mach number range of about 1.5 to 7.0. At speeds below 1.5 the problems considered here do not exist and system design requires only minor structural differences from those in surface-based equipment. At speeds above Mach 7 in atmospheric flight, the assumption of perfect gas flow begins to break down and the chemistry of the air must be considered. The nature of the aerodynamic effects on infrared systems is the same at these higher speeds, but the details of predicting these effects are much more complex and beyond the scope of this discussion. In this chapter, the ratio of specific heat for air is taken as 1.4 and the numerical coefficients are evaluated accordingly. In Section 24.4 the chemistry of the air is a factor and the limitation to Mach numbers below 7 no longer exists.

24.1.1. Symbols, Nomenclature and Units. The symbols, nomenclature and units used in this chapter are listed in Table 24-1.

Table 24-1. Symbols, Nomenclature and Units

Symbol	Nomenclature	Units
A_c	Area of collector	cm^2
A_d	Area of detector	cm^2
b_w	Thickness of window	cm
c_f	Local skin friction coefficient	—
c_p	Heat capacity; also, specific heat	$J\ g^{-1}\ K^{-1}$
D	Diameter	cm
E_λ	Spectral irradiance	$W\ cm^{-2}\mu m^{-1}$
F	Focal length	cm
H_i	Depth	cm
h	Heat transfer coefficient	$W\ cm^{-2}K^{-1}$
K	Empirical constant for loss in optical resolution by turbulence	$rad\ ft^{5/2}\ slug^{-1}$

Table 24-1. Symbols, Nomenclature and Units (*Continued*)

Symbol	Nomenclature	Units
k	Thermal conductivity	$W\ cm^{-1}K^{-1}$
k_{GD}	Gladstone-Dale constant	$ft^3\ slug^{-1}$
L	Radiance	$W\ cm^{-2}\ sr^{-1}$
l	Length	cm
M	Mach number	—
$M\lambda$	Spectral exitance	$W\ cm^{-2}\ \mu m^{-1}$
n	Index of refraction	—
P	Pressure	$N\ m^{-2}$
Pr	Prandtl number	—
q	Rate of heat transfer per unit area	$W\ cm^{-2}$
R	Radius of curvature; also, shock radius	cm; cm
Re	Reynolds number	—
r	Recovery factor	—
S	Apparent distance	cm
St	Stanton number	—
s.p.	Stagnation point	—
T	Temperature, absolute	K
t	Time	sec
U	Velocity in undisturbed stream far from body	$cm\ sec^{-1}$
u	Velocity	$cm\ sec^{-1}$
X	Distance	cm
x	Axial, or surface distance	cm
Y	Distance from wall	cm
β	Angle; also, velocity gradient	$rad;\ cm\ sec^{-1}\ cm^{-1}$
β'	Parameter	—
δ	Deflection angle; also, boundary layer thickness	rad; cm
ϵ_e	Equivalent total emissivity	—
θ	Angular station; also, shock angle	rad
λ	Wavelength	μm
μ	Viscosity	$N\ sec\ m^{-2}$
ν	Kinematic viscosity	$cm^2\ sec^{-1}$
ρ	Density	$g\ cm^{-3}$
σ	Stefan-Boltzmann constant; also, cone half-angle	$W\ cm^{-2}\ K^{-4}$; rad
τ	Time constant	sec
Φ	Flux	W
ϕ	Angle	rad
ψ	Entrance angle	rad
Ω	Field of view	rad
	Subscripts	
B	Blur-circle resolution limit	—

Table 24.1. Symbols, Nomenclature and Units (*Continued*)

Symbol	Nomenclature	Units
BB	Blackbody	—
D	Based upon diameter	—
d	Detector	—
e	External edge of boundary layer, effective	—
i	Initial; also, incompressible	—
o	Total or stagnation value	—
r	Recovery	—
s.p.	Stagnation point	—
t	Turbulent	—
w	Wall; also, window	—
x	Arbitrary distance	—
λ	Spectral quantity	—
∞	Free stream	—
1	Before air passes through shock wave	—
2	After air passes through shock wave	—

24.2. Infrared Windows

Infrared windows are basically protective coverings over the cavities or bays that house infrared or optical equipment. Generally, they are not an integral part of the functional scheme of the equipment and any change they make in radiation falling on the system produces a degradation of system performance. Satisfactory windows for ground-based equipment or low-speed aerial equipment can generally be designed by careful selection of a window material from those of Chapter 7 and then applying common optical and structural design techniques.

24.2.1. Window Requirements. The primary purpose of the window, whether it be a missile nose dome or a flat reconnaissance window well back along the aircraft fuselage, is to protect infrared equipment from the aerodynamic loads and severe heating rates encountered in supersonic flight. While performing this basic function, the window must meet the following additional requirements:

(1) The window must introduce only negligible distortion in the image formed by the optical system. This consideration nearly always results in a window which is a portion of a sphere or is made with one or more flat surfaces. Other shapes (cones, ogives, etc.) give large refraction errors and resolution losses unless the window is made very thin. (See Chapter 7.) However, windows that are too thin cannot carry the necessary structural loads.

(2) The window must transmit radiation in the appropriate wavelengths without gross attenuation. The usual window material selection is complicated in this case by the fact that many materials lose transparency as their temperature increases, particularly near the long-wavelength cutoff. (See Chapter 7.)

(3) The window must radiate only a negligible amount of energy in the wavelength region of system operation. Even though the window emissivity is low in these wavelengths (by virtue of requirement (2)) the window proximity to the sensing system

and the high window temperatures developed in supersonic flight make this a prime consideration. Window radiation can saturate the detector, and nonuniform radiation from the window can produce false signals which will mislead a tracking system or produce ghosts in a reconnaissance system.

(4) The window must not impose an unreasonable drag penalty on the vehicle.

(5) The window must withstand the mechanical and thermal loads induced by high-speed flight.

Other factors such as ease of fabrication, light weight, and ground handling loads must also be considered. Most of these requirements depend in whole or in part on the aerodynamic heating experienced in supersonic flight.

24.2.2. Window Heating

Aerodynamic Heat Transfer. The rate of aerodynamic heat transfer per unit area, q, of the wall (window) w can be written

$$q = h(T_r - T_w) \tag{24-1}$$

where h = heat-transfer coefficient

T_w = wall temperature

T_r = recovery temperature

The heat-transfer coefficient is a function of the wall geometry, the flight conditions, and to some extent, the wall temperature. The recovery temperature is equal to the temperature of the wall when there is no aerodynamic heat transfer. This temperature is usually defined by means of a recovery factor r:

$$T_r = T(1 + 0.2rM^2) \tag{24-2}$$

where M is the Mach number. If the wall (window) is assumed to be thermodynamically thin; i.e.,

$$\frac{b_w h}{k_w} \ll 1$$

where b_w = window thickness

k_w = thermal conductivity

and the radiation and conduction losses are negligible, then the aerodynamic heat-transfer rate can be equated to the rate of enthalpy increase within the wall, or

$$h(T_r - T_w) = \rho_w b_w c_{p_w} (dT_w/dt) \tag{24-3}$$

where ρ_w = window density

c_{p_w} = heat capacity of window

t = time

If there is a step change in the aerodynamic conditions, e.g., velocity (thus a step change in h and T_r), the solution is

$$\frac{T_r - T}{T_r - T_i} = \exp\left\{-\left(\frac{h}{\rho_w c_{p_w} b_w}\right) t\right\} \tag{24-4}$$

where T_i equals the initial temperature at $t = 0$.

At the higher Mach numbers (higher window temperatures), the radiation heat transfer plays a significant part. If this is to be considered, the heat-balance equation becomes

$$h(T_r - T_w) = \rho_w b_w c_{p_w} \frac{dT_w}{dt} + \sigma \epsilon_e(T) T_w^4 \qquad (24\text{-}5)$$

where $\epsilon_e(T)$ = equivalent total wall emissivity, which is a strong function of temperature because of its large spectral variation

σ = Stefan-Boltzmann constant

Equation (24-5) can be solved for a series of constant temperatures to yield a more accurate temperature history. The true equilibrium temperature $(dT/dt = 0)$ can also be found from Equation (24-5).

It is convenient to normalize the heat-transfer coefficient with respect to the local free-stream properties. The normalized heat-transfer coefficient is the Stanton number, St,

$$St = \frac{h}{\rho_e u_e c_p} \qquad (24\text{-}6)$$

where u is the velocity. (The subscript e indicates the flow properties of the external edge of the boundary layer.)

For a more rigorous treatment of aerodynamic heat transfer, see [24-1] and [24-2]. Details of the temperature distribution within windows, given the aerodynamically imposed boundary conditions, can be developed by methods described in [24-3] and [24-4].

Heat Transfer to a Flat Plate. An infrared systems engineer generally needs to know window temperatures. The process of finding these temperatures, T_w, is to solve Equation (24-5) using the Stanton number defined by Equation (24-6). Values of the Stanton number for various conditions can be found in Figures 24-1 or 24-2. Equation (24-5) is solved for a series of different values of T_w in the radiative term on an iterative basis.

The boundary-layer flow over a flat surface parallel to the flow direction can be either laminar or turbulent. In most flight situations where the flat window is more than a foot or so from the nose or leading edge of the vehicle, the flow is turbulent. Transition from laminar to turbulent flow takes place at Reynolds numbers, Re, of approximately 10^5 to 5×10^6, the lower value being more likely in an actual flight vehicle with the usual surface roughness. The Reynolds number is $Du\rho/\mu$, where D is the diameter of the tube in which flow occurs and μ is the viscosity.

In the laminar case, the results of Van Driest [24-5] can be used directly. These extensive heat-transfer calculations are summarized in Figure 24-1 as curves of $St\sqrt{Re}$ versus Mach number for various ratios of wall temperature, T_w, to free-stream temperature, T_∞.

Experimentally, the recovery factor, r, is given for laminar flow as $\sqrt{Pr}$, or about 0.848 as an average over normal flight conditions (Pr = Prandtl number, which is $c_p\mu/k$).

In the turbulent case, use is made of the Reynolds analogy, which relates the Stanton number to the local skin friction coefficient, c_f, through the equation

$$St = \frac{1}{s} \frac{c_f}{2} \qquad (24\text{-}7)$$

where $1/s$ is a measure of the accuracy of the analogy. For current purposes, assume $s = 1$ although in some cases s can be as low as 0.8. Extensive measurements of skin friction have been made, and a semi-empirical relation which holds generally for incompressible turbulent flow has been found which adequately predicts the local c_f.

$$c_{f_i} = 0.370 \ (\log_{10} Rex)^{2.584} \qquad (24\text{-}8)$$

where c_{f_i} = incompressible coefficient
x = some arbitrary distance

Flow compressibility has a marked influence on the friction coefficient. To account for this, one takes the ratio of the compressible to the incompressible friction coefficient for air to be [24-1]

$$\frac{c_f}{c_{f_i}} = \left[0.5 \left(1 + \frac{T_w}{T_e} \right) + 0.0394 M^2 \right]^{-0.648} \qquad (24\text{-}9)$$

where T_e is the temperature at the outer edge of the boundary layer. In the special case where the wall temperature is equal to the recovery temperature (insulated wall case)

$$\frac{c_f}{c_{f_i}} = \left[1 + 0.129 M^2 \right]^{-0.648} \qquad (24\text{-}10)$$

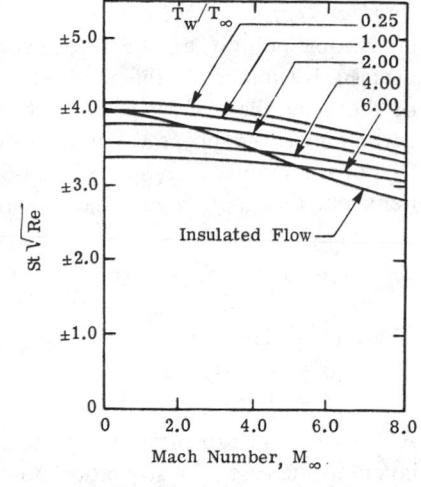

Fig. 24-1. Local heat-transfer coefficient. The coefficient is for a laminar boundary layer of a compressible fluid flowing along a flat plate, $T_\infty = 400°R$ [24-5].

The turbulent recovery factor, r, is approximately $(Pr)^{1/3}$, or about 0.896.

The insulated wall case serves as a handy reference for heat-transfer calculations. At $T_w = T_r$, there is no heat transfer; however, the Stanton number for the insulated case can be computed for these conditions. This corresponds rigorously to

$$St = \frac{dq}{dT} \ (\rho_e u_e c_p)^{-1} \qquad (24\text{-}11)$$

at $T_w = T_r$, and can be used to determine the heat-transfer coefficients at wall temperatures in the vicinity of T_r. Figure 24-2 shows the variation of both laminar and turbulent Stanton numbers as a function of Mach and Reynolds numbers.

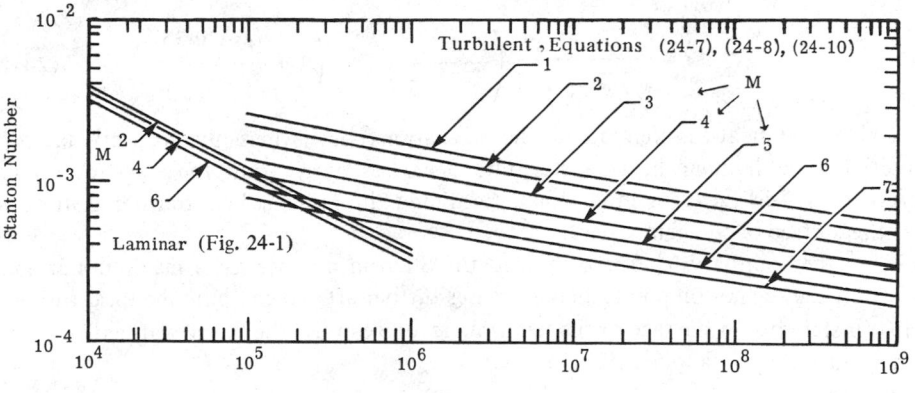

Fig. 24-2. Stanton number as a function of Reynolds number for insulated plate.

Heat Transfer to a Hemisphere. The aerodynamic flow to a hemispherical nose at the stagnation point (i.e., the foremost point on the hemisphere, regardless of angle of attack), is laminar. As the air accelerates in passing from the stagnation point to the base of the hemisphere, it can go through transition and the boundary layer generally becomes turbulent. This happens at Reynolds numbers around 4×10^5.

For the laminar case, Sibulkin [24-6] has shown that locally* the Stanton number on a hemisphere stagnation point can be written

$$St = 0.763 \left(\frac{c_p \mu}{k} \right)^{-0.6} \left(\frac{\beta D}{u} \right)^{1/2} \left(\frac{\rho u_\infty D}{\mu} \right)^{1/2} \tag{24-12}$$

where D = diameter of the hemisphere
β = velocity gradient
μ = viscosity

All of the fluid properties are evaluated at the local conditions just outside the boundary layer. In the case of a stagnation point (s.p.) this corresponds to stagnation temperature and pitot pressure.** This reduces Equation (24-12) into terms of free-stream flow:

$$St = 0.763 \left(\frac{\beta D}{u} \right)^{1/2} \left(\frac{\rho_\infty u D}{\mu_\infty} \right)^{-1/2} \left(\frac{c_p \mu}{k_\infty} \right)^{-0.6}_{\text{s.p.}} \left(\frac{\rho_o}{\rho_\infty} \right)^{1/2} \left(\frac{\mu_o}{\mu_\infty} \right)^{1/2} \tag{24-13}$$

where $(c_p \mu/k_\infty)_{\text{s.p.}}$ = stagnation Prandtl number (approximately 0.72 for normal flight conditions)
$\rho_\infty u D/\mu_\infty$ = free-stream Reynolds number based on hemisphere diameter
subscript o = total or stagnation value

From [24-1],

$$\beta \frac{D}{u} = \sqrt{8 \frac{\rho_\infty}{\rho_{02}}} \tag{24-14}$$

The subscript 02 denotes the total or stagnation value after the air passes through the shock wave. If it is assumed that $\mu \propto T^{0.76}$ and the perfect gas equations of [24-7] are used, then Equation (24-13) reduces to the general laminar Stanton number equation

$$St = \frac{9.37}{\sqrt{Re} \ \sqrt{u/D}} \left(\frac{M^2}{M^2 + 5} \right)^{0.25} \left(1 + \frac{M^2}{5} \right)^{0.38} \tag{24-15}$$

This gives the heat-transfer rate at the stagnation point where the recovery factor, r, equals 1. The laminar heat-transfer rate decreases away from the stagnation point. Reference [24-8] gives methods for determining the laminar heat-transfer distribution over the surface of the hemisphere.

When the boundary layer is turbulent, the Stanton number has a maximum at about $40°$ from the stagnation point, then decreases further aft. To compute the local turbulent heat-transfer rate on the face of the hemisphere, one can use the "local flat plate" method with good accuracy—about 10% error.

---

*A definition of locally just precedes Equation (24-16), below.
**The pressure measured in the stream flow outside the boundary layer by a pitot tube.

This approach calls for utilization of Equations (24-7), (24-8), and (24-9) to determine the local Stanton number. All of the flow properties are evaluated at "local" conditions; that is, at the pressure, velocity and temperature just outside the boundary layer at any angular station, θ. The Mach number distribution over the face of a hemisphere is insensitive to free-stream Mach number (Figure 24-3). The local pressure is given in [24-7].

$$\frac{P_e}{P_{02}} = \left(1 + \frac{M_e^2}{5}\right)^{-7/2} \qquad (24\text{-}16)$$

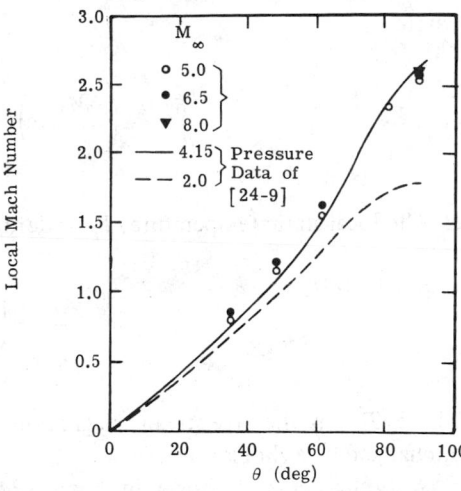

Fig. 24-3. Local Mach number on hemisphere.

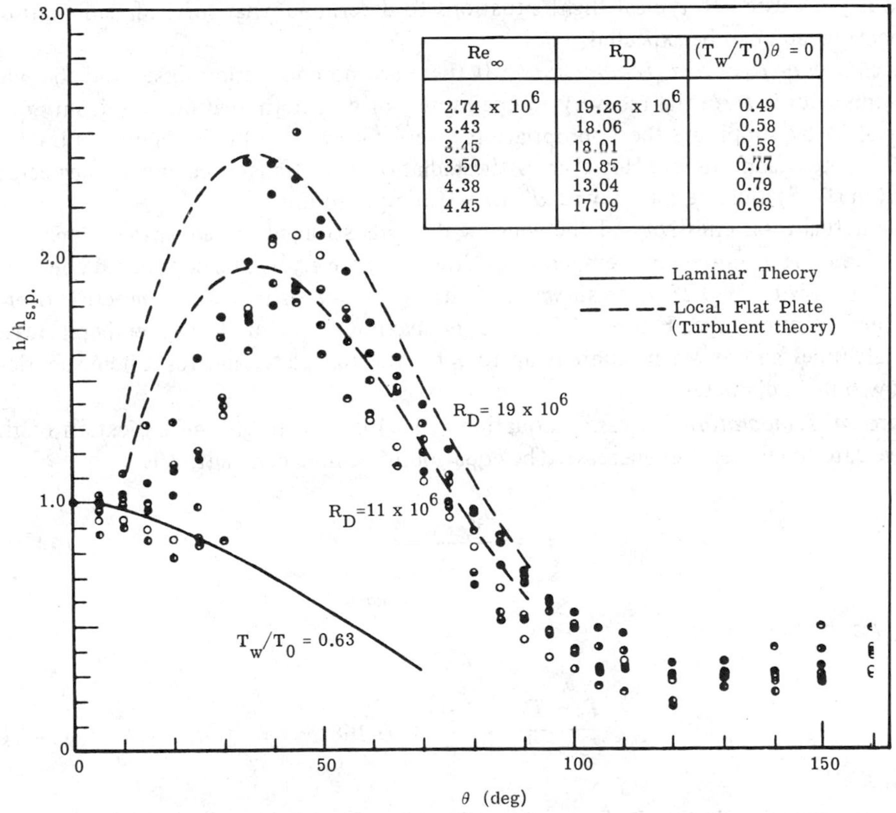

Re$_\infty$	R$_D$	$(T_w/T_0)\theta = 0$
2.74 x 10^6	19.26 x 10^6	0.49
3.43	18.06	0.58
3.45	18.01	0.58
3.50	10.85	0.77
4.38	13.04	0.79
4.45	17.09	0.69

——— Laminar Theory

– – – Local Flat Plate (Turbulent theory)

R$_D$= 19 x 10^6

R$_D$=11 x 10^6

T_w/T_0 = 0.63

Fig. 24-4. Theoretical and experimental heat-transfer distributions. Variation with θ of the ratio of a local heat-transfer coefficient to the value at the stagnation point (s.p.) at M_∞ = 2.0 and several free-stream Reynolds numbers [24-9].

where

$$\frac{P_{02}}{P_\infty} = \left(\frac{6M_1^2}{5}\right)^{7/2} \left(\frac{6}{7M_1^2 - 1}\right)^{5/2} \tag{24-17}$$

The local static temperature, T_e, is defined by

$$\frac{T_e}{T_{0_\infty}} = \left(1 + \frac{M_e^2}{5}\right)^{-1} \tag{24-18}$$

where T_{0_∞} is the free-stream stagnation temperature. (The stagnation temperature is unchanged by a shock wave.)

A typical case is shown in Figure 24-4, where both theoretical and experimental heat-transfer distributions are shown [24-9].

24.2.3. Typical Results. The following subsections apply some of the basic equations previously discussed to typical flight situations to determine what sorts of temperatures and heating rates can be expected.

Maximum or Recovery Temperature. If there are no conduction losses and the window emissivity is zero, the recovery temperature can be determined directly from Equation (24-2) by supplying the appropriate recovery factor: 0.848, for laminar flow and 0.896 for turbulent flow. However, if the radiation heat transfer cannot be neglected, Equation (24-5) must be solved, with $dT/dt = 0$ at equilibrium.

The actual total emissivity of the window depends strongly on the particular material selected and the instantaneous temperature. However, here ϵ_w will be considered constant. A total emissivity of 0.25, over all wavelengths, is characteristic of some materials over a large temperature range. Figure 24-5 shows the maximum or equilibrium temperatures at three altitudes and at Mach numbers up to 7 for the flat plate, and for a hemispherical window 6 in. in diameter.

Rate of Temperature Increase. Equation (24-4) is used to give an indication of the relative rate of temperature increase. The equation for a time constant, τ, is

$$\tau = \frac{\rho_w c_{p_w} b_w}{h} \tag{24-19}$$

and gives the time to reach

$$\frac{T_r - T}{T_r - T_i} = \frac{1}{e} \approx 0.368 \tag{24-20}$$

in the absence of radiation. In Figure 24-6 the generalized time constant is given for the conditions of Figure 24-5, and a typical material, MgF_2, 0.07 in. thick, is used to get a specific time constant.

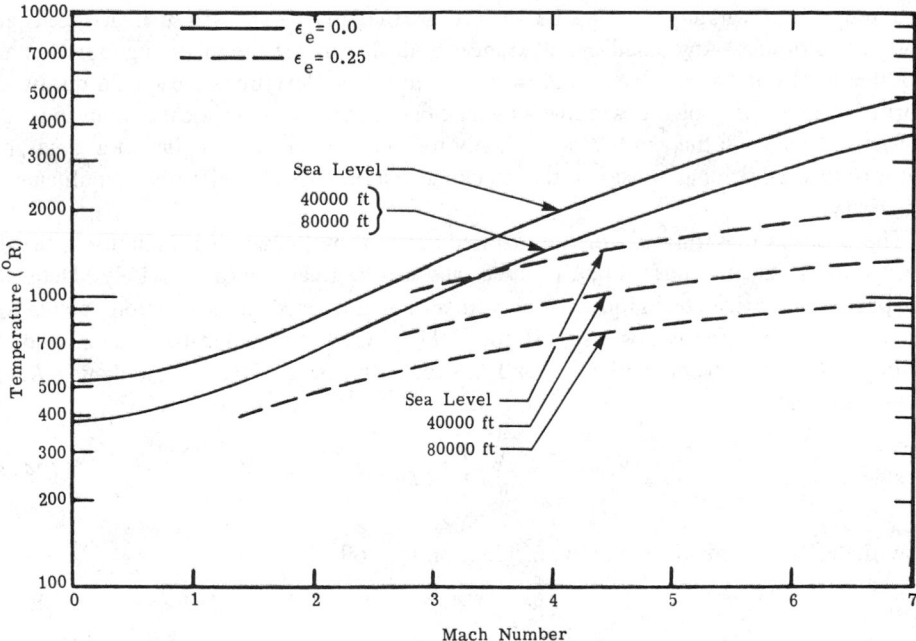

Fig. 24-5. Maximum temperature due to aerodynamic heating, turbulent flat plate, $x = 1$ ft.

24.2.4. Window Transmission and Radiation at Elevated Temperatures.

Aerodynamic heating can produce changes in the infrared characteristics of window materials. A major problem at elevated temperatures is a loss in transparency, particularly at the longer wavelengths. Transmission characteristics of materials at high temperatures are covered in Chapter 7. Any body, semitransparent or opaque, will radiate thermal energy in ever increasing amounts as the temperature rises. Two different types of effects are produced by a radiating window:

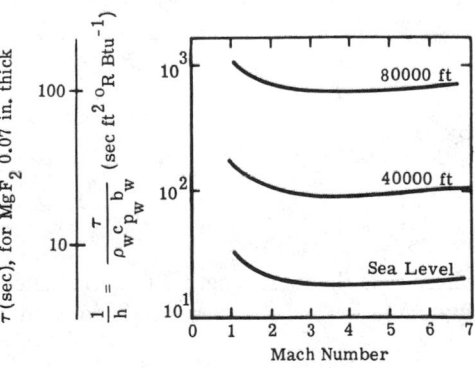

Fig. 24-6. Time constant for aerodynamic heating.

(1) Because of strong temperature gradients, or local "hot spots," false targets can appear.

(2) A large irradiance on the infrared detector, even though uniformly distributed, can saturate the detector and decrease its sensitivity.

The susceptibility of a particular guidance system to false targets associated with nonuniform temperature distributions is greatly dependent on the details of the discrimination and tracking methods employed. A window which is nearly hemispherical does not generally have large temperature gradients along its surface, even though the temperature level can be quite high. A pointed window, such as a cone, can have very high temperature gradients along its surface, particularly when used at high speeds and moderate altitudes. Any small protuberances optically distort the incoming radiation and produce local hot areas on the window. Generally, these deviations are spread out by the thermal conductivity of the window and are close enough to the optical system to be out of focus. A point on the window is not likely to appear as an exact replica of a target, but will produce additional noise in the tracking system, thereby effectively reducing its sensitivity.

The problem of saturation of the infrared detector by the window radiation is usually more critical than the false target problem and can be treated more precisely. Details of the particular optical techniques employed do not influence the saturation calculation, and a general analysis is possible. If the field of view is assumed to be small and the window is assumed to be isothermal and fills the entire field of view, the radiance, L, can be expressed as

$$L = \frac{1}{\pi} \int M_{\lambda BB} \epsilon \, d\lambda \qquad (24\text{-}21)$$

Now the effective radiant power, Φ, incident on the collector is

$$\Phi \doteq L A_c \pi (\Theta)^2 \qquad (24\text{-}22)$$

where Θ is the half-angle of the field of view. By neglecting the losses in radiant power due to any space filtering, refractive optical elements, etc., one can assume that the effective radiant power falling on the detector is equal to that falling on the collector. Therefore, the irradiance, E_d, at the detector is

$$E_d = \frac{\Phi}{A_d} = L \pi \Theta^2 \frac{A_c}{A_d} \qquad (24\text{-}23)$$

By geometry

$$\frac{A_c}{A_d} = \frac{1}{4} \left(\frac{D}{F} \right)^2 \Theta^2$$

where F is the focal length. The irradiance at the detector is a function of only the window radiation and the focal length to diameter ratio of the collector:

$$E_d = L \frac{\pi}{4} \left(\frac{F}{D} \right)^{-2} \qquad (24\text{-}24)$$

The emissivities of several window materials are shown in Figure 24-7. Others are covered in Chapter 7.

Figure 24-8 shows the detector effective spectral irradiance due to a typical window of MgF_2 at various temperatures.

In a conventional air-to-air missile application, the window temperature can reach several hundred degrees. Figure 24-8 shows that at the detector the background irradiance caused by the window can be many orders of magnitude greater than the irradiance from the target. Methods of calculating the radiance from windows having temperature gradients normal to the surface are discussed

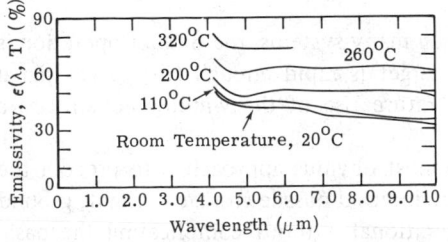

(a) Germanium (single crystal, 1.14 mm thick)

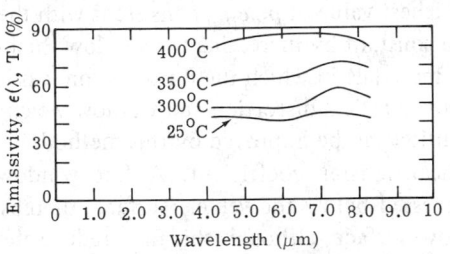

(b) Silicon (single crystal, 4.16 mm thick)

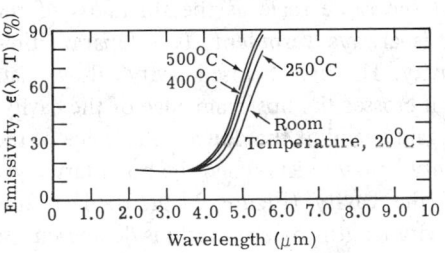

(c) Synthetic sapphire (4.11 mm thick)

Fig. 24-7. Emissivities of several window materials.

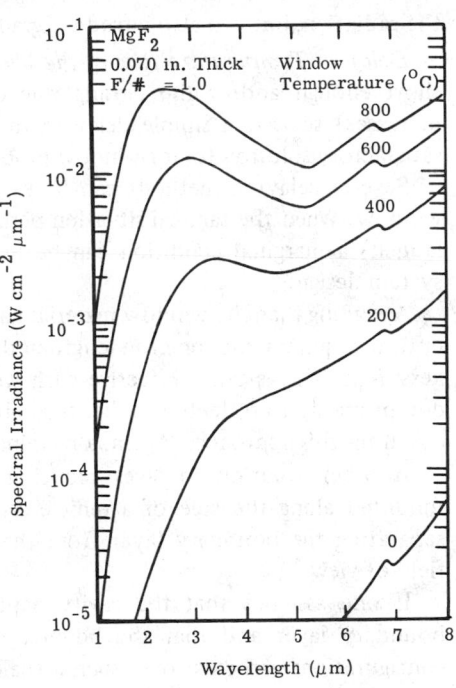

Fig. 24-8. Effective spectral irradiance at a detector. This is shown with an MgF_2 window versus wavelength, at various temperatures.

by Kiselev in Reference [24-10]. It should be noted that MgF_2 is a good infrared window material between 3 and 5 μm, and other materials would probably produce more of a saturation effect over the same temperature range.

24.2.5. Methods of Alleviating Hot-Window Problems

Optical-System Parameters. Several optical-system parameters have a considerable influence on the overall performance through the effects of window radiation. They include:

(1) Spectral region of sensitivity—the spectral distribution of the window radiation can alter the optimum wavelength limits from those indicated by target-detector-background considerations alone.

(2) Detector type—improved saturation characteristics can produce a greater signal-to-noise ratio than slightly greater detectivity in the presence of strong window radiation for the several detectors that can operate in the desired spectral region.

(3) Field of view—reduction of the field of view is important since the window irradiation on the cell and the cell area are reduced.

(4) Window location—for systems not requiring a nose installation, careful selection of the window location so as to avoid regions of high heat transfer (shock-wave impingement, boundary-layer reattachment, etc.) can reduce window radiation.

(5) Cooling the window—high conductivity windows with cooled mounts can reduce

the radiation, but can also introduce gradients.

Delay in Temperature Rise of the Window.   In many systems, the time of operation is short enough and/or the closing rate on the target is rapid enough (the target signal increases) so that a simple delay in the temperature rise of the window constitutes a satisfactory solution to hot-window problems.

Several delaying methods can be used. The most obvious approach is to precool the window. When the tactical situation permits (i.e., internal missile storage in flight, ground launch) a marginal condition can be made operational without complicating the basic system design.

Assuming that the window material has the highest value of $\rho_w c_{p_w}$ consistent with the optical requirements, one can lengthen the time constant by increasing the window thickness, b_w. This is fairly effective with window materials in which the transmission loss is due primarily to reflection rather than absorption. Optical distortion and window weight also limit this approach. Again, a marginal condition can be improved by this method.

Another solution involves reducing the heat-transfer coefficient. A flat window mounted along the side of a vehicle can be recessed below the vehicle surface, thereby separating the boundary layer from the window surface, although this may reduce the field of view.

If one assumes that the cavity depth is of the same scale as the thickness of the boundary layer and that the boundary layer is always turbulent, two separate flow configurations can exist over such a shallow cavity. The first, "closed cavity" flow exists when the boundary layer, initially separated as it crosses the upstream edge of the cavity, attaches to the cavity floor and then separates again ahead of the downstream edge of the cavity (Figure 24-9(a)). The second, "open cavity" flow exists when the boundary layer remains separated over the entire length of the cavity (Figure 24-9(b)). The flow configuration goes from closed to open as the cavity length-to-depth ratio is decreased. As the boundary layer becomes thicker with respect to the cavity depth, the transition from closed flow to open flow becomes more gradual and the flow configuration is not as sensitive to changes in length-to-depth ratio, l_e/H_i. Roughly, for a uniform depth cavity,

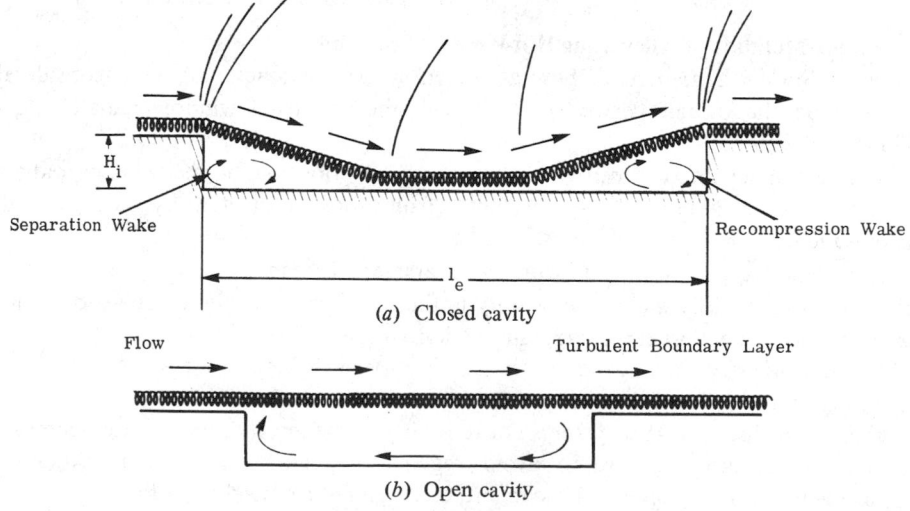

(a) Closed cavity

(b) Open cavity

Fig. 24-9. Flow configurations.

length-to-depth ratios less than about 10 will produce open cavity flow.

Theoretical treatment of aerodynamic heat transfer in separated flow has not proved satisfactory because of lack of available experimental data. For experimental data measured in a configuration similar to the one under consideration at moderate Mach numbers, see [24-11]. The heat-transfer coefficient, h, on the cavity floor, normalized with respect to the heat-transfer coefficient for an attached flow at the same body station, h_f, and Reynolds number, is shown in Figures 24-10 and 24-11.

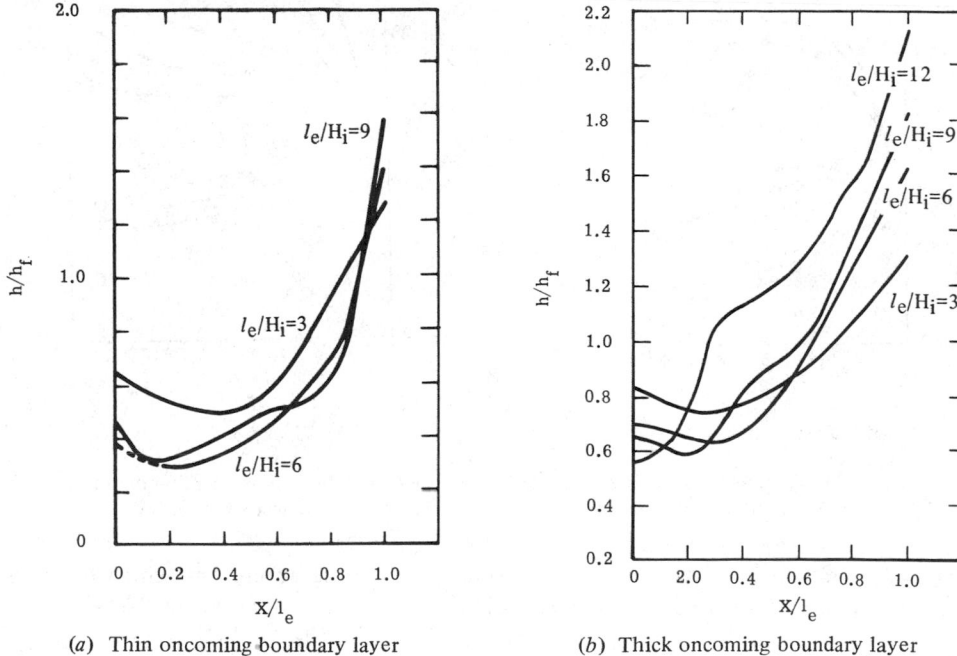

(a) Thin oncoming boundary layer   (b) Thick oncoming boundary layer

Fig. 24-10. Variation of heat-transfer distribution on floor of cutout as a function of l_e/H_i at $M_o = 2.9$ [24-11].

For all values of l_e/H_i the heat-transfer-coefficient ratio drops considerably below 1 in the separation wake, and rises above 1 in the recompression wake, reaching a maximum at the downstream edge of the cavity. The typical h/h_f distributions plotted in Figure 24-10 show that an increase in the boundary-layer thickness does not change the general character of the curves, but does significantly raise the average level of heat transfer. In Figure 24-11, the relative heat-transfer coefficient decreases with decreasing Mach number and a higher ratio of l_e/H_i produces a greater maximum heat-transfer coefficient.

24.3. Refraction by the Field of Flow

24.3.1. Index of Refraction of Air.

In the analysis of aerodynamic influences on electromagnetic radiation, one must consider the index of refraction of air, n. The major change in index of refraction of air takes place with a change in density, ρ. The variation of index of refraction for transparent substances as a function of density is [24-12]

$$\frac{n^2 - 1}{n^2 + 2} = k_{GD}\rho \qquad (24\text{-}25)$$

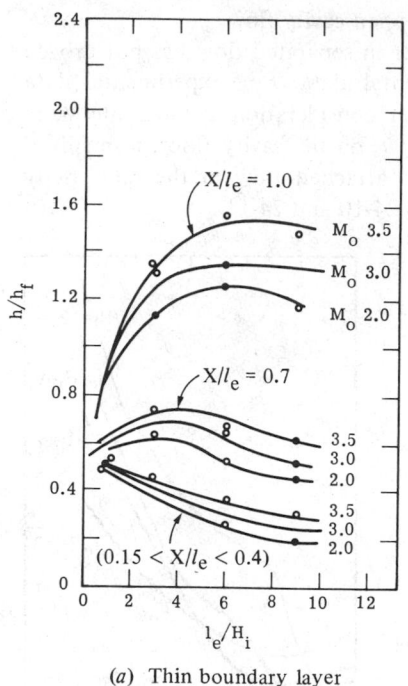

(a) Thin boundary layer

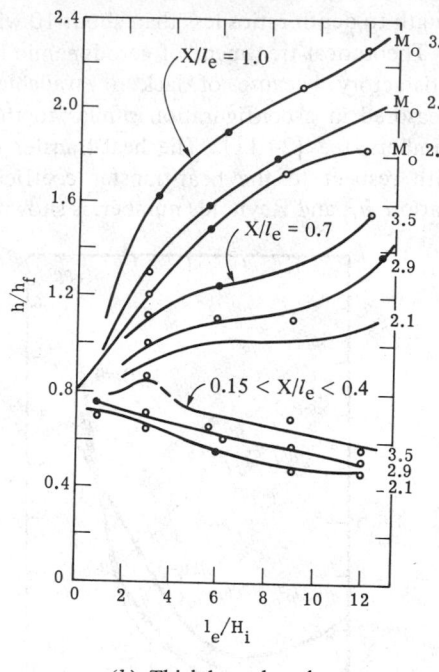

(b) Thick boundary layer

Fig. 24-11. Parametric map. It shows the heat-transfer ratio (minimum, maximum, and at $X/l_e = 0.7$) as functions of l_e/H_i and M_o. Note: points are averages over all tests.

where k_{GD} equals the Gladstone-Dale constant. This is the Lorenz law which for gases ($n \approx 1$) reduces to the simpler Gladstone-Dale law:

$$n - 1 = k_{GD}\rho \qquad (24\text{-}26)$$

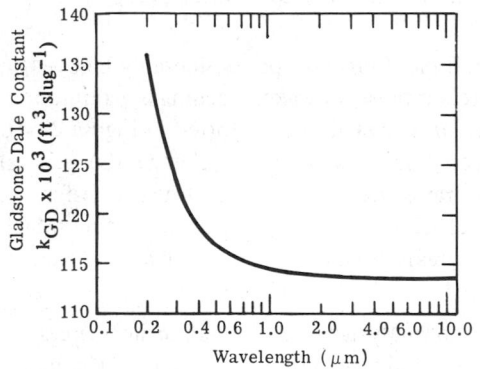

Fig. 24-12. Gladstone-Dale constant for air [24-13].

with virtually no loss in accuracy. Figure 24-12 shows the variation of k_{GD} in Equation (24-26) as a function of wavelength, as deduced from the data of [24-13]. Variations in index of refraction due to temperature effects on molecular structure and changes in atmospheric composition are neglected.

Figure 24-12 shows that the effects of refraction through nonuniformities in air are more pronounced in the visible portion of the spectrum than in the infrared. In the wavelength region beyond the near infrared ($\lambda > 1$ μm) the refractive index of air is nearly invariant with wavelength; thus one finds no dispersion and no chromatic aberration.

24.3.2. Shock Wave Effects. As a body travels through the atmosphere, the air must adjust to accommodate it. At supersonic velocities, the initial adjustment takes place across shock waves, which are discontinuities in the air properties.* Surfaces with sharp leading edges at moderate, positive angles of attack have these shock waves attached to the leading edge; relatively blunt bodies have shocks detached and standing some distance ahead of the body. The strength and orientation of these shock waves are determined by the flight Mach number and the geometry of the body.

The relation between the density ahead of the shock, ρ_1, and the density behind the shock, ρ_2, is

$$\frac{\rho_2}{\rho_1} = \frac{6(M_\infty \sin \theta)^2}{(M_\infty \sin \theta)^2 + 2} \tag{24-27}$$

Equation (24-27) shows that the density ratio (shock strength) increases with the shock angle, θ, and the free-stream Mach number, M, approaching the maximum value of 6 for a normal shock at M_∞.

No convenient, explicit relations exist for determining the shock angle. However, curves of shock angle as a function of Mach number and flow inclination are given in [24-7]. Figures 24-13 and 24-14 show representative curves of the functions for both wedge and conical flow.

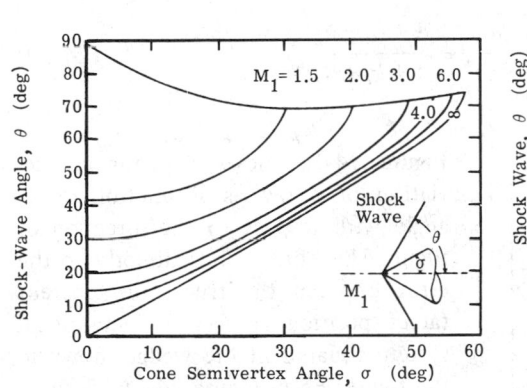

Fig. 24-13. Shock wave angle for conical flow.

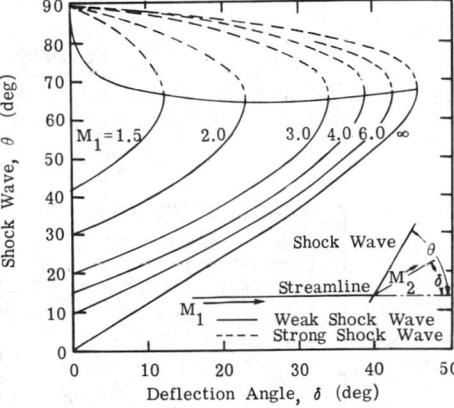

Fig. 24-14. Shock wave angle for wedge flow.

A shock wave produced by solid surfaces is exceptionally steady and, because of the density change across it, is a good refracting surface. Since a plane shock wave is not necessarily perpendicular to the optical axis or parallel to the window surface, the effects of refraction must be taken into account even in the simplest case. Using Snell's law, one can relate angle β_1, of the incident ray, and angle β_2, of the refracted ray, both measured from the normal to the shock plane, by Equation (24-28):

---

*Shock waves are not exactly discontinuities but do have a finite thickness. This thickness is about five molecular mean free paths, which allows translational and rotational equilibrium of the molecules to be established. At sea level the shock wave is about 1 μm thick.

$$\frac{\sin \beta_1}{\sin \beta_2} = \frac{1 + k_{GD}\rho_2}{1 + k_{GD}\rho_1} \qquad (24\text{-}28)$$

or alternatively,

$$\frac{\sin \beta_1}{\sin \beta_2} = \frac{1}{n_1} + \frac{n_1 - 1}{n_1}\frac{\rho_2}{\rho_1} \qquad (24\text{-}29)$$

where n_1 is the index value in the first medium. From Equation (24-29), the angular deviation of a ray passing through a plane shock wave can be calculated as a function of incidence angle. This is shown in Figure 24-15 for various density ratios and two altitudes. Figure 24-15 shows that the density ratio of 6 (the limiting value) at sea level is the worst possible refraction error. The angular deviation in this limiting case is seen to be about 1.5 mrad (300 sec of arc) at a 45° incidence angle. At the higher altitude, 50,000 ft, the maximum refraction error at 45° incidence is only 0.25 mil (50 sec), or about one-sixth of the sea-level value. The minimum density ratio shown is 2, which corresponds to a normal shock wave at about Mach 1.6. The deviation of the rays is only about one-fifth of the limiting value.

Another way of considering the variation of plane shock-wave refraction effects is to fix the body geometry and allow the Mach number to change, producing variations in density ratio and shock angle. Selecting a flow inclination of 10°, one gets

$$\frac{\cos (\theta + \phi_1)}{\cos (\theta + \phi_2)} = \frac{1 + k_{GD}\rho_2}{1 + k_{GD}\rho_1} \qquad (24\text{-}30)$$

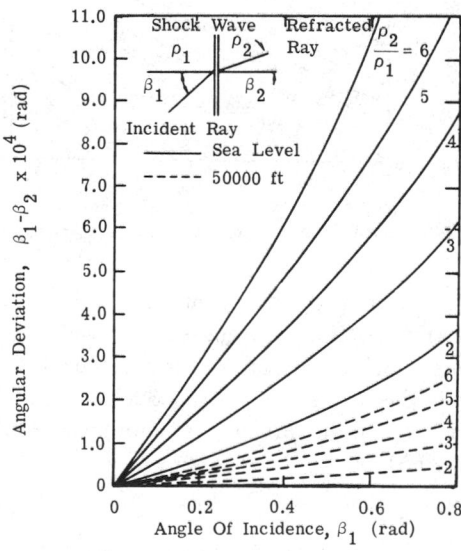

Figure 24-16 shows the angular deviation of a ray as a function of its angle with respect to the direction of flight. This corresponds directly to the error between the true and apparent target position as seen by an optical system situated in the wedge shown in the figure. From Figure 24-16 it can be seen that a target along the direction of flight will appear 0.2 mil (42 sec) too high at Mach 2, and about 1 mil too high at Mach 5. In the limiting case of a Mach number of infinity, the error along the flight direction is 3.2 mils. This error is less as the incident ray becomes more nearly perpendicular to the shock wave; and when it is exactly perpendicular, the error goes to zero.

Because of this shock wave, wide-angle photography suffers a distortion in

Fig. 24-15. Angular deviation of infrared radiation passing through plane shock waves.

the flight direction, even if the optical axis is perpendicular to it. If this distortion is evaluated as the difference in angular deviation of the rays at each edge of the field of view, the minimum distortion occurs when $\theta + \phi_1 = \pi/2$. At that point, the slopes of the curves of Figure 24-16 yield a distortion of about 0.3 mrad per radian field of view at Mach 2, and about 0.7 mrad per radian field of view at Mach 5. More explicitly, for example, a 1 rad, field of view instrument, in a $10°$ wedge flying at Mach 5, would show two objects, one at each limit of the field of view, too close together by at least 3.6 ft for every mile of range.

The curved shock wave occurs when the shock is detached from the leading edge or when the body itself presents a nonplane surface to the flow (e.g., curved cross-section, or body width not large compared to its thickness). The introduction of curvature into the density discontinuity produces a focusing effect which, except in special cases, is astigmatic. To understand the nature of the problem, consider first the conical shock wave influence on an incident plane wave of radiation approaching perpendicular to the cone axis, as shown in Figure 24-17. This curved shock acts as a lens and tends to focus incoming rays of parallel light. For the particular case of Figure 24-17, the focal length, F, is given by

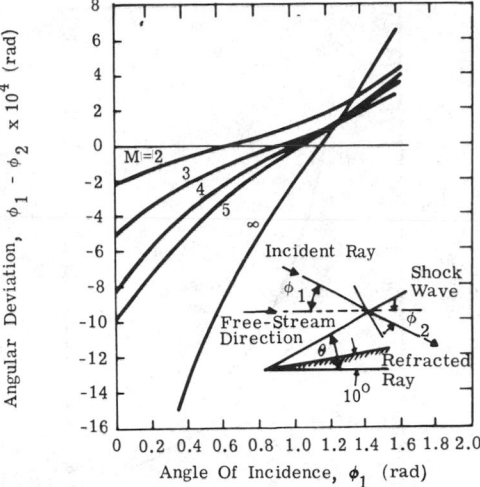

Fig. 24-16. Angular deviation of infrared radiation passing through plane shock waves produced by $10°$ wedge (at sea level).

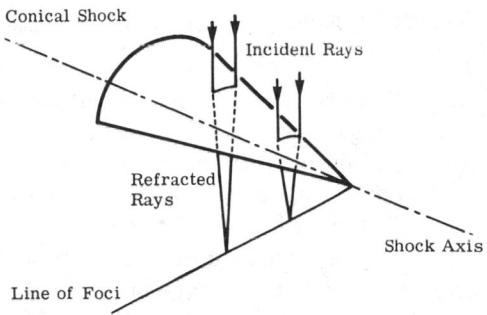

Fig. 24-17. Conical shock wave refraction of incident radiation.

$$\frac{F}{x} = \frac{n_1 \tan \theta}{n_2 - n_1} \tag{24-31}$$

where x = axial distance from the apex of the shock cone

θ = the shock angle

n_2 = index value in second medium

The locus of the "focal line" has been calculated for two cone half-angles, $10°$ and $20°$, and the variation with Mach number is shown in Figure 24-18. The shock lens becomes stronger (F/x less) with increasing cone angle and Mach number because of the increase in shock strength. The effect of this conical lens is to skew the focal plane of a

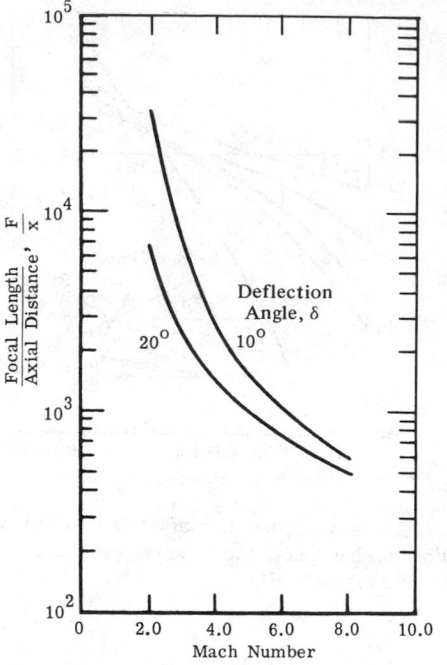

Fig. 24-18. Ratio of focal length to axial distance from apex for conical shock wave.

normal optical system whose axis is aligned with the incoming radiation. To estimate the importance of this effect, consider an $F/1$ optical system with a 15 cm aperture, 40 cm from the apex of a 10° half-angle cone. If the source is a line parallel to the cone axis and at an infinite distance, the change in the image at the focal plane can be calculated only on the basis of geometrical changes (i.e., the change in convergence angle of the rays). The magnitude of the defocusing due to the shock can then be compared to the diffraction limit of the simple optical system. Table 24-2 shows that the image focusing due to the conical shock wave at Mach 3, in an uncorrected system, would be about five times as large as the theoretical resolution limit and about ten times larger at Mach 5. Simply refocusing the system would bring the defocusing to within a factor of 2 of the theoretical resolution limit even at Mach 5. As the optical axis tilts toward the shock apex, the curvature normal to the incoming wave front becomes elliptical and then parabolic rather than circular, introducing further astigmatism. Also, as the optical path passes through regions near the apex of the shock cone, the radius of curvature of the shock decreases, producing a lens of greater power.

A more complicated optical mechanism exists in the case of flow about a conical body. This arises because the density of the air is not constant in the flow field between the shock wave and the body surface. A complicated integration process is required to accurately compute this type of effect; the approximate solutions are given in [24-14].

Table 24-2. Effect of Conical Shock Wave on $F/1$, 15 cm Optical System

	M = 3.0	M = 5.0
Theoretical resolution limit	2.4×10^{-5} rad	2.4×10^{-5} rad
Change (in focal length) due to shock	(fwd) 2.4×10^{-3} cm (aft) 1.7×10^{-3} cm	6.4×10^{-3} cm 4.6×10^{-3} cm
Angle subtended at image by 15 cm focus with shock	(fwd) 1.6×10^{-4} rad (aft) 1.1×10^{-4} rad	4.3×10^{-4} rad 3.1×10^{-4} rad
Maximum image angle subtended at optimum focus with shock	2.1×10^{-5} rad	5.8×10^{-5} rad

Typically, for a cone half-angle of 20° the contributions of the density gradients behind the shock wave are approximately 26% and 14% of the total error at Mach numbers of 2.0 and 3.5, respectively, when the ray is approximately perpendicular to the cone axis.

In the case of a hemisphere traveling at supersonic speeds, a nearly spherical detached shock forms ahead of the body and becomes part of the optical path. Assume, for example, a plane wave of radiation falling on the hemispherical shock along its axis of symmetry (Figure 24-19). Rays striking an oblique portion of the shock surface should be brought to some sort of focus on the axis of symmetry. However, the change in index of refraction across the shock is not constant but varies with the shock angle. It appears that each elemental annular ring, centered on the axis of symmetry, has its own focal length along the axis. Using the relations for the density change across the shock and Snell's law, one can compute this focal point as a function of the entrance angle, ψ. Figure 24-20 shows the results of such calculations. For Mach 3 flight at sea level, the focal length increases with ψ, but at Mach 5 it decreases with increasing ψ. Apparently, at some Mach number for sea-level flight the hemispherical shock behaves as a proper lens and has a single focal length. This Mach number appears to be about 3.26. The focal length becomes shorter as the Mach number increases because of the stronger shock. The simplified assumption of a spherical shock front obscures the detailed ray behavior due to the change of shock geometry with Mach number; however, this is of minor importance.

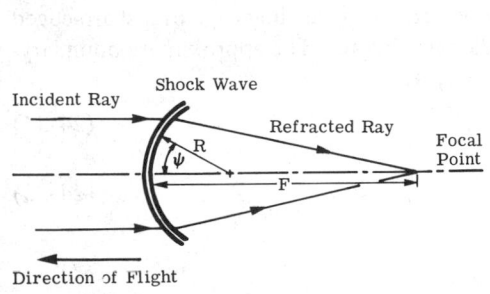

Fig. 24-19. Hemispherical shock wave refraction of incident radiation.

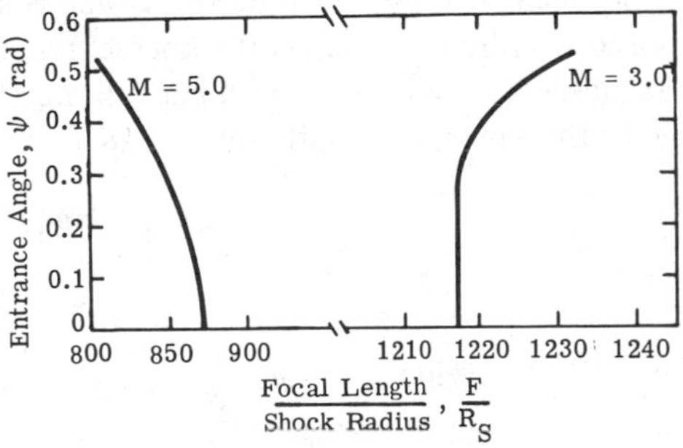

Fig. 24-20. Ratio of focal length, F, to shock radius, R_S, for hemispherical shock wave, $\lambda > 15$ μm, at sea level.

An additional optical element with a focal length equal to about a thousand times the shock radius is still significant. Consider, the same 15 cm, $F/1$ optical system behind a 25 cm diameter hemispherical shock. Again, the geometrical growth of the point's image in the focal plane is compared to the theoretical limit of resolution of the system. Table 24-3 shows the effects of the hemispherical shock wave at Mach 3 and Mach 5. The geometrical blurring of the image by the shock in the uncorrected focal plane is seen to be 50 to 100 times greater than the circle of confusion. By moving the focal plane a few tenths of a millimeter to obtain an optimum focus in the presence of the shock wave, one can lessen the geometrical spreading of the image to less than the original circle of confusion. If the optical system looks off the axis of the shock system, the situation is far more complex. The density changes lose their axial symmetry, and the intercepted section of the shock wave is no longer hemispherical but more nearly parabolic.

Table 24-3. Effect of Hemispherical Shock Wave on $F/1$, 15 cm Optical System

	M = 3.0 (rad)	M = 5.0 (rad)
Theoretical resolution limit	2.4×10^{-5}	2.4×10^{-5}
Angle subtended by image at 15 cm focus with shock	1.0×10^{-3}	1.4×10^{-3}
Angle subtended by image at optimum focus	1.5×10^{-5}	0.6×10^{-5}

24.3.3. Boundary-Layer Effects. The boundary layer can have any of several effects on the radiation passing through it. Steady-state effects can be similar to those found in the inviscid flow field, and the fluctuating nature of a turbulent boundary layer can have very serious optical consequences. Since the static pressure across a boundary layer is nearly constant, the variations in index of refraction are due entirely to the variation in static temperature. In the normal situation, n decreases from its free-stream value as the wall is approached. A minimum index of refraction would correspond to recovery temperature at the wall (Equation (24-2)), and this describes the "maximum effects" which can be expected (unless the wall is artificially heated by some means). It should be noted that boundary-layer thicknesses range from zero at the leading edge of a sharp-edged surface to several inches well back along a full-scale aircraft. The approximate boundary-layer thickness, δ, on a flat plate is given in [24-2] as

$$\delta_{\text{lam}} = 5 \sqrt{vx/u} \qquad (24\text{-}32)$$

$$\delta_{\text{turb}} = 0.37x \left(\frac{ux}{v}\right)^{-1/5} \qquad (24\text{-}33)$$

where u = velocity
v = kinematic viscosity
x = surface distance

It is seen that δ is proportional to $x^{1/2}$ in the laminar case and to $x^{4/5}$ in the turbulent case.

On the surface of a flat plate, two optical effects in the streamwise direction can be caused by this boundary-layer growth. First, the streamwise curvature of the boundary layer can cause a focusing effect, but some simple calculations show this to be negligible. At Mach 2 at sea level, for example, the focal length of the laminar boundary layer, only 30 cm from the leading edge, is greater than 10^{10} cm. The second effect is an angular deviation due to the locally nonparallel surfaces of the boundary layer (i.e., the surface at the wall and the outer edge). Under the conditions mentioned above, the prismlike deviation is about 1 μrad. Increased velocity and altitude both decrease the magnitude of this effect.

The curvature of a body and the resulting curvature of the boundary layer can cause another type of focusing to take place. Considering only curvature and refraction in a single plane, one can write the apparent distances of an image from the circularly curved interface between two media of differing index of refraction.

$$\frac{n_1}{S_1} + \frac{n_2}{S_2} = \frac{n_2 - n_1}{R_S} \tag{24-34}$$

where R_S = radius of curvature of the interface

S_1 and S_2 = the apparent distances as seen from within the two media

If the boundary layer is assumed to be thin compared to the radius of curvature, (i.e., the radius of curvature of the outer edge of the boundary is assumed equal to the inner radius), Equation (24-34) can be used to determine the refractive effects of the boundary layer. When Equation (24-34) is applied successively to thin layers of constant index of refraction, it is found that the indices of refraction at the wall and at the outer edge of the boundary layer will completely describe the refractive power of the boundary layer. By setting the original source distance at infinity, one makes the incoming rays parallel, and the resulting apparent source distance from the boundary layer is the effective focal length. Thus,

$$\frac{F}{R_S} = \frac{n_w}{(n_w - n_\infty)} \tag{24-35}$$

The index of refraction in the local free stream, n_∞, can be determined easily once flight conditions have been established; however, the index of refraction at the wall, n_w, cannot be precisely predicted unless some knowledge of the wall temperature is at hand. As an example, the maximum effect will be calculated (i.e., wall temperature equals recovery temperature). In general, the density of the boundary layer will be less at the wall than at the free-stream edge, and the minimum density possible is that associated with recovery temperature for a turbulent boundary layer. Recalling that the static pressure is constant across the boundary layer, one can use Equation (24-2) and the perfect gas law to find the density difference. With these results and the Gladstone-Dale law, one can rewrite Equation (24-35) as

$$\frac{F}{R_S} = -\left(\frac{1}{k_{GD}\rho_\infty}\right)(5r^{-1}M^{-2} + 1) - 5r^{-1}M^{-2} \tag{24-36}$$

From Equation (24-36) it can be seen that the focal length is negative, which means that the boundary layer acts as a concave lens and makes incoming parallel rays diverge. In addition, the larger the absolute value of the ratio F/R, the less the effect that the boundary layer has on optical performance. At low Mach numbers the boundary layer is not a very important optical element, and at Mach numbers above 5 or 6 the effect approaches a limit of

$$\frac{F}{R_S} = -\frac{1}{k_{GD}\rho_\infty} \tag{24-37}$$

Figure 24-21 shows the results of calculations based on Equation (24-36) as a plot of F/R_S versus local free-stream Mach number for a cylindrical body traveling with its axis at zero angle of attack and incident parallel rays perpendicular to the axis. The strength of the boundary layer as an optical element can be seen to be about the same as that of the hemispherical and conical shock waves previously discussed.

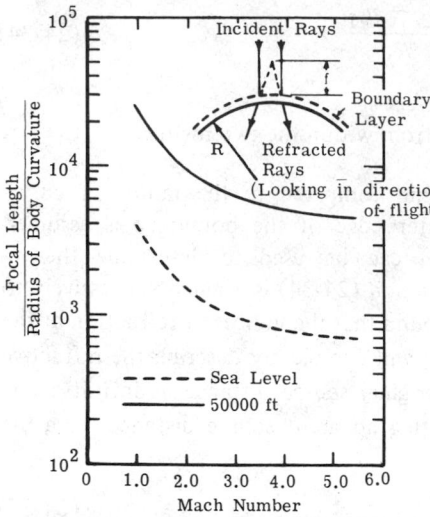

Fig. 24-21. Ratio of focal length to radius of boundary-layer curvature.

The example shown in Figure 24-21 is rather restricted. If the body under consideration is not cylindrical but is, for example, an ogive of circular cross-section, the increasing body radius causes the local flow properties to vary along the surface, and the variations in local free-stream Mach number and density may produce results somewhat different from those shown in Figure 24-21. In addition, the curvature in two directions in this case would create considerable astigmatism.

It should be emphasized that the foregoing discussion describes the behavior of the radiation just before it passes from the boundary layer into the protective window or optical device. The geometry and properties of the outermost optical element must be considered when the effects of the boundary layer are calculated for a particular configuration. For instance, if a thin protective window with the same radius of curvature as the boundary layer separates the boundary layer from an evacuated cavity, the effective focal length of the boundary layer-window combination is

$$\frac{F}{R_S} = -\frac{1}{k_{GD}\rho_\infty} \qquad (24\text{-}38)$$

regardless of the flight speed or window material.

24.3.4. Fluctuating-Boundary-Layer Effects. The turbulent boundary layer is characterized by random variations in velocity superimposed on the mean velocity distribution. Integrated aerodynamic effects of the turbulent boundary layer, such as heat transfer and friction drag, are fairly well known; but little is known about the structure of the turbulence itself, and no entirely satisfactory mathematical model is available.

Several optical effects are present because of the nonuniformities in the boundary layer. The perturbations of the index of refraction causes the incoming rays to be refracted to a varying extent throughout the field; the local disturbances not being constant with time. When an image is focused by an optical system, these disturbances give rise to

(1) Image dancing—random shifting in apparent object position.

(2) Scintillation—fluctuation in image intensity.

(3) Loss in resolution—initially parallel rays not coming to focus at the same point in the image plane.

Several attempts to determine the magnitude of these effects have been carried out [24-15, 24-16, 24-17, 24-18]. However, variations in experimental technique and measured quantities make it difficult to correlate the results. Some attempt to predict the loss in resolution for actual flight conditions was made [24-19] with the results noted below.

The scattering (loss in resolution) due to the turbulent boundary layer can be correlated on the basis of a parameter β', which is defined [24-18] as

$$\beta' = \frac{1}{\delta} \int_0^\delta \left[1 - \frac{\rho(y)}{\rho_\infty} \right] dy \qquad (24\text{-}39)$$

where δ = boundary-layer thickness

ρ_∞ = free-stream air density

$\rho(y)$ = local density as a function of distance from the wall (y)

To evaluate this integral, one must use some model of the boundary-layer profile. If one assumes that no heat is transferred to the wall, that the Prandtl number = 1, and that

$$\frac{u}{U} = \left(\frac{y}{\delta} \right)^{1/9}$$

the density ratio in air can be written

$$\frac{\rho}{\rho_\infty} = \left\{ 1 + 0.2 M_\infty^2 \left[1 - \left(\frac{y}{\delta} \right)^{2/9} \right] \right\}^{-1} \qquad (24\text{-}40)$$

In Figure 24-22, the Stine and Winovich parameter $\beta'/\rho_\infty \delta$ is shown. The boundary-layer thickness on a flat plate, δ, is approximately

$$\delta = 0.37 \frac{x}{(Re)^{1/5}} \qquad (24\text{-}41)$$

From the data of [24-16, 24-17, 24-18] and the analyses of [24-17] and [24-19], it appears that for the wind tunnel experiments, the empirical relation is

$$\theta_B \approx K B \rho_\infty \sqrt{\delta} \qquad (24\text{-}42)$$

where $K = 2.47$ rad ft$^{5/2}$ slug^{-1}

θ_B = resolution limit, i.e. the angular diameter of the aperture at the image plane through which 85% of the total energy from a point source (collimated beam) passes. [24-18]

Equation (24-42) is very approximate and should be considered only as a guide to the general magnitude of the image degradation. The difficulties in making wind tunnel measurements and the variations in technique among the various experimenters preclude an exact prediction of resolution loss. However, some valid general trends can be observed from Equation (24-41). First,

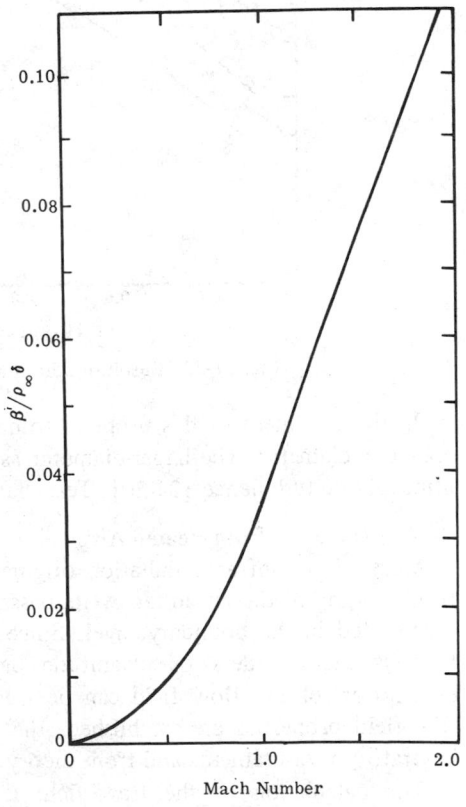

Fig. 24-22. Stine and Winovich scattering parameter versus Mach number for $u/U = (y/\delta)^{1/9}$.

the resolution limit is proportional to the free-stream density (and therefore altitude). Second, the resolution loss increases approximately as $x^{2.5}$, x being the distance from the aerodynamic leading edge of the vehicle. Figure 24-23 shows Equation (24-41) evaluated for several altitudes for a range of Mach numbers.

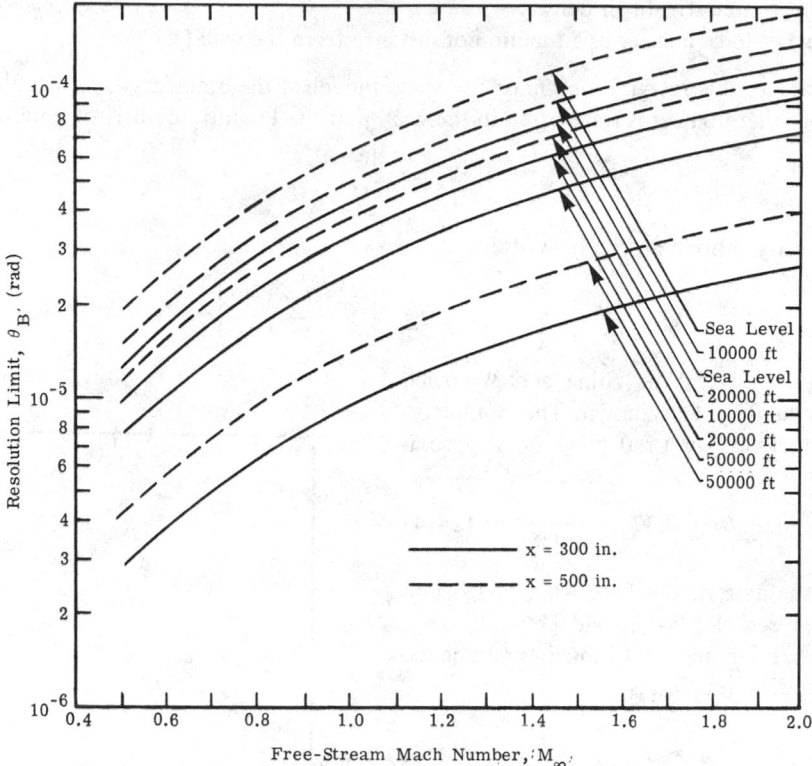

Fig. 24-23. Resolution limit as a function of M_∞ and body station.

In the treatment of this problem to date, no data have been used on the influence of aperture diameter. The larger-diameter astronomical telescopes are not as susceptible to atmospheric turbulence [24-20]. Tests have been made with apertures of 2 to 3 in.

24.4. Radiation from Heated Air

Most of the infrared radiation originating in the air surrounding the vehicle comes from heating of the air either by its passage through a strong shock wave or by its being accelerated in the boundary layer. Since the radiation energy associated with this flow field is many orders of magnitude below the kinetic and potential energies, the description of the flow field can be treated independent of the radiation. Once the flow-field properties are established, the radiation, in principle, can be estimated from laboratory measurements and from theory.

The calculation of the flow field is complex. Not only must the pressure and temperature be estimated at each point in the field from the vehicle out to the bow-shock wave, but also the nonequilibrium and chemical kinetic effects must be accounted for. These two considerations are vital if realistic estimates of air radiation are to be made, especially at high altitudes and high speeds. In particular, the molecular vibration-rotation relaxation times may, or may not, be long enough to provide a radiation-free area in the forward part of the flow field. The constituents of high-temperature air (in particular CO and NO which do not occur at ambient conditions) must be found to calculate a reasonable spectrum.

If the fluid properties are known, the radiation characteristics can be estimated from standard works on gaseous radiation such as References [24-21] and [24-22].

24.5. References and Bibliography

24.5.1. References

[24-1] C. C. Lin (ed.), *High Speed Aerodynamics and Jet Propulsion, Vol. V, Turbulent Flows and Heat Transfer,* Princeton University Press, Princeton, NJ, 1959.

[24-2] Herman Schlicting, *Boundary Layer Theory,* McGraw-Hill Book Company, New York, NY, 1968.

[24-3] P. J. Schneider, *Conduction Heat Transfer,* Addison-Wesley Publishing Company, Reading, MA, 1955.

[24-4] J. Fourier, *Analytical Theory of Heat,* Dover Publications, New York, NY, 1955.

[24-5] E. R. Van Driest, "Investigation of Laminar Boundary Layer in Compressible Fluids Using the Crocco Method," National Advisory Committee for Aeronautics (NACA), Washington, DC, Technical Note 2597, January 1952.

[24-6] M. J. Sibulkin, *Aeronautical Science,* Vol. 19, No. 570, 1952.

[24-7] National Advisory Committee for Aeronautics (NACA), Washington, DC, "Equations, Tables, and Charts for Compressible Flow," Technical Report 1135, 1947.

[24-8] H. A. Stine and K. Wanlass, "Theoretical and Experimental Investigation of Aerodynamic Heating and Isothermal Heat Transfer Parameters on a Hemispherical Nose with Laminar Boundary Layer at Supersonic Mach Numbers," National Advisory Committee for Aeronautics (NACA), Washington, DC, Technical Note 3344, 1957.

[24-9] I. E. Beckwith and J. S. Gallagher, "Heat Transfer and Recovery Temperatures on a Sphere with Laminar Transitional and Turbulent Boundary Layers at Mach Numbers of 2.00 and 4.15," National Advisory Committee for Aeronautics (NACA), Washington, DC, Technical Note 4125, 1957.

[24-10] K. A. Kieselev, "Computation of Radiance Emitted by a Nonuniformly Heated Flat Protective Window," *Soviet Journal of Optical Technology,* American Institute of Physics, New York, NY, Vol. 35, No. 4, July-August 1968, pp. 491-494.

[24-11] A. F. Charwat, C. F. Dewey, J. N. Roos, and J. A. Hitz, "An Investigation of Separated Flows, Part II, Flow in the Cavity and Heat Transfer," *Journal of the Aerospace Sciences,* Vol. 28, No. 7, July 1961.

[24-12] J. C. Slater and N. H. Frank, *Electromagnetism,* McGraw-Hill Book Company, New York, NY, 1969.

[24-13] S. L. Valley (ed.), *Handbook of Geophysics and Space Environments,* Air Force Cambridge Research Center, Cambridge, MA, 1965.

[24-14] H. A. Melkus, "The Behavior of a Light Ray Penetrating a Supersonic Flow Field," U.S. Air Force ARDC Report AFMDC-TR-59-39, October 1959.

[24-15] H. W. Liepmann, "Deflection and Diffusion of a Light Ray Passing Through a Boundary Layer," Douglas Aircraft Company, Santa Monica, CA, Report SM-14397, 1952.

[24-16] L. L. Baskins and L. E. Hamilton, "The Effect of Boundary Layer Thickness Upon the Optical Transmission Characteristics of a Supersonic Turbulent Boundary Layer," Northrop Aircraft Company, Beverly Hills, CA, Report NA 1-54-756, November 1954.

[24-17] C. J. Bartlett, "The Scattering of Light Rays in a Supersonic Turbulent Boundary Layer," M. S. Thesis, Massachusetts Institute of Technology, Department of Aeronautics and Astronautics, Cambridge, MA, June 1961.

[24-18] H. A. Stine and W. Winovich, "Light Diffusion Through High-Speed Turbulent Boundary Layers," National Advisory Committee for Aeronautics (NACA), Washington, DC, RM A56B21, May 1956.

[24-19] L. D. Lorah, Personal Files.

[24-20] A. H. Mikesell, *The Scintillation of Starlight*, U.S. Naval Observatory, Washington, DC, Second Series, Vol. XVII, Part IV, 1955.

[24-21] S. S. Penner, *Quantitative Molecular Spectroscopy and Gas Emissivities*, Addison-Wesley Publishing Company, Reading, MA, 1959.

[24-22] C. B. Ludwig, W. Malkus, J. E. Reardon, and J. A. L. Thomson, *Handbook of Infrared Radiation from Combustion Gases*, National Advisory Committee for Aeronautics (NACA), Washington, DC, SP-3080, 1973.

24.5.2. Bibliography

Bartlett, C. J., "The Scattering of Light Rays in a Supersonic Turbulent Boundary Layer," M. S. Thesis, Massachusetts Institute of Technology, Department of Aeronautics and Astronautics, Cambridge, MA, June 1961.

Baskins, L. L., and L. E. Hamilton, "The Effect of Boundary Layer Thickness Upon the Optical Transmission Characteristics of a Supersonic Turbulent Boundary Layer," Northrop Aircraft Company, Beverly Hills, CA, Report NA 1-54-756, November 1954.

Beckwith, I. E., and J. S. Gallagher, "Heat Transfer and Recovery Temperatures on a Sphere with Laminar Transitional and Turbulent Boundary Layers at Mach Numbers of 2.00 and 4.15," National Advisory Committee for Aeronautics (NACA), Washington, DC, Technical Note 4125, 1957.

Charwat, A. F., C. F. Dewey, J. N. Roos, and J. A. Hitz, "An Investigation of Separated Flows, Part II, Flow in the Cavity and Heat Transfer," *Journal of the Aerospace Sciences*, Vol. 28, No. 7, July 1961.

Fourier, J., *Analytical Theory of Heat*, Dover Publications, New York, NY, 1955.

Kieselev, K. A., "Computation of Radiance Emitted by a Nonuniformly Heated Flat Protective Window," *Soviet Journal of Optical Technology*, American Institute of Physics, New York, NY, Vol. 35, No. 4, July-August 1968, pp. 491-494.

Liepmann, H. W., "Deflection and Diffusion of a Light Ray Passing Through a Boundary Layer," Douglass Aircraft Company, Santa Monica, CA, Report SM-14397, 1952.

Lin, C. C. (ed.), *High Speed Aerodynamics and Jet Propulsion, Vol. V, Turbulent Flows and Heat Transfer*, Princeton University Press, Princeton, NJ, 1959.

Ludwig, C. B., W. Malkus, J. E. Reardon, and J. A. L. Thomson, *Handbook of Infrared Radiation from Combustion Gases*, National Advisory Committee for Aeronautics (NACA), Washington, DC, SP-3080, 1973.

Melkus, H. A., "The Behavior of a Light Ray Penetrating a Supersonic Flow Field," U.S. Air Force ARDC Report AFMDC-TR-59-39, October 1959.

Mikesell, A. H., *The Scintillation of Starlight*, U.S. Naval Observatory, Washington, DC, Second Series, Vol. XVII, Part IV, 1955.

National Advisory Committee for Aeronautics (NACA), Washington, DC, "Equations, Tables, and Charts for Compressible Flow," Technical Report 1135, 1947.

Penner, S. S., *Quantitative Molecular Spectroscopy and Gas Emissivities*, Addison-Wesley Publishing Company, Reading, MA, 1959.

Schlicting, Hermann, *Boundary Layer Theory*, McGraw-Hill Book Company, New York, NY, 1968.

Schneider, P. J., *Conduction Heat Transfer*, Addison-Wesley Publishing Company, Reading, MA, 1955.

Sibulkin, M. J., *Aeronautical Science*, Vol. 19, No. 570, 1952.

Slater, J. C., and N. H. Frank, *Electromagnetism*, Mc-Graw-Hill Book Company, New York, NY, 1969.

Stine, H. A., and K. Wanlass, "Theoretical and Experimental Investigation of Aerodynamic Heating and Isothermal Heat Transfer Parameters on a Hemispherical Nose with Laminar Boundary Layer at Supersonic Mach Numbers," National Advisory Committee for Aeronautics (NACA), Washington, DC, Technical Note 3344, 1957.

Stine, H. A., and W. Winovich, "Light Diffusion Through High-Speed Turbulent Boundary Layers," National Advisory Committee for Aeronautics (NACA), Washington, DC, RM A56B21, May 1956.

Valley, S. L. (ed.), *Handbook of Geophysics and Space Environments*, Air Force Cambridge Research Center, Cambridge, MA, 1965.

Van Driest, E. R., "Investigation of Laminar Boundary Layer in Compressible Fluids Using the Crocco Method," National Advisory Committee for Aeronautics (NACA), Washington, DC, Technical Note 2597, January 1952.

Chapter 25

PHYSICAL CONSTANTS AND
CONVERSION FACTORS

Jackson P. Livisay
Craig Mueller
Environmental Research Institute of Michigan

CONTENTS

TABLES AND FIGURES

25. Physical Constants and Conversion Factors

25.1. Introduction

The United States National Bureau of Standards (NBS) adopted the International System of Units (SI) in 1964 [25-1]. This system, which is proposed for international use by the International Bureau of Weights and Measures, was developed from the metric system. In 1972, the NBS published a translation of SI along with guidelines set forth by the International Committee on Weights and Measures [25-2]. The Task Group on Fundamental Constants of CODATA, the Committee on Data for Science and Technology, and the International Council of Scientific Unions (ICSU), recommended use of the International Physical Constants in 1973 [25-3].

25.1.1. Units and Physical Constants.
The defined values of the SI base units are given in Table 25-1. The adjusted values of physical constants are given in Table 25-2. Table 25-3 contains the angular definitions. Table 25-4 contains SI recommended prefixes. Some SI derived units are given in Table 25-5.

Table 25-1. SI Base Units [25-1]

Quantity	Unit	Symbol	Definition
Length	Meter	m	The length equal to 1,650,763.73 wavelengths in vacuo of the unperturbed transition $2p_{10} - 5d_5$ in ^{86}Kr
Mass	Kilogram	kg	The mass of the international kilogram at Sevres, France
Time	Second	sec	The duration of 9,192,631,770 periods of the radiation corresponding to the transition between two hyperfine levels of the ground state of the cesium-133 atom
Electric current	Ampere	A	The constant current which, if maintained in two, straight, parallel conductors of infinite length, with negligible circular cross-section, and placed 1 meter apart in vacuo, would produce between these conductors a force equal to 2×10^{-7} newton per meter of length
Thermodynamic temperature	Kelvin	K	The fraction 1/273.16 of the thermodynamic temperature of the triple point of water
Amount of substance	Mole	mol	The amount of substance containing the same number of atoms as 12 g of pure ^{12}C
Luminous intensity	Candela	cd	The luminous intensity, in the perpendicular direction of a surface of 1/600,000 square meter of a full radiator at the temperature of freezing platinum under a pressure of 101,325 newton per square meter

Table 25-2. (After [25-3])
Values of the 1973 International Physical Constants as
Recommended by CODATA-ICSU*

Quantity	Symbol	Value
Speed of light in vacuum	c	2.997924580(12)
Fine-structure constant, $[\mu_0 c^2/4\pi](e^2/\hbar c)$	α	7.2973506(60)
	α^{-1}	137.03604(11)
Electron charge	e	1.6021892(46)
		4.803242(14)
Planck constant	h	6.626176(36)
	$\hbar = h/2\pi$	1.0545887(57)
Avogadro constant	N	6.022045(31)
Atomic mass unit	u	1.6605655(86)
Electron rest mass	m_e	9.109534(47)
		5.110034 (14)†
$Nm_e = M_e$	M_e	5.4858026 (21)
Proton rest mass	m_p	1.6726485 (86)
		9.382796 (27)†
$Nm_p = M_p$	M_p	1.007276471(11)
Ratio of proton mass to electron mass	m_p/m_e	1836.15152(70)
Neutron rest mass	m_n	1.6749543 (86)
		9.395731 (27)†
$Nm_p = M_n$	M_n	1.008665012(37)
Electron charge to mass ratio	e/m_e	1.7588047(49)
		5.272764(15)
Magnetic flux quantum, $[c]^{-1}(hc/2e)$	Φ_0	2.0678506(54)
	h/e	4.135701(11)
		1.3795215(36)
Josephson frequency-voltage ratio	$2e/h$	4.835939(13)
Quantum of circulation	$h/2m_e$	3.6369455(60)
	h/m_e	7.273891(12)
Faraday constant	$\mathcal{F}$	9.648456(27)
		2.8925342(82)
Rydberg constant, $[\mu_0 c^2/4\pi]^2 (m_e e^4/4\pi\hbar^3 c)$	R_∞	1.097373177(83)
Bohr radius, $[\mu_0 c^2/4\pi]^{-1}(\hbar^2/m_e e^2) = \alpha/4\pi R_\infty$	a_0	5.2917706(44)
Classical electron radius, $[\mu_0 c^2/4\pi] (e^2/m_e c^2) = \alpha^3/4\pi R_\infty$	r_0	2.8179380(70)
Free electron g-factor, or electron magnetic moment in Bohr magnetons	$g_j/2 = \mu_e/\mu_B$	1.0011596567(35)
Free muon g-factor, or muon magnetic moment in units of $[c]$ $(e\hbar/2m_\mu c)$	$g_\mu/2$	1.00116616(31)
Bohr magneton, $[c]$ $(e\hbar/2m_e c)$	μ_B	9.274078(36)
Electron magnetic moment	μ_e	9.284832(36)
Gyromagnetic ratio of protons in H_2O	γ'_p	2.6751301(75)
	$\gamma'_p/2\pi$	4.257602(12)
γ'_p corrected for diamagnetism of H_2O	γ_p	2.6751987(75)
	$\gamma_p/2\pi$	4.257711(12)
Magnetic moment of protons in H_2O in Bohr magnetons	μ'_p/μ_B	1.52099322(10)

*Note that the numbers in parentheses are the one standard deviation uncertainties in the last digits of the quoted value computed on the basis of international consistency, that the unified atomic mass scale $^{12}C \equiv 12$ has been throughout, that u = atomic mass unit, C = coulomb, G = gauss, Hz = hertz = cycles/sec, J = joule, K = Kelvin (degrees Kelvin), T = tesla, $(10^4 G)$, V = volt, Wb = weber = $T \cdot m^2$, and W = watt. In the cases where formulas for constants are given (e.g., R_∞), the relations are written as the product of two factors. The second factor, in parentheses, is the expression to be used

Table 25-2. (After [25-3])
Values of the 1973 International Physical Constants as Recommended by CODATA-ICSU*((Continued))

Uncertainty (ppm)	Units		
	SI	cgs	MeV
0.004	10^8 m sec^{-1}	10^{10} cm sec^{-1}	–
0.82	10^{-3}	10^{-3}	–
0.82	–	–	–
2.9	10^{-19}C	10^{-20} emu	–
2.9	–	10^{-10} esu	–
5.4	10^{-34} J sec	10^{-27} erg sec	–
5.4	10^{-34} J sec	10^{-27} erg sec	–
5.1	10^{26} kmol^{-1}	10^{23} mol^{-1}	–
5.1	10^{-27} kg	10^{-24} g	–
5.1	10^{-31} kg	10^{-28} g	–
2.8†	–	–	10^{-1}
0.38	10^{-4} u	10^{-4} u	–
5.1	10^{-27} kg	10^{-24} g	–
2.8†	–	–	10^2
0.001	u	u	–
0.38	–	–	–
5.1	10^{-27} kg	10^{-24} g	–
2.8†	–	–	10^2
0.037	u	u	–
2.8	10^{11} C kg^{-1}	10^7 emu g^{-1}	–
2.8	–	10^{17} esu g$^{ l}$	
2.6	10^{-15} Wb	10^{-7} G cm^2	–
2.6	10^{-15} J sec C^{-1}	10^{-7} erg sec emu^{-1}	–
2.6	–	10^{-17} erg sec esu^{-1}	–
2.6	10^{14} Hz V^{-1}	–	–
1.6	10^{-4} J sec kg^{-1}	erg sec g^{-1}	–
1.6	10^{-4} J sec kg^{-1}	erg sec g^{-1}	–
2.8	10^7 C kmol^{-1}	10^4 C mol^{-1}	–
2.8	–	10^{14} esu mol^{-1}	–
0.075	10^7 m^{-1}	10^5 cm^{-1}	–
0.82	10^{-11} m	10^{-9} cm	–
2.5	10^{-15} m	10^{-13} cm	–
0.0035	–	–	–
0.31	–	–	–
3.9	10^{-24} J T^{-1}	10^{-21} erg G^{-1}	–
3.9	10^{-24} J T^{-1}	10^{-21} erg G^{-1}	–
2.8	10^8 rad sec^{-1} T^{-1}	10^4 rad sec^{-1} G^{-1}	–
2.8	10^7 Hz T^{-1}	10^3 Hz G^{-1}	–
2.8	10^8 rad sec^{-1} T^{-1}	10^4 rad sec^{-1} G^{-1}	–
2.8	10^7 Hz T^{-1}	10^3 Hz G^{-1}	–
0.066	10^{-3}	10^{-3}	–

when all quantities are expressed in cgs units, with the electron charge in electrostatic units. The first factor, in brackets, is to be included only if all quantities are expressed in SI units. We remind the reader that with the exception of the auxiliary constants which have been taken to be exact, the uncertainties of these constants are correlated, and therefore the general law of error propagation must be used in calculating additional quantities requiring two or more of these constants.

†Revised from the original table by one of its authors: Barry N. Taylor, National Bureau of Standards, Washington, DC.

Table 25-2. (After [25-3])
Values of the 1973 International Physical Constants as
Recommended by CODATA-ICSU* (*Continued*)

Quantity	Symbol	Value
Proton magnetic moment in Bohr magnetons	μ_p/μ_B	1.521032209(16)
Ratio of electron and proton magnetic moments	μ_e/μ_p	658.2106880(66)
Proton magnetic moment	μ_p	1.4106171(55)
Magnetic moment of protons in H_2O in nuclear magnetons	μ'_p/μ_N	2.7927740(11)
μ_p/μ_N corrected for diamagnetism of H_2O	μ_p/μ_N	2.7928456(11)
Nuclear magneton, [c] ($e\hbar/2m_pc$)	μ_N	5.050824(20)
Ratio of muon and proton magnetic moments	μ_μ/μ_p	3.1833402(72)
Muon magnetic moment	μ_μ	4.490474(18)
Ratio of muon mass to electron mass	m_μ/m_e	206.76865(47)
Muon rest mass	m_μ	1.883566(11)
	M_μ	0.11342920(26)
Compton wavelength of the electron, h/m_ec	λ_C	2.4263089(40)
	$\lambda_C/2\pi$	3.8615905(64)
Compton wavelength of the proton, h/m_pc	$\lambda_{C,p}$	1.3214099(22)
	$\lambda_{C,p}/2\pi$	2.1030892(36)
Compton wavelength of the neutron, h/m_nc	$\lambda_{C,n}$	1.3195909(22)
	$\lambda_{C,n}/2\pi$	2.1001941(35)
Standard volume of ideal gas	V_0	22.71081(71)
		22.41383(70)
Gas constant, V_0/T_0 (T_0 = 273.15 K)	R	8.31441(26)
		8.20568(26)
Boltzmann constant, R/N	k	1.380662(44)
Stefan-Boltzmann constant, $\pi^2k^4/60\hbar^3c^2$	σ	5.67032(71)
First radiation constant, $2\pi hc^2$	c_1	3.741832(20)
Second radiation constant, hc/k	c_2	1.438786(45)
Gravitational constant	G	6.6720(41)
kT for T = 300 K	kT	4.14199 (13)[†]
Energy-wave number conversion	hc	1.986478 (11)[†]
		1.2398520 (32)[†]
Energy-mass conversion	kg c^2	8.987551786 (72)[†]
		5.609545 (16)[†]
Energy-frequency conversion	1 eV h^{-1}[†]	2.4179696 (63)[†]

Table 25-2. (After [25-3])
Values of the 1973 International Physical Constants as Recommended by CODATA-ICSU* (Continued)

Uncertainty (ppm)	Units		
	SI	cgs	MeV
0.011	10^{-3}	10^{-3}	—
0.010	—	—	—
3.9	10^{-26} J T^{-1}	10^{-23} erg G^{-1}	—
0.38	—	—	—
0.38	—	—	—
3.9	10^{-27} J T^{-1}	10^{-24} erg G^{-1}	—
2.3	—	—	—
3.9	10^{-26} J T^{-1}	10^{-23} erg G^{-1}	—
2.3	—	—	—
5.6	10^{-28} kg	10^{-25} g	—
2.3	u	u	—
1.6	10^{-12} m	10^{-10} cm	—
1.6	10^{-13} m	10^{-11} cm	—
1.7	10^{-15} m	10^{-13} cm	—
1.7	10^{-16} m	10^{-14} cm	—
1.7	10^{-15} m	10^{-13} cm	—
1.7	10^{-16} m	10^{-16} cm	—
31	10^5 J kmol^{-1}	10^9 erg mol^{-1}	—
31	m^3 atm kmol^{-1}	10^3 cm^3 atm mol^{-1}	—
31	10^3 J kmol^{-1} K^{-1}	10^7 erg mol^{-1} K^{-1}	—
31	10^{-2} m^3 atm kmol^{-1} K^{-1}	10 cm^3 atm mol^{-1} K^{-1}	—
32	10^{-23} J K^{-1}	10^{-16} erg K^{-1}	—
125	10^{-8} W m^{-2} K^{-4}	10^{-5} erg sec^{-1} cm^{-2} K^{-4}	—
5.4	10^{-16} W m^2	10^{-5} erg cm^2 sec^{-1}	—
31	10^{-2} m K	cm K	—
615	10^{-11} N m^2 kg^{-2}	10^{-8} dyn cm^2 g^{-2}	—
32	10^{-21} J	10^{-14} erg	—
5.4	10^{-25} J m	10^{-16} erg cm	—
2.6[†]	—	10^{-4} eV cm	—
0.008	10^{16} J	10^{23} erg[†]	—
2.9[†]	—	—	10^{29}
2.6[†]	—	—	10^{14} Hz[†]

[†] Revised from the original table by one of its authors: **Barry** N. Taylor, National Bureau of Standards, Washington, DC.

Table 25-3. Angular Definitions [25-1]

Term	Symbol	Definition
Radian	rad	The plane angle subtended by an arc of a circle which is equal to the radius of the circle.
Steradian	sr	The solid angle subtended by a spherical surface area equal to the square of the radius of the sphere.

Table 25-4. SI Prefixes
[25-1, 25-2]

Factor	Prefix	Symbol
10^{18}	Exa	E
10^{15}	Peta	P
10^{12}	Tera	T
10^{9}	Giga	G
10^{6}	Mega	M
10^{3}	Kilo	k
10^{2}	Hecto	h
10	Deka	da
10^{-1}	Deci	d
10^{-2}	Centi	c
10^{-3}	Milli	m
10^{-6}	Micro	μ
10^{-9}	Nano	n
10^{-12}	Pico	p
10^{-15}	Femto	f
10^{-18}	Atto	a

Table 25-5. SI Derived Units [25-1]

Quantity	Unit	Symbol
Acceleration	Meter per second squared	$m\ sec^{-2}$
Activity (of a radioactive source)	1 per second	sec^{-1}
Angular acceleration	Radian per second squared	$rad\ sec^{-2}$
Angular velocity	Radian per second	$rad\ sec^{-1}$
Area	Square meter	m^2
Capacitance	Farad	F
Dynamic viscosity	Newton-second per square meter	$N\ sec\ m^{-2}$
Electric field strength	Volt per meter	$V\ m^{-1}$
Electric resistance	Ohm	Ω
Energy	Joule	J
Entropy	Joule per Kelvin	$J\ K^{-1}$
Force	Newton	N
Frequency	Hertz	Hz
Illuminance	Lux	lx
Inductance	Henry	H
Kinematic viscosity	Square meter per second	$m^2\ sec^{-1}$
Luminance	Candela per square meter	$cd\ m^{-2}$
Luminous flux	Lumen	lm
Magnetic field strength	Ampere per meter	$A\ m^{-1}$
Magnetic flux	Weber	Wb
Magnetic flux density	Tesla	T
Magnetomotive force	Ampere	A
Mass density (density)	Kilogram per cubic meter	$kg\ m^{-3}$
Potential difference, electro-motive force	Volt	V
Power	Watt	W
Pressure (mechanical stress)	Pascal	Pa
Quantity of electricity	Coulomb	C
Radiant intensity	Watt per steradian	$W\ sr^{-1}$
Specific heat capacity	Joule per kilogram Kelvin	$J\ kg^{-1}\ K^{-1}$
Speed, velocity	Meter per second	$m\ sec^{-1}$
Thermal conductivity	Watt per meter Kelvin	$W\ m^{-1}\ K^{-1}$
Volume	Cubic meter	m^3
Wave number	1 per meter	m^{-1}
Work, energy, quantity of heat	Joule	J

25.2. Conversion Factors

25.2.1. Length. The meter, m, defined in Table 25-1, is the base unit of length. Conversions from meters to other units are given in Table 25-6.

25.2.2. Area and Volume. These quantities are both derived from length. The base units are the square meter, m^2, and the cubic meter, m^3, respectively. Table 25-7 contains area conversions and Table 25-8 contains volume conversions.

Table 25-6. Length Conversions

Unit	Symbol (or Abbreviation)	Relationship		Conversion to Meter (m)	Reference
Fermi	—	—		10^{-15}	—
Angstrom	Å	—		10^{-10}	[25-3]
Micron	μm	10^4 Å	= 1 μm	10^{-6}	[25-3]
Inch	in.	2.54 cm	= 1 in.	0.0254*	[25-3]
Foot	ft	12 in.	= 1 ft	0.3048*	[25-3]
Yard	yd	3 ft (36 in.)	= 1 yd	0.9144*	[25-3]
Fathom	fath	2 yd (6 ft.)	= 1 fath	1.8288*	[25-3]
Link	—	7.92 in.	= 1 link	0.201168*	[25-1]
Rod	—	25 link (5.5 yd)	= 1 rod	5.0292	[25-1]
Chain	—	4 rod (22 yd)	= 1 chain	20.1168*	[25-1]
Furlong	fur	10 chain (220 yd)	= 1 fur	201.168*	[25-1]
Statute mile	mi	8 fur (1760 yd)	= 1 mi	1609.344	[25-1]
Nautical mile	nmi	1.15078 mi	= 1 nmi	1852.0*	[25-1]
Astronomical unit	AU	8.07775×10^7 nmi	= 1 AU	1.49600×10^{11}	[25-1]
Light year	light yr	6.3239×10^4 AU	= 1 light yr	9.46055×10^{15}	[25-1]
Parsec	pc	3.2562 light yr	= 1 pc	3.0857×10^{16}	[25-1]

*Defined value (1 in. = 2.54 cm exactly).

Table 25-7. Area Conversions

Unit	Symbol	Relationship		Conversion to Square Meter $(m^2)^*$	Reference
Barn	—	—		1.0×10^{-28}	[25-1]
Are	—	—		1.0×10^2	[25-1]
Hectare	—	—		1.0×10^4	[25-1]
Square inch	in.^{2}	—		6.4516×10^{-4}	[25-3]
Square foot	ft^2	144 in.^{2}	= 1 ft^2	9.290304×10^{-2}	[25-3]
Square yard	yd^2	9 ft^2	= 1 yd^2	8.3612736×10^{-1}	[25-3]
Acre	—	43,560 ft^2	= 1 acre	4.0468564224×10^3	[25-1]
Square statute mile	mi^2	640 acre	= 1 mi^2	$2.589988110336 \times 10^6$	[25-1]
Section	—	1 mi^2	= 1 section	$2.589988110336 \times 10^6$	[25-1]
Township	—	36 section	= 1 township	9.3239572×10^7	[25-1]

*Defined values.

Table 25-8. Volume Conversions

Unit	Symbol	Relationship		Conversion to Cubic Meter (m³)	Reference
Cubic centimeter	cm^3	–		10^{-6}	[25-3]
Liter	l	10^3 cm^3	= 1 l	10^{-3}	[25-3]
Cubic inch	in.^{3}	1.6387064×10 cm^3 = 1 in.^{3}		$1.6387064 \times 10^{-5*}$	[25-3]
Fluid ounce	fl oz	1.80469 in.^{3}	= 1 fl oz	2.9573530×10^{-5}	[25-3]
Pint (liquid)	pt	16 fl oz	= 1 pt	$4.73176473 \times 10^{-4*}$	[25-3]
Quart (liquid)	qt	2 pt	= 1 qt	$9.46352946 \times 10^{-4*}$	[25-3]
Gallon (liquid)	gal	4 qt	= 1 gal	$3.785411784 \times 10^{-3*}$	[25-3]
Cubic foot	ft^3	7.481 gal	= 1 ft^3	$2.8316846592 \times 10^{-2*}$	[25-3]
Cubic yard	yd^3	27 ft^3	= 1 yd^3	$7.6455485844 \times 10^{-1*}$	[25-3]
Acre foot	–	–		1.2334818×10^3	[25-1]

*Defined values.

25.2.3. Angle. The radian, rad, is the basic unit of a plane angle. Conversions are given in Table 25-9. The steradian, sr, is the basic unit of a solid angle.

Table 25-9. Angle Conversions [25-1]

Unit	Symbol	Relationship		Conversion to Radian (rad)
Milliradian	mrad	–		10^{-3}
Second	sec	–		$4.848136811 \times 10^{-6}$
Minute	min	60 sec	= 1 min	$2.908882087 \times 10^{-4}$
Degree	° or deg	60 min	= 1 deg	$1.745329252 \times 10^{-2}$
Quadrant	–	90 deg	= 1 quadrant	1.570796327
Centesimal second	centesimal sec	10^{-6} quadrant = 1 centesimal sec		$1.570796327 \times 10^{-6}$
Centesimal minute	centesimal min	10^{-4} quadrant = 1 centesimal min		$1.570796327 \times 10^{-4}$
Grad	–	10^{-2} quadrant = 1 grad		$1.570796327 \times 10^{-2}$
Circum-ference	–	4 quadrant = 2π rad = 1 circumference		6.283185308
Mil (military)	–	1/6400 = 1 mil circumference		$9.817477044 \times 10^{-4}$

25.2.4. Mass. The base unit of mass is the kilogram, kg. Conversions to other units are given in Table 25-10.

25.2.5. Density. Density is defined as mass per unit volume; the SI unit is kilogram per cubic meter, kg m^{-3}. Table 25-11 gives conversions to other units.

25.2.6. Time. The base unit of time is the second, sec. Conversions to other units are given in Table 25-12.

Table 25-10. Mass Conversions, Avoirdupois, Apothecaries, Troy

Unit	Symbol	Mass Conversion	Conversion to Kilograms (kg)	Reference
Tonne (i.e., metric ton)	t	–	10^3	[25-3]
Slug	–	–	1.45939029×10^1	[25-1]
Avoirdupois Grain	–	–	6.479891×10^{-5}*	[25-1]
Dram	dr	27.34375 grain = 1 dr	$1.771845195 \times 10^{-3}$	[25-3]
Ounce	oz	16 dr = 1 oz	$2.834952313 \times 10^{-2}$	[25-3]
Pound	lb	16 oz = 1 lb	4.5359237×10^{-1}*	[25-3]
Ton (short)	tn	2000 lb = 1 tn	9.0718474×10^2	[25-3]
Apothecaries Grain (see Avoirdupois)	–	–	–	–
Scruple	s ap	20 grain = 1 s ap	1.2959782×10^{-3}*	[25-1]
Dram	dr	3 s ap = 1 dr	3.8879346×10^{-3}	[25-1]
Ounce	oz	8 dr = 1 oz	$3.11034768 \times 10^{-2}$	[25-1]
Pound	lb	12 oz = 1 lb	$3.732317216 \times 10^{-1}$*	[25-1]
Troy Grain (see Avoirdupois)	–	–	–	–
Pennyweight	dwt	24 grain = 1 dwt	$1.555174384 \times 10^{-3}$*	[25-1]
Ounce	oz	20 dwt = 1 oz	$3.11034768 \times 10^{-2}$	[25-1]
Pound (see Apothecaries)	–	–	–	–

*Defined value.

Table 25-11. Density Conversions

Unit	Symbol	Conversion to Kilogram/Cubic Meter (kg m^{-3})	Reference
Gram/liter	g l^{-1}	1	[25-3]
Gram/cubic centimeter	g cm^{-3}	10^3	[25-3]
Gram/milliliter	g ml^{-1}	10^3	–
Pound mass/cubic inch	lbm in.$^{-3}$	2.7679905×10^4	[25-3]
Pound mass/cubic foot	lbm ft^{-3}	1.6018463×10^1	[25-3]
Slug/cubic foot	slug ft^{-3}	5.15379×10^2	[25-1]

Table 25-12. Time Conversion

Unit	Symbol	Relationship	Conversion to Mean Solar Second* (sec)	Reference
Second (sidereal)	sec	—	9.9726957×10^{-1}	[25-3]
Minute (sidereal)	min	60 sec (sidereal) = 1 min (sidereal)	5.9836174×10	[25-3]
Hour (sidereal)	h	60 min (sidereal) = 1 h (sidereal)	3.5901704×10^3	[25-3]
Day (sidereal)	d	24 h (sidereal) = 1 d (sidereal)	8.6164090×10^4	[25-3]
Year (sidereal)	yr	366.2564 d (sidereal) = 1 yr (sidereal)	3.1558150×10^7	[25-1]
Minute (mean solar)	—	60 sec (mean solar) = 1 min (mean solar)	6.0×10	[25-1]
Hour (mean solar)	—	60 min (mean solar) = 1 h (mean solar)	3.60×10^3	[25-1]
Day (mean solar)	—	24 h (mean solar) = 1 d (mean solar)	8.64×10^4	[25-1]
Month (mean calendar)	—	30.41667 d (mean solar) = 1 month (mean calendar)	2.628×10^6	[25-1]
Year (calendar)	—	365 d (mean solar) = 1 yr (calendar)	3.1536×10^7	[25-1]
Year (leap)		366 d (mean solar) = 1 yr (leap)	3.16224×10^7	[25-4]
Year (tropical)	—	365.24219 d (mean solar) = 1 yr (tropical)	3.1556926×10^7	[25-1]
Second		—	Consult *1977 American Ephemeris and Nautical Almanac*	[25-4]

*The unit of time (mean solar second) is based on the transition between two hyperfine levels of the ground state of the Cesium – 133 atom, with the value 9,192,631,770 cycles (Hz) as one second [25-3].

25.2.7. Velocity. Linear velocity is a vector quantity representing the time-rate of motion in a fixed direction. Angular velocity is a vector denoting the time-rate of angular motion about an axis; its units are radians per second, $\mathrm{rad\,sec}^{-1}$. Table 25-13 gives conversions. Mach number, a common velocity unit, is defined as the ratio of the given speed to that of sound at the same pressure and temperature.

25.2.8. Acceleration. The time-rate of change of the velocity vector in magnitude or direction is the definition of acceleration. The SI unit is meter per second squared, $\mathrm{m\,sec}^{-2}$. Table 25-14 gives conversions.

25.2.9. Force, Torque, Pressure, Work, Energy and Power. Force is a vector quantity defined as mass times acceleration ($\mathbf{f} = \mathbf{ma}$) or as the time-rate of change of momentum

Table 25-13. Speed Conversions

Unit	Symbol	Relationship	Conversion to Meter/Second (m sec^{-1})	Reference
Centimeter/second	cm sec^{-1}	–	10^{-2}	[25-3]
Kilometer/second	km sec^{-1}	–	10^3	[25-3]
Kilometer/hour	km h^{-1}	–	$2.7777777778 \times 10^{-1}$	[25-3]
Inch/second	in. sec^{-1}	–	2.54×10^{-2}*	[25-3]
Foot/second	ft sec^{-1}	12 in. sec^{-1} = 1 ft sec^{-1}	3.048×10^{-1}*	[25-3]
Statute mile/second	mi sec^{-1}	5280 ft sec^{-1} = 1 mi sec^{-1}	1.609344×10^3*	[25-1]
Statute mile/minute	mi min^{-1}	88 ft sec^{-1} = 1 mi min^{-1}	$2.68224 \times 10^*$	[25-1]
Statute mile/hour	mph	1.4666667 ft sec^{-1} = 1 mph	4.4704×10^{-1}*	[25-1]
Nautical mile/second	nmi sec^{-1}	6076.1033 ft sec^{-1} = 1 nmi sec^{-1}	1.852×10^3	[25-1]
Nautical mile/hour or knot (international)	nmi h^{-1}	1.6878099 ft sec^{-1} = 1 knot (international)	5.1444444×10^{-1}	[25-1]

*Defined value.

Table 25-14. Acceleration Conversions

Unit	Symbol	Conversion to Meter/Second Squared (m sec^{-2})	Reference
Meter/minute squared	m min^{-2}	2.777778×10^{-4}	[25-3]
Inch/second squared	in. sec^{-2}	2.54×10^{-2}*	[25-3]
Inch/minute squared	in. min^{-2}	7.0555556×10^{-6}*	[25-3]
Feet/second squared	ft sec^{-2}	3.048×10^{-1}*	[25-3]
Feet/minute squared	ft min^{-2}	8.4666667×10^{-5}	[25-3]
Statute mile/second squared	mi sec^{-2}	1.609344×10^3	[25-1]
Statute mile/minute squared	mi min^{-2}	4.47040×10^{-1}	[25-1]
Statute mile/hour squared	mi h^{-2}	1.241778×10^{-4}	[25-1]
Gal (Galileo)	–	10^{-2}*	[25-1]
Free fall standard	–	9.80665*	[25-1]

*Defined value.

($\mathbf{f} = d\mathbf{p}/dt$), where $\mathbf{p}$ is momentum. This allows for a change in mass. The SI unit of force is the newton, N, or kilogram-meter per second squared. Table 25-15 gives force conversions.

The torque about an origin of a force acting at a point is the product of the distance along a line between the point and the origin and the component of the force perpendicular to that line. The SI unit of the vector quantity torque is the newton-meter, N m. Table 25-16 gives torque conversions.

Table 25-15. Force Conversions [25-1]

Unit	Symbol	Conversion to Newton (N)
Dyne	–	10^{-5}
Kilogram force	kgf	9.80665*
Kip	–	4.448221615×10^3
Ounce force (avoirdupois)	ozf	2.7801385×10^{-1}
Pound force (avoirdupois)	lbf	4.448221615
Poundal	–	$1.382549543 \times 10^{-1}$

*Defined value.

Table 25-16. Torque Conversions [25-1]

Unit	Symbol	Conversion to Newton-Meter (N m)	Reference
Dyne-centimeter	dyne cm	1.0×10^{-7}	[25-1, 25-3]
Kilogram force-meter	kgf m	9.806650	[25-1]
Ounce force-inch	ozf in.	7.061552×10^{-3}	[25-1]
Pound force-inch	lbf in.	1.129848×10^{-1}	[25-1]
Pound force-foot	lbf ft	1.355818	[25-1]

Pressure is force per unit area normal to the surface. The SI unit of pressure is the torr which is the pressure of 1 mm Hg at $0°C$ and standard gravity. Pressure conversion factors are given in Table 25-17.

Work is usually thought of as force acting through a distance. Energy, classically, is of two types: potential and kinetic. Potential energy is the capability of doing work. Kinetic energy is the result of a mass in motion. Both work and energy have the joule, J, (newton-meter) as their base unit. Table 25-18 gives conversions for work and energy units.

Power is the time-rate at which work is done; the SI unit is watt, W. Table 25-19 gives power conversions.

Table 25-17. Pressure Conversions

Unit	Symbol	Conversion to Pascal (Pa) or Newton/Square-Meter ($N\ m^{-2}$)	Reference
Barye	–	10^{-1}	[25-1]
Millibar	mbar	10^2	[25-3]
Bar	–	10^5	[25-3]
Dyne/square centimeter	dyne cm^{-2}	10^{-1}	[25-3]
Atmosphere	atm	1.01325×10^5	[25-3]
Kilogram force/square meter	kgf m^{-2}	9.80665	[25-1]
Kilogram force/square centimeter	kgf cm^{-2}	9.80665×10^4	[25-1]
Pound force/square foot	lbf ft^{-2}	4.7880258×10	[25-1]
Pound force/square inch	psi, lbf in.$^{-2}$	6.8947572×10^3	[25-1]
Centimeter of water (4°C)	–	9.80638×10	[25-1]
Inch of water (60°F)	–	2.4884×10^2	[25-1]
Inch of water (6°C)	–	2.49082×10^2	[25-1]
Foot of water (4°C)	–	2.98898×10^3	[25-1]
Millimeter of mercury (0°C)	mm Hg	1.333224×10^2	[25-1]
Inch of mercury (60°F)	–	3.37685×10^3	[25-1]
Inch of mercury (0°C)	–	3.386389×10^3	[25-1]
Torr	–	1.333223×10^2	[25-3]

Table 25-18. Work and Energy Conversions

Unit	Symbol	Conversion to Joule (J)	Reference
Foot-pound force	ft lbf	1.3558179	[25-1]
Foot-poundal	–	4.2140110×10^{-2}	[25-1]
Ton (nuclear equivalent of TNT)	tn	4.20×10^9	[25-1]
British thermal unit (IST current)	Btu	1.055056×10^3*	[25-3]
Btu, IST before 1956	–	1.05504×10^3	[25-1]
Btu, mean	–	1.05587×10^3	[25-1]
Btu, thermochemical	–	1.054350×10^3	[25-1]
Btu ($39°$F)	–	1.05967×10^3	[25-1]
Btu ($60°$F)	–	1.05468×10^3	[25-1]
Calorie	cal	4.1868*	[25-3]
Calorie, mean	–	4.19002	[25-1]
Calorie, thermochemical	–	4.184**	[25-3]
Calorie ($15°$C)	–	4.18580	[25-1]
Calorie ($20°$C)	–	4.18190	[25-1]
Calorie kilogram	–	4.1868×10^3*	[25-3]
Calorie kilogram, mean	–	4.19002×10^3	[25-1]
Calorie kilogram, thermochemical	–	4.184×10^3**	[25-1]
Kilocalorie	kcal	4.1868×10^3*	[25-3]
Kilocalorie, mean	–	4.19002×10^3	[25-1]
Kilocalorie, thermochemical	–	4.184×10^3**	[25-3]
Electron volt	eV	$1.6021917 \times 10^{-19}$	[25-1]
Erg	–	1.00×10^{-7}**	[25-3]
Joule (International of 1948)	J	1.000165	[25-1]
Watt hour	W h	3.60×10^3**	[25-3]
Kilowatt hour	kW h	3.60×10^6**	[25-3]
Kilowatt hour (International of 1948)	–	3.60059×10^6	[25-1]

*International steam table.
**Defined value.

Table 25-19. Power Conversions

Unit	Symbol	Conversion to Watt (W)	Reference
Foot pound force/hour	ft lbf h^{-1}	3.7661610×10^{-4}	[25-1]
Foot pound force/minute	ft lbf min^{-1}	2.2596966×10^{-2}	[25-1]
Foot pound force/second	ft lbf sec^{-1}	1.3558179	[25-1]
Horsepower*	hp	7.4569987×10^2	[25-1]
Btu (thermochemical)/second	–	1.054350×10^3	[25-1]
Btu (thermochemical)/minute	–	1.757250×10	[25-1]
Calorie (thermochemical)/second	cal sec^{-1}	4.184**	[25-3]
Calorie (thermochemical)/minute	cal min^{-1}	6.9733333×10^{-2}	[25-3]
Kilocalorie (thermochemical)/minute	kcal min^{-1}	6.9733333×10	[25-3]
Kilocalorie (thermochemical)/second	kcal sec^{-1}	4.184×10^3**	[25-3]
Watt (International of 1948)	W	1.000165	[25-1]

*1 hp = 550 ft lbf sec^{-1}.
**Defined value.

25.2.10. Temperature. Temperature is a measure of the average, translational kinetic energy of the molecules of a substance; it has SI units of Kelvin, K. Figures 25-1 and 25-2 give conversions between common temperature scales.

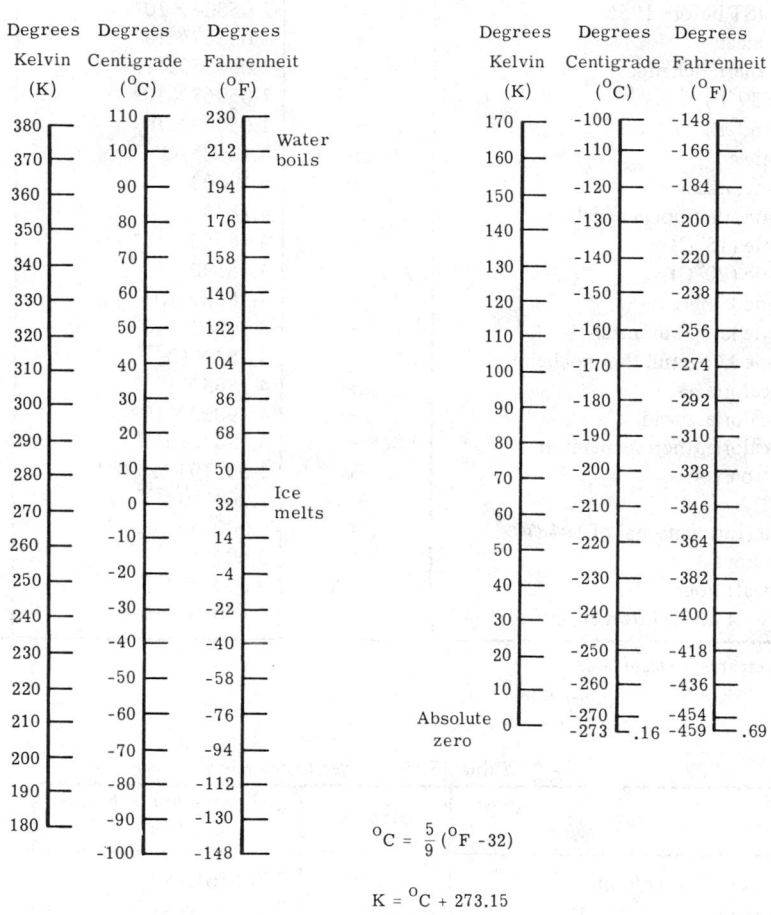

$$^{\circ}C = \frac{5}{9}(^{\circ}F - 32)$$

$$K = {}^{\circ}C + 273.15$$

Fig. 25-1. Temperature conversions for the Kelvin, Celsius, and Fahrenheit temperature scales.

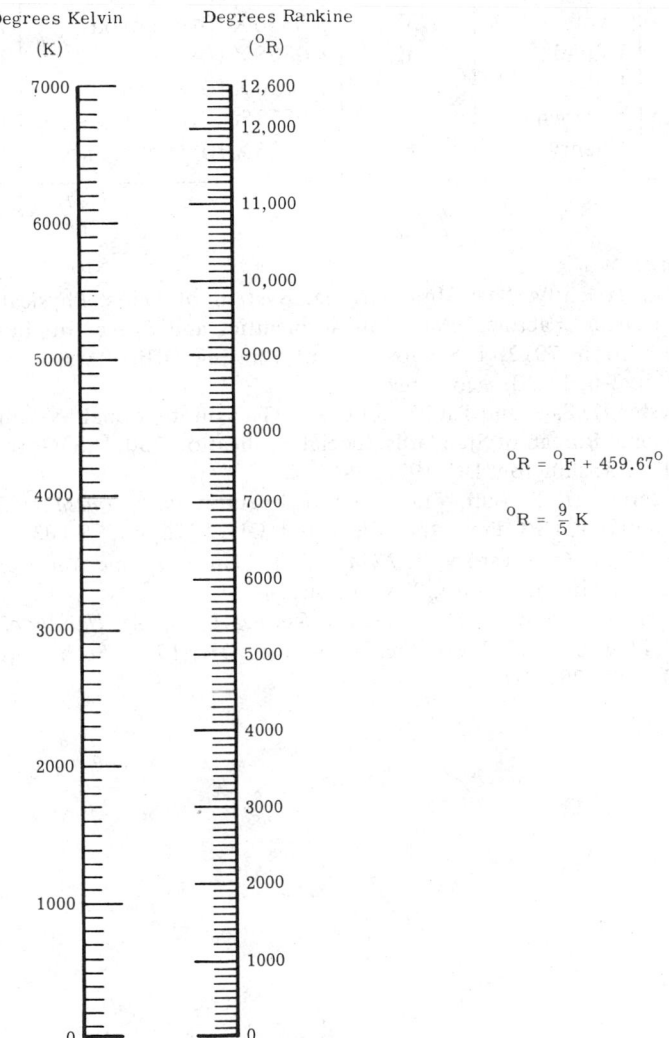

Fig. 25-2. Temperature conversions for the Kelvin and Rankine temperature scales.

25.2.11. Electrical Units. The ampere, A, defined in Table 25-1, is the standard of electric current. Table 25-20 gives conversions for electrical units.

Table 25-20. Electrical Unit Conversions

Quantity	System International	Symbol	Electrostatic	Electromagnetic
Charge	1 coulomb	C	2.9979×10^9 statcoulomb	10^{-1} abcoulomb
Current	1 ampere	A	2.9979×10^9 statampere	10^{-1} abampere
Potential	1 volt	V	3.3356×10^{-3} statvolt	10^8 abvolt
Capacity	1 farad	F	8.9878×10^{11} statfarad	10^{-9} abfarad
Resistance	1 ohm	Ω	1.1126×10^{-12} statohm	10^9 abohm
Conductance	1 siemen	S	8.987956×10^{11} statsiemen	10^{-9} absiemen
Inductance	1 henry	H	1.1126×10^{-12} stathenry	10^9 abhenry

25.3. References

[25-1]   G. A. Melchtly, "The International System of Units, Physical Constants and Conversion Factors," National Aeronautics and Space Administration Special Publication 7012, U.S. Government Printing Office, Washington, DC, 1973, pp. iii, 2-6, 11-20, second revision.

[25-2]   Chester H. Page and Paul Vigoureux, "The International System of Units (SI)," National Bureau of Standards Special Publication 330, U.S. Government Printing Office, Washington, DC, 1972.

[25-3]   Frederick D. Rossini, *Fundamental Measures and Constants for Science and Technology,* CRC Press, Inc., Cleveland, OH, 1976, pp. 98-103.

[25-4]   U.S. Naval Observatory, *1977 American Ephemeris and Nautical Almanac,* U.S. Nautical Almanac Office, Washington, DC, 1975.

[25-5]   Robert C. Weast, *Ph.D., Chemical Rubber Company Handbook of Chemistry and Physics,* CRC Press, Inc., Cleveland, OH, 1975, 56th edition pp. F 122, 285, 287, 294-316.

INDEX